# Calculus

## SINGLE VARIABLE

### Third Edition

**WILLIAM BRIGGS**
University of Colorado, Denver

**LYLE COCHRAN**
Whitworth University

**BERNARD GILLETT**
University of Colorado, Boulder

**ERIC SCHULZ**
Walla Walla Community College

| | |
|---|---|
| Director, Portfolio Management: | Deirdre Lynch |
| Executive Editor: | Jeff Weidenaar |
| Content Producer: | Rachel S. Reeve |
| Managing Producer: | Scott Disanno |
| Producer: | Stephanie Green |
| Manager, Courseware QA: | Mary Durnwald |
| Manager, Content Development: | Kristina Evans |
| Product Marketing Manager: | Emily Ockay |
| Field Marketing Manager: | Evan St. Cyr |
| Marketing Assistants: | Shannon McCormack, Erin Rush |
| Senior Author Support/Technology Specialist: | Joe Vetere |
| Manager, Rights and Permissions: | Gina Cheselka |
| Manufacturing Buyer: | Carol Melville, LSC Communications |
| Text and Cover Design, Production Coordination, Composition: | Cenveo® Publisher Services |
| Illustrations: | Network Graphics, Cenveo® Publisher Services |
| Cover Image: | Nutexzles/Moment/Getty Images |

**Library of Congress Cataloging-in-Publication Data**
Names: Briggs, William L., author. | Cochran, Lyle, author. | Gillett, Bernard, author. | Schulz, Eric P., author.
Title: Calculus / William Briggs (University of Colorado, Denver), Lyle Cochran (Whitworth University),
    Bernard Gillett (University of Colorado, Boulder), Eric Schulz (Walla Walla Community College).
Description: Third edition. | New York, NY : Pearson, [2019] | Includes index.
Identifiers: LCCN 2018002720 | ISBN 9780134765631 (hardcover) | ISBN 013476563X (hardcover)
Subjects:  LCSH: Calculus—Textbooks.
Classification: LCC QA303.2 .B749 2019 | DDC 515—dc23
LC record available at https://lccn.loc.gov/2018002720

1   18

ISBN 13: 978-0-13-476978-3
ISBN 10: 0-13-476978-3

*For Julie, Susan, Sally, Sue,*
*Katie, Jeremy, Elise, Mary, Claire, Katie, Chris, and Annie,*
*whose support, patience, and encouragement made this book possible.*

# Contents

# Preface

The third edition of *Calculus* supports a three-semester or four-quarter calculus sequence typically taken by students studying mathematics, engineering, the natural sciences, or economics. The third edition has the same goals as the first edition:

- to motivate the essential ideas of calculus with a lively narrative, demonstrating the utility of calculus with applications in diverse fields;

- to introduce new topics through concrete examples, applications, and analogies, appealing to students' intuition and geometric instincts to make calculus natural and believable; and

- once this intuitive foundation is established, to present generalizations and abstractions and to treat theoretical matters in a rigorous way.

The third edition both builds on the success of the previous two editions and addresses the feedback we have received. We have listened to and learned from the instructors who used the text. They have given us wise guidance about how to make the third edition an even more effective learning tool for students and a more powerful resource for instructors. Users of the text continue to tell us that it mirrors the course they teach—and, more important, that students actually read it! Of course, the third edition also benefits from our own experiences using the text, as well as from our experiences teaching mathematics at diverse institutions over the past 30 years.

## New to the Third Edition

### Exercises

The exercise sets are a major focus of the revision. In response to reviewer and instructor feedback, we've made some significant changes to the exercise sets by rearranging and relabeling exercises, modifying some exercises, and adding many new ones. Of the approximately 10,400 exercises appearing in this edition, 18% are new, and many of the exercises from the second edition were revised for this edition. We analyzed aggregated student usage and performance data from MyLab™ Math for the previous edition of this text. The results of this analysis helped us improve the quality and quantity of exercises that matter the most to instructors and students. We have also simplified the structure of the exercise sets from five parts to the following three:

1. **Getting Started** contains some of the former Review Questions but goes beyond those to include more conceptual exercises, along with new basic skills and short-answer exercises. Our goal in this section is to provide an excellent overall assessment of understanding of the key ideas of a section.

2. **Practice Exercises** consist primarily of exercises from the former Basic Skills, but they also include intermediate-level exercises from the former Further Explorations and Applications sections. Unlike previous editions, these exercises are not necessarily organized into groups corresponding to specific examples. For instance, instead of separating out Product Rule exercises from Quotient Rule exercises in Section 3.4, we

have merged these problems into one larger group of exercises. Consequently, specific instructions such as "Use the Product Rule to find the derivative of the following functions" and "Use the Quotient Rule to find the derivative of the given functions" have been replaced with the general instruction "Find the derivative of the following functions." With Product Rule and Quotient Rule exercises mixed together, students must first choose the correct method for evaluating derivatives before solving the problems.

3. **Explorations and Challenges** consist of more challenging problems and those that extend the content of the section.

We no longer have a section of the exercises called "Applications," but (somewhat ironically) in eliminating this section, we feel we are providing better coverage of applications because these exercises have been placed strategically *throughout the exercise sets*. Some are in Getting Started, most are in Practice Exercises, and some are in Explorations and Challenges. The applications nearly always have a boldface heading so that the topic of the application is readily apparent.

Regarding the boldface heads that precede exercises: These heads provide instructors with a quick way to discern the topic of a problem when creating assignments. We heard from users of earlier editions, however, that some of these heads provided too much guidance in how to solve a given problem. In this edition, therefore, we eliminated or reworded run-in heads that provided too much information about the solution method for a problem.

Finally, the **Chapter Review** exercises received a major revamp to provide more exercises (particularly intermediate-level problems) and more opportunities for students to choose a strategy of solution. More than 26% of the Chapter Review exercises are new.

## Content Changes

Below are noteworthy changes from the previous edition of the text. Many other detailed changes, not noted here, were made to improve the quality of the narrative and exercises. Bullet points with a ⚷ icon represent major content changes from the previous edition.

### Chapter 1 Functions

- Example 2 in Section 1.1 was modified with more emphasis on using algebraic techniques to determine the domain and range of a function. To better illustrate a common feature of limits, we replaced part (c) with a rational function that has a common factor in the numerator and denominator.

- Examples 7 and 8 in Section 1.1 from the second edition (2e) were moved forward in the narrative so that students get an intuitive feel for the composition of two functions using graphs and tables; compositions of functions using algebraic techniques follow.

- Example 10 in Section 1.1, illustrating the importance of secant lines, was made more relevant to students by using real data from a GPS watch during a hike. Corresponding exercises were also added.

- The Section 1.1 exercise set illustrates how the exercises have been reorganized and modified for the new edition. We have included a variety of short-answer, algebraic, graphical, and tabular exercises near the beginning of the exercise set, along with some new applied exercises.

- New applied exercises in biology, involving yeast growth, were added to Section 1.2 to illustrate the usefulness of linear functions.

- There is a 63% increase in the number of review exercises for Chapter 1. Most of the new exercises are intermediate-level exercises, providing more exhaustive coverage of the Chapter 1 content.

### Chapter 2 Limits

- Example 4 in Section 2.2 was revised, emphasizing an algebraic approach to a function with a jump discontinuity, rather than a graphical approach.

- Theorems 2.3 and 2.12 were modified, simplifying the notation to better connect with upcoming material.

- Example 7 in Section 2.3 was added to solidify the notions of left-, right-, and two-sided limits.

- Exercises were added to Section 2.5 to illustrate the similarities and differences between limits at infinity and infinite limits. We also included some easier exercises in Section 2.5 involving limits at infinity of functions containing square roots.

- Example 5 in Section 2.7 was added to demonstrate an epsilon-delta proof of a limit of a quadratic function.

- We added 17 epsilon-delta exercises to Section 2.7 to provide a greater variety of problems involving limits of quadratic, cubic, trigonometric, and absolute value functions.

### Chapter 3 Derivatives

- Chapter 3 now begins with a look back at average and instantaneous velocity, first encountered in Section 2.1, with a corresponding revised example in Section 3.1.

- ⚷ The derivative at a point and the derivative as a function are now treated separately in Sections 3.1 and 3.2.

- After defining the derivative at a point in Section 3.1 with a supporting example, we added a new subsection: Interpreting the Derivative (with two supporting examples).

- Several exercises were added to Section 3.3 that require students to use the Sum and Constant Rules, together with geometry, to evaluate derivatives.

- 🔑 The Power Rule for derivatives in Section 3.4 is stated for all real powers (later proved in Chapter 7). Example 4 in Section 3.4 includes two additional parts to highlight this change, and subsequent examples in upcoming sections rely on the more robust version of the Power Rule. The Power Rule for Rational Exponents in Section 3.8 was deleted because of this change.

- We combined the intermediate-level exercises in Section 3.4 involving the Product Rule and Quotient Rule together under one unified set of directions.

- In Section 3.7, we deleted references to Version 1 and Version 2 of the Chain Rule. Additionally, Chain Rule exercises involving repeated use of the rule were merged with the standard exercises.

- In Section 3.8, we added emphasis on simplifying derivative formulas for implicitly defined functions; see Examples 4 and 5.

- Example 3 in Section 3.9 was replaced; the new version shows how similar triangles are used in solving a related-rates problem.

### Chapter 4 Applications of the Derivative

- 🔑 The Mean Value Theorem (MVT) was moved from Section 4.6 to 4.2 so that the proof of Theorem 4.7 is not delayed. We added exercises to Section 4.2 that help students better understand the MVT geometrically.

- Example 5 in Section 4.5 was added to give guidance on a certain class of optimization problems.

- Example 3b in Section 4.7 was replaced to better drive home the need to simplify after applying l'Hôpital's Rule.

- Most of the intermediate exercises in Section 4.7 are no longer separated out by the type of indeterminate form, and we added some problems in which l'Hôpital's Rule does not apply.

- 🔑 Indefinite integrals of trigonometric functions with argument $ax$ (Table 4.9) were relocated to Section 5.5, where they are derived with the Substitution Rule.

- Example 5b in Section 4.9 was added to foreshadow a more complete treatment of the domain of an initial value problem found in Chapter 9.

- We added to Section 4.9 a significant number of intermediate antiderivative exercises that require some preliminary work (e.g., factoring, cancellation, expansion) before the antiderivatives can be determined.

### Chapter 5 Integration

- Examples 2 and 3 in Section 5.1 on approximating areas were replaced with a friendlier function where the grid points are more transparent; we return to these approximations in Section 5.3, where an exact result is given (Example 3b).

- Three properties of integrals (bounds on definite integrals) were added in Section 5.2 (Table 5.5); the last of these properties is used in the proof of the Fundamental Theorem (Section 5.3).

- Exercises were added to Sections 5.1 and 5.2 where students are required to evaluate Riemann sums using graphs or tables instead of formulas. These exercises will help students better understand the geometric meaning of Riemann sums.

- We added to Section 5.3 more exercises in which the integrand must be simplified before the integrals can be evaluated.

- A proof of Theorem 5.7 is now offered in Section 5.5.

- Table 5.6 lists the general integration formulas that were relocated from Section 4.9 to Section 5.5.

### Chapter 6 Applications of Integration

- Exercises requiring students to evaluate net change using graphs were added to Section 6.1.

- Exercises in Section 6.2 involving area calculations with respect to $x$ and $y$ are now combined under one unified set of directions (so that students must first determine the appropriate variable of integration).

- We increased the number of exercises in Sections 6.3 and 6.4 in which curves are revolved about lines other than the $x$- and $y$-axes. We also added introductory exercises that guide students, step by step, through the processes used to find volumes.

- A more gentle introduction to lifting problems (specifically, lifting a chain) was added in Section 6.7 and illustrated in Example 3, accompanied by additional exercises.

### Chapter 7 Logarithmic and Exponential Functions

- Exercises were added to Section 7.2 to give students practice at finding inverses of functions using the properties of exponential and logarithmic functions.

- New Getting Started application exercises (including a biology problem) were added to Section 7.2 to illustrate the usefulness of logarithmic and exponential functions.

- The introduction to exponential growth (Section 7.4) was rewritten to make a clear distinction between the relative growth rate (or percent change) of a quantity and the rate constant $k$. We revised the narrative so that the equation $y = y_0 e^{kt}$ applies to both growth and decay models. This revision resulted in a small change to the half-life formula.

- The variety of applied exercises in Section 7.4 was increased to further illustrate the utility of calculus in the study of exponential growth and decay.

### Chapter 8 Integration Techniques

- Table 8.1 now includes four standard trigonometric integrals that previously appeared in the section Trigonometric Integrals (8.3); these integrals are derived in Examples 1 and 2 in Section 8.1.

- 🔑 A new section (8.6) was added so that students can master integration techniques (that is, choose a strategy) apart from the context given in the previous five sections.

- In Section 8.5 we increased the number and variety of exercises where students must set up the appropriate form of the partial fraction decomposition of a rational function, including more with irreducible quadratic factors.

- A full derivation of Simpson's Rule was added to Section 8.8, accompanied by Example 7, additional figures, and an expanded exercise set.

• 🔑 The Comparison Test for improper integrals was added to Section 8.9, accompanied by Example 7, a two-part example. New exercises in Section 8.9 include doubly infinite improper integrals over infinite intervals.

## Chapter 9 Differential Equations

• 🔑 The chapter on differential equations that was available only online in the 2e was converted to a chapter of the text, replacing the single-section coverage found in the 2e.

• More attention was given to the domain of an initial value problem, resulting in the addition and revision of several examples and exercises throughout the chapter.

## Chapter 10 Sequences and Infinite Series

• 🔑 The second half of Chapter 10 was reordered: Comparison Tests (Section 10.5), Alternating Series (Section 10.6, which includes the topic of absolute convergence), The Ratio and Root Tests (Section 10.7), and Choosing a Convergence Test (Section 10.8; new section). We split the 2e section that covered the Comparison, Ratio, and Root Tests to avoid overwhelming students with too many tests at one time. Section 10.5 focuses entirely on the comparison tests; 39% of the exercises are new. The topic of alternating series now appears before the Ratio and Root Tests so that the latter tests may be stated in their more general form (they now apply to any series rather than only to series with positive terms). The final section (10.8) gives students an opportunity to master convergence tests after encountering each of them separately.

• The terminology associated with sequences (10.2) now includes *bounded above, bounded below,* and *bounded* (rather than only *bounded,* as found in earlier editions).

• Theorem 10.3 (Geometric Sequences) is now developed in the narrative rather than within an example, and an additional example (10.2.3) was added to reinforce the theorem and limit laws from Theorem 10.2.

• Example 5c in Section 10.2 uses mathematical induction to find the limit of a sequence defined recursively; this technique is reinforced in the exercise set.

• Example 3 in Section 10.3 was replaced with telescoping series that are not geometric and that require re-indexing.

• We increased the number and variety of exercises where the student must determine the appropriate series test necessary to determine convergence of a given series.

• We added some easier intermediate-level exercises to Section 10.6, where series are estimated using $n$th partial sums for a given value of $n$.

• Properties of Convergent Series (Theorem 10.8) was expanded (two more properties) and moved to Section 10.3 to better balance the material presented in Sections 10.3 and 10.4. Example 4 in Section 10.3 now has two parts to give students more exposure to the theorem.

## Chapter 11 Power Series

• Chapter 11 was revised to mesh with the changes made in Chapter 10.

• We included in Section 11.2 more exercises where the student must find the radius and interval of convergence.

• Example 2 in Section 11.3 was added to illustrate how to choose a different center for a series representation of a function when the original series for the function converges to the function on only part of its domain.

• We addressed an issue with the exercises in Section 11.2 of the previous edition by adding more exercises where the intervals of convergence either are closed or contain one, but not both, endpoints.

• We addressed an issue with exercises in the previous edition by adding many exercises that involve power series centered at locations other than 0.

## Chapter 12 Parametric and Polar Curves

• 🔑 The arc length of a two-dimensional curve described by parametric equations was added to Section 12.1, supported by two examples and additional exercises. Area and surfaces of revolution associated with parametric curves were also added to the exercises.

• In Example 3 in Section 12.2, we derive more general polar coordinate equations for circles.

• The arc length of a curve described in polar coordinates is given in Section 12.3.

## Chapter 13 Vectors and the Geometry of Space

• 🔑 The material from the 2e chapter Vectors and Vector-Valued Functions is now covered in this chapter and the following chapter.

• Example 5c in Section 13.1 was added to illustrate how to express a vector as a product of its magnitude and its direction.

• We increased the number of applied vector exercises in Section 13.1, starting with some easier exercises, resulting in a wider gradation of exercises.

• 🔑 We adopted a more traditional approach to lines and planes; these topics are now covered together in Section 13.5, followed by cylinders and quadric surfaces in Section 13.6. This arrangement gives students early exposure to all the basic three-dimensional objects that they will encounter throughout the remainder of the text.

• 🔑 A discussion of the distance from a point to a line was moved from the exercises into the narrative, supported with Example 3 in Section 13.5. Example 4 finds the point of intersection of two lines. Several related exercises were added to this section.

• In Section 13.6 there is a larger selection of exercises where the student must identify the quadric surface associated with a given equation. Exercises are also included where students design shapes using quadric surfaces.

## Chapter 14 Vector-Valued Functions

• More emphasis was placed on the surface(s) on which a space curve lies in Sections 14.1 and 14.3.

- We added exercises in Section 14.1 where students are asked to find the curve of intersection of two surfaces and where students must verify that a curve lies on a given surface.

- Example 3c in Section 14.3 was added to illustrate how a space curve can be mapped onto a sphere.

- Because the arc length of plane curves (described parametrically in Section 12.1 and with polar coordinates in Section 12.3) was moved to an earlier location in the text, Section 14.4 is now a shorter section.

### Chapter 15 Functions of Several Variables

- 🔑 Equations of planes and quadric surfaces were removed from this chapter and now appear in Chapter 13.

- The notation in Theorem 15.2 was simplified to match changes made to Theorem 2.3.

- Example 7 in Section 15.4 was added to illustrate how the Chain Rule is used to compute second partial derivatives.

- We added more challenging partial derivative exercises to Section 15.3 and more challenging Chain Rule exercises to Section 15.4.

- Example 7 in Section 15.5 was expanded to give students more practice finding equations of curves that lie on surfaces.

- Theorem 15.13 was added in Section 15.5; it's a three-dimensional version of Theorem 15.11.

- Example 7 in Section 15.7 was replaced with a more interesting example; the accompanying figure helps tell the story of maximum/minimum problems and can be used to preview Lagrange multipliers.

- We added to Section 15.7 some basic exercises that help students better understand the second derivative test for functions of two variables.

- 🔑 Example 1 in Section 15.8 was modified so that using Lagrange multipliers is the clear path to a solution, rather than eliminating one of the variables and using standard techniques. We also make it clear that care must be taken when using the method of Lagrange multipliers on sets that are not closed and bounded (absolute maximum and minimum values may not exist).

### Chapter 16 Multiple Integration

- Example 2 in Section 16.3 was modified because it was too similar to Example 1.

- More care was given to the notation used with polar, cylindrical, and spherical coordinates (see, for example, Theorem 16.3 and the development of integration in different coordinate systems).

- Example 3 in Section 16.4 was modified to make the integration a little more transparent and to show that changing variables to polar coordinates is permissible in more than just the $xy$-plane.

- More multiple integral exercises were added to Sections 16.1, 16.2, and 16.4, where integration by substitution or integration by parts is needed to evaluate the integrals.

- In Section 16.4 we added more exercises in which the integrals must first be evaluated with respect to $x$ or $y$ instead of $z$. We also included more exercises that require triple integrals to be expressed in several orderings.

### Chapter 17 Vector Calculus

- 🔑 Our approach to scalar line integrals was streamlined; Example 1 in Section 17.2 was modified to reflect this fact.

- We added basic exercises in Section 17.2 emphasizing the geometric meaning of line integrals in a vector field. A subset of exercises was added where line integrals are grouped so that the student must determine the type of line integral before evaluating the integral.

- Theorem 17.5 was added to Section 17.3; it addresses the converse of Theorem 17.4. We also promoted the area of a plane region by a line integral to theorem status (Theorem 17.8 in Section 17.4).

- More line integral exercises were added to Section 17.3 where the student must first find the potential function before evaluating the line integral over a conservative vector field using the Fundamental Theorem of Line Integrals.

- Example 3 in Section 17.7 was replaced to give an example of a surface whose bounding curve is not a plane curve and to provide an example that buttresses the claims made at the end of the section (that is, Two Final Notes on Stokes' Theorem).

- We added to Section 17.7 more challenging surface integrals that are evaluated using Stokes' Theorem.

## New to MyLab Math

- **Assignable Exercises** To better support students and instructors, we made the following changes to the assignable exercises:

  ○ Updated the solution processes in Help Me Solve This and View an Example to better match the techniques used in the text.

  ○ Added more Setup & Solve exercises to better mirror the types of responses that students are expected to provide on tests. We also added a parallel "standard" version of each Setup & Solve exercise to allow the instructor to determine which version to assign.

  ○ Added exercises corresponding to new exercises in the text.

○ Added exercises where MyLab Math users had identified gaps in coverage in the 2e.

○ Added extra practice exercises to each section (clearly labeled EXTRA). These "beyond the text" exercises are perfect for chapter reviews, quizzes, and tests.

○ Analyzed aggregated student usage and performance data from MyLab Math for the previous edition of this text. The results of this analysis helped improve the quality and quantity of exercises that matter the most to instructors and students.

• **Instructional Videos** For each section of the text, there is now a new full-lecture video. Many of these videos make use of Interactive Figures to enhance student understanding of concepts. To make it easier for students to navigate to the specific content they need, each lecture video is segmented into shorter clips (labeled Introduction, Example, or Summary). Both the full lectures and the video segments are assignable within MyLab Math. The videos were created by the following team: Matt Hudelson (Washington State University), Deb Carney and Rebecca Swanson (Colorado School of Mines), Greg Wisloski and Dan Radelet (Indiana University of Pennsylvania), and Nick Ormes (University of Denver).

• **Enhanced Interactive Figures** Incorporating functionality from several standard Interactive Figures makes Enhanced Interactive Figures mathematically richer and ideal for in-class demonstrations. Using a single figure, instructors can illustrate concepts that are difficult for students to visualize and can make important connections to key themes of calculus.

• **Enhanced Sample Assignments** These section-level assignments address gaps in pre-calculus skills with a personalized review of prerequisites, help keep skills fresh with spaced practice using key calculus concepts, and provide opportunities to work exercises without learning aids so students can check their understanding. They are assignable and editable.

• **Quick Quizzes** have been added to Learning Catalytics™ (an in-class assessment system) for every section of the text.

• **Maple™, Mathematica®, and Texas Instruments® Manuals and Projects** have all been updated to align with the latest software and hardware.

## Noteworthy Features

### Figures

Given the power of graphics software and the ease with which many students assimilate visual images, we devoted considerable time and deliberation to the figures in this text. Whenever possible, we let the figures communicate essential ideas using annotations reminiscent of an instructor's voice at the board. Readers will quickly find that the figures facilitate learning in new ways.

**Figure 6.40**

## Annotated Examples

Worked-out examples feature annotations in blue to guide students through the process of solving the example and to emphasize that each step in a mathematical argument must be rigorously justified. These annotations are designed to echo how instructors "talk through" examples in lecture. They also provide help for students who may struggle with the algebra and trigonometry steps within the solution process.

## Quick Checks

The narrative is interspersed with Quick Check questions that encourage students to do the calculus as they are reading about it. These questions resemble the kinds of questions instructors pose in class. Answers to the Quick Check questions are found at the end of the section in which they occur.

## Guided Projects

MyLab Math contains 78 Guided Projects that allow students to work in a directed, step-by-step fashion, with various objectives: to carry out extended calculations, to derive physical models, to explore related theoretical topics, or to investigate new applications of calculus. The Guided Projects vividly demonstrate the breadth of calculus and provide a wealth of mathematical excursions that go beyond the typical classroom experience. A list of related Guided Projects is included at the end of each chapter.

## Incorporating Technology

We believe that a calculus text should help students strengthen their analytical skills and demonstrate how technology can extend (not replace) those skills. Calculators and graphing utilities are additional tools in the kit, and students must learn when and when not to use them. Our goal is to accommodate the different policies regarding technology adopted by various instructors.

Throughout the text, exercises marked with **T** indicate that the use of technology—ranging from plotting a function with a graphing calculator to carrying out a calculation using a computer algebra system—may be needed. See page xx for information regarding our technology resource manuals covering Maple, Mathematica, and Texas Instruments graphing calculators.

## Text Versions

- **eBook with Interactive Figures**  The text is supported by a groundbreaking and award-winning electronic book created by Eric Schulz of Walla Walla Community College. This "live book" runs in Wolfram CDF Player (the free version of Mathematica) and contains the complete text of the print book plus interactive versions of approximately 700 figures. Instructors can use these interactive figures in the classroom to illustrate the important ideas of calculus, and students can explore them while they are reading the text. Our experience confirms that the interactive figures help build students' geometric intuition of calculus. The authors have written Interactive Figure Exercises that can be assigned via MyLab Math so that students can engage with the figures outside of class in a directed way. Available only within MyLab Math, the eBook provides instructors with powerful new teaching tools that expand and enrich the learning experience for students.

- **Other eBook Formats**  The text is also available in various stand-alone eBook formats. These are listed in the Pearson online catalog: **www.pearson.com**. MyLab Math also contains an HTML eBook that is screen-reader accessible.

- **Other Print Formats**  The text is also available in split editions (Single Variable [Chapters 1–12] and Multivariable [Chapters 10–17]) and in unbound (3-hole punched) formats. Again, see the Pearson online catalog for details: **www.pearson.com**.

# Acknowledgments

We would like to express our thanks to the following people, who made valuable contributions to this edition as it evolved through its many stages.

**Accuracy Checkers**

Jennifer Blue

Lisa Collette

Nathan Moyer

Tom Wegleitner

**Reviewers**

Siham Alfred, *Germanna Community College*

Darry Andrews, *Ohio State University*

Pichmony Anhaouy, *Langara College*

Raul Aparicio, *Blinn College*

Anthony Barcellos, *American River College*

Rajesh K. Barnwal, *Middle Tennessee State University*

Susan Barton, *Brigham Young University, Hawaii*

Aditya Baskota, *Kauai Community College*

Al Batten, *University of Colorado, Colorado Springs*

Jeffrey V. Berg, *Arapahoe Community College*

Martina Bode, *University of Illinois, Chicago*

Kaddour Boukaabar, *California University of Pennsylvania*

Paul W. Britt, *Our Lady of the Lake College*

Steve Brosnan, *Belmont Abbey College*

Brendan Burns Healy, *Tufts University*

MK Choy, *CUNY Hunter College*

Nathan P. Clements, *University of Wyoming*

Gregory M. Coldren, *Frederick Community College*

Brian Crimmel, *U.S. Coast Guard Academy*

Robert D. Crise, Jr, *Crafton Hills College*

Dibyajyoti Deb, *Oregon Institute of Technology*

Elena Dilai, *Monroe Community College*

Johnny I. Duke, *Georgia Highlands College*

Fedor Duzhin, *Nanyang Technological University*

Houssein El Turkey, *University of New Haven*

J. Robson Eby, *Blinn College*

Amy H. Erickson, *Georgia Gwinnett College*

Robert A. Farinelli, *College of Southern Maryland*

Rosemary Farley, *Manhattan College*

Lester A. French, Jr, *University of Maine, Augusta*

Pamela Frost, *Middlesex Community College*

Scott Gentile, *CUNY Hunter College*

Stephen V. Gilmore, *Johnson C. Smith University*

Joseph Hamelman, *Virginia Commonwealth University*

Jayne Ann Harderq, *Oral Roberts University*

Miriam Harris-Botzum, *Lehigh Carbon Community College*

Mako Haruta, *University of Hartford*

Ryan A. Hass, *Oregon State University*

Christy Hediger, *Lehigh Carbon Community College*

Joshua Hertz, *Northeastern University*

Peter Hocking, *Brunswick Community College*

Farhad Jafari, *University of Wyoming*

Yvette Janecek, *Blinn College*

Tom Jerardi, *Howard Community College*

Karen Jones, *Ivy Tech Community College*

Bir Kafle, *Purdue University Northwest*

Mike Kawai, *University of Colorado, Denver*

Mahmoud Khalili, *Oakton Community College*

Lynne E. Kowski, *Raritan Valley Community College*

Tatyana Kravchuk, *Northern Virginia Community College*

Lorraine Krusinski, *Brunswick High School*

Frederic L. Lang, *Howard Community College*

Robert Latta, *University of North Carolina, Charlotte*

Angelica Lyubenko, *University of Colorado, Denver*

Darren E. Mason, *Albion College*

John C. Medcalf, *American River College*

Elaine Merrill, *Brigham Young University, Hawaii*

Robert Miller, *Langara College*

Mishko Mitkovski, *Clemson University*

Carla A. Monticelli, *Camden County College*

Charles Moore, *Washington State University*

Humberto Munoz Barona, *Southern University*

Clark Musselman, *University of Washington, Bothell*

Glenn Newman, *Newbury College*

Daniela Nikolova-Popova, *Florida Atlantic University*

Janis M. Oldham, *North Carolina A&T State University*

Byungdo Park, *CUNY Hunter College*

Denise Race, *Eastfield College*

Hilary Risser, *Montana Tech*

Sylvester Roebuck, Jr, *Olive Harvey College*

John P. Roop, *North Carolina A&T State University*

Paul N. Runnion, *Missouri University of Science and Technology*

Kegan Samuel, *Middlesex Community College*

Steve Scarborough, *Oregon State University*

Vickey Sealy, *West Virginia University*

Ruth Seidel, *Blinn College*

Derek Smith, *Nashville State Community College*

James Talamo, *Ohio State University*

Yan Tang, *Embry-Riddle Aeronautical University*

Kye Taylor, *Tufts University*

Daniel Triolo, *Lake Sumter State College*

Cal Van Niewaal, *Coe College*

Robin Wilson, *California Polytechnic State University, Pomona*

Kurt Withey, *Northampton Community College*

Amy Yielding, *Eastern Oregon University*

Prudence York-Hammons, *Temple College*

Bradley R. Young, *Oakton Community College*

Pablo Zafra, *Kean University*

The following faculty members provided direction on the development of the MyLab Math course for this edition:

Heather Albrecht, *Blinn College*

Terry Barron, *Georgia Gwinnett College*

Jeffrey V. Berg, *Arapahoe Community College*

Joseph Bilello, *University of Colorado, Denver*

Mark Bollman, *Albion College*

Mike Bostick, *Central Wyoming College*

Laurie L. Boudreaux, *Nicholls State University*

Steve Brosnan, *Belmont Abbey College*

Debra S. Carney, *Colorado School of Mines*

Robert D. Crise, Jr, *Crafton Hills College*

Shannon Dingman, *University of Arkansas*

Deland Doscol, *North Georgia Technical College*

J. Robson Eby, *Blinn College*

Stephanie L. Fitch, *Missouri University of Science and Technology*

Dennis Garity, *Oregon State University*

Monica Geist, *Front Range Community College*

Brandie Gilchrist, *Central High School*

Roger M. Goldwyn, *Florida Atlantic University*

Maggie E. Habeeb, *California University of Pennsylvania*

Ryan A. Hass, *Oregon State University*

Danrun Huang, *St. Cloud State University*

Zhongming Huang, *Midland University*

Christa Johnson, *Guilford Technical Community College*

Bir Kafle, *Purdue University Northwest*

Semra Kilic-Bahi, *Colby-Sawyer College*

Kimberly Kinder, *Missouri University of Science and Technology*

Joseph Kotowski, *Oakton Community College*

Lynne E. Kowski, *Raritan Valley Community College*

Paula H. Krone, *Georgia Gwinnett College*

Daniel Lukac, *Miami University*

Jeffrey W. Lyons, *Nova Southeastern University*

James Magee, *Diablo Valley College*

Erum Marfani, *Frederick Community College*

Humberto Munoz Barona, *Southern University*

James A. Mendoza Alvarez, *University of Texas, Arlington*

Glenn Newman, *Newbury College*

Peter Nguyen, *Coker College*

Mike Nicholas, *Colorado School of Mines*

Seungly Oh, *Western New England University*

Denise Race, *Eastfield College*

Paul N. Runnion, *Missouri University of Science and Technology*

Polly Schulle, *Richland College*

Rebecca Swanson, *Colorado School of Mines*

William Sweet, *Blinn College*

M. Velentina Vega-Veglio, *William Paterson University*

Nick Wahle, *Cincinnati State Technical and Community College*

Patrick Wenkanaab, *Morgan State University*

Amanda Wheeler, *Amarillo College*

Diana White, *University of Colorado, Denver*

Gregory A. White, *Linn Benton Community College*

Joseph White, *Olympic College*

Robin T. Wilson, *California Polytechnic State University, Pomona*

Deborah A. Zankofski, *Prince George's Community College*

The following faculty were members of the Engineering Review Panel. This panel made recommendations to improve the text for engineering students.

Al Batten, *University of Colorado, Colorado Springs*

Josh Hertz, *Northeastern University*

Daniel Latta, *University of North Carolina, Charlotte*

Yan Tang, *Embry-Riddle Aeronautical University*

# MyLab Math Online Course
## for *Calculus, 3e*

(access code required)

MyLab™ Math is available to accompany Pearson's market-leading text offerings. To give students a consistent tone, voice, and teaching method, each text's flavor and approach are tightly integrated throughout the accompanying MyLab Math course, making learning the material as seamless as possible.

## PREPAREDNESS

One of the biggest challenges in calculus courses is making sure students are adequately prepared with the prerequisite skills needed to successfully complete their course work. MyLab Math supports students with just-in-time remediation and review of key concepts.

### Integrated Review Course

These special MyLab courses contain pre-made, assignable quizzes to assess the prerequisite skills needed for each chapter, plus personalized remediation for any gaps in skills that are identified. Each student, therefore, receives the appropriate level of help—no more, no less.

## DEVELOPING DEEPER UNDERSTANDING

MyLab Math provides content and tools that help students build a deeper understanding of course content than would otherwise be possible.

### eBook with Interactive Figures

The eBook includes approximately 700 figures that can be manipulated by students to provide a deeper geometric understanding of key concepts and examples as they read and learn new material. Students get unlimited access to the eBook within any MyLab Math course using that edition of the text. The authors have written Interactive Figure Exercises that can be assigned for homework so that students can engage with the figures outside of the classroom.

**pearson.com/mylab/math**

## Exercises with Immediate Feedback

The over 8000 homework and practice exercises for this text regenerate algorithmically to give students unlimited opportunity for practice and mastery. MyLab Math provides helpful feedback when students enter incorrect answers and includes the optional learning aids Help Me Solve This, View an Example, videos, and/or the eBook.

## NEW! Enhanced Sample Assignments

These section-level assignments include just-in-time review of prerequisites, help keep skills fresh with spaced practice of key concepts, and provide opportunities to work exercises without learning aids so students can check their understanding. They are assignable and editable within MyLab Math.

## Additional Conceptual Questions

Additional Conceptual Questions focus on deeper, theoretical understanding of the key concepts in calculus. These questions were written by faculty at Cornell University under an NSF grant and are also assignable through Learning Catalytics™.

## Setup & Solve Exercises

These exercises require students to show how they set up a problem, as well as the solution, thus better mirroring what is required on tests. This new type of exercise was widely praised by users of the second edition, so more were added to the third edition.

## ALL NEW! Instructional Videos

For each section of the text, there is now a new full-lecture video. Many of these videos make use of Interactive Figures to enhance student understanding of concepts. To make it easier for students to navigate to the content they need, each lecture video is segmented into shorter clips (labeled Introduction, Example, or Summary). Both the video lectures and the video segments are assignable within MyLab Math. The Guide to Video-Based Assignments makes it easy to assign videos for homework by showing which MyLab Math exercises correspond to each video.

**pearson.com/mylab/math**

## UPDATED! Technology Manuals (downloadable)

- Maple™ Manual and Projects by Kevin Reeves, East Texas Baptist University
- Mathematica® Manual and Projects by Todd Lee, Elon University
- TI-Graphing Calculator Manual by Elaine McDonald-Newman, Sonoma State University

These manuals cover *Maple* 2017, *Mathematica 11,* and the TI-84 Plus and TI-89, respectively. Each manual provides detailed guidance for integrating the software package or graphing calculator throughout the course, including syntax and commands. The projects include instructions and ready-made application files for Maple and Mathematica. The files can be downloaded from within MyLab Math.

## Student's Solutions Manuals (softcover and downloadable)

Single Variable Calculus (Chapters 1–12)
   ISBN: 0-13-477034-X | 978-0-13-477034-5
Multivariable Calculus (Chapters 10–17)
   ISBN: 0-13-476682-2 | 978-0-13-476682-9
Written by Mark Woodard (Furman University), the Student's Solutions Manual contains worked-out solutions to all the odd-numbered exercises. This manual is available in print and can be downloaded from within MyLab Math.

## SUPPORTING INSTRUCTION

MyLab Math comes from an experienced partner with educational expertise and an eye on the future. It provides resources to help you assess and improve student results at every turn and unparalleled flexibility to create a course tailored to you and your students.

## NEW! Enhanced Interactive Figures

Incorporating functionality from several standard Interactive Figures makes Enhanced Interactive Figures mathematically richer and ideal for in-class demonstrations. Using a single enhanced figure, instructors can illustrate concepts that are difficult for students to visualize and can make important connections to key themes of calculus.

## Learning Catalytics

Now included in all MyLab Math courses, this student response tool uses students' smartphones, tablets, or laptops to engage them in more interactive tasks and thinking during lecture. Learning Catalytics™ fosters student engagement and peer-to-peer learning with real-time analytics. Access pre-built exercises created specifically for calculus, including Quick Quiz exercises for each section of the text.

## PowerPoint Lecture Resources (downloadable)

Slides contain presentation resources such as key concepts, examples, definitions, figures, and tables from this text. They can be downloaded from within MyLab Math or from Pearson's online catalog at **www.pearson.com**.

## Comprehensive Gradebook

The gradebook includes enhanced reporting functionality, such as item analysis and a reporting dashboard to enable you to efficiently manage your course. Student performance data are presented at the class, section, and program levels in an accessible, visual manner so you'll have the information you need to keep your students on track.

## TestGen

TestGen® (**www.pearson.com/testgen**) enables instructors to build, edit, print, and administer tests using a computerized bank of questions developed to cover all the objectives of the text. TestGen is algorithmically based, allowing instructors to create multiple, but equivalent, versions of the same question or test with the click of a button. Instructors can also modify test bank questions and/or add new questions. The software and test bank are available for download from Pearson's online catalog, **www.pearson.com**. The questions are also assignable in MyLab Math.

## Instructor's Solutions Manual (downloadable)

Written by Mark Woodard (Furman University), the Instructor's Solutions Manual contains complete solutions to all the exercises in the text. It can be downloaded from within MyLab Math or from Pearson's online catalog, **www.pearson.com**.

## Instructor's Resource Guide (downloadable)

This resource includes Guided Projects that require students to make connections between concepts and applications from different parts of the calculus course. They are correlated to specific chapters of the text and can be assigned as individual or group work. The files can be downloaded from within MyLab Math or from Pearson's online catalog, **www.pearson.com**.

## Accessibility

Pearson works continuously to ensure that our products are as accessible as possible to all students. We are working toward achieving WCAG 2.0 Level AA and Section 508 standards, as expressed in the Pearson Guidelines for Accessible Educational Web Media, **www.pearson.com/mylab/math/accessibility**.

**pearson.com/mylab/math**

# Credits

**Chapter opener art:** Andrey Pavlov/123RF

## Chapter 1

**Page 23,** Rivas, E.M. et al (2014). A simple mathematical model that describes the growth of the area and the number of total and viable cells in yeast colonies. *Letters in Applied Microbiology*, 594(59). **Page 25,** Tucker, V. A. (2000, December). The Deep Fovea, Sideways Vision and Spiral Flight Paths in Raptors. *The Journal of Experimental Biology*, 203, 3745–3754. **Page 26,** Collings, B. J. (2007, January). Tennis (and Volleyball) Without Geometric Series. *The College Mathematics Journal*, 38(1). **Page 33,** Isaksen, D. C. (1996, September). How to Kick a Field Goal. *The College Mathematics Journal*, 27(4).

## Chapter 3

**Page 163,** Perloff, J. (2012). *Microeconomics*. Prentice Hall.

## Chapter 4

**Pages 205–206,** Yu, C. Chau, K.T. & Jiang J. Z. (2009). A flux-mnemonic permanent magnet brushless machine for wind power generation. *Journal of Applied Physics*, 105. **Pages 243,** Pennings, T. (2003, May). Do Dogs Know Calculus? *The College Mathematics Journal*, 34(6). **Page 245,** Apostol T. M. (1967). *Calculus*, Vol. 1. John Wiley & Sons; Dial, R. (2003). Energetic Savings and The Body Size Distributions of Gliding Mammals. *Evolutionary Ecology Research*, 5, 1151–1162. **Page 246,** Adam, J. A. (2011, June). Blood Vessel Branching: Beyond the Standard Calculus Problem. *Mathematics Magazine*, 84(3), 196–207. **Page 247,** Halmos, P. R. Problems For Mathematicians Young And Old. Copyright © Mathematical Association of America, 1991. All rights reserved; Dodge, W. & Viktora, S. (2002, November). Thinking out of the Box … Problem. *Mathematics Teacher*, 95(8).

## Chapter 5

**Page 316,** Barry, P. (2001, September). Integration from First Principles. *The College Mathematics Journal*, 32(4), 287–289. **Page 336,** Chen, H. (2005, December). Means Generated by an Integral. *Mathematics Magazine,* 78(5), 397–399; Tong, J. (2002, November). A Generalization of the Mean Value Theorem for Integrals. *The College Mathematics Journal*, 33(5).

## Chapter 6

**Page 362,** Zobitz, J. M. (2013, November). Forest Carbon Uptake and the Fundamental Theorem of Calculus. *The College Mathematics Journal*, 44(5), 421–424. **Page 372,** Cusick, W. L. (2008, April). Archimedean Quadrature Redux. *Mathematics Magazine,* 81(2), 83–95.

## Chapter 7

**Page 475,** Murray, I. W. & Wolf, B. O. (2012). Tissue Carbon Incorporation Rates and Diet-to-Tissue Discrimination in Ectotherms: Tortoises Are Really Slow. *Physiological and Biochemical Zoology*, 85(1). **Page 477,** Keller, J. (1973, September). A Theory of Competitive Running. *Physics Today*, 26(9); Schreiber, J. S. (2013). Motivating Calculus with Biology. *Mathematical Association of America*, Notes No. 81.

## Chapter 8

**Page 592,** Weidman, P. & Pinelis, I. (2004). Model equations for the Eiffel Tower profile: historical perspective and new results. *C. R. Mécanique*, 332, 571–584; Feuerman, M. et al. (1986, February). Problems. *Mathematics Magazine,* 59(1). **Page 596,** Galperin, G. & Ronsse, G. (2008, April). Lazy Student Integrals. *Mathematics Magazine,* 81(2), 152–154. **Pages 545–546,** Osler, J. T. (2003, May). Visual Proof of Two Integrals. *The College Mathematics Journal,* 34(3), 231–232.

## Chapter 10

**Page 682,** Fleron, J. F. (1999, January). Gabriel's Wedding Cake. *The College Mathematics Journal*, 30(1), 35–38; Selden, A. & Selden, J. (1993, November). Collegiate Mathematics Education Research: What Would That Be Like? *The College Mathematics Journal*, 24(5), 431–445. **Pages 682–683,** Chen, H. & Kennedy, C. (2012, May). Harmonic Series Meets Fibonacci Sequence. *The College Mathematics Journal*, 43(3), 237–243.

## Chapter 12

**Page 766,** Wagon, S. (2010). *Mathematica in Action*. Springer; created by Norton Starr, Amherst College. **Page 767,** Brannen, N. S. (2001, September). The Sun, the Moon, and Convexity. *The College Mathematics Journal*, 32(4), 268–272. **Page 774,** Fray, T. H. (1989). The butterfly curve. *American Mathematical Monthly*, 95(5), 442–443; revived in Wagon, S. & Packel, E. (1994). *Animating Calculus*. Freeman. **Page 778,** Wagon, S. & Packel, E. (1994). *Animating Calculus*. Freeman.

## Chapter 13

**Page 816,** Strang, G. (1991). Calculus. Wellesley-Cambridge Press. **Page 864,** Model Based 3D Tracking of an Articulated Hand, B. Stenger, P. R. S. Mendonça, R. Cipolla, CVPR, Vol. II, p. 310–315, December 2001. CVPR 2001: PROCEEDINGS OF THE 2001 IEEE COMPUTER SOCIETY CONFERENCE ON COMPUTER VISION AND PATTERN RECOGNITION by IEEE Computer Society Reproduced with permission of IEEE COMPUTER SOCIETY PRESS in the format Republish in a book via Copyright Clearance Center.

## Chapter 15

**Pages 929–930,** www.nfl.com. **Page 994,** Karim, R. (2014, December). Optimization of pectin isolation method from pineapple (ananas comosus l.) waste. *Carpathian Journal of Food Science and Technology*, 6(2), 116–122. **Page 996,** Rosenholtz, I. (1985, May). "The Only Critical Point in Town" Test. *Mathematics Magazine*, 58(3), 149–150 and Gillett, P. (1984). *Calculus and Analytical Geometry*, 2nd edition; Rosenholtz, I. (1987, February). Two mountains without a valley. *Mathematics Magazine*, 60(1); Math Horizons, Vol. 11, No. 4, April 2004.

## Chapter 16

**Page 1072,** Glaister, P. (1996). Golden Earrings. *Mathematical Gazette*, 80, 224–225.

# 1

# Functions

**Chapter Preview**   Mathematics is a language with an alphabet, a vocabulary, and many rules. Before beginning your calculus journey, you should be familiar with the elements of this language. Among these elements are algebra skills; the notation and terminology for various sets of real numbers; and the descriptions of lines, circles, and other basic sets in the coordinate plane. A review of this material is found in Appendix B, online at goo.gl/6DCbbM. This chapter begins with the fundamental concept of a function and then presents some of the functions needed for calculus: polynomials, rational functions, algebraic functions, and the trigonometric functions. (Logarithmic, exponential, and inverse functions are introduced in Chapter 7.) Before you begin studying calculus, it is important that you master the ideas in this chapter.

## 1.1 Review of Functions

Everywhere around us we see relationships among quantities, or **variables**. For example, the consumer price index changes in time and the temperature of the ocean varies with latitude. These relationships can often be expressed by mathematical objects called *functions*. Calculus is the study of functions, and because we use functions to describe the world around us, calculus is a universal language for human inquiry.

---

**DEFINITION**  **Function**

A **function** $f$ is a rule that assigns to each value $x$ in a set $D$ a *unique* value denoted $f(x)$. The set $D$ is the **domain** of the function. The **range** is the set of all values of $f(x)$ produced as $x$ varies over the entire domain (**Figure 1.1**).

---

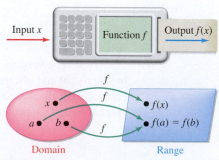

**Figure 1.1**

1

➤ If the domain is not specified, we take it to be the set of all values of $x$ for which $f$ is defined. We will see shortly that the domain and range of a function may be restricted by the context of the problem.

The **independent variable** is the variable associated with the domain; the **dependent variable** belongs to the range. The **graph** of a function $f$ is the set of all points $(x, y)$ in the $xy$-plane that satisfy the equation $y = f(x)$. The **argument** of a function is the expression on which the function works. For example, $x$ is the argument when we write $f(x)$. Similarly, 2 is the argument in $f(2)$ and $x^2 + 4$ is the argument in $f(x^2 + 4)$.

**QUICK CHECK 1**  If $f(x) = x^2 - 2x$, find $f(-1)$, $f(x^2)$, $f(t)$, and $f(p - 1)$. ◄

The requirement that a function assigns a *unique* value of the dependent variable to each value in the domain is expressed in the vertical line test (**Figure 1.2a**). For example, the outside temperature as it varies over the course of a day is a function of time (**Figure 1.2b**).

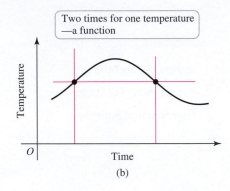

**Figure 1.2**

---

**Vertical Line Test**

A graph represents a function if and only if it passes the **vertical line test**: Every vertical line intersects the graph at most once. A graph that fails this test does not represent a function.

---

➤ A set of points or a graph that does *not* correspond to a function represents a **relation** between the variables. All functions are relations, but not all relations are functions.

**EXAMPLE 1**  **Identifying functions**  State whether each graph in **Figure 1.3** represents a function.

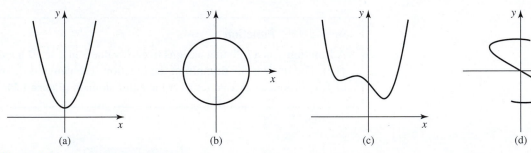

**Figure 1.3**

**SOLUTION**  The vertical line test indicates that only graphs (a) and (c) represent functions. In graphs (b) and (d), there are vertical lines that intersect the graph more than once. Equivalently, there are values of $x$ that correspond to more than one value of $y$. Therefore, graphs (b) and (d) do not pass the vertical line test and do not represent functions.

*Related Exercise 3* ◄

**EXAMPLE 2**  **Domain and range**  Determine the domain and range of each function.

**a.** $f(x) = x^2 + 1$     **b.** $g(x) = \sqrt{4 - x^2}$     **c.** $h(x) = \dfrac{x^2 - 3x + 2}{x - 1}$

**Figure 1.4**

**Figure 1.5**

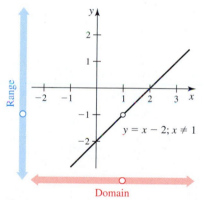

**Figure 1.6**

**SOLUTION**

**a.** Note that $f$ is defined for all values of $x$; therefore, its domain is the set of all real numbers, written $(-\infty, \infty)$ or $\mathbb{R}$. Because $x^2 \geq 0$ for all $x$, it follows that $x^2 + 1 \geq 1$, which implies that the range of $f$ is $[1, \infty)$. **Figure 1.4** shows the graph of $f$ along with its domain and range.

**b.** Functions involving square roots are defined provided the quantity under the root is nonnegative (additional restrictions may also apply). In this case, the function $g$ is defined provided $4 - x^2 \geq 0$, which means $x^2 \leq 4$, or $-2 \leq x \leq 2$. Therefore, the domain of $g$ is $[-2, 2]$. The graph of $g(x) = \sqrt{4 - x^2}$ is the upper half of a circle centered at the origin with radius 2 (**Figure 1.5**; see Appendix B, online at goo.gl/6DCbbM). From the graph we see that the range of $g$ is $[0, 2]$.

**c.** The function $h$ is defined for all values of $x \neq 1$, so its domain is $\{x: x \neq 1\}$. Factoring the numerator, we find that

$$h(x) = \frac{x^2 - 3x + 2}{x - 1} = \frac{(x - 1)(x - 2)}{x - 1} = x - 2, \text{ provided } x \neq 1.$$

The graph of $y = h(x)$, shown in **Figure 1.6**, is identical to the graph of the line $y = x - 2$ except that it has a hole at $(1, -1)$ because $h$ is undefined at $x = 1$. Therefore, the range of $h$ is $\{y: y \neq -1\}$.   *Related Exercises 23, 25* ◄

**EXAMPLE 3** **Domain and range in context** At time $t = 0$, a stone is thrown vertically upward from the ground at a speed of 30 m/s. Its height $h$ above the ground in meters (neglecting air resistance) is approximated by the function $f(t) = 30t - 5t^2$, where $t$ is measured in seconds. Find the domain and range of $f$ in the context of this particular problem.

**SOLUTION** Although $f$ is defined for all values of $t$, the only relevant times are between the time the stone is thrown ($t = 0$) and the time it strikes the ground, when $h = 0$. Solving the equation $h = 30t - 5t^2 = 0$, we find that

$$30t - 5t^2 = 0$$
$$5t(6 - t) = 0 \qquad \text{Factor.}$$
$$5t = 0 \quad \text{or} \quad 6 - t = 0 \qquad \text{Set each factor equal to 0.}$$
$$t = 0 \quad \text{or} \quad t = 6. \qquad \text{Solve.}$$

Therefore, the stone leaves the ground at $t = 0$ and returns to the ground at $t = 6$. An appropriate domain that fits the context of this problem is $\{t: 0 \leq t \leq 6\}$. The range consists of all values of $h = 30t - 5t^2$ as $t$ varies over $[0, 6]$. The largest value of $h$ occurs when the stone reaches its highest point at $t = 3$ (halfway through its flight), which is $h = f(3) = 45$. Therefore, the range is $[0, 45]$. These observations are confirmed by the graph of the height function (**Figure 1.7**). Note that this graph is *not* the trajectory of the stone; the stone moves vertically.

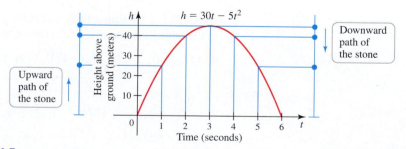

**Figure 1.7**   *Related Exercises 8–9* ◄

**QUICK CHECK 2** State the domain and range of $f(x) = (x^2 + 1)^{-1}$. ◄

## Composite Functions

Functions may be combined using sums $(f + g)$, differences $(f - g)$, products $(fg)$, or quotients $(f/g)$. The process called *composition* also produces new functions.

➤ In the composition $y = f(g(x))$, $f$ is the *outer function* and $g$ is the *inner function*.

> **DEFINITION Composite Functions**
>
> Given two functions $f$ and $g$, the composite function $f \circ g$ is defined by $(f \circ g)(x) = f(g(x))$. It is evaluated in two steps: $y = f(u)$, where $u = g(x)$. The domain of $f \circ g$ consists of all $x$ in the domain of $g$ such that $u = g(x)$ is in the domain of $f$ (**Figure 1.8**).

➤ Three different notations for intervals on the real number line will be used throughout the text:
- $[-2, 3)$ is an example of interval notation,
- $-2 \leq x < 3$ is inequality notation, and
- $\{x: -2 \leq x < 3\}$ is set notation.

(a)

(b)

**Figure 1.8**

**Figure 1.9**

**EXAMPLE 4   Using graphs to evaluate composite functions**  Use the graphs of $f$ and $g$ in **Figure 1.9** to find the following values.

**a.** $f(g(3))$     **b.** $g(f(3))$     **c.** $f(f(4))$     **d.** $f(g(f(8)))$

**SOLUTION**

**a.** The graphs indicate that $g(3) = 4$ and $f(4) = 8$, so $f(g(3)) = f(4) = 8$.

**b.** We see that $g(f(3)) = g(5) = 1$. Observe that $f(g(3)) \neq g(f(3))$.

**c.** In this case, $f(\underbrace{f(4)}_{8}) = f(8) = 6$.

**d.** Starting on the inside,

$$f(g(\underbrace{f(8)}_{6})) = f(\underbrace{g(6)}_{1}) = f(1) = 6.$$

*Related Exercise 15* ◀

**EXAMPLE 5   Using a table to evaluate composite functions**  Use the function values in the table to evaluate the following composite functions.

**a.** $(f \circ g)(0)$     **b.** $g(f(-1))$     **c.** $f(g(g(-1)))$

| $x$ | $-2$ | $-1$ | $0$ | $1$ | $2$ |
|---|---|---|---|---|---|
| $f(x)$ | $0$ | $1$ | $3$ | $4$ | $2$ |
| $g(x)$ | $-1$ | $0$ | $-2$ | $-3$ | $-4$ |

**SOLUTION**

**a.** Using the table, we see that $g(0) = -2$ and $f(-2) = 0$. Therefore, $(f \circ g)(0) = 0$.

**b.** Because $f(-1) = 1$ and $g(1) = -3$, it follows that $g(f(-1)) = -3$.

**c.** Starting with the inner function,

$$f(g(g(-1))) = f(\underbrace{g(0)}_{-2}) = f(-2) = 0.$$

*Related Exercise 16* ◀

**EXAMPLE 6**    **Composite functions and notation**  Let $f(x) = 3x^2 - x$ and $g(x) = 1/x$. Simplify the following expressions.

**a.** $f(5p + 1)$     **b.** $g(1/x)$     **c.** $f(g(x))$     **d.** $g(f(x))$

**SOLUTION**  In each case, the functions work on their arguments.

**a.** The argument of $f$ is $5p + 1$, so

$$f(5p + 1) = 3(5p + 1)^2 - (5p + 1) = 75p^2 + 25p + 2.$$

**b.** Because $g$ requires taking the reciprocal of the argument, we take the reciprocal of $1/x$ and find that $g(1/x) = 1/(1/x) = x$.

**c.** The argument of $f$ is $g(x)$, so

$$f(g(x)) = f\left(\frac{1}{x}\right) = 3\left(\frac{1}{x}\right)^2 - \left(\frac{1}{x}\right) = \frac{3}{x^2} - \frac{1}{x} = \frac{3 - x}{x^2}.$$

> ➤ Examples 6c and 6d demonstrate that, in general,
>
> $$f(g(x)) \neq g(f(x)).$$

**d.** The argument of $g$ is $f(x)$, so

$$g(f(x)) = g(3x^2 - x) = \frac{1}{3x^2 - x}.$$

*Related Exercises 33–37* ◄

**EXAMPLE 7**    **Working with composite functions**  Identify possible choices for the inner and outer functions in the following composite functions. Give the domain of the composite function.

**a.** $h(x) = \sqrt{9x - x^2}$     **b.** $h(x) = \dfrac{2}{(x^2 - 1)^3}$

**SOLUTION**

**a.** An obvious outer function is $f(x) = \sqrt{x}$, which works on the inner function $g(x) = 9x - x^2$. Therefore, $h$ can be expressed as $h = f \circ g$ or $h(x) = f(g(x))$. The domain of $f \circ g$ consists of all values of $x$ such that $9x - x^2 \geq 0$. Solving this inequality gives $\{x : 0 \leq x \leq 9\}$ as the domain of $f \circ g$.

> ➤ Techniques for solving inequalities are discussed in Appendix B, online at goo.gl/6DCbbM.

**b.** A good choice for an outer function is $f(x) = 2/x^3 = 2x^{-3}$, which works on the inner function $g(x) = x^2 - 1$. Therefore, $h$ can be expressed as $h = f \circ g$ or $h(x) = f(g(x))$. The domain of $f \circ g$ consists of all values of $g(x)$ such that $g(x) \neq 0$, which is $\{x : x \neq \pm 1\}$.

*Related Exercises 44–45* ◄

**EXAMPLE 8**    **More composite functions**  Given $f(x) = \sqrt[3]{x}$ and $g(x) = x^2 - x - 6$, find the following composite functions and their domains.

**a.** $g \circ f$     **b.** $g \circ g$

**SOLUTION**

**a.** We have

$$(g \circ f)(x) = g(f(x)) = g\left(\sqrt[3]{x}\right) = \left(\underbrace{\sqrt[3]{x}}_{f(x)}\right)^2 - \underbrace{\sqrt[3]{x}}_{f(x)} - 6 = x^{2/3} - x^{1/3} - 6.$$

Because the domains of $f$ and $g$ are $(-\infty, \infty)$, the domain of $f \circ g$ is also $(-\infty, \infty)$.

**b.** In this case, we have the composition of two polynomials:

$$(g \circ g)(x) = g(g(x))$$
$$= g(x^2 - x - 6)$$
$$= (\underbrace{x^2 - x - 6}_{g(x)})^2 - (\underbrace{x^2 - x - 6}_{g(x)}) - 6$$
$$= x^4 - 2x^3 - 12x^2 + 13x + 36.$$

**QUICK CHECK 3**  If $f(x) = x^2 + 1$ and $g(x) = x^2$, find $f \circ g$ and $g \circ f$. ◄

The domain of the composition of two polynomials is $(-\infty, \infty)$.

*Related Exercises 47–48* ◄

$$m_{\text{sec}} = \frac{f(x+h) - f(x)}{h}$$

**Figure 1.10**

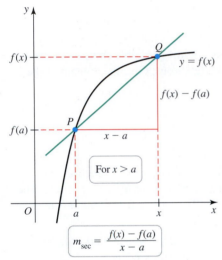

$$m_{\text{sec}} = \frac{f(x) - f(a)}{x - a}$$

**Figure 1.11**

▶ Treat $f(x + h)$ like the composition $f(g(x))$, where $x + h$ plays the role of $g(x)$. It may help to establish a pattern in your mind before evaluating $f(x + h)$. For instance, using the function in Example 9a, we have

$$f(x) = 3x^2 - x;$$
$$f(12) = 3 \cdot 12^2 - 12;$$
$$f(b) = 3b^2 - b;$$
$$f(\text{math}) = 3 \cdot \text{math}^2 - \text{math};$$

therefore,

$$f(x + h) = 3(x + h)^2 - (x + h).$$

▶ See the front papers of this text for a review of factoring formulas.

## Secant Lines and the Difference Quotient

As you will see shortly, slopes of lines and curves play a fundamental role in calculus. **Figure 1.10** shows two points $P(x, f(x))$ and $Q(x + h, f(x + h))$ on the graph of $y = f(x)$ in the case that $h > 0$. A line through any two points on a curve is called a **secant line**; its importance in the study of calculus is explained in Chapters 2 and 3. For now, we focus on the slope of the secant line through $P$ and $Q$, which is denoted $m_{\text{sec}}$ and is given by

$$m_{\text{sec}} = \frac{\text{change in } y}{\text{change in } x} = \frac{f(x+h) - f(x)}{(x+h) - x} = \frac{f(x+h) - f(x)}{h}.$$

The slope formula $\dfrac{f(x+h) - f(x)}{h}$ is also known as a **difference quotient**, and it can be expressed in several ways depending on how the coordinates of $P$ and $Q$ are labeled. For example, given the coordinates $P(a, f(a))$ and $Q(x, f(x))$ (**Figure 1.11**), the difference quotient is

$$m_{\text{sec}} = \frac{f(x) - f(a)}{x - a}.$$

We interpret the slope of the secant line in this form as the **average rate of change** of $f$ over the interval $[a, x]$.

### EXAMPLE 9   Working with difference quotients

**a.** Simplify the difference quotient $\dfrac{f(x+h) - f(x)}{h}$, for $f(x) = 3x^2 - x$.

**b.** Simplify the difference quotient $\dfrac{f(x) - f(a)}{x - a}$, for $f(x) = x^3$.

### SOLUTION

**a.** First note that $f(x + h) = 3(x + h)^2 - (x + h)$. We substitute this expression into the difference quotient and simplify:

$$\frac{f(x+h) - f(x)}{h} = \frac{\overbrace{3(x+h)^2 - (x+h)}^{f(x+h)} - \overbrace{(3x^2 - x)}^{f(x)}}{h}$$

$$= \frac{3(x^2 + 2xh + h^2) - (x+h) - (3x^2 - x)}{h} \qquad \text{Expand } (x+h)^2.$$

$$= \frac{3x^2 + 6xh + 3h^2 - x - h - 3x^2 + x}{h} \qquad \text{Distribute.}$$

$$= \frac{6xh + 3h^2 - h}{h} \qquad \text{Simplify.}$$

$$= \frac{h(6x + 3h - 1)}{h} = 6x + 3h - 1. \qquad \text{Factor and simplify.}$$

**b.** The factoring formula for the difference of perfect cubes is needed:

$$\frac{f(x) - f(a)}{x - a} = \frac{x^3 - a^3}{x - a}$$

$$= \frac{(x - a)(x^2 + ax + a^2)}{x - a} \qquad \text{Factoring formula}$$

$$= x^2 + ax + a^2. \qquad \text{Simplify.}$$

*Related Exercises 66, 72* ◀

**EXAMPLE 10  Interpreting the slope of the secant line**  The position of a hiker on a trail at various times $t$ is recorded by a GPS watch worn by the hiker. These data are then uploaded to a computer to produce the graph of the distance function $d = f(t)$ shown in Figure 1.12, where $d$ measures the distance traveled on the trail in miles and $t$ is the elapsed time in hours from the beginning of the hike.

**a.** Find the slope of the secant line that passes through the points on the graph corresponding to the trail segment between milepost 3 and milepost 5, and interpret the result.

**b.** Estimate the slope of the secant line that passes through points $A$ and $B$ in Figure 1.12, and compare it to the slope of the secant line found in part (a).

➤ Figure 1.12 contains actual GPS data collected in Rocky Mountain National Park. See Exercises 75–76 for another look at the data set.

**Figure 1.12**

**SOLUTION**

**a.** We see from the graph of $d = f(t)$ that 1.76 hours (about 1 hour and 46 minutes) has elapsed when the hiker arrives at milepost 3, while milepost 5 is reached 3.33 hours into the hike. This information is also expressed as $f(1.76) = 3$ and $f(3.33) = 5$. To find the slope of the secant line through these points, we compute the change in distance divided by the change in time:

$$m_{\text{sec}} = \frac{f(3.33) - f(1.76)}{3.33 - 1.76} = \frac{5 - 3}{3.33 - 1.76} \approx 1.3 \frac{\text{mi}}{\text{hr}}.$$

The units provide a clue about the physical meaning of the slope: It measures the average rate at which the distance changes per hour, which is the average speed of the hiker. In this case, the hiker walks with an average speed of approximately 1.3 mi/hr between mileposts 3 and 5.

**b.** From the graph we see that the coordinates of points $A$ and $B$ are approximately $(4.2, 5.3)$ and $(4.4, 5.8)$, respectively, which implies the hiker walks $5.8 - 5.3 = 0.5$ mi in $4.4 - 4.2 = 0.2$ hr. The slope of the secant line through $A$ and $B$ is

$$m_{\text{sec}} = \frac{\text{change in } d}{\text{change in } t} \approx \frac{0.5}{0.2} = 2.5 \frac{\text{mi}}{\text{hr}}.$$

For this segment of the trail, the hiker walks at an average speed of about 2.5 mi/hr, nearly twice as fast as the average speed computed in part (a). Expressed another way, steep sections of the distance curve yield steep secant lines, which correspond to faster average hiking speeds. Conversely, any secant line with slope equal to 0 corresponds to an average speed of 0. Looking one last time at Figure 1.12, we can identify the time intervals during which the hiker was resting alongside the trail—whenever the distance curve is horizontal, the hiker is not moving.

*Related Exercise 75* ◄

**QUICK CHECK 4**  Refer to Figure 1.12. Find the hiker's average speed during the first mile of the trail and then determine the hiker's average speed in the time interval from 3.9 to 4.1 hours. ◄

## Symmetry

The word *symmetry* has many meanings in mathematics. Here we consider symmetries of graphs and the relations they represent. Taking advantage of symmetry often saves time and leads to insights.

---

**DEFINITION**  **Symmetry in Graphs**

A graph is **symmetric with respect to the y-axis** if whenever the point $(x, y)$ is on the graph, the point $(-x, y)$ is also on the graph. This property means that the graph is unchanged when reflected across the y-axis (**Figure 1.13a**).

A graph is **symmetric with respect to the x-axis** if whenever the point $(x, y)$ is on the graph, the point $(x, -y)$ is also on the graph. This property means that the graph is unchanged when reflected across the x-axis (**Figure 1.13b**).

A graph is **symmetric with respect to the origin** if whenever the point $(x, y)$ is on the graph, the point $(-x, -y)$ is also on the graph (**Figure 1.13c**). Symmetry about both the x- and y-axes implies symmetry about the origin, but not vice versa.

---

(a)

(b)

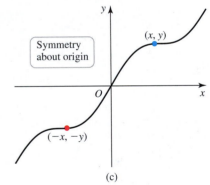

(c)

**Figure 1.13**

---

**DEFINITION**  **Symmetry in Functions**

An **even function** $f$ has the property that $f(-x) = f(x)$, for all $x$ in the domain. The graph of an even function is symmetric about the y-axis.

An **odd function** $f$ has the property that $f(-x) = -f(x)$, for all $x$ in the domain. The graph of an odd function is symmetric about the origin.

---

Polynomials consisting of only even powers of the variable (of the form $x^{2n}$, where $n$ is a nonnegative integer) are even functions. Polynomials consisting of only odd powers of the variable (of the form $x^{2n+1}$, where $n$ is a nonnegative integer) are odd functions.

**QUICK CHECK 5**  Explain why the graph of a nonzero function is never symmetric with respect to the x-axis. ◄

Even function: If $(x, y)$ is on the graph, then $(-x, y)$ is on the graph.

$y = x^4 - 2x^2 - 20$

$(-2, -12)$   $(2, -12)$

**Figure 1.14**

No symmetry: neither even nor odd function.

$y = x^3 - 3x + 1$

**Figure 1.15**

➤ The symmetry of compositions of even and odd functions is considered in Exercises 101–104.

**EXAMPLE 11** **Identifying symmetry in functions** Identify the symmetry, if any, in the following functions.

**a.** $f(x) = x^4 - 2x^2 - 20$    **b.** $g(x) = x^3 - 3x + 1$    **c.** $h(x) = \dfrac{1}{x^3 - x}$

**SOLUTION**

**a.** The function $f$ consists of only even powers of $x$ (where $20 = 20 \cdot 1 = 20x^0$ and $x^0$ is considered an even power). Therefore, $f$ is an even function (**Figure 1.14**). This fact is verified by showing that $f(-x) = f(x)$:

$$f(-x) = (-x)^4 - 2(-x)^2 - 20 = x^4 - 2x^2 - 20 = f(x).$$

**b.** The function $g$ consists of two odd powers and one even power (again, $1 = x^0$ is an even power). Therefore, we expect that $g$ has no symmetry about the $y$-axis or the origin (**Figure 1.15**). Note that

$$g(-x) = (-x)^3 - 3(-x) + 1 = -x^3 + 3x + 1,$$

so $g(-x)$ equals neither $g(x)$ nor $-g(x)$; therefore, $g$ has no symmetry.

**c.** In this case, $h$ is a composition of an odd function $f(x) = 1/x$ with an odd function $g(x) = x^3 - x$. Note that

$$h(-x) = \frac{1}{(-x)^3 - (-x)} = -\frac{1}{x^3 - x} = -h(x).$$

Because $h(-x) = -h(x)$, $h$ is an odd function (**Figure 1.16**).

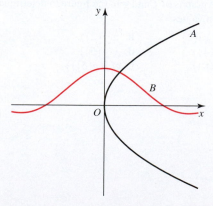

Odd function: If $(x, y)$ is on the graph, then $(-x, -y)$ is on the graph.

$y = \dfrac{1}{x^3 - x}$

$(-0.5, 2.67)$

$(1.5, 0.53)$

$(-1.5, -0.53)$

$(0.5, -2.67)$

**Figure 1.16**

*Related Exercises 79–81* ◄

---

## SECTION 1.1 EXERCISES

### Getting Started

1. Use the terms *domain*, *range*, *independent variable*, and *dependent variable* to explain how a function relates one variable to another variable.

2. Is the independent variable of a function associated with the domain or range? Is the dependent variable associated with the domain or range?

3. Decide whether graph $A$, graph $B$, or both represent functions.

**4.** The entire graph of $f$ is given. State the domain and range of $f$.

**5.** Which statement about a function is true? (i) For each value of $x$ in the domain, there corresponds one unique value of $y$ in the range; (ii) for each value of $y$ in the range, there corresponds one unique value of $x$ in the domain. Explain.

**6.** Determine the domain and range of $g(x) = \dfrac{x^2 - 1}{x - 1}$. Sketch a graph of $g$.

**7.** Determine the domain and range of $f(x) = 3x^2 - 10$.

**8.** **Throwing a stone** A stone is thrown vertically upward from the ground at a speed of 40 m/s at time $t = 0$. Its distance $d$ (in meters) above the ground (neglecting air resistance) is approximated by the function $f(t) = 40t - 5t^2$. Determine an appropriate domain for this function. Identify the independent and dependent variables.

**9.** **Water tower** A cylindrical water tower with a radius of 10 m and a height of 50 m is filled to a height of $h$ m. The volume $V$ of water (in cubic meters) is given by the function $g(h) = 100\pi h$. Identify the independent and dependent variables for this function, and then determine an appropriate domain.

**10.** Let $f(x) = 1/(x^3 + 1)$. Compute $f(2)$ and $f(y^2)$.

**11.** Let $f(x) = 2x + 1$ and $g(x) = 1/(x - 1)$. Simplify the expressions $f(g(1/2))$, $g(f(4))$, and $g(f(x))$.

**12.** Find functions $f$ and $g$ such that $f(g(x)) = (x^2 + 1)^5$. Find a different pair of functions $f$ and $g$ that also satisfy $f(g(x)) = (x^2 + 1)^5$.

**13.** Explain how to find the domain of $f \circ g$ if you know the domain and range of $f$ and $g$.

**14.** If $f(x) = \sqrt{x}$ and $g(x) = x^3 - 2$, simplify the expressions $(f \circ g)(3)$, $(f \circ f)(64)$, $(g \circ f)(x)$, and $(f \circ g)(x)$.

**15.** Use the graphs of $f$ and $g$ in the figure to determine the following function values.

    **a.** $(f \circ g)(2)$      **b.** $g(f(2))$
    **c.** $f(g(4))$      **d.** $g(f(5))$
    **e.** $f(f(8))$      **f.** $g(f(g(5)))$

**16.** Use the table to evaluate the given compositions.

| $x$ | $-1$ | $0$ | $1$ | $2$ | $3$ | $4$ |
|---|---|---|---|---|---|---|
| $f(x)$ | $3$ | $1$ | $0$ | $-1$ | $-3$ | $-1$ |
| $g(x)$ | $-1$ | $0$ | $2$ | $3$ | $4$ | $5$ |
| $h(x)$ | $0$ | $-1$ | $0$ | $3$ | $0$ | $4$ |

    **a.** $h(g(0))$      **b.** $g(f(4))$
    **c.** $h(h(0))$      **d.** $g(h(f(4)))$
    **e.** $f(f(f(1)))$      **f.** $h(h(h(0)))$
    **g.** $f(h(g(2)))$      **h.** $g(f(h(4)))$
    **i.** $g(g(g(1)))$      **j.** $f(f(h(3)))$

**17.** **Rising radiosonde** The National Weather Service releases approximately 70,000 radiosondes every year to collect data from the atmosphere. Attached to a balloon, a radiosonde rises at about 1000 ft/min until the balloon bursts in the upper atmosphere. Suppose a radiosonde is released from a point 6 ft above the ground and that 5 seconds later, it is 83 ft above the ground. Let $f(t)$ represent the height (in feet) that the radiosonde is above the ground $t$ seconds after it is released. Evaluate $\dfrac{f(5) - f(0)}{5 - 0}$ and interpret the meaning of this quotient.

**18.** **World record free fall** On October 14, 2012, Felix Baumgartner stepped off a balloon capsule at an altitude of 127,852.4 feet and began his free fall. It is claimed that Felix reached the speed of sound 34 seconds into his fall at an altitude of 109,731 feet and that he continued to fall at supersonic speed for 30 seconds until he was at an altitude of 75,330.4 feet. Let $f(t)$ equal the distance that Felix had fallen $t$ seconds after leaving his capsule. Calculate $f(0)$, $f(34)$, $f(64)$, and his average supersonic speed $\dfrac{f(64) - f(34)}{64 - 34}$ (in ft/s) over the time interval $[34, 64]$.
(*Source:* http://www.redbullstratos.com)

**19.** Suppose $f$ is an even function with $f(2) = 2$ and $g$ is an odd function with $g(2) = -2$. Evaluate $f(-2)$, $g(-2)$, $f(g(2))$, and $g(f(-2))$.

20. Complete the left half of the graph of $g$ if $g$ is an odd function.

21. State whether the functions represented by graphs $A$, $B$, and $C$ in the figure are even, odd, or neither.

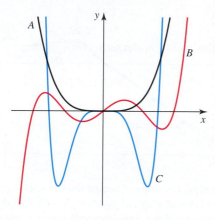

22. State whether the functions represented by graphs $A$, $B$, and $C$ in the figure are even, odd, or neither.

## Practice Exercises

**23–26. Domain and range** *State the domain and range of the function.*

23. $f(x) = \dfrac{x^2 - 5x + 6}{x - 2}$

24. $f(x) = \dfrac{x - 2}{2 - x}$

25. $f(x) = \sqrt{7 - x^2}$

26. $f(x) = -\sqrt{25 - x^2}$

**27–30. Domain** *State the domain of the function.*

27. $h(u) = \sqrt[3]{u - 1}$

28. $F(w) = \sqrt[4]{2 - w}$

29. $f(x) = (9 - x^2)^{3/2}$

30. $g(t) = \dfrac{1}{1 + t^2}$

**T 31. Launching a rocket** A small rocket is launched vertically upward from the edge of a cliff 80 ft above the ground at a speed of 96 ft/s. Its height (in feet) above the ground is given by $h(t) = -16t^2 + 96t + 80$, where $t$ represents time measured in seconds.

    **a.** Assuming the rocket is launched at $t = 0$, what is an appropriate domain for $h$?

    **b.** Graph $h$ and determine the time at which the rocket reaches its highest point. What is the height at that time?

32. **Draining a tank (Torricelli's law)** A cylindrical tank with a cross-sectional area of 10 m$^2$ is filled to a depth of 25 m with water. At $t = 0$ s, a drain in the bottom of the tank with an area of 1 m$^2$ is opened, allowing water to flow out of the tank. The depth of water in the tank (in meters) at time $t \geq 0$ is $d(t) = (5 - 0.22t)^2$.

    **a.** Check that $d(0) = 25$, as specified.

    **b.** At what time is the tank empty?

    **c.** What is an appropriate domain for $d$?

**33–42. Composite functions and notation** *Let $f(x) = x^2 - 4$, $g(x) = x^3$, and $F(x) = 1/(x - 3)$. Simplify or evaluate the following expressions.*

33. $g(1/z)$

34. $F(y^4)$

35. $F(g(y))$

36. $f(g(w))$

37. $g(f(u))$

38. $\dfrac{f(2 + h) - f(2)}{h}$

39. $F(F(x))$

40. $g(F(f(x)))$

41. $f(\sqrt{x + 4})$

42. $F\left(\dfrac{3x + 1}{x}\right)$

**43–46. Working with composite functions** *Find possible choices for outer and inner functions $f$ and $g$ such that the given function $h$ equals $f \circ g$.*

43. $h(x) = (x^3 - 5)^{10}$

44. $h(x) = \dfrac{2}{(x^6 + x^2 + 1)^2}$

45. $h(x) = \sqrt{x^4 + 2}$

46. $h(x) = \dfrac{1}{\sqrt{x^3 - 1}}$

**47–54. More composite functions** *Let $f(x) = |x|$, $g(x) = x^2 - 4$, $F(x) = \sqrt{x}$, and $G(x) = 1/(x - 2)$. Determine the following composite functions and give their domains.*

47. $f \circ g$

48. $g \circ f$

49. $f \circ G$

50. $f \circ g \circ G$

51. $G \circ g \circ f$

52. $g \circ F \circ F$

53. $g \circ g$

54. $G \circ G$

**55–60. Missing piece** *Let $g(x) = x^2 + 3$. Find a function $f$ that produces the given composition.*

55. $(f \circ g)(x) = x^2$

56. $(f \circ g)(x) = \dfrac{1}{x^2 + 3}$

57. $(f \circ g)(x) = x^4 + 6x^2 + 9$

58. $(f \circ g)(x) = x^4 + 6x^2 + 20$

59. $(g \circ f)(x) = x^4 + 3$

60. $(g \circ f)(x) = x^{2/3} + 3$

**61. Explain why or why not** Determine whether the following statements are true and give an explanation or counterexample.

  a. The range of $f(x) = 2x - 38$ is all real numbers.
  b. The relation $y = x^6 + 1$ is *not* a function because $y = 2$ for both $x = -1$ and $x = 1$.
  c. If $f(x) = x^{-1}$, then $f(1/x) = 1/f(x)$.
  d. In general, $f(f(x)) = (f(x))^2$.
  e. In general, $f(g(x)) = g(f(x))$.
  f. By definition, $f(g(x)) = (f \circ g)(x)$.
  g. If $f(x)$ is an even function, then $cf(ax)$ is an even function, where $a$ and $c$ are nonzero real numbers.
  h. If $f(x)$ is an odd function, then $f(x) + d$ is an odd function, where $d$ is a nonzero real number.
  i. If $f$ is both even *and* odd, then $f(x) = 0$ for all $x$.

**62–68. Working with difference quotients** *Simplify the difference quotient* $\dfrac{f(x + h) - f(x)}{h}$ *for the following functions.*

**62.** $f(x) = 10$

**63.** $f(x) = 3x$

**64.** $f(x) = 4x - 3$

**65.** $f(x) = x^2$

**66.** $f(x) = 2x^2 - 3x + 1$

**67.** $f(x) = \dfrac{2}{x}$

**68.** $f(x) = \dfrac{x}{x + 1}$

**69–74. Working with difference quotients** *Simplify the difference quotient* $\dfrac{f(x) - f(a)}{x - a}$ *for the following functions.*

**69.** $f(x) = x^2 + x$

**70.** $f(x) = 4 - 4x - x^2$

**71.** $f(x) = x^3 - 2x$

**72.** $f(x) = x^4$

**73.** $f(x) = -\dfrac{4}{x^2}$

**74.** $f(x) = \dfrac{1}{x} - x^2$

**75. GPS data** A GPS device tracks the elevation $E$ (in feet) of a hiker walking in the mountains. The elevation $t$ hours after beginning the hike is given in the figure.

  a. Find the slope of the secant line that passes through points $A$ and $B$. Interpret your answer as an average rate of change over the interval $1 \le t \le 3$.
  b. Repeat the procedure outlined in part (a) for the secant line that passes through points $P$ and $Q$.
  c. Notice that the curve in the figure is horizontal for an interval of time near $t = 5.5$ hr. Give a plausible explanation for the horizontal line segment.

**76. Elevation vs. Distance** The following graph, obtained from GPS data, shows the elevation of a hiker as a function of the distance $d$ from the starting point of the trail.

  a. Find the slope of the secant line that passes through points $A$ and $B$. Interpret your answer as an average rate of change over the interval $1 \le d \le 3$.
  b. Repeat the procedure outlined in part (a) for the secant line that passes through points $P$ and $Q$.
  c. Notice that the elevation function is nearly constant over the segment of the trail from mile $d = 4.5$ to mile $d = 5$. Give a plausible explanation for the horizontal line segment.

**77–78. Interpreting the slope of secant lines** *In each exercise, a function and an interval of its independent variable are given. The endpoints of the interval are associated with points P and Q on the graph of the function.*

  a. *Sketch a graph of the function and the secant line through P and Q.*
  b. *Find the slope of the secant line in part (a), and interpret your answer in terms of an average rate of change over the interval. Include units in your answer.*

**77.** After $t$ seconds, an object dropped from rest falls a distance $d = 16t^2$, where $d$ is measured in feet and $2 \le t \le 5$.

**78.** The volume $V$ of an ideal gas in cubic centimeters is given by $V = 2/p$, where $p$ is the pressure in atmospheres and $0.5 \le p \le 2$.

**79–86. Symmetry** *Determine whether the graphs of the following equations and functions are symmetric about the x-axis, the y-axis, or the origin. Check your work by graphing.*

**79.** $f(x) = x^4 + 5x^2 - 12$

**80.** $f(x) = 3x^5 + 2x^3 - x$

**81.** $f(x) = x^5 - x^3 - 2$

**82.** $f(x) = 2|x|$

**83.** $x^{2/3} + y^{2/3} = 1$

**84.** $x^3 - y^5 = 0$

**85.** $f(x) = x|x|$

**86.** $|x| + |y| = 1$

## Explorations and Challenges

**87. Composition of even and odd functions from graphs** Assume $f$ is an even function and $g$ is an odd function. Use the (incomplete) graphs of $f$ and $g$ in the figure to determine the following function values.

  a. $f(g(-2))$
  b. $g(f(-2))$
  c. $f(g(-4))$
  d. $g(f(5) - 8)$
  e. $g(g(-7))$
  f. $f(1 - f(8))$

**88. Composition of even and odd functions from tables** Assume $f$ is an even function, $g$ is an odd function, and both are defined at 0. Use the (incomplete) table to evaluate the given compositions.

| $x$ | 1 | 2 | 3 | 4 |
|---|---|---|---|---|
| $f(x)$ | 2 | $-1$ | 3 | $-4$ |
| $g(x)$ | $-3$ | $-1$ | $-4$ | $-2$ |

**a.** $f(g(-1))$      **b.** $g(f(-4))$
**c.** $f(g(-3))$      **d.** $f(g(-2))$
**e.** $g(g(-1))$      **f.** $f(g(0)-1)$
**g.** $f(g(g(-2)))$      **h.** $g(f(f(-4)))$
**i.** $g(g(g(-1)))$

**89. Absolute value graphs** Use the definition of absolute value (see Appendix B, online at goo.gl/6DCbbM) to graph the equation $|x| - |y| = 1$. Use a graphing utility to check your work.

**90. Graphing semicircles** Show that the graph of
$f(x) = 10 + \sqrt{-x^2 + 10x - 9}$ is the upper half of a circle. Then determine the domain and range of the function.

**91. Graphing semicircles** Show that the graph of
$g(x) = 2 - \sqrt{-x^2 + 6x + 16}$ is the lower half of a circle. Then determine the domain and range of the function.

**92. Even and odd at the origin**

   **a.** If $f(0)$ is defined and $f$ is an even function, is it necessarily true that $f(0) = 0$? Explain.

   **b.** If $f(0)$ is defined and $f$ is an odd function, is it necessarily true that $f(0) = 0$? Explain.

**T 93–96. Polynomial calculations** *Find a polynomial $f$ that satisfies the following properties. (Hint: Determine the degree of $f$; then substitute a polynomial of that degree and solve for its coefficients.)*

**93.** $f(f(x)) = 9x - 8$      **94.** $(f(x))^2 = 9x^2 - 12x + 4$

**95.** $f(f(x)) = x^4 - 12x^2 + 30$      **96.** $(f(x))^2 = x^4 - 12x^2 + 36$

**97–100. Difference quotients** *Simplify the difference quotients*
$$\frac{f(x+h) - f(x)}{h} \quad and \quad \frac{f(x) - f(a)}{x - a} \text{ by rationalizing the numerator.}$$

**97.** $f(x) = \sqrt{x}$      **98.** $f(x) = \sqrt{1 - 2x}$

**99.** $f(x) = -\dfrac{3}{\sqrt{x}}$      **100.** $f(x) = \sqrt{x^2 + 1}$

**101–104. Combining even and odd functions** *Let $E$ be an even function and $O$ be an odd function. Determine the symmetry, if any, of the following functions.*

**101.** $E + O$      **102.** $E \cdot O$

**103.** $O \circ E$      **104.** $E \circ O$

**QUICK CHECK ANSWERS**

**1.** $3, x^4 - 2x^2, t^2 - 2t, p^2 - 4p + 3$   **2.** Domain is all real numbers; range is $\{y : 0 < y \le 1\}$.   **3.** $(f \circ g)(x) = x^4 + 1$ and $(g \circ f)(x) = (x^2 + 1)^2$   **4.** Average speed $\approx 2.2$ mi/hr for first mile; average speed $= 0$ on $3.9 \le t \le 4.1$.   **5.** If the graph were symmetric with respect to the $x$-axis, it would not pass the vertical line test. ◀

# 1.2 Representing Functions

We consider four approaches to defining and representing functions: formulas, graphs, tables, and words.

## Using Formulas

The following list is a brief catalog of the families of functions that are introduced in this chapter and studied systematically throughout this text; they are all defined by *formulas*.

**1. Polynomials** are functions of the form

> ➤ One version of the Fundamental Theorem of Algebra states that a nonzero polynomial of degree $n$ has exactly $n$ (possibly complex) roots, counting each root up to its multiplicity.

$$p(x) = a_n x^n + a_{n-1} x^{n-1} + \cdots + a_1 x + a_0,$$

where the **coefficients** $a_0, a_1, \ldots, a_n$ are real numbers with $a_n \ne 0$ and the nonnegative integer $n$ is the **degree** of the polynomial. The domain of any polynomial is the set of all real numbers. An $n$th-degree polynomial can have as many as $n$ real **zeros** or **roots**— values of $x$ at which $p(x) = 0$; the zeros are points at which the graph of $p$ intersects the $x$-axis.

2. **Rational functions** are ratios of the form $f(x) = p(x)/q(x)$, where $p$ and $q$ are polynomials. Because division by zero is prohibited, the domain of a rational function is the set of all real numbers except those for which the denominator is zero.

3. **Algebraic functions** are constructed using the operations of algebra: addition, subtraction, multiplication, division, and roots. Examples of algebraic functions are $f(x) = \sqrt{2x^3 + 4}$ and $g(x) = x^{1/4}(x^3 + 2)$. In general, if an even root (square root, fourth root, and so forth) appears, then the domain does not contain points at which the quantity under the root is negative (and perhaps other points).

> ➤ Exponential and logarithmic functions, along with inverse trigonometric functions, are introduced in Chapter 7.

4. **Exponential functions** have the form $f(x) = b^x$, where the base $b \neq 1$ is a positive real number. Closely associated with exponential functions are **logarithmic functions** of the form $f(x) = \log_b x$, where $b > 0$ and $b \neq 1$. Exponential functions have a domain consisting of all real numbers. Logarithmic functions are defined for positive real numbers.

   The **natural exponential function** is $f(x) = e^x$, with base $b = e$, where $e \approx 2.71828\ldots$ is one of the fundamental constants of mathematics. Associated with the natural exponential function is the **natural logarithm function** $f(x) = \ln x$, which also has the base $b = e$.

5. The **trigonometric functions** are $\sin x$, $\cos x$, $\tan x$, $\cot x$, $\sec x$, and $\csc x$; they are fundamental to mathematics and many areas of application. Also important are their relatives, the **inverse trigonometric functions**.

6. Trigonometric, exponential, and logarithmic functions are a few examples of a large family called **transcendental functions**. **Figure 1.17** shows the organization of these functions, which are explored in detail in upcoming chapters.

**QUICK CHECK 1** Are all polynomials rational functions? Are all algebraic functions polynomials? ◄

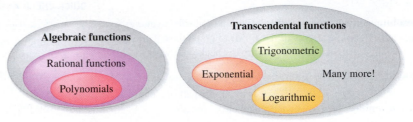

**Figure 1.17**

## Using Graphs

Although formulas are the most compact way to represent many functions, graphs often provide the most illuminating representations. Two of countless examples of functions and their graphs are shown in **Figure 1.18**. Much of this text is devoted to creating and analyzing graphs of functions.

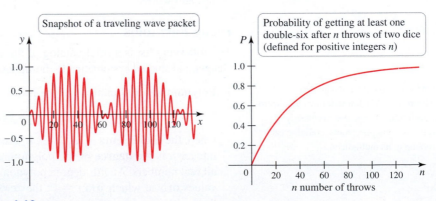

**Figure 1.18**

There are two approaches to graphing functions.

- Graphing calculators and graphing software are easy to use and powerful. Such **technology** easily produces graphs of most functions encountered in this text. We assume you know how to use a graphing utility.

- Graphing utilities, however, are not infallible. Therefore, you should also strive to master **analytical methods** (pencil-and-paper methods) in order to analyze functions and make accurate graphs by hand. Analytical methods rely heavily on calculus and are presented throughout this text.

**The important message is this**: Both technology and analytical methods are essential and must be used together in an integrated way to produce accurate graphs.

**Linear Functions**    One form of the equation of a line (see Appendix B, online at goo.gl/6DCbbM) is $y = mx + b$, where $m$ and $b$ are constants. Therefore, the function $f(x) = mx + b$ has a straight-line graph and is called a **linear function**.

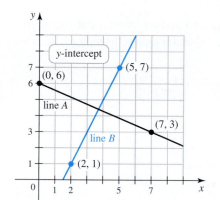

**Figure 1.19**

> In the solution to Example 1b, we used the point $(2, 1)$ to determine the value of $b$. Using the point $(5, 7)$—or, equivalently, the fact that $g(5) = 7$—leads to the same result.

**EXAMPLE 1    Linear functions and their graphs**    Determine the function represented by **(a)** line $A$ in **Figure 1.19** and **(b)** line $B$ in Figure 1.19.

**SOLUTION**

**a.** From the graph, we see that the $y$-intercept of line $A$ is $(0, 6)$. Using the points $(0, 6)$ and $(7, 3)$, we find that the slope of the line is

$$m = \frac{3 - 6}{7 - 0} = -\frac{3}{7}.$$

Therefore, the line is described by the function $f(x) = -\dfrac{3}{7}x + 6$.

**b.** Line $B$ passes through the points $(2, 1)$ and $(5, 7)$, so the slope of this line is

$$m = \frac{7 - 1}{5 - 2} = 2,$$

which implies that the line is described by a function of the form $g(x) = 2x + b$. To determine the value of $b$, note that $g(2) = 1$:

$$g(2) = 2 \cdot 2 + b = 1.$$

Solving for $b$, we find that $b = 1 - 4 = -3$. Therefore, the line is described by the function $g(x) = 2x - 3$.

*Related Exercises 3, 15–16* ◄

**EXAMPLE 2    Demand function for pizzas**    After studying sales for several months, the owner of a pizza chain knows that the number of two-topping pizzas sold in a week (called the *demand*) decreases as the price increases. Specifically, her data indicate that at a price of \$14 per pizza, an average of 400 pizzas are sold per week, while at a price of \$17 per pizza, an average of 250 pizzas are sold per week. Assume the demand $d$ is a *linear* function of the price $p$.

**a.** Find the constants $m$ and $b$ in the demand function $d = f(p) = mp + b$. Then graph $f$.

**b.** According to this model, how many pizzas (on average) are sold per week at a price of \$20?

**SOLUTION**

**a.** Two points on the graph of the demand function are given: $(p, d) = (14, 400)$ and $(17, 250)$. Therefore, the slope of the demand line is

$$m = \frac{400 - 250}{14 - 17} = -50 \text{ pizzas per dollar.}$$

The demand function $d = -50p + 1100$ is defined on the interval $0 \le p \le 22$.

$d = -50p + 1100$

**Figure 1.20**

➤ The units of the slope in Example 2 have meaning: For every dollar the price is reduced, an average of 50 more pizzas can be sold.

$(2, 3)$    $y = g(x)$
$(4, 3)$

**Figure 1.21**

$(2, 1)$

$y = \begin{cases} \dfrac{x^2 - 5x + 6}{x - 2} & \text{if } x \ne 2 \\ 1 & \text{if } x = 2 \end{cases}$

**Figure 1.22**

$y = |x|$

**Figure 1.23**

It follows that the equation of the linear demand function is

$$d - 250 = -50(p - 17).$$

Expressing $d$ as a function of $p$, we have $d = f(p) = -50p + 1100$ (**Figure 1.20**).

**b.** Using the demand function with a price of $20, the average number of pizzas that could be sold per week is $f(20) = 100$.

*Related Exercise 21* ◄

**Piecewise Functions**    A function may have different definitions on different parts of its domain. For example, income tax is levied in tax brackets that have different tax rates. Functions that have different definitions on different parts of their domain are called **piecewise functions**. If all the pieces are linear, the function is **piecewise linear**. Here are some examples.

**EXAMPLE 3**    **Defining a piecewise function**    The graph of a piecewise linear function $g$ is shown in **Figure 1.21**. Find a formula for the function.

**SOLUTION**    For $x < 2$, the graph is linear with a slope of 1 and a $y$-intercept of $(0, 0)$; its equation is $y = x$. For $x > 2$, the slope of the line is $-\frac{1}{2}$ and it passes through $(4, 3)$; an equation of this piece of the function is

$$y - 3 = -\frac{1}{2}(x - 4) \quad \text{or} \quad y = -\frac{1}{2}x + 5.$$

For $x = 2$, we have $g(2) = 3$. Therefore,

$$g(x) = \begin{cases} x & \text{if } x < 2 \\ 3 & \text{if } x = 2 \\ -\dfrac{1}{2}x + 5 & \text{if } x > 2. \end{cases}$$

*Related Exercises 25–26* ◄

**EXAMPLE 4**    **Graphing piecewise functions**    Graph the following functions.

**a.** $f(x) = \begin{cases} \dfrac{x^2 - 5x + 6}{x - 2} & \text{if } x \ne 2 \\ 1 & \text{if } x = 2 \end{cases}$

**b.** $f(x) = |x|$, the **absolute value** function

**SOLUTION**

**a.** The function $f$ is simplified by factoring and then canceling $x - 2$, assuming $x \ne 2$:

$$\frac{x^2 - 5x + 6}{x - 2} = \frac{(x - 2)(x - 3)}{x - 2} = x - 3.$$

Therefore, the graph of $f$ is identical to the graph of the line $y = x - 3$ when $x \ne 2$. We are given that $f(2) = 1$ (**Figure 1.22**).

**b.** The absolute value of a real number is defined as

$$f(x) = |x| = \begin{cases} x & \text{if } x \ge 0 \\ -x & \text{if } x < 0. \end{cases}$$

Graphing $y = -x$, for $x < 0$, and $y = x$, for $x \ge 0$, produces the graph in **Figure 1.23**.

*Related Exercises 29–30* ◄

**Power Functions**    Power functions are a special case of polynomials; they have the form $f(x) = x^n$, where $n$ is a positive integer. When $n$ is an even integer, the function values are nonnegative and the graph passes through the origin, opening upward (**Figure 1.24**). For odd integers, the power function $f(x) = x^n$ has values that are positive when $x$ is positive and negative when $x$ is negative (**Figure 1.25**).

**Figure 1.24**                    **Figure 1.25**

**QUICK CHECK 2** What is the range of $f(x) = x^7$? What is the range of $f(x) = x^8$? ◄

**Root Functions**    Root functions are a special case of algebraic functions; they have the form $f(x) = x^{1/n}$, where $n > 1$ is a positive integer. Notice that when $n$ is even (square roots, fourth roots, and so forth), the domain and range consist of nonnegative numbers. Their graphs begin steeply at the origin and flatten out as $x$ increases (**Figure 1.26**).

By contrast, the odd root functions (cube roots, fifth roots, and so forth) are defined for all real values of $x$, and their range is all real numbers. Their graphs pass through the origin, open upward for $x < 0$ and downward for $x > 0$, and flatten out as $x$ increases in magnitude (**Figure 1.27**).

➤ Recall that if $n$ is a positive integer, then $x^{1/n}$ is the $n$th root of $x$; that is, $f(x) = x^{1/n} = \sqrt[n]{x}$.

**QUICK CHECK 3** What are the domain and range of $f(x) = x^{1/7}$? What are the domain and range of $f(x) = x^{1/10}$? ◄

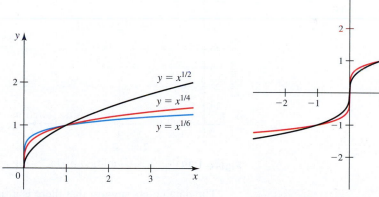

**Figure 1.26**                    **Figure 1.27**

**Rational Functions**    Rational functions appear frequently in this text, and much is said later about graphing rational functions. The following example illustrates how analysis and technology work together.

**EXAMPLE 5**    **Technology and analysis**  Consider the rational function

$$f(x) = \frac{3x^3 - x - 1}{x^3 + 2x^2 - 6}.$$

**a.** What is the domain of $f$?

**b.** Find the roots (zeros) of $f$.

**c.** Graph the function using a graphing utility.

$$y = \frac{3x^3 - x - 1}{x^3 + 2x^2 - 6}$$

**Figure 1.28**

> In Chapter 4, we show how calculus is used to locate the local maximum and local minimum values of a function.

**d.** At what points does the function have peaks and valleys?

**e.** How does $f$ behave as $x$ grows large in magnitude?

**SOLUTION**

**a.** The domain consists of all real numbers except those at which the denominator is zero. A graphing utility shows that the denominator has one real zero at $x \approx 1.34$ and therefore, the domain of $f$ is $\{x: x \neq 1.34\}$.

**b.** The roots of a rational function are the roots of the numerator, provided they are not also roots of the denominator. Using a graphing utility, the only real root of the numerator is $x \approx 0.85$.

**c.** After experimenting with the graphing window, a reasonable graph of $f$ is obtained (**Figure 1.28**). Because the denominator is zero at $x \approx 1.34$, the function becomes large in magnitude at nearby points, and $f$ has a *vertical asymptote*.

**d.** The function has two peaks (soon to be called *local maxima*), one near $x = -3.0$ and one near $x = 0.4$. The function also has two valleys (soon to be called *local minima*), one near $x = -0.3$ and one near $x = 2.6$.

**e.** By zooming out, it appears that as $x$ increases in the positive direction, the graph approaches the *horizontal asymptote* $y = 3$ from below, and as $x$ becomes large and negative, the graph approaches $y = 3$ from above.

*Related Exercises 35–36* ◄

## Using Tables

Sometimes functions do not originate as formulas or graphs; they may start as numbers or data. For example, suppose you do an experiment in which a marble is dropped into a cylinder filled with heavy oil and is allowed to fall freely. You measure the total distance $d$, in centimeters, that the marble falls at times $t = 0, 1, 2, 3, 4, 5, 6$, and 7 seconds after it is dropped (Table 1.1). The first step might be to plot the data points (**Figure 1.29**).

**Table 1.1**

| $t$ (s) | $d$ (cm) |
|---------|----------|
| 0 | 0 |
| 1 | 2 |
| 2 | 6 |
| 3 | 14 |
| 4 | 24 |
| 5 | 34 |
| 6 | 44 |
| 7 | 54 |

**Figure 1.29**

**Figure 1.30**

The data points suggest that there is a function $d = f(t)$ that gives the distance that the marble falls at *all* times of interest. Because the marble falls through the oil without abrupt changes, a smooth graph passing through the data points (**Figure 1.30**) is reasonable. Finding the function that best fits the data is a more difficult problem, which we discuss later in the text.

## Using Words

Using words may be the least mathematical way to define functions, but it is often the way in which functions originate. Once a function is defined in words, it can often be tabulated, graphed, or expressed as a formula.

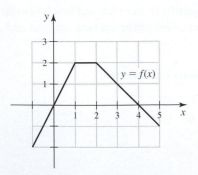

**Figure 1.31**

**EXAMPLE 6**   **A slope function**  Let $S$ be the **slope function** for a given function $f$. In words, this means that $S(x)$ is the slope of the curve $y = f(x)$ at the point $(x, f(x))$. Find and graph the slope function for the function $f$ in **Figure 1.31**.

$$S(x) = \begin{cases} 2 & \text{if } x < 1 \\ 0 & \text{if } 1 < x < 2 \\ -1 & \text{if } x > 2 \end{cases}$$

**Figure 1.32**

$$f(t) = \begin{cases} 2t & \text{if } 0 \le t \le 3 \\ 6 & \text{if } t > 3 \end{cases}$$

**Figure 1.33**

▶ Slope functions and area functions reappear in upcoming chapters and play an essential part in calculus.

**SOLUTION** For $x < 1$, the slope of $y = f(x)$ is 2. The slope is 0 for $1 < x < 2$, and the slope is $-1$ for $x > 2$. At $x = 1$ and $x = 2$, the graph of $f$ has a corner, so the slope is undefined at these points. Therefore, the domain of $S$ is the set of all real numbers except $x = 1$ and $x = 2$, and the slope function (**Figure 1.32**) is defined by the piecewise function

$$S(x) = \begin{cases} 2 & \text{if } x < 1 \\ 0 & \text{if } 1 < x < 2 \\ -1 & \text{if } x > 2. \end{cases}$$

*Related Exercises 47–48* ◀

**EXAMPLE 7  An area function** Let $A$ be an **area function** for a positive function $f$. In words, this means that $A(x)$ is the area of the region bounded by the graph of $f$ and the $t$-axis from $t = 0$ to $t = x$. Consider the function (**Figure 1.33**)

$$f(t) = \begin{cases} 2t & \text{if } 0 \le t \le 3 \\ 6 & \text{if } t > 3. \end{cases}$$

**a.** Find $A(2)$ and $A(5)$.

**b.** Find a piecewise formula for the area function for $f$.

**SOLUTION**

**a.** The value of $A(2)$ is the area of the shaded region between the graph of $f$ and the $t$-axis from $t = 0$ to $t = 2$ (**Figure 1.34a**). Using the formula for the area of a triangle,

$$A(2) = \frac{1}{2}(2)(4) = 4.$$

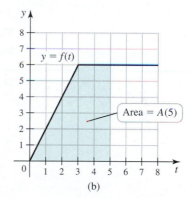

**Figure 1.34**

The value of $A(5)$ is the area of the shaded region between the graph of $f$ and the $t$-axis on the interval $[0, 5]$ (**Figure 1.34b**). This area equals the area of the triangle whose base is the interval $[0, 3]$ plus the area of the rectangle whose base is the interval $[3, 5]$:

$$A(5) = \underbrace{\frac{1}{2}(3)(6)}_{\substack{\text{area of the}\\\text{triangle}}} + \underbrace{(2)(6)}_{\substack{\text{area of the}\\\text{rectangle}}} = 21.$$

**b.** For $0 \le x \le 3$ (**Figure 1.35a**), $A(x)$ is the area of the triangle whose base is the interval $[0, x]$. Because the height of the triangle at $t = x$ is $f(x)$,

$$A(x) = \frac{1}{2}x\,f(x) = \frac{1}{2}x\underbrace{(2x)}_{f(x)} = x^2.$$

**Figure 1.35**

$$A(x) = \begin{cases} x^2 & \text{if } 0 \le x \le 3 \\ 6x - 9 & \text{if } x > 3 \end{cases}$$

**Figure 1.36**

For $x > 3$ (**Figure 1.35b**), $A(x)$ is the area of the triangle on the interval $[0, 3]$ plus the area of the rectangle on the interval $[3, x]$:

$$A(x) = \underbrace{\frac{1}{2}(3)(6)}_{\substack{\text{area of the} \\ \text{triangle}}} + \underbrace{(x - 3)(6)}_{\substack{\text{area of the} \\ \text{rectangle}}} = 6x - 9.$$

Therefore, the area function $A$ (**Figure 1.36**) has the piecewise definition

$$A(x) = \begin{cases} x^2 & \text{if } 0 \le x \le 3 \\ 6x - 9 & \text{if } x > 3. \end{cases}$$

*Related Exercises 51–52* ◄

## Transformations of Functions and Graphs

There are several ways to transform the graph of a function to produce graphs of new functions. Four transformations are common: *shifts* in the $x$- and $y$-directions and *scalings* in the $x$- and $y$-directions. These transformations, summarized in **Figures 1.37–1.42**, can save time in graphing and visualizing functions.

The graph of $y = f(x) + d$ is the graph of $y = f(x)$ shifted vertically by $d$ units: up if $d > 0$ and down if $d < 0$.

The graph of $y = f(x - b)$ is the graph of $y = f(x)$ shifted horizontally by $b$ units: right if $b > 0$ and left if $b < 0$.

**Figure 1.37**

**Figure 1.38**

For $c > 0$ and $c \neq 1$, the graph of $y = cf(x)$ is the graph of $y = f(x)$ scaled vertically by a factor of $c$: compressed if $0 < c < 1$ and stretched if $c > 1$.

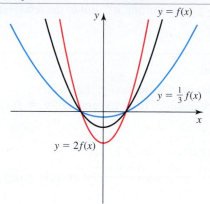

**Figure 1.39**

For $c < 0$, the graph of $y = cf(x)$ is the graph of $y = f(x)$ reflected across the $x$-axis and scaled vertically (if $c \neq -1$) by a factor of $|c|$: compressed if $0 < |c| < 1$ and stretched if $|c| > 1$.

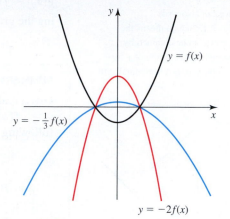

**Figure 1.40**

For $a > 0$ and $a \neq 1$, the graph of $y = f(ax)$ is the graph $y = f(x)$ scaled horizontally by a factor of $1/a$: compressed if $0 < 1/a < 1$ and stretched if $1/a > 1$.

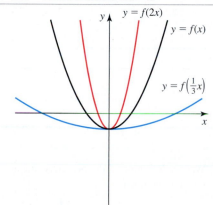

**Figure 1.41**

For $a < 0$, the graph of $y = f(ax)$ is the graph of $y = f(x)$ reflected across the $y$-axis and scaled horizontally (if $a \neq -1$) by a factor of $1/|a|$: compressed if $0 < 1/|a| < 1$ and stretched if $1/|a| > 1$.

**Figure 1.42**

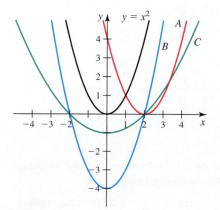

**Figure 1.43**

➤ You should verify that graph $C$ also corresponds to a horizontal stretch and a vertical shift. It has the equation $y = f(ax) - 1$, where $a = \frac{1}{2}$.

**EXAMPLE 8**  **Transforming parabolas**  The graphs $A$, $B$, and $C$ in **Figure 1.43** are obtained from the graph of $f(x) = x^2$ using shifts and scalings. Find the function that describes each graph.

**SOLUTION**

**a.** Graph $A$ is the graph of $f$ shifted to the right by 2 units. It represents the function

$$f(x - 2) = (x - 2)^2 = x^2 - 4x + 4.$$

**b.** Graph $B$ is the graph of $f$ shifted down by 4 units. It represents the function

$$f(x) - 4 = x^2 - 4.$$

**c.** Graph $C$ is a vertical scaling of the graph of $f$ shifted down by 1 unit. Therefore, it represents $cf(x) - 1 = cx^2 - 1$, for some value of $c$, with $0 < c < 1$ (because the graph is vertically compressed). Using the fact that graph $C$ passes through the points $(\pm 2, 0)$, we find that $c = \frac{1}{4}$. Therefore, the graph represents

$$y = \frac{1}{4} f(x) - 1 = \frac{1}{4} x^2 - 1.$$

*Related Exercises 55, 58* ◀

**QUICK CHECK 4** How do you modify the graph of $f(x) = 1/x$ to produce the graph of $g(x) = 1/(x + 4)$? ◄

➤ Note that we can also write $g(x) = 2\left|x + \frac{1}{2}\right|$, which means the graph of $g$ may also be obtained by vertically stretching the graph of $f$ by a factor of 2, followed by a horizontal shift.

Step 1: Horizontal compression
$y = |2x|$

$y = |2x + 1|$

Basic curve
$y = |x|$

Step 2: Horizontal shift
$y = \left|2\left(x + \frac{1}{2}\right)\right|$
$= |2x + 1|$

**Figure 1.44**

**EXAMPLE 9**  **Scaling and shifting**  Graph $g(x) = |2x + 1|$.

**SOLUTION**  We write the function as $g(x) = \left|2\left(x + \frac{1}{2}\right)\right|$. Letting $f(x) = |x|$, we have $g(x) = f\left(2\left(x + \frac{1}{2}\right)\right)$. Therefore, the graph of $g$ is obtained by horizontally compressing the graph of $f$ by a factor of $\frac{1}{2}$ and shifting it $\frac{1}{2}$ unit to the left (**Figure 1.44**).

*Related Exercise 64* ◄

---

**SUMMARY**  **Transformations**

Given real numbers $a$, $b$, $c$, and $d$ and a function $f$, the graph of $y = cf(a(x - b)) + d$ can be obtained from the graph of $y = f(x)$ in the following steps.

$$y = f(x) \xrightarrow{\substack{\textit{horizontal scaling by a factor of } 1/|a| \\ \textit{(and reflection across the y-axis if } a < 0)}} y = f(ax)$$

$$\xrightarrow{\substack{\textit{horizontal shift} \\ \textit{by b units}}} y = f(a(x - b))$$

$$\xrightarrow{\substack{\textit{vertical scaling by a factor of } |c| \textit{ (and} \\ \textit{reflection across the x-axis if } c < 0)}} y = cf(a(x - b))$$

$$\xrightarrow{\substack{\textit{vertical shift} \\ \textit{by d units}}} y = cf(a(x - b)) + d$$

---

## SECTION 1.2 EXERCISES

### Getting Started

1. Give four ways in which functions may be defined and represented.

2. What is the domain of a polynomial?

3. Determine the function $f$ represented by the graph of the line $y = f(x)$ in the figure.

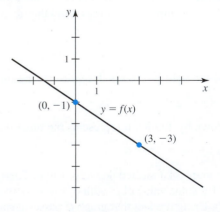

$(0, -1)$     $y = f(x)$

$(3, -3)$

4. Determine the linear function $g$ whose graph is parallel to the line $y = 2x + 1$ and passes through the point $(5, 0)$.

5. What is the domain of a rational function?

6. What is a piecewise linear function?

7. Write a definition of the piecewise linear function $y = f(x)$ that is given in the graph.

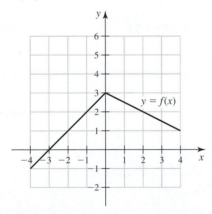

$y = f(x)$

8. The graph of $y = \sqrt{x}$ is shifted 2 units to the right and 3 units up. Write an equation for this transformed graph.

9. How do you obtain the graph of $y = f(x + 2)$ from the graph of $y = f(x)$?

10. How do you obtain the graph of $y = -3f(x)$ from the graph of $y = f(x)$?

11. How do you obtain the graph of $y = f(3x)$ from the graph of $y = f(x)$?

12. How do you obtain the graph of $y = 4(x + 3)^2 + 6$ from the graph of $y = x^2$?

13. The graphs of the functions $f$ and $g$ in the figure are obtained by vertical and horizontal shifts and scalings of $y = |x|$. Find formulas for $f$ and $g$. Verify your answers with a graphing utility.

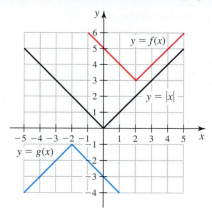

14. **Transformations** Use the graph of $f$ in the figure to plot the following functions.

    **a.** $y = -f(x)$      **b.** $y = f(x + 2)$
    **c.** $y = f(x - 2)$      **d.** $y = f(2x)$
    **e.** $y = f(x - 1) + 2$      **f.** $y = 2f(x)$

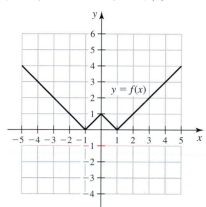

## Practice Exercises

15. **Graph of a linear function** Find and graph the linear function that passes through the points $(1, 3)$ and $(2, 5)$.

16. **Graph of a linear function** Find and graph the linear function that passes through the points $(2, -3)$ and $(5, 0)$.

17. **Linear function** Find the linear function whose graph passes though the point $(3, 2)$ and is parallel to the line $y = 3x + 8$.

18. **Linear function** Find the linear function whose graph passes though the point $(-1, 4)$ and is perpendicular to the line $y = \frac{1}{4}x - 7$.

**19–20. Yeast growth** *Consider a colony of yeast cells that has the shape of a cylinder. As the number of yeast cells increases, the cross-sectional area $A$ (in mm$^2$) of the colony increases but the height of the colony remains constant. If the colony starts from a single cell, the number of yeast cells (in millions) is approximated by the linear function $N(A) = C_s A$, where the constant $C_s$ is known as the cell-surface*

*coefficient. Use the given information to determine the cell-surface coefficient for each of the following colonies of yeast cells, and find the number of yeast cells in the colony when the cross-sectional area $A$ reaches 150 mm$^2$. (Source: Letters in Applied Microbiology, 594, 59, 2014)*

19. The scientific name of baker's or brewer's yeast (used in making bread, wine, and beer) is *Saccharomyces cerevisiae*. When the cross-sectional area of a colony of this yeast reaches 100 mm$^2$, there are 571 million yeast cells.

20. The yeast *Rhodotorula glutinis* is a laboratory contaminant. When the cross-sectional area of a colony reaches 100 mm$^2$, there are 226 million yeast cells.

21. **Demand function** Sales records indicate that if Blu-ray players are priced at \$250, then a large store sells an average of 12 units per day. If they are priced at \$200, then the store sells an average of 15 units per day. Find and graph the linear demand function for Blu-ray sales. For what prices is the demand function defined?

22. **Fundraiser** The Biology Club plans to have a fundraiser for which \$8 tickets will be sold. The cost of room rental and refreshments is \$175. Find and graph the function $p = f(n)$ that gives the profit from the fundraiser when $n$ tickets are sold. Notice that $f(0) = -\$175$; that is, the cost of room rental and refreshments must be paid regardless of how many tickets are sold. How many tickets must be sold for the fundraiser to break even (zero profit)?

23. **Bald eagle population** After DDT was banned and the Endangered Species Act was passed in 1973, the number of bald eagles in the United States increased dramatically. In the lower 48 states, the number of breeding pairs of bald eagles increased at a nearly linear rate from 1875 pairs in 1986 to 6471 pairs in 2000.

    **a.** Use the data points for 1986 and 2000 to find a linear function $p$ that models the number of breeding pairs from 1986 to 2000 $(0 \le t \le 14)$.

    **b.** Using the function in part (a), approximately how many breeding pairs were in the lower 48 states in 1995?

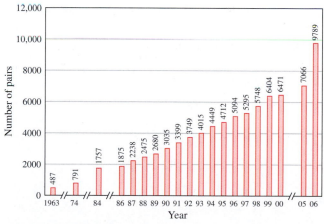

(*Source:* U.S. Fish and Wildlife Service)

24. **Taxicab fees** A taxicab ride costs \$3.50 plus \$2.50 per mile. Let $m$ be the distance (in miles) from the airport to a hotel. Find and graph the function $c(m)$ that represents the cost of taking a taxi from the airport to the hotel. Also determine how much it will cost if the hotel is 9 miles from the airport.

**25–26. Defining piecewise functions** *Write a definition of the function whose graph is given.*

**25.**

**26.**

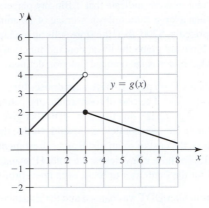

**27. Parking fees** Suppose that it costs 5¢ per minute to park at the airport, with the rate dropping to 3¢ per minute after 9 P.M. Find and graph the cost function $c(t)$ for values of $t$ satisfying $0 \le t \le 120$. Assume that $t$ is the number of minutes after 8 P.M.

**28. Taxicab fees** A taxicab ride costs $3.50 plus $2.50 per mile for the first 5 miles, with the rate dropping to $1.50 per mile after the fifth mile. Let $m$ be the distance (in miles) from the airport to a hotel. Find and graph the piecewise linear function $c(m)$ that represents the cost of taking a taxi from the airport to a hotel $m$ miles away.

**29–34. Piecewise linear functions** *Graph the following functions.*

**29.** $f(x) = \begin{cases} \dfrac{x^2 - x}{x - 1} & \text{if } x \ne 1 \\ 2 & \text{if } x = 1 \end{cases}$

**30.** $f(x) = \begin{cases} \dfrac{x^2 - x - 2}{x - 2} & \text{if } x \ne 2 \\ 4 & \text{if } x = 2 \end{cases}$

**31.** $f(x) = \begin{cases} 3x - 1 & \text{if } x \le 0 \\ -2x + 1 & \text{if } x > 0 \end{cases}$

**32.** $f(x) = \begin{cases} 3x - 1 & \text{if } x < 1 \\ x + 1 & \text{if } x \ge 1 \end{cases}$

**33.** $f(x) = \begin{cases} -2x - 1 & \text{if } x < -1 \\ 1 & \text{if } -1 \le x \le 1 \\ 2x - 1 & \text{if } x > 1 \end{cases}$

**34.** $f(x) = \begin{cases} 2x + 2 & \text{if } x < 0 \\ x + 2 & \text{if } 0 \le x \le 2 \\ 3 - \dfrac{x}{2} & \text{if } x > 2 \end{cases}$

**T 35–40. Graphs of functions**

a. *Use a graphing utility to produce a graph of the given function. Experiment with different windows to see how the graph changes on different scales. Sketch an accurate graph by hand after using the graphing utility.*
b. *Give the domain of the function.*
c. *Discuss interesting features of the function, such as peaks, valleys, and intercepts (as in Example 5).*

**35.** $f(x) = x^3 - 2x^2 + 6$    **36.** $f(x) = \sqrt[3]{2x^2 - 8}$

**37.** $g(x) = \left| \dfrac{x^2 - 4}{x + 3} \right|$    **38.** $f(x) = \dfrac{\sqrt{3x^2 - 12}}{x + 1}$

**39.** $f(x) = 3 - |2x - 1|$    **40.** $f(x) = \begin{cases} \dfrac{|x - 1|}{x - 1} & \text{if } x \ne 1 \\ 0 & \text{if } x = 1 \end{cases}$

**41. Features of a graph** Consider the graph of the function $f$ shown in the figure. Answer the following questions by referring to the points $A$–$I$.

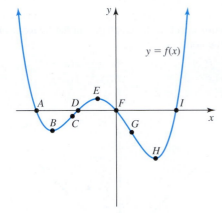

a. Which points correspond to the roots (zeros) of $f$?
b. Which points on the graph correspond to high points or peaks (soon to be called *local maximum* values of $f$)?
c. Which points on the graph correspond to low points or valleys (soon to be called *local minimum* values of $f$)?
d. As you move along the curve in the positive $x$-direction, at which point is the graph rising most rapidly?
e. As you move along the curve in the positive $x$-direction, at which point is the graph falling most rapidly?

**42. Features of a graph** Consider the graph of the function $g$ shown in the figure.

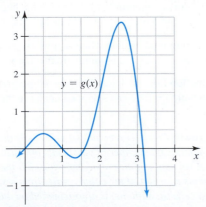

a. Give the approximate roots (zeros) of $g$.
b. Give the approximate coordinates of the high points or peaks (soon to be called *local maximum* values of $g$).
c. Give the approximate coordinates of the low points or valleys (soon to be called *local minimum* values of $g$).

**d.** Imagine moving along the curve in the positive *x*-direction on the interval $[0, 3]$. Give the approximate coordinates of the point at which the graph is rising most rapidly.

**e.** Imagine moving along the curve in the positive *x*-direction on the interval $[0, 3]$. Give the approximate coordinates of the point at which the graph is falling most rapidly.

**⊤ 43. Relative acuity of the human eye** The **fovea centralis** (or **fovea**) is responsible for the sharp central vision that humans use for reading and other detail-oriented eyesight. The relative acuity of a human eye, which measures the sharpness of vision, is modeled by the function

$$R(\theta) = \frac{0.568}{0.331|\theta| + 0.568},$$

where $\theta$ (in degrees) is the angular deviation of the line of sight from the center of the fovea (see figure).

**a.** Graph $R$, for $-15 \le \theta \le 15$.

**b.** For what value of $\theta$ is $R$ maximized? What does this fact indicate about our eyesight?

**c.** For what values of $\theta$ do we maintain at least 90% of our maximum relative acuity? (*Source: The Journal of Experimental Biology*, 203, 24, Dec 2000)

Fovea

**44–48. Slope functions** *Determine the slope function $S(x)$ for the following functions.*

**44.** $f(x) = 3$     **45.** $f(x) = 2x + 1$     **46.** $f(x) = |x|$

**47.** Use the figure for Exercise 7.

**48.** Use the figure for Exercise 26.

**49–52. Area functions** *Let $A(x)$ be the area of the region bounded by the t-axis and the graph of $y = f(t)$ from $t = 0$ to $t = x$. Consider the following functions and graphs.*

**a.** *Find $A(2)$.*    **b.** *Find $A(6)$.*    **c.** *Find a formula for $A(x)$.*

**49.** $f(t) = 6$

**50.** $f(t) = \dfrac{t}{2}$

**51.** $f(t) = \begin{cases} -2t + 8 & \text{if } t \le 3 \\ 2 & \text{if } t > 3 \end{cases}$

**52.** $f(t) = |t - 2| + 1$

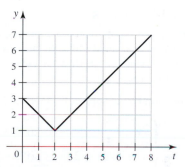

**53. Explain why or why not** Determine whether the following statements are true and give an explanation or counterexample.

**a.** All polynomials are rational functions, but not all rational functions are polynomials.

**b.** If $f$ is a linear polynomial, then $f \circ f$ is a quadratic polynomial.

**c.** If $f$ and $g$ are polynomials, then the degrees of $f \circ g$ and $g \circ f$ are equal.

**d.** The graph of $g(x) = f(x + 2)$ is the graph of $f$ shifted 2 units to the right.

**54. Shifting a graph** Use a shift to explain how the graph of $f(x) = \sqrt{-x^2 + 8x + 9}$ is obtained from the graph of $g(x) = \sqrt{25 - x^2}$. Sketch a graph of $f$.

**55. Transformations of $f(x) = x^2$** Use shifts and scalings to transform the graph of $f(x) = x^2$ into the graph of $g$. Use a graphing utility to check your work.

**a.** $g(x) = f(x - 3)$

**b.** $g(x) = f(2x - 4)$

**c.** $g(x) = -3f(x - 2) + 4$

**d.** $g(x) = 6f\left(\dfrac{x - 2}{3}\right) + 1$

**56. Transformations of $f(x) = \sqrt{x}$** Use shifts and scalings to transform the graph of $f(x) = \sqrt{x}$ into the graph of $g$. Use a graphing utility to check your work.

**a.** $g(x) = f(x + 4)$    **b.** $g(x) = 2f(2x - 1)$
**c.** $g(x) = \sqrt{x - 1}$    **d.** $g(x) = 3\sqrt{x - 1} - 5$

**T 57–64. Shifting and scaling** *Use shifts and scalings to graph the given functions. Then check your work with a graphing utility. Be sure to identify an original function on which the shifts and scalings are performed.*

**57.** $f(x) = (x - 2)^2 + 1$

**58.** $f(x) = x^2 - 2x + 3$
(*Hint:* Complete the square first.)

**59.** $g(x) = -3x^2$

**60.** $g(x) = 2x^3 - 1$

**61.** $g(x) = 2(x + 3)^2$

**62.** $p(x) = x^2 + 3x - 5$

**63.** $h(x) = -4x^2 - 4x + 12$

**64.** $h(x) = |3x - 6| + 1$

**65–67. Intersection problems** *Find the following points of intersection.*

**65.** The point(s) of intersection of the curves $y = 4\sqrt{2x}$ and $y = 2x^2$

**66.** The point(s) of intersection of the parabola $y = x^2 + 2$ and the line $y = x + 4$

**67.** The point(s) of intersection of the parabolas $y = x^2$ and $y = -x^2 + 8x$

## Explorations and Challenges

**68. Two semicircles** The entire graph of $f$ consists of the upper half of a circle of radius 2 centered at the origin and the lower half of a circle of radius 3 centered at $(5, 0)$. Find a piecewise function for $f$ and plot a graph of $f$.

**69. Piecewise function** Plot a graph of the function

$$f(x) = \begin{cases} \frac{3}{2}x & \text{if } 0 \leq x \leq 2 \\ 3 + \sqrt{x - 2} & \text{if } 2 < x \leq 6 \\ \sqrt{25 - (x - 6)^2} & \text{if } 6 < x \leq 11. \end{cases}$$

**70. Floor function** The floor function, or greatest integer function, $f(x) = \lfloor x \rfloor$, gives the greatest integer less than or equal to $x$. Graph the floor function for $-3 \leq x \leq 3$.

**71. Ceiling function** The ceiling function, or smallest integer function, $f(x) = \lceil x \rceil$, gives the smallest integer greater than or equal to $x$. Graph the ceiling function for $-3 \leq x \leq 3$.

**72. Sawtooth wave** Graph the sawtooth wave defined by

$$f(x) = \begin{cases} \vdots \\ x + 1 & \text{if } -1 \leq x < 0 \\ x & \text{if } 0 \leq x < 1 \\ x - 1 & \text{if } 1 \leq x < 2 \\ x - 2 & \text{if } 2 \leq x < 3 \\ \vdots \end{cases}$$

**73. Square wave** Graph the square wave defined by

$$f(x) = \begin{cases} 0 & \text{if } x < 0 \\ 1 & \text{if } 0 \leq x < 1 \\ 0 & \text{if } 1 \leq x < 2 \\ 1 & \text{if } 2 \leq x < 3 \\ \vdots \end{cases}$$

**74–76. Roots and powers** *Sketch a graph of the given pairs of functions. Be sure to draw the graphs accurately relative to each other.*

**74.** $y = x^4$ and $y = x^6$

**75.** $y = x^3$ and $y = x^7$

**76.** $y = x^{1/3}$ and $y = x^{1/5}$

**77. Tennis probabilities** Suppose the probability of a server winning any given point in a tennis match is a constant $p$, with $0 \leq p \leq 1$. Then the probability of the server winning a game when serving from deuce is

$$f(p) = \frac{p^2}{1 - 2p(1 - p)}.$$

**a.** Evaluate $f(0.75)$ and interpret the result.
**b.** Evaluate $f(0.25)$ and interpret the result.

(*Source: The College Mathematics Journal*, 38, 1, Jan 2007)

**78. Temperature scales**

**a.** Find the linear function $C = f(F)$ that gives the reading on the Celsius temperature scale corresponding to a reading on the Fahrenheit scale. Use the facts that $C = 0$ when $F = 32$ (freezing point) and $C = 100$ when $F = 212$ (boiling point).
**b.** At what temperature are the Celsius and Fahrenheit readings equal?

**79. Automobile lease vs. purchase** A car dealer offers a purchase option and a lease option on all new cars. Suppose you are interested in a car that can be bought outright for \$25,000 or leased for a start-up fee of \$1200 plus monthly payments of \$350.

**a.** Find the linear function $y = f(m)$ that gives the total amount you have paid on the lease option after $m$ months.
**b.** With the lease option, after a 48-month (4-year) term, the car has a residual value of \$10,000, which is the amount that you could pay to purchase the car. Assuming no other costs, should you lease or buy?

**T 80. Walking and rowing** Kelly has finished a picnic on an island that is 200 m off shore (see figure). She wants to return to a beach house that is 600 m from the point $P$ on the shore closest to the island. She plans to row a boat to a point on shore $x$ meters from $P$ and then jog along the (straight) shore to the house.

**a.** Let $d(x)$ be the total length of her trip as a function of $x$. Find and graph this function.

**b.** Suppose that Kelly can row at 2 m/s and jog at 4 m/s. Let $T(x)$ be the total time for her trip as a function of $x$. Find and graph $y = T(x)$.

**c.** Based on your graph in part (b), estimate the point on the shore at which Kelly should land to minimize the total time of her trip. What is that minimum time?

**T 81. Optimal boxes** Imagine a lidless box with height $h$ and a square base whose sides have length $x$. The box must have a volume of 125 ft³.

**a.** Find and graph the function $S(x)$ that gives the surface area of the box, for all values of $x > 0$.

**b.** Based on your graph in part (a), estimate the value of $x$ that produces the box with a minimum surface area.

**82. Composition of polynomials** Let $f$ be an $n$th-degree polynomial and let $g$ be an $m$th-degree polynomial. What is the degree of the following polynomials?

**a.** $f \cdot f$   **b.** $f \circ f$   **c.** $f \cdot g$   **d.** $f \circ g$

**83. Parabola vertex property** Prove that if a parabola crosses the $x$-axis twice, the $x$-coordinate of the vertex of the parabola is halfway between the $x$-intercepts.

**84. Parabola properties** Consider the general quadratic function $f(x) = ax^2 + bx + c$, with $a \neq 0$.

**a.** Find the coordinates of the vertex of the graph of the parabola $y = f(x)$ in terms of $a$, $b$, and $c$.

**b.** Find the conditions on $a$, $b$, and $c$ that guarantee that the graph of $f$ crosses the $x$-axis twice.

**T 85. Factorial function** The factorial function is defined for positive integers as $n! = n(n - 1)(n - 2) \cdots 3 \cdot 2 \cdot 1$.

**a.** Make a table of the factorial function, for $n = 1, 2, 3, 4$, and 5.

**b.** Graph these data points and then connect them with a smooth curve.

**c.** What is the least value of $n$ for which $n! > 10^6$?

**QUICK CHECK ANSWERS**

**1.** Yes; no   **2.** $(-\infty, \infty)$; $[0, \infty)$   **3.** Domain and range are $(-\infty, \infty)$. Domain and range are $[0, \infty)$.   **4.** Shift the graph of $f$ horizontally 4 units to the left. ◄

# 1.3 Trigonometric Functions

This section is a review of what you need to know in order to study the calculus of trigonometric functions.

## Radian Measure

Calculus typically requires that angles be measured in **radians** (rad). Working with a circle of radius $r$, the radian measure of an angle $\theta$ is the length of the arc associated with $\theta$, denoted $s$, divided by the radius of the circle $r$ (**Figure 1.45a**). Working on a unit circle ($r = 1$), the radian measure of an angle is simply the length of the arc associated with $\theta$ (**Figure 1.45b**). For example, the length of a full unit circle is $2\pi$; therefore, an angle with a radian measure of $\pi$ corresponds to a half circle ($\theta = 180°$) and an angle with a radian measure of $\pi/2$ corresponds to a quarter circle ($\theta = 90°$).

| Degrees | Radians |
|---------|---------|
| 0 | 0 |
| 30 | $\pi/6$ |
| 45 | $\pi/4$ |
| 60 | $\pi/3$ |
| 90 | $\pi/2$ |
| 120 | $2\pi/3$ |
| 135 | $3\pi/4$ |
| 150 | $5\pi/6$ |
| 180 | $\pi$ |

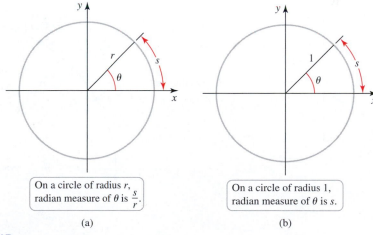

On a circle of radius $r$, radian measure of $\theta$ is $\dfrac{s}{r}$.

(a)

On a circle of radius 1, radian measure of $\theta$ is $s$.

(b)

**Figure 1.45**

**QUICK CHECK 1** What is the radian measure of a 270° angle? What is the degree measure of a $5\pi/4$-rad angle? ◄

## Trigonometric Functions

For acute angles, the trigonometric functions are defined as ratios of the lengths of the sides of a right triangle (**Figure 1.46**). To extend these definitions to include all angles, we work in an $xy$-coordinate system with a circle of radius $r$ centered at the origin. Suppose that $P(x, y)$ is a point on the circle. An angle $\theta$ is in **standard position** if its initial side is on the positive $x$-axis and its terminal side is the line segment $OP$ between the origin and $P$. An angle is positive if it is obtained by a counterclockwise rotation from the positive $x$-axis (**Figure 1.47**). When the right-triangle definitions of Figure 1.46 are used with the right triangle in Figure 1.47, the trigonometric functions may be expressed in terms of $x$, $y$, and the radius of the circle, $r = \sqrt{x^2 + y^2}$.

$$\sin \theta = \frac{\text{opp}}{\text{hyp}} \quad \cos \theta = \frac{\text{adj}}{\text{hyp}}$$

$$\tan \theta = \frac{\text{opp}}{\text{adj}} \quad \cot \theta = \frac{\text{adj}}{\text{opp}}$$

$$\sec \theta = \frac{\text{hyp}}{\text{adj}} \quad \csc \theta = \frac{\text{hyp}}{\text{opp}}$$

**Figure 1.46**

A positive angle $\theta$ results from a counterclockwise rotation.

**Figure 1.47**

➤ When working on a unit circle ($r = 1$), these definitions become

$$\sin \theta = y \qquad \cos \theta = x$$
$$\tan \theta = \frac{y}{x} \qquad \cot \theta = \frac{x}{y}$$
$$\sec \theta = \frac{1}{x} \qquad \csc \theta = \frac{1}{y}$$

---

**DEFINITION   Trigonometric Functions**

Let $P(x, y)$ be a point on a circle of radius $r$ associated with the angle $\theta$. Then

$$\sin \theta = \frac{y}{r}, \qquad \cos \theta = \frac{x}{r}, \qquad \tan \theta = \frac{y}{x},$$

$$\cot \theta = \frac{x}{y}, \qquad \sec \theta = \frac{r}{x}, \qquad \csc \theta = \frac{r}{y}.$$

---

➤ **Standard triangles**

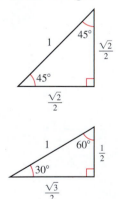

To evaluate the trigonometric functions at the standard angles (multiples of 30° and 45°), it is helpful to know the radian measure of those angles and the coordinates of the associated points on the unit circle (**Figure 1.48**).

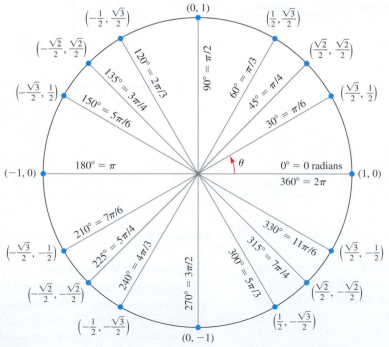

**Figure 1.48**

Combining the definitions of the trigonometric functions with the coordinates shown in Figure 1.48, we may evaluate these functions at any standard angle. For example,

$$\sin\frac{2\pi}{3} = \frac{\sqrt{3}}{2}, \qquad \cos\frac{5\pi}{6} = -\frac{\sqrt{3}}{2}, \qquad \tan\frac{7\pi}{6} = \frac{1}{\sqrt{3}}, \qquad \tan\frac{3\pi}{2} \text{ is undefined,}$$

$$\cot\frac{5\pi}{3} = -\frac{1}{\sqrt{3}}, \qquad \sec\frac{7\pi}{4} = \sqrt{2}, \qquad \csc\frac{3\pi}{2} = -1, \qquad \sec\frac{\pi}{2} \text{ is undefined.}$$

**EXAMPLE 1**    **Evaluating trigonometric functions**    Evaluate the following expressions.

**a.** $\sin(8\pi/3)$      **b.** $\csc(-11\pi/3)$

**SOLUTION**

**a.** The angle $\theta = 8\pi/3 = 2\pi + 2\pi/3$ corresponds to a *counterclockwise* revolution of one full circle ($2\pi$ radians) plus an additional $2\pi/3$ radians (**Figure 1.49**). Therefore, this angle has the same terminal side as the angle $2\pi/3$, and the corresponding point on the unit circle is $\left(-1/2, \sqrt{3}/2\right)$. It follows that $\sin(8\pi/3) = y = \sqrt{3}/2$.

**b.** The angle $\theta = -11\pi/3 = -2\pi - 5\pi/3$ corresponds to a *clockwise* revolution of one full circle ($2\pi$ radians) plus an additional $5\pi/3$ radians (**Figure 1.50**). Therefore, this angle has the same terminal side as the angle $\pi/3$. The coordinates of the corresponding point on the unit circle are $\left(1/2, \sqrt{3}/2\right)$, so $\csc(-11\pi/3) = 1/y = 2/\sqrt{3}$.

*Related Exercises 14, 18* ◀

**QUICK CHECK 2** Evaluate $\cos(11\pi/6)$ and $\sin(5\pi/4)$. ◀

## Trigonometric Identities

Trigonometric functions have a variety of properties, called identities, that are true for all angles in their domain. Here is a list of some commonly used identities.

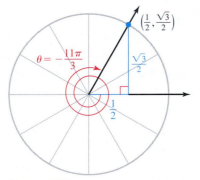

**Figure 1.49**

**Figure 1.50**

➤ In addition, to these identities, you should be familiar with the Law of Cosines and the Law of Sines. See Exercises 58 and 59.

➤ The half-angle identities for cosine and sine are also known as *Power-Reducing Identities*.

---

**Trigonometric Identities**

**Reciprocal Identities**

$$\tan\theta = \frac{\sin\theta}{\cos\theta} \qquad \cot\theta = \frac{1}{\tan\theta} = \frac{\cos\theta}{\sin\theta}$$

$$\csc\theta = \frac{1}{\sin\theta} \qquad \sec\theta = \frac{1}{\cos\theta}$$

**Pythagorean Identities**

$$\sin^2\theta + \cos^2\theta = 1 \qquad 1 + \cot^2\theta = \csc^2\theta \qquad \tan^2\theta + 1 = \sec^2\theta$$

**Double- and Half-Angle Identities**

$$\sin 2\theta = 2\sin\theta\cos\theta \qquad \cos 2\theta = \cos^2\theta - \sin^2\theta$$

$$\cos^2\theta = \frac{1 + \cos 2\theta}{2} \qquad \sin^2\theta = \frac{1 - \cos 2\theta}{2}$$

**QUICK CHECK 3** Use $\sin^2\theta + \cos^2\theta = 1$ to prove that $1 + \cot^2\theta = \csc^2\theta$. ◄

➤ By rationalizing the denominator, observe that $\dfrac{1}{\sqrt{2}} = \dfrac{1}{\sqrt{2}} \cdot \dfrac{\sqrt{2}}{\sqrt{2}} = \dfrac{\sqrt{2}}{2}$.

**EXAMPLE 2   Solving trigonometric equations**   Solve the following equations.

**a.** $\sqrt{2}\sin x + 1 = 0$     **b.** $\cos 2x = \sin 2x$, where $0 \le x < 2\pi$

**SOLUTION**

**a.** First, we solve for $\sin x$ to obtain $\sin x = -1/\sqrt{2} = -\sqrt{2}/2$. From the unit circle (Figure 1.48), we find that $\sin x = -\sqrt{2}/2$ if $x = 5\pi/4$ or $x = 7\pi/4$. Adding integer multiples of $2\pi$ produces additional solutions. Therefore, the set of all solutions is

$$x = \frac{5\pi}{4} + 2n\pi \quad \text{and} \quad x = \frac{7\pi}{4} + 2n\pi, \qquad \text{for } n = 0, \pm1, \pm2, \pm3, \dots.$$

**b.** Dividing both sides of the equation by $\cos 2x$ (assuming $\cos 2x \ne 0$), we obtain $\tan 2x = 1$. Letting $\theta = 2x$ gives us the equivalent equation $\tan\theta = 1$. This equation is satisfied by

$$\theta = \frac{\pi}{4}, \frac{5\pi}{4}, \frac{9\pi}{4}, \frac{13\pi}{4}, \frac{17\pi}{4}, \dots.$$

➤ Notice that the assumption $\cos 2x \ne 0$ is valid for these values of $x$.

Dividing by two and using the restriction $0 \le x < 2\pi$ gives the solutions

$$x = \frac{\theta}{2} = \frac{\pi}{8}, \frac{5\pi}{8}, \frac{9\pi}{8}, \text{ and } \frac{13\pi}{8}.$$

*Related Exercises 33, 38* ◄

## Graphs of the Trigonometric Functions

Trigonometric functions are examples of **periodic functions**: Their values repeat over every interval of some fixed length. A function $f$ is said to be periodic if $f(x + P) = f(x)$, for all $x$ in the domain, where the **period** $P$ is the smallest positive real number that has this property.

---

**Period of Trigonometric Functions**

The functions $\sin\theta$, $\cos\theta$, $\sec\theta$, and $\csc\theta$ have a period of $2\pi$:

$$\sin(\theta + 2\pi) = \sin\theta \qquad \cos(\theta + 2\pi) = \cos\theta$$
$$\sec(\theta + 2\pi) = \sec\theta \qquad \csc(\theta + 2\pi) = \csc\theta,$$

for all $\theta$ in the domain.

The functions $\tan\theta$ and $\cot\theta$ have a period of $\pi$:

$$\tan(\theta + \pi) = \tan\theta \qquad \cot(\theta + \pi) = \cot\theta,$$

for all $\theta$ in the domain.

---

The graph of $y = \sin\theta$ is shown in **Figure 1.51a**. Because $\csc\theta = 1/\sin\theta$, these two functions have the same sign, but $y = \csc\theta$ is undefined with vertical asymptotes at $\theta = 0, \pm\pi, \pm2\pi, \dots$. The functions $\cos\theta$ and $\sec\theta$ have a similar relationship (**Figure 1.51b**).

The graphs of $y = \sin\theta$ and its reciprocal, $y = \csc\theta$

The graphs of $y = \cos\theta$ and its reciprocal, $y = \sec\theta$

(a)     (b)

**Figure 1.51**

The graphs of $\tan \theta$ and $\cot \theta$ are shown in **Figure 1.52**. Each function has points, separated by $\pi$ units, at which it is undefined.

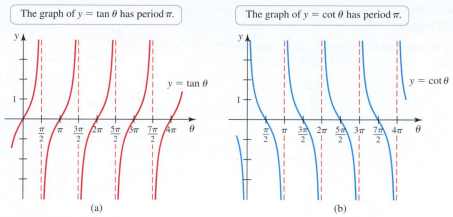

The graph of $y = \tan \theta$ has period $\pi$.

The graph of $y = \cot \theta$ has period $\pi$.

(a)                                        (b)

**Figure 1.52**

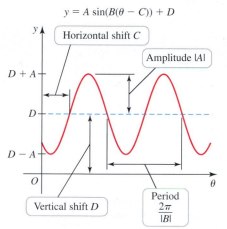

$y = A \sin(B(\theta - C)) + D$

Horizontal shift $C$

Amplitude $|A|$

Vertical shift $D$

Period $\dfrac{2\pi}{|B|}$

**Figure 1.53**

## Transforming Graphs

Many physical phenomena, such as the motion of waves and the rising and setting of the sun, can be modeled using trigonometric functions; the sine and cosine functions are especially useful. Using the transformation methods introduced in Section 1.2, we can show that the functions

$$y = A \sin (B(\theta - C)) + D \quad \text{and} \quad y = A \cos (B(\theta - C)) + D,$$

when compared to the graphs of $y = \sin \theta$ and $y = \cos \theta$, have a vertical scaling (or **amplitude**) of $|A|$, a period of $2\pi/|B|$, a horizontal shift (or **phase shift**) of $C$, and a **vertical shift** of $D$ (**Figure 1.53**).

For example, at latitude 40° north (Beijing, Madrid, Philadelphia), there are 12 hours of daylight on the equinoxes (approximately March 21 and September 21), with a maximum of 14.8 hours of daylight on the summer solstice (approximately June 21) and a minimum of 9.2 hours of daylight on the winter solstice (approximately December 21). Using this information, it can be shown that the function

$$D(t) = 2.8 \sin \left( \frac{2\pi}{365} (t - 81) \right) + 12$$

models the number of daylight hours $t$ days after January 1 (**Figure 1.54**; Exercise 70). The graph of this function is obtained from the graph of $y = \sin t$ by (1) a horizontal stretch by a factor of $365/2\pi$, (2) a horizontal shift of 81, (3) a vertical stretch by a factor of 2.8, and (4) a vertical shift of 12.

$y$ (hours)

Daylight function gives length of day throughout the year.

$y = 12$

$D(t) = 2.8 \sin \left( \dfrac{2\pi}{365}(t - 81) \right) + 12$

| Jan 1 | Mar 21 | June 21 | Sep 21 | Dec 21 | $t$ (days) |
| 0 | 81 | 173 | 265 | 356 | |

**Figure 1.54**

## SECTION 1.3 EXERCISES

### Getting Started

1. Define the six trigonometric functions in terms of the sides of a right triangle.

2. For the given angle $\theta$ corresponding to the point $P(-4, -3)$ in the figure, evaluate $\sin \theta$, $\cos \theta$, $\tan \theta$, $\cot \theta$, $\sec \theta$, and $\csc \theta$.

3. A projectile is launched at an angle of $\theta$ above the horizontal with an initial speed of $v$ ft/s and travels over level ground. The time of flight $t$ (the time it takes, in seconds, for the projectile to return to the ground) is approximated by the equation $t = \dfrac{v \sin \theta}{16}$. Determine the time of flight of a projectile if $\theta = \pi/6$ and $v = 96$.

4. A boat approaches a 50-ft-high lighthouse whose base is at sea level. Let $d$ be the distance between the boat and the base of the lighthouse. Let $L$ be the distance between the boat and the top of the lighthouse. Let $\theta$ be the angle of elevation between the boat and the top of the lighthouse.
   a. Express $d$ as a function of $\theta$.
   b. Express $L$ as a function of $\theta$.

5. How is the radian measure of an angle determined?

6. Explain what is meant by the period of a trigonometric function. What are the periods of the six trigonometric functions?

7. What are the three Pythagorean identities for the trigonometric functions?

8. Given that $\sin \theta = 1/\sqrt{5}$ and $\cos \theta = -2/\sqrt{5}$, use trigonometric identities to find the values of $\tan \theta$, $\cot \theta$, $\sec \theta$, and $\csc \theta$.

9. Solve the equation $\sin \theta = -1$, for $0 \le \theta < 2\pi$.

10. Solve the equation $\sin 2\theta = 1$, for $0 \le \theta < 2\pi$.

11. Where is the tangent function undefined?

12. What is the domain of the secant function?

### Practice Exercises

**13–28. Evaluating trigonometric functions** *Without using a calculator, evaluate the following expressions or state that the quantity is undefined.*

13. $\cos (2\pi/3)$
14. $\sin (2\pi/3)$
15. $\tan (-3\pi/4)$
16. $\tan (15\pi/4)$
17. $\cot (-13\pi/3)$
18. $\sec (7\pi/6)$

19. $\cot (-17\pi/3)$
20. $\sin (16\pi/3)$
21. $\cos 0$
22. $\sin (-\pi/2)$
23. $\cos (-\pi)$
24. $\tan 3\pi$
25. $\sec (5\pi/2)$
26. $\cot \pi$
27. $\cos (\pi/12)$ (*Hint:* Use a half-angle formula.)
28. $\sin (3\pi/8)$

**29–40. Solving trigonometric equations** *Solve the following equations.*

29. $\tan x = 1$
30. $2\theta \cos \theta + \theta = 0$
31. $\sin^2 \theta = \dfrac{1}{4}, 0 \le \theta < 2\pi$
32. $\cos^2 \theta = \dfrac{1}{2}, 0 \le \theta < 2\pi$
33. $\sqrt{2} \sin x - 1 = 0$
34. $\sin^2 \theta - 1 = 0$
35. $\sin \theta \cos \theta = 0, 0 \le \theta < 2\pi$
36. $\sin 3x = \dfrac{\sqrt{2}}{2}, 0 \le x < 2\pi$
37. $\cos 3x = \sin 3x, 0 \le x < 2\pi$
38. $\tan^2 2\theta = 1, 0 \le \theta < \pi$
**T** 39. $\sin 2\theta = \dfrac{1}{5}, 0 \le \theta \le \dfrac{\pi}{2}$
**T** 40. $\cos 3\theta = \dfrac{3}{7}, 0 \le \theta \le \pi$

**T 41–42. Projectile range** *A projectile is launched from the ground at an angle $\theta$ above the horizontal with an initial speed $v$ in ft/s. The range (the horizontal distance traveled by the projectile over level ground) is approximated by the equation $x = \dfrac{v^2}{32} \sin 2\theta$. Find all launch angles that satisfy the following conditions; express your answers in degrees.*

41. Initial speed of 150 ft/s; range of 400 ft
42. Initial speed of 160 ft/s; range of 350 ft

**43–48. Identities** *Prove the following identities.*

43. $\sec \theta = \dfrac{1}{\cos \theta}$
44. $\tan \theta = \dfrac{\sin \theta}{\cos \theta}$
45. $\tan^2 \theta + 1 = \sec^2 \theta$
46. $\dfrac{\sin \theta}{\csc \theta} + \dfrac{\cos \theta}{\sec \theta} = 1$
47. $\sec (\pi/2 - \theta) = \csc \theta$
48. $\sec (x + \pi) = -\sec x$

49. **Explain why or why not** Determine whether the following statements are true and give an explanation or counterexample.
   a. $\sin (a + b) = \sin a + \sin b$
   b. The equation $\cos \theta = 2$ has multiple solutions.
   c. The equation $\sin \theta = \frac{1}{2}$ has exactly one solution.
   d. The function $\sin (\pi x/12)$ has a period of 12.
   e. Of the six basic trigonometric functions, only tangent and cotangent have a range of $(-\infty, \infty)$.

**50–53. One function gives all six** *Given the following information about one trigonometric function, evaluate the other five functions.*

**50.** $\sin \theta = -\dfrac{4}{5}$ and $\pi < \theta < \dfrac{3\pi}{2}$

**51.** $\cos \theta = \dfrac{5}{13}$ and $0 < \theta < \dfrac{\pi}{2}$

**52.** $\sec \theta = \dfrac{5}{3}$ and $\dfrac{3\pi}{2} < \theta < 2\pi$

**53.** $\csc \theta = \dfrac{13}{12}$ and $0 < \theta < \dfrac{\pi}{2}$

**54–57. Amplitude and period** *Identify the amplitude and period of the following functions.*

**54.** $f(\theta) = 2 \sin 2\theta$        **55.** $g(\theta) = 3 \cos (\theta/3)$

**56.** $p(t) = 2.5 \sin \left( \dfrac{1}{2} (t - 3) \right)$   **57.** $q(x) = 3.6 \cos (\pi x/24)$

## Explorations and Challenges

**58. Law of Cosines** Use the figure to prove the Law of Cosines (which is a generalization of the Pythagorean theorem):
$c^2 = a^2 + b^2 - 2ab \cos \theta.$

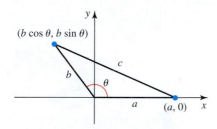

**59. Law of Sines** Use the figure to prove the Law of Sines:
$$\frac{\sin A}{a} = \frac{\sin B}{b} = \frac{\sin C}{c}.$$

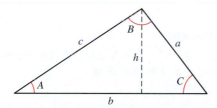

**T 60. Anchored sailboats** A sailboat named Ditl is anchored 200 feet north and 300 feet east of an observer standing on shore, while a second sailboat named Windborne is anchored 250 feet north and 100 feet west of the observer. Find the angle between the two sailboats as determined by the observer on shore.

**61. Area of a circular sector** Prove that the area of a sector of a circle of radius $r$ associated with a central angle $\theta$ (measured in radians) is $A = \frac{1}{2} r^2 \theta$.

**62–65. Graphing sine and cosine functions** *Beginning with the graphs of $y = \sin x$ or $y = \cos x$, use shifting and scaling transformations to sketch the graph of the following functions. Use a graphing utility to check your work.*

**62.** $f(x) = 3 \sin 2x$        **63.** $g(x) = -2 \cos (x/3)$

**64.** $p(x) = 3 \sin (2x - \pi/3) + 1$

**65.** $q(x) = 3.6 \cos (\pi x/24) + 2$

**66–67. Designer functions** *Design a sine function with the given properties.*

**66.** It has a period of 12 with a minimum value of $-4$ at $t = 0$ and a maximum value of 4 at $t = 6$.

**67.** It has a period of 24 with a minimum value of 10 at $t = 3$ and a maximum value of 16 at $t = 15$.

**T 68. Field goal attempt** During the 1950 Rose Bowl football game, The Ohio State University prepared to kick a field goal at a distance of 23 yd from the end line at point $A$ on the edge of the kicking region (see figure). But before the kick, Ohio State committed a penalty and the ball was backed up 5 yd to point $B$. After the game, the Ohio State coach claimed that his team deliberately committed a penalty to improve the kicking angle. Given that a successful kick must go between the goal posts $G_1$ and $G_2$, is $\angle G_1 B G_2$ greater than $\angle G_1 A G_2$? (In 1950, the goal posts were 23 ft 4 in apart, equidistant from the origin on the end line.) The boundaries of the kicking region are 53 ft 4 in apart and are equidistant from the $y$-axis. (*Source: The College Mathematics Journal, 27, 4, Sep 1996*)

**T 69. A surprising result** Earth is approximately circular in cross section, with a circumference at the equator of 24,882 miles. Suppose we use two ropes to create two concentric circles: one by wrapping a rope around the equator and another using a rope 38 ft longer (see figure). How much space is between the ropes?

**70. Daylight function for 40° N** Verify that the function

$$D(t) = 2.8 \sin\left(\frac{2\pi}{365}(t - 81)\right) + 12$$

has the following properties, where $t$ is measured in days and $D$ is the number of hours between sunrise and sunset.

a. It has a period of 365 days.
b. Its maximum and minimum values are 14.8 and 9.2, respectively, which occur approximately at $t = 172$ and $t = 355$, respectively (corresponding to the solstices).
c. $D(81) = 12$ and $D(264) \approx 12$ (corresponding to the equinoxes).

**71. Block on a spring** A light block hangs at rest from the end of a spring when it is pulled down 10 cm and released (see figure). Assume the block oscillates with an amplitude of 10 cm on either side of its rest position with a period of 1.5 s. Find a trigonometric function $d(t)$ that gives the displacement of the block $t$ seconds after it is released, where $d(t) > 0$ represents downward displacement.

Rest position, $d = 0$

$d(t) > 0$

**T 72. Viewing angles** An auditorium with a flat floor has a large flat-panel television on one wall. The lower edge of the television is 3 ft above the floor and the upper edge is 10 ft above the floor (see figure). Estimate the viewing angle $\theta$ at a distance $x = 10$ ft from the screen.

10 ft

3 ft

$\theta$

$x$

**73. Ladders** Two ladders of length $a$ lean against opposite walls of an alley with their feet touching (see figure). One ladder extends $h$ feet up the wall and makes a 75° angle with the ground. The other ladder extends $k$ feet up the opposite wall and makes a 45° angle with the ground. Find the width of the alley in terms of $h$. Assume the ground is horizontal and perpendicular to both walls.

$h$   $a$     $a$   $k$

75°      45°

**T 74. Pole in a corner** A pole of length $L$ is carried horizontally around a corner where a 3-ft-wide hallway meets a 4-ft-wide hallway. For $0 < \theta < \pi/2$, find the relationship between $L$ and $\theta$ at the moment when the pole simultaneously touches both walls and the corner $P$. Estimate $\theta$ when $L = 10$ ft.

Pole, length $L$

4 ft

$\theta$

$P$

3 ft

**QUICK CHECK ANSWERS**

**1.** $3\pi/2$; 225°  **2.** $\sqrt{3}/2$; $-\sqrt{2}/2$  **3.** Divide both sides of $\sin^2\theta + \cos^2\theta = 1$ by $\sin^2\theta$. ◄

---

## CHAPTER 1   REVIEW EXERCISES

**1. Explain why or why not** Determine whether the following statements are true and give an explanation or counterexample.

a. A function could have the property that $f(-x) = f(x)$, for all $x$.
b. $\cos(a + b) = \cos a + \cos b$, for all $a$ and $b$ in $[0, 2\pi]$.
c. If $f$ is a linear function of the form $f(x) = mx + b$, then $f(u + v) = f(u) + f(v)$, for all $u$ and $v$.
d. The function $f(x) = 1 - x$ has the property that $f(f(x)) = x$.
e. The set $\{x: |x + 3| > 4\}$ can be drawn on the number line without lifting your pencil.

**2.** **Functions** Decide whether graph $A$, graph $B$, or both represent functions.

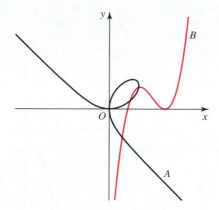

**3–5. Domain and range** *Determine the domain and range of the following functions.*

**3.** $f(w) = \dfrac{2w^2 - 3w - 2}{w - 2}$     **4.** $f(x) = x^5 + \sqrt{x}$

**5.** $h(z) = \sqrt{z^2 - 2z - 3}$

**6.** Suppose $f$ and $g$ are even functions with $f(2) = 2$ and $g(2) = -2$. Evaluate $f(g(2))$ and $g(f(-2))$.

**7.** **Composite functions** Let $f(x) = x^3$, $g(x) = \sin x$, and $h(x) = \sqrt{x}$.

   **a.** Evaluate $h(g(\pi/2))$.
   **b.** Find $h(f(x))$.
   **c.** Find $f(g(h(x)))$.
   **d.** Find the domain of $g \circ f$.
   **e.** Find the range of $f \circ g$.

**8.** **Composite functions** Find functions $f$ and $g$ such that $h = f \circ g$.

   **a.** $h(x) = \sin(x^2 + 1)$     **b.** $h(x) = (x^2 - 4)^{-3}$

**9–12. Simplifying difference quotients** *Evaluate and simplify the difference quotients* $\dfrac{f(x + h) - f(x)}{h}$ *and* $\dfrac{f(x) - f(a)}{x - a}$ *for each function.*

**9.** $f(x) = x^2 - 2x$     **10.** $f(x) = 4 - 5x$

**11.** $f(x) = x^3 + 2$     **12.** $f(x) = \dfrac{7}{x + 3}$

**13.** **Equations of lines** In each part below, find an equation of the line with the given properties.

   **a.** The line passing through the points $(2, -3)$ and $(4, 2)$
   **b.** The line with slope $\frac{3}{4}$ and $x$-intercept $(-4, 0)$
   **c.** The line with intercepts $(4, 0)$ and $(0, -2)$

**14.** **Population function** The population of a small town was 500 in 2018 and is growing at a rate of 24 people per year. Find and graph the linear population function $p(t)$ that gives the population of the town $t$ years after 2018. Then use this model to predict the population in 2033.

**15.** **Boiling-point function** Water boils at $212°$ F at sea level and at $200°$ F at an elevation of 6000 ft. Assume the boiling point $B$ varies linearly with altitude $a$. Find the function $B = f(a)$ that describes the dependence.

**16.** **Publishing costs** A small publisher plans to spend \$1000 for advertising a paperback book and estimates the printing cost is \$2.50 per book. The publisher will receive \$7 for each book sold.

   **a.** Find the function $C = f(x)$ that gives the cost of producing $x$ books.
   **b.** Find the function $R = g(x)$ that gives the revenue from selling $x$ books.
   **c.** Graph the cost and revenue functions; then find the number of books that must be sold for the publisher to break even.

**17.** **Graphing equations** Graph the following equations.

   **a.** $2x - 3y + 10 = 0$
   **b.** $y = x^2 + 2x - 3$
   **c.** $x^2 + 2x + y^2 + 4y + 1 = 0$
   **d.** $x^2 - 2x + y^2 - 8y + 5 = 0$

**18–20. Graphing functions** *Sketch a graph of each function.*

**18.** $f(x) = \begin{cases} 2x & \text{if } x \le 1 \\ 3 - x & \text{if } x > 1 \end{cases}$

**19.** $g(x) = \begin{cases} 4 - 2x & \text{if } x \le 1 \\ (x - 1)^2 + 2 & \text{if } x > 1 \end{cases}$

**20.** $h(x) = \begin{cases} \dfrac{3x^2 - 7x + 2}{x - 2} & \text{if } x \ne 2 \\ 6 & \text{if } x = 2 \end{cases}$

**21.** Let $f(t) = \begin{cases} t & \text{if } 0 \le t \le 4 \\ 8 - t & \text{if } 4 < t \le 8 \end{cases}$ and let $A(x)$ equal the area of the region bounded by the graph of $f$ and the $t$-axis from $t = 0$ to $t = x$, where $0 \le x \le 8$.

   **a.** Plot a graph of $f$.
   **b.** Find $A(2)$ and $A(6)$.
   **c.** Find a piecewise formula for $A(x)$.

**22.** **Piecewise linear functions** The parking costs in a city garage are \$2 for the first half hour and \$1 for each additional half hour. Graph the function $C = f(t)$ that gives the cost of parking for $t$ hours, where $0 \le t \le 3$.

**23.** **Graphing absolute value** Consider the function $f(x) = 2(x - |x|)$. Express the function in two pieces without using the absolute value. Then graph the function.

**24.** **Root functions** Graph the functions $f(x) = x^{1/3}$ and $g(x) = x^{1/4}$. Find all points where the two graphs intersect. For $x > 1$, is $f(x) > g(x)$ or is $g(x) > f(x)$?

**25.** **Root functions** Find the domain and range of the functions $f(x) = x^{1/7}$ and $g(x) = x^{1/4}$.

**26.** **Intersection points** Graph the equations $y = x^2$ and $x^2 + y^2 - 7y + 8 = 0$. At what point(s) do the curves intersect?

**27.** **Transformation of graphs** How is the graph of $y = x^2 + 6x - 3$ obtained from the graph of $y = x^2$?

**28. Shifting and scaling** The graph of $f$ is shown in the figure. Graph the following functions.

**a.** $f(x + 1)$   **b.** $2f(x - 1)$   **c.** $-f(x/2)$   **d.** $f(2(x - 1))$

**29. Symmetry** Identify the symmetry (if any) in the graphs of the following equations.

**a.** $y = \cos 3x$   **b.** $y = 3x^4 - 3x^2 + 1$   **c.** $y^2 - 4x^2 = 4$

**30. Degrees and radians**

**a.** Convert $135°$ to radian measure.

**b.** Convert $4\pi/5$ to degree measure.

**c.** What is the length of the arc on a circle of radius 10 associated with an angle of $4\pi/3$ (radians)?

**31–35. Solving equations** *Solve each equation.*

**31.** $1 - 2\sin^2\theta = 0, 0 \le \theta < 2\pi$

**32.** $\sin^2 2\theta = 1/2, -\pi/2 \le \theta \le \pi/2$

**33.** $4\cos^2 2\theta = 3, -\pi/2 \le \theta \le \pi/2$

**34.** $\sqrt{2}\sin 3\theta + 1 = 2, 0 \le \theta \le \pi$

**35.** $2\cos^2 2\theta - 3\cos 2\theta + 1 = 0, 0 \le \theta \le \pi$

**36. Graphing sine and cosine functions** Use shifts and scalings to graph the following functions, and identify the amplitude and period.

**a.** $f(x) = 4\cos(x/2)$     **b.** $g(\theta) = 2\sin(2\pi\theta/3)$

**c.** $h(\theta) = -\cos(2(\theta - \pi/4))$

**37. Designing functions** Find a trigonometric function $f$ that satisfies each set of properties. Answers are not unique.

**a.** It has a period of 6 with a minimum value of $-2$ at $t = 0$ and a maximum value of 2 at $t = 3$.

**b.** It has a period of 24 with a maximum value of 20 at $t = 6$ and a minimum value of 10 at $t = 18$.

**38. Graph to function** Find a trigonometric function $f$ represented by the graph in the figure.

**39. Matching** Match each function a–f with one of the graphs A–F.

**a.** $f(x) = -\sin x$        **b.** $f(x) = \cos 2x$

**c.** $f(x) = \tan(x/2)$      **d.** $f(x) = -\sec x$

**e.** $f(x) = \cot 2x$        **f.** $f(x) = \sin^2 x$

(A)                    (B)

(C)                    (D)

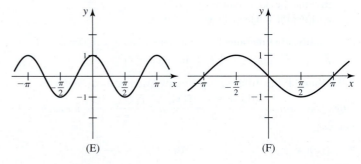

(E)                    (F)

**40–41. Intersection points** *Find the points at which the curves intersect on the given interval.*

**40.** $y = \sec x$ and $y = 2$ on $(-\pi/2, \pi/2)$

**41.** $y = \sin x$ and $y = -\dfrac{1}{2}$ on $(0, 2\pi)$

**42. Evaluating sine** Find the exact value of $\sin\dfrac{5\pi}{8}$.

**43. Evaluating cosine** Find the exact value of $\cos\dfrac{7\pi}{8}$.

**44–45. Identities** *Prove the following identities.*

**44.** $\dfrac{\sin\theta}{1 + \cos\theta} = \dfrac{1 - \cos\theta}{\sin\theta}$     **45.** $\tan 2\theta = \dfrac{2\tan\theta}{1 - \tan^2\theta}$

**46. Stereographic projections** A common way of displaying a sphere (such as Earth) on a plane (such as a map) is to use a *stereographic projection.* Here is the two-dimensional version of the method, which maps a circle to a line. Let $P$ be a point on the right half of a circle of radius $R$ identified by the angle $\varphi$. Find the function $x = F(\varphi)$ that gives the $x$-coordinate ($x \geq 0$) corresponding to $\varphi$ for $0 \leq \varphi \leq \pi$.

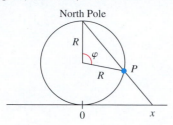

North Pole

**T 47. Sum of squared integers** Let $T(n) = 1^2 + 2^2 + \cdots + n^2$, where $n$ is a positive integer. It can be shown that

$$T(n) = \frac{n(n+1)(2n+1)}{6}.$$

   **a.** Make a table of $T(n)$, for $n = 1, 2, \ldots, 10$.
   **b.** How would you describe the domain of this function?
   **c.** What is the least value of $n$ for which $T(n) > 1000$?

**T 48. Sum of integers** Let $S(n) = 1 + 2 + \cdots + n$, where $n$ is a positive integer. It can be shown that $S(n) = n(n + 1)/2$.

   **a.** Make a table of $S(n)$, for $n = 1, 2, \ldots, 10$.
   **b.** How would you describe the domain of this function?
   **c.** What is the least value of $n$ for which $S(n) > 1000$?

**49. Little-known fact** The shortest day of the year occurs on the winter solstice (near December 21) and the longest day of the year occurs on the summer solstice (near June 21). However, the latest sunrise and the earliest sunset do not occur on the winter solstice, and the earliest sunrise and the latest sunset do not occur on the summer solstice. At latitude 40° north, the latest sunrise occurs on January 4 at 7:25 A.M. (14 days after the solstice), and the earliest sunset occurs on December 7 at 4:37 P.M. (14 days before the solstice). Similarly, the earliest sunrise occurs on July 2 at 4:30 A.M. (14 days after the solstice) and the latest sunset occurs on June 7 at 7:32 P.M. (14 days before the solstice). Using sine functions, devise a function $s(t)$ that gives the time of sunrise $t$ days after January 1 and a function $S(t)$ that gives the time of sunset $t$ days after January 1. Assume $s$ and $S$ are measured in minutes and that $s = 0$ and $S = 0$ correspond to 4:00 A.M. Graph the functions. Then graph the length of the day function $D(t) = S(t) - s(t)$ and show that the longest and shortest days occur on the solstices.

## Chapter 1 Guided Projects

*Applications of the material in this chapter and related topics can be found in the following Guided Projects. For additional information, see the Preface.*

- Problem-solving skills
- Constant rate problems
- Functions in action I

- Functions in action II
- Supply and demand
- Phase and amplitude

# 2

# Limits

**Chapter Preview** All of calculus is based on the idea of a *limit*. Not only are limits important in their own right but they also underlie the two fundamental operations of calculus: differentiation (calculating derivatives) and integration (evaluating integrals). Derivatives enable us to talk about the instantaneous rate of change of a function, which, in turn, leads to concepts such as velocity and acceleration, population growth rates, marginal cost, and flow rates. Integrals enable us to compute areas under curves, surface areas, and volumes. Because of the incredible reach of this single idea, it is essential to develop a solid understanding of limits. We first present limits intuitively by showing how they arise in computing instantaneous velocities and finding slopes of tangent lines. As the chapter progresses, we build more rigor into the definition of the limit and examine the different ways in which limits arise. The chapter concludes by introducing the important property of *continuity* and by giving the formal definition of a limit.

## 2.1 The Idea of Limits

This brief opening section illustrates how limits arise in two seemingly unrelated problems: finding the instantaneous velocity of a moving object and finding the slope of a line tangent to a curve. These two problems provide important insights into limits on an intuitive level. In the remainder of the chapter, we develop limits carefully and fill in the mathematical details.

### Average Velocity

Suppose you want to calculate your average velocity as you travel along a straight highway. If you pass milepost 100 at noon and milepost 130 at 12:30 P.M., you travel 30 miles in a half hour, so your **average velocity** over this time interval is $(30 \text{ mi})/(0.5 \text{ hr}) = 60 \text{ mi/hr}$. By contrast, even though your average velocity may be 60 mi/hr, it's almost certain that your **instantaneous velocity**, the speed indicated by the speedometer, varies from one moment to the next.

**EXAMPLE 1** **Average velocity** A rock is launched vertically upward from the ground with a speed of 96 ft/s. Neglecting air resistance, a well-known formula from physics states that the position of the rock after $t$ seconds is given by the function

$$s(t) = -16t^2 + 96t.$$

The position $s$ is measured in feet with $s = 0$ corresponding to the ground. Find the average velocity of the rock between each pair of times.

**a.** $t = 1$ s and $t = 3$ s  **b.** $t = 1$ s and $t = 2$ s

**SOLUTION** Figure 2.1 shows the position function of the rock on the time interval $0 \le t \le 3$. The graph is *not* the path of the rock. The rock travels up and down on a vertical line.

**Figure 2.1**

**a.** The average velocity of the rock over any time interval $[t_0, t_1]$ is the change in position divided by the elapsed time:

$$v_{av} = \frac{s(t_1) - s(t_0)}{t_1 - t_0}.$$

Therefore, the average velocity over the interval $[1, 3]$ is

$$v_{av} = \frac{s(3) - s(1)}{3 - 1} = \frac{144 \text{ ft} - 80 \text{ ft}}{3 \text{ s} - 1 \text{ s}} = \frac{64 \text{ ft}}{2 \text{ s}} = 32 \text{ ft/s}.$$

*Here is an important observation*: As shown in **Figure 2.2a**, the average velocity is simply the slope of the line joining the points $(1, s(1))$ and $(3, s(3))$ on the graph of the position function.

**QUICK CHECK 1** In Example 1, what is the average velocity between $t = 2$ and $t = 3$? ◄

**b.** The average velocity of the rock over the interval $[1, 2]$ is

$$v_{av} = \frac{s(2) - s(1)}{2 - 1} = \frac{128 \text{ ft} - 80 \text{ ft}}{2 \text{ s} - 1 \text{ s}} = \frac{48 \text{ ft}}{1 \text{ s}} = 48 \text{ ft/s}.$$

Again, the average velocity is the slope of the line joining the points $(1, s(1))$ and $(2, s(2))$ on the graph of the position function (**Figure 2.2b**).

**Figure 2.2**

*Related Exercises 13, 15* ◄

➤ See Section 1.1 for a discussion of secant
lines.

In Example 1, we computed slopes of lines passing through two points on a curve. Any such line joining two points on a curve is called a **secant line**. The slope of the secant line, denoted $m_{sec}$, for the position function in Example 1 on the interval $[t_0, t_1]$ is

$$m_{sec} = \frac{s(t_1) - s(t_0)}{t_1 - t_0}.$$

Example 1 demonstrates that the average velocity is the slope of a secant line on the graph of the position function; that is, $v_{av} = m_{sec}$ (**Figure 2.3**).

**Figure 2.3**

## Instantaneous Velocity

To compute the average velocity, we use the position of the object at *two* distinct points in time. How do we compute the instantaneous velocity at a *single* point in time? As illustrated in Example 2, the instantaneous velocity at a point $t = t_0$ is determined by computing average velocities over intervals $[t_0, t_1]$ that decrease in length. As $t_1$ approaches $t_0$, the average velocities typically approach a unique number, which is the instantaneous velocity. This single number is called a **limit**.

**QUICK CHECK 2** Explain the difference between average velocity and instantaneous velocity. ◄

**EXAMPLE 2   Instantaneous velocity** Estimate the instantaneous velocity of the rock in Example 1 at the *single* point $t = 1$.

**SOLUTION** We are interested in the instantaneous velocity at $t = 1$, so we compute the average velocity over smaller and smaller time intervals $[1, t]$ using the formula

$$v_{av} = \frac{s(t) - s(1)}{t - 1}.$$

**Table 2.1**

| Time interval | Average velocity |
|---|---|
| $[1, 2]$ | 48 ft/s |
| $[1, 1.5]$ | 56 ft/s |
| $[1, 1.1]$ | 62.4 ft/s |
| $[1, 1.01]$ | 63.84 ft/s |
| $[1, 1.001]$ | 63.984 ft/s |
| $[1, 1.0001]$ | 63.9984 ft/s |

Notice that these average velocities are also slopes of secant lines, several of which are shown in Table 2.1. For example, the average velocity on the interval $[1, 1.0001]$ is 63.9984 ft/s. Because this time interval is so short, the average velocity gives a good approximation to the instantaneous velocity at $t = 1$. We see that as $t$ approaches 1, the average velocities appear to approach 64 ft/s. In fact, we could make the average velocity as close to 64 ft/s as we like by taking $t$ sufficiently close to 1. Therefore, 64 ft/s is a reasonable estimate of the instantaneous velocity at $t = 1$.

*Related Exercises 17, 19* ◄

➤ The same instantaneous velocity is
obtained as $t$ approaches 1 from the left
(with $t < 1$) and as $t$ approaches 1 from
the right (with $t > 1$).

In language to be introduced in Section 2.2, we say that the limit of $v_{av}$ as $t$ approaches 1 equals the instantaneous velocity $v_{inst}$, which is 64 ft/s. This statement is illustrated in **Figure 2.4** and written compactly as

$$v_{inst} = \lim_{t \to 1} v_{av} = \lim_{t \to 1} \frac{s(t) - s(1)}{t - 1} = 64 \text{ ft/s}.$$

**Figure 2.4**

> We define tangent lines carefully in Section 3.1. For the moment, imagine zooming in on a point $P$ on a smooth curve. As you zoom in, the curve appears more and more like a line passing through $P$. This line is the *tangent line* at $P$. Because a smooth curve approaches a line as we zoom in on a point, a smooth curve is said to be *locally linear* at any given point.

## Slope of the Tangent Line

Several important conclusions follow from Examples 1 and 2. Each average velocity in Table 2.1 corresponds to the slope of a secant line on the graph of the position function (**Figure 2.5**). Just as the average velocities approach a limit as $t$ approaches 1, the slopes of the secant lines approach the same limit as $t$ approaches 1. Specifically, as $t$ approaches 1, two things happen:

1. The secant lines approach a unique line called the **tangent line**.

2. The slopes of the secant lines $m_{sec}$ approach the slope of the tangent line $m_{tan}$ at the point $(1, s(1))$. Therefore, the slope of the tangent line is also expressed as a limit:

$$m_{tan} = \lim_{t \to 1} m_{sec} = \lim_{t \to 1} \frac{s(t) - s(1)}{t - 1} = 64.$$

This limit is the same limit that defines the instantaneous velocity. Therefore, the instantaneous velocity at $t = 1$ is the slope of the line tangent to the position curve at $t = 1$.

**QUICK CHECK 3** In Figure 2.5, is $m_{tan}$ at $t = 2$ greater than or less than $m_{tan}$ at $t = 1$? ◄

**Figure 2.5**

The parallels between average and instantaneous velocities, on one hand, and between slopes of secant lines and tangent lines, on the other, illuminate the power behind the idea of a limit. As $t \to 1$, slopes of secant lines approach the slope of a tangent line. And as $t \to 1$, average velocities approach an instantaneous velocity. **Figure 2.6** summarizes these two parallel limit processes. These ideas lie at the foundation of what follows in the coming chapters.

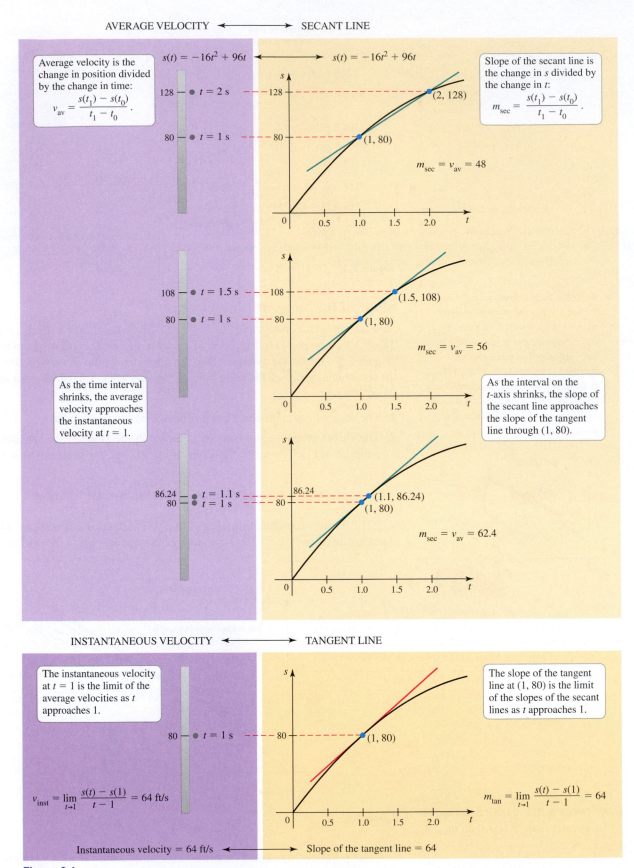

AVERAGE VELOCITY ◄————► SECANT LINE

$s(t) = -16t^2 + 96t$      $s(t) = -16t^2 + 96t$

Average velocity is the change in position divided by the change in time:
$$v_{av} = \frac{s(t_1) - s(t_0)}{t_1 - t_0}.$$

Slope of the secant line is the change in $s$ divided by the change in $t$:
$$m_{sec} = \frac{s(t_1) - s(t_0)}{t_1 - t_0}.$$

$m_{sec} = v_{av} = 48$

$m_{sec} = v_{av} = 56$

As the time interval shrinks, the average velocity approaches the instantaneous velocity at $t = 1$.

As the interval on the $t$-axis shrinks, the slope of the secant line approaches the slope of the tangent line through $(1, 80)$.

$m_{sec} = v_{av} = 62.4$

INSTANTANEOUS VELOCITY ◄————► TANGENT LINE

The instantaneous velocity at $t = 1$ is the limit of the average velocities as $t$ approaches 1.

The slope of the tangent line at $(1, 80)$ is the limit of the slopes of the secant lines as $t$ approaches 1.

$$v_{inst} = \lim_{t \to 1} \frac{s(t) - s(1)}{t - 1} = 64 \text{ ft/s}$$

$$m_{tan} = \lim_{t \to 1} \frac{s(t) - s(1)}{t - 1} = 64$$

Instantaneous velocity = 64 ft/s ◄————► Slope of the tangent line = 64

**Figure 2.6**

## SECTION 2.1 EXERCISES

### Getting Started

1. Suppose $s(t)$ is the position of an object moving along a line at time $t \geq 0$. What is the average velocity between the times $t = a$ and $t = b$?

2. Suppose $s(t)$ is the position of an object moving along a line at time $t \geq 0$. Describe a process for finding the instantaneous velocity at $t = a$.

3. The function $s(t)$ represents the position of an object at time $t$ moving along a line. Suppose $s(2) = 136$ and $s(3) = 156$. Find the average velocity of the object over the interval of time $[2, 3]$.

4. The function $s(t)$ represents the position of an object at time $t$ moving along a line. Suppose $s(1) = 84$ and $s(4) = 144$. Find the average velocity of the object over the interval of time $[1, 4]$.

5. The table gives the position $s(t)$ of an object moving along a line at time $t$, over a two-second interval. Find the average velocity of the object over the following intervals.

    **a.** $[0, 2]$     **b.** $[0, 1.5]$     **c.** $[0, 1]$     **d.** $[0, 0.5]$

    | $t$ | 0 | 0.5 | 1 | 1.5 | 2 |
    |-----|---|-----|---|-----|---|
    | $s(t)$ | 0 | 30 | 52 | 66 | 72 |

6. The graph gives the position $s(t)$ of an object moving along a line at time $t$, over a 2.5-second interval. Find the average velocity of the object over the following intervals.

    **a.** $[0.5, 2.5]$     **b.** $[0.5, 2]$     **c.** $[0.5, 1.5]$     **d.** $[0.5, 1]$

**T** 7. The following table gives the position $s(t)$ of an object moving along a line at time $t$. Determine the average velocities over the time intervals $[1, 1.01]$, $[1, 1.001]$, and $[1, 1.0001]$. Then make a conjecture about the value of the instantaneous velocity at $t = 1$.

| $t$ | 1 | 1.0001 | 1.001 | 1.01 |
|-----|---|--------|-------|------|
| $s(t)$ | 64 | 64.00479984 | 64.047984 | 64.4784 |

**T** 8. The following table gives the position $s(t)$ of an object moving along a line at time $t$. Determine the average velocities over the time intervals $[2, 2.01]$, $[2, 2.001]$, and $[2, 2.0001]$. Then make a conjecture about the value of the instantaneous velocity at $t = 2$.

| $t$ | 2 | 2.0001 | 2.001 | 2.01 |
|-----|---|--------|-------|------|
| $s(t)$ | 56 | 55.99959984 | 55.995984 | 55.9584 |

9. What is the slope of the secant line that passes through the points $(a, f(a))$ and $(b, f(b))$ on the graph of $f$?

10. Describe a process for finding the slope of the line tangent to the graph of $f$ at $(a, f(a))$.

11. Describe the parallels between finding the instantaneous velocity of an object at a point in time and finding the slope of the line tangent to the graph of a function at a point on the graph.

**T** 12. Given the function $f(x) = -16x^2 + 64x$, complete the following.

   **a.** Find the slopes of the secant lines that pass though the points $(x, f(x))$ and $(2, f(2))$, for $x = 1.5, 1.9, 1.99, 1.999$, and $1.9999$ (see figure).

   **b.** Make a conjecture about the value of the limit of the slopes of the secant lines that pass through $(x, f(x))$ and $(2, f(2))$ as $x$ approaches 2.

   **c.** What is the relationship between your answer to part (b) and the slope of the line tangent to the curve at $x = 2$ (see figure)?

   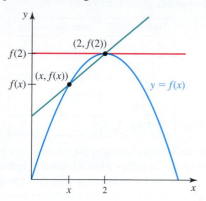

### Practice Exercises

**T** 13. **Average velocity** The position of an object moving vertically along a line is given by the function $s(t) = -16t^2 + 128t$. Find the average velocity of the object over the following intervals.

   **a.** $[1, 4]$     **b.** $[1, 3]$     **c.** $[1, 2]$

   **d.** $[1, 1 + h]$, where $h > 0$ is a real number

**T** 14. **Average velocity** The position of an object moving vertically along a line is given by the function $s(t) = -4.9t^2 + 30t + 20$. Find the average velocity of the object over the following intervals.

   **a.** $[0, 3]$     **b.** $[0, 2]$     **c.** $[0, 1]$

   **d.** $[0, h]$, where $h > 0$ is a real number

**T** 15. **Average velocity** Consider the position function $s(t) = -16t^2 + 100t$ representing the position of an object moving vertically along a line. Sketch a graph of $s$ with the secant line passing through $(0.5, s(0.5))$ and $(2, s(2))$. Determine the slope of the secant line and explain its relationship to the moving object.

16. **Average velocity** Consider the position function $s(t) = \sin \pi t$ representing the position of an object moving along a line on the end of a spring. Sketch a graph of $s$ together with the secant line passing through $(0, s(0))$ and $(0.5, s(0.5))$. Determine the slope of the secant line and explain its relationship to the moving object.

**T** 17. **Instantaneous velocity** Consider the position function $s(t) = -16t^2 + 128t$ (Exercise 13). Complete the following table with the appropriate average velocities. Then make a conjecture about the value of the instantaneous velocity at $t = 1$.

| Time interval | $[1, 2]$ | $[1, 1.5]$ | $[1, 1.1]$ | $[1, 1.01]$ | $[1, 1.001]$ |
|---------------|----------|------------|------------|-------------|--------------|
| Average velocity | | | | | |

**T 18. Instantaneous velocity** Consider the position function $s(t) = -4.9t^2 + 30t + 20$ (Exercise 14). Complete the following table with the appropriate average velocities. Then make a conjecture about the value of the instantaneous velocity at $t = 2$.

| Time interval | [2, 3] | [2, 2.5] | [2, 2.1] | [2, 2.01] | [2, 2.001] |
|---|---|---|---|---|---|
| Average velocity | | | | | |

**T 19. Instantaneous velocity** Consider the position function $s(t) = -16t^2 + 100t$. Complete the following table with the appropriate average velocities. Then make a conjecture about the value of the instantaneous velocity at $t = 3$.

| Time interval | Average velocity |
|---|---|
| [2, 3] | |
| [2.9, 3] | |
| [2.99, 3] | |
| [2.999, 3] | |
| [2.9999, 3] | |

**T 20. Instantaneous velocity** Consider the position function $s(t) = 3 \sin t$ that describes a block bouncing vertically on a spring. Complete the following table with the appropriate average velocities. Then make a conjecture about the value of the instantaneous velocity at $t = \pi/2$.

| Time interval | Average velocity |
|---|---|
| $[\pi/2, \pi]$ | |
| $[\pi/2, \pi/2 + 0.1]$ | |
| $[\pi/2, \pi/2 + 0.01]$ | |
| $[\pi/2, \pi/2 + 0.001]$ | |
| $[\pi/2, \pi/2 + 0.0001]$ | |

**T 21–24. Instantaneous velocity** *For the following position functions, make a table of average velocities similar to those in Exercises 19–20 and make a conjecture about the instantaneous velocity at the indicated time.*

**21.** $s(t) = -16t^2 + 80t + 60$ at $t = 3$

**22.** $s(t) = 20 \cos t$ at $t = \pi/2$

**23.** $s(t) = 40 \sin 2t$ at $t = 0$

**24.** $s(t) = 20/(t + 1)$ at $t = 0$

**T 25–28. Slopes of tangent lines** *For the following functions, make a table of slopes of secant lines and make a conjecture about the slope of the tangent line at the indicated point.*

**25.** $f(x) = 2x^2$ at $x = 2$

**26.** $f(x) = 3 \cos x$ at $x = \pi/2$

**27.** $f(x) = 1/(1 + x^2)$ at $x = -1$

**28.** $f(x) = x^3 - x$ at $x = 1$

### Explorations and Challenges

**T 29. Tangent lines with zero slope**

a. Graph the function $f(x) = x^2 - 4x + 3$.
b. Identify the point $(a, f(a))$ at which the function has a tangent line with zero slope.

c. Confirm your answer to part (b) by making a table of slopes of secant lines to approximate the slope of the tangent line at this point.

**T 30. Tangent lines with zero slope**

a. Graph the function $f(x) = 4 - x^2$.
b. Identify the point $(a, f(a))$ at which the function has a tangent line with zero slope.
c. Consider the point $(a, f(a))$ found in part (b). Is it true that the secant line between $(a - h, f(a - h))$ and $(a + h, f(a + h))$ has slope zero for any value of $h \neq 0$?

**T 31. Zero velocity** A projectile is fired vertically upward and has a position given by $s(t) = -16t^2 + 128t + 192$, for $0 \leq t \leq 9$.

a. Graph the position function, for $0 \leq t \leq 9$.
b. From the graph of the position function, identify the time at which the projectile has an instantaneous velocity of zero; call this time $t = a$.
c. Confirm your answer to part (b) by making a table of average velocities to approximate the instantaneous velocity at $t = a$.
d. For what values of $t$ on the interval $[0, 9]$ is the instantaneous velocity positive (the projectile moves upward)?
e. For what values of $t$ on the interval $[0, 9]$ is the instantaneous velocity negative (the projectile moves downward)?

**T 32. Impact speed** A rock is dropped off the edge of a cliff, and its distance $s$ (in feet) from the top of the cliff after $t$ seconds is $s(t) = 16t^2$. Assume the distance from the top of the cliff to the ground is 96 ft.

a. When will the rock strike the ground?
b. Make a table of average velocities and approximate the velocity at which the rock strikes the ground.

**T 33. Slope of tangent line** Given the function $f(x) = 1 - \cos x$ and the points $A(\pi/2, f(\pi/2))$, $B(\pi/2 + 0.05, f(\pi/2 + 0.05))$, $C(\pi/2 + 0.5, f(\pi/2 + 0.5))$, and $D(\pi, f(\pi))$ (see figure), find the slopes of the secant lines through $A$ and $D$, $A$ and $C$, and $A$ and $B$. Use your calculations to make a conjecture about the slope of the line tangent to the graph of $f$ at $x = \pi/2$.

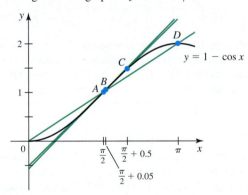

**QUICK CHECK ANSWERS**

**1.** 16 ft/s **2.** Average velocity is the velocity over an interval of time. Instantaneous velocity is the velocity at one point in time. **3.** Less than ◄

# 2.2 Definitions of Limits

Computing slopes of tangent lines and instantaneous velocities (Section 2.1) are just two of many important calculus problems that rely on limits. We now put these two problems aside until Chapter 3 and begin with a preliminary definition of the limit of a function.

> ➤ The terms *arbitrarily close* and *sufficiently close* will be made precise when rigorous definitions of limits are given in Section 2.7.

---

**DEFINITION** Limit of a Function (Preliminary)

Suppose the function $f$ is defined for all $x$ near $a$ except possibly at $a$. If $f(x)$ is arbitrarily close to $L$ (as close to $L$ as we like) for all $x$ sufficiently close (but not equal) to $a$, we write

$$\lim_{x \to a} f(x) = L$$

and say the limit of $f(x)$ as $x$ approaches $a$ equals $L$.

---

Informally, we say that $\lim_{x \to a} f(x) = L$ if $f(x)$ gets closer and closer to $L$ as $x$ gets closer and closer to $a$ from both sides of $a$. The value of $\lim_{x \to a} f(x)$ (if it exists) depends on the values of $f$ near $a$, but it does not depend on the value of $f(a)$. In some cases, the limit $\lim_{x \to a} f(x)$ equals $f(a)$. In other instances, $\lim_{x \to a} f(x)$ and $f(a)$ differ, or $f(a)$ may not even be defined.

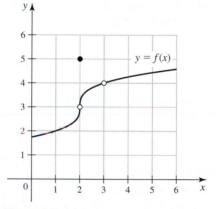

Figure 2.7

**EXAMPLE 1    Finding limits from a graph** Use the graph of $f$ (**Figure 2.7**) to determine the following values, if possible.

**a.** $f(1)$ and $\lim\limits_{x \to 1} f(x)$      **b.** $f(2)$ and $\lim\limits_{x \to 2} f(x)$      **c.** $f(3)$ and $\lim\limits_{x \to 3} f(x)$

**SOLUTION**

**a.** We see that $f(1) = 2$. As $x$ approaches 1 from either side, the values of $f(x)$ approach 2 (**Figure 2.8**). Therefore, $\lim\limits_{x \to 1} f(x) = 2$.

**b.** We see that $f(2) = 5$. However, as $x$ approaches 2 from either side, $f(x)$ approaches 3 because the points on the graph of $f$ approach the open circle at $(2, 3)$ (**Figure 2.9**). Therefore, $\lim\limits_{x \to 2} f(x) = 3$ even though $f(2) = 5$.

**c.** In this case, $f(3)$ is undefined. We see that $f(x)$ approaches 4 as $x$ approaches 3 from either side (**Figure 2.10**). Therefore, $\lim\limits_{x \to 3} f(x) = 4$ even though $f(3)$ does not exist.

Figure 2.8

Figure 2.9

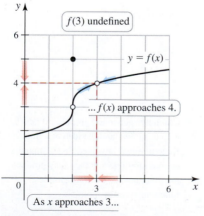

Figure 2.10

*Related Exercises 3–4* ◄

**QUICK CHECK 1** In Example 1, suppose we redefine the function at one point so that $f(1) = 1$. Does this change the value of $\lim_{x \to 1} f(x)$? ◄

In Example 1, we worked with the graph of a function to estimate limits. Let's now estimate limits using tabulated values of a function.

➤ In Example 2, we have not stated with certainty that $\lim_{x \to 1} f(x) = 0.5$. But this is a reasonable conjecture based on the numerical evidence. Methods for calculating limits precisely are introduced in Section 2.3.

**EXAMPLE 2** **Finding limits from a table** Create a table of values of $f(x) = \dfrac{\sqrt{x} - 1}{x - 1}$ corresponding to values of $x$ near 1. Then make a conjecture about the value of $\lim_{x \to 1} f(x)$.

**SOLUTION** Table 2.2 lists values of $f$ corresponding to values of $x$ approaching 1 from both sides. The numerical evidence suggests that $f(x)$ approaches 0.5 as $x$ approaches 1. Therefore, we make the conjecture that $\lim_{x \to 1} f(x) = 0.5$.

**Table 2.2**

|  |  |  |  |  | 1 |  |  |  |  |
|---|---|---|---|---|---|---|---|---|---|
| $x$ | 0.9 | 0.99 | 0.999 | 0.9999 | | 1.0001 | 1.001 | 1.01 | 1.1 |
| $f(x) = \dfrac{\sqrt{x} - 1}{x - 1}$ | 0.5131670 | 0.5012563 | 0.5001251 | 0.5000125 | | 0.4999875 | 0.4998751 | 0.4987562 | 0.4880885 |

*Related Exercises 7–8* ◄

## One-Sided Limits

The limit $\lim_{x \to a} f(x) = L$ is referred to as a *two-sided* limit because $f(x)$ approaches $L$ as $x$ approaches $a$ for values of $x$ less than $a$ *and* for values of $x$ greater than $a$. For some functions, it makes sense to examine *one-sided* limits called *right-sided* and *left-sided* limits.

➤ As with two-sided limits, the value of a one-sided limit (if it exists) depends on the values of $f(x)$ near $a$ but not on the value of $f(a)$.

---

**DEFINITION** **One-Sided Limits**

1. **Right-sided limit** Suppose $f$ is defined for all $x$ near $a$ with $x > a$. If $f(x)$ is arbitrarily close to $L$ for all $x$ sufficiently close to $a$ with $x > a$, we write

$$\lim_{x \to a^+} f(x) = L$$

and say the limit of $f(x)$ as $x$ approaches $a$ from the right equals $L$.

2. **Left-sided limit** Suppose $f$ is defined for all $x$ near $a$ with $x < a$. If $f(x)$ is arbitrarily close to $L$ for all $x$ sufficiently close to $a$ with $x < a$, we write

$$\lim_{x \to a^-} f(x) = L$$

and say the limit of $f(x)$ as $x$ approaches $a$ from the left equals $L$.

---

**EXAMPLE 3** **Examining limits graphically and numerically** Let $f(x) = \dfrac{x^3 - 8}{4(x - 2)}$.

Use tables and graphs to make a conjecture about the values of $\lim_{x \to 2^+} f(x)$, $\lim_{x \to 2^-} f(x)$, and $\lim_{x \to 2} f(x)$, if they exist.

➤ Computer-generated graphs and tables help us understand the idea of a limit. Keep in mind, however, that computers are not infallible and they may produce incorrect results, even for simple functions (see Example 5).

**SOLUTION** Figure 2.11a shows the graph of $f$ obtained with a graphing utility. The graph is misleading because $f(2)$ is undefined, which means there should be a hole in the graph at $(2, 3)$ (Figure 2.11b).

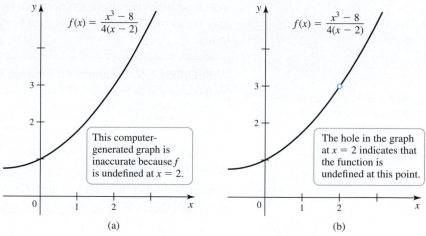

**Figure 2.11**

The graph in **Figure 2.12a** and the function values in Table 2.3 suggest that $f(x)$ approaches 3 as $x$ approaches 2 from the right. Therefore, we write the right-sided limit

$$\lim_{x \to 2^+} f(x) = 3,$$

which says the limit of $f(x)$ as $x$ approaches 2 from the right equals 3.

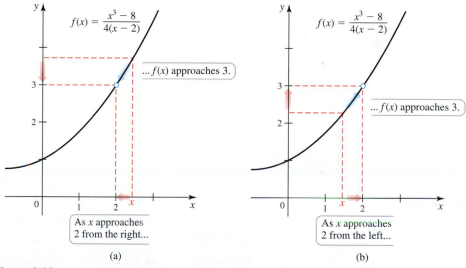

**Figure 2.12**

**Table 2.3**

| $x$ | 1.9 | 1.99 | 1.999 | 1.9999 | 2.0001 | 2.001 | 2.01 | 2.1 |
|---|---|---|---|---|---|---|---|---|
| $f(x) = \dfrac{x^3 - 8}{4(x - 2)}$ | 2.8525 | 2.985025 | 2.99850025 | 2.99985000 | 3.00015000 | 3.00150025 | 3.015025 | 3.1525 |

> ➤ Remember that the value of the limit does not depend on the value of $f(2)$. In Example 3, $\lim_{x \to 2} f(x) = 3$ despite the fact that $f(2)$ is undefined.

Similarly, **Figure 2.12b** and Table 2.3 suggest that as $x$ approaches 2 from the left, $f(x)$ approaches 3. So we write the left-sided limit

$$\lim_{x \to 2^-} f(x) = 3,$$

which says the limit of $f(x)$ as $x$ approaches 2 from the left equals 3. Because $f(x)$ approaches 3 as $x$ approaches 2 from either side, we write the two-sided limit $\lim_{x \to 2} f(x) = 3$.

*Related Exercises 27–28* ◀

Based on the previous example, you might wonder whether the limits $\lim_{x \to a^-} f(x)$, $\lim_{x \to a^+} f(x)$, and $\lim_{x \to a} f(x)$ always exist and are equal. The remaining examples demonstrate that these limits may have different values, and in some cases, one or more of these limits may not exist. The following theorem is useful when comparing one-sided and two-sided limits.

> **THEOREM 2.1  Relationship Between One-Sided and Two-Sided Limits**
> Assume $f$ is defined for all $x$ near $a$ except possibly at $a$. Then $\lim\limits_{x \to a} f(x) = L$ if and only if $\lim\limits_{x \to a^+} f(x) = L$ and $\lim\limits_{x \to a^-} f(x) = L$.

▶ Suppose $P$ and $Q$ are statements. We write $P$ *if and only if $Q$* when $P$ implies $Q$ and $Q$ implies $P$.

A proof of Theorem 2.1 is outlined in Exercise 56 of Section 2.7. Using this theorem, it follows that $\lim\limits_{x \to a} f(x) \neq L$ if either $\lim\limits_{x \to a^+} f(x) \neq L$ or $\lim\limits_{x \to a^-} f(x) \neq L$ (or both). Furthermore, if either $\lim\limits_{x \to a^+} f(x)$ or $\lim\limits_{x \to a^-} f(x)$ does not exist, then $\lim\limits_{x \to a} f(x)$ does not exist. We put these ideas to work in the next two examples.

**EXAMPLE 4  A function with a jump**  Sketch the graph of $g(x) = \dfrac{2x^2 - 6x + 4}{|x - 1|}$ and use the graph to find the values of $\lim\limits_{x \to 1^-} g(x)$, $\lim\limits_{x \to 1^+} g(x)$, and $\lim\limits_{x \to 1} g(x)$, if they exist.

**SOLUTION**  Sketching the graph of $g$ is straightforward if we first rewrite $g$ as a piecewise function. Factoring the numerator of $g$, we obtain

$$g(x) = \frac{2x^2 - 6x + 4}{|x - 1|} = \frac{2(x - 1)(x - 2)}{|x - 1|}.$$

Observe that $g$ is undefined at $x = 1$. For $x > 1$, $|x - 1| = x - 1$ and

$$g(x) = \frac{2(x - 1)(x - 2)}{x - 1} = 2x - 4.$$

Note that $x - 1 < 0$ when $x < 1$, which implies that $|x - 1| = -(x - 1)$ and

$$g(x) = \frac{2(x - 1)(x - 2)}{-(x - 1)} = -2x + 4.$$

**Figure 2.13**

Therefore, $g(x) = \begin{cases} 2x - 4 & \text{if } x > 1 \\ -2x + 4 & \text{if } x < 1 \end{cases}$ and the graph of $g$ consists of two linear pieces with a jump at $x = 1$ (**Figure 2.13**).

Examining the graph of $g$, we see that as $x$ approaches 1 from the left, $g(x)$ approaches 2. Therefore, $\lim\limits_{x \to 1^-} g(x) = 2$. As $x$ approaches 1 from the right, $g(x)$ approaches $-2$ and $\lim\limits_{x \to 1^+} g(x) = -2$. By Theorem 2.1, $\lim\limits_{x \to 1} g(x)$ does not exist because $\lim\limits_{x \to 1^-} g(x) \neq \lim\limits_{x \to 1^+} g(x)$.

*Related Exercises 19–20* ◀

**EXAMPLE 5  Some strange behavior**  Examine $\lim\limits_{x \to 0} \cos(1/x)$.

**SOLUTION**  From the first three values of $\cos(1/x)$ in Table 2.4, it is tempting to conclude that $\lim\limits_{x \to 0^+} \cos(1/x) = -1$. But this conclusion is not confirmed when we evaluate $\cos(1/x)$ for values of $x$ closer to 0.

**Table 2.4**

| $x$ | $\cos(1/x)$ |
| --- | --- |
| 0.001 | 0.56238 |
| 0.0001 | −0.95216 |
| 0.00001 | −0.99936 |
| 0.000001 | 0.93675 |
| 0.0000001 | −0.90727 |
| 0.00000001 | −0.36338 |

We might *incorrectly* conclude that $\cos(1/x)$ approaches $-1$ as $x$ approaches 0 from the right.

The behavior of $\cos(1/x)$ near 0 is better understood by letting $x = 1/(n\pi)$, where $n$ is a positive integer. By making this substitution, we can sample the function at discrete points that approach zero. In this case,

$$\cos\frac{1}{x} = \cos n\pi = \begin{cases} 1 & \text{if } n \text{ is even} \\ -1 & \text{if } n \text{ is odd.} \end{cases}$$

As $n$ increases, the values of $x = 1/(n\pi)$ approach zero, while the values of $\cos(1/x)$ oscillate between $-1$ and $1$ (**Figure 2.14**). Therefore, $\cos(1/x)$ does not approach a single number as $x$ approaches 0 from the right. We conclude that $\lim_{x\to 0^+} \cos(1/x)$ does *not* exist, which implies that $\lim_{x\to 0} \cos(1/x)$ does not exist.

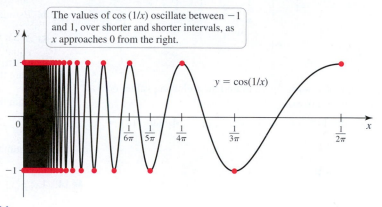

The values of cos (1/x) oscillate between −1 and 1, over shorter and shorter intervals, as x approaches 0 from the right.

$y = \cos(1/x)$

**QUICK CHECK 2** Why is the graph of $y = \cos(1/x)$ difficult to plot near $x = 0$, as suggested by Figure 2.14? ◄

**Figure 2.14**

*Related Exercise 43* ◄

Using tables and graphs to make conjectures for the values of limits worked well until Example 5. The limitation of technology in this example is not an isolated incident. For this reason, analytical techniques (paper-and-pencil methods) for finding limits are developed in the next section.

## SECTION 2.2 EXERCISES

### Getting Started

1. Explain the meaning of $\lim_{x\to a} f(x) = L$.

2. True or false: When $\lim_{x\to a} f(x)$ exists, it always equals $f(a)$. Explain.

3. Use the graph of $h$ in the figure to find the following values or state that they do not exist.

   **a.** $h(2)$   **b.** $\lim_{x\to 2} h(x)$   **c.** $h(4)$   **d.** $\lim_{x\to 4} h(x)$   **e.** $\lim_{x\to 5} h(x)$

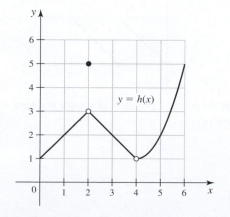

4. Use the graph of $g$ in the figure to find the following values or state that they do not exist.

   **a.** $g(0)$   **b.** $\lim_{x\to 0} g(x)$   **c.** $g(1)$   **d.** $\lim_{x\to 1} g(x)$

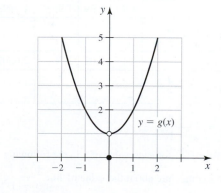

**5.** Use the graph of $f$ in the figure to find the following values or state that they do not exist.

   **a.** $f(1)$   **b.** $\lim\limits_{x \to 1} f(x)$   **c.** $f(0)$   **d.** $\lim\limits_{x \to 0} f(x)$

**6.** Use the graph of $f$ in the figure to find the following values or state that they do not exist.

   **a.** $f(2)$   **b.** $\lim\limits_{x \to 2} f(x)$   **c.** $\lim\limits_{x \to 4} f(x)$   **d.** $\lim\limits_{x \to 5} f(x)$

**T 7.** Let $f(x) = \dfrac{x^2 - 4}{x - 2}$.

   **a.** Calculate $f(x)$ for each value of $x$ in the following table.

   **b.** Make a conjecture about the value of $\lim\limits_{x \to 2} \dfrac{x^2 - 4}{x - 2}$.

| $x$ | 1.9 | 1.99 | 1.999 | 1.9999 |
|---|---|---|---|---|
| $f(x) = \dfrac{x^2 - 4}{x - 2}$ | | | | |
| $x$ | 2.1 | 2.01 | 2.001 | 2.0001 |
| $f(x) = \dfrac{x^2 - 4}{x - 2}$ | | | | |

**T 8.** Let $f(x) = \dfrac{x^3 - 1}{x - 1}$.

   **a.** Calculate $f(x)$ for each value of $x$ in the following table.

   **b.** Make a conjecture about the value of $\lim\limits_{x \to 1} \dfrac{x^3 - 1}{x - 1}$.

| $x$ | 0.9 | 0.99 | 0.999 | 0.9999 |
|---|---|---|---|---|
| $f(x) = \dfrac{x^3 - 1}{x - 1}$ | | | | |
| $x$ | 1.1 | 1.01 | 1.001 | 1.0001 |
| $f(x) = \dfrac{x^3 - 1}{x - 1}$ | | | | |

**T 9.** Let $g(t) = \dfrac{t - 9}{\sqrt{t} - 3}$.

   **a.** Make two tables, one showing values of $g$ for $t = 8.9,\ 8.99,$ and 8.999 and one showing values of $g$ for $t = 9.1,\ 9.01,$ and 9.001.

   **b.** Make a conjecture about the value of $\lim\limits_{t \to 9} \dfrac{t - 9}{\sqrt{t} - 3}$.

**T 10.** Let $f(x) = (1 + x)^{1/x}$.

   **a.** Make two tables, one showing values of $f$ for $x = 0.01,$ $0.001, 0.0001,$ and $0.00001$ and one showing values of $f$ for $x = -0.01,\ -0.001,\ -0.0001,$ and $-0.00001$. Round your answers to five digits.

   **b.** Estimate the value of $\lim\limits_{x \to 0} (1 + x)^{1/x}$.

   **c.** What mathematical constant does $\lim\limits_{x \to 0} (1 + x)^{1/x}$ appear to equal?

**11.** Explain the meaning of $\lim\limits_{x \to a^+} f(x) = L$.

**12.** Explain the meaning of $\lim\limits_{x \to a^-} f(x) = L$.

**13.** If $\lim\limits_{x \to a^-} f(x) = L$ and $\lim\limits_{x \to a^+} f(x) = M$, where $L$ and $M$ are finite real numbers, then how are $L$ and $M$ related if $\lim\limits_{x \to a} f(x)$ exists?

**T 14.** Let $g(x) = \dfrac{x^3 - 4x}{8|x - 2|}$.

   **a.** Calculate $g(x)$ for each value of $x$ in the following table.

   **b.** Make a conjecture about the values of $\lim\limits_{x \to 2^-} g(x),\ \lim\limits_{x \to 2^+} g(x),$ and $\lim\limits_{x \to 2} g(x)$ or state that they do not exist.

| $x$ | 1.9 | 1.99 | 1.999 | 1.9999 |
|---|---|---|---|---|
| $g(x) = \dfrac{x^3 - 4x}{8\|x - 2\|}$ | | | | |
| $x$ | 2.1 | 2.01 | 2.001 | 2.0001 |
| $g(x) = \dfrac{x^3 - 4x}{8\|x - 2\|}$ | | | | |

**15.** Use the graph of $f$ in the figure to find the following values or state that they do not exist. If a limit does not exist, explain why.

   **a.** $f(1)$   **b.** $\lim\limits_{x \to 1^-} f(x)$   **c.** $\lim\limits_{x \to 1^+} f(x)$   **d.** $\lim\limits_{x \to 1} f(x)$

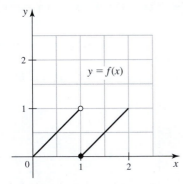

**16.** What are the potential problems of using a graphing utility to estimate $\lim\limits_{x \to a} f(x)$?

17. **Finding limits from a graph** Use the graph of $f$ in the figure to find the following values or state that they do not exist. If a limit does not exist, explain why.

    **a.** $f(1)$     **b.** $\lim\limits_{x \to 1^-} f(x)$     **c.** $\lim\limits_{x \to 1^+} f(x)$

    **d.** $\lim\limits_{x \to 1} f(x)$     **e.** $f(3)$     **f.** $\lim\limits_{x \to 3^-} f(x)$

    **g.** $\lim\limits_{x \to 3^+} f(x)$     **h.** $\lim\limits_{x \to 3} f(x)$     **i.** $f(2)$

    **j.** $\lim\limits_{x \to 2^-} f(x)$     **k.** $\lim\limits_{x \to 2^+} f(x)$     **l.** $\lim\limits_{x \to 2} f(x)$

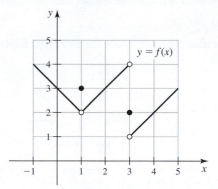

18. **One-sided and two-sided limits** Use the graph of $g$ in the figure to find the following values or state that they do not exist. If a limit does not exist, explain why.

    **a.** $g(2)$     **b.** $\lim\limits_{x \to 2^-} g(x)$     **c.** $\lim\limits_{x \to 2^+} g(x)$

    **d.** $\lim\limits_{x \to 2} g(x)$     **e.** $g(3)$     **f.** $\lim\limits_{x \to 3^-} g(x)$

    **g.** $\lim\limits_{x \to 3^+} g(x)$     **h.** $g(4)$     **i.** $\lim\limits_{x \to 4} g(x)$

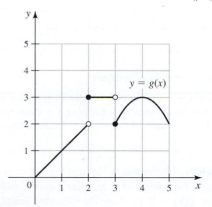

## Practice Exercises

**19–26. Evaluating limits graphically** *Sketch a graph of $f$ and use it to make a conjecture about the values of $f(a)$, $\lim\limits_{x \to a^-} f(x)$, $\lim\limits_{x \to a^+} f(x)$, and $\lim\limits_{x \to a} f(x)$ or state that they do not exist.*

19. $f(x) = \begin{cases} x^2 + 1 & \text{if } x \le -1 \\ 3 & \text{if } x > -1 \end{cases}; a = -1$

20. $f(x) = \begin{cases} 3 - x & \text{if } x < 2 \\ x - 1 & \text{if } x > 2 \end{cases}; a = 2$

21. $f(x) = \begin{cases} \sqrt{x} & \text{if } x < 4 \\ 3 & \text{if } x = 4; a = 4 \\ x + 1 & \text{if } x > 4 \end{cases}$

22. $f(x) = |x + 2| + 2; a = -2$

23. $f(x) = \dfrac{x^2 - 25}{x - 5}; a = 5$

**T** 24. $f(x) = \dfrac{x - 100}{\sqrt{x} - 10}; a = 100$

25. $f(x) = \dfrac{x^2 + x - 2}{x - 1}; a = 1$

26. $f(x) = \dfrac{1 - x^4}{x^2 - 1}; a = 1$

**T 27–32. Estimating limits graphically and numerically** *Use a graph of $f$ to estimate $\lim\limits_{x \to a} f(x)$ or to show that the limit does not exist. Evaluate $f(x)$ near $x = a$ to support your conjecture.*

27. $f(x) = \dfrac{x - 2}{\sin (x - 2)}; a = 2$

28. $f(x) = \dfrac{\tan^2 (\sin x)}{1 - \cos x}; a = 0$

29. $f(x) = \dfrac{1 - \cos (2x - 2)}{(x - 1)^2}; a = 1$

30. $f(x) = \dfrac{3 \sin x - 2 \cos x + 2}{x}; a = 0$

31. $f(x) = \dfrac{\sin (x + 1)}{|x + 1|}; a = -1$

32. $f(x) = \dfrac{x^3 - 4x^2 + 3x}{|x - 3|}; a = 3$

33. **Explain why or why not** Determine whether the following statements are true and give an explanation or counterexample.

    **a.** The value of $\lim\limits_{x \to 3} \dfrac{x^2 - 9}{x - 3}$ does not exist.

    **b.** The value of $\lim\limits_{x \to a} f(x)$ is always found by computing $f(a)$.

    **c.** The value of $\lim\limits_{x \to a} f(x)$ does not exist if $f(a)$ is undefined.

    **d.** $\lim\limits_{x \to 0} \sqrt{x} = 0$. (*Hint:* Graph $y = \sqrt{x}$.)

    **e.** $\lim\limits_{x \to \pi/2} \cot x = 0$. (*Hint:* Graph $y = \cot x$.)

34. **The Heaviside function** The Heaviside function is used in engineering applications to model flipping a switch. It is defined as

$$H(x) = \begin{cases} 0 & \text{if } x < 0 \\ 1 & \text{if } x \ge 0. \end{cases}$$

    **a.** Sketch a graph of $H$ on the interval $[-1, 1]$.

    **b.** Does $\lim\limits_{x \to 0} H(x)$ exist?

35. **Postage rates** Assume postage for sending a first-class letter in the United States is $0.47 for the first ounce (up to and including 1 oz) plus $0.21 for each additional ounce (up to and including each additional ounce).

    **a.** Graph the function $p = f(w)$ that gives the postage $p$ for sending a letter that weighs $w$ ounces, for $0 < w \le 3.5$.

    **b.** Evaluate $\lim\limits_{w \to 2.3} f(w)$.

    **c.** Does $\lim\limits_{w \to 3} f(w)$ exist? Explain.

**T 36–42. Calculator limits** *Estimate the following limits using graphs or tables.*

**36.** $\lim\limits_{h\to 0} \dfrac{\sin h}{h}$

**37.** $\lim\limits_{x\to \pi/2} \dfrac{\cot 3x}{\cos x}$

**38.** $\lim\limits_{x\to 1} \dfrac{18\left(\sqrt[3]{x}-1\right)}{x^3-1}$

**39.** $\lim\limits_{x\to 1} \dfrac{9\left(\sqrt{2x-x^4}-\sqrt[3]{x}\right)}{1-x^{3/4}}$

**40.** $\lim\limits_{x\to 0} \dfrac{1-\cos^5 2x}{1-\cos x}$

**41.** $\lim\limits_{x\to 8} \dfrac{(x^{4/3}-1)(x^{2/3}-4)}{x-8}$

**42.** $\lim\limits_{x\to 3} \dfrac{x^4-7x^3+15x^2-9x}{x-3}$

**T 43. Strange behavior near $x = 0$**

    **a.** Create a table of values of $\sin(1/x)$, for $x = 2/\pi, 2/(3\pi)$, $2/(5\pi), 2/(7\pi), 2/(9\pi)$, and $2/(11\pi)$. Describe the pattern of values you observe.

    **b.** Why does a graphing utility have difficulty plotting the graph of $y = \sin(1/x)$ near $x = 0$ (see figure)?

    **c.** What do you conclude about $\lim\limits_{x\to 0} \sin(1/x)$?

**T 44. Strange behavior near $x = 0$**

    **a.** Create a table of values of $\tan(3/x)$, for $x = 12/\pi, 12/(3\pi)$, $12/(5\pi), \ldots, 12/(11\pi)$. Describe the general pattern in the values you observe.

    **b.** Use a graphing utility to graph $y = \tan(3/x)$. Why do graphing utilities have difficulty plotting the graph near $x = 0$?

    **c.** What do you conclude about $\lim\limits_{x\to 0} \tan(3/x)$?

**45–49. Sketching graphs of functions** *Sketch the graph of a function with the given properties. You do not need to find a formula for the function.*

**45.** $f(2) = 1, \lim\limits_{x\to 2} f(x) = 3$

**46.** $f(1) = 0, f(2) = 4, f(3) = 6, \lim\limits_{x\to 2^-} f(x) = -3, \lim\limits_{x\to 2^+} f(x) = 5$

**47.** $g(1) = 0, g(2) = 1, g(3) = -2, \lim\limits_{x\to 2} g(x) = 0$,
$\lim\limits_{x\to 3^-} g(x) = -1, \lim\limits_{x\to 3^+} g(x) = -2$

**48.** $h(-1) = 2, \lim\limits_{x\to -1^-} h(x) = 0, \lim\limits_{x\to -1^+} h(x) = 3$,
$h(1) = \lim\limits_{x\to 1^-} h(x) = 1, \lim\limits_{x\to 1^+} h(x) = 4$

**49.** $p(0) = 2, \lim\limits_{x\to 0} p(x) = 0, \lim\limits_{x\to 2} p(x)$ does not exist,
$p(2) = \lim\limits_{x\to 2^+} p(x) = 1$

**50. A step function** Let $f(x) = \dfrac{|x|}{x}$, for $x \ne 0$.

    **a.** Sketch a graph of $f$ on the interval $[-2, 2]$.

    **b.** Does $\lim\limits_{x\to 0} f(x)$ exist? Explain your reasoning after first examining $\lim\limits_{x\to 0^-} f(x)$ and $\lim\limits_{x\to 0^+} f(x)$.

**51. The floor function** For any real number $x$, the *floor function* (or *greatest integer function*) $\lfloor x \rfloor$ is the greatest integer less than or equal to $x$ (see figure).

    **a.** Compute $\lim\limits_{x\to -1^-} \lfloor x \rfloor, \lim\limits_{x\to -1^+} \lfloor x \rfloor, \lim\limits_{x\to 2^-} \lfloor x \rfloor$, and $\lim\limits_{x\to 2^+} \lfloor x \rfloor$.

    **b.** Compute $\lim\limits_{x\to 2.3^-} \lfloor x \rfloor, \lim\limits_{x\to 2.3^+} \lfloor x \rfloor$, and $\lim\limits_{x\to 2.3} \lfloor x \rfloor$.

    **c.** For a given integer $a$, state the values of $\lim\limits_{x\to a^-} \lfloor x \rfloor$ and $\lim\limits_{x\to a^+} \lfloor x \rfloor$.

    **d.** In general, if $a$ is not an integer, state the values of $\lim\limits_{x\to a^-} \lfloor x \rfloor$ and $\lim\limits_{x\to a^+} \lfloor x \rfloor$.

    **e.** For what values of $a$ does $\lim\limits_{x\to a} \lfloor x \rfloor$ exist? Explain?

$y = \lfloor x \rfloor$

**52. The ceiling function** For any real number $x$, the *ceiling function* $\lceil x \rceil$ is the smallest integer greater than or equal to $x$.

    **a.** Graph the ceiling function $y = \lceil x \rceil$, for $-2 \le x \le 3$.

    **b.** Evaluate $\lim\limits_{x\to 2^-} \lceil x \rceil, \lim\limits_{x\to 1^+} \lceil x \rceil$, and $\lim\limits_{x\to 1.5} \lceil x \rceil$.

    **c.** For what values of $a$ does $\lim\limits_{x\to a} \lceil x \rceil$ exist? Explain.

### Explorations and Challenges

**53. Limits of even functions** A function $f$ is even if $f(-x) = f(x)$, for all $x$ in the domain of $f$. Suppose $f$ is even, with $\lim\limits_{x\to 2^+} f(x) = 5$ and $\lim\limits_{x\to 2^-} f(x) = 8$. Evaluate the following limits.

    **a.** $\lim\limits_{x\to -2^+} f(x)$     **b.** $\lim\limits_{x\to -2^-} f(x)$

**54. Limits of odd functions** A function $g$ is odd if $g(-x) = -g(x)$, for all $x$ in the domain of $g$. Suppose $g$ is odd, with $\lim\limits_{x\to 2^+} g(x) = 5$ and $\lim\limits_{x\to 2^-} g(x) = 8$. Evaluate the following limits.

    **a.** $\lim\limits_{x\to -2^+} g(x)$     **b.** $\lim\limits_{x\to -2^-} g(x)$

**T 55. Limits by graphs**

    **a.** Use a graphing utility to estimate $\lim\limits_{x\to 0} \dfrac{\tan 2x}{\sin x}, \lim\limits_{x\to 0} \dfrac{\tan 3x}{\sin x}$, and $\lim\limits_{x\to 0} \dfrac{\tan 4x}{\sin x}$.

    **b.** Make a conjecture about the value of $\lim\limits_{x\to 0} \dfrac{\tan px}{\sin x}$, for any real constant $p$.

**T 56. Limits by graphs** Graph $f(x) = \dfrac{\sin nx}{x}$, for $n = 1, 2, 3$, and $4$ (four graphs). Use the window $[-1, 1] \times [0, 5]$.

    **a.** Estimate $\lim\limits_{x\to 0} \dfrac{\sin x}{x}, \lim\limits_{x\to 0} \dfrac{\sin 2x}{x}, \lim\limits_{x\to 0} \dfrac{\sin 3x}{x}$, and $\lim\limits_{x\to 0} \dfrac{\sin 4x}{x}$.

    **b.** Make a conjecture about the value of $\lim\limits_{x\to 0} \dfrac{\sin px}{x}$, for any real constant $p$.

**T 57. Limits by graphs** Use a graphing utility to plot $y = \dfrac{\sin px}{\sin qx}$ for at least three different pairs of nonzero constants $p$ and $q$ of your choice. Estimate $\lim\limits_{x \to 0} \dfrac{\sin px}{\sin qx}$ in each case. Then use your work to make a conjecture about the value of $\lim\limits_{x \to 0} \dfrac{\sin px}{\sin qx}$ for any nonzero values of $p$ and $q$.

# 2.3 Techniques for Computing Limits

Graphical and numerical techniques for estimating limits, like those presented in the previous section, provide intuition about limits. These techniques, however, occasionally lead to incorrect results. Therefore, we turn our attention to analytical methods for evaluating limits precisely.

## Limits of Linear Functions

The graph of $f(x) = mx + b$ is a line with slope $m$ and $y$-intercept $b$. From **Figure 2.15**, we see that $f(x)$ approaches $f(a)$ as $x$ approaches $a$. Therefore, if $f$ is a linear function, we have $\lim\limits_{x \to a} f(x) = f(a)$. It follows that for linear functions, $\lim\limits_{x \to a} f(x)$ is found by direct substitution of $x = a$ into $f(x)$. This observation leads to the following theorem, which is proved in Exercise 39 of Section 2.7.

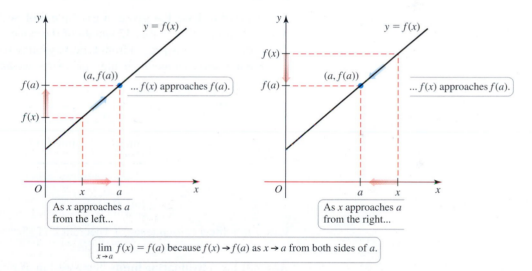

**Figure 2.15**

---

**THEOREM 2.2    Limits of Linear Functions**

Let $a$, $b$, and $m$ be real numbers. For linear functions $f(x) = mx + b$,

$$\lim_{x \to a} f(x) = f(a) = ma + b.$$

---

**EXAMPLE 1    Limits of linear functions**  Evaluate the following limits.

**a.** $\lim\limits_{x \to 3} f(x)$, where $f(x) = \tfrac{1}{2}x - 7$    **b.** $\lim\limits_{x \to 2} g(x)$, where $g(x) = 6$

**SOLUTION**

**a.** $\lim\limits_{x \to 3} f(x) = \lim\limits_{x \to 3} \left(\tfrac{1}{2}x - 7\right) = f(3) = -\tfrac{11}{2}$    **b.** $\lim\limits_{x \to 2} g(x) = \lim\limits_{x \to 2} 6 = g(2) = 6$

*Related Exercises 19, 22* ◄

## Limit Laws

The following limit laws greatly simplify the evaluation of many limits.

---

**THEOREM 2.3   Limit Laws**

Assume $\lim\limits_{x \to a} f(x)$ and $\lim\limits_{x \to a} g(x)$ exist. The following properties hold, where $c$ is a real number, and $n > 0$ is an integer.

1. **Sum** $\lim\limits_{x \to a} (f(x) + g(x)) = \lim\limits_{x \to a} f(x) + \lim\limits_{x \to a} g(x)$

2. **Difference** $\lim\limits_{x \to a} (f(x) - g(x)) = \lim\limits_{x \to a} f(x) - \lim\limits_{x \to a} g(x)$

3. **Constant multiple** $\lim\limits_{x \to a} (cf(x)) = c \lim\limits_{x \to a} f(x)$

4. **Product** $\lim\limits_{x \to a} (f(x)g(x)) = \left( \lim\limits_{x \to a} f(x) \right) \left( \lim\limits_{x \to a} g(x) \right)$

5. **Quotient** $\lim\limits_{x \to a} \left( \dfrac{f(x)}{g(x)} \right) = \dfrac{\lim\limits_{x \to a} f(x)}{\lim\limits_{x \to a} g(x)}$, provided $\lim\limits_{x \to a} g(x) \neq 0$

6. **Power** $\lim\limits_{x \to a} (f(x))^n = \left( \lim\limits_{x \to a} f(x) \right)^n$

7. **Root** $\lim\limits_{x \to a} (f(x))^{1/n} = \left( \lim\limits_{x \to a} f(x) \right)^{1/n}$, provided $f(x) > 0$, for $x$ near $a$, if $n$ is even

---

A proof of Law 1 is given in Example 6 of Section 2.7; the proofs of Laws 2 and 3 are asked for in Exercises 43 and 44 of the same section. Laws 4 and 5 are proved in Appendix A. Law 6 is proved from Law 4 as follows.

For a positive integer $n$, if $\lim\limits_{x \to a} f(x)$ exists, we have

$$\lim_{x \to a} (f(x))^n = \lim_{x \to a} (\underbrace{f(x)\, f(x) \cdots f(x)}_{n \text{ factors of } f(x)})$$

$$= \underbrace{\left( \lim_{x \to a} f(x) \right) \left( \lim_{x \to a} f(x) \right) \cdots \left( \lim_{x \to a} f(x) \right)}_{n \text{ factors of } \lim\limits_{x \to a} f(x)} \qquad \text{Repeated use of Law 4}$$

$$= \left( \lim_{x \to a} f(x) \right)^n.$$

Law 7 is a direct consequence of Theorem 2.11 (Section 2.6).

**EXAMPLE 2   Evaluating limits** Suppose $\lim\limits_{x \to 2} f(x) = 4$, $\lim\limits_{x \to 2} g(x) = 5$, and $\lim\limits_{x \to 2} h(x) = 8$. Use the limit laws in Theorem 2.3 to compute each limit.

**a.** $\lim\limits_{x \to 2} \dfrac{f(x) - g(x)}{h(x)}$     **b.** $\lim\limits_{x \to 2} (6f(x)g(x) + h(x))$     **c.** $\lim\limits_{x \to 2} (g(x))^3$

**SOLUTION**

**a.** $\lim\limits_{x \to 2} \dfrac{f(x) - g(x)}{h(x)} = \dfrac{\lim\limits_{x \to 2} (f(x) - g(x))}{\lim\limits_{x \to 2} h(x)}$     Law 5

$$= \frac{\lim\limits_{x \to 2} f(x) - \lim\limits_{x \to 2} g(x)}{\lim\limits_{x \to 2} h(x)} \qquad \text{Law 2}$$

$$= \frac{4 - 5}{8} = -\frac{1}{8}$$

**b.** $\lim\limits_{x\to 2} (6f(x)g(x) + h(x)) = \lim\limits_{x\to 2} (6f(x)g(x)) + \lim\limits_{x\to 2} h(x)$     Law 1

$\qquad\qquad\qquad\qquad = 6 \cdot \lim\limits_{x\to 2} (f(x)g(x)) + \lim\limits_{x\to 2} h(x)$     Law 3

$\qquad\qquad\qquad\qquad = 6\left(\lim\limits_{x\to 2} f(x)\right)\left(\lim\limits_{x\to 2} g(x)\right) + \lim\limits_{x\to 2} h(x)$     Law 4

$\qquad\qquad\qquad\qquad = 6 \cdot 4 \cdot 5 + 8 = 128$

**c.** $\lim\limits_{x\to 2} (g(x))^3 = \left(\lim\limits_{x\to 2} g(x)\right)^3 = 5^3 = 125$     Law 6

*Related Exercises 11–12* ◄

## Limits of Polynomial and Rational Functions

The limit laws are now used to find the limits of polynomial and rational functions. For example, to evaluate the limit of the polynomial $p(x) = 7x^3 + 3x^2 + 4x + 2$ at an arbitrary point $a$, we proceed as follows:

$$\lim_{x\to a} p(x) = \lim_{x\to a} (7x^3 + 3x^2 + 4x + 2)$$

$$= \lim_{x\to a} (7x^3) + \lim_{x\to a} (3x^2) + \lim_{x\to a} (4x + 2) \qquad \text{Law 1}$$

$$= 7 \lim_{x\to a} (x^3) + 3 \lim_{x\to a} (x^2) + \lim_{x\to a} (4x + 2) \qquad \text{Law 3}$$

$$= 7 \underbrace{\left(\lim_{x\to a} x\right)^3}_{a} + 3 \underbrace{\left(\lim_{x\to a} x\right)^2}_{a} + \underbrace{\lim_{x\to a} (4x + 2)}_{4a + 2} \qquad \text{Law 6}$$

$$= 7a^3 + 3a^2 + 4a + 2 = p(a). \qquad \text{Theorem 2.2}$$

As in the case of linear functions, the limit of a polynomial is found by direct substitution; that is, $\lim\limits_{x\to a} p(x) = p(a)$ (Exercise 105).

It is now a short step to evaluating limits of rational functions of the form $f(x) = p(x)/q(x)$, where $p$ and $q$ are polynomials. Applying Law 5, we have

$$\lim_{x\to a} \frac{p(x)}{q(x)} = \frac{\lim\limits_{x\to a} p(x)}{\lim\limits_{x\to a} q(x)} = \frac{p(a)}{q(a)}, \text{ provided } q(a) \neq 0,$$

which shows that limits of rational functions are also evaluated by direct substitution.

> ► The conditions under which direct substitution $\left(\lim\limits_{x\to a} f(x) = f(a)\right)$ can be used to evaluate a limit become clear in Section 2.6, when we discuss the important property of *continuity*.

---

**THEOREM 2.4**    **Limits of Polynomial and Rational Functions**

Assume $p$ and $q$ are polynomials and $a$ is a constant.

**a.** Polynomial functions:    $\lim\limits_{x\to a} p(x) = p(a)$

**b.** Rational functions:    $\lim\limits_{x\to a} \dfrac{p(x)}{q(x)} = \dfrac{p(a)}{q(a)}$, provided $q(a) \neq 0$

---

**QUICK CHECK 1**   Use Theorem 2.4 to evaluate $\lim\limits_{x\to 2} (2x^4 - 8x - 16)$ and $\lim\limits_{x\to -1} \dfrac{x - 1}{x}$. ◄

**EXAMPLE 3**   **Limit of a rational function**   Evaluate $\lim\limits_{x\to 2} \dfrac{3x^2 - 4x}{5x^3 - 36}$.

**SOLUTION**   Notice that the denominator of this function is nonzero at $x = 2$. Using Theorem 2.4b, we find that

$$\lim_{x\to 2} \frac{3x^2 - 4x}{5x^3 - 36} = \frac{3(2^2) - 4(2)}{5(2^3) - 36} = 1.$$

*Related Exercise 25* ◄

**QUICK CHECK 2**   Use Theorem 2.4 to compute $\lim\limits_{x\to 1} \dfrac{5x^4 - 3x^2 + 8x - 6}{x + 1}$. ◄

**EXAMPLE 4**   **An algebraic function**  Evaluate $\lim\limits_{x \to 2} \dfrac{\sqrt{2x^3 + 9} + 3x - 1}{4x + 1}$.

**SOLUTION**  Using Theorems 2.3 and 2.4, we have

$$\lim_{x \to 2} \frac{\sqrt{2x^3 + 9} + 3x - 1}{4x + 1} = \frac{\lim\limits_{x \to 2} \left( \sqrt{2x^3 + 9} + 3x - 1 \right)}{\lim\limits_{x \to 2} (4x + 1)} \qquad \text{Law 5}$$

$$= \frac{\sqrt{\lim\limits_{x \to 2} (2x^3 + 9)} + \lim\limits_{x \to 2} (3x - 1)}{\lim\limits_{x \to 2} (4x + 1)} \qquad \text{Laws 1 and 7}$$

$$= \frac{\sqrt{(2(2)^3 + 9)} + (3(2) - 1)}{(4(2) + 1)} \qquad \text{Theorem 2.4}$$

$$= \frac{\sqrt{25} + 5}{9} = \frac{10}{9}.$$

Notice that the limit at $x = 2$ equals the value of the function at $x = 2$.

*Related Exercises 26–27* ◄

## One-Sided Limits

Theorem 2.2, Limit Laws 1–6, and Theorem 2.4 also hold for left-sided and right-sided limits. In other words, these laws remain valid if we replace $\lim\limits_{x \to a}$ with $\lim\limits_{x \to a^+}$ or $\lim\limits_{x \to a^-}$. Law 7 must be modified slightly for one-sided limits, as shown in the next theorem.

---

**THEOREM 2.3 (CONTINUED)**   **Limit Laws for One-Sided Limits**
Laws 1–6 hold with $\lim\limits_{x \to a}$ replaced with $\lim\limits_{x \to a^+}$ or $\lim\limits_{x \to a^-}$. Law 7 is modified as follows. Assume $n > 0$ is an integer.

**7. Root**

**a.** $\lim\limits_{x \to a^+} (f(x))^{1/n} = \left( \lim\limits_{x \to a^+} f(x) \right)^{1/n}$, provided $f(x) \geq 0$, for $x$ near $a$ with
$$x > a, \text{ if } n \text{ is even}$$

**b.** $\lim\limits_{x \to a^-} (f(x))^{1/n} = \left( \lim\limits_{x \to a^-} f(x) \right)^{1/n}$, provided $f(x) \geq 0$, for $x$ near $a$ with
$$x < a, \text{ if } n \text{ is even}$$

---

**EXAMPLE 5**   **Calculating left- and right-sided limits**  Let

$$f(x) = \begin{cases} -2x + 4 & \text{if } x \leq 1 \\ \sqrt{x - 1} & \text{if } x > 1. \end{cases}$$

Find the values of $\lim\limits_{x \to 1^-} f(x)$, $\lim\limits_{x \to 1^+} f(x)$, and $\lim\limits_{x \to 1} f(x)$, or state that they do not exist.

**SOLUTION**  Notice that $f(x) = -2x + 4$, for $x \leq 1$. Therefore,

$$\lim_{x \to 1^-} f(x) = \lim_{x \to 1^-} (-2x + 4) = 2. \quad \text{Theorem 2.2}$$

For $x > 1$, note that $x - 1 > 0$; it follows that

$$\lim_{x \to 1^+} f(x) = \lim_{x \to 1^+} \sqrt{x - 1} = \sqrt{\lim_{x \to 1^+} (x - 1)} = 0. \quad \text{Law 7 for one-sided limits}$$

Because $\lim\limits_{x \to 1^-} f(x) \neq \lim\limits_{x \to 1^+} f(x)$, $\lim\limits_{x \to 1} f(x)$ does not exist by Theorem 2.1. The graph of $f$ (**Figure 2.16**) is consistent with these findings.

*Related Exercises 72–73* ◄

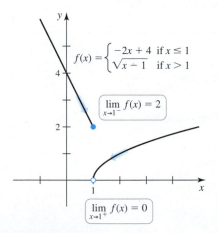

$$f(x) = \begin{cases} -2x + 4 & \text{if } x \leq 1 \\ \sqrt{x - 1} & \text{if } x > 1 \end{cases}$$

$$\lim_{x \to 1^-} f(x) = 2$$

$$\lim_{x \to 1^+} f(x) = 0$$

**Figure 2.16**

## Other Techniques

So far, we have evaluated limits by direct substitution. A more challenging problem is finding $\lim_{x \to a} f(x)$ when the limit exists, but $\lim_{x \to a} f(x) \neq f(a)$. Two typical cases are shown in **Figure 2.17**. In the first case, $f(a)$ is defined, but it is not equal to $\lim_{x \to a} f(x)$; in the second case, $f(a)$ is not defined at all. In both cases, direct substitution does not work—we need a new strategy. One way to evaluate a challenging limit is to replace it with an equivalent limit that *can* be evaluated by direct substitution. Example 6 illustrates two common scenarios.

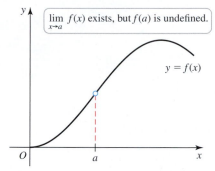

**Figure 2.17**

**EXAMPLE 6**   **Other techniques**  Evaluate the following limits.

**a.** $\displaystyle \lim_{x \to 2} \frac{x^2 - 6x + 8}{x^2 - 4}$          **b.** $\displaystyle \lim_{x \to 1} \frac{\sqrt{x} - 1}{x - 1}$

**SOLUTION**

**a. Factor and cancel**  This limit cannot be found by direct substitution because the denominator is zero when $x = 2$. Instead, the numerator and denominator are factored; then, assuming $x \neq 2$, we cancel like factors:

$$\frac{x^2 - 6x + 8}{x^2 - 4} = \frac{(x - 2)(x - 4)}{(x - 2)(x + 2)} = \frac{x - 4}{x + 2}.$$

> ➤ The argument used in Example 6a relies on the fact that in the limit process, $x$ approaches 2, but $x \neq 2$. Therefore, we may cancel like factors.

Because $\dfrac{x^2 - 6x + 8}{x^2 - 4} = \dfrac{x - 4}{x + 2}$ whenever $x \neq 2$, the two functions have the same limit as $x$ approaches 2 (**Figure 2.18**). Therefore,

$$\lim_{x \to 2} \frac{x^2 - 6x + 8}{x^2 - 4} = \lim_{x \to 2} \frac{x - 4}{x + 2} = \frac{2 - 4}{2 + 2} = -\frac{1}{2}.$$

**Figure 2.18**

> We multiply the given function by
> $$1 = \frac{\sqrt{x} + 1}{\sqrt{x} + 1}.$$

**b. Use conjugates** This limit was approximated numerically in Example 2 of Section 2.2; we made a conjecture that the value of the limit is $1/2$. Using direct substitution to verify this conjecture fails in this case, because the denominator is zero at $x = 1$. Instead, we first simplify the function by multiplying the numerator and denominator by the *algebraic conjugate* of the numerator. The conjugate of $\sqrt{x} - 1$ is $\sqrt{x} + 1$; therefore,

$$\frac{\sqrt{x} - 1}{x - 1} = \frac{(\sqrt{x} - 1)(\sqrt{x} + 1)}{(x - 1)(\sqrt{x} + 1)} \qquad \text{Rationalize the numerator; multiply by 1.}$$

$$= \frac{x + \sqrt{x} - \sqrt{x} - 1}{(x - 1)(\sqrt{x} + 1)} \qquad \text{Expand the numerator.}$$

$$= \frac{x - 1}{(x - 1)(\sqrt{x} + 1)} \qquad \text{Simplify.}$$

$$= \frac{1}{\sqrt{x} + 1}. \qquad \text{Cancel like factors assuming } x \neq 1.$$

The limit can now be evaluated:

$$\lim_{x \to 1} \frac{\sqrt{x} - 1}{x - 1} = \lim_{x \to 1} \frac{1}{\sqrt{x} + 1} = \frac{1}{1 + 1} = \frac{1}{2}.$$

*Related Exercises 34, 41* ◄

**QUICK CHECK 3** Evaluate
$$\lim_{x \to 5} \frac{x^2 - 7x + 10}{x - 5}. \blacktriangleleft$$

**EXAMPLE 7** **Finding limits** Let $f(x) = \dfrac{x^3 - 6x^2 + 8x}{\sqrt{x - 2}}$. Find the values of $\lim\limits_{x \to 2^-} f(x)$, $\lim\limits_{x \to 2^+} f(x)$, and $\lim\limits_{x \to 2} f(x)$, or state that they do not exist.

**SOLUTION** Because the denominator of $f$ is $\sqrt{x - 2}$, $f(x)$ is defined only when $x - 2 > 0$. Therefore, the domain of $f$ is $x > 2$ and it follows that $\lim\limits_{x \to 2^-} f(x)$ does not exist, which in turn implies that $\lim\limits_{x \to 2} f(x)$ does not exist (Theorem 2.1). To evaluate $\lim\limits_{x \to 2^+} f(x)$, factor the numerator of $f$ and simplify:

$$\lim_{x \to 2^+} f(x) = \lim_{x \to 2^+} \frac{x(x - 2)(x - 4)}{(x - 2)^{1/2}} \qquad \begin{array}{l}\text{Factor numerator;} \\ \sqrt{x - 2} = (x - 2)^{1/2}.\end{array}$$

$$= \lim_{x \to 2^+} x(x - 4)(x - 2)^{1/2} \qquad \begin{array}{l}\text{Simplify;} \\ (x - 2)/(x - 2)^{1/2} = (x - 2)^{1/2}.\end{array}$$

$$= \lim_{x \to 2^+} x \cdot \lim_{x \to 2^+} (x - 4) \cdot \lim_{x \to 2^+} (x - 2)^{1/2} \quad \text{Theorem 2.3}$$

$$= 2(-2)(0) = 0. \qquad \begin{array}{l}\text{Theorem 2.2, Law 7 for} \\ \text{one-sided limits}\end{array}$$

*Related Exercises 69–70* ◄

> The Squeeze Theorem is also called the Pinching Theorem or the Sandwich Theorem.

## The Squeeze Theorem

The *Squeeze Theorem* provides another useful method for calculating limits. Suppose the functions $f$ and $h$ have the same limit $L$ at $a$ and assume the function $g$ is trapped between $f$ and $h$ (**Figure 2.19**). The Squeeze Theorem says that $g$ must also have the limit $L$ at $a$. A proof of this theorem is assigned in Exercise 68 of Section 2.7.

---
**THEOREM 2.5** **The Squeeze Theorem**
Assume the functions $f$, $g$, and $h$ satisfy $f(x) \leq g(x) \leq h(x)$ for all values of $x$ near $a$, except possibly at $a$. If $\lim\limits_{x \to a} f(x) = \lim\limits_{x \to a} h(x) = L$, then $\lim\limits_{x \to a} g(x) = L$.

---

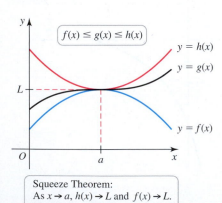

$f(x) \leq g(x) \leq h(x)$

$y = h(x)$

$y = g(x)$

$y = f(x)$

Squeeze Theorem:
As $x \to a$, $h(x) \to L$ and $f(x) \to L$.
Therefore, $g(x) \to L$.

**Figure 2.19**

**EXAMPLE 8** **Applying the Squeeze Theorem** Use the Squeeze Theorem to verify that $\lim\limits_{x \to 0} x^2 \sin(1/x) = 0$.

**SOLUTION** For any real number $\theta$, $-1 \leq \sin \theta \leq 1$. Letting $\theta = 1/x$ for $x \neq 0$, it follows that

$$-1 \leq \sin \frac{1}{x} \leq 1.$$

Noting that $x^2 > 0$ for $x \neq 0$, each term in this inequality is multiplied by $x^2$:

$$-x^2 \leq x^2 \sin \frac{1}{x} \leq x^2.$$

These inequalities are illustrated in **Figure 2.20**. Because $\lim\limits_{x\to 0} x^2 = \lim\limits_{x\to 0} (-x^2) = 0$, the Squeeze Theorem implies that $\lim\limits_{x\to 0} x^2 \sin (1/x) = 0$.

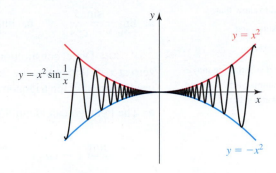

**QUICK CHECK 4** Suppose $f$ satisfies $1 \leq f(x) \leq 1 + \dfrac{x^2}{6}$ for all values of $x$ near zero. Find $\lim\limits_{x\to 0} f(x)$, if possible. ◄

**Figure 2.20**

*Related Exercises 81–82* ◄

## Trigonometric Limits

The Squeeze Theorem is used to evaluate two important limits that play a crucial role in establishing fundamental properties of the trigonometric functions in Section 2.6. These limits are

$$\lim_{x\to 0} \sin x = 0 \quad \text{and} \quad \lim_{x\to 0} \cos x = 1.$$

To verify these limits, the following inequalities, both valid on $-\pi/2 < x < \pi/2$, are used:

$$-|x| \leq \sin x \leq |x| \ \textbf{(Figure 2.21a)} \quad \text{and} \quad 1 - |x| \leq \cos x \leq 1 \ \textbf{(Figure 2.21b)}.$$

➤ See Exercise 104 for a geometric derivation of the inequalities illustrated in Figure 2.21.

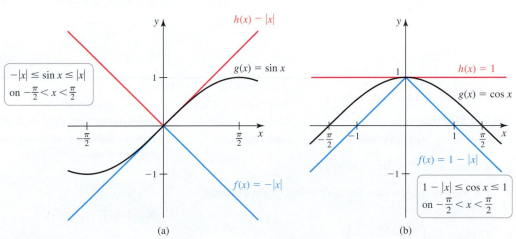

**Figure 2.21**

Let's begin with the inequality illustrated in Figure 2.21a. Letting $f(x) = -|x|$, $g(x) = \sin x$, and $h(x) = |x|$, we see that $g$ is trapped between $f$ and $h$ on $-\pi/2 < x < \pi/2$. Because $\lim\limits_{x\to 0} f(x) = \lim\limits_{x\to 0} h(x) = 0$ (Exercise 85), the Squeeze Theorem implies that $\lim\limits_{x\to 0} g(x) = \lim\limits_{x\to 0} \sin x = 0$.

To evaluate $\lim\limits_{x\to 0} \cos x$, let $f(x) = 1 - |x|$, $g(x) = \cos x$, and $h(x) = 1$, and notice that $g$ is again trapped between $f$ and $h$ on $-\pi/2 < x < \pi/2$ (Figure 2.21b). Because $\lim\limits_{x\to 0} f(x) = \lim\limits_{x\to 0} h(x) = 1$, the Squeeze Theorem implies that $\lim\limits_{x\to 0} g(x) = \lim\limits_{x\to 0} \cos x = 1$.

Having established that $\lim\limits_{x\to 0} \sin x = 0$ and $\lim\limits_{x\to 0} \cos x = 1$, we can evaluate more complicated limits involving trigonometric functions.

> ➤ Notice that direct substitution can be used to evaluate the important trigonometric limits just derived. That is, $\lim\limits_{x\to 0} \sin x = \sin 0 = 0$ and $\lim\limits_{x\to 0} \cos x = \cos 0 = 1$. In Section 2.6, these limits are used to show that
> $$\lim_{x\to a} \sin x = \sin a \text{ and}$$
> $$\lim_{x\to a} \cos x = \cos a.$$
> In other words, direct substitution may be used to evaluate limits of the sine and cosine functions for any value of $a$.

**EXAMPLE 9   Trigonometric limits**  Evaluate the following trigonometric limits.

**a.** $\lim\limits_{x\to 0} \dfrac{\sin^2 x}{1 - \cos x}$    **b.** $\lim\limits_{x\to 0} \dfrac{1 - \cos 2x}{\sin x}$

**SOLUTION**  Direct substitution does not work for either limit because the denominator is zero at $x = 0$ in both parts (a) and (b). Instead, trigonometric identities and limit laws are used to simplify the trigonometric function.

**a.** The Pythagorean identity $\sin^2 x + \cos^2 x = 1$ is used to simplify the function.

$$\lim_{x\to 0} \frac{\sin^2 x}{1 - \cos x} = \lim_{x\to 0} \frac{1 - \cos^2 x}{1 - \cos x} \qquad \sin^2 x = 1 - \cos^2 x$$

$$= \lim_{x\to 0} \frac{(1 - \cos x)(1 + \cos x)}{1 - \cos x} \qquad \text{Factor the numerator.}$$

$$= \lim_{x\to 0} (1 + \cos x) \qquad \text{Simplify.}$$

$$= \underbrace{\lim_{x\to 0} 1}_{1} + \underbrace{\lim_{x\to 0} \cos x}_{1} = 2 \qquad \text{Theorem 2.3 (Law 1)}$$

**b.** We use the identity $\cos 2x = \cos^2 x - \sin^2 x$ to simplify the function.

$$\lim_{x\to 0} \frac{1 - \cos 2x}{\sin x} = \lim_{x\to 0} \frac{1 - (\cos^2 x - \sin^2 x)}{\sin x} \qquad \cos 2x = \cos^2 x - \sin^2 x$$

$$= \lim_{x\to 0} \frac{1 - \cos^2 x + \sin^2 x}{\sin x} \qquad \text{Distribute.}$$

$$= \lim_{x\to 0} \frac{2 \sin^2 x}{\sin x} \qquad 1 - \cos^2 x = \sin^2 x$$

$$= 2 \underbrace{\lim_{x\to 0} \sin x}_{0} = 0 \qquad \text{Simplify; Theorem 2.3 (Law 3).}$$

*Related Exercises 60–61* ◀

---

# SECTION 2.3 EXERCISES

## Getting Started

1.  How is $\lim\limits_{x\to a} p(x)$ calculated if $p$ is a polynomial function?

2.  Evaluate $\lim\limits_{x\to 1} (x^3 + 3x^2 - 3x + 1)$.

3.  For what values of $a$ does $\lim\limits_{x\to a} r(x) = r(a)$ if $r$ is a rational function?

4.  Evaluate $\lim\limits_{x\to 4} \left( \dfrac{x^2 - 4x - 1}{3x - 1} \right)$.

5.  Explain why $\lim\limits_{x\to 3} \dfrac{x^2 - 7x + 12}{x - 3} = \lim\limits_{x\to 3} (x - 4)$ and then evaluate $\lim\limits_{x\to 3} \dfrac{x^2 - 7x + 12}{x - 3}$.

6.  Evaluate $\lim\limits_{x\to 5} \left( \dfrac{4x^2 - 100}{x - 5} \right)$.

**7–13.** *Assume $\lim\limits_{x\to 1} f(x) = 8$, $\lim\limits_{x\to 1} g(x) = 3$, and $\lim\limits_{x\to 1} h(x) = 2$. Compute the following limits and state the limit laws used to justify your computations.*

7.  $\lim\limits_{x\to 1} (4f(x))$

8.  $\lim\limits_{x\to 1} \dfrac{f(x)}{h(x)}$

9.  $\lim\limits_{x\to 1} (f(x) - g(x))$

10. $\lim\limits_{x\to 1} (f(x)h(x))$

11. $\lim\limits_{x\to 1} \dfrac{f(x)}{g(x) - h(x)}$

12. $\lim\limits_{x\to 1} \sqrt[3]{f(x)g(x) + 3}$

13. $\lim\limits_{x\to 1} (f(x))^{2/3}$

14. How are $\lim\limits_{x\to a^-} p(x)$ and $\lim\limits_{x\to a^+} p(x)$ calculated if $p$ is a polynomial function?

**15.** Suppose

$$g(x) = \begin{cases} 2x + 1 & \text{if } x \neq 0 \\ 5 & \text{if } x = 0. \end{cases}$$

Compute $g(0)$ and $\lim_{x \to 0} g(x)$.

**16.** Suppose

$$f(x) = \begin{cases} 4 & \text{if } x \leq 3 \\ x + 2 & \text{if } x > 3. \end{cases}$$

Compute $\lim_{x \to 3^-} f(x)$ and $\lim_{x \to 3^+} f(x)$. Then explain why $\lim_{x \to 3} f(x)$ does not exist.

**17.** Suppose $p$ and $q$ are polynomials. If $\lim_{x \to 0} \dfrac{p(x)}{q(x)} = 10$ and $q(0) = 2$, find $p(0)$.

**18.** Suppose $\lim_{x \to 2} f(x) = \lim_{x \to 2} h(x) = 5$. Find $\lim_{x \to 2} g(x)$, where $f(x) \leq g(x) \leq h(x)$, for all $x$.

### Practice Exercises

**19–70. Evaluating limits** *Find the following limits or state that they do not exist. Assume a, b, c, and k are fixed real numbers.*

**19.** $\lim_{x \to 4} (3x - 7)$

**20.** $\lim_{x \to 1} (-2x + 5)$

**21.** $\lim_{x \to -9} 5x$

**22.** $\lim_{x \to 6} 4$

**23.** $\lim_{x \to 1} (2x^3 - 3x^2 + 4x + 5)$

**24.** $\lim_{t \to -2} (t^2 + 5t + 7)$

**25.** $\lim_{x \to 1} \dfrac{5x^2 + 6x + 1}{8x - 4}$

**26.** $\lim_{t \to 3} \sqrt[3]{t^2 - 10}$

**27.** $\lim_{p \to 2} \dfrac{3p}{\sqrt{4p + 1} - 1}$

**28.** $\lim_{x \to 2} (x^2 - x)^5$

**29.** $\lim_{x \to 3} \dfrac{-5x}{\sqrt{4x - 3}}$

**30.** $\lim_{h \to 0} \dfrac{3}{\sqrt{16 + 3h} + 4}$

**31.** $\lim_{x \to 2} (5x - 6)^{3/2}$

**32.** $\lim_{h \to 0} \dfrac{100}{(10h - 1)^{11} + 2}$

**33.** $\lim_{x \to 1} \dfrac{x^2 - 1}{x - 1}$

**34.** $\lim_{x \to 3} \dfrac{x^2 - 2x - 3}{x - 3}$

**35.** $\lim_{x \to 4} \dfrac{x^2 - 16}{4 - x}$

**36.** $\lim_{t \to 2} \dfrac{3t^2 - 7t + 2}{2 - t}$

**37.** $\lim_{x \to b} \dfrac{(x - b)^{50} - x + b}{x - b}$

**38.** $\lim_{x \to -b} \dfrac{(x + b)^7 + (x + b)^{10}}{4(x + b)}$

**39.** $\lim_{x \to -1} \dfrac{(2x - 1)^2 - 9}{x + 1}$

**40.** $\lim_{h \to 0} \dfrac{\dfrac{1}{5 + h} - \dfrac{1}{5}}{h}$

**41.** $\lim_{x \to 9} \dfrac{\sqrt{x} - 3}{x - 9}$

**42.** $\lim_{w \to 1} \left( \dfrac{1}{w^2 - w} - \dfrac{1}{w - 1} \right)$

**43.** $\lim_{t \to 5} \left( \dfrac{1}{t^2 - 4t - 5} - \dfrac{1}{6(t - 5)} \right)$

**44.** $\lim_{t \to 3} \left( \left( 4t - \dfrac{2}{t - 3} \right)(6 + t - t^2) \right)$

**45.** $\lim_{x \to a} \dfrac{x - a}{\sqrt{x} - \sqrt{a}}, a > 0$

**46.** $\lim_{x \to a} \dfrac{x^2 - a^2}{\sqrt{x} - \sqrt{a}}, a > 0$

**47.** $\lim_{h \to 0} \dfrac{\sqrt{16 + h} - 4}{h}$

**48.** $\lim_{x \to c} \dfrac{x^2 - 2cx + c^2}{x - c}$

**49.** $\lim_{x \to 4} \dfrac{\dfrac{1}{x} - \dfrac{1}{4}}{x - 4}$

**50.** $\lim_{x \to 3} \dfrac{\dfrac{1}{x^2 + 2x} - \dfrac{1}{15}}{x - 3}$

**51.** $\lim_{x \to 1} \dfrac{\sqrt{10x - 9} - 1}{x - 1}$

**52.** $\lim_{x \to 2} \left( \dfrac{1}{x - 2} - \dfrac{2}{x^2 - 2x} \right)$

**53.** $\lim_{h \to 0} \dfrac{(5 + h)^2 - 25}{h}$

**54.** $\lim_{w \to -k} \dfrac{w^2 + 5kw + 4k^2}{w^2 + kw}, k \neq 0$

**55.** $\lim_{x \to 1} \dfrac{x - 1}{\sqrt{x} - 1}$

**56.** $\lim_{x \to 1} \dfrac{x - 1}{\sqrt{4x + 5} - 3}$

**57.** $\lim_{x \to 4} \dfrac{3(x - 4)\sqrt{x + 5}}{3 - \sqrt{x + 5}}$

**58.** $\lim_{x \to 0} \dfrac{x}{\sqrt{cx + 1} - 1}, c \neq 0$

**59.** $\lim_{x \to 0} x \cos x$

**60.** $\lim_{x \to 0} \dfrac{\sin 2x}{\sin x}$

**61.** $\lim_{x \to 0} \dfrac{1 - \cos x}{\cos^2 x - 3 \cos x + 2}$

**62.** $\lim_{x \to 0} \dfrac{\cos x - 1}{\cos^2 x - 1}$

**63.** $\lim_{x \to 0^-} \dfrac{x^2 - x}{|x|}$

**64.** $\lim_{w \to 3^-} \dfrac{|w - 3|}{w^2 - 7w + 12}$

**65.** $\lim_{t \to 2^+} \dfrac{|2t - 4|}{t^2 - 4}$

**66.** $\lim_{x \to -1} g(x)$, where $g(x) = \begin{cases} \dfrac{x^2 - 1}{x + 1} & \text{if } x < -1 \\ -2 & \text{if } x \geq -1 \end{cases}$

**67.** $\lim_{x \to 3} \dfrac{x - 3}{|x - 3|}$

**68.** $\lim_{x \to 5} \dfrac{|x - 5|}{x^2 - 25}$

**69.** $\lim_{x \to 1^-} \dfrac{x^3 + 1}{\sqrt{x} - 1}$

**70.** $\lim_{x \to 1^+} \dfrac{x - 1}{\sqrt{x^2 - 1}}$

**71. Explain why or why not** Determine whether the following statements are true and give an explanation or counterexample. Assume $a$ and $L$ are finite numbers.

**a.** If $\lim_{x \to a} f(x) = L$, then $f(a) = L$.

**b.** If $\lim_{x \to a^-} f(x) = L$, then $\lim_{x \to a^+} f(x) = L$.

**c.** If $\lim_{x \to a} f(x) = L$ and $\lim_{x \to a} g(x) = L$, then $f(a) = g(a)$.

**d.** The limit $\lim_{x \to a} \dfrac{f(x)}{g(x)}$ does not exist if $g(a) = 0$.

**e.** If $\lim_{x \to 1^+} \sqrt{f(x)} = \sqrt{\lim_{x \to 1^+} f(x)}$, it follows that

$$\lim_{x \to 1} \sqrt{f(x)} = \sqrt{\lim_{x \to 1} f(x)}.$$

**72. One-sided limits** Let

$$g(x) = \begin{cases} 5x - 15 & \text{if } x < 4 \\ \sqrt{6x + 1} & \text{if } x \geq 4. \end{cases}$$

Compute the following limits or state that they do not exist.

**a.** $\lim\limits_{x \to 4^-} g(x)$    **b.** $\lim\limits_{x \to 4^+} g(x)$    **c.** $\lim\limits_{x \to 4} g(x)$

**73. One-sided limits** Let

$$f(x) = \begin{cases} x^2 + 1 & \text{if } x < -1 \\ \sqrt{x + 1} & \text{if } x \geq -1. \end{cases}$$

Compute the following limits or state that they do not exist.

**a.** $\lim\limits_{x \to -1^-} f(x)$    **b.** $\lim\limits_{x \to -1^+} f(x)$    **c.** $\lim\limits_{x \to -1} f(x)$

**74. One-sided limits** Let

$$f(x) = \begin{cases} 0 & \text{if } x \leq -5 \\ \sqrt{25 - x^2} & \text{if } -5 < x < 5 \\ 3x & \text{if } x \geq 5. \end{cases}$$

Compute the following limits or state that they do not exist.

**a.** $\lim\limits_{x \to -5^-} f(x)$    **b.** $\lim\limits_{x \to -5^+} f(x)$    **c.** $\lim\limits_{x \to -5} f(x)$

**d.** $\lim\limits_{x \to 5^-} f(x)$    **e.** $\lim\limits_{x \to 5^+} f(x)$    **f.** $\lim\limits_{x \to 5} f(x)$

**75. One-sided limits**

**a.** Evaluate $\lim\limits_{x \to 2^+} \sqrt{x - 2}$.

**b.** Explain why $\lim\limits_{x \to 2^-} \sqrt{x - 2}$ does not exist.

**76. One-sided limits**

**a.** Evaluate $\lim\limits_{x \to 3^-} \sqrt{\dfrac{x - 3}{2 - x}}$.

**b.** Explain why $\lim\limits_{x \to 3^+} \sqrt{\dfrac{x - 3}{2 - x}}$ does not exist.

**T 77. Electric field** The magnitude of the electric field at a point $x$ meters from the midpoint of a 0.1-m line of charge is given by

$$E(x) = \frac{4.35}{x\sqrt{x^2 + 0.01}} \text{ (in units of newtons per coulomb, N/C).}$$

Evaluate $\lim\limits_{x \to 10} E(x)$.

**78. Torricelli's law** A cylindrical tank is filled with water to a depth of 9 meters. At $t = 0$, a drain in the bottom of the tank is opened, and water flows out of the tank. The depth of water in the tank (measured from the bottom of the tank) $t$ seconds after the drain is opened is approximated by $d(t) = (3 - 0.015t)^2$, for $0 \leq t \leq 200$. Evaluate and interpret $\lim\limits_{t \to 200^-} d(t)$.

**79. Limit of the radius of a cylinder** A right circular cylinder with a height of 10 cm and a surface area of $S$ cm$^2$ has a radius given by

$$r(S) = \frac{1}{2}\left(\sqrt{100 + \frac{2S}{\pi}} - 10\right).$$

Find $\lim\limits_{S \to 0^+} r(S)$ and interpret your result.

**80. A problem from relativity theory** Suppose a spaceship of length $L_0$ travels at a high speed $v$ relative to an observer. To the observer, the ship appears to have a smaller length given by the *Lorentz contraction formula*

$$L = L_0\sqrt{1 - \frac{v^2}{c^2}},$$

where $c$ is the speed of light.

**a.** What is the observed length $L$ of the ship if it is traveling at 50% of the speed of light?

**b.** What is the observed length $L$ of the ship if it is traveling at 75% of the speed of light?

**c.** In parts (a) and (b), what happens to $L$ as the speed of the ship increases?

**d.** Find $\lim\limits_{v \to c^-} L_0\sqrt{1 - \dfrac{v^2}{c^2}}$ and explain the significance of this limit.

**T 81. a.** Show that $-|x| \leq x \sin\dfrac{1}{x} \leq |x|$, for $x \neq 0$.

**b.** Illustrate the inequalities in part (a) with a graph.

**c.** Show that $\lim\limits_{x \to 0} x \sin\dfrac{1}{x} = 0$.

**T 82. A cosine limit** It can be shown that $1 - \dfrac{x^2}{2} \leq \cos x \leq 1$, for $x$ near 0.

**a.** Illustrate these inequalities with a graph.

**b.** Use these inequalities to evaluate $\lim\limits_{x \to 0} \cos x$.

**T 83. A sine limit** It can be shown that $1 - \dfrac{x^2}{6} \leq \dfrac{\sin x}{x} \leq 1$, for $x$ near 0.

**a.** Illustrate these inequalities with a graph.

**b.** Use these inequalities to evaluate $\lim\limits_{x \to 0} \dfrac{\sin x}{x}$.

**T 84. A secant limit**

**a.** Draw a graph to verify that $0 \leq x^2 \sec x^2 \leq x^4 + x^2$, for $x$ near 0.

**b.** Use the Squeeze Theorem to determine $\lim\limits_{x \to 0} x^2 \sec x^2$.

**85. Absolute value** Show that $\lim\limits_{x \to 0} |x| = 0$ by first evaluating $\lim\limits_{x \to 0^-} |x|$ and $\lim\limits_{x \to 0^+} |x|$. Recall that

$$|x| = \begin{cases} x & \text{if } x \geq 0 \\ -x & \text{if } x < 0. \end{cases}$$

**86. Absolute value limit** Show that $\lim\limits_{x \to a} |x| = |a|$, for any real number $a$. (*Hint:* Consider the cases $a < 0$, $a = 0$, and $a > 0$.)

**87. Finding a constant** Suppose

$$f(x) = \begin{cases} \dfrac{x^2 - 5x + 6}{x - 3} & \text{if } x \neq 3 \\ a & \text{if } x = 3. \end{cases}$$

Determine the value of the constant $a$ for which $\lim\limits_{x \to 3} f(x) = f(3)$.

**88. Finding a constant** Suppose

$$f(x) = \begin{cases} 3x + b & \text{if } x \leq 2 \\ x - 2 & \text{if } x > 2. \end{cases}$$

Determine the value of the constant $b$ for which $\lim\limits_{x \to 2} f(x)$ exists and state the value of the limit, if possible.

**89. Finding a constant** Suppose

$$g(x) = \begin{cases} x^2 - 5x & \text{if } x \leq -1 \\ ax^3 - 7 & \text{if } x > -1. \end{cases}$$

Determine the value of the constant $a$ for which $\lim\limits_{x \to -1} g(x)$ exists and state the value of the limit, if possible.

**90–94. Useful factorization formula** *Calculate the following limits using the factorization formula*

$$x^n - a^n = (x - a)(x^{n-1} + ax^{n-2} + a^2x^{n-3} + \cdots + a^{n-2}x + a^{n-1}),$$

*where n is a positive integer and a is a real number.*

**90.** $\displaystyle\lim_{x\to 2} \frac{x^5 - 32}{x - 2}$

**91.** $\displaystyle\lim_{x\to 1} \frac{x^6 - 1}{x - 1}$

**92.** $\displaystyle\lim_{x\to -1} \frac{x^7 + 1}{x + 1}$ (*Hint:* Use the formula for $x^7 - a^7$ with $a = -1$.)

**93.** $\displaystyle\lim_{x\to a} \frac{x^5 - a^5}{x - a}$

**94.** $\displaystyle\lim_{x\to a} \frac{x^n - a^n}{x - a}$, for any positive integer $n$

## Explorations and Challenges

**95. Even function limits** Suppose $f$ is an even function where $\displaystyle\lim_{x\to 1^-} f(x) = 5$ and $\displaystyle\lim_{x\to 1^+} f(x) = 6$. Find $\displaystyle\lim_{x\to -1^-} f(x)$ and $\displaystyle\lim_{x\to -1^+} f(x)$.

**96. Odd function limits** Suppose $g$ is an odd function where $\displaystyle\lim_{x\to 1^-} g(x) = 5$ and $\displaystyle\lim_{x\to 1^+} g(x) = 6$. Find $\displaystyle\lim_{x\to -1^-} g(x)$ and $\displaystyle\lim_{x\to -1^+} g(x)$.

**97.** Evaluate $\displaystyle\lim_{x\to 1} \frac{\sqrt[3]{x} - 1}{x - 1}$. (*Hint:* $x - 1 = (\sqrt[3]{x})^3 - 1^3$.)

**98.** Evaluate $\displaystyle\lim_{x\to 16} \frac{\sqrt[4]{x} - 2}{x - 16}$.

**99. Creating functions satisfying given limit conditions** Find functions $f$ and $g$ such that $\displaystyle\lim_{x\to 1} f(x) = 0$ and $\displaystyle\lim_{x\to 1} (f(x)g(x)) = 5$.

**100. Creating functions satisfying given limit conditions** Find a function $f$ satisfying $\displaystyle\lim_{x\to 1} \frac{f(x)}{x - 1} = 2$.

**101. Finding constants** Find constants $b$ and $c$ in the polynomial $p(x) = x^2 + bx + c$ such that $\displaystyle\lim_{x\to 2} \frac{p(x)}{x - 2} = 6$. Are the constants unique?

**102.** If $\displaystyle\lim_{x\to 1} f(x) = 4$, find $\displaystyle\lim_{x\to -1} f(x^2)$.

**103.** Suppose $g(x) = f(1 - x)$ for all $x$, $\displaystyle\lim_{x\to 1^+} f(x) = 4$, and $\displaystyle\lim_{x\to 1^-} f(x) = 6$. Find $\displaystyle\lim_{x\to 0^+} g(x)$ and $\displaystyle\lim_{x\to 0^-} g(x)$.

**104. Two trigonometric inequalities** Consider the angle $\theta$ in standard position in a unit circle where $0 \le \theta < \pi/2$ or $-\pi/2 < \theta < 0$ (use both figures).

   **a.** Show that $|AC| = |\sin\theta|$, for $-\pi/2 < \theta < \pi/2$. (*Hint:* Consider the cases $0 < \theta < \pi/2$ and $-\pi/2 < \theta < 0$ separately.)

   **b.** Show that $|\sin\theta| < |\theta|$, for $-\pi/2 < \theta < \pi/2$. (*Hint:* The length of arc $AB$ is $\theta$ if $0 \le \theta < \pi/2$, and is $-\theta$ if $-\pi/2 < \theta < 0$.)

   **c.** Conclude that $-|\theta| \le \sin\theta \le |\theta|$, for $-\pi/2 < \theta < \pi/2$.

   **d.** Show that $0 \le 1 - \cos\theta \le |\theta|$, for $-\pi/2 < \theta < \pi/2$.

   **e.** Show that $1 - |\theta| \le \cos\theta \le 1$, for $-\pi/2 < \theta < \pi/2$.

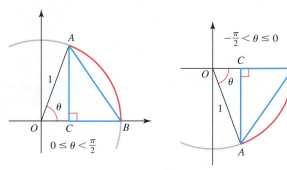

**105. Theorem 2.4a** Given the polynomial

$$p(x) = b_n x^n + b_{n-1} x^{n-1} + \cdots + b_1 x + b_0,$$

prove that $\displaystyle\lim_{x\to a} p(x) = p(a)$ for any value of $a$.

**QUICK CHECK ANSWERS**

**1.** $0, 2$    **2.** $2$    **3.** $3$    **4.** $1$ ◄

# 2.4 Infinite Limits

Two more limit scenarios are frequently encountered in calculus and are discussed in this section and the following section. An *infinite limit* occurs when function values increase or decrease without bound near a point. The other type of limit, known as a *limit at infinity*, occurs when the independent variable $x$ increases or decreases without bound. The ideas behind infinite limits and limits at infinity are quite different. Therefore, it is important to distinguish between these limits and the methods used to calculate them.

## An Overview

To illustrate the differences between infinite limits and limits at infinity, consider the values of $f(x) = 1/x^2$ in Table 2.5. As $x$ approaches 0 from either side, $f(x)$ grows larger and larger. Because $f(x)$ does not approach a finite number as $x$ approaches 0, $\displaystyle\lim_{x\to 0} f(x)$ does not exist.

Nevertheless, we use limit notation and write $\lim\limits_{x\to 0} f(x) = \infty$. The infinity symbol indicates that $f(x)$ grows arbitrarily large as $x$ approaches 0. This is an example of an *infinite limit*; in general, the *dependent variable* becomes arbitrarily large in magnitude as the *independent variable* approaches a finite number.

**Table 2.5**

| $x$ | $f(x) = 1/x^2$ |
|---|---|
| $\pm 0.1$ | 100 |
| $\pm 0.01$ | 10,000 |
| $\pm 0.001$ | 1,000,000 |
| $\downarrow$ | $\downarrow$ |
| 0 | $\infty$ |

$$\lim_{x\to 0} \frac{1}{x^2} = \infty$$

With *limits at infinity*, the opposite occurs: The *dependent variable* approaches a finite number as the *independent variable* becomes arbitrarily large in magnitude. In Table 2.6 we see that $f(x) = 1/x^2$ approaches 0 as $x$ increases. In this case, we write $\lim\limits_{x\to\infty} f(x) = 0$.

**Table 2.6**

| $x$ | $f(x) = 1/x^2$ |
|---|---|
| 10 | 0.01 |
| 100 | 0.0001 |
| 1000 | 0.000001 |
| $\downarrow$ | $\downarrow$ |
| $\infty$ | 0 |

$$\lim_{x\to -\infty} \frac{1}{x^2} = 0 \qquad \lim_{x\to\infty} \frac{1}{x^2} = 0$$

A general picture of these two limit scenarios—occurring with the same function—is shown in **Figure 2.22**.

**Figure 2.22**

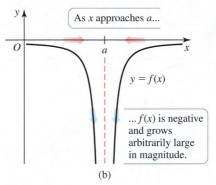

**Figure 2.23**

## Infinite Limits

The following definition of infinite limits is informal, but it is adequate for most functions encountered in this book. A precise definition is given in Section 2.7.

> **DEFINITION** **Infinite Limits**
>
> Suppose $f$ is defined for all $x$ near $a$. If $f(x)$ grows arbitrarily large for all $x$ sufficiently close (but not equal) to $a$ (**Figure 2.23a**), we write
>
> $$\lim_{x \to a} f(x) = \infty$$
>
> and say the limit of $f(x)$ as $x$ approaches $a$ is infinity.
>
>     If $f(x)$ is negative and grows arbitrarily large in magnitude for all $x$ sufficiently close (but not equal) to $a$ (**Figure 2.23b**), we write
>
> $$\lim_{x \to a} f(x) = -\infty$$
>
> and say the limit of $f(x)$ as $x$ approaches $a$ is negative infinity. *In both cases, the limit does not exist.*

**EXAMPLE 1** **Infinite limits** Analyze $\displaystyle\lim_{x \to 1} \frac{x}{(x^2 - 1)^2}$ and $\displaystyle\lim_{x \to -1} \frac{x}{(x^2 - 1)^2}$ using the graph of the function.

**Figure 2.24**

**SOLUTION** The graph of $f(x) = \dfrac{x}{(x^2 - 1)^2}$ (**Figure 2.24**) shows that as $x$ approaches 1 (from either side), the values of $f$ grow arbitrarily large. Therefore, the limit does not exist and we write

$$\lim_{x \to 1} \frac{x}{(x^2 - 1)^2} = \infty.$$

As $x$ approaches $-1$, the values of $f$ are negative and grow arbitrarily large in magnitude; therefore,

$$\lim_{x \to -1} \frac{x}{(x^2 - 1)^2} = -\infty.$$

*Related Exercise 6* ◀

    Example 1 illustrates *two-sided* infinite limits. As with finite limits, we also need to work with right-sided and left-sided infinite limits.

> **DEFINITION** **One-Sided Infinite Limits**
>
> Suppose $f$ is defined for all $x$ near $a$ with $x > a$. If $f(x)$ becomes arbitrarily large for all $x$ sufficiently close to $a$ with $x > a$, we write $\displaystyle\lim_{x \to a^+} f(x) = \infty$ (**Figure 2.25a**).
> The one-sided infinite limits $\displaystyle\lim_{x \to a^+} f(x) = -\infty$ (**Figure 2.25b**), $\displaystyle\lim_{x \to a^-} f(x) = \infty$ (**Figure 2.25c**), and $\displaystyle\lim_{x \to a^-} f(x) = -\infty$ (**Figure 2.25d**) are defined analogously.

**Figure 2.25**

In all the infinite limits illustrated in Figure 2.25, the line $x = a$ is called a *vertical asymptote*; it is a vertical line that is approached by the graph of $f$ as $x$ approaches $a$.

> **QUICK CHECK 1** Sketch the graph of a function and its vertical asymptote that satisfies the conditions $\lim\limits_{x \to 2^+} f(x) = -\infty$ and $\lim\limits_{x \to 2^-} f(x) = \infty$. ◀

> **DEFINITION  Vertical Asymptote**
>
> If $\lim\limits_{x \to a} f(x) = \pm\infty$, $\lim\limits_{x \to a^+} f(x) = \pm\infty$, or $\lim\limits_{x \to a^-} f(x) = \pm\infty$, the line $x = a$ is called a **vertical asymptote** of $f$.

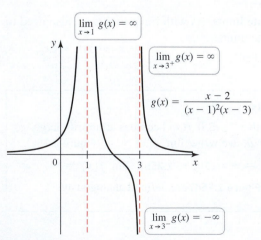

**Figure 2.26**

**EXAMPLE 2    Determining limits graphically** The vertical lines $x = 1$ and $x = 3$ are vertical asymptotes of the function $g(x) = \dfrac{x - 2}{(x - 1)^2 (x - 3)}$. Use Figure 2.26 to analyze the following limits.

**a.** $\lim\limits_{x \to 1} g(x)$    **b.** $\lim\limits_{x \to 3^-} g(x)$    **c.** $\lim\limits_{x \to 3} g(x)$

**SOLUTION**

**a.** The values of $g$ grow arbitrarily large as $x$ approaches 1 from either side. Therefore, $\lim\limits_{x \to 1} g(x) = \infty$.

**b.** The values of $g$ are negative and grow arbitrarily large in magnitude as $x$ approaches 3 from the left, so $\lim\limits_{x \to 3^-} g(x) = -\infty$.

**c.** Note that $\lim\limits_{x \to 3^+} g(x) = \infty$ and $\lim\limits_{x \to 3^-} g(x) = -\infty$. Because $g$ behaves differently as $x \to 3^-$ and as $x \to 3^+$, we do not write $\lim\limits_{x \to 3} g(x) = \infty$, nor do we write $\lim\limits_{x \to 3} g(x) = -\infty$. We simply say that the limit does not exist.

*Related Exercises 7–8* ◀

**Table 2.7**

| $x$ | $\dfrac{5 + x}{x}$ |
|---|---|
| 0.01 | $\dfrac{5.01}{0.01} = 501$ |
| 0.001 | $\dfrac{5.001}{0.001} = 5001$ |
| 0.0001 | $\dfrac{5.0001}{0.0001} = 50{,}001$ |
| $\downarrow$ | $\downarrow$ |
| $0^+$ | $\infty$ |

## Finding Infinite Limits Analytically

Many infinite limits are analyzed using a simple arithmetic property: The fraction $a/b$ grows arbitrarily large in magnitude if $b$ approaches 0 while $a$ remains nonzero and relatively constant. For example, consider the fraction $(5 + x)/x$ for values of $x$ approaching 0 from the right (Table 2.7).

We see that $\dfrac{5 + x}{x} \to \infty$ as $x \to 0^+$ because the numerator $5 + x$ approaches 5 while the denominator is positive and approaches 0. Therefore, we write $\lim\limits_{x \to 0^+} \dfrac{5 + x}{x} = \infty$. Similarly, $\lim\limits_{x \to 0^-} \dfrac{5 + x}{x} = -\infty$ because the numerator approaches 5 while the denominator approaches 0 through negative values.

**EXAMPLE 3**  **Determining limits analytically**  Analyze the following limits.

**a.** $\lim\limits_{x \to 3^+} \dfrac{2 - 5x}{x - 3}$      **b.** $\lim\limits_{x \to 3^-} \dfrac{2 - 5x}{x - 3}$

**SOLUTION**

**a.** As $x \to 3^+$, the numerator $2 - 5x$ approaches $2 - 5(3) = -13$ while the denominator $x - 3$ is positive and approaches 0. Therefore,

$$\lim\limits_{x \to 3^+} \underbrace{\frac{\overbrace{2 - 5x}^{\text{approaches } -13}}{x - 3}}_{\text{positive and approaches } 0} = -\infty.$$

**b.** As $x \to 3^-$, $2 - 5x$ approaches $2 - 5(3) = -13$ while $x - 3$ is negative and approaches 0. Therefore,

$$\lim\limits_{x \to 3^-} \underbrace{\frac{\overbrace{2 - 5x}^{\text{approaches } -13}}{x - 3}}_{\text{negative and approaches } 0} = \infty.$$

These limits imply that the given function has a vertical asymptote at $x = 3$; they also imply that the two-sided limit $\lim\limits_{x \to 3} \dfrac{2x - 5}{x - 3}$ does not exist.

*Related Exercises 21–22* ◄

**QUICK CHECK 2** Analyze $\lim\limits_{x \to 0^+} \dfrac{x - 5}{x}$ and $\lim\limits_{x \to 0^-} \dfrac{x - 5}{x}$ by determining the sign of the numerator and denominator. ◄

**EXAMPLE 4**  **Determining limits analytically**  Analyze $\lim\limits_{x \to -4^+} \dfrac{-x^3 + 5x^2 - 6x}{-x^3 - 4x^2}$.

**SOLUTION**  First we factor and simplify, assuming $x \neq 0$:

$$\frac{-x^3 + 5x^2 - 6x}{-x^3 - 4x^2} = \frac{-x(x - 2)(x - 3)}{-x^2(x + 4)} = \frac{(x - 2)(x - 3)}{x(x + 4)}.$$

➤ We can assume $x \neq 0$ because we are considering function values near $x = -4$.

As $x \to -4^+$, we find that

$$\lim\limits_{x \to -4^+} \frac{-x^3 + 5x^2 - 6x}{-x^3 - 4x^2} = \lim\limits_{x \to -4^+} \frac{\overbrace{(x - 2)(x - 3)}^{\text{approaches } 42}}{\underbrace{x(x + 4)}_{\substack{\text{negative and} \\ \text{approaches } 0}}} = -\infty.$$

This limit implies that the given function has a vertical asymptote at $x = -4$.

**QUICK CHECK 3** Verify that $x(x + 4) \to 0$ through negative values as $x \to -4^+$. ◄

*Related Exercises 28, 31* ◄

> Example 5 illustrates that $f(x)/g(x)$ might not grow arbitrarily large in magnitude if *both* $f(x)$ and $g(x)$ approach 0. Such limits are called *indeterminate forms*; they are examined in detail in Section 4.7.

> It is permissible to cancel the factor $x - 1$ in $\lim\limits_{x \to 1} \dfrac{(x-1)(x-3)}{(x-1)(x+1)}$ because $x$ approaches 1 but is not equal to 1. Therefore, $x - 1 \neq 0$.

**EXAMPLE 5**   **Location of vertical asymptotes**  Let $f(x) = \dfrac{x^2 - 4x + 3}{x^2 - 1}$. Determine the following limits and find the vertical asymptotes of $f$. Verify your work with a graphing utility.

**a.** $\lim\limits_{x \to 1} f(x)$     **b.** $\lim\limits_{x \to -1^-} f(x)$     **c.** $\lim\limits_{x \to -1^+} f(x)$

**SOLUTION**

**a.** Notice that as $x \to 1$, both the numerator and denominator of $f$ approach 0, and the function is undefined at $x = 1$. To compute $\lim\limits_{x \to 1} f(x)$, we first factor:

$$\lim_{x \to 1} f(x) = \lim_{x \to 1} \frac{x^2 - 4x + 3}{x^2 - 1}$$

$$= \lim_{x \to 1} \frac{(x-1)(x-3)}{(x-1)(x+1)} \qquad \text{Factor.}$$

$$= \lim_{x \to 1} \frac{(x-3)}{(x+1)} \qquad \text{Cancel like factors, } x \neq 1.$$

$$= \frac{1-3}{1+1} = -1. \qquad \text{Substitute } x = 1.$$

Therefore, $\lim\limits_{x \to 1} f(x) = -1$ (even though $f(1)$ is undefined). The line $x = 1$ is *not* a vertical asymptote of $f$.

**b.** In part (a) we showed that

$$f(x) = \frac{x^2 - 4x + 3}{x^2 - 1} = \frac{x-3}{x+1}, \quad \text{provided } x \neq 1.$$

We use this fact again. As $x$ approaches $-1$ from the left, the one-sided limit is

$$\lim_{x \to -1^-} f(x) = \lim_{x \to -1^-} \frac{\overbrace{x - 3}^{\text{approaches } -4}}{\underbrace{x + 1}_{\substack{\text{negative and} \\ \text{approaches } 0}}} = \infty.$$

**c.** As $x$ approaches $-1$ from the right, the one-sided limit is

$$\lim_{x \to -1^+} f(x) = \lim_{x \to -1^+} \frac{\overbrace{x - 3}^{\text{approaches } -4}}{\underbrace{x + 1}_{\substack{\text{positive and} \\ \text{approaches } 0}}} = -\infty.$$

The infinite limits $\lim\limits_{x \to -1^+} f(x) = -\infty$ and $\lim\limits_{x \to -1^-} f(x) = \infty$ each imply that the line $x = -1$ is a vertical asymptote of $f$. The graph of $f$ generated by a graphing utility *may* appear as shown in **Figure 2.27a**. If so, two corrections must be made. A hole should appear in the graph at $(1, -1)$ because $\lim\limits_{x \to 1} f(x) = -1$, but $f(1)$ is undefined.

It is also a good idea to replace the solid vertical line with a dashed line to emphasize that the vertical asymptote is not a part of the graph of $f$ (**Figure 2.27b**).

> ➤ Graphing utilities vary in how they display vertical asymptotes. The errors shown in Figure 2.27a do not occur on all graphing utilities.

Two versions of the graph of $y = \dfrac{x^2 - 4x + 3}{x^2 - 1}$

Calculator graph
(a)

Correct graph
(b)

**Figure 2.27**

**QUICK CHECK 4** The line $x = 2$ is not a vertical asymptote of $y = \dfrac{(x-1)(x-2)}{x-2}$. Why not? ◄

*Related Exercises 45–46* ◄

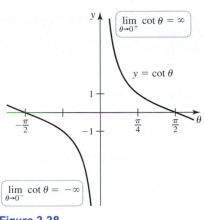

$\lim\limits_{\theta \to 0^+} \cot \theta = \infty$

$y = \cot \theta$

$\lim\limits_{\theta \to 0^-} \cot \theta = -\infty$

**Figure 2.28**

**EXAMPLE 6  Limits of trigonometric functions** Analyze the following limits.

**a.** $\lim\limits_{\theta \to 0^+} \cot \theta$  **b.** $\lim\limits_{\theta \to 0^-} \cot \theta$

**SOLUTION**

**a.** Recall that $\cot \theta = \cos \theta / \sin \theta$. Furthermore, it was shown in Section 2.3 that $\lim\limits_{\theta \to 0^+} \cos \theta = 1$ and that $\sin \theta$ is positive and approaches 0 as $\theta \to 0^+$. Therefore, as $\theta \to 0^+$, $\cot \theta$ becomes arbitrarily large and positive, which means $\lim\limits_{\theta \to 0^+} \cot \theta = \infty$. This limit is consistent with the graph of $\cot \theta$ (**Figure 2.28**), which has a vertical asymptote at $\theta = 0$.

**b.** In this case, $\lim\limits_{\theta \to 0^-} \cos \theta = 1$ and as $\theta \to 0^-$, $\sin \theta \to 0$ with $\sin \theta < 0$. Therefore, as $\theta \to 0^-$, $\cot \theta$ is negative and becomes arbitrarily large in magnitude. It follows that $\lim\limits_{\theta \to 0^-} \cot \theta = -\infty$, as confirmed by the graph of $\cot \theta$.

*Related Exercises 39–40* ◄

## SECTION 2.4 EXERCISES

### Getting Started

1. Explain the meaning of $\lim\limits_{x \to a^+} f(x) = -\infty$.

2. Explain the meaning of $\lim\limits_{x \to a} f(x) = \infty$.

3. What is a vertical asymptote?

4. Consider the function $F(x) = f(x)/g(x)$ with $g(a) = 0$. Does $F$ necessarily have a vertical asymptote at $x = a$? Explain your reasoning.

**T** 5. Compute the values of $f(x) = \dfrac{x+1}{(x-1)^2}$ in the following table and use them to determine $\lim\limits_{x \to 1} f(x)$.

| $x$ | $\dfrac{x+1}{(x-1)^2}$ | $x$ | $\dfrac{x+1}{(x-1)^2}$ |
|---|---|---|---|
| 1.1 | | 0.9 | |
| 1.01 | | 0.99 | |
| 1.001 | | 0.999 | |
| 1.0001 | | 0.9999 | |

6. Use the graph of $f(x) = \dfrac{x}{(x^2 - 2x - 3)^2}$ to determine $\lim\limits_{x \to -1} f(x)$ and $\lim\limits_{x \to 3} f(x)$.

$y = f(x)$

**7.** The graph of $f$ in the figure has vertical asymptotes at $x = 1$ and $x = 2$. Analyze the following limits.

   **a.** $\lim\limits_{x \to 1^-} f(x)$      **b.** $\lim\limits_{x \to 1^+} f(x)$      **c.** $\lim\limits_{x \to 1} f(x)$

   **d.** $\lim\limits_{x \to 2^-} f(x)$      **e.** $\lim\limits_{x \to 2^+} f(x)$      **f.** $\lim\limits_{x \to 2} f(x)$

**8.** The graph of $g$ in the figure has vertical asymptotes at $x = 2$ and $x = 4$. Analyze the following limits.

   **a.** $\lim\limits_{x \to 2^-} g(x)$      **b.** $\lim\limits_{x \to 2^+} g(x)$      **c.** $\lim\limits_{x \to 2} g(x)$

   **d.** $\lim\limits_{x \to 4^-} g(x)$      **e.** $\lim\limits_{x \to 4^+} g(x)$      **f.** $\lim\limits_{x \to 4} g(x)$

**9.** The graph of $h$ in the figure has vertical asymptotes at $x = -2$ and $x = 3$. Analyze the following limits.

   **a.** $\lim\limits_{x \to -2^-} h(x)$      **b.** $\lim\limits_{x \to -2^+} h(x)$      **c.** $\lim\limits_{x \to -2} h(x)$

   **d.** $\lim\limits_{x \to 3^-} h(x)$      **e.** $\lim\limits_{x \to 3^+} h(x)$      **f.** $\lim\limits_{x \to 3} h(x)$

**10.** The graph of $p$ in the figure has vertical asymptotes at $x = -2$ and $x = 3$. Analyze the following limits.

   **a.** $\lim\limits_{x \to -2^-} p(x)$      **b.** $\lim\limits_{x \to -2^+} p(x)$      **c.** $\lim\limits_{x \to -2} p(x)$

   **d.** $\lim\limits_{x \to 3^-} p(x)$      **e.** $\lim\limits_{x \to 3^+} p(x)$      **f.** $\lim\limits_{x \to 3} p(x)$

**T** **11.** Graph the function $f(x) = \dfrac{1}{x^2 - x}$ using a graphing utility with the window $[-1, 2] \times [-10, 10]$. Use your graph to determine the following limits.

   **a.** $\lim\limits_{x \to 0} f(x)$    **b.** $\lim\limits_{x \to 0^+} f(x)$    **c.** $\lim\limits_{x \to 1^-} f(x)$    **d.** $\lim\limits_{x \to 1^+} f(x)$

**T** **12. Analyzing infinite limits graphically** Graph the function $f(x) = x \cot x$ using a graphing utility. (Experiment with your choice of a graphing window.) Use your graph to determine the following limits.

   **a.** $\lim\limits_{x \to \pi^+} f(x)$    **b.** $\lim\limits_{x \to \pi^-} f(x)$    **c.** $\lim\limits_{x \to -\pi^+} f(x)$    **d.** $\lim\limits_{x \to -\pi^-} f(x)$

**13.** Suppose $f(x) \to 100$ and $g(x) \to 0$, with $g(x) < 0$ as $x \to 2$. Determine $\lim\limits_{x \to 2} \dfrac{f(x)}{g(x)}$.

**14.** Evaluate $\lim\limits_{x \to 3^-} \dfrac{1}{x - 3}$ and $\lim\limits_{x \to 3^+} \dfrac{1}{x - 3}$.

**15.** Verify that the function $f(x) = \dfrac{x^2 - 4x + 3}{x^2 - 3x + 2}$ is undefined at $x = 1$ and at $x = 2$. Does the graph of $f$ have vertical asymptotes at both these values of $x$? Explain.

**16.** Evaluate $\lim\limits_{x \to 0} \dfrac{x + 1}{1 - \cos x}$.

**17.** Sketch a possible graph of a function $f$, together with vertical asymptotes, satisfying all the following conditions on $[0, 4]$.

$$f(1) = 0, \quad f(3) \text{ is undefined}, \quad \lim\limits_{x \to 3} f(x) = 1,$$

$$\lim\limits_{x \to 0^+} f(x) = -\infty, \quad \lim\limits_{x \to 2} f(x) = \infty, \quad \lim\limits_{x \to 4^-} f(x) = \infty$$

**18.** Sketch a possible graph of a function $g$, together with vertical asymptotes, satisfying all the following conditions.

$$g(2) = 1, \quad g(5) = -1, \quad \lim\limits_{x \to 4} g(x) = -\infty,$$

$$\lim\limits_{x \to 7^-} g(x) = \infty, \quad \lim\limits_{x \to 7^+} g(x) = -\infty$$

**19.** Which of the following statements are correct? Choose all that apply.

   **a.** $\lim\limits_{x \to 1} \dfrac{1}{(x - 1)^2}$ does not exist    **b.** $\lim\limits_{x \to 1} \dfrac{1}{(x - 1)^2} = \infty$

   **c.** $\lim\limits_{x \to 1} \dfrac{1}{(x - 1)^2} = -\infty$

**20.** Which of the following statements are correct? Choose all that apply.

   **a.** $\lim\limits_{x \to 1^+} \dfrac{1}{1 - x}$ does not exist    **b.** $\lim\limits_{x \to 1^+} \dfrac{1}{1 - x} = \infty$

   **c.** $\lim\limits_{x \to 1^+} \dfrac{1}{1 - x} = -\infty$

### Practice Exercises

**21–44. Determining limits analytically** *Determine the following limits.*

**21. a.** $\lim\limits_{x \to 2^+} \dfrac{1}{x - 2}$    **b.** $\lim\limits_{x \to 2^-} \dfrac{1}{x - 2}$    **c.** $\lim\limits_{x \to 2} \dfrac{1}{x - 2}$

**22. a.** $\lim\limits_{x \to 3^+} \dfrac{2}{(x - 3)^3}$    **b.** $\lim\limits_{x \to 3^-} \dfrac{2}{(x - 3)^3}$    **c.** $\lim\limits_{x \to 3} \dfrac{2}{(x - 3)^3}$

**23.**
a. $\lim\limits_{x\to 4^+}\dfrac{x-5}{(x-4)^2}$
b. $\lim\limits_{x\to 4^-}\dfrac{x-5}{(x-4)^2}$
c. $\lim\limits_{x\to 4}\dfrac{x-5}{(x-4)^2}$

**24.**
a. $\lim\limits_{x\to 1^+}\dfrac{x}{|x-1|}$
b. $\lim\limits_{x\to 1^-}\dfrac{x}{|x-1|}$
c. $\lim\limits_{x\to 1}\dfrac{x}{|x-1|}$

**25.**
a. $\lim\limits_{z\to 3^+}\dfrac{(z-1)(z-2)}{(z-3)}$
b. $\lim\limits_{z\to 3^-}\dfrac{(z-1)(z-2)}{(z-3)}$

c. $\lim\limits_{z\to 3}\dfrac{(z-1)(z-2)}{(z-3)}$

**26.**
a. $\lim\limits_{x\to -2^+}\dfrac{(x-4)}{x(x+2)}$
b. $\lim\limits_{x\to -2^-}\dfrac{(x-4)}{x(x+2)}$
c. $\lim\limits_{x\to -2}\dfrac{(x-4)}{x(x+2)}$

**27.**
a. $\lim\limits_{x\to 2^+}\dfrac{x^2-4x+3}{(x-2)^2}$
b. $\lim\limits_{x\to 2^-}\dfrac{x^2-4x+3}{(x-2)^2}$

c. $\lim\limits_{x\to 2}\dfrac{x^2-4x+3}{(x-2)^2}$

**28.**
a. $\lim\limits_{t\to -2^+}\dfrac{t^3-5t^2+6t}{t^4-4t^2}$
b. $\lim\limits_{t\to -2^-}\dfrac{t^3-5t^2+6t}{t^4-4t^2}$

c. $\lim\limits_{t\to -2}\dfrac{t^3-5t^2+6t}{t^4-4t^2}$
d. $\lim\limits_{t\to 2}\dfrac{t^3-5t^2+6t}{t^4-4t^2}$

**29.**
a. $\lim\limits_{x\to 2^+}\dfrac{1}{\sqrt{x(x-2)}}$
b. $\lim\limits_{x\to 2^-}\dfrac{1}{\sqrt{x(x-2)}}$

c. $\lim\limits_{x\to 2}\dfrac{1}{\sqrt{x(x-2)}}$

**30.**
a. $\lim\limits_{x\to 1^+}\dfrac{x-3}{\sqrt{x^2-5x+4}}$
b. $\lim\limits_{x\to 1^-}\dfrac{x-3}{\sqrt{x^2-5x+4}}$

c. $\lim\limits_{x\to 1}\dfrac{x-3}{\sqrt{x^2-5x+4}}$

**31.**
a. $\lim\limits_{x\to 0}\dfrac{x-3}{x^4-9x^2}$
b. $\lim\limits_{x\to 3}\dfrac{x-3}{x^4-9x^2}$
c. $\lim\limits_{x\to -3}\dfrac{x-3}{x^4-9x^2}$

**32.**
a. $\lim\limits_{x\to 0}\dfrac{x-2}{x^5-4x^3}$
b. $\lim\limits_{x\to 2}\dfrac{x-2}{x^5-4x^3}$
c. $\lim\limits_{x\to -2}\dfrac{x-2}{x^5-4x^3}$

**33.** $\lim\limits_{x\to 0}\dfrac{x^3-5x^2}{x^2}$

**34.** $\lim\limits_{t\to 5}\dfrac{4t^2-100}{t-5}$

**35.** $\lim\limits_{x\to 1^+}\dfrac{x^2-5x+6}{x-1}$

**36.** $\lim\limits_{z\to 4}\dfrac{z-5}{(z^2-10z+24)^2}$

**37.** $\lim\limits_{x\to 6^+}\dfrac{x-7}{\sqrt{x-6}}$

**38.** $\lim\limits_{x\to 2^-}\dfrac{x-1}{\sqrt{(x-3)(x-2)}}$

**39.** $\lim\limits_{\theta\to 0^+}\csc\theta$

**40.** $\lim\limits_{x\to 0^-}\csc x$

**41.** $\lim\limits_{x\to 0^+}(-10\cot x)$

**42.** $\lim\limits_{\theta\to \pi/2^+}\dfrac{1}{3}\tan\theta$

**43.** $\lim\limits_{\theta\to 0}\dfrac{2+\sin\theta}{1-\cos^2\theta}$

**44.** $\lim\limits_{\theta\to 0^-}\dfrac{\sin\theta}{\cos^2\theta-1}$

**45. Location of vertical asymptotes** Analyze the following limits and find the vertical asymptotes of $f(x)=\dfrac{x-5}{x^2-25}$.

a. $\lim\limits_{x\to 5}f(x)$
b. $\lim\limits_{x\to -5^-}f(x)$
c. $\lim\limits_{x\to -5^+}f(x)$

**46. Location of vertical asymptotes** Analyze the following limits and find the vertical asymptotes of $f(x)=\dfrac{x+7}{x^4-49x^2}$.

a. $\lim\limits_{x\to 7^-}f(x)$
b. $\lim\limits_{x\to 7^+}f(x)$
c. $\lim\limits_{x\to -7}f(x)$
d. $\lim\limits_{x\to 0}f(x)$

**47–50. Finding vertical asymptotes** *Find all vertical asymptotes $x=a$ of the following functions. For each value of a, determine* $\lim\limits_{x\to a^+}f(x),\ \lim\limits_{x\to a^-}f(x),\ and\ \lim\limits_{x\to a}f(x).$

**47.** $f(x)=\dfrac{x^2-9x+14}{x^2-5x+6}$

**48.** $f(x)=\dfrac{\cos x}{x^2+2x}$

**49.** $f(x)=\dfrac{x+1}{x^3-4x^2+4x}$

**50.** $f(x)=\dfrac{x^3-10x^2+16x}{x^2-8x}$

**51. Checking your work graphically** Analyze the following limits. Then sketch a graph of $y=\tan x$ with the window $[-\pi,\pi]\times[-10,10]$ and use your graph to check your work.

a. $\lim\limits_{x\to \pi/2^+}\tan x$
b. $\lim\limits_{x\to \pi/2^-}\tan x$
c. $\lim\limits_{x\to -\pi/2^+}\tan x$
d. $\lim\limits_{x\to -\pi/2^-}\tan x$

**T 52. Checking your work graphically** Analyze the following limits. Then sketch a graph of $y=\sec x\tan x$ with the window $[-\pi,\pi]\times[-10,10]$ and use your graph to check your work.

a. $\lim\limits_{x\to \pi/2^+}\sec x\tan x$
b. $\lim\limits_{x\to \pi/2^-}\sec x\tan x$
c. $\lim\limits_{x\to -\pi/2^+}\sec x\tan x$
d. $\lim\limits_{x\to -\pi/2^-}\sec x\tan x$

**53. Explain why or why not** Determine whether the following statements are true and give an explanation or counterexample.

a. The line $x=1$ is a vertical asymptote of the function $f(x)=\dfrac{x^2-7x+6}{x^2-1}.$

b. The line $x=-1$ is a vertical asymptote of the function $f(x)=\dfrac{x^2-7x+6}{x^2-1}.$

c. If $g$ has a vertical asymptote at $x=1$ and $\lim\limits_{x\to 1^+}g(x)=\infty$, then $\lim\limits_{x\to 1^-}g(x)=\infty.$

**54. Matching** Match functions a–f with graphs A–F in the figure without using a graphing utility.

a. $f(x)=\dfrac{x}{x^2+1}$
b. $f(x)=\dfrac{x}{x^2-1}$
c. $f(x)=\dfrac{1}{x^2-1}$
d. $f(x)=\dfrac{x}{(x-1)^2}$
e. $f(x)=\dfrac{1}{(x-1)^2}$
f. $f(x)=\dfrac{x}{x+1}$

A.

B.

C.

D.

E.

F.

## Explorations and Challenges

**55. Finding a rational function** Find a rational function $r(x)$ such that $r(1)$ is undefined, $\lim_{x \to 1} r(x) = 0$, and $\lim_{x \to 2} r(x) = \infty$.

**56. Finding a function with vertical asymptotes** Find polynomials $p$ and $q$ such that $f = p/q$ is undefined at 1 and 2, but $f$ has a vertical asymptote only at 2. Sketch a graph of your function.

**57. Finding a function with infinite limits** Give a formula for a function $f$ that satisfies $\lim_{x \to 6^+} f(x) = \infty$ and $\lim_{x \to 6^-} f(x) = -\infty$.

**T 58–66. Asymptotes** *Use analytical methods and/or a graphing utility to identify the vertical asymptotes (if any) of the following functions.*

**58.** $f(x) = \dfrac{x^2 - 1}{x^4 - 1}$

**59.** $f(x) = \dfrac{x^2 - 3x + 2}{x^{10} - x^9}$

**60.** $g(x) = \cot\left(x - \dfrac{\pi}{2}\right)$, for $|x| \leq \pi$

**61.** $h(x) = \dfrac{\cos x}{(x + 1)^3}$

**62.** $p(x) = \sec\dfrac{\pi x}{2}$, for $|x| < 2$

**63.** $g(\theta) = \tan\dfrac{\pi\theta}{10}$

**64.** $q(s) = \dfrac{\pi}{s - \sin s}$

**65.** $f(x) = \dfrac{1}{\sqrt{x}\sec x}$

**66.** $g(x) = \dfrac{1}{\sqrt{x(x^2 - 1)}}$

**67. Limits with a parameter** Let $f(x) = \dfrac{x^2 - 7x + 12}{x - a}$.

    **a.** For what values of $a$, if any, does $\lim_{x \to a^+} f(x)$ equal a finite number?

    **b.** For what values of $a$, if any, does $\lim_{x \to a^+} f(x) = \infty$?

    **c.** For what values of $a$, if any, does $\lim_{x \to a^+} f(x) = -\infty$?

**68–69. Steep secant lines**

    **a.** *Given the graph of f in the following figures, find the slope of the secant line that passes through $(0, 0)$ and $(h, f(h))$ in terms of h, for $h > 0$ and $h < 0$.*

    **b.** *Analyze the limit of the slope of the secant line found in part (a) as $h \to 0^+$ and $h \to 0^-$. What does this tell you about the line tangent to the curve at $(0, 0)$?*

**68.** $f(x) = x^{1/3}$

**69.** $f(x) = x^{2/3}$

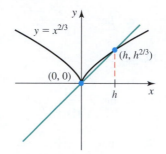

**QUICK CHECK ANSWERS**

**1.** Answers will vary, but all graphs should have a vertical asymptote at $x = 2$.   **2.** $-\infty; \infty$   **3.** As $x \to -4^+$, $x < 0$ and $(x + 4) > 0$, so $x(x + 4) \to 0$ through negative values.

**4.** $\lim_{x \to 2} \dfrac{(x - 1)(x - 2)}{x - 2} = \lim_{x \to 2} (x - 1) = 1$, which is not an infinite limit, so $x = 2$ is not a vertical asymptote. ◄

# 2.5 Limits at Infinity

Limits at infinity—as opposed to infinite limits—occur when the independent variable becomes large in magnitude. For this reason, limits at infinity determine what is called the *end behavior* of a function. An application of these limits is to determine whether a system (such as an ecosystem or a large oscillating structure) reaches a steady state as time increases.

## Limits at Infinity and Horizontal Asymptotes

Consider the function $f(x) = \dfrac{x}{\sqrt{x^2 + 1}}$ **(Figure 2.29)**, whose domain is $(-\infty, \infty)$. As $x$ becomes arbitrarily large (denoted $x \to \infty$), $f(x)$ approaches 1, and as $x$ becomes arbitrarily large in magnitude and negative (denoted $x \to -\infty$), $f(x)$ approaches $-1$. These limits are expressed as

$$\lim_{x \to \infty} f(x) = 1 \quad \text{and} \quad \lim_{x \to -\infty} f(x) = -1.$$

The graph of $f$ approaches the horizontal line $y = 1$ as $x \to \infty$, and it approaches the horizontal line $y = -1$ as $x \to -\infty$. These lines are called *horizontal asymptotes*.

**Figure 2.29**

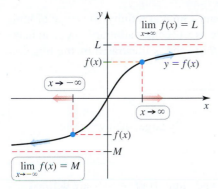

**Figure 2.30**

**QUICK CHECK 1** Evaluate $x/(x + 1)$ for $x = 10, 100,$ and $1000$. What is $\lim_{x \to \infty} \dfrac{x}{x + 1}$? ◄

➤ The limit laws of Theorem 2.3 and the Squeeze Theorem apply if $x \to a$ is replaced with $x \to \infty$ or $x \to -\infty$.

**Figure 2.31**

---

**DEFINITION** **Limits at Infinity and Horizontal Asymptotes**

If $f(x)$ becomes arbitrarily close to a finite number $L$ for all sufficiently large and positive $x$, then we write

$$\lim_{x \to \infty} f(x) = L.$$

We say the limit of $f(x)$ as $x$ approaches infinity is $L$. In this case, the line $y = L$ is a **horizontal asymptote** of $f$ (**Figure 2.30**). The limit at negative infinity, $\lim_{x \to -\infty} f(x) = M,$ is defined analogously. When this limit exists, $y = M$ is a horizontal asymptote.

---

**EXAMPLE 1** **Limits at infinity** Evaluate the following limits.

**a.** $\lim_{x \to -\infty} \left( 2 + \dfrac{10}{x^2} \right)$    **b.** $\lim_{x \to \infty} \left( 5 + \dfrac{\sin x}{\sqrt{x}} \right)$

**SOLUTION**

**a.** As $x$ becomes large and negative, $x^2$ becomes large and positive; in turn, $10/x^2$ approaches 0. By the limit laws of Theorem 2.3,

$$\lim_{x \to -\infty} \left( 2 + \frac{10}{x^2} \right) = \underbrace{\lim_{x \to -\infty} 2}_{\text{equals 2}} + \underbrace{\lim_{x \to -\infty} \left( \frac{10}{x^2} \right)}_{\text{equals 0}} = 2 + 0 = 2.$$

Therefore, the graph of $y = 2 + 10/x^2$ approaches the horizontal asymptote $y = 2$ as $x \to -\infty$ (**Figure 2.31**). Notice that $\lim_{x \to \infty} \left( 2 + \dfrac{10}{x^2} \right)$ is also equal to 2, which implies that the graph has a single horizontal asymptote.

**b.** The numerator of $\sin x / \sqrt{x}$ is bounded between $-1$ and $1$; therefore, for $x > 0$,

$$-\frac{1}{\sqrt{x}} \le \frac{\sin x}{\sqrt{x}} \le \frac{1}{\sqrt{x}}.$$

As $x \to \infty$, $\sqrt{x}$ becomes arbitrarily large, which means that

$$\lim_{x \to \infty} \frac{-1}{\sqrt{x}} = \lim_{x \to \infty} \frac{1}{\sqrt{x}} = 0.$$

It follows by the Squeeze Theorem (Theorem 2.5) that $\lim\limits_{x \to \infty} \dfrac{\sin x}{\sqrt{x}} = 0$.

Using the limit laws of Theorem 2.3,

$$\lim_{x \to \infty} \left( 5 + \frac{\sin x}{\sqrt{x}} \right) = \underbrace{\lim_{x \to \infty} 5}_{\text{equals } 5} + \underbrace{\lim_{x \to \infty} \frac{\sin x}{\sqrt{x}}}_{\text{equals } 0} = 5.$$

The graph of $y = 5 + \dfrac{\sin x}{\sqrt{x}}$ approaches the horizontal asymptote $y = 5$ as $x$ becomes

**Figure 2.32**

large (**Figure 2.32**). Note that the curve intersects its asymptote infinitely many times.

*Related Exercises 10, 19* ◀

## Infinite Limits at Infinity

It is possible for a limit to be *both* an infinite limit and a limit at infinity. This type of limit occurs if $f(x)$ becomes arbitrarily large in magnitude as $x$ becomes arbitrarily large in magnitude. Such a limit is called an *infinite limit at infinity* and is illustrated by the function $f(x) = x^3$ (**Figure 2.33**).

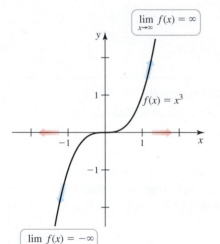

> **DEFINITION  Infinite Limits at Infinity**
>
> If $f(x)$ becomes arbitrarily large as $x$ becomes arbitrarily large, then we write
>
> $$\lim_{x \to \infty} f(x) = \infty.$$
>
> The limits $\lim\limits_{x \to \infty} f(x) = -\infty$, $\lim\limits_{x \to -\infty} f(x) = \infty$, and $\lim\limits_{x \to -\infty} f(x) = -\infty$ are defined similarly.

**Figure 2.33**

Infinite limits at infinity tell us about the behavior of polynomials for large-magnitude values of $x$. First, consider power functions $f(x) = x^n$, where $n$ is a positive integer. **Figure 2.34** shows that when $n$ is even, $\lim\limits_{x \to \pm\infty} x^n = \infty$, and when $n$ is odd, $\lim\limits_{x \to \infty} x^n = \infty$ and $\lim\limits_{x \to -\infty} x^n = -\infty$.

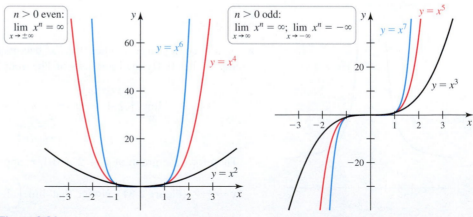

**Figure 2.34**

It follows that reciprocals of power functions $f(x) = 1/x^n = x^{-n}$, where $n$ is a positive integer, behave as follows:

$$\lim_{x \to \infty} \frac{1}{x^n} = \lim_{x \to \infty} x^{-n} = 0 \quad \text{and} \quad \lim_{x \to -\infty} \frac{1}{x^n} = \lim_{x \to -\infty} x^{-n} = 0.$$

**QUICK CHECK 2** Describe the behavior of $p(x) = -3x^3$ as $x \to \infty$ and as $x \to -\infty$. ◄

From here, it is a short step to finding the behavior of any polynomial as $x \to \pm\infty$. Let $p(x) = a_n x^n + a_{n-1}x^{n-1} + \cdots + a_2 x^2 + a_1 x + a_0$. We now write $p$ in the equivalent form

$$p(x) = x^n \left( a_n + \underbrace{\frac{a_{n-1}}{x}}_{\to 0} + \underbrace{\frac{a_{n-2}}{x^2}}_{\to 0} + \cdots + \underbrace{\frac{a_0}{x^n}}_{\to 0} \right).$$

Notice that as $x$ becomes large in magnitude, all the terms in $p$ except the first term approach zero. Therefore, as $x \to \pm\infty$, we see that $p(x) \approx a_n x^n$. This means that as $x \to \pm\infty$, the behavior of $p$ is determined by the term $a_n x^n$ with the highest power of $x$.

---

**THEOREM 2.6    Limits at Infinity of Powers and Polynomials**

Let $n$ be a positive integer and let $p$ be the polynomial
$p(x) = a_n x^n + a_{n-1}x^{n-1} + \cdots + a_2 x^2 + a_1 x + a_0$, where $a_n \neq 0$.

1. $\lim\limits_{x \to \pm\infty} x^n = \infty$ when $n$ is even.

2. $\lim\limits_{x \to \infty} x^n = \infty$ and $\lim\limits_{x \to -\infty} x^n = -\infty$ when $n$ is odd.

3. $\lim\limits_{x \to \pm\infty} \dfrac{1}{x^n} = \lim\limits_{x \to \pm\infty} x^{-n} = 0.$

4. $\lim\limits_{x \to \pm\infty} p(x) = \lim\limits_{x \to \pm\infty} a_n x^n = \pm\infty$, depending on the degree of the polynomial and the sign of the leading coefficient $a_n$.

---

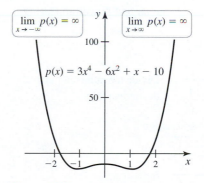

$\lim\limits_{x \to -\infty} p(x) = \infty$     $\lim\limits_{x \to \infty} p(x) = \infty$

$p(x) = 3x^4 - 6x^2 + x - 10$

**Figure 2.35**

$\lim\limits_{x \to -\infty} q(x) = \infty$

$q(x) = -2x^3 + 3x^2 - 12$

$\lim\limits_{x \to \infty} q(x) = -\infty$

**Figure 2.36**

**EXAMPLE 2    Limits at infinity** Determine the limits as $x \to \pm\infty$ of the following functions.

**a.** $p(x) = 3x^4 - 6x^2 + x - 10$    **b.** $q(x) = -2x^3 + 3x^2 - 12$

**SOLUTION**

**a.** We use the fact that the limit is determined by the behavior of the leading term:

$$\lim_{x \to \infty} (3x^4 - 6x^2 + x - 10) = \lim_{x \to \infty} 3\underbrace{x^4}_{\to \infty} = \infty.$$

Similarly,

$$\lim_{x \to -\infty} (3x^4 - 6x^2 + x - 10) = \lim_{x \to -\infty} 3\underbrace{x^4}_{\to \infty} = \infty.$$

Figure 2.35 illustrates these limits.

**b.** Noting that the leading coefficient is negative, we have

$$\lim_{x \to \infty} (-2x^3 + 3x^2 - 12) = \lim_{x \to \infty} (-2\underbrace{x^3}_{\to \infty}) = -\infty$$

$$\lim_{x \to -\infty} (-2x^3 + 3x^2 - 12) = \lim_{x \to -\infty} (-2\underbrace{x^3}_{\to -\infty}) = \infty.$$

The graph of $q$ (**Figure 2.36**) confirms these results.

*Related Exercises 21, 23* ◄

## End Behavior

The behavior of polynomials as $x \to \pm\infty$ is an example of what is often called *end behavior*. Having treated polynomials, we now turn to the end behavior of rational and algebraic functions.

**EXAMPLE 3  End behavior of rational functions** Use limits at infinity to determine the end behavior of the following rational functions.

**a.** $f(x) = \dfrac{3x + 2}{x^2 - 1}$    **b.** $g(x) = \dfrac{40x^4 + 4x^2 - 1}{10x^4 + 8x^2 + 1}$    **c.** $h(x) = \dfrac{x^3 - 2x + 1}{2x + 4}$

**SOLUTION**

➤ Recall that the *degree* of a polynomial is the highest power of $x$ that appears.

**a.** An effective approach for determining limits of rational functions at infinity is to divide both the numerator and denominator by $x^n$, where $n$ is the degree of the polynomial in the denominator. This strategy forces the terms corresponding to lower powers of $x$ to approach 0 in the limit. In this case, we divide by $x^2$:

$$\lim_{x \to \infty} \frac{3x + 2}{x^2 - 1} = \lim_{x \to \infty} \frac{\dfrac{3x + 2}{x^2}}{\dfrac{x^2 - 1}{x^2}} = \lim_{x \to \infty} \frac{\overbrace{\dfrac{3}{x} + \dfrac{2}{x^2}}^{\text{approaches } 0}}{\underbrace{1 - \dfrac{1}{x^2}}_{\text{approaches } 0}} = \frac{0}{1} = 0.$$

A similar calculation gives $\displaystyle\lim_{x \to -\infty} \frac{3x + 2}{x^2 - 1} = 0$; therefore, the graph of $f$ has the horizontal asymptote $y = 0$. You should confirm that the zeros of the denominator are $-1$ and $1$, which correspond to vertical asymptotes (**Figure 2.37**). In this example, the degree of the polynomial in the numerator is *less than* the degree of the polynomial in the denominator.

$f(x) = \dfrac{3x + 2}{x^2 - 1}$

$\displaystyle\lim_{x \to \infty} f(x) = 0$

$\displaystyle\lim_{x \to -\infty} f(x) = 0$

**Figure 2.37**

**b.** Again we divide both the numerator and denominator by the largest power appearing in the denominator, which is $x^4$:

$$\lim_{x \to \infty} \frac{40x^4 + 4x^2 - 1}{10x^4 + 8x^2 + 1} = \lim_{x \to \infty} \frac{\dfrac{40x^4}{x^4} + \dfrac{4x^2}{x^4} - \dfrac{1}{x^4}}{\dfrac{10x^4}{x^4} + \dfrac{8x^2}{x^4} + \dfrac{1}{x^4}} \qquad \text{\color{blue} Divide the numerator and denominator by } x^4.$$

$$= \lim_{x \to \infty} \frac{40 + \overbrace{\dfrac{4}{x^2}}^{\text{approaches } 0} - \overbrace{\dfrac{1}{x^4}}^{\text{approaches } 0}}{10 + \underbrace{\dfrac{8}{x^2}}_{\text{approaches } 0} + \underbrace{\dfrac{1}{x^4}}_{\text{approaches } 0}} \qquad \text{\color{blue} Simplify.}$$

$$= \frac{40 + 0 + 0}{10 + 0 + 0} = 4. \qquad \text{\color{blue} Evaluate limits.}$$

Using the same steps (dividing each term by $x^4$), it can be shown that $\displaystyle\lim_{x \to -\infty} \frac{40x^4 + 4x^2 - 1}{10x^4 + 8x^2 + 1} = 4$. This function has the horizontal asymptote $y = 4$ (**Figure 2.38**). Notice that the degree of the polynomial in the numerator *equals* the degree of the polynomial in the denominator.

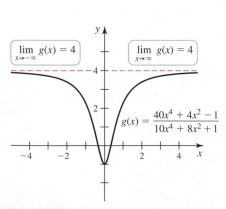

$\displaystyle\lim_{x \to -\infty} g(x) = 4$    $\displaystyle\lim_{x \to \infty} g(x) = 4$

$g(x) = \dfrac{40x^4 + 4x^2 - 1}{10x^4 + 8x^2 + 1}$

**Figure 2.38**

**c.** We divide the numerator and denominator by the largest power of $x$ appearing in the denominator, which is $x$, and then take the limit:

$$\lim_{x \to \infty} \frac{x^3 - 2x + 1}{2x + 4} = \lim_{x \to \infty} \frac{\dfrac{x^3}{x} - \dfrac{2x}{x} + \dfrac{1}{x}}{\dfrac{2x}{x} + \dfrac{4}{x}}$$    Divide the numerator and denominator by $x$.

$$= \lim_{x \to \infty} \frac{\overbrace{x^2}^{\text{arbitrarily large}} - \overbrace{2}^{\text{constant}} + \overbrace{\dfrac{1}{x}}^{\text{approaches 0}}}{\underbrace{2}_{\text{constant}} + \underbrace{\dfrac{4}{x}}_{\text{approaches 0}}}$$    Simplify.

$$= \infty.$$    Take limits.

$$\lim_{x \to \infty} h(x) = \infty$$

$$\lim_{x \to -\infty} h(x) = \infty$$

$$h(x) = \frac{x^3 - 2x + 1}{2x + 4}$$

**Figure 2.39**

As $x \to \infty$, all the terms in this function either approach zero or are constant—except the $x^2$-term in the numerator, which becomes arbitrarily large. Therefore, the limit of the function does not exist. Using a similar analysis, we find that $\lim\limits_{x \to -\infty} \dfrac{x^3 - 2x + 1}{2x + 4} = \infty$. These limits are not finite, so the graph of the function has no horizontal asymptote (**Figure 2.39**). There is, however, a vertical asymptote due to the fact that $x = -2$ is a zero of the denominator. In this case, the degree of the polynomial in the numerator is *greater than* the degree of the polynomial in the denominator.

*Related Exercises 38, 41, 43* ◄

A special case of end behavior arises with rational functions. As shown in the next example, if the graph of a function $f$ approaches a line (with finite and nonzero slope) as $x \to \pm\infty$, then that line is a **slant asymptote**, or **oblique asymptote**, of $f$.

**EXAMPLE 4** **Slant asymptotes** Determine the end behavior of the function $f(x) = \dfrac{2x^2 + 6x - 2}{x + 1}$.

**SOLUTION** We first divide the numerator and denominator by the largest power of $x$ appearing in the denominator, which is $x$:

$$\lim_{x \to \infty} \frac{2x^2 + 6x - 2}{x + 1} = \lim_{x \to \infty} \frac{\dfrac{2x^2}{x} + \dfrac{6x}{x} - \dfrac{2}{x}}{\dfrac{x}{x} + \dfrac{1}{x}}$$    Divide the numerator and denominator by $x$.

$$= \lim_{x \to \infty} \frac{\overbrace{2x}^{\text{arbitrarily large}} + \overbrace{6}^{\text{constant}} - \overbrace{\dfrac{2}{x}}^{\text{approaches 0}}}{\underbrace{1}_{\text{constant}} + \underbrace{\dfrac{1}{x}}_{\text{approaches 0}}}$$    Simplify.

$$= \infty.$$    Take limits.

A similar analysis shows that $\lim\limits_{x \to -\infty} \dfrac{2x^2 + 6x - 2}{x + 1} = -\infty$. Because these limits are not finite, $f$ has no horizontal asymptote.

However, there is more to be learned about the end behavior of this function. Using long division, the function $f$ is written

$$f(x) = \frac{2x^2 + 6x - 2}{x + 1} = \underbrace{2x + 4}_{\ell(x)} - \underbrace{\frac{6}{x + 1}}_{\substack{\text{approaches 0 as} \\ x \to \infty}}.$$

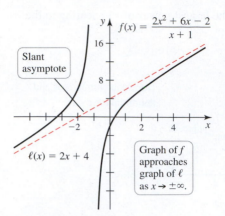

**Figure 2.40**

> More generally, a line $y = \ell(x)$ (with finite and nonzero slope) is a slant asymptote of a function $f$ if $\lim\limits_{x \to \infty} (f(x) - \ell(x)) = 0$ or $\lim\limits_{x \to -\infty} (f(x) - \ell(x)) = 0$.

**QUICK CHECK 3**  Use Theorem 2.7 to find the vertical and horizontal asymptotes of $y = \dfrac{10x}{3x - 1}$. ◄

As $x \to \infty$, the term $6/(x + 1)$ approaches 0, and we see that the function $f$ behaves like the linear function $\ell(x) = 2x + 4$. For this reason, the graph of $f$ approaches the graph of $\ell$ as $x \to \infty$ (**Figure 2.40**). A similar argument shows that the graph of $f$ also approaches the graph of $\ell$ as $x \to -\infty$. The line described by $\ell$ is a slant asymptote; it occurs with rational functions only when the degree of the polynomial in the numerator exceeds the degree of the polynomial in the denominator by exactly 1.

*Related Exercises 51–52* ◄

The conclusions reached in Examples 3 and 4 can be generalized for all rational functions. These results are summarized in Theorem 2.7; its proof is assigned in Exercise 80.

---

**THEOREM 2.7   End Behavior and Asymptotes of Rational Functions**

Suppose $f(x) = \dfrac{p(x)}{q(x)}$ is a rational function, where

$$p(x) = a_m x^m + a_{m-1} x^{m-1} + \cdots + a_2 x^2 + a_1 x + a_0 \quad \text{and}$$
$$q(x) = b_n x^n + b_{n-1} x^{n-1} + \cdots + b_2 x^2 + b_1 x + b_0,$$

with $a_m \neq 0$ and $b_n \neq 0$.

a. **Degree of numerator less than degree of denominator** If $m < n$, then $\lim\limits_{x \to \pm\infty} f(x) = 0$, and $y = 0$ is a horizontal asymptote of $f$.

b. **Degree of numerator equals degree of denominator** If $m = n$, then $\lim\limits_{x \to \pm\infty} f(x) = a_m/b_n$, and $y = a_m/b_n$ is a horizontal asymptote of $f$.

c. **Degree of numerator greater than degree of denominator** If $m > n$, then $\lim\limits_{x \to \pm\infty} f(x) = \infty$ or $-\infty$, and $f$ has no horizontal asymptote.

d. **Slant asymptote** If $m = n + 1$, then $\lim\limits_{x \to \pm\infty} f(x) = \infty$ or $-\infty$, and $f$ has no horizontal asymptote, but $f$ has a slant asymptote.

e. **Vertical asymptotes** Assuming $f$ is in reduced form ($p$ and $q$ share no common factors), vertical asymptotes occur at the zeros of $q$.

---

Although it isn't stated explicitly, Theorem 2.7 implies that a rational function can have at most one horizontal asymptote, and whenever there is a horizontal asymptote, $\lim\limits_{x \to \infty} \dfrac{p(x)}{q(x)} = \lim\limits_{x \to -\infty} \dfrac{p(x)}{q(x)}$. The same cannot be said of other functions, as the next examples show.

**EXAMPLE 5   End behavior of an algebraic function**  Use limits at infinity to determine the end behavior of $f(x) = \dfrac{10x^3 - 3x^2 + 8}{\sqrt{25x^6 + x^4 + 2}}$.

**SOLUTION**  The square root in the denominator forces us to revise the strategy used with rational functions. First, consider the limit as $x \to \infty$. The highest power of the polynomial in the denominator is 6. However, the polynomial is under a square root, so effectively, the term with the highest power in the denominator is $\sqrt{x^6} = x^3$. When we divide the numerator and denominator by $x^3$, for $x > 0$, the limit becomes

$$\lim_{x \to \infty} \frac{10x^3 - 3x^2 + 8}{\sqrt{25x^6 + x^4 + 2}} = \lim_{x \to \infty} \frac{\dfrac{10x^3}{x^3} - \dfrac{3x^2}{x^3} + \dfrac{8}{x^3}}{\sqrt{\dfrac{25x^6}{x^6} + \dfrac{x^4}{x^6} + \dfrac{2}{x^6}}} \qquad \text{Divide by } \sqrt{x^6} = x^3.$$

$$= \lim_{x \to \infty} \frac{10 - \overbrace{\dfrac{3}{x}}^{\text{approaches } 0} + \overbrace{\dfrac{8}{x^3}}^{\text{approaches } 0}}{\sqrt{25 + \underbrace{\dfrac{1}{x^2}}_{\text{approaches } 0} + \underbrace{\dfrac{2}{x^6}}_{\text{approaches } 0}}} \qquad \text{Simplify.}$$

▶ Recall that

$$\sqrt{x^2} = |x| = \begin{cases} x & \text{if } x \geq 0 \\ -x & \text{if } x < 0. \end{cases}$$

$$= \frac{10}{\sqrt{25}} = 2. \qquad \text{Evaluate limits.}$$

Therefore,

$$\sqrt{x^6} = |x^3| = \begin{cases} x^3 & \text{if } x \geq 0 \\ -x^3 & \text{if } x < 0. \end{cases}$$

As $x \to -\infty$, $x^3$ is negative, so we divide the numerator and denominator by $\sqrt{x^6} = -x^3$ (which is positive):

Because $x$ is negative as $x \to -\infty$, we have $\sqrt{x^6} = -x^3$.

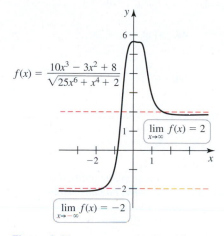

$f(x) = \dfrac{10x^3 - 3x^2 + 8}{\sqrt{25x^6 + x^4 + 2}}$

$\lim_{x \to \infty} f(x) = 2$

$\lim_{x \to -\infty} f(x) = -2$

**Figure 2.41**

$$\lim_{x \to -\infty} \frac{10x^3 - 3x^2 + 8}{\sqrt{25x^6 + x^4 + 2}} = \lim_{x \to -\infty} \frac{\dfrac{10x^3}{-x^3} - \dfrac{3x^2}{-x^3} + \dfrac{8}{-x^3}}{\sqrt{\dfrac{25x^6}{x^6} + \dfrac{x^4}{x^6} + \dfrac{2}{x^6}}} \qquad \text{Divide by } \sqrt{x^6} = -x^3 > 0.$$

$$= \lim_{x \to -\infty} \frac{-10 + \overbrace{\dfrac{3}{x}}^{\text{approaches } 0} - \overbrace{\dfrac{8}{x^3}}^{\text{approaches } 0}}{\sqrt{25 + \underbrace{\dfrac{1}{x^2}}_{\text{approaches } 0} + \underbrace{\dfrac{2}{x^6}}_{\text{approaches } 0}}} \qquad \text{Simplify.}$$

$$= -\frac{10}{\sqrt{25}} = -2. \qquad \text{Evaluate limits.}$$

The limits reveal two asymptotes, $y = 2$ and $y = -2$. In this particular case, the graph crosses both horizontal asymptotes (**Figure 2.41**).     *Related Exercises 46–47* ◄

## End Behavior of $\sin x$ and $\cos x$

Our future work requires knowing about the end behavior of $\sin x$ and $\cos x$. The values of both functions oscillate between $-1$ and $1$ as $x$ increases in magnitude. Therefore, $\lim_{x \to \pm\infty} \sin x$ and $\lim_{x \to \pm\infty} \cos x$ do not exist. However, both functions are bounded as $x \to \pm\infty$; that is, $|\sin x| \leq 1$ and $|\cos x| \leq 1$ for all $x$.

---

## SECTION 2.5 EXERCISES

### Getting Started

1. Explain the meaning of $\lim_{x \to -\infty} f(x) = 10$.

2. Evaluate $\lim_{x \to \infty} f(x)$ and $\lim_{x \to -\infty} f(x)$ using the figure.

$y = f(x)$

**3–13.** *Determine the following limits at infinity.*

3. $\lim_{x \to \infty} x^{12}$

4. $\lim_{x \to -\infty} 3x^{11}$

5. $\lim_{x \to \infty} x^{-6}$

6. $\lim_{x \to -\infty} x^{-11}$

7. $\lim_{t \to \infty} (-12t^{-5})$

8. $\lim_{x \to -\infty} 2x^{-8}$

9. $\lim_{x \to \infty} \left(3 + \dfrac{10}{x^2}\right)$

10. $\lim_{x \to \infty} \left(5 + \dfrac{1}{x} + \dfrac{10}{x^2}\right)$

11. $\lim_{x \to \infty} \dfrac{f(x)}{g(x)}$ if $f(x) \to 100{,}000$ and $g(x) \to \infty$ as $x \to \infty$

**12.** $\lim\limits_{x\to\infty} \dfrac{4x^2 + 2x + 3}{x^2}$

**13.** $\lim\limits_{x\to\infty} \dfrac{1-x}{2x}$, $\lim\limits_{x\to-\infty} \dfrac{1-x}{x^2}$, and $\lim\limits_{x\to\infty} \dfrac{1-x^2}{2x}$

**14.** Describe with a sketch the end behavior of $f(x) = \cos x$.

**15.** Suppose the function $g$ satisfies the inequality
$$3 - \frac{1}{x^2} \le g(x) \le 3 + \frac{1}{x^2},$$ for all nonzero values of $x$. Evaluate $\lim\limits_{x\to\infty} g(x)$ and $\lim\limits_{x\to-\infty} g(x)$.

**16.** The graph of $g$ has a vertical asymptote at $x = 2$ and horizontal asymptotes at $y = -1$ and $y = 3$ (see figure). Determine the following limits: $\lim\limits_{x\to-\infty} g(x)$, $\lim\limits_{x\to\infty} g(x)$, $\lim\limits_{x\to2^-} g(x)$, and $\lim\limits_{x\to2^+} g(x)$.

### Practice Exercises

**17–36. Limits at infinity** *Determine the following limits.*

**17.** $\lim\limits_{\theta\to\infty} \dfrac{\cos\theta}{\theta^2}$

**18.** $\lim\limits_{t\to\infty} \dfrac{5t^2 + t\sin t}{t^2}$

**19.** $\lim\limits_{x\to\infty} \dfrac{\cos x^5}{\sqrt{x}}$

**20.** $\lim\limits_{x\to-\infty} \left(5 + \dfrac{100}{x} + \dfrac{\sin^4 x^3}{x^2}\right)$

**21.** $\lim\limits_{x\to\infty} \left(3x^{12} - 9x^7\right)$

**22.** $\lim\limits_{x\to-\infty} \left(3x^7 + x^2\right)$

**23.** $\lim\limits_{x\to-\infty} \left(-3x^{16} + 2\right)$

**24.** $\lim\limits_{x\to-\infty} \left(2x^{-8} + 4x^3\right)$

**25.** $\lim\limits_{x\to\infty} \dfrac{14x^3 + 3x^2 - 2x}{21x^3 + x^2 + 2x + 1}$

**26.** $\lim\limits_{x\to\infty} \dfrac{9x^3 + x^2 - 5}{3x^4 + 4x^2}$

**27.** $\lim\limits_{x\to-\infty} \dfrac{3x^2 + 3x}{x + 1}$

**28.** $\lim\limits_{x\to\infty} \dfrac{x^4 + 7}{x^5 + x^2 - x}$

**29.** $\lim\limits_{w\to\infty} \dfrac{15w^2 + 3w + 1}{\sqrt{9w^4 + w^3}}$

**30.** $\lim\limits_{x\to-\infty} \dfrac{40x^4 + x^2 + 5x}{\sqrt{64x^8 + x^6}}$

**31.** $\lim\limits_{x\to-\infty} \dfrac{\sqrt{16x^2 + x}}{x}$

**32.** $\lim\limits_{x\to\infty} \dfrac{6x^2}{4x^2 + \sqrt{16x^4 + x^2}}$

**33.** $\lim\limits_{x\to\infty} \left(x^2 - \sqrt{x^4 + 3x^2}\right)$ (*Hint:* Multiply by $\dfrac{x^2 + \sqrt{x^4 + 3x^2}}{x^2 + \sqrt{x^4 + 3x^2}}$ first.)

**34.** $\lim\limits_{x\to-\infty} \left(x + \sqrt{x^2 - 5x}\right)$

**35.** $\lim\limits_{x\to\infty} \dfrac{\sin x}{x^2 + 1}$

**36.** $\lim\limits_{x\to-\infty} \dfrac{x^4 \cos x}{3x^6 + 7x^4}$

**37–50. Horizontal asymptotes** *Determine $\lim\limits_{x\to\infty} f(x)$ and $\lim\limits_{x\to-\infty} f(x)$ for the following functions. Then give the horizontal asymptotes of $f$ (if any).*

**37.** $f(x) = \dfrac{4x}{20x + 1}$

**38.** $f(x) = \dfrac{3x^2 - 7}{x^2 + 5x}$

**39.** $f(x) = \dfrac{6x^2 - 9x + 8}{3x^2 + 2}$

**40.** $f(x) = \dfrac{12x^8 - 3}{3x^8 - 2x^7}$

**41.** $f(x) = \dfrac{3x^3 - 7}{x^4 + 5x^2}$

**42.** $f(x) = \dfrac{2x + 1}{3x^4 - 2}$

**43.** $f(x) = \dfrac{40x^5 + x^2}{16x^4 - 2x}$

**44.** $f(x) = \dfrac{6x^2 + 1}{\sqrt{4x^4 + 3x + 1}}$

**45.** $f(x) = \dfrac{1}{2x^4 - \sqrt{4x^8 - 9x^4}}$

**46.** $f(x) = \dfrac{\sqrt{x^2 + 1}}{2x + 1}$

**47.** $f(x) = \dfrac{4x^3 + 1}{2x^3 + \sqrt{16x^6 + 1}}$

**48.** $f(x) = x - \sqrt{x^2 - 9x}$

**49.** $f(x) = \dfrac{\sqrt[3]{x^6 + 8}}{4x^2 + \sqrt{3x^4 + 1}}$

**50.** $f(x) = 4x\left(3x - \sqrt{9x^2 + 1}\right)$

**T** **51–56. Slant (oblique) asymptotes** *Complete the following steps for the given functions.*

  *a.* Find the slant asymptote of $f$.

  *b.* Find the vertical asymptotes of $f$ (if any).

  *c.* Graph $f$ and all of its asymptotes with a graphing utility. Then sketch a graph of the function by hand, correcting any errors appearing in the computer-generated graph.

**51.** $f(x) = \dfrac{x^2 - 3}{x + 6}$

**52.** $f(x) = \dfrac{x^2 - 1}{x + 2}$

**53.** $f(x) = \dfrac{x^2 - 2x + 5}{3x - 2}$

**54.** $f(x) = \dfrac{5x^2 - 4}{5x - 5}$

**55.** $f(x) = \dfrac{4x^3 + 4x^2 + 7x + 4}{x^2 + 1}$

**56.** $f(x) = \dfrac{3x^2 - 2x + 5}{3x + 4}$

**57.** **Explain why or why not** Determine whether the following statements are true and give an explanation or counterexample.

  *a.* The graph of a function can never cross one of its horizontal asymptotes.

  *b.* A rational function $f$ has both $\lim\limits_{x\to\infty} f(x) = L$ (where $L$ is finite) and $\lim\limits_{x\to-\infty} f(x) = \infty$.

  *c.* The graph of a function can have any number of vertical asymptotes but at most two horizontal asymptotes.

  *d.* $\lim\limits_{x\to\infty} \left(x^3 - x\right) = \lim\limits_{x\to\infty} x^3 - \lim\limits_{x\to\infty} x = \infty - \infty = 0$

**58–61. Steady states** *If a function f represents a system that varies in time, the existence of* $\lim\limits_{t \to \infty} f(t)$ *means that the system reaches a steady state (or equilibrium). For the following systems, determine whether a steady state exists and give the steady-state value.*

**58.** The population of a bacteria culture is given by $p(t) = \dfrac{2500}{t + 1}$.

**59.** The population of a culture of tumor cells is given by
$$p(t) = \frac{3500t}{t + 1}.$$

**60.** The population of a colony of squirrels is given by
$$p(t) = \frac{1500t^2}{2t^2 + 3}.$$

**61.** The amplitude of an oscillator is given by $a(t) = 2\left(\dfrac{t + \sin t}{t}\right)$.

**62–73. Horizontal and vertical asymptotes**

**a.** Analyze $\lim\limits_{x \to \infty} f(x)$ *and* $\lim\limits_{x \to -\infty} f(x)$, *and then identify any horizontal asymptotes.*

**b.** Find the vertical asymptotes. For each vertical asymptote $x = a$, analyze $\lim\limits_{x \to a^-} f(x)$ *and* $\lim\limits_{x \to a^+} f(x)$.

**62.** $f(x) = \dfrac{x^2 - 4x + 3}{x - 1}$

**63.** $f(x) = \dfrac{2x^3 + 10x^2 + 12x}{x^3 + 2x^2}$

**64.** $f(x) = \dfrac{\sqrt{16x^4 + 64x^2} + x^2}{2x^2 - 4}$

**65.** $f(x) = \dfrac{3x^4 + 3x^3 - 36x^2}{x^4 - 25x^2 + 144}$

**66.** $f(x) = x^2\left(4x^2 - \sqrt{16x^4 + 1}\right)$

**67.** $f(x) = \dfrac{x^2 - 9}{x(x - 3)}$

**68.** $f(x) = \dfrac{x^4 - 1}{x^2 - 1}$

**69.** $f(x) = \dfrac{\sqrt{x^2 + 2x + 6} - 3}{x - 1}$

**70.** $f(x) = \dfrac{|1 - x^2|}{x(x + 1)}$

**71.** $f(x) = \sqrt{|x|} - \sqrt{|x - 1|}$

**72.** $f(x) = \dfrac{2x}{\sqrt{x^2 - x - 2}}$

**73.** $f(x) = \dfrac{\cos x + 2\sqrt{x}}{\sqrt{x}}$.

**74–75. Sketching graphs** *Sketch a possible graph of a function f that satisfies all of the given conditions. Be sure to identify all vertical and horizontal asymptotes.*

**74.** $f(-1) = -2, f(1) = 2, f(0) = 0, \lim\limits_{x \to \infty} f(x) = 1,$
$\lim\limits_{x \to -\infty} f(x) = -1$

**75.** $\lim\limits_{x \to 0^+} f(x) = \infty, \lim\limits_{x \to 0^-} f(x) = -\infty, \lim\limits_{x \to \infty} f(x) = 1,$
$\lim\limits_{x \to -\infty} f(x) = -2$

## Explorations and Challenges

**76–79. Looking ahead to sequences** *A sequence is an infinite, ordered list of numbers that is often defined by a function. For example, the sequence* $\{2, 4, 6, 8, \ldots\}$ *is specified by the function* $f(n) = 2n$, *where* $n = 1, 2, 3, \ldots$. *The limit of such a sequence is* $\lim\limits_{n \to \infty} f(n)$, *provided the limit exists. All the limit laws for limits at infinity may be applied to limits of sequences. Find the limit of the following sequences or state that the limit does not exist.*

**76.** $\left\{4, 2, \dfrac{4}{3}, 1, \dfrac{4}{5}, \dfrac{2}{3}, \ldots\right\}$, which is defined by $f(n) = \dfrac{4}{n}$, for
$n = 1, 2, 3, \ldots$

**77.** $\left\{0, \dfrac{1}{2}, \dfrac{2}{3}, \dfrac{3}{4}, \ldots\right\}$, which is defined by $f(n) = \dfrac{n - 1}{n}$, for
$n = 1, 2, 3, \ldots$

**78.** $\left\{\dfrac{1}{2}, \dfrac{4}{3}, \dfrac{9}{4}, \dfrac{16}{5}, \ldots\right\}$, which is defined by $f(n) = \dfrac{n^2}{n + 1}$, for
$n = 1, 2, 3, \ldots$

**79.** $\left\{2, \dfrac{3}{4}, \dfrac{4}{9}, \dfrac{5}{16}, \ldots\right\}$, which is defined by $f(n) = \dfrac{n + 1}{n^2}$, for
$n = 1, 2, 3, \ldots$

**80. End behavior of rational functions** Suppose $f(x) = \dfrac{p(x)}{q(x)}$
is a rational function, where
$$p(x) = a_m x^m + a_{m-1} x^{m-1} + \cdots + a_2 x^2 + a_1 x + a_0,$$
$$q(x) = b_n x^n + b_{n-1} x^{n-1} + \cdots + b_2 x^2 + b_1 x + b_0, a_m \neq 0,$$
and $b_n \neq 0$.

**a.** Prove that if $m = n$, then $\lim\limits_{x \to \pm\infty} f(x) = \dfrac{a_m}{b_n}$.

**b.** Prove that if $m < n$, then $\lim\limits_{x \to \pm\infty} f(x) = 0$.

**81. Horizontal and slant asymptotes**

**a.** Is it possible for a rational function to have both slant and horizontal asymptotes? Explain.

**b.** Is it possible for an algebraic function to have two distinct slant asymptotes? Explain or give an example.

**QUICK CHECK ANSWERS**

**1.** $10/11, 100/101, 1000/1001; 1$    **2.** $p(x) \to -\infty$ as
$x \to \infty$ and $p(x) \to \infty$ as $x \to -\infty$    **3.** Horizontal asymptote is $y = \frac{10}{3}$; vertical asymptote is $x = \frac{1}{3}$. ◀

# 2.6 Continuity

The graphs of many functions encountered in this text contain no holes, jumps, or breaks. For example, if $L = f(t)$ represents the length (in inches) of a fish $t$ years after it is hatched, then the length of the fish changes gradually as $t$ increases. Consequently, the graph of $L = f(t)$ contains no breaks (**Figure 2.42a**). Some functions, however, do contain abrupt changes in their values. Consider a parking meter that accepts only quarters, where each quarter buys 15 minutes of parking. Letting $c(t)$ be the cost (in dollars) of parking for $t$ minutes, the graph of $c$ has breaks at integer multiples of 15 minutes (**Figure 2.42b**).

**Figure 2.42**

Informally, we say that a function $f$ is *continuous* at $a$ if the graph of $f$ does not have a hole or break at $a$ (that is, if the graph near $a$ can be drawn without lifting the pencil).

## Continuity at a Point

This informal description of continuity is sufficient for determining the continuity of simple functions, but it is not precise enough to deal with more complicated functions such as

$$h(x) = \begin{cases} x \sin \dfrac{1}{x} & \text{if } x \neq 0 \\ 0 & \text{if } x = 0. \end{cases}$$

It is difficult to determine whether the graph of $h$ has a break at 0 because it oscillates rapidly as $x$ approaches 0 (**Figure 2.43**). We need a better definition.

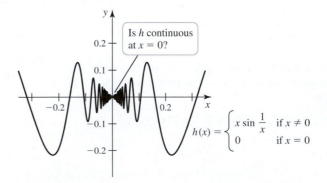

**Figure 2.43**

---

**DEFINITION**  **Continuity at a Point**

A function $f$ is **continuous** at $a$ if $\lim\limits_{x \to a} f(x) = f(a)$.

There is more to this definition than first appears. If $\lim\limits_{x \to a} f(x) = f(a)$, then $f(a)$ and $\lim\limits_{x \to a} f(x)$ must both exist, and they must be equal. The following checklist is helpful in determining whether a function is continuous at $a$.

---

**Continuity Checklist**

In order for $f$ to be continuous at $a$, the following three conditions must hold.

1. $f(a)$ is defined ($a$ is in the domain of $f$).
2. $\lim\limits_{x \to a} f(x)$ exists.
3. $\lim\limits_{x \to a} f(x) = f(a)$ (the value of $f$ equals the limit of $f$ at $a$).

---

If *any* item in the continuity checklist fails to hold, the function fails to be continuous at $a$. From this definition, we see that continuity has an important practical consequence:

> *If $f$ is continuous at $a$, then $\lim\limits_{x \to a} f(x) = f(a)$, and direct substitution may be used to evaluate $\lim\limits_{x \to a} f(x)$.*

Note that when $f$ is defined on an open interval containing $a$ (except possibly at $a$), we say that $f$ has a **discontinuity** at $a$ (or that $a$ is a **point of discontinuity**) if $f$ is not continuous at $a$.

**EXAMPLE 1  Points of discontinuity** Use the graph of $f$ in **Figure 2.44** to identify values of $x$ on the interval $(0, 7)$ at which $f$ has a discontinuity.

**SOLUTION** The function $f$ has discontinuities at $x = 1, 2, 3,$ and $5$ because the graph contains holes or breaks at these locations. The continuity checklist tells us why $f$ is not continuous at these points.

- $f(1)$ is not defined.
- $f(2) = 3$ and $\lim\limits_{x \to 2} f(x) = 1$. Therefore, $f(2)$ and $\lim\limits_{x \to 2} f(x)$ exist but are not equal.
- $\lim\limits_{x \to 3} f(x)$ does not exist because the left-sided limit $\lim\limits_{x \to 3^-} f(x) = 2$ differs from the right-sided limit $\lim\limits_{x \to 3^+} f(x) = 1$.
- Neither $\lim\limits_{x \to 5} f(x)$ nor $f(5)$ exists.

*Related Exercises 5–6* ◀

**EXAMPLE 2  Continuity at a point** Determine whether the following functions are continuous at $a$. Justify each answer using the continuity checklist.

**a.** $f(x) = \dfrac{3x^2 + 2x + 1}{x - 1};\quad a = 1$     **b.** $g(x) = \dfrac{3x^2 + 2x + 1}{x - 1};\quad a = 2$

**c.** $h(x) = \begin{cases} x \sin \dfrac{1}{x} & \text{if } x \neq 0 \\ 0 & \text{if } x = 0 \end{cases};\quad a = 0$

**SOLUTION**

**a.** The function $f$ is not continuous at 1 because $f(1)$ is undefined.

**b.** Because $g$ is a rational function and the denominator is nonzero at 2, it follows by Theorem 2.4 that $\lim\limits_{x \to 2} g(x) = g(2) = 17$. Therefore, $g$ is continuous at 2.

**c.** By definition, $h(0) = 0$. In Exercise 81 of Section 2.3, we used the Squeeze Theorem to show that $\lim\limits_{x \to 0} x \sin \dfrac{1}{x} = 0$. Therefore, $\lim\limits_{x \to 0} h(x) = h(0)$, which implies that $h$ is continuous at 0.

*Related Exercises 17, 22* ◀

---

*(left margin)*

**Figure 2.44**

➤ In Example 1, the discontinuities at $x = 1$ and $x = 2$ are called **removable discontinuities** because they can be removed by redefining the function at these points (in this case, $f(1) = 3$ and $f(2) = 1$). The discontinuity at $x = 3$ is called a **jump discontinuity**. The discontinuity at $x = 5$ is called an **infinite discontinuity**. These terms are discussed in Exercises 89–95.

**QUICK CHECK 1** For what values of $t$ in $(0, 60)$ does the graph of $y = c(t)$ in Figure 2.42b have a discontinuity? ◀

The following theorems make it easier to test various combinations of functions for continuity at a point.

---

**THEOREM 2.8    Continuity Rules**

If $f$ and $g$ are continuous at $a$, then the following functions are also continuous at $a$. Assume $c$ is a constant and $n > 0$ is an integer.

a. $f + g$                         b. $f - g$

c. $cf$                              d. $fg$

e. $f/g$, provided $g(a) \neq 0$       f. $(f(x))^n$

---

To prove the first result, note that if $f$ and $g$ are continuous at $a$, then $\lim_{x \to a} f(x) = f(a)$ and $\lim_{x \to a} g(x) = g(a)$. From the limit laws of Theorem 2.3, it follows that

$$\lim_{x \to a}(f(x) + g(x)) = f(a) + g(a).$$

Therefore, $f + g$ is continuous at $a$. Similar arguments lead to the continuity of differences, products, quotients, and powers of continuous functions. The next theorem is a direct consequence of Theorem 2.8.

---

**THEOREM 2.9    Polynomial and Rational Functions**

a. A polynomial function is continuous for all $x$.

b. A rational function (a function of the form $\dfrac{p}{q}$, where $p$ and $q$ are polynomials) is continuous for all $x$ for which $q(x) \neq 0$.

---

$f(x) = \dfrac{x}{x^2 - 7x + 12}$

Continuous everywhere except $x = 3$ and $x = 4$

**Figure 2.45**

**EXAMPLE 3    Applying the continuity theorems**  For what values of $x$ is the function

$$f(x) = \frac{x}{x^2 - 7x + 12} \quad \text{continuous?}$$

**SOLUTION**  Because $f$ is rational, Theorem 2.9b implies it is continuous for all $x$ at which the denominator is nonzero. The denominator factors as $(x - 3)(x - 4)$, so it is zero at $x = 3$ and $x = 4$. Therefore, $f$ is continuous for all $x$ except $x = 3$ and $x = 4$ (**Figure 2.45**).

*Related Exercises 26–27* ◄

The following theorem enables us to determine when a composition of two functions is continuous at a point. Its proof is informative and is outlined in Exercise 96.

---

**THEOREM 2.10    Continuity of Composite Functions at a Point**

If $g$ is continuous at $a$ and $f$ is continuous at $g(a)$, then the composite function $f \circ g$ is continuous at $a$.

---

Theorem 2.10 is useful because it allows us to conclude that the composition of two continuous functions is continuous at a point. For example, the composite function $\left(\dfrac{x}{x - 1}\right)^3$ is continuous for all $x \neq 1$. Furthermore, under the stated conditions on $f$ and $g$, the limit of $f \circ g$ is evaluated by direct substitution; that is,

$$\lim_{x \to a} f(g(x)) = f(g(a)).$$

**EXAMPLE 4  Limit of a composition**  Evaluate $\lim\limits_{x \to 0} \left( \dfrac{x^4 - 2x + 2}{x^6 + 2x^4 + 1} \right)^{10}$.

**SOLUTION**  The rational function $\dfrac{x^4 - 2x + 2}{x^6 + 2x^4 + 1}$ is continuous for all $x$ because its denominator is always positive (Theorem 2.9b). Therefore, $\left( \dfrac{x^4 - 2x + 2}{x^6 + 2x^4 + 1} \right)^{10}$, which is the composition of the continuous function $f(x) = x^{10}$ and a continuous rational function, is continuous for all $x$ by Theorem 2.10. By direct substitution,

$$\lim_{x \to 0} \left( \frac{x^4 - 2x + 2}{x^6 + 2x^4 + 1} \right)^{10} = \left( \frac{0^4 - 2 \cdot 0 + 2}{0^6 + 2 \cdot 0^4 + 1} \right)^{10} = 2^{10} = 1024.$$

*Related Exercises 31–32* ◄

Closely related to Theorem 2.10 are two useful results dealing with limits of composite functions. We present these results—one a more general version of the other—in a single theorem.

---

**THEOREM 2.11  Limits of Composite Functions**

1. If $g$ is continuous at $a$ and $f$ is continuous at $g(a)$, then

$$\lim_{x \to a} f(g(x)) = f\left( \lim_{x \to a} g(x) \right).$$

2. If $\lim\limits_{x \to a} g(x) = L$ and $f$ is continuous at $L$, then

$$\lim_{x \to a} f(g(x)) = f\left( \lim_{x \to a} g(x) \right).$$

---

**Proof:** The first statement follows directly from Theorem 2.10, which states that $\lim\limits_{x \to a} f(g(x)) = f(g(a))$. If $g$ is continuous at $a$, then $\lim\limits_{x \to a} g(x) = g(a)$, and it follows that

$$\lim_{x \to a} f(g(x)) = f(\underbrace{g(a)}_{\lim\limits_{x \to a} g(x)}) = f\left( \lim_{x \to a} g(x) \right).$$

The proof of the second statement (see Appendix A) relies on the formal definition of a limit, which is discussed in Section 2.7.   ◄

Both statements of Theorem 2.11 justify interchanging the order of a limit and a function evaluation. By the second statement, the inner function of the composition needn't be continuous at the point of interest, but it must have a limit at that point. Note also that $\lim\limits_{x \to a}$ can be replaced with $\lim\limits_{x \to a^+}$ or $\lim\limits_{x \to a^-}$ in statement (1) of Theorem 2.11, provided $g$ is *right-* or *left-continuous* (see p. 107) at $a$, respectively. In statement (2), $\lim\limits_{x \to a}$ can be replaced with $\lim\limits_{x \to \infty}$ or $\lim\limits_{x \to -\infty}$.

**EXAMPLE 5  Limits of a composite functions**  Evaluate the following limits.

**a.** $\lim\limits_{x \to -1} \sqrt{2x^2 - 1}$     **b.** $\lim\limits_{x \to 2} \cos\left( \dfrac{x^2 - 4}{x - 2} \right)$

**SOLUTION**

**a.** The inner function of the composite function $\sqrt{2x^2 - 1}$ is $g(x) = 2x^2 - 1$; it is continuous and positive at $-1$, and $g(-1) = 1$. Because $f(x) = \sqrt{x}$ is continuous at $g(-1) = 1$ (a consequence of Law 7, Theorem 2.3), we have, by the first statement of Theorem 2.11,

$$\lim_{x \to -1} \sqrt{2x^2 - 1} = \sqrt{\underbrace{\lim_{x \to -1} (2x^2 - 1)}_{1}} = \sqrt{1} = 1.$$

**b.** We show later in this section that $\cos x$ is continuous at all points of its domain. The inner function of the composite function $\cos\left(\dfrac{x^2 - 4}{x - 2}\right)$ is $\dfrac{x^2 - 4}{x - 2}$, which is not continuous at 2. However,

$$\lim_{x \to 2} \frac{x^2 - 4}{x - 2} = \lim_{x \to 2} \frac{(x - 2)(x + 2)}{x - 2} = \lim_{x \to 2} (x + 2) = 4.$$

Therefore, by the second statement of Theorem 2.11,

$$\lim_{x \to 2} \cos\left(\frac{x^2 - 4}{x - 2}\right) = \cos\left(\underbrace{\lim_{x \to 2} \frac{x^2 - 4}{x - 2}}_{4}\right) = \cos 4 \approx -0.654.$$

*Related Exercises 33–34* ◄

## Continuity on an Interval

A function is *continuous on an interval* if it is continuous at every point in that interval. Consider the functions $f$ and $g$ whose graphs are shown in **Figure 2.46**. Both these functions are continuous for all $x$ in $(a, b)$, but what about the endpoints? To answer this question, we introduce the ideas of *left-continuity* and *right-continuity*.

(a)

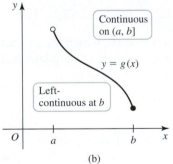

(b)

**Figure 2.46**

> **DEFINITION  Continuity at Endpoints**
>
> A function $f$ is **continuous from the right** (or **right-continuous**) at $a$ if $\lim\limits_{x \to a^+} f(x) = f(a)$, and $f$ is **continuous from the left** (or **left-continuous**) at $b$ if $\lim\limits_{x \to b^-} f(x) = f(b)$.

Combining the definitions of left-continuous and right-continuous with the definition of continuity at a point, we define what it means for a function to be continuous on an interval.

> **DEFINITION  Continuity on an Interval**
>
> A function $f$ is **continuous on an interval** $I$ if it is continuous at all points of $I$. If $I$ contains its endpoints, continuity on $I$ means continuous from the right or left at the endpoints.

To illustrate these definitions, consider again the functions in Figure 2.46. In **Figure 2.46a**, $f$ is continuous from the right at $a$ because $\lim\limits_{x \to a^+} f(x) = f(a)$, but it is not continuous from the left at $b$ because $f(b)$ is not defined. Therefore, $f$ is continuous on the interval $[a, b)$. The behavior of the function $g$ in **Figure 2.46b** is the opposite: It is continuous from the left at $b$, but it is not continuous from the right at $a$. Therefore, $g$ is continuous on $(a, b]$.

**QUICK CHECK 2** Modify the graphs of the functions $f$ and $g$ in Figure 2.46 to obtain functions that are continuous on $[a, b]$. ◄

**EXAMPLE 6   Intervals of continuity**  Determine the intervals of continuity for

$$f(x) = \begin{cases} x^2 + 1 & \text{if } x \le 0 \\ 3x + 5 & \text{if } x > 0. \end{cases}$$

**SOLUTION**  This piecewise function consists of two polynomials that describe a parabola and a line (**Figure 2.47**). By Theorem 2.9, $f$ is continuous for all $x \ne 0$. From its graph, it appears that $f$ is left-continuous at 0. This observation is verified by noting that

$$\lim_{x \to 0^-} f(x) = \lim_{x \to 0^-} (x^2 + 1) = 1,$$

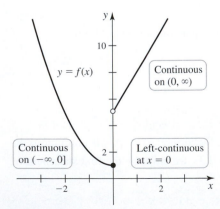

**Figure 2.47**

which means that $\lim\limits_{x \to 0^-} f(x) = f(0)$. However, because

$$\lim\limits_{x \to 0^+} f(x) = \lim\limits_{x \to 0^+} (3x + 5) = 5 \neq f(0),$$

we see that $f$ is not right-continuous at 0. Therefore, $f$ is continuous on $(-\infty, 0]$ and on $(0, \infty)$.

*Related Exercises 39–40* ◄

## Functions Involving Roots

Recall that Limit Law 7 of Theorem 2.3 states

$$\lim\limits_{x \to a} (f(x))^{1/n} = \left( \lim\limits_{x \to a} f(x) \right)^{1/n},$$

provided $f(x) > 0$, for $x$ near $a$, if $n$ is even. Therefore, if $n$ is odd and $f$ is continuous at $a$, then $(f(x))^{1/n}$ is continuous at $a$, because

$$\lim\limits_{x \to a} (f(x))^{1/n} = \left( \lim\limits_{x \to a} f(x) \right)^{1/n} = (f(a))^{1/n}.$$

When $n$ is even, the continuity of $(f(x))^{1/n}$ must be handled more carefully because this function is defined only when $f(x) \geq 0$. Exercise 73 of Section 2.7 establishes an important fact:

> *If $f$ is continuous at $a$, and $f(a) > 0$, then $f(x) > 0$ for all values of $x$ in some interval containing $a$.*

Combining this fact with Theorem 2.10 (the continuity of composite functions), it follows that $(f(x))^{1/n}$ is continuous at $a$ provided $f(a) > 0$. At points where $f(a) = 0$, the behavior of $(f(x))^{1/n}$ varies. Often we find that $(f(x))^{1/n}$ is left- or right-continuous at that point, or it may be continuous from both sides.

---

**THEOREM 2.12   Continuity of Functions with Roots**

Assume $n$ is a positive integer. If $n$ is an odd integer, then $(f(x))^{1/n}$ is continuous at all points at which $f$ is continuous.

If $n$ is even, then $(f(x))^{1/n}$ is continuous at all points $a$ at which $f$ is continuous and $f(a) > 0$.

---

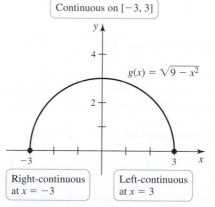

Continuous on $[-3, 3]$

$g(x) = \sqrt{9 - x^2}$

Right-continuous at $x = -3$

Left-continuous at $x = 3$

**Figure 2.48**

**EXAMPLE 7   Continuity with roots** For what values of $x$ are the following functions continuous?

**a.** $g(x) = \sqrt{9 - x^2}$

**b.** $f(x) = (x^2 - 2x + 4)^{2/3}$

**SOLUTION**

**a.** The graph of $g$ is the upper half of the circle $x^2 + y^2 = 9$ (which can be verified by solving $x^2 + y^2 = 9$ for $y$). From **Figure 2.48**, it appears that $g$ is continuous on $[-3, 3]$. To verify this fact, note that $g$ involves an even root ($n = 2$ in Theorem 2.12). If $-3 < x < 3$, then $9 - x^2 > 0$ and by Theorem 2.12, $g$ is continuous for all $x$ on $(-3, 3)$.

At the right endpoint, $\lim\limits_{x \to 3^-} \sqrt{9 - x^2} = 0 = g(3)$ by Limit Law 7 for one-sided limits, which implies that $g$ is left-continuous at 3. Similarly, $g$ is right-continuous at $-3$ because $\lim\limits_{x \to -3^+} \sqrt{9 - x^2} = 0 = g(-3)$. Therefore, $g$ is continuous on $[-3, 3]$.

**b.** The polynomial $x^2 - 2x + 4$ is continuous for all $x$ by Theorem 2.9a. Rewriting $f$ as $f(x) = ((x^2 - 2x + 4)^{1/3})^2$, we see that $f$ involves an odd root ($n = 3$ in Theorem 2.12). Therefore, $f$ is continuous for all $x$.

*Related Exercises 44–45* ◄

**QUICK CHECK 3** On what interval is $f(x) = x^{1/4}$ continuous? On what interval is $f(x) = x^{2/5}$ continuous? ◄

## Continuity of Trigonometric Functions

In Section 2.3, we used the Squeeze Theorem to show that $\lim\limits_{x \to 0} \sin x = 0$ and $\lim\limits_{x \to 0} \cos x = 1$. Because $\sin 0 = 0$ and $\cos 0 = 1$, these limits imply that $\sin x$ and $\cos x$ are continuous at 0. The graph of $y = \sin x$ (**Figure 2.49**) suggests that $\lim\limits_{x \to a} \sin x = \sin a$ for any value of $a$, which means that $\sin x$ is continuous on $(-\infty, \infty)$. The graph of $y = \cos x$ also indicates that $\cos x$ is continuous for all $x$. Exercise 99 outlines a proof of these results.

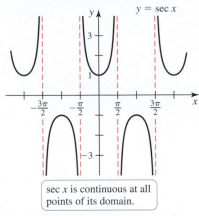

sec $x$ is continuous at all points of its domain.

**Figure 2.49**                    **Figure 2.50**

With these facts in hand, we appeal to Theorem 2.8e to discover that the remaining trigonometric functions are continuous on their domains. For example, because $\sec x = 1/\cos x$, the secant function is continuous for all $x$ for which $\cos x \neq 0$ (for all $x$ except odd multiples of $\pi/2$; **Figure 2.50**). Likewise, the tangent, cotangent, and cosecant functions are continuous at all points of their domains.

---

**THEOREM 2.13    Continuity of Trigonometric Functions**

The functions $\sin x$, $\cos x$, $\tan x$, $\cot x$, $\sec x$, and $\csc x$ are continuous at all points of their domains.

---

For each function listed in Theorem 2.13, we have $\lim\limits_{x \to a} f(x) = f(a)$, provided $a$ is in the domain of the function. This means that limits of these functions may be evaluated by direct substitution at points in the domain.

**EXAMPLE 8    Limits involving trigonometric functions** Evaluate $\lim\limits_{x \to 0} \dfrac{\cos^2 x - 1}{\cos x - 1}$.

**SOLUTION**    Both $\cos^2 x - 1$ and $\cos x - 1$ are continuous for all $x$ by Theorems 2.8 and 2.13. However, the ratio of these functions is continuous only when $\cos x - 1 \neq 0$, which occurs when $x$ is not an integer multiple of $2\pi$. Note that both the numerator and denominator of $\dfrac{\cos^2 x - 1}{\cos x - 1}$ approach 0 as $x \to 0$. To evaluate the limit, we factor and simplify:

$$\lim_{x \to 0} \frac{\cos^2 x - 1}{\cos x - 1} = \lim_{x \to 0} \frac{(\cos x - 1)(\cos x + 1)}{\cos x - 1} = \lim_{x \to 0} (\cos x + 1)$$

> Limits like the one in Example 8 are denoted 0/0 and are known as *indeterminate forms*, to be studied further in Section 4.7.

(where $\cos x - 1$ may be canceled because it is nonzero as $x$ approaches 0). The limit on the right is now evaluated using direct substitution:

$$\lim_{x \to 0} (\cos x + 1) = \cos 0 + 1 = 2.$$

*Related Exercises 62–63* ◄

We close this section with an important theorem that has both practical and theoretical uses.

## Intermediate Value Theorem

A common problem in mathematics is finding solutions to equations of the form $f(x) = L$. Before attempting to find values of $x$ satisfying this equation, it is worthwhile to determine whether a solution exists.

The existence of solutions is often established using a result known as the *Intermediate Value Theorem*. Given a function $f$ and a constant $L$, we assume $L$ lies strictly between $f(a)$ and $f(b)$. The Intermediate Value Theorem says that if $f$ is continuous on $[a, b]$, then the graph of $f$ must cross the horizontal line $y = L$ at least once (**Figure 2.51**). Although this theorem is easily illustrated, its proof is beyond the scope of this text.

Intermediate Value Theorem

There is *at least* one number $c$ in $(a, b)$ such that $f(c) = L$, where $L$ is between $f(a)$ and $f(b)$.

**Figure 2.51**

> **THEOREM 2.14    Intermediate Value Theorem**
> Suppose $f$ is continuous on the interval $[a, b]$ and $L$ is a number strictly between $f(a)$ and $f(b)$. Then there exists at least one number $c$ in $(a, b)$ satisfying $f(c) = L$.

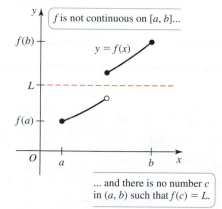

$f$ is not continuous on $[a, b]$...

... and there is no number $c$ in $(a, b)$ such that $f(c) = L$.

**Figure 2.52**

The importance of continuity in Theorem 2.14 is illustrated in **Figure 2.52**, where we see a function $f$ that is not continuous on $[a, b]$. For the value of $L$ shown in the figure, there is no value of $c$ in $(a, b)$ satisfying $f(c) = L$. The next example illustrates a practical application of the Intermediate Value Theorem.

**EXAMPLE 9    Finding an interest rate**  Suppose you invest $1000 in a special 5-year savings account with a fixed annual interest rate $r$, with monthly compounding. The amount of money $A$ in the account after 5 years (60 months) is

$$A(r) = 1000\left(1 + \frac{r}{12}\right)^{60}.$$ Your goal is to have $1400 in the account after 5 years.

**a.** Use the Intermediate Value Theorem to show there is a value of $r$ in $(0, 0.08)$—that is, an interest rate between 0% and 8%—for which $A(r) = 1400$.

**b.** Use a graphing utility to illustrate your explanation in part (a) and then estimate the interest rate required to reach your goal.

**SOLUTION**

**a.** As a polynomial in $r$ (of degree 60), $A(r) = 1000\left(1 + \frac{r}{12}\right)^{60}$ is continuous for all $r$.

Evaluating $A(r)$ at the endpoints of the interval $[0, 0.08]$, we have $A(0) = 1000$ and $A(0.08) \approx 1489.85$. Therefore,

$$A(0) < 1400 < A(0.08),$$

and it follows, by the Intermediate Value Theorem, that there is a value of $r$ in $(0, 0.08)$ for which $A(r) = 1400$.

**QUICK CHECK 4**  Does the equation $f(x) = x^3 + x + 1 = 0$ have a solution on the interval $[-1, 1]$? Explain. ◄

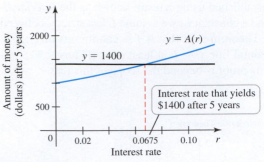

**Figure 2.53**

**b.** The graphs of $y = A(r)$ and the horizontal line $y = 1400$ are shown in **Figure 2.53**; it is evident that they intersect between $r = 0$ and $r = 0.08$. Solving $A(r) = 1400$ algebraically or using a root finder reveals that the curve and line intersect at $r \approx 0.0675$. Therefore, an interest rate of approximately 6.75% is required for the investment to be worth $1400 after 5 years.

*Related Exercises 65, 71* ◄

## SECTION 2.6 EXERCISES

### Getting Started

**1.** Which of the following functions are continuous for all values in their domain? Justify your answers.

  **a.** $a(t) =$ altitude of a skydiver $t$ seconds after jumping from a plane

  **b.** $n(t) =$ number of quarters needed to park legally in a metered parking space for $t$ minutes

  **c.** $T(t) =$ temperature $t$ minutes after midnight in Chicago on January 1

  **d.** $p(t) =$ number of points scored by a basketball player after $t$ minutes of a basketball game

**2.** Give the three conditions that must be satisfied by a function to be continuous at a point.

**3.** What does it mean for a function to be continuous on an interval?

**4.** We informally describe a function $f$ to be continuous at $a$ if its graph contains no holes or breaks at $a$. Explain why this is not an adequate definition of continuity.

**5–8.** *Determine the points on the interval $(0, 5)$ at which the following functions $f$ have discontinuities. At each point of discontinuity, state the conditions in the continuity checklist that are violated.*

**5.**

**6.**

**7.**

**8.**

**9.** Complete the following sentences in terms of a limit.

  **a.** A function is continuous from the left at $a$ if _____.
  **b.** A function is continuous from the right at $a$ if _____.

**10.** Evaluate $f(3)$ if $\lim\limits_{x \to 3^-} f(x) = 5$, $\lim\limits_{x \to 3^+} f(x) = 6$, and $f$ is right-continuous at $x = 3$.

**11–14.** *Determine the intervals of continuity for the following functions. At which endpoints of these intervals of continuity is $f$ continuous from the left or continuous from the right?*

**11.** The graph of Exercise 5        **12.** The graph of Exercise 6

**13.** The graph of Exercise 7        **14.** The graph of Exercise 8

**15.** What is the domain of $f(x) = \sqrt{1 - x^2}$? Where is $f$ continuous?

**16.** **Parking costs** Determine the intervals of continuity for the parking cost function $c$ introduced at the outset of this section (see figure). Consider $0 \le t \le 60$.

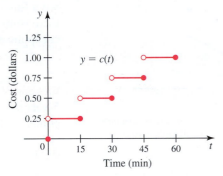

### Practice Exercises

**17–24. Continuity at a point** *Determine whether the following functions are continuous at $a$. Use the continuity checklist to justify your answer.*

**17.** $f(x) = \dfrac{2x^2 + 3x + 1}{x^2 + 5x}; a = -5$

**18.** $f(x) = \dfrac{2x^2 + 3x + 1}{x^2 + 5x}; a = 5$

**19.** $f(x) = \sqrt{x - 2}; a = 1$

**20.** $g(x) = \dfrac{1}{x - 3}; a = 3$

**21.** $f(x) = \begin{cases} \dfrac{x^2 - 1}{x - 1} & \text{if } x \neq 1 \\ 3 & \text{if } x = 1 \end{cases}; a = 1$

**22.** $f(x) = \begin{cases} \dfrac{x^2 - 4x + 3}{x - 3} & \text{if } x \neq 3 \\ 2 & \text{if } x = 3 \end{cases}; a = 3$

**23.** $f(x) = \dfrac{5x - 2}{x^2 - 9x + 20}; a = 4$

**24.** $f(x) = \begin{cases} \dfrac{x^2 + x}{x + 1} & \text{if } x \neq -1 \\ 2 & \text{if } x = -1 \end{cases}; a = -1$

**25–30. Continuity** *Determine the interval(s) on which the following functions are continuous.*

**25.** $p(x) = 4x^5 - 3x^2 + 1$

**26.** $g(x) = \dfrac{3x^2 - 6x + 7}{x^2 + x + 1}$

**27.** $f(x) = \dfrac{x^5 + 6x + 17}{x^2 - 9}$

**28.** $s(x) = \dfrac{x^2 - 4x + 3}{x^2 - 1}$

**29.** $f(x) = \dfrac{1}{x^2 - 4}$

**30.** $f(t) = \dfrac{t + 2}{t^2 - 4}$

**31–38. Limits** *Evaluate each limit and justify your answer.*

**31.** $\lim\limits_{x \to 0} (x^8 - 3x^6 - 1)^{40}$

**32.** $\lim\limits_{x \to 2} \left( \dfrac{3}{2x^5 - 4x^2 - 50} \right)^4$

**33.** $\lim\limits_{x \to 4} \sqrt{\dfrac{x^3 - 2x^2 - 8x}{x - 4}}$

**34.** $\lim\limits_{t \to 4} \dfrac{t - 4}{\sqrt{t} - 2}$

**35.** $\lim\limits_{x \to 1} \left( \dfrac{x + 5}{x + 2} \right)^4$

**36.** $\lim\limits_{x \to \infty} \left( \dfrac{2x + 1}{x} \right)^3$

**37.** $\lim\limits_{x \to 5} \left( \dfrac{\sqrt{x^2 - 16} - 3}{5x - 25} \right)^4$

**38.** $\lim\limits_{x \to 0} \left( \dfrac{x}{\sqrt{16x + 1} - 1} \right)^{1/3}$

**39–40. Intervals of continuity** *Complete the following steps for each function.*

**a.** *Use the continuity checklist to show that f is not continuous at the given value of a.*

**b.** *Determine whether f is continuous from the left or the right at a.*

**c.** *State the interval(s) of continuity.*

**39.** $f(x) = \begin{cases} 2x & \text{if } x < 1 \\ x^2 + 3x & \text{if } x \geq 1 \end{cases}; a = 1$

**40.** $f(x) = \begin{cases} x^3 + 4x + 1 & \text{if } x \leq 0 \\ 2x^3 & \text{if } x > 0 \end{cases}; a = 0$

**41–48. Functions with roots** *Determine the interval(s) on which the following functions are continuous. At which finite endpoints of the intervals of continuity is f continuous from the left or continuous from the right?*

**41.** $f(x) = \sqrt{5 - x}$

**42.** $f(x) = \sqrt{25 - x^2}$

**43.** $f(x) = \sqrt{2x^2 - 16}$

**44.** $f(x) = \sqrt{x^2 - 3x + 2}$

**45.** $f(x) = \sqrt[3]{x^2 - 2x - 3}$

**46.** $f(t) = (t^2 - 1)^{3/2}$

**47.** $f(x) = (2x - 3)^{2/3}$

**48.** $f(z) = (z - 1)^{3/4}$

**49–60.** *Evaluate each limit.*

**49.** $\lim\limits_{x \to 2} \sqrt{\dfrac{4x + 10}{2x - 2}}$

**50.** $\lim\limits_{x \to -1} \left( x^2 - 4 + \sqrt[3]{x^2 - 9} \right)$

**51.** $\lim\limits_{x \to \pi} \dfrac{\cos^2 x + 3\cos x + 2}{\cos x + 1}$

**52.** $\lim\limits_{x \to 3\pi/2} \dfrac{\sin^2 x + 6\sin x + 5}{\sin^2 x - 1}$

**53.** $\lim\limits_{x \to 3} \sqrt{x^2 + 7}$

**54.** $\lim\limits_{t \to 2} \dfrac{t^2 + 5}{1 + \sqrt{t^2 + 5}}$

**55.** $\lim\limits_{x \to \pi/2} \dfrac{\sin x - 1}{\sqrt{\sin x} - 1}$

**56.** $\lim\limits_{\theta \to 0} \dfrac{\dfrac{1}{2 + \sin \theta} - \dfrac{1}{2}}{\sin \theta}$

**57.** $\lim\limits_{x \to 0} \dfrac{\cos x - 1}{\sin^2 x}$

**58.** $\lim\limits_{x \to 0^+} \dfrac{1 - \cos^2 x}{\sin x}$

**59.** $\lim\limits_{t \to \infty} \dfrac{\sin t}{t^2}$

**60.** $\lim\limits_{x \to 0^+} \dfrac{\cos x}{x}$

**61–64. Continuity and limits with trigonometric functions** *Determine the interval(s) on which the following functions are continuous; then analyze the given limits.*

**61.** $f(x) = \csc x$; $\lim\limits_{x \to \pi/4} f(x)$; $\lim\limits_{x \to 2\pi^-} f(x)$

**62.** $f(x) = \sqrt{\sin x}$; $\lim\limits_{x \to \pi/2} f(x)$; $\lim\limits_{x \to 0^+} f(x)$

**63.** $f(x) = \dfrac{1 + \sin x}{\cos x}$; $\lim\limits_{x \to \pi/2^-} f(x)$; $\lim\limits_{x \to 4\pi/3} f(x)$

**64.** $f(x) = \dfrac{1}{2\cos x - 1}$; $\lim\limits_{x \to \pi/6} f(x)$

**T** **65–68. Applying the Intermediate Value Theorem**

**a.** *Use the Intermediate Value Theorem to show that the following equations have a solution on the given interval.*

**b.** *Use a graphing utility to find all the solutions to the equation on the given interval.*

**c.** *Illustrate your answers with an appropriate graph.*

**65.** $2x^3 + x - 2 = 0$; $(-1, 1)$

**66.** $\sqrt{x^4 + 25x^3 + 10} = 5$; $(0, 1)$

**67.** $x^3 - 5x^2 + 2x = -1$; $(-1, 5)$

**68.** $-x^5 - 4x^2 + 2\sqrt{x} + 5 = 0$; $(0, 3)$

**69.** **Explain why or why not** Determine whether the following statements are true and give an explanation or counterexample.

**a.** If a function is left-continuous and right-continuous at $a$, then it is continuous at $a$.

**b.** If a function is continuous at $a$, then it is left-continuous and right-continuous at $a$.

**c.** If $a < b$ and $f(a) \leq L \leq f(b)$, then there is some value of $c$ in $(a, b)$ for which $f(c) = L$.

**d.** Suppose $f$ is continuous on $[a, b]$. Then there is a point $c$ in $(a, b)$ such that $f(c) = (f(a) + f(b))/2$.

**T 70. Mortgage payments** You are shopping for a $250,000, 30-year (360-month) loan to buy a house. The monthly payment is given by

$$m(r) = \frac{250{,}000(r/12)}{1 - (1 + r/12)^{-360}},$$

where $r$ is the annual interest rate. Suppose banks are currently offering interest rates between 4% and 5%.

**a.** Show there is a value of $r$ in $(0.04, 0.05)$—an interest rate between 4% and 5%—that allows you to make monthly payments of $1300 per month.

**b.** Use a graph to illustrate your explanation to part (a). Then determine the interest rate you need for monthly payments of $1300.

**T 71. Interest rates** Suppose $5000 is invested in a savings account for 10 years (120 months), with an annual interest rate of $r$, compounded monthly. The amount of money in the account after 10 years is given by $A(r) = 5000(1 + r/12)^{120}$.

**a.** Show there is a value of $r$ in $(0, 0.08)$—an interest rate between 0% and 8%—that allows you to reach your savings goal of $7000 in 10 years.

**b.** Use a graph to illustrate your explanation in part (a). Then approximate the interest rate required to reach your goal.

**T 72. Investment problem** Assume you invest $250 at the end of each year for 10 years at an annual interest rate of $r$. The amount of money in your account after 10 years is given by

$$A(r) = \frac{250((1 + r)^{10} - 1)}{r}.$$ Assume your goal is to have $3500 in your account after 10 years.

**a.** Show that there is an interest rate $r$ in the interval $(0.01, 0.10)$—between 1% and 10%—that allows you to reach your financial goal.

**b.** Use a calculator to estimate the interest rate required to reach your financial goal.

**T 73.** Find an interval containing a solution to the equation $2x = \cos x$. Use a graphing utility to approximate the solution.

## Explorations and Challenges

**74. Continuity of the absolute value function** Prove that the absolute value function $|x|$ is continuous for all values of $x$. (*Hint:* Using the definition of the absolute value function, compute $\lim_{x \to 0^-} |x|$ and $\lim_{x \to 0^+} |x|$.)

**75–78. Continuity of functions with absolute values** *Use the continuity of the absolute value function (Exercise 74) to determine the interval(s) on which the following functions are continuous.*

**75.** $f(x) = |x^2 + 3x - 18|$       **76.** $g(x) = \left| \dfrac{x + 4}{x^2 - 4} \right|$

**77.** $h(x) = \left| \dfrac{1}{\sqrt{x} - 4} \right|$

**78.** $h(x) = |x^2 + 2x + 5| + \sqrt{x}$

**79. Pitfalls using technology** The graph of the *sawtooth function* $y = x - \lfloor x \rfloor$, where $\lfloor x \rfloor$ is the greatest integer function or floor function (Exercise 51, Section 2.2), was obtained using a graphing utility (see figure). Identify any inaccuracies appearing in the graph and then plot an accurate graph by hand.

$$y = x - \lfloor x \rfloor$$

**T 80. Pitfalls using technology** Graph the function $f(x) = \dfrac{\sin x}{x}$ using a graphing window of $[-\pi, \pi] \times [0, 2]$.

**a.** Sketch a copy of the graph obtained with your graphing device and describe any inaccuracies appearing in the graph.

**b.** Sketch an accurate graph of the function. Is $f$ continuous at 0?

**c.** What is the value of $\lim\limits_{x \to 0} \dfrac{\sin x}{x}$?

**81. Sketching functions**

**a.** Sketch the graph of a function that is not continuous at 1, but is defined at 1.

**b.** Sketch the graph of a function that is not continuous at 1, but has a limit at 1.

**82. An unknown constant** Determine the value of the constant $a$ for which the function

$$f(x) = \begin{cases} \dfrac{x^2 + 3x + 2}{x + 1} & \text{if } x \neq -1 \\ a & \text{if } x = -1 \end{cases}$$

is continuous at $-1$.

**83. An unknown constant** Let

$$g(x) = \begin{cases} x^2 + x & \text{if } x < 1 \\ a & \text{if } x = 1 \\ 3x + 5 & \text{if } x > 1. \end{cases}$$

**a.** Determine the value of $a$ for which $g$ is continuous from the left at 1.

**b.** Determine the value of $a$ for which $g$ is continuous from the right at 1.

**c.** Is there a value of $a$ for which $g$ is continuous at 1? Explain.

**T 84–85. Applying the Intermediate Value Theorem** *Use the Intermediate Value Theorem to verify that the following equations have three solutions on the given interval. Use a graphing utility to find the approximate roots.*

**84.** $x^3 + 10x^2 - 100x + 50 = 0; \ (-20, 10)$

**85.** $70x^3 - 87x^2 + 32x - 3 = 0; \ (0, 1)$

**86. Walk in the park** Suppose you park your car at a trailhead in a national park and begin a 2-hr hike to a lake at 7 A.M. on a Friday morning. On Sunday morning, you leave the lake at 7 A.M. and start the 2-hr hike back to your car. Assume the lake is 3 mi from your car. Let $f(t)$ be your distance from the car $t$ hours after 7 A.M. on Friday morning, and let $g(t)$ be your distance from the car $t$ hours after 7 A.M. on Sunday morning.

   **a.** Evaluate $f(0)$, $f(2)$, $g(0)$, and $g(2)$.

   **b.** Let $h(t) = f(t) - g(t)$. Find $h(0)$ and $h(2)$.

   **c.** Show that there is some point along the trail that you will pass at exactly the same time on both days.

**87. The monk and the mountain** A monk set out from a monastery in the valley at dawn. He walked all day up a winding path, stopping for lunch and taking a nap along the way. At dusk, he arrived at a temple on the mountaintop. The next day the monk made the return walk to the valley, leaving the temple at dawn, walking the same path for the entire day, and arriving at the monastery in the evening. Must there be one point along the path that the monk occupied at the same time of day on both the ascent and the descent? Explain. (*Hint:* The question can be answered without the Intermediate Value Theorem.) (*Source:* Arthur Koestler, *The Act of Creation*)

**88. Does continuity of $|f|$ imply continuity of $f$?** Let

$$g(x) = \begin{cases} 1 & \text{if } x \geq 0 \\ -1 & \text{if } x < 0. \end{cases}$$

   **a.** Write a formula for $|g(x)|$.

   **b.** Is $g$ continuous at $x = 0$? Explain.

   **c.** Is $|g|$ continuous at $x = 0$? Explain.

   **d.** For any function $f$, if $|f|$ is continuous at $a$, does it necessarily follow that $f$ is continuous at $a$? Explain.

**89–95. Classifying discontinuities** *The discontinuities in graphs (a) and (b) are* **removable discontinuities** *because they disappear if we define or redefine $f$ at $a$ such that $f(a) = \lim\limits_{x \to a} f(x)$. The function in graph (c) has a* **jump discontinuity** *because left and right limits exist at $a$ but are unequal. The discontinuity in graph (d) is an* **infinite discontinuity** *because the function has a vertical asymptote at $a$.*

(a)     (b)     (c)     (d)

**89.** Is the discontinuity at $a$ in graph (c) removable? Explain.

**90.** Is the discontinuity at $a$ in graph (d) removable? Explain.

**T 91–94.** *Classify the discontinuities in the following functions at the given points.*

**91.** $f(x) = \dfrac{x^2 - 7x + 10}{x - 2}; \ x = 2$

**92.** $g(x) = \begin{cases} \dfrac{x^2 - 1}{1 - x} & \text{if } x \neq 1 \\ 3 & \text{if } x = 1 \end{cases}; \ x = 1$

**93.** $h(x) = \dfrac{x^3 - 4x^2 + 4x}{x(x - 1)}; \ x = 0 \text{ and } x = 1$

**94.** $f(x) = \dfrac{|x - 2|}{x - 2}; \ x = 2$

**95. Do removable discontinuities exist?**

   **a.** Does the function $f(x) = x \sin(1/x)$ have a removable discontinuity at $x = 0$? Explain.

   **b.** Does the function $g(x) = \sin(1/x)$ have a removable discontinuity at $x = 0$? Explain.

**96. Continuity of composite functions** Prove Theorem 2.10: If $g$ is continuous at $a$ and $f$ is continuous at $g(a)$, then the composition $f \circ g$ is continuous at $a$. (*Hint:* Write the definition of continuity for $f$ and $g$ separately; then combine them to form the definition of continuity for $f \circ g$.)

**97. Continuity of compositions**

   **a.** Find functions $f$ and $g$ such that each function is continuous at 0 but $f \circ g$ is not continuous at 0.

   **b.** Explain why examples satisfying part (a) do not contradict Theorem 2.10.

**98. Violation of the Intermediate Value Theorem?** Let $f(x) = \dfrac{|x|}{x}$. Then $f(-2) = -1$ and $f(2) = 1$. Therefore, $f(-2) < 0 < f(2)$, but there is no value of $c$ between $-2$ and 2 for which $f(c) = 0$. Does this fact violate the Intermediate Value Theorem? Explain.

**99. Continuity of $\sin x$ and $\cos x$**

   **a.** Use the identity $\sin(a + h) = \sin a \cos h + \cos a \sin h$ with the fact that $\lim\limits_{x \to 0} \sin x = 0$ to prove that $\lim\limits_{x \to a} \sin x = \sin a$, thereby establishing that $\sin x$ is continuous for all $x$. (*Hint:* Let $h = x - a$ so that $x = a + h$ and note that $h \to 0$ as $x \to a$.)

   **b.** Use the identity $\cos(a + h) = \cos a \cos h - \sin a \sin h$ with the fact that $\lim\limits_{x \to 0} \cos x = 1$ to prove that $\lim\limits_{x \to a} \cos x = \cos a$.

**QUICK CHECK ANSWERS**

**1.** $t = 15, 30, 45$    **2.** Fill in the endpoints. **3.** $[0, \infty); (-\infty, \infty)$    **4.** The equation has a solution on the interval $[-1, 1]$ because $f$ is continuous on $[-1, 1]$ and $f(-1) < 0 < f(1)$. ◄

# 2.7 Precise Definitions of Limits

The limit definitions already encountered in this chapter are adequate for most elementary limits. However, some of the terminology used, such as *sufficiently close* and *arbitrarily large*, needs clarification. The goal of this section is to give limits a solid mathematical foundation by transforming the previous limit definitions into precise mathematical statements.

> ➤ The phrase *for all x near a* means for all
> $x$ in an open interval containing $a$.

> ➤ The Greek letters $\delta$ (delta) and $\varepsilon$ (epsilon)
> represent small positive numbers in the
> discussion of limits.

> ➤ The two conditions $|x - a| < \delta$
> and $x \neq a$ are written concisely as
> $0 < |x - a| < \delta$.

## Moving Toward a Precise Definition

Assume the function $f$ is defined for all $x$ near $a$, except possibly at $a$. Recall that $\lim_{x \to a} f(x) = L$ means that $f(x)$ is arbitrarily close to $L$ for all $x$ sufficiently close (but not equal) to $a$. This limit definition is made precise by observing that the distance between $f(x)$ and $L$ is $|f(x) - L|$ and that the distance between $x$ and $a$ is $|x - a|$. Therefore, we write $\lim_{x \to a} f(x) = L$ if we can make $|f(x) - L|$ arbitrarily small for any $x$, distinct from $a$, with $|x - a|$ sufficiently small. For instance, if we want $|f(x) - L|$ to be less than 0.1, then we must find a number $\delta > 0$ such that

$$|f(x) - L| < 0.1 \quad \text{whenever} \quad |x - a| < \delta \quad \text{and} \quad x \neq a.$$

If, instead, we want $|f(x) - L|$ to be less than 0.001, then we must find *another* number $\delta > 0$ such that

$$|f(x) - L| < 0.001 \quad \text{whenever} \quad 0 < |x - a| < \delta.$$

For the limit to exist, it must be true that for *any* $\varepsilon > 0$, we can always find a $\delta > 0$ such that

$$|f(x) - L| < \varepsilon \quad \text{whenever} \quad 0 < |x - a| < \delta.$$

**Figure 2.54**

> ➤ The founders of calculus, Isaac Newton
> (1642–1727) and Gottfried Leibniz
> (1646–1716), developed the core ideas
> of calculus without using a precise
> definition of a limit. It was not until the
> 19th century that a rigorous definition
> was introduced by Augustin-Louis
> Cauchy (1789–1857) and later refined by
> Karl Weierstrass (1815–1897).

**EXAMPLE 1  Determining values of $\delta$ from a graph**  Figure 2.54 shows the graph of a linear function $f$ with $\lim_{x \to 3} f(x) = 5$. For each value of $\varepsilon > 0$, determine a value of $\delta > 0$ satisfying the statement

$$|f(x) - 5| < \varepsilon \quad \text{whenever} \quad 0 < |x - 3| < \delta.$$

**a.** $\varepsilon = 1$    **b.** $\varepsilon = \frac{1}{2}$

**SOLUTION**

**a.** With $\varepsilon = 1$, we want $f(x)$ to be less than 1 unit from 5, which means $f(x)$ is between 4 and 6. To determine a corresponding value of $\delta$, draw the horizontal lines $y = 4$ and $y = 6$ (**Figure 2.55a**). Then sketch vertical lines passing through the points where the horizontal lines and the graph of $f$ intersect (**Figure 2.55b**). We see that the vertical

(a)

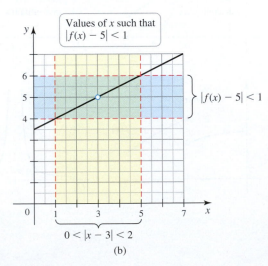

(b)

**Figure 2.55**

> Once an acceptable value of δ is found satisfying the statement
>
> $$|f(x) - L| < \varepsilon \quad \text{whenever}$$
>
> $$0 < |x - a| < \delta,$$
>
> any smaller positive value of δ also works.

lines intersect the x-axis at $x = 1$ and $x = 5$. Note that $f(x)$ is less than 1 unit from 5 on the y-axis if $x$ is within 2 units of 3 on the x-axis. So for $\varepsilon = 1$, we let $\delta = 2$ or any smaller positive value.

**b.** With $\varepsilon = \frac{1}{2}$, we want $f(x)$ to lie within a half-unit of 5, or equivalently, $f(x)$ must lie between 4.5 and 5.5. Proceeding as in part (a), we see that $f(x)$ is within a half-unit of 5 on the y-axis (**Figure 2.56a**) if $x$ is less than 1 unit from 3 (**Figure 2.56b**). So for $\varepsilon = \frac{1}{2}$, we let $\delta = 1$ or any smaller positive number.

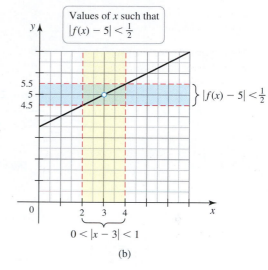

(a)     (b)

**Figure 2.56**

*Related Exercises 9–10* ◄

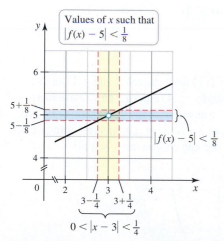

**Figure 2.57**

**QUICK CHECK 1** In Example 1, find a positive number δ satisfying the statement

$$|f(x) - 5| < \frac{1}{100}$$

whenever $0 < |x - 3| < \delta$. ◄

The idea of a limit, as illustrated in Example 1, may be described in terms of a contest between two people named Epp and Del. First, Epp picks a particular number $\varepsilon > 0$; then he challenges Del to find a corresponding value of $\delta > 0$ such that

$$|f(x) - 5| < \varepsilon \quad \text{whenever} \quad 0 < |x - 3| < \delta. \tag{1}$$

To illustrate, suppose Epp chooses $\varepsilon = 1$. From Example 1, we know that Del will satisfy (1) by choosing $0 < \delta \leq 2$. If Epp chooses $\varepsilon = \frac{1}{2}$, then (by Example 1) Del responds by letting $0 < \delta \leq 1$. If Epp lets $\varepsilon = \frac{1}{8}$, then Del chooses $0 < \delta \leq \frac{1}{4}$ (**Figure 2.57**). In fact, there is a pattern: For *any* $\varepsilon > 0$ that Epp chooses, no matter how small, Del will satisfy (1) by choosing a positive value of δ satisfying $0 < \delta \leq 2\varepsilon$. Del has discovered a mathematical relationship: If $0 < \delta \leq 2\varepsilon$ and $0 < |x - 3| < \delta$, then $|f(x) - 5| < \varepsilon$, for *any* $\varepsilon > 0$. This conversation illustrates the general procedure for proving that $\lim_{x \to a} f(x) = L$.

## A Precise Definition

Example 1 dealt with a linear function, but it points the way to a precise definition of a limit for any function. As shown in **Figure 2.58**, $\lim_{x \to a} f(x) = L$ means that for *any* positive number ε, there is another positive number δ such that

$$|f(x) - L| < \varepsilon \quad \text{whenever} \quad 0 < |x - a| < \delta.$$

In all limit proofs, the goal is to find a relationship between ε and δ that gives an admissible value of δ, in terms of ε only. This relationship must work for any positive value of ε.

**Figure 2.58**

---

**DEFINITION**  **Limit of a Function**

Assume $f(x)$ is defined for all $x$ in some open interval containing $a$, except possibly at $a$. We say **the limit of $f(x)$ as $x$ approaches $a$ is $L$**, written

$$\lim_{x \to a} f(x) = L,$$

if for *any* number $\varepsilon > 0$ there is a corresponding number $\delta > 0$ such that

$$|f(x) - L| < \varepsilon \quad \text{whenever} \quad 0 < |x - a| < \delta.$$

---

**EXAMPLE 2**  **Finding $\delta$ for a given $\varepsilon$ using a graphing utility**  Let $f(x) = x^3 - 6x^2 + 12x - 5$ and demonstrate that $\lim_{x \to 2} f(x) = 3$ as follows. For the given values of $\varepsilon$, use a graphing utility to find a value of $\delta > 0$ such that

$$|f(x) - 3| < \varepsilon \quad \text{whenever} \quad 0 < |x - 2| < \delta.$$

**a.** $\varepsilon = 1$    **b.** $\varepsilon = \frac{1}{2}$

> ➤ The value of $\delta$ in the precise definition of a limit depends only on $\varepsilon$.

> ➤ Definitions of the one-sided limits $\lim_{x \to a^+} f(x) = L$ and $\lim_{x \to a^-} f(x) = L$ are discussed in Exercises 51–55.

**SOLUTION**

**a.** The condition $|f(x) - 3| < \varepsilon = 1$ implies that $f(x)$ lies between 2 and 4. Using a graphing utility, we graph $f$ and the lines $y = 2$ and $y = 4$ (**Figure 2.59**). These lines intersect the graph of $f$ at $x = 1$ and at $x = 3$. We now sketch the vertical lines $x = 1$ and $x = 3$ and observe that $f(x)$ is within 1 unit of 3 whenever $x$ is within 1 unit of 2 on the $x$-axis (Figure 2.59). Therefore, with $\varepsilon = 1$, we can choose any $\delta$ with $0 < \delta \leq 1$.

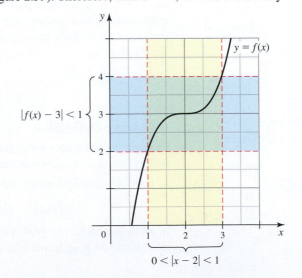

**Figure 2.59**

**b.** The condition $|f(x) - 3| < \varepsilon = \frac{1}{2}$ implies that $f(x)$ lies between 2.5 and 3.5 on the $y$-axis. We now find that the lines $y = 2.5$ and $y = 3.5$ intersect the graph of $f$ at $x \approx 1.21$ and $x \approx 2.79$ (**Figure 2.60**). Observe that if $x$ is less than 0.79 unit from 2 on the $x$-axis, then $f(x)$ is less than a half-unit from 3 on the $y$-axis. Therefore, with $\varepsilon = \frac{1}{2}$ we can choose any $\delta$ with $0 < \delta \leq 0.79$.

**Figure 2.60**

This procedure could be repeated for smaller and smaller values of $\varepsilon > 0$. For each value of $\varepsilon$, there exists a corresponding value of $\delta$, proving that the limit exists.

*Related Exercise 13* ◄

**QUICK CHECK 2** For the function $f$ given in Example 2, estimate a value of $\delta > 0$ satisfying $|f(x) - 3| < 0.25$ whenever $0 < |x - 2| < \delta$. ◄

The inequality $0 < |x - a| < \delta$ means that $x$ lies between $a - \delta$ and $a + \delta$ with $x \neq a$. We say that the interval $(a - \delta, a + \delta)$ is **symmetric about $a$** because $a$ is the midpoint of the interval. Symmetric intervals are convenient, but Example 3 demonstrates that we don't always get symmetric intervals without a bit of extra work.

**EXAMPLE 3** **Finding a symmetric interval** **Figure 2.61** shows the graph of $g$ with $\lim_{x \to 2} g(x) = 3$. For each value of $\varepsilon$, find the corresponding values of $\delta > 0$ that satisfy the condition

$$|g(x) - 3| < \varepsilon \quad \text{whenever} \quad 0 < |x - 2| < \delta.$$

**a.** $\varepsilon = 2$

**b.** $\varepsilon = 1$

**c.** For any given value of $\varepsilon$, make a conjecture about the corresponding values of $\delta$ that satisfy the limit condition.

**Figure 2.61**

**SOLUTION**

**a.** With $\varepsilon = 2$, we need a value of $\delta > 0$ such that $g(x)$ is within 2 units of 3, which means $g(x)$ is between 1 and 5, whenever $x$ is less than $\delta$ units from 2. The horizontal lines $y = 1$ and $y = 5$ intersect the graph of $g$ at $x = 1$ and $x = 6$. Therefore, $|g(x) - 3| < 2$ if $x$ lies in the interval $(1, 6)$ with $x \neq 2$ (**Figure 2.62a**). However, we want $x$ to lie in an interval that is *symmetric* about 2. We can guarantee that $|g(x) - 3| < 2$ in an interval symmetric about 2 only if $x$ is less than 1 unit away from 2, on either side of 2 (**Figure 2.62b**). Therefore, with $\varepsilon = 2$, we take $\delta = 1$ or any smaller positive number.

$|g(x) - 3| < 2$

$y = g(x)$

Values of $x$ such that $|g(x) - 3| < 2$

(a)

Symmetric interval $0 < |x - 2| < 1$ that guarantees $|g(x) - 3| < 2$

(b)

**Figure 2.62**

**b.** With $\varepsilon = 1$, $g(x)$ must lie between 2 and 4 (**Figure 2.63a**). This implies that $x$ must be within a half-unit to the left of 2 and within 2 units to the right of 2. Therefore, $|g(x) - 3| < 1$ provided $x$ lies in the interval $(1.5, 4)$. To obtain a symmetric interval about 2, we take $\delta = \frac{1}{2}$ or any smaller positive number. Then we are still guaranteed that $|g(x) - 3| < 1$ when $0 < |x - 2| < \frac{1}{2}$ (**Figure 2.63b**).

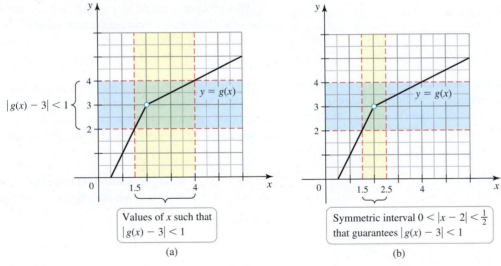

$|g(x) - 3| < 1$

$y = g(x)$

Values of $x$ such that $|g(x) - 3| < 1$

(a)

Symmetric interval $0 < |x - 2| < \frac{1}{2}$ that guarantees $|g(x) - 3| < 1$

(b)

**Figure 2.63**

**c.** From parts (a) and (b), it appears that if we choose $\delta \leq \varepsilon/2$, the limit condition is satisfied for any $\varepsilon > 0$.

*Related Exercises 15–16* ◄

In Examples 2 and 3, we showed that a limit exists by discovering a relationship between $\varepsilon$ and $\delta$ that satisfies the limit condition. We now generalize this procedure.

## Limit Proofs

We use the following two-step process to prove that $\lim\limits_{x \to a} f(x) = L$.

▶ The first step of the limit-proving process is the preliminary work of finding a candidate for δ. The second step verifies that the δ found in the first step actually works.

---

**Steps for proving that** $\lim\limits_{x \to a} f(x) = L$

1. **Find δ.** Let $\varepsilon$ be an arbitrary positive number. Use the inequality $|f(x) - L| < \varepsilon$ to find a condition of the form $|x - a| < \delta$, where $\delta$ depends only on the value of $\varepsilon$.

2. **Write a proof.** For any $\varepsilon > 0$, assume $0 < |x - a| < \delta$ and use the relationship between $\varepsilon$ and $\delta$ found in Step 1 to prove that $|f(x) - L| < \varepsilon$.

---

**EXAMPLE 4**  **Limit of a linear function**  Prove that $\lim\limits_{x \to 4} (4x - 15) = 1$ using the precise definition of a limit.

**SOLUTION**

*Step 1:*  *Find δ.* In this case, $a = 4$ and $L = 1$. Assuming $\varepsilon > 0$ is given, we use $|(4x - 15) - 1| < \varepsilon$ to find an inequality of the form $|x - 4| < \delta$. If $|(4x - 15) - 1| < \varepsilon$, then

$$|4x - 16| < \varepsilon$$

$$4|x - 4| < \varepsilon \qquad \text{Factor } 4x - 16.$$

$$|x - 4| < \frac{\varepsilon}{4}. \qquad \text{Divide by 4 and identify } \delta = \varepsilon/4.$$

We have shown that $|(4x - 15) - 1| < \varepsilon$ implies that $|x - 4| < \varepsilon/4$. Therefore, a plausible relationship between $\delta$ and $\varepsilon$ is $\delta = \varepsilon/4$. We now write the actual proof.

*Step 2:*  *Write a proof.* Let $\varepsilon > 0$ be given and assume $0 < |x - 4| < \delta$, where $\delta = \varepsilon/4$. The aim is to show that $|(4x - 15) - 1| < \varepsilon$ for all $x$ such that $0 < |x - 4| < \delta$. We simplify $|(4x - 15) - 1|$ and isolate the $|x - 4|$ term:

$$|(4x - 15) - 1| = |4x - 16|$$

$$= 4 \underbrace{|x - 4|}_{\text{less than } \delta = \varepsilon/4}$$

$$< 4\left(\frac{\varepsilon}{4}\right) = \varepsilon.$$

We have shown that for any $\varepsilon > 0$,

$$|f(x) - L| = |(4x - 15) - 1| < \varepsilon \quad \text{whenever} \quad 0 < |x - 4| < \delta,$$

provided $0 < \delta \le \varepsilon/4$. Therefore, $\lim\limits_{x \to 4} (4x - 15) = 1$.

*Related Exercises 19–20* ◀

**EXAMPLE 5**  **Limit of a quadratic function**  Prove that $\lim\limits_{x \to 5} x^2 = 25$ using the precise definition of a limit.

**SOLUTION**

*Step 1:*  *Find δ.* Given $\varepsilon > 0$, our task is to find an expression for $\delta > 0$ that depends only on $\varepsilon$, such that $|x^2 - 25| < \varepsilon$ whenever $0 < |x - 5| < \delta$. We begin by factoring $|x^2 - 25|$:

$$|x^2 - 25| = |(x + 5)(x - 5)| \qquad \text{Factor.}$$

$$= |x + 5||x - 5|. \qquad |ab| = |a||b|$$

Because the value of $\delta > 0$ in the inequality $0 < |x - 5| < \delta$ typically represents a small positive number, let's assume $\delta \le 1$ so that $|x - 5| < 1$, which implies that $-1 < x - 5 < 1$ or $4 < x < 6$. It follows that $x$ is positive, $|x + 5| < 11$, and

$$|x^2 - 25| = |x + 5||x - 5| < 11|x - 5|.$$

> The minimum value of $a$ and $b$ is denoted min $\{a, b\}$. If $x = $ min $\{a, b\}$, then $x$ is the smaller of $a$ and $b$. If $a = b$, then $x$ equals the common value of $a$ and $b$. In either case, $x \leq a$ and $x \leq b$.

Using this inequality, we have $|x^2 - 25| < \varepsilon$, provided $11|x - 5| < \varepsilon$ or $|x - 5| < \varepsilon/11$. Note that two restrictions have been placed on $|x - 5|$:

$$|x - 5| < 1 \quad \text{and} \quad |x - 5| < \frac{\varepsilon}{11}.$$

To ensure that both these inequalities are satisfied, let $\delta = $ min $\{1, \varepsilon/11\}$ so that $\delta$ equals the smaller of 1 and $\varepsilon/11$.

Step 2:   *Write a proof.* Let $\varepsilon > 0$ be given and assume $0 < |x - 5| < \delta$, where $\delta = $ min$\{1, \varepsilon/11\}$. By factoring $x^2 - 25$, we have

$$|x^2 - 25| = |x + 5||x - 5|.$$

Because $0 < |x - 5| < \delta$ and $\delta \leq \varepsilon/11$, we have $|x - 5| < \varepsilon/11$. It is also the case that $|x - 5| < 1$ because $\delta \leq 1$, which implies that $-1 < x - 5 < 1$ or $4 < x < 6$. Therefore, $|x + 5| < 11$ and

$$|x^2 - 25| = |x + 5||x - 5| < 11\left(\frac{\varepsilon}{11}\right) = \varepsilon.$$

We have shown that for any $\varepsilon > 0$, $|x^2 - 25| < \varepsilon$ whenever $0 < |x - 5| < \delta$, provided $0 < \delta = $ min$\{1, \varepsilon/11\}$. Therefore, $\lim_{x \to 5} x^2 = 25$.

*Related Exercises 27, 29* ◄

## Justifying Limit Laws

The precise definition of a limit is used to prove the limit laws in Theorem 2.3. Essential in several of these proofs is the **triangle inequality**, which states that

$$|x + y| \leq |x| + |y|, \quad \text{for all real numbers } x \text{ and } y.$$

**EXAMPLE 6   Proof of Limit Law 1** Prove that if $\lim_{x \to a} f(x)$ and $\lim_{x \to a} g(x)$ exist, then

$$\lim_{x \to a} (f(x) + g(x)) = \lim_{x \to a} f(x) + \lim_{x \to a} g(x).$$

> Because $\lim_{x \to a} f(x)$ exists, if there exists a $\delta > 0$ for any given $\varepsilon > 0$, then there also exists a $\delta > 0$ for any given $\frac{\varepsilon}{2}$.

**SOLUTION** Assume $\varepsilon > 0$ is given. Let $\lim_{x \to a} f(x) = L$, which implies there exists a $\delta_1 > 0$ such that

$$|f(x) - L| < \frac{\varepsilon}{2} \quad \text{whenever} \quad 0 < |x - a| < \delta_1.$$

Similarly, let $\lim_{x \to a} g(x) = M$, which implies there exists a $\delta_2 > 0$ such that

$$|g(x) - M| < \frac{\varepsilon}{2} \quad \text{whenever} \quad 0 < |x - a| < \delta_2.$$

Let $\delta = $ min $\{\delta_1, \delta_2\}$ and suppose $0 < |x - a| < \delta$. Because $\delta \leq \delta_1$, it follows that $0 < |x - a| < \delta_1$ and $|f(x) - L| < \varepsilon/2$. Similarly, because $\delta \leq \delta_2$, it follows that $0 < |x - a| < \delta_2$ and $|g(x) - M| < \varepsilon/2$. Therefore,

$$|(f(x) + g(x)) - (L + M)| = |(f(x) - L) + (g(x) - M)| \quad \text{\color{blue}Rearrange terms.}$$

$$\leq |f(x) - L| + |g(x) - M| \quad \text{\color{blue}Triangle inequality}$$

$$< \frac{\varepsilon}{2} + \frac{\varepsilon}{2} = \varepsilon.$$

> Proofs of other limit laws are outlined in Exercises 43–44.

We have shown that given any $\varepsilon > 0$, if $0 < |x - a| < \delta$, then $|(f(x) + g(x)) - (L + M)| < \varepsilon$, which implies that $\lim_{x \to a} (f(x) + g(x)) = L + M = \lim_{x \to a} f(x) + \lim_{x \to a} g(x)$.

*Related Exercises 43–44* ◄

## Infinite Limits

> Notice that for infinite limits, $N$ plays the role that $\varepsilon$ plays for regular limits. It sets a tolerance or bound for the function values $f(x)$.

In Section 2.4, we stated that $\lim_{x \to a} f(x) = \infty$ if $f(x)$ grows *arbitrarily large* as $x$ approaches $a$. More precisely, this means that for any positive number $N$ (no matter how large), $f(x)$ is larger than $N$ if $x$ is sufficiently close to $a$ but not equal to $a$.

> Precise definitions for $\lim\limits_{x \to a} f(x) = -\infty$,
> $\lim\limits_{x \to a^-} f(x) = -\infty$, $\lim\limits_{x \to a^+} f(x) = \infty$,
> $\lim\limits_{x \to a^+} f(x) = -\infty$, and $\lim\limits_{x \to a^-} f(x) = \infty$
> are given in Exercises 57–63.

---

**DEFINITION  Two-Sided Infinite Limit**

The **infinite limit** $\lim\limits_{x \to a} f(x) = \infty$ means that for any positive number $N$, there exists a corresponding $\delta > 0$ such that

$$f(x) > N \quad \text{whenever} \quad 0 < |x - a| < \delta.$$

---

As shown in **Figure 2.64**, to prove that $\lim\limits_{x \to a} f(x) = \infty$, we let $N$ represent *any* positive number. Then we find a value of $\delta > 0$, depending only on $N$, such that

$$f(x) > N \quad \text{whenever} \quad 0 < |x - a| < \delta.$$

This process is similar to the two-step process for finite limits.

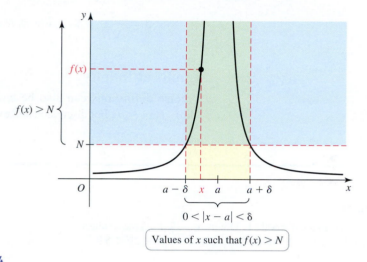

**Figure 2.64**

---

**Steps for proving that $\lim\limits_{x \to a} f(x) = \infty$**

1. **Find $\delta$.** Let $N$ be an arbitrary positive number. Use the statement $f(x) > N$ to find an inequality of the form $|x - a| < \delta$, where $\delta$ depends only on $N$.

2. **Write a proof.** For any $N > 0$, assume $0 < |x - a| < \delta$ and use the relationship between $N$ and $\delta$ found in Step 1 to prove that $f(x) > N$.

---

**EXAMPLE 7  An Infinite Limit Proof**  Let $f(x) = \dfrac{1}{(x - 2)^2}$. Prove that $\lim\limits_{x \to 2} f(x) = \infty$.

**SOLUTION**

*Step 1:  Find $\delta > 0$.* Assuming $N > 0$, we use the inequality $\dfrac{1}{(x - 2)^2} > N$ to find $\delta$, where $\delta$ depends only on $N$. Taking reciprocals of this inequality, it follows that

$$(x - 2)^2 < \frac{1}{N}$$

$$|x - 2| < \frac{1}{\sqrt{N}}. \quad \text{Take the square root of both sides.}$$

> Recall that $\sqrt{x^2} = |x|$.

The inequality $|x - 2| < \dfrac{1}{\sqrt{N}}$ has the form $|x - 2| < \delta$ if we let $\delta = \dfrac{1}{\sqrt{N}}$. We now write a proof based on this relationship between $\delta$ and $N$.

*Step 2:   Write a proof.* Suppose $N > 0$ is given. Let $\delta = \dfrac{1}{\sqrt{N}}$ and assume $0 < |x - 2| < \delta = \dfrac{1}{\sqrt{N}}$. Squaring both sides of the inequality $|x - 2| < \dfrac{1}{\sqrt{N}}$ and taking reciprocals, we have

$$(x - 2)^2 < \frac{1}{N} \quad \text{\color{blue}Square both sides.}$$

$$\frac{1}{(x - 2)^2} > N. \quad \text{\color{blue}Take reciprocals of both sides.}$$

**QUICK CHECK 3** In Example 7, if $N$ is increased by a factor of 100, how must $\delta$ change? ◄

We see that for any positive $N$, if $0 < |x - 2| < \delta = \dfrac{1}{\sqrt{N}}$, then $f(x) = \dfrac{1}{(x - 2)^2} > N$. It follows that $\lim\limits_{x \to 2} \dfrac{1}{(x - 2)^2} = \infty$. Note that because $\delta = \dfrac{1}{\sqrt{N}}$, $\delta$ decreases as $N$ increases.

*Related Exercises 45–46* ◄

### Limits at Infinity

Precise definitions can also be written for the limits at infinity $\lim\limits_{x \to \infty} f(x) = L$ and $\lim\limits_{x \to -\infty} f(x) = L$. For discussion and examples, see Exercises 64–65.

---

## SECTION 2.7 EXERCISES

### Getting Started

1. Suppose $x$ lies in the interval $(1, 3)$ with $x \neq 2$. Find the smallest positive value of $\delta$ such that the inequality $0 < |x - 2| < \delta$ is true.

2. Suppose $f(x)$ lies in the interval $(2, 6)$. What is the smallest value of $\varepsilon$ such that $|f(x) - 4| < \varepsilon$?

3. Which one of the following intervals is not symmetric about $x = 5$?

    **a.** $(1, 9)$     **b.** $(4, 6)$     **c.** $(3, 8)$     **d.** $(4.5, 5.5)$

4. Suppose $a$ is a constant and $\delta$ is a positive constant. Give a geometric description of the sets $\{x : |x - a| < \delta\}$ and $\{x : 0 < |x - a| < \delta\}$.

5. State the precise definition of $\lim\limits_{x \to a} f(x) = L$.

6. Interpret $|f(x) - L| < \varepsilon$ in words.

7. Suppose $|f(x) - 5| < 0.1$ whenever $0 < x < 5$. Find all values of $\delta > 0$ such that $|f(x) - 5| < 0.1$ whenever $0 < |x - 2| < \delta$.

8. Give the definition of $\lim\limits_{x \to a} f(x) = \infty$ and interpret it using pictures.

9. **Determining values of $\delta$ from a graph** The function $f$ in the figure satisfies $\lim\limits_{x \to 2} f(x) = 5$. Determine the largest value of $\delta > 0$ satisfying each statement.

    **a.** If $0 < |x - 2| < \delta$, then $|f(x) - 5| < 2$.

    **b.** If $0 < |x - 2| < \delta$, then $|f(x) - 5| < 1$.

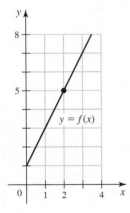

10. **Determining values of $\delta$ from a graph** The function $f$ in the figure satisfies $\lim\limits_{x \to 2} f(x) = 4$. Determine the largest value of $\delta > 0$ satisfying each statement.

    **a.** If $0 < |x - 2| < \delta$, then $|f(x) - 4| < 1$.

    **b.** If $0 < |x - 2| < \delta$, then $|f(x) - 4| < 1/2$.

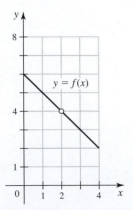

## Practice Exercises

**11. Determining values of δ from a graph** The function $f$ in the figure satisfies $\lim_{x \to 3} f(x) = 6$. Determine the largest value of $\delta > 0$ satisfying each statement.

  **a.** If $0 < |x - 3| < \delta$, then $|f(x) - 6| < 3$.

  **b.** If $0 < |x - 3| < \delta$, then $|f(x) - 6| < 1$.

**12. Determining values of δ from a graph** The function $f$ in the figure satisfies $\lim_{x \to 4} f(x) = 5$. Determine the largest value of $\delta > 0$ satisfying each statement.

  **a.** If $0 < |x - 4| < \delta$, then $|f(x) - 5| < 1$.

  **b.** If $0 < |x - 4| < \delta$, then $|f(x) - 5| < 0.5$.

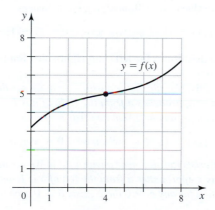

**T 13. Finding δ for a given ε using a graph** Let $f(x) = x^3 + 3$ and note that $\lim_{x \to 0} f(x) = 3$. For each value of $\varepsilon$, use a graphing utility to find all values of $\delta > 0$ such that $|f(x) - 3| < \varepsilon$ whenever $0 < |x - 0| < \delta$. Sketch graphs illustrating your work.

  **a.** $\varepsilon = 1$       **b.** $\varepsilon = 0.5$

**T 14. Finding δ for a given ε using a graph** Let $g(x) = 2x^3 - 12x^2 + 26x + 4$ and note that $\lim_{x \to 2} g(x) = 24$. For each value of $\varepsilon$, use a graphing utility to find all values of $\delta > 0$ such that $|g(x) - 24| < \varepsilon$ whenever $0 < |x - 2| < \delta$. Sketch graphs illustrating your work.

  **a.** $\varepsilon = 1$       **b.** $\varepsilon = 0.5$

**15. Finding a symmetric interval** The function $f$ in the figure satisfies $\lim_{x \to 2} f(x) = 3$. For each value of $\varepsilon$, find all values of $\delta > 0$ such that

$$|f(x) - 3| < \varepsilon \quad \text{whenever} \quad 0 < |x - 2| < \delta. \tag{2}$$

  **a.** $\varepsilon = 1$

  **b.** $\varepsilon = \frac{1}{2}$

  **c.** For any $\varepsilon > 0$, make a conjecture about the corresponding values of $\delta$ satisfying (2).

**16. Finding a symmetric interval** The function $f$ in the figure satisfies $\lim_{x \to 4} f(x) = 5$. For each value of $\varepsilon$, find all values of $\delta > 0$ such that

$$|f(x) - 5| < \varepsilon \quad \text{whenever} \quad 0 < |x - 4| < \delta. \tag{3}$$

  **a.** $\varepsilon = 2$       **b.** $\varepsilon = 1$

  **c.** For any $\varepsilon > 0$, make a conjecture about the corresponding values of $\delta$ satisfying (3).

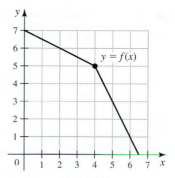

**T 17. Finding a symmetric interval** Let $f(x) = \dfrac{2x^2 - 2}{x - 1}$ and note that $\lim_{x \to 1} f(x) = 4$. For each value of $\varepsilon$, use a graphing utility to find all values of $\delta > 0$ such that $|f(x) - 4| < \varepsilon$ whenever $0 < |x - 1| < \delta$.

  **a.** $\varepsilon = 2$       **b.** $\varepsilon = 1$

  **c.** For any $\varepsilon > 0$, make a conjecture about the value of $\delta$ that satisfies the preceding inequality.

**T 18. Finding a symmetric interval** Let $f(x) = \begin{cases} \frac{1}{3}x + 1 & \text{if } x \le 3 \\ \frac{1}{2}x + \frac{1}{2} & \text{if } x > 3 \end{cases}$

and note that $\lim_{x \to 3} f(x) = 2$. For each value of $\varepsilon$, use a graphing utility to find all values of $\delta > 0$ such that $|f(x) - 2| < \varepsilon$ whenever $0 < |x - 3| < \delta$.

  **a.** $\varepsilon = \frac{1}{2}$       **b.** $\varepsilon = \frac{1}{4}$

  **c.** For any $\varepsilon > 0$, make a conjecture about the value of $\delta$ that satisfies the preceding inequality.

**19–42. Limit proofs** *Use the precise definition of a limit to prove the following limits. Specify a relationship between ε and δ that guarantees the limit exists.*

**19.** $\lim_{x \to 1} (8x + 5) = 13$       **20.** $\lim_{x \to 3} (-2x + 8) = 2$

**21.** $\lim\limits_{x \to 4} \dfrac{x^2 - 16}{x - 4} = 8$ (*Hint:* Factor and simplify.)

**22.** $\lim\limits_{x \to 3} \dfrac{x^2 - 7x + 12}{x - 3} = -1$

**23.** $\lim\limits_{x \to 0} |x| = 0$      **24.** $\lim\limits_{x \to 0} |5x| = 0$

**25.** $\lim\limits_{x \to 7} f(x) = 9$, where $f(x) = \begin{cases} 3x - 12 & \text{if } x \le 7 \\ x + 2 & \text{if } x > 7 \end{cases}$

**26.** $\lim\limits_{x \to 5} f(x) = 4$, where $f(x) = \begin{cases} 2x - 6 & \text{if } x \le 5 \\ -4x + 24 & \text{if } x > 5 \end{cases}$

**27.** $\lim\limits_{x \to 0} x^2 = 0$ (*Hint:* Use the identity $\sqrt{x^2} = |x|$.)

**28.** $\lim\limits_{x \to 3} (x - 3)^2 = 0$ (*Hint:* Use the identity $\sqrt{x^2} = |x|$.)

**29.** $\lim\limits_{x \to 2} (x^2 + 3x) = 10$      **30.** $\lim\limits_{x \to 4} (2x^2 - 4x + 1) = 17$

**31.** $\lim\limits_{x \to -3} |2x| = 6$ (*Hint:* Use the inequality $\big| |a| - |b| \big| \le |a - b|$, which holds for all constants $a$ and $b$ (see Exercise 74).)

**32.** $\lim\limits_{x \to 25} \sqrt{x} = 5$ (*Hint:* The factorization $x - 25 = (\sqrt{x} - 5)(\sqrt{x} + 5)$ implies that $\sqrt{x} - 5 = \dfrac{x - 25}{\sqrt{x} + 5}$.)

**33.** $\lim\limits_{x \to 3} \dfrac{1}{x} = \dfrac{1}{3}$ (*Hint:* As $x \to 3$, eventually the distance between $x$ and 3 is less than 1. Start by assuming $|x - 3| < 1$ and show $\dfrac{1}{|x|} < \dfrac{1}{2}$.)

**34.** $\lim\limits_{x \to 4} \dfrac{x - 4}{\sqrt{x} - 2} = 4$ (*Hint:* Multiply the numerator and denominator by $\sqrt{x} + 2$.)

**35.** $\lim\limits_{x \to 1/10} \dfrac{1}{x} = 10$ (*Hint:* To find $\delta$, you need to bound $x$ away from 0. So let $\left| x - \dfrac{1}{10} \right| < \dfrac{1}{20}$.)

**36.** $\lim\limits_{x \to 0} x \sin \dfrac{1}{x} = 0$

**37.** $\lim\limits_{x \to 0} (x^2 + x^4) = 0$ (*Hint:* You may use the fact that if $|x| < c$, then $x^2 < c^2$.)

**38.** $\lim\limits_{x \to a} b = b$, for any constants $a$ and $b$

**39.** $\lim\limits_{x \to a} (mx + b) = ma + b$, for any constants $a$, $b$, and $m$

**40.** $\lim\limits_{x \to 3} x^3 = 27$      **41.** $\lim\limits_{x \to 1} x^4 = 1$

**42.** $\lim\limits_{x \to 5} \dfrac{1}{x^2} = \dfrac{1}{25}$

**43.** **Proof of Limit Law 2** Suppose $\lim\limits_{x \to a} f(x) = L$ and $\lim\limits_{x \to a} g(x) = M$. Prove that $\lim\limits_{x \to a} (f(x) - g(x)) = L - M$.

**44.** **Proof of Limit Law 3** Suppose $\lim\limits_{x \to a} f(x) = L$. Prove that $\lim\limits_{x \to a} (cf(x)) = cL$, where $c$ is a constant.

**45–48. Limit proofs for infinite limits** *Use the precise definition of infinite limits to prove the following limits.*

**45.** $\lim\limits_{x \to 4} \dfrac{1}{(x - 4)^2} = \infty$      **46.** $\lim\limits_{x \to -1} \dfrac{1}{(x + 1)^4} = \infty$

**47.** $\lim\limits_{x \to 0} \left( \dfrac{1}{x^2} + 1 \right) = \infty$      **48.** $\lim\limits_{x \to 0} \left( \dfrac{1}{x^4} - \sin x \right) = \infty$

**49.** **Explain why or why not** Determine whether the following statements are true and give an explanation or counterexample. Assume $a$ and $L$ are finite numbers and assume $\lim\limits_{x \to a} f(x) = L$.

   **a.** For a given $\varepsilon > 0$, there is one value of $\delta > 0$ for which $|f(x) - L| < \varepsilon$ whenever $0 < |x - a| < \delta$.

   **b.** The limit $\lim\limits_{x \to a} f(x) = L$ means that given an arbitrary $\delta > 0$, we can always find an $\varepsilon > 0$ such that $|f(x) - L| < \varepsilon$ whenever $0 < |x - a| < \delta$.

   **c.** The limit $\lim\limits_{x \to a} f(x) = L$ means that for any arbitrary $\varepsilon > 0$, we can always find a $\delta > 0$ such that $|f(x) - L| < \varepsilon$ whenever $0 < |x - a| < \delta$.

   **d.** If $|x - a| < \delta$, then $a - \delta < x < a + \delta$.

**50.** **Finding $\delta$ algebraically** Let $f(x) = x^2 - 2x + 3$.

   **a.** For $\varepsilon = 0.25$, find the largest value of $\delta > 0$ satisfying the statement

   $$|f(x) - 2| < \varepsilon \quad \text{whenever} \quad 0 < |x - 1| < \delta.$$

   **b.** Verify that $\lim\limits_{x \to 1} f(x) = 2$ as follows. For any $\varepsilon > 0$, find the largest value of $\delta > 0$ satisfying the statement

   $$|f(x) - 2| < \varepsilon \quad \text{whenever} \quad 0 < |x - 1| < \delta.$$

**51–55. Precise definitions for left- and right-sided limits** *Use the following definitions.*

   *Assume $f$ exists for all $x$ near $a$ with $x > a$. We say that **the limit of $f(x)$ as $x$ approaches $a$ from the right of $a$ is $L$** and write $\lim\limits_{x \to a^+} f(x) = L$, if for any $\varepsilon > 0$ there exists $\delta > 0$ such that*

   $$|f(x) - L| < \varepsilon \quad \text{whenever} \quad 0 < x - a < \delta.$$

   *Assume $f$ exists for all $x$ near $a$ with $x < a$. We say that **the limit of $f(x)$ as $x$ approaches $a$ from the left of $a$ is $L$** and write $\lim\limits_{x \to a^-} f(x) = L$, if for any $\varepsilon > 0$ there exists $\delta > 0$ such that*

   $$|f(x) - L| < \varepsilon \quad \text{whenever} \quad 0 < a - x < \delta.$$

**51.** **Comparing definitions** Why is the last inequality in the definition of $\lim\limits_{x \to a} f(x) = L$, namely, $0 < |x - a| < \delta$, replaced with $0 < x - a < \delta$ in the definition of $\lim\limits_{x \to a^+} f(x) = L$?

**52.** **Comparing definitions** Why is the last inequality in the definition of $\lim\limits_{x \to a} f(x) = L$, namely, $0 < |x - a| < \delta$, replaced with $0 < a - x < \delta$ in the definition of $\lim\limits_{x \to a^-} f(x) = L$?

**53.** **One-sided limit proofs** Prove the following limits for

   $$f(x) = \begin{cases} 3x - 4 & \text{if } x < 0 \\ 2x - 4 & \text{if } x \ge 0. \end{cases}$$

   **a.** $\lim\limits_{x \to 0^+} f(x) = -4$   **b.** $\lim\limits_{x \to 0^-} f(x) = -4$   **c.** $\lim\limits_{x \to 0} f(x) = -4$

**54. Determining values of δ from a graph** The function $f$ in the figure satisfies $\lim\limits_{x \to 2^+} f(x) = 0$ and $\lim\limits_{x \to 2^-} f(x) = 1$. Determine all values of $\delta > 0$ that satisfy each statement.

a. $|f(x) - 0| < 2$ whenever $0 < x - 2 < \delta$

b. $|f(x) - 0| < 1$ whenever $0 < x - 2 < \delta$

c. $|f(x) - 1| < 2$ whenever $0 < 2 - x < \delta$

d. $|f(x) - 1| < 1$ whenever $0 < 2 - x < \delta$

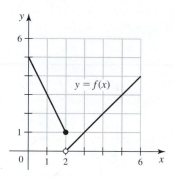

**55. One-sided limit proof** Prove that $\lim\limits_{x \to 0^+} \sqrt{x} = 0$.

## Explorations and Challenges

**56. The relationship between one-sided and two-sided limits** Prove the following statements to establish the fact that $\lim\limits_{x \to a} f(x) = L$ if and only if $\lim\limits_{x \to a^-} f(x) = L$ and $\lim\limits_{x \to a^+} f(x) = L$.

a. If $\lim\limits_{x \to a^-} f(x) = L$ and $\lim\limits_{x \to a^+} f(x) = L$, then $\lim\limits_{x \to a} f(x) = L$.

b. If $\lim\limits_{x \to a} f(x) = L$, then $\lim\limits_{x \to a^-} f(x) = L$ and $\lim\limits_{x \to a^+} f(x) = L$.

**57. Definition of one-sided infinite limits** We write $\lim\limits_{x \to a^+} f(x) = -\infty$ if for any negative number $N$, there exists $\delta > 0$ such that

$$f(x) < N \quad \text{whenever} \quad 0 < x - a < \delta.$$

a. Write an analogous formal definition for $\lim\limits_{x \to a^+} f(x) = \infty$.

b. Write an analogous formal definition for $\lim\limits_{x \to a^-} f(x) = -\infty$.

c. Write an analogous formal definition for $\lim\limits_{x \to a^-} f(x) = \infty$.

**58–59. One-sided infinite limits** *Use the definitions given in Exercise 57 to prove the following infinite limits.*

**58.** $\lim\limits_{x \to 1^+} \dfrac{1}{1 - x} = -\infty$    **59.** $\lim\limits_{x \to 1^-} \dfrac{1}{1 - x} = \infty$

**60–61. Definition of an infinite limit** *We write* $\lim\limits_{x \to a} f(x) = -\infty$ *if for any negative number $M$, there exists $\delta > 0$ such that*

$$f(x) < M \quad \text{whenever} \quad 0 < |x - a| < \delta.$$

*Use this definition to prove the following statements.*

**60.** $\lim\limits_{x \to 1} \dfrac{-2}{(x - 1)^2} = -\infty$    **61.** $\lim\limits_{x \to -2} \dfrac{-10}{(x + 2)^4} = -\infty$

**62.** Suppose $\lim\limits_{x \to a} f(x) = \infty$. Prove that $\lim\limits_{x \to a} (f(x) + c) = \infty$ for any constant $c$.

**63.** Suppose $\lim\limits_{x \to a} f(x) = \infty$ and $\lim\limits_{x \to a} g(x) = \infty$. Prove that $\lim\limits_{x \to a} (f(x) + g(x)) = \infty$.

**64–65. Definition of a limit at infinity** *The limit at infinity* $\lim\limits_{x \to \infty} f(x) = L$ *means that for any $\varepsilon > 0$, there exists $N > 0$ such that*

$$|f(x) - L| < \varepsilon \quad \text{whenever} \quad x > N.$$

*Use this definition to prove the following statements.*

**64.** $\lim\limits_{x \to \infty} \dfrac{10}{x} = 0$    **65.** $\lim\limits_{x \to \infty} \dfrac{2x + 1}{x} = 2$

**66–67. Definition of infinite limits at infinity** *We write* $\lim\limits_{x \to \infty} f(x) = \infty$ *if for any positive number $M$, there is a corresponding $N > 0$ such that*

$$f(x) > M \quad \text{whenever} \quad x > N.$$

*Use this definition to prove the following statements.*

**66.** $\lim\limits_{x \to \infty} \dfrac{x}{100} = \infty$    **67.** $\lim\limits_{x \to \infty} \dfrac{x^2 + x}{x} = \infty$

**68. Proof of the Squeeze Theorem** Assume the functions $f$, $g$, and $h$ satisfy the inequality $f(x) \leq g(x) \leq h(x)$, for all $x$ near $a$, except possibly at $a$. Prove that if $\lim\limits_{x \to a} f(x) = \lim\limits_{x \to a} h(x) = L$, then $\lim\limits_{x \to a} g(x) = L$.

**69. Limit proof** Suppose $f$ is defined for all $x$ near $a$, except possibly at $a$. Assume for any integer $N > 0$, there is another integer $M > 0$ such that $|f(x) - L| < 1/N$ whenever $|x - a| < 1/M$. Prove that $\lim\limits_{x \to a} f(x) = L$ using the precise definition of a limit.

**70–72. Proving that $\lim\limits_{x \to a} f(x) \neq L$** *Use the following definition for the nonexistence of a limit. Assume $f$ is defined for all $x$ near $a$, except possibly at $a$. We write $\lim\limits_{x \to a} f(x) \neq L$ if for some $\varepsilon > 0$, there is no value of $\delta > 0$ satisfying the condition*

$$|f(x) - L| < \varepsilon \quad \text{whenever} \quad 0 < |x - a| < \delta.$$

**70.** For the following function, note that $\lim\limits_{x \to 2} f(x) \neq 3$. Find all values of $\varepsilon > 0$ for which the preceding condition for nonexistence is satisfied.

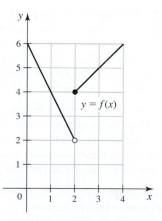

**71.** Prove that $\lim\limits_{x \to 0} \dfrac{|x|}{x}$ does not exist.

**72.** Let

$$f(x) = \begin{cases} 0 & \text{if } x \text{ is rational} \\ 1 & \text{if } x \text{ is irrational.} \end{cases}$$

Prove that $\lim\limits_{x \to a} f(x)$ does not exist for any value of $a$. (*Hint:* Assume $\lim\limits_{x \to a} f(x) = L$ for some values of $a$ and $L$, and let $\varepsilon = \dfrac{1}{2}$.)

**73. A continuity proof** Suppose $f$ is continuous at $a$ and defined for all $x$ near $a$. If $f(a) > 0$, show that there is a positive number $\delta > 0$ for which $f(x) > 0$ for all $x$ in $(a - \delta, a + \delta)$. (In other words, $f$ is positive for all $x$ in some interval containing $a$.)

**74.** Show that $\big| |a| - |b| \big| \le |a - b|$ for all constants $a$ and $b$. (*Hint:* Write $|a| = |(a - b) + b|$ and apply the triangle inequality to $|(a - b) + b|$.)

## CHAPTER 2    REVIEW EXERCISES

**1.    Explain why or why not** Determine whether the following statements are true and give an explanation or counterexample.

   **a.** The rational function $\dfrac{x - 1}{x^2 - 1}$ has vertical asymptotes at $x = -1$ and $x = 1$.

   **b.** Numerical or graphical methods always produce good estimates of $\lim\limits_{x \to a} f(x)$.

   **c.** The value of $\lim\limits_{x \to a} f(x)$, if it exists, is found by calculating $f(a)$.

   **d.** If $\lim\limits_{x \to a} f(x) = \infty$ or $\lim\limits_{x \to a} f(x) = -\infty$, then $\lim\limits_{x \to a} f(x)$ does not exist.

   **e.** If $\lim\limits_{x \to a} f(x)$ does not exist, then either $\lim\limits_{x \to a} f(x) = \infty$ or $\lim\limits_{x \to a} f(x) = -\infty$.

   **f.** The line $y = 2x + 1$ is a slant asymptote of the function $f(x) = \dfrac{2x^2 + x}{x - 3}$.

   **g.** If a function is continuous on the intervals $(a, b)$ and $[b, c)$, where $a < b < c$, then the function is also continuous on $(a, c)$.

   **h.** If $\lim\limits_{x \to a} f(x)$ can be calculated by direct substitution, then $f$ is continuous at $x = a$.

**2.    Average velocity** The height above the ground of a stone thrown upwards is given by $s(t)$, where $t$ is measured in seconds. After 1 second, the height of the stone is 48 feet above the ground, and after 1.5 seconds, the height of the stone is 60 feet above the ground. Evaluate $s(1)$ and $s(1.5)$, and then find the average velocity of the stone over the time interval $[1, 1.5]$.

**T 3.    Instantaneous velocity** A baseball is thrown upwards into the air; its distance above the ground after $t$ seconds is given by $s(t) = -16t^2 + 60t + 6$. Make a table of average velocities to make a conjecture about the instantaneous velocity of the baseball at $t = 1.5$ seconds after it was thrown into the air.

**4.    Estimating limits graphically** Use the graph of $f$ in the figure to evaluate the function or analyze the limit.

   **a.** $f(-1)$  **b.** $\lim\limits_{x \to -1^-} f(x)$  **c.** $\lim\limits_{x \to -1^+} f(x)$  **d.** $\lim\limits_{x \to -1} f(x)$

   **e.** $f(1)$  **f.** $\lim\limits_{x \to 1} f(x)$  **g.** $\lim\limits_{x \to 2} f(x)$  **h.** $\lim\limits_{x \to 3^-} f(x)$

   **i.** $\lim\limits_{x \to 3^+} f(x)$  **j.** $\lim\limits_{x \to 3} f(x)$

**5.    Points of discontinuity** Use the graph of $f$ in the figure to determine the values of $x$ in the interval $(-3, 5)$ at which $f$ fails to be continuous. Justify your answers using the continuity checklist.

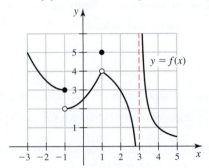

**T 6.    Computing a limit graphically and analytically**

   **a.** Graph $y = \dfrac{\sin 2\theta}{\sin \theta}$ with a graphing utility. Comment on any inaccuracies in the graph and then sketch an accurate graph of the function.

   **b.** Estimate $\lim\limits_{\theta \to 0} \dfrac{\sin 2\theta}{\sin \theta}$ using the graph in part (a).

   **c.** Verify your answer to part (b) by finding the value of $\lim\limits_{\theta \to 0} \dfrac{\sin 2\theta}{\sin \theta}$ analytically using the trigonometric identity $\sin 2\theta = 2 \sin \theta \cos \theta$.

**T 7.    Computing a limit numerically and analytically**

   **a.** Estimate $\lim\limits_{x \to \pi/4} \dfrac{\cos 2x}{\cos x - \sin x}$ by making a table of values of $\dfrac{\cos 2x}{\cos x - \sin x}$ for values of $x$ approaching $\pi/4$. Round your estimate to four digits.

   **b.** Use analytic methods to find the value of $\lim\limits_{x \to \pi/4} \dfrac{\cos 2x}{\cos x - \sin x}$.

**8. Snowboard rental** Suppose the rental cost for a snowboard is $25 for the first day (or any part of the first day) plus $15 for each additional day (or any part of a day).

   **a.** Graph the function $c = f(t)$ that gives the cost of renting a snowboard for $t$ days, for $0 \le t \le 5$.

   **b.** Evaluate $\lim_{t \to 2.9} f(t)$.

   **c.** Evaluate $\lim_{x \to 3^-} f(t)$ and $\lim_{t \to 3^+} f(t)$.

   **d.** Interpret the meaning of the limits in part (c).

   **e.** For what values of $t$ is $f$ continuous? Explain.

**9. Sketching a graph** Sketch the graph of a function $f$ with all the following properties.

$$\lim_{x \to -2^-} f(x) = \infty \qquad \lim_{x \to -2^+} f(x) = -\infty \qquad \lim_{x \to 0} f(x) = \infty$$

$$\lim_{x \to 3^-} f(x) = 2 \qquad \lim_{x \to 3^+} f(x) = 4 \qquad f(3) = 1$$

**10–45. Calculating limits** *Determine the following limits.*

**10.** $\lim_{x \to 1000} 18\pi^2$

**11.** $\lim_{x \to 1} \sqrt{5x + 6}$

**12.** $\lim_{h \to 0} \dfrac{\sqrt{5x + 5h} - \sqrt{5x}}{h}$, where $x > 0$ is constant

**13.** $\lim_{h \to 0} \dfrac{(h + 6)^2 + (h + 6) - 42}{h}$

**14.** $\lim_{x \to a} \dfrac{(3x + 1)^2 - (3a + 1)^2}{x - a}$, where $a$ is constant

**15.** $\lim_{x \to 1} \dfrac{x^3 - 7x^2 + 12x}{4 - x}$

**16.** $\lim_{x \to 4} \dfrac{x^3 - 7x^2 + 12x}{4 - x}$

**17.** $\lim_{x \to 1} \dfrac{1 - x^2}{x^2 - 8x + 7}$

**18.** $\lim_{x \to 3} \dfrac{\sqrt{3x + 16} - 5}{x - 3}$

**19.** $\lim_{x \to 3} \dfrac{1}{x - 3}\left(\dfrac{1}{\sqrt{x + 1}} - \dfrac{1}{2}\right)$

**20.** $\lim_{t \to 1/3} \dfrac{t - 1/3}{(3t - 1)^2}$

**21.** $\lim_{x \to 3} \dfrac{x^4 - 81}{x - 3}$

**22.** $\lim_{p \to 1} \dfrac{p^5 - 1}{p - 1}$

**23.** $\lim_{x \to 81} \dfrac{\sqrt[4]{x} - 3}{x - 81}$

**24.** $\lim_{\theta \to \pi/2} \dfrac{\sin^2 \theta - 5\sin \theta + 4}{\sin^2 \theta - 1}$

**25.** $\lim_{x \to \pi/2} \dfrac{\dfrac{1}{\sqrt{\sin x}} - 1}{x + \pi/2}$

**26.** $\lim_{x \to 1^+} \sqrt{\dfrac{x - 1}{x - 3}}$ and $\lim_{x \to 1^-} \sqrt{\dfrac{x - 1}{x - 3}}$

**27.** $\lim_{x \to 5} \dfrac{x - 7}{x(x - 5)^2}$

**28.** $\lim_{x \to -5^+} \dfrac{x - 5}{x + 5}$

**29.** $\lim_{x \to 3^-} \dfrac{x - 4}{x^2 - 3x}$

**30.** $\lim_{u \to 0^+} \dfrac{u - 1}{\sin u}$

**31.** $\lim_{x \to 1^+} \dfrac{4x^3 - 4x^2}{|x - 1|}$

**32.** $\lim_{x \to 2^-} \dfrac{|2x - 4|}{x^2 - 5x + 6}$

**33.** $\lim_{x \to 0^-} \dfrac{2}{\tan x}$

**34.** $\lim_{x \to \infty} \dfrac{4x^2 + 3x + 1}{\sqrt{8x^4 + 2}}$

**35.** $\lim_{x \to \infty} \dfrac{2x - 3}{4x + 10}$

**36.** $\lim_{x \to \infty} \dfrac{x^4 - 1}{x^5 + 2}$

**37.** $\lim_{x \to -\infty} \dfrac{3x + 1}{\sqrt{ax^2 + 2}}$, where $a$ is a positive constant

**38.** $\lim_{x \to \infty} \left(\sqrt{x^2 + ax} - \sqrt{x^2 - b}\right)$, where $a$ and $b$ are constants

**39.** $\lim_{x \to \infty} \left(\dfrac{1}{\sqrt{x^2 - ax} - \sqrt{x^2 - x}}\right)$, where $a$ is constant and $a \neq 1$

**40.** $\lim_{x \to -\infty} \dfrac{x^5}{\sqrt{x^{10} + 1} + \sqrt{3x^{10} + 1}}$

**41.** $\lim_{x \to -\infty} \dfrac{\sqrt{16x^6 - x}}{2x^3}$

**42.** $\lim_{x \to -\infty} (-3x^3 + 5)$

**43.** $\lim_{x \to -\infty} (|x - 1| + x)$ and $\lim_{x \to \infty} (|x - 1| + x)$

**44.** $\lim_{x \to -\infty} \dfrac{(|x - 2| + x)}{x}$ and $\lim_{x \to \infty} \dfrac{(|x - 2| + x)}{x}$

**45.** $\lim_{x \to \infty} \left(5 + \dfrac{\cos^4 x}{x^2 + x + 1}\right)$

**46.** Assume the function $g$ satisfies the inequality $1 \le g(x) \le \sin^2 x + 1$, for all values of $x$ near 0. Find $\lim_{x \to 0} g(x)$.

**T 47. An important limit**

   **a.** Use a graphing utility to illustrate the inequalities

   $$\cos x \le \dfrac{\sin x}{x} \le \dfrac{1}{\cos x}$$

   on $[-1, 1]$.

   **b.** Use part (a) to explain why $\lim_{x \to 0} \dfrac{\sin x}{x} = 1$.

**T 48. Finding vertical asymptotes** Let $f(x) = \dfrac{x^2 - 5x + 6}{x^2 - 2x}$.

   **a.** Analyze $\lim_{x \to 0^-} f(x)$, $\lim_{x \to 0^+} f(x)$, $\lim_{x \to 2^-} f(x)$, and $\lim_{x \to 2^+} f(x)$.

   **b.** Does the graph of $f$ have any vertical asymptotes? Explain.

   **c.** Graph $f$ using a graphing utility and then sketch the graph with paper and pencil, correcting any errors obtained with the graphing utility.

**49–54. End behavior** *Evaluate $\lim_{x \to \infty} f(x)$ and $\lim_{x \to -\infty} f(x)$.*

**49.** $f(x) = \dfrac{4x^3 + 1}{1 - x^3}$

**50.** $f(x) = \dfrac{x^6 + 1}{\sqrt{16x^{14} + 1}}$

**51.** $f(x) = \dfrac{12x^2}{\sqrt{16x^4 + 7}}$

**52.** $f(x) = \sqrt[3]{\dfrac{8x + 1}{x - 3}}$

**53.** $f(x) = \dfrac{3x^5 + 7x^3 + x}{4x^2 + 1}$

**54.** $f(x) = \dfrac{x + 1}{\sqrt{9x^2 + x}}$

**55–59. Slant asymptotes**

   **a.** *Analyze $\lim_{x \to \infty} f(x)$ and $\lim_{x \to -\infty} f(x)$ for each function.*

   **b.** *Determine whether $f$ has a slant asymptote. If so, write the equation of the slant asymptote.*

**55.** $f(x) = \dfrac{3x^2 + 5x + 7}{x + 1}$

**56.** $f(x) = \dfrac{9x^2 + 4}{(2x - 1)^2}$

**57.** $f(x) = \dfrac{1 + x - 2x^2 - x^3}{x^2 + 1}$     **58.** $f(x) = \dfrac{x(x + 2)^3}{3x^2 - 4x}$

**59.** $f(x) = \dfrac{4x^3 + x^2 + 7}{x^2 - x + 1}$

**60–63. Finding asymptotes** *Find all the asymptotes of the following functions.*

**60.** $f(x) = \dfrac{2x^2 + 6}{2x^2 + 3x - 2}$     **61.** $f(x) = \dfrac{\sqrt{x^6 + 1}}{x^3 - 1}$

**62.** $f(x) = \dfrac{2x^2 - 7}{x - 2}$     **63.** $f(x) = \dfrac{|x^3 + x|}{2x^2 + 1}$

**T 64.  Finding all asymptotes** Find all the asymptotes of $f(x) = \dfrac{x^2 + x + 3}{|x|}$. Plot a graph of $f$ together with its asymptotes. (*Hint:* Consider the cases $x = 0$, $x > 0$, and $x < 0$.)

**65–68. Continuity at a point** *Use the continuity checklist to determine whether the following functions are continuous at the given value of a.*

**65.** $f(x) = \dfrac{1}{x - 5}; a = 5$

**66.** $g(x) = \begin{cases} \dfrac{x^2 - 16}{x - 4} & \text{if } x \neq 4 \\ 9 & \text{if } x = 4 \end{cases}; a = 4$

**67.** $h(x) = \begin{cases} -2x + 14 & \text{if } x \leq 5 \\ \sqrt{x^2 - 9} & \text{if } x > 5 \end{cases}; a = 5$

**68.** $g(x) = \begin{cases} \dfrac{x^3 - 5x^2 + 6x}{x - 2} & \text{if } x \neq 2 \\ -2 & \text{if } x = 2 \end{cases}; a = 2$

**69–72. Continuity on intervals** *Find the intervals on which the following functions are continuous. Specify right- or left-continuity at the finite endpoints.*

**69.** $f(x) = \sqrt{x^2 - 5}$     **70.** $g(x) = \sqrt{x^2 - 5x + 6}$

**71.** $h(x) = \dfrac{2x}{x^3 - 25x}$     **72.** $g(x) = \cos\sqrt{x}$

**73.  Determining unknown constants** Let

$$g(x) = \begin{cases} 5x - 2 & \text{if } x < 1 \\ a & \text{if } x = 1 \\ ax^2 + bx & \text{if } x > 1. \end{cases}$$

Determine values of the constants $a$ and $b$, if possible, for which $g$ is continuous at $x = 1$.

**74.  Left- and right-continuity**

  **a.** Is $h(x) = \sqrt{x^2 - 9}$ left-continuous at $x = 3$? Explain.

  **b.** Is $h(x) = \sqrt{x^2 - 9}$ right-continuous at $x = 3$? Explain.

**75.** Sketch the graph of a function that is continuous on $(0, 1]$ and on $(1, 2)$ but is not continuous on $(0, 2)$.

**T 76–77. Intermediate Value Theorem**

  **a.** *Use the Intermediate Value Theorem to show that the equation has a solution in the given interval.*

  **b.** *Estimate a solution to the equation in the given interval using a root finder.*

**T 76.** $x^5 + 7x + 5 = 0; (-1, 0)$

**T 77.** $x = \cos x; \left(0, \dfrac{\pi}{2}\right)$

**78.** Suppose on a certain day, the low temperature was 32° at midnight, the high temperature was 65° at noon, and then the temperature dropped to 32° the following midnight. Prove there were two times during that day, which were 12 hours apart, when the temperatures were equal. (*Hint:* Let $T(t)$ equal the temperature $t$ hours after midnight and examine the function $f(t) = T(t) - T(t + 12)$, for $0 \leq t \leq 12$.)

**T 79.  Variable rectangles** Imagine the collection of rectangles with length $x$ and width $y$, such that their area is $xy = 100$.

  **a.** Show that the rectangle in this collection with length $x$ has perimeter $P(x) = 2x + 200/x$, where $x > 0$.

  **b.** Use the Intermediate Value Theorem to show that there is at least one rectangle with $2 \leq x \leq 30$ and perimeter 50.

  **c.** Estimate the length of the rectangle(s) with perimeter 50.

  **d.** Is there a rectangle in the collection with a perimeter of 30? Explain.

  **e.** Estimate the dimensions of the rectangle in the collection with the smallest possible perimeter.

**80–85. Limit proofs** *Use an appropriate limit definition to prove the following limits.*

**80.** $\lim\limits_{x \to 1} (5x - 2) = 3$     **81.** $\lim\limits_{x \to 5} \dfrac{x^2 - 25}{x - 5} = 10$

**82.** $\lim\limits_{x \to 3} f(x) = 5$, where $f(x) = \begin{cases} 3x - 4 & \text{if } x \leq 3 \\ -4x + 17 & \text{if } x > 3 \end{cases}$

**83.** $\lim\limits_{x \to 2} (3x^2 - 4) = 8$     **84.** $\lim\limits_{x \to 2^+} \sqrt{4x - 8} = 0$

**85.** $\lim\limits_{x \to 2} \dfrac{1}{(x - 2)^4} = \infty$

**86.  Limit proofs**

  **a.** Assume $|f(x)| \leq L$ for all $x$ near $a$ and $\lim\limits_{x \to a} g(x) = 0$. Give a formal proof that $\lim\limits_{x \to a} (f(x)g(x)) = 0$.

  **b.** Find a function $f$ for which $\lim\limits_{x \to 2} (f(x)(x - 2)) \neq 0$. Why doesn't this violate the result stated in part (a)?

  **c.** The Heaviside function is defined as

$$H(x) = \begin{cases} 0 & \text{if } x < 0 \\ 1 & \text{if } x \geq 0. \end{cases}$$

Explain why $\lim\limits_{x \to 0} (xH(x)) = 0$.

## Chapter 2 Guided Projects

*Applications of the material in this chapter and related topics can be found in the following Guided Projects. For additional information, see the Preface.*

- Fixed-point iteration
- Local linearity

# 3

# Derivatives

**Chapter Preview**  Now that you are familiar with limits, the door to calculus stands open. The first task is to introduce the fundamental concept of the *derivative*. Suppose a function $f$ represents a quantity of interest—for example, the variable cost of manufacturing an item, the population of a country, or the position of an orbiting satellite. The derivative of $f$ is another function, denoted $f'$, that gives the slope of the curve $y = f(x)$ as it changes with respect to $x$. Equivalently, the derivative of $f$ gives the *instantaneous rate of change* of $f$ with respect to the independent variable. We use limits not only to define the derivative but also to develop efficient rules for finding derivatives. The applications of the derivative—which we introduce along the way—are endless because almost everything around us is in a state of change, and derivatives describe change.

## 3.1 Introducing the Derivative

In this section, we return to the problem of finding the slope of a line tangent to a curve, introduced at the beginning of Chapter 2. This problem is important for several reasons.

- We identify the slope of the tangent line with the *instantaneous rate of change* of a function (**Figure 3.1**).
- The slopes of the tangent lines as they change along a curve are the values of a new function called the *derivative*.
- Looking farther ahead, if a curve represents the trajectory of a moving object, the tangent line at a point on the curve indicates the direction of motion at that point (**Figure 3.2**).

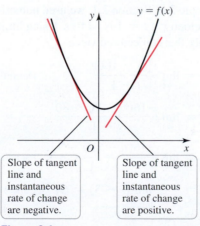

Slope of tangent line and instantaneous rate of change are negative.

Slope of tangent line and instantaneous rate of change are positive.

**Figure 3.1**

Trajectory of moving object

Tangents give direction of motion.

**Figure 3.2**

In Section 2.1, we used a limit to define the instantaneous velocity of an object that moves along a line. Recall that if $s(t)$ is the position of the object at time $t$, then the average velocity of the object over the time interval $[a, t]$ is

$$v_{\text{av}} = \frac{s(t) - s(a)}{t - a}.$$

The instantaneous velocity at time $t = a$ is the limit of the average velocity as $t \to a$:

$$v_{\text{inst}} = \lim_{t \to a} \frac{s(t) - s(a)}{t - a}.$$

We also learned that these quantities have important geometric interpretations. The average velocity is the slope of the secant line through the points $(a, s(a))$ and $(t, s(t))$ on the graph of $s$, and the instantaneous velocity is the slope of the tangent line through the point $(a, s(a))$ (**Figure 3.3**).

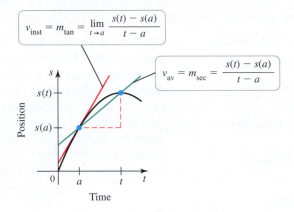

**Figure 3.3**

**EXAMPLE 1    Instantaneous velocity and tangent lines** A rock is launched vertically upward from the ground with an initial speed of 96 ft/s. The position of the rock in feet above the ground after $t$ seconds is given by the function $s(t) = -16t^2 + 96t$. Consider the point $P(1, 80)$ on the curve $y = s(t)$.

**a.** Find the instantaneous velocity of the rock 1 second after launch and find the slope of the line tangent to the graph of $s$ at $P$.

**b.** Find an equation of the tangent line in part (a).

**SOLUTION**

**a.** In Example 2 of Section 2.1, we used numerical evidence to estimate that the instantaneous velocity at $t = 1$ is 64 ft/s. Using limit techniques developed in Chapter 2, we can verify this conjectured value:

$$
\begin{aligned}
v_{\text{inst}} &= \lim_{t \to 1} \frac{s(t) - s(1)}{t - 1} && \text{\color{blue}{Definition of instantaneous velocity}} \\[2mm]
&= \lim_{t \to 1} \frac{-16t^2 + 96t - 80}{t - 1} && \text{\color{blue}{$s(t) = -16t^2 + 96t$; $s(1) = 80$}} \\[2mm]
&= \lim_{t \to 1} \frac{-16(t - 5)(t - 1)}{t - 1} && \text{\color{blue}{Factor the numerator.}} \\[2mm]
&= -16 \underbrace{\lim_{t \to 1} (t - 5)}_{-4} = 64. && \text{\color{blue}{Cancel factors ($t \neq 1$) and evaluate the limit.}}
\end{aligned}
$$

We see that the instantaneous velocity at $t = 1$ is 64 ft/s, which also equals the slope of the line tangent to the graph of $s$ at the point $P(1, 80)$.

**Figure 3.4**

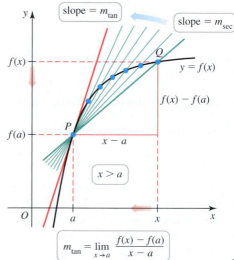

$$m_{tan} = \lim_{x \to a} \frac{f(x) - f(a)}{x - a}$$

**Figure 3.5**

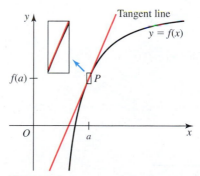

**Figure 3.6**

> ▶ The definition of $m_{sec}$ involves a *difference quotient*, introduced in Section 1.1.

> ▶ If $x$ and $y$ have physical units, then the average and instantaneous rates of change have units of (units of $y$)/(units of $x$). For example, if $y$ has units of meters and $x$ has units of seconds, the units of the rates of change are meters/second (m/s).

**b.** An equation of the line passing through $(1, 80)$ with slope 64 is $y - 80 = 64(t - 1)$ or $y = 64t + 16$ (**Figure 3.4**).

*Related Exercises 13–14* ◄

**QUICK CHECK 1** In Example 1, is the slope of the tangent line at $(2, 128)$ greater than or less than the slope at $(1, 80)$? ◄

The connection between the instantaneous rate of change of an object's position and the slope of a tangent line on the graph of the position function can be extended far beyond a discussion of velocity. In fact, the slope of a tangent line is one of the central concepts in calculus because it measures the instantaneous rate of change of a function. Whether a given function describes the position of an object, the population of a city, the concentration of a reactant in a chemical reaction, or the weight of a growing child, the slopes of tangent lines associated with these functions measure rates at which the quantities change.

## Tangent Lines and Rates of Change

Consider the curve $y = f(x)$ and a secant line intersecting the curve at the points $P(a, f(a))$ and $Q(x, f(x))$ (**Figure 3.5**). The difference $f(x) - f(a)$ is the change in the value of $f$ on the interval $[a, x]$, while $x - a$ is the change in $x$. As discussed in Chapters 1 and 2, the slope of the secant line $\overleftrightarrow{PQ}$ is

$$m_{sec} = \frac{f(x) - f(a)}{x - a},$$

and it gives the *average rate of change* in $f$ on the interval $[a, x]$.

Figure 3.5 also shows what happens as the variable point $x$ approaches the fixed point $a$. If the curve is smooth at $P(a, f(a))$—it has no kinks or corners—the secant lines approach a *unique* line that intersects the curve at $P$; this line is the *tangent line* at $P$. As $x$ approaches $a$, the slopes $m_{sec}$ of the secant lines approach a unique number $m_{tan}$ that we call the *slope of the tangent line*; that is,

$$m_{tan} = \lim_{x \to a} \frac{f(x) - f(a)}{x - a}.$$

The slope of the tangent line at $P$ is also called the *instantaneous rate of change* in $f$ at $a$ because it measures how quickly $f$ changes at $a$.

The tangent line has another geometric interpretation. As discussed in Section 2.1, if the curve $y = f(x)$ is smooth at a point $P(a, f(a))$, then the curve looks more like a line as we zoom in on $P$. The line that is approached as we zoom in on $P$ is the tangent line (**Figure 3.6**). A smooth curve has the property of *local linearity*, which means that if we look at a point on the curve locally (by zooming in), then the curve appears linear.

---

**DEFINITION** **Rate of Change and the Slope of the Tangent Line**

The **average rate of change** in $f$ on the interval $[a, x]$ is the slope of the corresponding secant line:

$$m_{sec} = \frac{f(x) - f(a)}{x - a}.$$

The **instantaneous rate of change** in $f$ at $a$ is

$$m_{tan} = \lim_{x \to a} \frac{f(x) - f(a)}{x - a}, \tag{1}$$

which is also the **slope of the tangent line** at $(a, f(a))$, provided this limit exists. The **tangent line** is the unique line through $(a, f(a))$ with slope $m_{tan}$. Its equation is

$$y - f(a) = m_{tan}(x - a).$$

---

**QUICK CHECK 2** Sketch the graph of a function $f$ near a point $a$. As in Figure 3.5, draw a secant line that passes through $(a, f(a))$ and a neighboring point $(x, f(x))$ with $x < a$. Use arrows to show how the secant lines approach the tangent line as $x$ approaches $a$. ◄

**EXAMPLE 2    Equation of a tangent line**  Find an equation of the line tangent to the graph of $f(x) = \dfrac{3}{x}$ at $\left(2, \dfrac{3}{2}\right)$.

**SOLUTION**  We use the definition of the slope of the tangent line with $a = 2$:

$$m_{\tan} = \lim_{x \to 2} \frac{f(x) - f(2)}{x - 2} \quad \text{Definition of slope of tangent line}$$

$$= \lim_{x \to 2} \frac{\dfrac{3}{x} - \dfrac{3}{2}}{x - 2} \quad f(x) = 3/x;\ f(2) = 3/2$$

$$= \lim_{x \to 2} \frac{\dfrac{6 - 3x}{2x}}{x - 2} \quad \text{Combine fractions with common denominator.}$$

$$= \lim_{x \to 2} \frac{-3(x - 2)}{2x(x - 2)} \quad \text{Simplify.}$$

$$= \lim_{x \to 2} \left(-\frac{3}{2x}\right) = -\frac{3}{4}. \quad \text{Cancel factors } (x \neq 2) \text{ and evaluate the limit.}$$

The tangent line has slope $m_{\tan} = -\dfrac{3}{4}$ and passes through the point $\left(2, \dfrac{3}{2}\right)$ (**Figure 3.7**). Its equation is $y - \dfrac{3}{2} = -\dfrac{3}{4}(x - 2)$ or $y = -\dfrac{3}{4}x + 3$. We could also say that the instantaneous rate of change in $f$ at $x = 2$ is $-3/4$. 

*Related Exercises 17–18* ◄

Slope of tangent line at $\left(2, \dfrac{3}{2}\right)$ is $m_{\tan} = -\dfrac{3}{4}$.

$y = \dfrac{3}{x}$

$\left(2, \dfrac{3}{2}\right)$

$y = -\dfrac{3}{4}x + 3$

**Figure 3.7**

An alternative formula for the slope of the tangent line is helpful for future work. Consider again the curve $y = f(x)$ and the secant line intersecting the curve at the points $P$ and $Q$. We now let $(a, f(a))$ and $(a + h, f(a + h))$ be the coordinates of $P$ and $Q$, respectively (**Figure 3.8**). The difference in the $x$-coordinates of $P$ and $Q$ is $(a + h) - a = h$. Note that $Q$ is located to the right of $P$ if $h > 0$ and to the left of $P$ if $h < 0$.

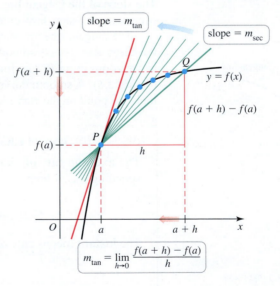

slope $= m_{\tan}$

slope $= m_{\sec}$

$y = f(x)$

$f(a + h)$

$f(a + h) - f(a)$

$f(a)$

$h$

$P$

$Q$

$O \quad a \quad a + h \quad x$

$$m_{\tan} = \lim_{h \to 0} \frac{f(a + h) - f(a)}{h}$$

**Figure 3.8**

The slope of the secant line $\overleftrightarrow{PQ}$ using the new notation is $m_{\sec} = \dfrac{f(a + h) - f(a)}{h}$. As $h$ approaches 0, the variable point $Q$ approaches $P$, and the slopes of the secant lines approach the slope of the tangent line. Therefore, the slope of the tangent line at $(a, f(a))$, which is also the instantaneous rate of change in $f$ at $a$, is

$$m_{\tan} = \lim_{h \to 0} \frac{f(a + h) - f(a)}{h}.$$

---

**ALTERNATIVE DEFINITION**  **Rate of Change and the Slope of the Tangent Line**

The **average rate of change** in $f$ on the interval $[a, a + h]$ is the slope of the corresponding secant line:

$$m_{\text{sec}} = \frac{f(a + h) - f(a)}{h}.$$

The **instantaneous rate of change** in $f$ at $a$ is

$$m_{\text{tan}} = \lim_{h \to 0} \frac{f(a + h) - f(a)}{h}, \tag{2}$$

which is also the **slope of the tangent line** at $(a, f(a))$, provided this limit exists.

---

**EXAMPLE 3**  **Equation of a tangent line**  Find an equation of the line tangent to the graph of $f(x) = x^3 + 4x$ at $(1, 5)$.

**SOLUTION**  We let $a = 1$ in definition (2) and first find $f(1 + h)$. After expanding and collecting terms, we have

$$f(1 + h) = (1 + h)^3 + 4(1 + h) = h^3 + 3h^2 + 7h + 5.$$

Substituting $f(1 + h)$ and $f(1) = 5$, the slope of the tangent line is

$$
\begin{aligned}
m_{\text{tan}} &= \lim_{h \to 0} \frac{f(1 + h) - f(1)}{h} && \text{Definition of } m_{\text{tan}} \\[2mm]
&= \lim_{h \to 0} \frac{(h^3 + 3h^2 + 7h + 5) - 5}{h} && \text{Substitute } f(1 + h) \text{ and } f(1) = 5. \\[2mm]
&= \lim_{h \to 0} \frac{h(h^2 + 3h + 7)}{h} && \text{Simplify.} \\[2mm]
&= \lim_{h \to 0} (h^2 + 3h + 7) && \text{Cancel } h, \text{ noting } h \neq 0. \\[2mm]
&= 7. && \text{Evaluate the limit.}
\end{aligned}
$$

The tangent line has slope $m_{\text{tan}} = 7$ and passes through the point $(1, 5)$ (**Figure 3.9**); its equation is $y - 5 = 7(x - 1)$ or $y = 7x - 2$. We could also say that the instantaneous rate of change in $f$ at $x = 1$ is 7.

*Related Exercises 23, 27* ◄

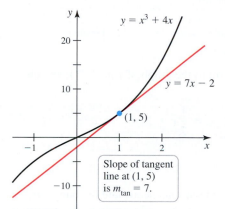

**Figure 3.9**

➤ By the definition of the limit as $h \to 0$, notice that $h$ approaches 0 but $h \neq 0$. Therefore, it is permissible to cancel $h$ from the numerator and denominator of

$$\frac{h(h^2 - 3h + 7)}{h}.$$

**QUICK CHECK 3**  Set up the calculation in Example 3 using definition (1) for the slope of the tangent line rather than definition (2). Does the calculation appear more difficult using definition (1)? ◄

## The Derivative

Computing the slope of the line tangent to the graph of a function $f$ at a given point $a$ gives us the instantaneous rate of change in $f$ at $a$. This information about the behavior of a function is so important that it has its own name and notation.

---

**DEFINITION**  **The Derivative of a Function at a Point**

The **derivative of $f$ at $a$**, denoted $f'(a)$, is given by either of the two following limits, provided the limits exist and $a$ is in the domain of $f$:

$$f'(a) = \lim_{x \to a} \frac{f(x) - f(a)}{x - a} \quad (1) \quad \text{or} \quad f'(a) = \lim_{h \to 0} \frac{f(a + h) - f(a)}{h}. \quad (2)$$

If $f'(a)$ exists, we say that $f$ is **differentiable** at $a$.

---

➤ The derivative notation $f'(a)$ is read *f prime of a*. A minor modification in the notation for the derivative is necessary when the name of the function changes. For example, given the function $y = g(x)$, its derivative at the point $a$ is $g'(a)$.

The limits that define the derivative of a function at a point are exactly the same limits used to compute the slope of a tangent line and the instantaneous rate of change of a function at a point. When you compute a derivative, remember that you are also finding a rate of change and the slope of a tangent line.

**Figure 3.10**

▶ Definition (2) could also be used to find $f'(2)$ in Example 4.

**QUICK CHECK 4** Verify that the derivative of the function $f$ in Example 4 at the point $(8, 5)$ is $f'(8) = 1/4$. Then find an equation of the line tangent to the graph of $f$ at the point $(8, 5)$. ◄

**EXAMPLE 4**  **Derivatives and tangent lines** Let $f(x) = \sqrt{2x} + 1$. Compute $f'(2)$, the derivative of $f$ at $x = 2$, and use the result to find an equation of the line tangent to the graph of $f$ at $(2, 3)$.

**SOLUTION** Definition (1) for $f'(a)$ is used here, with $a = 2$.

$$
\begin{aligned}
f'(2) &= \lim_{x \to 2} \frac{f(x) - f(2)}{x - 2} && \text{Definition of the derivative at a point} \\[2mm]
&= \lim_{x \to 2} \frac{\sqrt{2x} + 1 - 3}{x - 2} && \text{Substitute: } f(x) = \sqrt{2x} + 1 \text{ and } \\
& && f(2) = 3. \\[2mm]
&= \lim_{x \to 2} \frac{\sqrt{2x} - 2}{x - 2} \cdot \frac{\sqrt{2x} + 2}{\sqrt{2x} + 2} && \text{Multiply numerator and denominator} \\
& && \text{by } \sqrt{2x} + 2. \\[2mm]
&= \lim_{x \to 2} \frac{2x - 4}{(x - 2)(\sqrt{2x} + 2)} && \text{Simplify.} \\[2mm]
&= \lim_{x \to 2} \frac{2}{\sqrt{2x} + 2} = \frac{1}{2} && \text{Cancel factors } (x \neq 2) \text{ and evaluate} \\
& && \text{the limit.}
\end{aligned}
$$

Because $f'(2) = 1/2$ is the slope of the line tangent to the graph of $f$ at $(2, 3)$, an equation of the tangent line (**Figure 3.10**) is

$$
y - 3 = \frac{1}{2}(x - 2) \quad \text{or} \quad y = \frac{1}{2}x + 2.
$$

*Related Exercises 39–40* ◄

## Interpreting the Derivative

The derivative of a function $f$ at a point $a$ measures the instantaneous rate of change in $f$ at $a$. When the variables associated with a function represent physical quantities, the derivative takes on special meaning. For instance, suppose $T = g(d)$ describes the water temperature $T$ at depth $d$ in the ocean. Then $g'(d)$ measures the instantaneous rate of change in the water temperature at depth $d$. The following examples illustrate this idea.

**EXAMPLE 5**  **Instantaneous rate of change** Sound intensity $I$, measured in watts per square meter $(\text{W/m}^2)$ at a point $x$ meters from a sound source with acoustic power $P$, is given by $I(x) = \dfrac{P}{4\pi x^2}$. A typical sound system at a rock concert produces an acoustic power of about $P = 3$ W. Compute $I'(3)$ and interpret the result.

**SOLUTION** With $P = 3$, the sound intensity function is $I(x) = \dfrac{3}{4\pi x^2}$. A useful trick is to write $I$ as $I(x) = \dfrac{3}{4\pi} \cdot \dfrac{1}{x^2}$. Using definition (2) of the derivative at a point (p. 113), with $a = 3$, we have

$$
\begin{aligned}
I'(3) &= \lim_{h \to 0} \frac{I(3 + h) - I(3)}{h} && \text{Definition of } I'(3) \\
& && \text{Numerator units: } \text{W/m}^2; \text{ denominator units: m} \\[2mm]
&= \lim_{h \to 0} \frac{\dfrac{3}{4\pi} \cdot \dfrac{1}{(3 + h)^2} - \dfrac{3}{4\pi} \cdot \dfrac{1}{9}}{h} && \text{Evaluate } I(3 + h) \text{ and } I(3). \\[2mm]
&= \frac{3}{4\pi} \lim_{h \to 0} \frac{\dfrac{9 - (3 + h)^2}{9(3 + h)^2}}{h} && \text{Factor out } \dfrac{3}{4\pi} \text{ and simplify} \\
& && \text{remaining fraction.} \\[2mm]
&= \frac{3}{4\pi} \lim_{h \to 0} \frac{-h(6 + h)}{9(3 + h)^2} \cdot \frac{1}{h} && \text{Simplify.} \\[2mm]
&= -\frac{3}{4\pi} \lim_{h \to 0} \frac{6 + h}{9(3 + h)^2} && \text{Cancel } h; h \neq 0. \\[2mm]
&= -\frac{3}{4\pi} \cdot \frac{6}{81} = -\frac{1}{18\pi}. && \text{Evaluate limit.}
\end{aligned}
$$

**Figure 3.11**

The units associated with the numerator and denominator help to interpret this result. The change in sound intensity $I$ is measured in $W/m^2$ while the distance from the sound source is measured in meters (**Figure 3.11**). Therefore, the result $I'(3) = -1/(18\pi)$ means that the sound intensity decreases with an instantaneous rate of change of $1/(18\pi)$ $W/m^2$ *per meter* at a point 3 m from the sound source. Another useful interpretation is that when the distance from the source increases from 3 m to 4 m, the sound intensity decreases by about $1/(18\pi)$ $W/m^2$.

*Related Exercises 49–50* ◄

**EXAMPLE 6   Growth rates of Indian spotted owlets**   The Indian spotted owlet is a small owl that is indigenous to Southeast Asia. The body mass $m(t)$ (in grams) of an owl at an age of $t$ weeks is modeled by the graph in **Figure 3.12**. Estimate $m'(2)$ and state the physical meaning of this quantity. (*Source: ZooKeys, 132, 2011*)

**Figure 3.12**

**Figure 3.13**

**SOLUTION**   Recall that the derivative $m'(2)$ equals the slope of the line tangent to the graph of $y = m(t)$ at $t = 2$. One method for estimating $m'(2)$ is to sketch the line tangent to the curve at $t = 2$ and then estimate its slope (**Figure 3.13**). Searching for two convenient points on this line, we see that the tangent line passes through—or close to—the points $(1, 30)$ and $(2.5, 110)$. Therefore,

$$m'(2) \approx \frac{110 - 30}{2.5 - 1} = \frac{80 \text{ g}}{1.5 \text{ weeks}} \approx 53.3 \text{ g/week},$$

which means that the owl is growing at 53.3 g/week two weeks after birth.

*Related Exercises 53–54* ◄

➤ The graphical method used in Example 6 to estimate $m'(2)$ provides only a rough approximation to the value of the derivative: The answer depends on the accuracy with which the tangent line is sketched.

## SECTION 3.1 EXERCISES

### Getting Started

1. Use definition (1) (p. 111) for the slope of a tangent line to explain how slopes of secant lines approach the slope of the tangent line at a point.

2. Explain why the slope of a secant line can be interpreted as an average rate of change.

3. Explain why the slope of the tangent line can be interpreted as an instantaneous rate of change.

4. Explain the relationships among the slope of a tangent line, the instantaneous rate of change, and the value of the derivative at a point.

5. Given a function $f$ and a point $a$ in its domain, what does $f'(a)$ represent?

6. The following figure shows the graph of $f$ and a line tangent to the graph of $f$ at $x = 6$. Find $f(6)$ and $f'(6)$.

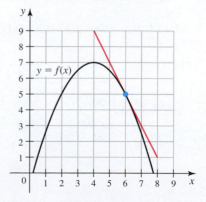

**7.** An equation of the line tangent to the graph of $f$ at the point $(2, 7)$ is $y = 4x - 1$. Find $f(2)$ and $f'(2)$.

**8.** An equation of the line tangent to the graph of $g$ at $x = 3$ is $y = 5x + 4$. Find $g(3)$ and $g'(3)$.

**9.** If $h(1) = 2$ and $h'(1) = 3$, find an equation of the line tangent to the graph of $h$ at $x = 1$.

**10.** If $f'(-2) = 7$, find an equation of the line tangent to the graph of $f$ at the point $(-2, 4)$.

**11.** Use definition (1) (p. 111) to find the slope of the line tangent to the graph of $f(x) = -5x + 1$ at the point $(1, -4)$.

**12.** Use definition (2) (p. 113) to find the slope of the line tangent to the graph of $f(x) = 5$ at $P(1, 5)$.

## Practice Exercises

**13–14. Velocity functions** *A projectile is fired vertically upward into the air; its position (in feet) above the ground after t seconds is given by the function s(t). For the following functions, use limits to determine the instantaneous velocity of the projectile at t = a seconds for the given value of a.*

**13.** $s(t) = -16t^2 + 100t; a = 1$

**14.** $s(t) = -16t^2 + 128t + 192; a = 2$

**15–20. Equations of tangent lines by definition (1)**

*a. Use definition (1) (p. 111) to find the slope of the line tangent to the graph of f at P.*
*b. Determine an equation of the tangent line at P.*
*c. Plot the graph of f and the tangent line at P.*

**15.** $f(x) = x^2 - 5; P(3, 4)$

**T 16.** $f(x) = -3x^2 - 5x + 1; P(1, -7)$

**17.** $f(x) = \dfrac{1}{x}; P(-1, -1)$

**T 18.** $f(x) = \dfrac{4}{x^2}; P(-1, 4)$

**T 19.** $f(x) = \sqrt{3x + 3}; P(2, 3)$

**T 20.** $f(x) = \dfrac{2}{\sqrt{x}}; P(4, 1)$

**21–32. Equations of tangent lines by definition (2)**

*a. Use definition (2) (p. 113) to find the slope of the line tangent to the graph of f at P.*
*b. Determine an equation of the tangent line at P.*

**21.** $f(x) = 2x + 1; P(0, 1)$      **22.** $f(x) = -7x; P(-1, 7)$

**23.** $f(x) = 3x^2 - 4x; P(1, -1)$  **24.** $f(x) = 8 - 2x^2; P(0, 8)$

**25.** $f(x) = x^2 - 4; P(2, 0)$    **26.** $f(x) = \dfrac{1}{x}; P(1, 1)$

**27.** $f(x) = x^3; P(1, 1)$        **28.** $f(x) = \dfrac{1}{2x + 1}; P(0, 1)$

**29.** $f(x) = \dfrac{1}{3 - 2x}; P\left(-1, \dfrac{1}{5}\right)$  **30.** $f(x) = \sqrt{x - 1}; P(2, 1)$

**31.** $f(x) = \sqrt{x + 3}; P(1, 2)$  **32.** $f(x) = \dfrac{x}{x + 1}; P(-2, 2)$

**33–42. Derivatives and tangent lines**
*a. For the following functions and values of a, find f'(a).*
*b. Determine an equation of the line tangent to the graph of f at the point (a, f(a)) for the given value of a.*

**33.** $f(x) = 8x; a = -3$          **34.** $f(x) = x^2; a = 3$

**35.** $f(x) = 4x^2 + 2x; a = -2$   **36.** $f(x) = 2x^3; a = 10$

**37.** $f(x) = \dfrac{1}{\sqrt{x}}; a = \dfrac{1}{4}$  **38.** $f(x) = \dfrac{1}{x^2}; a = 1$

**39.** $f(x) = \sqrt{2x + 1}; a = 4$  **40.** $f(x) = \sqrt{3x}; a = 12$

**41.** $f(x) = \dfrac{1}{x + 5}; a = 5$  **42.** $f(x) = \dfrac{1}{3x - 1}; a = 2$

**43–46. Derivative calculations** *Evaluate the derivative of the following functions at the given point.*

**43.** $f(t) = \dfrac{1}{t + 1}; a = 1$

**44.** $f(t) = t - t^2; a = 2$

**45.** $f(s) = 2\sqrt{s} - 1; a = 25$

**46.** $f(r) = \pi r^2; a = 3$

**47.** **Explain why or why not** Determine whether the following statements are true and give an explanation or counterexample.

   a. For linear functions, the slope of any secant line always equals the slope of any tangent line.
   b. The slope of the secant line passing through the points $P$ and $Q$ is less than the slope of the tangent line at $P$.
   c. Consider the graph of the parabola $f(x) = x^2$. For $a > 0$ and $h > 0$, the secant line through $(a, f(a))$ and $(a + h, f(a + h))$ always has a greater slope than the tangent line at $(a, f(a))$.

**48–51. Interpreting the derivative** *Find the derivative of each function at the given point and interpret the physical meaning of this quantity. Include units in your answer.*

**48.** When a faucet is turned on to fill a bathtub, the volume of water in gallons in the tub after $t$ minutes is $V(t) = 3t$. Find $V'(12)$.

**49.** An object dropped from rest falls $d(t) = 16t^2$ feet in $t$ seconds. Find $d'(4)$.

**50.** The gravitational force of attraction between two masses separated by a distance of $x$ meters is inversely proportional to the square of the distance between them, which implies that the force is described by the function $F(x) = k/x^2$, for some constant $k$, where $F(x)$ is measured in newtons. Find $F'(10)$, expressing your answer in terms of $k$.

**51.** Suppose the speed of a car approaching a stop sign is given by $v(t) = (t - 5)^2$, for $0 \le t \le 5$, where $t$ is measured in seconds and $v(t)$ is measured in meters per second. Find $v'(3)$.

**T 52.** **Population of Las Vegas** Let $p(t)$ represent the population of the Las Vegas metropolitan area $t$ years after 1970, as shown in the table and figure.

   a. Compute the average rate of growth of Las Vegas from 1970 to 1980.
   b. Explain why the average rate of growth calculated in part (a) is a good estimate of the instantaneous rate of growth of Las Vegas in 1975.

**c.** Compute the average rate of growth of Las Vegas from 1970 to 2000. Is the average rate of growth an overestimate or an underestimate of the instantaneous rate of growth of Las Vegas in 2000? Approximate the growth rate in 2000.

| Year | 1970 | 1980 | 1990 | 2000 | 2010 | 2020 |
|------|------|------|------|------|------|------|
| $t$ | 0 | 10 | 20 | 30 | 40 | 50 |
| $p(t)$ | 305,000 | 528,000 | 853,000 | 1,563,000 | 1,951,000 | 2,223,000 |

*Source:* U.S. Bureau of Census

**53. Owlet talons** Let $L(t)$ equal the average length (in mm) of the middle talon on an Indian spotted owlet that is $t$ weeks old, as shown in the figure.

**a.** Estimate $L'(1.5)$ and state the physical meaning of this quantity.
**b.** Estimate the value of $L'(a)$, for $a \geq 4$. What does this tell you about the talon lengths on these birds? (*Source: ZooKeys,* 132, 2011)

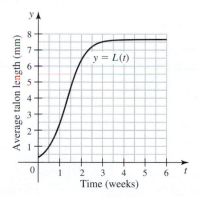

**54. Caffeine levels** Let $A(t)$ be the amount of caffeine (in mg) in the bloodstream $t$ hours after a cup of coffee has been consumed (see figure). Estimate the values of $A'(7)$ and $A'(15)$, rounding answers to the nearest whole number. Include units in your answers and interpret the physical meaning of these values.

**55.** Let $D(t)$ equal the number of daylight hours at a latitude of $40°$ N, $t$ days after January 1. Assuming $D(t)$ is approximated by a continuous function (see figure), estimate the values of $D'(60)$ and $D'(170)$. Include units in your answers and interpret your results.

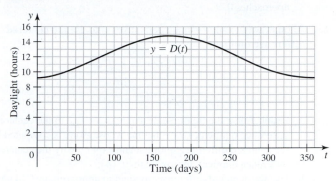

### Explorations and Challenges

**56–61. Find the function** *The following limits represent the slope of a curve $y = f(x)$ at the point $(a, f(a))$. Determine a possible function $f$ and number $a$; then calculate the limit.*

**56.** $\lim\limits_{x \to 1} \dfrac{3x^2 + 4x - 7}{x - 1}$

**57.** $\lim\limits_{x \to 2} \dfrac{5x^2 - 20}{x - 2}$

**58.** $\lim\limits_{x \to 2} \dfrac{\dfrac{1}{x + 1} - \dfrac{1}{3}}{x - 2}$

**59.** $\lim\limits_{h \to 0} \dfrac{(2 + h)^4 - 16}{h}$

**60.** $\lim\limits_{h \to 0} \dfrac{\sqrt{2 + h} - \sqrt{2}}{h}$

**61.** $\lim\limits_{h \to 0} \dfrac{|-1 + h| - 1}{h}$

**⊤ 62–65. Approximating derivatives** *Assuming the limit exists, the definition of the derivative $f'(a) = \lim\limits_{h \to 0} \dfrac{f(a + h) - f(a)}{h}$ implies that if $h$ is small, then an approximation to $f'(a)$ is given by*

$$f'(a) \approx \frac{f(a + h) - f(a)}{h}.$$

*If $h > 0$, then this approximation is called a **forward difference quotient**; if $h < 0$, it is a **backward difference quotient**. As shown in the following exercises, these formulas are used to approximate $f'$ at a point when $f$ is a complicated function or when $f$ is represented by a set of data points.*

**62.** Let $f(x) = \sqrt{x}$.

**a.** Find the exact value of $f'(4)$.

**b.** Show that $f'(4) \approx \dfrac{f(4 + h) - f(4)}{h} = \dfrac{\sqrt{4 + h} - 2}{h}$.

**c.** Complete columns 2 and 5 of the following table and describe how $\dfrac{\sqrt{4 + h} - 2}{h}$ behaves as $h$ approaches 0.

| $h$ | $(\sqrt{4 + h} - 2)/h$ | Error | $h$ | $(\sqrt{4 + h} - 2)/h$ | Error |
|-----|------------------------|-------|-----|------------------------|-------|
| 0.1 | | | −0.1 | | |
| 0.01 | | | −0.01 | | |
| 0.001 | | | −0.001 | | |
| 0.0001 | | | −0.0001 | | |

**d.** The accuracy of an approximation is measured by

$$\text{error} = |\text{exact value} - \text{approximate value}|.$$

Use the exact value of $f'(4)$ in part (a) to complete columns 3 and 6 in the table. Describe the behavior of the errors as $h$ approaches 0.

**63.** Another way to approximate derivatives is to use the **centered difference quotient:**

$$f'(a) \approx \frac{f(a + h) - f(a - h)}{2h}.$$

Again, consider $f(x) = \sqrt{x}$.

**a.** Graph $f$ near the point $(4, 2)$ and let $h = 1/2$ in the centered difference quotient. Draw the line whose slope is computed by the centered difference quotient and explain why the centered difference quotient approximates $f'(4)$.

**b.** Use the centered difference quotient to approximate $f'(4)$ by completing the table.

| $h$ | Approximation | Error |
|-----|---------------|-------|
| 0.1 | | |
| 0.01 | | |
| 0.001 | | |

**c.** Explain why it is not necessary to use negative values of $h$ in the table of part (b).

**d.** Compare the accuracy of the derivative estimates in part (b) with those found in Exercise 62.

**64.** The following table gives the distance $f(t)$ fallen by a smoke jumper $t$ seconds after she opens her chute.

**a.** Use the forward difference quotient with $h = 0.5$ to estimate the velocity of the smoke jumper at $t = 2$ seconds.

**b.** Repeat part (a) using the centered difference quotient.

| $t$ (seconds) | $f(t)$ (feet) |
|---------------|---------------|
| 0 | 0 |
| 0.5 | 4 |
| 1.0 | 15 |
| 1.5 | 33 |
| 2.0 | 55 |
| 2.5 | 81 |
| 3.0 | 109 |
| 3.5 | 138 |
| 4.0 | 169 |

**65.** The *error function* (denoted $\text{erf}(x)$) is an important function in statistics because it is related to the normal distribution. Its graph is shown in the figure, and values of $\text{erf}(x)$ at several points are shown in the table.

| $x$ | $\text{erf}(x)$ | $x$ | $\text{erf}(x)$ |
|------|-----------------|------|-----------------|
| 0.75 | 0.711156 | 1.05 | 0.862436 |
| 0.8 | 0.742101 | 1.1 | 0.880205 |
| 0.85 | 0.770668 | 1.15 | 0.896124 |
| 0.9 | 0.796908 | 1.2 | 0.910314 |
| 0.95 | 0.820891 | 1.25 | 0.922900 |
| 1.0 | 0.842701 | 1.3 | 0.934008 |

**a.** Use forward and centered difference quotients to find approximations to the derivative $\text{erf}'(1)$.

**b.** Given that $\text{erf}'(1) \approx 0.4151075$, compute the error in the approximations in part (a).

**QUICK CHECK ANSWERS**

**1.** The slope is less at $x = 2$.   **3.** Definition (1) requires factoring the numerator or long division to cancel $(x - 1)$.

**4.** $y - 5 = \dfrac{1}{4}(x - 8)$ ◄

# 3.2  The Derivative as a Function

In Section 3.1, we learned that the derivative of a function $f$ at a point $a$ is the slope of the line tangent to the graph of $f$ that passes through the point $(a, f(a))$. We now extend the concept of a derivative at a point to all (suitable) points in the domain of $f$ to create a new function called the derivative of $f$.

## The Derivative Function

So far we have computed the derivative of a function (or, equivalently, the slope of the tangent line) at one fixed point on a curve. If this point is moved along the curve, the tangent line also moves, and, in general, its slope changes (**Figure 3.14**). For this reason, the slope of the tangent line for the function $f$ is itself a function, called the *derivative* of $f$.

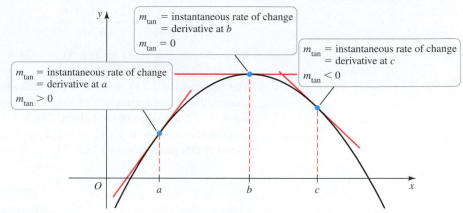

**Figure 3.14**

We let $f'$ (read $f$ *prime*) denote the derivative function for $f$, which means that $f'(a)$, when it exists, is the slope of the line tangent to the graph of $f$ at $(a, f(a))$. Using definition (2) for the derivative of $f$ at the point $a$ from Section 3.1, we have

$$f'(a) = \lim_{h \to 0} \frac{f(a + h) - f(a)}{h}.$$

We now take an important step. The derivative is a special function, but it works just like any other function. For example, if the graph of $f$ is smooth and 2 is in the domain of $f$, then $f'(2)$ is the slope of the line tangent to the graph of $f$ at the point $(2, f(2))$. Similarly, if $-2$ is in the domain of $f$, then $f'(-2)$ is the slope of the tangent line at the point $(-2, f(-2))$. In fact, if $x$ is *any* point in the domain of $f$, then $f'(x)$ is the slope of the tangent line at the point $(x, f(x))$. When we introduce a variable point $x$, the expression $f'(x)$ becomes the *derivative function*.

> ➤ To emphasize an important point, $f'(2)$ or $f'(-2)$ or $f'(a)$, for a real number $a$, are real numbers, whereas $f'$ and $f'(x)$ refer to the derivative *function*.

> ➤ The process of finding $f'$ is called *differentiation*, and to *differentiate* $f$ means to find $f'$.

---

**DEFINITION  The Derivative Function**

The **derivative** of $f$ is the function

$$f'(x) = \lim_{h \to 0} \frac{f(x + h) - f(x)}{h},$$

provided the limit exists and $x$ is in the domain of $f$. If $f'(x)$ exists, we say that $f$ is **differentiable** at $x$. If $f$ is differentiable at every point of an open interval $I$, we say that $f$ is differentiable on $I$.

---

Notice that the definition of $f'$ applies only at points in the domain of $f$. Therefore, the domain of $f'$ is no larger than the domain of $f$. If the limit in the definition of $f'$ fails to exist at some points, then the domain of $f'$ is a subset of the domain of $f$. Let's use this definition to compute a derivative function.

**EXAMPLE 1  Computing a derivative** Find the derivative of $f(x) = -x^2 + 6x$.

> ➤ Notice that this argument applies for $h > 0$ and for $h < 0$; that is, the limit as $h \to 0^+$ and the limit as $h \to 0^-$ are equal.

**SOLUTION**

$$f'(x) = \lim_{h \to 0} \frac{f(x + h) - f(x)}{h} \qquad \text{Definition of } f'(x)$$

$$= \lim_{h \to 0} \frac{\overbrace{-(x + h)^2 + 6(x + h)}^{f(x+h)} - \overbrace{(-x^2 + 6x)}^{f(x)}}{h} \qquad \text{Substitute.}$$

$$= \lim_{h \to 0} \frac{-(x^2 + 2xh + h^2) + 6x + 6h + x^2 - 6x}{h} \qquad \text{Expand the numerator.}$$

$$= \lim_{h \to 0} \frac{h(-2x - h + 6)}{h}$$

Simplify and factor out $h$.

$$= \lim_{h \to 0} (-2x - h + 6) = -2x + 6$$

Cancel $h$ and evaluate the limit.

The derivative is $f'(x) = -2x + 6$, which gives the slope of the tangent line (equivalently, the instantaneous rate of change) at *any* point on the curve. For example, at the point $(1, 5)$, the slope of the tangent line is $f'(1) = -2(1) + 6 = 4$. The slope of the tangent line at $(3, 9)$ is $f'(3) = -2(3) + 6 = 0$, which means the tangent line is horizontal at that point (**Figure 3.15**).

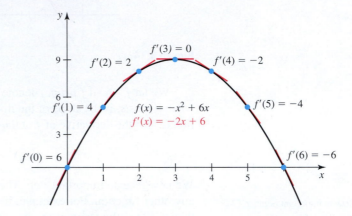

**QUICK CHECK 1** In Example 1, determine the slope of the tangent line at $x = 2$. ◄

**Figure 3.15**

*Related Exercises 23–24* ◄

**Figure 3.16**

➤ The notation $\dfrac{dy}{dx}$ is read *the derivative of y with respect to x or dy dx*. It does not mean $dy$ divided by $dx$, but it is a reminder of the limit of the quotient $\dfrac{\Delta y}{\Delta x}$.

➤ The derivative notation $dy/dx$ was introduced by Gottfried Wilhelm von Leibniz (1646–1716), one of the coinventors of calculus. His notation is used today in its original form. The notation used by Sir Isaac Newton (1642–1727), the other coinventor of calculus, is rarely used.

## Derivative Notation

For historical and practical reasons, several notations for the derivative are used. To see the origin of one notation, recall that the slope of the secant line through two points $P(x, f(x))$ and $Q(x + h, f(x + h))$ on the curve $y = f(x)$ is $\dfrac{f(x + h) - f(x)}{h}$. The quantity $h$ is the change in the $x$-coordinate in moving from $P$ to $Q$. A standard notation for change is the symbol $\Delta$ (uppercase Greek letter delta). So we replace $h$ with $\Delta x$ to represent the change in $x$. Similarly, $f(x + h) - f(x) = f(x + \Delta x) - f(x)$ is the change in $y$, denoted $\Delta y$ (**Figure 3.16**). Therefore, the slope of the secant line is

$$m_{\text{sec}} = \frac{f(x + \Delta x) - f(x)}{\Delta x} = \frac{\Delta y}{\Delta x}.$$

By letting $\Delta x \to 0$, the slope of the tangent line at $(x, f(x))$ is

$$f'(x) = \lim_{\Delta x \to 0} \frac{f(x + \Delta x) - f(x)}{\Delta x} = \lim_{\Delta x \to 0} \frac{\Delta y}{\Delta x} = \frac{dy}{dx}.$$

The new notation for the derivative is $\dfrac{dy}{dx}$; it reminds us that $f'(x)$ is the limit of $\dfrac{\Delta y}{\Delta x}$ as $\Delta x \to 0$.

In addition to the notation $f'(x)$ and $\dfrac{dy}{dx}$, other common ways of writing the derivative include

$$\frac{df}{dx}, \qquad \frac{d}{dx}(f(x)), \qquad D_x(f(x)), \quad \text{and} \quad y'(x).$$

The following notations represent the derivative of $f$ evaluated at $a$.

$$f'(a), \qquad y'(a), \qquad \left. \frac{df}{dx} \right|_{x=a}, \quad \text{and} \quad \left. \frac{dy}{dx} \right|_{x=a}$$

**QUICK CHECK 2** What are some other ways to write $f'(3)$, where $y = f(x)$? ◄

**EXAMPLE 2**  **A derivative calculation**  Let $y = f(x) = \sqrt{x}$.

➤ Example 2 gives the first of many derivative formulas to be presented in the text:

$$\frac{d}{dx}(\sqrt{x}) = \frac{1}{2\sqrt{x}}.$$

Remember this result. It will be used often.

**a.** Compute $\dfrac{dy}{dx}$.

**b.** Find an equation of the line tangent to the graph of $f$ at $(4, 2)$.

**SOLUTION**

**a.**
$$\frac{dy}{dx} = \lim_{h \to 0} \frac{f(x+h) - f(x)}{h} \qquad \text{Definition of } \frac{dy}{dx} = f'(x)$$

$$= \lim_{h \to 0} \frac{\sqrt{x+h} - \sqrt{x}}{h} \qquad \text{Substitute } f(x) = \sqrt{x}.$$

$$= \lim_{h \to 0} \frac{\left(\sqrt{x+h} - \sqrt{x}\right)\left(\sqrt{x+h} + \sqrt{x}\right)}{h\left(\sqrt{x+h} + \sqrt{x}\right)} \qquad \begin{array}{l}\text{Multiply the numerator and} \\ \text{denominator by } \sqrt{x+h} + \sqrt{x}.\end{array}$$

$$= \lim_{h \to 0} \frac{1}{\sqrt{x+h} + \sqrt{x}} = \frac{1}{2\sqrt{x}} \qquad \text{Simplify and evaluate the limit.}$$

Figure 3.17

**b.** The slope of the tangent line at $(4, 2)$ is

$$\left.\frac{dy}{dx}\right|_{x=4} = \frac{1}{2\sqrt{4}} = \frac{1}{4}.$$

Therefore, an equation of the tangent line (**Figure 3.17**) is

$$y - 2 = \frac{1}{4}(x - 4) \quad \text{or} \quad y = \frac{1}{4}x + 1.$$

*Related Exercises 37–38* ◄

**QUICK CHECK 3** In Example 2, do the slopes of the tangent lines increase or decrease as $x$ increases? Explain. ◄

If a function is given in terms of variables other than $x$ and $y$, we make an adjustment to the derivative definition. For example, if $y = g(t)$, we replace $f$ with $g$ and $x$ with $t$ to obtain the *derivative of g with respect to t*:

$$g'(t) = \lim_{h \to 0} \frac{g(t+h) - g(t)}{h}.$$

**QUICK CHECK 4** Express the derivative of $p = q(r)$ in three ways. ◄

Other notations for $g'(t)$ include $\dfrac{dg}{dt}, \dfrac{d}{dt}(g(t)), D_t(g(t))$, and $y'(t)$.

**EXAMPLE 3**  **Another derivative calculation**  Let $g(t) = 1/t^2$ and compute $g'(t)$.

**SOLUTION**

$$g'(t) = \lim_{h \to 0} \frac{g(t+h) - g(t)}{h} \qquad \text{Definition of } g'$$

$$= \lim_{h \to 0} \frac{1}{h}\left(\frac{1}{(t+h)^2} - \frac{1}{t^2}\right) \qquad \text{Substitute } g(t) = 1/t^2.$$

$$= \lim_{h \to 0} \frac{1}{h}\left(\frac{t^2 - (t+h)^2}{t^2(t+h)^2}\right) \qquad \text{Common denominator}$$

$$= \lim_{h \to 0} \frac{1}{h}\left(\frac{-2ht - h^2}{t^2(t+h)^2}\right) \qquad \text{Expand the numerator and simplify.}$$

$$= \lim_{h \to 0} \left(\frac{-2t - h}{t^2(t+h)^2}\right) \qquad h \neq 0; \text{ cancel } h.$$

$$= -\frac{2}{t^3} \qquad \text{Evaluate the limit.}$$

*Related Exercises 25–27* ◄

**Figure 3.18**

▶ In terms of limits at $x = -2$, we can
write $\lim\limits_{h \to 0^-} \dfrac{f(-2 + h) - f(-2)}{h} = -1$
and $\lim\limits_{h \to 0^+} \dfrac{f(-2 + h) - f(-2)}{h} = 1$.
Because the one-sided limits are not equal, $f'(-2)$ does not exist. The analogous one-sided limits at $x = 0$ are also unequal.

## Graphs of Derivatives

The function $f'$ is called the derivative of $f$ because it is *derived* from $f$. The following examples illustrate how to *derive* the graph of $f'$ from the graph of $f$.

**EXAMPLE 4   Graph of the derivative**   Sketch the graph of $f'$ from the graph of $f$ (**Figure 3.18**).

**SOLUTION**   The graph of $f$ consists of line segments, which are their own tangent lines. Therefore, the slope of the curve $y = f(x)$, for $x < -2$, is $-1$; that is, $f'(x) = -1$, for $x < -2$. Similarly, $f'(x) = 1$, for $-2 < x < 0$, and $f'(x) = -\frac{1}{2}$, for $x > 0$. **Figure 3.19** shows the graph of $f$ in black and the graph of $f'$ in red.

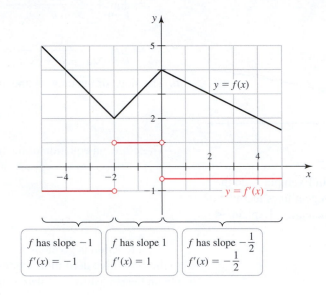

**Figure 3.19**

Notice that the slopes of the tangent lines change abruptly at $x = -2$ and $x = 0$. As a result, $f'(-2)$ and $f'(0)$ are undefined and the graph of the derivative has a discontinuity at these points.

*Related Exercises 17–18* ◀

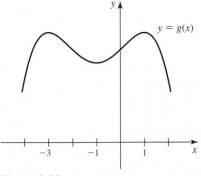

**Figure 3.20**

**QUICK CHECK 5**   Is it true that if $f(x) > 0$ at a point, then $f'(x) > 0$ at that point? Is it true that if $f'(x) > 0$ at a point, then $f(x) > 0$ at that point? Explain. ◀

**EXAMPLE 5   Graph of the derivative**   Sketch the graph of $g'$ using the graph of $g$ (**Figure 3.20**).

**SOLUTION**   Without an equation for $g$, the best we can do is to find the general shape of the graph of $g'$. Here are the key observations.

1. Note that the lines tangent to the graph of $g$ at $x = -3, -1$, and $1$ have a slope of $0$. Therefore,

$$g'(-3) = g'(-1) = g'(1) = 0,$$

which means the graph of $g'$ has $x$-intercepts at these points (**Figure 3.21a**).

2. For $x < -3$, the slopes of the tangent lines are positive and decrease to $0$ as $x$ approaches $-3$ from the left. Therefore, $g'(x)$ is positive for $x < -3$ and decreases to $0$ as $x$ approaches $-3$.

3. For $-3 < x < -1$, $g'(x)$ is negative; it initially decreases as $x$ increases and then increases to $0$ at $x = -1$. For $-1 < x < 1$, $g'(x)$ is positive; it initially increases as $x$ increases and then returns to $0$ at $x = 1$.

4. Finally, $g'(x)$ is negative and decreasing for $x > 1$. Because the slope of $g$ changes gradually, the graph of $g'$ is continuous with no jumps or breaks (**Figure 3.21b**).

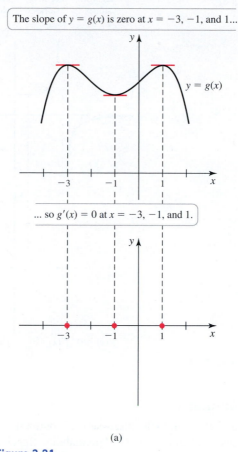

The slope of $y = g(x)$ is zero at $x = -3, -1$, and 1...

$y = g(x)$

... so $g'(x) = 0$ at $x = -3, -1$, and 1.

$y = g'(x)$

(a)

**Figure 3.21**

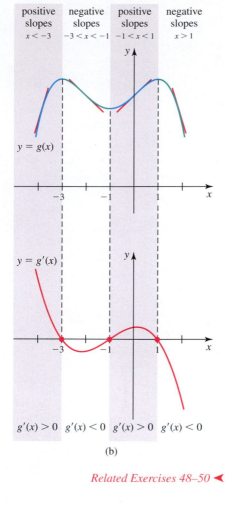

| positive slopes | negative slopes | positive slopes | negative slopes |
|---|---|---|---|
| $x < -3$ | $-3 < x < -1$ | $-1 < x < 1$ | $x > 1$ |

$y = g(x)$

$y = g'(x)$

$g'(x) > 0$  $g'(x) < 0$  $g'(x) > 0$  $g'(x) < 0$

(b)

*Related Exercises 48–50* ◀

$y = f(x)$

**Figure 3.22**

➤ Although it is the case in Example 6, a function and its derivative do not always share the same vertical asymptotes.

**EXAMPLE 6    Graphing the derivative with asymptotes**  The graph of the function $f$ is shown in **Figure 3.22**. Sketch a graph of its derivative.

**SOLUTION**  Identifying intervals on which the slopes of tangent lines are zero, positive, and negative, we make the following observations:

• A horizontal tangent line occurs at approximately $\left(-\frac{1}{3}, f\left(-\frac{1}{3}\right)\right)$. Therefore, $f'\left(-\frac{1}{3}\right) = 0$.

• On the interval $(-\infty, -1)$, slopes of tangent lines are negative and increase in magnitude without bound as we approach $-1$ from the left.

• On the interval $\left(-1, -\frac{1}{3}\right)$, slopes of tangent lines are negative and increase to zero at $-\frac{1}{3}$.

• On the interval $\left(-\frac{1}{3}, 1\right)$, slopes of tangent lines are positive and increase without bound as we approach 1 from the left.

• On the interval $(1, \infty)$, slopes of tangent lines are negative and increase to zero.

Assembling all this information, we obtain a graph of $f'$ shown in **Figure 3.23**. Notice that $f$ and $f'$ have the same vertical asymptotes. However, as we pass through $-1$, the sign of $f$ changes, but the sign of $f'$ does not change. As we pass through 1, the sign of $f$ does not change, but the sign of $f'$ does change.

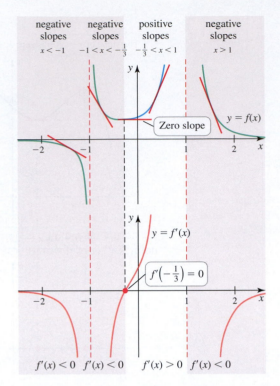

**Figure 3.23**

*Related Exercises 51–52* ◄

## Continuity

We now return to the discussion of continuity (Section 2.6) and investigate the relationship between continuity and differentiability. Specifically, we show that if a function is differentiable at a point, then it is also continuous at that point.

> **THEOREM 3.1   Differentiable Implies Continuous**
> If $f$ is differentiable at $a$, then $f$ is continuous at $a$.

**Proof:** Because $f$ is differentiable at a point $a$, we know that

$$f'(a) = \lim_{x \to a} \frac{f(x) - f(a)}{x - a}$$

exists. To show that $f$ is continuous at $a$, we must show that $\lim_{x \to a} f(x) = f(a)$. The key is the identity

> ▶ Expression (1) is an identity because it holds for all $x \neq a$, which can be seen by canceling $x - a$ and simplifying.

$$f(x) = \frac{f(x) - f(a)}{x - a}(x - a) + f(a), \quad \text{for } x \neq a. \tag{1}$$

Taking the limit as $x$ approaches $a$ on both sides of (1) and simplifying, we have

$$\lim_{x \to a} f(x) = \lim_{x \to a} \left( \frac{f(x) - f(a)}{x - a}(x - a) + f(a) \right) \qquad \text{\color{blue}Use identity.}$$

$$= \lim_{x \to a} \left( \underbrace{\frac{f(x) - f(a)}{x - a}}_{f'(a)} \right) \underbrace{\lim_{x \to a} (x - a)}_{0} + \underbrace{\lim_{x \to a} f(a)}_{f(a)} \qquad \text{\color{blue}Theorem 2.3}$$

$$= f'(a) \cdot 0 + f(a) \qquad \text{\color{blue}Evaluate limits.}$$

$$= f(a). \qquad \text{\color{blue}Simplify.}$$

**QUICK CHECK 6** Verify that the right-hand side of (1) equals $f(x)$ if $x \neq a$. ◄

Therefore, $\lim_{x \to a} f(x) = f(a)$, which means that $f$ is continuous at $a$. ◄

**Figure 3.24**

➤ The alternative version of Theorem 3.1 is called the *contrapositive* of the first statement of Theorem 3.1. A statement and its contrapositive are two equivalent ways of expressing the same statement. For example, the statement *If I live in Denver, then I live in Colorado* is logically equivalent to its contrapositive: *If I do not live in Colorado, then I do not live in Denver.*

➤ To avoid confusion about continuity and differentiability, it helps to think about the function $f(x) = |x|$: It is continuous everywhere but not differentiable at 0.

➤ Continuity requires that $\lim_{x \to a} (f(x) - f(a)) = 0$. Differentiability requires more: $\lim_{x \to a} \dfrac{f(x) - f(a)}{x - a}$ must exist.

➤ See Exercises 73–76 for a formal definition of a vertical tangent line.

Theorem 3.1 tells us that if $f$ is differentiable at a point, then it is necessarily continuous at that point. Therefore, if $f$ is *not* continuous at a point, then $f$ is *not* differentiable there (**Figure 3.24**). So Theorem 3.1 can be stated in another way.

---

**THEOREM 3.1 (ALTERNATIVE VERSION)   Not Continuous Implies Not Differentiable**
If $f$ is not continuous at $a$, then $f$ is not differentiable at $a$.

---

It is tempting to read more into Theorem 3.1 than what it actually states. If $f$ is continuous at a point, $f$ is *not* necessarily differentiable at that point. For example, consider the continuous function in **Figure 3.25** and note the **corner point** at $a$. Ignoring the portion of the graph for $x > a$, we might be tempted to conclude that $\ell_1$ is the line tangent to the curve at $a$. Ignoring the part of the graph for $x < a$, we might incorrectly conclude that $\ell_2$ is the line tangent to the curve at $a$. The slopes of $\ell_1$ and $\ell_2$ are not equal. Because of the abrupt change in the slope of the curve at $a$, $f$ is not differentiable at $a$: The limit that defines $f'$ does not exist at $a$.

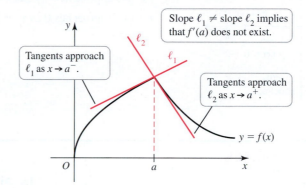

**Figure 3.25**

Another common situation occurs when the graph of a function $f$ has a vertical tangent line at $a$. In this case, $f'(a)$ is undefined because the slope of a vertical line is undefined. A vertical tangent line may occur at a sharp point on the curve called a **cusp** (for example, the function $f(x) = \sqrt{|x|}$ in **Figure 3.26a**). In other cases, a vertical tangent line may occur without a cusp (for example, the function $f(x) = \sqrt[3]{x}$ in **Figure 3.26b**).

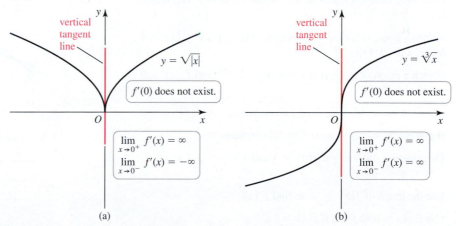

(a)                                          (b)

**Figure 3.26**

---

**When Is a Function Not Differentiable at a Point?**

A function $f$ is *not* differentiable at $a$ if at least one of the following conditions holds:

**a.** $f$ is not continuous at $a$ (Figure 3.24).

**b.** $f$ has a corner at $a$ (Figure 3.25).

**c.** $f$ has a vertical tangent at $a$ (Figure 3.26).

---

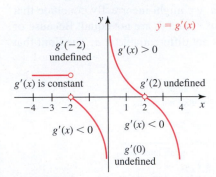

**Figure 3.27**

**Figure 3.28**

**EXAMPLE 7**   **Continuous and differentiable**  Consider the graph of $g$ in **Figure 3.27**.

**a.** Find the values of $x$ in the interval $(-4, 4)$ at which $g$ is not continuous.

**b.** Find the values of $x$ in the interval $(-4, 4)$ at which $g$ is not differentiable.

**c.** Sketch a graph of the derivative of $g$.

**SOLUTION**

**a.** The function $g$ fails to be continuous at $-2$ (where the one-sided limits are not equal) and at $2$ (where $g$ is not defined).

**b.** Because it is not continuous at $\pm 2$, $g$ is not differentiable at those points. Furthermore, $g$ is not differentiable at $0$, because the graph has a cusp at that point.

**c.** A sketch of the derivative (**Figure 3.28**) has the following features:

- $g'(x) > 0$, for $-4 < x < -2$ and $0 < x < 2$
- $g'(x) < 0$, for $-2 < x < 0$ and $2 < x < 4$
- $g'(x)$ approaches $-\infty$ as $x \to 0^-$ and as $x \to 4^-$, and $g'(x)$ approaches $\infty$ as $x \to 0^+$
- $g'(x)$ approaches $0$ as $x \to 2$ from either side, although $g'(2)$ does not exist.

*Related Exercises 53–54* ◀

---

## SECTION 3.2 EXERCISES

### Getting Started

1. For a given function $f$, what does $f'$ represent?

2. If $f'(x) = 3x + 2$, find the slope of the line tangent to the curve $y = f(x)$ at $x = 1, 2,$ and $3$.

3. Why is the notation $\dfrac{dy}{dx}$ used to represent the derivative?

4. Give three different notations for the derivative of $f$ with respect to $x$.

5. Sketch a graph of a function $f$, where $f(x) < 0$ and $f'(x) > 0$ for all $x$ in $(0, 1)$.

6. Sketch a graph of a function $f$, where $f(x) > 0$ and $f'(x) < 0$ for all $x$ in $(0, 2)$.

7. If $f$ is differentiable at $a$, must $f$ be continuous at $a$?

8. If $f$ is continuous at $a$, must $f$ be differentiable at $a$?

9. Describe the graph of $f$ if $f(0) = 1$ and $f'(x) = 3$, for $-\infty < x < \infty$.

10. Use the graph of $f(x) = |x|$ to find $f'(x)$.

11. Use limits to find $f'(x)$ if $f(x) = 7x$.

12. Use limits to find $f'(x)$ if $f(x) = -3x$.

13. Use limits to find $\dfrac{dy}{dx}$ if $y = x^2$. Then evaluate $\dfrac{dy}{dx}\Big|_{x=3}$ and $\dfrac{dy}{dx}\Big|_{x=-2}$.

14. The weight $w(x)$ (in pounds) of an Atlantic salmon can be estimated from its length $x$ (in inches). If $33 \le x \le 48$, the estimated weight is $w(x) = 1.5x - 35.8$. Use limits to find $w'(x)$ and interpret its meaning. (*Source:* www.atlanticsalmonfederation.org)

15. **Matching functions with derivatives**  Match graphs a–d of functions with graphs A–C of their derivatives.

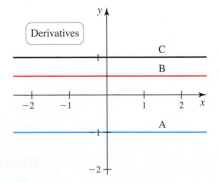

**16. Matching derivatives with functions** Match graphs a–d of derivative functions with possible graphs A–D of the corresponding functions.

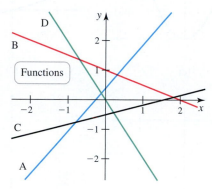

**17–18. Sketching derivatives** *Reproduce the graph of f and then sketch a graph of f ′ on the same axes.*

**17.**

**18.**

**19.** Use the graph of $f$ in the figure to do the following.

   **a.** Find the values of $x$ in $(-2, 2)$ at which $f$ is not continuous.

   **b.** Find the values of $x$ in $(-2, 2)$ at which $f$ is not differentiable.

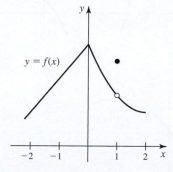

**20.** Use the graph of $g$ in the figure to do the following.

   **a.** Find the values of $x$ in $(-2, 2)$ at which $g$ is not continuous.

   **b.** Find the values of $x$ in $(-2, 2)$ at which $g$ is not differentiable.

## Practice Exercises

**21–30. Derivatives**

*a.* Use limits to find the derivative function $f'$ for the following functions $f$.

*b.* Evaluate $f'(a)$ for the given values of $a$.

**21.** $f(x) = 5x + 2;\ a = 1, 2$     **22.** $f(x) = 7;\ a = -1, 2$

**23.** $f(x) = 4x^2 + 1;\ a = 2, 4$    **24.** $f(x) = x^2 + 3x;\ a = -1, 4$

**25.** $f(x) = \dfrac{1}{x + 1};\ a = -\dfrac{1}{2}, 5$   **26.** $f(x) = \dfrac{x}{x + 2};\ a = -1, 0$

**27.** $f(t) = \dfrac{1}{\sqrt{t}};\ a = 9, \dfrac{1}{4}$     **28.** $f(w) = \sqrt{4w - 3};\ a = 1, 3$

**29.** $f(s) = 4s^3 + 3s;\ a = -3, -1$

**30.** $f(t) = 3t^4;\ a = -2, 2$

**31–32. Velocity functions** *A projectile is fired vertically upward into the air, and its position (in feet) above the ground after t seconds is given by the function s(t).*

*a.* For the following functions $s(t)$, find the instantaneous velocity function $v(t)$. (Recall that the velocity function $v$ is the derivative of the position function $s$.)

*b.* Determine the instantaneous velocity of the projectile at $t = 1$ and $t = 2$ seconds.

**31.** $s(t) = -16t^2 + 100t$     **32.** $s(t) = -16t^2 + 128t + 192$

**33.** Evaluate $\dfrac{dy}{dx}$ and $\dfrac{dy}{dx}\Big|_{x=2}$ if $y = \dfrac{x + 1}{x + 2}$.

**34.** Evaluate $\dfrac{ds}{dt}$ and $\dfrac{ds}{dt}\Big|_{t=-1}$ if $s = 11t^3 + t + 1$.

**35–40. Tangent lines**

*a.* Find the derivative function $f'$ for the following functions $f$.

*b.* Find an equation of the line tangent to the graph of $f$ at $(a, f(a))$ for the given value of $a$.

**35.** $f(x) = 3x^2 + 2x - 10;\ a = 1$

**36.** $f(x) = 5x^2 - 6x + 1;\ a = 2$

**37.** $f(x) = \sqrt{3x + 1};\ a = 8$     **38.** $f(x) = \sqrt{x + 2};\ a = 7$

**39.** $f(x) = \dfrac{2}{3x + 1};\ a = -1$     **40.** $f(x) = \dfrac{1}{x};\ a = -5$

**41. Power and energy** Energy is the capacity to do work, and power is the rate at which energy is used or consumed. Therefore, if $E(t)$ is the energy function for a system, then $P(t) = E'(t)$ is the power function. A unit of energy is the kilowatt-hour (1 kWh is the amount of energy needed to light ten 100-W light bulbs for an hour); the corresponding units for power are kilowatts. The following figure shows the energy demand function for a small community over a 25-hour period.

a. Estimate the power at $t = 10$ and $t = 20$ hr. Be sure to include units in your calculation.

b. At what times on the interval $[0, 25]$ is the power zero?

c. At what times on the interval $[0, 25]$ is the power a maximum?

**42. Slope of a line** Consider the line $f(x) = mx + b$, where $m$ and $b$ are constants. Show that $f'(x) = m$ for all $x$. Interpret this result.

**43. A derivative formula**

a. Use the definition of the derivative to determine
$$\frac{d}{dx}(ax^2 + bx + c),$$ where $a$, $b$, and $c$ are constants.

b. Let $f(x) = 4x^2 - 3x + 10$ and use part (a) to find $f'(x)$.

c. Use part (b) to find $f'(1)$.

**44. A derivative formula**

a. Use the definition of the derivative to determine
$$\frac{d}{dx}\left(\sqrt{ax + b}\right),$$ where $a$ and $b$ are constants.

b. Let $f(x) = \sqrt{5x + 9}$ and use part (a) to find $f'(x)$.

c. Use part (b) to find $f'(-1)$.

**45–46. Analyzing slopes** *Use the points A, B, C, D, and E in the following graphs to answer these questions.*

a. *At which points is the slope of the curve negative?*

b. *At which points is the slope of the curve positive?*

c. *Using A–E, list the slopes in decreasing order.*

**45.**

**46.**

**47. Matching functions with derivatives** Match the functions a–d in the first set of figures with the derivative functions A–D in the next set of figures.

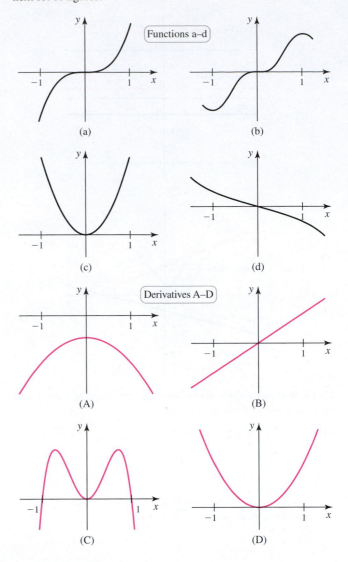

**48–52. Sketching derivatives** *Reproduce the graph of f and then plot a graph of f′ on the same axes.*

**48.**

**49.**

**50.**

**51.**

**52.**

**53. Where is the function continuous? Differentiable?** Use the graph of $f$ in the figure to do the following.

a. Find the values of $x$ in $(0, 3)$ at which $f$ is not continuous.
b. Find the values of $x$ in $(0, 3)$ at which $f$ is not differentiable.
c. Sketch a graph of $f'$.

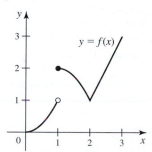

**54. Where is the function continuous? Differentiable?** Use the graph of $g$ in the figure to do the following.

a. Find the values of $x$ in $(0, 4)$ at which $g$ is not continuous.
b. Find the values of $x$ in $(0, 4)$ at which $g$ is not differentiable.
c. Sketch a graph of $g'$.

**55. Voltage on a capacitor** A capacitor is a device in an electrical circuit that stores charge. In one particular circuit, the charge on the capacitor $Q$ varies in time as shown in the figure.

a. At what time is the rate of change of the charge $Q'$ the greatest?
b. Is $Q'$ positive or negative for $t \geq 0$?

c. Is $Q'$ an increasing or decreasing function of time (or neither)?
d. Sketch the graph of $Q'$. You do not need a scale on the vertical axis.

**56. Logistic growth** A common model for population growth uses the logistic (or sigmoid) curve. Consider the logistic curve in the figure, where $P(t)$ is the population at time $t \geq 0$.

a. At approximately what time is the rate of growth $P'$ the greatest?
b. Is $P'$ positive or negative for $t \geq 0$?
c. Is $P'$ an increasing or decreasing function of time (or neither)?
d. Sketch the graph of $P'$. You do not need a scale on the vertical axis.

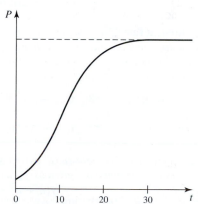

**57. Explain why or why not** Determine whether the following statements are true and give an explanation or counterexample.

a. If the function $f$ is differentiable for all values of $x$, then $f$ is continuous for all values of $x$.
b. The function $f(x) = |x + 1|$ is continuous for all $x$, but not differentiable for all $x$.
c. It is possible for the domain of $f$ to be $(a, b)$ and the domain of $f'$ to be $[a, b]$.

**Explorations and Challenges**

**58. Looking ahead: Derivative of $x^n$** Use the definition
$$f'(x) = \lim_{h \to 0} \frac{f(x + h) - f(x)}{h}$$ to find $f'(x)$ for the following functions.

a. $f(x) = x^2$
b. $f(x) = x^3$
c. $f(x) = x^4$
d. Based on your answers to parts (a)–(c), propose a formula for $f'(x)$ if $f(x) = x^n$, where $n$ is a positive integer.

**59. Determining the unknown constant** Let

$$f(x) = \begin{cases} 2x^2 & \text{if } x \leq 1 \\ ax - 2 & \text{if } x > 1. \end{cases}$$

Determine a value of $a$ (if possible) for which $f'$ is continuous at $x = 1$.

**60. Finding $f$ from $f'$** Sketch the graph of $f'(x) = 2$. Then sketch three possible graphs of $f$.

**61. Finding $f$ from $f'$** Sketch the graph of $f'(x) = x$. Then sketch a possible graph of $f$. Is more than one graph possible?

**62. Finding $f$ from $f'$** Create the graph of a continuous function $f$ such that

$$f'(x) = \begin{cases} 1 & \text{if } x < 0 \\ 0 & \text{if } 0 < x < 1 \\ -1 & \text{if } x > 1. \end{cases}$$

Is more than one graph possible?

**63–66. Normal lines** *A line perpendicular to another line or to a tangent line is often called a **normal line**. Find an equation of the line perpendicular to the line that is tangent to the following curves at the given point P.*

**63.** $y = 3x - 4$; $P(1, -1)$ **64.** $y = \sqrt{x}$; $P(4, 2)$

**65.** $y = \dfrac{2}{x}$; $P(1, 2)$ **66.** $y = x^2 - 3x$; $P(3, 0)$

**67–70. Aiming a tangent line** *Given the function $f$ and the point $Q$, find all points $P$ on the graph of $f$ such that the line tangent to $f$ at $P$ passes through $Q$. Check your work by graphing $f$ and the tangent lines.*

**67.** $f(x) = x^2 + 1$; $Q(3, 6)$ **68.** $f(x) = -x^2 + 4x - 3$; $Q(0, 6)$

**69.** $f(x) = \dfrac{1}{x}$; $Q(-2, 4)$ **70.** $f(x) = 3\sqrt{4x + 1}$; $Q(0, 5)$

**71–72. One-sided derivatives** *The **right-sided** and **left-sided** derivatives of a function at a point $a$ are given by $f'_+(a) =$*

$$\lim_{h \to 0^+} \frac{f(a + h) - f(a)}{h} \text{ and } f'_-(a) = \lim_{h \to 0^-} \frac{f(a + h) - f(a)}{h},$$

*respectively, provided these limits exist. The derivative $f'(a)$ exists if and only if $f'_+(a) = f'_-(a)$.*

**a.** *Sketch the following functions.*
**b.** *Compute $f'_+(a)$ and $f'_-(a)$ at the given point $a$.*
**c.** *Is $f$ continuous at $a$? Is $f$ differentiable at $a$?*

**71.** $f(x) = |x - 2|$; $a = 2$

**72.** $f(x) = \begin{cases} 4 - x^2 & \text{if } x \leq 1 \\ 2x + 1 & \text{if } x > 1 \end{cases}$; $a = 1$

**73–76. Vertical tangent lines** *If a function $f$ is continuous at $a$ and $\lim_{x \to a} |f'(x)| = \infty$, then the curve $y = f(x)$ has a vertical tangent line at $a$, and the equation of the tangent line is $x = a$. If $a$ is an endpoint of a domain, then the appropriate one-sided derivative (Exercises 71–72) is used. Use this information to answer the following questions.*

**T 73.** Graph the following functions and determine the location of the vertical tangent lines.

**a.** $f(x) = (x - 2)^{1/3}$ **b.** $f(x) = (x + 1)^{2/3}$
**c.** $f(x) = \sqrt{|x - 4|}$ **d.** $f(x) = x^{5/3} - 2x^{1/3}$

**74.** The preceding definition of a vertical tangent line includes four cases: $\lim_{x \to a^+} f'(x) = \pm \infty$ combined with $\lim_{x \to a^-} f'(x) = \pm \infty$ (for example, one case is $\lim_{x \to a^+} f'(x) = -\infty$ and $\lim_{x \to a^-} f'(x) = \infty$). Sketch a continuous function that has a vertical tangent line at $a$ in each of the four cases.

**75.** Verify that $f(x) = x^{1/3}$ has a vertical tangent line at $x = 0$.

**76.** Graph the following curves and determine the location of any vertical tangent lines.

**a.** $x^2 + y^2 = 9$ **b.** $x^2 + y^2 + 2x = 0$

**77. a.** Graph the function $f(x) = \begin{cases} x & \text{if } x \leq 0 \\ x + 1 & \text{if } x > 0. \end{cases}$

**b.** For $x < 0$, what is $f'(x)$?
**c.** For $x > 0$, what is $f'(x)$?
**d.** Graph $f'$ on its domain.
**e.** Is $f$ differentiable at 0? Explain.

**78.** Is $f(x) = \dfrac{x^2 - 5x + 6}{x - 2}$ differentiable at $x = 2$? Justify your answer.

**QUICK CHECK ANSWERS**

**1.** 2 **2.** $\dfrac{df}{dx}\Big|_{x=3}, \dfrac{dy}{dx}\Big|_{x=3}, y'(3)$ **3.** The slopes of the tangent lines decrease as $x$ increases because the values of $f'(x) = \dfrac{1}{2\sqrt{x}}$ decrease as $x$ increases. **4.** $\dfrac{dq}{dr}, \dfrac{dp}{dr}$, $D_r(q(r)), q'(r), p'(r)$ **5.** No; no ◄

# 3.3 Rules of Differentiation

If you always had to use limits to evaluate derivatives, as we did in Section 3.2, calculus would be a tedious affair. The goal of this chapter is to establish rules and formulas for quickly evaluating derivatives—not just for individual functions but for entire families of functions. By the end of the chapter, you will have learned many time-saving rules and formulas, all of which are listed in the endpapers of the text.

**Figure 3.29**

➤ The derivative of a constant function is 0 at every point because the values of a constant function do not change. This means the instantaneous rate of change is 0 at every point.

**QUICK CHECK 1** Find the values of $\frac{d}{dx}(11)$ and $\frac{d}{dx}(\pi)$. ◄

➤ Note that this factoring formula agrees with familiar factoring formulas for differences of squares and cubes:

$x^2 - a^2 = (x - a)(x + a)$

$x^3 - a^3 = (x - a)(x^2 + ax + a^2)$

## The Constant and Power Rules for Derivatives

The graph of the **constant function** $f(x) = c$ is a horizontal line with a slope of 0 at every point (**Figure 3.29**). It follows that $f'(x) = 0$ or, equivalently, $\frac{d}{dx}(c) = 0$ (Exercise 82). This observation leads to the *Constant Rule* for derivatives.

---

**THEOREM 3.2   Constant Rule**

If $c$ is a real number, then $\frac{d}{dx}(c) = 0$.

---

Next, consider power functions of the form $f(x) = x^n$, where $n$ is a nonnegative integer. If you completed Exercise 58 in Section 3.2, you used the limit definition of the derivative to discover that

$$\frac{d}{dx}(x^2) = 2x, \qquad \frac{d}{dx}(x^3) = 3x^2, \quad \text{and} \quad \frac{d}{dx}(x^4) = 4x^3.$$

In each case, the derivative of $x^n$ could be evaluated by placing the exponent $n$ in front of $x$ as a coefficient and decreasing the exponent by 1. Based on these observations, we state and prove the following theorem.

---

**THEOREM 3.3   Power Rule**

If $n$ is a nonnegative integer, then $\frac{d}{dx}(x^n) = nx^{n-1}$.

---

**Proof:** We let $f(x) = x^n$ and use the definition of the derivative in the form

$$f'(a) = \lim_{x \to a} \frac{f(x) - f(a)}{x - a}.$$

With $n = 1$ and $f(x) = x$, we have

$$f'(a) = \lim_{x \to a} \frac{f(x) - f(a)}{x - a} = \lim_{x \to a} \frac{x - a}{x - a} = 1,$$

as given by the Power Rule.

With $n \geq 2$ and $f(x) = x^n$, note that $f(x) - f(a) = x^n - a^n$. A factoring formula gives

$$x^n - a^n = (x - a)(x^{n-1} + x^{n-2}a + \cdots + xa^{n-2} + a^{n-1}).$$

Therefore,

$$f'(a) = \lim_{x \to a} \frac{x^n - a^n}{x - a} \qquad \text{Definition of } f'(a)$$

$$= \lim_{x \to a} \frac{(x - a)(x^{n-1} + x^{n-2}a + \cdots + xa^{n-2} + a^{n-1})}{x - a} \qquad \text{Factor } x^n - a^n.$$

$$= \lim_{x \to a} (x^{n-1} + x^{n-2}a + \cdots + xa^{n-2} + a^{n-1}) \qquad \text{Cancel common factors.}$$

$$= \underbrace{a^{n-1} + a^{n-2}\cdot a + \cdots + a\cdot a^{n-2} + a^{n-1}}_{n \text{ terms}} = na^{n-1}. \qquad \text{Evaluate the limit.}$$

Replacing $a$ with the variable $x$ in $f'(a) = na^{n-1}$, we obtain the result given in the Power Rule for $n \geq 2$. Finally, note that the Constant Rule is consistent with the Power Rule with $n = 0$. ◄

**EXAMPLE 1    Derivatives of power and constant functions**  Evaluate the following derivatives.

**a.** $\dfrac{d}{dx}(x^9)$    **b.** $\dfrac{d}{dx}(x)$    **c.** $\dfrac{d}{dx}(2^8)$

**SOLUTION**

**a.** $\dfrac{d}{dx}(x^9) = 9x^{9-1} = 9x^8$    Power Rule

**QUICK CHECK 2** Use the graph of $y = x$ to give a geometric explanation of why $\dfrac{d}{dx}(x) = 1$. ◄

**b.** $\dfrac{d}{dx}(x) = \dfrac{d}{dx}(x^1) = 1x^0 = 1$    Power Rule

**c.** You might be tempted to use the Power Rule here, but $2^8 = 256$ is a constant. So by the Constant Rule, $\dfrac{d}{dx}(2^8) = 0$.

*Related Exercises 19–22* ◄

## Constant Multiple Rule

Consider the problem of finding the derivative of a constant $c$ multiplied by a function $f$ (assuming $f'$ exists). We apply the definition of the derivative in the form

$$f'(x) = \lim_{h \to 0} \frac{f(x+h) - f(x)}{h}$$

to the function $cf$:

$$\frac{d}{dx}(cf(x)) = \lim_{h \to 0} \frac{cf(x+h) - cf(x)}{h} \qquad \text{Definition of the derivative of } cf$$

$$= \lim_{h \to 0} \frac{c(f(x+h) - f(x))}{h} \qquad \text{Factor out } c.$$

$$= c \lim_{h \to 0} \frac{f(x+h) - f(x)}{h} \qquad \text{Theorem 2.3}$$

$$= cf'(x). \qquad \text{Definition of } f'(x)$$

This calculation leads to the *Constant Multiple Rule* for derivatives.

➤ Theorem 3.4 says that the derivative of a constant multiplied by a function is the constant multiplied by the derivative of the function.

---

**THEOREM 3.4    Constant Multiple Rule**

If $f$ is differentiable at $x$ and $c$ is a constant, then

$$\frac{d}{dx}(cf(x)) = cf'(x).$$

---

**EXAMPLE 2    Derivatives of constant multiples of functions**  Evaluate the following derivatives.

**a.** $\dfrac{d}{dx}\left(-\dfrac{7x^{11}}{8}\right)$    **b.** $\dfrac{d}{dt}\left(\dfrac{3}{8}\sqrt{t}\right)$

**SOLUTION**

**a.** $\dfrac{d}{dx}\left(-\dfrac{7x^{11}}{8}\right) = -\dfrac{7}{8} \cdot \dfrac{d}{dx}(x^{11})$    Constant Multiple Rule

$$= -\frac{7}{8} \cdot 11x^{10} \qquad \text{Power Rule}$$

$$= -\frac{77}{8}x^{10} \qquad \text{Simplify.}$$

> In Example 2 of Section 3.2, we proved that $\frac{d}{dt}(\sqrt{t}) = \frac{1}{2\sqrt{t}}$.

**b.** $\dfrac{d}{dt}\left(\dfrac{3}{8}\sqrt{t}\right) = \dfrac{3}{8} \cdot \dfrac{d}{dt}(\sqrt{t})$    Constant Multiple Rule

$\qquad\qquad = \dfrac{3}{8} \cdot \dfrac{1}{2\sqrt{t}}$    $\dfrac{d}{dt}(\sqrt{t}) = \dfrac{1}{2\sqrt{t}}$

$\qquad\qquad = \dfrac{3}{16\sqrt{t}}$

*Related Exercises 23, 24, 28* ◄

## Sum Rule

Many functions are sums of simpler functions. Therefore, it is useful to establish a rule for calculating the derivative of the sum of two or more functions.

> In words, Theorem 3.5 states that the derivative of a sum is the sum of the derivatives.

---

**THEOREM 3.5    Sum Rule**

If $f$ and $g$ are differentiable at $x$, then

$$\frac{d}{dx}(f(x) + g(x)) = f'(x) + g'(x).$$

---

**Proof:** Let $F = f + g$, where $f$ and $g$ are differentiable at $x$, and use the definition of the derivative:

$$\frac{d}{dx}(f(x) + g(x)) = F'(x)$$

$$= \lim_{h \to 0} \frac{F(x + h) - F(x)}{h} \qquad \text{Definition of derivative}$$

$$= \lim_{h \to 0} \frac{(f(x + h) + g(x + h)) - (f(x) + g(x))}{h} \qquad F = f + g$$

$$= \lim_{h \to 0} \left(\frac{f(x + h) - f(x)}{h} + \frac{g(x + h) - g(x)}{h}\right) \qquad \text{Regroup.}$$

$$= \lim_{h \to 0} \frac{f(x + h) - f(x)}{h} + \lim_{h \to 0} \frac{g(x + h) - g(x)}{h} \qquad \text{Theorem 2.3}$$

$$= f'(x) + g'(x). \qquad \text{Definition of } f' \text{ and } g'$$ ◄

**QUICK CHECK 3** If $f(x) = x^2$ and $g(x) = 2x$, what is the derivative of $f(x) + g(x)$? ◄

The Sum Rule can be extended to three or more differentiable functions, $f_1, f_2, \ldots, f_n$, to obtain the **Generalized Sum Rule**:

$$\frac{d}{dx}(f_1(x) + f_2(x) + \cdots + f_n(x)) = f_1'(x) + f_2'(x) + \cdots + f_n'(x).$$

The difference of two functions $f - g$ can be rewritten as the sum $f + (-g)$. By combining the Sum Rule with the Constant Multiple Rule, the **Difference Rule** is established:

$$\frac{d}{dx}(f(x) - g(x)) = f'(x) - g'(x).$$

Let's put the Sum and Difference Rules to work on one of the more common problems: differentiating polynomials.

**EXAMPLE 3**  **Derivative of a polynomial**  Determine $\dfrac{d}{dw}(2w^3 + 9w^2 - 6w + 4)$.

**SOLUTION**

$$\frac{d}{dw}(2w^3 + 9w^2 - 6w + 4)$$

$$= \frac{d}{dw}(2w^3) + \frac{d}{dw}(9w^2) - \frac{d}{dw}(6w) + \frac{d}{dw}(4) \quad \text{Generalized Sum Rule and Difference Rule}$$

$$= 2\frac{d}{dw}(w^3) + 9\frac{d}{dw}(w^2) - 6\frac{d}{dw}(w) + \frac{d}{dw}(4) \quad \text{Constant Multiple Rule}$$

$$= 2 \cdot 3w^2 + 9 \cdot 2w - 6 \cdot 1 + 0 \quad \text{Power Rule and Constant Rule}$$

$$= 6w^2 + 18w - 6 \quad \text{Simplify.}$$

*Related Exercises 31–33* ◄

The technique used to differentiate the polynomial in Example 3 may be used for *any* polynomial. Much of the remainder of this chapter is devoted to discovering differentiation rules for rational, algebraic, and trigonometric functions.

$y = 2x^3 - 15x^2 + 24x$

Tangent line at $x \approx 0.697$ has slope 6.

Tangent line at $x \approx 4.303$ has slope 6.

**Figure 3.30**

**EXAMPLE 4**  **Slope of a tangent line**  Let $f(x) = 2x^3 - 15x^2 + 24x$. For what values of $x$ does the line tangent to the graph of $f$ have a slope of 6?

**SOLUTION**  The tangent line has a slope of 6 when

$$f'(x) = 6x^2 - 30x + 24 = 6.$$

Subtracting 6 from both sides of the equation and factoring, we have

$$6(x^2 - 5x + 3) = 0.$$

Using the quadratic formula, the roots are

$$x = \frac{5 - \sqrt{13}}{2} \approx 0.697 \quad \text{and} \quad x = \frac{5 + \sqrt{13}}{2} \approx 4.303.$$

Therefore, the slope of the curve at these points is 6 (**Figure 3.30**).

*Related Exercises 59–60* ◄

**QUICK CHECK 4**  Determine the point(s) at which $f(x) = x^3 - 12x$ has a horizontal tangent line. ◄

## Higher-Order Derivatives

➤ The prime notation, $f'$, $f''$, and $f'''$, is typically used only for the first, second, and third derivatives.

Because the derivative of a function $f$ is a function in its own right, we can take the derivative of $f'$. The result is the *second derivative of $f$*, denoted $f''$ (read $f$ *double prime*). The derivative of the second derivative is the *third derivative of $f$*, denoted $f'''$ or (read $f$ *triple prime*). In general, derivatives of order $n \geq 2$ are called *higher-order derivatives*.

> Parentheses are placed around $n$ to distinguish a derivative from a power. Therefore, $f^{(n)}$ is the $n$th derivative of $f$, and $f^n$ is the function $f$ raised to the $n$th power. By convention, $f^{(0)}$ is the function $f$ itself.

---

**DEFINITION** **Higher-Order Derivatives**

Assuming $y = f(x)$ can be differentiated as often as necessary, the **second derivative** of $f$ is

$$f''(x) = \frac{d}{dx}(f'(x)).$$

For integers $n \geq 1$, the **$n$th derivative** of $f$ is

$$f^{(n)}(x) = \frac{d}{dx}(f^{(n-1)}(x)).$$

---

> The notation $\dfrac{d^2 f}{dx^2}$ comes from $\dfrac{d}{dx}\left(\dfrac{df}{dx}\right)$ and is read $d$ 2 $f$ $dx$ squared.

Other common notations for the second derivative of $y = f(x)$ include $\dfrac{d^2 y}{dx^2}$ and $\dfrac{d^2 f}{dx^2}$; the notations $\dfrac{d^n y}{dx^n}$, $\dfrac{d^n f}{dx^n}$, and $y^{(n)}$ are used for the $n$th derivative of $f$.

**EXAMPLE 5** **Finding higher-order derivatives** Find the third derivative of the following functions.

**a.** $f(x) = 3x^3 - 5x + 12$　　　**b.** $y = 3t + 2t^{10}$

**SOLUTION**

**a.**

$$f'(x) = 9x^2 - 5$$

> In Example 5a, note that $f^{(4)}(x) = 0$, which means that all successive derivatives are also 0. In general, the $n$th derivative of an $n$th-degree polynomial is a constant, which implies that derivatives of order $k > n$ are 0.

$$f''(x) = \frac{d}{dx}(9x^2 - 5) = 18x$$

$$f'''(x) = 18$$

**b.** Here we use an alternative notation for higher-order derivatives:

$$\frac{dy}{dt} = \frac{d}{dt}(3t + 2t^{10}) = 3 + 20t^9$$

$$\frac{d^2 y}{dt^2} = \frac{d}{dt}(3 + 20t^9) = 180t^8$$

**QUICK CHECK 5** With $f(x) = x^5$, find $f^{(5)}(x)$ and $f^{(6)}(x)$. ◄

$$\frac{d^3 y}{dt^3} = \frac{d}{dt}(180t^8) = 1440t^7.$$

*Related Exercises 65–66* ◄

---

## SECTION 3.3 EXERCISES

### Getting Started

1. If the limit definition of a derivative can be used to find $f'$, then what is the purpose of using other rules to find $f'$?

2. In this section, we showed that the rule $\dfrac{d}{dx}(x^n) = nx^{n-1}$ is valid for what values of $n$?

3. Why is the derivative of a constant function zero?

4. How do you find the derivative of the sum of two functions $f + g$?

5. How do you find the derivative of a constant multiplied by a function?

6. How do you find the fifth derivative of a function?

7. Given that $f'(3) = 6$ and $g'(3) = -2$, find $(f + g)'(3)$.

8. If $f'(3) = 6$ and $g(x) = f(x) + x^2 + 1$, find $g'(3)$.

**9–11.** Let $F(x) = f(x) + g(x)$, $G(x) = f(x) - g(x)$, and $H(x) = 3f(x) + 2g(x)$, where the graphs of $f$ and $g$ are shown in the figure. Find each of the following.

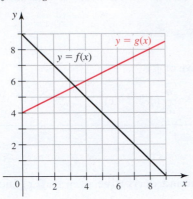

**9.** $F'(2)$    **10.** $G'(6)$    **11.** $H'(2)$

**12–14.** Use the table to find the following derivatives.

| $x$ | 1 | 2 | 3 | 4 | 5 |
|-----|---|---|---|---|---|
| $f'(x)$ | 3 | 5 | 2 | 1 | 4 |
| $g'(x)$ | 2 | 4 | 3 | 1 | 5 |

**12.** $\dfrac{d}{dx}(f(x) + g(x))\Big|_{x=1}$    **13.** $\dfrac{d}{dx}(1.5f(x))\Big|_{x=2}$

**14.** $\dfrac{d}{dx}(2x - 3g(x))\Big|_{x=4}$

**15.** If $f(t) = t^{10}$, find $f'(t)$, $f''(t)$, and $f'''(t)$.

**16.** Find an equation of the line tangent to the graph of $y = x^2$ at $x = 3$.

**17.** The line tangent to the graph of $f$ at $x = 5$ is $y = \dfrac{1}{10}x - 2$. Find $\dfrac{d}{dx}(4f(x))\Big|_{x=5}$.

**18.** The line tangent to the graph of $f$ at $x = 3$ is $y = 4x - 2$ and the line tangent to the graph of $g$ at $x = 3$ is $y = -5x + 1$. Find the values of $(f + g)(3)$ and $(f + g)'(3)$.

## Practice Exercises

**19–38. Derivatives** *Find the derivative of the following functions. See Example 2 of Section 3.2 for the derivative of $\sqrt{x}$.*

**19.** $y = x^5$    **20.** $f(t) = t$

**21.** $f(x) = 5$    **22.** $g(x) = 2^3$

**23.** $f(x) = 5x^3$    **24.** $g(w) = \dfrac{5}{6}w^{12}$

**25.** $h(t) = \dfrac{t^2}{2} + 1$    **26.** $f(v) = v^{100} + v + 10$

**27.** $p(x) = 8x$    **28.** $g(t) = 6\sqrt{t}$

**29.** $g(t) = 100t^2$    **30.** $f(s) = \dfrac{\sqrt{s}}{4}$

**31.** $f(x) = 3x^4 + 7x$    **32.** $g(x) = 6x^5 - \dfrac{5}{2}x^2 + x + 5$

**33.** $f(x) = 10x^4 - 32x + 4^3$    **34.** $f(t) = 6\sqrt{t} - 4t^3 + 9$

**35.** $g(w) = 2w^3 + 3w^2 + 10w$    **36.** $s(t) = 4\sqrt{t} - \dfrac{1}{4}t^4 + t + 1$

**37.** $f(x) = \begin{cases} x^2 + 1 & \text{if } x \le 0 \\ 2x^2 + x + 1 & \text{if } x > 0 \end{cases}$

**38.** $g(w) = \begin{cases} w & \text{if } w \le 1 \\ 2w^3 + 4w + 5 & \text{if } w > 1 \end{cases}$

**39. Height estimate** The distance an object falls (when released from rest, under the influence of Earth's gravity, and with no air resistance) is given by $d(t) = 16t^2$, where $d$ is measured in feet and $t$ is measured in seconds. A rock climber sits on a ledge on a vertical wall and carefully observes the time it takes for a small stone to fall from the ledge to the ground.

    **a.** Compute $d'(t)$. What units are associated with the derivative and what does it measure?

    **b.** If it takes 6 s for a stone to fall to the ground, how high is the ledge? How fast is the stone moving when it strikes the ground (in miles per hour)?

**T 40. Projectile trajectory** The position of a small rocket that is launched vertically upward is given by $s(t) = -5t^2 + 40t + 100$, for $0 \le t \le 10$, where $t$ is measured in seconds and $s$ is measured in meters above the ground.

    **a.** Find the rate of change in the position (instantaneous velocity) of the rocket, for $0 \le t \le 10$.

    **b.** At what time is the instantaneous velocity zero?

    **c.** At what time does the instantaneous velocity have the greatest magnitude, for $0 \le t \le 10$?

    **d.** Graph the position and instantaneous velocity, for $0 \le t \le 10$.

**41. City urbanization** City planners model the size of their city using the function $A(t) = -\dfrac{1}{50}t^2 + 2t + 20$, for $0 \le t \le 50$, where $A$ is measured in square miles and $t$ is the number of years after 2010.

    **a.** Compute $A'(t)$. What units are associated with this derivative and what does the derivative measure?

    **b.** How fast will the city be growing when it reaches a size of 38 mi²?

    **c.** Suppose the population density of the city remains constant from year to year at 1000 people/mi². Determine the growth rate of the population in 2030.

**42. Cell growth** When observations begin at $t = 0$, a cell culture has 1200 cells and continues to grow according to the function $p(t) = 1200 + 24t^4$, where $p$ is the number of cells and $t$ is measured in days.

    **a.** Compute $p'(t)$. What units are associated with the derivative and what does it measure?

    **b.** On the interval $[0, 4]$, when is the growth rate $p'(t)$ the least? When is it the greatest?

**43. Weight of Atlantic salmon** The weight $w(x)$ (in pounds) of an Atlantic salmon that is $x$ inches long can be estimated by the function

$$w(x) = \begin{cases} 0.4x - 5 & \text{if } 19 \le x \le 21 \\ 0.8x - 13.4 & \text{if } 21 < x \le 32 \\ 1.5x - 35.8 & \text{if } x > 32. \end{cases}$$

Calculate $w'(x)$ and explain the physical meaning of this derivative. (*Source*: www.atlanticsalmonfederation.org)

**44–54. Derivatives of products and quotients** *Find the derivative of the following functions by first expanding or simplifying the expression. Simplify your answers.*

**44.** $f(x) = \left(\sqrt{x} + 1\right)\left(\sqrt{x} - 1\right)$

**45.** $f(x) = (2x + 1)(3x^2 + 2)$

**46.** $g(r) = (5r^3 + 3r + 1)(r^2 + 3)$

**47.** $f(w) = \dfrac{w^3 - w}{w}$
**48.** $y = \dfrac{12s^3 - 8s^2 + 12s}{4s}$

**49.** $h(x) = (x^2 + 1)^2$
**50.** $h(x) = \sqrt{x}\left(\sqrt{x} - x^{3/2}\right)$

**51.** $g(x) = \dfrac{x^2 - 1}{x - 1}$
**52.** $h(x) = \dfrac{x^3 - 6x^2 + 8x}{x^2 - 2x}$

**53.** $y = \dfrac{x - a}{\sqrt{x} - \sqrt{a}}$, where $a$ is a positive constant

**54.** $y = \dfrac{x^2 - 2ax + a^2}{x - a}$, where $a$ is a constant

**T 55–58. Equations of tangent lines**

**a.** *Find an equation of the line tangent to the given curve at a.*
**b.** *Use a graphing utility to graph the curve and the tangent line on the same set of axes.*

**55.** $y = -3x^2 + 2$; $a = 1$

**56.** $y = x^3 - 4x^2 + 2x - 1$; $a = 2$

**57.** $y = \sqrt{x}$; $a = 4$
**58.** $y = \dfrac{1}{2}x^4 + x$; $a = 2$

**59. Finding slope locations** Let $f(x) = x^2 - 6x + 5$.

**a.** Find the values of $x$ for which the slope of the curve $y = f(x)$ is 0.
**b.** Find the values of $x$ for which the slope of the curve $y = f(x)$ is 2.

**60. Finding slope locations** Let $f(t) = t^3 - 27t + 5$.

**a.** Find the values of $t$ for which the slope of the curve $y = f(t)$ is 0.
**b.** Find the values of $t$ for which the slope of the curve $y = f(t)$ is 21.

**61. Finding slope locations** Let $f(x) = 2x^3 - 3x^2 - 12x + 4$.

**a.** Find all points on the graph of $f$ at which the tangent line is horizontal.
**b.** Find all points on the graph of $f$ at which the tangent line has slope 60.

**62. Finding slope locations** Let $f(x) = \sqrt{x}$.

**a.** Use $f'(x)$ to explain why the graph of $f$ has no horizontal tangent line.
**b.** Find all points on the graph of $f$ at which the tangent line has slope $\frac{1}{4}$.

**63. Finding slope locations** Let $f(x) = 4\sqrt{x} - x$.

**a.** Find all points on the graph of $f$ at which the tangent line is horizontal.
**b.** Find all points on the graph of $f$ at which the tangent line has slope $-\frac{1}{2}$.

**64–68. Higher-order derivatives** *Find $f'(x)$, $f''(x)$, and $f'''(x)$ for the following functions.*

**64.** $f(x) = 3x^3 + 5x^2 + 6x$
**65.** $f(x) = 5x^4 + 10x^3 + 3x + 6$

**66.** $f(x) = 3x^{12} + 4x^3$
**67.** $f(x) = \dfrac{x^2 - 7x - 8}{x + 1}$

**68.** $f(x) = \dfrac{1}{8}x^4 - 3x^2 + 1$

**69. Explain why or why not** Determine whether the following statements are true and give an explanation or counterexample.

**a.** $\dfrac{d}{dx}(10^5) = 5 \cdot 10^4$.

**b.** The slope of a line tangent to $f(x) = 4x + 1$ is never 0.

**c.** $\dfrac{d^n}{dx^n}(5x^3 + 2x + 5) = 0$, for any integer $n \ge 3$.

**70. Tangent lines** Suppose $f(3) = 1$ and $f'(3) = 4$. Let $g(x) = x^2 + f(x)$ and $h(x) = 3f(x)$.

**a.** Find an equation of the line tangent to $y = g(x)$ at $x = 3$.
**b.** Find an equation of the line tangent to $y = h(x)$ at $x = 3$.

**71. Derivatives from tangent lines** Suppose the line tangent to the graph of $f$ at $x = 2$ is $y = 4x + 1$ and suppose the line tangent to the graph of $g$ at $x = 2$ has slope 3 and passes through $(0, -2)$. Find an equation of the line tangent to the following curves at $x = 2$.

**a.** $y = f(x) + g(x)$
**b.** $y = f(x) - 2g(x)$
**c.** $y = 4f(x)$

**72. Tangent line** Find the equation of the line tangent to the curve $y = x + \sqrt{x}$ that has slope 2.

**73. Tangent line given** Determine the constants $b$ and $c$ such that the line tangent to $f(x) = x^2 + bx + c$ at $x = 1$ is $y = 4x + 2$.

**74–77. Derivatives from a graph** *Let $F = f + g$ and $G = 3f - g$, where the graphs of $f$ and $g$ are shown in the figure. Find the following derivatives.*

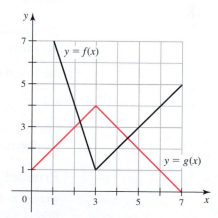

**74.** $F'(2)$
**75.** $G'(2)$
**76.** $F'(5)$
**77.** $G'(5)$

**78–81. Derivatives from limits** *The following limits represent $f'(a)$ for some function $f$ and some real number $a$.*

**a.** *Find a possible function $f$ and number $a$.*
**b.** *Evaluate the limit by computing $f'(a)$.*

**78.** $\displaystyle\lim_{x \to 1} \dfrac{x^{100} - 1}{x - 1}$
**79.** $\displaystyle\lim_{x \to 1} \dfrac{x^7 + x^6 - 2}{x - 1}$

**80.** $\lim\limits_{h \to 0} \dfrac{(1 + h)^8 + (1 + h)^3 - 2}{h}$

**81.** $\lim\limits_{h \to 0} \dfrac{\sqrt{9 + h} - \sqrt{9}}{h}$

### Explorations and Challenges

**82.** **Constant Rule proof** For the constant function $f(x) = c$, use the definition of the derivative to show that $f'(x) = 0$.

**83.** **Alternative proof of the Power Rule** The Binomial Theorem states that for any positive integer $n$,

$$(a + b)^n = a^n + na^{n-1}b + \frac{n(n-1)}{2 \cdot 1}a^{n-2}b^2$$
$$+ \frac{n(n-1)(n-2)}{3 \cdot 2 \cdot 1}a^{n-3}b^3 + \cdots + nab^{n-1} + b^n.$$

Use this formula and the definition

$$f'(x) = \lim\limits_{h \to 0} \frac{f(x + h) - f(x)}{h} \text{ to show that } \frac{d}{dx}(x^n) = nx^{n-1}, \text{ for}$$

any positive integer $n$.

**84.** **Power Rule for negative integers** Suppose $n$ is a negative integer and $f(x) = x^n$. Use the following steps to prove that $f'(a) = na^{n-1}$, which means the Power Rule for positive integers extends to all integers. This result is proved in Section 3.4 by a different method.

**a.** Assume $m = -n$, so that $m > 0$. Use the definition

$$f'(a) = \lim\limits_{x \to a} \frac{x^n - a^n}{x - a} = \lim\limits_{x \to a} \frac{x^{-m} - a^{-m}}{x - a}.$$

Simplify using the factoring rule (which is valid for $n > 0$)

$$x^n - a^n = (x - a)(x^{n-1} + x^{n-2}a + \cdots + xa^{n-2} + a^{n-1})$$

until it is possible to take the limit.

**b.** Use this result to find $\dfrac{d}{dx}(x^{-7})$ and $\dfrac{d}{dx}\left(\dfrac{1}{x^{10}}\right)$.

**85.** **Extending the Power Rule to $n = \dfrac{1}{2}, \dfrac{3}{2},$ and $\dfrac{5}{2}$** With Theorem 3.3 and Exercise 84, we have shown that the Power Rule, $\dfrac{d}{dx}(x^n) = nx^{n-1}$, applies to any integer $n$. Later in the chapter, we extend this rule so that it applies to any real number $n$.

**a.** Explain why the Power Rule is consistent with the formula $\dfrac{d}{dx}(\sqrt{x}) = \dfrac{1}{2\sqrt{x}}$.

**b.** Prove that the Power Rule holds for $n = \dfrac{3}{2}$. (*Hint:* Use the definition of the derivative: $\dfrac{d}{dx}(x^{3/2}) = \lim\limits_{h \to 0} \dfrac{(x + h)^{3/2} - x^{3/2}}{h}$.)

**c.** Prove that the Power Rule holds for $n = \dfrac{5}{2}$.

**d.** Propose a formula for $\dfrac{d}{dx}(x^{n/2})$ for any positive integer $n$.

**QUICK CHECK ANSWERS**

**1.** $\dfrac{d}{dx}(11) = 0$ and $\dfrac{d}{dx}(\pi) = 0$ because 11 and $\pi$ are constants.

**2.** The slope of the curve $y = x$ is 1 at any point; therefore, $\dfrac{d}{dx}(x) = 1$.    **3.** $2x + 2$    **4.** $x = 2$ and $x = -2$

**5.** $f^{(5)}(x) = 120, f^{(6)}(x) = 0$ ◄

# 3.4 The Product and Quotient Rules

The derivative of a sum of functions is the sum of the derivatives. So you might assume the derivative of a product of functions is the product of the derivatives. Consider, however, the functions $f(x) = x^3$ and $g(x) = x^4$. In this case, $\dfrac{d}{dx}(f(x)g(x)) = \dfrac{d}{dx}(x^7) = 7x^6$, but $f'(x)g'(x) = 3x^2 \cdot 4x^3 = 12x^5$. Therefore, $\dfrac{d}{dx}(f \cdot g) \neq f' \cdot g'$. Similarly, the derivative of a quotient is *not* the quotient of the derivatives. The purpose of this section is to develop rules for differentiating products and quotients of functions.

## Product Rule

Here is an anecdote that suggests the formula for the Product Rule. Imagine running along a road at a constant speed. Your speed is determined by two factors: the length of your stride and the number of strides you take each second. Therefore,

$$\text{running speed} = \text{stride length} \cdot \text{stride rate}.$$

For example, if your stride length is 3 ft per stride and you take 2 strides/s, then your speed is 6 ft/s.

Now suppose your stride length increases by 0.5 ft, from 3 to 3.5 ft. Then the change in speed is calculated as follows:

$$\textit{change} \text{ in speed} = \text{change in stride length} \cdot \text{stride rate}$$
$$= 0.5 \cdot 2 = 1 \text{ ft/s}.$$

Alternatively, suppose your stride length remains constant but your stride rate increases by 0.25 stride/s, from 2 to 2.25 strides/s. Then

$$change \text{ in speed} = \text{stride length} \cdot \text{change in stride rate}$$
$$= 3 \cdot 0.25 = 0.75 \text{ ft/s}.$$

If both your stride rate and stride length change simultaneously, we expect two contributions to the change in your running speed:

$$change \text{ in speed} = (\text{change in stride length} \cdot \text{stride rate})$$
$$+ (\text{stride length} \cdot \text{change in stride rate})$$
$$= 1 \text{ ft/s} + 0.75 \text{ ft/s} = 1.75 \text{ ft/s}.$$

This argument correctly suggests that the derivative (or rate of change) of a product of two functions has *two components*, as shown by the following rule.

> ▶ In words, Theorem 3.6 states that the derivative of the product of two functions equals the derivative of the first function multiplied by the second function, plus the first function multiplied by the derivative of the second function.

---

**THEOREM 3.6   Product Rule**

If $f$ and $g$ are differentiable at $x$, then

$$\frac{d}{dx}(f(x)g(x)) = f'(x)g(x) + f(x)g'(x).$$

---

**Proof:** We apply the definition of the derivative to the function $fg$:

$$\frac{d}{dx}(f(x)g(x)) = \lim_{h \to 0} \frac{f(x+h)g(x+h) - f(x)g(x)}{h}.$$

A useful tactic is to add $-f(x)g(x+h) + f(x)g(x+h)$ (which equals 0) to the numerator, so that

$$\frac{d}{dx}(f(x)g(x)) = \lim_{h \to 0} \frac{f(x+h)g(x+h) - f(x)g(x+h) + f(x)g(x+h) - f(x)g(x)}{h}.$$

The fraction is now split and the numerators are factored:

$$\frac{d}{dx}(f(x)g(x))$$

$$= \lim_{h \to 0} \frac{f(x+h)g(x+h) - f(x)g(x+h)}{h} + \lim_{h \to 0} \frac{f(x)g(x+h) - f(x)g(x)}{h}$$

> ▶ As $h \to 0$, $f(x)$ does not change in value because it is independent of $h$.

$$= \lim_{h \to 0} \left( \underbrace{\frac{f(x+h) - f(x)}{h}}_{\substack{\text{approaches } f'(x) \\ \text{as } h \to 0}} \cdot \underbrace{g(x+h)}_{\substack{\text{approaches} \\ g(x) \\ \text{as } h \to 0}} \right) + \lim_{h \to 0} \left( \underbrace{f(x)}_{\substack{\text{equals} \\ f(x) \text{ as} \\ h \to 0}} \cdot \underbrace{\frac{g(x+h) - g(x)}{h}}_{\substack{\text{approaches } g'(x) \\ \text{as } h \to 0}} \right)$$

$$= f'(x) \cdot g(x) + f(x) \cdot g'(x).$$

The continuity of $g$ is used to conclude that $\lim_{h \to 0} g(x+h) = g(x)$.   ◀

---

**EXAMPLE 1   Using the Product Rule**  Find and simplify the following derivatives.

**a.** $\dfrac{d}{dv}(v^2(2\sqrt{v} + 1))$     **b.** $\dfrac{d}{dx}((x^3 - 8)(x^2 + 4))$

> In Example 2 of Section 3.2, we proved
> that $\frac{d}{dv}(\sqrt{v}) = \frac{1}{2\sqrt{v}}$.

**SOLUTION**

**a.** $\dfrac{d}{dv}(v^2(2\sqrt{v}+1)) = \left(\dfrac{d}{dv}(v^2)\right)(2\sqrt{v}+1) + v^2\left(\dfrac{d}{dv}(2\sqrt{v}+1)\right)$    Product Rule

$$= 2v(2\sqrt{v}+1) + v^2\left(2\cdot\dfrac{1}{2\sqrt{v}}\right)$$    Evaluate the derivatives.

$$= 4v^{3/2} + 2v + v^{3/2} = 5v^{3/2} + 2v$$    Simplify.

**QUICK CHECK 1** Find the derivative of $f(x) = x^5$. Then find the same derivative using the Product Rule with $f(x) = x^2x^3$. ◄

**b.** $\dfrac{d}{dx}((x^3 - 8)(x^2 + 4)) = \underbrace{3x^2}\cdot(x^2+4) + (x^3-8)\cdot\underbrace{(2x)} = x\,(5x^3 + 12x - 16)$

$\phantom{xxxxxxxxxx}\underbrace{\phantom{xx}}_{\frac{d}{dx}(x^3-8)}\phantom{xxxxxxxxxxxxxx}\underbrace{\phantom{xx}}_{\frac{d}{dx}(x^2+4)}$

*Related Exercises 19–20* ◄

## Quotient Rule

Consider the quotient $q(x) = \dfrac{f(x)}{g(x)}$ and note that $f(x) = g(x)q(x)$. By the Product Rule, we have

$$f'(x) = g'(x)q(x) + g(x)q'(x).$$

Solving for $q'(x)$, we find that

$$q'(x) = \frac{f'(x) - g'(x)q(x)}{g(x)}.$$

Substituting $q(x) = \dfrac{f(x)}{g(x)}$ produces a rule for finding $q'(x)$:

$$q'(x) = \frac{f'(x) - g'(x)\dfrac{f(x)}{g(x)}}{g(x)}$$    Replace $q(x)$ with $\dfrac{f(x)}{g(x)}$.

$$= \frac{g(x)\left(f'(x) - g'(x)\dfrac{f(x)}{g(x)}\right)}{g(x)\cdot g(x)}$$    Multiply numerator and denominator by $g(x)$.

$$= \frac{g(x)f'(x) - f(x)g'(x)}{(g(x))^2}.$$    Simplify.

> In words, Theorem 3.7 states that the derivative of the quotient of two functions equals the denominator multiplied by the derivative of the numerator minus the numerator multiplied by the derivative of the denominator, all divided by the denominator squared. An easy way to remember the Quotient Rule is with
> $$\frac{LoD(Hi) - HiD(Lo)}{(Lo)^2}.$$

This calculation produces the correct result for the derivative of a quotient. However, there is one subtle point: How do we know that the derivative of $f/g$ exists in the first place? A complete proof of the Quotient Rule is outlined in Exercise 92.

---

**THEOREM 3.7  Quotient Rule**
If $f$ and $g$ are differentiable at $x$ and $g(x) \neq 0$, then the derivative of $f/g$ at $x$ exists and

$$\frac{d}{dx}\left(\frac{f(x)}{g(x)}\right) = \frac{g(x)f'(x) - f(x)g'(x)}{(g(x))^2}.$$

---

**EXAMPLE 2** **Using the Quotient Rule** Find and simplify the following derivatives.

**a.** $\dfrac{d}{dx}\left(\dfrac{x^2 + 3x + 4}{x^2 - 1}\right)$     **b.** $\dfrac{d}{dx}(2x^{-3})$

**SOLUTION**

> ➤ The Product and Quotient Rules are used on a regular basis throughout this text. It is a good idea to memorize these rules (along with the other derivative rules and formulas presented in this chapter) so that you can evaluate derivatives quickly.

**a.**
$$\dfrac{d}{dx}\left(\dfrac{x^2 + 3x + 4}{x^2 - 1}\right) = \dfrac{\overbrace{(x^2 - 1)(2x + 3)}^{\substack{(x^2 - 1)\,\cdot\,\text{the derivative}\\\text{of }(x^2 + 3x + 4)}} - \overbrace{(x^2 + 3x + 4)2x}^{\substack{(x^2 + 3x + 4)\,\cdot\,\text{the}\\\text{derivative of }(x^2 - 1)}}}{\underbrace{(x^2 - 1)^2}_{\substack{\text{the denominator}\\(x^2 - 1)\text{ squared}}}} \qquad \text{Quotient Rule}$$

$$= \dfrac{2x^3 - 2x + 3x^2 - 3 - 2x^3 - 6x^2 - 8x}{(x^2 - 1)^2} \qquad \text{Expand.}$$

$$= -\dfrac{3x^2 + 10x + 3}{(x^2 - 1)^2} \qquad \text{Simplify.}$$

**b.** We rewrite $2x^{-3}$ as $\dfrac{2}{x^3}$ and use the Quotient Rule:

$$\dfrac{d}{dx}\left(\dfrac{2}{x^3}\right) = \dfrac{x^3 \cdot 0 - 2 \cdot 3x^2}{(x^3)^2} = -\dfrac{6}{x^4} = -6x^{-4}.$$

*Related Exercises 22, 25* ◀

**QUICK CHECK 2** Find the derivative of $f(x) = x^5$. Then find the same derivative using the Quotient Rule with $f(x) = x^8/x^3$. ◀

**EXAMPLE 3** **Finding tangent lines** Find an equation of the line tangent to the graph of $f(x) = \dfrac{x^2 + 1}{x^2 - 4}$ at the point $(3, 2)$. Plot the curve and tangent line.

**SOLUTION** To find the slope of the tangent line, we compute $f'$ using the Quotient Rule:

$$f'(x) = \dfrac{(x^2 - 4)\,2x - (x^2 + 1)\,2x}{(x^2 - 4)^2} \qquad \text{Quotient Rule}$$

$$= \dfrac{2x^3 - 8x - 2x^3 - 2x}{(x^2 - 4)^2} = -\dfrac{10x}{(x^2 - 4)^2}. \qquad \text{Simplify.}$$

The slope of the tangent line at $(3, 2)$ is

$$m_{\text{tan}} = f'(3) = -\dfrac{10(3)}{(3^2 - 4)^2} = -\dfrac{6}{5}.$$

Therefore, an equation of the tangent line is

$$y - 2 = -\dfrac{6}{5}(x - 3), \quad \text{or} \quad y = -\dfrac{6}{5}x + \dfrac{28}{5}.$$

The graphs of $f$ and the tangent line are shown in **Figure 3.31**.

*Related Exercises 57–58* ◀

**Figure 3.31**

## Extending the Power Rule to Negative Integers

The Power Rule in Section 3.3 says that $\dfrac{d}{dx}(x^n) = nx^{n-1}$, for nonnegative integers $n$. Using the Quotient Rule, we show that the Power Rule also holds if $n$ is a negative integer. Assume $n$ is a negative integer and let $m = -n$, so that $m > 0$. Then

$$\frac{d}{dx}(x^n) = \frac{d}{dx}\left(\frac{1}{x^m}\right) \qquad\qquad x^n = \frac{1}{x^{-n}} = \frac{1}{x^m}$$

$$= \frac{x^m\left(\overbrace{\dfrac{d}{dx}(1)}^{\text{derivative of a constant is 0}}\right) - 1\left(\overbrace{\dfrac{d}{dx}x^m}^{\substack{\text{equals} \\ mx^{m-1}}}\right)}{(x^m)^2} \qquad \text{Quotient Rule}$$

$$= -\frac{mx^{m-1}}{x^{2m}} \qquad\qquad \text{Simplify.}$$

$$= -mx^{-m-1} \qquad\qquad \frac{x^{m-1}}{x^{2m}} = x^{m-1-2m}$$

$$= nx^{n-1}. \qquad\qquad \text{Replace } -m \text{ with } n.$$

This calculation leads to the first extension of the Power Rule; the rule now applies to all integers. In Theorem 3.8, we assert that, in fact, the Power Rule is valid for all real powers. A proof of this theorem appears in Chapter 7.

> ▶ In Theorem 3.8, it is necessary to restrict the domain for some values of $n$. If $n$ is irrational, then the Power Rule holds, provided that $x > 0$. If $n$ is a rational number of the form $p/q$, where $p$ and $q$ are integers with no common factors and $q$ is even, then the rule holds, provided that $x > 0$.

---

**THEOREM 3.8    Power Rule (general form)**
If $n$ is any real number, then

$$\frac{d}{dx}(x^n) = nx^{n-1}.$$

---

**QUICK CHECK 3** Find the derivative of $f(x) = 1/x^5$ in two different ways: using the Power Rule and using the Quotient Rule. ◄

**EXAMPLE 4    Using the Power Rule**  Find the following derivatives.

**a.** $\dfrac{d}{dx}\left(\dfrac{9}{x^5}\right)$ 

**b.** $\dfrac{d}{dt}\left(\dfrac{3t^{16} - 4}{t^6}\right)$

**c.** $\dfrac{d}{dz}\left(6\sqrt[3]{z}\right)$ 

**d.** $\dfrac{d}{dx}\left(\dfrac{3x^{5/2}}{2x^2 + 4}\right)$

**SOLUTION**

**a.** $\dfrac{d}{dx}\left(\dfrac{9}{x^5}\right) = \dfrac{d}{dx}(9x^{-5}) = 9(-5x^{-6}) = -45x^{-6} = -\dfrac{45}{x^6}$

**b.** This derivative can be evaluated by the Quotient Rule, but an alternative method is to rewrite the expression using negative powers:

$$\frac{3t^{16} - 4}{t^6} = \frac{3t^{16}}{t^6} - \frac{4}{t^6} = 3t^{10} - 4t^{-6}.$$

We now differentiate using the Power Rule:

$$\frac{d}{dt}\left(\frac{3t^{16} - 4}{t^6}\right) = \frac{d}{dt}(3t^{10} - 4t^{-6}) = 30t^9 + 24t^{-7}.$$

**c.** Express the cube root as a power and then apply the Constant Multiple and Power Rules:

$$\frac{d}{dz}(6\sqrt[3]{z}) = \frac{d}{dz}(6z^{1/3}) \qquad \sqrt[3]{z} = z^{1/3}$$

$$= 6 \cdot \frac{1}{3}z^{-2/3} \qquad \text{Constant Multiple and Power Rules}$$

$$= 2z^{-2/3}. \qquad \text{Simplify.}$$

**d.** The Quotient and Power Rules are required:

$$\frac{d}{dx}\left(\frac{3x^{5/2}}{2x^2 + 4}\right) = \frac{(2x^2 + 4) \cdot \frac{15}{2}x^{3/2} - 3x^{5/2} \cdot 4x}{(2x^2 + 4)^2} \qquad \text{Quotient and Power Rules}$$

$$= \frac{15x^{7/2} + 30x^{3/2} - 12x^{7/2}}{(2(x^2 + 2))^2} \qquad \text{Simplify.}$$

$$= \frac{3x^{7/2} + 30x^{3/2}}{4(x^2 + 2)^2} \qquad \text{Combine like terms.}$$

$$= \frac{3x^{3/2}(x^2 + 10)}{4(x^2 + 2)^2}. \qquad \text{Factor.}$$

*Related Exercises 37, 41, 49* ◄

## Rates of Change

Remember that the derivative has multiple uses and interpretations. The following example illustrates the derivative as a rate of change of a population. Specifically, the derivative tells us when the population is growing most rapidly and how the population behaves in the long run.

**EXAMPLE 5    Population growth rates** The population of a culture of cells increases and approaches a constant level (called the *steady state* or *carrying capacity*). The population is modeled by the function $p(t) = 400\left(\frac{t^2 + 1}{t^2 + 4}\right)$, where $t \geq 0$ is measured in hours (**Figure 3.32**)

**a.** Compute and graph the instantaneous growth rate of the population for $t \geq 0$.

**b.** At approximately what time is the instantaneous growth rate the greatest?

**c.** What is the steady-state population?

> Methods for determining exactly when the growth rate is a maximum are discussed in Chapter 4.

**SOLUTION**

**a.** The instantaneous growth rate is given by the derivative of the population function:

$$p'(t) = \frac{d}{dt}\left(400\left(\frac{t^2 + 1}{t^2 + 4}\right)\right)$$

$$= 400\frac{(t^2 + 4)(2t) - (t^2 + 1)(2t)}{(t^2 + 4)^2} \qquad \text{Quotient Rule}$$

$$= \frac{2400t}{(t^2 + 4)^2}. \qquad \text{Simplify.}$$

The growth rate has units of cells per hour; its graph is shown in Figure 3.32.

**b.** The growth rate $p'(t)$ has a maximum value at the point at which the population curve is steepest. Using a graphing utility, this point corresponds to $t \approx 1.15$ hr and the growth rate has a value of $p'(1.15) \approx 97$ cells/hr.

**c.** To determine whether the population approaches a fixed value after a long period of time (the steady-state population), we investigate the limit of the population function as $t \to \infty$. In this case, the steady-state population exists and is

$$\lim_{t \to \infty} p(t) = \lim_{t \to \infty} 400\left(\frac{t^2 + 1}{t^2 + 4}\right) = 400,$$

$$\underbrace{\phantom{\frac{t^2+1}{t^2+4}}}_{\text{approaches 1}}$$

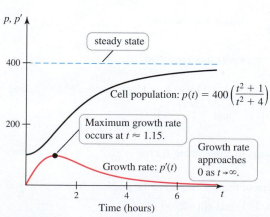

**Figure 3.32**

which is confirmed by the population curve in Figure 3.32. Notice that as the population approaches its steady state, the growth rate $p'$ approaches zero.

*Related Exercises 61–62* ◄

## Combining Derivative Rules

Some situations call for the use of multiple differentiation rules. This section concludes with one such example.

**EXAMPLE 6    Combining derivative rules**   Find the derivative of

$$y = \frac{4x(2x^3 - 3x^{-1})}{x^2 + 1}.$$

**SOLUTION**   In this case, we have the quotient of two functions, with a product in the numerator.

$$\frac{dy}{dx} = \frac{(x^2 + 1) \cdot \dfrac{d}{dx}(4x(2x^3 - 3x^{-1})) - (4x(2x^3 - 3x^{-1})) \cdot \dfrac{d}{dx}(x^2 + 1)}{(x^2 + 1)^2} \quad \text{Quotient Rule}$$

$$= \frac{(x^2 + 1)(4(2x^3 - 3x^{-1}) + 4x(6x^2 + 3x^{-2})) - (4x(2x^3 - 3x^{-1}))(2x)}{(x^2 + 1)^2}$$

Product Rule in the numerator

$$= \frac{8x(2x^4 + 4x^2 + 3)}{(x^2 + 1)^2} \quad \text{Simplify.}$$

*Related Exercises 43–44* ◄

---

## SECTION 3.4 EXERCISES

### Getting Started

1. How do you find the derivative of the product of two functions that are differentiable at a point?

2. How do you find the derivative of the quotient of two functions that are differentiable at a point?

3. Use the Product Rule to evaluate and simplify
   $$\frac{d}{dx}((x + 1)(3x + 2)).$$

4. Use the Product Rule to find $f'(1)$ given that $f(x) = x^4(1 + \sqrt{x})$.

5. Use the Quotient Rule to evaluate and simplify $\dfrac{d}{dx}\left(\dfrac{x - 1}{3x + 2}\right)$.

6. Use the Quotient Rule to find $g'(1)$ given that $g(x) = \dfrac{x^2}{x + 1}$.

**7–14.** *Find the derivative the following ways:*

a. *Using the Product Rule (Exercises 7–10) or the Quotient Rule (Exercises 11–14). Simplify your result.*

b. *By expanding the product first (Exercises 7–10) or by simplifying the quotient first (Exercises 11–14). Verify that your answer agrees with part (a).*

7. $f(x) = x(x - 1)$

8. $g(t) = (t + 1)(t^2 - t + 1)$

9. $f(x) = (x - 1)(3x + 4)$

10. $h(z) = (z^3 + 4z^2 + z)(z - 1)$

11. $f(t) = \dfrac{t^4 - 4t^2}{t^2}$

12. $g(s) = \dfrac{6s^3 - 15s^2 + 9s}{3s}$

13. $y = \dfrac{x^2 - a^2}{x - a}$, where $a$ is a constant

14. $y = \dfrac{ax^2 - x^3}{x - a}$, where $a$ is a constant

15. Given that $f(1) = 5$, $f'(1) = 4$, $g(1) = 2$, and $g'(1) = 3$, find
    $$\frac{d}{dx}(f(x)g(x))\Big|_{x=1} \text{ and } \frac{d}{dx}\left(\frac{f(x)}{g(x)}\right)\Big|_{x=1}.$$

16. Show two ways to differentiate $f(x) = \dfrac{1}{x^{10}}$.

17. Find the slope of the line tangent to the graph of $f(x) = \dfrac{x}{x + 6}$ at the point $(3, 1/3)$ and at $(-2, -1/2)$.

18. Given that $f(2) = 3$, $f'(2) = 4$, and $g(x) = xf(x)$, find $g(2)$ and $g'(2)$.

## Practice Exercises

**19–56. Derivatives** *Find and simplify the derivative of the following functions.*

19. $f(x) = 3x^4(2x^2 - 1)$

20. $g(x) = 6x - 2x(x^{10} - 3x^3)$

21. $f(x) = \dfrac{x}{x + 1}$

22. $f(x) = \dfrac{x^3 - 4x^2 + x}{x - 2}$

23. $f(t) = t^5(\sqrt{t} + 1)$

24. $g(w) = (w^8 + 3)(5w^2 + 3w + 1)$

25. $y = (3t - 1)(2t - 2)^{-1}$

26. $h(w) = \dfrac{w^2 - 1}{w^2 + 1}$

27. $g(w) = (w^3 + 4)(w^3 - 1)$

28. $s(t) = 4(3t^2 + 2t - 1)\sqrt{t}$

29. $h(x) = (x - 1)(x^3 + x^2 + x + 1)$

30. $f(x) = \left(1 + \dfrac{1}{x^2}\right)(x^2 + 1)$

31. $g(w) = \dfrac{w - 1}{\sqrt{w} + 1}$

32. $f(x) = \dfrac{2x + 1}{x - 1}$

33. $h(t) = \dfrac{t}{t^2 + 1}$

34. $f(t) = \dfrac{t + \sqrt{t} + 1}{t - \sqrt{t} - 1}$

35. $g(x) = \dfrac{x^4 + 1}{x^2 - 1}$

36. $y = (2\sqrt{x} - 1)(4x + 1)^{-1}$

37. $f(x) = 3x^{-9}$

38. $y = \dfrac{4}{p^3}$

39. $g(t) = 3t^2 + \dfrac{6}{t^7}$

40. $y = \dfrac{w^4 + 5w^2 + w}{w^2}$

41. $g(t) = \dfrac{t^3 + 3t^2 + t}{t^3}$

42. $p(x) = \dfrac{4x^3 + 3x + 1}{2x^5}$

43. $g(x) = \dfrac{x(3 - x)}{2x^2}$

44. $h(x) = \dfrac{(x - 1)(2x^2 - 1)}{(x^3 - 1)}$

45. $h(x) = \dfrac{4x}{(x^2 + x)(1 - x)}$

46. $h(x) = \dfrac{x + 1}{x^2(2x^3 + 1)}$

47. $g(w) = \dfrac{\sqrt{w} + w}{\sqrt{w} - w}$

48. $f(x) = \dfrac{4 - x^2}{x - 2}$

49. $h(w) = \dfrac{w^{5/3}}{w^{5/3} + 1}$

50. $g(x) = \dfrac{x^{4/3} - 1}{x^{4/3} + 1}$

51. $f(x) = 4x^2 - \dfrac{2x}{5x + 1}$

52. $f(w) = \dfrac{w + 2\sqrt{w} - 3}{\sqrt{w} + 3}$

53. $h(r) = \dfrac{2 - r - \sqrt{r}}{r + 1}$

54. $y = \dfrac{x - a}{\sqrt{x} - \sqrt{a}}$, where $a$ is a positive constant

55. $h(x) = (5x^7 + 5x)(6x^3 + 3x^2 + 3)$

56. $s(t) = (t + 1)(t + 2)(t + 3)$

**T 57–60. Equations of tangent lines**

**a.** *Find an equation of the line tangent to the given curve at a.*
**b.** *Use a graphing utility to graph the curve and the tangent line on the same set of axes.*

57. $y = \dfrac{x + 5}{x - 1}; a = 3$

58. $y = \dfrac{2x^2}{3x - 1}; a = 1$

59. $y = x(2x^{-2} + 1); a = -1$

60. $y = \dfrac{x - 2}{x + 1}; a = 1$

**T 61–62. Population growth** *Consider the following population functions.*

**a.** *Find the instantaneous growth rate of the population, for $t \geq 0$.*
**b.** *What is the instantaneous growth rate at $t = 5$?*
**c.** *Estimate the time when the instantaneous growth rate is greatest.*
**d.** *Evaluate and interpret $\lim\limits_{t \to \infty} p(t)$.*
**e.** *Use a graphing utility to graph the population and its growth rate.*

61. $p(t) = \dfrac{200t}{t + 2}$

62. $p(t) = 600\left(\dfrac{t^2 + 3}{t^2 + 9}\right)$

63. **Electrostatic force** The magnitude of the electrostatic force between two point charges $Q$ and $q$ of the same sign is given by $F(x) = \dfrac{kQq}{x^2}$, where $x$ is the distance (measured in meters) between the charges and $k = 9 \times 10^9$ N-m$^2$/C$^2$ is a physical constant (C stands for coulomb, the unit of charge; N stands for newton, the unit of force).

   **a.** Find the instantaneous rate of change of the force with respect to the distance between the charges.
   **b.** For two identical charges with $Q = q = 1$ C, what is the instantaneous rate of change of the force at a separation of $x = 0.001$ m?
   **c.** Does the magnitude of the instantaneous rate of change of the force increase or decrease with the separation? Explain.

64. **Gravitational force** The magnitude of the gravitational force between two objects of mass $M$ and $m$ is given by $F(x) = -\dfrac{GMm}{x^2}$, where $x$ is the distance between the centers of mass of the objects and $G = 6.7 \times 10^{-11}$ N-m$^2$/kg$^2$ is the gravitational constant (N stands for newton, the unit of force; the negative sign indicates an attractive force).

   **a.** Find the instantaneous rate of change of the force with respect to the distance between the objects.
   **b.** For two identical objects of mass $M = m = 0.1$ kg, what is the instantaneous rate of change of the force at a separation of $x = 0.01$ m?
   **c.** Does the magnitude of the instantaneous rate of change of the force increase or decrease with the separation? Explain.

65. **Explain why or why not** Determine whether the following statements are true and give an explanation or counterexample.

   **a.** $\dfrac{d}{dx}(\pi^5) = 5\pi^4$.

   **b.** The Quotient Rule must be used to evaluate $\dfrac{d}{dx}\left(\dfrac{x^2 + 3x + 2}{x}\right)$.

   **c.** $\dfrac{d}{dx}\left(\dfrac{1}{x^5}\right) = \dfrac{1}{5x^4}$.

**66–67. Higher-order derivatives** *Find $f'(x)$, $f''(x)$, and $f'''(x)$.*

66. $f(x) = \dfrac{1}{x}$

67. $f(x) = x^2(2 + x^{-3})$

**68–69. First and second derivatives** *Find $f'(x)$ and $f''(x)$.*

68. $f(x) = \dfrac{x}{x + 2}$

69. $f(x) = \dfrac{x^2 - 7x}{x + 1}$

**70. Tangent lines** Suppose $f(2) = 2$ and $f'(2) = 3$. Let $g(x) = x^2 f(x)$ and $h(x) = \dfrac{f(x)}{x - 3}$.

**a.** Find an equation of the line tangent to $y = g(x)$ at $x = 2$.

**b.** Find an equation of the line tangent to $y = h(x)$ at $x = 2$.

**T 71. The Witch of Agnesi** The graph of $y = \dfrac{a^3}{x^2 + a^2}$, where $a$ is a constant, is called the *witch of Agnesi* (named after the 18th-century Italian mathematician Maria Agnesi).

**a.** Let $a = 3$ and find an equation of the line tangent to

$$y = \frac{27}{x^2 + 9} \text{ at } x = 2.$$

**b.** Plot the function and the tangent line found in part (a).

**72–77. Derivatives from a table** *Use the following table to find the given derivatives.*

| $x$ | 1 | 2 | 3 | 4 |
|---|---|---|---|---|
| $f(x)$ | 5 | 4 | 3 | 2 |
| $f'(x)$ | 3 | 5 | 2 | 1 |
| $g(x)$ | 4 | 2 | 5 | 3 |
| $g'(x)$ | 2 | 4 | 3 | 1 |

**72.** $\dfrac{d}{dx}(f(x)g(x))\Big|_{x=1}$

**73.** $\dfrac{d}{dx}\left(\dfrac{f(x)}{g(x)}\right)\Big|_{x=2}$

**74.** $\dfrac{d}{dx}(xf(x))\Big|_{x=3}$

**75.** $\dfrac{d}{dx}\left(\dfrac{f(x)}{x+2}\right)\Big|_{x=4}$

**76.** $\dfrac{d}{dx}\left(\dfrac{xf(x)}{g(x)}\right)\Big|_{x=4}$

**77.** $\dfrac{d}{dx}\left(\dfrac{f(x)g(x)}{x}\right)\Big|_{x=4}$

**78–79. Flight formula for Indian spotted owlets** *The following table shows the average body mass $m(t)$ (in grams) and average wing chord length $w(t)$ (in millimeters), along with the derivatives $m'(t)$ and $w'(t)$, of $t$-week-old Indian spotted owlets. The **flight formula** function $f(t) = w(t)/m(t)$, which is the ratio of wing chord length to mass, is used to predict when these fledglings are old enough to fly. The values of $f$ are less than 1, but approach 1 as $t$ increases. When $f$ is close to 1, the fledglings are capable of flying, which is important for determining when rescued fledglings can be released back into the wild. (Source: ZooKeys, 132, 2011)*

| $t$ | $m(t)$ | $m'(t)$ | $w(t)$ | $w'(t)$ |
|---|---|---|---|---|
| 1 | 23.32 | 41.45 | 10.14 | 14.5 |
| 1.5 | 50.59 | 64.94 | 20.13 | 26.17 |
| 2 | 82.83 | 57.95 | 36.7 | 39.86 |
| 2.5 | 105.13 | 31.08 | 58.92 | 47.11 |
| 3 | 115.48 | 12.48 | 81.55 | 41.38 |
| 3.5 | 119.4 | 4.44 | 98.99 | 27.94 |
| 4 | 120.76 | 1.51 | 109.75 | 15.74 |
| 4.5 | 121.22 | 0.51 | 115.5 | 7.99 |
| 5 | 121.37 | 0.17 | 118.34 | 3.85 |
| 5.5 | 121.42 | 0.06 | 119.69 | 1.8 |
| 6 | 121.44 | 0.02 | 120.32 | 0.84 |
| 6.5 | 121.45 | 0.01 | 120.61 | 0.38 |

**78.** State the units associated with the following derivatives and state the physical meaning of each derivative.

**a.** $m'(t)$    **b.** $w'(t)$    **c.** $f'(t)$

**T 79.** Complete the following steps to examine the behavior of the flight formula.

**a.** Sketch an approximate graph of $y = f(t)$ by plotting and connecting the points $(1, f(1)), (1.5, f(1.5)), \ldots, (6.5, f(6.5))$ with a smooth curve.

**b.** For what value of $t$ does $f$ appear to be changing most rapidly? Round $t$ to the nearest whole number.

**c.** For the value of $t$ found in part (b), use the table and the Quotient Rule to find $f'(t)$. Describe what is happening to the bird at this stage in its life.

**d.** Use your graph of $f$ to predict what happens to $f'(t)$ as $t$ grows larger and confirm your prediction by evaluating $f'(6.5)$ using the Quotient Rule. Describe what is happening in the physical development of the fledglings as $t$ grows larger.

**80–81.** *Assume both the graphs of $f$ and $g$ pass through the point $(3, 2)$, $f'(3) = 5$, and $g'(3) = -10$. If $p(x) = f(x)g(x)$ and $q(x) = f(x)/g(x)$, find the following derivatives.*

**80.** $p'(3)$

**81.** $q'(3)$

**82.** Given that $f(1) = 2$ and $f'(1) = 2$, find the slope of the curve $y = xf(x)$ at the point $(1, 2)$.

**83–86. Derivatives from graphs** *Use the figure to find the following derivatives.*

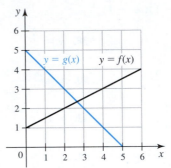

**83.** $\dfrac{d}{dx}(f(x)g(x))\Big|_{x=4}$

**84.** $\dfrac{d}{dx}\left(\dfrac{f(x)}{g(x)}\right)\Big|_{x=4}$

**85.** $\dfrac{d}{dx}(xg(x))\Big|_{x=2}$

**86.** $\dfrac{d}{dx}\left(\dfrac{x^2}{f(x)}\right)\Big|_{x=2}$

**87. Tangent lines** The line tangent to the curve $y = h(x)$ at $x = 4$ is $y = -3x + 14$. Find an equation of the line tangent to the following curves at $x = 4$.

**a.** $y = (x^2 - 3x)h(x)$    **b.** $y = h(x)/(x + 2)$

**88. Derivatives from tangent lines** Suppose the line tangent to the graph of $f$ at $x = 2$ is $y = 4x + 1$ and suppose $y = 3x - 2$ is the line tangent to the graph of $g$ at $x = 2$. Find an equation of the line tangent to the following curves at $x = 2$.

**a.** $y = f(x)g(x)$    **b.** $y = f(x)/g(x)$

## Explorations and Challenges

**89. Avoiding tedious work** Given that

$$q(x) = \frac{5x^8 + 6x^5 + 5x^4 + 3x^2 + 20x + 100}{10x^{10} + 8x^9 + 6x^5 + 6x^2 + 4x + 2}, \text{ find } q'(0)$$

without computing $q'(x)$. (*Hint:* Evaluate $f(0)$, $f'(0)$, $g(0)$, and $g'(0)$ where $f$ is the numerator of $q$ and $g$ is the denominator of $q$.)

**90.** Given that $p(x) = (10x^5 + 20x^3 + 100x^2 + 5x + 20) \cdot (10x^5 + 40x^3 + 20x^2 + 4x + 10)$, find $p'(0)$ without computing $p'(x)$.

**91. Means and tangents** Suppose $f$ is differentiable on an interval containing $a$ and $b$, and let $P(a, f(a))$ and $Q(b, f(b))$ be distinct points on the graph of $f$. Let $c$ be the $x$-coordinate of the point at which the lines tangent to the curve at $P$ and $Q$ intersect, assuming the tangent lines are not parallel (see figure).

a. If $f(x) = x^2$, show that $c = (a + b)/2$, the arithmetic mean of $a$ and $b$, for real numbers $a$ and $b$.
b. If $f(x) = \sqrt{x}$, show that $c = \sqrt{ab}$, the geometric mean of $a$ and $b$, for $a > 0$ and $b > 0$.
c. If $f(x) = 1/x$, show that $c = 2ab/(a + b)$, the harmonic mean of $a$ and $b$, for $a > 0$ and $b > 0$.
d. Find an expression for $c$ in terms of $a$ and $b$ for any (differentiable) function $f$ whenever $c$ exists.

**92. Proof of the Quotient Rule** Let $F = f/g$ be the quotient of two functions that are differentiable at $x$.

a. Use the definition of $F'$ to show that
$$\frac{d}{dx}\left(\frac{f(x)}{g(x)}\right) = \lim_{h \to 0} \frac{f(x + h)g(x) - f(x)g(x + h)}{hg(x + h)g(x)}.$$
b. Now add $-f(x)g(x) + f(x)g(x)$ (which equals 0) to the numerator in the preceding limit to obtain
$$\lim_{h \to 0} \frac{f(x + h)g(x) - f(x)g(x) + f(x)g(x) - f(x)g(x + h)}{hg(x + h)g(x)}.$$
Use this limit to obtain the Quotient Rule.
c. Explain why $F' = (f/g)'$ exists, whenever $g(x) \neq 0$.

**93. Product Rule for the second derivative** Assuming the first and second derivatives of $f$ and $g$ exist at $x$, find a formula for $d^2/dx^2(f(x)g(x))$.

**94. One of the Leibniz Rules** One of several Leibniz Rules in calculus deals with higher-order derivatives of products. Let $(fg)^{(n)}$ denote the $n$th derivative of the product $fg$, for $n \geq 1$.

a. Prove that $(fg)^{(2)} = f''g + 2f'g' + fg''$.
b. Prove that, in general,
$$(fg)^{(n)} = \sum_{k=0}^{n} \binom{n}{k} f^{(k)}g^{(n-k)},$$
where $\binom{n}{k} = \dfrac{n!}{k!(n - k)!}$ are the binomial coefficients.
c. Compare the result of (b) to the expansion of $(a + b)^n$.

**95. Product Rule for three functions** Assume $f$, $g$, and $h$ are differentiable at $x$.

a. Use the Product Rule (twice) to find a formula for $d/dx(f(x)g(x)h(x))$.
b. Use the formula in (a) to find $d/dx(\sqrt{x}(x - 1)(x + 3))$.

**QUICK CHECK ANSWERS**

**1.** $f'(x) = 5x^4$ by either method   **2.** $f'(x) = 5x^4$ by either method   **3.** $f'(x) = -5x^{-6}$ by either method ◄

# 3.5 Derivatives of Trigonometric Functions

> Results stated in this section assume angles are measured in *radians*.

From variations in market trends and ocean temperatures to daily fluctuations in tides and hormone levels, change is often cyclical or periodic. Trigonometric functions are well suited for describing such cyclical behavior. In this section, we investigate the derivatives of trigonometric functions and their many uses.

## Two Special Limits

Our principal goal is to determine derivative formulas for $\sin x$ and $\cos x$. To do this, we use two special limits.

**Table 3.1**

| $x$ | $\dfrac{\sin x}{x}$ |
|---|---|
| $\pm 0.1$ | 0.9983341665 |
| $\pm 0.01$ | 0.9999833334 |
| $\pm 0.001$ | 0.9999998333 |

---

**THEOREM 3.9   Trigonometric Limits**

$$\lim_{x \to 0} \frac{\sin x}{x} = 1 \qquad \lim_{x \to 0} \frac{\cos x - 1}{x} = 0$$

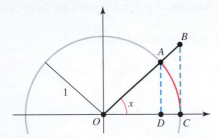

The graph of $y = \dfrac{\sin x}{x}$ suggests that $\lim\limits_{x \to 0} \dfrac{\sin x}{x} = 1$.

**Figure 3.33**

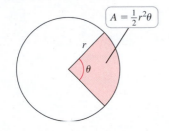

**Figure 3.34**

➤ Area of a sector of a circle of radius $r$ formed by a central angle $\theta$:

$$A = \tfrac{1}{2} r^2 \theta$$

➤ $\lim\limits_{x \to 0} \dfrac{\sin x}{x} = 1$ implies that if $|x|$ is small, then $\sin x \approx x$.

Note that these limits cannot be evaluated by direct substitution because in both cases, the numerator and denominator approach zero as $x \to 0$. We first examine numerical and graphical evidence supporting Theorem 3.9, and then we offer an analytic proof.

The values of $\dfrac{\sin x}{x}$, rounded to 10 digits, appear in Table 3.1. As $x$ approaches zero from both sides, it appears that $\dfrac{\sin x}{x}$ approaches 1. **Figure 3.33** shows a graph of $y = \dfrac{\sin x}{x}$, with a hole at $x = 0$, where the function is undefined. The graphical evidence also strongly suggests (but does not prove) that $\lim\limits_{x \to 0} \dfrac{\sin x}{x} = 1$. Similar evidence indicates that $\dfrac{\cos x - 1}{x}$ approaches 0 as $x$ approaches 0.

Using a geometric argument and the methods of Chapter 2, we now prove that $\lim\limits_{x \to 0} \dfrac{\sin x}{x} = 1$. The proof that $\lim\limits_{x \to 0} \dfrac{\cos x - 1}{x} = 0$ is found in Exercise 81.

**Proof:** Consider **Figure 3.34**, in which $\triangle OAD$, $\triangle OBC$, and the sector $OAC$ of the unit circle (with central angle $x$) are shown. Observe that with $0 < x < \pi/2$,

$$\text{area of } \triangle OAD < \text{area of sector } OAC < \text{area of } \triangle OBC. \tag{1}$$

Because the circle in Figure 3.34 is a unit circle, $OA = OC = 1$. It follows that $\sin x = \dfrac{AD}{OA} = AD$, $\cos x = \dfrac{OD}{OA} = OD$, and $\tan x = \dfrac{BC}{OC} = BC$. From these observations, we conclude that

• the area of $\triangle OAD = \dfrac{1}{2}(OD)(AD) = \dfrac{1}{2}\cos x \sin x$,

• the area of sector $OAC = \dfrac{1}{2} \cdot 1^2 \cdot x = \dfrac{x}{2}$, and

• the area of $\triangle OBC = \dfrac{1}{2}(OC)(BC) = \dfrac{1}{2}\tan x$.

Substituting these results into (1), we have

$$\frac{1}{2}\cos x \sin x < \frac{x}{2} < \frac{1}{2}\tan x.$$

Replacing $\tan x$ with $\dfrac{\sin x}{\cos x}$ and multiplying the inequalities by $\dfrac{2}{\sin x}$ (which is positive) leads to the inequalities

$$\cos x < \frac{x}{\sin x} < \frac{1}{\cos x}.$$

When we take reciprocals and reverse the inequalities, we have

$$\cos x < \frac{\sin x}{x} < \frac{1}{\cos x}, \tag{2}$$

for $0 < x < \pi/2$.

A similar argument shows that the inequalities in (2) also hold for $-\pi/2 < x < 0$. Taking the limit as $x \to 0$ in (2), we find that

$$\underbrace{\lim_{x \to 0} \cos x}_{1} \leq \lim_{x \to 0} \frac{\sin x}{x} \leq \underbrace{\lim_{x \to 0} \frac{1}{\cos x}}_{1}.$$

The Squeeze Theorem (Theorem 2.5) now implies that $\lim\limits_{x \to 0} \dfrac{\sin x}{x} = 1$.    ◄

**EXAMPLE 1**  **Calculating trigonometric limits**  Evaluate the following limits.

**a.** $\lim\limits_{x \to 0} \dfrac{\sin 4x}{x}$       **b.** $\lim\limits_{x \to 0} \dfrac{\sin 3x}{\sin 5x}$

**SOLUTION**

**a.** To use the fact that $\lim\limits_{x \to 0} \dfrac{\sin x}{x} = 1$, the argument of the sine function in the numerator must be the same as in the denominator. Multiplying and dividing $\dfrac{\sin 4x}{x}$ by 4, we evaluate the limit as follows:

$$\lim_{x \to 0} \frac{\sin 4x}{x} = \lim_{x \to 0} \frac{4 \sin 4x}{4x} \qquad \text{\color{blue}Multiply and divide by 4.}$$

$$= 4 \lim_{t \to 0} \underbrace{\frac{\sin t}{t}}_{1} \qquad \text{\color{blue}Factor out 4 and let } t = 4x; t \to 0 \text{ as } x \to 0.$$

$$= 4(1) = 4. \qquad \text{\color{blue}Theorem 3.9}$$

**b.** The first step is to divide the numerator and denominator of $\dfrac{\sin 3x}{\sin 5x}$ by $x$:

$$\frac{\sin 3x}{\sin 5x} = \frac{(\sin 3x)/x}{(\sin 5x)/x}.$$

As in part (a), we now divide and multiply $\dfrac{\sin 3x}{x}$ by 3 and divide and multiply $\dfrac{\sin 5x}{x}$ by 5. In the numerator, we let $t = 3x$, and in the denominator, we let $u = 5x$. In each case, $t \to 0$ and $u \to 0$ as $x \to 0$. Therefore,

$$\lim_{x \to 0} \frac{\sin 3x}{\sin 5x} = \lim_{x \to 0} \frac{\dfrac{3 \sin 3x}{3x}}{\dfrac{5 \sin 5x}{5x}} \qquad \text{\color{blue}Multiply and divide by 3 and 5.}$$

$$= \frac{3}{5} \frac{\lim\limits_{t \to 0} (\sin t)/t}{\lim\limits_{u \to 0} (\sin u)/u} \qquad \text{\color{blue}Let } t = 3x \text{ in numerator and } u = 5x \text{ in denominator.}$$

$$= \frac{3}{5} \cdot \frac{1}{1} = \frac{3}{5}. \qquad \text{\color{blue}Both limits equal 1.}$$

*Related Exercises 12–13* ◄

**QUICK CHECK 1**  Evaluate $\lim\limits_{x \to 0} \dfrac{\tan 2x}{x}$. ◄

We now use the important limits of Theorem 3.9 to establish the derivatives of $\sin x$ and $\cos x$.

## Derivatives of Sine and Cosine Functions

We start with the definition of the derivative,

$$f'(x) = \lim_{h \to 0} \frac{f(x + h) - f(x)}{h},$$

with $f(x) = \sin x$, and then appeal to the sine addition identity

$$\sin(x + h) = \sin x \cos h + \cos x \sin h.$$

Horizontal tangent lines for $f(x) = \sin x$ occur at the zeros of $f'(x) = \cos x$.

(a)

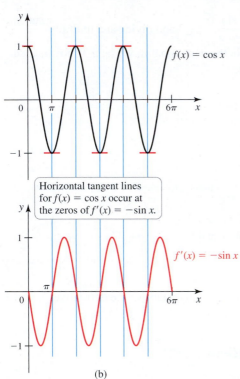

Horizontal tangent lines for $f(x) = \cos x$ occur at the zeros of $f'(x) = -\sin x$.

(b)

**Figure 3.35**

**QUICK CHECK 2** At what points on the interval $[0, 2\pi]$ does the graph of $f(x) = \sin x$ have tangent lines with positive slopes? At what points on the interval $[0, 2\pi]$ is $\cos x > 0$? Explain the connection. ◄

The derivative is

$$f'(x) = \lim_{h \to 0} \frac{\sin(x + h) - \sin x}{h} \qquad \text{Definition of derivative}$$

$$= \lim_{h \to 0} \frac{\sin x \cos h + \cos x \sin h - \sin x}{h} \qquad \text{Sine addition identity}$$

$$= \lim_{h \to 0} \frac{\sin x (\cos h - 1) + \cos x \sin h}{h} \qquad \text{Factor } \sin x.$$

$$= \lim_{h \to 0} \frac{\sin x (\cos h - 1)}{h} + \lim_{h \to 0} \frac{\cos x \sin h}{h} \qquad \text{Theorem 2.3}$$

$$= \sin x \left( \underbrace{\lim_{h \to 0} \frac{\cos h - 1}{h}}_{0} \right) + \cos x \left( \underbrace{\lim_{h \to 0} \frac{\sin h}{h}}_{1} \right) \qquad \begin{array}{l}\text{Both } \sin x \text{ and } \cos x \text{ are} \\ \text{independent of } h.\end{array}$$

$$= (\sin x)(0) + \cos x (1) \qquad \text{Theorem 3.9}$$

$$= \cos x. \qquad \text{Simplify.}$$

We have proved the important result that $\dfrac{d}{dx}(\sin x) = \cos x$.

The fact that $\dfrac{d}{dx}(\cos x) = -\sin x$ is proved in a similar way using a cosine addition identity (Exercise 83).

---

**THEOREM 3.10    Derivatives of Sine and Cosine**

$$\frac{d}{dx}(\sin x) = \cos x \qquad \frac{d}{dx}(\cos x) = -\sin x$$

---

From a geometric point of view, these derivative formulas make sense. Because $f(x) = \sin x$ is a periodic function, we expect its derivative to be periodic. Observe that the horizontal tangent lines on the graph of $f(x) = \sin x$ (**Figure 3.35a**) occur at the zeros of $f'(x) = \cos x$. Similarly, the horizontal tangent lines on the graph of $f(x) = \cos x$ occur at the zeros of $f'(x) = -\sin x$ (**Figure 3.35b**).

**EXAMPLE 2    Derivatives involving trigonometric functions** Calculate $dy/dx$ for the following functions.

**a.** $y = x^2 \cos x$    **b.** $y = \sin x - x \cos x$    **c.** $y = \dfrac{1 + \sin x}{1 - \sin x}$

**SOLUTION**

**a.** $\dfrac{dy}{dx} = \dfrac{d}{dx}(x^2 \cos x) = \overbrace{2x \cos x}^{\substack{\text{(derivative of} \\ x^2) \cdot \cos x}} + \overbrace{x^2(-\sin x)}^{\substack{x^2 \cdot \text{(derivative} \\ \text{of } \cos x)}}$    Product Rule

$\qquad\qquad = x(2 \cos x - x \sin x)$    Simplify.

**b.** $\dfrac{dy}{dx} = \dfrac{d}{dx}(\sin x) - \dfrac{d}{dx}(x \cos x)$    Difference Rule

$\qquad = \cos x - (\underbrace{(1) \cos x}_{\substack{\text{(derivative of } x) \\ \cdot \cos x}} + \underbrace{x(-\sin x)}_{\substack{x \cdot \text{(derivative} \\ \text{of } \cos x)}})$    Product Rule

$\qquad = x \sin x$    Simplify.

c. $\dfrac{dy}{dx} = \dfrac{(1 - \sin x)(\cos x) - (1 + \sin x)(-\cos x)}{(1 - \sin x)^2}$ — Quotient Rule

$= \dfrac{\cos x - \cos x \sin x + \cos x + \sin x \cos x}{(1 - \sin x)^2}$ — Expand.

$= \dfrac{2 \cos x}{(1 - \sin x)^2}$ — Simplify.

*Related Exercises 25, 27, 29* ◄

## Derivatives of Other Trigonometric Functions

The derivatives of $\tan x$, $\cot x$, $\sec x$, and $\csc x$ are obtained using the derivatives of $\sin x$ and $\cos x$ together with the Quotient Rule and trigonometric identities.

➤ Recall that $\tan x = \dfrac{\sin x}{\cos x}$, $\cot x = \dfrac{\cos x}{\sin x}$, $\sec x = \dfrac{1}{\cos x}$, and $\csc x = \dfrac{1}{\sin x}$.

**EXAMPLE 3** **Derivative of the tangent function** Calculate $\dfrac{d}{dx}(\tan x)$.

**SOLUTION** Using the identity $\tan x = \dfrac{\sin x}{\cos x}$ and the Quotient Rule, we have

$\dfrac{d}{dx}(\tan x) = \dfrac{d}{dx}\left(\dfrac{\sin x}{\cos x}\right)$

$= \dfrac{\cos x \cos x - \sin x(-\sin x)}{\cos^2 x}$ — Quotient Rule

$= \dfrac{\cos^2 x + \sin^2 x}{\cos^2 x}$ — Simplify numerator.

$= \dfrac{1}{\cos^2 x} = \sec^2 x.$ — $\cos^2 x + \sin^2 x = 1$

Therefore, $\dfrac{d}{dx}(\tan x) = \sec^2 x.$

*Related Exercises 52–54* ◄

➤ One way to remember Theorem 3.11 is to learn the derivatives of the sine, tangent, and secant functions. Then replace each function with its corresponding **cofunction** and put a negative sign on the right-hand side of the new derivative formula.

$\dfrac{d}{dx}(\sin x) = \cos x \quad \leftrightarrow$

$\dfrac{d}{dx}(\cos x) = -\sin x$

$\dfrac{d}{dx}(\tan x) = \sec^2 x \quad \leftrightarrow$

$\dfrac{d}{dx}(\cot x) = -\csc^2 x$

$\dfrac{d}{dx}(\sec x) = \sec x \tan x \quad \leftrightarrow$

$\dfrac{d}{dx}(\csc x) = -\csc x \cot x$

The derivatives of $\cot x$, $\sec x$, and $\csc x$ are given in Theorem 3.11 (Exercises 52–54).

**QUICK CHECK 3** The formulas for $\dfrac{d}{dx}(\cot x)$, $\dfrac{d}{dx}(\sec x)$, and $\dfrac{d}{dx}(\csc x)$ can be determined using the Quotient Rule. Why? ◄

---

**THEOREM 3.11** **Derivatives of the Trigonometric Functions**

$\dfrac{d}{dx}(\sin x) = \cos x \qquad \dfrac{d}{dx}(\cos x) = -\sin x$

$\dfrac{d}{dx}(\tan x) = \sec^2 x \qquad \dfrac{d}{dx}(\cot x) = -\csc^2 x$

$\dfrac{d}{dx}(\sec x) = \sec x \tan x \qquad \dfrac{d}{dx}(\csc x) = -\csc x \cot x$

**EXAMPLE 4** **Derivatives involving sec $x$ and csc $x$** Find the derivative of $y = \sec x \csc x$.

**SOLUTION**

$$\frac{dy}{dx} = \frac{d}{dx}(\sec x \cdot \csc x)$$

$$= \underbrace{\sec x \tan x \csc x}_{\text{derivative of } \sec x} + \sec x \underbrace{(-\csc x \cot x)}_{\text{derivative of } \csc x} \quad \text{Product Rule}$$

$$= \underbrace{\frac{1}{\cos x}}_{\sec x} \cdot \underbrace{\frac{\sin x}{\cos x}}_{\tan x} \cdot \underbrace{\frac{1}{\sin x}}_{\csc x} - \underbrace{\frac{1}{\cos x}}_{\sec x} \cdot \underbrace{\frac{1}{\sin x}}_{\csc x} \cdot \underbrace{\frac{\cos x}{\sin x}}_{\cot x} \quad \begin{array}{l}\text{Write functions in terms}\\\text{of } \sin x \text{ and } \cos x.\end{array}$$

$$= \frac{1}{\cos^2 x} - \frac{1}{\sin^2 x} \quad \text{Cancel and simplify.}$$

$$= \sec^2 x - \csc^2 x \quad \text{Definition of } \sec x \text{ and } \csc x$$

*Related Exercises 43–44* ◄

## Higher-Order Trigonometric Derivatives

Higher-order derivatives of the sine and cosine functions are important in many applications, particularly in problems that involve oscillations, vibrations, or waves. A few higher-order derivatives of $y = \sin x$ reveal a pattern.

$$\frac{dy}{dx} = \cos x \qquad\qquad \frac{d^2 y}{dx^2} = \frac{d}{dx}(\cos x) = -\sin x$$

$$\frac{d^3 y}{dx^3} = \frac{d}{dx}(-\sin x) = -\cos x \qquad\qquad \frac{d^4 y}{dx^4} = \frac{d}{dx}(-\cos x) = \sin x$$

We see that the higher-order derivatives of $\sin x$ cycle back periodically to $\pm \sin x$. In general, it can be shown that $\frac{d^{2n} y}{dx^{2n}} = (-1)^n \sin x$, with a similar result for $\cos x$ (Exercise 88). This cyclic behavior in the derivatives of $\sin x$ and $\cos x$ does not occur with the other trigonometric functions.

**QUICK CHECK 4** Find $\dfrac{d^2 y}{dx^2}$ and $\dfrac{d^4 y}{dx^4}$ when $y = \cos x$. Find $\dfrac{d^{40} y}{dx^{40}}$ and $\dfrac{d^{42} y}{dx^{42}}$ when $y = \sin x$. ◄

**EXAMPLE 5** **Second-order derivatives** Find the second derivative of $y = \csc x$.

**SOLUTION** By Theorem 3.11, $\dfrac{dy}{dx} = -\csc x \cot x$.

Applying the Product Rule gives the second derivative:

$$\frac{d^2 y}{dx^2} = \frac{d}{dx}(-\csc x \cot x)$$

$$= \left(\frac{d}{dx}(-\csc x)\right)\cot x - \csc x \frac{d}{dx}(\cot x) \quad \text{Product Rule}$$

$$= (\csc x \cot x)\cot x - \csc x(-\csc^2 x) \quad \text{Calculate derivatives.}$$

$$= \csc x(\cot^2 x + \csc^2 x). \quad \text{Factor.}$$

*Related Exercises 61–62* ◄

---

## SECTION 3.5 EXERCISES

### Getting Started

1. Why is it not possible to evaluate $\lim\limits_{x \to 0} \dfrac{\sin x}{x}$ by direct substitution?

2. How is $\lim\limits_{x \to 0} \dfrac{\sin x}{x}$ used in this section?

3. Explain why the Quotient Rule is used to determine the derivative of $\tan x$ and $\cot x$.

4. How can you use the derivatives $\dfrac{d}{dx}(\sin x) = \cos x$, $\dfrac{d}{dx}(\tan x) = \sec^2 x$, and $\dfrac{d}{dx}(\sec x) = \sec x \tan x$ to remember the derivatives of $\cos x$, $\cot x$, and $\csc x$?

5.  Let $f(x) = \sin x$. What is the value of $f'(\pi)$?

6.  Find the value of $\dfrac{d}{dx}(\tan x)\Big|_{x=\frac{\pi}{3}}$.

7.  Find an equation of the line tangent to the curve $y = \sin x$ at $x = 0$.

8.  Where does $\sin x$ have a horizontal tangent line? Where does $\cos x$ have a value of zero? Explain the connection between these two observations.

9.  Find $\dfrac{d^2}{dx^2}(\sin x + \cos x)$.    10. Find $\dfrac{d^2}{dx^2}(\sec x)$.

## Practice Exercises

**11–22. Trigonometric limits** *Use Theorem 3.9 to evaluate the following limits.*

11. $\displaystyle\lim_{x\to 0} \frac{\sin 3x}{x}$    12. $\displaystyle\lim_{x\to 0} \frac{\sin 5x}{3x}$

13. $\displaystyle\lim_{x\to 0} \frac{\sin 7x}{\sin 3x}$    14. $\displaystyle\lim_{x\to 0} \frac{\sin 3x}{\tan 4x}$

15. $\displaystyle\lim_{x\to 0} \frac{\tan 5x}{x}$    16. $\displaystyle\lim_{\theta\to 0} \frac{\cos^2\theta - 1}{\theta}$

17. $\displaystyle\lim_{x\to 0} \frac{\tan 7x}{\sin x}$    18. $\displaystyle\lim_{\theta\to 0} \frac{\sec\theta - 1}{\theta}$

19. $\displaystyle\lim_{x\to 2} \frac{\sin(x-2)}{x^2 - 4}$    20. $\displaystyle\lim_{x\to -3} \frac{\sin(x+3)}{x^2 + 8x + 15}$

21. $\displaystyle\lim_{x\to 0} \frac{\sin ax}{\sin bx}$, where $a$ and $b$ are constants with $b \ne 0$

22. $\displaystyle\lim_{x\to 0} \frac{\sin ax}{bx}$, where $a$ and $b$ are constants with $b \ne 0$

**23–51. Calculating derivatives** *Find the derivative of the following functions.*

23. $y = \sin x + \cos x$    24. $y = 5x^2 + \cos x$

25. $y = 3x^4 \sin x$    26. $y = \sin x + \dfrac{4\cos x}{x}$

27. $y = x \sin x$    28. $y = \dfrac{x}{\sin x + 1}$

29. $y = \dfrac{\cos x}{\sin x + 1}$    30. $y = \dfrac{\sin x}{1 + \cos x}$

31. $y = \sin x \cos x$

32. $y = \dfrac{a \sin x + b \cos x}{a \sin x - b \cos x}$; $a$ and $b$ are nonzero constants

33. $y = \cos^2 x$    34. $y = \dfrac{x \sin x}{1 + \cos x}$

35. $y = w^2 \sin w + 2w \cos w - 2 \sin w$

36. $y = -x^3 \cos x + 3x^2 \sin x + 6x \cos x - 6 \sin x$

37. $y = x \cos x \sin x$    38. $y = \dfrac{1}{2 + \sin x}$

39. $y = \dfrac{1 - \cos x}{1 + \cos x}$    40. $y = \dfrac{1 - \sin x}{1 + \sin x}$

41. $y = \dfrac{\sin x \cos x}{4 + \sin x \cos x}$    42. $y = \tan x + \cot x$

43. $y = \sec x + \csc x$    44. $y = \sec x \tan x$

45. $y = \sqrt{x} \csc x$    46. $y = \dfrac{\tan w}{1 + \tan w}$

47. $y = \dfrac{\cot x}{1 + \csc x}$    48. $y = \dfrac{\tan t}{1 + \sec t}$

49. $y = \tan x \sin x$    50. $y = \csc^2\theta - 1$

51. $y = x - \cos x \sin x$

**52–54. Verifying derivative formulas** *Verify the following derivative formulas using the Quotient Rule.*

52. $\dfrac{d}{dx}(\cot x) = -\csc^2 x$    53. $\dfrac{d}{dx}(\sec x) = \sec x \tan x$

54. $\dfrac{d}{dx}(\csc x) = -\csc x \cot x$

**T 55. Velocity of an oscillator** An object oscillates along a vertical line, and its position in centimeters is given by $y(t) = 30(\sin t - 1)$, where $t \ge 0$ is measured in seconds and $y$ is positive in the upward direction.

   a. Graph the position function, for $0 \le t \le 10$.
   b. Find the velocity of the oscillator, $v(t) = y'(t)$.
   c. Graph the velocity function, for $0 \le t \le 10$.
   d. At what times and positions is the velocity zero?
   e. At what times and positions is the velocity a maximum?
   f. The acceleration of the oscillator is $a(t) = v'(t)$. Find and graph the acceleration function.

**T 56. Resonance** An oscillator (such as a mass on a spring or a component in an electrical circuit) subject to external forces that have the same frequency as the oscillator itself may undergo motion called *resonance* (at least for short periods of time). The position function of an oscillator in resonance has the form $y(t) = At \sin t$, where $A$ is a constant.

   a. Graph the position function with $A = \frac{1}{2}$, for $0 \le t \le 20$. How does the amplitude of the oscillation (the height of the peaks) change as $t$ increases?
   b. Compute and graph the velocity of the object, $v(t) = y'(t)$ (with $A = \frac{1}{2}$), for $0 \le t \le 20$.
   c. Where do the zeros of the velocity function appear relative to the peaks and valleys of the position function?
   d. If the oscillator were a suspension bridge, explain why resonance could be catastrophic.

**57–64. Second derivatives** *Find $y''$ for the following functions.*

57. $y = x \sin x$    58. $y = x^2 \cos x$

59. $y = \dfrac{\sin x}{x}$    60. $y = \dfrac{x}{\cos x}$

61. $y = \cot x$    62. $y = \tan x$

63. $y = \sec x \csc x$    64. $y = \cos\theta \sin\theta$

65. **Explain why or why not** Determine whether the following statements are true and give an explanation or counterexample.

   a. $\dfrac{d}{dx}(\sin^2 x) = \cos^2 x$.    b. $\dfrac{d^2}{dx^2}(\sin x) = \sin x$.

   c. $\dfrac{d^4}{dx^4}(\cos x) = \cos x$.

   d. The function $\sec x$ is not differentiable at $x = \pi/2$.

**66–71. Trigonometric limits** *Evaluate the following limits or state that they do not exist. (Hint: Identify each limit as the derivative of a function at a point.)*

66. $\displaystyle\lim_{x\to \pi/2} \frac{\cos x}{x - \pi/2}$    67. $\displaystyle\lim_{x\to \pi/4} \frac{\tan x - 1}{x - \pi/4}$

**68.** $\displaystyle\lim_{h\to 0}\dfrac{\sin\left(\dfrac{\pi}{6}+h\right)-\dfrac{1}{2}}{h}$

**69.** $\displaystyle\lim_{h\to 0}\dfrac{\cos\left(\dfrac{\pi}{6}+h\right)-\dfrac{\sqrt{3}}{2}}{h}$

**70.** $\displaystyle\lim_{x\to\pi/4}\dfrac{\cot x-1}{x-\dfrac{\pi}{4}}$

**71.** $\displaystyle\lim_{h\to 0}\dfrac{\tan\left(\dfrac{5\pi}{6}+h\right)+\dfrac{1}{\sqrt{3}}}{h}$

**T 72–75. Equations of tangent lines**

**a.** *Find an equation of the line tangent to the following curves at the given value of x.*

**b.** *Use a graphing utility to plot the curve and the tangent line.*

**72.** $y=4\sin x\cos x;\ x=\dfrac{\pi}{3}$     **73.** $y=1+2\sin x;\ x=\dfrac{\pi}{6}$

**74.** $y=\csc x;\ x=\dfrac{\pi}{4}$     **75.** $y=\dfrac{\cos x}{1-\cos x};\ x=\dfrac{\pi}{3}$

**76. Locations of tangent lines**

   **a.** For what values of $x$ does $g(x)=x-\sin x$ have a horizontal tangent line?

   **b.** For what values of $x$ does $g(x)=x-\sin x$ have a slope of 1?

**77. Locations of horizontal tangent lines** For what values of $x$ does $f(x)=x-2\cos x$ have a horizontal tangent line?

**78. Matching** Match the graphs of the functions in a–d with the graphs of their derivatives in A–D.

Functions a–d

(a)

(b)

(c)

(d)

Derivatives A–D

(A)

(B)

(C)

(D)

**79. A differential equation** A differential equation is an equation involving an unknown function and its derivatives. Consider the differential equation $y''(t)+y(t)=0$.

   **a.** Show that $y=A\sin t$ satisfies the equation for any constant $A$.

   **b.** Show that $y=B\cos t$ satisfies the equation for any constant $B$.

   **c.** Show that $y=A\sin t+B\cos t$ satisfies the equation for any constants $A$ and $B$.

## Explorations and Challenges

**80. Using identities** Use the identity $\sin 2x=2\sin x\cos x$ to find $\dfrac{d}{dx}(\sin 2x)$. Then use the identity $\cos 2x=\cos^2 x-\sin^2 x$ to express the derivative of $\sin 2x$ in terms of $\cos 2x$.

**81. Proof of** $\displaystyle\lim_{x\to 0}\dfrac{\cos x-1}{x}=0$ Use the trigonometric identity $\cos^2 x+\sin^2 x=1$ to prove that $\displaystyle\lim_{x\to 0}\dfrac{\cos x-1}{x}=0$. (*Hint:* Begin by multiplying the numerator and denominator by $\cos x+1$.)

**82. Another method for proving** $\displaystyle\lim_{x\to 0}\dfrac{\cos x-1}{x}=0$ Use the half-angle formula $\sin^2 x=\dfrac{1-\cos 2x}{2}$ to prove that $\displaystyle\lim_{x\to 0}\dfrac{\cos x-1}{x}=0$.

**83. Proof of** $\dfrac{d}{dx}(\cos x)=-\sin x$ Use the definition of the derivative and the trigonometric identity

$$\cos(x+h)=\cos x\cos h-\sin x\sin h$$

to prove that $\dfrac{d}{dx}(\cos x)=-\sin x$.

**84. Continuity of a piecewise function** Let

$$f(x)=\begin{cases}\dfrac{3\sin x}{x} & \text{if } x\neq 0\\[2mm] a & \text{if } x=0.\end{cases}$$

For what values of $a$ is $f$ continuous?

**85. Continuity of a piecewise function** Let

$$g(x)=\begin{cases}\dfrac{1-\cos x}{2x} & \text{if } x\neq 0\\[2mm] a & \text{if } x=0.\end{cases}$$

For what values of $a$ is $g$ continuous?

**T 86. Computing limits with angles in degrees** Suppose your graphing calculator has two functions, one called $\sin x$, which calculates the sine of $x$ when $x$ is in radians, and the other called $s(x)$, which calculates the sine of $x$ when $x$ is in degrees.

   **a.** Explain why $s(x)=\sin\left(\dfrac{\pi}{180}x\right)$.

   **b.** Evaluate $\displaystyle\lim_{x\to 0}\dfrac{s(x)}{x}$. Verify your answer by estimating the limit on your calculator.

**87. Derivatives of $\sin^n x$** Calculate the following derivatives using the Product Rule.

   **a.** $\dfrac{d}{dx}(\sin^2 x)$     **b.** $\dfrac{d}{dx}(\sin^3 x)$     **c.** $\dfrac{d}{dx}(\sin^4 x)$

**d.** Based on your answers to parts (a)–(c), make a conjecture about $\dfrac{d}{dx}(\sin^n x)$, where $n$ is a positive integer. Then prove the result by induction.

**88.** Prove that
$$\frac{d^{2n}}{dx^{2n}}(\sin x) = (-1)^n \sin x \quad \text{and} \quad \frac{d^{2n}}{dx^{2n}}(\cos x) = (-1)^n \cos x.$$

**T 89–90. Difference quotients** *Suppose f is differentiable for all x and consider the function*

$$D(x) = \frac{f(x + 0.01) - f(x)}{0.01}.$$

*For the following functions, graph D on the given interval, and explain why the graph appears as it does. What is the relationship between the functions f and D?*

**89.** $f(x) = \sin x$ on $[-\pi, \pi]$    **90.** $f(x) = \dfrac{x^3}{3} + 1$ on $[-2, 2]$

# 3.6 Derivatives as Rates of Change

The theme of this section is the *derivative as a rate of change*. Observing the world around us, we see that almost everything is in a state of change: The size of the Internet is increasing; your blood pressure fluctuates; as supply increases, prices decrease; and the universe is expanding. This section explores a few of the many applications of this idea and demonstrates why calculus is called the mathematics of change.

## One-Dimensional Motion

> When describing the motion of objects, it is customary to use $t$ as the independent variable to represent time. Generally, motion is assumed to begin at $t = 0$.

Describing the motion of objects such as projectiles and planets was one of the challenges that led to the development of calculus in the 17th century. We begin by considering the motion of an object confined to one dimension; that is, the object moves along a line. This motion could be horizontal (for example, a car moving along a straight highway), or it could be vertical (such as a projectile launched vertically into the air).

Figure 3.36

**Position and Velocity** Suppose an object moves along a straight line and its location at time $t$ is given by the **position function** $s = f(t)$. All positions are measured relative to a reference point $s = 0$. The **displacement** of the object between $t = a$ and $t = a + \Delta t$ is $\Delta s = f(a + \Delta t) - f(a)$, where the elapsed time is $\Delta t$ units (**Figure 3.36**).

Recall from Section 2.1 that the *average velocity* of the object over the interval $[a, a + \Delta t]$ is the displacement $\Delta s$ of the object divided by the elapsed time $\Delta t$:

$$v_{av} = \frac{\Delta s}{\Delta t} = \frac{f(a + \Delta t) - f(a)}{\Delta t}.$$

The average velocity is the slope of the secant line passing through the points $P(a, f(a))$ and $Q(a + \Delta t, f(a + \Delta t))$ (**Figure 3.37**).

As $\Delta t$ approaches 0, the average velocity is calculated over smaller and smaller time intervals, and the limiting value of these average velocities, when it exists, is the *instantaneous velocity* at $a$. This is the same argument used to arrive at the derivative. The conclusion is that the instantaneous velocity at time $a$, denoted $v(a)$, is the derivative of the position function evaluated at $a$:

$$v(a) = \lim_{\Delta t \to 0} \frac{f(a + \Delta t) - f(a)}{\Delta t} = f'(a).$$

Equivalently, the instantaneous velocity at $a$ is the rate of change in the position function at $a$; it also equals the slope of the line tangent to the curve $s = f(t)$ at $P(a, f(a))$.

Figure 3.37

▶ Using the various derivative notations, the velocity is also written $v(t) = s'(t) = ds/dt$. If *average* or *instantaneous* is not specified, *velocity* is understood to mean instantaneous velocity.

**QUICK CHECK 1** Does the speedometer in your car measure average or instantaneous velocity? ◀

> **DEFINITION** Average and Instantaneous Velocity
>
> Let $s = f(t)$ be the position function of an object moving along a line. The **average velocity** of the object over the time interval $[a, a + \Delta t]$ is the slope of the secant line between $(a, f(a))$ and $(a + \Delta t, f(a + \Delta t))$:
>
> $$v_{av} = \frac{f(a + \Delta t) - f(a)}{\Delta t}.$$
>
> The **instantaneous velocity** at $a$ is the slope of the line tangent to the position curve, which is the derivative of the position function:
>
> $$v(a) = \lim_{\Delta t \to 0} \frac{f(a + \Delta t) - f(a)}{\Delta t} = f'(a).$$

**Figure 3.38**

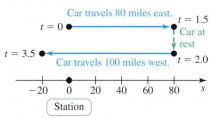

**Figure 3.39**

**EXAMPLE 1   Position and velocity of a patrol car**  Assume a police station is located along a straight east-west freeway. At noon ($t = 0$), a patrol car leaves the station heading east. The position function of the car $s = f(t)$ gives the location of the car in miles east ($s > 0$) or west ($s < 0$) of the station $t$ hours after noon (**Figure 3.38**).

a.  Describe the location of the patrol car during the first 3.5 hr of the trip.

b.  Calculate the displacement and average velocity of the car between 2:00 P.M. and 3:30 P.M. ($2 \le t \le 3.5$).

c.  At what time(s) is the instantaneous velocity greatest *as the car travels east*?

**SOLUTION**

a.  The graph of the position function indicates the car travels 80 miles east between $t = 0$ (noon) and $t = 1.5$ (1:30 P.M.). The car is at rest and its position does not change from $t = 1.5$ to $t = 2$ (that is, from 1:30 P.M. to 2:00 P.M.). Starting at $t = 2$, the car's distance from the station decreases, which means the car travels west, eventually ending up 20 miles west of the station at $t = 3.5$ (3:30 P.M.) (**Figure 3.39**).

b.  The position of the car at 3:30 P.M. is $f(3.5) = -20$ (Figure 3.38; the negative sign indicates the car is 20 miles *west* of the station), and the position of the car at 2:00 P.M. is $f(2) = 80$. Therefore, the displacement is

$$\Delta s = f(3.5) - f(2) = -20 \text{ mi} - 80 \text{ mi} = -100 \text{ mi}$$

during an elapsed time of $\Delta t = 3.5 - 2 = 1.5$ hr (the *negative* displacement indicates that the car moved 100 miles *west*). The average velocity is

$$v_{av} = \frac{\Delta s}{\Delta t} = \frac{-100 \text{ mi}}{1.5 \text{ hr}} \approx -66.7 \text{ mi/hr.}$$

c.  The greatest eastward instantaneous velocity corresponds to points at which the graph of the position function has the greatest positive slope. The greatest slope occurs between $t = 0.5$ and $t = 1$. During this time interval, the car also has a nearly constant velocity because the curve is approximately linear. We conclude that the eastward velocity is largest from 12:30 P.M. to 1:00 P.M.

*Related Exercises 11–12* ◀

**Speed and Acceleration**   When only the magnitude of the velocity is of interest, we use *speed*, which is the absolute value of the velocity:

$$\text{speed} = |v|.$$

For example, a car with an instantaneous velocity of $-30$ mi/hr has a speed of 30 mi/hr.

➤ Newton's First Law of Motion says that in the absence of external forces, a moving object has no acceleration, which means the magnitude and direction of the velocity are constant.

A more complete description of an object moving along a line includes its *acceleration*, which is the rate of change of the velocity; that is, acceleration is the derivative of the velocity function with respect to time $t$. If the acceleration is positive, the object's velocity increases; if it is negative, the object's velocity decreases. Because velocity is the derivative of the position function, acceleration is the second derivative of the position. Therefore,

$$a = \frac{dv}{dt} = \frac{d^2s}{dt^2}.$$

> **DEFINITION** Velocity, Speed, and Acceleration
>
> Suppose an object moves along a line with position $s = f(t)$. Then
>
> the **velocity** at time $t$ is $\qquad v = \dfrac{ds}{dt} = f'(t),$
>
> the **speed** at time $t$ is $\qquad |v| = |f'(t)|,$ and
>
> the **acceleration** at time $t$ is $\qquad a = \dfrac{dv}{dt} = \dfrac{d^2s}{dt^2} = f''(t).$

**QUICK CHECK 2** For an object moving along a line, is it possible for its velocity to increase while its speed decreases? Is it possible for its velocity to decrease while its speed increases? Give an example to support your answers. ◄

➤ The units of derivatives are consistent with the notation. If $s$ is measured in meters and $t$ is measured in seconds, the units of the velocity $\dfrac{ds}{dt}$ are m/s. The units of the acceleration $\dfrac{d^2s}{dt^2}$ are m/s².

**EXAMPLE 2** **Velocity and acceleration** Suppose the position (in feet) of an object moving horizontally along a line at time $t$ (in seconds) is $s = t^2 - 5t$, for $0 \le t \le 5$ (**Figure 3.40**). Assume positive values of $s$ correspond to positions to the right of $s = 0$.

**a.** Graph the velocity function on the interval $0 \le t \le 5$ and determine when the object is stationary, moving to the left, and moving to the right.

**b.** Graph the acceleration function on the interval $0 \le t \le 5$.

**c.** Describe the motion of the object.

**SOLUTION**

**a.** The velocity is $v = s'(t) = 2t - 5$. The object is stationary when $v = 2t - 5 = 0$, or at $t = 2.5$. Solving $v = 2t - 5 > 0$, the velocity is positive (motion to the right) for $\frac{5}{2} < t \le 5$. Similarly, the velocity is negative (motion to the left) for $0 \le t < \frac{5}{2}$. Though the velocity of the object is increasing at all times, its speed is decreasing for $0 \le t < \frac{5}{2}$, and then increasing for $\frac{5}{2} < t \le 5$. The graph of the velocity function (**Figure 3.41**) confirms these observations.

**b.** The acceleration is the derivative of the velocity or $a = v'(t) = s''(t) = 2$. This means that the acceleration is 2 ft/s², for $0 \le t \le 5$ (**Figure 3.42**).

**c.** Starting at an initial position of $s(0) = 0$, the object moves in the negative direction (to the left) with decreasing speed until it comes to rest momentarily at $s\left(\frac{5}{2}\right) = -\frac{25}{4}$. The object then moves in the positive direction (to the right) with increasing speed, reaching its initial position at $t = 5$. During this time interval, the acceleration is constant.

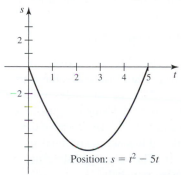

Position: $s = t^2 - 5t$

**Figure 3.40**

➤ Figure 3.40 gives the graph of the position function, not the path of the object. The motion is along a horizontal line.

**QUICK CHECK 3** Describe the velocity of an object that has a positive constant acceleration. Could an object have a positive acceleration and a decreasing speed? ◄

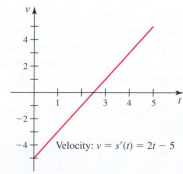

Velocity: $v = s'(t) = 2t - 5$

**Figure 3.41**

Acceleration: $a = s''(t) = 2$

**Figure 3.42**

*Related Exercises 15–16* ◄

➤ The acceleration due to Earth's gravitational field is denoted $g$. In metric units, $g \approx 9.8 \text{ m/s}^2$ on the surface of Earth; in the U.S. Customary System (USCS), $g \approx 32 \text{ ft/s}^2$.

➤ The position function in Example 3 is derived in Section 6.1. Once again we mention that the graph of the position function is not the path of the stone.

**Free Fall**  We now consider a problem in which an object moves vertically in Earth's gravitational field, assuming no other forces (such as air resistance) are at work.

**EXAMPLE 3   Motion in a gravitational field**  Suppose a stone is thrown vertically upward with an initial velocity of 64 ft/s from a bridge 96 ft above a river. By Newton's laws of motion, the position of the stone (measured as the height above the river) after $t$ seconds is

$$s(t) = -16t^2 + 64t + 96,$$

where $s = 0$ is the level of the river (**Figure 3.43a**).

**a.** Find the velocity and acceleration functions.

**b.** What is the highest point above the river reached by the stone?

**c.** With what velocity will the stone strike the river?

**SOLUTION**

**a.** The velocity of the stone is the derivative of the position function, and its acceleration is the derivative of the velocity function. Therefore,

$$v = \frac{ds}{dt} = -32t + 64 \quad \text{and} \quad a = \frac{dv}{dt} = -32.$$

**b.** When the stone reaches its high point, its velocity is zero (**Figure 3.43b**). Solving $v(t) = -32t + 64 = 0$ yields $t = 2$; therefore, the stone reaches its maximum height 2 seconds after it is thrown. Its height (in feet) at that instant is

$$s(2) = -16(2)^2 + 64(2) + 96 = 160.$$

**c.** To determine the velocity at which the stone strikes the river, we first determine *when* it reaches the river. The stone strikes the river when $s(t) = -16t^2 + 64t + 96 = 0$. Dividing both sides of the equation by $-16$, we obtain $t^2 - 4t - 6 = 0$. Using the quadratic formula, the solutions are $t \approx 5.16$ and $t \approx -1.16$. Because the stone is thrown at $t = 0$, only positive values of $t$ are of interest; therefore, the relevant root is $t \approx 5.16$. The velocity of the stone (in ft/s) when it strikes the river is approximately

$$v(5.16) = -32(5.16) + 64 = -101.12.$$

*Related Exercises 24–25* ◄

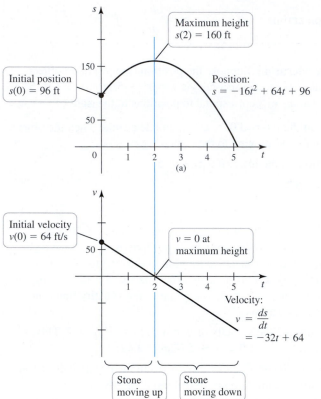

Figure 3.43

**QUICK CHECK 4** In Example 3, does the rock have a greater speed at $t = 1$ or $t = 3$? ◄

## Growth Models

Much of the change in the world around us can be classified as *growth*: Populations, prices, and computer networks all tend to increase in size. Modeling growth is important because it often leads to an understanding of underlying processes and allows for predictions.

We let $p = f(t)$ be the measure of a quantity of interest (for example, the population of a species or the consumer price index), where $t \geq 0$ represents time. The average growth rate of $p$ between time $t = a$ and a later time $t = a + \Delta t$ is the change $\Delta p$ divided by elapsed time $\Delta t$. Therefore, the **average growth rate** of $p$ on the interval $[a, a + \Delta t]$ is

$$\frac{\Delta p}{\Delta t} = \frac{f(a + \Delta t) - f(a)}{\Delta t}.$$

If we now let $\Delta t \to 0$, then $\dfrac{\Delta p}{\Delta t}$ approaches the derivative $\dfrac{dp}{dt}$, which is the **instantaneous growth rate** (or simply **growth rate**) of $p$ with respect to time:

$$\frac{dp}{dt} = \lim_{\Delta t \to 0} \frac{\Delta p}{\Delta t}.$$

Once again, we see the derivative appearing as an instantaneous rate of change. In the next example, a growth function and its derivative are approximated using real data.

**EXAMPLE 4   Internet growth**  The number of worldwide Internet users between 2000 and 2015 is shown in **Figure 3.44**. A reasonable fit to the data is given by the function $p(t) = 6t^2 + 98t + 431.2$, where $t$ is measured in years after 2000.

**a.** Use the function $p$ to approximate the average growth rate of Internet users from 2005 ($t = 5$) to 2010 ($t = 10$).

**b.** What was the instantaneous growth rate of Internet users in 2011?

**c.** Use a graphing utility to plot the growth rate $dp/dt$. What does the graph tell you about the growth rate between 2000 and 2015?

**d.** Assuming the growth function can be extended beyond 2015, what is the predicted number of Internet users in 2020 ($t = 20$)?

**SOLUTION**

**a.** The average growth rate over the interval $[5, 10]$ is

$$\frac{\Delta P}{\Delta t} = \frac{p(10) - p(5)}{10 - 5} \approx \frac{2011.2 - 1071.2}{5} = 188 \text{ million users/year.}$$

**b.** The growth rate at time $t$ is $p'(t) = 12t + 98$. In 2011 ($t = 11$), the growth rate was $p'(11) = 230$ million users per year.

**QUICK CHECK 5** Using the growth function in Example 4, compare the growth rates in 2001 and 2012. ◄

**c.** The graph of $p'$, for $0 \le t \le 15$, is shown in **Figure 3.45**. We see that the growth rate is positive and increasing, for $t \ge 0$.

**d.** The number of Internet users in 2020 is predicted to be $p(20) \approx 4791$ million, or about 4.8 billion users, which is approximately 62% of the world's population, assuming a projected population of 7.7 billion people in 2020.

**Figure 3.44**

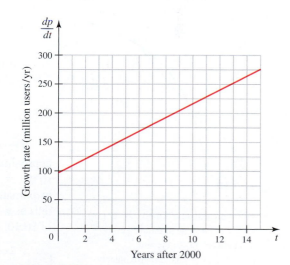

**Figure 3.45**

*Related Exercise 28* ◄

## Economics and Business

Our final examples illustrate how derivatives arise in economics and business. As you will see, the mathematics of derivatives is the same as it is in other applications. However, the vocabulary and interpretation are quite different.

**Figure 3.46**

➤ Although $x$ is a whole number of units, we treat it as a continuous variable, which is reasonable if $x$ is large.

**Figure 3.47**

➤ The average describes the past; the marginal describes the future.
    —Old saying

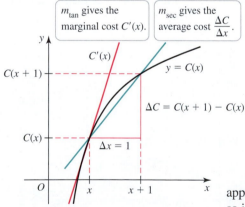

**Figure 3.48**

➤ The approximation $\Delta C / \Delta x \approx C'(x)$ says that the slope of the secant line between $(x, C(x))$ and $(x + 1, C(x + 1))$ is approximately equal to the slope of the tangent line at $(x, C(x))$. This approximation is good if the cost curve is nearly linear over a 1-unit interval.

**Average and Marginal Cost**   Imagine a company that manufactures large quantities of a product such as mousetraps, DVD players, or snowboards. Associated with the manufacturing process is a *cost function* $C(x)$ that gives the cost of manufacturing the first $x$ items of the product. A simple cost function might have the form $y = C(x) = 500 + 0.1x$, as shown in **Figure 3.46**. It includes a **fixed cost** of \$500 (setup costs and overhead) that is independent of the number of items produced. It also includes a **unit cost**, or **variable cost**, of \$0.10 per item produced. For example, the cost of producing the first 1000 items is $C(1000) = \$600$.

If the company produces $x$ items at a cost of $C(x)$, then the *average cost* is $\dfrac{C(x)}{x}$ per item. For the cost function $C(x) = 500 + 0.1x$, the average cost is

$$\frac{C(x)}{x} = \frac{500 + 0.1x}{x} = \frac{500}{x} + 0.1.$$

For example, the average cost of manufacturing the first 1000 items is

$$\frac{C(1000)}{1000} = \frac{\$600}{1000} = \$0.60/\text{unit}.$$

Plotting $C(x)/x$, we see that the average cost decreases as the number of items produced increases (**Figure 3.47**).

The average cost gives the cost of items already produced. But what about the cost of producing additional items? Once $x$ items have been produced, the cost of producing another $\Delta x$ items is $C(x + \Delta x) - C(x)$. Therefore, the average cost per item of producing those $\Delta x$ additional items is

$$\frac{C(x + \Delta x) - C(x)}{\Delta x} = \frac{\Delta C}{\Delta x}.$$

If we let $\Delta x \to 0$, we see that

$$\lim_{\Delta x \to 0} \frac{\Delta C}{\Delta x} = C'(x),$$

which is called the *marginal cost*. In reality, we cannot let $\Delta x \to 0$ because $\Delta x$ represents whole numbers of items. However, there is a useful interpretation of the marginal cost. Suppose $\Delta x = 1$. Then $\Delta C = C(x + 1) - C(x)$ is the cost to produce *one* additional item. In this case, we write

$$\frac{\Delta C}{\Delta x} = \frac{C(x + 1) - C(x)}{1}.$$

If the *slope* of the cost curve does not vary significantly near the point $x$, then—as shown in **Figure 3.48**—we have

$$\frac{\Delta C}{\Delta x} \approx \lim_{\Delta x \to 0} \frac{\Delta C}{\Delta x} = C'(x).$$

Therefore, the cost of producing one additional item, having already produced $x$ items, is approximated by the marginal cost $C'(x)$. In the preceding example, we have $C'(x) = 0.1$; so if $x = 1000$ items have been produced, then the cost of producing the 1001st item is approximately $C'(1000) = \$0.10$. With this simple linear cost function, the marginal cost tells us what we already know: The cost of producing one additional item is the variable cost of \$0.10. With more realistic cost functions, the marginal cost may be variable.

---

**DEFINITION   Average and Marginal Cost**

The **cost function** $C(x)$ gives the cost to produce the first $x$ items in a manufacturing process. The **average cost** to produce $x$ items is $\overline{C}(x) = C(x)/x$. The **marginal cost** $C'(x)$ is the approximate cost to produce one additional item after producing $x$ items.

Cost (dollars)

**Figure 3.49**

**EXAMPLE 5   Average and marginal costs** Suppose the cost of producing $x$ items is given by the function (**Figure 3.49**)

$$C(x) = -0.02x^2 + 50x + 100, \quad \text{for} \quad 0 \le x \le 1000.$$

**a.** Determine the average and marginal cost functions.

**b.** Determine the average and marginal cost for the first 100 items and interpret these values.

**c.** Determine the average and marginal cost for the first 900 items and interpret these values.

**SOLUTION**

**a.** The average cost is

$$\overline{C}(x) = \frac{C(x)}{x} = \frac{-0.02x^2 + 50x + 100}{x} = -0.02x + 50 + \frac{100}{x}$$

and the marginal cost is

$$\frac{dC}{dx} = -0.04x + 50.$$

The average cost decreases as the number of items produced increases (**Figure 3.50a**). The marginal cost decreases linearly with a slope of $-0.04$ (**Figure 3.50b**).

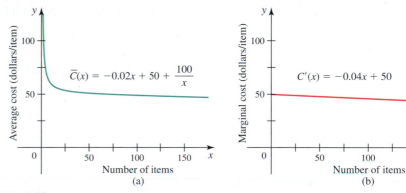

**Figure 3.50**

**b.** To produce $x = 100$ items, the average cost is

$$\overline{C}(100) = \frac{C(100)}{100} = \frac{-0.02(100)^2 + 50(100) + 100}{100} = \$49/\text{item}$$

and the marginal cost is

$$C'(100) = -0.04(100) + 50 = \$46/\text{item}.$$

These results mean that the average cost of producing the first 100 items is \$49 per item, but the cost of producing one additional item (the 101st item) is approximately \$46. Therefore, producing one more item is less expensive than the average cost of producing the first 100 items.

**c.** To produce $x = 900$ items, the average cost is

$$\overline{C}(900) = \frac{C(900)}{900} = \frac{-0.02(900)^2 + 50(900) + 100}{900} \approx \$32/\text{item}$$

and the marginal cost is

$$C'(900) = -0.04(900) + 50 = \$14/\text{item}.$$

**QUICK CHECK 6** In Example 5, what happens to the average cost as the number of items produced increases from $x = 1$ to $x = 100$? ◄

The comparison with part (b) is revealing. The average cost of producing 900 items has dropped to \$32 per item. More striking is that the marginal cost (the cost of producing the 901st item) has dropped to \$14.

*Related Exercises 29–30* ◄

**Figure 3.51**

**Elasticity in Economics**   Economists apply the term *elasticity* to prices, income, capital, labor, and other variables in systems with input and output. Elasticity describes how changes in the input to a system are related to changes in the output. Because elasticity involves change, it also involves derivatives.

A general rule is that as the price $p$ of an item increases, the number of sales of that item decreases. This relationship is expressed in a demand function. For example, suppose sales at a gas station have the linear demand function $D(p) = 1200 - 200p$ (**Figure 3.51**), where $D(p)$ is the number of gallons sold per day at a price $p$ (measured in dollars per gallon). According to this function, if gas sells at $3.60/gal, then the owner can expect to sell $D(3.6) = 480$ gallons. If the price is increased, sales decrease.

Suppose the price of a gallon of gasoline increases from \$3.60 to \$3.96 per gallon; call this change $\Delta p = \$0.36$. The resulting change in the number of gallons sold is $\Delta D = D(3.96) - D(3.60) = -72$. (The change is a decrease, so it is negative.) Comparisons of the variables are more meaningful if we work with percentages. Increasing the price from \$3.60 to \$3.96 per gallon is a percentage change of $\dfrac{\Delta p}{p} = \dfrac{\$0.36}{\$3.60} = 10\%$. Similarly, the corresponding percentage change in the gallons sold is $\dfrac{\Delta D}{D} = \dfrac{-72}{480} = -15\%$.

The *price elasticity of the demand* (or simply, *elasticity*) is the ratio of the percentage change in demand to the percentage change in price; that is, $E = \dfrac{\Delta D/D}{\Delta p/p}$. In the case of the gas demand function, the elasticity of this particular price change is $\dfrac{-15\%}{10\%} = -1.5$.

The elasticity is more useful when it is expressed as a function of the price. To do this, we consider small changes in $p$ and assume the corresponding changes in $D$ are also small. Using the definition of the derivative, the elasticity *function* is

$$E(p) = \lim_{\Delta p \to 0} \frac{\Delta D/D}{\Delta p/p} = \lim_{\Delta p \to 0} \frac{\Delta D}{\Delta p}\left(\frac{p}{D}\right) = \frac{dD}{dp}\frac{p}{D}.$$

Applying this definition to the gas demand function, we find that

$$E(p) = \frac{dD}{dp}\frac{p}{D}$$

$$= \underbrace{\frac{d}{dp}(1200 - 200p)}_{D}\,\underbrace{\frac{p}{1200 - 200p}}_{D} \qquad \text{Substitute } D = 1200 - 200p.$$

$$= -200\left(\frac{p}{1200 - 200p}\right) \qquad \text{Differentiate.}$$

$$= \frac{p}{p - 6}. \qquad \text{Simplify.}$$

Given a particular price, the elasticity is interpreted as the percentage change in the demand that results for every 1% change in the price. For example, in the gas demand case, with $p = \$3.60$, the elasticity is $E(3.6) = -1.5$; therefore, a 2% increase in the price results in a change of $-1.5 \cdot 2\% = -3\%$ (a decrease) in the number of gallons sold.

➤ Some texts define the elasticity as
$$E(p) = -\frac{dD}{dp}\frac{p}{D} \text{ to make } E(p) \text{ positive.}$$

---

**DEFINITION   Elasticity**

If the demand for a product varies with the price according to the function $D = f(p)$, then the **price elasticity of the demand** is $E(p) = \dfrac{dD}{dp}\dfrac{p}{D}$.

**EXAMPLE 6    Elasticity in pork prices**  The demand for processed pork in Canada is described by the function $D(p) = 286 - 20p$. (*Source:* J. Perloff, *Microeconomics*, Prentice Hall, 2012)

**a.** Compute and graph the price elasticity of the demand.

**b.** When $-\infty < E < -1$, the demand is said to be **elastic**. When $-1 < E < 0$, the demand is said to be **inelastic**. Interpret these terms.

**c.** For what prices is the demand for pork elastic? Inelastic?

**SOLUTION**

**a.** Substituting the demand function into the definition of elasticity, we find that

$$E(p) = \frac{dD}{dp}\frac{p}{D}$$

$$= \frac{d}{dp}\underbrace{(286 - 20p)}_{D}\underbrace{\frac{p}{286 - 20p}}_{D} \qquad \text{Substitute } D = 286 - 20p.$$

$$= -20\left(\frac{p}{286 - 20p}\right) \qquad \text{Differentiate.}$$

$$= -\frac{10p}{143 - 10p}. \qquad \text{Simplify.}$$

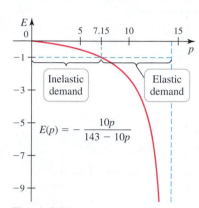

**Figure 3.52**

▶ When $E(p) = 0$, the demand for a good is said to be *perfectly inelastic*. In Example 6, this case occurs when $p = 0$.

Notice that the elasticity is undefined at $p = 14.3$, which is the price at which the demand reaches zero. (According to the model, no pork can be sold at prices above \$14.30.) Therefore, the domain of the elasticity function is $[0, 14.3)$, and on the interval $(0, 14.3)$, the elasticity is negative (**Figure 3.52**).

**b.** For prices with an elasticity in the interval $-\infty < E(p) < -1$, a $P\%$ increase in the price results in *more* than a $P\%$ decrease in the demand; this is the case of elastic (sensitive) demand. If a price has an elasticity in the interval $-1 < E(p) < 0$, a $P\%$ increase in the price results in *less* than a $P\%$ decrease in the demand; this is the case of inelastic (insensitive) demand.

**c.** Solving $E(p) = -\dfrac{10p}{143 - 10p} = -1$, we find that $E(p) < -1$, for $p > 7.15$.

For prices in this interval, the demand is elastic (Figure 3.52). For prices with $0 < p < 7.15$, the demand is inelastic.

*Related Exercises 33–34* ◀

---

## SECTION 3.6 EXERCISES

### Getting Started

**1.** Explain the difference between the average rate of change and the instantaneous rate of change of a function $f$.

**2.** Complete the following statement. If $\dfrac{dy}{dx}$ is large, then small changes in $x$ will result in relatively _____ changes in the value of $y$.

**3.** Complete the following statement. If $\dfrac{dy}{dx}$ is small, then small changes in $x$ will result in relatively _____ changes in the value of $y$.

**4.** Suppose the function $s(t)$ represents the position (in feet) of a stone $t$ seconds after it is thrown directly upward from 6 feet above Earth's surface.

   **a.** Is the stone's acceleration positive or negative when the stone is moving upward? Explain.

   **b.** What is the value of the instantaneous velocity $v(t)$ when the stone reaches its highest point? Explain.

**5.** Suppose $w(t)$ is the weight (in pounds) of a golden retriever puppy $t$ weeks after it is born. Interpret the meaning of $w'(15) = 1.75$.

**6.** What is the difference between the *velocity* and the *speed* of an object moving in a straight line?

**7.** Define the acceleration of an object moving in a straight line.

**8.** An object moving along a line has a constant negative acceleration. Describe the velocity of the object.

**9.** The speed of sound (in m/s) in dry air is approximated by the function $v(T) = 331 + 0.6T$, where $T$ is the air temperature (in degrees Celsius). Evaluate $v'(T)$ and interpret its meaning.

**10.** At noon, a city park manager starts filling a swimming pool. If $V(t)$ is the volume of water (in ft³) in the swimming pool $t$ hours after noon, then what does $dV/dt$ represent?

**11. Highway travel** A state patrol station is located on a straight north-south freeway. A patrol car leaves the station at 9:00 A.M. heading north with position function $s = f(t)$ that gives its location in miles $t$ hours after 9:00 A.M. (see figure). Assume $s$ is positive when the car is north of the patrol station.

**a.** Determine the average velocity of the car during the first 45 minutes of the trip.
**b.** Find the average velocity of the car over the interval $[0.25, 0.75]$. Is the average velocity a good estimate of the velocity at 9:30 A.M.?
**c.** Find the average velocity of the car over the interval $[1.75, 2.25]$. Estimate the velocity of the car at 11:00 A.M. and determine the direction in which the patrol car is moving.
**d.** Describe the motion of the patrol car relative to the patrol station between 9:00 A.M. and noon.

**12. Airline travel** The following figure shows the position function of an airliner on an out-and-back trip from Seattle to Minneapolis, where $s = f(t)$ is the number of ground miles from Seattle $t$ hours after take-off at 6:00 A.M. The plane returns to Seattle 8.5 hours later at 2:30 P.M.

**a.** Calculate the average velocity of the airliner during the first 1.5 hours of the trip ($0 \le t \le 1.5$).
**b.** Calculate the average velocity of the airliner between 1:30 P.M. and 2:30 P.M. ($7.5 \le t \le 8.5$).
**c.** At what time(s) is the velocity 0? Give a plausible explanation.
**d.** Determine the velocity of the airliner at noon ($t = 6$) and explain why the velocity is negative.

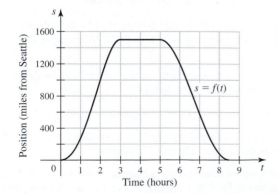

**13.** Suppose the average cost of producing 200 gas stoves is $70 per stove and the marginal cost at $x = 200$ is $65 per stove. Interpret these costs.

**14.** Explain why a decreasing demand function has a negative elasticity function.

### Practice Exercises

**15–20. Position, velocity, and acceleration** *Suppose the position of an object moving horizontally along a line after t seconds is given by the following functions $s = f(t)$, where s is measured in feet, with $s > 0$ corresponding to positions right of the origin.*

**a.** *Graph the position function.*
**b.** *Find and graph the velocity function. When is the object stationary, moving to the right, and moving to the left?*
**c.** *Determine the velocity and acceleration of the object at $t = 1$.*
**d.** *Determine the acceleration of the object when its velocity is zero.*
**e.** *On what intervals is the speed increasing?*

**15.** $f(t) = t^2 - 4t; 0 \le t \le 5$

**16.** $f(t) = -t^2 + 4t - 3; 0 \le t \le 5$

**17.** $f(t) = 2t^2 - 9t + 12; 0 \le t \le 3$

**18.** $f(t) = 18t - 3t^2; 0 \le t \le 8$

**19.** $f(t) = 2t^3 - 21t^2 + 60t; 0 \le t \le 6$

**20.** $f(t) = -6t^3 + 36t^2 - 54t; 0 \le t \le 4$

**21. A dropped stone on Earth** The height (in feet) of a stone dropped from a bridge 64 feet above a river at $t = 0$ seconds is given by $s(t) = -16t^2 + 64$. Find the velocity of the stone and its speed when it hits the water.

**22. A dropped stone on Mars** A stone is dropped off the edge of a 54-ft cliff on Mars, where the acceleration due to gravity is about 12 ft/s². The height (in feet) of the stone above the ground $t$ seconds after it is dropped is $s(t) = -6t^2 + 54$. Find the velocity of the stone and its speed when it hits the ground.

**23. Throwing a stone** Suppose a stone is thrown vertically upward from the edge of a cliff on Earth with an initial velocity of 32 ft/s from a height of 48 ft above the ground. The height (in feet) of the stone above the ground $t$ seconds after it is thrown is $s(t) = -16t^2 + 32t + 48$.

**a.** Determine the velocity $v$ of the stone after $t$ seconds.
**b.** When does the stone reach its highest point?
**c.** What is the height of the stone at the highest point?
**d.** When does the stone strike the ground?
**e.** With what velocity does the stone strike the ground?
**f.** On what intervals is the speed increasing?

**24.** Suppose a stone is thrown vertically upward from the edge of a cliff on Earth with an initial velocity of 19.6 m/s from a height of 24.5 m above the ground. The height (in meters) of the stone above the ground $t$ seconds after it is thrown is $s(t) = -4.9t^2 + 19.6t + 24.5$.

**a.** Determine the velocity $v$ of the stone after $t$ seconds.
**b.** When does the stone reach its highest point?
**c.** What is the height of the stone at the highest point?
**d.** When does the stone strike the ground?
**e.** With what velocity does the stone strike the ground?
**f.** On what intervals is the speed increasing?

**25.** Suppose a stone is thrown vertically upward from the edge of a cliff on Earth with an initial velocity of 64 ft/s from a height of 32 ft above the ground. The height (in feet) of the stone above the ground $t$ seconds after it is thrown is $s(t) = -16t^2 + 64t + 32$.

**a.** Determine the velocity $v$ of the stone after $t$ seconds.
**b.** When does the stone reach its highest point?

c. What is the height of the stone at the highest point?

d. When does the stone strike the ground?

e. With what velocity does the stone strike the ground?

f. On what intervals is the speed increasing?

26. **Maximum height** Suppose a baseball is thrown vertically upward from the ground with an initial velocity of $v_0$ ft/s. The approximate height of the ball (in feet) above the ground after $t$ seconds is given by $s(t) = -16t^2 + v_0 t$.

a. What is the height of the ball at its highest point?

b. With what velocity does the ball strike the ground?

27. **Initial velocity** Suppose a baseball is thrown vertically upward from the ground with an initial velocity of $v_0$ ft/s. Its height above the ground after $t$ seconds is given by $s(t) = -16t^2 + v_0 t$. Determine the initial velocity of the ball if it reaches a high point of 128 ft.

**T 28. Population growth in Washington** The population of the state of Washington (in millions) from 2010 ($t = 0$) to 2016 ($t = 6$) is modeled by the polynomial $p(t) = 0.0078t^2 + 0.028t + 6.73$.

a. Determine the average growth rate from 2010 to 2016.

b. What was the growth rate for Washington in 2011 ($t = 1$) and 2015 ($t = 5$)?

c. Use a graphing utility to graph $p'$ for $0 \le t \le 6$. What does this graph tell you about population growth in Washington during the period of time from 2010 to 2016?

**29–32. Average and marginal cost** *Consider the following cost functions.*

a. *Find the average cost and marginal cost functions.*

b. *Determine the average cost and the marginal cost when $x = a$.*

c. *Interpret the values obtained in part (b).*

29. $C(x) = 1000 + 0.1x, 0 \le x \le 5000, a = 2000$

30. $C(x) = 500 + 0.02x, 0 \le x \le 2000, a = 1000$

31. $C(x) = -0.01x^2 + 40x + 100, 0 \le x \le 1500, a = 1000$

32. $C(x) = -0.04x^2 + 100x + 800, 0 \le x \le 1000, a = 500$

33. **Demand and elasticity** Based on sales data over the past year, the owner of a DVD store devises the demand function $D(p) = 40 - 2p$, where $D(p)$ is the number of DVDs that can be sold in one day at a price of $p$ dollars.

a. According to the model, how many DVDs can be sold in a day at a price of $10?

b. According to the model, what is the maximum price that can be charged (above which no DVDs can be sold)?

c. Find the elasticity function for this demand function.

d. For what prices is the demand elastic? Inelastic?

e. If the price of DVDs is raised from $10.00 to $10.25, what is the exact percentage decrease in demand (using the demand function)?

f. If the price of DVDs is raised from $10.00 to $10.25, what is the approximate percentage decrease in demand (using the elasticity function)?

34. **Demand and elasticity** The economic advisor of a large tire store proposes the demand function $D(p) = \dfrac{1800}{p - 40}$, where $D(p)$ is the number of tires of one brand and size that can be sold in one day at a price $p$.

a. Recalling that the demand must be positive, what is the domain of this function?

b. According to the model, how many tires can be sold in a day at a price of $60 per tire?

c. Find the elasticity function on the domain of the demand function.

d. For what prices is the demand elastic? Inelastic?

e. If the price of tires is raised from $60 to $62, what is the approximate percentage decrease in demand (using the elasticity function)?

35. **Explain why or why not** Determine whether the following statements are true and give an explanation or counterexample.

a. If the acceleration of an object remains constant, then its velocity is constant.

b. If the acceleration of an object moving along a line is always 0, then its velocity is constant.

c. It is impossible for the instantaneous velocity at all times $a \le t \le b$ to equal the average velocity over the interval $a \le t \le b$.

d. A moving object can have negative acceleration and increasing speed.

36. **A feather dropped on the moon** On the moon, a feather will fall to the ground at the same rate as a heavy stone. Suppose a feather is dropped from a height of 40 m above the surface of the moon. Its height (in meters) above the ground after $t$ seconds is $s = 40 - 0.8t^2$. Determine the velocity and acceleration of the feather the moment it strikes the surface of the moon.

37. **Comparing velocities** A stone is thrown vertically into the air at an initial velocity of 96 ft/s. On Mars the height (in feet) of the stone above the ground after $t$ seconds is $s = 96t - 6t^2$, and on Earth it is $s = 96t - 16t^2$. How much higher will the stone travel on Mars than on Earth?

38. **Comparing velocities** Two stones are thrown vertically upward, each with an initial velocity of 48 ft/s at time $t = 0$. One stone is thrown from the edge of a bridge that is 32 feet above the ground, and the other stone is thrown from ground level. The height above the ground of the stone thrown from the bridge after $t$ seconds is $f(t) = -16t^2 + 48t + 32$, and the height of the stone thrown from the ground after $t$ seconds is $g(t) = -16t^2 + 48t$.

a. Show that the stones reach their high points at the same time.

b. How much higher does the stone thrown from the bridge go than the stone thrown from the ground?

c. When do the stones strike the ground and with what velocities?

39. **Matching heights** A stone is thrown with an initial velocity of 32 ft/s from the edge of a bridge that is 48 ft above the ground. The height of this stone above the ground $t$ seconds after it is thrown is $f(t) = -16t^2 + 32t + 48$. If a second stone is thrown from the ground, then its height above the ground after $t$ seconds is given by $g(t) = -16t^2 + v_0 t$, where $v_0$ is the initial velocity of the second stone. Determine the value of $v_0$ such that both stones reach the same high point.

40. **Velocity of a car** The graph shows the position $s = f(t)$ of a car $t$ hours after 5:00 P.M. relative to its starting point $s = 0$, where $s$ is measured in miles.

a. Describe the velocity of the car. Specifically, when is it speeding up and when is it slowing down?

b. At approximately what time is the car traveling the fastest? The slowest?

**c.** What is the approximate maximum velocity of the car? The approximate minimum velocity?

Time (hours)

**41. Velocity from position** The graph of $s = f(t)$ represents the position of an object moving along a line at time $t \geq 0$.

**a.** Assume the velocity of the object is 0 when $t = 0$. For what other values of $t$ is the velocity of the object zero?

**b.** When is the object moving in the positive direction and when is it moving in the negative direction?

**c.** Sketch a graph of the velocity function.

**d.** On what intervals is the speed increasing?

**42. Fish length** Assume the length $L$ (in centimeters) of a particular species of fish after $t$ years is modeled by the following graph.

**a.** What does $dL/dt$ represent and what happens to this derivative as $t$ increases?

**b.** What does the derivative tell you about how this species of fish grows?

**c.** Sketch a graph of $L'$ and $L''$.

**43–46. Average and marginal profit** *Let $C(x)$ represent the cost of producing $x$ items and $p(x)$ be the sale price per item if $x$ items are sold. The profit $P(x)$ of selling $x$ items is $P(x) = xp(x) - C(x)$ (revenue minus costs). The **average profit per item** when $x$ items are sold is $P(x)/x$ and the **marginal profit** is $dP/dx$. The marginal profit approximates the profit obtained by selling one more item, given that $x$*

items have already been sold. Consider the following cost functions $C$ and price functions $p$.

**a.** Find the profit function $P$.

**b.** Find the average profit function and the marginal profit function.

**c.** Find the average profit and the marginal profit if $x = a$ units are sold.

**d.** Interpret the meaning of the values obtained in part (c).

**43.** $C(x) = -0.02x^2 + 50x + 100$, $p(x) = 100$, $a = 500$

**44.** $C(x) = -0.02x^2 + 50x + 100$, $p(x) = 100 - 0.1x$, $a = 500$

**45.** $C(x) = -0.04x^2 + 100x + 800$, $p(x) = 200$, $a = 1000$

**46.** $C(x) = -0.04x^2 + 100x + 800$, $p(x) = 200 - 0.1x$, $a = 1000$

**47. U.S. population growth** The population $p(t)$ (in millions) of the United States $t$ years after the year 1900 is shown in the figure. Approximately when (in what year) was the U.S. population growing most slowly between 1925 and 2020? Estimate the growth rate in that year.

Years after 1900

**T 48. Average of marginal production** Economists use *production functions* to describe how the output of a system varies with respect to another variable such as labor or capital. For example, the production function $P(L) = 200L + 10L^2 - L^3$ gives the output of a system as a function of the number of laborers $L$. The *average product* $A(L)$ is the average output per laborer when $L$ laborers are working; that is, $A(L) = P(L)/L$. The *marginal product* $M(L)$ is the approximate change in output when one additional laborer is added to $L$ laborers; that is, $M(L) = dP/dL$.

**a.** For the given production function, compute and graph $P$, $A$, and $L$.

**b.** Suppose the peak of the average product curve occurs at $L = L_0$, so that $A'(L_0) = 0$. Show that for a general production function, $M(L_0) = A(L_0)$.

**T 49. Velocity of a marble** The position (in meters) of a marble, given an initial velocity and rolling up a long incline, is given by

$$s = \frac{100t}{t + 1},$$ where $t$ is measured in seconds and $s = 0$ is the starting point.

**a.** Graph the position function.

**b.** Find the velocity function for the marble.

**c.** Graph the velocity function and give a description of the motion of the marble.

**d.** At what time is the marble 80 m from its starting point?

**e.** At what time is the velocity 50 m/s?

**T 50. Tree growth** Let $b$ represent the base diameter of a conifer tree and let $h$ represent the height of the tree, where $b$ is measured in centimeters and $h$ is measured in meters. Assume the height is related to the base diameter by the function $h = 5.67 + 0.70b + 0.0067b^2$.

**a.** Graph the height function.

**b.** Plot and interpret the meaning of $dh/db$.

## Explorations and Challenges

**T 51. A different interpretation of marginal cost** Suppose a large company makes 25,000 gadgets per year in batches of $x$ items at a time. After analyzing setup costs to produce each batch and taking into account storage costs, planners have determined that the total cost $C(x)$ of producing 25,000 gadgets in batches of $x$ items at a time is given by

$$C(x) = 1{,}250{,}000 + \frac{125{,}000{,}000}{x} + 1.5x.$$

a. Determine the marginal cost and average cost functions. Graph and interpret these functions.
b. Determine the average cost and marginal cost when $x = 5000$.
c. The meaning of average cost and marginal cost here is different from earlier examples and exercises. Interpret the meaning of your answer in part (b).

**52. Diminishing returns** A cost function of the form $C(x) = \frac{1}{2}x^2$ reflects *diminishing returns to scale*. Find and graph the cost, average cost, and marginal cost functions. Interpret the graphs and explain the idea of diminishing returns.

**T 53. Revenue function** A store manager estimates that the demand for an energy drink decreases with increasing price according to the function $d(p) = \frac{100}{p^2 + 1}$, which means that at price $p$ (in dollars), $d(p)$ units can be sold. The revenue generated at price $p$ is $R(p) = p \cdot d(p)$ (price multiplied by number of units).

a. Find and graph the revenue function.
b. Find and graph the marginal revenue $R'(p)$.
c. From the graphs of $R$ and $R'$, estimate the price that should be charged to maximize the revenue.

**T 54. Fuel economy** Suppose you own a fuel-efficient hybrid automobile with a monitor on the dashboard that displays the mileage and gas consumption. The number of miles you can drive with $g$ gallons of gas remaining in the tank on a particular stretch of highway is given by $m(g) = 50g - 25.8g^2 + 12.5g^3 - 1.6g^4$, for $0 \le g \le 4$.

a. Graph and interpret the mileage function.
b. Graph and interpret the gas mileage $m(g)/g$.
c. Graph and interpret $dm/dg$.

**T 55. Spring oscillations** A spring hangs from the ceiling at equilibrium with a mass attached to its end. Suppose you pull downward on the mass and release it 10 inches below its equilibrium position with an upward push. The distance $x$ (in inches) of the mass from its equilibrium position after $t$ seconds is given by the function $x(t) = 10 \sin t - 10 \cos t$, where $x$ is positive when the mass is above the equilibrium position.

a. Graph and interpret this function.
b. Find $dx/dt$ and interpret the meaning of this derivative.
c. At what times is the velocity of the mass zero?
d. The function given here is a model for the motion of an object on a spring. In what ways is this model unrealistic?

**T 56. Power and energy** Power and energy are often used interchangeably, but they are quite different. **Energy** is what makes matter move or heat up. It is measured in units of **joules** or **Calories**, where 1 Cal = 4184 J. One hour of walking consumes roughly $10^6$ J, or 240 Cal. On the other hand, **power** is the rate at which energy is used, which is measured in **watts**, where 1 W = 1 J/s. Other useful units of power are **kilowatts** (1 kW = $10^3$ W) and **megawatts** (1 MW = $10^6$ W). If energy is used at a rate of 1 kW for one hour, the total amount of energy used is 1 **kilowatt-hour** (1 kWh = $3.6 \times 10^6$ J). Suppose the cumulative energy used in a large building over a 24-hr period is given by $E(t) = 100t + 4t^2 - \frac{t^3}{9}$ kWh, where $t = 0$ corresponds to midnight.

a. Graph the energy function.
b. The power is the rate of energy consumption; that is, $P(t) = E'(t)$. Find the power over the interval $0 \le t \le 24$.
c. Graph the power function and interpret the graph. What are the units of power in this case?

**57. A race** Jean and Juan run a one-lap race on a circular track. Their angular positions on the track during the race are given by the functions $\theta(t)$ and $\varphi(t)$, respectively, where $0 \le t \le 4$ and $t$ is measured in minutes (see figure). These angles are measured in radians, where $\theta = \varphi = 0$ represent the starting position and $\theta = \varphi = 2\pi$ represent the finish position. The angular velocities of the runners are $\theta'(t)$ and $\varphi'(t)$.

a. Compare in words the angular velocity of the two runners and the progress of the race.
b. Which runner has the greater average angular velocity?
c. Who wins the race?
d. Jean's position is given by $\theta(t) = \pi t^2/8$. What is her angular velocity at $t = 2$ and at what time is her angular velocity the greatest?
e. Juan's position is given by $\varphi(t) = \pi t(8 - t)/8$. What is his angular velocity at $t = 2$ and at what time is his angular velocity the greatest?

Circular track

**T 58. Flow from a tank** A cylindrical tank is full at time $t = 0$ when a valve in the bottom of the tank is opened. By Torricelli's law, the volume of water in the tank after $t$ hours is $V = 100(200 - t)^2$, measured in cubic meters.

a. Graph the volume function. What is the volume of water in the tank before the valve is opened?
b. How long does it take for the tank to empty?
c. Find the rate at which water flows from the tank and plot the flow rate function.
d. At what time is the magnitude of the flow rate a minimum? A maximum?

**59. Temperature distribution** A thin copper rod, 4 m in length, is heated at its midpoint, and the ends are held at a constant temperature of $0°$. When the temperature reaches equilibrium, the temperature profile is given by $T(x) = 40x(4 - x)$, where $0 \leq x \leq 4$ is the position along the rod. The **heat flux** at a point on the rod equals $-kT'(x)$, where $k > 0$ is a constant. If the heat flux is positive at a point, heat moves in the positive $x$-direction at that point, and if the heat flux is negative, heat moves in the negative $x$-direction.

   **a.** With $k = 1$, what is the heat flux at $x = 1$? At $x = 3$?
   **b.** For what values of $x$ is the heat flux negative? Positive?
   **c.** Explain the statement that heat flows out of the rod at its ends.

**T 60. Spring runoff** The flow of a small stream is monitored for 90 days between May 1 and August 1. The total water that flows past a gauging station is given by

$$V(t) = \begin{cases} \dfrac{4}{5}t^2 & \text{if } 0 \leq t < 45 \\ -\dfrac{4}{5}(t^2 - 180t + 4050) & \text{if } 45 \leq t < 90, \end{cases}$$

where $V$ is measured in cubic feet and $t$ is measured in days, with $t = 0$ corresponding to May 1.

   **a.** Graph the volume function.
   **b.** Find the flow rate function $V'(t)$ and graph it. What are the units of the flow rate?
   **c.** Describe the flow of the stream over the 3-month period. Specifically, when is the flow rate a maximum?

**QUICK CHECK ANSWERS**

**1.** Instantaneous velocity   **2.** Yes; yes   **3.** If an object has positive acceleration, then its velocity is increasing. If the velocity is negative but increasing, then the acceleration is positive and the speed is decreasing. For example, the velocity may increase from $-2$ m/s to $-1$ m/s to $0$ m/s.   **4.** $v(1) = 32$ ft/s and $v(3) = -32$ ft/s, so the speed is 32 ft/s at both times.   **5.** The growth rate in 2001 ($t = 1$) is 110 million users/year. It is less than half of the growth rate in 2012 ($t = 12$), which is 242 million users/year. **6.** As $x$ increases from 1 to 100, the average cost decreases from \$149.98/item to \$49/item. ◄

# 3.7 The Chain Rule

**QUICK CHECK 1** Explain why it is not practical to calculate $\dfrac{d}{dx}(5x + 4)^{100}$ by first expanding $(5x + 4)^{100}$. ◄

The differentiation rules presented so far allow us to find derivatives of many functions. However, these rules are inadequate for finding the derivatives of most *composite functions*. Here is a typical situation. If $f(x) = x^3$ and $g(x) = 5x + 4$, then their composition is $f(g(x)) = (5x + 4)^3$. One way to find the derivative is by expanding $(5x + 4)^3$ and differentiating the resulting polynomial. Unfortunately, this strategy becomes prohibitive for functions such as $(5x + 4)^{100}$. We need a better approach.

## Chain Rule Formulas

An efficient method for differentiating composite functions, called the *Chain Rule*, is motivated by the following example. Suppose Yancey, Uri, and Xan pick apples. Let $y$, $u$, and $x$ represent the number of apples picked in some period of time by Yancey, Uri, and Xan, respectively. Yancey picks apples three times faster than Uri, which means the rate at which Yancey picks apples with respect to Uri is $dy/du = 3$. Uri picks apples twice as fast as Xan, so $du/dx = 2$. Therefore, Yancey picks apples at a rate that is $3 \cdot 2 = 6$ times greater than Xan's rate, which means that $dy/dx = 6$ (**Figure 3.53**). Observe that

$$\frac{dy}{dx} = \frac{dy}{du} \cdot \frac{du}{dx} = 3 \cdot 2 = 6.$$

➤ Expressions such as $dy/dx$ should not be treated as fractions. Nevertheless, you can check symbolically that you have written the Chain Rule correctly by noting that $du$ appears in the "numerator" and "denominator." If it were "canceled," the Chain Rule would have $dy/dx$ on both sides.

The equation $\dfrac{dy}{dx} = \dfrac{dy}{du} \cdot \dfrac{du}{dx}$ is one form of the Chain Rule.

Alternatively, the Chain Rule may be expressed in terms of composite functions. Let $y = f(u)$ and $u = g(x)$, which means $y$ is related to $x$ through the composite function $y = f(u) = f(g(x))$. The derivative $\dfrac{dy}{dx}$ is now expressed as the product

$$\underbrace{\frac{d}{dx}\big(f(g(x))\big)}_{\frac{dy}{dx}} = \underbrace{f'(u)}_{\frac{dy}{du}} \cdot \underbrace{g'(x)}_{\frac{du}{dx}}.$$

3 times faster       2 times faster

$3 \times 2 = 6$ times faster

**Figure 3.53**

Replacing $u$ with $g(x)$ results in

$$\frac{d}{dx}\big(f(g(x))\big) = f'(g(x)) \cdot g'(x).$$

> ► The two equations in Theorem 3.12 differ only in notation (Leibniz notation for the derivative versus function notation). Mathematically, they are identical. The second equation states that the derivative of $y = f(g(x))$ is the derivative of $f$ evaluated at $g(x)$ multiplied by the derivative of $g$ evaluated at $x$.

> ► There may be several ways to choose an inner function $u = g(x)$ and an outer function $y = f(u)$. Nevertheless, we often refer to *the* inner and *the* outer function for the most obvious choices.

---

**THEOREM 3.12    The Chain Rule**

Suppose $y = f(u)$ is differentiable at $u = g(x)$ and $u = g(x)$ is differentiable at $x$. The composite function $y = f(g(x))$ is differentiable at $x$, and its derivative can be expressed in two equivalent ways.

$$\frac{dy}{dx} = \frac{dy}{du} \cdot \frac{du}{dx} \tag{1}$$

$$\frac{d}{dx}\big(f(g(x))\big) = f'(g(x)) \cdot g'(x) \tag{2}$$

---

A proof of the Chain Rule is given at the end of this section. For now, it's important to learn how to use it. With the composite function $f(g(x))$, we refer to $g$ as the *inner function* and to $f$ as the *outer function* of the composition. The key to using the Chain Rule is identifying the inner and outer functions. The following four steps outline the differentiation process, although you will soon find that the procedure can be streamlined.

---

**PROCEDURE    Using the Chain Rule**

Assume the differentiable function $y = f(g(x))$ is given.

1. Identify an outer function $f$ and an inner function $g$, and let $u = g(x)$.

2. Replace $g(x)$ with $u$ to express $y$ in terms of $u$:
$$y = f(\underbrace{g(x)}_{u}) \Rightarrow y = f(u).$$

3. Calculate the product $\dfrac{dy}{du} \cdot \dfrac{du}{dx}$.

4. Replace $u$ with $g(x)$ in $\dfrac{dy}{du}$ to obtain $\dfrac{dy}{dx}$.

---

**QUICK CHECK 2** Identify an inner function (call it $g$) of $y = (5x + 4)^3$. Let $u = g(x)$ and express the outer function $f$ in terms of $u$. ◄

---

**EXAMPLE 1    The Chain Rule** For each of the following composite functions, find the inner function $u = g(x)$ and the outer function $y = f(u)$. Use equation (1) of the Chain Rule to find $\dfrac{dy}{dx}$.

**a.** $y = (5x + 4)^{3/2}$    **b.** $y = \sin^3 x$    **c.** $y = \sin x^3$

**SOLUTION**

**a.** The inner function of $y = (5x + 4)^{3/2}$ is $u = 5x + 4$, and the outer function is $y = u^{3/2}$. Using equation (1) of the Chain Rule, we have

$$\begin{aligned}
\frac{dy}{dx} &= \frac{dy}{du} \cdot \frac{du}{dx} \\[2mm]
&= \frac{3}{2} u^{1/2} \cdot 5 \qquad {\color{blue} y = u^{3/2} \Rightarrow \frac{dy}{du} = \frac{3}{2}u^{1/2};\ u = 5x + 4 \Rightarrow \frac{du}{dx} = 5} \\[2mm]
&= \frac{3}{2}(5x + 4)^{1/2} \cdot 5 \qquad {\color{blue}\text{Replace } u \text{ with } 5x + 4.} \\[2mm]
&= \frac{15}{2}\sqrt{5x + 4}.
\end{aligned}$$

> ➤ With trigonometric functions, expressions such as $\sin^n x$ always mean $(\sin x)^n$, except when $n = -1$. In Example 1, $\sin^3 x = (\sin x)^3$.

**b.** Replacing the shorthand form $y = \sin^3 x$ with $y = (\sin x)^3$, we identify the inner function as $u = \sin x$. Letting $y = u^3$, we have

$$\frac{dy}{dx} = \frac{dy}{du} \cdot \frac{du}{dx} = 3u^2 \cdot \cos x = \underbrace{3 \sin^2 x}_{3u^2} \cos x.$$

**c.** Although $y = \sin x^3$ appears to be similar to the function $y = \sin^3 x$ in part (b), the inner function in this case is $u = x^3$ and the outer function is $y = \sin u$. Therefore,

$$\frac{dy}{dx} = \frac{dy}{du} \cdot \frac{du}{dx} = (\cos u) \cdot 3x^2 = 3x^2 \cos x^3.$$

*Related Exercises 13–15* ◄

When using equation (2) of the Chain Rule, $\dfrac{d}{dx}\big(f(g(x))\big) = f'(g(x)) \cdot g'(x)$, we identify an outer function $y = f(u)$ and an inner function $u = g(x)$. Then $\dfrac{d}{dx}\big(f(g(x))\big)$ is the product of $f'(u)$ evaluated at $u = g(x)$ and $g'(x)$.

**EXAMPLE 2    The Chain Rule** Use equation (2) of the Chain Rule to calculate the derivatives of the following functions.

**a.** $(6x^3 + 3x + 1)^{10}$    **b.** $\sqrt{5x^2 + 1}$    **c.** $\left(\dfrac{5t^2}{3t^2 + 2}\right)^3$

**SOLUTION**

**a.** The inner function of $(6x^3 + 3x + 1)^{10}$ is $g(x) = 6x^3 + 3x + 1$, and the outer function is $f(u) = u^{10}$. The derivative of the outer function is $f'(u) = 10u^9$, which, when evaluated at $g(x)$, is $10(6x^3 + 3x + 1)^9$. The derivative of the inner function is $g'(x) = 18x^2 + 3$. Multiplying the derivatives of the outer and inner functions, we have

$$\frac{d}{dx}\left((6x^3 + 3x + 1)^{10}\right) = \underbrace{10(6x^3 + 3x + 1)^9}_{f'(u) \text{ evaluated at } g(x)} \cdot \underbrace{(18x^2 + 3)}_{g'(x)}$$

$$= 30(6x^2 + 1)(6x^3 + 3x + 1)^9. \qquad \text{\color{blue}Factor and simplify.}$$

**b.** The inner function of $\sqrt{5x^2 + 1}$ is $g(x) = 5x^2 + 1$, and the outer function is $f(u) = \sqrt{u}$. The derivatives of these functions are $f'(u) = \dfrac{1}{2\sqrt{u}}$ and $g'(x) = 10x$. Therefore,

$$\frac{d}{dx}\sqrt{5x^2 + 1} = \underbrace{\frac{1}{2\sqrt{5x^2 + 1}}}_{\substack{f'(u) \text{ evaluated} \\ \text{at } g(x)}} \cdot \underbrace{10x}_{g'(x)} = \frac{5x}{\sqrt{5x^2 + 1}}.$$

**c.** The inner function of $\left(\dfrac{5t^2}{3t^2 + 2}\right)^3$ is $g(t) = \dfrac{5t^2}{3t^2 + 2}$. The outer function is $f(u) = u^3$, whose derivative is $f'(u) = 3u^2$. The derivative of the inner function requires the Quotient Rule. Applying the Chain Rule, we have

$$\frac{d}{dt}\left(\frac{5t^2}{3t^2 + 2}\right)^3 = \underbrace{3\left(\frac{5t^2}{3t^2 + 2}\right)^2}_{\substack{f'(u) \text{ evaluated} \\ \text{at } g(t)}} \cdot \underbrace{\frac{(3t^2 + 2)10t - 5t^2(6t)}{(3t^2 + 2)^2}}_{g'(t) \text{ by the Quotient Rule}} = \frac{1500t^5}{(3t^2 + 2)^4}.$$

*Related Exercises 26, 27, 31* ◄

The Chain Rule is also used to calculate the derivative of a composite function for a specific value of the variable. If $h(x) = f(g(x))$ and $a$ is a real number, then $h'(a) = f'(g(a))g'(a)$, provided the necessary derivatives exist. Therefore, $h'(a)$ is the derivative of $f$ evaluated at $g(a)$ multiplied by the derivative of $g$ evaluated at $a$.

**Table 3.2**

| $x$ | $f'(x)$ | $g(x)$ | $g'(x)$ |
|---|---|---|---|
| 1 | 5 | 2 | 3 |
| 2 | 7 | 1 | 4 |

**EXAMPLE 3**  **Calculating derivatives at a point** Let $h(x) = f(g(x))$. Use the values in Table 3.2 to calculate $h'(1)$ and $h'(2)$.

**SOLUTION** We use $h'(a) = f'(g(a))g'(a)$ with $a = 1$:

$$h'(1) = f'(g(1))g'(1) = f'(2)g'(1) = 7 \cdot 3 = 21.$$

With $a = 2$, we have

$$h'(2) = f'(g(2))g'(2) = f'(1)g'(2) = 5 \cdot 4 = 20.$$

*Related Exercises 23–24* ◄

**EXAMPLE 4**  **Applying the Chain Rule** A trail runner programs her GPS unit to record her altitude $a$ (in feet) every 10 minutes during a training run in the mountains; the resulting data are shown in Table 3.3. Meanwhile, at a nearby weather station, a weather probe records the atmospheric pressure $p$ (in hectopascals, or hPa) at various altitudes, shown in Table 3.4.

**Table 3.3**

| $t$ (minutes) | 0 | 10 | 20 | 30 | 40 | 50 | 60 | 70 | 80 |
|---|---|---|---|---|---|---|---|---|---|
| $a(t)$ (altitude) | 10,000 | 10,220 | 10,510 | 10,980 | 11,660 | 12,330 | 12,710 | 13,330 | 13,440 |

**Table 3.4**

| $a$ (altitude) | 5485 | 7795 | 10,260 | 11,330 | 12,330 | 13,330 | 14,330 | 15,830 | 16,230 |
|---|---|---|---|---|---|---|---|---|---|
| $p(a)$ (pressure) | 1000 | 925 | 840 | 821 | 793 | 765 | 738 | 700 | 690 |

Use the Chain Rule to estimate the rate of change in pressure per unit time experienced by the trail runner when she is 50 minutes into her run.

**SOLUTION** We seek the rate of change in the pressure $\dfrac{dp}{dt}$, which is given by the Chain Rule:

$$\frac{dp}{dt} = \frac{dp}{da}\frac{da}{dt}.$$

► The difference quotient $\dfrac{p(a + h) - p(a)}{h}$ is a *forward difference quotient* when $h > 0$ (see Exercises 62–65 in Section 3.1)

The runner is at an altitude of 12,330 feet 50 minutes into her run, so we must compute $dp/da$ when $a = 12,330$ and $da/dt$ when $t = 50$. These derivatives can be approximated using the following forward difference quotients.

$$\left.\frac{dp}{da}\right|_{a=12,330} \approx \frac{p(12,330 + 1000) - p(12,330)}{1000}$$

$$= \frac{765 - 793}{1000}$$

$$= -0.028 \ \frac{\text{hPa}}{\text{ft}}$$

$$\left.\frac{da}{dt}\right|_{t=50} \approx \frac{a(50 + 10) - a(50)}{10}$$

$$= \frac{12,710 - 12,330}{10}$$

$$= 38.0 \ \frac{\text{ft}}{\text{min}}$$

We now compute the rate of change of the pressure with respect to time:

$$\frac{dp}{dt} = \frac{dp}{da}\frac{da}{dt}$$

$$\approx -0.028 \ \frac{\text{hPa}}{\text{ft}} \cdot 38.0 \ \frac{\text{ft}}{\text{min}} = -1.06 \ \frac{\text{hPa}}{\text{min}}.$$

As expected, $dp/dt$ is negative because the pressure decreases with increasing altitude (Table 3.4) as the runner ascends the trail. Note also that the units are consistent.

*Related Exercises 72–73* ◄

## Chain Rule for Powers

The Chain Rule leads to a general derivative rule for powers of differentiable functions. In fact, we have already used it in several examples. Consider the function $f(x) = (g(x))^p$, where $p$ is a real number. Letting $f(u) = u^p$ be the outer function and $u = g(x)$ be the inner function, we obtain the Chain Rule for powers of functions.

---

> **THEOREM 3.13    Chain Rule for Powers**
> If $g$ is differentiable for all $x$ in its domain and $p$ is a real number, then
> $$\frac{d}{dx}\left((g(x))^p\right) = p(g(x))^{p-1}g'(x).$$

---

**EXAMPLE 5    Chain Rule for powers** Find $\dfrac{d}{dx}(\tan x + 10)^{21}$.

**SOLUTION** With $g(x) = \tan x + 10$, the Chain Rule gives

$$\frac{d}{dx}(\tan x + 10)^{21} = 21(\tan x + 10)^{20}\frac{d}{dx}(\tan x + 10)$$

$$= 21(\tan x + 10)^{20}\sec^2 x.$$

*Related Exercises 43–44* ◄

## The Composition of Three or More Functions

We can differentiate the composition of three or more functions by applying the Chain Rule repeatedly, as shown in the following example.

**EXAMPLE 6    Composition of three functions** Calculate the derivative of $\sin(\cos x^2)$.

**SOLUTION** The inner function of $\sin(\cos x^2)$ is $\cos x^2$. Because $\cos x^2$ is also a composition of two functions, the Chain Rule is used again to calculate $\dfrac{d}{dx}(\cos x^2)$, where $x^2$ is the inner function:

$$\frac{d}{dx}(\underbrace{\sin}_{\text{outer}}\,(\underbrace{\cos x^2}_{\text{inner}})) = \cos(\cos x^2)\frac{d}{dx}(\cos x^2) \qquad \text{Chain Rule}$$

$$= \cos(\cos x^2)\underbrace{(-\sin x^2)\cdot\frac{d}{dx}(x^2)}_{\frac{d}{dx}(\cos x^2)} \qquad \text{Chain Rule}$$

**QUICK CHECK 3** Let $y = \tan^{10}(x^5)$. Find $f$, $g$, and $h$ such that $y = f(u)$, where $u = g(v)$ and $v = h(x)$. ◄

$$= \cos(\cos x^2)\cdot(-\sin x^2)\cdot 2x \qquad \text{Differentiate } x^2.$$

$$= -2x\cos(\cos x^2)\sin x^2. \qquad \text{Simplify.}$$

*Related Exercises 49–50* ◄

The Chain Rule is often used together with other derivative rules. Example 7 illustrates how several differentiation rules are combined.

**EXAMPLE 7**   **Combining rules**   Find $\dfrac{d}{dx}\left(x^2\sqrt{x^2+1}\right)$.

**SOLUTION**   The given function is the product of $x^2$ and $\sqrt{x^2+1}$, and $\sqrt{x^2+1}$ is a composite function. We apply the Product Rule and then the Chain Rule:

$$\frac{d}{dx}\left(x^2\sqrt{x^2+1}\right) = \underbrace{\frac{d}{dx}(x^2)}_{2x}\cdot\sqrt{x^2+1} + \underbrace{x^2\cdot\frac{d}{dx}\left(\sqrt{x^2+1}\right)}_{\text{Use Chain Rule}} \qquad \text{Product Rule}$$

$$= 2x\sqrt{x^2+1} + x^2\cdot\frac{1}{2\sqrt{x^2+1}}\cdot 2x \qquad \text{Chain Rule}$$

$$= 2x\sqrt{x^2+1} + \frac{x^3}{\sqrt{x^2+1}} \qquad \text{Simplify.}$$

$$= \frac{3x^3+2x}{\sqrt{x^2+1}}. \qquad \text{Simplify.}$$

*Related Exercises 62–63* ◄

## Proof of the Chain Rule

Suppose $f$ is differentiable at $u = g(a)$, $g$ is differentiable at $a$, and $h(x) = f(g(x))$. By the definition of the derivative of $h$,

$$h'(a) = \lim_{x\to a}\frac{h(x)-h(a)}{x-a} = \lim_{x\to a}\frac{f(g(x))-f(g(a))}{x-a}. \qquad (3)$$

We assume $g(a) \neq g(x)$ for values of $x$ near $a$ but not equal to $a$. This assumption holds for most functions encountered in this text. For a proof of the Chain Rule without this assumption, see Exercise 101.

We multiply the right side of equation (3) by $\dfrac{g(x)-g(a)}{g(x)-g(a)}$, which equals 1, and let $v = g(x)$ and $u = g(a)$. The result is

$$h'(a) = \lim_{x\to a}\frac{f(g(x))-f(g(a))}{g(x)-g(a)}\cdot\frac{g(x)-g(a)}{x-a}$$

$$= \lim_{x\to a}\frac{f(v)-f(u)}{v-u}\cdot\frac{g(x)-g(a)}{x-a}.$$

By assumption, $g$ is differentiable at $a$; therefore, it is continuous at $a$. This means that $\lim_{x\to a} g(x) = g(a)$, so $v \to u$ as $x \to a$. Consequently,

$$h'(a) = \underbrace{\lim_{v\to u}\frac{f(v)-f(u)}{v-u}}_{f'(u)}\cdot\underbrace{\lim_{x\to a}\frac{g(x)-g(a)}{x-a}}_{g'(a)} = f'(u)g'(a).$$

Because $f$ and $g$ are differentiable at $u$ and $a$, respectively, the two limits in this expression exist; therefore, $h'(a)$ exists. Noting that $u = g(a)$, we have $h'(a) = f'(g(a))g'(a)$. Replacing $a$ with the variable $x$ gives the Chain Rule: $h'(x) = f'(g(x))g'(x)$.   ◄

## SECTION 3.7 EXERCISES

### Getting Started

1. Two equivalent forms of the Chain Rule for calculating the derivative of $y = f(g(x))$ are presented in this section. State both forms.

2. Identify the inner and outer functions in the composition $(x^2+10)^{-5}$.

3. Identify an inner function $u = g(x)$ and an outer function $y = f(u)$ of $y = (x^3+x+1)^4$. Then calculate $\dfrac{dy}{dx}$ using $\dfrac{dy}{dx} = \dfrac{dy}{du}\cdot\dfrac{du}{dx}$.

**4.** Identify an inner function $u = g(x)$ and an outer function $y = f(u)$ of $y = \sin(x^3 + 2x)$. Then calculate $\dfrac{dy}{dx}$ using $\dfrac{dy}{dx} = \dfrac{dy}{du} \cdot \dfrac{du}{dx}$.

**5.** The two composite functions $y = \cos^3 x$ and $y = \cos x^3$ look similar, but in fact are quite different. For each function, identify the inner function $u = g(x)$ and the outer function $y = f(u)$; then evaluate $\dfrac{dy}{dx}$ using the Chain Rule.

**6.** Let $h(x) = f(g(x))$, where $f$ and $g$ are differentiable on their domains. If $g(1) = 3$ and $g'(1) = 5$, what else do you need to know to calculate $h'(1)$?

**7.** Fill in the blanks. The derivative of $f(g(x))$ equals $f'$ evaluated at _____ multiplied by $g'$ evaluated at _____.

**8.** Evaluate the derivative of $y = (x^2 + 2x + 1)^2$ using $d/dx(f(g(x))) = f'(g(x)) \cdot g'(x)$.

**9.** Evaluate the derivative of $y = \sqrt{4x + 1}$ using $d/dx(f(g(x))) = f'(g(x)) \cdot g'(x)$.

**10.** Express $Q(x) = \cos^4(x^2 + 1)$ as the composition of three functions; that is, identify $f$, $g$, and $h$ such that $Q(x) = f(g(h(x)))$.

**11.** Given that $h(x) = f(g(x))$, find $h'(3)$ if $g(3) = 4$, $g'(3) = 5$, $f(4) = 9$, and $f'(4) = 10$.

**12.** Given that $h(x) = f(g(x))$, use the graphs of $f$ and $g$ to find $h'(4)$.

### Practice Exercises

**13–22.** *For each of the following composite functions, find an inner function $u = g(x)$ and an outer function $y = f(u)$ such that $y = f(g(x))$. Then calculate $dy/dx$.*

**13.** $y = (3x + 7)^{10}$

**14.** $y = (5x^2 + 11x)^{4/3}$

**15.** $y = \sin^5 x$

**16.** $y = \sin x^5$

**17.** $y = \sqrt{x^2 + 1}$

**18.** $y = \sqrt{7x - 1}$

**19.** $y = \sqrt{2 + \sin x}$

**20.** $y = (\cos x + \sin x)^2$

**21.** $y = \tan 5x^2$

**22.** $y = \sin \dfrac{x}{4}$

**23. Derivatives using tables** Let $h(x) = f(g(x))$ and $p(x) = g(f(x))$. Use the table to compute the following derivatives.

**a.** $h'(3)$  **b.** $h'(2)$  **c.** $p'(4)$  **d.** $p'(2)$  **e.** $h'(5)$

| $x$ | 1 | 2 | 3 | 4 | 5 |
|---|---|---|---|---|---|
| $f(x)$ | 0 | 3 | 5 | 1 | 0 |
| $f'(x)$ | 5 | 2 | $-5$ | $-8$ | $-10$ |
| $g(x)$ | 4 | 5 | 1 | 3 | 2 |
| $g'(x)$ | 2 | 10 | 20 | 15 | 20 |

**24. Derivatives using tables** Let $h(x) = f(g(x))$ and $k(x) = g(g(x))$. Use the table to compute the following derivatives.

**a.** $h'(1)$  **b.** $h'(2)$  **c.** $h'(3)$  **d.** $k'(3)$  **e.** $k'(1)$  **f.** $k'(5)$

| $x$ | 1 | 2 | 3 | 4 | 5 |
|---|---|---|---|---|---|
| $f'(x)$ | $-6$ | $-3$ | 8 | 7 | 2 |
| $g(x)$ | 4 | 1 | 5 | 2 | 3 |
| $g'(x)$ | 9 | 7 | 3 | $-1$ | $-5$ |

**25–70.** *Calculate the derivative of the following functions.*

**25.** $y = (3x^2 + 7x)^{10}$

**26.** $y = (x^2 + 2x + 7)^8$

**27.** $y = \sqrt{10x + 1}$

**28.** $y = \sqrt[3]{x^2 + 9}$

**29.** $y = 5(7x^3 + 1)^{-3}$

**30.** $y = \cos 5t$

**31.** $y = \sec(3x + 1)$

**32.** $y = \csc \sqrt{x}$

**33.** $y = \tan \sqrt{w}$

**34.** $y = \sin(\cos x)$

**35.** $y = \sin(4x^3 + 3x + 1)$

**36.** $y = \csc(t^2 + t)$

**37.** $y = (5x + 1)^{2/3}$

**38.** $y = x(x + 1)^{1/3}$

**39.** $y = \sqrt[4]{\dfrac{2x}{4x - 3}}$

**40.** $y = \cos^4 \theta + \sin^4 \theta$

**41.** $y = (\sec x + \tan x)^5$

**42.** $y = \sin(4 \cos z)$

**43.** $y = (2x^6 - 3x^3 + 3)^{25}$

**44.** $y = (\cos x + 2 \sin x)^8$

**45.** $y = (1 + 2 \tan u)^{4.5}$

**46.** $y = (1 - \sqrt{x})^4$

**47.** $y = \sqrt{1 + \cot^2 x}$

**48.** $y = \sqrt{(3x - 4)^2 + 3x}$

**49.** $y = (\sin^2 x + 1)^4$

**50.** $y = (\sin x^2 + 1)^4$

**51.** $y = \sin^5(\cos 3x)$

**52.** $y = \cos^{7/4}(4x^3)$

**53.** $y = \sqrt{x + \sqrt{x}}$

**54.** $y = \sqrt{x + \sqrt{x + \sqrt{x}}}$

**55.** $y = f(g(x^2))$, where $f$ and $g$ are differentiable for all real numbers

**56.** $y = (f(g(x^m)))^n$, where $f$ and $g$ are differentiable for all real numbers and $m$ and $n$ are constants

**57.** $y = \left(\dfrac{x}{x + 1}\right)^5$

**58.** $y = \left(\dfrac{x - 1}{x + 1}\right)^8$

**59.** $y = x(x^2 + 1)^3$

**60.** $y = \dfrac{x}{(x^2 + 1)^2}$

**61.** $y = \theta^2 \sec 5\theta$

**62.** $y = \left(\dfrac{3x}{4x + 2}\right)^5$

**63.** $y = ((x + 2)(x^2 + 1))^4$

**64.** $y = \left(\dfrac{\sin x}{\sin x + 1}\right)^2$

**65.** $y = \sqrt[5]{x^4 + \cos 2x}$

**66.** $y = \dfrac{x \cos x}{\sin x + 1}$

**67.** $y = (p + 3)^2 \sin p^2$

**68.** $y = (2z + 5)^{1.75} \tan z$

**69.** $y = \sqrt{f(x)}$, where $f$ is differentiable and nonnegative at $x$

**70.** $y = \sqrt[5]{f(x)g(x)}$, where $f$ and $g$ are differentiable at $x$

**71. Explain why or why not** Determine whether the following statements are true and give an explanation or counterexample.

**a.** The function $f(x) = x \sin x$ can be differentiated without using the Chain Rule.

**b.** The function $f(x) = (x^2 + 1)^{-12}$ should be differentiated using the Chain Rule.

**c.** The derivative of a product is *not* the product of the derivatives, but the derivative of a composition is a product of derivatives.

**d.** $\dfrac{d}{dx}P(Q(x)) = P'(x)Q'(x)$

**72. Changing temperature** The *lapse rate* is the rate at which the temperature in Earth's atmosphere decreases with altitude. For example, a lapse rate of 6.5° Celsius/km means the temperature *decreases* at a rate of 6.5°C per kilometer of altitude. The lapse rate varies with location and with other variables such as humidity. However, at a given time and location, the lapse rate is often nearly constant in the first 10 kilometers of the atmosphere. A radiosonde (weather balloon) is released from Earth's surface, and its altitude (measured in kilometers above sea level) at various times (measured in hours) is given in the table below.

| Time (hr) | 0 | 0.5 | 1 | 1.5 | 2 | 2.5 |
|---|---|---|---|---|---|---|
| Altitude (km) | 0.5 | 1.2 | 1.7 | 2.1 | 2.5 | 2.9 |

**a.** Assuming a lapse rate of 6.5°C/km, what is the approximate rate of change of the temperature with respect to time as the balloon rises 1.5 hours into the flight? Specify the units of your result and use a forward difference quotient when estimating the required derivative.

**b.** How does an increase in lapse rate change your answer in part (a)?

**c.** Is it necessary to know the actual temperature to carry out the calculation in part (a)? Explain.

**73. Applying the Chain Rule** Use the data in Tables 3.3 and 3.4 of Example 4 to estimate the rate of change in pressure with respect to time experienced by the runner when she is at an altitude of 13,330 ft. Make use of a forward difference quotient when estimating the required derivatives.

**74–77. Second derivatives** *Find* $\dfrac{d^2y}{dx^2}$ *for the following functions.*

**74.** $y = x \cos x^2$          **75.** $y = \sin x^2$

**76.** $y = \sqrt{x^2 + 2}$          **77.** $y = (x^2 + 1)^{-2}$

**78. Derivatives by different methods**

**a.** Calculate $\dfrac{d}{dx}(x^2 + x)^2$ using the Chain Rule. Simplify your answer.

**b.** Expand $(x^2 + x)^2$ first and then calculate the derivative. Verify that your answer agrees with part (a).

**79. Tangent lines** Determine an equation of the line tangent to the graph of $y = \dfrac{(x^2 - 1)^2}{x^3 - 6x - 1}$ at the point $(0, -1)$.

**T 80. Tangent lines** Determine equations of the lines tangent to the graph of $y = x\sqrt{5 - x^2}$ at the points $(1, 2)$ and $(-2, -2)$. Graph the function and the tangent lines.

**81. Tangent lines** Assume $f$ and $g$ are differentiable on their domains with $h(x) = f(g(x))$. Suppose the equation of the line tangent to the graph of $g$ at the point $(4, 7)$ is $y = 3x - 5$ and the equation of the line tangent to the graph of $f$ at $(7, 9)$ is $y = -2x + 23$.

**a.** Calculate $h(4)$ and $h'(4)$.

**b.** Determine an equation of the line tangent to the graph of $h$ at $(4, h(4))$.

**82. Tangent lines** Assume $f$ is a differentiable function whose graph passes through the point $(1, 4)$. Suppose $g(x) = f(x^2)$ and the line tangent to the graph of $f$ at $(1, 4)$ is $y = 3x + 1$. Find each of the following.

**a.** $g(1)$          **b.** $g'(x)$          **c.** $g'(1)$

**d.** An equation of the line tangent to the graph of $g$ when $x = 1$

**83. Tangent lines** Find the equation of the line tangent to $y = \sec 2x$ at $x = \dfrac{\pi}{6}$. Graph the function and the tangent line.

**84. Composition containing sin x** Suppose $f$ is differentiable on $[-2, 2]$ with $f'(0) = 3$ and $f'(1) = 5$. Let $g(x) = f(\sin x)$. Evaluate the following expressions.

**a.** $g'(0)$          **b.** $g'\left(\dfrac{\pi}{2}\right)$          **c.** $g'(\pi)$

**85. Composition containing sin x** Suppose $f$ is differentiable for all real numbers with $f(0) = -3$, $f(1) = 3$, $f'(0) = 3$, and $f'(1) = 5$. Let $g(x) = \sin(\pi f(x))$. Evaluate the following expressions.

**a.** $g'(0)$          **b.** $g'(1)$

**86–88. Vibrations of a spring** *Suppose an object of mass m is attached to the end of a spring hanging from the ceiling. The mass is at its equilibrium position y = 0 when the mass hangs at rest. Suppose you push the mass to a position $y_0$ units above its equilibrium position and release it. As the mass oscillates up and down (neglecting any friction in the system), the position y of the mass after t seconds is*

$$y = y_0 \cos\left(t\sqrt{\dfrac{k}{m}}\right), \tag{4}$$

*where k > 0 is a constant measuring the stiffness of the spring (the larger the value of k, the stiffer the spring) and y is positive in the upward direction.*

**86.** Use equation (4) to answer the following questions.

**a.** Find $dy/dt$, the velocity of the mass. Assume $k$ and $m$ are constant.

**b.** How would the velocity be affected if the experiment were repeated with four times the mass on the end of the spring?

**c.** How would the velocity be affected if the experiment were repeated with a spring having four times the stiffness ($k$ is increased by a factor of 4)?

**d.** Assume $y$ has units of meters, $t$ has units of seconds, $m$ has units of kg, and $k$ has units of kg/s². Show that the units of the velocity in part (a) are consistent.

**87.** Use equation (4) to answer the following questions.

**a.** Find the second derivative $\dfrac{d^2y}{dt^2}$.

**b.** Verify that $\dfrac{d^2y}{dt^2} = -\dfrac{k}{m}y$.

**88.** Use equation (4) to answer the following questions.

**a.** The *period T* is the time required by the mass to complete one oscillation. Show that $T = 2\pi\sqrt{\dfrac{m}{k}}$.

**b.** Assume $k$ is constant and calculate $\dfrac{dT}{dm}$.

**c.** Give a physical explanation of why $\dfrac{dT}{dm}$ is positive.

**T 89.** **Hours of daylight** The number of hours of daylight at any point on Earth fluctuates throughout the year. In the Northern Hemisphere, the shortest day is on the winter solstice and the longest day is on the summer solstice. At 40° north latitude, the length of a day is approximated by

$$D(t) = 12 - 3 \cos\left(\frac{2\pi(t + 10)}{365}\right),$$

where $D$ is measured in hours and $0 \le t \le 365$ is measured in days, with $t = 0$ corresponding to January 1.

a. Approximately how much daylight is there on March 1 ($t = 59$)?
b. Find the rate at which the daylight function changes.
c. Find the rate at which the daylight function changes on March 1. Convert your answer to units of min/day and explain what this result means.
d. Graph the function $y = D'(t)$ using a graphing utility.
e. At what times of the year is the length of day changing most rapidly? Least rapidly?

## Explorations and Challenges

**T 90.** **A mixing tank** A 500-liter (L) tank is filled with pure water. At time $t = 0$, a salt solution begins flowing into the tank at a rate of 5 L/min. At the same time, the (fully mixed) solution flows out of the tank at a rate of 5.5 L/min. The mass of salt in grams in the tank at any time $t \ge 0$ is given by

$$M(t) = 250(1000 - t)(1 - 10^{-30}(1000 - t)^{10})$$

and the volume of solution in the tank is given by

$$V(t) = 500 - 0.5t.$$

a. Graph the mass function and verify that $M(0) = 0$.
b. Graph the volume function and verify that the tank is empty when $t = 1000$ min.
c. The concentration of the salt solution in the tank (in g/L) is given by $C(t) = M(t)/V(t)$. Graph the concentration function and comment on its properties. Specifically, what are $C(0)$ and $\lim\limits_{t \to 1000^-} C(t)$?
d. Find the rate of change of the mass $M'(t)$, for $0 \le t \le 1000$.
e. Find the rate of change of the concentration $C'(t)$, for $0 \le t \le 1000$.
f. For what times is the concentration of the solution increasing? Decreasing?

**T 91.** **Power and energy** The total energy in megawatt-hr (MWh) used by a town is given by

$$E(t) = 400t + \frac{2400}{\pi} \sin\frac{\pi t}{12},$$

where $t \ge 0$ is measured in hours, with $t = 0$ corresponding to noon.

a. Find the power, or rate of energy consumption, $P(t) = E'(t)$ in units of megawatts (MW).
b. At what time of day is the rate of energy consumption a maximum? What is the power at that time of day?
c. At what time of day is the rate of energy consumption a minimum? What is the power at that time of day?
d. Sketch a graph of the power function reflecting the times when energy use is a minimum or a maximum.

**92.** **Deriving trigonometric identities**

a. Differentiate both sides of the identity $\cos 2t = \cos^2 t - \sin^2 t$ to prove that $\sin 2t = 2 \sin t \cos t$.
b. Verify that you obtain the same identity for $\sin 2t$ as in part (a) if you differentiate the identity $\cos 2t = 2 \cos^2 t - 1$.
c. Differentiate both sides of the identity $\sin 2t = 2 \sin t \cos t$ to prove that $\cos 2t = \cos^2 t - \sin^2 t$.

**93.** **Quotient Rule derivation** Suppose you forgot the Quotient Rule for calculating $\dfrac{d}{dx}\left(\dfrac{f(x)}{g(x)}\right)$. Use the Chain Rule and Product Rule with the identity $\dfrac{f(x)}{g(x)} = f(x)(g(x))^{-1}$ to derive the Quotient Rule.

**94.** **The Chain Rule for second derivatives**

a. Derive a formula for the second derivative, $\dfrac{d^2}{dx^2}(f(g(x)))$.
b. Use the formula in part (a) to calculate
$$\frac{d^2}{dx^2}(\sin(3x^4 + 5x^2 + 2)).$$

**T 95–98. Calculating limits** *The following limits are the derivatives of a composite function g at a point a.*

a. *Find a possible function g and number a.*
b. *Use the Chain Rule to find each limit. Verify your answer by using a calculator.*

**95.** $\lim\limits_{x \to 2} \dfrac{(x^2 - 3)^5 - 1}{x - 2}$

**96.** $\lim\limits_{x \to 0} \dfrac{\sqrt{4 + \sin x} - 2}{x}$

**97.** $\lim\limits_{h \to 0} \dfrac{\sin(\pi/2 + h)^2 - \sin(\pi^2/4)}{h}$

**98.** $\lim\limits_{h \to 0} \dfrac{\dfrac{1}{3((1 + h)^5 + 7)^{10}} - \dfrac{1}{3(8)^{10}}}{h}$

**99.** **Limit of a difference quotient** Assuming $f$ is differentiable for all $x$, simplify $\lim\limits_{x \to 5} \dfrac{f(x^2) - f(25)}{x - 5}$.

**100.** **Derivatives of even and odd functions** Recall that $f$ is even if $f(-x) = f(x)$, for all $x$ in the domain of $f$, and $f$ is odd if $f(-x) = -f(x)$, for all $x$ in the domain of $f$.

a. If $f$ is a differentiable, even function on its domain, determine whether $f'$ is even, odd, or neither.
b. If $f$ is a differentiable, odd function on its domain, determine whether $f'$ is even, odd, or neither.

**101.** **A general proof of the Chain Rule** Let $f$ and $g$ be differentiable functions with $h(x) = f(g(x))$. For a given constant $a$, let $u = g(a)$ and $v = g(x)$, and define

$$H(v) = \begin{cases} \dfrac{f(v) - f(u)}{v - u} - f'(u) & \text{if } v \ne u \\ 0 & \text{if } v = u. \end{cases}$$

a. Show that $\lim\limits_{v \to u} H(v) = 0$.
b. For any value of $u$, show that
$$f(v) - f(u) = (H(v) + f'(u))(v - u).$$
c. Show that
$$h'(a) = \lim\limits_{x \to a}\left((H(g(x)) + f'(g(a))) \cdot \frac{g(x) - g(a)}{x - a}\right).$$
d. Show that $h'(a) = f'(g(a))g'(a)$.

**QUICK CHECK ANSWERS**

**1.** The expansion of $(5x + 4)^{100}$ contains 101 terms. It would take too much time to calculate the expansion and the derivative.   **2.** The inner function is $u = 5x + 4$, and the outer function is $y = u^3$.   **3.** $f(u) = u^{10}$; $u = g(v) = \tan v$; $v = h(x) = x^5$ ◄

# 3.8 Implicit Differentiation

This chapter has been devoted to calculating derivatives of functions of the form $y = f(x)$, where $y$ is defined *explicitly* as a function of $x$. However, relations between variables are often expressed *implicitly*. For example, the equation of the unit circle $x^2 + y^2 = 1$ does not specify $y$ directly, but rather defines $y$ implicitly. This equation does not represent a single function because its graph fails the vertical line test (**Figure 3.54a**). If, however, the equation $x^2 + y^2 = 1$ is solved for $y$, then *two* functions emerge: $y = -\sqrt{1 - x^2}$ and $y = \sqrt{1 - x^2}$ (**Figure 3.54b**). Having identified two explicit functions that describe the circle, we can find their derivatives using the Chain Rule.

$$\text{If } y = \sqrt{1 - x^2}, \text{ then } \frac{dy}{dx} = -\frac{x}{\sqrt{1 - x^2}}. \tag{1}$$

$$\text{If } y = -\sqrt{1 - x^2}, \text{ then } \frac{dy}{dx} = \frac{x}{\sqrt{1 - x^2}}. \tag{2}$$

We use equation (1) to find the slope of the curve at any point on the upper half of the unit circle and equation (2) to find the slope of the curve at any point on the lower half of the circle.

**QUICK CHECK 1** The equation $x - y^2 = 0$ implicitly defines what two functions? ◄

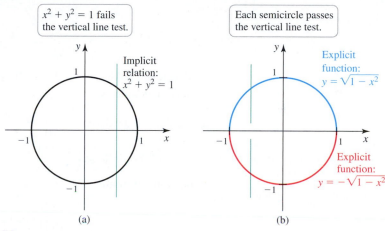

$x^2 + y^2 = 1$ fails the vertical line test.

Implicit relation: $x^2 + y^2 = 1$

Each semicircle passes the vertical line test.

Explicit function: $y = \sqrt{1 - x^2}$

Explicit function: $y = -\sqrt{1 - x^2}$

(a)          (b)

**Figure 3.54**

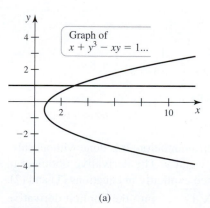

Graph of $x + y^3 - xy = 1$...

(a)

**Figure 3.55**

$y = f_1(x)$

$y = f_3(x)$

$y = f_2(x)$

... represents 3 functions.

(b)

Although it is straightforward to solve some implicit equations for $y$ (such as $x^2 + y^2 = 1$ or $x - y^2 = 0$), it is difficult or impossible to solve other equations for $y$. For example, the graph of $x + y^3 - xy = 1$ (**Figure 3.55a**) represents three functions: the upper half of a parabola $y = f_1(x)$, the lower half of a parabola $y = f_2(x)$, and the horizontal line $y = f_3(x)$ (**Figure 3.55b**). Solving for $y$ to obtain these three functions is challenging (Exercise 69), and even after solving for $y$, we would have to calculate derivatives for each of the three functions separately. The goal of this section is to find a *single* expression for the derivative *directly* from an equation without first solving for $y$. This technique, called **implicit differentiation**, is demonstrated through examples.

**EXAMPLE 1   Implicit differentiation**

**a.** Calculate $\dfrac{dy}{dx}$ directly from the equation for the unit circle $x^2 + y^2 = 1$.

**b.** Find the slope of the unit circle at $\left(\dfrac{1}{2}, \dfrac{\sqrt{3}}{2}\right)$ and $\left(\dfrac{1}{2}, -\dfrac{\sqrt{3}}{2}\right)$.

**SOLUTION**

**a.** To indicate the choice of $x$ as the independent variable, it is helpful to replace the variable $y$ with $y(x)$:

$$x^2 + (y(x))^2 = 1.   \text{Replace } y \text{ with } y(x).$$

We now take the derivative of each term in the equation *with respect to* $x$:

$$\underbrace{\frac{d}{dx}(x^2)}_{2x} + \underbrace{\frac{d}{dx}(y(x))^2}_{\text{Use the Chain Rule}} = \underbrace{\frac{d}{dx}(1)}_{0}.$$

By the Chain Rule, $\dfrac{d}{dx}(y(x))^2 = 2y(x)y'(x)$, or $\dfrac{d}{dx}(y^2) = 2y\dfrac{dy}{dx}$. Substituting this result, we have

$$2x + 2y\frac{dy}{dx} = 0.$$

The last step is to solve for $\dfrac{dy}{dx}$:

$$2y\frac{dy}{dx} = -2x   \text{Subtract } 2x \text{ from both sides.}$$

$$\frac{dy}{dx} = -\frac{x}{y}.   \text{Divide by } 2y \text{ and simplify.}$$

This result holds provided $y \neq 0$. At the points $(1, 0)$ and $(-1, 0)$, the circle has vertical tangent lines.

**b.** Notice that the derivative $\dfrac{dy}{dx} = -\dfrac{x}{y}$ depends on *both* $x$ and $y$. Therefore, to find the slope of the circle at $\left(\dfrac{1}{2}, \dfrac{\sqrt{3}}{2}\right)$, we substitute both $x = 1/2$ and $y = \sqrt{3}/2$ into the derivative formula. The result is

$$\left.\frac{dy}{dx}\right|_{\left(\frac{1}{2}, \frac{\sqrt{3}}{2}\right)} = -\frac{1/2}{\sqrt{3}/2} = -\frac{1}{\sqrt{3}}.$$

The slope of the curve at $\left(\dfrac{1}{2}, -\dfrac{\sqrt{3}}{2}\right)$ is

$$\left.\frac{dy}{dx}\right|_{\left(\frac{1}{2}, -\frac{\sqrt{3}}{2}\right)} = -\frac{1/2}{-\sqrt{3}/2} = \frac{1}{\sqrt{3}}.$$

The curve and tangent lines are shown in **Figure 3.56**.

*Related Exercises 13, 15* ◄

> Implicit differentiation usually produces an expression for $dy/dx$ in terms of both $x$ and $y$. The notation $\left.\dfrac{dy}{dx}\right|_{(a,\,b)}$ tells us to replace $x$ with $a$ and $y$ with $b$ in the expression for $dy/dx$.

**Figure 3.56**

Example 1 illustrates the technique of implicit differentiation. It is done without solving for $y$, and it produces $dy/dx$ in terms of both $x$ and $y$. The derivative obtained in Example 1 is consistent with the derivatives calculated explicitly in equations (1) and (2). For the upper half of the circle, substituting $y = \sqrt{1 - x^2}$ into the implicit derivative $dy/dx = -x/y$ gives

$$\frac{dy}{dx} = -\frac{x}{y} = -\frac{x}{\sqrt{1 - x^2}},$$

which agrees with equation (1). For the lower half of the circle, substituting $y = -\sqrt{1 - x^2}$ into $dy/dx = -x/y$ gives

$$\frac{dy}{dx} = -\frac{x}{y} = \frac{x}{\sqrt{1 - x^2}},$$

which is consistent with equation (2). Therefore, implicit differentiation gives a single unified derivative $dy/dx = -x/y$.

**EXAMPLE 2    Implicit differentiation**  Find $y'(x)$ when $\sin xy = x^2 + y$.

**SOLUTION**  It is impossible to solve this equation for $y$ in terms of $x$, so we differentiate implicitly. Differentiating both sides of the equation with respect to $x$, using the Chain Rule and the Product Rule on the left side, gives

$$(\cos xy)(y + xy') = 2x + y'.$$

We now solve for $y'$:

$$xy'\cos xy - y' = 2x - y\cos xy \qquad \text{Rearrange terms.}$$
$$y'(x\cos xy - 1) = 2x - y\cos xy \qquad \text{Factor on left side.}$$
$$y' = \frac{2x - y\cos xy}{x\cos xy - 1}. \qquad \text{Solve for } y'.$$

Notice that the final result gives $y'$ in terms of both $x$ and $y$.    *Related Exercises 31, 33* ◄

## Slopes of Tangent Lines

Derivatives obtained by implicit differentiation typically depend on $x$ and $y$. Therefore, the slope of a curve at a particular point $(a, b)$ requires both the $x$- and $y$-coordinates of the point. These coordinates are also needed to find an equation of the tangent line at that point.

**EXAMPLE 3    Finding tangent lines with implicit functions**  Find an equation of the line tangent to the curve $x^2 + xy - y^3 = 7$ at $(3, 2)$.

**SOLUTION**  We calculate the derivative of each term of the equation $x^2 + xy - y^3 = 7$ with respect to $x$:

$$\frac{d}{dx}(x^2) + \frac{d}{dx}(xy) - \frac{d}{dx}(y^3) = \frac{d}{dx}(7) \qquad \text{Differentiate each term.}$$

$$2x + \underbrace{y + xy'}_{\text{Product Rule}} - \underbrace{3y^2 y'}_{\text{Chain Rule}} = 0 \qquad \text{Calculate the derivatives.}$$

$$3y^2 y' - xy' = 2x + y \qquad \text{Group the terms containing } y'.$$

$$y' = \frac{2x + y}{3y^2 - x}. \qquad \text{Factor and solve for } y'.$$

To find the slope of the tangent line at $(3, 2)$, we substitute $x = 3$ and $y = 2$ into the derivative formula:

$$\left.\frac{dy}{dx}\right|_{(3, 2)} = \left.\frac{2x + y}{3y^2 - x}\right|_{(3, 2)} = \frac{8}{9}.$$

An equation of the line passing through $(3, 2)$ with slope $\frac{8}{9}$ is

$$y - 2 = \frac{8}{9}(x - 3) \quad \text{or} \quad y = \frac{8}{9}x - \frac{2}{3}.$$

**Figure 3.57** shows the graphs of the curve $x^2 + xy - y^3 = 7$ and the tangent line.
                                                    *Related Exercises 47–48* ◄

**EXAMPLE 4    Slope of a curve**  Find the slope of the curve $2(x + y)^{1/3} = y$ at the point $(4, 4)$.

**SOLUTION**  We begin by differentiating both sides of the given equation with respect to $x$:

$$\frac{2}{3}(x + y)^{-2/3}\left(1 + \frac{dy}{dx}\right) = \frac{dy}{dx} \qquad \begin{array}{l}\text{Implicit differentiation,}\\ \text{Chain Rule, Power Rule}\end{array}$$

$$\frac{2}{3}(x + y)^{-2/3} = \frac{dy}{dx} - \frac{2}{3}(x + y)^{-2/3}\frac{dy}{dx} \qquad \text{Expand and collect like terms.}$$

$$\frac{2}{3}(x + y)^{-2/3} = \frac{dy}{dx}\left(1 - \frac{2}{3}(x + y)^{-2/3}\right). \qquad \text{Factor out } \frac{dy}{dx}.$$

**QUICK CHECK 2**  Use implicit differentiation to find $\dfrac{dy}{dx}$ for $x - y^2 = 3$. ◄

**QUICK CHECK 3**  If a function is defined explicitly in the form $y = f(x)$, which coordinates are needed to find the slope of a tangent line—the $x$-coordinate, the $y$-coordinate, or both? ◄

➤ Because $y$ is a function of $x$, we have
$$\frac{d}{dx}(x) = 1 \quad \text{and}$$
$$\frac{d}{dx}(y) = y'.$$
To differentiate $y^3$ with respect to $x$, we need the Chain Rule.

**Figure 3.57**

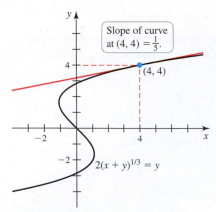

**Figure 3.58**

➤ The procedure outlined in Example 4 for simplifying $dy/dx$ is a useful trick. When computing derivatives using implicit differentiation, look back to the original equation to determine whether a substitution can be made to simplify the derivative. This technique is illustrated again in Example 5.

Solving for $dy/dx$, we find that

$$\frac{dy}{dx} = \frac{\frac{2}{3}(x+y)^{-2/3}}{1 - \frac{2}{3}(x+y)^{-2/3}}. \qquad \text{Divide by } 1 - \frac{2}{3}(x+y)^{-2/3}.$$

$$\frac{dy}{dx} = \frac{2}{3(x+y)^{2/3} - 2}. \qquad \text{Multiply by } 3(x+y)^{2/3} \text{ and simplify.}$$

At this stage, we could evaluate $dy/dx$ at the point $(4, 4)$ to determine the slope of the curve. However, the derivative can be simplified by employing a subtle maneuver. The original equation $2(x+y)^{1/3} = y$ implies that $(x+y)^{2/3} = y^2/4$. Therefore,

$$\frac{dy}{dx} = \frac{2}{3\underbrace{(x+y)^{2/3}}_{y^2/4} - 2} = \frac{8}{3y^2 - 8}.$$

With this simplified formula, finding the slope of the curve at the point $(4, 4)$ (**Figure 3.58**) requires only the $y$-coordinate:

$$\left.\frac{dy}{dx}\right|_{(4,4)} = \left.\frac{8}{3y^2 - 8}\right|_{y=4} = \frac{8}{3(4)^2 - 8} = \frac{1}{5}.$$

*Related Exercises 25–26* ◀

## Higher-Order Derivatives of Implicit Functions

In previous sections of this chapter, we found higher-order derivatives $\dfrac{d^n y}{dx^n}$ by first calculating $\dfrac{dy}{dx}, \dfrac{d^2 y}{dx^2}, \ldots,$ and $\dfrac{d^{n-1} y}{dx^{n-1}}$. The same approach is used with implicit differentiation.

**EXAMPLE 5**  **A second derivative**  Find $\dfrac{d^2 y}{dx^2}$ if $x^2 + y^2 = 1$.

**SOLUTION**  The first derivative $\dfrac{dy}{dx} = -\dfrac{x}{y}$ was computed in Example 1.

We now calculate the derivative of each side of this equation and simplify the right side:

$$\frac{d}{dx}\left(\frac{dy}{dx}\right) = \frac{d}{dx}\left(-\frac{x}{y}\right) \qquad \text{Take derivatives with respect to } x.$$

$$\frac{d^2 y}{dx^2} = -\frac{y \cdot 1 - x\dfrac{dy}{dx}}{y^2} \qquad \text{Quotient Rule}$$

$$= -\frac{y - x\left(-\dfrac{x}{y}\right)}{y^2} \qquad \text{Substitute for } \dfrac{dy}{dx}.$$

$$= -\frac{x^2 + y^2}{y^3} \qquad \text{Simplify.}$$

$$= -\frac{1}{y^3}. \qquad x^2 + y^2 = 1$$

In the last step of this calculation we used the original equation $x^2 + y^2 = 1$ to simplify the formula for $d^2 y/dx^2$.

*Related Exercises 51–52* ◀

## SECTION 3.8 EXERCISES

### Getting Started

1. For some equations, such as $x^2 + y^2 = 1$ or $x - y^2 = 0$, it is possible to solve for $y$ and then calculate $\dfrac{dy}{dx}$. Even in these cases, explain why implicit differentiation is usually a more efficient method for calculating the derivative.

2. Explain the differences between computing the derivatives of functions that are defined implicitly and explicitly.

3. Why are both the $x$-coordinate and the $y$-coordinate generally needed to find the slope of the tangent line at a point for an implicitly defined function?

4. Identify and correct the error in the following argument. Suppose $y^2 + 2y = 2x^3 - 7$. Differentiating both sides with respect to $x$ to find $\dfrac{dy}{dx}$, we have $2y + 2\dfrac{dy}{dx} = 6x^2$, which implies that $\dfrac{dy}{dx} = 3x^2 - y$.

**5–8.** Calculate $\dfrac{dy}{dx}$ using implicit differentiation.

5. $x = y^2$

6. $3x + 4y^3 = 7$

7. $\sin y + 2 = x$

8. $\sqrt{y} = \sqrt{x}$

9. Consider the curve defined by $2x - y + y^3 = 0$ (see figure).

   a. Find the coordinates of the $y$-intercepts of the curve.

   b. Verify that $\dfrac{dy}{dx} = \dfrac{2}{1 - 3y^2}$.

   c. Find the slope of the curve at each point where $x = 0$.

$2x - y + y^3 = 0$

10. Find the slope of the curve $x^2 + y^3 = 2$ at each point where $y = 1$ (see figure).

$x^2 + y^3 = 2$

11. Consider the curve $x = y^3$. Use implicit differentiation to verify that $\dfrac{dy}{dx} = \dfrac{1}{3y^2}$ and then find $\dfrac{d^2y}{dx^2}$.

12. Consider the curve $x = \sin y$. Use implicit differentiation to verify that $\dfrac{dy}{dx} = \sec y$ and then find $\dfrac{d^2y}{dx^2}$.

### Practice Exercises

**13–26. Implicit differentiation** *Carry out the following steps.*

a. *Use implicit differentiation to find* $\dfrac{dy}{dx}$.

b. *Find the slope of the curve at the given point.*

13. $x^4 + y^4 = 2$; $(1, -1)$

14. $y^2 + 1 = 2x$; $(1, 1)$

15. $y^2 = 4x$; $(1, 2)$

16. $y^2 + 3x = 8$; $\left(1, \sqrt{5}\right)$

17. $\sin y = 5x^4 - 5$; $(1, \pi)$

18. $\sqrt{x} - 2\sqrt{y} = 0$; $(4, 1)$

19. $\cos y = x$; $\left(0, \dfrac{\pi}{2}\right)$

20. $\tan xy = x + y$; $(0, 0)$

21. $xy = 7$; $(1, 7)$

22. $\dfrac{x}{y^2 + 1} = 1$; $(10, 3)$

23. $\sqrt[3]{x} + \sqrt[3]{y^4} = 2$; $(1, 1)$

24. $x^{2/3} + y^{2/3} = 2$; $(1, 1)$

25. $x\sqrt[3]{y} + y = 10$; $(1, 8)$

26. $(x + y)^{2/3} = y$; $(4, 4)$

**27–40. Implicit differentiation** *Use implicit differentiation to find* $\dfrac{dy}{dx}$.

27. $\sin x + \sin y = y$

28. $y = x \sin y$

29. $x + y = \cos y$

30. $x + 2y = \sqrt{y}$

31. $\sin xy = x + y$

32. $\tan (x + y) = 2y$

33. $\cos y^2 + x = y^2$

34. $y = \dfrac{x + 1}{y - 1}$

35. $x^3 = \dfrac{x + y}{x - y}$

36. $(xy + 1)^3 = x - y^2 + 8$

37. $6x^3 + 7y^3 = 13xy$

38. $\sin x \cos y = \sin x + \cos y$

39. $\sqrt{x^4 + y^2} = 5x + 2y^3$

40. $\sqrt{x + y^2} = \sin y$

41. **Cobb-Douglas production function** The output of an economic system $Q$, subject to two inputs, such as labor $L$ and capital $K$, is often modeled by the Cobb-Douglas production function $Q = cL^aK^b$. When $a + b = 1$, the case is called *constant returns to scale*. Suppose $Q = 1280$, $a = \dfrac{1}{3}$, $b = \dfrac{2}{3}$, and $c = 40$.

    a. Find the rate of change of capital with respect to labor, $dK/dL$.

    b. Evaluate the derivative in part (a) with $L = 8$ and $K = 64$.

42. **Surface area of a cone** The lateral surface area of a cone of radius $r$ and height $h$ (the surface area excluding the base) is $A = \pi r \sqrt{r^2 + h^2}$.

    a. Find $dr/dh$ for a cone with a lateral surface area of $A = 1500\pi$.

    b. Evaluate this derivative when $r = 30$ and $h = 40$.

**43. Volume of a spherical cap** Imagine slicing through a sphere with a plane (sheet of paper). The smaller piece produced is called a spherical cap. Its volume is $V = \pi h^2(3r - h)/3$, where $r$ is the radius of the sphere and $h$ is the thickness of the cap.

   **a.** Find $dr/dh$ for a spherical cap with a volume of $5\pi/3$.
   **b.** Evaluate this derivative when $r = 2$ and $h = 1$.

**44. Volume of a torus** The volume of a torus (doughnut or bagel) with an inner radius of $a$ and an outer radius of $b$ is $V = \pi^2(b + a)(b - a)^2/4$.

   **a.** Find $db/da$ for a torus with a volume of $64\pi^2$.
   **b.** Evaluate this derivative when $a = 6$ and $b = 10$.

**45–50. Tangent lines** *Carry out the following steps.*

*a. Verify that the given point lies on the curve.*
*b. Determine an equation of the line tangent to the curve at the given point.*

**45.** $\sin y + 5x = y^2;\ (0, 0)$    **46.** $x^3 + y^3 = 2xy;\ (1, 1)$

    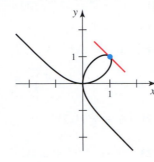

**47.** $x^2 + xy + y^2 = 7;\ (2, 1)$    **48.** $x^4 - x^2y + y^4 = 1;\ (-1, 1)$

    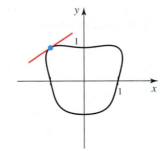

**49.** $\cos(x - y) + \sin y = \sqrt{2};$    **50.** $(x^2 + y^2)^2 = \dfrac{25}{4}xy^2;\ (1, 2)$
$\left(\dfrac{\pi}{2}, \dfrac{\pi}{4}\right)$

**51–56. Second derivatives** *Find $d^2y/dx^2$.*

**51.** $x + y^2 = 1$    **52.** $2x^2 + y^2 = 4$

**53.** $x + y = \sin y$    **54.** $x^4 + y^4 = 64$

**55.** $\cos y + x = y$    **56.** $\sin x + x^2y = 10$

**57. Explain why or why not** Determine whether the following statements are true and give an explanation or counterexample.

   **a.** For any equation containing the variables $x$ and $y$, the derivative $dy/dx$ can be found by first using algebra to rewrite the equation in the form $y = f(x)$.
   **b.** For the equation of a circle of radius $r$, $x^2 + y^2 = r^2$, we have $\dfrac{dy}{dx} = -\dfrac{x}{y}$, for $y \ne 0$ and any real number $r > 0$.
   **c.** If $x = 1$, then by implicit differentiation, $1 = 0$.
   **d.** If $xy = 1$, then $y' = 1/x$.

**58–59.** *Carry out the following steps.*

*a. Use implicit differentiation to find $\dfrac{dy}{dx}$.*
*b. Find the slope of the curve at the given point.*

**58.** $xy^{5/2} + x^{3/2}y = 12;\ (4, 1)$    **59.** $xy + x^{3/2}y^{-1/2} = 2;\ (1, 1)$

**T 60–62. Multiple tangent lines** *Complete the following steps.*

*a. Find equations of all lines tangent to the curve at the given value of x.*
*b. Graph the tangent lines on the given graph.*

**60.** $x + y^3 - y = 1;\ x = 1$    **61.** $x + y^2 - y = 1;\ x = 1$

    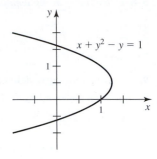

**62.** $4x^3 = y^2(4 - x);\ x = 2$
(cissoid of Diocles)

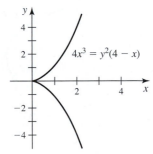

**63. Witch of Agnesi** Let $y(x^2 + 4) = 8$ (see figure).

   **a.** Use implicit differentiation to find $\dfrac{dy}{dx}$.
   **b.** Find equations of all lines tangent to the curve $y(x^2 + 4) = 8$ when $y = 1$.
   **c.** Solve the equation $y(x^2 + 4) = 8$ for $y$ to find an explicit expression for $y$ and then calculate $\dfrac{dy}{dx}$.
   **d.** Verify that the results of parts (a) and (c) are consistent.

**64. Vertical tangent lines**

    **a.** Determine the points where the curve $x + y^3 - y = 1$ has a vertical tangent line (see Exercise 60).

    **b.** Does the curve have any horizontal tangent lines? Explain.

**65. Vertical tangent lines**

    **a.** Determine the points where the curve $x + y^2 - y = 1$ has a vertical tangent line (Exercise 61).

    **b.** Does the curve have any horizontal tangent lines? Explain.

**T 66–67. Tangent lines for ellipses** *Find the equations of the vertical and horizontal tangent lines of the following ellipses.*

**66.** $x^2 + 4y^2 + 2xy = 12$

**67.** $9x^2 + y^2 - 36x + 6y + 36 = 0$

## Explorations and Challenges

**T 68–72. Identifying functions from an equation** *The following equations implicitly define one or more functions.*

    **a.** *Find $\dfrac{dy}{dx}$ using implicit differentiation.*

    **b.** *Solve the given equation for y to identify the implicitly defined functions $y = f_1(x), y = f_2(x), \ldots$.*

    **c.** *Use the functions found in part (b) to graph the given equation.*

**68.** $y^3 = ax^2$ (Neile's semicubical parabola)

**69.** $x + y^3 - xy = 1$ (*Hint:* Rewrite as $y^3 - 1 = xy - x$ and then factor both sides.)

**70.** $y^2 = \dfrac{x^2(4 - x)}{4 + x}$ (right strophoid)

**71.** $x^4 = 2(x^2 - y^2)$ (eight curve)

**72.** $y^2(x + 2) = x^2(6 - x)$ (trisectrix)

**T 73–78. Normal lines** *A **normal line** at a point P on a curve passes through P and is perpendicular to the line tangent to the curve at P (see figure). Use the following equations and graphs to determine an equation of the normal line at the given point. Illustrate your work by graphing the curve with the normal line.*

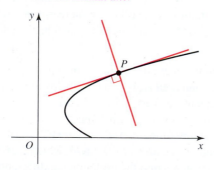

**73.** Exercise 45        **74.** Exercise 46

**75.** Exercise 47        **76.** Exercise 48

**77.** Exercise 49        **78.** Exercise 50

**T 79–82. Visualizing tangent and normal lines**

    **a.** *Determine an equation of the tangent line and the normal line at the given point $(x_0, y_0)$ on the following curves. (See instructions for Exercises 73–78.)*

    **b.** *Graph the tangent and normal lines on the given graph.*

**79.** $3x^3 + 7y^3 = 10y$; $(x_0, y_0) = (1, 1)$

**80.** $x^4 = 2x^2 + 2y^2$; $(x_0, y_0) = (2, 2)$ (kampyle of Eudoxus)

**81.** $(x^2 + y^2 - 2x)^2 = 2(x^2 + y^2)$; $(x_0, y_0) = (2, 2)$ (limaçon of Pascal)

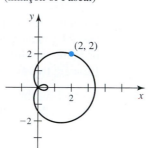

**82.** $(x^2 + y^2)^2 = \dfrac{25}{3}(x^2 - y^2)$; $(x_0, y_0) = (2, -1)$ (lemniscate of Bernoulli)

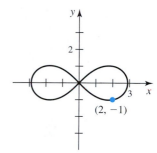

**83–85. Orthogonal trajectories** *Two curves are **orthogonal** to each other if their tangent lines are perpendicular at each point of intersection (recall that two lines are perpendicular to each other if their slopes are negative reciprocals). A family of curves forms **orthogonal trajectories** with another family of curves if each curve in one family is orthogonal to each curve in the other family. For example, the parabolas $y = cx^2$ form orthogonal trajectories with the family of ellipses $x^2 + 2y^2 = k$, where c and k are constants (see figure).*

    *Find $dy/dx$ for each equation of the following pairs. Use the derivatives to explain why the families of curves form orthogonal trajectories.*

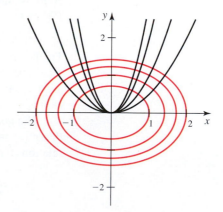

**83.** $y = mx$; $x^2 + y^2 = a^2$, where $m$ and $a$ are constants

**84.** $y = cx^2$; $x^2 + 2y^2 = k$, where $c$ and $k$ are constants

**85.** $xy = a$; $x^2 - y^2 = b$, where $a$ and $b$ are constants

**86. Finding slope** Find the slope of the curve $5\sqrt{x} - 10\sqrt{y} = \sin x$ at the point $(4\pi, \pi)$.

**87. A challenging derivative** Find $dy/dx$, where $(x^2 + y^2)(x^2 + y^2 + x) = 8xy^2$.

**88. A challenging derivative** Find $dy/dx$, where $\sqrt{3x^7 + y^2} = \sin^2 y + 100xy$.

**89. A challenging second derivative** Find $d^2y/dx^2$, where $\sqrt{y} + xy = 1$.

**T 90–92. Work carefully** *Proceed with caution when using implicit differentiation to find points at which a curve has a specified slope. For the following curves, find the points on the curve (if they exist) at which* the tangent line is horizontal or vertical. Once you have found possible points, make sure that they actually lie on the curve. Confirm your results with a graph.

**90.** $y^2 - 3xy = 2$

**91.** $x^2(3y^2 - 2y^3) = 4$

**92.** $x(1 - y^2) + y^3 = 0$

**QUICK CHECK ANSWERS**

**1.** $y = \sqrt{x}$ and $y = -\sqrt{x}$   **2.** $\dfrac{dy}{dx} = \dfrac{1}{2y}$

**3.** Only the $x$-coordinate is needed. ◄

# 3.9 Related Rates

We now return to the theme of derivatives as rates of change in problems in which the variables change with respect to *time*. The essential feature of these problems is that two or more variables, which are related in a known way, are themselves changing in time. Here are two examples illustrating this type of problem.

- An oil rig springs a leak and the oil spreads in a (roughly) circular patch around the rig. If the radius of the oil patch increases at a known rate, how fast is the area of the patch changing (Example 1)?
- Two airliners approach an airport with known speeds, one flying west and one flying north. How fast is the distance between the airliners changing (Example 2)?

In the first problem, the two related variables are the radius and the area of the oil patch. Both are changing in time. The second problem has three related variables: the positions of the two airliners and the distance between them. Again, the three variables change in time. The goal in both problems is to determine the rate of change of one of the variables at a specific moment of time—hence the name *related rates*.

We present a progression of examples in this section. After the first example, a general procedure is given for solving related-rate problems.

**Figure 3.59**

**EXAMPLE 1   Spreading oil** An oil rig springs a leak in calm seas, and the oil spreads in a circular patch around the rig. If the radius of the oil patch increases at a rate of 30 m/hr, how fast is the area of the patch increasing when the patch has a radius of 100 meters (**Figure 3.59**)?

**SOLUTION** Two variables change simultaneously: the radius of the circle and its area. The key relationship between the radius and area is $A = \pi r^2$. It helps to rewrite the basic relationship showing explicitly which quantities vary in time. In this case, we rewrite $A$ and $r$ as $A(t)$ and $r(t)$ to emphasize that they change with respect to $t$ (time). The general expression relating the radius and area at any time $t$ is $A(t) = \pi r(t)^2$.

The goal is to find the rate of change of the area of the circle, which is $A'(t)$, given that $r'(t) = 30$ m/hr. To introduce derivatives into the problem, we differentiate the area relation $A(t) = \pi r(t)^2$ with respect to $t$:

$$A'(t) = \frac{d}{dt}\left(\pi r(t)^2\right)$$

$$= \pi \frac{d}{dt}\left(r(t)^2\right)$$

$$= \pi \left(2r(t)\right) r'(t) \qquad \text{Chain Rule}$$

$$= 2\pi r(t)\, r'(t). \qquad \text{Simplify.}$$

Substituting the given values $r(t) = 100$ m and $r'(t) = 30$ m/hr, we have (including units)

$$A'(t) = 2\pi r(t)\, r'(t)$$

$$= 2\pi(100 \text{ m})\left(30\,\frac{\text{m}}{\text{hr}}\right)$$

$$= 6000\,\pi\,\frac{\text{m}^2}{\text{hr}}.$$

> It is important to remember that substitution of specific values of the variables occurs *after* differentiating.

**QUICK CHECK 1** In Example 1, what is the rate of change of the area when the radius is 200 m? 300 m? ◄

We see that the area of the oil spill increases at a rate of $6000\pi \approx 18{,}850$ m²/hr. Including units is a simple way to check your work. In this case, we expect an answer with units of area per unit time, so m²/hr makes sense. Notice that the rate of change of the area depends on the radius of the spill. As the radius increases, the rate of change of the area also increases.

*Related Exercises 5, 15* ◄

Using Example 1 as a template, we offer a general procedure for solving related-rate problems with the understanding that there are always variations that arise for individual problems.

---

**PROCEDURE**   **Steps for Related-Rate Problems**

1. Read the problem carefully, making a sketch to organize the given information. Identify the rates that are given and the rate that is to be determined.

2. Write one or more equations that express the basic relationships among the variables.

3. Introduce rates of change by differentiating the appropriate equation(s) with respect to time $t$.

4. Substitute known values and solve for the desired quantity.

5. Check that units are consistent and the answer is reasonable. (For example, does it have the correct sign?)

---

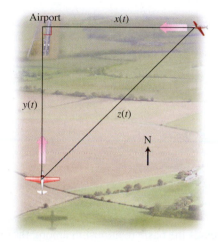

**Figure 3.60**

**EXAMPLE 2   Converging airplanes** Two small planes approach an airport, one flying due west at 120 mi/hr and the other flying due north at 150 mi/hr. Assuming they fly at the same constant elevation, how fast is the distance between the planes changing when the westbound plane is 180 miles from the airport and the northbound plane is 225 miles from the airport?

**SOLUTION** A sketch such as **Figure 3.60** helps us visualize the problem and organize the information. Let $x(t)$ and $y(t)$ denote the distance from the airport to the westbound and northbound planes, respectively. The paths of the two planes form the legs of a right triangle, and the distance between them, denoted $z(t)$, is the hypotenuse. By the Pythagorean theorem, $z^2 = x^2 + y^2$.

Our aim is to find $dz/dt$, the rate of change of the distance between the planes. We first differentiate both sides of $z^2 = x^2 + y^2$ with respect to $t$:

> In Example 1, we replaced $A$ and $r$ with $A(t)$ and $r(t)$, respectively, to remind us of the independent variable. After some practice, this replacement is not necessary.

$$\frac{d}{dt}(z^2) = \frac{d}{dt}(x^2 + y^2) \implies 2z\,\frac{dz}{dt} = 2x\,\frac{dx}{dt} + 2y\,\frac{dy}{dt}.$$

Notice that the Chain Rule is needed because $x$, $y$, and $z$ are functions of $t$. Solving for $dz/dt$ results in

> One could solve the equation $z^2 = x^2 + y^2$ for $z$, with the result
> $$z = \sqrt{x^2 + y^2},$$
> and then differentiate. However, it is easier to differentiate implicitly as shown in the example.

$$\frac{dz}{dt} = \frac{2x\,\frac{dx}{dt} + 2y\,\frac{dy}{dt}}{2z} = \frac{x\,\frac{dx}{dt} + y\,\frac{dy}{dt}}{z}.$$

This equation relates the unknown rate $dz/dt$ to the known quantities $x$, $y$, $z$, $dx/dt$, and $dy/dt$. For the westbound plane, $dx/dt = -120$ mi/hr (negative because the distance is decreasing), and for the northbound plane, $dy/dt = -150$ mi/hr. At the moment of interest, when $x = 180$ mi and $y = 225$ mi, the distance between the planes is

$$z = \sqrt{x^2 + y^2} = \sqrt{180^2 + 225^2} \approx 288 \text{ mi}.$$

Substituting these values gives

$$\frac{dz}{dt} = \frac{x\frac{dx}{dt} + y\frac{dy}{dt}}{z} \approx \frac{(180 \text{ mi})(-120 \text{ mi/hr}) + (225 \text{ mi})(-150 \text{ mi/hr})}{288 \text{ mi}}$$

$$\approx -192 \text{ mi/hr}.$$

Notice that $dz/dt < 0$, which means the distance between the planes is *decreasing* at a rate of about 192 mi/hr.

*Related Exercises 22–23* ◄

**QUICK CHECK 2** Assuming the same plane speeds as in Example 2, how fast is the distance between the planes changing if $x = 60$ mi and $y = 75$ mi? ◄

➤ Because coffee is draining out of the filter, the volume of coffee in the filter is decreasing, which implies that $dV/dt$ is negative.

**EXAMPLE 3    Morning coffee** Coffee is draining out of a conical filter at a rate of 2.25 in³/min. If the cone is 5 in tall and has a radius of 2 in, how fast is the coffee level dropping when the coffee is 3 in deep?

**SOLUTION** A sketch of the problem (**Figure 3.61a**) shows the three relevant variables: the volume $V$, the radius $r$, and the height $h$ of the coffee in the filter. The aim is to find the rate of change of the height $dh/dt$ at the instant that $h = 3$ in, given that $dV/dt = -2.25$ in³/min.

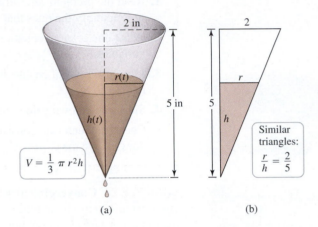

(a)                                (b)

**Figure 3.61**

The volume formula for a cone, $V = \dfrac{1}{3}\pi r^2 h$, expresses the basic relationship among the relevant variables. Using similar triangles (**Figure 3.61b**), we see that the ratio of the radius of the coffee in the cone to the height of the coffee is $2/5$ at all times; that is,

$$\frac{r}{h} = \frac{2}{5} \quad \text{or} \quad r = \frac{2}{5}h.$$

Substituting $r = \dfrac{2}{5}h$ into the volume formula gives $V$ in terms of $h$:

$$V = \frac{1}{3}\pi r^2 h = \frac{1}{3}\pi\left(\frac{2}{5}h\right)^2 h = \frac{4\pi}{75}h^3.$$

Rates of change are introduced by differentiating both sides of $V = \dfrac{4\pi}{75}h^3$ with respect to $t$:

$$\frac{dV}{dt} = \frac{4\pi}{25}h^2\frac{dh}{dt}.$$

Now we find $dh/dt$ at the instant when $h = 3$, given that $dV/dt = -2.25$ in$^3$/min. Solving for $dh/dt$ and substituting these values, we have

$$\frac{dh}{dt} = \frac{25 \, dV/dt}{4\pi h^2} \qquad \text{Solve for } \frac{dh}{dt}.$$

$$= \frac{25(-2.25 \text{ in}^3/\text{min})}{4\pi (3 \text{ in})^2} \approx -0.497 \, \frac{\text{in}}{\text{min}}. \qquad \text{Substitute for } \frac{dV}{dt} \text{ and } h.$$

At the instant that the coffee is 3 inches deep, the height decreases at a rate of $0.497$ in/min (almost half an inch per minute). Notice that the units work out consistently.

*Related Exercises 36–37* ◄

**QUICK CHECK 3** In Example 3, what is the rate of change of the height when $h = 2$ in? ◄

**Figure 3.62**

➤ The solution to Example 4 is reported in units of rad/s. Where did radians come from? Because a radian has no physical dimension (it is the ratio of an arc length to a radius), no unit appears. We write rad/s for clarity because $d\theta/dt$ is the rate of change of an angle.

**Figure 3.63**

➤ Recall that to convert radians to degrees, we use

$$\text{degrees} = \frac{180}{\pi} \cdot \text{radians}.$$

**EXAMPLE 4** **Observing a launch** An observer stands 200 meters from the launch site of a hot-air balloon at an elevation equal to the elevation of the launch site. The balloon rises vertically at a constant rate of 4 m/s. How fast is the angle of elevation of the balloon increasing 30 seconds after the launch? (The angle of elevation is the angle between the ground and the observer's line of sight to the balloon.)

**SOLUTION** Figure 3.62 shows the geometry of the launch. As the balloon rises, its distance from the ground $y$ and its angle of elevation $\theta$ change simultaneously. An equation expressing the relationship between these variables is $\tan \theta = y/200$. To find $d\theta/dt$, we differentiate both sides of this relationship using the Chain Rule:

$$\sec^2 \theta \, \frac{d\theta}{dt} = \frac{1}{200} \frac{dy}{dt}.$$

Next we solve for $\frac{d\theta}{dt}$:

$$\frac{d\theta}{dt} = \frac{dy/dt}{200 \sec^2 \theta} = \frac{(dy/dt) \cdot \cos^2 \theta}{200}.$$

The rate of change of the angle of elevation depends on the angle of elevation and the speed of the balloon. Thirty seconds after the launch, the balloon has risen $y = (4 \text{ m/s})(30 \text{ s}) = 120$ m. To complete the problem, we need the value of $\cos \theta$. Note that when $y = 120$ m, the distance between the observer and the balloon is

$$d = \sqrt{120^2 + 200^2} \approx 233.24 \text{ m}.$$

Therefore, $\cos \theta \approx 200/233.24 \approx 0.86$ (Figure 3.63), and the rate of change of the angle of elevation is

$$\frac{d\theta}{dt} = \frac{(dy/dt) \cdot \cos^2 \theta}{200} \approx \frac{(4 \text{ m/s})(0.86^2)}{200 \text{ m}} = 0.015 \text{ rad/s}.$$

At this instant, the balloon is rising at an angular rate of $0.015$ rad/s, or slightly less than $1°$/s, as seen by the observer.

*Related Exercise 46* ◄

**QUICK CHECK 4** In Example 4, notice that as the balloon rises (as $\theta$ increases), the rate of change of the angle of elevation decreases to zero. When does the maximum value of $\theta'(t)$ occur, and what is it? ◄

## SECTION 3.9 EXERCISES

### Getting Started

1. Give an example in which one dimension of a geometric figure changes and produces a corresponding change in the area or volume of the figure.

2. Charles' law states that for a fixed mass of gas under constant pressure, the volume $V$ and temperature $T$ of the gas (in kelvins) satisfy the equation $V = kT$, where $k$ is constant. Find an equation relating $dV/dt$ to $dT/dt$.

3. If two opposite sides of a rectangle increase in length, how must the other two opposite sides change if the area of the rectangle is to remain constant?

4. The temperature $F$ in degrees Fahrenheit is related to the temperature $C$ in degrees Celsius by the equation $F = \frac{9}{5}C + 32$.

   a. Find an equation relating $dF/dt$ to $dC/dt$.
   b. How fast is the temperature in an oven changing in degrees Fahrenheit per minute if it is rising at $10°$ Celsius per min?

5. A rectangular swimming pool 10 ft wide by 20 ft long and of uniform depth is being filled with water.

   a. If $t$ is elapsed time, $h$ is the height of the water, and $V$ is the volume of the water, find equations relating $V$ to $h$ and $dV/dt$ to $dh/dt$.
   b. At what rate is the volume of the water increasing if the water level is rising at $\frac{1}{4}$ ft/min?
   c. At what rate is the water level rising if the pool is filled at a rate of 10 ft³/min?

6. At all times, the length of a rectangle is twice the width $w$ of the rectangle as the area of the rectangle changes with respect to time $t$.

   a. Find an equation relating $A$ to $w$.
   b. Find an equation relating $dA/dt$ to $dw/dt$.

7. The volume $V$ of a sphere of radius $r$ changes over time $t$.

   a. Find an equation relating $dV/dt$ to $dr/dt$.
   b. At what rate is the volume changing if the radius increases at 2 in/min when the radius is 4 inches?
   c. At what rate is the radius changing if the volume increases at 10 in³/min when the radius is 5 inches?

8. At all times, the length of the long leg of a right triangle is 3 times the length $x$ of the short leg of the triangle. If the area of the triangle changes with respect to time $t$, find equations relating the area $A$ to $x$ and $dA/dt$ to $dx/dt$.

9. Assume $x$, $y$, and $z$ are functions of $t$ with $z = x + y^3$. Find $dz/dt$ when $dx/dt = -1$, $dy/dt = 5$, and $y = 2$.

10. Assume $w = x^2y^4$, where $x$ and $y$ are functions of $t$. Find $dw/dt$ when $x = 3$, $dx/dt = 2$, $dy/dt = 4$, and $y = 1$.

### Practice Exercises

11. **Expanding square** The sides of a square increase in length at a rate of 2 m/s.

    a. At what rate is the area of the square changing when the sides are 10 m long?
    b. At what rate is the area of the square changing when the sides are 20 m long?

12. **Shrinking square** The sides of a square decrease in length at a rate of 1 m/s.

    a. At what rate is the area of the square changing when the sides are 5 m long?
    b. At what rate are the lengths of the diagonals of the square changing?

13. **Expanding isosceles triangle** The legs of an isosceles right triangle increase in length at a rate of 2 m/s.

    a. At what rate is the area of the triangle changing when the legs are 2 m long?
    b. At what rate is the area of the triangle changing when the hypotenuse is 1 m long?
    c. At what rate is the length of the hypotenuse changing?

14. **Shrinking isosceles triangle** The hypotenuse of an isosceles right triangle decreases in length at a rate of 4 m/s.

    a. At what rate is the area of the triangle changing when the legs are 5 m long?
    b. At what rate are the lengths of the legs of the triangle changing?
    c. At what rate is the area of the triangle changing when the area is 4 m²?

15. **Expanding circle** The area of a circle increases at a rate of 1 cm²/s.

    a. How fast is the radius changing when the radius is 2 cm?
    b. How fast is the radius changing when the circumference is 2 cm?

16. **Expanding cube** The edges of a cube increase at a rate of 2 cm/s. How fast is the volume changing when the length of each edge is 50 cm?

17. **Shrinking circle** A circle has an initial radius of 50 ft when the radius begins decreasing at a rate of 2 ft/min. What is the rate of change of the area at the instant the radius is 10 ft?

18. **Shrinking cube** The volume of a cube decreases at a rate of 0.5 ft³/min. What is the rate of change of the side length when the side lengths are 12 ft?

19. **Balloons** A spherical balloon is inflated and its volume increases at a rate of 15 in³/min. What is the rate of change of its radius when the radius is 10 in?

20. **Expanding rectangle** A rectangle initially has dimensions 2 cm by 4 cm. All sides begin increasing in length at a rate of 1 cm/s. At what rate is the area of the rectangle increasing after 20 s?

21. **Melting snowball** A spherical snowball melts at a rate proportional to its surface area. Show that the rate of change of the radius is constant. (*Hint:* Surface area $= 4\pi r^2$.)

22. **Divergent paths** Two boats leave a port at the same time; one travels west at 20 mi/hr and the other travels south at 15 mi/hr.

    a. After 30 minutes, how far is each boat from port?
    b. At what rate is the distance between the boats changing 30 minutes after they leave the port?

23. **Time-lagged flights** An airliner passes over an airport at noon traveling 500 mi/hr due west. At 1:00 P.M., another airliner passes over the same airport at the same elevation traveling due north at 550 mi/hr. Assuming both airliners maintain their (equal) elevations, how fast is the distance between them changing at 2:30 P.M.?

24. **Flying a kite** Once Kate's kite reaches a height of 50 ft (above her hands), it rises no higher but drifts due east in a wind blowing 5 ft/s. How fast is the string running through Kate's hands at the moment when she has released 120 ft of string?

25. **Rope on a boat** A rope passing through a capstan on a dock is attached to a boat offshore. The rope is pulled in at a constant rate of 3 ft/s and the capstan is 5 ft vertically above the water. How fast is the boat traveling when it is 10 ft from the dock?

26. **Bug on a parabola** A bug is moving along the right side of the parabola $y = x^2$ at a rate such that its distance from the origin is increasing at 1 cm/min.

    **a.** At what rate is the $x$-coordinate of the bug increasing at the point $(2, 4)$?

    **b.** Use the equation $y = x^2$ to find an equation relating $\frac{dy}{dt}$ to $\frac{dx}{dt}$.

    **c.** At what rate is the $y$-coordinate of the bug increasing at the point $(2, 4)$?

27. **Balloons and motorcycles** A hot-air balloon is 150 ft above the ground when a motorcycle (traveling in a straight line on a horizontal road) passes directly beneath it going 40 mi/hr (58.67 ft/s). If the balloon rises vertically at a rate of 10 ft/s, what is the rate of change of the distance between the motorcycle and the balloon 10 seconds later?

28. **Baseball runners** Runners stand at first and second base in a baseball game. At the moment a ball is hit, the runner at first base runs to second base at 18 ft/s; simultaneously, the runner on second runs to third base at 20 ft/s. How fast is the distance between the runners changing 1 s after the ball is hit (see figure)? (*Hint:* The distance between consecutive bases is 90 ft and the bases lie at the corners of a square.)

29. **Fishing story** An angler hooks a trout and reels in his line at 4 in/s. Assume the tip of the fishing rod is 12 ft above the water and directly above the angler, and the fish is pulled horizontally directly toward the angler (see figure). Find the horizontal speed of the fish when it is 20 ft from the angler.

30. **Parabolic motion** An arrow is shot into the air and moves along the parabolic path $y = x(50 - x)$ (see figure). The horizontal

component of velocity is always 30 ft/s. What is the vertical component of velocity when (a) $x = 10$ and (b) $x = 40$?

31. **Draining a water heater** A water heater that has the shape of a right cylindrical tank with a radius of 1 ft and a height of 4 ft is being drained. How fast is water draining out of the tank (in ft³/min) if the water level is dropping at 6 in/min?

32. **Drinking a soda** At what rate (in in³/s) is soda being sucked out of a cylindrical glass that is 6 in tall and has a radius of 2 in? The depth of the soda decreases at a constant rate of 0.25 in/s.

33. **Piston compression** A piston is seated at the top of a cylindrical chamber with radius 5 cm when it starts moving into the chamber at a constant speed of 3 cm/s (see figure). What is the rate of change of the volume of the cylinder when the piston is 2 cm from the base of the chamber?

34. **Filling two pools** Two cylindrical swimming pools are being filled simultaneously at the same rate (in m³/min; see figure). The smaller pool has a radius of 5 m, and the water level rises at a rate of 0.5 m/min. The larger pool has a radius of 8 m. How fast is the water level rising in the larger pool?

35. **Growing sandpile** Sand falls from an overhead bin and accumulates in a conical pile with a radius that is always three times its height. Suppose the height of the pile increases at a rate of 2 cm/s when the pile is 12 cm high. At what rate is the sand leaving the bin at that instant?

**36. Draining a tank** An inverted conical water tank with a height of 12 ft and a radius of 6 ft is drained through a hole in the vertex at a rate of 2 ft³/s (see figure). What is the rate of change of the water depth when the water depth is 3 ft? (*Hint:* Use similar triangles.)

6 ft

12 ft

Outflow 2 ft³/s

**37. Draining a cone** Water is drained out of an inverted cone that has the same dimensions as the cone depicted in Exercise 36. If the water level drops at 1 ft/min, at what rate is water (in ft³/min) draining from the tank when the water depth is 6 ft?

**38. Two tanks** A conical tank with an upper radius of 4 m and a height of 5 m drains into a cylindrical tank with a radius of 4 m and a height of 5 m (see figure). If the water level in the conical tank drops at a rate of 0.5 m/min, at what rate does the water level in the cylindrical tank rise when the water level in the conical tank is 3 m? 1 m?

4 m

5 m

4 m

5 m

**39. Filling a hemispherical tank** A hemispherical tank with a radius of 10 m is filled from an inflow pipe at a rate of 3 m³/min (see figure). How fast is the water level rising when the water level is 5 m from the bottom of the tank? (*Hint:* The volume of a cap of thickness $h$ sliced from a sphere of radius $r$ is $\pi h^2(3r - h)/3$.)

Inflow 3 m³/min

10 m

**40. Surface area of hemispherical tank** For the situation described in Exercise 39, what is the rate of change of the area of the exposed surface of the water when the water is 5 m deep?

**41. Ladder against the wall** A 13-ft ladder is leaning against a vertical wall (see figure) when Jack begins pulling the foot of the ladder away from the wall at a rate of 0.5 ft/s. How fast is the top of the ladder sliding down the wall when the foot of the ladder is 5 ft from the wall?

13 ft

**42. Ladder against the wall again** A 12-ft ladder is leaning against a vertical wall when Jack begins pulling the foot of the ladder away from the wall at a rate of 0.2 ft/s. What is the configuration of the ladder at the instant when the vertical speed of the top of the ladder equals the horizontal speed of the foot of the ladder?

**43. Moving shadow** A 5-ft-tall woman walks at 8 ft/s toward a streetlight that is 20 ft above the ground. What is the rate of change of the length of her shadow when she is 15 ft from the streetlight? At what rate is the tip of her shadow moving?

**44. Another moving shadow** A landscape light at ground level lights up the side of a tall building that is 15 feet from the light. A 6-ft-tall man starts walking (on flat terrain) from the light directly toward the building. How fast is he walking when he is 9 feet from the light if his shadow on the building is shrinking at 2 ft/s at that instant?

**45. Watching an elevator** An observer is 20 m above the ground floor of a large hotel atrium looking at a glass-enclosed elevator shaft that is 20 m horizontally from the observer (see figure). The angle of elevation of the elevator is the angle that the observer's line of sight makes with the horizontal (it may be positive or negative). Assuming the elevator rises at a rate of 5 m/s, what is the rate of change of the angle of elevation when the elevator is 10 m above the ground? When the elevator is 40 m above the ground?

Elevator

$\theta > 0$
$\theta < 0$   Observer

20 m

20 m

**T 46. Observing a launch** An observer stands 300 ft from the launch site of a hot-air balloon at an elevation equal to the elevation of the launch site. The balloon is launched vertically and maintains a constant upward velocity of 20 ft/s. What is the rate of change of the angle of elevation of the balloon when it is 400 ft from the ground? (*Hint:* The angle of elevation is the angle $\theta$ between the observer's line of sight to the balloon and the ground.)

**T 47. Viewing angle** The bottom of a large theater screen is 3 ft above your eye level and the top of the screen is 10 ft above your eye level. Assume you walk away from the screen (perpendicular to the screen) at a rate of 3 ft/s while looking at the screen. What is the rate of change of the viewing angle $\theta$ when you are 30 ft from the wall on which the screen hangs, assuming the floor is horizontal (see figure)?

**48. Altitude of a jet** A jet ascends at a 10° angle from the horizontal with an airspeed of 550 mi/hr (its speed along its line of flight is 550 mi/hr). How fast is the altitude of the jet increasing? If the sun is directly overhead, how fast is the shadow of the jet moving on the ground?

**49. Rate of dive of a submarine** A surface ship is moving (horizontally) in a straight line at 10 km/hr. At the same time, an enemy submarine maintains a position directly below the ship while diving at an angle that is 20° below the horizontal. How fast is the submarine's altitude decreasing?

**50. Revolving light beam** A lighthouse stands 500 m off a straight shore, and the focused beam of its light revolves (at a constant rate) four times each minute. As shown in the figure, $P$ is the point on shore closest to the lighthouse and $Q$ is a point on the shore 200 m from $P$. What is the speed of the beam along the shore when it strikes the point $Q$? Describe how the speed of the beam along the shore varies with the distance between $P$ and $Q$. Neglect the height of the lighthouse.

**51. Filming a race** A camera is set up at the starting line of a drag race 50 ft from a dragster at the starting line (camera 1 in the figure). Two seconds after the start of the race, the dragster has traveled 100 ft and the camera is turning at 0.75 rad/s while filming the dragster.

**a.** What is the speed of the dragster at this point?

**b.** A second camera (camera 2 in the figure) filming the dragster is located on the starting line 100 ft away from the dragster at the start of the race. How fast is this camera turning 2 s after the start of the race?

**52. Fishing reel** An angler hooks a trout and begins turning her circular reel at 1.5 rev/s. Assume the radius of the reel (and the fishing line on it) is 2 inches.

**a.** Let $R$ equal the number of revolutions the angler has turned her reel and suppose $L$ is the amount of line that she has reeled in. Find an equation for $L$ as a function of $R$.

**b.** How fast is she reeling in her fishing line?

**53. Wind energy** The kinetic energy $E$ (in joules) of a mass in motion satisfies the equation $E = \frac{1}{2}mv^2$, where mass $m$ is measured in kg and velocity $v$ is measured in m/s.

**a.** Power $P$ is defined to be $dE/dt$, the rate of change in energy with respect to time. Power is measured in units of watts (W), where 1 W = 1 joule/s. If the velocity $v$ is constant, use implicit differentiation to find an equation for power $P$ in terms of the derivative $dm/dt$.

**b.** Wind turbines use kinetic energy in the wind to create electrical power. In this case, the derivative $dm/dt$ is called the **mass flow rate** and it satisfies the equation $\frac{dm}{dt} = \rho A v$, where $\rho$ is the density of the air in kg/m³, $A$ is the sweep area in m² of the wind turbine (see figure), and $v$ is the velocity of the wind in m/s. Show that $P = \frac{1}{2}\rho A v^3$.

**c.** Suppose a blade on a small wind turbine has a length of 3 m. Find the available power $P$ if the wind is blowing at 10 m/s. (*Hint:* Use $\rho = 1.23$ kg/m³ for the density of air. The density of air varies, but this is a reasonable average value.)

**d.** Wind turbines convert only a small percentage of the available wind power into electricity. Assume the wind turbine described in this exercise converts only 25% of the available wind power into electricity. How much electrical power is produced?

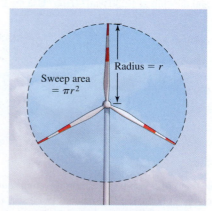

**54. Boyle's law** Robert Boyle (1627–1691) found that for a given quantity of gas at a constant temperature, the pressure $P$ (in kPa) and volume $V$ of the gas (in m$^3$) are accurately approximated by the equation $V = k/P$, where $k > 0$ is constant. Suppose the volume of an expanding gas is increasing at a rate of 0.15 m$^3$/min when the volume $V = 0.5$ m$^3$ and the pressure is $P = 50$ kPa. At what rate is pressure changing at this moment?

### Explorations and Challenges

**55. Clock hands** The hands of the clock in the tower of the Houses of Parliament in London are approximately 3 m and 2.5 m in length. How fast is the distance between the tips of the hands changing at 9:00? (*Hint:* Use the Law of Cosines.)

**56. Divergent paths** Two boats leave a port at the same time, one traveling west at 20 mi/hr and the other traveling southwest (45° south of west) at 15 mi/hr. After 30 minutes, how far apart are the boats and at what rate is the distance between them changing? (*Hint:* Use the Law of Cosines.)

**57. Filling a pool** A swimming pool is 50 m long and 20 m wide. Its depth decreases linearly along the length from 3 m to 1 m (see figure). It was initially empty and is being filled at a rate of 1 m$^3$/min. How fast is the water level rising 250 min after the filling begins? How long will it take to fill the pool?

Inflow 1 m$^3$/min

50 m

20 m

1 m

3 m

**58. Disappearing triangle** An equilateral triangle initially has sides of length 20 ft when each vertex moves toward the midpoint of the opposite side at a rate of 1.5 ft/min. Assuming the triangle remains equilateral, what is the rate of change of the area of the triangle at the instant the triangle disappears?

**T 59. Oblique tracking** A port and a radar station are 2 mi apart on a straight shore running east and west (see figure). A ship leaves the port at noon traveling northeast at a rate of 15 mi/hr. If the ship maintains its speed and course, what is the rate of change of the tracking angle $\theta$ between the shore and the line between the radar station and the ship at 12:30 P.M.? (*Hint:* Use the Law of Sines.)

N

Northeast course

$s$

$C$

$\theta$

$A$  45°  $B$

Port        Radar station

|← 2 mi →|

**T 60. Oblique tracking** A ship leaves port traveling southwest at a rate of 12 mi/hr. At noon, the ship reaches its closest approach to a radar station, which is on the shore 1.5 mi from the port. If the

ship maintains its speed and course, what is the rate of change of the tracking angle $\theta$ between the radar station and the ship at 1:30 P.M. (see figure)? (*Hint:* Use the Law of Sines.)

N

Radar station  |← 1.5 mi →|  Port

$\theta$      45°

Southwest course

Point of closest approach to radar station

**61. Navigation** A boat leaves a port traveling due east at 12 mi/hr. At the same time, another boat leaves the same port traveling northeast at 15 mi/hr. The angle $\theta$ of the line between the boats is measured relative to due north (see figure). What is the rate of change of this angle 30 min after the boats leave the port? 2 hr after the boats leave the port?

N

15 mi/hr

$\theta$

Port

12 mi/hr

**62. Watching a Ferris wheel** An observer stands 20 m from the bottom of a 10-m-tall Ferris wheel on a line that is perpendicular to the face of the Ferris wheel. The wheel revolves at a rate of $\pi$ rad/min and the observer's line of sight with a specific seat on the wheel makes an angle $\theta$ with the ground (see figure). Forty seconds after that seat leaves the lowest point on the wheel, what is the rate of change of $\theta$? Assume the observer's eyes are level with the bottom of the wheel.

$y$

$\pi$ rad/min

10 m

$\theta$

20 m

$x$

**63. Draining a trough** A trough in the shape of a half cylinder has length 5 m and radius 1 m. The trough is full of water when a valve is opened, and water flows out of the bottom of the trough at a rate of 1.5 m$^3$/hr (see figure). (*Hint:* The area of a sector of a circle of radius $r$ subtended by an angle $\theta$ is $r^2\theta/2$.)

**a.** How fast is the water level changing when the water level is 0.5 m from the bottom of the trough?

**b.** What is the rate of change of the surface area of the water when the water is 0.5 m deep?

Outflow
1.5 m³/hr

**64. Searchlight—wide beam** A revolving searchlight, which is 100 m from the nearest point on a straight highway, casts a horizontal beam along a highway (see figure). The beam leaves the spotlight at an angle of $\pi/16$ rad and revolves at a rate $\pi/6$ rad/s. Let $w$ be the width of the beam as it sweeps along the highway

and let $\theta$ be the angle that the center of the beam makes with the perpendicular to the highway. What is the rate of change of $w$ when $\theta = \pi/3$? Neglect the height of the searchlight.

$\theta$ is the angle between the center of the beam and the line perpendicular to the highway.

100 m

$\frac{\pi}{16}$ rad

$w$

**QUICK CHECK ANSWERS**

**1.** $12{,}000\pi$ m²/hr, $18{,}000\pi$ m²/hr   **2.** $-192$ mi/hr
**3.** $-1.12$ in/min   **4.** $t = 0, \theta = 0, \theta'(0) = 0.02$ rad/s ◄

# CHAPTER 3   REVIEW EXERCISES

**1. Explain why or why not** Determine whether the following statements are true and give an explanation or counterexample.

   **a.** The function $f(x) = |2x + 1|$ is continuous for all $x$; therefore, it is differentiable for all $x$.

   **b.** If $\dfrac{d}{dx}(f(x)) = \dfrac{d}{dx}(g(x))$, then $f = g$.

   **c.** For any function $f$, $\dfrac{d}{dx}|f(x)| = |f'(x)|$.

   **d.** The value of $f'(a)$ fails to exist only if the curve $y = f(x)$ has a vertical tangent line at $x = a$.

   **e.** An object can have negative acceleration and increasing speed.

**2–4.** *Evaluate the derivative of each of the following functions using a limit definition of the derivative. Check your work by evaluating the derivatives using the derivative rules given in this chapter.*

**2.** $f(x) = x^2 + 2x + 9$       **3.** $g(x) = \dfrac{1}{x^2 + 5}$

**4.** $h(t) = \sqrt{3t + 5}$

**5–8.** *Use differentiation to verify each equation.*

**5.** $\dfrac{d}{dx}(\tan^3 x - 3\tan x + 3x) = 3\tan^4 x$

**6.** $\dfrac{d}{dx}\left(\dfrac{x}{\sqrt{1 - x^2}}\right) = \dfrac{1}{(1 - x^2)^{3/2}}$

**7.** $\dfrac{d}{d\theta}(\sec^3\theta - 3\sec\theta) = \tan^3\theta\sec\theta$

**8.** $\dfrac{d}{dv}(6v - 4)(v + 1)^{3/2} = 15v\sqrt{v + 1}$

**9–46.** *Evaluate and simplify $y'$.*

**9.** $y = \dfrac{2}{3}x^3 + \pi x^2 + 7x + 1$   **10.** $y = (t + 1)^{10}(t + 2)$

**11.** $y = \sqrt{2 + \sin w + \cos w}$   **12.** $y = 2x^{\sqrt{2}}$

**13.** $y = (\tan w + 10)^5$   **14.** $y = (2x - 3)x^{3/2}$

**15.** $y = (1 + x^4)^{3/2}$   **16.** $y = 2x\sqrt{x^2 - 2x + 2}$

**17.** $y = 5t^2\sin t$   **18.** $y = 5x + \sin^3 x + \sin x^3$

**19.** $y = \dfrac{v^{100}}{v + 1}$   **20.** $y = \dfrac{\cos u + u}{\cos u - u}$

**21.** $y = \dfrac{\sec 2w}{\sec 2w + 1}$   **22.** $y = \left(\dfrac{\sin x}{\cos x + 1}\right)^{1/3}$

**23.** $y = \sin^{10}(5t^2 + 1)$   **24.** $y = \sqrt{10 + \cot^2 x}$

**25.** $y = (5t^2 + 10)^{100}$   **26.** $y = \dfrac{\csc 2\theta}{1 + \csc 2\theta}$

**27.** $y = 4\tan(\theta^2 + 3\theta + 2)$   **28.** $y = \csc^5 3x$

**29.** $y = \dfrac{4u^2 + u}{8u + 1}$   **30.** $y = \left(\dfrac{3t^2 - 1}{3t^2 + 1}\right)^{-3}$

**31.** $y = \tan(\sin\theta)$   **32.** $y = \left(\dfrac{v}{v + 1}\right)^{4/3}$

**33.** $y = \sin\sqrt{\cos^2 x + 1}$   **34.** $y = \sin(\sin(3\theta + 2))$

**35.** $y = -x^2\cos x + 2x\sin x + 2\cos x$

**36.** $y = \sin^{10}\theta^{10}$

**37.** $y = \dfrac{5}{12}(x + 1)^{12/5} - \dfrac{5}{7}(x + 1)^{7/5}$

**38.** $y = |2x - 6|$

**39.** $y = \sin\left(\dfrac{x}{x + 1}\right)$   **40.** $y = \left(\dfrac{v}{3v^2 + 2v + 1}\right)^{1/3}$

**41.** $y = \dfrac{\tan^2 w}{\sec w + 1}$   **42.** $y = \dfrac{(2u + 1)^{20}}{(2u + 1)^{20} + 1}$

**43.** $x = \cos(x - y)$

**44.** $xy^4 + x^4y = 1$

**45.** $y\sqrt{x^2 + y^2} = 15$

**46.** $\sin x \cos(y - 1) = \dfrac{1}{2}$

**47.** Evaluate $\dfrac{d}{dx}(x \sec x)\Big|_{x=\frac{\pi}{4}}$.

**48.** Find $f'(1)$ when $f(\theta) = \sin(\pi\theta)$.

**49–52. Higher-order derivatives** *Find and simplify* $y''$.

**49.** $y = (x - 3)\sqrt{x + 2}$

**50.** $y = \dfrac{3x - 1}{x + 1}$

**51.** $xy + y^2 = 1$

**52.** $x + \sin y = y$

**53–56. Tangent lines** *Find an equation of the line tangent to each of the following curves at the given point.*

**53.** $y = 3x^3 + \sin x;\ (0, 0)$

**54.** $y = \dfrac{4x}{x^2 + 3};\ (1, 1)$

**55.** $y + \sqrt{xy} = 6;\ (1, 4)$

**56.** $x^2y + y^3 = 5;\ (2, 1)$

**57–60. Derivative formulas** *Evaluate the following derivatives. Express your answers in terms of* $f$, $g$, $f'$, *and* $g'$.

**57.** $\dfrac{d}{dx}(x^2 f(x))$

**58.** $\dfrac{d}{dx}\sqrt{\dfrac{f(x)}{g(x)}},\ \dfrac{f(x)}{g(x)} \geq 0$

**59.** $\dfrac{d}{dx}\left(\dfrac{xf(x)}{g(x)}\right)$

**60.** $\dfrac{d}{dx}f\left(\sqrt{g(x)}\right),\ g(x) \geq 0$

**61. Matching functions and derivatives** Match the functions a–d with the derivatives A–D.

(a)

(b)

(c)

(d)

(A)      (B)

(C)      (D)

**62.** Sketch a graph of $f'$ for the function $f$ shown in the figure.

**63.** Sketch a graph of $g'$ for the function $g$ shown in the figure.

**64.** Use the given graphs of $f$ and $g$ to find each derivative.

  **a.** $\dfrac{d}{dx}(5f(x) + 3g(x))\Big|_{x=1}$      **b.** $\dfrac{d}{dx}(f(x)g(x))\Big|_{x=1}$

  **c.** $\dfrac{d}{dx}\left(\dfrac{f(x)}{g(x)}\right)\Big|_{x=3}$      **d.** $\dfrac{d}{dx}(f(f(x)))\Big|_{x=4}$

  **e.** $\dfrac{d}{dx}(g(f(x)))\Big|_{x=1}$

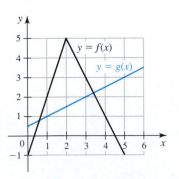

**65. Finding derivatives from a table** Find the values of the following derivatives using the table.

| $x$ | 1 | 3 | 5 | 7 | 9 |
|---|---|---|---|---|---|
| $f(x)$ | 3 | 1 | 9 | 7 | 5 |
| $f'(x)$ | 7 | 9 | 5 | 1 | 3 |
| $g(x)$ | 9 | 7 | 5 | 3 | 1 |
| $g'(x)$ | 5 | 9 | 3 | 1 | 7 |

**a.** $\dfrac{d}{dx}(f(x) + 2g(x))\Big|_{x=3}$    **b.** $\dfrac{d}{dx}\left(\dfrac{f(x)}{g(x)}\right)\Big|_{x=1}$

**c.** $\dfrac{d}{dx}(f(x)g(x))\Big|_{x=3}$    **d.** $\dfrac{d}{dx}(f(x)^3)\Big|_{x=5}$

**66–67. Derivatives from a graph** *If possible, evaluate the following derivatives using the graphs of f and f′.*

**66. a.** $\dfrac{d}{dx}(xf(x))\Big|_{x=2}$    **b.** $\dfrac{d}{dx}(f(x^2))\Big|_{x=1}$

**c.** $\dfrac{d}{dx}(f(f(x)))\Big|_{x=1}$

**67. a.** $\dfrac{d}{dx}\left(\dfrac{x^3}{f(x)}\right)\Big|_{x=2}$    **b.** $\dfrac{d}{dx}(\sqrt{f(x)})\Big|_{x=1}$

**c.** $\dfrac{d}{dx}(\cos(\pi f(x)))\Big|_{x=3}$

**68–69.** *The line tangent to $y = f(x)$ at $x = 3$ is $y = 4x - 10$ and the line tangent to $y = g(x)$ at $x = 5$ is $y = 6x - 27$.*

**68.** Compute $f(3)$, $f'(3)$, $g(5)$, and $g'(5)$.

**69.** Find an equation of the line tangent to the graph of $y = f(g(x))$ at $x = 5$.

**70. Horizontal motion** The position of an object moving horizontally after $t$ seconds is given by the function $s = 27t - t^3$, for $t > 0$, where $s$ is measured in feet, with $s > 0$ corresponding to positions right of the origin.

**a.** When is the object stationary, moving to the right, and moving to the left?

**b.** Determine the velocity and acceleration of the object at $t = 2$.

**c.** Determine the acceleration of the object when its velocity is zero.

**d.** On what time intervals is the speed decreasing?

**71. Projectile on Mars** Suppose an object is fired vertically upward from the ground on Mars with an initial velocity of 96 ft/s. The height $s$ (in feet) of the object above the ground after $t$ seconds is given by $s = 96t - 6t^2$.

**a.** Determine the instantaneous velocity of the object at $t = 1$.

**b.** When will the object have an instantaneous velocity of 12 ft/s?

**c.** What is the height of the object at the highest point of its trajectory?

**d.** With what speed does the object strike the ground?

**72. Growth rate of bacteria** Suppose the following graph represents the number of bacteria in a culture $t$ hours after the start of an experiment.

**a.** At approximately what time is the instantaneous growth rate the greatest, for $0 \leq t \leq 36$? Estimate the growth rate at this time.

**b.** At approximately what time in the interval $0 \leq t \leq 36$ is the instantaneous growth rate the least? Estimate the instantaneous growth rate at this time.

**c.** What is the average growth rate over the interval $0 \leq t \leq 36$?

**73. Population of the United States** The population of the United States (in millions) by decade is given in the table, where $t$ is the number of years after 1910. These data are plotted and fitted with a smooth curve $y = p(t)$ in the figure.

**a.** Compute the average rate of population growth from 1950 to 1960.

**b.** Explain why the average rate of growth from 1950 to 1960 is a good approximation to the (instantaneous) rate of growth in 1955.

**c.** Estimate the instantaneous rate of growth in 1985.

| Year | 1910 | 1920 | 1930 | 1940 | 1950 | 1960 |
|---|---|---|---|---|---|---|
| $t$ | 0 | 10 | 20 | 30 | 40 | 50 |
| $p(t)$ | 92 | 106 | 123 | 132 | 152 | 179 |

| Year | 1970 | 1980 | 1990 | 2000 | 2010 | 2020 |
|---|---|---|---|---|---|---|
| $t$ | 60 | 70 | 80 | 90 | 100 | 110 |
| $p(t)$ | 203 | 227 | 249 | 281 | 309 | 335 |

**74. Velocity of a skydiver** Assume the graph represents the distance (in meters) that a skydiver has fallen $t$ seconds after jumping out of a plane.

a. Estimate the velocity of the skydiver at $t = 15$.
b. Estimate the velocity of the skydiver at $t = 70$.
c. Estimate the average velocity of the skydiver between $t = 20$ and $t = 90$.
d. Sketch a graph of the velocity function, for $0 \le t \le 120$.
e. What significant event occurred at $t = 30$?

**75.** For what value(s) of $x$ is the line tangent to the curve $y = x\sqrt{6 - x}$ horizontal? Vertical?

**76–77. Limits** *The following limits represent the derivative of a function $f$ at a point $a$. Find a possible $f$ and $a$, and then evaluate the limit.*

**76.** $\displaystyle \lim_{h \to 0} \frac{\sin^2\left(\dfrac{\pi}{4} + h\right) - \dfrac{1}{2}}{h}$

**77.** $\displaystyle \lim_{x \to 5} \frac{\tan\left(\pi\sqrt{3x - 11}\right)}{x - 5}$

**T 78. Velocity of a probe** A small probe is launched vertically from the ground. After it reaches its high point, a parachute deploys and the probe descends to Earth. The height of the probe above ground is
$$s(t) = \frac{300t - 50t^2}{t^3 + 2}, \text{ for } 0 \le t \le 6.$$

a. Graph the height function and describe the motion of the probe.
b. Find the velocity of the probe.
c. Graph the velocity function and determine the approximate time at which the velocity is a maximum.

**79. Marginal and average cost** Suppose the cost of producing $x$ lawn mowers is $C(x) = -0.02x^2 + 400x + 5000$.

a. Determine the average and marginal costs for $x = 3000$ lawn mowers.
b. Interpret the meaning of your results in part (a).

**80. Marginal and average cost** Assume $C(x) = -0.0001x^3 + 0.05x^2 + 60x + 800$ is the cost of making $x$ fly rods.

a. Determine the average and marginal costs for $x = 400$ fly rods.
b. Interpret the meaning of your results in part (a).

**T 81. Population growth** Suppose $p(t) = -1.7t^3 + 72t^2 + 7200t + 80,000$ is the population of a city $t$ years after 1950.

a. Determine the average rate of growth of the city from 1950 to 2000.
b. What was the rate of growth of the city in 1990?

**T 82. Position of a piston** The distance between the head of a piston and the end of a cylindrical chamber is given by $x(t) = \dfrac{8t}{t + 1}$ cm, for $t \ge 0$ (measured in seconds). The radius of the cylinder is 4 cm.

a. Find the volume of the chamber, for $t \ge 0$.
b. Find the rate of change of the volume $V'(t)$, for $t \ge 0$.
c. Graph the derivative of the volume function. On what intervals is the volume increasing? Decreasing?

**83. Boat rates** Two boats leave a dock at the same time. One boat travels south at 30 mi/hr and the other travels east at 40 mi/hr. After half an hour, how fast is the distance between the boats increasing?

**84. Rate of inflation of a balloon** A spherical balloon is inflated at a rate of 10 cm³/min. At what rate is the diameter of the balloon increasing when the balloon has a diameter of 5 cm?

**85. Rate of descent of a hot-air balloon** A rope is attached to the bottom of a hot-air balloon that is floating above a flat field. If the angle of the rope to the ground remains 65° and the rope is pulled in at 5 ft/s, how quickly is the elevation of the balloon changing?

**86. Filling a tank** Water flows into a conical tank at a rate of 2 ft³/min. If the radius of the top of the tank is 4 ft and the height is 6 ft, determine how quickly the water level is rising when the water is 2 ft deep in the tank.

**87. Angle of elevation** A jet flying at 450 mi/hr and traveling in a straight line at a constant elevation of 500 ft passes directly over a spectator at an air show. How quickly is the angle of elevation (between the ground and the line from the spectator to the jet) changing 2 seconds later?

**T 88. Viewing angle** A man whose eye level is 6 ft above the ground walks toward a billboard at a rate of 2 ft/s. The bottom of the billboard is 10 ft above the ground, and it is 15 ft high. The man's viewing angle is the angle formed by the lines between the man's eyes and the top and bottom of the billboard. At what rate is the viewing angle changing when the man is 30 ft from the billboard?

**89. Shadow length** A street light is fastened to the top of a 15-ft-high pole. If a 5-ft-tall women walks away from the pole in a straight line over level ground at a rate of 3 ft/s, how fast is the length of her shadow changing when she is 10 ft away from the pole?

**90. A parabola property** Let $f(x) = x^2$.

    **a.** Show that $\dfrac{f(x) - f(y)}{x - y} = f'\left(\dfrac{x + y}{2}\right)$, for all $x \neq y$.

    **b.** Is this property true for $f(x) = ax^2$, where $a$ is a nonzero real number?

    **c.** Give a geometrical interpretation of this property.

    **d.** Is this property true for $f(x) = ax^3$?

## Chapter 3 Guided Projects

*Applications of the material in this chapter and related topics can be found in the following Guided Projects. For additional information, see the Preface.*

- Numerical differentiation
- Elasticity in economics

# Applications of the Derivative

**Chapter Preview**   Much of the previous chapter was devoted to the basic mechanics of derivatives: evaluating them and interpreting them as rates of change. We now apply derivatives to a variety of mathematical questions, many of which concern properties of functions and their graphs. One outcome of this work is a set of analytical curve-sketching methods that produce accurate graphs of functions. Equally important, derivatives allow us to formulate and solve a wealth of practical problems. For example, an asteroid passes perilously close to Earth: At what point along their trajectories is the distance separating them smallest, and what is the minimum distance? An economist has a mathematical model that relates the demand for a product to its price: What price maximizes the revenue? In this chapter, we develop the tools needed to answer such questions. In addition, we begin an ongoing discussion about approximating functions, we present an important result called the Mean Value Theorem, and we work with a powerful method that enables us to evaluate a new kind of limit. The chapter concludes with two important topics: a numerical approach to approximating roots of functions, called Newton's method, and a preview of integral calculus, which is the subject of Chapter 5.

## 4.1 Maxima and Minima

With a working understanding of derivatives, we now undertake one of the fundamental tasks of calculus: analyzing the behavior of functions and producing accurate graphs of them. An important question associated with any function concerns its maximum and minimum values: On a given interval (perhaps the entire domain), where does the function assume its largest and smallest values? Questions about maximum and minimum values take on added significance when a function represents a practical quantity, such as the profits of a company, the surface area of a container, or the speed of a space vehicle.

### Absolute Maxima and Minima

Imagine taking a long hike through varying terrain from west to east. Your elevation changes as you walk over hills, through valleys, and across plains, and you reach several high and low points along the journey. Analogously, when we examine a function $f$ over an interval on the $x$-axis, its values increase and decrease, reaching high points and low points (**Figure 4.1**). You can view our study of functions in this chapter as an exploratory hike along the $x$-axis.

Figure 4.1

> **DEFINITION  Absolute Maximum and Minimum**
>
> Let $f$ be defined on a set $D$ containing $c$. If $f(c) \geq f(x)$ for every $x$ in $D$, then $f(c)$ is an **absolute maximum** value of $f$ on $D$. If $f(c) \leq f(x)$ for every $x$ in $D$, then $f(c)$ is an **absolute minimum** value of $f$ on $D$. An **absolute extreme value** is either an absolute maximum value or an absolute minimum value.

The existence and location of absolute extreme values depend on both the function and the interval of interest. **Figure 4.2** shows various cases for the function $f(x) = 4 - x^2$. Notice that if the interval of interest is not closed, a function might not attain absolute extreme values (Figure 4.2a, c, and d).

> Absolute maximum and minimum values are also called *global* maximum and minimum values. The plural of *maximum* is *maxima*; the plural of *minimum* is *minima*.

$y = 4 - x^2$ on $(-\infty, \infty)$

Absolute max of 4 at $x = 0$

No absolute min

(a) Absolute max only

$y = 4 - x^2$ on $[0, 2]$

Absolute max of 4 at $x = 0$

Absolute min of 0 at $x = 2$

(b) Absolute max and min

$y = 4 - x^2$ on $(0, 2]$

No absolute max

Absolute min of 0 at $x = 2$

(c) Absolute min only

$y = 4 - x^2$ on $(0, 2)$

No absolute max

No absolute min

(d) No absolute max or min

**Figure 4.2**

However, defining a function on a closed interval is not enough to guarantee the existence of absolute extreme values. Both functions in **Figure 4.3** are defined at every point of a closed interval, but neither function attains an absolute maximum—the discontinuity in each function prevents it from happening.

No absolute maximum value

$$y = \begin{cases} x & \text{if } 0 \le x < 1 \\ 0 & \text{if } x = 1 \end{cases}$$

Absolute minimum value of 0 at $x = 0$ and $x = 1$

(a)

No absolute maximum value

Absolute minimum

$y = f(x)$

(b)

**Figure 4.3**

It turns out that *two* conditions ensure the existence of absolute maximum and minimum values on an interval: The function must be continuous on the interval, and the interval must be closed and bounded.

> ➤ The proof of the Extreme Value Theorem relies on properties of the real numbers, found in advanced books.

---

**THEOREM 4.1   Extreme Value Theorem**

A function that is continuous on a closed interval $[a, b]$ has an absolute maximum value and an absolute minimum value on that interval.

---

**QUICK CHECK 1** Sketch the graph of a function that is continuous on an interval but does not have an absolute minimum value. Sketch the graph of a function that is defined on a closed interval but does not have an absolute minimum value. ◄

**EXAMPLE 1   Locating absolute maximum and minimum values**   For the functions in **Figure 4.4**, identify the location of the absolute maximum value and the absolute minimum value on the interval $[a, b]$. Do the functions meet the conditions of the Extreme Value Theorem?

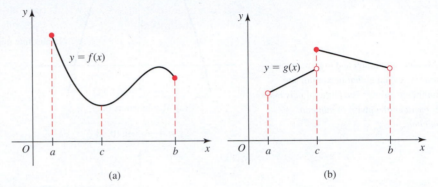

**Figure 4.4**

**SOLUTION**

**a.** The function $f$ is continuous on the closed interval $[a, b]$, so the Extreme Value Theorem guarantees an absolute maximum (which occurs at $a$) and an absolute minimum (which occurs at $c$).

**b.** The function $g$ does not satisfy the conditions of the Extreme Value Theorem because it is not continuous, and it is defined only on the open interval $(a, b)$. It does not have an absolute minimum value. It does, however, have an absolute maximum at $c$. Therefore, a function may violate the conditions of the Extreme Value Theorem and still have an absolute maximum or minimum (or both).

*Related Exercises 11–14* ◄

## Local Maxima and Minima

**Figure 4.5** shows a function $f$ defined on the interval $[a, b]$. It has an absolute minimum at the endpoint $b$ and an absolute maximum at the interior point $s$. In addition, the function has

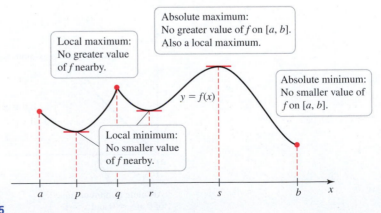

**Figure 4.5**

special behavior at $q$, where its value is greatest *among values at nearby points*, and at $p$ and $r$, where its value is least *among values at nearby points*. A point at which a function takes on the maximum or minimum value among values at nearby points is important.

> ➤ Local maximum and minimum values are also called *relative maximum* and *minimum values*. Local extrema (plural) and *local extremum* (singular) refer to either local maxima or local minima.

---

**DEFINITION**   **Local Maximum and Minimum Values**

Suppose $c$ is an interior point of some interval $I$ on which $f$ is defined. If $f(c) \geq f(x)$ for all $x$ in $I$, then $f(c)$ is a **local maximum** value of $f$. If $f(c) \leq f(x)$ for all $x$ in $I$, then $f(c)$ is a **local minimum** value of $f$.

---

In this text, we adopt the convention that local maximum values and local minimum values occur only at interior points of the interval(s) of interest. For example, in Figure 4.5, the minimum value that occurs at the endpoint $b$ is not a local minimum. However, it is the absolute minimum of the function on $[a, b]$.

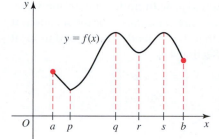

**Figure 4.6**

**EXAMPLE 2**   **Locating various maxima and minima**   **Figure 4.6** shows the graph of a function defined on $[a, b]$. Identify the location of the various maxima and minima using the terms *absolute* and *local*.

**SOLUTION**   The function $f$ is continuous on a closed interval; by Theorem 4.1, it has absolute maximum and minimum values on $[a, b]$. The function has a local minimum value and its absolute minimum value at $p$. It has another local minimum value at $r$. The absolute maximum value of $f$ occurs at both $q$ and $s$ (which also correspond to local maximum values).

*Related Exercises 15–18* ◀

**Critical Points**   Another look at Figure 4.6 shows that local maxima and minima occur at points in the open interval $(a, b)$ where the derivative is zero ($x = q$, $r$, and $s$) and at points where the derivative fails to exist ($x = p$). We now make this observation precise.

    **Figure 4.7** illustrates a function that is differentiable at $c$ with a local maximum at $c$. For $x$ near $c$ with $x < c$, the secant lines through $(x, f(x))$ and $(c, f(c))$ have nonnegative slopes. For $x$ near $c$ with $x > c$, the secant lines through $(x, f(x))$ and $(c, f(c))$ have nonpositive slopes. As $x \rightarrow c$, the slopes of these secant lines approach the slope of the tangent line at $(c, f(c))$. These observations imply that the slope of the tangent line must be both nonnegative and nonpositive, which happens only if $f'(c) = 0$. Similar reasoning leads to the same conclusion for a function with a local minimum at $c$: if $f'(c)$ exists, then $f'(c)$ must be zero. This argument is an outline of the proof (Exercise 85) of the following theorem.

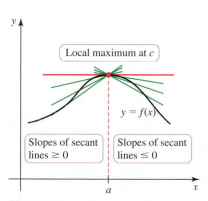

**Figure 4.7**

---

**THEOREM 4.2**   **Local Extreme Value Theorem**

If $f$ has a local maximum or minimum value at $c$ and $f'(c)$ exists, then $f'(c) = 0$.

---

> ➤ Theorem 4.2, often attributed to Fermat, is one of the clearest examples in mathematics of a necessary, but not sufficient, condition. A local maximum (or minimum) at $c$ necessarily implies a critical point at $c$, but a critical point at $c$ is not sufficient to imply a local maximum (or minimum)

Local extrema can also occur at points $c$ where $f'(c)$ does not exist. **Figure 4.8** shows two such cases, one in which $c$ is a point of discontinuity and one in which $f$ has a corner point at $c$. Because local extrema may occur at points $c$ where $f'(c) = 0$ or where $f'(c)$ does not exist, we make the following definition.

---

**DEFINITION**   **Critical Point**

An interior point $c$ of the domain of $f$ at which $f'(c) = 0$ or $f'(c)$ fails to exist is called a **critical point** of $f$.

---

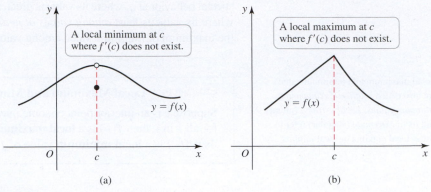

**Figure 4.8**

Note that the converse of Theorem 4.2 is not necessarily true. It is possible that $f'(c) = 0$ at a point without a local maximum or local minimum value occurring there (**Figure 4.9a**). It is also possible that $f'(c)$ fails to exist, with no local extreme value occurring at $c$ (**Figure 4.9b**). Therefore, critical points are *candidates* for the location of local extreme values, but you must determine whether they actually correspond to local maxima or minima. This procedure is discussed in Section 4.3.

**Figure 4.9**

**Figure 4.10**

**EXAMPLE 3** **Locating critical points** Find the critical points of $f(x) = \dfrac{x}{x^2 + 1}$.

**SOLUTION** Note that $f$ is differentiable on its domain, which is $(-\infty, \infty)$. By the Quotient Rule,

$$f'(x) = \frac{(x^2 + 1) - 2x^2}{(x^2 + 1)^2} = \frac{1 - x^2}{(x^2 + 1)^2}.$$

Setting $f'(x) = 0$ and noting that $x^2 + 1 > 0$ for all $x$, the critical points satisfy the equation $1 - x^2 = 0$. Therefore, the critical points are $x = 1$ and $x = -1$. The graph of $f$ (**Figure 4.10**) shows that $f$ has a local (and absolute) maximum at $\left(1, \frac{1}{2}\right)$ and a local (and absolute) minimum at $\left(-1, -\frac{1}{2}\right)$.

*Related Exercises 30–31* ◄

**QUICK CHECK 2** Consider the function $f(x) = x^3$. Where is the critical point of $f$? Does $f$ have a local maximum or minimum at the critical point? ◄

## Locating Absolute Maxima and Minima

Theorem 4.1 guarantees the existence of absolute extreme values of a continuous function on a closed interval $[a, b]$, but it doesn't say where these values are located. Two observations lead to a procedure for locating absolute extreme values.

• An absolute extreme value in the interior of an interval is also a local extreme value, and we know that local extreme values occur at the critical points of $f$.

• Absolute extreme values may also occur at the endpoints of the interval of interest.

These two facts suggest the following procedure for locating the absolute extreme values of a continuous function on a closed interval.

> **PROCEDURE** **Locating Absolute Extreme Values on a Closed Interval**
>
> Assume the function $f$ is continuous on the closed interval $[a, b]$.
>
> 1. Locate the critical points $c$ in $(a, b)$, where $f'(c) = 0$ or $f'(c)$ does not exist. These points are candidates for absolute maxima and minima.
>
> 2. Evaluate $f$ at the critical points and at the endpoints of $[a, b]$.
>
> 3. Choose the largest and smallest values of $f$ from Step 2 for the absolute maximum and minimum values, respectively.

Note that the preceding procedure box does not address the case in which $f$ is continuous on an open interval. If the interval of interest is an open interval, then absolute extreme values—if they exist—occur at interior points.

**EXAMPLE 4** **Absolute extreme values** Find the absolute maximum and minimum values of the following functions.

**a.** $f(x) = x^4 - 2x^3$ on the interval $[-2, 2]$

**b.** $g(x) = x^{2/3}(2 - x)$ on the interval $[-1, 2]$

**SOLUTION**

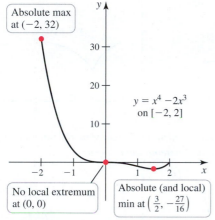

Absolute max at $(-2, 32)$

$y = x^4 - 2x^3$ on $[-2, 2]$

No local extremum at $(0, 0)$

Absolute (and local) min at $\left(\frac{3}{2}, -\frac{27}{16}\right)$

**Figure 4.11**

**a.** Because $f$ is a polynomial, its derivative exists everywhere. So if $f$ has critical points, they are points at which $f'(x) = 0$. Computing $f'$ and setting it equal to zero, we have

$$f'(x) = 4x^3 - 6x^2 = 2x^2 (2x - 3) = 0.$$

Solving this equation gives the critical points $x = 0$ and $x = \frac{3}{2}$, both of which lie in the interval $[-2, 2]$; these points and the endpoints are *candidates* for the location of absolute extrema. Evaluating $f$ at each of these points, we have

$$f(-2) = 32, \quad f(0) = 0, \quad f\left(\tfrac{3}{2}\right) = -\tfrac{27}{16}, \quad \text{and} \quad f(2) = 0.$$

The largest of these function values is $f(-2) = 32$, which is the absolute maximum of $f$ on $[-2, 2]$. The smallest of these values is $f\left(\frac{3}{2}\right) = -\frac{27}{16}$, which is the absolute minimum of $f$ on $[-2, 2]$. The graph of $f$ (**Figure 4.11**) shows that the critical point $x = 0$ corresponds to neither a local maximum nor a local minimum.

**b.** Differentiating $g(x) = x^{2/3}(2 - x) = 2x^{2/3} - x^{5/3}$, we have

$$g'(x) = \frac{4}{3}x^{-1/3} - \frac{5}{3}x^{2/3} = \frac{4 - 5x}{3x^{1/3}}.$$

Absolute max at $(-1, 3)$

$y = x^{2/3}(2 - x)$ on $[-1, 2]$

Local max at $\left(\frac{4}{5}, 1.03\right)$

Absolute min at $(2, 0)$

Absolute (and local) min at $(0, 0)$

**Figure 4.12**

Because $g'(0)$ is undefined and 0 is in the domain of $g$, $x = 0$ is a critical point. In addition, $g'(x) = 0$ when $4 - 5x = 0$, so $x = \frac{4}{5}$ is also a critical point. These two critical points and the endpoints are *candidates* for the location of absolute extrema. The next step is to evaluate $g$ at the critical points and endpoints:

$$g(-1) = 3, \quad g(0) = 0, \quad g\left(\tfrac{4}{5}\right) \approx 1.03, \quad \text{and} \quad g(2) = 0.$$

The largest of these function values is $g(-1) = 3$, which is the absolute maximum value of $g$ on $[-1, 2]$. The least of these values is 0, which occurs twice. Therefore, $g$ has its absolute minimum value on $[-1, 2]$ at the critical point $x = 0$ and the endpoint $x = 2$ (**Figure 4.12**).

*Related Exercises 44, 50* ◀

We now apply these ideas to a practical situation.

**EXAMPLE 5** **Trajectory high point** A stone is launched vertically upward from a bridge 80 ft above the ground at a speed of 64 ft/s. Its height above the ground $t$ seconds after the launch is given by

➤ The derivation of the position function for an object moving in a gravitational field is given in Section 6.1.

$$f(t) = -16t^2 + 64t + 80, \quad \text{for } 0 \leq t \leq 5.$$

When does the stone reach its maximum height?

**SOLUTION** We must evaluate the height function at the critical points and at the endpoints. The critical points satisfy the equation

$$f'(t) = -32t + 64 = -32(t - 2) = 0,$$

so the only critical point is $t = 2$. We now evaluate $f$ at the endpoints and at the critical point:

$$f(0) = 80, \quad f(2) = 144, \quad \text{and} \quad f(5) = 0.$$

On the interval $[0, 5]$, the absolute maximum occurs at $t = 2$, at which time the stone reaches a height of 144 ft. Because $f'(t)$ is the velocity of the stone, the maximum height occurs at the instant the velocity is zero.

*Related Exercise 67* ◄

# SECTION 4.1 EXERCISES

## Getting Started

1. What does it mean for a function to have an absolute extreme value at a point $c$ of an interval $[a, b]$?

2. What are local maximum and minimum values of a function?

3. What conditions must be met to ensure that a function has an absolute maximum value and an absolute minimum value on an interval?

4. Sketch the graph of a function that is continuous on an open interval $(a, b)$ but has neither an absolute maximum nor an absolute minimum value on $(a, b)$.

5. Sketch the graph of a function that has an absolute maximum, a local minimum, but no absolute minimum on $[0, 3]$.

6. What is a critical point of a function?

7. Sketch the graph of a function $f$ that has a local maximum value at a point $c$ where $f'(c) = 0$.

8. Sketch the graph of a function $f$ that has a local minimum value at a point $c$ where $f'(c)$ is undefined.

9. How do you determine the absolute maximum and minimum values of a continuous function on a closed interval?

10. Explain how a function can have an absolute minimum value at an endpoint of an interval.

**11–14.** *Use the following graphs to identify the points (if any) on the interval $[a, b]$ at which the function has an absolute maximum or an absolute minimum value.*

11. **12.**

13. **14.**

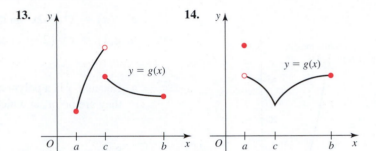

**15–18.** *Use the following graphs to identify the points on the interval $[a, b]$ at which local and absolute extreme values occur.*

15. **16.**

17. **18.**

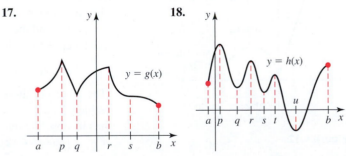

**19–22.** *Sketch the graph of a continuous function $f$ on $[0, 4]$ satisfying the given properties.*

19. $f'(x) = 0$ for $x = 1$ and $2$; $f$ has an absolute maximum at $x = 4$; $f$ has an absolute minimum at $x = 0$; and $f$ has a local minimum at $x = 2$.

20. $f'(x) = 0$ for $x = 1, 2,$ and $3$; $f$ has an absolute minimum at $x = 1$; $f$ has no local extremum at $x = 2$; and $f$ has an absolute maximum at $x = 3$.

**21.** $f'(1)$ and $f'(3)$ are undefined; $f'(2) = 0$; $f$ has a local maximum at $x = 1$; $f$ has a local minimum at $x = 2$; $f$ has an absolute maximum at $x = 3$; and $f$ has an absolute minimum at $x = 4$.

**22.** $f'(x) = 0$ at $x = 1$ and $3$; $f'(2)$ is undefined; $f$ has an absolute maximum at $x = 2$; $f$ has neither a local maximum nor a local minimum at $x = 1$; and $f$ has an absolute minimum at $x = 3$.

## Practice Exercises

**23–40. Locating critical points** *Find the critical points of the following functions. Assume $a$ is a nonzero constant.*

**23.** $f(x) = 3x^2 - 4x + 2$

**24.** $f(x) = \dfrac{1}{8}x^3 - \dfrac{1}{2}x$

**25.** $f(x) = \dfrac{x^3}{3} - 9x$

**26.** $f(x) = \dfrac{x^4}{4} - \dfrac{x^3}{3} - 3x^2 + 10$

**27.** $f(x) = 3x^3 + \dfrac{3x^2}{2} - 2x$

**28.** $f(x) = \dfrac{4x^5}{5} - 3x^3 + 5$

**29.** $f(x) = x^3 - 4a^2x$

**30.** $f(x) = \dfrac{(x + 1)^2}{x^2 + 1}$

**31.** $f(t) = \dfrac{4t}{4t^3 + 1}$

**32.** $f(x) = 12x^5 - 20x^3$

**33.** $f(x) = 2\sqrt{x} - x$ on $[0, 4]$

**34.** $f(x) = \sin x \cos x$

**35.** $f(x) = x^2\sqrt{x + 5}$

**36.** $f(t) = (x - 6)\sqrt{x}$ on $[0, 4]$

**37.** $f(x) = x\sqrt{x - a}$

**38.** $f(x) = \dfrac{x}{\sqrt{x - a}}$

**39.** $f(t) = \dfrac{1}{5}t^5 - a^4t$

**40.** $f(x) = x^3 - 3ax^2 + 3a^2x - a^3$

**41–62. Absolute maxima and minima** *Determine the location and value of the absolute extreme values of $f$ on the given interval, if they exist.*

**41.** $f(x) = x^2 - 10$ on $[-2, 3]$

**42.** $f(x) = (x + 1)^{4/3}$ on $[-9, 7]$

**43.** $f(x) = x^3 - 3x^2$ on $[-1, 3]$

**44.** $f(x) = x^4 - 4x^3 + 4x^2$ on $[-1, 3]$

**45.** $f(x) = 3x^5 - 25x^3 + 60x$ on $[-2, 3]$

**46.** $f(x) = \sin 2x - \sqrt{3}x$ on $[0, \pi]$

**47.** $f(x) = \cos^2 x$ on $[0, \pi]$

**48.** $f(x) = \dfrac{x}{(x^2 + 3)^2}$ on $[-2, 2]$

**49.** $f(x) = \sin 3x$ on $[-\pi/4, \pi/3]$

**50.** $f(x) = 3x^{2/3} - x$ on $[0, 27]$

**51.** $f(x) = x + \sin 2x$ on $[0, \pi/2]$

**52.** $f(x) = x\sqrt{2 - x^2}$ on $[-\sqrt{2}, \sqrt{2}]$

**53.** $f(x) = 2x^3 - 15x^2 + 24x$ on $[0, 5]$

**54.** $f(x) = 4x^3 - 21x^2 + 36x$ on $[1, 3]$

**55.** $f(x) = \dfrac{4x^3}{3} + 5x^2 - 6x$ on $[-4, 1]$

**56.** $f(x) = 2x^6 - 15x^4 + 24x^2$ on $[-2, 2]$

**57.** $f(x) = \dfrac{x}{(x^2 + 9)^5}$ on $[-2, 2]$

**58.** $f(x) = x^{1/2}\left(\dfrac{x^2}{5} - 4\right)$ on $[0, 4]$

**59.** $f(x) = \sec x$ on $\left[-\dfrac{\pi}{4}, \dfrac{\pi}{4}\right]$

**60.** $f(x) = x^{1/3}(x + 4)$ on $[-27, 27]$

**61.** $f(x) = x^{2/3}(4 - x^2)$ on $[-2, 2]$

**62.** $f(t) = \dfrac{3t}{t^2 + 1}$ on $[-2, 2]$

**63. Efficiency of wind turbines** A wind turbine converts wind energy into electrical power. Let $v_1$ equal the upstream velocity of the wind before it encounters the wind turbine, and let $v_2$ equal the downstream velocity of the wind after it passes through the area swept out by the turbine blades.

**a.** Assuming that $v_1 > 0$, give a physical explanation to show that $0 \le \dfrac{v_2}{v_1} \le 1$.

**b.** The amount of power extracted from the wind depends on the ratio $r = \dfrac{v_2}{v_1}$, the ratio of the downstream velocity to upstream velocity. Let $R(r)$ equal the fraction of power that is extracted from the total available power in the wind stream, for a given value of $r$. In about 1920, the German physicist Albert Betz showed that $R(r) = \dfrac{1}{2}(1 + r)(1 - r^2)$, where $0 \le r \le 1$ (a derivation of $R$ is outlined in Exercise 64). Calculate $R(1)$ and explain how you could have arrived at this value without using the formula for $R$. Give a physical explanation of why it is unlikely or impossible for it to be the case that $r = 1$.

**c.** Calculate $R(0)$ and give a physical explanation of why it is unlikely or impossible for it to be the case that $r = 0$.

**d.** The maximum value of $R$ is called the *Betz limit*. It represents the theoretical maximum amount of power that can be extracted from the wind. Find this value and explain its physical meaning.

**64. Derivation of wind turbine formula** A derivation of the function $R$ in Exercise 63, based on three equations from physics, is outlined here. Consider again the figure given in Exercise 63, where $v_1$ equals the upstream velocity of the wind just before the wind stream encounters the wind turbine, and $v_2$ equals the downstream velocity of the wind just after the wind stream passes through the area swept out by the turbine blades. An equation for the power extracted by the rotor blades, based on conservation of momentum, is $P = v^2\rho A(v_1 - v_2)$, where $v$ is the velocity of the wind (in m/s) as it passes through the turbine blades, $\rho$ is the density of air (in kg/m$^3$), and $A$ is the area (in m$^2$) of the circular region swept out by the rotor blades.

a. Another expression for the power extracted by the rotor blades, based on conservation of energy, is $P = \dfrac{1}{2}\rho vA(v_1{}^2 - v_2{}^2)$. Equate the two power equations and solve for $v$.

b. Show that $P = \dfrac{\rho A}{4}(v_1 + v_2)(v_1{}^2 - v_2{}^2)$.

c. If the wind were to pass through the same area $A$ without being disturbed by rotor blades, the amount of available power would be $P_0 = \dfrac{\rho Av_1{}^3}{2}$. Let $r = \dfrac{v_2}{v_1}$ and simplify the ratio $\dfrac{P}{P_0}$ to obtain the function $R(r)$ given in Exercise 63. (*Source: Journal of Applied Physics*, 105, 2009)

65. **Maximum distance from origin** Suppose the position of an object moving horizontally after $t$ seconds is given by the function $s(t) = 32t - t^4$, where $0 \le t \le 3$ and $s$ is measured in feet, with $s > 0$ corresponding to positions to the right of the origin. When is the object farthest to the right?

66. **Minimum-surface-area box** All boxes with a square base and a volume of 50 ft$^3$ have a surface area given by $S(x) = 2x^2 + \dfrac{200}{x}$, where $x$ is the length of the sides of the base. Find the absolute minimum of the surface area function on the interval $(0, \infty)$. What are the dimensions of the box with minimum surface area?

67. **Trajectory high point** A stone is launched vertically upward from a cliff 192 ft above the ground at a speed of 64 ft/s. Its height above the ground $t$ seconds after the launch is given by $s = -16t^2 + 64t + 192$, for $0 \le t \le 6$. When does the stone reach its maximum height?

68. **Maximizing revenue** A sales analyst determines that the revenue from sales of fruit smoothies is given by $R(x) = -60x^2 + 300x$, where $x$ is the price in dollars charged per item, for $0 \le x \le 5$.

a. Find the critical points of the revenue function.
b. Determine the absolute maximum value of the revenue function and the price that maximizes the revenue.

69. **Maximizing profit** Suppose a tour guide has a bus that holds a maximum of 100 people. Assume his profit (in dollars) for taking $n$ people on a city tour is $P(n) = n(50 - 0.5n) - 100$. (Although $P$ is defined only for positive integers, treat it as a continuous function.)

a. How many people should the guide take on a tour to maximize the profit?
b. Suppose the bus holds a maximum of 45 people. How many people should be taken on a tour to maximize the profit?

70. **Minimizing rectangle perimeters** All rectangles with an area of 64 have a perimeter given by $P(x) = 2x + \dfrac{128}{x}$, where $x$ is the length of one side of the rectangle. Find the absolute minimum value of the perimeter function on the interval $(0, \infty)$. What are the dimensions of the rectangle with minimum perimeter?

71. **Explain why or why not** Determine whether the following statements are true and give an explanation or counterexample.

a. The function $f(x) = \sqrt{x}$ has a local maximum on the interval $[0, \infty)$.
b. If a function has an absolute maximum on a closed interval, then the function must be continuous on that interval.
c. A function $f$ has the property that $f'(2) = 0$. Therefore, $f$ has a local extreme value at $x = 2$.
d. Absolute extreme values of a function on a closed interval always occur at a critical point or an endpoint of the interval.

## Explorations and Challenges

**T 72–73. Absolute maxima and minima**

a. *Find the critical points of f on the given interval.*
b. *Determine the absolute extreme values of f on the given interval. Use a graphing utility to confirm your conclusions.*

72. $f(x) = \dfrac{x}{\sqrt{x - 4}}$ on $[6, 12]$    73. $f(x) = |2x - x^2|$ on $[-2, 3]$

**T 74–77. Critical points and extreme values**

a. *Find the critical points of the following functions on the given interval. Use a root finder, if necessary.*
b. *Use a graphing utility to determine whether the critical points correspond to local maxima, local minima, or neither.*
c. *Find the absolute maximum and minimum values on the given interval, if they exist.*

74. $f(x) = 6x^4 - 16x^3 - 45x^2 + 54x + 23$ on $[-5, 5]$

75. $f(\theta) = 2\sin\theta + \cos\theta$ on $[-2\pi, 2\pi]$

76. $g(x) = (x - 3)^{5/3}(x + 2)$ on $[-4, 4]$

77. $h(x) = \dfrac{5 - x}{x^2 + 2x - 3}$ on $[-10, 10]$

**T 78–79. Absolute value functions** *Graph the following functions and determine the local and absolute extreme values on the given interval.*

78. $f(x) = |x - 3| + |x + 2|$ on $[-4, 4]$

79. $g(x) = |x - 3| - 2|x + 1|$ on $[-2, 3]$

**T 80. Dancing on a parabola** Two people, $A$ and $B$, walk along the parabola $y = x^2$ in such a way that the line segment $L$ between them is always perpendicular to the line tangent to the parabola at $A$'s position. The goal of this exercise is to determine the positions of $A$ and $B$ when $L$ has minimum length. Assume the coordinates of $A$ are $(a, a^2)$.

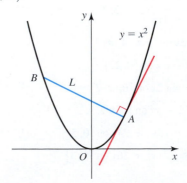

**a.** Find the slope of the line tangent to the parabola at $A$, and find the slope of the line that is perpendicular to the tangent line at $A$.

**b.** Find the equation of the line joining $A$ and $B$.

**c.** Find the position of $B$ on the parabola.

**d.** Write the function $F(a)$ that gives the *square* of the distance between $A$ and $B$ as it varies with $a$. (The square of the distance is minimized at the same point that the distance is minimized; it is easier to work with the square of the distance.)

**e.** Find the critical point of $F$ on the interval $a > 0$.

**f.** Evaluate $F$ at the critical point and verify that it corresponds to an absolute minimum. What are the positions of $A$ and $B$ that minimize the length of $L$? What is the minimum length?

**g.** Graph the function $F$ to check your work.

**T 81. Every second counts** You must get from a point $P$ on the straight shore of a lake to a stranded swimmer who is 50 m from a point $Q$ on the shore that is 50 m from you (see figure). Assuming that you can swim at a speed of 2 m/s and run at a speed of 4 m/s, the goal of this exercise is to determine the point along the shore, $x$ meters from $Q$, where you should stop running and start swimming to reach the swimmer in the minimum time.

**a.** Find the function $T$ that gives the travel time as a function of $x$, where $0 \le x \le 50$.

**b.** Find the critical point of $T$ on $(0, 50)$.

**c.** Evaluate $T$ at the critical point and the endpoints ($x = 0$ and $x = 50$) to verify that the critical point corresponds to an absolute minimum. What is the minimum travel time?

**d.** Graph the function $T$ to check your work.

**82. Extreme values of parabolas** Consider the function $f(x) = ax^2 + bx + c$, with $a \ne 0$. Explain geometrically why $f$ has exactly one absolute extreme value on $(-\infty, \infty)$. Find the critical point to determine the value of $x$ at which $f$ has an extreme value.

**83. Values of related functions** Suppose $f$ is differentiable on $(-\infty, \infty)$ and assume it has a local extreme value at the point $x = 2$, where $f(2) = 0$. Let $g(x) = xf(x) + 1$ and let $h(x) = xf(x) + x + 1$, for all values of $x$.

**a.** Evaluate $g(2)$, $h(2)$, $g'(2)$, and $h'(2)$.

**b.** Does either $g$ or $h$ have a local extreme value at $x = 2$? Explain.

**T 84. A family of double-humped functions** Consider the functions

$$f(x) = \frac{x}{(x^2 + 1)^n}, \text{ where } n \text{ is a positive integer.}$$

**a.** Show that these functions are odd for all positive integers $n$.

**b.** Show that the critical points of these functions are

$$x = \pm\frac{1}{\sqrt{2n - 1}}, \text{ for all positive integers } n. \text{ (Start with the special cases } n = 1 \text{ and } n = 2.)$$

**c.** Show that as $n$ increases, the absolute maximum values of these functions decrease.

**d.** Use a graphing utility to verify your conclusions.

**85. Proof of the Local Extreme Value Theorem** Prove Theorem 4.2 for a local maximum: If $f$ has a local maximum value at the point $c$ and $f'(c)$ exists, then $f'(c) = 0$. Use the following steps.

**a.** Suppose $f$ has a local maximum at $c$. What is the sign of $f(x) - f(c)$ if $x$ is near $c$ and $x > c$? What is the sign of $f(x) - f(c)$ if $x$ is near $c$ and $x < c$?

**b.** If $f'(c)$ exists, then it is defined by $\lim\limits_{x \to c} \dfrac{f(x) - f(c)}{x - c}$. Examine this limit as $x \to c^+$ and conclude that $f'(c) \le 0$.

**c.** Examine the limit in part (b) as $x \to c^-$ and conclude that $f'(c) \ge 0$.

**d.** Combine parts (b) and (c) to conclude that $f'(c) = 0$.

**86. Even and odd functions**

**a.** Suppose a nonconstant even function $f$ has a local minimum at $c$. Does $f$ have a local maximum or minimum at $-c$? Explain. (An even function satisfies $f(-x) = f(x)$.)

**b.** Suppose a nonconstant odd function $f$ has a local minimum at $c$. Does $f$ have a local maximum or minimum at $-c$? Explain. (An odd function satisfies $f(-x) = -f(x)$.)

**QUICK CHECK ANSWERS**

**1.** The continuous function $f(x) = x$ does not have an absolute minimum on the open interval $(0, 1)$. The function $f(x) = -x$ on $\left[0, \frac{1}{2}\right)$ and $f(x) = 0$ on $\left[\frac{1}{2}, 1\right]$ does not have an absolute minimum on $[0, 1]$.   **2.** The critical point is $x = 0$. Although $f'(0) = 0$, the function has neither a local maximum nor minimum at $x = 0$. ◄

# 4.2 Mean Value Theorem

In Section 4.1, we learned how to find the absolute extrema of a function. This information is needed to produce an accurate graph of a function (Sections 4.3 and 4.4) and to solve optimization problems (Section 4.5). The procedures used to solve these types of problems depend on several results developed over the next two sections.

We begin with the *Mean Value Theorem*, a cornerstone in the theoretical framework of calculus. Several critical theorems rely on the Mean Value Theorem; this theorem also appears in practical applications. The proof of the Mean Value Theorem relies on a preliminary result known as Rolle's Theorem.

**Figure 4.13**

> ➤ The Extreme Value Theorem, discussed in Section 4.1, states that a function that is continuous on a closed bounded interval attains its absolute maximum and minimum values on that interval.

## Rolle's Theorem

Consider a function $f$ that is continuous on a closed interval $[a, b]$ and differentiable on the open interval $(a, b)$. Furthermore, assume $f$ has the special property that $f(a) = f(b)$ (**Figure 4.13**). The statement of Rolle's Theorem is not surprising: It says that somewhere between $a$ and $b$, there is at least one point at which $f$ has a horizontal tangent line.

---

**THEOREM 4.3   Rolle's Theorem**

Let $f$ be continuous on a closed interval $[a, b]$ and differentiable on $(a, b)$ with $f(a) = f(b)$. There is at least one point $c$ in $(a, b)$ such that $f'(c) = 0$.

---

**Proof:** The function $f$ satisfies the conditions of Theorem 4.1 (Extreme Value Theorem); therefore, it attains its absolute maximum and minimum values on $[a, b]$. Those values are attained either at an endpoint or at an interior point $c$.

*Case 1:* First suppose $f$ attains both its absolute maximum and minimum values at the endpoints. Because $f(a) = f(b)$, the maximum and minimum values are equal, and it follows that $f$ is a constant function on $[a, b]$. Therefore, $f'(x) = 0$ for all $x$ in $(a, b)$, and the conclusion of the theorem holds.

*Case 2:* Assume at least one of the absolute extreme values of $f$ does not occur at an endpoint. Then $f$ must attain an absolute extreme value at an interior point of $[a, b]$; therefore, $f$ must have either a local maximum or a local minimum at a point $c$ in $(a, b)$. We know from Theorem 4.2 that at a local extremum, the derivative is zero. Therefore, $f'(c) = 0$ for at least one point $c$ of $(a, b)$, and again the conclusion of the theorem holds.   ◀

Why does Rolle's Theorem require continuity? A function that is not continuous on $[a, b]$ may have identical values at both endpoints and still not have a horizontal tangent line at any point on the interval (**Figure 4.14a**). Similarly, a function that is continuous on $[a, b]$ but not differentiable at a point of $(a, b)$ may also fail to have a horizontal tangent line (**Figure 4.14b**).

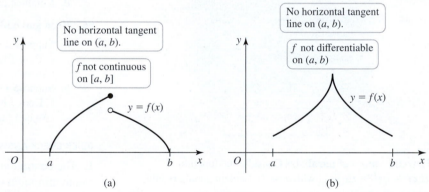

(a)                                           (b)

**Figure 4.14**

**QUICK CHECK 1** Where on the interval $[0, 4]$ does $f(x) = 4x - x^2$ have a horizontal tangent line? ◀

**EXAMPLE 1   Verifying Rolle's Theorem** Find an interval $I$ on which Rolle's Theorem applies to $f(x) = x^3 - 7x^2 + 10x$. Then find all points $c$ in $I$ at which $f'(c) = 0$.

**SOLUTION** Because $f$ is a polynomial, it is everywhere continuous and differentiable. We need an interval $[a, b]$ with the property that $f(a) = f(b)$. Noting that $f(x) = x(x - 2)(x - 5)$, we choose the interval $[0, 5]$, because $f(0) = f(5) = 0$ (other intervals are possible). The goal is to find points $c$ in the interval $(0, 5)$ at which $f'(c) = 0$, which amounts to the familiar task of finding the critical points of $f$. The critical points satisfy

$$f'(x) = 3x^2 - 14x + 10 = 0.$$

Using the quadratic formula, the roots are

$$x = \frac{7 \pm \sqrt{19}}{3}, \quad \text{or} \quad x \approx 0.88 \quad \text{and} \quad x \approx 3.79.$$

**Figure 4.15**

As shown in **Figure 4.15**, the graph of $f$ has two points at which the tangent line is horizontal.

*Related Exercises 11, 16* ◀

## Mean Value Theorem

These lines are parallel and their slopes are equal, that is...

Secant line: slope $= \dfrac{f(b) - f(a)}{b - a}$

Tangent line: slope $= f'(c)$

$y = f(x)$

$\dfrac{f(b) - f(a)}{b - a} = f'(c)$

**Figure 4.16**

The Mean Value Theorem is easily understood with the aid of a picture. **Figure 4.16** shows a function $f$ differentiable on $(a, b)$ with a secant line passing through $(a, f(a))$ and $(b, f(b))$; the slope of the secant line is the average rate of change of $f$ over $[a, b]$. The Mean Value Theorem claims that there exists a point $c$ in $(a, b)$ at which the slope of the tangent line at $c$ is equal to the slope of the secant line. In other words, we can find a point on the graph of $f$ where the tangent line is parallel to the secant line.

> **THEOREM 4.4   Mean Value Theorem**
> If $f$ is continuous on the closed interval $[a, b]$ and differentiable on $(a, b)$, then there is at least one point $c$ in $(a, b)$ such that
> $$\frac{f(b) - f(a)}{b - a} = f'(c).$$

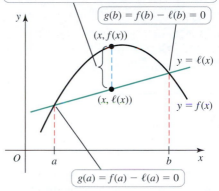

$g(x) = f(x) - \ell(x)$: The difference between the $y$-coordinates of the points $(x, \ell(x))$ and $(x, f(x))$

$g(b) = f(b) - \ell(b) = 0$

$(x, f(x))$

$y = \ell(x)$

$(x, \ell(x))$

$y = f(x)$

$g(a) = f(a) - \ell(a) = 0$

**Figure 4.17**

➤ The proofs of Rolle's Theorem and the Mean Value Theorem are nonconstructive: The theorems claim that a certain point exists, but their proofs do not say how to find it.

**Proof:** The strategy of the proof is to use the function $f$ of the Mean Value Theorem to form a new function $g$ that satisfies Rolle's Theorem. Notice that the continuity and differentiability conditions of Rolle's Theorem and the Mean Value Theorem are the same. We devise $g$ so that it satisfies the conditions $g(a) = g(b) = 0$.

As shown in **Figure 4.17**, the secant line passing through $(a, f(a))$ and $(b, f(b))$ is described by a function $\ell$. We now define a new function $g$ that measures the difference between the function values of $f$ and $\ell$. This function is simply $g(x) = f(x) - \ell(x)$. Because $f$ and $\ell$ are continuous on $[a, b]$ and differentiable on $(a, b)$, it follows that $g$ is also continuous on $[a, b]$ and differentiable on $(a, b)$. Furthermore, because the graphs of $f$ and $\ell$ intersect at $x = a$ and $x = b$, we have $g(a) = f(a) - \ell(a) = 0$ and $g(b) = f(b) - \ell(b) = 0$.

We now have a function $g$ that satisfies the conditions of Rolle's Theorem. By that theorem, we are guaranteed the existence of at least one point $c$ in the interval $(a, b)$ such that $g'(c) = 0$. By the definition of $g$, this condition implies that $f'(c) - \ell'(c) = 0$, or $f'(c) = \ell'(c)$.

We are almost finished. What is $\ell'(c)$? It is just the slope of the secant line, which is
$$\frac{f(b) - f(a)}{b - a}.$$
Therefore, $f'(c) = \ell'(c)$ implies that
$$\frac{f(b) - f(a)}{b - a} = f'(c). \qquad \blacktriangleleft$$

**QUICK CHECK 2** Sketch the graph of a function that illustrates why the continuity condition of the Mean Value Theorem is needed. Sketch the graph of a function that illustrates why the differentiability condition of the Mean Value Theorem is needed. ◄

The following situation offers an interpretation of the Mean Value Theorem. Imagine driving for 2 hours to a town 100 miles away. While your average speed is $100 \text{ mi}/2 \text{ hr} = 50 \text{ mi/hr}$, your instantaneous speed (measured by the speedometer) almost certainly varies. The Mean Value Theorem says that at some point during the trip, your instantaneous speed equals your average speed, which is 50 mi/hr. In Example 2, we apply these ideas to the science of weather forecasting.

**EXAMPLE 2   Mean Value Theorem in action** The *lapse rate* is the rate at which the temperature $T$ decreases in the atmosphere with respect to increasing altitude $z$. It is typically reported in units of °C/km and is defined by $\gamma = -dT/dz$. When the lapse rate rises above 7°C/km in a certain layer of the atmosphere, it indicates favorable conditions for thunderstorm and tornado formation, provided other atmospheric conditions are also present.

Suppose the temperature at $z = 2.9$ km is $T = 7.6$°C and the temperature at $z = 5.6$ km is $T = -14.3$°C. Assume also that the temperature function is continuous and differentiable at all altitudes of interest. What can a meteorologist conclude from these data?

**SOLUTION** Figure 4.18 shows the two data points plotted on a graph of altitude and temperature. The slope of the line joining these points is

$$\frac{-14.3°C - 7.6°C}{5.6 \text{ km} - 2.9 \text{ km}} = -8.1°C/\text{km},$$

> Meteorologists look for "steep" lapse rates in the layer of the atmosphere where the pressure is between 700 and 500 hPa (hectopascals). This range of pressure typically corresponds to altitudes between 3 km and 5.5 km. The data in Example 2 were recorded in Denver at nearly the same time a tornado struck 50 mi to the north.

which means, on average, the temperature is decreasing at 8.1°C/km in the layer of air between 2.9 km and 5.6 km. With only two data points, we cannot know the entire temperature profile. The Mean Value Theorem, however, guarantees that there is at least one altitude at which $dT/dz = -8.1°C/\text{km}$. At each such altitude, the lapse rate is $\gamma = -dT/dz = 8.1°C/\text{km}$. Because this lapse rate is above the 7°C/km threshold associated with unstable weather, the meteorologist might expect an increased likelihood of severe storms.

**Figure 4.18**

*Related Exercises 19, 40* ◀

**EXAMPLE 3   Verifying the Mean Value Theorem** Determine whether the function $f(x) = 2x^3 - 3x + 1$ satisfies the conditions of the Mean Value Theorem on the interval $[-2, 2]$. If so, find the point(s) guaranteed to exist by the theorem.

**SOLUTION** The polynomial $f$ is everywhere continuous and differentiable, so it satisfies the conditions of the Mean Value Theorem. The average rate of change of the function on the interval $[-2, 2]$ is

$$\frac{f(2) - f(-2)}{2 - (-2)} = \frac{11 - (-9)}{4} = 5.$$

The goal is to find points in $(-2, 2)$ at which the line tangent to the curve has a slope of 5—that is, to find points at which $f'(x) = 5$. Differentiating $f$, this condition becomes

$$f'(x) = 6x^2 - 3 = 5 \quad \text{or} \quad x^2 = \frac{4}{3}.$$

**Figure 4.19**

Therefore, the points guaranteed to exist by the Mean Value Theorem are $x = \pm 2/\sqrt{3} \approx \pm 1.15$. The tangent lines have slope 5 at the corresponding points on the curve (**Figure 4.19**).

*Related Exercises 21–22* ◀

## Consequences of the Mean Value Theorem

We close with two results that follow from the Mean Value Theorem.

We already know that the derivative of a constant function is zero; that is, if $f(x) = C$, then $f'(x) = 0$ (Theorem 3.2). Theorem 4.5 states the converse of this result.

---

**THEOREM 4.5   Zero Derivative Implies Constant Function**
If $f$ is differentiable and $f'(x) = 0$ at all points of an open interval $I$, then $f$ is a constant function on $I$.

---

**Proof:** Suppose $f'(x) = 0$ on $[a, b]$, where $a$ and $b$ are distinct points of $I$. By the Mean Value Theorem, there exists a point $c$ in $(a, b)$ such that

$$\frac{f(b) - f(a)}{b - a} = \underbrace{f'(c) = 0}_{\substack{f'(x) = 0 \text{ for} \\ \text{all } x \text{ in } I}}.$$

Multiplying both sides of this equation by $b - a \neq 0$, it follows that $f(b) = f(a)$, and this is true for every pair of points $a$ and $b$ in $I$. If $f(b) = f(a)$ for every pair of points in an interval, then $f$ is a constant function on that interval. ◄

Theorem 4.6 builds on the conclusion of Theorem 4.5.

---

**THEOREM 4.6  Functions with Equal Derivatives Differ by a Constant**
If two functions have the property that $f'(x) = g'(x)$, for all $x$ of an open interval $I$, then $f(x) - g(x) = C$ on $I$, where $C$ is a constant; that is, $f$ and $g$ differ by a constant.

---

**Proof:** The fact that $f'(x) = g'(x)$ on $I$ implies that $f'(x) - g'(x) = 0$ on $I$. Recall that the derivative of a difference of two functions equals the difference of the derivatives, so we can write

$$f'(x) - g'(x) = (f - g)'(x) = 0.$$

**QUICK CHECK 3** Give two distinct linear functions $f$ and $g$ that satisfy $f'(x) = g'(x)$; that is, the lines have equal slopes. Show that $f$ and $g$ differ by a constant. ◄

Now we have a function $f - g$ whose derivative is zero on $I$. By Theorem 4.5, $f(x) - g(x) = C$, for all $x$ in $I$, where $C$ is a constant; that is, $f$ and $g$ differ by a constant. ◄

The utility of Theorems 4.5 and 4.6 will become apparent in Section 4.9, where we establish a pivotal result that has far-reaching consequences.

---

## SECTION 4.2 EXERCISES

### Getting Started

1. Explain Rolle's Theorem with a sketch.

2. Draw the graph of a function for which the conclusion of Rolle's Theorem does not hold.

3. Explain why Rolle's Theorem cannot be applied to the function $f(x) = |x|$ on the interval $[-a, a]$ for any $a > 0$.

4. Explain the Mean Value Theorem with a sketch.

**5–7.** *For each function $f$ and interval $[a, b]$, a graph of $f$ is given along with the secant line that passes though the graph of $f$ at $x = a$ and $x = b$.*

**a.** *Use the graph to make a conjecture about the value(s) of $c$ satisfying the equation $\dfrac{f(b) - f(a)}{b - a} = f'(c)$.*

**b.** *Verify your answer to part (a) by solving the equation $\dfrac{f(b) - f(a)}{b - a} = f'(c)$ for $c$.*

5.  $f(x) = \dfrac{x^2}{4} + 1; [-2, 4]$

6.  $f(x) = 2\sqrt{x}; [0, 4]$

**7.** $f(x) = \dfrac{x^5}{16}; [-2, 2]$

$y = \dfrac{x^5}{16}$

**8.** At what points $c$ does the conclusion of the Mean Value Theorem hold for $f(x) = x^3$ on the interval $[-10, 10]$?

**9.** Draw the graph of a function for which the conclusion of the Mean Value Theorem does not hold.

**10.** Let $f(x) = x^{2/3}$. Show that there is no value of $c$ in the interval $(-1, 8)$ for which $f'(c) = \dfrac{f(8) - f(-1)}{8 - (-1)}$ and explain why this does not violate the Mean Value Theorem.

## Practice Exercises

**11–18. Rolle's Theorem** *Determine whether Rolle's Theorem applies to the following functions on the given interval. If so, find the point(s) guaranteed to exist by Rolle's Theorem.*

**11.** $f(x) = x(x - 1)^2; [0, 1]$     **12.** $f(x) = \sin 2x; [0, \pi/2]$

**13.** $f(x) = \cos 4x; [\pi/8, 3\pi/8]$  **14.** $f(x) = 1 - |x|; [-1, 1]$

**15.** $f(x) = 1 - x^{2/3}; [-1, 1]$

**16.** $f(x) = x^3 - 2x^2 - 8x; [-2, 4]$

**17.** $g(x) = x^3 - x^2 - 5x - 3; [-1, 3]$

**18.** $h(x) = \sqrt{x}; [0, a]$, where $a > 0$

**19. Lapse rates in the atmosphere** Refer to Example 2. Concurrent measurements indicate that at an elevation of 6.1 km, the temperature is $-10.3°C$, and at an elevation of 3.2 km, the temperature is $8.0°C$. Based on the Mean Value Theorem, can you conclude that the lapse rate exceeds the threshold value of $7°C/km$ at some intermediate elevation? Explain.

**20. Drag racer acceleration** The fastest drag racers can reach a speed of 330 mi/hr over a quarter-mile strip in 4.45 seconds (from a standing start). Complete the following sentence about such a drag racer: At some point during the race, the maximum acceleration of the drag racer is at least _____ mi/hr/s.

**21–32. Mean Value Theorem** *Consider the following functions on the given interval $[a, b]$.*

**a.** *Determine whether the Mean Value Theorem applies to the following functions on the given interval $[a, b]$.*
**b.** *If so, find the point(s) that are guaranteed to exist by the Mean Value Theorem.*

**21.** $f(x) = 7 - x^2; [-1, 2]$     **22.** $f(x) = x^3 - 2x^2; [0, 1]$

**23.** $f(x) = \begin{cases} -2x & \text{if } x < 0 \\ x & \text{if } x \geq 0 \end{cases}; [-1, 1]$

**24.** $f(x) = \dfrac{1}{(x-1)^2}; [0, 2]$   **25.** $f(x) = \sqrt{x}; [1, 4]$

**26.** $f(x) = |x - 1|; [-1, 4]$   **T 27.** $f(x) = \sin x; \left[-\dfrac{\pi}{2}, \dfrac{\pi}{2}\right]$

**T 28.** $f(x) = \tan x; \left[0, \dfrac{\pi}{4}\right]$   **29.** $f(x) = x^{-1/3}; \left[\dfrac{1}{8}, 8\right]$

**30.** $f(x) = x + \dfrac{1}{x}; [1, 3]$   **31.** $f(x) = 2x^{1/3}; [-8, 8]$

**32.** $f(x) = \dfrac{x}{x+2}; [-1, 2]$

**33. Explain why or why not** Determine whether the following statements are true and give an explanation or counterexample.

**a.** The continuous function $f(x) = 1 - |x|$ satisfies the conditions of the Mean Value Theorem on the interval $[-1, 1]$.
**b.** Two differentiable functions that differ by a constant always have the same derivative.
**c.** If $f'(x) = 0$, then $f(x) = 10$.

**34.** Without evaluating derivatives, determine which of the functions $f(x) = \sin^2 x$, $g(x) = -\cos^2 x$, $h(x) = 2\sin^2 x$, and $p(x) = 1/\csc^2 x$ have the same derivative.

**35.** Without evaluating derivatives, determine which of the functions $g(x) = 2x^{10}$, $h(x) = x^{10} + 2$, and $p(x) = x^{10} - \sin 2$ have the same derivative as $f(x) = x^{10}$.

**36.** Find all functions $f$ whose derivative is $f'(x) = x + 1$.

**37. Mean Value Theorem and graphs** By visual inspection, locate all points on the interval $(-4, 4)$ at which the slope of the tangent line equals the average rate of change of the function on the interval $[-4, 4]$.

**38–39. Mean Value Theorem and graphs** *Find all points on the interval $(1, 3)$ at which the slope of the tangent line equals the average rate of change of $f$ on $[1, 3]$. Reconcile your results with the Mean Value Theorem.*

**38.**

$y = f(x)$

**39.**

$y = f(x)$

**40. Avalanche forecasting** Avalanche forecasters measure the *temperature gradient* $\dfrac{dT}{dh}$, which is the rate at which the temperature in a snowpack $T$ changes with respect to its depth $h$. A large temperature gradient may lead to a weak layer in the snowpack. When these weak layers collapse, avalanches occur. Avalanche forecasters use the following rule of thumb: If $\dfrac{dT}{dh}$ exceeds 10°C/m anywhere in the snowpack, conditions are favorable for weak-layer formation, and the risk of avalanche increases. Assume the temperature function is continuous and differentiable.

   **a.** An avalanche forecaster digs a snow pit and takes two temperature measurements. At the surface ($h = 0$), the temperature is −16°C. At a depth of 1.1 m, the temperature is −2°C. Using the Mean Value Theorem, what can he conclude about the temperature gradient? Is the formation of a weak layer likely?
   **b.** One mile away, a skier finds that the temperature at a depth of 1.4 m is −1°C, and at the surface it is −12°C. What can be concluded about the temperature gradient? Is the formation of a weak layer in her location likely?
   **c.** Because snow is an excellent insulator, the temperature of snow-covered ground is near 0°C. Furthermore, the surface temperature of snow in a particular area does not vary much from one location to the next. Explain why a weak layer is more likely to form in places where the snowpack is not too deep.
   **d.** The term *isothermal* is used to describe the situation where all layers of the snowpack are at the same temperature (typically near the freezing point). Is a weak layer likely to form in isothermal snow? Explain.

**41. Mean Value Theorem and the police** A state patrol officer saw a car start from rest at a highway on-ramp. She radioed ahead to a patrol officer 30 mi along the highway. When the car reached the location of the second officer 28 min later, it was clocked going 60 mi/hr. The driver of the car was given a ticket for exceeding the 60-mi/hr speed limit. Why can the officer conclude that the driver exceeded the speed limit?

**42. Mean Value Theorem and the police again** Compare carefully to Exercise 41. A state patrol officer saw a car start from rest at a highway on-ramp. She radioed ahead to another officer 30 mi along the highway. When the car reached the location of the second officer 30 min later, it was clocked going 60 mi/hr. Can the patrol officer conclude that the driver exceeded the speed limit?

**43. Running pace** Explain why if a runner completes a 6.2-mi (10-km) race in 32 min, then he must have been running at exactly 11 mi/hr at least twice in the race. Assume the runner's speed at the finish line is zero.

**44. Mean Value Theorem for linear functions** Interpret the Mean Value Theorem when it is applied to any linear function.

**45. Mean Value Theorem for quadratic functions** Consider the quadratic function $f(x) = Ax^2 + Bx + C$, where $A$, $B$, and $C$ are real numbers with $A \neq 0$. Show that when the Mean Value Theorem is applied to $f$ on the interval $[a, b]$, the number $c$ guaranteed by the theorem is the midpoint of the interval.

**46. Means**
   **a.** Show that the point $c$ guaranteed to exist by the Mean Value Theorem for $f(x) = x^2$ on $[a, b]$ is the arithmetic mean of $a$ and $b$; that is, $c = \dfrac{a + b}{2}$.
   **b.** Show that the point $c$ guaranteed to exist by the Mean Value Theorem for $f(x) = \dfrac{1}{x}$ on $[a, b]$, where $0 < a < b$, is the geometric mean of $a$ and $b$; that is, $c = \sqrt{ab}$.

**47. Equal derivatives** Verify that the functions $f(x) = \tan^2 x$ and $g(x) = \sec^2 x$ have the same derivative. What can you say about the difference $f − g$? Explain.

**48. Equal derivatives** Verify that the functions $f(x) = \sin^2 x$ and $g(x) = -\cos^2 x$ have the same derivative. What can you say about the difference $f − g$? Explain.

**49. 100-m speed** The Jamaican sprinter Usain Bolt set a world record of 9.58 s in the 100-meter dash in the summer of 2009. Did his speed ever exceed 37 km/hr during the race? Explain.

**Explorations and Challenges**

**50.** Suppose $f'(x) > 1$, for all $x > 0$, and $f(0) = 0$. Show that $f(x) > x$, for all $x > 0$.

**51.** Suppose $f'(x) < 2$, for all $x \geq 2$, and $f(2) = 7$. Show that $f(4) < 11$.

**52.** Use the Mean Value Theorem to prove that $1 + \dfrac{a}{2} > \sqrt{1 + a}$, for $a > 0$. (*Hint:* For a given value of $a > 0$, let $f(x) = \sqrt{1 + x}$ on $[0, a]$ and use the fact that $\sqrt{1 + c} > 1$, for $c > 0$.)

**53.** Prove the following statements.
   **a.** $|\sin a − \sin b| \leq |a − b|$, for any real numbers $a$ and $b$
   **b.** $|\sin a| \leq |a|$, for any real number $a$

**54. Condition for nondifferentiability** Suppose $f'(x) < 0 < f''(x)$, for $x < a$, and $f'(x) > 0 > f''(x)$, for $x > a$. Prove that $f$ is not differentiable at $a$. (*Hint:* Assume $f$ is differentiable at $a$, and apply the Mean Value Theorem to $f'$.) More generally, show that if $f'$ and $f''$ change sign at the same point, then $f$ is not differentiable at that point.

**55. Generalized Mean Value Theorem** Suppose the functions $f$ and $g$ are continuous on $[a, b]$ and differentiable on $(a, b)$, where $g(a) \neq g(b)$. Then there is a point $c$ in $(a, b)$ at which

$$\frac{f(b) - f(a)}{g(b) - g(a)} = \frac{f'(c)}{g'(c)}.$$

This result is known as the **Generalized (or Cauchy's) Mean Value Theorem**.

**a.** If $g(x) = x$, then show that the Generalized Mean Value Theorem reduces to the Mean Value Theorem.

**b.** Suppose $f(x) = x^2 - 1$, $g(x) = 4x + 2$, and $[a, b] = [0, 1]$. Find a value of $c$ satisfying the Generalized Mean Value Theorem.

**QUICK CHECK ANSWERS**

**1.** $x = 2$   **2.** The functions shown in Figure 4.14 provide examples.   **3.** The graphs of $f(x) = 3x$ and $g(x) = 3x + 2$ have the same slope. Note that $f(x) - g(x) = -2$, a constant. ◄

# 4.3 What Derivatives Tell Us

In Section 4.1, we saw that the derivative is a tool for finding critical points, which are related to local maxima and minima. As we show in this section, derivatives (first *and* second derivatives) tell us much more about the behavior of functions.

## Increasing and Decreasing Functions

We have used the terms *increasing* and *decreasing* informally in earlier sections to describe a function or its graph. For example, the graph in **Figure 4.20a** rises as $x$ increases, so the corresponding function is increasing. In **Figure 4.20b**, the graph falls as $x$ increases, so the corresponding function is decreasing. The following definition makes these ideas precise.

➤ A function is called **monotonic** if it is either increasing or decreasing. We can make a further distinction by defining **nondecreasing** ($f(x_2) \geq f(x_1)$ whenever $x_2 > x_1$) and **nonincreasing** ($f(x_2) \leq f(x_1)$ whenever $x_2 > x_1$).

---

**DEFINITION  Increasing and Decreasing Functions**

Suppose a function $f$ is defined on an interval $I$. We say that $f$ is **increasing** on $I$ if $f(x_2) > f(x_1)$ whenever $x_1$ and $x_2$ are in $I$ and $x_2 > x_1$. We say that $f$ is **decreasing** on $I$ if $f(x_2) < f(x_1)$ whenever $x_1$ and $x_2$ are in $I$ and $x_2 > x_1$.

---

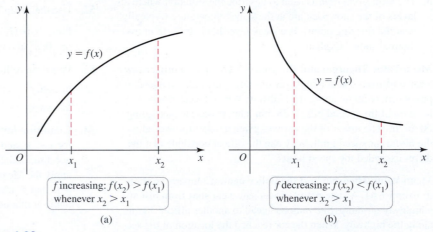

**Figure 4.20**

**Intervals of Increase and Decrease**   The graph of a function $f$ gives us an idea of the intervals on which $f$ is increasing and decreasing. But how do we determine those intervals precisely? This question is answered by making a connection to the derivative.

Recall that the derivative of a function gives the slopes of tangent lines. If the derivative is positive on an interval, the tangent lines on that interval have positive slopes, and the function is increasing on the interval (**Figure 4.21a**). Said differently, positive derivatives on an interval imply positive rates of change on the interval, which, in turn, indicate an increase in function values.

Similarly, if the derivative is negative on an interval, the tangent lines on that interval have negative slopes, and the function is decreasing on that interval (**Figure 4.21b**). These observations lead to Theorem 4.7, whose proof relies on the Mean Value Theorem.

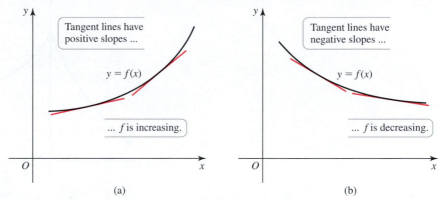

**Figure 4.21**

> The converse of Theorem 4.7 may not be true. According to the definition, $f(x) = x^3$ is increasing on $(-\infty, \infty)$ but it is not true that $f'(x) > 0$ on $(-\infty, \infty)$ (because $f'(0) = 0$).

---

**THEOREM 4.7    Test for Intervals of Increase and Decrease**
Suppose $f$ is continuous on an interval $I$ and differentiable at all interior points of $I$. If $f'(x) > 0$ at all interior points of $I$, then $f$ is increasing on $I$. If $f'(x) < 0$ at all interior points of $I$, then $f$ is decreasing on $I$.

---

**Proof:** Let $a$ and $b$ be any two distinct points in the interval $I$ with $b > a$. By the Mean Value Theorem,

$$\frac{f(b) - f(a)}{b - a} = f'(c),$$

for some $c$ between $a$ and $b$. Equivalently,

$$f(b) - f(a) = f'(c)(b - a).$$

Notice that $b - a > 0$ by assumption. So if $f'(c) > 0$, then $f(b) - f(a) > 0$. Therefore, for all $a$ and $b$ in $I$ with $b > a$, we have $f(b) > f(a)$, which implies that $f$ is increasing on $I$. Similarly, if $f'(c) < 0$, then $f(b) - f(a) < 0$ or $f(b) < f(a)$. It follows that $f$ is decreasing on $I$. ◄

**QUICK CHECK 1** Explain why a positive derivative on an interval implies that the function is increasing on the interval. ◄

**EXAMPLE 1    Sketching a function**  Sketch a graph of a function $f$ that is continuous on $(-\infty, \infty)$ and satisfies the following conditions.

1. $f' > 0$ on $(-\infty, 0)$, $(4, 6)$, and $(6, \infty)$.     2. $f' < 0$ on $(0, 4)$.
3. $f'(0)$ is undefined.     4. $f'(4) = f'(6) = 0$.

**SOLUTION**  By condition (1), $f$ is increasing on the intervals $(-\infty, 0)$, $(4, 6)$, and $(6, \infty)$. By condition (2), $f$ is decreasing on $(0, 4)$. Continuity of $f$ and condition (3) imply that $f$ has a cusp or corner at $x = 0$, and by condition (4), the graph has a horizontal tangent line at $x = 4$ and $x = 6$. It is useful to summarize these results using a *sign graph* (**Figure 4.22**).

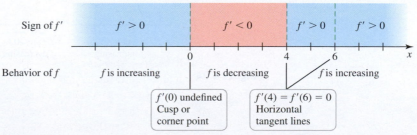

**Figure 4.22**

One possible graph satisfying these conditions is shown in **Figure 4.23**. Notice that the graph has a cusp at $x = 0$. Furthermore, although $f'(4) = f'(6) = 0$, $f$ has a local minimum at $x = 4$ but no local extremum at $x = 6$.

**Figure 4.23**

*Related Exercises 9–10* ◄

**EXAMPLE 2    Intervals of increase and decrease**    Find the intervals on which the function $f(x) = 2x^3 + 3x^2 + 1$ is increasing and the intervals on which it is decreasing.

**SOLUTION**    Note that $f'(x) = 6x^2 + 6x = 6x(x + 1)$. To find the intervals of increase, we first solve $6x(x + 1) = 0$ and determine that the critical points are $x = 0$ and $x = -1$. If $f'$ changes sign, then it does so at these points and nowhere else; that is, $f'$ has the same sign throughout each of the intervals $(-\infty, -1)$, $(-1, 0)$, and $(0, \infty)$. Evaluating $f'$ at selected points of each interval determines the sign of $f'$ on that interval.

> ➤ Appendix B, online at goo.gl/6DCbbM, shows how to solve inequalities using test values.

- At $x = -2$, $f'(-2) = 12 > 0$, so $f' > 0$ and $f$ is increasing on $(-\infty, -1)$.
- At $x = -\frac{1}{2}$, $f'\left(-\frac{1}{2}\right) = -\frac{3}{2} < 0$, so $f' < 0$ and $f$ is decreasing on $(-1, 0)$.
- At $x = 1$, $f'(1) = 12 > 0$, so $f' > 0$ and $f$ is increasing on $(0, \infty)$.

The graph of $f$ has a horizontal tangent line at $x = -1$ and $x = 0$. **Figure 4.24** shows the graph of $f$ superimposed on the graph of $f'$, confirming our conclusions.

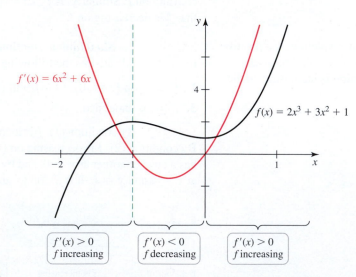

**Figure 4.24**

*Related Exercises 22–27* ◄

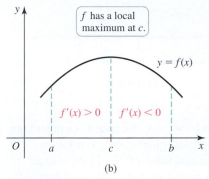

**Figure 4.25**

**QUICK CHECK 2** Sketch a function $f$ that is differentiable on $(-\infty, \infty)$ with the following properties: (i) $x = 0$ and $x = 2$ are critical points; (ii) $f$ is increasing on $(-\infty, 2)$; (iii) $f$ is decreasing on $(2, \infty)$. ◄

## Identifying Local Maxima and Minima

Using what we know about increasing and decreasing functions, we can now identify local extrema. Suppose $x = c$ is a critical point of $f$, where $f'(c) = 0$. Suppose also that $f'$ changes sign at $c$ with $f'(x) < 0$ on an interval $(a, c)$ to the left of $c$ and $f'(x) > 0$ on an interval $(c, b)$ to the right of $c$. In this case $f'$ is decreasing to the left of $c$ and increasing to the right of $c$, which means that $f$ has a local minimum at $c$, as shown in **Figure 4.25a**.

Similarly, suppose $f'$ changes sign at $c$ with $f'(x) > 0$ on an interval $(a, c)$ to the left of $c$ and $f'(x) < 0$ on an interval $(c, b)$ to the right of $c$. Then $f$ is increasing to the left of $c$ and decreasing to the right of $c$, so $f$ has a local maximum at $c$, as shown in **Figure 4.25b**.

**Figure 4.26** shows typical features of a function on an interval $[a, b]$. At local maxima or minima ($c_2$, $c_3$, and $c_4$), $f'$ changes sign. Although $c_1$ and $c_5$ are critical points, the sign of $f'$ is the same on both sides near these points, so there is no local maximum or minimum at these points. As emphasized earlier, *critical points do not always correspond to local extreme values.*

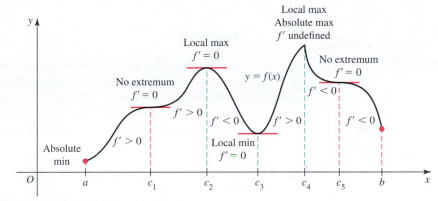

**Figure 4.26**

**First Derivative Test** The observations used to interpret Figure 4.26 are summarized in a powerful test for identifying local maxima and minima.

---

**THEOREM 4.8 First Derivative Test**
Assume $f$ is continuous on an interval that contains a critical point $c$, and assume $f$ is differentiable on an interval containing $c$, except perhaps at $c$ itself.

- If $f'$ changes sign from positive to negative as $x$ increases through $c$, then $f$ has a **local maximum** at $c$.

- If $f'$ changes sign from negative to positive as $x$ increases through $c$, then $f$ has a **local minimum** at $c$.

- If $f'$ is positive on both sides near $c$ or negative on both sides near $c$, then $f$ has no local extreme value at $c$.

---

**Proof:** Suppose $f'(x) > 0$ on an interval $(a, c)$. By Theorem 4.7, we know that $f$ is increasing on $(a, c)$, which implies that $f(x) < f(c)$ for all $x$ in $(a, c)$. Similarly, suppose $f'(x) < 0$ on an interval $(c, b)$. This time Theorem 4.7 says that $f$ is decreasing on $(c, b)$, which implies that $f(x) < f(c)$ for all $x$ in $(c, b)$. Therefore, $f(x) \leq f(c)$ for all $x$ in $(a, b)$ and $f$ has a local maximum at $c$. The proofs of the other two cases are similar. ◄

**EXAMPLE 3 Using the First Derivative Test** Consider the function

$$f(x) = 3x^4 - 4x^3 - 6x^2 + 12x + 1.$$

**a.** Find the intervals on which $f$ is increasing and those on which it is decreasing.

**b.** Identify the local extrema of $f$.

**SOLUTION**

**a.** Differentiating $f$, we find that

$$f'(x) = 12x^3 - 12x^2 - 12x + 12$$
$$= 12(x^3 - x^2 - x + 1)$$
$$= 12(x + 1)(x - 1)^2.$$

Solving $f'(x) = 0$ gives the critical points $x = -1$ and $x = 1$. The critical points determine the intervals $(-\infty, -1)$, $(-1, 1)$, and $(1, \infty)$ on which $f'$ does not change sign. Choosing a test point in each interval, a sign graph of $f'$ is constructed (**Figure 4.27**) that summarizes the behavior of $f$.

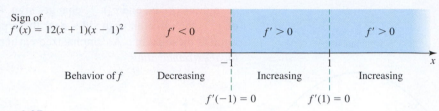

Sign of
$f'(x) = 12(x + 1)(x - 1)^2$

| | $f' < 0$ | $f' > 0$ | $f' > 0$ | |
|---|---|---|---|---|

Behavior of $f$      Decreasing      Increasing      Increasing

$$f'(-1) = 0 \qquad f'(1) = 0$$

**Figure 4.27**

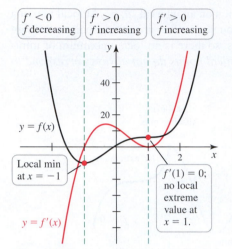

| $f' < 0$ | $f' > 0$ | $f' > 0$ |
|---|---|---|
| $f$ decreasing | $f$ increasing | $f$ increasing |

$y = f(x)$

Local min at $x = -1$

$f'(1) = 0$; no local extreme value at $x = 1$.

$y = f'(x)$

**Figure 4.28**

**b.** Note that $f$ is a polynomial, so it is continuous on $(-\infty, \infty)$. Because $f'$ changes sign from negative to positive as $x$ passes through the critical point $x = -1$, it follows by the First Derivative Test that $f$ has a local minimum value of $f(-1) = -10$ at $x = -1$. Observe that $f'$ is positive on both sides near $x = 1$, so $f$ does not have a local extreme value at $x = 1$ (**Figure 4.28**).

*Related Exercises 45–46* ◄

**EXAMPLE 4   Extreme values**  Find the local extrema of the function $g(x) = x^{2/3}(2 - x)$.

**SOLUTION**  In Example 4b of Section 4.1, we found that

$$g'(x) = \frac{4}{3}x^{-1/3} - \frac{5}{3}x^{2/3} = \frac{4 - 5x}{3x^{1/3}}$$

and that the critical points of $g$ are $x = 0$ and $x = \frac{4}{5}$. These two critical points are *candidates* for local extrema, and Theorem 4.8 is used to classify each as a local maximum, local minimum, or neither.

On the interval $(-\infty, 0)$, the numerator of $g'$ is positive and the denominator is negative (**Figure 4.29**). Therefore, $g'(x) < 0$ on $(-\infty, 0)$. On the interval $\left(0, \frac{4}{5}\right)$, the numerator of $g'$ is positive, as is the denominator. Therefore, $g'(x) > 0$ on $\left(0, \frac{4}{5}\right)$. We see that as $x$ passes through 0, $g'$ changes sign from negative to positive, which means $g$ has a local minimum at 0. A similar argument shows that $g'$ changes sign from positive to negative as $x$ passes through $\frac{4}{5}$, so $g$ has a local maximum at $\frac{4}{5}$. These observations are confirmed by the graphs of $g$ and $g'$ (**Figure 4.30**).

**QUICK CHECK 3** Explain how the First Derivative Test determines whether $f(x) = x^2$ has a local maximum or local minimum at the critical point $x = 0$. ◄

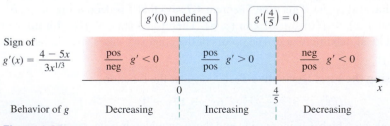

Sign of
$g'(x) = \dfrac{4 - 5x}{3x^{1/3}}$

| $\dfrac{\text{pos}}{\text{neg}}$ $g' < 0$ | $\dfrac{\text{pos}}{\text{pos}}$ $g' > 0$ | $\dfrac{\text{neg}}{\text{pos}}$ $g' < 0$ |
|---|---|---|

$g'(0)$ undefined        $g'\left(\frac{4}{5}\right) = 0$

Behavior of $g$      Decreasing      Increasing      Decreasing

**Figure 4.29**

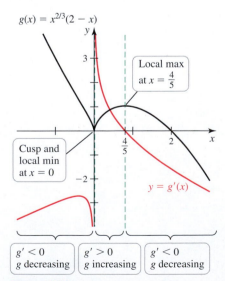

$g(x) = x^{2/3}(2 - x)$

Local max at $x = \frac{4}{5}$

Cusp and local min at $x = 0$

$y = g'(x)$

| $g' < 0$ | $g' > 0$ | $g' < 0$ |
|---|---|---|
| $g$ decreasing | $g$ increasing | $g$ decreasing |

**Figure 4.30**

*Related Exercise 47* ◄

**Absolute Extreme Values on Any Interval**   Theorem 4.1 guarantees the existence of absolute extreme values only on closed intervals. What can be said about absolute extrema on intervals that are not closed? The following theorem provides a valuable test.

The graph of $f$ cannot bend downward to fall below $f(c)$ because that creates additional local extreme values.

A single local min on $I$

$y = f(x)$

**Figure 4.31**

---

> **THEOREM 4.9   One Local Extremum Implies Absolute Extremum**
> Suppose $f$ is continuous on an interval $I$ that contains exactly one local extremum at $c$.
>
> • If a local maximum occurs at $c$, then $f(c)$ is the absolute maximum of $f$ on $I$.
> • If a local minimum occurs at $c$, then $f(c)$ is the absolute minimum of $f$ on $I$.

The proof of Theorem 4.9 is beyond the scope of this text, although **Figure 4.31** illustrates why the theorem is plausible. Assume $f$ has exactly one local minimum on $I$ at $c$. Notice that there can be no other point on the graph at which $f$ has a value less than $f(c)$. If such a point did exist, the graph would have to bend downward to drop below $f(c)$, which, by the continuity of $f$, cannot happen as it implies additional local extreme values on $I$. A similar argument applies to a solitary local maximum.

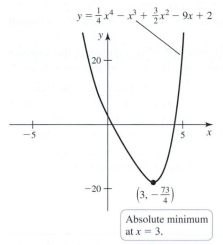

$y = \frac{1}{4}x^4 - x^3 + \frac{3}{2}x^2 - 9x + 2$

$\left(3, -\frac{73}{4}\right)$

Absolute minimum at $x = 3$.

**Figure 4.32**

**EXAMPLE 5   Finding an absolute extremum**   Verify that $f(x) = \frac{1}{4}x^4 - x^3 + \frac{3}{2}x^2 - 9x + 2$ has an absolute extreme value on its domain.

**SOLUTION**   As a polynomial, $f$ is differentiable on its domain $(-\infty, \infty)$ with

$$f'(x) = x^3 - 3x^2 + 3x - 9 = (x - 3)(x^2 + 3).$$

Solving $f'(x) = 0$ and noting that $x^2 + 3 > 0$ for all $x$ gives the single critical point $x = 3$. It may be verified that $f'(x) < 0$ for $x < 3$ and $f'(x) > 0$ for $x > 3$. Therefore, by Theorem 4.8, $f$ has a local minimum at $x = 3$. Because it is the only local extremum on $(-\infty, \infty)$, it follows from Theorem 4.9 that the absolute minimum value of $f$ occurs at $x = 3$, where $f(3) = -\frac{73}{4}$ (**Figure 4.32**).

*Related Exercises 49–51* ◄

## Concavity and Inflection Points

Just as the first derivative is related to the slope of tangent lines, the second derivative also has geometric meaning. Consider $f(x) = \sin x$, shown in **Figure 4.33**. Its graph bends upward for $-\pi < x < 0$, reflecting the fact that the slopes of the tangent lines increase as $x$ increases. It follows that the first derivative is increasing for $-\pi < x < 0$. A function with the property that $f'$ is increasing on an interval is *concave up* on that interval.

Similarly, $f(x) = \sin x$ bends downward for $0 < x < \pi$ because it has a decreasing first derivative on that interval. A function with the property that $f'$ is decreasing as $x$ increases on an interval is *concave down* on that interval.

Here is another useful characterization of concavity. If a function is concave up at a point (any point in $(-\pi, 0)$, Figure 4.33), then its graph near that point lies *above* the tangent line at that point. Similarly, if a function is concave down at a point (any point in $(0, \pi)$, Figure 4.33), then its graph near that point lies *below* the tangent line at that point (Exercise 102).

Finally, imagine a function $f$ that changes concavity (from up to down, or vice versa) at a point $c$ in the domain of $f$. For example, $f(x) = \sin x$ in Figure 4.33 changes from concave up to concave down as $x$ passes through $x = 0$. A point on the graph of $f$ at which $f$ changes concavity is called an *inflection point*.

$f'$ increases and slopes of tangent lines increase as $x$ increases

$y = \sin x$

Concave down

Concave up

$f'$ decreases and slopes of tangent lines decrease as $x$ increases

**Figure 4.33**

---

> **DEFINITION   Concavity and Inflection Point**
> Let $f$ be differentiable on an open interval $I$. If $f'$ is increasing on $I$, then $f$ is **concave up** on $I$. If $f'$ is decreasing on $I$, then $f$ is **concave down** on $I$.
> If $f$ is continuous at $c$ and $f$ changes concavity at $c$ (from up to down, or vice versa), then $f$ has an **inflection point** at $c$.

Applying Theorem 4.7 to $f'$ leads to a test for concavity in terms of the second derivative. Specifically, if $f'' > 0$ on an interval $I$, then $f'$ is increasing on $I$ and $f$ is concave up on $I$. Similarly, if $f'' < 0$ on $I$, then $f$ is concave down on $I$. In addition, if the values of $f''$ change sign at a point $c$ (from positive to negative, or vice versa), then the concavity of $f$ changes at $c$ and $f$ has an inflection point at $c$ (**Figure 4.34a**). We now have a useful interpretation of the second derivative: It measures *concavity*.

---

**THEOREM 4.10    Test for Concavity**
Suppose $f''$ exists on an open interval $I$.

- If $f'' > 0$ on $I$, then $f$ is concave up on $I$.
- If $f'' < 0$ on $I$, then $f$ is concave down on $I$.
- If $c$ is a point of $I$ at which $f''$ changes sign at $c$ (from positive to negative, or vice versa), then $f$ has an inflection point at $c$.

---

There are a few important but subtle points here. The fact that $f''(c) = 0$ does not necessarily imply that $f$ has an inflection point at $c$. A good example is $f(x) = x^4$. Although $f''(0) = 0$, the concavity does not change at $x = 0$ (a similar function is shown in **Figure 4.34b**).

Typically, if $f$ has an inflection point at $c$, then $f''(c) = 0$, reflecting the smooth change in concavity. However, an inflection point may also occur at a point where $f''$ does not exist. For example, the function $f(x) = x^{1/3}$ has a vertical tangent line and an inflection point at $x = 0$ (a similar function is shown in **Figure 4.34c**). Finally, note that the function shown in **Figure 4.34d**, with behavior similar to that of $f(x) = x^{2/3}$, does not have an inflection point at $c$ despite the fact that $f''(c)$ does not exist. In summary, if $f''(c) = 0$ or $f''(c)$ does not exist, then $(c, f(c))$ is a candidate for an inflection point. To be certain an inflection point occurs at $c$, we must show that the concavity of $f$ changes at $c$.

**QUICK CHECK 4** Verify that the function $f(x) = x^4$ is concave up for $x > 0$ and for $x < 0$. Is $x = 0$ an inflection point? Explain. ◄

**Figure 4.34**

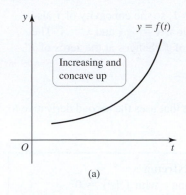

Increasing and
concave up

$y = f(t)$

(a)

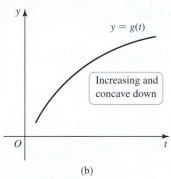

$y = g(t)$

Increasing and
concave down

(b)

**Figure 4.35**

**EXAMPLE 6    Interpreting concavity** Sketch a function satisfying each set of conditions on some interval.

**a.** $f'(t) > 0$ and $f''(t) > 0$          **b.** $g'(t) > 0$ and $g''(t) < 0$

**c.** Which of the functions, $f$ or $g$, could describe a population that increases and approaches a steady state as $t \to \infty$?

**SOLUTION**

**a.** **Figure 4.35a** shows the graph of a function that is increasing $(f'(t) > 0)$ and concave up $(f''(t) > 0)$.

**b.** **Figure 4.35b** shows the graph of a function that is increasing $(g'(t) > 0)$ and concave down $(g''(t) < 0)$.

**c.** Because $f$ increases at an *increasing* rate, the graph of $f$ could not approach a horizontal asymptote, so $f$ could not describe a population that approaches a steady state. On the other hand, $g$ increases at a *decreasing* rate, so its graph could approach a horizontal asymptote, depending on the rate at which $g$ increases.

*Related Exercises 53–56* ◄

**EXAMPLE 7    Detecting concavity** Identify the intervals on which the function $f(x) = 3x^4 - 4x^3 - 6x^2 + 12x + 1$ is concave up or concave down. Then locate the inflection points.

**SOLUTION** This function was considered in Example 3, where we found that

$$f'(x) = 12(x + 1)(x - 1)^2.$$

It follows that

$$f''(x) = 12(x - 1)(3x + 1).$$

We see that $f''(x) = 0$ at $x = 1$ and $x = -\frac{1}{3}$. These points are *candidates* for inflection points; to be certain that they are inflection points, we must determine whether the concavity changes at these points. The sign graph in **Figure 4.36** shows the following:

- $f''(x) > 0$ and $f$ is concave up on $\left(-\infty, -\frac{1}{3}\right)$ and $(1, \infty)$.
- $f''(x) < 0$ and $f$ is concave down on $\left(-\frac{1}{3}, 1\right)$.

**Figure 4.36**

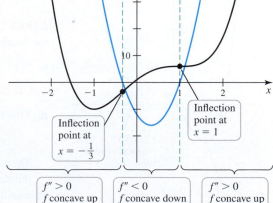

**Figure 4.37**

We see that the sign of $f''$ changes at $x = -\frac{1}{3}$ and at $x = 1$, so the concavity of $f$ also changes at these points. Therefore, inflection points occur at $x = -\frac{1}{3}$ and $x = 1$. The graphs of $f$ and $f''$ (**Figure 4.37**) show that the concavity of $f$ changes at the zeros of $f''$.

*Related Exercises 58–59* ◄

**Second Derivative Test**    It is now a short step to a test that uses the second derivative to identify local maxima and minima.

> **THEOREM 4.11    Second Derivative Test for Local Extrema**
> Suppose $f''$ is continuous on an open interval containing $c$ with $f'(c) = 0$.
>
> • If $f''(c) > 0$, then $f$ has a local minimum at $c$ (**Figure 4.38a**).
>
> • If $f''(c) < 0$, then $f$ has a local maximum at $c$ (**Figure 4.38b**).
>
> • If $f''(c) = 0$, then the test is inconclusive; $f$ may have a local maximum, a local minimum, or neither at $c$.

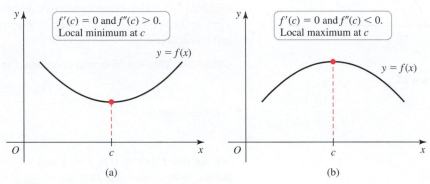

**Figure 4.38**

**QUICK CHECK 5** Sketch a graph of a function with $f'(x) > 0$ and $f''(x) > 0$ on an interval. Sketch a graph of a function with $f'(x) < 0$ and $f''(x) < 0$ on an interval. ◄

**Proof:** Assume $f''(c) > 0$. Because $f''$ is continuous on an interval containing $c$, it follows that $f'' > 0$ on some open interval $I$ containing $c$ and that $f'$ is increasing on $I$. Because $f'(c) = 0$, it follows that $f'$ changes sign at $c$ from negative to positive, which, by the First Derivative Test, implies that $f$ has a local minimum at $c$. The proofs of the second and third statements are similar.    ◄

**EXAMPLE 8    The Second Derivative Test**    Use the Second Derivative Test to locate the local extrema of the following functions.

**a.** $f(x) = 3x^4 - 4x^3 - 6x^2 + 12x + 1$ on $[-2, 2]$        **b.** $f(x) = \sin^2 x$

**SOLUTION**

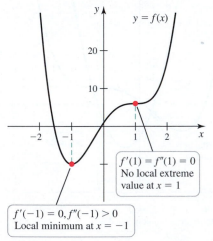

**Figure 4.39**

**a.** This function was considered in Examples 3 and 7, where we found that

$$f'(x) = 12(x + 1)(x - 1)^2 \quad \text{and} \quad f''(x) = 12(x - 1)(3x + 1).$$

Therefore, the critical points of $f$ are $x = -1$ and $x = 1$. Evaluating $f''$ at the critical points, we find that $f''(-1) = 48 > 0$. By the Second Derivative Test, $f$ has a local minimum at $x = -1$. At the other critical point, $f''(1) = 0$, so the test is inconclusive. You can check that the first derivative does not change sign at $x = 1$, which means $f$ does not have a local maximum or minimum at $x = 1$ (**Figure 4.39**).

**b.** Using the Chain Rule and a trigonometric identity, we have $f'(x) = 2 \sin x \cos x = \sin 2x$ and $f''(x) = 2 \cos 2x$. The critical points occur when $f'(x) = \sin 2x = 0$, or when $x = 0, \pm\pi/2, \pm\pi, \ldots$. To apply the Second Derivative Test, we evaluate $f''$ at the critical points:

- $f''(0) = 2 > 0$, so $f$ has a local minimum at $x = 0$.
- $f''(\pm\pi/2) = -2 < 0$, so $f$ has a local maximum at $x = \pm\pi/2$.
- $f''(\pm\pi) = 2 > 0$, so $f$ has a local minimum at $x = \pm\pi$.

This pattern continues, and we see that $f$ has alternating local maxima and minima, evenly spaced every $\pi/2$ units (**Figure 4.40**).

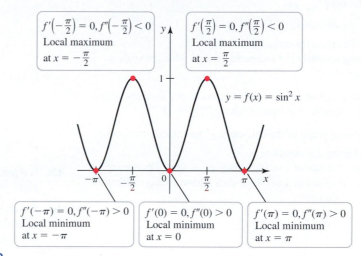

**Figure 4.40**

*Related Exercises 70, 72* ◄

## Recap of Derivative Properties

This section has demonstrated that the first and second derivatives of a function provide valuable information about its graph. The relationships among a function's derivatives and its extreme values and concavity are summarized in **Figure 4.41**.

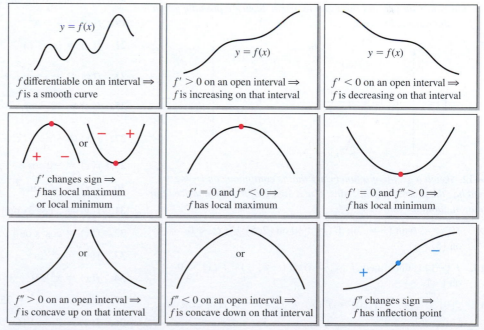

**Figure 4.41**

## SECTION 4.3 EXERCISES

### Getting Started

1.  Explain how the first derivative of a function determines where the function is increasing or decreasing.

2.  Explain how to apply the First Derivative Test.

3.  Suppose the derivative of $f$ is $f'(x) = x - 3$.
    **a.** Find the critical points of $f$.
    **b.** On what intervals is $f$ increasing and on what intervals is $f$ decreasing?

4.  Suppose the derivative of $f$ is $f'(x) = (x - 1)(x - 2)$.
    **a.** Find the critical points of $f$.
    **b.** On what intervals is $f$ increasing and on what intervals is $f$ decreasing?

5.  Sketch the graph of a function that has neither a local maximum nor a local minimum at a point where $f'(x) = 0$.

6.  The following graph of the derivative $g'$ has exactly two roots.
    **a.** Find the critical points of $g$.
    **b.** For what values of $x$ in $(0, 3)$ is $g$ increasing? Decreasing?
    **c.** For what values of $x$ in $(0, 3)$ does $g$ have a local maximum? A local minimum?

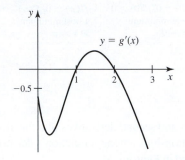

**7–8.** *The following figures give the graph of the derivative of a continuous function $f$ that passes through the origin. Sketch a possible graph of $f$ on the same set of axes.*

7.

8.

**9–12.** *Sketch a graph of a function $f$ that is continuous on $(-\infty, \infty)$ and has the following properties. Use a sign graph to summarize information about the function.*

9.  $f'(x) < 0$ on $(-\infty, 2)$; $f'(x) > 0$ on $(2, 5)$; $f'(x) < 0$ on $(5, \infty)$

10. $f'(-1)$ is undefined; $f'(x) > 0$ on $(-\infty, -1)$; $f'(x) < 0$ on $(-1, \infty)$

11. $f(0) = f(4) = f'(0) = f'(2) = f'(4) = 0$; $f(x) \geq 0$ on $(-\infty, \infty)$

12. $f'(-2) = f'(2) = f'(6) = 0$; $f'(x) \geq 0$ on $(-\infty, \infty)$

13. Suppose $g''(x) = 2 - x$.
    **a.** On what intervals is $g$ concave up and on what intervals is $g$ concave down?
    **b.** State the inflection points of $g$.

14. The following graph of $g''$ has exactly three $x$-intercepts.
    **a.** For what values of $x$ in $(-4, 3)$ is the graph of $g$ concave up? Concave down?
    **b.** State the inflection points of $g$ that lie in $(-4, 3)$.

15. Is it possible for a function to satisfy $f(x) > 0$, $f'(x) > 0$, and $f''(x) < 0$ on an interval? Explain.

16. Sketch a graph of a function that changes from concave up to concave down as $x$ increases. Describe how the second derivative of this function changes.

17. Give a function that does not have an inflection point at a point where $f''(x) = 0$.

18. Suppose $f$ is continuous on an interval containing a critical point $c$ and $f''(c) = 0$. How do you determine whether $f$ has a local extreme value at $x = c$?

### Practice Exercises

**19–40. Increasing and decreasing functions** *Find the intervals on which $f$ is increasing and the intervals on which it is decreasing.*

19. $f(x) = 4 - x^2$

20. $f(x) = x^2 - 16$

21. $f(x) = (x - 1)^2$

22. $f(x) = x^3 + 4x$

23. $f(x) = \dfrac{x^3}{3} - \dfrac{5x^2}{2} + 4x$

24. $f(x) = -\dfrac{x^3}{3} + \dfrac{x^2}{2} + 2x$

25. $f(x) = 12 + x - x^2$

26. $f(x) = x^4 - 4x^3 + 4x^2$

27. $f(x) = -\dfrac{x^4}{4} + x^3 - x^2$

28. $f(x) = 2x^5 - \dfrac{15x^4}{4} + \dfrac{5x^3}{3}$

29. $f(x) = -2\cos x - x$ on $[0, 2\pi]$

30. $f(x) = \sqrt{2}\sin x - x$ on $[0, 2\pi]$

31. $f(x) = 3\cos 3x$ on $[-\pi, \pi]$

32. $f(x) = \cos^2 x$ on $[-\pi, \pi]$

33. $f(x) = x^{2/3}(x^2 - 4)$

34. $f(x) = x^2\sqrt{9 - x^2}$ on $(-3, 3)$

35. $f(x) = -12x^5 + 75x^4 - 80x^3$

36. $f(x) = 3x^4 - 16x^3 + 24x^2$

37. $f(x) = -2x^4 + x^2 + 10$

**38.** $f(x) = \dfrac{x^4}{4} - \dfrac{8x^3}{3} + \dfrac{15x^2}{2} + 8$

**39.** $f(x) = \sin x - x \cos x$ on $(0, 2\pi)$

**40.** $f(x) = x^2 \sin x - 2 \sin x + 2x \cos x$ on $(0, 2\pi)$

**41–48. First Derivative Test**

**a.** Locate the critical points of $f$.

**b.** Use the First Derivative Test to locate the local maximum and minimum values.

**c.** Identify the absolute maximum and minimum values of the function on the given interval (when they exist).

**41.** $f(x) = x^2 + 3$ on $[-3, 2]$

**42.** $f(x) = -x^2 - x + 2$ on $[-4, 4]$

**43.** $f(x) = x\sqrt{4 - x^2}$ on $[-2, 2]$

**44.** $f(x) = 2x^3 + 3x^2 - 12x + 1$ on $[-2, 4]$

**45.** $f(x) = -x^3 + 9x$ on $[-4, 3]$

**46.** $f(x) = 2x^5 - 5x^4 - 10x^3 + 4$ on $[-2, 4]$

**47.** $f(x) = x^{2/3}(x - 5)$ on $[-5, 5]$

**48.** $f(x) = \dfrac{x^2}{x^2 - 1}$ on $[-4, 4]$

**49–52. Absolute extreme values** *Verify that the following functions satisfy the conditions of Theorem 4.9 on their domains. Then find the location and value of the absolute extrema guaranteed by the theorem.*

**49.** $f(x) = -3x^2 + 2x - 5$     **50.** $f(x) = 4x + 1/\sqrt{x}$

**51.** $A(r) = 24/r + 2\pi r^2, r > 0$

**52.** $f(x) = x\sqrt{3 - x}$

**53–56. Sketching curves** *Sketch a graph of a function $f$ that is continuous on $(-\infty, \infty)$ and has the following properties.*

**53.** $f'(x) > 0, f''(x) > 0$

**54.** $f'(x) < 0$ and $f''(x) > 0$ on $(-\infty, 0)$; $f'(x) > 0$ and $f''(x) > 0$ on $(0, \infty)$

**55.** $f'(x) < 0$ and $f''(x) < 0$ on $(-\infty, 0)$; $f'(x) < 0$ and $f''(x) > 0$ on $(0, \infty)$

**56.** $f'(x) < 0$ and $f''(x) > 0$ on $(-\infty, 0)$; $f'(x) < 0$ and $f''(x) < 0$ on $(0, \infty)$

**57–68. Concavity** *Determine the intervals on which the following functions are concave up or concave down. Identify any inflection points.*

**57.** $f(x) = x^4 - 2x^3 + 1$    **58.** $f(x) = -x^4 - 2x^3 + 12x^2$

**59.** $f(x) = 5x^4 - 20x^3 + 10$    **60.** $f(x) = \dfrac{1}{1 + x^2}$

**61.** $g(t) = \dfrac{t - 2}{t + 3}$     **62.** $g(x) = \sqrt[3]{x - 4}$

**63.** $f(\theta) = \theta \sin \theta + 2 \cos \theta$ on $(0, 2\pi)$

**64.** $f(x) = x^2 + 4 \sin x$ on $[0, \pi]$

**65.** $f(x) = 4(x + 1)^{5/2}(3x - 4)$ on $[-1, \infty)$

**66.** $h(t) = 2 + \cos 2t$ on $[0, \pi]$

**67.** $g(t) = 3t^5 - 30t^4 + 80t^3 + 100$

**68.** $f(x) = 2x^4 + 8x^3 + 12x^2 - x - 2$

**69–84. Second Derivative Test** *Locate the critical points of the following functions. Then use the Second Derivative Test to determine (if possible) whether they correspond to local maxima or local minima.*

**69.** $f(x) = x^3 - 3x^2$     **70.** $f(x) = 6x^2 - x^3$

**71.** $f(x) = 4 - x^2$     **72.** $f(x) = x^3 - \dfrac{3}{2}x^2 - 36x$

**73.** $f(x) = 2x^3 - 3x^2 + 12$    **74.** $p(x) = \dfrac{x - 4}{x^2 + 20}$

**75.** $f(x) = 3x^4 - 4x^3 + 2$    **76.** $g(x) = \dfrac{x^4}{2 - 12x^2}$

**T 77.** $f(x) = x^3 + \cos x$    **78.** $f(x) = \sqrt{x}\left(\dfrac{12}{7}x^3 - 4x^2\right)$

**79.** $p(t) = 2t^3 + 3t^2 - 36t$

**80.** $f(x) = \dfrac{x^4}{4} - \dfrac{5x^3}{3} - 4x^2 + 48x$

**81.** $h(x) = (x + a)^4$; $a$ constant

**82.** $f(x) = x^3 - 13x^2 - 9x$

**83.** $f(\theta) = 2 \sin 3\theta - 3\theta$ on $[0, \pi]$

**84.** $f(x) = 2x^{-3} - x^{-2}$

**85.** **Explain why or why not** Determine whether the following statements are true and give an explanation or counterexample.

**a.** If $f'(x) > 0$ and $f''(x) < 0$ on an interval, then $f$ is increasing at a decreasing rate on the interval.

**b.** If $f'(c) > 0$ and $f''(c) = 0$, then $f$ has a local maximum at $c$.

**c.** Two functions that differ by an additive constant both increase and decrease on the same intervals.

**d.** If $f$ and $g$ increase on an interval, then the product $fg$ also increases on that interval.

**e.** There exists a function $f$ that is continuous on $(-\infty, \infty)$ with exactly three critical points, all of which correspond to local maxima.

**86–87. Functions from derivatives** *Consider the following graphs of $f'$ and $f''$. On the same set of axes, sketch the graph of a possible function $f$. The graphs of $f$ are not unique.*

**86.**

**87.**

**88. Is it possible?** Determine whether the following properties can be satisfied by a function that is continuous on $(-\infty, \infty)$. If such a function is possible, provide an example or a sketch of the function. If such a function is not possible, explain why.

   **a.** A function $f$ is concave down and positive everywhere.
   **b.** A function $f$ is increasing and concave down everywhere.
   **c.** A function $f$ has exactly two local extrema and three inflection points.
   **d.** A function $f$ has exactly four zeros and two local extrema.

**89. Matching derivatives and functions** The following figures show the graphs of three functions (graphs a–c). Match each function with its first derivative (graphs A–C) *and* its second derivative (graphs i–iii).

(a)

(b)

(c)

(A)

(B)

(C)

(i)

(ii)

(iii)

**90. Graphical analysis** The figure shows the graphs of $f$, $f'$, and $f''$. Which curve is which?

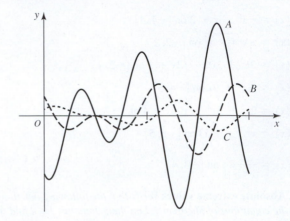

**91. Sketching graphs** Sketch the graph of a function $f$ that is continuous on $[a, b]$ such that $f$, $f'$, and $f''$ have the signs indicated in the following table on $[a, b]$. There are eight different cases lettered A–H that correspond to eight different graphs.

| Case | A | B | C | D | E | F | G | H |
|------|---|---|---|---|---|---|---|---|
| $f$  | + | + | + | + | − | − | − | − |
| $f'$ | + | + | − | − | + | + | − | − |
| $f''$| + | − | + | − | + | − | + | − |

**92–95. Designer functions** *Sketch the graph of a function $f$ that is continuous on $(-\infty, \infty)$ and satisfies the following sets of conditions.*

**92.** $f''(x) > 0$ on $(-\infty, -2)$; $f''(-2) = 0$; $f'(-1) = f'(1) = 0$; $f''(2) = 0$; $f'(3) = 0$; $f''(x) > 0$ on $(4, \infty)$

**93.** $f(-2) = f''(-1) = 0$; $f'\left(-\frac{3}{2}\right) = 0$; $f(0) = f'(0) = 0$; $f(1) = f'(1) = 0$

**94.** $f'(x) > 0$, for all $x$ in the domain of $f'$; $f'(-2)$ and $f'(1)$ do not exist; $f''(0) = 0$

**95.** $f''(x) > 0$ on $(-\infty, -2)$; $f''(x) < 0$ on $(-2, 1)$; $f''(x) > 0$ on $(1, 3)$; $f''(x) < 0$ on $(3, \infty)$

**T 96. Graph carefully** Graph the function $f(x) = 60x^5 - 901x^3 + 27x$ in the window $[-4, 4] \times [-10{,}000, 10{,}000]$. How many extreme values do you see? Locate *all* the extreme values by analyzing $f'$.

**97. Interpreting the derivative** The graph of $f'$ on the interval $[-3, 2]$ is shown in the figure.

   **a.** On what interval(s) is $f$ increasing? Decreasing?
   **b.** Find the critical points of $f$. Which critical points correspond to local maxima? Local minima? Neither?
   **c.** At what point(s) does $f$ have an inflection point?
   **d.** On what interval(s) is $f$ concave up? Concave down?

e. Sketch the graph of $f''$.
f. Sketch one possible graph of $f$.

**98. Concavity of parabolas** Consider the general parabola described by the function $f(x) = ax^2 + bx + c$. For what values of $a$, $b$, and $c$ is $f$ concave up? For what values of $a$, $b$, and $c$ is $f$ concave down?

## Explorations and Challenges

**T** **99. Demand functions and elasticity** Economists use *demand functions* to describe how much of a commodity can be sold at varying prices. For example, the demand function $D(p) = 500 - 10p$ says that at a price of $p = 10$, a quantity of $D(10) = 400$ units of the commodity can be sold. The elasticity $E = \dfrac{dD}{dp}\dfrac{p}{D}$ of the demand gives the approximate percent change in the demand for every 1% change in the price. (See Section 3.6 or the Guided Project *Elasticity in Economics* for more on demand functions and elasticity.)

a. Compute the elasticity of the demand function $D(p) = 500 - 10p$.
b. If the price is \$12 and increases by 4.5%, what is the approximate percent change in the demand?
c. Show that for the linear demand function $D(p) = a - bp$, where $a$ and $b$ are positive real numbers, the elasticity is a decreasing function, for $p \geq 0$ and $p \neq a/b$.
d. Show that the demand function $D(p) = a/p^b$, where $a$ and $b$ are positive real numbers, has a constant elasticity for all positive prices.

**100. Population models** A typical population curve is shown in the figure. The population is small at $t = 0$ and increases toward a steady-state level called the *carrying capacity*. Explain why the maximum growth rate occurs at an inflection point of the population curve.

**101. Population models** The population of a species is given by the function $P(t) = \dfrac{Kt^2}{t^2 + b}$, where $t \geq 0$ is measured in years and $K$ and $b$ are positive real numbers.

a. With $K = 300$ and $b = 30$, what is $\lim\limits_{t \to \infty} P(t)$, the carrying capacity of the population?
b. With $K = 300$ and $b = 30$, when does the maximum growth rate occur?
c. For arbitrary positive values of $K$ and $b$, when does the maximum growth rate occur (in terms of $K$ and $b$)?

**102. Tangent lines and concavity** Give an argument to support the claim that if a function is concave up at a point, then the tangent line at that point lies below the curve near that point.

**103. General quartic** Show that the general quartic (fourth-degree) polynomial $f(x) = x^4 + ax^3 + bx^2 + cx + d$, where $a$, $b$, $c$, and $d$ are real numbers, has either zero or two inflection points, and the latter case occurs provided $b < \dfrac{3a^2}{8}$.

**104. First Derivative Test is not exhaustive** Sketch the graph of a (simple) nonconstant function $f$ that has a local maximum at $x = 1$, with $f'(1) = 0$, where $f'$ does not change sign from positive to negative as $x$ increases through 1. Why can't the First Derivative Test be used to classify the critical point at $x = 1$ as a local maximum? How could the test be rephrased to account for such a critical point?

**105. Properties of cubics** Consider the general cubic polynomial $f(x) = x^3 + ax^2 + bx + c$, where $a$, $b$, and $c$ are real numbers.

a. Prove that $f$ has exactly one local maximum and one local minimum provided $a^2 > 3b$.
b. Prove that $f$ has no extreme values if $a^2 < 3b$.

**QUICK CHECK ANSWERS**

**1.** Positive derivatives on an interval mean the curve is rising on the interval, which means the function is increasing on the interval.    **2.**

**3.** $f'(x) < 0$ on $(-\infty, 0)$ and $f'(x) > 0$ on $(0, \infty)$. Therefore, $f$ has a local minimum at $x = 0$ by the First Derivative Test.    **4.** $f''(x) = 12x^2$, so $f''(x) > 0$ for $x < 0$ and for $x > 0$. There is no inflection point at $x = 0$ because the second derivative does not change sign.    **5.** The first curve should be rising and concave up. The second curve should be falling and concave down.

# 4.4 Graphing Functions

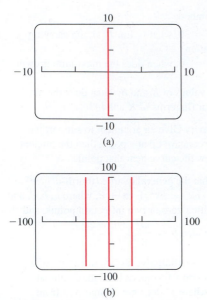

(a)

(b)

**Figure 4.42**

**QUICK CHECK 1** Graph $f(x) = x^3/3 - 400x$ using various windows on a graphing calculator. Find a window that gives a better graph of $f$ than those in Figure 4.42. ◄

We have now collected the tools required for a comprehensive approach to graphing functions. These *analytical methods* are indispensable, even with the availability of powerful graphing utilities, as illustrated by the following example.

## Calculators and Analysis

Suppose you want to graph the harmless-looking function $f(x) = x^3/3 - 400x$. The result of plotting $f$ using a graphing calculator with a default window of $[-10, 10] \times [-10, 10]$ is shown in **Figure 4.42a**; one vertical line appears on the screen. Zooming out to the window $[-100, 100] \times [-100, 100]$ produces three vertical lines (**Figure 4.42b**); it is still difficult to understand the behavior of the function using only this graph. Expanding the window even more to $[-1000, 1000] \times [-1000, 1000]$ is no better. So what do we do?

The function $f(x) = x^3/3 - 400x$ has a reasonable graph (see Example 2), but it cannot be found automatically by letting technology do all the work. Here is the message of this section: Graphing utilities are valuable for exploring functions, producing preliminary graphs, and checking your work. But they should not be relied on exclusively because they cannot explain *why* a graph has its shape. Rather, graphing utilities should be used in an interactive way with the analytical methods presented in this chapter.

## Graphing Guidelines

The following set of guidelines need not be followed exactly for every function, and you will find that several steps can often be done at once. Depending on the specific problem, some of the steps are best done analytically, while other steps can be done with a graphing utility. Experiment with both approaches and try to find a good balance. We also present a schematic record-keeping procedure to keep track of discoveries as they are made.

---

**Graphing Guidelines for $y = f(x)$**

1. **Identify the domain or interval of interest.** On what interval(s) should the function be graphed? It may be the domain of the function or some subset of the domain.

2. **Exploit symmetry.** Take advantage of symmetry. For example, is the function *even* $(f(-x) = f(x))$, *odd* $(f(-x) = -f(x))$, or neither?

3. **Find the first and second derivatives.** They are needed to determine extreme values, concavity, inflection points, and the intervals on which $f$ is increasing or decreasing. Computing derivatives—particularly second derivatives—may not be practical, so some functions may need to be graphed without complete derivative information.

4. **Find critical points and possible inflection points.** Determine points at which $f'(x) = 0$ or $f'$ is undefined. Determine points at which $f''(x) = 0$ or $f''$ is undefined.

5. **Find intervals on which the function is increasing/decreasing and concave up/down.** The first derivative determines the intervals on which $f$ is increasing or decreasing. The second derivative determines the intervals on which the function is concave up or concave down.

6. **Identify extreme values and inflection points.** Use either the First or Second Derivative Test to classify the critical points. Both $x$- and $y$-coordinates of maxima, minima, and inflection points are needed for graphing.

7. **Locate all asymptotes and determine end behavior.** Vertical asymptotes often occur at zeros of denominators. Horizontal asymptotes require examining limits as $x \to \pm\infty$; these limits determine end behavior. Slant asymptotes occur with rational functions in which the degree of the numerator is one more than the degree of the denominator.

8. **Find the intercepts.** The $y$-intercept of the graph is found by setting $x = 0$. The $x$-intercepts are found by solving $f(x) = 0$; they are the real zeros (or roots) of $f$.

9. **Choose an appropriate graphing window and plot a graph.** Use the results of the previous steps to graph the function. If you use graphing software, check for consistency with your analytical work. Is your graph *complete*—that is, does it show all the essential details of the function?

> Limits at infinity determine the end behavior of many functions, though a modification is required for some functions. For example, the end behavior of $f(x) = \dfrac{1+x}{x+\sqrt{x}}$ is found by computing $\lim\limits_{x\to 0^+} f(x)$ and $\lim\limits_{x\to\infty} f(x)$ because the domain of $f$ is $(0, \infty)$.

**EXAMPLE 1   A warm-up** Given the following information about the first and second derivatives of a function $f$ that is continuous on $(-\infty, \infty)$, summarize the information using a sign graph, and then sketch a possible graph of $f$.

$$f' < 0, f'' > 0 \text{ on } (-\infty, 0) \quad f' > 0, f'' > 0 \text{ on } (0, 1) \quad f' > 0, f'' < 0 \text{ on } (1, 2)$$
$$f' < 0, f'' < 0 \text{ on } (2, 3) \qquad f' < 0, f'' > 0 \text{ on } (3, 4) \quad f' > 0, f'' > 0 \text{ on } (4, \infty)$$

**SOLUTION** Figure 4.43 uses the given information to determine the behavior of $f$ and its graph. For example, on the interval $(-\infty, 0)$, $f$ is decreasing and concave up, so we sketch a segment of a curve with these properties on this interval. Continuing in this manner, we obtain a useful summary of the properties of $f$.

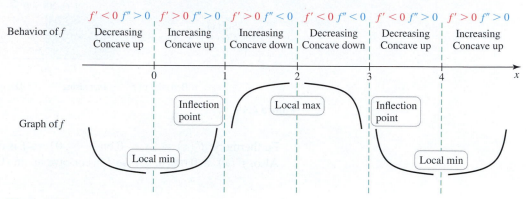

**Figure 4.43**

Assembling the information shown in Figure 4.43, a possible graph of $f$ is produced (Figure 4.44). Notice that derivative information is not sufficient to determine the $y$-coordinates of points on the curve.

**QUICK CHECK 2** Explain why the functions $f$ and $f + C$, where $C$ is a constant, have the same derivative properties. ◄

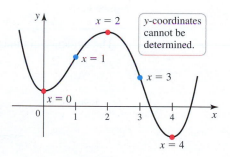

**Figure 4.44**

*Related Exercises 7–8* ◄

**EXAMPLE 2   A deceptive polynomial** Use the graphing guidelines to graph
$$f(x) = \frac{x^3}{3} - 400x \text{ on its domain.}$$

**SOLUTION**

1. **Domain** The domain of any polynomial is $(-\infty, \infty)$.

2. **Symmetry** Because $f$ consists of odd powers of the variable, it is an odd function. Its graph is symmetric about the origin.

> Notice that the first derivative of an odd polynomial is an even polynomial and the second derivative is an odd polynomial (assuming the original polynomial is of degree 3 or greater).

3. **Derivatives** The derivatives of $f$ are
$$f'(x) = x^2 - 400 \quad \text{and} \quad f''(x) = 2x.$$

4. **Critical points and possible inflection points** Solving $f'(x) = 0$, we find that the critical points are $x = \pm 20$. Solving $f''(x) = 0$, we see that a possible inflection point occurs at $x = 0$.

➤ See Appendix B, online at
goo.gl/6DCbbM, for solving
inequalities using test values.

**5. Increasing/decreasing and concavity** Note that

$$f'(x) = x^2 - 400 = (x - 20)(x + 20).$$

Solving the inequality $f'(x) < 0$, we find that $f$ is decreasing on the interval $(-20, 20)$. Solving the inequality $f'(x) > 0$ reveals that $f$ is increasing on the intervals $(-\infty, -20)$ and $(20, \infty)$ (**Figure 4.45**). By the First Derivative Test, we have enough information to conclude that $f$ has a local maximum at $x = -20$ and a local minimum at $x = 20$.

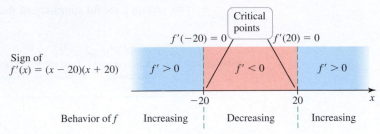

**Figure 4.45**

Furthermore, $f''(x) = 2x < 0$ on $(-\infty, 0)$, so $f$ is concave down on this interval. Also, $f''(x) > 0$ on $(0, \infty)$, so $f$ is concave up on $(0, \infty)$ (**Figure 4.46**).

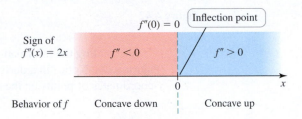

**Figure 4.46**

The evidence obtained so far is summarized in **Figure 4.47**.

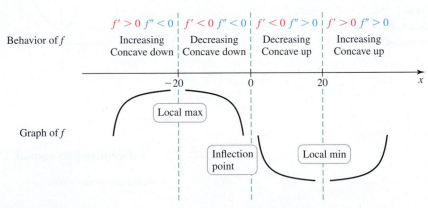

**Figure 4.47**

**6. Extreme values and inflection points** In this case, the Second Derivative Test is easily applied and it confirms what we have already learned. Because $f''(-20) < 0$ and $f''(20) > 0$, $f$ has a local maximum at $x = -20$ and a local minimum at $x = 20$. The corresponding function values are $f(-20) = 16{,}000/3 = 5333\frac{1}{3}$ and $f(20) = -f(-20) = -5333\frac{1}{3}$. Finally, we see that $f''$ changes sign at $x = 0$, making $(0, 0)$ an inflection point.

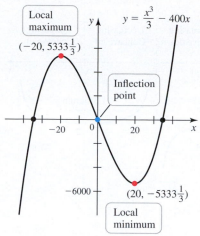

**Figure 4.48**

> Sign charts and sign graphs (Table 4.1 and Figure 4.49) must be constructed carefully when vertical asymptotes are involved: The sign of $f'$ and $f''$ may or may not change at an asymptote.

7. **Asymptotes and end behavior** Polynomials have neither vertical nor horizontal asymptotes. Because the highest-power term in the polynomial is $x^3$ (an odd power) and the leading coefficient is positive, we have the end behavior

$$\lim_{x \to \infty} f(x) = \infty \quad \text{and} \quad \lim_{x \to -\infty} f(x) = -\infty.$$

8. **Intercepts** The $y$-intercept is $(0, 0)$. We solve the equation $f(x) = 0$ to find the $x$-intercepts:

$$\frac{x^3}{3} - 400x = x\left(\frac{x^2}{3} - 400\right) = 0.$$

The roots of this equation are $x = 0$ and $x = \pm\sqrt{1200} \approx \pm 34.6$.

9. **Graph the function** Using the information found in Steps 1–8, we choose the graphing window $[-40, 40] \times [-6000, 6000]$ and produce the graph shown in **Figure 4.48**. Notice that the symmetry detected in Step 2 is evident in this graph.

*Related Exercises 17–18* ◄

**EXAMPLE 3** **The surprises of a rational function** Use the graphing guidelines to graph $f(x) = \dfrac{10x^3}{x^2 - 1}$ on its domain.

**SOLUTION**

1. **Domain** The zeros of the denominator are $x = \pm 1$, so the domain is $\{x : x \neq \pm 1\}$.

2. **Symmetry** This function consists of an odd function divided by an even function. The product or quotient of an even function and an odd function is odd. Therefore, the graph is symmetric about the origin.

3. **Derivatives** The Quotient Rule is used to find the first and second derivatives:

$$f'(x) = \frac{10x^2(x^2 - 3)}{(x^2 - 1)^2} \quad \text{and} \quad f''(x) = \frac{20x(x^2 + 3)}{(x^2 - 1)^3}.$$

4. **Critical points and possible inflection points** The solutions of $f'(x) = 0$ occur where the numerator equals 0, provided the denominator is nonzero at those points. Solving $10x^2(x^2 - 3) = 0$ gives the critical points $x = 0$ and $x = \pm\sqrt{3}$. The solutions of $f''(x) = 0$ are found by solving $20x(x^2 + 3) = 0$; we see that the only possible inflection point occurs at $x = 0$.

5. **Increasing/decreasing and concavity** To find the sign of $f'$, first note that the denominator of $f'$ is nonnegative, as is the factor $10x^2$ in the numerator. So the sign of $f'$ is determined by the sign of the factor $x^2 - 3$, which is negative on $(-\sqrt{3}, \sqrt{3})$ (excluding $x = \pm 1$) and positive on $(-\infty, -\sqrt{3})$ and $(\sqrt{3}, \infty)$. Therefore, $f$ is decreasing on $(-\sqrt{3}, \sqrt{3})$ (excluding $x = \pm 1$) and increasing on $(-\infty, -\sqrt{3})$ and $(\sqrt{3}, \infty)$.

The sign of $f''$ is a bit trickier. Because $x^2 + 3$ is positive, the sign of $f''$ is determined by the sign of $20x$ in the numerator and $(x^2 - 1)^3$ in the denominator. When $20x$ and $(x^2 - 1)^3$ have the same sign, $f''(x) > 0$; when $20x$ and $(x^2 - 1)^3$ have opposite signs, $f''(x) < 0$ (Table 4.1). The results of this analysis are shown in **Figure 4.49**.

**Table 4.1**

| | $20x$ | $x^2 + 3$ | $(x^2 - 1)^3$ | Sign of $f''$ |
|---|---|---|---|---|
| $(-\infty, -1)$ | $-$ | $+$ | $+$ | $-$ |
| $(-1, 0)$ | $-$ | $+$ | $-$ | $+$ |
| $(0, 1)$ | $+$ | $+$ | $-$ | $-$ |
| $(1, \infty)$ | $+$ | $+$ | $+$ | $+$ |

**Figure 4.49**

6. **Extreme values and inflection points** The First Derivative Test is easily applied by looking at Figure 4.49. The function is increasing on $(-\infty, -\sqrt{3})$ and decreasing on $(-\sqrt{3}, -1)$; therefore, $f$ has a local maximum at $x = -\sqrt{3}$, where $f(-\sqrt{3}) = -15\sqrt{3}$. Similarly, $f$ has a local minimum at $x = \sqrt{3}$, where $f(\sqrt{3}) = 15\sqrt{3}$. (These results could also be obtained with the Second Derivative Test.) There is no local extreme value at the critical point $x = 0$, only a horizontal tangent line.

Using Table 4.1 from Step 5, we see that $f''$ changes sign at $x = \pm 1$ and at $x = 0$. The points $x = \pm 1$ are not in the domain of $f$, so they cannot correspond to inflection points. However, there is an inflection point at $(0, 0)$.

7. **Asymptotes and end behavior** Recall from Section 2.4 that zeros of the denominator, which in this case are $x = \pm 1$, are candidates for vertical asymptotes. Checking the behavior of $f$ on either side of $x = \pm 1$, we find

$$\lim_{x \to -1^-} f(x) = -\infty, \qquad \lim_{x \to -1^+} f(x) = \infty,$$
$$\lim_{x \to 1^-} f(x) = -\infty, \qquad \lim_{x \to 1^+} f(x) = \infty.$$

It follows that $f$ has vertical asymptotes at $x = \pm 1$. The degree of the numerator is greater than the degree of the denominator, so there is no horizontal asymptote. Using long division, it can be shown that

$$f(x) = 10x + \frac{10x}{x^2 - 1}.$$

Therefore, as $x \to \pm\infty$, the graph of $f$ approaches the line $y = 10x$. This line is a slant asymptote (Section 2.5).

8. **Intercepts** The zeros of a rational function coincide with the zeros of the numerator, provided those points are not also zeros of the denominator. In this case, the zeros of $f$ satisfy $10x^3 = 0$, or $x = 0$ (which is not a zero of the denominator). Therefore, $(0, 0)$ is both the $x$-intercept and the $y$-intercept.

9. **Graphing** We now assemble an accurate graph of $f$, as shown in **Figure 4.50**. A window of $[-3, 3] \times [-40, 40]$ gives a complete graph of the function. Notice that the symmetry about the origin deduced in Step 2 is apparent in the graph.

*Related Exercises 30–31* ◄

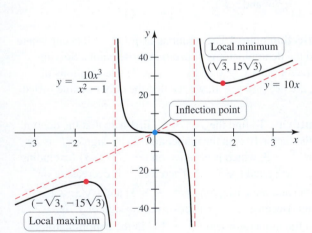

$$y = \frac{10x^3}{x^2 - 1}$$

**Figure 4.50**

**QUICK CHECK 3** Verify that the function $f$ in Example 3 is symmetric about the origin by showing that $f(-x) = -f(x)$. ◄

In the next example, we show how the guidelines may be streamlined to some extent.

**EXAMPLE 4** **Roots and cusps** Graph $f(x) = \frac{1}{8} x^{2/3}(9x^2 - 8x - 16)$ on its domain.

**SOLUTION** The domain of $f$ is $(-\infty, \infty)$. The polynomial factor in $f$ consists of both even and odd powers, so $f$ has no special symmetry. Computing the first derivative is straightforward if you first expand $f$ as a sum of three terms:

$$f'(x) = \frac{d}{dx}\left(\frac{9x^{8/3}}{8} - x^{5/3} - 2x^{2/3}\right) \quad \text{Expand } f.$$

$$= 3x^{5/3} - \frac{5}{3}x^{2/3} - \frac{4}{3}x^{-1/3} \quad \text{Differentiate.}$$

$$= \frac{(x-1)(9x+4)}{3x^{1/3}}. \quad \text{Simplify.}$$

**Table 4.2**

| | $3x^{1/3}$ | $9x+4$ | $x-1$ | Sign of $f'$ |
|---|---|---|---|---|
| $(-\infty, -\frac{4}{9})$ | $-$ | $-$ | $-$ | $-$ |
| $(-\frac{4}{9}, 0)$ | $-$ | $+$ | $-$ | $+$ |
| $(0, 1)$ | $+$ | $+$ | $-$ | $-$ |
| $(1, \infty)$ | $+$ | $+$ | $+$ | $+$ |

The critical points are now identified: $f'$ is undefined at $x = 0$ (because $x^{-1/3}$ is undefined there) and $f'(x) = 0$ at $x = 1$ and $x = -\frac{4}{9}$. So we have three critical points to analyze. Table 4.2 tracks the signs of the factors in the numerator and denominator of $f'$ and shows the sign of $f'$ on the relevant intervals; this information is recorded in **Figure 4.51**.

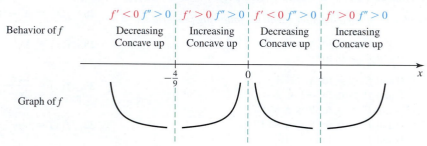

**Figure 4.51**

➤ To show that $f''(x) > 0$ (for $x \neq 0$), analyze its numerator and denominator.
- The graph of $45x^2 - 10x + 4$ is a parabola that opens upward (its leading coefficient is positive) with no $x$-intercepts (verified by the quadratic formula). Therefore, the numerator is always positive.
- The denominator is nonnegative because $9x^{4/3} = 9(x^{1/3})^4$ (even powers are nonnegative).

We use the second line in the calculation of $f'$ to compute the second derivative:

$$f''(x) = \frac{d}{dx}\left(3x^{5/3} - \frac{5}{3}x^{2/3} - \frac{4}{3}x^{-1/3}\right)$$

$$= 5x^{2/3} - \frac{10}{9}x^{-1/3} + \frac{4}{9}x^{-4/3} \quad \text{Differentiate.}$$

$$= \frac{45x^2 - 10x + 4}{9x^{4/3}}. \quad \text{Simplify.}$$

Solving $f''(x) = 0$, we discover that $f''(x) > 0$, for all $x$ except $x = 0$, where it is undefined. Therefore, $f$ is concave up on $(-\infty, 0)$ and $(0, \infty)$ (Figure 4.51).

By the Second Derivative Test, because $f''(x) > 0$, for $x \neq 0$, the critical points $x = -\frac{4}{9}$ and $x = 1$ correspond to local minima; their $y$-coordinates are $f(-\frac{4}{9}) \approx -0.78$ and $f(1) = -\frac{15}{8} = -1.875$.

What about the third critical point, $x = 0$? Note that $f(0) = 0$, and $f$ is increasing just to the left of 0 and decreasing just to the right. By the First Derivative Test, $f$ has a local maximum at $x = 0$. Furthermore, $f'(x) \to \infty$ as $x \to 0^-$ and $f'(x) \to -\infty$ as $x \to 0^+$, so the graph of $f$ has a cusp at $x = 0$.

As $x \to \pm\infty$, $f$ is dominated by its highest-power term, which is $9x^{8/3}/8$. This term becomes large and positive as $x \to \pm\infty$; therefore, $f$ has no absolute maximum. Its absolute minimum occurs at $x = 1$ because, comparing the two local minima, $f(1) < f(-\frac{4}{9})$.

The roots of $f$ satisfy $\frac{1}{8}x^{2/3}(9x^2 - 8x - 16) = 0$, which gives $x = 0$ and

$$x = \frac{4}{9}(1 \pm \sqrt{10}) \approx -0.96 \quad \text{and} \quad 1.85. \quad \text{Use the quadratic formula.}$$

With the information gathered in this analysis, we obtain the graph shown in **Figure 4.52**.

$$y = \frac{x^{2/3}(9x^2 - 8x - 16)}{8}$$

Local maximum

$(0, 0)$

$(-\frac{4}{9}, -0.78)$

Local minimum

$(1, -\frac{15}{8})$   Absolute minimum

**Figure 4.52**

*Related Exercises 38–40* ◄

## SECTION 4.4 EXERCISES

### Getting Started

1. Why is it important to determine the domain of $f$ before graphing $f$?

2. Explain why it is useful to know about symmetry in a function.

3. Can the graph of a polynomial have vertical or horizontal asymptotes? Explain.

4. Where are the vertical asymptotes of a rational function located?

5. How do you find the absolute maximum and minimum values of a function that is continuous on a closed interval?

6. Describe the possible end behavior of a polynomial.

**7–8.** *Sketch a graph of a function $f$ with the following properties.*

7. $f' < 0$ and $f'' < 0$, for $x < 3$
   $f' < 0$ and $f'' > 0$, for $x > 3$

8. $f' < 0$ and $f'' < 0$, for $x < -1$
   $f' < 0$ and $f'' > 0$, for $-1 < x < 2$
   $f' > 0$ and $f'' > 0$, for $2 < x < 8$
   $f' > 0$ and $f'' < 0$, for $8 < x < 10$
   $f' > 0$ and $f'' > 0$, for $x > 10$

**9–12.** *Sketch a continuous function $f$ on some interval that has the properties described. Answers will vary.*

9. The function $f$ has one inflection point but no local extrema.

10. The function $f$ has three real zeros and exactly two local minima.

11. The function $f$ satisfies $f'(-2) = 2$, $f'(0) = 0$, $f'(1) = -3$, and $f'(4) = 1$.

12. The function $f$ has the same finite limit as $x \to \pm\infty$ and has exactly one local minimum and one local maximum.

### Practice Exercises

13. Let $f(x) = (x - 3)(x + 3)^2$.
    a. Verify that $f'(x) = 3(x - 1)(x + 3)$ and $f''(x) = 6(x + 1)$.
    b. Find the critical points and possible inflection points of $f$.
    c. Find the intervals on which $f$ is increasing or decreasing.
    d. Determine the intervals on which $f$ is concave up or concave down.
    e. Identify the local extreme values and inflection points of $f$.
    f. State the $x$- and $y$-intercepts of the graph of $f$.
    g. Use your work in parts (a) through (f) to sketch a graph of $f$.

14. If $f(x) = \dfrac{1}{3x^4 + 5}$, it can be shown that $f'(x) = -\dfrac{12x^3}{(3x^4 + 5)^2}$

    and $f''(x) = \dfrac{180x^2(x^2 + 1)(x + 1)(x - 1)}{(3x^4 + 5)^3}$. Use these functions

    to complete the following steps.

    a. Find the critical points and possible inflection points of $f$.
    b. Find the intervals on which $f$ is increasing or decreasing.
    c. Determine the intervals on which $f$ is concave up or concave down.
    d. Identify the local extreme values and inflection points of $f$.
    e. State the $x$- and $y$-intercepts of the graph of $f$.
    f. Find the asymptotes of $f$.
    g. Use your work in parts (a) through (f) to sketch a graph of $f$.

**15–46. Graphing functions** *Use the guidelines of this section to make a complete graph of $f$.*

15. $f(x) = x^2 - 6x$

16. $f(x) = x - x^2$

17. $f(x) = x^3 - 6x^2 + 9x$

18. $f(x) = 3x - x^3$

19. $f(x) = x^4 - 6x^2$

20. $f(x) = x^4 + 4x^3$

21. $f(x) = (x - 6)(x + 6)^2$

22. $f(x) = 27(x - 2)^2(x + 2)$

23. $f(x) = x^3 - 6x^2 - 135x$

24. $f(x) = x^4 + 8x^3 - 270x^2 + 1$

25. $f(x) = x^3 - 3x^2 - 144x - 140$

26. $f(x) = x^3 - 147x + 286$

27. $f(x) = x - 2\sqrt{x}$

28. $f(x) = 3\sqrt{x} - x^{3/2}$

29. $f(x) = \dfrac{3x}{x^2 - 1}$

30. $f(x) = \dfrac{2x - 3}{2x - 8}$

31. $f(x) = \dfrac{x^2}{x - 2}$

32. $f(x) = \dfrac{x^2}{x^2 - 4}$

33. $f(x) = \dfrac{x^2 + 12}{2x + 1}$

34. $f(x) = \dfrac{4x}{x^2 + 3}$

35. $f(x) = \sqrt{x}(x - 3)$

36. $f(x) = x^{1/3}(4 - x)$

37. $f(x) = x + 2\cos x$ on $[-2\pi, 2\pi]$

38. $f(x) = x - 3x^{2/3}$

39. $f(x) = x - 3x^{1/3}$

40. $f(x) = 2 - 2x^{2/3} + x^{4/3}$

41. $f(x) = \sin x - x$ on $[0, 2\pi]$

42. $f(x) = x\sqrt{x + 3}$

43. $f(x) = \dfrac{1 + x}{\sqrt{x}}$

44. $f(x) = x + \cos 2x$ on $[0, \pi]$

45. $f(x) = x + \tan x$ on $\left(-\dfrac{\pi}{2}, \dfrac{\pi}{2}\right)$

46. $f(x) = \dfrac{3}{x^2} - \dfrac{54}{x^4}$

**47–48.** *Use the graphs of $f'$ and $f''$ to find the critical points and inflection points of $f$, the intervals on which $f$ is increasing or decreasing, and the intervals of concavity. Then graph $f$ assuming $f(0) = 0$.*

47.

**48.**

**T 49–54. Graphing with technology** *Make a complete graph of the following functions. A graphing utility is useful in locating intercepts, local extreme values, and inflection points.*

**49.** $f(x) = \frac{1}{3}x^3 - 2x^2 - 5x + 2$

**50.** $f(x) = \frac{1}{15}x^3 - x + 1$

**51.** $f(x) = 3x^4 + 4x^3 - 12x^2$

**52.** $f(x) = x^3 - 33x^2 + 216x - 2$

**53.** $f(x) = \frac{3x - 5}{x^2 - 1}$      **54.** $f(x) = x^{1/3}(x - 2)^2$

**55. Explain why or why not** Determine whether the following statements are true and give an explanation or counterexample.

**a.** If the zeros of $f'$ are $-3$, $1$, and $4$, then the local extrema of $f$ are located at these points.

**b.** If the zeros of $f''$ are $-2$ and $4$, then the inflection points of $f$ are located at these points.

**c.** If the zeros of the denominator of $f$ are $-3$ and $4$, then $f$ has vertical asymptotes at these points.

**d.** If a rational function has a finite limit as $x \to \infty$, then it must have a finite limit as $x \to -\infty$.

**56–59. Functions from derivatives** *Use the derivative $f'$ to determine the x-coordinates of the local maxima and minima of $f$, and the intervals on which $f$ is increasing or decreasing. Sketch a possible graph of $f$ ($f$ is not unique).*

**56.** $f'(x) = (x - 1)(x + 2)(x + 4)$

**57.** $f'(x) = 10 \sin 2x$ on $[-2\pi, 2\pi]$

**58.** $f'(x) = \frac{1}{6}(x + 1)(x - 2)^2(x - 3)$

**59.** $f'(x) = x^2(x + 2)(x - 1)$

### Explorations and Challenges

**60. Derivative information** Suppose a continuous function $f$ is concave up on $(-\infty, 0)$ and $(0, \infty)$. Assume $f$ has a local maximum at $x = 0$. What, if anything, do you know about $f'(0)$? Explain with an illustration.

**61. Height functions** The figure shows six containers, each of which is filled from the top. Assume water is poured into the containers at a constant rate and each container is filled in 10 s. Assume also that the horizontal cross sections of the containers are always circles. Let $h(t)$ be the depth of water in the container at time $t$, for $0 \le t \le 10$.

**a.** For each container, sketch a graph of the function $y = h(t)$, for $0 \le t \le 10$.

**b.** Explain why $h$ is an increasing function.

**c.** Describe the concavity of the function. Identify inflection points when they occur.

**d.** For each container, where does $h'$ (the derivative of $h$) have an absolute maximum on $[0, 10]$?

**T 62. A pursuit curve** A man stands 1 mi east of a crossroads. At noon, a dog starts walking north from the crossroads at 1 mi/hr (see figure). At the same instant, the man starts walking and at all times walks directly toward the dog at $s > 1$ mi/hr. The path in the xy-plane followed by the man as he pursues the dog is given by the function

$$y = f(x) = \frac{s}{2}\left(\frac{x^{(s+1)/s}}{s + 1} - \frac{x^{(s-1)/s}}{s - 1}\right) + \frac{s}{s^2 - 1}.$$

Select various values of $s > 1$ and graph this pursuit curve. Comment on the changes in the curve as $s$ increases.

**T 63–66. Combining technology with analytical methods** *Use a graphing utility together with analytical methods to create a complete graph of the following functions. Be sure to find and label the intercepts, local extrema, inflection points, and asymptotes, and find the intervals on which the function is increasing or decreasing, and the intervals on which the function is concave up or concave down.*

**63.** $f(x) = \frac{x \sin x}{x^2 + 1}$ on $[-2\pi, 2\pi]$

**64.** $f(x) = 3\sqrt[4]{x} - \sqrt{x} - 2$

**65.** $f(x) = 3x^4 - 44x^3 + 60x^2$ (*Hint:* Two different graphing windows may be needed.)

**66.** $f(x) = \frac{\sqrt{4x^2 + 1}}{x^2 + 1}$

**T 67–71. Special curves** *The following classical curves have been studied by generations of mathematicians. Use analytical methods (including implicit differentiation) and a graphing utility to graph the curves. Include as much detail as possible.*

**67.** $x^{2/3} + y^{2/3} = 1$ (Astroid or hypocycloid with four cusps)

**68.** $y = \dfrac{8}{x^2 + 4}$ (Witch of Agnesi)

**69.** $x^3 + y^3 = 3xy$ (Folium of Descartes)

**70.** $y^2 = x^3(1 - x)$ (Pear curve)

**71.** $x^4 - x^2 + y^2 = 0$ (Figure-8 curve)

**T 72. Elliptic curves** The equation $y^2 = x^3 - ax + 3$, where $a$ is a parameter, defines a well-known family of *elliptic curves*.

  **a.** Plot a graph of the curve when $a = 3$.

  **b.** Plot a graph of the curve when $a = 4$.

  **c.** By experimentation, determine the approximate value of $a$ $(3 < a < 4)$ at which the graph separates into two curves.

**T 73. Lamé curves** The equation $|y/a|^n + |x/a|^n = 1$, where $n$ and $a$ are positive real numbers, defines the family of Lamé curves. Make a complete graph of this function with $a = 1$, for $n = \frac{2}{3}$, 1, 2, and 3. Describe the progression that you observe as $n$ increases.

**QUICK CHECK ANSWERS**

**1.** Make the window larger in the $y$-direction.

**2.** Notice that $f$ and $f + C$ have the same derivatives.

**3.** $f(-x) = \dfrac{10(-x)^3}{(-x)^2 - 1} = -\dfrac{10x^3}{x^2 - 1} = -f(x)$ ◄

# 4.5 Optimization Problems

The theme of this section is *optimization*, a topic arising in many disciplines that rely on mathematics. A structural engineer may seek the dimensions of a beam that maximize strength for a specified cost. A packaging designer may seek the dimensions of a container that maximize the volume of the container for a given surface area. Airline strategists need to find the best allocation of airliners among several hubs to minimize fuel costs and maximize passenger miles. In all these examples, the challenge is to find an *efficient* way to carry out a task, where "efficient" could mean least expensive, most profitable, least time consuming, or, as you will see, many other measures.

To introduce the ideas behind optimization problems, think about pairs of nonnegative real numbers $x$ and $y$ between 0 and 20 with the property that their sum is 20, that is, $x + y = 20$. Of all possible pairs, which has the greatest product?

Table 4.3 displays a few cases showing how the product of two nonnegative numbers varies while their sum remains constant. The condition that $x + y = 20$ is called a **constraint**: It tells us to consider only (nonnegative) values of $x$ and $y$ satisfying this equation.

The quantity that we wish to maximize (or minimize in other cases) is called the **objective function**; in this case, the objective function is the product $P = xy$. From Table 4.3, it appears that the product is greatest if both $x$ and $y$ are near the middle of the interval $[0, 20]$.

This simple problem has all the essential features of optimization problems. At their heart, optimization problems take the following form:

*What is the maximum (minimum) value of an objective function subject to the given constraint(s)?*

For the problem at hand, this question would be stated as, "What pair of nonnegative numbers maximizes $P = xy$ subject to the constraint $x + y = 20$?" The first step is to use the constraint to express the objective function $P = xy$ in terms of a single variable. In this case, the constraint is

$$x + y = 20, \quad \text{or} \quad y = 20 - x.$$

Substituting for $y$, the objective function becomes

$$P = xy = x(20 - x) = 20x - x^2,$$

which is a function of the single variable $x$. Notice that the values of $x$ lie in the interval $0 \le x \le 20$ with $P(0) = P(20) = 0$.

**Table 4.3**

| $x$ | $y$ | $x + y$ | $P = xy$ |
|-----|-----|---------|----------|
| 1 | 19 | 20 | 19 |
| 5.5 | 14.5 | 20 | 79.75 |
| 9 | 11 | 20 | 99 |
| 13 | 7 | 20 | 91 |
| 18 | 2 | 20 | 36 |

➤ In this problem, it is just as easy to eliminate $x$ as $y$. In other problems, eliminating one variable may result in less work than eliminating other variables.

**Figure 4.53**

**QUICK CHECK 1** Verify that in the previous example, the same result is obtained if the constraint $x + y = 20$ is used to eliminate $x$ rather than $y$. ◄

Constraint: $4x + y = 400$

**Figure 4.54**

➤ Recall from Section 4.1 that, when they exist, the absolute extreme values occur at critical points or endpoints.

**QUICK CHECK 2** Find the objective function in Example 1 (in terms of $x$) (i) if there is no interior fence and (ii) if there is one interior fence that forms a right angle with the barn, as in Figure 4.54. ◄

To maximize $P$, we first find the critical points by solving

$$P'(x) = 20 - 2x = 0$$

to obtain the solution $x = 10$. To find the absolute maximum value of $P$ on the interval $[0, 20]$, we check the endpoints and the critical points. Because $P(0) = P(20) = 0$ and $P(10) = 100$, we conclude that $P$ has its absolute maximum value at $x = 10$. By the constraint $x + y = 20$, the numbers with the greatest product are $x = y = 10$, and their product is $P = 100$.

**Figure 4.53** summarizes this problem. We see the constraint line $x + y = 20$ in the $xy$-plane. Above the line is the objective function $P = xy$. As $x$ and $y$ vary along the constraint line, the objective function changes, reaching a maximum value of 100 when $x = y = 10$.

Most optimization problems have the same basic structure as the preceding example: There is an objective function, which may involve several variables, and one or more constraints. The methods of calculus (Sections 4.1 and 4.3) are used to find the minimum or maximum values of the objective function.

**EXAMPLE 1**  **Rancher's dilemma**  A rancher has 400 ft of fence for constructing a rectangular corral. One side of the corral will be formed by a barn and requires no fence. Three exterior fences and two interior fences partition the corral into three rectangular regions as shown in **Figure 4.54**. What are the dimensions of the corral that maximize the enclosed area? What is the area of that corral?

**SOLUTION**  We first sketch the corral (Figure 4.54), where $x \geq 0$ is the width and $y \geq 0$ is the length of the corral. The amount of fence required is $4x + y$, so the constraint is $4x + y = 400$, or $y = 400 - 4x$.

The objective function to be maximized is the area of the corral, $A = xy$. Using $y = 400 - 4x$, we eliminate $y$ and express $A$ as a function of $x$:

$$A = xy = x(400 - 4x) = 400x - 4x^2.$$

Notice that the width of the corral must be at least $x = 0$, and it cannot exceed $x = 100$ (because 400 ft of fence are available). Therefore, we maximize $A(x) = 400x - 4x^2$, for $0 \leq x \leq 100$. The critical points of the objective function satisfy

$$A'(x) = 400 - 8x = 0,$$

which has the solution $x = 50$. To find the absolute maximum value of $A$, we check the endpoints of $[0, 100]$ and the critical point $x = 50$. Because $A(0) = A(100) = 0$ and $A(50) = 10,000$, the absolute maximum value of $A$ occurs when $x = 50$. Using the constraint, the optimal length of the corral is $y = 400 - 4(50) = 200$. Therefore, the maximum area of 10,000 ft$^2$ is achieved with dimensions $x = 50$ ft and $y = 200$ ft. The objective function $A$ is shown in **Figure 4.55**.

**Figure 4.55**

*Related Exercises 12, 16* ◄

Objective function: $V = w^2h$
Constraint: $2w + h = 64$

**Figure 4.56**

Optimal solution at $w = \frac{64}{3}$

Objective function:
$V = 64w^2 - 2w^3$

**Figure 4.57**

**QUICK CHECK 3** Find the objective function in Example 2 (in terms of $w$) if the constraint is that the sum of length and width and height cannot exceed 108 in. ◄

**Figure 4.58**

➤ To show that the chord length of a circle is $2r \sin(\theta/2)$, draw a line from the center of the circle to the midpoint of the chord (Figure 4.59). This line bisects the angle $\theta$. Using a right triangle, half the length of the chord is $r \sin(\theta/2)$.

**EXAMPLE 2  Airline regulations** Suppose an airline policy states that all baggage must be box-shaped with a sum of length, width, and height not exceeding 64 in. What are the dimensions and volume of a square-based box with the greatest volume under these conditions?

**SOLUTION** We sketch a square-based box whose length and width are $w$ and whose height is $h$ (**Figure 4.56**). By the airline policy, the constraint is $2w + h = 64$. The objective function is the volume, $V = w^2h$. Either $w$ or $h$ may be eliminated from the objective function; the constraint $h = 64 - 2w$ implies that the volume is

$$V = w^2h = w^2(64 - 2w) = 64w^2 - 2w^3.$$

The objective function has now been expressed in terms of a single variable. Notice that $w$ is nonnegative and cannot exceed 32, so the domain of $V$ is $0 \le w \le 32$. The critical points satisfy

$$V'(w) = 128w - 6w^2 = 2w(64 - 3w) = 0,$$

which has roots $w = 0$ and $w = \frac{64}{3}$. By the First (or Second) Derivative Test, $w = \frac{64}{3}$ corresponds to a local maximum. At the endpoints, $V(0) = V(32) = 0$. Therefore, the volume function has an absolute maximum of $V(64/3) \approx 9709$ in³. The dimensions of the optimal box are $w = 64/3$ in and $h = 64 - 2w = 64/3$ in, so the optimal box is a cube. A graph of the volume function is shown in **Figure 4.57**.

*Related Exercises 19–20* ◄

**Optimization Guidelines** With two examples providing some insight, we present a procedure for solving optimization problems. These guidelines provide a general framework, but the details may vary depending on the problem.

**Guidelines for Optimization Problems**
1. Read the problem carefully, identify the variables, and organize the given information with a picture.
2. Identify the objective function (the function to be optimized). Write it in terms of the variables of the problem.
3. Identify the constraint(s). Write them in terms of the variables of the problem.
4. Use the constraint(s) to eliminate all but one independent variable of the objective function.
5. With the objective function expressed in terms of a single variable, find the interval of interest for that variable.
6. Use methods of calculus to find the absolute maximum or minimum value of the objective function on the interval of interest. If necessary, check the endpoints.

**EXAMPLE 3  Walking and swimming** Suppose you are standing on the shore of a circular pond with a radius of 1 mile and you want to get to a point on the shore directly opposite your position (on the other end of a diameter). You plan to swim at 2 mi/hr from your current position to another point $P$ on the shore and then walk at 3 mi/hr along the shore to the terminal point (**Figure 4.58**). How should you choose $P$ to minimize the total time for the trip?

**SOLUTION** As shown in Figure 4.58, the initial point is chosen arbitrarily, and the terminal point is at the other end of a diameter. The easiest way to describe the transition point $P$ is to refer to the central angle $\theta$. If $\theta = 0$, then the entire trip is done by walking; if $\theta = \pi$, the entire trip is done by swimming. So the interval of interest is $0 \le \theta \le \pi$.

The objective function is the total travel time as it varies with $\theta$. For each leg of the trip (swim and walk), the travel time is the distance traveled divided by the speed. We need a few facts from circular geometry. The length of the swimming leg is the length of the chord of the circle corresponding to the angle $\theta$. For a circle of radius $r$, this chord

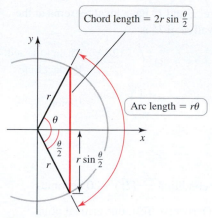

**Figure 4.59**

▶ You can check two special cases: If the entire trip is done walking, the travel time is $(\pi \text{ mi})/(3 \text{ mi/hr}) \approx 1.05$ hr. If the entire trip is done swimming, the travel time is $(2 \text{ mi})/(2 \text{ mi/hr}) = 1$ hr.

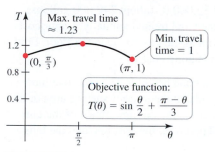

**Figure 4.60**

length is given by $2r \sin (\theta/2)$ (**Figure 4.59**). So the time for the swimming leg (with $r = 1$ and a speed of 2 mi/hr) is

$$\text{time} = \frac{\text{distance}}{\text{rate}} = \frac{2 \sin (\theta/2)}{2} = \sin \frac{\theta}{2}.$$

The length of the walking leg is the length of the arc of the circle corresponding to the angle $\pi - \theta$. For a circle of radius $r$, the arc length corresponding to an angle $\theta$ is $r\theta$ (**Figure 4.59**). Therefore, the time for the walking leg (with an angle $\pi - \theta$, $r = 1$, and a speed of 3 mi/hr) is

$$\text{time} = \frac{\text{distance}}{\text{rate}} = \frac{\pi - \theta}{3}.$$

The total travel time for the trip (in hours) is the objective function

$$T(\theta) = \sin \frac{\theta}{2} + \frac{\pi - \theta}{3}, \quad \text{for} \quad 0 \le \theta \le \pi.$$

We now analyze the objective function. The critical points of $T$ satisfy

$$\frac{dT}{d\theta} = \frac{1}{2} \cos \frac{\theta}{2} - \frac{1}{3} = 0 \quad \text{or} \quad \cos \frac{\theta}{2} = \frac{2}{3}.$$

Using a calculator, the only solution in the interval $[0, \pi]$ is $\theta \approx 1.68$ rad $\approx 96°$, which is the critical point.

Evaluating the objective function at the critical point and at the endpoints, we find that $T(1.68) \approx 1.23$ hr, $T(0) = \pi/3 \approx 1.05$ hr, and $T(\pi) = 1$ hr. We conclude that the minimum travel time is $T(\pi) = 1$ hr when the entire trip is done swimming. The *maximum* travel time, corresponding to $\theta \approx 96°$, is $T \approx 1.23$ hr.

The objective function is shown in **Figure 4.60**. In general, the maximum and minimum travel times depend on the walking and swimming speeds (Exercise 26).

*Related Exercises 26–27* ◀

**EXAMPLE 4    Ladder over the fence** An 8-foot-tall fence runs parallel to the side of a house 3 feet away (**Figure 4.61a**). What is the length of the shortest ladder that clears the fence and reaches the house? Assume the vertical wall of the house and the horizontal ground have infinite extent (see Exercise 31 for more realistic assumptions).

**SOLUTION** Let's first ask why we expect a minimum ladder length. You could put the foot of the ladder far from the fence so that it clears the fence at a shallow angle, but the ladder would be long. Or you could put the foot of the ladder close to the fence so that it clears the fence at a steep angle, but again, the ladder would be long. Somewhere between these extremes is a ladder position that minimizes the ladder length.

The objective function in this problem is the ladder length $L$. The position of the ladder is specified by $x$, the distance between the foot of the ladder and the fence (**Figure 4.61b**). The goal is to express $L$ as a function of $x$, where $x > 0$.

The Pythagorean theorem gives the relationship

$$L^2 = (x + 3)^2 + b^2,$$

where $b$ is the height of the top of the ladder above the ground. Similar triangles give the constraint $8/x = b/(3 + x)$. We now solve the constraint equation for $b$ and substitute to express $L^2$ in terms of $x$:

$$L^2 = (x + 3)^2 + \underbrace{\left( \frac{8(x + 3)}{x} \right)^2}_{b} = (x + 3)^2 \left( 1 + \frac{64}{x^2} \right).$$

(a)

(b)

**Figure 4.61**

At this juncture, we could find the critical points of $L$ by first solving the preceding equation for $L$ and then solving $L' = 0$. However, the solution is simplified considerably if we note that $L$ is a nonnegative function. Therefore, $L$ and $L^2$ have local extrema at the same points, so we choose to minimize $L^2$. The derivative of $L^2$ is

$$\frac{d}{dx}\left((x+3)^2\left(1+\frac{64}{x^2}\right)\right) = 2(x+3)\left(1+\frac{64}{x^2}\right) + (x+3)^2\left(-\frac{128}{x^3}\right) \qquad \text{Chain Rule and Product Rule}$$

$$= 2(x+3)\left(1+\frac{64}{x^2} - (x+3)\frac{64}{x^3}\right) \qquad \text{Factor.}$$

$$= \frac{2(x+3)(x^3 - 192)}{x^3}. \qquad \text{Simplify.}$$

Because $x > 0$, we have $x + 3 \neq 0$; therefore, the condition $\dfrac{d}{dx}(L^2) = 0$ becomes $x^3 - 192 = 0$, or $x = 4\sqrt[3]{3} \approx 5.77$. By the First Derivative Test, this critical point corresponds to a local minimum. By Theorem 4.11, this solitary local minimum is also the absolute minimum on the interval $(0, \infty)$. Therefore, the minimum ladder length occurs when the foot of the ladder is approximately 5.77 ft from the fence. We find that $L^2(5.77) \approx 224.77$ and the minimum ladder length is $\sqrt{224.77} \approx 15$ ft.

*Related Exercises 30–31* ◄

**EXAMPLE 5**  **Water tower**  A water storage tank in a small community is built in the shape of a right circular cylinder with a capacity of 32,000 ft³ (about 240,000 gallons). The interior wall and floor of the tank must be cleaned and treated annually. Labor costs for cleaning the wall are twice as high per square foot as the cost to clean the floor. Find the radius and height of the tank that minimize the cleaning cost.

**SOLUTION**  Figure 4.62 shows a water tank with radius $r$ and height $h$. The objective function describes the cost to clean the tank, which depends on the interior surface area of the tank. The area of the tank's circular floor is $\pi r^2$. Because the circumference of the tank is $2\pi r$, the lateral surface area of the wall is $2\pi rh$. Therefore, the cost to clean the tank is proportional to

$$C = 2 \cdot \underbrace{2\pi rh}_{\text{area of wall}} + \underbrace{\pi r^2}_{\text{area of floor}}.$$

It costs twice as much per square foot to clean the wall, so the area of the wall is multiplied by 2.

To find a constraint equation, we start with the volume of a cylinder $V = \pi r^2 h$ and replace $V$ with 32,000 to obtain $\pi r^2 h = 32{,}000$, or $h = 32{,}000/(\pi r^2)$. Substituting this expression for $h$ into the cost equation, we obtain a function of the radius:

$$C(r) = 4\pi r \frac{32{,}000}{\pi r^2} + \pi r^2 = \frac{128{,}000}{r} + \pi r^2.$$

To minimize $C$, we find its critical points on the interval $0 < r < \infty$:

$$C'(r) = -\frac{128{,}000}{r^2} + 2\pi r = 0 \qquad \text{Set the derivative equal to 0.}$$

$$2\pi r = \frac{128{,}000}{r^2} \qquad \text{Rearrange equation.}$$

$$r^3 = \frac{64{,}000}{\pi} \qquad \text{Multiply by } r^2 (r > 0) \text{ and divide by } 2\pi.$$

$$r = \frac{40}{\sqrt[3]{\pi}} \approx 27.3 \text{ ft.} \qquad \text{Solve for } r.$$

**Figure 4.62**

➤ We don't need to know the actual cost to clean the tank. Whatever it costs (per square foot) to clean the floor, it costs twice as much to clean the wall, which implies that the objective function in Example 5 is a constant multiple of the actual cleaning cost. Because $\dfrac{d}{dx}(kf(x)) = k\dfrac{d}{dx}(f(x))$, both $f(x)$ and $kf(x)$ share the same critical points, so either function yields the solution.

This solitary critical point corresponds to a local minimum (confirm that $C''(r) > 0$ and use the Second Derivative Test), so by Theorem 4.9, it also corresponds to the absolute minimum. Because $h = 32{,}000/(\pi r^2)$, the dimensions of the tank that minimize the cleaning cost are

$$r \approx 27.3 \text{ ft} \quad \text{and} \quad h = 32{,}000/(\pi r^2) \approx 13.7 \text{ ft.}$$

A bit of algebra shows that $r$ is exactly twice as large as $h$, so the optimal tank is 4 times as wide as it is high.

*Related Exercises 21, 35, 42* ◄

## SECTION 4.5 EXERCISES

### Getting Started

1. Fill in the blanks: The goal of an optimization problem is to find the maximum or minimum value of the _____ function subject to the _____ .

2. If the objective function involves more than one independent variable, how are the extra variables eliminated?

3. Suppose the objective function is $Q = x^2y$ and you know that $x + y = 10$. Write the objective function first in terms of $x$ and then in terms of $y$.

4. Suppose you wish to minimize a continuous objective function on a closed interval, but you find that it has only a single local maximum. Where should you look for the solution to the problem?

5. Suppose the objective function $P = xy$ is subject to the constraint $10x + y = 100$, where $x$ and $y$ are real numbers.
   a. Eliminate the variable $y$ from the objective function so that $P$ is expressed as a function of one variable $x$.
   b. Find the absolute maximum value of $P$ subject to the given constraint.

6. Suppose $S = x + 2y$ is an objective function subject to the constraint $xy = 50$, for $x > 0$ and $y > 0$.
   a. Eliminate the variable $y$ from the objective function so that $S$ is expressed as a function of one variable $x$.
   b. Find the absolute minimum value of $S$ subject to the given constraint.

7. What two nonnegative real numbers with a sum of 23 have the largest possible product?

8. What two nonnegative real numbers $a$ and $b$ whose sum is 23 maximize $a^2 + b^2$? Minimize $a^2 + b^2$?

9. What two positive real numbers whose product is 50 have the smallest possible sum?

10. Find numbers $x$ and $y$ satisfying the equation $3x + y = 12$ such that the product of $x$ and $y$ is as large as possible.

### Practice Exercises

11. **Maximum-area rectangles** Of all rectangles with a perimeter of 10, which one has the maximum area? (Give the dimensions.)

12. **Maximum-area rectangles** Of all rectangles with a fixed perimeter of $P$, which one has the maximum area? (Give the dimensions in terms of $P$.)

13. **Minimum-perimeter rectangles** Of all rectangles of area 100, which one has the minimum perimeter?

14. **Minimum-perimeter rectangles** Of all rectangles with a fixed area $A$, which one has the minimum perimeter? (Give the dimensions in terms of $A$.)

15. **Minimum sum** Find positive numbers $x$ and $y$ satisfying the equation $xy = 12$ such that the sum $2x + y$ is as small as possible.

16. **Pen problems**
   a. A rectangular pen is built with one side against a barn. Two hundred meters of fencing are used for the other three sides of the pen. What dimensions maximize the area of the pen?

   b. A rancher plans to make four identical and adjacent rectangular pens against a barn, each with an area of 100 m$^2$ (see figure). What are the dimensions of each pen that minimize the amount of fence that must be used?

17. **Rectangles beneath a semicircle** A rectangle is constructed with its base on the diameter of a semicircle with radius 5 and its two other vertices on the semicircle. What are the dimensions of the rectangle with maximum area?

18. **Rectangles beneath a parabola** A rectangle is constructed with its base on the $x$-axis and two of its vertices on the parabola $y = 48 - x^2$. What are the dimensions of the rectangle with the maximum area? What is the area?

19. **Minimum-surface-area box** Of all boxes with a square base and a volume of 8 m$^3$, which one has the minimum surface area? (Give its dimensions.)

20. **Maximum-volume box** Suppose an airline policy states that all baggage must be box-shaped with a sum of length, width, and height not exceeding 108 in. What are the dimensions and volume of a square-based box with the greatest volume under these conditions?

21. **Shipping crates** A square-based, box-shaped shipping crate is designed to have a volume of 16 ft$^3$. The material used to make the base costs twice as much (per square foot) as the material in the sides, and the material used to make the top costs half as much (per square foot) as the material in the sides. What are the dimensions of the crate that minimize the cost of materials?

22. **Closest point on a line** What point on the line $y = 3x + 4$ is closest to the origin?

**T** 23. **Closest point on a curve** What point on the parabola $y = 1 - x^2$ is closest to the point $(1, 1)$?

24. **Minimum distance** Find the point $P$ on the curve $y = x^2$ that is closest to the point $(18, 0)$. What is the least distance between $P$ and $(18, 0)$? (*Hint:* Use synthetic division.)

25. **Minimum distance** Find the point $P$ on the line $y = 3x$ that is closest to the point $(50, 0)$. What is the least distance between $P$ and $(50, 0)$?

26. **Walking and swimming** A man wishes to get from an initial point on the shore of a circular pond with radius 1 mi to a point on the shore directly opposite (on the other end of the diameter). He plans to swim from the initial point to another point on the shore and then walk along the shore to the terminal point.
   a. If he swims at 2 mi/hr and walks at 4 mi/hr, what are the maximum and minimum times for the trip?
   b. If he swims at 2 mi/hr and walks at 1.5 mi/hr, what are the maximum and minimum times for the trip?
   c. If he swims at 2 mi/hr, what is the minimum walking speed for which it is quickest to walk the entire distance?

**27. Walking and rowing** A boat on the ocean is 4 mi from the nearest point on a straight shoreline; that point is 6 mi from a restaurant on the shore (see figure). A woman plans to row the boat straight to a point on the shore and then walk along the shore to the restaurant.

    **a.** If she walks at 3 mi/hr and rows at 2 mi/hr, at which point on the shore should she land to minimize the total travel time?

    **b.** If she walks at 3 mi/hr, what is the minimum speed at which she must row so that the quickest way to the restaurant is to row directly (with no walking)?

**28. Laying cable** An island is 3.5 mi from the nearest point on a straight shoreline; that point is 8 mi from a power station (see figure). A utility company plans to lay electrical cable underwater from the island to the shore and then underground along the shore to the power station. Assume it costs $2400/mi to lay underwater cable and $1200/mi to lay underground cable. At what point should the underwater cable meet the shore in order to minimize the cost of the project?

**29. Laying cable again** Solve the problem in Exercise 28, but this time minimize the cost with respect to the smaller angle $\theta$ between the underwater cable and the shore. (You should get the same answer.)

**30. Shortest ladder** A 10-ft-tall fence runs parallel to the wall of a house at a distance of 4 ft. Find the length of the shortest ladder that extends from the ground to the house without touching the fence. Assume the vertical wall of the house and the horizontal ground have infinite extent.

**31. Shortest ladder—more realistic** An 8-ft-tall fence runs parallel to the wall of a house at a distance of 5 ft. Find the length of the shortest ladder that extends from the ground to the house without touching the fence. Assume the vertical wall of the house is 20 ft high and the horizontal ground extends 20 ft from the fence.

**32. Circle and square** A piece of wire of length 60 is cut, and the resulting two pieces are formed to make a circle and a square. Where should the wire be cut to (a) maximize and (b) minimize the combined area of the circle and the square?

**33. Maximum-volume cone** A cone is constructed by cutting a sector from a circular sheet of metal with radius 20. The cut sheet is then folded up and welded (see figure). Find the radius and height of the cone with maximum volume that can be formed in this way.

**34. Slant height and cones** Among all right circular cones with a slant height of 3, what are the dimensions (radius and height) that maximize the volume of the cone? The slant height of a cone is the distance from the outer edge of the base to the vertex.

**35. Optimal soda can**

    **a. Classical problem** Find the radius and height of a cylindrical soda can with a volume of 354 cm$^3$ that minimize the surface area.

    **b. Real problem** Compare your answer in part (a) to a real soda can, which has a volume of 354 cm$^3$, a radius of 3.1 cm, and a height of 12.0 cm, to conclude that real soda cans do not seem to have an optimal design. Then use the fact that real soda cans have a double thickness in their top and bottom surfaces to find the radius and height that minimize the surface area of a real can (the surface areas of the top and bottom are now twice their values in part (a)). Are these dimensions closer to the dimensions of a real soda can?

**36. Covering a marble** Imagine a flat-bottomed cylindrical pot with a circular cross section of radius 4. A marble with radius $0 < r < 4$ is placed in the bottom of the pot. What is the radius of the marble that requires the most water to cover it completely?

**37. Optimal garden** A rectangular flower garden with an area of 30 m$^2$ is surrounded by a grass border 1 m wide on two sides and 2 m wide on the other two sides (see figure). What dimensions of the garden minimize the combined area of the garden and borders?

**38. Rectangles beneath a line**

    **a.** A rectangle is constructed with one side on the positive $x$-axis, one side on the positive $y$-axis, and the vertex opposite the origin on the line $y = 10 - 2x$. What dimensions maximize the area of the rectangle? What is the maximum area?

    **b.** Is it possible to construct a rectangle with a greater area than that found in part (a) by placing one side of the rectangle on the line $y = 10 - 2x$ and the two vertices not on that line on the positive $x$- and $y$-axes? Find the dimensions of the rectangle of maximum area that can be constructed in this way.

**39. Designing a box** Two squares of length $x$ are cut out of adjacent corners of an $18'' \times 18''$ piece of cardboard and two rectangles of length $9''$ and width $x$ are cut out of the other two corners of the cardboard (see figure). The resulting piece of cardboard is then folded along the dashed lines to form an enclosed box.

Find the dimensions and volume of the largest box that can be formed in this way.

40. **Folded boxes**

   a. Squares with sides of length $x$ are cut out of each corner of a rectangular piece of cardboard measuring 5 ft by 8 ft. The resulting piece of cardboard is then folded into a box without a lid. Find the volume of the largest box that can be formed in this way.

   b. Squares with sides of length $x$ are cut out of each corner of a square piece of cardboard with sides of length $\ell$. Find the volume of the largest open box that can be formed in this way.

41. **Norman window** A window consists of rectangular pane of glass surmounted by a semicircular pane of glass (see figure). If the perimeter of the window is 20 feet, determine the dimensions of the window that maximize the area of the window.

42. **Light transmission** A window consists of a rectangular pane of clear glass surmounted by a semicircular pane of tinted glass. The clear glass transmits twice as much light per unit of surface area as the tinted glass. Of all such windows with a fixed perimeter $P$, what are the dimensions of the window that transmits the most light?

43. **Kepler's wine barrel** Several mathematical stories originated with the second wedding of the mathematician and astronomer Johannes Kepler. Here is one: While shopping for wine for his wedding, Kepler noticed that the price of a barrel of wine (here assumed to be a cylinder) was determined solely by the length $d$ of a dipstick that was inserted diagonally through a centered hole in the top of the barrel to the edge of the base of the barrel (see figure). Kepler realized that this measurement does not determine the volume of the barrel and that for a fixed value of $d$, the volume varies with the radius $r$ and height $h$ of the barrel. For a fixed value of $d$, what is the ratio $r/h$ that maximizes the volume of the barrel?

44. **Maximizing profit** Suppose you own a tour bus and you book groups of 20 to 70 people for a day tour. The cost per person is $30 minus $0.25 for every ticket sold. If gas and other miscellaneous costs are $200, how many tickets should you sell to maximize your profit? Treat the number of tickets as a nonnegative real number.

45. **Maximum-volume cylinder in a sphere** Find the dimensions of the right circular cylinder of maximum volume that can be placed inside of a sphere of radius $R$.

46. **Cone in a sphere** Find the height $h$, radius $r$, and volume of a right circular cone with maximum volume that is inscribed in a sphere of radius $R$.

47. **Cone in a cone** A right circular cone is inscribed inside a larger right circular cone with a volume of 150 cm³. The axes of the cones coincide, and the vertex of the inner cone touches the center of the base of the outer cone. Find the ratio of the heights of the cones that maximizes the volume of the inner cone.

48. **Do dogs know calculus?** A mathematician stands on a beach with his dog at point $A$. He throws a tennis ball so that it hits the water at point $B$. The dog, wanting to get to the tennis ball as quickly as possible, runs along the straight beach line to point $D$ and then swims from point $D$ to point $B$ to retrieve his ball. Assume $C$ is the point on the edge of the beach closest to the tennis ball (see figure).

   a. Assume the dog runs at speed $r$ and swims at speed $s$, where $r > s$ and both are measured in meters per second. Also assume the lengths of $BC$, $CD$, and $AC$ are $x$, $y$, and $z$, respectively. Find a function $T(y)$ representing the total time it takes for the dog to get to the ball.

   b. Verify that the value of $y$ that minimizes the time it takes to retrieve the ball is $y = \dfrac{x}{\sqrt{r/s + 1}\sqrt{r/s - 1}}$.

   c. If the dog runs at 8 m/s and swims at 1 m/s, what ratio $y/x$ produces the fastest retrieving time?

   d. A dog named Elvis who runs at 6.4 m/s and swims at 0.910 m/s was found to use an average ratio of $y/x$ of 0.144 to retrieve his ball. Does Elvis appear to know calculus?

   (*Source*: T. Pennings, *Do Dogs Know Calculus? The College Mathematics Journal*, 34, 3, May 2003)

49. **Travel costs** A simple model for travel costs involves the cost of gasoline and the cost of a driver. Specifically, assume gasoline costs $p$/gallon and the vehicle gets $g$ miles per gallon. Also assume the driver earns $w$/hour.

   a. A plausible function to describe how gas mileage (in mi/gal) varies with speed is $g(v) = v(85 - v)/60$. Evaluate $g(0)$, $g(40)$, and $g(60)$ and explain why these values are reasonable.

   b. At what speed does the gas mileage function have its maximum?

c. Explain why $C(v) = Lp/g(v) + Lw/v$ gives the cost of the trip in dollars, where $L$ is the length of the trip and $v$ is the constant speed. Show that the dimensions are consistent.

d. Let $L = 400$ mi, $p = \$4/\text{gal}$, and $w = \$20/\text{hr}$. At what (constant) speed should the vehicle be driven to minimize the cost of the trip?

e. Should the optimal speed be increased or decreased (compared with part (d)) if $L$ is increased from 400 mi to 500 mi? Explain.

f. Should the optimal speed be increased or decreased (compared with part (d)) if $p$ is increased from $\$4/\text{gal}$ to $\$4.20/\text{gal}$? Explain.

g. Should the optimal speed be increased or decreased (compared with part (d)) if $w$ is decreased from $\$20/\text{hr}$ to $\$15/\text{hr}$? Explain.

50. **Suspension system** A load must be suspended 6 m below a high ceiling using cables attached to two supports that are 2 m apart (see figure). How far below the ceiling ($x$ in the figure) should the cables be joined to minimize the total length of cable used?

51. **Viewing angles** An auditorium with a flat floor has a large screen on one wall. The lower edge of the screen is 3 ft above eye level and the upper edge of the screen is 10 ft above eye level (see figure). How far from the screen should you stand to maximize your viewing angle? (*Hint:* Use the identity
$$\tan(\alpha - \beta) = \frac{\tan\alpha - \tan\beta}{1 + \tan\alpha\tan\beta}.)$$

**T 52. Basketball shot** A basketball is shot with an initial velocity of $v$ ft/s at an angle of $45°$ to the floor. The center of the basketball is 8 ft above the floor at a horizontal distance of 18 feet from the center of the basketball hoop when it is released. The height $h$ (in feet) of the center of the basketball after it has traveled a horizontal distance of $x$ feet is modeled by the function
$$h(x) = -\frac{32x^2}{v^2} + x + 8 \text{ (see figure)}.$$

a. Find the initial velocity $v$ if the center of the basketball passes the center of the hoop that is located 10 ft above the floor. Assume the ball does not hit the front of the hoop (otherwise it might not pass though the center of the hoop). The validity of this assumption is explored in the remainder of this exercise.

b. During the flight of the basketball, show that the distance $s$ from the center of the basketball to the front of the hoop is
$$s = \sqrt{(x - 17.25)^2 + \left(-\frac{4x^2}{81} + x - 2\right)^2}.$$

(*Hint:* The diameter of the basketball hoop is 18 inches.)

c. Determine whether the assumption that the basketball does not hit the front of the hoop in part (a) is valid. Use the fact that the diameter of a women's basketball is about 9.23 inches. (*Hint:* The ball will hit the front of the hoop if, during its flight, the distance from the center of the ball to the front of the hoop is less than the radius of the basketball.)

d. A men's basketball has a diameter of about 9.5 inches. Would this larger ball lead to a different conclusion than in part (c)?

53. **Light sources** The intensity of a light source at a distance is directly proportional to the strength of the source and inversely proportional to the square of the distance from the source. Two light sources, one twice as strong as the other, are 12 m apart. At what point on the line segment joining the sources is the intensity the weakest?

54. **Snell's Law** Suppose a light source at $A$ is in a medium in which light travels at a speed $v_1$ and that point $B$ is in a medium in which light travels at a speed $v_2$ (see figure). Using Fermat's Principle, which states that light travels along the path that requires the minimum travel time (Exercise 55), show that the path taken between points $A$ and $B$ satisfies $(\sin\theta_1)/v_1 = (\sin\theta_2)/v_2$.

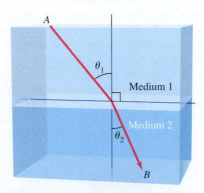

55. **Fermat's Principle**

a. Two poles of heights $m$ and $n$ are separated by a horizontal distance $d$. A rope is stretched from the top of one pole to the ground and then to the top of the other pole. Show that the configuration that requires the least amount of rope occurs when $\theta_1 = \theta_2$ (see figure).

(a)

**b.** Fermat's Principle states that when light travels between two points in the same medium (at a constant speed), it travels on the path that minimizes the travel time. Show that when light from a source $A$ reflects off a surface and is received at point $B$, the angle of incidence equals the angle of reflection, or $\theta_1 = \theta_2$ (see figure).

(b)

**56. Crankshaft** A crank of radius $r$ rotates with an angular frequency $\omega$. It is connected to a piston by a connecting rod of length $L$ (see figure). The acceleration of the piston varies with the position of the crank according to the function

$$a(\theta) = \omega^2 r\left(\cos\theta + \frac{r\cos 2\theta}{L}\right).$$

For $\omega = 1$, $L = 2$, and $r = 1$, find the values of $\theta$, with $0 \le \theta \le 2\pi$, for which the acceleration of the piston is a maximum and minimum.

**57. Making silos** A grain silo consists of a cylindrical concrete tower surmounted by a metal hemispherical dome. The metal in the dome costs 1.5 times as much as the concrete (per unit of surface area). If the volume of the silo is 750 m³, what are the dimensions of the silo (radius and height of the cylindrical tower) that minimize the cost of the materials? Assume the silo has no floor and no flat ceiling under the dome.

**58. Searchlight problem—narrow beam** A searchlight is 100 m from the nearest point on a straight highway (see figure). As it rotates, the searchlight casts a horizontal beam that intersects the highway in a point. If the light revolves at a rate of $\pi/6$ rad/s, find the rate at which the beam sweeps along the highway as a function of $\theta$. For what value of $\theta$ is this rate maximized?

**59. Minimizing related functions** Complete each of the following parts.
**a.** What value of $x$ minimizes $f(x) = (x - a_1)^2 + (x - a_2)^2$, for constants $a_1$ and $a_2$?
**b.** What value of $x$ minimizes $f(x) = (x - a_1)^2 + (x - a_2)^2 + (x - a_3)^2$, for constants $a_1$, $a_2$, and $a_3$?
**c.** Based on the answers to parts (a) and (b), make a conjecture about the values of $x$ that minimize $f(x) = (x - a_1)^2 + (x - a_2)^2 + \cdots + (x - a_n)^2$, for a positive integer $n$ and constants $a_1, a_2, \ldots, a_n$. Use calculus to verify your conjecture.

(*Source:* T. Apostol, *Calculus*, Vol. 1, John Wiley and Sons, 1967)

#### Explorations and Challenges

**T 60. Gliding mammals** Many species of small mammals (such as flying squirrels and marsupial gliders) have the ability to walk and glide. Recent research suggests that these animals choose the most energy-efficient means of travel. According to one empirical model, the energy required for a glider with body mass $m$ to walk a horizontal distance $D$ is $8.46Dm^{2/3}$ (where $m$ is measured in grams, $D$ is measured in meters, and energy is measured in microliters of oxygen consumed in respiration). The energy cost of climbing to a height $D\tan\theta$ and gliding a horizontal distance $D$ at an angle of $\theta$ is modeled by $1.36mD\tan\theta$ (where $\theta = 0$ represents horizontal flight and $\theta > 45°$ represents controlled falling). Therefore, the function

$$S(m, \theta) = 8.46m^{2/3} - 1.36m\tan\theta$$

gives the energy difference per horizontal meter traveled between walking and gliding: If $S > 0$ for given values of $m$ and $\theta$, then it is more costly to walk than to glide.

**a.** For what glide angles is it more efficient for a 200-gram animal to glide rather than walk?
**b.** Find an equation that relates $\theta$ to $m$ in the case that walking and gliding are equally efficient. Does the angle $\theta$ increase or decrease as the mass $m$ increases?
**c.** To make gliding more efficient than walking, do larger gliders have a larger or smaller selection of glide angles than smaller gliders?
**d.** Let $\theta = 25°$ (a typical glide angle). Graph $S$ as a function of $m$, for $0 \le m \le 3000$. For what values of $m$ is gliding more efficient?
**e.** For $\theta = 25°$, what value of $m$ (call it $m^*$) maximizes $S$?
**f.** Does $m^*$, as defined in part (e), increase or decrease with increasing $\theta$? That is, as a glider reduces its glide angle, does its optimal size become larger or smaller?
**g.** Assuming Dumbo is a gliding elephant whose weight is 1 metric ton ($10^6$ g), what glide angle would Dumbo use to be more efficient at gliding than walking?

(*Source:* R. Dial, *Energetic savings and the body size distribution of gliding mammals, Evolutionary Ecology Research*, 5, 2003)

**61. Metal rain gutters** A rain gutter is made from sheets of metal 9 in wide. The gutters have a 3-in base and two 3-in sides, folded up at an angle $\theta$ (see figure). What angle $\theta$ maximizes the cross-sectional area of the gutter?

3 in       3 in
         $\theta$
     3 in

Cross-sectional area

**62. Crease-length problem** A rectangular sheet of paper of width $a$ and length $b$, where $0 < a < b$, is folded by taking one corner of the sheet and placing it at a point $P$ on the opposite long side of the sheet (see figure). The fold is then flattened to form a crease across the sheet. Assuming that the fold is made so that there is no flap extending beyond the original sheet, find the point $P$ that produces the crease of minimum length. What is the length of that crease?

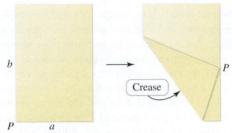

Crease

**T 63. Watching a Ferris wheel** An observer stands 20 m from the bottom of a Ferris wheel on a line that is perpendicular to the face of the wheel, with her eyes at the level of the bottom of the wheel. The wheel revolves at a rate of $\pi$ rad/min, and the observer's line of sight to a specific seat on the Ferris wheel makes an angle $\theta$ with the horizontal (see figure). At what time during a full revolution is $\theta$ changing most rapidly?

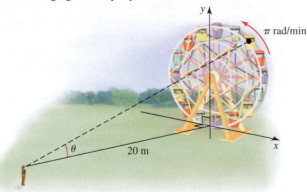

$\pi$ rad/min

$\theta$       20 m

**64. Blood flow** The resistance to blood flow in the circulatory system is one measure of how hard the heart works to pump blood through blood vessels. Lower resistance may correspond to a healthier, higher-efficiency circulatory system. Consider a smaller straight blood vessel of radius $r_2$ that branches off a larger straight blood vessel of radius $r_1$ at an angle $\theta$ (see figure with given lengths $\ell_1$, $\ell_2$, $\ell_3$, and $\ell_4$). Using Poiseuille's Law, it can be shown that the total resistance $T$ to blood flowing along the path from points $A$ to $B$ to $D$ is $T = k\left(\dfrac{\ell_1}{r_1^4} + \dfrac{\ell_2}{r_2^4}\right)$, where $k > 0$ is a constant.

**a.** Show that $T = k\left(\dfrac{\ell_4 - \ell_3 \cot\theta}{r_1^4} + \dfrac{\ell_3 \csc\theta}{r_2^4}\right)$, assuming that the line segment from $A$ to $C$ is perpendicular to the line segment from $C$ to $D$ and $0 < \theta < \dfrac{\pi}{2}$.

**b.** Show that the total resistance $T$ is minimized when $\cos\theta = \left(\dfrac{r_2}{r_1}\right)^4$.

**c.** If the radius of the smaller vessel is 85% of the radius of the larger vessel, then find the value of $\theta$ that minimizes $T$. State your answer in degrees. (*Source: Blood Vessel Branching: Beyond the Standard Calculus Problem, Mathematics Magazine, 84, 2011*)

**65. Sum of isosceles distances**

**a.** An isosceles triangle has a base of length 4 and two sides of length $2\sqrt{2}$. Let $P$ be a point on the perpendicular bisector of the base. Find the location $P$ that minimizes the sum of the distances between $P$ and the three vertices.

**b.** Assume in part (a) that the height of the isosceles triangle is $h > 0$ and its base has length 4. Show that the location of $P$ that gives a minimum solution is independent of $h$ for $h \geq \dfrac{2}{\sqrt{3}}$.

**66. Cylinder and cones (Putnam Exam 1938)** Right circular cones of height $h$ and radius $r$ are attached to each end of a right circular cylinder of height $h$ and radius $r$, forming a double-pointed object. For a given surface area $A$, what are the dimensions $r$ and $h$ that maximize the volume of the object?

**67. Slowest shortcut** Suppose you are standing in a field near a straight section of railroad tracks just as the locomotive of a train passes the point nearest to you, which is $1/4$ mi away. The train, with length $1/3$ mi, is traveling at 20 mi/hr. If you start running in a straight line across the field, how slowly can you run and still catch the train? In which direction should you run?

**68. Rectangles in triangles** Find the dimensions and area of the rectangle of maximum area that can be inscribed in the following figures.

**a.** A right triangle with a given hypotenuse length $L$
**b.** An equilateral triangle with a given side length $L$
**c.** A right triangle with a given area $A$
**d.** An arbitrary triangle with a given area $A$ (The result applies to any triangle, but first consider triangles for which all the angles are less than or equal to 90°.)

**69. Cylinder in a cone** A right circular cylinder is placed inside a cone of radius $R$ and height $H$ so that the base of the cylinder lies on the base of the cone.

**a.** Find the dimensions of the cylinder with maximum volume. Specifically, show that the volume of the maximum-volume cylinder is $4/9$ the volume of the cone.

**b.** Find the dimensions of the cylinder with maximum lateral surface area (area of the curved surface).

**70. Another pen problem** A rancher is building a horse pen on the corner of her property using 1000 ft of fencing. Because of the unusual shape of her property, the pen must be built in the shape of a trapezoid (see figure).

   **a.** Determine the lengths of the sides that maximize the area of the pen.

   **b.** Suppose there is already a fence along the side of the property opposite the side of length $y$. Find the lengths of the sides that maximize the area of the pen, using 1000 ft of fencing.

**71. Minimum-length roads** A house is located at each corner of a square with side lengths of 1 mi. What is the length of the shortest road system with straight roads that connects all of the houses by roads (that is, a road system that allows one to drive from any house to any other house)? (*Hint:* Place two points inside the square at which roads meet.) (*Source:* Paul Halmos, *Problems for Mathematicians Young and Old*, MAA, 1991)

**72. The arbelos** An arbelos is the region enclosed by three mutually tangent semicircles; it is the region inside the larger semicircle and outside the two smaller semicircles (see figure).

   **a.** Given an arbelos in which the diameter of the largest circle is 1, what positions of point $B$ maximize the area of the arbelos?

   **b.** Show that the area of the arbelos is the area of a circle whose diameter is the distance $BD$ in the figure.

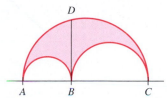

**73. Proximity questions** Find the point on the graph of $y = \sqrt{x}$ that is nearest the point $(p, 0)$ if (i) $p > \dfrac{1}{2}$; and (ii) $0 < p < \dfrac{1}{2}$. Express the answer in terms of $p$.

**74. Turning a corner with a pole**

   **a.** What is the length of the longest pole that can be carried horizontally around a corner at which a 3-ft corridor and a 4-ft corridor meet at right angles?

   **b.** What is the length of the longest pole that can be carried horizontally around a corner at which a corridor that is $a$ ft wide and a corridor that is $b$ ft wide meet at right angles?

   **c.** What is the length of the longest pole that can be carried horizontally around a corner at which a corridor that is $a = 5$ ft wide and a corridor that is $b = 5$ ft wide meet at an angle of $120°$?

   **d.** What is the length of the longest pole that can be carried around a corner at which a corridor that is $a$ ft wide and a corridor that is $b$ feet wide meet at right angles, assuming there is an 8-foot ceiling and that you may tilt the pole at any angle?

**75. Tree notch (Putnam Exam 1938, rephrased)** A notch is cut in a cylindrical vertical tree trunk (see figure). The notch penetrates to the axis of the cylinder and is bounded by two half-planes that intersect on a diameter $D$ of the tree. The angle between the two half-planes is $\theta$. Prove that for a given tree and fixed angle $\theta$, the volume of the notch is minimized by taking the bounding planes at equal angles to the horizontal plane that also passes through $D$.

**76. Circle in a triangle** What are the radius and area of the circle of maximum area that can be inscribed in an isosceles triangle whose two equal sides have length 1?

**T 77. A challenging pen problem** A farmer uses 200 meters of fencing to build two triangular pens against a barn (see figure); the pens are constructed with three sides and a diagonal dividing fence. What dimensions maximize the area of the pen?

**78. Folded boxes** Squares with sides of length $x$ are cut out of each corner of a rectangular piece of cardboard with sides of length $\ell$ and $L$. The resulting piece of cardboard is then folded into a box without a lid. Holding $\ell$ fixed, find the size of the corner squares $x$ that maximizes the volume of the box as $L \to \infty$. (*Source: Mathematics Teacher*, Nov 2002)

**QUICK CHECK ANSWERS**

**2.** (i) $A = 400x - 2x^2$ (ii) $A = 400x - 3x^2$
**3.** $V = 108w^2 - 2w^3$ ◄

# 4.6 Linear Approximation and Differentials

Imagine plotting a smooth curve with a graphing utility. Now pick a point $P$ on the curve, draw the line tangent to the curve at $P$, and zoom in on it several times. As you successively enlarge the curve near $P$, it looks more and more like the tangent line (**Figure 4.63a**). This fundamental observation—that smooth curves appear straighter on smaller scales—is called *local linearity*; it is the basis of many important mathematical ideas, one of which is *linear approximation*.

Now consider a curve with a corner or cusp at a point $Q$ (**Figure 4.63b**). No amount of magnification "straightens out" the curve or removes the corner at $Q$. The different behavior at $P$ and $Q$ is related to the idea of differentiability: The function in Figure 4.63a is differentiable at $P$, whereas the function in Figure 4.63b is not differentiable at $Q$. One of the requirements for the techniques presented in this section is that the function be differentiable at the point in question.

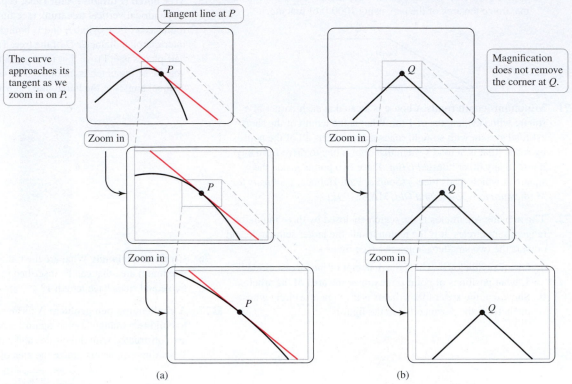

Figure 4.63

## Linear Approximation

Figure 4.63a suggests that when we zoom in on the graph of a smooth function at a point $P$, the curve approaches its tangent line at $P$. This fact is the key to understanding linear approximation. The idea is to use the line tangent to the curve at $P$ to approximate the value of the function at points near $P$. Here's how it works.

Assume $f$ is differentiable on an interval containing the point $a$. The slope of the line tangent to the curve at the point $(a, f(a))$ is $f'(a)$. Therefore, an equation of the tangent line is

$$y - f(a) = f'(a)(x - a) \quad \text{or} \quad y = \underbrace{f(a) + f'(a)(x - a)}_{L(x)}.$$

This tangent line represents a new function $L$ that we call the *linear approximation* to $f$ at the point $a$ (**Figure 4.64**). If $f$ and $f'$ are easy to evaluate at $a$, then the value of $f$ at points near $a$ is easily approximated using the linear approximation $L$. That is,

$$f(x) \approx L(x) = f(a) + f'(a)(x - a).$$

This approximation improves as $x$ approaches $a$.

Figure 4.64

**QUICK CHECK 1** Sketch the graph of a function $f$ that is concave up on an interval containing the point $a$. Sketch the linear approximation to $f$ at $a$. Is the graph of the linear approximation above or below the graph of $f$? ◄

---

**DEFINITION** **Linear Approximation to $f$ at $a$**

Suppose $f$ is differentiable on an interval $I$ containing the point $a$. The **linear approximation** to $f$ at $a$ is the linear function

$$L(x) = f(a) + f'(a)(x - a), \quad \text{for } x \text{ in } I.$$

**EXAMPLE 1    Useful driving math**  Suppose you are driving along a highway at a constant speed and you record the number of seconds it takes to travel between two consecutive mile markers. If it takes 60 seconds to travel one mile, then your average speed is 1 mi/60 s or 60 mi/hr. Now suppose you travel one mile in $60 + x$ seconds; for example, if it takes 62 seconds, then $x = 2$, and if it takes 57 seconds, then $x = -3$. In this case, your average speed over one mile is 1 mi/$(60 + x)$ s. Because there are 3600 s in 1 hr, the function

> ➤ In Example 1, notice that when $x$ is positive, you are driving slower than 60 mi/hr; when $x$ is negative, you are driving faster than 60 mi/hr.

$$s(x) = \frac{3600}{60 + x} = 3600(60 + x)^{-1}$$

gives your average speed in mi/hr if you travel one mile in $x$ seconds more or less than 60 seconds. For example, if you travel one mile in 62 seconds, then $x = 2$ and your average speed is $s(2) \approx 58.06$ mi/hr. If you travel one mile in 57 seconds, then $x = -3$ and your average speed is $s(-3) \approx 63.16$ mi/hr. Because you don't want to use a calculator while driving, you need an easy approximation to this function. Use linear approximation to derive such a formula.

**SOLUTION**  The idea is to find the linear approximation to $s$ at the point 0. We first use the Chain Rule to compute

$$s'(x) = -3600(60 + x)^{-2},$$

and then note that $s(0) = 60$ and $s'(0) = -3600 \cdot 60^{-2} = -1$. Using the linear approximation formula, we find that

$$s(x) \approx L(x) = s(0) + s'(0)(x - 0) = 60 - x.$$

**QUICK CHECK 2**  In Example 1, suppose you travel one mile in 75 seconds. What is the average speed given by the linear approximation formula? What is the exact average speed? Explain the discrepancy between the two values. ◄

For example, if you travel one mile in 62 seconds, then $x = 2$ and your average speed is approximately $L(2) = 58$ mi/hr, which is very close to the exact value given previously. If you travel one mile in 57 seconds, then $x = -3$ and your average speed is approximately $L(-3) = 63$ mi/hr, which again is close to the exact value.

*Related Exercises 13–14* ◄

**EXAMPLE 2    Linear approximations and errors**

**a.** Find the linear approximation to $f(x) = \sqrt{x}$ at $x = 1$ and use it to approximate $\sqrt{1.1}$.

**b.** Use linear approximation to estimate the value of $\sqrt{0.1}$.

**SOLUTION**

**a.** We construct the linear approximation

$$L(x) = f(a) + f'(a)(x - a),$$

where $f(x) = \sqrt{x}$, $f'(x) = 1/(2\sqrt{x})$, and $a = 1$. Noting that $f(a) = f(1) = 1$ and $f'(a) = f'(1) = \frac{1}{2}$, we have

$$L(x) = 1 + \frac{1}{2}(x - 1) = \frac{1}{2}(x + 1),$$

which describes the line tangent to the curve at the point $(1, 1)$ **(Figure 4.65)**. Because $x = 1.1$ is near $x = 1$, we approximate $\sqrt{1.1}$ by $L(1.1)$:

$$\sqrt{1.1} \approx L(1.1) = \frac{1}{2}(1.1 + 1) = 1.05.$$

The exact value is $f(1.1) = \sqrt{1.1} = 1.0488 \ldots$ ; therefore, the linear approximation has an error of about 0.0012. Furthermore, our approximation is an *overestimate* because the tangent line lies above the graph of $f$. In Table 4.4, we see several approximations to $\sqrt{x}$ for $x$ near 1 and the associated errors $|L(x) - \sqrt{x}|$. Clearly, the errors decrease as $x$ approaches 1.

Linear approximation:
$L(x) = \frac{1}{2}(x + 1)$

$f(x) = \sqrt{x}$

$(1, 1)$

$L(1.1) = 1.05$
$\sqrt{1.1} = 1.0488\ldots$

The error in the approximation decreases as $x \to 1$.

**Figure 4.65**

**Table 4.4**

| $x$ | $L(x)$ | Exact $\sqrt{x}$ | Error |
|-----|--------|------------------|-------|
| 1.2 | 1.1 | $1.0954\ldots$ | $4.6 \times 10^{-3}$ |
| 1.1 | 1.05 | $1.0488\ldots$ | $1.2 \times 10^{-3}$ |
| 1.01 | 1.005 | $1.0049\ldots$ | $1.2 \times 10^{-5}$ |
| 1.001 | 1.0005 | $1.0005\ldots$ | $1.2 \times 10^{-7}$ |

**b.** If the linear approximation $L(x) = \frac{1}{2}(x + 1)$ obtained in part (a) is used to approximate $\sqrt{0.1}$, we have

$$\sqrt{0.1} \approx L(0.1) = \frac{1}{2}(0.1 + 1) = 0.55.$$

A calculator gives $\sqrt{0.1} = 0.3162\ldots$, which shows that the approximation is well off the mark. The error arises because the tangent line through $(1, 1)$ is not close to the curve at $x = 0.1$ (Figure 4.65). For this reason, we seek a different value of $a$, with the requirement that it is near $x = 0.1$, and both $f(a)$ and $f'(a)$ are easily computed. It is tempting to try $a = 0$, but $f'(0)$ is undefined. One choice that works well is $a = \frac{9}{100} = 0.09$. Using the linear approximation $L(x) = f(a) + f'(a)(x - a)$, we have

> ► We choose $a = \frac{9}{100}$ because it is close to 0.1 and its square root is easy to evaluate.

$$\sqrt{0.1} \approx L(0.1) = \overset{f(a)}{\sqrt{\frac{9}{100}}} + \overset{f'(a)}{\frac{1}{2\sqrt{9/100}}} \overset{(x-a)}{\left(\frac{1}{10} - \frac{9}{100}\right)}$$

$$= \frac{3}{10} + \frac{10}{6}\left(\frac{1}{100}\right)$$

$$= \frac{19}{60} \approx 0.3167.$$

**QUICK CHECK 3** Suppose you want to use linear approximation to estimate $\sqrt{0.18}$. What is a good choice for $a$? ◄

This approximation agrees with the exact value to three decimal places.

*Related Exercises 31–32* ◄

Linear approximation:
$L(x) = x$
$f(x) = \sin x$

**Figure 4.66**

**EXAMPLE 3   Linear approximation for the sine function**   Find the linear approximation to $f(x) = \sin x$ at $x = 0$ and use it to approximate $\sin 2.5°$.

**SOLUTION**   We first construct a linear approximation $L(x) = f(a) + f'(a)(x - a)$, where $f(x) = \sin x$ and $a = 0$. Noting that $f(0) = 0$ and $f'(0) = \cos(0) = 1$, we have

$$L(x) = 0 + 1(x - 0) = x.$$

Again, the linear approximation is the line tangent to the curve at the point $(0, 0)$ (Figure 4.66). Before using $L(x)$ to approximate $\sin 2.5°$, we convert to radian measure (the derivative formulas for trigonometric functions require angles in radians):

$$2.5° = 2.5°\left(\frac{\pi}{180°}\right) = \frac{\pi}{72} \approx 0.04363 \text{ rad}.$$

Therefore, $\sin 2.5° \approx L(0.04363) = 0.04363$. A calculator gives $\sin 2.5° \approx 0.04362$, so the approximation is accurate to four decimal places.

*Related Exercises 38, 46* ◄

In Examples 2 and 3, we used a calculator to check the accuracy of our approximations. This raises a question: Why bother with linear approximation when a calculator does a better job? There are some good answers to that question.

Linear approximation is actually just the first step in the process of *polynomial approximation*. While linear approximation does a decent job of estimating function values when $x$ is near $a$, we can generally do better with higher-degree polynomials. These ideas are explored further in Chapter 11.

Linear approximation also allows us to discover simple approximations to complicated functions. In Example 3, we found the *small-angle approximation* to the sine *function*: $\sin x \approx x$ for $x$ near 0.

**QUICK CHECK 4** Explain why the linear approximation to $f(x) = \cos x$ at $x = 0$ is $L(x) = 1$. ◄

## Linear Approximation and Concavity

Additional insight into linear approximation is gained by bringing concavity into the picture. Figure 4.67a shows the graph of a function $f$ and its linear approximation (tangent line) at the point $(a, f(a))$. In this particular case, $f$ is concave up on an interval containing $a$, and the graph of $L$ lies below the graph of $f$ near $a$. As a result, the linear approximation evaluated at a point near $a$ is less than the exact value of $f$ at that point. In other words, the linear approximation *underestimates* values of $f$ near $a$.

The contrasting case is shown in **Figure 4.67b**, where we see graphs of $f$ and $L$ when $f$ is concave down on an interval containing $a$. Now the graph of $L$ lies above the graph of $f$, which means the linear approximation *overestimates* values of $f$ near $a$.

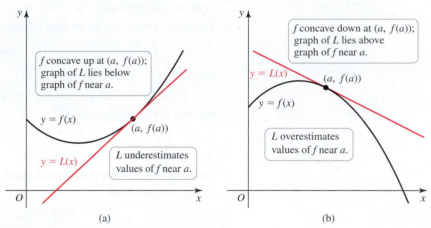

**Figure 4.67**

We can make another observation related to the degree of concavity (also called *curvature*). A large value of $|f''(a)|$ (large curvature) means that near $(a, f(a))$, the slope of the curve changes rapidly and the graph of $f$ separates quickly from the tangent line. A small value of $|f''(a)|$ (small curvature) means the slope of the curve changes slowly and the curve is relatively flat near $(a, f(a))$; therefore, the curve remains close to the tangent line. As a result, absolute errors in linear approximation are larger when $|f''(a)|$ is large.

**EXAMPLE 4    Linear approximation and concavity**

**a.** Find the linear approximation to $f(x) = x^{1/3}$ at $x = 1$ and $x = 27$.

**b.** Use the linear approximations of part (a) to approximate $\sqrt[3]{2}$ and $\sqrt[3]{26}$.

**c.** Are the approximations in part (b) overestimates or underestimates?

**d.** Compute the error in each approximation of part (b). Which error is greater? Explain.

**SOLUTION**

**a.** Note that

$$f(1) = 1, \quad f(27) = 3, \quad f'(x) = \frac{1}{3x^{2/3}}, \quad f'(1) = \frac{1}{3}, \quad \text{and} \quad f'(27) = \frac{1}{27}.$$

Therefore, the linear approximation at $x = 1$ is

$$L_1(x) = 1 + \frac{1}{3}(x - 1) = \frac{1}{3}x + \frac{2}{3},$$

and the linear approximation at $x = 27$ is

$$L_2(x) = 3 + \frac{1}{27}(x - 27) = \frac{1}{27}x + 2.$$

**b.** Using the results of part (a), we find that

$$\sqrt[3]{2} \approx L_1(2) = \frac{1}{3} \cdot 2 + \frac{2}{3} = \frac{4}{3} \approx 1.333$$

and

$$\sqrt[3]{26} \approx L_2(26) = \frac{1}{27} \cdot 26 + 2 \approx 2.963.$$

(a)

(b)

**Figure 4.68**

**c.** Figure 4.68 shows the graphs of $f$ and the linear approximations $L_1$ and $L_2$ at $x = 1$ and $x = 27$, respectively (note the different scales on the $x$-axes). We see that $f$ is concave down at both points, which is confirmed by the fact that

$$f''(x) = -\frac{2}{9}x^{-5/3} < 0, \quad \text{for} \quad x > 0.$$

As a result, the linear approximations lie above the graph of $f$ and both approximations are overestimates.

**d.** The errors in the two linear approximations are

$$\left| L_1(2) - 2^{1/3} \right| \approx 0.073 \quad \text{and} \quad \left| L_2(26) - 26^{1/3} \right| \approx 0.00047.$$

Because $\left| f''(1) \right| \approx 0.22$ and $\left| f''(27) \right| \approx 0.00091$, the curvature of $f$ is greater at $x = 1$ than at $x = 27$, explaining why the approximation of $\sqrt[3]{26}$ is more accurate than the approximation of $\sqrt[3]{2}$.

*Related Exercises 47–50* ◄

**A Variation on Linear Approximation**   Linear approximation says that a function $f$ can be approximated as

$$f(x) \approx f(a) + f'(a)(x - a),$$

where $a$ is fixed and $x$ is a nearby point. We first rewrite this expression as

$$\underbrace{f(x) - f(a)}_{\Delta y} \approx f'(a)\underbrace{(x - a)}_{\Delta x}.$$

It is customary to use the notation $\Delta$ (capital Greek delta) to denote a change. The factor $x - a$ is the change in the $x$-coordinate between $a$ and a nearby point $x$. Similarly, $f(x) - f(a)$ is the corresponding change in the $y$-coordinate (Figure 4.69). So we write this approximation as

$$\Delta y \approx f'(a)\, \Delta x.$$

In other words, a change in $y$ (the function value) can be approximated by the corresponding change in $x$ magnified or diminished by a factor of $f'(a)$. This interpretation states the familiar fact that $f'(a)$ is the rate of change of $y$ with respect to $x$.

**Figure 4.69**

---

**Relationship Between $\Delta x$ and $\Delta y$**

Suppose $f$ is differentiable on an interval $I$ containing the point $a$. The change in the value of $f$ between two points $a$ and $a + \Delta x$ is approximately

$$\Delta y \approx f'(a)\, \Delta x,$$

where $a + \Delta x$ is in $I$.

## EXAMPLE 5 Estimating changes with linear approximations

**a.** Approximate the change in $y = f(x) = x^9 - 2x + 1$ when $x$ changes from 1.00 to 1.05.

**b.** Approximate the change in the surface area of a spherical hot-air balloon when the radius decreases from 4 m to 3.9 m.

**SOLUTION**

**a.** The change in $y$ is $\Delta y \approx f'(a)\,\Delta x$, where $a = 1$, $\Delta x = 0.05$, and $f'(x) = 9x^8 - 2$. Substituting these values, we find that

$$\Delta y \approx f'(a)\,\Delta x = f'(1) \cdot 0.05 = 7 \cdot 0.05 = 0.35.$$

If $x$ increases from 1.00 to 1.05, then $y$ increases by approximately 0.35.

> Notice that the units in these calculations are consistent. If $r$ has units of meters (m), $S'$ has units of $m^2/m = m$, so $\Delta S$ has units of $m^2$, as it should.

**b.** The surface area of a sphere is $S = 4\pi r^2$, so the change in the surface area when the radius changes by $\Delta r$ is $\Delta S \approx S'(a)\,\Delta r$. Substituting $S'(r) = 8\pi r$, $a = 4$, and $\Delta r = -0.1$, the approximate change in the surface area is

$$\Delta S \approx S'(a)\,\Delta r = S'(4) \cdot (-0.1) = 32\pi \cdot (-0.1) \approx -10.05.$$

The change in surface area is approximately $-10.05$ $m^2$; it is negative, reflecting a decrease.

*Related Exercises 55, 57* ◄

**QUICK CHECK 5** Given that the volume of a sphere is $V = 4\pi r^3/3$, find an expression for the approximate change in the volume when the radius changes from $a$ to $a + \Delta r$. ◄

---

**SUMMARY** Uses of Linear Approximation

- To approximate $f$ near $x = a$, use

$$f(x) \approx L(x) = f(a) + f'(a)(x - a).$$

- To approximate the change $\Delta y$ in the dependent variable when the independent variable $x$ changes from $a$ to $a + \Delta x$, use

$$\Delta y \approx f'(a)\,\Delta x.$$

---

## Differentials

We now introduce an important concept that allows us to distinguish two related quantities:

- the change in the function $y = f(x)$ as $x$ changes from $a$ to $a + \Delta x$ (which we call $\Delta y$, as before), and

- the change in the linear approximation $y = L(x)$ as $x$ changes from $a$ to $a + \Delta x$ (which we call the *differential dy*).

Consider a function $y = f(x)$ differentiable on an interval containing $a$. If the $x$-coordinate changes from $a$ to $a + \Delta x$, then the corresponding change in the function is *exactly*

$$\Delta y = f(a + \Delta x) - f(a).$$

Using the linear approximation $L(x) = f(a) + f'(a)(x - a)$, the change in $L$ as $x$ changes from $a$ to $a + \Delta x$ is

$$\Delta L = L(a + \Delta x) - L(a)$$
$$= \underbrace{(f(a) + f'(a)(a + \Delta x - a))}_{L(a + \Delta x)} - \underbrace{(f(a) + f'(a)(a - a))}_{L(a)}$$
$$= f'(a)\,\Delta x.$$

To distinguish $\Delta y$ and $\Delta L$, we define two new variables called *differentials*. The differential $dx$ is simply $\Delta x$; the differential $dy$ is the change in the linear approximation, which is $\Delta L = f'(a)\,\Delta x$. Using this notation,

$$\Delta L = \underbrace{dy}_{\text{same as }\Delta L} = f'(a)\,\Delta x = f'(a)\,\underbrace{dx}_{\text{same as }\Delta x}.$$

Therefore, at the point $a$, we have $dy = f'(a)\,dx$. More generally, we replace the fixed point $a$ with a variable point $x$ and write

$$dy = f'(x)\,dx.$$

---

**DEFINITION  Differentials**

Let $f$ be differentiable on an interval containing $x$. A small change in $x$ is denoted by the **differential** $dx$. The corresponding change in $f$ is approximated by the **differential** $dy = f'(x)\,dx$; that is,

$$\Delta y = f(x + dx) - f(x) \approx dy = f'(x)\,dx.$$

---

> As one of the two co-inventors of calculus, Gottfried Leibniz relied on the idea of differentials in his development of calculus. Leibniz's notation for differentials is essentially the same as the notation we use today. An Irish philosopher of the day, Bishop Berkeley, called differentials "the ghost of departed quantities."

**Figure 4.70** shows that if $\Delta x = dx$ is small, then the change in $f$, which is $\Delta y$, is well approximated by the change in the linear approximation, which is $dy$. Furthermore, the approximation $\Delta y \approx dy$ improves as $dx$ approaches 0. The notation for differentials is consistent with the notation for the derivative: If we divide both sides of $dy = f'(x)\,dx$ by $dx$, we have (symbolically)

$$\frac{dy}{dx} = \frac{f'(x)\,dx}{dx} = f'(x).$$

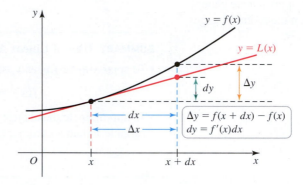

**Figure 4.70**

**EXAMPLE 6  Differentials as change** Use the notation of differentials to write the approximate change in $f(x) = 3\cos^2 x$ given a small change $dx$.

> Recall that $\sin 2x = 2 \sin x \cos x$.

**SOLUTION** With $f(x) = 3\cos^2 x$, we have $f'(x) = -6 \cos x \sin x = -3 \sin 2x$. Therefore,

$$dy = f'(x)\,dx = -3 \sin 2x\,dx.$$

The interpretation is that a small change $dx$ in the independent variable $x$ produces an approximate change in the dependent variable of $dy = -3 \sin 2x\,dx$. For example, if $x$ increases from $x = \pi/4$ to $x = \pi/4 + 0.1$, then $dx = 0.1$ and

$$dy = -3 \sin (\pi/2)(0.1) = -0.3.$$

The approximate change in the function is $-0.3$, which means a decrease of approximately 0.3.

*Related Exercises 60, 66* ◄

---

## SECTION 4.6 EXERCISES

### Getting Started

1. Sketch the graph of a smooth function $f$ and label a point $P(a, f(a))$ on the curve. Draw the line that represents the linear approximation to $f$ at $P$.

2. Suppose you find the linear approximation to a differentiable function at a local maximum of that function. Describe the graph of the linear approximation.

3. How is linear approximation used to approximate the value of a function $f$ near a point at which $f$ and $f'$ are easily evaluated?

4. How can linear approximation be used to approximate the change in $y = f(x)$ given a change in $x$?

5. Suppose $f$ is differentiable on $(-\infty, \infty)$, $f(1) = 2$, and $f'(1) = 3$. Find the linear approximation to $f$ at $x = 1$ and use it to approximate $f(1.1)$.

**6.** Suppose $f$ is differentiable on $(-\infty, \infty)$ and the equation of the line tangent to the graph of $f$ at $x = 2$ is $y = 5x - 3$. Use the linear approximation to $f$ at $x = 2$ to approximate $f(2.01)$.

**7.** Use linear approximation to estimate $f(3.85)$ given that $f(4) = 3$ and $f'(4) = 2$.

**8.** Use linear approximation to estimate $f(5.1)$ given that $f(5) = 10$ and $f'(5) = -2$.

**9.** Given a function $f$ that is differentiable on its domain, write and explain the relationship between the differentials $dx$ and $dy$.

**10.** Does the differential $dy$ represent the change in $f$ or the change in the linear approximation to $f$? Explain.

**11.** Suppose $f$ is differentiable on $(-\infty, \infty)$ and $f(5.01) - f(5) = 0.25$. Use linear approximation to estimate the value of $f'(5)$.

**12.** Suppose $f$ is differentiable on $(-\infty, \infty)$, $f(5.99) = 7$, and $f(6) = 7.002$. Use linear approximation to estimate the value of $f'(6)$.

## Practice Exercises

**13–14. Estimating speed** *Use the linear approximation given in Example 1 to answer the following questions.*

**13.** If you travel one mile in 59 seconds, what is your approximate average speed? What is your exact speed?

**14.** If you travel one mile in 63 seconds, what is your approximate average speed? What is your exact speed?

**15–18. Estimating time** *Suppose you want to travel D miles at a constant speed of $(60 + x)$ mi/hr, where x could be positive or negative. The time in minutes required to travel D miles is $T(x) = 60D(60 + x)^{-1}$.*

**15.** Show that the linear approximation to $T$ at the point $x = 0$ is
$$T(x) \approx L(x) = D\left(1 - \frac{x}{60}\right).$$

**16.** Use the result of Exercise 15 to approximate the amount of time it takes to drive 45 miles at 62 mi/hr. What is the exact time required?

**17.** Use the result of Exercise 15 to approximate the amount of time it takes to drive 80 miles at 57 mi/hr. What is the exact time required?

**18.** Use the result of Exercise 15 to approximate the amount of time it takes to drive 93 miles at 63 mi/hr. What is the exact time required?

**19–24. Linear approximation** *Find the linear approximation to the following functions at the given point a.*

**19.** $f(x) = 4x^2 + x; a = 1$     **20.** $f(x) = x^3 - 5x + 3; a = 2$

**21.** $g(t) = \sqrt{2t + 9}; a = -4$     **22.** $h(w) = \sqrt{5w - 1}; a = 1$

**23.** $f(x) = \sqrt[3]{x}; a = 8$     **24.** $f(x) = 9(4x + 11)^{2/3}; a = 4$

**T 25–36. Linear approximation**

**a.** Write the equation of the line that represents the linear approximation to the following functions at the given point a.
**b.** Use the linear approximation to estimate the given quantity.
**c.** Compute the percent error in your approximation, $100|\text{approximation} - \text{exact}|/|\text{exact}|$, where the exact value is given by a calculator.

**25.** $f(x) = 12 - x^2; a = 2; f(2.1)$

**26.** $f(x) = \sin x; a = \pi/4; f(0.75)$

**27.** $f(x) = 1/(1 + x); a = 0; f(-0.1)$

**28.** $f(x) = x/(x + 1); a = 1; f(1.1)$

**29.** $f(x) = \cos x; a = 0; f(-0.01)$

**30.** $f(x) = x^{-3}; a = 1; f(1.05)$

**31.** $f(x) = (8 + x)^{-1/3}; a = 0; f(-0.1)$

**32.** $f(x) = \sqrt[4]{x}; a = 81; f(85)$

**33.** $f(x) = 1/(x + 1); a = 0; 1/1.1$

**34.** $f(x) = \cos x; a = \pi/4; \cos 0.8$

**35.** $f(x) = \sqrt[3]{64 + x}; a = 0; \sqrt[3]{62.5}$

**36.** $f(x) = \tan x; a = 0; \tan 3°$

**37–46. Estimations with linear approximation** *Use linear approximations to estimate the following quantities. Choose a value of a to produce a small error.*

**37.** $1/203$     **38.** $\tan(-2°)$     **39.** $\sqrt{146}$     **40.** $\sqrt[3]{65}$

**41.** $1/1.05$     **42.** $\sqrt{5/29}$     **43.** $\sin(\pi/4 + 0.1)$

**44.** $1/\sqrt{119}$     **45.** $1/\sqrt[3]{510}$     **46.** $\cos 31°$

**47–50. Linear approximation and concavity** *Carry out the following steps for the given functions f and points a.*

**a.** Find the linear approximation L to the function f at the point a.
**b.** Graph f and L on the same set of axes.
**c.** Based on the graphs in part (b), state whether linear approximations to f near x = a are underestimates or overestimates.
**d.** Compute $f''(a)$ to confirm your conclusions in part (c).

**47.** $f(x) = \frac{2}{x}; a = 1$     **48.** $f(x) = 5 - x^2; a = 2$

**49.** $f(x) = 1/\sqrt{x}; a = 1$     **50.** $f(x) = \sqrt{2}\cos x; a = \frac{\pi}{4}$

**T 51. Error in driving speed** Consider again the average speed $s(x)$ and its linear approximation $L(x)$ discussed in Example 1. The error in using $L(x)$ to approximate $s(x)$ is given by $E(x) = |L(x) - s(x)|$. Use a graphing utility to determine the (approximate) values of x for which $E(x) \le 1$. What does your answer say about the accuracy of the average speeds estimated by $L(x)$ over this interval?

**52. Ideal Gas Law** The pressure P, temperature T, and volume V of an ideal gas are related by $PV = nRT$, where n is the number of moles of the gas and R is the universal gas constant. For the purposes of this exercise, let $nR = 1$; therefore, $P = T/V$.

**a.** Suppose the volume is held constant and the temperature increases by $\Delta T = 0.05$. What is the approximate change in the pressure? Does the pressure increase or decrease?
**b.** Suppose the temperature is held constant and the volume increases by $\Delta V = 0.1$. What is the approximate change in the pressure? Does the pressure increase or decrease?
**c.** Suppose the pressure is held constant and the volume increases by $\Delta V = 0.1$. What is the approximate change in the temperature? Does the temperature increase or decrease?

**53. Explain why or why not** Determine whether the following statements are true and give an explanation or counterexample.

a. The linear approximation to $f(x) = x^2$ at $x = 0$ is $L(x) = 0$.
b. Linear approximation at $x = 0$ provides a good approximation to $f(x) = |x|$.
c. If $f(x) = mx + b$, then the linear approximation to $f$ at any point is $L(x) = f(x)$.

**54. Time function** Show that the function $T(x) = 60D(60 + x)^{-1}$ gives the time in minutes required to drive $D$ miles at $60 + x$ miles per hour.

**55–58. Approximating changes**

**55.** Approximate the change in the volume of a sphere when its radius changes from $r = 5$ ft to $r = 5.1$ ft $\left(V(r) = \frac{4}{3}\pi r^3\right)$.

**56.** Approximate the change in the volume of a right circular cone of fixed height $h = 4$ m when its radius increases from $r = 3$ m to $r = 3.05$ m $\left(V(r) = \frac{1}{3}\pi r^2 h\right)$.

**57.** Approximate the change in the lateral surface area (excluding the area of the base) of a right circular cone of fixed height $h = 6$ m when its radius decreases from $r = 10$ m to $r = 9.9$ m $\left(S = \pi r\sqrt{r^2 + h^2}\right)$.

**58.** Approximate the change in the magnitude of the electrostatic force between two charges when the distance between them increases from $r = 20$ m to $r = 21$ m $\left(F(r) = 0.01/r^2\right)$.

**59–66. Differentials** *Consider the following functions and express the relationship between a small change in x and the corresponding change in y in the form $dy = f'(x)dx$.*

**59.** $f(x) = 2x + 1$

**60.** $f(x) = \sin^2 x$

**61.** $f(x) = \dfrac{1}{x^3}$

**62.** $f(x) = \sqrt{x^2 + 1}$

**63.** $f(x) = 2 - a\cos x$, $a$ constant

**64.** $f(x) = \dfrac{x + 4}{4 - x}$

**65.** $f(x) = 3x^3 - 4x$

**66.** $f(x) = \tan x$

## Explorations and Challenges

**T 67. Errors in approximations** Suppose $f(x) = \sqrt[3]{x}$ is to be approximated near $x = 8$. Find the linear approximation to $f$ at 8. Then complete the following table, showing the errors in various approximations. Use a calculator to obtain the exact values. The percent error is $100|\text{approximation} - \text{exact}|/|\text{exact}|$. Comment on the behavior of the errors as $x$ approaches 8.

| $x$ | Linear approx. | Exact value | Percent error |
|---|---|---|---|
| 8.1 | | | |
| 8.01 | | | |
| 8.001 | | | |
| 8.0001 | | | |
| 7.9999 | | | |
| 7.999 | | | |
| 7.99 | | | |
| 7.9 | | | |

**T 68. Errors in approximations** Suppose $f(x) = 1/(x + 1)$ is to be approximated near $x = 0$. Find the linear approximation to $f$ at 0. Then complete the following table, showing the errors in various approximations. Use a calculator to obtain the exact values. The percent error is $100|\text{approximation} - \text{exact}|/|\text{exact}|$. Comment on the behavior of the errors as $x$ approaches 0.

| $x$ | Linear approx. | Exact value | Percent error |
|---|---|---|---|
| 0.1 | | | |
| 0.01 | | | |
| 0.001 | | | |
| 0.0001 | | | |
| −0.0001 | | | |
| −0.001 | | | |
| −0.01 | | | |
| −0.1 | | | |

**69. Linear approximation and the second derivative** Draw the graph of a function $f$ such that $f(1) = f'(1) = f''(1) = 1$. Draw the linear approximation to the function at the point $(1, 1)$. Now draw the graph of another function $g$ such that $g(1) = g'(1) = 1$ and $g''(1) = 10$. (It is not possible to represent the second derivative exactly, but your graphs should reflect the fact that $f''(1)$ is relatively small compared to $g''(1)$.) Now suppose linear approximations are used to approximate $f(1.1)$ and $g(1.1)$.

a. Which function has the more accurate linear approximation near $x = 1$ and why?
b. Explain why the error in the linear approximation to $f$ near a point $a$ is proportional to the magnitude of $f''(a)$.

**QUICK CHECK ANSWERS**

1. The linear approximation lies below the graph of $f$ for $x$ near $a$. 2. $L(15) = 45$, $s(15) = 48$; $x = 15$ is not close to 0. 3. $a = 0.16$ 4. Note that $f(0) = 1$ and $f'(0) = 0$, so $L(x) = 1$ (this is the line tangent to $y = \cos x$ at $(0, 1)$). 5. $\Delta V \approx 4\pi a^2 \Delta r$ ◀

# 4.7 L'Hôpital's Rule

The study of limits in Chapter 2 was thorough but not exhaustive. Some limits, called *indeterminate forms*, cannot generally be evaluated using the techniques presented in Chapter 2. These limits tend to be the more interesting limits that arise in practice. A powerful result called *l'Hôpital's Rule* enables us to evaluate such limits with relative ease.

Here is how indeterminate forms arise. If $f$ is a *continuous* function at a point $a$, then we know that $\lim_{x \to a} f(x) = f(a)$, allowing the limit to be evaluated by computing $f(a)$. But there are many limits that cannot be evaluated by substitution. In fact, we encountered such a limit in Section 3.5:

$$\lim_{x \to 0} \frac{\sin x}{x} = 1.$$

If we attempt to substitute $x = 0$ into $(\sin x)/x$, we get $0/0$, which has no meaning. Yet we proved that $(\sin x)/x$ has the limit 1 at $x = 0$ (Theorem 3.9). This limit is an example of an *indeterminate form*; specifically, $\lim_{x \to 0} \frac{\sin x}{x}$ has the form $0/0$ because the numerator and denominator both approach 0 as $x \to 0$.

The meaning of an *indeterminate form* is further illustrated by $\lim_{x \to \infty} \frac{ax}{x + 1}$, where $a \neq 0$. This limit has the indeterminate form $\infty/\infty$ (meaning that the numerator and denominator become arbitrarily large in magnitude as $x \to \infty$), but the actual value of the limit is $\lim_{x \to \infty} \frac{ax}{x + 1} = a \lim_{x \to \infty} \frac{x}{x + 1} = a$. In general, a limit with the form $\infty/\infty$ or $0/0$ can have *any* value—which is why these limits must be handled carefully.

## L' Hôpital's Rule for the Form 0/0

Consider a function of the form $f(x)/g(x)$ and assume $\lim_{x \to a} f(x) = \lim_{x \to a} g(x) = 0$. Then the limit $\lim_{x \to a} \frac{f(x)}{g(x)}$ has the indeterminate form $0/0$. We first state l'Hôpital's Rule and then prove a special case.

> ▶ The notations $0/0$ and $\infty/\infty$ are merely symbols used to describe various types of indeterminate forms. The notation $0/0$ does not imply division by 0.

> ▶ Guillaume François l'Hôpital (lo-pee-tal) (1661–1704) is credited with writing the first calculus textbook. Much of the material in his book, including l'Hôpital's Rule, was provided by the Swiss mathematician Johann Bernoulli (1667–1748).

---

**THEOREM 4.12  L'Hôpital's Rule**

Suppose $f$ and $g$ are differentiable on an open interval $I$ containing $a$ with $g'(x) \neq 0$ on $I$ when $x \neq a$. If $\lim_{x \to a} f(x) = \lim_{x \to a} g(x) = 0$, then

$$\lim_{x \to a} \frac{f(x)}{g(x)} = \lim_{x \to a} \frac{f'(x)}{g'(x)},$$

provided the limit on the right exists (or is $\pm \infty$). The rule also applies if $x \to a$ is replaced with $x \to \pm \infty$, $x \to a^+$, or $x \to a^-$.

---

**Proof (special case):** The proof of this theorem relies on the Generalized Mean Value Theorem (Exercise 55 of Section 4.2). We prove a special case of the theorem in which we assume $f'$ and $g'$ are continuous at $a$, $f(a) = g(a) = 0$, and $g'(a) \neq 0$. We have

$$\lim_{x \to a} \frac{f'(x)}{g'(x)} = \frac{f'(a)}{g'(a)} \qquad \text{Continuity of } f' \text{ and } g'$$

$$= \frac{\displaystyle\lim_{x \to a} \frac{f(x) - f(a)}{x - a}}{\displaystyle\lim_{x \to a} \frac{g(x) - g(a)}{x - a}} \qquad \text{Definition of } f'(a) \text{ and } g'(a)$$

▶ The definition of the derivative provides an example of an indeterminate form. Assuming $f$ is differentiable at $x$,

$$f'(x) = \lim_{h \to 0} \frac{f(x + h) - f(x)}{h}$$

has the form $0/0$.

$$\frac{f(x)}{g(x)} = \frac{f'(x)}{g'(x)} = \frac{4}{2} = 2$$

**Figure 4.71**

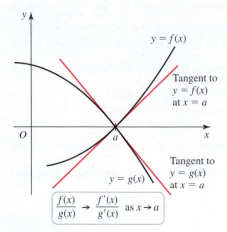

$$\frac{f(x)}{g(x)} \to \frac{f'(x)}{g'(x)} \quad \text{as } x \to a$$

**Figure 4.72**

**QUICK CHECK 1** Which of the following functions lead to an indeterminate form as $x \to 0$: $f(x) = x^2/(x + 2)$, $g(x) = (\tan 3x)/x$, or $h(x) = (1 - \cos x)/x^2$? ◀

$$= \lim_{x \to a} \frac{\dfrac{f(x) - f(a)}{x - a}}{\dfrac{g(x) - g(a)}{x - a}} \qquad \text{Limit of a quotient, } g'(a) \neq 0$$

$$= \lim_{x \to a} \frac{f(x) - f(a)}{g(x) - g(a)} \qquad \text{Cancel } x - a.$$

$$= \lim_{x \to a} \frac{f(x)}{g(x)}. \qquad f(a) = g(a) = 0 \qquad \blacktriangleleft$$

The geometry of l'Hôpital's Rule offers some insight. First consider two *linear* functions, $f$ and $g$, whose graphs both pass through the point $(a, 0)$ with slopes 4 and 2, respectively; this means that

$$f(x) = 4(x - a) \quad \text{and} \quad g(x) = 2(x - a).$$

Furthermore, $f(a) = g(a) = 0$, $f'(x) = 4$, and $g'(x) = 2$ (**Figure 4.71**).

Looking at the quotient $f/g$, we see that

$$\frac{f(x)}{g(x)} = \frac{4(x - a)}{2(x - a)} = \frac{4}{2} = \frac{f'(x)}{g'(x)}.$$

This argument may be generalized, and we find that for any linear functions $f$ and $g$ with $f(a) = g(a) = 0$,

$$\lim_{x \to a} \frac{f(x)}{g(x)} = \lim_{x \to a} \frac{f'(x)}{g'(x)},$$

provided $g'(a) \neq 0$.

If $f$ and $g$ are not linear functions, we replace them with their linear approximations at $a$ (**Figure 4.72**). Zooming in on the point $a$, the curves are close to their respective tangent lines $y = f'(a)(x - a)$ and $y = g'(a)(x - a)$, which have slopes $f'(a)$ and $g'(a) \neq 0$, respectively. Near $x = a$, we have

$$\frac{f(x)}{g(x)} \approx \frac{f'(a)(x - a)}{g'(a)(x - a)} = \frac{f'(a)}{g'(a)}.$$

Therefore, the ratio of the functions is well approximated by the ratio of the derivatives. In the limit as $x \to a$, we again have

$$\lim_{x \to a} \frac{f(x)}{g(x)} = \lim_{x \to a} \frac{f'(x)}{g'(x)}.$$

**EXAMPLE 1   Using l'Hôpital's Rule** Evaluate the following limits.

**a.** $\displaystyle \lim_{x \to 1} \frac{x^3 + x^2 - 2x}{x - 1}$     **b.** $\displaystyle \lim_{x \to 0} \frac{\sqrt{9 + 3x} - 3}{x}$

**SOLUTION**

▶ The limit in Example 1a can also be evaluated by factoring the numerator and canceling $(x - 1)$:

$$\lim_{x \to 1} \frac{x^3 + x^2 - 2x}{x - 1}$$
$$= \lim_{x \to 1} \frac{x(x - 1)(x + 2)}{x - 1}$$
$$= \lim_{x \to 1} x(x + 2) = 3.$$

**a.** Direct substitution of $x = 1$ into $\dfrac{x^3 + x^2 - 2x}{x - 1}$ produces the indeterminate form $0/0$.

Applying l'Hôpital's Rule with $f(x) = x^3 + x^2 - 2x$ and $g(x) = x - 1$ gives

$$\lim_{x \to 1} \frac{x^3 + x^2 - 2x}{x - 1} = \lim_{x \to 1} \frac{f'(x)}{g'(x)} = \lim_{x \to 1} \frac{3x^2 + 2x - 2}{1} = 3.$$

**b.** Substituting $x = 0$ into this function produces the indeterminate form $0/0$. Let $f(x) = \sqrt{9 + 3x} - 3$ and $g(x) = x$, and note that $f'(x) = \dfrac{3}{2\sqrt{9 + 3x}}$ and $g'(x) = 1$. Applying l'Hôpital's Rule, we have

$$\lim_{x \to 0} \underbrace{\frac{\sqrt{9 + 3x} - 3}{x}}_{f/g} = \lim_{x \to 0} \underbrace{\frac{\dfrac{3}{2\sqrt{9 + 3x}}}{1}}_{f'/g'} = \frac{1}{2}.$$

*Related Exercises 11–12* ◄

L'Hôpital's Rule requires evaluating $\lim\limits_{x \to a} f'(x)/g'(x)$. It may happen that this second limit is another indeterminate form to which l'Hôpital's Rule may again be applied.

**EXAMPLE 2  L'Hôpital's Rule repeated**　Evaluate the following limits.

**a.** $\lim\limits_{x \to 0} \dfrac{\sec x - 1}{x^2}$　　　**b.** $\lim\limits_{x \to 2} \dfrac{x^3 - 3x^2 + 4}{x^4 - 4x^3 + 7x^2 - 12x + 12}$

**SOLUTION**

**a.** This limit has the indeterminate form $0/0$. Applying l'Hôpital's Rule, we have

$$\lim_{x \to 0} \frac{\sec x - 1}{x^2} = \lim_{x \to 0} \frac{\sec x \tan x}{2x},$$

which is another limit of the form $0/0$. Therefore, we apply l'Hôpital's Rule again:

> The second application of l'Hôpital's Rule in Example 2a is convenient, though not necessary. One could also write
> $$\lim_{x \to 0} \frac{\sec x \tan x}{2x} = \lim_{x \to 0} \frac{1}{\cos x} \cdot \frac{\sin x}{\cos x} \cdot \frac{1}{2x}$$
> $$= \frac{1}{2} \lim_{x \to 0} \underbrace{\frac{\sin x}{x}}_{1} \cdot \underbrace{\lim_{x \to 0} \frac{1}{\cos^2 x}}_{1} = \frac{1}{2}.$$

$$\lim_{x \to 0} \frac{\sec x - 1}{x^2} = \lim_{x \to 0} \frac{\sec x \tan x}{2x} \qquad \text{\color{blue}L'Hôpital's Rule}$$

$$= \lim_{x \to 0} \frac{(\sec x \tan x)\tan x + \sec x (\sec^2 x)}{2} \qquad \text{\color{blue}L'Hôpital's Rule again; Product Rule}$$

$$= \lim_{x \to 0} \frac{\sec x \underbrace{\tan^2 x}_{\text{approaches 0}} + \underbrace{\sec^3 x}_{\text{approaches 1}}}{2} = \frac{1}{2}. \qquad \text{\color{blue}Evaluate limit.}$$

**b.** Evaluating the numerator and denominator at $x = 2$, we see that this limit has the form $0/0$. Applying l'Hôpital's Rule twice, we have

$$\lim_{x \to 2} \frac{x^3 - 3x^2 + 4}{x^4 - 4x^3 + 7x^2 - 12x + 12} = \underbrace{\lim_{x \to 2} \frac{3x^2 - 6x}{4x^3 - 12x^2 + 14x - 12}}_{\text{limit of the form } 0/0} \qquad \text{\color{blue}L'Hôpital's Rule}$$

$$= \lim_{x \to 2} \frac{6x - 6}{12x^2 - 24x + 14} \qquad \text{\color{blue}L'Hôpital's Rule again}$$

$$= \frac{3}{7}. \qquad \text{\color{blue}Evaluate limit.}$$

It is easy to overlook a crucial step in this computation: After applying l'Hôpital's Rule the first time, you *must* establish that the new limit is an indeterminate form before applying l'Hôpital's Rule a second time.

*Related Exercises 30, 31* ◄

## Indeterminate Form $\infty/\infty$

L'Hôpital's Rule also applies directly to limits of the form $\lim\limits_{x \to a} f(x)/g(x)$, where $\lim\limits_{x \to a} f(x) = \pm\infty$ and $\lim\limits_{x \to a} g(x) = \pm\infty$; this indeterminate form is denoted $\infty/\infty$. The proof of this result is found in advanced books.

> **THEOREM 4.13** L'Hôpital's Rule ($\infty/\infty$)
>
> Suppose $f$ and $g$ are differentiable on an open interval $I$ containing $a$, with $g'(x) \neq 0$ on $I$ when $x \neq a$. If $\lim\limits_{x \to a} f(x) = \pm\infty$ and $\lim\limits_{x \to a} g(x) = \pm\infty$, then
>
> $$\lim_{x \to a} \frac{f(x)}{g(x)} = \lim_{x \to a} \frac{f'(x)}{g'(x)},$$
>
> provided the limit on the right exists (or is $\pm\infty$). The rule also applies for $x \to \pm\infty$, $x \to a^+$, or $x \to a^-$.

**QUICK CHECK 2** Which of the following functions lead to an indeterminate form as $x \to \infty$: $f(x) = (\sin x)/x$, $g(x) = (3x^2 + 4)/x^2$, or $h(x) = (x - 1)/x^3$? ◄

**EXAMPLE 3** L'Hôpital's Rule for $\infty/\infty$ Evaluate the following limits.

**a.** $\lim\limits_{x \to \infty} \dfrac{4x^3 - 6x^2 + 1}{2x^3 - 10x + 3}$  **b.** $\lim\limits_{x \to 0^+} \dfrac{1 - \cot(\sin x)}{\cot x}$

**SOLUTION**

➤ As shown in Section 2.5, the limit in Example 3a could also be evaluated by first dividing the numerator and denominator by $x^3$ or by using Theorem 2.7.

**a.** This limit has the indeterminate form $\infty/\infty$ because both the numerator and the denominator approach $\infty$ as $x \to \infty$. Applying l'Hôpital's Rule three times, we have

$$\underbrace{\lim_{x \to \infty} \frac{4x^3 - 6x^2 + 1}{2x^3 - 10x + 3}}_{\infty/\infty} = \underbrace{\lim_{x \to \infty} \frac{12x^2 - 12x}{6x^2 - 10}}_{\infty/\infty} = \underbrace{\lim_{x \to \infty} \frac{24x - 12}{12x}}_{\infty/\infty} = \lim_{x \to \infty} \frac{24}{12} = 2.$$

**b.** In this limit, the numerator approaches $-\infty$ and the denominator approaches $\infty$ as $x \to 0^+$. L'Hôpital's Rule gives us

➤ In the solution to Example 3b, notice that we simplify $(\csc^2(\sin x) \cos x)/(-\csc^2 x)$ before using L'Hôpital's Rule again. This step is important.

$$\lim_{x \to 0^+} \frac{1 - \cot(\sin x)}{\cot x} = \lim_{x \to 0^+} \frac{\csc^2(\sin x) \cos x}{-\csc^2 x} \quad \text{L'Hôpital's Rule}$$

$$= \lim_{x \to 0^+} \frac{-\sin^2 x \cos x}{\sin^2(\sin x)}. \quad \text{Simplify.}$$

The new limit has the form $0/0$, so we apply L'Hôpital's Rule again. However, before doing so, we use limit laws to ease the computations involved:

$$\lim_{x \to 0^+} \underbrace{\frac{-\sin^2 x \cos x}{\sin^2(\sin x)}}_{0/0} = -\left( \underbrace{\lim_{x \to 0^+} \frac{\sin x}{\sin(\sin x)}}_{\text{Apply l'Hôpital's Rule.}} \right)^2 \cdot \underbrace{\lim_{x \to 0^+} \cos x}_{\text{approaches 1}} \quad \text{Limit laws}$$

$$= -\left( \lim_{x \to 0^+} \frac{\cos x}{\cos(\sin x) \cos x} \right)^2 \cdot 1 \quad \text{L'Hôpital's Rule}$$

$$= -1 \cdot 1 = -1. \quad \text{Evaluate limits.}$$

Note that we are justified in using the limit laws because both limits in the product exist.

*Related Exercises 18, 39* ◄

### Related Indeterminate Forms: $0 \cdot \infty$ and $\infty - \infty$

We now consider limits of the form $\lim\limits_{x \to a} f(x)g(x)$, where $\lim\limits_{x \to a} f(x) = 0$ and $\lim\limits_{x \to a} g(x) = \pm\infty$; such limits are denoted $0 \cdot \infty$. *L'Hôpital's Rule cannot be directly applied to limits of this form.* Furthermore, it's risky to jump to conclusions about such limits. Suppose $f(x) = x$ and $g(x) = \dfrac{1}{x^2}$, in which case $\lim\limits_{x \to 0} f(x) = 0$, $\lim\limits_{x \to 0} g(x) = \infty$, and $\lim\limits_{x \to 0} f(x)g(x) = \lim\limits_{x \to 0} \dfrac{1}{x}$ does not exist. On the other hand, if $f(x) = x$ and $g(x) = \dfrac{1}{\sqrt{x}}$, we have $\lim\limits_{x \to 0^+} f(x) = 0$, $\lim\limits_{x \to 0^+} g(x) = \infty$, and $\lim\limits_{x \to 0^+} f(x)g(x) = \lim\limits_{x \to 0^+} \sqrt{x} = 0$. So a limit of the form $0 \cdot \infty$, in which the two functions compete with each other, may have any value or may not exist. The following example illustrates how this indeterminate form can be recast in the form $0/0$ or $\infty/\infty$.

**EXAMPLE 4**  **L'Hôpital's Rule for $0 \cdot \infty$**  Evaluate $\lim_{x \to \infty} x^2 \sin\left(\dfrac{1}{4x^2}\right)$.

**SOLUTION**  This limit has the form $0 \cdot \infty$. A common technique that converts this form to either $0/0$ or $\infty/\infty$ is to *divide by the reciprocal*. We rewrite the limit and apply l'Hôpital's Rule:

$$\underbrace{\lim_{x \to \infty} x^2 \sin\left(\frac{1}{4x^2}\right)}_{0 \cdot \infty \text{ form}} = \underbrace{\lim_{x \to \infty} \frac{\sin\left(\dfrac{1}{4x^2}\right)}{\dfrac{1}{x^2}}}_{\text{recast in } 0/0 \text{ form}} \qquad x^2 = \frac{1}{1/x^2}$$

$$= \lim_{x \to \infty} \frac{\cos\left(\dfrac{1}{4x^2}\right)\dfrac{1}{4}(-2x^{-3})}{-2x^{-3}} \qquad \text{L'Hôpital's Rule}$$

$$= \frac{1}{4} \lim_{x \to \infty} \cos\left(\frac{1}{4x^2}\right) \qquad \text{Simplify.}$$

$$= \frac{1}{4}. \qquad \frac{1}{4x^2} \to 0, \cos 0 = 1$$

*Related Exercise 41* ◄

**QUICK CHECK 3**  What is the form of the limit $\lim_{x \to \pi/2^-} (x - \pi/2)(\tan x)$? Write it in the form $0/0$. ◄

Limits of the form $\lim_{x \to a} (f(x) - g(x))$, where $\lim_{x \to a} f(x) = \infty$ and $\lim_{x \to a} g(x) = \infty$, are indeterminate forms that we denote $\infty - \infty$. L'Hôpital's Rule cannot be applied directly to an $\infty - \infty$ form. It must first be expressed in the form $0/0$ or $\infty/\infty$. With the $\infty - \infty$ form, it is easy to reach erroneous conclusions. For example, if $f(x) = 3x + 5$ and $g(x) = 3x$, then

$$\lim_{x \to \infty} ((3x + 5) - (3x)) = 5.$$

However, if $f(x) = 3x$ and $g(x) = 2x$, then

$$\lim_{x \to \infty} (3x - 2x) = \lim_{x \to \infty} x = \infty.$$

These examples show again why indeterminate forms are deceptive. Before proceeding, we introduce another useful technique.

Occasionally, it helps to convert a limit as $x \to \infty$ to a limit as $t \to 0^+$ (or vice versa) by a *change of variables*. To evaluate $\lim_{x \to \infty} f(x)$, we define $t = 1/x$ and note that as $x \to \infty, t \to 0^+$. Then

$$\lim_{x \to \infty} f(x) = \lim_{t \to 0^+} f\left(\frac{1}{t}\right).$$

This idea is illustrated in the next example.

**EXAMPLE 5**  **L'Hôpital's Rule for $\infty - \infty$**  Evaluate $\lim_{x \to \infty} \left(x - \sqrt{x^2 - 3x}\right)$.

**SOLUTION**  As $x \to \infty$, both terms in the difference $x - \sqrt{x^2 - 3x}$ approach $\infty$ and this limit has the form $\infty - \infty$. We first factor $x$ from the expression and form a quotient:

$$\lim_{x \to \infty} \left(x - \sqrt{x^2 - 3x}\right) = \lim_{x \to \infty} \left(x - \sqrt{x^2(1 - 3/x)}\right) \qquad \text{Factor } x^2 \text{ under square root.}$$

$$= \lim_{x \to \infty} x(1 - \sqrt{1 - 3/x}) \qquad x > 0, \text{ so } \sqrt{x^2} = x$$

$$= \lim_{x \to \infty} \frac{1 - \sqrt{1 - 3/x}}{1/x}. \qquad \begin{array}{l} \text{Write } 0 \cdot \infty \text{ form as } 0/0 \text{ form;} \\ x = \dfrac{1}{1/x}. \end{array}$$

This new limit has the form $0/0$, and l'Hôpital's Rule may be applied. One way to proceed is to use the change of variables $t = 1/x$:

$$\lim_{x \to \infty} \frac{1 - \sqrt{1 - 3/x}}{1/x} = \lim_{t \to 0^+} \frac{1 - \sqrt{1 - 3t}}{t} \qquad \text{Let } t = 1/x; \text{ replace } \lim_{x \to \infty} \text{ with } \lim_{t \to 0^+}.$$

$$= \lim_{t \to 0^+} \frac{\dfrac{3}{2\sqrt{1 - 3t}}}{1} \qquad \text{L'Hôpital's Rule}$$

$$= \frac{3}{2}. \qquad \text{Evaluate limit.}$$

*Related Exercises 49–50* ◄

## Pitfalls in Using l'Hôpital's Rule

We close with a list of common pitfalls to watch out for when using l'Hôpital's Rule.

**1.** L'Hôpital's Rule says $\displaystyle\lim_{x \to a} \frac{f(x)}{g(x)} = \lim_{x \to a} \frac{f'(x)}{g'(x)}$, not

$$\lim_{x \to a} \frac{f(x)}{g(x)} = \lim_{x \to a} \left( \frac{f(x)}{g(x)} \right)' \quad \text{or} \quad \lim_{x \to a} \frac{f(x)}{g(x)} = \lim_{x \to a} \left( \frac{1}{g(x)} \right)' f'(x).$$

In other words, you should evaluate $f'(x)$ and $g'(x)$, form their quotient, and then take the limit. Don't confuse l'Hôpital's Rule with the Quotient Rule.

**2.** Be sure that the given limit involves the indeterminate form $0/0$ or $\infty/\infty$ before applying l'Hôpital's Rule. For example, consider the following erroneous use of l'Hôpital's Rule:

$$\lim_{x \to 0} \frac{1 - \sin x}{\cos x} = \lim_{x \to 0} \frac{\cos x}{\sin x},$$

which does not exist. The original limit is not an indeterminate form in the first place. This limit should be evaluated by direct substitution:

$$\lim_{x \to 0} \frac{1 - \sin x}{\cos x} = \frac{1 - \sin 0}{1} = 1.$$

**3.** When using l'Hôpital's Rule repeatedly, be sure to simplify expressions as much as possible at each step and evaluate the limit as soon as the new limit is no longer an indeterminate form.

**4.** Repeated use of l'Hôpital's Rule occasionally leads to unending cycles, in which case other methods must be used. For example, limits of the form $\displaystyle\lim_{x \to \infty} \frac{\sqrt{ax + 1}}{\sqrt{bx + 1}}$, where $a$ and $b$ are positive real numbers, lead to such behavior (Exercise 67).

**5.** Be sure that the limit produced by l'Hôpital's Rule exists. Consider $\displaystyle\lim_{x \to \infty} \frac{3x + \cos x}{x}$, which has the form $\infty/\infty$. Applying l'Hôpital's Rule, we have

$$\lim_{x \to \infty} \frac{3x + \cos x}{x} = \lim_{x \to \infty} \frac{3 - \sin x}{1}.$$

It is tempting to conclude that because the limit on the right side does not exist, the original limit also does not exist. In fact, the original limit has a value of 3 (divide numerator and denominator by $x$). To reach a conclusion from l'Hôpital's Rule, the limit produced by l'Hôpital's Rule must exist (or be $\pm \infty$).

## SECTION 4.7 EXERCISES

### Getting Started

1. Explain with examples what is meant by the indeterminate form $0/0$.

2. Why are special methods such as l'Hôpital's Rule, rather than substitution, needed to evaluate indeterminate forms?

3. Explain the steps used to apply l'Hôpital's Rule to a limit of the form $0/0$.

4. Give examples of each of the following.
   a. A limit with the indeterminate form $0/0$ that equals 3.
   b. A limit with the indeterminate form $0/0$ that equals 4.

5. Give examples of each of the following.
   a. A limit with the indeterminate form $0 \cdot \infty$ that equals 1.
   b. A limit with the indeterminate form $0 \cdot \infty$ that equals 2.

6. Which of the following limits can be evaluated without l'Hôpital's Rule? Evaluate each limit.

   a. $\displaystyle\lim_{x \to 0} \frac{\sin x}{x^3 + 2x + 1}$
   b. $\displaystyle\lim_{x \to 0} \frac{\sin x}{x^3 + 2x}$

7. Explain how to convert a limit of the form $0 \cdot \infty$ to a limit of the form $0/0$ or $\infty/\infty$.

8. Give an example of a limit of the form $\infty/\infty$ as $x \to 0$.

9. The limit $\displaystyle\lim_{x \to 1^-} (x - 1) \tan \frac{\pi x}{2}$ has the indeterminate form $0 \cdot \infty$. Convert the limit to the indeterminate form $0/0$ and then evaluate the limit.

10. Evaluate $\displaystyle\lim_{x \to 2} \frac{x^3 - 3x^2 + 2x}{x - 2}$ using l'Hôpital's Rule and then check your work by evaluating the limit using an appropriate Chapter 2 method.

### Practice Exercises

**11–59. Limits** *Evaluate the following limits. Use l'Hôpital's Rule when it is convenient and applicable.*

11. $\displaystyle\lim_{x \to 2} \frac{x^2 - 2x}{x^2 - 6x + 8}$

12. $\displaystyle\lim_{x \to -1} \frac{x^4 + x^3 + 2x + 2}{x + 1}$

13. $\displaystyle\lim_{x \to 1} \frac{x^2 + 2x}{x + 3}$

14. $\displaystyle\lim_{x \to 1} \frac{\cos \pi x + 1}{\sqrt{x} - 1}$

15. $\displaystyle\lim_{x \to 1} \frac{4 \sin \pi x}{\cos \pi x + x}$

16. $\displaystyle\lim_{x \to \pi} \frac{4 \sin x}{x + \cos x - \pi + 1}$

17. $\displaystyle\lim_{x \to \infty} \frac{3x^4 - x^2}{6x^4 + 12}$

18. $\displaystyle\lim_{x \to \infty} \frac{4x^3 - 2x^2 + 6}{\pi x^3 + 4}$

19. $\displaystyle\lim_{x \to 1} \frac{x - 1}{\sqrt{x} - 1}$

20. $\displaystyle\lim_{x \to 0^+} \frac{x - 3\sqrt{x}}{x - \sqrt{x}}$

21. $\displaystyle\lim_{x \to 0} \frac{3 \sin 4x}{5x}$

22. $\displaystyle\lim_{x \to 2\pi} \frac{x \sin x + x^2 - 4\pi^2}{x - 2\pi}$

23. $\displaystyle\lim_{u \to \pi/4} \frac{\tan u - \cot u}{u - \pi/4}$

24. $\displaystyle\lim_{z \to 0} \frac{\tan 4z}{\tan 7z}$

25. $\displaystyle\lim_{x \to 0} \frac{1 - \cos 3x}{8x^2}$

26. $\displaystyle\lim_{x \to 0} \frac{\sin^2 3x}{x^2}$

27. $\displaystyle\lim_{x \to \pi} \frac{\cos x + 1}{(x - \pi)^2}$

28. $\displaystyle\lim_{x \to 0} \frac{\cos x - x^2 - 1}{3x^2 + x^4}$

29. $\displaystyle\lim_{x \to \pi/2^-} \frac{\tan x}{3/(2x - \pi)}$

30. $\displaystyle\lim_{x \to \pi^-} \frac{\csc x + x}{\tan \frac{x}{2}}$

31. $\displaystyle\lim_{x \to 0} \frac{\tan x - x}{x^3}$

32. $\displaystyle\lim_{x \to 0} \frac{\sin x - x}{7x^3}$

33. $\displaystyle\lim_{x \to -1} \frac{x^3 - x^2 - 5x - 3}{x^4 + 2x^3 - x^2 - 4x - 2}$

34. $\displaystyle\lim_{x \to 1} \frac{x^n - 1}{x - 1}$, $n$ is a positive integer

35. $\displaystyle\lim_{v \to 3} \frac{v - 1 - \sqrt{v^2 - 5}}{v - 3}$

36. $\displaystyle\lim_{y \to 2} \frac{y^2 + y - 6}{\sqrt{8 - y^2} - y}$

37. $\displaystyle\lim_{x \to 2} \frac{x^2 - 4x + 4}{\sin^2 \pi x}$

38. $\displaystyle\lim_{x \to 2} \frac{\sqrt[3]{3x + 2} - 2}{x - 2}$

39. $\displaystyle\lim_{x \to \pi/2^-} \frac{1 + 2 \sec x}{1 + \tan x}$

40. $\displaystyle\lim_{x \to 1^+} \left( \frac{1}{x - 1} - \frac{1}{\sqrt{x - 1}} \right)$

41. $\displaystyle\lim_{x \to 0} x \csc x$

42. $\displaystyle\lim_{x \to 1^-} (1 - x) \tan \frac{\pi x}{2}$

43. $\displaystyle\lim_{x \to 0} \csc 6x \sin 7x$

44. $\displaystyle\lim_{x \to \infty} \frac{3}{x} \csc \frac{5}{x}$

45. $\displaystyle\lim_{x \to \pi/2^-} \left( \frac{\pi}{2} - x \right) \sec x$

46. $\displaystyle\lim_{x \to 0^+} (\sin x) \sqrt{\frac{1 - x}{x}}$

47. $\displaystyle\lim_{x \to 0^+} \left( \cot x - \frac{1}{x} \right)$

48. $\displaystyle\lim_{\theta \to \pi/2^-} (\tan \theta - \sec \theta)$

49. $\displaystyle\lim_{x \to \infty} \left( x^2 - \sqrt{x^4 + 16x^2} \right)$

50. $\displaystyle\lim_{x \to \infty} \left( x - \sqrt{x^2 + 4x} \right)$

51. $\displaystyle\lim_{x \to \infty} \left( \sqrt{x - 2} - \sqrt{x - 4} \right)$

52. $\displaystyle\lim_{x \to \infty} \left( \sqrt{x^2 - 1} - \sqrt[3]{x^3 - 1} \right)$

53. $\displaystyle\lim_{n \to \infty} \frac{1 + 2 + \cdots + n}{n^2}$ $\left( Hint: \text{Use } 1 + 2 + \cdots + n = \frac{n(n + 1)}{2}. \right)$

54. $\displaystyle\lim_{\theta \to \infty} \frac{\sin 2\theta - \theta^3}{30^3}$

55. $\displaystyle\lim_{x \to 6} \frac{\sqrt[5]{5x + 2} - 2}{1/x - 1/6}$

56. $\displaystyle\lim_{x \to \pi/2} (\pi - 2x) \tan x$

57. $\displaystyle\lim_{x \to \infty} x^3 \left( \frac{1}{x} - \sin \frac{1}{x} \right)$

58. $\displaystyle\lim_{x \to 0} \left( \frac{1}{x} - \frac{\sin x}{x^2} \right)$

59. $\displaystyle\lim_{x \to \pi/2^-} \frac{\sec x - 1}{\tan x}$

60. **An optics limit** The theory of interference of coherent oscillators requires the limit $\displaystyle\lim_{\delta \to 2m\pi} \frac{\sin^2(N\delta/2)}{\sin^2(\delta/2)}$, where $N$ is a positive integer and $m$ is any integer. Show that the value of this limit is $N^2$.

61. **Explain why or why not** Determine whether the following statements are true and give an explanation or counterexample.

   a. By l'Hôpital's Rule, $\displaystyle\lim_{x \to 2} \frac{x - 2}{x^2 - 1} = \lim_{x \to 2} \frac{1}{2x} = \frac{1}{4}$.

   b. $\displaystyle\lim_{x \to 0} x \sin x = \lim_{x \to 0} f(x)g(x) = \lim_{x \to 0} f'(x) \lim_{x \to 0} g'(x) = \left( \lim_{x \to 0} 1 \right) \left( \lim_{x \to 0} \cos x \right) = 1$.

**c.** $\lim\limits_{x\to 2}\dfrac{x^3-2x^2+x-2}{x^2-1}=\lim\limits_{x\to 2}\dfrac{3x^2-4x+1}{2x}.$

**d.** $\lim\limits_{x\to 2}\dfrac{x^3-2x^2+x-2}{x^2-4}=\lim\limits_{x\to 2}\dfrac{3x^2-4x+1}{2x}.$

**62–63. Two methods** *Evaluate the following limits in two different ways: Use the methods of Chapter 2 and use l'Hôpital's Rule.*

**62.** $\lim\limits_{x\to\infty}\dfrac{2x^3-x^2+1}{5x^3+2x}$

**63.** $\lim\limits_{x\to 0}\dfrac{1-\cos^2 x}{1-\cos x}$

### Explorations and Challenges

**64–65. More limits** *Evaluate the following limits.*

**64.** $\lim\limits_{n\to\infty}\left(\cot\dfrac{1}{n}-n\right)$

**65.** $\lim\limits_{n\to\infty}\left(n\cot\dfrac{1}{n}-n^2\right)$

**66.** Let $a$ and $b$ be positive real numbers. Evaluate
$\lim\limits_{x\to\infty}\left(ax-\sqrt{a^2x^2-bx}\right)$ in terms of $a$ and $b$.

**67.** **L'Hôpital loops** Consider the limit $\lim\limits_{x\to\infty}\dfrac{\sqrt{ax+b}}{\sqrt{cx+d}}$, where $a$, $b$, $c$, and $d$ are positive real numbers. Show that l'Hôpital's Rule fails for this limit. Find the limit using another method.

**68.** Evaluate one of the limits l'Hôpital used in his own textbook in about 1700:
$$\lim\limits_{x\to a}\dfrac{\sqrt{2a^3x-x^4}-a\sqrt[3]{a^2x}}{a-\sqrt[4]{ax^3}},\text{ where }a\text{ is a real number.}$$

**69.** **A geometric limit** Let $f(\theta)$ be the area of the triangle $ABP$ (see figure), and let $g(\theta)$ be the area of the region between the chord $PB$ and the arc $PB$. Evaluate $\lim\limits_{\theta\to 0}\dfrac{g(\theta)}{f(\theta)}$.

**QUICK CHECK ANSWERS**

**1.** $g$ and $h$    **2.** $g$ and $h$    **3.** $0\cdot\infty$; $(x-\pi/2)/\cot x$ ◄

---

► Newton's method is attributed to Sir Isaac Newton, who devised the method in 1669. However, similar methods were known prior to Newton's time. A special case of Newton's method for approximating square roots is called the Babylonian method and was probably invented by Greek mathematicians.

# 4.8 Newton's Method

A common problem that arises in mathematics is finding the *roots*, or *zeros*, of a function. The roots of a function are the values of $x$ that satisfy the equation $f(x)=0$. Equivalently, they correspond to the $x$-intercepts of the graph of $f$. You have already seen an important example of a root-finding problem. To find the critical points of a function $f$, we must solve the equation $f'(x)=0$; that is, we find the roots of $f'$. Newton's method, which we discuss in this section, is one of the most effective methods for *approximating* the roots of a function.

## Why Approximate?

A little background about roots of functions explains why a method is needed to approximate roots. If you are given a linear function, such as $f(x)=2x-9$, you know how to use algebraic methods to solve $f(x)=0$ and find the single root $x=\frac{9}{2}$. Similarly, given the quadratic function $f(x)=x^2-6x-72$, you know how to factor or use the quadratic formula to discover that the roots are $x=12$ and $x=-6$. It turns out that formulas also exist for finding the roots of cubic (third-degree) and quartic (fourth-degree) polynomials. Methods such as factoring and algebra are called *analytical methods*; when they work, they give the roots of a function *exactly* in terms of arithmetic operations and radicals.

Here is an important fact: Apart from the functions we have listed—polynomials of degree four or less—analytical methods do not give the roots of most functions. To be sure, there are special cases in which analytical methods work. For example, you should verify that the single real root of $f(x)=\sin x-x$ is $x=0$ and that the two real roots of $f(x)=x^{10}-1$ are $x=1$ and $x=-1$. But in general, the roots of even relatively simple functions such as $f(x)=\cos x-x$ cannot be found exactly using analytical methods.

When analytical methods do not work, which is the majority of cases, we need another approach. That approach is to approximate roots using *numerical methods*, such as Newton's method.

### Deriving Newton's Method

Newton's method is most easily derived geometrically. Assume $r$ is a root of $f$ that we wish to approximate; this means that $f(r) = 0$. We also assume $f$ is differentiable on some interval containing $r$. Suppose $x_0$ is an initial approximation to $r$ that is generally obtained by some preliminary analysis. A better approximation to $r$ is often obtained by carrying out the following two steps:

• Draw a line tangent to the curve $y = f(x)$ at the point $(x_0, f(x_0))$.

• Find the point $(x_1, 0)$ at which the tangent line intersects the $x$-axis, and let $x_1$ be the new approximation to $r$.

For the curve shown in **Figure 4.73a**, $x_1$ is a better approximation to the root $r$ than $x_0$.

> Sequences are the subject of Chapter 10. An infinite, ordered list of numbers
> $$\{x_1, x_2, x_3, \ldots\}$$
> is a sequence, and if its values approach a number $r$, we say that the sequence *converges* to $r$. If a sequence fails to approach a single number, the sequence *diverges*.

(a)          (b)          (c)

**Figure 4.73**

**Figure 4.74**

To improve the approximation $x_1$, we repeat the two-step process, using $x_1$ to determine the next estimate $x_2$ (**Figure 4.73b**). Then we use $x_2$ to obtain $x_3$ (**Figure 4.73c**), and so forth. Continuing in this fashion, we obtain a *sequence* of approximations $\{x_1, x_2, x_3, \ldots\}$ that ideally get closer and closer, or *converge*, to the root $r$. Several steps of Newton's method and the convergence of the approximations to the root are shown in **Figure 4.74**.

All that remains is to find a formula that captures the process just described. Assume we have computed the $n$th approximation $x_n$ to the root $r$ and we want to obtain the next approximation $x_{n+1}$. We first draw the line tangent to the curve at the point $(x_n, f(x_n))$; its slope is $m = f'(x_n)$. Using the point-slope form of the equation of a line, an equation of the tangent line at the point $(x_n, f(x_n))$ is

$$y - f(x_n) = \underbrace{f'(x_n)}_{m}(x - x_n).$$

> Recall that the point-slope form of the equation of a line with slope $m$ passing through $(x_n, y_n)$ is
> $$y - y_n = m(x - x_n).$$

We find the point at which this line intersects the $x$-axis by setting $y = 0$ in the equation of the line and solving for $x$. This value of $x$ becomes the new approximation $x_{n+1}$:

$$\underbrace{0}_{\text{set } y \text{ to } 0} - f(x_n) = f'(x_n)(\underbrace{x}_{\substack{\text{becomes} \\ x_{n+1}}} - x_n).$$

Solving for $x$ and calling it $x_{n+1}$, we find that

$$\underbrace{x_{n+1}}_{\substack{\text{new} \\ \text{approximation}}} = \underbrace{x_n}_{\substack{\text{current} \\ \text{approximation}}} - \frac{f(x_n)}{f'(x_n)}, \quad \text{provided } f'(x_n) \neq 0.$$

**QUICK CHECK 1** Verify that setting $y = 0$ in the equation $y - f(x_n) = f'(x_n)(x - x_n)$ and solving for $x$ gives the formula for Newton's method. ◄

We have derived the general step of Newton's method for approximating roots of a function $f$. This step is repeated for $n = 0, 1, 2, \ldots$, until a termination condition is met (to be discussed).

---

> **PROCEDURE**   **Newton's Method for Approximating Roots of $f(x) = 0$**
>
> **1.** Choose an initial approximation $x_0$ as close to a root as possible.
>
> **2.** For $n = 0, 1, 2, \ldots$,
>
> $$x_{n+1} = x_n - \frac{f(x_n)}{f'(x_n)},$$
>
> provided $f'(x_n) \neq 0$.
>
> **3.** End the calculations when a termination condition is met.

➤ Newton's method is an example of a repetitive loop calculation called an *iteration*. The most efficient way to implement the method is with a calculator or computer. The method is also included in many software packages.

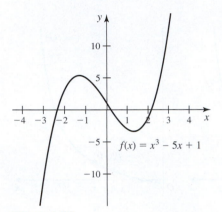

**Figure 4.75**

**EXAMPLE 1**   **Applying Newton's method**  Approximate the roots of $f(x) = x^3 - 5x + 1$ (**Figure 4.75**) using seven steps of Newton's method. Use $x_0 = -3$, $x_0 = 1$, and $x_0 = 4$ as initial approximations.

**SOLUTION**  Noting that $f'(x) = 3x^2 - 5$, Newton's method takes the form

$$x_{n+1} = x_n - \frac{\overbrace{x_n^3 - 5x_n + 1}^{f(x_n)}}{\underbrace{3x_n^2 - 5}_{f'(x_n)}} = \frac{2x_n^3 - 1}{3x_n^2 - 5},$$

where $n = 0, 1, 2, \ldots$, and $x_0$ is specified. With an initial approximation of $x_0 = -3$, the first approximation is

$$x_1 = \frac{2x_0^3 - 1}{3x_0^2 - 5} = \frac{2(-3)^3 - 1}{3(-3)^2 - 5} = -2.5.$$

The second approximation is

$$x_2 = \frac{2x_1^3 - 1}{3x_1^2 - 5} = \frac{2(-2.5)^3 - 1}{3(-2.5)^2 - 5} \approx -2.345455.$$

**Table 4.5**

| $n$ | $x_n$ | $x_n$ | $x_n$ |
|---|---|---|---|
| 0 | $-3$ | 1 | 4 |
| 1 | $-2.500000$ | $-0.500000$ | 2.953488 |
| 2 | $-2.345455$ | 0.294118 | 2.386813 |
| 3 | $-2.330203$ | 0.200215 | 2.166534 |
| 4 | $-2.330059$ | 0.201639 | 2.129453 |
| 5 | $-2.330059$ | 0.201640 | 2.128420 |
| 6 | $-2.330059$ | 0.201640 | 2.128419 |
| 7 | $-2.330059$ | 0.201640 | 2.128419 |

Continuing in this fashion, we generate the first seven approximations shown in Table 4.5. The approximations generated from the initial approximations $x_0 = 1$ and $x_0 = 4$ are also shown in the table.

Notice that with the initial approximation $x_0 = -3$ (second column), the resulting sequence of approximations settles on the value $-2.330059$ after four iterations, and then there are no further changes in these digits. A similar behavior is seen with the initial approximations $x_0 = 1$ and $x_0 = 4$. Based on this evidence, we conclude that $-2.330059$, $0.201640$, and $2.128419$ are approximations to the roots of $f$ with at least six digits (to the right of the decimal point) of accuracy.

A graph of $f$ (**Figure 4.76**) confirms that $f$ has three real roots and that the Newton approximations to the three roots are reasonable. The figure also shows the first three Newton approximations at each root.

*Related Exercises 13–14* ◄

➤ The numbers in Table 4.5 were computed with 16 decimal digits of precision. The results are displayed with 6 digits to the right of the decimal point.

**QUICK CHECK 2**  What happens if you apply Newton's method to the function $f(x) = x$? ◄

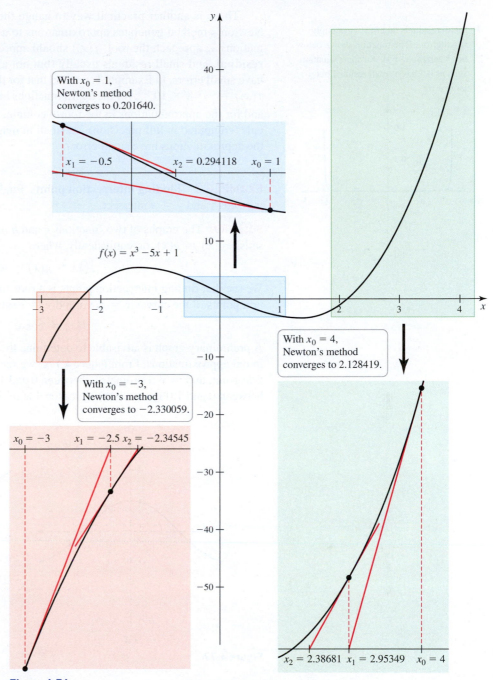

With $x_0 = 1$, Newton's method converges to 0.201640.

$x_1 = -0.5$    $x_2 = 0.294118$    $x_0 = 1$

$f(x) = x^3 - 5x + 1$

With $x_0 = 4$, Newton's method converges to 2.128419.

With $x_0 = -3$, Newton's method converges to $-2.330059$.

$x_0 = -3$    $x_1 = -2.5$    $x_2 = -2.34545$

$x_2 = 2.38681$    $x_1 = 2.95349$    $x_0 = 4$

**Figure 4.76**

## When Do You Stop?

Example 1 raises an important question and gives a practical answer: How many Newton approximations should you compute? Ideally, we would like to compute the **error** in $x_n$ as an approximation to the root $r$, which is the quantity $|x_n - r|$. Unfortunately, we don't know $r$ in practice; it is the quantity that we are trying to approximate. So we need a practical way to estimate the error.

In the second column of Table 4.5, we see that $x_4$ and $x_5$ agree in their seven digits, $-2.330059$. A general rule of thumb is that if two successive approximations agree to, say, seven digits, then those common digits are accurate (as an approximation to the root). So if you want $p$ digits of accuracy in your approximation, you should compute either until two successive approximations agree to $p$ digits or until some maximum number of iterations is exceeded (in which case Newton's method has failed to find an approximation of the root with the desired accuracy).

➤ If you write a program for Newton's method, it is a good idea to specify a maximum number of iterations as an escape clause in case the method does not converge.

> ➤ Small residuals do not always imply small errors: The function shown below has a zero at $x = 0$. An approximation such as 0.5 has a small residual but a large error.

There is another practical way to gauge the accuracy of approximations. Because Newton's method generates approximations to a root of $f$, it follows that as the approximations $x_n$ approach the root, $f(x_n)$ should approach zero. The quantity $f(x_n)$ is called a **residual**, and small residuals usually (but not always) suggest that the approximations have small errors. In Example 1, we find that for the approximations in the second column, $f(x_7) = -1.8 \times 10^{-15}$; for the approximations in the third column, $f(x_7) = 1.1 \times 10^{-16}$; and for the approximations in the fourth column, $f(x_7) = -1.8 \times 10^{-15}$. All these residuals (computed in full precision) are small in magnitude, giving us added confidence that the approximations have small errors.

**EXAMPLE 2    Finding intersection points**  Find the points at which the curves $y = \cos x$ and $y = x$ intersect.

**SOLUTION**  The graphs of two functions $g$ and $h$ intersect at points whose $x$-coordinates satisfy $g(x) = h(x)$, or, equivalently, where

$$f(x) = g(x) - h(x) = 0.$$

We see that finding intersection points is a root-finding problem. In this case, the intersection points of the curves $y = \cos x$ and $y = x$ satisfy

$$f(x) = \cos x - x = 0.$$

A preliminary graph is advisable to determine the number of intersection points and good initial approximations. From **Figure 4.77a**, we see that the two curves have one intersection point, and its $x$-coordinate is between 0 and 1. Equivalently, the function $f$ has a zero between 0 and 1 (**Figure 4.77b**). A reasonable initial approximation is $x_0 = 0.5$.

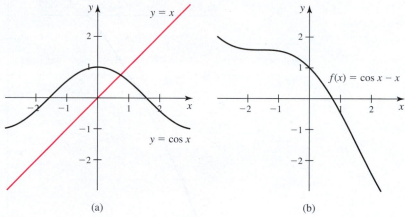

(a)                                                    (b)

**Figure 4.77**

**Table 4.6**

| $n$ | $x_n$ | Residual |
|---|---|---|
| 0 | 0.5 | 0.377583 |
| 1 | 0.755222 | $-0.0271033$ |
| 2 | 0.739142 | $-0.0000946154$ |
| 3 | 0.739085 | $-1.18098 \times 10^{-9}$ |
| 4 | 0.739085 | 0 |
| 5 | 0.739085 | 0 |

Newton's method takes the form

$$x_{n+1} = x_n - \overbrace{\frac{\overbrace{\cos x_n - x_n}^{f(x_n)}}{\underbrace{-\sin x_n - 1}_{f'(x_n)}}} = \frac{x_n \sin x_n + \cos x_n}{\sin x_n + 1}.$$

The results of Newton's method, using an initial approximation of $x_0 = 0.5$, are shown in Table 4.6.

We see that after four iterations, the approximations agree to six digits, so we take 0.739085 as the approximation to the root. Furthermore, the residuals, shown in the last column and computed with full precision, are essentially zero, which confirms the accuracy of the approximation. Therefore, the intersection point is approximately $(0.739085, 0.739085)$ (because the point lies on the line $y = x$).

*Related Exercises 27, 32* ◄

**Figure 4.78**

**Table 4.7**

| $n$ | $x_n$ | $x_n$ |
|---|---|---|
| 0 | $-1.000000$ | $1.000000$ |
| 1 | $-0.800000$ | $0.800000$ |
| 2 | $-0.706030$ | $0.668421$ |
| 3 | $-0.684424$ | $0.614669$ |
| 4 | $-0.683353$ | $0.607971$ |
| 5 | $-0.683350$ | $0.607886$ |
| 6 | $-0.683350$ | $0.607886$ |

> ➤ A more thorough analysis of the rate at which Newton's method converges and the ways in which it fails to converge is presented in a course in numerical analysis.
>
> Newton's method is widely used because in general, it has a remarkable rate of convergence; the number of digits of accuracy roughly doubles with each iteration.

**EXAMPLE 3** **Finding Local Extrema** Find the $x$-coordinate of the local maxima and local minima of the function $f(x) = x^5 - x^4 + x^2 - x$ on the interval $(-\infty, \infty)$.

**SOLUTION** A graph of the function provides some guidance. **Figure 4.78** shows that $f$ has a single local maximum on the interval $[-1, 0]$ and a single local minimum on the interval $[0, 1]$.

To locate the local extrema, we must find the critical points by solving

$$f'(x) = 5x^4 - 4x^3 + 2x - 1 = 0.$$

To this equation, we apply Newton's method. The results of the calculations, using initial approximations of $x_0 = -1$ and $x_0 = 1$, are shown in Table 4.7.

Newton's method finds the two critical points quickly, and they are consistent with the graph of $f$. We conclude that the local maximum occurs at $x \approx -0.683350$ and the local minimum occurs at $x \approx 0.607886$.

*Related Exercises 41–42* ◄

## Pitfalls of Newton's Method

Given a good initial approximation, Newton's method usually converges to a root. In addition, when the method converges, it usually does so quickly. However, when Newton's method fails, it does so in curious and spectacular ways. The formula for Newton's method suggests one way in which the method could encounter difficulties: The term $f'(x_n)$ appears in a denominator, so if at any step $f'(x_n) = 0$, then the method breaks down. Furthermore, if $f'(x_n)$ is close to zero at any step, then the method may converge slowly or may fail to converge. The following example shows three ways in which Newton's method may go awry.

**EXAMPLE 4** **Difficulties with Newton's method** Find the root of $f(x) = \dfrac{8x^2}{3x^2 + 1}$ using Newton's method with initial approximations of $x_0 = 1$, $x_0 = 0.15$, and $x_0 = 1.1$.

**SOLUTION** Notice that $f$ has the single root $x = 0$. So the point of the example is not to find the root, but to investigate the performance of Newton's method. Computing $f'$ and doing a few steps of algebra show that the formula for Newton's method is

$$x_{n+1} = x_n - \frac{f(x_n)}{f'(x_n)} = \frac{x_n}{2}\left(1 - 3x_n^2\right).$$

The results of five iterations of Newton's method are displayed in Table 4.8, and they tell three different stories.

**Table 4.8**

| $n$ | $x_n$ | $x_n$ | $x_n$ |
|---|---|---|---|
| 0 | 1 | 0.15 | 1.1 |
| 1 | $-1$ | 0.0699375 | $-1.4465$ |
| 2 | 1 | 0.0344556 | 3.81665 |
| 3 | $-1$ | 0.0171665 | $-81.4865$ |
| 4 | 1 | 0.00857564 | $8.11572 \times 10^5$ |
| 5 | $-1$ | 0.00428687 | $-8.01692 \times 10^{17}$ |

The approximations generated using $x_0 = 1$ (second column) get stuck in a cycle that alternates between $+1$ and $-1$. The geometry underlying this rare occurrence is illustrated in **Figure 4.79**.

**Figure 4.79**

The approximations generated using $x_0 = 0.15$ (third column) actually converge to the root 0, but they converge slowly (**Figure 4.80**). Notice that the error is reduced by a factor of approximately 2 with each step. Newton's method usually has a faster rate of error reduction. The slow convergence is due to the fact that both $f$ and $f'$ have zeros at 0. As mentioned earlier, if the approximations $x_n$ approach a zero of $f'$, the rate of convergence is often compromised.

The approximations generated using $x_0 = 1.1$ (fourth column) increase in magnitude quickly and do not converge to a finite value, even though this initial approximation seems reasonable. The geometry of this case is shown in **Figure 4.81**.

**Figure 4.80**

**Figure 4.81**

The three cases in this example illustrate several ways in which Newton's method may fail to converge at its usual rate: The approximations may cycle or wander, they may converge slowly, or they may diverge (often at a rapid rate).

*Related Exercises 50–51* ◀

# SECTION 4.8 EXERCISES

## Getting Started

1. Give a geometric explanation of Newton's method.

2. Explain how the iteration formula for Newton's method works.

3. A graph of $f$ and the lines tangent to $f$ at $x = 1, 2,$ and 3 are given. If $x_0 = 3$, find the values of $x_1, x_2,$ and $x_3$ that are obtained by applying Newton's method.

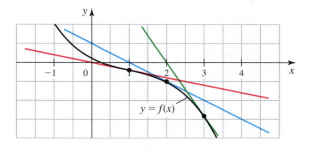

4. A graph of $f$ and the lines tangent to $f$ at $x = -3, 2,$ and 3 are given. If $x_0 = 3$, find the values of $x_1, x_2,$ and $x_3$ that are obtained by applying Newton's method.

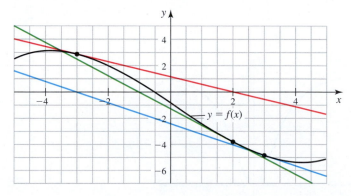

**5.** Let $f(x) = 2x^3 - 6x^2 + 4x$. Use Newton's method to find $x_1$ given that $x_0 = 1.4$. Use the graph of $f$ (see figure) and an appropriate tangent line to illustrate how $x_1$ is obtained from $x_0$.

**6.** The function $f(x) = \dfrac{4x}{x^2 + 4}$ is differentiable and has a local maximum at $x = 2$, where $f'(2) = 0$ (see figure).

  **a.** Use a graphical explanation to show that Newton's method applied to $f(x) = 0$ fails to produce a value $x_1$ for the initial approximation $x_0 = 2$.

  **b.** Use the formula for Newton's method to show that Newton's method fails to produce a value $x_1$ with the initial approximation $x_0 = 2$.

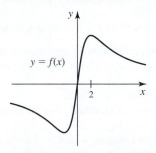

**7.** How do you decide when to terminate Newton's method?

**8.** Give the formula for Newton's method for the function $f(x) = x^2 - 5$.

**T 9–12.** *Write the formula for Newton's method and use the given initial approximation to compute the approximations $x_1$ and $x_2$.*

**9.** $f(x) = x^2 - 6; x_0 = 3$   **10.** $f(x) = x^2 - 2x - 3; x_0 = 2$

**11.** $f(x) = 2 - \tan x; x_0 = 1$   **12.** $f(x) = x^3 - 2; x_0 = 2$

### Practice Exercises

**T 13–19. Finding roots with Newton's method** *For the given function $f$ and initial approximation $x_0$, use Newton's method to approximate a root of $f$. Stop calculating approximations when two successive approximations agree to five digits to the right of the decimal point after rounding. Show your work by making a table similar to that in Example 1.*

**13.** $f(x) = x^2 - 10; x_0 = 3$

**14.** $f(x) = x^3 + x^2 + 1; x_0 = -1.5$

**15.** $f(x) = \sin x + x - 1; x_0 = 0.5$

**16.** $f(x) = x^2 - \cos x; x_0 = 1.5$

**17.** $f(x) = \tan x - 2x; x_0 = 1.2$

**18.** $f(x) = 3 \sin x - 2; x_0 = 1$

**19.** $f(x) = x^2 - \sqrt{x} - 1; x_0 = 2$

**T 20–26. Finding all roots** *Use Newton's method to find all the roots of the following functions. Use preliminary analysis and graphing to determine good initial approximations.*

**20.** $f(x) = \cos x - \dfrac{x}{7}$   **21.** $f(x) = \cos 2x - x^2 + 2x$

**22.** $f(x) = \dfrac{x}{6} - \sec x$ on $[0, 8]$

**23.** $f(x) = 2 - \sin x - \sqrt{x}$   **24.** $f(x) = \dfrac{x^5}{5} - \dfrac{x^3}{4} - \dfrac{1}{20}$

**25.** $f(x) = \cos 3x + \sin 5x + x$

**26.** $f(x) = x^2(x - 100) + 1$

**T 27–32. Finding intersection points** *Use Newton's method to approximate all the intersection points of the following pairs of curves. Some preliminary graphing or analysis may help in choosing good initial approximations.*

**27.** $y = \sin x$ and $y = \dfrac{x}{2}$

**28.** $y = -x^3$ and $y = x + 1$

**29.** $y = \dfrac{1}{x}$ and $y = 4 - x^2$

**30.** $y = x^3$ and $y = x^2 + 1$

**31.** $y = 4\sqrt{x}$ and $y = x^2 + 1$

**32.** $y = \sin x$ and $y = \sqrt{x} - 1$

**T 33.** **Retirement account** Assume you invest $10,000 at the end of each year for 30 years at an annual interest rate of $r$. The amount of money in your account after 30 years is
$$A = \frac{10{,}000\left((1 + r)^{30} - 1\right)}{r}.$$ Assume you want $1,000,000 in your account after 30 years.

  **a.** Show that the minimum value of $r$ required to meet your investment needs satisfies the equation
  $1{,}000{,}000r - 10{,}000(1 + r)^{30} + 10{,}000 = 0.$

  **b.** Apply Newton's method to solve the equation in part (a) to find the interest rate required to meet your investment goal.

**T 34.** **Investment problem** A one-time investment of $2500 is deposited in a 5-year savings account paying a fixed annual interest rate $r$, with monthly compounding. The amount of money in the account after 5 years is $A(r) = 2500\left(1 + \dfrac{r}{12}\right)^{60}$.

  **a.** Use Newton's method to find the value of $r$ if the goal is to have $3200 in the account after 5 years.

  **b.** Verify your answer to part (a) algebraically.

**T 35.** **Mortgage payments.** The monthly payment on a $200,000, 30-year (360-month) home loan is given by
$$m(r) = \frac{200{,}000(r/12)}{1 - (1 + r/12)^{-360}},$$
where $r$ is the annual interest rate. Use Newton's method to determine the interest rate $r$ that enables you to make monthly payments of $1200 per month.

**T 36.** **The sinc function** The sinc function, $\text{sinc}(x) = \dfrac{\sin x}{x}$ for $x \neq 0$, $\text{sinc}(0) = 1$, appears frequently in signal-processing applications.

    **a.** Graph the sinc function on $[-2\pi, 2\pi]$.

    **b.** Locate the first local minimum and the first local maximum of $\text{sinc}(x)$, for $x > 0$.

**T 37–40. Estimating roots** *The values of various roots can be approximated using Newton's method. For example, to approximate the value of $\sqrt[3]{10}$, we let $x = \sqrt[3]{10}$ and cube both sides of the equation to obtain $x^3 = 10$, or $x^3 - 10 = 0$. Therefore, $\sqrt[3]{10}$ is a root of $p(x) = x^3 - 10$, which we can approximate by applying Newton's method. Approximate each value of r by first finding a polynomial with integer coefficients that has a root r. Use an appropriate value of $x_0$ and stop calculating approximations when two successive approximations agree to five digits to the right of the decimal point after rounding.*

**37.** $r = 7^{1/4}$                   **38.** $r = 2^{1/3}$

**39.** $r = (-9)^{1/3}$          **40.** $r = (-67)^{1/5}$

**T 41–44. Newton's method and curve sketching** *Use Newton's method to find approximate answers to the following questions.*

**41.** Where is the first local minimum of $f(x) = \dfrac{\cos x}{x}$ on the interval $(0, \infty)$ located?

**42.** Where are all the local extrema of $f(x) = 3x^4 + 8x^3 + 12x^2 + 48x$ located?

**43.** Where are the inflection points of $f(x) = \dfrac{9}{5}x^5 - \dfrac{15}{2}x^4 + \dfrac{7}{3}x^3 + 30x^2 + 1$ located?

**44.** Where is the first local maximum of $f(x) = 2x + 5\cos x$ on the interval $(0, \infty)$ located?

**45.** **Explain why or why not** Determine whether the following statements are true and give an explanation or counterexample.

    **a.** Newton's method is an example of a numerical method for approximating the roots of a function.

    **b.** Newton's method gives better approximations to the roots of a quadratic equation than the quadratic formula.

    **c.** Newton's method always finds an approximate root of a function.

**T 46–49. Fixed points** *An important question about many functions concerns the existence and location of fixed points. A **fixed point** of f is a value of x that satisfies the equation $f(x) = x$; it corresponds to a point at which the graph of f intersects the line $y = x$. Find all the fixed points of the following functions. Use preliminary analysis and graphing to determine good initial approximations.*

**46.** $f(x) = x^3 - 3x^2 + x + 1$    **47.** $f(x) = \cos x$

**48.** $f(x) = \tan\dfrac{x}{2}$ on $(-\pi, \pi)$    **49.** $f(x) = 2x\cos x$ on $[0, 2]$

### Explorations and Challenges

**T 50–51. Pitfalls of Newton's method** *Let $f(x) = \dfrac{x}{1 + x^2}$, which has just one root, $r = 0$.*

**50.** Use the initial approximation $x_0 = 1/\sqrt{3}$ to complete the following steps.

    **a.** Use Newton's method to find the exact values of $x_1$ and $x_2$.

    **b.** State the values of $x_3, x_4, x_5, \ldots$ without performing any additional calculations.

    **c.** Use a graph of f to illustrate why Newton's method produces the values found in part (b).

    **d.** Why does Newton's method fail to approximate the root $r = 0$ if $x_0 = 1/\sqrt{3}$?

**51.** Use the initial approximation $x_0 = 2$ to complete the following steps.

    **a.** Use Newton's method to find the values of $x_1$, $x_2$, and $x_3$.

    **b.** Use a graph of f to illustrate why Newton's method produces the values $x_1$ and $x_2$ found in part (a).

    **c.** Why does Newton's method fail to approximate the root $r = 0$ if $x_0 = 2$?

**T 52.** **Approximating square roots** Let $a > 0$ be given and suppose we want to approximate $\sqrt{a}$ using Newton's method.

    **a.** Explain why the square root problem is equivalent to finding the positive root of $f(x) = x^2 - a$.

    **b.** Show that Newton's method applied to this function takes the form (sometimes called the Babylonian method)

$$x_{n+1} = \frac{1}{2}\left(x_n + \frac{a}{x_n}\right), \text{ for } n = 0, 1, 2, \ldots.$$

    **c.** How would you choose initial approximations to approximate $\sqrt{13}$ and $\sqrt{73}$?

    **d.** Approximate $\sqrt{13}$ and $\sqrt{73}$ with at least ten significant digits.

**T 53.** **Approximating reciprocals** To approximate the reciprocal of a number $a$ without using division, we can apply Newton's method to the function $f(x) = \dfrac{1}{x} - a$.

    **a.** Verify that Newton's method gives the formula $x_{n+1} = (2 - ax_n)x_n$.

    **b.** Apply Newton's method with $a = 7$ using a starting value of your choice. Compute an approximation with eight digits of accuracy. What number does Newton's method approximate in this case?

**T 54.** **Modified Newton's method** The function $f$ has a root of *multiplicity* 2 at $r$ if $f(r) = f'(r) = 0$ and $f''(r) \neq 0$. In this case, a slight modification of Newton's method, known as the *modified* (or *accelerated*) Newton's method, is given by the formula

$$x_{n+1} = x_n - \frac{2f(x_n)}{f'(x_n)}, \text{ for } n = 0, 1, 2, \ldots.$$

This modified form generally increases the rate of convergence.

    **a.** Verify that 1.5 is a root of multiplicity 2 of the function $f(x) = \sin \pi x + 1$.

    **b.** Apply Newton's method and the modified Newton's method using $x_0 = 1.25$ to find the value of $x_3$ in each case. Compare the accuracy of these values of $x_3$.

    **c.** Consider the function $f(x) = \dfrac{8x^2}{3x^2 + 1}$ given in Example 4. Use the modified Newton's method to find the value of $x_3$ using $x_0 = 0.15$. Compare this value to the value of $x_3$ found in Example 4 with $x_0 = 0.15$.

**T 55.** **An eigenvalue problem** A certain kind of differential equation (see Chapter 9) leads to the root-finding problem $\tan \pi\lambda = \lambda$, where the roots $\lambda$ are called **eigenvalues**. Find the first three positive eigenvalues of this problem.

**T 56. Fixed points of quadratics and quartics** Let $f(x) = ax(1 - x)$, where $a$ is a real number and $0 \leq x \leq 1$. Recall that the fixed point of a function is a value of $x$ such that $f(x) = x$ (Exercises 46–49).

    **a.** Without using a calculator, find the values of $a$, with $0 < a \leq 4$, such that $f$ has a fixed point. Give the fixed point in terms of $a$.

    **b.** Consider the polynomial $g(x) = f(f(x))$. Write $g$ in terms of $a$ and powers of $x$. What is its degree?

    **c.** Graph $g$ for $a = 2, 3$, and $4$.

    **d.** Find the number and location of the fixed points of $g$ for $a = 2, 3$, and $4$ on the interval $0 \leq x \leq 1$.

**T 57. Basins of attraction** Suppose $f$ has a real root $r$ and Newton's method is used to approximate $r$ with an initial approximation $x_0$. The **basin of attraction** of $r$ is the set of initial approximations that produce a sequence that converges to $r$. Points near $r$ are often in the basin of attraction of $r$—but not always. Sometimes an initial approximation $x_0$ may produce a sequence that doesn't converge, and sometimes an initial approximation $x_0$

may produce a sequence that converges to a distant root. Let $f(x) = (x + 2)(x + 1)(x - 3)$, which has roots $x = -2, -1$, and $3$. Use Newton's method with initial approximations on the interval $[-4, 4]$ and determine (approximately) the basin of each root.

**T 58.** The functions $f(x) = (x - 1)^2$ and $g(x) = x^2 - 1$ both have a root at $x = 1$. Apply Newton's method to both functions with an initial approximation $x_0 = 2$. Compare the rates at which the method converges in these cases, and give an explanation.

**QUICK CHECK ANSWERS**

**1.** $0 - f(x_n) = f'(x_n)(x - x_n) \implies -\dfrac{f(x_n)}{f'(x_n)} = x - x_n \implies$

$x = x_n - \dfrac{f(x_n)}{f'(x_n)}$    **2.** Newton's method will find the root

$x = 0$ exactly in one step. ◄

# 4.9 Antiderivatives

The goal of differentiation is to find the derivative $f'$ of a given function $f$. The reverse process, called *antidifferentiation*, is equally important: Given a function $f$, we look for an *antiderivative* function $F$ whose derivative is $f$; that is, a function $F$ such that $F' = f$.

---

> **DEFINITION Antiderivative**
>
> A function $F$ is an **antiderivative** of $f$ on an interval $I$ provided $F'(x) = f(x)$, for all $x$ in $I$.

---

In this section, we revisit derivative formulas developed in previous chapters to discover corresponding antiderivative formulas.

## Thinking Backward

Consider the derivative formula $\dfrac{d}{dx}(x) = 1$. It implies that an antiderivative of $f(x) = 1$ is $F(x) = x$ because $F'(x) = f(x)$. Using the same logic, we can write

$$\frac{d}{dx}(x^2) = 2x \quad \implies \quad \text{an antiderivative of } f(x) = 2x \text{ is } F(x) = x^2 \text{ and}$$

$$\frac{d}{dx}(\sin x) = \cos x \quad \implies \quad \text{an antiderivative of } f(x) = \cos x \text{ is } F(x) = \sin x.$$

**QUICK CHECK 1** Verify by differentiation that $x^4$ is an antiderivative of $4x^3$. ◄

Each of these proposed antiderivative formulas is easily checked by showing that $F' = f$.

    An immediate question arises: Does a function have more than one antiderivative? To answer this question, let's focus on $f(x) = 1$ and the antiderivative $F(x) = x$. Because the derivative of a constant $C$ is zero, we see that $F(x) = x + C$ is also an antiderivative of $f(x) = 1$, which is easy to check:

$$F'(x) = \frac{d}{dx}(x + C) = 1 = f(x).$$

Therefore, $f(x) = 1$ actually has an infinite number of antiderivatives. For the same reason, any function of the form $F(x) = x^2 + C$ is an antiderivative of $f(x) = 2x$, and any function of the form $F(x) = \sin x + C$ is an antiderivative of $f(x) = \cos x$, where $C$ is an arbitrary constant.

We might ask whether there are still *more* antiderivatives of a given function. The following theorem provides the answer.

---

**THEOREM 4.14    The Family of Antiderivatives**
Let $F$ be any antiderivative of $f$ on an interval $I$. Then *all* the antiderivatives of $f$ on $I$ have the form $F + C$, where $C$ is an arbitrary constant.

---

**Proof:** Suppose $F$ and $G$ are antiderivatives of $f$ on an interval $I$. Then $F' = f$ and $G' = f$, which implies that $F' = G'$ on $I$. Theorem 4.6 states that functions with equal derivatives differ by a constant. Therefore, $G = F + C$, and all antiderivatives of $f$ have the form $F + C$, where $C$ is an arbitrary constant.    ◄

Theorem 4.14 says that while there are infinitely many antiderivatives of a function, they are all of one family, namely, those functions of the form $F + C$. Because the antiderivatives of a particular function differ by a constant, the antiderivatives are vertical translations of one another (**Figure 4.82**).

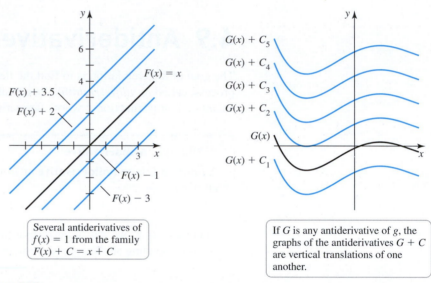

Several antiderivatives of $f(x) = 1$ from the family $F(x) + C = x + C$

If $G$ is any antiderivative of $g$, the graphs of the antiderivatives $G + C$ are vertical translations of one another.

**Figure 4.82**

**EXAMPLE 1    Finding antiderivatives**  Use what you know about derivatives to find all antiderivatives of the following functions.

**a.** $f(x) = 3x^2$    **b.** $f(x) = -\dfrac{9}{x^{10}}$    **c.** $f(t) = \sin t$

**SOLUTION**

**a.** Note that $\dfrac{d}{dx}(x^3) = 3x^2$. Therefore, an antiderivative of $f(x) = 3x^2$ is $x^3$. By Theorem 4.14, the complete family of antiderivatives is $F(x) = x^3 + C$, where $C$ is an arbitrary constant.

**b.** Because $\dfrac{d}{dx}(x^{-9}) = -9x^{-10} = -\dfrac{9}{x^{10}}$, all antiderivatives of $f$ are of the form
$F(x) = x^{-9} + C$, where $C$ is an arbitrary constant.

**c.** Recall that $\dfrac{d}{dt}(\cos t) = -\sin t$. We seek a function whose derivative is $\sin t$, not $-\sin t$. Observing that $\dfrac{d}{dt}(-\cos t) = \sin t$, it follows that the antiderivatives of $\sin t$ are $F(t) = -\cos t + C$, where $C$ is an arbitrary constant.

*Related Exercises 12–13, 17* ◄

**QUICK CHECK 2** Find the family of antiderivatives for each of $f(x) = \dfrac{1}{2\sqrt{x}}$, $g(x) = 4x^3$, and $h(x) = \sec^2 x$. ◄

## Indefinite Integrals

The notation $\dfrac{d}{dx}(f(x))$ means *take the derivative of $f(x)$* with respect to $x$. We need analogous notation for antiderivatives. For historical reasons that become apparent in the next chapter, the notation that means *find the antiderivatives of $f$* is the **indefinite integral** $\int f(x)\,dx$. Every time an indefinite integral sign $\int$ appears, it is followed by a function called the **integrand**, which in turn is followed by the differential $dx$. For now, $dx$ simply means that $x$ is the independent variable, or the **variable of integration**. The notation $\int f(x)\,dx$ represents *all* the antiderivatives of $f$. When the integrand is a function of a variable different from $x$—say, $g(t)$—we write $\int g(t)\,dt$ to represent the antiderivatives of $g$.

Using this new notation, the three results of Example 1 are written

$$\int 3x^2\,dx = x^3 + C, \quad \int\left(-\frac{9}{x^{10}}\right)dx = x^{-9} + C, \quad \text{and} \quad \int \sin t\,dt = -\cos t + C,$$

where $C$ is an arbitrary constant called a **constant of integration**. The derivative formulas presented earlier in the text may be written in terms of indefinite integrals. We begin with the Power Rule.

> ➤ Notice that if $p = -1$ in Theorem 4.15, then $F(x)$ is undefined. The antiderivative of $f(x) = x^{-1}$ is discussed in Chapter 7. The case $p = 0$ says that $\int 1\,dx = x + C$.

---

**THEOREM 4.15   Power Rule for Indefinite Integrals**

$$\int x^p\,dx = \frac{x^{p+1}}{p+1} + C,$$

where $p \neq -1$ is a real number and $C$ is an arbitrary constant.

---

**Proof:** The theorem says that the antiderivatives of $f(x) = x^p$ have the form $F(x) = \dfrac{x^{p+1}}{p+1} + C$. Differentiating $F$, we verify that $F'(x) = f(x)$, provided $p \neq -1$:

> ➤ Any indefinite integral calculation can be checked by differentiation: The derivative of the alleged indefinite integral must equal the integrand.

$$F'(x) = \frac{d}{dx}\left(\frac{x^{p+1}}{p+1} + C\right)$$

$$= \frac{d}{dx}\left(\frac{x^{p+1}}{p+1}\right) + \underbrace{\frac{d}{dx}(C)}_{0}$$

$$= \frac{(p+1)x^{(p+1)-1}}{p+1} + 0 = x^p. \qquad \blacktriangleleft$$

Theorems 3.4 and 3.5 (Section 3.3) state the Constant Multiple and Sum Rules for derivatives. Here are the corresponding antiderivative rules, which are proved by differentiation.

---

**THEOREM 4.16   Constant Multiple and Sum Rules**

**Constant Multiple Rule:** $\displaystyle\int cf(x)\,dx = c\int f(x)\,dx$, for real numbers $c$

**Sum Rule:** $\displaystyle\int (f(x) + g(x))\,dx = \int f(x)\,dx + \int g(x)\,dx$

---

The following example shows how Theorems 4.15 and 4.16 are used.

**EXAMPLE 2   Indefinite integrals**   Determine the following indefinite integrals.

**a.** $\displaystyle\int (3x^5 + 2 - 5\sqrt{x})\,dx$   **b.** $\displaystyle\int\left(\frac{4x^{19} - 5x^{-8}}{x^2}\right)dx$   **c.** $\displaystyle\int (z^2 + 1)(2z - 5)\,dz$

> $\int dx$ means $\int 1\, dx$, which is the indefinite integral of the constant function $f(x) = 1$, so $\int dx = x + C$.

> Each indefinite integral in Example 2a produces an arbitrary constant, all of which may be combined in one arbitrary constant called $C$.

**SOLUTION**

**a.** $\displaystyle \int (3x^5 + 2 - 5\sqrt{x})\, dx = \int 3x^5\, dx + \int 2\, dx - \int 5x^{1/2}\, dx$    Sum Rule

$\displaystyle = 3\int x^5\, dx + 2\int dx - 5\int x^{1/2}\, dx$    Constant Multiple Rule

$\displaystyle = 3 \cdot \frac{x^6}{6} + 2 \cdot x - 5 \cdot \frac{x^{3/2}}{3/2} + C$    Power Rule

$\displaystyle = \frac{x^6}{2} + 2x - \frac{10}{3}x^{3/2} + C$    Simplify.

**b.** $\displaystyle \int \left( \frac{4x^{19} - 5x^{-8}}{x^2} \right) dx = \int (4x^{17} - 5x^{-10})\, dx$    Simplify the integrand.

$\displaystyle = 4\int x^{17}\, dx - 5\int x^{-10}\, dx$    Sum and Constant Multiple Rules

$\displaystyle = 4 \cdot \frac{x^{18}}{18} - 5 \cdot \frac{x^{-9}}{-9} + C$    Power Rule

$\displaystyle = \frac{2x^{18}}{9} + \frac{5x^{-9}}{9} + C$    Simplify.

> Examples 2b and 2c show that in general, the indefinite integral of a product or quotient is not the product or quotient of indefinite integrals.

**c.** $\displaystyle \int (z^2 + 1)(2z - 5)\, dz = \int (2z^3 - 5z^2 + 2z - 5)\, dz$    Expand integrand.

$\displaystyle = \frac{1}{2}z^4 - \frac{5}{3}z^3 + z^2 - 5z + C$    Integrate each term.

All these results should be checked by differentiation.

*Related Exercises 24, 25, 31, 35* ◄

## Indefinite Integrals of Trigonometric Functions

We used a familiar derivative formula in Example 1c to find the antiderivative of $\sin x$. Our goal in this section is to write the other derivative results for trigonometric functions as indefinite integrals.

**EXAMPLE 3**    **Indefinite integrals of trigonometric functions**    Evaluate the following indefinite integrals.

**a.** $\displaystyle \int \sec^2 x\, dx$        **b.** $\displaystyle \int (2x + 3\cos x)\, dx$        **c.** $\displaystyle \int \frac{\sin x}{\cos^2 x}\, dx$

> Remember the words that go with antiderivatives and indefinite integrals. The statement $\dfrac{d}{dx}(\tan x) = \sec^2 x$ says that $\tan x$ can be differentiated to get $\sec^2 x$. Therefore,
> $$\int \sec^2 x\, dx = \tan x + C.$$

**SOLUTION**

**a.** The derivative result $\dfrac{d}{dx}(\tan x) = \sec^2 x$ is reversed to produce the indefinite integral

$$\int \sec^2 x\, dx = \tan x + C.$$

**b.** We first split the integral into two integrals using Theorem 4.16:

$$\int (2x + 3\cos x)\, dx = 2\int x\, dx + 3\int \cos x\, dx.$$    Sum and Constant Multiple Rules

The first of these new integrals is handled by the Power Rule, and the second integral is evaluated by reversing the derivative result $\dfrac{d}{dx}(\sin x) = \cos x$:

$$2\int x\, dx + 3\int \cos x\, dx = 2 \cdot \frac{x^2}{2} + 3\sin x + C$$    Power Rule; $\dfrac{d}{dx}(\sin x) = \cos x \Rightarrow$ $\int \cos x\, dx = \sin x + C$

$$= x^2 + 3\sin x + C.$$    Simplify.

**c.** When we rewrite the integrand, a familiar derivative formula emerges:

$$\int \frac{\sin x}{\cos^2 x} \, dx = \int \underbrace{\frac{1}{\cos x}}_{\sec x} \cdot \underbrace{\frac{\sin x}{\cos x}}_{\tan x} \, dx$$

$$= \int \sec x \tan x \, dx \qquad \text{Rewrite the integrand.}$$

$$= \sec x + C.$$

$$\frac{d}{dx}(\sec x) = \sec x \tan x \Rightarrow$$
$$\int \sec x \tan x \, dx = \sec x + C$$

*Related Exercises 40, 41, 47* ◄

The ideas illustrated in Example 3 are used to obtain the integrals in Table 4.9, where we assume $C$ is an arbitrary constant. Example 4 provides additional integrals involving trigonometric functions.

**Table 4.9  Indefinite Integrals of Trigonometric Functions**

**1.** $\dfrac{d}{dx}(\sin x) = \cos x \quad \Rightarrow \quad \displaystyle\int \cos x \, dx = \sin x + C$

**2.** $\dfrac{d}{dx}(\cos x) = -\sin x \quad \Rightarrow \quad \displaystyle\int \sin x \, dx = -\cos x + C$

**3.** $\dfrac{d}{dx}(\tan x) = \sec^2 x \quad \Rightarrow \quad \displaystyle\int \sec^2 x \, dx = \tan x + C$

**4.** $\dfrac{d}{dx}(\cot x) = -\csc^2 x \quad \Rightarrow \quad \displaystyle\int \csc^2 x \, dx = -\cot x + C$

**5.** $\dfrac{d}{dx}(\sec x) = \sec x \tan x \quad \Rightarrow \quad \displaystyle\int \sec x \tan x \, dx = \sec x + C$

**6.** $\dfrac{d}{dx}(\csc x) = -\csc x \cot x \quad \Rightarrow \quad \displaystyle\int \csc x \cot x \, dx = -\csc x + C$

**QUICK CHECK 3** Use differentiation to verify result 6 in Table 4.9:

$$\int \csc x \cot x \, dx = -\csc x + C. \blacktriangleleft$$

**EXAMPLE 4  Indefinite integrals involving trigonometric functions** Determine the following indefinite integrals.

**a.** $\displaystyle\int \left( \frac{2}{\pi} \sin x - 2 \csc^2 x \right) dx$     **b.** $\displaystyle\int \frac{4 \cos x + \sin^2 x}{\sin^2 x} \, dx$

**SOLUTION**

**a.** Splitting up the integral (Theorem 4.16) and then using Table 4.9, we have

$$\int \left( \frac{2}{\pi} \sin x - 2 \csc^2 x \right) dx = \frac{2}{\pi} \int \sin x \, dx - 2 \int \csc^2 x \, dx \qquad \begin{array}{l}\text{Sum and Constant} \\ \text{Multiple Rules}\end{array}$$

$$= \frac{2}{\pi}(-\cos x) - 2(-\cot x) + C \qquad \begin{array}{l}\text{Results (2) and (4),} \\ \text{Table 4.16}\end{array}$$

$$= 2 \cot x - \frac{2}{\pi} \cos x + C. \qquad \text{Simplify.}$$

**b.** Again, we split up the integral and then rewrite the first term in the integrand so that it matches result (6) in Table 4.9:

$$\int \frac{4 \cos x + \sin^2 x}{\sin^2 x} \, dx = \int \left( \underbrace{\frac{4}{\sin x} \cdot \frac{\cos x}{\sin x}}_{4 \csc x \quad \cot x} + \underbrace{\frac{\sin^2 x}{\sin^2 x}}_{1} \right) dx$$

$$= 4 \int \csc x \cot x \, dx + \int dx \qquad \begin{array}{l}\text{Rewrite the integrand;} \\ \text{Theorem 4.16.}\end{array}$$

$$= -4 \csc x + x + C. \qquad \begin{array}{l}\text{Result (6), Table 4.9;} \\ \displaystyle\int dx = x + C\end{array}$$

*Related Exercises 45, 50* ◄

## Introduction to Differential Equations

An equation involving an unknown function and its derivatives is called a **differential equation**. Here is an example to get us started.

Suppose you know that the derivative of a function $f$ satisfies the equation

$$f'(x) = 2x + 10.$$

**QUICK CHECK 4** Explain why an antiderivative of $f'$ is $f$. ◄

To find a function $f$ that satisfies this equation, we note that the solutions are antiderivatives of $2x + 10$, which are $x^2 + 10x + C$, where $C$ is an arbitrary constant. So we have found an infinite number of solutions, all of the form $f(x) = x^2 + 10x + C$.

Now consider a more general differential equation of the form $f'(x) = g(x)$, where $g$ is given and $f$ is unknown. The solution $f$ consists of the antiderivatives of $g$, which involve an arbitrary constant. In most practical cases, the differential equation is accompanied by an **initial condition** that allows us to determine the arbitrary constant. Therefore, we consider problems of the form

$$f'(x) = g(x), \quad \text{where } g \text{ is given, and} \qquad \text{\color{blue}Differential equation}$$
$$f(a) = b, \qquad \text{where } a \text{ and } b \text{ are given.} \quad \text{\color{blue}Initial condition}$$

A differential equation coupled with an initial condition is called an **initial value problem**.

**EXAMPLE 5    Initial value problems**  Solve the following initial value problems.

**a.** $f'(x) = x^2 - 2x; f(1) = \dfrac{1}{3}$    **b.** $f'(t) = 1 - \dfrac{1}{t^2}; f(1) = 0$

**c.** Solve the differential equation in part (b) using the initial condition $f(-1) = -2$.

**SOLUTION**

**a.** The solution $f$ is an antiderivative of $x^2 - 2x$. Therefore,

$$f(x) = \frac{x^3}{3} - x^2 + C,$$

where $C$ is an arbitrary constant. We have determined that the solution is a member of a family of functions, all of which differ by a constant. This family of functions, called the **general solution**, is shown in **Figure 4.83**, where we see curves for various choices of $C$.

Using the initial condition $f(1) = \frac{1}{3}$, we must find the particular function in the general solution whose graph passes through the point $\left(1, \frac{1}{3}\right)$. Imposing the condition $f(1) = \frac{1}{3}$, we reason as follows:

$$f(x) = \frac{x^3}{3} - x^2 + C \qquad \text{\color{blue}General solution}$$

$$f(1) = \frac{1}{3} - 1 + C \qquad \text{\color{blue}Substitute } x = 1.$$

$$\frac{1}{3} = \frac{1}{3} - 1 + C \qquad \text{\color{blue}}f(1) = \tfrac{1}{3}$$

$$C = 1. \qquad \text{\color{blue}Solve for } C.$$

Therefore, the solution to the initial value problem is

$$f(x) = \frac{x^3}{3} - x^2 + 1,$$

which is just one of the curves in the family shown in Figure 4.83.

**b.** By the Power Rule (Theorem 4.15), an antiderivative of $1 - 1/t^2$ is $t + 1/t$. Therefore, the family of functions

$$f(t) = t + \frac{1}{t} + C$$

represents the general solution of the differential equation. **Figure 4.84** shows several members of this family for various values of $C$; notice that for each value of $C$,

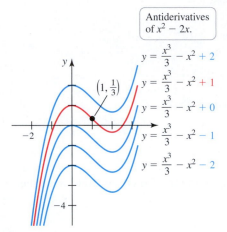

Antiderivatives of $x^2 - 2x$.

$y = \dfrac{x^3}{3} - x^2 + 2$

$y = \dfrac{x^3}{3} - x^2 + 1$

$y = \dfrac{x^3}{3} - x^2 + 0$

$y = \dfrac{x^3}{3} - x^2 - 1$

$y = \dfrac{x^3}{3} - x^2 - 2$

**Figure 4.83**

➤ It is advisable to check that the solution satisfies the original problem: $f'(x) = x^2 - 2x$ and $f(1) = \frac{1}{3} - 1 + 1 = \frac{1}{3}$.

the graph of $f$ has two branches. The initial condition $f(1) = 0$ determines the value of $C$:

$$f(t) = t + \frac{1}{t} + C \qquad \text{General solution}$$

$$f(1) = 1 + 1 + C \qquad \text{Substitute } t = 1.$$

$$0 = 2 + C \qquad\qquad f(1) = 0$$

$$C = -2 \qquad\qquad\quad \text{Solve for } C.$$

> ➤ Because $f(t) = t + \dfrac{1}{t} - 2$ is undefined at $t = 0$, it cannot be considered a solution to Example 5b over its entire domain. As discussed in Chapter 9, a solution to an initial value problem is a differentiable function defined on a single interval.

Substituting $C = -2$ into the general solution gives $f(t) = t + \dfrac{1}{t} - 2$. However, only the right branch of the curve $y = t + \dfrac{1}{t} - 2$ passes through the point $(1, 0)$ specified by the initial condition (Figure 4.84, quadrant I). Restricting the domain of $f$ to reflect this fact, we find that the solution to the initial value problem is

$$f(t) = t + \frac{1}{t} - 2, \ t > 0.$$

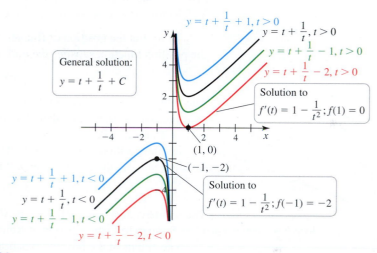

**Figure 4.84**

c. Using the general solution from part (b) and the initial condition $f(-1) = -2$ leads to $f(-1) = -1 - 1 + C = -2$, which implies that $C = 0$. Substituting this value into the general solution yields $f(t) = t + \dfrac{1}{t}$, but just as we saw in part (b), an additional restriction is needed because only the left branch of the curve $y = t + \dfrac{1}{t}$ passes through the point $(-1, -2)$ (Figure 4.84, quadrant III). The solution to this initial value problem is

$$f(t) = t + \frac{1}{t}, \ t < 0.$$

*Related Exercises 62, 68* ◀

**QUICK CHECK 5** Position is an antiderivative of velocity. But there are infinitely many antiderivatives that differ by a constant. Explain how two objects can have the same velocity function but two different position functions. ◀

> ➤ The convention with motion problems is to assume that motion begins at $t = 0$. This means that initial conditions are specified at $t = 0$.

## Motion Problems Revisited

Antiderivatives allow us to revisit the topic of one-dimensional motion introduced in Section 3.6. Suppose the position of an object that moves along a line relative to an origin is $s(t)$, where $t \geq 0$ measures elapsed time. The velocity of the object is $v(t) = s'(t)$, which may now be read in terms of antiderivatives: *The position function is an antiderivative of the velocity.* If we are given the velocity function of an object and its position at a particular time, we can determine its position at all future times by solving an initial value problem.

We also know that the acceleration $a(t)$ of an object moving in one dimension is the rate of change of the velocity, which means $a(t) = v'(t)$. In antiderivative terms, this says that the velocity is an antiderivative of the acceleration. Therefore, if we are given the acceleration of an object and its velocity at a particular time, we can determine its velocity at all times. These ideas lie at the heart of modeling the motion of objects.

---

**Initial Value Problems for Velocity and Position**

Suppose an object moves along a line with a (known) velocity $v(t)$, for $t \geq 0$. Then its position is found by solving the initial value problem

$$s'(t) = v(t), \quad s(0) = s_0, \text{ where } s_0 \text{ is the (known) initial position.}$$

If the (known) acceleration of the object $a(t)$ is given, then its velocity is found by solving the initial value problem

$$v'(t) = a(t), \quad v(0) = v_0, \text{ where } v_0 \text{ is the (known) initial velocity.}$$

---

**EXAMPLE 6   A race** Runner A begins at the point $s(0) = 0$ and runs on a straight and level road with velocity $v(t) = 2t$. Runner B begins with a head start at the point $S(0) = 8$ and runs with velocity $V(t) = 2$. Find the positions of the runners for $t \geq 0$ and determine who is ahead at $t = 6$ time units.

**SOLUTION** Let the position of Runner A be $s(t)$, with an initial position $s(0) = 0$. Then the position function satisfies the initial value problem

$$s'(t) = 2t, \quad s(0) = 0.$$

The solution is an antiderivative of $s'(t) = 2t$, which has the form $s(t) = t^2 + C$. Substituting $s(0) = 0$, we find that $C = 0$. Therefore, the position of Runner A is given by $s(t) = t^2$, for $t \geq 0$.

Let the position of Runner B be $S(t)$, with an initial position $S(0) = 8$. This position function satisfies the initial value problem

$$S'(t) = 2, \quad S(0) = 8.$$

The antiderivatives of $S'(t) = 2$ are $S(t) = 2t + C$. Substituting $S(0) = 8$ implies that $C = 8$. Therefore, the position of Runner B is given by $S(t) = 2t + 8$, for $t \geq 0$.

The graphs of the position functions are shown in **Figure 4.85**. Runner B begins with a head start but is overtaken when $s(t) = S(t)$, or when $t^2 = 2t + 8$. The solutions of this equation are $t = 4$ and $t = -2$. Only the positive solution is relevant because the race takes place for $t \geq 0$, so Runner A overtakes Runner B at $t = 4$, when $s = S = 16$. When $t = 6$, Runner A has the lead.

*Related Exercises 87–88* ◀

**Figure 4.85**

**EXAMPLE 7   Motion with gravity** Neglecting air resistance, the motion of an object moving vertically near Earth's surface is determined by the acceleration due to gravity, which is approximately 9.8 m/s². Suppose a stone is thrown vertically upward at $t = 0$ with a velocity of 40 m/s from the edge of a cliff that is 100 m above a river.

**a.** Find the velocity $v(t)$ of the object, for $t \geq 0$.

**b.** Find the position $s(t)$ of the object, for $t \geq 0$.

**c.** Find the maximum height of the object above the river.

**d.** With what speed does the object strike the river?

**SOLUTION** We establish a coordinate system in which the positive $s$-axis points vertically upward with $s = 0$ corresponding to the river (**Figure 4.86**). Let $s(t)$ be the position of the stone measured relative to the river, for $t \geq 0$. The initial velocity of the stone is $v(0) = 40$ m/s and the initial position of the stone is $s(0) = 100$ m.

**a.** The acceleration due to gravity points in the *negative* $s$-direction. Therefore, the initial value problem governing the motion of the object is

$$\text{acceleration} = v'(t) = -9.8, v(0) = 40.$$

**Figure 4.86**

▶ The acceleration due to gravity at Earth's surface is approximately $g = 9.8 \text{ m/s}^2$, or $g = 32 \text{ ft/s}^2$. It varies (even at sea level) from about 9.8640 at the poles to 9.7982 at the equator. The equation $v'(t) = -g$ is an instance of Newton's Second Law of Motion and assumes no other forces (such as air resistance) are present.

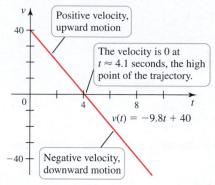

**Figure 4.87**

The antiderivatives of $-9.8$ are $v(t) = -9.8t + C$. The initial condition $v(0) = 40$ gives $C = 40$. Therefore, the velocity of the stone is

$$v(t) = -9.8t + 40.$$

As shown in **Figure 4.87**, the velocity decreases from its initial value $v(0) = 40$ until it reaches zero at the high point of the trajectory. This point is reached when

$$v(t) = -9.8t + 40 = 0$$

or when $t \approx 4.1$ s. For $t > 4.1$, the velocity is negative and increases in magnitude as the stone falls to Earth.

**b.** Knowing the velocity function of the stone, we can determine its position. The position function satisfies the initial value problem

$$v(t) = s'(t) = -9.8t + 40, \quad s(0) = 100.$$

The antiderivatives of $-9.8t + 40$ are

$$s(t) = -4.9t^2 + 40t + C.$$

The initial condition $s(0) = 100$ implies $C = 100$, so the position function of the stone is

$$s(t) = -4.9t^2 + 40t + 100,$$

as shown in **Figure 4.88**. The parabolic graph of the position function is not the actual trajectory of the stone; the stone moves vertically along the $s$-axis.

**c.** The position function of the stone increases for $0 < t < 4.1$. At $t \approx 4.1$, the stone reaches a high point of $s(4.1) \approx 182$ m.

**d.** For $t > 4.1$, the position function decreases, and the stone strikes the river when $s(t) = 0$. The roots of this equation are $t \approx 10.2$ and $t \approx -2.0$. Only the first root is relevant because the motion starts at $t = 0$. Therefore, the stone strikes the ground at $t \approx 10.2$ s. Its speed (in m/s) at this instant is $|v(10.2)| \approx |-60| = 60$.

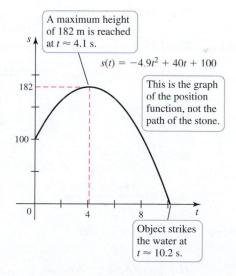

**Figure 4.88**

*Related Exercises 89–90* ◀

## SECTION 4.9 EXERCISES

### Getting Started

**1.** Fill in the blanks with either of the words *the derivative* or *an antiderivative*: If $F'(x) = f(x)$, then $f$ is _____ of $F$, and $F$ is _____ of $f$.

**2.** Describe the set of antiderivatives of $f(x) = 0$.

**3.** Describe the set of antiderivatives of $f(x) = 1$.

**4.** Why do two different antiderivatives of a function differ by a constant?

**5.** Give the antiderivatives of $x^p$. For what values of $p$ does your answer apply?

**6.** Give the antiderivatives of $a \csc^2 x$, where $a$ is a constant.

**7.** Give the antiderivatives of $1/x^2$.

**8.** Evaluate $\int a \cos x \, dx$ and $\int a \sin x \, dx$, where $a$ is a constant.

**9.** If $F(x) = x^2 - 3x + C$ and $F(-1) = 4$, what is the value of $C$?

**10.** For a given function $f$, explain the steps used to solve the initial value problem $F'(t) = f(t)$, $F(0) = 10$.

## Practice Exercises

**11–22. Finding antiderivatives** *Find all the antiderivatives of the following functions. Check your work by taking derivatives.*

**11.** $f(x) = 5x^4$

**12.** $g(x) = 11x^{10}$

**13.** $f(x) = 2 \sin x + 1$

**14.** $g(x) = -4 \cos x - x$

**15.** $p(x) = 3 \sec^2 x$

**16.** $q(s) = \csc^2 s$

**17.** $f(y) = -\dfrac{2}{y^3}$

**18.** $h(z) = -6z^{-7}$

**19.** $f(x) = \dfrac{1}{\sqrt{x}}$

**20.** $h(y) = \sqrt[3]{y}$

**21.** $f(x) = \dfrac{7}{2}x^{5/2}$

**22.** $f(t) = \pi$

**23–54. Indefinite integrals** *Determine the following indefinite integrals. Check your work by differentiation.*

**23.** $\int (3x^5 - 5x^9) \, dx$

**24.** $\int (3u^{-2} - 4u^2 + 1) \, du$

**25.** $\int \left(4\sqrt{x} - \dfrac{4}{\sqrt{x}}\right) dx$

**26.** $\int \left(\dfrac{5}{t^2} + 4t^2\right) dt$

**27.** $\int (5s + 3)^2 \, ds$

**28.** $\int 5m(12m^3 - 10m) \, dm$

**29.** $\int (3x^{1/3} + 4x^{-1/3} + 6) \, dx$

**30.** $\int 5\sqrt[4]{x} \, dx$

**31.** $\int (3x + 1)(4 - x) \, dx$

**32.** $\int (4z^{1/3} - z^{-1/3}) \, dz$

**33.** $\int \left(\dfrac{3}{x^4} + 2 - \dfrac{3}{x^2}\right) dx$

**34.** $\int \sqrt[5]{r^2} \, dr$

**35.** $\int \dfrac{4x^4 - 6x^2}{x} \, dx$

**36.** $\int \dfrac{12t^8 - t}{t^{3/2}} \, dt$

**37.** $\int \dfrac{x^2 - 36}{x - 6} \, dx$

**38.** $\int \dfrac{y^3 - 9y^2 + 20y}{y - 4} \, dy$

**39.** $\int (\csc^2 \theta + 2\theta^2 - 3\theta) \, d\theta$

**40.** $\int (\csc^2 \theta + 1) \, d\theta$

**41.** $\int \dfrac{2 + 3 \cos y}{\sin^2 y} \, dy$

**42.** $\int \sin t(4 \csc t - \cot t) \, dt$

**43.** $\int (\sec^2 x - 1) \, dx$

**44.** $\int \dfrac{\sec^3 v - \sec^2 v}{\sec v - 1} \, dv$

**45.** $\int (\sec^2 \theta + \sec \theta \tan \theta) \, d\theta$

**46.** $\int \dfrac{\sin \theta - 1}{\cos^2 \theta} \, d\theta$

**47.** $\int (3t^2 + 2 \csc^2 t) \, dt$

**48.** $\int \csc x(\cot x - \csc x) \, dx$

**49.** $\int \sec \theta(\tan \theta + \sec \theta + \cos \theta) \, d\theta$

**50.** $\int \dfrac{\csc^3 x + 1}{\csc x} \, dx$

**51.** $\int \dfrac{\tan x + \sec x}{\sec x} \, dx$

**52.** $\int \left(\sqrt[3]{x^2} + \sqrt{x^3}\right) dx$

**53.** $\int \sqrt{x} \left(2x^6 - 4\sqrt[3]{x}\right) dx$

**54.** $\int \dfrac{16 \cos^2 w - 81 \sin^2 w}{4 \cos w - 9 \sin w} \, dw$

**55–60. Particular antiderivatives** *For the following functions $f$, find the antiderivative $F$ that satisfies the given condition.*

**55.** $f(x) = x^5 - 2x^2 + 1$; $F(0) = 1$

**56.** $f(x) = 4\sqrt{x} + 6$; $F(1) = 8$

**57.** $f(x) = 8x^3 + \sin x$; $F(0) = 2$

**58.** $f(t) = \sec^2 t$; $F(\pi/4) = 1$, $-\pi/2 < t < \pi/2$

**59.** $f(v) = \sec v \tan v$; $F(0) = 2$, $-\pi/2 < v < \pi/2$

**60.** $f(\theta) = 2 \sin \theta - 4 \cos \theta$; $F\left(\dfrac{\pi}{4}\right) = 2$

**61–68. Solving initial value problems** *Find the solution of the following initial value problems.*

**61.** $f'(x) = 2x - 3$; $f(0) = 4$

**62.** $g'(x) = 7x^6 - 4x^3 + 12$; $g(1) = 24$

**63.** $g'(x) = 7x\left(x^6 - \dfrac{1}{7}\right)$; $g(1) = 2$

**64.** $h'(t) = 1 + 6 \sin t$; $h\left(\dfrac{\pi}{3}\right) = -3$

**65.** $f'(u) = 4(\cos u - \sin u)$; $f\left(\dfrac{\pi}{2}\right) = 0$

**66.** $f'(x) = \sin x + \cos x + 1$; $f(\pi) = 3$

**67.** $y'(\theta) = \dfrac{\sqrt{2} \cos^3 \theta + 1}{\cos^2 \theta}$; $y\left(\dfrac{\pi}{4}\right) = 3$, $-\pi/2 < \theta < \pi/2$

**68.** $v'(x) = 4x^{1/3} + 2x^{-1/3}$; $v(8) = 40$, $x > 0$

**T** **69–72. Graphing general solutions** *Graph several functions that satisfy each of the following differential equations. Then find and graph the particular function that satisfies the given initial condition.*

**69.** $f'(x) = 2x - 5$; $f(0) = 4$

**70.** $f'(x) = 3x^2 - 1$; $f(1) = 2$

**71.** $f'(x) = 3x + \sin x$; $f(0) = 3$

**72.** $f'(x) = \cos x - \sin x + 2$; $f(0) = 1$

**73–78. Velocity to position** *Given the following velocity functions of an object moving along a line, find the position function with the given initial position.*

**73.** $v(t) = 2t + 4$; $s(0) = 0$    **74.** $v(t) = 5$; $s(0) = 4$

**75.** $v(t) = 2\sqrt{t}$; $s(0) = 1$    **76.** $v(t) = 2 \cos t$; $s(0) = 0$

**77.** $v(t) = 6t^2 + 4t - 10$; $s(0) = 0$

**78.** $v(t) = 4t + \sin t$; $s(0) = 0$

**79–84. Acceleration to position** *Given the following acceleration functions of an object moving along a line, find the position function with the given initial velocity and position.*

**79.** $a(t) = -32$; $v(0) = 20$, $s(0) = 0$

**80.** $a(t) = 4$; $v(0) = -3$, $s(0) = 2$

**81.** $a(t) = 0.2t$; $v(0) = 0$, $s(0) = 1$

**82.** $a(t) = 2 \cos t; v(0) = 1, s(0) = 0$

**83.** $a(t) = 2 + 3 \sin t; v(0) = 1, s(0) = 10$

**84.** $a(t) = -9.8; v(0) = 5, s(0) = 0$

**85.** A car starting at rest accelerates at $16 \text{ ft/s}^2$ for 5 seconds on a straight road. How far does it travel during this time?

**86.** A car is moving at 60 mi/hr (88 ft/s) on a straight road when the driver steps on the brake pedal and begins decelerating at a constant rate of $10 \text{ ft/s}^2$ for 3 seconds. How far did the car go during this 3-second interval?

**87–88. Races** *The velocity function and initial position of Runners A and B are given. Analyze the race that results by graphing the position functions of the runners and finding the time and positions (if any) at which they first pass each other.*

**87.** A: $v(t) = \sin t; s(0) = 0$    B: $V(t) = \cos t; S(0) = 0$

**88.** A: $v(t) = 2t; s(0) = 0$    B: $V(t) = 3\sqrt{t}; S(0) = 0$

**⊤ 89–92. Motion with gravity** *Consider the following descriptions of the vertical motion of an object subject only to the acceleration due to gravity. Begin with the acceleration equation $a(t) = v'(t) = -g$, where $g = 9.8 \text{ m/s}^2$.*

*a. Find the velocity of the object for all relevant times.*
*b. Find the position of the object for all relevant times.*
*c. Find the time when the object reaches its highest point. What is the height?*
*d. Find the time when the object strikes the ground.*

**89.** A softball is popped up vertically (from the ground) with a velocity of 30 m/s.

**90.** A stone is thrown vertically upward with a velocity of 30 m/s from the edge of a cliff 200 m above a river.

**91.** A payload is released at an elevation of 400 m from a hot-air balloon that is rising at a rate of 10 m/s.

**92.** A payload is dropped at an elevation of 400 m from a hot-air balloon that is descending at a rate of 10 m/s.

**93. Explain why or why not** Determine whether the following statements are true and give an explanation or counterexample.

a. $F(x) = x^3 - 4x + 100$ and $G(x) = x^3 - 4x - 100$ are antiderivatives of the same function.
b. If $F'(x) = f(x)$, then $f$ is an antiderivative of $F$.
c. If $F'(x) = f(x)$, then $\int f(x)\, dx = F(x) + C$.
d. $f(x) = x^3 + 3$ and $g(x) = x^3 - 4$ are derivatives of the same function.
e. If $F'(x) = G'(x)$, then $F(x) = G(x)$.

**Explorations and Challenges**

**94–97. Functions from higher derivatives** *Find the function F that satisfies the following differential equations and initial conditions.*

**94.** $F''(x) = 1; F'(0) = 3, F(0) = 4$

**95.** $F''(x) = \cos x; F'(0) = 3, F(\pi) = 4$

**96.** $F'''(x) = 4x; F''(0) = 0, F'(0) = 1, F(0) = 3$

**97.** $F'''(x) = 672x^5 + 24x; F''(0) = 0, F'(0) = 2, F(0) = 1$

**98. Mass on a spring** A mass oscillates up and down on the end of a spring. Find its position $s$ relative to the equilibrium position if its acceleration is $a(t) = 2 \sin t$ and its initial velocity and position are $v(0) = 3$ and $s(0) = 0$, respectively.

**99. Flow rate** A large tank is filled with water when an outflow valve is opened at $t = 0$. Water flows out at a rate, in gal/min, given by $Q'(t) = 0.1(100 - t^2)$, for $0 \le t \le 10$.

a. Find the amount of water $Q(t)$ that has flowed out of the tank after $t$ minutes, given the initial condition $Q(0) = 0$.
b. Graph the flow function $Q$, for $0 \le t \le 10$.
c. How much water flows out of the tank in 10 min?

**100. General head start problem** Suppose object A is located at $s = 0$ at time $t = 0$ and starts moving along the $s$-axis with a velocity given by $v(t) = 2at$, where $a > 0$. Object B is located at $s = c > 0$ at $t = 0$ and starts moving along the $s$-axis with a constant velocity given by $V(t) = b > 0$. Show that A always overtakes B at time

$$t = \frac{b + \sqrt{b^2 + 4ac}}{2a}.$$

**101–104. Verifying indefinite integrals** *Verify the following indefinite integrals by differentiation. These integrals are derived in later chapters.*

**101.** $\int \dfrac{\cos \sqrt{x}}{\sqrt{x}}\, dx = 2 \sin \sqrt{x} + C$

**102.** $\int \dfrac{x}{\sqrt{x^2 + 1}}\, dx = \sqrt{x^2 + 1} + C$

**103.** $\int x^2 \cos x^3\, dx = \dfrac{1}{3} \sin x^3 + C$

**104.** $\int \dfrac{x}{(x^2 - 1)^2}\, dx = -\dfrac{1}{2(x^2 - 1)} + C$

**QUICK CHECK ANSWERS**

**1.** $d/dx(x^4) = 4x^3$ **2.** $\sqrt{x} + C$, $x^4 + C$, $\tan x + C$
**3.** $d/dx(-\csc x + C) = \csc x \cot x$ **4.** One function that can be differentiated to get $f'$ is $f$. Therefore, $f$ is an antiderivative of $f'$. **5.** The two position functions involve two different initial positions; they differ by a constant. ◄

## CHAPTER 4    REVIEW EXERCISES

**1. Explain why or why not** Determine whether the following statements are true and give an explanation or counterexample.

a. If $f'(c) = 0$, then $f$ has a local minimum or maximum at $c$.
b. If $f''(c) = 0$, then $f$ has an inflection point at $c$.
c. $F(x) = x^2 + 10$ and $G(x) = x^2 - 100$ are antiderivatives of the same function.
d. Between two local minima of a function continuous on $(-\infty, \infty)$, there must be a local maximum.
e. The linear approximation to $f(x) = \sin x$ at $x = 0$ is $L(x) = x$.
f. If $\lim\limits_{x \to \infty} f(x) = \infty$ and $\lim\limits_{x \to \infty} g(x) = \infty$, then
$$\lim\limits_{x \to \infty} (f(x) - g(x)) = 0.$$

**2. Locating extrema** Consider the graph of a function $f$ on the interval $[-3, 3]$.

    **a.** Give the approximate coordinates of the local maxima and minima of $f$.

    **b.** Give the approximate coordinates of the absolute maximum and minimum values of $f$ (if they exist).

    **c.** Give the approximate coordinates of the inflection point(s) of $f$.

    **d.** Give the approximate coordinates of the zero(s) of $f$.

    **e.** On what intervals (approximately) is $f$ concave up?

    **f.** On what intervals (approximately) is $f$ concave down?

**3–4. Designer functions** *Sketch the graph of a function continuous on the given interval that satisfies the following conditions.*

**3.** $f$ is continuous on the interval $[-4, 4]$; $f'(x) = 0$ for $x = -2, 0,$ and $3$; $f$ has an absolute minimum at $x = 3$; $f$ has a local minimum at $x = -2$; $f$ has a local maximum at $x = 0$; $f$ has an absolute maximum at $x = -4$.

**4.** $f$ is continuous on $(-\infty, \infty)$; $f'(x) < 0$ and $f''(x) < 0$ on $(-\infty, 0)$; $f'(x) > 0$ and $f''(x) > 0$ on $(0, \infty)$.

**5.** Use the graphs of $f'$ and $f''$ to complete the following steps.

    **a.** Find the critical points of $f$ and determine where $f$ is increasing and where it is decreasing.

    **b.** Determine the locations of the inflection points of $f$ and the intervals on which $f$ is concave up or concave down.

    **c.** Determine where $f$ has local maxima and minima.

    **d.** Plot a possible graph of $f$.

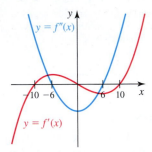

**6–12. Critical points** *Find the critical points of the following functions on the given intervals. Identify the absolute maximum and absolute minimum values (if they exist).*

**6.** $f(x) = x^3 - 6x^2$ on $[-1, 5]$

**7.** $f(x) = 3x^4 - 6x^2 + 9$ on $[-2, 2]$

**8.** $g(x) = x^4 - 50x^2$ on $[-1, 5]$

**9.** $f(x) = 2x^3 - 3x^2 - 36x + 12$ on $(-\infty, \infty)$

**10.** $f(x) = \sin 2x + 3$ on $[-\pi, \pi]$

**11.** $g(x) = -\dfrac{1}{2}\sin x + \dfrac{1}{2}\sin x \cos x$ on $[0, 2\pi]$

**12.** $f(x) = 4x^{1/2} - x^{5/2}$ on $[0, 4]$

**T 13. Absolute values** Consider the function $f(x) = |x - 2| + |x + 3|$ on $[-4, 4]$. Graph $f$, identify the critical points, and give the coordinates of the local and absolute extreme values.

**T 14–16.** *Use $f'$ and $f''$ to complete parts (a) and (b).*

    **a.** Find the intervals on which $f$ is increasing and the intervals on which it is decreasing.

    **b.** Find the intervals on which $f$ is concave up and the intervals on which it is concave down.

**14.** $f(x) = 9(x - 1)^{5/3}(x + 3)$   **15.** $f(x) = \dfrac{x^9}{9} + 3x^5 - 16x$

**16.** $f(x) = x\sqrt{x + 9}$

**17. Inflection points** Does $f(x) = 2x^5 - 10x^4 + 20x^3 + x + 1$ have any inflection points? If so, identify them.

**T 18.** Does $f(x) = \dfrac{x^6}{2} + \dfrac{5x^4}{4} - 15x^2$ have any inflection points? If so, identify them.

**19.** Identify the critical points and the inflection points of $f(x) = (x - a)(x + a)^3$, for $a > 0$.

**20–28. Curve sketching** *Use the guidelines given in Section 4.4 to make a complete graph of the following functions on their domains or on the given interval. Use a graphing utility to check your work.*

**20.** $f(x) = x(x + 4)^3$       **21.** $f(x) = \dfrac{x^4}{2} - 3x^2 + 4x + 1$

**22.** $f(x) = \dfrac{3x}{x^2 + 2}$

**23.** $f(x) = 4\cos(\pi(x - 1))$ on $[0, 2]$

**24.** $f(x) = \dfrac{x^2 + 3}{x - 1}$       **25.** $f(x) = \dfrac{10x^2}{x^2 + 3}$

**26.** $f(x) = x\sqrt[3]{x + 4}$ (*Hint:* In addition to using the graphing guidelines outlined in Section 4.4, describe the line tangent to the graph of $f$ at $x = -4$ by examining $\lim\limits_{x \to -4} f'(x)$.)

**27.** $f(x) = \sqrt[3]{x} - \sqrt{x} + 2$     **28.** $f(x) = \dfrac{x^2 + 12}{x - 2}$

**29. Optimal popcorn box** A small popcorn box is created from a $12'' \times 12''$ sheet of paperboard by first cutting out four shaded rectangles, each of length $x$ and width $x/2$ (see figure). The remaining paperboard is folded along the solid lines to form a box. What dimensions of the box maximize the volume of the box?

**30. Minimizing time** Hannah is standing on the edge of an island, 1 mile from a straight shoreline (see figure). She wants to return to her beach house that is 2 miles from the point $P$ on shore that is closest to where she is standing. Given that she can swim at 2 mi/hr and jog at 6 mi/hr, find the point at which she should come ashore to minimize the total time of her trip. If she starts swimming at noon, can she make it home before 1 P.M.?

**T 31. Minimizing sound intensity** Two sound speakers are 100 m apart and one speaker is three times as loud as the other speaker. At what point on a line segment between the speakers is the sound intensity the weakest? (*Hint:* Sound intensity is directly proportional to the sound level and inversely proportional to the square of the distance from the sound source.)

**32. Hockey problem** A hockey player skates on a line that is perpendicular to the goal line. If the line on which he is skating is 10 feet to the left of the center of the hockey goal and if the hockey goal is 6 ft wide, then where (what value of $x$) should he shoot the puck to maximize the angle $\theta$ on goal (see figure)? (*Hint:* Use the identity $\tan(\alpha - \beta) = \dfrac{\tan \alpha - \tan \beta}{1 + \tan \alpha \tan \beta}$.)

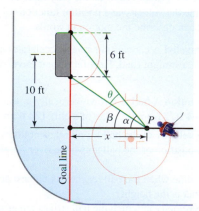

**33. Optimization** A right triangle has legs of length $h$ and $r$ and a hypotenuse of length 4 (see figure). It is revolved about the leg of length $h$ to sweep out a right circular cone. What values of $h$ and $r$ maximize the volume of the cone? (Volume of a cone $= \dfrac{1}{3}\pi r^2 h$.)

**T 34. Rectangles beneath a curve** A rectangle is constructed with one side on the positive $x$-axis, one side on the positive $y$-axis, and the vertex opposite the origin on the curve $y = \cos x$, for $0 < x < \pi/2$. Approximate the dimensions of the rectangle that maximize the area of the rectangle. What is the area?

**35. Maximum printable area** A rectangular page in a text (with width $x$ and length $y$) has an area of 98 in², top and bottom margins

set at 1 in, and left and right margins set at $1/2$ in. The printable area of the page is the rectangle that lies within the margins. What are the dimensions of the page that maximize the printable area?

**36. Nearest point** What point on the graph of $f(x) = 5/2 - x^2$ is closest to the origin? (*Hint:* You can minimize the square of the distance.)

**37. Maximum area** A line segment of length 10 joins the points $(0, p)$ and $(q, 0)$ to form a triangle in the first quadrant. Find the values of $p$ and $q$ that maximize the area of the triangle.

**38. Minimum painting surface** A metal cistern in the shape of a right circular cylinder with volume $V = 50$ m³ needs to be painted each year to reduce corrosion. The paint is applied only to surfaces exposed to the elements (the outside cylinder wall and the circular top). Find the dimensions $r$ and $h$ of the cylinder that minimize the area of the painted surfaces.

**39–40. Linear approximation**

a. *Find the linear approximation to $f$ at the given point $a$.*
b. *Use your answer from part (a) to estimate the given function value. Does your approximation underestimate or overestimate the exact function value?*

**39.** $f(x) = x^{2/3}; a = 27; f(29)$    **40.** $f(x) = \sqrt{x}; a = 4; f(4.1)$

**41–42. Estimations with linear approximation** *Use linear approximation to estimate the following quantities. Choose a value of a to produce a small error.*

**41.** $\dfrac{1}{4.2^2}$              **42.** $\sqrt{9.05}$

**43. Change in elevation** The elevation $h$ (in feet above the ground) of a stone dropped from a height of 1000 ft is modeled by the equation $h(t) = 1000 - 16t^2$, where $t$ is measured in seconds and air resistance is neglected. Approximate the change in elevation over the interval $5 \le t \le 5.7$ (recall that $\Delta h \approx h'(a)\,\Delta t$).

**44. Growth rate of bamboo** Bamboo belongs to the grass family and is one of the fastest growing plants in the world.

a. A bamboo shoot was 500 cm tall at 10:00 A.M. and 515 cm tall at 3:00 P.M. Compute the average growth rate of the bamboo shoot in cm/hr over the period of time from 10:00 A.M. to 3:00 P.M.
b. Based on the Mean Value Theorem, what can you conclude about the instantaneous growth rate of bamboo measured in *millimeters per second* between 10:00 A.M. and 3:00 P.M.?

**45. Mean Value Theorem** For the function $f(x) = \sqrt{10x}$ and the interval $[a, b] = [0, 10]$, use the graph to make a conjecture about the value of $c$ for which $\dfrac{f(b) - f(a)}{b - a} = f'(c)$. Then verify your conjectured value by solving the equation $\dfrac{f(b) - f(a)}{b - a} = f'(c)$ for $c$.

**46. Mean Value Theorem** Explain why the Mean Value Theorem does not apply to the function $f(x) = |x|$ on $[a, b] = [-1, 2]$.

**47. Mean Value Theorem** The population of a culture of cells grows according to the function $P(t) = \dfrac{100t}{t + 1}$, where $t \geq 0$ is measured in weeks.

   **a.** What is the average rate of change in the population over the interval $[0, 8]$?

   **b.** At what point of the interval $[0, 8]$ is the instantaneous rate of change equal to the average rate of change?

**T 48. Newton's method** Use Newton's method to approximate the roots of $f(x) = \sin x - x^2 + 1$ to six digits. Make a table showing the first five approximations for each root using an initial estimate of your choice.

**T 49. Newton's method** Use Newton's method to approximate the roots of $f(x) = 3x^3 - 4x^2 + 1$ to six digits.

**T 50. Newton's method** Use Newton's method to approximate the $x$-coordinates of the inflection points of $f(x) = 2x^5 - 6x^3 - 4x + 2$ to six digits.

**51–53. Two methods** *Evaluate the following limits in two different ways: with and without l'Hôpital's Rule.*

**51.** $\displaystyle \lim_{x \to \infty} \dfrac{2x^5 - x + 1}{5x^6 + x}$

**52.** $\displaystyle \lim_{x \to \infty} \dfrac{4x^4 - \sqrt{x}}{2x^4 + x^{-1}}$

**53.** $\displaystyle \lim_{x \to 1} \dfrac{\sqrt{3x + 1} - 2}{x - 1}$

**54–61. Limits** *Evaluate the following limits. Use l'Hôpital's Rule when needed.*

**54.** $\displaystyle \lim_{t \to 2} \dfrac{t^3 - t^2 - 2t}{t^2 - 4}$

**55.** $\displaystyle \lim_{t \to 0} \dfrac{1 - \cos 6t}{2t}$

**56.** $\displaystyle \lim_{x \to \infty} \dfrac{5x^2 + 2x - 5}{\sqrt{x^4 - 1}}$

**57.** $\displaystyle \lim_{\theta \to 0} \dfrac{3 \sin^2 2\theta}{\theta^2}$

**58.** $\displaystyle \lim_{x \to \infty} \left( \sqrt{x^2 + x + 1} - \sqrt{x^2 - x} \right)$

**59.** $\displaystyle \lim_{\theta \to 0} 2\theta \cot 3\theta$

**60.** $\displaystyle \lim_{\theta \to 0} \dfrac{3 \sin 8\theta}{8 \sin 3\theta}$

**61.** $\displaystyle \lim_{x \to 1} \dfrac{x^4 - x^3 - 3x^2 + 5x - 2}{x^3 + x^2 - 5x + 3}$

**62–71. Indefinite integrals** *Determine the following indefinite integrals.*

**62.** $\displaystyle \int (x^8 - 3x^3 + 1)\, dx$

**63.** $\displaystyle \int (2x + 1)^2\, dx$

**64.** $\displaystyle \int \dfrac{x^3 - x^2}{x}\, dx$

**65.** $\displaystyle \int \left( \dfrac{1}{x^2} - \dfrac{2}{x^{5/2}} \right) dx$

**66.** $\displaystyle \int \dfrac{x^4 - 2\sqrt{x} + 2}{x^2}\, dx$

**67.** $\displaystyle \int (1 + 3 \cos \theta)\, d\theta$

**68.** $\displaystyle \int 2 \sec^2 \theta\, d\theta$

**69.** $\displaystyle \int \dfrac{dx}{1 - \sin^2 x}$

**70.** $\displaystyle \int \dfrac{1 + \tan \theta}{\sec \theta}\, d\theta$

**71.** $\displaystyle \int \left( \sqrt[4]{x^3} + \sqrt{x^5} \right) dx$

**72–75. Functions from derivatives** *Find the function $f$ with the following properties.*

**72.** $f'(x) = 3x^2 - 1$; $f(0) = 10$

**73.** $f'(t) = \sin t + 2t$; $f(0) = 5$

**74.** $f'(t) = t^2 + t^{-2}$; $f(1) = 1$, for $t > 0$

**75.** $f'(x) = \cos^2 \dfrac{x}{2}$; $f(0) = 1$

**T 76. Motion along a line** Two objects move along the $x$-axis with position functions $x_1(t) = 2 \sin t$ and $x_2(t) = \sin\left( t - \dfrac{\pi}{2} \right)$. At what times on the interval $[0, 2\pi]$ are the objects closest to each other and farthest from each other?

**77. Vertical motion with gravity** A rocket is launched vertically upward with an initial velocity of 120 m/s from a platform that is 125 m above the ground. Assume the only force at work is gravity. Determine the velocity and position functions of the rocket, for $t \geq 0$. Then describe the motion in words.

**78. Distance traveled** A car starting at rest accelerates at 20 ft/s$^2$ for 5 seconds on a straight road. How far does it travel during this time?

**79. Projectile motion** A ball is thrown vertically upward with a velocity of 64 ft/s from the edge of a cliff 128 ft above a river.

   **a.** Find the velocity of the ball for all relevant times.

   **b.** Find the position of the ball above the river for all relevant times.

   **c.** Find the time when the ball reaches its highest point above the river. What is the height?

   **d.** Find the velocity at which the ball strikes the ground.

**80. Cosine limits** Let $n$ be a positive integer. Evaluate the following limits.

   **a.** $\displaystyle \lim_{x \to 0} \dfrac{1 - \cos x^n}{x^{2n}}$

   **b.** $\displaystyle \lim_{x \to 0} \dfrac{1 - \cos^n x}{x^2}$

## Chapter 4 Guided Projects

*Applications of the material in this chapter and related topics can be found in the following Guided Projects. For additional information, see the Preface.*

- Ice cream, geometry, and calculus
- Newton's method

# 5

# Integration

**Chapter Preview**  We are now at a critical point in the calculus story. Many would argue that this chapter is the cornerstone of calculus because it explains the relationship between the two processes of calculus: differentiation and integration. We begin by explaining why finding the area of regions bounded by the graphs of functions is such an important problem in calculus. Then you will see how antiderivatives lead to definite integrals, which are used to solve the area problem. But there is more to the story. You will also see the remarkable connection between derivatives and integrals, which is expressed in the Fundamental Theorem of Calculus. In this chapter, we develop key properties of definite integrals, investigate a few of their many applications, and present the first of several powerful techniques for evaluating definite integrals.

## 5.1 Approximating Areas under Curves

The derivative of a function is associated with rates of change and slopes of tangent lines. We also know that antiderivatives (or indefinite integrals) reverse the derivative operation. **Figure 5.1** summarizes our current understanding and raises the question: What is the geometric meaning of the integral? The following example reveals a clue.

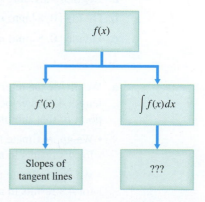

**Figure 5.1**

### Area under a Velocity Curve

Consider an object moving along a line with a known position function. You learned in previous chapters that the slope of the line tangent to the graph of the position function at a certain time gives the velocity $v$ at that time. We now turn the situation around. If we know the velocity function of a moving object, what can we learn about its position function?

> Recall from Section 3.6 that the *displacement* of an object moving along a line is given by

> final position − initial position.

> If the velocity of an object is positive, its displacement equals the distance traveled.

Imagine a car traveling at a constant velocity of 60 mi/hr along a straight highway over a two-hour period. The graph of the velocity function $v = 60$ on the interval $0 \le t \le 2$ is a horizontal line (**Figure 5.2**). The displacement of the car between $t = 0$ and $t = 2$ is found by a familiar formula:

$$\text{displacement} = \text{rate} \cdot \text{time}$$
$$= 60 \text{ mi/hr} \cdot 2 \text{ hr} = 120 \text{ mi}.$$

This product is the area of the rectangle formed by the velocity curve and the *t*-axis between $t = 0$ and $t = 2$ (**Figure 5.3**). In this case (constant positive velocity), we see that the area between the velocity curve and the *t*-axis is the displacement of the moving object.

> The side lengths of the rectangle in Figure 5.3 have units mi/hr and hr. Therefore, the units of the area are mi/hr · hr = mi, which is a unit of displacement.

**QUICK CHECK 1** What is the displacement of an object that travels at a constant velocity of 10 mi/hr for a half hour, 20 mi/hr for the next half hour, and 30 mi/hr for the next hour? ◄

**Figure 5.2**

**Figure 5.3**

Because objects do not necessarily move at a constant velocity, we first extend these ideas to positive velocities that *change* over an interval of time. One strategy is to divide the time interval into many subintervals and approximate the velocity on each subinterval with a constant velocity. Then the displacements on each subinterval are calculated and summed. This strategy produces only an approximation to the displacement; however, this approximation generally improves as the number of subintervals increases.

**EXAMPLE 1**    **Approximating the displacement**  Suppose the velocity in m/s of an object moving along a line is given by the function $v = t^2$, where $0 \le t \le 8$. Approximate the displacement of the object by dividing the time interval $[0, 8]$ into $n$ subintervals of equal length. On each subinterval, approximate the velocity with a constant equal to the value of $v$ evaluated at the midpoint of the subinterval.

**a.** Begin by dividing $[0, 8]$ into $n = 2$ subintervals: $[0, 4]$ and $[4, 8]$.

**b.** Divide $[0, 8]$ into $n = 4$ subintervals: $[0, 2]$, $[2, 4]$, $[4, 6]$, and $[6, 8]$.

**c.** Divide $[0, 8]$ into $n = 8$ subintervals of equal length.

**SOLUTION**

**a.** We divide the interval $[0, 8]$ into $n = 2$ subintervals, $[0, 4]$ and $[4, 8]$, each with length 4. The velocity on each subinterval is approximated by evaluating $v$ at the midpoint of that subinterval (**Figure 5.4a**).

 • We approximate the velocity on $[0, 4]$ by $v(2) = 2^2 = 4$ m/s. Traveling at 4 m/s for 4 s results in a displacement of 4 m/s · 4 s = 16 m.

 • We approximate the velocity on $[4, 8]$ by $v(6) = 6^2 = 36$ m/s. Traveling at 36 m/s for 4 s results in a displacement of 36 m/s · 4 s = 144 m.

Therefore, an approximation to the displacement over the entire interval $[0, 8]$ is

$$(v(2) \cdot 4 \text{ s}) + (v(6) \cdot 4 \text{ s}) = (4 \text{ m/s} \cdot 4 \text{ s}) + (36 \text{ m/s} \cdot 4 \text{ s}) = 160 \text{ m}.$$

**b.** With $n = 4$ (**Figure 5.4b**), each subinterval has length 2. The approximate displacement over the entire interval is

$$\underbrace{(1 \text{ m/s} \cdot 2 \text{ s})}_{v(1)} + \underbrace{(9 \text{ m/s} \cdot 2 \text{ s})}_{v(3)} + \underbrace{(25 \text{ m/s} \cdot 2 \text{ s})}_{v(5)} + \underbrace{(49 \text{ m/s} \cdot 2 \text{ s})}_{v(7)} = 168 \text{ m}.$$

**c.** With $n = 8$ subintervals (**Figure 5.4c**), the approximation to the displacement is 170 m. In each case, the approximate displacement is the sum of the areas of the rectangles under the velocity curve.

> The midpoint of each subinterval is used to approximate the velocity over that subinterval.

$n = 2$
(a)

$n = 4$
(b)

$n = 8$
(c)

**Figure 5.4**

*Related Exercises 3, 15–16* ◀

**QUICK CHECK 2** In Example 1, if we used $n = 32$ subintervals of equal length, what would be the length of each subinterval? Find the midpoint of the first and last subinterval. ◀

The progression in Example 1 may be continued. Larger values of $n$ mean more rectangles; in general, more rectangles give a better fit to the region under the curve (**Figure 5.5**). With the help of a calculator, we can generate the approximations in Table 5.1 using $n = 1, 2, 4, 8, 16, 32,$ and $64$ subintervals. Observe that as $n$ increases, the approximations appear to approach a limit of approximately 170.7 m. The limit is the exact displacement, which is represented by the area of the region under the velocity curve. This strategy of taking limits of sums is developed fully in Section 5.2.

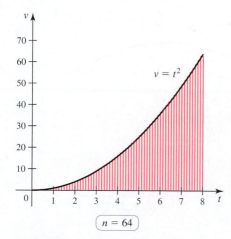
$n = 64$

**Figure 5.5**

**Table 5.1** Approximations to the area under the velocity curve $v = t^2$ on $[0, 8]$

| Number of subintervals | Length of each subinterval | Approximate displacement (area under curve) |
|---|---|---|
| 1 | 8 s | 128.0 m |
| 2 | 4 s | 160.0 m |
| 4 | 2 s | 168.0 m |
| 8 | 1 s | 170.0 m |
| 16 | 0.5 s | 170.5 m |
| 32 | 0.25 s | 170.625 m |
| 64 | 0.125 s | 170.65625 m |

➤ The language "the area of the region bounded by the graph of a function" is often abbreviated as "the area under the curve."

**Figure 5.6**

## Approximating Areas by Riemann Sums

We wouldn't spend much time investigating areas under curves if the idea applied only to computing displacements from velocity curves. However, the problem of finding areas under curves arises frequently and turns out to be immensely important—as you will see in the next two chapters. For this reason, we now develop a systematic method for approximating areas under curves. Consider a function $f$ that is continuous and nonnegative on an interval $[a, b]$. The goal is to approximate the area of the region $R$ bounded by the graph of $f$ and the $x$-axis from $x = a$ to $x = b$ (**Figure 5.6**). We begin by dividing the interval $[a, b]$ into $n$ subintervals of equal length,

$$[x_0, x_1], [x_1, x_2], \ldots, [x_{n-1}, x_n],$$

where $a = x_0$ and $b = x_n$ (**Figure 5.7**). The length of each subinterval, denoted $\Delta x$, is found by dividing the length of the entire interval by $n$:

$$\Delta x = \frac{b - a}{n}.$$

**Figure 5.7**

---

**DEFINITION   Regular Partition**

Suppose $[a, b]$ is a closed interval containing $n$ subintervals

$$[x_0, x_1], [x_1, x_2], \ldots, [x_{n-1}, x_n]$$

of equal length $\Delta x = \dfrac{b - a}{n}$, with $a = x_0$ and $b = x_n$. The endpoints $x_0, x_1, x_2, \ldots,$ $x_{n-1}, x_n$ of the subintervals are called **grid points**, and they create a **regular partition** of the interval $[a, b]$. In general, the $k$th grid point is

$$x_k = a + k\Delta x, \text{ for } k = 0, 1, 2, \ldots, n.$$

---

**QUICK CHECK 3** If the interval $[1, 9]$ is partitioned into 4 subintervals of equal length, what is $\Delta x$? List the grid points $x_0, x_1, x_2, x_3,$ and $x_4$. ◄

In the $k$th subinterval $[x_{k-1}, x_k]$, we choose any point $x_k^*$ and build a rectangle whose height is $f(x_k^*)$, the value of $f$ at $x_k^*$ (**Figure 5.8**). The area of the rectangle on the $k$th subinterval is

$$\text{height} \cdot \text{base} = f(x_k^*)\Delta x, \qquad \text{where } k = 1, 2, \ldots, n.$$

Summing the areas of the rectangles in Figure 5.8, we obtain an approximation to the area of $R$, which is called a **Riemann sum**:

$$f(x_1^*)\Delta x + f(x_2^*)\Delta x + \cdots + f(x_n^*)\Delta x.$$

Three notable Riemann sums are the *left*, *right*, and *midpoint Riemann sums*.

➤ Although the idea of integration was developed in the 17th century, it was almost 200 years later that the German mathematician Bernhard Riemann (1826–1866) worked on the mathematical theory underlying integration.

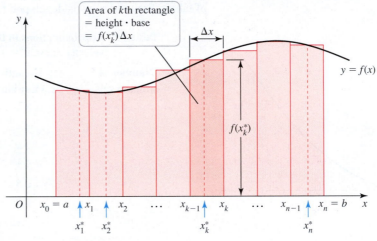

**Figure 5.8**

---

**DEFINITION   Riemann Sum**

Suppose $f$ is defined on a closed interval $[a, b]$, which is divided into $n$ subintervals of equal length $\Delta x$. If $x_k^*$ is any point in the $k$th subinterval $[x_{k-1}, x_k]$, for $k = 1, 2, \ldots, n,$ then

$$f(x_1^*)\Delta x + f(x_2^*)\Delta x + \cdots + f(x_n^*)\Delta x$$

is called a **Riemann sum** for $f$ on $[a, b]$. This sum is called

• a **left Riemann sum** if $x_k^*$ is the left endpoint of $[x_{k-1}, x_k]$ (**Figure 5.9**);

• a **right Riemann sum** if $x_k^*$ is the right endpoint of $[x_{k-1}, x_k]$ (**Figure 5.10**); and

• a **midpoint Riemann sum** if $x_k^*$ is the midpoint of $[x_{k-1}, x_k]$ (**Figure 5.11**), for $k = 1, 2, \ldots, n.$

**Figure 5.9**

**Figure 5.10**

**Figure 5.11**

**EXAMPLE 2   Left and right Riemann sums**  Let $R$ be the region bounded by the graph of $f(x) = 3\sqrt{x}$ and the $x$-axis between $x = 4$ and $x = 16$.

**a.** Approximate the area of $R$ using a left Riemann sum with $n = 6$ subintervals. Illustrate the sum with the appropriate rectangles.

**b.** Approximate the area of $R$ using a right Riemann sum with $n = 6$ subintervals. Illustrate the sum with the appropriate rectangles.

**c.** Do the area approximations in parts (a) and (b) underestimate or overestimate the actual area under the curve?

**SOLUTION**  Dividing the interval $[a, b] = [4, 16]$ into $n = 6$ subintervals means that the length of each subinterval is

$$\Delta x = \frac{b - a}{n} = \frac{16 - 4}{6} = 2;$$

therefore the grid points are 4, 6, 8, 10, 12, 14, and 16.

**a.** To find the left Riemann sum, we set $x_1^*, x_2^*, \ldots, x_6^*$ equal to the left endpoints of the six subintervals. The heights of the rectangles are $f(x_k^*)$, for $k = 1, 2, \ldots, 6$.

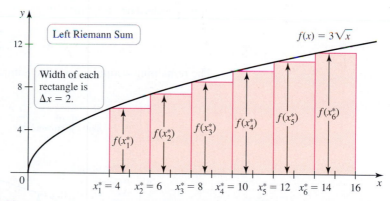

**Figure 5.12**

The resulting left Riemann sum (**Figure 5.12**) is

$$f(x_1^*)\Delta x + f(x_2^*)\Delta x + \cdots + f(x_6^*)\Delta x$$
$$= f(4) \cdot 2 + f(6) \cdot 2 + f(8) \cdot 2 + f(10) \cdot 2 + f(12) \cdot 2 + f(14) \cdot 2$$
$$= 3\sqrt{4} \cdot 2 + 3\sqrt{6} \cdot 2 + 3\sqrt{8} \cdot 2 + 3\sqrt{10} \cdot 2 + 3\sqrt{12} \cdot 2 + 3\sqrt{14} \cdot 2$$
$$\approx 105.876.$$

**b.** In a right Riemann sum, the right endpoints are used for $x_1^*, x_2^*, \ldots, x_6^*$, and the heights of the rectangles are $f(x_k^*)$, for $k = 1, 2, \ldots, 6$.

**Figure 5.13**

The resulting right Riemann sum (**Figure 5.13**) is

$$f(x_1^*)\Delta x + f(x_2^*)\Delta x + \cdots + f(x_6^*)\Delta x$$
$$= f(6)\cdot 2 + f(8)\cdot 2 + f(10)\cdot 2 + f(12)\cdot 2 + f(14)\cdot 2 + f(16)\cdot 2$$
$$= 3\sqrt{6}\cdot 2 + 3\sqrt{8}\cdot 2 + 3\sqrt{10}\cdot 2 + 3\sqrt{12}\cdot 2 + 3\sqrt{14}\cdot 2 + 3\sqrt{16}\cdot 2$$
$$\approx 117.876.$$

**QUICK CHECK 4** If the function in Example 2 is replaced with $f(x) = 1/x$, does the left Riemann sum or the right Riemann sum overestimate the area under the curve? ◄

c. Looking at the graphs, we see that the left Riemann sum in part (a) underestimates the actual area of $R$, whereas the right Riemann sum in part (b) overestimates the area of $R$. Therefore, the area of $R$ is between 105.876 and 117.876. These approximations improve as the number of rectangles increases.

*Related Exercises 23–24, 29* ◄

**EXAMPLE 3  A midpoint Riemann sum** Let $R$ be the region bounded by the graph of $f(x) = 3\sqrt{x}$ and the $x$-axis between $x = 4$ and $x = 16$. Approximate the area of $R$ using a midpoint Riemann sum with $n = 6$ subintervals. Illustrate the sum with the appropriate rectangles.

**SOLUTION** The grid points and the length of the subintervals are the same as in Example 2. To find the midpoint Riemann sum, we set $x_1^*, x_2^*, \ldots, x_6^*$ equal to the midpoints of the subintervals. The midpoint of the first subinterval is the average of $x_0$ and $x_1$, which is

$$x_1^* = \frac{x_0 + x_1}{2} = \frac{4 + 6}{2} = 5.$$

The remaining midpoints are also computed by averaging the two nearest grid points.

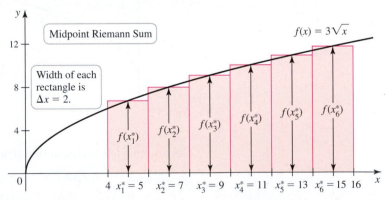

**Figure 5.14**

The resulting midpoint Riemann sum (**Figure 5.14**) is

$$f(x_1^*)\Delta x + f(x_2^*)\Delta x + \cdots + f(x_6^*)\Delta x$$
$$= f(5)\cdot 2 + f(7)\cdot 2 + f(9)\cdot 2 + f(11)\cdot 2 + f(13)\cdot 2 + f(15)\cdot 2$$
$$= 3\sqrt{5}\cdot 2 + 3\sqrt{7}\cdot 2 + 3\sqrt{9}\cdot 2 + 3\sqrt{11}\cdot 2 + 3\sqrt{13}\cdot 2 + 3\sqrt{15}\cdot 2$$
$$\approx 112.062.$$

Comparing the midpoint Riemann sum (Figure 5.14) with the left and right Riemann sums (Figures 5.12 and 5.13) suggests that the midpoint sum is a more accurate estimate of the area under the curve. Indeed, in Section 5.3, we will learn that the exact area under the curve is 112.

*Related Exercises 33–34, 39* ◄

**EXAMPLE 4   Riemann sums from tables** Estimate the area $A$ under the graph of $f$ on the interval $[0, 2]$ using left and right Riemann sums with $n = 4$, where $f$ is continuous but known only at the points in Table 5.2.

**SOLUTION** With $n = 4$ subintervals on the interval $[0, 2]$, $\Delta x = 2/4 = 0.5$. Using the left endpoint of each subinterval, the left Riemann sum is

$$A \approx f(0)\Delta x + f(0.5)\Delta x + f(1.0)\Delta x + f(1.5)\Delta x$$
$$= 1 \cdot 0.5 + 3 \cdot 0.5 + 4.5 \cdot 0.5 + 5.5 \cdot 0.5 = 7.0.$$

Using the right endpoint of each subinterval, the right Riemann sum is

$$A \approx f(0.5)\Delta x + f(1.0)\Delta x + f(1.5)\Delta x + f(2.0)\Delta x$$
$$= 3 \cdot 0.5 + 4.5 \cdot 0.5 + 5.5 \cdot 0.5 + 6.0 \cdot 0.5 = 9.5.$$

With only five function values, these estimates of the area are necessarily crude. Better estimates are obtained by using more subintervals and more function values.

*Related Exercises 43–44* ◄

**Table 5.2**

| $x$ | $f(x)$ |
|-----|--------|
| 0 | 1 |
| 0.5 | 3 |
| 1.0 | 4.5 |
| 1.5 | 5.5 |
| 2.0 | 6.0 |

## Sigma (Summation) Notation

Working with Riemann sums is cumbersome with large numbers of subintervals. Therefore, we pause for a moment to introduce some notation that simplifies our work.

**Sigma** (or **summation**) **notation** is used to express sums in a compact way. For example, the sum $1 + 2 + 3 + \cdots + 10$ is represented in sigma notation as $\sum_{k=1}^{10} k$. Here is how the notation works. The symbol $\Sigma$ (*sigma*, the Greek capital S) stands for *sum*. The **index** $k$ takes on all integer values from the lower limit ($k = 1$) to the upper limit ($k = 10$). The expression that immediately follows $\Sigma$ (the **summand**) is evaluated for each value of $k$, and the resulting values are summed. Here are some examples.

$$\sum_{k=1}^{99} k = 1 + 2 + 3 + \cdots + 99 = 4950 \qquad \sum_{k=1}^{n} k = 1 + 2 + \cdots + n$$

$$\sum_{k=0}^{3} k^2 = 0^2 + 1^2 + 2^2 + 3^2 = 14 \qquad \sum_{k=1}^{4} (2k + 1) = 3 + 5 + 7 + 9 = 24$$

$$\sum_{k=-1}^{2} (k^2 + k) = ((-1)^2 + (-1)) + (0^2 + 0) + (1^2 + 1) + (2^2 + 2) = 8$$

The index in a sum is a *dummy variable*. It is internal to the sum, so it does not matter what symbol you choose as an index. For example,

$$\sum_{k=1}^{99} k = \sum_{n=1}^{99} n = \sum_{p=1}^{99} p.$$

Two properties of sums and sigma notation are useful in upcoming work. Suppose $\{a_1, a_2, \ldots, a_n\}$ and $\{b_1, b_2, \ldots, b_n\}$ are two sets of real numbers, and suppose $c$ is a real number. Then we can factor multiplicative constants out of a sum:

*Constant Multiple Rule*   $$\sum_{k=1}^{n} ca_k = c \sum_{k=1}^{n} a_k.$$

We can also split a sum into two sums:

*Addition Rule*   $$\sum_{k=1}^{n} (a_k + b_k) = \sum_{k=1}^{n} a_k + \sum_{k=1}^{n} b_k.$$

In the coming examples and exercises, the following formulas for sums of powers of integers are essential.

> ➤ Formulas for $\displaystyle\sum_{k=1}^{n} k^p$, where $p$ is a positive integer, have been known for centuries. The formulas for $p = 0, 1, 2,$ and 3 are relatively simple. The formulas become complicated as $p$ increases.

**THEOREM 5.1   Sums of Powers of Integers**

Let $n$ be a positive integer and $c$ a real number.

$$\sum_{k=1}^{n} c = cn \qquad\qquad \sum_{k=1}^{n} k = \frac{n(n+1)}{2}$$

$$\sum_{k=1}^{n} k^2 = \frac{n(n+1)(2n+1)}{6} \qquad \sum_{k=1}^{n} k^3 = \frac{n^2(n+1)^2}{4}$$

## Riemann Sums Using Sigma Notation

With sigma notation, a Riemann sum has the convenient compact form

$$f(x_1^*)\Delta x + f(x_2^*)\Delta x + \cdots + f(x_n^*)\Delta x = \sum_{k=1}^{n} f(x_k^*)\Delta x.$$

To express left, right, and midpoint Riemann sums in sigma notation, we must identify the points $x_k^*$.

- For left Riemann sums, the left endpoints of the subintervals are $x_k^* = a + (k-1)\Delta x$, for $k = 1, \ldots, n$.
- For right Riemann sums, the right endpoints of the subintervals are $x_k^* = a + k\Delta x$, for $k = 1, \ldots, n$.
- For midpoint Riemann sums, the midpoints of the subintervals are $x_k^* = a + (k - \frac{1}{2})\Delta x$, for $k = 1, \ldots, n$.

The three Riemann sums are written compactly as follows.

> ➤ For the left Riemann sum,
> $x_1^* = a + 0 \cdot \Delta x, x_2^* = a + 1 \cdot \Delta x,$
> $x_3^* = a + 2 \cdot \Delta x,$
> and in general, $x_k^* = a + (k-1)\Delta x$, for $k = 1, \ldots, n.$
>
> For the right Riemann sum,
> $x_1^* = a + 1 \cdot \Delta x, x_2^* = a + 2 \cdot \Delta x,$
> $x_3^* = a + 3 \cdot \Delta x,$
> and in general, $x_k^* = a + k\Delta x$, for $k = 1, \ldots, n.$
>
> For the midpoint Riemann sum,
> $x_1^* = a + \frac{1}{2}\Delta x, x_2^* = a + \frac{3}{2}\Delta x,$
> and in general,
> $x_k^* = a + (k - \frac{1}{2})\Delta x = \dfrac{x_k + x_{k-1}}{2},$
> for $k = 1, \ldots, n.$

**DEFINITION   Left, Right, and Midpoint Riemann Sums in Sigma Notation**

Suppose $f$ is defined on a closed interval $[a, b]$, which is divided into $n$ subintervals of equal length $\Delta x$. If $x_k^*$ is a point in the $k$th subinterval $[x_{k-1}, x_k]$, for

$k = 1, 2, \ldots, n$, then the **Riemann sum** for $f$ on $[a, b]$ is $\displaystyle\sum_{k=1}^{n} f(x_k^*)\Delta x$. Three cases arise in practice.

- $\displaystyle\sum_{k=1}^{n} f(x_k^*)\Delta x$ is a **left Riemann sum** if $x_k^* = a + (k-1)\Delta x$.

- $\displaystyle\sum_{k=1}^{n} f(x_k^*)\Delta x$ is a **right Riemann sum** if $x_k^* = a + k\Delta x$.

- $\displaystyle\sum_{k=1}^{n} f(x_k^*)\Delta x$ is a **midpoint Riemann sum** if $x_k^* = a + (k - \frac{1}{2})\Delta x$.

**EXAMPLE 5   Calculating Riemann sums**  Evaluate the left, right, and midpoint Riemann sums for $f(x) = x^3 + 1$ between $a = 0$ and $b = 2$ using $n = 50$ subintervals. Make a conjecture about the exact area of the region under the curve (**Figure 5.15**).

**SOLUTION**  With $n = 50$, the length of each subinterval is

$$\Delta x = \frac{b - a}{n} = \frac{2 - 0}{50} = \frac{1}{25} = 0.04.$$

The value of $x_k^*$ for the left Riemann sum is

$$x_k^* = a + (k-1)\Delta x = 0 + 0.04(k-1) = 0.04k - 0.04,$$

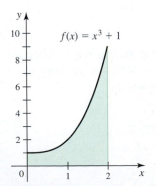

**Figure 5.15**

$f(x) = x^3 + 1$

for $k = 1, 2, \ldots, 50$. Therefore, the left Riemann sum, evaluated with a calculator, is

$$\sum_{k=1}^{n} f(x_k^*)\Delta x = \sum_{k=1}^{50} f(0.04k - 0.04)0.04 = 5.8416.$$

To evaluate the right Riemann sum, we let $x_k^* = a + k\Delta x = 0.04k$ and find that

$$\sum_{k=1}^{n} f(x_k^*)\Delta x = \sum_{k=1}^{50} f(0.04k)0.04 = 6.1616.$$

For the midpoint Riemann sum, we let

$$x_k^* = a + \left(k - \frac{1}{2}\right)\Delta x = 0 + 0.04\left(k - \frac{1}{2}\right) = 0.04k - 0.02.$$

The value of the sum is

$$\sum_{k=1}^{n} f(x_k^*)\,\Delta x = \sum_{k=1}^{50} f(0.04k - 0.02)0.04 = 5.9992.$$

Because $f$ is increasing on $[0, 2]$, the left Riemann sum underestimates the area of the shaded region in Figure 5.15, and the right Riemann sum overestimates the area. Therefore, the exact area lies between 5.8416 and 6.1616. The midpoint Riemann sum usually gives the best estimate for increasing or decreasing functions.

Table 5.3 shows the left, right, and midpoint Riemann sum approximations for values of $n$ up to 200. All three sets of approximations approach a value near 6, which is a reasonable estimate of the area under the curve. In Section 5.2, we show rigorously that the limit of all three Riemann sums as $n \to \infty$ is 6.

**ALTERNATIVE SOLUTION**  It is worth examining another approach to Example 5. Consider the right Riemann sum given previously:

$$\sum_{k=1}^{n} f(x_k^*)\,\Delta x = \sum_{k=1}^{50} f(0.04k)0.04.$$

Rather than evaluating this sum with a calculator, we note that $f(0.04k) = (0.04k)^3 + 1$ and use the properties of sums:

$$\sum_{k=1}^{n} f(x_k^*)\Delta x = \sum_{k=1}^{50} \underbrace{((0.04k)^3 + 1)}_{f(x_k^*)}\underbrace{0.04}_{\Delta x}$$

$$= \sum_{k=1}^{50} (0.04k)^3\, 0.04 + \sum_{k=1}^{50} 1\cdot 0.04 \qquad \sum(a_k + b_k) = \sum a_k + \sum b_k$$

$$= (0.04)^4 \sum_{k=1}^{50} k^3 + 0.04 \sum_{k=1}^{50} 1. \qquad \sum ca_k = c\sum a_k$$

Using the summation formulas for powers of integers in Theorem 5.1, we find that

$$\sum_{k=1}^{50} 1 = 50 \quad \text{and} \quad \sum_{k=1}^{50} k^3 = \frac{50^2 \cdot 51^2}{4}.$$

Substituting the values of these sums into the right Riemann sum yields

$$\sum_{k=1}^{50} f(x_k^*)\,\Delta x = \frac{3851}{625} = 6.1616,$$

confirming the result given by a calculator. The idea of evaluating Riemann sums for *arbitrary* values of $n$ is used in Section 5.2, where we evaluate the limit of the Riemann sum as $n \to \infty$.

*Related Exercises 51–52* ◄

**Table 5.3   Left, right, and midpoint Riemann sum approximations**

| $n$ | $L_n$ | $R_n$ | $M_n$ |
|---|---|---|---|
| 20 | 5.61 | 6.41 | 5.995 |
| 40 | 5.8025 | 6.2025 | 5.99875 |
| 60 | 5.86778 | 6.13444 | 5.99944 |
| 80 | 5.90063 | 6.10063 | 5.99969 |
| 100 | 5.9204 | 6.0804 | 5.9998 |
| 120 | 5.93361 | 6.06694 | 5.99986 |
| 140 | 5.94306 | 6.05735 | 5.9999 |
| 160 | 5.95016 | 6.05016 | 5.99992 |
| 180 | 5.95568 | 6.04457 | 5.99994 |
| 200 | 5.9601 | 6.0401 | 5.99995 |

## SECTION 5.1 EXERCISES

### Getting Started

**1.** Suppose an object moves along a line at 15 m/s, for $0 \leq t < 2$, and at 25 m/s, for $2 \leq t \leq 5$, where $t$ is measured in seconds. Sketch the graph of the velocity function and find the displacement of the object, for $0 \leq t \leq 5$.

**2.** Given the graph of the positive velocity of an object moving along a line, what is the geometrical representation of its displacement over a time interval $[a, b]$?

**3.** The velocity in ft/s of an object moving along a line is given by $v = f(t)$ on the interval $0 \leq t \leq 8$ (see figure), where $t$ is measured in seconds.

   **a.** Divide the interval $[0, 8]$ into $n = 2$ subintervals, $[0, 4]$ and $[4, 8]$. On each subinterval, assume the object moves at a constant velocity equal to the value of $v$ evaluated at the midpoint of the subinterval, and use these approximations to estimate the displacement of the object on $[0, 8]$ (see part (a) of the figure).

   **b.** Repeat part (a) for $n = 4$ subintervals (see part (b) of the figure).

(a)

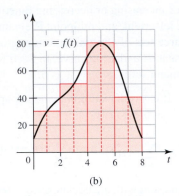

(b)

**4.** The velocity in ft/s of an object moving along a line is given by $v = f(t)$ on the interval $0 \leq t \leq 6$ (see figure), where $t$ is measured in seconds.

   **a.** Divide the interval $[0, 6]$ into $n = 3$ subintervals, $[0, 2]$, $[2, 4]$, and $[4, 6]$. On each subinterval, assume the object moves at a constant velocity equal to the value of $v$ evaluated at the left endpoint of the subinterval, and use these approximations to estimate the displacement of the object on $[0, 6]$ (see part (a) of the figure).

   **b.** Repeat part (a) for $n = 6$ subintervals (see part (b) of the figure).

(a)

(b)

**5.** The velocity in ft/s of an object moving along a line is given by $v = f(t)$ on the interval $0 \leq t \leq 6$ (see figure), where $t$ is measured in seconds.

   **a.** Divide the interval $[0, 6]$ into $n = 3$ subintervals, $[0, 2]$, $[2, 4]$, and $[4, 6]$. On each subinterval, assume the object moves at a constant velocity equal to the value of $v$ evaluated at the right endpoint of the subinterval, and use these approximations to estimate the displacement of the object on $[0, 6]$ (see part (a) of the figure).

   **b.** Repeat part (a) for $n = 6$ subintervals (see part (b) of the figure).

(a)

(b)

**6.** The velocity in ft/s of an object moving along a line is given by $v = f(t)$ on the interval $0 \leq t \leq 10$ (see figure), where $t$ is measured in seconds.

   **a.** Divide the interval $[0, 10]$ into $n = 5$ subintervals. On each subinterval, assume the object moves at a constant velocity equal to the value of $v$ evaluated at the left endpoint of the subinterval, and use these approximations to estimate the displacement of the object on $[0, 10]$.

   **b.** Repeat part (a) using the right endpoints to estimate the displacement on $[0, 10]$.

   **c.** Repeat part (a) using the midpoints of each subinterval to estimate the displacement on $[0, 10]$.

**7.** Suppose you want to approximate the area of the region bounded by the graph of $f(x) = \cos x$ and the $x$-axis between $x = 0$ and $x = \dfrac{\pi}{2}$. Explain a possible strategy.

**8.** Explain how Riemann sum approximations to the area of a region under a curve change as the number of subintervals increases.

**9.** **Approximating area from a graph** Approximate the area of the region bounded by the graph (see figure) and the $x$-axis by dividing the interval $[1, 7]$ into $n = 6$ subintervals. Use a left and right Riemann sum to obtain two different approximations.

**10. Approximating area from a graph** Approximate the area of the region bounded by the graph (see figure) and the $x$-axis by dividing the interval $[0, 6]$ into $n = 3$ subintervals. Use a left, right, and midpoint Riemann sum to obtain three different approximations.

**11.** Suppose the interval $[1, 3]$ is partitioned into $n = 4$ subintervals. What is the subinterval length $\Delta x$? List the grid points $x_0, x_1, x_2, x_3,$ and $x_4$. Which points are used for the left, right, and midpoint Riemann sums?

**12.** Suppose the interval $[2, 6]$ is partitioned into $n = 4$ subintervals with grid points $x_0 = 2, x_1 = 3, x_2 = 4, x_3 = 5,$ and $x_4 = 6$. Write, but do not evaluate, the left, right, and midpoint Riemann sums for $f(x) = x^2$.

**13.** Does a right Riemann sum underestimate or overestimate the area of the region under the graph of a function that is positive and decreasing on an interval $[a, b]$? Explain.

**14.** Does a left Riemann sum underestimate or overestimate the area of the region under the graph of a function that is positive and increasing on an interval $[a, b]$? Explain.

## Practice Exercises

**15. Approximating displacement** The velocity in ft/s of an object moving along a line is given by $v = 3t^2 + 1$ on the interval $0 \le t \le 4$, where $t$ is measured in seconds.

**a.** Divide the interval $[0, 4]$ into $n = 4$ subintervals, $[0, 1]$, $[1, 2], [2, 3],$ and $[3, 4]$. On each subinterval, assume the object moves at a constant velocity equal to $v$ evaluated at the midpoint of the subinterval, and use these approximations to estimate the displacement of the object on $[0, 4]$ (see part (a) of the figure).

**b.** Repeat part (a) for $n = 8$ subintervals (see part (b) of the figure).

(a)                  (b)

**T 16. Approximating displacement** The velocity in ft/s of an object moving along a line is given by $v = \sqrt{10t}$ on the interval $1 \le t \le 7$, where $t$ is measured in seconds.

**a.** Divide the interval $[1, 7]$ into $n = 3$ subintervals, $[1, 3]$, $[3, 5],$ and $[5, 7]$. On each subinterval, assume the object moves at a constant velocity equal to $v$ evaluated at the midpoint of the subinterval, and use these approximations to estimate the displacement of the object on $[1, 7]$ (see part (a) of the figure).

**b.** Repeat part (a) for $n = 6$ subintervals (see part (b) of the figure).

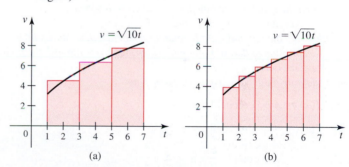

(a)                  (b)

**17–22. Approximating displacement** *The velocity of an object is given by the following functions on a specified interval. Approximate the displacement of the object on this interval by subdividing the interval into n subintervals. Use the left endpoint of each subinterval to compute the height of the rectangles.*

**17.** $v = 2t + 1$ (m/s), for $0 \le t \le 8; n = 2$

**18.** $v = t^3 + 1$ (m/s), for $0 \le t \le 3; n = 3$

**19.** $v = \dfrac{1}{2t + 1}$ (m/s), for $0 \le t \le 8; n = 4$

**20.** $v = \dfrac{t^2}{2} + 4$ (ft/s), for $0 \le t \le 12; n = 6$

**T 21.** $v = 4\sqrt{t + 1}$ (mi/hr), for $0 \le t \le 15; n = 5$

**22.** $v = \dfrac{t + 3}{6}$ (m/s), for $0 \le t \le 4; n = 4$

**23–24. Left and right Riemann sums** *Use the figures to calculate the left and right Riemann sums for f on the given interval and for the given value of n.*

**23.** $f(x) = x + 1$ on $[1, 6]$; $n = 5$

**24.** $f(x) = \dfrac{1}{x}$ on $[1, 5]$; $n = 4$

**25–32. Left and right Riemann sums** *Complete the following steps for the given function, interval, and value of n.*

*a.* Sketch the graph of the function on the given interval.
*b.* Calculate $\Delta x$ and the grid points $x_0, x_1, \ldots, x_n$.
*c.* Illustrate the left and right Riemann sums. Then determine which Riemann sum underestimates and which sum overestimates the area under the curve.
*d.* Calculate the left and right Riemann sums.

**25.** $f(x) = x + 1$ on $[0, 4]$; $n = 4$

**26.** $f(x) = 9 - x$ on $[3, 8]$; $n = 5$

**T 27.** $f(x) = \cos x$ on $\left[0, \dfrac{\pi}{2}\right]$; $n = 4$

**28.** $f(x) = \sin(\pi x/6)$ on $[0, 3]$; $n = 3$

**29.** $f(x) = x^2 - 1$ on $[2, 4]$; $n = 4$

**30.** $f(x) = 2x^2$ on $[1, 6]$; $n = 5$

**T 31.** $f(x) = \sqrt{x}$ on $[0, 3]$; $n = 6$

**T 32.** $f(x) = 2^x$ on $[0, 1]$; $n = 4$

**33. A midpoint Riemann sum** Approximate the area of the region bounded by the graph of $f(x) = 100 - x^2$ and the $x$-axis on $[0, 10]$ with $n = 5$ subintervals. Use the midpoint of each subinterval to determine the height of each rectangle (see figure).

**T 34. A midpoint Riemann sum** Approximate the area of the region bounded by the graph of $f(t) = \cos \dfrac{t}{2}$ and the $t$-axis on $[0, \pi]$ with $n = 4$ subintervals. Use the midpoint of each subinterval to determine the height of each rectangle (see figure).

**35. Free fall** On October 14, 2012, Felix Baumgartner stepped off a balloon capsule at an altitude of almost 39 km above Earth's surface and began his free fall. His velocity in m/s during the fall is given in the figure. It is claimed that Felix reached the speed of sound 34 seconds into his fall and that he continued to fall at supersonic speed for 30 seconds. (*Source:* http://www.redbullstratos.com)

*a.* Divide the interval $[34, 64]$ into $n = 5$ subintervals with the grid points $x_0 = 34, x_1 = 40, x_2 = 46, x_3 = 52, x_4 = 58$, and $x_5 = 64$. Use left and right Riemann sums to estimate how far Felix fell while traveling at supersonic speed.
*b.* It is claimed that the actual distance that Felix fell at supersonic speed was approximately 10,485 m. Which estimate in part (a) produced the more accurate estimate?
*c.* How could you obtain more accurate estimates of the total distance fallen than those found in part (a)?

**36. Free fall** Use geometry and the figure given in Exercise 35 to estimate how far Felix fell in the first 20 seconds of his free fall.

**37–42. Midpoint Riemann sums** *Complete the following steps for the given function, interval, and value of n.*

a. *Sketch the graph of the function on the given interval.*
b. *Calculate $\Delta x$ and the grid points $x_0, x_1, \ldots, x_n$.*
c. *Illustrate the midpoint Riemann sum by sketching the appropriate rectangles.*
d. *Calculate the midpoint Riemann sum.*

**37.** $f(x) = 2x + 1$ on $[0, 4]$; $n = 4$

**T 38.** $f(x) = 2 \cos (\pi x/2)$ on $[0, 1]$; $n = 6$

**T 39.** $f(x) = \sqrt{x}$ on $[1, 3]$; $n = 4$

**40.** $f(x) = x^2$ on $[0, 4]$; $n = 4$

**41.** $f(x) = \dfrac{1}{x}$ on $[1, 6]$; $n = 5$

**42.** $f(x) = 4 - x$ on $[-1, 4]$; $n = 5$

**43–44. Riemann sums from tables** *Evaluate the left and right Riemann sums for f over the given interval for the given value of n.*

**43.** $[0, 2]$; $n = 4$

| $x$ | 0 | 0.5 | 1 | 1.5 | 2 |
|---|---|---|---|---|---|
| $f(x)$ | 5 | 3 | 2 | 1 | 1 |

**44.** $[1, 5]$; $n = 8$

| $x$ | 1 | 1.5 | 2 | 2.5 | 3 | 3.5 | 4 | 4.5 | 5 |
|---|---|---|---|---|---|---|---|---|---|
| $f(x)$ | 0 | 2 | 3 | 2 | 2 | 1 | 0 | 2 | 3 |

**45. Displacement from a table of velocities** The velocities (in mi/hr) of an automobile moving along a straight highway over a two-hour period are given in the following table.

| $t$ (hr) | 0 | 0.25 | 0.5 | 0.75 | 1 | 1.25 | 1.5 | 1.75 | 2 |
|---|---|---|---|---|---|---|---|---|---|
| $v$ (mi/hr) | 50 | 50 | 60 | 60 | 55 | 65 | 50 | 60 | 70 |

a. Sketch a smooth curve passing through the data points.
b. Find the midpoint Riemann sum approximation to the displacement on $[0, 2]$ with $n = 2$ and $n = 4$ subintervals.

**46. Displacement from a table of velocities** The velocities (in m/s) of an automobile moving along a straight freeway over a four-second period are given in the following table.

| $t$ (s) | 0 | 0.5 | 1 | 1.5 | 2 | 2.5 | 3 | 3.5 | 4 |
|---|---|---|---|---|---|---|---|---|---|
| $v$ (m/s) | 20 | 25 | 30 | 35 | 30 | 30 | 35 | 40 | 40 |

a. Sketch a smooth curve passing through the data points.
b. Find the midpoint Riemann sum approximation to the displacement on $[0, 4]$ with $n = 2$ and $n = 4$ subintervals.

**47. Sigma notation** Express the following sums using sigma notation. (Answers are not unique.)

a. $1 + 2 + 3 + 4 + 5$
b. $4 + 5 + 6 + 7 + 8 + 9$
c. $1^2 + 2^2 + 3^2 + 4^2$
d. $1 + \dfrac{1}{2} + \dfrac{1}{3} + \dfrac{1}{4}$

**48. Sigma notation** Express the following sums using sigma notation. (Answers are not unique.)

a. $1 + 3 + 5 + 7 + \cdots + 99$
b. $4 + 9 + 14 + \cdots + 44$
c. $3 + 8 + 13 + \cdots + 63$
d. $\dfrac{1}{1 \cdot 2} + \dfrac{1}{2 \cdot 3} + \dfrac{1}{3 \cdot 4} + \cdots + \dfrac{1}{49 \cdot 50}$

**49. Sigma notation** Evaluate the following expressions.

a. $\displaystyle\sum_{k=1}^{10} k$
b. $\displaystyle\sum_{k=1}^{6} (2k + 1)$
c. $\displaystyle\sum_{k=1}^{4} k^2$
d. $\displaystyle\sum_{n=1}^{5} (1 + n^2)$
e. $\displaystyle\sum_{m=1}^{3} \dfrac{2m + 2}{3}$
f. $\displaystyle\sum_{j=1}^{3} (3j - 4)$
g. $\displaystyle\sum_{p=1}^{5} (2p + p^2)$
h. $\displaystyle\sum_{n=0}^{4} \sin \dfrac{n\pi}{2}$

**T 50. Evaluating sums** Evaluate the following expressions by two methods. (i) Use Theorem 5.1. (ii) Use a calculator.

a. $\displaystyle\sum_{k=1}^{45} k$
b. $\displaystyle\sum_{k=1}^{45} (5k - 1)$
c. $\displaystyle\sum_{k=1}^{75} 2k^2$
d. $\displaystyle\sum_{n=1}^{50} (1 + n^2)$
e. $\displaystyle\sum_{m=1}^{75} \dfrac{2m + 2}{3}$
f. $\displaystyle\sum_{j=1}^{20} (3j - 4)$
g. $\displaystyle\sum_{p=1}^{35} (2p + p^2)$
h. $\displaystyle\sum_{n=0}^{40} (n^2 + 3n - 1)$

**T 51–54. Riemann sums for larger values of n** *Complete the following steps for the given function f and interval.*

a. *For the given value of n, use sigma notation to write the left, right, and midpoint Riemann sums. Then evaluate each sum using a calculator.*
b. *Based on the approximations found in part (a), estimate the area of the region bounded by the graph of f and the x-axis on the interval.*

**51.** $f(x) = 3\sqrt{x}$ on $[0, 4]$; $n = 40$

**52.** $f(x) = x^2 + 1$ on $[-1, 1]$; $n = 50$

**53.** $f(x) = x^2 - 1$ on $[2, 5]$; $n = 75$

**54.** $f(x) = \cos 2x$ on $\left[0, \dfrac{\pi}{4}\right]$; $n = 60$

**T 55–58. Approximating areas with a calculator** *Use a calculator and right Riemann sums to approximate the area of the given region. Present your calculations in a table showing the approximations for $n = 10, 30, 60,$ and $80$ subintervals. Make a conjecture about the limit of Riemann sums as $n \to \infty$.*

**55.** The region bounded by the graph of $f(x) = 12 - 3x^2$ and the x-axis on the interval $[-1, 1]$.

**56.** The region bounded by the graph of $f(x) = 3x^2 + 1$ and the x-axis on the interval $[-1, 1]$.

**57.** The region bounded by the graph of $f(x) = \dfrac{1 - \cos x}{2}$ and the x-axis on the interval $[-\pi, \pi]$.

**58.** The region bounded by the graph of $f(x) = \sin x + \cos x$ and the x-axis on the interval $[0, \pi/2]$.

**59. Explain why or why not** Determine whether the following statements are true and give an explanation or counterexample.

a. Consider the linear function $f(x) = 2x + 5$ and the region bounded by its graph and the x-axis on the interval $[3, 6]$. Suppose the area of this region is approximated using midpoint

Riemann sums. Then the approximations give the exact area of the region for any number of subintervals.

**b.** A left Riemann sum always overestimates the area of a region bounded by a positive increasing function and the $x$-axis on an interval $[a, b]$.

**c.** For an increasing or decreasing nonconstant function on an interval $[a, b]$ and a given value of $n$, the value of the midpoint Riemann sum always lies between the values of the left and right Riemann sums.

**T 60. Riemann sums for a semicircle** Let $f(x) = \sqrt{1 - x^2}$.

**a.** Show that the graph of $f$ is the upper half of a circle of radius 1 centered at the origin.

**b.** Estimate the area between the graph of $f$ and the $x$-axis on the interval $[-1, 1]$ using a midpoint Riemann sum with $n = 25$.

**c.** Repeat part (b) using $n = 75$ rectangles.

**d.** What happens to the midpoint Riemann sums on $[-1, 1]$ as $n \to \infty$?

**T 61–64. Sigma notation for Riemann sums** *Use sigma notation to write the following Riemann sums. Then evaluate each Riemann sum using Theorem 5.1 or a calculator.*

**61.** The right Riemann sum for $f(x) = x + 1$ on $[0, 4]$ with $n = 50$

**62.** The left Riemann sum for $f(x) = \dfrac{3}{x}$ on $[1, 3]$ with $n = 30$

**63.** The midpoint Riemann sum for $f(x) = x^3$ on $[3, 11]$ with $n = 32$

**64.** The midpoint Riemann sum for $f(x) = 1 + \cos \pi x$ on $[0, 2]$ with $n = 50$

**65–68. Identifying Riemann sums** *Fill in the blanks with an interval and a value of n.*

**65.** $\displaystyle\sum_{k=1}^{4} f(1 + k) \cdot 1$ is a right Riemann sum for $f$ on the interval

$[\underline{\quad}, \underline{\quad}]$ with $n = \underline{\quad}$.

**66.** $\displaystyle\sum_{k=1}^{4} f(2 + k) \cdot 1$ is a right Riemann sum for $f$ on the interval

$[\underline{\quad}, \underline{\quad}]$ with $n = \underline{\quad}$.

**67.** $\displaystyle\sum_{k=1}^{4} f(1.5 + k) \cdot 1$ is a midpoint Riemann sum for $f$ on the interval

$[\underline{\quad}, \underline{\quad}]$ with $n = \underline{\quad}$.

**68.** $\displaystyle\sum_{k=1}^{8} f\left(1.5 + \dfrac{k}{2}\right) \cdot \dfrac{1}{2}$ is a left Riemann sum for $f$ on the interval

$[\underline{\quad}, \underline{\quad}]$ with $n = \underline{\quad}$.

**69. Approximating areas** Estimate the area of the region bounded by the graph of $f(x) = x^2 + 2$ and the $x$-axis on $[0, 2]$ in the following ways.

**a.** Divide $[0, 2]$ into $n = 4$ subintervals and approximate the area of the region using a left Riemann sum. Illustrate the solution geometrically.

**b.** Divide $[0, 2]$ into $n = 4$ subintervals and approximate the area of the region using a midpoint Riemann sum. Illustrate the solution geometrically.

**c.** Divide $[0, 2]$ into $n = 4$ subintervals and approximate the area of the region using a right Riemann sum. Illustrate the solution geometrically.

**70. Displacement from a velocity graph** Consider the velocity function for an object moving along a line (see figure).

**a.** Describe the motion of the object over the interval $[0, 6]$.

**b.** Use geometry to find the displacement of the object between $t = 0$ and $t = 3$.

**c.** Use geometry to find the displacement of the object between $t = 3$ and $t = 5$.

**d.** Assuming the velocity remains 30 m/s, for $t \geq 4$, find the function that gives the displacement between $t = 0$ and any time $t \geq 4$.

**71. Displacement from a velocity graph** Consider the velocity function for an object moving along a line (see figure).

**a.** Describe the motion of the object over the interval $[0, 6]$.

**b.** Use geometry to find the displacement of the object between $t = 0$ and $t = 2$.

**c.** Use geometry to find the displacement of the object between $t = 2$ and $t = 5$.

**d.** Assuming the velocity remains 10 m/s, for $t \geq 5$, find the function that gives the displacement between $t = 0$ and any time $t \geq 5$.

**72. Flow rates** Suppose a gauge at the outflow of a reservoir measures the flow rate of water in units of $\text{ft}^3/\text{hr}$. In Chapter 6, we show that the total amount of water that flows out of the reservoir is the area under the flow rate curve. Consider the flow rate function shown in the figure.

**a.** Find the amount of water (in units of $\text{ft}^3$) that flows out of the reservoir over the interval $[0, 4]$.

**b.** Find the amount of water that flows out of the reservoir over the interval $[8, 10]$.

**c.** Does more water flow out of the reservoir over the interval $[0, 4]$ or $[4, 6]$?

**d.** Show that the units of your answer are consistent with the units of the variables on the axes.

73. **Mass from density** A thin 10-cm rod is made of an alloy whose density varies along its length according to the function shown in the figure. Assume density is measured in units of g/cm. In Chapter 6, we show that the mass of the rod is the area under the density curve.

    a. Find the mass of the left half of the rod ($0 \le x \le 5$).
    b. Find the mass of the right half of the rod ($5 \le x \le 10$).
    c. Find the mass of the entire rod ($0 \le x \le 10$).
    d. Find the point along the rod at which it will balance (called the center of mass).

Length (cm)

**74–75. Displacement from velocity** *The following functions describe the velocity of a car (in mi/hr) moving along a straight highway for a 3-hr interval. In each case, find the function that gives the displacement of the car over the interval* $[0, t]$, *where* $0 \le t \le 3$.

74. $v(t) = \begin{cases} 40 & \text{if } 0 \le t \le 1.5 \\ 50 & \text{if } 1.5 < t \le 3 \end{cases}$

Time (hr)

75. $v(t) = \begin{cases} 30 & \text{if } 0 \le t \le 2 \\ 50 & \text{if } 2 < t \le 2.5 \\ 44 & \text{if } 2.5 < t \le 3 \end{cases}$

Time (hr)

**T 76–77. Functions with absolute value** *Use a calculator and the method of your choice to approximate the area of the following regions. Present your calculations in a table, showing approximations using* $n = 16, 32,$ *and* $64$ *subintervals. Make a conjecture about the limits of the approximations.*

76. The region bounded by the graph of $f(x) = |25 - x^2|$ and the $x$-axis on the interval $[0, 10]$.

77. The region bounded by the graph of $f(x) = |1 - x^3|$ and the $x$-axis on the interval $[-1, 2]$.

### Explorations and Challenges

78. **Riemann sums for constant functions** Let $f(x) = c$, where $c > 0$, be a constant function on $[a, b]$. Prove that any Riemann sum for any value of $n$ gives the exact area of the region between the graph of $f$ and the $x$-axis on $[a, b]$.

79. **Riemann sums for linear functions** Assume the linear function $f(x) = mx + c$ is positive on the interval $[a, b]$. Prove that the midpoint Riemann sum with any value of $n$ gives the exact area of the region between the graph of $f$ and the $x$-axis on $[a, b]$.

80. **Shape of the graph for left Riemann sums** Suppose a left Riemann sum is used to approximate the area of the region bounded by the graph of a positive function and the $x$-axis on the interval $[a, b]$. Fill in the following table to indicate whether the resulting approximation underestimates or overestimates the exact area in the four cases shown. Use a sketch to explain your reasoning in each case.

|  | Increasing on $[a, b]$ | Decreasing on $[a, b]$ |
|---|---|---|
| Concave up on $[a, b]$ |  |  |
| Concave down on $[a, b]$ |  |  |

81. **Shape of the graph for right Riemann sums** Suppose a right Riemann sum is used to approximate the area of the region bounded by the graph of a positive function and the $x$-axis on the interval $[a, b]$. Fill in the following table to indicate whether the resulting approximation underestimates or overestimates the exact area in the four cases shown. Use a sketch to explain your reasoning in each case.

|  | Increasing on $[a, b]$ | Decreasing on $[a, b]$ |
|---|---|---|
| Concave up on $[a, b]$ |  |  |
| Concave down on $[a, b]$ |  |  |

**QUICK CHECK ANSWERS**

1. 45 mi   2. 0.25, 0.125, 7.875   3. $\Delta x = 2$; $\{1, 3, 5, 7, 9\}$
4. The left sum overestimates the area. ◄

# 5.2 Definite Integrals

We introduced Riemann sums in Section 5.1 as a way to approximate the area of a region bounded by a curve $y = f(x)$ and the $x$-axis on an interval $[a, b]$. In that discussion, we assumed $f$ to be nonnegative on the interval. Our next task is to discover the geometric meaning of Riemann sums when $f$ is negative on some or all of $[a, b]$. Once this matter is settled, we proceed to the main event of this section, which is to define the *definite integral*. With definite integrals, the approximations given by Riemann sums become exact.

## Net Area

How do we interpret Riemann sums when $f$ is negative at some or all points of $[a, b]$? The answer follows directly from the Riemann sum definition.

**EXAMPLE 1    Interpreting Riemann sums**  Evaluate and interpret the following Riemann sums for $f(x) = 1 - x^2$ on the interval $[a, b]$ with $n$ equally spaced subintervals.

**a.** A midpoint Riemann sum with $[a, b] = [1, 3]$ and $n = 4$

**b.** A left Riemann sum with $[a, b] = [0, 3]$ and $n = 6$

**SOLUTION**

**a.** The length of each subinterval is $\Delta x = \dfrac{b - a}{n} = \dfrac{3 - 1}{4} = 0.5$. So the grid points are

$$x_0 = 1, \quad x_1 = 1.5, \quad x_2 = 2, \quad \text{and} \quad x_3 = 2.5, \quad \text{and} \quad x_4 = 3.$$

To compute the midpoint Riemann sum, we evaluate $f$ at the midpoints of the subintervals, which are

$$x_1^* = 1.25, \quad x_2^* = 1.75, \quad x_3^* = 2.25, \quad \text{and} \quad x_4^* = 2.75.$$

The resulting midpoint Riemann sum is

$$\sum_{k=1}^{n} f(x_k^*)\,\Delta x = \sum_{k=1}^{4} f(x_k^*)(0.5)$$

$$= f(1.25)(0.5) + f(1.75)(0.5) + f(2.25)(0.5) + f(2.75)(0.5)$$

$$= (-0.5625 - 2.0625 - 4.0625 - 6.5625)0.5$$

$$= -6.625.$$

All values of $f(x_k^*)$ are negative, so the Riemann sum is also negative. Because area is always a nonnegative quantity, this Riemann sum does not approximate the area of the region between the curve and the $x$-axis on $[1, 3]$. Notice, however, that the values of $f(x_k^*)$ are the *negative* of the heights of the corresponding rectangles (**Figure 5.16**). Therefore, the Riemann sum approximates the *negative* of the area of the region bounded by the curve.

The midpoint Riemann sum for $f(x) = 1 - x^2$ on $[1, 3]$ is $-6.625$.

$f(x) = 1 - x^2$

**Figure 5.16**

**b.** The length of each subinterval is $\Delta x = \dfrac{b - a}{n} = \dfrac{3 - 0}{6} = 0.5$, and the grid points are

$$x_0 = 0, \quad x_1 = 0.5, \quad x_2 = 1, \quad x_3 = 1.5, \quad x_4 = 2, \quad x_5 = 2.5, \quad \text{and} \quad x_6 = 3.$$

To calculate the left Riemann sum, we set $x_1^*, x_2^*, \ldots, x_6^*$ equal to the left endpoints of the subintervals:

$$x_1^* = 0, \quad x_2^* = 0.5, \quad x_3^* = 1, \quad x_4^* = 1.5, \quad x_5^* = 2, \quad \text{and} \quad x_6^* = 2.5.$$

The resulting left Riemann sum is

$$\sum_{k=1}^{n} f(x_k^*)\Delta x = \sum_{k=1}^{6} f(x_k^*)(0.5)$$

$$= \underbrace{(f(0) + f(0.5) + f(1)}_{\text{nonnegative contribution}} + \underbrace{f(1.5) + f(2) + f(2.5))}_{\text{negative contribution}}0.5$$

$$= (1 + 0.75 + 0 - 1.25 - 3 - 5.25)0.5$$

$$= -3.875.$$

The left Riemann sum on [0, 1.5] is 0.875.

The left Riemann sum on [1.5, 3] is −4.75.

The resulting left Riemann sum on [0, 3] is −3.875.

$f(x) = 1 - x^2$

**Figure 5.17**

In this case, the values of $f(x_k^*)$ are nonnegative for $k = 1, 2,$ and 3, and negative for $k = 4, 5,$ and 6 (**Figure 5.17**). Where $f$ is positive, we get positive contributions to the Riemann sum, and where $f$ is negative, we get negative contributions to the sum.

*Related Exercises 17, 22* ◄

Let's recap what we learned in Example 1. On intervals where $f(x) < 0$, Riemann sums approximate the *negative* of the area of the region bounded by the curve (**Figure 5.18**).

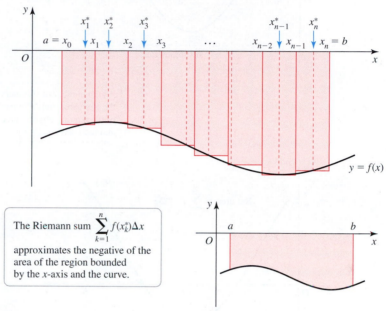

The Riemann sum $\sum_{k=1}^{n} f(x_k^*)\Delta x$ approximates the negative of the area of the region bounded by the $x$-axis and the curve.

**Figure 5.18**

In the more general case that $f$ is positive on part of $[a, b]$, we get positive contributions to the sum where $f$ is positive and negative contributions to the sum where $f$ is negative. In this case, Riemann sums approximate the area of the regions that lie above the $x$-axis *minus* the area of the regions that lie *below* the $x$-axis (**Figure 5.19**). This difference between the positive and negative contributions is called the *net area*; it can be positive, negative, or zero.

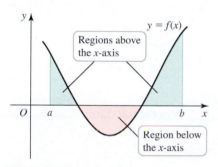

**Figure 5.19**

➤ Net area suggests the difference between positive and negative contributions much like net change or net profit. Some texts use the term **signed area** for net area.

**QUICK CHECK 1** Suppose $f(x) = -5$. What is the net area of the region bounded by the graph of $f$ and the $x$-axis on the interval $[1, 5]$? Make a sketch of the function and the region. ◄

**QUICK CHECK 2** Sketch a continuous function $f$ that is positive over the interval $[0, 1)$ and negative over the interval $(1, 2]$, such that the net area of the region bounded by the graph of $f$ and the $x$-axis on $[0, 2]$ is zero. ◄

**DEFINITION Net Area**

Consider the region $R$ bounded by the graph of a continuous function $f$ and the $x$-axis between $x = a$ and $x = b$. The **net area** of $R$ is the sum of the areas of the parts of $R$ that lie above the $x$-axis *minus* the sum of the areas of the parts of $R$ that lie below the $x$-axis on $[a, b]$.

## The Definite Integral

Riemann sums for $f$ on $[a, b]$ give *approximations* to the net area of the region bounded by the graph of $f$ and the $x$-axis between $x = a$ and $x = b$, where $a < b$. How can we make these approximations exact? If $f$ is continuous on $[a, b]$, it is reasonable to expect the Riemann sum approximations to approach the exact value of the net area as the number of subintervals $n \to \infty$ and as the length of the subintervals $\Delta x \to 0$ (**Figure 5.20**). In terms of limits, we write

$$\text{net area} = \lim_{n \to \infty} \sum_{k=1}^{n} f(x_k^*)\Delta x.$$

The Riemann sums we have used so far involve regular partitions in which the subintervals have the same length $\Delta x$. We now introduce partitions of $[a, b]$ in which the

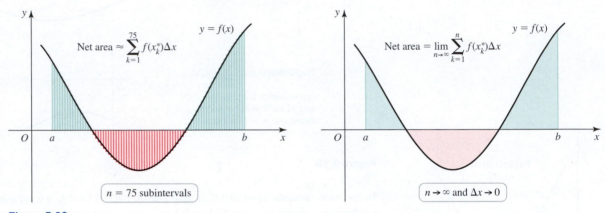

**Figure 5.20**

lengths of the subintervals are not necessarily equal. A **general partition** of $[a, b]$ consists of the $n$ subintervals

$$[x_0, x_1], [x_1, x_2], \ldots, [x_{n-1}, x_n],$$

where $x_0 = a$ and $x_n = b$. The length of the $k$th subinterval is $\Delta x_k = x_k - x_{k-1}$, for $k = 1, \ldots, n$. We let $x_k^*$ be any point in the subinterval $[x_{k-1}, x_k]$. This general partition is used to define the *general Riemann sum*.

---

**DEFINITION   General Riemann Sum**

Suppose $[x_0, x_1], [x_1, x_2], \ldots, [x_{n-1}, x_n]$ are subintervals of $[a, b]$ with

$$a = x_0 < x_1 < x_2 < \cdots < x_{n-1} < x_n = b.$$

Let $\Delta x_k$ be the length of the subinterval $[x_{k-1}, x_k]$ and let $x_k^*$ be any point in $[x_{k-1}, x_k]$, for $k = 1, 2, \ldots, n$.

If $f$ is defined on $[a, b]$, the sum

$$\sum_{k=1}^{n} f(x_k^*) \Delta x_k = f(x_1^*) \Delta x_1 + f(x_2^*) \Delta x_2 + \cdots + f(x_n^*) \Delta x_n$$

is called a **general Riemann sum for $f$ on $[a, b]$**.

As was the case for regular Riemann sums, if we choose $x_k^*$ to be the left endpoint of $[x_{k-1}, x_k]$, for $k = 1, 2, \ldots, n$, then the general Riemann sum is a left Riemann sum. Similarly, if we choose $x_k^*$ to be the right endpoint $[x_{k-1}, x_k]$, for $k = 1, 2, \ldots, n$, then the general Riemann sum is a right Riemann sum, and if we choose $x_k^*$ to be the midpoint of the interval $[x_{k-1}, x_k]$, for $k = 1, 2, \ldots, n$, then the general Riemann sum is a midpoint Riemann sum.

Now consider the limit of $\sum_{k=1}^{n} f(x_k^*) \Delta x_k$ as $n \to \infty$ and as *all* the $\Delta x_k \to 0$. We let $\Delta$ denote the largest value of $\Delta x_k$; that is, $\Delta = \max\{\Delta x_1, \Delta x_2, \ldots, \Delta x_n\}$. Observe that if $\Delta \to 0$, then $\Delta x_k \to 0$, for $k = 1, 2, \ldots, n$. For the limit $\lim_{\Delta \to 0} \sum_{k=1}^{n} f(x_k^*) \Delta x_k$ to exist, it must have the same value over all general partitions of $[a, b]$ and for all choices of $x_k^*$ on a partition.

> Note that $\Delta \to 0$ forces all $\Delta x_k \to 0$, which forces $n \to \infty$. Therefore, it suffices to write $\Delta \to 0$ in the limit.

---

**DEFINITION  Definite Integral**

A function $f$ defined on $[a, b]$ is **integrable** on $[a, b]$ if $\lim_{\Delta \to 0} \sum_{k=1}^{n} f(x_k^*) \Delta x_k$ exists and is unique over all partitions of $[a, b]$ and all choices of $x_k^*$ on a partition. This limit is the **definite integral of $f$ from $a$ to $b$**, which we write

$$\int_a^b f(x)\, dx = \lim_{\Delta \to 0} \sum_{k=1}^{n} f(x_k^*) \Delta x_k.$$

---

When the limit defining the definite integral of $f$ exists, it equals the net area of the region bounded by the graph of $f$ and the $x$-axis on $[a, b]$. It is imperative to remember that the indefinite integral $\int f(x)\, dx$ is a family of functions of $x$ (the antiderivatives of $f$) and that the definite integral $\int_a^b f(x)\, dx$ is a real number (the net area of a region).

**Notation**   The notation for the definite integral requires some explanation. There is a direct match between the notation on either side of the equation in the definition (**Figure 5.21**). In the limit as $\Delta \to 0$, the finite sum, denoted $\sum$, becomes a sum with an infinite number of terms, denoted $\int$. The integral sign $\int$ is an elongated $S$ for sum. The **limits of integration**, $a$ and $b$, and the limits of summation also match: The lower limit in the sum, $k = 1$, corresponds to the left endpoint of the interval, $x = a$, and the upper limit in the sum, $k = n$, corresponds to the right endpoint of the interval, $x = b$. The function under the integral sign is called the **integrand**. Finally, the differential $dx$ in the integral (which corresponds to $\Delta x_k$ in the sum) is an essential part of the notation; it tells us that the **variable of integration** is $x$.

The variable of integration is a dummy variable that is completely internal to the integral. It does not matter what the variable of integration is called, as long as it does not conflict with other variables that are in use. Therefore, the integrals in **Figure 5.22** all have the same meaning.

Figure 5.21

> For Leibniz, who introduced this notation in 1675, $dx$ represented the width of an infinitesimally thin rectangle and $f(x)\, dx$ represented the area of such a rectangle. He used $\int_a^b f(x)\, dx$ to denote the sum of these areas from $a$ to $b$.

Figure 5.22

The strategy of slicing a region into smaller parts, summing the results from the parts, and taking a limit is used repeatedly in calculus and its applications. We call this strategy the **slice-and-sum method**. It often results in a Riemann sum whose limit is a definite integral.

## Evaluating Definite Integrals

> ➤ A function $f$ is bounded on an interval $I$ if there is a number $M$ such that $|f(x)| < M$ for all $x$ in $I$.

Most of the functions encountered in this text are integrable (see Exercise 91 for an exception). In fact, if $f$ is continuous on $[a, b]$ or if $f$ is bounded on $[a, b]$ with a finite number of discontinuities, then $f$ is integrable on $[a, b]$. The proof of this result lies beyond the scope of this text.

---

**THEOREM 5.2  Integrable Functions**

If $f$ is continuous on $[a, b]$ or bounded on $[a, b]$ with a finite number of discontinuities, then $f$ is integrable on $[a, b]$.

---

**QUICK CHECK 3** Graph $f(x) = x$ and use geometry to evaluate $\int_{-1}^{1} x \, dx$. ◄

When $f$ is continuous on $[a, b]$, we have seen that the definite integral $\int_a^b f(x) \, dx$ is the net area bounded by the graph of $f$ and the $x$-axis on $[a, b]$. **Figure 5.23** illustrates how the idea of net area carries over to piecewise continuous functions (Exercises 84–88).

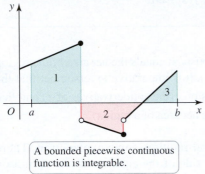

Net area $= \int_a^b f(x) \, dx$

$=$ area above $x$-axis (Regions 1 and 3)
$-$ area below $x$-axis (Region 2)

A bounded piecewise continuous function is integrable.

**Figure 5.23**

**EXAMPLE 2  Identifying the limit of a sum**  Assume

$$\lim_{\Delta \to 0} \sum_{k=1}^{n} (3x_k^{*2} + 2x_k^* + 1)\Delta x_k$$

is the limit of a Riemann sum for a function $f$ on $[1, 3]$. Identify the function $f$ and express the limit as a definite integral. What does the definite integral represent geometrically?

**SOLUTION**  By comparing the sum $\sum_{k=1}^{n} (3x_k^{*2} + 2x_k^* + 1)\Delta x_k$ to the general Riemann sum $\sum_{k=1}^{n} f(x_k^*)\Delta x_k$, we see that $f(x) = 3x^2 + 2x + 1$. Because $f$ is a polynomial, it is continuous on $[1, 3]$ and is, therefore, integrable on $[1, 3]$. It follows that

$$\lim_{\Delta \to 0} \sum_{k=1}^{n} (3x_k^{*2} + 2x_k^* + 1)\Delta x_k = \int_1^3 (3x^2 + 2x + 1) \, dx.$$

Because $f$ is positive on $[1, 3]$, the definite integral $\int_1^3 (3x^2 + 2x + 1) \, dx$ is the area of the region bounded by the curve $y = 3x^2 + 2x + 1$ and the $x$-axis on $[1, 3]$ (**Figure 5.24**).

$$\lim_{\Delta \to 0} \sum_{k=1}^{n} (3x_k^{*2} + 2x_k^* + 1)\Delta x_k = \int_1^3 (3x^2 + 2x + 1) dx$$

**Figure 5.24**

*Related Exercises 33–34* ◄

**EXAMPLE 3  Evaluating definite integrals using geometry**  Use familiar area formulas to evaluate the following definite integrals.

**a.** $\displaystyle\int_2^4 (2x + 3) \, dx$   **b.** $\displaystyle\int_1^6 (2x - 6) \, dx$   **c.** $\displaystyle\int_3^4 \sqrt{1 - (x - 3)^2} \, dx$

**Figure 5.25**

▶ **A trapezoid and its area** When $a = 0$, we get the area of a triangle. When $a = b$, we get the area of a rectangle.

**SOLUTION** To evaluate these definite integrals geometrically, a sketch of the corresponding region is essential.

**a.** The definite integral $\int_2^4 (2x + 3)\, dx$ is the area of the trapezoid bounded by the $x$-axis and the line $y = 2x + 3$ from $x = 2$ to $x = 4$ (**Figure 5.25**). The width of its base is 2 and the lengths of its two parallel sides are $f(2) = 7$ and $f(4) = 11$. Using the area formula for a trapezoid, we have

$$\int_2^4 (2x + 3)\, dx = \frac{1}{2} \cdot 2(11 + 7) = 18.$$

**b.** A sketch shows that the regions bounded by the line $y = 2x - 6$ and the $x$-axis are triangles (**Figure 5.26**). The area of the triangle on the interval $[1, 3]$ is $\frac{1}{2} \cdot 2 \cdot 4 = 4$. Similarly, the area of the triangle on $[3, 6]$ is $\frac{1}{2} \cdot 3 \cdot 6 = 9$. The definite integral is the net area of the entire region, which is the area of the triangle above the $x$-axis minus the area of the triangle below the $x$-axis:

$$\int_1^6 (2x - 6)\, dx = \text{net area} = 9 - 4 = 5.$$

**Figure 5.26**

Area of shaded region $= \frac{1}{4}\pi(1)^2 = \frac{1}{4}\pi$

**Figure 5.27**

**c.** We first let $y = \sqrt{1 - (x - 3)^2}$ and observe that $y \geq 0$ when $2 \leq x \leq 4$. Squaring both sides leads to the equation $(x - 3)^2 + y^2 = 1$, whose graph is a circle of radius 1 centered at $(3, 0)$. Because $y \geq 0$, the graph of $y = \sqrt{1 - (x - 3)^2}$ is the upper half of the circle. It follows that the integral $\int_3^4 \sqrt{1 - (x - 3)^2}\, dx$ is the area of a quarter circle of radius 1 (**Figure 5.27**). Therefore,

$$\int_3^4 \sqrt{1 - (x - 3)^2}\, dx = \frac{1}{4}\pi(1)^2 = \frac{\pi}{4}.$$

*Related Exercises 37–38, 41* ◄

**QUICK CHECK 4** Let $f(x) = 5$ and use geometry to evaluate $\int_1^3 f(x)\, dx$. What is the value of $\int_a^b c\, dx$, where $c$ is a real number? ◄

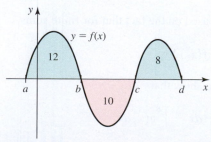

**Figure 5.28**

**EXAMPLE 4  Definite integrals from graphs** Figure 5.28 shows the graph of a function $f$ with the areas of the regions bounded by its graph and the $x$-axis given. Find the values of the following definite integrals.

**a.** $\displaystyle\int_a^b f(x)\, dx$    **b.** $\displaystyle\int_b^c f(x)\, dx$    **c.** $\displaystyle\int_a^c f(x)\, dx$    **d.** $\displaystyle\int_b^d f(x)\, dx$

**SOLUTION**

**a.** Because $f(x) \geq 0$ on $[a, b]$, the value of the definite integral is the area of the region between the graph and the $x$-axis on $[a, b]$; that is, $\int_a^b f(x)\, dx = 12$.

**b.** Because $f(x) \leq 0$ on $[b, c]$, the value of the definite integral is the negative of the area of the corresponding region; that is, $\int_b^c f(x)\, dx = -10$.

**c.** The value of the definite integral is the area of the region on $[a, b]$ (where $f(x) \geq 0$) minus the area of the region on $[b, c]$ (where $f(x) \leq 0$). Therefore, $\int_a^c f(x)\, dx = 12 - 10 = 2$.

**d.** Reasoning as in part (c), we have $\int_b^d f(x)\, dx = -10 + 8 = -2$.

*Related Exercises 57–60* ◄

## Properties of Definite Integrals

Recall that the definite integral $\int_a^b f(x)\,dx$ was defined assuming $a < b$. There are, however, occasions when it is necessary to reverse the limits of integration. If $f$ is integrable on $[a, b]$, we define

$$\int_b^a f(x)\,dx = -\int_a^b f(x)\,dx.$$

In other words, reversing the limits of integration changes the sign of the integral.

Another fundamental property of integrals is that if we integrate from a point to itself, then the length of the interval of integration is zero, which means the definite integral is also zero.

---

**DEFINITION**  **Reversing Limits and Identical Limits of Integration**

Suppose $f$ is integrable on $[a, b]$.

**1.** $\displaystyle\int_b^a f(x)\,dx = -\int_a^b f(x)\,dx$         **2.** $\displaystyle\int_a^a f(x)\,dx = 0$

---

**QUICK CHECK 5** Evaluate $\int_a^b f(x)\,dx + \int_b^a f(x)\,dx$ assuming $f$ is integrable on $[a, b]$. ◄

**Integral of a Sum**   Definite integrals possess other properties that often simplify their evaluation. Assume $f$ and $g$ are integrable on $[a, b]$. The first property states that their sum $f + g$ is integrable on $[a, b]$ and the integral of their sum is the sum of their integrals:

$$\int_a^b (f(x) + g(x))\,dx = \int_a^b f(x)\,dx + \int_a^b g(x)\,dx.$$

We prove this property assuming $f$ and $g$ are continuous. In this case, $f + g$ is continuous and, therefore, integrable. We then have

$$\int_a^b (f(x) + g(x))\,dx = \lim_{\Delta \to 0} \sum_{k=1}^n (f(x_k^*) + g(x_k^*))\Delta x_k \qquad \text{\color{blue}Definition of definite integral}$$

$$= \lim_{\Delta \to 0} \left( \sum_{k=1}^n f(x_k^*)\Delta x_k + \sum_{k=1}^n g(x_k^*)\Delta x_k \right) \qquad \text{\color{blue}Addition rule for finite sums}$$

$$= \lim_{\Delta \to 0} \sum_{k=1}^n f(x_k^*)\Delta x_k + \lim_{\Delta \to 0} \sum_{k=1}^n g(x_k^*)\Delta x_k \qquad \text{\color{blue}Limit of a sum}$$

$$= \int_a^b f(x)\,dx + \int_a^b g(x)\,dx. \qquad \text{\color{blue}Definition of definite integral}$$

**Constants in Integrals**   Another property of definite integrals is that constants can be factored out of the integral. If $f$ is integrable on $[a, b]$ and $c$ is a constant, then $cf$ is integrable on $[a, b]$ and

$$\int_a^b cf(x)\,dx = c\int_a^b f(x)\,dx.$$

The justification for this property (Exercise 89) is based on the fact that for finite sums,

$$\sum_{k=1}^n cf(x_k^*)\Delta x_k = c\sum_{k=1}^n f(x_k^*)\Delta x_k.$$

Combining the previous two properties, we can also show that

$$\int_a^b (f(x) - g(x))\,dx = \int_a^b f(x)\,dx - \int_a^b g(x)\,dx;$$

that is, the integral of the difference of two functions is the difference of the integrals.

**Integrals over Subintervals**   If the point $p$ lies between $a$ and $b$, then the integral on $[a, b]$ may be split into two integrals. As shown in **Figure 5.29**, we have the property

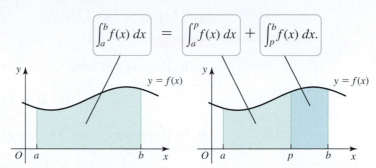

Figure 5.29

It is surprising that this property also holds when $p$ lies outside the interval $[a, b]$. For example, if $a < b < p$ and $f$ is integrable on $[a, p]$, then it follows (**Figure 5.30**) that

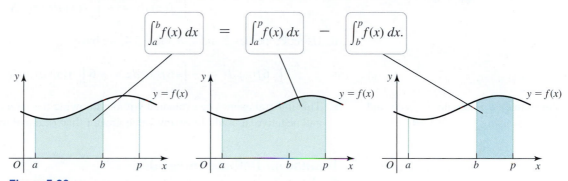

Figure 5.30

Because $\int_p^b f(x)\, dx = -\int_b^p f(x)\, dx$, we have the original property:

$$\int_a^b f(x)\, dx = \int_a^p f(x)\, dx + \int_p^b f(x)\, dx.$$

**Integrals of Absolute Values**   Finally, how do we interpret $\int_a^b |f(x)|\, dx$, the integral of the absolute value of an integrable function? The graphs $f$ and $|f|$ are shown in **Figure 5.31**. The integral $\int_a^b |f(x)|\, dx$ gives the area of regions $R_1^*$ and $R_2$. But $R_1$ and $R_1^*$ have the same area; therefore, $\int_a^b |f(x)|\, dx$ also gives the area of $R_1$ and $R_2$. The conclusion is that $\int_a^b |f(x)|\, dx$ is the area of the entire region (above and below the $x$-axis) that lies between the graph of $f$ and the $x$-axis on $[a, b]$.

All these properties will be used frequently in upcoming work. It's worth collecting them in one table (Table 5.4).

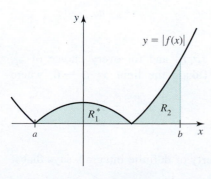

$$\int_a^b |f(x)|\, dx = \text{area of } R_1^* + \text{area of } R_2$$
$$= \text{area of } R_1 + \text{area of } R_2$$

**Figure 5.31**

**Table 5.4  Properties of definite integrals**

Let $f$ and $g$ be integrable functions on an interval that contains $a$, $b$, and $p$.

**1.** $\int_a^a f(x)\, dx = 0$   Definition

**2.** $\int_b^a f(x)\, dx = -\int_a^b f(x)\, dx$   Definition

**3.** $\int_a^b (f(x) \pm g(x))\, dx = \int_a^b f(x)\, dx \pm \int_a^b g(x)\, dx$

**4.** $\int_a^b c f(x)\, dx = c \int_a^b f(x)\, dx$, for any constant $c$

**5.** $\int_a^b f(x)\, dx = \int_a^p f(x)\, dx + \int_p^b f(x)\, dx$

**6.** The function $|f|$ is integrable on $[a, b]$, and $\int_a^b |f(x)|\, dx$ is the sum of the areas of the regions bounded by the graph of $f$ and the $x$-axis on $[a, b]$.

**EXAMPLE 5**  **Properties of integrals**  Assume $\int_0^5 f(x)\,dx = 3$ and $\int_0^7 f(x)\,dx = -10$. Evaluate the following integrals, if possible.

**a.** $\int_0^7 2f(x)\,dx$  **b.** $\int_5^7 f(x)\,dx$  **c.** $\int_5^0 f(x)\,dx$  **d.** $\int_7^0 6f(x)\,dx$  **e.** $\int_0^7 |f(x)|\,dx$

**SOLUTION**

**a.** By Property 4 of Table 5.4,

$$\int_0^7 2f(x)\,dx = 2\int_0^7 f(x)\,dx = 2 \cdot (-10) = -20.$$

**b.** By Property 5 of Table 5.4, $\int_0^7 f(x)\,dx = \int_0^5 f(x)\,dx + \int_5^7 f(x)\,dx$. Therefore,

$$\int_5^7 f(x)\,dx = \int_0^7 f(x)\,dx - \int_0^5 f(x)\,dx = -10 - 3 = -13.$$

**c.** By Property 2 of Table 5.4,

$$\int_5^0 f(x)\,dx = -\int_0^5 f(x)\,dx = -3.$$

**d.** Using Properties 2 and 4 of Table 5.4, we have

$$\int_7^0 6f(x)\,dx = -\int_0^7 6f(x)\,dx = -6\int_0^7 f(x)\,dx = (-6)(-10) = 60.$$

**QUICK CHECK 6** Evaluate $\int_{-1}^2 x\,dx$ and $\int_{-1}^2 |x|\,dx$ using geometry. ◀

**e.** This integral cannot be evaluated without knowing the intervals on which $f$ is positive and negative. It could have any value greater than or equal to 10.

*Related Exercises 49–50* ◀

## Bounds on Definite Integrals

We conclude our discussion of properties of definite integrals with three results that are helpful for upcoming theoretical work. We assume $f$ and $g$ are continuous on $[a, b]$, where $b > a$.

**Nonnegative Integrand**  If $f(x) \geq 0$ on $[a, b]$, it is geometrically apparent that $\int_a^b f(x)\,dx \geq 0$ (**Figure 5.32**). To prove this result, suppose $m$ is the absolute minimum value of $f$ on $[a, b]$, guaranteed to exist by Theorem 4.1 (note that $m \geq 0$). Working with a general partition of $[a, b]$, observe that

$\int_a^b f(x)\,dx$ = net area under the curve $y = f(x)$; when $f(x) \geq 0$, net area $\geq 0$.

**Figure 5.32**

$$0 \leq m(b - a) = m\sum_{k=1}^{n} \Delta x_k \qquad \sum_{k=1}^{n} \Delta x_k = (b - a)$$

$$= \sum_{k=1}^{n} m\,\Delta x_k \qquad \text{Property of finite sums, Section 5.1}$$

$$\leq \underbrace{\sum_{k=1}^{n} f(x_k^*)\,\Delta x_k}_{\substack{\text{General Riemann} \\ \text{sum for } f \text{ on } [a, b].}} \qquad m \leq f(x_k^*) \text{ for all } x_k^* \text{ in } [a, b]$$

We have shown that for any general partition of $[a, b]$ and for every choice of $x_k^*$, the general Riemann sum for $f$ is nonnegative. Taking the limit as $\Delta \to 0$, where $\Delta = \max\{\Delta x_1, \Delta x_2, \ldots, \Delta x_n\}$, we have

$$\int_a^b f(x)\,dx \geq 0.$$

If $f(x) \geq g(x)$ on $[a, b]$, then $\int_a^b f(x)\,dx \geq \int_a^b g(x)\,dx$.

**Figure 5.33**

**Comparing Definite Integrals**  A related property of definite integrals says that if $f(x) \geq g(x)$ on $[a, b]$ (**Figure 5.33**), then

$$\int_a^b f(x)\,dx \geq \int_a^b g(x)\,dx.$$

This property follows from the previous result and Property 3 of Table 5.4 (Exercise 94).

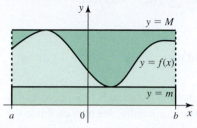

Figure 5.34

**Lower and Upper Bounds**    Because $f$ is continuous on $[a, b]$, it attains an absolute minimum value $m$ and an absolute maximum value $M$ on $[a, b]$ (Theorem 4.1). Our final property (**Figure 5.34**) claims that

$$m(b - a) \leq \int_a^b f(x)\,dx \leq M(b - a).$$

To prove this result, once again working with a general partition of $[a, b]$, notice that

$$m(b - a) = m\sum_{k=1}^{n}\Delta x_k \qquad \sum_{k=1}^{n}\Delta x_k = b - a$$

$$= \sum_{k=1}^{n}m\Delta x_k \qquad \text{Property of finite sums}$$

$$\leq \sum_{k=1}^{n}f(x_k^*)\Delta x_k \qquad m \leq f(x_k^*) \text{ for all } x_k^* \text{ in } [a, b]$$

$$\leq \sum_{k=1}^{n}M\Delta x_k \qquad f(x_k^*) \leq M \text{ for all } x_k^* \text{ in } [a, b]$$

$$= M\sum_{k=1}^{n}\Delta x_k \qquad \text{Property of finite sums}$$

$$= M(b - a). \qquad \sum_{k=1}^{n}\Delta x_k = b - a$$

We have shown that $m(b - a) \leq \sum_{k=1}^{n}f(x_k^*)\Delta x_k \leq M(b - a)$ for any partition of $[a, b]$ and for every choice of $x_k^*$. Letting $\Delta \to 0$, we obtain

$$m(b - a) \leq \int_a^b f(x)\,dx \leq M(b - a).$$

Table 5.5 summarizes our findings.

▶ Although we proved Properties 3, 4, 7, 8, and 9 of Tables 5.4 and 5.5 for continuous functions $f$ and $g$, these properties hold when $f$ and $g$ are integrable.

**Table 5.5   Additional properties of definite integrals**

Let $f$ and $g$ be integrable functions on $[a, b]$, where $b > a$.

**7.** If $f(x) \geq 0$ on $[a, b]$, then $\int_a^b f(x)\,dx \geq 0$.

**8.** If $f(x) \geq g(x)$ on $[a, b]$, then $\int_a^b f(x)\,dx \geq \int_a^b g(x)\,dx$.

**9.** If $m \leq f(x) \leq M$, then $m(b - a) \leq \int_a^b f(x)\,dx \leq M(b - a)$.

In the next section, Property 9 is used to prove the central result of this chapter, the Fundamental Theorem of Calculus.

## Evaluating Definite Integrals Using Limits

In Example 3, we used area formulas for trapezoids, triangles, and circles to evaluate definite integrals. Regions bounded by more general functions have curved boundaries for which conventional geometrical methods do not work. At this point in our discussion, the only way to handle such integrals is to appeal to the definition of the definite integral and the summation formulas given in Theorem 5.1.

We know that if $f$ is integrable on $[a, b]$, then $\int_a^b f(x)\,dx = \lim_{\Delta \to 0} \sum_{k=1}^{n} f(x_k^*)\Delta x_k$, for any partition of $[a, b]$ and any points $x_k^*$. To simplify these calculations, we use equally spaced grid points and right Riemann sums. That is, for each value of $n$, we let $\Delta x_k = \Delta x = \dfrac{b-a}{n}$ and $x_k^* = a + k\,\Delta x$, for $k = 1, 2, \ldots, n$. Then as $n \to \infty$ and $\Delta \to 0$,

$$\int_a^b f(x)\,dx = \lim_{\Delta \to 0} \sum_{k=1}^n f(x_k^*)\Delta x_k = \lim_{n \to \infty} \sum_{k=1}^n f(a + k\Delta x)\Delta x.$$

$$x_k^* = a + k\Delta x = \frac{2k}{n}$$
$$k = 1, \ldots, n$$

**EXAMPLE 6** **Evaluating definite integrals** Find the value of $\int_0^2 (x^3 + 1)\,dx$ by evaluating a right Riemann sum and letting $n \to \infty$.

**SOLUTION** Based on approximations found in Example 5, Section 5.1, we conjectured that the value of this integral is 6. To verify this conjecture, we now evaluate the integral exactly. The interval $[a, b] = [0, 2]$ is divided into $n$ subintervals of length $\Delta x = \dfrac{b-a}{n} = \dfrac{2}{n}$, which produces the grid points

$$x_k^* = a + k\Delta x = 0 + k \cdot \frac{2}{n} = \frac{2k}{n}, \qquad \text{for } k = 1, 2, \ldots, n.$$

Letting $f(x) = x^3 + 1$, the right Riemann sum is

> An analogous calculation could be done using left Riemann sums or midpoint Riemann sums.

$$\sum_{k=1}^n f(x_k^*)\Delta x = \sum_{k=1}^n \left( \left( \frac{2k}{n} \right)^3 + 1 \right) \frac{2}{n}$$

$$= \frac{2}{n} \sum_{k=1}^n \left( \frac{8k^3}{n^3} + 1 \right) \qquad \sum_{k=1}^n ca_k = c\sum_{k=1}^n a_k$$

$$= \frac{2}{n} \left( \frac{8}{n^3} \sum_{k=1}^n k^3 + \sum_{k=1}^n 1 \right) \qquad \sum_{k=1}^n (ca_k + b_k) = c\sum_{k=1}^n a_k + \sum_{k=1}^n b_k$$

$$= \frac{2}{n} \left( \frac{8}{n^3} \left( \frac{n^2(n+1)^2}{4} \right) + n \right) \qquad \sum_{k=1}^n k^3 = \frac{n^2(n+1)^2}{4} \text{ and } \sum_{k=1}^n 1 = n;$$
$$\qquad\qquad\qquad\qquad\qquad\qquad\qquad \text{Theorem 5.1}$$

$$= \frac{4(n^2 + 2n + 1)}{n^2} + 2. \qquad \text{Simplify.}$$

Now we evaluate $\int_0^2 (x^3 + 1)\,dx$ by letting $n \to \infty$ in the Riemann sum:

$$\int_0^2 (x^3 + 1)\,dx = \lim_{n \to \infty} \sum_{k=1}^n f(x_k^*)\Delta x$$

$$= \lim_{n \to \infty} \left( \frac{4(n^2 + 2n + 1)}{n^2} + 2 \right)$$

$$= 4 \underbrace{\lim_{n \to \infty} \left( \frac{n^2 + 2n + 1}{n^2} \right)}_{1} + \lim_{n \to \infty} 2$$

$$= 4(1) + 2 = 6.$$

Therefore, $\int_0^2 (x^3 + 1)\,dx = 6$, confirming our conjecture in Example 5, Section 5.1.

*Related Exercises 78, 80* ◄

The Riemann sum calculations in Example 6 are tedious even if $f$ is a simple function. For polynomials of degree 4 and higher, the calculations are more challenging, and for rational and transcendental functions, advanced mathematical results are needed. The next section introduces more efficient methods for evaluating definite integrals.

## SECTION 5.2 EXERCISES

### Getting Started

1. What does net area measure?

2. Under what conditions does the net area of a region (bounded by a continuous function) equal the area of a region? When does the net area of a region differ from the area of a region?

3. Use the graph of $y = f(x)$ to estimate $\int_0^6 f(x)\,dx$ using a left and right Riemann sum with $n = 6$.

4. Use the graph of $y = g(x)$ to estimate $\int_2^{10} g(x)\,dx$ using a left, right, and midpoint Riemann sum with $n = 4$.

5. Suppose $f$ is continuous on $[2, 8]$. Use the table of values of $f$ to estimate $\int_2^8 f(x)\,dx$ using a left, right, and midpoint Riemann sum with $n = 3$.

| $x$ | 2 | 3 | 4 | 5 | 6 | 7 | 8 |
|-----|---|----|----|----|----|----|----|
| $f(x)$ | 2 | −2 | −3 | −4 | −5 | −2 | −1 |

6. Suppose $g$ is continuous on $[1, 9]$. Use the table of values of $g$ to estimate $\int_1^9 g(x)\,dx$ using a left, right, and midpoint Riemann sum with $n = 4$.

| $x$ | 1 | 2 | 3 | 4 | 5 | 6 | 7 | 8 | 9 |
|-----|----|----|----|---|---|---|---|---|----|
| $g(x)$ | −5 | −4 | −3 | 0 | 2 | 4 | 5 | 0 | −1 |

7. Sketch a graph of $y = 2$ on $[-1, 4]$ and use geometry to find the exact value of $\int_{-1}^4 2\,dx$.

8. Sketch a graph of $y = -3$ on $[1, 5]$ and use geometry to find the exact value of $\int_1^5 (-3)\,dx$.

9. Sketch a graph of $y = 2x$ on $[-1, 2]$ and use geometry to find the exact value of $\int_{-1}^2 2x\,dx$.

10. Suppose $\int_1^3 f(x)\,dx = 10$ and $\int_1^3 g(x)\,dx = -20$. Evaluate $\int_1^3 (2f(x) - 4g(x))\,dx$ and $\int_3^1 (2f(x) - 4g(x))\,dx$.

11. Use graphs to evaluate $\int_0^{2\pi} \sin x\,dx$ and $\int_0^{2\pi} \cos x\,dx$.

12. Explain how the notation for Riemann sums, $\sum_{k=1}^n f(x_k^*)\Delta x_k$, corresponds to the notation for the definite integral, $\int_a^b f(x)\,dx$.

13. Give a geometrical explanation of why $\int_a^a f(x)\,dx = 0$.

14. Use Table 5.4 to rewrite $\int_1^6 (2x^3 - 4x)\,dx$ as the difference of two integrals.

15. Use geometry to find a formula for $\int_0^a x\,dx$, in terms of a constant $a > 0$.

16. If $f$ is continuous on $[a, b]$ and $\int_a^b |f(x)|\,dx = 0$, what can you conclude about $f$?

### Practice Exercises

**17–20. Approximating net area** *The following functions are negative on the given interval.*

a. *Sketch the function on the interval.*
b. *Approximate the net area bounded by the graph of f and the x-axis on the interval using a left, right, and midpoint Riemann sum with $n = 4$.*

17. $f(x) = -2x - 1$ on $[0, 4]$    **T** 18. $f(x) = -4 - x^3$ on $[3, 7]$

**T** 19. $f(x) = \sin 2x$ on $\left[\dfrac{\pi}{2}, \pi\right]$    **T** 20. $f(x) = x^3 - 1$ on $[-2, 0]$

**T 21–24. Approximating net area** *The following functions are positive and negative on the given interval.*

a. *Sketch the function on the interval.*
b. *Approximate the net area bounded by the graph of f and the x-axis on the interval using a left, right, and midpoint Riemann sum with $n = 4$.*
c. *Use the sketch in part (a) to show which intervals of $[a, b]$ make positive and negative contributions to the net area.*

21. $f(x) = 4 - 2x$ on $[0, 4]$    22. $f(x) = 8 - 2x^2$ on $[0, 4]$

23. $f(x) = \sin 2x$ on $\left[0, \dfrac{3\pi}{4}\right]$    24. $f(x) = x^3$ on $[-1, 2]$

**25–28. Area versus net area** *Graph the following functions. Then use geometry (not Riemann sums) to find the area and the net area of the region described.*

25. The region between the graph of $y = -3x$ and the x-axis, for $-2 \le x \le 2$

26. The region between the graph of $y = 4x - 8$ and the x-axis, for $-4 \le x \le 8$

27. The region between the graph of $y = 1 - |x|$ and the x-axis, for $-2 \le x \le 2$

28. The region between the graph of $y = 3x - 6$ and the x-axis, for $0 \le x \le 6$

**T 29–32. Approximating definite integrals** *Complete the following steps for the given integral and the given value of n.*

a. *Sketch the graph of the integrand on the interval of integration.*
b. *Calculate $\Delta x$ and the grid points $x_0, x_1, \ldots, x_n$, assuming a regular partition.*
c. *Calculate the left and right Riemann sums for the given value of n.*
d. *Determine which Riemann sum (left or right) underestimates the value of the definite integral and which overestimates the value of the definite integral.*

**29.** $\int_3^6 (1 - 2x)\, dx;\ n = 6$    **30.** $\int_0^2 (x^2 - 2)\, dx;\ n = 4$

**31.** $\int_1^7 \frac{1}{x}\, dx;\ n = 6$    **32.** $\int_0^{\pi/2} \cos x\, dx;\ n = 4$

**33–36. Identifying definite integrals as limits of sums** *Consider the following limits of Riemann sums for a function f on $[a, b]$. Identify f and express the limit as a definite integral.*

**33.** $\displaystyle\lim_{\Delta \to 0} \sum_{k=1}^{n} (x_k^{*2} + 1)\Delta x_k$ on $[0, 2]$

**34.** $\displaystyle\lim_{\Delta \to 0} \sum_{k=1}^{n} (4 - x_k^{*2})\Delta x_k$ on $[-2, 2]$

**35.** $\displaystyle\lim_{\Delta \to 0} \sum_{k=1}^{n} x_k^{*}(\cos x_k^{*})\Delta x_k$ on $[1, 2]$

**36.** $\displaystyle\lim_{\Delta \to 0} \sum_{k=1}^{n} |x_k^{*2} - 1|\Delta x_k$ on $[-2, 2]$

**37–44. Definite integrals** *Use geometry (not Riemann sums) to evaluate the following definite integrals. Sketch a graph of the integrand, show the region in question, and interpret your result.*

**37.** $\int_0^4 (8 - 2x)\, dx$    **38.** $\int_{-4}^2 (2x + 4)\, dx$

**39.** $\int_{-1}^2 (-|x|)\, dx$    **40.** $\int_0^2 (1 - x)\, dx$

**41.** $\int_0^4 \sqrt{16 - x^2}\, dx$    **42.** $\int_{-1}^3 \sqrt{4 - (x - 1)^2}\, dx$

**43.** $\int_0^4 f(x)\, dx$, where $f(x) = \begin{cases} 5 & \text{if } x \le 2 \\ 3x - 1 & \text{if } x > 2 \end{cases}$

**44.** $\int_1^{10} g(x)\, dx$, where $g(x) = \begin{cases} 4x & \text{if } 0 \le x \le 2 \\ -8x + 16 & \text{if } 2 < x \le 3 \\ -8 & \text{if } x > 3 \end{cases}$

**45–48.** *The accompanying figure shows four regions bounded by the graph of $y = x \sin x$: $R_1, R_2, R_3,$ and $R_4$, whose areas are $1$, $\pi - 1$, $\pi + 1$, and $2\pi - 1$, respectively. (We verify these results later in the text.) Use this information to evaluate the following integrals.*

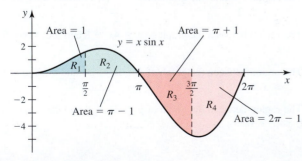

**45.** $\int_0^{\pi} x \sin x\, dx$    **46.** $\int_0^{3\pi/2} x \sin x\, dx$

**47.** $\int_0^{2\pi} x \sin x\, dx$    **48.** $\int_{\pi/2}^{2\pi} x \sin x\, dx$

**49. Properties of integrals** Use only the fact that $\int_0^4 3x(4 - x)\, dx = 32$, and the definitions and properties of integrals, to evaluate the following integrals, if possible.

a. $\int_4^0 3x(4 - x)\, dx$    b. $\int_0^4 x(x - 4)\, dx$

c. $\int_4^0 6x(4 - x)\, dx$    d. $\int_0^8 3x(4 - x)\, dx$

**50. Properties of integrals** Suppose $\int_1^4 f(x)\, dx = 8$ and $\int_1^6 f(x)\, dx = 5$. Evaluate the following integrals.

a. $\int_1^4 (-3f(x))\, dx$    b. $\int_1^4 3f(x)\, dx$

c. $\int_6^4 12f(x)\, dx$    d. $\int_4^6 3f(x)\, dx$

**51. Properties of integrals** Suppose $\int_0^3 f(x)\, dx = 2$, $\int_3^6 f(x)\, dx = -5$, and $\int_3^6 g(x)\, dx = 1$. Evaluate the following integrals.

a. $\int_0^3 5f(x)\, dx$    b. $\int_3^6 (-3g(x))\, dx$

c. $\int_3^6 (3f(x) - g(x))\, dx$    d. $\int_6^3 (f(x) + 2g(x))\, dx$

**52. Properties of integrals** Suppose $f(x) \ge 0$ on $[0, 2]$, $f(x) \le 0$ on $[2, 5]$, $\int_0^2 f(x)\, dx = 6$, and $\int_2^5 f(x)\, dx = -8$. Evaluate the following integrals.

a. $\int_0^5 f(x)\, dx$    b. $\int_0^5 |f(x)|\, dx$

c. $\int_2^5 4|f(x)|\, dx$    d. $\int_0^5 (f(x) + |f(x)|)\, dx$

**53. Properties of integrals** Consider two functions $f$ and $g$ on $[1, 6]$ such that $\int_1^6 f(x)\, dx = 10$, $\int_1^6 g(x)\, dx = 5$, $\int_4^6 f(x)\, dx = 5$, and $\int_1^4 g(x)\, dx = 2$. Evaluate the following integrals.

a. $\int_1^4 3f(x)\, dx$    b. $\int_1^6 (f(x) - g(x))\, dx$

c. $\int_1^4 (f(x) - g(x))\, dx$    d. $\int_4^6 (g(x) - f(x))\, dx$

e. $\int_4^6 8g(x)\, dx$    f. $\int_4^1 2f(x)\, dx$

**54.** Suppose $f$ is continuous on $[1, 5]$ and $2 \le f(x) \le 3$, for all $x$ in $[1, 5]$. Find lower and upper bounds for $\int_1^5 f(x)\, dx$.

**55–56. Using properties of integrals** *Use the value of the first integral I to evaluate the two given integrals.*

**55.** $I = \int_0^1 (x^3 - 2x)\, dx = -\dfrac{3}{4}$

a. $\int_0^1 (4x - 2x^3)\, dx$    b. $\int_1^0 (2x - x^3)\, dx$

**56.** $I = \int_0^{\pi/2} (\cos \theta - 2 \sin \theta)\, d\theta = -1$

a. $\int_0^{\pi/2} (2 \sin \theta - \cos \theta)\, d\theta$    b. $\int_{\pi/2}^0 (4 \cos \theta - 8 \sin \theta)\, d\theta$

**57–64. Definite integrals from graphs** *The figure shows the areas of regions bounded by the graph of f and the x-axis. Evaluate the following integrals.*

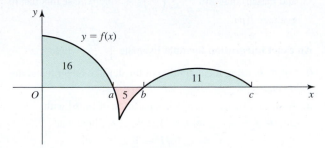

**57.** $\int_0^a f(x)\, dx$

**58.** $\int_0^b f(x)\, dx$

**59.** $\int_a^c f(x)\, dx$

**60.** $\int_0^c f(x)\, dx$

**61.** $\int_0^c |f(x)|\, dx$

**62.** $\int_0^c (2|f(x)| + 3f(x))\, dx$

**63.** $\int_a^0 f(x)\, dx$

**64.** $\int_c^0 |f(x)|\, dx$

**65.** Use geometry and properties of integrals to evaluate
$\int_0^1 (2x + \sqrt{1 - x^2} + 1)\, dx$.

**66.** Use geometry and properties of integrals to evaluate
$\int_1^5 (|x - 2| + \sqrt{-x^2 + 6x - 5})\, dx$.

**67. Explain why or why not** Determine whether the following statements are true and give an explanation or counterexample.

**a.** If $f$ is a constant function on the interval $[a, b]$, then the right and left Riemann sums give the exact value of $\int_a^b f(x)\, dx$, for any positive integer $n$.

**b.** If $f$ is a linear function on the interval $[a, b]$, then a midpoint Riemann sum gives the exact value of $\int_a^b f(x)\, dx$, for any positive integer $n$.

**c.** $\int_0^{2\pi/a} \sin ax\, dx = \int_0^{2\pi/a} \cos ax\, dx = 0$ (*Hint:* Graph the functions and use properties of trigonometric functions.)

**d.** If $\int_a^b f(x)\, dx = \int_b^a f(x)\, dx$, then $f$ is a constant function.

**e.** Property 4 of Table 5.4 implies that $\int_a^b xf(x)\, dx = x\int_a^b f(x)\, dx$.

**T 68–70. Approximating definite integrals with a calculator** *Consider the following definite integrals.*

**a.** *Write the left and right Riemann sums in sigma notation for an arbitrary value of n.*

**b.** *Evaluate each sum using a calculator with n = 20, 50, and 100. Use these values to estimate the value of the integral.*

**68.** $\int_4^9 3\sqrt{x}\, dx$

**69.** $\int_0^1 (x^2 + 1)\, dx$

**70.** $\int_{-1}^1 \pi \cos\left(\frac{\pi x}{2}\right) dx$

**T 71–74. Midpoint Riemann sums with a calculator** *Consider the following definite integrals.*

**a.** *Write the midpoint Riemann sum in sigma notation for an arbitrary value of n.*

**b.** *Evaluate each sum using a calculator with n = 20, 50, and 100. Use these values to estimate the value of the integral.*

**71.** $\int_1^4 2\sqrt{x}\, dx$

**72.** $\int_{-1}^2 \sin\left(\frac{\pi x}{4}\right) dx$

**73.** $\int_0^4 (4x - x^2)\, dx$

**74.** $\int_0^3 \frac{2}{\sqrt{x + 1}}\, dx$

**75–81. Limits of sums** *Use the definition of the definite integral to evaluate the following definite integrals. Use right Riemann sums and Theorem 5.1.*

**75.** $\int_0^2 (2x + 1)\, dx$

**76.** $\int_1^5 (1 - x)\, dx$

**77.** $\int_3^7 (4x + 6)\, dx$

**78.** $\int_0^2 (x^2 - 1)\, dx$

**79.** $\int_1^4 (x^2 - 1)\, dx$

**80.** $\int_0^2 (x^3 + x + 1)\, dx$

**81.** $\int_0^1 (4x^3 + 3x^2)\, dx$

### Explorations and Challenges

**82–83. Area by geometry** *Use geometry to evaluate the following integrals.*

**82.** $\int_1^6 |2x - 4|\, dx$

**83.** $\int_{-6}^4 \sqrt{24 - 2x - x^2}\, dx$

**84. Integrating piecewise continuous functions** Suppose $f$ is continuous on the intervals $[a, p]$ and $[p, b]$, where $a < p < b$, with a finite jump at $p$. Form a uniform partition on the interval $[a, p]$ with $n$ grid points and another uniform partition on the interval $[p, b]$ with $m$ grid points, where $p$ is a grid point of both partitions. Write a Riemann sum for $\int_a^b f(x)\, dx$ and separate it into two pieces for $[a, p]$ and $[p, b]$. Explain why $\int_a^b f(x)\, dx = \int_a^p f(x)\, dx + \int_p^b f(x)\, dx$.

**85–86. Integrating piecewise continuous functions** *Use geometry and the result of Exercise 84 to evaluate the following integrals.*

**85.** $\int_0^{10} f(x)\, dx$, where $f(x) = \begin{cases} 2 & \text{if } 0 \le x \le 5 \\ 3 & \text{if } 5 < x \le 10 \end{cases}$

**86.** $\int_1^6 f(x)\, dx$, where $f(x) = \begin{cases} 2x & \text{if } 1 \le x < 4 \\ 10 - 2x & \text{if } 4 \le x \le 6 \end{cases}$

**87–88. Integrating piecewise continuous functions** *Recall that the floor function $\lfloor x \rfloor$ is the greatest integer less than or equal to $x$ and that the ceiling function $\lceil x \rceil$ is the least integer greater than or equal to $x$. Use the result of Exercise 84 and the graphs to evaluate the following integrals.*

**87.** $\int_1^5 x\lfloor x \rfloor\, dx$

**88.** $\int_0^4 \dfrac{x}{\lceil x \rceil}\,dx$

$y = \dfrac{x}{\lceil x \rceil}$

**89. Constants in integrals** Use the definition of the definite integral to justify the property $\int_a^b cf(x)\,dx = c\int_a^b f(x)\,dx$, where $f$ is continuous and $c$ is a real number.

**90. Zero net area** Assuming $0 < c < d$, find the value of $b$ (in terms of $c$ and $d$) for which $\int_c^d (x + b)\,dx = 0$.

**91. A nonintegrable function** Consider the function defined on $[0, 1]$ such that $f(x) = 1$ if $x$ is a rational number and $f(x) = 0$ if $x$ is irrational. This function has an infinite number of discontinuities, and the integral $\int_0^1 f(x)\,dx$ does not exist. Show that the right, left, and midpoint Riemann sums on *regular* partitions with $n$ subintervals equal 1 for all $n$. (*Hint:* Between any two real numbers lie a rational and an irrational number.)

**92. Powers of $x$ by Riemann sums** Consider the integral $I(p) = \int_0^1 x^p\,dx$, where $p$ is a positive integer.

**a.** Write the left Riemann sum for the integral with $n$ subintervals.

**b.** It is a fact (proved by the 17th-century mathematicians Fermat and Pascal) that $\displaystyle \lim_{n\to\infty} \frac{1}{n}\sum_{k=0}^{n-1}\left(\frac{k}{n}\right)^p = \frac{1}{p+1}$. Use this fact to evaluate $I(p)$.

**93. An exact integration formula** Evaluate $\int_a^b \dfrac{dx}{x^2}$, where $0 < a < b$, using the definition of the definite integral and the following steps.

**a.** Assume $\{x_0, x_1, \ldots, x_n\}$ is a partition of $[a, b]$ with $\Delta x_k = x_k - x_{k-1}$, for $k = 1, 2, \ldots, n$. Show that
$$x_{k-1} \le \sqrt{x_{k-1}x_k} \le x_k, \text{ for } k = 1, 2, \ldots, n.$$

**b.** Show that $\dfrac{1}{x_{k-1}} - \dfrac{1}{x_k} = \dfrac{\Delta x_k}{x_{k-1}x_k}$, for $k = 1, 2, \ldots, n$.

**c.** Simplify the general Riemann sum for $\int_a^b \dfrac{dx}{x^2}$ using $x_k^* = \sqrt{x_{k-1}x_k}$.

**d.** Conclude that $\int_a^b \dfrac{dx}{x^2} = \dfrac{1}{a} - \dfrac{1}{b}$.

(*Source: The College Mathematics Journal*, 32, 4, Sep 2001)

**94.** Use Property 3 of Table 5.4 and Property 7 of Table 5.5 to prove Property 8 of Table 5.5.

**QUICK CHECK ANSWERS**

**1.** $-20$    **2.** $f(x) = 1 - x$ is one possibility.    **3.** $0$
**4.** $10; c(b - a)$    **5.** $0$    **6.** $\frac{3}{2}; \frac{5}{2}$ ◄

# 5.3 Fundamental Theorem of Calculus

Evaluating definite integrals using limits of Riemann sums, as described in Section 5.2, is usually not possible or practical. Fortunately, there is a powerful and practical method for evaluating definite integrals, which is developed in this section. Along the way, we discover the inverse relationship between differentiation and integration, expressed in the most important result of calculus, the Fundamental Theorem of Calculus. The first step in this process is to introduce *area functions* (first seen in Section 1.2).

## Area Functions

The concept of an area function is crucial to the discussion about the connection between derivatives and integrals. We start with a continuous function $y = f(t)$ defined for $t \ge a$, where $a$ is a fixed number. The *area function* for $f$ with left endpoint $a$ is denoted $A(x)$; it gives the net area of the region bounded by the graph of $f$ and the $t$-axis between $t = a$ and $t = x$ (**Figure 5.35**). The net area of this region is also given by the definite integral.

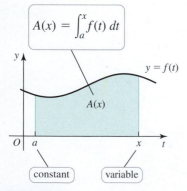

$A(x) = \int_a^x f(t)\,dt$

$y = f(t)$

$A(x)$

constant    variable

**Figure 5.35**

Independent variable of the area function

$$A(x) = \int_a^x f(t)\,dt$$

Variable of integration (dummy variable)

> ▶ Suppose you want to devise a function $A$ whose value is the sum of the first $n$ positive integers, $1 + 2 + \cdots + n$. The most compact way to write this function is $A(n) = \sum_{k=1}^{n} k$, for $n \geq 1$. In this function, $n$ is the independent variable of the function $A$, and $k$, which is internal to the sum, is a dummy variable. This function involving a sum is analogous to an area function involving an integral.

Notice that $x$ is the upper limit of the integral *and* the independent variable of the area function: As $x$ changes, so does the net area under the curve. Because the symbol $x$ is already in use as the independent variable for $A$, we must choose another symbol for the variable of integration. Any symbol—except $x$—can be used because it is a *dummy variable*; we have chosen $t$ as the integration variable.

**Figure 5.36** gives a general view of how an area function is generated. Suppose $f$ is a continuous function and $a$ is a fixed number. Now choose a point $b > a$. The net area of the region between the graph of $f$ and the $t$-axis on the interval $[a, b]$ is $A(b)$. Moving the right endpoint to $(c, 0)$ or $(d, 0)$ produces different regions with net areas $A(c)$ and $A(d)$, respectively. In general, if $x > a$ is a variable point, then $A(x) = \int_a^x f(t)\, dt$ is the net area of the region between the graph of $f$ and the $t$-axis on the interval $[a, x]$.

> ▶ Notice that $t$ is the independent variable when we plot $f$, and $x$ is the independent variable when we plot $A$.

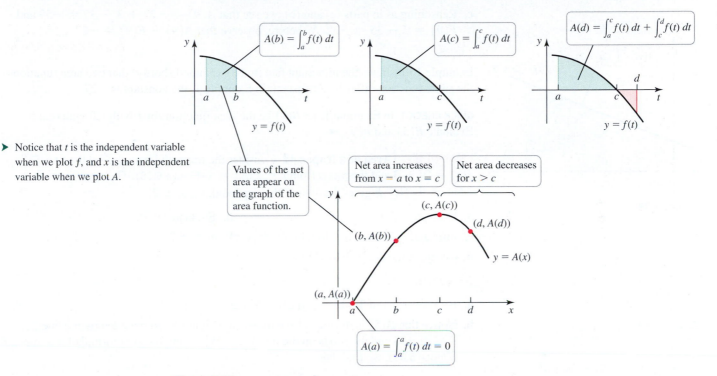

**Figure 5.36**

Figure 5.36 shows how $A(x)$ varies with respect to $x$. Notice that $A(a) = \int_a^a f(t)\, dt = 0$. Then for $x > a$, the net area increases for $x < c$, at which point $f(c) = 0$. For $x > c$, the function $f$ is negative, which produces a negative contribution to the area function. As a result, the area function decreases for $x > c$.

---

**DEFINITION  Area Function**

Let $f$ be a continuous function, for $t \geq a$. The **area function for $f$ with left endpoint $a$** is

$$A(x) = \int_a^x f(t)\, dt,$$

where $x \geq a$. The area function gives the net area of the region bounded by the graph of $f$ and the $t$-axis on the interval $[a, x]$.

---

The following two examples illustrate the idea of area functions.

**Figure 5.37**

**EXAMPLE 1  Comparing area functions** The graph of $f$ is shown in **Figure 5.37** with areas of various regions marked. Let $A(x) = \int_{-1}^x f(t)\, dt$ and $F(x) = \int_3^x f(t)\, dt$ be two area functions for $f$ (note the different left endpoints). Evaluate the following area functions.

**a.** $A(3)$ and $F(3)$         **b.** $A(5)$ and $F(5)$         **c.** $A(9)$ and $F(9)$

**SOLUTION**

**a.** The value of $A(3) = \int_{-1}^{3} f(t)\,dt$ is the net area of the region bounded by the graph of $f$ and the $t$-axis on the interval $[-1, 3]$. Using the graph of $f$, we see that $A(3) = -27$ (because this region has an area of 27 and lies below the $t$-axis). On the other hand, $F(3) = \int_{3}^{3} f(t)\,dt = 0$ by Property 1 of Table 5.4. Notice that $A(3) - F(3) = -27$.

**b.** The value of $A(5) = \int_{-1}^{5} f(t)\,dt$ is found by subtracting the area of the region that lies below the $t$-axis on $[-1, 3]$ from the area of the region that lies above the $t$-axis on $[3, 5]$. Therefore, $A(5) = 3 - 27 = -24$. Similarly, $F(5)$ is the net area of the region bounded by the graph of $f$ and the $t$-axis on the interval $[3, 5]$; therefore, $F(5) = 3$. Notice that $A(5) - F(5) = -27$.

**c.** Reasoning as in parts (a) and (b), we see that $A(9) = -27 + 3 - 35 = -59$ and $F(9) = 3 - 35 = -32$. As before, observe that $A(9) - F(9) = -27$.

*Related Exercises 13–14* ◄

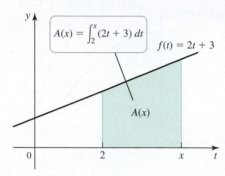

Figure 5.38

Example 1 illustrates the important fact (to be explained shortly) that two area functions of the same function differ by a constant; in Example 1, the constant is $-27$.

**QUICK CHECK 1** In Example 1, let $B(x)$ be the area function for $f$ with left endpoint 5. Evaluate $B(5)$ and $B(9)$. ◄

**EXAMPLE 2   Area of a trapezoid** Consider the trapezoid bounded by the line $f(t) = 2t + 3$ and the $t$-axis from $t = 2$ to $t = x$ (**Figure 5.38**). The area function $A(x) = \int_{2}^{x} f(t)\,dt$ gives the area of the trapezoid, for $x \geq 2$.

**a.** Evaluate $A(2)$.                    **b.** Evaluate $A(5)$.

**c.** Find and graph the area function $y = A(x)$, for $x \geq 2$.

**d.** Compare the derivative of $A$ to $f$.

Figure 5.39

**SOLUTION**

**a.** By Property 1 of Table 5.4, $A(2) = \int_{2}^{2} (2t + 3)\,dt = 0$.

**b.** Notice that $A(5)$ is the area of the trapezoid (Figure 5.38) bounded by the line $y = 2t + 3$ and the $t$-axis on the interval $[2, 5]$. Using the area formula for a trapezoid (**Figure 5.39**), we find that

$$A(5) = \int_{2}^{5} (2t + 3)\,dt = \frac{1}{2}\underbrace{(5 - 2)}_{\substack{\text{distance between}\\\text{parallel sides}}} \cdot \underbrace{(f(2) + f(5))}_{\substack{\text{sum of parallel}\\\text{side lengths}}} = \frac{1}{2} \cdot 3(7 + 13) = 30.$$

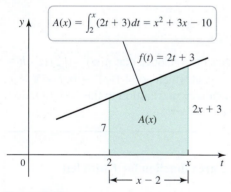

Figure 5.40

**c.** Now the right endpoint of the base is a variable $x \geq 2$ (**Figure 5.40**). The distance between the parallel sides of the trapezoid is $x - 2$. By the area formula for a trapezoid, the area of this trapezoid, for any $x \geq 2$, is

$$A(x) = \frac{1}{2}\underbrace{(x - 2)}_{\substack{\text{distance between}\\\text{parallel sides}}} \cdot \underbrace{(f(2) + f(x))}_{\substack{\text{sum of parallel}\\\text{side lengths}}}$$

$$= \frac{1}{2}(x - 2)(7 + 2x + 3)$$

$$= (x - 2)(x + 5)$$

$$= x^2 + 3x - 10.$$

Expressing the area function in terms of an integral with a variable upper limit, we have

$$A(x) = \int_{2}^{x} (2t + 3)\,dt = x^2 + 3x - 10, \text{ for } x \geq 2.$$

Figure 5.41

Because the line $f(t) = 2t + 3$ is above the $t$-axis, for $t \geq 2$, the area function $A(x) = x^2 + 3x - 10$ is an increasing function of $x$ with $A(2) = 0$ (**Figure 5.41**).

➤ Recall that if $A'(x) = f(x)$, then $f$ is the derivative of $A$; equivalently, $A$ is an antiderivative of $f$.

**QUICK CHECK 2** Verify that the area function in Example 2c gives the correct area when $x = 6$ and $x = 10$. ◄

**d.** Differentiating the area function, we find that

$$A'(x) = \frac{d}{dx}(x^2 + 3x - 10) = 2x + 3 = f(x).$$

Therefore, $A'(x) = f(x)$, or equivalently, the area function $A$ is an antiderivative of $f$. We soon show that this relationship is not an accident; it is the first part of the Fundamental Theorem of Calculus.

*Related Exercises 18–19* ◄

## Fundamental Theorem of Calculus

Example 2 suggests that the area function $A$ for a linear function $f$ is an antiderivative of $f$; that is, $A'(x) = f(x)$. Our goal is to show that this conjecture holds for more general functions. Let's start with an intuitive argument; a formal proof is given at the end of the section.

Assume $f$ is a continuous function defined on an interval $[a, b]$. As before, $A(x) = \int_a^x f(t)\, dt$ is the area function for $f$ with a left endpoint $a$: It gives the net area of the region bounded by the graph of $f$ and the $t$-axis on the interval $[a, x]$, for $x \geq a$. Figure 5.42 is the key to the argument.

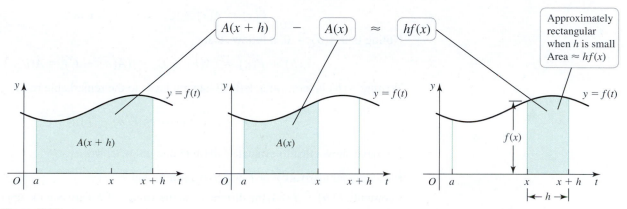

**Figure 5.42**

Note that with $h > 0$, $A(x + h)$ is the net area of the region whose base is the interval $[a, x + h]$ and $A(x)$ is the net area of the region whose base is the interval $[a, x]$. So the difference $A(x + h) - A(x)$ is the net area of the region whose base is the interval $[x, x + h]$. If $h$ is small, the region in question is nearly rectangular with a base of length $h$ and a height $f(x)$. Therefore, the net area of this region is

$$A(x + h) - A(x) \approx hf(x).$$

Dividing by $h$, we have

$$\frac{A(x + h) - A(x)}{h} \approx f(x).$$

➤ Recall that

$$f'(x) = \lim_{h \to 0} \frac{f(x + h) - f(x)}{h}.$$

If the function $f$ is replaced with $A$, then

$$A'(x) = \lim_{h \to 0} \frac{A(x + h) - A(x)}{h}.$$

An analogous argument can be made with $h < 0$. Now observe that as $h$ tends to zero, this approximation improves. In the limit as $h \to 0$, we have

$$\underbrace{\lim_{h \to 0} \frac{A(x + h) - A(x)}{h}}_{A'(x)} = \underbrace{\lim_{h \to 0} f(x)}_{f(x)}.$$

We see that indeed $A'(x) = f(x)$. Because $A(x) = \int_a^x f(t)\, dt$, the result can also be written

$$A'(x) = \frac{d}{dx}\underbrace{\int_a^x f(t)\, dt}_{A(x)} = f(x),$$

which says that the derivative of the integral of $f$ is $f$. This conclusion is the first part of the Fundamental Theorem of Calculus.

---

**THEOREM 5.3 (PART 1)   Fundamental Theorem of Calculus**

If $f$ is continuous on $[a, b]$, then the area function

$$A(x) = \int_a^x f(t)\, dt, \quad \text{for} \quad a \le x \le b,$$

is continuous on $[a, b]$ and differentiable on $(a, b)$. The area function satisfies $A'(x) = f(x)$. Equivalently,

$$A'(x) = \frac{d}{dx} \int_a^x f(t)\, dt = f(x),$$

which means that the area function of $f$ is an antiderivative of $f$ on $[a, b]$.

---

Given that $A$ is an antiderivative of $f$ on $[a, b]$, it is one short step to a powerful method for evaluating definite integrals. Remember (Section 4.9) that any two antiderivatives of $f$ differ by a constant. Assuming $F$ is any other antiderivative of $f$ on $[a, b]$, we have

$$F(x) = A(x) + C, \quad \text{for} \quad a \le x \le b.$$

Noting that $A(a) = 0$, it follows that

$$F(b) - F(a) = (A(b) + C) - (A(a) + C) = A(b).$$

Writing $A(b)$ in terms of a definite integral leads to the remarkable result

$$A(b) = \int_a^b f(x)\, dx = F(b) - F(a).$$

We have shown that to evaluate a definite integral of $f$, we

• find any antiderivative of $f$, which we call $F$; and

• compute $F(b) - F(a)$, the difference in the values of $F$ between the upper and lower limits of integration.

This process comprises the second part of the Fundamental Theorem of Calculus.

---

**THEOREM 5.3 (PART 2)   Fundamental Theorem of Calculus**

If $f$ is continuous on $[a, b]$ and $F$ is any antiderivative of $f$ on $[a, b]$, then

$$\int_a^b f(x)\, dx = F(b) - F(a).$$

---

It is customary and convenient to denote the difference $F(b) - F(a)$ by $F(x)\big|_a^b$. Using this shorthand, Part 2 of the Fundamental Theorem is summarized in **Figure 5.43**.

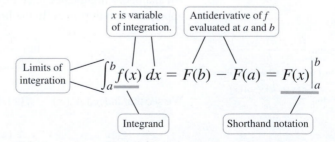

**Figure 5.43**

**QUICK CHECK 3** Evaluate $\left( \dfrac{x}{x + 1} \right)\Big|_1^2$.  ◄

**The Inverse Relationship Between Differentiation and Integration**    It is worth pausing to observe that the two parts of the Fundamental Theorem express the inverse relationship between differentiation and integration. Part 1 of the Fundamental Theorem says

$$\frac{d}{dx} \int_a^x f(t) \, dt = f(x),$$

or the derivative of the integral of $f$ is $f$ itself.

Noting that $f$ is an antiderivative of $f'$, Part 2 of the Fundamental Theorem says

$$\int_a^b f'(x) \, dx = f(b) - f(a),$$

**QUICK CHECK 4** Explain why $f$ is an antiderivative of $f'$. ◄

or the definite integral of the derivative of $f$ is given in terms of $f$ evaluated at two points. In other words, the integral "undoes" the derivative.

This last relationship is important because it expresses the integral as an *accumulation* operation. Suppose we know the rate of change of $f$ (which is $f'$) on an interval $[a, b]$. The Fundamental Theorem says that we can integrate (that is, sum or accumulate) the rate of change over that interval, and the result is simply the difference in $f$ evaluated at the endpoints. You will see this accumulation property used many times in the next chapter. Now let's use the Fundamental Theorem to evaluate definite integrals.

**EXAMPLE 3    Evaluating definite integrals**    Evaluate the following definite integrals using the Fundamental Theorem of Calculus, Part 2. Interpret each result geometrically.

**a.** $\int_0^{10} (60x - 6x^2) \, dx$    **b.** $\int_4^{16} 3\sqrt{x} \, dx$    **c.** $\int_0^{2\pi} 3 \sin x \, dx$    **d.** $\int_{1/16}^{1/4} \frac{\sqrt{t} - 2t}{t} \, dt$

**SOLUTION**

**a.** Using the antiderivative rules of Section 4.9, an antiderivative of $f(x) = 60x - 6x^2$ is $F(x) = 30x^2 - 2x^3$. By the Fundamental Theorem, the value of the definite integral is

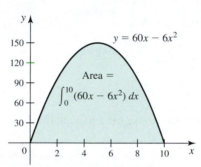

Figure 5.44

$$\int_0^{10} (60x - 6x^2) \, dx = \underbrace{(30x^2 - 2x^3)}_{F(x)} \Big|_0^{10} \qquad \text{Fundamental Theorem}$$

$$= \underbrace{(30 \cdot 10^2 - 2 \cdot 10^3)}_{F(10)} - \underbrace{(30 \cdot 0^2 - 2 \cdot 0^3)}_{F(0)} \qquad \begin{array}{l}\text{Evaluate at } x = 10 \\ \text{and } x = 0.\end{array}$$

$$= (3000 - 2000) - 0$$

$$= 1000. \qquad \text{Simplify.}$$

Because $f(x) = 60x - 6x^2$ is positive on $[0, 10]$, the definite integral $\int_0^{10} (60x - 6x^2) \, dx$ is the area of the region between the graph of $f$ and the $x$-axis on the interval $[0, 10]$ (**Figure 5.44**).

**b.** Because $f(x) = 3\sqrt{x}$ is positive on $[4, 16]$ (**Figure 5.45**), the definite integral $\int_4^{16} 3\sqrt{x} \, dx$ equals the area of the region under the graph of $f$ on $[4, 16]$. In Example 3 of Section 5.1, we used a midpoint Riemann sum to show that the area is approximately 112.062. Using the Fundamental Theorem, we can compute the exact area. Noting that an antiderivative of $x^{1/2}$ is $\frac{2}{3}x^{3/2}$, we have

Figure 5.45

$$\int_4^{16} 3\sqrt{x} \, dx = 3 \int_4^{16} x^{1/2} \, dx \qquad \text{Property 4, Table 5.4}$$

$$= 3 \cdot \frac{2}{3} x^{3/2} \Big|_4^{16} \qquad \text{Fundamental Theorem}$$

$$= 2(16^{3/2} - 4^{3/2}) \qquad \text{Evaluate at } x = 16 \text{ and } x = 4.$$

$$= 2(64 - 8) = 112. \qquad \text{Simplify.}$$

**c.** As shown in **Figure 5.46**, the region bounded by the graph of $f(x) = 3 \sin x$ and the $x$-axis on $[0, 2\pi]$ consists of two parts, one above the $x$-axis and one below the $x$-axis. By the symmetry of $f$, these two regions have the same area, so the definite integral

Figure 5.46

➤ When evaluating definite integrals, factor out multiplicative constants when possible, as we did in parts (b) and (c) of Example 3. To illustrate: It is better to write

$$-3 \cos x \Big|_0^{2\pi} = -3(\cos 2\pi - \cos 0) \text{ than}$$

$$-3 \cos x \Big|_0^{2\pi} = (-3 \cos 2\pi) - (-3 \cos 0).$$

over $[0, 2\pi]$, which gives the net area of the entire region, is zero. Let's confirm this fact. An antiderivative of $f(x) = 3 \sin x$ is $-3 \cos x$. Therefore, the value of the definite integral is

$$\int_0^{2\pi} 3 \sin x \, dx = -3 \cos x \Big|_0^{2\pi} \quad \text{Fundamental Theorem}$$

$$= -3(\cos 2\pi - \cos 0) \quad \text{Substitute.}$$

$$= -3(1 - 1) = 0. \quad \text{Simplify.}$$

**d.** Although the variable of integration is $t$, rather than $x$, we proceed as before, after simplifying the integrand:

$$f(t) = \frac{\sqrt{t} - 2t}{t} = \frac{1}{\sqrt{t}} - 2.$$

➤ We know that

$$\frac{d}{dt}(t^{1/2}) = \frac{1}{2}t^{-1/2}.$$

Therefore, $\int \frac{1}{2}t^{-1/2} dt = t^{1/2} + C$

and $\int \frac{dt}{\sqrt{t}} = \int t^{-1/2} dt = 2t^{1/2} + C.$

Finding antiderivatives with respect to $t$ and applying the Fundamental Theorem, we have

$$\int_{1/16}^{1/4} \frac{\sqrt{t} - 2t}{t} \, dt = \int_{1/16}^{1/4} (t^{-1/2} - 2) \, dt \quad \begin{array}{l}\text{Simplify the} \\ \text{integrand.}\end{array}$$

$$= (2t^{1/2} - 2t) \Big|_{1/16}^{1/4} \quad \begin{array}{l}\text{Fundamental} \\ \text{Theorem}\end{array}$$

$$= \left(2\left(\frac{1}{4}\right)^{1/2} - \frac{1}{2}\right) - \left(2\left(\frac{1}{16}\right)^{1/2} - \frac{1}{8}\right) \quad \text{Evaluate.}$$

$$= 1 - \frac{1}{2} - \frac{1}{2} + \frac{1}{8} \quad \text{Simplify.}$$

$$= \frac{1}{8}.$$

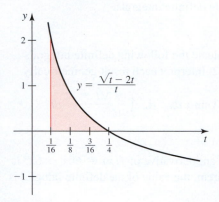

**Figure 5.47**

The definite integral is positive because the graph of $f$ lies above the $t$-axis on the interval of integration $\left[\frac{1}{16}, \frac{1}{4}\right]$ (**Figure 5.47**).

*Related Exercises 23–24, 26, 36* ◄

**EXAMPLE 4  Net areas and definite integrals** The graph of $f(x) = 6x(x + 1)(x - 2)$ is shown in **Figure 5.48**. The region $R_1$ is bounded by the curve and the $x$-axis on the interval $[-1, 0]$, and $R_2$ is bounded by the curve and the $x$-axis on the interval $[0, 2]$.

**a.** Find the *net area* of the region between the curve and the $x$-axis on $[-1, 2]$.

**b.** Find the *area* of the region between the curve and the $x$-axis on $[-1, 2]$.

**SOLUTION**

**a.** The net area of the region is given by a definite integral. The integrand $f$ is first expanded to find an antiderivative:

$$\int_{-1}^{2} f(x) \, dx = \int_{-1}^{2} (6x^3 - 6x^2 - 12x) \, dx. \quad \text{Expand } f.$$

$$= \left(\frac{3}{2}x^4 - 2x^3 - 6x^2\right)\Big|_{-1}^{2} \quad \text{Fundamental Theorem}$$

$$= -\frac{27}{2}. \quad \text{Simplify.}$$

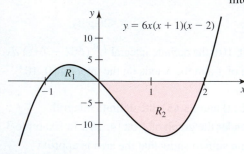

**Figure 5.48**

The net area of the region between the curve and the $x$-axis on $[-1, 2]$ is $-\frac{27}{2}$, which is the area of $R_1$ *minus* the area of $R_2$ (Figure 5.48). Because $R_2$ has a larger area than $R_1$, the net area is negative.

**b.** The region $R_1$ lies above the $x$-axis, so its area is

$$\int_{-1}^{0} (6x^3 - 6x^2 - 12x) \, dx = \left(\frac{3}{2}x^4 - 2x^3 - 6x^2\right)\Big|_{-1}^{0} = \frac{5}{2}.$$

The region $R_2$ lies below the $x$-axis, so its net area is negative:

$$\int_0^2 (6x^3 - 6x^2 - 12x)\, dx = \left(\frac{3}{2}x^4 - 2x^3 - 6x^2\right)\Big|_0^2 = -16.$$

Therefore, the *area* of $R_2$ is $-(-16) = 16$. The combined area of $R_1$ and $R_2$ is $\frac{5}{2} + 16 = \frac{37}{2}$. We could also find the area of this region directly by evaluating $\int_{-1}^2 |f(x)|\, dx$.

*Related Exercises 66, 68* ◄

Examples 3 and 4 make use of Part 2 of the Fundamental Theorem, which is the most potent tool for evaluating definite integrals. The remaining examples illustrate the use of the equally important Part 1 of the Fundamental Theorem.

**EXAMPLE 5   Derivatives of integrals**  Use Part 1 of the Fundamental Theorem to simplify the following expressions.

**a.** $\dfrac{d}{dx}\displaystyle\int_1^x \sin^2 t\, dt$    **b.** $\dfrac{d}{dx}\displaystyle\int_x^5 \sqrt{t^2 + 1}\, dt$    **c.** $\dfrac{d}{dx}\displaystyle\int_0^{x^2} \cos t^2\, dt$

**SOLUTION**

**a.** Using Part 1 of the Fundamental Theorem, we see that

$$\frac{d}{dx}\int_1^x \sin^2 t\, dt = \sin^2 x.$$

**b.** To apply Part 1 of the Fundamental Theorem, the variable must appear in the upper limit. Therefore, we use the fact that $\int_a^b f(t)\, dt = -\int_b^a f(t)\, dt$ and then apply the Fundamental Theorem:

$$\frac{d}{dx}\int_x^5 \sqrt{t^2 + 1}\, dt = -\frac{d}{dx}\int_5^x \sqrt{t^2 + 1}\, dt = -\sqrt{x^2 + 1}.$$

**c.** The upper limit of the integral is not $x$, but a function of $x$. Therefore, the function to be differentiated is a composite function, which requires the Chain Rule. We let $u = x^2$ to produce

$$y = g(u) = \int_0^u \cos t^2\, dt.$$

By the Chain Rule,

$$\frac{d}{dx}\int_0^{x^2} \cos t^2\, dt = \frac{dy}{dx} = \frac{dy}{du}\frac{du}{dx} \qquad \text{\color{blue}{Chain Rule}}$$

➤ Example 5c illustrates one case of Leibniz's Rule:

$$\frac{d}{dx}\int_a^{g(x)} f(t)\, dt = f(g(x))g'(x).$$

$$= \left(\frac{d}{du}\int_0^u \cos t^2\, dt\right)(2x) \qquad \text{\color{blue}{Substitute for $g$; note that $u'(x) = 2x$.}}$$

$$= (\cos u^2)(2x) \qquad \text{\color{blue}{Fundamental Theorem}}$$

$$= 2x \cos x^4. \qquad \text{\color{blue}{Substitute $u = x^2$.}}$$

*Related Exercises 71, 73, 76* ◄

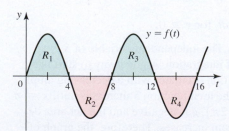

**Figure 5.49**

**EXAMPLE 6   Working with area functions**  Consider the function $f$ shown in **Figure 5.49** and its area function $A(x) = \int_0^x f(t)\, dt$, for $0 \le x \le 17$. Assume the four regions $R_1, R_2, R_3$, and $R_4$ have the same area. Based on the graph of $f$, do the following.

**a.** Find the zeros of $A$ on $[0, 17]$.

**b.** Find the points on $[0, 17]$ at which $A$ has a local maximum or a local minimum.

**c.** Sketch a graph of $A$, for $0 \le x \le 17$.

**SOLUTION**

**a.** The area function $A(x) = \int_0^x f(t)\, dt$ gives the net area bounded by the graph of $f$ and the $t$-axis on the interval $[0, x]$ (**Figure 5.50a**). Therefore, $A(0) = \int_0^0 f(t)\, dt = 0$. Because $R_1$ and $R_2$ have the same area but lie on opposite sides of the $t$-axis, it follows that $A(8) = \int_0^8 f(t)\, dt = 0$. Similarly, $A(16) = \int_0^{16} f(t)\, dt = 0$. Therefore, the zeros of $A$ are $x = 0, 8$, and 16.

> ➤ Recall that local extrema occur only at interior points of the domain.

**b.** Observe that the function $f$ is positive, for $0 < t < 4$, which implies that $A(x)$ increases as $x$ increases from 0 to 4 (**Figure 5.50b**). Then as $x$ increases from 4 to 8, $A(x)$ decreases because $f$ is negative, for $4 < t < 8$ (**Figure 5.50c**). Similarly, $A(x)$ increases as $x$ increases from $x = 8$ to $x = 12$ (**Figure 5.50d**) and decreases from $x = 12$ to $x = 16$. By the First Derivative Test, $A$ has local minima at $x = 8$ and $x = 16$ and local maxima at $x = 4$ and $x = 12$ (**Figure 5.50e**).

**c.** Combining the observations in parts (a) and (b) leads to a qualitative sketch of $A$ (Figure 5.50e). Note that $A(x) \geq 0$, for all $x \geq 0$. It is not possible to determine function values ($y$-coordinates) on the graph of $A$.

$A(x) = \int_0^x f(t)\, dt =$ net area over $[0, x]$.

(a)

The net area $A(x)$ increases as $x$ increases from 0 to 4.

(b)

The net area decreases as $x$ increases from 4 to 8.

(c)

The net area $A(x)$ increases as $x$ increases from 8 to 12.

(d)

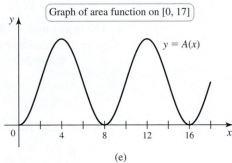

Graph of area function on [0, 17]

$y = A(x)$

(e)

**Figure 5.50**

*Related Exercises 69–70* ◀

**EXAMPLE 7  The sine integral function**  Let

$$g(t) = \begin{cases} \dfrac{\sin t}{t} & \text{if } t > 0 \\ 1 & \text{if } t = 0. \end{cases}$$

Graph the *sine integral function* $S(x) = \int_0^x g(t)\, dt$, for $x \geq 0$.

**SOLUTION**  Notice that $S$ is an area function for $g$. The independent variable of $S$ is $x$, and $t$ has been chosen as the (dummy) variable of integration. A good way to start is by graphing the integrand $g$ (**Figure 5.51a**). The function oscillates with a decreasing amplitude with $g(0) = 1$. Beginning with $S(0) = 0$, the area function $S$ increases until $x = \pi$ because $g$ is positive on $(0, \pi)$. However, on $(\pi, 2\pi)$, $g$ is negative and the net area decreases. On $(2\pi, 3\pi)$, $g$ is positive again, so $S$ again increases. Therefore, the graph of $S$ has alternating local maxima and minima. Because the amplitude of $g$ decreases, each maximum of $S$ is less than the previous maximum, and each minimum of $S$ is greater than the previous minimum (**Figure 5.51b**). Determining the exact value of $S$ at these maxima and minima is difficult.

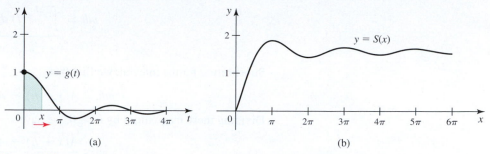

**Figure 5.51**

Appealing to Part 1 of the Fundamental Theorem, we find that

$$S'(x) = \frac{d}{dx} \int_0^x g(t)\, dt = \frac{\sin x}{x}, \text{ for } x > 0.$$

> Note that
>
> $$\lim_{x \to \infty} S'(x) = \lim_{x \to \infty} g(x) = 0.$$

As anticipated, the derivative of $S$ changes sign at integer multiples of $\pi$. Specifically, $S'$ is positive and $S$ increases on the intervals $(0, \pi), (2\pi, 3\pi), \ldots,$ $(2n\pi, (2n + 1)\pi), \ldots,$ while $S'$ is negative and $S$ decreases on the remaining intervals. Clearly, $S$ has local maxima at $x = \pi, 3\pi, 5\pi, \ldots,$ and it has local minima at $x = 2\pi, 4\pi, 6\pi, \ldots.$

One more observation is helpful. It can be shown that although $S$ oscillates for increasing $x$, its graph gradually flattens out and approaches a horizontal asymptote. (Finding the exact value of this horizontal asymptote is challenging; see Exercise 113.) Assembling all these observations, the graph of the sine integral function emerges (Figure 5.51b).

*Related Exercises 99–100* ◄

We conclude this section with a formal proof of the Fundamental Theorem of Calculus.

**Proof of the Fundamental Theorem:** Let $f$ be continuous on $[a, b]$ and let $A$ be the area function for $f$ with left endpoint $a$. The first step is to prove that $A$ is differentiable on $(a, b)$ and $A'(x) = f(x)$, which is Part 1 of the Fundamental Theorem. The proof of Part 2 then follows.

**Step 1.** We assume $a < x < b$ and use the definition of the derivative,

$$A'(x) = \lim_{h \to 0} \frac{A(x + h) - A(x)}{h}.$$

First assume $h > 0$. Using **Figure 5.52** and Property 5 of Table 5.4, we have

$$A(x + h) - A(x) = \int_a^{x+h} f(t)\, dt - \int_a^x f(t)\, dt = \int_x^{x+h} f(t)\, dt.$$

That is, $A(x + h) - A(x)$ is the net area of the region bounded by the curve on the interval $[x, x + h]$.

**Figure 5.52**

> ▶ The quantities $m$ and $M$ exist for any $h > 0$; however, their values depend on $h$.

Let $m$ and $M$ be the minimum and maximum values of $f$ on $[x, x + h]$, respectively, which exist by the continuity of $f$. Note that $m \leq f(t) \leq M$ on $[x, x + h]$ (which is an interval of length $h$). It follows, by Property 9 of Table 5.5, that

$$mh \leq \underbrace{\int_x^{x+h} f(t)\, dt}_{A(x+h)\,-\,A(x)} \leq Mh.$$

Substituting for the integral, we find that

$$mh \leq A(x + h) - A(x) \leq Mh.$$

Dividing these inequalities by $h > 0$, we have

$$m \leq \frac{A(x + h) - A(x)}{h} \leq M.$$

The case $h < 0$ is handled similarly and leads to the same conclusion.

We now take the limit as $h \to 0$ across these inequalities. As $h \to 0$, $m$ and $M$ approach $f(x)$, because $f$ is continuous at $x$. At the same time, as $h \to 0$, the quotient that is sandwiched between $m$ and $M$ approaches $A'(x)$:

$$\underbrace{\lim_{h \to 0} m}_{f(x)} \leq \underbrace{\lim_{h \to 0} \frac{A(x + h) - A(x)}{h}}_{A'(x)} \leq \underbrace{\lim_{h \to 0} M}_{f(x)}.$$

By the Squeeze Theorem (Theorem 2.5), we conclude that $A'(x)$ exists and $A$ is differentiable for $a < x < b$. Furthermore, $A'(x) = f(x)$. Finally, because $A$ is differentiable on $(a, b)$, $A$ is continuous on $(a, b)$ by Theorem 3.1. Exercise 116 shows that $A$ is also right- and left-continuous at the endpoints $a$ and $b$, respectively.

> ▶ Once again we use an important fact: Two antiderivatives of the same function differ by a constant.

**Step 2.** Having established that the area function $A$ is an antiderivative of $f$, we know that $F(x) = A(x) + C$, where $F$ is any antiderivative of $f$ and $C$ is a constant. Noting that $A(a) = 0$, it follows that

$$F(b) - F(a) = (A(b) + C) - (A(a) + C) = A(b).$$

Writing $A(b)$ in terms of a definite integral, we have

$$A(b) = \int_a^b f(x)\, dx = F(b) - F(a),$$

which is Part 2 of the Fundamental Theorem.    ◀

## SECTION 5.3 EXERCISES

### Getting Started

1. Suppose $A$ is an area function of $f$. What is the relationship between $f$ and $A$?

2. Suppose $F$ is an antiderivative of $f$ and $A$ is an area function of $f$. What is the relationship between $F$ and $A$?

3. Explain mathematically how the Fundamental Theorem of Calculus is used to evaluate definite integrals.

4. Let $f(x) = c$, where $c$ is a positive constant. Explain why an area function of $f$ is an increasing function.

5. The linear function $f(x) = 3 - x$ is decreasing on the interval $[0, 3]$. Is its area function for $f$ (with left endpoint 0) increasing or decreasing on the interval $[0, 3]$? Draw a picture and explain.

6. Evaluate $\int_0^2 3x^2\, dx$ and $\int_{-2}^2 3x^2\, dx$.

7. Explain in words and express mathematically the inverse relationship between differentiation and integration as given by Part 1 of the Fundamental Theorem of Calculus.

8. Why can the constant of integration be omitted from the antiderivative when evaluating a definite integral?

9. Evaluate $\dfrac{d}{dx} \int_a^x f(t)\, dt$ and $\dfrac{d}{dx} \int_a^b f(t)\, dt$, where $a$ and $b$ are constants.

10. Explain why $\int_a^b f'(x)\, dx = f(b) - f(a)$.

11. Evaluate $\int_3^8 f'(t)\, dt$, where $f'$ is continuous on $[3, 8]$, $f(3) = 4$, and $f(8) = 20$.

12. Evaluate $\int_2^7 3\, dx$ using the Fundamental Theorem of Calculus. Check your work by evaluating the integral using geometry.

## Practice Exercises

**13. Area functions** The graph of $f$ is shown in the figure. Let $A(x) = \int_{-2}^{x} f(t)\, dt$ and $F(x) = \int_{4}^{x} f(t)\, dt$ be two area functions for $f$. Evaluate the following area functions.

   **a.** $A(-2)$   **b.** $F(8)$   **c.** $A(4)$   **d.** $F(4)$   **e.** $A(8)$

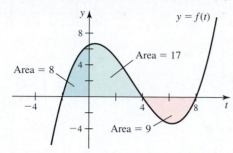

**14. Area functions** The graph of $f$ is shown in the figure. Let $A(x) = \int_{0}^{x} f(t)\, dt$ and $F(x) = \int_{2}^{x} f(t)\, dt$ be two area functions for $f$. Evaluate the following area functions.

   **a.** $A(2)$   **b.** $F(5)$   **c.** $A(0)$   **d.** $F(8)$
   **e.** $A(8)$   **f.** $A(5)$   **g.** $F(2)$

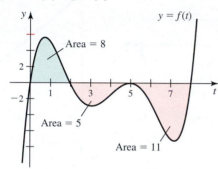

**15–16. Area functions for constant functions** *Consider the following functions $f$ and real numbers $a$ (see figure).*

**a.** Find and graph the area function $A(x) = \int_{a}^{x} f(t)\, dt$ for $f$.
**b.** Verify that $A'(x) = f(x)$.

**15.** $f(t) = 5, a = 0$        **16.** $f(t) = 5, a = -5$

**17. Area functions for the same linear function** Let $f(t) = t$ and consider the two area functions $A(x) = \int_{0}^{x} f(t)\, dt$ and $F(x) = \int_{2}^{x} f(t)\, dt$.

   **a.** Evaluate $A(2)$ and $A(4)$. Then use geometry to find an expression for $A(x)$, for $x \geq 0$.
   **b.** Evaluate $F(4)$ and $F(6)$. Then use geometry to find an expression for $F(x)$, for $x \geq 2$.
   **c.** Show that $A(x) - F(x)$ is a constant and that $A'(x) = F'(x) = f(x)$.

**18. Area functions for the same linear function** Let $f(t) = 2t - 2$ and consider the two area functions $A(x) = \int_{1}^{x} f(t)\, dt$ and $F(x) = \int_{4}^{x} f(t)\, dt$.

   **a.** Evaluate $A(2)$ and $A(3)$. Then use geometry to find an expression for $A(x)$, for $x \geq 1$.
   **b.** Evaluate $F(5)$ and $F(6)$. Then use geometry to find an expression for $F(x)$, for $x \geq 4$.
   **c.** Show that $A(x) - F(x)$ is a constant and that $A'(x) = F'(x) = f(x)$.

**19–22. Area functions for linear functions** *Consider the following functions $f$ and real numbers $a$ (see figure).*

**a.** Find and graph the area function $A(x) = \int_{a}^{x} f(t)\, dt$.
**b.** Verify that $A'(x) = f(x)$.

**19.** $f(t) = t + 5, a = -5$     **20.** $f(t) = 2t + 5, a = 0$

**21.** $f(t) = 3t + 1, a = 2$     **22.** $f(t) = 4t + 2, a = 0$

**23–24. Definite integrals** *Evaluate the following integrals using the Fundamental Theorem of Calculus. Explain why your result is consistent with the figure.*

**23.** $\int_{0}^{1} (x^2 - 2x + 3)\, dx$     **24.** $\int_{-\pi/4}^{7\pi/4} (\sin x + \cos x)\, dx$

 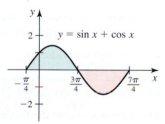

**25–28. Definite integrals** *Evaluate the following integrals using the Fundamental Theorem of Calculus. Sketch the graph of the integrand and shade the region whose net area you have found.*

**25.** $\int_{-2}^{3} (x^2 - x - 6)\, dx$     **26.** $\int_{0}^{1} (x - \sqrt{x})\, dx$

**27.** $\int_{0}^{5} (x^2 - 9)\, dx$     **28.** $\int_{1/2}^{2} \left(1 - \frac{1}{x^2}\right) dx$

**29–60. Definite integrals** *Evaluate the following integrals using the Fundamental Theorem of Calculus.*

**29.** $\int_{0}^{2} 4x^3\, dx$   **30.** $\int_{0}^{2} (3x^2 + 2x)\, dx$   **31.** $\int_{1}^{8} 8x^{1/3}\, dx$

**32.** $\int_{1}^{16} x^{-5/4}\, dx$   **33.** $\int_{0}^{1} (x + \sqrt{x})\, dx$   **34.** $\int_{0}^{\pi/4} 2\cos x\, dx$

**35.** $\int_{1}^{9} \frac{2}{\sqrt{x}}\, dx$   **36.** $\int_{4}^{9} \frac{2 + \sqrt{t}}{\sqrt{t}}\, dt$   **37.** $\int_{-2}^{2} (x^2 - 4)\, dx$

**38.** $\int_{0}^{\pi/4} (\sin x + \cos x)\, dx$   **39.** $\int_{1/2}^{1} (t^{-3} - 8)\, dt$

**40.** $\int_{0}^{4} t(t - 2)(t - 4)\, dt$

**41.** $\displaystyle\int_1^4 (1 - x)(x - 4)\, dx$

**42.** $\displaystyle\int_2^0 t(t + 1)\, dt$

**43.** $\displaystyle\int_{-2}^{-1} x^{-3}\, dx$

**44.** $\displaystyle\int_0^\pi (1 - \sin x)\, dx$

**45.** $\displaystyle\int_0^{\pi/4} \sec^2 \theta\, d\theta$

**46.** $\displaystyle\int_{-\pi/2}^{\pi/2} (\cos x - 1)\, dx$

**47.** $\displaystyle\int_1^2 \frac{3}{w^2}\, dw$

**48.** $\displaystyle\int_4^9 \frac{x - \sqrt{x}}{x^3}\, dx$

**49.** $\displaystyle\int_1^8 \sqrt[3]{y}\, dy$

**50.** $\displaystyle\frac{1}{2}\int_1^4 \frac{x^2 - 1}{x^2}\, dx$

**51.** $\displaystyle\int_1^4 \frac{x - 2}{\sqrt{x}}\, dx$

**52.** $\displaystyle\int_1^2 \frac{2s - 4}{s^3}\, ds$

**53.** $\displaystyle\int_0^{\pi/3} \sec x \tan x\, dx$

**54.** $\displaystyle\int_{\pi/4}^{\pi/2} \csc^2 \theta\, d\theta$

**55.** $\displaystyle\int_{\pi/4}^{3\pi/4} (\cot^2 x + 1)\, dx$

**56.** $\displaystyle\int_1^2 \frac{x^2 + 6x + 8}{x^4 + 2x^3}\, dx$

**57.** $\displaystyle\int_{\pi/4}^{\pi/2} \csc x \cot x\, dx$

**58.** $\displaystyle\int_0^{\pi/4} \sec x(\sec x + \cos x)\, dx$

**59.** $\displaystyle\int_0^\pi f(x)\, dx$, where $f(x) = \begin{cases} \sin x + 1 & \text{if } x \le \pi/2 \\ 2\cos x + 2 & \text{if } x > \pi/2 \end{cases}$

**60.** $\displaystyle\int_1^3 g(x)\, dx$, where $g(x) = \begin{cases} 3x^2 + 4x + 1 & \text{if } x \le 2 \\ 2x + 5 & \text{if } x > 2 \end{cases}$

**61–64. Area** *Find (i) the net area and (ii) the area of the following regions. Graph the function and indicate the region in question.*

**61.** The region bounded by $y = x^{1/2}$ and the $x$-axis between $x = 1$ and $x = 4$

**62.** The region above the $x$-axis bounded by $y = 4 - x^2$

**63.** The region below the $x$-axis bounded by $y = x^4 - 16$

**64.** The region bounded by $y = 6 \cos x$ and the $x$-axis between $x = -\pi/2$ and $x = \pi$

**65–70. Areas of regions** *Find the area of the region bounded by the graph of f and the x-axis on the given interval.*

**65.** $f(x) = x^2 - 25$ on $[2, 4]$    **66.** $f(x) = x^3 - 1$ on $[-1, 2]$

**67.** $f(x) = \dfrac{1}{x^5}$ on $[-2, -1]$

**68.** $f(x) = x(x + 1)(x - 2)$ on $[-1, 2]$

**69.** $f(x) = \sin x$ on $\left[-\dfrac{\pi}{4}, \dfrac{3\pi}{4}\right]$    **70.** $f(x) = \cos x$ on $\left[\dfrac{\pi}{2}, \pi\right]$

**71–84. Derivatives of integrals** *Simplify the following expressions.*

**71.** $\displaystyle\frac{d}{dx}\int_3^x (t^2 + t + 1)\, dt$

**72.** $\displaystyle\frac{d}{dx}\int_0^x \sin^2 t\, dt$

**73.** $\displaystyle\frac{d}{dx}\int_x^1 \sqrt{t^4 + 1}\, dt$

**74.** $\displaystyle\frac{d}{dx}\int_x^0 \frac{dp}{p^2 + 1}$

**75.** $\displaystyle\frac{d}{dx}\int_2^{x^3} \frac{dp}{p^2}$

**76.** $\displaystyle\frac{d}{dx}\int_0^{x^2} \frac{dt}{t^2 + 4}$

**77.** $\displaystyle\frac{d}{dx}\int_0^{\cos x} (t^4 + 6)\, dt$

**78.** $\displaystyle\frac{d}{dx}\int_x^1 \cos^3 t\, dt$

**79.** $\displaystyle\frac{d}{dz}\int_{\sin z}^{10} \frac{dt}{t^4 + 1}$

**80.** $\displaystyle\frac{d}{dy}\int_{y^3}^{10} \sqrt{x^6 + 1}\, dx$

**81.** $\displaystyle\frac{d}{dt}\left(\int_1^t \frac{3}{x}\, dx - \int_{t^2}^1 \frac{3}{x}\, dx\right)$

**82.** $\displaystyle\frac{d}{dt}\left(\int_0^t \frac{dx}{1 + x^2} + \int_1^{1/t} \frac{dx}{1 + x^2}\right)$

**83.** $\displaystyle\frac{d}{dx}\int_{-x}^x \sqrt{1 + t^2}\, dt$

*(Hint:* $\displaystyle\int_{-x}^x \sqrt{1 + t^2}\, dt = \int_{-x}^0 \sqrt{1 + t^2}\, dt + \int_0^x \sqrt{1 + t^2}\, dt.)$

**84.** $\displaystyle\frac{d}{dx}\int_x^{x^2} \sin t^2\, dt$

**85. Matching functions with area functions** Match the functions $f$, whose graphs are given in a–d, with the area functions $A(x) = \displaystyle\int_0^x f(t)\, dt$, whose graphs are given in A–D.

(a)

(b)

(c)

(d)

(A)

(B)

(C)

(D)

**86–89. Working with area functions** *Consider the function f and its graph.*

**a.** *Estimate the zeros of the area function $A(x) = \int_0^x f(t)\,dt$, for $0 \le x \le 10$.*

**b.** *Estimate the points (if any) at which A has a local maximum or minimum.*

**c.** *Sketch a graph of A, for $0 \le x \le 10$, without a scale on the y-axis.*

**86.**

**87.**

**88.**

**89.**

**90. Area functions from graphs** The graph of $f$ is given in the figure. Let $A(x) = \int_0^x f(t)\,dt$ and evaluate $A(1), A(2), A(4),$ and $A(6)$.

**91. Area functions from graphs** The graph of $f$ is given in the figure. Let $A(x) = \int_0^x f(t)\,dt$ and evaluate $A(2), A(5), A(8),$ and $A(12)$.

**92–96. Working with area functions** *Consider the function f and the points a, b, and c.*

**a.** *Find the area function $A(x) = \int_a^x f(t)\,dt$ using the Fundamental Theorem.*

**b.** *Graph f and A.*

**c.** *Evaluate $A(b)$ and $A(c)$. Interpret the results using the graphs of part (b).*

**92.** $f(x) = \sin x; a = 0, b = \dfrac{\pi}{2}, c = \pi$

**93.** $f(x) = \cos \pi x; a = 0, b = \dfrac{1}{2}, c = 1$

**94.** $f(x) = -12x(x - 1)(x - 2); a = 0, b = 1, c = 2$

**95.** $f(x) = \cos x; a = 0, b = \dfrac{\pi}{2}, c = \pi$

**96.** $f(x) = \dfrac{1}{x^2}; a = 1, b = 2, c = 4$

**97.** Find the critical points of the function

$$f(x) = \int_{-1}^x t^2(t - 3)(t - 4)\,dt,$$

and determine the intervals on which $f$ is increasing or decreasing.

**98.** Determine the intervals on which the function

$$g(x) = \int_x^0 \frac{t}{t^2 + 1}\,dt \text{ is concave up or concave down.}$$

**T 99–100. Functions defined by integrals** *Consider the function g, which is given in terms of a definite integral with a variable upper limit.*

**a.** *Graph the integrand.*

**b.** *Calculate $g'(x)$.*

**c.** *Graph g, showing all your work and reasoning.*

**99.** $g(x) = \int_0^x \sin^2 t\,dt$

**100.** $g(x) = \int_0^x \sin(\pi t^2)\,dt$ (a Fresnel integral)

**T 101–104. Areas of regions** *Find the area of the region R bounded by the graph of f and the x-axis on the given interval. Graph f and show the region R.*

**101.** $f(x) = 2 - |x|$ on $[-2, 4]$

**102.** $f(x) = 16 - x^4$ on $[-2, 2]$

**103.** $f(x) = x^4 - 4$ on $[1, 4]$

**104.** $f(x) = x^2(x - 2)$ on $[-1, 3]$

**105. Explain why or why not** Determine whether the following statements are true and give an explanation or counterexample.

**a.** Suppose $f$ is a positive decreasing function, for $x > 0$. Then the area function $A(x) = \int_0^x f(t)\,dt$ is an increasing function of $x$.

**b.** Suppose $f$ is a negative increasing function, for $x > 0$. Then the area function $A(x) = \int_0^x f(t)\,dt$ is a decreasing function of $x$.

**c.** The functions $p(x) = \sin 3x$ and $q(x) = 4 \sin 3x$ are antiderivatives of the same function.

**d.** If $A(x) = 3x^2 - x - 3$ is an area function for $f$, then $B(x) = 3x^2 - x$ is also an area function for $f$.

## Explorations and Challenges

**106.** Evaluate $\displaystyle\lim_{x \to 2} \frac{\int_2^x \sqrt{t^2 + t + 3}\,dt}{x^2 - 4}$.

**107. Maximum net area** What value of $b > -1$ maximizes the integral $\int_{-1}^b x^2(3 - x)\,dx$?

**108. Maximum net area** Graph the function $f(x) = 8 + 2x - x^2$ and determine the values of $a$ and $b$ that maximize the value of the integral $\int_a^b f(x)\,dx$.

**109. Zero net area** Consider the function $f(x) = x^2 - 4x$.

    **a.** Graph $f$ on the interval $x \geq 0$.

    **b.** For what value of $b > 0$ is $\int_0^b f(x)\,dx = 0$?

    **c.** In general, for the function $f(x) = x^2 - ax$, where $a > 0$, for what value of $b > 0$ (as a function of $a$) is $\int_0^b f(x)\,dx = 0$?

**110. Cubic zero net area** Consider the graph of the cubic $y = x(x - a)(x - b)$, where $0 < a < b$. Verify that the graph bounds a region above the $x$-axis, for $0 < x < a$, and bounds a region below the $x$-axis, for $a < x < b$. What is the relationship between $a$ and $b$ if the areas of these two regions are equal?

**111. An integral equation** Use the Fundamental Theorem of Calculus, Part 1, to find the function $f$ that satisfies the equation

$$\int_0^x f(t)\,dt = 2\cos x + 3x - 2.$$

Verify the result by substitution into the equation.

**112. Max/min of area functions** Suppose $f$ is continuous on $[0, \infty)$ and $A(x)$ is the net area of the region bounded by the graph of $f$ and the $t$-axis on $[0, x]$. Show that the local maxima and minima of $A$ occur at the zeros of $f$. Verify this fact with the function $f(x) = x^2 - 10x$.

**T 113. Asymptote of sine integral** Use a calculator to approximate

$$\lim_{x \to \infty} S(x) = \lim_{x \to \infty} \int_0^x \frac{\sin t}{t}\,dt,$$

where $S$ is the sine integral function (see Example 7). Explain your reasoning.

**114. Sine integral** Show that the sine integral $S(x) = \int_0^x \frac{\sin t}{t}\,dt$ satisfies the (differential) equation $xS'(x) + 2S''(x) + xS'''(x) = 0$.

**115. Fresnel integral** Show that the Fresnel integral $S(x) = \int_0^x \sin t^2\,dt$ satisfies the (differential) equation $(S'(x))^2 + \left(\dfrac{S''(x)}{2x}\right)^2 = 1$.

**116. Continuity at the endpoints** Assume $f$ is continuous on $[a, b]$ and let $A$ be the area function for $f$ with left endpoint $a$. Let $m^*$ and $M^*$ be the absolute minimum and maximum values of $f$ on $[a, b]$, respectively.

    **a.** Prove that $m^*(x - a) \leq A(x) \leq M^*(x - a)$, for all $x$ in $[a, b]$. Use this result and the Squeeze Theorem to show that $A$ is continuous from the right at $x = a$.

    **b.** Prove that $m^*(b - x) \leq A(b) - A(x) \leq M^*(b - x)$, for all $x$ in $[a, b]$. Use this result to show that $A$ is continuous from the left at $x = b$.

**117. Discrete version of the Fundamental Theorem** In this exercise, we work with a discrete problem and show why the relationship $\int_a^b f'(x)\,dx = f(b) - f(a)$ makes sense. Suppose we have a set of equally spaced grid points

$$\{a = x_0 < x_1 < x_2 < \cdots < x_{n-1} < x_n = b\},$$

where the distance between any two grid points is $\Delta x$. Suppose also that at each grid point $x_k$, a function value $f(x_k)$ is defined, for $k = 0, \ldots, n$.

    **a.** We now replace the integral with a sum and replace the derivative with a difference quotient. Explain why $\int_a^b f'(x)\,dx$ is analogous to $\displaystyle\sum_{k=1}^n \underbrace{\frac{f(x_k) - f(x_{k-1})}{\Delta x}}_{\approx f'(x_k)} \Delta x.$

    **b.** Simplify the sum in part (a) and show that it is equal to $f(b) - f(a)$.

    **c.** Explain the correspondence between the integral relationship and the summation relationship.

**QUICK CHECK ANSWERS**

**1.** $0, -35$  **2.** $A(6) = 44$; $A(10) = 120$  **3.** $\frac{2}{3} - \frac{1}{2} = \frac{1}{6}$
**4.** If $f$ is differentiated, we get $f'$. Therefore, $f$ is an antiderivative of $f'$. ◄

# 5.4 Working with Integrals

With the Fundamental Theorem of Calculus in hand, we may begin an investigation of integration and its applications. In this section, we discuss the role of symmetry in integrals, we use the slice-and-sum strategy to define the average value of a function, and we explore a theoretical result called the Mean Value Theorem for Integrals.

## Integrating Even and Odd Functions

Symmetry appears throughout mathematics in many different forms, and its use often leads to insights and efficiencies. Here we use the symmetry of a function to simplify integral calculations.

Section 1.1 introduced the symmetry of even and odd functions. An **even function** satisfies the property $f(-x) = f(x)$, which means that its graph is symmetric about the $y$-axis (**Figure 5.53a**). Examples of even functions are $f(x) = \cos x$ and $f(x) = x^n$, where $n$ is an even integer. An **odd function** satisfies the property $f(-x) = -f(x)$, which means that its graph is symmetric about the origin (**Figure 5.53b**). Examples of odd functions are $f(x) = \sin x$ and $f(x) = x^n$, where $n$ is an odd integer.

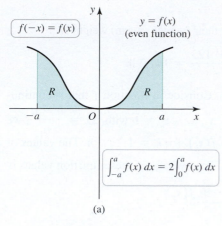

$f(-x) = f(x)$
$y = f(x)$ (even function)

$$\int_{-a}^{a} f(x)\, dx = 2\int_{0}^{a} f(x)\, dx$$

(a)

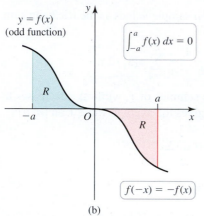

$y = f(x)$ (odd function)

$$\int_{-a}^{a} f(x)\, dx = 0$$

$f(-x) = -f(x)$

(b)

**Figure 5.53**

**QUICK CHECK 1** If $f$ and $g$ are both even functions, is the product $fg$ even or odd? Use the facts that $f(-x) = f(x)$ and $g(-x) = g(x)$. ◄

Special things happen when we integrate even and odd functions on intervals centered at the origin. First suppose $f$ is an even function and consider $\int_{-a}^{a} f(x)\, dx$. From Figure 5.53a, we see that the integral of $f$ on $[-a, 0]$ equals the integral of $f$ on $[0, a]$. Therefore, the integral on $[-a, a]$ is twice the integral on $[0, a]$, or

$$\int_{-a}^{a} f(x)\, dx = 2\int_{0}^{a} f(x)\, dx.$$

On the other hand, suppose $f$ is an odd function and consider $\int_{-a}^{a} f(x)\, dx$. As shown in Figure 5.53b, the integral on the interval $[-a, 0]$ is the negative of the integral on $[0, a]$. Therefore, the integral on $[-a, a]$ is zero, or

$$\int_{-a}^{a} f(x)\, dx = 0.$$

We summarize these results in the following theorem.

---

**THEOREM 5.4    Integrals of Even and Odd Functions**

Let $a$ be a positive real number and let $f$ be an integrable function on the interval $[-a, a]$.

- If $f$ is even, $\int_{-a}^{a} f(x)\, dx = 2\int_{0}^{a} f(x)\, dx$.
- If $f$ is odd, $\int_{-a}^{a} f(x)\, dx = 0$.

---

The following example shows how symmetry can simplify integration.

**EXAMPLE 1    Integrating symmetric functions** Evaluate the following integrals using symmetry arguments.

**a.** $\displaystyle\int_{-2}^{2} (x^4 - 3x^3)\, dx$    **b.** $\displaystyle\int_{-\pi/2}^{\pi/2} (\cos x - 4\sin^3 x)\, dx$

**SOLUTION**

**a.** Note that $x^4 - 3x^3$ is neither odd nor even, so Theorem 5.4 cannot be applied directly. However, we can split the integral and then use symmetry:

$$\int_{-2}^{2} (x^4 - 3x^3)\, dx = \int_{-2}^{2} x^4\, dx - 3\underbrace{\int_{-2}^{2} x^3\, dx}_{0} \quad \text{Properties 3 and 4 of Table 5.4}$$

$$= 2\int_{0}^{2} x^4\, dx - 0 \qquad x^4 \text{ is even; } x^3 \text{ is odd.}$$

$$= 2\left(\frac{x^5}{5}\right)\Big|_{0}^{2} \qquad \text{Fundamental Theorem}$$

$$= 2\left(\frac{32}{5}\right) = \frac{64}{5}. \qquad \text{Simplify.}$$

Notice how the odd-powered term of the integrand is eliminated by symmetry. Integration of the even-powered term is simplified because the lower limit is zero.

> There are a couple of ways to see that $\sin^3 x$ is an odd function. Its graph is symmetric about the origin, indicating that $\sin^3(-x) = -\sin^3 x$. Or by analogy, take an odd power of $x$ and raise it to an odd power. For example, $(x^5)^3 = x^{15}$, which is odd. See Exercises 49–52 for direct proofs of symmetry in composite functions.

**b.** The $\cos x$ term is an even function, so it can be integrated on the interval $[0, \pi/2]$. What about $\sin^3 x$? It is an odd function raised to an odd power, which results in an odd function; its integral on $[-\pi/2, \pi/2]$ is zero. Therefore,

$$\int_{-\pi/2}^{\pi/2} (\cos x - 4\sin^3 x)\, dx = 2\int_{0}^{\pi/2} \cos x\, dx - 0 \quad \text{Symmetry}$$

$$= 2\sin x \Big|_{0}^{\pi/2} \qquad \text{Fundamental Theorem}$$

$$= 2(1 - 0) = 2. \qquad \text{Simplify.}$$

*Related Exercises 15–16* ◄

## Average Value of a Function

If five people weigh 155, 143, 180, 105, and 123 lb, their average (mean) weight is

$$\frac{155 + 143 + 180 + 105 + 123}{5} = 141.2 \text{ lb}.$$

This idea generalizes quite naturally to functions. Consider a function $f$ that is continuous on $[a, b]$. Using a regular partition $x_0 = a, x_1, x_2, \ldots, x_n = b$ with $\Delta x = \frac{b - a}{n}$, we select a point $x_k^*$ in each subinterval and compute $f(x_k^*)$, for $k = 1, \ldots, n$. The values of $f(x_k^*)$ may be viewed as a sampling of $f$ on $[a, b]$. The average of these function values is

$$\frac{f(x_1^*) + f(x_2^*) + \cdots + f(x_n^*)}{n}.$$

Noting that $n = \frac{b - a}{\Delta x}$, we write the average of the $n$ sample values as the Riemann sum

$$\frac{f(x_1^*) + f(x_2^*) + \cdots + f(x_n^*)}{(b - a)/\Delta x} = \frac{1}{b - a} \sum_{k=1}^{n} f(x_k^*) \Delta x.$$

Now suppose we increase $n$, taking more and more samples of $f$, while $\Delta x$ decreases to zero. The limit of this sum is a definite integral that gives the average value $\bar{f}$ on $[a, b]$:

$$\bar{f} = \frac{1}{b - a} \lim_{n \to \infty} \sum_{k=1}^{n} f(x_k^*) \Delta x$$

$$= \frac{1}{b - a} \int_a^b f(x) \, dx.$$

This definition of the average value of a function is analogous to the definition of the average of a finite set of numbers.

---

**DEFINITION  Average Value of a Function**

The average value of an integrable function $f$ on the interval $[a, b]$ is

$$\bar{f} = \frac{1}{b - a} \int_a^b f(x) \, dx.$$

---

The average value of a function $f$ on an interval $[a, b]$ has a clear geometrical interpretation. Multiplying both sides of the definition of average value by $(b - a)$, we have

$$\underbrace{(b - a)\bar{f}}_{\substack{\text{net area of} \\ \text{rectangle}}} = \underbrace{\int_a^b f(x) \, dx.}_{\substack{\text{net area of region} \\ \text{bounded by curve}}}$$

We see that $|\bar{f}|$ is the height of a rectangle with base $[a, b]$, and that rectangle has the same net area as the region bounded by the graph of $f$ on the interval $[a, b]$ (**Figure 5.54**). Note that $\bar{f}$ may be zero or negative.

**QUICK CHECK 2** What is the average value of a constant function on an interval? What is the average value of an odd function on an interval $[-a, a]$? ◄

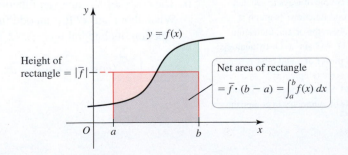

**Figure 5.54**

**EXAMPLE 2    Average elevation**  A hiking trail has an elevation given by

$$f(x) = 60x^3 - 650x^2 + 1200x + 4500,$$

where $f$ is measured in feet above sea level and $x$ represents horizontal distance along the trail in miles, with $0 \le x \le 5$. What is the average elevation of the trail?

**Figure 5.55**

**SOLUTION**  The trail ranges between elevations of about 2000 and 5000 ft (**Figure 5.55**). If we let the endpoints of the trail correspond to the horizontal distances $a = 0$ and $b = 5$, the average elevation of the trail in feet is

$$\bar{f} = \frac{1}{5} \int_0^5 (60x^3 - 650x^2 + 1200x + 4500) \, dx$$

$$= \frac{1}{5} \left( 60 \frac{x^4}{4} - 650 \frac{x^3}{3} + 1200 \frac{x^2}{2} + 4500x \right) \Big|_0^5 \qquad \text{Fundamental Theorem}$$

$$= 3958 \tfrac{1}{3}. \qquad \text{Simplify.}$$

The average elevation of the trail is slightly less than 3960 ft.

*Related Exercises 26, 34* ◀

## Mean Value Theorem for Integrals

The average value of a function brings us close to an important theoretical result. The Mean Value Theorem for Integrals says that if $f$ is continuous on $[a, b]$, then there is at least one point $c$ in the interval $(a, b)$ such that $f(c)$ equals the average value of $f$ on $(a, b)$. In other words, the horizontal line $y = \bar{f}$ intersects the graph of $f$ for some point $c$ in $(a, b)$ (**Figure 5.56**). If $f$ were not continuous, such a point might not exist.

➤ Compare this statement to that of the Mean Value Theorem for Derivatives: There is at least one point $c$ in $(a, b)$ such that $f'(c)$ equals the average slope of $f$.

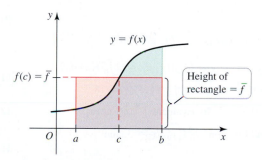

**Figure 5.56**

➤ Theorem 5.5 guarantees a point $c$ in the open interval $(a, b)$ at which $f$ equals its average value. However, $f$ may also equal its average value at an endpoint of that interval.

---

**THEOREM 5.5    Mean Value Theorem for Integrals**
Let $f$ be continuous on the interval $[a, b]$. There exists a point $c$ in $(a, b)$ such that

$$f(c) = \bar{f} = \frac{1}{b - a} \int_a^b f(t) \, dt.$$

---

**Proof:**  We begin by letting $F(x) = \int_a^x f(t) \, dt$ and noting that $F$ is continuous on $[a, b]$ and differentiable on $(a, b)$ (by Theorem 5.3, Part 1). We now apply the Mean Value Theorem for derivatives (Theorem 4.4) to $F$ and conclude that there exists at least one point $c$ in $(a, b)$ such that

$$\underbrace{F'(c)}_{f(c)} = \frac{F(b) - F(a)}{b - a}.$$

By Theorem 5.3, Part 1, we know that $F'(c) = f(c)$, and by Theorem 5.3, Part 2, we know that

$$F(b) - F(a) = \int_a^b f(t) \, dt.$$

> A more general form of the Mean Value Theorem states that if $f$ and $g$ are continuous on $[a, b]$ with $g(x) \geq 0$ on $[a, b]$, then there exists a number $c$ in $(a, b)$ such that
>
> $$\int_a^b f(x)g(x)\, dx = f(c)\int_a^b g(x)\, dx.$$

**QUICK CHECK 3** Explain why $f(x) = 0$ for at least one point of $(a, b)$ if $f$ is continuous and $\int_a^b f(x)\, dx = 0$. ◄

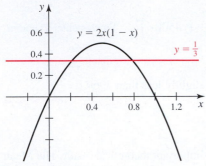

**Figure 5.57**

Combining these observations, we have

$$f(c) = \frac{1}{b - a}\int_a^b f(t)\, dt,$$

where $c$ is a point in $(a, b)$.

◄

**EXAMPLE 3**  **Average value equals function value**  Find the point(s) on the interval $(0, 1)$ at which $f(x) = 2x(1 - x)$ equals its average value on $[0, 1]$.

**SOLUTION**  The average value of $f$ on $[0, 1]$ is

$$\bar{f} = \frac{1}{1 - 0}\int_0^1 2x(1 - x)\, dx = \left(x^2 - \frac{2}{3}x^3\right)\Big|_0^1 = \frac{1}{3}.$$

We must find the points on $(0, 1)$ at which $f(x) = \frac{1}{3}$ (**Figure 5.57**). Using the quadratic formula, the two solutions of $f(x) = 2x(1 - x) = \frac{1}{3}$ are

$$\frac{1 - \sqrt{1/3}}{2} \approx 0.211 \quad \text{and} \quad \frac{1 + \sqrt{1/3}}{2} \approx 0.789.$$

These two points are located symmetrically on either side of $x = \frac{1}{2}$. The two solutions, 0.211 and 0.789, are the same for $f(x) = ax(1 - x)$ for any nonzero value of $a$ (Exercise 53).

*Related Exercises 41–42* ◄

---

## SECTION 5.4 EXERCISES

### Getting Started

1. If $f$ is an odd function, why is $\int_{-a}^a f(x)\, dx = 0$?

2. If $f$ is an even function, why is $\int_{-a}^a f(x)\, dx = 2\int_0^a f(x)\, dx$?

3. Suppose $f$ is an even function and $\int_{-8}^8 f(x)\, dx = 18$.

   **a.** Evaluate $\int_0^8 f(x)\, dx$.     **b.** Evaluate $\int_{-8}^8 xf(x)\, dx$.

4. Suppose $f$ is an odd function, $\int_0^4 f(x)\, dx = 3$, and $\int_0^8 f(x)\, dx = 9$.

   **a.** Evaluate $\int_{-4}^8 f(x)\, dx$.     **b.** Evaluate $\int_{-8}^4 f(x)\, dx$.

5. Use symmetry to explain why

   $$\int_{-4}^4 (5x^4 + 3x^3 + 2x^2 + x + 1)\, dx = 2\int_0^4 (5x^4 + 2x^2 + 1)\, dx.$$

6. Use symmetry to fill in the blanks:
   $\int_{-\pi}^\pi (\sin x + \cos x)\, dx = \underline{\quad} \int_0^\pi \underline{\quad}\, dx.$

7. Is $x^{12}$ an even or odd function? Is $\sin x^2$ an even or odd function?

8. Explain how to find the average value of a function on an interval $[a, b]$ and why this definition is analogous to the definition of the average of a set of numbers.

9. Explain the statement that a continuous function on an interval $[a, b]$ equals its average value at some point on $(a, b)$.

10. Sketch the function $y = x$ on the interval $[0, 2]$ and let $R$ be the region bounded by $y = x$ and the $x$-axis on $[0, 2]$. Now sketch a rectangle in the first quadrant whose base is $[0, 2]$ and whose area equals the area of $R$.

### Practice Exercises

**11–24. Symmetry in integrals** *Use symmetry to evaluate the following integrals.*

11. $\int_{-2}^2 x^9\, dx$

12. $\int_{-200}^{200} 2x^5\, dx$

13. $\int_{-2}^2 (3x^8 - 2)\, dx$

14. $\int_{-\pi/4}^{\pi/4} \cos x\, dx$

15. $\int_{-2}^2 (x^2 + x^3)\, dx$

16. $\int_{-\pi}^\pi t^2 \sin t\, dt$

17. $\int_{-2}^2 (x^9 - 3x^5 + 2x^2 - 10)\, dx$

18. $\int_{-\pi/2}^{\pi/2} 5 \sin \theta\, d\theta$

19. $\int_{-\pi/4}^{\pi/4} \sin^5 t\, dt$

20. $\int_{-1}^1 (1 - |x|)\, dx$

21. $\int_{-\pi/4}^{\pi/4} \sec^2 x\, dx$

22. $\int_{-\pi/4}^{\pi/4} \tan \theta\, d\theta$

23. $\int_{-2}^2 \dfrac{x^3 - 4x}{x^2 + 1}\, dx$

24. $\int_{-2}^2 (1 - |x|^3)\, dx$

**25–32. Average values** *Find the average value of the following functions on the given interval. Draw a graph of the function and indicate the average value.*

25. $f(x) = x^3$ on $[-1, 1]$

26. $f(x) = x^2 + 1$ on $[-2, 2]$

27. $f(x) = \dfrac{1}{\sqrt{x}}$ on $[1, 4]$

28. $f(x) = \dfrac{1}{x^3}$ on $[1, 2]$

29. $f(x) = \cos x$ on $\left[-\dfrac{\pi}{2}, \dfrac{\pi}{2}\right]$

30. $f(x) = x(1 - x)$ on $[0, 1]$

**31.** $f(x) = x^n$ on $[0, 1]$, for any positive integer $n$

**32.** $f(x) = x^{1/n}$ on $[0, 1]$, for any positive integer $n$

**33. Average distance on a parabola** What is the average distance between the parabola $y = 30x(20 - x)$ and the $x$-axis on the interval $[0, 20]$?

**T 34. Average elevation** The elevation of a path is given by $f(x) = x^3 - 5x^2 + 30$, where $x$ measures horizontal distance. Draw a graph of the elevation function and find its average value, for $0 \le x \le 4$.

**35. Average velocity** The velocity in m/s of an object moving along a line over the time interval $[0, 6]$ is $v(t) = t^2 + 3t$. Find the average velocity of the object over this time interval.

**36. Average velocity** A rock is launched vertically upward from the ground with a speed of 64 ft/s. The height of the rock (in ft) above the ground after $t$ seconds is given by the function $s(t) = -16t^2 + 64t$. Find its average velocity during its flight.

**37. Average height of an arch** The height of an arch above the ground is given by the function $y = 10 \sin x$, for $0 \le x \le \pi$. What is the average height of the arch above the ground?

**38. Average height of a wave** The surface of a water wave is described by $y = 5(1 + \cos x)$, for $-\pi \le x \le \pi$, where $y = 0$ corresponds to a trough of the wave (see figure). Find the average height of the wave above the trough on $[-\pi, \pi]$.

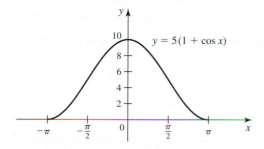

**39–44. Mean Value Theorem for Integrals** *Find or approximate all points at which the given function equals its average value on the given interval.*

**39.** $f(x) = 8 - 2x$ on $[0, 4]$   **T 40.** $f(x) = \cos x$ on $\left[-\frac{\pi}{2}, \frac{\pi}{2}\right]$

**41.** $f(x) = 1 - \dfrac{x^2}{a^2}$ on $[0, a]$, where $a$ is a positive real number

**T 42.** $f(x) = \dfrac{\pi}{4} \sin x$ on $[0, \pi]$   **43.** $f(x) = 1 - |x|$ on $[-1, 1]$

**44.** $f(x) = \dfrac{1}{x^2}$ on $[1, 4]$

**45. Explain why or why not** Determine whether the following statements are true and give an explanation or counterexample.

**a.** If $f$ is symmetric about the line $x = 2$, then $\int_0^4 f(x)\,dx = 2\int_0^2 f(x)\,dx$.

**b.** If $f$ has the property $f(a + x) = -f(a - x)$, for all $x$, where $a$ is constant, then $\int_{a-2}^{a+2} f(x)\,dx = 0$.

**c.** The average value of a linear function on an interval $[a, b]$ is the function value at the midpoint of $[a, b]$.

**d.** Consider the function $f(x) = x(a - x)$ on the interval $[0, a]$, for $a > 0$. Its average value on $[0, a]$ is $\frac{1}{2}$ of its maximum value.

**46. Planetary orbits** The planets orbit the Sun in elliptical orbits with the Sun at one focus (see Section 12.4 for more on ellipses). The equation of an ellipse whose dimensions are $2a$ in the $x$-direction and $2b$ in the $y$-direction is $\dfrac{x^2}{a^2} + \dfrac{y^2}{b^2} = 1$.

**a.** Let $d^2$ denote the square of the distance from a planet to the center of the ellipse at $(0, 0)$. Integrate over the interval $[-a, a]$ to show that the average value of $d^2$ is $\dfrac{a^2 + 2b^2}{3}$.

**b.** Show that in the case of a circle ($a = b = R$), the average value in part (a) is $R^2$.

**c.** Assuming $0 < b < a$, the coordinates of the Sun are $\left(\sqrt{a^2 - b^2}, 0\right)$. Let $D^2$ denote the square of the distance from the planet to the Sun. Integrate over the interval $[-a, a]$ to show that the average value of $D^2$ is $\dfrac{4a^2 - b^2}{3}$.

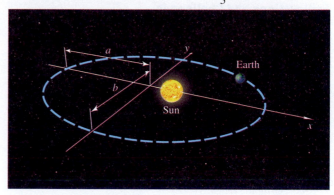

**47. Gateway Arch** The Gateway Arch in St. Louis is 630 ft high and has a 630-ft base. Its shape can be modeled by the parabola

$$y = 630\left(1 - \left(\frac{x}{315}\right)^2\right).$$

Find the average height of the arch above the ground.

**T 48. Comparing a sine and a quadratic function** Consider the functions $f(x) = \sin x$ and $g(x) = \dfrac{4}{\pi^2} x(\pi - x)$.

**a.** Carefully graph $f$ and $g$ on the same set of axes. Verify that both functions have a single local maximum on the interval $[0, \pi]$ and that they have the same maximum value on $[0, \pi]$.

**b.** On the interval $[0, \pi]$, which is true: $f(x) \ge g(x)$, $g(x) \ge f(x)$, or neither?

**c.** Compute and compare the average values of $f$ and $g$ on $[0, \pi]$.

**49–52. Symmetry of composite functions** *Prove that the integrand is either even or odd. Then give the value of the integral or show how it can be simplified. Assume $f$ and $g$ are even functions and $p$ and $q$ are odd functions.*

**49.** $\displaystyle\int_{-a}^{a} f(g(x))\, dx$

**50.** $\displaystyle\int_{-a}^{a} f(p(x))\, dx$

**51.** $\displaystyle\int_{-a}^{a} p(g(x))\, dx$

**52.** $\displaystyle\int_{-a}^{a} p(q(x))\, dx$

**53. Average value with a parameter** Consider the function $f(x) = ax(1 - x)$ on the interval $[0, 1]$, where $a$ is a positive real number.

   **a.** Find the average value of $f$ as a function of $a$.

   **b.** Find the points at which the value of $f$ equals its average value, and prove that they are independent of $a$.

## Explorations and Challenges

**54. Alternative definitions of means** Consider the function

$$f(t) = \frac{\int_{a}^{b} x^{t+1}\, dx}{\int_{a}^{b} x^{t}\, dx}.$$

Show that the following means can be defined in terms of $f$.

   **a.** Arithmetic mean: $f(0) = \dfrac{a + b}{2}$

   **b.** Geometric mean: $f\left(-\dfrac{3}{2}\right) = \sqrt{ab}$

   **c.** Harmonic mean: $f(-3) = \dfrac{2ab}{a + b}$

(*Source: Mathematics Magazine*, 78, 5, Dec 2005)

**55. Problems of antiquity** Several calculus problems were solved by Greek mathematicians long before the discovery of calculus. The following problems were solved by Archimedes using methods that predated calculus by 2000 years.

   **a.** Show that the area of a segment of a parabola is $4/3$ that of its inscribed triangle of greatest area. In other words, the area bounded by the parabola $y = a^2 - x^2$ and the $x$-axis is $4/3$ the area of the triangle with vertices $(\pm a, 0)$ and $(0, a^2)$. Assume $a > 0$ is unspecified.

   **b.** Show that the area bounded by the parabola $y = a^2 - x^2$ and the $x$-axis is $2/3$ the area of the rectangle with vertices $(\pm a, 0)$ and $(\pm a, a^2)$. Assume $a > 0$ is unspecified.

**56. Average value of the derivative** Suppose $f'$ is a continuous function for all real numbers. Show that the average value of the derivative on an interval $[a, b]$ is $\overline{f'} = \dfrac{f(b) - f(a)}{b - a}$. Interpret this result in terms of secant lines.

**57. Symmetry of powers** Fill in the following table with either **even** or **odd**, and prove each result. Assume $n$ is a nonnegative integer and $f^n$ means the $n$th power of $f$.

|  | $f$ is even | $f$ is odd |
|---|---|---|
| $n$ is even | $f^n$ is _____ | $f^n$ is _____ |
| $n$ is odd | $f^n$ is _____ | $f^n$ is _____ |

**58. Bounds on an integral** Suppose $f$ is continuous on $[a, b]$ with $f''(x) > 0$ on the interval. It can be shown that

$$(b - a)f\left(\frac{a + b}{2}\right) \leq \int_{a}^{b} f(x)\, dx \leq (b - a)\frac{f(a) + f(b)}{2}.$$

   **a.** Assuming $f$ is nonnegative on $[a, b]$, draw a figure to illustrate the geometric meaning of these inequalities. Discuss your conclusions.

   **b.** Divide these inequalities by $(b - a)$ and interpret the resulting inequalities in terms of the average value of $f$ on $[a, b]$.

**59. Generalizing the Mean Value Theorem for Integrals** Suppose $f$ and $g$ are continuous on $[a, b]$ and let

$$h(x) = (x - b)\int_{a}^{x} f(t)\, dt + (x - a)\int_{x}^{b} g(t)\, dt.$$

   **a.** Use Rolle's Theorem to show that there is a number $c$ in $(a, b)$ such that

$$\int_{a}^{c} f(t)\, dt + \int_{c}^{b} g(t)\, dt = f(c)(b - c) + g(c)(c - a),$$

which is a generalization of the Mean Value Theorem for Integrals.

   **b.** Show that there is a number $c$ in $(a, b)$ such that $\int_{a}^{c} f(t)\, dt = f(c)(b - c)$.

   **c.** Use a sketch to interpret part (b) geometrically.

   **d.** Use the result of part (a) to give an alternative proof of the Mean Value Theorem for Integrals.

(*Source: The College Mathematics Journal*, 33, 5, Nov 2002)

**60. Evaluating a sine integral by Riemann sums** Consider the integral $I = \int_{0}^{\pi/2} \sin x\, dx$.

   **a.** Write the left Riemann sum for $I$ with $n$ subintervals.

   **b.** Show that $\displaystyle\lim_{\theta \to 0} \theta\left(\frac{\cos\theta + \sin\theta - 1}{2(1 - \cos\theta)}\right) = 1$.

   **c.** It is a fact that $\displaystyle\sum_{k=0}^{n-1} \sin\left(\frac{\pi k}{2n}\right) = \frac{\cos\left(\dfrac{\pi}{2n}\right) + \sin\left(\dfrac{\pi}{2n}\right) - 1}{2\left(1 - \cos\left(\dfrac{\pi}{2n}\right)\right)}$.

   Use this fact and part (b) to evaluate $I$ by taking the limit of the Riemann sum as $n \to \infty$.

## QUICK CHECK ANSWERS

**1.** $f(-x)g(-x) = f(x)g(x)$; therefore, $fg$ is even.

**2.** The average value is the constant; the average value is 0.

**3.** The average value is zero on the interval; by the Mean Value Theorem for Integrals, $f(x) = 0$ at some point on the interval. ◄

# 5.5 Substitution Rule

Given just about any differentiable function, with enough know-how and persistence, you can compute its derivative. But the same cannot be said of antiderivatives. Many functions, even relatively simple ones, do not have antiderivatives that can be expressed in terms of familiar functions. Examples are $\sin x^2$ and $(\sin x)/x$. The immediate goal of this section is to enlarge the family of functions for which we can find antiderivatives. This campaign resumes in Chapter 8, where additional integration methods are developed.

## Indefinite Integrals

One way to find new antiderivative rules is to start with familiar derivative rules and work backward. When applied to the Chain Rule, this strategy leads to the Substitution Rule. For example, consider the indefinite integral $\int \cos 2x \, dx$. The closest familiar integral related to this problem is

> ➤ We assume $C$ is an arbitrary constant without stating so each time it appears.

$$\int \cos x \, dx = \sin x + C,$$

which is true because

$$\frac{d}{dx}(\sin x + C) = \cos x.$$

Therefore, we might *incorrectly* conclude that the indefinite integral of $\cos 2x$ is $\sin 2x + C$. However, by the Chain Rule,

$$\frac{d}{dx}(\sin 2x + C) = 2 \cos 2x \neq \cos 2x.$$

Note that $\sin 2x$ fails to be an antiderivative of $\cos 2x$ by a multiplicative factor of 2. A small adjustment corrects this problem. Let's try $\frac{1}{2} \sin 2x$:

$$\frac{d}{dx}\left(\frac{1}{2} \sin 2x\right) = \frac{1}{2} \cdot 2 \cos 2x = \cos 2x.$$

It works! So we have

$$\int \cos 2x \, dx = \frac{1}{2} \sin 2x + C.$$

The trial-and-error approach of the previous example is impractical for complicated integrals. To develop a systematic method, consider a composite function $F(g(x))$, where $F$ is an antiderivative of $f$; that is, $F' = f$. Using the Chain Rule to differentiate the composite function $F(g(x))$, we find that

$$\frac{d}{dx}(F(g(x))) = F'(g(x))g'(x) = \underbrace{f(g(x))}g'(x).$$

This equation says that $F(g(x))$ is an antiderivative of $f(g(x))g'(x)$, which is written

$$\int f(g(x))g'(x) \, dx = F(g(x)) + C, \tag{1}$$

where $F$ is any antiderivative of $f$.

> ➤ You can call the new variable anything you want because it is just another variable of integration. Typically, $u$ is the standard choice for the new variable.

Why is this approach called the *Substitution Rule* (or *Change of Variables Rule*)? In the composite function $f(g(x))$ in equation (1), we identify the inner function as $u = g(x)$, which implies that $du = g'(x) \, dx$. Making this identification, the integral in equation (1) is written

$$\int \underbrace{f(g(x))}_{f(u)} \underbrace{g'(x) \, dx}_{du} = \int f(u) \, du = F(u) + C.$$

We see that the integral $\int f(g(x))g'(x) \, dx$ with respect to $x$ is replaced with a new integral $\int f(u) \, du$ with respect to the new variable $u$. In other words, we have substituted the new variable $u$ for the old variable $x$. Of course, if the new integral with respect to $u$ is no easier to find than the original integral, then the change of variables has not helped. The Substitution Rule requires plenty of practice until certain patterns become familiar.

---

**THEOREM 5.6   Substitution Rule for Indefinite Integrals**

Let $u = g(x)$, where $g$ is differentiable on an interval, and let $f$ be continuous on the corresponding range of $g$. On that interval,

$$\int f(g(x))g'(x)\,dx = \int f(u)\,du.$$

---

In practice, Theorem 5.6 is applied using the following procedure.

---

**PROCEDURE   Substitution Rule (Change of Variables)**

1. Given an indefinite integral involving a composite function $f(g(x))$, identify an inner function $u = g(x)$ such that a constant multiple of $g'(x)$ appears in the integrand.

2. Substitute $u = g(x)$ and $du = g'(x)\,dx$ in the integral.

3. Evaluate the new indefinite integral with respect to $u$.

4. Write the result in terms of $x$ using $u = g(x)$.

*Disclaimer: Not all integrals yield to the Substitution Rule.*

---

**EXAMPLE 1   Perfect substitutions**  Use the Substitution Rule to find the following indefinite integrals. Check your work by differentiating.

**a.** $\displaystyle\int 2(2x + 1)^3\,dx$     **b.** $\displaystyle\int 2x \cos x^2\,dx$

**SOLUTION**

**a.** We identify $u = 2x + 1$ as the inner function of the composite function $(2x + 1)^3$. Therefore, we choose the new variable $u = 2x + 1$, which implies that $\dfrac{du}{dx} = 2$, or $du = 2\,dx$. Notice that $du = 2\,dx$ appears as a factor in the integrand. The change of variables looks like this:

$$\int \underbrace{(2x + 1)^3}_{u^3} \cdot \underbrace{2\,dx}_{du} = \int u^3\,du \qquad \text{Substitute } u = 2x + 1, du = 2\,dx.$$

$$= \frac{u^4}{4} + C \qquad \text{Antiderivative}$$

$$= \frac{(2x + 1)^4}{4} + C. \quad \text{Replace } u \text{ with } 2x + 1.$$

▶ Use the Chain Rule to check that
$$\frac{d}{dx}\left(\frac{(2x + 1)^4}{4} + C\right) = 2(2x + 1)^3.$$

Notice that the final step uses $u = 2x + 1$ to return to the original variable.

**b.** The composite function $\cos x^2$ has the inner function $u = x^2$, which implies that $du = 2x\,dx$. The change of variables appears as

$$\int \underbrace{(\cos x^2)}_{\cos u}\,\underbrace{2x\,dx}_{du} = \int \cos u\,du \quad \text{Substitute } u = x^2, du = 2x\,dx.$$

$$= \sin u + C \quad \text{Antiderivative}$$

$$= \sin x^2 + C. \quad \text{Replace } u \text{ with } x^2.$$

To check the result, we compute $\dfrac{d}{dx}(\sin x^2 + C) = (\cos x^2)\,2x = 2x \cos x^2$.

*Related Exercises 17, 20–21* ◀

**QUICK CHECK 1** Find a new variable $u$ so that $\int 4x^3(x^4 + 5)^{10}\,dx = \int u^{10}\,du$. ◀

Most substitutions are not perfect. The remaining examples show more typical situations that require introducing a constant factor.

**EXAMPLE 2** **Introducing a constant** Find the following indefinite integrals.

**a.** $\int x^4(x^5 + 6)^9\, dx$ **b.** $\int \cos^3 x \sin x\, dx$

**SOLUTION**

**a.** The inner function of the composite function $(x^5 + 6)^9$ is $x^5 + 6$, and its derivative $5x^4$ also appears in the integrand (up to a multiplicative factor). Therefore, we use the substitution $u = x^5 + 6$, which implies that $du = 5x^4\, dx$, or $x^4\, dx = \dfrac{1}{5}\, du$. By the Substitution Rule,

$$\int \underbrace{(x^5 + 6)^9}_{u^9}\underbrace{x^4\, dx}_{\frac{1}{5}\, du} = \int u^9 \cdot \frac{1}{5}\, du \qquad \text{Substitute } u = x^5 + 6,\ du = 5x^4\, dx \Rightarrow x^4\, dx = \frac{1}{5}\, du.$$

$$= \frac{1}{5}\int u^9\, du \qquad \int cf(x)\, dx = c\int f(x)\, dx$$

$$= \frac{1}{5}\cdot\frac{u^{10}}{10} + C \qquad \text{Antiderivative}$$

$$= \frac{1}{50}(x^5 + 6)^{10} + C. \qquad \text{Replace } u \text{ with } x^5 + 6.$$

**b.** The integrand can be written as $(\cos x)^3 \sin x$. The inner function in the composition $(\cos x)^3$ is $\cos x$, which suggests the substitution $u = \cos x$. Note that $du = -\sin x\, dx$ or $\sin x\, dx = -du$. The change of variables appears as

$$\int \underbrace{\cos^3 x}_{u^3}\underbrace{\sin x\, dx}_{-du} = -\int u^3\, du \qquad \text{Substitute } u = \cos x,\ du = -\sin x\, dx.$$

$$= -\frac{u^4}{4} + C \qquad \text{Antiderivative}$$

$$= -\frac{\cos^4 x}{4} + C. \qquad \text{Replace } u \text{ with } \cos x.$$

*Related Exercises 23–24* ◀

Sometimes the choice for a $u$-substitution is not so obvious *or* more than one $u$-substitution works. The following example illustrates both of these points.

**EXAMPLE 3** **Variations on the substitution method** Find $\displaystyle\int \frac{x}{\sqrt{x+1}}\, dx$.

**SOLUTION**

**Substitution 1** The composite function $\sqrt{x+1}$ suggests the new variable $u = x + 1$. You might doubt whether this choice will work because $du = dx$, which leaves the $x$ in the numerator of the integrand unaccounted for. But let's proceed. Letting $u = x + 1$, we have $x = u - 1$, $du = dx$, and

$$\int \frac{x}{\sqrt{x+1}}\, dx = \int \frac{u-1}{\sqrt{u}}\, du \qquad \text{Substitute } u = x + 1,\ du = dx.$$

$$= \int\left(\sqrt{u} - \frac{1}{\sqrt{u}}\right) du \qquad \text{Rewrite integrand.}$$

$$= \int (u^{1/2} - u^{-1/2})\, du. \qquad \text{Fractional powers}$$

**QUICK CHECK 2** In Example 2a, explain why the same substitution would not work as well for the integral $\int x^3(x^5 + 6)^9\, dx$. ◀

We integrate each term individually and then return to the original variable $x$:

$$\int (u^{1/2} - u^{-1/2})\, du = \frac{2}{3} u^{3/2} - 2u^{1/2} + C \qquad \text{Antiderivatives}$$

$$= \frac{2}{3}(x+1)^{3/2} - 2(x+1)^{1/2} + C \qquad \text{Replace } u \text{ with } x + 1.$$

$$= \frac{2}{3}(x+1)^{1/2}(x-2) + C. \qquad \begin{array}{l}\text{Factor out } (x+1)^{1/2}\\ \text{and simplify.}\end{array}$$

▶ In Substitution 2, you could also use the fact that

$$u'(x) = \frac{1}{2\sqrt{x+1}},$$

which implies

$$du = \frac{1}{2\sqrt{x+1}}\, dx.$$

**Substitution 2**   Another possible substitution is $u = \sqrt{x+1}$. Now $u^2 = x+1$, $x = u^2 - 1$, and $dx = 2u\, du$. Making these substitutions leads to

$$\int \frac{x}{\sqrt{x+1}}\, dx = \int \frac{u^2-1}{u}\, 2u\, du \qquad \text{Substitute } u = \sqrt{x+1},\, x = u^2 - 1.$$

$$= 2\int (u^2 - 1)\, du \qquad \text{Simplify the integrand.}$$

$$= 2\left(\frac{u^3}{3} - u\right) + C \qquad \text{Antiderivatives}$$

$$= \frac{2}{3}(x+1)^{3/2} - 2(x+1)^{1/2} + C \qquad \text{Replace } u \text{ with } \sqrt{x+1}.$$

$$= \frac{2}{3}(x+1)^{1/2}(x-2) + C. \qquad \text{Factor out } (x+1)^{1/2} \text{ and simplify.}$$

Observe that the same indefinite integral is found using either substitution.

*Related Exercises 74–75* ◄

## General Formulas for Indefinite Integrals

Integrals of the form $\int f(ax)\, dx$ occur frequently in the remainder of this text, so our aim here is to generalize the integral formulas first introduced in Section 4.9. We encountered an integral of the form $\int f(ax)\, dx$ in the opening of this section, where we used trial and error to discover that $\int \cos 2x\, dx = \frac{1}{2}\sin 2x + C$. Notice that 2 could be replaced by any nonzero constant $a$ to produce the more general result

$$\int \cos ax\, dx = \frac{1}{a}\sin ax + C.$$

Let's verify this result without resorting to trial and error by using substitution. Letting $u = ax$, we have $du = a\, dx$, or $dx = \frac{1}{a}\, du$. By the Substitution Rule,

$$\int \cos ax\, dx = \int \cos u \cdot \frac{1}{a}\, du \qquad \text{Substitute } u = ax,\, dx = \frac{1}{a}\, du.$$

$$= \frac{1}{a}\int \cos u\, du \qquad \int cf(x)\, dx = c\int f(x)\, dx$$

$$= \frac{1}{a}\sin u + C \qquad \text{Antiderivative}$$

$$= \frac{1}{a}\sin ax + C. \qquad \text{Replace } u \text{ with } ax.$$

**QUICK CHECK 3** Evaluate $\int \cos 6x\, dx$ without using the substitution method. ◄

Now that we have established a general result for $\int \cos ax\, dx$, we can use it to evaluate integrals such as $\int \cos 6x\, dx$ without resorting to the substitution method. Table 5.6 lists additional general formulas for standard trigonometric integrals, where we assume $a \ne 0$ is a real number. The derivations of results (2)–(6) are similar to the derivation of result (1) just given. Note that all these integration formulas can be verified by differentiation.

**Table 5.6** General Integration Formulas

---

**1.** $\int \cos ax \, dx = \dfrac{1}{a} \sin ax + C$ 
 **2.** $\int \sin ax \, dx = -\dfrac{1}{a} \cos ax + C$

**3.** $\int \sec^2 ax \, dx = \dfrac{1}{a} \tan ax + C$ 
 **4.** $\int \csc^2 ax \, dx = -\dfrac{1}{a} \cot ax + C$

**5.** $\int \sec ax \tan ax \, dx = \dfrac{1}{a} \sec ax + C$ 
 **6.** $\int \csc ax \cot ax = -\dfrac{1}{a} \csc ax + C$

---

## Definite Integrals

The Substitution Rule is also used for definite integrals; in fact, there are two ways to proceed.

- You may use the Substitution Rule to find an antiderivative $F$ and then use the Fundamental Theorem to evaluate $F(b) - F(a)$.

- Alternatively, once you have changed variables from $x$ to $u$, you also may change the limits of integration and complete the integration with respect to $u$. Specifically, if $u = g(x)$, the lower limit $x = a$ is replaced with $u = g(a)$ and the upper limit $x = b$ is replaced with $u = g(b)$.

The second option tends to be more efficient, and we use it whenever possible. This approach is summarized in the following theorem, which is then applied to several definite integrals.

> **THEOREM 5.7   Substitution Rule for Definite Integrals**
> Let $u = g(x)$, where $g'$ is continuous on $[a, b]$, and let $f$ be continuous on the range of $g$. Then
> $$\int_a^b f(g(x))g'(x) \, dx = \int_{g(a)}^{g(b)} f(u) \, du. \tag{2}$$

**Proof:** We apply Part 2 of the Fundamental Theorem of Calculus to both sides of equation (2). Let $F$ be an antiderivative of $f$. Then, by equation (1) (p. 337), we know that $F(g(x))$ is an antiderivative of $f(g(x))g'(x)$, which implies that

$$\int_a^b f(g(x))g'(x) \, dx = F(g(x))\Big|_a^b = F(g(b)) - F(g(a)). \qquad \text{Fundamental Theorem, Part 2}$$

Applying the Fundamental Theorem to the right side of equation (2) leads to the same result:

$$\int_{g(a)}^{g(b)} f(u) \, du = F(u)\Big|_{g(a)}^{g(b)} = F(g(b)) - F(g(a)). \qquad ◄$$

**EXAMPLE 4   Definite integrals** Evaluate the following integrals.

**a.** $\displaystyle\int_0^2 \dfrac{dx}{(x + 3)^3}$ 
 **b.** $\displaystyle\int_{-1}^2 \dfrac{x^2}{(x^3 + 2)^3} \, dx$ 
 **c.** $\displaystyle\int_0^{\pi/2} \sin^4 x \cos x \, dx$

**SOLUTION**

➤ When the integrand has the form $f(ax + b)$, the substitution $u = ax + b$ is often effective.

**a.** Let the new variable be $u = x + 3$ and then $du = dx$. Because we have changed the variable of integration from $x$ to $u$, the limits of integration must also be expressed in terms of $u$. In this case,

$$x = 0 \text{ implies } u = 0 + 3 = 3, \quad \text{Lower limit}$$
$$x = 2 \text{ implies } u = 2 + 3 = 5. \quad \text{Upper limit}$$

The entire integration is carried out as follows:

$$\int_0^2 \frac{dx}{(x+3)^3} = \int_3^5 u^{-3}\,du \qquad \text{Substitute } u = x + 3, du = dx.$$

$$= -\frac{u^{-2}}{2}\Big|_3^5 \qquad \text{Fundamental Theorem}$$

$$= -\frac{1}{2}(5^{-2} - 3^{-2}) = \frac{8}{225}. \qquad \text{Simplify.}$$

**b.** Notice that a multiple of the derivative of $x^3 + 2$ appears in the numerator; therefore, we let $u = x^3 + 2$, which implies that $du = 3x^2\,dx$, or $x^2\,dx = \frac{1}{3}\,du$. We also change the limits of integration:

$$x = -1 \text{ implies } u = -1 + 2 = 1, \quad \text{Lower limit}$$
$$x = 2 \quad \text{implies } u = 2^3 + 2 = 10. \quad \text{Upper limit}$$

Changing variables, we have

$$\int_{-1}^2 \frac{x^2}{(x^3+2)^3}\,dx = \frac{1}{3}\int_1^{10} u^{-3}\,du \qquad \text{Substitute } u = x^3 + 2, du = 3x^2\,dx.$$

$$= \frac{1}{3}\left(-\frac{u^{-2}}{2}\right)\Big|_1^{10} \qquad \text{Fundamental Theorem}$$

$$= \frac{1}{3}\left(-\frac{1}{200} - \left(-\frac{1}{2}\right)\right) \qquad \text{Simplify.}$$

$$= \frac{33}{200}.$$

**c.** Let $u = \sin x$, which implies that $du = \cos x\,dx$. The lower limit of integration becomes $u = 0$ and the upper limit becomes $u = 1$. Changing variables, we have

$$\int_0^{\pi/2} \sin^4 x \cos x\,dx = \int_0^1 u^4\,du \qquad u = \sin x, du = \cos x\,dx$$

$$= \frac{u^5}{5}\Big|_0^1 = \frac{1}{5}. \qquad \text{Fundamental Theorem}$$

*Related Exercises 44, 47, 60* ◄

The Substitution Rule enables us to find two standard integrals that appear frequently in practice, $\int \sin^2 x\,dx$ and $\int \cos^2 x\,dx$. These integrals are handled using the identities

$$\sin^2 x = \frac{1 - \cos 2x}{2} \quad \text{and} \quad \cos^2 x = \frac{1 + \cos 2x}{2}.$$

**EXAMPLE 5**  **Integral of $\cos^2\theta$**  Evaluate $\displaystyle\int_0^{\pi/2} \cos^2\theta\,d\theta$.

**SOLUTION**  Working with the indefinite integral first, we use the identity for $\cos^2\theta$:

$$\int \cos^2\theta\,d\theta = \int \frac{1 + \cos 2\theta}{2}\,d\theta = \frac{1}{2}\int d\theta + \frac{1}{2}\int \cos 2\theta\,d\theta.$$

► See Exercise 108 for a generalization of Example 5. Trigonometric integrals involving powers of $\sin x$ and $\cos x$ are explored in greater detail in Section 8.3.

Result (1) of Table 5.6 is used for the second integral, and we have

$$\int \cos^2\theta\,d\theta = \frac{1}{2}\int d\theta + \frac{1}{2}\underbrace{\int \cos 2\theta\,d\theta}_{\frac{1}{2}\sin 2\theta\,+\,C}$$

$$= \frac{\theta}{2} + \frac{1}{4}\sin 2\theta + C. \qquad \text{Evaluate integrals;  Table 5.6.}$$

Using the Fundamental Theorem of Calculus, the value of the definite integral is

$$\int_0^{\pi/2} \cos^2 \theta \, d\theta = \left( \frac{\theta}{2} + \frac{1}{4} \sin 2\theta \right)\Big|_0^{\pi/2}$$

$$= \left( \frac{\pi}{4} + \frac{1}{4} \sin \pi \right) - \left( 0 + \frac{1}{4} \sin 0 \right) = \frac{\pi}{4}.$$

*Related Exercises 83, 87* ◄

## Geometry of Substitution

The Substitution Rule has a geometric interpretation. To keep matters simple, consider the integral $\int_0^2 2(2x + 1) \, dx$. The graph of the integrand $y = 2(2x + 1)$ on the interval $[0, 2]$ is shown in Figure 5.58a, along with the region $R$ whose area is given by the integral. The change of variables $u = 2x + 1$, $du = 2 \, dx$, $u(0) = 1$, and $u(2) = 5$ leads to the new integral

$$\int_0^2 2(2x + 1) \, dx = \int_1^5 u \, du.$$

Figure 5.58b shows the graph of the new integrand $y = u$ on the interval $[1, 5]$ and the region $R'$ whose area is given by the new integral. You can check that the areas of $R$ and $R'$ are equal. An analogous interpretation may be given to more complicated integrands and substitutions.

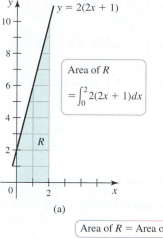

**Figure 5.58**

**QUICK CHECK 4** Changes of variables occur frequently in mathematics. For example, suppose you want to solve the equation $x^4 - 13x^2 + 36 = 0$. If you use the substitution $u = x^2$, what is the new equation that must be solved for $u$? What are the roots of the original equation? ◄

# SECTION 5.5 EXERCISES

## Getting Started

1. On which derivative rule is the Substitution Rule based?

2. Why is the Substitution Rule referred to as a change of variables?

3. The composite function $f(g(x))$ consists of an inner function $g$ and an outer function $f$. If an integrand includes $f(g(x))$, which function is often a likely choice for a new variable $u$?

4. Find a suitable substitution for evaluating $\int \tan x \sec^2 x \, dx$ and explain your choice.

5. When using a change of variables $u = g(x)$ to evaluate the definite integral $\int_a^b f(g(x))g'(x) \, dx$, how are the limits of integration transformed?

6. If the change of variables $u = x^2 - 4$ is used to evaluate the definite integral $\int_2^4 f(x) \, dx$, what are the new limits of integration?

**7–10.** *Use the given substitution to evaluate the following indefinite integrals. Check your answer by differentiating.*

7. $\int 2x(x^2 + 1)^4 \, dx$, $u = x^2 + 1$

8. $\int 8x \cos (4x^2 + 3) \, dx$, $u = 4x^2 + 3$

9. $\int \sin^3 x \cos x \, dx$, $u = \sin x$

10. $\int (6x + 1)\sqrt{3x^2 + x} \, dx$, $u = 3x^2 + x$

**11–14.** *Use a substitution of the form $u = ax + b$ to evaluate the following indefinite integrals.*

11. $\int (x + 1)^{12} \, dx$

12. $\int \sin 10x \, dx$

13. $\int \sqrt{2x + 1} \, dx$

14. $\int \cos (2x + 5) \, dx$

15. Use Table 5.6 to evaluate the following indefinite integrals.

a. $\int \csc^2 10x \, dx$

b. $\int \sec 5x \tan 5x \, dx$

c. $\int \sin 7x \, dx$

d. $\int \cos \frac{x}{7} \, dx$

16. Use Table 5.6 to evaluate the following definite integrals.

a. $\int_0^{2\pi} \sin \frac{x}{2} \, dx$

b. $\int_0^{\pi/40} \cos 20x \, dx$

c. $\int_{\pi/8}^{\pi/4} \csc 2x \cot 2x \, dx$

d. $\int_0^{\pi/16} \sec^2 4x \, dx$

## Practice Exercises

**17–40. Indefinite integrals** *Use a change of variables or Table 5.6 to evaluate the following indefinite integrals. Check your work by differentiating.*

17. $\int 2x(x^2 - 1)^{99} \, dx$

18. $\int x \sin x^2 \, dx$

**19.** $\int \dfrac{2x^2}{\sqrt{1 - 4x^3}}\, dx$

**20.** $\int \dfrac{(\sqrt{x} + 1)^4}{2\sqrt{x}}\, dx$

**21.** $\int (x^2 + x)^{10}(2x + 1)\, dx$

**22.** $\int \dfrac{1}{(10x - 3)^2}\, dx$

**23.** $\int x^3(x^4 + 16)^6\, dx$

**24.** $\int \sin^{10}\theta \cos\theta\, d\theta$

**25.** $\int \dfrac{x}{\sqrt{4 - 9x^2}}\, dx$

**26.** $\int x^9 \sin x^{10}\, dx$

**27.** $\int (x^6 - 3x^2)^4(x^5 - x)\, dx$

**28.** $\int \dfrac{x}{(1 + 4x^2)^3}\, dx$

**29.** $\int x \csc x^2 \cot x^2\, dx$

**30.** $\int \sec 4w \tan 4w\, dw$

**31.** $\int \sec^2(10x + 7)\, dx$

**32.** $\int \csc^4 x \cot x\, dx$

**33.** $\int t^3 \sin t^4 \cos t^4\, dt$

**34.** $\int (\sin^5 x + 3\sin^3 x - \sin x)\cos x\, dx$

**35.** $\int \dfrac{\csc^2 x}{\cot^3 x}\, dx$

**36.** $\int (x^{3/2} + 8)^5 \sqrt{x}\, dx$

**37.** $\int \sin x \sec^8 x\, dx$

**38.** $\int \dfrac{\cos\sqrt{t}}{\sqrt{t}}\, dt$

**39.** $\int (\sec w + 3)^9 \sec w \tan w\, dw$

**40.** $\int (\sec^5 x + \sec^3 x)\tan x\, dx$

**41–70. Definite integrals** *Use a change of variables or Table 5.6 to evaluate the following definite integrals.*

**41.** $\displaystyle\int_0^{\pi/8} \cos 2x\, dx$

**42.** $\displaystyle\int_{\pi/16}^{\pi/8} 8 \csc^2 4x\, dx$

**43.** $\displaystyle\int_0^1 2x(4 - x^2)\, dx$

**44.** $\displaystyle\int_0^2 \dfrac{2x}{(x^2 + 1)^2}\, dx$

**45.** $\displaystyle\int_0^{\pi/4} \dfrac{\sec^2 x}{\sqrt{\tan x + 3}}\, dx$

**46.** $\displaystyle\int_{-2\pi}^{2\pi} \cos\dfrac{\theta}{8}\, d\theta$

**47.** $\displaystyle\int_0^{\pi/2} \sin^2\theta \cos\theta\, d\theta$

**48.** $\displaystyle\int_0^{\pi/4} \dfrac{\sin x}{\cos^2 x}\, dx$

**49.** $\displaystyle\int_0^{\pi/4} (\sec^3 x + \sec^2 x)\tan x\, dx$

**50.** $\displaystyle\int_{\pi/12}^{\pi/6} \csc^4 3x \cot 3x\, dx$

**51.** $\displaystyle\int_{-\pi/12}^{\pi/8} \sec^2 2y\, dy$

**52.** $\displaystyle\int_0^4 \dfrac{p}{\sqrt{9 + p^2}}\, dp$

**53.** $\displaystyle\int_{\pi/8}^{\pi/4} (\sin^3 2x + \sin 2x)\cos 2x\, dx$

**54.** $\displaystyle\int_0^{\pi/4} \dfrac{\sin\theta}{\cos^3\theta}\, d\theta$

**55.** $\displaystyle\int_2^6 \dfrac{x}{\sqrt{2x - 3}}\, dx$

**56.** $\displaystyle\int_0^1 \dfrac{v^3 + 1}{\sqrt{v^4 + 4v + 4}}\, dv$

**57.** $\displaystyle\int_0^{\sqrt{3}} \dfrac{w}{\sqrt{w^2 + 1}}\, dw$

**58.** $\displaystyle\int_0^4 3\sqrt{2t + 1}\, dt$

**59.** $\displaystyle\int_0^1 (2p + 1)^3\, dp$

**60.** $\displaystyle\int_0^2 \dfrac{81q}{(2q^2 + 1)^3}\, dq$

**61.** $\displaystyle\int_0^1 x\sqrt{1 - x^2}\, dx$

**62.** $\displaystyle\int_1^3 \dfrac{(1 + 4/x)^2}{x^2}\, dx$

**63.** $\displaystyle\int_2^3 \dfrac{x}{\sqrt[3]{x^2 - 1}}\, dx$

**64.** $\displaystyle\int_0^{6/5} \dfrac{x}{(25x^2 + 36)^2}\, dx$

**65.** $\displaystyle\int_0^2 x^3\sqrt{16 - x^4}\, dx$

**66.** $\displaystyle\int_{-1}^1 (x - 1)(x^2 - 2x)^7\, dx$

**67.** $\displaystyle\int_0^1 x(x + 1)^{10}(x - 1)^{10}\, dx$

**68.** $\displaystyle\int_{\pi/8}^{\pi/4} \csc^2 2x \sqrt{1 + \cot 2x}\, dx$

**69.** $\displaystyle\int_1^2 \dfrac{4}{9x^2 + 6x + 1}\, dx$

**70.** $\displaystyle\int_0^{\pi/2} (2 + \sin x)\sin x \cos x\, dx$

**71. Average velocity** An object moves in one dimension with a velocity in m/s given by $v(t) = 8\sin\pi t + 2t$. Find its average velocity over the time interval from $t = 0$ to $t = 10$, where $t$ is measured in seconds.

**T 72. Periodic motion** An object moves along a line with a velocity in m/s given by $v(t) = 8\cos\dfrac{\pi t}{6}$. Its initial position is $s(0) = 0$.

  **a.** Graph the velocity function.
  **b.** As discussed in Chapter 6, the position of the object is given by
  $$s(t) = \int_0^t v(y)\, dy,\ \text{for } t \ge 0.$$
  Find the position function, for $t \ge 0$.
  **c.** What is the period of the motion—that is, starting at any point, how long does it take the object to return to that position?

**73. Population models** The population of a culture of bacteria has a growth rate given by $p'(t) = \dfrac{200}{(t + 1)^r}$ bacteria per hour, for $t \ge 0$, where $r > 1$ is a real number. In Chapter 6 it is shown that the increase in the population over the time interval $[0, t]$ is given by $\int_0^t p'(s)\, ds$. (Note that the growth rate decreases in time, reflecting competition for space and food.)

  **a.** Using the population model with $r = 2$, what is the increase in the population over the time interval $0 \le t \le 4$?
  **b.** Using the population model with $r = 3$, what is the increase in the population over the time interval $0 \le t \le 6$?
  **c.** Let $\Delta P$ be the increase in the population over a fixed time interval $[0, T]$. For fixed $T$, does $\Delta P$ increase or decrease with the parameter $r$? Explain.
  **d.** A lab technician measures an increase in the population of 350 bacteria over the 10-hr period $[0, 10]$. Estimate the value of $r$ that best fits this data point.
  **e.** Looking ahead: Use the population model in part (b) to find the increase in population over the time interval $[0, T]$, for any $T > 0$. If the culture is allowed to grow indefinitely ($T \to \infty$), does the bacteria population increase without bound? Or does it approach a finite limit?

**74–82. Variations on the substitution method** *Evaluate the following integrals.*

**74.** $\int \dfrac{x}{(x - 2)^3}\, dx$

**75.** $\int \dfrac{x}{\sqrt{x - 4}}\, dx$

**76.** $\int \dfrac{y^2}{(y + 1)^4}\, dy$

**77.** $\int \dfrac{x}{\sqrt[3]{x + 4}}\, dx$

**78.** $\displaystyle\int \frac{2x}{\sqrt{3x+2}}\,dx$

**79.** $\displaystyle\int x\sqrt[3]{2x+1}\,dx$

**80.** $\displaystyle\int (z+1)\sqrt{3z+2}\,dz$

**81.** $\displaystyle\int x(x+10)^9 dx$

**82.** $\displaystyle\int x^3(x^2+1)^{10}\,dx$

**83–90. Integrals with $\sin^2 x$ and $\cos^2 x$** *Evaluate the following integrals.*

**83.** $\displaystyle\int_{-\pi}^{\pi}\cos^2 x\,dx$

**84.** $\displaystyle\int \sin^2 x\,dx$

**85.** $\displaystyle\int \sin^2\left(\theta+\frac{\pi}{6}\right)d\theta$

**86.** $\displaystyle\int_0^{\pi/4}\cos^2 8\theta\,d\theta$

**87.** $\displaystyle\int_{-\pi/4}^{\pi/4}\sin^2 2\theta\,d\theta$

**88.** $\displaystyle\int x\cos^2 x^2\,dx$

**89.** $\displaystyle\int_0^{\pi/6}\frac{\sin 2y}{(\sin^2 y+2)^2}\,dy$ (*Hint:* $\sin 2y=2\sin y\cos y$.)

**90.** $\displaystyle\int_0^{\pi/2}\sin^4\theta\,d\theta$

**91. Explain why or why not** Determine whether the following statements are true and give an explanation or counterexample. Assume $f$, $f'$, and $f''$ are continuous functions for all real numbers.

**a.** $\displaystyle\int f(x)f'(x)\,dx = \frac{1}{2}(f(x))^2 + C.$

**b.** $\displaystyle\int (f(x))^n f'(x)\,dx = \frac{1}{n+1}(f(x))^{n+1} + C, n\neq -1.$

**c.** $\displaystyle\int \sin 2x\,dx = 2\int \sin x\,dx.$

**d.** $\displaystyle\int (x^2+1)^9 dx = \frac{(x^2+1)^{10}}{10}+C.$

**e.** $\displaystyle\int_a^b f'(x)f''(x)\,dx = f'(b)-f'(a).$

**92–94. Areas of regions** *Find the area of the following regions.*

**92.** The region bounded by the graph of $f(x)=\dfrac{x}{\sqrt{x^2-9}}$ and the $x$-axis between $x=4$ and $x=5$

**93.** The region bounded by the graph of $f(x)=x\sin x^2$ and the $x$-axis between $x=0$ and $x=\sqrt{\pi}$

**94.** The region bounded by the graph of $f(x)=(x-4)^4$ and the $x$-axis between $x=2$ and $x=6$

### Explorations and Challenges

**95. Morphing parabolas** The family of parabolas $y=\dfrac{1}{a}-\dfrac{x^2}{a^3}$, where $a>0$, has the property that for $x\ge 0$, the $x$-intercept is $(a,0)$ and the $y$-intercept is $(0,1/a)$. Let $A(a)$ be the area of the region in the first quadrant bounded by the parabola and the $x$-axis. Find $A(a)$ and determine whether it is an increasing, decreasing, or constant function of $a$.

**96. Substitutions** Suppose $f$ is an even function with $\int_0^8 f(x)\,dx=9$. Evaluate each integral.

**a.** $\displaystyle\int_{-1}^1 xf(x^2)\,dx.$

**b.** $\displaystyle\int_{-2}^2 x^2 f(x^3)\,dx.$

**97. Substitutions** Suppose $p$ is a nonzero real number and $f$ is an odd function with $\int_0^1 f(x)\,dx=\pi$. Evaluate each integral.

**a.** $\displaystyle\int_0^{\pi/(2p)}(\cos px)f(\sin px)\,dx$

**b.** $\displaystyle\int_{-\pi/2}^{\pi/2}(\cos x)f(\sin x)\,dx$

**98. Average distance on a triangle** Consider the right triangle with vertices $(0,0)$, $(0,b)$, and $(a,0)$, where $a>0$ and $b>0$. Show that the average vertical distance from points on the $x$-axis to the hypotenuse is $b/2$, for all $a>0$.

**T 99. Average value of sine functions** Use a graphing utility to verify that the functions $f(x)=\sin kx$ have a period of $2\pi/k$, where $k=1,2,3,\dots$. Equivalently, the first "hump" of $f(x)=\sin kx$ occurs on the interval $[0,\pi/k]$. Verify that the average value of the first hump of $f(x)=\sin kx$ is independent of $k$. What is the average value?

**100. Equal areas** The area of the shaded region under the curve $y=2\sin 2x$ in part (a) of the figure equals the area of the shaded region under the curve $y=\sin x$ in part (b) of the figure. Explain why this is true without computing areas.

(a)

(b)

**101. Equal areas** The area of the shaded region under the curve $y=\dfrac{(\sqrt{x}-1)^2}{2\sqrt{x}}$ on the interval $[4,9]$ in part (a) of the following figure equals the area of the shaded region under the curve $y=x^2$ on the interval $[1,2]$ in part (b) of the figure. Without computing areas, explain why.

(a)

(b)

**102–104. General results** *Evaluate the following integrals in which the function $f$ is unspecified. Note that $f^{(p)}$ is the pth derivative of $f$ and $f^p$ is the pth power of $f$. Assume $f$ and its derivatives are continuous for all real numbers.*

**102.** $\displaystyle\int (5f^3(x)+7f^2(x)+f(x))f'(x)\,dx$

**103.** $\displaystyle\int_1^2 (5f^3(x)+7f^2(x)+f(x))f'(x)\,dx$, where $f(1)=4$, $f(2)=5$

**104.** $\displaystyle\int (f^{(p)}(x))^n f^{(p+1)}(x)\,dx$, where $p$ is a positive integer, $n\neq -1$

**105–107. More than one way** *Occasionally, two different substitutions do the job. Use each substitution to evaluate the following integrals.*

**105.** $\int_0^1 x\sqrt{x + a}\, dx; a > 0\ (u = \sqrt{x + a}\text{ and } u = x + a)$

**106.** $\int_0^1 x\sqrt[p]{x + a}\, dx; a > 0\ (u = \sqrt[p]{x + a}\text{ and } u = x + a)$

**107.** $\int \sec^3 \theta \tan \theta\, d\theta\ (u = \cos \theta\text{ and } u = \sec \theta)$

**108. $\sin^2 ax$ and $\cos^2 ax$ integrals** Use the Substitution Rule to prove that

$$\int \sin^2 ax\, dx = \frac{x}{2} - \frac{\sin (2ax)}{4a} + C \text{ and}$$

$$\int \cos^2 ax\, dx = \frac{x}{2} + \frac{\sin (2ax)}{4a} + C.$$

**109. Integral of $\sin^2 x \cos^2 x$** Consider the integral $I = \int \sin^2 x \cos^2 x\, dx$.

   **a.** Find $I$ using the identity $\sin 2x = 2 \sin x \cos x$.

   **b.** Find $I$ using the identity $\cos^2 x = 1 - \sin^2 x$.

   **c.** Confirm that the results in parts (a) and (b) are consistent and compare the work involved in the two methods.

**110. Substitution: shift** Perhaps the simplest change of variables is the shift or translation given by $u = x + c$, where $c$ is a real number.

   **a.** Prove that shifting a function does not change the net area under the curve, in the sense that

$$\int_a^b f(x + c)\, dx = \int_{a+c}^{b+c} f(u)\, du.$$

   **b.** Draw a picture to illustrate this change of variables in the case where $f(x) = \sin x$, $a = 0$, $b = \pi$, and $c = \pi/2$.

**111. Substitution: scaling** Another change of variables that can be interpreted geometrically is the scaling $u = cx$, where $c$ is a real number. Prove and interpret the fact that

$$\int_a^b f(cx)\, dx = \frac{1}{c} \int_{ac}^{bc} f(u)\, du.$$

Draw a picture to illustrate this change of variables in the case where $f(x) = \sin x$, $a = 0$, $b = \pi$, and $c = 1/2$.

**112–115. Multiple substitutions** *If necessary, use two or more substitutions to find the following integrals.*

**112.** $\int x \sin^4 x^2 \cos x^2\, dx$ (*Hint:* Begin with $u = x^2$, and then use $v = \sin u$.)

**113.** $\int \dfrac{dx}{\sqrt{1 + \sqrt{1 + x}}}$ (*Hint:* Begin with $u = \sqrt{1 + x}$.)

**114.** $\int \tan^{10} 4x \sec^2 4x\, dx$ (*Hint:* Begin with $u = 4x$.)

**115.** $\int_0^{\pi/2} \dfrac{\cos \theta \sin \theta}{\sqrt{\cos^2 \theta + 16}}\, d\theta$ (*Hint:* Begin with $u = \cos \theta$.)

**QUICK CHECK ANSWERS**

**1.** $u = x^4 + 5$   **2.** With $u = x^5 + 6$, we have $du = 5x^4$, and $x^4$ does not appear in the integrand.   **3.** $\dfrac{1}{6} \sin 6x + C$

**4.** New equation: $u^2 - 13u + 36 = 0$; roots: $x = \pm 2, \pm 3$ ◄

## CHAPTER 5    REVIEW EXERCISES

**1.** **Explain why or why not** Determine whether the following statements are true and give an explanation or counterexample. Assume $f$ and $f'$ are continuous functions for all real numbers.

   **a.** If $A(x) = \int_a^x f(t)\, dt$ and $f(t) = 2t - 3$, then $A$ is a quadratic function.

   **b.** Given an area function $A(x) = \int_a^x f(t)\, dt$ and an antiderivative $F$ of $f$, it follows that $A'(x) = F(x)$.

   **c.** $\int_a^b f'(x)\, dx = f(b) - f(a)$.

   **d.** If $f$ is continuous on $[a, b]$ and $\int_a^b |f(x)|\, dx = 0$, then $f(x) = 0$ on $[a, b]$.

   **e.** If the average value of $f$ on $[a, b]$ is zero, then $f(x) = 0$ on $[a, b]$.

   **f.** $\int_a^b (2f(x) - 3g(x))\, dx = 2\int_a^b f(x)\, dx + 3\int_b^a g(x)\, dx$.

   **g.** $\int f'(g(x))g'(x)\, dx = f(g(x)) + C$.

**2.** The velocity in ft/s of an object moving along a line is given by the function $v = f(t)$ on the interval $0 \le t \le 12$, where $t$ is in seconds (see figure). Estimate the displacement of the object on $[0, 12]$ using left, right, and midpoint Riemann sums with $n = 3$ subintervals and a regular partition.

**3.** **Ascent rate of a scuba diver** Divers who ascend too quickly in the water risk *decompression illness*. A common recommendation for a maximum rate of ascent is 30 feet/minute with a 5-minute safety stop 15 feet below the surface of the water. Suppose a diver ascends to the surface in 8 minutes according to the velocity function

$$v(t) = \begin{cases} 30 & \text{if } 0 \le t \le 2 \\ 0 & \text{if } 2 < t \le 7 \\ 15 & \text{if } 7 < t \le 8. \end{cases}$$

   **a.** Graph the velocity function $v$.

   **b.** Compute the area under the velocity curve.

   **c.** Interpret the physical meaning of the area under the velocity curve.

**4.** Use the tabulated values of $f$ to estimate the value of $\int_0^6 f(x)\,dx$ by evaluating the left, right, and midpoint Riemann sums using a regular partition with $n = 3$ subintervals.

| $x$ | 0 | 1 | 2 | 3 | 4 | 5 | 6 |
|-----|----|----|----|----|----|----|----|
| $f(x)$ | 20 | 22 | 26 | 30 | 34 | 38 | 40 |

**5.** Estimate $\int_1^4 \sqrt{4x + 1}\,dx$ by evaluating the left, right, and midpoint Riemann sums using a regular partition with $n = 6$ subintervals.

**6.** **Evaluating Riemann sums** Consider the function $f(x) = 3x + 4$ on the interval $[3, 7]$. Show that the midpoint Riemann sum with $n = 4$ gives the exact area of the region bounded by the graph. Assume a regular partition.

**T 7.** **Estimating a definite integral** Use a calculator and midpoint Riemann sums to approximate $\int_1^{25} \sqrt{2x - 1}\,dx$. Present your calculations in a table showing the approximations for $n = 10$, 30, and 60 subintervals, assuming a regular partition. Make a conjecture about the exact value of the integral and verify your conjecture using the Fundamental Theorem of Calculus.

**8.** Suppose the expression $\lim\limits_{\Delta \to 0} \sum\limits_{k=1}^{n} (x_k^{*3} + x_k^*)\Delta x_k$ is the limit of a Riemann sum of a function $f$ on $[3, 8]$. Identify a possible function $f$ and express the limit as a definite integral.

**9.** **Integration by Riemann sums** Consider the integral $\int_1^4 (3x - 2)\,dx$.

   **a.** Evaluate the right Riemann sum for the integral with $n = 3$.

   **b.** Use summation notation to express the right Riemann sum in terms of a positive integer $n$.

   **c.** Evaluate the definite integral by taking the limit as $n \to \infty$ of the Riemann sum in part (b).

   **d.** Confirm the result from part (c) by graphing $y = 3x - 2$ and using geometry to evaluate the integral. Then evaluate $\int_1^4 (3x - 2)\,dx$ with the Fundamental Theorem of Calculus.

**10–13. Limit definition of the definite integral** *Use the limit definition of the definite integral with right Riemann sums and a regular partition to evaluate the following definite integrals. Use the Fundamental Theorem of Calculus to check your answer.*

**10.** $\int_0^1 (4x - 2)\,dx$      **11.** $\int_0^2 (x^2 - 4)\,dx$

**12.** $\int_1^2 (3x^2 + x)\,dx$      **13.** $\int_0^4 (x^3 - x)\,dx$

**14.** **Sum to integral** Evaluate the following limit by identifying the integral that it represents:

$$\lim_{n \to \infty} \sum_{k=1}^{n} \left(\left(\frac{4k}{n}\right)^5 + 1\right)\frac{4}{n}.$$

**15.** **Symmetry properties** Suppose $\int_0^4 f(x)\,dx = 10$ and $\int_0^4 g(x)\,dx = 20$. Furthermore, suppose $f$ is an even function and $g$ is an odd function. Evaluate the following integrals.

   **a.** $\int_{-4}^4 f(x)\,dx$      **b.** $\int_{-4}^4 3g(x)\,dx$

   **c.** $\int_{-4}^4 (4f(x) - 3g(x))\,dx$    **d.** $\int_0^1 8xf(4x^2)\,dx$

   **e.** $\int_{-2}^2 3xf(x)\,dx$

**16.** **Properties of integrals** The figure shows the areas of regions bounded by the graph of $f$ and the $x$-axis. Evaluate the following integrals.

   **a.** $\int_a^c f(x)\,dx$      **b.** $\int_b^d f(x)\,dx$

   **c.** $\int_c^b 2f(x)\,dx$      **d.** $\int_a^d 4f(x)\,dx$

   **e.** $\int_a^b 3f(x)\,dx$      **f.** $\int_b^d 2f(x)\,dx$

   **g.** $\int_a^c |f(x)|\,dx$      **h.** $\int_d^a |f(x)|\,dx$

**17–22. Properties of integrals** *Suppose $\int_1^4 f(x)\,dx = 6$, $\int_1^4 g(x)\,dx = 4$, and $\int_3^4 f(x)\,dx = 2$. Evaluate the following integrals or state that there is not enough information.*

**17.** $\int_1^4 3f(x)\,dx$      **18.** $-\int_4^1 2f(x)\,dx$

**19.** $\int_1^4 (3f(x) - 2g(x))\,dx$      **20.** $\int_1^4 f(x)g(x)\,dx$

**21.** $\int_1^3 \frac{f(x)}{g(x)}\,dx$      **22.** $\int_4^1 (f(x) - g(x))\,dx$

**23.** **Area by geometry** Use geometry to evaluate the following definite integrals, where the graph of $f$ is given in the figure.

   **a.** $\int_0^4 f(x)\,dx$      **b.** $\int_6^4 f(x)\,dx$

   **c.** $\int_5^7 f(x)\,dx$      **d.** $\int_0^7 f(x)\,dx$

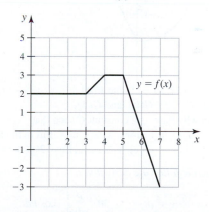

**24.** Use geometry to find the displacement of an object moving along a line for the time intervals (i) $0 \le t \le 5$, (ii) $3 \le t \le 7$, and (iii) $0 \le t \le 8$, where the graph of its velocity $v = g(t)$ is given in the figure.

**25–26.** *Use geometry and properties of integrals to evaluate the following definite integrals.*

**25.** $\int_0^4 \sqrt{8x - x^2}\, dx.$ (*Hint:* Complete the square.)

**26.** $\int_4^0 \left(2x + \sqrt{16 - x^2}\right) dx$ (*Hint:* Write the integral as a sum of two integrals.)

**27. Identifying functions** Match the graphs $A$, $B$, and $C$ in the figure with the functions $f(x)$, $f'(x)$, and $\int_0^x f(t)\, dt$.

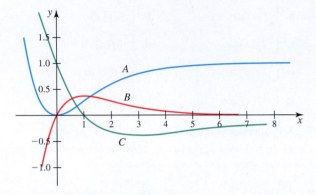

**28. Area functions** Consider the graph of the function $f$ in the figure and let $F(x) = \int_0^x f(t)\, dt$ and $G(x) = \int_1^x f(t)\, dt$. Assume the graph consists of a line segment from $(0, -2)$ to $(2, 2)$ and two quarter circles of radius 2.

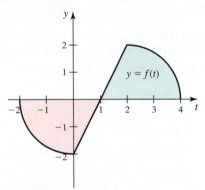

$y = f(t)$

**a.** Evaluate $F(2)$, $F(-2)$, and $F(4)$.
**b.** Evaluate $G(-2)$, $G(0)$, and $G(4)$.
**c.** Explain why there is a constant $C$ such that $F(x) = G(x) + C$, for $-2 \le x \le 4$. Fill in the blank with a number: $F(x) = G(x) +$ _____ , for $-2 \le x \le 4$.

**29–34.** *Evaluate the following derivatives.*

**29.** $\dfrac{d}{dx} \int_7^x \sqrt{1 + t^4 + t^6}\, dt$

**30.** $\dfrac{d}{dx} \int_3^{x^2} \cos t\, dt$

**31.** $\dfrac{d}{dx} \int_x^5 \sin w^6\, dw$

**32.** $\dfrac{d}{dx} \int_{x^2}^5 \sin w^6\, dw$

**33.** $\dfrac{d}{dx} \int_{-x}^x \dfrac{dt}{t^{10} + 1}$

**34.** $\dfrac{d}{dx} \int_{x^2}^{7x} \sin^3 t\, dt$

**35.** Find the intervals on which $f(x) = \int_x^1 (t - 3)(t - 6)^{11}\, dt$ is increasing and the intervals on which it is decreasing.

**36. Area function by geometry** Use geometry to find the area $A(x)$ that is bounded by the graph of $f(t) = 2t - 4$ and the $t$-axis between the point $(2, 0)$ and the variable point $(x, 0)$, where $x \ge 2$. Verify that $A'(x) = f(x)$.

**37.** Given that $F' = f$, use the substitution method to show that $\int f(ax + b)\, dx = \dfrac{1}{a} F(ax + b) + C$, for nonzero constants $a$ and $b$.

**38–80. Evaluating integrals** *Evaluate the following integrals.*

**38.** $\int_1^5 dx$

**39.** $\int_{-2}^2 (3x^4 - 2x + 1)\, dx$

**40.** $\int_0^1 (4x^{21} - 2x^{16} + 1)\, dx$

**41.** $\int_0^1 (9x^8 - 7x^6)\, dx$

**42.** $\int \dfrac{\sqrt{x} + 1}{\sqrt{x}}\, dx$

**43.** $\int_0^1 \sqrt{x}\left(\sqrt{x} + 1\right) dx$

**44.** $\int (3x + 1)(3x^2 + 2x + 1)^3\, dx$

**45.** $\int_{\pi/6}^{\pi/3} (\sec^2 t + \csc^2 t)\, dt$

**46.** $\int_{\pi/12}^{\pi/9} (\csc 3x \cot 3x + \sec 3x \tan 3x)\, dx$

**47.** $\int_{\sqrt{2}}^{\sqrt{10}} \dfrac{x}{\sqrt{x^2 - 1}}\, dx$

**48.** $\int_1^4 \dfrac{\sqrt{v} + v}{v}\, dv$

**49.** $\int \dfrac{\cos x}{\sin^{7/4} x}\, dx$

**50.** $\int_1^4 \dfrac{dx}{\sqrt{x}\left(1 + \sqrt{x}\right)^2}$

**51.** $\int x^2 \cos x^3\, dx$

**52.** $\int \dfrac{\cos 3t}{(1 + \sin 3t)^2}\, dt$

**53.** $\int \dfrac{\sin 14w}{(16 + \sin^2 7w)^2}\, dw$

**54.** $\int \sqrt{1 + \tan 2t}\, \sec^2 2t\, dt$

**55.** $\int (\sin^3 3x + \sin^2 3x) \cos 3x\, dx$

**56.** $\int \cos 3x\, dx$

**57.** $\int_0^2 (2x + 1)^3\, dx$

**58.** $\int \sin z \sin (\cos z)\, dz$

**59.** $\int_0^1 \dfrac{6x}{(4 - x^2)^{3/2}}\, dx$

**60.** $\int \dfrac{y^2}{(y^3 + 27)^2}\, dy$

**61.** $\int \dfrac{x^2 + 2x - 2}{(x^3 + 3x^2 - 6x)^2}\, dx$

**62.** $\int y^2(3y^3 + 1)^4\, dy$

**63.** $\int_0^{2\pi} \cos^2 \dfrac{x}{6}\, dx$

**64.** $\int x \sin x^2 \cos^8 x^2\, dx$

**65.** $\int_0^\pi \sin^2 5\theta\, d\theta$

**66.** $\int_0^\pi (1 - \cos^2 3\theta)\, d\theta$

**67.** $\int_{-5}^5 \dfrac{w^3}{\sqrt{w^{50} + w^{20} + 1}}\, dw$ (*Hint:* Use symmetry.)

**68.** $\int_{-3}^3 (511x^{17} + 302x^{13} + 117x^9 + 303x^3 + x^2)\, dx$

**69.** $\int \dfrac{1}{x^2} \sin \dfrac{1}{x}\, dx$

**70.** $\int_1^4 \dfrac{1 + x^{3/2}}{x^{1/2}}\, dx$

**71.** $\displaystyle\int_0^1 \frac{\sqrt{x}}{(\sqrt{x}+1)^4}\,dx$

**72.** $\displaystyle\int \frac{(\cos\sqrt{x})\sqrt{\sin\sqrt{x}}}{\sqrt{x}}\,dx$

**73.** $\displaystyle\int x(x+3)^{10}\,dx$

**74.** $\displaystyle\int x^7\sqrt{x^4+1}\,dx$

**75.** $\displaystyle\int_0^3 \frac{x}{\sqrt{25-x^2}}\,dx$

**76.** $\displaystyle\int \frac{x}{(x+1)^3}\,dx$

**77.** $\displaystyle\int_{-10}^{10} \frac{x}{\sqrt{200-x^2}}\,dx$

**78.** $\displaystyle\int_{-\pi/2}^{\pi/2} (\cos 2x + \cos x \sin x - 3\sin x^5)\,dx$

**79.** $\displaystyle\int_0^4 f(x)\,dx$ for $f(x) = \begin{cases} 2x+1 & \text{if } x \le 3 \\ 3x^2 + 2x - 8 & \text{if } x > 3 \end{cases}$

**80.** $\displaystyle\int_0^5 |2x-8|\,dx$

**81–84. Area of regions** *Compute the area of the region bounded by the graph of f and the x-axis on the given interval. You may find it useful to sketch the region.*

**81.** $f(x) = 16 - x^2$ on $[-4, 4]$

**82.** $f(x) = x^3 - x$ on $[-1, 0]$

**83.** $f(x) = 2\sin\dfrac{x}{4}$ on $[0, 2\pi]$

**84.** $f(x) = \cos\pi x$ on $[0, 2]$

**85–86. Area versus net area** *Find (i) the net area and (ii) the area of the region bounded by the graph of f and the x-axis on the given interval. You may find it useful to sketch the region.*

**85.** $f(x) = x^4 - x^2$ on $[-1, 1]$

**86.** $f(x) = x^2 - x$ on $[0, 3]$

**87.** Suppose $\int_1^2 f(x)\,dx = 10$. Evaluate the following integrals.

  **a.** $\displaystyle\int_{-1}^{-\sqrt{2}} x f(x^2)\,dx$      **b.** $\displaystyle\int_1^4 \frac{f(\sqrt{x})}{\sqrt{x}}\,dx$

**88. Root mean square** The root mean square (or RMS) is another measure of average value, often used with oscillating functions (for example, sine and cosine functions that describe the current, voltage, or power in an alternating circuit). The RMS of a function $f$ on the interval $[0, T]$ is

$$\overline{f}_{\text{RMS}} = \sqrt{\frac{1}{T}\int_0^T f(t)^2\,dt}.$$

Compute the RMS of $f(t) = A\sin(\omega t)$, where $A$ and $\omega$ are positive constants and $T$ is any integer multiple of the period of $f$, which is $2\pi/\omega$.

**89. Displacement from velocity** A particle moves along a line with a velocity given by $v(t) = 5\sin\pi t$, starting with an initial position $s(0) = 0$. Find the displacement of the particle between $t = 0$ and $t = 2$, which is given by $s(t) = \int_0^2 v(t)\,dt$. Find the distance traveled by the particle during this interval, which is $\int_0^2 |v(t)|\,dt$.

**90. Velocity to displacement** An object travels on the x-axis with a velocity given by $v(t) = 2t + 5$, for $0 \le t \le 4$.

  **a.** How far does the object travel, for $0 \le t \le 4$?

  **b.** What is the average velocity of the object on the interval $[0, 4]$?

  **c.** True or false: The object would travel as far as in part (a) if it traveled at its average velocity (a constant), for $0 \le t \le 4$.

**91.** Find the average value of $f(x) = \cos 3x$ on $\left[0, \dfrac{\pi}{6}\right]$.

**92. Average height** A baseball is launched into the outfield on a parabolic trajectory given by $y = 0.01x(200 - x)$. Find the average height of the baseball over the horizontal extent of its flight.

**93. Average values** Integration is not needed.

  **a.** Find the average value of $f$ shown in the figure on the interval $[1, 6]$ and then find the point(s) $c$ in $(1, 6)$ guaranteed to exist by the Mean Value Theorem for Integrals.

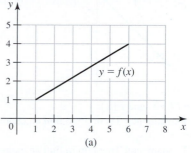

(a)

  **b.** Find the average value of $f$ shown in the figure on the interval $[2, 6]$ and then find the point(s) $c$ in $(2, 6)$ guaranteed to exist by the Mean Value Theorem for Integrals.

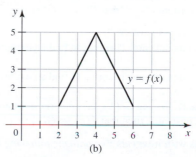

(b)

**94. An unknown function** The function $f$ satisfies the equation $3x^4 - 48 = \int_2^x f(t)\,dt$. Find $f$ and check your answer by substitution.

**95. An unknown function** Assume $f'$ is a continuous on $[2, 4]$, $\int_1^2 f'(2x)\,dx = 10$, and $f(2) = 4$. Evaluate $f(4)$.

**96. Function defined by an integral** Let $H(x) = \int_0^x \sqrt{4 - t^2}\,dt$, for $-2 \le x \le 2$.

  **a.** Evaluate $H(0)$.

  **b.** Evaluate $H'(1)$.

  **c.** Evaluate $H'(2)$.

  **d.** Use geometry to evaluate $H(2)$.

  **e.** Find the value of $s$ such that $H(x) = sH(-x)$.

**97. Function defined by an integral** Make a graph of the function

$$f(x) = \int_1^x \frac{dt}{t}, \text{ for } x \ge 1.$$ Be sure to include all of the evidence you used to arrive at the graph.

**98. Change of variables** Use the change of variables $u^3 = x^2 - 1$ to evaluate the integral $\int_1^3 x\sqrt[3]{x^2 - 1}\,dx$.

**99–100. Area functions and the Fundamental Theorem** *Consider the function*

$$f(t) = \begin{cases} t & \text{if } -2 \le t < 0 \\ t^2/2 & \text{if } 0 \le t \le 2 \end{cases}$$

*and its graph shown below. Let* $F(x) = \int_{-1}^{x} f(t)\, dt$ *and* $G(x) = \int_{-2}^{x} f(t)\, dt$.

**99. a.** Evaluate $F(-2)$ and $F(2)$.
   **b.** Use the Fundamental Theorem to find an expression for $F'(x)$, for $-2 \le x < 0$.
   **c.** Use the Fundamental Theorem to find an expression for $F'(x)$, for $0 \le x \le 2$.
   **d.** Evaluate $F'(-1)$ and $F'(1)$. Interpret these values.
   **e.** Evaluate $F''(-1)$ and $F''(1)$.
   **f.** Find a constant $C$ such that $F(x) = G(x) + C$.

**100. a.** Evaluate $G(-1)$ and $G(1)$.
   **b.** Use the Fundamental Theorem to find an expression for $G'(x)$, for $-2 \le x < 0$.
   **c.** Use the Fundamental Theorem to find an expression for $G'(x)$, for $0 \le x \le 2$.
   **d.** Evaluate $G'(0)$ and $G'(1)$. Interpret these values.
   **e.** Find a constant $C$ such that $F(x) = G(x) + C$.

**101–102. Limits with integrals** *Evaluate the following limits.*

**101.** $\displaystyle \lim_{x \to 1} \frac{\int_1^x \cos \pi t^2 \, dt}{x - 1}$

**102.** $\displaystyle \lim_{x \to 1} \frac{\int_1^{x^2} \cos t^3 \, dt}{x - 1}$

**103.** Evaluate $\displaystyle \int_0^3 x \lceil x \rceil \, dx$.

**104. Area with a parameter** Let $a > 0$ be a real number and consider the family of functions $f(x) = \sin ax$ on the interval $[0, \pi/a]$.
   **a.** Graph $f$, for $a = 1, 2$, and 3.
   **b.** Let $g(a)$ be the area of the region bounded by the graph of $f$ and the $x$-axis on the interval $[0, \pi/a]$. Graph $g$ for $0 < a < \infty$. Is $g$ an increasing function, a decreasing function, or neither?

**105.** Evaluate $\int \sec^8 (\tan x^2) \sin (\tan x^2)\, x \sec^2 x^2 \, dx$.

**T 106. Area function properties** Consider the function $f(t) = t^2 - 5t + 4$ and the area function $A(x) = \int_0^x f(t)\, dt$.
   **a.** Graph $f$ on the interval $[0, 6]$.
   **b.** Compute and graph $A$ on the interval $[0, 6]$.
   **c.** Show that the local extrema of $A$ occur at the zeros of $f$.
   **d.** Give a geometric and analytical explanation for the observation in part (c).
   **e.** Find the approximate zeros of $A$, other than 0, and call them $x_1$ and $x_2$, where $x_1 < x_2$.
   **f.** Find $b$ such that the area bounded by the graph of $f$ and the $t$-axis on the interval $[0, x_1]$ equals the area bounded by the graph of $f$ and the $t$-axis on the interval $[x_1, b]$.
   **g.** If $f$ is an integrable function and $A(x) = \int_a^x f(t)\, dt$, is it always true that the local extrema of $A$ occur at the zeros of $f$? Explain.

**107. Function defined by an integral** Let $f(x) = \int_0^x (t - 1)^{15} (t - 2)^9 \, dt$.
   **a.** Find the intervals on which $f$ is increasing and the intervals on which $f$ is decreasing.
   **b.** Find the intervals on which $f$ is concave up and the intervals on which $f$ is concave down.
   **c.** For what values of $x$ does $f$ have local minima? Local maxima?
   **d.** Where are the inflection points of $f$?

**108. Unit area sine curve** Find the value of $c$ such that the region bounded by $y = c \sin x$ and the $x$-axis on the interval $[0, \pi]$ has area 1.

**109. Equivalent equations** Explain why, if a function $u$ satisfies the equation $u(x) + 2 \int_0^x u(t)\, dt = 10$, then it also satisfies the equation $u'(x) + 2u(x) = 0$. Is it true that if $u$ satisfies the second equation, then it satisfies the first equation?

**110. Unit area cubic** Find the value of $c > 0$ such that the region bounded by the cubic $y = x(x - c)^2$ and the $x$-axis on the interval $[0, c]$ has area 1.

# Chapter 5 Guided Projects

*Applications of the material in this chapter and related topics can be found in the following Guided Projects. For additional information, see the Preface.*

- Limits of sums
- Distribution of wealth

- Symmetry in integrals

# 6

# Applications of Integration

**Chapter Preview** Now that we have some basic techniques for evaluating integrals, we turn our attention to the uses of integration, which are virtually endless. We first illustrate the general rule that if the rate of change of a quantity is known, then integration can be used to determine the net change or the future value of that quantity over a certain time interval. Next, we explore some rich geometric applications of integration: computing the area of regions bounded by several curves, the volume and surface area of three-dimensional solids, and the length of curves. We end the chapter with an examination of a variety of physical applications of integration, such as finding the work done by a variable force and computing the total force exerted by water behind a dam. All these applications are unified by their use of the *slice-and-sum* strategy.

## 6.1 Velocity and Net Change

In previous chapters, we established the relationship between the position and velocity of an object moving along a line. With integration, we can now say much more about this relationship. Once we relate velocity and position through integration, we can make analogous observations about a variety of other practical problems, which include fluid flow, population growth, manufacturing costs, and production and consumption of natural resources. The ideas in this section come directly from the Fundamental Theorem of Calculus, and they are among the most powerful applications of calculus.

### Velocity, Position, and Displacement

Suppose you are driving along a straight highway and your position relative to a reference point or origin is $s(t)$ for times $t \geq 0$. Your *displacement* over a time interval $[a, b]$ is the change in the position $s(b) - s(a)$ (**Figure 6.1**). If $s(b) > s(a)$, then your displacement is positive; when $s(b) < s(a)$, your displacement is negative.

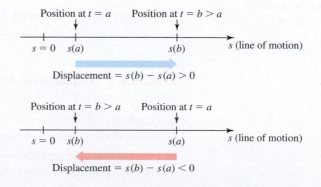

**Figure 6.1**

Now assume $v(t)$ is the velocity of the object at a particular time $t$. Recall from Chapter 3 that $v(t) = s'(t)$, which means that $s$ is an antiderivative of $v$. From the Fundamental Theorem of Calculus, it follows that

$$\int_a^b v(t)\, dt = \int_a^b s'(t)\, dt = s(b) - s(a) = \text{displacement}.$$

We see that the definite integral $\int_a^b v(t)\, dt$ is the displacement (change in position) between times $t = a$ and $t = b$. Equivalently, the displacement over the time interval $[a, b]$ is the net area under the velocity curve over $[a, b]$ (**Figure 6.2a**).

Not to be confused with the displacement is the *distance traveled* over a time interval, which is the total distance traveled by the object, independent of the direction of motion. If the velocity is positive, the object moves in the positive direction and the displacement equals the distance traveled. However, if the velocity changes sign, then the displacement and the distance traveled are not generally equal.

**QUICK CHECK 1** A police officer leaves his station on a north-south freeway at 9 A.M., traveling north (the positive direction) for 40 mi between 9 A.M. and 10 A.M. From 10 A.M. to 11 A.M., he travels south to a point 20 mi south of the station. What are the distance traveled and the displacement between 9 A.M. and 11 A.M.? ◄

To compute the distance traveled, we need the magnitude, but not the sign, of the velocity. The magnitude of the velocity $|v(t)|$ is called the *speed*. The distance traveled over a small time interval $dt$ is $|v(t)|\, dt$ (speed multiplied by elapsed time). Summing these distances, the distance traveled over the time interval $[a, b]$ is the integral of the speed; that is,

$$\text{distance traveled} = \int_a^b |v(t)|\, dt.$$

As shown in **Figure 6.2b**, integrating the speed produces the area (not net area) bounded by the velocity curve and the $t$-axis, which corresponds to the distance traveled. The distance traveled is always nonnegative.

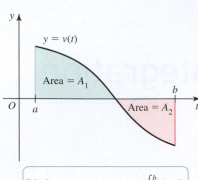

$$\text{Displacement} = A_1 - A_2 = \int_a^b v(t)\, dt$$

(a)

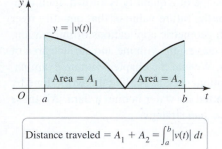

$$\text{Distance traveled} = A_1 + A_2 = \int_a^b |v(t)|\, dt$$

(b)

**Figure 6.2**

**QUICK CHECK 2** Describe a possible motion of an object along a line, for $0 \le t \le 5$, for which the displacement and the distance traveled are different. ◄

---

**DEFINITION  Position, Velocity, Displacement, and Distance**

1. The **position** of an object moving along a line at time $t$, denoted $s(t)$, is the location of the object relative to the origin.

2. The **velocity** of an object at time $t$ is $v(t) = s'(t)$.

3. The **displacement** of the object between $t = a$ and $t = b > a$ is

$$s(b) - s(a) = \int_a^b v(t)\, dt.$$

4. The **distance traveled** by the object between $t = a$ and $t = b > a$ is

$$\int_a^b |v(t)|\, dt,$$

where $|v(t)|$ is the **speed** of the object at time $t$.

---

**EXAMPLE 1  Displacement from velocity** A jogger runs along a straight road with velocity (in mi/hr) $v(t) = 2t^2 - 8t + 6$, for $0 \le t \le 3$, where $t$ is measured in hours.

a. Graph the velocity function over the interval $[0, 3]$. Determine when the jogger moves in the positive direction and when she moves in the negative direction.

b. Find the displacement of the jogger (in miles) on the time intervals $[0, 1]$, $[1, 3]$, and $[0, 3]$. Interpret these results.

c. Find the distance traveled over the interval $[0, 3]$.

(a)

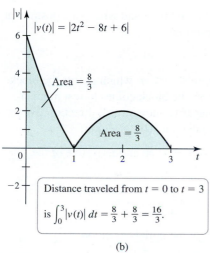

(b)

**Figure 6.3**

▶ Note that $t$ is the independent variable of the position function. Therefore, another (dummy) variable, in this case $x$, must be used as the variable of integration.

▶ Theorem 6.1 is a consequence (actually a statement) of the Fundamental Theorem of Calculus.

**SOLUTION**

**a.** By solving $v(t) = 2t^2 - 8t + 6 = 2(t - 1)(t - 3) = 0$, we find that the velocity is zero at $t = 1$ and $t = 3$; these values are the $t$-intercepts of the graph of $v$, which is an upward-opening parabola with a $v$-intercept of 6 (**Figure 6.3a**). The velocity is positive on the interval $0 \leq t < 1$, which means the jogger moves in the positive $s$ direction. For $1 < t < 3$, the velocity is negative and the jogger moves in the negative $s$ direction.

**b.** The displacement (in miles) over the interval $[0, 1]$ is

$$s(1) - s(0) = \int_0^1 v(t) \, dt$$

$$= \int_0^1 (2t^2 - 8t + 6) \, dt \qquad \text{Substitute for } v.$$

$$= \left( \frac{2}{3} t^3 - 4t^2 + 6t \right) \Big|_0^1 = \frac{8}{3}. \qquad \text{Evaluate integral.}$$

A similar calculation shows that the displacement over the interval $[1, 3]$ is

$$s(3) - s(1) = \int_1^3 v(t) \, dt = -\frac{8}{3}.$$

Over the interval $[0, 3]$, the displacement is $\frac{8}{3} + \left( -\frac{8}{3} \right) = 0$, which means the jogger returns to the starting point after three hours.

**c.** From part (b), we can deduce the total distance traveled by the jogger. On the interval $[0, 1]$, the distance traveled is $\frac{8}{3}$ mi; on the interval $[1, 3]$, the distance traveled is also $\frac{8}{3}$ mi. Therefore, the distance traveled on $[0, 3]$ is $\frac{16}{3}$ mi. Alternatively (**Figure 6.3b**), we can integrate the speed and get the same result:

$$\int_0^3 |v(t)| \, dt = \int_0^1 (2t^2 - 8t + 6) \, dt + \int_1^3 (-(2t^2 - 8t + 6)) \, dt \quad \text{Definition of } |v(t)|$$

$$= \left( \frac{2}{3} t^3 - 4t^2 + 6t \right) \Big|_0^1 + \left( -\frac{2}{3} t^3 + 4t^2 - 6t \right) \Big|_1^3 \quad \text{Evaluate integrals.}$$

$$= \frac{16}{3}. \qquad \text{Simplify.}$$

*Related Exercises 7–8* ◀

## Future Value of the Position Function

To find the displacement of an object, we do not need to know its initial position. For example, whether an object moves from $s = -20$ to $s = -10$ or from $s = 50$ to $s = 60$, its displacement is 10 units. What happens if we are interested in the actual *position* of the object at some future time?

Suppose we know the velocity of an object and its initial position $s(0)$. The goal is to find the position $s(t)$ at some future time $t \geq 0$. The Fundamental Theorem of Calculus gives us the answer directly. Because the position $s$ is an antiderivative of the velocity $v$, we have

$$\int_0^t v(x) \, dx = \int_0^t s'(x) \, dx = s(x) \Big|_0^t = s(t) - s(0).$$

Rearranging this expression leads to the following result.

---

**THEOREM 6.1 Position from Velocity**

Given the velocity $v(t)$ of an object moving along a line and its initial position $s(0)$, the position function of the object for future times $t \geq 0$ is

$$\underbrace{s(t)}_{\substack{\text{position} \\ \text{at } t}} = \underbrace{s(0)}_{\substack{\text{initial} \\ \text{position}}} + \underbrace{\int_0^t v(x) \, dx}_{\substack{\text{displacement} \\ \text{over } [0, t]}}.$$

---

**QUICK CHECK 3** Is the position $s(t)$ a number or a function? For fixed times $t = a$ and $t = b$, is the displacement $s(b) - s(a)$ a number or a function? ◄

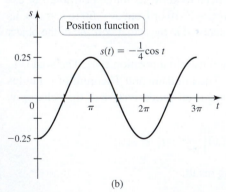

**Figure 6.4**

**Figure 6.5**

➤ It is worth repeating that to find the displacement, we need to know only the velocity. To find the position, we must know both the velocity and the initial position $s(0)$.

Theorem 6.1 says that to find the position $s(t)$, we add the displacement over the interval $[0, t]$ to the initial position $s(0)$.

There are two *equivalent* ways to determine the position function:

• Using antiderivatives (Section 4.9)

• Using Theorem 6.1

The latter method is usually more efficient, but either method produces the same result. The following example illustrates both approaches.

**EXAMPLE 2    Position from velocity** A block hangs at rest from a massless spring at the origin ($s = 0$). At $t = 0$, the block is pulled downward $\frac{1}{4}$ m to its initial position $s(0) = -\frac{1}{4}$ and released (**Figure 6.4**). Its velocity (in m/s) is given by $v(t) = \frac{1}{4} \sin t$, for $t \geq 0$. Assume the upward direction is positive.

**a.** Find the position of the block, for $t \geq 0$.

**b.** Graph the position function, for $0 \leq t \leq 3\pi$.

**c.** When does the block move through the origin for the first time?

**d.** When does the block reach its highest point for the first time and what is its position at that time? When does the block return to its lowest point?

**SOLUTION**

**a.** The velocity function (**Figure 6.5a**) is positive for $0 < t < \pi$, which means the block moves in the positive (upward) direction. At $t = \pi$, the block comes to rest momentarily; for $\pi < t < 2\pi$, the block moves in the negative (downward) direction. We let $s(t)$ be the position at time $t \geq 0$ with the initial position $s(0) = -\frac{1}{4}$ m.

**Method 1: Using antiderivatives** Because the position is an antiderivative of the velocity, we have

$$s(t) = \int v(t)\, dt = \int \frac{1}{4} \sin t\, dt = -\frac{1}{4} \cos t + C.$$

To determine the arbitrary constant $C$, we substitute the initial condition $s(0) = -\frac{1}{4}$ into the expression for $s(t)$:

$$-\frac{1}{4} = -\frac{1}{4} \cos 0 + C.$$

Solving for $C$, we find that $C = 0$. Therefore, the position for any time $t \geq 0$ is

$$s(t) = -\frac{1}{4} \cos t.$$

**Method 2: Using Theorem 6.1** Alternatively, we may use the relationship

$$s(t) = s(0) + \int_0^t v(x)\, dx.$$

Substituting $v(x) = \frac{1}{4} \sin x$ and $s(0) = -\frac{1}{4}$, the position function is

$$s(t) = \underbrace{-\frac{1}{4}}_{s(0)} + \int_0^t \underbrace{\frac{1}{4} \sin x}_{v(x)}\, dx$$

$$= -\frac{1}{4} - \left( \frac{1}{4} \cos x \right) \Big|_0^t \qquad \text{Evaluate integral.}$$

$$= -\frac{1}{4} - \frac{1}{4} (\cos t - 1) \qquad \text{Simplify.}$$

$$= -\frac{1}{4} \cos t. \qquad \text{Simplify.}$$

**b.** The graph of the position function is shown in **Figure 6.5b**. We see that $s(0) = -\frac{1}{4}$ m, as prescribed.

**c.** The block initially moves in the positive $s$ direction (upward), reaching the origin ($s = 0$) when $s(t) = -\frac{1}{4} \cos t = 0$. So the block arrives at the origin for the first time when $t = \pi/2$.

**d.** The block moves in the positive direction and reaches its high point for the first time when $t = \pi$; the position at that moment is $s(\pi) = \frac{1}{4}$ m. The block then reverses direction and moves in the negative (downward) direction, reaching its low point at $t = 2\pi$. This motion repeats every $2\pi$ seconds.

*Related Exercises 17, 20* ◀

**QUICK CHECK 4** Without doing further calculations, what are the displacement and distance traveled by the block in Example 2 over the interval $[0, 2\pi]$? ◀

▶ The terminal velocity of an object depends on its density, shape, and size and on the medium through which it falls. Estimates for humans in free fall in the lower atmosphere vary from 120 mi/hr (54 m/s) to 180 mi/hr (80 m/s).

**EXAMPLE 3   Skydiving**  Suppose a skydiver leaps from a hovering helicopter and falls in a straight line. Assume he reaches a terminal velocity of 80 m/s immediately at $t = 0$ and falls for 19 seconds, at which time he opens his parachute. The velocity decreases linearly to 6 m/s over a two-second period and then remains constant until he reaches the ground at $t = 40$ s. The motion is described by the velocity function

$$v(t) = \begin{cases} 80 & \text{if } 0 \le t < 19 \\ 783 - 37t & \text{if } 19 \le t < 21 \\ 6 & \text{if } 21 \le t \le 40. \end{cases}$$

Determine the height above the ground from which the skydiver jumped.

**SOLUTION**  We let the position of the skydiver increase *downward* with the origin ($s = 0$) corresponding to the position of the helicopter. The velocity is positive, so the distance traveled by the skydiver equals the displacement, which is

$$\int_0^{40} |v(t)| \, dt = \int_0^{19} 80 \, dt + \int_{19}^{21} (783 - 37t) \, dt + \int_{21}^{40} 6 \, dt$$

$$= 80t \Big|_0^{19} + \left(783t - \frac{37t^2}{2}\right)\Big|_{19}^{21} + 6t \Big|_{21}^{40} \qquad \text{Fundamental Theorem}$$

$$= 1720. \qquad \text{Evaluate and simplify.}$$

The skydiver jumped from 1720 m above the ground. Notice that the displacement of the skydiver is the area under the velocity curve (**Figure 6.6**).

**QUICK CHECK 5** Suppose (unrealistically) in Example 3 that the velocity of the skydiver is 80 m/s, for $0 \le t < 20$, and then it changes instantaneously to 6 m/s, for $20 \le t \le 40$. Sketch the velocity function and, without integrating, find the distance the skydiver falls in 40 s. ◀

$$\text{Distance traveled} = \int_0^{40} |v(t)| \, dt = \int_0^{19} 80 \, dt + \int_{19}^{21} (783 - 37t) \, dt + \int_{21}^{40} 6 \, dt$$

**Figure 6.6**

*Related Exercises 27–28* ◀

## Acceleration

Because the acceleration of an object moving along a line is given by $a(t) = v'(t)$, the relationship between velocity and acceleration is the same as the relationship between position and velocity. Given the acceleration of an object, the change in velocity over an interval $[a, b]$ is

$$\text{change in velocity} = v(b) - v(a) = \int_a^b v'(t)\, dt = \int_a^b a(t)\, dt.$$

Furthermore, if we know the acceleration and initial velocity $v(0)$, then we can also find the velocity at future times.

> ➤ Theorem 6.2 is a consequence of the Fundamental Theorem of Calculus.

> **THEOREM 6.2    Velocity from Acceleration**
> Given the acceleration $a(t)$ of an object moving along a line and its initial velocity $v(0)$, the velocity of the object for future times $t \geq 0$ is
>
> $$v(t) = v(0) + \int_0^t a(x)\, dx.$$

**Figure 6.7**

**Figure 6.8**

**EXAMPLE 4    Motion in a gravitational field**  An artillery shell is fired directly upward with an initial velocity of 300 m/s from a point 30 m above the ground (**Figure 6.7**). Assume only the force of gravity acts on the shell and it produces an acceleration of 9.8 m/s². Find the velocity of the shell while it is in the air.

**SOLUTION**  We let the positive direction be upward with the origin ($s = 0$) corresponding to the ground. The initial velocity of the shell is $v(0) = 300$ m/s. The acceleration due to gravity is downward; therefore, $a(t) = -9.8$ m/s². Integrating the acceleration, the velocity is

$$v(t) = \underbrace{v(0)}_{300\ \text{m/s}} + \int_0^t \underbrace{a(x)}_{-9.8\ \text{m/s}^2}\, dx = 300 + \int_0^t (-9.8)\, dx = 300 - 9.8t.$$

The velocity decreases from its initial value of 300 m/s, reaching zero at the high point of the trajectory when $v(t) = 300 - 9.8t = 0$, or at $t \approx 30.6$ s (**Figure 6.8**). At this point, the velocity becomes negative, and the shell begins its descent to Earth.

Knowing the velocity function, you could now find the position function using the methods of Example 2.

*Related Exercises 30–31* ◀

## Net Change and Future Value

Everything we have said about velocity, position, and displacement carries over to more general situations. Suppose you are interested in some quantity $Q$ that changes over *time*; $Q$ may represent the amount of water in a reservoir, the population of a cell culture, or the amount of a resource that is consumed or produced. If you are given the rate $Q'$ at which $Q$ changes, then integration allows you to calculate either the net change in the quantity $Q$ or the future value of $Q$.

We argue just as we did for velocity and position: Because $Q(t)$ is an antiderivative of $Q'(t)$, the Fundamental Theorem of Calculus tells us that

$$\int_a^b Q'(t)\, dt = Q(b) - Q(a) = \text{net change in } Q \text{ over } [a, b].$$

Geometrically, the net change in $Q$ over the time interval $[a, b]$ is the net area under the graph of $Q'$ over $[a, b]$. We interpret the product $Q'(t)\, dt$ as a change in $Q$ over a small increment of

> Note that the units in the integral are consistent. For example, if $Q'$ has units of gallons/second, and $t$ and $x$ have units of seconds, then $Q'(x)\,dx$ has units of (gallons/second)(seconds) = gallons, which are the units of $Q$.

time. Integrating $Q'(t)$ accumulates, or adds up, these small changes over the interval $[a, b]$. The result is the net change in $Q$ between $t = a$ and $t = b$. We see that accumulating the rate of change of a quantity over the interval gives the net change in that quantity over the interval.

Alternatively, suppose we are given both the rate of change $Q'$ and the initial value $Q(0)$. Integrating over the interval $[0, t]$, where $t \geq 0$, we have

$$\int_0^t Q'(x)\,dx = Q(t) - Q(0).$$

Rearranging this equation, we write the value of $Q$ at any future time $t \geq 0$ as

$$\underbrace{Q(t)}_{\substack{\text{future} \\ \text{value}}} = \underbrace{Q(0)}_{\substack{\text{initial} \\ \text{value}}} + \underbrace{\int_0^t Q'(x)\,dx.}_{\substack{\text{net change} \\ \text{over } [0, t]}}$$

> At the risk of being repetitious, Theorem 6.3 is also a consequence of the Fundamental Theorem of Calculus. We assume $Q'$ is an integrable function.

---

**THEOREM 6.3    Net Change and Future Value**

Suppose a quantity $Q$ changes over time at a known rate $Q'$. Then the **net change** in $Q$ between $t = a$ and $t = b > a$ is

$$\underbrace{Q(b) - Q(a)}_{\text{net change in } Q} = \int_a^b Q'(t)\,dt.$$

Given the initial value $Q(0)$, the **future value** of $Q$ at time $t \geq 0$ is

$$Q(t) = Q(0) + \int_0^t Q'(x)\,dx.$$

---

The correspondences between velocity–displacement problems and more general problems are shown in Table 6.1.

**Table 6.1**

| Velocity–Displacement Problems | General Problems |
| --- | --- |
| Position $s(t)$ | Quantity $Q(t)$ (such as volume or population) |
| Velocity: $s'(t) = v(t)$ | Rate of change: $Q'(t)$ |
| Displacement: $s(b) - s(a) = \int_a^b v(t)\,dt$ | Net change: $Q(b) - Q(a) = \int_a^b Q'(t)\,dt$ |
| Future position: $s(t) = s(0) + \int_0^t v(x)\,dx$ | Future value of $Q$: $Q(t) = Q(0) + \int_0^t Q'(x)\,dx$ |

We now consider two general applications that involve integrating rates of change.

**EXAMPLE 5    Emptying a tank**  Imagine an open cylindrical tank with radius $R$ and height $H$ filled to the top with water. At time $t = 0$, a circular drain of radius $r$ in the bottom of the tank is opened and water flows out of the tank (**Figure 6.9**). Using a result known as **Torricelli's law** (proposed in 1643), it can be shown that under ideal conditions the rate of change of the volume of water in the tank at time $t \geq 0$ is

$$V'(t) = \frac{\pi r^4 g}{R^2} t - \pi r^2 \sqrt{2gH},$$

**Figure 6.9**

where $g = 980 \text{ cm/s}^2$ is the acceleration due to gravity. In the specific case that $r = 5$ cm, $R = 50$ cm, and $H = 100$ cm, we have $V'(t) = \pi(245t - 3500\sqrt{10})$, where $t$ is measured in seconds and $V(0) = 250{,}000\pi \text{ cm}^3$. Find the function that gives the volume of water remaining in the tank until the tank is empty.

**Figure 6.10**

**SOLUTION** As shown in **Figure 6.10**, the rate of change of the volume is negative, reflecting the decreasing volume of water in the tank. Knowing the initial volume $V(0)$ and the rate of change $V'(t)$, Theorem 6.3 gives the volume of water in the tank:

$$V(t) = V(0) + \int_0^t V'(x)\, dx \qquad \text{Theorem 6.3}$$

$$= \underbrace{250{,}000\pi}_{V(0)} + \int_0^t \underbrace{(\pi(245x - 3500\sqrt{10}))}_{V'(x)}\, dx \quad \text{Substitute.}$$

$$= 250{,}000\pi + \pi\left(\frac{245x^2}{2} - 3500\sqrt{10}x\right)\Big|_0^t \quad \begin{array}{l}\text{Fundamental}\\ \text{Theorem}\end{array}$$

$$= \pi(122.5t^2 - 3500\sqrt{10}t + 250{,}000) \qquad \text{Simplify.}$$

The graph of the volume function (Figure 6.10) shows that the volume decreases until the volume reaches zero at $t \approx 45.2$ seconds—the same time at which the rate of change equals zero.

*Related Exercise 44* ◄

**EXAMPLE 6  Production costs** A book publisher estimates that the marginal cost of producing a particular title (in dollars/book) is given by

$$C'(x) = 12 - 0.0002x,$$

where $0 \le x \le 50{,}000$ is the number of books printed. What is the cost of producing the 12,001st through the 15,000th book?

**SOLUTION** Recall from Section 3.6 that the cost function $C(x)$ is the cost required to produce $x$ units of a product. The marginal cost $C'(x)$ is the approximate cost of producing one additional unit after $x$ units have already been produced. The cost of producing books $x = 12{,}001$ through $x = 15{,}000$ is the cost of producing 15,000 books minus the cost of producing the first 12,000 books. Therefore, the cost in dollars of producing books 12,001 through 15,000 is

> ► Although $x$ is a positive integer (the number of books produced), we treat it as a continuous variable in this example.

$$C(15{,}000) - C(12{,}000) = \int_{12{,}000}^{15{,}000} C'(x)\, dx$$

$$= \int_{12{,}000}^{15{,}000} (12 - 0.0002x)\, dx \quad \text{Substitute for } C'(x).$$

$$= (12x - 0.0001x^2)\Big|_{12{,}000}^{15{,}000} \quad \text{Fundamental Theorem}$$

$$= 27{,}900. \qquad \text{Simplify.}$$

*Related Exercises 51–52* ◄

> **QUICK CHECK 6** Is the cost of increasing production from 9000 books to 12,000 books in Example 6 more or less than the cost of increasing production from 12,000 books to 15,000 books? Explain. ◄

---

## SECTION 6.1 EXERCISES

### Getting Started

1. Explain the meaning of position, displacement, and distance traveled as they apply to an object moving along a line.

2. Suppose the velocity of an object moving along a line is positive. Are displacement and distance traveled equal? Explain.

3. Given the velocity function $v$ of an object moving along a line, explain how definite integrals can be used to find the displacement of the object.

4. Explain how to use definite integrals to find the net change in a quantity, given the rate of change of that quantity.

5. Given the rate of change of a quantity $Q$ and its initial value $Q(0)$, explain how to find the value of $Q$ at a future time $t \ge 0$.

6. What is the result of integrating a population growth rate between times $t = a$ and $t = b$, where $b > a$?

7. **Displacement and distance from velocity** Consider the graph shown in the figure, which gives the velocity of an object moving along a line. Assume time is measured in hours and distance is measured in miles. The areas of three regions bounded by the velocity curve and the $t$-axis are also given.

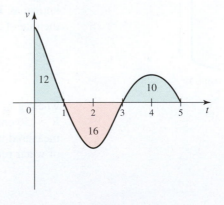

**a.** On what intervals is the object moving in the positive direction?

**b.** What is the displacement of the object over the interval $[0, 3]$?

**c.** What is the total distance traveled by the object over the interval $[1, 5]$?

**d.** What is the displacement of the object over the interval $[0, 5]$?

**e.** Describe the position of the object relative to its initial position after 5 hours.

**8. Displacement and distance from velocity** Consider the velocity function of an object moving along a line (see figure). Assume time is measured in seconds and distance is measured in meters. The areas of four regions bounded by the velocity curve and the $t$-axis are also given.

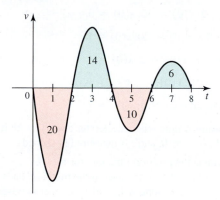

**a.** On what intervals is the object moving in the negative direction?

**b.** What is the displacement of the object over the interval $[2, 6]$?

**c.** What is the total distance traveled by the object over the interval $[0, 6]$?

**d.** What is the displacement of the object over the interval $[0, 8]$?

**e.** Describe the position of the object relative to its initial position after 8 seconds.

**9–10. Velocity graphs** *The figures show velocity functions for motion along a line. Assume the motion begins with an initial position of $s(0) = 0$. Determine the following.*

**a.** *The displacement between $t = 0$ and $t = 5$*

**b.** *The distance traveled between $t = 0$ and $t = 5$*

**c.** *The position at $t = 5$*

**d.** *A piecewise function for $s(t)$*

**9.**

**10.**

**Practice Exercises**

**11. Distance traveled and displacement** Suppose an object moves along a line with velocity (in m/s) $v(t) = 3 \sin 2t$, for $0 \le t \le 2\pi$, where $t$ is measured in seconds (see figure).

**a.** Find the distance traveled by the object on the time interval $[0, \pi/2]$.

**b.** Find the displacement of the object on the time intervals $[0, \pi/2], [0, \pi], [0, 3\pi/2]$, and $[0, 2\pi]$. (*Hint:* Use your answer to part (a) together with the symmetry of the graph to find the displacement values.)

**c.** Find the distance traveled by the object on the time interval $[0, 2\pi]$.

**12. Distance traveled and displacement** Suppose an object moves along a line with velocity (in ft/s) $v(t) = 6 - 2t$, for $0 \le t \le 6$, where $t$ is measured in seconds.

**a.** Graph the velocity function on the interval $0 \le t \le 6$. Determine when the motion is in the positive direction and when it is in the negative direction on $0 \le t \le 6$.

**b.** Find the displacement of the object on the interval $0 \le t \le 6$.

**c.** Find the distance traveled by the object on the interval $0 \le t \le 6$.

**13–16. Displacement from velocity** *Consider an object moving along a line with the given velocity v. Assume time t is measured in seconds and velocities have units of m/s.*

**a.** *Determine when the motion is in the positive direction and when it is in the negative direction.*

**b.** *Find the displacement over the given interval.*

**c.** *Find the distance traveled over the given interval.*

**13.** $v(t) = 3t^2 - 6t$ on $[0, 3]$

**14.** $v(t) = 4t^3 - 24t^2 + 20t$ on $[0, 5]$

**15.** $v(t) = 3t^2 - 18t + 24$ on $[0, 5]$

**16.** $v(t) = 4 \cos \pi t$ on $[0, 2]$

**17–22. Position from velocity** *Consider an object moving along a line with the given velocity v and initial position.*

**a.** *Determine the position function, for $t \ge 0$, using the antiderivative method.*

**b.** *Determine the position function, for $t \ge 0$, using the Fundamental Theorem of Calculus (Theorem 6.1). Check for agreement with the answer to part (a).*

**17.** $v(t) = \sin t$ on $[0, 2\pi]$; $s(0) = 1$

**18.** $v(t) = -t^3 + 3t^2 - 2t$ on $[0, 3]$; $s(0) = 4$

**19.** $v(t) = 6 - 2t$ on $[0, 5]$; $s(0) = 0$

**20.** $v(t) = 3 \sin \pi t$ on $[0, 4]$; $s(0) = 1$

**21.** $v(t) = 9 - t^2$ on $[0, 4]$; $s(0) = -2$

**22.** $v(t) = \dfrac{1}{(t + 1)^2}$ on $[0, 4]$; $s(0) = -2$

**T 23. Oscillating motion** A mass hanging from a spring is set in motion, and its ensuing velocity is given by $v(t) = 2\pi \cos \pi t$, for $t \geq 0$. Assume the positive direction is upward and $s(0) = 0$.

  **a.** Determine the position function, for $t \geq 0$.
  **b.** Graph the position function on the interval $[0, 4]$.
  **c.** At what times does the mass reach its low point the first three times?
  **d.** At what times does the mass reach its high point the first three times?

**24. Cycling distance** A cyclist rides down a long straight road with a velocity (in m/min) given by $v(t) = 400 - 20t$, for $0 \leq t \leq 10$, where $t$ is measured in minutes.

  **a.** How far does the cyclist travel in the first 5 min?
  **b.** How far does the cyclist travel in the first 10 min?
  **c.** How far has the cyclist traveled when her velocity is 250 m/min?

**25. Flying into a headwind** The velocity (in mi/hr) of an airplane flying into a headwind is given by $v(t) = 30(16 - t^2)$, for $0 \leq t \leq 3$. Assume $s(0) = 0$ and $t$ is measured in hours.

  **a.** Determine the position function, for $0 \leq t \leq 3$.
  **b.** How far does the airplane travel in the first 2 hr?
  **c.** How far has the airplane traveled at the instant its velocity reaches 400 mi/hr?

**26. Day hike** The velocity (in mi/hr) of a hiker walking along a straight trail is given by $v(t) = 3 \sin^2 \dfrac{\pi t}{2}$, for $0 \leq t \leq 4$.

  Assume $s(0) = 0$ and $t$ is measured in hours.

  **a.** Determine the position function, for $0 \leq t \leq 4$.
  (*Hint:* $\sin^2 t = \dfrac{1 - \cos 2t}{2}$.)

  **b.** What is the distance traveled by the hiker in the first 15 min of the hike?
  **c.** What is the hiker's position at $t = 3$?

**27. Piecewise velocity** The velocity of a (fast) automobile on a straight highway is given by the function

$$v(t) = \begin{cases} 3t & \text{if } 0 \leq t < 20 \\ 60 & \text{if } 20 \leq t < 45 \\ 240 - 4t & \text{if } t \geq 45, \end{cases}$$

where $t$ is measured in seconds and $v$ has units of m/s.

  **a.** Graph the velocity function, for $0 \leq t \leq 70$. When is the velocity a maximum? When is the velocity zero?
  **b.** What is the distance traveled by the automobile in the first 30 s?
  **c.** What is the distance traveled by the automobile in the first 60 s?
  **d.** What is the position of the automobile when $t = 75$?

**28. Probe speed** A data collection probe is dropped from a stationary balloon, and it falls with a velocity (in m/s) given by $v(t) = 9.8t$,

neglecting air resistance. After 10 s, a chute deploys and the probe immediately slows to a constant speed of 10 m/s, which it maintains until it enters the ocean.

  **a.** Graph the velocity function.
  **b.** How far does the probe fall in the first 30 s after it is released?
  **c.** If the probe was released from an altitude of 3 km, when does it enter the ocean?

**29–34. Position and velocity from acceleration** *Find the position and velocity of an object moving along a straight line with the given acceleration, initial velocity, and initial position. Use the Fundamental Theorem of Calculus (Theorems 6.1 and 6.2).*

**29.** $a(t) = -32$; $v(0) = 70$; $s(0) = 10$

**30.** $a(t) = -32$; $v(0) = 50$; $s(0) = 0$

**31.** $a(t) = -9.8$; $v(0) = 20$; $s(0) = 0$

**32.** $a(t) = \sin \pi t$; $v(0) = 2$; $s(0) = 1$

**33.** $a(t) = -0.01t$; $v(0) = 10$; $s(0) = 0$

**34.** $a(t) = \cos 2t$; $v(0) = 5$; $s(0) = 7$

**35. Acceleration** A drag racer accelerates at $a(t) = 88$ ft/s². Assume $v(0) = 0$, $s(0) = 0$, and $t$ is measured in seconds.

  **a.** Determine the position function, for $t \geq 0$.
  **b.** How far does the racer travel in the first 4 seconds?
  **c.** At this rate, how long will it take the racer to travel $\frac{1}{4}$ mi?
  **d.** How long does it take the racer to travel 300 ft?
  **e.** How far has the racer traveled when it reaches a speed of 178 ft/s?

**36. Deceleration** A car slows down with an acceleration of $a(t) = -15$ ft/s². Assume $v(0) = 60$ ft/s, $s(0) = 0$, and $t$ is measured in seconds.

  **a.** Determine the position function, for $t \geq 0$.
  **b.** How far does the car travel in the time it takes to come to rest?

**37. Approaching a station** At $t = 0$, a train approaching a station begins decelerating from a speed of 80 mi/hr according to the acceleration function $a(t) = -1280(1 + 8t)^{-3}$, where $t \geq 0$ is measured in hours. How far does the train travel between $t = 0$ and $t = 0.2$? Between $t = 0.2$ and $t = 0.4$? The units of acceleration are mi/hr².

**38–40. Population growth**

**38.** Starting with an initial value of $P(0) = 55$, the population of a prairie dog community grows at a rate of $P'(t) = 20 - t/5$ (prairie dogs/month), for $0 \leq t \leq 200$, where $t$ is measured in months.

  **a.** What is the population 6 months later?
  **b.** Find the population $P(t)$, for $0 \leq t \leq 200$.

**39.** When records were first kept ($t = 0$), the population of a rural town was 250 people. During the following years, the population grew at a rate of $P'(t) = 30(1 + \sqrt{t})$, where $t$ is measured in years.

  **a.** What is the population after 20 years?
  **b.** Find the population $P(t)$ at any time $t \geq 0$.

**40.** The population of a community of foxes is observed to fluctuate on a 10-year cycle due to variations in the availability of prey. When population measurements began ($t = 0$), the population was 35 foxes. The growth rate in units of foxes/year was observed to be

$$P'(t) = 5 + 10 \sin \frac{\pi t}{5}.$$

  **a.** What is the population 15 years later? 35 years later?
  **b.** Find the population $P(t)$ at any time $t \geq 0$.

**41. Oil production** An oil refinery produces oil at a variable rate given by

$$Q'(t) = \begin{cases} 800 & \text{if } 0 \leq t < 30 \\ 2600 - 60t & \text{if } 30 \leq t < 40 \\ 200 & \text{if } t \geq 40, \end{cases}$$

where $t$ is measured in days and $Q$ is measured in barrels.

  **a.** How many barrels are produced in the first 35 days?
  **b.** How many barrels are produced in the first 50 days?
  **c.** Without using integration, determine the number of barrels produced over the interval $[60, 80]$.

**42. Flow rates in the Spokane River** The daily discharge of the Spokane River as it flows through Spokane, Washington, in April and June is modeled by the functions

$$r_1(t) = 0.25t^2 + 37.46t + 722.47 \text{ (April) and}$$

$$r_2(t) = 0.90t^2 - 69.06t + 2053.12 \text{ (June),}$$

where the discharge is measured in millions of cubic feet per day, and $t = 0$ corresponds to the beginning of the first day of the month (see figure).

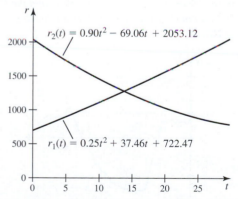

  **a.** Determine the total amount of water that flows through Spokane in April (30 days).
  **b.** Determine the total amount of water that flows through Spokane in June (30 days).
  **c.** The Spokane River flows out of Lake Coeur d'Alene, which contains approximately 0.67 mi³ of water. Determine the percentage of Lake Coeur d'Alene's volume that flows through Spokane in April and June.

**43. Two runners** At noon ($t = 0$), Alicia starts running along a long straight road at 4 mi/hr. Her velocity decreases according to the function $v(t) = 4/(t + 1)^2$, for $t \geq 0$. At noon, Boris also starts running along the same road with a 1-mi head start on Alicia; his velocity is given by $u(t) = 2/(t + 1)^2$, for $t \geq 0$. Assume $t$ is measured in hours.

  **a.** Find the position functions for Alicia and Boris, where $s = 0$ corresponds to Alicia's starting point.
  **b.** When, if ever, does Alicia overtake Boris?
  **c.** How much of a head start would Boris need to stay ahead of Alicia indefinitely?

**44. Filling a tank** A 2000-liter cistern is empty when water begins flowing into it (at $t = 0$) at a rate (in L/min) given by $Q'(t) = 3\sqrt{t}$, where $t$ is measured in minutes.

  **a.** How much water flows into the cistern in 1 hour?
  **b.** Find the function that gives the amount of water in the tank at any time $t \geq 0$.
  **c.** When will the tank be full?

**T 45. Filling a reservoir** A reservoir with a capacity of 2500 m³ is filled with a single inflow pipe. The reservoir is empty when the inflow pipe is opened at $t = 0$. Letting $Q(t)$ be the amount of water in the reservoir at time $t$, the flow rate of water into the reservoir (in m³/hr) oscillates on a 24-hr cycle (see figure) and is given by

$$Q'(t) = 20\left(1 + \cos \frac{\pi t}{12}\right).$$

  **a.** How much water flows into the reservoir in the first 2 hr?
  **b.** Find the function that gives the amount of water in the reservoir over the interval $[0, t]$, where $t \geq 0$.
  **c.** When is the reservoir full?

**46. Blood flow** A typical human heart pumps 70 mL of blood (the stroke volume) with each beat. Assuming a heart rate of 60 beats/min (1 beat/s), a reasonable model for the outflow rate of the heart is $V'(t) = 70(1 + \sin 2\pi t)$, where $V(t)$ is the amount of blood (in milliliters) pumped over the interval $[0, t]$, $V(0) = 0$, and $t$ is measured in seconds.

  **a.** Verify that the amount of blood pumped over a one-second interval is 70 mL.
  **b.** Find the function that gives the total blood pumped between $t = 0$ and a future time $t > 0$.
  **c.** What is the cardiac output over a period of 1 min? (Use calculus; then check your answer with algebra.)

**47. Air flow in the lungs** A simple model (with different parameters for different people) for the flow of air in and out of the lungs is

$$V'(t) = -\frac{\pi}{2} \sin \frac{\pi t}{2},$$

where $V(t)$ (measured in liters) is the volume of air in the lungs at time $t \geq 0$, $t$ is measured in seconds, and $t = 0$ corresponds to a time at which the lungs are full and exhalation begins. Only a fraction of the air in the lungs is exchanged with each breath. The amount that is exchanged is called the *tidal volume*.

a. Find the volume function $V$ assuming $V(0) = 6$ L.
b. What is the breathing rate in breaths/min?
c. What is the tidal volume and what is the total capacity of the lungs?

48. **Oscillating growth rates** Some species have growth rates that oscillate with an (approximately) constant period $P$. Consider the growth rate function

$$N'(t) = r + A \sin \frac{2\pi t}{P},$$

where $A$ and $r$ are constants with units of individuals/yr, and $t$ is measured in years. A species becomes extinct if its population ever reaches 0 after $t = 0$.

a. Suppose $P = 10$, $A = 20$, and $r = 0$. If the initial population is $N(0) = 10$, does the population ever become extinct? Explain.
b. Suppose $P = 10$, $A = 20$, and $r = 0$. If the initial population is $N(0) = 100$, does the population ever become extinct? Explain.
c. Suppose $P = 10$, $A = 50$, and $r = 5$. If the initial population is $N(0) = 10$, does the population ever become extinct? Explain.
d. Suppose $P = 10$, $A = 50$, and $r = -5$. Find the initial population $N(0)$ needed to ensure that the population never becomes extinct.

49. **Power and energy** The terms *power* and *energy* are often used interchangeably, but they are quite different. **Energy** is what makes matter move or heat up and is measured in units of **joules** (J) or **Calories** (Cal), where 1 Cal = 4184 J. One hour of walking consumes roughly $10^6$ J, or 250 Cal. On the other hand, **power** is the rate at which energy is used and is measured in **watts** (W; 1 W = 1 J/s). Other useful units of power are **kilowatts** (1 kW = $10^3$ W) and **megawatts** (1 MW = $10^6$ W). If energy is used at a rate of 1 kW for 1 hr, the total amount of energy used is 1 **kilowatt-hour** (kWh), which is $3.6 \times 10^6$ J.

Suppose the power function of a large city over a 24-hr period is given by

$$P(t) = E'(t) = 300 - 200 \sin \frac{\pi t}{12},$$

where $P$ is measured in megawatts and $t = 0$ corresponds to 6:00 P.M. (see figure).

a. How much energy is consumed by this city in a typical 24-hr period? Express the answer in megawatt-hours and in joules.
b. Burning 1 kg of coal produces about 450 kWh of energy. How many kilograms of coal are required to meet the energy needs of the city for 1 day? For 1 year?
c. Fission of 1 gram of uranium-235 (U-235) produces about 16,000 kWh of energy. How many grams of uranium are needed to meet the energy needs of the city for 1 day? For 1 year?
d. A typical wind turbine generates electrical power at a rate of about 200 kW. Approximately how many wind turbines are needed to meet the average energy needs of the city?

50. **Carbon uptake** An important process in the study of global warming and greenhouse gases is the *net ecosystem exchange*, which is the rate at which carbon leaves an ecosystem and enters the atmosphere in a particular geographic region. Let $N(t)$ equal the net ecosystem exchange on an average July day in a high-altitude coniferous forest, where $N(t)$ is measured in grams of carbon per square meter per hour and $t$ is the number of hours past midnight so that $0 \leq t \leq 24$ (see figure). Negative values of $N$ correspond to times when the amount of carbon in the atmosphere decreases, and positive values of $N$ occur when the amount of carbon in the atmosphere increases.

a. Trees and other plants help reduce carbon emissions in the atmosphere by using photosynthesis to absorb carbon dioxide and release oxygen. Give a possible explanation why $N$ is negative on the interval $5 < t < 17$.
b. The *cumulative net carbon uptake*, $\int_0^{24} N(t)\, dt$, is the net change in the amount of carbon in the atmosphere over a 24-hour period in the coniferous forest. Use a midpoint Riemann sum with $n = 12$ subintervals to estimate the cumulative net carbon uptake in the coniferous forest and interpret the result. (*Source: Forest Carbon Uptake and the Fundamental Theorem of Calculus, The College Mathematics Journal, 44, 5, Nov 2013*)

**51–54. Marginal cost** *Consider the following marginal cost functions.*
a. *Find the additional cost incurred in dollars when production is increased from 100 units to 150 units.*
b. *Find the additional cost incurred in dollars when production is increased from 500 units to 550 units.*

51. $C'(x) = 2000 - 0.5x$          52. $C'(x) = 200 - 0.05x$

53. $C'(x) = 300 + 10x - 0.01x^2$

54. $C'(x) = 3000 - x - 0.001x^2$

**55. Explain why or why not** Determine whether the following statements are true and give an explanation or counterexample.

  **a.** The distance traveled by an object moving along a line is the same as the displacement of the object.

  **b.** When the velocity is positive on an interval, the displacement and the distance traveled on that interval are equal.

  **c.** Consider a tank that is filled and drained at a flow rate of
  $$V'(t) = 1 - \frac{t^2}{100} \text{ (gal/min), for } t \geq 0,$$ where $t$ is measured in minutes. It follows that the volume of water in the tank increases for 10 min and then decreases until the tank is empty.

  **d.** A particular marginal cost function has the property that it is positive and decreasing. The cost of increasing production from $A$ units to $2A$ units is greater than the cost of increasing production from $2A$ units to $3A$ units.

**56–59. Equivalent constant velocity** *Consider the following velocity functions. In each case, complete the sentence: The same distance could have been traveled over the given time period at a <u>constant</u> velocity of _____.*

**56.** $v(t) = 2t + 6$, for $0 \leq t \leq 8$

**57.** $v(t) = 1 - \dfrac{t^2}{16}$, for $0 \leq t \leq 4$

**58.** $v(t) = 2 \sin t$, for $0 \leq t \leq \pi$

**59.** $v(t) = t(25 - t^2)^{1/2}$, for $0 \leq t \leq 5$

## Explorations and Challenges

**60. Where do they meet?** Kelly started at noon ($t = 0$) riding a bike from Niwot to Berthoud, a distance of 20 km, with velocity
$$v(t) = \frac{15}{(t + 1)^2}$$ (decreasing because of fatigue). Sandy started at noon ($t = 0$) riding a bike in the opposite direction from Berthoud to Niwot with velocity $u(t) = \dfrac{20}{(t + 1)^2}$ (also decreasing because of fatigue). Assume distance is measured in kilometers and time is measured in hours.

  **a.** Make a graph of Kelly's distance from Niwot as a function of time.

  **b.** Make a graph of Sandy's distance from Berthoud as a function of time.

  **c.** When do they meet? How far has each person traveled when they meet?

  **d.** More generally, if the riders' speeds are $v(t) = \dfrac{A}{(t + 1)^2}$ and $u(t) = \dfrac{B}{(t + 1)^2}$ and the distance between the towns is $D$, what conditions on $A$, $B$, and $D$ must be met to ensure that the riders will pass each other?

  **e.** Looking ahead: With the velocity functions given in part (d), make a conjecture about the maximum distance each person can ride (given unlimited time).

**61. Bike race** Theo and Sasha start at the same place on a straight road, riding bikes with the following velocities (measured in mi/hr). Assume $t$ is measured in hours.

  Theo: $v_T(t) = 10$, for $t \geq 0$

  Sasha: $v_S(t) = 15t$, for $0 \leq t \leq 1$, and $v_S(t) = 15$, for $t > 1$

  **a.** Graph the velocity function for both riders.

  **b.** If the riders ride for 1 hr, who rides farther? Interpret your answer geometrically using the graphs of part (a).

  **c.** If the riders ride for 2 hr, who rides farther? Interpret your answer geometrically using the graphs of part (a).

  **d.** Which rider arrives first at the 10-, 15-, and 20-mile markers of the race? Interpret your answers geometrically using the graphs of part (a).

  **e.** Suppose Sasha gives Theo a head start of 0.2 mi and the riders ride for 20 mi. Who wins the race?

  **f.** Suppose Sasha gives Theo a head start of 0.2 hr and the riders ride for 20 mi. Who wins the race?

**62. Variable gravity** At Earth's surface, the acceleration due to gravity is approximately $g = 9.8 \text{ m/s}^2$ (with local variations). However, the acceleration decreases with distance from the surface according to Newton's law of gravitation. At a distance of $y$ meters from Earth's surface, the acceleration is given by
$$a(y) = -\frac{g}{(1 + y/R)^2},$$
where $R = 6.4 \times 10^6$ m is the radius of Earth.

  **a.** Suppose a projectile is launched upward with an initial velocity of $v_0$ m/s. Let $v(t)$ be its velocity and $y(t)$ its height (in meters) above the surface $t$ seconds after the launch. Neglecting forces such as air resistance, explain why $\dfrac{dv}{dt} = a(y)$ and $\dfrac{dy}{dt} = v(t)$.

  **b.** Use the Chain Rule to show that $\dfrac{dv}{dt} = \dfrac{1}{2}\dfrac{d}{dy}(v^2)$.

  **c.** Show that the equation of motion for the projectile is $\dfrac{1}{2}\dfrac{d}{dy}(v^2) = a(y)$, where $a(y)$ is given previously.

  **d.** Integrate both sides of the equation in part (c) with respect to $y$ using the fact that when $y = 0$, $v = v_0$. Show that
  $$\frac{1}{2}(v^2 - v_0^2) = gR\left(\frac{1}{1 + y/R} - 1\right).$$

  **e.** When the projectile reaches its maximum height, $v = 0$. Use this fact to determine that the maximum height is
  $$y_{max} = \frac{Rv_0^2}{2gR - v_0^2}.$$

  **f.** Graph $y_{max}$ as a function of $v_0$. What is the maximum height when $v_0 = 500$ m/s, 1500 m/s, and 5 km/s?

  **g.** Show that the value of $v_0$ needed to put the projectile into orbit (called the escape velocity) is $\sqrt{2gR}$.

**63.** Suppose $f$ and $g$ have continuous derivatives on an interval $[a, b]$. Prove that if $f(a) = g(a)$ and $f(b) = g(b)$, then
$$\int_a^b f'(x)\, dx = \int_a^b g'(x)\, dx.$$

**64.** Use Exercise 63 to prove that if two runners start and finish at the same time and place, then *regardless of the velocities at which they run*, their displacements are equal.

**65.** Use Exercise 63 to prove that if two trails start at the same place and finish at the same place, then *regardless of the ups and downs of the trails*, they have the same net change in elevation.

**66.** Without evaluating integrals, prove that

$$\int_0^2 \frac{d}{dx}(12 \sin \pi x^2)\, dx = \int_0^2 \frac{d}{dx}(x^{10}(2-x)^3)\, dx.$$

**QUICK CHECK ANSWERS**

**1.** Displacement $= -20$ mi (20 mi south); distance traveled $= 100$ mi   **2.** Suppose the object moves in the positive direction, for $0 \le t \le 3$, and then moves in the negative direction, for $3 < t \le 5$.   **3.** A function; a number

**4.** Displacement $= 0$; distance traveled $= 1$   **5.** 1720 m
**6.** The production cost would increase more between 9000 and 12,000 books than between 12,000 and 15,000 books. Graph $C'$ and look at the area under the curve. ◀

# 6.2 Regions Between Curves

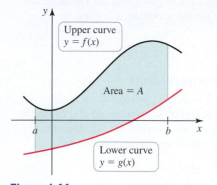

Figure 6.11

In this section, the method for finding the area of a region bounded by a single curve is generalized to regions bounded by two or more curves. Consider two functions $f$ and $g$ continuous on an interval $[a, b]$ on which $f(x) \ge g(x)$ (**Figure 6.11**). The goal is to find the area $A$ of the region bounded by the two curves and the vertical lines $x = a$ and $x = b$.

Once again, we rely on the *slice-and-sum* strategy (Section 5.2) for finding areas by Riemann sums. The interval $[a, b]$ is partitioned into $n$ subintervals using uniformly spaced grid points separated by a distance $\Delta x = (b - a)/n$ (**Figure 6.12**). On each subinterval, we build a rectangle extending from the lower curve to the upper curve. On the $k$th subinterval, a point $x_k^*$ is chosen, and the height of the corresponding rectangle is taken to be $f(x_k^*) - g(x_k^*)$. Therefore, the area of the $k$th rectangle is $(f(x_k^*) - g(x_k^*))\Delta x$ (**Figure 6.13**). Summing the areas of the $n$ rectangles gives an approximation to the area of the region between the curves:

$$A \approx \sum_{k=1}^{n} (f(x_k^*) - g(x_k^*))\Delta x.$$

**Figure 6.12**            **Figure 6.13**

As the number of grid points increases, $\Delta x$ approaches zero and these sums approach the area of the region between the curves; that is,

$$A = \lim_{n\to\infty} \sum_{k=1}^{n} (f(x_k^*) - g(x_k^*))\Delta x.$$

The limit of these Riemann sums is a definite integral of the function $f - g$.

> It is helpful to interpret the area formula: $f(x) - g(x)$ is the height of a rectangle and $dx$ represents its width. We sum (integrate) the areas of the rectangles $(f(x) - g(x))\, dx$ to obtain the area of the region.

**QUICK CHECK 1** In the area formula for a region between two curves, verify that if the lower curve is $g(x) = 0$, the formula becomes the usual formula for the area of the region bounded by $y = f(x)$ and the $x$-axis. ◄

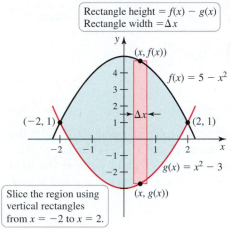

Rectangle height = $f(x) - g(x)$
Rectangle width = $\Delta x$

Slice the region using vertical rectangles from $x = -2$ to $x = 2$.

**Figure 6.14**

**QUICK CHECK 2** Interpret the area formula when it is written in the form $A = \int_a^b f(x)\, dx - \int_a^b g(x)\, dx$, where $f(x) \geq g(x) \geq 0$ on $[a, b]$. ◄

---

**DEFINITION Area of a Region Between Two Curves**

Suppose $f$ and $g$ are continuous functions with $f(x) \geq g(x)$ on the interval $[a, b]$. The area of the region bounded by the graphs of $f$ and $g$ on $[a, b]$ is

$$A = \int_a^b (f(x) - g(x))\, dx.$$

**EXAMPLE 1 Area between curves** Find the area of the region bounded by the graphs of $f(x) = 5 - x^2$ and $g(x) = x^2 - 3$ (**Figure 6.14**).

**SOLUTION** A key step in the solution of many area problems is finding the intersection points of the boundary curves, which often determine the limits of integration. The intersection points of these two curves satisfy the equation $5 - x^2 = x^2 - 3$. The solutions of this equation are $x = -2$ and $x = 2$, which become the lower and upper limits of integration, respectively. The graph of $f$ is the upper curve and the graph of $g$ is the lower curve on the interval $[-2, 2]$. Therefore, the area of the region is

$$A = \int_{-2}^{2} (\underbrace{(5 - x^2)}_{f(x)} - \underbrace{(x^2 - 3)}_{g(x)})\, dx \quad \text{Substitute for } f \text{ and } g.$$

$$= 2\int_0^2 (8 - 2x^2)\, dx \quad \text{Simplify and use symmetry.}$$

$$= 2\left(8x - \frac{2}{3}x^3\right)\Big|_0^2 \quad \text{Fundamental Theorem}$$

$$= \frac{64}{3}. \quad \text{Simplify.}$$

Notice how the symmetry of the problem simplifies the integration. Additionally, note that the area formula $A = \int_a^b (f(x) - g(x))\, dx$ is valid even if one or both curves lie below the $x$-axis, as long as $f(x) \geq g(x)$ on $[a, b]$.

*Related Exercises 9–10* ◄

**EXAMPLE 2 Compound region** Find the area of the region bounded by the graphs of $f(x) = -x^2 + 3x + 6$ and $g(x) = |2x|$ (**Figure 6.15a**).

**SOLUTION** The lower boundary of the region consists of two different branches of the absolute value function. In situations like this, the region is divided into two (or more) subregions whose areas are found independently and then summed; these subregions are labeled $R_1$ and $R_2$ (**Figure 6.15b**). By the definition of absolute value,

$$g(x) = |2x| = \begin{cases} 2x & \text{if } x \geq 0 \\ -2x & \text{if } x < 0. \end{cases}$$

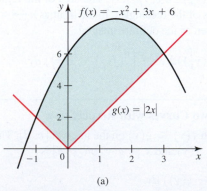

**Figure 6.15**

▶ The solution $x = 6$ corresponds to the intersection of the parabola $y = -x^2 + 3x + 6$ and the line $y = -2x$ in the fourth quadrant, not shown in Figure 6.15 because $g(x) = -2x$ only when $x < 0$.

The left intersection point of $f$ and $g$ satisfies $-2x = -x^2 + 3x + 6$, or $x^2 - 5x - 6 = 0$. Solving for $x$, we find that $(x + 1)(x - 6) = 0$, which implies $x = -1$ or $x = 6$; only the first solution is relevant. The right intersection point of $f$ and $g$ satisfies $2x = -x^2 + 3x + 6$; you should verify that the relevant solution in this case is $x = 3$.

Given these points of intersection, we see that the region $R_1$ is bounded by $y = -x^2 + 3x + 6$ and $y = -2x$ on the interval $[-1, 0]$. Similarly, region $R_2$ is bounded by $y = -x^2 + 3x + 6$ and $y = 2x$ on $[0, 3]$ (Figure 6.15b). Therefore,

$$A = \underbrace{\int_{-1}^{0}((-x^2 + 3x + 6) - (-2x))\, dx}_{\text{area of region } R_1} + \underbrace{\int_{0}^{3}((-x^2 + 3x + 6) - 2x)\, dx}_{\text{area of region } R_2}$$

$$= \int_{-1}^{0}(-x^2 + 5x + 6)\, dx + \int_{0}^{3}(-x^2 + x + 6)\, dx \qquad \text{Simplify.}$$

$$= \left(-\frac{x^3}{3} + \frac{5}{2}x^2 + 6x\right)\Big|_{-1}^{0} + \left(-\frac{x^3}{3} + \frac{1}{2}x^2 + 6x\right)\Big|_{0}^{3} \qquad \text{Fundamental Theorem}$$

$$= 0 - \left(\frac{1}{3} + \frac{5}{2} - 6\right) + \left(-9 + \frac{9}{2} + 18\right) - 0 = \frac{50}{3}. \qquad \text{Simplify.}$$

*Related Exercises 15–16* ◀

## Integrating with Respect to $y$

There are occasions when it is convenient to reverse the roles of $x$ and $y$. Consider the regions shown in **Figure 6.16** that are bounded by the graphs of $x = f(y)$ and $x = g(y)$, where $f(y) \geq g(y)$, for $c \leq y \leq d$ (which implies that the graph of $f$ lies to the right of the graph of $g$). The lower and upper boundaries of the regions are $y = c$ and $y = d$, respectively.

**Figure 6.16**

Area of $k$th rectangle $= (f(y_k^*) - g(y_k^*))\, \Delta y$

Area of region: $A \approx \sum_{k=1}^{n} (f(y_k^*) - g(y_k^*))\, \Delta y$

**Figure 6.17**

In cases such as these, we treat $y$ as the independent variable and divide the interval $[c, d]$ into $n$ subintervals of width $\Delta y = (d - c)/n$ (**Figure 6.17**). On the $k$th subinterval, a point $y_k^*$ is selected, and we construct a rectangle that extends from the left curve to the right curve. The $k$th rectangle has length $f(y_k^*) - g(y_k^*)$, so the area of the $k$th rectangle is $(f(y_k^*) - g(y_k^*))\Delta y$. The area of the region is approximated by the sum of the areas of the rectangles. In the limit as $n \to \infty$ and $\Delta y \to 0$, the area of the region is given as the definite integral

$$A = \lim_{n \to \infty} \sum_{k=1}^{n} (f(y_k^*) - g(y_k^*))\Delta y = \int_{c}^{d}(f(y) - g(y))\, dy.$$

▶ This area formula is analogous to the one given on p. 365; it is now expressed with respect to the $y$-axis. In this case, $f(y) - g(y)$ is the length of a rectangle and $dy$ represents its width. We sum (integrate) the areas of the rectangles $(f(y) - g(y))\, dy$ to obtain the area of the region.

> **DEFINITION  Area of a Region Between Two Curves with Respect to $y$**
>
> Suppose $f$ and $g$ are continuous functions with $f(y) \geq g(y)$ on the interval $[c, d]$. The area of the region bounded by the graphs $x = f(y)$ and $x = g(y)$ on $[c, d]$ is
>
> $$A = \int_{c}^{d}(f(y) - g(y))\, dy.$$

**EXAMPLE 3**  **Integrating with respect to *y***  Find the area of the region *R* bounded by the graphs of $y = x^3$, $y = x + 6$, and the *x*-axis.

**SOLUTION**  The area of this region could be found by integrating with respect to *x*. But this approach requires splitting the region into two pieces (**Figure 6.18**). Alternatively, we can view *y* as the independent variable, express the bounding curves as functions of *y*, and make horizontal slices parallel to the *x*-axis (**Figure 6.19**).

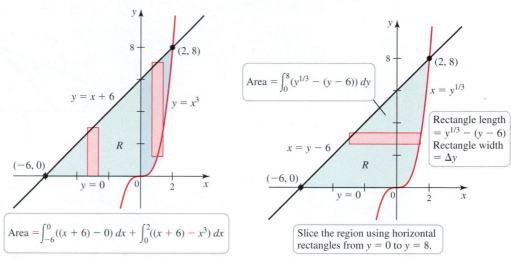

Area $= \int_{-6}^{0}((x + 6) - 0)\,dx + \int_{0}^{2}((x + 6) - x^3)\,dx$

**Figure 6.18**

Area $= \int_{0}^{8}(y^{1/3} - (y - 6))\,dy$

Rectangle length $= y^{1/3} - (y - 6)$
Rectangle width $= \Delta y$

Slice the region using horizontal rectangles from $y = 0$ to $y = 8$.

**Figure 6.19**

▶ You may use synthetic division or a root finder to factor the cubic polynomial in Example 3. Then the quadratic formula shows that the equation

$$y^2 - 10y + 27 = 0$$

has no real roots.

Solving for *x* in terms of *y*, the right curve $y = x^3$ becomes $x = f(y) = y^{1/3}$. The left curve $y = x + 6$ becomes $x = g(y) = y - 6$. The intersection point of the curves satisfies the equation $y^{1/3} = y - 6$, or $y = (y - 6)^3$. Expanding this equation gives the cubic equation

$$y^3 - 18y^2 + 107y - 216 = (y - 8)(y^2 - 10y + 27) = 0,$$

whose only real root is $y = 8$. As shown in Figure 6.19, the areas of the slices through the region are summed from $y = 0$ to $y = 8$. Therefore, the area of the region is given by

$$\int_{0}^{8}(y^{1/3} - (y - 6))\,dy = \left(\frac{3}{4}y^{4/3} - \frac{y^2}{2} + 6y\right)\Big|_{0}^{8} \quad \text{Fundamental Theorem}$$

$$= \left(\frac{3}{4}\cdot 16 - 32 + 48\right) - 0 = 28. \quad \text{Simplify.}$$

*Related Exercises 19–20* ◀

**QUICK CHECK 3**  The region *R* is bounded by the curve $y = \sqrt{x}$, the line $y = x - 2$, and the *x*-axis. Express the area of *R* in terms of (a) integral(s) with respect to *x* and (b) integral(s) with respect to *y*. ◀

Area of region *R*
$= \int_{0}^{4}((y + 4) - y^{3/2})\,dy$

$y = x^{2/3}, x = y^{3/2}$
*R*
$y = x - 4,$
$x = y + 4$
$(8, 4)$

**Figure 6.20**

**EXAMPLE 4**  **Calculus and geometry**  Find the area of the region *R* in the first quadrant bounded by the curves $y = x^{2/3}$ and $y = x - 4$ (**Figure 6.20**).

**SOLUTION**  Slicing the region vertically and integrating with respect to *x* requires two integrals. Slicing the region horizontally requires a single integral with respect to *y*. The second approach appears to involve less work.

Slicing horizontally, the right bounding curve is $x = y + 4$ and the left bounding curve is $x = y^{3/2}$. The two curves intersect at $(8, 4)$, so the limits of integration are $y = 0$ and $y = 4$. The area of *R* is

$$\int_{0}^{4}(\underbrace{(y + 4)}_{\text{right curve}} - \underbrace{y^{3/2}}_{\text{left curve}})\,dy = \left(\frac{y^2}{2} + 4y - \frac{2}{5}y^{5/2}\right)\Big|_{0}^{4} = \frac{56}{5}.$$

Can this area be found using a different approach? Sometimes it helps to use geometry. Notice that the region *R* can be formed by taking the entire region under the curve

➤ To find the point of intersection in Example 4, solve $y^{3/2} = y + 4$ by first squaring both sides of the equation.

$y = x^{2/3}$ on the interval $[0, 8]$ and then removing a triangle whose base is the interval $[4, 8]$ (**Figure 6.21**). The area of the region $R_1$ under the curve $y = x^{2/3}$ is

$$\int_0^8 x^{2/3}\, dx = \frac{3}{5}x^{5/3}\Big|_0^8 = \frac{96}{5}.$$

The triangle $R_2$ has a base of length 4 and a height of 4, so its area is $\frac{1}{2}\cdot 4\cdot 4 = 8$. Therefore, the area of $R$ is $\frac{96}{5} - 8 = \frac{56}{5}$, which agrees with the first calculation.

Area of region $R$ = (area of region $R_1$) − (area of region $R_2$)

**QUICK CHECK 4** An alternative way to determine the area of the region in Example 3 (Figure 6.18) is to compute $18 + \int_0^2 (x + 6 - x^3)\, dx$. Why? ◄

**Figure 6.21**

*Related Exercises 34–36* ◄

## SECTION 6.2 EXERCISES

### Getting Started

1. Set up a sum of two integrals that equals the area of the shaded region bounded by the graphs of the functions $f$ and $g$ on $[a, c]$ (see figure). Assume the curves intersect at $x = b$.

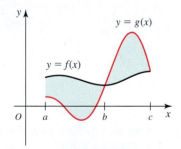

2. Set up an integral that equals the area of the region (see figure) in the following two ways. Do not evaluate the integrals.
   a. Using integration with respect to $x$.
   b. Using integration with respect to $y$.

3. Make a sketch to show a case in which the area bounded by two curves is most easily found by integrating with respect to $x$.

4. Make a sketch to show a case in which the area bounded by two curves is most easily found by integrating with respect to $y$.

5. Find the area of the region (see figure) in two ways.
   a. Using integration with respect to $x$
   b. Using geometry

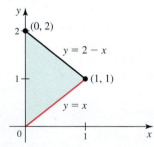

6. Find the area of the region (see figure) in two ways.
   a. By integrating with respect to $y$
   b. Using geometry

**7.** Express the area of the shaded region in Exercise 5 as the sum of two integrals with respect to $y$. Do not evaluate the integrals.

**8.** Express the area of the shaded region in Exercise 6 as the sum of two integrals with respect to $x$. Do not evaluate the integrals.

## Practice Exercises

**9–30. Finding area** *Determine the area of the shaded region in the following figures.*

**9.**

**10.**

**11.**

**12.**

**13.**

**14.**

**15.**

**16.**

**17.**

**18.**

**19.**

**20.**

**21.**

**22.**

**23.**

**24.**

**25.**

**26.**

**27.**

**28.**

**29.**

**30.**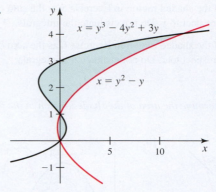

**31–32. Two approaches** *Express the area of the following shaded regions in terms of (a) one or more integrals with respect to x and (b) one or more integrals with respect to y. You do not need to evaluate the integrals.*

**31.** **32.**

**33.** **Area between velocity curves** Two runners, starting at the same location, run along a straight road for 1 hour. The velocity of one runner is $v_1(t) = 7t$ and the velocity of the other runner is $v_2(t) = 10\sqrt{t}$. Assume $t$ is measured in hours and the velocities $v_1(t)$ and $v_2(t)$ are measured in km/hr. Determine the area between the curves $y = v_1(t)$ and $y = v_2(t)$, for $0 \le t \le 1$. Interpret the physical meaning of this area.

**34–36. Calculus and geometry** *For the given regions $R_1$ and $R_2$, complete the following steps.*

*a. Find the area of region $R_1$.*

*b. Find the area of region $R_2$ using geometry and the answer to part (a).*

**34.** $R_1$ is the region in the first quadrant bounded by the $y$-axis and the curves $y = 2x^2$ and $y = 3 - x$; $R_2$ is the region in the first quadrant bounded by the $x$-axis and the curves $y = 2x^2$ and $y = 3 - x$ (see figure).

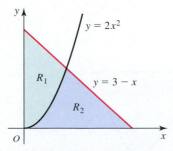

**T 35.** $R_1$ is the region in the first quadrant bounded by the line $x = 1$ and the curve $y = 6x(2 - x^2)^2$; $R_2$ is the region in the first quadrant bounded the curve $y = 6x(2 - x^2)^2$ and the line $y = 6x$.

**36.** $R_1$ is the region in the first quadrant bounded by the coordinate axes and the curve $x = \cos y$; $R_2$ is the region bounded by the lines $y = \dfrac{\pi}{2}$ and $x = 1$, and the curve $x = \cos y$.

**37–62. Regions between curves** *Find the area of the region described in the following exercises.*

**37.** The region bounded by $y = 4x + 4$, $y = 6x + 6$, and $x = 4$

**38.** The region bounded by $y = \cos x$ and $y = \sin x$ on the interval $\left[\dfrac{\pi}{4}, \dfrac{5\pi}{4}\right]$

**39.** The region bounded by $y = 2x^2$ and $y = x^2 + 4$

**40.** The region bounded by $y = 6x$ and $y = 3x^2 - 6x$

**41.** The region bounded by $y = 2x$ and $y = 2x^2 - 4$

**42.** The region bounded by $y = 24\sqrt{x}$ and $y = 3x^2$

**43.** The region bounded by $y = x$, $y = 1/x^2$, $y = 0$, and $x = 2$

**44.** The region in the first quadrant on the interval $[0, 2]$ bounded by $y = 4x - x^2$ and $y = 4x - 4$

**45.** The region bounded by $y = 2 - |x|$ and $y = x^2$

**46.** The region bounded by $y = x^3$ and $y = 9x$

**47.** The region bounded by $y = |x - 3|$ and $y = \dfrac{x}{2}$

**48.** The region bounded by $y = 3x - x^3$ and $y = -x$

**49.** The region in the first quadrant bounded by $y = x^{2/3}$ and $y = 4$

**50.** The region in the first quadrant bounded by $y = 2$ and $y = 2\sin x$ on the interval $[0, \pi/2]$

**51.** The region bounded by $y = x/4$ and $x = y^3$

**T 52.** The region below the line $y = 2$ and above the curve $y = \sec^2 x$ on the interval $[0, \pi/4]$

**T 53.** The region between the line $y = x$ and the curve $y = 2x\sqrt{1 - x^2}$ in the first quadrant

**54.** The region bounded by $x = y^2 - 4$ and $y = \dfrac{x}{3}$

**55.** The region bounded by $y = \sqrt{x}$, $y = 2x - 15$, and $y = 0$

**56.** The region bounded by $y = 2$ and $y = x^{2/3} + 1$

**57.** The region bounded by $y = x^2 - 2x + 1$ and $y = 5x - 9$

**58.** The region bounded by $x = y(y - 1)$ and $x = -y(y - 1)$

**59.** The region bounded by $x = y(y - 1)$ and $y = \dfrac{x}{3}$

**60.** The region bounded by $y = \sin x$ and $y = x(x - \pi)$ on the interval $[0, \pi]$

**61.** The region bounded by $y = 1 - \cos x$ and $y = \sin x$ on the interval $[\pi/2, 2\pi]$

**62.** The region in the first quadrant bounded by $y = x^{-2}$, $y = 8x$, and $y = x/8$

**63–64. Complicated regions** *Find the area of the shaded regions in the following figures.*

**63.**

**64.**

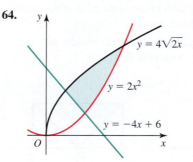

**65. Explain why or why not** Determine whether the following statements are true and give an explanation or counterexample.

   **a.** The area of the region bounded by $y = x$ and $x = y^2$ can be found only by integrating with respect to $x$.

   **b.** The area of the region between $y = \sin x$ and $y = \cos x$ on the interval $[0, \pi/2]$ is $\int_0^{\pi/2}(\cos x - \sin x)\,dx$.

   **c.** $\displaystyle\int_0^1 (x - x^2)\,dx = \int_0^1 (\sqrt{y} - y)\,dy$.

## Explorations and Challenges

**66. Differences of even functions** Assume $f$ and $g$ are even, integrable functions on $[-a, a]$, where $a > 1$. Suppose $f(x) > g(x) > 0$ on $[-a, a]$ and that the area bounded by the graphs of $f$ and $g$ on $[-a, a]$ is 10. What is the value of $\int_0^{\sqrt{a}} x(f(x^2) - g(x^2))\,dx$?

**67. Area of a curve defined implicitly** Determine the area of the shaded region bounded by the curve $x^2 = y^4(1 - y^3)$ (see figure).

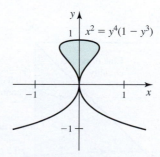

**68–71. Roots and powers** *Find the area of the following regions, expressing your results in terms of the positive integer $n \geq 2$.*

**68.** The region bounded by $f(x) = x$ and $g(x) = x^n$, for $x \geq 0$

**69.** The region bounded by $f(x) = x$ and $g(x) = x^{1/n}$, for $x \geq 0$

**70.** The region bounded by $f(x) = x^{1/n}$ and $g(x) = x^n$, for $x \geq 0$

**71.** Let $A_n$ be the area of the region bounded by $f(x) = x^{1/n}$ and $g(x) = x^n$ on the interval $[0, 1]$, where $n$ is a positive integer. Evaluate $\lim\limits_{n \to \infty} A_n$ and interpret the result.

**72–73. Bisecting regions** *For each region R, find the horizontal line $y = k$ that divides R into two subregions of equal area.*

**72.** *R* is the region bounded by $y = 1 - x$, the *x*-axis, and the *y*-axis.

**73.** *R* is the region bounded by $y = 1 - |x - 1|$ and the *x*-axis.

**74. Geometric probability** Suppose a dartboard occupies the square $\{(x, y): 0 \leq |x| \leq 1, 0 \leq |y| \leq 1\}$. A dart is thrown randomly at the board many times (meaning it is equally likely to land at any point in the square). What fraction of the dart throws land closer to the edge of the board than to the center? Equivalently, what is the probability that the dart lands closer to the edge of the board than to the center? Proceed as follows.

**a.** Argue that by symmetry it is necessary to consider only one quarter of the board, say the region $R$: $\{(x,y): |x| \leq y \leq 1\}$.

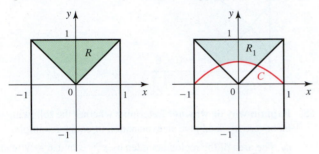

**b.** Find the curve $C$ in this region that is equidistant from the center of the board and the top edge of the board (see figure).

**c.** The probability that the dart lands closer to the edge of the board than to the center is the ratio of the area of the region $R_1$ above $C$ to the area of the entire region $R$. Compute this probability.

**T 75. Lorenz curves and the Gini index** A **Lorenz curve** is given by $y = L(x)$, where $0 \leq x \leq 1$ represents the lowest fraction of the population of a society in terms of wealth, and $0 \leq y \leq 1$ represents the fraction of the total wealth that is owned by that fraction of the society. For example, the Lorenz curve in the figure shows that $L(0.5) = 0.2$, which means that the lowest 0.5 (50%) of the society owns 0.2 (20%) of the wealth. (See Guided Project *Distribution of Wealth* for more on Lorenz curves.)

**a.** A Lorenz curve $y = L(x)$ is accompanied by the line $y = x$, called the **line of perfect equality**. Explain why this line is given this name.

**b.** Explain why a Lorenz curve satisfies the conditions $L(0) = 0$, $L(1) = 1$, $L(x) \leq x$, and $L'(x) \geq 0$ on $[0, 1]$.

**c.** Graph the Lorenz curves $L(x) = x^p$ corresponding to $p = 1.1$, 1.5, 2, 3, and 4. Which value of $p$ corresponds to the *most* equitable distribution of wealth (closest to the line of perfect equality)? Which value of $p$ corresponds to the *least* equitable distribution of wealth? Explain.

**d.** The information in the Lorenz curve is often summarized in a single measure called the **Gini index**, which is defined as follows. Let $A$ be the area of the region between $y = x$ and $y = L(x)$ (see figure) and let $B$ be the area of the region between $y = L(x)$ and the *x*-axis. Then the Gini index is

$$G = \frac{A}{A + B}.$$ Show that $G = 2A = 1 - 2\int_0^1 L(x)\, dx$.

**e.** Compute the Gini index for the cases $L(x) = x^p$ and $p = 1.1$, 1.5, 2, 3, and 4.

**f.** What is the smallest interval $[a, b]$ on which values of the Gini index lie for $L(x) = x^p$ with $p \geq 1$? Which endpoints of $[a, b]$ correspond to the least and most equitable distribution of wealth?

**g.** Consider the Lorenz curve described by $L(x) = \dfrac{5x^2}{6} + \dfrac{x}{6}$. Show that it satisfies the conditions $L(0) = 0, L(1) = 1$, and $L'(x) \geq 0$ on $[0, 1]$. Find the Gini index for this function.

**76. Equal area properties for parabolas** Consider the parabola $y = x^2$. Let $P, Q$, and $R$ be points on the parabola with $R$ between $P$ and $Q$ on the curve. Let $\ell_P, \ell_Q$, and $\ell_R$ be the lines tangent to the parabola at $P, Q$, and $R$, respectively (see figure). Let $P'$ be the intersection point of $\ell_Q$ and $\ell_R$, let $Q'$ be the intersection point of $\ell_P$ and $\ell_R$, and let $R'$ be the intersection point of $\ell_P$ and $\ell_Q$. Prove that Area $\triangle PQR = 2 \cdot$ Area $\triangle P'Q'R'$ in the following cases. (In fact, the property holds for any three points on any parabola.) (*Source: Mathematics Magazine*, 81, 2, Apr 2008)

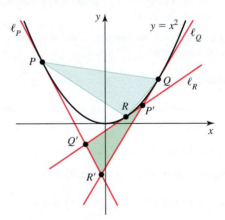

**a.** $P(-a, a^2), Q(a, a^2)$, and $R(0, 0)$, where $a$ is a positive real number

**b.** $P(-a, a^2), Q(b, b^2)$, and $R(0, 0)$, where $a$ and $b$ are positive real numbers

**c.** $P(-a, a^2), Q(b, b^2)$, and $R$ is any point between $P$ and $Q$ on the curve

**T 77. Roots and powers** Consider the functions $f(x) = x^n$ and $g(x) = x^{1/n}$, where $n \geq 2$ is a positive integer.

**a.** Graph $f$ and $g$ for $n = 2, 3$, and 4, for $x \geq 0$.

**b.** Give a geometric interpretation of the area function
$A_n(x) = \int_0^x (f(s) - g(s))\, ds$, for $n = 2, 3, 4, \ldots$ and $x > 0$.

**c.** Find the positive root of $A_n(x) = 0$ in terms of $n$. Does the root increase or decrease with $n$?

**T 78. Shifting sines** Consider the functions $f(x) = a \sin 2x$ and
$g(x) = \dfrac{\sin x}{a}$, where $a > 0$ is a real number.

**a.** Graph the two functions on the interval $[0, \frac{\pi}{2}]$, for $a = \frac{1}{2}, 1$, and 2.

**b.** Show that the curves have an intersection point $x^*$ (other than $x = 0$) on $[0, \frac{\pi}{2}]$ that satisfies $\cos x^* = \frac{1}{2a^2}$, provided $a > 1/\sqrt{2}$.

**c.** Find the area of the region between the two curves on $[0, x^*]$ when $a = 1$.

**d.** Show that as $a \to 1/\sqrt{2}^+$, the area of the region between the two curves on $[0, x^*]$ approaches zero.

# 6.3 Volume by Slicing

We have seen that integration is used to compute the area of two-dimensional regions bounded by curves. Integrals are also used to find the volume of three-dimensional regions (or solids). Once again, the slice-and-sum method is the key to solving these problems.

## General Slicing Method

Consider a solid object that extends in the $x$-direction from $x = a$ to $x = b$. Imagine making a vertical cut through the solid, perpendicular to the $x$-axis at a particular point $x$, and suppose the area of the cross section created by the cut is given by a known integrable function $A$ (**Figure 6.22**).

To find the volume of this solid, we first divide $[a, b]$ into $n$ subintervals of length $\Delta x = (b - a)/n$. The endpoints of the subintervals are the grid points $x_0 = a, x_1, x_2, \ldots, x_n = b$. We now make vertical cuts through the solid perpendicular to the $x$-axis at each grid point, which produces $n$ slices of thickness $\Delta x$. (Imagine cutting a loaf of bread to create $n$ slices of equal width.) On each subinterval, an arbitrary point $x_k^*$ is identified. The $k$th slice through the solid has a thickness $\Delta x$, and we take $A(x_k^*)$ as a representative cross-sectional area of the slice. Therefore, the volume of the $k$th slice is approximately $A(x_k^*)\Delta x$ (**Figure 6.23**). Summing the volumes of the slices, the approximate volume of the solid is

$$V \approx \sum_{k=1}^{n} A(x_k^*)\Delta x.$$

Cross section with area $A(x)$

**Figure 6.22**

Cross-sectional area $= A(x_k^*)$

Volume of $k$th slice $\approx A(x_k^*)\Delta x$

**Figure 6.23**

As the number of slices increases ($n \to \infty$) and the thickness of each slice goes to zero ($\Delta x \to 0$), the exact volume $V$ is obtained in terms of a definite integral (**Figure 6.24**):

$$V = \lim_{n \to \infty} \sum_{k=1}^{n} A(x_k^*) \Delta x = \int_a^b A(x)\, dx.$$

Increase the number of slices.

$$\text{Volume} = \lim_{n \to \infty} \sum_{k=1}^{n} A(x_k^*) \Delta x$$
$$= \int_a^b A(x)\, dx$$

$n \to \infty$

**Figure 6.24**

We summarize the important general slicing method, which is also the basis of other volume formulas to follow.

> ➤ The factors in this volume integral have meaning: $A(x)$ is the cross-sectional area of a slice and $dx$ represents its thickness. Summing (integrating) the volumes of the slices $A(x)\, dx$ gives the volume of the solid.

**QUICK CHECK 1** Why is the volume, as given by the general slicing method, equal to the average value of the area function $A$ on $[a, b]$ multiplied by $b - a$? ◄

---

**General Slicing Method**

Suppose a solid object extends from $x = a$ to $x = b$, and the cross section of the solid perpendicular to the $x$-axis has an area given by a function $A$ that is integrable on $[a, b]$. The volume of the solid is

$$V = \int_a^b A(x)\, dx.$$

---

**EXAMPLE 1   Volume of a "parabolic cube"** Let $R$ be the region in the first quadrant bounded by the coordinate axes and the curve $y = 1 - x^2$. A solid has a base $R$, and cross sections through the solid perpendicular to the base and parallel to the $y$-axis are squares (**Figure 6.25a**). Find the volume of the solid.

**SOLUTION** Focus on a cross section through the solid at a point $x$, where $0 \le x \le 1$. That cross section is a square with sides of length $1 - x^2$. Therefore, the area of a typical cross section is $A(x) = (1 - x^2)^2$. Using the general slicing method, the volume of the solid is

$$
\begin{aligned}
V &= \int_0^1 A(x)\, dx & \text{General slicing method} \\
&= \int_0^1 (1 - x^2)^2\, dx & \text{Substitute for } A(x). \\
&= \int_0^1 (1 - 2x^2 + x^4)\, dx & \text{Expand integrand.} \\
&= \frac{8}{15}. & \text{Evaluate.}
\end{aligned}
$$

The actual solid with a square cross section is shown in **Figure 6.25b**.

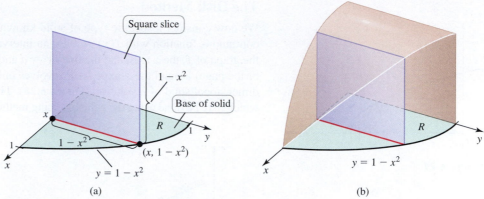

**Figure 6.25**

*Related Exercises 11–12* ◄

**EXAMPLE 2**  **Volume of a "parabolic hemisphere"**  A solid has a base that is bounded by the curves $y = x^2$ and $y = 2 - x^2$ in the $xy$-plane. Cross sections through the solid perpendicular to the base and parallel to the $y$-axis are semicircular disks. Find the volume of the solid.

**SOLUTION**  Because a typical cross section perpendicular to the $x$-axis is a semicircular disk (**Figure 6.26**), the area of a cross section is $\frac{1}{2}\pi r^2$, where $r$ is the radius of the cross section.

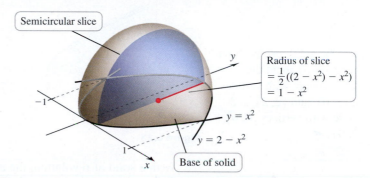

**Figure 6.26**

The key observation is that this radius is one-half of the distance between the upper bounding curve $y = 2 - x^2$ and the lower bounding curve $y = x^2$. So the radius at the point $x$ is

$$r = \frac{1}{2}\left((2 - x^2) - x^2\right) = 1 - x^2.$$

This means that the area of the semicircular cross section at the point $x$ is

$$A(x) = \frac{1}{2}\pi r^2 = \frac{\pi}{2}(1 - x^2)^2.$$

The intersection points of the two bounding curves satisfy $2 - x^2 = x^2$, which has solutions $x = \pm 1$. Therefore, the cross sections lie between $x = -1$ and $x = 1$. Integrating the cross-sectional areas, the volume of the solid is

$$
\begin{aligned}
V &= \int_{-1}^{1} A(x)\, dx && \text{General slicing method} \\
&= \int_{-1}^{1} \frac{\pi}{2}(1 - x^2)^2\, dx && \text{Substitute for } A(x). \\
&= \frac{\pi}{2}\int_{-1}^{1}(1 - 2x^2 + x^4)\, dx && \text{Expand integrand.} \\
&= \frac{8\pi}{15}. && \text{Evaluate.}
\end{aligned}
$$

**QUICK CHECK 2**  In Example 2, what is the cross-sectional area function $A(x)$ if cross sections perpendicular to the base are squares rather than semicircles? ◄

*Related Exercise 15* ◄

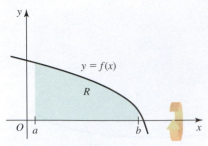

**Figure 6.27**

## The Disk Method

We now consider a specific type of solid known as a **solid of revolution**. Suppose $f$ is a continuous function with $f(x) \geq 0$ on an interval $[a, b]$. Let $R$ be the region bounded by the graph of $f$, the $x$-axis, and the lines $x = a$ and $x = b$ (**Figure 6.27**). Now revolve $R$ (out of the page) around the $x$-axis. As $R$ revolves once about the $x$-axis, it sweeps out a three-dimensional solid of revolution (**Figure 6.28**). The goal is to find the volume of this solid, which may be done using the general slicing method.

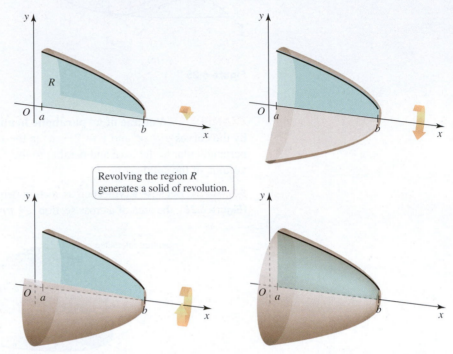

Revolving the region $R$ generates a solid of revolution.

**QUICK CHECK 3** What solid results when the region $R$ is revolved about the $x$-axis if (a) $R$ is a square with vertices $(0, 0)$, $(0, 2)$, $(2, 0)$, and $(2, 2)$, and (b) $R$ is a triangle with vertices $(0, 0)$, $(0, 2)$, and $(2, 0)$? ◄

**Figure 6.28**

With a solid of revolution, the cross-sectional area function has a special form because all cross sections perpendicular to the $x$-axis are *circular disks* with radius $f(x)$ (**Figure 6.29**). Therefore, the cross section at the point $x$, where $a \leq x \leq b$, has area

$$A(x) = \pi(\text{radius})^2 = \pi f(x)^2.$$

By the general slicing method, the volume of the solid is

$$V = \int_a^b A(x)\, dx = \int_a^b \pi f(x)^2\, dx.$$

Because each slice through the solid is a circular disk, the resulting method is called the *disk method*.

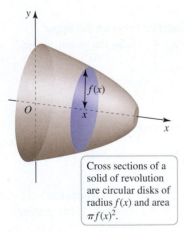

Cross sections of a solid of revolution are circular disks of radius $f(x)$ and area $\pi f(x)^2$.

**Figure 6.29**

---

**Disk Method about the $x$-Axis**

Let $f$ be continuous with $f(x) \geq 0$ on the interval $[a, b]$. If the region $R$ bounded by the graph of $f$, the $x$-axis, and the lines $x = a$ and $x = b$ is revolved about the $x$-axis, the volume of the resulting solid of revolution is

$$V = \int_a^b \pi \underbrace{f(x)^2}_{\substack{\text{disk} \\ \text{radius}}} dx.$$

---

**EXAMPLE 3**   **Disk method at work**   Let $R$ be the region bounded by the curve $f(x) = (x + 1)^2$, the $x$-axis, and the lines $x = 0$ and $x = 2$ (**Figure 6.30a**). Find the volume of the solid of revolution obtained by revolving $R$ about the $x$-axis.

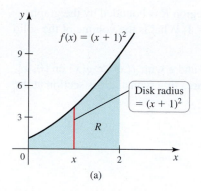

$f(x) = (x + 1)^2$

Disk radius
$= (x + 1)^2$

$R$

(a)

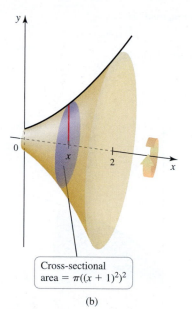

Cross-sectional
area $= \pi((x + 1)^2)^2$

(b)

**Figure 6.30**

$r_o$

$r_i$

Cross-sectional
area $= \pi r_o^2 - \pi r_i^2$

$(x, f(x))$

$(x, g(x))$

Cross-sectional area
$= \pi(f(x)^2 - g(x)^2)$

**Figure 6.32**

**SOLUTION** When the region $R$ is revolved about the $x$-axis, it generates a solid of revolution (**Figure 6.30**b). A cross section perpendicular to the $x$-axis at the point $0 \le x \le 2$ is a circular disk of radius $f(x)$. Therefore, a typical cross section has area

$$A(x) = \pi f(x)^2 = \pi((x + 1)^2)^2.$$

Integrating these cross-sectional areas between $x = 0$ and $x = 2$ gives the volume of the solid:

$$V = \int_0^2 A(x)\, dx = \int_0^2 \pi(\underbrace{(x + 1)^2}_{\text{disk radius}})^2\, dx \quad \text{Substitute for } A(x).$$

$$= \int_0^2 \pi(x + 1)^4\, dx \quad \text{Simplify.}$$

$$= \pi \left. \frac{u^5}{5} \right|_1^3 = \frac{242\,\pi}{5}. \quad \text{Let } u = x + 1 \text{ and evaluate.}$$

*Related Exercises 17, 19* ◀

### Washer Method

A slight variation on the disk method enables us to compute the volume of more exotic solids of revolution. Suppose $R$ is the region bounded by the graphs of $f$ and $g$ between $x = a$ and $x = b$, where $f(x) \ge g(x) \ge 0$ (**Figure 6.31**). If $R$ is revolved about the $x$-axis to generate a solid of revolution, the resulting solid generally has a hole through it.

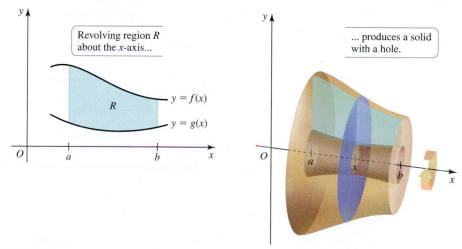

Revolving region $R$
about the $x$-axis...

$y = f(x)$

$R$

$y = g(x)$

... produces a solid
with a hole.

**Figure 6.31**

Once again we apply the general slicing method. In this case, a cross section through the solid perpendicular to the $x$-axis is a circular *washer* with an outer radius of $r_o = f(x)$ and an inner radius of $r_i = g(x)$, where $a \le x \le b$. The area of the cross section is the area of the entire disk minus the area of the hole, or

$$A(x) = \pi(r_o^2 - r_i^2) = \pi(f(x)^2 - g(x)^2)$$

(**Figure 6.32**). The general slicing method gives the volume of the solid.

---

**Washer Method about the $x$-Axis**

Let $f$ and $g$ be continuous functions with $f(x) \ge g(x) \ge 0$ on $[a, b]$. Let $R$ be the region bounded by $y = f(x)$, $y = g(x)$, and the lines $x = a$ and $x = b$. When $R$ is revolved about the $x$-axis, the volume of the resulting solid of revolution is

$$V = \int_a^b \pi(\underbrace{f(x)^2}_{\substack{\text{outer} \\ \text{radius}}} - \underbrace{g(x)^2}_{\substack{\text{inner} \\ \text{radius}}})\, dx.$$

---

**QUICK CHECK 4** Show that when $g(x) = 0$ in the washer method, the result is the disk method. ◄

> The washer method is really two applications of the disk method. We compute the volume of the entire solid without the hole (by the disk method) and then subtract the volume of the hole (also computed by the disk method).

**EXAMPLE 4  Volume by the washer method** The region $R$ is bounded by the graphs of $f(x) = \sqrt{x}$ and $g(x) = x^2$ between $x = 0$ and $x = 1$. What is the volume of the solid that results when $R$ is revolved about the $x$-axis?

**SOLUTION** The region $R$ is bounded by the graphs of $f$ and $g$ with $f(x) \geq g(x)$ on $[0, 1]$, so the washer method is applicable (**Figure 6.33**). The area of a typical cross section at the point $x$ is

$$A(x) = \pi(\,f(x)^2 - g(x)^2) = \pi((\underbrace{\sqrt{x}}_{\text{outer radius}})^2 - (\underbrace{x^2}_{\text{inner radius}})^2) = \pi(x - x^4).$$

Therefore, the volume of the solid is

$$V = \int_0^1 \pi(x - x^4)\,dx \qquad \text{Washer method}$$

$$= \pi\left(\frac{x^2}{2} - \frac{x^5}{5}\right)\Big|_0^1 = \frac{3\pi}{10}. \qquad \text{Fundamental Theorem}$$

> Ignoring the factor of $\pi$, the integrand in the washer method integral is $f(x)^2 - g(x)^2$, which is not equal to $(f(x) - g(x))^2$.

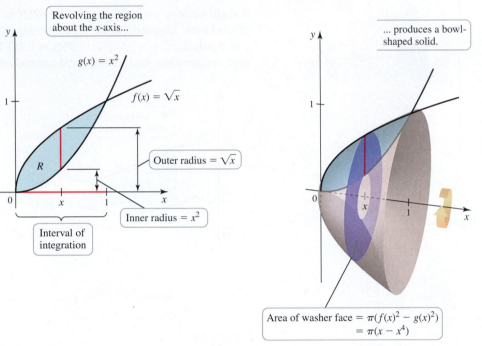

Revolving the region about the $x$-axis...

$g(x) = x^2$

$f(x) = \sqrt{x}$

Outer radius $= \sqrt{x}$

$R$

Inner radius $= x^2$

Interval of integration

... produces a bowl-shaped solid.

Area of washer face $= \pi(f(x)^2 - g(x)^2)$
$= \pi(x - x^4)$

**Figure 6.33**

*Related Exercises 21–22* ◄

**QUICK CHECK 5** Suppose the region in Example 4 is revolved about the line $y = -1$ instead of the $x$-axis. (a) What is the inner radius of a typical washer? (b) What is the outer radius of a typical washer? ◄

## Revolving about the $y$-Axis

Everything you learned about revolving regions about the $x$-axis applies to revolving regions about the $y$-axis. Consider a region $R$ bounded by the curve $x = p(y)$ on the right, the curve $x = q(y)$ on the left, and the horizontal lines $y = c$ and $y = d$ (**Figure 6.34a**).

To find the volume of the solid generated when $R$ is revolved about the $y$-axis, we use the general slicing method—now with respect to the $y$-axis (**Figure 6.34b**). The area of a typical cross section is $A(y) = \pi(p(y)^2 - q(y)^2)$, where $c \leq y \leq d$. As before, integrating these cross-sectional areas of the solid gives the volume.

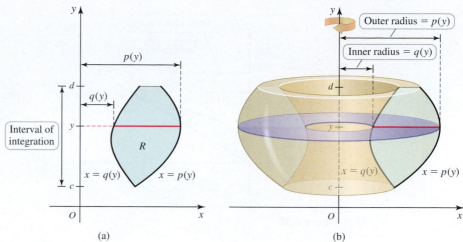

**Figure 6.34**

> The disk/washer method about the *y*-axis is the disk/washer method about the *x*-axis with *x* replaced with *y*.

---

**Disk and Washer Methods about the *y*-Axis**

Let $p$ and $q$ be continuous functions with $p(y) \geq q(y) \geq 0$ on $[c, d]$. Let $R$ be the region bounded by $x = p(y)$, $x = q(y)$, and the lines $y = c$ and $y = d$. When $R$ is revolved about the *y*-axis, the volume of the resulting solid of revolution is given by

$$V = \int_c^d \pi(\underbrace{p(y)^2}_{\substack{\text{outer} \\ \text{radius}}} - \underbrace{q(y)^2}_{\substack{\text{inner} \\ \text{radius}}}) \, dy.$$

If $q(y) = 0$, the disk method results:

$$V = \int_c^d \pi \underbrace{p(y)^2}_{\substack{\text{disk} \\ \text{radius}}} \, dy.$$

---

**EXAMPLE 5    Which solid has greater volume?**  Let $R$ be the region in the first quadrant bounded by the graphs of $x = y^3$ and $x = 4y$. Which is greater, the volume of the solid generated when $R$ is revolved about the *x*-axis or the *y*-axis?

**SOLUTION**  Solving $y^3 = 4y$, or equivalently, $y(y^2 - 4) = 0$, we find that the bounding curves of $R$ intersect at the points $(0, 0)$ and $(8, 2)$. When the region $R$ (**Figure 6.35a**) is revolved about the *y*-axis, it generates a funnel with a curved inner surface (**Figure 6.35b**). Washer-shaped cross sections perpendicular to the *y*-axis extend from $y = 0$ to $y = 2$. The outer radius of the cross section at the point $y$ is determined by the line $x = p(y) = 4y$. The inner radius of the cross section at the point $y$ is determined by the curve $x = q(y) = y^3$. Applying the washer method, the volume of this solid is

$$V = \int_0^2 \pi(p(y)^2 - q(y)^2) \, dy \qquad \text{Washer method}$$

$$= \int_0^2 \pi(16y^2 - y^6) \, dy \qquad \text{Substitute for } p \text{ and } q.$$

$$= \pi\left(\frac{16}{3}y^3 - \frac{y^7}{7}\right)\Big|_0^2 \qquad \text{Fundamental Theorem}$$

$$= \frac{512\pi}{21}. \qquad \text{Evaluate.}$$

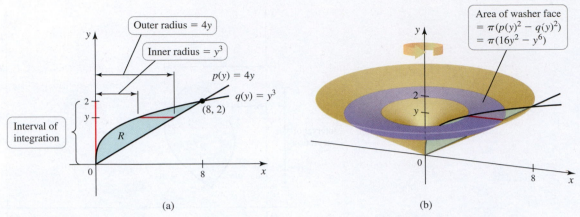

**Figure 6.35**

When the region $R$ is revolved about the $x$-axis, it generates a different funnel (**Figure 6.36**). Vertical slices through the solid between $x = 0$ and $x = 8$ produce washers. The outer radius of the washer at the point $x$ is determined by the curve $x = y^3$, or $y = f(x) = x^{1/3}$. The inner radius is determined by $x = 4y$, or $y = g(x) = x/4$. The volume of the resulting solid is

$$V = \int_0^8 \pi(f(x)^2 - g(x)^2)\, dx \quad \text{Washer method}$$

$$= \int_0^8 \pi\left(x^{2/3} - \frac{x^2}{16}\right) dx \quad \text{Substitute for } f \text{ and } g.$$

$$= \pi\left(\frac{3}{5}x^{5/3} - \frac{x^3}{48}\right)\Big|_0^8 \quad \text{Fundamental Theorem}$$

$$= \frac{128\pi}{15}. \quad \text{Evaluate.}$$

We see that revolving the region about the $y$-axis produces a solid of greater volume.

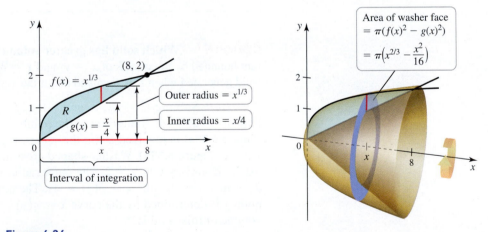

**QUICK CHECK 6** The region in the first quadrant bounded by $y = x$ and $y = x^3$ is revolved about the $y$-axis. Give the integral for the volume of the solid that is generated. ◄

**Figure 6.36**

*Related Exercises 43–44* ◄

The disk and washer methods may be generalized to handle situations in which a region $R$ is revolved about a line parallel to one of the coordinate axes. The next example discusses three such cases.

**EXAMPLE 6** **Revolving about other lines** Let $f(x) = \sqrt{x} + 1$ and $g(x) = x^2 + 1$.

**a.** Find the volume of the solid generated when the region $R_1$ bounded by the graph of $f$ and the line $y = 2$ on the interval $[0, 1]$ is revolved about the line $y = 2$.

**b.** Find the volume of the solid generated when the region $R_2$ bounded by the graphs of $f$ and $g$ on the interval $[0, 1]$ is revolved about the line $y = -1$.

**c.** Find the volume of the solid generated when the region $R_2$ bounded by the graphs of $f$ and $g$ on the interval $[0, 1]$ is revolved about the line $x = 2$.

**SOLUTION**

**a.** Figure 6.37a shows the region $R_1$ and the axis of revolution. Applying the disk method, we see that a disk located at a point $x$ has a radius of $2 - f(x) = 2 - (\sqrt{x} + 1) = 1 - \sqrt{x}$. Therefore, the volume of the solid generated when $R_1$ is revolved about $y = 2$ is

$$\int_0^1 \pi \underbrace{(1 - \sqrt{x})^2}_{\substack{\text{disk} \\ \text{radius}}} dx = \pi \int_0^1 (1 - 2\sqrt{x} + x)\, dx = \frac{\pi}{6}.$$

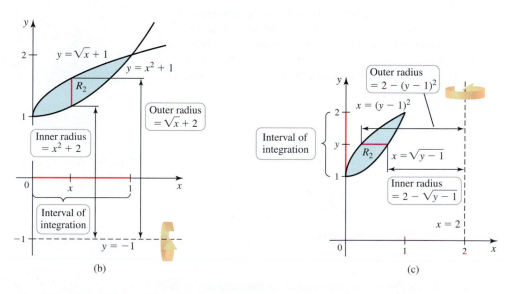

**Figure 6.37**

**b.** When the graph of $f$ is revolved about $y = -1$, it sweeps out a solid of revolution whose radius at a point $x$ is $f(x) + 1 = \sqrt{x} + 2$. Similarly, when the graph of $g$ is revolved about $y = -1$, it sweeps out a solid of revolution whose radius at a point $x$ is $g(x) + 1 = x^2 + 2$ (Figure 6.37b). Using the washer method, the volume of the solid generated when $R_2$ is revolved about $y = -1$ is

$$\int_0^1 \pi (\underbrace{(\sqrt{x} + 2)^2}_{\substack{\text{outer} \\ \text{radius}}} - \underbrace{(x^2 + 2)^2}_{\substack{\text{inner} \\ \text{radius}}})\, dx \qquad \text{Washer method}$$

$$= \pi \int_0^1 (-x^4 - 4x^2 + x + 4\sqrt{x})\, dx \qquad \text{Simplify integrand.}$$

$$= \frac{49\pi}{30}. \qquad \text{Evaluate integral.}$$

**c.** When the region $R_2$ is revolved about the line $x = 2$, we use the washer method and integrate in the $y$-direction. First note that the graph of $f$ is described by $y = \sqrt{x} + 1$, or equivalently, $x = (y - 1)^2$, for $y \geq 1$. Also, the graph of $g$ is described by $y = x^2 + 1$, or equivalently, $x = \sqrt{y - 1}$ for $y \geq 1$ (Figure 6.37c). When the graph of $f$ is revolved about the line $x = 2$, the radius of a typical disk at a point $y$ is $2 - (y - 1)^2$. Similarly, when the graph of $g$ is revolved about $x = 2$, the radius of a typical disk at a point $y$ is $2 - \sqrt{y - 1}$. Finally, observe that the extent of the region $R_2$ in the $y$-direction is the interval $1 \leq y \leq 2$.

Applying the washer method, simplifying the integrand, and integrating powers of $y$, the volume of the solid of revolution is

$$\int_1^2 \pi(\underbrace{(2-(y-1)^2)^2}_{\substack{\text{outer} \\ \text{radius}}} - \underbrace{(2-\sqrt{y-1})^2}_{\substack{\text{inner} \\ \text{radius}}}) \, dy = \frac{31\pi}{30}.$$

*Related Exercises 45, 48* ◄

## SECTION 6.3 EXERCISES

### Getting Started

1. Suppose a cut is made through a solid object perpendicular to the $x$-axis at a particular point $x$. Explain the meaning of $A(x)$.

2. A solid has a circular base; cross sections perpendicular to the base are squares. What method should be used to find the volume of the solid?

3. Consider a solid whose base is the region in the first quadrant bounded by the curve $y = \sqrt{3-x}$ and the line $x = 2$, and whose cross sections through the solid perpendicular to the $x$-axis are squares.

   a. Find an expression for the area $A(x)$ of a cross section of the solid at a point $x$ in $[0, 2]$.
   b. Write an integral for the volume of the solid.

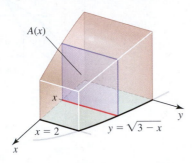

4. Why is the disk method a special case of the general slicing method?

5. Let $R$ be the region bounded by the curve $y = \sqrt{\cos x}$ and the $x$-axis on $[0, \pi/2]$. A solid of revolution is obtained by revolving $R$ about the $x$-axis (see figures).

   a. Find an expression for the radius of a cross section of the solid of revolution at a point $x$ in $[0, \pi/2]$.
   b. Find an expression for the area $A(x)$ of a cross section of the solid at a point $x$ in $[0, \pi/2]$.
   c. Write an integral for the volume of the solid.

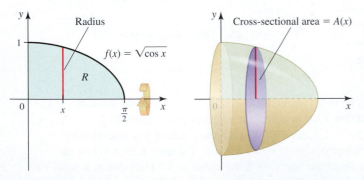

6. Let $R$ be the region bounded by the curve $x = \cos y$ and the $x$-axis on $[0, 1]$. A solid of revolution is obtained by revolving $R$ about the $y$-axis (see figures).

   a. Find an expression for the radius of a cross section of the solid at a point $y$ in $[0, \pi/2]$.
   b. Find an expression for the area $A(y)$ of a cross section of the solid at a point $y$ in $[0, \pi/2]$.
   c. Write an integral for the volume of the solid.

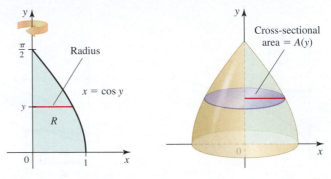

**7–10.** *Use the region $R$ that is bounded by the graphs of $y = 1 + \sqrt{x}$, $x = 4$, and $y = 1$ to complete the exercises.*

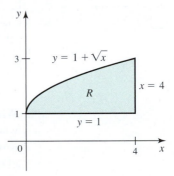

7. Region $R$ is revolved about the $x$-axis to form a solid of revolution whose cross sections are washers.

   a. What is the outer radius of a cross section of the solid at a point $x$ in $[0, 4]$?
   b. What is the inner radius of a cross section of the solid at a point $x$ in $[0, 4]$?
   c. What is the area $A(x)$ of a cross section of the solid at a point $x$ in $[0, 4]$?
   d. Write an integral for the volume of the solid.

8. Region $R$ is revolved about the $y$-axis to form a solid of revolution whose cross sections are washers.

   a. What is the outer radius of a cross section of the solid at a point $y$ in $[1, 3]$?
   b. What is the inner radius of a cross section of the solid at a point $y$ in $[1, 3]$?
   c. What is the area $A(y)$ of a cross section of the solid at a point $y$ in $[1, 3]$?
   d. Write an integral for the volume of the solid.

**9.** Region $R$ is revolved about the line $y = 1$ to form a solid of revolution.

    **a.** What is the radius of a cross section of the solid at a point $x$ in $[0, 4]$?

    **b.** What is the area $A(x)$ of a cross section of the solid at a point $x$ in $[0, 4]$?

    **c.** Write an integral for the volume of the solid.

**10.** Region $R$ is revolved about the line $x = 4$ to form a solid of revolution.

    **a.** What is the radius of a cross section of the solid at a point $y$ in $[1, 3]$?

    **b.** What is the area $A(y)$ of a cross section of the solid at a point $y$ in $[1, 3]$?

    **c.** Write an integral for the volume of the solid.

## Practice Exercises

**11–16. General slicing method** *Use the general slicing method to find the volume of the following solids.*

**11.** The solid whose base is the region bounded by the semicircle $y = \sqrt{1 - x^2}$ and the $x$-axis, and whose cross sections through the solid perpendicular to the $x$-axis are squares

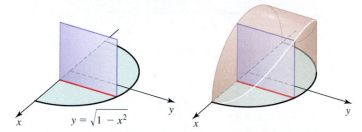

**12.** The solid whose base is the region bounded by the curves $y = x^2$ and $y = 2 - x^2$, and whose cross sections through the solid perpendicular to the $x$-axis are squares

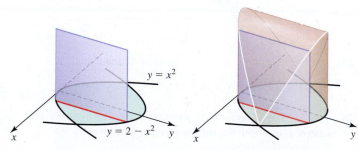

**13.** The solid whose base is the region bounded by the curve $y = \sqrt{\cos x}$ and the $x$-axis on $[-\pi/2, \pi/2]$, and whose cross sections through the solid perpendicular to the $x$-axis are isosceles right triangles with a horizontal leg in the $xy$-plane and a vertical leg above the $x$-axis

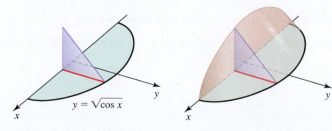

**14.** The solid with a semicircular base of radius 5 whose cross sections perpendicular to the base and parallel to the diameter are squares

**15.** The solid whose base is the triangle with vertices $(0, 0)$, $(2, 0)$, and $(0, 2)$, and whose cross sections perpendicular to the base and parallel to the $y$-axis are semicircles

**16.** The solid whose base is the region bounded by $y = x^2$ and the line $y = 1$, and whose cross sections perpendicular to the base and parallel to the $y$-axis are squares

**17–40. Solids of revolution** *Let $R$ be the region bounded by the following curves. Find the volume of the solid generated when $R$ is revolved about the given axis.*

**17.** $y = 2x$, $y = 0$, and $x = 3$; about the $x$-axis (Verify that your answer agrees with the volume formula for a cone.)

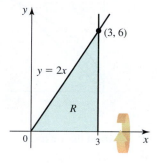

**18.** $y = 2 - 2x$, $y = 0$, and $x = 0$; about the $x$-axis (Verify that your answer agrees with the volume formula for a cone.)

**19.** $y = 4 - x^2$, $y = 0$, and $x = 0$; about the $x$-axis

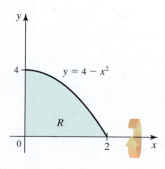

**20.** $y = \dfrac{1}{x^2}$, $y = 0$, $x = 1$, and $x = 4$; about the $x$-axis

**21.** $y = x$ and $y = 2\sqrt{x}$; about the $x$-axis

**22.** $y = x$ and $y = \sqrt[4]{x}$; about the $x$-axis

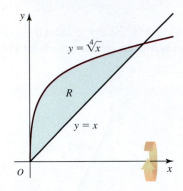

**23.** $y = \sin x$, $y = 1 - \sin x$, $x = \pi/6$, and $x = 5\pi/6$; about the $x$-axis

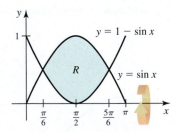

**24.** $y = x$, $y = x + 2$, $x = 0$, and $x = 4$; about the $x$-axis

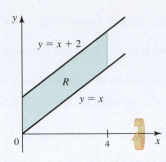

**25.** $y = x^3$, $y = 0$, and $x = 1$; about the $y$-axis

**26.** $y = x$, $y = 2x$, and $y = 6$; about the $y$-axis

**27.** $y = \cos x$ on $[0, \pi/2]$, $y = 0$, and $x = 0$; about the $x$-axis
(*Hint:* Recall that $\cos^2 x = \dfrac{1 + \cos 2x}{2}$.)

**28.** $y = \sqrt{25 - x^2}$ and $y = 0$; about the $x$-axis (Verify that your answer agrees with the volume formula for a sphere.)

**29.** $y = \sin x$ on $[0, \pi]$ and $y = 0$; about the $x$-axis
(*Hint:* Recall that $\sin^2 x = \dfrac{1 - \cos 2x}{2}$.)

**30.** $y = \sec x$, $y = 0$, $x = 0$, and $x = \frac{\pi}{4}$; about the $x$-axis

**31.** $y = 2 + \sqrt{x}$, $y = 2 - \sqrt{x}$, and $x = 4$; about the $y$-axis

**32.** $y = \sqrt{\sin x}$, $y = 1$, and $x = 0$; about the $x$-axis

**33.** $y = \sin x$ and $y = \sqrt{\sin x}$, for $0 \le x \le \frac{\pi}{2}$; about the $x$-axis

**34.** $y = |x|$ and $y = 2 - x^2$; about the $x$-axis

**35.** $y = \sqrt{x}$, $y = 0$, and $x = 4$; about the $y$-axis

**36.** $y = 4 - x^2$, $x = 2$, and $y = 4$; about the $y$-axis

**37.** $y = x^{-3/2}$, $y = 1$, $x = 1$, and $x = 6$; about the $x$-axis

**38.** $y = x^2$, $y = 2 - x$, and $y = 0$; about the $y$-axis

**39.** $y = x^{1/3}$, $y = 4 - x^{1/3}$, and $x = 0$; about the $y$-axis

**40.** $y = x^{-1}$, $y = 0$, $x = 1$, and $x = p > 0$; about the $x$-axis (Is the volume bounded as $p \to \infty$?)

**41–44. Which is greater?** *For the following regions R, determine which is greater: the volume of the solid generated when R is revolved about the x-axis or about the y-axis.*

**41.** $R$ is bounded by $y = 2x$, the $x$-axis, and $x = 5$.

**42.** $R$ is bounded by $y = 4 - 2x$, the $x$-axis, and the $y$-axis.

**43.** $R$ is bounded by $y = 1 - x^3$, the $x$-axis, and the $y$-axis.

**44.** $R$ is bounded by $y = x^2$ and $y = \sqrt{8x}$.

**45–55. Revolution about other axes** *Let R be the region bounded by the following curves. Find the volume of the solid generated when R is revolved about the given line.*

**45.** $y = 1 - \sqrt{x}$, $y = 1$, and $x = 1$; about $y = 1$

**46.** $y = \sqrt{x} + 2$, $y = 2$, and $x = 1$; about $y = 2$

**47.** $x = 2 - \sec y$, $x = 2$, $y = \frac{\pi}{3}$, and $y = 0$; about $x = 2$

**48.** $y = 1 - \sqrt{x}$, $x = 1$, and $y = 1$; about $x = 1$

**49.** $x = 0$, $y = \sqrt{x}$, and $y = 1$; about $y = 1$

**50.** $x = 0$, $y = \sqrt{x}$, and $y = 2$; about $x = 4$

**51.** $y = 2 \sin x$ and $y = 0$ on $[0, \pi]$; about $y = -2$

**52.** $y = x^2/4$ and the $y$-axis on the interval $0 \le y \le 1$; about $x = -1$

**53.** $y = \sin x$ and $y = 1 - \sin x$ on the interval $\frac{\pi}{6} \le x \le \frac{5\pi}{6}$; about $y = -1$

**54.** $y = x$ and $y = 1 + \frac{x}{2}$; about $y = 3$

**55.** $y = 2 - x$ and $y = 2 - 2x$; about $x = 3$

**56. Comparing volumes** The region $R$ is bounded by the graph of $f(x) = 2x(2 - x)$ and the $x$-axis. Which is greater, the volume of the solid generated when $R$ is revolved about the line $y = 2$ or the volume of the solid generated when $R$ is revolved about the line $y = 0$? Use integration to justify your answer.

**57. Explain why or why not** Determine whether the following statements are true and give an explanation or counterexample.

  **a.** A pyramid is a solid of revolution.
  **b.** The volume of a hemisphere can be computed using the disk method.
  **c.** Let $R_1$ be the region bounded by $y = \cos x$ and the $x$-axis on $[-\pi/2, \pi/2]$. Let $R_2$ be the region bounded by $y = \sin x$ and the $x$-axis on $[0, \pi]$. The volumes of the solids generated when $R_1$ and $R_2$ are revolved about the $x$-axis are equal.

### Explorations and Challenges

**58.** Use calculus to find the volume of a tetrahedron (pyramid with four triangular faces), all of whose edges have length 4.

**59. Fermat's volume calculation (1636)** Let $R$ be the region bounded by the curve $y = \sqrt{x + a}$ (with $a > 0$), the $y$-axis, and the $x$-axis. Let $S$ be the solid generated by rotating $R$ about the $y$-axis. Let $T$ be the inscribed cone that has the same circular base as $S$ and height $\sqrt{a}$. Show that $\frac{\text{volume}(S)}{\text{volume}(T)} = \frac{8}{5}$.

**60. Solid from a piecewise function** Let

$$f(x) = \begin{cases} x & \text{if } 0 \le x \le 2 \\ 2x - 2 & \text{if } 2 < x \le 5 \\ -2x + 18 & \text{if } 5 < x \le 6. \end{cases}$$

Find the volume of the solid formed when the region bounded by the graph of $f$, the $x$-axis, and the line $x = 6$ is revolved about the $x$-axis.

**61. Solids from integrals** Sketch a solid of revolution whose volume by the disk method is given by the following integrals. Indicate the function that generates the solid. Solutions are not unique.

a. $\displaystyle\int_0^\pi \pi \sin^2 x \, dx$

b. $\displaystyle\int_0^2 \pi(x^2 + 2x + 1) \, dx$

**62. Volume of a cup** A 6-inch-tall plastic cup is shaped like a surface obtained by rotating a line segment in the first quadrant about the $x$-axis. Given that the radius of the base of the cup is 1 inch, the radius of the top of the cup is 2 inches, and the cup is filled to the brim with water, use integration to approximate the volume of the water in the cup.

**63. Estimating volume** Suppose the region bounded by the curve $y = f(x)$ from $x = 0$ to $x = 4$ (see figure) is revolved about the $x$-axis to form a solid of revolution. Use left, right, and midpoint Riemann sums, with $n = 4$ subintervals of equal length, to estimate the volume of the solid of revolution.

**T 64. Volume of a wooden object** A solid wooden object turned on a lathe has a length of 50 cm and the diameters (measured in cm) shown in the figure. (A lathe is a tool that spins and cuts a block of wood so that it has circular cross sections.) Use left Riemann sums with uniformly spaced grid points to estimate the volume of the object.

**65. Cylinder, cone, hemisphere** A right circular cylinder with height $R$ and radius $R$ has a volume of $V_C = \pi R^3$ (height = radius).

a. Find the volume of the cone that is inscribed in the cylinder with the same base as the cylinder and height $R$. Express the volume in terms of $V_C$.

b. Find the volume of the hemisphere that is inscribed in the cylinder with the same base as the cylinder. Express the volume in terms of $V_C$.

**66. Water in a bowl** A hemispherical bowl of radius 8 inches is filled to a depth of $h$ inches, where $0 \le h \le 8$. Find the volume of water in the bowl as a function of $h$. (Check the special cases $h = 0$ and $h = 8$.)

**T 67. A torus (doughnut)** Find the volume of the torus formed when the circle of radius 2 centered at $(3, 0)$ is revolved about the $y$-axis. Use geometry to evaluate the integral.

**68. Which is greater?** Let $R$ be the region bounded by $y = x^2$ and $y = \sqrt{x}$. Use integration to determine which is greater: the volume of the solid generated when $R$ is revolved about the $x$-axis or about the line $y = 1$.

**69. Cavalieri's principle** *Cavalieri's principle* states that if two solids with equal altitudes have the same cross-sectional areas at every height, then they have equal volumes (see figure).

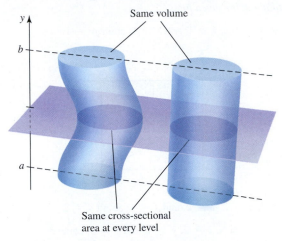

a. Use the general slicing method to justify Cavalieri's principle.

b. Use Cavalieri's principle to find the volume of a circular cylinder of radius $r$ and height $h$ whose axis is at an angle of $\frac{\pi}{4}$ to the base (see figure).

**70. Limiting volume** Consider the region $R$ in the first quadrant bounded by $y = x^{1/n}$ and $y = x^n$, where $n > 1$ is a positive number.

   **a.** Find the volume $V(n)$ of the solid generated when $R$ is revolved about the $x$-axis. Express your answer in terms of $n$.

   **b.** Evaluate $\lim_{n \to \infty} V(n)$. Interpret this limit geometrically.

**QUICK CHECK ANSWERS**

**1.** The average value of $A$ on $[a, b]$ is $\overline{A} = \dfrac{1}{b-a}\displaystyle\int_a^b A(x)\,dx$.
Therefore, $V = (b - a)\overline{A}$.   **2.** $A(x) = (2 - 2x^2)^2$

**3. (a)** A cylinder with height 2 and radius 2; **(b)** a cone with height 2 and base radius 2   **4.** When $g(x) = 0$, the washer

method $V = \displaystyle\int_a^b \pi(f(x)^2 - g(x)^2)\,dx$ reduces to the disk

method $V = \displaystyle\int_a^b \pi(f(x)^2)\,dx$.   **5. (a)** Inner radius $= x^2 + 1$;

**(b)** outer radius $= \sqrt{x} + 1$   **6.** $\displaystyle\int_0^1 \pi(y^{2/3} - y^2)\,dy$ ◄

# 6.4 Volume by Shells

You can solve many challenging volume problems using the disk/washer method. There are, however, some volume problems that are difficult to solve with this method. For this reason, we extend our discussion of volume problems to the *shell method*, which—like the disk/washer method—is used to compute the volume of solids of revolution.

➤ Why another method? Suppose $R$ is the region in the first quadrant bounded by the graph of $y = x^2 - x^3$ and the $x$-axis (**Figure 6.38**). When $R$ is revolved about the $y$-axis, the resulting solid has a volume that is difficult to compute using the washer method. The volume is much easier to compute using the shell method.

## Cylindrical Shells

Let $R$ be a region bounded by the graph of $f$, the $x$-axis, and the lines $x = a$ and $x = b$, where $0 \le a < b$ and $f(x) \ge 0$ on $[a, b]$. When $R$ is revolved about the $y$-axis, a solid is generated (**Figure 6.39**) whose volume is computed with the slice-and-sum strategy.

**Figure 6.38**

**Figure 6.39**

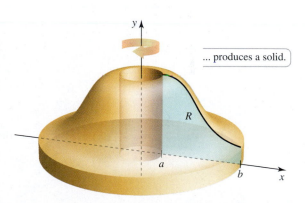

We divide $[a, b]$ into $n$ subintervals of length $\Delta x = (b - a)/n$ and identify an arbitrary point $x_k^*$ on the $k$th subinterval, for $k = 1, \ldots, n$. Now observe the rectangle built on the $k$th subinterval with a height of $f(x_k^*)$ and a width $\Delta x$ (**Figure 6.40**). As it revolves about the $y$-axis, this rectangle sweeps out a thin *cylindrical shell*.

**Figure 6.40**

When the $k$th cylindrical shell is unwrapped (**Figure 6.41**), it approximates a thin rectangular slab. The approximate length of the slab is the circumference of a circle with radius $x_k^*$, which is $2\pi x_k^*$. The height of the slab is the height of the original rectangle $f(x_k^*)$ and its thickness is $\Delta x$; therefore, the volume of the $k$th shell is approximately

$$\underbrace{2\pi x_k^*}_{\text{length}} \cdot \underbrace{f(x_k^*)}_{\text{height}} \cdot \underbrace{\Delta x}_{\text{thickness}} = 2\pi x_k^* f(x_k^*)\Delta x.$$

Summing the volumes of the $n$ cylindrical shells gives an approximation to the volume of the entire solid:

$$V \approx \sum_{k=1}^{n} 2\pi x_k^* f(x_k^*)\Delta x.$$

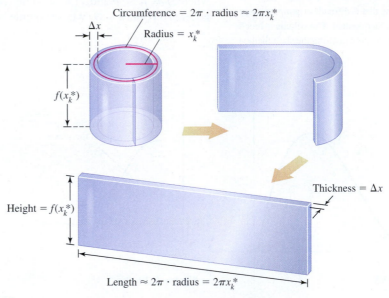

**Figure 6.41**

As $n$ increases and $\Delta x$ approaches 0 (**Figure 6.42**), we obtain the exact volume of the solid as a definite integral:

$$V = \lim_{n \to \infty} \sum_{k=1}^{n} \underbrace{2\pi\, x_k^*}_{\substack{\text{shell} \\ \text{circumference}}} \overbrace{f(x_k^*)}^{\substack{\text{shell} \\ \text{height}}} \underbrace{\Delta x}_{\substack{\text{shell} \\ \text{thickness}}} = \int_a^b 2\pi x f(x)\, dx.$$

➤ Rather than memorizing, think of the meaning of the factors in this formula: $f(x)$ is the height of a single cylindrical shell, $2\pi x$ is the circumference of the shell, and $dx$ corresponds to the thickness of a shell. Therefore, $2\pi x\, f(x)\, dx$ represents the volume of a single shell, and we sum the volumes from $x = a$ to $x = b$. Notice that the integrand for the shell method is the function $A(x)$ that gives the surface area of the shell of radius $x$, for $a \leq x \leq b$.

Increase the number of shells.

$$\text{Volume} = \lim_{n \to \infty} \sum_{k=1}^{n} 2\pi x_k^* f(x_k^*)\Delta x$$
$$= \int_a^b 2\pi x f(x)\, dx$$

$n \to \infty$

**Figure 6.42**

Before doing examples, we generalize this method as we did for the disk method. Suppose the region $R$ is bounded by two curves, $y = f(x)$ and $y = g(x)$, where $f(x) \geq g(x)$ on $[a, b]$ (**Figure 6.43**). What is the volume of the solid generated when $R$ is revolved about the $y$-axis?

Shell height $= f(x_k^*) - g(x_k^*)$

Shell radius $- x_k^*$

Volume of $k$th shell $\approx 2\pi x_k^*(f(x_k^*) - g(x_k^*))\Delta x$

**Figure 6.43**

The situation is similar to the case we just considered. A typical rectangle in $R$ sweeps out a cylindrical shell, but now the height of the $k$th shell is $f(x_k^*) - g(x_k^*)$, for $k = 1, \ldots, n$. As before, we take the radius of the $k$th shell to be $x_k^*$, which means the volume of the $k$th shell is approximated by $2\pi x_k^*(f(x_k^*) - g(x_k^*))\Delta x$ (Figure 6.43). Summing the volumes of all the shells gives an approximation to the volume of the entire solid:

$$V \approx \sum_{k=1}^{n} \underbrace{2\pi x_k^*}_{\substack{\text{shell} \\ \text{circumference}}} \underbrace{(f(x_k^*) - g(x_k^*))}_{\substack{\text{shell} \\ \text{height}}}\Delta x.$$

Taking the limit as $n \to \infty$ (which implies that $\Delta x \to 0$), the exact volume is the definite integral

$$V = \lim_{n \to \infty} \sum_{k=1}^{n} 2\pi x_k^*(f(x_k^*) - g(x_k^*))\Delta x = \int_a^b 2\pi x(f(x) - g(x))\, dx.$$

We now have the formula for the shell method.

> ➤ An analogous formula for the shell method when $R$ is revolved about the $x$-axis is obtained by reversing the roles of $x$ and $y$:
>
> $$V = \int_c^d 2\pi y(p(y) - q(y))\, dy.$$
>
> We assume $R$ is bounded by the curves $x = p(y)$ and $x = q(y)$ where $p(y) \geq q(y)$ on $[c, d]$.

---

**Volume by the Shell Method**

Let $f$ and $g$ be continuous functions with $f(x) \geq g(x)$ on $[a, b]$. If $R$ is the region bounded by the curves $y = f(x)$ and $y = g(x)$ between the lines $x = a$ and $x = b$, the volume of the solid generated when $R$ is revolved about the $y$-axis is

$$V = \int_a^b 2\pi x \underbrace{(f(x) - g(x))}_{\substack{\text{shell} \\ \text{circumference height}}}\, dx.$$

---

Figure 6.44

**EXAMPLE 1    A sine bowl**  Let $R$ be the region bounded by the graph of $f(x) = \sin x^2$, the $x$-axis, and the vertical line $x = \sqrt{\pi/2}$ (**Figure 6.44**). Find the volume of the solid generated when $R$ is revolved about the $y$-axis.

**SOLUTION**  Revolving $R$ about the $y$-axis produces a bowl-shaped region (**Figure 6.45**). The radius of a typical cylindrical shell is $x$ and its height is $f(x) = \sin x^2$. Therefore, the volume by the shell method is

$$V = \int_a^b \underbrace{2\pi x}_{\substack{\text{shell} \\ \text{circumference}}} \underbrace{f(x)}_{\substack{\text{shell} \\ \text{height}}}\, dx = \int_0^{\sqrt{\pi/2}} 2\pi x \sin x^2\, dx.$$

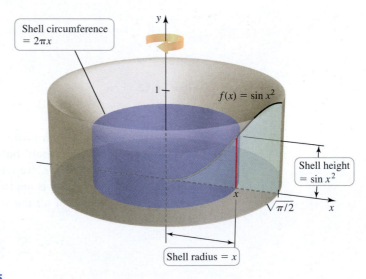

Figure 6.45

➤ When computing volumes using the shell method, it is best to sketch the region $R$ in the $xy$-plane and draw a slice through the region that generates a typical shell.

Now we make the change of variables $u = x^2$, which means that $du = 2x\,dx$. The lower limit $x = 0$ becomes $u = 0$, and the upper limit $x = \sqrt{\pi/2}$ becomes $u = \pi/2$. The volume of the solid is

$$V = \int_0^{\sqrt{\pi/2}} \underbrace{2\pi x}_{\substack{\text{shell}\\\text{circumference}}}\ \underbrace{\sin x^2}_{\substack{\text{shell}\\\text{height}}}\,dx = \pi \int_0^{\pi/2} \sin u\,du \qquad \color{blue}{u = x^2,\, du = 2x\,dx}$$

$$= \pi(-\cos u)\Big|_0^{\pi/2} \qquad \color{blue}{\text{Fundamental Theorem}}$$

$$= \pi(0 - (-1)) = \pi. \qquad \color{blue}{\text{Simplify.}}$$

Related Exercises 10, 17 ◄

**QUICK CHECK 1** The triangle bounded by the $x$-axis, the line $y = 2x$, and the line $x = 1$ is revolved about the $y$-axis. Give an integral that equals the volume of the resulting solid using the shell method. ◄

**EXAMPLE 2** **Shells about the $x$-axis** Let $R$ be the region in the first quadrant bounded by the graph of $y = \sqrt{x - 2}$ and the line $y = 2$. Find the volume of the solid generated when $R$ is revolved about the $x$-axis.

**SOLUTION** The revolution is about the $x$-axis, so the integration in the shell method is with respect to $y$. A typical shell runs parallel to the $x$-axis and has radius $y$, where $0 \le y \le 2$; the shells extend from the $y$-axis to the curve $y = \sqrt{x - 2}$ (**Figure 6.46**). Solving $y = \sqrt{x - 2}$ for $x$, we have $x = y^2 + 2$, which is the height of the shell at the point $y$ (**Figure 6.47**). Integrating with respect to $y$, the volume of the solid is

$$V = \int_0^2 \underbrace{2\pi y}_{\substack{\text{shell}\\\text{circumference}}}\ \underbrace{(y^2 + 2)}_{\substack{\text{shell}\\\text{height}}}\,dy = 2\pi \int_0^2 (y^3 + 2y)\,dy = 16\pi.$$

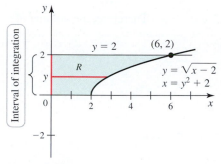

**Figure 6.46**

➤ In Example 2, we could use the disk/washer method to compute the volume, but notice that this approach requires splitting the region into two subregions. A better approach is to use the shell method and integrate along the $y$-axis.

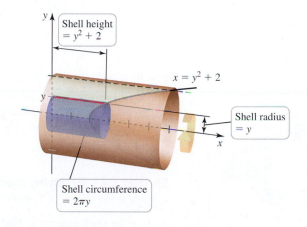

**Figure 6.47**

Related Exercises 13, 21 ◄

**EXAMPLE 3** **Volume of a drilled sphere** A cylindrical hole with radius $r$ is drilled symmetrically through the center of a sphere with radius $a$, where $0 \le r \le a$. What is the volume of the remaining material?

**SOLUTION** The $y$-axis is chosen to coincide with the axis of the cylindrical hole. We let $R$ be the region in the $xy$-plane bounded above by $f(x) = \sqrt{a^2 - x^2}$, the upper half of a circle of radius $a$, and bounded below by $g(x) = -\sqrt{a^2 - x^2}$, the lower half of a circle of radius $a$, for $r \le x \le a$ (**Figure 6.48a**). Slices are taken perpendicular to the $x$-axis from $x = r$ to $x = a$. When a slice is revolved about the $y$-axis, it sweeps out a cylindrical shell that is concentric with the hole through the sphere (**Figure 6.48b**). The radius of a

**Figure 6.48**

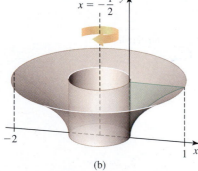

typical shell is $x$ and its height is $f(x) - g(x) = 2\sqrt{a^2 - x^2}$. Therefore, the volume of the material that remains in the sphere is

$$
V = \int_r^a \underbrace{2\pi x}_{\substack{\text{shell} \\ \text{circumference}}} \underbrace{(2\sqrt{a^2 - x^2})}_{\substack{\text{shell} \\ \text{height}}} dx
$$

$$
= -2\pi \int_{a^2-r^2}^{0} \sqrt{u}\, du \quad u = a^2 - x^2,\, du = -2x\, dx
$$

$$
= 2\pi \left( \frac{2}{3} u^{3/2} \right) \Big|_0^{a^2-r^2} \quad \textcolor{blue}{\text{Fundamental Theorem}}
$$

$$
= \frac{4\pi}{3} (a^2 - r^2)^{3/2}. \quad \textcolor{blue}{\text{Simplify.}}
$$

It is important to check the result by examining special cases. In the case that $r = a$ (the radius of the hole equals the radius of the sphere), our calculation gives a volume of 0, which is correct. In the case that $r = 0$ (no hole in the sphere), our calculation gives the correct volume of a sphere, $\frac{4}{3}\pi a^3$.

*Related Exercises 62–63* ◄

**EXAMPLE 4    Revolving about other lines** Let $R$ be the region bounded by the curve $y = \sqrt{x}$, the line $y = 1$, and the $y$-axis (**Figure 6.49a**).

**a.** Use the shell method to find the volume of the solid generated when $R$ is revolved about the line $x = -\frac{1}{2}$ (**Figure 6.49b**).

**b.** Use the disk/washer method to find the volume of the solid generated when $R$ is revolved about the line $y = 1$ (**Figure 6.49c**).

**SOLUTION**

**a.** Using the shell method, we must imagine taking slices through $R$ parallel to the $y$-axis. A typical slice through $R$ at a point $x$, where $0 \leq x \leq 1$, has height $1 - \sqrt{x}$. When that slice is revolved about the line $x = -\frac{1}{2}$, it sweeps out a cylindrical shell with a radius of $x + \frac{1}{2}$ and a height of $1 - \sqrt{x}$ (**Figure 6.50**). A slight modification of the standard shell method gives the volume of the solid:

**Figure 6.49**

$$
\int_0^1 \underbrace{2\pi \left( x + \frac{1}{2} \right)}_{\substack{\text{shell} \\ \text{circumference}}} \underbrace{(1 - \sqrt{x})}_{\substack{\text{shell} \\ \text{height}}} dx = 2\pi \int_0^1 \left( x - x^{3/2} + \frac{1}{2} - \frac{x^{1/2}}{2} \right) dx \quad \textcolor{blue}{\substack{\text{Expand} \\ \text{integrand.}}}
$$

$$
= 2\pi \left( \frac{1}{2} x^2 - \frac{2}{5} x^{5/2} + \frac{1}{2} x - \frac{1}{3} x^{3/2} \right) \Big|_0^1 = \frac{8\pi}{15}. \quad \textcolor{blue}{\substack{\text{Evaluate} \\ \text{integral.}}}
$$

➤ If we instead revolved about the *y*-axis
($x = 0$), the radius of the shell would be *x*.
Because we are revolving about the line
$x = -\frac{1}{2}$, the radius of the shell is $x + \frac{1}{2}$.

**Figure 6.50**

**b.** Using the disk/washer method, we take slices through *R* parallel to the *y*-axis. Consider a typical slice at a point *x*, where $0 \le x \le 1$. Its length, now measured with respect to the line $y = 1$, is $1 - \sqrt{x}$. When that slice is revolved about the line $y = 1$, it sweeps out a disk of radius $1 - \sqrt{x}$ (**Figure 6.51**). By the disk/washer method, the volume of the solid is

➤ The disk/washer method can also be
used for part (a), and the shell method
can also be used for part (b).

$$\int_0^1 \pi \underbrace{(1 - \sqrt{x})^2}_{\text{disk radius}} dx = \pi \int_0^1 (1 - 2\sqrt{x} + x)\, dx \quad \text{\color{blue}Expand integrand.}$$

$$= \pi \left( x - \frac{4}{3}x^{3/2} + \frac{1}{2}x^2 \right)\Big|_0^1 \quad \text{\color{blue}Evaluate integral.}$$

$$= \frac{\pi}{6}.$$

**Figure 6.51**

**QUICK CHECK 2** Write the volume integral in Example 4b in the case that *R* is revolved about the line $y = -5$. ◄

*Related Exercises 32–33* ◄

## Restoring Order

After working with slices, disks, washers, and shells, you may feel somewhat overwhelmed. How do you choose a method, and which method is best?

Notice that the disk method is just a special case of the washer method. So for solids of revolution, the choice is between the washer method and the shell method. In *principle*, either method can be used. In *practice*, one method usually produces an integral that is easier to evaluate than the other method. The following table summarizes these methods.

| SUMMARY    Disk/Washer and Shell Methods | |
|---|---|
| **Integration with respect to $x$**  | **Disk/washer method about the $x$-axis** <br> Disks/washers are *perpendicular* to the $x$-axis. <br><br> $$\int_a^b \pi \underbrace{(f(x)^2}_{\substack{\text{outer} \\ \text{radius}}} - \underbrace{g(x)^2)}_{\substack{\text{inner} \\ \text{radius}}}\, dx$$ |
|  | **Shell method about the $y$-axis** <br> Shells are *parallel* to the $y$-axis. <br><br> $$\int_a^b 2\pi \underbrace{x}_{\substack{\text{shell} \\ \text{circumference}}} \underbrace{(f(x) - g(x))}_{\substack{\text{shell} \\ \text{height}}}\, dx$$ |
| **Integration with respect to $y$**  | **Disk/washer method about the $y$-axis** <br> Disks/washers are *perpendicular* to the $y$-axis. <br><br> $$\int_c^d \pi \underbrace{(p(y)^2}_{\substack{\text{outer} \\ \text{radius}}} - \underbrace{q(y)^2)}_{\substack{\text{inner} \\ \text{radius}}}\, dy$$ |
| | **Shell method about the $x$-axis** <br> Shells are *parallel* to the $x$-axis. <br><br> $$\int_c^d 2\pi \underbrace{y}_{\substack{\text{shell} \\ \text{circumference}}} \underbrace{(p(y) - q(y))}_{\substack{\text{shell} \\ \text{height}}}\, dy$$ |

**Figure 6.52**

The following example shows that while two methods may be used on the same problem, one of them may be preferable.

**EXAMPLE 5    Volume by which method?** The region $R$ is bounded by the graphs of $f(x) = 2x - x^2$ and $g(x) = x$ on the interval $[0, 1]$ (**Figure 6.52**). Use the washer method and the shell method to find the volume of the solid formed when $R$ is revolved about the $x$-axis.

**SOLUTION** Solving $f(x) = g(x)$, we find that the curves intersect at the points $(0, 0)$ and $(1, 1)$. Using the washer method, the upper bounding curve is the graph of $f$, the lower bounding curve is the graph of $g$, and a typical washer is perpendicular to the $x$-axis (**Figure 6.53**). Therefore, the volume is

$$V = \int_0^1 \pi \left( \underbrace{(2x - x^2)^2}_{\substack{\text{outer}\\\text{radius}}} - \underbrace{x^2}_{\substack{\text{inner}\\\text{radius}}} \right) dx \qquad \text{Washer method}$$

$$= \pi \int_0^1 (x^4 - 4x^3 + 3x^2)\, dx \qquad \text{Expand integrand.}$$

$$= \pi \left( \frac{x^5}{5} - x^4 + x^3 \right)\Big|_0^1 = \frac{\pi}{5}. \qquad \text{Evaluate integral.}$$

▶ To solve $y = 2x - x^2$ for $x$, write the equation as $x^2 - 2x + y = 0$ and complete the square or use the quadratic formula.

The shell method requires expressing the bounding curves in the form $x = p(y)$ for the right curve and $x = q(y)$ for the left curve. The right curve is $x = y$. Solving $y = 2x - x^2$ for $x$, we find that $x = 1 - \sqrt{1 - y}$ describes the left curve. A typical shell is parallel to the $x$-axis (**Figure 6.54**). Therefore, the volume is

$$V = \int_0^1 \underbrace{2\pi y}_{\substack{\text{shell}\\\text{circumference}}} \underbrace{(y - (1 - \sqrt{1 - y}))}_{\substack{\text{shell}\\\text{height}}} dy.$$

This integral equals $\frac{\pi}{5}$, but it is more difficult to evaluate than the integral required by the washer method. In this case, the washer method is preferable. Of course, the shell method may be preferable for other problems.

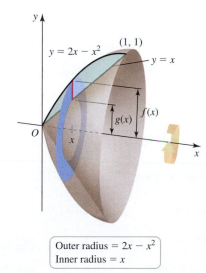

Outer radius = $2x - x^2$
Inner radius = $x$

**Figure 6.53**

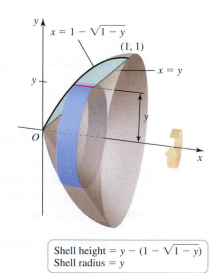

Shell height = $y - (1 - \sqrt{1 - y})$
Shell radius = $y$

**Figure 6.54**

**QUICK CHECK 3** Suppose the region in Example 5 is revolved about the $y$-axis. Which method (washer or shell) leads to an easier integral? ◀

*Related Exercises 49–50* ◀

---

## SECTION 6.4 EXERCISES

### Getting Started

1. Assume $f$ and $g$ are continuous, with $f(x) \geq g(x) \geq 0$ on $[a, b]$. The region bounded by the graphs of $f$ and $g$ and the lines $x = a$ and $x = b$ is revolved about the $y$-axis. Write the integral given by the shell method that equals the volume of the resulting solid.

2. Fill in the blanks: A region $R$ is revolved about the $y$-axis. The volume of the resulting solid could (in principle) be found by using the disk/washer method and integrating with respect to _____ or using the shell method and integrating with respect to _____.

3. Fill in the blanks: A region $R$ is revolved about the $x$-axis. The volume of the resulting solid could (in principle) be found by using the disk/washer method and integrating with respect to _____ or using the shell method and integrating with respect to _____.

4. Look again at the region $R$ in Figure 6.38 (p. 387). Explain why it would be difficult to use the washer method to find the volume of the solid of revolution that results when $R$ is revolved about the $y$-axis.

**5.** Let $R$ be the region in the first quadrant bounded above by the curve $y = 2 - x^2$ and bounded below by the line $y = x$. Suppose the shell method is used to determine the volume of the solid generated by revolving $R$ about the $y$-axis.

   **a.** What is the radius of a cylindrical shell at a point $x$ in $[0, 2]$?

   **b.** What is the height of a cylindrical shell at a point $x$ in $[0, 2]$?

   **c.** Write an integral for the volume of the solid using the shell method.

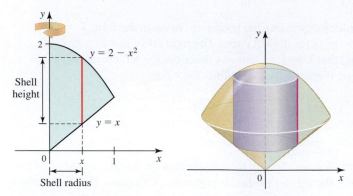

**6–8.** *Let $R$ be the region bounded by the curves $y = 2 - \sqrt{x}, y = 2$, and $x = 4$ in the first quadrant.*

**6.** Suppose the shell method is used to determine the volume of the solid generated by revolving $R$ about the $x$-axis.

   **a.** What is the radius of a cylindrical shell at a point $y$ in $[0, 2]$?

   **b.** What is the height of a cylindrical shell at a point $y$ in $[0, 2]$?

   **c.** Write an integral for the volume of the solid using the shell method.

**7.** Suppose the shell method is used to determine the volume of the solid generated by revolving $R$ about the line $y = 2$.

   **a.** What is the radius of a cylindrical shell at a point $y$ in $[0, 2]$?

   **b.** What is the height of a cylindrical shell at a point $y$ in $[0, 2]$?

   **c.** Write an integral for the volume of the solid using the shell method.

**8.** Suppose the shell method is used to determine the volume of the solid generated by revolving $R$ about the line $x = 4$.

   **a.** What is the radius of a cylindrical shell at a point $x$ in $[0, 4]$?

   **b.** What is the height of a cylindrical shell at a point $x$ in $[0, 4]$?

   **c.** Write an integral for the volume of the solid using the shell method.

## Practice Exercises

**9–30. Shell method** *Let $R$ be the region bounded by the following curves. Use the shell method to find the volume of the solid generated when $R$ is revolved about indicated axis.*

**9.** $y = x - x^2$ and $y = 0$; about the $y$-axis

**10.** $y = (1 + x^2)^{-2}, y = 0, x = 0$, and $x = 1$; about the $y$-axis

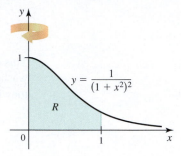

**11.** $y = 3x, y = 3$, and $x = 0$; about the $y$-axis (Use integration and check your answer using the volume formula for a cone.)

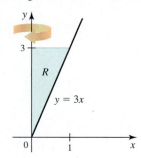

**12.** $y = -x^2 + 4x + 2$ and $y = x^2 - 6x + 10$; about the $y$-axis

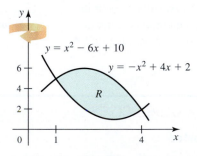

**13.** $y = \sqrt{x}, y = 0$, and $x = 4$; about the $x$-axis

**14.** $y = \sqrt{x}$, $y = 2 - x$, and $y = 0$; about the $x$-axis

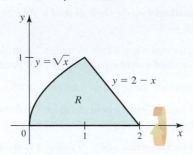

**15.** $y = 4 - x$, $y = 2$, and $x = 0$; about the $x$-axis

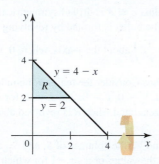

**16.** $x = \dfrac{1}{2\sqrt{y}}$, $x = \dfrac{1}{2}$ and $y = \dfrac{1}{4}$; about the $x$-axis

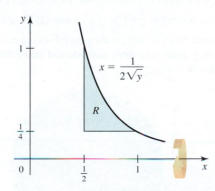

**17.** $y = \cos x^2$ and $y = 0$, for $0 \le x \le \sqrt{\pi/2}$; about the $y$-axis

**18.** $y = 6 - x$, $y = 0$, $x = 2$, and $x = 4$; about the $y$-axis

**19.** $y = 1 - x^2$, $x = 0$, and $y = 0$, in the first quadrant; about the $y$-axis

**20.** $y = \sqrt{x}$, $y = 0$, and $x = 1$; about the $y$-axis

**21.** $x = y^2$, $x = 0$, and $y = 3$; about the $x$-axis

**22.** $y = x^3$, $y = 1$, and $x = 0$; about the $x$-axis

**23.** $y = x$, $y = 2 - x$, and $y = 0$; about the $x$-axis

**24.** $y = \sqrt{4 - 2x^2}$, $y = 0$, and $x = 0$, in the first quadrant; about the $y$-axis

**T 25.** $y = \dfrac{1}{(x^2 + 1)^2}$, $y = 0$, $x = 1$, and $x = 2$; about the $y$-axis

**26.** $x = y^2$, $x = 4$, and $y = 0$; about the $x$-axis

**T 27.** $y = 2x^{-3/2}$, $y = 2$, $y = 16$, and $x = 0$; about the $x$-axis

**T 28.** $y = (1 - x)^{-1/2}$, $y = 1$, $y = 2$, and $x = 0$; about the $y$-axis

**29.** $y = x^3 - x^8 + 1$ and $y = 1$; about the $y$-axis

**T 30.** $y = \sqrt{50 - 2x^2}$, in the first quadrant; about the $x$-axis

**31–34. Shell and washer methods** *Let R be the region bounded by the following curves. Use both the shell method and the washer method to find the volume of the solid generated when R is revolved about the indicated axis.*

**31.** $y = x$ and $y = x^{1/3}$, in the first quadrant; about the $x$-axis

**32.** $y = \dfrac{1}{x + 1}$ and $y = 1 - \dfrac{x}{3}$; about the $x$-axis

**33.** $y = (x - 2)^3 - 2$, $x = 0$, and $y = 25$; about the $y$-axis

**34.** $y = 8$, $y = 2x + 2$, $x = 0$, and $x = 2$; about the $y$-axis

**35–40. Shell method about other lines** *Let R be the region bounded by $y = x^2$, $x = 1$, and $y = 0$. Use the shell method to find the volume of the solid generated when R is revolved about the following lines.*

**35.** $x = -2$      **36.** $x = 1$      **37.** $x = 2$

**38.** $y = 1$      **39.** $y = -2$      **40.** $y = 2$

**41–44. Shell and washer methods about other lines** *Use both the shell method and the washer method to find the volume of the solid that is generated when the region in the first quadrant bounded by $y = x^2$, $y = 1$, and $x = 0$ is revolved about the following lines.*

**41.** $y = -2$    **42.** $x = -1$    **43.** $y = 6$    **44.** $x = 2$

**45. Volume of a sphere** Let $R$ be the region bounded by the upper half of the circle $x^2 + y^2 = r^2$ and the $x$-axis. A sphere of radius $r$ is obtained by revolving $R$ about the $x$-axis.

   **a.** Use the shell method to verify that the volume of a sphere of radius $r$ is $\frac{4}{3}\pi r^3$.

   **b.** Repeat part (a) using the disk method.

**46. Comparing American and rugby union footballs** An ellipse centered at the origin is described by the equation $\dfrac{x^2}{a^2} + \dfrac{y^2}{b^2} = 1$, where $a$ and $b$ are positive constants. If the upper half of such an ellipse is revolved about the $x$-axis, the resulting surface is an *ellipsoid*.

   **a.** Use the washer method to find the volume of an ellipsoid (in terms of $a$ and $b$). Check your work using the shell method.

   **b.** Both American and rugby union footballs have the shape of ellipsoids. The maximum regulation size of a rugby union football corresponds to parameters of $a = 6$ in and $b = 3.82$ in, and the maximum regulation size of an American football corresponds to parameters of $a = 5.62$ in and $b = 3.38$ in. Find the volume of each football.

   **c.** Fill in the blank: At their maximum regulation sizes, the volume of a rugby union football has approximately _____ times the volume of an American football.

**47. A torus (doughnut)** A torus is formed when a circle of radius 2 centered at $(3, 0)$ is revolved about the $y$-axis.

    **a.** Use the shell method to write an integral for the volume of the torus.

    **b.** Use the washer method to write an integral for the volume of the torus.

    **c.** Find the volume of the torus by evaluating one of the two integrals obtained in parts (a) and (b). (*Hint:* Both integrals can be evaluated without using the Fundamental Theorem of Calculus.)

**48. A cone by two methods** Verify that the volume of a right circular cone with a base radius of $r$ and a height of $h$ is $\pi r^2 h/3$. Use the region bounded by the line $y = rx/h$, the $x$-axis, and the line $x = h$, where the region is rotated around the $x$-axis. Then (a) use the disk method and integrate with respect to $x$, and (b) use the shell method and integrate with respect to $y$.

**49–58. Choose your method** *Let R be the region bounded by the following curves. Use the method of your choice to find the volume of the solid generated when R is revolved about the given axis.*

**49.** $y = x - x^4$ and $y = 0$; about the $x$-axis

**50.** $y = x - x^4$ and $y = 0$; about the $y$-axis

**51.** $y = x^2$ and $y = 2 - x^2$; about the $x$-axis

**52.** $y = \sin x$ and $y = 1 - \sin x$, for $\pi/6 \le x \le 5\pi/6$; about the $x$-axis

**53.** $y = x, y = 2x + 2, x = 2$, and $x = 6$; about the $x$-axis

**54.** $y = x^3, y = 0$, and $x = 2$; about the $x$-axis

**55.** $y = 4 - x^2$ and $y = 2 - x$, in the first quadrant; about the $y$-axis

**56.** $y = 2, y = 2x + 2$, and $x = 6$; about the $y$-axis

**57.** $y = x^2, y = 2 - x$, and $x = 0$, in the first quadrant; about the $y$-axis

**58.** $y = \sqrt{x}$, the $x$-axis, and $x = 4$; about the $x$-axis

**59. Explain why or why not** Determine whether the following statements are true and give an explanation or counterexample.

    **a.** When using the shell method, the axis of the cylindrical shells is parallel to the axis of revolution.

    **b.** If a region is revolved about the $y$-axis, then the shell method must be used.

    **c.** If a region is revolved about the $x$-axis, then in principle, it is possible to use the disk/washer method and integrate with respect to $x$ or to use the shell method and integrate with respect to $y$.

## Explorations and Challenges

**60–64. Shell method** *Use the shell method to find the volume of the following solids.*

**60.** The solid formed when a hole of radius 2 is drilled symmetrically along the axis of a right circular cylinder of height 6 and radius 4

**61.** A right circular cone of radius 3 and height 8

**62.** The solid formed when a hole of radius 3 is drilled symmetrically through the center of a sphere of radius 6

**63.** The solid formed when a hole of radius 3 is drilled symmetrically along the axis of a right circular cone of radius 6 and height 9

**64.** A hole of radius $r \le R$ is drilled symmetrically along the axis of a bullet. The bullet is formed by revolving the parabola $y = 6\left(1 - \dfrac{x^2}{R^2}\right)$ about the $y$-axis, where $0 \le x \le R$.

**T 65. Equal volumes** Consider the region $R$ bounded by the curves $y = ax^2 + 1, y = 0, x = 0$, and $x = 1$, for $a \ge -1$. Let $S_1$ and $S_2$ be solids generated when $R$ is revolved about the $x$- and $y$-axes, respectively.

    **a.** Find $V_1$ and $V_2$, the volumes of $S_1$ and $S_2$, as functions of $a$.

    **b.** What are the values of $a \ge -1$ for which $V_1(a) = V_2(a)$?

**66. A spherical cap** Consider the cap of thickness $h$ that has been sliced from a sphere of radius $r$ (see figure). Verify that the volume of the cap is $\dfrac{\pi}{3}h^2(3r - h)$ using (a) the washer method, (b) the shell method, and (c) the general slicing method. Check for consistency among the three methods and check the special cases $h = r$ and $h = 0$.

**67. Change of variables** Suppose $f(x) > 0$, for all $x$, and $\int_0^4 f(x)\,dx = 10$. Let $R$ be the region in the first quadrant bounded by the coordinate axes, $y = f(x^2)$, and $x = 2$. Find the volume of the solid generated by revolving $R$ around the $y$-axis.

**68. Equal integrals** Without evaluating integrals, explain the following equalities. (*Hint:* Draw pictures.)

    **a.** $\pi \displaystyle\int_0^4 (8 - 2x)^2\,dx = 2\pi \int_0^8 y\left(4 - \dfrac{y}{2}\right)dy$

    **b.** $\displaystyle\int_0^2 (25 - (x^2 + 1)^2)\,dx = 2\int_1^5 y\sqrt{y - 1}\,dy$

**69. Volumes without calculus** Solve the following problems with *and* without calculus. A good picture helps.

    **a.** A cube with side length $r$ is inscribed in a sphere, which is inscribed in a right circular cone, which is inscribed in a right circular cylinder. The side length (slant height) of the cone is equal to its diameter. What is the volume of the cylinder?

**b.** A cube is inscribed in a right circular cone with a radius of 1 and a height of 3. What is the volume of the cube?

**c.** A cylindrical hole 10 in long is drilled symmetrically through the center of a sphere. How much material is left in the sphere? (Enough information *is* given).

**70. Wedge from a tree** Imagine a cylindrical tree of radius $a$. A wedge is cut from the tree by making two cuts: one in a horizontal plane $P$ perpendicular to the axis of the cylinder, and one that makes an angle $\theta$ with $P$, intersecting $P$ along a diameter of the tree (see figure). What is the volume of the wedge?

**71. Different axes of revolution** Suppose $R$ is the region bounded by $y = f(x)$ and $y = g(x)$ on the interval $[a, b]$, where $f(x) \geq g(x) \geq 0$.

**a.** Show that if $R$ is revolved about the horizontal line $y = y_0$ that lies below $R$, then by the washer method, the volume of the resulting solid is

$$V = \int_a^b \pi\left((f(x) - y_0)^2 - (g(x) - y_0)^2\right) dx.$$

**b.** How is this formula changed if the line $y = y_0$ lies above $R$?

**72. Different axes of revolution** Suppose $R$ is the region bounded by $y = f(x)$ and $y = g(x)$ on the interval $[a, b]$, where $f(x) \geq g(x)$.

**a.** Show that if $R$ is revolved about the vertical line $x = x_0$, where $x_0 < a$, then by the shell method, the volume of the resulting solid is $V = \int_a^b 2\pi(x - x_0)(f(x) - g(x)) dx$.

**b.** How is this formula changed if $x_0 > b$?

**QUICK CHECK ANSWERS**

**1.** $\int_0^1 2\pi x(2x)\, dx$  **2.** $V = \int_0^1 \pi(36 - (\sqrt{x} + 5)^2)\, dx$
**3.** The shell method is easier. ◄

# 6.5 Length of Curves

The space station orbits Earth in an elliptical path. How far does it travel in one orbit? A baseball slugger launches a home run into the upper deck and the sportscaster claims it landed 480 feet from home plate. But how far did the ball actually travel along its flight path? These questions deal with the length of trajectories or, more generally, with *arc length*. As you will see, their answers can be found by integration.

There are two common ways to formulate problems about arc length: The curve may be given explicitly in the form $y = f(x)$ or it may be defined *parametrically*. In this section, we deal with the first case. Parametric curves and the associated arc length problem are discussed in Section 12.1.

## Arc Length for $y = f(x)$

Suppose a curve is given by $y = f(x)$, where $f$ is a function with a continuous first derivative on the interval $[a, b]$. The goal is to determine how far you would travel if you walked along the curve from $(a, f(a))$ to $(b, f(b))$. This distance is the arc length, which we denote $L$.

As shown in **Figure 6.55**, we divide $[a, b]$ into $n$ subintervals of length $\Delta x = (b - a)/n$, where $x_k$ is the right endpoint of the $k$th subinterval, for $k = 1, \ldots, n$.

The length of the red polygonal line from $(a, f(a))$ to $(b, f(b))$ approximates $L$.

**Figure 6.55**

➤ More generally, we may choose any point in the $k$th subinterval and $\Delta x$ may vary from one subinterval to the next. Using right endpoints, as we do here, simplifies the discussion and leads to the same result.

Joining the corresponding points on the curve by line segments, we obtain a polygonal line with $n$ line segments. If $n$ is large and $\Delta x$ is small, the length of the polygonal line is a good approximation to the length of the actual curve. The strategy is to find the length of the polygonal line and then let $n$ increase, while $\Delta x$ goes to zero, to get the exact length of the curve.

Consider the $k$th subinterval $[x_{k-1}, x_k]$ and the line segment between the points $(x_{k-1}, f(x_{k-1}))$ and $(x_k, f(x_k))$. We let the change in the $y$-coordinate between these points be

$$\Delta y_k = f(x_k) - f(x_{k-1}).$$

➤ Notice that $\Delta x$ is the same for each subinterval, but $\Delta y_k$ depends on the subinterval.

The $k$th line segment is the hypotenuse of a right triangle with sides of length $\Delta x$ and $|\Delta y_k| = |f(x_k) - f(x_{k-1})|$. The length of each line segment is

$$\sqrt{(\Delta x)^2 + |\Delta y_k|^2}, \quad \text{for} \quad k = 1, 2, \ldots, n.$$

Summing these lengths, we obtain the length of the polygonal line, which approximates the length $L$ of the curve:

$$L \approx \sum_{k=1}^{n} \sqrt{(\Delta x)^2 + |\Delta y_k|^2}.$$

Mean Value Theorem

In previous applications of the integral, we would, at this point, take the limit as $n \to \infty$ and $\Delta x \to 0$ to obtain a definite integral. However, because of the presence of the $\Delta y_k$ term, we must complete one additional step before taking a limit. Notice that the slope of the line segment on the $k$th subinterval is $\Delta y_k / \Delta x$ (rise over run). By the Mean Value Theorem (see the margin figure and Section 4.2), this slope equals $f'(x_k^*)$ for some point $x_k^*$ on the $k$th subinterval. Therefore,

$$
\begin{aligned}
L &\approx \sum_{k=1}^{n} \sqrt{(\Delta x)^2 + |\Delta y_k|^2} \\
&= \sum_{k=1}^{n} \sqrt{(\Delta x)^2 \left(1 + \left(\frac{\Delta y_k}{\Delta x}\right)^2\right)} && \text{Factor out } (\Delta x)^2. \\
&= \sum_{k=1}^{n} \sqrt{1 + \left(\frac{\Delta y_k}{\Delta x}\right)^2}\, \Delta x && \text{Bring } \Delta x \text{ out of the square root.} \\
&= \sum_{k=1}^{n} \sqrt{1 + f'(x_k^*)^2}\, \Delta x. && \text{Mean Value Theorem}
\end{aligned}
$$

Now we have a Riemann sum. As $n$ increases and as $\Delta x$ approaches zero, the sum approaches a definite integral, which is also the exact length of the curve. We have

$$L = \lim_{n \to \infty} \sum_{k=1}^{n} \sqrt{1 + f'(x_k^*)^2}\, \Delta x = \int_a^b \sqrt{1 + f'(x)^2}\, dx.$$

➤ Note that $1 + f'(x)^2$ is positive, so the square root in the integrand is defined whenever $f'$ exists. To ensure that $\sqrt{1 + f'(x)^2}$ is integrable on $[a, b]$, we require that $f'$ be continuous.

**QUICK CHECK 1** What does the arc length formula give for the length of the line $y = x$ between $x = 0$ and $x = a$, where $a \geq 0$? ◄

---

**DEFINITION  Arc Length for $y = f(x)$**

Let $f$ have a continuous first derivative on the interval $[a, b]$. The length of the curve from $(a, f(a))$ to $(b, f(b))$ is

$$L = \int_a^b \sqrt{1 + f'(x)^2}\, dx.$$

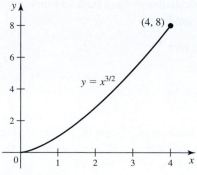

**Figure 6.56**

**EXAMPLE 1**   **Arc length**   Find the length of the curve $f(x) = x^{3/2}$ between $x = 0$ and $x = 4$ (**Figure 6.56**).

**SOLUTION**   Notice that $f'(x) = \frac{3}{2} x^{1/2}$, which is continuous on the interval $[0, 4]$. Using the arc length formula, we have

$$L = \int_a^b \sqrt{1 + f'(x)^2}\, dx = \int_0^4 \sqrt{1 + \left(\frac{3}{2} x^{1/2}\right)^2}\, dx \qquad \text{Substitute for } f'(x).$$

$$= \int_0^4 \sqrt{1 + \frac{9}{4} x}\, dx \qquad \text{Simplify.}$$

$$= \frac{4}{9} \int_1^{10} \sqrt{u}\, du \qquad u = 1 + \frac{9x}{4}, du = \frac{9}{4} dx$$

$$= \frac{4}{9} \left(\frac{2}{3} u^{3/2}\right)\Big|_1^{10} \qquad \text{Fundamental Theorem}$$

$$= \frac{8}{27} \left(10^{3/2} - 1\right). \qquad \text{Simplify.}$$

The length of the curve is $\frac{8}{27} \left(10^{3/2} - 1\right) \approx 9.1$ units.

*Related Exercise 11* ◄

**Figure 6.57**

**EXAMPLE 2**   **Arc length calculation**   Find the length of the curve $f(x) = x^3 + \dfrac{1}{12x}$ on the interval $\left[\frac{1}{2}, 2\right]$ (**Figure 6.57**).

**SOLUTION**   We first calculate $f'(x) = 3x^2 - \dfrac{1}{12x^2}$ and $f'(x)^2 = 9x^4 - \dfrac{1}{2} + \dfrac{1}{144x^4}$. The length of the curve on $\left[\frac{1}{2}, 2\right]$ is

$$L = \int_{1/2}^2 \sqrt{1 + f'(x)^2}\, dx = \int_{1/2}^2 \sqrt{1 + \left(9x^4 - \frac{1}{2} + \frac{1}{144x^4}\right)}\, dx \qquad \text{Substitute.}$$

$$= \int_{1/2}^2 \sqrt{\left(3x^2 + \frac{1}{12x^2}\right)^2}\, dx \qquad \text{Factor.}$$

$$= \int_{1/2}^2 \left(3x^2 + \frac{1}{12x^2}\right) dx \qquad \text{Simplify.}$$

$$= \left(x^3 - \frac{1}{12x}\right)\Big|_{1/2}^2 = 8. \qquad \begin{array}{l}\text{Evaluate the}\\\text{integral.}\end{array}$$

*Related Exercise 10* ◄

**EXAMPLE 3**   **Looking ahead**   Consider the segment of the parabola $f(x) = x^2$ on the interval $[0, 2]$.

**a.** Write the integral for the length of the curve.

**b.** Use a calculator to evaluate the integral.

**SOLUTION**

**a.** Noting that $f'(x) = 2x$, the arc length integral is

$$\int_0^2 \sqrt{1 + f'(x)^2}\, dx = \int_0^2 \sqrt{1 + 4x^2}\, dx.$$

**b.** Using integration techniques presented so far, this integral cannot be evaluated (the required method is given in Section 8.4). This is typical of arc length integrals—even simple functions can lead to arc length integrals that are difficult to evaluate analytically. Without an analytical method, we may use numerical integration to *approximate*

> ► When relying on technology, it is a good idea to check whether an answer is plausible. In Example 3, we found that the arc length of $y = x^2$ on $[0, 2]$ is approximately 4.647. The straight-line distance between $(0, 0)$ and $(2, 4)$ is $\sqrt{20} \approx 4.472$, so our answer is reasonable.

the value of a definite integral (Section 8.8). Many calculators have built-in functions for this purpose. For this integral, the approximate arc length is

$$\int_0^2 \sqrt{1 + 4x^2}\, dx \approx 4.647.$$

*Related Exercises 21, 24* ◄

## Arc Length for $x = g(y)$

Sometimes it is advantageous to describe a curve as a function of $y$—that is, $x = g(y)$. The arc length formula in this case is derived exactly as in the case of $y = f(x)$, switching the roles of $x$ and $y$. The result is the following arc length formula.

**QUICK CHECK 2** What does the arc length formula give for the length of the line $x = y$ between $y = c$ and $y = d$, where $d \geq c$? Is the result consistent with the result given by the Pythagorean theorem? ◄

> **DEFINITION   Arc Length for $x = g(y)$**
>
> Let $x = g(y)$ have a continuous first derivative on the interval $[c, d]$. The length of the curve from $(g(c), c)$ to $(g(d), d)$ is
>
> $$L = \int_c^d \sqrt{1 + g'(y)^2}\, dy.$$

**Figure 6.58**

**EXAMPLE 4    Arc length**  Find the length of the curve $y = f(x) = x^{2/3}$ between $x = 0$ and $x = 8$ (**Figure 6.58**).

**SOLUTION**  The derivative of $f(x) = x^{2/3}$ is $f'(x) = \frac{2}{3} x^{-1/3}$, which is undefined at $x = 0$. Therefore, the arc length formula with respect to $x$ cannot be used, yet the curve certainly appears to have a well-defined length.

The key is to describe the curve with $y$ as the independent variable. Solving $y = x^{2/3}$ for $x$, we have $x = g(y) = \pm y^{3/2}$. Notice that when $x = 8$, $y = 8^{2/3} = 4$, which says that we should use the positive branch of $\pm y^{3/2}$. Therefore, finding the length of the curve $y = f(x) = x^{2/3}$ from $x = 0$ to $x = 8$ is equivalent to finding the length of the curve $x = g(y) = y^{3/2}$ from $y = 0$ to $y = 4$. This is precisely the problem solved in Example 1. The arc length is $\frac{8}{27}(10^{3/2} - 1) \approx 9.1$ units.

*Related Exercises 18–19* ◄

**QUICK CHECK 3** Write the integral for the length of the curve $x = \sin y$ on the interval $0 \leq y \leq \pi$. ◄

## SECTION 6.5 EXERCISES

### Getting Started

1. Explain the steps required to find the length of a curve $y = f(x)$ between $x = a$ and $x = b$.

2. Explain the steps required to find the length of a curve $x = g(y)$ between $y = c$ and $y = d$.

**3–6. Setting up arc length integrals** *Write and simplify, but do not evaluate, an integral with respect to x that gives the length of the following curves on the given interval.*

3. $y = x^3 + 2$ on $[-2, 5]$

4. $y = 2 \cos 3x$ on $[-\pi, \pi]$

5. $y = \dfrac{1}{x^2}$ on $[1, 2]$

6. $y = \sin \pi x$ on $[0, 1]$

7. Find the arc length of the line $y = 2x + 1$ on $[1, 5]$ using calculus and verify your answer using geometry.

8. Find the arc length of the line $y = 4 - 3x$ on $[-3, 2]$ using calculus and verify your answer using geometry.

### Practice Exercises

**9–20. Arc length calculations** *Find the arc length of the following curves on the given interval.*

9. $y = -8x - 3$ on $[-2, 6]$ (Use calculus.)

10. $y = \dfrac{x^3}{3} + \dfrac{1}{4x}$ on $[1, 5]$       11. $y = \dfrac{x^{3/2}}{3}$ on $[0, 60]$

12. $y = \dfrac{3}{10} x^{1/3} - \dfrac{3}{2} x^{5/3}$ on $[1, 3]$

13. $y = \dfrac{(x^2 + 2)^{3/2}}{3}$ on $[0, 1]$       14. $y = \dfrac{x^{3/2}}{3} - x^{1/2}$ on $[4, 16]$

**15.** $y = \dfrac{x^4}{4} + \dfrac{1}{8x^2}$ on $[1, 2]$     **16.** $y = \dfrac{2}{3}x^{3/2} - \dfrac{1}{2}x^{1/2}$ on $[1, 9]$

**17.** $x = \dfrac{(y^2 - 2)^{3/2}}{3}$, for $3 \le y \le 6$

**18.** $x = \dfrac{9}{4}y^{2/3} - \dfrac{1}{8}y^{4/3}$, for $1 \le y \le 2$

**19.** $x = 2y - 4$, for $-3 \le y \le 4$ (Use calculus, but check your work using geometry.)

**20.** $x = \dfrac{y^5}{5} + \dfrac{1}{12y^3}$, for $2 \le y \le 4$

**T 21–30. Arc length by calculator**

**a.** *Write and simplify the integral that gives the arc length of the following curves on the given interval.*

**b.** *If necessary, use technology to evaluate or approximate the integral.*

**21.** $y = x^2$, for $-1 \le x \le 1$     **22.** $y = \sin x$, for $0 \le x \le \pi$

**23.** $y = \tan x$, for $0 \le x \le \pi/4$   **24.** $y = \dfrac{x^3}{3}$, for $-1 \le x \le 1$

**25.** $x = \sqrt{y - 2}$, for $3 \le y \le 4$

**26.** $y = \dfrac{8}{x^2}$, for $1 \le x \le 4$

**27.** $y = \cos 2x$, for $0 \le x \le \pi$

**28.** $y = 4x - x^2$, for $0 \le x \le 4$

**29.** $y = \dfrac{1}{x}$, for $1 \le x \le 10$

**30.** $x = \dfrac{1}{y^2 + 1}$, for $-5 \le y \le 5$

**T 31. Golden Gate cables** The profile of the cables on a suspension bridge may be modeled by a parabola. The central span of the Golden Gate Bridge (see figure) is 1280 m long and 152 m high. The parabola $y = 0.00037x^2$ gives a good fit to the shape of the cables, where $|x| \le 640$, and $x$ and $y$ are measured in meters. Approximate the length of the cables that stretch between the tops of the two towers.

**T 32. Gateway Arch** The shape of the Gateway Arch in St. Louis (with a height and a base length of 630 ft) is modeled by the function

$$y = 630\left(1 - \left(\dfrac{x}{315}\right)^2\right),$$ where $|x| \le 315$, and $x$ and $y$ are measured in feet (see figure). Estimate the length of the Gateway Arch.

**33. Explain why or why not** Determine whether the following statements are true and give an explanation or counterexample.

**a.** $\displaystyle\int_a^b \sqrt{1 + f'(x)^2}\, dx = \int_a^b (1 + f'(x))\, dx$

**b.** Assuming $f'$ is continuous on the interval $[a, b]$, the length of the curve $y = f(x)$ on $[a, b]$ is the area under the curve $y = \sqrt{1 + f'(x)^2}$ on $[a, b]$.

**c.** Arc length may be negative if $f(x) < 0$ on part of the interval in question.

**34. Arc length for a line** Consider the segment of the line $y = mx + c$ on the interval $[a, b]$. Use the arc length formula to show that the length of the line segment is $(b - a)\sqrt{1 + m^2}$. Verify this result by computing the length of the line segment using the distance formula.

**35. Functions from arc length** What differentiable functions have an arc length on the interval $[a, b]$ given by the following integrals? Note that the answers are not unique. Give a family of functions that satisfy the conditions.

**a.** $\displaystyle\int_a^b \sqrt{1 + 16x^4}\, dx$     **b.** $\displaystyle\int_a^b \sqrt{1 + 36 \cos^2 2x}\, dx$

**36. Function from arc length** Find a curve that passes through the point $(1, 5)$ and has an arc length on the interval $[2, 6]$ given by $\displaystyle\int_2^6 \sqrt{1 + 16x^{-6}}\, dx$.

**T 37. Cosine vs. parabola** Which curve has the greater length on the interval $[-1, 1]$, $y = 1 - x^2$ or $y = \cos\dfrac{\pi x}{2}$?

**38. Function defined as an integral** Write the integral that gives the length of the curve $y = f(x) = \int_0^x \sin t\, dt$ on the interval $[0, \pi]$.

**Explorations and Challenges**

**39. Lengths of related curves** Suppose the graph of $f$ on the interval $[a, b]$ has length $L$, where $f'$ is continuous on $[a, b]$. Evaluate the following integrals in terms of $L$.

**a.** $\displaystyle\int_{a/2}^{b/2} \sqrt{1 + f'(2x)^2}\, dx$

**b.** $\displaystyle\int_{a/c}^{b/c} \sqrt{1 + f'(cx)^2}\, dx$ if $c \neq 0$

**40. Lengths of symmetric curves** Suppose a curve is described by $y = f(x)$ on the interval $[-b, b]$, where $f'$ is continuous on $[-b, b]$. Show that if $f$ is odd *or* $f$ is even, then the length of the curve $y = f(x)$ from $x = -b$ to $x = b$ is twice the length of the curve from $x = 0$ to $x = b$. Use a geometric argument and prove it using integration.

**41. A family of algebraic functions**

   **a.** Show that the arc length integral for the function
$$f(x) = ax^n + \frac{1}{4an(n-2)x^{n-2}}, \text{ where } a \text{ and } n \text{ are positive}$$
real numbers with $n \neq 2$, may be integrated using methods you already know.

   **b.** Verify that the arc length of the curve $y = f(x)$ on the interval $[1, 2]$ is

$$a(2^n - 1) + \frac{1 - 2^{2-n}}{4an(n-2)}.$$

**T 42. Bernoulli's "parabolas"** Johann Bernoulli (1667−1748) evaluated the arc length of curves of the form $y = x^{(2n+1)/2n}$, where $n$ is a positive integer, on the interval $[0, a]$.

   **a.** Write the arc length integral.

   **b.** Make the change of variables $u^2 = 1 + \left(\frac{2n+1}{2n}\right)^2 x^{1/n}$ to obtain a new integral with respect to $u$.

   **c.** Use the Binomial Theorem to expand this integrand and evaluate the integral.

   **d.** The case $n = 1$ ($y = x^{3/2}$) was done in Example 1. With $a = 1$, compute the arc length in the cases $n = 2$ and $n = 3$. Does the arc length increase or decrease with $n$?

   **e.** Graph the arc length of the curves for $a = 1$ as a function of $n$ (treat $n$ as a real number when graphing).

**QUICK CHECK ANSWERS**

**1.** $\sqrt{2}a$ (the length of the line segment joining the points)
**2.** $\sqrt{2}(d - c)$ (the length of the line segment joining the points)   **3.** $L = \int_0^\pi \sqrt{1 + \cos^2 y}\, dy$ ◄

# 6.6 Surface Area

In Sections 6.3 and 6.4, we introduced solids of revolution and presented methods for computing the volume of such solids. We now consider a related problem: computing the *area* of the surface of a solid of revolution. Surface area calculations are important in aerodynamics (computing the lift on an airplane wing) and biology (computing transport rates across cell membranes), to name just two applications. Here is an interesting observation: A surface area problem is "between" a volume problem (which is three-dimensional) and an arc length problem (which is one-dimensional). For this reason, you will see ideas that appear in both volume and arc length calculations as we develop the surface area integral.

## Some Preliminary Calculations

Consider a curve $y = f(x)$ on an interval $[a, b]$, where $f$ is a nonnegative function with a continuous first derivative on $[a, b]$. Now imagine revolving the curve about the $x$-axis to generate a *surface of revolution* (**Figure 6.59**). Our objective is to find the area of this surface.

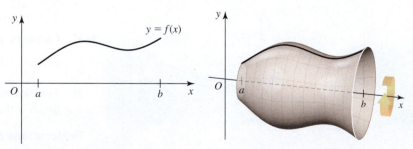

**Figure 6.59**

Before tackling this problem, we consider a preliminary problem upon which we build a general surface area formula. First consider the graph of $f(x) = rx/h$ on the interval $[0, h]$, where $h > 0$ and $r > 0$. When this line segment is revolved about the $x$-axis, it generates the surface of a cone of radius $r$ and height $h$ (**Figure 6.60**). A formula from geometry states that the surface area of a right circular cone of radius $r$ and height $h$ (excluding the base) is $\pi r \sqrt{r^2 + h^2} = \pi r \ell$, where $\ell$ is the slant height of the cone (the length of the slanted "edge" of the cone).

One way to derive the formula for the surface area of a cone (not including the base) is to cut the cone on a line from its base to its vertex. When the cone is unfolded, it forms a sector of a circular disk of radius $\ell$ with a curved edge of length $2\pi r$. This sector is a fraction
$$\frac{2\pi r}{2\pi\ell} = \frac{r}{\ell}$$ of a full circular disk of
radius $\ell$. So the area of the sector, which is also the surface area of the cone, is
$$\pi\ell^2\cdot\frac{r}{\ell} = \pi r\ell.$$

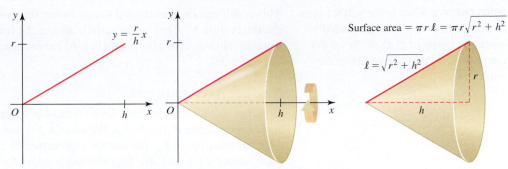

**Figure 6.60**

Curved edge length $= 2\pi r$

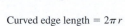

Curved edge length $= 2\pi r$

**QUICK CHECK 1** Which is greater, the surface area of a cone of height 10 and radius 20 or the surface area of a cone of height 20 and radius 10 (excluding the bases)? ◄

Surface area of frustum:
$$S = \pi(f(b) + f(a))\ell$$
$$= \pi(r_2 + r_1)\ell$$

With this result, we can solve a preliminary problem that will be useful. Consider the linear function $f(x) = cx$ on the interval $[a, b]$, where $0 < a < b$ and $c > 0$. When this line segment is revolved about the $x$-axis, it generates a *frustum of a cone* (a cone whose top has been sliced off). The goal is to find $S$, the surface area of the frustum. **Figure 6.61** shows that $S$ is the difference between the surface area $S_b$ of the cone that extends over the interval $[0, b]$ and the surface area $S_a$ of the cone that extends over the interval $[0, a]$.

| Surface area of large cone | $-$ | Surface area of small cone | $=$ | Surface area of frustum |

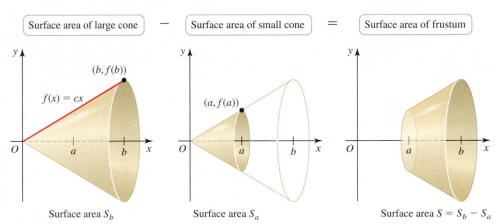

**Figure 6.61**

Notice that the radius of the cone on $[0, b]$ is $r = f(b) = cb$, and its height is $h = b$. Therefore, this cone has surface area
$$S_b = \pi r\sqrt{r^2 + h^2} = \pi(bc)\sqrt{(bc)^2 + b^2} = \pi b^2 c\sqrt{c^2 + 1}.$$

Similarly, the cone on $[0, a]$ has radius $r = f(a) = ca$ and height $h = a$, so its surface area is
$$S_a = \pi(ac)\sqrt{(ac)^2 + a^2} = \pi a^2 c\sqrt{c^2 + 1}.$$

The difference of the surface areas $S_b - S_a$ is the surface area $S$ of the frustum on $[a, b]$:
$$S = S_b - S_a = \pi b^2 c\sqrt{c^2 + 1} - \pi a^2 c\sqrt{c^2 + 1}$$
$$= \pi c(b^2 - a^2)\sqrt{c^2 + 1}.$$

A slightly different form of this surface area formula will be useful. Observe that the line segment from $(a, f(a))$ to $(b, f(b))$ (which is the slant height of the frustum in Figure 6.61) has length
$$\ell = \sqrt{(b - a)^2 + (bc - ac)^2} = (b - a)\sqrt{c^2 + 1}.$$

Therefore, the surface area of the frustum can also be written
$$S = \pi c(b^2 - a^2)\sqrt{c^2 + 1}$$
$$= \pi c(b + a)(b - a)\sqrt{c^2 + 1} \quad \text{Factor } b^2 - a^2.$$
$$= \pi\underbrace{(cb + ca)}_{f(b)\quad f(a)}\underbrace{(b - a)\sqrt{c^2 + 1}}_{\ell} \quad \text{Expand } c(b + a).$$
$$= \pi(f(b) + f(a))\ell.$$

**QUICK CHECK 2** What is the surface area of the frustum of a cone generated when the graph of $f(x) = 3x$ on the interval $[2, 5]$ is revolved about the $x$-axis? ◄

This result can be generalized to *any* linear function $g(x) = cx + d$ that is positive on the interval $[a, b]$. That is, the surface area of the frustum generated by revolving the line segment between $(a, g(a))$ and $(b, g(b))$ about the $x$-axis is given by $\pi(g(b) + g(a))\ell$ (Exercise 38).

### Surface Area Formula

With the surface area formula for a frustum of a cone, we now derive a general area formula for a surface of revolution. We assume $f$ is a nonnegative function with a continuous first derivative on $[a, b]$. The surface is generated by revolving the graph of $f$ on the interval $[a, b]$ about the $x$-axis. We begin by subdividing the interval $[a, b]$ into $n$ subintervals of equal length $\Delta x = \dfrac{b - a}{n}$. The grid points in this partition are

$$x_0 = a, x_1, x_2, \ldots, x_{n-1}, x_n = b.$$

Now consider the $k$th subinterval $[x_{k-1}, x_k]$ and the line segment between the points $(x_{k-1}, f(x_{k-1}))$ and $(x_k, f(x_k))$ (**Figure 6.62**). We let the change in the $y$-coordinates between these points be $\Delta y_k = f(x_k) - f(x_{k-1})$.

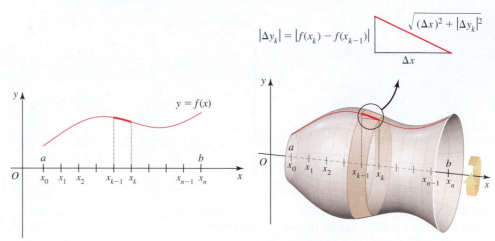

**Figure 6.62**                    **Figure 6.63**

When this line segment is revolved about the $x$-axis, it generates a frustum of a cone (**Figure 6.63**). The slant height of this frustum is the length of the hypotenuse of a right triangle whose sides have lengths $\Delta x$ and $|\Delta y_k|$. Therefore, the slant height of the $k$th frustum is

$$\sqrt{(\Delta x)^2 + |\Delta y_k|^2} = \sqrt{(\Delta x)^2 + (\Delta y_k)^2}$$

and its surface area is

$$S_k = \pi(f(x_k) + f(x_{k-1}))\sqrt{(\Delta x)^2 + (\Delta y_k)^2}.$$

It follows that the area $S$ of the entire surface of revolution is approximately the sum of the surface areas of the individual frustums $S_k$, for $k = 1, \ldots, n$; that is,

$$S \approx \sum_{k=1}^{n} S_k = \sum_{k=1}^{n} \pi(f(x_k) + f(x_{k-1}))\sqrt{(\Delta x)^2 + (\Delta y_k)^2}.$$

➤ Notice that $f$ is assumed to be differentiable on $[a, b]$; therefore, it satisfies the conditions of the Mean Value Theorem. Recall that a similar argument was used to derive the arc length formula in Section 6.5.

We would like to identify this sum as a Riemann sum. However, one more step is required to put it in the correct form. We apply the Mean Value Theorem on the $k$th subinterval $[x_{k-1}, x_k]$ and observe that

$$\frac{f(x_k) - f(x_{k-1})}{\Delta x} = f'(x_k^*),$$

for some number $x_k^*$ in the interval $(x_{k-1}, x_k)$, for $k = 1, \ldots, n$. It follows that $\Delta y_k = f(x_k) - f(x_{k-1}) = f'(x_k^*)\Delta x$.

We now replace $\Delta y_k$ with $f'(x_k^*)\Delta x$ in the expression for the approximate surface area. The result is

$$S \approx \sum_{k=1}^{n} S_k = \sum_{k=1}^{n} \pi(f(x_k) + f(x_{k-1}))\sqrt{(\Delta x)^2 + (\Delta y_k)^2}$$

$$= \sum_{k=1}^{n} \pi(f(x_k) + f(x_{k-1}))\sqrt{(\Delta x)^2(1 + f'(x_k^*)^2)} \qquad \text{Mean Value Theorem}$$

$$= \sum_{k=1}^{n} \pi(f(x_k) + f(x_{k-1}))\sqrt{1 + f'(x_k^*)^2}\,\Delta x. \qquad \text{Factor out } \Delta x.$$

When $\Delta x$ is small, we have $x_{k-1} \approx x_k \approx x_k^*$, and by the continuity of $f$, it follows that $f(x_{k-1}) \approx f(x_k) \approx f(x_k^*)$, for $k = 1, \ldots, n$. These observations allow us to write

$$S \approx \sum_{k=1}^{n} \pi(f(x_k^*) + f(x_k^*))\sqrt{1 + f'(x_k^*)^2}\,\Delta x$$

$$= \sum_{k=1}^{n} 2\pi f(x_k^*)\sqrt{1 + f'(x_k^*)^2}\,\Delta x.$$

This approximation to $S$, which has the form of a Riemann sum, improves as the number of subintervals increases and as the length of the subintervals approaches 0. Specifically, as $n \to \infty$ and as $\Delta x \to 0$, we obtain an integral for the surface area:

$$S = \lim_{n \to \infty} \sum_{k=1}^{n} 2\pi f(x_k^*)\sqrt{1 + f'(x_k^*)^2}\,\Delta x$$

$$= \int_a^b 2\pi f(x)\sqrt{1 + f'(x)^2}\,dx.$$

**QUICK CHECK 3** Let $f(x) = c$, where $c > 0$. What surface is generated when the graph of $f$ on $[a, b]$ is revolved about the $x$-axis? Without using calculus, what is the area of the surface? ◄

> **DEFINITION  Area of a Surface of Revolution**
>
> Let $f$ be a nonnegative function with a continuous first derivative on the interval $[a, b]$. The area of the surface generated when the graph of $f$ on the interval $[a, b]$ is revolved about the $x$-axis is
>
> $$S = \int_a^b 2\pi f(x)\sqrt{1 + f'(x)^2}\,dx.$$

**EXAMPLE 1  Using the surface area formula** The graph of $f(x) = 2\sqrt{x}$ on the interval $[1, 3]$ is revolved about the $x$-axis. What is the area of the surface generated (**Figure 6.64**)?

**SOLUTION** Noting that $f'(x) = \dfrac{1}{\sqrt{x}}$, the surface area formula gives

$$S = \int_a^b 2\pi f(x)\sqrt{1 + f'(x)^2}\,dx$$

$$= 2\pi \int_1^3 2\sqrt{x}\sqrt{1 + \frac{1}{x}}\,dx \qquad \text{Substitute for } f \text{ and } f'.$$

$$= 4\pi \int_1^3 \sqrt{x + 1}\,dx \qquad \text{Simplify.}$$

$$= \frac{8\pi}{3}(x + 1)^{3/2}\Big|_1^3 = \frac{16\pi}{3}(4 - \sqrt{2}). \qquad \text{Integrate and simplify.}$$

*Related Exercise 9* ◄

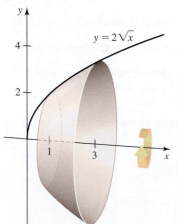

$y = 2\sqrt{x}$

**Figure 6.64**

**EXAMPLE 2  Surface area of a spherical zone** A spherical zone is produced when a sphere of radius $a$ is sliced by two parallel planes. In this example, we compute the surface area of the spherical zone that results when the first plane is oriented vertically and cuts the sphere in half while the second plane lies $h$ units to the right, where $0 \le h < a$

(**Figure 6.65a**). Show that the area of this spherical zone of width $h$ cut from a sphere of radius $a$ is $2\pi a h$, and use the result to show that the surface area of the sphere is $4\pi a^2$.

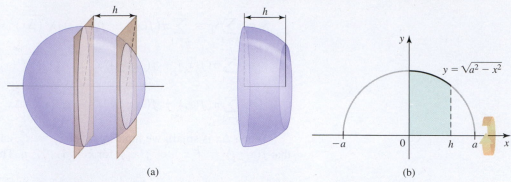

**Figure 6.65**

**SOLUTION** To generate the spherical zone, we revolve the curve $f(x) = \sqrt{a^2 - x^2}$ on the interval $[0, h]$ about the $x$-axis (**Figure 6.65b**). Noting that $f'(x) = -x(a^2 - x^2)^{-1/2}$, the surface area of the spherical zone of width $h$ is

$$S = \int_a^b 2\pi f(x)\sqrt{1 + f'(x)^2}\, dx$$

$$= 2\pi \int_0^h \sqrt{a^2 - x^2}\sqrt{1 + (-x(a^2 - x^2)^{-1/2})^2}\, dx \qquad \text{Substitute for } f \text{ and } f'.$$

$$= 2\pi \int_0^h \sqrt{a^2 - x^2}\sqrt{\frac{a^2}{a^2 - x^2}}\, dx \qquad\qquad \text{Simplify.}$$

$$= 2\pi \int_0^h a\, dx = 2\pi a h. \qquad\qquad\qquad \text{Simplify and integrate.}$$

To find the surface area of the entire sphere, it is tempting to integrate over the interval $[-a, a]$. However, $f$ is not differentiable at $\pm a$, so we cannot use the surface area formula on p. 407. Instead, we use the formula for the area of the spherical zone and let $h$ approach $a$ from the left to find the area of the associated hemisphere:

$$\text{Area of zone} = 2\pi a h, \text{ so}$$

$$\text{Area of hemisphere} = \lim_{h \to a^-} 2\pi a h = 2\pi a^2.$$

Therefore, the surface area of a sphere of radius $a$ is $2(2\pi a^2) = 4\pi a^2$.

*Related Exercise 13* ◄

**EXAMPLE 3  Painting a funnel** The curved surface of a funnel is generated by revolving the graph of $y = f(x) = x^3 + \dfrac{1}{12x}$ on the interval $[1, 2]$ about the $x$-axis (**Figure 6.66**).

Approximately what volume of paint is needed to cover the outside of the funnel with a layer of paint 0.05 cm thick? Assume $x$ and $y$ are measured in centimeters.

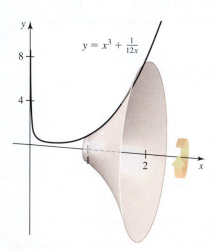

**Figure 6.66**

**SOLUTION** Note that $f'(x) = 3x^2 - \dfrac{1}{12x^2}$. Therefore, the surface area of the funnel in cm$^2$ is

$$S = \int_a^b 2\pi f(x)\sqrt{1 + f'(x)^2}\, dx$$

$$= 2\pi \int_1^2 \left(x^3 + \frac{1}{12x}\right)\sqrt{1 + \left(3x^2 - \frac{1}{12x^2}\right)^2}\, dx \qquad \text{Substitute for } f \text{ and } f'.$$

$$= 2\pi \int_1^2 \left(x^3 + \frac{1}{12x}\right)\sqrt{\left(3x^2 + \frac{1}{12x^2}\right)^2}\, dx \qquad \text{Expand and factor under square root.}$$

$$= 2\pi \int_1^2 \left(x^3 + \frac{1}{12x}\right)\left(3x^2 + \frac{1}{12x^2}\right) dx \qquad \text{Simplify.}$$

$$= \frac{12{,}289}{192}\pi. \qquad\qquad\qquad\qquad\qquad\qquad \text{Evaluate integral.}$$

Because the paint layer is 0.05 cm thick, the volume of paint needed is approximately

$$\left(\frac{12{,}289\pi}{192} \text{ cm}^2\right)(0.05 \text{ cm}) \approx 10.1 \text{ cm}^3.$$

*Related Exercises 21–22* ◄

The derivation that led to the surface area integral may be used when a curve is revolved about the *y*-axis (rather than the *x*-axis). The result is the same integral, where *x* is replaced with *y*. For example, if the curve $x = g(y)$ on the interval $[c, d]$ is revolved about the *y*-axis, the area of the surface generated is

$$S = \int_c^d 2\pi g(y)\sqrt{1 + g'(y)^2}\, dy.$$

To use this integral, we must first describe the given curve as a differentiable function of *y*.

**EXAMPLE 4  Change of perspective**  Consider the curve defined by the equation $4y^3 - 10x\sqrt{y} + 5 = 0$, for $1 \le y \le 2$. Find the area of the surface generated when this curve is revolved about the *y*-axis (**Figure 6.67**).

**SOLUTION**  We solve the equation $4y^3 - 10x\sqrt{y} + 5 = 0$ for *x* so that the curve is expressed as a function of *y*:

$$4y^3 - 10x\sqrt{y} + 5 = 0$$

$$10x\sqrt{y} = 4y^3 + 5 \qquad \text{Rearrange terms.}$$

$$x = g(y) = \frac{2}{5}y^{5/2} + \frac{1}{2}y^{-1/2}. \quad \text{Divide by } 10\sqrt{y}.$$

Note that $g'(y) = y^{3/2} - \frac{1}{4}y^{-3/2}$ and that the interval of integration on the *y*-axis is $[1, 2]$. The area of the surface is

$$S = \int_c^d 2\pi g(y)\sqrt{1 + g'(y)^2}\, dy$$

$$= 2\pi \int_1^2 \left(\frac{2}{5}y^{5/2} + \frac{1}{2}y^{-1/2}\right)\sqrt{1 + \left(y^{3/2} - \frac{1}{4}y^{-3/2}\right)^2}\, dy \quad \text{Substitute for } g \text{ and } g'.$$

$$= 2\pi \int_1^2 \left(\frac{2}{5}y^{5/2} + \frac{1}{2}y^{-1/2}\right)\sqrt{\left(y^{3/2} + \frac{1}{4}y^{-3/2}\right)^2}\, dy \qquad \text{Expand and factor.}$$

$$= 2\pi \int_1^2 \left(\frac{2}{5}y^4 + \frac{3}{5}y + \frac{1}{8}y^{-2}\right) dy \qquad \text{Simplify.}$$

$$= 2\pi \left(\frac{2}{25}y^5 + \frac{3}{10}y^2 - \frac{1}{8y}\right)\Big|_1^2 = \frac{1377}{200}\pi. \qquad \text{Integrate and evaluate.}$$

*Related Exercise 34* ◄

**Figure 6.67**

The graph shows $4y^3 - 10x\sqrt{y} + 5 = 0$.

---

## SECTION 6.6 EXERCISES

### Getting Started

1. What is the area of the curved surface of a right circular cone of radius 3 and height 4?

2. A frustum of a cone is generated by revolving the graph of $y = 4x$ on the interval $[2, 6]$ about the *x*-axis. What is the area of the surface of the frustum?

3. Assume *f* is a nonnegative function with a continuous first derivative on $[a, b]$. The curve $y = f(x)$ on $[a, b]$ is revolved about the *x*-axis. Explain how to find the area of the surface that is generated.

4. Assume *g* is a nonnegative function with a continuous first derivative on $[c, d]$. The curve $x = g(y)$ on $[c, d]$ is revolved about the *y*-axis. Explain how to find the area of the surface that is generated.

5. A surface is generated by revolving the line $f(x) = 2 - x$, for $0 \le x \le 2$, about the *x*-axis. Find the area of the resulting surface in the following ways.

   a. Using calculus
   b. Using geometry, after first determining the shape and dimensions of the surface

**6.** A surface is generated by revolving the line $x = 3$, for $0 \le y \le 8$, about the $y$-axis. Find the area of the resulting surface in the following ways.

    **a.** Using calculus with $g(y) = 3$

    **b.** Using geometry, after first determining the shape and dimensions of the surface

## Practice Exercises

**7–20. Computing surface areas** *Find the area of the surface generated when the given curve is revolved about the given axis.*

**7.** $y = 3x + 4$, for $0 \le x \le 6$; about the $x$-axis

**8.** $y = 12 - 3x$, for $1 \le x \le 3$; about the $x$-axis

**9.** $y = 8\sqrt{x}$, for $9 \le x \le 20$; about the $x$-axis

**10.** $y = x^3$, for $0 \le x \le 1$; about the $x$-axis

**11.** $y = (3x)^{1/3}$, for $0 \le x \le \dfrac{8}{3}$; about the $y$-axis

**12.** $y = \dfrac{x^2}{4}$, for $2 \le x \le 4$; about the $y$-axis

**13.** $y = \sqrt{1 - x^2}$, for $-\dfrac{1}{2} \le x \le \dfrac{1}{2}$; about the $x$-axis

**14.** $x = \sqrt{-y^2 + 6y - 8}$, for $3 \le y \le \dfrac{7}{2}$; about the $y$-axis

**15.** $y = 4x - 1$, for $1 \le x \le 4$; about the $y$-axis (*Hint:* Integrate with respect to $y$.)

**16.** $y = \sqrt{4x + 6}$, for $0 \le x \le 5$; about the $x$-axis

**17.** $y = \sqrt{-x^2 + 6x - 5}$, for $2 \le x \le 3$; about the $x$-axis

**18.** $y = \sqrt{5x - x^2}$, for $1 \le x \le 4$; about the $x$-axis

**19.** $x = \sqrt{12y - y^2}$, for $2 \le y \le 10$; about the $y$-axis

**20.** $y = 1 + \sqrt{1 - x^2}$ between the points $(1, 1)$ and $\left(\dfrac{\sqrt{3}}{2}, \dfrac{3}{2}\right)$; about the $y$-axis

**21–22. Painting surfaces** *A 1.5-mm layer of paint is applied to one side of the following surfaces. Find the approximate volume of paint needed. Assume $x$ and $y$ are measured in meters.*

**21.** The spherical zone generated when the curve $y = \sqrt{8x - x^2}$ on the interval $1 \le x \le 7$ is revolved about the $x$-axis

**22.** The spherical zone generated when the upper portion of the circle $x^2 + y^2 = 100$ on the interval $-8 \le x \le 8$ is revolved about the $x$-axis

**23. Explain why or why not** Determine whether the following statements are true and give an explanation or counterexample.

    **a.** If the curve $y = f(x)$ on the interval $[a, b]$ is revolved about the $y$-axis, the area of the surface generated is

$$\int_{f(a)}^{f(b)} 2\pi f(y) \sqrt{1 + f'(y)^2}\, dy.$$

    **b.** If $f$ is not one-to-one on the interval $[a, b]$, then the area of the surface generated when the graph of $f$ on $[a, b]$ is revolved about the $x$-axis is not defined.

    **c.** Let $f(x) = 12x^2$. The area of the surface generated when the graph of $f$ on $[-4, 4]$ is revolved about the $x$-axis is twice the area of the surface generated when the graph of $f$ on $[0, 4]$ is revolved about the $x$-axis.

    **d.** Let $f(x) = 12x^2$. The area of the surface generated when the graph of $f$ on $[-4, 4]$ is revolved about the $y$-axis is twice the area of the surface generated when the graph of $f$ on $[0, 4]$ is revolved about the $y$-axis.

**T 24–28. Surface area using technology** *Consider the following curves on the given intervals.*

    **a.** *Write the integral that gives the area of the surface generated when the curve is revolved about the given axis.*

    **b.** *Use a calculator or software to approximate the surface area.*

**24.** $y = x^5$, for $0 \le x \le 1$; about the $x$-axis

**25.** $y = \cos x$, for $0 \le x \le \dfrac{\pi}{2}$; about the $x$-axis

**26.** $y = (2x + 3)^2$, for $0 \le x \le 1$; about the $y$-axis

**27.** $y = \tan x$, for $0 \le x \le \dfrac{\pi}{4}$; about the $x$-axis

**28.** $y = \sqrt{\sin \pi x}$, for $0 \le x \le \dfrac{1}{2}$; about the $x$-axis

**29. Revolving an astroid** Consider the upper half of the astroid described by $x^{2/3} + y^{2/3} = a^{2/3}$, where $a > 0$ and $|x| \le a$. Find the area of the surface generated when this curve is revolved about the $x$-axis. Note that the function describing the curve is not differentiable at 0. However, the surface area integral can be evaluated using symmetry and methods you know.

**30. Cones and cylinders** The volume of a cone of radius $r$ and height $h$ is one-third the volume of a cylinder with the same radius and height. Does the surface area of a cone of radius $r$ and height $h$ equal one-third the surface area of a cylinder with the same radius and height? If not, find the correct relationship. Exclude the bases of the cone and cylinder.

## Explorations and Challenges

**31–35. Challenging surface area calculations** *Find the area of the surface generated when the given curve is revolved about the given axis.*

**31.** $y = x^{3/2} - \dfrac{x^{1/2}}{3}$, for $1 \le x \le 2$; about the $x$-axis

**32.** $y = \dfrac{x^4}{8} + \dfrac{1}{4x^2}$, for $1 \le x \le 2$; about the $x$-axis

**33.** $y = \dfrac{x^3}{3} + \dfrac{1}{4x}$, for $\dfrac{1}{2} \le x \le 2$; about the $x$-axis

**34.** $y = \dfrac{1}{2}\left(1 - \sqrt{1 - 4x^2}\right)$ between the points $\left(\dfrac{\sqrt{3}}{4}, \dfrac{1}{4}\right)$ and $\left(\dfrac{1}{2}, \dfrac{1}{2}\right)$; about the $y$-axis

**35.** $x = 4y^{3/2} - \dfrac{y^{1/2}}{12}$, for $1 \le y \le 4$; about the $y$-axis

**36. Surface area of an ellipsoid** If the top half of the ellipse $\dfrac{x^2}{a^2} + \dfrac{y^2}{b^2} = 1$ is revolved about the $x$-axis, the result is an *ellipsoid* whose axis along the $x$-axis has length $2a$, whose axis along the $y$-axis has length $2b$, and whose axis perpendicular to the $xy$-plane has length $2b$. We assume that $0 < b < a$ (see figure). Use the following steps to find the surface area $S$ of this ellipsoid.

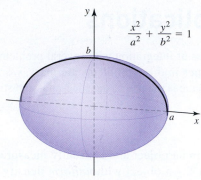

$$\frac{x^2}{a^2} + \frac{y^2}{b^2} = 1$$

**a.** Use the surface area formula to show that

$$S = \frac{4\pi b}{a} \int_0^a \sqrt{a^2 - c^2x^2}\, dx, \text{ where } c^2 = 1 - \frac{b^2}{a^2}.$$

**b.** Use the change of variables $u = cx$ to show that

$$S = \frac{4\pi b}{\sqrt{a^2 - b^2}} \int_0^{\sqrt{a^2-b^2}} \sqrt{a^2 - u^2}\, du.$$

**T c.** Use the integral formula in part (b) together with a calculator or software to approximate the surface area of an ellipse if $a = 4$ and $b = 3$.

**T d.** The surface area of the ellipsoid can be estimated by the Knud Thomsen formula

$$S \approx 4\pi \left( \frac{2(ab)^p + b^{2p}}{3} \right)^{1/p},$$

where $p = 1.6075$. Use this formula to estimate the surface area of an ellipse, where $a = 4$ and $b = 3$. Compare this approximation with the result found in part (c).

**e.** Use part (a) to show that if $a = b$, then $S = 4\pi a^2$, which is the surface area of a sphere of radius $a$.

**f.** How accurate is the formula in part (d) if $a = b$?

**37. Zones of a sphere** Suppose a sphere of radius $r$ is sliced by two horizontal planes $h$ units apart (see figure). Show that the surface area of the resulting zone on the sphere is $2\pi rh$, independent of the location of the cutting planes.

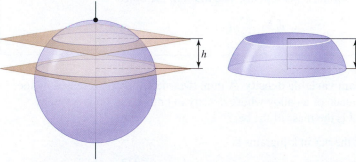

**38. Surface area of a frustum** Show that the surface area of the frustum of a cone generated by revolving the line segment between $(a, g(a))$ and $(b, g(b))$ about the x-axis is $\pi(g(b) + g(a))\ell$, for any linear function $g(x) = cx + d$ that is positive on the interval $[a, b]$, where $\ell$ is the slant height of the frustum.

**39. Scaling surface area** Let $f$ be a nonnegative function with a continuous first derivative on $[a, b]$ and suppose $g(x) = c\, f(x)$ and $h(x) = f(cx)$, where $c > 0$. When the curve $y = f(x)$ on $[a, b]$ is revolved about the x-axis, the area of the resulting surface is $A$. Evaluate the following integrals in terms of $A$ and $c$.

**a.** $\int_a^b 2\pi g(x) \sqrt{c^2 + g'(x)^2}\, dx$

**b.** $\int_{a/c}^{b/c} 2\pi h(x) \sqrt{c^2 + h'(x)^2}\, dx$

**40. Surface plus cylinder** Suppose $f$ is a nonnegative function with a continuous first derivative on $[a, b]$. Let $L$ equal the length of the graph of $f$ on $[a, b]$ and let $S$ be the area of the surface generated by revolving the graph of $f$ on $[a, b]$ about the x-axis. For a positive constant $C$, assume the curve $y = f(x) + C$ is revolved about the x-axis. Show that the area of the resulting surface equals the sum of $S$ and the surface area of a right circular cylinder of radius $C$ and height $L$.

**41. Surface-area-to-volume ratio (SAV)** In the design of solid objects (both artificial and natural), the ratio of the surface area to the volume of the object is important. Animals typically generate heat at a rate proportional to their volume and lose heat at a rate proportional to their surface area. Therefore, animals with a low SAV ratio tend to retain heat whereas animals with a high SAV ratio (such as children and hummingbirds) lose heat relatively quickly.

**a.** What is the SAV ratio of a cube with side lengths $R$?

**b.** What is the SAV ratio of a ball with radius $R$?

**c.** For a fixed constant $h > 0$, consider a ball of radius $\sqrt[3]{h}/4$ and an ellipsoid with a long axis of length $2\sqrt[3]{h}$ and short axes that are each half the length of the long axis. Show that the ball and ellipsoid have the same volume. (*Hint:* The volume of an ellipsoid is $\frac{4}{3}\pi ABC$, where the axes have lengths $2A$, $2B$, and $2C$).

**T d.** Create a table to compare SAV ratios of the balls and ellipsoids of equal volume described in part (c) using values of $h = 1.1, 5, 10,$ and $20$. (*Hint:* Find the surface area of the ellipsoids using the integral in part (b) of Exercise 36 together with a calculator or other technology.)

**e.** Among all ellipsoids of a fixed volume, which one would you choose for the shape of an animal if the goal is to minimize heat loss?

**QUICK CHECK ANSWERS**

**1.** The surface area of the first cone $(200\sqrt{5}\pi)$ is twice as great as the surface area of the second cone $(100\sqrt{5}\pi)$.
**2.** The surface area is $63\sqrt{10}\pi$. **3.** The surface is a cylinder of radius $c$ and height $b - a$. The area of the curved surface is $2\pi c(b - a)$. ◄

# 6.7 Physical Applications

We conclude this chapter on applications of integration with several problems from physics and engineering. The physical themes in these problems are mass, work, force, and pressure. The common mathematical theme is the use of the slice-and-sum strategy, which always leads to a definite integral.

➤ In Chapter 16, we return to mass calculations for two- and three-dimensional objects (plates and solids).

$x = a$                           $x = b$

**Figure 6.68**

**QUICK CHECK 1** In Figure 6.68, suppose $a = 0$, $b = 3$, and the density of the rod in g/cm is $\rho(x) = (4 - x)$. (a) Where is the rod lightest and heaviest? (b) What is the density at the middle of the bar? ◄

Mass of $k$th subinterval:
$m_k \approx \rho(x_k^*)\Delta x$

$x = a$        $x_k^*$        $x = b$

**Figure 6.69**

➤ Note that the units of the integral work out as they should: $\rho$ has units of mass per length and $dx$ has units of length, so $\rho(x)\,dx$ has units of mass.

## Density and Mass

Density is the concentration of mass in an object and is usually measured in units of mass per volume (for example, g/cm$^3$). An object with *uniform* density satisfies the basic relationship

$$\text{mass} = \text{density} \cdot \text{volume}.$$

When the density of an object *varies*, this formula no longer holds, and we must appeal to calculus.

In this section, we introduce mass calculations for thin objects that can be viewed as line segments (such as wires or thin bars). The bar shown in **Figure 6.68** has a density $\rho$ that varies along its length. For one-dimensional objects, we use *linear density* with units of mass per length (for example, g/cm). What is the mass of such an object?

We begin by dividing the bar, represented by the interval $a \leq x \leq b$, into $n$ subintervals of equal length $\Delta x = (b - a)/n$ (**Figure 6.69**). Let $x_k^*$ be any point in the $k$th subinterval, for $k = 1, \ldots, n$. The mass of the $k$th segment of the bar $m_k$ is approximately the density at $x_k^*$ multiplied by the length of the interval, or $m_k \approx \rho(x_k^*)\Delta x$. So the approximate mass of the entire bar is

$$\sum_{k=1}^{n} m_k \approx \sum_{k=1}^{n} \underbrace{\rho(x_k^*)\Delta x}_{m_k}.$$

The exact mass is obtained by taking the limit as $n \to \infty$ and as $\Delta x \to 0$, which produces a definite integral.

---

**DEFINITION  Mass of a One-Dimensional Object**

Suppose a thin bar or wire is represented by the interval $a \leq x \leq b$ with a density function $\rho$ (with units of mass per length). The **mass** of the object is

$$m = \int_a^b \rho(x)\,dx.$$

---

**QUICK CHECK 2** A thin bar occupies the interval $0 \leq x \leq 2$ and has a density in kg/m of $\rho(x) = (1 + x^2)$. Using the minimum value of the density, what is a lower bound for the mass of the object? Using the maximum value of the density, what is an upper bound for the mass of the object? ◄

**EXAMPLE 1   Mass from variable density**  A thin, two-meter bar, represented by the interval $0 \leq x \leq 2$, is made of an alloy whose density in units of kg/m is given by $\rho(x) = (1 + x^2)$. What is the mass of the bar?

**SOLUTION** The mass of the bar in kilograms is

$$m = \int_a^b \rho(x)\,dx = \int_0^2 (1 + x^2)\,dx = \left(x + \frac{x^3}{3}\right)\Big|_0^2 = \frac{14}{3}.$$

*Related Exercises 14–15* ◄

## Work

Work can be described as the change in energy when a force causes a displacement of an object. When you carry a basket of laundry up a flight of stairs or push a stalled car, you apply a force that results in the displacement of an object, and work is done. If a *constant* force $F$ displaces an object a distance $d$ in the direction of the force, the work done is the force multiplied by the distance:

$$\text{work} = \text{force} \cdot \text{distance}.$$

**Figure 6.70**

**QUICK CHECK 3** Explain why the sum of the work over $n$ subintervals is only an approximation of the total work. ◄

It is easiest to use metric units for force and work. A newton (N) is the force required to give a 1-kg mass an acceleration of $1 \text{ m/s}^2$. A joule (J) is 1 newton-meter (N-m), the work done by a 1-N force acting over a distance of 1 m.

Calculus enters the picture with *variable* forces. Suppose an object is moved along the $x$-axis by a variable force $F$ that is directed along the $x$-axis (**Figure 6.70**). How much work is done in moving the object between $x = a$ and $x = b$? Once again, we use the slice-and-sum strategy.

The interval $[a, b]$ is divided into $n$ subintervals of equal length $\Delta x = (b - a)/n$. We let $x_k^*$ be any point in the $k$th subinterval, for $k = 1, \ldots, n$. On that subinterval, the force is approximately constant with a value of $F(x_k^*)$. Therefore, the work done in moving the object across the $k$th subinterval is approximately $F(x_k^*)\Delta x$ (force · distance). Summing the work done over each of the $n$ subintervals, the total work over the interval $[a, b]$ is approximately

$$W \approx \sum_{k=1}^{n} F(x_k^*)\Delta x.$$

This approximation becomes exact when we take the limit as $n \to \infty$ and $\Delta x \to 0$. The total work done is the integral of the force over the interval $[a, b]$ (or, equivalently, the net area under the force curve in Figure 6.70).

---

**DEFINITION  Work**

The work done by a variable force $F$ moving an object along a line from $x = a$ to $x = b$ in the direction of the force is

$$W = \int_a^b F(x)\, dx.$$

---

An application of force and work that is easy to visualize is the stretching and compression of a spring. Suppose an object is attached to a spring on a frictionless horizontal surface; the object slides back and forth under the influence of the spring. We say that the spring is at *equilibrium* when it is neither compressed nor stretched. It is convenient to let $x$ be the position of the object, where $x = 0$ is the equilibrium position (**Figure 6.71**).

**Figure 6.71**

**Figure 6.72**

According to **Hooke's law**, the force required to keep the spring in a compressed or stretched position $x$ units from the equilibrium position is $F(x) = kx$, where the positive spring constant $k$ measures the stiffness of the spring. Note that to stretch the spring to a position $x > 0$, a force $F > 0$ (in the positive direction) is required. To compress the spring to a position $x < 0$, a force $F < 0$ (in the negative direction) is required (**Figure 6.72**). In other words, the force required to displace the spring is always in the direction of the displacement.

**EXAMPLE 2   Compressing a spring**  Suppose a force of 10 N is required to stretch a spring 0.1 m from its equilibrium position and hold it in that position.

**a.** Assuming the spring obeys Hooke's law, find the spring constant $k$.

**b.** How much work is needed to *compress* the spring 0.5 m from its equilibrium position?

**c.** How much work is needed to *stretch* the spring 0.25 m from its equilibrium position?

**d.** How much additional work is required to stretch the spring 0.25 m if it has already been stretched 0.1 m from its equilibrium position?

**SOLUTION**

▶ Hooke's law was proposed by the English scientist Robert Hooke (1635–1703), who also coined the biological term *cell*. Hooke's law works well for springs made of many common materials. However, some springs obey more complicated spring laws (see Exercise 59).

**a.** The fact that a force of 10 N is required to keep the spring stretched at $x = 0.1$ m means (by Hooke's law) that $F(0.1) = k(0.1\text{ m}) = 10$ N. Solving for the spring constant, we find that $k = 100$ N/m. Therefore, Hooke's law for this spring is $F(x) = 100x$.

**b.** The work in joules required to compress the spring from $x = 0$ to $x = -0.5$ is

$$W = \int_a^b F(x)\,dx = \int_0^{-0.5} 100x\,dx = 50x^2\Big|_0^{-0.5} = 12.5.$$

▶ Notice again that the units in the integral are consistent. If $F$ has units of N and $x$ has units of m, then $W$ has units of $F\,dx$, or N-m, which are the units of work (1 N-m = 1 J).

**c.** The work in joules required to stretch the spring from $x = 0$ to $x = 0.25$ is

$$W = \int_a^b F(x)\,dx = \int_0^{0.25} 100x\,dx = 50x^2\Big|_0^{0.25} = 3.125.$$

**d.** The work in joules required to stretch the spring from $x = 0.1$ to $x = 0.35$ is

**QUICK CHECK 4**  In Example 2, explain why more work is needed in part (d) than in part (c), even though the displacement is the same. ◀

$$W = \int_a^b F(x)\,dx = \int_{0.1}^{0.35} 100x\,dx = 50x^2\Big|_{0.1}^{0.35} = 5.625.$$

Comparing parts (c) and (d), we see that more work is required to stretch the spring 0.25 m starting at $x = 0.1$ than starting at $x = 0$.

*Related Exercises 23–24* ◀

**Figure 6.73**

▶ The units of $\rho$, $\Delta y$, $g$, and $L - y_k^*$ are kg/m, m, m/s², and m, respectively. When multiplied, they yield $\dfrac{\text{kg m}^2}{\text{s}^2}$, which are the units of a joule.

**Lifting Problems**  Another common work problem arises when the motion is vertical and the force is due to gravity. The gravitational force exerted on an object with mass $m$ (measured in kg) is $F = mg$, where $g = 9.8$ m/s² is the acceleration due to gravity near the surface of Earth. The work in joules required to lift an object of mass $m$ a vertical distance of $y$ meters is

$$\text{work} = \text{force} \cdot \text{distance} = mgy.$$

This type of problem becomes interesting when the object being lifted is a rope, a chain, or a cable. In these situations, different parts of the object are lifted different distances, so integration is necessary.

Imagine a chain of length $L$ meters with a constant density of $\rho$ kg/m hanging vertically from a scaffolding platform at a construction site. To compute the work done in lifting the chain to the platform, we introduce a coordinate system where $y = 0$ corresponds to the bottom of the chain and $y = L$ corresponds to the top of the chain (**Figure 6.73**). We then divide the interval $[0, L]$ into $n$ subintervals of equal length $\Delta y$ and choose a point $y_k^*$ from each subinterval $[y_{k-1}, y_k]$, for $k = 1, \ldots, n$.

The mass of each segment of the chain is $m = \rho \Delta y$, and the point $y_k^*$ in the $k$th segment is lifted a distance of $L - y_k^*$. Therefore, the work required to lift the $k$th segment is approximately

$$W_k = \underbrace{\rho \Delta y g}_{\text{force}} \cdot \underbrace{(L - y_k^*)}_{\text{distance}},$$

so the total work required to lift the chain is

$$W \approx \sum_{k=1}^{n} \rho g (L - y_k^*) \Delta y.$$

As the length of each segment $\Delta y$ tends to zero and the number of segments tends to infinity, we obtain a definite integral for the total work in the limit:

$$W = \lim_{n\to\infty} \sum_{k=1}^{n} \rho g (L - y_k^*)\Delta y = \int_0^L \rho g (L - y)\,dy. \qquad (1)$$

The function $L - y$ in equation (1) measures the distance through which a point $y$ on the chain moves. Recognize that this function changes depending on the location of the origin relative to the chain, as illustrated in Example 3 (and Quick Check 6).

**EXAMPLE 3   Lifting a chain and bucket**  A ten-meter chain with density of $1.5\ \text{kg}/\text{m}$ hangs from a platform at a construction site that is 11 meters above the ground (**Figure 6.74a**).

**a.** Compute the work required to lift the chain to the platform.

**b.** Several packages of nails are placed in a one-meter-tall bucket that rests on the ground; the mass of the bucket and nails together is 15 kg, and the chain is attached to the bucket (**Figure 6.74b**). How much work is required to lift the bucket to the platform?

**SOLUTION**

**a.** We use a coordinate system with $y = 0$ placed at the bottom of the chain. By equation (1), the work in joules required to lift the chain to the platform is

$$W = \int_0^L \rho g (L - y)\,dy = 1.5g \int_0^{10}(10 - y)\,dy \quad \text{\small $L = 10$ m; $\rho = 1.5\ \text{kg}/\text{m}$}$$

$$= 1.5g\left( 10y - \frac{y^2}{2}\right)\Big|_0^{10} \quad \text{\small Integrate.}$$

$$= 1.5g\left( 100 - \frac{100}{2}\right) \quad \text{\small Evaluate.}$$

$$= 735. \quad \text{\small Simplify; $g = 9.8\ \text{m}/\text{s}$.}$$

Had we instead chosen a coordinate system with $y = 0$ corresponding to the ground, the work integral would be

$$W = \int_1^{11} 1.5g(11 - y)\,dy$$

because the chain would then lie in the interval $[1, 11]$, and a point $y$ on the chain would move a distance of $(11 - y)$ m before reaching the platform. You can verify that the value of this definite integral is also 735 joules.

**b.** This time we use a coordinate system with $y = 0$ placed on the ground at the bottom of the bucket; we compute the work required to lift the chain and bucket separately. As shown in part (a), the work needed to lift the chain is 735 joules. Integration is not necessary to compute the work needed to lift the bucket (see Quick Check 5). We simply note that the bucket has a mass of 15 kg and every point in the bucket moves a distance of 11 m, so the work in joules required to lift it is

$$W = mgy = 15(9.8)(11) = 1617.$$

The work required to lift both the chain and the bucket is $735 + 1617 = 2352$ joules.

*Related Exercises 31–32* ◄

**QUICK CHECK 5**  In Example 3b, the bucket occupies the interval $[0, 1]$ and the chain occupies the interval $[1, 11]$ (Figure 6.74b). Why is integration used to compute the work needed to lift the chain but not to compute the work needed to lift the bucket? ◄

**Figure 6.74**

**QUICK CHECK 6** Set up and evaluate the work integral in Example 3a using a coordinate system with $y = 0$ placed at the top of the chain and with a positive $y$-axis that points downward. ◄

▶ The choice of a coordinate system is somewhat arbitrary and may depend on the geometry of the problem. You can let the $y$-axis point upward or downward, and there are usually several logical choices for the location of $y = 0$. You should experiment with different coordinate systems.

**Pumping Problems**   The chains and cables encountered in the lifting problems from the previous pages can be modeled as one-dimensional objects. An extra layer of complexity is added when investigating the work required to pump fluid from a tank, although the same principles remain in play.

Suppose a fluid such as water is pumped out of a tank to a height $h$ above the bottom of the tank. How much work is required, assuming the tank is full of water? Three key observations lead to the solution.

• Water from different levels of the tank is lifted different vertical distances, requiring different amounts of work.

• Two equal volumes of water from the same horizontal plane are lifted the same distance and require the same amount of work.

• A volume $V$ of water has mass $\rho V$, where $\rho = 1 \text{ g/cm}^3 = 1000 \text{ kg/m}^3$ is the density of water.

To solve this problem, we let the $y$-axis point upward with $y = 0$ at the bottom of the tank. The body of water that must be lifted extends from $y = 0$ to $y = b$ (which *may* be the top of the tank). The level to which the water must be raised is $y = h$, where $h \geq b$ (**Figure 6.75**). We now slice the water into $n$ horizontal layers, each having thickness $\Delta y$. The $k$th layer occupying the interval $[y_{k-1}, y_k]$, for $k = 1, \ldots, n$, is approximately $y_k^*$ units above the bottom of the tank, where $y_k^*$ is any point in $[y_{k-1}, y_k]$.

**Figure 6.75**

The cross-sectional area of the $k$th layer at $y_k^*$, denoted $A(y_k^*)$, is determined by the shape of the tank; the solution depends on being able to find $A$ for all values of $y$. Because the volume of the $k$th layer is approximately $A(y_k^*)\Delta y$, the force on the $k$th layer (its weight) is

$$F_k = mg \approx \underbrace{\underbrace{A(y_k^*)\Delta y}_{\text{volume}} \cdot \underbrace{\rho}_{\text{density}}}_{\text{mass}} \cdot g.$$

To reach the level $y = h$, the $k$th layer is lifted an approximate distance $(h - y_k^*)$ (Figure 6.75). So the work in lifting the $k$th layer to a height $h$ is approximately

$$W_k = \underbrace{A(y_k^*)\Delta y \rho g}_{\text{force}} \cdot \underbrace{(h - y_k^*)}_{\text{distance}}.$$

Summing the work required to lift all the layers to a height $h$, the total work is

$$W \approx \sum_{k=1}^{n} W_k = \sum_{k=1}^{n} A(y_k^*)\rho g(h - y_k^*)\Delta y.$$

This approximation becomes more accurate as the width of the layers $\Delta y$ tends to zero and the number of layers tends to infinity. In this limit, we obtain a definite integral from $y = 0$ to $y = b$. The total work required to empty the tank is

$$W = \lim_{n \to \infty} \sum_{k=1}^{n} A(y_k^*)\rho g(h - y_k^*)\Delta y = \int_0^b \rho g A(y)(h - y)\, dy.$$

This derivation assumes the *bottom* of the tank is at $y = 0$, in which case the distance that the slice at level $y$ must be lifted is $D(y) = h - y$. If you choose a different location for the origin, the function $D$ will be different. Here is a general procedure for any choice of origin.

> Notice that the work integral for pumping problems reduces to the work integral for lifting problems given in equation (1) on p. 415 when we take $A(y) = 1$.

---

**PROCEDURE**    **Solving Pumping Problems**

1. Draw a $y$-axis in the vertical direction (parallel to gravity) and choose a convenient origin. Assume the interval $[a, b]$ corresponds to the vertical extent of the fluid.

2. For $a \le y \le b$, find the cross-sectional area $A(y)$ of the horizontal slices and the distance $D(y)$ the slices must be lifted.

3. The work required to lift the water is

$$W = \int_a^b \rho g A(y) D(y)\, dy.$$

---

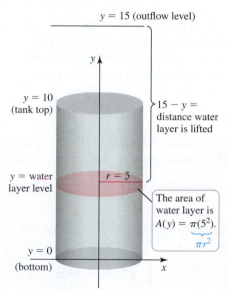

$y = 15$ (outflow level)

$y = 10$ (tank top)

$15 - y =$ distance water layer is lifted

$y =$ water layer level

$r = 5$

The area of water layer is $A(y) = \pi(5^2)$.
$\pi r^2$

$y = 0$ (bottom)

**Figure 6.76**

> Recall that $g \approx 9.8 \text{ m/s}^2$. You should verify that the units are consistent in this calculation: The units of $\rho$, $g$, $A(y)$, $D(y)$, and $dy$ are kg/m$^3$, m/s$^2$, m$^2$, m, and m, respectively. The resulting units of $W$ are kg m$^2$/s$^2$, or J. A more convenient unit for large amounts of work and energy is the kilowatt-hour, which is 3.6 million joules.

**QUICK CHECK 7** In Example 4, how would the integral change if the outflow pipe were at the top of the tank? ◄

We now use this procedure to solve two pumping problems.

**EXAMPLE 4**    **Pumping water** How much work is needed to pump all the water out of a cylindrical tank with a height of 10 m and a radius of 5 m? The water is pumped to an outflow pipe 15 m above the bottom of the tank.

**SOLUTION** Figure 6.76 shows the cylindrical tank filled to capacity and the outflow 15 m above the bottom of the tank. We let $y = 0$ represent the bottom of the tank and $y = 10$ represent the top of the tank. In this case, all horizontal slices are circular disks of radius $r = 5$ m. Therefore, for $0 \le y \le 10$, the cross-sectional area is

$$A(y) = \pi r^2 = \pi 5^2 = 25\pi.$$

Note that the water is pumped to a level $h = 15$ m above the bottom of the tank, so the lifting distance is $D(y) = 15 - y$. The resulting work integral is

$$W = \int_0^{10} \rho g \underbrace{A(y)}_{25\pi} \underbrace{D(y)}_{15 - y}\, dy = 25\pi\rho g \int_0^{10} (15 - y)\, dy.$$

Substituting $\rho = 1000 \text{ kg/m}^3$ and $g = 9.8 \text{ m/s}^2$, the total work in joules is

$$W = 25\pi\rho g \int_0^{10} (15 - y)\, dy$$

$$= 25\pi \underbrace{(1000)}_{\rho}\underbrace{(9.8)}_{g}\left( 15y - \frac{1}{2}y^2 \right)\Big|_0^{10}$$

$$\approx 7.7 \times 10^7.$$

The work required to pump the water out of the tank is approximately 77 million joules.

*Related Exercises 36–37* ◄

**EXAMPLE 5    Pumping gasoline**   A cylindrical tank with a length of 10 m and a radius of 5 m is on its side and half full of gasoline (**Figure 6.77**). How much work is required to empty the tank through an outlet pipe at the top of the tank? (The density of gasoline is $\rho = 737 \text{ kg/m}^3$.)

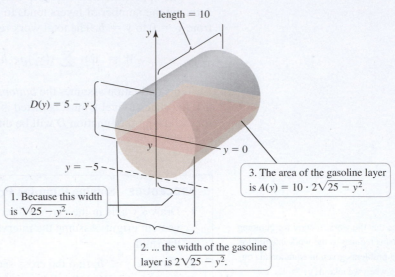

length = 10

$D(y) = 5 - y$

$y = -5$

$y = 0$

1. Because this width is $\sqrt{25 - y^2}$...

2. ... the width of the gasoline layer is $2\sqrt{25 - y^2}$.

3. The area of the gasoline layer is $A(y) = 10 \cdot 2\sqrt{25 - y^2}$.

**Figure 6.77**

> The equation of the right side of the circle is $x = \sqrt{25 - y^2}$.

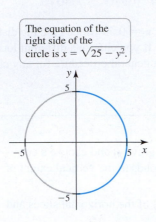

> Again, there are several choices for the location of the origin. The location in this example makes $A(y)$ easy to compute.

**SOLUTION**   In this problem, we choose a different origin by letting $y = 0$ and $y = -5$ correspond to the center and the bottom of the tank, respectively. For $-5 \leq y \leq 0$, a horizontal layer of gasoline located at a depth $y$ is a rectangle with a length of 10 and a width of $2\sqrt{25 - y^2}$ (Figure 6.77). Therefore, the cross-sectional area of the layer at depth $y$ is

$$A(y) = 20\sqrt{25 - y^2}.$$

The distance the layer at level $y$ must be lifted to reach the top of the tank is $D(y) = 5 - y$, where $-5 \leq y \leq 0$; note that $5 \leq D(y) \leq 10$. The resulting work integral is

$$W = \underbrace{737}_{\rho}\underbrace{(9.8)}_{g}\int_{-5}^{0}\underbrace{20\sqrt{25 - y^2}}_{A(y)}\underbrace{(5 - y)}_{D(y)}\,dy = 144{,}452\int_{-5}^{0}\sqrt{25 - y^2}\,(5 - y)\,dy.$$

This integral is evaluated by splitting the integrand into two pieces and recognizing that one piece is the area of a quarter circle of radius 5:

$$\int_{-5}^{0}\sqrt{25 - y^2}\,(5 - y)\,dy = 5\underbrace{\int_{-5}^{0}\sqrt{25 - y^2}\,dy}_{\text{area of quarter circle}} - \underbrace{\int_{-5}^{0}y\sqrt{25 - y^2}\,dy}_{\text{let } u = 25 - y^2;\ du = -2y\,dy}$$

$$= 5\cdot\frac{25\pi}{4} + \frac{1}{2}\int_{0}^{25}\sqrt{u}\,du$$

$$= \frac{125\pi}{4} + \frac{1}{3}u^{3/2}\Big|_{0}^{25} = \frac{375\pi + 500}{12}.$$

Multiplying this result by $20\,\rho g = 144{,}452$, we find that the work required is approximately 20.2 million joules.

*Related Exercises 42–44* ◄

## Force and Pressure

Another application of integration deals with the force exerted on a surface by a body of water. Again, we need a few physical principles.

   Pressure is a force per unit area, measured in units such as newtons per square meter ($\text{N/m}^2$). For example, the pressure of the atmosphere on the surface of Earth is about 14 lb/in² (approximately 100 kilopascals, or $10^5 \text{ N/m}^2$). As another example, if you stood on the bottom of a swimming pool, you would feel pressure due to the weight (force) of the column of water above your head. If your head is flat and has surface area $A$ m$^2$ and it is $h$ meters below the surface, then the column of water above your head has volume $Ah$ m$^3$. That column of water exerts a force (its weight)

$$F = \text{mass} \cdot \text{acceleration} = \underbrace{\text{volume} \cdot \text{density}}_{\text{mass}} \cdot g = Ah\rho g,$$

where $\rho$ is the density of water and $g$ is the acceleration due to gravity. Therefore, the pressure on your head is the force divided by the surface area of your head:

$$\text{pressure} = \frac{\text{force}}{A} = \frac{Ah\rho g}{A} = \rho gh.$$

This pressure is called **hydrostatic pressure** (meaning the pressure of *water at rest*), and it has the following important property: *It has the same magnitude in all directions.* Specifically, the hydrostatic pressure on a vertical wall of the swimming pool at a depth $h$ is also $\rho gh$. This is the only fact needed to find the total force on vertical walls such as dams and swimming pools. We assume the water completely covers the face of the dam.

The first step in finding the force on the face of the dam is to introduce a coordinate system. We choose a $y$-axis pointing upward with $y = 0$ corresponding to the base of the dam and $y = a$ corresponding to the top of the dam (**Figure 6.78**). Because the pressure varies with depth ($y$-direction), the dam is sliced horizontally into $n$ strips of equal thickness $\Delta y$. The $k$th strip corresponds to the interval $[y_{k-1}, y_k]$, and we let $y_k^*$ be any point in that interval. The depth of that strip is approximately $h = a - y_k^*$, so the hydrostatic pressure on that strip is approximately $\rho g(a - y_k^*)$.

The crux of any dam problem is finding the width of the strips as a function of $y$, which we denote $w(y)$. Each dam has its own width function; however, once the width function is known, the solution follows directly. The approximate area of the $k$th strip is its width multiplied by its thickness, or $w(y_k^*)\Delta y$. The force on the $k$th strip (which is the area of the strip multiplied by the pressure) is approximately

$$F_k = \underbrace{\rho g(a - y_k^*)}_{\text{pressure}}\underbrace{w(y_k^*)\Delta y}_{\text{area of strip}}.$$

Summing the forces over the $n$ strips, the total force is

$$F \approx \sum_{k=1}^{n} F_k = \sum_{k=1}^{n} \rho g(a - y_k^*)w(y_k^*)\Delta y.$$

To find the exact force, we let the thickness of the strips tend to zero and the number of strips tend to infinity, which produces a definite integral. The limits of integration correspond to the base ($y = 0$) and top ($y = a$) of the dam. Therefore, the total force on the dam is

$$F = \lim_{n \to \infty} \sum_{k=1}^{n} \rho g(a - y_k^*)w(y_k^*)\Delta y = \int_0^a \rho g(a - y)w(y)\,dy.$$

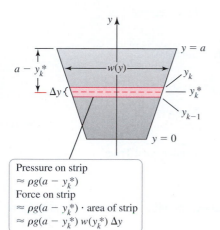

Pressure on strip
$\approx \rho g(a - y_k^*)$
Force on strip
$\approx \rho g(a - y_k^*) \cdot$ area of strip
$\approx \rho g(a - y_k^*)\, w(y_k^*)\, \Delta y$

**Figure 6.78**

➤ We have chosen $y = 0$ to be the base of the dam. Depending on the geometry of the problem, it may be more convenient (less computation) to let $y = 0$ be at the top of the dam. Note that the depth function $D(y) = a - y$ varies with the chosen coordinate system. Experiment with different choices.

**PROCEDURE   Solving Force-on-Dam Problems**

1. Draw a $y$-axis on the face of the dam in the vertical direction and choose a convenient origin (often taken to be the base of the dam).

2. Find the width function $w(y)$ for each value of $y$ on the face of the dam.

3. If the base of the dam is at $y = 0$ and the top of the dam is at $y = a$, then the total force on the dam is

$$F = \int_0^a \rho g\underbrace{(a - y)}_{\text{depth}}\underbrace{w(y)}_{\text{width}}\,dy.$$

**Figure 6.79**

**EXAMPLE 6   Force on a dam**   A large vertical dam in the shape of a symmetric trapezoid has a height of 30 m, a width of 20 m at its base, and a width of 40 m at the top (**Figure 6.79**). What is the total force on the face of the dam when the reservoir is full?

**SOLUTION**  We place the origin at the center of the base of the dam (**Figure 6.80**). The right slanted edge of the dam is a segment of the line that passes through the points $(10, 0)$ and $(20, 30)$. An equation of that line is

$$y - 0 = \frac{30}{10}(x - 10) \quad \text{or} \quad y = 3x - 30 \quad \text{or} \quad x = \frac{1}{3}(y + 30).$$

▶ You should check the width function: $w(0) = 20$ (the width of the dam at its base) and $w(30) = 40$ (the width of the dam at its top).

Notice that at a depth of $y$, where $0 \le y \le 30$, the width of the dam is

$$w(y) = 2x = \frac{2}{3}(y + 30).$$

**Figure 6.80**

Using $\rho = 1000 \text{ kg/m}^3$ and $g = 9.8 \text{ m/s}^2$, the total force on the dam (in newtons) is

$$F = \int_0^a \rho g(a - y)w(y)\, dy \qquad \text{Force integral}$$

$$= \rho g \int_0^{30} \underbrace{(30 - y)}_{a-y} \underbrace{\frac{2}{3}(y + 30)}_{w(y)}\, dy \qquad \text{Substitute.}$$

$$= \frac{2}{3}\rho g \int_0^{30} (900 - y^2)\, dy \qquad \text{Simplify.}$$

$$= \frac{2}{3}\rho g \left( 900y - \frac{y^3}{3} \right) \Big|_0^{30} \qquad \text{Fundamental Theorem}$$

$$\approx 1.18 \times 10^8.$$

The force of $1.18 \times 10^8$ N on the dam amounts to about 26 million pounds, or 13,000 tons.

*Related Exercises 46–47* ◀

---

## SECTION 6.7 EXERCISES

### Getting Started

1.  Suppose a 1-m cylindrical bar has a constant density of 1 g/cm for its left half and a constant density of 2 g/cm for its right half. What is its mass?

2.  Explain how to find the mass of a one-dimensional object with a variable density $\rho$.

3.  How much work is required to move an object from $x = 0$ to $x = 5$ (measured in meters) in the presence of a constant force of 5 N acting along the $x$-axis?

4.  Why is integration used to find the work done by a variable force?

5.  Why is integration used to find the work required to pump water out of a tank?

6.  Why is integration used to find the total force on the face of a dam?

7.  What is the pressure on a horizontal surface with an area of 2 m$^2$ that is 4 m underwater?

8.  Explain why you integrate in the vertical direction (parallel to the acceleration due to gravity) rather than the horizontal direction to find the force on the face of a dam.

**9–12.** *Consider the cylindrical tank in Example 4 that has a height of 10 m and a radius of 5 m. Recall that if the tank is full of water, then $\int_0^{10} 25\pi \rho g(15 - y)\, dy$ equals the work required to pump all the water out of the tank through an outflow pipe that is 15 m above the bottom of the tank. Revise this work integral for the following scenarios. (Do not evaluate the integrals.)*

9.  The work required to empty the top half of the tank

10.  The work required to empty the tank if it is half full

11.  The work required to empty the tank through an outflow pipe at the top of the tank

12.  The work required to empty the tank if the water in the tank is only 3 m deep and the outflow pipe is at the top of the tank

### Practice Exercises

**13–20. Mass of one-dimensional objects** *Find the mass of the following thin bars with the given density function.*

13.  $\rho(x) = 1 + \sin x$, for $0 \le x \le \pi$

14.  $\rho(x) = 1 + x^3$, for $0 \le x \le 1$

15. $\rho(x) = 2 - \dfrac{x}{2}$, for $0 \leq x \leq 2$

16. $\rho(x) = 1 + 3 \sin x$, for $0 \leq x \leq \pi$

17. $\rho(x) = x\sqrt{2 - x^2}$, for $0 \leq x \leq 1$

18. $\rho(x) = \begin{cases} 1 & \text{if } 0 \leq x \leq 2 \\ 2 & \text{if } 2 < x \leq 3 \end{cases}$

19. $\rho(x) = \begin{cases} 1 & \text{if } 0 \leq x \leq 2 \\ 1 + x & \text{if } 2 < x \leq 4 \end{cases}$

20. $\rho(x) = \begin{cases} x^2 & \text{if } 0 \leq x \leq 1 \\ x(2 - x) & \text{if } 1 < x \leq 2 \end{cases}$

21. **Work from force** How much work is required to move an object from $x = 0$ to $x = 3$ (measured in meters) in the presence of a force (in N) given by $F(x) = 2x$ acting along the $x$-axis?

22. **Work from force** How much work is required to move an object from $x = 1$ to $x = 3$ (measured in meters) in the presence of a force (in N) given by $F(x) = \dfrac{2}{x^2}$ acting along the $x$-axis?

23. **Compressing and stretching a spring** Suppose a force of 30 N is required to stretch and hold a spring 0.2 m from its equilibrium position.

    a. Assuming the spring obeys Hooke's law, find the spring constant $k$.

    b. How much work is required to compress the spring 0.4 m from its equilibrium position?

    c. How much work is required to stretch the spring 0.3 m from its equilibrium position?

    d. How much additional work is required to stretch the spring 0.2 m if it has already been stretched 0.2 m from its equilibrium position?

24. **Compressing and stretching a spring** Suppose a force of 15 N is required to stretch and hold a spring 0.25 m from its equilibrium position.

    a. Assuming the spring obeys Hooke's law, find the spring constant $k$.

    b. How much work is required to compress the spring 0.2 m from its equilibrium position?

    c. How much additional work is required to stretch the spring 0.3 m if it has already been stretched 0.25 m from its equilibrium position?

25. **Work done by a spring** A spring on a horizontal surface can be stretched and held 0.5 m from its equilibrium position with a force of 50 N.

    a. How much work is done in stretching the spring 1.5 m from its equilibrium position?

    b. How much work is done in compressing the spring 0.5 m from its equilibrium position?

26. **Shock absorber** A heavy-duty shock absorber is compressed 2 cm from its equilibrium position by a mass of 500 kg. How much work is required to compress the shock absorber 4 cm from its equilibrium position? (A mass of 500 kg exerts a force (in newtons) of $500g$, where $g = 9.8 \text{ m/s}^2$.)

27. **Calculating work for different springs** Calculate the work required to stretch the following springs 0.5 m from their equilibrium positions. Assume Hooke's law is obeyed.

    a. A spring that requires a force of 50 N to be stretched 0.2 m from its equilibrium position

    b. A spring that requires 50 J of work to be stretched 0.2 m from its equilibrium position

28. **Calculating work for different springs** Calculate the work required to stretch the following springs 0.4 m from their equilibrium positions. Assume Hooke's law is obeyed.

    a. A spring that requires a force of 50 N to be stretched 0.1 m from its equilibrium position

    b. A spring that requires 2 J of work to be stretched 0.1 m from its equilibrium position

29. **Calculating work for different springs** Calculate the work required to stretch the following springs 1.25 m from their equilibrium positions. Assume Hooke's law is obeyed.

    a. A spring that requires 100 J of work to be stretched 0.5 m from its equilibrium position

    b. A spring that requires a force of 250 N to be stretched 0.5 m from its equilibrium position

30. **Work function** A spring has a restoring force given by $F(x) = 25x$. Let $W(x)$ be the work required to stretch the spring from its equilibrium position $(x = 0)$ to a variable distance $x$. Find and graph the work function. Compare the work required to stretch the spring $x$ units from equilibrium to the work required to compress the spring $x$ units from equilibrium.

31. **Winding a chain** A 30-m-long chain hangs vertically from a cylinder attached to a winch. Assume there is no friction in the system and the chain has a density of 5 kg/m.

    a. How much work is required to wind the entire chain onto the cylinder using the winch?

    b. How much work is required to wind the chain onto the cylinder if a 50-kg block is attached to the end of the chain?

32. **Coiling a rope** A 60-m-long, 9.4-mm-diameter rope hangs freely from a ledge. The density of the rope is 55 g/m. How much work is needed to pull the entire rope to the ledge?

33. **Winding part of a chain** A 20-m-long, 50-kg chain hangs vertically from a cylinder attached to a winch. How much work is needed to wind the upper half of the chain onto the winch?

34. **Leaky bucket** A 1-kg bucket resting on the ground contains 3 kg of water. How much work is required to raise the bucket vertically a distance of 10 m if water leaks out of the bucket at a constant rate of $\frac{1}{5}$ kg/m? Assume the weight of the rope used to raise the bucket is negligible. (*Hint:* Use the definition of work, $W = \int_a^b F(y)\, dy$, where $F$ is the variable force required to lift an object along a vertical line from $y = a$ to $y = b$.)

35. **Emptying a swimming pool** A swimming pool has the shape of a box with a base that measures 25 m by 15 m and a uniform depth of 2.5 m. How much work is required to pump the water out of the pool (to the level of the top of the pool) when it is full?

36. **Emptying a cylindrical tank** A cylindrical water tank has height 8 m and radius 2 m (see figure).

    a. If the tank is full of water, how much work is required to pump the water to the level of the top of the tank and out of the tank?

**b.** Is it true that it takes half as much work to pump the water out of the tank when it is half full as when it is full? Explain.

**37. Emptying a half-full cylindrical tank** Suppose the water tank in Exercise 36 is half full of water. Determine the work required to empty the tank by pumping the water to a level 2 m above the top of the tank.

**38. Emptying a partially filled swimming pool** If the water in the swimming pool in Exercise 35 is 2 m deep, then how much work is required to pump all the water to a level 3 m above the bottom of the pool?

**39. Emptying a conical tank** A water tank is shaped like an inverted cone with height 6 m and base radius 1.5 m (see figure).

**a.** If the tank is full, how much work is required to pump the water to the level of the top of the tank and out of the tank?

**b.** Is it true that it takes half as much work to pump the water out of the tank when it is filled to half its depth as when it is full? Explain.

**40. Upper and lower half** A cylinder with height 8 m and radius 3 m is filled with water and must be emptied through an outlet pipe 2 m above the top of the cylinder.

**a.** Compute the work required to empty the water in the top half of the tank.

**b.** Compute the work required to empty the (equal amount of) water in the lower half of the tank.

**c.** Interpret the results of parts (a) and (b).

**41. Filling a spherical tank** A spherical water tank with an inner radius of 8 m has its lowest point 2 m above the ground. It is filled by a pipe that feeds the tank at its lowest point (see figure). Neglecting the volume of the inflow pipe, how much work is required to fill the tank if it is initially empty?

**42. Emptying a water trough** A water trough has a semicircular cross section with a radius of 0.25 m and a length of 3 m (see figure).

**a.** How much work is required to pump the water out of the trough (to the level of the top of the trough) when it is full?

**b.** If the length is doubled, is the required work doubled? Explain.

**c.** If the radius is doubled, is the required work doubled? Explain.

**43. Emptying a water trough** A cattle trough has a trapezoidal cross section with a height of 1 m and horizontal sides of length $\frac{1}{2}$ m and 1 m. Assume the length of the trough is 10 m (see figure).

**a.** How much work is required to pump the water out of the trough (to the level of the top of the trough) when it is full?

**b.** If the length is doubled, is the required work doubled? Explain.

**44. Pumping water** Suppose the tank in Example 5 is full of water (rather than half full of gasoline). Determine the work required to pump all the water to an outlet pipe 15 m above the bottom of the tank.

**45. Emptying a conical tank** An inverted cone (base above the vertex) is 2 m high and has a base radius of $\frac{1}{2}$ m. If the tank is full, how much work is required to pump the water to a level 1 m above the top of the tank?

**46–49. Force on dams** *The following figures show the shapes and dimensions of small dams. Assuming the water level is at the top of the dam, find the total force on the face of the dam.*

**46.**

**47.**

**48.**

**49.**

**50. Parabolic dam** The lower edge of a dam is defined by the parabola $y = x^2/16$ (see figure). Use a coordinate system with $y = 0$ at the bottom of the dam to determine the total force of the water on the dam. Lengths are measured in meters. Assume the water level is at the top of the dam.

**51. Force on a triangular plate** A plate shaped like an isosceles triangle with a height of 1 m is placed on a vertical wall 1 m below the surface of a pool filled with water (see figure). Compute the force on the plate.

**52–54. Force on a window** *A diving pool that is* 4 m *deep and full of water has a viewing window on one of its vertical walls. Find the force on the following windows.*

**52.** The window is a square, 0.5 m on a side, with the lower edge of the window on the bottom of the pool.

**53.** The window is a square, 0.5 m on a side, with the lower edge of the window 1 m from the bottom of the pool.

**54.** The window is circular, with a radius of 0.5 m, tangent to the bottom of the pool.

**55. Force on a building** A large building shaped like a box is 50 m high with a face that is 80 m wide. A strong wind blows directly at the face of the building, exerting a pressure of 150 N/m² at the ground and increasing with height according to $P(y) = 150 + 2y$, where $y$ is the height above the ground. Calculate the total force on the building, which is a measure of the resistance that must be included in the design of the building.

**56. Force on the end of a tank** Determine the force on a circular end of the tank in Figure 6.77 if the tank is full of gasoline. The density of gasoline is $\rho = 737$ kg/m³.

**57. Explain why or why not** Determine whether the following statements are true and give an explanation or counterexample.

  **a.** The mass of a thin wire is the length of the wire times its average density over its length.

**b.** The work required to stretch a linear spring (that obeys Hooke's law) 100 cm from equilibrium is the same as the work required to compress it 100 cm from equilibrium.

**c.** The work required to lift a 10-kg object vertically 10 m is the same as the work required to lift a 20-kg object vertically 5 m.

**d.** The total force on a 10-ft² region on the (horizontal) floor of a pool is the same as the total force on a 10-ft² region on a (vertical) wall of the pool.

### Explorations and Challenges

**58. Mass of two bars** Two bars of length $L$ have densities $\rho_1(x) = 4(x + 1)^{-2}$ and $\rho_2(x) = 6(x + 1)^{-3}$, for $0 \le x \le L$.

  **a.** For what values of $L$ is bar 1 heavier than bar 2?

  **b.** As the lengths of the bars increase, do their masses increase without bound? Explain.

**59. A nonlinear spring** Hooke's law is applicable to idealized (linear) springs that are not stretched or compressed too far from their equilibrium positions. Consider a nonlinear spring whose restoring force is given by $F(x) = 16x - 0.1x^3$, for $|x| \le 7$.

  **a.** Graph the restoring force and interpret it.

  **b.** How much work is done in stretching the spring from its equilibrium position ($x = 0$) to $x = 1.5$?

  **c.** How much work is done in compressing the spring from its equilibrium position ($x = 0$) to $x = -2$?

**60. A vertical spring** A 10-kg mass is attached to a spring that hangs vertically and is stretched 2 m from the equilibrium position of the spring. Assume a linear spring with $F(x) = kx$.

  **a.** How much work is required to compress the spring and lift the mass 0.5 m?

  **b.** How much work is required to stretch the spring and lower the mass 0.5 m?

**61. Leaky cement bucket** A 350 kg-bucket containing 4650 kg of cement is resting on the ground when a crane begins lifting it at a constant rate of 5 m/min. As the crane raises the bucket, cement leaks out of the bucket at a constant rate of 100 kg/min. How much work is required to lift the bucket a distance of 30 m if we ignore the weight of the crane cable attached to the bucket?

**62. Emptying a real swimming pool** A swimming pool is 20 m long and 10 m wide, with a bottom that slopes uniformly from a depth of 1 m at one end to a depth of 2 m at the other end (see figure). Assuming the pool is full, how much work is required to pump the water to a level 0.2 m above the top of the pool?

**63. Drinking juice** A glass has circular cross sections that taper (linearly) from a radius of 5 cm at the top of the glass to a radius of 4 cm at the bottom. The glass is 15 cm high and full of orange juice. How much work is required to drink all the juice through a straw if your mouth is 5 cm above the top of the glass? Assume the density of orange juice equals the density of water.

**64. Lifting a pendulum** A body of mass $m$ is suspended by a rod of length $L$ that pivots without friction (see figure). The mass is slowly lifted along a circular arc to a height $h$.

   **a.** Assuming the only force acting on the mass is the gravitational force, show that the component of this force acting along the arc of motion is $F = mg \sin \theta$.

   **b.** Noting that an element of length along the path of the pendulum is $ds = L \, d\theta$, evaluate an integral in $\theta$ to show that the work done in lifting the mass to a height $h$ is $mgh$.

**65. Critical depth** A large tank has a plastic window on one wall that is designed to withstand a force of 90,000 N. The square window is 2 m on a side, and its lower edge is 1 m from the bottom of the tank.

   **a.** If the tank is filled to a depth of 4 m, will the window withstand the resulting force?

   **b.** What is the maximum depth to which the tank can be filled without the window failing?

**66. Orientation and force** A plate shaped like an equilateral triangle 1 m on a side is placed on a vertical wall 1 m below the surface of a pool filled with water. On which plate in the figure is the force greater? Try to anticipate the answer and then compute the force on each plate.

**67. Orientation and force** A square plate 1 m on a side is placed on a vertical wall 1 m below the surface of a pool filled with water. On which plate in the figure is the force greater? Try to anticipate the answer and then compute the force on each plate.

**68. Work by two different integrals** A rigid body with a mass of 2 kg moves along a line due to a force that produces a position function $x(t) = 4t^2$, where $x$ is measured in meters and $t$ is measured in seconds. Find the work done during the first 5 s in two ways.

   **a.** Note that $x''(t) = 8$; then use Newton's second law ($F = ma = mx''(t)$) to evaluate the work integral $W = \int_{x_0}^{x_f} F(x) \, dx$, where $x_0$ and $x_f$ are the initial and final positions, respectively.

   **b.** Change variables in the work integral and integrate with respect to $t$. Be sure your answer agrees with part (a).

**69. Work in a gravitational field** For large distances from the surface of Earth, the gravitational force is given by $F(x) = \dfrac{GMm}{(x + R)^2}$, where $G = 6.7 \times 10^{-11}$ N-m$^2$/kg$^2$ is the gravitational constant, $M = 6 \times 10^{24}$ kg is the mass of Earth, $m$ is the mass of the object in the gravitational field, $R = 6.378 \times 10^6$ m is the radius of Earth, and $x \geq 0$ is the distance above the surface of Earth (in meters).

   **a.** How much work is required to launch a rocket with a mass of 500 kg in a vertical flight path to a height of 2500 km (from Earth's surface)?

   **b.** Find the work required to launch the rocket to a height of $x$ kilometers, for $x > 0$.

   **c.** How much work is required to reach outer space ($x \to \infty$)?

   **d.** Equate the work in part (c) to the initial kinetic energy of the rocket, $\frac{1}{2}mv^2$, to compute the escape velocity of the rocket.

**70. Buoyancy** Archimedes' principle says that the buoyant force exerted on an object that is (partially or totally) submerged in water is equal to the weight of the water displaced by the object (see figure). Let $\rho_w = 1$ g/cm$^3 = 1000$ kg/m$^3$ be the density of water and let $\rho$ be the density of an object in water. Let $f = \rho/\rho_w$. If $0 < f \leq 1$, then the object floats with a fraction $f$ of its volume submerged; if $f > 1$, then the object sinks. Consider a cubical box with sides 2 m long floating in water with one-half of its volume submerged ($\rho = \rho_w/2$). Find the force required to fully submerge the box (so its top surface is at the water level). (See the Guided Project *Buoyancy and Archimedes' Principle* for further explorations of buoyancy problems.)

**QUICK CHECK ANSWERS**

**1. a.** The bar is heaviest at the left end and lightest at the right end. **b.** $\rho = 2.5$ g/cm. **2.** Minimum mass = 2 kg; maximum mass = 10 kg **3.** We assume the force is constant over each subinterval, when, in fact, it varies over each subinterval. **4.** The restoring force of the spring increases as the spring is stretched ($f(x) = 100x$). Greater restoring forces are encountered on the interval $[0.1, 0.35]$ than on the interval $[0, 0.25]$. **5.** The chain is a flexible object, and different points on the chain are lifted different distances as it is pulled to the platform. The bucket, however, is a rigid object, and every point in the bucket is lifted 11 m when it goes from resting on the ground to resting on the platform (in other words, it can be treated as a point mass). **6.** $\int_0^{10} 1.5gy \, dy = 735$ joules. **7.** The factor $(15 - y)$ in the integral is replaced with $(10 - y)$. ◄

## CHAPTER 6   REVIEW EXERCISES

1. **Explain why or why not** Determine whether the following statements are true and give an explanation or counterexample.

   a. A region $R$ is revolved about the $y$-axis to generate a solid $S$. To find the volume of $S$, you could either use the disk/washer method and integrate with respect to $y$ or use the shell method and integrate with respect to $x$.

   b. Given only the velocity of an object moving on a line, it is possible to find its displacement, but not its position.

   c. If water flows into a tank at a constant rate (for example, 6 gal/min), the volume of water in the tank increases according to a linear function of time.

**2–3. Displacement, distance, and position** *Consider an object moving along a line with the following velocities and initial positions. Assume time t is measured in seconds and velocities have units of m/s.*

a. *Over the given interval, determine when the object is moving in the positive direction and when it is moving in the negative direction.*

b. *Find the displacement over the given interval.*

c. *Find the distance traveled over the given interval.*

d. *Determine the position function $s(t)$ using the Fundamental Theorem of Calculus (Theorem 6.1). Check your answer by finding the position function using the antiderivative method.*

2. $v(t) = 1 - 2\cos(\pi t/3)$, for $0 \le t \le 3$; $s(0) = 0$

3. $v(t) = 12t^2 - 30t + 12$, for $0 \le t \le 3$; $s(0) = 1$

4. **Displacement from velocity** The velocity of an object moving along a line is given by $v(t) = 20\cos \pi t$ (in ft/s). What is the displacement of the object after 1.5 s?

5. **Position, displacement, and distance** A projectile is launched vertically from the ground at $t = 0$, and its velocity in flight (in m/s) is given by $v(t) = 20 - 10t$. Find the position, displacement, and distance traveled after $t$ seconds, for $0 \le t \le 4$.

6. **Deceleration** At $t = 0$, a car begins decelerating from a velocity of 80 ft/s at a constant rate of 5 ft/s². Find its position function assuming $s(0) = 0$.

7. **An oscillator** The acceleration of an object moving along a line is given by $a(t) = 2\sin \dfrac{\pi t}{4}$. The initial velocity and position are $v(0) = -\dfrac{8}{\pi}$ and $s(0) = 0$.

   a. Find the velocity and position for $t \ge 0$.

   b. What are the minimum and maximum values of $s$?

   c. Find the average velocity and average position over the interval $[0, 8]$.

8. **A race** Starting at the same point on a straight road, Anna and Benny begin running with velocities (in mi/hr) given by $v_A(t) = 2t + 1$ and $v_B(t) = 4 - t$, respectively.

   a. Graph the velocity functions, for $0 \le t \le 4$.

   b. If the runners run for 1 hr, who runs farther? Interpret your conclusion geometrically using the graph in part (a).

   c. If the runners run for 6 mi, who wins the race? Interpret your conclusion geometrically using the graph in part (a).

9. **Fuel consumption** A small plane in flight consumes fuel at a rate (in gal/min) given by

   $$R'(t) = \begin{cases} 4t^{1/3} & \text{if } 0 \le t \le 8 \quad \text{(take-off)} \\ 2 & \text{if } t > 8 \quad \text{(cruising)}. \end{cases}$$

   a. Find a function $R$ that gives the total fuel consumed, for $0 \le t \le 8$.

   b. Find a function $R$ that gives the total fuel consumed, for $t \ge 0$.

   c. If the fuel tank capacity is 150 gal, when does the fuel run out?

**T 10. Decreasing velocity** A projectile is fired upward, and its velocity (in m/s) is given by $v(t) = \dfrac{200}{\sqrt{t + 1}}$, for $t \ge 0$.

   a. Graph the velocity function, for $t \ge 0$.

   b. Find and graph the position function for the projectile, for $t \ge 0$, assuming $s(0) = 0$.

   c. Given unlimited time, can the projectile travel 2500 m? If so, at what time does the distance traveled equal 2500 m?

11. Explain why $\int_0^2 x^2\,dx = \int_0^4 (2 - \sqrt{y})\,dy$ without evaluating the integrals.

**12–23. Areas of regions** *Determine the area of the given region.*

12.

13.

14.

15. The region bounded by $y = x$, $y = \sqrt{x - 1}$, $y = 2$, and $y = 0$

16. The region bounded by $y = \sin\dfrac{x}{2}$ and $y = \cos\dfrac{x}{2}$ on $\left[\dfrac{-3\pi}{2}, \dfrac{5\pi}{2}\right]$

**T 17.** The region in the first quadrant bounded by $y = 4x$ and $y = x\sqrt{25 - x^2}$

**18.** The region bounded by $y = 6 - 3x^2$ and $y = 6x - 3$

**19.** The region bounded by $y = x^2$, $y = 2x^2 - 4x$, and $y = 0$

**20.** The region in the first quadrant bounded by the curve $\sqrt{x} + \sqrt{y} = 1$

**21.** The region in the first quadrant bounded by $y = x/6$ and $y = 1 - |x/2 - 1|$

**22.** The region in the first quadrant bounded by $y = x^p$ and $y = \sqrt[p]{x}$, where $p = 100$ and $p = 1000$ (two area calculations)

**23.** The region bounded by $y = 2x^2 - 6x + 5$ and $y = 1$

**24.** Determine the area of the region bounded by the curves $x = y^2$ and $x = (2 - y^2)^2$ (see figure).

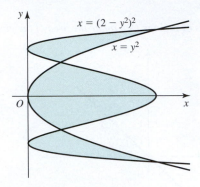

**25–31. Multiple regions** *The regions $R_1$, $R_2$, and $R_3$ (see figure) are formed by the graphs of $y = 2\sqrt{x}$, $y = 3 - x$, and $x = 3$.*

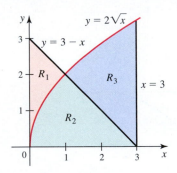

**25.** Find the area of each of the regions $R_1$, $R_2$, and $R_3$.

**26.** Find the volume of the solid obtained by revolving region $R_1$ about the $x$-axis.

**27.** Find the volume of the solid obtained by revolving region $R_1$ about the $y$-axis.

**28.** Find the volume of the solid obtained by revolving region $R_2$ about the $y$-axis.

**29.** Find the volume of the solid obtained by revolving region $R_2$ about the $x$-axis.

**30.** Use the disk method to find an integral, or sum of integrals, that equals the volume of the solid obtained by revolving region $R_3$ about the line $x = 3$. Do not evaluate the integral.

**31.** Use the shell method to find an integral, or sum of integrals, that equals the volume of the solid obtained by revolving region $R_3$ about the line $x = 3$. Do not evaluate the integral.

**32. Area and volume** The region $R$ is bounded by the curves $x = y^2 + 2$, $y = x - 4$, and $y = 0$ (see figure).

   **a.** Write a single integral that gives the area of $R$.
   **b.** Write a single integral that gives the volume of the solid generated when $R$ is revolved about the $x$-axis.
   **c.** Write a single integral that gives the volume of the solid generated when $R$ is revolved about the $y$-axis.
   **d.** Suppose $S$ is a solid whose base is $R$ and whose cross sections perpendicular to $R$ and parallel to the $x$-axis are semicircles. Write a single integral that gives the volume of $S$.

**33–36. Area and volume** *Let $R$ be the region in the first quadrant bounded by the graph of $f(x) = \begin{cases} 1 & \text{if } 0 \le x \le 1 \\ 2 - \sqrt{x} & \text{if } 1 < x \le 4. \end{cases}$*

**33.** Find the area of the region $R$.

**34.** Find the volume of the solid generated when $R$ is revolved about the $x$-axis.

**35.** Find the volume of the solid generated when $R$ is revolved about the $y$-axis.

**T 36.** Find the area of the surface generated when the curve $y = f(x)$, for $0 \le x \le 4$, is revolved about the $x$-axis.

**T 37.** Find the area of the shaded regions $R_1$ and $R_2$ shown in the figure.

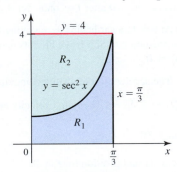

**38. An area function** Let $R(x)$ be the area of the shaded region between the graphs of $y = f(t)$ and $y = g(t)$ on the interval $[a, x]$ (see figure).

   **a.** Sketch a plausible graph of $R$, for $a \le x \le c$.
   **b.** Give expressions for $R(x)$ and $R'(x)$, for $a \le x \le c$.

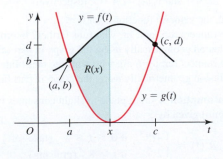

**39. An area function** Consider the functions $y = \dfrac{x^2}{a}$ and $y = \sqrt{\dfrac{x}{a}}$, where $a > 0$. Find $A(a)$, the area of the region between the curves.

**40. Two methods** The region $R$ in the first quadrant bounded by the parabola $y = 4 - x^2$ and coordinate axes is revolved about the $y$-axis to produce a dome-shaped solid. Find the volume of the solid in the following ways:

**a.** Apply the disk method and integrate with respect to $y$.
**b.** Apply the shell method and integrate with respect to $x$.

**41–55. Volumes of solids** *Choose the general slicing method, the disk/washer method, or the shell method to answer the following questions.*

**41.** What is the volume of the solid whose base is the region in the first quadrant bounded by $y = \sqrt{x}$, $y = 2 - x$, and the $x$-axis, and whose cross sections perpendicular to the base and parallel to the $y$-axis are squares?

**42.** What is the volume of the solid whose base is the region in the first quadrant bounded by $y = \sqrt{x}$, $y = 2 - x$, and the $x$-axis, and whose cross sections perpendicular to the base and parallel to the $y$-axis are semicircles?

**43.** What is the volume of the solid whose base is the region in the first quadrant bounded by $y = \sqrt{x}$, $y = 2 - x$, and the $y$-axis, and whose cross sections perpendicular to the base and parallel to the $x$-axis are squares?

**44.** The region bounded by the curves $y = -x^2 + 2x + 2$ and $y = 2x^2 - 4x + 2$ is revolved about the $x$-axis. What is the volume of the solid that is generated?

**45.** The region bounded by the curve $y = 1 + \sqrt{x}$, the curve $y = 1 - \sqrt{x}$, and the line $x = 1$ is revolved about the $y$-axis. Find the volume of the resulting solid by (a) integrating with respect to $x$ and (b) integrating with respect to $y$. Be sure your answers agree.

**46.** The region bounded by the curves $y = x + 1$, $y = 12/x$, and $y = 1$ is revolved about the $x$-axis. What is the volume of the solid that is generated?

**47.** The region bounded by $y = 1/x^3$, $y = 0$, $x = 2$, and $x = 4$ is revolved about the $y$-axis. What is the volume of the solid that is generated?

**48.** The region bounded by the curves $y = \sec x$ and $y = 2$, for $0 \le x \le \frac{\pi}{3}$, is revolved about the $x$-axis. What is the volume of the solid that is generated?

**49.** The region bounded by $y = (1 - x^2)^{-1/2}$ and the $x$-axis over the interval $[0, \sqrt{3}/2]$ is revolved about the $y$-axis. What is the volume of the solid that is generated?

**50.** The region bounded by the graph of $y = 4 - x^2$ and the $x$-axis on the interval $[-2, 2]$ is revolved about the line $x = -2$. What is the volume of the solid that is generated?

**51.** The region bounded by the graphs of $y = (x - 2)^2$ and $y = 4$ is revolved about the line $y = 4$. What is the volume of the resulting solid?

**52.** The region bounded by the graphs of $y = 6x$ and $y = x^2 + 5$ is revolved about the line $y = -1$ and the line $x = -1$. Find the volumes of the resulting solids. Which one is greater?

**53.** The region bounded by the graphs of $y = 2x$, $y = 6 - x$, and $y = 0$ is revolved about the line $y = -2$ and the line $x = -2$. Find the volumes of the resulting solids. Which one is greater?

**54.** The region bounded by $y = 4/\sqrt{x + 1}$, $y = 1$, and $x = 0$ is revolved about the $y$-axis. What is the volume of the solid that is generated?

**55.** The region in the first quadrant bounded by $y = \sec x$, $y = 2$, and $x = 0$ is revolved about the $x$-axis. What is the volume of the solid that is generated?

**56. Comparing volumes** Let $R$ be the region in the first quadrant bounded by the $y$-axis and the graphs of $y = 3x$ and $y = 4 - x^2$. Answer the following questions by finding the volumes of the described solids.

**a.** Which is greater, the volume of the solid obtained by revolving $R$ about the $x$-axis or the line $y = 5$?
**b.** Which is greater, the volume of the solid obtained by revolving $R$ about the $y$-axis or the line $x = 1$?

**57. Comparing volumes** Let $R$ be the region bounded by the graph of $f(x) = cx(1 - x)$ and the $x$-axis on $[0, 1]$. Find the positive value of $c$ such that the volume of the solid generated by revolving $R$ about the $x$-axis equals the volume of the solid generated by revolving $R$ about the $y$-axis.

**58–60. Arc length** *Find the length of the following curves.*

**58.** $y = 2x + 4$ on $[-2, 2]$ (Use calculus.)

**59.** $y = x^{1/2} - x^{3/2}/3$ on $[1, 3]$

**60.** $y = \dfrac{x^3}{6} + \dfrac{1}{2x}$ on $[1, 2]$

**T 61–65. Arc length by calculator** *Write and simplify the integral that gives the arc length of the following curves on the given interval. Then use technology to approximate the integral.*

**61.** $y = (x + 1)^2$ on $[2, 4]$

**62.** $y = \sin \pi x$ on $[0, 1]$

**63.** $y = \sqrt{x^4 + x^2}$ on $[0, 1]$

**64.** $y = \dfrac{x^3}{3} + \dfrac{x^2}{2}$ on $[0, 2]$

**65.** $y = \cot x$ on $\left[-\dfrac{\pi}{2}, -\dfrac{\pi}{4}\right]$

**66. Surface area and volume** Let $f(x) = \frac{1}{3}x^3$ and let $R$ be the region bounded by the graph of $f$ and the $x$-axis on the interval $[0, 2]$.

**a.** Find the area of the surface generated when the graph of $f$ on $[0, 2]$ is revolved about the $x$-axis.
**b.** Find the volume of the solid generated when $R$ is revolved about the $y$-axis.
**c.** Find the volume of the solid generated when $R$ is revolved about the $x$-axis.

**67. Surface area and volume** Let $f(x) = \sqrt{3x - x^2}$ and let $R$ be the region bounded by the graph of $f$ and the $x$-axis on the interval $[0, 3]$.

**a.** Find the area of the surface generated when the graph of $f$ on $[0, 3]$ is revolved about the $x$-axis.
**b.** Find the volume of the solid generated when $R$ is revolved about the $x$-axis.

**68. Surface area of a cone** Find the surface area of a cone (excluding the base) with radius 4 and height 8 using integration and a surface area integral.

**T 69. Surface area and more** Let $f(x) = \dfrac{x^4}{2} + \dfrac{1}{16x^2}$ and let $R$ be the region bounded by the graph of $f$ and the $x$-axis on the interval $[1, 2]$.

    **a.** Find the area of the surface generated when the graph of $f$ on $[1, 2]$ is revolved about the $x$-axis. Use technology to evaluate or approximate the integral.

    **b.** Find the length of the curve $y = f(x)$ on $[1, 2]$.

    **c.** Find the volume of the solid generated when $R$ is revolved about the $x$-axis. Use technology to evaluate or approximate the integral.

**70–72. Variable density in one dimension** *Find the mass of the following thin bars.*

**70.** A bar on the interval $0 \le x \le 9$ with a density (in g/m) given by $\rho(x) = 3 + 2\sqrt{x}$

**71.** A 3-m bar with a density (in g/m) of $\rho(x) = 150 - \sqrt{3x}$, for $0 \le x \le 3$

**72.** A bar on the interval $0 \le x \le 6$ with a density

$$\rho(x) = \begin{cases} 1 & \text{if } 0 \le x < 2 \\ 2 & \text{if } 2 \le x < 4 \\ 4 & \text{if } 4 \le x \le 6 \end{cases}$$

**73. Spring work**

    **a.** It takes 50 J of work to stretch a spring 0.2 m from its equilibrium position. How much work is needed to stretch it an additional 0.5 m?

    **b.** It takes 50 N of force to stretch a spring 0.2 m from its equilibrium position. How much work is needed to stretch it an additional 0.5 m?

**74. Leaky bucket** A 1-kg bucket resting on the ground initially contains 6 kg of water. The bucket is lifted from the ground at a constant rate, and while it is rising, water leaks out of the bucket at a constant rate of $\frac{1}{4}$ kg/m. How much work is required to lift the bucket a vertical distance of 8 m? Assume the weight of the rope used to raise the bucket is negligible.

**75. Lifting problem** A 10-m, 20-kg chain hangs vertically from a cylinder attached to a winch.

    **a.** How much work is required to wind the entire chain onto the cylinder using the winch?

    **b.** How much work is required to wind the upper 4 m of the chain onto the cylinder using the winch?

**76. Lifting problem** A 4-kg mass is attached to the bottom of a 5-m, 15-kg chain. If the chain hangs from a platform, how much work is required to pull the chain and the mass onto the platform?

**77. Pumping water** A water tank has the shape of a box that is 2 m wide, 4 m long, and 6 m high.

    **a.** If the tank is full, how much work is required to pump the water to the level of the top of the tank?

    **b.** If the water in the tank is 2 m deep, how much work is required to pump the water to a level of 1 m above the top of the tank?

**78. Pumping water** A cylindrical water tank has a height of 6 m and a radius of 4 m. How much work is required to empty the full tank by pumping the water to an outflow pipe at the top of the tank?

**79. Pumping water** A water tank that is full of water has the shape of an inverted cone with a height of 6 m and a radius of 4 m (see figure). Assume the water is pumped out to the level of the top of the tank.

    **a.** How much work is required to pump out the water?

    **b.** Which requires more work, pumping out the top 3 m of water or the bottom 3 m of water?

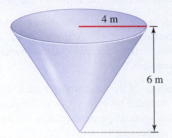

**80. Pumping water** A water tank that has the shape of a sphere with a radius of 2 m is half full of water. How much work is required to pump the water to the level of the top of the tank?

**81. Pumping water** A tank has the shape of the surface obtained by revolving the curve $y = x^2$, for $0 \le x \le 3$, about the $y$-axis. (Assume $x$ and $y$ are in meters.) If the tank is full of water, how much work is required to pump the water to an outflow pipe 1 m above the top of the tank?

**82–84. Fluid forces** *Suppose the following plates are placed on a vertical wall so that the top of the plate is 2 m below the surface of a pool that is filled with water. Compute the force on each plate.*

**82.** A square plate with side length of 1 m, oriented so that the top and bottom sides of the square are horizontal

**83.** A plate shaped like an isosceles triangle with a height of 1 m and a base of length $\dfrac{1}{2}$ m (see figure)

**84.** A circular plate with a radius of 2 m

**85. Force on a dam** Find the total force on the face of a semicircular dam with a radius of 20 m when its reservoir is full of water. The diameter of the semicircle is the top of the dam.

86. **Equal area property for parabolas** Let $f(x) = ax^2 + bx + c$ be an arbitrary quadratic function and choose two points $x = p$ and $x = q$. Let $L_1$ be the line tangent to the graph of $f$ at the point $(p, f(p))$ and let $L_2$ be the line tangent to the graph at the point $(q, f(q))$. Let $x = s$ be the vertical line through the intersection point of $L_1$ and $L_2$. Finally, let $R_1$ be the region bounded by $y = f(x)$, $L_1$, and the vertical line $x = s$, and let $R_2$ be the region bounded by $y = f(x)$, $L_2$, and the vertical line $x = s$. Prove that the area of $R_1$ equals the area of $R_2$.

## Chapter 6 Guided Projects

*Applications of the material in this chapter and related topics can be found in the following Guided Projects. For additional information, see the Preface.*

- Geometric probability
- Mathematics of the CD player
- Designing a water clock
- Buoyancy and Archimedes' principle
- Dipstick problems

# 7

# Logarithmic and Exponential Functions

**Chapter Preview**  In Chapter 1, we presented a catalog of elementary functions that included polynomial, algebraic, and trigonometric functions. Until now, we have studied the calculus of these three families of functions, exploring their applications along the way. However, many areas of mathematics rely on other functions, such as exponential and logarithmic functions, which are known to you from algebra. It turns out that an exponential function of the form $b^x$, where $b > 0$, is the *inverse function* of the logarithmic function $\log_b x$ (and vice versa). In light of this fact, the chapter opens with a general discussion of inverse functions. The *natural logarithmic function* is introduced next, along with its inverse, the *natural exponential function*. We then develop the various derivatives and integrals associated with these functions. Inverse functions are again put to work to define the inverse trigonometric functions and to produce their derivative and integral properties. With exponential and logarithmic functions in the picture, we can resume the discussion of l'Hôpital's Rule that began in Chapter 4. New indeterminate forms are explored, leading to a ranking of functions by their growth rates. Finally, the chapter concludes with a study of *hyperbolic functions* and their inverses; these functions are related to exponential, logarithmic and trigonometric functions. Throughout the chapter, we emphasize many practical applications of each of these families of functions.

## 7.1 Inverse Functions

From your study of algebra, you know that when a function has an *inverse function*, the two functions are related in special ways. Roughly speaking, the action of one function undoes the action of the other. In this section, we begin with a review of inverse functions: when they exist, how to find them, and how to graph them. We then investigate the relationship between the derivative of a function and the derivative of its inverse function. With this background, we devote much of the rest of the chapter to developing new functions that arise as the inverses of familiar functions.

### Existence of Inverse Functions

Consider the linear function $f(x) = 2x$, which takes any value of $x$ and doubles it. The function that reverses this process by taking any value of $f(x) = 2x$ and mapping it back to $x$ is called the *inverse function* of $f$, denoted $f^{-1}$. In this case, the inverse function is $f^{-1}(x) = x/2$. The effect of applying these two functions in succession looks like this:

$$x \overset{f}{\longrightarrow} 2x \overset{f^{-1}}{\longrightarrow} x.$$

We now generalize this idea.

> **DEFINITION** **Inverse Function**
>
> Given a function $f$, its inverse (if it exists) is a function $f^{-1}$ such that whenever $y = f(x)$, then $f^{-1}(y) = x$ (**Figure 7.1**).

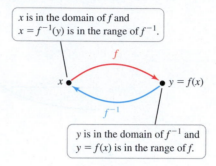

**Figure 7.1**

**QUICK CHECK 1** What is the inverse of $f(x) = \frac{1}{3}x$? What is the inverse of $f(x) = x - 7$? ◄

Because the inverse "undoes" the original function, if we start with a value of $x$, apply $f$ to it, and then apply $f^{-1}$ to the result, we recover the original value of $x$; that is,

$$f^{-1}(f(x)) = x.$$

Similarly, if we apply $f^{-1}$ to a value of $y$ and then apply $f$ to the result, we recover the original value of $y$; that is,

$$f(f^{-1}(y)) = y.$$

> ➤ The notation $f^{-1}$ for the inverse can be confusing. The inverse is not the reciprocal; that is, $f^{-1}(x)$ is not $1/f(x) = (f(x))^{-1}$. We adopt the common convention of using simply *inverse* to mean *inverse function*.

**One-to-One Functions**  We have defined the inverse of a function but said nothing about when it exists. To ensure that $f$ has an inverse on a domain, $f$ must be *one-to-one* on that domain. This property means that every output of the function $f$ corresponds to exactly one input. The one-to-one property is checked graphically by using the *horizontal line test*.

> ➤ The vertical line test determines whether $f$ is a function. The horizontal line test determines whether $f$ is one-to-one.

> **DEFINITION** **One-to-One Functions and the Horizontal Line Test**
>
> A function $f$ is **one-to-one** on a domain $D$ if each value of $f(x)$ corresponds to exactly one value of $x$ in $D$. More precisely, $f$ is one-to-one on $D$ if $f(x_1) \neq f(x_2)$ whenever $x_1 \neq x_2$, for $x_1$ and $x_2$ in $D$. The **horizontal line test** says that every horizontal line intersects the graph of a one-to-one function at most once (**Figure 7.2**).

**Figure 7.2**

For example, in **Figure 7.3**, some horizontal lines intersect the graph of $f(x) = x^2$ twice. Therefore, $f$ does not have an inverse function on the interval $(-\infty, \infty)$. However, if the domain of $f$ is restricted to one of the intervals $(-\infty, 0]$ or $[0, \infty)$, then the graph of $f$ passes the horizontal line test, and $f$ is one-to-one on these intervals.

**Figure 7.3**

**EXAMPLE 1   One-to-one functions**  Determine the (largest possible) intervals on which the function $f(x) = 2x^2 - x^4$ (**Figure 7.4**) is one-to-one.

**Figure 7.4**

**SOLUTION**  The function is not one-to-one on the entire real line because it fails the horizontal line test. However, on the intervals $(-\infty, -1]$, $[-1, 0]$, $[0, 1]$, and $[1, \infty)$, $f$ is one-to-one. The function is also one-to-one on any subinterval of these four intervals.

*Related Exercises 5–6* ◄

**Conditions for the Existence of Inverse Functions**    Figure 7.5a illustrates the actions of a one-to-one function $f$ and its inverse $f^{-1}$. We see that $f$ maps a value of $x$ to a unique value of $y$. In turn, $f^{-1}$ maps that value of $y$ back to the original value of $x$. This procedure cannot be carried out if $f$ is not one-to-one (**Figure 7.5b**).

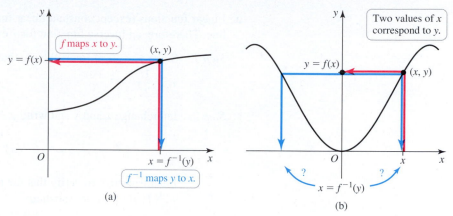

(a)                     (b)

**Figure 7.5**

> The statement that a one-to-one function has an inverse may be plausible based on its graph. However, the proof of this theorem is fairly technical and is omitted.

---

**THEOREM 7.1**   **Existence of Inverse Functions**

Let $f$ be a one-to-one function on a domain $D$ with a range $R$. Then $f$ has a unique inverse $f^{-1}$ with domain $R$ and range $D$ such that

$$f^{-1}(f(x)) = x \quad \text{and} \quad f(f^{-1}(y)) = y,$$

where $x$ is in $D$ and $y$ is in $R$.

---

**QUICK CHECK 2**   The function that gives degrees Fahrenheit in terms of degrees Celsius is $F = \dfrac{9}{5}C + 32$. Why does this function have an inverse? ◄

**EXAMPLE 2**   **Does an inverse exist?**   Determine the largest intervals on which $f(x) = x^2 - 1$ has an inverse function.

**SOLUTION**   On the interval $(-\infty, \infty)$ the function does not pass the horizontal line test and is not one-to-one (**Figure 7.6**). However, if the domain of $f$ is restricted to the interval $(-\infty, 0]$ or $[0, \infty)$, then $f$ is one-to-one and an inverse exists.

*Related Exercises 20, 22* ◄

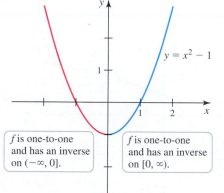

$y = x^2 - 1$

$f$ is one-to-one and has an inverse on $(-\infty, 0]$.

$f$ is one-to-one and has an inverse on $[0, \infty)$.

**Figure 7.6**

**Finding Inverse Functions**   The crux of finding an inverse for a function $f$ is solving the equation $y = f(x)$ for $x$ in terms of $y$. If it is possible to do so, then we have found a relationship of the form $x = f^{-1}(y)$. Interchanging $x$ and $y$ in $x = f^{-1}(y)$ so that $x$ is the independent variable (which is the customary role for $x$), the inverse has the form $y = f^{-1}(x)$. Notice that if $f$ is not one-to-one, this process leads to more than one inverse function.

> Once you find a formula for $f^{-1}$, you can check your work by verifying that $f^{-1}(f(x)) = x$ and $f(f^{-1}(x)) = x$.

> A constant function (whose graph is a horizontal line) fails the horizontal line test and does not have an inverse.

---

**PROCEDURE**   **Finding an Inverse Function**

Suppose $f$ is one-to-one on an interval $I$. To find $f^{-1}$, use the following steps.

1. Solve $y = f(x)$ for $x$. If necessary, restrict the resulting function so that $x$ lies in $I$.

2. Interchange $x$ and $y$ and write $y = f^{-1}(x)$.

---

**EXAMPLE 3**   **Finding inverse functions**  Find the inverse(s) of the following functions. Restrict the domain of $f$ if necessary.

**a.** $f(x) = 2x + 6$          **b.** $f(x) = x^2 - 1$

**SOLUTION**

**a.** Linear functions (except constant linear functions) are one-to-one on the entire real line. Therefore, an inverse function for $f$ exists for all values of $x$.

*Step 1:*  Solve $y = f(x)$ for $x$: We see that $y = 2x + 6$ implies that $2x = y - 6$, or $x = \dfrac{1}{2}y - 3$.

*Step 2:*  Interchange $x$ and $y$ and write $y = f^{-1}(x)$:

$$y = f^{-1}(x) = \frac{1}{2}x - 3.$$

It is instructive to verify that the inverse relations $f(f^{-1}(x)) = x$ and $f^{-1}(f(x)) = x$ are satisfied:

$$f(f^{-1}(x)) = f\left(\frac{1}{2}x - 3\right) = \underbrace{2\left(\frac{1}{2}x - 3\right) + 6}_{f(x)\,=\,2x\,+\,6} = x - 6 + 6 = x, \text{ and}$$

$$f^{-1}(f(x)) = f^{-1}(2x + 6) = \underbrace{\frac{1}{2}(2x + 6) - 3}_{f^{-1}(x)\,=\,\frac{1}{2}x\,-\,3} = x + 3 - 3 = x.$$

**b.** As shown in Example 2, the function $f(x) = x^2 - 1$ is not one-to-one on the entire real line; however, it is one-to-one on $(-\infty, 0]$ and on $[0, \infty)$ **(Figure 7.7a)**. If we restrict our attention to either of these intervals, then an inverse function can be found.

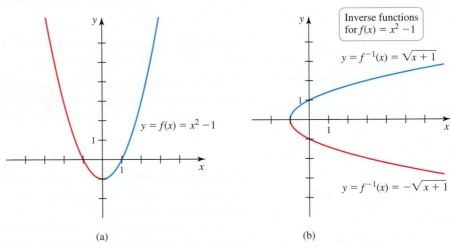

(a)                                      (b)

**Figure 7.7**

*Step 1:*  Solve $y = f(x)$ for $x$:

$$y = x^2 - 1$$
$$x^2 = y + 1$$
$$x = \begin{cases} \sqrt{y + 1} \\ -\sqrt{y + 1}. \end{cases}$$

Each branch of the square root corresponds to an inverse function.

*Step 2:* Interchange $x$ and $y$ and write $y = f^{-1}(x)$:

$$y = f^{-1}(x) = \sqrt{x+1} \quad \text{or} \quad y = f^{-1}(x) = -\sqrt{x+1}.$$

The interpretation of this result is important. Taking the positive branch of the square root, the inverse function $y = f^{-1}(x) = \sqrt{x+1}$ gives positive values of $y$; it corresponds to the branch of $f(x) = x^2 - 1$ on the interval $[0, \infty)$ (**Figure 7.7b**). The negative branch of the square root, $y = f^{-1}(x) = -\sqrt{x+1}$, is another inverse function that gives negative values of $y$; it corresponds to the branch of $f(x) = x^2 - 1$ on the interval $(-\infty, 0]$.

**QUICK CHECK 3** On what interval(s) does the function $f(x) = x^3$ have an inverse? ◄

*Related Exercises 31–32* ◄

## Graphing Inverse Functions

The graphs of a function and its inverse have a special relationship, which is illustrated in the following example.

**EXAMPLE 4    Graphing inverse functions** Plot $f$ and $f^{-1}$ on the same coordinate axes.

**a.** $f(x) = 2x + 6$      **b.** $f(x) = \sqrt{x-1}$

**SOLUTION**

**a.** The inverse of $f(x) = 2x + 6$, found in Example 3, is

$$y = f^{-1}(x) = \frac{1}{2}x - 3.$$

The graphs of $f$ and $f^{-1}$ are shown in **Figure 7.8**. Notice that both $f$ and $f^{-1}$ are increasing linear functions that intersect at $(-6, -6)$.

The function $f(x) = 2x + 6$ and its inverse $f^{-1}(x) = \frac{x}{2} - 3$ are symmetric about the line $y = x$.

The function $f(x) = \sqrt{x-1}$ $(x \geq 1)$ and its inverse $f^{-1}(x) = x^2 + 1$ $(x \geq 0)$ are symmetric about $y = x$.

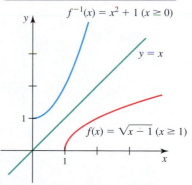

**Figure 7.8**          **Figure 7.9**

**b.** The domain of $f(x) = \sqrt{x-1}$ is $[1, \infty)$, and its range is $[0, \infty)$. On this domain, $f$ is one-to-one and has an inverse. It can be found in two steps:

*Step 1:* Solve $y = \sqrt{x-1}$ for $x$:

$$y^2 = x - 1 \quad \text{or} \quad x = y^2 + 1.$$

*Step 2:* Interchange $x$ and $y$ and write $y = f^{-1}(x)$:

$$y = f^{-1}(x) = x^2 + 1.$$

The graphs of $f$ and $f^{-1}$ are shown in **Figure 7.9**; notice that the domain of $f^{-1}$ (which is $x \geq 0$) corresponds to the range of $f$ (which is $y \geq 0$).

*Related Exercises 24–25* ◄

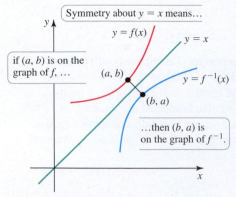

**Figure 7.10**

Looking closely at the graphs in Figure 7.8 and Figure 7.9, you see a symmetry that always occurs when a function and its inverse are plotted on the same set of axes. In each figure, one curve is the reflection of the other curve across the line $y = x$. These curves are *symmetric about the line $y = x$*, which means that the point $(a, b)$ is on one curve whenever the point $(b, a)$ is on the other curve (**Figure 7.10**).

The explanation for the symmetry comes directly from the definition of the inverse. Suppose that the point $(a, b)$ is on the graph of $y = f(x)$, which means that $b = f(a)$. By the definition of the inverse function, we know that $a = f^{-1}(b)$, which means that the point $(b, a)$ is on the graph of $y = f^{-1}(x)$. This argument applies to all relevant points $(a, b)$, so whenever $(a, b)$ is on the graph of $f$, $(b, a)$ is on the graph of $f^{-1}$. As a consequence, the graphs are symmetric about the line $y = x$.

Now suppose a function $f$ is continuous and one-to-one on an interval $I$. Reflecting the graph of $f$ through the line $y = x$ generates the graph of $f^{-1}$. The reflection process introduces no discontinuities in the graph of $f^{-1}$, so it is plausible (and indeed, true) that $f^{-1}$ is continuous on the interval corresponding to $I$. We state this fact without a formal proof.

> **THEOREM 7.2   Continuity of Inverse Functions**
> If a function $f$ is continuous on an interval $I$ and has an inverse on $I$, then its inverse $f^{-1}$ is also continuous (on the interval consisting of the points $f(x)$, where $x$ is in $I$).

## Derivatives of Inverse Functions

Here is an important question that bears on upcoming work: Given a differentiable function $f$ that is one-to-one on an interval, how do we evaluate the derivative of $f^{-1}$? The key to finding the derivative of the inverse function lies in the symmetry of the graphs of $f$ and $f^{-1}$.

**EXAMPLE 5    Linear functions, inverses, and derivatives**  Consider the general linear function $y = f(x) = mx + b$, where $m \neq 0$ and $b$ are constants.

**a.** Write the inverse of $f$ in the form $y = f^{-1}(x)$.

**b.** Find the derivative of the inverse $\dfrac{d}{dx}\left(f^{-1}(x)\right)$.

**c.** Consider the specific case $f(x) = 2x - 12$. Graph $f$ and $f^{-1}$, and find the slope of each line.

**SOLUTION**

**a.** Solving $y = mx + b$ for $x$, we find that $mx = y - b$, or

$$x = \frac{y}{m} - \frac{b}{m}.$$

Writing this function in the form $y = f^{-1}(x)$ (by reversing the roles of $x$ and $y$), we have

$$y = f^{-1}(x) = \frac{x}{m} - \frac{b}{m},$$

which describes a line with slope $1/m$.

**b.** The derivative of $f^{-1}$ is

$$(f^{-1})'(x) = \frac{1}{m}.$$

Notice that $f'(x) = m$, so the derivative of $f^{-1}$ is the reciprocal of $f'$.

c. In the case that $f(x) = 2x - 12$, we have $f^{-1}(x) = x/2 + 6$. The graphs of these two lines are symmetric about the line $y = x$ (**Figure 7.11**). Furthermore, the slope of the line $y = f(x)$ is 2 and the slope of $y = f^{-1}(x)$ is $\frac{1}{2}$; that is, the slopes (and therefore the derivatives) are reciprocals of each other.

**Figure 7.11**

*Related Exercises 39–40* ◀

The reciprocal property obeyed by $f'$ and $(f^{-1})'$ in Example 5 holds for all differentiable functions with inverses. **Figure 7.12** shows the graphs of a typical one-to-one function and its inverse. It also shows a pair of symmetric points—$(x_0, y_0)$ on the graph of $f$ and $(y_0, x_0)$ on the graph of $f^{-1}$—along with the tangent lines at these points.

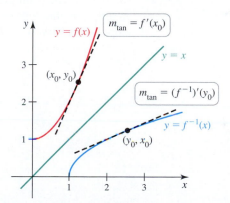

**Figure 7.12**

Notice that as the lines tangent to the graph of $f$ get steeper (as $x$ increases), the corresponding lines tangent to the graph of $f^{-1}$ get less steep. The next theorem makes this relationship precise.

---

**THEOREM 7.3  Derivative of the Inverse Function**

Let $f$ be differentiable and have an inverse on an interval $I$. If $x_0$ is a point of $I$ at which $f'(x_0) \neq 0$, then $f^{-1}$ is differentiable at $y_0 = f(x_0)$ and

$$(f^{-1})'(y_0) = \frac{1}{f'(x_0)}, \quad \text{where} \quad y_0 = f(x_0).$$

---

▶ The result of Theorem 7.3 is also written in the form

$$(f^{-1})'(f(x_0)) = \frac{1}{f'(x_0)}$$

or

$$(f^{-1})'(y_0) = \frac{1}{f'(f^{-1}(y_0))}.$$

To understand this theorem, suppose that $(x_0, y_0)$ is a point on the graph of $f$, which means that $(y_0, x_0)$ is the corresponding point on the graph of $f^{-1}$. Then the slope of the line tangent to the graph of $f^{-1}$ at the point $(y_0, x_0)$ is the reciprocal of the slope of the line tangent to the graph of $f$ at the point $(x_0, y_0)$. Importantly, the theorem says that we can evaluate the derivative of the inverse function without finding the inverse function itself.

**Proof:** Before doing a short calculation, we note two facts:

- At a point $x_0$ where $f$ is differentiable, $y_0 = f(x_0)$ and $x_0 = f^{-1}(y_0)$.
- Because $f$ is differentiable at $x_0$, $f$ is continuous at $x_0$ (Theorem 3.1), which implies that $f^{-1}$ is also continuous at $y_0$ (Theorem 7.2). Therefore, as $y \to y_0$, $x \to x_0$.

Using the definition of the derivative, we have

$$(f^{-1})'(y_0) = \lim_{y \to y_0} \frac{f^{-1}(y) - f^{-1}(y_0)}{y - y_0} \qquad \text{Definition of derivative of } f^{-1}$$

$$= \lim_{x \to x_0} \frac{x - x_0}{f(x) - f(x_0)} \qquad y = f(x) \text{ and } x = f^{-1}(y); x \to x_0 \text{ as } y \to y_0$$

$$= \lim_{x \to x_0} \frac{1}{\dfrac{f(x) - f(x_0)}{x - x_0}} \qquad \frac{a}{b} = \frac{1}{b/a}$$

$$= \frac{1}{f'(x_0)}. \qquad \text{Definition of derivative of } f$$

**QUICK CHECK 4** Sketch the graphs of $f(x) = x^3$ and $f^{-1}(x) = x^{1/3}$. Then verify that Theorem 7.3 holds at the point $(1, 1)$. ◄

We have shown that $(f^{-1})'(y_0)$ exists ($f^{-1}$ is differentiable at $y_0$) and that it equals the reciprocal of $f'(x_0)$. ◄

**EXAMPLE 6** **Derivative of an inverse function** The function $f(x) = \sqrt{x} + x^2 + 1$ is one-to-one, for $x \geq 0$, and has an inverse on that interval. Find the slope of the curve $y = f^{-1}(x)$ at the point $(3, 1)$.

**SOLUTION** The point $(1, 3)$ is on the graph of $f$; therefore, $(3, 1)$ is on the graph of $f^{-1}$. In this case, the slope of the curve $y = f^{-1}(x)$ at the point $(3, 1)$ is the reciprocal of the slope of the curve $y = f(x)$ at $(1, 3)$ (**Figure 7.13**).

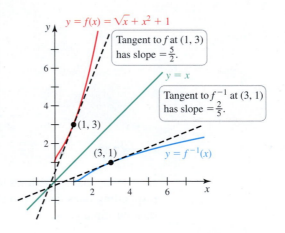

**Figure 7.13**

Note that $f'(x) = \dfrac{1}{2\sqrt{x}} + 2x$, which means that $f'(1) = \dfrac{1}{2} + 2 = \dfrac{5}{2}$. Therefore,

$$(f^{-1})'(3) = \frac{1}{f'(1)} = \frac{1}{5/2} = \frac{2}{5}.$$

Observe that it is not necessary to find a formula for $f^{-1}$ to evaluate its derivative at a point.

*Related Exercises 47–48* ◄

**EXAMPLE 7  Derivatives of an inverse function**  Use the values of a one-to-one differentiable function $f$ in Table 7.1 to compute the indicated derivatives or state that the derivative cannot be determined.

**Table 7.1**

| $x$ | $-1$ | $0$ | $1$ | $2$ | $3$ |
|---|---|---|---|---|---|
| $f(x)$ | $2$ | $3$ | $5$ | $6$ | $7$ |
| $f'(x)$ | $1/2$ | $2$ | $3/2$ | $1$ | $2/3$ |

**a.** $(f^{-1})'(5)$        **b.** $(f^{-1})'(2)$        **c.** $(f^{-1})'(1)$

**SOLUTION**  We use the relationship $(f^{-1})'(y_0) = \dfrac{1}{f'(x_0)}$, where $y_0 = f(x_0)$.

**a.** In this case, $y_0 = f(x_0) = 5$. Using Table 7.1, we see that $x_0 = 1$ and $f'(1) = \dfrac{3}{2}$.
Therefore, $(f^{-1})'(5) = \dfrac{1}{f'(1)} = \dfrac{2}{3}$.

**b.** In this case, $y_0 = f(x_0) = 2$, which implies that $x_0 = -1$ and $f'(-1) = \dfrac{1}{2}$.
Therefore, $(f^{-1})'(2) = \dfrac{1}{f'(-1)} = 2$.

**c.** With $y_0 = f(x_0) = 1$, Table 7.1 does not supply a value of $x_0$. Therefore, neither $f'(x_0)$ nor $(f^{-1})'(1)$ can be determined.

*Related Exercises 53–54* ◄

---

## SECTION 7.1 EXERCISES

### Getting Started

1. Explain why a function that is not one-to-one on an interval $I$ cannot have an inverse function on $I$.

2. Give an example of a function that is one-to-one on the entire real number line.

3. Sketch a graph of a function that is one-to-one on the interval $(-\infty, 0]$ but is not one-to-one on $(-\infty, \infty)$.

4. Sketch a graph of a function that is one-to-one on the intervals $(-\infty, -2]$ and $[-2, \infty)$ but is not one-to-one on $(-\infty, \infty)$.

5. Find three intervals on which $f$ is one-to-one, making each interval as large as possible.

6. Find four intervals on which $f$ is one-to-one, making each interval as large as possible.

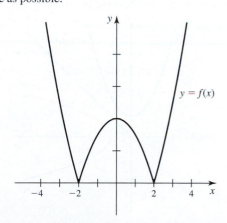

7. Find the inverse of the function $f(x) = 2x$. Verify that $f(f^{-1}(x)) = x$ and $f^{-1}(f(x)) = x$.

8. Find the inverse of the function $f(x) = \sqrt{x}$. Verify that $f(f^{-1}(x)) = x$ and $f^{-1}(f(x)) = x$.

**9.** Sketch the graph of the inverse of $f$.

**10.** Sketch the graph of the inverse of $f$.

**11–12.** *The parabola* $y = x^2 + 1$ *consists of two one-to-one functions,* $g_1(x)$ *and* $g_2(x)$. *Complete each exercise and confirm that your answers are consistent with the graphs displayed in the figure.*

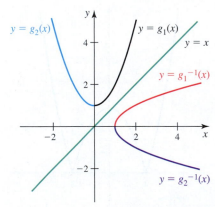

**11.** Find formulas for $g_1(x)$ and $g_1^{-1}(x)$. State the domain and range of each function.

**12.** Find formulas for $g_2(x)$ and $g_2^{-1}(x)$. State the domain and range of each function.

**13.** Use the graph of $f$ to find $f^{-1}(2)$, $f^{-1}(9)$, and $f^{-1}(12)$.

**14.** Explain how to find $(f^{-1})'(y_0)$ given that $y_0 = f(x_0)$.

**15.** Suppose $f$ is a one-to-one function with $f(2) = 8$ and $f'(2) = 4$. What is the value of $(f^{-1})'(8)$?

**16.** Suppose $f$ is a one-to-one function and $y = 3x - 2$ is the line tangent to the graph of $f$ at the point $(1, 3)$. Evaluate $(f^{-1})'(3)$.

## Practice Exercises

**17–22. Where do inverses exist?** *Use analytical and/or graphical methods to determine the largest possible sets of points on which the following functions have an inverse.*

**17.** $f(x) = 3x + 4$                **18.** $f(x) = |2x + 1|$

**19.** $f(x) = 1/(x - 5)$           **20.** $f(x) = -(6 - x)^2$

**21.** $f(x) = 1/x^2$                **22.** $f(x) = x^2 - 2x + 8$
                                      (*Hint:* Complete the square.)

**23–28. Graphing inverse functions** *Find the inverse function (on the given interval, if specified) and graph both $f$ and $f^{-1}$ on the same set of axes. Check your work by looking for the required symmetry in the graphs.*

**23.** $f(x) = 8 - 4x$              **24.** $f(x) = 3x + 5$

**25.** $f(x) = \sqrt{x + 2}$        **26.** $f(x) = \sqrt{3 - x}$

**27.** $f(x) = (x - 2)^2 - 1$, for $x \geq 2$

**28.** $f(x) = x^2 + 4$, for $x \geq 0$

**29–36. Finding inverse functions**

**a.** Find the inverse of each function (on the given interval, if specified) and write it in the form $y = f^{-1}(x)$.

**b.** Verify the relationships $f(f^{-1}(x)) = x$ and $f^{-1}(f(x)) = x$.

**29.** $f(x) = 6 - 4x$              **30.** $f(x) = 3x^3$

**31.** $f(x) = (x - 7)^{5/3}$       **32.** $f(x) = x^4 - 16$, for $x \leq 0$

**33.** $f(x) = 1 - \dfrac{1}{x}$    **34.** $f(x) = 2/(x^2 + 1)$, for $x \geq 0$

**35.** $f(x) = 6/(x^2 - 9)$, for $x > 3$

**36.** $f(x) = x/(x - 2)$, for $x > 2$

**37. Splitting up curves** The unit circle $x^2 + y^2 = 1$ consists of four one-to-one functions, $f_1(x)$, $f_2(x)$, $f_3(x)$, and $f_4(x)$ (see figure).

  **a.** Find the domain and a formula for each function.
  **b.** Find the inverse of each function and write it as $y = f^{-1}(x)$.

**38. Splitting up curves** The equation $y^4 = 4x^2$ is associated with four one-to-one functions, $f_1(x)$, $f_2(x)$, $f_3(x)$, and $f_4(x)$ (see figure).

  **a.** Find the domain and a formula for each function.
  **b.** Find the inverse of each function and write it as $y = f^{-1}(x)$.

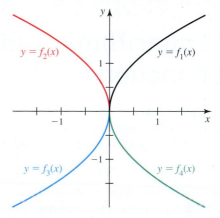

**39–44. Derivatives of inverse functions at a point** *Find the derivative of the inverse of the following functions at the specified point on the graph of the inverse function. You do not need to find $f^{-1}$.*

**39.** $f(x) = 3x + 4$; $(16, 4)$     **40.** $f(x) = \frac{1}{2}x + 8$; $(10, 4)$

**41.** $f(x) = -5x + 4$; $(-1, 1)$

**42.** $f(x) = x^2 + 1$, for $x \geq 0$; $(5, 2)$

**43.** $f(x) = \tan x$; $(1, \pi/4)$

**44.** $f(x) = x^2 - 2x - 3$, for $x \leq 1$; $(12, -3)$

**45–48. Slopes of tangent lines** *Given the function $f$, find the slope of the line tangent to the graph of $f^{-1}$ at the specified point on the graph of $f^{-1}$.*

**45.** $f(x) = \sqrt{x}$; $(2, 4)$     **46.** $f(x) = x^3$; $(8, 2)$

**47.** $f(x) = (x + 2)^2$, for $x \geq -2$; $(36, 4)$

**48.** $f(x) = -x^2 + 8$, for $x \geq 0$; $(7, 1)$

**49–52. Derivatives and inverse functions**

**49.** Find $(f^{-1})'(3)$, where $f(x) = x^3 + x + 1$.

**50.** Suppose the slope of the curve $y = f(x)$ at $(7, 4)$ is $\frac{2}{3}$. Find the slope of the curve $y = f^{-1}(x)$ at $(4, 7)$.

**51.** Suppose the slope of the curve $y = f^{-1}(x)$ at $(4, 7)$ is $\frac{4}{5}$. Find $f'(7)$.

**52.** Suppose the slope of the curve $y = f(x)$ at $(4, 7)$ is $\frac{1}{5}$. Find $(f^{-1})'(7)$.

**53–54. Derivatives of inverse functions from a table** *Use the following tables to determine the indicated derivatives or state that the derivative cannot be determined.*

**53.**

| $x$ | $-2$ | $-1$ | $0$ | $1$ | $2$ |
|---|---|---|---|---|---|
| $f(x)$ | 2 | 3 | 4 | 6 | 7 |
| $f'(x)$ | 1 | 1/2 | 2 | 3/2 | 1 |

  **a.** $(f^{-1})'(4)$   **b.** $(f^{-1})'(6)$   **c.** $(f^{-1})'(1)$   **d.** $f'(1)$

**54.**

| $x$ | $-4$ | $-2$ | $0$ | $2$ | $4$ |
|---|---|---|---|---|---|
| $f(x)$ | 0 | 1 | 2 | 3 | 4 |
| $f'(x)$ | 5 | 4 | 3 | 2 | 1 |

  **a.** $f'(f(0))$   **b.** $(f^{-1})'(0)$   **c.** $(f^{-1})'(1)$   **d.** $(f^{-1})'(f(4))$

**55. Explain why or why not** Determine whether the following statements are true and give an explanation or counterexample.

  **a.** If $f(x) = x^2 + 1$, then $f^{-1}(x) = 1/(x^2 + 1)$.
  **b.** If $f(x) = 1/x$, then $f^{-1}(x) = 1/x$.
  **c.** When restricted to the largest possible intervals, the function $f(x) = x^3 + x$ has three different inverses.
  **d.** When restricted to the largest possible intervals, a tenth-degree polynomial could have ten different inverses.
  **e.** If $f(x) = 1/x$, then $(f^{-1})'(x) = -1/x^2$.

**T 56. Piecewise linear function** Consider the function $f(x) = |x| - 2|x - 1|$.

  **a.** Find the largest possible intervals on which $f$ is one-to-one.
  **b.** Find explicit formulas for the inverse of $f$ on the intervals in part (a).

**57–60. Finding all inverses** *Find all the inverses associated with the following functions and state their domains.*

**57.** $f(x) = (x + 1)^3$      **58.** $f(x) = (x - 4)^2$

**59.** $f(x) = 2/(x^2 + 2)$      **60.** $f(x) = 2x/(x + 2)$

**61–68. Derivatives of inverse functions** *Consider the following functions (on the given interval, if specified). Find the inverse function, express it as a function of $x$, and find the derivative of the inverse function.*

**61.** $f(x) = 3x - 4$      **62.** $f(x) = |x + 2|$, for $x \leq -2$

**63.** $f(x) = x^2 - 4$, for $x \geq 0$    **64.** $f(x) = \dfrac{x}{x + 5}$

**65.** $f(x) = \sqrt{x + 2}$      **66.** $f(x) = x^{2/3}$, for $x \geq 0$

**67.** $f(x) = x^{-1/2}$      **68.** $f(x) = x^3 + 3$

**69–72. Geometry functions** *Each function describes the volume $V$ or surface area $S$ of a three-dimensional solid in terms of its radius. Find the inverse of each function that gives the radius in terms of $V$ or $S$. Assume that $r$, $V$, and $S$ are nonnegative. Express your answer in the form $r = f^{-1}(S)$ or $r = f^{-1}(V)$.*

**69.** Sphere: $V = \frac{4}{3}\pi r^3$      **70.** Sphere: $S = 4\pi r^2$

**71.** Cylinder with height 10: $V = 10\pi r^2$

**72.** Cone with height 12: $V = 4\pi r^2$

## Explorations and Challenges

**T 73. Inverses of a quartic** Consider the quartic polynomial
$y = f(x) = x^4 - x^2$.

    **a.** Graph $f$ and find the largest intervals on which it is one-to-one. The goal is to find the inverse function on each of these intervals.

    **b.** Make the substitution $u = x^2$ to solve the equation $y = f(x)$ for $x$ in terms of $y$. Be sure you have included all possible solutions.

    **c.** Write each inverse function in the form $y = f^{-1}(x)$ for each of the intervals found in part (a).

**74. Inverse of composite functions**

    **a.** Let $g(x) = 2x + 3$ and $h(x) = x^3$. Consider the composite function $f(x) = g(h(x))$. Find $f^{-1}$ directly and then express it in terms of $g^{-1}$ and $h^{-1}$.

    **b.** Let $g(x) = x^2 + 1$ and $h(x) = \sqrt{x}$. Consider the composite function $f(x) = g(h(x))$. Find $f^{-1}$ directly and then express it in terms of $g^{-1}$ and $h^{-1}$.

    **c.** Explain why if $g$ and $h$ are one-to-one, the inverse of $f(x) = g(h(x))$ exists.

**75–76. Inverses of (some) cubics** *Finding the inverse of a cubic polynomial is equivalent to solving a cubic equation. A special case that is simpler than the general case is the cubic $f(x) = x^3 + ax$. Find the inverse of the following cubics using the substitution (known as Vieta's substitution) $x = z - a/(3z)$. Be sure to determine where the function is one-to-one.*

**75.** $f(x) = x^3 + 2x$          **76.** $f(x) = x^3 + 4x - 1$

**77. Tangents and inverses** Suppose $y = L(x) = ax + b$ (with $a \neq 0$) is the equation of the line tangent to the graph of a one-to-one function $f$ at $(x_0, y_0)$. Also, suppose that $y = M(x) = cx + d$ is the equation of the line tangent to the graph of $f^{-1}$ at $(y_0, x_0)$.

    **a.** Express $a$ and $b$ in terms of $x_0$ and $y_0$.

    **b.** Express $c$ in terms of $a$, and $d$ in terms of $a$, $x_0$ and $y_0$.

    **c.** Prove that $L^{-1}(x) = M(x)$.

**QUICK CHECK ANSWERS**

**1.** $f^{-1}(x) = 3x$; $f^{-1}(x) = x + 7$   **2.** For every Fahrenheit temperature, there is exactly one Celsius temperature, and vice versa. The given relation is also a linear function. It is one-to-one, so it has an inverse function.   **3.** The function $f(x) = x^3$ is one-to-one on $(-\infty, \infty)$, so it has an inverse for all values of $x$.   **4.** $f'(1) = 3$, $(f^{-1})'(1) = \frac{1}{3}$ ◄

# 7.2 The Natural Logarithmic and Exponential Functions

► Logarithms were invented around 1600 for computational purposes by the Scotsman John Napier and the Englishman Henry Briggs. Unfortunately, the word *logarithm*, derived from the Greek for "reasoning (*logos*) with numbers (*arithmos*)," doesn't help with the meaning of the word. **When you see *logarithm*, you should think *exponent*.**

Your understanding of logarithms and exponentials as algebraic operations is important, and it will be put to use in the coming pages. For example, you may recall the following relationship that will be used frequently: If $b$ denotes a *base* with $b > 0$ and $b \neq 1$, then

$$y = b^x \quad \text{if and only if} \quad x = \log_b y.$$

However, to do calculus with logarithms and exponentials, we must view them not just as operations, but as functions. Once we define logarithmic and exponential *functions*, many important questions follow.

• What are the domains of $b^x$ and $\log_b x$?

• How do we assign meaning to expressions such as $2^\pi$ or $\log_3 \pi$?

• Are these functions continuous on their domains?

• What are their derivatives?

• What new integrals can be evaluated using these functions?

It all begins with the *natural logarithmic function*, which is defined in terms of a definite integral, after which we use the theory of inverse functions (Section 7.1) to develop the *natural exponential function*. Our objective in this section is to place these important functions on a solid foundation by presenting a rigorous development of their properties.

    Before embarking on this program, we offer a roadmap to help guide you through the section. We carry out the following three steps.

**1.** We first define the natural logarithmic function, denoted $\ln x$, in terms of an integral, and then derive the properties of $\ln x$ directly from this new definition.

**2.** Next, the natural exponential function $e^x$ is introduced as the inverse of $\ln x$, and the properties of $e^x$ are developed by appealing to this inverse relationship. We also present derivative and integral formulas associated with these functions.

**3.** Finally, we define the general exponential function $b^x$ in terms of $e^x$ so that two crucial properties of the natural logarithmic and exponential functions can be extended to all real numbers. One of these properties is used to derive a limit definition of $e$ that is used to approximate $e$.

After establishing the properties of the natural logarithmic and exponential functions, we conclude the section with derivative and integral formulas associated with $e^x$, and we present the technique of logarithmic differentiation.

## Step 1: The Natural Logarithm

Our aim is to develop the properties of the natural logarithm using definite integrals. We begin with the following definition.

---

**DEFINITION** **The Natural Logarithm**

The **natural logarithm** of a number $x > 0$ is $\ln x = \displaystyle\int_1^x \frac{1}{t}\, dt.$

---

All the familiar geometric and algebraic properties of the natural logarithmic function follow directly from this integral definition.

### Properties of the Natural Logarithm

**Domain, range, and sign** Because the natural logarithm is defined as a definite integral, its value is the net area under the curve $y = 1/t$ between $t = 1$ and $t = x$. The integrand is undefined at $t = 0$, so the domain of $\ln x$ is $(0, \infty)$. On the interval $(1, \infty)$, $\ln x$ is positive because the net area of the region under the curve is positive (**Figure 7.14a**). On $(0, 1)$, we have $\displaystyle\int_1^x \frac{1}{t}\, dt = -\int_x^1 \frac{1}{t}\, dt$, which implies $\ln x$ is negative (**Figure 7.14b**). As expected, when $x = 1$, we have $\ln 1 = \displaystyle\int_1^1 \frac{1}{t}\, dt = 0$. The net area interpretation of $\ln x$ also implies that the range of $\ln x$ is $(-\infty, \infty)$. (See Exercise 112 for an outline of a proof.)

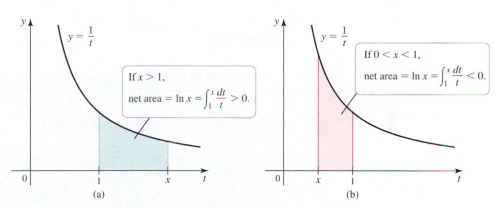

**Figure 7.14**

**Derivative** The derivative of the natural logarithm follows immediately from its definition and the Fundamental Theorem of Calculus:

$$\frac{d}{dx}(\ln x) = \frac{d}{dx}\int_1^x \frac{dt}{t} = \frac{1}{x}, \qquad \text{for } x > 0.$$

We have two important consequences:

- Because its derivative is defined for $x > 0$, $\ln x$ is a differentiable function for $x > 0$, which means it is continuous on its domain (Theorem 3.1).

- Because $1/x > 0$ for $x > 0$, $\ln x$ is strictly increasing and one-to-one on its domain; therefore, it has a well-defined inverse.

The Chain Rule allows us to extend the derivative property to all nonzero real numbers (Exercise 110). By differentiating $\ln(-x)$ for $x < 0$, we find that

$$\frac{d}{dx}(\ln|x|) = \frac{1}{x}, \text{ for } x \neq 0.$$

> By the Fundamental Theorem of Calculus,
>
> $$\frac{d}{dx}\int_a^x f(t)\, dt = f(x).$$

**QUICK CHECK 1** What is the domain of $\ln |x|$? ◄

More generally, by the Chain Rule,

$$\frac{d}{dx}(\ln |u(x)|) = \frac{d}{du}(\ln |u(x)|)u'(x) = \frac{u'(x)}{u(x)}.$$

**Graph of ln x** As noted before, $\ln x$ is continuous and strictly increasing for $x > 0$. The second derivative, $\dfrac{d^2}{dx^2}(\ln x) = -\dfrac{1}{x^2}$, is negative for $x > 0$, which implies the graph of $\ln x$ is concave down for $x > 0$. As demonstrated in Exercise 112,

$$\lim_{x \to \infty} \ln x = \infty \quad \text{and} \quad \lim_{x \to 0^+} \ln x = -\infty.$$

This information, coupled with the fact that $\ln 1 = 0$, gives the graph of $y = \ln x$ (**Figure 7.15**). The graphs of $y = \ln x$, $y = \ln |x|$, and their derivatives are shown in **Figure 7.16**.

$\dfrac{d}{dx}(\ln x) > 0 \Rightarrow \ln x$ is increasing for $x > 0$

$\lim_{x \to \infty} \ln x = \infty$

$y = \ln x$

$(1, 0)$

$\dfrac{d^2}{dx^2}(\ln x) < 0 \Rightarrow$ concave down

$\lim_{x \to 0^+} \ln x = -\infty$

**Figure 7.15**

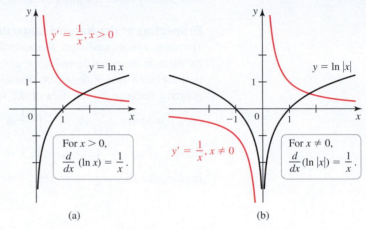

$y' = \dfrac{1}{x}, x > 0$

$y = \ln x$

For $x > 0$,

$\dfrac{d}{dx}(\ln x) = \dfrac{1}{x}.$

$y' = \dfrac{1}{x}, x \neq 0$

$y = \ln |x|$

For $x \neq 0$,

$\dfrac{d}{dx}(\ln |x|) = \dfrac{1}{x}.$

(a)                        (b)

**Figure 7.16**

**Logarithm of a product** The familiar logarithm property

$$\ln xy = \ln x + \ln y, \qquad \text{for } x > 0, \quad y > 0,$$

may be proved using the integral definition:

$$\ln xy = \int_1^{xy} \frac{dt}{t} \qquad \text{Definition of } \ln xy$$

$$= \int_1^x \frac{dt}{t} + \int_x^{xy} \frac{dt}{t} \qquad \text{Additive property of integrals}$$

$$= \int_1^x \frac{dt}{t} + \int_1^y \frac{du}{u} \qquad \text{Substitute } u = t/x \text{ in second integral.}$$

$$= \ln x + \ln y. \qquad \text{Definition of the natural logarithm}$$

**Logarithm of a quotient** Assuming $x > 0$ and $y > 0$, the product property and a bit of algebra give

$$\ln x = \ln \left( y \cdot \frac{x}{y} \right) = \ln y + \ln \frac{x}{y}.$$

Solving for $\ln (x/y)$, we have

$$\ln \frac{x}{y} = \ln x - \ln y,$$

which is the quotient property for logarithms. (Also see Exercise 72.)

**Logarithm of a power**   Assuming $x > 0$ and $p$ is an integer, we have

$$\ln x^p = \int_1^{x^p} \frac{dt}{t} \qquad \text{Definition of } \ln x^p$$

$$= p \int_1^x \frac{du}{u} \qquad \text{Let } t = u^p;\ dt = pu^{p-1}\, du.$$

$$= p \ln x. \qquad \text{By definition, } \ln x = \int_1^x \frac{du}{u}.$$

This argument relies on the Power Rule $(dt = pu^{p-1}\, du)$, which we proved only for integer exponents in Sections 3.3 and 3.4. Later in this section, we prove that $\ln x^p = p \ln x$ for all real values of $p$.

**Integrals**   Because $\dfrac{d}{dx}(\ln |x|) = \dfrac{1}{x}$, we have

$$\int \frac{1}{x}\, dx = \ln |x| + C.$$

We have shown that the following properties of $\ln x$ follow from its integral definition.

---

**THEOREM 7.4**   **Properties of the Natural Logarithm**

1. The domain and range of $\ln x$ are $(0, \infty)$ and $(-\infty, \infty)$, respectively.
2. $\ln xy = \ln x + \ln y, \quad$ for $x > 0,\ y > 0$
3. $\ln (x/y) = \ln x - \ln y, \quad$ for $x > 0,\ y > 0$
4. $\ln x^p = p \ln x, \quad$ for $x > 0$ and $p$ an integer
5. $\dfrac{d}{dx}(\ln |x|) = \dfrac{1}{x}, \quad$ for $x \neq 0$
6. $\dfrac{d}{dx}(\ln |u(x)|) = \dfrac{u'(x)}{u(x)}, \quad$ for $u(x) \neq 0$
7. $\displaystyle\int \frac{1}{x}\, dx = \ln |x| + C$

---

**EXAMPLE 1**   **Derivatives involving ln $x$**   Find $\dfrac{dy}{dx}$ for the following functions.

**a.** $y = \ln 4x$      **b.** $y = x \ln x$      **c.** $y = \ln |\sec x|$      **d.** $y = \dfrac{\ln x^2}{x^2}$

**SOLUTION**

**a.** Using the Chain Rule,

$$\frac{dy}{dx} = \frac{d}{dx}(\ln 4x) = \frac{1}{4x}\cdot 4 = \frac{1}{x}.$$

An alternative method uses a property of logarithms before differentiating:

$$\frac{d}{dx}(\ln 4x) = \frac{d}{dx}(\ln 4 + \ln x) \qquad \ln xy = \ln x + \ln y$$

> Because $\ln x$ and $\ln 4x$ differ by a constant ($\ln 4x = \ln x + \ln 4$), the derivatives of $\ln x$ and $\ln 4x$ are equal.

$$= 0 + \frac{1}{x} = \frac{1}{x}. \qquad \ln 4 \text{ is a constant.}$$

**b.** By the Product Rule,

$$\frac{dy}{dx} = \frac{d}{dx}(x \ln x) = 1 \cdot \ln x + x \cdot \frac{1}{x} = \ln x + 1.$$

**c.** Using property 6 of Theorem 7.4,

$$\frac{dy}{dx} = \frac{1}{\sec x}\left(\frac{d}{dx}(\sec x)\right) = \frac{1}{\sec x}(\sec x \tan x) = \tan x.$$

**QUICK CHECK 2** Find $\dfrac{d}{dx}(\ln x^p)$, where $x > 0$ and $p$ is an integer, in two ways: (1) using the Chain Rule and (2) by first using a property of logarithms. ◄

**d.** The Quotient Rule and Chain Rule give

$$\frac{dy}{dx} = \frac{x^2\left(\dfrac{1}{x^2} \cdot 2x\right) - (\ln x^2)\,2x}{(x^2)^2} = \frac{2x - 2x\ln x^2}{x^4} = \frac{2(1 - \ln x^2)}{x^3}.$$

*Related Exercises 17–18, 21–22* ◄

**EXAMPLE 2**  **Integrals with ln $x$**  Evaluate $\displaystyle\int_0^4 \frac{x}{x^2 + 9}\,dx$.

**SOLUTION**

$$\int_0^4 \frac{x}{x^2 + 9}\,dx = \frac{1}{2}\int_9^{25} \frac{du}{u} \qquad \text{Let } u = x^2 + 9;\ du = 2x\,dx.$$

$$= \frac{1}{2}\ln|u|\,\Big|_9^{25} \qquad \text{Fundamental Theorem}$$

$$= \frac{1}{2}(\ln 25 - \ln 9) \qquad \text{Evaluate.}$$

$$= \ln\frac{5}{3} \qquad \text{Properties of logarithms}$$

*Related Exercise 44* ◄

## Step 2: The Exponential Function

We have established that $f(x) = \ln x$ is a continuous, increasing function on the interval $(0, \infty)$. Therefore, it is one-to-one and its inverse function exists on $(0, \infty)$. We denote the inverse function $f^{-1}(x) = \exp(x)$. Its graph is obtained by reflecting the graph of $f(x) = \ln x$ about the line $y = x$ (**Figure 7.17**). The domain of $\exp(x)$ is $(-\infty, \infty)$ because the range of $\ln x$ is $(-\infty, \infty)$, and the range of $\exp(x)$ is $(0, \infty)$ because the domain of $\ln x$ is $(0, \infty)$.

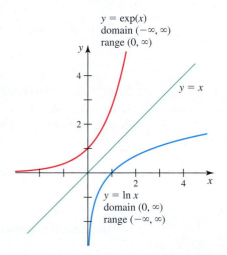

**Figure 7.17**

The usual relationships between a function and its inverse also hold:

• $y = \exp(x)$ if and only if $x = \ln y$

• $\exp(\ln x) = x$, for $x > 0$, and $\ln(\exp(x)) = x$, for all $x$.

We now appeal to the properties of $\ln x$ and use the inverse relations between $\ln x$ and $\exp(x)$ to show that $\exp(x)$ satisfies the expected properties of any exponential function. For example, if $x_1 = \ln y_1$ and $x_2 = \ln y_2$, then it follows that $y_1 = \exp(x_1)$, $y_2 = \exp(x_2)$, and

$$\exp(x_1 + x_2) = \exp\underbrace{(\ln y_1 + \ln y_2)}_{\ln y_1 y_2} \qquad \text{Substitute } x_1 = \ln y_1, x_2 = \ln y_2.$$

$$= \exp(\ln y_1 y_2) \qquad \qquad \text{Properties of logarithms}$$

$$= y_1 y_2 \qquad \qquad \qquad \text{Inverse property of } \exp(x) \text{ and } \ln x$$

$$= \exp(x_1)\exp(x_2). \qquad y_1 = \exp(x_1), y_2 = \exp(x_2)$$

Therefore, $\exp(x)$ satisfies the property of exponential functions $b^{x_1+x_2} = b^{x_1}b^{x_2}$. Similar arguments show that $\exp(x)$ satisfies other characteristic properties of exponential functions (Exercise 111):

$$\exp(0) = 1,$$

$$\exp(x_1 - x_2) = \frac{\exp(x_1)}{\exp(x_2)}, \quad \text{and}$$

$$(\exp(x))^p = \exp(px), \quad \text{for integers } p.$$

Suspecting that $\exp(x)$ is an exponential function, we proceed to identify its base. Let's consider the real number $\exp(1)$, and with a bit of forethought, call it $e$. The inverse relationship between $\ln x$ and $\exp(x)$ implies that

$$\text{if } e = \exp(1), \text{ then} \quad \ln e = \ln(\exp(1)) = 1.$$

Using the fact that $\ln e = 1$ and the integral definition of $\ln x$, we now formally define $e$.

> ▶ The constant $e$ was identified and named by the Swiss mathematician Leonhard Euler (1707–1783) (pronounced "oiler").

---

**DEFINITION  The Number $e$**

The number $e$ is the real number that satisfies $\ln e = \displaystyle\int_1^e \frac{dt}{t} = 1$.

---

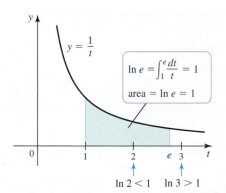

$$y = \frac{1}{t}$$

$$\ln e = \int_1^e \frac{dt}{t} = 1$$

$$\text{area} = \ln e = 1$$

$$\ln 2 < 1 \quad \ln 3 > 1$$

**Figure 7.18**

The number $e$ has the property that the area of the region bounded by the graph of $y = \dfrac{1}{t}$ and the $t$-axis on the interval $[1, e]$ is 1 (**Figure 7.18**). Note that $\ln 2 < 1$ and $\ln 3 > 1$ (Exercise 113). Because $\ln x$ is continuous on its domain, the Intermediate Value Theorem ensures that there is a number $e$ with $2 < e < 3$ such that $\ln e = 1$.

We can now show that indeed $\exp(x)$ is the exponential function $e^x$. Assume that $p$ is an integer and note that $e^p > 0$. By property 4 of Theorem 7.4 we have

$$\ln e^p = p \underbrace{\ln e}_{1} = p.$$

Using the inverse relationship between $\ln x$ and $\exp(x)$, we also know that

$$\ln \exp(p) = p.$$

Equating these two expressions for $p$, we conclude that $\ln e^p = \ln \exp(p)$. Because $\ln x$ is a one-to-one function, it follows that

$$e^p = \exp(p), \quad \text{for integers } p,$$

> ▶ Because $\exp(x)$ is the exponential function with base $e$, we conclude that its inverse $\ln x$ is the logarithmic function with base $e$. That is
>
> $$\ln x = \log_e x.$$

and we conclude that $\exp(x)$ is the exponential function with base $e$. We already know how to evaluate $e^x$ when $x$ is rational. For example, $e^3 = e \cdot e \cdot e$, $e^{-2} = \dfrac{1}{e \cdot e}$, and $e^{1/2} = \sqrt{e}$. But how do we evaluate $e^x$ when $x$ is irrational? We proceed as follows. The function $x = \ln y$ is defined for $y > 0$ and its range is all real numbers. Therefore, the domain of its inverse $y = \exp(x)$ is all real numbers; that is, $\exp(x)$ is defined for all real numbers. We now define $e^x$ to be $\exp(x)$ when $x$ is irrational.

---

**DEFINITION** **The Exponential Function**

For real numbers $x$, $y = e^x = \exp(x)$, where $x = \ln y$.

---

We may now dispense with the notation $\exp(x)$ and use $e^x$ as the inverse of $\ln x$. The usual inverse relationships between $e^x$ and $\ln x$ hold, and the properties of $\exp(x)$ can now be written for $e^x$.

---

**THEOREM 7.5** **Properties of $e^x$**

The exponential function $e^x$ satisfies the following properties, all of which result from the integral definition of $\ln x$. Let $x$ and $y$ be real numbers.

1. $e^{x+y} = e^x e^y$
2. $e^{x-y} = e^x / e^y$
3. $(e^x)^p = e^{xp}$, where $p$ is an integer
4. $\ln(e^x) = x$, for all $x$
5. $e^{\ln x} = x$, for $x > 0$

---

## Step 3: General Exponential Functions

It is now a short step to define the exponential function $b^x$ for positive bases with $b \neq 1$ and for all real numbers $x$. By properties 3 and 5 of Theorem 7.5, if $x$ is an integer, then

$$b^x = \underbrace{\left(e^{\ln b}\right)}_{b}{}^x = e^{x \ln b};$$

this important relationship expresses $b^x$ in terms of $e^x$. Because $e^x$ is defined for all real $x$, we use this relationship to define $b^x$ for all real $x$.

---

**DEFINITION** **Exponential Functions with General Bases**

Let $b$ be a positive real number with $b \neq 1$. Then for all real $x$,

$$b^x = e^{x \ln b}.$$

---

This definition comes with an immediate and important consequence. We use the definition of $b^x$ to write

$$x^p = e^{p \ln x}, \text{ for } x > 0 \text{ and } p \text{ real}.$$

> Knowing that $\ln x^p = p \ln x$ for real $p$, we can also extend property 3 of Theorem 7.5 to real numbers. For real $x$ and $y$, we take the natural logarithm of both sides of $z = (e^x)^y$, which gives $\ln z = y \ln e^x = xy$, or $z = e^{xy}$. Therefore, $(e^x)^y = e^{xy}$.

Taking the natural logarithm of both sides and using the inverse relationship between $e^x$ and $\ln x$, we find that

$$\ln x^p = \ln e^{p \ln x} = p \ln x, \text{ for } x > 0 \text{ and } p \text{ real}.$$

In this way, we extend property 4 of Theorem 7.4 to real powers.

**QUICK CHECK 3** Simplify $e^{\ln 2x}$, $\ln(e^{2x})$, $e^{2 \ln x}$, and $\ln(2e^x)$. ◄

**Approximating e**   We have shown that the number $e$ serves as a base for both $\ln x$ and $e^x$, but how do we approximate its value? Recall that the derivative of $\ln x$ at $x = 1$ is 1. By the definition of the derivative, it follows that

> ➤ Because $\dfrac{d}{dx}(\ln x) = \dfrac{1}{x}$,
> $$\frac{d}{dx}(\ln x)\Big|_{x=1} = \frac{1}{1} = 1.$$

$$
\begin{aligned}
1 = \frac{d}{dx}(\ln x)\Big|_{x=1} &= \lim_{h \to 0} \frac{\ln(1+h) - \ln 1}{h} && \text{Derivative of } \ln x \text{ at } x = 1 \\
&= \lim_{h \to 0} \frac{\ln(1+h)}{h} && \ln 1 = 0 \\
&= \lim_{h \to 0} \ln(1+h)^{1/h}. && p \ln x = \ln x^p
\end{aligned}
$$

> ➤ Here we rely on Theorem 2.11 of Section 2.6. If $f$ is continuous at $g(a)$ and $\lim_{x \to a} g(x)$ exists, then $\lim_{x \to a} f(g(x)) = f(\lim_{x \to a} g(x))$.

The natural logarithm is continuous for $x > 0$, so it is permissible to interchange the order of $\lim_{h \to 0}$ and the evaluation of $\ln(1+h)^{1/h}$. The result is that

$$\ln\left(\lim_{h \to 0}(1+h)^{1/h}\right) = 1.$$

$$\underbrace{\phantom{\lim_{h \to 0}(1+h)^{1/h}}}_{e}$$

Observe that the limit within the brackets is $e$ because $\ln e = 1$ and only one number satisfies this equation. Therefore, we have isolated $e$ as a limit:

$$e = \lim_{h \to 0}(1+h)^{1/h}.$$

It is evident from Table 7.2 that $(1+h)^{1/h} \to 2.718282\ldots$ as $h \to 0$.

**Table 7.2**

| $h$ | $(1+h)^{1/h}$ | $h$ | $(1+h)^{1/h}$ |
|---|---|---|---|
| $10^{-1}$ | 2.593742 | $-10^{-1}$ | 2.867972 |
| $10^{-2}$ | 2.704814 | $-10^{-2}$ | 2.731999 |
| $10^{-3}$ | 2.716924 | $-10^{-3}$ | 2.719642 |
| $10^{-4}$ | 2.718146 | $-10^{-4}$ | 2.718418 |
| $10^{-5}$ | 2.718268 | $-10^{-5}$ | 2.718295 |
| $10^{-6}$ | 2.718280 | $-10^{-6}$ | 2.718283 |
| $10^{-7}$ | 2.718282 | $-10^{-7}$ | 2.718282 |

The value of this limit is $e$, and it has been computed to millions of digits. A better approximation,

$$e \approx 2.718281828459045,$$

is obtained by methods introduced in Chapter 11.

## Derivatives and Integrals

The derivative of the exponential function follows directly from Theorem 7.3 (derivatives of inverse functions) or by using the Chain Rule. Taking the latter course, we observe that $\ln(e^x) = x$ and then differentiate both sides with respect to $x$:

$$\frac{d}{dx}(\ln e^x) = \frac{d}{dx}(x)$$

$$\underbrace{\phantom{\frac{d}{dx}(\ln e^x)}}_{1}$$

$$\frac{1}{e^x}\frac{d}{dx}(e^x) = 1 \qquad\qquad \frac{d}{dx}(\ln u(x)) = \frac{u'(x)}{u(x)} \;\;\text{(Chain Rule)}$$

$$\frac{d}{dx}(e^x) = e^x. \qquad\qquad \text{Solve for } \frac{d}{dx}(e^x).$$

**Figure 7.19**

➤ As shown in Example 5a, the integral formula in Theorem 7.6 can be generalized:

$$\int e^{ax}\, dx = \frac{1}{a} e^{ax} + C.$$

**QUICK CHECK 4** What is the slope of the curve $y = e^x$ at $x = \ln 2$? What is the area of the region bounded by the graph of $y = e^x$ and the $x$-axis between $x = 0$ and $x = \ln 2$? ◄

We obtain the remarkable result that the exponential function is its own derivative, which implies that the line tangent to the graph of $y = e^x$ at $(0, 1)$ has slope 1 (**Figure 7.19**). It immediately follows that $e^x$ is its own antiderivative up to an arbitrary constant; that is,

$$\int e^x\, dx = e^x + C.$$

Extending these results using the Chain Rule, we have the following theorem.

---

**THEOREM 7.6    Derivative and Integral of the Exponential Function**
For real numbers $x$,

$$\frac{d}{dx}\left(e^{u(x)}\right) = e^{u(x)}u'(x) \quad \text{and} \quad \int e^x\, dx = e^x + C.$$

---

**EXAMPLE 3    Derivatives involving exponential functions** Evaluate the following derivatives.

**a.** $\dfrac{d}{dx}\left(3e^{2x} - 4e^x + e^{-3x}\right)$    **b.** $\dfrac{d}{dt}\left(\dfrac{e^t}{e^{2t} - 1}\right)$    **c.** $\dfrac{d}{dx}\left(e^{\cos \pi x}\right)\Big|_{x=1/2}$

**SOLUTION**

**a.** $\dfrac{d}{dx}\left(3e^{2x} - 4e^x + e^{-3x}\right) = 3\dfrac{d}{dx}\left(e^{2x}\right) - 4\dfrac{d}{dx}\left(e^x\right) + \dfrac{d}{dx}\left(e^{-3x}\right)$    Sum and Constant Multiple Rules

$$= 3 \cdot 2 \cdot e^{2x} - 4e^x + (-3)e^{-3x}$$    Chain Rule

$$= 6e^{2x} - 4e^x - 3e^{-3x}$$    Simplify.

**b.** $\dfrac{d}{dt}\left(\dfrac{e^t}{e^{2t} - 1}\right) = \dfrac{(e^{2t} - 1)e^t - e^t \cdot 2e^{2t}}{(e^{2t} - 1)^2} = -\dfrac{e^{3t} + e^t}{(e^{2t} - 1)^2}$    Quotient Rule

**c.** First note that by the Chain Rule, we have

$$\frac{d}{dx}\left(e^{\cos \pi x}\right) = -\pi \sin \pi x \cdot e^{\cos \pi x}.$$

Therefore,

$$\frac{d}{dx}\left(e^{\cos \pi x}\right)\Big|_{x=1/2} = -\pi \underbrace{\sin\left(\pi/2\right)}_{1} \cdot \underbrace{e^{\cos\left(\pi/2\right)}}_{e^0 = 1} = -\pi.$$

*Related Exercises 33, 35, 37* ◄

**EXAMPLE 4    Finding tangent lines**

**a.** Write an equation of the line tangent to the graph of $f(x) = 2x - \dfrac{e^x}{2}$ at the point $\left(0, -\frac{1}{2}\right)$.

**b.** Find the point(s) on the graph of $f$ where the tangent line is horizontal.

**SOLUTION**

**a.** To find the slope of the tangent line at $\left(0, -\frac{1}{2}\right)$, we first calculate $f'(x)$:

$$f'(x) = \frac{d}{dx}\left(2x - \frac{e^x}{2}\right)$$

$$= \frac{d}{dx}(2x) - \frac{d}{dx}\left(\frac{1}{2}e^x\right)$$    Difference Rule

$$= 2 - \frac{1}{2}e^x.$$    Evaluate derivatives.

It follows that the slope of the tangent line at $\left(0, -\frac{1}{2}\right)$ is

$$f'(0) = 2 - \frac{1}{2}e^0 = \frac{3}{2}.$$

**Figure 7.20** shows the tangent line passing through $\left(0, -\frac{1}{2}\right)$; it has the equation

$$y - \left(-\frac{1}{2}\right) = \frac{3}{2}(x - 0) \quad \text{or} \quad y = \frac{3}{2}x - \frac{1}{2}.$$

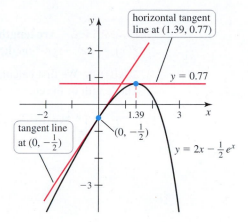

**Figure 7.20**

**b.** Because the slope of a horizontal tangent line is 0, our goal is to solve $f'(x) = 2 - \frac{1}{2}e^x = 0$. We multiply both sides of this equation by 2 and rearrange to arrive at the equation $e^x = 4$. Taking the natural logarithm of both sides, we find that $x = \ln 4$. Therefore, $f'(x) = 0$ at $x = \ln 4 \approx 1.39$, and $f$ has a horizontal tangent at $(\ln 4, f(\ln 4)) \approx (1.39, 0.77)$ (Figure 7.20).

*Related Exercises 39–40* ◄

**EXAMPLE 5**    **Integrals with $e^x$**    Evaluate the following integrals.

**a.** $\displaystyle\int e^{10x}\, dx$    **b.** $\displaystyle\int \frac{e^x}{1 + e^x}\, dx$

**SOLUTION**

**a.** We let $u = 10x$, which implies $du = 10\, dx$, or $dx = \dfrac{1}{10}\, du$:

$$\int \underbrace{e^{10x}}_{e^u}\, \underbrace{dx}_{\frac{1}{10}\, du} = \int e^u \frac{1}{10}\, du \qquad u = 10x,\ du = 10\, dx$$

$$= \frac{1}{10} \int e^u\, du \qquad \int cf(x)\, dx = c\int f(x)\, dx$$

$$= \frac{1}{10} e^u + C \qquad \text{Antiderivative}$$

$$= \frac{1}{10} e^{10x} + C. \qquad \text{Replace } u \text{ with } 10x.$$

The procedure used here can be generalized by replacing 10 with a nonzero constant $a$ to obtain

$$\int e^{ax} dx = \frac{1}{a} e^{ax} + C.$$

**b.** The change of variables $u = 1 + e^x$ implies $du = e^x \, dx$:

$$\int \underbrace{\frac{1}{1 + e^x}}_{u} \underbrace{e^x \, dx}_{du} = \int \frac{1}{u} \, du \qquad u = 1 + e^x, du = e^x \, dx$$

$$= \ln |u| + C \qquad \text{Antiderivative of } u^{-1}$$

$$= \ln (1 + e^x) + C. \quad \text{Replace } u \text{ by } 1 + e^x.$$

Note that the absolute value may be removed from $\ln |u|$ because $1 + e^x > 0$, for all $x$.

*Related Exercises 52, 60* ◄

**EXAMPLE 6    Arc length of an exponential curve**  Find the length of the curve $f(x) = 2e^x + \frac{1}{8}e^{-x}$ on the interval $[0, \ln 2]$.

**SOLUTION**  We first calculate $f'(x) = 2e^x - \frac{1}{8}e^{-x}$ and $f'(x)^2 = 4e^{2x} - \frac{1}{2} + \frac{1}{64}e^{-2x}$. The length of the curve on the interval $[0, \ln 2]$ is

$$L = \int_0^{\ln 2} \sqrt{1 + f'(x)^2} \, dx = \int_0^{\ln 2} \sqrt{1 + \left(4e^{2x} - \frac{1}{2} + \frac{1}{64}e^{-2x}\right)} \, dx$$

$$= \int_0^{\ln 2} \sqrt{4e^{2x} + \frac{1}{2} + \frac{1}{64}e^{-2x}} \, dx \quad \text{Simplify.}$$

$$= \int_0^{\ln 2} \sqrt{\left(2e^x + \frac{1}{8}e^{-x}\right)^2} \, dx \qquad \text{Factor.}$$

$$= \int_0^{\ln 2} \left(2e^x + \frac{1}{8}e^{-x}\right) dx \qquad \text{Simplify.}$$

$$= \left(2e^x - \frac{1}{8}e^{-x}\right)\Big|_0^{\ln 2} = \frac{33}{16}. \qquad \text{Evaluate the integral.}$$

*Related Exercises 61–62* ◄

## Logarithmic Differentiation

Products, quotients, and powers of functions are usually differentiated using the derivative rules of the same name (perhaps combined with the Chain Rule). There are times, however, when the direct computation of a derivative is tedious. Consider the function

$$f(x) = \frac{(x^3 - 1)^4 \sqrt{3x - 1}}{x^2 + 4}.$$

> ► The properties of logarithms needed for logarithmic differentiation (where $x > 0$, $y > 0$, and $z$ is any real number):
>
> **1.** $\ln xy = \ln x + \ln y$
>
> **2.** $\ln (x/y) = \ln x - \ln y$
>
> **3.** $\ln x^z = z \ln x$
>
> All three properties are used in Example 7.

We would need the Quotient, Product, and Chain Rules just to compute $f'(x)$, and simplifying the result would require additional work. The properties of logarithms developed in this section are useful for differentiating such functions.

**EXAMPLE 7    Logarithmic differentiation**  Let $f(x) = \dfrac{(x^2 + 1)^4 e^x}{x^2 + 4}$ and compute $f'(x)$.

**SOLUTION**  We begin by taking the natural logarithm of both sides and simplifying the result:

> ► In the event that $f(x) \leq 0$ for some values of $x$, $\ln f(x)$ is not defined. In that case, we generally find the derivative of $|y| = |f(x)|$.

$$\ln f(x) = \ln \left(\frac{(x^2 + 1)^4 e^x}{x^2 + 4}\right)$$

$$= \ln (x^2 + 1)^4 + \ln e^x - \ln (x^2 + 4) \quad \ln xy = \ln x + \ln y$$

$$= 4 \ln (x^2 + 1) + x - \ln (x^2 + 4). \quad \ln x^y = y \ln x; \ln e^x = x$$

We now differentiate both sides using the Chain Rule; specifically, the derivative of the left side is $\dfrac{d}{dx}(\ln f(x)) = \dfrac{f'(x)}{f(x)}$. Therefore,

$$\frac{f'(x)}{f(x)} = 4 \cdot \frac{1}{x^2 + 1} \cdot 2x + 1 - \frac{1}{x^2 + 4} \cdot 2x.$$

Solving for $f'(x)$, we have

$$f'(x) = f(x)\left(\frac{8x}{x^2 + 1} + 1 - \frac{2x}{x^2 + 4}\right).$$

Finally, we replace $f(x)$ with the original function:

$$f'(x) = \frac{(x^2 + 1)^4 e^x}{x^2 + 4}\left(\frac{8x}{x^2 + 1} - \frac{2x}{x^2 + 4} + 1\right).$$

*Related Exercises 63, 67* ◄

Logarithmic differentiation also provides a method for finding derivatives of tower functions, which are functions of the form $g(x)^{h(x)}$. For example, the derivative of $f(x) = x^x$ is computed as follows, assuming $x > 0$:

$$f(x) = x^x$$

$\ln f(x) = \ln x^x = x \ln x$    Take logarithms of both sides; use properties.

$\dfrac{1}{f(x)} f'(x) = 1 \cdot \ln x + x \cdot \dfrac{1}{x}$    Differentiate both sides.

$f'(x) = f(x)(\ln x + 1)$    Solve for $f'(x)$ and simplify.

$f'(x) = x^x(\ln x + 1)$.    Replace $f(x)$ with $x^x$.

## SECTION 7.2 EXERCISES

### Getting Started

1. What are the domain and range of $\ln x$?

2. Give a geometrical interpretation of the function $\ln x = \displaystyle\int_1^x \frac{dt}{t}$.

3. Differentiate both sides of $x = e^y$ with respect to $x$ to show that $\dfrac{d}{dx}(\ln x) = \dfrac{1}{x}$, for $x > 0$.

4. Sketch the graph of $f(x) = \ln |x|$ and explain how the graph shows that $f'(x) = 1/x$.

5. Show that $\dfrac{d}{dx}(\ln kx) = \dfrac{d}{dx}(\ln x)$, where $x > 0$ and $k$ is a positive real number.

6. Explain the general procedure of logarithmic differentiation.

**7–10. Finding inverse functions** *Find the inverse $f^{-1}(x)$ of each function.*

7. $f(x) = e^{2x+6}$

8. $f(x) = 4e^{5x}$

9. $f(x) = \ln(3x + 1)$

10. $f(x) = 1/(e^x + 1)$

**11–14. Solving equations** *Solve the following equations.*

11. $\ln x = -1$

12. $\ln y = 3$

13. $e^{3x+1} = 10$

14. $e^{-2x} = 6$

T **15. Using inverse relations** One hundred grams of a particular radioactive substance decays according to the function $m(t) = 100e^{-t/650}$, where $t > 0$ measures time in years. When does the mass reach 50 grams?

T **16. Mass of juvenile desert tortoises** In a study conducted at the University of New Mexico, it was found that the mass $m(t)$ (in grams) of juvenile desert tortoises $t$ days after a switch to a particular diet is accurately described by the function $m(t) = m_0 e^{0.004t}$, where $m_0$ is the mass of the tortoise at the time of the diet switch. According to this function, how long does it take a juvenile desert tortoise to reach a mass of 150 g if it had a mass of 64 g when its diet was switched? (Based on *Physiological and Biochemical Zoology*, 85, 1, 2012)

### Practice Exercises

**17–38. Derivatives** *Find the following derivatives.*

17. $\dfrac{d}{dx}(\ln 7x)$

18. $\dfrac{d}{dx}(x^2 \ln x)$

19. $\dfrac{d}{dx}(\ln x^2)$

20. $\dfrac{d}{dx}(\ln 2x^8)$

21. $\dfrac{d}{dx}(\ln |\sin x|)$

22. $\dfrac{d}{dx}\left(\dfrac{\ln x^2}{x}\right)$

23. $\dfrac{d}{dx}\left(\ln\left(\dfrac{x + 1}{x - 1}\right)\right)$

24. $\dfrac{d}{dx}(e^x \ln x)$

25. $\dfrac{d}{dx}((x^2 + 1)\ln x)$

26. $\dfrac{d}{dx}(\ln |x^2 - 1|)$

27. $\dfrac{d}{dx}(\ln(\ln x))$

28. $\dfrac{d}{dx}(\ln(\cos^2 x))$

29. $\dfrac{d}{dx}\left(\dfrac{\ln x}{\ln x + 1}\right)$

30. $\dfrac{d}{dx}\left(\dfrac{\ln x}{x}\right)$

**31.** $\dfrac{d}{dx}\left(\dfrac{e^x}{e^x + 1}\right)$

**32.** $\dfrac{d}{dx}\left(\dfrac{2e^x - 1}{2e^x + 1}\right)$

**33.** $\dfrac{d}{dx}(9e^{-x} - 5e^{2x} - 6e^x)$

**34.** $\dfrac{d}{dx}(xe^{-x} - e^{2x})$

**35.** $\dfrac{d}{dx}\left(\dfrac{e^{2x}}{e^{-x} + 2}\right)$

**36.** $\dfrac{d}{dx}(\cot e^x)$

**37.** $\dfrac{d}{dx}(e^{\sin 2x})\Big|_{x=\pi/4}$

**38.** $\dfrac{d}{dx}(\ln(e^{2x} + 1))\Big|_{x=\ln 2}$

**39–40. Equations of tangent lines** *Find an equation of the line tangent to the following curves at the point* $(a, f(a))$.

**39.** $y = \dfrac{e^x}{4} - x;\ a = 0$

**40.** $y = 2e^x - 1;\ a = \ln 3$

**41–60. Integrals** *Evaluate the following integrals. Include absolute values only when needed.*

**41.** $\displaystyle\int \dfrac{3}{x - 10}\,dx$

**42.** $\displaystyle\int \dfrac{dx}{4x - 3}$

**43.** $\displaystyle\int \left(\dfrac{2}{x - 4} - \dfrac{3}{2x + 1}\right) dx$

**44.** $\displaystyle\int \dfrac{x^2}{2x^3 + 1}\,dx$

**45.** $\displaystyle\int_0^3 \dfrac{2x - 1}{x + 1}\,dx$

**46.** $\displaystyle\int \dfrac{\sec^2 x}{\tan x}\,dx$

**47.** $\displaystyle\int_3^4 \dfrac{dx}{2x \ln x \ln^3 (\ln x)}$

**48.** $\displaystyle\int_0^{\pi/2} \dfrac{\sin x}{1 + \cos x}\,dx$

**49.** $\displaystyle\int \dfrac{dx}{x \ln x \ln (\ln x)}$

**50.** $\displaystyle\int_0^1 \dfrac{y \ln^4 (y^2 + 1)}{y^2 + 1}\,dy$

**51.** $\displaystyle\int \dfrac{e^{2x} - e^{-2x}}{2}\,dx$

**52.** $\displaystyle\int 3e^{-4t}\,dt$

**53.** $\displaystyle\int (e^{2x} + 1)\,dx$

**54.** $\dfrac{1}{2}\displaystyle\int_0^{\ln 2} e^x dx$

**55.** $\displaystyle\int_0^{\ln 3} e^x(e^{3x} + e^{2x} + e^x)\,dx$

**56.** $\displaystyle\int (2e^{-10z} + 3e^{5z})\,dz$

**57.** $\displaystyle\int \dfrac{e^x + e^{-x}}{e^x - e^{-x}}\,dx$

**58.** $\displaystyle\int \dfrac{e^{\sin x}}{\sec x}\,dx$

**59.** $\displaystyle\int \dfrac{e^{\sqrt{x}}}{\sqrt{x}}\,dx$

**60.** $\displaystyle\int_{-2}^2 \dfrac{e^{z/2}}{e^{z/2} + 1}\,dz$

**61–62. Arc length** *Find the length of the following curves.*

**61.** $x = 2e^{\sqrt{2}y} + \dfrac{1}{16}e^{-\sqrt{2}y},\ \text{for } 0 \le y \le \dfrac{\ln 2}{\sqrt{2}}$

**62.** $y = \dfrac{1}{2}(e^x + e^{-x}),\ \text{for } -\ln 2 \le x \le \ln 2$

**63–70. Logarithmic differentiation** *Use logarithmic differentiation to evaluate* $f'(x)$.

**63.** $f(x) = \dfrac{(x + 1)^{10}}{(2x - 4)^8}$

**64.** $f(x) = x^2 \cos x$

**65.** $f(x) = x^{\ln x}$

**66.** $f(x) = \dfrac{\tan^{10} x}{(5x + 3)^6}$

**67.** $f(x) = \dfrac{(x + 1)^{3/2}(x - 4)^{5/2}}{(5x + 3)^{2/3}}$

**68.** $f(x) = \dfrac{x^8 \cos^3 x}{\sqrt{x - 1}}$

**69.** $f(x) = (\sin x)^{\tan x}$

**70.** $f(x) = \left(1 + \dfrac{1}{x}\right)^{2x}$

**71. Explain why or why not** Determine whether the following statements are true and give an explanation or counterexample. Assume $x > 0$ and $y > 0$.

**a.** $\ln xy = \ln x + \ln y$.

**b.** $\ln 0 = 1$.

**c.** $\ln (x + y) = \ln x + \ln y$.

**d.** $\dfrac{d}{dx}(e^3) = e^3$.

**e.** $\dfrac{d}{dx}(e^x) = xe^{x-1}$.

**f.** $\dfrac{d^n}{dx^n}(e^{3x}) = 3^n e^{3x}$, for any integer $n \ge 1$.

**g.** The area between the curve $y = 1/x$ and the $x$-axis on the interval $[1, e]$ is 1.

**72. Logarithm properties** Use the integral definition of the natural logarithm to prove that $\ln (x/y) = \ln x - \ln y$.

**73. Looking ahead: Integrals of tan $x$ and cot $x$** Use a change of variables to verify each integral.

**a.** $\displaystyle\int \tan x\,dx = -\ln |\cos x| + C = \ln |\sec x| + C$

**b.** $\displaystyle\int \cot x\,dx = \ln |\sin x| + C$

**T 74. Behavior at the origin** Using calculus and accurate sketches, explain how the graphs of $f(x) = x^p \ln x$ differ as $x \to 0^+$ for $p = \frac{1}{2}, 1,$ and 2.

**75. Average value** What is the average value of $f(x) = 1/x$ on the interval $[1, p]$ for $p > 1$? What is the average value of $f$ as $p \to \infty$?

**76–97. Miscellaneous derivatives and integrals.** *Evaluate the following derivatives and integrals.*

**76.** $\dfrac{d}{dx}(x^{2x})$

**77.** $\dfrac{d}{dx}(e^{-10x^2})$

**78.** $\dfrac{d}{dx}(x^{\tan x})$

**79.** $\dfrac{d}{dx}(\ln \sqrt{10x})$

**80.** $\dfrac{d}{dx}(x^e + e^x)$

**81.** $\dfrac{d}{dx}\left(\dfrac{(x^2 + 1)(x - 3)}{(x + 2)^3}\right)$

**82.** $\dfrac{d}{dx}(\ln (\sec^4 x \tan^2 x))$

**83.** $\dfrac{d}{dx}(\sin (\sin (e^x)))$

**84.** $\dfrac{d}{dx}(\sin^2 (e^{3x+1}))$

**85.** $\dfrac{d}{dx}\left(\dfrac{xe^x}{x + 1}\right)$

**86.** $\dfrac{d}{dx}\left(\left(\dfrac{e^x}{x + 1}\right)^8\right)$

**87.** $\displaystyle\int x^2 e^{x^3}\,dx$

**88.** $\displaystyle\int_0^\pi e^{\sin x} \cos x\,dx$

**89.** $\displaystyle\int_1^{2e} \dfrac{e^{\ln x}}{x}\,dx$

**90.** $\displaystyle\int \dfrac{\sin (\ln x)}{4x}\,dx$

**91.** $\displaystyle\int_1^{e^2} \dfrac{\ln^5 x}{x}\,dx$

**92.** $\displaystyle\int \dfrac{\ln^2 x + 2 \ln x - 1}{x}\,dx$

**93.** $\int \dfrac{x}{x-2}\,dx$ (*Hint*: Let $u = x - 2$.)

**94.** $\int_0^{\ln 4} \dfrac{e^x}{3 + 2e^x}\,dx$

**95.** $\int_0^{\pi/6} \dfrac{\sin 2y}{\sin^2 y + 2}\,dy$ (*Hint*: $\sin 2y = 2\sin y \cos y$.)

**96.** $\int \dfrac{e^{2x}}{e^{2x} + 1}\,dx$ **97.** $\int \dfrac{dt}{t \ln t^2}$

**98–99. First and second derivative analysis**

**a.** Find the critical points of $f$.
**b.** Use the First Derivative Test to locate the local maximum and minimum values.
**c.** Determine the intervals on which $f$ is concave up or concave down. Identify any inflection points.

**98.** $f(x) = x^4 e^{-x}$ **99.** $f(x) = xe^{-x^2/2}$

**100–101. Linear approximation**

**a.** Find the linear approximation $L$ to the function $f$ at the point $a$.
**b.** Graph $f$ and $L$ on the same set of axes.
**c.** Use the linear approximation to estimate the given quantity.

**100.** $f(x) = \ln(1 + x); a = 0; \ln 1.9$

**101.** $f(x) = e^{-x}; a = \ln 2; 1/e$

**102. Solid of revolution** The region bounded by the graphs of $x = 0, x = \sqrt{\ln y}$, and $x = \sqrt{2 - \ln y}$ in the first quadrant is revolved about the $y$-axis. What is the volume of the resulting solid?

## Explorations and Challenges

**103. Probability as an integral** Two points $P$ and $Q$ are chosen randomly, one on each of two adjacent sides of a unit square (see figure). What is the probability that the area of the triangle formed by the sides of the square and the line segment $PQ$ is less than one-fourth the area of the square? Begin by showing that $x$ and $y$ must satisfy $xy < \frac{1}{2}$ in order for the area condition to be met.

Then argue that the required probability is $\dfrac{1}{2} + \displaystyle\int_{1/2}^1 \dfrac{dx}{2x}$ and evaluate the integral.

**104–107. Logistic growth** *Scientists often use the logistic growth function* $P(t) = \dfrac{P_0 K}{P_0 + (K - P_0)e^{-r_0 t}}$ *to model population growth,*

*where $P_0$ is the initial population at time $t = 0$, $K$ is the* **carrying capacity,** *and $r_0$ is the base growth rate. The carrying capacity is a theoretical upper bound on the total population that the surrounding*

*environment can support. The figure shows the sigmoid (S-shaped) curve associated with a typical logistic model.*

**T 104. Population crash** The logistic model can be used for situations in which the initial population $P_0$ is above the carrying capacity $K$. For example, consider a deer population of 1500 on an island where a fire has reduced the carrying capacity to 1000 deer.

**a.** Assuming a base growth rate of $r_0 = 0.1$ and an initial population of $P(0) = 1500$, write a logistic growth function for the deer population and graph it. Based on the graph, what happens to the deer population in the long run?
**b.** How fast (in deer per year) is the population declining immediately after the fire at $t = 0$?
**c.** How long does it take the deer population to decline to 1200 deer?

**T 105. Gone fishing** When a reservoir is created by a new dam, 50 fish are introduced into the reservoir, which has an estimated carrying capacity of 8000 fish. A logistic model of the fish population is $P(t) = \dfrac{400{,}000}{50 + 7950e^{-0.5t}}$, where $t$ is measured in years.

**a.** Graph $P$ using a graphing utility. Experiment with different windows until you produce an S-shaped curve characteristic of the logistic model. What window works well for this function?
**b.** How long does it take the population to reach 5000 fish? How long does it take the population to reach 90% of the carrying capacity?
**c.** How fast (in fish per year) is the population growing at $t = 0$? At $t = 5$?
**d.** Graph $P'$ and use the graph to estimate the year in which the population is growing fastest.

**T 106. World population (part 1)** The population of the world reached 6 billion in 1999 ($t = 0$). Assume Earth's carrying capacity is 15 billion and the base growth rate is $r_0 = 0.025$ per year.

**a.** Write a logistic growth function for the world's population (in billions), and graph your equation on the interval $0 \le t \le 200$ using a graphing utility.
**b.** What will the population be in the year 2020? When will it reach 12 billion?

**107. World population (part 2)** The *relative growth rate* $r$ of a function $f$ measures the rate of change of the function compared to its value at a particular point. It is computed as $r(t) = f'(t)/f(t)$.

**a.** Confirm that the relative growth rate in 1999 ($t = 0$) for the logistic model in Exercise 106 is $r(0) = P'(0)/P(0) = 0.015$. This means the world's population was growing at 1.5% per year in 1999.

**b.** Compute the relative growth rate of the world's population in 2010 and 2020. What appears to be happening to the relative growth rate as time increases?

**c.** Evaluate $\lim_{t \to \infty} r(t) = \lim_{t \to \infty} \dfrac{P'(t)}{P(t)}$, where $P(t)$ is the logistic growth function from Exercise 106. What does your answer say about populations that follow a logistic growth pattern?

**108. Snowplow problem** With snow on the ground and falling at a constant rate, a snowplow began plowing down a long straight road at noon. The plow traveled twice as far in the first hour as it did in the second hour. At what time did the snow start falling? Assume the plowing rate is inversely proportional to the depth of the snow.

**109. Depletion of natural resources** Suppose that $r(t) = r_0 e^{-kt}$, with $k > 0$, is the rate at which a nation extracts oil, where $r_0 = 10^7$ barrels/yr is the current rate of extraction. Suppose also that the estimate of the total oil reserve is $2 \times 10^9$ barrels.

**a.** Find $Q(t)$, the total amount of oil extracted by the nation after $t$ years.

**b.** Evaluate $\lim_{t \to \infty} Q(t)$ and explain the meaning of this limit.

**c.** Find the minimum decay constant $k$ for which the total oil reserves will last forever.

**d.** Suppose $r_0 = 2 \times 10^7$ barrels/yr and the decay constant $k$ is the minimum value found in part (c). How long will the total oil reserves last?

**110. Derivative of $\ln |x|$** Differentiate $\ln x$ for $x > 0$ and differentiate $\ln(-x)$ for $x < 0$ to conclude that $\dfrac{d}{dx}(\ln |x|) = \dfrac{1}{x}$.

**111. Properties of $e^x$** Use the inverse relations between $\ln x$ and $e^x$ ($\exp(x)$) and the properties of $\ln x$ to prove the following properties.

**a.** $\exp(0) = 1$

**b.** $\exp(x - y) = \dfrac{\exp(x)}{\exp(y)}$

**c.** $(\exp(x))^p = \exp(px)$, where $p$ is an integer

**112. $\ln x$ is unbounded** Use the following argument to show that $\lim_{x \to \infty} \ln x = \infty$ and $\lim_{x \to 0^+} \ln x = -\infty$.

**a.** Make a sketch of the function $f(x) = 1/x$ on the interval $[1, 2]$. Explain why the area of the region bounded by $y = f(x)$ and the $x$-axis on $[1, 2]$ is $\ln 2$.

**b.** Construct a rectangle over the interval $[1, 2]$ with height $\frac{1}{2}$. Explain why $\ln 2 > \frac{1}{2}$.

**c.** Show that $\ln 2^n > n/2$ and $\ln 2^{-n} < -n/2$.

**d.** Conclude that $\lim_{x \to \infty} \ln x = \infty$ and $\lim_{x \to 0^+} \ln x = -\infty$.

**113. Bounds on $e$** Use a left Riemann sum with at least $n = 2$ subintervals of equal length to approximate $\ln 2 = \displaystyle\int_1^2 \dfrac{dt}{t}$ and show that $\ln 2 < 1$. Use a right Riemann sum with $n = 7$ subintervals of equal length to approximate $\ln 3 = \displaystyle\int_1^3 \dfrac{dt}{t}$ and show that $\ln 3 > 1$.

**114. Alternative proof of product property** Assume that $y > 0$ is fixed and that $x > 0$. Show that $\dfrac{d}{dx}(\ln xy) = \dfrac{d}{dx}(\ln x)$. Recall that if two functions have the same derivative, then they differ by an additive constant. Set $x = 1$ to evaluate the constant and prove that $\ln xy = \ln x + \ln y$.

**115. Harmonic sum** In Chapter 10, we will encounter the harmonic sum $1 + \dfrac{1}{2} + \dfrac{1}{3} + \cdots + \dfrac{1}{n}$. Use a left Riemann sum to approximate $\displaystyle\int_1^{n+1} \dfrac{dx}{x}$ (with unit spacing between the grid points) to show that $1 + \dfrac{1}{2} + \dfrac{1}{3} + \cdots + \dfrac{1}{n} > \ln(n + 1)$. Use this fact to conclude that $\lim_{n \to \infty} \left(1 + \dfrac{1}{2} + \dfrac{1}{3} + \cdots + \dfrac{1}{n}\right)$ does not exist.

**T 116. Tangency question** It is easily verified that the graphs of $y = x^2$ and $y = e^x$ have no point of intersection (for $x > 0$), and the graphs of $y = x^3$ and $y = e^x$ have two points of intersection. It follows that for some real number $2 < p < 3$, the graphs of $y = x^p$ and $y = e^x$ have exactly one point of intersection (for $x > 0$). Using analytical and/or graphical methods, determine $p$ and the coordinates of the single point of intersection.

**117. Property of exponents** Prove that for real numbers $x$, $a$, and $b \neq 0$, $\left(\dfrac{a}{b}\right)^x = \dfrac{a^x}{b^x}$.

**QUICK CHECK ANSWERS**

1. $\{x : x \neq 0\}$   2. $p/x$   3. $2x, 2x, x^2, x + \ln 2$
4. Slope $= 2$; area $= 1$ ◄

# 7.3 Logarithmic and Exponential Functions with Other Bases

In Section 7.2, we briefly introduced the general exponential function $b^x$ for theoretical purposes, defining it in terms of the base $e$. It turns out that situations occasionally arise in which it is more convenient to work with bases other than $e$. In this section, we develop the properties of the function $b^x$ and its inverse, which is the logarithmic function $\log_b x$, where the base $b$ is a positive number with $b \neq 1$. Derivatives and integrals associated with these important functions follow from their relationship to the natural exponential and logarithmic functions. We also return to the Power Rule $\left(\dfrac{d}{dx}(x^p) = px^{p-1}\right)$ from Chapter 3, now armed with the tools to show that it holds for all real powers $p$.

**QUICK CHECK 1** Is it possible to raise a positive number $b$ to a power and obtain a negative number? Is it possible to obtain zero? ◄

## General Exponential Functions

Recall from Section 7.2 that the general exponential function $b^x$ is defined for all real numbers $x$ in terms of the base $e$:

$$b^x = e^{x \ln b}, \text{ where the base } b \text{ is positive, with } b \neq 1.$$

We begin by establishing the properties of this family of functions.

### Properties of $f(x) = b^x$

1. Because $f(x) = b^x$ is defined for all real numbers, the domain of $f$ is $(-\infty, \infty)$. Because $b^x = e^{x \ln b}$ and the range of $e^x$ is $(0, \infty)$, the range of $f$ is $(0, \infty)$.

2. For all $b > 0$, $b^0 = e^{0 \ln b} = 1$. Therefore, $f(0) = 1$.

3. If $b > 1$, then $\ln b > 0$ and $b^x = e^{x \ln b}$ is an increasing function (**Figure 7.21**). For example, if $b = 2$, then $2^x > 2^y$ whenever $x > y$.

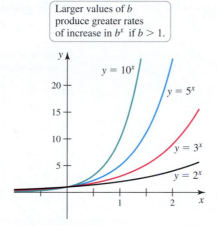

Larger values of $b$ produce greater rates of increase in $b^x$ if $b > 1$.

**Figure 7.21**

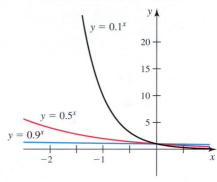

Smaller values of $b$ produce greater rates of decrease in $b^x$ if $0 < b < 1$.

**Figure 7.22**

> The proof of the property
> $$\left(\frac{a}{b}\right)^x = \frac{a^x}{b^x}$$
> was considered in Exercise 117 of Section 7.2.

**QUICK CHECK 2** Explain why $f(x) = (1/3)^x$ is a decreasing function. ◄

4. If $0 < b < 1$, then $\ln b < 0$ and $b^x = e^{x \ln b}$ is a decreasing function (**Figure 7.22**).

For example, if $b = \frac{1}{2}$, then

$$f(x) = \left(\frac{1}{2}\right)^x = \frac{1}{2^x} = 2^{-x},$$

and because $2^x$ increases with $x$, $2^{-x}$ decreases with $x$.

The graphs of $b^x$ for various values of $b$ are horizontal scalings of the graph of $y = e^x$ (recall from Section 1.2 that the graph of $y = f(ax)$ is a horizontal scaling of the graph of $y = f(x)$ and note that $b^x = e^{x \ln b}$).

## General Logarithmic Functions

Everything we learned about inverse functions is now applied to the exponential function $f(x) = b^x$. For any $b > 0$, with $b \neq 1$, this function is one-to-one on the interval $(-\infty, \infty)$. Therefore, it has an inverse.

---

**DEFINITION  Logarithmic Function Base $b$**

For any base $b > 0$, with $b \neq 1$, the **logarithmic function base $b$**, denoted $\log_b x$, is the inverse of the exponential function $b^x$.

---

The inverse relationship between logarithmic and exponential functions may be stated concisely in several ways. First, we have

$$y = \log_b x \quad \text{if and only if} \quad b^y = x.$$

Combining these two conditions results in two important relations.

---

**Inverse Relations for Exponential and Logarithmic Functions**

For any base $b > 0$, with $b \neq 1$, the following inverse relations hold.

**I1.** $b^{\log_b x} = x$, for $x > 0$

**I2.** $\log_b b^x = x$, for all $x$

---

The graph of the logarithmic function is generated using the symmetry of the graphs of a function and its inverse. **Figure 7.23** shows how the graph of $y = b^x$, for $b > 1$, is reflected across the line $y = x$ to obtain the graph of $y = \log_b x$.

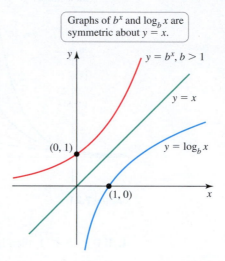

Graphs of $b^x$ and $\log_b x$ are symmetric about $y = x$.

$y = b^x, b > 1$

$y = x$

$(0, 1)$

$y = \log_b x$

$(1, 0)$

**Figure 7.23**

The graphs of $y = \log_b x$ are shown (**Figure 7.24**) for several bases $b > 1$. Logarithms with bases $0 < b < 1$, although well defined, are generally not used (and they can be expressed in terms of bases with $b > 1$).

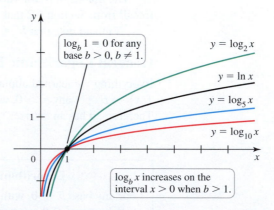

$\log_b 1 = 0$ for any base $b > 0, b \neq 1$.

$y = \log_2 x$

$y = \ln x$

$y = \log_5 x$

$y = \log_{10} x$

$\log_b x$ increases on the interval $x > 0$ when $b > 1$.

**Figure 7.24**

Logarithmic functions with base $b > 0$ satisfy properties that parallel the properties of the exponential functions given earlier.

**Properties of $\log_b x$**

1. Because the range of $b^x$ is $(0, \infty)$, the domain of $\log_b x$ is $(0, \infty)$.

2. The domain of $b^x$ is $(-\infty, \infty)$, which implies that the range of $\log_b x$ is $(-\infty, \infty)$.

3. Because $b^0 = 1$, it follows that $\log_b 1 = 0$.

4. If $b > 1$, then $\log_b x$ is an increasing function of $x$. For example, if $b = e$, then $\ln x > \ln y$ whenever $x > y$ (Figure 7.24).

**QUICK CHECK 3** What is the domain of $f(x) = \log_b(x^2)$? What is the range of $f(x) = \log_b(x^2)$? ◄

**EXAMPLE 1  Using inverse relations** One thousand grams of a particular radioactive substance decays according to the function $m(t) = 1000e^{-t/850}$, where $t \geq 0$ measures time in years and $m$ is measured in grams.

a. When does the mass of the substance reach the safe level, deemed to be 1 g?

b. Write the mass function with a base of $\frac{1}{2}$.

**SOLUTION**

▶ Provided the arguments are positive, we can take the $\log_b$ of both sides of an equation and produce an equivalent equation.

a. Setting $m(t) = 1$, we solve $1000e^{-t/850} = 1$ by dividing both sides by 1000 and taking the natural logarithm of both sides:

$$\ln\left(e^{-t/850}\right) = \ln\frac{1}{1000}.$$

This equation is simplified by calculating $\ln(1/1000) \approx -6.908$ and observing that $\ln\left(e^{-t/850}\right) = -\dfrac{t}{850}$ (inverse property I2). Therefore,

$$-\frac{t}{850} \approx -6.908,$$

which implies that $t \approx (-850)(-6.908) \approx 5872$ years.

b. We seek a function $p(t)$ such that $m(t) = 1000\,e^{-t/850} = 1000\left(\dfrac{1}{2}\right)^{p(t)}$. Canceling 1000 and taking the natural logarithm of both sides, we proceed as follows:

$$\ln\left(e^{-t/850}\right) = \ln\left(\frac{1}{2}\right)^{p(t)}$$

$$-\frac{t}{850} = p(t)\ln\frac{1}{2} = -p(t)\ln 2 \quad \text{Properties of logarithms}$$

$$p(t) = \frac{t}{850\ln 2}. \quad \text{Solve for } p(t).$$

The mass function can be written $m(t) = 1000\left(\frac{1}{2}\right)^{t/(850\ln 2)}$, which says that when $t$ increases by $850\ln 2 \approx 589$ years, the mass decreases by a factor of $\frac{1}{2}$.

*Related Exercises 15, 18* ◄

## The Derivative of $b^x$

A rule similar to $\dfrac{d}{dx}(e^x) = e^x$ exists for computing the derivative of $b^x$, where $b > 0$. Because $b^x = e^{x\ln b}$, the derivative of $b^x$ is

$$\frac{d}{dx}(b^x) = \frac{d}{dx}(e^{x\ln b}) = \underbrace{e^{x\ln b}}_{b^x}\cdot \ln b. \quad \text{Chain Rule}$$

Noting that $e^{x \ln b} = b^x$ results in the following theorem.

➤ Check that when $b = e$, Theorem 7.7 becomes

$$\frac{d}{dx}(e^x) = e^x$$

> **THEOREM 7.7   Derivative of $b^x$**
> If $b > 0$ and $b \neq 1$, then for all $x$,
>
> $$\frac{d}{dx}(b^x) = b^x \ln b.$$

Notice that when $b > 1$, $\ln b > 0$ and the graph of $y = b^x$ has tangent lines with positive slopes for all $x$. When $0 < b < 1$, $\ln b < 0$ and the graph of $y = b^x$ has tangent lines with negative slopes for all $x$. In either case, the tangent line at $(0, 1)$ has slope $\ln b$ (**Figure 7.25**).

Slope of line tangent to $y = b^x$ at $x = 0$ is $\ln b$.

**Figure 7.25**

**EXAMPLE 2   Derivatives with $b^x$**   Find the derivative of the following functions.

**a.** $f(x) = 3^x$          **b.** $g(t) = 108 \cdot 2^{t/12}$

**SOLUTION**

**a.** Using Theorem 7.7, $f'(x) = 3^x \ln 3$.

**b.**
$$g'(t) = 108 \frac{d}{dt}(2^{t/12}) \qquad \text{Constant Multiple Rule}$$

$$= 108 \cdot \ln 2 \cdot 2^{t/12} \underbrace{\frac{d}{dt}\left(\frac{t}{12}\right)}_{1/12} \qquad \text{Chain Rule}$$

$$= 9 \ln 2 \cdot 2^{t/12} \qquad \text{Simplify.}$$

*Related Exercises 23–25* ◀

**EXAMPLE 3   An exponential model**   Table 7.3 and **Figure 7.26** show how the incidence of Down syndrome in newborn infants increases with the age of the mother.

**Table 7.3**

| Mother's Age | Incidence of Down Syndrome | Decimal Equivalent |
|---|---|---|
| 30 | 1 in 900 | 0.00111 |
| 35 | 1 in 400 | 0.00250 |
| 36 | 1 in 300 | 0.00333 |
| 37 | 1 in 230 | 0.00435 |
| 38 | 1 in 180 | 0.00556 |
| 39 | 1 in 135 | 0.00741 |
| 40 | 1 in 105 | 0.00952 |
| 42 | 1 in 60 | 0.01667 |
| 44 | 1 in 35 | 0.02857 |
| 46 | 1 in 20 | 0.05000 |
| 48 | 1 in 16 | 0.06250 |
| 49 | 1 in 12 | 0.08333 |

(Based on E.G. Hook and A. Lindsjo, *The American Journal of Human Genetics*, 30, Jan 1978)

**Figure 7.26**

▶ The model in Example 3 was created using a method called *exponential regression*. The parameters $A$ and $B$ are chosen so that the function $P(a) = A \cdot B^a$ fits the data as closely as possible.

The data can be modeled with the exponential function $P(a) = \dfrac{1}{1{,}613{,}000} 1.2733^a$,

where $a$ is the age of the mother (in years) and $P(a)$ is the incidence (number of children with Down syndrome per total births).

**a.** According to the model, at what age is the incidence of Down syndrome equal to 0.01 (that is, 1 in 100)?

**b.** Compute $P'(a)$.

**c.** Find $P'(35)$ and $P'(46)$, and interpret each.

**SOLUTION**

**a.** We let $P(a) = 0.01$ and solve for $a$:

$$0.01 = \frac{1}{1{,}613{,}000} 1.2733^a$$

$$\ln 16{,}130 = \ln (1.2733^a) \quad \text{Multiply both sides by 1,613,000 and take logarithms of both sides.}$$

$$\ln 16{,}130 = a \ln 1.2733 \quad \text{Property of logarithms}$$

$$a = \frac{\ln 16{,}130}{\ln 1.2733} \approx 40 \text{ (years old)}. \quad \text{Solve for } a.$$

**b.** $P'(a) = \dfrac{1}{1{,}613{,}000} \dfrac{d}{da}(1.2733^a)$

$$= \frac{1}{1{,}613{,}000} 1.2733^a \ln 1.2733$$

$$\approx \frac{1}{6{,}676{,}000} 1.2733^a$$

**c.** The derivative measures the rate of change of the incidence with respect to age. For a 35-year-old woman,

$$P'(35) = \frac{1}{6{,}676{,}000} 1.2733^{35} \approx 0.0007,$$

which means the incidence increases at a rate of about 0.0007/year. By age 46, the rate of change is

$$P'(46) = \frac{1}{6{,}676{,}000} 1.2733^{46} \approx 0.01,$$

which is a significant increase over the rate of change of the incidence at age 35.

*Related Exercise 35* ◀

**QUICK CHECK 4** Suppose $A = 500(1.045)^t$. Compute $\dfrac{dA}{dt}$. ◀

## Integral of $b^x$

The fact that $\dfrac{d}{dx}(b^x) = b^x \ln b$ leads immediately to the following indefinite integral result.

---

**THEOREM 7.8   Indefinite integral of $b^x$**

For $b > 0$ and $b \neq 1$, $\displaystyle\int b^x \, dx = \frac{1}{\ln b} b^x + C.$

---

**EXAMPLE 4    Integrals involving exponentials with other bases** Evaluate the following integrals.

**a.** $\displaystyle\int x\,3^{x^2}\,dx$    **b.** $\displaystyle\int_1^4 \frac{6^{-\sqrt{x}}}{\sqrt{x}}\,dx$

**SOLUTION**

**a.**
$$\int x\,3^{x^2}\,dx = \frac{1}{2}\int 3^u\,du \qquad u = x^2,\, du = 2x\,dx$$

$$= \frac{1}{2}\frac{1}{\ln 3}\,3^u + C \qquad \text{Integrate.}$$

$$= \frac{1}{2\ln 3}\,3^{x^2} + C \qquad \text{Substitute } u = x^2.$$

**b.**
$$\int_1^4 \frac{6^{-\sqrt{x}}}{\sqrt{x}}\,dx = -2\int_{-1}^{-2} 6^u\,du \qquad u = -\sqrt{x},\, du = -\frac{1}{2\sqrt{x}}\,dx$$

$$= -\frac{2}{\ln 6}\,6^u\Big|_{-1}^{-2} \qquad \text{Fundamental Theorem}$$

$$= \frac{5}{18\ln 6} \qquad \text{Simplify.}$$

*Related Exercises 40, 43* ◄

## The General Power Rule

In Theorem 3.8 of Section 3.4, we claimed that the Power Rule for derivatives, $\dfrac{d}{dx}(x^p) = px^{p-1}$, is valid for real powers $p$. We now have the tools to prove this important result.

---

**THEOREM 7.9    General Power Rule**
For real numbers $p$ and for $x > 0$,

$$\frac{d}{dx}(x^p) = px^{p-1}.$$

Furthermore, if $u$ is a positive differentiable function on its domain, then

$$\frac{d}{dx}(u(x)^p) = p(u(x))^{p-1}\cdot u'(x).$$

---

**Proof:** For $x > 0$ and real numbers $p$, the derivative of $x^p$ is computed as follows:

$$\frac{d}{dx}(x^p) = \frac{d}{dx}\left(e^{p\ln x}\right) \qquad x^p = e^{p\ln x}$$

$$= \underbrace{e^{p\ln x}}_{x^p}\cdot\frac{p}{x} \qquad \text{Chain Rule; } \frac{d}{dx}(p\ln x) = p\cdot\frac{1}{x}$$

$$= x^p\cdot\frac{p}{x} \qquad e^{p\ln x} = x^p$$

$$= px^{p-1}. \qquad \text{Simplify.}$$

We see that $\dfrac{d}{dx}(x^p) = px^{p-1}$, for all real powers $p$. The second part of the General Power Rule follows from the Chain Rule. ◄

**EXAMPLE 5  Computing derivatives** Find the derivative of the following functions.

**a.** $y = x^\pi$         **b.** $y = \pi^x$         **c.** $y = (x^2 + 4)^e$

> Recall that power functions have the variable in the base, while exponential functions have the variable in the exponent.

**SOLUTION**

**a.** With $y = x^\pi$, we have a power function with an irrational exponent; by the General Power Rule,

$$\frac{dy}{dx} = \pi x^{\pi - 1}, \quad \text{for } x > 0.$$

**b.** Here we have an exponential function with base $b = \pi$. By Theorem 7.7,

$$\frac{dy}{dx} = \pi^x \ln \pi.$$

**c.** The Chain Rule and General Power Rule are required:

$$\frac{dy}{dx} = e(x^2 + 4)^{e-1} \cdot 2x = 2ex\,(x^2 + 4)^{e-1}.$$

Because $x^2 + 4 > 0$ for all $x$, the result is valid for all $x$.

*Related Exercises 45–46* ◄

Functions of the form $f(x) = (g(x))^{h(x)}$, where both $g$ and $h$ are nonconstant functions, are neither exponential functions nor power functions (they are sometimes called *tower functions*). To compute their derivatives, we use the identity $b^x = e^{x \ln b}$ to rewrite $f$ with base $e$:

$$f(x) = (g(x))^{h(x)} = e^{h(x) \ln g(x)}.$$

> Recall that the derivative of the tower function $f(x) = x^x$ was determined using logarithmic differentiation in Section 7.2. Examples 6 and 7 offer an alternative method for differentiating such functions. Either method is acceptable; in fact, they are equivalent.

This function carries the restriction $g(x) > 0$. The derivative of $f$ is then computed using the methods developed in this section.

**EXAMPLE 6  Tower function** Let $f(x) = x^{\sin x}$, for $x > 0$.

**a.** Find $f'(x)$.         **b.** Evaluate $f'\left(\dfrac{\pi}{2}\right)$.

**SOLUTION**

**a.** The key step is to use $b^x = e^{x \ln b}$ to write $f$ in the form

$$f(x) = x^{\sin x} = e^{(\sin x)\ln x}.$$

We now differentiate:

$$f'(x) = e^{\sin x \ln x} \frac{d}{dx}((\sin x)\ln x) \qquad \text{Chain Rule}$$

$$= \underbrace{e^{(\sin x)\ln x}}_{x^{\sin x}}\left((\cos x)\ln x + \frac{\sin x}{x}\right) \qquad \text{Product Rule}$$

$$= x^{\sin x}\left((\cos x)\ln x + \frac{\sin x}{x}\right).$$

**b.** Letting $x = \dfrac{\pi}{2}$, we find that

$$f'\left(\frac{\pi}{2}\right) = \left(\frac{\pi}{2}\right)^{\sin(\pi/2)}\left(\underbrace{\left(\cos\frac{\pi}{2}\right)}_{0}\ln\frac{\pi}{2} + \underbrace{\frac{\sin(\pi/2)}{\pi/2}}_{2/\pi}\right) \qquad \text{Substitute } x = \frac{\pi}{2}.$$

$$= \frac{\pi}{2}\left(0 + \frac{2}{\pi}\right) = 1.$$

*Related Exercises 51–52* ◄

**EXAMPLE 7  Finding a horizontal tangent line**  Determine whether the graph of $f(x) = x^x$, for $x > 0$, has any horizontal tangent lines.

**SOLUTION**  A horizontal tangent occurs when $f'(x) = 0$. To find the derivative, we first write $f(x) = x^x = e^{x \ln x}$:

$$\frac{d}{dx}(x^x) = \frac{d}{dx}(e^{x \ln x})$$

$$= \underbrace{e^{x \ln x}}_{x^x}\left(1 \cdot \ln x + x \cdot \frac{1}{x}\right) \quad \text{Chain Rule; Product Rule}$$

$$= x^x(\ln x + 1). \qquad \text{Simplify; } e^{x \ln x} = x^x.$$

The equation $f'(x) = 0$ implies that $x^x = 0$ or $\ln x + 1 = 0$. The first equation has no solution because $x^x = e^{x \ln x} > 0$, for all $x > 0$. We solve the second equation, $\ln x + 1 = 0$, as follows:

$$\ln x = -1$$

$$e^{\ln x} = e^{-1} \quad \text{Exponentiate both sides.}$$

$$x = \frac{1}{e}. \quad e^{\ln x} = x$$

Therefore, the graph of $f(x) = x^x$ has a single horizontal tangent at $(e^{-1}, f(e^{-1})) \approx (0.368, 0.692)$. Using the First Derivative Test and Theorem 4.9, we could verify what the graph of $f$ shows (**Figure 7.27**): The absolute minimum of $f$ occurs at $x = 1/e$.

*Related Exercises 69–70* ◀

**Figure 7.27**

## Derivatives of General Logarithmic Functions

The general exponential function $f(x) = b^x$ is one-to-one when $b > 0$ with $b \neq 1$. The inverse function $f^{-1}(x) = \log_b x$ is the logarithmic function with base $b$. To find the derivative of the logarithmic function, we begin with the inverse relationship

$$y = \log_b x \iff x = b^y.$$

Differentiating both sides of $x = b^y$ with respect to $x$, we obtain

$$1 = b^y \ln b \cdot \frac{dy}{dx} \quad \text{Implicit differentiation}$$

$$\frac{dy}{dx} = \frac{1}{b^y \ln b} \quad \text{Solve for } \frac{dy}{dx}.$$

$$\frac{dy}{dx} = \frac{1}{x \ln b}. \quad b^y = x$$

> ▶ An alternative proof of Theorem 7.10 uses the change-of-base formula
>
> $\log_b x = \dfrac{\ln x}{\ln b}$ (see Exercise 112 for a proof of this formula). Differentiating both sides of this equation gives the same result.

---

**THEOREM 7.10  Derivative of $\log_b x$**

If $b > 0$ and $b \neq 1$, then

$$\frac{d}{dx}(\log_b x) = \frac{1}{x \ln b}, \quad \text{for } x > 0 \quad \text{and} \quad \frac{d}{dx}(\log_b |x|) = \frac{1}{x \ln b}, \quad \text{for } x \neq 0.$$

---

**QUICK CHECK 5**  Compute $dy/dx$ for $y = \log_3 x$. ◀

**EXAMPLE 8  Derivatives with general logarithms**  Compute the derivative of the following functions.

**a.** $f(x) = \log_5(2x + 1)$          **b.** $T(n) = n \log_2 n$

**SOLUTION**

> The function in Example 8b is used in computer science as an estimate on the computing time needed to carry out a *sorting algorithm* on a list of *n* items.

**a.** We use Theorem 7.10 with the Chain Rule assuming $2x + 1 > 0$:

$$f'(x) = \frac{1}{(2x+1)\ln 5} \cdot 2 = \frac{2}{\ln 5} \cdot \frac{1}{2x+1}. \quad \text{Chain Rule}$$

**b.**

$$T'(n) = \log_2 n + n \cdot \frac{1}{n \ln 2} = \log_2 n + \frac{1}{\ln 2} \quad \text{Product Rule}$$

We can change bases and write the result in base *e*:

$$T'(n) = \frac{\ln n}{\ln 2} + \frac{1}{\ln 2} = \frac{\ln n + 1}{\ln 2}.$$

*Related Exercises 63, 65* ◄

**QUICK CHECK 6** Show that the derivative computed in Example 8b can be expressed in base 2 as $T'(n) = \log_2 (en)$. ◄

---

## SECTION 7.3 EXERCISES

### Getting Started

1. State the derivative rule for the exponential function $f(x) = b^x$. How does it differ from the derivative formula for $e^x$?

2. State the derivative rule for the logarithmic function $f(x) = \log_b x$. How does it differ from the derivative formula for $\ln x$?

3. Explain why $b^x = e^{x \ln b}$.

4. Express the function $f(x) = g(x)^{h(x)}$ in terms of the natural logarithmic and natural exponential functions (base *e*).

5. Evaluate $\dfrac{d}{dx}(7^x)$.

6. What is the inverse function of $b^x$, and what are its domain and range?

7. Express $3^x$, $x^\pi$, and $x^{\sin x}$ using the base *e*.

8. Evaluate $\int 4^x \, dx$.

9. Evaluate $\dfrac{d}{dx}(x^e)$ and $\dfrac{d}{dx}(e^x)$.

10. Evaluate $\int x^\pi \, dx$ and $\int \pi^x \, dx$.

### Practice Exercises

**11–22. Solving equations** *Solve the following equations.*

11. $\log_{10} x = 3$

12. $\log_5 x = -1$

13. $\log_8 x = \frac{1}{3}$

14. $\log_b 125 = 3$

15. $7^x = 21$

16. $2^x = 55$

17. $3^{3x-4} = 15$

18. $5^{3x} = 29$

19. $10^{x^2-4} = 1$

20. $3^{x^2-5x-5} = \frac{1}{3}$

21. $2^{|x|} = 16$

22. $9^x + 3^{x+1} - 18 = 0$

**23–34. Derivatives of $b^x$** *Find the derivatives of the following functions.*

23. $y = 8^x$

24. $y = 5^{3t}$

25. $y = 5 \cdot 4^x$

26. $y = 4^{-x} \sin x$

27. $y = x^3 \cdot 3^x$

28. $P = \dfrac{40}{1 + 2^{-t}}$

29. $A = 250(1.045)^{4t}$

30. $y = \ln 10^x$

31. $y = 2^{3 + \sin x}$

32. $y = 10^{\ln 2x}$

33. $y = \dfrac{2^x}{2^x + 1}$

34. $y = 10^x(\ln 10^x - 1)$

35. **Exponential model** The following table shows the *time of useful consciousness* at various altitudes in the situation where a pressurized airplane suddenly loses pressure. The change in pressure drastically reduces available oxygen, and hypoxia sets in. The upper value of each time interval is roughly modeled by $T = 10 \cdot 2^{-0.274a}$, where $T$ measures time in minutes and $a$ is the altitude over 22,000 in thousands of feet ($a = 0$ corresponds to 22,000 ft).

| Altitude (in ft) | Time of Useful Consciousness |
|---|---|
| 22,000 | 5 to 10 min |
| 25,000 | 3 to 5 min |
| 28,000 | 2.5 to 3 min |
| 30,000 | 1 to 2 min |
| 35,000 | 30 to 60 s |
| 40,000 | 15 to 20 s |
| 45,000 | 9 to 15 s |

a. A Learjet flying at 38,000 ft ($a = 16$) suddenly loses pressure when the seal on a window fails. According to this model, how long do the pilot and passengers have to deploy oxygen masks before they become incapacitated?

b. What is the average rate of change of $T$ with respect to $a$ over the interval from 24,000 to 30,000 ft (include units)?

c. Find the instantaneous rate of change $dT/da$, compute it at 30,000 ft, and interpret its meaning.

36. **Magnitude of an earthquake** The energy (in joules) released by an earthquake of magnitude $M$ is given by the equation $E = 25,000 \cdot 10^{1.5M}$. (This equation can be solved for $M$ to define the magnitude of a given earthquake; it is a refinement of the original Richter scale created by Charles Richter in 1935.)

a. Compute the energy released by earthquakes of magnitude 1, 2, 3, 4, and 5. Plot the points on a graph and join them with a smooth curve.

b. Compute $dE/dM$ and evaluate it for $M = 3$. What does this derivative mean? ($M$ has no units, so the units of the derivative are J per change in magnitude.)

**37. Diagnostic scanning** Iodine-123 is a radioactive isotope used in medicine to test the function of the thyroid gland. If a 350-microcurie ($\mu$Ci) dose of iodine-123 is administered to a patient, the quantity $Q$ left in the body after $t$ hours is approximately $Q = 350(\frac{1}{2})^{t/13.1}$.

   **a.** How long does it take for the level of iodine-123 to drop to 10 $\mu$Ci?

   **b.** Find the rate of change of the quantity of iodine-123 at 12 hours, 1 day, and 2 days. What do your answers say about the rate at which iodine decreases as time increases?

**38–44. Integrals with general bases** *Evaluate the following integrals.*

**38.** $\displaystyle\int 2^{3x} \, dx$

**39.** $\displaystyle\int_{-1}^{1} 10^x \, dx$

**40.** $\displaystyle\int_0^{\pi/2} 4^{\sin x} \cos x \, dx$

**41.** $\displaystyle\int_1^2 (1 + \ln x) x^x \, dx$

**42.** $\displaystyle\int_{1/3}^{1/2} \frac{10^{1/p}}{p^2} \, dp$

**43.** $\displaystyle\int x^2 6^{x^3 + 8} \, dx$

**44.** $\displaystyle\int \frac{4^{\cot x}}{\sin^2 x} \, dx$

**45–50. General Power Rule** *Use the General Power Rule where appropriate to find the derivative of the following functions.*

**45.** $g(y) = e^y \cdot y^e$

**46.** $f(x) = 2x^{\sqrt{2}}$

**47.** $s(t) = \cos 2^t$

**48.** $y = \ln (x^3 + 1)^\pi$

**49.** $f(x) = (2x - 3)x^{3/2}$

**50.** $y = \tan x^{0.74}$

**51–56. Derivatives of tower functions (or $g^h$)** *Find the derivative of each function and evaluate the derivative at the given value of a.*

**51.** $f(x) = x^{\cos x}$; $a = \pi/2$

**52.** $g(x) = x^{\ln x}$; $a = e$

**53.** $h(x) = x^{\sqrt{x}}$; $a = 4$

**54.** $f(x) = (x^2 + 1)^x$; $a = 1$

**55.** $f(x) = (\sin x)^{\ln x}$; $a = \pi/2$   **56.** $f(x) = (\tan x)^{x-1}$; $a = \pi/4$

**57–68. Derivatives** *Evaluate the derivatives of the following functions.*

**57.** $f(x) = (2x)^{4x}$

**58.** $f(x) = x^{\sqrt{3}}$

**59.** $h(x) = 2^{(x^2)}$

**60.** $h(t) = (\sin t)^{\sqrt{t}}$

**61.** $H(x) = (x + 1)^{2x}$

**62.** $p(x) = x^{-\ln x}$

**63.** $y = 4 \log_3 (x^2 - 1)$

**64.** $y = \log_{10} x$

**65.** $y = (\cos x) \ln (\cos^2 x)$

**66.** $y = \log_8 |\tan x|$

**67.** $y = \dfrac{1}{\log_4 x}$

**68.** $y = \log_2 (\log_2 x)$

**69–72. Tangent lines**

**69.** Find an equation of the line tangent to $y = x^{\sin x}$ at the point $x = 1$.

**70.** Determine whether the graph of $y = x^{\sqrt{x}}$ has any horizontal tangent lines.

**71.** The graph of $y = (x^2)^x$ has two horizontal tangent lines. Find equations for both of them.

**72.** The graph of $y = x^{\ln x}$ has one horizontal tangent line. Find an equation for it.

**73. Explain why or why not** Determine whether the following statements are true and give an explanation or counterexample.

   **a.** The derivative of $\log_2 9$ is $1/(9 \ln 2)$.
   **b.** $\ln (x + 1) + \ln (x - 1) = \ln (x^2 - 1)$, for all $x$.
   **c.** The exponential function $2^{x+1}$ can be written in base $e$ as $e^{2 \ln (x+1)}$.
   **d.** $\dfrac{d}{dx}(\sqrt{2}^x) = x\sqrt{2}^{x-1}$.
   **e.** $\dfrac{d}{dx}(x^{\sqrt{2}}) = \sqrt{2}x^{\sqrt{2}-1}$.

**74. Graphs of exponential functions** The following figure shows the graphs of $y = 2^x$, $y = 3^x$, $y = 2^{-x}$, and $y = 3^{-x}$. Match each curve with the correct function.

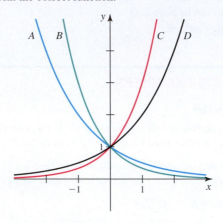

**75. Graphs of logarithmic functions** The following figure shows the graphs of $y = \log_2 x$, $y = \log_4 x$, and $y = \log_{10} x$. Match each curve with the correct function.

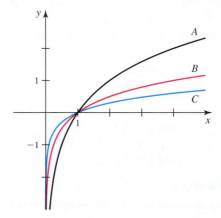

**76. Graphs of modified logarithmic functions** Without using a graphing utility, sketch the graph of $y = \log_2 x$. Then on the same set of axes, sketch the graphs of $y = \log_2 (x - 1)$, $y = \log_2 x^2$, $y = (\log_2 x)^2$, and $y = \log_2 x + 1$.

**77–80. Higher-order derivatives** *Find the following higher-order derivatives.*

**77.** $\dfrac{d^2}{dx^2} (\log_{10} x)$

**78.** $\dfrac{d^3}{dx^3} (x^{4.2}) \Big|_{x=1}$

**79.** $\dfrac{d^3}{dx^3} (x^2 \ln x)$

**80.** $\dfrac{d^n}{dx^n} (2^x)$

**81–82. Derivatives by different methods** *Calculate the derivative of the following functions (i) using the fact that $b^x = e^{x \ln b}$ and (ii) by using logarithmic differentiation. Verify that both answers are the same.*

**81.** $y = 1.05^x$

**82.** $y = (x^2 + 1)^x$

**83–88. Derivatives of logarithmic functions** *Use the properties of logarithms to simplify the following functions before computing $f'(x)$.*

**83.** $f(x) = \ln(3x + 1)^4$

**84.** $f(x) = \ln\dfrac{2x}{(x^2 + 1)^3}$

**85.** $f(x) = \log_{10}\sqrt{10x}$

**86.** $f(x) = \log_2\dfrac{8}{\sqrt{x + 1}}$

**87.** $f(x) = \ln\dfrac{(2x - 1)(x + 2)^3}{(1 - 4x)^2}$

**88.** $f(x) = \ln(\sec^4 x \tan^2 x)$

**T 89. Tangent line** Find the equation of the line tangent to $y = 2^{\sin x}$ at $x = \pi/2$. Graph the function and the tangent line.

**T 90. Horizontal tangents** The graph of $y = \cos x \cdot \ln \cos^2 x$ has seven horizontal tangent lines on the interval $[0, 2\pi]$. Find the approximate $x$-coordinates of all points at which these tangent lines occur.

**91–98. Derivatives of tower functions** *Compute the following derivatives. Use logarithmic differentiation where appropriate.*

**91.** $\dfrac{d}{dx}(x^{10x})$

**92.** $\dfrac{d}{dx}(2x)^{2x}$

**93.** $\dfrac{d}{dx}(x^{\cos x})$

**94.** $\dfrac{d}{dx}(x^2 + 1)^{-x}$

**95.** $\dfrac{d}{dx}\left(1 + \dfrac{1}{x}\right)^x$

**96.** $\dfrac{d}{dx}(1 + x^2)^{\sin x}$

**97.** $\dfrac{d}{dx}(x^{(x^{10})})$

**98.** $\dfrac{d}{dx}(\ln x)^{x^2}$

**99–103. Miscellaneous integrals** *Evaluate the following integrals.*

**99.** $\displaystyle\int 3^{-2x}\, dx$

**100.** $\displaystyle\int_0^5 5^{5x}\, dx$

**101.** $\displaystyle\int x^2 10^{x^3}\, dx$

**102.** $\displaystyle\int_0^\pi 2^{\sin x}\cos x\, dx$

**103.** $\displaystyle\int_1^{2e}\dfrac{3^{\ln x}}{x}\, dx$

### Explorations and Challenges

**T 104. Triple intersection** Graph the functions $f(x) = x^3$, $g(x) = 3^x$, and $h(x) = x^x$ and find their common intersection point (exactly).

**105–108. Calculating limits exactly** *Use the definition of the derivative to evaluate the following limits.*

**105.** $\displaystyle\lim_{x \to e}\dfrac{\ln x - 1}{x - e}$

**106.** $\displaystyle\lim_{h \to 0}\dfrac{\ln(e^8 + h) - 8}{h}$

**107.** $\displaystyle\lim_{h \to 0}\dfrac{(3 + h)^{3+h} - 27}{h}$

**108.** $\displaystyle\lim_{x \to 2}\dfrac{5^x - 25}{x - 2}$

**109. Derivative of $u(x)^{v(x)}$** Use logarithmic differentiation to prove that

$$\frac{d}{dx}(u(x)^{v(x)}) = u(x)^{v(x)}\left(\frac{dv}{dx}\ln u(x) + \frac{v(x)}{u(x)}\frac{du}{dx}\right).$$

**T 110. Tangency question** It is easily verified that the graphs of $y = 1.1^x$ and $y = x$ have two points of intersection, while the graphs of $y = 2^x$ and $y = x$ have no point of intersection. It follows that for some real number $1 < p < 2$, the graphs of $y = p^x$ and $y = x$ have exactly one point of intersection. Using analytical and/or graphical methods, determine $p$ and the coordinates of the single point of intersection.

**111. Nice property** Prove that $(\log_b c)(\log_c b) = 1$, for $b > 0$ and $c > 0$, $b \neq 1$, and $c \neq 1$.

**112. Change of base** Use the relationship $y = \log_b x \Leftrightarrow b^y = x$ to prove $\log_b x = \dfrac{\ln x}{\ln b}$.

**QUICK CHECK ANSWERS**

**1.** $b^x$ is always positive (and never zero) for all $x$ and for positive bases $b$.  **2.** Because $(1/3)^x = 1/3^x$ and $3^x$ increases as $x$ increases, it follows that $(1/3)^x$ decreases as $x$ increases.  **3.** $\{x: x \neq 0\}$, $\mathbb{R}$

**4.** $\dfrac{dA}{dt} = 500(1.045)^t \cdot \ln 1.045 \approx 22(1.045)^t$

**5.** $\dfrac{dy}{dx} = \dfrac{1}{x \ln 3}$

**6.** $T'(n) = \log_2 n + \dfrac{1}{\ln 2} = \log_2 n + \dfrac{1}{\dfrac{\log_2 2}{\log_2 e}}$

$= \log_2 n + \log_2 e = \log_2(en)$ ◄

# 7.4 Exponential Models

The uses of exponential functions are wide-ranging. In this section, you will see them applied to problems in finance, medicine, ecology, biology, economics, pharmacokinetics, anthropology, and physics.

## Exponential Growth

Imagine a quantity that grows in the following manner: The percent change in the quantity over the span of one year—or over any unit of time—remains constant. For example, suppose the value of a real estate investment increases at 7% per year, or the number of yeast cells in a batch of bread dough grows at 1.3% per minute. In both cases, we can create a simple function that models the growth of the quantity.

Let's focus on the real estate investment growing at 7% per year; we assume its value at time $t = 0$ is $y_0$. To compute its value one year later, we simply take 7% of $y_0$ and add it to $y_0$:

$$\text{value at } t = 1: \ y_0 + 0.07y_0 = y_0(1 + 0.07) = y_0(1.07).$$

The previous calculation shows that multiplication of a quantity by 1.07 is equivalent to increasing the quantity by 7%. Therefore, the value of the real estate investment after two years is

$$\text{value at } t = 2: \ \underbrace{y_0(1.07)}_{\text{value at } t = 1}(1.07) = y_0(1.07)^2.$$

Already a pattern is emerging. To find the value of the investment at $t = 3$ years, we multiply the value of the investment at $t = 2$ years by 1.07 to arrive at

$$\text{value at } t = 3: \ \underbrace{y_0(1.07)^2}_{\text{value at } t = 2}(1.07) = y_0(1.07)^3.$$

Building on this pattern, we find that the value of the investment at any time $t$ is given by the exponential function

$$y(t) = y_0(1.07)^t. \tag{1}$$

While the exponential function in equation (1) with base 1.07 adequately describes the value of the real estate investment at any time $t$, it is generally easier to work with functions expressed in base $e$. Recall from Section 7.2 that $b^x = e^{x \ln b}$. Therefore, equation (1) can be written as

$$y(t) = y_0(1.07)^t = y_0 e^{t \ln 1.07} \approx y_0 e^{0.0677t}. \quad \text{\color{blue}In } 1.07 \approx 0.0677$$

It is evident that we could carry out this procedure for any quantity whose percent change from one time unit to the next remains constant; the result would be

$$y(t) = y_0 e^{kt},$$

where the constant $k$, called the **rate constant**, is determined by the constant percent change in the quantity over one time unit. When $k > 0$, the function $y(t) = y_0 e^{kt}$ is called an *exponential growth function* (**Figure 7.28**), and it has a special property.

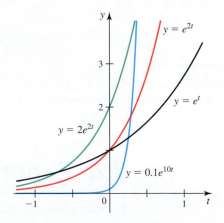

**Figure 7.28**

Notice that the derivative of $y$ is

$$y'(t) = y_0 k e^{kt} = k(\underbrace{y_0 e^{kt}}_{y(t)}); \tag{2}$$

that is, $y'(t) = ky(t)$. Equation (2) says that the rate of change of an exponential growth function is proportional to its value. If $y(t)$ represents a population, then $y'(t)$ is the (**absolute**) **growth rate** of the population, with units such as people/month or cells/hr. Because the growth rate is proportional to the value of the function, the larger the population, the faster its growth.

Another way to talk about growth rates is to use the **relative growth rate**, which is the growth rate divided by the current value of the quantity, or $y'(t)/y(t)$. When equation (2) is written in the form $y'(t)/y(t) = k$, it has another interpretation. It says that *a quantity that grows exponentially has a constant relative growth rate*. Constant relative growth rate, or constant percent change, is the hallmark of exponential growth. Returning to equation (1), we see that a constant percent change of 7% per year is equivalent to a rate constant of $k = \ln 1.07 \approx 0.0677$. Example 1 illustrates the difference between exponential growth and linear growth.

> ➤ The derivative $\dfrac{dy}{dt}$ is the *absolute* growth rate, but it is usually called the *growth rate*.

> ➤ A consumer price index that increases at a constant rate of 4% per year increases exponentially. A currency that is devalued at a constant rate of 3% per month decreases exponentially. By contrast, linear growth is characterized by constant absolute growth rates, such as 500 people per year or $400 per month.

**EXAMPLE 1    Linear versus exponential growth**  Suppose the population of the town of Pine is given by $P(t) = 1500 + 125t$, while the population of the town of Spruce is given by $S(t) = 1500e^{0.1t}$, where $t \geq 0$ is measured in years. Find the growth rate and the relative growth rate of each town.

**SOLUTION**  Note that Pine grows according to a linear function, while Spruce grows exponentially (**Figure 7.29**). The growth rate of Pine is $P'(t) = 125$ people/year, which is constant for all times. The growth rate of Spruce is

$$S'(t) = 0.1\underbrace{(1500e^{0.1t})}_{S(t)} = 0.1S(t),$$

showing that the growth rate is proportional to the population. The relative growth rate of Pine is $\dfrac{P'(t)}{P(t)} = \dfrac{125}{1500 + 125t}$, which decreases in time. The relative growth rate of Spruce is

$$\frac{S'(t)}{S(t)} = \frac{0.1 \cdot 1500e^{0.1t}}{1500e^{0.1t}} = 0.1,$$

which is constant for all times. In summary, the linear population function has a *constant absolute growth rate* and the exponential population function has a *constant relative growth rate*.

**Figure 7.29**

*Related Exercises 13–14* ◄

**QUICK CHECK 1** Population $A$ increases at a constant rate of 4%/yr. Population $B$ increases at a constant rate of 500 people/yr. Which population exhibits exponential growth? What kind of growth is exhibited by the other population? ◄

The rate constant $k$ in $y(t) = y_0 e^{kt}$ determines the growth rate of the exponential function. For problems that involve time, the units of $k$ are time$^{-1}$; for example, if $t$ is measured in months, the units of $k$ are month$^{-1}$. In this way, the exponent $kt$ is dimensionless (without units).

Unless there is good reason to do otherwise, it is customary to take $t = 0$ as the reference point for time. Notice that with $y(t) = y_0 e^{kt}$, we have $y(0) = y_0$. Therefore, $y_0$ has a simple meaning: It is the *initial value* of the quantity of interest. In the examples that follow, two pieces of information are typically given: the initial value and clues for determining the rate constant $k$. The initial value and the rate constant determine an exponential growth function completely.

> ➤ The unit time$^{-1}$ is read *per unit time*. For example, month$^{-1}$ is read *per month*.

---

**Exponential Growth Functions**

Exponential growth is described by functions of the form $y(t) = y_0 e^{kt}$. The **initial value** of $y$ at $t = 0$ is $y(0) = y_0$, and the **rate constant** $k > 0$ determines the rate of growth. Exponential growth is characterized by a constant relative growth rate.

---

➤ Note that the initial value $y_0$ appears on both sides of this equation. It may be canceled, meaning that the doubling time is independent of the initial condition: *The doubling time is constant for all t.*

Because exponential growth is characterized by a constant relative growth rate, the time required for a quantity to double (a 100% increase) is constant. Therefore, one way to describe an exponentially growing quantity is to give its *doubling time*. To compute the time it takes the function $y(t) = y_0 e^{kt}$ to double in value, say from $y_0$ to $2y_0$, we find the value of $t$ that satisfies

$$2y_0 = y_0 e^{kt}.$$

Canceling $y_0$ from the equation $2y_0 = y_0 e^{kt}$ leaves the equation $2 = e^{kt}$. Taking logarithms of both sides, we have $\ln 2 = \ln e^{kt}$, or $\ln 2 = kt$, which has the solution $t = \dfrac{\ln 2}{k}$. We denote this doubling time $T_2$ so that $T_2 = \dfrac{\ln 2}{k}$. If $y$ increases exponentially, the time it takes to double from 100 to 200 is the same as the time it takes to double from 1000 to 2000.

---

**DEFINITION  Doubling Time**

The quantity described by the function $y(t) = y_0 e^{kt}$, for $k > 0$, has a constant

**doubling time** of $T_2 = \dfrac{\ln 2}{k}$, with the same units as $t$.

---

**QUICK CHECK 2** Verify that the time needed for $y(t) = y_0 e^{kt}$ to double from $y_0$ to $2y_0$ is the same as the time needed to double from $2y_0$ to $4y_0$. ◄

➤ **World population**

| | |
|---|---|
| 1804 | 1 billion |
| 1927 | 2 billion |
| 1960 | 3 billion |
| 1974 | 4 billion |
| 1987 | 5 billion |
| 1999 | 6 billion |
| 2012 | 7 billion |
| 2050 | 9 billion (proj.) |

**EXAMPLE 2  World population** Human population growth rates vary geographically and fluctuate over time. The overall growth rate for world population peaked at an annual rate of 2.1% per year in the 1960s. Assume a world population of 6.0 billion in 1999 $(t = 0)$ and 7.4 billion in 2017 $(t = 18)$.

**a.** Find an exponential growth function for the world population that fits the two data points.

**b.** Find the doubling time for the world population using the model in part (a).

**c.** Find the (absolute) growth rate $y'(t)$ and graph it, for $0 \le t \le 50$.

**d.** According to the growth model, how fast is the population expected to be growing in 2020 $(t = 21)$?

**SOLUTION**

**a.** Let $y(t)$ be world population measured in billions of people $t$ years after 1999. We use the growth function $y(t) = y_0 e^{kt}$, where $y_0$ and $k$ must be determined. The initial value is $y_0 = 6$ (billion). To determine the rate constant $k$, we use the fact that $y(18) = 7.4$. Substituting $t = 18$ into the growth function with $y_0 = 6$ implies

$$y(18) = 6e^{18k} = 7.4.$$

Solving for $k$ yields the rate constant $k = \dfrac{\ln(7.4/6)}{18} \approx 0.01165 \text{ year}^{-1}$.

Therefore, the growth function is

$$y(t) = 6e^{0.01165t}.$$

**b.** The doubling time of the population is

$$T_2 = \frac{\ln 2}{k} \approx \frac{\ln 2}{0.01165} \approx 59.5 \text{ years.}$$

**c.** Working with the growth function $y(t) = 6e^{0.01165t}$, we find that

$$y'(t) = 6(0.01165)e^{0.01165t} = 0.0699e^{0.01165t},$$

which has units of *billions of people/year*. As shown in **Figure 7.30** the growth rate itself increases exponentially.

**Figure 7.30**

**d.** In 2020 ($t = 21$), the growth rate is expected to be

$$y'(21) = 0.0699e^{(0.01165)(21)} \approx 0.0893 \text{ billion people/year},$$

or roughly 89 million people/year.

*Related Exercises 16–17* ◄

**QUICK CHECK 3** Assume $y(t) = 100e^{0.05t}$. By (exactly) what percentage does $y$ increase when $t$ increases by 1 unit? ◄

➤ Converted to a daily rate (dividing by 365), the world population in 2020 is expected to increase at a rate of roughly 244,600 people per day.

➤ The concept of continuous compounding is explained in Exercise 61 of Section 7.6.

**A Financial Model**   Exponential functions are used in many financial applications, several of which are explored in the exercises. For now, consider a simple savings account in which an initial deposit earns interest that is reinvested in the account. Interest payments are made on a regular basis (for example, annually, monthly, daily), or interest may be compounded continuously. In all cases, the balance in the account increases exponentially at a rate that can be determined from the advertised **annual percentage yield** (or **APY**) of the account. Assuming no additional deposits are made, the balance in the account is given by the exponential growth function $y(t) = y_0 e^{kt}$, where $y_0$ is the initial deposit, $t$ is measured in years, and $k$ is determined by the annual percentage yield.

**EXAMPLE 3**   **Compounding**   The APY of a savings account is the percentage increase in the balance over the course of a year. Suppose you deposit $500 in a savings account that has an APY of 6.18% per year. Assume the interest rate remains constant and no additional deposits or withdrawals are made. How long will it take the balance to reach $2500?

**SOLUTION**   Because the balance grows by a fixed percentage every year, it grows exponentially. Letting $y(t)$ be the balance $t$ years after the initial deposit of $y_0 = \$500$, we have $y(t) = y_0 e^{kt}$, where the rate constant $k$ must be determined. Note that if the initial balance is $y_0$, one year later the balance is 6.18% more, or

➤ If the balance increases by 6.18% in one year, it increases by a factor of 1.0618 in one year.

$$y(1) = 1.0618y_0 = y_0 e^k.$$

Solving for $k$, we find that the rate constant is

$$k = \ln 1.0618 \approx 0.060 \text{ yr}^{-1}.$$

Therefore, the balance at any time $t \geq 0$ is $y(t) = 500e^{0.060t}$. To determine the time required for the balance to reach $2500, we solve the equation

$$y(t) = 500e^{0.060t} = 2500.$$

Dividing by 500 and taking the natural logarithm of both sides yields

$$0.060t = \ln 5.$$

The balance reaches \$2500 in $t = (\ln 5)/0.060 \approx 26.8$ yr.

*Related Exercises 18–19* ◄

**Resource Consumption**  Among the many resources that people use, energy is certainly one of the most important. The basic unit of energy is the **joule** (J), roughly the energy needed to lift a 0.1-kg object (say an orange) 1 m. The *rate* at which energy is consumed is called **power.** The basic unit of power is the **watt** (W), where 1 W = 1 J/s. If you turn on a 100-W lightbulb for 1 min, the bulb consumes energy at a rate of 100 J/s, and it uses a total of 100 J/s · 60 s = 6000 J of energy.

A more useful measure of energy for large quantities is the **kilowatt-hour** (kWh). A kilowatt is 1000 W, or 1000 J/s. So if you consume energy at the rate of 1 kW for 1 hr (3600 s), you use a total of 1000 J/s · 3600 s = $3.6 \times 10^6$ J, which is 1 kWh. A person running for one hour consumes roughly 1 kWh of energy. A typical house uses on the order of 1000 kWh of energy in a month.

Assume the total energy used (by a person, machine, or city) is given by the function $E(t)$. Because the power $P(t)$ is the rate at which energy is used, we have $P(t) = E'(t)$. Using the ideas of Section 6.1, the total amount of energy used between the times $t = a$ and $t = b$ is

$$\text{total energy used} = \int_a^b E'(t)\,dt = \int_a^b P(t)\,dt.$$

We see that energy is the area under the power curve. With this background, we can investigate a situation in which the rate of energy consumption increases exponentially.

**EXAMPLE 4**  **Energy consumption**  At the beginning of 2010, the rate of energy consumption for the city of Denver was 7000 megawatts (MW), where 1 MW = $10^6$ W. That rate is expected to increase at an annual growth rate of 2% per year.

**a.** Find the function that gives the power or rate of energy consumption for all times after the beginning of 2010.

**b.** Find the total amount of energy used during 2018.

**c.** Find the function that gives the total (cumulative) amount of energy used by the city between 2010 and any time $t \geq 0$.

**SOLUTION**

**a.** Let $t \geq 0$ be the number of years after the beginning of 2010 and let $P(t)$ be the power function that gives the rate of energy consumption at time $t$. Because $P$ increases at a constant rate of 2% per year, it increases exponentially. Therefore, $P(t) = P_0 e^{kt}$, where $P_0 = 7000$ MW. We determine $k$ as before by setting $t = 1$; after one year the power is

$$P(1) = P_0 e^k = 1.02 P_0.  \quad \text{Increase by 2\%, or by a factor of 1.02.}$$

Canceling $P_0$ and solving for $k$, we find that $k = \ln 1.02 \approx 0.0198$. Therefore, the power function (**Figure 7.31**) is

$$P(t) = 7000 e^{0.0198t}, \quad \text{for } t \geq 0.$$

**b.** The entire year 2018 corresponds to the interval $8 \leq t \leq 9$. Substituting $P(t) = 7000 e^{0.0198t}$ reveals that the total energy used in 2018 was

$$\int_8^9 P(t)\,dt = \int_8^9 7000 e^{0.0198t}\,dt \quad \text{Substitute for } P(t).$$

$$= \frac{7000}{0.0198} e^{0.0198t} \Big|_8^9 \quad \text{Fundamental Theorem}$$

$$\approx 8283. \quad \text{Evaluate.}$$

Because the units of $P$ are MW and $t$ is measured in years, the units of energy are MW-yr. To convert to MWh, we multiply by 8760 hr/yr to get the total energy of about $7.26 \times 10^7$ MWh (or $7.26 \times 10^{10}$ kWh).

**Figure 7.31**

► It is a common mistake to assume that if the annual growth rate is 2% per year, then $k = 2\% = 0.02$ year$^{-1}$. The rate constant $k$ must be calculated, as it is in Example 4a, to give $k \approx 0.0198$. For larger growth rates, the difference between $k$ and the actual growth rate is greater.

**Figure 7.32**

**c.** The total energy used between $t = 0$ and any future time $t$ is given by the future value formula (Section 6.1):

$$E(t) = E(0) + \int_0^t E'(s)\, ds = E(0) + \int_0^t P(s)\, ds.$$

Assuming $t = 0$ corresponds to the beginning of 2010, we take $E(0) = 0$. Substituting again for the power function $P$, the total energy (in MW-yr) at time $t$ is

$$E(t) = E(0) + \int_0^t P(s)\, ds$$

$$= 0 + \int_0^t 7000e^{0.0198s}\, ds \quad \text{Substitute for } P(s) \text{ and } E(0).$$

$$= \frac{7000}{0.0198} e^{0.0198s}\Big|_0^t \quad \text{Fundamental Theorem}$$

$$\approx 353{,}535(e^{0.0198t} - 1). \quad \text{Evaluate.}$$

As shown in **Figure 7.32**, when the rate of energy consumption increases exponentially, the total amount of energy consumed also increases exponentially.

*Related Exercises 24, 26* ◀

## Exponential Decay

Everything you have learned about exponential growth carries over directly to exponential decay. A function that decreases exponentially has the form $y(t) = y_0 e^{kt}$, where $y_0 = y(0)$ is the initial value and $k < 0$ is the rate constant.

Exponential decay is characterized by a constant relative decay rate and by a constant *half-life*. For example, radioactive plutonium has a half-life of 24,000 years. An initial sample of 1 mg decays to 0.5 mg after 24,000 years and to 0.25 mg after 48,000 years. To compute the half-life, denoted $T_{1/2}$, we determine the time required for the quantity $y(t) = y_0 e^{kt}$ to reach one-half of its current value; that is, we solve $y_0/2 = y_0 e^{kt}$ for $t$. Canceling $y_0$ and taking logarithms of both sides, we find that

$$\frac{1}{2} = e^{kt} \quad \Rightarrow \quad \ln\frac{1}{2} = kt \quad \Rightarrow \quad t = \frac{\ln(1/2)}{k}.$$

**QUICK CHECK 4** If a quantity decreases by a factor of 8 every 30 years, what is its half-life? ◀

➤ Because $\ln(1/2) < 0$ and $k < 0$, it follows that $T_{1/2} > 0$, as it should be.

---

**Exponential Decay Functions**

Exponential decay is described by functions of the form $y(t) = y_0 e^{kt}$. The initial value of $y$ is $y(0) = y_0$, and the rate constant $k < 0$ determines the rate of decay. Exponential decay is characterized by a constant relative decay rate. The constant **half-life** is $T_{1/2} = \dfrac{\ln(1/2)}{k}$, with the same units as $t$.

---

**Radiometric Dating**  A powerful method for estimating the age of ancient objects (for example, fossils, bones, meteorites, and cave paintings) relies on the radioactive decay of certain elements. A common version of radiometric dating uses the carbon isotope C-14, which is present in all living matter. When a living organism dies, it ceases to replace C-14, and the C-14 that is present decays with a half-life of about $T_{1/2} = 5730$ years. Comparing the C-14 in a living organism to the amount in a dead sample provides an estimate of its age.

**EXAMPLE 5**  **Radiometric dating**  Researchers determine that a fossilized bone has 30% of the C-14 of a live bone. Estimate the age of the bone. Assume a half-life for C-14 of 5730 years.

**SOLUTION**  The exponential decay function $y(t) = y_0 e^{kt}$ represents the amount of C-14 in the bone $t$ years after the animal died. By the half-life formula, $T_{1/2} = \ln(1/2)/k$. Substituting $T_{1/2} = 5730$ yr, the rate constant is

$$k = \frac{\ln(1/2)}{T_{1/2}} = \frac{\ln(1/2)}{5730 \text{ yr}} \approx -0.000121 \text{ yr}^{-1}.$$

Assume the amount of C-14 in a living bone is $y_0$. Over $t$ years, the amount of C-14 in the fossilized bone decays to 30% of its initial value, or $0.3y_0$. Using the decay function, we have

$$0.3y_0 = y_0 e^{-0.000121t}.$$

Solving for $t$, the age of the bone (in years) is

$$t = \frac{\ln 0.3}{-0.000121} \approx 9950.$$

*Related Exercises 34–35* ◄

**Pharmacokinetics** Pharmacokinetics describes the processes by which drugs are assimilated by the body. The elimination of most drugs from the body may be modeled by an exponential decay function with a known half-life (alcohol is a notable exception). The simplest models assume an entire drug dose is immediately absorbed into the blood. This assumption is a bit of an idealization; more refined mathematical models account for the absorption process.

➤ **Approximate half-lives of common drugs**

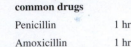

| | |
|---|---|
| Penicillin | 1 hr |
| Amoxicillin | 1 hr |
| Nicotine | 2 hr |
| Morphine | 3 hr |
| Tetracycline | 9 hr |
| Digitalis | 33 hr |
| Phenobarbitol | 2–6 days |

**EXAMPLE 6    Pharmacokinetics** An exponential decay function $y(t) = y_0 e^{kt}$ models the amount of drug in the blood $t$ hr after an initial dose of $y_0 = 100$ mg is administered. Assume the half-life of the drug is 16 hours.

**a.** Find the exponential decay function that governs the amount of drug in the blood.

**b.** How much time is required for the drug to reach 1% of the initial dose (1 mg)?

**c.** If a second 100-mg dose is given 12 hr after the first dose, how much time is required for the drug level to reach 1 mg?

**SOLUTION**

**a.** Knowing that the half-life is 16 hr, the rate constant is

$$k = \frac{\ln(1/2)}{T_{1/2}} = \frac{\ln(1/2)}{16 \text{ hr}} \approx -0.0433 \text{ hr}^{-1}.$$

Therefore, the decay function is $y(t) = 100 e^{-0.0433t}$.

**b.** The time required for the drug to reach 1 mg is the solution of

$$100 e^{-0.0433t} = 1.$$

Solving for $t$, we have

$$t = \frac{\ln 0.01}{-0.0433 \text{ hr}^{-1}} \approx 106 \text{ hr}.$$

It takes more than 4 days for the drug to be reduced to 1% of the initial dose.

**c.** Using the exponential decay function of part (a), the amount of drug in the blood after 12 hr is

$$y(12) = 100 e^{-0.0433 \cdot 12} \approx 59.5 \text{ mg}.$$

The second 100-mg dose given after 12 hr increases the amount of drug (assuming instantaneous absorption) to 159.5 mg. This amount becomes the new initial value for another exponential decay process (**Figure 7.33**). Measuring $t$ from the time of the second dose, the amount of drug in the blood is

$$y(t) = 159.5 e^{-0.0433t}.$$

The amount of drug reaches 1 mg when

$$y(t) = 159.5 e^{-0.0433t} = 1,$$

which implies that

$$t = \frac{-\ln 159.5}{-0.0433 \text{ hr}^{-1}} = 117.1 \text{ hr}.$$

$$\ln \frac{1}{159.5} = -\ln 159.5$$

**Figure 7.33**

Approximately 117 hr after the second dose (or 129 hr after the first dose), the amount of drug reaches 1 mg.

*Related Exercises 29, 38* ◄

## SECTION 7.4 EXERCISES

### Getting Started

1.  In terms of relative growth rate, what is the defining property of exponential growth?

2.  Give two pieces of information that may be used to formulate an exponential growth or decay function.

3.  Explain the meaning of doubling time.

4.  Explain the meaning of half-life.

5.  How are the rate constant and the doubling time related?

6.  How are the rate constant and the half-life related?

7.  Suppose a quantity described by the function $y(t) = y_0 e^{kt}$, where $t$ is measured in years, has a doubling time of 20 years. Find the rate constant $k$.

8.  Suppose a quantity is described by the function $y(t) = 30,000e^{-0.05t}$, where $t$ is measured in years. Find the half-life of the quantity.

9.  Give two examples of processes that are modeled by exponential growth.

10. Give two examples of processes that are modeled by exponential decay.

11. Because of the absence of predators, the number of rabbits on a small island increases at a rate of 11% per month. If $y(t)$ equals the number of rabbits on the island $t$ months from now, find the rate constant $k$ for the growth function $y(t) = y_0 e^{kt}$.

12. After the introduction of foxes on an island, the number of rabbits on the island decreases by 4.5% per month. If $y(t)$ equals the number of rabbits on the island $t$ months after foxes were introduced, find the rate constant $k$ for the exponential decay function $y(t) = y_0 e^{kt}$.

### Practice Exercises

**13–14. Absolute and relative growth rates** *Two functions f and g are given. Show that the growth rate of the linear function is constant and the relative growth rate of the exponential function is constant.*

13. $f(t) = 100 + 10.5t, \quad g(t) = 100e^{t/10}$

14. $f(t) = 2200 + 400t, \quad g(t) = 400 \cdot 2^{t/20}$

**15–20. Designing exponential growth functions** *Complete the following steps for the given situation.*

a. *Find the rate constant k and use it to devise an exponential growth function that fits the given data.*
b. *Answer the accompanying question.*

15. **Population** The population of a town with a 2016 population of 90,000 grows at a rate of 2.4%/yr. In what year will the population reach 120,000?

16. **Population** The population of Clark County, Nevada, was about 2.115 million in 2015. Assuming an annual growth rate of 1.5%/yr, what will the county population be in 2025?

17. **Population** The current population of a town is 50,000 and is growing exponentially. If the population is projected to be 55,000 in 10 years, then what will be the population 20 years from now?

18. **Savings account** An initial deposit of $1500 is placed in a savings account with an APY of 3.1%. How long will it take until the balance of the account is $2500? Assume the interest rate remains constant and no additional deposits or withdrawals are made.

19. **Rising costs** Between 2010 and 2016, the average rate of inflation was about 1.6%/yr. If a cart of groceries cost $100 in 2010, what will it cost in 2025, assuming the rate of inflation remains constant at 1.6%?

20. **Cell growth** The number of cells in a tumor doubles every 6 weeks starting with 8 cells. After how many weeks does the tumor have 1500 cells?

21. **Determining APY** Suppose $1000 is deposited in a savings account that increases exponentially. Determine the APY if the account increases to $1200 in 5 years. Assume the interest rate remains constant and no additional deposits or withdrawals are made.

22. **Tortoise growth** In a study conducted at University of New Mexico, it was found that the mass (weight) of juvenile desert tortoises exhibited exponential growth after a diet switch. One of these tortoises had a mass of about 64 g at the time of the diet switch, and 33 days later the mass was 73 g. How many days after the diet switch did the tortoise have a mass of 100 g? (*Source: Physiological and Biochemical Zoology, 85, 1, 2012*)

23. **Projection sensitivity** According to the 2014 national population projections published by the U.S. Census Bureau, the U.S. population is projected to be 334.4 million in 2020 with an estimated growth rate of 0.79%/yr.

    a. Based on these figures, find the doubling time and the projected population in 2050. Assume the growth rate remains constant.
    b. Suppose the actual growth rate is instead 0.7%. What are the resulting doubling time and projected 2050 population?

24. **Energy consumption** On the first day of the year ($t = 0$), a city uses electricity at a rate of 2000 MW. That rate is projected to increase at a rate of 1.3% per year.

    a. Based on these figures, find an exponential growth function for the power (rate of electricity use) for the city.
    b. Find the total energy (in MW-yr) used by the city over four full years beginning at $t = 0$.
    c. Find a function that gives the total energy used (in MW-yr) between $t = 0$ and any future time $t > 0$.

25. **Population of Texas** Texas was the third fastest growing state in the United States in 2016. Texas grew from 25.1 million in 2010 to 26.47 million in 2016. Use an exponential growth model to predict the population of Texas in 2025.

26. **Oil consumption** Starting in 2018 ($t = 0$), the rate at which oil is consumed by a small country increases at a rate of 1.5%/yr, starting with an initial rate of 1.2 million barrels/yr.

    a. How much oil is consumed over the course of the year 2018 (between $t = 0$ and $t = 1$)?
    b. Find the function that gives the amount of oil consumed between $t = 0$ and any future time $t$.
    c. How many years after 2018 will the amount of oil consumed since 2018 reach 10 million barrels?

**27–30. Designing exponential decay functions** *Devise an exponential decay function that fits the following data; then answer the accompanying questions. Be sure to identify the reference point ($t = 0$) and units of time.*

27. **Crime rate** The homicide rate decreases at a rate of 3%/yr in a city that had 800 homicides/yr in 2018. At this rate, when will the homicide rate reach 600 homicides/yr?

28. **Drug metabolism** A drug is eliminated from the body at a rate of 15%/hr. After how many hours does the amount of drug reach 10% of the initial dose?

29. **Valium metabolism** The drug Valium is eliminated from the bloodstream with a half-life of 36 hr. Suppose a patient receives an initial dose of 20 mg of Valium at midnight. How much Valium is in the patient's blood at noon the next day? When will the Valium concentration reach 10% of its initial level?

30. **China's population** China's one-child policy was implemented with a goal of reducing China's population to 700 million by 2050 (from 1.2 billion in 2000). Suppose China's population then declined at a rate of 0.5%/yr. Would this rate of decline be sufficient to meet the goal?

31. **Population of West Virginia** The population of West Virginia decreased from about 1.853 million in 2010 to 1.831 million in 2016. Use an exponential model to predict the population in 2025. Explain why an exponential (decay) model might not be an appropriate long-term model of the population of West Virginia.

32. **Depreciation of equipment** A large die-casting machine used to make automobile engine blocks is purchased for $2.5 million. For tax purposes, the value of the machine can be depreciated by 6.8% of its current value each year.

    a. What is the value of the machine after 10 years?
    b. After how many years is the value of the machine 10% of its original value?

33. **Atmospheric pressure** The pressure of Earth's atmosphere at sea level is approximately 1000 millibars and decreases exponentially with elevation. At an elevation of 30,000 ft (approximately the altitude of Mt. Everest), the pressure is one-third the sea-level pressure. At what elevation is the pressure half the sea-level pressure? At what elevation is it 1% of the sea-level pressure?

34. **Carbon dating** The half-life of C-14 is about 5730 yr.

    a. Archaeologists find a piece of cloth painted with organic dyes. Analysis of the dye in the cloth shows that only 77% of the C-14 originally in the dye remains. When was the cloth painted?
    b. A well-preserved piece of wood found at an archaeological site has 6.2% of the C-14 that it had when it was alive. Estimate when the wood was cut.

35. **Uranium dating** Uranium-238 (U-238) has a half-life of 4.5 billion years. Geologists find a rock containing a mixture of U-238 and lead, and they determine that 85% of the original U-238 remains; the other 15% has decayed into lead. How old is the rock?

36. **Radioiodine treatment** Roughly 12,000 Americans are diagnosed with thyroid cancer every year, which accounts for 1% of all cancer cases. It occurs in women three times as frequently as in men. Fortunately, thyroid cancer can be treated successfully in many cases with radioactive iodine, or I-131. This unstable form of iodine has a half-life of 8 days and is given in small doses measured in millicuries.

    a. Suppose a patient is given an initial dose of 100 millicuries. Find the function that gives the amount of I-131 in the body after $t \geq 0$ days.
    b. How long does it take the amount of I-131 to reach 10% of the initial dose?
    c. Finding the initial dose to give a particular patient is a critical calculation. How does the time to reach 10% of the initial dose change if the initial dose is increased by 5%?

**37–38. Caffeine** *After an individual drinks a beverage containing caffeine, the amount of caffeine in the bloodstream can be modeled by an exponential decay function, with a half-life that depends on several factors, including age and body weight. For the sake of simplicity, assume the caffeine in the following drinks immediately enters the bloodstream upon consumption.*

37. An individual consumes an energy drink that contains caffeine. If 80% of the caffeine from the energy drink is still in the bloodstream 2 hours later, find the half-life of caffeine for this individual.

38. An individual consumes two cups of coffee, each containing 90 mg of caffeine, two hours apart. Assume the half-life of caffeine for this individual is 5.7 hours.

    a. Determine the amount of caffeine in the bloodstream 1 hour after drinking the first cup of coffee.
    b. Determine the amount of caffeine in the bloodstream 1 hour after drinking the second cup of coffee.

**39–40. LED lighting** *LED (light-emitting diode) bulbs are rapidly decreasing in cost, and they are more energy efficient than standard incandescent light bulbs and CFL (compact fluorescent light) bulbs. By some estimates, LED bulbs last more than 40 times longer than incandescent bulbs and more than 8 times longer than CFL bulbs. Haitz's law, which is explored in the following two exercises, predicts that over time, LED bulbs will exponentially increase in efficiency and exponentially decrease in cost.*

39. Haitz's law predicts that the cost per lumen of an LED bulb decreases by a factor of 10 every 10 years. This means that 10 years from now, the cost of an LED bulb will be 1/10 of its current cost. Predict the cost of a particular LED bulb in 2021 if it costs 4 dollars in 2018.

40. LED packages are assemblies that house LED chips. Haitz's law predicts that the amount of light produced by an LED package increases by a factor of 20 every 10 years. Compared to its current light output, how much more light will an LED package produce 3 years from now?

41. **Tumor growth** Suppose the cells of a tumor are idealized as spheres, each with a radius of 5 $\mu$m (micrometers). The number of cells has a doubling time of 35 days. Approximately how long will it take a single cell to grow into a multi-celled spherical tumor with a volume of 0.5 cm$^3$ (1 cm = 10,000 $\mu$m)? Assume the tumor spheres are tightly packed.

42. **Tripling time** A quantity increases according to the exponential function $y(t) = y_0 e^{kt}$. What is the tripling time for the quantity? What is the time required for the quantity to increase $p$-fold?

43. **Explain why or why not** Determine whether the following statements are true and give an explanation or counterexample.

    a. A quantity that increases at 6%/yr obeys the growth function $y(t) = y_0 e^{0.06t}$.
    b. If a quantity increases by 10%/yr, it increases by 30% over 3 years.

**c.** A quantity decreases by one-third every month. Therefore, it decreases exponentially.

**d.** If the rate constant of an exponential growth function is increased, its doubling time is decreased.

**e.** If a quantity increases exponentially, the time required to increase by a factor of 10 remains constant for all time.

## Explorations and Challenges

**T 44. A running model** A model for the startup of a runner in a short race results in the velocity function $v(t) = a(1 - e^{-t/c})$, where $a$ and $c$ are positive constants, $t$ is measured in seconds, and $v$ has units of m/s. (*Source:* Joe Keller, *A Theory of Competitive Running*, *Physics Today*, 26, Sep 1973)

**a.** Graph the velocity function for $a = 12$ and $c = 2$. What is the runner's maximum velocity?

**b.** Using the velocity in part (a) and assuming $s(0) = 0$, find the position function $s(t)$, for $t \geq 0$.

**c.** Graph the position function and estimate the time required to run 100 m.

**T 45. Chemotherapy** In an experimental study at Dartmouth College, mice with tumors were treated with the chemotherapeutic drug Cisplatin. Before treatment, the tumors consisted entirely of *clonogenic* cells that divide rapidly, causing the tumors to double in size every 2.9 days. Immediately after treatment, 99% of the cells in the tumor became *quiescent* cells which do not divide and lose 50% of their volume every 5.7 days. For a particular mouse, assume the tumor size is 0.5 cm$^3$ at the time of treatment.

**a.** Find an exponential decay function $V_1(t)$ that equals the total volume of the quiescent cells in the tumor $t$ days after treatment.

**b.** Find an exponential growth function $V_2(t)$ that equals the total volume of the clonogenic cells in the tumor $t$ days after treatment.

**c.** Use parts (a) and (b) to find a function $V(t)$ that equals the volume of the tumor $t$ days after treatment.

**d.** Plot a graph of $V(t)$ for $0 \leq t \leq 15$. What happens to the size of the tumor, assuming there are no follow-up treatments with Cisplatin?

**e.** In cases where more than one chemotherapy treatment is required, it is often best to give a second treatment just before the tumor starts growing again. For the mice in this exercise, when should the second treatment be given? (*Source*: *Motivating Calculus with Biology*, MAA Notes, 81, 2013)

**46. Overtaking** City A has a current population of 500,000 people and grows at a rate of 3%/yr. City B has a current population of 300,000 and grows at a rate of 5%/yr.

**a.** When will the cities have the same population?

**b.** Suppose City C has a current population of $y_0 < 500,000$ and a growth rate of $p > 3\%$/yr. What is the relationship between $y_0$ and $p$ such that Cities A and C have the same population in 10 years?

**T 47. A slowing race** Starting at the same time and place, Abe and Bob race, running at velocities $u(t) = \dfrac{4}{t + 1}$ mi/hr and $v(t) = 4e^{-t/2}$ mi/hr, respectively, for $t \geq 0$.

**a.** Who is ahead after $t = 5$ hr? After $t = 10$ hr?

**b.** Find and graph the position functions of both runners. Which runner can run only a finite distance in an unlimited amount of time?

**48. Rule of 70** Bankers use the Rule of 70, which says that if an account increases at a fixed rate of $p\%$/yr, its doubling time is approximately $70/p$. Use linear approximation to explain why and when this is true.

**49. Compounded inflation** The U.S. government reports the rate of inflation (as measured by the consumer price index) both monthly and annually. Suppose for a particular month, the *monthly* rate of inflation is reported as 0.8%. Assuming this rate remains constant, what is the corresponding *annual* rate of inflation? Is the annual rate 12 times the monthly rate? Explain.

**50. Acceleration, velocity, position** Suppose the acceleration of an object moving along a line is given by $a(t) = -kv(t)$, where $k$ is a positive constant and $v$ is the object's velocity. Assume the initial velocity and position are given by $v(0) = 10$ and $s(0) = 0$, respectively.

**a.** Use $a(t) = v'(t)$ to find the velocity of the object as a function of time.

**b.** Use $v(t) = s'(t)$ to find the position of the object as a function of time.

**c.** Use the fact that $\dfrac{dv}{dt} = \dfrac{dv}{ds}\dfrac{ds}{dt}$ (by the Chain Rule) to find the velocity as a function of position.

**T 51. Air resistance** (adapted from Putnam Exam, 1939) An object moves freely in a straight line, acted on by air resistance, which is proportional to its speed; this means its acceleration is $a(t) = -kv(t)$. The velocity of the object decreases from 1000 ft/s to 900 ft/s over a distance of 1200 ft. Approximate the time required for this deceleration to occur. (Exercise 50 may be useful.)

**52. General relative growth rates** Define the relative growth rate of the function $f$ over the time interval $T$ to be the relative change in $f$ over an interval of length $T$:

$$R_T = \frac{f(t + T) - f(t)}{f(t)}.$$

Show that for the exponential function $y(t) = y_0 e^{kt}$, the relative growth rate $R_T$, for fixed $T$, is constant for all $t$.

**53. Equivalent growth functions** The same exponential growth function can be written in the forms $y(t) = y_0 e^{kt}$, $y(t) = y_0(1 + r)^t$, and $y(t) = y_0 2^{t/T_2}$. Write $k$ as a function of $r$, $r$ as a function of $T_2$, and $T_2$ as a function of $k$.

**54. Geometric means** A quantity grows exponentially according to $y(t) = y_0 e^{kt}$. What is the relationship among $m$, $n$, and $p$ such that $y(p) = \sqrt{y(m)y(n)}$?

**55. Constant doubling time** Prove that the doubling time for an exponentially increasing quantity is constant for all time.

**QUICK CHECK ANSWERS**

**1.** Population $A$ grows exponentially; population $B$ grows linearly.   **3.** The function $100e^{0.05t}$ increases by a factor of 1.0513, or by 5.13%, in 1 unit of time.   **4.** 10 years ◄

# 7.5 Inverse Trigonometric Functions

We used the idea of an inverse function to relate the natural logarithmic function to the natural exponential function. We now carry out a similar procedure with trigonometric functions. Our goal is to develop the inverses of the sine and cosine in detail. The inverses of the other four trigonometric functions then follow in an analogous way. When we investigate the calculus of inverse trigonometric functions, we discover several new and important derivatives and integrals.

## Inverse Sine and Cosine

So far, we have asked this question: Given an angle $x$, what is $\sin x$ or $\cos x$? Now we ask the opposite question: Given a number $y$, what is the angle $x$ such that $\sin x = y$? Or what is the angle $x$ such that $\cos x = y$? These are inverse questions.

There are a few things to notice right away. First, these questions do not make sense if $|y| > 1$, because $-1 \le \sin x \le 1$ and $-1 \le \cos x \le 1$. Next, let's select an acceptable value of $y$, say, $y = \frac{1}{2}$, and find the angle $x$ that satisfies $\sin x = y = \frac{1}{2}$. It is apparent that infinitely many angles satisfy $\sin x = \frac{1}{2}$; all angles of the form $\pi/6 \pm 2n\pi$ and $5\pi/6 \pm 2n\pi$, where $n$ is an integer, answer the inverse question (**Figure 7.34**). A similar situation occurs with the cosine function.

These inverse questions do not have unique answers because $\sin x$ and $\cos x$ are not one-to-one on their domains. To define their inverses, these functions are restricted to intervals on which they are one-to-one. For the sine function, the standard choice is $[-\pi/2, \pi/2]$; for cosine, it is $[0, \pi]$ (**Figure 7.35**). Now when we ask for the angle $x$ on the interval $[-\pi/2, \pi/2]$ such that $\sin x = \frac{1}{2}$, there is one answer: $x = \pi/6$. When we ask for the angle $x$ on the interval $[0, \pi]$ such that $\cos x = -\frac{1}{2}$, there is one answer: $x = 2\pi/3$.

We define the **inverse sine**, or **arcsine**, denoted $y = \sin^{-1} x$ or $y = \arcsin x$, such that $y$ is the angle whose sine is $x$, with the provision that $y$ lies in the interval $[-\pi/2, \pi/2]$. Similarly, we define the **inverse cosine**, or **arccosine**, denoted $y = \cos^{-1} x$ or $y = \arccos x$, such that $y$ is the angle whose cosine is $x$, with the provision that $y$ lies in the interval $[0, \pi]$.

Infinitely many values of $x$ satisfy $\sin x = \frac{1}{2}$.

**Figure 7.34**

➤ The notation for the inverse trigonometric functions invites confusion: $\sin^{-1} x$ and $\cos^{-1} x$ do not mean the reciprocals of $\sin x$ and $\cos x$. The expression $\sin^{-1} x$ should be read "*angle whose sine is x*," and $\cos^{-1} x$ should be read "*angle whose cosine is x*." The values of $\sin^{-1} x$ and $\cos^{-1} x$ are angles.

Restrict the domain of $y = \sin x$ to $\left[-\frac{\pi}{2}, \frac{\pi}{2}\right]$.

Restrict the domain of $y = \cos x$ to $[0, \pi]$.

**Figure 7.35**

---

**DEFINITION  Inverse Sine and Cosine**

$y = \sin^{-1} x$ is the value of $y$ such that $x = \sin y$, where $-\pi/2 \le y \le \pi/2$.
$y = \cos^{-1} x$ is the value of $y$ such that $x = \cos y$, where $0 \le y \le \pi$. The domain of both $\sin^{-1} x$ and $\cos^{-1} x$ is $\{x : -1 \le x \le 1\}$.

Recall that any invertible function and its inverse satisfy the properties

$$f(f^{-1}(y)) = y \quad \text{and} \quad f^{-1}(f(x)) = x.$$

These properties apply to the inverse sine and cosine, as long as we observe the restrictions on the domains. Here is what we can say:

- $\sin(\sin^{-1}x) = x$ and $\cos(\cos^{-1}x) = x$, for $-1 \le x \le 1$.
- $\sin^{-1}(\sin y) = y$, for $-\pi/2 \le y \le \pi/2$.
- $\cos^{-1}(\cos y) = y$, for $0 \le y \le \pi$.

**QUICK CHECK 1** Explain why $\sin^{-1}(\sin 0) = 0$, but $\sin^{-1}(\sin 2\pi) \ne 2\pi$. ◄

**EXAMPLE 1** **Working with inverse sine and cosine** Evaluate the following expressions.

**a.** $\sin^{-1}(\sqrt{3}/2)$  **b.** $\cos^{-1}(-\sqrt{3}/2)$  **c.** $\cos^{-1}(\cos 3\pi)$  **d.** $\sin^{-1}\left(\sin\dfrac{3\pi}{4}\right)$

**SOLUTION**

**a.** $\sin^{-1}(\sqrt{3}/2) = \pi/3$ because $\sin(\pi/3) = \sqrt{3}/2$ and $\pi/3$ is in the interval $[-\pi/2, \pi/2]$.

**b.** $\cos^{-1}(-\sqrt{3}/2) = 5\pi/6$ because $\cos(5\pi/6) = -\sqrt{3}/2$ and $5\pi/6$ is in the interval $[0, \pi]$.

**c.** It's tempting to conclude that $\cos^{-1}(\cos 3\pi) = 3\pi$, but the result of an inverse cosine operation must lie in the interval $[0, \pi]$. Because $\cos(3\pi) = -1$ and $\cos^{-1}(-1) = \pi$, we have

$$\cos^{-1}(\underbrace{\cos 3\pi}_{-1}) = \cos^{-1}(-1) = \pi.$$

**d.** $\sin^{-1}\left(\underbrace{\sin\dfrac{3\pi}{4}}_{\sqrt{2}/2}\right) = \sin^{-1}\dfrac{\sqrt{2}}{2} = \dfrac{\pi}{4}$

*Related Exercises 21–24* ◄

**Graphs and Properties** Recall from Section 7.1 that the graph of the inverse $f^{-1}$ is obtained by reflecting the graph of $f$ about the identity line $y = x$. This operation produces the graphs of the inverse sine (**Figure 7.36**) and inverse cosine (**Figure 7.37**). The graphs make it easy to compare the domain and range of each function and its inverse.

The graphs of $y = \sin x$ and $y = \sin^{-1}x$ are symmetric about the line $y = x$.

Range of $\sin x$ = Domain of $\sin^{-1}x$ $[-1, 1]$

$\left[-\dfrac{\pi}{2}, \dfrac{\pi}{2}\right]$

Restricted domain of $\sin x$ = Range of $\sin^{-1}x$

**Figure 7.36**

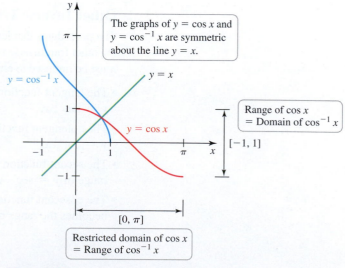

The graphs of $y = \cos x$ and $y = \cos^{-1}x$ are symmetric about the line $y = x$.

Range of $\cos x$ = Domain of $\cos^{-1}x$ $[-1, 1]$

$[0, \pi]$

Restricted domain of $\cos x$ = Range of $\cos^{-1}x$

**Figure 7.37**

**EXAMPLE 2    Right-triangle relationships**

**a.** Suppose $\theta = \sin^{-1}(2/5)$. Find $\cos\theta$ and $\tan\theta$.

**b.** Find an alternative form for $\cot\left(\cos^{-1}(x/4)\right)$ in terms of $x$.

**SOLUTION**

$\sin\theta = \dfrac{2}{5}$

$\cos\theta = \dfrac{\sqrt{21}}{5}$

$\tan\theta = \dfrac{2}{\sqrt{21}}$

**Figure 7.38**

**a.** Relationships between the trigonometric functions and their inverses are often simplified using a right-triangle sketch. Notice that $0 < \theta < \pi/2$. The right triangle in **Figure 7.38** satisfies the relationship $\sin\theta = \frac{2}{5}$, or, equivalently, $\theta = \sin^{-1}\frac{2}{5}$. We label the angle $\theta$ and the lengths of two sides; we then see the length of the third side is $\sqrt{21}$ (by the Pythagorean theorem). Now it is easy to read directly from the triangle:

$$\cos\theta = \frac{\sqrt{21}}{5} \quad \text{and} \quad \tan\theta = \frac{2}{\sqrt{21}}.$$

$\cos\theta = \dfrac{x}{4}$

**Figure 7.39**

**b.** We draw a right triangle with an angle $\theta$ satisfying $\cos\theta = x/4$, or, equivalently, $\theta = \cos^{-1}(x/4)$, where $0 < x < 4$ (**Figure 7.39**). The length of the third side of the triangle is $\sqrt{16 - x^2}$. It now follows that

$$\cot\left(\underbrace{\cos^{-1}\frac{x}{4}}_{\theta}\right) = \frac{x}{\sqrt{16 - x^2}}, \text{ for } 0 < x < 4 \text{ and } \theta \text{ in the first quadrant.}$$

A similar argument with $\theta$ in the second quadrant shows this relationship is valid for $|x| < 4$.

*Related Exercises 25–26* ◄

**EXAMPLE 3    A useful identity**  Use right triangles to explain why $\cos^{-1}x + \sin^{-1}x = \pi/2$.

**SOLUTION**  We draw a right triangle in a unit circle and label the acute angles $\theta$ and $\varphi$ (**Figure 7.40**). These angles satisfy $\cos\theta = x$, or $\theta = \cos^{-1}x$, and $\sin\varphi = x$, or $\varphi = \sin^{-1}x$. Because $\theta$ and $\varphi$ are complementary angles, we have

$$\frac{\pi}{2} = \theta + \varphi = \cos^{-1}x + \sin^{-1}x.$$

This result holds for $0 \le x \le 1$. An analogous argument extends the property to $-1 \le x \le 1$.

*Related Exercises 31–32* ◄

**Figure 7.40**

$\sin\varphi = x \Rightarrow \varphi = \sin^{-1}x$

$\cos\theta = x \Rightarrow \theta = \cos^{-1}x$

## Other Inverse Trigonometric Functions

The procedures that led to the inverse sine and inverse cosine functions can be used to obtain the other four inverse trigonometric functions. Each of these functions carries a restriction that must be imposed to ensure that an inverse exists.

- The tangent function is one-to-one on $(-\pi/2, \pi/2)$, which becomes the range of $y = \tan^{-1}x$.
- The cotangent function is one-to-one on $(0, \pi)$, which becomes the range of $y = \cot^{-1}x$.
- The secant function is one-to-one on $[0, \pi]$, excluding $x = \pi/2$; this set becomes the range of $y = \sec^{-1}x$.
- The cosecant function is one-to-one on $[-\pi/2, \pi/2]$, excluding $x = 0$; this set becomes the range of $y = \csc^{-1}x$.

The inverse tangent, cotangent, secant, and cosecant are defined as follows.

> ➤ Tables, books, and computer algebra systems differ on the definition of the inverses of the secant, cosecant, and cotangent. For example, in some books, $\sec^{-1} x$ is defined to lie in the interval $[-\pi, -\pi/2]$ when $x < 0$.

---

**DEFINITION** **Other Inverse Trigonometric Functions**

$y = \tan^{-1} x$ is the value of $y$ such that $x = \tan y$, where $-\pi/2 < y < \pi/2$.

$y = \cot^{-1} x$ is the value of $y$ such that $x = \cot y$, where $0 < y < \pi$.

The domain of both $\tan^{-1} x$ and $\cot^{-1} x$ is $\{x : -\infty < x < \infty\}$.

$y = \sec^{-1} x$ is the value of $y$ such that $x = \sec y$, where $0 \le y \le \pi$, with $y \ne \pi/2$.

$y = \csc^{-1} x$ is the value of $y$ such that $x = \csc y$, where $-\pi/2 \le y \le \pi/2$, with $y \ne 0$.

The domain of both $\sec^{-1} x$ and $\csc^{-1} x$ is $\{x : |x| \ge 1\}$.

---

The graphs of these inverse functions are obtained by reflecting the graphs of the original trigonometric functions about the line $y = x$ (**Figures 7.41–7.44**). The inverse secant and cosecant are somewhat irregular. The domain of the secant function (Figure 7.43)

**Figure 7.41**

**Figure 7.42**

**Figure 7.43**

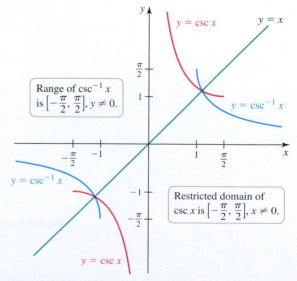

**Figure 7.44**

is restricted to the set $[0, \pi]$, excluding $x = \pi/2$, where the secant has a vertical asymptote. This asymptote splits the range of the secant into two disjoint intervals $(-\infty, -1]$ and $[1, \infty)$, which, in turn, splits the domain of the inverse secant into the same two intervals. A similar situation occurs with the cosecant.

**EXAMPLE 4   Working with inverse trigonometric functions** Evaluate or simplify the following expressions.

**a.** $\tan^{-1}(-1/\sqrt{3})$      **b.** $\sec^{-1}(-2)$      **c.** $\sin(\tan^{-1}x)$

**SOLUTION**

**a.** The result of an inverse tangent operation must lie in the interval $(-\pi/2, \pi/2)$. Therefore,

$$\tan^{-1}\left(-\frac{1}{\sqrt{3}}\right) = -\frac{\pi}{6} \quad \text{because} \quad \tan\left(-\frac{\pi}{6}\right) = -\frac{1}{\sqrt{3}}.$$

**b.** The result of an inverse secant operation when $x \le -1$ must lie in the interval $(\pi/2, \pi]$. Therefore,

$$\sec^{-1}(-2) = \frac{2\pi}{3} \quad \text{because} \quad \sec\frac{2\pi}{3} = -2.$$

**c.** Figure 7.45 shows a right triangle with the relationship $x = \tan\theta$ or $\theta = \tan^{-1}x$, in the case that $0 \le \theta < \pi/2$. We see that

$$\sin\underbrace{(\tan^{-1}x)}_{\theta} = \frac{x}{\sqrt{1 + x^2}}.$$

The same result follows if $-\pi/2 < \theta < 0$, in which case $x < 0$ and $\sin\theta < 0$.

*Related Exercises 34–35, 41* ◄

Figure 7.45

**QUICK CHECK 2** Evaluate $\sec^{-1}1$ and $\tan^{-1}1$. ◄

## Inverse Sine and Its Derivative

Recall that $y = \sin^{-1}x$ is the value of $y$ such that $x = \sin y$, where $-\pi/2 \le y \le \pi/2$. The domain of $\sin^{-1}x$ is $\{x: -1 \le x \le 1\}$ (Figure 7.46).

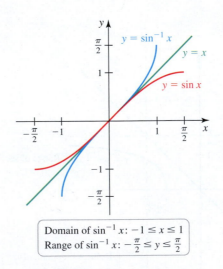

Figure 7.46

The derivative of $y = \sin^{-1} x$ follows by differentiating both sides of $x = \sin y$ with respect to $x$, simplifying, and solving for $dy/dx$:

$$x = \sin y \qquad \textcolor{blue}{y = \sin^{-1} x \iff x = \sin y}$$

$$\frac{d}{dx}(x) = \frac{d}{dx}(\sin y) \qquad \textcolor{blue}{\text{Differentiate with respect to } x.}$$

$$1 = (\cos y)\frac{dy}{dx} \qquad \textcolor{blue}{\text{Chain Rule on the right side}}$$

$$\frac{dy}{dx} = \frac{1}{\cos y}. \qquad \textcolor{blue}{\text{Solve for } \frac{dy}{dx}.}$$

The identity $\sin^2 y + \cos^2 y = 1$ is used to express this derivative in terms of $x$. Solving for $\cos y$ yields

$$\cos y = \pm \sqrt{1 - \underbrace{\sin^2 y}_{x^2}} \qquad \textcolor{blue}{x = \sin y \implies x^2 = \sin^2 y}$$

$$= \pm \sqrt{1 - x^2}.$$

Because $y$ is restricted to the interval $-\pi/2 \le y \le \pi/2$, we have $\cos y \ge 0$. Therefore, we choose the positive branch of the square root, and it follows that

$$\frac{dy}{dx} = \frac{d}{dx}(\sin^{-1} x) = \frac{1}{\sqrt{1 - x^2}}.$$

This result is consistent with the graph of $f(x) = \sin^{-1} x$ (**Figure 7.47**).

**Figure 7.47**

---

**THEOREM 7.11  Derivative of Inverse Sine**

$$\frac{d}{dx}(\sin^{-1} x) = \frac{1}{\sqrt{1 - x^2}}, \qquad \text{for } -1 < x < 1$$

---

**QUICK CHECK 3** Is $f(x) = \sin^{-1} x$ an even or odd function? Is $f'(x)$ an even or odd function? ◄

**EXAMPLE 5  Derivatives involving the inverse sine**  Compute the following derivatives.

**a.** $\dfrac{d}{dx}(\sin^{-1}(x^2 - 1))$ \qquad **b.** $\dfrac{d}{dx}(\cos(\sin^{-1} x))$

**SOLUTION**  We apply the Chain Rule for both derivatives.

**a.** $\dfrac{d}{dx}(\sin^{-1}(\underbrace{x^2-1}_{u})) = \underbrace{\dfrac{1}{\sqrt{1-(x^2-1)^2}}}_{\text{derivative of } \sin^{-1} u \atop \text{evaluated at } u = x^2-1} \cdot \underbrace{2x}_{u'(x)} = \dfrac{2x}{\sqrt{2x^2-x^4}}$

> The result in Example 5b could have been obtained by noting that $\cos(\sin^{-1}x) = \sqrt{1-x^2}$ and differentiating this expression (Exercise 108).

**b.** $\dfrac{d}{dx}(\cos(\underbrace{\sin^{-1}x}_{u})) = \underbrace{-\sin(\sin^{-1}x)}_{\text{the derivative of the} \atop \text{outer function } \cos u \atop \text{evaluated at } u = \sin^{-1}x} \cdot \underbrace{\dfrac{1}{\sqrt{1-x^2}}}_{\text{the derivative of the} \atop \text{inner function } \sin^{-1}x} = -\dfrac{x}{\sqrt{1-x^2}}$

This result is valid for $-1 < x < 1$, where $\sin(\sin^{-1}x) = x$.

*Related Exercises 49, 51* ◄

## Derivatives of Inverse Tangent and Inverse Secant

The derivatives of the inverse tangent and inverse secant are derived using a method similar to that used for the inverse sine. Once these three derivative results are known, the derivatives of the inverse cosine, cotangent, and cosecant follow immediately.

**Inverse Tangent**   Recall that $y = \tan^{-1}x$ is the value of $y$ such that $x = \tan y$, where $-\pi/2 < y < \pi/2$. The domain of $y = \tan^{-1}x$ is $\{x: -\infty < x < \infty\}$ (**Figure 7.48**).

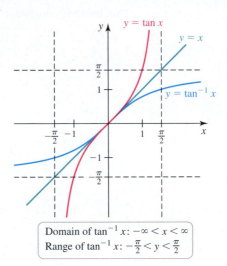

Domain of $\tan^{-1}x$: $-\infty < x < \infty$
Range of $\tan^{-1}x$: $-\dfrac{\pi}{2} < y < \dfrac{\pi}{2}$

**Figure 7.48**

To find $\dfrac{dy}{dx}$, we differentiate both sides of $x = \tan y$ with respect to $x$ and simplify:

$$x = \tan y \qquad y = \tan^{-1}x \iff x = \tan y$$

$$\dfrac{d}{dx}(x) = \dfrac{d}{dx}(\tan y) \quad \text{Differentiate with respect to } x.$$

$$1 = \sec^2 y \cdot \dfrac{dy}{dx} \quad \text{Chain Rule}$$

$$\dfrac{dy}{dx} = \dfrac{1}{\sec^2 y}. \quad \text{Solve for } \dfrac{dy}{dx}.$$

To express this derivative in terms of $x$, we combine the trigonometric identity $\sec^2 y = 1 + \tan^2 y$ with $x = \tan y$ to obtain $\sec^2 y = 1 + x^2$. Substituting this result into the expression for $dy/dx$, it follows that

$$\dfrac{dy}{dx} = \dfrac{d}{dx}(\tan^{-1}x) = \dfrac{1}{1+x^2}.$$

The graphs of the inverse tangent and its derivative (**Figure 7.49**) are informative. Letting $f(x) = \tan^{-1} x$ and $f'(x) = \dfrac{1}{1 + x^2}$, we see that $f'(0) = 1$, which is the maximum value of the derivative; that is, $\tan^{-1} x$ has its maximum slope at $x = 0$. As $x \to \infty$, $f'(x)$ approaches zero; likewise, as $x \to -\infty$, $f'(x)$ approaches zero.

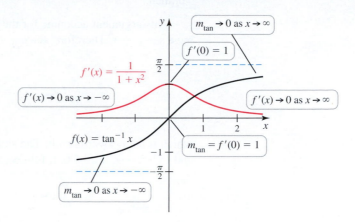

**QUICK CHECK 4** How do the slopes of the lines tangent to the graph of $y = \tan^{-1} x$ behave as $x \to \infty$? ◄

**Figure 7.49**

**Inverse Secant** Recall that $y = \sec^{-1} x$ is the value of $y$ such that $x = \sec y$, where $0 \le y \le \pi$, with $y \ne \pi/2$. The domain of $y = \sec^{-1} x$ is $\{x : |x| \ge 1\}$ (**Figure 7.50**).

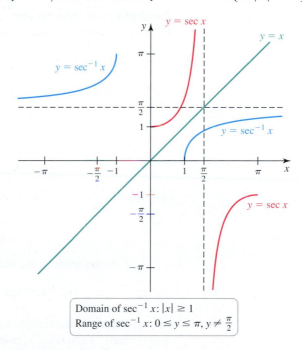

**Figure 7.50**

The derivative of the inverse secant presents a new twist. Let $y = \sec^{-1} x$, or $x = \sec y$, and then differentiate both sides of $x = \sec y$ with respect to $x$:

$$1 = \sec y \tan y \, \frac{dy}{dx}.$$

Solving for $\dfrac{dy}{dx}$ produces

$$\frac{dy}{dx} = \frac{d}{dx}(\sec^{-1} x) = \frac{1}{\sec y \tan y}.$$

The final step is to express $\sec y \tan y$ in terms of $x$ by using the identity $\sec^2 y = 1 + \tan^2 y$. Solving this equation for $\tan y$, we have

$$\tan y = \pm \sqrt{\underbrace{\sec^2 y}_{x^2} - 1} = \pm \sqrt{x^2 - 1}.$$

Two cases must be examined to resolve the sign on the square root:

• By the definition of $y = \sec^{-1} x$, if $x \geq 1$, then $0 \leq y < \pi/2$ and $\tan y > 0$. In this case, we choose the positive branch and take $\tan y = \sqrt{x^2 - 1}$.

• However, if $x \leq -1$, then $\pi/2 < y \leq \pi$ and $\tan y < 0$. Now we choose the negative branch.

This argument accounts for the $\tan y$ factor in the derivative. For the $\sec y$ factor, we have $\sec y = x$. Therefore, the derivative of the inverse secant is

$$\frac{d}{dx}(\sec^{-1} x) = \begin{cases} \dfrac{1}{x\sqrt{x^2 - 1}} & \text{if } x > 1 \\[2mm] -\dfrac{1}{x\sqrt{x^2 - 1}} & \text{if } x < -1, \end{cases}$$

which is an awkward result. The absolute value helps here: Recall that $|x| = x$, if $x > 0$, and $|x| = -x$, if $x < 0$. It follows that

$$\frac{d}{dx}(\sec^{-1} x) = \frac{1}{|x|\sqrt{x^2 - 1}}, \text{ for } |x| > 1.$$

We see that the slope of the inverse secant function is always positive, which is consistent with this derivative result (Figure 7.50).

**Derivatives of Other Inverse Trigonometric Functions**   The hard work is complete. The derivative of the inverse cosine results from the identity

> ▶ This identity was proved in Example 3 of this section.

$$\cos^{-1} x + \sin^{-1} x = \frac{\pi}{2}.$$

Differentiating both sides of this equation with respect to $x$, we find that

$$\frac{d}{dx}(\cos^{-1} x) + \underbrace{\frac{d}{dx}(\sin^{-1} x)}_{1/\sqrt{1-x^2}} = \underbrace{\frac{d}{dx}\left(\frac{\pi}{2}\right)}_{0}.$$

Solving for $\dfrac{d}{dx}(\cos^{-1} x)$ reveals that the required derivative is

$$\frac{d}{dx}(\cos^{-1} x) = -\frac{1}{\sqrt{1 - x^2}}.$$

In a similar manner, the analogous identities

$$\cot^{-1} x + \tan^{-1} x = \frac{\pi}{2} \quad \text{and} \quad \csc^{-1} x + \sec^{-1} x = \frac{\pi}{2}$$

are used to show that the derivatives of $\cot^{-1} x$ and $\csc^{-1} x$ are the negative of the derivatives of $\tan^{-1} x$ and $\sec^{-1} x$, respectively (Exercise 106).

> **QUICK CHECK 5** Summarize how the derivatives of inverse trigonometric functions are related to the derivatives of the corresponding inverse cofunctions (for example, inverse tangent and inverse cotangent). ◀

---

**THEOREM 7.12   Derivatives of Inverse Trigonometric Functions**

$$\frac{d}{dx}(\sin^{-1} x) = \frac{1}{\sqrt{1 - x^2}} \qquad \frac{d}{dx}(\cos^{-1} x) = -\frac{1}{\sqrt{1 - x^2}}, \quad \text{for } -1 < x < 1$$

$$\frac{d}{dx}(\tan^{-1} x) = \frac{1}{1 + x^2} \qquad \frac{d}{dx}(\cot^{-1} x) = -\frac{1}{1 + x^2}, \quad \text{for } -\infty < x < \infty$$

$$\frac{d}{dx}(\sec^{-1} x) = \frac{1}{|x|\sqrt{x^2 - 1}} \qquad \frac{d}{dx}(\csc^{-1} x) = -\frac{1}{|x|\sqrt{x^2 - 1}}, \quad \text{for } |x| > 1$$

**EXAMPLE 6**   **Derivatives of inverse trigonometric functions**

**a.** Evaluate $f'(2\sqrt{3})$, where $f(x) = x \tan^{-1}(x/2)$.

**b.** Find an equation of the line tangent to the graph of $g(x) = \sec^{-1} 2x$ at the point $(1, \pi/3)$.

**SOLUTION**

**a.** $f'(x) = 1 \cdot \tan^{-1}\dfrac{x}{2} + x \underbrace{\dfrac{1}{1 + (x/2)^2} \cdot \dfrac{1}{2}}_{\frac{d}{dx}(\tan^{-1}(x/2))}$    Product Rule and Chain Rule

$= \tan^{-1}\dfrac{x}{2} + \dfrac{2x}{4 + x^2}$    Simplify.

We evaluate $f'$ at $x = 2\sqrt{3}$ and note that $\tan^{-1}\sqrt{3} = \pi/3$:

$$f'(2\sqrt{3}) = \tan^{-1}\sqrt{3} + \frac{2(2\sqrt{3})}{4 + (2\sqrt{3})^2} = \frac{\pi}{3} + \frac{\sqrt{3}}{4}.$$

**b.** The slope of the tangent line at $(1, \pi/3)$ is $g'(1)$. Using the Chain Rule, we have

$$g'(x) = \frac{d}{dx}(\sec^{-1} 2x) = \frac{2}{|2x|\sqrt{4x^2 - 1}} = \frac{1}{|x|\sqrt{4x^2 - 1}}.$$

It follows that $g'(1) = 1/\sqrt{3}$. An equation of the tangent line is

$$\left(y - \frac{\pi}{3}\right) = \frac{1}{\sqrt{3}}(x - 1) \quad \text{or} \quad y = \frac{1}{\sqrt{3}}x + \frac{\pi}{3} - \frac{1}{\sqrt{3}}.$$

*Related Exercises 55, 58* ◄

**EXAMPLE 7**   **Shadows in a ballpark**   As the sun descends behind the 150-ft grandstand of a baseball stadium, the shadow of the stadium moves across the field (**Figure 7.51**). Let $\ell$ be the line segment between the edge of the shadow and the sun, and let $\theta$ be the angle of elevation of the sun—the angle between $\ell$ and the horizontal. The length of the shadow $s$ is the distance between the edge of the shadow and the base of the grandstand.

**a.** Express $\theta$ as a function of the shadow length $s$.

**b.** Compute $d\theta/ds$ when $s = 200$ ft and explain what this rate of change measures.

**Figure 7.51**

**SOLUTION**

**a.** The tangent of $\theta$ is

$$\tan \theta = \frac{150}{s},$$

where $s > 0$. Taking the inverse tangent of both sides of this equation, we find that

$$\theta = \tan^{-1}\frac{150}{s}.$$

**Figure 7.52** illustrates how the sun's angle of elevation $\theta$ approaches $\pi/2$ as the length of the shadow approaches zero ($\theta = \pi/2$ means the sun is overhead). As the shadow length increases, $\theta$ decreases and approaches zero.

**b.** Using the Chain Rule, we have

$$\frac{d\theta}{ds} = \frac{1}{1 + (150/s)^2}\frac{d}{ds}\left(\frac{150}{s}\right)$$    Chain Rule; $\frac{d}{du}(\tan^{-1} u) = \frac{1}{1 + u^2}$

$$= \frac{1}{1 + (150/s)^2}\left(-\frac{150}{s^2}\right)$$    Evaluate the derivative.

$$= -\frac{150}{s^2 + 22{,}500}.$$    Simplify.

As the shadow length increases, the angle of elevation decreases.

$\theta = \tan^{-1}\dfrac{150}{s}$

Angle of elevation

$s$ (feet)

**Figure 7.52**

**QUICK CHECK 6** Example 7 makes the claim that $d\theta/ds = -0.0024$ rad/ft is equivalent to $-0.138°$/ft. Verify this claim. ◄

Notice that $d\theta/ds$ is negative for all values of $s$, which means that as shadows lengthen, the angle of elevation decreases (Figure 7.52). At $s = 200$ ft, we have

$$\frac{d\theta}{ds}\bigg|_{s=200} = -\frac{150}{200^2 + 150^2} = -0.0024 \frac{\text{rad}}{\text{ft}}.$$

When the length of the shadow is $s = 200$ ft, the angle of elevation is changing at a rate of $-0.0024$ rad/ft, or $-0.138°$/ft.

*Related Exercises 73–74* ◄

## Integrals Involving Inverse Trigonometric Functions

It is now a straightforward matter to write the results of Theorem 7.12 in terms of indefinite integrals. Notice that only three (rather than six) new integrals arise in this way. However, they are important integrals that are used frequently in upcoming chapters.

We generalize the integrals using the following observation applied to the inverse sine derivative. Because

**QUICK CHECK 7** Why do the derivatives of the six inverse trigonometric functions lead to only three independent indefinite integrals? ◄

$$\frac{d}{dx}(\sin^{-1} x) = \frac{1}{\sqrt{1 - x^2}}, \quad \text{for } |x| < 1,$$

the Chain Rule gives

$$\frac{d}{dx}\left(\sin^{-1}\frac{x}{a}\right) = \frac{1}{\sqrt{1 - (x/a)^2}} \cdot \frac{1}{a} = \frac{1}{\sqrt{a^2 - x^2}}, \quad \text{for } |x| < a,$$

where $a > 0$ is a constant. Writing this result as an indefinite integral, we have

$$\int \frac{dx}{\sqrt{a^2 - x^2}} = \sin^{-1}\frac{x}{a} + C.$$

A similar calculation with the inverse tangent and inverse secant derivatives gives the following indefinite integrals.

---

**THEOREM 7.13  Integrals Involving Inverse Trigonometric Functions**

1. $\displaystyle\int \frac{dx}{\sqrt{a^2 - x^2}} = \sin^{-1}\frac{x}{a} + C$

2. $\displaystyle\int \frac{dx}{a^2 + x^2} = \frac{1}{a}\tan^{-1}\frac{x}{a} + C$

3. $\displaystyle\int \frac{dx}{x\sqrt{x^2 - a^2}} = \frac{1}{a}\sec^{-1}\left|\frac{x}{a}\right| + C, \, a > 0$

---

**EXAMPLE 8  Evaluating integrals** Evaluate the following integrals.

**a.** $\displaystyle\int \frac{4}{\sqrt{9 - x^2}}\, dx$    **b.** $\displaystyle\int \frac{dx}{16x^2 + 1}$    **c.** $\displaystyle\int_{5/\sqrt{3}}^{5} \frac{dx}{x\sqrt{4x^2 - 25}}$

**SOLUTION**

**a.** Setting $a = 3$ in the first result of Theorem 7.13, we have

$$\int \frac{4}{\sqrt{9 - x^2}}\, dx = 4\int \frac{dx}{\sqrt{3^2 - x^2}} = 4\sin^{-1}\frac{x}{3} + C.$$

**b.** An algebra step is needed to put this integral in a form that matches Theorem 7.13. We first write

$$\int \frac{dx}{16x^2 + 1} = \frac{1}{16}\int \frac{dx}{x^2 + \frac{1}{16}} = \frac{1}{16}\int \frac{dx}{x^2 + \left(\frac{1}{4}\right)^2}.$$

Setting $a = \frac{1}{4}$ in part 2 of Theorem 7.13 gives

$$\int \frac{dx}{16x^2 + 1} = \frac{1}{16}\int \frac{dx}{x^2 + (\frac{1}{4})^2} = \frac{1}{16} \cdot 4 \tan^{-1} 4x + C = \frac{1}{4}\tan^{-1} 4x + C.$$

**c.** A change of variables puts the integral in the form of the third integral in Theorem 7.13. Letting $u = 2x$, we have

$$\int_{5/\sqrt{3}}^{5} \frac{dx}{x\sqrt{4x^2 - 25}} = \int_{10/\sqrt{3}}^{10} \frac{\frac{1}{2}\,du}{\frac{1}{2}\,u\sqrt{u^2 - 25}} \qquad u = 2x,\ du = 2\,dx$$

$$= \int_{10/\sqrt{3}}^{10} \frac{du}{u\sqrt{u^2 - 25}} \qquad \text{Simplify.}$$

$$= \frac{1}{5}\sec^{-1}\frac{u}{5}\bigg|_{10/\sqrt{3}}^{10} \qquad \text{Theorem 7.13, } a = 5$$

$$= \frac{1}{5}\left(\sec^{-1} 2 - \sec^{-1}\frac{2}{\sqrt{3}}\right) \qquad \text{Evaluate.}$$

$$= \frac{1}{5}\left(\frac{\pi}{3} - \frac{\pi}{6}\right) = \frac{\pi}{30}. \qquad \text{Simplify.}$$

*Related Exercises 77–79* ◄

**EXAMPLE 9** **Circumference of a circle** Confirm that the circumference of a circle of radius $a$ is $2\pi a$.

**SOLUTION** The upper half of a circle of radius $a$ centered at $(0, 0)$ is given by the function $f(x) = \sqrt{a^2 - x^2}$, for $|x| \le a$ (**Figure 7.53**). So we might consider using the arc length formula on the interval $[-a, a]$ to find the length of a semicircle. However, the circle has vertical tangent lines at $x = \pm a$ and $f'(\pm a)$ is undefined, which prevents us from using the arc length formula. An alternative approach is to use symmetry and avoid the points $x = \pm a$. For example, let's compute the length of one-eighth of the circle on the interval $[0, a/\sqrt{2}]$ (Figure 7.53).

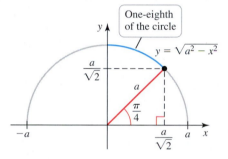

**Figure 7.53**

We first determine that $f'(x) = -\dfrac{x}{\sqrt{a^2 - x^2}}$, which is continuous on $[0, a/\sqrt{2}]$.

The length of one-eighth of the circle is

$$\int_0^{a/\sqrt{2}} \sqrt{1 + f'(x)^2}\,dx = \int_0^{a/\sqrt{2}} \sqrt{1 + \left(-\frac{x}{\sqrt{a^2 - x^2}}\right)^2}\,dx$$

$$= \int_0^{a/\sqrt{2}} \sqrt{\frac{a^2}{a^2 - x^2}}\,dx \qquad \text{Simplify.}$$

$$= a\int_0^{a/\sqrt{2}} \frac{dx}{\sqrt{a^2 - x^2}} \qquad \text{Simplify; } a > 0$$

$$= a\,\sin^{-1}\frac{x}{a}\bigg|_0^{a/\sqrt{2}} \qquad \text{Integrate.}$$

$$= a\left(\sin^{-1}\frac{1}{\sqrt{2}} - 0\right) \qquad \text{Evaluate.}$$

$$= \frac{\pi a}{4}. \qquad \text{Simplify.}$$

▶ The arc length integral for the semicircle on $[-a, a]$ is an example of an *improper integral*, a topic considered in Section 8.9.

It follows that the circumference of the full circle is $8(\pi a/4) = 2\pi a$ units.

*Related Exercises 81, 83* ◄

## SECTION 7.5 EXERCISES

### Getting Started

1. Explain why the domain of the sine function must be restricted in order to define its inverse function.

2. Why do values of $\cos^{-1} x$ lie in the interval $[0, \pi]$?

3. Evaluate $\cos^{-1}(\cos(5\pi/4))$.

4. Evaluate $\sin^{-1}(\sin(11\pi/6))$.

5. Is it true that $\tan(\tan^{-1} x) = x$ for all $x$? Is it true that $\tan^{-1}(\tan x) = x$ for all $x$?

6. Sketch the graphs of $y = \cos x$ and $y = \cos^{-1} x$ on the same set of axes.

7. The function $\tan x$ is undefined at $x = \pm \pi/2$. How does this fact appear in the graph of $y = \tan^{-1} x$?

8. State the domain and range of $\sec^{-1} x$.

9. State the derivative formulas for $\sin^{-1} x$, $\tan^{-1} x$, and $\sec^{-1} x$.

10. What is the slope of the line tangent to the graph of $y = \sin^{-1} x$ at $x = 0$?

11. What is the slope of the line tangent to the graph of $y = \tan^{-1} x$ at $x = -2$?

12. How are the derivatives of $\sin^{-1} x$ and $\cos^{-1} x$ related?

13. Find the slope of the curve $y = \sin^{-1} x$ at $\left(\dfrac{1}{2}, \dfrac{\pi}{6}\right)$ without calculating the derivative of $\sin^{-1} x$.

14. Find the slope of the curve $y = \tan^{-1} x$ at $\left(1, \dfrac{\pi}{4}\right)$ without calculating the derivative of $\tan^{-1} x$.

15. Evaluate $\displaystyle\int \frac{a}{\sqrt{1 - x^2}}\, dx$, where $a$ is a constant.

16. Evaluate $\displaystyle\int_0^{1/2} \frac{dx}{1 + 4x^2}$ using the substitution $u = 2x$.

### Practice Exercises

**17–24. Inverse sines and cosines** *Without using a calculator, evaluate the following expressions, or state that the quantity is undefined.*

17. $\sin^{-1} 1$

18. $\cos^{-1}(-1)$

19. $\sin^{-1}\left(-\frac{1}{2}\right)$

20. $\cos^{-1} 2$

21. $\cos^{-1}\left(-\frac{1}{2}\right)$

22. $\sin^{-1}(-1)$

23. $\cos(\cos^{-1}(-1))$

24. $\cos^{-1}(\cos(7\pi/6))$

**25–30. Right-triangle relationships** *Draw a right triangle to simplify the given expressions. Assume $x > 0$.*

25. $\cos(\sin^{-1} x)$

26. $\cos(\sin^{-1}(x/3))$

27. $\sin(\cos^{-1}(x/2))$

28. $\sin^{-1}(\cos \theta)$, for $0 \le \theta \le \pi/2$

29. $\sin(2\cos^{-1} x)$   (*Hint: Use $\sin 2\theta = 2 \sin \theta \cos \theta$.*)

30. $\cos(2\sin^{-1} x)$   (*Hint: Use $\cos 2\theta = \cos^2 \theta - \sin^2 \theta$.*)

**31–32. Identities** *Prove the following identities.*

31. $\cos^{-1} x + \cos^{-1}(-x) = \pi$

32. $\sin^{-1} y + \sin^{-1}(-y) = 0$

**33–40. Evaluating inverse trigonometric functions** *Without using a calculator, evaluate or simplify the following expressions.*

33. $\tan^{-1} \sqrt{3}$

34. $\cot^{-1}(-1/\sqrt{3})$

35. $\sec^{-1} 2$

36. $\csc^{-1}(-1)$

37. $\tan^{-1}(\tan(\pi/4))$

38. $\tan^{-1}(\tan(3\pi/4))$

39. $\csc^{-1}(\sec 2)$

40. $\tan(\tan^{-1} 1)$

**41–46. Right-triangle relationships** *Draw a right triangle to simplify the given expressions. Assume $x > 0$.*

41. $\cos(\tan^{-1} x)$

42. $\tan(\cos^{-1} x)$

43. $\sec(\tan^{-1} x)$

44. $\sin(\sec^{-1} 3x)$

45. $\sec(\csc^{-1} 2x)$

46. $\tan(\csc^{-1}(x/2))$

**47–48. Right-triangle pictures** *Express $\theta$ in terms of $x$ using the inverse sine, inverse tangent, and inverse secant functions.*

47.

48.
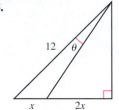

**49–72. Derivatives** *Evaluate the derivatives of the following functions.*

49. $f(x) = \sin^{-1} 2x$

50. $f(x) = x \sin^{-1} x$

51. $f(w) = \cos(\sin^{-1} 2w)$

52. $f(x) = \sin^{-1}(\ln x)$

53. $f(x) = \sin^{-1}(e^{-2x})$

54. $f(x) = \sin^{-1}(e^{\sin x})$

55. $f(y) = \tan^{-1}(2y^2 - 4)$

56. $g(z) = \tan^{-1}(1/z)$

57. $f(z) = \cot^{-1} \sqrt{z}$

58. $f(x) = \sec^{-1} \sqrt{x}$

59. $f(x) = \cos^{-1}(1/x)$

60. $f(t) = (\cos^{-1} t)^2$

61. $f(u) = \csc^{-1}(2u + 1)$

62. $f(t) = \ln(\tan^{-1} t)$

63. $f(y) = \cot^{-1}(1/(y^2 + 1))$

64. $f(w) = \sin(\sec^{-1} 2w)$

65. $f(x) = \sec^{-1}(\ln x)$

66. $f(x) = \tan^{-1}(e^{4x})$

67. $f(x) = \csc^{-1}(\tan e^x)$

68. $f(x) = \sin(\tan^{-1} \ln x)$

69. $f(s) = \cot^{-1}(e^s)$

70. $f(x) = 1/(\tan^{-1}(x^2 + 4))$

71. $f(w) = w^2 - \tan^{-1} w^2$

72. $f(t) = \ln(\sin^{-1} t^2)$

**T** 73. **Angular size** A boat sails directly toward a 150-m skyscraper that stands on the edge of a harbor. The angular size $\theta$ of the building is the angle formed by lines from the top and bottom of the building to the observer (see figure).

   a. What is the rate of change of the angular size $d\theta/dx$ when the boat is 500 m from the building?

**b.** Graph $d\theta/dx$ as a function of $x$ and determine the point at which the angular size changes most rapidly.

**T 74. Angle of elevation** A small plane, moving at 70 m/s, flies horizontally on a line 400 m directly above an observer. Let $\theta$ be the angle of elevation of the plane (see figure).

**a.** What is the rate of change of the angle of elevation $d\theta/dx$ when the plane is 500 m past the observer?

**b.** Graph $d\theta/dx$ as a function of $x$ and determine the point at which $\theta$ changes most rapidly.

**T 75. Towing a boat** A boat is towed toward a dock by a cable attached to a winch that stands 10 ft above the water level (see figure). Let $\theta$ be the angle of elevation of the winch, and let $\ell$ be the length of the cable as the boat is towed toward the dock.

**a.** Show that the rate of change of $\theta$ with respect to $\ell$ is

$$\frac{d\theta}{d\ell} = -\frac{10}{\ell\sqrt{\ell^2 - 100}}.$$

**b.** Compute $\dfrac{d\theta}{d\ell}$ when $\ell = 50, 20,$ and 11 ft.

**c.** Find $\lim\limits_{\ell \to 10^+} \dfrac{d\theta}{d\ell}$, and explain what happens as the last foot of cable is reeled in (note that the boat is at the dock when $\ell = 10$).

**d.** It is evident from the figure that $\theta$ increases as the boat is towed to the dock. Why, then, is $d\theta/d\ell$ negative?

**76. Hockey problem** A hockey player states on a line that is perpendicular to the goal line (see figure).

**a.** Express the angle $\theta$ on goal as a function of $x$.

**b.** Determine the distance $x$ that maximizes the angle $\theta$.

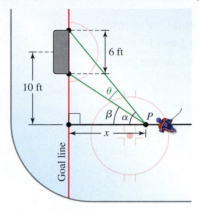

**77–90. Integrals involving inverse trigonometric functions** *Evaluate the following integrals.*

**77.** $\displaystyle\int \frac{6}{\sqrt{25 - x^2}}\, dx$

**78.** $\displaystyle\int \frac{3}{4 + v^2}\, dv$

**79.** $\displaystyle\int \frac{dx}{x\sqrt{x^2 - 100}}$

**80.** $\displaystyle\int \frac{2}{16z^2 + 25}\, dz$

**81.** $\displaystyle\int_0^3 \frac{dx}{\sqrt{36 - x^2}}$

**82.** $\displaystyle\int_2^{2\sqrt{3}} \frac{5}{x^2 + 4}\, dx$

**83.** $\displaystyle\int_0^{3/2} \frac{dx}{\sqrt{36 - 4x^2}}$

**84.** $\displaystyle\int_0^{5/4} \frac{3}{64x^2 + 100}\, dx$

**85.** $\displaystyle\int_{-\ln\sqrt{3}}^0 \frac{e^x}{1 + e^{2x}}\, dx$

**86.** $\displaystyle\int_1^{\sqrt{e}} \frac{dx}{x\sqrt{1 - \ln^2 x}}$

**87.** $\displaystyle\int \frac{dy}{y^2 - 4y + 5}$

**88.** $\displaystyle\int \frac{dx}{(x + 3)\sqrt{x^2 + 6x}}$

**89.** $\displaystyle\int \frac{e^x}{e^{2x} + 4}\, dx$

**90.** $\displaystyle\int \frac{dx}{x^3 + x^{-1}}$

**91. Explain why or why not** Determine whether the following statements are true and give an explanation or counterexample.

**a.** $\dfrac{\sin^{-1} x}{\cos^{-1} x} = \tan^{-1} x.$

**b.** $\cos^{-1}(\cos(15\pi/16)) = 15\pi/16.$

**c.** $\sin^{-1} x = 1/\sin x.$

**d.** $\dfrac{d}{dx}(\sin^{-1} x + \cos^{-1} x) = 0.$

**e.** $\dfrac{d}{dx}(\tan^{-1} x) = \sec^2 x.$

**f.** The lines tangent to the graph of $y = \sin^{-1} x$ on the interval $[-1, 1]$ have a minimum slope of 1.

**g.** The lines tangent to the graph of $y = \sin x$ on the interval $[-\pi/2, \pi/2]$ have a maximum slope of 1.

**92–95. One function gives all six** *Given the following information about one trigonometric function, evaluate the other five functions.*

**92.** $\sin\theta = -\dfrac{4}{5}$ and $\pi < \theta < 3\pi/2$

**93.** $\cos\theta = \dfrac{5}{13}$ and $0 < \theta < \pi/2$

**94.** $\sec \theta = \dfrac{5}{3}$ and $3\pi/2 < \theta < 2\pi$

**95.** $\csc \theta = \dfrac{13}{12}$ and $0 < \theta < \pi/2$

## Explorations and Challenges

**T 96–99. Graphing $f$ and $f'$**

a. *Graph $f$ with a graphing utility.*
b. *Compute and graph $f'$.*
c. *Verify that the zeros of $f'$ correspond to points at which $f$ has a horizontal tangent line.*

**96.** $f(x) = (x - 1)\sin^{-1} x$ on $[-1, 1]$

**97.** $f(x) = (x^2 - 1)\sin^{-1} x$ on $[-1, 1]$

**98.** $f(x) = (\sec^{-1} x)/x$ on $[1, \infty)$

**99.** $f(x) = e^{-x}\tan^{-1} x$ on $[0, \infty)$

**T 100. Graphing with inverse trigonometric functions**

a. Graph the function $f(x) = \dfrac{\tan^{-1} x}{x^2 + 1}$.

b. Compute and graph $f'$, and then approximate the roots of $f'(x)$.

c. Verify that the zeros of $f'$ correspond to points at which $f$ has a horizontal tangent line.

**101. Angle to a particle (part 1)** A particle travels clockwise on a circular path of diameter $D$, monitored by a sensor on the circle at point $P$; the other endpoint of the diameter on which the sensor lies is $Q$ (see figure). Let $\theta$ be the angle between the diameter $PQ$ and the line from the sensor to the particle. Let $c$ be the length of the chord from the particle's position to $Q$.

a. Calculate $\dfrac{d\theta}{dc}$.

b. Evaluate $\left.\dfrac{d\theta}{dc}\right|_{c=0}$

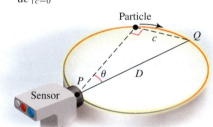

**102. Angle to a particle (part 2)** The figure in Exercise 101 shows the particle traveling away from the sensor, which may have influenced your solution (we expect you used the inverse sine function). Suppose instead that the particle approaches the sensor (see figure). How would this change the solution? Explain the differences in the two answers.

**103. Tracking a dive** A biologist standing at the bottom of an 80-ft vertical cliff watches a peregrine falcon dive from the top of the cliff at a 45° angle from the horizontal (see figure).

a. Express the angle of elevation $\theta$ from the biologist to the falcon as a function of the height $h$ of the bird above the ground. (*Hint:* The vertical distance between the top of the cliff and the falcon is $80 - h$.)

b. What is the rate of change of $\theta$ with respect to the bird's height when it is 60 ft above the ground?

**104. Derivative of the inverse sine** Find the derivative of the inverse sine using Theorem 7.3.

**105. Derivative of the inverse cosine** Find the derivative of the inverse cosine using Theorem 7.3.

**106. Derivative of $\cot^{-1} x$ and $\csc^{-1} x$** Use a trigonometric identity to show that the derivatives of the inverse cotangent and inverse cosecant differ from the derivatives of the inverse tangent and inverse secant, respectively, by a multiplicative factor of $-1$.

**107–110. Identity proofs** *Prove the following identities and give the values of $x$ for which they are true.*

**107.** $\cos\left(2\sin^{-1} x\right) = 1 - 2x^2$

**108.** $\cos\left(\sin^{-1} x\right) = \sqrt{1 - x^2}$

**109.** $\sin\left(2\sin^{-1} x\right) = 2x\sqrt{1 - x^2}$

**110.** $\tan\left(2\tan^{-1} x\right) = \dfrac{2x}{1 - x^2}$

**QUICK CHECK ANSWERS**

**1.** $\sin^{-1}(\sin 0) = \sin^{-1} 0 = 0$ and $\sin^{-1}(\sin 2\pi) = \sin^{-1} 0 = 0$    **2.** $0, \pi/4$    **3.** $f(x) = \sin^{-1} x$ is odd, while $f'(x) = 1/\sqrt{1 - x^2}$ is even.    **4.** The slopes of the tangent lines approach 0.    **5.** One is the negative of the other.    **6.** Recall that $1° = \pi/180$ rad. So 0.0024 rad/ft is equivalent to 0.138°/ft.    **7.** Because $\sin^{-1} x$ and $\cos^{-1} x$ differ by a constant, they are both antiderivatives of $(1 - x^2)^{-1/2}$. A similar argument applies to $\tan^{-1} x$ and $\cot^{-1} x$, and $\sec^{-1} x$ and $\csc^{-1} x$. ◄

# 7.6 L'Hôpital's Rule and Growth Rates of Functions

We first encountered l'Hôpital's Rule in Section 4.7, where it was applied directly to the indeterminate forms $0/0$ and $\infty/\infty$. We may now combine properties of the exponential function with l'Hôpital's Rule to evaluate limits with the indeterminate forms that we denote $1^{\infty}$, $0^0$, and $\infty^0$. This technique greatly expands our limit-taking capabilities and it offers some surprising results. A valuable outcome of this section is a ranking of functions according to their growth rates as $x \to \infty$. For example, which function grows faster as $x \to \infty$, $1.001^x$ or $x^{100}$? Rankings such as this should become part of your mathematical intuition; they are used often in upcoming chapters.

## A Brief Review

In Section 4.7, you learned that limits of the indeterminate forms $0/0$ and $\infty/\infty$ may be evaluated by applying l'Hôpital's Rule. Specifically, l'Hôpital's Rule says that under suitable conditions on $f$ and $g$, if $\lim\limits_{x \to a} f(x) = \lim\limits_{x \to a} g(x) = 0$, or $\lim\limits_{x \to a} f(x) = \pm\infty$ and $\lim\limits_{x \to a} g(x) = \pm\infty$, then

$$\lim_{x \to a} \frac{f(x)}{g(x)} = \lim_{x \to a} \frac{f'(x)}{g'(x)},$$

provided the limit on the right exists (or is $\pm\infty$). The rule also applies if $x \to a$ is replaced with $x \to \pm\infty$, $x \to a^+$, or $x \to a^-$. Example 1 provides a review of these ideas using the transcendental functions introduced in this chapter. We also review the method used to analyze limits of the form $0 \cdot \infty$.

**EXAMPLE 1** **L'Hôpital's Rule revisited** Evaluate the following limits.

**a.** $\lim\limits_{x \to 0} \dfrac{e^{2x} - 2x - 1}{x \sin x}$     **b.** $\lim\limits_{x \to \infty} 3^x \left(\pi/2 - \tan^{-1} x\right)$     **c.** $\lim\limits_{x \to \infty} \dfrac{x}{\ln x}$

**SOLUTION**

**a.** Evaluating the numerator and denominator of the function at 0, we see that this limit has the indeterminate form $0/0$. Applying l'Hôpital's Rule results in another limit of the form $0/0$, so we use the rule again:

$$\lim_{x \to 0} \frac{e^{2x} - 2x - 1}{x \sin x} = \lim_{x \to 0} \underbrace{\frac{2e^{2x} - 2}{\sin x + x \cos x}}_{\text{limit form } 0/0} \qquad \text{L'Hôpital's Rule}$$

$$= \lim_{x \to 0} \frac{4e^{2x}}{2 \cos x - x \sin x} \qquad \text{L'Hôpital's Rule again}$$

$$= \frac{4}{2} = 2. \qquad \text{Evaluate limit.}$$

> The fact that $f(x) = \tan x$ has a vertical asymptote at $x = \dfrac{\pi}{2}$ implies that its inverse $f^{-1}(x) = \tan^{-1} x$ has a horizontal asymptote of $y = \dfrac{\pi}{2}$, which in turn implies that $\lim\limits_{x \to \infty} \tan^{-1} x = \dfrac{\pi}{2}$. A similar argument leads to $\lim\limits_{x \to -\infty} \tan^{-1} x = -\dfrac{\pi}{2}$, as shown in the figure.

**b.** Because $\lim\limits_{x \to \infty} \tan^{-1} x = \dfrac{\pi}{2}$ (see margin note), the limit form is $0 \cdot \infty$, which may be converted to the form $0/0$ so that l'Hôpital's Rule can be applied:

$$\lim_{x \to \infty} 3^x(\pi/2 - \tan^{-1} x) = \lim_{x \to \infty} \frac{\pi/2 - \tan^{-1} x}{3^{-x}} \qquad 3^x = \frac{1}{3^{-x}}$$

$$= \lim_{x \to \infty} \frac{-1/(1 + x^2)}{-3^{-x} \ln 3} \qquad \text{L'Hôpital's Rule for } 0/0 \text{ form}$$

$$= \lim_{x \to \infty} \frac{3^x}{(1 + x^2) \ln 3}. \qquad \text{Simplify.}$$

$$\lim_{x \to \pi/2^-} \tan x = \infty$$

$$\lim_{x \to -\infty} \tan^{-1} x = -\frac{\pi}{2}$$

$$\lim_{x \to \infty} \tan^{-1} x = \frac{\pi}{2}$$

$$\lim_{x \to -\pi/2^+} \tan x = -\infty$$

The new limit is of the form $\infty/\infty$. Applying l'Hôpital's Rule two more times, we have

$$\underbrace{\lim_{x\to\infty}\frac{3^x}{(1+x^2)\ln 3}}_{\infty/\infty}=\underbrace{\lim_{x\to\infty}\frac{3^x\ln 3}{2x\ln 3}}_{\text{l'Hôpital's Rule}}=\underbrace{\lim_{x\to\infty}\frac{3^x}{2x}}_{\text{Simplify; }\infty/\infty}=\underbrace{\lim_{x\to\infty}\frac{3^x\ln 3}{2}}_{\text{l'Hôpital's Rule}}=\infty.$$

**c.** This time we have a limit of the form $\infty/\infty$, to which we apply l'Hôpital's Rule:

$$\lim_{x\to\infty}\frac{x}{\ln x}=\lim_{x\to\infty}\frac{1}{1/x}\qquad\text{L'Hôpital's Rule}$$

$$=\lim_{x\to\infty}x=\infty.\qquad\text{Simplify and evaluate.}$$

*Related Exercises 7, 14, 20* ◄

## Indeterminate Forms $1^\infty$, $0^0$, and $\infty^0$

The indeterminate forms $1^\infty$, $0^0$, and $\infty^0$ all arise in limits of the form $\lim_{x\to a}f(x)^{g(x)}$. However, l'Hôpital's Rule cannot be applied directly to these indeterminate forms. They must first be expressed in the form $0/0$ or $\infty/\infty$. Here is how we proceed.

The inverse relationship between $\ln x$ and $e^x$ says that $f^g=e^{g\ln f}$, so we first write

$$\lim_{x\to a}f(x)^{g(x)}=\lim_{x\to a}e^{g(x)\ln f(x)}.$$

By the continuity of the exponential function, we switch the order of the limit and the exponential function (Theorem 2.11); therefore,

$$\lim_{x\to a}f(x)^{g(x)}=\lim_{x\to a}e^{g(x)\ln f(x)}=e^{\lim_{x\to a}g(x)\ln f(x)},$$

provided $\lim_{x\to a}g(x)\ln f(x)$ exists or is $\pm\infty$. Therefore, $\lim_{x\to a}f(x)^{g(x)}$ is evaluated using the following two steps.

> ► Notice the following:
> For $1^\infty$, $L$ has the form $\infty\cdot\ln 1=\infty\cdot 0$.
> For $0^0$, $L$ has the form $0\cdot\ln 0=0\cdot\infty$.
> For $\infty^0$, $L$ has the form $0\cdot\ln\infty=0\cdot\infty$.

**QUICK CHECK 1** Explain why a limit of the form $0^\infty$ is not an indeterminate form. ◄

---

**PROCEDURE**   **Indeterminate forms $1^\infty$, $0^0$, and $\infty^0$**

Assume $\lim_{x\to a}f(x)^{g(x)}$ has the indeterminate form $1^\infty$, $0^0$, or $\infty^0$.

1. Analyze $L=\lim_{x\to a}g(x)\ln f(x)$. This limit can be put in the form $0/0$ or $\infty/\infty$, both of which are handled by l'Hôpital's Rule.

2. When $L$ is finite, $\lim_{x\to a}f(x)^{g(x)}=e^L$. If $L=\infty$ or $-\infty$, then $\lim_{x\to a}f(x)^{g(x)}=\infty$ or $\lim_{x\to a}f(x)^{g(x)}=0$, respectively.

---

**EXAMPLE 2**   **Indeterminate forms**  Evaluate the following limits.

**a.** $\lim_{x\to 0^+}x^x$    **b.** $\lim_{x\to\infty}\left(1+\frac{1}{x}\right)^x$    **c.** $\lim_{x\to 0^+}(\csc x)^x$

**SOLUTION**

**a.** This limit has the form $0^0$. Using the given two-step procedure, we note that $x^x=e^{x\ln x}$ and first evaluate

$$L=\lim_{x\to 0^+}x\ln x.$$

This limit has the form $0\cdot\infty$, which may be put in the form $\infty/\infty$ so that l'Hôpital's Rule can be applied:

$$L=\lim_{x\to 0^+}x\ln x=\lim_{x\to 0^+}\frac{\ln x}{1/x}\qquad x=\frac{1}{1/x}$$

$$=\lim_{x\to 0^+}\frac{1/x}{-1/x^2}\qquad\text{L'Hôpital's Rule for }\infty/\infty\text{ form}$$

$$=\lim_{x\to 0^+}(-x)=0.\qquad\text{Simplify and evaluate the limit.}$$

**Figure 7.54**

▶ The limit in Example 2b is often given as a definition of $e$. It is a special case of the more general limit

$$\lim_{x \to \infty} \left(1 + \frac{a}{x}\right)^x = e^a.$$

See Exercise 67.

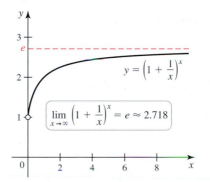

**Figure 7.55**

The second step is to exponentiate the limit:

$$\lim_{x \to 0^+} x^x = e^L = e^0 = 1.$$

We conclude that $\lim_{x \to 0^+} x^x = 1$ (**Figure 7.54**).

**b.** This limit has the form $1^\infty$. Noting that $(1 + 1/x)^x = e^{x \ln (1 + 1/x)}$, the first step is to evaluate

$$L = \lim_{x \to \infty} x \ln \left(1 + \frac{1}{x}\right),$$

which has the form $0 \cdot \infty$. Proceeding as in part (a), we have

$$L = \lim_{x \to \infty} x \ln \left(1 + \frac{1}{x}\right) = \lim_{x \to \infty} \frac{\ln (1 + 1/x)}{1/x} \qquad x = \frac{1}{1/x}$$

$$= \lim_{x \to \infty} \frac{\dfrac{1}{1 + 1/x} \cdot \left(-\dfrac{1}{x^2}\right)}{\left(-\dfrac{1}{x^2}\right)} \qquad \text{L'Hôpital's Rule for 0/0 form}$$

$$= \lim_{x \to \infty} \frac{1}{1 + 1/x} = 1 \qquad \text{Simplify and evaluate.}$$

The second step is to exponentiate the limit:

$$\lim_{x \to \infty} \left(1 + \frac{1}{x}\right)^x = e^L = e^1 = e.$$

The function $y = (1 + 1/x)^x$ (**Figure 7.55**) has a horizontal asymptote $y = e \approx 2.71828$.

**c.** Note that $\lim_{x \to 0^+} \csc x = \infty$, so the limit has the form $\infty^0$. Using the two-step procedure with $(\csc x)^x = e^{x \ln \csc x}$, the first step is to evaluate

$$L = \lim_{x \to 0^+} x \ln \csc x = \lim_{x \to 0^+} \frac{\ln \csc x}{x^{-1}},$$

which has the form $\infty / \infty$. Applying l'Hôpital's Rule, we find that

$$L = \lim_{x \to 0^+} \frac{\ln \csc x}{x^{-1}} = \lim_{x \to 0^+} \frac{\dfrac{1}{\csc x}(-\csc x \cot x)}{-x^{-2}} = \lim_{x \to 0^+} \frac{\cot x}{x^{-2}}.$$

The resulting limit is another indeterminate form. However, some simplification reveals that

$$L = \lim_{x \to 0^+} \frac{\cot x}{x^{-2}} = \lim_{x \to 0^+} \frac{x^2 \cos x}{\sin x} = \underbrace{\lim_{x \to 0^+} \frac{x}{\sin x}}_{1} \cdot \underbrace{\lim_{x \to 0^+} x}_{0} \cdot \underbrace{\lim_{x \to 0^+} \cos x}_{1} = 0.$$

Therefore, $L = 0$ and $\lim_{x \to 0^+} (\csc x)^x = e^L = e^0 = 1$.

*Related Exercises 21, 23–24* ◀

## Growth Rates of Functions

An important use of l'Hôpital's Rule is to compare the growth rates of functions. Here are two questions—one practical and one theoretical—that demonstrate the importance of comparative growth rates of functions.

▶ Models of epidemics produce more complicated functions than the one given here, but they have the same general features.

• A particular theory for modeling the spread of an epidemic predicts that the number of infected people $t$ days after the start of the epidemic is given by the function

$$N(t) = 2.5t^2 e^{-0.01t} = 2.5 \frac{t^2}{e^{0.01t}}.$$

*Question:* In the long run (as $t \to \infty$), does the epidemic spread or does it die out?

▶ The Prime Number Theorem was proved simultaneously (two different proofs) in 1896 by Jacques Hadamard and Charles de la Vallée Poussin, relying on fundamental ideas contributed by Riemann.

• A prime number is an integer $p \geq 2$ that has only two divisors, 1 and itself. The first few prime numbers are 2, 3, 5, 7, and 11. A celebrated theorem states that the number of prime numbers less than $x$ is approximately

$$P(x) = \frac{x}{\ln x}, \qquad \text{for large values of } x.$$

*Question:* According to this function, is the number of prime numbers infinite?

These two questions involve a comparison of two functions. In the first question, if $t^2$ grows faster than $e^{0.01t}$ as $t \to \infty$, then $\lim_{t \to \infty} N(t) = \infty$ and the epidemic grows. If $e^{0.01t}$ grows faster than $t^2$ as $t \to \infty$, then $\lim_{t \to \infty} N(t) = 0$ and the epidemic eventually dies out. We will explain what is meant by *grows faster than* in a moment.

In the second question, the comparison is between $x$ and $\ln x$. In Example 1c, we found that $\lim_{x \to \infty} P(x) = \infty$; therefore, we conclude that the number of prime numbers is infinite.

Our goal is to obtain a ranking of the following families of functions based on their growth rates:

▶ Another function with a large growth rate is the factorial function, defined for integers as $f(n) = n! = n(n-1) \cdots 2 \cdot 1$. See Exercise 66.

**QUICK CHECK 2** Before proceeding, use your intuition and rank these classes of functions in order of their growth rates. ◄

• $mx$, where $m > 0$ (represents linear functions)
• $x^p$, where $p > 0$ (represents polynomials and algebraic functions)
• $x^x$ (sometimes called a *superexponential* or *tower function*)
• $\ln x$ (represents logarithmic functions)
• $\ln^q x$, where $q > 0$ (represents powers of logarithmic functions)
• $x^p \ln x$, where $p > 0$ (a combination of powers and logarithms)
• $e^x$ (represents exponential functions).

We need to be precise about growth rates and what it means for $f$ to grow faster than $g$ as $x \to \infty$. We work with the following definitions.

---

**DEFINITION  Growth Rates of Functions (as $x \to \infty$)**

Suppose $f$ and $g$ are functions with $\lim_{x \to \infty} f(x) = \lim_{x \to \infty} g(x) = \infty$. Then **$f$ grows faster than $g$** as $x \to \infty$ if

$$\lim_{x \to \infty} \frac{g(x)}{f(x)} = 0 \quad \text{or, equivalently, if} \quad \lim_{x \to \infty} \frac{f(x)}{g(x)} = \infty.$$

The functions $f$ and $g$ have **comparable growth rates** if

$$\lim_{x \to \infty} \frac{f(x)}{g(x)} = M,$$

where $0 < M < \infty$ ($M$ is positive and finite).

---

Figure 7.56

The idea of growth rates is illustrated nicely with graphs. **Figure 7.56** shows a family of linear functions of the form $y = mx$, where $m > 0$, and powers of $x$ of the form $y = x^p$, where $p > 1$. We see that powers of $x$ grow faster (their curves rise at a greater rate) than the linear functions as $x \to \infty$.

$y = b^x$ for $b = 2$, 2.25, 2.5, 2.75, 3

> Exponential functions grow faster than powers of $x$ as $x \to \infty$.

$y = x^p$ for $p = 2$, 2.25, 2.5, 2.75, 3

**Figure 7.57**

**QUICK CHECK 3** Compare the growth rates of $f(x) = x^2$ and $g(x) = x^3$ as $x \to \infty$. Compare the growth rates of $f(x) = x^2$ and $g(x) = 10x^2$ as $x \to \infty$. ◀

**Figure 7.57** shows that exponential functions of the form $y = b^x$, where $b > 1$, grow faster than powers of $x$ of the form $y = x^p$, where $p > 0$, as $x \to \infty$ (Example 4).

We now begin a systematic comparison of growth rates. Note that a growth rate limit involves an indeterminate form $\infty / \infty$, so l'Hôpital's Rule is always in the picture.

**EXAMPLE 3** **Powers of $x$ vs. powers of ln $x$** Compare the growth rates as $x \to \infty$ of the following pairs of functions.

**a.** $f(x) = \ln x$ and $g(x) = x^p$, where $p > 0$

**b.** $f(x) = \ln^q x$ and $g(x) = x^p$, where $p > 0$ and $q > 0$

**SOLUTION**

**a.** The limit of the ratio of the two functions is

$$\lim_{x \to \infty} \frac{\ln x}{x^p} = \lim_{x \to \infty} \frac{1/x}{px^{p-1}} \quad \text{L'Hôpital's Rule}$$

$$= \lim_{x \to \infty} \frac{1}{px^p} \quad \text{Simplify.}$$

$$= 0. \quad \text{Evaluate the limit.}$$

We see that any positive power of $x$ grows faster than $\ln x$.

**b.** We compare $\ln^q x$ and $x^p$ by observing that

$$\lim_{x \to \infty} \frac{\ln^q x}{x^p} = \lim_{x \to \infty} \left( \frac{\ln x}{x^{p/q}} \right)^q = \left( \underbrace{\lim_{x \to \infty} \frac{\ln x}{x^{p/q}}}_{0} \right)^q.$$

By part (a), $\lim_{x \to \infty} \dfrac{\ln x}{x^{p/q}} = 0$ (because $p/q > 0$). Therefore, $\lim_{x \to \infty} \dfrac{\ln^q x}{x^p} = 0$ (because $q > 0$). We conclude that positive powers of $x$ grow faster than positive powers of $\ln x$.

*Related Exercises 36, 38* ◀

**EXAMPLE 4** **Powers of $x$ vs. exponentials** Compare the rates of growth of $f(x) = x^p$ and $g(x) = e^x$ as $x \to \infty$, where $p$ is a positive real number.

**SOLUTION** The goal is to evaluate $\lim_{x \to \infty} \dfrac{x^p}{e^x}$ for $p > 0$. This comparison is most easily done using Example 3 and a change of variables. We let $x = \ln t$ and note that as $x \to \infty$, we also have $t \to \infty$. With this substitution, $x^p = \ln^p t$ and $e^x = e^{\ln t} = t$. Therefore,

$$\lim_{x \to \infty} \frac{x^p}{e^x} = \lim_{t \to \infty} \frac{\ln^p t}{t} = 0. \quad \text{Example 3}$$

A similar argument applies when comparing the growth rates of $f(x) = x^p$ and $h(x) = b^x$, for $b > 1$ (Exercise 63). We see that increasing exponential functions grow faster than positive powers of $x$ (Figure 7.57).

*Related Exercises 31, 37* ◀

These examples, together with the comparison of exponential functions $b^x$ and the superexponential $x^x$ (Exercise 68), establish a ranking of growth rates.

---

**THEOREM 7.14** **Ranking Growth Rates as $x \to \infty$**

Let $f \ll g$ mean that $g$ grows faster than $f$ as $x \to \infty$. With positive real numbers $p, q, r,$ and $s$ and $b > 1$,

$$\ln^q x \ll x^p \ll x^p \ln^r x \ll x^{p+s} \ll b^x \ll x^x.$$

---

You should try to build these relative growth rates into your intuition. They are useful in future chapters (Chapter 10 on sequences, in particular), and they can be used to evaluate limits at infinity quickly.

## SECTION 7.6 EXERCISES

### Getting Started

1. Explain why the form $1^\infty$ is indeterminate and cannot be evaluated by substitution. Explain how the competing functions behave.

2. Give the two-step method for attacking an indeterminate limit of the form $\lim\limits_{x \to a} f(x)^{g(x)}$.

3. In terms of limits, what does it mean for $f$ to grow faster than $g$ as $x \to \infty$?

4. In terms of limits, what does it mean for the rates of growth of $f$ and $g$ to be comparable as $x \to \infty$?

5. Rank the functions $x^3$, $\ln x$, $x^x$, and $2^x$ in order of increasing growth rates as $x \to \infty$.

6. Rank the functions $x^{100}$, $\ln x^{10}$, $x^x$, and $10^x$ in order of increasing growth rates as $x \to \infty$.

### Practice Exercises

**7–20. A brief review** *Evaluate the following limits.*

7. $\lim\limits_{x \to 1} \dfrac{\ln x}{4x - x^2 - 3}$

8. $\lim\limits_{x \to 0} \dfrac{e^x - 1}{x^2 + 3x}$

9. $\lim\limits_{x \to e} \dfrac{\ln x - 1}{x - e}$

10. $\lim\limits_{x \to 1} \dfrac{4 \tan^{-1} x - \pi}{x - 1}$

11. $\lim\limits_{x \to 0} \dfrac{e^x - \sin x - 1}{x^4 + 8x^3 + 12x^2}$

12. $\lim\limits_{x \to 0} \dfrac{e^x - x - 1}{5x^2}$

13. $\lim\limits_{x \to \infty} \dfrac{\log_2 x}{\log_3 x}$

14. $\lim\limits_{x \to 0^+} \dfrac{1 - \ln x}{1 + \ln x}$

15. $\lim\limits_{x \to \infty} \dfrac{\ln (3x + 5)}{\ln (7x + 3) + 1}$

16. $\lim\limits_{x \to \infty} \dfrac{\ln (3x + 5e^x)}{\ln (7x + 3e^{2x})}$

17. $\lim\limits_{x \to \infty} \dfrac{x^2 - \ln (2/x)}{3x^2 + 2x}$

18. $\lim\limits_{x \to \infty} e^x \cot^{-1} x$

19. $\lim\limits_{x \to 0^+} (\sin^{-1} x) \cot x$

20. $\lim\limits_{x \to 1^-} \tan \left( \dfrac{\pi x}{2} \right) \ln x$

**21–30. $1^\infty$, $0^0$, $\infty^0$ forms** *Evaluate the following limits or explain why they do not exist. Check your results by graphing.*

21. $\lim\limits_{x \to 0} (1 + 2x)^{3/x}$

22. $\lim\limits_{x \to 0} (1 + 4x)^{3/x}$

23. $\lim\limits_{\theta \to \pi/2^-} (\tan \theta)^{\cos \theta}$

24. $\lim\limits_{\theta \to 0^+} (\sin \theta)^{\tan \theta}$

25. $\lim\limits_{x \to 0^+} (1 + x)^{\cot x}$

26. $\lim\limits_{x \to \infty} \left( 1 + \dfrac{1}{x} \right)^{\ln x}$

27. $\lim\limits_{x \to 0^+} (\tan x)^x$

28. $\lim\limits_{z \to \infty} \left( 1 + \dfrac{10}{z^2} \right)^{z^2}$

29. $\lim\limits_{x \to 0} (x + \cos x)^{1/x}$

30. $\lim\limits_{x \to 0^+} \left( \dfrac{1}{3} \cdot 3^x + \dfrac{2}{3} \cdot 2^x \right)^{1/x}$

**31–42. Comparing growth rates** *Use limit methods to determine which of the two given functions grows faster or state that they have comparable growth rates.*

31. $x^{10}$; $e^{0.01x}$

32. $x^2 \ln x$; $\ln^2 x$

33. $\ln x^{20}$; $\ln x$

34. $\ln x$; $\ln (\ln x)$

35. $100^x$; $x^x$

36. $x^2 \ln x$; $x^3$

37. $x^{20}$; $1.00001^x$

38. $x^{10} \ln^{10} x$; $x^{11}$

39. $x^x$; $(x/2)^x$

40. $\ln \sqrt{x}$; $\ln^2 x$

41. $e^{x^2}$; $e^{10x}$

42. $e^{x^2}$; $x^{x/10}$

43. **Explain why or why not** Determine whether the following statements are true and give an explanation or counterexample.

   a. $\lim\limits_{x \to 0^+} x^{1/x}$ is an indeterminate form.

   b. The number 1 raised to any fixed power is 1. Therefore, because $(1 + x) \to 1$ as $x \to 0$, $(1 + x)^{1/x} \to 1$ as $x \to 0$.

   c. The functions $\ln x^{100}$ and $\ln x$ have comparable growth rates as $x \to \infty$.

   d. The function $e^x$ grows faster than $2^x$ as $x \to \infty$.

**44–52. Miscellaneous limits by any means** *Use analytical methods to evaluate the following limits.*

44. $\lim\limits_{x \to 0^+} x^{1/\ln x}$

45. $\lim\limits_{x \to 0} \left( \dfrac{\sin x}{x} \right)^{1/x^2}$

46. $\lim\limits_{x \to 0^+} (\cot x)^x$

47. $\lim\limits_{x \to \infty} x^{1/x}$

48. $\lim\limits_{x \to \infty} \left( \dfrac{1}{2x} \right)^{3/x}$

49. $\lim\limits_{x \to 1^+} (\sqrt{x} - 1)^{\sin \pi x}$

50. $\lim\limits_{x \to 2^+} (\ln (x^2 - 3))^{x^3 - 2x - 4}$

51. $\lim\limits_{x \to 1^+} \left( \dfrac{1}{x - 1} - \dfrac{1}{\ln x} \right)$

52. $\lim\limits_{x \to 1^-} \left( \dfrac{x}{x - 1} - \dfrac{x}{\ln x} \right)$

**T** 53. **It may take time** The ranking of growth rates given in the text applies for $x \to \infty$. However, these rates may not be evident for small values of $x$. For example, an exponential grows faster than any power of $x$. However, for $1 < x < 19,800$, $x^2$ is greater than $e^{x/1000}$. For the following pairs of functions, estimate the point at which the faster-growing function overtakes the slower-growing function (for the last time).

   a. $\ln^3 x$ and $x^{0.3}$
   b. $2^{x/100}$ and $x^3$
   c. $x^{x/100}$ and $e^x$
   d. $\ln^{10} x$ and $e^{x/10}$

**54–57. Limits with parameters** *Evaluate the following limits in terms of the parameters $a$ and $b$, which are positive real numbers. In each case, graph the function for specific values of the parameters to check your results.*

54. $\lim\limits_{x \to 0} (1 + ax)^{b/x}$

55. $\lim\limits_{x \to 0^+} (a^x - b^x)^x$, $a > b > 0$

56. $\lim\limits_{x \to 0^+} (a^x - b^x)^{1/x}$, $a > b > 0$

57. $\lim\limits_{x \to 0} \dfrac{a^x - b^x}{x}$

58. **Avoiding l'Hôpital's Rule** Let $L = \lim\limits_{x \to 0} \dfrac{x - \sin x}{x^3}$.

   a. Compute $L$ using l'Hôpital's Rule.

   b. Now compute $L$ using the following interesting detour. First use the sine triplication formula $\sin 3x = 3 \sin x - 4 \sin^3 x$ to replace $\sin x$ by $3 \sin (x/3) - 4 \sin^3 (x/3)$ in the expression for $L$.

   c. Let $t = x/3$ and note that as $x \to 0$, $t \to 0$. Use $\lim\limits_{t \to 0} \dfrac{\sin t}{t} = 1$ to show that $L = \dfrac{L}{9} + \dfrac{4}{27}$.

   d. Solve for $L$ and check for agreement with part (a). (Source: *Teaching Mathematics and its Applications*, Fabio Cavallini, 3, 1988)

**59. Two methods**

    **a.** The derivative of a function $f$ at a point $a$ is given by $\lim\limits_{h \to 0} \dfrac{3^h - 1}{h}$. Find a possible $f$ and $a$, and evaluate the limit.

    **b.** Evaluate $\lim\limits_{h \to 0} \dfrac{3^h - 1}{h}$ using l'Hôpital's Rule.

**60. Two methods**

    **a.** The derivative of a function $f$ at a point $a$ is given by $\lim\limits_{x \to 0} \dfrac{e^x - 1}{x}$. Find a possible $f$ and $a$, and evaluate the limit.

    **b.** Evaluate $\lim\limits_{x \to 0} \dfrac{e^x - 1}{x}$ using l'Hôpital's Rule.

**61. Compound interest** Suppose you make a deposit of $\$P$ into a savings account that earns interest at a rate of $100r\,\%$ per year.

    **a.** Show that if interest is compounded once per year, then the balance after $t$ years is $B(t) = P(1 + r)^t$.

    **b.** If interest is compounded $m$ times per year, then the balance after $t$ years is $B(t) = P(1 + r/m)^{mt}$. For example, $m = 12$ corresponds to monthly compounding, and the interest rate for each month is $r/12$. In the limit $m \to \infty$, the compounding is said to be *continuous*. Show that with continuous compounding, the balance after $t$ years is $B(t) = Pe^{rt}$.

### Explorations and Challenges

**T 62. Algorithm complexity** The complexity of a computer algorithm is the number of operations or steps the algorithm needs to complete its task assuming there are $n$ pieces of input (for example, the number of steps needed to put $n$ numbers in ascending order). Four algorithms for doing the same task have complexities of A: $n^{3/2}$, B: $n \log_2 n$, C: $n(\log_2 n)^2$, and D: $\sqrt{n} \log_2 n$. Rank the algorithms in order of increasing efficiency for large values of $n$. Graph the complexities as they vary with $n$ and comment on your observations.

**63. Exponential functions and powers** Show that any exponential function $b^x$, for $b > 1$, grows faster than $x^p$, for $p > 0$.

**64. Exponentials with different bases** Show that $f(x) = a^x$ grows faster than $g(x) = b^x$ as $x \to \infty$ if $1 < b < a$.

**65. Logs with different bases** Show that $f(x) = \log_a x$ and $g(x) = \log_b x$, where $a > 1$ and $b > 1$, grow at a comparable rate as $x \to \infty$.

**T 66. Factorial growth rate** The factorial function is defined for positive integers as $n! = n(n - 1)(n - 2) \cdots 3 \cdot 2 \cdot 1$. For example, $5! = 5 \cdot 4 \cdot 3 \cdot 2 \cdot 1 = 120$. A valuable result that gives good approximations to $n!$ for large values of $n$ is *Stirling's formula*, $n! \approx \sqrt{2\pi n}\, n^n e^{-n}$. Use this formula and a calculator to determine where the factorial function appears in the ranking of growth rates given in Theorem 7.14. (See the Guided Project *Stirling's formula and $n!$*.)

**67. Exponential limit** Prove that $\lim\limits_{x \to \infty} \left(1 + \dfrac{a}{x}\right)^x = e^a$, for $a \neq 0$.

**68. Exponentials vs. super exponentials** Show that $x^x$ grows faster than $b^x$ as $x \to \infty$ for $b > 1$.

**69. Exponential growth rates**

    **a.** For what values of $b > 0$ does $b^x$ grow faster than $e^x$ as $x \to \infty$?

    **b.** Compare the growth rates of $e^x$ and $e^{ax}$ as $x \to \infty$, for $a > 0$.

**T 70. A fascinating function** Consider the function $f(x) = (ab^x + (1 - a)c^x)^{1/x}$, where $a$, $b$, and $c$ are positive real numbers with $0 < a < 1$.

    **a.** Graph $f$ for several sets of $(a, b, c)$. Verify that in all cases $f$ is an increasing function with a single inflection point, for all $x$.

    **b.** Use analytical methods to determine $\lim\limits_{x \to 0} f(x)$ in terms of $a$, $b$, and $c$.

    **c.** Show that $\lim\limits_{x \to \infty} f(x) = \max\{b, c\}$ and $\lim\limits_{x \to -\infty} f(x) = \min\{b, c\}$, for any $0 < a < 1$.

    **d.** Estimate the location of the inflection point of $f$.

**QUICK CHECK ANSWERS**

**1.** The form $0^\infty$ (for example, $\lim\limits_{x \to 0^+} x^{1/x}$) is not indeterminate, because as the base goes to zero, raising it to larger and larger powers drives the entire function to zero. **3.** $x^3$ grows faster than $x^2$ as $x \to \infty$, whereas $x^2$ and $10x^2$ have comparable growth rates as $x \to \infty$. ◀

# 7.7 Hyperbolic Functions

In this section, we introduce a new family of functions called the *hyperbolic* functions, which are closely related to both trigonometric functions and exponential functions. Hyperbolic functions find widespread use in fluid dynamics, projectile motion, architecture, and electrical engineering, to name just a few areas. Hyperbolic functions are also important in the development of many theoretical results in mathematics.

## Relationship Between Trigonometric and Hyperbolic Functions

The trigonometric functions defined in Chapter 1 are based on relationships involving a circle—for this reason, trigonometric functions are also known as *circular* functions. Specifically, $\cos t$ and $\sin t$ are equal to the $x$- and $y$-coordinates, respectively, of the point $P(x, y)$ on the unit circle that corresponds to an angle of $t$ radians (**Figure 7.58**). We can also regard $t$ as the length of the arc from $(1, 0)$ to the point $P(x, y)$.

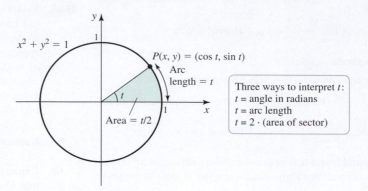

**Figure 7.58**

> Recall that the area of a circular sector of radius $r$ and angle $\theta$ is $A = \frac{1}{2}r^2\theta$. With $r = 1$ and $\theta = t$, we have $A = \frac{1}{2}t$, which implies $t = 2A$.

There is yet another way to interpret the number $t$, and it is this third interpretation that links the trigonometric and hyperbolic functions. Observe that $t$ is twice the area of the circular sector in Figure 7.58. The functions $\cos t$ and $\sin t$ are still defined as the $x$- and $y$-coordinates of the point $P$, but now we associate $P$ with a sector whose area is one-half of $t$.

The *hyperbolic cosine* and *hyperbolic sine* are defined in an analogous fashion using the hyperbola $x^2 - y^2 = 1$ instead of the circle $x^2 + y^2 = 1$. Consider the region bounded by the $x$-axis, the right branch of the unit hyperbola $x^2 - y^2 = 1$, and a line segment from the origin to a point $P(x, y)$ on the hyperbola (**Figure 7.59**); let $t$ equal twice the area of this region.

**Figure 7.59**

The hyperbolic cosine of $t$, denoted $\cosh t$, is the $x$-coordinate of $P$, and the hyperbolic sine of $t$, denoted $\sinh t$, is the $y$-coordinate of $P$. Expressing $x$ and $y$ in terms of $t$ leads to the standard definitions of the hyperbolic functions. We accomplish this task by writing $t$, which is an area, as an integral that depends on the coordinates of $P$. In Exercise 112, we ask you to carry out the calculations to show that

$$x = \cosh t = \frac{e^t + e^{-t}}{2} \quad \text{and} \quad y = \sinh t = \frac{e^t - e^{-t}}{2}.$$

Everything that follows in this section is based on these two definitions.

## Definitions, Identities, and Graphs of the Hyperbolic Functions

Once the hyperbolic cosine and hyperbolic sine are defined, the four remaining hyperbolic functions follow in a manner analogous to the trigonometric functions.

> There is no universally accepted pronunciation of the names of the hyperbolic functions. In the United States, *cohsh x* (long *oh* sound) and *sinch x* are common choices for $\cosh x$ and $\sinh x$. The pronunciations *tanch x*, *cotanch x*, *seech x* or *sech x*, and *coseech x* or *cosech x* are used for the other functions. International pronunciations vary as well.

---

**DEFINITION  Hyperbolic Functions**

**Hyperbolic cosine**

$$\cosh x = \frac{e^x + e^{-x}}{2}$$

**Hyperbolic sine**

$$\sinh x = \frac{e^x - e^{-x}}{2}$$

**Hyperbolic tangent**

$$\tanh x = \frac{\sinh x}{\cosh x} = \frac{e^x - e^{-x}}{e^x + e^{-x}}$$

**Hyperbolic cotangent**

$$\coth x = \frac{\cosh x}{\sinh x} = \frac{e^x + e^{-x}}{e^x - e^{-x}}$$

**Hyperbolic secant**

$$\operatorname{sech} x = \frac{1}{\cosh x} = \frac{2}{e^x + e^{-x}}$$

**Hyperbolic cosecant**

$$\operatorname{csch} x = \frac{1}{\sinh x} = \frac{2}{e^x - e^{-x}}$$

The hyperbolic functions satisfy many important identities. Let's begin with the fundamental identity for hyperbolic functions, which is analogous to the familiar trigonometric identity $\cos^2 x + \sin^2 x = 1$:

$$\cosh^2 x - \sinh^2 x = 1.$$

> ➤ The fundamental identity for hyperbolic functions can also be understood in terms of the geometric definition of the hyperbolic functions. Because the point $P(\cosh t, \sinh t)$ is on the hyperbola $x^2 - y^2 = 1$, the coordinates of $P$ satisfy the equation of the hyperbola, which leads immediately to
> $$\cosh^2 t - \sinh^2 t = 1.$$

This identity is verified by appealing to the definitions:

$$\cosh^2 x - \sinh^2 x = \left(\frac{e^x + e^{-x}}{2}\right)^2 - \left(\frac{e^x - e^{-x}}{2}\right)^2 \qquad \text{Definition of } \cosh x \text{ and } \sinh x$$

$$= \frac{e^{2x} + 2 + e^{-2x} - (e^{2x} - 2 + e^{-2x})}{4} \qquad \text{Expand and combine fractions.}$$

$$= \frac{4}{4} = 1. \qquad \text{Simplify.}$$

**EXAMPLE 1    Deriving hyperbolic identities**

**a.** Use the fundamental identity $\cosh^2 x - \sinh^2 x = 1$ to prove that $1 - \tanh^2 x = \text{sech}^2 x$.

**b.** Derive the identity $\sinh 2x = 2 \sinh x \cosh x$.

**SOLUTION**

**a.** Dividing both sides of the fundamental identity $\cosh^2 x - \sinh^2 x = 1$ by $\cosh^2 x$ leads to the desired result:

$$\cosh^2 x - \sinh^2 x = 1 \qquad \text{Fundamental identity}$$

$$\frac{\cosh^2 x}{\cosh^2 x} - \underbrace{\frac{\sinh^2 x}{\cosh^2 x}}_{\tanh^2 x} = \underbrace{\frac{1}{\cosh^2 x}}_{\text{sech}^2 x} \qquad \text{Divide both sides by } \cosh^2 x.$$

$$1 - \tanh^2 x = \text{sech}^2 x. \qquad \text{Identify functions.}$$

**b.** Using the definition of the hyperbolic sine, we have

$$\sinh 2x = \frac{e^{2x} - e^{-2x}}{2} \qquad \text{Definition of } \sinh$$

$$= \frac{(e^x - e^{-x})(e^x + e^{-x})}{2} \qquad \text{Factor; difference of perfect squares}$$

$$= 2 \sinh x \cosh x. \qquad \text{Identify functions.}$$

*Related Exercises 16–18* ◄

The identities in Example 1 are just two of many useful hyperbolic identities, some of which we list next.

---

**Hyperbolic Identities**

$$\cosh^2 x - \sinh^2 x = 1 \qquad\qquad \cosh(-x) = \cosh x$$

$$1 - \tanh^2 x = \text{sech}^2 x \qquad\qquad \sinh(-x) = -\sinh x$$

$$\coth^2 x - 1 = \text{csch}^2 x \qquad\qquad \tanh(-x) = -\tanh x$$

$$\cosh(x + y) = \cosh x \cosh y + \sinh x \sinh y$$

$$\sinh(x + y) = \sinh x \cosh y + \cosh x \sinh y$$

$$\cosh 2x = \cosh^2 x + \sinh^2 x \qquad \sinh 2x = 2 \sinh x \cosh x$$

$$\cosh^2 x = \frac{\cosh 2x + 1}{2} \qquad\qquad \sinh^2 x = \frac{\cosh 2x - 1}{2}$$

---

Graphs of the hyperbolic functions are relatively easy to produce because they are based on the familiar graphs of $e^x$ and $e^{-x}$. Recall that $\lim\limits_{x \to \infty} e^{-x} = 0$ and that $\lim\limits_{x \to -\infty} e^x = 0$. With these facts in mind, we see that the graph of $\cosh x$ (**Figure 7.60**) approaches the graph of $y = \dfrac{1}{2} e^x$ as $x \to \infty$ because $\cosh x = \dfrac{e^x + e^{-x}}{2} \approx \dfrac{e^x}{2}$ for large values of $x$.

**Figure 7.60**

A similar argument shows that as $x \to -\infty$, $\cosh x$ approaches $y = \dfrac{1}{2} e^{-x}$. Note also that $\cosh x$ is an even function:

$$\cosh(-x) = \frac{e^{-x} + e^{-(-x)}}{2} = \frac{e^x + e^{-x}}{2} = \cosh x.$$

Finally, $\cosh 0 = \dfrac{e^0 + e^0}{2} = 1$, so its $y$-intercept is $(0, 1)$. The behavior of $\sinh x$, an odd function also shown in Figure 7.60, can be explained in much the same way.

The graphs of the other four hyperbolic functions are shown in **Figure 7.61**. As a consequence of their definitions, we see that the domain of $\cosh x$, $\sinh x$, $\tanh x$, and $\text{sech}\, x$ is $(-\infty, \infty)$, whereas the domain of $\coth x$ and $\text{csch}\, x$ is the set of all real numbers excluding 0.

**QUICK CHECK 1** Use the definition of the hyperbolic sine to show that $\sinh x$ is an odd function. ◄

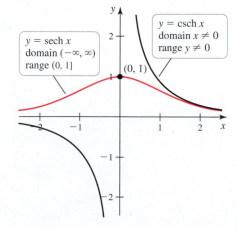

**QUICK CHECK 2** Explain why the graph of $\tanh x$ has the horizontal asymptotes $y = 1$ and $y = -1$. ◄

**Figure 7.61**

## Derivatives and Integrals of Hyperbolic Functions

Because the hyperbolic functions are defined in terms of $e^x$ and $e^{-x}$, computing their derivatives is straightforward. The derivatives of the hyperbolic functions are given in Theorem 7.15—reversing these formulas produces corresponding integral formulas.

▶ The identities, derivative formulas, and integral formulas for the hyperbolic functions are similar to the corresponding formulas for the trigonometric functions, which makes them easy to remember. However, be aware of some subtle differences in the signs associated with these formulas. For instance,

$$d/dx(\cos x) = -\sin x,$$

whereas

$$d/dx(\cosh x) = \sinh x.$$

---

**THEOREM 7.15  Derivative and Integral Formulas**

1. $\dfrac{d}{dx}(\cosh x) = \sinh x \quad \Rightarrow \quad \displaystyle\int \sinh x \, dx = \cosh x + C$

2. $\dfrac{d}{dx}(\sinh x) = \cosh x \quad \Rightarrow \quad \displaystyle\int \cosh x \, dx = \sinh x + C$

3. $\dfrac{d}{dx}(\tanh x) = \text{sech}^2 x \quad \Rightarrow \quad \displaystyle\int \text{sech}^2 x \, dx = \tanh x + C$

4. $\dfrac{d}{dx}(\coth x) = -\text{csch}^2 x \quad \Rightarrow \quad \displaystyle\int \text{csch}^2 x \, dx = -\coth x + C$

5. $\dfrac{d}{dx}(\text{sech } x) = -\text{sech } x \tanh x \quad \Rightarrow \quad \displaystyle\int \text{sech } x \tanh x \, dx = -\text{sech } x + C$

6. $\dfrac{d}{dx}(\text{csch } x) = -\text{csch } x \coth x \quad \Rightarrow \quad \displaystyle\int \text{csch } x \coth x \, dx = -\text{csch } x + C$

---

**Proof:** Using the definitions of $\cosh x$ and $\sinh x$, we have

$$\frac{d}{dx}(\cosh x) = \frac{d}{dx}\left(\frac{e^x + e^{-x}}{2}\right) = \frac{e^x - e^{-x}}{2} = \sinh x \quad \text{and}$$

$$\frac{d}{dx}(\sinh x) = \frac{d}{dx}\left(\frac{e^x - e^{-x}}{2}\right) = \frac{e^x + e^{-x}}{2} = \cosh x.$$

To prove formula (3), we begin with $\tanh x = \sinh x / \cosh x$ and then apply the Quotient Rule:

$$\frac{d}{dx}(\tanh x) = \frac{d}{dx}\left(\frac{\sinh x}{\cosh x}\right) \qquad \text{Definition of } \tanh x$$

$$= \frac{(\cosh x)\cosh x - (\sinh x)\sinh x}{\cosh^2 x} \qquad \text{Quotient Rule}$$

$$= \frac{1}{\cosh^2 x} \qquad \cosh^2 x - \sinh^2 x = 1$$

$$= \text{sech}^2 x. \qquad \text{sech } x = 1/\cosh x$$

The proofs of the remaining derivative formulas are assigned in Exercises 19–21. The integral formulas are a direct consequence of their corresponding derivative formulas. ◀

**EXAMPLE 2  Derivatives and integrals of hyperbolic functions** Evaluate the following derivatives and integrals.

a. $\dfrac{d}{dx}(\text{sech } 3x)$

b. $\dfrac{d^2}{dx^2}(\text{sech } 3x)$

c. $\displaystyle\int \frac{\text{csch}^2 \sqrt{x}}{\sqrt{x}} \, dx$

d. $\displaystyle\int_0^{\ln 3} \sinh^3 x \cosh x \, dx$

**SOLUTION**

a. Combining formula (5) of Theorem 7.15 with the Chain Rule gives

$$\frac{d}{dx}(\text{sech } 3x) = -3 \text{ sech } 3x \tanh 3x.$$

**b.** Applying the Product Rule and Chain Rule to the result of part (a), we have

$$\frac{d^2}{dx^2}(\text{sech } 3x) = \frac{d}{dx}(-3 \text{ sech } 3x \tanh 3x)$$

$$= \underbrace{\frac{d}{dx}(-3 \text{ sech } 3x)}_{9 \text{ sech } 3x \tanh 3x} \cdot \tanh 3x + (-3 \text{ sech } 3x) \cdot \underbrace{\frac{d}{dx}(\tanh 3x)}_{3 \text{ sech}^2 3x} \qquad \text{Product Rule}$$

$$= 9 \text{ sech } 3x \tanh^2 3x - 9 \text{ sech}^3 3x \qquad\qquad\qquad \text{Chain Rule}$$

$$= 9 \text{ sech } 3x(\tanh^2 3x - \text{ sech}^2 3x). \qquad\qquad\qquad \text{Simplify.}$$

**c.** The integrand suggests the substitution $u = \sqrt{x}$:

$$\int \frac{\text{csch}^2 \sqrt{x}}{\sqrt{x}} dx = 2 \int \text{csch}^2 u \, du \qquad \text{Let } u = \sqrt{x}; du = \frac{1}{2\sqrt{x}} dx.$$

$$= -2 \coth u + C \qquad \text{Formula (4), Theorem 7.15}$$

$$= -2 \coth \sqrt{x} + C. \quad u = \sqrt{x}$$

**d.** The derivative formula $\dfrac{d}{dx}(\sinh x) = \cosh x$ suggests the substitution $u = \sinh x$:

$$\int_0^{\ln 3} \sinh^3 x \cosh x \, dx = \int_0^{4/3} u^3 \, du. \quad \text{Let } u = \sinh x; du = \cosh x \, dx.$$

The new limits of integration are determined by the calculations

$$x = 0 \quad \Rightarrow \quad u = \sinh 0 = 0 \quad \text{and}$$

$$x = \ln 3 \quad \Rightarrow \quad u = \sinh(\ln 3) = \frac{e^{\ln 3} - e^{-\ln 3}}{2} = \frac{3 - 1/3}{2} = \frac{4}{3}.$$

We now evaluate the integral in the variable $u$:

$$\int_0^{4/3} u^3 \, du = \frac{1}{4} u^4 \Big|_0^{4/3}$$

**QUICK CHECK 3** Find both the derivative and indefinite integral of $f(x) = 4 \cosh 2x$. ◄

$$= \frac{1}{4}\left(\left(\frac{4}{3}\right)^4 - 0^4\right) = \frac{64}{81}.$$

*Related Exercises 22, 28, 43, 44* ◄

Theorem 7.16 presents integral formulas for the four hyperbolic functions not covered in Theorem 7.15.

---

**THEOREM 7.16  Integrals of Hyperbolic Functions**

**1.** $\displaystyle\int \tanh x \, dx = \ln \cosh x + C$     **2.** $\displaystyle\int \coth x \, dx = \ln|\sinh x| + C$

**3.** $\displaystyle\int \text{sech } x \, dx = \tan^{-1}(\sinh x) + C$     **4.** $\displaystyle\int \text{csch } x \, dx = \ln|\tanh(x/2)| + C$

---

**Proof:** Formula (1) is derived by first writing $\tanh x$ in terms of $\sinh x$ and $\cosh x$:

$$\int \tanh x \, dx = \int \frac{\sinh x}{\cosh x} dx \qquad \text{Definition of } \tanh x$$

$$= \int \frac{1}{u} du \qquad \text{Let } u = \cosh x; du = \sinh x \, dx.$$

$$= \ln|u| + C \qquad \text{Evaluate integral.}$$

$$= \ln \cosh x + C. \quad u = \cosh x > 0$$

Formula (2) is derived in a similar fashion (Exercise 60). The more challenging proofs of formulas (3) and (4) are considered in Exercises 107 and 108. ◄

**EXAMPLE 3** **Integrals involving hyperbolic functions** Determine the indefinite integral $\int x \coth x^2 \, dx$.

**SOLUTION** The integrand suggests the substitution $u = x^2$:

$$\int x \coth x^2 \, dx = \frac{1}{2} \int \coth u \, du \qquad \text{Let } u = x^2; \, du = 2x \, dx.$$

$$= \frac{1}{2} \ln|\sinh u| + C \qquad \text{Evaluate integral; use Theorem 7.16.}$$

$$= \frac{1}{2} \ln(\sinh x^2) + C. \qquad u = x^2; \sinh x^2 \geq 0$$

**QUICK CHECK 4** Determine the indefinite integral $\int \operatorname{csch} 2x \, dx$. ◄

*Related Exercises 45, 58* ◄

## Inverse Hyperbolic Functions

At present, we don't have the tools for evaluating an integral such as $\int \dfrac{dx}{\sqrt{x^2 + 4}}$. By studying inverse hyperbolic functions, we can discover new integration formulas. Inverse hyperbolic functions are also useful for solving equations involving hyperbolic functions.

Figures 7.60 and 7.61 show that the functions $\sinh x$, $\tanh x$, $\coth x$, and $\operatorname{csch} x$ are all one-to-one on their respective domains. This observation implies that each of these functions has a well-defined inverse. However, the function $y = \cosh x$ is not one-to-one on $(-\infty, \infty)$, so its inverse, denoted $y = \cosh^{-1} x$, exists only if we restrict the domain of $\cosh x$. Specifically, when $y = \cosh x$ is restricted to the interval $[0, \infty)$, it is one-to-one, and its inverse is defined as follows:

$$y = \cosh^{-1} x \quad \text{if and only if} \quad x = \cosh y, \text{ for } x \geq 1 \text{ and } 0 \leq y < \infty.$$

**Figure 7.62a** shows the graph of $y = \cosh^{-1} x$, obtained by reflecting the graph of $y = \cosh x$ on $[0, \infty)$ over the line $y = x$. The definitions and graphs of the other five inverse hyperbolic functions are also shown in **Figure 7.62**. Notice that the domain of $y = \operatorname{sech} x$ (**Figure 7.62d**) must be restricted to $[0, \infty)$ to ensure the existence of its inverse.

**Figure 7.62**

Because hyperbolic functions are defined in terms of exponential functions, we can find explicit formulas for their inverses in terms of logarithms. For example, let's start with the definition of the inverse hyperbolic sine. For all real $x$ and $y$, we have

$$y = \sinh^{-1} x \quad \Leftrightarrow \quad x = \sinh y.$$

Following the procedure outlined in Section 7.1, we solve

$$x = \sinh y = \frac{e^y - e^{-y}}{2}$$

for $y$ to give a formula for $\sinh^{-1} x$:

$$x = \frac{e^y - e^{-y}}{2} \quad \Rightarrow \quad e^y - 2x - e^{-y} = 0 \qquad \text{Rearrange equation.}$$

$$\Rightarrow \quad (e^y)^2 - 2xe^y - 1 = 0. \quad \text{Multiply by } e^y.$$

At this stage, we recognize a quadratic equation in $e^y$ and solve for $e^y$ using the quadratic formula, with $a = 1$, $b = -2x$, and $c = -1$:

$$e^y = \frac{2x \pm \sqrt{4x^2 + 4}}{2} = x \pm \sqrt{x^2 + 1} = \underbrace{x + \sqrt{x^2 + 1}}_{\text{choose positive root}}.$$

Because $e^y > 0$ and $\sqrt{x^2 + 1} > x$, the positive root must be chosen. We now solve for $y$ by taking the natural logarithm of both sides:

$$e^y = x + \sqrt{x^2 + 1} \quad \Rightarrow \quad y = \ln(x + \sqrt{x^2 + 1}).$$

Therefore, the formula we seek is $\sinh^{-1} x = \ln(x + \sqrt{x^2 + 1})$.

A similar procedure can be carried out for the other inverse hyperbolic functions (Exercise 110). Theorem 7.17 lists the results of these calculations.

➤ Most calculators allow for the direct evaluation of the hyperbolic sine, cosine, and tangent, along with their inverses, but are not programmed to evaluate $\operatorname{sech}^{-1} x$, $\operatorname{csch}^{-1} x$, and $\coth^{-1} x$. The formulas in Theorem 7.17 are useful for evaluating these functions on a calculator.

---

**THEOREM 7.17**   **Inverses of the Hyperbolic Functions Expressed as Logarithms**

$$\cosh^{-1} x = \ln\left(x + \sqrt{x^2 - 1}\right) \ (x \ge 1) \qquad \operatorname{sech}^{-1} x = \cosh^{-1} \frac{1}{x} \ (0 < x \le 1)$$

$$\sinh^{-1} x = \ln\left(x + \sqrt{x^2 + 1}\right) \qquad\qquad \operatorname{csch}^{-1} x = \sinh^{-1} \frac{1}{x} \ (x \ne 0)$$

$$\tanh^{-1} x = \frac{1}{2} \ln\left(\frac{1 + x}{1 - x}\right) (|x| < 1) \qquad \coth^{-1} x = \tanh^{-1} \frac{1}{x} \ (|x| > 1)$$

---

Notice that the formulas in Theorem 7.17 for the inverse hyperbolic secant, cosecant, and cotangent are given in terms of the inverses of their corresponding reciprocal functions. Justification for these formulas follows from the definitions of the inverse functions. For example, from the definition of $\operatorname{csch}^{-1} x$, we have

$$y = \operatorname{csch}^{-1} x \quad \Leftrightarrow \quad x = \operatorname{csch} y \quad \Leftrightarrow \quad 1/x = \sinh y.$$

Applying the inverse hyperbolic sine to both sides of $1/x = \sinh y$ yields

$$\sinh^{-1}(1/x) = \underbrace{\sinh^{-1}(\sinh y)}_{y} \quad \text{or} \quad y = \operatorname{csch}^{-1} x = \sinh^{-1}(1/x).$$

Similar derivations yield the other two formulas.

**EXAMPLE 4**   **Points of intersection**   Find the points at which the curves $y = \cosh x$ and $y = \frac{5}{3}$ intersect (**Figure 7.63**).

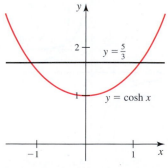

**Figure 7.63**

**QUICK CHECK 5** Use the results of Example 4 to write an integral for the area of the region bounded by $y = \cosh x$ and $y = \frac{5}{3}$ (Figure 7.63) and then evaluate the integral. ◄

**SOLUTION** The $x$-coordinates of the points of intersection satisfy the equation $\cosh x = \frac{5}{3}$, which is solved by applying $\cosh^{-1}$ to both sides of the equation. However, evaluating $\cosh^{-1}(\cosh x)$ requires care—in Exercise 105, you are asked to show that $\cosh^{-1}(\cosh x) = |x|$. With this fact, the points of intersection can be found:

$$\cosh x = \frac{5}{3} \qquad \text{Set equations equal to one another.}$$

$$\cosh^{-1}(\cosh x) = \cosh^{-1}\frac{5}{3} \qquad \text{Apply } \cosh^{-1} \text{ to both sides.}$$

$$|x| = \ln\left(\frac{5}{3} + \sqrt{\left(\frac{5}{3}\right)^2 - 1}\right) \qquad \text{Simplify; use Theorem 7.17.}$$

$$x = \pm \ln 3. \qquad \text{Simplify.}$$

The points of intersection lie on the line $y = \frac{5}{3}$, so the points are $\left(-\ln 3, \frac{5}{3}\right)$ and $\left(\ln 3, \frac{5}{3}\right)$.

*Related Exercises 61–62* ◄

## Derivatives of the Inverse Hyperbolic Functions and Related Integral Formulas

The derivatives of the inverse hyperbolic functions can be computed directly from the logarithmic formulas given in Theorem 7.17. However, it is more efficient to use the definitions in Figure 7.62.

Recall that the inverse hyperbolic sine is defined by

$$y = \sinh^{-1} x \quad \Leftrightarrow \quad x = \sinh y.$$

We differentiate both sides of $x = \sinh y$ with respect to $x$ and solve for $dy/dx$:

$$x = \sinh y \qquad\qquad y = \sinh^{-1} x \;\Leftrightarrow\; x = \sinh y$$

$$1 = (\cosh y)\frac{dy}{dx} \qquad \text{Use implicit differentiation.}$$

$$\frac{dy}{dx} = \frac{1}{\cosh y} \qquad \text{Solve for } dy/dx.$$

$$\frac{dy}{dx} = \frac{1}{\pm\sqrt{\sinh^2 y + 1}} \qquad \cosh^2 y - \sinh^2 y = 1$$

$$\frac{dy}{dx} = \frac{1}{\sqrt{x^2 + 1}}. \qquad x = \sinh y$$

In the last step, the positive root is chosen because $\cosh y > 0$ for all $y$.

The derivatives of the other inverse hyperbolic functions, listed in Theorem 7.18, are derived in a similar way (Exercise 106).

---

**THEOREM 7.18** **Derivatives of the Inverse Hyperbolic Functions**

$$\frac{d}{dx}(\cosh^{-1} x) = \frac{1}{\sqrt{x^2 - 1}} \; (x > 1) \qquad \frac{d}{dx}(\sinh^{-1} x) = \frac{1}{\sqrt{x^2 + 1}}$$

$$\frac{d}{dx}(\tanh^{-1} x) = \frac{1}{1 - x^2} \; (|x| < 1) \qquad \frac{d}{dx}(\coth^{-1} x) = \frac{1}{1 - x^2} \; (|x| > 1)$$

$$\frac{d}{dx}(\operatorname{sech}^{-1} x) = -\frac{1}{x\sqrt{1 - x^2}} \; (0 < x < 1) \qquad \frac{d}{dx}(\operatorname{csch}^{-1} x) = -\frac{1}{|x|\sqrt{1 + x^2}} \; (x \neq 0)$$

---

The restrictions associated with the formulas in Theorem 7.18 are a direct consequence of the domains of the inverse functions (Figure 7.62). Note that the derivative of both $\tanh^{-1} x$ and $\coth^{-1} x$ is $1/(1 - x^2)$, although this result is valid on different domains

$(|x| < 1$ for $\tanh^{-1} x$ and $|x| > 1$ for $\coth^{-1} x$). These facts have a bearing on formula (3) in the next theorem, which is a reversal of the derivative formulas in Theorem 7.18. Here we list integral results, where $a$ is a positive constant; each formula can be verified by differentiation.

> ➤ The integrals in Theorem 7.19 appear again in Chapter 8 in terms of logarithms and with fewer restrictions on the variable of integration.

---

**THEOREM 7.19    Integral Formulas**

1. $\displaystyle\int \frac{dx}{\sqrt{x^2 - a^2}} = \cosh^{-1} \frac{x}{a} + C,$ for $x > a$

2. $\displaystyle\int \frac{dx}{\sqrt{x^2 + a^2}} = \sinh^{-1} \frac{x}{a} + C,$ for all $x$

3. $\displaystyle\int \frac{dx}{a^2 - x^2} = \begin{cases} \dfrac{1}{a} \tanh^{-1} \dfrac{x}{a} + C, & \text{for } |x| < a \\[2mm] \dfrac{1}{a} \coth^{-1} \dfrac{x}{a} + C, & \text{for } |x| > a \end{cases}$

4. $\displaystyle\int \frac{dx}{x\sqrt{a^2 - x^2}} = -\frac{1}{a} \operatorname{sech}^{-1} \frac{x}{a} + C,$ for $0 < x < a$

5. $\displaystyle\int \frac{dx}{x\sqrt{a^2 + x^2}} = -\frac{1}{a} \operatorname{csch}^{-1} \frac{|x|}{a} + C,$ for $x \neq 0$

---

**EXAMPLE 5    Derivatives of inverse hyperbolic functions** Compute $dy/dx$ for each function.

**a.** $y = \tanh^{-1} 3x$     **b.** $y = x^2 \sinh^{-1} x$

**SOLUTION**

**a.** Using the Chain Rule, we have

$$\frac{dy}{dx} = \frac{d}{dx}(\tanh^{-1} 3x) = \frac{1}{1 - (3x)^2} \cdot 3 = \frac{3}{1 - 9x^2}.$$

> ➤ The function $3/(1 - 9x^2)$ in the solution to Example 5a is defined for all $x \neq \pm 1/3$. However, the derivative formula $dy/dx = 3/(1 - 9x^2)$ is valid only on $-1/3 < x < 1/3$ because $\tanh^{-1} 3x$ is defined only on $-1/3 < x < 1/3$. The result of computing $\frac{d}{dx}(\coth^{-1} 3x)$ is the same, but valid on $(-\infty, -1/3) \cup (1/3, \infty)$.

**b.** $\dfrac{dy}{dx} = 2x \sinh^{-1} x + x^2 \cdot \dfrac{1}{\sqrt{x^2 + 1}}$     Product Rule; Theorem 7.18

$\qquad = x\left(\dfrac{2\sqrt{x^2 + 1} \cdot \sinh^{-1} x + x}{\sqrt{x^2 + 1}}\right)$     Simplify.

*Related Exercises 31, 35* ◄

**EXAMPLE 6    Integral computations**

**a.** Compute the area of the region bounded by $y = 1/\sqrt{x^2 + 16}$ over the interval $[0, 3]$.

**b.** Evaluate $\displaystyle\int_9^{25} \frac{dx}{\sqrt{x}(4 - x)}$.

**SOLUTION**

**a.** The region in question is shown in **Figure 7.64**, and its area is given by $\displaystyle\int_0^3 \frac{dx}{\sqrt{x^2 + 16}}$.

Using formula (2) in Theorem 7.19 with $a = 4$, we have

$$\int_0^3 \frac{dx}{\sqrt{x^2 + 16}} = \sinh^{-1} \frac{x}{4}\Big|_0^3 \qquad \text{Theorem 7.19}$$

$$= \sinh^{-1} \frac{3}{4} - \sinh^{-1} 0 \quad \text{Evaluate.}$$

$$= \sinh^{-1} \frac{3}{4}. \qquad \sinh^{-1} 0 = 0$$

**Figure 7.64**

A calculator gives an approximate result of $\sinh^{-1}(3/4) \approx 0.693$. The exact result can be written in terms of logarithms using Theorem 7.17:

$$\sinh^{-1}(3/4) = \ln(3/4 + \sqrt{(3/4)^2 + 1}) = \ln 2.$$

**b.** The integral doesn't match any of the formulas in Theorem 7.19, so we use the substitution $u = \sqrt{x}$:

$$\int_9^{25} \frac{dx}{\sqrt{x}(4 - x)} = 2\int_3^5 \frac{du}{4 - u^2}. \quad \text{Let } u = \sqrt{x}; du = \frac{dx}{2\sqrt{x}}.$$

The new integral now matches formula (3), with $a = 2$. We conclude that

$$2\int_3^5 \frac{du}{4 - u^2} = 2 \cdot \frac{1}{2} \coth^{-1} \frac{u}{2}\Big|_3^5 \qquad \int \frac{dx}{a^2 - x^2} = \frac{1}{a} \coth^{-1} \frac{x}{a} + C$$

$$= \coth^{-1} \frac{5}{2} - \coth^{-1} \frac{3}{2}. \quad \text{Evaluate.}$$

The antiderivative involving $\coth^{-1} x$ was chosen because the interval of integration $(3 \le u \le 5)$ satisfies $|u| > a = 2$. Theorem 7.17 is used to express the result in numerical form in case your calculator cannot evaluate $\coth^{-1} x$:

$$\coth^{-1} \frac{5}{2} - \coth^{-1} \frac{3}{2} = \tanh^{-1} \frac{2}{5} - \tanh^{-1} \frac{2}{3} \approx -0.381.$$

*Related Exercises 47, 63* ◄

**QUICK CHECK 6** Evaluate $\int_0^1 \frac{du}{4 - u^2}$. ◄

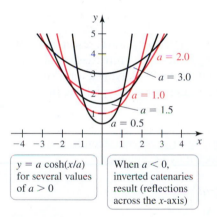

$y = a \cosh(x/a)$ for several values of $a > 0$

When $a < 0$, inverted catenaries result (reflections across the $x$-axis)

**Figure 7.65**

➤ A Tyrolean traverse is used to pass over difficult terrain, such as a chasm between two cliffs or a raging river. A rope is strung between two anchor points, the climber clips onto the rope and then traverses the gap by pulling on the rope.

## Applications of Hyperbolic Functions

This section concludes with a brief look at two applied problems associated with hyperbolic functions. Additional applications are presented in the exercises.

**The Catenary** When a free-hanging rope or flexible cable supporting only its own weight is attached to two points of equal height, it takes the shape of a curve known as a *catenary*. You can see catenaries in telephone wires, ropes strung across chasms for Tyrolean traverses (Example 7), and spider webs.

The equation for a general catenary is $y = a \cosh(x/a)$, where $a \ne 0$ is a real number. When $a < 0$, the curve is called an inverted catenary, sometimes used in the design of arches. **Figure 7.65** illustrates catenaries for several values of $a$.

**EXAMPLE 7** **Length of a catenary** A climber anchors a rope at two points of equal height, separated by a distance of 100 ft, in order to perform a *Tyrolean traverse*. The rope follows the catenary $f(x) = 200 \cosh(x/200)$ over the interval $[-50, 50]$ (**Figure 7.66**). Find the length of the unweighted rope between the two anchor points.

**Figure 7.66**

> Using the principles of vector analysis introduced in Chapter 13, the tension in the rope and the forces acting upon the anchors in Example 7 can be computed. This is crucial information for anyone setting up a Tyrolean traverse; the *sag angle* (Exercise 72) figures into these calculations. Similar calculations are important for catenary lifelines used in construction and for rigging camera shots in Hollywood movies.

**SOLUTION** Recall from Section 6.5 that the arc length of the curve $y = f(x)$ over the interval $[a, b]$ is $L = \int_a^b \sqrt{1 + f'(x)^2}\, dx$. Also note that

$$f'(x) = 200 \sinh\left(\frac{x}{200}\right) \cdot \frac{1}{200} = \sinh\frac{x}{200}.$$

Therefore, the length of the rope is

$$L = \int_{-50}^{50} \sqrt{1 + \sinh^2\left(\frac{x}{200}\right)}\, dx \qquad \text{Arc length formula}$$

$$= 2\int_0^{50} \sqrt{1 + \sinh^2\left(\frac{x}{200}\right)}\, dx \qquad \text{Use symmetry.}$$

$$= 400\int_0^{1/4} \sqrt{1 + \sinh^2 u}\, du \qquad \text{Let } u = \frac{x}{200}.$$

$$= 400\int_0^{1/4} \cosh u\, du \qquad 1 + \sinh^2 u = \cosh^2 u$$

$$= 400 \sinh u\Big|_0^{1/4} \qquad \text{Evaluate integral.}$$

$$= 400\left(\sinh\frac{1}{4} - \sinh 0\right) \qquad \text{Simplify.}$$

$$\approx 101 \text{ ft.} \qquad \text{Evaluate.}$$

*Related Exercises 69–70* ◄

**Velocity of a Wave**    To describe the characteristics of a traveling wave, researchers formulate a *wave equation* that reflects the known (or hypothesized) properties of the wave and that often takes the form of a differential equation (Chapter 9). Solving a wave equation produces additional information about the wave, and it turns out that hyperbolic functions may arise naturally in this context.

**EXAMPLE 8    Velocity of an ocean wave**    The velocity $v$ (in meters/second) of an idealized surface wave traveling on the ocean is modeled by the equation

$$v = \sqrt{\frac{g\lambda}{2\pi} \tanh\left(\frac{2\pi d}{\lambda}\right)},$$

where $g = 9.8 \text{ m/s}^2$ is the acceleration due to gravity, $\lambda$ is the wavelength measured in meters from crest to crest, and $d$ is the depth of the undisturbed water, also measured in meters (**Figure 7.67**).

**Figure 7.67**

> In fluid dynamics, *water depth* is often discussed in terms of the depth-to-wavelength ratio $d/\lambda$, not the actual depth of the water. Three classifications are generally used:

*shallow water*: $d/\lambda < 0.05$

*intermediate depth*: $0.05 < d/\lambda < 0.5$

*deep water*: $d/\lambda > 0.5$

**a.** A sea kayaker observes several waves that pass beneath her kayak, and she estimates that $\lambda = 12$ m and $v = 4$ m/s. How deep is the water in which she is kayaking?

**b.** The *deep-water* equation for wave velocity is $v = \sqrt{\dfrac{g\lambda}{2\pi}}$, which is an approximation to the velocity formula given above. Waves are said to be in deep water if the depth-to-wavelength ratio $d/\lambda$ is greater than $\dfrac{1}{2}$. Explain why $v = \sqrt{\dfrac{g\lambda}{2\pi}}$ is a good approximation when $d/\lambda > \dfrac{1}{2}$.

**SOLUTION**

**a.** We substitute $\lambda = 12$ and $v = 4$ into the velocity equation and solve for $d$.

$$4 = \sqrt{\frac{g \cdot 12}{2\pi} \tanh\left(\frac{2\pi d}{12}\right)} \;\Rightarrow\; 16 = \frac{6g}{\pi} \tanh\left(\frac{\pi d}{6}\right) \qquad \text{Square both sides.}$$

$$\Rightarrow\; \frac{8\pi}{3g} = \tanh\left(\frac{\pi d}{6}\right) \qquad \text{Multiply by } \frac{\pi}{6g}.$$

In order to extract $d$ from the argument of tanh, we apply $\tanh^{-1}$ to both sides of the equation and then use the property $\tanh^{-1}(\tanh x) = x$, for all $x$.

$$\tanh^{-1}\left(\frac{8\pi}{3g}\right) = \tanh^{-1}\left(\tanh\left(\frac{\pi d}{6}\right)\right) \qquad \text{Apply } \tanh^{-1} \text{ to both sides.}$$

$$\tanh^{-1}\left(\frac{8\pi}{29.4}\right) = \frac{\pi d}{6} \qquad \text{Simplify; } 3g = 29.4.$$

$$d = \frac{6}{\pi}\tanh^{-1}\left(\frac{8\pi}{29.4}\right) \approx 2.4 \text{ m} \qquad \text{Solve for } d.$$

Therefore, the kayaker is in water that is about 2.4 m deep.

**b.** Recall that $y = \tanh x$ is an increasing function ($dy/dx = \text{sech}^2 x > 0$) whose values approach 1 as $x \to \infty$ (**Figure 7.68**). Also notice that when $\dfrac{d}{\lambda} = \dfrac{1}{2}$,

$$\tanh\left(\frac{2\pi d}{\lambda}\right) = \tanh \pi \approx 0.996, \text{ which is nearly equal to 1. These facts imply that}$$

whenever $\dfrac{d}{\lambda} > \dfrac{1}{2}$, we can replace $\tanh\left(\dfrac{2\pi d}{\lambda}\right)$ with 1 in the velocity formula, resulting

in the deep-water velocity function $v = \sqrt{\dfrac{g\lambda}{2\pi}}$.

*Related Exercises 73–74* ◄

$\tanh x \to 1$ as $x \to \infty$

$y = \tanh x$

**Figure 7.68**

**QUICK CHECK 7** Explain why longer waves travel faster than shorter waves in deep water. ◄

---

# SECTION 7.7 EXERCISES

## Getting Started

1. State the definitions of the hyperbolic cosine and hyperbolic sine functions.

2. Sketch the graphs of $y = \cosh x$, $y = \sinh x$, and $y = \tanh x$ (include asymptotes), and state whether each function is even, odd, or neither.

3. What is the fundamental identity for hyperbolic functions?

4. How are the derivative formulas for the hyperbolic functions and the trigonometric functions alike? How are they different?

5. Express $\sinh^{-1} x$ in terms of logarithms.

6. What is the domain of $\text{sech}^{-1} x$? How is $\text{sech}^{-1} x$ defined in terms of the inverse hyperbolic cosine?

7. A calculator has a built-in $\sinh^{-1} x$ function, but no $\text{csch}^{-1} x$ function. How do you evaluate $\text{csch}^{-1} 5$ on such a calculator?

8. On what interval is the formula $\dfrac{d}{dx}(\tanh^{-1} x) = \dfrac{1}{1-x^2}$ valid?

9. When evaluating the definite integral $\displaystyle\int_6^8 \frac{dx}{16-x^2}$, why must you choose the antiderivative $\dfrac{1}{4}\coth^{-1}\dfrac{x}{4}$ rather than $\dfrac{1}{4}\tanh^{-1}\dfrac{x}{4}$?

10. How does the graph of the catenary $y = a\cosh\dfrac{x}{a}$ change as $a > 0$ increases?

## Practice Exercises

**11–15. Identities** *Prove each identity using the definitions of the hyperbolic functions.*

11. $\tanh x = \dfrac{e^{2x}-1}{e^{2x}+1}$

12. $\tanh(-x) = -\tanh x$

13. $\cosh 2x = \cosh^2 x + \sinh^2 x$ (*Hint:* Begin with the right side of the equation.)

14. $2\sinh(\ln(\sec x)) = \sin x \tan x$

15. $\cosh x + \sinh x = e^x$

**16–18. Identities** *Use the given identity to prove the related identity.*

16. Use the fundamental identity $\cosh^2 x - \sinh^2 x = 1$ to prove the identity $\coth^2 x - 1 = \text{csch}^2 x$.

17. Use the identity $\cosh 2x = \cosh^2 x + \sinh^2 x$ to prove the identities $\cosh^2 x = \dfrac{\cosh 2x + 1}{2}$ and $\sinh^2 x = \dfrac{\cosh 2x - 1}{2}$.

18. Use the identity $\cosh(x+y) = \cosh x \cosh y + \sinh x \sinh y$ to prove the identity $\cosh 2x = \cosh^2 x + \sinh^2 x$.

**19–21. Derivative formulas** *Derive the following derivative formulas given that $\dfrac{d}{dx}(\cosh x) = \sinh x$ and $\dfrac{d}{dx}(\sinh x) = \cosh x$.*

19. $\dfrac{d}{dx}(\coth x) = -\text{csch}^2 x$

20. $\dfrac{d}{dx}(\text{sech } x) = -\text{sech } x \tanh x$

21. $\dfrac{d}{dx}(\text{csch } x) = -\text{csch } x \coth x$

**22–36. Derivatives** *Find the derivatives of the following functions.*

22. $f(x) = \sinh 4x$

23. $f(x) = \cosh^2 x$

24. $f(x) = -\sinh^3 4x$

25. $f(x) = \tanh^2 x$

26. $f(x) = \sqrt{\coth 3x}$

27. $f(x) = \ln \text{sech } 2x$

28. $f(x) = x \tanh x$

29. $f(x) = x^2 \cosh^2 3x$

30. $f(x) = \dfrac{x}{\text{csch } x}$

31. $f(x) = \cosh^{-1} 4x$

**32.** $f(t) = 2 \tanh^{-1} \sqrt{t}$     **33.** $f(v) = \sinh^{-1} v^2$

**34.** $f(x) = \operatorname{csch}^{-1} \dfrac{2}{x}$     **35.** $f(x) = x \sinh^{-1} x - \sqrt{x^2 + 1}$

**36.** $f(u) = \sinh^{-1}(\tan u)$

**37–56. Integrals** *Evaluate each integral.*

**37.** $\displaystyle\int \cosh 2x \, dx$     **38.** $\displaystyle\int \operatorname{sech}^2 w \tanh w \, dw$

**39.** $\displaystyle\int \frac{\sinh x}{1 + \cosh x} \, dx$     **40.** $\displaystyle\int \coth^2 x \operatorname{csch}^2 x \, dx$

**41.** $\displaystyle\int \tanh^2 x \, dx$ *(Hint: Use an identity.)*

**42.** $\displaystyle\int \sinh^2 z \, dz$ *(Hint: Use an identity.)*

**43.** $\displaystyle\int_0^1 \cosh^3 3x \sinh 3x \, dx$     **44.** $\displaystyle\int_0^4 \frac{\operatorname{sech}^2 \sqrt{x}}{\sqrt{x}} \, dx$

**45.** $\displaystyle\int_0^{\ln 2} \tanh x \, dx$     **46.** $\displaystyle\int_{\ln 2}^{\ln 3} \operatorname{csch} y \, dy$

**47.** $\displaystyle\int \frac{dx}{8 - x^2}, x > 2\sqrt{2}$     **48.** $\displaystyle\int \frac{dx}{\sqrt{x^2 - 16}}, x > 4$

**49.** $\displaystyle\int \frac{e^x}{36 - e^{2x}} \, dx, x < \ln 6$     **50.** $\displaystyle\int \frac{dx}{x\sqrt{16 + x^2}}$

**51.** $\displaystyle\int \frac{dx}{x\sqrt{4 - x^8}}$     **52.** $\displaystyle\int \frac{dx}{x\sqrt{1 + x^4}}$

**53.** $\displaystyle\int \frac{\cosh z}{\sinh^2 z} \, dz$     **54.** $\displaystyle\int \frac{\cos \theta}{9 - \sin^2 \theta} \, d\theta$

**55.** $\displaystyle\int_{5/12}^{3/4} \frac{\sinh^{-1} x}{\sqrt{x^2 + 1}} \, dx$

**56.** $\displaystyle\int_{25}^{225} \frac{dx}{\sqrt{x^2 + 25x}}$ *(Hint:* $\sqrt{x^2 + 25x} = \sqrt{x}\sqrt{x + 25}$.*)*

**57–58. Two ways** *Evaluate the following integrals two ways.*

**a.** *Simplify the integrand first and then integrate.*

**b.** *Change variables (let $u = \ln x$), integrate, and then simplify your answer. Verify that both methods give the same answer.*

**57.** $\displaystyle\int \frac{\sinh (\ln x)}{x} \, dx$     **58.** $\displaystyle\int_1^{\sqrt{3}} \frac{\operatorname{sech} (\ln x)}{x} \, dx$

**T 59. Visual approximation**

    **a.** Use a graphing utility to sketch the graph of $y = \coth x$ and then explain why $\int_5^{10} \coth x \, dx \approx 5$.

    **b.** Evaluate $\int_5^{10} \coth x \, dx$ analytically and use a calculator to arrive at a decimal approximation to the answer. How large is the error in the approximation in part (a)?

**60. Integral proof** Prove the formula $\int \coth x \, dx = \ln |\sinh x| + C$ of Theorem 7.16.

**61–62. Points of intersection and area**

**a.** *Sketch the graphs of the functions $f$ and $g$ and find the $x$-coordinate of the points at which they intersect.*

**b.** *Compute the area of the region described.*

**61.** $f(x) = \operatorname{sech} x, g(x) = \tanh x$; the region bounded by the graphs of $f$, $g$, and the $y$-axis

**62.** $f(x) = \sinh x, g(x) = \tanh x$; the region bounded by the graphs of $f$, $g$, and $x = \ln 3$

**63–68. Definite integrals** *Evaluate the following definite integrals. Use Theorem 7.17 to express your answer in terms of logarithms.*

**63.** $\displaystyle\int_1^{e^2} \frac{dx}{x\sqrt{\ln^2 x + 1}}$     **64.** $\displaystyle\int_5^{3\sqrt{5}} \frac{dx}{\sqrt{x^2 - 9}}$

**65.** $\displaystyle\int_{-2}^2 \frac{dt}{t^2 - 9}$     **66.** $\displaystyle\int_{1/6}^{1/4} \frac{dt}{t\sqrt{1 - 4t^2}}$

**67.** $\displaystyle\int_{1/8}^1 \frac{dx}{x\sqrt{1 + x^{2/3}}}$     **68.** $\displaystyle\int_{\ln 5}^{\ln 9} \frac{\cosh x}{4 - \sinh^2 x} \, dx$

**69. Catenary arch** The portion of the curve $y = \dfrac{17}{15} - \cosh x$ that lies above the $x$-axis forms a catenary arch. Find the average height of the arch above the $x$-axis.

**70. Length of a catenary** Show that the arc length of the catenary $y = \cosh x$ over the interval $[0, a]$ is $L = \sinh a$.

**T 71. Power lines** A power line is attached at the same height to two utility poles that are separated by a distance of 100 ft; the power line follows the curve $f(x) = a \cosh \dfrac{x}{a}$. Use the following steps to find the value of $a$ that produces a sag of 10 ft midway between the poles. Use a coordinate system that places the poles at $x = \pm 50$.

    **a.** Show that $a$ satisfies the equation $\cosh \dfrac{50}{a} - 1 = \dfrac{10}{a}$.

    **b.** Let $t = \dfrac{10}{a}$, confirm that the equation in part (a) reduces to $\cosh 5t - 1 = t$, and solve for $t$ using a graphing utility. Report your answer accurate to two decimal places.

    **c.** Use your answer in part (b) to find $a$, and then compute the length of the power line.

**72. Sag angle** Imagine a climber clipping onto the rope described in Example 7 and pulling himself to the rope's midpoint. Because the rope is supporting the weight of the climber, it no longer takes the shape of the catenary $y = 200 \cosh \dfrac{x}{200}$. Instead, the rope (nearly) forms two sides of an isosceles triangle. Compute the *sag angle* $\theta$ illustrated in the figure, assuming the rope does not stretch when weighted. Recall from Example 7 that the length of the rope is 101 ft.

**T 73. Wavelength** The velocity of a surface wave on the ocean is given by $v = \sqrt{\dfrac{g\lambda}{2\pi} \tanh\left(\dfrac{2\pi d}{\lambda}\right)}$ (Example 8). Use a graphing utility or root finder to approximate the wavelength $\lambda$ of an ocean wave traveling at $v = 7$ m/s in water that is $d = 10$ m deep.

**74. Wave velocity** Use Exercise 73 to do the following calculations.

   **a.** Find the velocity of a wave where $\lambda = 50$ m and $d = 20$ m.

   **b.** Determine the depth of the water if a wave with $\lambda = 15$ m is traveling at $v = 4.5$ m/s.

**75. Shallow-water velocity equation**

   **a.** Confirm that the linear approximation to $f(x) = \tanh x$ at $a = 0$ is $L(x) = x$.

   **b.** Recall that the velocity of a surface wave on the ocean is $v = \sqrt{\dfrac{g\lambda}{2\pi} \tanh\left(\dfrac{2\pi d}{\lambda}\right)}$. In fluid dynamics, *shallow water* refers to water where the depth-to-wavelength ratio $\dfrac{d}{\lambda} < 0.05$.

Use your answer to part (a) to explain why the approximate shallow-water velocity equation is $v = \sqrt{gd}$.

   **c.** Use the shallow-water velocity equation to explain why waves tend to slow down as they approach the shore.

**76. Tsunamis** A tsunami is an ocean wave often caused by earthquakes on the ocean floor; these waves typically have long wavelengths, ranging from 150 to 1000 km. Imagine a tsunami traveling across the Pacific Ocean, which is the deepest ocean in the world, with an average depth of about 4000 m. Explain why the *shallow-water velocity equation* (Exercise 75) applies to tsunamis even though the actual depth of the water is large. What does the shallow-water equation say about the speed of a tsunami in the Pacific Ocean (use $d = 4000$ m)?

**77. Explain why or why not** Determine whether the following statements are true and give an explanation or counterexample.

   **a.** $\dfrac{d}{dx}(\sinh(\ln 3)) = \dfrac{\cosh(\ln 3)}{3}.$

   **b.** $\dfrac{d}{dx}(\sinh x) = \cosh x$ and $\dfrac{d}{dx}(\cosh x) = -\sinh x.$

   **c.** $\ln(1 + \sqrt{2}) = -\ln(-1 + \sqrt{2}).$

   **d.** $\displaystyle\int_0^1 \dfrac{dx}{4 - x^2} = \dfrac{1}{2}\left(\coth^{-1}\dfrac{1}{2} - \coth^{-1}0\right).$

**T 78. Evaluating hyperbolic functions** Use a calculator to evaluate each expression or state that the value does not exist. Report answers accurate to four decimal places to the right of the decimal point.

   **a.** $\coth 4$              **b.** $\tanh^{-1} 2$

   **c.** $\operatorname{csch}^{-1} 5$         **d.** $\operatorname{csch} x \Big|_{1/2}^{2}$

   **e.** $\ln\left|\tanh\dfrac{x}{2}\right|\Big|_1^{10}$     **f.** $\tan^{-1}(\sinh x)\Big|_{-3}^{3}$

   **g.** $\dfrac{1}{4}\coth^{-1}\dfrac{x}{4}\Big|_{20}^{36}$

**79. Evaluating hyperbolic functions** Evaluate each expression without using a calculator or state that the value does not exist. Simplify answers as much as possible.

   **a.** $\cosh 0$            **b.** $\tanh 0$

   **c.** $\operatorname{csch} 0$          **d.** $\operatorname{sech}(\sinh 0)$

   **e.** $\coth(\ln 5)$        **f.** $\sinh(2\ln 3)$

   **g.** $\cosh^2 1$          **h.** $\operatorname{sech}^{-1}(\ln 3)$

   **i.** $\cosh^{-1}\dfrac{17}{8}$       **j.** $\sinh^{-1}\dfrac{e^2 - 1}{2e}$

**80. Confirming a graph** The graph of $f(x) = \sinh x$ is shown in Figure 7.60. Use calculus to find the intervals on which $f$ is increasing or decreasing, and to find the intervals on which $f$ is concave up or concave down to confirm that the graph is correct.

**81. Critical points** Find the critical points of the function $f(x) = \sinh^2 x \cosh x$.

**T 82. Critical points**

   **a.** Show that the critical points of $f(x) = \dfrac{\cosh x}{x}$ satisfy $x = \coth x$.

   **b.** Use a root finder to approximate the critical points of $f$.

**83. Points of inflection** Find the $x$-coordinate of the point(s) of inflection of $f(x) = \tanh^2 x$.

**84. Points of inflection** Find the $x$-coordinate of the point(s) of inflection of $f(x) = \operatorname{sech} x$. Report exact answers in terms of logarithms (use Theorem 7.17).

**85. Area of region** Find the area of the region bounded by $y = \operatorname{sech} x$, $x = 1$, and the unit circle (see figure).

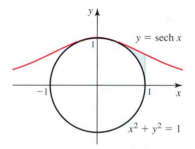

**86. Solid of revolution** Compute the volume of the solid of revolution that results when the region in Exercise 85 is revolved about the $x$-axis.

**87. L'Hôpital loophole** Explain why l'Hôpital's Rule fails when applied to the limit $\displaystyle\lim_{x\to\infty} \dfrac{\sinh x}{\cosh x}$ and then find the limit another way.

**88–91. Limits** *Use l'Hôpital's Rule to evaluate the following limits.*

**88.** $\displaystyle\lim_{x\to\infty} \dfrac{1 - \coth x}{1 - \tanh x}$      **89.** $\displaystyle\lim_{x\to 0} \dfrac{\tanh^{-1} x}{\tan(\pi x/2)}$

**90.** $\displaystyle\lim_{x\to 1^-} \dfrac{\tanh^{-1} x}{\tan(\pi x/2)}$    **91.** $\displaystyle\lim_{x\to 0^+} (\tanh x)^x$

**T 92. Slant asymptote** The linear function $\ell(x) = mx + b$, for finite $m \neq 0$, is a slant asymptote of $f(x)$ if $\displaystyle\lim_{x\to\infty}(f(x) - \ell(x)) = 0$.

   **a.** Use a graphing utility to make a sketch that shows $\ell(x) = x$ is a slant asymptote of $f(x) = x \tanh x$. Does $f$ have any other slant asymptotes?

   **b.** Provide an intuitive argument showing that $f(x) = x \tanh x$ behaves like $\ell(x) = x$ as $x$ gets large.

   **c.** Prove that $\ell(x) = x$ is a slant asymptote of $f$ by confirming $\displaystyle\lim_{x\to\infty}(x \tanh x - x) = 0$.

## Explorations and Challenges

**93. Kiln design** Find the volume interior to the inverted catenary kiln (an oven used to fire pottery) shown in the figure.

$y = 3 - \cosh x$

(0, 0)

6

**T 94. Newton's method** Use Newton's method to find all local extreme values of $f(x) = x \operatorname{sech} x$.

**95. Falling body** When an object falling from rest encounters air resistance proportional to the square of its velocity, the distance it falls (in meters) after $t$ seconds is given by

$$d(t) = \frac{m}{k} \ln\left(\cosh\left(\sqrt{\frac{kg}{m}}\, t\right)\right),$$ where $m$ is the mass of the object

in kilograms, $g = 9.8 \text{ m/s}^2$ is the acceleration due to gravity, and $k$ is a physical constant.

   **a.** A BASE jumper ($m = 75$ kg) leaps from a tall cliff and performs a ten-second delay (she free-falls for 10 s and then opens her chute). How far does she fall in 10 s? Assume $k = 0.2$.

   **b.** How long does it take for her to fall the first 100 m? The second 100 m? What is her average velocity over each of these intervals?

**96. Velocity of falling body** Refer to Exercise 95, which gives the position function for a falling body. Use $m = 75$ kg and $k = 0.2$.

   **a.** Confirm that the BASE jumper's velocity $t$ seconds after jumping is $v(t) = d'(t) = \sqrt{\dfrac{mg}{k}} \tanh\left(\sqrt{\dfrac{kg}{m}}\, t\right)$.

   **b.** How fast is the BASE jumper falling ten seconds after jumping?

   **c.** How long does it take for the BASE jumper to reach a speed of 45 m/s (roughly 100 mi/hr)?

**97. Terminal velocity** Refer to Exercises 95 and 96.

   **a.** Compute a jumper's *terminal velocity*, which is defined as

$$\lim_{t\to\infty} v(t) = \lim_{t\to\infty} \sqrt{\frac{mg}{k}} \tanh\left(\sqrt{\frac{kg}{m}}\, t\right).$$

   **b.** Find the terminal velocity for the jumper in Exercise 96 ($m = 75$ kg and $k = 0.2$).

   **c.** How long does it take any falling object to reach a speed equal to 95% of its terminal velocity? Leave your answer in terms of $k$, $g$, and $m$.

   **d.** How tall must a cliff be so that the BASE jumper ($m = 75$ kg and $k = 0.2$) reaches 95% of terminal velocity? Assume the jumper needs at least 300 m at the end of free fall to deploy the chute and land safely.

**98. Acceleration of a falling body**

   **a.** Find the acceleration $a(t) = v'(t)$ of a falling body whose velocity is given in part (a) of Exercise 96.

   **b.** Compute $\lim_{t\to\infty} a(t)$. Explain your answer as it relates to terminal velocity (Exercise 97).

**99. Differential equations** Hyperbolic functions are useful in solving differential equations (Chapter 9). Show that the functions $y = A \sinh kx$ and $y = B \cosh kx$, where $A$, $B$, and $k$ are constants, satisfy the equation $y''(x) - k^2 y(x) = 0$.

**100. Surface area of a catenoid** When the catenary $y = a \cosh \dfrac{x}{a}$ is revolved about the $x$-axis, it sweeps out a surface of revolution called a *catenoid*. Find the area of the surface generated when $y = \cosh x$ on $[-\ln 2, \ln 2]$ is rotated about the $x$-axis.

**101–104. Proving identities** *Prove the following identities.*

**101.** $\sinh(\cosh^{-1} x) = \sqrt{x^2 - 1}$, for $x \geq 1$

**102.** $\cosh(\sinh^{-1} x) = \sqrt{x^2 + 1}$, for all $x$

**103.** $\cosh(x + y) = \cosh x \cosh y + \sinh x \sinh y$

**104.** $\sinh(x + y) = \sinh x \cosh y + \cosh x \sinh y$

**105. Inverse identity** Show that $\cosh^{-1}(\cosh x) = |x|$ by using the formula $\cosh^{-1} t = \ln\left(t + \sqrt{t^2 - 1}\right)$ and considering the cases $x \geq 0$ and $x < 0$.

**106. Theorem 7.18**

   **a.** The definition of the inverse hyperbolic cosine is $y = \cosh^{-1} x \Leftrightarrow x = \cosh y$, for $x \geq 1, 0 \leq y < \infty$. Use implicit differentiation to show that $\dfrac{d}{dx}(\cosh^{-1} x) = \dfrac{1}{\sqrt{x^2 - 1}}$.

   **b.** Differentiate $\sinh^{-1} x = \ln\left(x + \sqrt{x^2 + 1}\right)$ to show that $\dfrac{d}{dx}(\sinh^{-1} x) = \dfrac{1}{\sqrt{x^2 + 1}}$.

**107. Many formulas** There are several ways to express the indefinite integral of $\operatorname{sech} x$.

   **a.** Show that $\int \operatorname{sech} x \, dx = \tan^{-1}(\sinh x) + C$ (Theorem 7.16). (*Hint:* Write $\operatorname{sech} x = \dfrac{1}{\cosh x} = \dfrac{\cosh x}{\cosh^2 x} = \dfrac{\cosh x}{1 + \sinh^2 x}$ and then make a change of variables.)

   **b.** Show that $\int \operatorname{sech} x \, dx = \sin^{-1}(\tanh x) + C$. (*Hint:* Show that $\operatorname{sech} x = \dfrac{\operatorname{sech}^2 x}{\sqrt{1 - \tanh^2 x}}$ and then make a change of variables.)

   **c.** Verify that $\int \operatorname{sech} x \, dx = 2 \tan^{-1} e^x + C$ by proving $\dfrac{d}{dx}(2 \tan^{-1} e^x) = \operatorname{sech} x$.

**108. Integral formula** Carry out the following steps to derive the formula $\int \operatorname{csch} x \, dx = \ln\left|\tanh \dfrac{x}{2}\right| + C$ (Theorem 7.16).

   **a.** Change variables with the substitution $u = \dfrac{x}{2}$ to show that $\int \operatorname{csch} x \, dx = \int \dfrac{2 \, du}{\sinh 2u}$.

   **b.** Use the identity for $\sinh 2u$ to show that $\dfrac{2}{\sinh 2u} = \dfrac{\operatorname{sech}^2 u}{\tanh u}$.

   **c.** Change variables again to determine $\int \dfrac{\operatorname{sech}^2 u}{\tanh u} \, du$, and then express your answer in terms of $x$.

**109. Arc length** Use the result of Exercise 108 to find the arc length of the curve $f(x) = \ln\left|\tanh \dfrac{x}{2}\right|$ on $[\ln 2, \ln 8]$.

**110. Inverse hyperbolic tangent** Recall that the inverse hyperbolic tangent is defined as $y = \tanh^{-1} x \Leftrightarrow x = \tanh y$, for $-1 < x < 1$ and all real $y$. Solve $x = \tanh y$ for $y$ to express the formula for $\tanh^{-1} x$ in terms of logarithms.

**111. Integral family** Use the substitution $u = x^r$ to show that
$$\int \frac{dx}{x\sqrt{1-x^{2r}}} = -\frac{1}{r}\operatorname{sech}^{-1} x^r + C, \text{ for } r > 0 \text{ and } 0 < x < 1.$$

**112. Definitions of hyperbolic sine and cosine** Complete the following steps to prove that when the $x$- and $y$-coordinates of a point on the hyperbola $x^2 - y^2 = 1$ are defined as $\cosh t$ and $\sinh t$, respectively, where $t$ is twice the area of the shaded region in the figure, $x$ and $y$ can be expressed as
$$x = \cosh t = \frac{e^t + e^{-t}}{2} \quad \text{and} \quad y = \sinh t = \frac{e^t - e^{-t}}{2}.$$

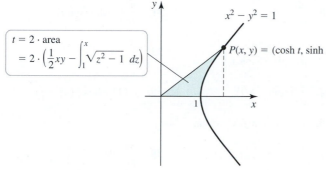

**a.** Explain why twice the area of the shaded region is given by
$$t = 2 \cdot \left(\frac{1}{2}xy - \int_1^x \sqrt{z^2 - 1}\, dz\right)$$
$$= x\sqrt{x^2 - 1} - 2\int_1^x \sqrt{z^2 - 1}\, dz.$$

**b.** In Chapter 8, the formula for the integral in part (a) is derived:
$$\int \sqrt{z^2 - 1}\, dz = \frac{z}{2}\sqrt{z^2 - 1} - \frac{1}{2}\ln\left|z + \sqrt{z^2 - 1}\right| + C.$$

Evaluate this integral on the interval $[1, x]$, explain why the absolute value can be dropped, and combine the result with part (a) to show that
$$t = \ln\left(x + \sqrt{x^2 - 1}\right).$$

**c.** Solve the final equation from part (b) for $x$ to show that
$$x = \frac{e^t + e^{-t}}{2}.$$

**d.** Use the fact that $y = \sqrt{x^2 - 1}$, in combination with part (c), to show that $y = \dfrac{e^t - e^{-t}}{2}$.

**QUICK CHECK ANSWERS**

**1.** $\sinh(-x) = \dfrac{e^{-x} - e^{-(-x)}}{2} = -\dfrac{e^x - e^{-x}}{2} = -\sinh x$

**2.** Because $\tanh x = \dfrac{e^x - e^{-x}}{e^x + e^{-x}}$ and $\lim\limits_{x \to \infty} e^{-x} = 0$, $\tanh x \approx \dfrac{e^x}{e^x} = 1$ for large $x$, which implies that $y = 1$ is a horizontal asymptote. A similar argument shows that $\tanh x \to -1$ as $x \to -\infty$, which means that $y = -1$ is also a horizontal asymptote. **3.** $\dfrac{d}{dx}(4 \cosh 2x) = 8 \sinh 2x$; $\int 4 \cosh 2x\, dx = 2 \sinh 2x + C$ **4.** $\dfrac{1}{2}\ln|\tanh x| + C$

**5.** Area $= 2\displaystyle\int_0^{\ln 3}\left(\frac{5}{3} - \cosh x\right) dx$

$$= \frac{2}{3}(5 \ln 3 - 4) \approx 0.995$$

**6.** $\displaystyle\int_0^1 \frac{du}{4 - u^2} = \frac{1}{2}\tanh^{-1}\frac{1}{2} \approx 0.275$ **7.** The deep-water velocity formula is $v = \sqrt{\dfrac{g\lambda}{2\pi}}$, which is an increasing function of the wavelength $\lambda$. Therefore, larger values of $\lambda$ correspond to faster waves. ◀

# CHAPTER 7 REVIEW EXERCISES

**1. Explain why or why not** Determine whether the following statements are true and give an explanation or counterexample.

**a.** If $f(x) = 2/x$, then $f^{-1}(x) = 2/x$.

**b.** $\ln xy = (\ln x)(\ln y)$.

**c.** The slope of a line tangent to $f(x) = e^x$ is never 0.

**d.** The function $y = Ae^{0.1t}$ increases by 10% when $t$ increases by one unit.

**e.** $\dfrac{d}{dx}(b^x) = b^x$ for exactly one positive value of $b$.

**f.** $\lim\limits_{x \to 0^+} \sin x^{\ln x}$ is an indeterminate form.

**g.** The domain of $\tan^{-1} x$ is $\{x : |x| < \pi/2\}$.

**h.** $\pi^{3x} = e^{\pi \ln 3x}$.

**i.** $\sinh(\ln x) = \dfrac{x^2 - 1}{2x}$, for $x > 0$.

**2–3. Properties of logarithms and exponentials** *Use properties of logarithms and exponentials, not a calculator, for the following exercises.*

**2.** Solve the equation $48 = 6e^{4k}$ for $k$.

**3.** Solve the equation $\log_{10} x^2 + 3 \log_{10} x = \log_{10} 32$ for $x$. Does the answer depend on the base of the log in the equation?

**4.   Graphs of logarithmic and exponential functions** The figure shows the graphs of $y = 2^x$, $y = 3^{-x}$, and $y = -\ln x$. Match each curve with the correct function.

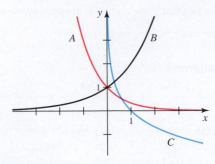

**T 5–6. Existence of inverses** *Determine the largest intervals on which the following functions have an inverse.*

**5.**   $f(x) = x^3 - 3x^2$        **6.**   $g(t) = 2 \sin (t/3)$

**T 7–8. Finding inverses** *Find the inverse on the specified interval and express it in the form $y = f^{-1}(x)$. Then graph $f$ and $f^{-1}$.*

**7.**   $f(x) = x^2 - 4x + 5$, for $x \geq 2$

**8.**   $f(x) = 1/x^2$, for $x > 0$

**9–14. Finding inverse functions** *Find the inverse $f^{-1}(x)$ of each function (on the given interval, if specified).*

**9.**   $f(x) = \dfrac{5 - 2x}{3}$        **10.**   $f(x) = \ln (x^2 + 1)$, for $x \geq 0$

**11.**   $f(x) = 10^{-2x}$        **12.**   $f(x) = \log_{10} 4x$

**13.**   $f(x) = e^x/(e^x + 2)$        **14.**   $f(x) = x/(x - 2)$, for $x > 2$

**15–17. Derivative of the inverse at a point** *Consider the following functions. In each case, without finding the inverse, evaluate the derivative of the inverse at the given point.*

**15.**   $f(x) = \cos x$, $0 \leq x \leq \pi$, at $f(\pi/4)$

**16.**   $f(x) = 1/(x + 1)$, $x > -1$, at $f(0)$

**17.**   $f(x) = \sqrt{x^3 + x - 1}$ at $y = 3$

**T 18.   A function and its inverse function** The function $f(x) = \dfrac{x}{x + 1}$

is one-to-one for $x > -1$ and has an inverse on that interval.

**a.** Graph $f$, for $x > -1$.
**b.** Find the inverse function $f^{-1}$ corresponding to the function graphed in part (a). Graph $f^{-1}$ on the same set of axes as in part (a).
**c.** Evaluate the derivative of $f^{-1}$ at the point $\left(\frac{1}{2}, 1\right)$.
**d.** Sketch the tangent lines on the graphs of $f$ and $f^{-1}$ at $\left(1, \frac{1}{2}\right)$ and $\left(\frac{1}{2}, 1\right)$, respectively.

**19–22. Derivatives of inverse functions** *Consider the following functions (on the given interval, if specified). Find the inverse function, express it as a function of $x$, and find the derivative of the inverse function.*

**19.**   $f(x) = 6 - 3x$        **20.**   $f(x) = 3^{2x} - 1$

**21.**   $f(x) = \sqrt{4x - 1}$        **22.**   $f(x) = x^2 + 2$, $x \leq 0$

**23.   Derivative of the inverse in two ways** Let $f(x) = \sin x$, $f^{-1}(x) = \sin^{-1} x$, and $(x_0, y_0) = (\pi/4, 1/\sqrt{2})$.

**a.** Evaluate $(f^{-1})'(1/\sqrt{2})$ using Theorem 7.3.
**b.** Evaluate $(f^{-1})'(1/\sqrt{2})$ directly by differentiating $f^{-1}$. Check for agreement with part (a).

**24–40. Evaluating derivatives** *Evaluate and simplify the following derivatives.*

**24.**   $\dfrac{d}{dx}\left(xe^{-10x}\right)$        **25.**   $\dfrac{d}{dx}\left(x \ln^2 x\right)$

**26.**   $\dfrac{d}{dw}\left(e^{-w} \ln w\right)$        **27.**   $\dfrac{d}{dx}\left(2^{x^2 - x}\right)$

**28.**   $\dfrac{d}{dx}\left(\log_3 (x + 8)\right)$        **29.**   $\dfrac{d}{dx}\left(\sin^{-1} \dfrac{1}{x}\right)$

**30.**   $\dfrac{d}{dx}\left(x^{\sin 5x}\right)$        **31.**   $\dfrac{d}{dx}\left(\cosh t \sinh t\right)$

**32.**   $\dfrac{d}{dx}\left(\dfrac{\sinh x}{1 + \sinh x}\right)$        **33.**   $\dfrac{d}{dx}\left(\sin^{-1} e^x\right)$

**34.**   $\dfrac{d}{dt}\left(\sinh^{-1} \sqrt{t}\right)$        **35.**   $f'(1)$ when $f(x) = x^{1/x}$

**36.**   $f'(1)$ when $f(x) = \tan^{-1}(4x^2)$

**37.**   $\dfrac{d}{dx}\left(x \sec^{-1} x\right)\Big|_{x = \frac{2}{\sqrt{3}}}$        **38.**   $\dfrac{d}{dx}\left(\tan^{-1} e^{-x}\right)\Big|_{x = 0}$

**39.**   $\dfrac{d}{dx}\left(x^2 \sinh x\right)$        **40.**   $\dfrac{d}{dx}\left(\cosh x^2\right)$

**41–46. Inverse sines and cosines** *Evaluate or simplify the following expressions.*

**41.**   $\cos^{-1} \dfrac{\sqrt{3}}{2}$        **42.**   $\cos^{-1}\left(-\dfrac{\sqrt{3}}{2}\right)$        **43.**   $\sin^{-1}\left(\sin\left(-\dfrac{5\pi}{6}\right)\right)$

**44.**   $\cos^{-1}\left(\cos \dfrac{5\pi}{4}\right)$        **45.**   $\sin (\sin^{-1} x)$        **46.**   $\cos^{-1}(\sin 3\pi)$

**47.   Right triangles** Given that $\theta = \sin^{-1} \dfrac{12}{13}$, evaluate $\cos \theta$, $\tan \theta$, $\cot \theta$, $\sec \theta$, and $\csc \theta$.

**48–53. Right-triangle relationships** *Draw a right triangle to simplify the given expression. Assume $x > 0$ and $0 \leq \theta \leq \pi/2$.*

**48.**   $\tan (\sec^{-1} (x/2))$        **49.**   $\cot^{-1} (\tan \theta)$

**50.**   $\csc^{-1} (\sec \theta)$        **51.**   $\sin^{-1} x + \sin^{-1} (-x)$

**52.**   $\sec (\csc^{-1} 3x)$        **53.**   $\cot (\sin^{-1} (2x - 1))$

**54–67. Integrals** *Evaluate the following integrals.*

**54.**   $\displaystyle \int \dfrac{e^x}{4e^x + 6}\, dx$        **55.**   $\displaystyle \int_{e^2}^{e^8} \dfrac{dx}{x \ln x}$

**56.**   $\displaystyle \int_1^4 \dfrac{10^{\sqrt{x}}}{\sqrt{x}}\, dx$        **57.**   $\displaystyle \int \dfrac{x + 4}{x^2 + 8x + 25}\, dx$

**58.**   $\displaystyle \int \dfrac{dx}{\sqrt{49 - 4x^2}}$        **59.**   $\displaystyle \int \dfrac{3}{2x^2 + 1}\, dx$

**60.**   $\displaystyle \int \dfrac{dt}{2t\sqrt{t^2 - 4}}$        **61.**   $\displaystyle \int_{-1}^{2\sqrt{3}-1} \dfrac{dx}{x^2 + 2x + 5}$

**62.**   $\displaystyle \int \dfrac{(\ln \ln x)^4}{x \ln x}\, dx$        **63.**   $\displaystyle \int_{-2/3}^{2/\sqrt{3}} \dfrac{dx}{\sqrt{16 - 9x^2}}$

**64.** $\displaystyle\int_{\ln 2}^{\ln 3} \coth s \, ds$

**65.** $\displaystyle\int \frac{dx}{\sqrt{x^2 - 9}}, x > 3$

**66.** $\displaystyle\int \frac{e^x}{\sqrt{e^{2x} + 4}} \, dx$

**67.** $\displaystyle\int_0^1 \frac{x^2}{9 - x^6} \, dx$

**68–69. Arc length** *Find the length of the following curves.*

**68.** $y = \ln\left(x - \sqrt{x^2 - 1}\right)$, for $1 \leq x \leq \sqrt{2}$

**69.** $y = 3 \ln x - \dfrac{x^2}{24}$, for $1 \leq x \leq 6$

**70–72. Volume problems** *Compute the volume of the following solids of revolution.*

**70.** The region bounded by the curve $y = e^{-x^2}$ and the $x$-axis on the interval $\left[0, \sqrt{\ln 3}\right]$ is revolved about the $y$-axis.

**71.** The region bounded by the curve $y = \dfrac{2}{1 + x^2}$ and the $x$-axis on the interval $[0, 4]$ is revolved about the $y$-axis.

**72.** The region bounded by the curve $y = (4 - x^2)^{-1/4}$ and the $x$-axis on the interval $[0, 1]$ is revolved about the $x$-axis.

**T 73.** **An exponential bike ride** Tom and Sue took a bike ride, both starting at the same time and position. Tom started riding at 20 mi/hr, and his velocity decreased according to the function $v(t) = 20e^{-2t}$, for $t \geq 0$. Sue started riding at 15 mi/hr, and her velocity decreased according to the function $u(t) = 15e^{-t}$, for $t \geq 0$.

    **a.** Find and graph the position functions of Tom and Sue.
    **b.** Find the times at which the riders had the same position at the same time.
    **c.** Who ultimately took the lead and remained in the lead?

**74.** **Population growth** The population of a large city grows exponentially with a current population of 1.3 million and a predicted population of 1.45 million 10 years from now.

    **a.** Use an exponential model to estimate the population in 20 years. Assume the annual growth rate is constant.
    **b.** Find the doubling time of the population.

**75.** **Caffeine** An adult consumes an espresso containing 75 mg of caffeine. If the caffeine has a half-life of 5.5 hours, when will the amount of caffeine in her bloodstream equal 30 mg?

**76.** **Two cups of coffee** A college student consumed two cups of coffee, each containing 80 mg of caffeine, 90 minutes apart. Suppose that just before he consumed his second cup of coffee, 65 mg of caffeine was still in his system from the first cup of coffee. How much caffeine remains in his system 7 hours after he drank his first cup of coffee?

**77.** **Moore's Law** In 1965, Gordon Moore observed that the number of transistors that could be placed on an integrated circuit was approximately doubling each year, and he predicted that this trend would continue for another decade. In 1975, Moore revised the doubling time to every two years, and this prediction became known as *Moore's Law*.

    **a.** In 1979, Intel introduced the Intel 8088 processor; each of its integrated circuits contained 29,000 transistors. Use Moore's revised doubling time to find a function $y(t)$ that approximates the number of transistors on an integrated circuit $t$ years after 1979.

    **b.** In 2000, the Pentium 4 integrated circuit was introduced; it contained 42 million transistors. Compare this value to the value obtained using the function found in part (a).

**78.** **Radioactive decay** The mass of radioactive material in a sample has decreased by 30% since the decay began. Assuming a half-life of 1500 years, how long ago did the decay begin?

**79.** **Population growth** Growing from an initial population of 150,000 at a constant annual growth rate of 4%/yr, how long will it take a city to reach a population of 1 million?

**80.** **Savings account** A savings account advertises an annual percentage yield (APY) of 5.4%, which means that the balance in the account increases at an annual growth rate of 5.4%/yr.

    **a.** Find the balance in the account for $t \geq 0$ with an initial deposit of $1500, assuming the APY remains fixed and no additional deposits or withdrawals are made.
    **b.** What is the doubling time of the balance?
    **c.** After how many years does the balance reach $5000?

**81–82. Curve sketching** *Use the graphing techniques of Section 4.4 to graph the following functions on their domains. Identify local extreme points, inflection points, concavity, and end behavior. Use a graphing utility only to check your work.*

**81.** $f(x) = e^x(x^2 - x)$

**82.** $f(x) = \ln x - \ln^2 x$

**T 83.** **Arc length** Find the length of the curve $y = \ln x$ between $x = 1$ and $x = b > 1$ given that

$$\int \frac{\sqrt{x^2 + a^2}}{x} \, dx = \sqrt{x^2 + a^2} - a \ln\left(\frac{a + \sqrt{x^2 + a^2}}{x}\right) + C.$$

Use any means to approximate the value of $b$ for which the curve has length 2.

**84–89. Limits** *Evaluate the following limits.*

**84.** $\displaystyle\lim_{x \to 0} \frac{e^{-2x} - 1 + 2x}{x^2}$

**85.** $\displaystyle\lim_{x \to \pi/2^-} (\sin x)^{\tan x}$

**86.** $\displaystyle\lim_{x \to \infty} \frac{\ln^3 x}{\sqrt{x}}$

**87.** $\displaystyle\lim_{x \to \infty} \ln\left(\frac{x + 1}{x - 1}\right)$

**88.** $\displaystyle\lim_{x \to \infty} \left(\frac{2}{\pi} \tan^{-1} x\right)^x$

**89.** $\displaystyle\lim_{x \to \infty} \left(1 - \frac{3}{x}\right)^x$

**90–97. Comparing growth rates** *Determine which of the two functions grows faster or state that they have comparable growth rates.*

**90.** $x^{100}$ and $1.1^x$

**91.** $x^{1/2}$ and $x^{1/3}$

**92.** $\ln x$ and $\log_{10} x$

**93.** $\sqrt{x}$ and $\ln^{10} x$

**94.** $10x$ and $\ln x^2$

**95.** $e^x$ and $3^x$

**96.** $\sqrt{x^6 + 10}$ and $x^3$

**97.** $2^x$ and $4^{x/2}$

**98.** **Logs of logs** Compare the growth rates of $\ln x$, $\ln(\ln x)$, and $\ln(\ln(\ln x))$.

**T 99.** **Two limits with exponentials** Evaluate $\displaystyle\lim_{x \to 0^+} \frac{x}{\sqrt{1 - e^{-x^2}}}$ and $\displaystyle\lim_{x \to 0^+} \frac{x^2}{1 - e^{-x^2}}$, and confirm your result by graphing.

**100.** **Geometric mean** Prove that $\displaystyle\lim_{r \to 0} \left(\frac{a^r + b^r + c^r}{3}\right)^{1/r} = \sqrt[3]{abc}$, where $a$, $b$, and $c$ are positive real numbers.

**101. Towers of exponents** The functions

$$f(x) = (x^x)^x \quad \text{and} \quad g(x) = x^{(x^x)}$$

are different functions. For example, $f(3) = 19{,}683$ and

$g(3) \approx 7.6 \times 10^{12}$. Determine whether $\lim\limits_{x \to 0^+} f(x)$ and $\lim\limits_{x \to 0^+} g(x)$
are indeterminate forms and evaluate the limits.

**T 102. A family of super-exponential functions** Let $f(x) = (a + x)^x$,
where $a > 0$.

a. What is the domain of $f$ (in terms of $a$)?
b. Describe the end behavior of $f$ (near the left boundary of its
   domain and as $x \to \infty$).
c. Compute $f'$. Then graph $f$ and $f'$ for $a = 0.5, 1, 2,$ and $3$.
d. Show that $f$ has a single local minimum at the point $z$ that
   satisfies $(z + a) \ln(z + a) + z = 0$.
e. Describe how $z$ (found in part (d)) varies as $a$ increases.
   Describe how $f(z)$ varies as $a$ increases.

**103. Limits for $e$** Consider the function $g(x) = (1 + 1/x)^{x+a}$.
Show that if $0 \le a < \frac{1}{2}$, then $g(x) \to e$ from *below* as $x \to \infty$;
if $\frac{1}{2} \le a < 1$, then $g(x) \to e$ from *above* as $x \to \infty$.

**104. Arc length of family of exponential functions**

a. Show that the arc length integral for the function

$$f(x) = Ae^{ax} + \frac{1}{4Aa^2} e^{-ax}, \text{ where } a > 0 \text{ and } A > 0, \text{ may be}$$

   integrated using methods you already know.
b. Verify that the arc length of the curve $y = f(x)$ on the interval
   $[0, \ln 2]$ is

$$A(2^a - 1) - \frac{1}{4a^2 A}(2^{-a} - 1).$$

**T 105. Log-normal probability distribution** A commonly used distribution in probability and statistics is the log-normal distribution.
(If the logarithm of a variable has a normal distribution, then the
variable itself has a log-normal distribution.) The distribution
function is

$$f(x) = \frac{1}{x\sigma\sqrt{2\pi}} e^{-\ln^2 x/(2\sigma^2)}, \text{ for } x > 0,$$

where $\ln x$ has zero mean and standard deviation $\sigma > 0$.

a. Graph $f$ for $\sigma = \frac{1}{2}, 1,$ and $2$. Based on your graphs, does
   $\lim\limits_{x \to 0^+} f(x)$ appear to exist?
b. Evaluate $\lim\limits_{x \to 0^+} f(x)$. (*Hint:* Let $x = e^y$.)
c. Show that $f$ has a single local maximum at $x^* = e^{-\sigma^2}$.
d. Evaluate $f(x^*)$ and express the result as a function of $\sigma$.
e. For what value of $\sigma > 0$ in part (d) does $f(x^*)$ have a
   minimum?

**T 106. A rush hour function** The function $f(t) = \sin(e^{a \cos t}) + 1$ is
a periodic function with the property that the number of local
extrema on the interval $[0, 2\pi]$ is determined by the parameter
$a > 0$. For some values of $a$, the function could be used to model
traffic flow (two rush hours per day) or ocean tides (two low tides
and two high tides per day).

a. Graph $f$ on the interval $[0, 2\pi]$ for $a = 0.3, 1.3,$ and $1.7$. Confirm that in these cases, $f$ has one, three, and five local extrema
   on $(0, 2\pi)$, respectively.
b. Prove that the period of $f$ is $2\pi$, for all $a > 0$.
c. Prove that $f$ has extreme values at $t = 0, \pi,$ and $2\pi$, for all
   $a > 0$.

d. Prove that if $0 < a < \ln(\pi/2)$, then $f$ has one local minimum on $(0, 2\pi)$.
e. Prove that if $\ln(\pi/2) < a < \ln(3\pi/2)$, then $f$ has one local
   minimum and two local maxima on $(0, 2\pi)$.
f. Prove that if $a > \ln(3\pi/2)$, then $f$ has at least five local extreme points on $(0, 2\pi)$.

**107. Viewing angles** An auditorium with a flat floor has a large
screen on one wall. The lower edge of the screen is 3 ft above
eye level and the upper edge of the screen is 10 ft above eye
level (see figure).

a. Express $\theta$ as a function of $x$.
b. Maximize the function in part (a) to determine how far you
   should stand from the screen to maximize your viewing
   angle $\theta$.

**T 108. Blood testing** Suppose that a blood test for a disease must be
given to a population of $N$ people, where $N$ is large. At most,
$N$ individual blood tests must be done. The following strategy
reduces the number of tests. Suppose 100 people are selected
from the population and their blood samples are pooled. One test
determines whether any of the 100 people test positive. If the test
is positive, those 100 people are tested individually, making 101
tests necessary. However, if the pooled sample tests negative, then
100 people have been tested with one test. This procedure is then
repeated. Probability theory shows that if the group size is $x$ (for
example, $x = 100$, as described here), then the average number of
blood tests required to test $N$ people is $N(1 - q^x + 1/x)$, where
$q$ is the probability that any one person tests negative. What group
size $x$ minimizes the average number of tests in the case that
$N = 10{,}000$ and $q = 0.95$? Assume that $x$ is a nonnegative real
number.

**T 109. Bungee jumper** A woman attached to a bungee cord jumps from
a bridge that is 30 m above a river. Her height in meters above
the river $t$ seconds after the jump is $y(t) = 15(1 + e^{-t} \cos t)$, for
$t \ge 0$.

a. Determine her velocity at $t = 1$ and $t = 3$.
b. Use a graphing utility to determine when she is moving downward and when she is moving upward during the first 10 s.
c. Use a graphing utility to estimate the maximum upward
   velocity.

**T 110. Cell population** The population of a culture of cells after $t$ days is
approximated by the function $P(t) = \dfrac{1600}{1 + 7e^{-0.02t}}$, for $t \ge 0$.

a. Graph the population function.
b. What is the average growth rate during the first 10 days?
c. Looking at the graph, when does the growth rate appear to be
   a maximum?

**d.** Differentiate the population function to determine the growth rate function $P'(t)$.

**e.** Graph the growth rate. When is it a maximum and what is the population at the time that the growth rate is a maximum?

**111. Derivatives of hyperbolic functions** Compute the following derivatives.

**a.** $\dfrac{d^6}{dx^6}(\cosh x)$          **b.** $\dfrac{d}{dx}(x \operatorname{sech} x)$

**112. Area of region** Find the area of the region bounded by the curves $f(x) = 8 \operatorname{sech}^2 x$ and $g(x) = \cosh x$.

**113. Linear approximation** Find the linear approximation to $f(x) = \cosh x$ at $a = \ln 3$ and then use it to approximate the value of $\cosh 1$.

**114. Limit** Evaluate $\lim\limits_{x \to \infty} (\tanh x)^x$.

## Chapter 7 Guided Projects

*Applications of the material in this chapter and related topics can be found in the following Guided Projects. For additional information, see the Preface.*

- Means and tangent lines
- Inverse sine from geometry
- Optimizing fuel use
- Enzyme kinetics

- Oscillators
- Pharmacokinetics–drug metabolism
- Acid, noise, and earthquakes
- Landing an airliner

- Hyperbolic functions
- Atmospheric $CO_2$

# 8

# Integration Techniques

**Chapter Preview**  In this chapter, we return to integration methods and present a variety of new strategies that supplement the substitution (or change of variables) method. The new techniques introduced here are integration by parts, trigonometric substitution, and partial fractions. Taken altogether, these *analytical methods* (pencil-and-paper methods) greatly enlarge the collection of integrals that we can evaluate. Nevertheless, it is important to recognize that these methods are limited because many integrals do not yield to them. For this reason, we also introduce table-based methods, which are used to evaluate many indefinite integrals, and computer-based methods for approximating definite integrals. The discussion then turns to integrals that have either infinite integrands or infinite intervals of integration. These integrals, called *improper integrals*, offer surprising results and have many practical applications.

## 8.1 Basic Approaches

Before plunging into new integration techniques, we devote this section to two practical goals. The first is to review what you learned about the substitution method in Section 5.5. The other is to introduce several basic simplifying procedures that are worth keeping in mind when evaluating integrals.

➤ Table 8.1 is similar to Table 4.9 in Section 4.9. It is a subset of the table of integrals at the back of the text.

Table 8.1 includes the basic integration formulas presented earlier in the text. To this list we add four standard trigonometric integrals (formulas 14–17), which are derived in Examples 1 and 2 using substitution.

**Table 8.1  Basic Integration Formulas**

**1.** $\int k \, dx = kx + C, k \text{ real}$

**2.** $\int x^p \, dx = \dfrac{x^{p+1}}{p+1} + C, p \neq -1 \text{ real}$

**3.** $\int \cos ax \, dx = \dfrac{1}{a} \sin ax + C$

**4.** $\int \sin ax \, dx = -\dfrac{1}{a} \cos ax + C$

**5.** $\int \sec^2 ax \, dx = \dfrac{1}{a} \tan ax + C$

**6.** $\int \csc^2 ax \, dx = -\dfrac{1}{a} \cot ax + C$

**7.** $\int \sec ax \tan ax \, dx = \dfrac{1}{a} \sec ax + C$

**8.** $\int \csc ax \cot ax \, dx = -\dfrac{1}{a} \csc ax + C$

**9.** $\int e^{ax} \, dx = \dfrac{1}{a} e^{ax} + C$

**10.** $\int \dfrac{dx}{x} = \ln |x| + C$

**11.** $\int \dfrac{dx}{a^2 + x^2} = \dfrac{1}{a} \tan^{-1} \dfrac{x}{a} + C$

**12.** $\int \dfrac{dx}{\sqrt{a^2 - x^2}} = \sin^{-1} \dfrac{x}{a} + C, a > 0$

**13.** $\int \dfrac{dx}{x\sqrt{x^2 - a^2}} = \dfrac{1}{a} \sec^{-1} \left| \dfrac{x}{a} \right| + C, a > 0$

**14.** $\int \tan ax \, dx = \dfrac{1}{a} \ln |\sec ax| + C$

**15.** $\int \cot ax \, dx = \dfrac{1}{a} \ln |\sin ax| + C$

**16.** $\int \sec ax \, dx = \dfrac{1}{a} \ln |\sec ax + \tan ax| + C$

**17.** $\int \csc ax \, dx = -\dfrac{1}{a} \ln |\csc ax + \cot ax| + C$

In Section 4.9, we introduced formulas for $\int \cos x \, dx$ and $\int \sin x \, dx$ and then derived more general results listed in Table 5.6 of Section 5.5. Missing from that table of integrals are formulas for the antiderivatives of $\tan x$, $\cot x$, $\sec x$, and $\csc x$. Examples 1 and 2 (and their related exercises) fill in these holes.

**EXAMPLE 1**  **Substitution review**  Derive integral formula 14 in Table 8.1:

$$\int \tan ax \, dx = \frac{1}{a} \ln|\sec ax| + C.$$

**SOLUTION**  We begin with a simpler form of the integral, $\int \tan x \, dx$, and express $\tan x$ in terms of $\sin x$ and $\cos x$ to prepare for a standard substitution:

$$\int \tan x \, dx = \int \frac{\sin x}{\cos x} dx$$

$$= -\int \frac{1}{u} \, du \qquad\qquad u = \cos x; \, du = -\sin x \, dx$$

$$= -\ln|u| + C = -\ln|\cos x| + C.$$

Using the properties of logarithms, the integral can also be written

$$\int \tan x \, dx = -\ln|\cos x| + C = -\ln|1/\sec x| + C$$

$$= -(\ln 1 - \ln|\sec x|) = \ln|\sec x| + C.$$

**QUICK CHECK 1** What change of variable would you use for the integral $\int \dfrac{\sec^2 x}{\tan^3 x} dx$? ◄

With this result in hand, the more general formula given in Table 8.1 is easily derived by using the substitution $u = ax$. The derivation of formula 15 in Table 8.1 is similar (Exercise 66).

*Related Exercise 66* ◄

**EXAMPLE 2**  **Multiplication by 1**  Derive integral formula 16 in Table 8.1:

$$\int \sec ax \, dx = \frac{1}{a} \ln|\sec ax + \tan ax| + C.$$

**SOLUTION**  As in Example 1, it is easier to begin with a simpler form of the integral, $\int \sec x \, dx$. The key to evaluating this integral is admittedly not obvious, and the trick works only on special integrals. The idea is to multiply the integrand by 1, but the challenge is finding the appropriate representation of 1. In this case, we use

$$1 = \frac{\sec x + \tan x}{\sec x + \tan x}.$$

The integral is evaluated as follows:

$$\int \sec x \, dx = \int \sec x \cdot \underbrace{\frac{\sec x + \tan x}{\sec x + \tan x}}_{1} dx \quad \text{Multiply integrand by 1.}$$

$$= \int \frac{\sec^2 x + \sec x \tan x}{\sec x + \tan x} dx \quad \text{Expand numerator.}$$

$$= \int \frac{du}{u} \qquad\qquad u = \sec x + \tan x; \, du = (\sec^2 x + \sec x \tan x) dx$$

$$= \ln|u| + C \qquad\qquad \text{Integrate.}$$

$$= \ln|\sec x + \tan x| + C. \qquad u = \sec x + \tan x$$

Using the substitution $u = ax$ with $\int \sec ax \, dx$ leads to the more general result in Table 8.1. The derivation of formula 17 in Table 8.1 employs the same technique used here (Exercise 67).

*Related Exercise 67* ◄

**EXAMPLE 3**  **Subtle substitution**  Evaluate $\int \dfrac{dx}{e^x + e^{-x}}$.

**SOLUTION**  In this case, we see nothing in Table 8.1 that resembles the given integral. In a spirit of trial and error, we try the technique used in Example 2 and multiply the numerator and denominator of the integrand by $e^x$:

$$\int \frac{dx}{e^x + e^{-x}} = \int \frac{e^x}{e^{2x} + 1}\, dx. \qquad e^x \cdot e^x = e^{2x}$$

This form of the integrand suggests the substitution $u = e^x$, which implies that $du = e^x\, dx$. Making these substitutions, the integral becomes

$$\int \frac{e^x}{e^{2x} + 1}\, dx = \int \frac{du}{u^2 + 1} \qquad \text{Substitute } u = e^x, du = e^x\, dx.$$

$$= \tan^{-1} u + C \qquad \text{Table 8.1}$$

$$= \tan^{-1} e^x + C. \quad u = e^x$$

*Related Exercises 45–46* ◄

**EXAMPLE 4**  **Split up fractions**  Evaluate $\int \dfrac{\cos x + \sin^3 x}{\sec x}\, dx$.

**SOLUTION**  Don't overlook the opportunity to split a fraction into two or more fractions. In this case, the integrand is simplified in a useful way:

$$\int \frac{\cos x + \sin^3 x}{\sec x}\, dx = \int \frac{\cos x}{\sec x}\, dx + \int \frac{\sin^3 x}{\sec x}\, dx \qquad \text{Split fraction.}$$

$$= \int \cos^2 x\, dx + \int \sin^3 x \cos x\, dx. \quad \sec x = \frac{1}{\cos x}$$

> **Half-angle formulas**
>
> $\cos^2 x = \dfrac{1 + \cos 2x}{2}$
>
> $\sin^2 x = \dfrac{1 - \cos 2x}{2}$

The first of the resulting integrals is evaluated using a half-angle formula (Example 5 of Section 5.5). In the second integral, the substitution $u = \sin x$ is used:

$$\int \frac{\cos x + \sin^3 x}{\sec x}\, dx = \int \cos^2 x\, dx + \int \sin^3 x \cos x\, dx$$

$$= \int \frac{1 + \cos 2x}{2}\, dx + \int \sin^3 x \cos x\, dx \quad \text{Half-angle formula}$$

**QUICK CHECK 2**  Explain how to simplify the integrand of $\int \dfrac{x^3 + \sqrt{x}}{x^{3/2}}\, dx$ before integrating. ◄

$$= \frac{1}{2} \int dx + \frac{1}{2} \int \cos 2x\, dx + \int u^3\, du \qquad u = \sin x, du = \cos x\, dx$$

$$= \frac{x}{2} + \frac{1}{4} \sin 2x + \frac{1}{4} \sin^4 x + C. \qquad \text{Evaluate integrals.}$$

*Related Exercises 22, 25* ◄

**EXAMPLE 5**  **Division with rational functions**  Evaluate $\int \dfrac{x^2 + 2x - 1}{x + 4}\, dx$.

**SOLUTION**  When integrating rational functions (polynomials in the numerator and denominator), check to see whether the function is *improper* (the degree of the numerator is greater than or equal to the degree of the denominator). In this example, we have an improper rational function, and long division is used to simplify it. The integration is done as follows:

> $\phantom{x + 4\,} x - 2$
> $x + 4 \overline{)\, x^2 + 2x - 1}$
> $\phantom{x + 4)\,} \underline{x^2 + 4x}$
> $\phantom{x + 4)x^2} -2x - 1$
> $\phantom{x + 4)x^2} \underline{-2x - 8}$
> $\phantom{x + 4)x^2 - 2x} 7$

$$\int \frac{x^2 + 2x - 1}{x + 4}\, dx = \int (x - 2)\, dx + \int \frac{7}{x + 4}\, dx \qquad \text{Long division}$$

$$= \frac{x^2}{2} - 2x + 7 \ln |x + 4| + C. \qquad \text{Evaluate integrals.}$$

*Related Exercises 34–35* ◄

**QUICK CHECK 3** Explain how to simplify the integrand of $\int \dfrac{x+1}{x-1}\, dx$ before integrating. ◄

**EXAMPLE 6** **Complete the square** Evaluate $\int \dfrac{dx}{\sqrt{-x^2 - 8x - 7}}$.

**SOLUTION** We don't see an integral in Table 8.1 that looks like the given integral, so some preliminary work is needed. In this case, the key is to complete the square on the polynomial in the denominator. We find that

$$\begin{aligned}
-x^2 - 8x - 7 &= -(x^2 + 8x + 7) \\
&= -(x^2 + 8x \underbrace{+ 16 - 16}_{\text{add and subtract 16}} + 7) \quad \text{Complete the square.} \\
&= -((x+4)^2 - 9) \quad \text{Factor and combine terms.} \\
&= 9 - (x+4)^2. \quad \text{Rearrange terms.}
\end{aligned}$$

After a change of variables, the integral is recognizable:

$$\begin{aligned}
\int \frac{dx}{\sqrt{-7 - 8x - x^2}} &= \int \frac{dx}{\sqrt{9 - (x+4)^2}} \quad \text{Complete the square.} \\
&= \int \frac{du}{\sqrt{9 - u^2}} \quad u = x+4, du = dx \\
&= \sin^{-1} \frac{u}{3} + C \quad \text{Table 8.1} \\
&= \sin^{-1}\left(\frac{x+4}{3}\right) + C. \quad \text{Replace } u \text{ with } x+4.
\end{aligned}$$

*Related Exercises 31, 37* ◄

**QUICK CHECK 4** Express $x^2 + 6x + 16$ in terms of a perfect square. ◄

The techniques illustrated in this section are designed to transform or simplify an integrand before you apply a specific method. In fact, these ideas may help you recognize the best method to use. Keep them in mind as you learn new integration methods and improve your integration skills.

## SECTION 8.1 EXERCISES

### Getting Started

1. What change of variables would you use for the integral $\int (4 - 7x)^{-6}\, dx$?

2. Evaluate $\int (\sec x + 1)^2\, dx$. (*Hint:* Expand $(\sec x + 1)^2$ first.)

3. What trigonometric identity is useful in evaluating $\int \sin^2 x\, dx$?

4. Let $f(x) = \dfrac{4x^3 + x^2 + 4x + 2}{x^2 + 1}$. Use long division to show that $f(x) = 4x + 1 + \dfrac{1}{x^2 + 1}$ and use this result to evaluate $\int f(x)\, dx$.

5. Describe a first step in integrating $\int \dfrac{10}{x^2 - 4x - 9}\, dx$.

6. Evaluate $\int \dfrac{2x+1}{x^2+1}\, dx$ using the following steps.
   a. Fill in the blanks: By splitting the integrand into two fractions, we have $\int \dfrac{2x+1}{x^2+1}\, dx = \int \underline{\quad}\, dx + \int \underline{\quad}\, dx$.
   b. Evaluate the two integrals on the right side of the equation in part (a).

### Practice Exercises

**7–64. Integration review** *Evaluate the following integrals.*

7. $\int \dfrac{dx}{(3 - 5x)^4}$

8. $\int (9x - 2)^{-3}\, dx$

9. $\int_0^{3\pi/8} \sin\left(2x - \dfrac{\pi}{4}\right) dx$

10. $\int e^{3 - 4x}\, dx$

11. $\int \dfrac{\ln 2x}{x}\, dx$

12. $\int_{-5}^0 \dfrac{dx}{\sqrt{4 - x}}$

13. $\int \dfrac{e^x}{e^x + 1}\, dx$

14. $\int_0^1 x\, 3^{x^2 + 1}\, dx$

15. $\int_1^{e^2} \dfrac{\ln^2(x^2)}{x}\, dx$

16. $\int_0^1 \dfrac{t^2}{1 + t^6}\, dt$

17. $\int_1^2 s(s - 1)^9\, ds$

18. $\int_3^7 (t - 6)\sqrt{t - 3}\, dt$

19. $\int \dfrac{(\ln w - 1)^7 \ln w}{w}\, dw$

20. $\int e^x(1 + e^x)^9(1 - e^x)\, dx$

21. $\int \dfrac{x+2}{x^2+4}\, dx$

22. $\int \dfrac{\sin x + 1}{\cos x}\, dx$

**23.** $\displaystyle\int e^x \csc\left(3e^x + 4\right) dx$

**24.** $\displaystyle\int_0^9 \frac{x^{5/2} - x^{1/2}}{x^{3/2}} dx$

**25.** $\displaystyle\int_0^{\pi/4} \frac{\sec\theta + \csc\theta}{\sec\theta\,\csc\theta} d\theta$

**26.** $\displaystyle\int \frac{4 + e^{-2x}}{e^{3x}} dx$

**27.** $\displaystyle\int \frac{2 - 3x}{\sqrt{1 - x^2}} dx$

**28.** $\displaystyle\int \frac{3x + 1}{\sqrt{4 - x^2}} dx$

**29.** $\displaystyle\int_{\pi/4}^{\pi/2} \sqrt{1 + \cot^2 x}\, dx$

**30.** $\displaystyle\int_{e^{\pi/4}}^{e^{\pi/3}} \frac{\cot(\ln x)}{x} dx$

**31.** $\displaystyle\int \frac{dx}{x^2 - 2x + 10}$

**32.** $\displaystyle\int_0^2 \frac{x}{x^2 + 4x + 8} dx$

**33.** $\displaystyle\int \frac{x^3 + 2x^2 + 5x + 3}{x^2 + x + 2} dx$

**34.** $\displaystyle\int_2^4 \frac{x^2 + 2}{x - 1} dx$

**35.** $\displaystyle\int_0^1 \frac{t^4 + t^3 + t^2 + t + 1}{t^2 + 1} dt$

**36.** $\displaystyle\int \frac{t^3 - 2}{t + 1} dt$

**37.** $\displaystyle\int \frac{d\theta}{\sqrt{27 - 6\theta - \theta^2}}$

**38.** $\displaystyle\int \frac{x}{x^4 + 2x^2 + 1} dx$

**39.** $\displaystyle\int \frac{d\theta}{1 + \sin\theta}$

**40.** $\displaystyle\int \frac{1 - x}{1 - \sqrt{x}} dx$

**41.** $\displaystyle\int \frac{dx}{\sec x - 1}$

**42.** $\displaystyle\int \frac{d\theta}{1 - \csc\theta}$

**43.** $\displaystyle\int \frac{\cosh 3x}{1 + \sinh 3x} dx$

**44.** $\displaystyle\int_0^{\sqrt{3}} \frac{6x^3}{\sqrt{x^2 + 1}} dx$

**45.** $\displaystyle\int \frac{e^x}{e^x - 2e^{-x}} dx$

**46.** $\displaystyle\int \frac{e^{2z}}{e^{2z} - 4e^{-z}} dz$

**47.** $\displaystyle\int \frac{dx}{x^{-1} + 1}$

**48.** $\displaystyle\int \frac{dy}{y^{-1} + y^{-3}}$

**49.** $\displaystyle\int \sqrt{9 + \sqrt{t + 1}}\, dt$

**50.** $\displaystyle\int_4^9 \frac{dx}{1 - \sqrt{x}}$

**51.** $\displaystyle\int_{-1}^0 \frac{x}{x^2 + 2x + 2} dx$

**52.** $\displaystyle\int_{\pi/6}^{\pi/2} \frac{dy}{\sin y}$

**53.** $\displaystyle\int e^x \sec\left(e^x + 1\right) dx$

**54.** $\displaystyle\int_0^1 \sqrt{1 + \sqrt{x}}\, dx$

**55.** $\displaystyle\int \sin x \sin 2x\, dx$

**56.** $\displaystyle\int_0^{\pi/2} \sqrt{1 + \cos 2x}\, dx$

**57.** $\displaystyle\int \frac{dx}{x^{1/2} + x^{3/2}}$

**58.** $\displaystyle\int_0^1 \frac{dp}{4 - \sqrt{p}}$

**59.** $\displaystyle\int \frac{x - 2}{x^2 + 6x + 13} dx$

**60.** $\displaystyle\int_0^{\pi/4} 3\sqrt{1 + \sin 2x}\, dx$

**61.** $\displaystyle\int \frac{e^x}{e^{2x} + 2e^x + 1} dx$

**62.** $\displaystyle\int \frac{-x^5 - x^4 - 2x^3 + 4x + 3}{x^2 + x + 1} dx$

**63.** $\displaystyle\int_1^3 \frac{2}{x^2 + 2x + 1} dx$

**64.** $\displaystyle\int_0^2 \frac{2}{s^3 + 3s^2 + 3s + 1} ds$

**65. Explain why or why not** Determine whether the following statements are true and give an explanation or counterexample.

    **a.** $\displaystyle\int \frac{3}{x^2 + 4} dx = \int \frac{3}{x^2} dx + \int \frac{3}{4} dx.$

    **b.** Long division simplifies the evaluation of the integral
$\displaystyle\int \frac{x^3 + 2}{3x^4 + x} dx.$

    **c.** $\displaystyle\int \frac{1}{\sin x + 1} dx = \ln|\sin x + 1| + C.$

    **d.** $\displaystyle\int \frac{dx}{e^x} = \ln e^x + C.$

**66–67. Integrals of cot x and csc x**

**66.** Use a change of variables to prove that $\int \cot x\, dx = \ln|\sin x| + C.$ (*Hint:* See Example 1.)

**67.** Prove that $\int \csc x\, dx = -\ln|\csc x + \cot x| + C.$ (*Hint:* See Example 2.)

**68. Different methods**

    **a.** Evaluate $\int \cot x \csc^2 x\, dx$ using the substitution $u = \cot x$.

    **b.** Evaluate $\int \cot x \csc^2 x\, dx$ using the substitution $u = \csc x$.

    **c.** Reconcile the results in parts (a) and (b).

**69. Different substitutions**

    **a.** Evaluate $\int \tan x \sec^2 x\, dx$ using the substitution $u = \tan x$.

    **b.** Evaluate $\int \tan x \sec^2 x\, dx$ using the substitution $u = \sec x$.

    **c.** Reconcile the results in parts (a) and (b).

**70. Different methods** Let $I = \displaystyle\int \frac{x + 2}{x + 4} dx.$

    **a.** Evaluate $I$ after first performing long division on the integrand.

    **b.** Evaluate $I$ without performing long division on the integrand.

    **c.** Reconcile the results in parts (a) and (b).

**71. Different methods** Let $I = \displaystyle\int \frac{x^2}{x + 1} dx.$

    **a.** Evaluate $I$ using the substitution $u = x + 1$.

    **b.** Evaluate $I$ after first performing long division on the integrand.

    **c.** Reconcile the results in parts (a) and (b).

**72. Area of a region between curves** Find the area of the entire region bounded by the curves $y = \dfrac{x^3}{x^2 + 1}$ and $y = \dfrac{8x}{x^2 + 1}$.

**73. Area of a region between curves** Find the area of the region bounded by the curves $y = \dfrac{x^2}{x^3 - 3x}$ and $y = \dfrac{1}{x^3 - 3x}$ on the interval $[2, 4]$.

**74. Volume of a solid** Consider the region $R$ bounded by the graph of $f(x) = \dfrac{1}{x + 2}$ and the $x$-axis on the interval $[0, 3]$. Find the volume of the solid formed when $R$ is revolved about the $y$-axis.

**75. Volume of a solid** Consider the region $R$ bounded by the graph of $f(x) = \sqrt{x^2 + 1}$ and the $x$-axis on the interval $[0, 2]$. Find the volume of the solid formed when $R$ is revolved about the $y$-axis.

**Explorations and Challenges**

**76. Different substitutions**

    **a.** Show that $\displaystyle\int \frac{dx}{\sqrt{x - x^2}} = \sin^{-1}(2x - 1) + C$ using either $u = 2x - 1$ or $u = x - \dfrac{1}{2}$.

**b.** Show that $\int \dfrac{dx}{\sqrt{x - x^2}} = 2\sin^{-1}\sqrt{x} + C$ using $u = \sqrt{x}$.

**c.** Prove the identity $2\sin^{-1}\sqrt{x} - \sin^{-1}(2x - 1) = \dfrac{\pi}{2}$.

*(Source: The College Mathematics Journal, 32, 5, Nov 2001)*

**77. Surface area** Let $f(x) = \sqrt{x + 1}$. Find the area of the surface generated when the region bounded by the graph of $f$ and the $x$-axis on the interval $[0, 1]$ is revolved about the $x$-axis.

**78. Surface area** Find the area of the surface generated when the region bounded by the graph of $y = e^x + \dfrac{1}{4}e^{-x}$ and the $x$-axis on the interval $[0, \ln 2]$ is revolved about the $x$-axis.

**79. Arc length** Find the length of the curve $y = x^{5/4}$ on the interval $[0, 1]$. (*Hint:* Write the arc length integral and let $u^2 = 1 + \left(\dfrac{5}{4}\right)^2 \sqrt{x}$.)

**80. Skydiving** A skydiver in free fall subject to gravitational acceleration and air resistance has a velocity given by

$$v(t) = v_T\left(\frac{e^{at} - 1}{e^{at} + 1}\right),$$ where $v_T$ is the terminal velocity and $a > 0$ is a physical constant. Find the distance that the skydiver falls after $t$ seconds, which is $d(t) = \int_0^t v(y)\, dy$.

**QUICK CHECK ANSWERS**

**1.** Let $u = \tan x$.    **2.** Write the integrand as $x^{3/2} + x^{-1}$.

**3.** Use long division to write the integrand as $1 + \dfrac{2}{x - 1}$.

**4.** $(x + 3)^2 + 7$ ◄

# 8.2 Integration by Parts

The Substitution Rule (Section 5.5) emerges when we reverse the Chain Rule for derivatives. In this section, we use a similar strategy and reverse the Product Rule for derivatives. The result is an integration technique called *integration by parts*. To illustrate the importance of integration by parts, consider the indefinite integrals

$$\int e^x\, dx = e^x + C \quad \text{and} \quad \int xe^x\, dx.$$

The first integral is an elementary integral that we have already encountered. The second integral is only slightly different—and yet, the appearance of the product $xe^x$ in the integrand makes this integral (at the moment) impossible to evaluate. Integration by parts is ideally suited for evaluating integrals of *products* of functions.

## Integration by Parts for Indefinite Integrals

Given two differentiable functions $u$ and $v$, the Product Rule states that

$$\frac{d}{dx}\left(u(x)v(x)\right) = u'(x)v(x) + u(x)v'(x).$$

By integrating both sides, we can write this rule in terms of an indefinite integral:

$$u(x)v(x) = \int \left(u'(x)v(x) + u(x)v'(x)\right) dx.$$

Rearranging this expression in the form

$$\int u(x)\underbrace{v'(x)\, dx}_{dv} = u(x)v(x) - \int v(x)\underbrace{u'(x)\, dx}_{du}$$

leads to the basic relationship for *integration by parts*. It is expressed compactly by noting that $du = u'(x)\, dx$ and $dv = v'(x)\, dx$. Suppressing the independent variable $x$, we have

$$\int u\, dv = uv - \int v\, du.$$

The integral $\int u\, dv$ is viewed as the given integral, and we use integration by parts to express it in terms of a new integral $\int v\, du$. The technique is successful if the new integral can be evaluated.

> **Integration by Parts**
>
> Suppose $u$ and $v$ are differentiable functions. Then
>
> $$\int u\, dv = uv - \int v\, du.$$

▶ The integration by parts calculation may be done without including the constant of integration—as long as it is included in the final result.

▶ The arrows in the table show how to combine factors in the integration by parts formula. The first arrow indicates the product $uv$; the second arrow indicates the integrand $v\,du$.

▶ To make the table, first write the functions in the original integral:

$$u = \underline{\hspace{1cm}}, dv = \underline{\hspace{1cm}}.$$

Then find the functions in the new integral by differentiating $u$ and integrating $dv$:

$$du = \underline{\hspace{1cm}}, v = \underline{\hspace{1cm}}.$$

**QUICK CHECK 1** What are the best choices for $u$ and $dv$ in evaluating $\int x \cos x\,dx$? ◀

**EXAMPLE 1    Integration by parts** Evaluate $\int xe^x\,dx$.

**SOLUTION** The presence of *products* in the integrand often suggests integration by parts. We split the product $xe^x$ into two factors, one of which must be identified as $u$ and the other as $dv$ (the latter always includes the differential $dx$). Powers of $x$ are *often* good choices for $u$. The choice for $dv$ should be easy to integrate. In this case, the choices $u = x$ and $dv = e^x\,dx$ are advisable. It follows that $du = dx$. The relationship $dv = e^x\,dx$ means that $v$ is an antiderivative of $e^x$, which implies $v = e^x$. A table is helpful for organizing these calculations.

| Functions in original integral | $u = x$ | $dv = e^x\,dx$ |
|---|---|---|
| Functions in new integral | $du = dx$ | $v = e^x$ |

The integration by parts rule is now applied:

$$\int \underbrace{x}_{u}\ \underbrace{e^x\,dx}_{dv} = \underbrace{x}_{u}\ \underbrace{e^x}_{v} - \int \underbrace{e^x}_{v}\ \underbrace{dx}_{du}.$$

The original integral $\int xe^x\,dx$ has been replaced with the integral of $e^x$, which is easier to evaluate: $\int e^x\,dx = e^x + C$. The entire procedure looks like this:

$$\int xe^x\,dx = xe^x - \int e^x\,dx \quad \text{Integration by parts}$$
$$= xe^x - e^x + C. \quad \text{Evaluate the new integral.}$$

*Related Exercises 11–12* ◀

**EXAMPLE 2    Integration by parts** Evaluate $\int x \sin x\,dx$.

**SOLUTION** Remembering that powers of $x$ are often a good choice for $u$, we form the following table.

| $u = x$ | $dv = \sin x\,dx$ |
|---|---|
| $du = dx$ | $v = -\cos x$ |

Applying integration by parts, we have

$$\int \underbrace{x}_{u}\ \underbrace{\sin x\,dx}_{dv} = \underbrace{x}_{u}\ \underbrace{(-\cos x)}_{v} - \int \underbrace{(-\cos x)}_{v}\ \underbrace{dx}_{du} \quad \text{Integration by parts}$$
$$= -x \cos x + \sin x + C. \quad \text{Evaluate } \int \cos x\,dx = \sin x.$$

*Related Exercises 10, 16* ◀

In general, integration by parts works when we can easily integrate the choice for $dv$ and when the new integral is easier to evaluate than the original. Integration by parts is often used for integrals of the form $\int x^n f(x)\,dx$, where $n$ is a positive integer. Such integrals generally require the repeated use of integration by parts, as shown in the following example.

**EXAMPLE 3    Repeated use of integration by parts**

**a.** Evaluate $\int x^2 e^x\,dx$.

**b.** How would you evaluate $\int x^n e^x\,dx$, where $n$ is a positive integer?

**SOLUTION**

**a.** The factor $x^2$ is a good choice for $u$, leaving $dv = e^x\,dx$. We then have

$$\int \underbrace{x^2}_{u}\ \underbrace{e^x\,dx}_{dv} = \underbrace{x^2}_{u}\ \underbrace{e^x}_{v} - \int \underbrace{e^x}_{v}\ \underbrace{2x\,dx}_{du}.$$

Notice that the new integral on the right side is simpler than the original integral because the power of $x$ has been reduced by one. In fact, the new integral was evaluated in Example 1. Therefore, after using integration by parts twice, we have

$$\int x^2 e^x\, dx = x^2 e^x - 2\int xe^x\, dx \qquad \text{Integration by parts}$$

$$= x^2 e^x - 2(xe^x - e^x) + C \quad \text{Result of Example 1}$$

$$= e^x (x^2 - 2x + 2) + C. \qquad \text{Simplify.}$$

**b.** We now let $u = x^n$ and $dv = e^x\, dx$. The integration takes the form

$$\int x^n e^x\, dx = x^n e^x - n\int x^{n-1} e^x\, dx.$$

We see that integration by parts reduces the power of the variable in the integrand. The integral in part (a) with $n = 2$ requires two uses of integration by parts. You can probably anticipate that evaluating the integral $\int x^n e^x\, dx$ requires $n$ applications of integration by parts to reach the integral $\int e^x\, dx$, which is easily evaluated.

*Related Exercises 24–25* ◄

➤ An integral identity in which the power of a variable is reduced is called a **reduction formula.** Other examples of reduction formulas are explored in Exercises 50–53.

**EXAMPLE 4    Repeated use of integration by parts** Evaluate $\int e^{2x} \sin x\, dx$.

**SOLUTION**  The integrand consists of a product, which suggests integration by parts. In this case, there is no obvious choice for $u$ and $dv$, so let's try the following choices.

➤ In Example 4, we could also use $u = \sin x$ and $dv = e^{2x}\, dx$. In general, some trial and error may be required when using integration by parts. Effective choices come with practice.

The integral then becomes

$$\int e^{2x} \sin x\, dx = -e^{2x} \cos x + 2\int e^{2x} \cos x\, dx. \tag{1}$$

The original integral has been expressed in terms of a new integral, $\int e^{2x} \cos x\, dx$, which is no easier to evaluate than the original integral. It is tempting to start over with a new choice of $u$ and $dv$, but a little persistence pays off. Suppose we evaluate $\int e^{2x} \cos x\, dx$ using integration by parts with the following choices.

➤ When using integration by parts, the acronym LIPET may help. If the integrand is the product of two or more functions, choose $u$ to be the first function type that appears in the list

**L**ogarithmic, **I**nverse trigonometric, **P**olynomial, **E**xponential, **T**rigonometric.

| $u = e^{2x}$ | $dv = \cos x\, dx$ |
|---|---|
| $du = 2e^{2x}\, dx$ | $v = \sin x$ |

Integrating by parts, we have

$$\int e^{2x} \cos x\, dx = e^{2x} \sin x - 2\int e^{2x} \sin x\, dx. \tag{2}$$

Now observe that equation (2) contains the original integral, $\int e^{2x} \sin x\, dx$. Substituting the result of equation (2) into equation (1), we find that

$$\int e^{2x} \sin x\, dx = -e^{2x} \cos x + 2\int e^{2x} \cos x\, dx$$

$$= -e^{2x} \cos x + 2\left(e^{2x} \sin x - 2\int e^{2x} \sin x\, dx\right) \quad \text{Substitute for } \int e^{2x} \cos x\, dx.$$

$$= -e^{2x} \cos x + 2e^{2x} \sin x - 4\int e^{2x} \sin x\, dx. \quad \text{Simplify.}$$

➤ To solve for $\int e^{2x} \sin x\, dx$ in the equation $\int e^{2x} \sin x\, dx = -e^{2x} \cos x + 2e^{2x} \sin x - 4\int e^{2x} \sin x\, dx$, add $4\int e^{2x} \sin x\, dx$ to both sides of the equation and then divide both sides by 5.

Now it is a matter of solving for $\int e^{2x} \sin x\, dx$ and including the constant of integration:

$$\int e^{2x} \sin x\, dx = \frac{1}{5} e^{2x} (2 \sin x - \cos x) + C.$$

*Related Exercises 27–28* ◄

## Integration by Parts for Definite Integrals

Integration by parts with definite integrals presents two options. You can use the method outlined in Examples 1–4 to find an antiderivative and then evaluate it at the upper and lower limits of integration. Alternatively, the limits of integration can be incorporated directly into the integration by parts process. With the second approach, integration by parts for definite integrals has the following form.

> ▶ Integration by parts for definite integrals still has the form
>
> $$\int u \, dv = uv - \int v \, du.$$
>
> However, both definite integrals must be written with respect to $x$.

---

**Integration by Parts for Definite Integrals**

Let $u$ and $v$ be differentiable. Then

$$\int_a^b u(x)v'(x) \, dx = u(x)v(x) \Big|_a^b - \int_a^b v(x)u'(x) \, dx.$$

---

**EXAMPLE 5**   **A definite integral**   Evaluate $\int_1^2 \ln x \, dx$.

**SOLUTION**   This example is instructive because the integrand does not appear to be a product. The key is to view the integrand as the product $(\ln x)(1 \, dx)$. Then the following choices are plausible.

| $u = \ln x$ | $dv = dx$ |
|---|---|
| $du = \dfrac{1}{x} \, dx$ | $v = x$ |

Using integration by parts, we have

$$\int_1^2 \underbrace{\ln x}_{u} \underbrace{dx}_{dv} = \left( (\ln x) \, x \right) \Big|_1^2 - \int_1^2 \underbrace{x}_{v} \underbrace{\frac{1}{x} \, dx}_{du} \qquad \text{Integration by parts}$$

$$= (x \ln x - x) \Big|_1^2 \qquad \text{Integrate and simplify.}$$

$$= (2 \ln 2 - 2) - (0 - 1) \qquad \text{Evaluate.}$$

$$= 2 \ln 2 - 1 \approx 0.386. \qquad \text{Simplify.}$$

*Related Exercises 33–34* ◀

In Example 5, we evaluated a definite integral of $\ln x$. The corresponding indefinite integral can be added to our list of integration formulas.

**QUICK CHECK 2** Verify by differentiation that $\int \ln x \, dx = x \ln x - x + C$. ◀

---

**Integral of ln x**

$$\int \ln x \, dx = x \ln x - x + C$$

---

We now apply integration by parts to a familiar geometry problem.

**EXAMPLE 6**   **Solids of revolution**   Let $R$ be the region bounded by $y = \ln x$, the $x$-axis, and the line $x = e$ (**Figure 8.1**). Find the volume of the solid that is generated when the region $R$ is revolved about the $x$-axis.

**SOLUTION**   Revolving $R$ about the $x$-axis generates a solid whose volume is computed with the disk method (Section 6.3). Its volume is

$$V = \int_1^e \pi (\ln x)^2 \, dx.$$

**Figure 8.1**

► Recall that if $f(x) \geq 0$ on $[a, b]$ and the region bounded by the graph of $f$ and the $x$-axis on $[a, b]$ is revolved about the $x$-axis, then the volume of the solid generated is $V = \int_a^b \pi f(x)^2 \, dx$.

We integrate by parts with the following assignments.

The integration is carried out as follows, using the indefinite integral of $\ln x$ just given:

$$V = \int_1^e \pi (\ln x)^2 \, dx \qquad \text{Disk method}$$

$$= \pi \left( \underbrace{(\ln x)^2}_{u} \underbrace{x}_{v} \Big|_1^e - \int_1^e \underbrace{x}_{v} \underbrace{\frac{2 \ln x}{x}}_{du} \, dx \right) \qquad \text{Integration by parts}$$

$$= \pi \left( x(\ln x)^2 \Big|_1^e - 2 \int_1^e \ln x \, dx \right) \qquad \text{Simplify.}$$

**QUICK CHECK 3** How many times do you need to integrate by parts to reduce $\int_1^e (\ln x)^6 \, dx$ to an integral of $\ln x$? ◄

$$= \pi \left( x(\ln x)^2 \Big|_1^e - 2(x \ln x - x) \Big|_1^e \right) \qquad \int \ln x \, dx = x \ln x - x + C$$

$$= \pi (e(\ln e)^2 - 2e \ln e + 2e - 2) \qquad \text{Evaluate and simplify.}$$

$$= \pi (e - 2) \approx 2.257. \qquad \text{Simplify.}$$

*Related Exercises 45, 47* ◄

## SECTION 8.2 EXERCISES

### Getting Started

1. On which derivative rule is integration by parts based?

2. Use integration by parts to evaluate $\int x \cos x \, dx$ with $u = x$ and $dv = \cos x \, dx$.

3. Use integration by parts to evaluate $\int x \ln x \, dx$ with $u = \ln x$ and $dv = x \, dx$.

4. How is integration by parts used to evaluate a definite integral?

5. What type of integrand is a good candidate for integration by parts?

6. How would you choose $dv$ when evaluating $\int x^n e^{ax} \, dx$ using integration by parts?

**7–8.** *Use a substitution to reduce the following integrals to $\int \ln u \, du$. Then evaluate the resulting integral using the formula for $\int \ln x \, dx$.*

7. $\int (\sec^2 x) \ln (\tan x + 2) \, dx$  8. $\int (\cos x) \ln (\sin x) \, dx$

### Practice Exercises

**9–40. Integration by parts** *Evaluate the following integrals using integration by parts.*

9. $\int x \cos 5x \, dx$

10. $\int x \sin 2x \, dx$

11. $\int t e^{6t} \, dt$

12. $\int 2x e^{3x} \, dx$

13. $\int x \ln 10x \, dx$

14. $\int s e^{-2s} \, ds$

15. $\int (2w + 4) \cos 2w \, dw$

16. $\int \theta \sec^2 \theta \, d\theta$

17. $\int x \, 3^x \, dx$

18. $\int x^9 \ln x \, dx$

19. $\int \frac{\ln x}{x^{10}} \, dx$

20. $\int \sin^{-1} x \, dx$

21. $\int x \sin x \cos x \, dx$

22. $\int e^{2x} \sin e^x \, dx$

23. $\int x^2 \sin 2x \, dx$

24. $\int x^2 e^{4x} \, dx$

25. $\int t^2 e^{-t} \, dt$

26. $\int t^3 \sin t \, dt$

27. $\int e^x \cos x \, dx$

28. $\int e^{3x} \cos 2x \, dx$

29. $\int e^{-x} \sin 4x \, dx$

30. $\int e^{-2\theta} \sin 6\theta \, d\theta$

31. $\int e^{3x} \sin e^x \, dx$

32. $\int_0^1 x^2 \, 2^x \, dx$

33. $\int_0^\pi x \sin x \, dx$

34. $\int_1^e \ln 2x \, dx$

35. $\int_0^{\pi/2} x \cos 2x \, dx$

36. $\int_0^{\ln 2} x e^x \, dx$

37. $\int_1^{e^2} x^2 \ln x \, dx$

38. $\int x^2 \ln^2 x \, dx$

39. $\int_0^1 \sin^{-1} y \, dy$

40. $\int e^{\sqrt{x}} \, dx$

41. Evaluate the integral in part (a) and then use this result to evaluate the integral in part (b).

   a. $\int \tan^{-1} x \, dx$    b. $\int x \tan^{-1} x^2 \, dx$

**42–47. Volumes of solids** *Find the volume of the solid that is generated when the given region is revolved as described.*

42. The region bounded by $f(x) = \ln x$, $y = 1$, and the coordinate axes is revolved about the $x$-axis.

**43.** The region bounded by $f(x) = e^{-x}$, $x = \ln 2$, and the coordinate axes is revolved about the $y$-axis.

**44.** The region bounded by $f(x) = \sin x$ and the $x$-axis on $[0, \pi]$ is revolved about the $y$-axis.

**45.** The region bounded by $g(x) = \sqrt{\ln x}$ and the $x$-axis on $[1, e]$ is revolved about the $x$-axis.

**46.** The region bounded by $f(x) = e^{-x}$ and the $x$-axis on $[0, \ln 2]$ is revolved about the line $x = \ln 2$.

**47.** The region bounded by $f(x) = x \ln x$ and the $x$-axis on $[1, e^2]$ is revolved about the $x$-axis.

**48. Integral of $\sec^3 x$** Use integration by parts to show that
$$\int \sec^3 x \, dx = \frac{1}{2} \sec x \tan x + \frac{1}{2} \int \sec x \, dx.$$

**49. Explain why or why not** Determine whether the following statements are true and give an explanation or counterexample.

  **a.** $\int uv' \, dx = \left( \int u \, dx \right)\left( \int v' \, dx \right).$

  **b.** $\int uv' \, dx = uv - \int vu' \, dx.$

  **c.** $\int v \, du = uv - \int u \, dv.$

**50–53. Reduction formulas** *Use integration by parts to derive the following reduction formulas.*

**50.** $\int x^n e^{ax} \, dx = \dfrac{x^n e^{ax}}{a} - \dfrac{n}{a} \int x^{n-1} e^{ax} \, dx$, for $a \neq 0$

**51.** $\int x^n \cos ax \, dx = \dfrac{x^n \sin ax}{a} - \dfrac{n}{a} \int x^{n-1} \sin ax \, dx$, for $a \neq 0$

**52.** $\int x^n \sin ax \, dx = -\dfrac{x^n \cos ax}{a} + \dfrac{n}{a} \int x^{n-1} \cos ax \, dx$, for $a \neq 0$

**53.** $\int \ln^n x \, dx = x \ln^n x - n \int \ln^{n-1} x \, dx$

**54–57. Applying reduction formulas** *Use the reduction formulas in Exercises 50–53 to evaluate the following integrals.*

**54.** $\int x^2 e^{3x} \, dx$

**55.** $\int x^2 \cos 5x \, dx$

**56.** $\int x^3 \sin x \, dx$

**57.** $\int_1^e \ln^3 x \, dx$

**58. Two methods** Evaluate $\int_0^{\pi/3} \sin x \ln(\cos x) \, dx$ in the following two ways.

  **a.** Use integration by parts.      **b.** Use substitution.

**59. Two methods**

  **a.** Evaluate $\displaystyle\int \frac{x}{\sqrt{x+1}} \, dx$ using integration by parts.

  **b.** Evaluate $\displaystyle\int \frac{x}{\sqrt{x+1}} \, dx$ using substitution.

  **c.** Verify that your answers to parts (a) and (b) are consistent.

**60. Two methods**

  **a.** Evaluate $\int x \ln x^2 \, dx$ using the substitution $u = x^2$ and evaluating $\int \ln u \, du$.

  **b.** Evaluate $\int x \ln x^2 \, dx$ using integration by parts.

  **c.** Verify that your answers to parts (a) and (b) are consistent.

**61. Logarithm base $b$** Prove that $\displaystyle\int \log_b x \, dx = \frac{1}{\ln b}(x \ln x - x) + C.$

**62. Two integration methods** Evaluate $\int \sin x \cos x \, dx$ using integration by parts. Then evaluate the integral using a substitution. Reconcile your answers.

**63. Combining two integration methods** Evaluate $\int \cos \sqrt{x} \, dx$ using a substitution followed by integration by parts.

**64. Combining two integration methods** Evaluate $\int_0^{\pi^2/4} \sin \sqrt{x} \, dx$ using a substitution followed by integration by parts.

**65. An identity** Show that if $f$ has continuous derivatives on $[a, b]$ and $f'(a) = f'(b) = 0$, then
$$\int_a^b x f''(x) \, dx = f(a) - f(b).$$

**66. Integrating derivatives** Use integration by parts to show that if $f'$ is continuous on $[a, b]$, then
$$\int_a^b f(x) f'(x) \, dx = \frac{1}{2}(f(b)^2 - f(a)^2).$$

**67. Function defined as an integral** Find the arc length of the function $f(x) = \int_e^x \sqrt{\ln^2 t - 1} \, dt$ on $[e, e^3]$.

**68. Log integrals** Use integration by parts to show that for $m \neq -1$,
$$\int x^m \ln x \, dx = \frac{x^{m+1}}{m+1}\left( \ln x - \frac{1}{m+1} \right) + C$$
and for $m = -1$,
$$\int \frac{\ln x}{x} \, dx = \frac{1}{2} \ln^2 x + C.$$

**69. Comparing volumes** Let $R$ be the region bounded by $y = \sin x$ and the $x$-axis on the interval $[0, \pi]$. Which is greater, the volume of the solid generated when $R$ is revolved about the $x$-axis or the volume of the solid generated when $R$ is revolved about the $y$-axis?

**70. A useful integral**

  **a.** Use integration by parts to show that if $f'$ is continuous, then
$$\int x f'(x) \, dx = x f(x) - \int f(x) \, dx.$$

  **b.** Use part (a) to evaluate $\int x e^{3x} \, dx$.

**71. Solid of revolution** Find the volume of the solid generated when the region bounded by $y = \cos x$ and the $x$-axis on the interval $[0, \pi/2]$ is revolved about the $y$-axis.

### Explorations and Challenges

**72. Between the sine and inverse sine** Find the area of the region bounded by the curves $y = \sin x$ and $y = \sin^{-1} x$ on the interval $[0, 1/2]$.

**73. Two useful exponential integrals** Use integration by parts to derive the following formulas for real numbers $a$ and $b$.
$$\int e^{ax} \sin bx \, dx = \frac{e^{ax}(a \sin bx - b \cos bx)}{a^2 + b^2} + C$$
$$\int e^{ax} \cos bx \, dx = \frac{e^{ax}(a \cos bx + b \sin bx)}{a^2 + b^2} + C$$

**74. Integrating inverse functions** Assume $f$ has an inverse on its domain.

**a.** Let $y = f^{-1}(x)$ and show that
$$\int f^{-1}(x)\, dx = \int y f'(y)\, dy.$$

**b.** Use part (a) to show that
$$\int f^{-1}(x)\, dx = yf(y) - \int f(y)\, dy.$$

**c.** Use the result of part (b) to evaluate $\int \ln x\, dx$ (express the result in terms of $x$).

**d.** Use the result of part (b) to evaluate $\int \sin^{-1} x\, dx$.

**e.** Use the result of part (b) to evaluate $\int \tan^{-1} x\, dx$.

**T 75. Oscillator displacements** Suppose a mass on a spring that is slowed by friction has the position function $s(t) = e^{-t}\sin t$.

**a.** Graph the position function. At what times does the oscillator pass through the position $s = 0$?

**b.** Find the average value of the position on the interval $[0, \pi]$.

**c.** Generalize part (b) and find the average value of the position on the interval $[n\pi, (n + 1)\pi]$, for $n = 0, 1, 2, \ldots$.

**d.** Let $a_n$ be the absolute value of the average position on the interval $[n\pi, (n + 1)\pi]$, for $n = 0, 1, 2, \ldots$. Describe the pattern in the numbers $a_0, a_1, a_2, \ldots$.

**76. Find the error** Suppose you evaluate $\int \dfrac{dx}{x}$ using integration by parts. With $u = 1/x$ and $dv = dx$, you find that $du = -1/x^2\, dx$, $v = x$, and
$$\int \frac{dx}{x} = \left(\frac{1}{x}\right)x - \int x\left(-\frac{1}{x^2}\right)dx = 1 + \int \frac{dx}{x}.$$
You conclude that $0 = 1$. Explain the problem with the calculation.

**77. Tabular integration** Consider the integral $\int f(x)g(x)\, dx$, where $f$ can be differentiated repeatedly and $g$ can be integrated repeatedly. Let $G_k$ represent the result of calculating $k$ indefinite integrals of $g$, where the constants of integration are omitted.

**a.** Show that integration by parts, when applied to $\int f(x)g(x)\, dx$ with the choices $u = f(x)$ and $dv = g(x)\, dx$, leads to $\int f(x)g(x)\, dx = f(x)G_1(x) - \int f'(x)G_1(x)\, dx$. This formula can be remembered by utilizing the following table, where a right arrow represents a product of functions on the right side of the integration by parts formula, and a left arrow represents the *integral* of a product of functions (also appearing on the right side of the formula). Explain the significance of the signs associated with the arrows.

| f and its derivatives | g and its integrals |
|---|---|
| $f(x)$ | $g(x)$ |
| $f'(x)$ | $G_1(x)$ |

**b.** Perform integration by parts again on $\int f'(x)G_1(x)\, dx$ (from part (a)) with the choices $u = f'(x)$ and $dv = G_1(x)\, dx$ to show that $\int f(x)g(x)\, dx = f(x)G_1(x) - f'(x)G_2(x) + \int f''(x)G_2(x)\, dx$. Explain the connection between this integral formula and the following table, paying close attention to the signs attached to the arrows.

| f and its derivatives | g and its integrals |
|---|---|
| $f(x)$ | $g(x)$ |
| $f'(x)$ | $G_1(x)$ |
| $f''(x)$ | $G_2(x)$ |

**c.** Continue the pattern established in parts (a) and (b) and integrate by parts a third time. Write the integral formula that results from three applications of integration by parts, and construct the associated *tabular integration* table (include signs of the arrows).

**d.** The tabular integration table from part (c) is easily extended to allow for as many steps as necessary in the process of integration by parts. Evaluate $\int x^2 e^{x/2}\, dx$ by constructing an appropriate table, and explain why the process terminates after four rows of the table have been filled in.

**e.** Use tabular integration to evaluate $\int x^3 \cos x\, dx$. How many rows of the table are necessary? Why?

**f.** Explain why tabular integration is particularly suited to integrals of the form $\int p_n(x)g(x)\, dx$, where $p_n$ is a polynomial of degree $n > 0$ (and where, as before, we assume $g$ is easily integrated as many times as necessary).

**78. Practice with tabular integration** Evaluate the following integrals using tabular integration (refer to Exercise 77).

**a.** $\displaystyle\int x^4 e^x\, dx$

**b.** $\displaystyle\int 7xe^{3x}\, dx$

**c.** $\displaystyle\int_{-1}^{0} 2x^2\sqrt{x + 1}\, dx$

**d.** $\displaystyle\int (x^3 - 2x)\sin 2x\, dx$

**e.** $\displaystyle\int \frac{2x^2 - 3x}{(x - 1)^3}\, dx$

**f.** $\displaystyle\int \frac{x^2 + 3x + 4}{\sqrt[3]{2x + 1}}\, dx$

**g.** Why doesn't tabular integration work well when applied to
$$\int \frac{x}{\sqrt{1 - x^2}}\, dx?$$
Evaluate this integral using a different method.

**79. Tabular integration extended** Refer to Exercise 77.

**a.** The following table shows the method of tabular integration applied to $\int e^x\cos x\, dx$. Use the table to express $\int e^x\cos x\, dx$ in terms of the sum of functions and an indefinite integral.

| f and its derivatives | g and its integrals |
|---|---|
| $e^x$ | $\cos x$ |
| $e^x$ | $\sin x$ |
| $e^x$ | $-\cos x$ |

**b.** Solve the equation in part (a) for $\int e^x\cos x\, dx$.

**c.** Evaluate $\int e^{-2x}\sin 3x\, dx$ by applying the idea from parts (a) and (b).

**80. An identity** Show that if $f$ and $g$ have continuous second derivatives and $f(0) = f(1) = g(0) = g(1) = 0$, then
$$\int_0^1 f''(x)g(x)\, dx = \int_0^1 f(x)g''(x)\, dx.$$

**81. Possible and impossible integrals** Let $I_n = \int x^n e^{-x^2}\, dx$, where $n$ is a nonnegative integer.

**a.** $I_0 = \int e^{-x^2}\, dx$ cannot be expressed in terms of elementary functions. Evaluate $I_1$.

**b.** Use integration by parts to evaluate $I_3$.

**c.** Use integration by parts and the result of part (b) to evaluate $I_5$.

**d.** Show that, in general, if $n$ is odd, then $I_n = -\frac{1}{2} e^{-x^2} p_{n-1}(x)$, where $p_{n-1}$ is a polynomial of degree $n - 1$.

**e.** Argue that if $n$ is even, then $I_n$ cannot be expressed in terms of elementary functions.

**82. A family of exponentials** The curves $y = xe^{-ax}$ are shown in the figure for $a = 1, 2,$ and 3.

**a.** Find the area of the region bounded by $y = xe^{-x}$ and the $x$-axis on the interval $[0, 4]$.

**b.** Find the area of the region bounded by $y = xe^{-ax}$ and the $x$-axis on the interval $[0, 4]$, where $a > 0$.

**c.** Find the area of the region bounded by $y = xe^{-ax}$ and the $x$-axis on the interval $[0, b]$. Because this area depends on $a$ and $b$, we call it $A(a, b)$.

**d.** Use part (c) to show that $A(1, \ln b) = 4A\left(2, \frac{\ln b}{2}\right)$.

**e.** Does this pattern continue? Is it true that $A(1, \ln b) = a^2 A(a, (\ln b)/a)$?

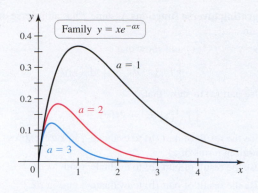

Family $y = xe^{-ax}$

$a = 1$

$a = 2$

$a = 3$

**QUICK CHECK ANSWERS**

**1.** Let $u = x$ and $dv = \cos x \, dx$.

**2.** $\frac{d}{dx}(x \ln x - x + C) = \ln x$

**3.** Integration by parts must be applied five times. ◄

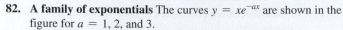

# 8.3 Trigonometric Integrals

At the moment, our inventory of integrals involving trigonometric functions is rather limited. For example, we can integrate the basic trigonometric functions found in Table 8.1, but have not yet encountered integrals such as $\int \cos^5 x \, dx$ and $\int \cos^2 x \sin^4 x \, dx$. The goal of this section is to develop techniques for evaluating integrals involving powers of trigonometric functions. These techniques are indispensable when we encounter *trigonometric substitutions* in the next section.

## Integrating Powers of sin $x$ or cos $x$

Two strategies are used when evaluating integrals of the form $\int \sin^m x \, dx$ or $\int \cos^n x \, dx$, where $m$ and $n$ are positive integers. Both strategies use trigonometric identities to recast the integrand, as shown in the first example.

> Some of the techniques described in this section also work for negative powers of trigonometric functions.

**EXAMPLE 1    Powers of sine or cosine**  Evaluate the following integrals.

**a.** $\int \cos^5 x \, dx$          **b.** $\int \sin^4 x \, dx$

**SOLUTION**

> Pythagorean identities:
> $$\cos^2 x + \sin^2 x = 1$$
> $$1 + \tan^2 x = \sec^2 x$$
> $$\cot^2 x + 1 = \csc^2 x$$

**a.** Integrals involving odd powers of $\cos x$ (or $\sin x$) are most easily evaluated by splitting off a single factor of $\cos x$ (or $\sin x$). In this case, we rewrite $\cos^5 x$ as $\cos^4 x \cdot \cos x$. Notice that $\cos^4 x$ can be written in terms of $\sin x$ using the identity $\cos^2 x = 1 - \sin^2 x$. The result is an integrand that readily yields to the substitution $u = \sin x$:

$$\int \cos^5 x \, dx = \int \cos^4 x \cos x \, dx \qquad \text{Split off } \cos x.$$

$$= \int (1 - \sin^2 x)^2 \cos x \, dx \qquad \text{Pythagorean identity}$$

$$= \int (1 - u^2)^2 \, du \qquad \text{Let } u = \sin x; \, du = \cos x \, dx.$$

$$= \int (1 - 2u^2 + u^4) \, du \qquad \text{Expand.}$$

$$= u - \frac{2}{3} u^3 + \frac{1}{5} u^5 + C \qquad \text{Integrate.}$$

$$= \sin x - \frac{2}{3} \sin^3 x + \frac{1}{5} \sin^5 x + C. \qquad \text{Replace } u \text{ with } \sin x.$$

> ➤ The half-angle formulas for $\sin^2 x$ and $\cos^2 x$ are easily confused. Use the phrase "sine is minus" to remember that a minus sign is associated with the half-angle formula for $\sin^2 x$, whereas a positive sign is used for $\cos^2 x$.

**b.** With even positive powers of $\sin x$ or $\cos x$, we use the half-angle formulas

$$\sin^2 x = \frac{1 - \cos 2x}{2} \quad \text{and} \quad \cos^2 x = \frac{1 + \cos 2x}{2}$$

to reduce the powers in the integrand:

$$\int \sin^4 x \, dx = \int \underbrace{\left(\frac{1 - \cos 2x}{2}\right)^2}_{\sin^2 x} dx \qquad \text{Half-angle formula}$$

$$= \frac{1}{4} \int (1 - 2\cos 2x + \cos^2 2x) \, dx. \quad \text{Expand the integrand.}$$

Using the half-angle formula for $\cos^2 2x$, the evaluation may be completed:

$$\int \sin^4 x \, dx = \frac{1}{4} \int \left(1 - 2\cos 2x + \underbrace{\frac{1 + \cos 4x}{2}}_{\cos^2 2x}\right) dx \quad \text{Half-angle formula}$$

$$= \frac{1}{4} \int \left(\frac{3}{2} - 2\cos 2x + \frac{1}{2}\cos 4x\right) dx \qquad \text{Simplify.}$$

$$= \frac{3x}{8} - \frac{1}{4}\sin 2x + \frac{1}{32}\sin 4x + C. \qquad \text{Evaluate the integrals.}$$

*Related Exercises 9–11* ◄

**QUICK CHECK 1** Evaluate $\int \sin^3 x \, dx$ by splitting off a factor of $\sin x$, rewriting $\sin^2 x$ in terms of $\cos x$, and using an appropriate $u$-substitution. ◄

## Integrating Products of Powers of $\sin x$ and $\cos x$

We now consider integrals of the form $\int \sin^m x \cos^n x \, dx$. If $m$ is an odd, positive integer, we split off a factor of $\sin x$ and write the remaining even power of $\sin x$ in terms of cosine functions. This step prepares the integrand for the substitution $u = \cos x$, and the resulting integral is readily evaluated. A similar strategy is used when $n$ is an odd, positive integer.

If both $m$ and $n$ are even positive integers, the half-angle formulas are used to transform the integrand into a polynomial in $\cos 2x$, each of whose terms can be integrated, as shown in Example 2.

**EXAMPLE 2** **Products of sine and cosine** Evaluate the following integrals.

**a.** $\int \sin^4 x \cos^2 x \, dx$      **b.** $\int \sin^3 x \cos^{-2} x \, dx$

**SOLUTION**

**a.** When both powers are even positive integers, the half-angle formulas are used:

$$\int \sin^4 x \cos^2 x \, dx = \int \underbrace{\left(\frac{1 - \cos 2x}{2}\right)^2}_{\sin^2 x}\underbrace{\left(\frac{1 + \cos 2x}{2}\right)}_{\cos^2 x} dx \qquad \text{Half-angle formulas}$$

$$= \frac{1}{8} \int (1 - \cos 2x - \cos^2 2x + \cos^3 2x) \, dx. \quad \text{Expand.}$$

The third term in the integrand is rewritten with a half-angle formula. For the last term, a factor of $\cos 2x$ is split off, and the resulting even power of $\cos 2x$ is written in terms of $\sin 2x$ to prepare for a $u$-substitution:

$$\int \sin^4 x \cos^2 x \, dx =$$

$$\frac{1}{8} \int \left(1 - \cos 2x - \left(\overbrace{\frac{1 + \cos 4x}{2}}^{\cos^2 2x}\right)\right) dx + \frac{1}{8} \int \overbrace{(1 - \sin^2 2x)}^{\cos^2 2x} \cdot \cos 2x \, dx.$$

Finally, the integrals are evaluated, using the substitution $u = \sin 2x$ for the second integral. After simplification, we find that

$$\int \sin^4 x \cos^2 x \, dx = \frac{1}{16}x - \frac{1}{64}\sin 4x - \frac{1}{48}\sin^3 2x + C.$$

**b.** When at least one power is odd and positive, the following approach works:

$$\int \sin^3 x \cos^{-2} x \, dx = \int \sin^2 x \cos^{-2} x \cdot \sin x \, dx \qquad \text{Split off } \sin x.$$

$$= \int (1 - \cos^2 x) \cos^{-2} x \cdot \sin x \, dx \qquad \text{Pythagorean identity}$$

$$= -\int (1 - u^2) u^{-2} \, du \qquad u = \cos x; \; du = -\sin x \, dx$$

$$= \int (1 - u^{-2}) \, du = u + \frac{1}{u} + C \qquad \text{Evaluate the integral.}$$

$$= \cos x + \sec x + C. \qquad \text{Replace } u \text{ with } \cos x.$$

*Related Exercises 15–17* ◀

**QUICK CHECK 2** What strategy would you use to evaluate $\int \sin^3 x \cos^3 x \, dx$? ◀

Table 8.2 summarizes the techniques used to evaluate integrals of the form $\int \sin^m x \cos^n x \, dx$.

**Table 8.2**

| $\int \sin^m x \cos^n x \, dx$ | Strategy |
|---|---|
| $m$ odd and positive, $n$ real | Split off $\sin x$, rewrite the resulting even power of $\sin x$ in terms of $\cos x$, and then use $u = \cos x$. |
| $n$ odd and positive, $m$ real | Split off $\cos x$, rewrite the resulting even power of $\cos x$ in terms of $\sin x$, and then use $u = \sin x$. |
| $m$ and $n$ both even, nonnegative integers | Use half-angle formulas to transform the integrand into a polynomial in $\cos 2x$, and apply the preceding strategies once again to powers of $\cos 2x$ greater than 1. |

➤ If both $m$ and $n$ are odd, you may split off $\sin x$ or $\cos x$, though in practice, one choice may be more efficient than the other.

## Reduction Formulas

Evaluating an integral such as $\int \sin^8 x \, dx$ using the method of Example 1b is tedious, at best. For this reason, *reduction formulas* have been developed to ease the workload. A reduction formula equates an integral involving a power of a function with another integral in which the power is reduced; several reduction formulas were encountered in Exercises 50–53 of Section 8.2. Here are some frequently used reduction formulas for trigonometric integrals.

---

**Reduction Formulas**

Assume $n$ is a positive integer.

**1.** $\displaystyle\int \sin^n x \, dx = -\frac{\sin^{n-1} x \cos x}{n} + \frac{n-1}{n} \int \sin^{n-2} x \, dx$

**2.** $\displaystyle\int \cos^n x \, dx = \frac{\cos^{n-1} x \sin x}{n} + \frac{n-1}{n} \int \cos^{n-2} x \, dx$

**3.** $\displaystyle\int \tan^n x \, dx = \frac{\tan^{n-1} x}{n-1} - \int \tan^{n-2} x \, dx, \quad n \neq 1$

**4.** $\displaystyle\int \sec^n x \, dx = \frac{\sec^{n-2} x \tan x}{n-1} + \frac{n-2}{n-1} \int \sec^{n-2} x \, dx, \quad n \neq 1$

---

Formulas 1, 3, and 4 are derived in Exercises 72–74. The derivation of formula 2 is similar to that of formula 1.

**EXAMPLE 3** **Powers of tan $x$** Evaluate $\int \tan^4 x \, dx$.

**SOLUTION** Reduction formula 3 gives

$$\int \tan^4 x \, dx = \frac{1}{3} \tan^3 x - \underbrace{\int \tan^2 x \, dx}_{\text{use (3) again}}$$

$$= \frac{1}{3} \tan^3 x - (\tan x - \int \underbrace{\tan^0 x}_{=1} \, dx)$$

$$= \frac{1}{3} \tan^3 x - \tan x + x + C.$$

An alternative solution uses the identity $\tan^2 x = \sec^2 x - 1$:

$$\int \tan^4 x \, dx = \int \tan^2 x \underbrace{(\sec^2 x - 1)}_{\tan^2 x} \, dx$$

$$= \int \tan^2 x \sec^2 x \, dx - \int \tan^2 x \, dx.$$

The substitution $u = \tan x$, $du = \sec^2 x \, dx$ is used in the first integral, and the identity $\tan^2 x = \sec^2 x - 1$ is used again in the second integral:

$$\int \tan^4 x \, dx = \int \underbrace{\tan^2 x}_{u^2} \underbrace{\sec^2 x \, dx}_{du} - \int \tan^2 x \, dx$$

$$= \int u^2 \, du - \int (\sec^2 x - 1) \, dx \qquad \text{Substitution and identity}$$

$$= \frac{u^3}{3} - \tan x + x + C \qquad \text{Evaluate integrals.}$$

$$= \frac{1}{3} \tan^3 x - \tan x + x + C. \qquad u = \tan x$$

*Related Exercises 30–31* ◄

Note that for odd powers of $\tan x$ and $\sec x$, the use of reduction formula 3 or 4 will eventually lead to $\int \tan x \, dx$ or $\int \sec x \, dx$. Table 8.1 in Section 8.1 gives these integrals, along with the integrals of $\cot x$ and $\csc x$.

### Integrating Products of Powers of $\tan x$ and $\sec x$

Integrals of the form $\int \tan^m x \sec^n x \, dx$ are evaluated using methods analogous to those used for $\int \sin^m x \cos^n x \, dx$. For example, if $n$ is an even positive integer, we split off a factor of $\sec^2 x$ and write the remaining even power of $\sec x$ in terms of $\tan x$. This step prepares the integral for the substitution $u = \tan x$. If $m$ is odd and positive, we split off a factor of $\sec x \tan x$ (the derivative of $\sec x$), which prepares the integral for the substitution $u = \sec x$. If $m$ is even and $n$ is odd, the integrand is expressed as a polynomial in $\sec x$, each of whose terms is handled by a reduction formula. Example 4 illustrates these techniques.

**EXAMPLE 4** **Products of $\tan x$ and $\sec x$** Evaluate the following integrals.

**a.** $\int \tan^3 x \sec^4 x \, dx$ **b.** $\int \tan^2 x \sec x \, dx$

**SOLUTION**

**a.** With an even power of $\sec x$, we split off a factor of $\sec^2 x$ and prepare the integral for the substitution $u = \tan x$:

$$\int \tan^3 x \sec^4 x \, dx = \int \tan^3 x \sec^2 x \cdot \sec^2 x \, dx$$

$$= \int \tan^3 x \, (\tan^2 x + 1) \cdot \sec^2 x \, dx \qquad \sec^2 x = \tan^2 x + 1$$

$$= \int u^3 \, (u^2 + 1) \, du \qquad u = \tan x; \, du = \sec^2 x \, dx$$

$$= \frac{1}{6} \tan^6 x + \frac{1}{4} \tan^4 x + C. \qquad \text{Evaluate; } u = \tan x.$$

▶ In Example 4a, the two methods produce results that look different but are equivalent. This is common when evaluating trigonometric integrals. For instance, evaluate $\int \sin^4 x \, dx$ using reduction formula 1, and compare your answer to

$$\frac{3x}{8} - \frac{1}{4}\sin 2x + \frac{1}{32}\sin 4x + C,$$

the solution found in Example 1b.

Because the integrand also has an odd power of $\tan x$, an alternative solution is to split off a factor of $\sec x \tan x$ and prepare the integral for the substitution $u = \sec x$:

$$\int \tan^3 x \sec^4 x \, dx = \int \underbrace{\tan^2 x}_{\sec^2 x - 1} \sec^3 x \cdot \sec x \tan x \, dx$$

$$= \int (\sec^2 x - 1) \sec^3 x \cdot \sec x \tan x \, dx$$

$$= \int (u^2 - 1) u^3 \, du \qquad \begin{array}{l} u = \sec x; \\ du = \sec x \tan x \, dx \end{array}$$

$$= \frac{1}{6} \sec^6 x - \frac{1}{4} \sec^4 x + C. \qquad \text{Evaluate; } u = \sec x.$$

The apparent difference in the two solutions given here is reconciled by using the identity $1 + \tan^2 x = \sec^2 x$ to transform the second result into the first, the only difference being an additive constant, which is part of $C$.

**b.** In this case, we write the even power of $\tan x$ in terms of $\sec x$:

$$\int \tan^2 x \sec x \, dx = \int (\sec^2 x - 1) \sec x \, dx \qquad \tan^2 x = \sec^2 x - 1$$

$$= \int \sec^3 x \, dx - \int \sec x \, dx \qquad \text{Expand.}$$

$$\overbrace{= \frac{1}{2} \sec x \tan x + \frac{1}{2} \int \sec x \, dx}^{\text{reduction formula 4}} - \int \sec x \, dx \qquad \text{Use reduction formula.}$$

$$= \frac{1}{2} \sec x \tan x - \frac{1}{2} \ln |\sec x + \tan x| + C. \qquad \begin{array}{l}\text{Add secant integrals; use} \\ \text{Table 8.1 in Section 8.1.}\end{array}$$

*Related Exercises 33–35* ◀

Table 8.3 summarizes the methods used to integrate $\int \tan^m x \sec^n x \, dx$. Analogous techniques are used for $\int \cot^m x \csc^n x \, dx$.

**Table 8.3**

| $\int \tan^m x \sec^n x \, dx$ | Strategy |
|---|---|
| $n$ even and positive, $m$ real | Split off $\sec^2 x$, rewrite the remaining even power of $\sec x$ in terms of $\tan x$, and use $u = \tan x$. |
| $m$ odd and positive, $n$ real | Split off $\sec x \tan x$, rewrite the remaining even power of $\tan x$ in terms of $\sec x$, and use $u = \sec x$. |
| $m$ even and positive, $n$ odd and positive | Rewrite the even power of $\tan x$ in terms of $\sec x$ to produce a polynomial in $\sec x$; apply reduction formula 4 to each term. |

## SECTION 8.3 EXERCISES

### Getting Started

1. State the half-angle identities used to integrate $\sin^2 x$ and $\cos^2 x$.

2. State the three Pythagorean identities.

3. Describe the method used to integrate $\sin^3 x$.

4. Describe the method used to integrate $\sin^m x \cos^n x$, for $m$ even and $n$ odd.

5. What is a reduction formula?

6. How would you evaluate $\int \cos^2 x \sin^3 x \, dx$?

7. How would you evaluate $\int \tan^{10} x \sec^2 x \, dx$?

8. How would you evaluate $\int \sec^{12} x \tan x \, dx$?

### Practice Exercises

**9–61. Trigonometric integrals** *Evaluate the following integrals.*

9. $\int \cos^3 x \, dx$

10. $\int \sin^3 x \, dx$

11. $\int \sin^2 3x \, dx$

12. $\int \cos^4 2\theta \, d\theta$

13. $\int \sin^5 x \, dx$

14. $\int \cos^3 20x \, dx$

15. $\int \sin^3 x \cos^2 x \, dx$

16. $\int \sin^2 \theta \cos^5 \theta \, d\theta$

17. $\int \cos^3 x \sqrt{\sin x} \, dx$

18. $\int \sin^3 \theta \cos^{-2} \theta \, d\theta$

19. $\int_0^{\pi/3} \sin^5 x \cos^{-2} x \, dx$

20. $\int \sin^{-3/2} x \cos^3 x \, dx$

**21.** $\displaystyle\int_0^{\pi/2} \cos^3 x \sqrt{\sin^3 x}\, dx$

**22.** $\displaystyle\int_{\pi/4}^{\pi/2} \sin^2 2x \cos^3 2x\, dx$

**23.** $\displaystyle\int \sin^2 x \cos^2 x\, dx$

**24.** $\displaystyle\int \sin^3 x \cos^5 x\, dx$

**25.** $\displaystyle\int \sin^2 x \cos^4 x\, dx$

**26.** $\displaystyle\int \sin^3 x \cos^{3/2} x\, dx$

**27.** $\displaystyle\int \tan^2 x\, dx$

**28.** $\displaystyle\int 6 \sec^4 x\, dx$

**29.** $\displaystyle\int \cot^4 x\, dx$

**30.** $\displaystyle\int \tan^3 \theta\, d\theta$

**31.** $\displaystyle\int 20 \tan^6 x\, dx$

**32.** $\displaystyle\int \cot^5 3x\, dx$

**33.** $\displaystyle\int 10 \tan^9 x \sec^2 x\, dx$

**34.** $\displaystyle\int \tan^9 x \sec^4 x\, dx$

**35.** $\displaystyle\int \tan x \sec^3 x\, dx$

**36.** $\displaystyle\int \tan 4x \sec^{3/2} 4x\, dx$

**37.** $\displaystyle\int \frac{\sec^4 (\ln \theta)}{\theta}\, d\theta$

**38.** $\displaystyle\int \tan^5 \theta \sec^4 \theta\, d\theta$

**39.** $\displaystyle\int_{-\pi/3}^{\pi/3} \sqrt{\sec^2 \theta - 1}\, d\theta$

**40.** $\displaystyle\int_0^{\pi/6} \tan^5 2x \sec 2x\, dx$

**41.** $\displaystyle\int_0^{\pi/4} \sec^7 x \sin x\, dx$

**42.** $\displaystyle\int \sqrt{\tan x} \sec^4 x\, dx$

**43.** $\displaystyle\int \tan^3 4x\, dx$

**44.** $\displaystyle\int \frac{\sec^2 x}{\tan^5 x}\, dx$

**45.** $\displaystyle\int \sec^2 x \tan^{1/2} x\, dx$

**46.** $\displaystyle\int \sec^{-2} x \tan^3 x\, dx$

**47.** $\displaystyle\int \frac{\csc^4 x}{\cot^2 x}\, dx$

**48.** $\displaystyle\int \csc^{10} x \cot x\, dx$

**49.** $\displaystyle\int_{\pi/20}^{\pi/10} \csc^2 5w \cot^4 5w\, dw$

**50.** $\displaystyle\int \csc^{10} x \cot^3 x\, dx$

**51.** $\displaystyle\int (\csc^2 x + \csc^4 x)\, dx$

**52.** $\displaystyle\int_0^{\pi/8} (\tan 2x + \tan^3 2x)\, dx$

**53.** $\displaystyle\int_0^{\pi/4} \sec^4 \theta\, d\theta$

**54.** $\displaystyle\int_0^{\sqrt{\pi/2}} x \sin^3(x^2)\, dx$

**55.** $\displaystyle\int_{\pi/6}^{\pi/3} \cot^3 \theta\, d\theta$

**56.** $\displaystyle\int_0^{\pi/4} \tan^3 \theta \sec^2 \theta\, d\theta$

**57.** $\displaystyle\int_0^{\pi} (1 - \cos 2x)^{3/2}\, dx$

**58.** $\displaystyle\int_{-\pi/4}^{\pi/4} \sqrt{1 + \cos 4x}\, dx$

**59.** $\displaystyle\int_0^{\pi/2} \sqrt{1 - \cos 2x}\, dx$

**60.** $\displaystyle\int_0^{\pi/8} \sqrt{1 - \cos 8x}\, dx$

**61.** $\displaystyle\int_0^{\pi/4} (1 + \cos 4x)^{3/2}\, dx$

**62. Arc length** Find the length of the curve $y = \ln(\sec x)$, for $0 \le x \le \pi/4$.

**63. Explain why or why not** Determine whether the following statements are true and give an explanation or counterexample.

   **a.** If $m$ is a positive integer, then $\int_0^\pi \cos^{2m+1} x\, dx = 0$.

   **b.** If $m$ is a positive integer, then $\int_0^\pi \sin^m x\, dx = 0$.

**64. Sine football** Find the volume of the solid generated when the region bounded by $y = \sin x$ and the $x$-axis on the interval $[0, \pi]$ is revolved about the $x$-axis.

**65. Volume** Find the volume of the solid generated when the region bounded by $y = \sin^2 x \cos^{3/2} x$ and the $x$-axis on the interval $[0, \pi/2]$ is revolved about the $x$-axis.

**66. Particle position** A particle moves along a line with a velocity (in m/s) given by $v(t) = \sec^4 \dfrac{\pi t}{12}$, for $0 \le t \le 5$, where $t$ is measured in seconds. Determine the position function $s(t)$, for $0 \le t \le 5$. Assume $s(0) = 0$.

**67–70. Integrals of the form $\int \sin mx \cos nx\, dx$** Use the following three identities to evaluate the given integrals.

$$\sin mx \sin nx = \frac{1}{2}\left(\cos\left((m - n)x\right) - \cos\left((m + n)x\right)\right)$$

$$\sin mx \cos nx = \frac{1}{2}\left(\sin\left((m - n)x\right) + \sin\left((m + n)x\right)\right)$$

$$\cos mx \cos nx = \frac{1}{2}\left(\cos\left((m - n)x\right) + \cos\left((m + n)x\right)\right)$$

**67.** $\displaystyle\int \sin 3x \cos 7x\, dx$

**68.** $\displaystyle\int \sin 5x \sin 7x\, dx$

**69.** $\displaystyle\int \sin 3x \sin 2x\, dx$

**70.** $\displaystyle\int \cos x \cos 2x\, dx$

**Explorations and Challenges**

**71.** Prove the following **orthogonality relations** (which are used to generate *Fourier series*). Assume $m$ and $n$ are integers with $m \ne n$.

   **a.** $\displaystyle\int_0^\pi \sin mx \sin nx\, dx = 0$    **b.** $\displaystyle\int_0^\pi \cos mx \cos nx\, dx = 0$

   **c.** $\displaystyle\int_0^\pi \sin mx \cos nx\, dx = 0$, for $|m + n|$ even

**72. A sine reduction formula** Use integration by parts to obtain the reduction formula for positive integers $n$:

$$\int \sin^n x\, dx = -\sin^{n-1} x \cos x + (n - 1)\int \sin^{n-2} x \cos^2 x\, dx.$$

Then use an identity to obtain the reduction formula

$$\int \sin^n x\, dx = -\frac{\sin^{n-1} x \cos x}{n} + \frac{n - 1}{n}\int \sin^{n-2} x\, dx.$$

Use this reduction formula to evaluate $\int \sin^6 x\, dx$.

**73. A tangent reduction formula** Prove that for positive integers $n \ne 1$,

$$\int \tan^n x\, dx = \frac{\tan^{n-1} x}{n - 1} - \int \tan^{n-2} x\, dx.$$

Use the formula to evaluate $\int_0^{\pi/4} \tan^3 x\, dx$.

**74. A secant reduction formula** Prove that for positive integers $n \ne 1$,

$$\int \sec^n x\, dx = \frac{\sec^{n-2} x \tan x}{n - 1} + \frac{n - 2}{n - 1}\int \sec^{n-2} x\, dx.$$

(*Hint:* Integrate by parts with $u = \sec^{n-2} x$ and $dv = \sec^2 x\, dx$.)

**75. Exploring powers of sine and cosine**

  **a.** Graph the functions $f_1(x) = \sin^2 x$ and $f_2(x) = \sin^2 2x$ on the interval $[0, \pi]$. Find the area under these curves on $[0, \pi]$.

  **b.** Graph a few more of the functions $f_n(x) = \sin^2 nx$ on the interval $[0, \pi]$, where $n$ is a positive integer. Find the area under these curves on $[0, \pi]$. Comment on your observations.

  **c.** Prove that $\int_0^\pi \sin^2 nx \, dx$ has the same value for all positive integers $n$.

  **d.** Does the conclusion of part (c) hold if sine is replaced by cosine?

  **e.** Repeat parts (a), (b), and (c) with $\sin^2 x$ replaced by $\sin^4 x$. Comment on your observations.

**QUICK CHECK ANSWERS**

**1.** $\frac{1}{3} \cos^3 x - \cos x + C$   **2.** Write $\int \sin^3 x \cos^3 x \, dx = \int \sin^2 x \cos^3 x \sin x \, dx = \int (1 - \cos^2 x) \cos^3 x \sin x \, dx$. Then use the substitution $u = \cos x$. Or begin by writing $\int \sin^3 x \cos^3 x \, dx = \int \sin^3 x \cos^2 x \cos x \, dx$. ◄

# 8.4 Trigonometric Substitutions

In Example 3 of Section 6.5, we wrote the arc length integral for the segment of the parabola $y = x^2$ on the interval $[0, 2]$ as

$$\int_0^2 \sqrt{1 + 4x^2} \, dx = \int_0^2 2\sqrt{\tfrac{1}{4} + x^2} \, dx.$$

At the time, we did not have the analytical methods needed to evaluate this integral. The difficulty with $\int_0^2 \sqrt{1 + 4x^2} \, dx$ is that the square root of a sum (or difference) of two squares is not easily simplified. On the other hand, the square root of a product of two squares is easily simplified: $\sqrt{A^2 B^2} = |AB|$. If we could somehow replace $1 + 4x^2$ with a product of squares, this integral might be easier to evaluate. The goal of this section is to introduce techniques that transform sums of squares $a^2 + x^2$ (and the difference of squares $a^2 - x^2$ and $x^2 - a^2$) into products of squares.

Integrals similar to the arc length integral for the parabola arise in many different situations. For example, electrostatic, magnetic, and gravitational forces obey an inverse square law (their strength is proportional to $1/r^2$, where $r$ is a distance). Computing these force fields in two dimensions leads to integrals such as $\displaystyle\int \frac{dx}{\sqrt{x^2 + a^2}}$ or $\displaystyle\int \frac{dx}{(x^2 + a^2)^{3/2}}$.

It turns out that integrals containing the terms $a^2 \pm x^2$ or $x^2 - a^2$, where $a$ is a constant, can be simplified using somewhat unexpected substitutions involving trigonometric functions. The new integrals produced by these substitutions are often trigonometric integrals of the variety studied in the preceding section.

## Integrals Involving $a^2 - x^2$

> ► To understand how a difference of squares is rewritten as a product of squares, think of the Pythagorean Theorem: $a^2 + b^2 = c^2$. A rearrangement of this theorem leads to the standard substitution for integrals involving the difference of squares $a^2 - x^2$. The term $\sqrt{a^2 - x^2}$ is the length of one side of a right triangle whose hypotenuse has length $a$ and whose other side has length $x$. Labeling one acute angle $\theta$, we see that the substitution $x = a \sin \theta$ transforms $a^2 - x^2$ into $a^2 \cos^2 \theta$.

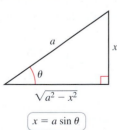

$$x = a \sin \theta$$

Suppose you are faced with an integral whose integrand contains the term $a^2 - x^2$, where $a$ is a positive constant. Observe what happens when $x$ is replaced with $a \sin \theta$:

$$
\begin{aligned}
a^2 - x^2 &= a^2 - (a \sin \theta)^2 \quad &\text{Replace } x \text{ with } a \sin \theta.\\
&= a^2 - a^2 \sin^2 \theta \quad &\text{Simplify.}\\
&= a^2 (1 - \sin^2 \theta) \quad &\text{Factor.}\\
&= a^2 \cos^2 \theta. \quad &1 - \sin^2 \theta = \cos^2 \theta
\end{aligned}
$$

This calculation shows that the change of variables $x = a \sin \theta$ turns the difference $a^2 - x^2$ into the product $a^2 \cos^2 \theta$. The resulting integral—now with respect to $\theta$—is often easier to evaluate than the original integral. The details of this procedure are spelled out in the following examples.

**QUICK CHECK 1** Use a substitution of the form $x = a \sin \theta$ to transform $9 - x^2$ into a product. ◄

**EXAMPLE 1**   **Area of a circle**  Verify that the area of a circle of radius $a$ is $\pi a^2$.

**SOLUTION**  The function $f(x) = \sqrt{a^2 - x^2}$ describes the upper half of a circle centered at the origin with radius $a$ (**Figure 8.2**). The region under this curve on the interval $[0, a]$ is a quarter-circle. Therefore, the area of the full circle is $4 \int_0^a \sqrt{a^2 - x^2} \, dx$.

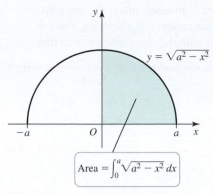

Area $= \int_0^a \sqrt{a^2 - x^2}\, dx$

**Figure 8.2**

$x = a \sin \theta \Rightarrow \theta = \sin^{-1} \dfrac{x}{a}$
$-\dfrac{\pi}{2} \le \theta \le \dfrac{\pi}{2}$

**Figure 8.3**

▶ The key identities for integrating $\sin^2 \theta$ and $\cos^2 \theta$ are

$$\sin^2 \theta = \frac{1 - \cos 2\theta}{2} \quad \text{and}$$

$$\cos^2 \theta = \frac{1 + \cos 2\theta}{2}.$$

Because the integrand contains the expression $a^2 - x^2$, we use the trigonometric substitution $x = a \sin \theta$, which can also be written $\theta = \sin^{-1}(x/a)$. This substitution is well defined only when the angle $\theta$ is restricted to the interval $-\dfrac{\pi}{2} \le \theta \le \dfrac{\pi}{2}$, which is the range of $\sin^{-1}(x/a)$ (**Figure 8.3**). As with all substitutions, the differential associated with the substitution must be computed:

$$x = a \sin \theta \quad \text{implies that} \quad dx = a \cos \theta\, d\theta.$$

Notice that the new variable $\theta$ plays the role of an angle. Replacing $x$ with $a \sin \theta$ in the integrand, we have

$$
\begin{aligned}
\sqrt{a^2 - x^2} &= \sqrt{a^2 - (a \sin \theta)^2} && \text{Replace } x \text{ with } a \sin \theta. \\
&= \sqrt{a^2 (1 - \sin^2 \theta)} && \text{Factor.} \\
&= \sqrt{a^2 \cos^2 \theta} && 1 - \sin^2 \theta = \cos^2 \theta \\
&= |a \cos \theta| && \sqrt{x^2} = |x| \\
&= a \cos \theta. && a > 0, \cos \theta \ge 0, \text{ for } -\tfrac{\pi}{2} \le \theta \le \tfrac{\pi}{2}
\end{aligned}
$$

We also change the limits of integration: When $x = 0$, $\theta = \sin^{-1} 0 = 0$; when $x = a$, $\theta = \sin^{-1}(a/a) = \sin^{-1} 1 = \pi/2$. Making these substitutions, we evaluate the integral as follows:

$$
\begin{aligned}
4 \int_0^a \sqrt{a^2 - x^2}\, dx &= 4 \int_0^{\pi/2} \underbrace{a \cos \theta}_{\substack{\text{integrand} \\ \text{simplified}}} \cdot \underbrace{a \cos \theta\, d\theta}_{dx} && x = a \sin \theta,\ dx = a \cos \theta\, d\theta \\
&= 4a^2 \int_0^{\pi/2} \cos^2 \theta\, d\theta && \text{Simplify.} \\
&= 4a^2 \left( \frac{\theta}{2} + \frac{\sin 2\theta}{4} \right) \Big|_0^{\pi/2} && \cos^2 \theta = \frac{1 + \cos 2\theta}{2} \\
&= 4a^2 \left( \left( \frac{\pi}{4} + 0 \right) - (0 + 0) \right) = \pi a^2. && \text{Simplify.}
\end{aligned}
$$

A similar calculation (Exercise 58) gives the area of an ellipse.

*Related Exercises 7, 9* ◀

**EXAMPLE 2**   **Sine substitution**   Evaluate $\displaystyle \int \frac{dx}{(16 - x^2)^{3/2}}$.

**SOLUTION**   The factor $16 - x^2$ has the form $a^2 - x^2$ with $a = 4$, so we use the substitution $x = 4 \sin \theta$. It follows that $dx = 4 \cos \theta\, d\theta$. We now simplify $(16 - x^2)^{3/2}$:

$$
\begin{aligned}
(16 - x^2)^{3/2} &= (16 - (4 \sin \theta)^2)^{3/2} && \text{Substitute } x = 4 \sin \theta. \\
&= (16 (1 - \sin^2 \theta))^{3/2} && \text{Factor.} \\
&= (16 \cos^2 \theta)^{3/2} && 1 - \sin^2 \theta = \cos^2 \theta \\
&= 64 \cos^3 \theta. && \text{Simplify.}
\end{aligned}
$$

Replacing the factors $(16 - x^2)^{3/2}$ and $dx$ of the original integral with appropriate expressions in $\theta$, we have

$$
\begin{aligned}
\int \frac{\overbrace{dx}^{4 \cos \theta\, d\theta}}{\underbrace{(16 - x^2)^{3/2}}_{64 \cos^3 \theta}} &= \int \frac{4 \cos \theta}{64 \cos^3 \theta}\, d\theta \\
&= \frac{1}{16} \int \frac{d\theta}{\cos^2 \theta} \\
&= \frac{1}{16} \int \sec^2 \theta\, d\theta && \text{Simplify.} \\
&= \frac{1}{16} \tan \theta + C. && \text{Evaluate the integral.}
\end{aligned}
$$

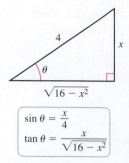

$$\sin \theta = \frac{x}{4}$$
$$\tan \theta = \frac{x}{\sqrt{16 - x^2}}$$

**Figure 8.4**

The final step is to express this result in terms of $x$. In many integrals, this step is most easily done with a reference triangle showing the relationship between $x$ and $\theta$. **Figure 8.4** shows a right triangle with an angle $\theta$ and with the sides labeled such that $x = 4 \sin \theta$ (or $\sin \theta = x/4$). Using this triangle, we see that $\tan \theta = \dfrac{x}{\sqrt{16 - x^2}}$, which implies that

$$\int \frac{dx}{(16 - x^2)^{3/2}} = \frac{1}{16} \tan \theta + C = \frac{x}{16\sqrt{16 - x^2}} + C.$$

*Related Exercises 8, 11* ◀

## Integrals Involving $a^2 + x^2$ or $x^2 - a^2$

The additional trigonometric substitutions involving tangent and secant use a procedure similar to that used for the sine substitution. **Figure 8.5** and Table 8.4 summarize the three basic trigonometric substitutions for real numbers $a > 0$.

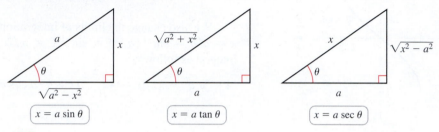

$$x = a \sin \theta \qquad x = a \tan \theta \qquad x = a \sec \theta$$

**Figure 8.5**

$$x = a \tan \theta \Rightarrow \theta = \tan^{-1} \frac{x}{a}$$
$$-\frac{\pi}{2} < \theta < \frac{\pi}{2}$$

**Figure 8.6**

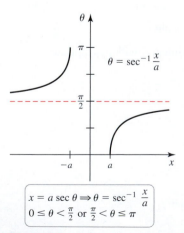

$$x = a \sec \theta \Rightarrow \theta = \sec^{-1} \frac{x}{a}$$
$$0 \le \theta < \frac{\pi}{2} \text{ or } \frac{\pi}{2} < \theta \le \pi$$

**Figure 8.7**

**Table 8.4**

| **The Integral Contains . . .** | **Corresponding Substitution** | **Useful Identity** |
|---|---|---|
| $a^2 - x^2$ | $x = a \sin \theta, -\dfrac{\pi}{2} \le \theta \le \dfrac{\pi}{2}$, for $\lvert x \rvert \le a$ | $a^2 - a^2 \sin^2 \theta = a^2 \cos^2 \theta$ |
| $a^2 + x^2$ | $x = a \tan \theta, -\dfrac{\pi}{2} < \theta < \dfrac{\pi}{2}$ | $a^2 + a^2 \tan^2 \theta = a^2 \sec^2 \theta$ |
| $x^2 - a^2$ | $x = a \sec \theta, \begin{cases} 0 \le \theta < \dfrac{\pi}{2}, \text{ for } x \ge a \\ \dfrac{\pi}{2} < \theta \le \pi, \text{ for } x \le -a \end{cases}$ | $a^2 \sec^2 \theta - a^2 = a^2 \tan^2 \theta$ |

In order for the tangent substitution $x = a \tan \theta$ to be well defined, the angle $\theta$ must be restricted to the interval $-\pi/2 < \theta < \pi/2$, which is the range of $\tan^{-1}(x/a)$ (**Figure 8.6**). On this interval, $\sec \theta > 0$ and with $a > 0$, it is valid to write

$$\sqrt{a^2 + x^2} = \sqrt{a^2 + (a \tan \theta)^2} = \sqrt{a^2 \underbrace{(1 + \tan^2 \theta)}_{\sec^2 \theta}} = a \sec \theta.$$

With the secant substitution, there is a technicality. As discussed in Section 1.4, $\theta = \sec^{-1}(x/a)$ is defined for $x \ge a$, in which case $0 \le \theta < \pi/2$, *and* for $x \le -a$, in which case $\pi/2 < \theta \le \pi$ (**Figure 8.7**). These restrictions on $\theta$ must be treated carefully when simplifying integrands with a factor of $\sqrt{x^2 - a^2}$. Because $\tan \theta$ is positive in the first quadrant but negative in the second, we have

$$\sqrt{x^2 - a^2} = \sqrt{a^2 \underbrace{(\sec^2 \theta - 1)}_{\tan^2 \theta}} = \lvert a \tan \theta \rvert = \begin{cases} a \tan \theta & \text{if } 0 \le \theta < \dfrac{\pi}{2} \\ -a \tan \theta & \text{if } \dfrac{\pi}{2} < \theta \le \pi. \end{cases}$$

**QUICK CHECK 2** Suggest a trigonometric substitution for each integral, and then write the new integral that results after carrying out the substitution.

**a.** $\displaystyle\int \frac{x^2}{\sqrt{x^2+9}}\,dx$

**b.** $\displaystyle\int \frac{3}{\sqrt{x^2-16}}\,dx, \ x>4$ ◄

> Because we are evaluating a definite integral in Example 3, we could change the limits of integration to $\theta=0$ and $\theta=\tan^{-1}4$. However, $\tan^{-1}4$ is not a standard angle, so it is easier to express the antiderivative in terms of $x$ and use the original limits of integration.

When evaluating a definite integral, you should check the limits of integration to see which of these two cases applies. For indefinite integrals, a piecewise formula is often needed, unless a restriction on the variable is given in the problem (see Exercises 79–80).

**EXAMPLE 3  Arc length of a parabola** Evaluate $\int_0^2 \sqrt{1+4x^2}\,dx$, the arc length of the segment of the parabola $y=x^2$ on $[0,2]$.

**SOLUTION** Removing a factor of 4 from the square root, we have

$$\int_0^2 \sqrt{1+4x^2}\,dx = 2\int_0^2 \sqrt{\tfrac14 + x^2}\,dx = 2\int_0^2 \sqrt{\left(\tfrac12\right)^2 + x^2}\,dx.$$

The integrand contains the expression $a^2+x^2$, with $a=\tfrac12$, which suggests the substitution $x=\tfrac12\tan\theta$. It follows that $dx=\tfrac12\sec^2\theta\,d\theta$, and

$$\sqrt{\left(\tfrac12\right)^2 + x^2} = \sqrt{\left(\tfrac12\right)^2 + \left(\tfrac12\tan\theta\right)^2} = \frac12\underbrace{\sqrt{1+\tan^2\theta}}_{\sec^2\theta} = \frac12\sec\theta.$$

Setting aside the limits of integration for the moment, we compute the antiderivative:

$$2\int \sqrt{\left(\tfrac12\right)^2 + x^2}\,dx = 2\int \tfrac12\sec\theta\ \underbrace{\tfrac12\sec^2\theta\,d\theta}_{dx} \qquad \begin{aligned}&x=\tfrac12\tan\theta,\\ &dx=\tfrac12\sec^2\theta\,d\theta\end{aligned}$$

$$= \frac12\int \sec^3\theta\,d\theta \qquad\qquad \text{Simplify.}$$

$$= \frac14\left(\sec\theta\tan\theta + \ln\left|\sec\theta + \tan\theta\right|\right) + C. \qquad \begin{aligned}&\text{Reduction formula 4,}\\ &\text{Section 8.3}\end{aligned}$$

Using a reference triangle (**Figure 8.8**), we express the antiderivative in terms of the original variable $x$ and evaluate the definite integral:

$$2\int_0^2 \sqrt{\left(\tfrac12\right)^2 + x^2}\,dx = \frac14\left(\underbrace{\sqrt{1+4x^2}}_{\sec\theta}\underbrace{2x}_{\tan\theta} + \ln\left|\underbrace{\sqrt{1+4x^2}}_{\sec\theta} + \underbrace{2x}_{\tan\theta}\right|\right)\Bigg|_0^2$$

$$= \frac14\left(4\sqrt{17} + \ln\left(\sqrt{17}+4\right)\right) \approx 4.65.$$

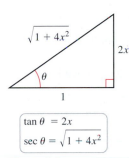

$\sqrt{1+4x^2}$

$2x$

$\theta$

$1$

$\tan\theta = 2x$

$\sec\theta = \sqrt{1+4x^2}$

**Figure 8.8**

*Related Exercises 41, 77* ◄

**QUICK CHECK 3** The integral $\displaystyle\int \frac{dx}{a^2+x^2} = \frac1a\tan^{-1}\frac{x}{a} + C$ is given in Section 7.5. Verify this result with the appropriate trigonometric substitution. ◄

**EXAMPLE 4  Another tangent substitution** Evaluate $\displaystyle\int \frac{dx}{(1+x^2)^2}$.

**SOLUTION** The factor $1+x^2$ suggests the substitution $x=\tan\theta$. It follows that $\theta=\tan^{-1}x$, $dx=\sec^2\theta\,d\theta$, and

$$(1+x^2)^2 = (1+\tan^2\theta)^2 = \sec^4\theta.$$
$$\underbrace{\phantom{(1+\tan^2\theta)}}_{\sec^2\theta}$$

Substituting these factors leads to

$$\int \frac{dx}{(1+x^2)^2} = \int \frac{\sec^2\theta}{\sec^4\theta}\,d\theta \qquad x=\tan\theta,\ dx=\sec^2\theta\,d\theta$$

$$= \int \cos^2\theta\,d\theta \qquad\qquad \text{Simplify.}$$

$$= \frac{\theta}{2} + \frac{\sin2\theta}{4} + C. \qquad \text{Integrate } \cos^2\theta = \frac{1+\cos2\theta}{2}.$$

$$\sin \theta = \frac{x}{\sqrt{1 + x^2}}$$
$$\cos \theta = \frac{1}{\sqrt{1 + x^2}}$$

**Figure 8.9**

The final step is to return to the original variable $x$. The first term $\theta/2$ is replaced with $\frac{1}{2} \tan^{-1} x$. The second term involving $\sin 2\theta$ requires the identity $\sin 2\theta = 2 \sin \theta \cos \theta$. The reference triangle (**Figure 8.9**) tells us that

$$\frac{1}{4} \sin 2\theta = \frac{1}{2} \sin \theta \cos \theta = \frac{1}{2} \cdot \frac{x}{\sqrt{1 + x^2}} \cdot \frac{1}{\sqrt{1 + x^2}} = \frac{x}{2(1 + x^2)}.$$

The integration can now be completed:

$$\int \frac{dx}{(1 + x^2)^2} = \frac{\theta}{2} + \frac{\sin 2\theta}{4} + C$$

$$= \frac{1}{2} \tan^{-1} x + \frac{x}{2(1 + x^2)} + C.$$

*Related Exercises 28–29* ◀

**EXAMPLE 5**   **Multiple approaches**  Evaluate the integral $\displaystyle\int \frac{dx}{\sqrt{x^2 + 4}}$.

**SOLUTION**  Our goal is to show that several different methods lead to the same end.

<u>Solution 1:</u> The term $x^2 + 4$ suggests the substitution $x = 2 \tan \theta$, which implies that $dx = 2 \sec^2 \theta \, d\theta$ and

$$\sqrt{x^2 + 4} = \sqrt{4 \tan^2 \theta + 4} = \sqrt{4(\tan^2 \theta + 1)} = 2\sqrt{\sec^2 \theta} = 2 \sec \theta.$$

Making these substitutions, the integral becomes

$$\int \frac{dx}{\sqrt{x^2 + 4}} = \int \frac{2 \sec^2 \theta}{2 \sec \theta} \, d\theta = \int \sec \theta \, d\theta = \ln |\sec \theta + \tan \theta| + C.$$

To express the indefinite integral in terms of $x$, notice that with $x = 2 \tan \theta$, we have

$$\tan \theta = \frac{x}{2} \quad \text{and} \quad \sec \theta = \sqrt{\tan^2 \theta + 1} = \frac{1}{2}\sqrt{x^2 + 4}.$$

Therefore,

$$\int \frac{dx}{\sqrt{x^2 + 4}} = \ln |\sec \theta + \tan \theta| + C$$

$$= \ln \left| \frac{1}{2}\sqrt{x^2 + 4} + \frac{x}{2} \right| + C \qquad \text{Substitute for } \sec \theta \text{ and } \tan \theta.$$

$$= \ln \left( \frac{1}{2}\left(\sqrt{x^2 + 4} + x\right) \right) + C \qquad \text{Factor; } \sqrt{x^2 + 4} + x > 0.$$

$$= \ln \frac{1}{2} + \ln\left(\sqrt{x^2 + 4} + x\right) + C \qquad \ln ab = \ln a + \ln b$$

$$= \ln\left(\sqrt{x^2 + 4} + x\right) + C. \qquad \text{Absorb constant in } C.$$

<u>Solution 2:</u> Using Theorem 7.19 of Section 7.7, we see that

$$\int \frac{dx}{\sqrt{x^2 + 4}} = \sinh^{-1} \frac{x}{2} + C.$$

By Theorem 7.17 of Section 7.7, we also know that

$$\sinh^{-1} \frac{x}{2} = \ln\left( \frac{x}{2} + \sqrt{\left(\frac{x}{2}\right)^2 + 1} \right) = \ln\left( \frac{1}{2}\left(\sqrt{x^2 + 4} + x\right) \right),$$

which leads to the same result as in Solution 1.

<u>Solution 3:</u> Yet another approach is to use the substitution $x = 2 \sinh t$, which implies that $dx = 2 \cosh t \, dt$ and

$$\sqrt{x^2 + 4} = \sqrt{4 \sinh^2 t + 4} = \sqrt{4(\sinh^2 t + 1)} = 2\sqrt{\cosh^2 t} = 2 \cosh t.$$

The original integral now becomes

$$\int \frac{dx}{\sqrt{x^2 + 4}} = \int \frac{2 \cosh t}{2 \cosh t} \, dt = \int dt = t + C.$$

Because $x = 2 \sinh t$, we have $t = \sinh^{-1} \frac{x}{2}$, which, by Theorem 7.17, leads to the result found in Solution 2.

This example shows that some integrals may be evaluated by more than one method. With practice, you will learn to identify the best method for a given integral.

*Related Exercises 44, 49* ◄

> ➤ Recall that to complete the square with $x^2 + bx + c$, you add and subtract $(b/2)^2$ to and from the expression, and then factor to form a perfect square. You could also make the single substitution $x + 2 = 3 \sec \theta$ in Example 6.

**EXAMPLE 6    A secant substitution**  Evaluate $\int_1^4 \frac{\sqrt{x^2 + 4x - 5}}{x + 2} \, dx$.

**SOLUTION**  This example illustrates a useful preliminary step first encountered in Section 8.1. The integrand does not contain any of the patterns in Table 8.4 that suggest a trigonometric substitution. Completing the square does, however, lead to one of those patterns. Noting that $x^2 + 4x - 5 = (x + 2)^2 - 9$, we change variables with $u = x + 2$ and write the integral as

$$\int_1^4 \frac{\sqrt{x^2 + 4x - 5}}{x + 2} \, dx = \int_1^4 \frac{\sqrt{(x + 2)^2 - 9}}{x + 2} \, dx \quad \text{Complete the square.}$$

$$= \int_3^6 \frac{\sqrt{u^2 - 9}}{u} \, du. \quad \begin{array}{l} u = x + 2, du = dx \\ \text{Change limits of integration.} \end{array}$$

> ➤ The substitution $u = 3 \sec \theta$ can be rewritten as $\theta = \sec^{-1}(u/3)$. Because $u \geq 3$ in the integral $\int_3^6 \frac{\sqrt{u^2 - 9}}{u} \, du$, we have $0 \leq \theta < \frac{\pi}{2}$.

This new integral calls for the secant substitution $u = 3 \sec \theta$ (where $0 \leq \theta < \pi/2$), which implies that $du = 3 \sec \theta \tan \theta \, d\theta$ and $\sqrt{u^2 - 9} = 3 \tan \theta$. We also change the limits of integration: When $u = 3$, $\theta = 0$, and when $u = 6$, $\theta = \pi/3$. The complete integration can now be done:

$$\int_1^4 \frac{\sqrt{x^2 + 4x - 5}}{x + 2} \, dx = \int_3^6 \frac{\sqrt{u^2 - 9}}{u} \, du \quad u = x + 2, du = dx$$

$$= \int_0^{\pi/3} \frac{3 \tan \theta}{3 \sec \theta} \, 3 \sec \theta \tan \theta \, d\theta \quad u = 3 \sec \theta, du = 3 \sec \theta \tan \theta \, d\theta$$

$$= 3 \int_0^{\pi/3} \tan^2 \theta \, d\theta \quad \text{Simplify.}$$

$$= 3 \int_0^{\pi/3} (\sec^2 \theta - 1) \, d\theta \quad \tan^2 \theta = \sec^2 \theta - 1$$

$$= 3 (\tan \theta - \theta) \Big|_0^{\pi/3} \quad \text{Evaluate the integral.}$$

$$= 3\sqrt{3} - \pi. \quad \text{Simplify.}$$

*Related Exercises 40, 60* ◄

## SECTION 8.4 EXERCISES

### Getting Started

1. What change of variables is suggested by an integral containing $\sqrt{x^2 - 9}$?

2. What change of variables is suggested by an integral containing $\sqrt{x^2 + 36}$?

3. What change of variables is suggested by an integral containing $\sqrt{100 - x^2}$?

4. If $x = 4 \tan \theta$, express $\sin \theta$ in terms of $x$.

5. Using the trigonometric substitution $x = 2 \sin \theta$, for $-\frac{\pi}{2} \leq \theta \leq \frac{\pi}{2}$, express $\cot \theta$ in terms of $x$.

6. Using the trigonometric substitution $x = 8 \sec \theta$, $x \geq 8$ and $0 < \theta \leq \frac{\pi}{2}$, express $\tan \theta$ in terms of $x$.

### Practice Exercises

**7–56. Trigonometric substitutions** *Evaluate the following integrals using trigonometric substitution.*

7. $\int_0^{5/2} \frac{dx}{\sqrt{25 - x^2}}$  (*Hint:* Check your answer without using trigonometric substitution.)

**8.** $\displaystyle\int_0^{3/2} \frac{dx}{(9 - x^2)^{3/2}}$

**9.** $\displaystyle\int_5^{5\sqrt{3}} \sqrt{100 - x^2}\, dx$

**10.** $\displaystyle\int_0^{\sqrt{2}} \frac{x^2}{\sqrt{4 - x^2}}\, dx$

**11.** $\displaystyle\int_{1/2}^{\sqrt{3}/2} \frac{x^2}{\sqrt{1 - x^2}}\, dx$

**12.** $\displaystyle\int_{1/2}^{1} \frac{\sqrt{1 - x^2}}{x^2}\, dx$

**13.** $\displaystyle\int \frac{dx}{\sqrt{16 - x^2}}$

**14.** $\displaystyle\int \sqrt{36 - t^2}\, dt$

**15.** $\displaystyle\int \frac{dx}{x^2\sqrt{x^2 + 9}}$

**16.** $\displaystyle\int \frac{x^2}{(25 + x^2)^2}\, dx$

**17.** $\displaystyle\int_0^2 \frac{x^2}{x^2 + 4}\, dx$

**18.** $\displaystyle\int \frac{dx}{(1 + x^2)^{3/2}}$

**19.** $\displaystyle\int \frac{dx}{\sqrt{x^2 - 81}}, x > 9$

**20.** $\displaystyle\int \frac{dx}{\sqrt{x^2 - 49}}, x > 7$

**21.** $\displaystyle\int \sqrt{64 - x^2}\, dx$

**22.** $\displaystyle\int \frac{dt}{t^2\sqrt{9 - t^2}}$

**23.** $\displaystyle\int \frac{dx}{(25 - x^2)^{3/2}}$

**24.** $\displaystyle\int \frac{\sqrt{9 - x^2}}{x^2}\, dx$

**25.** $\displaystyle\int \frac{\sqrt{9 - x^2}}{x}\, dx$

**26.** $\displaystyle\int_{\sqrt{2}}^2 \frac{\sqrt{x^2 - 1}}{x}\, dx$

**27.** $\displaystyle\int_0^{1/3} \frac{dx}{(9x^2 + 1)^{3/2}}$

**28.** $\displaystyle\int_0^6 \frac{z^2}{(z^2 + 36)^2}\, dz$

**29.** $\displaystyle\int \frac{dx}{(4 + x^2)^2}$

**30.** $\displaystyle\int x^3\sqrt{1 - x^2}\, dx$

**31.** $\displaystyle\int \frac{x^2}{\sqrt{16 - x^2}}\, dx$

**32.** $\displaystyle\int \frac{dx}{(x^2 - 36)^{3/2}}, x > 6$

**33.** $\displaystyle\int \frac{\sqrt{x^2 - 9}}{x}\, dx, x > 3$

**34.** $\displaystyle\int \frac{dx}{x^3\sqrt{x^2 - 1}}, x > 1$

**35.** $\displaystyle\int \frac{dx}{x(x^2 - 1)^{3/2}}, x > 1$

**36.** $\displaystyle\int_{8\sqrt{2}}^{16} \frac{dx}{\sqrt{x^2 - 64}}$

**37.** $\displaystyle\int_{1/\sqrt{3}}^1 \frac{dx}{x^2\sqrt{1 + x^2}}$

**38.** $\displaystyle\int_1^{\sqrt{2}} \frac{dx}{x^2\sqrt{4 - x^2}}$

**39.** $\displaystyle\int \frac{x^2}{(100 - x^2)^{3/2}}\, dx$

**40.** $\displaystyle\int_{10/\sqrt{3}}^{10} \frac{dy}{\sqrt{y^2 - 25}}$

**41.** $\displaystyle\int \frac{dx}{(1 + 4x^2)^{3/2}}$

**42.** $\displaystyle\int \frac{dx}{x^2\sqrt{9x^2 - 1}}, x > \frac{1}{3}$

**43.** $\displaystyle\int_0^{4/\sqrt{3}} \frac{dx}{\sqrt{x^2 + 16}}$

**44.** $\displaystyle\int \frac{dx}{\sqrt{16 + 4x^2}}$

**45.** $\displaystyle\int \frac{x^3}{(81 - x^2)^2}\, dx$

**46.** $\displaystyle\int \frac{dx}{\sqrt{1 - 2x^2}}$

**47.** $\displaystyle\int_{4/\sqrt{3}}^4 \frac{dx}{x^2(x^2 - 4)}$

**48.** $\displaystyle\int \sqrt{9 - 4x^2}\, dx$

**49.** $\displaystyle\int_0^{1/\sqrt{3}} \sqrt{x^2 + 1}\, dx$

**50.** $\displaystyle\int (36 - 9x^2)^{-3/2}\, dx$

**51.** $\displaystyle\int \frac{x^2}{\sqrt{4 + x^2}}\, dx$

**52.** $\displaystyle\int \frac{\sqrt{4x^2 - 1}}{x^2}\, dx, x > \frac{1}{2}$

**53.** $\displaystyle\int \frac{\sqrt{9x^2 - 25}}{x^3}\, dx, x > \frac{5}{3}$

**54.** $\displaystyle\int \frac{y^4}{1 + y^2}\, dy$

**55.** $\displaystyle\int \frac{dx}{x^3\sqrt{x^2 - 100}}, x > 10$

**56.** $\displaystyle\int \frac{x^3}{(x^2 - 16)^{3/2}}\, dx, x < -4$

**57.** **Explain why or why not** Determine whether the following statements are true and give an explanation or counterexample.

   **a.** If $x = 4\tan\theta$, then $\csc\theta = 4/x$.

   **b.** The integral $\displaystyle\int_1^2 \sqrt{1 - x^2}\, dx$ does not have a finite real value.

   **c.** The integral $\displaystyle\int_1^2 \sqrt{x^2 - 1}\, dx$ does not have a finite real value.

   **d.** The integral $\displaystyle\int \frac{dx}{x^2 + 4x + 9}$ cannot be evaluated using a trigonometric substitution.

**58.** **Area of an ellipse** The upper half of the ellipse centered at the origin with axes of length $2a$ and $2b$ is described by $y = \dfrac{b}{a}\sqrt{a^2 - x^2}$ (see figure). Find the area of the ellipse in terms of $a$ and $b$.

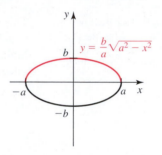

**59.** **Area of a segment of a circle** Use two approaches to show that the area of a cap (or segment) of a circle of radius $r$ subtended by an angle $\theta$ (see figure) is given by

$$A_{\text{seg}} = \frac{1}{2}r^2(\theta - \sin\theta).$$

   **a.** Find the area using geometry (no calculus).
   **b.** Find the area using calculus.

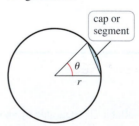

cap or segment

**60–69. Completing the square** *Evaluate the following integrals.*

**60.** $\displaystyle\int \frac{dx}{x^2 - 6x + 34}$

**61.** $\displaystyle\int \frac{dx}{\sqrt{3 - 2x - x^2}}$

**62.** $\displaystyle\int \frac{du}{2u^2 - 12u + 36}$

**63.** $\displaystyle\int \frac{dx}{x^2 + 6x + 18}$

**64.** $\displaystyle\int \frac{x^2 - 2x + 1}{\sqrt{x^2 - 2x + 10}}\, dx$

**65.** $\displaystyle\int_{1/2}^{(\sqrt{2}+3)/(2\sqrt{2})} \frac{dx}{8x^2 - 8x + 11}$

**66.** $\displaystyle\int_1^4 \frac{dt}{t^2 - 2t + 10}$    **67.** $\displaystyle\int \frac{x^2 - 8x + 16}{(9 + 8x - x^2)^{3/2}} \, dx$

**68.** $\displaystyle\int \frac{dx}{\sqrt{(x-1)(3-x)}}$    **69.** $\displaystyle\int_{2+\sqrt{2}}^4 \frac{dx}{\sqrt{(x-1)(x-3)}}$

**T 70–74. Using the integral of $\sec^3 u$** *By reduction formula 4 in Section 8.3,*

$$\int \sec^3 u \, du = \frac{1}{2}(\sec u \tan u + \ln |\sec u + \tan u|) + C.$$

*Graph the following functions and find the area under the curve on the given interval.*

**70.** $f(x) = (4 + x^2)^{1/2}, [0, 2]$    **71.** $f(x) = (9 - x^2)^{-2}, \left[0, \dfrac{3}{2}\right]$

**72.** $f(x) = (x^2 - 25)^{1/2}, [5, 10]$

**73.** $f(x) = \sqrt{x^2 - 9}/x, [3, 6]$

**74.** $f(x) = \dfrac{1}{x\sqrt{x^2 - 36}}, \left[\dfrac{12}{\sqrt{3}}, 12\right]$

**75.** Find the area of the region bounded by the curve $f(x) = (16 + x^2)^{-3/2}$ and the $x$-axis on the interval $[0, 3]$.

### Explorations and Challenges

**76. Area and volume** Consider the function $f(x) = (9 + x^2)^{-1/2}$ and the region $R$ on the interval $[0, 4]$ (see figure).

   **a.** Find the area of $R$.
   **b.** Find the volume of the solid generated when $R$ is revolved about the $x$-axis.
   **c.** Find the volume of the solid generated when $R$ is revolved about the $y$-axis.

**77. Arc length of a parabola** Find the length of the curve $y = ax^2$ from $x = 0$ to $x = 10$, where $a > 0$ is a real number.

**78. Computing areas** On the interval $[0, 2]$, the graphs of $f(x) = x^2/3$ and $g(x) = x^2(9 - x^2)^{-1/2}$ have similar shapes.

   **a.** Find the area of the region bounded by the graph of $f$ and the $x$-axis on the interval $[0, 2]$.
   **b.** Find the area of the region bounded by the graph of $g$ and the $x$-axis on the interval $[0, 2]$.
   **c.** Which region has greater area?

**T 79–80. Care with the secant substitution** *Recall that the substitution $x = a \sec \theta$ implies either $x \geq a$ (in which case $0 \leq \theta < \pi/2$ and $\tan \theta \geq 0$) or $x \leq -a$ (in which case $\pi/2 < \theta \leq \pi$ and $\tan \theta \leq 0$).*

**79.** Show that $\displaystyle\int \frac{dx}{x\sqrt{x^2 - 1}} = \begin{cases} \sec^{-1} x + C & \text{if } x > 1 \\ -\sec^{-1} x + C & \text{if } x < -1. \end{cases}$

**80.** Evaluate for $\displaystyle\int \frac{\sqrt{x^2 - 1}}{x^3} \, dx$, for $x > 1$ and for $x < -1$.

**81. Electric field due to a line of charge** A total charge of $Q$ is distributed uniformly on a line segment of length $2L$ along the $y$-axis (see figure). The $x$-component of the electric field at a point $(a, 0)$ is given by

$$E_x(a) = \frac{kQa}{2L} \int_{-L}^L \frac{dy}{(a^2 + y^2)^{3/2}},$$

where $k$ is a physical constant and $a > 0$.

   **a.** Confirm that $E_x(a) = \dfrac{kQ}{a\sqrt{a^2 + L^2}}$.
   **b.** Letting $\rho = Q/2L$ be the charge density on the line segment, show that if $L \to \infty$, then $E_x(a) = 2k\rho/a$.

(See the Guided Project *Electric field integrals* for a derivation of this and other similar integrals.)

**82. Magnetic field due to current in a straight wire** A long straight wire of length $2L$ on the $y$-axis carries a current $I$. According to the Biot-Savart Law, the magnitude of the magnetic field due to the current at a point $(a, 0)$ is given by

$$B(a) = \frac{\mu_0 I}{4\pi} \int_{-L}^L \frac{\sin \theta}{r^2} \, dy,$$

where $\mu_0$ is a physical constant, $a > 0$, and $\theta$, $r$, and $y$ are related as shown in the figure.

   **a.** Show that the magnitude of the magnetic field at $(a, 0)$ is
$$B(a) = \frac{\mu_0 I L}{2\pi a \sqrt{a^2 + L^2}}.$$
   **b.** What is the magnitude of the magnetic field at $(a, 0)$ due to an infinitely long wire $(L \to \infty)$?

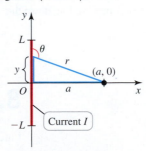

**83. Visual proof** Let $F(x) = \int_0^x \sqrt{a^2 - t^2} \, dt$. The figure shows that $F(x) = $ area of sector $OAB$ + area of triangle $OBC$.

   **a.** Use the figure to prove that
$$F(x) = \frac{a^2 \sin^{-1}(x/a)}{2} + \frac{x\sqrt{a^2 - x^2}}{2}.$$

**b.** Conclude that

$$\int \sqrt{a^2 - x^2}\, dx = \frac{a^2 \sin^{-1}(x/a)}{2} + \frac{x\sqrt{a^2 - x^2}}{2} + C.$$

(*Source: The College Mathematics Journal*, 34, 3, May 2003)

**T 84. Maximum path length of a projectile** (Adapted from Putnam Exam 1940) A projectile is launched from the ground with an initial speed $V$ at an angle $\theta$ from the horizontal. Assume the $x$-axis is the horizontal ground and $y$ is the height above the ground. Neglecting air resistance and letting $g$ be the acceleration due to gravity, it can be shown that the trajectory of the projectile is given by

$$y = -\frac{1}{2}kx^2 + y_{\max}, \quad \text{where } k = \frac{g}{(V\cos\theta)^2}$$

and $y_{\max} = \dfrac{(V\sin\theta)^2}{2g}$.

**a.** Note that the high point of the trajectory occurs at $(0, y_{\max})$. If the projectile is on the ground at $(-a, 0)$ and $(a, 0)$, what is $a$?

**b.** Show that the length of the trajectory (arc length) is

$$2\int_0^a \sqrt{1 + k^2 x^2}\, dx.$$

**c.** Evaluate the arc length integral and express your result in terms of $V$, $g$, and $\theta$.

**d.** For a fixed value of $V$ and $g$, show that the launch angle $\theta$ that maximizes the length of the trajectory satisfies $(\sin\theta)\ln(\sec\theta + \tan\theta) = 1$.

**e.** Use a graphing utility to approximate the optimal launch angle.

**85. Fastest descent time** The cycloid is the curve traced by a point on the rim of a rolling wheel. Imagine a wire shaped like an inverted cycloid (see figure). A bead sliding down this wire without friction has some remarkable properties. Among all wire shapes, the cycloid is the shape that produces the fastest descent time (see the Guided Project *The amazing cycloid* for more about the *brachistochrone property*). It can be shown that the descent time between any two points $0 \le a < b \le \pi$ on the curve is

$$\text{descent time} = \int_a^b \sqrt{\frac{1 - \cos t}{g(\cos a - \cos t)}}\, dt,$$

where $g$ is the acceleration due to gravity, $t = 0$ corresponds to the top of the wire, and $t = \pi$ corresponds to the lowest point on the wire.

**a.** Find the descent time on the interval $[a, b]$ by making the substitution $u = \cos t$.

**b.** Show that when $b = \pi$, the descent time is the same for all values of $a$; that is, the descent time to the bottom of the wire is the same for all starting points.

**86. Clever substitution** Evaluate $\displaystyle\int \frac{dx}{1 + \sin x + \cos x}$ using the substitution $x = 2\tan^{-1}\theta$. The identities $\sin x = 2\sin\dfrac{x}{2}\cos\dfrac{x}{2}$ and $\cos x = \cos^2\dfrac{x}{2} - \sin^2\dfrac{x}{2}$ are helpful.

**QUICK CHECK ANSWERS**

**1.** Use $x = 3\sin\theta$ to obtain $9\cos^2\theta$.    **2.** (a) Use $x = 3\tan\theta$; $\int 9\tan^2\theta\sec\theta\, d\theta$. (b) Use $x = 4\sec\theta$; $\int 3\sec\theta\, d\theta$.    **3.** Let $x = a\tan\theta$, so that $dx = a\sec^2\theta\, d\theta$. The new integral is $\displaystyle\int \frac{a\sec^2\theta\, d\theta}{a^2(1 + \tan^2\theta)} = \frac{1}{a}\int d\theta = \frac{1}{a}\theta + C = \frac{1}{a}\tan^{-1}\frac{x}{a} + C.$ ◄

# 8.5 Partial Fractions

▶ Recall that a rational function has the form $p/q$, where $p$ and $q$ are polynomials.

Later in this chapter, we will see that finding the velocity of a skydiver requires evaluating an integral of the form $\displaystyle\int \frac{dv}{a - bv^2}$ and finding the population of a species that is limited in size involves an integral of the form $\displaystyle\int \frac{dP}{aP(1 - bP)}$, where $a$ and $b$ are constants in both cases. These integrals have the common feature that their integrands are rational functions. Similar integrals result from modeling mechanical and electrical networks. The goal of this section is to introduce the *method of partial fractions* for integrating rational functions. When combined with standard and trigonometric substitutions (Section 8.4), this method allows us (in principle) to integrate any rational function.

## Method of Partial Fractions

Given a function such as

$$f(x) = \frac{1}{x - 2} + \frac{2}{x + 4},$$

it is a straightforward task to find a common denominator and write the equivalent expression

$$f(x) = \frac{(x + 4) + 2(x - 2)}{(x - 2)(x + 4)} = \frac{3x}{(x - 2)(x + 4)} = \frac{3x}{x^2 + 2x - 8}.$$

The purpose of partial fractions is to reverse this process. Given a rational function that is difficult to integrate, the method of partial fractions produces an equivalent function that is much easier to integrate.

| **Rational function** | | **Partial fraction decomposition** |
|---|---|---|
| $\dfrac{3x}{x^2 + 2x - 8}$ | $\xrightarrow[\text{partial fractions}]{\textit{method of}}$ | $\dfrac{1}{x - 2} + \dfrac{2}{x + 4}$ |
| **Difficult to integrate** | | **Easy to integrate** |
| $\displaystyle\int \frac{3x}{x^2 + 2x - 8}\, dx$ | | $\displaystyle\int \left(\frac{1}{x - 2} + \frac{2}{x + 4}\right) dx$ |

**QUICK CHECK 1** Find an antiderivative of $f(x) = \dfrac{1}{x - 2} + \dfrac{2}{x + 4}$. ◄

**The Key Idea** Working with the same function, $f(x) = \dfrac{3x}{(x - 2)(x + 4)}$, our objective is to write it in the form

$$\frac{A}{x - 2} + \frac{B}{x + 4},$$

➤ Notice that the numerator of the original rational function does not affect the form of the partial fraction decomposition. The constants $A$ and $B$ are called *undetermined coefficients*.

where $A$ and $B$ are constants to be determined. This expression is called the **partial fraction decomposition** of the original function; in this case, it has two terms, one for each factor in the denominator of the original function.

The constants $A$ and $B$ are determined using the condition that the original function $f$ and its partial fraction decomposition must be equal for all values of $x$ in the domain of $f$; that is,

$$\frac{3x}{(x - 2)(x + 4)} = \frac{A}{x - 2} + \frac{B}{x + 4}. \tag{1}$$

➤ This step requires that $x \neq 2$ and $x \neq -4$; both values are outside the domain of $f$.

Multiplying both sides of equation (1) by $(x - 2)(x + 4)$ gives

$$3x = A(x + 4) + B(x - 2).$$

Collecting like powers of $x$ results in

$$3x = (A + B)x + (4A - 2B). \tag{2}$$

If equation (2) is to hold for all values of $x$, then

- the coefficients of $x^1$ on both sides of the equation must be equal, and
- the coefficients of $x^0$ (that is, the constants) on both sides of the equation must be equal.

$$3x + 0 = (A + B)x + (4A - 2B)$$

These observations lead to two equations for $A$ and $B$.

Equate coefficients of $x^1$:    $3 = A + B$

Equate coefficients of $x^0$:    $0 = 4A - 2B$

The first equation says that $A = 3 - B$. Substituting $A = 3 - B$ into the second equation gives the equation $0 = 4(3 - B) - 2B$. Solving for $B$, we find that $6B = 12$, or $B = 2$. The value of $A$ now follows; we have $A = 3 - B = 1$.

Substituting these values of $A$ and $B$ into equation (1), the partial fraction decomposition is

$$\frac{3x}{(x-2)(x+4)} = \frac{1}{x-2} + \frac{2}{x+4}.$$

> Like an ordinary fraction, a rational function is said to be in **reduced form** if the numerator and denominator have no common factors, and it is said to be **proper** if the degree of the numerator is less than the degree of the denominator.

## Simple Linear Factors

The previous calculation illustrates the method of partial fractions with **simple linear factors**, meaning the denominator of the original function consists only of linear factors of the form $(x-r)$, which appear to the first power and no higher power. Here is the general procedure for this case.

> For the sake of simplicity, the factors of the denominator $q$ in Step 1 are written with a leading coefficient of 1. The same procedure may be applied to factors whose leading coefficients are not equal to 1 as long as no factor is a constant multiple of any other factor. For example, $(x+2)(2x+4) = 2(x+2)^2$ should be recognized as a repeated linear factor (p. 550), not two simple linear factors. See Example 2.

**QUICK CHECK 2** If the denominator of a reduced proper rational function is $(x-1)(x+5)(x-10)$, what is the general form of its partial fraction decomposition? ◄

---

**PROCEDURE**  **Partial Fractions with Simple Linear Factors**

Suppose $f(x) = p(x)/q(x)$, where $p$ and $q$ are polynomials with no common factors and with the degree of $p$ less than the degree of $q$. Assume $q$ is the product of simple linear factors. The partial fraction decomposition is obtained as follows.

**Step 1.**  **Factor the denominator** $q$ in the form $(x-r_1)(x-r_2)\cdots(x-r_n)$, where $r_1, \ldots, r_n$ are distinct real numbers.

**Step 2.**  **Partial fraction decomposition** Form the partial fraction decomposition by writing

$$\frac{p(x)}{q(x)} = \frac{A_1}{(x-r_1)} + \frac{A_2}{(x-r_2)} + \cdots + \frac{A_n}{(x-r_n)}.$$

**Step 3.**  **Clear denominators** Multiply both sides of the equation in Step 2 by $q(x) = (x-r_1)(x-r_2)\cdots(x-r_n)$, which produces conditions for $A_1, \ldots, A_n$.

**Step 4.**  **Solve for coefficients** Equate like powers of $x$ in Step 3 to solve for the undetermined coefficients $A_1, \ldots, A_n$.

---

### EXAMPLE 1    Integrating with partial fractions

**a.** Find the partial fraction decomposition for $f(x) = \dfrac{3x^2 + 7x - 2}{x^3 - x^2 - 2x}$.

**b.** Evaluate $\displaystyle\int f(x)\, dx$.

**SOLUTION**

**a.** The partial fraction decomposition is done in four steps.

*Step 1:*  Factoring the denominator, we find that

$$x^3 - x^2 - 2x = x(x+1)(x-2),$$

in which only simple linear factors appear.

> You can call the undetermined coefficients $A_1, A_2, A_3, \ldots$ or $A, B, C, \ldots$. The latter choice avoids subscripts.

*Step 2:*  The partial fraction decomposition has one term for each factor in the denominator:

$$\frac{3x^2 + 7x - 2}{x(x+1)(x-2)} = \frac{A}{x} + \frac{B}{x+1} + \frac{C}{x-2}. \tag{3}$$

The goal is to find the undetermined coefficients $A$, $B$, and $C$.

*Step 3:*  We multiply both sides of equation (3) by $x(x+1)(x-2)$:

$$3x^2 + 7x - 2 = A(x+1)(x-2) + Bx(x-2) + Cx(x+1)$$
$$= (A+B+C)x^2 + (-A - 2B + C)x - 2A.$$

*Step 4:*   We now equate coefficients of $x^2$, $x^1$, and $x^0$ on both sides of the equation in Step 3.

$$
\begin{aligned}
\text{Equate coefficients of } x^2: &\quad A + B + C = 3 \\
\text{Equate coefficients of } x^1: &\quad -A - 2B + C = 7 \\
\text{Equate coefficients of } x^0: &\quad -2A = -2
\end{aligned}
$$

The third equation implies that $A = 1$, which is substituted into the first two equations to give

$$
B + C = 2 \quad \text{and} \quad -2B + C = 8.
$$

Solving for $B$ and $C$, we conclude that $A = 1$, $B = -2$, and $C = 4$. Substituting the values of $A$, $B$, and $C$ into equation (3), the partial fraction decomposition is

$$
f(x) = \frac{1}{x} - \frac{2}{x+1} + \frac{4}{x-2}.
$$

**b.** Integration is now straightforward:

$$
\begin{aligned}
\int \frac{3x^2 + 7x - 2}{x^3 - x^2 - 2x}\, dx &= \int \left( \frac{1}{x} - \frac{2}{x+1} + \frac{4}{x-2} \right) dx && \text{\color{blue} Partial fractions} \\
&= \ln|x| - 2\ln|x+1| + 4\ln|x-2| + K && \text{\color{blue} Integrate; arbitrary constant } K. \\
&= \ln \frac{|x|(x-2)^4}{(x+1)^2} + K. && \text{\color{blue} Properties of logarithms}
\end{aligned}
$$

*Related Exercises 30, 35* ◄

**A Shortcut (Convenient Values)**   Solving for more than three unknown coefficients in a partial fraction decomposition may be difficult. In the case of simple linear factors, a shortcut saves work. In Example 1, Step 3 led to the equation

$$
3x^2 + 7x - 2 = A(x+1)(x-2) + Bx(x-2) + Cx(x+1).
$$

> ➤ In cases other than simple linear factors, the shortcut can be used to determine some, but not all the coefficients, which reduces the work required to find the remaining coefficients. A modified shortcut can be utilized to find all the coefficients; see the margin note next to Example 3.

Because this equation holds for *all* values of $x$, it must hold for any particular value of $x$. By choosing values of $x$ judiciously, it is easy to solve for $A$, $B$, and $C$. For example, setting $x = 0$ in this equation results in $-2 = -2A$, or $A = 1$. Setting $x = -1$ results in $-6 = 3B$, or $B = -2$, and setting $x = 2$ results in $24 = 6C$, or $C = 4$. In each case, we choose a value of $x$ that eliminates all but one term on the right side of the equation.

**EXAMPLE 2**   **Using the shortcut**

**a.** Find the partial fraction decomposition for $f(x) = \dfrac{9x^2 + 2x - 1}{(x-1)(2x^2 + 7x - 4)}$.

**b.** Evaluate $\displaystyle\int_2^{14} f(x)\, dx$.

**SOLUTION**

**a.** We use four steps to obtain the partial fraction decomposition.

*Step 1:*   Factoring the denominator of $f$ results in $(x-1)(2x-1)(x+4)$, so the integrand has three simple linear factors.

*Step 2:*   We form the partial fraction decomposition with one term for each factor in the denominator:

$$
\frac{9x^2 + 2x - 1}{(x-1)(2x-1)(x+4)} = \frac{A}{x-1} + \frac{B}{2x-1} + \frac{C}{x+4}. \tag{4}
$$

The goal is to find the undetermined coefficients $A$, $B$, and $C$.

*Step 3:*   We now multiply both sides of equation (4) by $(x-1)(2x-1)(x+4)$:

$$
9x^2 + 2x - 1 = A(2x-1)(x+4) + B(x-1)(x+4) + C(x-1)(2x-1). \tag{5}
$$

*Step 4:* The shortcut is now used to determine $A$, $B$, and $C$. Substituting $x = 1$, $1/2$, and $-4$ in equation (5) allows us to solve directly for the coefficients:

$$\text{Letting } x = 1 \Rightarrow 10 = 5A + 0 \cdot B + 0 \cdot C \Rightarrow A = 2;$$

$$\text{Letting } x = 1/2 \Rightarrow 9/4 = 0 \cdot A - 9B/4 + 0 \cdot C \Rightarrow B = -1;$$

$$\text{Letting } x = -4 \Rightarrow 135 = 0 \cdot A + 0 \cdot B + 45C \Rightarrow C = 3.$$

Substituting the values of $A$, $B$, and $C$ into equation (4) gives the partial fraction decomposition

$$f(x) = \frac{2}{x - 1} - \frac{1}{2x - 1} + \frac{3}{x + 4}.$$

**b.** We now carry out the integration.

$$\int_2^{14} f(x)\, dx = \int_2^{14}\left(\frac{2}{x - 1} - \frac{1}{2x - 1} + \frac{3}{x + 4}\right) dx \qquad \text{Partial fractions}$$

$$= \left(2\ln|x - 1| - \frac{1}{2}\ln|2x - 1| + 3\ln|x + 4|\right)\Bigg|_2^{14} \qquad \begin{array}{l}\text{Integrate;}\\ u = 2x - 1,\\ du = 2dx.\end{array}$$

$$= 2\ln 13 - \frac{1}{2}\ln 27 + 3\ln 18 - \left(2\underbrace{\ln 1}_{0} - \frac{1}{2}\ln 3 + 3\ln 6\right) \qquad \text{Evaluate.}$$

$$= \ln 13^2 + \underbrace{\ln 18^3 - \ln 6^3}_{\ln 27} - \frac{1}{2}\underbrace{(\ln 27 - \ln 3)}_{\ln 3} \qquad \text{Log properties}$$

$$= \ln 1521 \approx 7.327 \qquad \text{Simplify.}$$

*Related Exercises 27–28* ◀

## Repeated Linear Factors

▶ *Simple* means the factor is raised to the first power; *repeated* means the factor is raised to an integer power greater than 1.

The preceding discussion relies on the assumption that the denominator of the rational function can be factored into simple linear factors of the form $(x - r)$. But what about denominators such as $x^2(x - 3)$ or $(x + 2)^2(x - 4)^3$, in which linear factors are raised to integer powers greater than 1? In these cases we have *repeated linear factors*, and a modification to the previous procedure must be made.

Here is the modification: Suppose the factor $(x - r)^m$ appears in the denominator, where $m > 1$ is an integer. Then there must be a partial fraction for each power of $(x - r)$ up to and including the $m$th power. For example, if $x^2(x - 3)^4$ appears in the denominator, then the partial fraction decomposition includes the terms

▶ Think of $x^2$ as the repeated linear factor $(x - 0)^2$.

$$\frac{A}{x} + \frac{B}{x^2} + \frac{C}{(x - 3)} + \frac{D}{(x - 3)^2} + \frac{E}{(x - 3)^3} + \frac{F}{(x - 3)^4}.$$

The rest of the partial fraction procedure remains the same, although the amount of work increases as the number of coefficients increases.

---

**PROCEDURE**    **Partial Fractions for Repeated Linear Factors**

Suppose the repeated linear factor $(x - r)^m$ appears in the denominator of a proper rational function in reduced form. The partial fraction decomposition has a partial fraction for each power of $(x - r)$ up to and including the $m$th power; that is, the partial fraction decomposition contains the sum

$$\frac{A_1}{(x - r)} + \frac{A_2}{(x - r)^2} + \frac{A_3}{(x - r)^3} + \cdots + \frac{A_m}{(x - r)^m},$$

where $A_1, \ldots, A_m$ are constants to be determined.

---

**EXAMPLE 3**    **Integrating with repeated linear factors**   Evaluate $\int f(x)\, dx$, where

$$f(x) = \frac{5x^2 - 3x + 2}{x^3 - 2x^2}.$$

**SOLUTION** The denominator factors as $x^3 - 2x^2 = x^2(x - 2)$, so it has one simple linear factor $(x - 2)$ and one repeated linear factor $x^2$. The partial fraction decomposition has the form

$$\frac{5x^2 - 3x + 2}{x^2(x - 2)} = \frac{A}{x} + \frac{B}{x^2} + \frac{C}{(x - 2)}.$$

Multiplying both sides of the partial fraction decomposition by $x^2(x - 2)$, we find

$$5x^2 - 3x + 2 = Ax(x - 2) + B(x - 2) + Cx^2$$
$$= (A + C)x^2 + (-2A + B)x - 2B.$$

The coefficients $A$, $B$, and $C$ are determined by equating the coefficients of $x^2$, $x^1$, and $x^0$.

> The shortcut can be used to obtain two of the three coefficients easily. Choosing $x = 0$ allows $B$ to be determined. Choosing $x = 2$ determines $C$. To find $A$, any other value of $x$ may be substituted.

| | |
|---|---|
| Equate coefficients of $x^2$: | $A + C = 5$ |
| Equate coefficients of $x^1$: | $-2A + B = -3$ |
| Equate coefficients of $x^0$: | $-2B = 2$ |

Solving these three equations in three unknowns results in the solution $A = 1$, $B = -1$, and $C = 4$. When $A$, $B$, and $C$ are substituted, the partial fraction decomposition is

$$f(x) = \frac{1}{x} - \frac{1}{x^2} + \frac{4}{x - 2}.$$

Integration is now straightforward:

$$\int \frac{5x^2 - 3x + 2}{x^3 - 2x^2}\, dx = \int \left( \frac{1}{x} - \frac{1}{x^2} + \frac{4}{x - 2} \right) dx \qquad \text{Partial fractions}$$

**QUICK CHECK 3** State the form of the partial fraction decomposition of the reduced proper rational function $p(x)/q(x)$ if $q(x) = x^2(x - 3)^2(x - 1)$. ◄

$$= \ln |x| + \frac{1}{x} + 4 \ln |x - 2| + K \quad \text{Integrate; arbitrary constant } K.$$

$$= \frac{1}{x} + \ln \left( |x|(x - 2)^4 \right) + K. \qquad \text{Properties of logarithms}$$

*Related Exercises 40–41* ◄

## Irreducible Quadratic Factors

By the Fundamental Theorem of Algebra, we know that a polynomial with real-valued coefficients can be written as the product of linear factors of the form $x - r$ and *irreducible quadratic factors* of the form $ax^2 + bx + c$, where $r$, $a$, $b$, and $c$ are real numbers. By irreducible, we mean that $ax^2 + bx + c$ cannot be factored over the real numbers. For example, the polynomial

> The quadratic $ax^2 + bx + c$ has no real roots and cannot be factored over the real numbers if $b^2 - 4ac < 0$.

$$x^9 + 4x^8 + 6x^7 + 34x^6 + 64x^5 - 84x^4 - 287x^3 - 500x^2 - 354x - 180$$

factors as

$$\underbrace{(x - 2)}_{\substack{\text{linear} \\ \text{factor}}}\underbrace{(x + 3)^2}_{\substack{\text{repeated} \\ \text{linear} \\ \text{factor}}}\underbrace{(x^2 - 2x + 10)}_{\substack{\text{irreducible} \\ \text{quadratic} \\ \text{factor}}}\underbrace{(x^2 + x + 1)^2}_{\substack{\text{repeated} \\ \text{irreducible} \\ \text{quadratic factor}}}.$$

In this factored form, we see linear factors (simple and repeated) and irreducible quadratic factors (simple and repeated).

With irreducible quadratic factors, two cases must be considered: simple and repeated factors. Simple quadratic factors are examined in the following examples, and repeated quadratic factors (which generally involve long computations) are explored in the exercises.

---

**PROCEDURE**   **Partial Fractions with Simple Irreducible Quadratic Factors**

Suppose a simple irreducible factor $ax^2 + bx + c$ appears in the denominator of a proper rational function in reduced form. The partial fraction decomposition contains a term of the form

$$\frac{Ax + B}{ax^2 + bx + c},$$

where $A$ and $B$ are unknown coefficients to be determined.

---

**EXAMPLE 4**   **Setting up partial fractions**   Give the appropriate form of the partial fraction decomposition for the following functions.

**a.** $\dfrac{x^2 + 1}{x^4 - 4x^3 - 32x^2}$    **b.** $\dfrac{10}{(x - 2)^2(x^2 + 2x + 2)}$

**SOLUTION**

**a.** The denominator factors as $x^2(x^2 - 4x - 32) = x^2(x - 8)(x + 4)$. Therefore, $x$ is a repeated linear factor, and $(x - 8)$ and $(x + 4)$ are simple linear factors. The required form of the decomposition is

$$\frac{A}{x} + \frac{B}{x^2} + \frac{C}{x - 8} + \frac{D}{x + 4}.$$

We see that the factor $x^2 - 4x - 32$ is quadratic, but it can be factored as $(x - 8)(x + 4)$, so it is not irreducible.

> In Example 4b, the factor $(x - 2)^2$ cannot be treated as an irreducible quadratic factor; it is a repeated linear factor.

**b.** The denominator is already fully factored. The quadratic factor $x^2 + 2x + 2$ cannot be factored using real numbers; therefore, it is irreducible. The form of the decomposition is

$$\frac{A}{x - 2} + \frac{B}{(x - 2)^2} + \frac{Cx + D}{x^2 + 2x + 2}.$$

*Related Exercises 8, 10* ◄

**EXAMPLE 5**   **Integrating with quadratic factors**   Evaluate

$$\int \frac{7x^2 - 13x + 13}{(x - 2)(x^2 - 2x + 3)}\, dx.$$

**SOLUTION**   Note that the quadratic factor is irreducible. The appropriate form of the partial fraction decomposition is

$$\frac{7x^2 - 13x + 13}{(x - 2)(x^2 - 2x + 3)} = \frac{A}{x - 2} + \frac{Bx + C}{x^2 - 2x + 3}.$$

Multiplying both sides of this equation by $(x - 2)(x^2 - 2x + 3)$ leads to

$$\begin{aligned}
7x^2 - 13x + 13 &= A(x^2 - 2x + 3) + (Bx + C)(x - 2) \\
&= (A + B)x^2 + (-2A - 2B + C)x + (3A - 2C).
\end{aligned}$$

Equating coefficients of equal powers of $x$ results in the equations

$$A + B = 7, \quad -2A - 2B + C = -13, \quad \text{and} \quad 3A - 2C = 13.$$

Solving this system of equations gives $A = 5$, $B = 2$, and $C = 1$; therefore, the original integral can be written as

$$\int \frac{7x^2 - 13x + 13}{(x - 2)(x^2 - 2x + 3)}\, dx = \int \frac{5}{x - 2}\, dx + \int \frac{2x + 1}{x^2 - 2x + 3}\, dx.$$

Let's work on the second (more difficult) integral. The substitution $u = x^2 - 2x + 3$ would work if $du = (2x - 2)\,dx$ appeared in the numerator. For this reason, we write the numerator as $2x + 1 = (2x - 2) + 3$ and split the integral:

$$\int \frac{2x + 1}{x^2 - 2x + 3}\,dx = \int \frac{2x - 2}{x^2 - 2x + 3}\,dx + \int \frac{3}{x^2 - 2x + 3}\,dx.$$

Assembling all the pieces, we have

$$\int \frac{7x^2 - 13x + 13}{(x - 2)(x^2 - 2x + 3)}\,dx$$

$$= \int \frac{5}{x - 2}\,dx + \int \underbrace{\frac{2x - 2}{x^2 - 2x + 3}}_{\text{let } u = x^2 - 2x + 3}\,dx + \int \frac{3}{\underbrace{x^2 - 2x + 3}_{(x - 1)^2 + 2}}\,dx \qquad \text{Complete the square in the third integral.}$$

$$= 5 \ln |x - 2| + \ln |x^2 - 2x + 3| + \frac{3}{\sqrt{2}} \tan^{-1}\left(\frac{x - 1}{\sqrt{2}}\right) + K \qquad \text{Integrate.}$$

$$= \ln \left| (x - 2)^5 (x^2 - 2x + 3) \right| + \frac{3}{\sqrt{2}} \tan^{-1}\left(\frac{x - 1}{\sqrt{2}}\right) + K. \qquad \text{Property of logarithms}$$

To evaluate the last integral $\displaystyle\int \frac{3}{x^2 - 2x + 3}\,dx$, we completed the square in the denominator and used the substitution $u = x - 1$ to produce $3\displaystyle\int \frac{du}{u^2 + 2}$, which is a standard form.

*Related Exercises 53, 55* ◄

**Long Division** The preceding discussion of partial fraction decomposition assumes $f(x) = p(x)/q(x)$ is a proper rational function. If this is not the case and we are faced with an improper rational function $f$, we divide the denominator into the numerator and express $f$ in two parts. One part will be a polynomial, and the other will be a proper rational function. For example, given the function

$$f(x) = \frac{2x^3 + 11x^2 + 28x + 33}{x^2 - x - 6},$$

we perform long division.

$$
\begin{array}{r}
2x + 13 \phantom{xxxxxxxxx} \\
x^2 - x - 6 \overline{\smash{)}\ 2x^3 + 11x^2 + 28x + 33} \\
\underline{2x^3 - 2x^2 - 12x \phantom{xxxxxx}} \\
13x^2 + 40x + 33 \\
\underline{13x^2 - 13x - 78} \\
53x + 111
\end{array}
$$

It follows that

$$f(x) = \underbrace{2x + 13}_{\substack{\text{polynomial;} \\ \text{easy to} \\ \text{integrate}}} + \underbrace{\frac{53x + 111}{x^2 - x - 6}}_{\substack{\text{apply partial fraction} \\ \text{decomposition}}}.$$

The first piece is easily integrated, and the second piece now qualifies for the methods described in this section.

**QUICK CHECK 4** What is the result of simplifying $\dfrac{x}{x + 1}$ by long division? ◄

**SUMMARY**  **Partial Fraction Decompositions**

Let $f(x) = p(x)/q(x)$ be a proper rational function in reduced form. Assume the denominator $q$ has been factored completely over the real numbers and $m$ is a positive integer.

1.  **Simple linear factor** A factor $x - r$ in the denominator requires the partial fraction $\dfrac{A}{x - r}$.

2.  **Repeated linear factor** A factor $(x - r)^m$ with $m > 1$ in the denominator requires the partial fractions

$$\frac{A_1}{(x - r)} + \frac{A_2}{(x - r)^2} + \frac{A_3}{(x - r)^3} + \cdots + \frac{A_m}{(x - r)^m}.$$

3.  **Simple irreducible quadratic factor** An irreducible factor $ax^2 + bx + c$ in the denominator requires the partial fraction

$$\frac{Ax + B}{ax^2 + bx + c}.$$

4.  **Repeated irreducible quadratic factor** (See Exercises 61–64.) An irreducible factor $(ax^2 + bx + c)^m$ with $m > 1$ in the denominator requires the partial fractions

$$\frac{A_1x + B_1}{ax^2 + bx + c} + \frac{A_2x + B_2}{(ax^2 + bx + c)^2} + \cdots + \frac{A_mx + B_m}{(ax^2 + bx + c)^m}.$$

## SECTION 8.5 EXERCISES

### Getting Started

1.  What kinds of functions can be integrated using partial fraction decomposition?

2.  Give an example of each of the following.

    a.  A simple linear factor
    b.  A repeated linear factor
    c.  A simple irreducible quadratic factor
    d.  A repeated irreducible quadratic factor

3.  What term(s) should appear in the partial fraction decomposition of a proper rational function with each of the following?

    a.  A factor of $x - 3$ in the denominator
    b.  A factor of $(x - 4)^3$ in the denominator
    c.  A factor of $x^2 + 2x + 6$ in the denominator

4.  What is the first step in integrating $\dfrac{x^2 + 2x - 3}{x + 1}$?

**5–16.** *Set up the appropriate form of the partial fraction decomposition for the following expressions. Do not find the values of the unknown constants.*

5.  $\dfrac{4x}{x^2 - 9x + 20}$

6.  $\dfrac{4x + 1}{4x^2 - 1}$

7.  $\dfrac{x + 3}{(x - 5)^2}$

8.  $\dfrac{2}{x^3 - 2x^2 + x}$

9.  $\dfrac{4}{x^5 - 5x^3 + 4x}$

10. $\dfrac{20x}{(x - 1)^2(x^2 + 1)}$

11. $\dfrac{1}{x(x^2 + 1)}$

12. $\dfrac{2x^2 + 3}{(x^2 - 8x + 16)(x^2 + 3x + 4)}$

13. $\dfrac{x^4 + 12x^2}{(x - 2)^2(x^2 + x + 2)^2}$

14. $\dfrac{6x^4 - 4x^3 + 15x^2 - 5x + 7}{(x - 2)(2x^2 + 3)^2}$

15. $\dfrac{x}{(x^4 - 16)^2}$

16. $\dfrac{x^2 + 2x + 6}{x^3(x^2 + x + 1)^2}$

### Practice Exercises

**17–22.** *Give the partial fraction decomposition for the following expressions.*

17. $\dfrac{5x - 7}{x^2 - 3x + 2}$

18. $\dfrac{11x - 10}{x^2 - x}$

19. $\dfrac{6}{x^2 - 2x - 8}$

20. $\dfrac{x^2 - 4x + 11}{(x - 3)(x - 1)(x + 1)}$

21. $\dfrac{2x^2 + 5x + 6}{x^2 + 3x + 2}$  (*Hint:* Use long division first.)

22. $\dfrac{x^4 + 2x^3 + x}{x^2 - 1}$

**23–64. Integration** *Evaluate the following integrals.*

23. $\displaystyle\int \frac{3}{(x - 1)(x + 2)}\, dx$

24. $\displaystyle\int \frac{8}{(x - 2)(x + 6)}\, dx$

25. $\displaystyle\int \frac{6}{x^2 - 1}\, dx$

26. $\displaystyle\int_0^1 \frac{dt}{t^2 - 9}$

**27.** $\displaystyle\int \frac{8x-5}{3x^2-5x+2}\,dx$

**28.** $\displaystyle\int_1^2 \frac{7x-2}{3x^2-2x}\,dx$

**29.** $\displaystyle\int_{-1}^2 \frac{5x}{x^2-x-6}\,dx$

**30.** $\displaystyle\int \frac{21x^2}{x^3-x^2-12x}\,dx$

**31.** $\displaystyle\int \frac{6x^2}{x^4-5x^2+4}\,dx$

**32.** $\displaystyle\int \frac{4x-2}{x^3-x}\,dx$

**33.** $\displaystyle\int \frac{3x^2+4x-6}{x^2-3x+2}\,dx$

**34.** $\displaystyle\int \frac{2z^3+z^2-6z+7}{z^2+z-6}\,dz$

**35.** $\displaystyle\int \frac{x^2+12x-4}{x^3-4x}\,dx$

**36.** $\displaystyle\int \frac{z^2+20z-15}{z^3+4z^2-5z}\,dz$

**37.** $\displaystyle\int \frac{dx}{x^4-10x^2+9}$

**38.** $\displaystyle\int_0^5 \frac{2}{x^2-4x-32}\,dx$

**39.** $\displaystyle\int \frac{81}{x^3-9x^2}\,dx$

**40.** $\displaystyle\int \frac{16x^2}{(x-6)(x+2)^2}\,dx$

**41.** $\displaystyle\int_{-1}^1 \frac{x}{(x+3)^2}\,dx$

**42.** $\displaystyle\int \frac{dx}{x^3-2x^2-4x+8}$

**43.** $\displaystyle\int \frac{2}{x^3+x^2}\,dx$

**44.** $\displaystyle\int_1^2 \frac{2}{t^3(t+1)}\,dt$

**45.** $\displaystyle\int \frac{x-5}{x^2(x+1)}\,dx$

**46.** $\displaystyle\int \frac{x^2}{(x-2)^3}\,dx$

**47.** $\displaystyle\int \frac{x^3-10x^2+27x}{x^2-10x+25}\,dx$

**48.** $\displaystyle\int \frac{x^3+2}{x^3-2x^2+x}\,dx$

**49.** $\displaystyle\int \frac{x^2-4}{x^3-2x^2+x}\,dx$

**50.** $\displaystyle\int \frac{8(x^2+4)}{x(x^2+8)}\,dx$

**51.** $\displaystyle\int \frac{x^2+x+2}{(x+1)(x^2+1)}\,dx$

**52.** $\displaystyle\int \frac{x^2+3x+2}{x(x^2+2x+2)}\,dx$

**53.** $\displaystyle\int \frac{2x^2+5x+5}{(x+1)(x^2+2x+2)}\,dx$

**54.** $\displaystyle\int \frac{z+1}{z(z^2+4)}\,dz$

**55.** $\displaystyle\int \frac{20x}{(x-1)(x^2+4x+5)}\,dx$

**56.** $\displaystyle\int \frac{2x+1}{x^2+4}\,dx$

**57.** $\displaystyle\int \frac{x^3+5x}{(x^2+3)^2}\,dx$

**58.** $\displaystyle\int \frac{x^4+1}{x^3+9x}\,dx$

**59.** $\displaystyle\int \frac{x^3+6x^2+12x+6}{(x^2+6x+10)^2}\,dx$

**60.** $\displaystyle\int \frac{dy}{(y^2+1)(y^2+2)}$

**61.** $\displaystyle\int \frac{2}{x(x^2+1)^2}\,dx$

**62.** $\displaystyle\int \frac{dx}{(x+1)(x^2+2x+2)^2}$

**63.** $\displaystyle\int \frac{9x^2+x+21}{(3x^2+7)^2}\,dx$

**64.** $\displaystyle\int \frac{9x^5+6x^3}{(3x^2+1)^3}\,dx$

**65. Explain why or why not** Determine whether the following statements are true and give an explanation or counterexample.

    **a.** To evaluate $\displaystyle\int \frac{4x^6}{x^4+3x^2}\,dx$, the first step is to find the partial fraction decomposition of the integrand.

    **b.** The easiest way to evaluate $\displaystyle\int \frac{6x+1}{3x^2+x}\,dx$ is with a partial fraction decomposition of the integrand.

    **c.** The rational function $f(x) = \dfrac{1}{x^2-13x+42}$ has an irreducible quadratic denominator.

    **d.** The rational function $f(x) = \dfrac{1}{x^2-13x+43}$ has an irreducible quadratic denominator.

**T** **66–68. Areas of regions** *Find the area of the following regions.*

**66.** The region bounded by the curve $y = \dfrac{x-x^2}{(x+1)(x^2+1)}$ and the $x$-axis from $x=0$ to $x=1$

**67.** The region bounded by the curve $y = \dfrac{10}{x^2-2x-24}$, the $x$-axis, and the lines $x=-2$ and $x=2$

**68.** The region in the first quadrant bounded by the curves $y = \dfrac{3x^2+2x+1}{x(x^2+x+1)}$, $y = \dfrac{2}{x}$, and $x=2$

**69–72. Volumes of solids** *Find the volume of the following solids.*

**69.** The region bounded by $y = \dfrac{x}{x+1}$, the $x$-axis, and $x=4$ is revolved about the $x$-axis.

**70.** The region bounded by $y = \dfrac{1}{x^2(x^2+2)^2}$, $y=0$, $x=1$, and $x=2$ is revolved about the $y$-axis.

**71.** The region bounded by $y = \dfrac{1}{\sqrt{x(3-x)}}$, $y=0$, $x=1$, and $x=2$ is revolved about the $x$-axis.

**72.** The region bounded by $y = \dfrac{3x^2+25}{x^2(x^2+25)}$, $y=0$, $x=5$, and $x=10$ is revolved about the $y$-axis.

**73. Two methods** Evaluate $\displaystyle\int \frac{dx}{x^2-1}$, for $x>1$, in two ways: using partial fractions and a trigonometric substitution. Reconcile your two answers.

**T** **74–75. Finding constants with a computer algebra system** *Give the appropriate form of the partial fraction decomposition of the expression, and then use a computer algebra system to find the unknown constants.*

**74.** $\dfrac{3x^2+2x+1}{(x+1)^3(x^2+x+1)^2}$

**75.** $\dfrac{x^4+3x^2+1}{x(x^2+1)^2(x^2+x+4)^2}$

**76–83. Preliminary steps** *The following integrals require a preliminary step such as a change of variables before using the method of partial fractions. Evaluate these integrals.*

**76.** $\displaystyle\int \frac{\cos\theta}{(\sin^3\theta - 4\sin\theta)}\,d\theta$

**77.** $\displaystyle\int \frac{e^x}{(e^x-1)(e^x+2)}\,dx$

**78.** $\displaystyle\int \frac{dy}{y(\sqrt{a}-\sqrt{y})}$, $a>0$ (*Hint:* Let $u = \sqrt{y}$.)

**79.** $\displaystyle\int \frac{\sec t}{1+\sin t}\,dt$

**80.** $\displaystyle\int \sqrt{e^x+1}\,dx$ (*Hint:* Let $u = \sqrt{e^x+1}$.)

**81.** $\displaystyle\int \frac{(e^{3x}+e^{2x}+e^x)}{(e^{2x}+1)^2}\,dx$

**82.** $\displaystyle\int \frac{dx}{x\sqrt{1+2x}}$

**83.** $\displaystyle\int \frac{dx}{1 + e^x}$

**84. What's wrong?** Why are there no constants $A$ and $B$ satisfying

$$\frac{x^2}{(x - 4)(x + 5)} = \frac{A}{x - 4} + \frac{B}{x + 5}?$$

**85. Another form of** $\displaystyle\int \sec x\, dx$

  **a.** Verify the identity $\sec x = \dfrac{\cos x}{1 - \sin^2 x}$.

  **b.** Use the identity in part (a) to verify that

$$\int \sec x\, dx = \frac{1}{2} \ln \left| \frac{1 + \sin x}{1 - \sin x} \right| + C.$$

  (*Source: The College Mathematics Journal, 32, 5, Nov 2001*)

### Explorations and Challenges

**86. Partial fractions from an integral** A computer algebra system was used to obtain the following result.

$$\int \frac{2x^4 + 14x^3 + 14x^2 + 51x + 19}{x^5 + x^4 + 3x^3 + 3x^2 - 4x - 4}\, dx$$

$$= 5 \ln (x - 1) - 3 \ln (x + 1) - \frac{3}{x + 1} + \frac{1}{2} \tan^{-1} \frac{x}{2} + C$$

Use this fact to find the partial fraction decomposition of the integrand on the left side of the equation.

**87–92.** *An integrand with trigonometric functions in the numerator and denominator can often be converted to a rational function using the substitution $u = \tan (x/2)$ or, equivalently, $x = 2 \tan^{-1} u$. The following relations are used in making this change of variables.*

$$A\!: dx = \frac{2}{1 + u^2}\, du \quad B\!: \sin x = \frac{2u}{1 + u^2} \quad C\!: \cos x = \frac{1 - u^2}{1 + u^2}$$

**87.** Verify relation $A$ by differentiating $x = 2 \tan^{-1} u$. Verify relations $B$ and $C$ using a right-triangle diagram and the double-angle formulas

$$\sin x = 2 \sin \frac{x}{2} \cos \frac{x}{2} \quad \text{and} \quad \cos x = 2 \cos^2 \frac{x}{2} - 1.$$

**88.** Evaluate $\displaystyle\int \frac{dx}{2 + \cos x}$.  **89.** Evaluate $\displaystyle\int \frac{dx}{1 - \cos x}$.

**90.** Evaluate $\displaystyle\int \frac{dx}{1 + \sin x + \cos x}$.

**91.** Evaluate $\displaystyle\int_0^{\pi/2} \frac{d\theta}{\cos \theta + \sin \theta}$.  **92.** Evaluate $\displaystyle\int_0^{\pi/3} \frac{\sin \theta}{1 - \sin \theta}\, d\theta$.

**93. Three start-ups** Three cars, $A$, $B$, and $C$, start from rest and accelerate along a line according to the following velocity functions:

$$v_A(t) = \frac{88t}{t + 1}, \quad v_B(t) = \frac{88t^2}{(t + 1)^2}, \quad \text{and} \quad v_C(t) = \frac{88t^2}{t^2 + 1}.$$

  **a.** Which car travels farthest on the interval $0 \le t \le 1$?
  **b.** Which car travels farthest on the interval $0 \le t \le 5$?
  **c.** Find the position functions for each car assuming each car starts at the origin.
  **d.** Which car ultimately gains the lead and remains in front?

**T 94. Skydiving** A skydiver has a downward velocity given by

$$v(t) = V_T \left( \frac{1 - e^{-2gt/V_T}}{1 + e^{-2gt/V_T}} \right),$$

where $t = 0$ is the instant the skydiver starts falling, $g = 9.8 \text{ m/s}^2$ is the acceleration due to gravity, and $V_T$ is the terminal velocity of the skydiver.

  **a.** Evaluate $v(0)$ and $\displaystyle\lim_{t \to \infty} v(t)$ and interpret these results.
  **b.** Graph the velocity function.
  **c.** Verify by integration that the position function is given by

$$s(t) = V_T t + \frac{V_T^2}{g} \ln \left( \frac{1 + e^{-2gt/V_T}}{2} \right),$$

  where $s'(t) = v(t)$ and $s(0) = 0$.
  **d.** Graph the position function.
  (See the Guided Project *Terminal velocity* for more details on free fall and terminal velocity.)

**95. $\pi < 22/7$** One of the earliest approximations to $\pi$ is $22/7$. Verify that $0 < \displaystyle\int_0^1 \frac{x^4(1 - x)^4}{1 + x^2}\, dx = \frac{22}{7} - \pi$. Why can you conclude that $\pi < 22/7$?

**96. Challenge** Show that with the change of variables $u = \sqrt{\tan x}$, the integral $\int \sqrt{\tan x}\, dx$ can be converted to an integral amenable to partial fractions. Evaluate $\int_0^{\pi/4} \sqrt{\tan x}\, dx$.

**QUICK CHECK ANSWERS**

**1.** $\ln |x - 2| + 2 \ln |x + 4| = \ln |(x - 2)(x + 4)^2|$
**2.** $A/(x - 1) + B/(x + 5) + C/(x - 10)$
**3.** $A/x + B/x^2 + C/(x - 3) + D/(x - 3)^2 + E/(x - 1)$
**4.** $1 - \dfrac{1}{x + 1}$ ◄

# 8.6 Integration Strategies

In the previous five sections, we introduced various methods for evaluating integrals, though always in the context of learning a specific technique of integration. For instance, in Section 8.3, we focused on integrals involving trigonometric functions. In this section, the chief goal is to illustrate how to attack a generic integration problem when the appropriate method is not obvious.

## Integration Strategies

As a starting point, we assume you know all the integrals listed in Table 8.1 at the beginning of this chapter, and that you are familiar with the reduction formulas presented in

Sections 8.2 and 8.3. Your instructor may also require that you know the formulas presented in Section 7.7 that involve hyperbolic functions.

We noted in Section 5.5 that computing the derivative of a (differentiable) function usually requires only the proper application of the derivative rules from Chapter 3. Finding the antiderivative of a function is, generally speaking, more difficult. In what follows, we offer a collection of strategies to use when confronted with an unfamiliar integral, illustrated with brief examples.

**Rewrite the integrand**  Several techniques can be used to transform an integrand. We can split up fractions, use long division or trigonometric identities, complete the square in an integrand that includes a quadratic function, perform algebraic manipulations (such as multiplying numerator and denominator by the same expression), and cancel common factors.

**EXAMPLE 1** **Rewrite the integrand**  Evaluate $\int \dfrac{\sin x + 1}{\cos^2 x}\, dx$.

**SOLUTION**  Split the fraction and use trigonometric identities to rewrite the integral as the sum of two integrals:

$$\int \frac{\sin x + 1}{\cos^2 x}\, dx = \int \left( \frac{1}{\cos x} \cdot \frac{\sin x}{\cos x} + \frac{1}{\cos^2 x} \right) dx = \int \sec x \tan x\, dx + \int \sec^2 x\, dx.$$

Both integrals on the right are now easily integrated using Table 8.1.

*Related Exercises 4, 11* ◄

> **QUICK CHECK 1** Use Table 8.1 (p. 520) to complete the process of evaluating $\int \dfrac{\sin x + 1}{\cos^2 x}\, dx$ given in Example 1. ◄

**Try substitution**  As a general rule, if an integrand contains the function $f(x)$ and its derivative $f'(x)$, the substitution $u = f(x)$ often results in an easier integral. For instance, $\int \dfrac{\sin (1/x)}{x^2}\, dx$ suggests the substitution $u = 1/x$ because $\dfrac{du}{dx} = -\dfrac{1}{x^2}$ appears in the integrand (up to a constant multiplicative factor). When attempting to evaluate an integral using the substitution method, it is important to understand that there is no list of "rules" one must follow—any substitution is fair game. On occasion, a creative substitution does the trick.

> **QUICK CHECK 2** Use the substitution $u = \dfrac{1}{x}$ to evaluate $\int \dfrac{\sin (1/x)}{x^2}\, dx$. ◄

**EXAMPLE 2** **Substitution after rewriting the integrand**  Evaluate $\int \dfrac{x^2}{\sqrt{4 - x^6}}\, dx$.

**SOLUTION**  Expressing the denominator of the integrand as $\sqrt{4 - (x^3)^2}$ looks promising, particularly because a constant multiple of the derivative of $x^3$ appears in the numerator. Therefore, we try the substitution $u = x^3$:

$$\int \frac{x^2}{\sqrt{4 - x^6}}\, dx = \int \frac{x^2}{\sqrt{4 - (x^3)^2}}\, dx = \frac{1}{3} \int \frac{du}{\sqrt{4 - u^2}}. \qquad \text{Let } u = x^3;\ du = 3x^2\, dx.$$

> **QUICK CHECK 3** Complete the evaluation of $\int \dfrac{x^2}{\sqrt{4 - x^6}}\, dx$ in Example 2; make sure to express your answer in terms of $x$. ◄

Observe that $\int \dfrac{du}{\sqrt{4 - u^2}}$ is a standard integral, found in Table 8.1.

*Related Exercises 14, 43* ◄

**Recognize a common pattern present in the integrand**  Here we list certain patterns to look for in the integrand and the associated integration techniques.

1. Integrating the product of two functions often calls for integration by parts (Section 8.2), particularly if the factors of the product are a polynomial and a transcendental function (a logarithmic, exponential, trigonometric, or inverse trigonometric function). In some cases, integration by parts is successful when the integrand is a single function (for example, $\int \ln x\, dx$ or $\int \sin^{-1} x\, dx$).

2. Powers and products of $\sin x$ and $\cos x$, $\tan x$ and $\sec x$, or $\cot x$ and $\csc x$ are integrated using identities, reduction formulas, or the substitutions given in Tables 8.2 and 8.3 of Section 8.3.

3. Integrals containing the expression $a^2 - x^2$, $a^2 + x^2$, or $x^2 - a^2$ may yield to trigonometric substitution. Use the substitutions given in Table 8.4 of Section 8.4 to tackle such integrals, but don't overlook a simpler solution. For example, $\int \dfrac{x}{1 - x^2}\, dx$ is most easily evaluated using the substitution $u = 1 - x^2$ rather than $x = \sin\theta$.

4. Use partial fraction decomposition (Section 8.5) to integrate rational functions. Remember to first cancel common factors and to apply long division if the rational function is not proper.

**EXAMPLE 3**    **Identify a strategy**  Suggest a technique of integration for each of the following integrals based on patterns in the integrand.

**a.** $\displaystyle\int \frac{4 - 3x^2}{x(x^2 - 4)}\, dx$    **b.** $\displaystyle\int xe^{\sqrt{1+x^2}}\, dx$    **c.** $\displaystyle\int \ln(1 + x^2)\, dx$

**SOLUTION**

**a.** The integrand is a rational function whose denominator factors easily, so partial fraction decomposition is an obvious choice. However, it is worth considering other options that may save time. Expanding the denominator, we recognize the integral admits an almost perfect $u$-substitution:

$$\int \frac{4 - 3x^2}{x(x^2 - 4)}\, dx = \int \frac{4 - 3x^2}{x^3 - 4x}\, dx. \qquad \begin{array}{l}\text{Let } u = x^3 - 4x;\, du = (3x^2 - 4)\,dx,\\ \text{or } -du = (4 - 3x^2)\,dx.\end{array}$$

This substitution leads to $-\displaystyle\int \frac{du}{u}$, which is a standard integral.

**b.** The integrand suggests several techniques. The substitution $u = 1 + x^2$ looks promising given that $du/dx = 2x$ appears in the integrand (up to a constant multiple). This choice leads to $\dfrac{1}{2}\displaystyle\int e^{\sqrt{u}}\, du$, which requires a crafty substitution followed by integration by parts.

   Given that the integrand is a product of two functions, we might try integration by parts as a first step. We leave it to the reader to conclude that this path is not fruitful.

   Finally, the presence of $1 + x^2$ in the integrand suggests the substitution $x = \tan\theta$. This choice leads to

$$\int xe^{\sqrt{1+x^2}}\, dx = \int \tan\theta \cdot e^{\sqrt{1+\tan^2\theta}} \sec^2\theta\, d\theta \quad x = \tan\theta,\, dx = \sec^2\theta\, d\theta$$

$$= \int \sec\theta \cdot e^{\sec\theta} \sec\theta \tan\theta\, d\theta \quad \begin{array}{l}\text{Rewrite the integrand to prepare}\\ \text{for a substitution.}\end{array}$$

$$= \int ue^u\, du. \qquad \text{Let } u = \sec\theta;\, du = \sec\theta \tan\theta\, d\theta.$$

> The substitution $u = \sqrt{1 + x^2}$ is yet another option for the integral in Example 3b; this choice leads immediately (though not transparently) to $\int ue^u\, du$. It takes a lot of practice to find creative solutions such as this one.

We encountered the last integral in Example 1 of Section 8.2 and evaluated it using integration by parts.

**c.** Because $1 + x^2$ is present in the integrand, we could try the substitution $x = \tan\theta$, which yields $\int \ln(\sec^2\theta) \sec^2\theta\, d\theta$. Using integration by parts on this new integral eventually leads to the answer, but it is more efficient to use integration by parts from the beginning. We try the following choices for $u$ and $dv$.

| $u = \ln(1 + x^2)$ | $dv = dx$ |
|---|---|
| $du = \dfrac{2x}{1 + x^2}\, dx$ | $v = x$ |

The integral then becomes

$$\int \ln(1 + x^2)\, dx = x\ln(1 + x^2) - 2\int \frac{x^2}{1 + x^2}\, dx,$$

and the new integral is handled with long division followed by routine integration.

*Related Exercises 55, 70, 96* ◄

The integrals in Example 3 bring to light two important concepts. First, recognize that more than one technique may work for a given integral. In practice, one method is usually more efficient than the others, so be ready to try another method if your first choice does not work. Second, to determine which choice is best, you may need to carry out the first few steps of several techniques in order to recognize the best approach.

**Be persistent**   Be prepared to experience dead ends when attempting to evaluate an integral. Look again at the integral given in Example 1. Instead of splitting the fraction, suppose we used the identity $\cos^2 x = 1 - \sin^2 x$:

$$\int \frac{\sin x + 1}{\cos^2 x}\, dx = \int \frac{\sin x + 1}{1 - \sin^2 x}\, dx = \int \frac{\sin x + 1}{(1 + \sin x)(1 - \sin x)}\, dx = \int \frac{1}{1 - \sin x}\, dx.$$

No error has been committed in the previous calculation, but the final integral is not easily integrated in its present form. In situations such as this one, start again using a different approach.

**EXAMPLE 4**   **Integration by parts?**   Evaluate $\int x e^{\sqrt{x}}\, dx$.

**SOLUTION**   The integrand is the product of a power of $x$ and a transcendental function; integrals of this form often yield to integration by parts. We try the following choices for $u$ and $dv$.

| $u = e^{\sqrt{x}}$ | $dv = x\, dx$ |
|---|---|
| $du = \dfrac{e^{\sqrt{x}}}{2\sqrt{x}}\, dx$ | $v = \dfrac{x^2}{2}$ |

The integral then becomes

$$\int x e^{\sqrt{x}}\, dx = \frac{x^2 e^{\sqrt{x}}}{2} - \frac{1}{4}\int x^{3/2} e^{\sqrt{x}}\, dx.$$

> The integral that results in Example 4 when applying integration by parts, $\int x^{3/2}\, e^{\sqrt{x}}\, dx$, can be evaluated. For instance, the substitution $u = e^{\sqrt{x}}$ leads to $2 \int \ln^4 u\, du$, and the substitution $u = \sqrt{x}$ leads to $2 \int u^4 e^u\, du$, both of which can be handled by reduction formulas given in the exercise set of Section 8.2. However, the second solution, using the substitution $u = \sqrt{x}$, is more efficient.

The new integral appears to be more difficult than the original. Because integration by parts was not successful, we return to the original integral. Let's try the substitution $u = \sqrt{x}$, which implies that $du = \dfrac{1}{2\sqrt{x}}\, dx$. At this point, there are two paths we might follow.

*Option 1:* Our substitution would succeed if $2\sqrt{x}$ appeared in the denominator of the integrand. Using the technique of multiplication by 1, we multiply the numerator and denominator of the integrand by $2\sqrt{x}$ to prepare for the substitution:

$$\int x e^{\sqrt{x}}\, dx = \int \frac{2\sqrt{x} \cdot x e^{\sqrt{x}}}{2\sqrt{x}}\, dx \quad \text{Multiply by 1.}$$

$$= 2\int u^3 e^u\, du. \qquad u = \sqrt{x};\, du = \frac{dx}{2\sqrt{x}};\, \sqrt{x} \cdot x = (\sqrt{x})^3 = u^3$$

*Option 2:* Because $u = \sqrt{x}$, we have $x = u^2$, which implies that $dx = 2u\, du$. Making these substitutions in the integral leads to

$$\int x e^{\sqrt{x}}\, dx = \int u^2 e^u \cdot 2u\, du \quad u = \sqrt{x};\, x = u^2;\, dx = 2u\, du$$

$$= 2\int u^3 e^u\, du. \qquad \text{Simplify.}$$

In both cases we arrive at the same integral, $2\int u^3 e^u\, du$, which is handled by applying repeated integration by parts (see tabular integration, Exercises 77–79, Section 8.2) or by using a reduction formula (Exercise 50, Section 8.2).

*Related Exercise 98* ◄

**Multiple techniques**   More than one technique may be necessary for complicated integrals. We conclude this section with a full evaluation of an integral that requires several techniques of integration.

**EXAMPLE 5    Multiple techniques needed**  Find the area of the surface of revolution that results when the curve $f(x) = e^x$ on $[0, \ln 2]$ is revolved about the $x$-axis.

**SOLUTION**  Recall from Section 6.6 that the integral $\int_a^b 2\pi f(x)\sqrt{1 + f'(x)^2}\, dx$ gives the area of the surface generated when $y = f(x)$ on $[a, b]$ is revolved about the $x$-axis. Because $f'(x) = e^x$, the area of the surface is

$$\int_0^{\ln 2} 2\pi e^x \sqrt{1 + (e^x)^2}\, dx.$$

The presence of $e^x$ and its derivative in the integrand suggests the substitution $u = e^x$:

$$\int_0^{\ln 2} 2\pi e^x \sqrt{1 + (e^x)^2}\, dx = 2\pi \int_1^2 \sqrt{1 + u^2}\, du \qquad \begin{array}{l} u = e^x; \, du = e^x\, dx \\ x = 0 \Rightarrow u = 1; \, x = \ln 2 \Rightarrow u = 2 \end{array}$$

The new integrand contains $1 + u^2$, so we try the trigonometric substitution $u = \tan\theta$. Setting the definite integral aside for the moment, we first focus on evaluating the indefinite integral:

$$2\pi \int \sqrt{1 + u^2}\, du = 2\pi \int \sqrt{1 + \tan^2\theta}\, \sec^2\theta\, d\theta \qquad \text{Let } u = \tan\theta; \, du = \sec^2\theta\, d\theta.$$

$$= 2\pi \int \sec^3\theta\, d\theta. \qquad 1 + \tan^2\theta = \sec^2\theta$$

To complete the solution, we rely on the secant reduction formula from Section 8.3 and return to the variable $u$:

$$2\pi \int \sec^3\theta\, d\theta = 2\pi \left( \frac{1}{2} \sec\theta \tan\theta + \frac{1}{2} \int \sec\theta\, d\theta \right) \qquad \text{\textcolor{blue}{Secant reduction formula}}$$

$$= \pi\left(\sec\theta \tan\theta + \ln|\sec\theta + \tan\theta|\right) + C \qquad \text{\textcolor{blue}{Evaluate }} \int \sec\theta\, d\theta.$$

$$= \pi\left(u\sqrt{1 + u^2} + \ln|u + \sqrt{1 + u^2}|\right) + C. \qquad \text{\textcolor{blue}{$\tan\theta = u$; $\sec\theta = \sqrt{1 + u^2}$}}$$

Evaluating the antiderivative using limits of integration from the integral in $u$ gives the area of the surface:

$$\text{Area} = \pi\left(u\sqrt{1 + u^2} + \ln\left(u + \sqrt{1 + u^2}\right)\right)\Big|_1^2$$

$$= \pi\left(2\sqrt{5} - \sqrt{2} + \ln\frac{2 + \sqrt{5}}{1 + \sqrt{2}}\right) \approx 11.37.$$

*Related Exercise 87* ◄

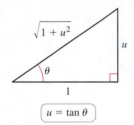

$u = \tan\theta$

---

# SECTION 8.6 EXERCISES

## Getting Started

**1–6. Choosing an integration strategy** *Identify a technique of integration for evaluating the following integrals. If necessary, explain how to first simplify the integrand before applying the suggested technique of integration. You do not need to evaluate the integrals.*

**1.** $\displaystyle\int 4x \sin 5x\, dx$

**2.** $\displaystyle\int (1 + \tan x)\sec^2 x\, dx$

**3.** $\displaystyle\int \frac{x^3}{\sqrt{64 - x^2}}\, dx$

**4.** $\displaystyle\int \frac{\tan^2 x + 1}{\tan x}\, dx$

**5.** $\displaystyle\int \frac{5x^2 + 18x + 20}{(2x + 3)(x^2 + 4x + 8)}\, dx$

**6.** $\displaystyle\int \frac{\cos^5 x \sin^4 x}{1 - \sin^2 x}\, dx$

## Practice Exercises

**7–84.** *Evaluate the following integrals.*

**7.** $\displaystyle\int_0^{\pi/2} \frac{\sin\theta}{1 + \cos^2\theta}\, d\theta$

**8.** $\displaystyle\int \cos^2 10x\, dx$

**9.** $\displaystyle\int_4^6 \frac{dx}{\sqrt{8x - x^2}}$

**10.** $\displaystyle\int \sin^9 x \cos^3 x\, dx$

**11.** $\displaystyle\int_0^{\pi/4} (\sec x - \cos x)^2\, dx$

**12.** $\displaystyle\int \frac{e^x}{\sqrt{1 - e^{2x}}}\, dx$

**13.** $\displaystyle\int \frac{dx}{e^x\sqrt{1 - e^{2x}}}$

**14.** $\displaystyle\int \frac{x^{-2} + x^{-3}}{x^{-1} + 16x^{-3}}\, dx$

**15.** $\displaystyle\int_1^4 \frac{2^{\sqrt{x}}}{\sqrt{x}}\, dx$

**16.** $\displaystyle\int \frac{dx}{x^4 - 1}$

**17.** $\displaystyle\int_1^2 w^3 e^{w^2}\, dw$

**18.** $\displaystyle\int_5^6 x(x-5)^{10}\, dx$

**19.** $\displaystyle\int_0^{\pi/2} \sin^7 x\, dx$

**20.** $\displaystyle\int_1^3 \frac{dt}{\sqrt{t}(t+1)}$

**21.** $\displaystyle\int x^9 \ln 3x\, dx$

**22.** $\displaystyle\int \frac{dx}{(x-a)(x-b)},\, a\neq b$

**23.** $\displaystyle\int \frac{\sin x}{\cos^2 x + \cos x}\, dx$

**24.** $\displaystyle\int \frac{3w^5 + 2w^4 - 12w^3 - 12w - 32}{w^3 - 4w}\, dw$

**25.** $\displaystyle\int \frac{dx}{x\sqrt{1-x^2}}$

**26.** $\displaystyle\int_{1/e}^1 \frac{dx}{x(\ln^2 x + 2\ln x + 2)}$

**27.** $\displaystyle\int \sin^4 \frac{x}{2}\, dx$

**28.** $\displaystyle\int \frac{3x^2 + 2x + 3}{x^4 + 2x^2 + 1}\, dx$

**29.** $\displaystyle\int \frac{2\cos x + \cot x}{1 + \sin x}\, dx$

**30.** $\displaystyle\int_{5/2}^{5\sqrt{3}/2} \frac{dv}{v^2\sqrt{25-v^2}}$

**31.** $\displaystyle\int \sqrt{36-9x^2}\, dx$

**32.** $\displaystyle\int \frac{dx}{\sqrt{36x^2-25}},\, x>5/6$

**33.** $\displaystyle\int \frac{e^x}{a^2 + e^{2x}}\, dx,\, a\neq 0$

**34.** $\displaystyle\int_0^{\pi/9} \frac{\sin 3x}{\cos 3x + 1}\, dx$

**35.** $\displaystyle\int_0^{\pi/4} (\tan^2\theta + \tan\theta + 1)\sec^2\theta\, d\theta$

**36.** $\displaystyle\int x\, 10^x\, dx$

**37.** $\displaystyle\int_0^{\pi/6} \frac{dx}{1-\sin 2x}$

**38.** $\displaystyle\int_{\pi/6}^{\pi/2} \cos x \ln(\sin x)\, dx$

**39.** $\displaystyle\int \sin x \ln(\sin x)\, dx$

**40.** $\displaystyle\int \sin 2x \ln(\sin x)\, dx$

**41.** $\displaystyle\int \cot^{3/2} x \csc^4 x\, dx$

**42.** $\displaystyle\int_0^{1/2} \frac{\sin^{-1} x}{\sqrt{1-x^2}}\, dx$

**43.** $\displaystyle\int \frac{x^9}{\sqrt{1-x^{20}}}\, dx$

**44.** $\displaystyle\int \frac{dx}{x^3 - x^2}$

**45.** $\displaystyle\int_0^{\ln 2} \frac{1}{(1+e^x)^2}\, dx$

**46.** $\displaystyle\int \frac{dx}{e^{2x}+1}$

**47.** $\displaystyle\int \frac{2x^3 + x^2 - 2x - 4}{x^2 - x - 2}\, dx$

**48.** $\displaystyle\int \frac{\sqrt{16-x^2}}{x^2}\, dx$

**49.** $\displaystyle\int \tan^3 x \sec^9 x\, dx$

**50.** $\displaystyle\int \tan^7 x \sec^4 x\, dx$

**51.** $\displaystyle\int_0^{\pi/3} \tan x \sec^{7/4} x\, dx$

**52.** $\displaystyle\int t^2 e^{3t}\, dt$

**53.** $\displaystyle\int e^x \cot^3 e^x\, dx$

**54.** $\displaystyle\int \frac{2x^2 + 3x + 26}{(x-2)(x^2+16)}\, dx$

**55.** $\displaystyle\int \frac{3x^2 + 3x + 1}{x^3 + x}\, dx$

**56.** $\displaystyle\int_\pi^{3\pi/2} \sin 2x\, e^{\sin^2 x}\, dx$

**57.** $\displaystyle\int \sin\sqrt{x}\, dx$

**58.** $\displaystyle\int w^2 \tan^{-1} w\, dw$

**59.** $\displaystyle\int \frac{dx}{x^4 + x^2}$

**60.** $\displaystyle\int_0^{\pi/2} e^{-3x}\cos x\, dx$

**61.** $\displaystyle\int_0^{\sqrt{2}/2} e^{\sin^{-1}x}\, dx$

**62.** $\displaystyle\int_0^{\pi/2} \sqrt{1+\cos\theta}\, d\theta$

**63.** $\displaystyle\int x^a \ln x\, dx,\, a\neq -1$

**64.** $\displaystyle\int \frac{\ln ax}{x}\, dx,\, a\neq 0$

**65.** $\displaystyle\int_0^{1/6} \frac{dx}{\sqrt{1-9x^2}}$

**66.** $\displaystyle\int \frac{x}{\sqrt{1-9x^2}}\, dx$

**67.** $\displaystyle\int \frac{x^2}{\sqrt{1-9x^2}}\, dx$

**68.** $\displaystyle\int \frac{e^x}{e^{2x} + 2e^x + 17}\, dx$

**69.** $\displaystyle\int \frac{dx}{1 - x^2 + \sqrt{1-x^2}}$

**70.** $\displaystyle\int \ln(x^2 + a^2)\, dx,\, a\neq 0$

**71.** $\displaystyle\int \frac{1-\cos x}{1+\cos x}\, dx$

**72.** $\displaystyle\int x^2 \sinh x\, dx$

**73.** $\displaystyle\int_9^{16} \sqrt{1+\sqrt{x}}\, dx$

**74.** $\displaystyle\int \frac{e^{3x}}{e^x - 1}\, dx$

**75.** $\displaystyle\int_1^3 \frac{\tan^{-1}\sqrt{x}}{x^{1/2} + x^{3/2}}\, dx$

**76.** $\displaystyle\int \frac{x}{x^2 + 6x + 18}\, dx$

**77.** $\displaystyle\int \cos^{-1} x\, dx$

**78.** $\displaystyle\int (\cos^{-1} x)^2\, dx$

**79.** $\displaystyle\int \frac{\sin^{-1} x}{x^2}\, dx$

**80.** $\displaystyle\int_{-2}^{-1} \sqrt{-4x-x^2}\, dx$

**81.** $\displaystyle\int \frac{x^4 + 2x^3 + 5x^2 + 2x + 1}{x^5 + 2x^3 + x}\, dx$

**82.** $\displaystyle\int \frac{dx}{1+\tan x}$

**83.** $\displaystyle\int e^x \sin^{998}(e^x)\cos^3(e^x)\, dx$

**84.** $\displaystyle\int \frac{\tan\theta + \tan^3\theta}{(1+\tan\theta)^{50}}\, d\theta$

**85. Explain why or why not** Determine whether the following statements are true and give an explanation or counterexample.

a. More than one integration method can be used to evaluate
$$\int \frac{dx}{1-x^2}.$$

b. Using the substitution $u = \sqrt[3]{x}$ in $\displaystyle\int \sin\sqrt[3]{x}\, dx$ leads to
$$\int 3u^2 \sin u\, du.$$

c. The most efficient way to evaluate $\displaystyle\int \tan 3x \sec^2 3x\, dx$ is to first rewrite the integrand in terms of $\sin 3x$ and $\cos 3x$.

d. Using the substitution $u = \tan x$ in $\displaystyle\int \frac{\tan^2 x}{\tan x - 1}\, dx$ leads to
$$\int \frac{u^2}{u-1}\, du.$$

**86. Area** Find the area of the region bounded by the curves
$$y = \frac{x}{x^2 - 2x + 2},\quad y = \frac{2}{x^2 - 2x + 2},\quad \text{and } x = 0.$$

**87. Surface area** Find the area of the surface generated when the curve $f(x) = \sin x$ on $[0, \pi/2]$ is revolved about the $x$-axis.

**88. Volume** Find the volume of the solid obtained by revolving the region bounded by the curve $y = \dfrac{x}{\sqrt{4 - x^2}}$ on $[0, 1]$ about the $y$-axis.

**89. Volume** Find the volume of the solid obtained by revolving the region bounded by the curve $y = \dfrac{1}{1 - \sin x}$ on $[0, \pi/4]$ about the $x$-axis.

**90. Work** Let $R$ be the region in the first quadrant bounded by the curve $y = \sqrt{x^4 - 4}$, and the lines $y = 0$ and $y = 2$. Suppose a tank that is full of water has the shape of a solid of revolution obtained by revolving region $R$ about the $y$-axis. How much work is required to pump all the water to the top of the tank? Assume $x$ and $y$ are in meters.

**91. Work** Let $R$ be the region in the first quadrant bounded by the curve $y = \sec^{-1} x$ and the line $y = \pi/3$. Suppose a tank that is full of water has the shape of a solid of revolution obtained by revolving region $R$ about the $y$-axis. How much work is required to pump all the water to the top of the tank? Assume $x$ and $y$ are in meters.

**Explorations and Challenges**

**92–98.** *Evaluate the following integrals.*

**92.** $\displaystyle\int_{1}^{\sqrt[3]{2}} y^8 e^{y^3}\, dy$

**93.** $\displaystyle\int \dfrac{68}{e^{2x} + 2e^x + 17}\, dx$

**94.** $\displaystyle\int \dfrac{dt}{t^3 + 1}$

**95.** $\displaystyle\int \dfrac{dx}{1 - \tan^2 x}$

**96.** $\displaystyle\int e^{\sqrt[3]{x}}\, dx$

**97.** $\displaystyle\int \tan^{-1} \sqrt[3]{x}\, dx$

**98.** $\displaystyle\int e^{\sqrt{\sin x}} \cos x\, dx$

**99. Surface area** Find the area of the surface generated when the curve $f(x) = \tan x$ on $[0, \pi/4]$ is revolved about the $x$-axis.

**QUICK CHECK ANSWERS**

**1.** $\sec x + \tan x + C$   **2.** $\cos(1/x) + C$

**3.** $\dfrac{1}{3} \sin^{-1} \dfrac{x^3}{2} + C$ ◄

# 8.7 Other Methods of Integration

The integration methods studied so far—various substitutions, integration by parts, and partial fractions—are examples of *analytical methods*; they are done with pencil and paper, and they give exact results. While many important integrals can be evaluated with analytical methods, many more integrals lie beyond their reach. For example, the following integrals cannot be evaluated in terms of familiar functions:

$$\int e^{x^2}\, dx, \quad \int \sin x^2\, dx, \quad \int \frac{\sin x}{x}\, dx, \quad \int \frac{e^{-x}}{x}\, dx, \quad \text{and} \quad \int \ln(\ln x)\, dx.$$

The next two sections survey alternative strategies for evaluating integrals when standard analytical methods do not work. These strategies fall into three categories.

**Figure 8.10**

1. **Tables of integrals** The endpapers of this text contain a table of many standard integrals. Because these integrals were evaluated analytically, using tables is considered an analytical method. Tables of integrals also contain reduction formulas like those discussed in Sections 8.2 and 8.3.

2. **Symbolic methods** Computer algebra systems have sophisticated algorithms to evaluate difficult integrals. Many definite and indefinite integrals can be evaluated exactly using these symbolic methods.

3. **Numerical methods** The value of a definite integral can be approximated accurately using numerical methods introduced in the next section. *Numerical* means that these methods compute numbers rather than manipulate symbols. Computers and calculators often have built-in functions to carry out numerical calculations.

Figure 8.10 is a chart of the various integration strategies and how they are related.

## Using Tables of Integrals

Given a specific integral, you *may* be able to find the identical integral in a table of integrals. More often, some preliminary work is needed to convert the given integral into one that appears in a table. Most tables give only indefinite integrals, although some tables include special definite integrals. The following examples illustrate various ways in which tables of integrals are used.

▶ A short table of integrals is found at the end of the text. Longer tables of integrals are found online and in venerable collections such as the *CRC Mathematical Tables* and *Handbook of Mathematical Functions*, by Abramowitz and Stegun.

**EXAMPLE 1**   **Using tables of integrals**  Evaluate the integral $\int \dfrac{dx}{x\sqrt{2x-9}}$.

**SOLUTION** It is worth noting that this integral may be evaluated with the change of variables $u^2 = 2x - 9$. Alternatively, a table of integrals includes the integral

$$\int \frac{dx}{x\sqrt{ax-b}} = \frac{2}{\sqrt{b}}\tan^{-1}\sqrt{\frac{ax-b}{b}} + C, \quad \text{where} \quad b > 0,$$

which matches the given integral. Letting $a = 2$ and $b = 9$, we find that

$$\int \frac{dx}{x\sqrt{2x-9}} = \frac{2}{\sqrt{9}}\tan^{-1}\sqrt{\frac{2x-9}{9}} + C = \frac{2}{3}\tan^{-1}\frac{\sqrt{2x-9}}{3} + C.$$

*Related Exercises 9–10* ◀

▶ Letting $u^2 = 2x - 9$, we have $u\,du = dx$ and $x = \frac{1}{2}(u^2 + 9)$. Therefore,

$$\int \frac{dx}{x\sqrt{2x-9}} = 2\int \frac{du}{u^2+9}.$$

**EXAMPLE 2**   **Preliminary work**  Evaluate $\int \sqrt{x^2 + 6x}\,dx$.

**SOLUTION** Most tables of integrals do not include this integral. The nearest integral you are likely to find is $\int \sqrt{x^2 \pm a^2}\,dx$. The given integral can be put into this form by completing the square and using a substitution:

$$x^2 + 6x = x^2 + 6x + 9 - 9 = (x+3)^2 - 9.$$

With the change of variables $u = x + 3$, the evaluation appears as follows:

$$\int \sqrt{x^2 + 6x}\,dx = \int \sqrt{(x+3)^2 - 9}\,dx \qquad \text{Complete the square.}$$

$$= \int \sqrt{u^2 - 9}\,du \qquad u = x + 3,\ du = dx$$

$$= \frac{u}{2}\sqrt{u^2 - 9} - \frac{9}{2}\ln\left|u + \sqrt{u^2 - 9}\right| + C \qquad \text{Table of integrals}$$

$$= \frac{x+3}{2}\sqrt{(x+3)^2 - 9} - \frac{9}{2}\ln\left|x + 3 + \sqrt{(x+3)^2 - 9}\right| + C$$

$$\text{Replace } u \text{ with } x + 3.$$

$$= \frac{x+3}{2}\sqrt{x^2 + 6x} - \frac{9}{2}\ln\left|x + 3 + \sqrt{x^2 + 6x}\right| + C. \quad \text{Simplify.}$$

*Related Exercises 31–33* ◀

**Figure 8.11**

**QUICK CHECK 1** Use the result of Example 3 to evaluate

$$\int_0^{\pi/2} \frac{dx}{1 + \sin x}.$$ ◀

**EXAMPLE 3**   **Using tables of integrals for area**  Find the area of the region bounded by the curve $y = \dfrac{1}{1 + \sin x}$ and the $x$-axis between $x = 0$ and $x = \pi$.

**SOLUTION** The region in question (**Figure 8.11**) lies entirely above the $x$-axis, so its area is $\int_0^\pi \dfrac{dx}{1 + \sin x}$. A matching integral in a table of integrals is

$$\int \frac{dx}{1 + \sin ax} = -\frac{1}{a}\tan\left(\frac{\pi}{4} - \frac{ax}{2}\right) + C.$$

Evaluating the definite integral with $a = 1$, we have

$$\int_0^\pi \frac{dx}{1 + \sin x} = \left.-\tan\left(\frac{\pi}{4} - \frac{x}{2}\right)\right|_0^\pi = -\tan\left(-\frac{\pi}{4}\right) - \left(-\tan\frac{\pi}{4}\right) = 2.$$

*Related Exercises 45–46* ◀

## Symbolic Methods

Computer algebra systems evaluate many integrals exactly using symbolic methods, and they approximate many definite integrals using numerical methods. Different software packages may produce different results for the same indefinite integral, but ultimately, they must agree. The discussion that follows does not rely on one particular computer algebra system. Rather, it illustrates results from different systems and shows some of the idiosyncrasies of using a computer algebra system.

**QUICK CHECK 2** Using one computer algebra system, it was found that $\int \sin x \cos x \, dx = \frac{1}{2} \sin^2 x + C$; using another computer algebra system, it was found that $\int \sin x \cos x \, dx = -\frac{1}{2} \cos^2 x + C$. Reconcile the two answers. ◄

> Some computer algebra systems do not include the constant of integration after evaluating an indefinite integral. But it should always be included when reporting a result.

**EXAMPLE 4**  **Apparent discrepancies**  Evaluate $\int \dfrac{dx}{\sqrt{e^x + 1}}$ using tables and a computer algebra system.

**SOLUTION**  Using one particular computer algebra system, we find that

$$\int \frac{dx}{\sqrt{e^x + 1}} = -2 \tanh^{-1} \sqrt{e^x + 1} + C,$$

> Recall that the *hyperbolic tangent* is defined as
> $$\tanh x = \frac{e^x - e^{-x}}{e^x + e^{-x}}.$$
> Its inverse is the *inverse hyperbolic tangent*, written $\tanh^{-1} x$.

where $\tanh^{-1}$ is the *inverse hyperbolic tangent* function (Section 7.7). However, we can obtain a result in terms of more familiar functions by first using the substitution $u = e^x$, which implies that $du = e^x \, dx$ or $dx = du/e^x = du/u$. The integral becomes

$$\int \frac{dx}{\sqrt{e^x + 1}} = \int \frac{du}{u\sqrt{u + 1}}.$$

Using a computer algebra system again, we obtain

$$\int \frac{dx}{\sqrt{e^x + 1}} = \int \frac{du}{u\sqrt{u + 1}} = \ln\left(\sqrt{1 + u} - 1\right) - \ln\left(\sqrt{1 + u} + 1\right)$$

$$= \ln\left(\sqrt{1 + e^x} - 1\right) - \ln\left(\sqrt{1 + e^x} + 1\right).$$

> Some computer algebra systems use $\log x$ for $\ln x$.

A table of integrals leads to a third equivalent form of the integral:

$$\int \frac{dx}{\sqrt{e^x + 1}} = \int \frac{du}{u\sqrt{u + 1}} = \ln\left(\frac{\sqrt{u + 1} - 1}{\sqrt{u + 1} + 1}\right) + C$$

$$= \ln\left(\frac{\sqrt{e^x + 1} - 1}{\sqrt{e^x + 1} + 1}\right) + C.$$

Often the difference between two results is a few steps of algebra or a trigonometric identity. In this case, the final two results are reconciled using logarithm properties. This example illustrates that some computer algebra systems do not include constants of integration and may omit absolute values when logarithms appear. It is important for the user to determine whether integration constants and absolute values are needed.

*Related Exercises 49–50* ◄

**QUICK CHECK 3** Using partial fractions, we know that $\int \dfrac{dx}{x(x + 1)} = \ln\left|\dfrac{x}{x + 1}\right| + C$.

Using a computer algebra system, we find that $\int \dfrac{dx}{x(x + 1)} = \ln x - \ln(x + 1)$. What is wrong with the result from the computer algebra system? ◄

**EXAMPLE 5** **Symbolic versus numerical integration** Use a computer algebra system to evaluate $\int_0^1 \sin x^2 \, dx$.

**SOLUTION** Sometimes a computer algebra system gives the exact value of an integral in terms of an unfamiliar function, or it may not be able to evaluate the integral exactly. For example, one particular computer algebra system returns the result

$$\int_0^1 \sin x^2 \, dx = \sqrt{\frac{\pi}{2}} \, S\left(\sqrt{\frac{2}{\pi}}\right),$$

where $S$ is known as the *Fresnel integral function* $\left(S(x) = \int_0^x \sin\left(\frac{\pi t^2}{2}\right) dt\right)$. However, if the computer algebra system is instructed to approximate the integral, the result is

$$\int_0^1 \sin x^2 \, dx \approx 0.3102683017,$$

which is an excellent approximation.

*Related Exercises 61, 63* ◄

---

## SECTION 8.7 EXERCISES

### Getting Started

1. Give some examples of analytical methods for evaluating integrals.

2. Does a computer algebra system give an exact result for an indefinite integral? Explain.

3. Why might an integral found in a table differ from the same integral evaluated by a computer algebra system?

4. Is a reduction formula an analytical method or a numerical method? Explain.

5. Evaluate $\int e^x \cos^3(e^x) \, dx$ using tables after performing the substitution $u = e^x$.

6. Evaluate $\int \cos x \sqrt{100 - \sin^2 x} \, dx$ using tables after performing the substitution $u = \sin x$.

### Practice Exercises

**7–40. Table look-up integrals** *Use a table of integrals to evaluate the following indefinite integrals. Some of the integrals require preliminary work, such as completing the square or changing variables, before they can be found in a table.*

7. $\int \cos^{-1} x \, dx$

8. $\int \sin 3x \cos 2x \, dx$

9. $\int \dfrac{dx}{\sqrt{x^2 + 16}}$

10. $\int \dfrac{dx}{\sqrt{x^2 - 25}}$

11. $\int \dfrac{3u}{2u + 7} \, du$

12. $\int \dfrac{dy}{y(2y + 9)}$

13. $\int \dfrac{dx}{1 - \cos 4x}$

14. $\int \dfrac{dx}{x\sqrt{81 - x^2}}$

15. $\int \dfrac{x}{\sqrt{4x + 1}} \, dx$

16. $\int t\sqrt{4t + 12} \, dt$

17. $\int \dfrac{dx}{\sqrt{9x^2 - 100}}, x > \dfrac{10}{3}$

18. $\int \dfrac{dx}{225 - 16x^2}$

19. $\int \dfrac{e^x}{\sqrt{e^{2x} + 4}} \, dx$

20. $\int \dfrac{\sqrt{\ln^2 x + 4}}{x} \, dx$

21. $\int \dfrac{\cos x}{\sin^2 x + 2 \sin x} \, dx$

22. $\int \dfrac{\cos^{-1}\sqrt{x}}{\sqrt{x}} \, dx$

23. $\int \dfrac{(\ln x)\sin^{-1}(\ln x)}{x} \, dx$

24. $\int \dfrac{dt}{\sqrt{1 + 4e^t}}$

25. $\int \dfrac{dx}{(16 + 9x^2)^{3/2}}$

26. $\int \sqrt{4x^2 - 9} \, dx, x > \dfrac{3}{2}$

27. $\int \dfrac{dx}{x\sqrt{144 - x^2}}$

28. $\int \dfrac{dv}{v(v^2 + 8)}$

29. $\int \ln^2 x \, dx$

30. $\int x^2 e^{5x} \, dx$

31. $\int \sqrt{x^2 + 10x} \, dx, x > 0$

32. $\int \sqrt{x^2 - 8x} \, dx, x > 8$

33. $\int \dfrac{dx}{x^2 + 2x + 10}$

34. $\int \sqrt{x^2 - 4x + 8} \, dx$

35. $\int \dfrac{dx}{x(x^{10} + 1)}$

36. $\int \dfrac{dt}{t(t^8 - 256)}$

37. $\int \dfrac{dx}{\sqrt{x^2 - 6x}}, x > 6$

38. $\int \dfrac{dx}{\sqrt{x^2 + 10x}}, x > 0$

39. $\int \dfrac{\tan^{-1} x^3}{x^4} \, dx$

40. $\int \dfrac{e^{3t}}{\sqrt{4 + e^{2t}}} \, dt$

**41–48. Geometry problems** *Use a table of integrals to solve the following problems.*

41. Find the length of the curve $y = \dfrac{x^2}{4}$ on the interval $[0, 8]$.

42. Find the length of the curve $y = x^{3/2} + 8$ on the interval $[0, 2]$.

43. Find the length of the curve $y = e^x$ on the interval $[0, \ln 2]$.

**44.** The region bounded by the graph of $y = x^2\sqrt{\ln x}$ and the $x$-axis on the interval $[1, e]$ is revolved about the $x$-axis. What is the volume of the solid that is formed?

**45.** The region bounded by the graph of $y = 1/\sqrt{x + 4}$ and the $x$-axis on the interval $[0, 12]$ is revolved about the $y$-axis. What is the volume of the solid that is formed?

**46.** Find the area of the region bounded by the graph of

$$y = \frac{1}{\sqrt{x^2 - 2x + 2}}$$ and the $x$-axis between $x = 0$ and $x = 3$.

**47.** The region bounded by the graphs of $y = \pi/2$, $y = \sin^{-1} x$, and the $y$-axis is revolved about the $y$-axis. What is the volume of the solid that is formed?

**48.** The graphs of $f(x) = \dfrac{2}{x^2 + 1}$ and $g(x) = \dfrac{7}{4\sqrt{x^2 + 1}}$ are shown in the figure. Which is greater, the average value of $f$ or the average value of $g$ on the interval $[-1, 1]$?

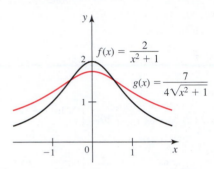

**T** **49–63. Integrating with a CAS** *Use a computer algebra system to evaluate the following integrals. Find both an exact result and an approximate result for each definite integral. Assume a is a positive real number.*

**49.** $\displaystyle\int \frac{x}{\sqrt{2x + 3}}\, dx$

**50.** $\displaystyle\int \sqrt{4x^2 + 36}\, dx$

**51.** $\displaystyle\int \tan^2 3x\, dx$

**52.** $\displaystyle\int_0^{\pi/2} \cos^6 x\, dx$

**53.** $\displaystyle\int_0^4 (9 + x^2)^{3/2}\, dx$

**54.** $\displaystyle\int (a^2 - t^2)^{-2}\, dt$

**55.** $\displaystyle\int \frac{(x^2 - a^2)^{3/2}}{x}\, dx$

**56.** $\displaystyle\int \frac{dx}{x(a^2 - x^2)^2}$

**57.** $\displaystyle\int_0^{\pi/2} \frac{dt}{1 + \tan^2 t}$

**58.** $\displaystyle\int_0^{2\pi} \frac{dt}{(4 + 2\sin t)^2}$

**59.** $\displaystyle\int (a^2 - x^2)^{3/2}\, dx$

**60.** $\displaystyle\int (y^2 + a^2)^{-5/2}\, dy$

**61.** $\displaystyle\int_0^1 (\ln x)\ln(1 + x)\, dx$

**62.** $\displaystyle\int_0^1 x^{31} e^{x^2}\, dx$

**63.** $\displaystyle\int_1^2 \frac{x^{19}}{\sqrt{x^4 - 1}}\, dx$

**64.** Using a computer algebra system, it was determined that

$$\int x(x + 1)^8\, dx = \frac{x^{10}}{10} + \frac{8x^9}{9} + \frac{7x^8}{2} + 8x^7 + \frac{35x^6}{3} + \frac{56x^5}{5} +$$

$$7x^4 + \frac{8x^3}{3} + \frac{x^2}{2} + C.$$ Use integration by substitution to evaluate

$$\int x(x + 1)^8\, dx.$$

**65–68. Reduction formulas** *Use the reduction formulas in a table of integrals to evaluate the following integrals.*

**65.** $\displaystyle\int x^3 e^{2x}\, dx$

**66.** $\displaystyle\int p^2 e^{-3p}\, dp$

**67.** $\displaystyle\int \tan^4 3y\, dy$

**68.** $\displaystyle\int \sec^4 4t\, dt$

**T** **69–70. Using a computer algebra system** *Use a computer algebra system to solve the following problems.*

**69.** Find the exact area of the region bounded by the curves

$$y = \sqrt{x + \sqrt{x}} \text{ and } y = \frac{2}{\sqrt{1 + \sqrt{x}}}$$ in the first quadrant.

**70.** Find the approximate area of the surface generated when the curve $y = 1 + \sin x + \cos x$, for $0 \le x \le \pi$, is revolved about the $x$-axis.

**71–74. Deriving formulas** *Evaluate the following integrals. Assume a and b are real numbers and n is a positive integer.*

**71.** $\displaystyle\int \frac{x}{ax + b}\, dx$ (*Hint:* Let $u = ax + b$.)

**72.** $\displaystyle\int \frac{x}{\sqrt{ax + b}}\, dx$ (*Hint:* Let $u^2 = ax + b$.)

**73.** $\displaystyle\int x(ax + b)^n\, dx$ (*Hint:* Let $u = ax + b$.)

**74.** $\displaystyle\int x^n \sin^{-1} x\, dx$ (*Hint:* Try integration by parts.)

**75. Explain why or why not** Determine whether the following statements are true and give an explanation or counterexample.

  **a.** It is possible for a computer algebra system to give the result
$$\int \frac{dx}{x(x - 1)} = \ln(x - 1) - \ln x$$ and a table of integrals to give the result $\displaystyle\int \frac{dx}{x(x - 1)} = \ln\left|\frac{x - 1}{x}\right| + C.$

  **b.** A computer algebra system working in symbolic mode could give the result $\displaystyle\int_0^1 x^8\, dx = \frac{1}{9}$, and a computer algebra system working in approximate (numerical) mode could give the result $\int_0^1 x^8\, dx = 0.11111111$.

## Explorations and Challenges

**76. Apparent discrepancy** Three different computer algebra systems give the following results:

$$\int \frac{dx}{x\sqrt{x^4 - 1}} = \frac{1}{2}\cos^{-1}\sqrt{x^{-4}} = \frac{1}{2}\cos^{-1} x^{-2} = \frac{1}{2}\tan^{-1}\sqrt{x^4 - 1}.$$

Explain how all three can be correct.

**77. Reconciling results** Using one computer algebra system, it was found that $\displaystyle\int \frac{dx}{1 + \sin x} = \frac{\sin x - 1}{\cos x}$, and using another computer algebra system, it was found that

$$\int \frac{dx}{1 + \sin x} = \frac{2\sin(x/2)}{\cos(x/2) + \sin(x/2)}.$$ Reconcile the two answers.

**78. Apparent discrepancy** Resolve the apparent discrepancy

between $\displaystyle\int \frac{dx}{x(x - 1)(x + 2)} = \frac{1}{6}\ln\frac{(x - 1)^2|x + 2|}{|x|^3} + C$ and

$$\int \frac{dx}{x(x - 1)(x + 2)} = \frac{\ln|x - 1|}{3} + \frac{\ln|x + 2|}{6} - \frac{\ln|x|}{2} + C.$$

**T 79–82. Double table look-up** *The following integrals may require more than one table look-up. Evaluate the integrals using a table of integrals, and then check your answer with a computer algebra system.*

**79.** $\int x \sin^{-1} 2x \, dx$

**80.** $\int 4x \cos^{-1} 10x \, dx$

**81.** $\int \dfrac{\tan^{-1} x}{x^2} \, dx$

**82.** $\int \dfrac{\sin^{-1} ax}{x^2} \, dx, \ a > 0$

**83. Evaluating an integral without the Fundamental Theorem of Calculus** Evaluate $\int_0^{\pi/4} \ln\,(1 + \tan x)\,dx$ using the following steps.

   **a.** If $f$ is integrable on $[0, b]$, use substitution to show that
$$\int_0^b f(x)\,dx = \int_0^{b/2}(f(x) + f(b - x))\,dx.$$

   **b.** Use part (a) and the identity $\tan\,(\alpha + \beta) = \dfrac{\tan \alpha + \tan \beta}{1 - \tan \alpha \tan \beta}$ to evaluate $\int_0^{\pi/4} \ln\,(1 + \tan x)\,dx$.

   *(Source: The College Mathematics Journal, 33, 4, Sep 2004)*

**84. Two integration approaches** Evaluate $\int \cos\,(\ln x)\,dx$ two different ways:

   **a.** Use tables after first using the substitution $u = \ln x$.

   **b.** Use integration by parts twice to verify your answer to part (a).

**T 85. Period of a pendulum** Consider a pendulum of length $L$ meters that swings under the influence of gravity alone. Suppose the pendulum starts swinging with an initial displacement of $\theta_0$ radians (see figure). The period (time to complete one full cycle) is given by

$$T = \frac{4}{\omega} \int_0^{\pi/2} \frac{d\varphi}{\sqrt{1 - k^2 \sin^2 \varphi}},$$

where $\omega^2 = \dfrac{g}{L}$, $g = 9.8$ m/s$^2$ is the acceleration due to gravity, and $k^2 = \sin^2 \dfrac{\theta_0}{2}$. Assume $L = 9.8$ m, which means $\omega = 1\text{s}^{-1}$.

   **a.** Use a computer algebra system to find the period of the pendulum for $\theta_0 = 0.1, 0.2, \ldots, 0.9, 1.0$ rad.

   **b.** For small values of $\theta_0$, the period should be approximately $2\pi$ s. For what values of $\theta_0$ are your computed values within 10% of $2\pi$ (relative error less than 0.1)?

**86. Arc length of a parabola** Let $L(c)$ be the length of the parabola $f(x) = x^2$ from $x = 0$ to $x = c$, where $c \geq 0$ is a constant.

   **a.** Find an expression for $L$.

   **b.** Is $L$ concave up or concave down on $[0, \infty)$?

   **c.** Show that as $c$ becomes large and positive, the arc length function increases as $c^2$; that is, $L(c) \approx kc^2$, where $k$ is a constant.

**T 87. Powers of sine and cosine** It can be shown that

$$\int_0^{\pi/2} \sin^n x \, dx = \int_0^{\pi/2} \cos^n x \, dx =$$

$$\begin{cases} \dfrac{1 \cdot 3 \cdot 5 \cdots (n - 1)}{2 \cdot 4 \cdot 6 \cdots n} \cdot \dfrac{\pi}{2} & \text{if } n \geq 2 \text{ is an even integer} \\[2mm] \dfrac{2 \cdot 4 \cdot 6 \cdots (n - 1)}{3 \cdot 5 \cdot 7 \cdots n} & \text{if } n \geq 3 \text{ is an odd integer.} \end{cases}$$

   **a.** Use a computer algebra system to confirm this result for $n = 2, 3, 4$, and 5.

   **b.** Evaluate the integrals with $n = 10$ and confirm the result.

   **c.** Using graphing and/or symbolic computation, determine whether the values of the integrals increase or decrease as $n$ increases.

**T 88. A remarkable integral** It is a fact that $\int_0^{\pi/2} \dfrac{dx}{1 + \tan^m x} = \dfrac{\pi}{4}$, for *all* real numbers $m$.

   **a.** Graph the integrand for $m = -2, -3/2, -1, -1/2, 0, 1/2, 1, 3/2$, and 2, and explain geometrically how the area under the curve on the interval $[0, \pi/2]$ remains constant as $m$ varies.

   **b.** Use a computer algebra system to confirm that the integral is constant for all $m$.

**QUICK CHECK ANSWERS**

**1.** 1  **2.** Because $\sin^2 x = 1 - \cos^2 x$, the two results differ by a constant, which can be absorbed in the arbitrary constant $C$.
**3.** The second result agrees with the first for $x > 0$ after using $\ln a - \ln b = \ln\,(a/b)$. The second result should have absolute values and an arbitrary constant. ◄

# 8.8 Numerical Integration

Situations arise in which the analytical methods we have developed so far cannot be used to evaluate a definite integral. For example, an integrand may not have an obvious antiderivative (such as $\cos x^2$ and $1/\ln x$), or perhaps the value of the integrand is known only at a finite set of points, which makes finding an antiderivative impossible.

When analytical methods fail, we often turn to *numerical methods*, which are typically done on a calculator or computer. These methods do not produce exact values of definite integrals, but they provide approximations that are generally quite accurate. Many calculators, software packages, and computer algebra systems have built-in numerical integration methods. In this section, we explore some of these methods.

## Absolute and Relative Error

Because numerical methods do not typically produce exact results, we should be concerned about the accuracy of approximations, which leads to the ideas of *absolute* and *relative error.*

> **DEFINITION** **Absolute and Relative Error**
>
> Suppose $c$ is a computed numerical solution to a problem having an exact solution $x$. There are two common measures of the error in $c$ as an approximation to $x$:
>
> $$\textbf{absolute error} = |c - x|$$
>
> and
>
> $$\textbf{relative error} = \frac{|c - x|}{|x|} \quad (\text{if } x \neq 0).$$

➤ Because the exact solution is usually not known, the goal in practice is to estimate the maximum size of the error.

**EXAMPLE 1** **Absolute and relative error** The ancient Greeks used $\frac{22}{7}$ to approximate the value of $\pi$. Determine the absolute and relative error in this approximation to $\pi$.

**SOLUTION** Letting $c = \frac{22}{7}$ be the approximate value of $x = \pi$, we find that

$$\text{absolute error} = \left| \frac{22}{7} - \pi \right| \approx 0.00126$$

and

$$\text{relative error} = \frac{|22/7 - \pi|}{|\pi|} \approx 0.000402 \approx 0.04\%.$$

*Related Exercises 11–14* ◄

## Midpoint Rule

Many numerical integration methods are based on the ideas that underlie Riemann sums; these methods approximate the net area of regions bounded by curves. A typical problem is shown in **Figure 8.12**, where we see a function $f$ defined on an interval $[a, b]$. The goal is to

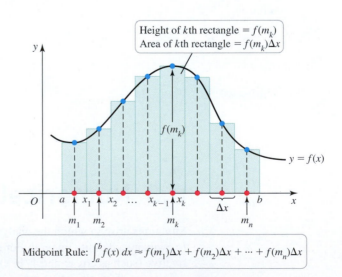

Midpoint Rule: $\int_a^b f(x)\, dx \approx f(m_1)\Delta x + f(m_2)\Delta x + \cdots + f(m_n)\Delta x$

**Figure 8.12**

approximate the value of $\int_a^b f(x)\,dx$. As with Riemann sums, we first partition the interval $[a, b]$ into $n$ subintervals of equal length $\Delta x = (b - a)/n$. This partition establishes $n + 1$ grid points

$$x_0 = a, \quad x_1 = a + \Delta x, \quad x_2 = a + 2\Delta x, \ldots, \quad x_k = a + k\Delta x, \ldots, \quad x_n = b.$$

The $k$th subinterval is $[x_{k-1}, x_k]$, for $k = 1, 2, \ldots, n$.

The Midpoint Rule approximates the region under the curve using rectangles. The bases of the rectangles have width $\Delta x$. The height of the $k$th rectangle is $f(m_k)$, where $m_k = (x_{k-1} + x_k)/2$ is the midpoint of the $k$th subinterval (Figure 8.12). Therefore, the net area of the $k$th rectangle is $f(m_k)\Delta x$.

Let $M(n)$ be the Midpoint Rule approximation to the integral using $n$ rectangles. Summing the net areas of the rectangles, we have

$$\int_a^b f(x)\,dx \approx M(n)$$

$$= f(m_1)\Delta x + f(m_2)\Delta x + \cdots + f(m_n)\Delta x$$

$$= \sum_{k=1}^{n} f\left(\frac{x_{k-1} + x_k}{2}\right)\Delta x.$$

> If $f(m_k) < 0$ for some $k$, then the net area of the corresponding rectangle is negative, which makes a negative contribution to the approximation (Section 5.2).

Just as with Riemann sums, the Midpoint Rule approximations to $\int_a^b f(x)\,dx$ generally improve as $n$ increases.

> The Midpoint Rule is a midpoint Riemann sum, introduced in Section 5.1.

---

**DEFINITION  Midpoint Rule**

Suppose $f$ is defined and integrable on $[a, b]$. The **Midpoint Rule approximation** to $\int_a^b f(x)\,dx$ using $n$ equally spaced subintervals on $[a, b]$ is

$$M(n) = f(m_1)\Delta x + f(m_2)\Delta x + \cdots + f(m_n)\Delta x$$

$$= \sum_{k=1}^{n} f\left(\frac{x_{k-1} + x_k}{2}\right)\Delta x,$$

where $\Delta x = (b - a)/n, x_0 = a, x_k = a + k\Delta x$, and $m_k = (x_{k-1} + x_k)/2$ is the midpoint of $[x_{k-1}, x_k]$, for $k = 1, \ldots, n$.

---

**QUICK CHECK 1** To apply the Midpoint Rule on the interval $[3, 11]$ with $n = 4$, at what points must the integrand be evaluated? ◄

**Figure 8.13**

**EXAMPLE 2  Applying the Midpoint Rule**  Approximate $\int_2^4 x^2\,dx$ using the Midpoint Rule with $n = 4$ and $n = 8$ subintervals.

**SOLUTION**  With $a = 2, b = 4$, and $n = 4$ subintervals, the length of each subinterval is $\Delta x = (b - a)/n = 2/4 = 0.5$. The grid points are

$$x_0 = 2, \quad x_1 = 2.5, \quad x_2 = 3, \quad x_3 = 3.5, \quad \text{and} \quad x_4 = 4.$$

The integrand must be evaluated at the midpoints (**Figure 8.13**)

$$m_1 = 2.25, \quad m_2 = 2.75, \quad m_3 = 3.25, \quad \text{and} \quad m_4 = 3.75.$$

With $f(x) = x^2$ and $n = 4$, the Midpoint Rule approximation is

$$M(4) = f(m_1)\Delta x + f(m_2)\Delta x + f(m_3)\Delta x + f(m_4)\Delta x$$

$$= (m_1^2 + m_2^2 + m_3^2 + m_4^2)\Delta x$$

$$= (2.25^2 + 2.75^2 + 3.25^2 + 3.75^2) \cdot 0.5$$

$$= 18.625.$$

The exact area of the region is $\frac{56}{3}$, so this Midpoint Rule approximation has an absolute error of

$$|18.625 - 56/3| \approx 0.0417$$

and a relative error of

$$\left|\frac{18.625 - 56/3}{56/3}\right| \approx 0.00223 = 0.223\%.$$

Using $n = 8$ subintervals, the midpoint approximation is

$$M(8) = \sum_{k=1}^{8} f(m_k)\Delta x = 18.65625,$$

which has an absolute error of about 0.0104 and a relative error of about 0.0558%. We see that increasing $n$ and using more rectangles decreases the error in the approximations.

*Related Exercises 15–16* ◄

## The Trapezoid Rule

Another numerical method for estimating $\int_a^b f(x)\,dx$ is the Trapezoid Rule, which uses the same partition of the interval $[a, b]$ described for the Midpoint Rule. Instead of approximating the region under the curve by rectangles, the Trapezoid Rule uses (what else?) trapezoids. The bases of the trapezoids have length $\Delta x$. The sides of the $k$th trapezoid have lengths $f(x_{k-1})$ and $f(x_k)$, for $k = 1, 2, \ldots, n$ (**Figure 8.14**). Therefore, the net area of the $k$th trapezoid is $\dfrac{f(x_{k-1}) + f(x_k)}{2}\Delta x$.

Area of a trapezoid

$$\text{Area} = h\left(\frac{a+b}{2}\right)$$

➤ This derivation of the Trapezoid Rule assumes $f$ is nonnegative on $[a, b]$. However, the same argument can be used if $f$ is negative on all or part of $[a, b]$. In fact, the argument illustrates how negative contributions to the net area arise when $f$ is negative.

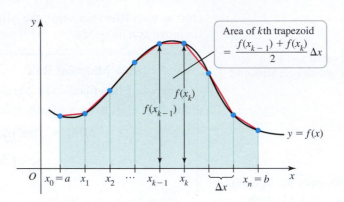

Trapezoid Rule: $\int_a^b f(x)dx \approx \left(\frac{1}{2}f(x_0) + f(x_1) + \cdots + f(x_{n-1}) + \frac{1}{2}f(x_n)\right)\Delta x$

**Figure 8.14**

Letting $T(n)$ be the Trapezoid Rule approximation to the integral using $n$ subintervals, we have

$$\int_a^b f(x)\,dx \approx T(n)$$

$$= \underbrace{\frac{f(x_0) + f(x_1)}{2}\Delta x}_{\text{area of first trapezoid}} + \underbrace{\frac{f(x_1) + f(x_2)}{2}\Delta x}_{\text{area of second trapezoid}} + \cdots + \underbrace{\frac{f(x_{n-1}) + f(x_n)}{2}\Delta x}_{\text{area of }n\text{th trapezoid}}$$

$$= \left(\frac{f(x_0)}{2} + \underbrace{\frac{f(x_1)}{2} + \frac{f(x_1)}{2}}_{f(x_1)} + \cdots + \underbrace{\frac{f(x_{n-1})}{2} + \frac{f(x_{n-1})}{2}}_{f(x_{n-1})} + \frac{f(x_n)}{2}\right)\Delta x$$

$$= \left(\frac{f(x_0)}{2} + \underbrace{f(x_1) + \cdots + f(x_{n-1})}_{\displaystyle\sum_{k=1}^{n-1} f(x_k)} + \frac{f(x_n)}{2}\right)\Delta x.$$

As with the Midpoint Rule, the Trapezoid Rule approximations generally improve as $n$ increases.

QUICK CHECK 2 Does the Trapezoid
Rule underestimate or overestimate
the value of $\int_0^4 x^2\,dx$? ◄

**DEFINITION** **Trapezoid Rule**

Suppose $f$ is defined and integrable on $[a, b]$. The **Trapezoid Rule approximation**
to $\int_a^b f(x)\,dx$ using $n$ equally spaced subintervals on $[a, b]$ is

$$T(n) = \left(\frac{1}{2} f(x_0) + \sum_{k=1}^{n-1} f(x_k) + \frac{1}{2} f(x_n)\right)\Delta x,$$

where $\Delta x = (b - a)/n$ and $x_k = a + k\Delta x$, for $k = 0, 1, \ldots, n$.

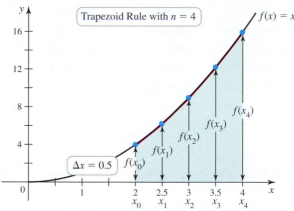

**Figure 8.15**

**EXAMPLE 3** **Applying the Trapezoid Rule** Approximate $\int_2^4 x^2\,dx$
using the Trapezoid Rule with $n = 4$ subintervals.

**SOLUTION** As in Example 2, the grid points are

$$x_0 = 2, \quad x_1 = 2.5, \quad x_2 = 3, \quad x_3 = 3.5, \quad \text{and} \quad x_4 = 4.$$

With $f(x) = x^2$ and $n = 4$, the Trapezoid Rule approximation is

$$
\begin{aligned}
T(4) &= \tfrac{1}{2} f(x_0)\Delta x + f(x_1)\Delta x + f(x_2)\Delta x + f(x_3)\Delta x + \tfrac{1}{2} f(x_4)\Delta x \\
&= \left(\tfrac{1}{2} x_0^2 + x_1^2 + x_2^2 + x_3^2 + \tfrac{1}{2} x_4^2\right)\Delta x \\
&= \left(\tfrac{1}{2}\cdot 2^2 + 2.5^2 + 3^2 + 3.5^2 + \tfrac{1}{2}\cdot 4^2\right)\cdot 0.5 \\
&= 18.75.
\end{aligned}
$$

**Figure 8.15** shows the approximation with $n = 4$ trapezoids. The
exact area of the region is $56/3$, so the Trapezoid Rule approximation
has an absolute error of about $0.0833$ and a relative error of approximately $0.00446$, or $0.446\%$. Increasing $n$ decreases this error.

*Related Exercises 19–20* ◄

**EXAMPLE 4** **Errors in the Midpoint and Trapezoid Rules** Given that

$$\int_0^1 xe^{-x}\,dx = 1 - 2e^{-1},$$

find the absolute errors in the Midpoint Rule and Trapezoid Rule approximations to the
integral with $n = 4, 8, 16, 32, 64,$ and $128$ subintervals.

**SOLUTION** Because the exact value of the integral is known (which does *not* happen in
practice), we can compute the error in various approximations. For example, if $n = 16$,
then

$$\Delta x = \frac{1}{16} \quad \text{and} \quad x_k = \frac{k}{16}, \quad \text{for } k = 0, 1, \ldots, n.$$

Using sigma notation and a calculator, we have

$$M(16) = \sum_{k=1}^{16} f\!\left(\overbrace{\frac{(k-1)/16 + k/16}{2}}^{x_{k-1} \quad x_k}\right)\overbrace{\frac{1}{16}}^{\Delta x} = \sum_{k=1}^{16} f\!\left(\frac{2k-1}{32}\right)\frac{1}{16} \approx 0.26440383609318$$

and

$$T(16) = \left(\frac{1}{2} \underbrace{f(0)}_{x_0 \,\equiv\, a} + \sum_{k=1}^{\overbrace{15}^{n-1}} \underbrace{f(k/16)}_{x_k} + \frac{1}{2} \underbrace{f(1)}_{x_{16} \,=\, b}\right)\frac{1}{16} \approx 0.26391564480235.$$

The absolute error in the Midpoint Rule approximation with $n = 16$ is
$|M(16) - (1 - 2e^{-1})| \approx 0.000163$. The absolute error in the Trapezoid Rule approxi-
mation with $n = 16$ is $|T(16) - (1 - 2e^{-1})| \approx 0.000325$.

The Midpoint Rule and Trapezoid Rule approximations to the integral, together with the associated absolute errors, are shown in Table 8.5 for various values of $n$. Notice that as $n$ increases, the errors in both methods decrease, as expected. With $n = 128$ subintervals, the approximations $M(128)$ and $T(128)$ agree to four decimal places. Based on these approximations, a good approximation to the integral is 0.2642. The way in which the errors decrease is also worth noting. If you look carefully at both error columns in Table 8.5, you will see that each time $n$ is doubled (or $\Delta x$ is halved), the error decreases by a factor of approximately 4.

**QUICK CHECK 3** Compute the approximate factor by which the error decreases in Table 8.5 between $T(16)$ and $T(32)$, and between $T(32)$ and $T(64)$. ◄

**Table 8.5**

| $n$ | $M(n)$ | $T(n)$ | Error in $M(n)$ | Error in $T(n)$ |
|---|---|---|---|---|
| 4 | 0.26683456310319 | 0.25904504019141 | 0.00259 | 0.00520 |
| 8 | 0.26489148795740 | 0.26293980164730 | 0.000650 | 0.00130 |
| 16 | 0.26440383609318 | 0.26391564480235 | 0.000163 | 0.000325 |
| 32 | 0.26428180513718 | 0.26415974044777 | 0.0000407 | 0.0000814 |
| 64 | 0.26425129001915 | 0.26422077279247 | 0.0000102 | 0.0000203 |
| 128 | 0.26424366077837 | 0.26423603140581 | 0.00000254 | 0.00000509 |

*Related Exercises 31–32* ◄

**Table 8.6**

| Year | World Crude Oil Production (billions of barrels/yr) |
|---|---|
| 1995 | 21.9 |
| 1996 | 22.3 |
| 1997 | 23.0 |
| 1998 | 23.7 |
| 1999 | 24.5 |
| 2000 | 23.7 |
| 2001 | 25.2 |
| 2002 | 24.8 |
| 2003 | 24.5 |
| 2004 | 25.2 |
| 2005 | 25.9 |
| 2006 | 26.3 |
| 2007 | 27.0 |
| 2008 | 26.9 |
| 2009 | 26.4 |
| 2010 | 27.0 |
| 2011 | 27.0 |

(*Source:* U.S. Energy Information Administration)

We now apply the Midpoint and Trapezoid Rules to a problem with real data.

**EXAMPLE 5** **World oil production** Table 8.6 and **Figure 8.16** show data for the rate of world crude oil production (in billions of barrels/yr) over a 16-year period. If the rate of oil production is given by the (assumed to be integrable) function $R$, then the total amount of oil produced in billions of barrels over the time period $a \le t \le b$ is $Q = \int_a^b R(t)\, dt$ (Section 6.1). Use the Midpoint and Trapezoid Rules to approximate the total oil produced between 1995 and 2011.

**SOLUTION** For convenience, let $t = 0$ represent 1995 and $t = 16$ represent 2011. We let $R(t)$ be the rate of oil production in the year corresponding to $t$ (for example, $R(6) = 25.2$ is the rate in 2001). The goal is to approximate $Q = \int_0^{16} R(t)\, dt$. If we use $n = 4$ subintervals, then $\Delta t = 4$ yr. The resulting Midpoint and Trapezoid Rule approximations (in billions of barrels) are

$$Q \approx M(4) = (R(2) + R(6) + R(10) + R(14))\Delta t$$
$$= (23.0 + 25.2 + 25.9 + 26.4)4$$
$$= 402.0$$

and

$$Q \approx T(4) = \left(\frac{1}{2}R(0) + R(4) + R(8) + R(12) + \frac{1}{2}R(16)\right)\Delta t$$
$$= \left(\frac{1}{2}\cdot 21.9 + 24.5 + 24.5 + 27.0 + \frac{1}{2}\cdot 27.0\right)4$$
$$= 401.8.$$

The two methods give reasonable agreement. Using $n = 8$ subintervals, with $\Delta t = 2$ yr, similar calculations give the approximations

$$Q \approx M(8) = 399.8 \quad \text{and} \quad Q \approx T(8) = 401.9.$$

The given data do not allow us to compute the next Midpoint Rule approximation $M(16)$. However, we can compute the next Trapezoid Rule approximation $T(16)$, and here is a good way to do it. If $T(n)$ and $M(n)$ are known, then the next Trapezoid Rule approximation is (Exercise 76)

$$T(2n) = \frac{T(n) + M(n)}{2}.$$

**Figure 8.16**
(*Source:* U.S. Energy Information Administration)

Using this identity, we find that

$$T(16) = \frac{T(8) + M(8)}{2} = \frac{401.9 + 399.8}{2} \approx 400.9.$$

Based on these calculations, the best approximation to the total oil produced between 1995 and 2011 is 400.9 billion barrels.  *Related Exercises 37–40* ◄

The Midpoint and Trapezoid Rules, as well as left and right Riemann sums, can be applied to problems in which data are given on a nonuniform grid (that is, the lengths of the subintervals vary). In the case of the Trapezoid Rule, the net areas of the approximating trapezoids must be computed individually and then summed, as shown in the next example.

**EXAMPLE 6**  **Net change in sea level**  Table 8.7 lists rates of change $s'(t)$ in global sea level $s(t)$ in various years from 1995 ($t = 0$) to 2011 ($t = 16$), with rates of change reported in mm/yr.

**Table 8.7**

| $t$ (years from 1995) | 0 (1995) | 3 (1998) | 5 (2000) | 7 (2002) | 8 (2003) | 12 (2007) | 14 (2009) | 16 (2011) |
|---|---|---|---|---|---|---|---|---|
| $s'(t)$ (mm/yr) | 0.51 | 5.19 | 4.39 | 2.21 | 5.24 | 0.63 | 4.19 | 2.38 |

(*Source:* Collecte Localisation Satellites/Centre national d'études spatiales/Legos)

> The rate of change in sea level varies from one location on Earth to the next; sea level also varies seasonally and is influenced by ocean currents. The data in Table 8.7 reflect approximate rates of change at the beginning of each year listed, averaged over the entire globe.

a. Assuming $s'$ is continuous on $[0, 16]$, explain how a definite integral can be used to find the net change in sea level from 1995 to 2011; then write the definite integral.

b. Use the data in Table 8.7 and generalize the Trapezoid Rule to estimate the value of the integral from part (a).

**SOLUTION**

a. The net change in any quantity $Q$ over the interval $[a, b]$ is $Q(b) - Q(a)$ (Section 6.1). When the rate of change $Q'$ is known, the net change in $Q$ is found by integrating $Q'$ over the same interval; that is,

$$\text{net change in } Q = Q(b) - Q(a) = \int_a^b Q'(t)\,dt. \quad \text{Fundamental Theorem}$$

Therefore, the net change in sea level from 1995 to 2011 is $\int_0^{16} s'(t)\,dt$.

b. The values from Table 8.7 are plotted in **Figure 8.17**, accompanied by seven trapezoids whose area approximates $\int_0^{16} s'(t)\,dt$. Notice that the grid points (the $t$-values in Table 8.7) do not form a regular partition of the interval $[0, 16]$. Therefore, we must generalize the standard Trapezoid Rule and compute the area of each trapezoid separately.

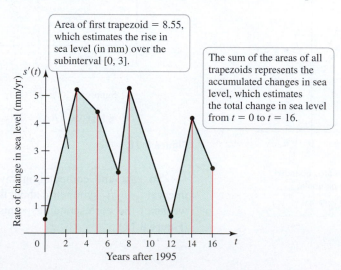

Area of first trapezoid = 8.55, which estimates the rise in sea level (in mm) over the subinterval [0, 3].

The sum of the areas of all trapezoids represents the accumulated changes in sea level, which estimates the total change in sea level from $t = 0$ to $t = 16$.

**Figure 8.17**

Focusing on the first trapezoid over the subinterval $[0, 3]$, we find that its area is

$$\underbrace{\text{area of first trapezoid}}_{A = \frac{1}{2}(b_1 + b_2)h} = \frac{1}{2} \cdot \underbrace{(s'(0) + s'(3))}_{\text{measured in mm/yr}} \cdot \underset{\text{yr}}{3} = \frac{1}{2} \cdot (0.51 + 5.19) \cdot 3 = 8.55.$$

Because $s'$ is measured in mm/yr and $t$ is measured in yr, the area of this trapezoid (8.55) is interpreted as the approximate net change in sea level from 1995 to 1998, measured in mm. As we add new trapezoid areas to the ongoing sum that approximates $\int_0^{16} s'(t)\, dt$, the changes in sea level accumulate, resulting in the total change in sea level on $[0, 16]$. The sum of the areas of all seven trapezoids is

$$\underbrace{\frac{1}{2}(s'(0) + s'(3)) \cdot 3}_{\text{first trapezoid}} + \underbrace{\frac{1}{2}(s'(3) + s'(5)) \cdot 2}_{\text{second trapezoid...}} + \frac{1}{2}(s'(5) + s'(7)) \cdot 2 + \frac{1}{2}(s'(7) + s'(8)) \cdot 1$$

$$+ \frac{1}{2}(s'(8) + s'(12)) \cdot 4 + \frac{1}{2}(s'(12) + s'(14)) \cdot 2 + \underbrace{\frac{1}{2}(s'(14) + s'(16)) \cdot 2}_{\text{... last trapezoid}} = 51.585.$$

We conclude that an estimate of the rise in sea level from 1995 to 2011 is 51.585 mm.

*Related Exercises 41, 43* ◄

## Simpson's Rule

An improvement over the Midpoint Rule and the Trapezoid Rule results when the graph of $f$ is approximated with curves rather than line segments. Returning to the partition of the interval $[a, b]$ used by the Midpoint and Trapezoid Rules, suppose we work with three neighboring points on the curve $y = f(x)$, say $(x_0, f(x_0))$, $(x_1, f(x_1))$, and $(x_2, f(x_2))$. These three points determine a parabola, and it is easy to find the net area bounded by a parabola on the interval $[x_0, x_2]$. When this idea is applied to every group of three consecutive points along the interval of integration, the result is *Simpson's Rule* (**Figure 8.18**). Our immediate goal is to determine a formula for this rule.

Simpson's Rule: $\int_a^b f(x)\, dx$

$\approx \left( f(x_0) + 4f(x_1) + 2f(x_2) + 4f(x_3) + \cdots + 2f(x_{n-2}) + 4f(x_{n-1}) + f(x_n) \right) \dfrac{\Delta x}{3}$

**Figure 8.18**

$p(x) = Ax^2 + Bx + C$

$y = f(x)$

$(-h, y_0)$     $(h, y_2)$

$(0, y_1)$

Net area bounded by parabola:
$\dfrac{h}{3}(y_0 + 4y_1 + y_2)$

**Figure 8.19**

To ease the computation involved in deriving Simpson's Rule, we consider an interval $[-h, h]$ symmetric about the origin with grid points $-h$, $0$, and $h$, where $h = \Delta x$; we label the corresponding points on the graph of $f$ as $(-h, y_0)$, $(0, y_1)$, and $(h, y_2)$ (**Figure 8.19**).

If these points are to lie on the parabola $p(x) = Ax^2 + Bx + C$, they must satisfy the following conditions:

$$p(-h) = A(-h)^2 + B(-h) + C = y_0, \quad \text{or} \quad Ah^2 - Bh + C = y_0;$$
$$p(0) = A(0)^2 + B(0) + C = y_1, \quad \text{or} \quad C = y_1; \text{ and}$$
$$p(h) = A(h)^2 + B(h) + C = y_2, \quad \text{or} \quad Ah^2 + Bh + C = y_2.$$

Solving this system of equations for $A$, $B$, and $C$ yields

$$A = \frac{y_0 - 2y_1 + y_2}{2h^2}, \quad B = \frac{y_2 - y_0}{2h}, \quad \text{and } C = y_1.$$

Therefore, the net area bounded by the parabola is given by

> Recall that if $f(x)$ is an even function and $g(x)$ is odd, then
> $$\int_{-h}^{h} f(x)\,dx = 2\int_{0}^{h} f(x)\,dx \quad \text{and}$$
> $$\int_{-h}^{h} g(x)\,dx = 0.$$

$$\int_{-h}^{h}\left(\underbrace{\frac{y_0 - 2y_1 + y_2}{2h^2}x^2}_{A} + \overbrace{\underbrace{\frac{y_2 - y_0}{2h}x}_{B}}^{\text{odd function}} + \underbrace{y_1}_{C}\right)dx = 2\int_{0}^{h}\left(\frac{y_0 - 2y_1 + y_2}{2h^2}x^2 + y_1\right)dx \qquad \text{Use symmetry.}$$

$$= 2\left(\frac{y_0 - 2y_1 + y_2}{6h^2}x^3 + y_1 x\right)\Big|_0^h \qquad \text{Integrate.}$$

> In the event that the three points $(-h, y_0)$, $(0, y_1)$, and $(h, y_2)$ are collinear, there is no parabola that passes through all the points. However, it can be shown that the net area bounded by the line segment through these points is also given by $\frac{h}{3}(y_0 + 4y_1 + y_2)$.

$$= 2\left(\frac{(y_0 - 2y_1 + y_2)h}{6} + y_1 h\right) \qquad \text{Evaluate.}$$

$$= \frac{h}{3}(y_0 + 4y_1 + y_2). \qquad \text{Simplify.}$$

Notice that a horizontal shift of the parabola shown in Figure 8.19 does not change its associated net area, which implies that the net area bounded by the parabola on $[x_0, x_2]$ shown in Figure 8.18 is also given by

$$\frac{h}{3}(y_0 + 4y_1 + y_2), \quad \text{or} \quad (f(x_0) + 4f(x_1) + f(x_2))\frac{\Delta x}{3}.$$

Similarly, the net area bounded by the parabola on the interval $[x_2, x_4]$ is

$$(f(x_2) + 4f(x_3) + f(x_4))\frac{\Delta x}{3}.$$

When we sum the net area of all such parabolas, we arrive at an approximation to the net area bounded by the graph of $f$ on $[a, b]$. With $n$ subintervals, Simpson's Rule is denoted $S(n)$ and is given by

$$S(n) = \underbrace{(f(x_0) + 4f(x_1) + f(x_2))\frac{\Delta x}{3}}_{\text{net area bounded by first parabola}} + \underbrace{(f(x_2) + 4f(x_3) + f(x_4))\frac{\Delta x}{3}}_{\text{net area bounded by second parabola}} + \cdots$$

$$+ \underbrace{(f(x_{n-2}) + 4f(x_{n-1}) + f(x_n))\frac{\Delta x}{3}}_{\text{net area bounded by }n\text{th parabola}}$$

$$= (f(x_0) + 4f(x_1) + 2f(x_2) + 4f(x_3) + \cdots + 2f(x_{n-2}) + 4f(x_{n-1}) + f(x_n))\frac{\Delta x}{3}.$$

Notice that apart from the first and last terms, the coefficients alternate between 4 and 2; to apply this rule, *n must be an even integer.*

> **DEFINITION  Simpson's Rule**
>
> Suppose $f$ is defined and integrable on $[a, b]$ and $n \geq 2$ is an even integer. The **Simpson's Rule approximation** to $\int_a^b f(x)\, dx$ using $n$ equally spaced subintervals on $[a, b]$ is
>
> $$S(n) = (f(x_0) + 4f(x_1) + 2f(x_2) + 4f(x_3) + \cdots + 4f(x_{n-1}) + f(x_n))\frac{\Delta x}{3},$$
>
> where $n$ is an even integer, $\Delta x = (b - a)/n$, and $x_k = a + k\Delta x$, for $k = 0, 1, \ldots, n$.

**EXAMPLE 7   Applying Simpson's Rule**  Approximate $\int_{\pi/2}^{5\pi/2} \dfrac{\sin x}{x}\, dx$ using Simpson's Rule with $n = 4$ subintervals.

**SOLUTION**  We have $\Delta x = \dfrac{5\pi/2 - \pi/2}{4} = \pi/2$, and therefore the grid points are

$$x_0 = \pi/2, \ x_1 = \pi, \ x_2 = 3\pi/2, \ x_3 = 2\pi, \quad \text{and} \quad x_4 = 5\pi/2.$$

With $f(x) = \dfrac{\sin x}{x}$ and $n = 4$, the approximation given by Simpson's Rule is

$$S(4) = (f(x_0) + 4f(x_1) + 2f(x_2) + 4f(x_3) + f(x_4))\frac{\pi/2}{3}$$

$$= \left(\frac{\sin x_0}{x_0} + \frac{4\sin x_1}{x_1} + \frac{2\sin x_2}{x_2} + \frac{4\sin x_3}{x_3} + \frac{\sin x_4}{x_4}\right)\frac{\pi}{6}$$

$$= \left(\frac{1}{\pi/2} + \frac{4(0)}{\pi} + \frac{2(-1)}{3\pi/2} + \frac{4(0)}{2\pi} + \frac{1}{5\pi/2}\right)\frac{\pi}{6}$$

$$= \left(\frac{2}{\pi} - \frac{4}{3\pi} + \frac{2}{5\pi}\right)\frac{\pi}{6}$$

$$= \frac{8}{45} \approx 0.178.$$

$f(x) = \dfrac{\sin x}{x}$

Simpson's Rule with $n = 4$

**Figure 8.20**

**QUICK CHECK 4** Write the Simpson's Rule formula for $\int_0^3 g(x)\, dx$ with $n = 6$ subintervals. ◄

**Figure 8.20** shows the approximation with two parabolas. To perform error analysis on our approximation, we need to look ahead to Theorem 8.1 (p. 577); this theorem involves finding an upper bound of $|f^{(4)}(x)|$ on the interval $\left[\frac{\pi}{2}, \frac{5\pi}{2}\right]$.

*Related Exercises 23, 26* ◄

You can use the formula for Simpson's Rule, but here is an easier way. If you already have the Trapezoid Rule approximations $T(n)$ and $T(2n)$, the next Simpson's Rule approximation follows immediately with a simple calculation (Exercise 78):

$$S(2n) = \frac{4T(2n) - T(n)}{3}.$$

**EXAMPLE 8   Errors in the Trapezoid Rule and Simpson's Rule**  Given that $\int_0^1 xe^{-x}\, dx = 1 - 2e^{-1}$, find the absolute errors in the Trapezoid Rule and Simpson's Rule approximations to the integral with $n = 8, 16, 32, 64,$ and $128$ subintervals.

**SOLUTION**  Because the shortcut formula for Simpson's Rule is based on values generated by the Trapezoid Rule, it is best to calculate the Trapezoid Rule approximations first. The second column of Table 8.8 shows the Trapezoid Rule approximations computed in Example 4. Having a column of Trapezoid Rule approximations, we can readily find the corresponding Simpson's Rule approximations. For example, if $n = 4$, we have

$$S(8) = \frac{4T(8) - T(4)}{3} \approx 0.26423805546593.$$

The table also shows the absolute errors in the approximations. The Simpson's Rule errors decrease more rapidly than the Trapezoid Rule errors. By careful inspection, you will see that the Simpson's Rule errors decrease with a clear pattern: Each time $n$ is doubled (or $\Delta x$ is halved), the errors decrease by a factor of approximately 16, which makes Simpson's Rule a more accurate method.

**Table 8.8**

| $n$ | $T(n)$ | $S(n)$ | Error in $T(n)$ | Error in $S(n)$ |
|---|---|---|---|---|
| 4 | 0.25904504019141 | | 0.00520 | |
| 8 | 0.26293980164730 | 0.26423805546593 | 0.00130 | 0.00000306 |
| 16 | 0.26391564480235 | 0.26424092585404 | 0.000325 | 0.000000192 |
| 32 | 0.26415974044777 | 0.26424110566291 | 0.0000814 | 0.0000000120 |
| 64 | 0.26422077279247 | 0.26424111690738 | 0.0000203 | 0.000000000750 |
| 128 | 0.26423603140581 | 0.26424111761026 | 0.00000509 | 0.0000000000469 |

> **QUICK CHECK 5** Compute the approximate factor by which the error decreases in Table 8.8 between $S(16)$ and $S(32)$ and between $S(32)$ and $S(64)$. ◄

*Related Exercises 45–48* ◄

## Errors in Numerical Integration

A detailed analysis of the errors in the three methods we have discussed goes beyond the scope of the text. We state without proof the standard error theorems for the methods and note that Examples 2, 3, 4, and 8 are consistent with these results.

> ➤ Because $\Delta x = \dfrac{b - a}{n}$, the error bounds in Theorem 8.1 can also be written as
> $$E_M \le \frac{k(b - a)^3}{24n^2},$$
> $$E_T \le \frac{k(b - a)^3}{12n^2}, \text{ and}$$
> $$E_S \le \frac{K(b - a)^5}{180n^4}.$$

---

**THEOREM 8.1    Errors in Numerical Integration**

Assume $f''$ is continuous on the interval $[a, b]$ and $k$ is a real number such that $|f''(x)| \le k$, for all $x$ in $[a, b]$. The absolute errors in approximating the integral $\int_a^b f(x)\,dx$ by the Midpoint Rule and Trapezoid Rule with $n$ subintervals satisfy the inequalities

$$E_M \le \frac{k(b - a)}{24}(\Delta x)^2 \quad \text{and} \quad E_T \le \frac{k(b - a)}{12}(\Delta x)^2,$$

respectively, where $\Delta x = (b - a)/n$.

Assume $f^{(4)}$ is continuous on the interval $[a, b]$ and $K$ is a real number such that $|f^{(4)}(x)| \le K$ on $[a, b]$. The absolute error in approximating the integral $\int_a^b f(x)\,dx$ by Simpson's Rule with $n$ subintervals satisfies the inequality

$$E_S \le \frac{K(b - a)}{180}(\Delta x)^4.$$

---

> **QUICK CHECK 6** In Example 7, we used Simpson's Rule with $n = 4$ to find that $\int_{\pi/2}^{5\pi/2} \dfrac{\sin x}{x}\,dx \approx 0.178$. Use Theorem 8.1 to find an upper bound on $E_S$, the absolute error in this approximation, given that $|f^{(4)}(x)| \le 0.2$ on $\left[\frac{\pi}{2}, \frac{5\pi}{2}\right]$. ◄

The absolute errors associated with the Midpoint Rule and Trapezoid Rule are proportional to $(\Delta x)^2$. So if $\Delta x$ is reduced by a factor of 2, the errors decrease roughly by a factor of 4, as shown in Example 4. Simpson's Rule is a more accurate method; its error is proportional to $(\Delta x)^4$, which means that if $\Delta x$ is reduced by a factor of 2, the errors decrease roughly by a factor of 16, as shown in Example 8. Computing both the Trapezoid Rule and Simpson's Rule together, as shown in Example 8, is a powerful method that produces accurate approximations with relatively little work.

# SECTION 8.8 EXERCISES

## Getting Started

1. If the interval $[4, 18]$ is partitioned into $n = 28$ subintervals of equal length, what is $\Delta x$?

2. Explain geometrically how the Midpoint Rule is used to approximate a definite integral.

3. Explain geometrically how the Trapezoid Rule is used to approximate a definite integral.

4. If the Midpoint Rule is used on the interval $[-1, 11]$ with $n = 3$ subintervals, at what $x$-coordinates is the integrand evaluated?

5–8. *Compute the following estimates of $\int_0^8 f(x)\, dx$ using the graph in the figure.*

5. $M(4)$     6. $T(4)$     7. $S(4)$     8. $S(8)$

9. If the Trapezoid Rule is used on the interval $[-1, 9]$ with $n = 5$ subintervals, at what $x$-coordinates is the integrand evaluated?

10. Suppose two Trapezoidal Rule approximations of $\int_a^b f(x)\, dx$ are $T(2) = 6$ and $T(4) = 5.1$. Find the Simpson's Rule approximation $S(4)$.

**T** 11–14. *Compute the absolute and relative errors in using $c$ to approximate $x$.*

11. $x = \pi; c = 3.14$

12. $x = \sqrt{2}; c = 1.414$

13. $x = e; c = 2.72$

14. $x = e; c = 2.718$

## Practice Exercises

**T** 15–18. **Midpoint Rule approximations** *Find the indicated Midpoint Rule approximations to the following integrals.*

15. $\int_2^{10} 2x^2\, dx$ using $n = 1, 2,$ and 4 subintervals

16. $\int_1^9 x^3\, dx$ using $n = 1, 2,$ and 4 subintervals

17. $\int_0^1 \sin \pi x\, dx$ using $n = 6$ subintervals

18. $\int_0^1 e^{-x}\, dx$ using $n = 8$ subintervals

**T** 19–22. **Trapezoid Rule approximations** *Find the indicated Trapezoid Rule approximations to the following integrals.*

19. $\int_2^{10} 2x^2\, dx$ using $n = 2, 4,$ and 8 subintervals

20. $\int_1^9 x^3\, dx$ using $n = 2, 4,$ and 8 subintervals

21. $\int_0^1 \sin \pi x\, dx$ using $n = 6$ subintervals

22. $\int_0^1 e^{-x}\, dx$ using $n = 8$ subintervals

**T** 23–26. **Simpson's Rule approximations** *Find the indicated Simpson's Rule approximations to the following integrals.*

23. $\int_0^\pi \sqrt{\sin x}\, dx$ using $n = 4$ and $n = 6$ subintervals

24. $\int_4^8 \sqrt{x}\, dx$ using $n = 4$ and $n = 8$ subintervals

25. $\int_{-2}^3 e^{-x^2}\, dx$ using $n = 10$ subintervals

26. $\int_2^4 \cos \sqrt{x}\, dx$ using $n = 8$ subintervals

**T** 27. **Midpoint Rule, Trapezoid Rule, and relative error** Find the Midpoint and Trapezoid Rule approximations to $\int_0^1 \sin \pi x\, dx$ using $n = 25$ subintervals. Compute the relative error of each approximation.

**T** 28. **Midpoint Rule, Trapezoid Rule, and relative error** Find the Midpoint and Trapezoid Rule approximations to $\int_0^1 e^{-x}\, dx$ using $n = 50$ subintervals. Compute the relative error of each approximation.

**T** 29–34. **Comparing the Midpoint and Trapezoid Rules** *Apply the Midpoint and Trapezoid Rules to the following integrals. Make a table similar to Table 8.5 showing the approximations and errors for $n = 4, 8, 16,$ and 32. The exact values of the integrals are given for computing the error.*

29. $\int_1^5 (3x^2 - 2x)\, dx = 100$     30. $\int_{-2}^6 \left(\frac{x^3}{16} - x\right) dx = 4$

31. $\int_0^{\pi/4} 3 \sin 2x\, dx = \frac{3}{2}$     32. $\int_1^e \ln x\, dx = 1$

33. $\int_0^\pi \sin x \cos 3x\, dx = 0$     34. $\int_0^8 e^{-2x}\, dx = \dfrac{1 - e^{-16}}{2}$

35–36. **River flow rates** *The following figure shows the discharge rate $r(t)$ of the Snoqualmie River near Carnation, Washington, starting on a February day when the air temperature was rising. The variable $t$ is the number of hours after midnight, $r(t)$ is given in millions of cubic feet per hour, and $\int_0^{24} r(t)\, dt$ equals the total amount of water that flows by the town of Carnation over a 24-hour period. Estimate $\int_0^{24} r(t)\, dt$ using the Trapezoidal Rule and Simpson's Rule with the following values of $n$.*

35. $n = 4$     36. $n = 6$

**T 37–40. Temperature data** *Hourly temperature data for Boulder, Colorado; San Francisco, California; Nantucket, Massachusetts; and Duluth, Minnesota, over a 12-hr period on the same day of January are shown in the figure. Assume these data are taken from a continuous temperature function $T(t)$. The average temperature (in °F) over the 12-hr period is* $\overline{T} = \dfrac{1}{12}\int_0^{12} T(t)\,dt.$

Hours after 6:00 A.M.

| t | 0 | 1 | 2 | 3 | 4 | 5 | 6 | 7 | 8 | 9 | 10 | 11 | 12 |
|---|---|---|---|---|---|---|---|---|---|---|----|----|----|
| B | 47 | 50 | 46 | 45 | 48 | 52 | 54 | 61 | 62 | 63 | 63 | 59 | 55 |
| SF | 41 | 44 | 46 | 48 | 52 | 53 | 53 | 53 | 51 | 51 | 49 | 47 | 47 |
| N | 35 | 34 | 34 | 36 | 36 | 37 | 37 | 36 | 35 | 35 | 34 | 33 | 32 |
| D | 9 | 11 | 11 | 12 | 14 | 15 | 17 | 19 | 20 | 22 | 24 | 24 | 25 |

**37.** Find an accurate approximation to the average temperature over the 12-hr period for Boulder. State your method.

**38.** Find an accurate approximation to the average temperature over the 12-hr period for San Francisco. State your method.

**39.** Find an accurate approximation to the average temperature over the 12-hr period for Nantucket. State your method.

**40.** Find an accurate approximation to the average temperature over the 12-hr period for Duluth. State your method.

**41–44. Nonuniform grids** *Use the indicated methods to solve the following problems with nonuniform grids.*

**41.** A curling iron is plugged into an outlet at time $t = 0$. Its temperature $T$ in degrees Fahrenheit, assumed to be a continuous function that is strictly increasing and concave down on $0 \le t \le 120$, is given at various times (in seconds) in the table.

| t (seconds) | 0 | 20 | 45 | 60 | 90 | 110 | 120 |
|---|---|---|---|---|---|---|---|
| T(t) (°F) | 70 | 130 | 200 | 239 | 311 | 355 | 375 |

**a.** Approximate $\dfrac{1}{120}\int_0^{120} T(t)\,dt$ in three ways: using a left Riemann sum, a right Riemann sum, and the Trapezoid Rule. Interpret the value of $\dfrac{1}{120}\int_0^{120} T(t)\,dt$ in the context of this problem.

**b.** Which of the estimates in part (a) overestimate the value of $\dfrac{1}{120}\int_0^{120} T(t)\,dt$? Which underestimate it? Justify your answers with a simple sketch of the sums you computed.

**c.** Evaluate and interpret $\dfrac{1}{120}\int_0^{120} T'(t)\,dt$ in the context of this problem.

**42. Approximating integrals** The function $f$ is twice differentiable on $(-\infty, \infty)$. Values of $f$ at various points on $[0, 20]$ are given in the table.

| x | 0 | 4 | 7 | 12 | 14 | 18 | 20 |
|---|---|---|---|---|---|---|---|
| f(x) | 3 | 0 | -2 | -1 | 2 | 4 | 7 |

**a.** Approximate $\int_0^{20} f(x)\,dx$ in three ways: using a left Riemann sum, a right Riemann sum, and the Trapezoid Rule.

**b.** A scatterplot of the data in the table is provided in the figure. Use the scatterplot to illustrate each of the approximations in part (a) by sketching appropriate rectangles for the Riemann sums and by sketching trapezoids for the Trapezoid Rule approximation.

**c.** Evaluate $\int_4^{12} (3f'(x) + 2)\,dx$.

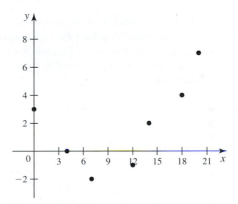

**43.** A hot-air balloon is launched from an elevation of 5400 ft above sea level. As it rises, the vertical velocity is computed using a device (called a *variometer*) that measures the change in atmospheric pressure. The vertical velocities at selected times are shown in the table (with units of ft/min).

| t (min) | 0 | 1 | 1.5 | 3 | 3.5 | 4 | 5 |
|---|---|---|---|---|---|---|---|
| Velocity (ft/min) | 0 | 100 | 120 | 150 | 110 | 90 | 80 |

**a.** Use the Trapezoid Rule to estimate the elevation of the balloon after five minutes. Remember that the balloon starts at an elevation of 5400 ft.

**b.** Use a right Riemann sum to estimate the elevation of the balloon after five minutes.

**c.** A polynomial that fits the data reasonably well is

$$g(t) = 3.49t^3 - 43.21t^2 + 142.43t - 1.75.$$

Estimate the elevation of the balloon after five minutes using this polynomial.

**44.** A piece of wood paneling must be cut in the shape shown in the figure. The coordinates of several points on its curved surface are also shown (with units of inches).

**a.** Estimate the surface area of the paneling using the Trapezoid Rule.

**b.** Estimate the surface area of the paneling using a left Riemann sum.

c. Could two identical pieces be cut from a 9-in by 9-in piece of wood? Answer carefully.

**T 45–48. Trapezoid Rule and Simpson's Rule** *Consider the following integrals and the given values of n.*

*a. Find the Trapezoid Rule approximations to the integral using n and 2n subintervals.*

*b. Find the Simpson's Rule approximation to the integral using 2n subintervals. It is easiest to obtain Simpson's Rule approximations from the Trapezoid Rule approximations, as in Example 8.*

*c. Compute the absolute errors in the Trapezoid Rule and Simpson's Rule with 2n subintervals.*

**45.** $\int_0^1 e^{2x}\,dx;\ n = 25$  **46.** $\int_0^2 x^4\,dx;\ n = 30$

**47.** $\int_1^e \frac{dx}{x};\ n = 50$  **48.** $\int_0^{\pi/4} \frac{dx}{1 + x^2};\ n = 64$

**T 49–52. Simpson's Rule** *Apply Simpson's Rule to the following integrals. It is easiest to obtain the Simpson's Rule approximations from the Trapezoid Rule approximations, as in Example 8. Make a table similar to Table 8.8 showing the approximations and errors for n = 4, 8, 16, and 32. The exact values of the integrals are given for computing the error.*

**49.** $\int_0^4 (3x^5 - 8x^3)\,dx = 1536$  **50.** $\int_1^e \ln x\,dx = 1$

**51.** $\int_0^{\pi} e^{-t}\sin t\,dt = \frac{1}{2}(e^{-\pi} + 1)$

**52.** $\int_0^6 3e^{-3x}\,dx = 1 - e^{-18} \approx 1.000000$

**53. Explain why or why not** Determine whether the following statements are true and give an explanation or counterexample.

a. Suppose $\int_a^b f(x)\,dx$ is approximated with Simpson's Rule using $n = 18$ subintervals, where $|f^{(4)}(x)| \le 1$ on $[a, b]$. The absolute error $E_S$ in approximating the integral satisfies $E_S \le \frac{(\Delta x)^5}{10}$.

b. If the number of subintervals used in the Midpoint Rule is increased by a factor of 3, the error is expected to decrease by a factor of 8.

c. If the number of subintervals used in the Trapezoid Rule is increased by a factor of 4, the error is expected to decrease by a factor of 16.

**T 54–57. Comparing the Midpoint and Trapezoid Rules** *Compare the errors in the Midpoint and Trapezoid Rules with n = 4, 8, 16, and 32 subintervals when they are applied to the following integrals (with their exact values given).*

**54.** $\int_0^{\pi/2} \sin^6 x\,dx = \frac{5\pi}{32}$  **55.** $\int_0^{\pi/2} \cos^9 x\,dx = \frac{128}{315}$

**56.** $\int_0^1 (8x^7 - 7x^8)\,dx = \frac{2}{9}$

**57.** $\int_0^{\pi} \ln(5 + 3\cos x)\,dx = \pi \ln \frac{9}{2}$

**T 58–61. Using Simpson's Rule** *Approximate the following integrals using Simpson's Rule. Experiment with values of n to ensure that the error is less than $10^{-3}$.*

**58.** $\int_0^{2\pi} \frac{dx}{(5 + 3\sin x)^2} = \frac{5\pi}{32}$

**59.** $\int_0^{\pi} \frac{4\cos x}{5 - 4\cos x}\,dx = \frac{2\pi}{3}$

**60.** $\int_0^{\pi} \ln(2 + \cos x)\,dx = \pi \ln\left(\frac{2 + \sqrt{3}}{2}\right)$

**61.** $\int_0^{\pi} \sin 6x \cos 3x\,dx = \frac{4}{9}$

**T 62. Period of a pendulum** A standard pendulum of length $L$ that swings under the influence of gravity alone (no resistance) has a period of

$$T = \frac{4}{\omega}\int_0^{\pi/2} \frac{d\varphi}{\sqrt{1 - k^2 \sin^2 \varphi}},$$

where $\omega^2 = g/L$, $k^2 = \sin^2(\theta_0/2)$, $g = 9.8\ \text{m/s}^2$ is the acceleration due to gravity, and $\theta_0$ is the initial angle from which the pendulum is released (in radians). Use numerical integration to approximate the period of a pendulum with $L = 1$ m that is released from an angle of $\theta_0 = \pi/4$ rad.

**T 63. Normal distribution of heights** The heights of U.S. men are normally distributed with a mean of 69 in and a standard deviation of 3 in. This means that the fraction of men with a height between $a$ and $b$ (with $a < b$) inches is given by the integral

$$\frac{1}{3\sqrt{2\pi}}\int_a^b e^{-((x-69)/3)^2/2}\,dx.$$

What percentage of American men are between 66 and 72 inches tall? Use the method of your choice, and experiment with the number of subintervals until you obtain successive approximations that differ by less than $10^{-3}$.

**T 64. Normal distribution of movie lengths** A study revealed that the lengths of U.S. movies are normally distributed with a mean of 110 minutes and a standard deviation of 22 minutes. This means that the fraction of movies with lengths between $a$ and $b$ minutes (with $a < b$) is given by the integral

$$\frac{1}{22\sqrt{2\pi}}\int_a^b e^{-((x-110)/22)^2/2}\,dx.$$

What percentage of U.S. movies are between 1 hr and 1.5 hr long (60–90 min)?

**T 65. U.S. oil produced and imported** The figure shows the rate at which U.S. oil was produced and imported between 1920 and 2005 in units of millions of barrels per day. The total amount of oil produced or imported is given by the area of the region under the corresponding curve. Be careful with units because both days and years are used in this data set.

a. Use numerical integration to estimate the amount of U.S. oil produced between 1940 and 2000. Use the method of your choice, and experiment with values of $n$.

**b.** Use numerical integration to estimate the amount of oil imported between 1940 and 2000. Use the method of your choice, and experiment with values of $n$.

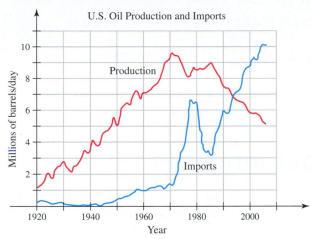

U.S. Oil Production and Imports

(*Source:* U.S. Energy Information Administration)

**T 66–71. Estimating error** *Refer to Theorem 8.1 in the following exercises.*

**66.** Let $f(x) = \cos x^2$.

**a.** Find a Midpoint Rule approximation to $\int_{-1}^{1} \cos x^2 \, dx$ using $n = 30$ subintervals.
**b.** Calculate $f''(x)$.
**c.** Explain why $|f''(x)| \le 6$ on $[-1, 1]$.
**d.** Use Theorem 8.1 to find an upper bound on the absolute error in the estimate found in part (a).

**67.** Let $f(x) = \sqrt{x^3 + 1}$.

**a.** Find a Midpoint Rule approximation to $\int_{1}^{6} \sqrt{x^3 + 1} \, dx$ using $n = 50$ subintervals.
**b.** Calculate $f''(x)$.
**c.** Use the fact that $f''$ is decreasing and positive on $[1, 6]$ to show that $|f''(x)| \le 15/(8\sqrt{2})$ on $[1, 6]$.
**d.** Use Theorem 8.1 to find an upper bound on the absolute error in the estimate found in part (a).

**68.** Let $f(x) = e^{x^2}$.

**a.** Find a Trapezoid Rule approximation to $\int_{0}^{1} e^{x^2} \, dx$ using $n = 50$ subintervals.
**b.** Calculate $f''(x)$.
**c.** Explain why $|f''(x)| < 18$ on $[0, 1]$, given that $e < 3$.
**d.** Use Theorem 8.1 to find an upper bound on the absolute error in the estimate found in part (a).

**69.** Let $f(x) = \sin e^x$.

**a.** Find a Trapezoid Rule approximation to $\int_{0}^{1} \sin e^x \, dx$ using $n = 40$ subintervals.
**b.** Calculate $f''(x)$.
**c.** Explain why $|f''(x)| < 6$ on $[0, 1]$, given that $e < 3$. (*Hint:* Graph $f''$.)
**d.** Find an upper bound on the absolute error in the estimate found in part (a) using Theorem 8.1.

**70.** Let $f(x) = e^{-x^2}$.

**a.** Find a Simpson's Rule approximation to $\int_{0}^{3} e^{-x^2} \, dx$ using $n = 30$ subintervals.

**b.** Calculate $f^{(4)}(x)$.
**c.** Find an upper bound on the absolute error in the estimate found in part (a) using Theorem 8.1. (*Hint:* Use a graph to find an upper bound for $|f^{(4)}(x)|$ on $[0, 3]$.)

**71.** Let $f(x) = \sqrt{\sin x}$.

**a.** Find a Simpson's Rule approximation to $\int_{1}^{2} \sqrt{\sin x} \, dx$ using $n = 20$ subintervals.
**b.** Find an upper bound on the absolute error in the estimate found in part (a) using Theorem 8.1. (*Hint:* Use the fact that $|f^{(4)}(x)| \le 1$ on $[1, 2]$.)

## Explorations and Challenges

**72. Exact Trapezoid Rule** Prove that the Trapezoid Rule is exact (no error) when approximating the definite integral of a linear function.

**T 73. Arc length of an ellipse** The length of an ellipse with axes of length $2a$ and $2b$ is

$$\int_{0}^{2\pi} \sqrt{a^2 \cos^2 t + b^2 \sin^2 t} \, dt.$$

Use numerical integration, and experiment with different values of $n$ to approximate the length of an ellipse with $a = 4$ and $b = 8$.

**T 74. Sine integral** The theory of diffraction produces the sine integral function $\text{Si}(x) = \int_{0}^{x} \frac{\sin t}{t} \, dt$. Use the Midpoint Rule to approximate $\text{Si}(1)$ and $\text{Si}(10)$. (Recall that $\lim_{x \to 0} \frac{\sin x}{x} = 1$.) Experiment with the number of subintervals until you obtain approximations that have an error less than $10^{-3}$. A rule of thumb is that if two successive approximations differ by less than $10^{-3}$, then the error is usually less than $10^{-3}$.

**75. Exact Simpson's Rule**

**a.** Use Simpson's Rule to approximate $\int_{0}^{4} x^3 \, dx$ using two subintervals ($n = 2$); compare the approximation to the value of the integral.
**b.** Use Simpson's Rule to approximate $\int_{0}^{4} x^3 \, dx$ using four subintervals ($n = 4$); compare the approximation to the value of the integral.
**c.** Use the error bound associated with Simpson's Rule given in Theorem 8.1 to explain why the approximations in parts (a) and (b) give the exact value of the integral.
**d.** Use Theorem 8.1 to explain why a Simpson's Rule approximation using any (even) number of subintervals gives the exact value of $\int_{a}^{b} f(x) \, dx$, where $f(x)$ is a polynomial of degree 3 or less.

**76. Shortcut for the Trapezoid Rule** Given a Midpoint Rule approximation $M(n)$ and a Trapezoid Rule approximation $T(n)$ for a continuous function on $[a, b]$ with $n$ subintervals, show that

$$T(2n) = \frac{T(n) + M(n)}{2}.$$

**77. Trapezoid Rule and concavity** Suppose $f$ is positive and its first two derivatives are continuous on $[a, b]$. If $f''$ is positive on $[a, b]$, then is a Trapezoid Rule estimate of $\int_{a}^{b} f(x) \, dx$ an underestimate or overestimate of the integral? Justify your answer using Theorem 8.1 and an illustration.

**78. Shortcut for Simpson's Rule** Using the notation of the text, prove that $S(2n) = \dfrac{4T(2n) - T(n)}{3}$, for $n \geq 1$.

**T 79. Another Simpson's Rule formula** Another Simpson's Rule formula is $S(2n) = \dfrac{2M(n) + T(n)}{3}$, for $n \geq 1$. Use this rule to estimate $\displaystyle\int_1^e \frac{1}{x} \, dx$ using $n = 10$ subintervals.

# 8.9 Improper Integrals

The definite integrals we have encountered so far involve finite-valued functions and finite intervals of integration. In this section, you will see that definite integrals can sometimes be evaluated when these conditions are not met. Here is an example. The energy required to launch a rocket from the surface of Earth ($R = 6370$ km from the center of Earth) to an altitude $H$ is given by an integral of the form $\int_R^{R+H} k/x^2 \, dx$, where $k$ is a constant that includes the mass of the rocket, the mass of Earth, and the gravitational constant. This integral may be evaluated for any finite altitude $H > 0$. Now suppose the aim is to launch the rocket to an arbitrarily large altitude $H$ so that it escapes Earth's gravitational field. The energy required is given by the preceding integral as $H \to \infty$, which we write $\int_R^{\infty} k/x^2 \, dx$. This integral is an example of an *improper integral*, and it has a finite value (which explains why it is possible to launch rockets to outer space). For historical reasons, the term *improper integral* is used for cases in which

• the interval of integration is infinite, or

• the integrand is unbounded on the interval of integration.

In this section, we explore improper integrals and their many uses.

### Infinite Intervals

A simple example illustrates what can happen when we integrate a function over an infinite interval. Consider the integral $\displaystyle\int_1^b \frac{1}{x^2} \, dx$, for any real number $b > 1$. As shown in Figure 8.21, this integral gives the area of the region bounded by the curve $y = x^{-2}$ and the $x$-axis between $x = 1$ and $x = b$. In fact, the value of the integral is

$$\int_1^b \frac{1}{x^2} \, dx = -\frac{1}{x}\Big|_1^b = 1 - \frac{1}{b}.$$

For example, if $b = 2$, the area under the curve is $\frac{1}{2}$; if $b = 3$, the area under the curve is $\frac{2}{3}$. In general, as $b$ increases, the area under the curve increases.

Now let's ask what happens to the area as $b$ becomes arbitrarily large. Letting $b \to \infty$, the area of the region under the curve is

$$\lim_{b \to \infty} \left(1 - \frac{1}{b}\right) = 1.$$

We have discovered, surprising as it may seem, a curve of *infinite* length that bounds a region with *finite* area (1 square unit).

We express this result as

$$\int_1^{\infty} \frac{1}{x^2} \, dx = 1,$$

which is an improper integral because $\infty$ appears in the upper limit. In general, to evaluate $\int_a^{\infty} f(x) \, dx$, we first integrate over a finite interval $[a, b]$ and then let $b \to \infty$. Similar procedures are used to evaluate $\int_{-\infty}^{b} f(x) \, dx$ and $\int_{-\infty}^{\infty} f(x) \, dx$.

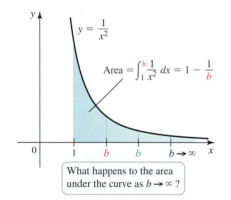

$y = \dfrac{1}{x^2}$

Area $= \displaystyle\int_1^b \frac{1}{x^2} \, dx = 1 - \frac{1}{b}$

What happens to the area under the curve as $b \to \infty$?

**Figure 8.21**

**DEFINITION  Improper Integrals over Infinite Intervals**

**1.** If $f$ is continuous on $[a, \infty)$, then

$$\int_a^\infty f(x)\,dx = \lim_{b \to \infty} \int_a^b f(x)\,dx.$$

**2.** If $f$ is continuous on $(-\infty, b]$, then

$$\int_{-\infty}^b f(x)\,dx = \lim_{a \to -\infty} \int_a^b f(x)\,dx.$$

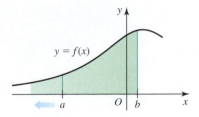

> Doubly infinite integrals (case 3 in the definition) must be evaluated as two independent limits and not as
>
> $$\int_{-\infty}^\infty f(x)\,dx = \lim_{b \to \infty} \int_{-b}^b f(x)\,dx.$$

**3.** If $f$ is continuous on $(-\infty, \infty)$, then

$$\int_{-\infty}^\infty f(x)\,dx = \lim_{a \to -\infty} \int_a^c f(x)\,dx$$

$$+ \lim_{b \to \infty} \int_c^b f(x)\,dx,$$

where $c$ is any real number. It can be shown that the choice of $c$ does not affect the value or convergence of the original integral.

If the limits in cases 1–3 exist, then the improper integrals **converge**; otherwise, they **diverge**.

**EXAMPLE 1  Infinite intervals**  Evaluate each integral.

**a.** $\displaystyle\int_0^\infty e^{-x}\,dx$    **b.** $\displaystyle\int_{-\infty}^\infty \frac{dx}{1 + x^2}$

**SOLUTION**

**a.** Using the definition of the improper integral, we have

$$\int_0^\infty e^{-x}\,dx = \lim_{b \to \infty} \int_0^b e^{-x}\,dx \qquad \text{\color{blue}Definition of improper integral}$$

$$= \lim_{b \to \infty} \left(-e^{-x}\right)\Big|_0^b \qquad \text{\color{blue}Evaluate the integral.}$$

$$= \lim_{b \to \infty} \left(1 - e^{-b}\right) \qquad \text{\color{blue}Simplify.}$$

$$= \left(1 - \underbrace{\lim_{b \to \infty} \frac{1}{e^b}}_{\text{equals 0}}\right) = 1. \qquad \text{\color{blue}Evaluate the limit; } e^{-b} = \frac{1}{e^b}.$$

In this case, the limit exists, so the integral converges and the region under the curve has a finite area of 1 (**Figure 8.22**).

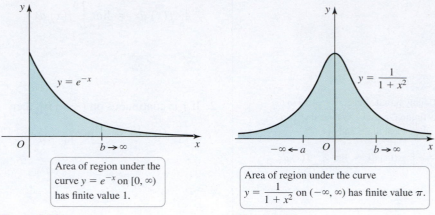

Area of region under the curve $y = e^{-x}$ on $[0, \infty)$ has finite value 1.

**Figure 8.22**

Area of region under the curve $y = \dfrac{1}{1 + x^2}$ on $(-\infty, \infty)$ has finite value $\pi$.

**Figure 8.23**

▶ Recall that

$$\int \frac{dx}{a^2 + x^2} = \frac{1}{a}\tan^{-1}\frac{x}{a} + C.$$

The graph of $y = \tan^{-1} x$ shows that

$$\lim_{x \to \infty} \tan^{-1} x = \frac{\pi}{2} \text{ and } \lim_{x \to -\infty} \tan^{-1} x = -\frac{\pi}{2}.$$

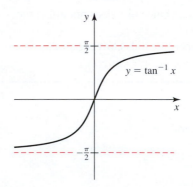

**QUICK CHECK 1** The function $f(x) = 1 + x^{-1}$ decreases to 1 as $x \to \infty$. Does $\int_1^\infty f(x)\, dx$ exist? ◀

**b.** Using the definition of the improper integral, we choose $c = 0$ and write

$$\int_{-\infty}^\infty \frac{dx}{1 + x^2} = \lim_{a \to -\infty} \int_a^c \frac{dx}{1 + x^2} + \lim_{b \to \infty} \int_c^b \frac{dx}{1 + x^2} \qquad \text{Definition of improper integral}$$

$$= \lim_{a \to -\infty} \tan^{-1} x \Big|_a^0 + \lim_{b \to \infty} \tan^{-1} x \Big|_0^b \qquad \text{Evaluate integral; } c = 0.$$

$$= \lim_{a \to -\infty} (0 - \tan^{-1} a) + \lim_{b \to \infty} (\tan^{-1} b - 0) \qquad \text{Simplify.}$$

$$= \frac{\pi}{2} + \frac{\pi}{2} = \pi. \qquad \text{Evaluate limits.}$$

The same result is obtained with any value of the intermediate point $c$; therefore, the value of the integral is $\pi$ (**Figure 8.23**).

*Related Exercises 10, 15* ◀

**EXAMPLE 2  The family $f(x) = 1/x^p$** Consider the family of functions $f(x) = 1/x^p$, where $p$ is a real number. For what values of $p$ does $\int_1^\infty f(x)\, dx$ converge?

**SOLUTION** For $p > 0$, the functions $f(x) = 1/x^p$ approach zero as $x \to \infty$, with larger values of $p$ giving greater rates of decrease (**Figure 8.24**). Assuming $p \neq 1$, the integral is evaluated as follows:

$$\int_1^\infty \frac{1}{x^p}\, dx = \lim_{b \to \infty} \int_1^b x^{-p}\, dx \qquad \text{Definition of improper integral}$$

$$= \frac{1}{1 - p} \lim_{b \to \infty} \left( x^{1-p} \Big|_1^b \right) \qquad \text{Evaluate the integral on a finite interval.}$$

$$= \frac{1}{1 - p} \lim_{b \to \infty} (b^{1-p} - 1). \qquad \text{Simplify.}$$

It is easiest to consider three cases.

*Case 1:* If $p > 1$, then $p - 1 > 0$, and $b^{1-p} = 1/b^{p-1}$ approaches 0 as $b \to \infty$. Therefore, the integral converges and its value is

$$\int_1^\infty \frac{1}{x^p}\, dx = \frac{1}{1 - p} \lim_{b \to \infty} \underbrace{(b^{1-p} - 1)}_{\text{approaches } 0} = \frac{1}{p - 1}.$$

▶ Recall that for $p \neq 1$,

$$\int \frac{1}{x^p}\, dx = \int x^{-p}\, dx$$
$$= \frac{x^{-p+1}}{-p + 1} + C$$
$$= \frac{x^{1-p}}{1 - p} + C.$$

▶ Example 2 is important in the study of infinite series in Chapter 10. It shows that a continuous function $f$ must do more than simply decrease to zero for its integral on $[a, \infty)$ to converge; it must decrease to zero *sufficiently fast*.

$$\int_1^\infty \frac{dx}{x^p} = \frac{1}{p-1}, \text{ if } p > 1.$$

**Figure 8.24**

**QUICK CHECK 2** Use the result of Example 2 to evaluate $\int_1^\infty \frac{1}{x^4} \, dx.$ ◄

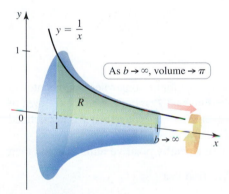

**Figure 8.25**

▶ The integral in Example 3b can be evaluated directly by using the substitution $u = x^2$ and then consulting a table of integrals.

▶ The solid in Examples 3a and 3b is called *Gabriel's horn* or *Torricelli's trumpet.* We have shown that—quite remarkably—it has finite volume and infinite surface area.

*Case 2:*   If $p < 1$, then $1 - p > 0$, and the integral diverges:

$$\int_1^\infty \frac{1}{x^p} \, dx = \frac{1}{1-p} \underbrace{\lim_{b \to \infty} (b^{1-p} - 1)}_{\text{arbitrarily large}} = \infty.$$

*Case 3:*   If $p = 1$, then $\int_1^\infty \frac{1}{x} \, dx = \lim_{b \to \infty} \ln b = \infty$, so the integral diverges.

In summary, $\int_1^\infty \frac{1}{x^p} \, dx = \frac{1}{p-1}$ if $p > 1$, and the integral diverges if $p \leq 1$.

*Related Exercises 8–9* ◄

**EXAMPLE 3**   **Solids of revolution**   Let $R$ be the region bounded by the graph of $y = x^{-1}$ and the $x$-axis, for $x \geq 1$.

**a.** What is the volume of the solid generated when $R$ is revolved about the $x$-axis?

**b.** What is the surface area of the solid generated when $R$ is revolved about the $x$-axis?

**c.** What is the volume of the solid generated when $R$ is revolved about the $y$-axis?

**SOLUTION**

**a.** The region $R$ and the corresponding solid of revolution are shown in **Figure 8.25**. We use the disk method (Section 6.3) over the interval $[1, b]$ and then let $b \to \infty$:

$$\text{Volume} = \int_1^\infty \pi (f(x))^2 \, dx \qquad \text{Disk method}$$

$$= \pi \lim_{b \to \infty} \int_1^b \frac{1}{x^2} \, dx \qquad \text{Definition of improper integral}$$

$$= \pi \lim_{b \to \infty} \left( 1 - \frac{1}{b} \right) = \pi. \qquad \text{Evaluate the integral.}$$

The improper integral exists, and the solid has a volume of $\pi$ cubic units.

**b.** Using the results of Section 6.6, the area of the surface generated on the interval $[1, b]$, where $b > 1$, is

$$\int_1^b 2\pi f(x) \sqrt{1 + f'(x)^2} \, dx.$$

The area of the surface generated on the interval $[1, \infty)$ is found by letting $b \to \infty$:

$$\text{Surface area} = 2\pi \lim_{b \to \infty} \int_1^b f(x) \sqrt{1 + f'(x)^2} \, dx \qquad \text{Surface area formula; let } b \to \infty.$$

$$= 2\pi \lim_{b \to \infty} \int_1^b \frac{1}{x} \sqrt{1 + \left( -\frac{1}{x^2} \right)^2} \, dx \qquad \text{Substitute } f \text{ and } f'.$$

$$= 2\pi \lim_{b \to \infty} \int_1^b \frac{1}{x^3} \sqrt{1 + x^4} \, dx. \qquad \text{Simplify.}$$

Notice that on the interval of integration $x \geq 1$, we have $\sqrt{1 + x^4} > \sqrt{x^4} = x^2$, which means that

$$\frac{1}{x^3} \sqrt{1 + x^4} > \frac{x^2}{x^3} = \frac{1}{x}.$$

Therefore, for all $b$ with $1 < b < \infty$,

$$\text{Surface area} = 2\pi \int_1^b \frac{1}{x^3} \sqrt{1 + x^4} \, dx > 2\pi \int_1^b \frac{1}{x} \, dx.$$

Because $2\pi \lim_{b \to \infty} \int_1^b \frac{1}{x} \, dx = \infty$ (by Example 2), the preceding inequality implies that

$2\pi \lim_{b \to \infty} \int_1^b \frac{1}{x^3} \sqrt{1 + x^4} \, dx = \infty$. Therefore, the integral diverges and the surface area of the solid is infinite.

➤ Recall that if $f(x) > 0$ on $[a, b]$ and the region bounded by the graph of $f$ and the $x$-axis on $[a, b]$ is revolved about the $y$-axis, the volume of the solid generated is

$$V = \int_a^b 2\pi x f(x)\, dx.$$

**c.** The region in question and the corresponding solid of revolution are shown in **Figure 8.26**. Using the shell method (Section 6.4) on the interval $[1, b)$ and letting $b \to \infty$, the volume is given by

$$
\begin{aligned}
\text{Volume} &= \int_1^\infty 2\pi x f(x)\, dx && \text{\color{blue}Shell method} \\
&= 2\pi \int_1^\infty 1\, dx && \text{\color{blue}$f(x) = x^{-1}$} \\
&= 2\pi \lim_{b \to \infty} \int_1^b 1\, dx && \text{\color{blue}Definition of improper integral} \\
&= 2\pi \lim_{b \to \infty} (b - 1) && \text{\color{blue}Evaluate the integral over a finite interval.} \\
&= \infty. && \text{\color{blue}The improper integral diverges.}
\end{aligned}
$$

Revolving the region $R$ about the $y$-axis, the volume of the resulting solid is infinite.

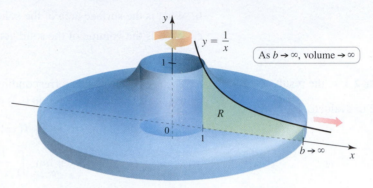

**Figure 8.26**

*Related Exercises 65–66* ◄

## Unbounded Integrands

Improper integrals also occur when the integrand becomes infinite somewhere on the interval of integration. Consider the function $f(x) = 1/\sqrt{x}$ (**Figure 8.27**). Let's examine the area of the region bounded by the graph of $f$ between $x = 0$ and $x = 1$. Notice that $f$ is not defined at $x = 0$, and it increases without bound as $x \to 0^+$.

The idea here is to replace the lower limit 0 with a nearby positive number $c$ and then consider the integral $\int_c^1 \dfrac{dx}{\sqrt{x}}$, where $0 < c < 1$. We find that

$$\int_c^1 \frac{dx}{\sqrt{x}} = 2\sqrt{x}\,\Big|_c^1 = 2(1 - \sqrt{c}).$$

To find the area of the region under the curve over the interval $(0, 1]$, we let $c \to 0^+$. The resulting area, which we denote $\int_0^1 \dfrac{dx}{\sqrt{x}}$, is

$$\lim_{c \to 0^+} \int_c^1 \frac{dx}{\sqrt{x}} = \lim_{c \to 0^+} 2(1 - \sqrt{c}) = 2.$$

Once again, we have a surprising result: Although the region in question has a boundary curve with infinite length, the area of the region is finite.

What happens to the area under the curve as $c \to 0^+$?

**Figure 8.27**

➤ The functions $f(x) = 1/x^p$ are unbounded at $x = 0$, for $p > 0$. It can be shown (Exercise 94) that

$$\int_0^1 \frac{dx}{x^p} = \frac{1}{1 - p},$$

provided $p < 1$. Otherwise, the integral diverges.

**QUICK CHECK 3** Explain why the one-sided limit $c \to 0^+$ (instead of a two-sided limit) must be used in the previous calculation. ◄

The preceding example shows that if a function is unbounded at a point $p$, it may be possible to integrate that function over an interval that contains $p$. The point $p$ may occur at either endpoint or at an interior point of the interval of integration.

**DEFINITION** Improper Integrals with an Unbounded Integrand

**1.** Suppose $f$ is continuous on $(a, b]$ with
$\lim\limits_{x \to a^+} f(x) = \pm \infty$. Then

$$\int_a^b f(x)\, dx = \lim_{c \to a^+} \int_c^b f(x)\, dx.$$

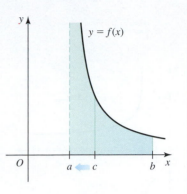

**2.** Suppose $f$ is continuous on $[a, b)$ with
$\lim\limits_{x \to b^-} f(x) = \pm \infty$. Then

$$\int_a^b f(x)\, dx = \lim_{c \to b^-} \int_a^c f(x)\, dx.$$

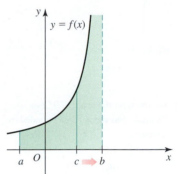

**3.** Suppose $f$ is continuous on $[a, b]$ except at
the interior point $p$ where $f$ is unbounded.
Then

$$\int_a^b f(x)\, dx = \lim_{c \to p^-} \int_a^c f(x)\, dx + \lim_{d \to p^+} \int_d^b f(x)\, dx.$$

If the limits in cases 1–3 exist, then
the improper integrals **converge**;
otherwise, they **diverge**.

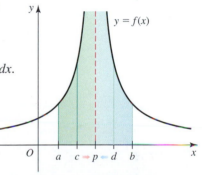

▶ Recall that
$$\int \frac{dx}{\sqrt{a^2 - x^2}} = \sin^{-1}\frac{x}{a} + C.$$

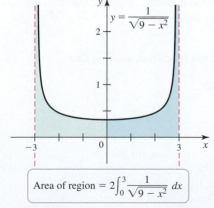

**Figure 8.28**

**EXAMPLE 4** **Infinite integrand** Find the area of the region $R$ between the graph of
$f(x) = \dfrac{1}{\sqrt{9 - x^2}}$ and the $x$-axis on the interval $(-3, 3)$ (if it exists).

**SOLUTION** The integrand is even and has vertical asymptotes at $x = \pm 3$ (**Figure 8.28**).
By symmetry, the area of $R$ is given by

$$\int_{-3}^{3} \frac{1}{\sqrt{9 - x^2}}\, dx = 2\int_{0}^{3} \frac{1}{\sqrt{9 - x^2}}\, dx,$$

assuming these improper integrals exist. Because the integrand is unbounded at $x = 3$,
we replace the upper limit with $c$, evaluate the resulting integral, and then let $c \to 3^-$:

$$2\int_0^3 \frac{dx}{\sqrt{9 - x^2}} = 2 \lim_{c \to 3^-} \int_0^c \frac{dx}{\sqrt{9 - x^2}} \qquad \text{Definition of improper integral}$$

$$= 2 \lim_{c \to 3^-} \sin^{-1}\frac{x}{3}\Big|_0^c \qquad \text{Evaluate the integral.}$$

$$= 2 \lim_{c \to 3^-} \left( \underbrace{\sin^{-1}\frac{c}{3}}_{\text{approaches } \pi/2} - \underbrace{\sin^{-1} 0}_{\text{equals } 0} \right). \qquad \text{Simplify.}$$

Note that as $c \to 3^-$, $\sin^{-1}(c/3) \to \sin^{-1} 1 = \pi/2$. Therefore, the area of $R$ is

$$2 \int_0^3 \frac{1}{\sqrt{9 - x^2}} \, dx = 2\left(\frac{\pi}{2} - 0\right) = \pi.$$

*Related Exercises 37–38* ◄

**EXAMPLE 5   Infinite integrand at an interior point**  Evaluate $\int_1^{10} \frac{dx}{(x-2)^{1/3}}$.

**SOLUTION**  The integrand is unbounded at $x = 2$, which is an interior point of the interval of integration (**Figure 8.29**). We split the interval into two subintervals and evaluate an improper integral on each subinterval:

$$\int_1^{10} \frac{dx}{(x-2)^{1/3}} = \lim_{c \to 2^-} \int_1^c \frac{dx}{(x-2)^{1/3}} + \lim_{d \to 2^+} \int_d^{10} \frac{dx}{(x-2)^{1/3}} \qquad \text{Definition of improper integral}$$

$$= \lim_{c \to 2^-} \frac{3}{2} (x-2)^{2/3} \Big|_1^c + \lim_{d \to 2^+} \frac{3}{2} (x-2)^{2/3} \Big|_d^{10} \qquad \text{Evaluate integrals.}$$

$$= \frac{3}{2} \left( \lim_{c \to 2^-} (c-2)^{2/3} - (1-2)^{2/3} \right)$$

$$+ \frac{3}{2} \left( (10-2)^{2/3} - \lim_{d \to 2^+} (d-2)^{2/3} \right) \qquad \text{Simplify.}$$

$$= \frac{3}{2} \left( 0 - (-1)^{2/3} + 8^{2/3} - 0 \right) = \frac{9}{2}. \qquad \text{Evaluate limits.}$$

*Related Exercises 48–49* ◄

**Figure 8.29**

➤ We interpret the result of 9/2 in Example 5 as the net area bounded by the curve $y = 1/(x-2)^{1/3}$ over the interval $[1, 10]$.

The following example illustrates one of many practical uses of improper integrals.

**EXAMPLE 6   Bioavailability**  The most efficient way to deliver a drug to its intended target site is to administer it intravenously (directly into the blood). If a drug is administered any other way (for example, by injection, orally, by nasal inhalant, or by skin patch), then some of the drug is typically lost to absorption before it gets to the blood. By definition, the bioavailability of a drug measures the effectiveness of a nonintravenous method compared to the intravenous method. The bioavailability of intravenous dosing is 100%.

Let the functions $C_i(t)$ and $C_o(t)$ give the concentration of a drug in the blood, for times $t \geq 0$, using intravenous and oral dosing, respectively. (These functions can be determined through clinical experiments.) Assuming the same amount of drug is initially administered by both methods, the bioavailability for an oral dose is defined to be

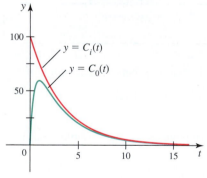

$$F = \frac{\text{AUC}_o}{\text{AUC}_i} = \frac{\int_0^\infty C_o(t) \, dt}{\int_0^\infty C_i(t) \, dt},$$

where AUC is used in the pharmacology literature to mean *area under the curve*.

Suppose the concentration of a certain drug in the blood in mg/L when given intravenously is $C_i(t) = 100e^{-0.3t}$, where $t \geq 0$ is measured in hours. Suppose also that the concentration of the same drug when delivered orally is $C_o(t) = 90\left(e^{-0.3t} - e^{-2.5t}\right)$ (**Figure 8.30**). Find the bioavailability of the drug.

**Figure 8.30**

**SOLUTION**  Evaluating the integrals of the concentration functions, we find that

$$\text{AUC}_i = \int_0^\infty C_i(t) \, dt = \int_0^\infty 100e^{-0.3t} \, dt$$

$$= \lim_{b \to \infty} \int_0^b 100e^{-0.3t} \, dt \qquad \text{Improper integral}$$

$$= \lim_{b \to \infty} \frac{1000}{3} \left( 1 - \underbrace{e^{-0.3b}}_{\text{approaches zero}} \right) \qquad \text{Evaluate the integral.}$$

$$= \frac{1000}{3}. \qquad \text{Evaluate the limit.}$$

Similarly,

$$\text{AUC}_o = \int_0^\infty C_o(t)\, dt = \int_0^\infty 90(e^{-0.3t} - e^{-2.5t})\, dt$$

$$= \lim_{b \to \infty} \int_0^b 90(e^{-0.3t} - e^{-2.5t})\, dt \qquad \text{Improper integral}$$

$$= \lim_{b \to \infty} \left( 300(1 - \underbrace{e^{-0.3b}}_{\substack{\text{approaches} \\ \text{zero}}}) - 36(1 - \underbrace{e^{-2.5b}}_{\substack{\text{approaches} \\ \text{zero}}}) \right) \qquad \text{Evaluate the integral.}$$

$$= 264. \qquad \text{Evaluate the limit.}$$

Therefore, the bioavailability is $F = 264/(1000/3) = 0.792$, which means oral administration of the drug is roughly 80% as effective as intravenous dosing. Notice that $F$ is the ratio of the areas under the two curves on the interval $[0, \infty)$.

*Related Exercises 61–62* ◄

## The Comparison Test

> ▶ The integral $\int_a^\infty e^{-x^2}\, dx$ plays an essential role in probability theory, and it is important to determine whether it converges—see Example 7a.

There are some cases in which we cannot determine whether a given improper integral converges, simply because it is impossible to compute an antiderivative using the catalog of integrals in this text. For example, the antiderivative of $g(x) = e^{-x^2}$ is not expressible in terms of elementary functions, so the procedures used earlier in this section fail to determine whether $\int_a^\infty e^{-x^2}\, dx$ converges, for any real number $a$. Theorem 8.2 is helpful here.

---

**THEOREM 8.2  Comparison Test for Improper Integrals**
Suppose the functions $f$ and $g$ are continuous on the interval $[a, \infty)$, with $f(x) \ge g(x) \ge 0$, for $x \ge a$.

1. If $\int_a^\infty f(x)\, dx$ converges, then $\int_a^\infty g(x)\, dx$ converges.

2. If $\int_a^\infty g(x)\, dx$ diverges, then $\int_a^\infty f(x)\, dx$ diverges.

---

We omit a formal proof, although **Figure 8.31** helps to explain statement (1) of Theorem 8.2. If the area beneath the graph of $f$ is finite on $[a, \infty)$ and if $f(x) \ge g(x) \ge 0$, then the area under the graph of $g$ on $[a, \infty)$ must also be finite. Statement (2) follows from (1) because it is the contrapositive of (1). Analogous theorems apply to improper integrals of the form $\int_{-\infty}^b f(x)\, dx$ and to integrals over finite intervals with an unbounded integrand.

If the area under the graph of $f$ on $[a, \infty)$ is finite . . .

$f(x) \ge g(x)$ on $[a, \infty)$

. . . then the area under the graph of $g$ on $[a, \infty)$ is also finite.

**Figure 8.31**

**EXAMPLE 7  Comparison Test**  Use Theorem 8.2 to determine whether the following integrals converge.

**a.** $\displaystyle\int_1^\infty e^{-x^2}\, dx$      **b.** $\displaystyle\int_1^\infty \frac{1}{\sqrt[3]{x^2 - 0.5}}\, dx$

**SOLUTION**

**a.** Before invoking Theorem 8.2, it helps to predict whether a given improper integral converges or diverges so we can decide whether to apply statement (1) or (2) of the Comparison Test. In Example 1a, we learned that $\int_0^\infty e^{-x}\, dx$ converges, which implies that $\int_1^\infty e^{-x}\, dx$ also converges (Quick Check 4). Because $e^{-x^2}$ approaches 0 more rapidly than $e^{-x}$ as $x \to \infty$ ($x^2 > x$ for large values of $x$), we expect that $\int_1^\infty e^{-x^2}\, dx$ also converges. So we let $f(x) = e^{-x}$ and $g(x) = e^{-x^2}$ in statement (1); our goal is to prove that $f(x) \ge g(x)$ on $[1, \infty)$ (note that both functions are positive for all $x$). Because $x \le x^2$ and $e^x$ is increasing on $[1, \infty)$, we have $e^x \le e^{x^2}$ on $[1, \infty)$, which implies that

$$e^{-x} \ge e^{-x^2} \text{ on } [1, \infty). \quad \text{Figure 8.32}$$

Therefore, by statement (1) of Theorem 8.2, $\int_1^\infty e^{-x^2}\, dx$ converges.

**QUICK CHECK 4**  Show that $\displaystyle\int_1^\infty e^{-x}\, dx = \frac{1}{e}.$ ◄

$e^{-x} \ge e^{-x^2}$ on $[1, \infty)$

**Figure 8.32**

**b.** One challenge in using Theorem 8.2 is choosing an appropriate improper integral for comparison to the given improper integral. In this case, we see that for large values of $x$,

$$\frac{1}{\sqrt[3]{x^2 - 0.5}} \approx \frac{1}{\sqrt[3]{x^2}} = \frac{1}{x^{2/3}}, \text{ which suggests that we compare } \int_1^\infty \frac{1}{\sqrt[3]{x^2 - 0.5}} \, dx \text{ to}$$

$\int_1^\infty \frac{1}{x^{2/3}} \, dx$. In Example 2 we showed that $\int_1^\infty \frac{1}{x^p} \, dx$ diverges if $p \le 1$, so we

expect that $\int_1^\infty \frac{1}{\sqrt[3]{x^2 - 0.5}} \, dx$ also diverges. Therefore, we let $f(x) = \sqrt[3]{x^2 - 0.5}$ and

$g(x) = x^{2/3}$ in statement (2) of Theorem 8.2; our goal is to show that $f(x) \ge g(x)$ on $[1, \infty)$. Note that $\sqrt[3]{x^2 - 0.5} < \sqrt[3]{x^2} = x^{2/3}$ on $[1, \infty)$, which implies that

$$\frac{1}{\sqrt[3]{x^2 - 0.5}} > \frac{1}{x^{2/3}} \text{ on } [1, \infty).$$

By statement (2) of Theorem 8.2, $\int_1^\infty \frac{1}{\sqrt[3]{x^2 - 0.5}} \, dx$ diverges.

*Related Exercises 77, 79* ◄

We close this section with two important points. First, for improper integrals of the form $\int_a^\infty g(x) \, dx$, where $g$ satisfies the conditions of Theorem 8.2, one need not find a function $f$ that satisfies the theorem's conditions on the entire interval $[a, \infty)$. Rather, it suffices to find a function $f$ such that

$$f(x) \ge g(x) \text{ on } [c, \infty), \text{ where } c > a, \text{ and } \int_c^\infty f(x) \, dx \text{ converges.}$$

These two conditions are enough to conclude that $\int_a^\infty g(x) \, dx$ converges because $g(x)$ is bounded on $[a, c]$ and therefore $\int_a^c g(x) \, dx$ is finite. For example, even though $e^{-x}$ is not greater than or equal to $e^{-x^2}$ on $[0, \infty)$ (Figure 8.32), we can use the result of Example 7a to conclude that $\int_0^\infty e^{-x^2} \, dx$ (with lower limit of 0) converges—adding the finite number $\int_0^1 e^{-x^2} \, dx$ to the convergent integral $\int_1^\infty e^{-x^2} \, dx$ does not affect whether it converges. Similar principles can be applied to the other forms of improper integrals encountered in this section.

The second point is that we never discovered the value of the integral in Example 7a. In future chapters, we develop methods that provide accurate approximations to the value of a convergent integral, and in Section 16.3, we show that $\int_0^\infty e^{-x^2} \, dx = \frac{\sqrt{\pi}}{2}$.

# SECTION 8.9 EXERCISES

## Getting Started

**1.** What are the two general ways in which an improper integral may occur?

**2.** Evaluate $\int_2^\infty \frac{dx}{x^3}$ after writing the expression as a limit.

**3.** Rewrite $\int_2^\infty \frac{dx}{x^{1/5}}$ as a limit and then show that the integral diverges.

**4.** Evaluate $\int_0^1 \frac{dx}{x^{1/5}}$ after writing the integral as a limit.

**5.** Write $\lim_{a \to -\infty} \int_a^0 f(x) \, dx + \lim_{b \to \infty} \int_0^b f(x) \, dx$ as an improper integral.

**6.** For what values of $p$ does $\int_1^\infty x^{-p} \, dx$ converge?

## Practice Exercises

**7–58. Improper integrals** *Evaluate the following integrals or state that they diverge.*

**7.** $\int_3^\infty \frac{dx}{x^2}$

**8.** $\int_2^\infty \frac{dx}{x}$

**9.** $\int_2^\infty \frac{dx}{\sqrt{x}}$

**10.** $\int_0^\infty e^{-2x} \, dx$

**11.** $\int_0^\infty e^{-ax} \, dx, a > 0$

**12.** $\int_{-\infty}^{-1} \frac{dx}{\sqrt[3]{x}}$

**13.** $\int_0^\infty \cos x \, dx$

**14.** $\int_{-\infty}^{-1} \frac{dx}{x^3}$

**15.** $\int_{-\infty}^\infty \frac{dx}{x^2 + 100}$

**16.** $\int_{-\infty}^\infty \frac{dx}{x^2 + a^2}, a > 0$

**17.** $\int_7^\infty \dfrac{dx}{(x+1)^{1/3}}$

**18.** $\int_2^\infty \dfrac{dx}{(x+2)^2}$

**19.** $\int_1^\infty \dfrac{3x^2+1}{x^3+x}\,dx$

**20.** $\int_1^\infty 2^{-x}\,dx$

**21.** $\int_2^\infty \dfrac{\cos(\pi/x)}{x^2}\,dx$

**22.** $\int_{-\infty}^{-2} \dfrac{1}{z^2}\sin\dfrac{\pi}{z}\,dz$

**23.** $\int_0^\infty \dfrac{e^u}{e^{2u}+1}\,du$

**24.** $\int_{-\infty}^a \sqrt{e^x}\,dx,\ a>0$

**25.** $\int_{-\infty}^\infty \dfrac{e^{3x}}{1+e^{6x}}\,dx$

**26.** $\int_{-\infty}^\infty \dfrac{x}{(x^2+1)^2}\,dx$

**27.** $\int_{-\infty}^\infty xe^{-x^2}\,dx$

**28.** $\int_1^\infty \dfrac{\tan^{-1}s}{s^2+1}\,ds$

**29.** $\int_{-\infty}^\infty \dfrac{(\tan^{-1}t)^2}{t^2+1}\,dt$

**30.** $\int_{-\infty}^0 e^x\,dx$

**31.** $\int_1^\infty \dfrac{dv}{v(v+1)}$

**32.** $\int_1^\infty \dfrac{dx}{x^2(x-1)}$

**33.** $\int_2^\infty \dfrac{dy}{y\ln y}$

**34.** $\int_{-\infty}^{-4/\pi} \dfrac{1}{x^2}\sec^2\!\Big(\dfrac{1}{x}\Big)\,dx$

**35.** $\int_{-\infty}^0 \dfrac{dx}{\sqrt[3]{2-x}}$

**36.** $\int_{e^2}^\infty \dfrac{dx}{x\ln^p x},\ p>1$

**37.** $\int_0^8 \dfrac{dx}{\sqrt[3]{x}}$

**38.** $\int_1^2 \dfrac{dx}{\sqrt{x-1}}$

**39.** $\int_0^{\pi/2} \tan\theta\,d\theta$

**40.** $\int_{-3}^1 \dfrac{dx}{(2x+6)^{2/3}}$

**41.** $\int_0^{\pi/2} \sec x\tan x\,dx$

**42.** $\int_3^4 \dfrac{dz}{(z-3)^{3/2}}$

**43.** $\int_0^1 \dfrac{e^{\sqrt{x}}}{\sqrt{x}}\,dx$

**44.** $\int_0^{\ln 3} \dfrac{e^y}{(e^y-1)^{2/3}}\,dy$

**45.** $\int_0^1 \dfrac{x^3}{x^4-1}\,dx$

**46.** $\int_1^\infty \dfrac{dx}{\sqrt[3]{x-1}}$

**47.** $\int_0^{10} \dfrac{dx}{\sqrt[4]{10-x}}$

**48.** $\int_1^{11} \dfrac{dx}{(x-3)^{2/3}}$

**49.** $\int_0^2 \dfrac{dx}{(x-1)^2}$

**50.** $\int_0^9 \dfrac{dx}{(x-1)^{1/3}}$

**51.** $\int_{-2}^2 \dfrac{dp}{\sqrt{4-p^2}}$

**52.** $\int_0^\infty xe^{-x}\,dx$

**53.** $\int_0^1 \ln x\,dx$

**54.** $\int_1^\infty \dfrac{\ln x}{x^2}\,dx$

**55.** $\int_0^{\ln 2} \dfrac{e^x}{\sqrt{e^{2x}-1}}\,dx$

**56.** $\int_0^1 \dfrac{dx}{x+\sqrt{x}}\,dx$

**57.** $\int_{-\infty}^\infty e^{-|x|}\,dx$

**58.** $\int_{-\infty}^\infty \dfrac{dx}{x^2+2x+5}$

**59. Perpetual annuity** Imagine that today you deposit $B in a savings account that earns interest at a rate of p% per year compounded continuously (see Section 7.4). The goal is to draw an income of $I per year from the account forever. The amount of money that must be deposited is $B = I\int_0^\infty e^{-rt}\,dt$, where $r = p/100$. Suppose you find an account that earns 12% interest annually and you wish to have an income from the account of $5000 per year. How much must you deposit today?

**60. Draining a pool** Water is drained from a swimming pool at a rate given by $R(t) = 100e^{-0.05t}$ gal/hr. If the drain is left open indefinitely, how much water drains from the pool?

**61. Bioavailability** When a drug is given intravenously, the concentration of the drug in the blood is $C_i(t) = 250e^{-0.08t}$, for $t \ge 0$. When the same drug is given orally, the concentration of the drug in the blood is $C_o(t) = 200(e^{-0.08t} - e^{-1.8t})$, for $t \ge 0$. Compute the bioavailability of the drug.

**62. Electronic chips** Suppose the probability that a particular computer chip fails after $a$ hours of operation is $0.00005\int_a^\infty e^{-0.00005t}\,dt$.

   **a.** Find the probability that the computer chip fails after 15,000 hr of operation.

   **b.** Of the chips that are still operating after 15,000 hr, what fraction of these will operate for at least another 15,000 hr?

   **c.** Evaluate $0.00005\int_0^\infty e^{-0.00005t}\,dt$ and interpret its meaning.

**63. Average lifetime** The average time until a computer chip fails (see Exercise 62) is $0.00005\int_0^\infty te^{-0.00005t}\,dt$. Find this value.

**64. Maximum distance** An object moves on a line with velocity

$$v(t) = \frac{10}{(t+1)^2}\text{ mi/hr, for }t \ge 0\text{, where }t\text{ is measured in hours.}$$

What is the maximum distance the object can travel?

**65–76. Volumes** *Find the volume of the described solid of revolution or state that it does not exist.*

**65.** The region bounded by $f(x) = x^{-2}$ and the $x$-axis on the interval $[1, \infty)$ is revolved about the $x$-axis.

**66.** The region bounded by $f(x) = (x^2+1)^{-1/2}$ and the $x$-axis on the interval $[2, \infty)$ is revolved about the $x$-axis.

**67.** The region bounded by $f(x) = \sqrt{\dfrac{x+1}{x^3}}$ and the $x$-axis on the interval $[1, \infty)$ is revolved about the $x$-axis.

**68.** The region bounded by $f(x) = (x+1)^{-3}$ and the $x$-axis on the interval $[0, \infty)$ is revolved about the $y$-axis.

**69.** The region bounded by $f(x) = \dfrac{1}{\sqrt{x}\ln x}$ and the $x$-axis on the interval $[2, \infty)$ is revolved about the $x$-axis.

**70.** The region bounded by $f(x) = \dfrac{\sqrt{x}}{\sqrt[3]{x^2+1}}$ and the $x$-axis on the interval $[0, \infty)$ is revolved about the $x$-axis.

**71.** The region bounded by $f(x) = (x-1)^{-1/4}$ and the $x$-axis on the interval $(1, 2]$ is revolved about the $x$-axis.

**72.** The region bounded by $f(x) = (x+1)^{-3/2}$ and the $x$-axis on the interval $(-1, 1]$ is revolved about the line $y = -1$.

**73.** The region bounded by $f(x) = \tan x$ and the $x$-axis on the interval $[0, \pi/2)$ is revolved about the $x$-axis.

**74.** The region bounded by $f(x) = -\ln x$ and the $x$-axis on the interval $(0, 1]$ is revolved about the $x$-axis.

**75.** The region bounded by $f(x) = (4-x)^{-1/3}$ and the $x$-axis on the interval $[0, 4)$ is revolved about the $y$-axis.

**76.** The region bounded by $f(x) = (x^2-1)^{-1/4}$ and the $x$-axis on the interval $(1, 2]$ is revolved about the $y$-axis.

**77–86. Comparison Test** *Determine whether the following integrals converge or diverge.*

**77.** $\displaystyle\int_1^\infty \frac{dx}{x^3 + 1}$

**78.** $\displaystyle\int_0^\infty \frac{dx}{e^x + x + 1}$

**79.** $\displaystyle\int_3^\infty \frac{dx}{\ln x}$ *(Hint: $\ln x \le x$.)*

**80.** $\displaystyle\int_2^\infty \frac{x^3}{x^4 - x - 1}\,dx$

**81.** $\displaystyle\int_1^\infty \frac{\sin^2 x}{x^2}\,dx$

**82.** $\displaystyle\int_1^\infty \frac{1}{e^x(1 + x^2)}\,dx$

**83.** $\displaystyle\int_1^\infty \frac{2 + \cos x}{\sqrt{x}}\,dx$

**84.** $\displaystyle\int_1^\infty \frac{2 + \cos x}{x^2}\,dx$

**85.** $\displaystyle\int_0^1 \frac{dx}{\sqrt{x^{1/3} + x}}$

**86.** $\displaystyle\int_0^1 \frac{\sin x + 1}{x^5}\,dx$

**87. Explain why or why not** Determine whether the following statements are true and give an explanation or counterexample.

    **a.** If $f$ is continuous and $0 < f(x) < g(x)$ on the interval $[0, \infty)$, and $\displaystyle\int_0^\infty g(x)\,dx = M < \infty$, then $\displaystyle\int_0^\infty f(x)\,dx$ exists.

    **b.** If $\displaystyle\lim_{x\to\infty} f(x) = 1$, then $\displaystyle\int_0^\infty f(x)\,dx$ exists.

    **c.** If $\displaystyle\int_0^1 x^{-p}\,dx$ exists, then $\displaystyle\int_0^1 x^{-q}\,dx$ exists, where $q > p$.

    **d.** If $\displaystyle\int_1^\infty x^{-p}\,dx$ exists, then $\displaystyle\int_1^\infty x^{-q}\,dx$ exists, where $q > p$.

    **e.** $\displaystyle\int_1^\infty \frac{dx}{x^{3p+2}}$ exists, for $p > -\frac{1}{3}$.

**88. Incorrect calculation**

    **a.** What is wrong with this calculation?

$$\int_{-1}^1 \frac{dx}{x} = \ln|x|\,\Big|_{-1}^1 = \ln 1 - \ln 1 = 0$$

    **b.** Evaluate $\displaystyle\int_{-1}^1 \frac{dx}{x}$ or show that the integral does not exist.

**89. Area between curves** Let $R$ be the region bounded by the graphs of $y = e^{-ax}$ and $y = e^{-bx}$, for $x \ge 0$, where $a > b > 0$. Find the area of $R$ in terms of $a$ and $b$.

**90. Area between curves** Let $R$ be the region bounded by the graphs of $y = x^{-p}$ and $y = x^{-q}$, for $x \ge 1$, where $q > p > 1$. Find the area of $R$ in terms of $p$ and $q$.

**T 91. Regions bounded by exponentials** Let $a > 0$ and let $R$ be the region bounded by the graph of $y = e^{-ax}$ and the $x$-axis on the interval $[b, \infty)$.

    **a.** Find $A(a, b)$, the area of $R$ as a function of $a$ and $b$.

    **b.** Find the relationship $b = g(a)$ such that $A(a, b) = 2$.

    **c.** What is the minimum value of $b$ (call it $b^*$) such that when $b > b^*$, $A(a, b) = 2$ for some value of $a > 0$?

## Explorations and Challenges

**92–93. Improper integrals with infinite intervals and unbounded integrands** *For a real number a, suppose* $\displaystyle\lim_{x\to a^+} f(x) = -\infty$ *or* $\displaystyle\lim_{x\to a^+} f(x) = \infty$. *In these cases, the integral* $\displaystyle\int_a^\infty f(x)\,dx$ *is improper for two reasons:* $\infty$ *appears in the upper limit and $f$ is unbounded at* $x = a$. *It can be shown that* $\displaystyle\int_a^\infty f(x)\,dx = \int_a^c f(x)\,dx + \int_c^\infty f(x)\,dx$, *for any $c > a$. Use this result to evaluate the following improper integrals.*

**92.** $\displaystyle\int_0^\infty x^{-x}(\ln x + 1)\,dx$

**93.** $\displaystyle\int_1^\infty \frac{dx}{x\sqrt{x - 1}}$

**94. The family** $f(x) = \dfrac{1}{x^p}$ **revisited** Consider the family of functions $f(x) = \dfrac{1}{x^p}$, where $p$ is a real number. For what values of $p$ does the integral $\displaystyle\int_0^1 f(x)\,dx$ exist? What is its value?

**T 95–98. Numerical methods** *Use numerical methods or a calculator to approximate the following integrals as closely as possible. The exact value of each integral is given.*

**95.** $\displaystyle\int_0^{\pi/2} \ln(\sin x)\,dx = \int_0^{\pi/2} \ln(\cos x)\,dx = -\frac{\pi \ln 2}{2}$

**96.** $\displaystyle\int_0^\infty \frac{\sin^2 x}{x^2}\,dx = \frac{\pi}{2}$

**97.** $\displaystyle\int_0^\infty \ln\left(\frac{e^x + 1}{e^x - 1}\right)dx = \frac{\pi^2}{4}$

**98.** $\displaystyle\int_0^1 \frac{\ln x}{1 + x}\,dx = -\frac{\pi^2}{12}$

**99. Decaying oscillations** Let $a > 0$ and $b$ be real numbers. Use integration to confirm the following identities. (See Exercise 73 of Section 8.2.)

    **a.** $\displaystyle\int_0^\infty e^{-ax} \cos bx\,dx = \frac{a}{a^2 + b^2}$

    **b.** $\displaystyle\int_0^\infty e^{-ax} \sin bx\,dx = \frac{b}{a^2 + b^2}$

**100. The Eiffel Tower property** Let $R$ be the region between the curves $y = e^{-cx}$ and $y = -e^{-cx}$ on the interval $[a, \infty)$, where $a \ge 0$ and $c > 0$. The center of mass of $R$ is located at $(\bar{x}, 0)$, where $\bar{x} = \dfrac{\int_a^\infty x e^{-cx}\,dx}{\int_a^\infty e^{-cx}\,dx}$. (The profile of the Eiffel Tower is modeled by the two exponential curves; see the Guided Project *The exponential Eiffel Tower*.)

    **a.** For $a = 0$ and $c = 2$, sketch the curves that define $R$ and find the center of mass of $R$. Indicate the location of the center of mass.

    **b.** With $a = 0$ and $c = 2$, find equations of the lines tangent to the curves at the points corresponding to $x = 0$.

    **c.** Show that the tangent lines intersect at the center of mass.

    **d.** Show that this same property holds for any $a \ge 0$ and any $c > 0$; that is, the tangent lines to the curves $y = \pm e^{-cx}$ at $x = a$ intersect at the center of mass of $R$.

*(Source: P. Weidman and I. Pinelis, Comptes Rendu, Mechanique, 332, 571–584, 2004)*

**101. Many methods needed** Show that $\displaystyle\int_0^\infty \frac{\sqrt{x}\,\ln x}{(1 + x)^2}\,dx = \pi$ in the following steps.

    **a.** Integrate by parts with $u = \sqrt{x}\,\ln x$.

    **b.** Change variables by letting $y = 1/x$.

    **c.** Show that $\displaystyle\int_0^1 \frac{\ln x}{\sqrt{x}(1 + x)}\,dx = -\int_1^\infty \frac{\ln x}{\sqrt{x}(1 + x)}\,dx$ (and that both integrals converge). Conclude that $\displaystyle\int_0^\infty \frac{\ln x}{\sqrt{x}(1 + x)}\,dx = 0$.

    **d.** Evaluate the remaining integral using the change of variables $z = \sqrt{x}$.

*(Source: Mathematics Magazine 59, 1, Feb 1986)*

**102–106. Laplace transforms** *A powerful tool in solving problems in engineering and physics is the Laplace transform. Given a function* $f(t)$, *the Laplace transform is a new function* $F(s)$ *defined by*

$$F(s) = \int_0^\infty e^{-st} f(t)\, dt,$$

*where we assume s is a positive real number. For example, to find the Laplace transform of* $f(t) = e^{-t}$, *the following improper integral is evaluated using integration by parts:*

$$F(s) = \int_0^\infty e^{-st} e^{-t}\, dt = \int_0^\infty e^{-(s+1)t}\, dt = \frac{1}{s+1}.$$

*Verify the following Laplace transforms, where a is a real number.*

**102.** $f(t) = 1 \to F(s) = \dfrac{1}{s}$    **103.** $f(t) = e^{at} \to F(s) = \dfrac{1}{s-a}$

**104.** $f(t) = t \to F(s) = \dfrac{1}{s^2}$

**105.** $f(t) = \sin at \to F(s) = \dfrac{a}{s^2 + a^2}$

**106.** $f(t) = \cos at \to F(s) = \dfrac{s}{s^2 + a^2}$

**107. Improper integrals** Evaluate the following improper integrals (Putnam Exam, 1939).

**a.** $\displaystyle\int_1^3 \frac{dx}{\sqrt{(x-1)(3-x)}}$    **b.** $\displaystyle\int_1^\infty \frac{dx}{e^{x+1} + e^{3-x}}$

**108. Draining a tank** Water is drained from a 3000-gal tank at a rate that starts at 100 gal/hr and decreases continuously by 5%/hr. If the drain is left open indefinitely, how much water drains from the tank? Can a full tank be emptied at this rate?

**109. Escape velocity and black holes** The work required to launch an object from the surface of Earth to outer space is given by $W = \int_R^\infty F(x)\, dx$, where $R = 6370$ km is the approximate radius of Earth, $F(x) = \dfrac{GMm}{x^2}$ is the gravitational force between Earth and the object, $G$ is the gravitational constant, $M$ is the mass of Earth, $m$ is the mass of the object, and $GM = 4 \times 10^{14}\,\text{m}^3/\text{s}^2$.

**a.** Find the work required to launch an object in terms of $m$.
**b.** What escape velocity $v_e$ is required to give the object a kinetic energy $\frac{1}{2} m v_e^2$ equal to $W$?
**c.** The French scientist Laplace anticipated the existence of black holes in the 18th century with the following argument: If a body has an escape velocity that equals or exceeds the speed

of light, $c = 300{,}000$ km/s, then light cannot escape the body and it cannot be seen. Show that such a body has a radius $R \le 2GM/c^2$. For Earth to be a black hole, what would its radius need to be?

**110. Adding a proton to a nucleus** The nucleus of an atom is positively charged because it consists of positively charged protons and uncharged neutrons. To bring a free proton toward a nucleus, a repulsive force $F(r) = kqQ/r^2$ must be overcome, where $q = 1.6 \times 10^{-19}$ C (coulombs) is the charge on the proton, $k = 9 \times 10^9\,\text{N} \cdot \text{m}^2/\text{C}^2$, $Q$ is the charge on the nucleus, and $r$ is the distance between the center of the nucleus and the proton. Find the work required to bring a free proton (assumed to be a point mass) from a large distance $(r \to \infty)$ to the edge of a nucleus that has a charge $Q = 50q$ and a radius of $6 \times 10^{-11}$ m.

**111. Gamma function** The gamma function is defined by $\Gamma(p) = \int_0^\infty x^{p-1} e^{-x}\, dx$, for $p$ not equal to zero or a negative integer.

**a.** Use the reduction formula

$$\int_0^\infty x^p e^{-x}\, dx = p \int_0^\infty x^{p-1} e^{-x}\, dx \quad \text{for } p = 1, 2, 3, \ldots$$

to show that $\Gamma(p+1) = p!$ ($p$ factorial).

**b.** Use the substitution $x = u^2$ and the fact that $\displaystyle\int_0^\infty e^{-u^2}\, du = \frac{\sqrt{\pi}}{2}$ to show that $\Gamma\!\left(\dfrac{1}{2}\right) = \sqrt{\pi}$.

**T 112. Gaussians** An important function in statistics is the Gaussian (or normal distribution, or bell-shaped curve), $f(x) = e^{-ax^2}$.

**a.** Graph the Gaussian for $a = 0.5$, 1, and 2.
**b.** Given that $\displaystyle\int_{-\infty}^\infty e^{-ax^2}\, dx = \sqrt{\frac{\pi}{a}}$, compute the area under the curves in part (a).
**c.** Complete the square to evaluate $\displaystyle\int_{-\infty}^\infty e^{-(ax^2 + bx + c)}\, dx$, where $a > 0$, $b$, and $c$ are real numbers.

**QUICK CHECK ANSWERS**

**1.** The integral diverges. $\displaystyle\lim_{b \to \infty} \int_1^b (1 + x^{-1})\, dx =$

$\displaystyle\lim_{b \to \infty} (x + \ln x)\Big|_1^b$ does not exist.    **2.** $\frac{1}{3}$    **3.** $c$ must approach 0 through values in the interval of integration $(0, 1)$. Therefore, $c \to 0^+$.    **4.** $\displaystyle\int_1^\infty e^{-x}\, dx = \lim_{b \to \infty}\left(-e^{-x}\Big|_1^b\right) = e^{-1}$ ◄

---

## CHAPTER 8   REVIEW EXERCISES

**1.** **Explain why or why not** Determine whether the following statements are true and give an explanation or counterexample.

**a.** The integral $\displaystyle\int x^2 e^{2x}\, dx$ can be evaluated using integration by parts.

**b.** To evaluate the integral $\displaystyle\int \frac{dx}{\sqrt{x^2 - 100}}$ analytically, it is best to use partial fractions.

**c.** One computer algebra system produces $\displaystyle\int 2 \sin x \cos x\, dx = \sin^2 x$. Another computer algebra system

produces $\displaystyle\int 2 \sin x \cos x\, dx = -\cos^2 x$. One computer algebra system is wrong (apart from a missing constant of integration).

**d.** $\displaystyle\int 2 \sin x \cos x\, dx = -\frac{1}{2}\cos 2x + C$.

**e.** The best approach to evaluating $\displaystyle\int \frac{x^3 + 1}{3x^2}\, dx$ is to use the change of variables $u = x^3 + 1$.

**2–74. Integration techniques** *Use the methods introduced in Sections 8.1 through 8.5 to evaluate the following integrals.*

**2.** $\displaystyle\int \cos\left(\frac{x}{2} + \frac{\pi}{3}\right) dx$

**3.** $\displaystyle\int \frac{3x}{\sqrt{x+4}} \, dx$

**4.** $\displaystyle\int_{-1}^{\ln 2} \frac{3t}{e^t} \, dt$

**5.** $\displaystyle\int \frac{x}{2\sqrt{x+2}} \, dx$

**6.** $\displaystyle\int \frac{2 - \sin 2\theta}{\cos^2 2\theta} \, d\theta$

**7.** $\displaystyle\int_{-2}^{1} \frac{3}{x^2 + 4x + 13} \, dx$

**8.** $\displaystyle\int_{\pi}^{2\pi} \cot \frac{x}{3} \, dx$

**9.** $\displaystyle\int_{0}^{\pi/4} \cos^5 2x \sin^2 2x \, dx$

**10.** $\displaystyle\int \frac{x^3 + 3x^2 + 1}{x^3 + 1} \, dx$

**11.** $\displaystyle\int \frac{\sqrt{t-1}}{2t} \, dt$

**12.** $\displaystyle\int \frac{8x + 5}{2x^2 + 3x + 1} \, dx$

**13.** $\displaystyle\int_{0}^{\pi} e^{3x} \sin 6x \, dx$

**14.** $\displaystyle\int_{2}^{3} (6w - 10)e^{3w} \, dw$

**15.** $\displaystyle\int_{1}^{2} \frac{3x^5 + 48x^3 + 3x^2 + 16}{x^3 + 16x} \, dx$

**16.** $\displaystyle\int \frac{x^6 + 2x^4 + x^2 + 1}{(x^2 + 1)^2} \, dx$

**17.** $\displaystyle\int \frac{2x^2 + 7x + 4}{x^3 + 2x^2 + 2x} \, dx$

**18.** $\displaystyle\int_{0}^{\sqrt{2}} \frac{x+1}{3x^2 + 6} \, dx$

**19.** $\displaystyle\int \frac{x^2 + 4x + 7}{(x+3)^3} \, dx$

**20.** $\displaystyle\int \frac{\sqrt{1-x^2}}{x} \, dx$

**21.** $\displaystyle\int_{\sqrt{2}}^{2} \frac{\sqrt{x^2-1}}{x} \, dx$

**22.** $\displaystyle\int \tan^3 5\theta \, d\theta$

**23.** $\displaystyle\int \frac{\sin^4 t}{\cos^6 t} \, dt$

**24.** $\displaystyle\int \frac{dx}{\sqrt{18x - x^2}}$

**25.** $\displaystyle\int_{-3/2}^{-1} \frac{dx}{4x^2 + 12x + 10}$

**26.** $\displaystyle\int \tan 10x \, dx$

**27.** $\displaystyle\int \frac{dw}{(w+1)^2 \sqrt{w^2 + 2w - 8}}$

**28.** $\displaystyle\int \frac{\sin^3 x}{\cos^5 x} \, dx$

**29.** $\displaystyle\int \frac{\cos^4 x}{\sin^6 x} \, dx$

**30.** $\displaystyle\int x \tan^{-1} 7x \, dx$

**31.** $\displaystyle\int x \sinh 2x \, dx$

**32.** $\displaystyle\int \csc^2 6x \cot 6x \, dx$

**33.** $\displaystyle\int \tan^3 3\theta \sec^3 3\theta \, d\theta$

**34.** $\displaystyle\int \frac{w^3}{\sqrt{36 - 9w^2}} \, dw$

**35.** $\displaystyle\int \frac{x^3}{\sqrt{4x^2 + 16}} \, dx$

**36.** $\displaystyle\int_{-1/2}^{1/2} \frac{u^2 + 1}{u^2 - 1} \, du$

**37.** $\displaystyle\int \frac{3x^3 + 4x^2 + 6x}{(x+1)^2(x^2+4)} \, dx$

**38.** $\displaystyle\int_{\pi/4}^{\pi/2} x \csc^2 x \, dx$

**39.** $\displaystyle\int \frac{dt}{2 + e^t}$

**40.** $\displaystyle\int \frac{x^2 - 4}{x + 4} \, dx$

**41.** $\displaystyle\int \frac{d\theta}{1 + \cos 4\theta}$

**42.** $\displaystyle\int x^2 \cos x \, dx$

**43.** $\displaystyle\int e^x \sin x \, dx$

**44.** $\displaystyle\int_{1}^{e} x^2 \ln x \, dx$

**45.** $\displaystyle\int \frac{2x^3 + 5x^2 + 13x + 9}{x^2(x^2 + 4x + 9)} \, dx$

**46.** $\displaystyle\int \frac{x^3 + 4x^2 + 12x + 4}{(x^2 + 4x + 10)^2} \, dx$

**47.** $\displaystyle\int \cos^2 4\theta \, d\theta$

**48.** $\displaystyle\int \sin 3x \cos^6 3x \, dx$

**49.** $\displaystyle\int \sec^{49} 2z \tan 2z \, dz$

**50.** $\displaystyle\int_{0}^{\pi/6} \cos^4 3x \, dx$

**51.** $\displaystyle\int_{0}^{\pi/4} \sin^5 4\theta \, d\theta$

**52.** $\displaystyle\int \tan^4 2u \, du$

**53.** $\displaystyle\int \frac{dx}{\sqrt{x} + \sqrt[3]{x}}$ (*Hint:* Let $u = x^6$.)

**54.** $\displaystyle\int \frac{dx}{\sqrt{9x^2 - 25}}, \, x > \frac{5}{3}$

**55.** $\displaystyle\int \frac{dy}{y^2 \sqrt{18 - 2y^2}}$

**56.** $\displaystyle\int_{0}^{\sqrt{3}/2} \frac{x^2}{(1 - x^2)^{3/2}} \, dx$

**57.** $\displaystyle\int_{0}^{\sqrt{3}/2} \frac{4}{9 + 4x^2} \, dx$

**58.** $\displaystyle\int \frac{(1 - u^2)^{5/2}}{u^8} \, du$

**59.** $\displaystyle\int \operatorname{sech}^2 x \sinh x \, dx$

**60.** $\displaystyle\int x^2 \cosh x \, dx$

**61.** $\displaystyle\int_{0}^{\ln(\sqrt{3}+2)} \frac{\cosh x}{\sqrt{4 - \sinh^2 x}} \, dx$

**62.** $\displaystyle\int \sinh^{-1} x \, dx$

**63.** $\displaystyle\int \frac{dx}{x^2 - 2x - 15}$

**64.** $\displaystyle\int \frac{dx}{x^3 - 2x^2}$

**65.** $\displaystyle\int_{0}^{1} \frac{dy}{(y+1)(y^2+1)}$

**66.** $\displaystyle\int_{0}^{\infty} \frac{6x}{1 + x^6} \, dx$

**67.** $\displaystyle\int_{0}^{2} \frac{dx}{\sqrt[3]{|x-1|}}$

**68.** $\displaystyle\int_{-1}^{1} \frac{dx}{x^2 + 2x + 5}$

**69.** $\displaystyle\int \frac{dx}{x^2 - x - 2}$

**70.** $\displaystyle\int \frac{3x^2 + x - 3}{x^2 - 1} \, dx$

**71.** $\displaystyle\int \frac{2x^2 - 4x}{x^2 - 4} \, dx$

**72.** $\displaystyle\int_{1/12}^{1/4} \frac{dx}{\sqrt{x}(1 + 4x)}$

**73.** $\displaystyle\int \frac{e^{2t}}{(1 + e^{4t})^{3/2}} \, dt$

**74.** $\displaystyle\int \frac{dx}{\sqrt{\sqrt{\sqrt{1 + \sqrt{x}}}}}$

**75.** Evaluate the integral in part (a) and then use this result to evaluate the integral in part (b).

  **a.** $\displaystyle\int e^x \sec e^x \tan e^x \, dx$    **b.** $\displaystyle\int e^{2x} \sec e^x \tan e^x \, dx$

**76–81. Table of integrals** *Use a table of integrals to evaluate the following integrals.*

**76.** $\displaystyle\int x(2x + 3)^5 \, dx$

**77.** $\displaystyle\int \frac{dx}{x\sqrt{4x - 6}}$

**78.** $\displaystyle\int_0^{\pi/2} \frac{d\theta}{1 + \sin 2\theta}$

**79.** $\displaystyle\int \sec^5 x \, dx$

**80.** $\displaystyle\int_0^1 \frac{e^x}{e^{2x}\sqrt{16 - e^{2x}}} \, dx$

**81.** $\displaystyle\int_0^{\pi/2} \cos x \ln^3(1 + \sin x) \, dx$

**82–88. Improper integrals** *Evaluate the following integrals or show that the integral diverges.*

**82.** $\displaystyle\int_{-\infty}^{-1} \frac{dx}{(x - 1)^4}$

**83.** $\displaystyle\int_0^{\infty} xe^{-x} \, dx$

**84.** $\displaystyle\int_0^{\pi} \sec^2 x \, dx$

**85.** $\displaystyle\int_0^3 \frac{dx}{\sqrt{9 - x^2}}$

**86.** $\displaystyle\int_{-\infty}^{\infty} \frac{x^3}{1 + x^8} \, dx$

**87.** $\displaystyle\int_1^{\infty} \frac{dx}{x(x - 1)^{1/3}}$ (*Hint:* First write the integral as the sum of two improper integrals.)

**88.** $\displaystyle\int_{-\infty}^{\infty} \frac{2}{x^2 + 2x + 2} \, dx$

**89–91. Comparison Test** *Determine whether the following integrals converge or diverge.*

**89.** $\displaystyle\int_1^{\infty} \frac{dx}{x^5 + x^4 + x^3 + 1}$

**90.** $\displaystyle\int_0^1 \frac{dx}{\sqrt{x + \sin x}}$

**91.** $\displaystyle\int_3^{\infty} \frac{x^3}{\sqrt{x^7 - 1}} \, dx$

**92. Integral with a parameter** For what values of $p$ does the integral $\displaystyle\int_2^{\infty} \frac{dx}{x \ln^p x}$ converge and what is its value (in terms of $p$)?

**T 93–94. Approximations** *Use a computer algebra system to approximate the value of the following integrals.*

**93.** $\displaystyle\int_{-1}^1 e^{-2x^2} \, dx$

**94.** $\displaystyle\int_1^{\sqrt{e}} x^3 \ln^3 x \, dx$

**T 95–98. Numerical integration** *Estimate the following integrals using the Midpoint Rule $M(n)$, the Trapezoidal Rule $T(n)$, and Simpson's Rule $S(n)$ for the given values of n.*

**95.** $\displaystyle\int_0^8 f(x) \, dx$; $n = 4$ (see figure)

**96.** $\displaystyle\int_1^3 \frac{dx}{x^3 + x + 1}$; $n = 4$

**97.** $\displaystyle\int_0^1 \tan x^2 \, dx$; $n = 40$

**98.** $\displaystyle\int_1^8 e^{\sin x} \, dx$; $n = 60$

**99. Improper integrals by numerical methods** Use the Trapezoid Rule (Section 8.8) to approximate $\int_0^R e^{-x^2} \, dx$ with $R = 2$, 4, and 8. For each value of $R$, take $n = 4$, 8, 16, and 32, and compare approximations with successive values of $n$. Use these approximations to approximate $I = \int_0^{\infty} e^{-x^2} \, dx$.

**100. Comparing areas** Show that the area of the region bounded by the graph of $y = ae^{-ax}$ and the $x$-axis on the interval $[0, \infty)$ is the same for all values of $a > 0$.

**101. Comparing volumes** Let $R$ be the region bounded by the graph of $y = \sin x$ and the $x$-axis on the interval $[0, \pi]$. Which is greater, the volume of the solid generated when $R$ is revolved about the $x$-axis or about the $y$-axis?

**102–105. Volumes** *The region R is bounded by the curve $y = \ln x$ and the $x$-axis on the interval $[1, e]$. Find the volume of the solid that is generated when R is revolved in the following ways.*

**102.** About the $y$-axis

**103.** About the $x$-axis

**104.** About the line $y = 1$

**105.** About the line $x = 1$

**106. Arc length** Find the length of the curve

$$y = \frac{x}{2}\sqrt{3 - x^2} + \frac{3}{2}\sin^{-1}\frac{x}{\sqrt{3}} \text{ from } x = 0 \text{ to } x = 1.$$

**T 107. Zero log integral** It is evident from the graph of $y = \ln x$ that for every real number $a$ with $0 < a < 1$, there is a unique real number $b = g(a)$ with $b > 1$, such that $\int_a^b \ln x \, dx = 0$ (the net area bounded by the graph of $y = \ln x$ on $[a, b]$ is 0).

  **a.** Approximate $b = g\left(\dfrac{1}{2}\right)$.

  **b.** Approximate $b = g\left(\dfrac{1}{3}\right)$.

  **c.** Find the equation satisfied by all pairs of numbers $(a, b)$ such that $b = g(a)$.

  **d.** Is $g$ an increasing or decreasing function of $a$? Explain.

**108. Arc length** Find the length of the curve $y = \ln x$ on the interval $[1, e^2]$.

**109. Average velocity** Find the average velocity of a projectile whose velocity over the interval $0 \le t \le \pi$ is given by $v(t) = 10 \sin 3t$.

**110. Comparing distances** Suppose two cars started at the same time and place ($t = 0$ and $s = 0$). The velocity of car A (in mi/hr) is given by $u(t) = 40/(t + 1)$ and the velocity of car B (in mi/hr) is given by $v(t) = 40e^{-t/2}$.

  **a.** After $t = 2$ hr, which car has traveled farther?

  **b.** After $t = 3$ hr, which car has traveled farther?

  **c.** If allowed to travel indefinitely ($t \to \infty$), which car will travel a finite distance?

**111. Traffic flow** When data from a traffic study are fitted to a curve, the flow rate of cars past a point on a highway is approximated by $R(t) = 800te^{-t/2}$ cars/hr. How many cars pass the measuring site during the time interval $0 \le t \le 4$?

**T 112. Comparing integrals** Graph the functions $f(x) = \pm\dfrac{1}{x^2}$, $g(x) = \dfrac{\cos x}{x^2}$, and $h(x) = \dfrac{\cos^2 x}{x^2}$. Without evaluating integrals and knowing that $\int_1^{\infty} f(x) \, dx$ has a finite value, determine whether $\int_1^{\infty} g(x) \, dx$ and $\int_1^{\infty} h(x) \, dx$ have finite values.

**113. A family of logarithm integrals** Let $I(p) = \int_1^e \frac{\ln x}{x^p} \, dx$, where $p$ is a real number.

a. Find an expression for $I(p)$, for all real values of $p$.
b. Evaluate $\lim_{p \to \infty} I(p)$ and $\lim_{p \to -\infty} I(p)$.
c. For what value of $p$ is $I(p) = 1$?

**T 114. Arc length of the natural logarithm** Consider the curve $y = \ln x$.

a. Find the length of the curve from $x = 1$ to $x = a$ and call it $L(a)$. (*Hint:* The change of variables $u = \sqrt{x^2 + 1}$ allows evaluation by partial fractions.)
b. Graph $L(a)$.
c. As $a$ increases, $L(a)$ increases as what power of $a$?

**T 115. Best approximation** Let $I = \int_0^1 \frac{x^2 - x}{\ln x} \, dx$. Use any method you choose to find a good approximation to $I$. You may use the facts that $\lim_{x \to 0^+} \frac{x^2 - x}{\ln x} = 0$ and $\lim_{x \to 1^-} \frac{x^2 - x}{\ln x} = 1$.

**T 116. Numerical integration** Use a calculator to determine the integer $n$ that satisfies $\int_0^{1/2} \frac{\ln(1 + 2x)}{x} \, dx = \frac{\pi^2}{n}$.

**T 117. Numerical integration** Use a calculator to determine the integer $n$ that satisfies $\int_0^1 \frac{\sin^{-1} x}{x} \, dx = \frac{\pi \ln 2}{n}$.

**118. Two worthy integrals**

a. Let $I(a) = \int_0^\infty \frac{dx}{(1 + x^a)(1 + x^2)}$, where $a$ is a real number. Evaluate $I(a)$ and show that its value is independent of $a$. (*Hint:* Split the integral into two integrals over $[0, 1]$ and $[1, \infty)$; then use a change of variables to convert the second integral into an integral over $[0, 1]$.)
b. Let $f$ be any positive continuous function on $[0, \pi/2]$. Evaluate $\int_0^{\pi/2} \frac{f(\cos x)}{f(\cos x) + f(\sin x)} \, dx$. (*Hint:* Use the identity $\cos\left(\frac{\pi}{2} - x\right) = \sin x$.)

(*Source: Mathematics Magazine* 81, 2, Apr 2008)

**T 119. Comparing volumes** Let $R$ be the region bounded by $y = \ln x$, the $x$-axis, and the line $x = a$, where $a > 1$.

a. Find the volume $V_1(a)$ of the solid generated when $R$ is revolved about the $x$-axis (as a function of $a$).
b. Find the volume $V_2(a)$ of the solid generated when $R$ is revolved about the $y$-axis (as a function of $a$).
c. Graph $V_1$ and $V_2$. For what values of $a > 1$ is $V_1(a) > V_2(a)$?

**120. Equal volumes**

a. Let $R$ be the region bounded by the graph of $f(x) = x^{-p}$ and the $x$-axis, for $x \geq 1$. Let $V_1$ and $V_2$ be the volumes of the solids generated when $R$ is revolved about the $x$-axis and the

$y$-axis, respectively, if they exist. For what values of $p$ (if any) is $V_1 = V_2$?
b. Repeat part (a) on the interval $(0, 1]$.

**121. Equal volumes** Let $R_1$ be the region bounded by the graph of $y = e^{-ax}$ and the $x$-axis on the interval $[0, b]$, where $a > 0$ and $b > 0$. Let $R_2$ be the region bounded by the graph of $y = e^{-ax}$ and the $x$-axis on the interval $[b, \infty)$. Let $V_1$ and $V_2$ be the volumes of the solids generated when $R_1$ and $R_2$ are revolved about the $x$-axis. Find and graph the relationship between $a$ and $b$ for which $V_1 = V_2$.

**122. Comparing areas** The region $R_1$ is bounded by the graph of $y = \tan x$ and the $x$-axis on the interval $[0, \pi/3]$. The region $R_2$ is bounded by the graph of $y = \sec x$ and the $x$-axis on the interval $[0, \pi/6]$. Which region has the greater area?

**123. Region between curves** Find the area of the region bounded by the graphs of $y = \tan x$ and $y = \sec x$ on the interval $[0, \pi/4]$.

**124. Mercator map projection** The Mercator map projection was proposed by the Flemish geographer Gerardus Mercator (1512–1594). The stretching factor of the Mercator map as a function of the latitude $\theta$ is given by the function

$$G(\theta) = \int_0^\theta \sec x \, dx.$$

Graph $G$, for $0 \leq \theta < \pi/2$. (See the Guided Project *Mercator projections* for a derivation of this integral.)

**125. Wallis products** Complete the following steps to prove a well-known formula discovered by the 17th-century English mathematician John Wallis.

a. Use a reduction formula to show that

$$\int_0^\pi \sin^m x \, dx = \frac{m-1}{m} \int_0^\pi \sin^{m-2} x \, dx,$$

for any integer $m \geq 2$.
b. Show that $\int_0^\pi \sin^{2n} x \, dx = \frac{2n-1}{2n} \cdot \frac{2n-3}{2n-2} \cdots \frac{1}{2} \cdot \pi$, for any integer $n \geq 1$. The product on the right is called a *Wallis product*.
c. Show that $\int_0^\pi \sin^{2n+1} x \, dx = \frac{2n}{2n+1} \cdot \frac{2n-2}{2n-1} \cdots \frac{2}{3} \cdot 2$, for any integer $n \geq 1$. The product on the right is another example of a Wallis product.
d. Use the inequality $0 \leq \sin x \leq 1$ on $[0, \pi]$ to show that $0 \leq \int_0^\pi \sin^{m+1} x \, dx \leq \int_0^\pi \sin^m x \, dx$, for any integer $m \geq 0$.
e. Use part (d) to show that

$$\frac{2n}{2n+1} \cdot \frac{2n-2}{2n-1} \cdots \frac{2}{3} \cdot 2 \leq \frac{2n-1}{2n} \cdot \frac{2n-3}{2n-2} \cdots \frac{1}{2} \cdot \pi$$
$$\leq \frac{2n-2}{2n-1} \cdot \frac{2n-4}{2n-3} \cdots \frac{2}{3} \cdot 2,$$

for any integer $n \geq 2$.
f. Show that $\frac{2n}{2n+1} \cdot \frac{\pi}{2} \leq \frac{(2n)^2}{2n+1} \cdot \frac{(2n-2)^2}{(2n-1)^2} \cdots \frac{2^2}{3^2} \leq \frac{\pi}{2}$, for any integer $n \geq 2$.
g. Use part (f) to conclude that

$$\frac{\pi}{2} = \lim_{n \to \infty} \left( \frac{2}{1} \cdot \frac{2}{3} \cdot \frac{4}{3} \cdot \frac{4}{5} \cdots \frac{2n}{2n-1} \cdot \frac{2n}{2n+1} \right).$$

This is known as the *Wallis formula*.

# Chapter 8 Guided Projects

*Applications of the material in this chapter and related topics can be found in the following Guided Projects. For additional information, see the Preface.*

- Simpson's Rule
- How long will your iPod last?

- Mercator projections

# 9

# Differential Equations

**Chapter Preview**  If you wanted to demonstrate the utility of mathematics to a skeptic, perhaps the most convincing way would be to talk about *differential equations*. This vast subject lies at the heart of mathematical modeling and is used in engineering, physics, chemistry, biology, geophysics, economics and finance, and health sciences. Its many applications in these areas include analyzing the stability of buildings and bridges, simulating planet and satellite orbits, describing chemical reactions, modeling populations and epidemics, predicting weather, locating oil reserves, forecasting financial markets, producing medical images, and simulating drug kinetics. Differential equations rely heavily on calculus and are usually studied in advanced courses that follow calculus. Nevertheless, you have now seen enough calculus to take a brief tour of this rich and powerful subject.

## 9.1 Basic Ideas

If you studied Section 4.9 or 6.1, then you saw a preview of differential equations. There you learned that if you are given the derivative of a function (for example, a velocity or some other rate of change), you can find the function itself by integration. This process amounts to solving a differential equation.

A differential equation involves an unknown function $y$ and its derivatives. The unknown in a differential equation is not a number (as in an algebraic equation), but rather a *function*. Examples of differential equations are

> Common choices for the independent variable in a differential equation are $x$ and $t$, with $t$ being used for time-dependent problems.

$$\text{(A) } \frac{dy}{dx} + 4y = \cos x, \quad \text{(B) } \frac{d^2y}{dx^2} + 16y = 0, \quad \text{and} \quad \text{(C) } y'(t) = 0.1y(100 - y).$$

In each case, the goal is to find functions $y$ that satisfy the equation. Just to be clear about what we mean by a solution, consider equation (B). If we substitute $y = \cos 4x$ and $y'' = -16 \cos 4x$ into this equation, we find that

$$\underbrace{-16 \cos 4x}_{y''} + \underbrace{16 \cos 4x}_{16y} = 0,$$

which implies that $y = \cos 4x$ is a solution of the equation. You should verify that $y = C \cos 4x$ is also a solution, for any real number $C$ (as is $y = C \sin 4x$). Already we see that one differential equation can have many solutions.

Let's begin with a brief discussion of the terminology associated with differential equations. The **order** of a differential equation is the highest order that appears on a derivative in the equation. Of the three differential equations just given, (A) and (C) are first order, and (B) is second order. A differential equation is **linear** if the unknown

➤ A *linear* differential equation cannot have terms such as $y^2$, $yy'$, or $\sin y$, where $y$ is the unknown function.

function $y$ and its derivatives appear only to the first power and are not composed with other functions. Furthermore, a linear equation cannot have products or quotients of $y$ and its derivatives. Of the equations just given, (A) and (B) are linear, but (C) is **nonlinear** (because the right side contains $y^2$). The general **first-order linear differential equation** has the form

$$\frac{dy}{dt} + p(t)y = q(t),$$

where $p$ and $q$ are given functions of $t$. Notice that $y$ and $y'$ appear to the first power and not in products or compositions that involve $y$ or $y'$, which makes the equation linear.

Solving a first-order differential equation requires integration—you must "undo" the derivative $y'$ in order to find $y$. Integration introduces an arbitrary constant, so the most general solution of a first-order differential equation typically involves one arbitrary constant. Similarly, the most general solution of a second-order differential equation involves two arbitrary constants, and for an $n$th-order differential equation, the most general solution involves $n$ arbitrary constants. The most general family of functions that solves a differential equation, including the appropriate number of arbitrary constants, is called (not surprisingly) the **general solution**.

➤ To keep matters simple, we will use *general solution* to refer to the most general family of solutions of a differential equation. However, some nonlinear equations may have isolated solutions that are not included in this family of solutions. For example, you should check that for real numbers $C$, the functions $y = 1/(C - t)$ satisfy the equation $y'(t) = y^2$. Therefore, we call $y = 1/(C - t)$ the general solution of the equation, even though it doesn't include the solution $y = 0$.

A differential equation is often accompanied by **initial conditions** that specify the values of $y$, and possibly its derivatives, at a particular point. In general, an $n$th-order equation requires $n$ initial conditions, which can be used to determine the $n$ arbitrary constants in the general solution. A differential equation, together with the appropriate number of initial conditions, is called an **initial value problem**. A typical first-order initial value problem has the form

$$y'(t) = F(t, y) \quad \text{Differential equation}$$
$$y(a) = b, \quad \text{Initial condition}$$

where $a$ and $b$ are given, and $F$ is a given expression that involves $t$ and/or $y$. A **solution** to the first-order initial value problem $y'(t) = F(t, y)$, $y(a) = b$ is a function $y(t)$ that satisfies the differential equation on an interval $I$ and whose graph includes the point $(a, b)$. We refer to $I$ as the **domain** of the solution.

As we see in upcoming examples, the domain of a solution to an initial value problem is the natural domain of the function $y$, possibly restricted by the properties of the differential equation, the initial condition(s), or the physical context of the differential equation.

## EXAMPLE 1   Verifying solutions

**a.** Show by substitution that the function $y(t) = Ce^{2.5t}$ is a solution of the differential equation $y'(t) = 2.5y(t)$, where $C$ is an arbitrary constant.

**b.** Show by substitution that the function $y(t) = 3.2e^{2.5t}$ satisfies the initial value problem

$$y'(t) = 2.5y(t) \quad \text{Differential equation}$$
$$y(0) = 3.2. \quad \text{Initial condition}$$

### SOLUTION

**a.** We differentiate $y(t) = Ce^{2.5t}$ to get $y'(t) = 2.5Ce^{2.5t}$. Substituting into the differential equation, we find that

$$y'(t) = \underbrace{2.5Ce^{2.5t}}_{y'(t)} = 2.5\underbrace{Ce^{2.5t}}_{y(t)} = 2.5y(t).$$

In other words, the function $y(t) = Ce^{2.5t}$ satisfies the equation $y'(t) = 2.5y(t)$, for any value of $C$. Therefore, $y(t) = Ce^{2.5t}$ is a family of solutions of the differential equation.

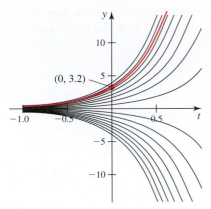

(0, 3.2)

**Figure 9.1**

**b.** By part (a) with $C = 3.2$, the function $y(t) = 3.2e^{2.5t}$ satisfies the differential equation $y'(t) = 2.5y(t)$. We can also check that this function satisfies the initial condition $y(0) = 3.2$:

$$y(0) = 3.2e^{2.5 \cdot 0} = 3.2 \cdot e^0 = 3.2.$$

Therefore, $y(t) = 3.2e^{2.5t}$ is a solution of the initial value problem, and its domain is $(-\infty, \infty)$ because $y(t)$ satisfies the differential equation for every real value of $t$. **Figure 9.1** shows the family of curves $y = Ce^{2.5t}$ with several different values of the constant $C$. It also shows the function $y(t) = 3.2e^{2.5t}$ highlighted in red, which is the solution of the initial value problem.

*Related Exercises 7–8* ◀

**EXAMPLE 2**    **General solutions**  Find the general solution of the following differential equations.

**a.** $y'(t) = 5 \cos t + 6 \sin 3t$
**b.** $y''(t) = 10t^3 - 144t^7 + 12t$

**SOLUTION**

**a.** The solution of the equation consists of the antiderivatives of $5 \cos t + 6 \sin 3t$. Taking the indefinite integral of both sides of the equation, we have

$$\int y'(t)\,dt = \int (5 \cos t + 6 \sin 3t)\,dt \quad \text{Integrate both sides with respect to } t.$$

$$y(t) = 5 \sin t - 2 \cos 3t + C, \quad \text{Evaluate integrals.}$$

where $C$ is an arbitrary constant. The function $y(t) = 5 \sin t - 2 \cos 3t + C$ is the general solution of the differential equation.

**b.** In this second-order equation, we are given $y''(t)$ in terms of the independent variable $t$. Taking the indefinite integral of both sides of the equation yields

$$\int y''(t)\,dt = \int (10t^3 - 144t^7 + 12t)\,dt \quad \text{Integrate both sides with respect to } t.$$

$$y'(t) = \frac{5}{2}t^4 - 18t^8 + 6t^2 + C_1. \quad \text{Evaluate integrals.}$$

Integrating once gives $y'(t)$ and introduces an arbitrary constant that we call $C_1$. We now integrate again:

$$\int y'(t)\,dt = \int \left( \frac{5}{2}t^4 - 18t^8 + 6t^2 + C_1 \right) dt \quad \text{Integrate both sides with respect to } t.$$

$$y(t) = \frac{1}{2}t^5 - 2t^9 + 2t^3 + C_1 t + C_2. \quad \text{Evaluate integrals.}$$

**QUICK CHECK 1** What are the orders of the equations in Example 2? Are they linear or nonlinear? ◀

This function, which involves two arbitrary constants, is the general solution of the differential equation.

*Related Exercises 22–23* ◀

**EXAMPLE 3**    **An initial value problem**  Solve the initial value problem

$$y'(t) = 10e^{-t/2}, \quad y(0) = 4.$$

**SOLUTION**  The general solution is found by taking the indefinite integral of both sides of the differential equation with respect to $t$:

$$\int y'(t)\,dt = \int 10e^{-t/2}\,dt \quad \text{Integrate both sides with respect to } t.$$

$$y(t) = -20e^{-t/2} + C. \quad \text{Evaluate integrals.}$$

**Figure 9.2**

**QUICK CHECK 2** What is the solution of the initial value problem in Example 3 with the initial condition $y(0) = 16$? ◄

We have found the general solution, which involves one arbitrary constant. To determine its value, we use the initial condition by substituting $t = 0$ and $y = 4$ into the general solution:

$$\underbrace{y(0)}_{4} = -20e^{-0/2} + C = -20 + C,$$

which implies that $4 = -20 + C$ or $C = 24$. Therefore, the solution of the initial value problem is $y(t) = -20e^{-t/2} + 24$ (**Figure 9.2**). You should check that this function satisfies the initial condition and that it satisfies the differential equation on $(-\infty, \infty)$; therefore, the domain of the solution is all real numbers.

*Related Exercises 33–34* ◄

In Example 3, the interval on which the solution $y(t) = -20e^{t/2} + 24$ satisfies the differential equation matches the natural domain of the function, which is $(-\infty, \infty)$. In such cases it is not necessary to state the domain of the solution. However, in Example 4, we see that the differential equation and the initial condition may restrict the natural domain of the solution function.

**EXAMPLE 4** **Determining the domain** Solve each initial value problem and determine the domain of the solution.

**a.** $y'(t) = 1 + \dfrac{2}{\sqrt[3]{t}}, y(1) = 4$      **b.** $y'(x) = \csc 2x \cot 2x, y(\pi/4) = 0$

**SOLUTION**

**a.** Integrating both sides of the differential equation with respect to $t$ gives the general solution:

$$\int y'(t)\, dt = \int (1 + 2t^{-1/3})\, dt \quad \text{Integrate both sides with respect to } t.$$

$$y(t) = t + 3t^{2/3} + C. \quad \text{Evaluate integrals.}$$

The initial condition $y(1) = 4$ determines the value of $C$:

$$y(1) = \underbrace{1 + 3 \cdot 1^{2/3}}_{4} + C = 4 + C = 4,$$

Domain of $y(t) = t + 3t^{2/3}$ taken alone: $(-\infty, \infty)$

Considered as a solution to an initial value problem, the domain of $y(t)$ is $(0, \infty)$.

**Figure 9.3**

**QUICK CHECK 3** Solve the initial value problem in Example 4a with an initial condition of $y(-1) = 4$. What is the domain of the solution? ◄

which implies that $C = 0$. Therefore, $y(t) = t + 3t^{2/3}$ is the solution to the initial value problem. When treated as a stand-alone function, the domain of $y(t) = t + 3t^{2/3}$ is all real numbers, as shown in **Figure 9.3**. However, when it is regarded as a solution to the differential equation $y'(t) = 1 + \dfrac{2}{\sqrt[3]{t}}$, the domain of $y(t)$ cannot include $t = 0$ because $y'(t)$ is undefined at 0. Only the right branch of $y(t) = t + 3t^{2/3}$ satisfies both the differential equation and the initial condition $y(1) = 4$, which implies that the domain of the solution to the initial value problem is $(0, \infty)$.

**b.** To find the general solution, we integrate both sides of the equation with respect to $x$ using the substitution $u = 2x$:

$$\int y'(x)\, dx = \int \csc 2x \cot 2x\, dx \quad \text{Integrate with respect to } x.$$

$$y(x) = \frac{1}{2} \int \csc u \cot u\, du \quad u = 2x;\, du = 2\, dx$$

$$= -\frac{1}{2} \csc u + C = -\frac{1}{2} \csc 2x + C. \quad \text{Evaluate integral.}$$

The initial condition $y(\pi/4) = 0$ implies that

$$y\left(\frac{\pi}{4}\right) = -\frac{1}{2} \csc \underbrace{\frac{\pi}{2}}_{1} + C = 0 \quad \text{or} \quad C = \frac{1}{2},$$

$$y(x) = \tfrac{1}{2} - \tfrac{1}{2}\csc 2x$$

Domain: $\left(0, \tfrac{\pi}{2}\right)$

**Figure 9.4**

> The term *initial condition* originates with equations in which the independent variable is *time*. The initial condition $y(0) = a$ gives the initial state of the system. In such cases, the solution is usually graphed only for $t \geq 0$, and the domain of the solution is determined by the physical constraints of the problem.

so the solution to the initial value problem is $y(x) = \dfrac{1}{2} - \dfrac{1}{2}\csc 2x$. The graph of $y$ has infinitely many branches (**Figure 9.4**); we choose the branch that satisfies the initial condition, which implies that the domain of the solution is $(0, \pi/2)$.

*Related Exercises 35–36* ◄

## Differential Equations in Action

We close this section with three examples of differential equations that are used to model particular physical systems. The first example of one-dimensional motion in a gravitational field was introduced in Example 7 of Section 4.9; it is useful to revisit this problem using the language of differential equations. The equations in Examples 6 and 7 reappear later in the chapter when we show how to solve them. These examples illustrate how the domain of a solution is determined by the physical context of the problem.

**EXAMPLE 5 Motion in a gravitational field** A stone is launched vertically upward with a velocity of $v_0$ m/s from a point $s_0$ meters above the ground, where $v_0 > 0$ and $s_0 \geq 0$. Assume the stone is launched at time $t = 0$ and that $s(t)$ is the position of the stone at time $t \geq 0$; the positive $s$-axis points upward with the origin at the ground. By Newton's Second Law of Motion, assuming no air resistance, the position of the stone is governed by the differential equation $s''(t) = -g$, where $g = 9.8$ m/s$^2$ is the acceleration due to gravity (in the downward direction).

**a.** Find the position $s(t)$ of the stone for all times at which the stone is above the ground.

**b.** At what time does the stone reach its highest point and what is its height above the ground?

**c.** Does the stone go higher if it is launched at $v(0) = v_0 = 39.2$ m/s from the ground ($s_0 = 0$) or at $v_0 = 19.6$ m/s from a height of $s_0 = 50$ m?

**SOLUTION**

**a.** Integrating both sides of the differential equation $s''(t) = -9.8$ gives the velocity $v(t)$:

$$\int s''(t)\,dt = -\int 9.8\,dt \qquad \text{Integrate both sides.}$$

$$s'(t) = v(t) = -9.8t + C_1. \qquad \text{Evaluate integrals.}$$

To evaluate the constant $C_1$, we use the initial condition $v(0) = v_0$, finding that $v(0) = -9.8 \cdot 0 + C_1 = C_1 = v_0$. Therefore, $C_1 = v_0$ and the velocity is $v(t) = s'(t) = -9.8t + v_0$.

Integrating both sides of this velocity equation gives the position function:

$$\int s'(t)\,dt = \int (-9.8t + v_0)\,dt \qquad \text{Integrate both sides.}$$

$$s(t) = -4.9t^2 + v_0 t + C_2. \qquad \text{Evaluate integrals.}$$

We now use the initial condition $s(0) = s_0$ to evaluate $C_2$, finding that

$$s(0) = -4.9 \cdot 0^2 + v_0 \cdot 0 + C_2 = C_2 = s_0.$$

Therefore, $C_2 = s_0$, and the position function is $s(t) = -4.9t^2 + v_0 t + s_0$, where $v_0$ and $s_0$ are given. This function is valid while the stone is in flight. To find the domain of the solution, which depends on $v_0$ and $s_0$, we solve $s(t) \geq 0$.

> To find the time at which the stone reaches its highest point, we could instead locate the local maximum of the position function, which also requires solving $s'(t) = v(t) = 0$.

**b.** The stone reaches its highest point when $v(t) = 0$. Solving $v(t) = -9.8t + v_0 = 0$, we find that the stone reaches its highest point when $t = v_0/9.8$, measured in seconds. So the position at the highest point is

$$s_{\max} = s\left(\frac{v_0}{9.8}\right) = -4.9\left(\frac{v_0}{9.8}\right)^2 + v_0\left(\frac{v_0}{9.8}\right) + s_0 = \frac{v_0^2}{19.6} + s_0.$$

**Figure 9.5**

➤ The curves in Figure 9.5 are not the trajectories of the stones. The motion is one-dimensional because the stones travel along a vertical line.

**c.** Now it is a matter of substituting the given values of $s_0$ and $v_0$. In the first case, with $v_0 = 39.2$ and $s_0 = 0$, we have $s_{max} = 78.4$ m. In the second case, with $v_0 = 19.6$ and $s_0 = 50$, we have $s_{max} = 69.6$ m. The position functions in the two cases are shown in **Figure 9.5**. We see that the stone goes higher with $v_0 = 39.2$ and $s_0 = 0$.

*Related Exercises 43–44* ◄

**QUICK CHECK 4** Suppose the initial conditions in Example 5a are $v_0 = 14.7$ m/s and $s_0 = 49$ m. Write the position function $s(t)$, and state its domain. At what time will the stone reach its maximum height? What is the maximum height at that time? ◄

**EXAMPLE 6  A harvesting model**  A simple model of a harvested resource (for example, timber or fish) assumes a competition between the harvesting and the natural growth of the resource. This process may be described by the differential equation

$$\underbrace{p'(t)}_{\substack{\text{rate of} \\ \text{change}}} = \underbrace{rp(t)}_{\substack{\text{natural} \\ \text{growth rate}}} - \underbrace{H}_{\text{harvesting}}, \quad \text{for} \quad t \geq 0,$$

where $p(t)$ is the amount (or population) of the resource at time $t \geq 0$, $r > 0$ is the natural growth rate of the resource, and $H > 0$ is the constant harvesting rate. An initial condition $p(0) = p_0$ is also specified to create an initial value problem. Notice that the rate of change $p'(t)$ has a positive contribution from the natural growth rate and a negative contribution from the harvesting term.

**a.** For given constants $p_0$, $r$, and $H$, verify that the function

$$p(t) = \left(p_0 - \frac{H}{r}\right)e^{rt} + \frac{H}{r}$$

is a solution of the initial value problem.

**b.** Let $p_0 = 1000$ and $r = 0.1$. Graph the solutions for the harvesting rates $H = 50, 90, 130,$ and $170$. Describe and interpret the four curves.

**c.** What value of $H$ gives a constant value of $p$, for all $t \geq 0$?

**SOLUTION**

**a.** Differentiating the given solution, we find that

$$p'(t) = \left(p_0 - \frac{H}{r}\right)re^{rt} = (rp_0 - H)e^{rt}.$$

Simplifying the right side of the differential equation, we find that

$$rp(t) - H = r\left(\underbrace{\left(p_0 - \frac{H}{r}\right)e^{rt} + \frac{H}{r}}_{p(t)}\right) - H = (rp_0 - H)e^{rt}.$$

Therefore, the left and right sides of the equation $p'(t) = rp(t) - H$ are equal, so the equation is satisfied by the given function. You can verify that the given solution also satisfies $p(0) = p_0$, which means $p$ satisfies the initial value problem.

**b.** Letting $p_0 = 1000$ and $r = 0.1$, the function

$$p(t) = (1000 - 10H)e^{0.1t} + 10H$$

is graphed in **Figure 9.6**, for $H = 50, 90, 130,$ and $170$. We see that for small values of $H$ ($H = 50$ and $H = 90$), the amount of the resource increases with time. On the other hand, for large values of $H$ ($H = 130$ and $H = 170$), the amount of the resource decreases with time, eventually reaching zero. The model predicts that if the harvesting rate is too large, the resource will eventually disappear. The domain of each solution depends on $p_0$, $r$, and $H$. For example, with $p_0 = 1000$, $r = 0.1$, and $H = 170$, we find that the domain is approximately $[0, 8.9]$ (bottom curve in Figure 9.6).

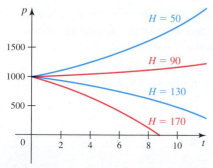

**Figure 9.6**

**c.** The solution

$$p(t) = (1000 - 10H)e^{0.1t} + 10H$$

is constant (independent of $t$) when $1000 - 10H = 0$, which implies $H = 100$. In this case, the solution is

$$p(t) = \underbrace{(1000 - 10H)}_{0} e^{0.1t} + 10H = 1000.$$

Therefore, if the harvesting rate is $H = 100$, then the harvesting exactly balances the natural growth of the resource, and $p$ is constant. This solution is called an *equilibrium solution*. For $H > 100$, the amount of resource decreases in time, and for $H < 100$, it increases in time.

*Related Exercises 45–46* ◀

▶ Evangelista Torricelli was an Italian mathematician and physicist who lived from 1608 to 1647. He is credited with inventing the barometer.

**Figure 9.7**

**EXAMPLE 7**  **Flow from a tank**  Imagine a large cylindrical tank with cross-sectional area $A$. The bottom of the tank has a circular drain with cross-sectional area $a$. Assume the tank is initially filled with water to a height (in meters) of $h(0) = H$ (**Figure 9.7**). According to Torricelli's law, the height of the water $t$ seconds after the drain is opened is described by the differential equation

$$h'(t) = -k\sqrt{h}, \quad \text{where } t \geq 0, \, k = \frac{a}{A}\sqrt{2g},$$

and $g = 9.8 \text{ m/s}^2$ is the acceleration due to gravity.

**a.** According to the differential equation, is $h$ an increasing or decreasing function of $t$, for $t \geq 0$?

**b.** Verify by substitution that the solution of the initial value problem is

$$h(t) = \left(\sqrt{H} - \frac{kt}{2}\right)^2.$$

**c.** Graph the solution for $H = 1.44$ m, $A = 1 \text{ m}^2$, and $a = 0.05 \text{ m}^2$.

**d.** After how many seconds is the tank in part (c) empty?

**SOLUTION**

**a.** Because $k > 0$, the differential equation implies that $h'(t) < 0$, for $t \geq 0$. Therefore, the height of the water decreases in time, consistent with the fact that the tank is being drained.

**b.** We first check the initial condition. Substituting $t = 0$ into the proposed solution, we see that

$$h(0) = \left(\sqrt{H} - \frac{k \cdot 0}{2}\right)^2 = (\sqrt{H})^2 = H.$$

Differentiating the proposed solution, we have

$$h'(t) = 2\underbrace{\left(\sqrt{H} - \frac{kt}{2}\right)}_{\sqrt{h(t)}}\left(-\frac{k}{2}\right) = -k\sqrt{h}.$$

Therefore, $h$ satisfies the initial condition and the differential equation.

**c.** With the given values of the parameters,

$$k = \frac{a}{A}\sqrt{2g} = \frac{0.05 \text{ m}^2}{1 \text{ m}^2}\sqrt{2 \cdot 9.8 \text{ m/s}^2} \approx 0.22 \text{ m}^{1/2}/\text{s},$$

**Figure 9.8**

**QUICK CHECK 5** In Example 7, if the height function were given by $h(t) = (4.2 - 0.14t)^2$, at what time would the tank be empty? What does your answer say about the domain of this solution? ◄

and the solution becomes

$$h(t) = \left(\sqrt{H} - \frac{kt}{2}\right)^2 \approx (\sqrt{1.44} - 0.11t)^2 = (1.2 - 0.11t)^2.$$

The graph of the solution (**Figure 9.8**) shows the height of the water decreasing from $h(0) = 1.44$ to zero at approximately $t = 11$ s.

**d.** Solving the equation

$$h(t) = (1.2 - 0.11t)^2 = 0,$$

we find that the tank is empty at $t \approx 10.9$ s. This calculation determines the domain of the solution, which is approximately $[0, 10.9]$.

*Related Exercises 47–48* ◄

**Final Note** Throughout this section, we found solutions to initial value problems without worrying about whether there might be other solutions. Once we find a solution to an initial value problem, how can we be sure there aren't other solutions? More generally, given a particular initial value problem, how do we know whether a solution exists and whether it is unique?

These theoretical questions have answers, and they are provided by powerful *existence and uniqueness theorems*. These theorems and their proofs are quite technical and are handled in advanced courses. Here is an informal statement of an existence and uniqueness theorem for a particular class of initial value problems encountered in this chapter:

The solution of the general first-order initial value problem

$$y'(t) = F(t, y), y(a) = b$$

exists and is unique in some region that contains the point $(a, b)$ provided $F$ is a "well-behaved" function in that region.

The technical challenges arise in defining *well-behaved* in the most general way possible. The initial value problems we consider in this chapter satisfy the conditions of this theorem, and can be assumed to have unique solutions.

---

## SECTION 9.1 EXERCISES

### Getting Started

1. Consider the differential equation $y'(t) + 9y(t) = 10$.

   a. How many arbitrary constants appear in the general solution of the differential equation?
   b. Is the differential equation linear or nonlinear?

2. If the general solution of a differential equation is $y(t) = Ce^{-3t} + 10$, what is the solution that satisfies the initial condition $y(0) = 5$?

3. Does the function $y(t) = 2t$ satisfy the differential equation $y'''(t) + y'(t) = 2$?

4. Does the function $y(t) = 6e^{-3t}$ satisfy the initial value problem $y'(t) - 3y(t) = 0, y(0) = 6$?

5. The solution to the initial value problem $y'(t) = 2 \sec t \tan t$, $y(\pi) = -2$ is $y(t) = 2 \sec t$. What is the domain of this solution? (*Hint:* Sketch a graph of $y$).

6. Explain why the graph of the solution to the initial value problem
   $$y'(t) = \frac{t^2}{1 - t}, y(-1) = \ln 2$$
   cannot cross the line $t = 1$.

### Practice Exercises

**7–16. Verifying general solutions** *Verify that the given function is a solution of the differential equation that follows it. Assume $C, C_1, C_2,$ and $C_3$ are arbitrary constants.*

7. $y(t) = Ce^{-5t}; y'(t) + 5y(t) = 0$

8. $y(t) = Ct^3; ty'(t) - 3y(t) = 0$

9. $y(t) = C_1 \sin 4t + C_2 \cos 4t; y''(t) + 16y(t) = 0$

10. $y(x) = C_1 e^{-x} + C_2 e^x; y''(x) - y(x) = 0$

11. $u(t) = Ce^{1/(4t^4)}; u'(t) + \frac{1}{t^5} u(t) = 0$

12. $u(t) = C_1 e^t + C_2 te^t; u''(t) - 2u'(t) + u(t) = 0$

13. $g(x) = C_1 e^{-2x} + C_2 xe^{-2x} + 2; g''(x) + 4g'(x) + 4g(x) = 8$

14. $u(t) = C_1 t^2 + C_2 t^3; t^2 u''(t) - 4tu'(t) + 6u(t) = 0$

15. $u(t) = C_1 t^5 + C_2 t^{-4} - t^3; t^2 u''(t) - 20u(t) = 14t^3$

16. $z(t) = C_1 e^{-t} + C_2 e^{2t} + C_3 e^{-3t} - e^t;$
    $z'''(t) + 2z''(t) - 5z'(t) - 6z(t) = 8e^t$

**17–20. Verifying solutions of initial value problems** *Verify that the given function y is a solution of the initial value problem that follows it.*

**17.** $y(t) = 16e^{2t} - 10$; $y'(t) - 2y(t) = 20$, $y(0) = 6$

**18.** $y(t) = 8t^6 - 3$; $ty'(t) - 6y(t) = 18$, $y(1) = 5$

**19.** $y(t) = -3 \cos 3t$; $y''(t) + 9y(t) = 0$, $y(0) = -3$, $y'(0) = 0$

**20.** $y(x) = \frac{1}{4}(e^{2x} - e^{-2x})$; $y''(x) - 4y(x) = 0$, $y(0) = 0$, $y'(0) = 1$

**21–32. Finding general solutions** *Find the general solution of each differential equation. Use* $C, C_1, C_2, \ldots$ *to denote arbitrary constants.*

**21.** $y'(t) = 3 + e^{-2t}$

**22.** $y'(t) = 12t^5 - 20t^4 + 2 - 6t^{-2}$

**23.** $y'(x) = 4 \tan 2x - 3 \cos x$

**24.** $p'(x) = \frac{16}{x^9} - 5 + 14x^6$

**25.** $y''(t) = 60t^4 - 4 + 12t^{-3}$

**26.** $y''(t) = 15e^{3t} + \sin 4t$

**27.** $u''(x) = 55x^9 + 36x^7 - 21x^5 + 10x^{-3}$

**28.** $v''(x) = \frac{5}{x^2}$        **29.** $u'(x) = \frac{2(x-1)}{x^2 + 4}$

**30.** $y'(t) = t \ln t + 1$     **31.** $y''(x) = \frac{x}{(1 - x^2)^{3/2}}$

**32.** $v'(t) = \frac{4}{t^2 - 4}$

**33–42. Solving initial value problems** *Solve the following initial value problems.*

**33.** $y'(t) = 1 + e^t$, $y(0) = 4$

**34.** $y'(t) = \sin t + \cos 2t$, $y(0) = 4$

**35.** $y'(x) = 3x^2 - 3x^{-4}$, $y(1) = 0$

**36.** $y'(x) = 4 \sec^2 2x$, $y(0) = 8$

**37.** $y''(t) = 12t - 20t^3$, $y(0) = 1$, $y'(0) = 0$

**38.** $u''(x) = 4e^{2x} - 8e^{-2x}$, $u(0) = 1$, $u'(0) = 3$

**39.** $y''(t) = te^t$, $y(0) = 0$, $y'(0) = 1$

**40.** $y'(t) = te^t$, $y(0) = -1$

**41.** $u'(x) = \frac{1}{x^2 + 16} - 4$, $u(0) = 2$

**42.** $p'(x) = \frac{2}{x^2 + x}$, $p(1) = 0$

**43–44. Motion in a gravitational field** *An object is fired vertically upward with an initial velocity* $v(0) = v_0$ *from an initial position* $s(0) = s_0$.

**a.** *For the following values of* $v_0$ *and* $s_0$, *find the position and velocity functions for all times at which the object is above the ground* $(s = 0)$.

**b.** *Find the time at which the highest point of the trajectory is reached and the height of the object at that time.*

**43.** $v_0 = 29.4$ m/s, $s_0 = 30$ m

**44.** $v_0 = 49$ m/s, $s_0 = 60$ m

**45–46. Harvesting problems** *Consider the harvesting problem in Example 6.*

**45.** If $r = 0.05$ and $p_0 = 1500$, for what values of $H$ is the amount of the resource increasing? For what value of $H$ is the amount of the resource constant? If $H = 100$, when does the resource vanish?

**46.** If $r = 0.05$ and $H = 500$, for what values of $p_0$ is the amount of the resource decreasing? For what value of $p_0$ is the amount of the resource constant? If $p_0 = 9000$, when does the resource vanish?

**T 47–48. Draining tanks** *Consider the tank problem in Example 7. For the following parameter values, find the water height function. Then determine the approximate time at which the tank is first empty and graph the solution.*

**47.** $H = 1.96$ m, $A = 1.5$ m$^2$, $a = 0.3$ m$^2$

**48.** $H = 2.25$ m, $A = 2$ m$^2$, $a = 0.5$ m$^2$

**49.** **Explain why or why not** Determine whether the following statements are true and give an explanation or counterexample.

    **a.** The general solution of the differential equation $y'(t) = 1$ is $y(t) = t$.

    **b.** The differential equation $y''(t) - y(t)y'(t) = 0$ is second order and linear.

    **c.** To find the solution of an initial value problem, we usually begin by finding a general solution of the differential equation.

## Explorations and Challenges

**50.** **A second-order equation** Consider the differential equation $y''(t) - k^2 y(t) = 0$, where $k > 0$ is a real number.

    **a.** Verify by substitution that when $k = 1$, a solution of the equation is $y(t) = C_1 e^t + C_2 e^{-t}$. You may assume this function is the general solution.

    **b.** Verify by substitution that when $k = 2$, the general solution of the equation is $y(t) = C_1 e^{2t} + C_2 e^{-2t}$.

    **c.** Give the general solution of the equation for arbitrary $k > 0$ and verify your conjecture.

    **d.** For a positive real number $k$, verify that the general solution of the equation may also be expressed in the form $y(t) = C_1 \cosh kt + C_2 \sinh kt$, where cosh and sinh are the hyperbolic cosine and hyperbolic sine, respectively (Section 7.3).

**51.** **Another second-order equation** Consider the differential equation $y''(t) + k^2 y(t) = 0$, where $k$ is a positive real number.

    **a.** Verify by substitution that when $k = 1$, a solution of the equation is $y(t) = C_1 \sin t + C_2 \cos t$. You may assume this function is the general solution.

    **b.** Verify by substitution that when $k = 2$, the general solution of the equation is $y(t) = C_1 \sin 2t + C_2 \cos 2t$.

    **c.** Give the general solution of the equation for arbitrary $k > 0$ and verify your conjecture.

**52–56.** *In this section, several models are presented and the solution of the associated differential equation is given. Later in the chapter, we present methods for solving these differential equations.*

**T 52.** **Drug infusion** The delivery of a drug (such as an antibiotic) through an intravenous line may be modeled by the differential equation $m'(t) + km(t) = I$, where $m(t)$ is the mass of the drug in the blood at time $t \geq 0$, $k$ is a constant that describes the rate at which the drug is absorbed, and $I$ is the infusion rate.

**a.** Show by substitution that if the initial mass of drug in the blood is zero ($m(0) = 0$), then the solution of the initial value problem is $m(t) = \dfrac{I}{k}(1 - e^{-kt})$.

**b.** Graph the solution for $I = 10$ mg/hr and $k = 0.05$ hr$^{-1}$.

**c.** Evaluate $\displaystyle\lim_{t \to \infty} m(t)$, the steady-state drug level, and verify the result using the graph in part (b).

**T 53.** **Logistic population growth** Widely used models for population growth involve the *logistic equation* $P'(t) = rP\left(1 - \dfrac{P}{K}\right)$, where $P(t)$ is the population, for $t \geq 0$, and $r > 0$ and $K > 0$ are given constants.

**a.** Verify by substitution that the general solution of the equation is $P(t) = \dfrac{K}{1 + Ce^{-rt}}$, where $C$ is an arbitrary constant.

**b.** Find the value of $C$ that corresponds to the initial condition $P(0) = 50$.

**c.** Graph the solution for $P(0) = 50$, $r = 0.1$, and $K = 300$.

**d.** Find $\displaystyle\lim_{t \to \infty} P(t)$ and check that the result is consistent with the graph in part (c).

**T 54.** **Free fall** One possible model that describes the free fall of an object in a gravitational field subject to air resistance uses the equation $v'(t) = g - bv$, where $v(t)$ is the velocity of the object for $t \geq 0$, $g = 9.8$ m/s$^2$ is the acceleration due to gravity, and $b > 0$ is a constant that involves the mass of the object and the air resistance.

**a.** Verify by substitution that a solution of the equation, subject to the initial condition $v(0) = 0$, is $v(t) = \dfrac{g}{b}(1 - e^{-bt})$.

**b.** Graph the solution with $b = 0.1$ s$^{-1}$.

**c.** Using the graph in part (b), estimate the terminal velocity $\displaystyle\lim_{t \to \infty} v(t)$.

**T 55.** **Chemical rate equations** The reaction of certain chemical compounds can be modeled using a differential equation of the form

$y'(t) = -ky^n(t)$, where $y(t)$ is the concentration of the compound, for $t \geq 0$, $k > 0$ is a constant that determines the speed of the reaction, and $n$ is a positive integer called the *order* of the reaction. Assume the initial concentration of the compound is $y(0) = y_0 > 0$.

**a.** Consider a first-order reaction ($n = 1$) and show that the solution of the initial value problem is $y(t) = y_0 e^{-kt}$.

**b.** Consider a second-order reaction ($n = 2$) and show that the solution of the initial value problem is $y(t) = \dfrac{y_0}{y_0 kt + 1}$.

**c.** Let $y_0 = 1$ and $k = 0.1$. Graph the first-order and second-order solutions found in parts (a) and (b). Compare the two reactions.

**T 56.** **Tumor growth** The growth of cancer tumors may be modeled by the Gompertz growth equation. Let $M(t)$ be the mass of a tumor, for $t \geq 0$. The relevant initial value problem is

$$\frac{dM}{dt} = -rM(t)\ln\left(\frac{M(t)}{K}\right), \quad M(0) = M_0,$$

where $r$ and $K$ are positive constants and $0 < M_0 < K$.

**a.** Show by substitution that the solution of the initial value problem is

$$M(t) = K\left(\frac{M_0}{K}\right)^{\exp(-rt)}.$$

**b.** Graph the solution for $M_0 = 100$ and $r = 0.05$.

**c.** Using the graph in part (b), estimate $\displaystyle\lim_{t \to \infty} M(t)$, the limiting size of the tumor.

**QUICK CHECK ANSWERS**

**1.** The first equation is first order and linear. The second equation is second order and linear.   **2.** $y(t) = -20e^{-t/2} + 36$
**3.** $y(t) = t + 3t^{2/3} + 2$; domain is $(-\infty, 0)$
**4.** $s(t) = -4.9t^2 + 14.7t + 49$; domain is $[0, 5]$; stone reaches maximum height of 60.025 m at $t = 1.5$ s
**5.** The tank is empty at $t = 30$ s; the domain is $[0, 30]$. ◄

# 9.2 Direction Fields and Euler's Method

The goal of this chapter is to present methods for finding solutions of various kinds of differential equations. However, before taking up that task, we spend a few pages investigating a remarkable fact: It is possible to visualize and draw approximate graphs of the solutions of a differential equation without ever solving the equation. You might wonder how one can graph a function without knowing a formula for it. It turns out that the differential equation itself contains enough information to draw accurate graphs of its solutions. The tool that makes this visualization possible and allows us to explore the geometry of a differential equation is called the *direction field* (or *slope field*).

## Direction Fields

We work with first-order differential equations of the form

$$\frac{dy}{dt} = f(t, y),$$

where the notation $f(t, y)$ means an expression involving the independent variable $t$ and/or the unknown solution $y$. If a solution of this equation is displayed in the $ty$-plane, then the

**Figure 9.9**

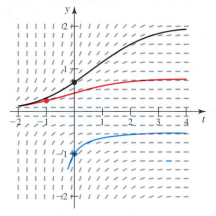

**Figure 9.10**

➤ If the function $f$ in the differential equation is even slightly complicated, drawing the direction field by hand is tedious. It's best to use a calculator or software. Examples 1 and 2 show some basic steps in plotting fairly simple direction fields by hand.

differential equation simply says that at each point $(t, y)$ of the solution curve, the slope of the curve is $y'(t) = f(t, y)$ (**Figure 9.9**). A **direction field** is a picture that shows the slope of the solution at selected points in the $ty$-plane.

For example, consider the equation $y'(t) = f(t, y) = y^2 e^{-t}$. We choose a regular grid of points in the $ty$-plane and at each point $P(t, y)$ we make a small line segment with slope $y^2 e^{-t}$. The line segment at a point $P$ gives the slope of the solution curve that passes through $P$ (**Figure 9.10**). For instance, we see that along the $t$-axis ($y = 0$), the slopes of the line segments are $f(t, 0) = 0$, which means the line segments are horizontal. Along the $y$-axis ($t = 0$), the slopes of the line segments are $f(0, y) = y^2$, which means the slopes of the line segments increase as we move up or down the $y$-axis.

Now suppose an initial condition $y(0) = \frac{2}{3}$ is given. We start at the point $(0, \frac{2}{3})$ in the $ty$-plane and sketch a curve that follows the flow of the direction field (black curve in Figure 9.10). At each point of the solution curve, the slope matches the direction field. Different initial conditions ($y(-1) = \frac{1}{3}$ and $y(0) = -1$ in Figure 9.10) give different solution curves. The collection of solution curves for several different initial conditions is a representation of the general solution of the equation.

**EXAMPLE 1    Direction field for a linear differential equation**  **Figure 9.11** shows the direction field for the equation $y'(t) = y - 2$, for $t \geq 0$ and $y \geq 0$. For what initial conditions at $t = 0$ are the solutions constant? Increasing? Decreasing?

**SOLUTION**  The direction field has horizontal line segments (slope zero) for $y = 2$. Therefore, $y'(t) = 0$ when $y = 2$, for all $t \geq 0$. These horizontal line segments correspond to a solution that is constant in time; that is, if the initial condition is $y(0) = 2$, then the solution is $y(t) = 2$, for all $t \geq 0$.

We also see that the direction field has line segments with positive slopes above the line $y = 2$ (with increasing slopes as you move away from $y = 2$). Therefore, $y'(t) > 0$ when $y > 2$, and solutions are increasing in this region.

Similarly, the direction field has line segments with negative slopes below the line $y = 2$ (with increasingly negative slopes as you move away from $y = 2$). Therefore, $y'(t) < 0$ when $y < 2$, and solutions are decreasing in this region.

Combining these observations, we see that if the initial condition satisfies $y(0) > 2$, the resulting solution is increasing, for $t \geq 0$. If the initial condition satisfies $y(0) < 2$, the resulting solution is decreasing, for $t \geq 0$. **Figure 9.12** shows the solution curves with initial conditions $y(0) = 2.25$, $y(0) = 2$, and $y(0) = 1.75$.

**QUICK CHECK 1**  Assuming solutions are unique (at most one solution curve passes through each point), explain why a solution curve cannot cross the line $y = 2$ in Example 1. ◄

**Figure 9.11**

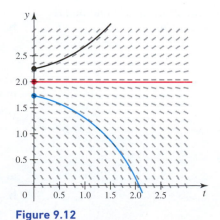

**Figure 9.12**

*Related Exercises 12, 17* ◄

➤ A differential equation in which the function $f$ is independent of $t$ is said to be **autonomous**.

For a differential equation of the form $y'(t) = f(y)$ (that is, the function $f$ depends only on $y$), the following steps are useful in sketching the direction field. Notice that because the direction field depends only on $y$, it has the same slope on any given horizontal line. A detailed direction field is usually not required.

---

**PROCEDURE    Sketching a Direction Field by Hand for $y'(t) = f(y)$**

1. Find the values of $y$ for which $f(y) = 0$. For example, suppose $f(a) = 0$. Then we have $y'(t) = 0$ whenever $y = a$, and the direction field at all points $(t, a)$ consists of horizontal line segments. If the initial condition is $y(0) = a$, then the solution is $y(t) = a$, for all $t \geq 0$. Such a constant solution is called an **equilibrium solution**.

2. Find the values of $y$ for which $f(y) > 0$. For example, suppose $f(b) > 0$. Then $y'(t) > 0$ whenever $y = b$. It follows that the direction field at all points $(t, b)$ has line segments with positive slopes, and the solution is increasing at those points.

3. Find the values of $y$ for which $f(y) < 0$. For example, suppose $f(c) < 0$. Then $y'(t) < 0$ whenever $y = c$. It follows that the direction field at all points $(t, c)$ has line segments with negative slopes and the solution is decreasing at those points.

---

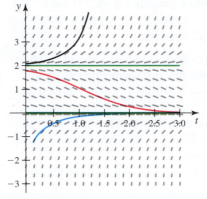

**Figure 9.13**

**QUICK CHECK 2** In Example 2, is the solution to the equation increasing or decreasing if the initial condition is $y(0) = 2.01$? Is it increasing or decreasing if the initial condition is $y(1) = -1$? ◄

**EXAMPLE 2    Direction field for a simple nonlinear equation** Consider the differential equation $y'(t) = y(y - 2)$, for $t \geq 0$.

**a.** For what initial conditions $y(0) = a$ is the resulting solution constant? Increasing? Decreasing?

**b.** Sketch the direction field for the equation.

**SOLUTION**

**a.** We follow the steps given in the procedure.

1. Letting $f(y) = y(y - 2)$, we see that $f(y) = 0$ when $y = 0$ or $y = 2$. Therefore, the direction field has horizontal line segments when $y = 0$ and $y = 2$. As a result, the constant functions $y(t) = 0$ and $y(t) = 2$, for $t \geq 0$, are equilibrium solutions (**Figure 9.13**).

2. The solutions of the inequality $f(y) = y(y - 2) > 0$ are $y < 0$ or $y > 2$. Therefore, below the line $y = 0$ and above the line $y = 2$, the direction field has positive slopes and the solutions are increasing in these regions.

3. The solution of the inequality $f(y) = y(y - 2) < 0$ is $0 < y < 2$. Therefore, between the lines $y = 0$ and $y = 2$, the direction field has negative slopes and the solutions are decreasing in this region.

**b.** The direction field is shown in Figure 9.13 with several representative solutions.

*Related Exercises 18, 20* ◄

**EXAMPLE 3    Direction field for the logistic equation** The logistic equation is commonly used to model populations with a stable equilibrium solution (called the *carrying capacity*). Consider the logistic equation

$$\frac{dP}{dt} = 0.1P\left(1 - \frac{P}{300}\right), \text{ for } t \geq 0.$$

**a.** Sketch the direction field of the equation.

**b.** Sketch solution curves corresponding to the initial conditions $P(0) = 50$, $P(0) = 150$, and $P(0) = 350$.

**c.** Find and interpret $\lim\limits_{t \to \infty} P(t)$.

**SOLUTION**

➤ The constant solutions $P = 0$ and $P = 300$ are equilibrium solutions. The solution $P = 0$ is an *unstable* equilibrium because nearby solution curves move away from $P = 0$. By contrast, the solution $P = 300$ is a *stable equilibrium* because nearby solution curves are attracted to $P = 300$.

**a.** We follow the steps in the summary box for sketching the direction field. Because $P$ represents a population, we assume $P \geq 0$.

1. Notice that $P'(t) = 0$ when $P = 0$ or $P = 300$. Therefore, if the initial population is either $P = 0$ or $P = 300$, then $P'(t) = 0$, for all $t \geq 0$, and the solution is

Solution curves in the direction field
for $\dfrac{dP}{dt} = 0.1P\left(1 - \dfrac{P}{300}\right)$

**Figure 9.14**

constant. For this reason we expect the direction field to show horizontal lines (with zero slope) at $P = 0$ and $P = 300$.

2. The equation implies that $P'(t) > 0$ provided $0 < P < 300$. Therefore, the direction field has positive slopes and the solutions are increasing, for $t \geq 0$ and $0 < P < 300$.

3. The equation also implies that $P'(t) < 0$ provided $P > 300$ (it was assumed $P \geq 0$). Therefore, the direction field has negative slopes and the solutions are decreasing, for $t \geq 0$ and $P > 300$.

b. **Figure 9.14** shows the direction field with three solution curves corresponding to the three different initial conditions.

c. The horizontal line $P = 300$ corresponds to the carrying capacity of the population. We see that if the initial population is positive and less than 300, the resulting solution increases to the carrying capacity from below. If the initial population is greater than 300, the resulting solution decreases to the carrying capacity from above.

*Related Exercises 21–22* ◄

**QUICK CHECK 3** According to Figure 9.14, for what approximate value of $P$ is the growth rate of the solution the greatest? ◄

## Euler's Method

➤ Euler proposed his method for finding approximate solutions to differential equations 200 years before digital computers were invented.

**Figure 9.15**

➤ See Exercise 45 for setting up Euler's method on a more general interval $[a, b]$.

➤ The argument used to derive the first step of Euler's method is really an application of linear approximation (Section 4.6). We draw a line tangent to the curve at the point $(t_0, u_0)$. The point on that line corresponding to $t = t_1$ is $(t_1, u_1)$, where $u_1$ is the Euler approximation to $y(t_1)$.

**Figure 9.16**

Direction fields are useful for at least two reasons. As shown in previous examples, a direction field provides valuable qualitative information about the solutions of a differential equation *without solving the equation*. In addition, it turns out that direction fields are the basis for many computer-based methods for approximating solutions of a differential equation. The computer begins with the initial condition and advances the solution in small steps, always following the direction field at each time step. The simplest method that uses this idea is called *Euler's method*.

Suppose we wish to approximate the solution to the initial value problem $y'(t) = f(t, y), y(0) = A$ on an interval $[0, T]$. We begin by dividing the interval $[0, T]$ into $N$ **time steps** of equal length $\Delta t = \dfrac{T}{N}$. In so doing, we create a set of grid points (**Figure 9.15**)

$$t_0 = 0, t_1 = \Delta t, t_2 = 2\Delta t, \ldots, t_k = k\Delta t, \ldots, t_N = N\Delta t = T.$$

The exact solution of the initial value problem at the grid points is $y(t_k)$, for $k = 0, 1, 2, \ldots, N$, which is generally unknown unless we are able to solve the original differential equation. The goal is to compute a set of *approximations* to the exact solution at the grid points, which we denote $u_k$, for $k = 0, 1, 2, \ldots, N$; that is, $u_k \approx y(t_k)$.

The initial condition says that $u_0 = y(0) = A$ (exactly). We now make one step forward in time of length $\Delta t$ and compute an approximation $u_1$ to $y(t_1)$. The key observation is that, according to the direction field, the solution at the point $(t_0, u_0)$ has slope $f(t_0, u_0)$. We obtain $u_1$ from $u_0$ by drawing a line segment starting at $(t_0, u_0)$ with horizontal extent $\Delta t$ and slope $f(t_0, u_0)$. The other endpoint of the line segment is $(t_1, u_1)$ (**Figure 9.16**). Applying the slope formula to the two points $(t_0, u_0)$ and $(t_1, u_1)$, we have

$$f(t_0, u_0) = \frac{u_1 - u_0}{t_1 - t_0}.$$

Solving for $u_1$ and noting that $t_1 - t_0 = \Delta t$, we have

$$u_1 = u_0 + f(t_0, u_0)\Delta t.$$

This basic *Euler step* is now repeated over each time step until we reach $t = T$. That is, having computed $u_1$, we apply the same argument to obtain $u_2$. From $u_2$, we compute $u_3$. In general, $u_{k+1}$ is computed from $u_k$, for $k = 0, 1, 2, \ldots, N - 1$. Hand calculations with Euler's method quickly become laborious. The method is usually carried out on a calculator or with a computer program. It is also included in many software packages.

> **PROCEDURE**   Euler's Method for $y'(t) = f(t, y), y(0) = A$ on $[0, T]$
>
> 1. Choose either a time step $\Delta t$ or a positive integer $N$ such that $\Delta t = \dfrac{T}{N}$ and $t_k = k\Delta t$, for $k = 0, 1, 2, \ldots, N - 1$.
> 2. Let $u_0 = y(0) = A$.
> 3. For $k = 0, 1, 2, \ldots, N - 1$, compute
>
> $$u_{k+1} = u_k + f(t_k, u_k)\Delta t.$$
>
> Each $u_k$ is an approximation to the exact solution $y(t_k)$.

**EXAMPLE 4   Using Euler's method**  Find an approximate solution to the initial value problem $y'(t) = t - \dfrac{y}{2}, y(0) = 1$, on the interval $[0, 2]$. Use the time steps $\Delta t = 0.2$ $(N = 10)$ and $\Delta t = 0.1$ $(N = 20)$. Which time step gives a better approximation to the exact solution, which is $y(t) = 5e^{-t/2} + 2t - 4$?

**SOLUTION**  With a time step of $\Delta t = 0.2$, the grid points on the interval $[0, 2]$ are

$$t_0 = 0.0, t_1 = 0.2, t_2 = 0.4, \ldots, t_{10} = 2.0.$$

We identify $f(t, y) = t - \dfrac{y}{2}$ and let $u_k$ be the Euler approximation to $y(t_k)$. Euler's method takes the form

$$u_0 = y(0) = 1, \quad u_{k+1} = u_k + f(t_k, u_k)\Delta t = u_k + \left(t_k - \frac{u_k}{2}\right)\Delta t,$$

where $k = 0, 1, 2, \ldots, 9$. For example, the value of the approximation $u_1$ is given by

$$u_1 = u_0 + f(t_0, u_0)\Delta t = u_0 + \left(t_0 - \frac{u_0}{2}\right)\Delta t = 1 + \left(0 - \frac{1}{2}\right) \cdot 0.2 = 0.900,$$

and the value of $u_2$ is given by

$$u_2 = u_1 + f(t_1, u_1)\Delta t = u_1 + \left(t_1 - \frac{u_1}{2}\right)\Delta t = 0.9 + \left(0.2 - \frac{0.9}{2}\right) \cdot 0.2 = 0.850.$$

A similar procedure is used with $\Delta t = 0.1$. In this case, $N = 20$ time steps are needed to cover the interval $[0, 2]$. The results of the two calculations are shown in **Figure 9.17**, where the exact solution appears as a solid curve and the Euler approximations are shown as points. From these graphs, it appears that the time step $\Delta t = 0.1$ gives better approximations to the solution.

A more detailed account of these calculations is given in Table 9.1, which shows the numerical values of the Euler approximations for $\Delta t = 0.2$ and $\Delta t = 0.1$. Notice that the approximations with $\Delta t = 0.1$ are tabulated at *every other* time step so that they may be compared to the $\Delta t = 0.2$ approximations.

**Figure 9.17**

> Because computers produce small errors at each time step, taking a large number of time steps may eventually lead to an unacceptable accumulation of errors. When more accuracy is needed, it may be best to use other methods that require more work per time step, but also give more accurate results.

**QUICK CHECK 4** Notice that the errors in Table 9.1 increase in time for both time steps. Give a possible explanation for this increase in the errors. ◄

**Table 9.1**

| $t_k$ | $u_k(\Delta t = 0.2)$ | $u_k(\Delta t = 0.1)$ | $e_k(\Delta t = 0.2)$ | $e_k(\Delta t = 0.1)$ |
|---|---|---|---|---|
| 0.0 | 1.000 | 1.000 | 0.000 | 0.000 |
| 0.2 | 0.900 | 0.913 | 0.0242 | 0.0117 |
| 0.4 | 0.850 | 0.873 | 0.0437 | 0.0211 |
| 0.6 | 0.845 | 0.875 | 0.0591 | 0.0286 |
| 0.8 | 0.881 | 0.917 | 0.0711 | 0.0345 |
| 1.0 | 0.952 | 0.994 | 0.0802 | 0.0390 |
| 1.2 | 1.057 | 1.102 | 0.0869 | 0.0423 |
| 1.4 | 1.191 | 1.238 | 0.0914 | 0.0446 |
| 1.6 | 1.352 | 1.401 | 0.0943 | 0.0460 |
| 1.8 | 1.537 | 1.586 | 0.0957 | 0.0468 |
| 2.0 | 1.743 | 1.792 | 0.0960 | 0.0470 |

How accurate are these approximations? Although it does not generally happen in practice, we can determine the solution of this particular initial value problem exactly. (You can check that the solution is $y(t) = 5e^{-t/2} + 2t - 4$.) We investigate the accuracy of the Euler approximations by computing the *error*, $e_k = |u_k - y(t_k)|$, at each grid point. The error simply measures the difference between the exact solution and the corresponding approximations. The last two columns of Table 9.1 show the errors associated with the approximations. We see that at every grid point, the approximations with $\Delta t = 0.1$ have errors with roughly half the magnitude of the errors with $\Delta t = 0.2$.

This pattern is typical of Euler's method. If we focus on one point in time, halving the time step roughly halves the errors. However, nothing is free: Halving the time step also requires twice as many time steps and twice the amount of computational work to cover the same time interval.

*Related Exercises 29, 32* ◄

**Final Note**    Euler's method is the simplest of a vast collection of *numerical methods* for approximating solutions of differential equations (often studied in courses on *numerical analysis*). As we have seen, Euler's method uses linear approximation; that is, the method follows the direction field using line segments. This idea works well provided the direction field varies smoothly and slowly. In less well behaved cases, Euler's method may encounter difficulties. More robust and accurate methods do a better job of following the direction field (for example, by using parabolas or higher-degree polynomials instead of linear approximation). While these refined methods are generally more accurate than Euler's method, they often require more computational work per time step. As with Euler's method, all methods have the property that their accuracy improves as the time step decreases. The upshot is that there are often trade-offs in choosing a method to approximate the solution of a differential equation. However, Euler's method is a good place to start and may be adequate.

## SECTION 9.2 EXERCISES

### Getting Started

1. Explain how to sketch the direction field of the equation $y'(t) = f(t, y)$, where $f$ is given.

2. Consider the differential equation $y'(t) = t^2 - 3y^2$ and the solution curve that passes through the point $(3, 1)$. What is the slope of the curve at $(3, 1)$?

3. Consider the initial value problem $y'(t) = t^2 - 3y^2$, $y(3) = 1$. What is the approximation to $y(3.1)$ given by Euler's method with a time step of $\Delta t = 0.1$?

4. Give a geometrical explanation of how Euler's method works.

5. **Matching direction fields** Match equations a–d with direction fields A–D.

   **a.** $y'(t) = \dfrac{t}{2}$            **b.** $y'(t) = \dfrac{y}{2}$

   **c.** $y'(t) = \dfrac{t^2 + y^2}{2}$    **d.** $y'(t) = \dfrac{y}{t}$

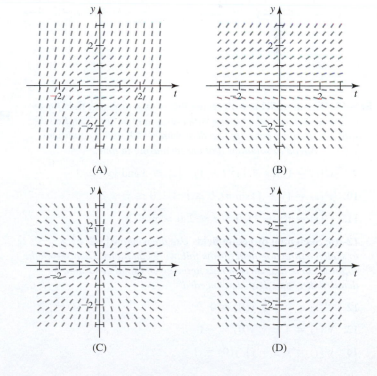

(A)             (B)

(C)             (D)

**6.  Identifying direction fields** Which of the differential equations a–d corresponds to the following direction field? Explain your reasoning.

**a.** $y'(t) = 0.5(y + 1)(t - 1)$

**b.** $y'(t) = -0.5(y + 1)(t - 1)$

**c.** $y'(t) = 0.5(y - 1)(t + 1)$

**d.** $y'(t) = -0.5(y - 1)(t + 1)$

## Practice Exercises

**7–8. Direction fields** *A differential equation and its direction field are shown in the following figures. Sketch a graph of the solution curve that passes through the given initial conditions.*

**7.** $y'(t) = \dfrac{t^2}{y^2 + 1}$, $y(0) = -2$  **8.**  $y'(t) = \dfrac{\sin t}{y}$, $y(-2) = -2$

and $y(-2) = 0$.   and $y(-2) = 2$

**T 9–11. Direction fields with technology** *Plot a direction field for the following differential equation with a graphing utility. Then find the solutions that are constant and determine which initial conditions $y(0) = A$ lead to solutions that are increasing in time.*

**9.** $y'(t) = 0.5(y + 1)^2(t - 1)^2$, $|t| \leq 3$ and $|y| \leq 3$

**10.** $y'(t) = (y - 1)\sin \pi t$, $0 \leq t \leq 2$, $0 \leq y \leq 2$

**11.** $y'(t) = t(y - 1)$, $0 \leq t \leq 2$, $0 \leq y \leq 2$

**12–16. Sketching direction fields** *Use the window $[-2, 2] \times [-2, 2]$ to sketch a direction field for the following equations. Then sketch the solution curve that corresponds to the given initial condition. A detailed direction field is not needed.*

**12.** $y'(t) = y - 3$, $y(0) = 1$

**13.** $y'(t) = 4 - y$, $y(0) = -1$

**14.** $y'(t) = y(2 - y)$, $y(0) = 1$

**15.** $y'(x) = \sin x$, $y(-2) = 2$

**16.** $y'(x) = \sin y$, $y(-2) = \dfrac{1}{2}$

**17–20. Increasing and decreasing solutions** *Consider the following differential equations. A detailed direction field is not needed.*

**a.** *Find the solutions that are constant, for all $t \geq 0$ (the equilibrium solutions).*

**b.** *In what regions are solutions increasing? Decreasing?*

**c.** *Which initial conditions $y(0) = A$ lead to solutions that are increasing in time? Decreasing?*

**d.** *Sketch the direction field and verify that it is consistent with parts (a)–(c).*

**17.** $y'(t) = (y - 1)(1 + y)$   **18.**  $y'(t) = (y - 2)(y + 1)$

**19.** $y'(t) = \cos y$, for $|y| \leq \pi$   **20.**  $y'(t) = y(y + 3)(4 - y)$

**21–24. Logistic equations** *Consider the following logistic equations. In each case, sketch the direction field, draw the solution curve for each initial condition, and find the equilibrium solutions. A detailed direction field is not needed. Assume $t \geq 0$ and $P \geq 0$.*

**21.** $P'(t) = 0.05P\left(1 - \dfrac{P}{500}\right)$; $P(0) = 100$, $P(0) = 400$, $P(0) = 700$

**22.** $P'(t) = 0.1P\left(1 - \dfrac{P}{1200}\right)$; $P(0) = 600$, $P(0) = 800$, $P(0) = 1600$

**23.** $P'(t) = 0.02P\left(4 - \dfrac{P}{800}\right)$; $P(0) = 1600$, $P(0) = 2400$, $P(0) = 4000$

**24.** $P'(t) = 0.05P - 0.001P^2$; $P(0) = 10$, $P(0) = 40$, $P(0) = 80$

**25–28. Two steps of Euler's method** *For the following initial value problems, compute the first two approximations $u_1$ and $u_2$ given by Euler's method using the given time step.*

**25.** $y'(t) = 2y$, $y(0) = 2$; $\Delta t = 0.5$

**26.** $y'(t) = -y$, $y(0) = -1$; $\Delta t = 0.2$

**27.** $y'(t) = 2 - y$, $y(0) = 1$; $\Delta t = 0.1$

**28.** $y'(t) = t + y$, $y(0) = 4$; $\Delta t = 0.5$

**T 29–32. Errors in Euler's method** *Consider the following initial value problems.*

**a.** *Find the approximations to $y(0.2)$ and $y(0.4)$ using Euler's method with time steps of $\Delta t = 0.2, 0.1, 0.05$, and $0.025$.*

**b.** *Using the exact solution given, compute the errors in the Euler approximations at $t = 0.2$ and $t = 0.4$.*

**c.** *Which time step results in the more accurate approximation? Explain your observations.*

**d.** *In general, how does halving the time step affect the error at $t = 0.2$ and $t = 0.4$?*

**29.** $y'(t) = -y$, $y(0) = 1$; $y(t) = e^{-t}$

**30.** $y'(t) = \dfrac{y}{2}$, $y(0) = 2$; $y(t) = 2e^{t/2}$

**31.** $y'(t) = 4 - y$, $y(0) = 3$; $y(t) = 4 - e^{-t}$

**32.** $y'(t) = 2t + 1$, $y(0) = 0$; $y(t) = t^2 + t$

**T** **33–36. Computing Euler approximations** *Use a calculator or computer program to carry out the following steps.*

**a.** *Approximate the value of $y(T)$ using Euler's method with the given time step on the interval $[0, T]$.*

**b.** *Using the exact solution (also given), find the error in the approximation to $y(T)$ (only at the right endpoint of the time interval).*

**c.** *Repeating parts (a) and (b) using half the time step used in those calculations, again find an approximation to $y(T)$.*

**d.** *Compare the errors in the approximations to $y(T)$.*

**33.** $y'(t) = -2y, y(0) = 1; \Delta t = 0.2, T = 2; y(t) = e^{-2t}$

**34.** $y'(t) = 6 - 2y, y(0) = -1; \Delta t = 0.2, T = 3;$
$y(t) = 3 - 4e^{-2t}$

**35.** $y'(t) = t - y, y(0) = 4; \Delta t = 0.2, T = 4;$
$y(t) = 5e^{-t} + t - 1$

**36.** $y'(t) = \dfrac{t}{y}, y(0) = 4; \Delta t = 0.1, T = 2; y(t) = \sqrt{t^2 + 16}$

**37.** **Explain why or why not** Determine whether the following statements are true and give an explanation or counterexample.

   **a.** A direction field allows you to visualize the solution of a differential equation, but it does not give exact values of the solution at particular points.

   **b.** Euler's method is used to compute exact values of the solution of an initial value problem.

**38–43. Equilibrium solutions** *A differential equation of the form $y'(t) = f(y)$ is said to be **autonomous** (the function $f$ depends only on $y$). The constant function $y = y_0$ is an **equilibrium solution** of the equation provided $f(y_0) = 0$ (because then $y'(t) = 0$ and the solution remains constant for all $t$). Note that equilibrium solutions correspond to horizontal lines in the direction field. Note also that for autonomous equations, the direction field is independent of $t$. Carry out the following analysis on the given equations.*

**a.** *Find the equilibrium solutions.*

**b.** *Sketch the direction field, for $t \geq 0$.*

**c.** *Sketch the solution curve that corresponds to the initial condition $y(0) = 1$.*

**38.** $y'(t) = 2y + 4$        **39.** $y'(t) = 6 - 2y$

**40.** $y'(t) = y(2 - y)$      **41.** $y'(t) = y(y - 3)$

**42.** $y'(t) = \sin y$         **43.** $y'(t) = y(y - 3)(y + 2)$

## Explorations and Challenges

**44.** **Direction field analysis** Consider the first-order initial value problem $y'(t) = ay + b, y(0) = A$, for $t \geq 0$, where $a$, $b$, and $A$ are real numbers.

   **a.** Explain why $y = -b/a$ is an equilibrium solution and corresponds to a horizontal line in the direction field.

   **b.** Draw a representative direction field in the case that $a > 0$. Show that if $A > -b/a$, then the solution increases for $t \geq 0$, and that if $A < -b/a$, then the solution decreases for $t \geq 0$.

   **c.** Draw a representative direction field in the case that $a < 0$. Show that if $A > -b/a$, then the solution decreases for $t \geq 0$, and that if $A < -b/a$, then the solution increases for $t \geq 0$.

**45.** **Euler's method on more general grids** Suppose the solution of the initial value problem $y'(t) = f(t, y), y(a) = A$ is to be approximated on the interval $[a, b]$.

   **a.** If $N + 1$ grid points are used (including the endpoints), what is the time step $\Delta t$?

   **b.** Write the first step of Euler's method to compute $u_1$.

   **c.** Write the general step of Euler's method that applies, for $k = 0, 1, \ldots, N - 1$.

**46–48. Analyzing models** *The following models were discussed in Section 9.1 and reappear in later sections of this chapter. In each case, carry out the indicated analysis using direction fields.*

**46.** **Drug infusion** The delivery of a drug (such as an antibiotic) through an intravenous line may be modeled by the differential equation $m'(t) + km(t) = I$, where $m(t)$ is the mass of the drug in the blood at time $t \geq 0$, $k$ is a constant that describes the rate at which the drug is absorbed, and $I$ is the infusion rate. Let $I = 10$ mg/hr and $k = 0.05$ hr$^{-1}$.

   **a.** Draw the direction field, for $0 \leq t \leq 100, 0 \leq y \leq 600$.

   **b.** For what initial values $m(0) = A$ are solutions increasing? Decreasing?

   **c.** What is the equilibrium solution?

**47.** **Free fall** A model that describes the free fall of an object in a gravitational field subject to air resistance uses the equation $v'(t) = g - bv$, where $v(t)$ is the velocity of the object, for $t \geq 0$, $g = 9.8$ m/s$^2$ is the acceleration due to gravity, and $b > 0$ is a constant that involves the mass of the object and the air resistance. Let $b = 0.1$ s$^{-1}$.

   **a.** Draw the direction field for $0 \leq t \leq 60, 0 \leq y \leq 150$.

   **b.** For what initial values $v(0) = A$ are solutions increasing? Decreasing?

   **c.** What is the equilibrium solution?

**48.** **Chemical rate equations** Consider the chemical rate equations $y'(t) = -ky(t)$ and $y'(t) = -ky^2(t)$, where $y(t)$ is the concentration of the compound for $t \geq 0$, and $k > 0$ is a constant that determines the speed of the reaction. Assume the initial concentration of the compound is $y(0) = y_0 > 0$.

   **a.** Let $k = 0.3$ and make a sketch of the direction fields for both equations. What is the equilibrium solution in both cases?

   **b.** According to the direction fields, which reaction approaches its equilibrium solution faster?

**49.** **Convergence of Euler's method** Suppose Euler's method is applied to the initial value problem $y'(t) = ay, y(0) = 1$, which has the exact solution $y(t) = e^{at}$. For this exercise, let $h$ denote the time step (rather than $\Delta t$). The grid points are then given by $t_k = kh$. We let $u_k$ be the Euler approximation to the exact solution $y(t_k)$, for $k = 0, 1, 2, \ldots$.

   **a.** Show that Euler's method applied to this problem can be written $u_0 = 1, u_{k+1} = (1 + ah)u_k$, for $k = 0, 1, 2, \ldots$.

   **b.** Show by substitution that $u_k = (1 + ah)^k$ is a solution of the equations in part (a), for $k = 0, 1, 2, \ldots$.

   **c.** Recall from Section 4.7 that $\lim\limits_{h \to 0} (1 + ah)^{1/h} = e^a$. Use this fact to show that as the time step goes to zero ($h \to 0$, with $t_k = kh$ fixed), the approximations given by Euler's method approach the exact solution of the initial value problem; that is, $\lim\limits_{h \to 0} u_k = \lim\limits_{h \to 0} (1 + ah)^k = y(t_k) = e^{at_k}$.

**50. Stability of Euler's method** Consider the initial value problem $y'(t) = -ay$, $y(0) = 1$, where $a > 0$; it has the exact solution $y(t) = e^{-at}$, which is a decreasing function.

**a.** Show that Euler's method applied to this problem with time step $h$ can be written $u_0 = 1$, $u_{k+1} = (1 - ah)u_k$, for $k = 0, 1, 2, \ldots$.

**b.** Show by substitution that $u_k = (1 - ah)^k$ is a solution of the equations in part (a), for $k = 0, 1, 2, \ldots$.

**c.** Explain why, as $k$ increases, the Euler approximations $u_k = (1 - ah)^k$ decrease in magnitude only if $|1 - ah| < 1$.

**d.** Show that the inequality in part (c) implies that the time step must satisfy the condition $0 < h < \dfrac{2}{a}$. If the time step does not satisfy this condition, then Euler's method is *unstable* and produces approximations that actually increase in time.

**QUICK CHECK ANSWERS**

**1.** To cross the line $y = 2$, the solution must have a slope different from zero when $y = 2$. However, according to the direction field, a solution on the line $y = 2$ has zero slope.　**2.** The solutions originating at both initial conditions are increasing.　**3.** The direction field is steepest when $P = 150$.　**4.** Each step of Euler's method introduces an error. With each successive step of the calculation, the errors could accumulate (or propagate). ◄

# 9.3 Separable Differential Equations

Sketching solutions of a differential equation using its direction field is a powerful technique, and it provides a wealth of information about the solutions. However, valuable as they are, direction fields do not produce the actual solutions of a differential equation. In this section, we examine methods that lead to the solutions of certain differential equations in terms of an algebraic expression (often called an *analytical solution*). The equations we consider are first order and belong to a class called *separable equations*.

## Method of Solution

> If the equation has the form $y'(t) = f(t)$ (that is, the right side depends only on $t$), then solving the equation amounts to finding the antiderivatives of $f$, a problem discussed in Section 4.9.

The most general first-order differential equation has the form $y'(t) = F(t, y)$, where $F(t, y)$ is an expression that may involve both the independent variable $t$ and the unknown function $y$. We have a *chance* of solving such an equation if it can be written in the form

$$g(y)y'(t) = h(t).$$

In the equation $g(y)y'(t) = h(t)$, the factor $g(y)$ involves only $y$, and $h(t)$ involves only $t$; that is, the variables have been separated. An equation that can be written in this form is said to be **separable**.

In general, we solve a separable differential equation by integrating both sides of the equation with respect to $t$:

$$\int g(y)\underbrace{y'(t)\,dt}_{dy} = \int h(t)\,dt \qquad \text{Integrate both sides with respect to } t.$$

$$\int g(y)\,dy = \int h(t)\,dt. \qquad \text{Change variables on left; } dy = y'(t)\,dt.$$

The fact that $dy = y'(t)\,dt$ on the left side of the equation leaves us with two integrals to evaluate, one with respect to $y$ and one with respect to $t$. Finding a solution depends on evaluating these integrals.

**QUICK CHECK 1** Which of the following equations are separable? (A) $y'(t) = y + t$, (B) $y'(t) = \dfrac{ty}{t + 1}$, and (C) $y'(x) = e^{x+y}$ ◄

**EXAMPLE 1** **A separable equation** Find a function that satisfies the following initial value problem.

$$y'(t) = y^2 e^{-t}, \quad y(0) = \tfrac{1}{2}$$

**SOLUTION** The equation is written in separable form by dividing both sides of the equation by $y^2$ to give $\dfrac{y'(t)}{y^2} = e^{-t}$. We now integrate both sides of the equation with respect to $t$ and evaluate the resulting integrals:

$$\int \frac{1}{y^2} y'(t)\, dt = \int e^{-t}\, dt$$

$$\int \frac{dy}{y^2} = \int e^{-t}\, dt \qquad \text{Change variables on left side.}$$

$$-\frac{1}{y} = -e^{-t} + C. \qquad \text{Evaluate integrals.}$$

Solving for $y$ gives the general solution

$$y(t) = \frac{1}{e^{-t} - C}.$$

> ➤ In practice, the change of variable on the left side is often omitted, and we go directly to the second step, which is to integrate the left side with respect to $y$ and the right side with respect to $t$.

> ➤ Notice that each integration produces a constant of integration. The two constants of integration may be combined into one.

The initial condition $y(0) = \frac{1}{2}$ implies that

$$y(0) = \frac{1}{e^0 - C} = \frac{1}{1 - C} = \frac{1}{2}.$$

It follows that $C = -1$, so the solution to the initial value problem is

$$y(t) = \frac{1}{e^{-t} + 1}, \text{ for all real } t.$$

The solution (**Figure 9.18**) passes through $\left(0, \frac{1}{2}\right)$ and increases to approach the asymptote $y = 1$ because $\displaystyle\lim_{t\to\infty} \frac{1}{e^{-t} + 1} = 1.$

Related Exercises 17, 20 ◀

**Figure 9.18**

**QUICK CHECK 2** Write

$$y'(t) = (t^2 + 1)/y^3$$

in separated form. ◀

**EXAMPLE 2**    **Another separable equation**   Find the solutions of the equation $y'(x) = e^{-y} \sin x$ subject to the three different initial conditions

$$y(0) = 1, \quad y\!\left(\frac{\pi}{2}\right) = \frac{1}{2}, \quad \text{and} \quad y(\pi) = \ln 2.$$

**SOLUTION** Writing the equation in the form $e^y y'(x) = \sin x$, we see that it is separable. Integrating both sides with respect to $x$, we have

$$\int e^y y'(x)\, dx = \int \sin x\, dx$$

$$\int e^y\, dy = \int \sin x\, dx \qquad \text{Change variables on left side.}$$

$$e^y = -\cos x + C. \qquad \text{Evaluate integrals.}$$

The general solution $y$ is found by taking logarithms of both sides of this equation:

$$y(x) = \ln\,(C - \cos x).$$

The three initial conditions are now used to evaluate the constant $C$ for the three solutions:

$$y(0) = 1 \Rightarrow 1 = \ln\,(C - \cos 0) = \ln\,(C - 1) \Rightarrow e = C - 1 \Rightarrow C = e + 1,$$

$$y\!\left(\frac{\pi}{2}\right) = \frac{1}{2} \Rightarrow \frac{1}{2} = \ln\!\left(C - \cos\frac{\pi}{2}\right) = \ln C \Rightarrow C = e^{1/2}, \text{ and}$$

$$y(\pi) = \ln 2 \Rightarrow \ln 2 = \ln\,(C - \cos \pi) = \ln\,(C + 1) \Rightarrow 2 = C + 1 \Rightarrow C = 1.$$

Substituting these values of $C$ into the general solution gives the solutions of the three initial value problems (**Figure 9.19**):

$$y_1(x) = \ln\,(e + 1 - \cos x), \tag{1}$$
$$y_2(x) = \ln\,(e^{1/2} - \cos x), \text{ and} \tag{2}$$
$$y_3(x) = \ln\,(1 - \cos x). \tag{3}$$

**Figure 9.19**

**QUICK CHECK 3** Find the value of the constant $C$ in Example 2 with the initial condition $y(\pi) = 0$, and then explain why the domain of the corresponding solution is $(\pi/2, 3\pi/2)$. ◄

Solutions (1) and (2) have a domain of all real numbers (both $e + 1 - \cos x$ and $e^{1/2} - \cos x$ are greater than 0, for all $x$). The graph of solution (3) has a vertical asymptote at $x = 0, \pm2\pi, \pm4\pi, \ldots$ because $1 - \cos x = 0$ at those points and $\lim\limits_{x \to 0^+} \ln x = -\infty$. Therefore, we choose the branch of the graph that passes through the initial condition $(\pi, \ln 2)$ (Figure 9.19) and conclude that the domain of $y_3(x)$ is $(0, 2\pi)$.

*Related Exercises 26–27* ◄

Even if we can evaluate the integrals necessary to solve a separable equation, the solution may not be easily expressed in an explicit form. Here is an example of a solution that is best left in implicit form.

**EXAMPLE 3**    **An implicit solution** Find and graph the solution of the initial value problem

$$(\cos y)\, y'(t) = \sin^2 t \cos t, \quad y(0) = \frac{\pi}{6}.$$

**SOLUTION** The equation is already in separated form. Integrating both sides with respect to $t$, we have

> ▶ For the integral on the right side, we use the substitution $u = \sin t$. The integral becomes $\int u^2\, du = \frac{1}{3}u^3 + C$.

$$\int \cos y\, dy = \int \sin^2 t \cos t\, dt \quad \text{Integrate both sides.}$$

$$\sin y = \frac{1}{3}\sin^3 t + C. \quad \text{Evaluate integrals.}$$

When imposing the initial condition in this case, it is best to leave the general solution in implicit form. Substituting $t = 0$ and $y = \dfrac{\pi}{6}$ into the general solution, we find that

> ▶ Care must be used when graphing and interpreting implicit solutions. The graph of
> $$\sin y = \frac{1}{3}\sin^3 t + \frac{1}{2}$$
> is a family of an infinite number of curves.

$$\sin\frac{\pi}{6} = \frac{1}{3}\sin^3 0 + C \quad \text{or} \quad C = \frac{1}{2}.$$

Therefore, the solution of the initial value problem is

$$\sin y = \frac{1}{3}\sin^3 t + \frac{1}{2}.$$

In order to graph the solution in this implicit form, it is easiest to use graphing software. The result is shown in **Figure 9.20**.

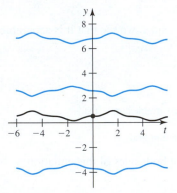

You must choose the curve that satisfies the initial condition, as shown in Figure 9.20. The domain of this solution is $(-\infty, \infty)$.

**Figure 9.20**

*Related Exercises 33, 36* ◄

**QUICK CHECK 4** Find the value of the constant $C$ in Example 3 with the initial condition $y\!\left(\dfrac{\pi}{6}\right) = 0$. ◄

## Logistic Equation Revisited

> ▶ The derivation of the logistic equation is discussed in Section 9.5.

In Exercise 53 of Section 9.1, we introduced the logistic equation, which is commonly used for modeling populations, epidemics, and the spread of rumors. In Section 9.2, we investigated the direction field associated with the logistic equation. It turns out that the logistic equation is a separable equation, so we now have the tools needed to solve it.

**EXAMPLE 4** **Logistic population growth** Assume 50 fruit flies are in a large jar at the beginning of an experiment. Let $P(t)$ be the number of fruit flies in the jar $t$ days later. At first, the population grows exponentially, but due to limited space and food supply, the growth rate decreases and the population is prevented from growing without bound. This experiment is modeled by the *logistic equation*

$$\frac{dP}{dt} = 0.1P\left(1 - \frac{P}{300}\right), \text{ for } t \geq 0,$$

together with the initial condition $P(0) = 50$. Solve this initial value problem.

**SOLUTION** We see that the equation is separable by writing it in the form

$$\frac{1}{P\left(1 - \dfrac{P}{300}\right)} \cdot \frac{dP}{dt} = 0.1.$$

Integrating both sides with respect to $t$ leads to the equation

$$\int \frac{dP}{P\left(1 - \dfrac{P}{300}\right)} = \int 0.1 \, dt. \tag{4}$$

The integral on the right side of equation (4) is $\int 0.1 \, dt = 0.1t + C$.

Because the integrand on the left side is a rational function in $P$, we use partial fractions. You should verify that

$$\frac{1}{P\left(1 - \dfrac{P}{300}\right)} = \frac{300}{P(300 - P)} = \frac{1}{P} + \frac{1}{300 - P}$$

and therefore,

$$\int \frac{1}{P\left(1 - \dfrac{P}{300}\right)} \, dP = \int \left(\frac{1}{P} + \frac{1}{300 - P}\right) dP = \ln\left|\frac{P}{300 - P}\right| + C.$$

After integration, equation (4) becomes

$$\ln\left|\frac{P}{300 - P}\right| = 0.1t + C. \tag{5}$$

The next step is to solve for $P$, which is tangled up inside the logarithm. We first exponentiate both sides of equation (5) to obtain

$$\left|\frac{P}{300 - P}\right| = e^C \cdot e^{0.1t}.$$

We can remove the absolute value on the left side of equation (5) by writing

$$\frac{P}{300 - P} = \pm \, e^C \cdot e^{0.1t}.$$

> There are not many times in mathematics when we can redefine a constant in the middle of a calculation. When working with arbitrary constants, it may be possible, if it is done carefully.

At this point, a useful trick simplifies matters. Because $C$ is an arbitrary constant, $\pm e^C$ is also an arbitrary constant, so we rename $\pm e^C$ as $C$. We now have

$$\frac{P}{300 - P} = Ce^{0.1t}. \tag{6}$$

Solving equation (6) for $P$ and replacing $1/C$ by $C$ gives the general solution

$$P(t) = \frac{300}{1 + Ce^{-0.1t}}.$$

➤ We could also use the initial condition in equation (6) to solve for C.

**Figure 9.21**

**Figure 9.21** shows the general solution, with curves corresponding to several different values of C. Using the initial condition $P(0) = 50$, we find the value of C for our specific problem is $C = 5$. It follows that the solution of the initial value problem is

$$P(t) = \frac{300}{1 + 5e^{-0.1t}}, \text{ for } t \geq 0.$$

Figure 9.21 also shows this particular solution (in red) among the curves in the general solution. A significant feature of this model is that, for $0 < P(0) < 300$, the population increases, but not without bound. Instead, it approaches an **equilibrium**, or **steady-state**, solution with a value of

$$\lim_{t \to \infty} P(t) = \lim_{t \to \infty} \frac{300}{1 + 5e^{-0.1t}} = 300,$$

which is the maximum population that the environment (space and food supply) can sustain. This equilibrium population is called the **carrying capacity**. Notice that all the curves in the general solution approach the carrying capacity as $t$ increases.

*Related Exercises 39–40* ◄

---

## SECTION 9.3 EXERCISES

### Getting Started

1. What is a separable first-order differential equation?

2. Is the equation $t^2 y'(t) = \dfrac{t + 4}{y^2}$ separable?

3. Is the equation $y'(t) = 2y - t$ separable?

4. Explain how to solve a separable differential equation of the form $g(t)y'(t) = h(t)$.

### Practice Exercises

**5–16. Solving separable equations** *Find the general solution of the following equations. Express the solution explicitly as a function of the independent variable.*

5. $t^{-3} y'(t) = 1$

6. $e^{4t} y'(t) = 5$

7. $\dfrac{dy}{dt} = \dfrac{3t^2}{y}$

8. $\dfrac{dy}{dx} = y(x^2 + 1)$

9. $y'(t) = e^{y/2} \sin t$

10. $x^2 \dfrac{dw}{dx} = \sqrt{w}(3x + 1)$

11. $x^2 y'(x) = y^2$

12. $(t^2 + 1)^3 yy'(t) = t(y^2 + 4)$

13. $y'(t)\csc t = -\dfrac{y^3}{2}$

14. $y'(t)e^{t/2} = y^2 + 4$

15. $u'(x) = e^{2x - u}$

16. $xu'(x) = u^2 - 4$

**17–32. Solving initial value problems** *Determine whether the following equations are separable. If so, solve the initial value problem.*

17. $2yy'(t) = 3t^2, y(0) = 9$

18. $y'(t) = e^{ty}, y(0) = 1$

19. $\dfrac{dy}{dt} = ty + 2, y(1) = 2$

20. $y'(t) = y(4t^3 + 1), y(0) = 4$

21. $y'(t) = ye^t, y(0) = -1$

22. $y'(x) = y \cos x, y(0) = 3$

23. $\dfrac{dy}{dx} = e^{x-y}, y(0) = \ln 3$

24. $y'(t) = \cos^2 y, y(1) = \dfrac{\pi}{4}$

25. $y'(t) = \dfrac{\ln^3 t}{te^y}, y(1) = 0$

26. $ty'(t) = y(y + 1), y(3) = 1$

27. $y'(t) = \dfrac{\sec^2 t}{2y}, y(\pi/4) = 1$

28. $y'(t) = \dfrac{y + 3}{5t + 6}, y(2) = 0$

29. $y'(t) = \dfrac{t}{y}, y(1) = 2$

30. $y'(t) = y^3 \sin t, y(0) = 1$

31. $ty'(t) = 1, y(1) = 2$

32. $(\sec t)y'(t) = 1, y(0) = 1$

**T 33–38. Solutions in implicit form** *Solve the following initial value problems and leave the solution in implicit form. Use graphing software to plot the solution. If the implicit solution describes more than one function, be sure to indicate which function corresponds to the solution of the initial value problem.*

33. $y'(t) = \dfrac{2t^2}{y^2 - 1}, y(0) = 0$

34. $y'(x) = \dfrac{1 + x}{2 - y}, y(1) = 1$

35. $u'(x) = \csc u \cos \dfrac{x}{2}, u(\pi) = \dfrac{\pi}{2}$

36. $yy'(x) = \dfrac{2x}{(2 + y^2)^2}, y(1) = -1$

37. $y'(x)\sqrt{y + 4} = \sqrt{x + 1}, y(3) = 5$

38. $z'(x) = \dfrac{z^2 + 4}{x^2 + 16}, z(4) = 2$

**T 39. Logistic equation for a population** A community of hares on an island has a population of 50 when observations begin (at $t = 0$). The population is modeled by the initial value problem

$$\dfrac{dP}{dt} = 0.08P\left(1 - \dfrac{P}{200}\right), P(0) = 50.$$

a. Find and graph the solution of the initial value problem, for $t \geq 0$.
b. What is the steady-state population?

**T 40. Logistic equation for an epidemic** When an infected person is introduced into a closed and otherwise healthy community, the number of people who contract the disease (in the absence of any intervention) may be modeled by the logistic equation

$$\dfrac{dP}{dt} = kP\left(1 - \dfrac{P}{A}\right), P(0) = P_0,$$

where $k$ is a positive infection rate, $A$ is the number of people in the community, and $P_0$ is the number of infected people at $t = 0$. The model also assumes no recovery.

**a.** Find the solution of the initial value problem, for $t \geq 0$, in terms of $k$, $A$, and $P_0$.

**b.** Graph the solution in the case that $k = 0.025$, $A = 300$, and $P_0 = 1$.

**c.** For a fixed value of $k$ and $A$, describe the long-term behavior of the solutions, for any $P_0$ with $0 < P_0 < A$.

**41. Explain why or why not** Determine whether the following statements are true and give an explanation or counterexample.

**a.** The equation $u'(x) = (x^2 u^7)^{-1}$ is separable.

**b.** The general solution of the separable equation

$$y'(t) = \frac{t}{y^7 + 10y^4}$$ can be expressed explicitly with $y$ in

terms of $t$.

**c.** The general solution of the equation $yy'(x) = xe^{-y}$ can be found using integration by parts.

**42–43. Implicit solutions for separable equations** *For the following separable equations, carry out the indicated analysis.*

**a.** *Find the general solution of the equation.*

**b.** *Find the value of the arbitrary constant associated with each initial condition. (Each initial condition requires a different constant.)*

**c.** *Use the graph of the general solution that is provided to sketch the solution curve for each initial condition.*

**42.** $y'(t) = \dfrac{t^2}{y^2 + 1}$; $y(-1) = 1$, $y(0) = 0$, $y(-1) = -1$

**43.** $e^{-y/2} y'(x) = 4x \sin x^2 - x$; $y(0) = 0$,

$$y(0) = \ln\left(\frac{1}{4}\right), y\left(\sqrt{\frac{\pi}{2}}\right) = 0$$

**44. Orthogonal trajectories** Two curves are orthogonal to each other if their tangent lines are perpendicular at each point of intersection. A family of curves forms **orthogonal trajectories** with another family of curves if each curve in one family is orthogonal to each curve in the other family. Use the following steps to find the orthogonal trajectories of the family of ellipses $2x^2 + y^2 = a^2$.

**a.** Apply implicit differentiation to $2x^2 + y^2 = a^2$ to show that $\dfrac{dy}{dx} = \dfrac{-2x}{y}$.

**b.** The family of trajectories orthogonal to $2x^2 + y^2 = a^2$ satisfies the differential equation $\dfrac{dy}{dx} = \dfrac{y}{2x}$. Why?

**c.** Solve the differential equation in part (b) to verify that $y^2 = e^C|x|$ and then explain why it follows that $y^2 = kx$, where $k$ is an arbitrary constant. Therefore, the family of parabolas $y^2 = kx$ forms the orthogonal trajectories of the family of ellipses $2x^2 + y^2 = a^2$.

**45. Orthogonal trajectories** Use the method in Exercise 44 to find the orthogonal trajectories for the family of circles $x^2 + y^2 = a^2$.

**T 46. Logistic equation for spread of rumors** Sociologists model the spread of rumors using logistic equations. The key assumption is that at any given time, a fraction $y$ of the population, where $0 \leq y \leq 1$, knows the rumor, while the remaining fraction $1 - y$ does not. Furthermore, the rumor spreads by interactions between those who know the rumor and those who do not. The number of such interactions is proportional to $y(1 - y)$. Therefore, the equation that describes the spread of the rumor is $y'(t) = ky(1 - y)$, for $t \geq 0$, where $k$ is a positive real number and $t$ is measured in weeks. The number of people who initially know the rumor is $y(0) = y_0$, where $0 \leq y_0 \leq 1$.

**a.** Solve this initial value problem and give the solution in terms of $k$ and $y_0$.

**b.** Assume $k = 0.3$ weeks$^{-1}$ and graph the solution for $y_0 = 0.1$ and $y_0 = 0.7$.

**c.** Describe and interpret the long-term behavior of the rumor function, for any $0 \leq y_0 \leq 1$.

**T 47. Free fall** An object in free fall may be modeled by assuming the only forces at work are the gravitational force and air resistance. By Newton's Second Law of Motion (mass · acceleration = the sum of external forces), the velocity of the object satisfies the differential equation

$$\underbrace{m}_{\text{mass}} \cdot \underbrace{v'(t)}_{\text{acceleration}} = \underbrace{mg + f(v)}_{\text{external forces}},$$

where $f$ is a function that models the air resistance (assuming the positive direction is downward). One common assumption (often used for motion in air) is that $f(v) = -kv^2$, for $t \geq 0$, where $k > 0$ is a drag coefficient.

**a.** Show that the equation can be written in the form $v'(t) = g - av^2$, where $a = \dfrac{k}{m}$.

**b.** For what (positive) value of $v$ is $v'(t) = 0$? (This equilibrium solution is called the *terminal velocity*.)

**c.** Find the solution of this separable equation assuming $v(0) = 0$ and $0 < v^2 < \dfrac{g}{a}$.

**d.** Graph the solution found in part (c) with $g = 9.8$ m/s$^2$, $m = 1$, and $k = 0.1$, and verify that the terminal velocity agrees with the value found in part (b).

**T 48. Free fall** Using the background given in Exercise 47, assume the resistance is given by $f(v) = -Rv$, for $t \geq 0$, where $R > 0$ is a drag coefficient (an assumption often made for a heavy medium such as water or oil).

a. Show that the equation can be written in the form
$$v'(t) = g - bv, \text{ where } b = \frac{R}{m}.$$

b. For what value of $v$ is $v'(t) = 0$? (This equilibrium solution is called the terminal velocity.)

c. Find the solution of this separable equation assuming $v(0) = 0$ and $0 < v < \dfrac{g}{b}$.

d. Graph the solution found in part (c) with $g = 9.8 \text{ m/s}^2$, $m = 1$, and $R = 0.1$, and verify that the terminal velocity agrees with the value found in part (b).

**T 49. Torricelli's law** An open cylindrical tank initially filled with water drains through a hole in the bottom of the tank according to Torricelli's law (see figure). If $h(t)$ is the depth of water in the tank for $t \geq 0$ s, then Torricelli's law implies $h'(t) = -k\sqrt{h}$, where $k$ is a constant that includes $g = 9.8 \text{ m/s}^2$, the radius of the tank, and the radius of the drain. Assume the initial depth of the water is $h(0) = H$ m.

a. Find the solution of the initial value problem.
b. Find the solution in the case that $k = 0.1$ and $H = 0.5$ m.
c. In part (b), how long does it take for the tank to drain?
d. Graph the solution in part (b) and check that it is consistent with part (c).

$h(t)$

**T 50. Chemical rate equations** Let $y(t)$ be the concentration of a substance in a chemical reaction (typical units are moles/liter). The change in the concentration, under appropriate conditions, is modeled by the equation $\dfrac{dy}{dt} = -ky^n$, for $t \geq 0$, where $k > 0$ is a rate constant and the positive integer $n$ is the order of the reaction.

a. Show that for a first-order reaction ($n = 1$), the concentration obeys an exponential decay law.
b. Solve the initial value problem for a second-order reaction ($n = 2$) assuming $y(0) = y_0$.
c. Graph the concentration for a first-order and second-order reaction with $k = 0.1$ and $y_0 = 1$.

**Explorations and Challenges**

**T 51. Tumor growth** The Gompertz growth equation is often used to model the growth of tumors. Let $M(t)$ be the mass of a tumor at time $t \geq 0$. The relevant initial value problem is
$$\frac{dM}{dt} = -rM \ln\left(\frac{M}{K}\right), M(0) = M_0,$$
where $r$ and $K$ are positive constants and $0 < M_0 < K$.

a. Graph the growth rate function $R(M) = -rM \ln\left(\dfrac{M}{K}\right)$ (which equals $M'(t)$) assuming $r = 1$ and $K = 4$. For what values of $M$ is the growth rate positive? For what value of $M$ is the growth rate a maximum?

b. Solve the initial value problem and graph the solution for $r = 1$, $K = 4$, and $M_0 = 1$. Describe the growth pattern of the tumor. Is the growth unbounded? If not, what is the limiting size of the tumor?

**T 52. Technology for an initial value problem** Solve $y'(t) = ye^t \cos^3 4t$, $y(0) = 1$, and plot the solution for $0 \leq t \leq \pi$.

**53. Blowup in finite time** Consider the initial value problem $y'(t) = y^{n+1}$, $y(0) = y_0$, where $n$ is a positive integer.

a. Solve the initial value problem with $n = 1$ and $y_0 = 1$.
b. Solve the initial value problem with $n = 2$ and $y_0 = 1/\sqrt{2}$.
c. Solve the problem for positive integers $n$ and $y_0 = n^{-1/n}$. How do solutions behave as $t \to 1^-$?

**T 54. Analysis of a separable equation** Consider the differential equation $yy'(t) = \dfrac{1}{2}e^t + t$ and carry out the following analysis.

a. Find the general solution of the equation and express it explicitly as a function of $t$ in two cases: $y > 0$ and $y < 0$.
b. Find the solutions that satisfy the initial conditions $y(-1) = 1$ and $y(-1) = 2$.
c. Graph the solutions in part (b) and describe their behavior as $t$ increases.
d. Find the solutions that satisfy the initial conditions $y(-1) = -1$ and $y(-1) = -2$.
e. Graph the solutions in part (d) and describe their behavior as $t$ increases.

**QUICK CHECK ANSWERS**

**1.** B and C are separable.   **2.** $y^3 y'(t) = t^2 + 1$   **3.** The initial condition $y(\pi) = 0$ implies that $C = 0$; the solution $y(x) = \ln(-\cos x)$ is defined only when $-\cos x > 0$, which occurs on the intervals ... $\left(-\frac{3\pi}{2}, -\frac{\pi}{2}\right), \left(\frac{\pi}{2}, \frac{3\pi}{2}\right), \left(\frac{5\pi}{2}, \frac{7\pi}{2}\right), \ldots$ The interval containing the initial condition, $\left(\frac{\pi}{2}, \frac{3\pi}{2}\right)$, is the domain.   **4.** $C = -\frac{1}{24}$ ◄

# 9.4 Special First-Order Linear Differential Equations

> ➤ The exponential growth and decay problems studied in Section 7.4 appear again in this section, but now with a differential equations perspective.

We now focus on a special class of differential equations with so many interesting applications that they warrant special attention. All the equations we study in this section are first order and linear.

## Method of Solution

Consider the first-order linear equation $y'(t) = ky + b$, where $k \neq 0$ and $b$ are real numbers. By varying the values of $k$ and $b$, this versatile equation may be used to model a wide variety of phenomena. Specifically, the terms of the equation have the following general meaning:

$$\underbrace{y'(t)}_{\substack{\text{rate of change} \\ \text{of } y}} = \underbrace{ky(t)}_{\substack{\text{natural growth or} \\ \text{decay rate of } y}} + \underbrace{b.}_{\substack{\text{growth or decay} \\ \text{rate due to} \\ \text{external effects}}}$$

> In the most general first-order linear equation, $k$ and $b$ are functions of $t$. This general first-order linear equation is not separable. See Exercises 45–48 for this more challenging case.

For example, if $y$ represents the number of fish in a hatchery, then $ky(t)$ (with $k > 0$) models exponential growth in the fish population, in the absence of other factors, and $b < 0$ is the harvesting rate at which the population is depleted. As another example, if $y$ represents the amount of a drug in the blood, then $ky(t)$ (with $k < 0$) models exponential decay of the drug through the kidneys, and $b > 0$ is the rate at which the drug is added to the blood intravenously. Because $k$ and $b$ are constants, the equation is separable and we can give an explicit solution.

To solve this equation, we begin by dividing both sides of $y'(t) = ky + b$ by $ky + b$ to express it in separated form:

$$\frac{y'(t)}{ky + b} = 1.$$

We now integrate both sides of this equation with respect to $t$ and observe that $dy = y'(t)\,dt$, which gives

$$\int \frac{dy}{ky + b} = \int dt \qquad \text{Integrate both sides of the equation.}$$

$$\frac{1}{k} \ln |ky + b| = t + C. \qquad \text{Evaluate integrals.}$$

For the moment, we assume $ky + b > 0$, or $y > -b/k$, so the absolute value may be removed. Multiplying through by $k$ and exponentiating both sides of the equation, we have

$$ky + b = e^{kt + kC} = e^{kt} \cdot \underbrace{e^{kC}}_{\text{redefine as } C} = Ce^{kt}.$$

Notice that we use the standard practice of redefining the arbitrary constant $C$ as we solve for $y$: If $C$ is arbitrary, then $e^{kC}$ and $C/k$ are also arbitrary. We now solve for the general solution:

$$y(t) = Ce^{kt} - \frac{b}{k}.$$

We can also show that if $ky + b < 0$, or $y < -b/k$, then the same solution results (Exercise 36).

---

**SUMMARY**  **Solution of a First-Order Linear Differential Equation**

The general solution of the first-order linear equation $y'(t) = ky + b$, where $k \neq 0$ and $b$ are real numbers, is

$$y(t) = Ce^{kt} - \frac{b}{k},$$

where $C$ is an arbitrary constant. Given an initial condition, the value of $C$ may be determined.

---

**QUICK CHECK 1** Verify by substitution that $y(t) = Ce^{kt} - b/k$ is a solution of $y'(t) = ky + b$, for real numbers $b$ and $k \neq 0$. ◄

**EXAMPLE 1   An initial value problem for drug dosing** A drug is administered to a patient through an intravenous line at a rate of 6 mg/hr. The drug has a half-life that corresponds to a rate constant of $k = 0.03$ hr$^{-1}$. Let $y(t)$ be the amount of drug in the blood, for $t \geq 0$. Solve the initial value problem that governs the process,

$$y'(t) = -0.03y + 6, \quad y(0) = 0,$$

and interpret the solution.

**SOLUTION** The equation has the form $y'(t) = ky + b$, where $k = -0.03$ and $b = 6$. Therefore, the general solution is

$$y(t) = Ce^{-0.03t} + 200.$$

To determine the value of $C$ for this particular problem, we substitute $y(0) = 0$ into the general solution. The result is that $y(0) = C + 200 = 0$, which implies that $C = -200$. Therefore, the solution of the initial value problem is

$$y(t) = -200e^{-0.03t} + 200 = 200(1 - e^{-0.03t}), \text{ for } t \geq 0.$$

The graph of the solution (**Figure 9.22**) reveals an important fact: The amount of drug in the blood increases, but it approaches a steady-state level of

$$\lim_{t \to \infty} y(t) = \lim_{t \to \infty} (200(1 - e^{-0.03t})) = 200 \text{ mg}.$$

A doctor can obtain practical information from this solution. For example, after 100 hours, the drug level reaches 95% of the steady state.

*Related Exercise 31* ◀

**Figure 9.22**

Graph showing Steady state: $y = 200$ and $y = 200(1 - e^{-0.03t})$, with $t$-axis from 0 to 140 and $y$-axis up to 200.

**QUICK CHECK 2** If the rate constant in Example 1 were 0.3 instead of 0.03, would the steady-state level of the drug change? If so, to what value? ◀

➤ The idea of stable and unstable equilibrium solutions can be illustrated using a hemispherical bowl and a small ball. When the ball rests at the bottom of the bowl, it is at rest in an equilibrium state. If the ball is moved away from the equilibrium state, it returns to that state.

Stable

By contrast, when the ball rests on top of the inverted bowl, it is at rest in an equilibrium state. However, if the ball is moved away from the equilibrium state, it does not return to that state.

Unstable

**EXAMPLE 2   Direction field analysis** Use direction fields to analyze the behavior of the solutions of the following equations, where $k > 0$ and $b$ is nonzero. Assume $t \geq 0$.

**a.** $y'(t) = -ky + b$          **b.** $y'(t) = ky + b$

**SOLUTION**

**a.** First notice that $y'(t) = 0$ when $y = b/k$. Therefore, the direction field consists of horizontal line segments when $y = b/k$. These horizontal line segments correspond to the constant *equilibrium* solution $y(t) = \dfrac{b}{k}$. Depending on the sign of $b$, the constant solution could be positive or negative. If $-ky + b > 0$, or equivalently, $y < b/k$, then $y'(t) > 0$, and solutions are increasing in this region. Similarly, if $-ky + b < 0$, or equivalently, $y > b/k$, then $y'(t) < 0$, and solutions are decreasing in this region. **Figure 9.23** shows a typical direction field in the case that $b > 0$. Notice that the solution curves are attracted to the equilibrium solution. For this reason, the equilibrium is said to be *stable*.

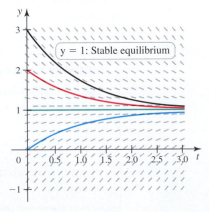

$y = 1$: Stable equilibrium

**Figure 9.23**

**b.** The analysis is similar to that in part (a). In this case, we have an equilibrium solution at $y = -b/k$, which may be positive or negative depending on the sign of $b$. If

**Figure 9.24**

$ky + b > 0$, or equivalently, $y > -b/k$, then $y'(t) > 0$, and solutions are increasing in this region. Similarly, if $ky + b < 0$, or equivalently, $y < -b/k$, then $y'(t) < 0$, and solutions are decreasing in this region. **Figure 9.24** shows a direction field for $b < 0$. Now the solution curves move away from the equilibrium solution, and the equilibrium is *unstable*.

*Related Exercises 17–18* ◄

**QUICK CHECK 3** What is the equilibrium solution of the equation $y'(t) = 2y - 4$? Is it stable or unstable? ◄

We give a qualitative summary of the important ideas introduced in Example 2.

---

**SUMMARY    Equilibrium Solutions**

The differential equation $y'(t) = f(y)$ has a (constant) **equilibrium** solution $y = a$ when $f(a) = 0$. The equilibrium is **stable** if initial conditions near $y = a$ produce solutions that approach $y = a$ as $t \to \infty$. The equilibrium is **unstable** if initial conditions near $y = a$ produce solutions that do not approach $y = a$ as $t \to \infty$.

---

**EXAMPLE 3    Paying off a loan** Suppose you borrow \$60,000 with a monthly interest rate of 0.5% and plan to pay it back with monthly payments of \$600. The balance in the loan is described approximately by the initial value problem

$$B'(t) = \underbrace{0.005B}_{\text{interest}} - \underbrace{600}_{\substack{\text{monthly} \\ \text{payments}}}, B(0) = 60,000,$$

where $B(t)$ is the balance in the loan after $t$ months. Notice that the interest increases the loan balance, while the monthly payments decrease the loan balance.

**a.** Find and graph the loan balance function.

**b.** After approximately how many months does the loan balance reach zero?

**SOLUTION**

► Loan payments are an example of a discrete process (interest is assessed and payments are made each month). However, they may be modeled as a continuous process using a differential equation because the time interval between payments is small compared to the length of the entire loan process. Discrete processes are often modeled using *difference equations*.

**a.** The differential equation has the form $y'(t) = ky + b$, where $k = 0.005$ month$^{-1}$ and $b = -\$600/\text{month}$. Using the summary box, the general solution is

$$B(t) = Ce^{kt} - \frac{b}{k} = Ce^{0.005t} + 120,000.$$

The initial condition implies that

$$B(0) = C + 120,000 = 60,000 \Rightarrow C = -60,000.$$

Therefore, the solution of the initial value problem is

$$B(t) = Ce^{kt} - \frac{b}{k} = 120,000 - 60,000e^{0.005t}.$$

**b.** The graph (**Figure 9.25**) shows the loan balance decreasing and reaching zero at $t \approx 139$ months (which implies an approximate domain of $[0, 139]$). This fact can be confirmed by solving $B(t) = 0$ algebraically.

*Related Exercises 23–26* ◄

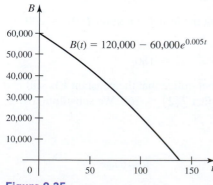

**Figure 9.25**

## Newton's Law of Cooling

Imagine taking a fired bowl out of a hot pottery kiln and putting it on a rack to cool at room temperature. Your intuition tells you that because the temperature of the bowl is greater than the temperature of the room, the pot cools and its temperature approaches the temperature of the room. (We assume the room is sufficiently large that the heating of the room by the bowl is negligible.)

It turns out that this process can be described approximately using a first-order differential equation similar to those studied in this section. That equation is often called

Newton's Law of Cooling, and it is based on the familiar observation that *heat flows from hot to cold*. The solution of the equation gives the temperature of the bowl at all times after it is removed from the kiln.

We let $t = 0$ be the time at which the bowl is removed from the kiln. The temperature of the bowl at any time $t \geq 0$ is $T(t)$, and $T(0) = T_0$ is the temperature of the bowl as it comes out of the kiln. We also let $A$ be the temperature of the room or the ambient temperature. Both $T_0$ and $A$ are assumed to be known.

Newton's Law of Cooling says that the rate at which the temperature changes at any time is proportional to the temperature *difference* between the bowl and the room at that time; that is,

$$\frac{dT}{dt} = -k(T(t) - A),$$

where $k > 0$ is a constant determined by the thermal properties of the bowl. Notice that the equation makes sense.

- If $T(t) > A$ (the bowl is hotter than the room), then $\dfrac{dT}{dt} < 0$, and the temperature of the bowl decreases (cooling).

- If $T(t) < A$ (the bowl is colder than the room), then $\dfrac{dT}{dt} > 0$, and the temperature of the bowl increases (heating).

We see that Newton's Law of Cooling amounts to a first-order differential equation that we know how to solve. The equation has the form $T'(t) = -kT + b$, where $b = kA$. This equation was studied earlier in the section; its general solution is

$$T(t) = Ce^{-kt} + A.$$

When we use the initial condition $T(0) = T_0$ to determine $C$, we find that

$$T(0) = C + A = T_0 \quad \Rightarrow \quad C = T_0 - A.$$

Therefore, the solution of the initial value problem is

$$T(t) = (T_0 - A)e^{-kt} + A.$$

**QUICK CHECK 4** Verify that the solution of the initial value problem satisfies $T(0) = T_0$. What is the solution of the problem if $T_0 = A$? ◄

Newton's Law of Cooling models the cooling process well when the object is a good conductor of heat and when the temperature is fairly uniform throughout the object.

**EXAMPLE 4**  **Cooling a bowl**  A bowl is removed from a pottery kiln at a temperature of 200°C and placed on a rack in a room with an ambient temperature of 20°C. Two minutes after the bowl is removed, its temperature is 160°C. Find the temperature of the bowl, for all $t \geq 0$.

**SOLUTION**  Letting $A = 20$, the general solution of the cooling equation is

$$T(t) = Ce^{-kt} + 20.$$

As always, the arbitrary constant is determined using the initial condition $T(0) = 200$. Substituting this condition we find that

$$T(0) = C + 20 = 200 \quad \Rightarrow \quad C = 180.$$

> ► The value of the thermal constant $k$ is known for common materials. Example 4 illustrates one way to estimate the constant experimentally.

The solution at this point is $T(t) = 180e^{-kt} + 20$, but notice that the constant $k$ is still unknown. It is determined using the additional fact that $T(2) = 160$. We substitute this condition into the solution and solve for $k$:

$$T(2) = 180e^{-2k} + 20 = 160 \qquad \text{Substitute } t = 2.$$
$$180e^{-2k} = 140 \qquad \text{Rearrange.}$$
$$e^{-2k} = \frac{140}{180} = \frac{7}{9} \qquad \text{Rearrange.}$$
$$k = -\frac{1}{2}\ln\frac{7}{9} \approx 0.126. \qquad \text{Solve for } k.$$

**Figure 9.26**

Therefore, the solution for $t \geq 0$ is

$$T(t) = 180e^{-kt} + 20 \approx 180e^{-0.126t} + 20.$$

The graph (**Figure 9.26**) confirms that $T(0) = 200$ and that $T(2) = 160$. Notice also that $\lim\limits_{t \to \infty} T(t) = 20$, meaning that the temperature of the bowl approaches the ambient temperature as $t \to \infty$. Equivalently, the solution $T = 20$ is a stable equilibrium of the system.

*Related Exercises 27, 30* ◄

**QUICK CHECK 5** In general, what is the equilibrium temperature for any Newton cooling problem? Is it a stable or unstable equilibrium? ◄

---

## SECTION 9.4 EXERCISES

### Getting Started

1. The general solution of a first-order linear differential equation is $y(t) = Ce^{-10t} - 13$. What solution satisfies the initial condition $y(0) = 4$?

2. What is the general solution of the equation $y'(t) = 3y - 12$?

3. What is the general solution of the equation $y'(t) = -4y + 6$?

4. What is the equilibrium solution of the equation $y'(t) = 3y - 9$? Is it stable or unstable?

### Practice Exercises

**5–10. First-order linear equations** *Find the general solution of the following equations.*

5. $y'(t) = 3y - 4$

6. $y'(x) = -y + 2$

7. $y'(x) + 2y = -4$

8. $y'(x) = 2y + 6$

9. $u'(t) + 12u = 15$

10. $v'(y) - \dfrac{v}{2} = 14$

**11–16. Initial value problems** *Solve the following initial value problems.*

11. $y'(t) = 3y - 6,\ y(0) = 9$

12. $y'(x) = -y + 2,\ y(0) = -2$

13. $y'(t) - 2y = 8,\ y(0) = 0$

14. $u'(x) = 2u + 6,\ u(1) = 6$

15. $y'(t) - 3y = 12,\ y(1) = 4$

16. $z'(t) + \dfrac{z}{2} = 6,\ z(-1) = 0$

**17–22. Stability of equilibrium points** *Find the equilibrium solution of the following equations, make a sketch of the direction field, for $t \geq 0$, and determine whether the equilibrium solution is stable. The direction field needs to indicate only whether solutions are increasing or decreasing on either side of the equilibrium solution.*

17. $y'(t) = 12y - 18$

18. $y'(t) = -6y + 12$

19. $y'(t) = -\dfrac{y}{3} - 1$

20. $y'(t) - \dfrac{y}{4} - 1 = 0$

21. $u'(t) + 7u + 21 = 0$

22. $u'(t) - 4u = 3$

**23–26. Loan problems** *The following initial value problems model the payoff of a loan. In each case, solve the initial value problem, for $t \geq 0$, graph the solution, and determine the first month in which the loan balance is zero.*

23. $B'(t) = 0.005B - 500,\ B(0) = 50{,}000$

24. $B'(t) = 0.01B - 750,\ B(0) = 45{,}000$

25. $B'(t) = 0.0075B - 1500,\ B(0) = 100{,}000$

26. $B'(t) = 0.004B - 800,\ B(0) = 40{,}000$

**27–30. Newton's Law of Cooling** *Solve the differential equation for Newton's Law of Cooling to find the temperature function in the following cases. Then answer any additional questions.*

27. A cup of coffee has a temperature of 90°C when it is poured and allowed to cool in a room with a temperature of 25°C. One minute after the coffee is poured, its temperature is 85°C. How long must you wait until the coffee is cool enough to drink, say 30°C?

28. An iron rod is removed from a blacksmith's forge at a temperature of 900°C. Assume $k = 0.02$ and the rod cools in a room with a temperature of 30°C. When does the temperature of the rod reach 100°C?

29. A glass of milk is moved from a refrigerator with a temperature of 5°C to a room with a temperature of 20°C. One minute later the milk has warmed to a temperature of 7°C. After how many minutes does the milk have a temperature that is 90% of the ambient temperature?

30. A pot of boiling soup (100°C) is put in a cellar with a temperature of 10°C. After 30 minutes, the soup has cooled to 80°C. When will the temperature of the soup reach 30°C?

**T** **31. Intravenous drug dosing** The amount of drug in the blood of a patient (in milligrams) administered via an intravenous line is governed by the initial value problem $y'(t) = -0.02y + 3$, $y(0) = 0$, where $t$ is measured in hours.

    **a.** Find and graph the solution of the initial value problem.

    **b.** What is the steady-state level of the drug?

    **c.** When does the drug level reach 90% of the steady-state value?

**T 32. Fish harvesting** A fish hatchery has 500 fish at $t = 0$, when harvesting begins at a rate of $b > 0$ fish/year. The fish population is modeled by the initial value problem $y'(t) = 0.01y - b$, $y(0) = 500$, where $t$ is measured in years.

   **a.** Find the fish population, for $t \geq 0$, in terms of the harvesting rate $b$.

   **b.** Graph the solution in the case that $b = 40$ fish/year. Describe the solution.

   **c.** Graph the solution in the case that $b = 60$ fish/year. Describe the solution.

**33. Optimal harvesting rate** Let $y(t)$ be the population of a species that is being harvested, for $t \geq 0$. Consider the harvesting model $y'(t) = 0.008y - h$, $y(0) = y_0$, where $h$ is the annual harvesting rate, $y_0$ is the initial population of the species, and $t$ is measured in years.

   **a.** If $y_0 = 2000$, what harvesting rate should be used to maintain a constant population of $y = 2000$, for $t \geq 0$?

   **b.** If the harvesting rate is $h = 200$/year, what initial population ensures a constant population?

**T 34. Endowment model** An endowment is an investment account in which the balance ideally remains constant and withdrawals are made on the interest earned by the account. Such an account may be modeled by the initial value problem $B'(t) = rB - m$, for $t \geq 0$, with $B(0) = B_0$. The constant $r > 0$ reflects the annual interest rate, $m > 0$ is the annual rate of withdrawal, $B_0$ is the initial balance in the account, and $t$ is measured in years.

   **a.** Solve the initial value problem with $r = 0.05$, $m = \$1000$/year, and $B_0 = \$15,000$. Does the balance in the account increase or decrease?

   **b.** If $r = 0.05$ and $B_0 = \$50,000$, what is the annual withdrawal rate $m$ that ensures a constant balance in the account? What is the constant balance?

**35. Explain why or why not** Determine whether the following statements are true and give an explanation or counterexample.

   **a.** The general solution of $y'(t) = 2y - 18$ is $y(t) = 2e^{2t} + 9$.

   **b.** If $k > 0$ and $b > 0$, then $y(t) = 0$ is never a solution of $y'(t) = ky - b$.

   **c.** The equation $y'(t) = ty(t) + 3$ is separable and can be solved using the methods of this section.

   **d.** According to Newton's Law of Cooling, the temperature of a hot object will reach the ambient temperature after a finite amount of time.

**36. Case 2 of the general solution** Solve the equation $y'(t) = ky + b$ in the case that $ky + b < 0$ and verify that the general solution is $y(t) = Ce^{kt} - \dfrac{b}{k}$.

**37. A bad loan** Consider a loan repayment plan described by the initial value problem

$$B'(t) = 0.03B - 600, \quad B(0) = 40,000,$$

where the amount borrowed is $B(0) = \$40,000$, the monthly payments are $\$600$, and $B(t)$ is the unpaid balance in the loan.

   **a.** Find the solution of the initial value problem and explain why $B$ is an increasing function.

   **b.** What is the most that you can borrow under the terms of this loan without going further into debt each month?

   **c.** Now consider the more general loan repayment plan described by the initial value problem

$$B'(t) = rB - m, \quad B(0) = B_0,$$

   where $r > 0$ reflects the interest rate, $m > 0$ is the monthly payment, and $B_0 > 0$ is the amount borrowed. In terms of $m$ and $r$, what is the maximum amount $B_0$ that can be borrowed without going further into debt each month?

**38. Cooling time** Suppose an object with an initial temperature of $T_0 > 0$ is put in surroundings with an ambient temperature of $A$, where $A < \dfrac{T_0}{2}$. Let $t_{1/2}$ be the time required for the object to cool to $\dfrac{T_0}{2}$.

   **a.** Show that $t_{1/2} = -\dfrac{1}{k} \ln\left( \dfrac{T_0 - 2A}{2(T_0 - A)} \right)$.

   **b.** Does $t_{1/2}$ increase or decrease as $k$ increases? Explain.

   **c.** Why is the condition $A < \dfrac{T_0}{2}$ needed?

## Explorations and Challenges

**39–42. Special equations** *A special class of first-order linear equations have the form* $a(t)y'(t) + a'(t)y(t) = f(t)$, *where $a$ and $f$ are given functions of $t$. Notice that the left side of this equation can be written as the derivative of a product, so the equation has the form*

$$a(t)y'(t) + a'(t)y(t) = \frac{d}{dt}(a(t)y(t)) = f(t).$$

*Therefore, the equation can be solved by integrating both sides with respect to $t$. Use this idea to solve the following initial value problems.*

**39.** $ty'(t) + y = 1 + t, y(1) = 4$

**40.** $t^3 y'(t) + 3t^2 y = \dfrac{1 + t}{t}, y(1) = 6$

**41.** $e^{-t} y'(t) - e^{-t} y = e^{2t}, y(0) = 4$

**42.** $(t^2 + 1)y'(t) + 2ty = 3t^2, y(2) = 8$

**43. Change of variables in a Bernoulli equation** The equation $y'(t) + ay = by^p$, where $a$, $b$, and $p$ are real numbers, is called a *Bernoulli equation*. Unless $p = 1$, the equation is nonlinear and would appear to be difficult to solve—except for a small miracle. Through the change of variables $v(t) = (y(t))^{1-p}$, the equation can be made linear. Carry out the following steps.

   **a.** Letting $v = y^{1-p}$, show that $y'(t) = \dfrac{y(t)^p}{1 - p} v'(t)$.

   **b.** Substitute this expression for $y'(t)$ into the differential equation and simplify to obtain the new (linear) equation $v'(t) = a(1 - p)v = b(1 - p)$, which can be solved using the methods of this section. The solution $y$ of the original equation can then be found from $v$.

**44. Solving Bernoulli equations** Use the method outlined in Exercise 43 to solve the following Bernoulli equations.

   **a.** $y'(t) + y = 2y^2$

   **b.** $y'(t) - 2y = 3y^{-1}$

   **c.** $y'(t) + y = \sqrt{y}$

**45–48. General first-order linear equations** *Consider the general first-order linear equation* $y'(t) + a(t)y(t) = f(t)$. *This equation can be solved, in principle, by defining the* **integrating factor** $p(t) = \exp\left(\int a(t)\,dt\right)$. *Here is how the integrating factor works. Multiply both sides of the equation by p (which is always positive) and show that the left side becomes an exact derivative. Therefore, the equation becomes*

$$p(t)(y'(t) + a(t)y(t)) = \frac{d}{dt}(p(t)y(t)) = p(t)f(t).$$

*Now integrate both sides of the equation with respect to t to obtain the solution. Use this method to solve the following initial value problems. Begin by computing the required integrating factor.*

**45.** $y'(t) + \dfrac{1}{t}y(t) = 0,\ y(1) = 6$

**46.** $y'(t) + \dfrac{3}{t}y(t) = 1 - 2t,\ y(2) = 0$

**47.** $y'(t) + \dfrac{2t}{t^2 + 1}y(t) = 1 + 3t^2,\ y(1) = 4$

**48.** $y'(t) + 2ty(t) = 3t,\ y(0) = 1$

**QUICK CHECK ANSWERS**

**1.** $y'(t) = Cke^{kt}$, while $ky + b = k(Ce^{kt} - b/k) + b = Cke^{kt}$.    **2.** The steady-state drug level would be $y = 20$.
**3.** The equilibrium solution $y = 2$ is unstable.
**4.** $T(0) = (T_0 - A) + A = T_0$. If $T_0 = A$, $T(t) = A$ for all $t \geq 0$.    **5.** The ambient temperature is a stable equilibrium. ◄

# 9.5 Modeling with Differential Equations

Many examples and exercises of this chapter have illustrated the use of differential equations to model various real-world problems. In this concluding section, we focus on three specific applications and explore some of the ideas involved in formulating mathematical models. The first application is the modeling of populations, examples of which we have already encountered. Next we derive the differential equation that governs a mixed-tank reaction. Finally, we introduce and analyze a well-known two-species ecosystem model.

## Population Models

So far, we have seen two examples of differential equations that model population growth. Letting $P(t)$ be the population of a species at time $t \geq 0$, both equations have the general form $P'(t) = f(P)$, where $f(P)$ is a function that depends only on the population, and $r$ and $K$ are constants.

$$\text{Exponential growth: } P'(t) = f(P) = rP$$

$$\text{Logistic growth: } P'(t) = f(P) = rP\left(1 - \frac{P}{K}\right)$$

The **growth rate function** $f$ specifies the rate of growth of the population and is chosen to give the best description of the population. **Figure 9.27** shows a graph of the growth rate functions for the exponential and logistic models. Note that population $P$ is the variable on the horizontal axis and the growth rate function, which defines $P'$, is on the vertical axis. In both cases, the growth rate function is nonnegative, so both models describe populations that increase. Population values for which $f(P) = 0$ correspond to equilibrium solutions.

For the exponential model, the growth rate function increases linearly with the population size, implying that the larger the population, the larger the growth rate. Therefore, with this model, populations increase (unrealistically) without bound (**Figure 9.28**).

The growth rate function for the logistic model has zeros at $P = 0$ and $P = K$ (equilibrium solutions) and has a local maximum at $P = K/2$. As a result, the population increases slowly at first, approaches a maximum growth rate, and then grows more slowly as it approaches the **carrying capacity** $P = K$ (Figure 9.28). The important feature of this model is that the population is bounded in size, reflecting the effects of overcrowding or a shortage of resources.

**Figure 9.27**

**Figure 9.28**

An important observation about the logistic model is that when the population is small compared to the carrying capacity (often written $P \ll K$), the population grows exponentially with a rate constant $r$. We see this fact in the growth rate function:

$$f(P) = rP\left(1 - \underbrace{\frac{P}{K}}_{\text{smal}}\right) \approx rP.$$

**QUICK CHECK 1** Explain why the maximum growth rate for the logistic equation occurs at $P = K/2$. ◄

This fact is also evident in Figure 9.28, where the population curves are nearly identical for small values of $t$. Therefore, $r$ may be interpreted as the natural growth rate of the species in ideal conditions (unlimited space and resources).

**EXAMPLE 1**    **Designing a logistic model**  Wildlife biologists observe a prairie dog community for several years. When observations begin, there are 8 prairie dogs; after one year the population reaches 20 prairie dogs. After 10 years, the population has leveled out at approximately 200 prairie dogs. Assuming a logistic growth model applies to this community, find a function that models the population.

**SOLUTION**  The biologists' measurements suggest that the initial population of the community is $P_0 = 8$ and the carrying capacity is $K = 200$. Using the logistic equation, the resulting initial value problem is

$$P'(t) = rP\left(1 - \frac{P}{200}\right), \quad P(0) = P_0 = 8.$$

Using the methods of Section 9.3, the solution of this problem (Exercise 33) is

$$P(t) = \frac{200}{24e^{-rt} + 1}.$$

Carrying capacity: $P = 200$

$P(t) = \dfrac{200}{24e^{-0.981t} + 1}$, for $t \geq 0$

$P(0) = 8, P(1) = 20$

**Figure 9.29**

Notice that the natural growth rate $r$ is still undetermined; it is estimated using the fact that the population after one year is 20 prairie dogs. Substituting $P(1) = 20$ into the solution, we have

$$P(1) = \frac{200}{24e^{-r} + 1} = 20 \quad \color{blue}{\text{Substitute } t = 1 \text{ and } P = 20.}$$

$$e^{-r} = \frac{3}{8} \quad \color{blue}{\text{Simplify.}}$$

$$r = -\ln\frac{3}{8} \approx 0.981. \quad \color{blue}{\text{Take logarithms of both sides.}}$$

Substituting this value of $r$, we obtain the population function shown in **Figure 9.29**. Notice that the initial condition $P(0) = 8$ is satisfied, $P(1) \approx 20$, and the population approaches the carrying capacity of 200 prairie dogs.

*Related Exercises 17–18* ◄

**EXAMPLE 2**    **Gompertz growth model**  Models of tumor growth often use the Gompertz equation

$$M'(t) = -rM \ln\left(\frac{M}{K}\right),$$

where $M(t)$ is the mass of the tumor at time $t \geq 0$, and $r$ and $K$ are positive constants.

**a.** Graph the growth rate function for the Gompertz model with $M > 0$, discuss its features, and compare it to the logistic growth rate function.

**b.** Find the general solution of the Gompertz equation with positive values of $r$, $K$, and $M(0) = M_0$, assuming $0 < M_0 < K$.

**c.** Graph the solution in part (b) when $r = 0.5$, $K = 10$, and $M_0 = 0.01$.

**Figure 9.30**

> Recall that $\exp(u)$ is another way to write $e^u$.

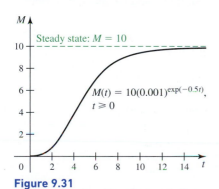

Steady state: $M = 10$

$M(t) = 10(0.001)^{\exp(-0.5t)}$,
$t \geq 0$

**Figure 9.31**

**SOLUTION**

**a.** The growth rate function $f(M) = -rM \ln\left(\dfrac{M}{K}\right)$ is a skewed version of the logistic growth rate function (**Figure 9.30**). It is left as an exercise (Exercise 32) to show that the Gompertz model has a maximum growth rate of $rK/e$ when $M = K/e$ (compared to the logistic growth rate function, which has a maximum of $rK/4$ when $P = K/2$).

**b.** The Gompertz equation is separable and is solved using the methods of Section 9.3:

$$\frac{M'(t)}{M \ln\left(\dfrac{M}{K}\right)} = -r \qquad \text{Write equation in separated form.}$$

$$\int \frac{dM}{M \ln\left(\dfrac{M}{K}\right)} = -\int r\, dt \qquad \text{Integrate both sides; } M'(t)\, dt = dM.$$

$$\ln\left|\ln\left(\dfrac{M}{K}\right)\right| = -rt + C \qquad \text{Integrate with } u = \ln\left(\dfrac{M}{K}\right) \text{ on left side.}$$

$$\ln\left(\dfrac{M}{K}\right) = Ce^{-rt} \qquad \text{Exponentiate both sides; relabel } e^{\pm C} \text{ as } C.$$

$$M(t) = K \exp\left(Ce^{-rt}\right). \qquad \text{Exponentiate both sides.}$$

We now have the general solution. The initial condition $M(0) = M_0$ implies that $M_0 = Ke^C$. Solving for $C$, we find that $C = \ln \dfrac{M_0}{K}$.

Substituting this value of $C$ into the general solution gives the solution of the initial value problem (Exercise 19):

$$M(t) = K\left(\frac{M_0}{K}\right)^{\exp(-rt)}.$$

You should check that this unusual solution satisfies the initial condition $M(0) = M_0$. Furthermore, the solution has a steady state given by $\lim\limits_{t \to \infty} M(t) = K$ (Exercise 34).

**c.** The solution given in part (b) with $r = 0.5$, $K = 10$, and $M_0 = 0.01$ is

$$M(t) = 10(0.001)^{\exp(-0.5t)};$$

its graph is shown in **Figure 9.31**. Notice that the solution approaches the steady-state mass $M = K = 10$.

*Related Exercises 20–22* ◄

### Stirred Tank Reactions

In their many forms and variations, models of stirred tank reactions are used to simulate industrial and manufacturing processes. They are also adapted so they can be applied to physiological problems, such as the assimilation of drugs by systems of organs. The differential equations governing these reactions can be derived from first principles, and we have the tools to solve them.

A stirred tank reaction takes place in a large tank that is initially filled with a solution of a soluble substance, such as salt or sugar. The solution has a known initial concentration of the substance, measured in grams per liter (g/L). The tank is filled by an inflow pipe at a known rate of $R$ liters per second (L/s) with a solution of the same substance that has a known concentration of $C_i$ g/L. The tank also has an outflow pipe that allows solution to leave the tank at a rate equal to the inflow rate of $R$ (L/s). Therefore, at all times the volume of solution, denoted $V$ and measured in liters, is constant. The configuration of the tank and the names of the various parameters are shown in **Figure 9.32**.

Imagine that at time $t = 0$, the inflow and outflow pipes are opened and solution begins flowing into and out of the tank. We assume at all times the tank is thoroughly stirred, so the solution in the tank has a uniform—but changing—concentration. The goal is to find the mass of the substance (salt or sugar) in the tank at all times.

Stirring device

Inflow rate $R$
Inflow concentration $C_i$

Volume $V$

Outflow rate $R$

**Figure 9.32**

**QUICK CHECK 2** Suppose the tank is filled with a salt solution that initially has a concentration of 50 g/L and the inflow pipe carries pure water (concentration of 0 g/L). If the stirred tank reaction runs for a long time, what is the eventual concentration of the salt solution in the tank? ◄

A key fact in modeling the stirred tank reaction is that

$$\text{concentration} = \frac{\text{mass}}{\text{volume}} \quad \text{or} \quad \text{mass} = \text{concentration} \cdot \text{volume}.$$

Let $m(t)$ be the mass of the substance in the tank at time $t \geq 0$, with $m(0) = m_0$ given. Assuming the mass $m(t)$ is known at some time $t$, we ask how it changes in a small time interval $[t, t + \Delta t]$ to give a new mass $m(t + \Delta t)$. The task is to account for all the mass that flows into and out of the tank during this time interval. Here is how the mass changes:

$$\underbrace{m(t + \Delta t)}_{\substack{\text{mass at end} \\ \text{of interval}}} \approx \underbrace{m(t)}_{\substack{\text{current} \\ \text{mass}}} + \underbrace{\text{mass that flows in}}_{C_i R \Delta t} - \underbrace{\text{mass that flows out}}_{\frac{m(t)}{V} R \Delta t}.$$

The crux of the modeling process is to determine the inflow and outflow terms in this equation. Consider the inflow first. Solution flows in at a rate $R$ L/s, so the volume of solution that flows into the tank in $\Delta t$ seconds is $R \Delta t$ liters (check that the units work out). The solution that flows into the tank has a concentration of $C_i$ g/L; therefore, the mass of substance that flows into the tank in time $\Delta t$ is

$$\underbrace{C_i}_{\substack{\text{concentration} \\ g/L}} \cdot \underbrace{R}_{\substack{\text{inflow rate} \\ L/s}} \cdot \underbrace{\Delta t}_{\substack{\text{time} \\ \text{interval}}} \quad \text{(grams)}.$$

(Remember that mass = concentration · volume; check the units.)

Now let's look at the outflow term. At time $t$, the mass of substance in the tank is $m(t)$, so the concentration of the solution is $m(t)/V$ g/L. As with the inflow, the volume of solution that flows out of the tank in $\Delta t$ seconds is $R \Delta t$ liters, and the mass of substance that flows out of the tank in time $\Delta t$ is

$$\underbrace{\frac{m(t)}{V}}_{\substack{\text{concentration} \\ g/L}} \cdot \underbrace{R}_{\substack{\text{outflow rate} \\ L/s}} \cdot \underbrace{\Delta t}_{\substack{\text{time} \\ \text{interval}}} \quad \text{(grams)}.$$

We now substitute these quantities into the mass change equation:

$$m(t + \Delta t) \approx m(t) + \underbrace{C_i R \Delta t}_{\text{inflow}} - \underbrace{\frac{m(t)}{V} R \Delta t}_{\text{outflow}}.$$

This equation is an approximation because the mass of substance changes during the time interval $[t, t + \Delta t]$. However, the approximation improves as the length of the time interval $\Delta t$ decreases. We divide through the mass change equation by $\Delta t$:

$$\underbrace{\frac{m(t + \Delta t) - m(t)}{\Delta t}}_{\to\, m'(t) \text{ as } \Delta t \to 0} \approx C_i R - \frac{m(t)}{V} R.$$

Observe that the left side of the equation approaches the derivative $m'(t)$ as $\Delta t$ approaches zero. The result is a differential equation that governs the mass of the substance in the stirred tank. We have a familiar, linear first-order initial value problem to solve:

$$m'(t) = -\frac{R}{V} m(t) + C_i R, \quad m(0) = m_0.$$

The solution of this equation is analyzed in Exercise 35.

**EXAMPLE 3**   **A stirred tank**   A 1000-L tank is filled with a brine (salt) solution with an initial concentration of 5 g/L. Brine solution with a concentration of 25 g/L flows into the tank at a rate of 8 L/s, while thoroughly mixed solution flows out of the tank at 8 L/s.

**a.** Find the mass of salt in the tank, for $t \geq 0$.

**b.** Find the concentration of the solution in the tank, for $t \geq 0$.

**SOLUTION**

**a.** We are given the initial concentration of the solution in the tank. To find the initial mass of salt in the tank, we multiply the concentration by the volume:

$$m_0 = 1000 \text{ L} \cdot 5 \frac{g}{L} = 5000 \text{ g}.$$

The inflow concentration is $C_i = 25$ g/L and inflow rate is $R = 8$ L/s. Therefore, the initial value problem for the reaction is

$$m'(t) = -\frac{8}{1000} m(t) + 25 \cdot 8$$

$$= -0.008m(t) + 200, \quad m(0) = 5000.$$

This is an equation of the form $y'(t) = ky + b$, which was discussed in Section 9.4. Letting $k = -0.008$ and $b = 200$, the general solution is

$$m(t) = Ce^{kt} - \frac{b}{k} = Ce^{-0.008t} - \frac{200}{(-0.008)} = Ce^{-0.008t} + 25{,}000.$$

The initial condition $m(0) = 5000$, when substituted into the general solution, implies that

$$5000 = C + 25{,}000 \quad \Rightarrow \quad C = -20{,}000.$$

Therefore, the solution of the initial value problem is

$$m(t) = 25{,}000 - 20{,}000e^{-0.008t}, \text{ for } t \geq 0.$$

The graph of $m$ (**Figure 9.33**) indicates that the mass of salt in the tank approaches 25,000 g as $t$ increases. This mass corresponds to a concentration of 25,000 g/1000 L = 25 g/L, which is the concentration of the inflow solution. As time increases, the original solution in the tank is replaced by the inflow solution.

**b.** The concentration is found by dividing the mass function by the volume of the tank. Therefore, the concentration function is

$$C(t) = 25 - 20e^{-0.008t}, \text{ for } t \geq 0.$$

*Related Exercises 24–25* ◄

**Figure 9.33**

**Figure 9.34**

➤ The original predator-prey model is attributed to the Belgian mathematician Pierre François Verhulst (1804–1849). The model was further developed independently by the American biophysicist Alfred Lotka and the Italian mathematician Vito Volterra, who used it to study shark populations. These equations are also called the Lotka-Volterra equations.

## Predator-Prey Models

Perhaps the best-known graph in wildlife ecology shows 100 years of data, collected by the Hudson Bay Company, of populations of Canadian lynx and snowshoe hare (**Figure 9.34**). The striking features of these graphs are the cyclic fluctuations of the two populations and the fact that the hare population is out of phase with the lynx population. In general, two species may interact in a competitive way, in a cooperative way, or, as in the case of the lynx-hare pair, as predator and prey. In this section, we investigate the fundamental model for describing predator-prey interactions.

Our task is to consider a system consisting of two species—one a predator and one a prey—and to devise a *pair* of differential equations that describes their interactions and whose solutions give their populations. To be specific, let the predator be foxes, whose population at time $t \geq 0$ is $F(t)$, and let the prey be hares, whose population at time $t \geq 0$ is $H(t)$. Here are the assumptions that underlie the model.

• When the hare population (prey) is sparse, the fox population decreases exponentially, while encounters between hares and foxes increase the fox population (the hares are the food supply).

• When the fox population (predators) is sparse, the hare population increases exponentially, while encounters between hares and foxes deplete the hare population (the foxes eat the hares).

Here is a set of differential equations that incorporate these assumptions.

$$\underbrace{F'(t)}_{\substack{\text{rate of change} \\ \text{of fox population}}} = \underbrace{-aF(t)}_{\substack{\text{natural decay} \\ \text{of foxes}}} + \underbrace{bF(t)H(t)}_{\substack{\text{increase in foxes} \\ \text{due to fox-hare} \\ \text{encounters}}}$$

$$\underbrace{H'(t)}_{\substack{\text{rate of change} \\ \text{of hare population}}} = \underbrace{cH(t)}_{\substack{\text{natural growth} \\ \text{of hares}}} - \underbrace{dF(t)H(t)}_{\substack{\text{decrease in hares} \\ \text{due to fox-hare} \\ \text{encounters}}}$$

In these equations, $a$, $b$, $c$, and $d$ are positive real numbers.

Notice that in the first equation, the rate of change of the fox population decreases with the size of the fox population and increases with the number of fox-hare interactions. We assume the number of fox-hare interactions is proportional to the product of the fox and hare populations. In the second equation, the rate of change of the hare population increases with the size of the hare population and decreases with the number of fox-hare interactions.

We have no methods for solving such a pair of equations; indeed, finding an analytical solution is challenging (Exercise 39). Fortunately, we can resort to a familiar tool to study the solutions: direction fields. However, in this case, because there are two unknown solutions, the direction field is plotted in the $FH$-plane.

Let's first rewrite the governing equations more compactly.

$$F'(t) = -aF + bFH = F(-a + bH)$$
$$H'(t) = cH - dFH = H(c - dF)$$

As with the direction fields we studied earlier, we look for conditions for which the derivatives are zero, positive, and negative. Because $F$ and $H$ are populations, we assume they have nonnegative values.

The points in the $FH$-plane at which $F'(t) = H'(t) = 0$ are special because they correspond to equilibrium solutions. You should verify that there are two such points: $(F, H) = (0, 0)$ and $(F, H) = \left(\dfrac{c}{d}, \dfrac{a}{b}\right)$. If the two populations have either of these initial values, then they remain constant for all time.

Recalling that $F > 0$, the condition $F'(t) = F(-a + bH) > 0$ is satisfied when $-a + bH > 0$—or, equivalently, when $H > \dfrac{a}{b}$. Using similar reasoning, the condition $F'(t) < 0$ is satisfied when $0 < H < \dfrac{a}{b}$. Repeating this process on the second equation gives $H'(t) = H(c - dF) > 0$ when $0 < F < \dfrac{c}{d}$ and $H'(t) < 0$ when $F > \dfrac{c}{d}$.

**Figure 9.35** summarizes everything we have learned so far.

We see that the vertical line $F = \dfrac{c}{d}$ and the horizontal line $H = \dfrac{a}{b}$ divide the first quadrant of the $FH$-plane into four regions. In each region, the derivatives of $F$ and $H$ have particular signs. For example, in the region nearest to the origin, we have $F' < 0$ and $H' > 0$, which means that $F$ is decreasing and $H$ is increasing in this region. Therefore, we mark this region with an arrow that points in the direction of decreasing $F$ and increasing $H$. All solution curves move in the negative $F$-direction and positive $H$-direction in this region. Similar arguments explain the arrows in the other three regions of Figure 9.35.

If we stand back and look at Figure 9.35, we can see the general "flow" of the solution curves. They circulate around the equilibrium point in the clockwise direction. While it is not evident from this analysis, it can be shown that the solution curves are actually closed curves; that is, they close on themselves. Therefore, if we choose an initial population of foxes and hares, corresponding to a single point in the $FH$-plane, the resulting solution curve eventually returns to the same point. In the process, both $F$ and $H$ oscillate in a cyclic fashion—as seen in the Hudson Bay data. **Figure 9.36** shows two solution curves superimposed on the direction field.

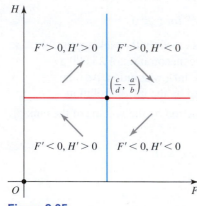

**Figure 9.35**

➤ Notice that when we plot solutions in the $FH$-plane, the independent variable $t$ does not appear explicitly in the graph. That is, the graph in the $FH$-plane is different from a graph of $F$ as a function of $t$ or of $H$ as a function of $t$. As a result, it is not possible to determine the period of the oscillations from the graph in the $FH$-plane.

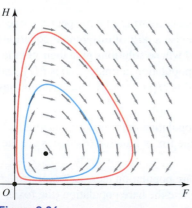

**Figure 9.36**

**QUICK CHECK 3** Explain why a closed solution curve in the *FH*-plane represents fox and hare populations that oscillate in a cyclic way. ◄

➤ The variables in population models are often scaled to some reference quantity. For example, *F* and *H* may be measured in hundreds of individuals, so that $F = 3$ might mean 300 foxes.

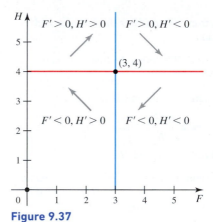

**Figure 9.37**

**EXAMPLE 4** **A predator-prey model** Consider the predator-prey model given by the equations

$$F'(t) = -12F + 3FH,$$
$$H'(t) = 15H - 5FH.$$

**a.** Find the lines on which $F' = 0$ or $H' = 0$, and the equilibrium points of the system.

**b.** Make a sketch of four regions in the first quadrant of the *FH*-plane and indicate the directions in which the solution curves move in each region.

**c.** Sketch a representative solution curve in the *FH*-plane.

**SOLUTION**

**a.** Using the first equation and solving $F' = 0$ gives the condition

$$-12F + 3FH = 3F(-4 + H) = 0,$$

which implies that $F' = 0$ when $F = 0$ or when $H = 4$. Using the second equation and solving $H' = 0$ implies that

$$15H - 5FH = 5H(3 - F) = 0.$$

Therefore, $H' = 0$ when $H = 0$ or $F = 3$. The equilibrium points occur when $F' = H' = 0$ (simultaneously). These conditions are satisfied at the points $(0, 0)$ and $(F, H) = (3, 4)$. Therefore, the system has two equilibrium points. These observations are recorded in **Figure 9.37**.

**b.** The horizontal line $H = 4$ and the vertical line $F = 3$ divide the first quadrant of the *FH*-plane into four regions. The condition

$$F' = -12F + 3FH = 3F(-4 + H) > 0$$

implies that $H > 4$ (recall that $F > 0$). It follows that $F' < 0$ when $0 < H < 4$. Similarly,

$$H' = 15H - 5FH = 5H(3 - F) > 0$$

when $0 < F < 3$, and $H' < 0$ when $F > 3$. These observations are also shown in Figure 9.37, and from them, we can see that the solution curves circulate around the equilibrium point $(3, 4)$ in the clockwise direction.

**c.** **Figure 9.38** shows the direction field in detail with a typical solution curve superimposed on the direction field.

**Figure 9.38**

**Figure 9.39**

A final view of the solutions is obtained by using a numerical method, such as Euler's method, to approximate the solutions of the predator-prey equations. **Figure 9.39** shows the fox and hare populations, now graphed as functions of time. The cyclic behavior is evident, and the period of the oscillations is also seen to be approximately 0.5 time units.

*Related Exercises 27–30* ◄

## SECTION 9.5 EXERCISES

### Getting Started

1. Explain how the growth rate function determines the solution of a population model.

2. What is a carrying capacity? Mathematically, how does it appear on the graph of a population function?

3. Explain how the growth rate function can be decreasing while the population function is increasing.

4. Explain how a stirred tank reaction works.

5. Is the differential equation that describes a stirred tank reaction (as developed in this section) linear or nonlinear? What is its order?

6. What are the assumptions underlying the predator-prey model discussed in this section?

7. Describe the solution curves in a predator-prey model in the $FH$-plane.

8. Describe the behavior of the two populations in a predator-prey model as functions of time.

**9–14. Growth rate functions** *Make a sketch of the population function P (as a function of time) that results from the following growth rate functions. Assume the population at time $t = 0$ begins at some positive value.*

**9.**

**10.**

**11.**

**12.**

**13.**

**14.**

### Practice Exercises

**T 15–16. Solving logistic equations** *Write a logistic equation with the following parameter values. Then solve the initial value problem and graph the solution. Let r be the natural growth rate, K the carrying capacity, and $P_0$ the initial population.*

15. $r = 0.2, K = 300, P_0 = 50$

16. $r = 0.4, K = 5500, P_0 = 100$

**T 17–18. Designing logistic functions** *Use the method of Example 1 to find a logistic function that describes the following populations. Graph the population function.*

17. The population increases from 200 to 600 in the first year and eventually levels off at 2000.

18. The population increases from 50 to 60 in the first month and eventually levels off at 150.

19. **General Gompertz solution** Solve the initial value problem

$$M'(t) = -rM \ln\left(\frac{M}{K}\right), M(0) = M_0$$

with arbitrary positive values of $r$, $K$, and $M_0$.

**T 20–22. Solving the Gompertz equation** *Solve the Gompertz equation in Exercise 19 with the given values of r, K, and $M_0$. Then graph the solution to be sure that $M(0)$ and $\lim_{t \to \infty} M(t)$ are correct.*

20. $r = 0.1, K = 500, M_0 = 50$

21. $r = 0.05, K = 1200, M_0 = 90$

22. $r = 0.6, K = 5500, M_0 = 20$

**23–26. Stirred tank reactions** *For each of the following stirred tank reactions, carry out the following analysis.*

**a.** *Write an initial value problem for the mass of the substance.*
**b.** *Solve the initial value problem.*

23. A 500-L tank is initially filled with pure water. A copper sulfate solution with a concentration of 20 g/L flows into the tank at a rate of 4 L/min. The thoroughly mixed solution is drained from the tank at a rate of 4 L/min.

24. A 1500-L tank is initially filled with a solution that contains 3000 g of salt. A salt solution with a concentration of 20 g/L flows into the tank at a rate of 3 L/min. The thoroughly mixed solution is drained from the tank at a rate of 3 L/min.

25. A 2000-L tank is initially filled with a sugar solution with a concentration of 40 g/L. A sugar solution with a concentration of 10 g/L flows into the tank at a rate of 10 L/min. The thoroughly mixed solution is drained from the tank at a rate of 10 L/min.

26. A one-million-liter pond is contaminated by a chemical pollutant with a concentration of 20 g/L. The source of the pollutant is removed, and pure water is allowed to flow into the pond at a rate of 1200 L/hr. Assuming the pond is thoroughly mixed and drained at a rate of 1200 L/hr, how long does it take to reduce the concentration of the solution in the pond to 10% of the initial value?

**27–30. Predator-prey models** *Consider the following pairs of differential equations that model a predator-prey system with populations x and y. In each case, carry out the following steps.*

**a.** *Identify which equation corresponds to the predator and which corresponds to the prey.*

**b.** *Find the lines along which $x'(t) = 0$. Find the lines along which $y'(t) = 0$.*

**c.** *Find the equilibrium points for the system.*

**d.** *Identify the four regions in the first quadrant of the xy-plane in which $x'$ and $y'$ are positive or negative.*

**e.** *Sketch a representative solution curve in the xy-plane and indicate the direction in which the solution evolves.*

**27.** $x'(t) = -3x + 6xy,\ y'(t) = y - 4xy$

**28.** $x'(t) = 2x - 4xy,\ y'(t) = -y + 2xy$

**29.** $x'(t) = -3x + xy,\ y'(t) = 2y - xy$

**30.** $x'(t) = 2x - xy,\ y'(t) = -y + xy$

**31. Explain why or why not** Determine whether the following statements are true and give an explanation or counterexample.

    **a.** If the growth rate function for a population model is positive, then the population is increasing.

    **b.** The solution of a stirred tank initial value problem always approaches a constant as $t \to \infty$.

    **c.** In the predator-prey models discussed in this section, if the initial predator population is zero and the initial prey population is positive, then the prey population increases without bound.

## Explorations and Challenges

**32. Growth rate functions**

    **a.** Show that the logistic growth rate function $f(P) = rP\left(1 - \dfrac{P}{K}\right)$ has a maximum value of $\dfrac{rK}{4}$ at the point $P = \dfrac{K}{2}$.

    **b.** Show that the Gompertz growth rate function
$$f(M) = -rM \ln\left(\frac{M}{K}\right)$$ has a maximum value of $\dfrac{rK}{e}$ at the point $M = \dfrac{K}{e}$.

**33. Solution of the logistic equation** Use separation of variables to show that the solution of the initial value problem
$$P'(t) = rP\left(1 - \frac{P}{K}\right), \quad P(0) = P_0$$
is $P(t) = \dfrac{K}{\left(\dfrac{K}{P_0} - 1\right)e^{-rt} + 1}$.

**34. Properties of the Gompertz solution** Verify that the function
$$M(t) = K\left(\frac{M_0}{K}\right)^{\exp(-rt)}$$
satisfies the properties $M(0) = M_0$ and $\lim\limits_{t\to\infty} M(t) = K$.

**35. Properties of stirred tank solutions**

    **a.** Show that for general positive values of $R$, $V$, $C_i$, and $m_0$, the solution of the initial value problem
$$m'(t) = -\frac{R}{V}m(t) + C_i R, \quad m(0) = m_0$$
is $m(t) = (m_0 - C_i V)e^{-Rt/V} + C_i V$.

    **b.** Verify that $m(0) = m_0$.

    **c.** Evaluate $\lim\limits_{t\to\infty} m(t)$ and give a physical interpretation of the result.

    **d.** Suppose $m_0$ and $V$ are fixed. Describe the effect of increasing $R$ on the graph of the solution.

**36. A physiological model** A common assumption in modeling drug assimilation is that the blood volume in a person is a single compartment that behaves like a stirred tank. Suppose the blood volume is a four-liter tank that initially has a zero concentration of a particular drug. At time $t = 0$, an intravenous line is inserted into a vein (into the tank) that carries a drug solution with a concentration of 500 mg/L. The inflow rate is 0.06 L/min. Assume the drug is quickly mixed thoroughly in the blood and that the volume of blood remains constant.

    **a.** Write an initial value problem that models the mass of the drug in the blood, for $t \ge 0$.

    **b.** Solve the initial value problem, and graph both the mass of the drug and the concentration of the drug.

    **c.** What is the steady-state mass of the drug in the blood?

    **d.** After how many minutes does the drug mass reach 90% of its steady-state level?

**37. RC circuit equation** Suppose a battery with voltage $V$ is connected in series to a capacitor (a charge storage device) with capacitance $C$ and a resistor with resistance $R$. As the charge $Q$ in the capacitor increases, the current $I$ across the capacitor decreases according to the following initial value problems. Solve each initial value problem and interpret the solution. Assume $C$, $R$, and $V$ are given.

    **a.** $I'(t) + \dfrac{1}{RC}I(t) = 0,\ I(0) = \dfrac{V}{R}$

    **b.** $Q'(t) + \dfrac{1}{RC}Q(t) = \dfrac{V}{R},\ Q(0) = 0$

**38. U.S. population projections** According to the U.S. Census Bureau, the nation's population (to the nearest million) was 296 million in 2005 and 321 million in 2015. The Bureau also projects a 2050 population of 398 million. To construct a logistic model, both the growth rate and the carrying capacity must be estimated. There are several ways to estimate these parameters. Here is one approach:

    **a.** Assume $t = 0$ corresponds to 2005 and that the population growth is exponential for the first ten years; that is, between 2005 and 2015, the population is given by $P(t) = P(0)e^{rt}$. Estimate the growth rate $r$ using this assumption.

    **b.** Write the solution of the logistic equation with the value of $r$ found in part (a). Use the projected value $P(45) = 398$ million to find a value of the carrying capacity $K$.

    **c.** According to the logistic model determined in parts (a) and (b), when will the U.S. population reach 95% of its carrying capacity?

    **d.** Estimations of this kind must be made and interpreted carefully. Suppose the projected population for 2050 is 410 million rather than 398 million. What is the value of the carrying capacity in this case?

e. Repeat part (d) assuming the projected population for 2050 is 380 million rather than 398 million. What is the value of the carrying capacity in this case?

f. Comment on the sensitivity of the carrying capacity to the 35-year population projection.

**T 39. Analytical solution of the predator-prey equations** The solution of the predator-prey equations

$$x'(t) = -ax + bxy, \quad y'(t) = cy - dxy$$

can be viewed as parametric equations that describe the solution curves. Assume $a$, $b$, $c$, and $d$ are positive constants and consider solutions in the first quadrant.

a. Recalling that $\dfrac{dy}{dx} = \dfrac{y'(t)}{x'(t)}$, divide the first equation by the second equation to obtain a separable differential equation in terms of $x$ and $y$.

b. Show that the general solution can be written in the implicit form $e^{dx+by} = Cx^c y^a$, where $C$ is an arbitrary constant.

c. Let $a = 0.8$, $b = 0.4$, $c = 0.9$, and $d = 0.3$. Plot the solution curves for $C = 1.5$, 2, and 2.5, and confirm that they are, in fact, closed curves. Use the graphing window $[0, 9] \times [0, 9]$.

**QUICK CHECK ANSWERS**

1. The graph of the growth rate function is a parabola with zeros (intercepts on the horizontal axis) at $P = 0$ and $P = K$. The vertex of a parabola occurs at the midpoint of the interval between the zeros, or in this case, at $P = K/2$.   **2.** 0 g/L

3. Trace a closed solution curve in the $FH$-plane, and watch the values of either variable, $F$ or $H$. As you traverse the curve once, the values of each variable increase, then decrease (or vice versa), and return to their starting values. ◀

---

## CHAPTER 9   REVIEW EXERCISES

1. **Explain why or why not** Determine whether the following statements are true and give an explanation or counterexample.

   a. The differential equation $y' + 2y = t$ is first-order, linear, and separable.

   b. The differential equation $y'y = 2t^2$ is first-order, linear, and separable.

   c. The function $y = t + 1/t$, for $t > 0$, satisfies the initial value problem $ty' + y = 2t$, $y(1) = 2$.

   d. The direction field for the differential equation $y'(t) = t + y(t)$ is plotted in the $ty$-plane.

   e. Euler's method gives the exact solution to the initial value problem $y' = ty^2$, $y(0) = 3$ on the interval $[0, a]$ provided $a$ is not too large.

**2–10. General solutions** *Use the method of your choice to find the general solution of the following differential equations.*

2. $y'(t) + 3y = 0$

3. $y'(t) + 2y = 6$

4. $p'(x) = 4p + 8$

5. $y'(t) = 2ty$

6. $y'(t) = \sqrt{\dfrac{y}{t}}$

7. $y'(t) = \dfrac{y}{t^2 + 1}$

8. $y'(x) = \dfrac{\sin x}{2y}$

9. $y'(t) = (2t + 1)(y^2 + 1)$

10. $z'(t) = \dfrac{tz}{t^2 + 1}$

**11–18. Solving initial value problems** *Use the method of your choice to find the solution of the following initial value problems.*

11. $y'(t) = 2t + \cos t$, $y(0) = 1$

12. $y'(t) = -3y + 9$, $y(0) = 4$

13. $Q'(t) = Q - 8$, $Q(1) = 0$

14. $y'(x) = \dfrac{x}{y}$, $y(2) = 4$

15. $u'(t) = \left(\dfrac{u}{t}\right)^{1/3}$, $u(1) = 8$

16. $y'(x) = 4x \csc y$, $y(0) = \dfrac{\pi}{2}$

17. $t(t^2 + 1)s'(t) = s$, $s(1) = 1$

18. $\theta'(x) = 4x \cos^2 \theta$, $\theta(0) = \dfrac{\pi}{4}$

19. **Direction fields** Consider the direction field for the equation $y' = y(2 - y)$ shown in the figure and initial conditions of the form $y(0) = A$.

   a. Sketch a solution on the direction field with the initial condition $y(0) = 1$.

   b. Sketch a solution on the direction field with the initial condition $y(0) = 3$.

   c. For what values of $A$ are the corresponding solutions increasing, for $t \geq 0$?

   d. For what values of $A$ are the corresponding solutions decreasing, for $t \geq 0$?

   e. Identify the equilibrium solutions for the differential equation.

**20. Direction fields** The direction field for the equation
$y'(t) = t - y$, for $|t| \leq 4$ and $|y| \leq 4$, is shown in the figure.

   **a.** Use the direction field to sketch the solution curve that passes through the point $(0, 1/2)$.

   **b.** Use the direction field to sketch the solution curve that passes through the point $(0, -1/2)$.

   **c.** In what region of the $ty$-plane are solutions increasing? Decreasing?

   **d.** Complete the following sentence. The solution of the differential equation with the initial condition $y(0) = A$, where $A$ is a real number, approaches the line _____ as $t \to \infty$.

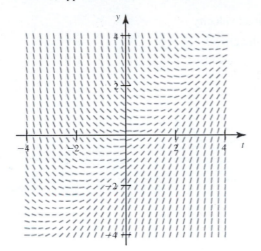

**21. Euler's method** Consider the initial value problem $y'(t) = \dfrac{1}{2y}$, $y(0) = 1$.

   **a.** Use Euler's method with $\Delta t = 0.1$ to compute approximations to $y(0.1)$ and $y(0.2)$.

   **b.** Use Euler's method with $\Delta t = 0.05$ to compute approximations to $y(0.1)$ and $y(0.2)$.

   **c.** The exact solution of this initial value problem is $y = \sqrt{t + 1}$, for $t > -1$. Compute the errors on the approximations to $y(0.2)$ found in parts (a) and (b). Which approximation gives the smaller error?

**22–25. Equilibrium solutions** *Find the equilibrium solutions of the following equations and determine whether each solution is stable or unstable.*

**22.** $y'(t) = y(2 - y)$       **23.** $y'(t) = y(3 + y)(y - 5)$

**24.** $y'(t) = \sin 2y$, for $|y| < \pi$   **25.** $y'(t) = y^3 - y^2 - 2y$

**26. Logistic growth** The population of a rabbit community is governed by the initial value problem

$$P'(t) = 0.2P\left(1 - \frac{P}{1200}\right), P(0) = 50.$$

   **a.** Find the equilibrium solutions.

   **b.** Find the population, for all times $t \geq 0$.

   **c.** What is the carrying capacity of the population?

   **d.** What is the population when the growth rate is a maximum?

**27. Logistic growth parameters** A cell culture has a population of 20 when a nutrient solution is added at $t = 0$. After 20 hours, the cell population is 80 and the carrying capacity of the culture is estimated to be 1600 cells.

   **a.** Use the population data at $t = 0$ and $t = 20$ to find the natural growth rate of the population.

   **b.** Find and solve the logistic equation for the cell population.

   **c.** After how many hours does the population reach half of the carrying capacity?

**28. Logistic growth in India** The population of India was 435 million in 1960 ($t = 0$) and 487 million in 1965 ($t = 5$). The projected population for 2050 is 1.57 billion.

   **a.** Assume the population increased exponentially between 1960 and 1965, and use the populations in these years to determine the natural growth rate in a logistic model.

   **b.** Use the solution of the logistic equation and the 2050 projected population to determine the carrying capacity.

   **c.** Based on the values of $r$ and $K$ found in parts (a) and (b), write the logistic growth function for India's population (measured in millions of people).

   **d.** In approximately what year does the population of India first exceed 2 billion people?

   **e.** Discuss some possible shortcomings of this model. Why might the carrying capacity be either greater than or less than the value predicted by the model?

**29. Stirred tank reaction** A 100-L tank is filled with pure water when an inflow pipe is opened and a sugar solution with a concentration of 20 gm/L flows into the tank at a rate of 0.5 L/min. The solution is thoroughly mixed and flows out of the tank at a rate of 0.5 L/min.

   **a.** Find the mass of sugar in the tank at all times after the inflow pipe is opened.

   **b.** What is the steady-state mass of sugar in the tank?

   **c.** At what time does the mass of sugar reach 95% of its steady-state level?

**30. Newton's Law of Cooling** A cup of coffee is removed from a microwave oven with a temperature of 80°C and allowed to cool in a room with a temperature of 25°C. Five minutes later, the temperature of the coffee is 60°C.

   **a.** Find the rate constant $k$ for the cooling process.

   **b.** Find the temperature of the coffee, for $t \geq 0$.

   **c.** When does the temperature of the coffee reach 50°C?

**31. A predator-prey model** Consider the predator-prey model

$$x'(t) = -4x + 2xy, \, y'(t) = 5y - xy.$$

   **a.** Does $x$ represent the population of the predator species or the prey species?

   **b.** Find the lines along which $x'(t) = 0$. Find the lines along which $y'(t) = 0$.

   **c.** Find the equilibrium points for the system.

   **d.** Identify the four regions in the first quadrant of the $xy$-plane in which $x'$ and $y'$ are positive or negative.

   **e.** Sketch a typical solution curve in the $xy$-plane. In which direction does the solution evolve?

**32. A first-order equation** Consider the equation
$t^2 y'(t) + 2ty(t) = e^{-t}$.

   **a.** Show that the left side of the equation can be written as the derivative of a single term.

   **b.** Integrate both sides of the equation to obtain the general solution.

   **c.** Find the solution that satisfies the condition $y(1) = 0$.

**33. A second-order equation** Consider the equation
$t^2 y''(t) + 2ty'(t) - 12y(t) = 0$.

**a.** Look for solutions of the form $y(t) = t^p$, where $p$ is to be determined. Substitute this trial solution into the equation and find two values of $p$ that give solutions; call them $p_1$ and $p_2$.

**b.** Assuming the general solution of the equation is $y(t) = C_1 t^{p_1} + C_2 t^{p_2}$, find the solution that satisfies the conditions $y(1) = 0, y'(1) = 7$.

## Chapter 9 Guided Projects

*Applications of the material in this chapter and related topics can be found in the following Guided Projects. For additional information, see the Preface.*

- Cooling coffee
- Euler's method for differential equations
- Predator-prey models
- Period of the pendulum

- Terminal velocity
- Logistic growth
- A pursuit problem

# 10

# Sequences and Infinite Series

**Chapter Preview** This chapter covers topics that lie at the foundation of calculus—indeed, at the foundation of mathematics. The first task is to make a clear distinction between a *sequence* and an *infinite series*. A sequence is an ordered *list* of numbers, $a_1, a_2, \ldots$, while an infinite series is a *sum* of numbers, $a_1 + a_2 + \cdots$. The idea of convergence to a limit is important for both sequences and series, but convergence is analyzed differently in the two cases. To determine limits of sequences, we use the same tools used for limits of functions at infinity. The analysis of infinite series requires new methods, which are introduced in this chapter. The study of infinite series begins with *geometric series*, which have theoretical importance and are used to answer many practical questions (When is your auto loan paid off? How much antibiotic is in your blood if you take three pills per day?). We then present several tests that are used to determine whether a given series converges. In the final section of the chapter, we offer guidelines to help choose an appropriate convergence test for a given series.

## 10.1 An Overview

To understand sequences and series, you must understand how they differ and how they are related. The purposes of this opening section are to introduce sequences and series in concrete terms, and to illustrate both their differences and their relationships with each other.

### Examples of Sequences

> ➤ In keeping with common practice, the terms *series* and *infinite series* are used interchangeably throughout this chapter.

Consider the following *list* of numbers:

$$\{1, 4, 7, 10, 13, 16, \ldots\}.$$

Each number in the list is obtained by adding 3 to the previous number in the list. With this rule, we could extend the list indefinitely.

This list is an example of a *sequence*, where each number in the sequence is called a **term** of the sequence. We denote sequences in any of the following forms:

> ➤ The dots ( . . . , an ellipsis) after the last number of a sequence mean that the list continues indefinitely.

$$\{a_1, a_2, a_3, \ldots, a_n, \ldots\}, \quad \{a_n\}_{n=1}^{\infty}, \quad \text{or} \quad \{a_n\}.$$

The subscript $n$ that appears in $a_n$ is called an **index**, and it indicates the order of terms in the sequence. The choice of a starting index is arbitrary, but sequences usually begin with $n = 0$ or $n = 1$.

The sequence $\{1, 4, 7, 10, \ldots\}$ can be defined in two ways. First, we have the rule that each term of the sequence is 3 more than the previous term; that is, $a_2 = a_1 + 3$, $a_3 = a_2 + 3$, $a_4 = a_3 + 3$, and so forth. In general, we see that

$$a_1 = 1 \quad \text{and} \quad a_{n+1} = a_n + 3, \quad \text{for } n = 1, 2, 3, \ldots.$$

This way of defining a sequence is called a *recurrence relation* (or an *implicit formula*). It specifies the initial term of the sequence (in this case, $a_1 = 1$) and gives a general rule for computing the next term of the sequence from previous terms. For example, if you know $a_{100}$, the recurrence relation can be used to find $a_{101}$.

Suppose instead you want to find $a_{147}$ directly without computing the first 146 terms of the sequence. The first four terms of the sequence can be written

$$a_1 = 1 + (3 \cdot 0), \quad a_2 = 1 + (3 \cdot 1), \quad a_3 = 1 + (3 \cdot 2), \quad a_4 = 1 + (3 \cdot 3).$$

Observe the pattern: The $n$th term of the sequence is 1 plus 3 multiplied by $n - 1$, or

$$a_n = 1 + 3(n - 1) = 3n - 2, \quad \text{for } n = 1, 2, 3, \ldots.$$

> **QUICK CHECK 1** Find $a_{10}$ for the sequence $\{1, 4, 7, 10, \ldots\}$ using the recurrence relation and then again using the explicit formula for the $n$th term. ◄

With this *explicit formula*, the $n$th term of the sequence is determined directly from the value of $n$. For example, with $n = 147$,

$$a_{147} = 3 \cdot \underset{n}{147} - \underset{n}{2} = 439.$$

> ➤ When defined by an explicit formula $a_n = f(n)$, it is evident that sequences are functions. The domain is generally a subset of the nonnegative integers, and one real number $a_n$ is assigned to each integer $n$ in the domain.

---

**DEFINITION Sequence**

A **sequence** $\{a_n\}$ is an ordered list of numbers of the form

$$\{a_1, a_2, a_3, \ldots, a_n, \ldots\}.$$

A sequence may be generated by a **recurrence relation** of the form $a_{n+1} = f(a_n)$, for $n = 1, 2, 3, \ldots$, where $a_1$ is given. A sequence may also be defined with an **explicit formula** of the form $a_n = f(n)$, for $n = 1, 2, 3, \ldots$.

---

**EXAMPLE 1 Explicit formulas** Use the explicit formula for $\{a_n\}_{n=1}^{\infty}$ to write the first four terms of each sequence. Sketch a graph of the sequence.

**a.** $a_n = \dfrac{1}{2^n}$  **b.** $a_n = \dfrac{(-1)^n n}{n^2 + 1}$

**SOLUTION**

**a.** Substituting $n = 1, 2, 3, 4, \ldots$ into the explicit formula $a_n = \dfrac{1}{2^n}$, we find that the terms of the sequence are

$$\left\{ \frac{1}{2}, \frac{1}{2^2}, \frac{1}{2^3}, \frac{1}{2^4}, \ldots \right\} = \left\{ \frac{1}{2}, \frac{1}{4}, \frac{1}{8}, \frac{1}{16}, \ldots \right\}.$$

The graph of a sequence is the graph of a function that is defined only on a set of integers. In this case, we plot the coordinate pairs $(n, a_n)$, for $n = 1, 2, 3, \ldots$, resulting in a graph consisting of individual points. The graph of the sequence $a_n = \dfrac{1}{2^n}$ suggests that the terms of this sequence approach 0 as $n$ increases (**Figure 10.1**).

**b.** Substituting $n = 1, 2, 3, 4, \ldots$ into the explicit formula, the terms of the sequence are

$$\left\{ \frac{(-1)^1(1)}{1^2 + 1}, \frac{(-1)^2 2}{2^2 + 1}, \frac{(-1)^3 3}{3^2 + 1}, \frac{(-1)^4 4}{4^2 + 1}, \ldots \right\} = \left\{ -\frac{1}{2}, \frac{2}{5}, -\frac{3}{10}, \frac{4}{17}, \ldots \right\}.$$

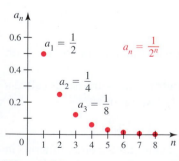

**Figure 10.1**

> ➤ The "switch" $(-1)^n$ is used frequently to alternate the signs of the terms of sequences and series.

**Figure 10.2**

From the graph (**Figure 10.2**), we see that the terms of the sequence alternate in sign and appear to approach 0 as $n$ increases.

*Related Exercises 13, 16* ◄

**EXAMPLE 2**   **Recurrence relations**  Use the recurrence relation for $\{a_n\}_{n=1}^{\infty}$ to write the first four terms of the sequences

$$a_{n+1} = 2a_n + 1,\, a_1 = 1 \quad \text{and} \quad a_{n+1} = 2a_n + 1,\, a_1 = -1.$$

**SOLUTION**  Notice that the recurrence relation is the same for the two sequences; only the first term differs. The first four terms of each sequence are as follows.

| $n$ | $a_n$ with $a_1 = 1$ | $a_n$ with $a_1 = -1$ |
|---|---|---|
| 1 | $a_1 = 1$ (given) | $a_1 = -1$ (given) |
| 2 | $a_2 = 2a_1 + 1 = 2 \cdot 1 + 1 = 3$ | $a_2 = 2a_1 + 1 = 2(-1) + 1 = -1$ |
| 3 | $a_3 = 2a_2 + 1 = 2 \cdot 3 + 1 = 7$ | $a_3 = 2a_2 + 1 = 2(-1) + 1 = -1$ |
| 4 | $a_4 = 2a_3 + 1 = 2 \cdot 7 + 1 = 15$ | $a_4 = 2a_3 + 1 = 2(-1) + 1 = -1$ |

**QUICK CHECK 2** Find an explicit formula for the sequence $\{1, 3, 7, 15, \dots\}$ (Example 2). ◄

We see that the terms of the first sequence increase without bound, while all terms of the second sequence are $-1$. Clearly, the initial term of the sequence may determine the behavior of the entire sequence.

*Related Exercises 21–22* ◄

**EXAMPLE 3**   **Working with sequences**  Consider the following sequences.

**a.** $\{a_n\} = \{-2, 5, 12, 19, \dots\}$    **b.** $\{b_n\} = \{3, 6, 12, 24, 48, \dots\}$

**(i)**  Find the next two terms of the sequence.

**(ii)** Find a recurrence relation that generates the sequence.

**(iii)** Find an explicit formula for the $n$th term of the sequence.

**SOLUTION**

**a. (i)**  Each term is obtained by adding 7 to its predecessor. The next two terms are $19 + 7 = 26$ and $26 + 7 = 33$.

➤ In Example 3, we chose the starting index $n = 0$. Other choices are possible.

**(ii)** Because each term is seven more than its predecessor, a recurrence relation is

$$a_{n+1} = a_n + 7,\, a_0 = -2, \quad \text{for } n = 0, 1, 2, \dots.$$

**(iii)** Notice that $a_0 = -2$, $a_1 = -2 + (1 \cdot 7)$, and $a_2 = -2 + (2 \cdot 7)$, so an explicit formula is

$$a_n = 7n - 2, \quad \text{for } n = 0, 1, 2, \dots.$$

**b. (i)**  Each term is obtained by multiplying its predecessor by 2. The next two terms are $48 \cdot 2 = 96$ and $96 \cdot 2 = 192$.

**(ii)** Because each term is two times its predecessor, a recurrence relation is

$$b_{n+1} = 2b_n,\, a_0 = 3, \quad \text{for } n = 0, 1, 2, \dots.$$

**(iii)** To obtain an explicit formula, note that $a_0 = 3$, $a_1 = 3(2^1)$, and $a_2 = 3(2^2)$. In general,

$$a_n = 3(2^n), \quad \text{for } n = 0, 1, 2, \dots.$$

*Related Exercises 27–28* ◄

## Limit of a Sequence

Perhaps the most important question about a sequence is this: If you go farther and farther out in the sequence, $a_{100}, \dots, a_{10,000}, \dots, a_{100,000}, \dots$, how do the terms of the sequence behave? Do they approach a specific number, and if so, what is that number? Or do they grow in magnitude without bound? Or do they wander around with or without a pattern?

The long-term behavior of a sequence is described by its **limit**. The limit of a sequence is defined rigorously in the next section. For now, we work with an informal definition.

---

**DEFINITION  Limit of a Sequence**

If the terms of a sequence $\{a_n\}$ approach a unique number $L$ as $n$ increases—that is, if $a_n$ can be made arbitrarily close to $L$ by taking $n$ sufficiently large—then we say $\lim\limits_{n\to\infty} a_n = L$ exists, and the sequence **converges** to $L$. If the terms of the sequence do not approach a single number as $n$ increases, the sequence has no limit, and the sequence **diverges**.

---

**EXAMPLE 4  Limits of sequences**  Write the first four terms of each sequence. If you believe the sequence converges, make a conjecture about its limit. If the sequence appears to diverge, explain why.

**a.** $\left\{\dfrac{(-1)^n}{n^2 + 1}\right\}_{n=1}^{\infty}$                    Explicit formula

**b.** $\{\cos n\pi\}_{n=1}^{\infty}$                    Explicit formula

**c.** $\{a_n\}_{n=1}^{\infty}$, where $a_{n+1} = -2a_n$, $a_1 = 1$    Recurrence relation

**SOLUTION**

**a.** Beginning with $n = 1$, the first four terms of the sequence are

$$\left\{\frac{(-1)^1}{1^2 + 1}, \frac{(-1)^2}{2^2 + 1}, \frac{(-1)^3}{3^2 + 1}, \frac{(-1)^4}{4^2 + 1}, \ldots\right\} = \left\{-\frac{1}{2}, \frac{1}{5}, -\frac{1}{10}, \frac{1}{17}, \ldots\right\}.$$

The terms decrease in magnitude and approach zero with alternating signs. The limit appears to be 0 (**Figure 10.3**).

**b.** The first four terms of the sequence are

$$\{\cos \pi, \cos 2\pi, \cos 3\pi, \cos 4\pi, \ldots\} = \{-1, 1, -1, 1, \ldots\}.$$

In this case, the terms of the sequence alternate between $-1$ and $+1$, and never approach a single value. Therefore, the sequence diverges (**Figure 10.4**).

**c.** The first four terms of the sequence are

$$\{1, -2a_1, -2a_2, -2a_3, \ldots\} = \{1, -2, 4, -8, \ldots\}.$$

Because the magnitudes of the terms increase without bound, the sequence diverges (**Figure 10.5**).

**Figure 10.3**

**Figure 10.4**

**Figure 10.5**

*Related Exercises 35–36* ◄

**EXAMPLE 5**    **Limit of a sequence**   Enumerate and graph the terms of the following sequence, and make a conjecture about its limit.

$$a_n = \frac{4n^3}{n^3 + 1}, \quad \text{for } n = 1, 2, 3, \ldots. \quad \text{Explicit formula}$$

**SOLUTION**   The first 14 terms of the sequence $\{a_n\}$ are tabulated in Table 10.1 and graphed in **Figure 10.6**. The terms appear to approach 4.

**Table 10.1**

| $n$ | $a_n$ | $n$ | $a_n$ |
|---|---|---|---|
| 1 | 2.000 | 8 | 3.992 |
| 2 | 3.556 | 9 | 3.995 |
| 3 | 3.857 | 10 | 3.996 |
| 4 | 3.938 | 11 | 3.997 |
| 5 | 3.968 | 12 | 3.998 |
| 6 | 3.982 | 13 | 3.998 |
| 7 | 3.988 | 14 | 3.999 |

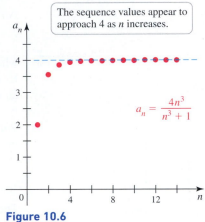

The sequence values appear to approach 4 as $n$ increases.

$a_n = \dfrac{4n^3}{n^3 + 1}$

**Figure 10.6**

*Related Exercise 45* ◄

The height of each bounce of the basketball is 0.8 of the height of the previous bounce.

20 ft

16 ft

12.8 ft

10.24 ft

Time

**Figure 10.7**

**EXAMPLE 6**    **A bouncing ball**   A basketball tossed straight up in the air reaches a high point and falls to the floor. Each time the ball bounces on the floor it rebounds to 0.8 of its previous height. Let $h_n$ be the high point after the $n$th bounce, with the initial height being $h_0 = 20$ ft.

**a.** Find a recurrence relation and an explicit formula for the sequence $\{h_n\}$.

**b.** What is the high point after the 10th bounce? After the 20th bounce?

**c.** Speculate on the limit of the sequence $\{h_n\}$.

**SOLUTION**

**a.** We first write and graph the heights of the ball for several bounces using the rule that each height is 0.8 of the previous height (**Figure 10.7**). For example, we have

$$h_0 = 20 \text{ ft,}$$
$$h_1 = 0.8 \, h_0 = 16 \text{ ft,}$$
$$h_2 = 0.8 \, h_1 = 0.8^2 \, h_0 = 12.80 \text{ ft,}$$
$$h_3 = 0.8 \, h_2 = 0.8^3 \, h_0 = 10.24 \text{ ft, and}$$
$$h_4 = 0.8 \, h_3 = 0.8^4 \, h_0 \approx 8.19 \text{ ft.}$$

Each number in the list is 0.8 of the previous number. Therefore, the recurrence relation for the sequence of heights is

$$h_{n+1} = 0.8 \, h_n, \quad h_0 = 20, \quad \text{for } n = 0, 1, 2, 3, \ldots.$$

To find an explicit formula for the $n$th term, note that

$$h_1 = h_0 \cdot 0.8, \quad h_2 = h_0 \cdot 0.8^2, \quad h_3 = h_0 \cdot 0.8^3, \quad \text{and} \quad h_4 = h_0 \cdot 0.8^4.$$

In general, we have

$$h_n = h_0 \cdot 0.8^n = 20 \cdot 0.8^n, \quad \text{for } n = 0, 1, 2, 3, \ldots,$$

which is an explicit formula for the terms of the sequence.

**b.** Using the explicit formula for the sequence, we see that after $n = 10$ bounces, the next height is

$$h_{10} = 20 \cdot 0.8^{10} \approx 2.15 \text{ ft.}$$

**Figure 10.8**

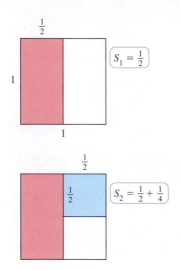

$S_1 = \frac{1}{2}$

$S_2 = \frac{1}{2} + \frac{1}{4}$

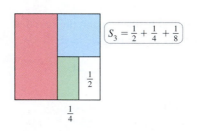

$S_3 = \frac{1}{2} + \frac{1}{4} + \frac{1}{8}$

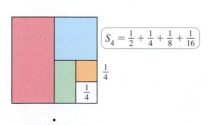

$S_4 = \frac{1}{2} + \frac{1}{4} + \frac{1}{8} + \frac{1}{16}$

$\vdots$

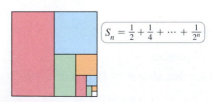

$S_n = \frac{1}{2} + \frac{1}{4} + \cdots + \frac{1}{2^n}$

**Figure 10.9**

After $n = 20$ bounces, the next height is

$$h_{20} = 20 \cdot 0.8^{20} \approx 0.23 \text{ ft.}$$

**c.** The terms of the sequence (**Figure 10.8**) appear to decrease and approach 0. A reasonable conjecture is that $\lim\limits_{n \to \infty} h_n = 0$.

*Related Exercises 57–60* ◄

## Infinite Series and the Sequence of Partial Sums

An infinite series can be viewed as a *sum* of an infinite set of numbers; it has the form

$$a_1 + a_2 + \cdots + a_n + \cdots,$$

where the terms of the series, $a_1, a_2, \ldots$, are real numbers. We first answer the question: How is it possible to sum an infinite set of numbers and produce a finite number? Here is an informative example.

Consider a unit square (sides of length 1) that is subdivided as shown in **Figure 10.9**. We let $S_n$ be the area of the colored region in the $n$th figure of the progression. The area of the colored region in the first figure is

$$S_1 = \frac{1}{2} \cdot 1 = \frac{1}{2}. \quad \frac{1}{2} = \frac{2^1 - 1}{2^1}$$

The area of the colored region in the second figure is $S_1$ plus the area of the smaller blue square, which is $\frac{1}{2} \cdot \frac{1}{2} = \frac{1}{4}$. Therefore,

$$S_2 = \frac{1}{2} + \frac{1}{4} = \frac{3}{4}. \quad \frac{3}{4} = \frac{2^2 - 1}{2^2}$$

The area of the colored region in the third figure is $S_2$ plus the area of the smaller green rectangle, which is $\frac{1}{2} \cdot \frac{1}{4} = \frac{1}{8}$. Therefore,

$$S_3 = \frac{1}{2} + \frac{1}{4} + \frac{1}{8} = \frac{7}{8}. \quad \frac{7}{8} = \frac{2^3 - 1}{2^3}$$

Continuing in this manner, we find that

$$S_n = \frac{1}{2} + \frac{1}{4} + \frac{1}{8} + \cdots + \frac{1}{2^n} = \frac{2^n - 1}{2^n}.$$

If this process is continued indefinitely, the area of the colored region $S_n$ approaches the area of the unit square, which is 1. So it is plausible that

$$\lim_{n \to \infty} S_n = \underbrace{\frac{1}{2} + \frac{1}{4} + \frac{1}{8} + \cdots}_{\text{sum continues indefinitely}} = 1.$$

The explicit formula $S_n = \dfrac{2^n - 1}{2^n}$ can be analyzed to verify our assertion that $\lim\limits_{n \to \infty} S_n = 1$; we turn to that task in Section 10.2.

This example shows that it is possible to sum an infinite set of numbers and obtain a finite number—in this case, the sum is 1. The sequence $\{S_n\}$ generated in this example is extremely important. It is called a *sequence of partial sums*, and its limit is the value of the infinite series $\frac{1}{2} + \frac{1}{4} + \frac{1}{8} + \cdots$. The idea of a sequence of partial sums is illustrated by the decimal expansion of 1.

**EXAMPLE 7   Working with series**   Consider the infinite series

$$0.9 + 0.09 + 0.009 + 0.0009 + \cdots,$$

where each term of the sum is $\frac{1}{10}$ of the previous term.

**a.** Find the sum of the first one, two, three, and four terms of the series.

**b.** What value would you assign to the infinite series $0.9 + 0.09 + 0.009 + \cdots$?

**SOLUTION**

**a.** Let $S_n$ denote the sum of the first $n$ terms of the given series. Then

$$S_1 = 0.9,$$
$$S_2 = 0.9 + 0.09 = 0.99,$$
$$S_3 = 0.9 + 0.09 + 0.009 = 0.999, \text{ and}$$
$$S_4 = 0.9 + 0.09 + 0.009 + 0.0009 = 0.9999.$$

**b.** The sums $S_1, S_2, \ldots, S_n$ form a sequence $\{S_n\}$, which is a sequence of partial sums. As more and more terms are included, the values of $S_n$ approach 1. Therefore, a reasonable conjecture for the value of the series is 1:

$$\underbrace{0.9 + \underbrace{0.09 + \underbrace{0.009 + 0.0009}_{S_3 = 0.999} }_{S_2 = 0.99}}_{S_1 = 0.9} + \cdots = 1.$$

*Related Exercises 61–62* ◄

**QUICK CHECK 3** Reasoning as in Example 7, what is the value of $0.3 + 0.03 + 0.003 + \cdots$? ◄

The $n$th term of the sequence of partial sums is

$$S_n = \underbrace{0.9 + 0.09 + 0.009 + \cdots + 0.0\ldots09}_{n \text{ terms}} = \sum_{k=1}^{n} 9 \cdot 0.1^k.$$

► Recall the summation notation introduced in Chapter 5: $\sum_{k=1}^{n} a_k$ means $a_1 + a_2 + \cdots + a_n$.

We observed that $\lim_{n \to \infty} S_n = 1$. For this reason, we write

$$\lim_{n \to \infty} S_n = \lim_{n \to \infty} \underbrace{\sum_{k=1}^{n} 9 \cdot 0.1^k}_{S_n} = \underbrace{\sum_{k=1}^{\infty} 9 \cdot 0.1^k}_{\text{new object}} = 1.$$

By letting $n \to \infty$, a new mathematical object $\sum_{k=1}^{\infty} 9 \cdot 0.1^k$ is created. It is an infinite series, and its value is the *limit* of the sequence of partial sums.

► The term *series* is used for historical reasons. When you see *series*, you should think *sum*.

---

**DEFINITION   Infinite Series**

Given a sequence $\{a_1, a_2, a_3, \ldots\}$, the sum of its terms

$$a_1 + a_2 + a_3 + \cdots = \sum_{k=1}^{\infty} a_k$$

is called an **infinite series**. The **sequence of partial sums** $\{S_n\}$ associated with this series has the terms

$$S_1 = a_1$$
$$S_2 = a_1 + a_2$$
$$S_3 = a_1 + a_2 + a_3$$
$$\vdots$$
$$S_n = a_1 + a_2 + a_3 + \cdots + a_n = \sum_{k=1}^{n} a_k, \quad \text{for } n = 1, 2, 3, \ldots.$$

If the sequence of partial sums $\{S_n\}$ has a limit $L$, the infinite series **converges** to that limit, and we write

$$\sum_{k=1}^{\infty} a_k = \lim_{n \to \infty} \underbrace{\sum_{k=1}^{n} a_k}_{S_n} = \lim_{n \to \infty} S_n = L.$$

**QUICK CHECK 4** Do the series $\sum_{k=1}^{\infty} 1$ and $\sum_{k=1}^{\infty} k$ converge or diverge? ◄

If the sequence of partial sums diverges, the infinite series also **diverges**.

---

**EXAMPLE 8**   **Sequence of partial sums**  Consider the infinite series

$$\sum_{k=1}^{\infty} \frac{1}{k(k+1)}.$$

**a.** Find the first four terms of the sequence of partial sums.

**b.** Find an expression for $S_n$ and make a conjecture about the value of the series.

**SOLUTION**

**a.** The sequence of partial sums can be evaluated explicitly:

$$S_1 = \sum_{k=1}^{1} \frac{1}{k(k+1)} = \frac{1}{2},$$

$$S_2 = \sum_{k=1}^{2} \frac{1}{k(k+1)} = \frac{1}{2} + \frac{1}{6} = \frac{2}{3},$$

$$S_3 = \sum_{k=1}^{3} \frac{1}{k(k+1)} = \frac{1}{2} + \frac{1}{6} + \frac{1}{12} = \frac{3}{4}, \text{ and}$$

$$S_4 = \sum_{k=1}^{4} \frac{1}{k(k+1)} = \frac{1}{2} + \frac{1}{6} + \frac{1}{12} + \frac{1}{20} = \frac{4}{5}.$$

**Figure 10.10**

**b.** Based on the pattern in the sequence of partial sums, a reasonable conjecture is that

$$S_n = \frac{n}{n+1}, \text{ for } n = 1, 2, 3, \ldots, \text{ which produces the sequence } \left\{ \frac{1}{2}, \frac{2}{3}, \frac{3}{4}, \frac{4}{5}, \frac{5}{6}, \ldots \right\}.$$

It appears that the values of $S_n$ approach 1 (**Figure 10.10**), so a reasonable conjecture is

that $\displaystyle\lim_{n \to \infty} S_n = \lim_{n \to \infty} \frac{n}{n+1} = 1$. Therefore, we claim that

$$\sum_{k=1}^{\infty} \frac{1}{k(k+1)} = \lim_{n \to \infty} S_n = 1.$$

*Related Exercise 67* ◄

**QUICK CHECK 5** Find the first four terms of the sequence of partial sums for the series $\displaystyle\sum_{k=1}^{\infty} (-1)^k k$. Does the series converge or diverge? ◄

## Summary

This section features three key ideas to keep in mind.

- A *sequence* $\{a_1, a_2, \ldots, a_n, \ldots\}$ is an ordered *list* of numbers.

- An *infinite series* $\displaystyle\sum_{k=1}^{\infty} a_k = a_1 + a_2 + a_3 + \cdots$ is a *sum* of numbers.

- A *sequence of partial sums* $\{S_1, S_2, S_3, \ldots\}$, where $S_n = a_1 + a_2 + \cdots + a_n$, is used to evaluate the series $\displaystyle\sum_{k=1}^{\infty} a_k$.

For sequences, we ask about the behavior of the individual terms as we go out farther and farther in the list; that is, we ask about $\displaystyle\lim_{n \to \infty} a_n$. For infinite series, we examine the sequence of partial sums related to the series. If the sequence of partial sums $\{S_n\}$ has a limit, then the infinite series $\displaystyle\sum_{k=1}^{\infty} a_k$ converges to that limit. If the sequence of partial sums does not have a limit, the infinite series diverges.

Table 10.2 shows the correspondences between sequences/series and functions, and between summation and integration. For a sequence, the index $n$ plays the role of the independent variable and takes on integer values; the terms of the sequence $\{a_n\}$ correspond to the dependent variable.

With sequences $\{a_n\}$, the idea of accumulation corresponds to summation, whereas with functions, accumulation corresponds to integration. A finite sum is analogous to integrating a function over a finite interval. An infinite series is analogous to integrating a function over an infinite interval.

**Table 10.2**

|  | Sequences/Series | Functions |
|---|---|---|
| Independent variable | $n$ | $x$ |
| Dependent variable | $a_n$ | $f(x)$ |
| Domain | Integers<br>e.g., $n = 1, 2, 3, \ldots$ | Real numbers<br>e.g., $\{x : x \geq 1\}$ |
| Accumulation | Sums | Integrals |
| Accumulation over a finite interval | $\displaystyle\sum_{k=1}^{n} a_k$ | $\displaystyle\int_1^n f(x)\,dx$ |
| Accumulation over an infinite interval | $\displaystyle\sum_{k=1}^{\infty} a_k$ | $\displaystyle\int_1^{\infty} f(x)\,dx$ |

## SECTION 10.1 EXERCISES

### Getting Started

| $n$ | $a_n$ |
|---|---|
| 1 | 2.59374246 |
| 2 | 2.70481383 |
| 3 | 2.71692393 |
| 4 | 2.71814593 |
| 5 | 2.71826824 |
| 6 | 2.71828047 |
| 7 | 2.71828169 |
| 8 | 2.71828179 |
| 9 | 2.71828204 |
| 10 | 2.71828203 |

1. Define *sequence* and give an example.

2. Suppose the sequence $\{a_n\}$ is defined by the explicit formula $a_n = 1/n$, for $n = 1, 2, 3, \ldots$. Write out the first five terms of the sequence.

3. Suppose the sequence $\{a_n\}$ is defined by the recurrence relation $a_{n+1} = na_n$, for $n = 1, 2, 3, \ldots$, where $a_1 = 1$. Write out the first five terms of the sequence.

4. Find two different explicit formulas for the sequence $\{-1, 1, -1, 1, -1, 1, \ldots\}$.

5. Find two different explicit formulas for the sequence $\{1, -2, 3, -4, -5, \ldots\}$.

6. The first ten terms of the sequence $\{2\tan^{-1} 10^n\}_{n=1}^{\infty}$ are rounded to 8 digits right of the decimal point (see table). Make a conjecture about the limit of the sequence.

| $n$ | $a_n$ |
|---|---|
| 1 | 2.94225535 |
| 2 | 3.12159332 |
| 3 | 3.13959265 |
| 4 | 3.14139265 |
| 5 | 3.14157265 |
| 6 | 3.14159065 |
| 7 | 3.14159245 |
| 8 | 3.14159263 |
| 9 | 3.14159265 |
| 10 | 3.14159265 |

7. The first ten terms of the sequence $\left\{\left(1 + \dfrac{1}{10^n}\right)^{10^n}\right\}_{n=1}^{\infty}$ are rounded to 8 digits right of the decimal point (see table). Make a conjecture about the limit of the sequence.

8. Does the sequence $\{1, 2, 1, 2, 1, 2, \ldots\}$ converge? Explain.

9. The terms of a sequence of partial sums are defined by $S_n = \displaystyle\sum_{k=1}^{n} k^2$, for $n = 1, 2, 3, \ldots$. Evaluate the first four terms of the sequence.

10. Given the series $\displaystyle\sum_{k=1}^{\infty} k$, evaluate the first four terms of its sequence of partial sums $S_n = \displaystyle\sum_{k=1}^{n} k$.

11. Use sigma notation to write an infinite series whose first four successive partial sums are 10, 20, 30, and 40.

12. Consider the infinite series $\displaystyle\sum_{k=1}^{\infty} \dfrac{1}{k}$. Evaluate the first four terms of the sequence of partial sums.

### Practice Exercises

**13–20. Explicit formulas** *Write the first four terms of the sequence* $\{a_n\}_{n=1}^{\infty}$.

13. $a_n = \dfrac{1}{10^n}$

14. $a_n = 3n + 1$

15. $a_n = \dfrac{(-1)^n}{2^n}$

16. $a_n = 2 + (-1)^n$

**17.**  $a_n = \dfrac{2^{n+1}}{2^n + 1}$

**18.**  $a_n = n + \dfrac{1}{n}$

**19.**  $a_n = 1 + \sin \dfrac{\pi n}{2}$

**20.**  $a_n = n!$ (*Hint:* Recall that $n! = n(n-1)(n-2) \cdots 2 \cdot 1$.)

**21–26. Recurrence relations** *Write the first four terms of the sequence* $\{a_n\}$ *defined by the following recurrence relations.*

**21.**  $a_{n+1} = 2a_n;\ a_1 = 2$

**22.**  $a_{n+1} = \dfrac{a_n}{2};\ a_1 = 32$

**23.**  $a_{n+1} = 3a_n - 12;\ a_1 = 10$

**24.**  $a_{n+1} = a_n^2 - 1;\ a_1 = 1$

**25.**  $a_{n+1} = \dfrac{1}{1 + a_n};\ a_0 = 1$

**26.**  $a_{n+1} = a_n + a_{n-1};\ a_1 = 1, a_0 = 1$

**27–34. Working with sequences** *Several terms of a sequence* $\{a_n\}_{n=1}^{\infty}$ *are given.*

**a.** *Find the next two terms of the sequence.*
**b.** *Find a recurrence relation that generates the sequence (supply the initial value of the index and the first term of the sequence).*
**c.** *Find an explicit formula for the nth term of the sequence.*

**27.**  $\left\{ 1, \dfrac{1}{2}, \dfrac{1}{4}, \dfrac{1}{8}, \dfrac{1}{16}, \ldots \right\}$

**28.**  $\{2, 5, 8, 11, \ldots\}$

**29.**  $\{1, 2, 4, 8, 16, \ldots\}$

**30.**  $\{64, 32, 16, 8, 4, \ldots\}$

**31.**  $\{1, 3, 9, 27, 81, \ldots\}$

**32.**  $\{1, 4, 9, 16, 25, \ldots\}$

**33.**  $\{-5, 5, -5, 5, \ldots\}$

**34.**  $\{1, 0, 1, 0, 1, 0, 1, \ldots\}$

**35–44. Limits of sequences** *Write the terms* $a_1, a_2, a_3,$ *and* $a_4$ *of the following sequences. If the sequence appears to converge, make a conjecture about its limit. If the sequence diverges, explain why.*

**35.**  $a_n = 10^n - 1;\ n = 1, 2, 3, \ldots$

**36.**  $a_{n+1} = \dfrac{10}{a_n};\ a_1 = 1$

**37.**  $a_n = \dfrac{1}{10^n};\ n = 1, 2, 3, \ldots$

**38.**  $a_{n+1} = \dfrac{a_n}{10};\ a_0 = 1$

**39.**  $a_n = 3 + \cos \pi n;\ n = 1, 2, 3, \ldots$

**40.**  $a_n = 1 - 10^{-n};\ n = 1, 2, 3, \ldots$

**41.**  $a_{n+1} = 1 + \dfrac{a_n}{2};\ a_0 = 2$

**42.**  $a_{n+1} = 1 - \dfrac{a_n}{2};\ a_0 = \dfrac{2}{3}$

**T 43.**  $a_{n+1} = \dfrac{a_n}{11} + 50;\ a_0 = 50$

**44.**  $a_{n+1} = 10a_n - 1;\ a_0 = 0$

**T 45–48. Explicit formulas for sequences** *Consider the formulas for the following sequences* $\{a_n\}_{n=1}^{\infty}$. *Make a table with at least ten terms and determine a plausible limit of the sequence or state that the sequence diverges.*

**45.**  $a_n = \dfrac{5^n}{5^n + 1}$

**46.**  $a_n = 2^n \sin(2^{-n})$

**47.**  $a_n = n^2 + n$

**48.**  $a_n = \dfrac{100n - 1}{10n}$

**49–50. Limits from graphs** *Consider the following sequences.*

**a.** *Find the first four terms of the sequence.*
**b.** *Based on part (a) and the figure, determine a plausible limit of the sequence.*

**49.**  $a_n = 2 + 2^{-n};\ n = 1, 2, 3, \ldots$

**50.**  $a_n = \dfrac{n^2}{n^2 - 1};\ n = 2, 3, 4, \ldots$

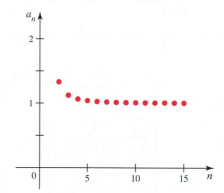

**T 51–56. Recurrence relations** *Consider the following recurrence relations. Make a table with at least ten terms and determine a plausible limit of the sequence or state that the sequence diverges.*

**51.**  $a_{n+1} = \dfrac{1}{2}a_n + 2;\ a_1 = 3$

**52.**  $a_n = \dfrac{1}{4}a_{n-1} - 3;\ a_0 = 1$

**53.**  $a_{n+1} = 4a_n + 1;\ a_0 = 1$

**54.**  $a_{n+1} = \dfrac{a_n}{10} + 3;\ a_0 = 10$

**55.**  $a_{n+1} = \dfrac{1}{2}\sqrt{a_n} + 3;\ a_1 = 8$

**56.**  $a_{n+1} = \sqrt{8a_n + 9};\ a_1 = 10$

**57–60. Heights of bouncing balls** *A ball is thrown upward to a height of $h_0$ meters. After each bounce, the ball rebounds to a fraction $r$ of its previous height. Let $h_n$ be the height after the nth bounce. Consider the following values of $h_0$ and $r$.*

*a. Find the first four terms of the sequence of heights $\{h_n\}$.*
*b. Find an explicit formula for the nth term of the sequence $\{h_n\}$.*

**57.** $h_0 = 20, \ r = 0.5$       **58.** $h_0 = 10, \ r = 0.9$

**59.** $h_0 = 30, \ r = 0.25$       **60.** $h_0 = 20, \ r = 0.75$

**61–66. Sequences of partial sums** *For the following infinite series, find the first four terms of the sequence of partial sums. Then make a conjecture about the value of the infinite series or state that the series diverges.*

**61.** $0.3 + 0.03 + 0.003 + \cdots$

**62.** $0.6 + 0.06 + 0.006 + \cdots$

**63.** $4 + 0.9 + 0.09 + 0.009 + \cdots$

**64.** $\displaystyle\sum_{k=1}^{\infty} 10^k$       **65.** $\displaystyle\sum_{k=1}^{\infty} \frac{6}{10^k}$

**66.** $\displaystyle\sum_{k=1}^{\infty} \cos \pi k$

**67–70. Formulas for sequences of partial sums** *Consider the following infinite series.*

*a. Find the first four partial sums $S_1, S_2, S_3,$ and $S_4$ of the series.*
*b. Find a formula for the nth partial sum $S_n$ of the infinite series. Use this formula to find the next four partial sums $S_5, S_6, S_7,$ and $S_8$ of the infinite series.*
*c. Make a conjecture for the value of the series.*

**67.** $\displaystyle\sum_{k=1}^{\infty} \frac{2}{(2k-1)(2k+1)}$       **68.** $\displaystyle\sum_{k=1}^{\infty} \frac{1}{2^k}$

**69.** $\displaystyle\sum_{k=1}^{\infty} 90(0.1)^k$       **70.** $\displaystyle\sum_{k=1}^{\infty} \frac{2}{3^{k-1}}$

**71. Explain why or why not** Determine whether the following statements are true and give an explanation or counterexample.

  **a.** The sequence of partial sums for the series $1 + 2 + 3 + \cdots$ is $\{1, 3, 6, 10, \ldots\}$.
  **b.** If a sequence of positive numbers converges, then the sequence is decreasing.
  **c.** If the terms of the sequence $\{a_n\}$ are positive and increasing, then the sequence of partial sums for the series $\displaystyle\sum_{k=1}^{\infty} a_k$ diverges.

**72–75. Practical sequences** *Consider the following situations that generate a sequence.*

*a. Write out the first five terms of the sequence.*
*b. Find an explicit formula for the terms of the sequence.*
*c. Find a recurrence relation that generates the sequence.*
*d. Using a calculator or a graphing utility, estimate the limit of the sequence or state that it does not exist.*

**72. Population growth** When a biologist begins a study, a colony of prairie dogs has a population of 250. Regular measurements reveal that each month the prairie dog population increases by 3%. Let $p_n$ be the population (rounded to whole numbers) at the end of the nth month, where the initial population is $p_0 = 250$.

**73. Radioactive decay** A material transmutes 50% of its mass to another element every 10 years due to radioactive decay. Let $M_n$ be the mass of the radioactive material at the end of the nth decade, where the initial mass of the material is $M_0 = 20$ g.

**74. Consumer Price Index** The Consumer Price Index (the CPI is a measure of the U.S. cost of living) is given a base value of 100 in the year 1984. Assume the CPI has increased by an average of 3% per year since 1984. Let $c_n$ be the CPI $n$ years after 1984, where $c_0 = 100$.

**75. Drug elimination** Jack took a 200-mg dose of a pain killer at midnight. Every hour, 5% of the drug is washed out of his bloodstream. Let $d_n$ be the amount of drug in Jack's blood $n$ hours after the drug was taken, where $d_0 = 200$ mg.

### Explorations and Challenges

**76–77. Distance traveled by bouncing balls** *A ball is thrown upward to a height of $h_0$ meters. After each bounce, the ball rebounds to a fraction $r$ of its previous height. Let $h_n$ be the height after the nth bounce and let $S_n$ be the total distance the ball has traveled at the moment of the nth bounce.*

*a. Find the first four terms of the sequence $\{S_n\}$.*
*b. Make a table of 20 terms of the sequence $\{S_n\}$ and determine a plausible value for the limit of $\{S_n\}$.*

**76.** $h_0 = 20, r = 0.5$       **77.** $h_0 = 20, r = 0.75$

**78. A square root finder** A well-known method for approximating $\sqrt{c}$ for positive real numbers $c$ consists of the following recurrence relation (based on Newton's method; see Section 4.8). Let $a_0 = c$ and

$$a_{n+1} = \frac{1}{2}\left(a_n + \frac{c}{a_n}\right), \quad \text{for } n = 0, 1, 2, 3, \ldots.$$

  **a.** Use this recurrence relation to approximate $\sqrt{10}$. How many terms of the sequence are needed to approximate $\sqrt{10}$ with an error less than 0.01? How many terms of the sequence are needed to approximate $\sqrt{10}$ with an error less than 0.0001? (To compute the error, assume a calculator gives the exact value.)
  **b.** Use this recurrence relation to approximate $\sqrt{c}$, for $c = 2, 3, \ldots, 10$. Make a table showing the number of terms of the sequence needed to approximate $\sqrt{c}$ with an error less than 0.01.

**79–80. Fixed-point iteration** *A method for estimating a solution to the equation $x = f(x)$, known as **fixed-point iteration**, is based on the following recurrence relation. Let $x_0 = c$ and $x_{n+1} = f(x_n)$, for $n = 1, 2, 3, \ldots$ and a real number $c$. If the sequence $\{x_n\}_{n=0}^{\infty}$ converges to $L$, then $L$ is a solution to the equation $x = f(x)$ and $L$ is called a fixed point of $f$. To estimate $L$ with $p$ digits of accuracy to the right of the decimal point, we can compute the terms of the sequence $\{x_n\}_{n=0}^{\infty}$ until two successive values agree to $p$ digits of accuracy. Use fixed-point iteration to find a solution to the following equations with $p = 3$ digits of accuracy using the given value of $x_0$.*

**79.** $x = \cos x; \ x_0 = 0.8$       **80.** $x = \dfrac{\sqrt{x^3 + 1}}{20}; \ x_0 = 5$

**QUICK CHECK ANSWERS**

**1.** $a_{10} = 28$   **2.** $a_n = 2^n - 1, \ n = 1, 2, 3, \ldots$
**3.** $0.33333 \ldots = \frac{1}{3}$   **4.** Both diverge.   **5.** $S_1 = -1, S_2 = 1,$ $S_3 = -2, S_4 = 2$; the series diverges. ◄

# 10.2 Sequences

The previous section sets the stage for an in-depth investigation of sequences and infinite series. This section is devoted to sequences, and the remainder of the chapter deals with series.

## Limit of a Sequence and Limit Laws

A fundamental question about sequences concerns the behavior of the terms as we go out farther and farther in the sequence. For example, in the sequence

$$\{a_n\}_{n=0}^{\infty} = \left\{\frac{1}{n^2 + 1}\right\}_{n=0}^{\infty} = \left\{1, \frac{1}{2}, \frac{1}{5}, \frac{1}{10}, \dots\right\},$$

the terms remain positive and decrease to 0. We say that this sequence converges and its limit is 0, written $\lim\limits_{n\to\infty} a_n = 0$. Similarly, the terms of the sequence

$$\{b_n\}_{n=1}^{\infty} = \left\{(-1)^n \frac{n(n+1)}{2}\right\}_{n=1}^{\infty} = \{-1, 3, -6, 10, \dots\}$$

increase in magnitude and do not approach a unique value as $n$ increases. In this case, we say that the sequence diverges.

Limits of sequences are really no different from limits at infinity of functions except that the variable $n$ assumes only integer values as $n \to \infty$. This idea works as follows.

Given a sequence $\{a_n\}$, we define a function $f$ such that $f(n) = a_n$ for all indices $n$. For example, if $a_n = n/(n+1)$, then we let $f(x) = x/(x+1)$. By the methods of Section 2.5, we know that $\lim\limits_{x\to\infty} f(x) = 1$; because the terms of the sequence lie on the graph of $f$, it follows that $\lim\limits_{n\to\infty} a_n = 1$ (**Figure 10.11**). This reasoning is the basis of the following theorem.

$$\lim_{x\to\infty} f(x) = 1 \Rightarrow \lim_{n\to\infty} a_n = 1$$

$$f(x) = \frac{x}{x+1} \qquad a_n = \frac{n}{n+1}$$

**Figure 10.11**

▶ The converse of Theorem 10.1 is not true. For example, if $a_n = \cos 2\pi n$, then $\lim\limits_{n\to\infty} a_n = 1$, but $\lim\limits_{x\to\infty} \cos 2\pi x$ does not exist.

---

**THEOREM 10.1  Limits of Sequences from Limits of Functions**
Suppose $f$ is a function such that $f(n) = a_n$, for positive integers $n$. If $\lim\limits_{x\to\infty} f(x) = L$, then the limit of the sequence $\{a_n\}$ is also $L$, where $L$ may be $\pm\infty$.

---

Because of the correspondence between limits of sequences and limits of functions at infinity, we have the following properties that are analogous to those for functions given in Theorem 2.3.

▶ The limit of a sequence $\{a_n\}$ is determined by the terms in the *tail* of the sequence—the terms with large values of $n$. If the sequences $\{a_n\}$ and $\{b_n\}$ differ in their first 100 terms but have identical terms for $n > 100$, then they have the same limit. For this reason, the initial index of a sequence (for example, $n = 0$ or $n = 1$) is often not specified.

---

**THEOREM 10.2  Limit Laws for Sequences**
Assume the sequences $\{a_n\}$ and $\{b_n\}$ have limits $A$ and $B$, respectively. Then

**1.** $\lim\limits_{n\to\infty} (a_n \pm b_n) = A \pm B$

**2.** $\lim\limits_{n\to\infty} ca_n = cA$, where $c$ is a real number

**3.** $\lim\limits_{n\to\infty} a_n b_n = AB$

**4.** $\lim\limits_{n\to\infty} \dfrac{a_n}{b_n} = \dfrac{A}{B}$, provided $B \neq 0$.

---

**EXAMPLE 1  Limits of sequences**  Determine the limits of the following sequences.

**a.** $a_n = \dfrac{3n^3}{n^3 + 1}$      **b.** $b_n = \left(\dfrac{n+5}{n}\right)^n$      **c.** $c_n = n^{1/n}$

**SOLUTION**

**a.** A function with the property that $f(n) = a_n$ is $f(x) = \dfrac{3x^3}{x^3 + 1}$. Dividing numerator and denominator by $x^3$ (or appealing to Theorem 2.7), we find that $\lim\limits_{x \to \infty} f(x) = 3$. (Alternatively, we can apply l'Hôpital's Rule and obtain the same result.) We conclude that $\lim\limits_{n \to \infty} a_n = 3$.

**b.** The limit

$$\lim_{n \to \infty} b_n = \lim_{n \to \infty} \left( \frac{n + 5}{n} \right)^n = \lim_{n \to \infty} \left( 1 + \frac{5}{n} \right)^n$$

has the indeterminate form $1^\infty$. Recall that for this limit (Section 7.6), we first evaluate

$$L = \lim_{n \to \infty} \ln \left( 1 + \frac{5}{n} \right)^n = \lim_{n \to \infty} n \ln \left( 1 + \frac{5}{n} \right),$$

> It is customary to treat $n$ as a continuous variable and differentiate with respect to $n$, rather than write the sequence as a function of $x$, as was done in Example 1a.

and then, if $L$ exists, $\lim\limits_{n \to \infty} b_n = e^L$. Using l'Hôpital's Rule for the indeterminate form $0/0$, we have

$$L = \lim_{n \to \infty} n \ln \left( 1 + \frac{5}{n} \right) = \lim_{n \to \infty} \frac{\ln \left( 1 + (5/n) \right)}{1/n} \qquad \textcolor{blue}{\text{Indeterminate form } 0/0}$$

> For a review of l'Hôpital's Rule, see Section 7.6, where we showed that
> $$\lim_{x \to \infty} \left( 1 + \frac{a}{x} \right)^x = e^a.$$

$$= \lim_{n \to \infty} \frac{\dfrac{1}{1 + (5/n)} \left( -\dfrac{5}{n^2} \right)}{-1/n^2} \qquad \textcolor{blue}{\text{L'Hôpital's Rule}}$$

$$= \lim_{n \to \infty} \frac{5}{1 + (5/n)} = 5. \qquad \textcolor{blue}{\text{Simplify; } 5/n \to 0 \text{ as } n \to \infty.}$$

Because $\lim\limits_{n \to \infty} b_n = e^L = e^5$, we have $\lim\limits_{n \to \infty} \left( \dfrac{5 + n}{n} \right)^n = e^5$.

**c.** The limit has the indeterminate form $\infty^0$, so we first evaluate $L = \lim\limits_{n \to \infty} \ln n^{1/n} = \lim\limits_{n \to \infty} \dfrac{\ln n}{n}$; if $L$ exists, then $\lim\limits_{n \to \infty} c_n = e^L$. Using either l'Hôpital's Rule or the relative growth rates in Section 7.6, we find that $L = 0$. Therefore, $\lim\limits_{n \to \infty} c_n = e^0 = 1$.

*Related Exercises 14, 38, 39* ◄

## Terminology for Sequences

We now introduce some terminology for sequences that is similar to that used for functions. The following terms are used to describe sequences $\{a_n\}$.

> Because an increasing sequence is, by definition, nondecreasing, it is also monotonic. Similarly, a decreasing sequence is monotonic.

---

**DEFINITIONS** Terminology for Sequences

- $\{a_n\}$ is **increasing** if $a_{n+1} > a_n$; for example, $\{0, 1, 2, 3, \ldots\}$.
- $\{a_n\}$ is **nondecreasing** if $a_{n+1} \geq a_n$; for example, $\{1, 1, 2, 2, 3, 3, \ldots\}$.
- $\{a_n\}$ is **decreasing** if $a_{n+1} < a_n$; for example, $\{2, 1, 0, -1, \ldots\}$.
- $\{a_n\}$ is **nonincreasing** if $a_{n+1} \leq a_n$; for example, $\{0, -1, -1, -2, -2, \ldots\}$.
- $\{a_n\}$ is **monotonic** if it is either nonincreasing or nondecreasing (it moves in one direction).
- $\{a_n\}$ is **bounded above** if there is a number $M$ such that $a_n \leq M$, for all relevant values of $n$, and $\{a_n\}$ is **bounded below** if there is a number $N$ such that $a_n \geq N$, for all relevant values of $n$.
- If $\{a_n\}$ is bounded above and bounded below, then we say that $\{a_n\}$ is a **bounded** sequence.

---

For example, the sequence

$$\{a_n\} = \left\{1 - \frac{1}{n}\right\}_{n=1}^{\infty} = \left\{0, \frac{1}{2}, \frac{2}{3}, \frac{3}{4}, \dots\right\}$$

satisfies $0 \le a_n \le 1$, for $n \ge 1$, and its terms are increasing in size. Therefore, the sequence is bounded below, bounded above, and increasing; it is also monotonic (**Figure 10.12**). The sequence

$$\{b_n\} = \left\{1 + \frac{1}{n}\right\}_{n=1}^{\infty} = \left\{2, \frac{3}{2}, \frac{4}{3}, \frac{5}{4}, \dots\right\}$$

satisfies $1 \le b_n \le 2$, for $n \ge 1$, and its terms are decreasing in size. Therefore, the sequence is bounded below, bounded above, and decreasing; it is also monotonic (Figure 10.12). Because $\{a_n\}$ and $\{b_n\}$ are bounded below and above, both are bounded sequences.

**QUICK CHECK 1** Classify the following sequences as bounded, monotonic, or neither.

**a.** $\left\{\frac{1}{2}, \frac{3}{4}, \frac{7}{8}, \frac{15}{16}, \dots\right\}$

**b.** $\left\{1, -\frac{1}{2}, \frac{1}{4}, -\frac{1}{8}, \frac{1}{16}, \dots\right\}$

**c.** $\{1, -2, 3, -4, 5, \dots\}$

**d.** $\{1, 1, 1, 1, \dots\}$ ◄

**Figure 10.12**

**Figure 10.13**

**Figure 10.14**

**EXAMPLE 2** **Limits of sequences and graphing** Compare and contrast the behavior of $\{a_n\}$ and $\{b_n\}$ as $n \to \infty$.

**a.** $a_n = \dfrac{n^{3/2}}{n^{3/2} + 1}$  **b.** $b_n = \dfrac{(-1)^n n^{3/2}}{n^{3/2} + 1}$

**SOLUTION**

**a.** The terms of $\{a_n\}$ are positive, increasing, and bounded (**Figure 10.13**). Dividing the numerator and denominator of $a_n$ by $n^{3/2}$, we see that

$$\lim_{n\to\infty} a_n = \lim_{n\to\infty} \frac{n^{3/2}}{n^{3/2} + 1} = \lim_{n\to\infty} \frac{1}{1 + \underbrace{\frac{1}{n^{3/2}}}_{\text{approaches } 0 \text{ as } n \to \infty}} = 1.$$

**b.** The terms of the bounded sequence $\{b_n\}$ alternate in sign. Using the result of part (a), it follows that the even terms form an increasing sequence that approaches 1, and the odd terms form a decreasing sequence that approaches $-1$ (**Figure 10.14**). Therefore, the sequence diverges, illustrating the fact that the presence of $(-1)^n$ may significantly alter the behavior of a sequence.

*Related Exercises 55, 60* ◄

## Geometric Sequences

**Geometric sequences** have the property that each term is obtained by multiplying the previous term by a fixed constant, called the **ratio**. They have the form $\{r^n\}$ or $\{ar^n\}$, where the ratio $r$ and $a \ne 0$ are real numbers. The value of $r$ determines the behavior of the sequence.

When a number less than 1 in magnitude is raised to increasing powers, the resulting numbers decrease to zero. For example, the geometric sequence $\{0.75^n\}$, where $r = 0.75$, converges to zero and is monotonic (**Figure 10.15**). If we choose $r = -0.75$, we have $\{(-0.75)^n\} = \{(-1)^n 0.75^n\}$. Observe that the factor $(-1)^n$ oscillates between 1 and $-1$, while $0.75^n$ decreases to zero as $n$ increases. Therefore, the sequence oscillates and converges to zero (**Figure 10.16**).

**Figure 10.15**                              **Figure 10.16**

When a number greater than 1 in magnitude is raised to increasing powers, the resulting numbers increase in magnitude. For example, letting $r = 1.15$ results in the sequence $\{1.15^n\}$, whose terms are positive and increase without bound. In this case, the sequence diverges and is monotonic (**Figure 10.17**). Choosing $r = -1.15$, we have $\{(-1.15)^n\} = \{(-1)^n 1.15^n\}$. As before, the factor $(-1)^n$ oscillates between 1 and $-1$, while $1.15^n$ increases without bound as $n$ increases. The terms of the sequence increase in magnitude without bound and alternate in sign. In this case, the sequence oscillates and diverges (**Figure 10.18**).

**Figure 10.17**                              **Figure 10.18**

**QUICK CHECK 2** Describe the behavior of $\{r^n\}$ in the cases $r = -1$ and $r = 1$. ◀

The preceding discussion and Quick Check 2 are summarized in the following theorem.

**THEOREM 10.3   Geometric Sequences**

Let $r$ be a real number. Then

$$\lim_{n \to \infty} r^n = \begin{cases} 0 & \text{if } |r| < 1 \\ 1 & \text{if } r = 1 \\ \text{does not exist} & \text{if } r \le -1 \text{ or } r > 1. \end{cases}$$

If $r > 0$, then $\{r^n\}$ is a monotonic sequence. If $r < 0$, then $\{r^n\}$ oscillates.

**EXAMPLE 3   Using Limit Laws**  Determine the limits of the following sequences.

**a.** $a_n = 5(0.6)^n - \dfrac{1}{3^n}$       **b.** $b_n = \dfrac{2n^2 + n}{2^n(3n^2 - 4)}$

**SOLUTION**

**a.** Using Laws 1 and 2 from Theorem 10.2, we have

$$\lim_{n\to\infty} a_n = \lim_{n\to\infty} \left( 5(0.6)^n - \frac{1}{3^n} \right) = 5 \lim_{n\to\infty} (0.6)^n - \lim_{n\to\infty} \left( \frac{1}{3} \right)^n.$$

Both $\{(0.6)^n\}$ and $\{(1/3)^n\}$ are geometric sequences with $|r| < 1$, which implies that

$$\lim_{n\to\infty} a_n = 5 \underbrace{\lim_{n\to\infty} (0.6)^n}_{0} - \underbrace{\lim_{n\to\infty} \left( \frac{1}{3} \right)^n}_{0} = 5 \cdot 0 - 0 = 0. \quad \text{\color{blue}Theorem 10.3}$$

**b.** We write the limit of the sequence as a product of limits:

$$\lim_{n\to\infty} b_n = \lim_{n\to\infty} \frac{2n^2 + n}{2^n(3n^2 - 4)} = \lim_{n\to\infty} \left( \frac{1}{2} \right)^n \cdot \lim_{n\to\infty} \frac{2n^2 + n}{3n^2 - 4}. \quad \text{\color{blue}Law 3, Theorem 10.2}$$

The first of the new limits involves a geometric sequence with $r = 1/2$, and the second limit yields to Theorem 2.7:

$$\lim_{n\to\infty} b_n = \underbrace{\lim_{n\to\infty} \left( \frac{1}{2} \right)^n}_{0} \cdot \underbrace{\lim_{n\to\infty} \frac{2n^2 + n}{3n^2 - 4}}_{2/3} = 0 \cdot \frac{2}{3} = 0. \quad \text{\color{blue}Evaluate limits.}$$

*Related Exercises 26–27* ◄

The previous examples show that a sequence may display any of the following behaviors:

• It may converge to a single value, which is the limit of the sequence.

• Its terms may increase in magnitude without bound (either with one sign or with mixed signs), in which case the sequence diverges.

• Its terms may remain bounded but settle into an oscillating pattern in which the terms approach two or more values; in this case, the sequence diverges.

   Not illustrated in the preceding examples is one other type of behavior: The terms of a sequence may remain bounded, but wander chaotically forever without a pattern. In this case, the sequence also diverges (see the Guided Project *Chaos!*).

## The Squeeze Theorem

We cite two theorems that are used to evaluate limits and to establish that limits exist. The first theorem is a direct analog of the Squeeze Theorem in Section 2.3.

**Figure 10.19**

> **THEOREM 10.4   Squeeze Theorem for Sequences**
> Let $\{a_n\}$, $\{b_n\}$, and $\{c_n\}$ be sequences with $a_n \le b_n \le c_n$, for all integers $n$ greater than some index $N$. If $\lim\limits_{n\to\infty} a_n = \lim\limits_{n\to\infty} c_n = L$, then $\lim\limits_{n\to\infty} b_n = L$ (**Figure 10.19**).

**EXAMPLE 4   Squeeze Theorem**  Find the limit of the sequence $b_n = \dfrac{\cos n}{n^2 + 1}$.

**SOLUTION**  The goal is to find two sequences $\{a_n\}$ and $\{c_n\}$ whose terms lie below and above the terms of the given sequence $\{b_n\}$. Note that $-1 \le \cos n \le 1$, for all $n$. Therefore,

$$\underbrace{-\frac{1}{n^2 + 1}}_{a_n} \le \underbrace{\frac{\cos n}{n^2 + 1}}_{b_n} \le \underbrace{\frac{1}{n^2 + 1}}_{c_n}.$$

**Figure 10.20**

Letting $a_n = -\dfrac{1}{n^2 + 1}$ and $c_n = \dfrac{1}{n^2 + 1}$, we have $a_n \leq b_n \leq c_n$, for $n \geq 1$. Furthermore, $\lim\limits_{n \to \infty} a_n = \lim\limits_{n \to \infty} c_n = 0$. By the Squeeze Theorem, $\lim\limits_{n \to \infty} b_n = 0$ (**Figure 10.20**).

*Related Exercises 65, 67* ◄

## Bounded Monotonic Sequence Theorem

Suppose you pour a cup of hot coffee and put it on your desk to cool. Assume every minute you measure the temperature of the coffee to create a sequence of temperature readings $\{T_1, T_2, T_3, \ldots\}$. This sequence has two notable properties: First, the terms of the sequence are decreasing (because the coffee is cooling); and second, the sequence is bounded below (because the temperature of the coffee cannot be less than the temperature of the surrounding room). In fact, if the measurements continue indefinitely, the sequence of temperatures converges to the temperature of the room. This example illustrates an important theorem that characterizes convergent sequences in terms of boundedness and monotonicity. The theorem is easy to believe, but its proof is beyond the scope of this text.

> **THEOREM 10.5   Bounded Monotonic Sequence**
> A bounded monotonic sequence converges.

> ➤ **Some optional terminology:** $M$ is called an *upper bound* of the sequence in Figure 10.21a, and $N$ is a *lower bound* of the sequence in Figure 10.21b. The number $M^*$ is the *least upper bound* of a sequence (or a set) if it is the smallest of all the upper bounds. It is a fundamental property of the real numbers that if a sequence (or a nonempty set) is bounded above, then it has a least upper bound. It can be shown that an increasing sequence that is bounded above converges to its least upper bound. Similarly, a decreasing sequence that is bounded below converges to its greatest lower bound.

**Figure 10.21** shows the two cases of this theorem. In the first case, we see a nondecreasing sequence, all of whose terms are less than $M$. It must converge to a limit less than or equal to $M$. Similarly, a nonincreasing sequence, all of whose terms are greater than $N$, must converge to a limit greater than or equal to $N$.

**Figure 10.21**

## An Application: Recurrence Relations

> ➤ Most drugs decay exponentially in the bloodstream and have a characteristic half-life assuming the drug absorbs quickly into the blood.

**EXAMPLE 5   Sequences for drug doses** Suppose your doctor prescribes a 100-mg dose of an antibiotic to be taken every 12 hours. Furthermore, the drug is known to have a half-life of 12 hours; that is, every 12 hours half of the drug in your blood is eliminated.

**a.** Find the sequence that gives the amount of drug in your blood immediately after each dose.

**b.** Use a graph to propose the limit of this sequence; that is, in the long run, how much drug do you have in your blood?

**c.** Find the limit of the sequence directly.

**SOLUTION**

**a.** Let $d_n$ be the amount of drug in the blood immediately following the $n$th dose, where $n = 1, 2, 3, \ldots$ and $d_1 = 100$ mg. We want to write a recurrence relation that gives the amount of drug in the blood after the $(n + 1)$st dose $(d_{n+1})$ in terms of the

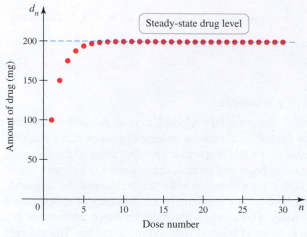

**Figure 10.22**

▶ *Mathematical induction* is used to prove statements involving natural numbers. In the first step, we show that the statement holds for $n = 1$. In the second step, we show that if the statement is true for $n = k$, then the statement is also true for $n = k + 1$, where $k$ is any natural number greater than 1. If the proof of the second step is successful, it implies that the statement holds for any natural number.

**QUICK CHECK 3** If a drug has the same half-life as in Example 5, (i) how would the steady-state level of drug in the blood change if the regular dose were 150 mg instead of 100 mg? (ii) How would the steady-state level change if the dosing interval were 6 hr instead of 12 hr? ◀

amount of drug after the $n$th dose ($d_n$). In the 12 hours between the $n$th dose and the $(n + 1)$st dose, half of the drug in the blood is eliminated *and* another 100 mg of drug is added. So we have

$$d_{n+1} = 0.5\,d_n + 100, \qquad \text{for } n = 1, 2, 3, \ldots, \text{ with } d_1 = 100,$$

which is the recurrence relation for the sequence $\{d_n\}$.

**b.** We see from **Figure 10.22** that after about 10 doses (5 days) the amount of antibiotic in the blood is close to 200 mg, and— importantly for your body—it never exceeds 200 mg.

**c.** Figure 10.22 gives evidence that the terms of the sequence are increasing and bounded above by 200. To prove these facts, note that $d_1 = 100 < 200$. Now suppose $d_k < 200$, for any $k > 1$. The recurrence relation in part (a) implies that $d_{k+1} = 0.5d_k + 100$. Because $d_k < 200$, we have

$$d_{k+1} = 0.5\,\underbrace{d_k}_{<200} + 100 < 0.5(200) + 100 = 200.$$

We conclude by mathematical induction that $d_k < 200$, for all $k$. To prove that $\{d_k\}$ is increasing (and therefore monotonic), we need to show that $d_{k+1} - d_k > 0$. Note that

$$d_{k+1} - d_k = \underbrace{0.5d_k + 100}_{\text{recurrence relation}} - d_k = 100 - 0.5d_k > 0;$$

the final inequality follows from the fact that $d_k < 200$. Because $\{d_k\}$ is increasing, it is bounded below by its first term, which is $d_1 = 100$, and therefore $\{d_k\}$ is bounded. By the Bounded Monotonic Sequence Theorem, the sequence has a limit; therefore, $\lim_{n\to\infty} d_n = L$ and $\lim_{n\to\infty} d_{n+1} = L$. We now take the limit of both sides of the recurrence relation:

$$d_{n+1} = 0.5\,d_n + 100 \qquad \text{\color{blue}Recurrence relation}$$
$$\underbrace{\lim_{n\to\infty} d_{n+1}}_{L} = 0.5\,\underbrace{\lim_{n\to\infty} d_n}_{L} + \lim_{n\to\infty} 100 \qquad \text{\color{blue}Limits of both sides}$$
$$L = 0.5L + 100. \qquad \text{\color{blue}Substitute } L.$$

Solving for $L$, the steady-state drug level is $L = 200$.

*Related Exercises 71, 89* ◀

## Growth Rates of Sequences

All the hard work we did in Section 7.6 to establish the relative growth rates of functions is now applied to sequences. Here is the question: Given two nondecreasing sequences of positive terms $\{a_n\}$ and $\{b_n\}$, which sequence grows faster as $n \to \infty$? As with functions, to compare growth rates, we evaluate $\lim_{n\to\infty} a_n/b_n$. If $\lim_{n\to\infty} a_n/b_n = 0$, then $\{b_n\}$ grows faster than $\{a_n\}$. If $\lim_{n\to\infty} a_n/b_n = \infty$, then $\{a_n\}$ grows faster than $\{b_n\}$.

Using the results of Section 7.6, we immediately arrive at the following ranking of growth rates of sequences as $n \to \infty$, with positive real numbers $p, q, r, s$, and $b > 1$:

$$\{\ln^q n\} \ll \{n^p\} \ll \{n^p \ln^r n\} \ll \{n^{p+s}\} \ll \{b^n\} \ll \{n^n\}.$$

▶ $0! = 1$ (by definition)
$1! = 1$
$2! = 2 \cdot 1! = 2$
$3! = 3 \cdot 2! = 6$
$4! = 4 \cdot 3! = 24$
$5! = 5 \cdot 4! = 120$
$6! = 6 \cdot 5! = 720$

As before, the notation $\{a_n\} \ll \{b_n\}$ means $\{b_n\}$ *grows faster* than $\{a_n\}$ as $n \to \infty$. Another important sequence that should be added to the list is the **factorial sequence** $\{n!\}$, where $n! = n(n - 1)(n - 2) \cdots 2 \cdot 1$. Where does the factorial sequence $\{n!\}$ appear in the list? The following argument provides some intuition. Notice that

$$n^n = \underbrace{n \cdot n \cdot n \cdots n}_{n \text{ factors}}, \qquad \text{whereas}$$

$$n! = \underbrace{n \cdot (n - 1) \cdot (n - 2) \cdots 2 \cdot 1}_{n \text{ factors}}.$$

The $n$th term of both sequences involves the product of $n$ factors; however, the factors of $n!$ decrease, while the factors of $n^n$ are the same. Based on this observation, we claim that $\{n^n\}$ grows faster than $\{n!\}$, and we have the ordering $\{n!\} \ll \{n^n\}$. But where does $\{n!\}$ appear in the list relative to $\{b^n\}$? Again, some intuition is gained by noting that

$$b^n = \underbrace{b \cdot b \cdot b \cdots b}_{n \text{ factors}}, \qquad \text{whereas}$$

$$n! = \underbrace{n \cdot (n-1) \cdot (n-2) \cdots 2 \cdot 1}_{n \text{ factors}}.$$

The $n$th term of both sequences involves a product of $n$ factors; however, the factors of $b^n$ remain constant as $n$ increases, while the factors of $n!$ increase with $n$. So we claim that $\{n!\}$ grows faster than $\{b^n\}$. This conjecture is supported by computation, although the outcome of the race may not be immediately evident if $b$ is large (Exercise 111).

---

**THEOREM 10.6  Growth Rates of Sequences**
The following sequences are ordered according to increasing growth rates as

$n \to \infty$; that is, if $\{a_n\}$ appears before $\{b_n\}$ in the list, then $\displaystyle\lim_{n \to \infty} \frac{a_n}{b_n} = 0$ and

$$\lim_{n \to \infty} \frac{b_n}{a_n} = \infty:$$

$$\{\ln^q n\} \ll \{n^p\} \ll \{n^p \ln^r n\} \ll \{n^{p+s}\} \ll \{b^n\} \ll \{n!\} \ll \{n^n\}.$$

The ordering applies for positive real numbers $p, q, r, s,$ and $b > 1$.

---

**QUICK CHECK 4** Which sequence grows faster: $\{\ln n\}$ or $\{n^{1.1}\}$? What is

$$\lim_{n \to \infty} \frac{n^{1,000,000}}{e^n}? \blacktriangleleft$$

It is worth noting that the rankings in Theorem 10.6 do not change if a sequence is multiplied by a positive constant (Exercise 110).

**EXAMPLE 6  Convergence and growth rates** Compare growth rates of sequences to determine whether the following sequences converge.

**a.** $\left\{ \dfrac{\ln n^{10}}{0.00001n} \right\}$   **b.** $\left\{ \dfrac{n^8 \ln n}{n^{8.001}} \right\}$   **c.** $\left\{ \dfrac{n!}{10^n} \right\}$

**SOLUTION**

**a.** Because $\ln n^{10} = 10 \ln n$, the sequence in the numerator is a constant multiple of the sequence $\{\ln n\}$. Similarly, the sequence in the denominator is a constant multiple of the sequence $\{n\}$. By Theorem 10.6, $\{n\}$ grows faster than $\{\ln n\}$ as $n \to \infty$; therefore, the sequence $\left\{ \dfrac{\ln n^{10}}{0.00001n} \right\}$ converges to zero.

**b.** The sequence in the numerator is $\{n^p \ln^r n\}$ of Theorem 10.6 with $p = 8$ and $r = 1$. The sequence in the denominator is $\{n^{p+s}\}$ of Theorem 10.6 with $p = 8$ and $s = 0.001$. Because $\{n^{p+s}\}$ grows faster than $\{n^p \ln^r n\}$ as $n \to \infty$, we conclude that $\left\{ \dfrac{n^8 \ln n}{n^{8.001}} \right\}$ converges to zero.

**c.** Using Theorem 10.6, we see that $n!$ grows faster than any exponential function as $n \to \infty$. Therefore, $\displaystyle\lim_{n \to \infty} \frac{n!}{10^n} = \infty$, and the sequence diverges. **Figure 10.23** gives a visual comparison of the growth rates of $\{n!\}$ and $\{10^n\}$. Because these sequences grow so quickly, we plot the logarithm of the terms. The exponential sequence $\{10^n\}$ dominates the factorial sequence $\{n!\}$ until $n = 25$ terms. At that point, the factorial sequence overtakes the exponential sequence.

**Figure 10.23**

At $n = 25$, $n! > 10^n$

$\{10^n\}$

$\{n!\}$

*Related Exercises 76, 78* ◀

## Formal Definition of a Limit of a Sequence

As with limits of functions, there is a formal definition of the limit of a sequence.

> **DEFINITION  Limit of a Sequence**
>
> The sequence $\{a_n\}$ converges to $L$ provided the terms of $a_n$ can be made arbitrarily close to $L$ by taking $n$ sufficiently large. More precisely, $\{a_n\}$ has the unique limit $L$ if, given any $\varepsilon > 0$, it is possible to find a positive integer $N$ (depending only on $\varepsilon$) such that
>
> $$|a_n - L| < \varepsilon \qquad \text{whenever } n > N.$$
>
> If the **limit of a sequence** is $L$, we say the sequence **converges** to $L$, written
>
> $$\lim_{n \to \infty} a_n = L.$$
>
> A sequence that does not converge is said to **diverge**.

When $n > N$,
$|a_n - L| < \varepsilon$

**Figure 10.24**

The formal definition of the limit of a convergent sequence is interpreted in much the same way as the limit at infinity of a function. Given a small tolerance $\varepsilon > 0$, how far out in the sequence must you go so that all succeeding terms are within $\varepsilon$ of the limit $L$ (**Figure 10.24**)? Given *any* value of $\varepsilon > 0$ (no matter how small), you must find a value of $N$ such that all terms beyond $a_N$ are within $\varepsilon$ of $L$.

**EXAMPLE 7  Limits using the formal definition**  Consider the claim that

$$\lim_{n \to \infty} a_n = \lim_{n \to \infty} \frac{n}{n - 1} = 1.$$

**a.** Given $\varepsilon = 0.01$, find a value of $N$ that satisfies the conditions of the limit definition.

**b.** Prove that $\lim_{n \to \infty} a_n = 1$.

**SOLUTION**

**a.** We must find an integer $N$ such that $|a_n - 1| < \varepsilon = 0.01$ whenever $n > N$. This condition can be written

$$|a_n - 1| = \left| \frac{n}{n - 1} - 1 \right| = \left| \frac{1}{n - 1} \right| < 0.01.$$

Noting that $n > 1$, the absolute value can be removed. The condition on $n$ becomes $n - 1 > 1/0.01 = 100$, or $n > 101$. Therefore, we take $N = 101$ or any larger number. This means that $|a_n - 1| < 0.01$ whenever $n > 101$.

**b.** Given *any* $\varepsilon > 0$, we must find a value of $N$ (depending on $\varepsilon$) that guarantees $|a_n - 1| = \left| \dfrac{n}{n - 1} - 1 \right| < \varepsilon$ whenever $n > N$. For $n > 1$, the inequality $\left| \dfrac{n}{n - 1} - 1 \right| < \varepsilon$ implies that

$$\left| \frac{n}{n - 1} - 1 \right| = \frac{1}{n - 1} < \varepsilon.$$

> ➤ In general, $1/\varepsilon + 1$ is not an integer, so $N$ should be the least integer greater than $1/\varepsilon + 1$ or any larger integer.

Solving for $n$, we find that $\dfrac{1}{n - 1} < \varepsilon$ or $n - 1 > \dfrac{1}{\varepsilon}$ or $n > \dfrac{1}{\varepsilon} + 1$. Therefore, given a tolerance $\varepsilon > 0$, we must look beyond $a_N$ in the sequence, where $N \geq \dfrac{1}{\varepsilon} + 1$, to be sure that the terms of the sequence are within $\varepsilon$ of the limit 1. Because we can provide a value of $N$ for *any* $\varepsilon > 0$, the limit exists and equals 1.

*Related Exercise 105* ◄

## SECTION 10.2 EXERCISES

### Getting Started

1. Give an example of a nonincreasing sequence with a limit.

2. Give an example of a nondecreasing sequence without a limit.

3. Give an example of a bounded sequence that has a limit.

4. Give an example of a bounded sequence without a limit.

5. For what values of $r$ does the sequence $\{r^n\}$ converge? Diverge?

**6–9.** *Determine whether the following sequences converge or diverge, and state whether they are monotonic or whether they oscillate. Give the limit when the sequence converges.*

6. $\{0.2^n\}$

7. $\{1.00001^n\}$

8. $\{(-2.5)^n\}$

9. $\{(-0.7)^n\}$

10. Find the limit of the sequence $\{a_n\}$ if $1 - \dfrac{1}{n} < a_n < 1 + \dfrac{1}{n}$, for every integer $n \geq 1$.

11. Compare the growth rates of $\{n^{100}\}$ and $\{e^{n/100}\}$ as $n \to \infty$.

12. Use Theorem 10.6 to evaluate $\lim\limits_{n \to \infty} \dfrac{n^{100}}{n^n}$.

### Practice Exercises

**13–52. Limits of sequences** *Find the limit of the following sequences or determine that the sequence diverges.*

13. $\left\{ \dfrac{n^3}{n^4 + 1} \right\}$

14. $\left\{ \dfrac{n^{12}}{3n^{12} + 4} \right\}$

15. $\left\{ \dfrac{3n^3 - 1}{2n^3 + 1} \right\}$

16. $\left\{ \dfrac{n^5 + 3n}{10n^3 + n} \right\}$

17. $\left\{ \tan^{-1}\left( \dfrac{10n}{10n + 4} \right) \right\}$

18. $\left\{ \cot\left( \dfrac{n\pi}{2n + 2} \right) \right\}$

19. $\left\{ 1 + \cos\dfrac{1}{n} \right\}$

20. $\{ \ln(n^3 + 1) - \ln(3n^3 + 10n) \}$

21. $\left\{ \ln \sin\dfrac{1}{n} + \ln n \right\}$

22. $\left\{ \dfrac{k}{\sqrt{9k^2 + 1}} \right\}$

23. $\left\{ \dfrac{\sqrt{4n^4 + 3n}}{8n^2 + 1} \right\}$

24. $\left\{ \dfrac{2e^n + 1}{e^n} \right\}$

25. $\left\{ \dfrac{\ln n^2}{\ln 3n} \right\}$

26. $\{ 5(-1.01)^n \}$

27. $\{ 2^{n+1} 3^{-n} \}$

28. $\{ 100(-0.003)^n \}$

29. $\{ (0.5)^n + 3(0.75)^n \}$

30. $\left\{ \dfrac{e^n + \pi^n}{e^n} \right\}$

31. $\left\{ \dfrac{3^{n+1} + 3}{3^n} \right\}$

32. $\left\{ \dfrac{3^n}{3^n + 4^n} \right\}$

33. $\left\{ \dfrac{(n+1)!}{n!} \right\}$

34. $\left\{ \dfrac{(2n)! \, n^2}{(2n + 2)!} \right\}$

35. $\{ \tan^{-1} n \}$

36. $\left\{ \dfrac{e^{n/10}}{2^n} \right\}$

37. $\{ \sqrt{n^2 + 1} - n \}$

38. $\{ n^{2/n} \}$

39. $\left\{ \left( 1 + \dfrac{2}{n} \right)^n \right\}$

40. $\left\{ \dfrac{\ln n}{n^{1.1}} \right\}$

41. $\{ \sqrt[n]{e^{3n+4}} \}$

42. $\left\{ \left( \dfrac{n}{n + 5} \right)^n \right\}$

43. $\left\{ \sqrt{\left( 1 + \dfrac{1}{2n} \right)^n} \right\}$

44. $\left\{ \left( 1 + \dfrac{4}{n} \right)^{3n} \right\}$

45. $\left\{ \dfrac{n}{e^n + 3n} \right\}$

46. $\left\{ \dfrac{\ln(1/n)}{n} \right\}$

47. $\left\{ \left( \dfrac{1}{n} \right)^{1/n} \right\}$

48. $\left\{ \left( 1 - \dfrac{4}{n} \right)^n \right\}$

49. $\{ b_n \}$, where $b_n = \begin{cases} \dfrac{n}{n + 1} & \text{if } n \leq 5000 \\ ne^{-n} & \text{if } n > 5000 \end{cases}$

50. $\left\{ n\left( 1 - \cos\dfrac{1}{n} \right) \right\}$

51. $\left\{ n \sin\dfrac{6}{n} \right\}$

52. $\left\{ \dfrac{n^8 + n^7}{n^7 + n^8 \ln n} \right\}$

53. Plot a graph of the sequence $\{a_n\}$, for $a_n = \sin\dfrac{n\pi}{2}$. Then determine the limit of the sequence or explain why the sequence diverges.

54. Plot a graph of the sequence $\{a_n\}$, for $a_n = \dfrac{(-1)^n n}{n + 1}$. Then determine the limit of the sequence or explain why the sequence diverges.

**55–70. More sequences** *Find the limit of the following sequences or determine that the sequence diverges.*

55. $\left\{ \dfrac{(-1)^n}{2^n} \right\}$

56. $\left\{ \dfrac{(-1)^n}{n} \right\}$

57. $a_n = (-1)^n \sqrt[n]{n}$

58. $\left\{ \dfrac{\cos(n\pi/2)}{\sqrt{n}} \right\}$

59. $\left\{ \dfrac{n \sin^3(n\pi/2)}{n + 1} \right\}$

60. $\left\{ \dfrac{(-1)^{n+1} n^2}{2n^3 + n} \right\}$

61. $a_n = e^{-n} \cos n$

62. $a_n = \dfrac{e^{-n}}{2 \sin(e^{-n})}$

63. $\left\{ \dfrac{\tan^{-1} n}{n} \right\}$

64. $\left\{ \dfrac{(2n^3 + n) \tan^{-1} n}{n^3 + 4} \right\}$

65. $\left\{ \dfrac{\cos n}{n} \right\}$

66. $\left\{ \dfrac{\sin 6n}{5n} \right\}$

67. $\left\{ \dfrac{\sin n}{2^n} \right\}$

68. $\left\{ \int_1^n x^{-2} \, dx \right\}$

69. $\left\{ \dfrac{75^{n-1}}{99^n} + \dfrac{5^n \sin n}{8^n} \right\}$

70. $\left\{ \cos(0.99^n) + \dfrac{7^n + 9^n}{63^n} \right\}$

**T 71. Periodic dosing** Many people take aspirin on a regular basis as a preventive measure for heart disease. Suppose a person takes 80 mg of aspirin every 24 hours. Assume aspirin has a half-life of 24 hours; that is, every 24 hours, half of the drug in the blood is eliminated.

   **a.** Find a recurrence relation for the sequence $\{d_n\}$ that gives the amount of drug in the blood after the $n$th dose, where $d_1 = 80$.

   **b.** Use a calculator to estimate this limit. In the long run, how much drug is in the person's blood?

   **c.** Assuming the sequence has a limit, confirm the result of part (b) by finding the limit of $\{d_n\}$ directly.

**T 72. A car loan** Marie takes out a $20,000 loan for a new car. The loan has an annual interest rate of 6% or, equivalently, a monthly interest rate of 0.5%. Each month, the bank adds interest to the loan balance (the interest is always 0.5% of the current balance), and then Marie makes a $200 payment to reduce the loan balance. Let $B_n$ be the loan balance immediately after the $n$th payment, where $B_0 = \$20,000$.

   **a.** Write the first five terms of the sequence $\{B_n\}$.

   **b.** Find a recurrence relation that generates the sequence $\{B_n\}$.

   **c.** Determine how many months are needed to reduce the loan balance to zero.

**T 73. A savings plan** James begins a savings plan in which he deposits $100 at the beginning of each month into an account that earns 9% interest annually or, equivalently, 0.75% per month. To be clear, on the first day of each month, the bank adds 0.75% of the current balance as interest, and then James deposits $100. Let $B_n$ be the balance in the account after the $n$th payment, where $B_0 = \$0$.

   **a.** Write the first five terms of the sequence $\{B_n\}$.

   **b.** Find a recurrence relation that generates the sequence $\{B_n\}$.

   **c.** How many months are needed to reach a balance of $5000?

**T 74. Diluting a solution** A tank is filled with 100 L of a 40% alcohol solution (by volume). You repeatedly perform the following operation: Remove 2 L of the solution from the tank and replace them with 2 L of 10% alcohol solution.

   **a.** Let $C_n$ be the concentration of the solution in the tank after the $n$th replacement, where $C_0 = 40\%$. Write the first five terms of the sequence $\{C_n\}$.

   **b.** After how many replacements does the alcohol concentration reach 15%?

   **c.** Determine the limiting (steady-state) concentration of the solution that is approached after many replacements.

**75–82. Growth rates of sequences** *Use Theorem 10.6 to find the limit of the following sequences or state that they diverge.*

**75.** $\left\{\dfrac{n!}{n^n}\right\}$     **76.** $\left\{\dfrac{3^n}{n!}\right\}$     **77.** $\left\{\dfrac{n^{10}}{\ln^{20} n}\right\}$

**78.** $\left\{\dfrac{n^{10}}{\ln^{1000} n}\right\}$     **79.** $\left\{\dfrac{n^{1000}}{2^n}\right\}$     **80.** $a_n = \dfrac{4^n + 5n!}{n! + 2^n}$

**81.** $a_n = \dfrac{6^n + 3^n}{6^n + n^{100}}$     **82.** $a_n = \dfrac{7^n}{n^7 5^n}$

**83. Explain why or why not** Determine whether the following statements are true and give an explanation or counterexample.

   **a.** If $\lim\limits_{n\to\infty} a_n = 1$ and $\lim\limits_{n\to\infty} b_n = 3$, then $\lim\limits_{n\to\infty} \dfrac{b_n}{a_n} = 3$.

   **b.** If $\lim\limits_{n\to\infty} a_n = 0$ and $\lim\limits_{n\to\infty} b_n = \infty$, then $\lim\limits_{n\to\infty} a_n b_n = 0$.

   **c.** The convergent sequences $\{a_n\}$ and $\{b_n\}$ differ in their first 100 terms, but $a_n = b_n$ for $n > 100$. It follows that $\lim\limits_{n\to\infty} a_n = \lim\limits_{n\to\infty} b_n$.

   **d.** If $\{a_n\} = \left\{1, \dfrac{1}{2}, \dfrac{1}{3}, \dfrac{1}{4}, \dfrac{1}{5}, \dots\right\}$ and $\{b_n\} = \left\{1, 0, \dfrac{1}{2}, 0, \dfrac{1}{3}, 0, \dfrac{1}{4}, 0, \dots\right\}$, then $\lim\limits_{n\to\infty} a_n = \lim\limits_{n\to\infty} b_n$.

   **e.** If the sequence $\{a_n\}$ converges, then the sequence $\{(-1)^n a_n\}$ converges.

   **f.** If the sequence $\{a_n\}$ diverges, then the sequence $\{0.000001 a_n\}$ diverges.

**T 84–87. Sequences by recurrence relations** *The following sequences, defined by a recurrence relation, are monotonic and bounded, and therefore converge by Theorem 10.5.*

   **a.** *Examine the first three terms of the sequence to determine whether the sequence is nondecreasing or nonincreasing.*

   **b.** *Use analytical methods to find the limit of the sequence.*

**84.** $a_{n+1} = \dfrac{1}{2}a_n + 2;\ a_0 = 1$

**85.** $a_{n+1} = 2a_n(1 - a_n);\ a_0 = 0.3$

**86.** $a_{n+1} = \dfrac{1}{2}\left(a_n + \dfrac{2}{a_n}\right);\ a_0 = 2$

**T 87.** $a_{n+1} = \sqrt{2 + a_n};\ a_0 = 3$

**T 88. Fish harvesting** A fishery manager knows that her fish population naturally increases at a rate of 1.5% per month, while 80 fish are harvested each month. Let $F_n$ be the fish population after the $n$th month, where $F_0 = 4000$ fish.

   **a.** Write out the first five terms of the sequence $\{F_n\}$.

   **b.** Find a recurrence relation that generates the sequence $\{F_n\}$.

   **c.** Does the fish population decrease or increase in the long run?

   **d.** Determine whether the fish population decreases or increases in the long run if the initial population is 5500 fish.

   **e.** Determine the initial fish population $F_0$ below which the population decreases.

**T 89. Drug dosing** A patient takes 75 mg of a medication every 12 hours; 60% of the medication in the blood is eliminated every 12 hours.

   **a.** Let $d_n$ equal the amount of medication (in mg) in the bloodstream after $n$ doses, where $d_1 = 75$. Find a recurrence relation for $d_n$.

   **b.** Show that $\{d_n\}$ is monotonic and bounded, and therefore converges.

   **c.** Find the limit of the sequence. What is the physical meaning of this limit?

**T 90. Sleep model** After many nights of observation, you notice that if you oversleep one night, you tend to undersleep the following night, and vice versa. This pattern of compensation is described by the relationship

$$x_{n+1} = \frac{1}{2}(x_n + x_{n-1}), \quad \text{for } n = 1, 2, 3, \dots,$$

where $x_n$ is the number of hours of sleep you get on the $n$th night, and $x_0 = 7$ and $x_1 = 6$ are the number of hours of sleep on the first two nights, respectively.

   **a.** Write out the first six terms of the sequence $\{x_n\}$ and confirm that the terms alternately increase and decrease.

**b.** Show that the explicit formula

$$x_n = \frac{19}{3} + \frac{2}{3}\left(-\frac{1}{2}\right)^n, \quad \text{for } n \geq 0,$$

generates the terms of the sequence in part (a).

**c.** Assume the limit of the sequence exists. What is the limit of the sequence?

**T 91. Calculator algorithm** The CORDIC (COordinate Rotation DIgital Calculation) algorithm is used by most calculators to evaluate trigonometric and logarithmic functions. An important number in the CORDIC algorithm, called the *aggregate constant*, is given by the infinite product $\prod_{n=0}^{\infty} \frac{2^n}{\sqrt{1 + 2^{2n}}}$, where $\prod_{n=0}^{N} a_n$ is the *partial product* $a_0 \cdot a_1 \cdots a_N$. This infinite product is the limit of the sequence

$$\left\{ \prod_{n=0}^{0} \frac{2^n}{\sqrt{1 + 2^{2n}}}, \prod_{n=0}^{1} \frac{2^n}{\sqrt{1 + 2^{2n}}}, \prod_{n=0}^{2} \frac{2^n}{\sqrt{1 + 2^{2n}}}, \cdots \right\}.$$

Estimate the value of the aggregate constant. (See the Guided Project *CORDIC algorithms: How your calculator works*.)

**92. A sequence of products** Find the limit of the sequence

$$\{a_n\}_{n=2}^{\infty} = \left\{ \left(1 - \frac{1}{2}\right)\left(1 - \frac{1}{3}\right)\cdots\left(1 - \frac{1}{n}\right) \right\}_{n=2}^{\infty}.$$

## Explorations and Challenges

**93.** Suppose the sequence $\{a_n\}_{n=0}^{\infty}$ is defined by the recurrence relation $a_{n+1} = \frac{1}{3}a_n + 6; a_0 = 3$.

**a.** Prove that the sequence is increasing and bounded.
**b.** Explain why $\{a_n\}_{n=0}^{\infty}$ converges and find the limit.

**94.** Suppose the sequence $\{a_n\}_{n=0}^{\infty}$ is defined by the recurrence relation $a_{n+1} = \sqrt{a_n + 20}; a_0 = 6$.

**a.** Prove that the sequence is decreasing and bounded.
**b.** Explain why $\{a_n\}_{n=0}^{\infty}$ converges and find the limit.

**T 95. Repeated square roots** Consider the sequence defined by $a_{n+1} = \sqrt{2 + a_n}, a_0 = \sqrt{2}$, for $n = 0, 1, 2, 3, \ldots$.

**a.** Evaluate the first four terms of the sequence $\{a_n\}$. State the exact values first, and then the approximate values.
**b.** Show that the sequence is increasing and bounded.
**c.** Assuming the limit exists, use the method of Example 5 to determine the limit exactly.

**T 96. Towers of powers** For a positive real number $p$, the tower of exponents $p^{p^{p^{\cdot^{\cdot^{\cdot}}}}}$ continues indefinitely but the expression is ambiguous. The tower could be built from the top as the limit of the sequence $\{p^p, (p^p)^p, ((p^p)^p)^p, \ldots\}$, in which case the sequence is defined recursively as

$$a_{n+1} = a_n^p \text{ (building from the top)}, \tag{1}$$

where $a_1 = p^p$. The tower could also be built from the bottom as the limit of the sequence $\{p^p, p^{(p^p)}, p^{(p^{(p^p)})}, \ldots\}$, in which case the sequence is defined recursively as

$$a_{n+1} = p^{a_n} \text{ (building from the bottom)}, \tag{2}$$

where again $a_1 = p^p$.

**a.** Estimate the value of the tower with $p = 0.5$ by building from the top. That is, use tables to estimate the limit of the sequence defined recursively by (1) with $p = 0.5$. Estimate the maximum value of $p > 0$ for which the sequence has a limit.

**b.** Estimate the value of the tower with $p = 1.2$ by building from the bottom. That is, use tables to estimate the limit of the sequence defined recursively by (2) with $p = 1.2$. Estimate the maximum value of $p > 1$ for which the sequence has a limit.

**T 97. Fibonacci sequence** The famous Fibonacci sequence was proposed by Leonardo Pisano, also known as Fibonacci, in about A.D. 1200 as a model for the growth of rabbit populations. It is given by the recurrence relation $f_{n+1} = f_n + f_{n-1}$, for $n = 1, 2, 3, \ldots$, where $f_0 = 1$ and $f_1 = 1$. Each term of the sequence is the sum of its two predecessors.

**a.** Write out the first ten terms of the sequence.
**b.** Is the sequence bounded?
**c.** Estimate or determine $\varphi = \lim_{n \to \infty} \frac{f_{n+1}}{f_n}$, the ratio of the successive terms of the sequence. Provide evidence that $\varphi = \frac{1 + \sqrt{5}}{2}$, a number known as the *golden mean*.
**d.** Use induction to verify the remarkable result that

$$f_n = \frac{1}{\sqrt{5}}\left(\varphi^n - (-1)^n \varphi^{-n}\right).$$

**98. Arithmetic-geometric mean** Pick two positive numbers $a_0$ and $b_0$ with $a_0 > b_0$ and write out the first few terms of the two sequences $\{a_n\}$ and $\{b_n\}$:

$$a_{n+1} = \frac{a_n + b_n}{2}, \quad b_{n+1} = \sqrt{a_n b_n}, \quad \text{for } n = 0, 1, 2, \ldots.$$

Recall that the arithmetic mean $A = (p + q)/2$ and the geometric mean $G = \sqrt{pq}$ of two positive numbers $p$ and $q$ satisfy $A \geq G$.

**a.** Show that $a_n > b_n$ for all $n$.
**b.** Show that $\{a_n\}$ is a decreasing sequence and $\{b_n\}$ is an increasing sequence.
**c.** Conclude that $\{a_n\}$ and $\{b_n\}$ converge.
**d.** Show that $a_{n+1} - b_{n+1} < \frac{a_n - b_n}{2}$ and conclude that $\lim_{n \to \infty} a_n = \lim_{n \to \infty} b_n$. The common value of these limits is called the arithmetic-geometric mean of $a_0$ and $b_0$, denoted $AGM(a_0, b_0)$.
**e.** Estimate $AGM(12, 20)$. Estimate Gauss' constant $\frac{1}{AGM(1, \sqrt{2})}$.

**T 99. Continued fractions** The expression

$$1 + \cfrac{1}{1 + \cfrac{1}{1 + \cfrac{1}{1 + \cfrac{1}{1 + \ddots}}}},$$

where the process continues indefinitely, is called a *continued fraction*.

**a.** Show that this expression can be built in steps using the recurrence relation $a_0 = 1, a_{n+1} = 1 + \frac{1}{a_n}$ for $n = 0, 1, 2, 3, \ldots$. Explain why the value of the expression can be interpreted as $\lim_{n \to \infty} a_n$, provided the limit exists.
**b.** Evaluate the first five terms of the sequence $\{a_n\}$.
**c.** Using computation and/or graphing, estimate the limit of the sequence.

**d.** Assuming the limit exists, use the method of Example 5 to determine the limit exactly. Compare your estimate with $\dfrac{1 + \sqrt{5}}{2}$, a number known as the *golden mean*.

**e.** Assuming the limit exists, use the same ideas to determine the value of

$$a + \cfrac{b}{a + \cfrac{b}{a + \cfrac{b}{a + \cfrac{b}{a + \cfrac{b}{\ddots}}}}},$$

where $a$ and $b$ are positive real numbers.

**100. The hailstone sequence** Here is a fascinating (unsolved) problem known as the hailstone problem (or the Ulam Conjecture or the Collatz Conjecture). It involves sequences in two different ways. First, choose a positive integer $N$ and call it $a_0$. This is the *seed* of a sequence. The rest of the sequence is generated as follows: For $n = 0, 1, 2, \ldots$

$$a_{n+1} = \begin{cases} a_n/2 & \text{if } a_n \text{ is even} \\ 3a_n + 1 & \text{if } a_n \text{ is odd.} \end{cases}$$

However, if $a_n = 1$ for any $n$, then the sequence terminates.

**a.** Compute the sequence that results from the seeds $N = 2, 3, 4, \ldots, 10$. You should verify that in all these cases, the sequence eventually terminates. The hailstone conjecture (still unproved) states that for all positive integers $N$, the sequence terminates after a finite number of terms.

**b.** Now define the hailstone sequence $\{H_k\}$, which is the number of terms needed for the sequence $\{a_n\}$ to terminate starting with a seed of $k$. Verify that $H_2 = 1$, $H_3 = 7$, and $H_4 = 2$.

**c.** Plot as many terms of the hailstone sequence as is feasible. How did the sequence get its name? Does the conjecture appear to be true?

**101–106. Formal proofs of limits** *Use the formal definition of the limit of a sequence to prove the following limits.*

**101.** $\displaystyle\lim_{n \to \infty} \frac{1}{n} = 0$

**102.** $\displaystyle\lim_{n \to \infty} \frac{1}{n^2} = 0$

**103.** $\displaystyle\lim_{n \to \infty} \frac{3n^2}{4n^2 + 1} = \frac{3}{4}$

**104.** $\displaystyle\lim_{n \to \infty} b^{-n} = 0$, for $b > 1$

**105.** $\displaystyle\lim_{n \to \infty} \frac{cn}{bn + 1} = \frac{c}{b}$, for real numbers $b > 0$ and $c > 0$

**106.** $\displaystyle\lim_{n \to \infty} \frac{n}{n^2 + 1} = 0$

**T 107. Comparing sequences with a parameter** For what values of $a$ does the sequence $\{n!\}$ grow faster than the sequence $\{n^{an}\}$? (*Hint:* Stirling's formula is useful: $n! \approx \sqrt{2\pi n}\, n^n e^{-n}$, for large values of $n$.)

**108–109. Reindexing** *Express each sequence $\{a_n\}_{n=1}^{\infty}$ as an equivalent sequence of the form $\{b_n\}_{n=3}^{\infty}$.*

**108.** $\{2n + 1\}_{n=1}^{\infty}$

**109.** $\{n^2 + 6n - 9\}_{n=1}^{\infty}$

**110.** Prove that if $\{a_n\} \ll \{b_n\}$ (as used in Theorem 10.6), then $\{ca_n\} \ll \{db_n\}$, where $c$ and $d$ are positive real numbers.

**T 111. Crossover point** The sequence $\{n!\}$ ultimately grows faster than the sequence $\{b^n\}$, for any $b > 1$, as $n \to \infty$. However, $b^n$ is generally greater than $n!$ for small values of $n$. Use a calculator to determine the smallest value of $n$ such that $n! > b^n$ for each of the cases $b = 2$, $b = e$, and $b = 10$.

**QUICK CHECK ANSWERS**

**1. a.** Bounded, monotonic; **b.** Bounded, not monotonic; **c.** Not bounded, not monotonic; **d.** Bounded, monotonic (both nonincreasing and nondecreasing) **2.** If $r = -1$, the sequence is $\{-1, 1, -1, 1, \ldots\}$, the terms alternate in sign, and the sequence diverges. If $r = 1$, the sequence is $\{1, 1, 1, 1, \ldots\}$, the terms are constant, and the sequence converges to 1. **3.** Both changes would increase the steady-state level of drug. **4.** $\{n^{1.1}\}$ grows faster; the limit is 0. ◄

# 10.3 Infinite Series

> ▶ The sequence of partial sums may be visualized nicely as follows:
> $$\underbrace{\overbrace{\underbrace{a_1 + a_2}_{S_1} + a_3}^{S_2} + a_4}_{S_3} + \cdots$$

When the initial value of the index is not 1, the $n$th partial sum is still the sum of the first $n$ terms of the series. For example, the $n$th partial sum of $\displaystyle\sum_{k=3}^{\infty} a_k$ is

$$S_n = \underbrace{a_3 + a_4 + \cdots + a_{n+2}}_{n \text{ terms}}.$$

We begin our discussion of infinite series with *geometric series*. These series arise more frequently than any other infinite series, they are used in many practical problems, and they illustrate all the essential features of infinite series in general. First let's summarize some important ideas from Section 10.1.

Recall that every infinite series $\displaystyle\sum_{k=1}^{\infty} a_k$ has a sequence of partial sums:

$$S_1 = a_1, \qquad S_2 = a_1 + a_2, \qquad S_3 = a_1 + a_2 + a_3,$$

and in general, $S_n = \displaystyle\sum_{k=1}^{n} a_k$, for $n = 1, 2, 3, \ldots$.

If the sequence of partial sums $\{S_n\}$ converges—that is, if $\displaystyle\lim_{n \to \infty} S_n = L$—then the value of the infinite series is also $L$. If the sequence of partial sums diverges, then the infinite series also diverges.

**QUICK CHECK 1** Write the $n$th partial sum of the series $\displaystyle\sum_{k=0}^{\infty} b_n$. ◄

In summary, to evaluate an infinite series, it is necessary to determine a formula for the sequence of partial sums $\{S_n\}$ and then find its limit. This procedure can be carried out with the series that we discuss in this section: geometric series and telescoping series.

## Geometric Sums and Series

➤ Geometric *sequences* have the form $\{r^k\}$ or $\{ar^k\}$. Geometric *sums* and *series* have the form $\sum r^k$ or $\sum ar^k$.

As a preliminary step to geometric series, we study geometric sums, which are *finite sums* in which each term in the sum is a constant multiple of the previous term. A **geometric sum** with $n$ terms has the form

$$S_n = a + ar + ar^2 + \cdots + ar^{n-1} = \sum_{k=0}^{n-1} ar^k,$$

**QUICK CHECK 2** Which of the following sums are not geometric sums?

**a.** $\displaystyle\sum_{k=0}^{10} \left(\frac{1}{2}\right)^k$    **b.** $\displaystyle\sum_{k=0}^{20} \frac{1}{k}$

**c.** $\displaystyle\sum_{k=0}^{30} (2k + 1)$ ◄

where $a \neq 0$ and $r$ are real numbers; $r$ is called the **ratio** of the sum and $a$ is its first term. For example, the geometric sum with $r = 0.1$, $a = 0.9$, and $n = 4$ is

$$0.9 + 0.09 + 0.009 + 0.0009 = 0.9(1 + 0.1 + 0.01 + 0.001)$$

$$= \sum_{k=0}^{3} 0.9(0.1^k).$$

Our goal is to find a formula for the value of the geometric sum

$$S_n = a + ar + ar^2 + \cdots + ar^{n-1}, \tag{1}$$

for any values of $a \neq 0$, $r$, and the positive integer $n$. Doing so requires a clever maneuver. The first step is to multiply both sides of equation (1) by the ratio $r$:

$$rS_n = r(a + ar + ar^2 + ar^3 + \cdots + ar^{n-1})$$
$$= ar + ar^2 + ar^3 + \cdots + ar^{n-1} + ar^n. \tag{2}$$

➤ The notation $\displaystyle\sum_{k=0}^{\infty} ar^k$ appears to have an undefined first term when $r = 0$. The notation is understood to mean $a + ar + ar^2 + \cdots$ and therefore, the series has a value of $a$ when $r = 0$.

We now subtract equation (2) from equation (1). Notice how most of the terms on the right sides of these equations cancel, leaving

$$S_n - rS_n = a - ar^n.$$

Assuming $r \neq 1$ and solving for $S_n$ results in a general formula for the value of a geometric sum:

$$S_n = a\frac{1 - r^n}{1 - r}. \tag{3}$$

**QUICK CHECK 3** Verify that the geometric sum formula gives the correct result for the sums $1 + \frac{1}{2}$ and $\frac{1}{2} + \frac{1}{4} + \frac{1}{8}$. ◄

Having dealt with geometric sums, it is a short step to *geometric series*. We simply note that the geometric sums $S_n = \displaystyle\sum_{k=0}^{n-1} ar^k$ form the sequence of partial sums for the geometric series $\displaystyle\sum_{k=0}^{\infty} ar^k$. The value of the geometric series is the limit of its sequence of partial sums (provided that limit exists). Using equation (3), we have

$$\underbrace{\sum_{k=0}^{\infty} ar^k}_{\text{geometric series}} = \lim_{n\to\infty} \underbrace{\sum_{k=0}^{n-1} ar^k}_{\text{geometric sum } S_n} = \lim_{n\to\infty} a\frac{1 - r^n}{1 - r}.$$

To compute this limit, we must examine the behavior of $r^n$ as $n \to \infty$. Recall from our work with geometric sequences (Section 10.2) that

$$\lim_{n\to\infty} r^n = \begin{cases} 0 & \text{if } |r| < 1 \\ 1 & \text{if } r = 1 \\ \text{does not exist} & \text{if } r \leq -1 \text{ or } r > 1. \end{cases}$$

**Case 1:** $|r| < 1$  Because $\displaystyle\lim_{n\to\infty} r^n = 0$, we have

$$\lim_{n\to\infty} S_n = \lim_{n\to\infty} a\frac{1 - r^n}{1 - r} = a\frac{1 - \overset{0}{\overbrace{\lim_{n\to\infty} r^n}}}{1 - r} = \frac{a}{1 - r}.$$

In the case that $|r| < 1$, the geometric series *converges* to $\dfrac{a}{1 - r}$.

**Case 2:** $|r| > 1$ In this case, $\lim\limits_{n \to \infty} r^n$ does not exist, so $\lim\limits_{n \to \infty} S_n$ does not exist and the series *diverges*.

**Case 3:** $|r| = 1$ If $r = 1$, then the geometric series is $\sum\limits_{k=0}^{\infty} a = a + a + a + \cdots$, which

diverges. If $r = -1$, the geometric series is $a \sum\limits_{k=0}^{\infty} (-1)^k = a - a + a - \cdots$, which also

diverges (because the sequence of partial sums oscillates between 0 and $a$). We summarize these results in Theorem 10.7.

> **QUICK CHECK 4** Evaluate
> $\frac{1}{2} + \frac{1}{4} + \frac{1}{8} + \frac{1}{16} + \cdots$. ◄

---

**THEOREM 10.7  Geometric Series**

Let $a \neq 0$ and $r$ be real numbers. If $|r| < 1$, then $\sum\limits_{k=0}^{\infty} ar^k = \dfrac{a}{1 - r}$. If $|r| \geq 1$,

then the series diverges.

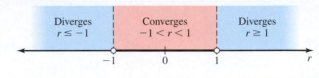

---

> **QUICK CHECK 5** Explain why $\sum\limits_{k=0}^{\infty} 0.2^k$
>
> converges and why $\sum\limits_{k=0}^{\infty} 2^k$ diverges. ◄

**EXAMPLE 1   Geometric series** Evaluate the following geometric series or state that the series diverges.

**a.** $\sum\limits_{k=0}^{\infty} 1.1^k$     **b.** $\sum\limits_{k=0}^{\infty} e^{-k}$     **c.** $\sum\limits_{k=2}^{\infty} 3(-0.75)^k$

**SOLUTION**

**a.** The ratio of this geometric series is $r = 1.1$. Because $|r| \geq 1$, the series diverges.

**b.** Note that $e^{-k} = \dfrac{1}{e^k} = \left(\dfrac{1}{e}\right)^k$. Therefore, the ratio of the series is $r = \dfrac{1}{e}$, and its first

term is $a = 1$. Because $|r| < 1$, the series converges and its value is

$$\sum_{k=0}^{\infty} e^{-k} = \sum_{k=0}^{\infty} \left(\frac{1}{e}\right)^k = \frac{1}{1 - (1/e)} = \frac{e}{e - 1} \approx 1.582.$$

> ➤ The series in Example 1c is called an *alternating series* because the terms alternate in sign. Such series are discussed in detail in Section 10.6.

**c.** Writing out the first few terms of the series is helpful:

$$\sum_{k=2}^{\infty} 3(-0.75)^k = \underbrace{3(-0.75)^2}_{a} + \underbrace{3(-0.75)^3}_{ar} + \underbrace{3(-0.75)^4}_{ar^2} + \cdots.$$

We see that the first term of the series is $a = 3(-0.75)^2$ and that the ratio of the series is $r = -0.75$. Because $|r| < 1$, the series converges, and its value is

$$\sum_{k=2}^{\infty} 3(-0.75)^k = \underbrace{\frac{3(-0.75)^2}{1 - (-0.75)}}_{\frac{a}{1 - r}} = \frac{27}{28}.$$

*Related Exercises 24, 27, 29* ◄

**EXAMPLE 2   Decimal expansions as geometric series** Write $1.0\overline{35} = 1.0353535\ldots$ as a geometric series and express its value as a fraction.

**SOLUTION** Notice that the decimal part of this number is a convergent geometric series with $a = 0.035$ and $r = 0.01$:

$$1.0353535\ldots = 1 + \underbrace{0.035 + 0.00035 + 0.0000035 + \cdots}_{\text{geometric series with } a = 0.035 \text{ and } r = 0.01}.$$

Evaluating the series, we have

$$1.0353535\ldots = 1 + \frac{a}{1-r} = 1 + \frac{0.035}{1-0.01} = 1 + \frac{35}{990} = \frac{205}{198}.$$

*Related Exercises 49–50* ◄

## Telescoping Series

With geometric series, we carried out the entire evaluation process by finding a formula for the sequence of partial sums and evaluating the limit of the sequence. Not many infinite series can be subjected to this sort of analysis. With another class of series, called **telescoping series**, it can also be done. Here is an example.

**EXAMPLE 3**  **Telescoping series**  Evaluate the following series.

**a.** $\displaystyle\sum_{k=1}^{\infty}\left(\cos\frac{1}{k} - \cos\frac{1}{k+1}\right)$    **b.** $\displaystyle\sum_{k=3}^{\infty}\frac{1}{(k-2)(k-1)}$

**SOLUTION**

**a.** The $n$th term of the sequence of partial sums is

$$S_n = \sum_{k=1}^{n}\left(\cos\frac{1}{k} - \cos\frac{1}{k+1}\right)$$

$$= \left(\cos 1 - \cos\frac{1}{2}\right) + \left(\cos\frac{1}{2} - \cos\frac{1}{3}\right) + \cdots + \left(\cos\frac{1}{n} - \cos\frac{1}{n+1}\right)$$

$$= \cos 1 + \underbrace{\left(-\cos\frac{1}{2} + \cos\frac{1}{2}\right)}_{0} + \cdots + \underbrace{\left(-\cos\frac{1}{n} + \cos\frac{1}{n}\right)}_{0} - \cos\frac{1}{n+1} \qquad \text{Regroup terms.}$$

$$= \cos 1 - \cos\frac{1}{n+1}. \qquad \text{Simplify.}$$

Observe that the interior terms of the sum cancel (or telescope), leaving a simple expression for $S_n$. Taking the limit, we find that

$$\sum_{k=1}^{\infty}\left(\cos\frac{1}{k} - \cos\frac{1}{k+1}\right) = \lim_{n\to\infty} S_n$$

$$= \lim_{n\to\infty}\left(\cos 1 - \underbrace{\cos\frac{1}{n+1}}_{\to\,\cos 0\,=\,1}\right) = \cos 1 - 1 \approx -0.46.$$

**b.** Because the initial value of the index is $k=3$, the $n$th partial sum $S_n$ involves $n$ terms that run from $k=3$ to $k=n+2$, which is somewhat awkward. The $n$th partial sums are most easily written when the initial value of the index is $k=1$. We *reindex* the original sum by using the substitution $j=k-2$, which implies that $j=1$ when $k=3$ and that $k-1=j+1$:

$$\sum_{k=3}^{\infty}\frac{1}{(k-2)(k-1)} = \sum_{j=1}^{\infty}\frac{1}{j(j+1)} \qquad \text{Let } j=k-2.$$

$$= \sum_{k=1}^{\infty}\frac{1}{k(k+1)}.$$

The variable used as an index does not affect the value of a series, so it is permissible to return to the index $k$.

► See Section 8.5 for a review of partial fractions.

Using the method of partial fractions, the sequence of partial sums is

$$S_n = \sum_{k=1}^{n}\frac{1}{k(k+1)} = \sum_{k=1}^{n}\left(\frac{1}{k} - \frac{1}{k+1}\right).$$

Writing out this sum, we see that

$$S_n = \left(1 - \frac{1}{2}\right) + \left(\frac{1}{2} - \frac{1}{3}\right) + \left(\frac{1}{3} - \frac{1}{4}\right) + \cdots + \left(\frac{1}{n} - \frac{1}{n+1}\right)$$

$$= 1 + \underbrace{\left(-\frac{1}{2} + \frac{1}{2}\right)}_{0} + \underbrace{\left(-\frac{1}{3} + \frac{1}{3}\right)}_{0} + \cdots + \underbrace{\left(-\frac{1}{n} + \frac{1}{n}\right)}_{0} - \frac{1}{n+1}$$

$$= 1 - \frac{1}{n+1}.$$

Again, the sum telescopes and all the interior terms cancel. The result is a simple formula for the $n$th term of the sequence of partial sums. The value of the series is

$$\sum_{k=1}^{\infty} \frac{1}{k(k+1)} = \lim_{n \to \infty} S_n = \lim_{n \to \infty} \left(1 - \frac{1}{n+1}\right) = 1.$$

*Related Exercises 59, 69* ◄

## Properties of Convergent Series

We close this section with several properties that are useful in upcoming work. The notation $\Sigma a_k$, without initial and final values of $k$, is used to refer to a general infinite series whose terms may be positive or negative (or both).

---

**THEOREM 10.8  Properties of Convergent Series**

1. Suppose $\Sigma a_k$ converges to $A$ and $c$ is a real number. The series $\Sigma ca_k$ converges, and $\Sigma ca_k = c\Sigma a_k = cA$.

2. Suppose $\Sigma a_k$ diverges. Then $\Sigma ca_k$ also diverges, for any real number $c \neq 0$.

3. Suppose $\Sigma a_k$ converges to $A$ and $\Sigma b_k$ converges to $B$. The series $\Sigma(a_k \pm b_k)$ converges, and $\Sigma(a_k \pm b_k) = \Sigma a_k \pm \Sigma b_k = A \pm B$.

4. Suppose $\Sigma a_k$ diverges and $\Sigma b_k$ converges. Then $\Sigma(a_k \pm b_k)$ diverges.

5. If $M$ is a positive integer, then $\sum_{k=1}^{\infty} a_k$ and $\sum_{k=M}^{\infty} a_k$ either both converge or both diverge. In general, *whether* a series converges does not depend on a finite number of terms added to or removed from the series. However, the *value* of a convergent series does change if nonzero terms are added or removed.

---

> The **leading terms** of an infinite series are those at the beginning with a small index. The **tail** of an infinite series consists of the terms at the "end" of the series with a large and increasing index. The convergence or divergence of an infinite series depends on the tail of the series, while the value of a convergent series is determined primarily by the leading terms.

**Proof:** These properties are proved using properties of finite sums and limits of sequences. To prove Property 1, assume $\sum_{k=1}^{\infty} a_k$ converges to $A$ and note that

$$\sum_{k=1}^{\infty} ca_k = \lim_{n \to \infty} \sum_{k=1}^{n} ca_k \qquad \text{Definition of infinite series}$$

$$= \lim_{n \to \infty} c \sum_{k=1}^{n} a_k \qquad \text{Property of finite sums}$$

$$= c \lim_{n \to \infty} \sum_{k=1}^{n} a_k \qquad \text{Property of limits}$$

$$= c \sum_{k=1}^{\infty} a_k \qquad \text{Definition of infinite series}$$

$$= cA. \qquad \text{Value of the series}$$

The proofs of Properties 2 and 3 are similar (Exercises 96–97); we offer Property 4 with no proof.

Property 5 follows by noting that for finite sums with $1 < M < n$,

$$\sum_{k=M}^{n} a_k = \sum_{k=1}^{n} a_k - \sum_{k=1}^{M-1} a_k.$$

Letting $n \to \infty$ in this equation and assuming the value of the infinite series is $\sum_{k=1}^{\infty} a_k = A$, it follows that

$$\sum_{k=M}^{\infty} a_k = \underbrace{\sum_{k=1}^{\infty} a_k}_{A} - \underbrace{\sum_{k=1}^{M-1} a_k}_{\text{finite number}}.$$

**QUICK CHECK 6** Explain why if $\sum_{k=1}^{\infty} a_k$ converges, then the series $\sum_{k=5}^{\infty} a_k$ (with a different starting index) also converges. Do the two series have the same value? ◄

Because the right side has a finite value, $\sum_{k=M}^{\infty} a_k$ converges. Similarly, if $\sum_{k=M}^{\infty} a_k$ converges, then $\sum_{k=1}^{\infty} a_k$ converges. By an analogous argument, if one of these series diverges, then the other series diverges. ◄

**EXAMPLE 4** **Using properties of series** Evaluate the following infinite series.

**a.** $\sum_{k=2}^{\infty} \left( \frac{1}{3^k} - \frac{2}{3^{k+2}} \right)$    **b.** $\sum_{k=1}^{\infty} \left( 5\left(\frac{2}{3}\right)^k - \frac{7^{k-1}}{6^k} \right)$

**SOLUTION**

**a.** As long as both $\sum_{k=2}^{\infty} \frac{1}{3^k}$ and $\sum_{k=2}^{\infty} \frac{2}{3^{k+1}}$ converge, we can use Properties 1 and 3 of Theorem 10.8 to write

$$\sum_{k=2}^{\infty} \left( \frac{1}{3^k} - \frac{2}{3^{k+1}} \right) = \sum_{k=2}^{\infty} \frac{1}{3^k} - 2\sum_{k=2}^{\infty} \frac{1}{3^{k+1}}.$$

Both of the new series are geometric, and the ratio of each series is $r = \frac{1}{3} < 1$, so each series converges. The first term of $\sum_{k=2}^{\infty} \frac{1}{3^k}$ is $a = \frac{1}{9}$ and the first term of $\sum_{k=2}^{\infty} \frac{1}{3^{k+1}}$ is $a = \frac{1}{27}$, so we have

$$\sum_{k=2}^{\infty} \left( \frac{1}{3^k} - \frac{2}{3^{k+1}} \right) = \sum_{k=2}^{\infty} \frac{1}{3^k} - 2\sum_{k=2}^{\infty} \frac{1}{3^{k+1}} = \frac{\frac{1}{9}}{1 - \frac{1}{3}} - 2\frac{\frac{1}{27}}{1 - \frac{1}{3}} = \frac{1}{9} \cdot \frac{3}{2} - \frac{2}{27} \cdot \frac{3}{2} = \frac{1}{18}.$$

**b.** We examine the two series $\sum_{k=1}^{\infty} 5\left(\frac{2}{3}\right)^k$ and $\sum_{k=1}^{\infty} \frac{7^{k-1}}{6^k}$ individually. The first series is a geometric series and is evaluated using the methods of Section 10.3. Its first few terms are

$$\sum_{k=1}^{\infty} 5\left(\frac{2}{3}\right)^k = 5\left(\frac{2}{3}\right) + 5\left(\frac{2}{3}\right)^2 + 5\left(\frac{2}{3}\right)^3 + \cdots.$$

The first term of the series is $a = 5\left(\frac{2}{3}\right)$ and the ratio is $r = \frac{2}{3} < 1$; therefore,

$$\sum_{k=1}^{\infty} 5\left(\frac{2}{3}\right)^k = \frac{a}{1-r} = \frac{5\left(\frac{2}{3}\right)}{1 - \frac{2}{3}} = 10.$$

Writing out the first few terms of the second series, we see that it, too, is a geometric series:

$$\sum_{k=1}^{\infty} \frac{7^{k-1}}{6^k} = \frac{1}{6} + \frac{7}{6^2} + \frac{7^2}{6^3} + \cdots.$$

**QUICK CHECK 7** For a series with positive terms, explain why the sequence of partial sums $\{S_n\}$ is an increasing sequence. ◄

Because the ratio is $r = \frac{7}{6} > 1$, $\sum_{k=1}^{\infty} \frac{7^{k-1}}{6^k}$ diverges. By Property 4 of Theorem 10.8, the original series also diverges.

*Related Exercises 81–82* ◄

## SECTION 10.3 EXERCISES

### Getting Started

1. What is meant by the *ratio* of a geometric series?

2. What is the difference between a geometric sum and a geometric series?

3. Does a geometric series always have a finite value?

4. Does a geometric sum always have a finite value?

5. Find the first term $a$ and the ratio $r$ of each geometric series.

   a. $\displaystyle\sum_{k=0}^{\infty} \frac{2}{3}\left(\frac{1}{5}\right)^k$
   
   b. $\displaystyle\sum_{k=2}^{\infty} \frac{1}{3}\left(-\frac{1}{3}\right)^k$

6. What is the condition for convergence of the geometric series
   $$\sum_{k=0}^{\infty} ar^k?$$

7. Find a formula for the $n$th partial sum $S_n$ of $\displaystyle\sum_{k=1}^{\infty}\left(\frac{1}{k+3} - \frac{1}{k+4}\right)$.
   Use your formula to find the sum of the first 36 terms of the series.

8. Reindex the series $\displaystyle\sum_{k=5}^{\infty} \frac{3}{4k^2 - 63}$ so that it starts at $k = 1$.

### Practice Exercises

**9–15. Geometric sums** *Evaluate each geometric sum.*

9. $\displaystyle\sum_{k=0}^{8} 3^k$

**T** 10. $\displaystyle\sum_{k=0}^{10} \left(\frac{1}{4}\right)^k$

**T** 11. $\displaystyle\sum_{k=0}^{20} \left(\frac{2}{5}\right)^{2k}$

12. $\displaystyle\sum_{k=4}^{12} 2^k$

**T** 13. $\displaystyle\sum_{k=0}^{9} \left(-\frac{3}{4}\right)^k$

14. $\displaystyle\sum_{k=0}^{6} \pi^k$

**T** 15. $\dfrac{1}{4} + \dfrac{1}{12} + \dfrac{1}{36} + \dfrac{1}{108} + \cdots + \dfrac{1}{2916}$

**T** **16–17. Periodic savings** *Suppose you deposit m dollars at the beginning of every month in a savings account that earns a monthly interest rate of r, which is the annual interest rate divided by 12 (for example, if the annual interest rate is 2.4%, r = 0.024/12 = 0.002). For an initial investment of m dollars, the amount of money in your account at the beginning of the second month is the sum of your second deposit and your initial deposit plus interest, or m + m(1 + r). Continuing in this fashion, it can be shown that the amount of money in your account after n months is $A_n = m + m(1 + r) + \cdots + m(1 + r)^{n-1}$. Use geometric sums to determine the amount of money in your savings account after 5 years (60 months) using the given monthly deposit and interest rate.*

16. Monthly deposits of $100 at an annual interest rate of 1.8%

17. Monthly deposits of $250 at a monthly interest rate of 0.2%

**18–20. Evaluating geometric series two ways** *Evaluate each geometric series two ways.*

a. *Find the $n$th partial sum $S_n$ of the series and evaluate $\lim_{n\to\infty} S_n$.*
b. *Evaluate the series using Theorem 10.7.*

18. $\displaystyle\sum_{k=0}^{\infty} \left(\frac{2}{5}\right)^k$

19. $\displaystyle\sum_{k=0}^{\infty} \left(-\frac{2}{7}\right)^k$

20. $2 + \dfrac{4}{3} + \dfrac{8}{9} + \dfrac{16}{27} + \cdots$

**21–42. Geometric series** *Evaluate each geometric series or state that it diverges.*

21. $\displaystyle\sum_{k=0}^{\infty} \left(\frac{1}{4}\right)^k$

22. $\displaystyle\sum_{k=0}^{\infty} \left(\frac{3}{5}\right)^k$

23. $\displaystyle\sum_{k=0}^{\infty} \left(-\frac{9}{10}\right)^k$

24. $\displaystyle\sum_{k=1}^{\infty} \left(-\frac{2}{3}\right)^k$

25. $\displaystyle\sum_{k=0}^{\infty} 0.9^k$

26. $1 + \dfrac{2}{7} + \dfrac{2^2}{7^2} + \dfrac{2^3}{7^3} + \cdots$

27. $1 + 1.01 + 1.01^2 + 1.01^3 + \cdots$

28. $1 + \dfrac{1}{\pi} + \dfrac{1}{\pi^2} + \dfrac{1}{\pi^3} + \cdots$

29. $\displaystyle\sum_{k=1}^{\infty} e^{-2k}$

30. $\displaystyle\sum_{m=2}^{\infty} \frac{5}{2^m}$

31. $\displaystyle\sum_{k=1}^{\infty} 2^{-3k}$

32. $\displaystyle\sum_{k=3}^{\infty} \frac{3 \cdot 4^k}{7^k}$

33. $\displaystyle\sum_{k=4}^{\infty} \frac{1}{5^k}$

34. $\displaystyle\sum_{k=0}^{\infty} \left(\frac{4}{3}\right)^{-k}$

35. $\displaystyle\sum_{k=0}^{\infty} 3(-\pi)^{-k}$

36. $\displaystyle\sum_{k=1}^{\infty} (-e)^{-k}$

37. $1 + \dfrac{e}{\pi} + \dfrac{e^2}{\pi^2} + \dfrac{e^3}{\pi^3} + \cdots$

38. $\dfrac{1}{16} + \dfrac{3}{64} + \dfrac{9}{256} + \dfrac{27}{1024} + \cdots$

39. $\displaystyle\sum_{k=2}^{\infty} (-0.15)^k$

40. $\displaystyle\sum_{k=1}^{\infty} 3\left(-\frac{1}{8}\right)^{3k}$

41. $\displaystyle\sum_{k=1}^{\infty} \frac{4}{12^k}$

42. $\displaystyle\sum_{k=2}^{\infty} 3e^{-k}$

**43–44. Periodic doses** *Suppose you take a dose of m mg of a particular medication once per day. Assume f equals the fraction of the medication that remains in your blood one day later. Just after taking another dose of medication on the second day, the amount of medication in your blood equals the sum of the second dose and the fraction of the first dose remaining in your blood, which is m + mf. Continuing in this fashion, the amount of medication in your blood just after your nth dose is $A_n = m + mf + \cdots + mf^{n-1}$. For the given values of f and m, calculate $A_5, A_{10}, A_{30}$, and $\lim_{n\to\infty} A_n$. Interpret the meaning of the limit $\lim_{n\to\infty} A_n$.*

43. $f = 0.25, m = 200$ mg

44. $f = 0.4, m = 150$ mg

45. **Periodic doses** Suppose you take 200 mg of an antibiotic every 6 hr. The half-life of the drug (the time it takes for half of the drug to be eliminated from your blood) is 6 hr. Use infinite series to find the long-term (steady-state) amount of antibiotic in your blood. You may assume the steady-state amount is finite.

**46–53. Decimal expansions** *Write each repeating decimal first as a geometric series and then as a fraction (a ratio of two integers).*

46. $0.\overline{6} = 0.666\ldots$

47. $0.\overline{3} = 0.333\ldots$

48. $0.\overline{09} = 0.090909\ldots$

49. $0.\overline{037} = 0.037037\ldots$

**50.** $1.\overline{25} = 1.252525\ldots$

**51.** $0.\overline{456} = 0.456456456\ldots$

**52.** $5.12\overline{83} = 5.12838383\ldots$

**53.** $0.00\overline{952} = 0.00952952\ldots$

**54–69. Telescoping series** *For the following telescoping series, find a formula for the nth term of the sequence of partial sums* $\{S_n\}$. *Then evaluate* $\lim_{n \to \infty} S_n$ *to obtain the value of the series or state that the series diverges.*

**54.** $\sum_{k=1}^{\infty} \left( \dfrac{1}{k+2} - \dfrac{1}{k+3} \right)$

**55.** $\sum_{k=1}^{\infty} \left( \dfrac{1}{k+1} - \dfrac{1}{k+2} \right)$

**56.** $\sum_{k=1}^{\infty} \dfrac{20}{25k^2 + 15k - 4}$

**57.** $\sum_{k=1}^{\infty} \dfrac{1}{(k+6)(k+7)}$

**58.** $\sum_{k=-3}^{\infty} \dfrac{10}{4k^2 + 32k + 63}$

**59.** $\sum_{k=3}^{\infty} \dfrac{4}{(4k-3)(4k+1)}$

**60.** $\sum_{k=0}^{\infty} \dfrac{1}{(3k+1)(3k+4)}$

**61.** $\sum_{k=1}^{\infty} \ln \dfrac{k+1}{k}$

**62.** $\sum_{k=3}^{\infty} \dfrac{2}{(2k-1)(2k+1)}$

**63.** $\sum_{k=1}^{\infty} \dfrac{1}{(k+p)(k+p+1)}$, where $p$ is a positive integer

**64.** $\sum_{k=1}^{\infty} \dfrac{1}{(ak+1)(ak+a+1)}$, where $a$ is a positive integer

**65.** $\sum_{k=1}^{\infty} \left( \dfrac{1}{\sqrt{k+1}} - \dfrac{1}{\sqrt{k+3}} \right)$

**66.** $\sum_{k=1}^{\infty} \dfrac{6}{k^2 + 2k}$

**67.** $\sum_{k=1}^{\infty} \dfrac{3}{k^2 + 5k + 4}$

**68.** $\sum_{k=1}^{\infty} (\sqrt{k+1} - \sqrt{k})$

**69.** $\sum_{k=1}^{\infty} (\tan^{-1}(k+1) - \tan^{-1}k)$

**70. Evaluating an infinite series two ways** Evaluate the series
$$\sum_{k=1}^{\infty} \left( \dfrac{1}{2^k} - \dfrac{1}{2^{k+1}} \right) \text{ two ways.}$$

  **a.** Use a telescoping series argument.
  **b.** Use a geometric series argument with Theorem 10.8.

**71. Evaluating an infinite series two ways** Evaluate the series
$$\sum_{k=1}^{\infty} \left( \dfrac{4}{3^k} - \dfrac{4}{3^{k+1}} \right) \text{ two ways.}$$

  **a.** Use a telescoping series argument.
  **b.** Use a geometric series argument with Theorem 10.8.

**72–86. Evaluating series** *Evaluate each series or state that it diverges.*

**72.** $\sum_{k=1}^{\infty} \dfrac{1}{9k^2 + 39k + 40}$

**73.** $\sum_{k=0}^{\infty} \dfrac{1}{16k^2 + 8k - 3}$

**74.** $\sum_{k=0}^{\infty} \left( \sin \left( \dfrac{(k+1)\pi}{2k+1} \right) - \sin \left( \dfrac{k\pi}{2k-1} \right) \right)$

**75.** $\sum_{k=0}^{\infty} \left( \dfrac{1}{4} \right)^k 5^{3-k}$

**76.** $\sum_{k=2}^{\infty} \left( \dfrac{3}{8} \right)^{3k}$

**77.** $\sum_{k=1}^{\infty} \dfrac{(-2)^k}{3^{k+1}}$

**78.** $\sum_{k=1}^{\infty} \left( \sin^{-1} \left( \dfrac{1}{k} \right) - \sin^{-1} \left( \dfrac{1}{k+1} \right) \right)$

**79.** $\sum_{k=2}^{\infty} \dfrac{\ln((k+1)k^{-1})}{(\ln k)\ln(k+1)}$

**80.** $\sum_{k=1}^{\infty} \dfrac{\pi^k}{e^{k+1}}$

**81.** $\sum_{k=0}^{\infty} \left( 3 \left( \dfrac{2}{5} \right)^k - 2 \left( \dfrac{5}{7} \right)^k \right)$

**82.** $\sum_{k=1}^{\infty} \left( 2 \left( \dfrac{3}{5} \right)^k + 3 \left( \dfrac{4}{9} \right)^k \right)$

**83.** $\sum_{k=1}^{\infty} \left( \dfrac{1}{3} \left( \dfrac{5}{6} \right)^k + \dfrac{3}{5} \left( \dfrac{7}{9} \right)^k \right)$

**84.** $\sum_{k=0}^{\infty} \left( \dfrac{1}{2}(0.2)^k + \dfrac{3}{2}(0.8)^k \right)$

**85.** $\sum_{k=1}^{\infty} \left( \left( \dfrac{1}{6} \right)^k + \left( \dfrac{1}{3} \right)^{k-1} \right)$

**86.** $\sum_{k=0}^{\infty} \dfrac{2 - 3^k}{6^k}$

**87. Explain why or why not** Determine whether the following statements are true and give an explanation or counterexample.

  **a.** If $\sum_{k=1}^{\infty} a_k$ converges, then $\sum_{k=10}^{\infty} a_k$ converges.

  **b.** If $\sum_{k=1}^{\infty} a_k$ diverges, then $\sum_{k=10}^{\infty} a_k$ diverges.

  **c.** If $\sum a_k$ converges, then $\sum (a_k + 0.0001)$ converges.

  **d.** If $\sum p^k$ diverges, then $\sum (p + 0.001)^k$ diverges, for a fixed real number $p$.

  **e.** $\sum_{k=1}^{\infty} \left( \dfrac{\pi}{e} \right)^{-k}$ is a convergent geometric series.

  **f.** If the series $\sum_{k=1}^{\infty} a^k$ converges and $|a| < |b|$, then the series $\sum_{k=1}^{\infty} b^k$ converges.

  **g.** Viewed as a function of $r$, the series $1 + r + r^2 + r^3 + \cdots$ takes on all values in the interval $\left( \dfrac{1}{2}, \infty \right)$.

**88–89. Binary numbers** *Humans use the ten digits 0 through 9 to form base-10 or decimal numbers, whereas computers calculate and store numbers internally as binary numbers—numbers consisting entirely of 0's and 1's. For this exercise, we consider binary numbers that have the form* $0.b_1 b_2 b_3 \ldots$, *where each of the digits* $b_1, b_2, b_3, \ldots$ *is either 0 or 1. The base-10 representation of the binary number* $0.b_1 b_2 b_3 \ldots$ *is the infinite series* $\dfrac{b_1}{2^1} + \dfrac{b_2}{2^2} + \dfrac{b_3}{2^3} + \cdots$.

**88.** Verify that the base-10 representation of the binary number
  $0.010101\ldots$ (which can also be written as $0.\overline{01}$) is $\dfrac{1}{3}$.

**89.** Computers can store only a finite number of digits and therefore numbers with nonterminating digits must be rounded or truncated before they can be used and stored by a computer.

  **a.** Find the base-10 representation of the binary number $0.\overline{0011}$.
  **b.** Suppose a computer rounds the binary number $0.\overline{0011}$ to $b = 0.0011001100110011$ before storing it. Find the approximate base-10 representation of $b$, rounding your answer to 8 digits to the right of the decimal place.

**90.** For what value of $r$ does $1 + r + r^2 + r^3 + \cdots = 10$?

**T 91.** For what value of $r$ does $\sum_{k=3}^{\infty} r^{2k} = 10$?

**92. Value of a series**

a. Evaluate the series

$$\sum_{k=1}^{\infty} \frac{3^k}{(3^{k+1} - 1)(3^k - 1)}.$$

(*Hint:* Find constants $b$ and $c$ so that

$$\frac{3^k}{(3^{k+1} - 1)(3^k - 1)} = \frac{b}{3^k - 1} + \frac{c}{3^{k+1} - 1}.)$$

b. For what values of $a$ does the series

$$\sum_{k=1}^{\infty} \frac{a^k}{(a^{k+1} - 1)(a^k - 1)}$$

converge, and in those cases, what is its value?

**93–95. Loans** *Suppose you borrow P dollars from a bank at a monthly interest rate of r and you make monthly payments of M dollars/month to pay off the loan (for example, if your borrow $30,000 at a rate of 1.5% per month, then P = 30,000 and r = 0.015). Each month, the bank first adds interest to the loan balance and then subtracts your monthly payment to determine the new balance on your loan. So if $A_0$ is your original loan balance and $A_n$ equals your loan balance after n payments, then your balance after n payments is given by the recursive formula $A_n = A_{n-1}(1 + r) - M, A_0 = P$.*

**93.** Use the recursive formula and geometric sums to obtain the explicit formula $A_n = \left( A_0 - \dfrac{M}{r} \right)(1 + r)^n + \dfrac{M}{r}.$

**T 94. Car loan** Suppose you borrow $20,000 for a new car at a monthly interest rate of 0.75%. If you make payments of $600/month, after how many months will the loan balance be zero?

**T 95. House loan** Suppose you take out a home mortgage for $180,000 at a monthly interest rate of 0.5%. If you make payments of $1000/month, after how many months will the loan balance be zero?

## Explorations and Challenges

**96. Properties proof** Use the ideas in the proof of Property 1 of Theorem 10.8 to prove Property 3 of Theorem 10.8.

**97. Property of divergent series** Prove Property 2 of Theorem 10.8: If $\Sigma a_k$ diverges, then $\Sigma c a_k$ also diverges, for any real number $c \neq 0$.

**98. Zeno's paradox** The Greek philosopher Zeno of Elea (who lived about 450 B.C.) invented many paradoxes, the most famous of which tells of a race between the swift warrior Achilles and a tortoise. Zeno argued as follows.

*The slower when running will never be overtaken by the quicker; for that which is pursuing must first reach the point from which that which is fleeing started, so that the slower must necessarily always be some distance ahead.*

In other words, giving the tortoise a head start ensures that Achilles will never overtake the tortoise because every time Achilles reaches the point where the tortoise was, the tortoise has moved ahead. Resolve this paradox by assuming Achilles gives the tortoise a 1-mi head start and runs 5 mi/hr to the tortoise's 1 mi/hr. How far does Achilles run before he overtakes the tortoise, and how long does it take?

**99. Archimedes' quadrature of the parabola** The Greeks solved several calculus problems almost 2000 years before the discovery of calculus. One example is Archimedes' calculation of the area

of the region $R$ bounded by a segment of a parabola, which he did using the "method of exhaustion." As shown in the figure, the idea was to fill $R$ with an infinite sequence of triangles. Archimedes began with an isosceles triangle inscribed in the parabola, with area $A_1$, and proceeded in stages, with the number of new triangles doubling at each stage. He was able to show (the key to the solution) that at each stage, the area of a new triangle is 1/8 of the area of a triangle at the previous stage; for example, $A_2 = 1/8 \, A_1$, and so forth. Show, as Archimedes did, that the area of $R$ is 4/3 times the area of $A_1$.

**100. Double glass** An insulated window consists of two parallel panes of glass with a small spacing between them. Suppose each pane reflects a fraction $p$ of the incoming light and transmits the remaining light. Considering all reflections of light between the panes, what fraction of the incoming light is ultimately transmitted by the window? Assume the amount of incoming light is 1.

**101. Snowflake island fractal** The fractal called the *snowflake island* (or *Koch island*) is constructed as follows: Let $I_0$ be an equilateral triangle with sides of length 1. The figure $I_1$ is obtained by replacing the middle third of each side of $I_0$ with a new outward equilateral triangle with sides of length 1/3 (see figure). The process is repeated, where $I_{n+1}$ is obtained by replacing the middle third of each side of $I_n$ with a new outward equilateral triangle with sides of length $\dfrac{1}{3^{n+1}}$. The limiting figure as $n \to \infty$ is called the snowflake island.

a. Let $L_n$ be the perimeter of $I_n$. Show that $\lim_{n \to \infty} L_n = \infty$.

b. Let $A_n$ be the area of $I_n$. Find $\lim_{n \to \infty} A_n$. It exists!

**102. Bubbles** Imagine a stack of hemispherical soap bubbles with decreasing radii $r_1 = 1, r_2, r_3, \ldots$ (see figure). Let $h_n$ be the distance between the diameters of bubble $n$ and bubble $n + 1$, and let $H_n$ be the total height of the stack with $n$ bubbles.

**a.** Use the Pythagorean theorem to show that in a stack with $n$ bubbles, $h_1^2 = r_1^2 - r_2^2, h_2^2 = r_2^2 - r_3^2$, and so forth. Note that for the last bubble $h_n = r_n$.

**b.** Use part (a) to show that the height of a stack with $n$ bubbles is

$$H_n = \sqrt{r_1^2 - r_2^2} + \sqrt{r_2^2 - r_3^2} + \cdots + \sqrt{r_{n-1}^2 - r_n^2} + r_n.$$

**c.** The height of a stack of bubbles depends on how the radii decrease. Suppose $r_1 = 1, r_2 = a, r_3 = a^2, \ldots, r_n = a^{n-1}$, where $0 < a < 1$ is a fixed real number. In terms of $a$, find the height $H_n$ of a stack with $n$ bubbles.

**d.** Suppose the stack in part (c) is extended indefinitely ($n \to \infty$). In terms of $a$, how high would the stack be?

**e.** Challenge problem: Fix $n$ and determine the sequence of radii $r_1, r_2, r_3, \ldots, r_n$ that maximizes $H_n$, the height of the stack with $n$ bubbles.

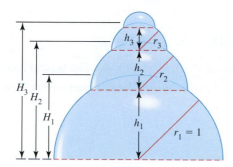

**103. Remainder term** Consider the geometric series $S = \sum\limits_{k=0}^{\infty} r^k$, which has the value $\dfrac{1}{1-r}$ provided $|r| < 1$. Let $S_n = \sum\limits_{k=0}^{n-1} r^k = \dfrac{1-r^n}{1-r}$ be the sum of the first $n$ terms. The magnitude of the remainder $R_n$ is the error in approximating $S$ by $S_n$. Show that

$$R_n = S - S_n = \frac{r^n}{1-r}.$$

**104–107. Comparing remainder terms** *Use Exercise 103 to determine how many terms of each series are needed so that the partial sum is within $10^{-6}$ of the value of the series (that is, to ensure $|R_n| < 10^{-6}$).*

**104. a.** $\sum\limits_{k=0}^{\infty} 0.6^k$      **b.** $\sum\limits_{k=0}^{\infty} 0.15^k$

**105. a.** $\sum\limits_{k=0}^{\infty} (-0.8)^k$      **b.** $\sum\limits_{k=0}^{\infty} 0.2^k$

**106. a.** $\sum\limits_{k=0}^{\infty} 0.72^k$      **b.** $\sum\limits_{k=0}^{\infty} (-0.25)^k$

**107. a.** $\sum\limits_{k=0}^{\infty} \left(\dfrac{1}{\pi}\right)^k$      **b.** $\sum\limits_{k=0}^{\infty} \left(\dfrac{1}{e}\right)^k$

**108. Function defined by a series** Suppose a function $f$ is defined by the geometric series $f(x) = \sum\limits_{k=0}^{\infty} x^k$.

**a.** Evaluate $f(0), f(0.2), f(0.5), f(1)$, and $f(1.5)$, if possible.

**b.** What is the domain of $f$?

**109. Function defined by a series** Suppose a function $f$ is defined by the geometric series $f(x) = \sum\limits_{k=0}^{\infty} (-1)^k x^k$.

**a.** Evaluate $f(0), f(0.2), f(0.5), f(1)$, and $f(1.5)$, if possible.

**b.** What is the domain of $f$?

**110. Function defined by a series** Suppose a function $f$ is defined by the geometric series $f(x) = \sum\limits_{k=0}^{\infty} x^{2k}$.

**a.** Evaluate $f(0), f(0.2), f(0.5), f(1)$, and $f(1.5)$, if possible.

**b.** What is the domain of $f$?

**111. Series in an equation** For what values of $x$ does the geometric series

$$f(x) = \sum_{k=0}^{\infty} \left(\frac{1}{1+x}\right)^k$$

converge? Solve $f(x) = 3$.

**QUICK CHECK ANSWERS**

**1.** $S_n = b_0 + b_1 + \cdots + b_{n-1}$    **2.** b and c    **3.** Using the formula, the values are $\frac{3}{2}$ and $\frac{7}{8}$.    **4.** 1    **5.** The first converges because $|r| = 0.2 < 1$; the second diverges because $|r| = 2 > 1$.    **6.** Removing a finite number of terms does not change whether the series converges. It generally changes the value of the series.    **7.** Given the $n$th term of the sequence of partial sums $S_n$, the next term is obtained by adding a positive number. So $S_{n+1} > S_n$, which means the sequence is increasing. ◄

# 10.4 The Divergence and Integral Tests

With geometric series and telescoping series, the sequence of partial sums can be found and its limit can be evaluated (when it exists). Unfortunately, it is difficult or impossible to find an explicit formula for the sequence of partial sums for most infinite series. Therefore, it is difficult to obtain the exact value of most convergent series.

In light of these observations, we now shift our focus and ask a simple *yes* or *no* question: Given an infinite series, does it converge? If the answer is *no*, the series diverges and there are no more questions to ask. If the answer is *yes*, the series converges and it may be possible to estimate its value.

## The Divergence Test

One of the simplest and most useful tests determines whether an infinite series *diverges*. Though our focus in this section and the next is on series with positive terms, the Divergence Test applies to series with arbitrary terms.

---

**THEOREM 10.9  Divergence Test**

If $\sum a_k$ converges, then $\lim_{k \to \infty} a_k = 0$. Equivalently, if $\lim_{k \to \infty} a_k \neq 0$, then the series diverges.

---

*Important note:* Theorem 10.9 cannot be used to conclude that a series converges.

**Proof:** Let $\{S_n\}$ be the sequence of partial sums for the series $\sum a_k$. Assuming the series converges, it has a finite value, call it $S$, where

$$S = \lim_{n \to \infty} S_n = \lim_{n \to \infty} S_{n-1}.$$

Note that $S_n - S_{n-1} = a_n$. Therefore,

$$\lim_{n \to \infty} a_n = \lim_{n \to \infty} (S_n - S_{n-1}) = S - S = 0;$$

that is, $\lim_{n \to \infty} a_n = 0$ (which implies $\lim_{k \to \infty} a_k = 0$; **Figure 10.25**). The second part of the test follows immediately because it is the *contrapositive* of the first part (see margin note).  ◄

> ➤ If the statement *if p, then q* is true, then its contrapositive, *if (not q), then (not p)*, is also true. However its converse, *if q, then p*, is not necessarily true. Try it out on the true statement, *if I live in Paris, then I live in France.*

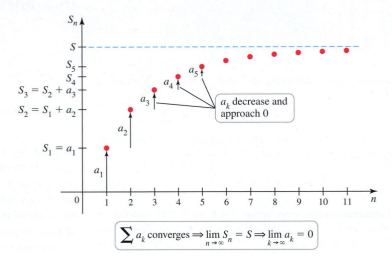

**Figure 10.25**

**EXAMPLE 1  Using the Divergence Test**  Determine whether the following series diverge or state that the Divergence Test is inconclusive.

**a.** $\displaystyle\sum_{k=0}^{\infty} \frac{k}{k+1}$    **b.** $\displaystyle\sum_{k=1}^{\infty} \frac{1 + 3^k}{2^k}$    **c.** $\displaystyle\sum_{k=1}^{\infty} \frac{1}{k}$    **d.** $\displaystyle\sum_{k=1}^{\infty} \frac{1}{k^2}$

**SOLUTION**  By the Divergence Test, if $\lim_{k \to \infty} a_k \neq 0$, then the series $\sum a_k$ diverges.

**a.** $\displaystyle\lim_{k \to \infty} a_k = \lim_{k \to \infty} \frac{k}{k+1} = 1 \neq 0$

The terms of the series do not approach zero, so the series diverges by the Divergence Test.

**b.** $\lim_{k \to \infty} a_k = \lim_{k \to \infty} \dfrac{1 + 3^k}{2^k}$

$$= \lim_{k \to \infty} \left( \underbrace{2^{-k}}_{\to 0} + \underbrace{\left( \dfrac{3}{2} \right)^k}_{\to \infty} \right) \quad \text{Simplify.}$$

$$= \infty$$

In this case, $\lim_{k \to \infty} a_k \neq 0$, so the corresponding series $\sum_{k=1}^{\infty} \dfrac{1 + 3^k}{2^k}$ diverges by the Divergence Test.

**c.** $\lim_{k \to \infty} a_k = \lim_{k \to \infty} \dfrac{1}{k} = 0$

In this case, the terms of the series approach zero, so the Divergence Test is inconclusive. Remember, the Divergence Test cannot be used to prove that a series converges.

**QUICK CHECK 1** Apply the Divergence Test to the geometric series $\sum r^k$. For what values of $r$ does the series diverge? ◄

**d.** $\lim_{k \to \infty} a_k = \lim_{k \to \infty} \dfrac{1}{k^2} = 0$

As in part (c), the terms of the series approach 0, so the Divergence Test is inconclusive.

*Related Exercises 9–10* ◄

To summarize: If the terms $a_k$ of a given series do *not* approach zero as $k \to \infty$, then the series diverges. Unfortunately, the test is easy to misuse. It's tempting to conclude that if the terms of the series approach zero, then the series converges. However, look again at the series in Examples 1c and 1d. Although it is true that $\lim_{k \to \infty} a_k = 0$ for both series, we will soon discover that one of them converges while the other diverges. We cannot tell which behavior to expect based only on the observation that $\lim_{k \to \infty} a_k = 0$.

## The Harmonic Series

We now look at an example with a surprising result. Consider the infinite series

$$\sum_{k=1}^{\infty} \frac{1}{k} = 1 + \frac{1}{2} + \frac{1}{3} + \frac{1}{4} + \frac{1}{5} + \cdots,$$

a famous series known as the **harmonic series**. Does it converge? As explained in Example 1c, this question cannot be answered by the Divergence Test, despite the fact that $\lim_{k \to \infty} \dfrac{1}{k} = 0$. Suppose instead you try to answer the convergence question by writing out the terms of the sequence of partial sums:

$$S_1 = 1, \qquad\qquad S_2 = 1 + \frac{1}{2} = \frac{3}{2},$$

$$S_3 = 1 + \frac{1}{2} + \frac{1}{3} = \frac{11}{6}, \qquad S_4 = 1 + \frac{1}{2} + \frac{1}{3} + \frac{1}{4} = \frac{25}{12},$$

and in general,

$$S_n = \sum_{k=1}^{n} \frac{1}{k} = 1 + \frac{1}{2} + \frac{1}{3} + \frac{1}{4} + \cdots + \frac{1}{n}.$$

There is no obvious pattern in this sequence, and in fact, no simple explicit formula for $S_n$ exists, so we analyze the sequence numerically. Have a look at the first 200 terms of the sequence of partial sums shown in **Figure 10.26**. What do you think—does the series converge? The terms of the sequence of partial sums increase, but at a decreasing rate. They could approach a limit or they could increase without bound.

Computing additional terms of the sequence of partial sums does not provide conclusive evidence. Table 10.3 shows that the sum of the first million terms is less than 15; the sum of the first $10^{40}$ terms—an unimaginably large number of terms—is less than 100. This is a case in which computation alone is not sufficient to determine whether a series converges. We need another way to determine whether the series converges.

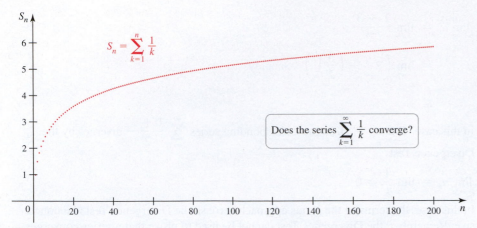

**Figure 10.26**

**Table 10.3**

| $n$ | $S_n$ | $n$ | $S_n$ |
|---|---|---|---|
| $10^3$ | $\approx 7.49$ | $10^{10}$ | $\approx 23.60$ |
| $10^4$ | $\approx 9.79$ | $10^{20}$ | $\approx 46.63$ |
| $10^5$ | $\approx 12.09$ | $10^{30}$ | $\approx 69.65$ |
| $10^6$ | $\approx 14.39$ | $10^{40}$ | $\approx 92.68$ |

Does the series $\displaystyle\sum_{k=1}^{\infty} \frac{1}{k}$ converge?

**Figure 10.27**

The sum of the areas of the rectangles is greater than the area under the curve $y = 1/x$ from $x = 1$ to $x = n + 1$:
$$S_n > \int_1^{n+1} \frac{dx}{x}.$$

➤ Recall that $\displaystyle\int \frac{dx}{x} = \ln |x| + C$. In Section 8.9, we showed that $\displaystyle\int_1^{\infty} \frac{dx}{x^p}$ diverges for $p \leq 1$. Therefore, $\displaystyle\int_1^{\infty} \frac{dx}{x}$ diverges.

Observe that the $n$th term of the sequence of partial sums,

$$S_n = \sum_{k=1}^{n} \frac{1}{k} = 1 + \frac{1}{2} + \frac{1}{3} + \frac{1}{4} + \cdots + \frac{1}{n},$$

is represented geometrically by a left Riemann sum of the function $y = \dfrac{1}{x}$ on the interval $[1, n + 1]$ (**Figure 10.27**). This fact follows by noticing that the areas of the rectangles, from left to right, are $1, \dfrac{1}{2}, \ldots,$ and $\dfrac{1}{n}$. Comparing the sum of the areas of these $n$ rectangles with the area under the curve, we see that $S_n > \displaystyle\int_1^{n+1} \frac{dx}{x}$. We know that $\displaystyle\int_1^{n+1} \frac{dx}{x} = \ln (n + 1)$ increases without bound as $n$ increases. Because $S_n$ exceeds $\displaystyle\int_1^{n+1} \frac{dx}{x}$, $S_n$ also increases without bound; therefore, $\displaystyle\lim_{n \to \infty} S_n = \infty$ and the harmonic series $\displaystyle\sum_{k=1}^{\infty} \frac{1}{k}$ diverges. This argument justifies the following theorem.

---

**THEOREM 10.10   Harmonic Series**

The harmonic series $\displaystyle\sum_{k=1}^{\infty} \frac{1}{k} = 1 + \frac{1}{2} + \frac{1}{3} + \frac{1}{4} + \frac{1}{5} + \cdots$ diverges—even though the terms of the series approach zero.

---

The ideas used to demonstrate that the harmonic series diverges are now used to prove a new and powerful convergence test. This test and those presented in Section 10.5 apply only to series with positive terms.

## The Integral Test

The fact that infinite series are sums and that integrals are limits of sums suggests a connection between series and integrals. The Integral Test exploits this connection.

---

**THEOREM 10.11   Integral Test**

Suppose $f$ is a continuous, positive, decreasing function, for $x \geq 1$, and let $a_k = f(k)$, for $k = 1, 2, 3, \ldots$. Then

$$\sum_{k=1}^{\infty} a_k \quad \text{and} \quad \int_1^{\infty} f(x)\, dx$$

either both converge or both diverge. In the case of convergence, the value of the integral is *not* equal to the value of the series.

---

➤ The Integral Test also applies if the terms of the series $a_k$ are decreasing for $k > N$ for some finite number $N > 1$. The proof can be modified to account for this situation.

**Proof:** By comparing the shaded regions in **Figure 10.28**, it follows that

$$\sum_{k=2}^{n} a_k < \int_{1}^{n} f(x)\, dx < \sum_{k=1}^{n-1} a_k. \tag{1}$$

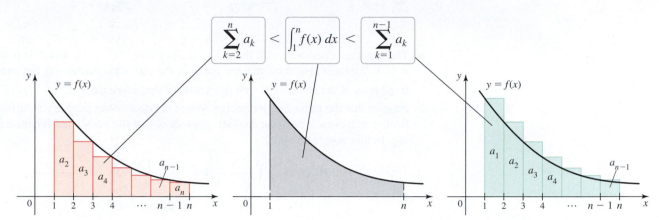

**Figure 10.28**

The proof must demonstrate two results: If the improper integral $\int_{1}^{\infty} f(x)\, dx$ has a finite value, then the infinite series converges, *and* if the infinite series converges, then the improper integral has a finite value. First suppose the improper integral $\int_{1}^{\infty} f(x)\, dx$ has a finite value, say $I$. We have

$$\sum_{k=1}^{n} a_k = a_1 + \sum_{k=2}^{n} a_k \qquad \text{Separate the first term of the series.}$$

$$< a_1 + \int_{1}^{n} f(x)\, dx \quad \text{Left inequality in expression (1)}$$

$$< a_1 + \int_{1}^{\infty} f(x)\, dx \quad f \text{ is positive, so } \int_{1}^{n} f(x)\, dx < \int_{1}^{\infty} f(x)\, dx.$$

$$= a_1 + I.$$

> In this proof, we rely twice on the Bounded Monotonic Sequence Theorem of Section 10.2: A bounded monotonic sequence converges.

This argument implies that the terms of the sequence of partial sums $S_n = \sum_{k=1}^{n} a_k$ are bounded above by $a_1 + I$. Because $\{S_n\}$ is also increasing (the series consists of positive terms), the sequence of partial sums converges, which means the series $\sum_{k=1}^{\infty} a_k$ converges (to a value less than or equal to $a_1 + I$).

Now suppose the infinite series $\sum_{k=1}^{\infty} a_k$ converges and has a value $S$. We have

$$\int_{1}^{n} f(x)\, dx < \sum_{k=1}^{n-1} a_k \quad \text{Right inequality in expression (1)}$$

$$< \sum_{k=1}^{\infty} a_k \quad \text{Terms } a_k \text{ are positive.}$$

$$= S. \qquad \text{Value of infinite series}$$

> An extended version of this proof can be used to show that, in fact,
> $$\int_{1}^{\infty} f(x)\, dx < \sum_{k=1}^{\infty} a_k \text{ (strict inequality)}$$
> in all cases.

We see that the sequence $\left\{ \int_{1}^{n} f(x)\, dx \right\}$ is increasing (because $f(x) > 0$) and bounded above by a fixed number $S$. Therefore, the improper integral $\int_{1}^{\infty} f(x)\, dx = \lim_{n \to \infty} \int_{1}^{n} f(x)\, dx$ has a finite value (less than or equal to $S$).

We have shown that if $\int_{1}^{\infty} f(x)\, dx$ is finite, then $\sum a_k$ converges, and vice versa. The same inequalities imply that $\int_{1}^{\infty} f(x)\, dx$ and $\sum a_k$ also diverge together. ◄

The Integral Test is used to determine *whether* a series converges or diverges. For this reason, adding or subtracting a few terms in the series *or* changing the lower limit of integration to another finite point does not change the outcome of the test. Therefore, the test depends on neither the lower index of the series nor the lower limit of the integral.

**EXAMPLE 2**  **Applying the Integral Test**  Determine whether the following series converge.

**a.** $\displaystyle\sum_{k=1}^{\infty} \frac{k}{k^2 + 1}$     **b.** $\displaystyle\sum_{k=3}^{\infty} \frac{1}{\sqrt{2k - 5}}$     **c.** $\displaystyle\sum_{k=0}^{\infty} \frac{1}{k^2 + 4}$

**SOLUTION**

**a.** The function associated with this series is $f(x) = x/(x^2 + 1)$, which is positive, for $x \geq 1$. We must also show that the terms of the series are decreasing beyond some fixed term of the series. The first few terms of the series are $\left\{\frac{1}{2}, \frac{2}{5}, \frac{3}{10}, \frac{4}{17}, \ldots\right\}$, and it appears that the terms are decreasing. When the decreasing property is difficult to confirm, one approach is to use derivatives to show that the associated function is decreasing. In this case, we have

$$f'(x) = \underbrace{\frac{d}{dx}\left(\frac{x}{x^2 + 1}\right) = \frac{x^2 + 1 - 2x^2}{(x^2 + 1)^2}}_{\text{Quotient Rule}} = \frac{1 - x^2}{(x^2 + 1)^2}.$$

For $x > 1$, $f'(x) < 0$, which implies that the function and the terms of the series are decreasing. The integral that determines convergence is

$$\int_1^{\infty} \frac{x}{x^2 + 1}\, dx = \lim_{b \to \infty} \int_1^b \frac{x}{x^2 + 1}\, dx \qquad \text{Definition of improper integral}$$

$$= \lim_{b \to \infty} \frac{1}{2} \ln\left(x^2 + 1\right)\Big|_1^b \qquad \text{Evaluate integral.}$$

$$= \frac{1}{2} \lim_{b \to \infty} \left(\ln\left(b^2 + 1\right) - \ln 2\right) \quad \text{Simplify.}$$

$$= \infty. \qquad \lim_{b \to \infty} \ln\left(b^2 + 1\right) = \infty$$

Because the integral diverges, the series diverges.

**b.** The Integral Test may be modified to accommodate initial indices other than $k = 1$. The terms of this series decrease, for $k \geq 3$. In this case, the relevant integral is

$$\int_3^{\infty} \frac{dx}{\sqrt{2x - 5}} = \lim_{b \to \infty} \int_3^b \frac{dx}{\sqrt{2x - 5}} \qquad \text{Definition of improper integral}$$

$$= \lim_{b \to \infty} \sqrt{2x - 5}\,\Big|_3^b \qquad \text{Evaluate integral.}$$

$$= \infty. \qquad \lim_{b \to \infty} \sqrt{2b - 5} = \infty$$

Because the integral diverges, the series also diverges.

**c.** The terms of the series are positive and decreasing, for $k \geq 0$. The relevant integral is

$$\int_0^{\infty} \frac{dx}{x^2 + 4} = \lim_{b \to \infty} \int_0^b \frac{dx}{x^2 + 4} \qquad \text{Definition of improper integral}$$

$$= \lim_{b \to \infty} \frac{1}{2} \tan^{-1}\frac{x}{2}\Big|_0^b \qquad \text{Evaluate integral.}$$

$$= \frac{1}{2} \underbrace{\lim_{b \to \infty} \tan^{-1}\frac{b}{2}}_{\frac{\pi}{2}} - \tan^{-1} 0 \quad \text{Simplify.}$$

$$= \frac{\pi}{4}. \qquad \tan^{-1} x \to \frac{\pi}{2}, \text{ as } x \to \infty.$$

Because the integral is finite (equivalently, it converges), the infinite series also converges $\left(\text{but not to } \dfrac{\pi}{4}\right)$.

*Related Exercises 17, 20* ◄

## The *p*-series

The Integral Test is used to prove Theorem 10.12, which addresses the convergence of an entire family of infinite series known as the *p-series*.

> **THEOREM 10.12    Convergence of the *p*-series**
>
> The ***p*-series** $\sum\limits_{k=1}^{\infty} \dfrac{1}{k^p}$ converges for $p > 1$ and diverges for $p \leq 1$.

**QUICK CHECK 2** Which of the following series are *p*-series, and which series converge?

a. $\sum\limits_{k=1}^{\infty} k^{-0.8}$    b. $\sum\limits_{k=1}^{\infty} 2^{-k}$    c. $\sum\limits_{k=10}^{\infty} k^{-4}$ ◄

**Proof:** To apply the Integral Test, observe that the terms of the given series are positive and decreasing, for $p > 0$. The function associated with the series is $f(x) = \dfrac{1}{x^p}$. The relevant integral is $\int_1^{\infty} \dfrac{dx}{x^p}$. Appealing to Example 2 in Section 8.9, recall that this improper integral converges for $p > 1$ and diverges for $p \leq 1$. Therefore, by the Integral Test, the *p*-series $\sum\limits_{k=1}^{\infty} \dfrac{1}{k^p}$ converges for $p > 1$ and diverges for $0 < p \leq 1$. For $p \leq 0$, the series diverges by the Divergence Test.     ◄

**EXAMPLE 3    Using the *p*-series test**  Determine whether the following series converge or diverge.

a. $\sum\limits_{k=1}^{\infty} k^{-3}$    b. $\sum\limits_{k=1}^{\infty} \dfrac{1}{\sqrt[4]{k^3}}$    c. $\sum\limits_{k=4}^{\infty} \dfrac{1}{(k-1)^2}$

**SOLUTION**

a. Because $\sum\limits_{k=1}^{\infty} k^{-3} = \sum\limits_{k=1}^{\infty} \dfrac{1}{k^3}$ is a *p*-series with $p = 3$, it converges by Theorem 10.12.

b. This series is a *p*-series with $p = \frac{3}{4}$. By Theorem 10.12, it diverges.

c. The series

$$\sum_{k=4}^{\infty} \frac{1}{(k-1)^2} = \sum_{k=3}^{\infty} \frac{1}{k^2} = \frac{1}{3^2} + \frac{1}{4^2} + \frac{1}{5^2} + \cdots$$

is a convergent *p*-series ($p = 2$) without the first two terms. Recall that adding or removing a finite number of terms does not affect the convergence of a series (Property 5, Theorem 10.8). Therefore, the given series converges.

*Related Exercises 23–24* ◄

## Estimating the Value of Infinite Series

The Integral Test is powerful in its own right, but it comes with an added bonus. It can be used to estimate the value of a convergent series with positive terms. We define the **remainder** to be the error in approximating a convergent series by the sum of its first *n* terms; that is,

$$R_n = \underbrace{\sum_{k=1}^{\infty} a_k}_{\substack{\text{value of} \\ \text{series}}} - \underbrace{\sum_{k=1}^{n} a_k}_{\substack{\text{approximation based} \\ \text{on first } n \text{ terms}}} = a_{n+1} + a_{n+2} + a_{n+3} + \cdots.$$

**QUICK CHECK 3** If $\sum a_k$ is a convergent series of positive terms, why is $R_n > 0$? ◄

The remainder consists of the *tail* of the series—those terms beyond $a_n$. For series with positive terms, the remainder is positive.

We now argue much as we did in the proof of the Integral Test. Let $f$ be a continuous, positive, decreasing function such that $f(k) = a_k$, for all relevant $k$. From **Figure 10.29**, we see that $\int_{n+1}^{\infty} f(x)\, dx < R_n$.

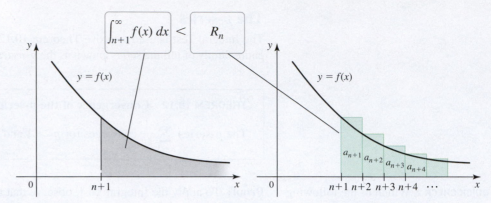

**Figure 10.29**

Similarly, **Figure 10.30** shows that $R_n < \int_n^\infty f(x)\, dx$.

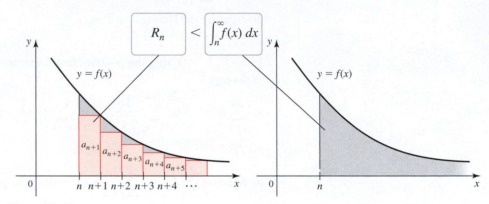

**Figure 10.30**

Combining these two inequalities, the remainder is squeezed between two integrals:

$$\int_{n+1}^\infty f(x)\, dx < R_n < \int_n^\infty f(x)\, dx. \qquad (2)$$

If the integrals can be evaluated, this result provides an estimate of the remainder.

There is, however, another equally useful way to express this result. Notice that the value of the series is

$$S = \sum_{k=1}^\infty a_k = \underbrace{\sum_{k=1}^n a_k}_{S_n} + R_n,$$

which is the sum of the first $n$ terms $S_n$ and the remainder $R_n$. Adding $S_n$ to each term of (2), we have

$$\underbrace{S_n + \int_{n+1}^\infty f(x)\, dx}_{L_n} < \underbrace{\sum_{k=1}^\infty a_k}_{S_n + R_n = S} < \underbrace{S_n + \int_n^\infty f(x)\, dx}_{U_n}.$$

These inequalities can be abbreviated as $L_n < S < U_n$, where $S$ is the exact value of the series, and $L_n$ and $U_n$ are lower and upper bounds for $S$, respectively. If the integrals in these bounds can be evaluated, it is straightforward to compute $S_n$ (by summing the first $n$ terms of the series) and to compute both $L_n$ and $U_n$.

---

**THEOREM 10.13   Estimating Series with Positive Terms**

Let $f$ be a continuous, positive, decreasing function, for $x \geq 1$, and let $a_k = f(k)$, for $k = 1, 2, 3, \ldots$. Let $S = \sum_{k=1}^{\infty} a_k$ be a convergent series and let $S_n = \sum_{k=1}^{n} a_k$ be the sum of the first $n$ terms of the series. The remainder $R_n = S - S_n$ satisfies

$$R_n < \int_n^{\infty} f(x)\, dx.$$

Furthermore, the exact value of the series is bounded as follows:

$$L_n = S_n + \int_{n+1}^{\infty} f(x)\, dx \ < \ \sum_{k=1}^{\infty} a_k \ < \ S_n + \int_n^{\infty} f(x)\, dx = U_n.$$

---

**EXAMPLE 4   Approximating a $p$-series**

**a.** How many terms of the convergent $p$-series $\sum_{k=1}^{\infty} \dfrac{1}{k^2}$ must be summed to obtain an approximation that is within $10^{-3}$ of the exact value of the series?

**b.** Find an approximation to the series using 50 terms of the series.

**SOLUTION** The function associated with this series is $f(x) = 1/x^2$.

**a.** Using the bound on the remainder, we have

$$R_n < \int_n^{\infty} f(x)\, dx = \int_n^{\infty} \frac{dx}{x^2} = \frac{1}{n}.$$

To ensure that $R_n < 10^{-3}$, we must choose $n$ so that $1/n < 10^{-3}$, which implies that $n > 1000$. In other words, we must sum at least 1001 terms of the series to be sure that the remainder is less than $10^{-3}$.

**b.** Using the bounds on the series, we have $L_n < S < U_n$, where $S$ is the exact value of the series, and

$$L_n = S_n + \int_{n+1}^{\infty} \frac{dx}{x^2} = S_n + \frac{1}{n+1} \quad \text{and} \quad U_n = S_n + \int_n^{\infty} \frac{dx}{x^2} = S_n + \frac{1}{n}.$$

> ▶ The values of $p$-series with even values of $p$ are generally known. For example, with $p = 2$, the series converges to $\pi^2/6$ (a proof is outlined in Exercise 68); with $p = 4$, the series converges to $\pi^4/90$. The values of $p$-series with odd values of $p$ are not known.

Therefore, the series is bounded as follows:

$$S_n + \frac{1}{n+1} < S < S_n + \frac{1}{n},$$

where $S_n$ is the sum of the first $n$ terms. Using a calculator to sum the first 50 terms of the series, we find that $S_{50} \approx 1.625133$. The exact value of the series is in the interval

$$S_{50} + \frac{1}{50+1} < S < S_{50} + \frac{1}{50},$$

or $1.644741 < S < 1.645133$. Taking the average of these two bounds as our approximation of $S$, we find that $S \approx 1.644937$. This estimate is better than simply using $S_{50}$. **Figure 10.31a** shows the lower and upper bounds, $L_n$ and $U_n$, respectively, for $n = 1, 2, \ldots, 50$. **Figure 10.31b** shows these bounds on an enlarged scale for $n = 50, 51, \ldots, 100$. These figures illustrate how the exact value of the series is squeezed into a narrowing interval as $n$ increases.

(a)    (b)

**Figure 10.31**

*Related Exercises 41, 43* ◀

# SECTION 10.4 EXERCISES

## Getting Started

1. If we know that $\lim_{k \to \infty} a_k = 1$, then what can we say about $\sum_{k=1}^{\infty} a_k$?

2. Is it true that if the terms of a series of positive terms decrease to zero, then the series converges? Explain using an example.

3. If we know that $\sum_{k=1}^{\infty} a_k = 10{,}000$, then what can we say about $\lim_{k \to \infty} a_k$?

4. For what values of $p$ does the series $\sum_{k=1}^{\infty} \dfrac{1}{k^p}$ converge? For what values of $p$ does it diverge?

5. For what values of $p$ does the series $\sum_{k=10}^{\infty} \dfrac{1}{k^p}$ converge (initial index is 10)? For what values of $p$ does it diverge?

6. Explain why the sequence of partial sums for a series of positive terms is an increasing sequence.

7. Define the remainder of an infinite series.

8. If a series of positive terms converges, does it follow that the remainder $R_n$ must decrease to zero as $n \to \infty$? Explain.

## Practice Exercises

**9–16. Divergence Test** *Use the Divergence Test to determine whether the following series diverge or state that the test is inconclusive.*

9. $\sum_{k=0}^{\infty} \dfrac{k}{2k+1}$

10. $\sum_{k=1}^{\infty} \dfrac{k}{k^2+1}$

11. $\sum_{k=0}^{\infty} \dfrac{1}{1000+k}$

12. $\sum_{k=1}^{\infty} \dfrac{k^3}{k^3+1}$

13. $\sum_{k=2}^{\infty} \dfrac{k}{\ln k}$

14. $\sum_{k=1}^{\infty} \dfrac{k^2}{2^k}$

15. $\sum_{k=2}^{\infty} \dfrac{\sqrt{k}}{\ln^{10} k}$

16. $\sum_{k=1}^{\infty} \dfrac{k^3}{k!}$

**17–22. Integral Test** *Use the Integral Test to determine whether the following series converge after showing that the conditions of the Integral Test are satisfied.*

17. $\sum_{k=0}^{\infty} \dfrac{1}{\sqrt{k+8}}$

18. $\sum_{k=1}^{\infty} \dfrac{1}{(2k+4)^2}$

19. $\sum_{k=1}^{\infty} \dfrac{1}{\sqrt[3]{5k+3}}$

20. $\sum_{k=1}^{\infty} \dfrac{e^k}{1+e^{2k}}$

21. $\sum_{k=1}^{\infty} ke^{-2k^2}$

22. $\sum_{k=2}^{\infty} \dfrac{1}{k(\ln k)^2}$

**23–38. Divergence, Integral, and p-series Tests** *Use the Divergence Test, the Integral Test, or the p-series test to determine whether the following series converge.*

23. $\sum_{k=1}^{\infty} k^{-1/5}$

24. $\sum_{k=1}^{\infty} \dfrac{1}{\sqrt{k^3}}$

25. $\sum_{k=1}^{\infty} \dfrac{k^3}{e^{k^4}}$

26. $\sum_{k=1}^{\infty} \dfrac{k^2}{k^3+5}$

27. $\sum_{k=1}^{\infty} k^{1/k}$

28. $\sum_{k=1}^{\infty} \dfrac{\sqrt{k^2+1}}{k}$

29. $\sum_{k=1}^{\infty} \dfrac{1}{k^{10}}$

30. $\sum_{k=2}^{\infty} \dfrac{k^e}{k^\pi}$

31. $\sum_{k=3}^{\infty} \dfrac{1}{(k-2)^4}$

32. $\sum_{k=1}^{\infty} 2k^{-3/2}$

33. $\sum_{k=1}^{\infty} \dfrac{k}{e^k}$

34. $\sum_{k=3}^{\infty} \dfrac{1}{k(\ln k)\ln \ln k}$

35. $\sum_{k=1}^{\infty} \left(\dfrac{k}{k+10}\right)^k$

36. $\sum_{k=1}^{\infty} \dfrac{k}{(k^2+1)^3}$

37. $\sum_{k=1}^{\infty} \dfrac{1}{\sqrt[3]{k}}$

38. $\sum_{k=1}^{\infty} \dfrac{1}{\sqrt[3]{27k^2}}$

**T 39–40. Lower and upper bounds of a series** *For each convergent series and given value of n, use Theorem 10.13 to complete the following.*

a. *Use $S_n$ to estimate the sum of the series.*
b. *Find an upper bound for the remainder $R_n$.*
c. *Find lower and upper bounds ($L_n$ and $U_n$, respectively) for the exact value of the series.*

**39.** $\displaystyle\sum_{k=1}^{\infty} \frac{1}{k^7}; n = 2$  **40.** $\displaystyle\sum_{k=1}^{\infty} \frac{1}{k^3}; n = 5$

**T 41–44. Remainders and estimates** *Consider the following convergent series.*

a. *Find an upper bound for the remainder in terms of n.*
b. *Find how many terms are needed to ensure that the remainder is less than $10^{-3}$.*
c. *Find lower and upper bounds ($L_n$ and $U_n$, respectively) on the exact value of the series.*
d. *Find an interval in which the value of the series must lie if you approximate it using ten terms of the series.*

**41.** $\displaystyle\sum_{k=1}^{\infty} \frac{1}{k^6}$  **42.** $\displaystyle\sum_{k=2}^{\infty} \frac{1}{k(\ln k)^2}$

**43.** $\displaystyle\sum_{k=1}^{\infty} \frac{1}{3^k}$  **44.** $\displaystyle\sum_{k=1}^{\infty} ke^{-k^2}$

**T 45.** Estimate the series $\displaystyle\sum_{k=1}^{\infty} \frac{1}{k^7}$ to within $10^{-4}$ of its exact value.

**T 46.** Estimate the series $\displaystyle\sum_{k=1}^{\infty} \frac{1}{(3k + 2)^2}$ to within $10^{-3}$ of its exact value.

**47. Explain why or why not** Determine whether the following statements are true and give an explanation or counterexample.

a. The sum $\displaystyle\sum_{k=1}^{\infty} \frac{1}{3^k}$ is a *p*-series.

b. The sum $\displaystyle\sum_{k=3}^{\infty} \frac{1}{\sqrt{k - 2}}$ is a *p*-series.

c. Suppose $f$ is a continuous, positive, decreasing function, for $x \geq 1$, and $a_k = f(k)$, for $k = 1, 2, 3, \ldots$. If $\displaystyle\sum_{k=1}^{\infty} a_k$ converges to $L$, then $\displaystyle\int_1^{\infty} f(x)\,dx$ converges to $L$.

d. Every partial sum $S_n$ of the series $\displaystyle\sum_{k=1}^{\infty} \frac{1}{k^2}$ underestimates the exact value of $\displaystyle\sum_{k=1}^{\infty} \frac{1}{k^2}$.

e. If $\sum k^{-p}$ converges, then $\sum k^{-p+0.001}$ converges.

f. If $\displaystyle\lim_{k\to\infty} a_k = 0$, then $\sum a_k$ converges.

**48–63. Choose your test** *Determine whether the following series converge or diverge using the properties and tests introduced in Sections 10.3 and 10.4.*

**48.** $\displaystyle\sum_{k=1}^{\infty} \ln\left(\frac{2k^6}{1 + k^6}\right)$  **49.** $\displaystyle\sum_{k=2}^{\infty} \frac{1}{e^k}$

**50.** $\displaystyle\sum_{k=1}^{\infty} \sqrt{\frac{k + 1}{k}}$  **51.** $\displaystyle\sum_{k=1}^{\infty} \frac{1}{(3k + 1)(3k + 4)}$

**52.** $\displaystyle\sum_{k=0}^{\infty} \frac{10}{k^2 + 9}$  **53.** $\displaystyle\sum_{k=1}^{\infty} \frac{k}{\sqrt{k^2 + 1}}$

**54.** $\displaystyle\sum_{k=1}^{\infty} \frac{2^k + 3^k}{4^k}$  **55.** $\displaystyle\sum_{k=3}^{\infty} \frac{4}{k\sqrt{\ln k}}$

**56.** $\displaystyle\sum_{k=1}^{\infty} \frac{1}{k^2 + 7k + 12}$  **57.** $\displaystyle\sum_{k=1}^{\infty} \left(\frac{5}{6}\right)^{-k}$

**58.** $\displaystyle\sum_{k=1}^{\infty} \left(\frac{k + 6}{k}\right)^k$  **59.** $\displaystyle\sum_{k=4}^{\infty} \frac{3}{(k - 3)^4}$

**60.** $\displaystyle\sum_{k=1}^{\infty} \frac{3^k}{k^2 + 1}$  **61.** $\displaystyle\frac{2}{4^2} + \frac{2}{5^2} + \frac{2}{6^2} + \cdots$

**62.** $\displaystyle\frac{1}{3} + \frac{1}{5} + \frac{1}{7} + \cdots$  **63.** $\displaystyle\sum_{k=1}^{\infty} \frac{3^{k+2}}{5^k}$

**64. Log *p*-series** Consider the series $\displaystyle\sum_{k=2}^{\infty} \frac{1}{k(\ln k)^p}$, where $p$ is a real number.

a. Use the Integral Test to determine the values of $p$ for which this series converges.
b. Does this series converge faster for $p = 2$ or $p = 3$? Explain.

**65. Loglog *p*-series** Consider the series $\displaystyle\sum_{k=2}^{\infty} \frac{1}{k(\ln k)(\ln \ln k)^p}$, where $p$ is a real number.

a. For what values of $p$ does this series converge?
b. Which of the following series converges faster? Explain.

$$\sum_{k=2}^{\infty} \frac{1}{k(\ln k)^2} \quad \text{or} \quad \sum_{k=3}^{\infty} \frac{1}{k(\ln k)(\ln \ln k)^2}?$$

## Explorations and Challenges

**66. Prime numbers** The prime numbers are those positive integers that are divisible by only 1 and themselves (for example, 2, 3, 5, 7, 11, 13, ... ). A celebrated theorem states that the sequence of prime numbers $\{p_k\}$ satisfies $\displaystyle\lim_{k\to\infty} \frac{p_k}{k \ln k} = 1$. Show that $\displaystyle\sum_{k=2}^{\infty} \frac{1}{k \ln k}$ diverges, which implies that the series $\displaystyle\sum_{k=1}^{\infty} \frac{1}{p_k}$ diverges.

**T 67. The zeta function** The Riemann zeta function is the subject of extensive research and is associated with several renowned unsolved problems. It is defined by $\displaystyle\zeta(x) = \sum_{k=1}^{\infty} \frac{1}{k^x}$. When $x$ is a real number, the zeta function becomes a *p*-series. For even positive integers $p$, the value of $\zeta(p)$ is known exactly. For example,

$$\sum_{k=1}^{\infty} \frac{1}{k^2} = \frac{\pi^2}{6}, \quad \sum_{k=1}^{\infty} \frac{1}{k^4} = \frac{\pi^4}{90}, \quad \text{and} \quad \sum_{k=1}^{\infty} \frac{1}{k^6} = \frac{\pi^6}{945}, \ldots.$$

Use the estimation techniques described in the text to approximate $\zeta(3)$ and $\zeta(5)$ (whose values are not known exactly) with a remainder less than $10^{-3}$.

**68. Showing that** $\displaystyle\sum_{k=1}^{\infty} \frac{1}{k^2} = \frac{\pi^2}{6}$ In 1734, Leonhard Euler informally proved that $\displaystyle\sum_{k=1}^{\infty} \frac{1}{k^2} = \frac{\pi^2}{6}$. An elegant proof is outlined here that uses the inequality

$$\cot^2 x < \frac{1}{x^2} < 1 + \cot^2 x \ \left(\text{provided } 0 < x < \frac{\pi}{2}\right)$$

and the identity

$$\sum_{k=1}^{n}\cot^2 k\theta = \frac{n(2n-1)}{3}, \text{ for } n = 1, 2, 3, \ldots, \text{ where } \theta = \frac{\pi}{2n+1}.$$

**a.** Show that $\displaystyle\sum_{k=1}^{n}\cot^2 k\theta < \frac{1}{\theta^2}\sum_{k=1}^{n}\frac{1}{k^2} < n + \sum_{k=1}^{n}\cot^2 k\theta.$

**b.** Use the inequality in part (a) to show that

$$\frac{n(2n-1)\pi^2}{3(2n+1)^2} < \sum_{k=1}^{n}\frac{1}{k^2} < \frac{n(2n+2)\pi^2}{3(2n+1)^2}.$$

**c.** Use the Squeeze Theorem to conclude that $\displaystyle\sum_{k=1}^{\infty}\frac{1}{k^2} = \frac{\pi^2}{6}.$

(*Source: The College Mathematics Journal, 24, 5, Nov 1993*)

**69. Reciprocals of odd squares** Assume $\displaystyle\sum_{k=1}^{\infty}\frac{1}{k^2} = \frac{\pi^2}{6}$ (Exercises 67 and 68) and the terms of this series may be rearranged without changing the value of the series. Determine the sum of the reciprocals of the squares of the odd positive integers.

**70. Gabriel's wedding cake** Consider a wedding cake of infinite height, each layer of which is a right circular cylinder of height 1. The bottom layer of the cake has a radius of 1, the second layer has a radius of $1/2$, the third layer has a radius of $1/3$, and the $n$th layer has a radius of $1/n$ (see figure).

**a.** Find the area of the lateral (vertical) sides of the wedding cake.
**b.** Determine the volume of the cake. (*Hint:* Use the result of Exercise 68.)
**c.** Comment on your answers to parts (a) and (b).

(*Source: The College Mathematics Journal, 30, 1, Jan 1999*)

**71. A divergence proof** Give an argument similar to that given in the text for the harmonic series to show that $\displaystyle\sum_{k=1}^{\infty}\frac{1}{\sqrt{k}}$ diverges.

**T 72. Shifted $p$-series** Consider the sequence $\{F_n\}$ defined by

$$F_n = \sum_{k=1}^{\infty}\frac{1}{k(k+n)},$$

for $n = 0, 1, 2, \ldots$. When $n = 0$, the series is a $p$-series, and we have $F_0 = \dfrac{\pi^2}{6}$ (Exercises 67 and 68).

**a.** Explain why $\{F_n\}$ is a decreasing sequence.
**b.** Plot $\{F_n\}$, for $n = 1, 2, \ldots, 20$.
**c.** Based on your experiments, make a conjecture about $\lim_{n\to\infty} F_n$.

**73. A sequence of sums** Consider the sequence $\{x_n\}$ defined for $n = 1, 2, 3, \ldots$ by

$$x_n = \sum_{k=n+1}^{2n}\frac{1}{k} = \frac{1}{n+1} + \frac{1}{n+2} + \cdots + \frac{1}{2n}.$$

**a.** Write out the terms $x_1, x_2, x_3$.
**b.** Show that $\dfrac{1}{2} \le x_n < 1$, for $n = 1, 2, 3, \ldots$.
**c.** Show that $x_n$ is the right Riemann sum for $\displaystyle\int_1^2 \frac{dx}{x}$ using $n$ subintervals.
**d.** Conclude that $\lim_{n\to\infty} x_n = \ln 2$.

**T 74. The harmonic series and Euler's constant**
**a.** Sketch the function $f(x) = 1/x$ on the interval $[1, n+1]$, where $n$ is a positive integer. Use this graph to verify that

$$\ln(n+1) < 1 + \frac{1}{2} + \frac{1}{3} + \cdots + \frac{1}{n} < 1 + \ln n.$$

**b.** Let $S_n$ be the sum of the first $n$ terms of the harmonic series, so part (a) says $\ln(n+1) < S_n < 1 + \ln n$. Define the new sequence $\{E_n\}$ by

$$E_n = S_n - \ln(n+1), \quad \text{for } n = 1, 2, 3, \ldots.$$

Show that $E_n > 0$ for $n = 1, 2, 3, \ldots$.
**c.** Using a figure similar to that used in part (a), show that

$$\frac{1}{n+1} > \ln(n+2) - \ln(n+1).$$

**d.** Use parts (a) and (c) to show that $\{E_n\}$ is an increasing sequence ($E_{n+1} > E_n$).
**e.** Use part (a) to show that $\{E_n\}$ is bounded above by 1.
**f.** Conclude from parts (d) and (e) that $\{E_n\}$ has a limit less than or equal to 1. This limit is known as **Euler's constant** and is denoted $\gamma$ (the Greek lowercase gamma).
**g.** By computing terms of $\{E_n\}$, estimate the value of $\gamma$ and compare it to the value $\gamma \approx 0.5772$. (It has been conjectured that $\gamma$ is irrational.)
**h.** The preceding arguments show that the sum of the first $n$ terms of the harmonic series satisfy $S_n \approx 0.5772 + \ln(n+1)$. How many terms must be summed for the sum to exceed 10?

**75. The harmonic series and the Fibonacci sequence** The Fibonacci sequence $\{1, 1, 2, 3, 5, 8, 13, \ldots\}$ is generated by the recurrence relation

$$f_{n+1} = f_n + f_{n-1}, \text{ for } n = 1, 2, 3, \ldots, \text{ where } f_0 = 1, f_1 = 1.$$

**a.** It can be shown that the sequence of ratios of successive terms of the sequence $\left\{\dfrac{f_{n+1}}{f_n}\right\}$ has a limit $\varphi$. Divide both sides of the recurrence relation by $f_n$, take the limit as $n \to \infty$, and show that $\varphi = \lim_{n\to\infty}\dfrac{f_{n+1}}{f_n} = \dfrac{1+\sqrt{5}}{2} \approx 1.618$.

**b.** Show that $\lim_{n\to\infty}\dfrac{f_{n-1}}{f_{n+1}} = 1 - \dfrac{1}{\varphi} \approx 0.382$.

**c.** Now consider the harmonic series and group terms as follows:

$$\sum_{k=1}^{\infty}\frac{1}{k} = 1 + \frac{1}{2} + \frac{1}{3} + \underbrace{\left(\frac{1}{4} + \frac{1}{5}\right)}_{2 \text{ terms}} + \underbrace{\left(\frac{1}{6} + \frac{1}{7} + \frac{1}{8}\right)}_{3 \text{ terms}}$$

$$+ \underbrace{\left(\frac{1}{9} + \cdots + \frac{1}{13}\right)}_{5 \text{ terms}} + \cdots.$$

With the Fibonacci sequence in mind, show that

$$\sum_{k=1}^{\infty} \frac{1}{k} \geq 1 + \frac{1}{2} + \frac{1}{3} + \frac{2}{5} + \frac{3}{8} + \frac{5}{13} + \cdots = 1 + \sum_{k=1}^{\infty} \frac{f_{k-1}}{f_{k+1}}.$$

**d.** Use part (b) to conclude that the harmonic series diverges.

(*Source: The College Mathematics Journal, 43, May 2012*)

**76. Stacking dominoes** Consider a set of identical dominoes that are 2 in long. The dominoes are stacked on top of each other with their long edges aligned so that each domino overhangs the one beneath it *as far as possible* (see figure).

**a.** If there are $n$ dominoes in the stack, what is the *greatest* distance that the top domino can be made to overhang the bottom domino? (*Hint:* Put the $n$th domino beneath the previous $n - 1$ dominoes.)

**b.** If we allow for infinitely many dominoes in the stack, what is the greatest distance that the top domino can be made to overhang the bottom domino?

**QUICK CHECK ANSWERS**

**1.** The series diverges for $|r| \geq 1$.   **2. a.** Divergent $p$-series
**b.** Convergent geometric series   **c.** Convergent $p$-series
**3.** The remainder is $R_n = a_{n+1} + a_{n+2} + \cdots$, which consists of positive numbers. ◀

# 10.5 Comparison Tests

The Integral Test from the previous section is easily applied to a series such as $\displaystyle\sum_{k=1}^{\infty} \frac{4}{4 + k^2}$ because $\displaystyle\int_1^{\infty} \frac{4}{4 + x^2}\, dx$ is a standard integral. However, using the Integral Test to determine whether $\displaystyle\sum_{k=1}^{\infty} \frac{5k^4 - 2k^2 + 3}{2k^6 - k + 5}$ converges is decidedly more difficult. For this reason, we develop two additional convergence tests, called *comparison tests*, and we use one of them in Example 2a to investigate the convergence of $\displaystyle\sum_{k=1}^{\infty} \frac{5k^4 - 2k^2 + 3}{2k^6 - k + 5}$. As with the Integral Test, both comparison tests apply only to series with positive terms.

## The Comparison Test

Comparison tests use known series to test the convergence of unknown series. The first test is the Direct Comparison Test or simply the Comparison Test.

> ➤ Whether a series converges depends on the behavior of terms in the tail (large values of the index). So the inequalities $a_k \leq b_k$ and $b_k \leq a_k$ need not hold for all terms of the series. They must hold for all $k > N$ for some positive integer $N$.

---

**THEOREM 10.14   Comparison Test**

Let $\sum a_k$ and $\sum b_k$ be series with positive terms.

**1.** If $a_k \leq b_k$ and $\sum b_k$ converges, then $\sum a_k$ converges.
**2.** If $b_k \leq a_k$ and $\sum b_k$ diverges, then $\sum a_k$ diverges.

---

**Proof:** Assume $\sum b_k$ converges, which means that $\sum b_k$ has a finite value $B$. The sequence of partial sums for $\sum a_k$ satisfies

$$S_n = \sum_{k=1}^{n} a_k \leq \sum_{k=1}^{n} b_k \qquad a_k \leq b_k$$

$$< \sum_{k=1}^{\infty} b_k \qquad \text{Positive terms are added to a finite sum.}$$

$$= B. \qquad \text{Value of series}$$

Therefore, the sequence of partial sums for $\sum a_k$ is increasing and bounded above by $B$. By the Bounded Monotonic Sequence Theorem (Theorem 10.5), the sequence of partial sums of $\sum a_k$ has a limit, which implies that $\sum a_k$ converges. The second case of the theorem is proved in a similar way. ◀

The Comparison Test can be illustrated with graphs of sequences of partial sums. Consider the series

$$\sum_{k=1}^{\infty} a_k = \sum_{k=1}^{\infty} \frac{1}{k^2 + 10} \quad \text{and} \quad \sum_{k=1}^{\infty} b_k = \sum_{k=1}^{\infty} \frac{1}{k^2}.$$

Because $\frac{1}{k^2 + 10} < \frac{1}{k^2}$, it follows that $a_k < b_k$, for $k \geq 1$. Furthermore, $\sum b_k$ is a convergent $p$-series. By the Comparison Test, we conclude that $\sum a_k$ also converges (**Figure 10.32**). The second case of the Comparison Test is illustrated with the series

$$\sum_{k=4}^{\infty} a_k = \sum_{k=4}^{\infty} \frac{1}{\sqrt{k - 3}} \quad \text{and} \quad \sum_{k=4}^{\infty} b_k = \sum_{k=4}^{\infty} \frac{1}{\sqrt{k}}.$$

Now $\frac{1}{\sqrt{k}} < \frac{1}{\sqrt{k - 3}}$, for $k \geq 4$. Therefore, $b_k < a_k$, for $k \geq 4$. Because $\sum b_k$ is a divergent $p$-series, by the Comparison Test, $\sum a_k$ also diverges. **Figure 10.33** shows that the sequence of partial sums for $\sum a_k$ lies above the sequence of partial sums for $\sum b_k$. Because the sequence of partial sums for $\sum b_k$ diverges, the sequence of partial sums for $\sum a_k$ also diverges.

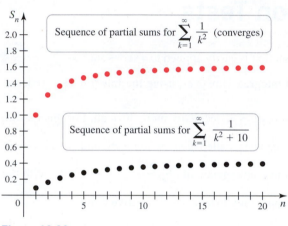

**Figure 10.32**

**Figure 10.33**

The key in using the Comparison Test is finding an appropriate comparison series. Plenty of practice will enable you to spot patterns and choose good comparison series.

**EXAMPLE 1**    **Using the Comparison Test**  Determine whether the following series converge.

**a.** $\displaystyle\sum_{k=1}^{\infty} \frac{k^3}{2k^4 - 1}$            **b.** $\displaystyle\sum_{k=2}^{\infty} \frac{\ln k}{k^3}$

**SOLUTION**  In using comparison tests, it's helpful to get a feel for how the terms of the given series behave.

▶ If $\sum a_k$ diverges, then $\sum c a_k$ also diverges for any constant $c \neq 0$ (Exercise 97 of Section 10.3).

**a.** As we go farther and farther out in this series ($k \to \infty$), the terms behave like

$$\frac{k^3}{2k^4 - 1} \approx \frac{k^3}{2k^4} = \frac{1}{2k}.$$

So a reasonable choice for a comparison series is the divergent series $\sum \frac{1}{2k}$. We must now show that the terms of the given series are *greater* than the terms of the comparison series. It is done by noting that $2k^4 - 1 < 2k^4$. Inverting both sides, we have

$$\frac{1}{2k^4 - 1} > \frac{1}{2k^4}, \quad \text{which implies that} \quad \frac{k^3}{2k^4 - 1} > \frac{k^3}{2k^4} = \frac{1}{2k}.$$

Because $\sum \frac{1}{2k}$ diverges, case (2) of the Comparison Test implies that the given series also diverges.

**b.** We note that $\ln k < k$, for $k \geq 2$, and then divide by $k^3$:

$$\frac{\ln k}{k^3} < \frac{k}{k^3} = \frac{1}{k^2}.$$

**QUICK CHECK 1** Explain why it is difficult to use the divergent series $\Sigma 1/k$ as a comparison series to test $\Sigma 1/(k+1)$. ◄

Therefore, an appropriate comparison series is the convergent $p$-series $\sum \frac{1}{k^2}$. Because $\sum \frac{1}{k^2}$ converges, the given series converges.

*Related Exercises 13, 19* ◄

## The Limit Comparison Test

The Comparison Test should be tried if there is an obvious comparison series and the necessary inequality is easily established. Notice, however, that if the series in Example 1a were

$$\sum_{k=1}^{\infty} \frac{k^3}{2k^4 + 10} \text{ instead of } \sum_{k=1}^{\infty} \frac{k^3}{2k^4 - 1}, \text{ then the comparison to the series } \sum \frac{1}{2k} \text{ would not}$$

work. Rather than fiddling with inequalities, it is often easier to use a more refined test called the *Limit Comparison Test*.

**QUICK CHECK 2** For case (1) of the Limit Comparison Test, we must have $0 < L < \infty$. Why can either $a_k$ or $b_k$ be chosen as the known comparison series? That is, why can $L$ be the limit of either $a_k/b_k$ or $b_k/a_k$? ◄

---

**THEOREM 10.15　Limit Comparison Test**

Let $\Sigma a_k$ and $\Sigma b_k$ be series with positive terms and let

$$\lim_{k \to \infty} \frac{a_k}{b_k} = L.$$

1. If $0 < L < \infty$ (that is, $L$ is a finite positive number), then $\Sigma a_k$ and $\Sigma b_k$ either both converge or both diverge.
2. If $L = 0$ and $\Sigma b_k$ converges, then $\Sigma a_k$ converges.
3. If $L = \infty$ and $\Sigma b_k$ diverges, then $\Sigma a_k$ diverges.

---

> ➤ Recall that $|x| < a$ is equivalent to $-a < x < a$.

**Proof (Case 1):** Recall the definition of $\displaystyle\lim_{k \to \infty} \frac{a_k}{b_k} = L$: Given any $\varepsilon > 0$, $\left|\dfrac{a_k}{b_k} - L\right| < \varepsilon$ provided $k$ is sufficiently large. In this case, let's take $\varepsilon = L/2$. It then follows that for sufficiently large $k$, $\left|\dfrac{a_k}{b_k} - L\right| < \dfrac{L}{2}$, or (removing the absolute value) $-\dfrac{L}{2} < \dfrac{a_k}{b_k} - L < \dfrac{L}{2}$.
Adding $L$ to all terms in these inequalities, we have

$$\frac{L}{2} < \frac{a_k}{b_k} < \frac{3L}{2}.$$

These inequalities imply that for sufficiently large $k$,

$$\frac{Lb_k}{2} < a_k < \frac{3Lb_k}{2}.$$

We see that the terms of $\Sigma a_k$ are sandwiched between multiples of the terms of $\Sigma b_k$. By the Comparison Test, it follows that the two series converge or diverge together. Cases (2) and (3) ($L = 0$ and $L = \infty$, respectively) are treated in Exercise 65. ◄

**EXAMPLE 2　Using the Limit Comparison Test** Determine whether the following series converge.

**a.** $\displaystyle\sum_{k=1}^{\infty} \frac{5k^4 - 2k^2 + 3}{2k^6 - k + 5}$　　**b.** $\displaystyle\sum_{k=1}^{\infty} \frac{\ln k}{k^2}$.

**SOLUTION** In both cases, we must find a comparison series whose terms behave like the terms of the given series as $k \to \infty$.

**a.** As $k \to \infty$, a rational function behaves like the ratio of the leading (highest-power) terms. In this case, as $k \to \infty$,

$$\frac{5k^4 - 2k^2 + 3}{2k^6 - k + 5} \approx \frac{5k^4}{2k^6} = \frac{5}{2k^2}.$$

Therefore, a reasonable comparison series is the convergent $p$-series $\sum_{k=1}^{\infty} \frac{1}{k^2}$ (the factor of $5/2$ does not affect whether the given series converges). Having chosen a comparison series, we compute the limit $L$:

$$L = \lim_{k \to \infty} \frac{(5k^4 - 2k^2 + 3)/(2k^6 - k + 5)}{1/k^2} \qquad \text{Ratio of terms of series}$$

$$= \lim_{k \to \infty} \frac{k^2(5k^4 - 2k^2 + 3)}{2k^6 - k + 5} \qquad \text{Simplify.}$$

$$= \lim_{k \to \infty} \frac{5k^6 - 2k^4 + 3k^2}{2k^6 - k + 5} = \frac{5}{2}. \qquad \text{Simplify and evaluate the limit.}$$

We see that $0 < L < \infty$; therefore, the given series converges.

**b.** Why is this series interesting? We know that $\sum_{k=1}^{\infty} \frac{1}{k^2}$ converges and that $\sum_{k=1}^{\infty} \frac{1}{k}$ diverges. The given series $\sum_{k=1}^{\infty} \frac{\ln k}{k^2}$ is "between" these two series. This observation suggests that we use either $\sum_{k=1}^{\infty} \frac{1}{k^2}$ or $\sum_{k=1}^{\infty} \frac{1}{k}$ as a comparison series. In the first case, letting $a_k = \ln k/k^2$ and $b_k = 1/k^2$, we find that

$$L = \lim_{k \to \infty} \frac{a_k}{b_k} = \lim_{k \to \infty} \frac{\ln k/k^2}{1/k^2} = \lim_{k \to \infty} \ln k = \infty.$$

Case (3) of the Limit Comparison Test does not apply here because the comparison series $\sum_{k=1}^{\infty} \frac{1}{k^2}$ converges; case (3) is conclusive only when the comparison series *diverges*.

If, instead, we use the comparison series $\sum b_k = \sum \frac{1}{k}$, then

$$L = \lim_{k \to \infty} \frac{a_k}{b_k} = \lim_{k \to \infty} \frac{\ln k/k^2}{1/k} = \lim_{k \to \infty} \frac{\ln k}{k} = 0.$$

Case (2) of the Limit Comparison Test does not apply here because the comparison series $\sum_{k=1}^{\infty} \frac{1}{k}$ diverges; case (2) is conclusive only when the comparison series *converges*.

With a bit more cunning, the Limit Comparison Test becomes conclusive. A series that lies "between" $\sum_{k=1}^{\infty} \frac{1}{k^2}$ and $\sum_{k=1}^{\infty} \frac{1}{k}$ is the convergent $p$-series $\sum_{k=1}^{\infty} \frac{1}{k^{3/2}}$; we try it as a comparison series. Letting $a_k = \ln k/k^2$ and $b_k = 1/k^{3/2}$, we find that

$$L = \lim_{k \to \infty} \frac{a_k}{b_k} = \lim_{k \to \infty} \frac{\ln k/k^2}{1/k^{3/2}} = \lim_{k \to \infty} \frac{\ln k}{\sqrt{k}} = 0.$$

(This limit is evaluated using l'Hôpital's Rule or by recalling that $\ln k$ grows more slowly than any positive power of $k$.) Now case (2) of the Limit Comparison Test applies; the comparison series $\sum \frac{1}{k^{3/2}}$ converges, so the given series converges.

*Related Exercises 11, 30* ◄

## SECTION 10.5 EXERCISES

### Getting Started

1. Explain how the Limit Comparison Test works.

2. Explain why, with a series of positive terms, the sequence of partial sums is an increasing sequence.

3. What comparison series would you use with the Comparison Test to determine whether $\displaystyle\sum_{k=1}^{\infty} \frac{1}{k^2 + 1}$ converges?

4. Can the Comparison Test, with the comparison series $\displaystyle\sum_{k=1}^{\infty} \frac{1}{k^4}$, be used to show that the series $\displaystyle\sum_{k=1}^{\infty} \frac{1}{k^4 - 1}$ converges?

5. What comparison series would you use with the Comparison Test to determine whether $\displaystyle\sum_{k=1}^{\infty} \frac{2^k}{3^k + 1}$ converges?

6. Determine whether $\displaystyle\sum_{k=2}^{\infty} \frac{1}{\sqrt{k} - 1}$ converges using the Comparison Test with the comparison series $\displaystyle\sum_{k=1}^{\infty} \frac{1}{\sqrt{k}}$.

7. What comparison series would you use with the Limit Comparison Test to determine whether $\displaystyle\sum_{k=1}^{\infty} \frac{k^2 + k + 5}{k^3 + 3k + 1}$ converges?

8. Determine whether $\displaystyle\sum_{k=1}^{\infty} \frac{k + 1}{3k^3 + 2}$ converges using the Limit Comparison Test with the comparison series $\displaystyle\sum_{k=1}^{\infty} \frac{1}{k^2}$.

### Practice Exercises

**9–36. Comparison tests** *Use the Comparison Test or the Limit Comparison Test to determine whether the following series converge.*

9. $\displaystyle\sum_{k=1}^{\infty} \frac{1}{k^2 + 4}$

10. $\displaystyle\sum_{k=1}^{\infty} \frac{k^2 + k - 1}{k^4 + 4k^2 - 3}$

11. $\displaystyle\sum_{k=1}^{\infty} \frac{k^2 - 1}{k^3 + 4}$

12. $\displaystyle\sum_{k=1}^{\infty} \frac{0.0001}{k + 4}$

13. $\displaystyle\sum_{k=1}^{\infty} \frac{\sqrt{k}}{k^2 + 3}$

14. $\displaystyle\sum_{k=1}^{\infty} \frac{1}{5^k + 3}$

15. $\displaystyle\sum_{k=1}^{\infty} \frac{4^k}{5^k - 3}$

16. $\displaystyle\sum_{k=1}^{\infty} \frac{\sin^2 k}{k^2}$

17. $\displaystyle\sum_{k=1}^{\infty} \frac{1}{k^{3/2} + 1}$

18. $\displaystyle\sum_{k=1}^{\infty} \frac{1}{k \cdot 10^k}$

19. $\displaystyle\sum_{k=4}^{\infty} \frac{1 + \cos^2 k}{k - 3}$

20. $\displaystyle\sum_{k=1}^{\infty} \frac{k^2 + k + 2}{6^k(k^2 + 1)}$

21. $\displaystyle\sum_{k=1}^{\infty} \frac{(3k^3 + 4)(7k^2 + 1)}{(2k^3 + 1)(4k^3 - 1)}$

22. $\displaystyle\sum_{k=1}^{\infty} \sqrt{\frac{k}{k^3 + 1}}$

23. $\displaystyle\sum_{k=1}^{\infty} \frac{\sin(1/k)}{k^2}$

24. $\displaystyle\sum_{k=1}^{\infty} \frac{1}{3^k - 2^k}$

25. $\displaystyle\sum_{k=1}^{\infty} \frac{1}{2k - \sqrt{k}}$

26. $\displaystyle\sum_{k=1}^{\infty} \frac{1}{k\sqrt{k} + 2}$

27. $\displaystyle\sum_{k=1}^{\infty} \frac{2 + (-1)^k}{k^2}$

28. $\displaystyle\sum_{k=1}^{\infty} \frac{2 + \sin k}{k}$

29. $\displaystyle\sum_{k=1}^{\infty} \frac{\sqrt[3]{k^2 + 1}}{\sqrt{k^3 + 2}}$

30. $\displaystyle\sum_{k=3}^{\infty} \frac{1}{(k \ln k)^2}$

31. $\displaystyle\sum_{k=1}^{\infty} \frac{20}{\sqrt[3]{k} + \sqrt{k}}$

32. $\displaystyle\sum_{k=3}^{\infty} \frac{\sqrt{\ln k}}{k}$

33. $\displaystyle\sum_{k=1}^{\infty} \frac{1}{2^{\ln k}}$

34. $\displaystyle\sum_{k=2}^{\infty} \frac{\ln^2 k}{k^4}$

**T** 35. $\displaystyle\sum_{k=1}^{\infty} \frac{1}{4^{\ln k}}$

36. $\displaystyle\sum_{k=1}^{\infty} \frac{5}{\pi^k + e^k - 2}$

37. **Explain why or why not** Determine whether the following statements are true and give an explanation or counterexample.

   a. Suppose $0 < a_k < b_k$. If $\Sigma a_k$ converges, then $\Sigma b_k$ converges.

   b. Suppose $0 < a_k < b_k$. If $\Sigma a_k$ diverges, then $\Sigma b_k$ diverges.

   c. Suppose $0 < b_k < c_k < a_k$. If $\Sigma a_k$ converges, then $\Sigma b_k$ and $\Sigma c_k$ converge.

   d. When applying the Limit Comparison Test, an appropriate comparison series for $\displaystyle\sum_{k=1}^{\infty} \frac{k^2 + 2k + 1}{k^5 + 5k + 7}$ is $\displaystyle\sum_{k=1}^{\infty} \frac{1}{k^3}$.

**38–39. Examining a series two ways** *Determine whether the follow series converge using either the Comparison Test or the Limit Comparison Test. Then use another method to check your answer.*

38. $\displaystyle\sum_{k=1}^{\infty} \frac{e^{2k}}{e^{4k} + 1}$

39. $\displaystyle\sum_{k=1}^{\infty} \frac{1}{k^2 + 2k + 1}$

**40–62. Choose your test** *Use the test of your choice to determine whether the following series converge.*

40. $\left(\dfrac{1}{2}\right)^2 + \left(\dfrac{2}{3}\right)^3 + \left(\dfrac{3}{4}\right)^4 + \cdots$

41. $\displaystyle\sum_{k=1}^{\infty} \left(1 + \frac{2}{k}\right)^k$

42. $\displaystyle\sum_{k=1}^{\infty} \frac{e^{2k} + k}{e^{5k} - k^2}$

43. $\displaystyle\sum_{k=1}^{\infty} \frac{k^2 + 2k + 1}{3k^2 + 1}$

44. $\displaystyle\sum_{k=1}^{\infty} \frac{2^k}{e^k - 1}$

45. $\displaystyle\sum_{k=3}^{\infty} \frac{1}{\ln k}$

46. $\displaystyle\sum_{k=1}^{\infty} \frac{1}{5^k - 1}$

47. $\displaystyle\sum_{k=1}^{\infty} \frac{1}{\sqrt{k^3 - k + 1}}$

48. $\displaystyle\sum_{k=3}^{\infty} \frac{1}{5^k - 3^k}$

49. $\displaystyle\sum_{k=1}^{\infty} \left(\frac{1}{k} + 2^{-k}\right)$

50. $\displaystyle\sum_{k=2}^{\infty} \frac{5 \ln k}{k}$

51. $\displaystyle\sum_{k=1}^{\infty} \frac{k^8}{k^{11} + 3}$

52. $\displaystyle\sum_{k=2}^{\infty} \frac{1}{k^2 \ln k}$

53. $\displaystyle\sum_{k=1}^{\infty} \frac{1}{k^{1+p}}, p > 0$

54. $\displaystyle\sum_{k=1}^{\infty} \frac{\tan^{-1} k}{k^2}$

**55.** $\displaystyle\sum_{k=1}^{\infty} \ln\left(\frac{k+2}{k+1}\right)$

**56.** $\displaystyle\sum_{k=1}^{\infty} k^{-1/k}$

**57.** $\displaystyle\sum_{k=2}^{\infty} \frac{1}{k^{\ln k}}$

**58.** $\displaystyle\sum_{k=1}^{\infty} \sin^2 \frac{1}{k}$

**59.** $\displaystyle\sum_{k=1}^{\infty} \tan \frac{1}{k}$

**60.** $\displaystyle\sum_{k=1}^{\infty} \sqrt{\frac{1}{2^{2k}} + \frac{1}{k^2}}$

**61.** $\dfrac{1}{1\cdot 3} + \dfrac{1}{3\cdot 5} + \dfrac{1}{5\cdot 7} + \cdots$

**62.** $\dfrac{1}{2^2} + \dfrac{2}{3^2} + \dfrac{3}{4^2} + \cdots$

**63. Series of squares** Prove that if $\sum a_k$ is a convergent series of positive terms, then the series $\sum a_k^2$ also converges.

**64. Two sine series** Determine whether the following series converge.

**a.** $\displaystyle\sum_{k=1}^{\infty} \sin \frac{1}{k}$

**b.** $\displaystyle\sum_{k=1}^{\infty} \frac{1}{k} \sin \frac{1}{k}$

### Explorations and Challenges

**65. Limit Comparison Test proof** Use the proof of case (1) of the Limit Comparison Test (Theorem 10.15) to prove cases (2) and (3).

**66. Infinite products** An infinite product $P = a_1 a_2 a_3 \ldots$, which is denoted $\displaystyle\prod_{k=1}^{\infty} a_k$, is the limit of the *sequence of partial products* $\{a_1, a_1 a_2, a_1 a_2 a_3, \ldots\}$. Assume $a_k > 0$ for all $k$ and $L$ is a finite constant.

**a.** Evaluate $\displaystyle\prod_{k=1}^{\infty}\left(\frac{k}{k+1}\right) = \frac{1}{2}\cdot\frac{2}{3}\cdot\frac{3}{4}\cdot\frac{4}{5}\cdots$.

**b.** Show that if $\displaystyle\sum_{k=1}^{\infty} \ln a_k = L$, then $\displaystyle\prod_{k=1}^{\infty} a_k = e^L$.

**c.** Use the result of part (b) to evaluate
$$\prod_{k=0}^{\infty} e^{1/2^k} = e \cdot e^{1/2} \cdot e^{1/4} \cdot e^{1/8}\cdots.$$

**67. An early limit** Working in the early 1600s, the mathematicians Wallis, Pascal, and Fermat wanted to calculate the area of the region under the curve $y = x^p$ between $x = 0$ and $x = 1$, where $p$ is a positive integer. Using arguments that predated the Fundamental Theorem of Calculus, they were able to prove that
$$\lim_{n\to\infty} \frac{1}{n}\sum_{k=0}^{n-1}\left(\frac{k}{n}\right)^p = \frac{1}{p+1}.$$
Use what you know about Riemann sums and integrals to verify this limit.

**QUICK CHECK ANSWERS**

**1.** To use the Comparison Test, we would need to show that $1/(k+1) > 1/k$, which is not true.   **2.** If $\displaystyle\lim_{k\to\infty} \frac{a_k}{b_k} = L$ for $0 < L < \infty$, then $\displaystyle\lim_{k\to\infty} \frac{b_k}{a_k} = \frac{1}{L}$, where $0 < 1/L < \infty$. ◄

# 10.6 Alternating Series

Our previous discussion focused on infinite series with positive terms, which is certainly an important part of the entire subject. But there are many interesting series with terms of mixed sign. For example, the series
$$1 + \frac{1}{2} - \frac{1}{3} - \frac{1}{4} + \frac{1}{5} + \frac{1}{6} - \frac{1}{7} - \frac{1}{8} + \cdots$$
has the pattern that two positive terms are followed by two negative terms, and vice versa. Clearly, infinite series could have endless sign patterns, so we need to restrict our attention.

Fortunately, the simplest sign pattern is also the most important. We consider **alternating series** in which the signs strictly alternate, as in the series
$$\sum_{k=1}^{\infty} \frac{(-1)^{k+1}}{k} = 1 - \frac{1}{2} + \frac{1}{3} - \frac{1}{4} + \frac{1}{5} - \frac{1}{6} + \frac{1}{7} - \frac{1}{8} + \cdots.$$

The factor $(-1)^{k+1}$ (or $(-1)^k$) has the pattern $\{\ldots, 1, -1, 1, -1, \ldots\}$ and provides the alternating signs.

## Alternating Harmonic Series

Let's see what is different about alternating series by working with the series $\displaystyle\sum_{k=1}^{\infty} \frac{(-1)^{k+1}}{k}$, which is called the **alternating harmonic series**. Recall that this series *without* the alternating signs, $\displaystyle\sum_{k=1}^{\infty} \frac{1}{k}$, is the *divergent* harmonic series. So an immediate question is whether the presence of alternating signs affects the convergence of a series.

$$S_n = \sum_{k=1}^{n} \frac{(-1)^{k+1}}{k}$$

Sequence of partial sums for the alternating harmonic series

**Figure 10.34**

We investigate this question by looking at the sequence of partial sums for the series. In this case, the first four terms of the sequence of partial sums are

$$S_1 = 1$$

$$S_2 = 1 - \frac{1}{2} = \frac{1}{2}$$

$$S_3 = 1 - \frac{1}{2} + \frac{1}{3} = \frac{5}{6}$$

$$S_4 = 1 - \frac{1}{2} + \frac{1}{3} - \frac{1}{4} = \frac{7}{12}.$$

Plotting the first 30 terms of the sequence of partial sums results in **Figure 10.34**, which has several noteworthy features.

- The terms of the sequence of partial sums appear to converge to a limit; if they do, it means that, while the harmonic series diverges, the *alternating* harmonic series converges. We will soon learn that taking a divergent series with positive terms and making it an alternating series *may* turn it into a convergent series.

- For series with *positive* terms, the sequence of partial sums is necessarily an increasing sequence. Because the terms of an alternating series alternate in sign, the sequence of partial sums is not increasing (Figure 10.34).

- For the alternating harmonic series, the odd terms of the sequence of partial sums form a decreasing sequence and the even terms form an increasing sequence. As a result, the limit of the sequence of partial sums lies between any two consecutive terms of the sequence.

**QUICK CHECK 1** Write out the first few terms of the sequence of partial sums for the alternating series $1 - 2 + 3 - 4 + 5 - 6 + \cdots$. Does this series appear to converge or diverge? ◄

### Alternating Series Test

We now consider alternating series in general, which are written $\sum(-1)^{k+1}a_k$, where $a_k > 0$. With the exception of the Divergence Test, none of the convergence tests for series with positive terms applies to alternating series. Fortunately, one test works for most alternating series—and it is easy to use.

> ➤ Recall that the Divergence Test of Section 10.4 applies to all series: If the terms of *any* series (including an alternating series) do not tend to zero, then the series diverges.

> ➤ Depending on the sign of the first term of the series, an alternating series may be written with $(-1)^k$ or $(-1)^{k+1}$. Theorem 10.16 also applies to series of the form $\sum(-1)^k a_k$.

---

**THEOREM 10.16  Alternating Series Test**
The alternating series $\sum(-1)^{k+1}a_k$ converges provided

1. the terms of the series are nonincreasing in magnitude ($0 < a_{k+1} \le a_k$, for $k$ greater than some index $N$) and

2. $\lim\limits_{k\to\infty} a_k = 0.$

---

There is potential for confusion here. *For series of positive terms,* $\lim\limits_{k\to\infty} a_k = 0$ ***does not** imply convergence. For alternating series with nonincreasing terms,* $\lim\limits_{k\to\infty} a_k = 0$ ***does** imply convergence.*

**Proof:** The proof is short and instructive; it relies on **Figure 10.35**. We consider an alternating series in the form

$$\sum_{k=1}^{\infty}(-1)^{k+1}a_k = a_1 - a_2 + a_3 - a_4 + \cdots.$$

Because the terms of the series are nonincreasing in magnitude, the even terms of the sequence of partial sums $\{S_{2k}\} = \{S_2, S_4, \ldots\}$ form a nondecreasing sequence that is bounded above by $S_1$. By the Bounded Monotonic Sequence Theorem (Section 10.2), this sequence has a limit; call it $L$. Similarly, the odd terms of the sequence of partial sums $\{S_{2k-1}\} = \{S_1, S_3, \ldots\}$ form a nonincreasing sequence that is bounded below by $S_2$. By

**Figure 10.35**

the Bounded Monotonic Sequence Theorem, this sequence also has a limit; call it $L'$. At the moment, we cannot conclude that $L = L'$. However, notice that $S_{2k} = S_{2k-1} - a_{2k}$. By the condition that $\lim\limits_{k\to\infty} a_k = 0$, it follows that

$$\underbrace{\lim_{k\to\infty} S_{2k}}_{L} = \underbrace{\lim_{k\to\infty} S_{2k-1}}_{L'} - \underbrace{\lim_{k\to\infty} a_{2k}}_{0},$$

or $L = L'$. Therefore, the sequence of partial sums converges to a (unique) limit, and the corresponding alternating series converges to that limit.  ◄

Now we can confirm that the alternating harmonic series $\sum\limits_{k=1}^{\infty} \dfrac{(-1)^{k+1}}{k}$ converges. This fact follows immediately from the Alternating Series Test because the terms $a_k = \dfrac{1}{k}$ decrease and $\lim\limits_{k\to\infty} a_k = 0$.

> $\sum\limits_{k=1}^{\infty} \dfrac{1}{k}$ $\quad\sum\limits_{k=1}^{\infty} \dfrac{(-1)^{k+1}}{k}$
> • Diverges $\qquad$ • Converges
> • Partial sums $\quad$ • Partial sums
> $\quad$ increase. $\qquad$ bound the
> $\qquad\qquad\qquad$ series above
> $\qquad\qquad\qquad$ and below.

---

**THEOREM 10.17  Alternating Harmonic Series**

The alternating harmonic series $\sum\limits_{k=1}^{\infty} \dfrac{(-1)^{k+1}}{k}$ converges (even though the harmonic series $\sum\limits_{k=1}^{\infty} \dfrac{1}{k}$ diverges).

---

**QUICK CHECK 2** Explain why the value of a convergent alternating series, with terms that are nonincreasing in magnitude, is trapped between successive terms of the sequence of partial sums. ◄

**EXAMPLE 1  Alternating Series Test** Determine whether the following series converge or diverge.

**a.** $\sum\limits_{k=1}^{\infty} \dfrac{(-1)^{k+1}}{k^2}$ $\qquad$ **b.** $2 - \dfrac{3}{2} + \dfrac{4}{3} - \dfrac{5}{4} + \cdots$ $\qquad$ **c.** $\sum\limits_{k=2}^{\infty} \dfrac{(-1)^k \ln k}{k}$

**SOLUTION**

**a.** The terms of this series decrease in magnitude, for $k \geq 1$. Furthermore,

$$\lim_{k\to\infty} a_k = \lim_{k\to\infty} \frac{1}{k^2} = 0.$$

Therefore, the series converges.

**b.** Using sigma notation, this series is written as $\sum\limits_{k=1}^{\infty} (-1)^{k+1} \dfrac{k+1}{k}$. Because $\lim\limits_{k\to\infty} \dfrac{k+1}{k} = 1$, the sequence $\left\{ (-1)^{k+1} \dfrac{k+1}{k} \right\}$ oscillates and diverges. Therefore, $\sum\limits_{k=1}^{\infty} (-1)^{k+1} \dfrac{k+1}{k}$ diverges by the Divergence Test.

**c.** The first step is to show that the terms decrease in magnitude after some fixed term of the series. One way to proceed is to look at the function $f(x) = \dfrac{\ln x}{x}$, which generates the terms of the series. By the Quotient Rule, $f'(x) = \dfrac{1 - \ln x}{x^2}$. The fact that $f'(x) < 0$, for $x > e$, implies that the terms $\dfrac{\ln k}{k}$ decrease, for $k \geq 3$. As long as the terms of the series decrease for all $k$ greater than some fixed integer, the first condition of the test is met. Furthermore, using l'Hôpital's Rule or the fact that $\{\ln k\}$ increases more slowly than $\{k\}$ (Section 10.2), we see that

$$\lim_{k \to \infty} a_k = \lim_{k \to \infty} \frac{\ln k}{k} = 0.$$

The conditions of the Alternating Series Test are met and the series converges.

*Related Exercises 14, 15, 18* ◄

## Remainders in Alternating Series

Recall that if a series converges to a value $S$, then the remainder is $R_n = S - S_n$, where $S_n$ is the sum of the first $n$ terms of the series. The magnitude of the remainder is the *absolute error* in approximating $S$ by $S_n$.

An upper bound on the magnitude of the remainder in an alternating series arises from the following observation: When the terms are nonincreasing in magnitude, the value of the series is always trapped between successive terms of the sequence of partial sums. Therefore, as shown in **Figure 10.36**,

$$|R_n| = |S - S_n| \leq |S_{n+1} - S_n| = a_{n+1}.$$

This argument justifies the following theorem.

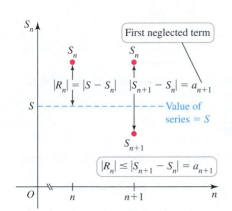

**Figure 10.36**

---

**THEOREM 10.18**    **Remainder in Alternating Series**

Let $\displaystyle\sum_{k=1}^{\infty} (-1)^{k+1} a_k$ be a convergent alternating series with terms that are nonincreasing in magnitude. Let $R_n = S - S_n$ be the remainder in approximating the value of that series by the sum of its first $n$ terms. Then $|R_n| \leq a_{n+1}$. In other words, the magnitude of the remainder is less than or equal to the magnitude of the first neglected term.

---

**EXAMPLE 2**    **Remainder in an alternating series**

**a.** In turns out that $\ln 2 = 1 - \dfrac{1}{2} + \dfrac{1}{3} - \dfrac{1}{4} + \cdots = \displaystyle\sum_{k=1}^{\infty} \dfrac{(-1)^{k+1}}{k}$. How many terms of the series are required to approximate $\ln 2$ with an error less than $10^{-6}$? The exact value of the series is given but is not needed to answer the question.

**b.** Consider the series $-1 + \dfrac{1}{2!} - \dfrac{1}{3!} + \dfrac{1}{4!} - \cdots = \displaystyle\sum_{k=1}^{\infty} \dfrac{(-1)^k}{k!}$. Find an upper bound for the magnitude of the error in approximating the value of the series (which is $e^{-1} - 1$) with $n = 9$ terms.

**SOLUTION** Notice that both series meet the conditions of Theorem 10.18.

**a.** The series is expressed as the sum of the first $n$ terms plus the remainder:

$$\sum_{k=1}^{\infty} \frac{(-1)^{k+1}}{k} = \underbrace{1 - \frac{1}{2} + \frac{1}{3} - \frac{1}{4} + \cdots + \frac{(-1)^{n+1}}{n}}_{S_n \,=\, \text{the sum of the first } n \text{ terms}} + \overbrace{\underbrace{\frac{(-1)^{n+2}}{n+1}}_{\substack{|R_n| = |S - S_n| \text{ is less} \\ \text{than the magnitude} \\ \text{of this term}}} + \cdots}^{R_n}.$$

In magnitude, the remainder is less than or equal to the magnitude of the $(n + 1)$st term:

$$|R_n| = |S - S_n| \le a_{n+1} = \frac{1}{n + 1}.$$

To ensure that the error is less than $10^{-6}$, we require that

$$a_{n+1} = \frac{1}{n + 1} < 10^{-6}, \quad \text{or} \quad n + 1 > 10^6.$$

Therefore, it takes 1 million terms of the series to approximate ln 2 with an error less than $10^{-6}$.

**b.** The series may be expressed as the sum of the first nine terms plus the remainder:

$$\sum_{k=1}^{\infty} \frac{(-1)^k}{k!} = \underbrace{-1 + \frac{1}{2!} - \frac{1}{3!} + \cdots - \frac{1}{9!}}_{S_9 \, = \, \text{sum of first 9 terms}} + \overbrace{\frac{1}{10!}}^{R_9} - \cdots.$$

$|R_9| = |S - S_9|$ is less than this term

The error committed when using the first nine terms to approximate the value of the series satisfies

$$|R_9| = |S - S_9| \le a_{10} = \frac{1}{10!} \approx 2.8 \times 10^{-7}.$$

Therefore, the error is no greater than $2.8 \times 10^{-7}$. As a check, the difference between the sum of the first nine terms, $\sum_{k=1}^{9} \frac{(-1)^k}{k!} \approx -0.632120811$, and the exact value, $S = e^{-1} - 1 \approx -0.632120559$, is approximately $2.5 \times 10^{-7}$. Therefore, the actual error satisfies the bound given by Theorem 10.18.

*Related Exercises 32–33* ◀

**QUICK CHECK 3** Compare and comment on the speed of convergence of the two series in Example 2. Why does one series converge more rapidly than the other? ◀

## Absolute and Conditional Convergence

In this final segment, we introduce some terminology that is needed in Chapter 11. We now let the notation $\sum a_k$ denote any series—a series of positive terms, an alternating series, or even a more general infinite series.

Look again at the convergent alternating harmonic series $\sum (-1)^{k+1}/k$. The corresponding series of positive terms, $\sum 1/k$, is the divergent harmonic series. In contrast, we saw in Example 1a that the alternating series $\sum (-1)^{k+1}/k^2$ converges, and the corresponding $p$-series of positive terms $\sum 1/k^2$ also converges. These examples illustrate that removing the alternating signs in a convergent series *may* or *may not* result in a convergent series. The terminology that we now introduce distinguishes between these cases.

---

**DEFINITION  Absolute and Conditional Convergence**

If $\sum |a_k|$ converges, then $\sum a_k$ **converges absolutely**.

If $\sum |a_k|$ diverges and $\sum a_k$ converges, then $\sum a_k$ **converges conditionally**.

---

The series $\sum (-1)^{k+1}/k^2$ is an example of an absolutely convergent series because the series of absolute values,

$$\sum_{k=1}^{\infty} \left| \frac{(-1)^{k+1}}{k^2} \right| = \sum_{k=1}^{\infty} \frac{1}{k^2},$$

is a convergent $p$-series. In this case, removing the alternating signs in the series does *not* affect its convergence.

On the other hand, the convergent alternating harmonic series $\sum (-1)^{k+1}/k$ has the property that the corresponding series of absolute values,

$$\sum_{k=1}^{\infty} \left| \frac{(-1)^{k+1}}{k} \right| = \sum_{k=1}^{\infty} \frac{1}{k},$$

does *not* converge. In this case, removing the alternating signs in the series *does* affect convergence, so this series does not converge absolutely. Instead, we say it *converges conditionally*. A convergent series (such as $\sum (-1)^{k+1}/k$) may not converge absolutely. However, if a series converges absolutely, then it converges.

> **THEOREM 10.19   Absolute Convergence Implies Convergence**
> If $\sum |a_k|$ converges, then $\sum a_k$ converges (absolute convergence implies convergence). Equivalently, if $\sum a_k$ diverges, then $\sum |a_k|$ diverges.

**Proof:** Because $|a_k| = a_k$ or $|a_k| = -a_k$, it follows that $0 \le a_k + |a_k| \le 2|a_k|$. By assumption, $\sum |a_k|$ converges, which, in turn, implies that $2\sum |a_k|$ converges. Using the Comparison Test and the inequality $0 \le a_k + |a_k| \le 2|a_k|$, it follows that $\sum (a_k + |a_k|)$ converges. Now note that

$$\sum a_k = \sum (a_k + |a_k| - |a_k|) = \underbrace{\sum (a_k + |a_k|)}_{\text{converges}} - \underbrace{\sum |a_k|}_{\text{converges}}.$$

We see that $\sum a_k$ is the sum of two convergent series, so it also converges. The second statement of the theorem is logically equivalent to the first statement. ◄

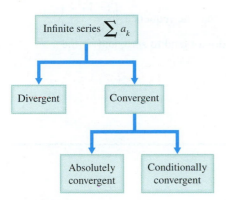

**Figure 10.37**

**Figure 10.37** gives an overview of absolute and conditional convergence. It shows the universe of all infinite series, split first according to whether they converge or diverge. Convergent series are further divided between absolutely and conditionally convergent series. Here are a few more consequences of these definitions.

- The distinction between absolute and conditional convergence is relevant only for series of mixed sign, which includes alternating series. If a series of positive terms converges, it converges absolutely; conditional convergence does not apply.

- To test for absolute convergence, we test the series $\sum |a_k|$, which is a series of positive terms. Therefore, the convergence tests of Sections 10.4 and 10.5 (for positive-term series) are used to determine absolute convergence.

**QUICK CHECK 4** Explain why a convergent series of positive terms converges absolutely. ◄

**EXAMPLE 3   Absolute and conditional convergence**   Determine whether the following series diverge, converge absolutely, or converge conditionally.

**a.** $\displaystyle\sum_{k=1}^{\infty} \frac{(-1)^{k+1}}{\sqrt{k}}$     **b.** $\displaystyle\sum_{k=1}^{\infty} \frac{(-1)^{k+1}}{\sqrt{k^3}}$     **c.** $\displaystyle\sum_{k=1}^{\infty} \frac{\sin k}{k^2}$     **d.** $\displaystyle\sum_{k=1}^{\infty} \frac{(-1)^k k}{k+1}$

**SOLUTION**

**a.** We examine the series of absolute values,

$$\sum_{k=1}^{\infty} \left| \frac{(-1)^{k+1}}{\sqrt{k}} \right| = \sum_{k=1}^{\infty} \frac{1}{\sqrt{k}},$$

which is a divergent *p*-series (with $p = \frac{1}{2} < 1$). Therefore, the given alternating series does not converge absolutely. To determine whether the series converges conditionally, we look at the original series—with alternating signs. The magnitude of the terms of this series decrease with $\lim_{k \to \infty} 1/\sqrt{k} = 0$, so by the Alternating Series Test, the series converges. Because this series converges, but not absolutely, it converges conditionally.

**b.** To assess absolute convergence, we look at the series of absolute values,

$$\sum_{k=1}^{\infty} \left| \frac{(-1)^{k+1}}{\sqrt{k^3}} \right| = \sum_{k=1}^{\infty} \frac{1}{k^{3/2}},$$

which is a convergent $p$-series (with $p = \frac{3}{2} > 1$). Therefore, the original alternating series converges absolutely (and by Theorem 10.19 it converges).

**c.** The terms of this series do not strictly alternate in sign (the first few signs are $+ + + - - -$), so the Alternating Series Test does not apply. Because $|\sin k| \leq 1$, the terms of the series of absolute values satisfy

$$\left| \frac{\sin k}{k^2} \right| = \frac{|\sin k|}{k^2} \leq \frac{1}{k^2}.$$

The series $\sum \dfrac{1}{k^2}$ is a convergent $p$-series. Therefore, by the Comparison Test, the series $\sum \left| \dfrac{\sin k}{k^2} \right|$ converges, which implies that the series $\sum \dfrac{\sin k}{k^2}$ converges absolutely (and by Theorem 10.19 it converges).

**d.** Notice that $\lim\limits_{k \to \infty} k/(k+1) = 1$, which implies that the sequence $\left\{ \dfrac{(-1)^k k}{k+1} \right\}$ oscillates and diverges. The terms of the series do not tend to zero, and by the Divergence Test, the series diverges.

*Related Exercises 45, 51, 53* ◄

---

## SECTION 10.6 EXERCISES

### Getting Started

1. Explain why the sequence of partial sums for an alternating series is not an increasing sequence.

2. Describe how to apply the Alternating Series Test.

3. Why does the value of a converging alternating series with terms that are nonincreasing in magnitude lie between any two consecutive terms of its sequence of partial sums?

4. Suppose an alternating series with terms that are nonincreasing in magnitude converges to a value $L$. Explain how to estimate the remainder that occurs when the series is terminated after $n$ terms.

5. Explain why the magnitude of the remainder in an alternating series (with terms that are nonincreasing in magnitude) is less than or equal to the magnitude of the first neglected term.

6. Give an example of a convergent alternating series that fails to converge absolutely.

7. Is it possible for a series of positive terms to converge conditionally? Explain.

8. Does the geometric series $\sum\limits_{k=0}^{\infty} (-0.5)^k$ converge absolutely?

9. Is it possible for an alternating series to converge absolutely but not conditionally?

10. Determine the values of $p$ for which the series $\sum\limits_{k=0}^{\infty} \dfrac{(-1)^k}{k^p}$ converges conditionally.

### Practice Exercises

**11–27. Alternating Series Test** *Determine whether the following series converge.*

11. $\displaystyle\sum_{k=0}^{\infty} \frac{(-1)^k}{2k+1}$

12. $\displaystyle\sum_{k=1}^{\infty} \frac{(-1)^k}{\sqrt{k}}$

13. $\displaystyle\sum_{k=1}^{\infty} \frac{(-1)^k k}{3k+2}$

14. $\displaystyle\sum_{k=1}^{\infty} (-1)^k \left( 1 + \frac{1}{k} \right)^k$

15. $\displaystyle\sum_{k=1}^{\infty} \frac{(-1)^{k+1}}{k^3}$

16. $\displaystyle\sum_{k=0}^{\infty} \frac{(-1)^k}{k^2+10}$

17. $\displaystyle\sum_{k=1}^{\infty} (-1)^{k+1} \frac{k^2}{k^3+1}$

18. $\displaystyle\sum_{k=2}^{\infty} (-1)^k \frac{\ln k}{k^2}$

19. $\displaystyle\sum_{k=2}^{\infty} (-1)^k \frac{k^2-1}{k^2+3}$

20. $\displaystyle\sum_{k=0}^{\infty} \left( -\frac{1}{5} \right)^k$

21. $\displaystyle\sum_{k=2}^{\infty} (-1)^k \left( 1 + \frac{1}{k} \right)$

22. $\displaystyle\sum_{k=1}^{\infty} \frac{\cos \pi k}{k^2}$

23. $\displaystyle\sum_{k=1}^{\infty} (-1)^k \frac{k^{11} + 2k^5 + 1}{4k(k^{10}+1)}$

24. $\displaystyle\sum_{k=2}^{\infty} \frac{(-1)^k}{k \ln^2 k}$

25. $\displaystyle\sum_{k=1}^{\infty} (-1)^{k+1} k^{1/k}$

26. $\displaystyle\sum_{k=1}^{\infty} (-1)^k k \sin \frac{1}{k}$

27. $\displaystyle\sum_{k=0}^{\infty} \frac{(-1)^k}{\sqrt{k^2+4}}$

**T** **28–32. Estimating errors in partial sums** *For each of the following convergent alternating series, evaluate the nth partial sum for the given value of n. Then use Theorem 10.18 to find an upper bound for the error $|S - S_n|$ in using the nth partial sum $S_n$ to estimate the value of the series S.*

**28.** $\displaystyle\sum_{k=1}^{\infty} \frac{(-1)^{k+1}}{k^3 + 1}$; $n = 3$

**29.** $\displaystyle\sum_{k=1}^{\infty} \frac{(-1)^k}{k^4}$; $n = 4$

**30.** $\displaystyle\sum_{k=0}^{\infty} \frac{(-1)^k}{(3k+1)^4}$; $n = 2$

**31.** $\displaystyle\sum_{k=0}^{\infty} \frac{(-1)^k}{k^4 + k^2 + 1}$; $n = 5$

**32.** $\displaystyle\sum_{k=0}^{\infty} \frac{(-1)^k}{(2k+1)!}$; $n = 3$

**T** **33–38. Remainders in alternating series** *Determine how many terms of the following convergent series must be summed to be sure that the remainder is less than $10^{-4}$ in magnitude. Although you do not need it, the exact value of the series is given in each case.*

**33.** $\ln 2 = \displaystyle\sum_{k=1}^{\infty} \frac{(-1)^{k+1}}{k}$

**34.** $\dfrac{1}{e} = \displaystyle\sum_{k=0}^{\infty} \frac{(-1)^k}{k!}$

**35.** $\dfrac{\pi}{4} = \displaystyle\sum_{k=0}^{\infty} \frac{(-1)^k}{2k+1}$

**36.** $\dfrac{\pi^2}{12} = \displaystyle\sum_{k=1}^{\infty} \frac{(-1)^{k+1}}{k^2}$

**37.** $\dfrac{7\pi^4}{720} = \displaystyle\sum_{k=1}^{\infty} \frac{(-1)^{k+1}}{k^4}$

**38.** $\dfrac{\pi^3}{32} = \displaystyle\sum_{k=0}^{\infty} \frac{(-1)^k}{(2k+1)^3}$

**T** **39–44. Estimating infinite series** *Estimate the value of the following convergent series with an absolute error less than $10^{-3}$.*

**39.** $\displaystyle\sum_{k=1}^{\infty} \frac{(-1)^k}{k^5}$

**40.** $\displaystyle\sum_{k=1}^{\infty} \frac{(-1)^k}{(2k+1)^3}$

**41.** $\displaystyle\sum_{k=1}^{\infty} \frac{(-1)^k k}{k^2 + 1}$

**42.** $\displaystyle\sum_{k=1}^{\infty} \frac{(-1)^k k}{k^4 + 1}$

**43.** $\displaystyle\sum_{k=1}^{\infty} \frac{(-1)^k}{k^k}$

**44.** $\displaystyle\sum_{k=1}^{\infty} \frac{(-1)^{k+1}}{(2k+1)!}$

**45–63. Absolute and conditional convergence** *Determine whether the following series converge absolutely, converge conditionally, or diverge.*

**45.** $\displaystyle\sum_{k=1}^{\infty} \frac{(-1)^k}{k^{2/3}}$

**46.** $\displaystyle\sum_{k=0}^{\infty} \frac{(-1)^k 2k}{\sqrt{k^2 + 9}}$

**47.** $\displaystyle\sum_{k=1}^{\infty} \frac{(-1)^{k+1}}{(k+1)!}$

**48.** $\displaystyle\sum_{k=1}^{\infty} \left(-\frac{1}{3}\right)^k$

**49.** $\displaystyle\sum_{k=1}^{\infty} \left(\frac{3}{4}\right)^k$

**50.** $\displaystyle\sum_{k=1}^{\infty} \frac{(-1)^{k+1}}{k^{0.99}}$

**51.** $\displaystyle\sum_{k=1}^{\infty} \frac{\cos k}{k^3}$

**52.** $\displaystyle\sum_{k=1}^{\infty} \frac{(-1)^k k^2}{\sqrt{k^6 + 1}}$

**53.** $\displaystyle\sum_{k=1}^{\infty} (-1)^k \tan^{-1} k$

**54.** $\displaystyle\sum_{k=1}^{\infty} (-1)^k e^{-k}$

**55.** $\displaystyle\sum_{k=1}^{\infty} \frac{(-1)^{k+1}}{2\sqrt{k} - 1}$

**56.** $\displaystyle\sum_{k=2}^{\infty} \frac{(-1)^k}{3^k - 1}$

**57.** $\displaystyle\sum_{k=1}^{\infty} \frac{(-1)^k k}{2k+1}$

**58.** $\displaystyle\sum_{k=3}^{\infty} \frac{(-1)^k}{\ln k}$

**59.** $\displaystyle\sum_{k=1}^{\infty} \frac{(-1)^k \tan^{-1} k}{k^3}$

**60.** $\displaystyle\sum_{k=1}^{\infty} (-1)^k \frac{k^2 + 1}{3k^4 + 3}$

**61.** $\displaystyle\sum_{k=2}^{\infty} (-1)^k \frac{k^2 + 1}{k^3 - 1}$

**62.** $\displaystyle\sum_{k=1}^{\infty} \frac{\sin k}{3^k + 4^k}$

**63.** $\displaystyle\sum_{k=1}^{\infty} (-1)^{k+1} \frac{k!}{k^k}$ (*Hint:* Show that $\dfrac{k!}{k^k} \leq \dfrac{2}{k^2}$, for $k \geq 3$.)

**64. Alternating Series Test** Show that the series
$$\frac{1}{3} - \frac{2}{5} + \frac{3}{7} - \frac{4}{9} + \cdots = \sum_{k=1}^{\infty} (-1)^{k+1} \frac{k}{2k+1}$$
diverges. Which condition of the Alternating Series Test is not satisfied?

**65. Explain why or why not** Determine whether the following statements are true and give an explanation or counterexample.

  **a.** A series that converges must converge absolutely.
  **b.** A series that converges absolutely must converge.
  **c.** A series that converges conditionally must converge.
  **d.** If $\sum a_k$ diverges, then $\sum |a_k|$ diverges.
  **e.** If $\sum a_k^2$ converges, then $\sum a_k$ converges.
  **f.** If $a_k > 0$ and $\sum a_k$ converges, then $\sum a_k^2$ converges.
  **g.** If $\sum a_k$ converges conditionally, then $\sum |a_k|$ diverges.

## Explorations and Challenges

**66. Rearranging series** It can be proved that if a series converges absolutely, then its terms may be summed in any order without changing the value of the series. However, if a series converges conditionally, then the value of the series depends on the order of summation. For example, the (conditionally convergent) alternating harmonic series has the value
$$1 - \frac{1}{2} + \frac{1}{3} - \frac{1}{4} + \cdots = \ln 2.$$
Show that by rearranging the terms (so the sign pattern is $++-$),
$$1 + \frac{1}{3} - \frac{1}{2} + \frac{1}{5} + \frac{1}{7} - \frac{1}{4} + \cdots = \frac{3}{2} \ln 2.$$

**67. Alternating p-series** Given that $\displaystyle\sum_{k=1}^{\infty} \frac{1}{k^2} = \frac{\pi^2}{6}$, show that $\displaystyle\sum_{k=1}^{\infty} \frac{(-1)^{k+1}}{k^2} = \frac{\pi^2}{12}$. (Assume the result of Exercise 66.)

**68. Alternating p-series** Given that $\displaystyle\sum_{k=1}^{\infty} \frac{1}{k^4} = \frac{\pi^4}{90}$, show that $\displaystyle\sum_{k=1}^{\infty} \frac{(-1)^{k+1}}{k^4} = \frac{7\pi^4}{720}$. (Assume the result of Exercise 66.)

**69. A fallacy** Explain the fallacy in the following argument. Let $x = 1 + \dfrac{1}{3} + \dfrac{1}{5} + \dfrac{1}{7} + \cdots$ and $y = \dfrac{1}{2} + \dfrac{1}{4} + \dfrac{1}{6} + \dfrac{1}{8} + \cdots$. It follows that $2y = x + y$, which implies that $x = y$. On the other hand,
$$x - y = \left(1 - \frac{1}{2}\right) + \left(\frac{1}{3} - \frac{1}{4}\right) + \left(\frac{1}{5} - \frac{1}{6}\right) + \cdots > 0$$
$$\qquad\quad \underbrace{\qquad}_{>0} \qquad \underbrace{\qquad}_{>0} \qquad \underbrace{\qquad}_{>0}$$
is a sum of positive terms, so $x > y$. Therefore, we have shown that $x = y$ and $x > y$.

**70. Conditions of the Alternating Series Test** It can be shown that if the sequence $\{a_{2n}\} = \{a_2, a_4, a_6, \ldots\}$ and the sequence $\{a_{2n-1}\} = \{a_1, a_3, a_5, \ldots\}$ both converge to $L$, then the sequence $\{a_n\} = \{a_1, a_2, a_3, \ldots\}$ converges to $L$. It is also the case that if $\{a_{2n}\}$ or $\{a_{2n-1}\}$ diverges, then $\{a_n\}$ diverges. Use these results in this exercise. Consider the alternating series

$$\sum_{k=1}^{\infty} (-1)^{k+1} a_k, \text{ where } a_k = \begin{cases} \dfrac{4}{k+1} & \text{if } k \text{ is odd} \\ \dfrac{2}{k} & \text{if } k \text{ is even.} \end{cases}$$

**a.** Show that $\{a_n\}$ converges to 0.

**b.** Show that $S_{2n} = \displaystyle\sum_{k=1}^{n} \dfrac{1}{k}$ and explain why $\displaystyle\lim_{k \to \infty} S_{2n} = \infty$.

**c.** Explain why the series $\displaystyle\sum_{k=1}^{\infty} (-1)^{k+1} a_k$ diverges even though $\{a_n\}$ converges to 0. Explain why this result does not contradict the Alternating Series Test.

**71. A diverging alternating series** Consider the alternating series

$$\sum_{k=1}^{\infty} (-1)^{k+1} a_k = \frac{1}{1^2} - \frac{1}{1} + \frac{1}{2^2} - \frac{1}{2} + \frac{1}{3^2} - \frac{1}{3} + \cdots.$$

**a.** Show that the individual terms of the series converge to 0. (*Hint:* See Exercise 70.)

**b.** Find a formula for $S_{2n}$, the sum of the first $2n$ terms of the series.

**c.** Explain why the alternating series diverges even though individual terms of the series converge to 0. Explain why this result does not contradict the Alternating Series Test.

**QUICK CHECK ANSWERS**

**1.** $1, -1, 2, -2, 3, -3, \ldots$; series diverges.  **2.** The even terms of the sequence of partial sums approach the value of the series from one side; the odd terms of the sequence of partial sums approach the value of the series from the other side.  **3.** The second series with $k!$ in the denominator converges much more quickly than the first series because $k!$ increases much faster than $k$ as $k \to \infty$.  **4.** If a series has positive terms, the series of absolute values is the same as the series itself. ◄

# 10.7 The Ratio and Root Tests

We now consider two additional convergence tests for general series: the Ratio Test and the Root Test. Both tests are used frequently throughout the next chapter. As in previous sections, these tests determine *whether* an infinite series converges, but they do not establish the value of the series.

## The Ratio Test

The tests we have developed so far to determine whether a series converges are powerful, but they all have limitations. For example, the Integral Test requires evaluating integrals, and it is difficult to use with a series such as $\displaystyle\sum_{k=1}^{\infty} \frac{1}{k!}$ because integrating $1/k!$ is problematic. Further, the Integral Test and the comparison tests work only with series that have positive terms.

The Ratio Test, which we introduce in the next theorem, significantly enlarges the set of infinite series that we can analyze. It can be applied to any series, though its success depends on evaluating the limit of the ratio of successive terms in a series. The conclusions reached by the Ratio Test rely on the idea of absolute convergence and Theorem 10.19 from Section 10.6. We assume $a_k \neq 0$, for all $k$, so that the ratio $a_{k+1}/a_k$ is defined.

> **THEOREM 10.20  Ratio Test**
>
> Let $\sum a_k$ be an infinite series, and let $r = \displaystyle\lim_{k \to \infty} \left| \frac{a_{k+1}}{a_k} \right|$.
>
> **1.** If $r < 1$, the series converges absolutely, and therefore it converges (by Theorem 10.19).
>
> **2.** If $r > 1$ (including $r = \infty$), the series diverges.
>
> **3.** If $r = 1$, the test is inconclusive.

> See Appendix A for a formal proof of Theorem 10.20.

**Proof (outline):** The idea behind the proof provides insight. Let's assume the limit $r \geq 0$ exists. Then as $k$ gets large and the ratio $|a_{k+1}/a_k|$ approaches $r$, we have $|a_{k+1}| \approx r|a_k|$. Therefore, as one goes farther and farther out in the series, it behaves like

$$|a_k| + |a_{k+1}| + |a_{k+2}| + \cdots \approx |a_k| + r|a_k| + r^2|a_k| + r^3|a_k| + \cdots$$
$$= |a_k|(1 + r + r^2 + r^3 + \cdots).$$

The tail of the series, which determines whether the series converges, behaves like a geometric series with ratio $r$. We know that if $r < 1$, the geometric series converges, and if $r > 1$, the series diverges, which is the conclusion of the Ratio Test. ◄

**EXAMPLE 1   Using the Ratio Test**  Use the Ratio Test to determine whether the following series converge.

**a.** $\displaystyle\sum_{k=1}^{\infty} \frac{10^k}{k!}$     **b.** $\displaystyle\sum_{k=1}^{\infty} \frac{(-1)^k k^k}{k!}$     **c.** $\displaystyle\sum_{k=1}^{\infty} (-1)^{k+1} e^{-k}(k^2 + 4)$

**SOLUTION**  In each case, the limit of the ratio of successive terms is determined.

> Recall that
> $$k! = k \cdot (k-1) \cdots 2 \cdot 1.$$
> Therefore,
> $$(k+1)! = (k+1)\underbrace{k(k-1)\cdots 1}_{k!}$$
> $$= (k+1)k!.$$

**a.** In this case, we have a series with positive terms. Therefore,

$$r = \lim_{k \to \infty} \left| \frac{a_{k+1}}{a_k} \right| = \lim_{k \to \infty} \frac{10^{k+1}/(k+1)!}{10^k/k!} \qquad \text{Substitute } a_{k+1} \text{ and } a_k.$$

$$= \lim_{k \to \infty} \frac{10^{k+1}}{10^k} \cdot \frac{k!}{(k+1)k!} \qquad \text{Invert and multiply.}$$

$$= \lim_{k \to \infty} \frac{10}{k+1} = 0 \qquad \text{Simplify and evaluate the limit.}$$

Because $r = 0 < 1$, the series converges by the Ratio Test.

**b.** $\displaystyle r = \lim_{k \to \infty} \left| \frac{a_{k+1}}{a_k} \right| = \lim_{k \to \infty} \left| \frac{(-1)^{k+1}(k+1)^{k+1}/(k+1)!}{(-1)^k k^k/k!} \right| \qquad \text{Substitute } a_{k+1} \text{ and } a_k.$

$$= \lim_{k \to \infty} \left( \frac{k+1}{k} \right)^k \qquad \text{Simplify.}$$

> Recall from Section 4.7 that
> $$\lim_{k \to \infty} \left(1 + \frac{1}{k}\right)^k = e \approx 2.718.$$

$$= \lim_{k \to \infty} \left(1 + \frac{1}{k}\right)^k = e \qquad \text{Simplify and evaluate the limit.}$$

Because $r = e > 1$, the series diverges by the Ratio Test. Alternatively, we could have noted that $\lim\limits_{k \to \infty} k^k/k! = \infty$ (Theorem 10.6) and used the Divergence Test to reach the same conclusion.

**c.** $\displaystyle r = \lim_{k \to \infty} \left| \frac{a_{k+1}}{a_k} \right| = \lim_{k \to \infty} \frac{e^{-(k+1)}((k+1)^2 + 4)}{e^{-k}(k^2 + 4)} \qquad \text{Substitute } a_{k+1} \text{ and } a_k.$

$$= \lim_{k \to \infty} \frac{e^{-k}e^{-1}(k^2 + 2k + 5)}{e^{-k}(k^2 + 4)} \qquad \text{Simplify.}$$

$$= e^{-1} \lim_{k \to \infty} \underbrace{\frac{k^2 + 2k + 5}{k^2 + 4}}_{1} \qquad \text{Simplify.}$$

$$= e^{-1}$$

**QUICK CHECK 1** Evaluate $10!/9!$, $(k+2)!/k!$, and $k!/(k+1)!$ ◄

Because $e^{-1} = \dfrac{1}{e} < 1$, the series converges absolutely by the Ratio Test, which implies that it converges.

*Related Exercises 10, 14* ◄

The Ratio Test is conclusive for many series. Nevertheless, observe what happens

▶ In Section 10.8, we offer guidelines to help you to decide which convergence test is best suited for a given series.

when the Ratio Test is applied to the harmonic series $\displaystyle\sum_{k=1}^{\infty}\frac{1}{k}$:

$$r = \lim_{k\to\infty}\left|\frac{a_{k+1}}{a_k}\right| = \lim_{k\to\infty}\frac{1/(k+1)}{1/k} = \lim_{k\to\infty}\frac{k}{k+1} = 1,$$

which means the test is inconclusive. We know the harmonic series diverges, yet the Ratio Test cannot be used to establish this fact. Like all the convergence tests encountered in this chapter, the Ratio Test works only for certain classes of series.

**QUICK CHECK 2** Verify that the Ratio Test is inconclusive for $\displaystyle\sum_{k=1}^{\infty}\frac{1}{k^2}$. What test could be applied to show that $\displaystyle\sum_{k=1}^{\infty}\frac{1}{k^2}$ converges? ◄

## The Root Test

The Root Test is similar to the Ratio Test and is well suited to series whose general term involves powers.

---

**THEOREM 10.21**    **Root Test**

Let $\sum a_k$ be an infinite series, and let $\rho = \displaystyle\lim_{k\to\infty}\sqrt[k]{|a_k|}$.

1. If $\rho < 1$, the series converges absolutely, and therefore it converges (by Theorem 10.19).
2. If $\rho > 1$ (including $\rho = \infty$), the series diverges.
3. If $\rho = 1$, the test is inconclusive.

---

**Proof (outline):** Assume the limit $\rho$ exists. If $k$ is large, we have $\rho \approx \sqrt[k]{|a_k|}$ or $|a_k| \approx \rho^k$. For large values of $k$, the tail of the series, which determines whether a series converges, behaves like

$$|a_k| + |a_{k+1}| + |a_{k+2}| + \cdots \approx \rho^k + \rho^{k+1} + \rho^{k+2} + \cdots.$$

▶ See Appendix A for a formal proof of Theorem 10.21.

Therefore, the tail of the series is approximately a geometric series with ratio $\rho$. If $\rho < 1$, the geometric series converges, and if $\rho > 1$, the series diverges, which is the conclusion of the Root Test. ◄

**EXAMPLE 2**    **Using the Root Test**    Use the Root Test to determine whether the following series converge.

**a.** $\displaystyle\sum_{k=1}^{\infty}\left(\frac{3-4k^2}{7k^2+6}\right)^k$      **b.** $\displaystyle\sum_{k=1}^{\infty}\frac{(-2)^k}{k^{10}}$

**SOLUTION**

**a.** The required limit is

$$\rho = \lim_{k\to\infty}\sqrt[k]{\left|\left(\frac{3-4k^2}{7k^2+6}\right)^k\right|} = \lim_{k\to\infty}\frac{4k^2-3}{7k^2+6} = \frac{4}{7}. \qquad \color{blue}{|3-4k^2| = 4k^2-3, \text{ for } k > 1}$$

Because $\rho < 1$, the series converges absolutely by the Root Test (and therefore it converges).

**b.** In this case,

$$\rho = \lim_{k\to\infty}\sqrt[k]{\left|\frac{(-2)^k}{k^{10}}\right|} = \lim_{k\to\infty}\frac{2}{k^{10/k}} = \lim_{k\to\infty}\frac{2}{(k^{1/k})^{10}} = 2. \qquad \color{blue}{\lim_{k\to\infty}k^{1/k} = 1}$$

Because $\rho > 1$, the series diverges by the Root Test.

     We could have used the Ratio Test for both series in this example, but the Root Test is easier to apply in each case. In part (b), the Divergence Test leads to the same conclusion.

*Related Exercises 17, 32* ◄

## SECTION 10.7 EXERCISES

### Getting Started

1. Explain how the Ratio Test works.

2. Explain how the Root Test works.

3. Evaluate $1000!/998!$ without a calculator.

4. Evaluate $(100!)^2/(99!)^2$ without a calculator.

5. Simplify $\dfrac{k!}{(k+2)!}$, for any integer $k \geq 0$.

6. Simplify $\dfrac{(2k+1)!}{(2k-1)!}$, for any integer $k \geq 1$.

7. What test is advisable if a series involves a factorial term?

8. Can the value of a series be determined using the Root Test or the Ratio Test?

### Practice Exercises

**9–30. The Ratio and Root Tests** *Use the Ratio Test or the Root Test to determine whether the following series converge absolutely or diverge.*

9. $\displaystyle\sum_{k=1}^{\infty} \frac{(-1)^k}{k!}$

10. $\displaystyle\sum_{k=1}^{\infty} \frac{(-2)^k}{k!}$

11. $\displaystyle\sum_{k=1}^{\infty} (-1)^{k+1} \left( \frac{10k^3 + k}{9k^3 + k + 1} \right)^k$

12. $\displaystyle\sum_{k=1}^{\infty} \left( -\frac{2k}{k+1} \right)^k$

13. $\displaystyle\sum_{k=1}^{\infty} \frac{k^2}{4^k}$

14. $\displaystyle\sum_{k=1}^{\infty} \frac{k^k}{2^k}$

15. $\displaystyle\sum_{k=1}^{\infty} (-1)^{k+1} k e^{-k}$

16. $\displaystyle\sum_{k=1}^{\infty} (-1)^k \frac{k!}{k^k}$

17. $\displaystyle\sum_{k=1}^{\infty} \frac{(-7)^k}{k^2}$

18. $\displaystyle\sum_{k=1}^{\infty} \left( 1 + \frac{3}{k} \right)^{k^2}$

19. $\displaystyle\sum_{k=1}^{\infty} \frac{2^k}{k^{99}}$

20. $\displaystyle\sum_{k=1}^{\infty} \frac{(-1)^k k!}{k^6}$

21. $\displaystyle\sum_{k=1}^{\infty} \frac{(k!)^2}{(2k)!}$

22. $2 + \dfrac{4}{16} + \dfrac{8}{81} + \dfrac{16}{256} + \cdots$

23. $\displaystyle\sum_{k=1}^{\infty} \left( \frac{k}{k+1} \right)^{2k^2}$

24. $\displaystyle\sum_{k=1}^{\infty} (-1)^{k+1} \frac{3^{k^2}}{k^k}$

25. $\displaystyle\sum_{k=1}^{\infty} (-1)^{k+1} \left( \frac{3k^2 + 4k}{2k^2 + 1} \right)^k$

26. $\displaystyle\sum_{k=1}^{\infty} \left( \frac{1}{\ln(k+1)} \right)^k$

27. $1 + \left( \dfrac{1}{2} \right)^2 + \left( \dfrac{1}{3} \right)^3 + \left( \dfrac{1}{4} \right)^4 + \cdots$

28. $\displaystyle\sum_{k=1}^{\infty} \frac{k}{e^k}$

29. $\displaystyle\sum_{k=1}^{\infty} \frac{(-1)^{k+1} k^{2k}}{k! k!}$

30. $\displaystyle\sum_{k=1}^{\infty} \left( \frac{k}{1 - 5k} \right)^k$

31. **Explain why or why not** Determine whether the following statements are true and give an explanation or counterexample.

a. $n! n! = (2n)!$, for all positive integers $n$.

b. $\dfrac{(2n)!}{(2n-1)!} = 2n$.

c. If $\displaystyle\lim_{k \to \infty} \sqrt[k]{|a_k|} = \frac{1}{4}$, then $\displaystyle\sum 10 a_k$ converges absolutely.

d. The Ratio Test is always inconclusive when applied to $\displaystyle\sum a_k$, where $a_k$ is a nonzero rational function of $k$.

**32–49. Choose your test** *Use the test of your choice to determine whether the following series converge absolutely, converge conditionally, or diverge.*

32. $\displaystyle\sum_{k=1}^{\infty} \left( \frac{4k^3 + 7}{9k^2 - 1} \right)^k$

33. $\displaystyle\sum_{k=1}^{\infty} (-1)^k \frac{2^{k+1}}{9^{k-1}}$

34. $\displaystyle\sum_{k=1}^{\infty} \frac{k!}{(2k+1)!}$

35. $\displaystyle\sum_{k=1}^{\infty} \frac{(2k+1)!}{(k!)^2}$

36. $\displaystyle\sum_{k=1}^{\infty} \left( \frac{k^2}{2k^2 + 1} \right)^k$

37. $\displaystyle\sum_{k=1}^{\infty} \frac{k^{100}}{(k+1)!}$

38. $\displaystyle\sum_{k=1}^{\infty} k^3 \sin(1/k^3)$

39. $\displaystyle\sum_{k=1}^{\infty} \frac{(-1)^k k^3}{\sqrt{k^8 + 1}}$

40. $\displaystyle\sum_{k=1}^{\infty} \frac{1}{(1+p)^k}, p > 0$

41. $\displaystyle\sum_{k=1}^{\infty} \left( \sqrt[k]{k} - 1 \right)^{2k}$

42. $\displaystyle\sum_{k=1}^{\infty} \frac{(k!)^3}{(3k)!}$

43. $\displaystyle\sum_{k=1}^{\infty} \frac{2^k k!}{k^k}$

44. $\displaystyle\sum_{k=1}^{\infty} \left( 1 - \frac{1}{k} \right)^{k^2}$

45. $\displaystyle\sum_{k=1}^{\infty} \frac{(-1)^k}{k^{0.99}}$

46. $\displaystyle\sum_{k=2}^{\infty} 100 k^{-k}$

47. $\dfrac{1}{1!} + \dfrac{4}{2!} + \dfrac{9}{3!} + \dfrac{16}{4!} + \cdots$

48. $\displaystyle\sum_{k=1}^{\infty} \frac{(-1)^k}{\tan^{-1} k}$

49. $\displaystyle\sum_{k=1}^{\infty} \frac{(-1)^k}{\sqrt{k^{3/2} + k}}$

### Explorations and Challenges

**50–57. Convergence parameter** *Find the values of the parameter $p > 0$ for which the following series converge.*

50. $\displaystyle\sum_{k=2}^{\infty} \frac{1}{(\ln k)^p}$

51. $\displaystyle\sum_{k=2}^{\infty} \frac{\ln k}{k^p}$

52. $\displaystyle\sum_{k=2}^{\infty} \frac{1}{k(\ln k)(\ln \ln k)^p}$

53. $\displaystyle\sum_{k=2}^{\infty} \left( \frac{\ln k}{k} \right)^p$

54. $\displaystyle\sum_{k=0}^{\infty} \frac{k! p^k}{(k+1)^k}$

(*Hint:* Stirling's formula is useful: $k! \approx \sqrt{2\pi k}\, k^k e^{-k}$, for large $k$.)

55. $\displaystyle\sum_{k=1}^{\infty} \frac{k p^k}{k+1}$

56. $\displaystyle\sum_{k=1}^{\infty} \ln \left( \frac{k}{k+1} \right)^p$

57. $\displaystyle\sum_{k=1}^{\infty} \left( 1 - \frac{p}{k} \right)^k$

**58–63. A glimpse ahead to power series** *Use the Ratio Test or the Root Test to determine the values of x for which each series converges.*

**58.** $\displaystyle\sum_{k=1}^{\infty} \frac{x^k}{k!}$  **59.** $\displaystyle\sum_{k=1}^{\infty} x^k$  **60.** $\displaystyle\sum_{k=1}^{\infty} \frac{x^k}{k}$

**61.** $\displaystyle\sum_{k=1}^{\infty} \frac{x^k}{k^2}$  **62.** $\displaystyle\sum_{k=1}^{\infty} \frac{x^{2k}}{k^2}$  **63.** $\displaystyle\sum_{k=1}^{\infty} \frac{x^k}{2^k}$

**64. Stirling's formula** Complete the following steps to find the values of $p > 0$ for which the series $\displaystyle\sum_{k=1}^{\infty} \frac{1 \cdot 3 \cdot 5 \cdots (2k-1)}{p^k k!}$ converges.

**a.** Use the Ratio Test to show that $\displaystyle\sum_{k=1}^{\infty} \frac{1 \cdot 3 \cdot 5 \cdots (2k-1)}{p^k k!}$ converges for $p > 2$.

**b.** Use Stirling's formula, $k! \approx \sqrt{2\pi k}\, k^k e^{-k}$ for large $k$, to determine whether the series converges with $p = 2$.

(*Hint:* $1 \cdot 3 \cdot 5 \cdots (2k-1) = \dfrac{1 \cdot 2 \cdot 3 \cdot 4 \cdot 5 \cdot 6 \cdots (2k-1)2k}{2 \cdot 4 \cdot 6 \cdots 2k}$.)

(See the Guided Project *Stirling's formula and n!* for more on this topic.)

**QUICK CHECK ANSWERS**

**1.** 10; $(k+2)(k+1)$; $1/(k+1)$  **2.** The Integral Test or $p$-series with $p = 2$ ◄

# 10.8 Choosing a Convergence Test

In previous sections, we presented methods for determining whether an infinite series converges, and in some cases we discovered that it is possible to determine the exact value of a convergent series. In this section, we step back and focus on general strategies for classifying infinite series and deciding which tests are appropriate for a specific series.

## Series Strategies

It is helpful if you are already familiar with the convergence tests summarized in Table 10.4. With these tests in mind, we explore various strategies and illustrate them with examples. These strategies can be applied sequentially, more or less in the order in which they are given.

**Categorize the Series** The first step in approaching an infinite series is to note some of the distinctive properties that help to classify it. Is it a series with positive terms? An alternating series? Or something more general? Does the general term consist of rational functions, exponential functions, or factorials? Can the series be simplified in any obvious ways before applying convergence tests? Here is a quick suggestive example.

**EXAMPLE 1** **Categorize the series** Discuss the series $\displaystyle\sum_{k=1}^{\infty} \frac{2^k + \cos(\pi k)\sqrt{k}}{3^{k+1}}$.

**SOLUTION** A good first step in analyzing this series would be to split it into two series that can be examined separately. We begin by writing

$$\sum_{k=1}^{\infty} \frac{2^k + \cos(\pi k)\sqrt{k}}{3^{k+1}} = \underbrace{\sum_{k=1}^{\infty} \frac{2^k}{3^{k+1}}}_{\text{geometric series}} + \underbrace{\sum_{k=1}^{\infty} \frac{\cos(\pi k)\sqrt{k}}{3^{k+1}}}_{\text{alternating series}}.$$

**QUICK CHECK 1** Evaluate the geometric series in Example 1. Does the alternating series in Example 1 converge? ◄

The general term of the first series is the ratio of exponential functions, so this series is a geometric series that can be evaluated explicitly. Because $\cos(\pi k) = (-1)^k$, we see that the second series is an alternating series, whose convergence can also be tested. Notice that for the original series to converge, both the geometric series and the alternating series must converge.

*Related Exercise 8* ◄

**Apply the Divergence Test** Almost without exception, the Divergence Test is the easiest test to apply because it requires nothing more than taking a limit. Furthermore, it can lead to a quick conclusion, namely that the given series diverges.

**Table 10.4** Special Series and Convergence Tests

| Series or Test | Form of Series | Condition for Convergence | Condition for Divergence | Comments |
|---|---|---|---|---|
| Geometric series | $\displaystyle\sum_{k=0}^{\infty} ar^k, a \neq 0$ | $\|r\| < 1$ | $\|r\| \geq 1$ | If $\|r\| < 1$, then $\displaystyle\sum_{k=0}^{\infty} ar^k = \frac{a}{1-r}$. |
| Divergence Test | $\displaystyle\sum_{k=1}^{\infty} a_k$ | Does not apply | $\displaystyle\lim_{k \to \infty} a_k \neq 0$ | Cannot be used to prove convergence |
| Integral Test | $\displaystyle\sum_{k=1}^{\infty} a_k$, where $a_k = f(k)$ and $f$ is continuous, positive, and decreasing | $\displaystyle\int_1^{\infty} f(x)\, dx$ converges. | $\displaystyle\int_1^{\infty} f(x)\, dx$ diverges. | The value of the integral is not the value of the series. |
| $p$-series | $\displaystyle\sum_{k=1}^{\infty} \frac{1}{k^p}$ | $p > 1$ | $p \leq 1$ | Useful for comparison tests |
| Ratio Test | $\displaystyle\sum_{k=1}^{\infty} a_k$ | $\displaystyle\lim_{k \to \infty} \left\| \frac{a_{k+1}}{a_k} \right\| < 1$ | $\displaystyle\lim_{k \to \infty} \left\| \frac{a_{k+1}}{a_k} \right\| > 1$ | Inconclusive if $\displaystyle\lim_{k \to \infty} \left\| \frac{a_{k+1}}{a_k} \right\| = 1$ |
| Root Test | $\displaystyle\sum_{k=1}^{\infty} a_k$ | $\displaystyle\lim_{k \to \infty} \sqrt[k]{\|a_k\|} < 1$ | $\displaystyle\lim_{k \to \infty} \sqrt[k]{\|a_k\|} > 1$ | Inconclusive if $\displaystyle\lim_{k \to \infty} \sqrt[k]{\|a_k\|} = 1$ |
| Comparison Test | $\displaystyle\sum_{k=1}^{\infty} a_k$, where $a_k > 0$ | $a_k \leq b_k$ and $\displaystyle\sum_{k=1}^{\infty} b_k$ converges. | $b_k \leq a_k$ and $\displaystyle\sum_{k=1}^{\infty} b_k$ diverges. | $\displaystyle\sum_{k=1}^{\infty} a_k$ is given; you supply $\displaystyle\sum_{k=1}^{\infty} b_k$. |
| Limit Comparison Test | $\displaystyle\sum_{k=1}^{\infty} a_k$, where $a_k > 0, b_k > 0$ | $0 \leq \displaystyle\lim_{k \to \infty} \frac{a_k}{b_k} < \infty$ and $\displaystyle\sum_{k=1}^{\infty} b_k$ converges. | $\displaystyle\lim_{k \to \infty} \frac{a_k}{b_k} > 0$ and $\displaystyle\sum_{k=1}^{\infty} b_k$ diverges. | $\displaystyle\sum_{k=1}^{\infty} a_k$ is given; you supply $\displaystyle\sum_{k=1}^{\infty} b_k$. |
| Alternating Series Test | $\displaystyle\sum_{k=1}^{\infty} (-1)^k a_k$, where $a_k > 0$ | $\displaystyle\lim_{k \to \infty} a_k = 0$ and $0 < a_{k+1} \leq a_k$ | $\displaystyle\lim_{k \to \infty} a_k \neq 0$ | Remainder $R_n$ satisfies $\|R_n\| \leq a_{n+1}$ |
| Absolute Convergence | $\displaystyle\sum_{k=1}^{\infty} a_k, a_k$ arbitrary | $\displaystyle\sum_{k=1}^{\infty} \|a_k\|$ converges. | | Applies to arbitrary series |

**EXAMPLE 2**   **Divergence Test**   Does the series $\displaystyle\sum_{k=1}^{\infty} \left(1 - \frac{10}{k}\right)^k$ converge or diverge?

**SOLUTION** The terms of this series alternate in sign for $k < 10$, and then remain positive and close to zero for $k > 10$. Applying the Divergence Test, we evaluate the limit of the general term using l'Hôpital's Rule and find that

$$\lim_{k \to \infty} \left(1 - \frac{10}{k}\right)^k = e^{-10} \neq 0.$$

**QUICK CHECK 2** Show the steps in evaluating the limit in Example 2. ◄

Because the limit of the general term is non-zero, the series diverges (very slowly).

*Related Exercise 18* ◄

**Identify Special Series**   We have studied several special families of series that are readily identified and analyzed for convergence. Specifically, you should look for geometric series, $p$-series, and (less obvious and less common) telescoping series.

**EXAMPLE 3** **Special series** Does the series $\displaystyle\sum_{k=4}^{\infty} \frac{1}{\sqrt[4]{k^2 - 6k + 9}}$ converge or diverge?

**SOLUTION** The Divergence Test is inconclusive for this series of positive terms. One approach is to rewrite the series in the following steps:

$$\sum_{k=4}^{\infty} \frac{1}{\sqrt[4]{k^2 - 6k + 9}} = \sum_{k=4}^{\infty} \frac{1}{\sqrt[4]{(k-3)^2}} \qquad \text{Factor denominator.}$$

$$= \sum_{j=1}^{\infty} \frac{1}{\sqrt[4]{j^2}} \qquad \text{Shift index; let } j = k - 3.$$

$$= \sum_{k=1}^{\infty} \frac{1}{\sqrt{k}}. \qquad \text{Simplify.}$$

We now identify this series as a $p$-series with $p = \dfrac{1}{2} < 1$; therefore, it diverges.

*Related Exercise 24* ◄

**Integral, Ratio, and Root Tests** You have probably evaluated enough integrals by now to guess whether the general term of a series can be integrated easily. There are also some immediate clues: If a series involves factorials (something involving $k!$) or terms of the form $f(k)^k$, the Integral Test is not likely to help. On the other hand, series with terms of the form $k!$, $k^k$, or $a^k$, where $a$ is a constant, often submit to the Ratio Test or the Root Test.

> ➤ When using the Integral Test, make sure the integrand satisfies the conditions given in Theorem 10.11: It must be continuous, positive, and decreasing on $[k, \infty)$ for some finite number $k$. In Example 4, $f(x) = x^2 e^{-2x}$ satisfies these conditions on $[1, \infty)$.

**EXAMPLE 4** **More than one test** Does the series $\displaystyle\sum_{k=1}^{\infty} k^2 e^{-2k}$ converge or diverge?

**SOLUTION** The general term of this series of positive terms can be integrated using integration by parts, which reveals that $\displaystyle\int_1^{\infty} x^2 e^{-2x}\, dx = \frac{5e^{-2}}{4} < \infty$. Because the integral converges, the infinite series also converges. However, it may be easier to apply the Ratio Test. We find that

$$\lim_{k\to\infty} \frac{(k+1)^2 e^{-2(k+1)}}{k^2 e^{-2k}} = e^{-2} \underbrace{\lim_{k\to\infty}\left(\frac{k+1}{k}\right)^2}_{1} = e^{-2} \cdot 1 \approx 0.135 < 1.$$

> **QUICK CHECK 3** Verify the limit in the Ratio Test in Example 4. ◄

Because the ratio of successive terms in less than 1, the series converges.

*Related Exercise 25* ◄

**Rational Terms** If a series involves a rational function, or ratios of roots or algebraic functions, then a comparison test is a likely candidate. When the ordinary Comparison Test does not work easily, the more powerful Limit Comparison Test may lead to a conclusion. As shown in the following example, analyzing the end behavior (as $k \to \infty$) of the general term is often the trick needed to find a good comparison series.

**EXAMPLE 5** **Comparison tests** Does the series $\displaystyle\sum_{k=2}^{\infty} \sqrt[3]{\frac{k^2 - 1}{k^4 + 4}}$ converge or diverge?

**SOLUTION** The Divergence Test is inconclusive. Ruling out the Integral, Ratio, and Root Tests as impractical, we are left with a comparison test. The end behavior of the general term is determined by working with the leading terms of the numerator and denominator. We find that for large values of $k$,

$$\sqrt[3]{\frac{k^2 - 1}{k^4 + 4}} \approx \sqrt[3]{\frac{k^2}{k^4}} = \frac{1}{k^{2/3}}.$$

Therefore, a good comparison series is $\displaystyle\sum k^{-2/3}$, which is a divergent $p$-series. The ordinary Comparison Test is difficult to apply in this case, so we turn to the Limit

Comparison Test. Taking the limit of the ratio of terms of the given series and the comparison series, we find that

$$L = \lim_{k \to \infty} \frac{\sqrt[3]{\dfrac{k^2 - 1}{k^4 + 4}}}{k^{-2/3}} = \lim_{k \to \infty} \sqrt[3]{\frac{k^2 - 1}{k^4 + 4} \cdot k^{2/3}} = \lim_{k \to \infty} \sqrt[3]{\frac{k^4 - k^2}{k^4 + 4}} = \sqrt[3]{\lim_{k \to \infty} \frac{1 - k^{-2}}{1 + 4k^{-4}}} = 1.$$

Because $0 < L < \infty$ and the comparison series diverges, the given series also diverges.

*Related Exercise 49* ◄

The guidelines presented in this section are a good start, but in the end, convergence tests are mastered through practice. It's your turn.

## SECTION 10.8 EXERCISES

### Getting Started

**1–10. Choosing convergence tests** *Identify a convergence test for each of the following series. If necessary, explain how to simplify or rewrite the series before applying the convergence test. You do not need to carry out the convergence test.*

**1.** $\displaystyle\sum_{k=1}^{\infty} (-1)^k \left( 2 + \frac{1}{k^2} \right)^k$

**2.** $\displaystyle\sum_{k=3}^{\infty} \frac{2}{k^2 - k - 2}$

**3.** $\displaystyle\sum_{k=3}^{\infty} \frac{2k^2}{k^2 - k - 2}$

**4.** $\displaystyle\sum_{k=3}^{\infty} \frac{1}{k \ln^7 k}$

**5.** $\displaystyle\sum_{k=10}^{\infty} \frac{1}{(k - 9)^5}$

**6.** $\displaystyle\sum_{k=10}^{\infty} \frac{100^k}{k! \, k^2}$

**7.** $\displaystyle\sum_{k=1}^{\infty} \frac{k^2}{k^4 + k^3 + 1}$

**8.** $\displaystyle\sum_{k=1}^{\infty} \frac{(-3)^k}{4^{k+1}}$

**9.** $\displaystyle\sum_{k=1}^{\infty} \frac{(-1)^{k+1}}{\sqrt{2^k + \ln k}}$

**10.** $\displaystyle\sum_{k=1}^{\infty} (\tan^{-1} 2k - \tan^{-1}(2k - 2))$

### Practice Exercises

**11–86. Applying convergence tests** *Determine whether the following series converge. Justify your answers.*

**11.** $\displaystyle\sum_{k=1}^{\infty} \frac{2k^4 + k}{4k^4 - 8k}$

**12.** $\displaystyle\sum_{k=1}^{\infty} 7^{-2k}$

**13.** $\displaystyle\sum_{k=3}^{\infty} \frac{5}{2 + \ln k}$

**14.** $\displaystyle\sum_{k=1}^{\infty} \frac{7k^2 - k - 2}{4k^4 - 3k + 1}$

**15.** $\displaystyle\sum_{k=1}^{\infty} \frac{(-7)^k}{k!}$

**16.** $\displaystyle\sum_{k=1}^{\infty} \frac{7^k}{k! + 10}$

**17.** $\displaystyle\sum_{k=1}^{\infty} \frac{(-k)^3}{3k^3 + 2}$

**18.** $\displaystyle\sum_{k=1}^{\infty} \left( 1 + \frac{a}{k} \right)^k$, $a$ is real

**19.** $\displaystyle\sum_{k=0}^{\infty} \frac{3^{k+4}}{5^{k-2}}$

**20.** $\displaystyle\sum_{j=1}^{\infty} \frac{4j^{10}}{j^{11} + 1}$

**21.** $\displaystyle\sum_{k=1}^{\infty} \frac{(-1)^k k}{k^3 + 1}$

**22.** $\displaystyle\sum_{k=1}^{\infty} \left( \frac{e + 1}{\pi} \right)^k$

**23.** $\displaystyle\sum_{k=1}^{\infty} \frac{k^5}{5^k}$

**24.** $\displaystyle\sum_{k=1}^{\infty} \frac{4}{(k + 3)^3}$

**25.** $\displaystyle\sum_{k=1}^{\infty} \frac{1}{\sqrt{k} e^{\sqrt{k}}}$

**26.** $\displaystyle\sum_{k=1}^{\infty} \frac{5 + \sin k}{\sqrt{k}}$

**27.** $\displaystyle\sum_{k=1}^{\infty} \frac{3 + \cos 5k}{k^3}$

**28.** $\displaystyle\sum_{k=3}^{\infty} \frac{(-1)^k \ln k}{k^{1/3}}$

**29.** $\displaystyle\sum_{k=1}^{\infty} \frac{10^k + 1}{k^{10}}$

**30.** $\displaystyle\sum_{k=3}^{\infty} \frac{1}{k^3 \ln k}$

**31.** $\displaystyle\sum_{j=1}^{\infty} \frac{5}{j^2 + 4}$

**32.** $\displaystyle\sum_{k=1}^{\infty} \frac{k^k}{(k + 2)^k}$

**33.** $\displaystyle\sum_{k=3}^{\infty} \frac{1}{k^{1/3} \ln k}$

**34.** $\displaystyle\sum_{k=1}^{\infty} \frac{(-1)^k 5k^2}{\sqrt{3k^5 + 1}}$

**35.** $\displaystyle\sum_{k=1}^{\infty} \frac{2^k 3^k}{k^k}$

**36.** $\displaystyle\sum_{k=1}^{\infty} (-1)^{k+1} \frac{2k + 1}{k!}$

**37.** $\displaystyle\sum_{k=1}^{\infty} (-1)^k \left( \frac{5k}{3k + 7} \right)^k$

**38.** $\displaystyle\sum_{k=1}^{\infty} \frac{2^k (k - 1)!}{k!}$

**39.** $\displaystyle\sum_{k=1}^{\infty} \frac{5^k (k!)^2}{(2k)!}$

**40.** $\displaystyle\sum_{j=1}^{\infty} \frac{\cos ((2j + 1)\pi)}{j^2 + 1}$

**41.** $\displaystyle\sum_{k=1}^{\infty} \frac{2^k}{3^k - 2^k}$

**42.** $\displaystyle\sum_{k=1}^{\infty} \frac{2 + (-1)^k}{2^k}$

**43.** $\displaystyle\sum_{k=1}^{\infty} \cos \frac{(3k - 1)\pi}{3}$

**44.** $\displaystyle\sum_{k=10}^{\infty} \frac{1}{\sqrt{k(k - 9)}}$

**45.** $\displaystyle\sum_{k=1}^{\infty} \frac{k^4}{e^{k^5}}$

**46.** $\displaystyle\sum_{k=1}^{\infty} \frac{1}{(k + 1)! - k!}$

**47.** $\displaystyle\sum_{k=1}^{\infty} \frac{(4k)!}{(k!)^4}$

**48.** $\dfrac{2}{3} + \dfrac{3}{8} + \dfrac{4}{15} + \dfrac{5}{24} + \dfrac{6}{35} + \cdots$

**49.** $\displaystyle\sum_{k=1}^{\infty} \frac{\sqrt[5]{k}}{\sqrt[5]{k^7 + 1}}$

**50.** $\displaystyle\sum_{k=1}^{\infty} e^{-k^3}$

**51.** $\displaystyle\sum_{k=1}^{\infty} \frac{7^k + 11^k}{11^k}$

**52.** $\displaystyle\sum_{k=1}^{\infty} \frac{7^k + 11^k}{13^k}$

**53.** $\displaystyle\sum_{k=1}^{\infty} \sin \frac{1}{k^9}$

**54.** $\displaystyle\sum_{j=1}^{\infty} j^9 \sin \frac{1}{j^9}$

**55.** $\displaystyle\sum_{k=1}^{\infty} \cos \frac{1}{k^9}$

**56.** $\displaystyle\sum_{k=1}^{\infty} \left(\frac{k}{k+10}\right)^{-k^2}$

**57.** $\displaystyle\sum_{k=1}^{\infty} 5^{1-2k}$

**58.** $\displaystyle\sum_{k=1}^{\infty} \frac{(-1)^k}{\sqrt{\ln(k+2)}}$

**59.** $\displaystyle\sum_{k=1}^{\infty} \frac{k!}{k^k + 3}$

**60.** $\displaystyle\sum_{k=1}^{\infty} \left(\frac{1}{k^2} + \frac{1}{k^5}\right)$

**61.** $\displaystyle\sum_{k=1}^{\infty} \frac{1}{\ln(e^k + 1)}$

**62.** $\displaystyle\sum_{k=0}^{\infty} \left(\frac{\tan^{-1} k}{\pi}\right)^k$

**63.** $\displaystyle\sum_{k=1}^{\infty} \left(\frac{k+a}{k}\right)^{k^2}, a > 0$

**64.** $\displaystyle\frac{1}{1\cdot 4} + \frac{1}{2\cdot 7} + \frac{1}{3\cdot 10} + \frac{1}{4\cdot 13} + \cdots$

**65.** $\displaystyle\sum_{k=1}^{\infty} \left(\cos\frac{1}{k} - \cos\frac{1}{k+1}\right)$

**66.** $\displaystyle\sum_{k=1}^{\infty} \frac{4^{k^2}}{k!}$

**67.** $\displaystyle\sum_{j=1}^{\infty} \frac{\cot^{-1}\frac{1}{j}}{2^j}$

**68.** $\displaystyle\sum_{k=2}^{\infty} \frac{\sqrt{k+1} - \sqrt{k}}{\sqrt{k^2 - k}}$

**69.** $\displaystyle\sum_{k=1}^{\infty} \left(1 + \frac{1}{2k}\right)^k$

**70.** $\displaystyle\sum_{k=0}^{\infty} e^{k/100}$

**71.** $\displaystyle\sum_{k=1}^{\infty} \frac{\ln^2 k}{k^{3/2}}$

**72.** $\displaystyle\sum_{k=1}^{\infty} (-1)^k \cos\frac{1}{k}$

**73.** $\displaystyle\sum_{k=0}^{\infty} k^2 \cdot 1.001^{-k}$

**74.** $\displaystyle\sum_{k=0}^{\infty} k \cdot 0.999^{-k}$

**75.** $\displaystyle\sum_{k=1}^{\infty} \left(-\frac{1}{k}\right)^k$

**76.** $\displaystyle\sum_{k=3}^{\infty} \frac{2}{k(2-k)}$

**77.** $\displaystyle\sum_{k=0}^{\infty} \frac{3k}{\sqrt[4]{k^4 + 3}}$

**78.** $\displaystyle\sum_{k=0}^{\infty} k \cdot 3^{-k^2}$

**79.** $\displaystyle\sum_{k=1}^{\infty} \tan^{-1}\frac{1}{\sqrt{k}}$

**80.** $\displaystyle\sum_{k=1}^{\infty} \left(\frac{3k^8 - 2}{3k^9 + 2}\right)^k$

**81.** $\displaystyle\sum_{k=1}^{\infty} \left(\frac{1}{\sqrt{k+2}} - \frac{1}{\sqrt{k}}\right)$

**82.** $\displaystyle\sum_{k=0}^{\infty} \left(\left(\frac{2}{3}\right)^{k+1} - \left(\frac{3}{2}\right)^{-k+1}\right)$

**83.** $\displaystyle\sum_{j=2}^{\infty} \frac{1}{j\ln^{10} j}$

**84.** $\displaystyle\sum_{k=1}^{\infty} \tan^{-1}\frac{1}{k^3}$

**85.** $\displaystyle\frac{1}{2\cdot 3} + \frac{1}{4\cdot 5} + \frac{1}{6\cdot 7} + \frac{1}{8\cdot 9} + \cdots$

**86.** $\displaystyle\sum_{k=1}^{\infty} \frac{a^k}{k!}, a \neq 0$

**87.** **Explain why or why not** Determine whether the following statements are true and give an explanation or counterexample.

    **a.** If the Limit Comparison Test can be applied successfully to a given series with a certain comparison series, the Comparison Test also works with the same comparison series.

    **b.** The series $\displaystyle\sum_{k=3}^{\infty} \frac{1}{k\ln^p k}$ converges for the same values of $p$ as the series $\displaystyle\sum_{k=3}^{\infty} \frac{1}{k^p}$.

    **c.** Both the Ratio Test and the Root Test can be applied conclusively to a geometric series.

    **d.** The Alternating Series Test can be used to show that some series diverge.

### Explorations and Challenges

**88–93. A few more series** *Determine whether the following series converge. Justify your answers.*

**88.** $\displaystyle\sum_{k=1}^{\infty} \left(\sqrt{k^4 + 1} - k^2\right)$

**89.** $\displaystyle\sum_{k=1}^{\infty} \frac{k}{1 + 2 + \cdots + k}$

**90.** $\displaystyle\sum_{k=1}^{\infty} \left(\frac{1^3 + 2^3 + 3^3 + \cdots + k^3}{k^5}\right)$

**91.** $\displaystyle\sum_{k=0}^{\infty} \ln\left(\frac{2k+1}{2k+4}\right)$

**92.** $\displaystyle\sum_{k=1}^{\infty} \frac{1}{10^{\ln k}}$

**93.** $\displaystyle\sum_{k=1}^{\infty} \frac{1}{2^{\ln k} + 2}$

**QUICK CHECK ANSWERS**

**1.** $2/3$; yes ◄

---

### CHAPTER 10   REVIEW EXERCISES

**1.** **Explain why or why not** Determine whether the following statements are true and give an explanation or counterexample.

    **a.** The terms of the sequence $\{a_n\}$ increase in magnitude, so the limit of the sequence does not exist.

    **b.** The terms of the series $\displaystyle\sum \frac{1}{\sqrt{k}}$ approach zero, so the series converges.

    **c.** The terms of the sequence of partial sums of the series $\displaystyle\sum a_k$ approach $\frac{5}{2}$, so the infinite series converges to $\frac{5}{2}$.

    **d.** An alternating series that converges absolutely must converge conditionally.

    **e.** The sequence $a_n = \dfrac{n^2}{n^2 + 1}$ converges.

    **f.** The sequence $a_n = \dfrac{(-1)^n n^2}{n^2 + 1}$ converges.

    **g.** The series $\displaystyle\sum_{k=1}^{\infty} \frac{k^2}{k^2 + 1}$ converges.

    **h.** The sequence of partial sums associated with the series $\displaystyle\sum_{k=1}^{\infty} \frac{1}{k^2 + 1}$ converges.

**2.** **Related sequences**

   **a.** Find the limit of the sequence $\left\{ \dfrac{12n^3 + 4n}{6n^3 + 5} \right\}$.

   **b.** Explain why the sequence $\left\{ (-1)^n \dfrac{12n^3 + 4n}{6n^3 + 5} \right\}$ diverges.

**3.** **Geometric sums** Evaluate the geometric sums $\displaystyle\sum_{k=0}^{9} (0.2)^k$ and $\displaystyle\sum_{k=2}^{9} (0.2)^k$.

**4.** **A savings plan** Suppose you open a savings account by depositing \$100. The account earns interest at an annual rate of 3% per year (0.25% per month). At the end of each month, you earn interest on the current balance, and then you deposit \$100. Let $B_n$ be the balance at the beginning of the $n$th month, where $B_0 = \$100$.

   **a.** Find a recurrence relation for the sequence $\{B_n\}$.

   **b.** Find an explicit formula that gives $B_n$, for $n = 0, 1, 2, 3, \ldots$.

   **c.** Find $B_{30}$, the balance at the beginning of the 30th month.

**5.** **Partial sums** Let $S_n$ be the $n$th partial sum of $\displaystyle\sum_{k=1}^{\infty} a_k = 8$. Find $\displaystyle\lim_{k\to\infty} a_k$ and $\displaystyle\lim_{n\to\infty} S_n$.

**6.** Find the value of $r$ for which $\displaystyle\sum_{k=0}^{\infty} 3r^k = 6$.

**7.** Give an example (if possible) of a sequence $\{a_k\}$ that converges, while the series $\displaystyle\sum_{k=1}^{\infty} a_k$ diverges.

**8.** Give an example (if possible) of a series $\displaystyle\sum_{k=1}^{\infty} a_k$ that converges, while the sequence $\{a_k\}$ diverges.

**9.** **Sequences versus series**

   **a.** Find the limit of the sequence $\left\{ \left( -\dfrac{4}{5} \right)^k \right\}$.

   **b.** Evaluate $\displaystyle\sum_{k=0}^{\infty} \left( -\dfrac{4}{5} \right)^k$.

**10.** **Sequences versus series**

   **a.** Find the limit of the sequence $\left\{ \dfrac{1}{k} - \dfrac{1}{k+1} \right\}$.

   **b.** Evaluate $\displaystyle\sum_{k=1}^{\infty} \left( \dfrac{1}{k} - \dfrac{1}{k+1} \right)$.

**11.** **a.** Does the sequence $\left\{ \dfrac{k}{k+1} \right\}$ converge? Why or why not?

   **b.** Does the series $\displaystyle\sum_{k=1}^{\infty} \dfrac{k}{k+1}$ converge? Why or why not?

**12–24. Limits of sequences** *Evaluate the limit of the sequence or state that it does not exist.*

**12.** $a_n = \dfrac{3n^3 + 4n}{6n^3 + 5}$

**13.** $a_n = (-1)^n \dfrac{3n^3 + 4n}{6n^3 + 5}$

**14.** $a_n = \dfrac{(2n+2)!}{(2n)!\, n^2}$

**15.** $a_n = \dfrac{2^n + 5^{n+1}}{5^n}$

**16.** $a_n = \dfrac{n^2 + 4}{\sqrt{4n^4 + 1}}$

**17.** $a_n = \dfrac{8^n}{n!}$

**18.** $a_n = \left( 1 + \dfrac{3}{n} \right)^{2n}$

**19.** $a_n = \dfrac{(3n^2 + 2n + 1)\sin n}{4n^3 + n}$ (*Hint:* Use the Squeeze Theorem.)

**20.** $a_n = n - \sqrt{n^2 - 1}$

**21.** $a_n = \left( \dfrac{1}{n} \right)^{1/\ln n}$

**22.** $a_n = \sin \dfrac{\pi n}{6}$

**23.** $a_n = \dfrac{(-1)^n}{0.9^n}$

**24.** $a_n = \dfrac{\pi}{2} - \tan^{-1} n$

**25–26. Recursively defined sequences** *The following sequences* $\{a_n\}_{n=0}^{\infty}$ *are defined by a recurrence relation. Assume each sequence is monotonic and bounded.*

   **a.** *Find the first five terms* $a_0, a_1, \ldots, a_4$ *of each sequence.*

   **b.** *Determine the limit of each sequence.*

**25.** $a_{n+1} = \dfrac{1}{2} a_n + 8; \; a_0 = 80$

**T 26.** $a_{n+1} = \sqrt{\dfrac{1}{3} a_n + 34}; \; a_0 = 81$

**27–37. Evaluating series** *Evaluate the following infinite series or state that the series diverges.*

**27.** $\displaystyle\sum_{k=1}^{\infty} 3(1.001)^k$

**28.** $\displaystyle\sum_{k=1}^{\infty} \left( \dfrac{9}{10} \right)^k$

**29.** $\displaystyle\sum_{k=0}^{\infty} \left( \left( \dfrac{1}{3} \right)^k + \left( \dfrac{4}{3} \right)^k \right)$

**30.** $\displaystyle\sum_{k=1}^{\infty} \dfrac{3^k - 5^k}{10^k}$

**31.** $\displaystyle\sum_{k=1}^{\infty} \ln \left( \dfrac{2k+1}{2k-1} \right)$

**32.** $\displaystyle\sum_{k=0}^{\infty} \left( -\dfrac{1}{5} \right)^k$

**33.** $\displaystyle\sum_{k=0}^{\infty} (\tan^{-1}(k+2) - \tan^{-1} k)$

**34.** $\displaystyle\sum_{k=2}^{\infty} \left( \dfrac{1}{\sqrt{k}} - \dfrac{1}{\sqrt{k-1}} \right)$

**35.** $\displaystyle\sum_{k=1}^{\infty} \dfrac{9}{(3k-2)(3k+1)}$

**36.** $\displaystyle\sum_{k=1}^{\infty} 4^{-3k}$

**37.** $\displaystyle\sum_{k=1}^{\infty} \dfrac{2^k}{3^{k+2}}$

**38.** Express $0.29292929\ldots$ as a ratio of two integers.

**39.** Express $0.314141414\ldots$ as a ratio of two integers.

**40.** **Finding steady states through sequences** Suppose you take 100 mg of aspirin once per day. Assume the aspirin has a half-life of one day; that is, every day, half of the aspirin in your blood is eliminated. Assume $d_n$ is the amount of aspirin in your blood after the $n$th dose, where $d_1 = 100$.

   **a.** Find a recurrence relation for the sequence $\{d_n\}$.

   **b.** Assuming the sequence $\{d_n\}$ converges, find the long-term (steady-state) amount of aspirin in your blood.

**41.** **Finding steady states using infinite series** Solve Exercise 40 by expressing the amount of aspirin in your blood as a geometric series and evaluating the series.

**42–76. Convergence or divergence** *Use a convergence test of your choice to determine whether the following series converge.*

**42.** $\displaystyle\sum_{k=1}^{\infty} \frac{k}{k^{5/2}+1}$

**43.** $\displaystyle\sum_{k=1}^{\infty} k^{-2/3}$

**44.** $\displaystyle\sum_{j=1}^{\infty} \frac{2j^4+j^3+1}{3j^6+4}$

**45.** $\displaystyle\sum_{k=1}^{\infty} \frac{2k^4+k^3+1}{3k^4+4}$

**46.** $\displaystyle\sum_{k=1}^{\infty} \frac{k^{k/2}2^k}{\sqrt{k!}}$

**47.** $\displaystyle\sum_{k=1}^{\infty} \frac{7+\sin k}{k^2}$

**48.** $\displaystyle\sum_{k=1}^{\infty} \frac{2k^4+1}{\sqrt{9k^8+2}}$

**49.** $\displaystyle\sum_{k=1}^{\infty} \frac{k^4}{\sqrt{9k^{12}+2}}$

**50.** $\displaystyle\sum_{k=1}^{\infty} \sqrt{\frac{k!}{k^k}}$

**51.** $\displaystyle\sum_{k=1}^{\infty} \frac{2^k}{e^k}$

**52.** $\displaystyle\sum_{k=1}^{\infty} \left(\frac{k}{k+3}\right)^{2k}$

**53.** $\displaystyle\sum_{k=1}^{\infty} \left(\frac{k}{2k+3}\right)^{k}$

**54.** $\displaystyle\sum_{k=1}^{\infty} \sin \frac{1}{k^4}$

**55.** $\displaystyle\sum_{k=1}^{\infty} \frac{k!}{e^k k^k}$

**56.** $\displaystyle\sum_{k=1}^{\infty} \frac{1}{\sqrt{k}\sqrt{k+1}}$

**57.** $\displaystyle\sum_{k=1}^{\infty} \frac{5^k}{2^{2k+1}}$

**58.** $\displaystyle\sum_{k=1}^{\infty} ke^{-k}$

**59.** $\displaystyle\sum_{j=0}^{\infty} \frac{2\cdot 4^j}{(2j+1)!}$

**60.** $\displaystyle\sum_{k=0}^{\infty} \frac{9^k}{(2k)!}$

**61.** $\displaystyle\sum_{k=3}^{\infty} \frac{\ln k}{k^{3/2}}$

**62.** $\displaystyle\sum_{k=1}^{\infty} \frac{k^2}{(2k-1)!}$

**63.** $\displaystyle\sum_{k=1}^{\infty} \frac{3}{2+e^k}$

**64.** $\displaystyle\sum_{k=1}^{\infty} k \sin \frac{1}{k}$

**65.** $\displaystyle\sum_{k=1}^{\infty} \frac{\sqrt[k]{k}}{k^3}$

**66.** $\displaystyle\sum_{k=1}^{\infty} \frac{1}{1+\ln k}$

**67.** $\displaystyle\sum_{k=1}^{\infty} k^5 e^{-k}$

**68.** $\displaystyle\sum_{k=4}^{\infty} \frac{2}{k^2-10}$

**69.** $\displaystyle\sum_{k=1}^{\infty} \left(1-\cos\frac{1}{k}\right)$

**70.** $\displaystyle\sum_{k=1}^{\infty} \left(\frac{\pi}{2}-\tan^{-1}k\right)$

**71.** $\displaystyle\sum_{k=1}^{\infty} \left(1-\cos\frac{1}{k}\right)^2$

**72.** $\displaystyle\sum_{k=1}^{\infty} \left(\frac{\pi}{2}-\tan^{-1}k\right)^2$

**73.** $\displaystyle\sum_{k=1}^{\infty} \frac{\coth k}{k}$

**74.** $\displaystyle\sum_{k=1}^{\infty} \frac{1}{\sinh k}$

**75.** $\displaystyle\sum_{k=1}^{\infty} \tanh k$

**76.** $\displaystyle\sum_{k=0}^{\infty} \operatorname{sech} k$

**77–87. Absolute or conditional convergence** *Determine whether the following series converge absolutely, converge conditionally, or diverge.*

**77.** $\displaystyle\sum_{k=1}^{\infty} \frac{(-1)^{k+1}}{k^{3/7}}$

**78.** $\displaystyle\sum_{k=1}^{\infty} \frac{(-1)^k}{k^{5/2}}$

**79.** $\displaystyle\sum_{k=2}^{\infty} \frac{(-1)^k}{k^2-1}$

**80.** $\displaystyle\sum_{j=1}^{\infty} \frac{(-1)^j}{\ln(j+3)}$

**81.** $\displaystyle\sum_{k=1}^{\infty} \frac{(-1)^{k+1}(k^2+4)}{2k^2+1}$

**82.** $\displaystyle\sum_{k=1}^{\infty} \frac{(-1)^k}{\sqrt{k^2+1}}$

**83.** $\displaystyle\sum_{k=1}^{\infty} (-1)^k k e^{-k}$

**84.** $\displaystyle\sum_{k=0}^{\infty} \frac{(-1)^k}{e^k+e^{-k}}$

**85.** $\displaystyle\sum_{k=1}^{\infty} \frac{(-1)^{k+1}10^k}{k!}$

**86.** $\displaystyle\sum_{k=2}^{\infty} \frac{(-1)^k}{k \ln k}$

**87.** $\displaystyle\sum_{k=1}^{\infty} \frac{(-2)^{k+1}}{k^2}$

**88. Logarithmic *p*-series** Show that the series $\displaystyle\sum_{k=2}^{\infty} \frac{1}{k(\ln k)^p}$ converges provided $p > 1$.

**T 89–90. Lower and upper bounds of a series** *For each convergent series and given value of n, complete the following.*

**a.** Use the nth partial sum $S_n$ to estimate the value of the series.
**b.** Find an upper bound for the remainder $R_n$.
**c.** Find lower and upper bounds ($L_n$ and $U_n$, respectively) for the exact value of the series.

**89.** $\displaystyle\sum_{k=1}^{\infty} \frac{1}{k^5}; n = 5$

**90.** $\displaystyle\sum_{k=1}^{\infty} \frac{1}{k^2+1}; n = 20$

**91.** Estimate the value of the series $\displaystyle\sum_{k=1}^{\infty} \frac{1}{(2k+5)^3}$ to within $10^{-4}$ of its exact value.

**92.** Estimate the value of the series $\displaystyle\sum_{k=1}^{\infty} \frac{(-1)^{k+1}}{(2k+5)^3}$ using the fifth partial sum $S_5$ and then find an upper bound for the error $|S - S_5|$.

**T 93. Error in a finite alternating sum** How many terms of the series $\displaystyle\sum_{k=1}^{\infty} \frac{(-1)^{k+1}}{k^4}$ must be summed to ensure that the approximation error is less than $10^{-8}$?

**94. Equations involving series** Solve the following equations for $x$.

**a.** $\displaystyle\sum_{k=0}^{\infty} e^{kx} = 2$

**b.** $\displaystyle\sum_{k=0}^{\infty} (3x)^k = 4$

**95. Building a tunnel—first scenario** A crew of workers is constructing a tunnel through a mountain. Understandably, the rate of construction decreases because rocks and earth must be removed a greater distance as the tunnel gets longer. Suppose each week the crew digs 0.95 of the distance it dug the previous week. In the first week, the crew constructed 100 m of tunnel.

**a.** How far does the crew dig in 10 weeks? 20 weeks? N weeks?
**b.** What is the longest tunnel the crew can build at this rate?

**96. Building a tunnel—second scenario** As in Exercise 95, a crew of workers is constructing a tunnel. The time required to dig 100 m increases by 10% each week, starting with 1 week to dig the first 100 m. Can the crew complete a 1.5-km (1500-m) tunnel in 30 weeks? Explain.

**97. Pages of circles** On page 1 of a book, there is one circle of radius 1. On page 2, there are two circles of radius $1/2$. On page $n$, there are $2^{n-1}$ circles of radius $2^{-n+1}$.

**a.** What is the sum of the areas of the circles on page $n$ of the book?
**b.** Assuming the book continues indefinitely ($n \to \infty$), what is the sum of the areas of all the circles in the book?

**T 98. Max sine sequence** Let $a_n = \max\{\sin 1, \sin 2, \ldots, \sin n\}$, for $n = 1, 2, 3, \ldots$, where $\max\{\ldots\}$ denotes the maximum element of the set. Does $\{a_n\}$ converge? If so, make a conjecture about the limit.

**99. Sierpinski triangle** The fractal called the *Sierpinski triangle* is the limit of a sequence of figures. Starting with the equilateral triangle with sides of length 1, an inverted equilateral triangle with sides of length $1/2$ is removed. Then, three inverted equilateral triangles with sides of length $1/4$ are removed from this figure (see figure). The process continues in this way. Let $T_n$ be the total area of the removed triangles after stage $n$ of the process. The area of an equilateral triangle with side length $L$ is $A = \dfrac{\sqrt{3}L^2}{4}$.

    **a.** Find $T_1$ and $T_2$, the total area of the removed triangles after stages 1 and 2, respectively.

    **b.** Find $T_n$, for $n = 1, 2, 3, \ldots$.

    **c.** Find $\displaystyle\lim_{n \to \infty} T_n$.

    **d.** What is the area of the original triangle that remains as $n \to \infty$?

    Initial stage          First stage          Second stage

**100. Multiplier effect** Imagine that the government of a small community decides to give a total of $\$W$, distributed equally, to all its citizens. Suppose each month each citizen saves a fraction $p$ of his or her new wealth and spends the remaining $1 - p$ in the community. Assume no money leaves or enters the community, and all the spent money is redistributed throughout the community.

    **a.** If this cycle of saving and spending continues for many months, how much money is ultimately spent? Specifically, by what factor is the initial investment of $\$W$ increased? Economists refer to this increase in the investment as the *multiplier effect*.

    **b.** Evaluate the limits $p \to 0$ and $p \to 1$, and interpret their meanings. (See the Guided Project *Economic stimulus packages* for more on stimulus packages.)

**101. Bouncing ball for time** Suppose a rubber ball, when dropped from a given height, returns to a fraction $p$ of that height. In the absence of air resistance, a ball dropped from a height $h$ requires $\sqrt{\dfrac{2h}{g}}$ seconds to fall to the ground, where $g \approx 9.8 \text{ m/s}^2$ is the acceleration due to gravity. The time taken to bounce *up* to a given height equals the time to fall from that height to the ground. How long does it take for a ball dropped from 10 m to come to rest?

## Chapter 10 Guided Projects

*Applications of the material in this chapter and related topics can be found in the following Guided Projects. For additional information, see the Preface.*

- Chaos!
- Financial matters
- Periodic drug dosing
- Economic stimulus packages

- The mathematics of loans
- Archimedes' approximation to $\pi$
- Exact values of infinite series
- Conditional convergence in a crystal lattice

# 11

# Power Series

**Chapter Preview**   Until now, you have worked with infinite series consisting of real numbers. In this chapter, we make a seemingly small, but significant, change by considering infinite series whose terms include powers of a variable. With this change, an infinite series becomes a *power series*. One of the most fundamental ideas in all of calculus is that functions can be represented by power series. As a first step toward this result, we look at approximating functions using polynomials. The transition from polynomials to power series is then straightforward, and we learn how to represent the familiar functions of mathematics in terms of power series called *Taylor series*. The remainder of the chapter is devoted to the properties and many uses of Taylor series.

# 11.1 Approximating Functions with Polynomials

Power series provide a way to represent familiar functions and to define new functions. For this reason, power series—like sets and functions—are among the most fundamental entities in mathematics.

## What Is a Power Series?

A *power series* is an infinite series of the form

$$\sum_{k=0}^{\infty} c_k x^k = \underbrace{c_0 + c_1 x + c_2 x^2 + \cdots + c_n x^n}_{n\text{th-degree polynomial}} + \underbrace{c_{n+1} x^{n+1} + \cdots,}_{\text{terms continue}}$$

or, more generally,

$$\sum_{k=0}^{\infty} c_k (x - a)^k = \underbrace{c_0 + c_1(x - a) + \cdots + c_n(x - a)^n}_{n\text{th-degree polynomial}} + \underbrace{c_{n+1}(x - a)^{n+1} + \cdots,}_{\text{terms continue}}$$

where the *center* of the series $a$ and the coefficients $c_k$ are constants. This type of series is called a power series because it consists of powers of $x$ or $(x - a)$.

Viewed another way, a power series is built up from polynomials of increasing degree, as shown in the following progression.

$$\text{Degree 0: } c_0$$
$$\text{Degree 1: } c_0 + c_1 x$$
$$\text{Degree 2: } c_0 + c_1 x + c_2 x^2$$
$$\vdots \qquad \vdots \qquad \vdots$$
$$\text{Degree } n: c_0 + c_1 x + c_2 x^2 + \cdots + c_n x^n = \sum_{k=0}^{n} c_k x^k$$

$$\left.\begin{array}{c}\\\\\\\\\\\\\end{array}\right\} \text{Polynomials}$$

$$\vdots \qquad \vdots \qquad \vdots$$
$$c_0 + c_1 x + c_2 x^2 + \cdots + c_n x^n + \cdots = \sum_{k=0}^{\infty} c_k x^k$$

$$\left.\begin{array}{c}\\\\\end{array}\right\} \text{Power series}$$

According to this perspective, a power series is a "super-polynomial." Therefore, we begin our exploration of power series by using polynomials to approximate functions.

## Polynomial Approximation

An important observation motivates our work. To evaluate a polynomial $\left(\text{such as } f(x) = x^8 - 4x^5 + \frac{1}{2}\right)$, all we need is arithmetic—addition, subtraction, multiplication, and division. However, algebraic functions $\left(\text{say, } f(x) = \sqrt[3]{x^4 - 1}\right)$ and the trigonometric, logarithmic, and exponential functions usually cannot be evaluated exactly using arithmetic. Therefore, it makes practical sense to use the simplest of functions, polynomials, to approximate more complicated functions.

## Linear and Quadratic Approximation

In Section 4.6, you learned that if a function $f$ is differentiable at a point $a$, then it can be approximated near $a$ by its tangent line, which is the linear approximation to $f$ at the point $a$. The linear approximation at $a$ is given by

$$y - f(a) = f'(a)(x - a) \quad \text{or} \quad y = f(a) + f'(a)(x - a).$$

Because the linear approximation is a first-degree polynomial, we name it $p_1$:

$$p_1(x) = f(a) + f'(a)(x - a).$$

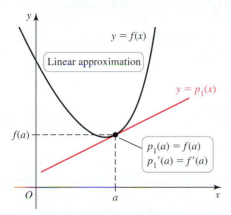

**Figure 11.1**

This polynomial has some important properties: It matches $f$ in *value* and in *slope* at $a$. In other words (**Figure 11.1**),

$$p_1(a) = f(a) \quad \text{and} \quad p_1'(a) = f'(a).$$

Linear approximation works well if $f$ has a fairly constant slope near $a$. However, if $f$ has a lot of curvature near $a$, then the tangent line may not provide an accurate approximation. To remedy this situation, we create a quadratic approximating polynomial by adding one new term to the linear polynomial. Denoting this new polynomial $p_2$, we let

$$p_2(x) = \underbrace{f(a) + f'(a)(x - a)}_{p_1(x)} + \underbrace{c_2(x - a)^2}_{\text{quadratic term}}.$$

The new term consists of a coefficient $c_2$ that must be determined and a quadratic factor $(x - a)^2$.

To determine $c_2$ and to ensure that $p_2$ is a good approximation to $f$ near the point $a$, we require that $p_2$ agree with $f$ in value, slope, and concavity at $a$; that is, $p_2$ must satisfy the matching conditions

$$p_2(a) = f(a), \quad p_2'(a) = f'(a), \quad \text{and} \quad p_2''(a) = f''(a),$$

where we assume $f$ and its first and second derivatives exist at $a$ (**Figure 11.2**).

Substituting $x = a$ into $p_2$, we see immediately that $p_2(a) = f(a)$, so the first matching condition is met. Differentiating $p_2$ once, we have

$$p_2'(x) = f'(a) + 2c_2(x - a).$$

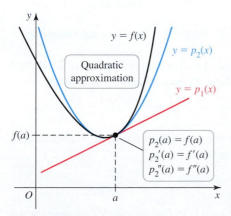

**Figure 11.2**

▶ Matching concavity (second derivatives) ensures that the graph of $p_2$ bends in the same direction as the graph of $f$ at $a$.

So $p_2'(a) = f'(a)$, and the second matching condition is also met. Because $p_2''(a) = 2c_2$, the third matching condition is

$$p_2''(a) = 2c_2 = f''(a).$$

It follows that $c_2 = \frac{1}{2}f''(a)$; therefore, the quadratic approximating polynomial is

$$p_2(x) = \underbrace{f(a) + f'(a)(x - a)}_{p_1(x)} + \frac{f''(a)}{2}(x - a)^2.$$

**EXAMPLE 1** **Linear and quadratic approximations for ln $x$**

**a.** Find the linear approximation to $f(x) = \ln x$ at $x = 1$.

**b.** Find the quadratic approximation to $f(x) = \ln x$ at $x = 1$.

**c.** Use these approximations to estimate ln 1.05.

**SOLUTION**

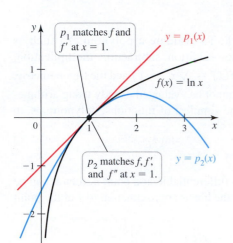

$p_1$ matches $f$ and $f'$ at $x = 1$.

$y = p_1(x)$

$f(x) = \ln x$

$p_2$ matches $f, f',$ and $f''$ at $x = 1$.

$y = p_2(x)$

**Figure 11.3**

**a.** Note that $f(1) = 0$, $f'(x) = 1/x$, and $f'(1) = 1$. Therefore, the linear approximation to $f(x) = \ln x$ at $x = 1$ is

$$p_1(x) = f(1) + f'(1)(x - 1) = 0 + 1(x - 1) = x - 1.$$

As shown in **Figure 11.3**, $p_1$ matches $f$ in value $(p_1(1) = f(1))$ and in slope $(p_1'(1) = f'(1))$ at $x = 1$.

**b.** We first compute $f''(x) = -1/x^2$ and $f''(1) = -1$. Building on the linear approximation found in part (a), the quadratic approximation is

$$p_2(x) = \underbrace{x - 1}_{p_1(x)} + \underbrace{\frac{1}{2}f''(1)}_{c_2}(x - 1)^2$$

$$= x - 1 - \frac{1}{2}(x - 1)^2.$$

Because $p_2$ matches $f$ in value, slope, and concavity at $x = 1$, it provides a better approximation to $f$ near $x = 1$ (Figure 11.3).

**c.** To approximate ln 1.05, we substitute $x = 1.05$ into each polynomial approximation:

$$p_1(1.05) = 1.05 - 1 = 0.05 \text{ and} \qquad \text{Linear approximation}$$

$$p_2(1.05) = 1.05 - 1 - \frac{1}{2}(1.05 - 1)^2 = 0.04875. \quad \text{Quadratic approximation}$$

The value of ln 1.05 given by a calculator, rounded to five decimal places, is 0.04879, showing the improvement in quadratic approximation over linear approximation.

*Related Exercises 10, 13* ◄

We now extend the idea of linear and quadratic approximation to obtain higher-degree polynomials that generally provide better approximations.

## Taylor Polynomials

➤ Building on ideas that were already circulating in the early 18th century, Brook Taylor (1685–1731) published Taylor's Theorem in 1715. He is also credited with discovering integration by parts.

Assume $f$ and its first $n$ derivatives exist at $a$; our goal is to find an $n$th-degree polynomial that approximates the values of $f$ near $a$. The first step is to use $p_2$ to obtain a cubic polynomial $p_3$ of the form

$$p_3(x) = p_2(x) + c_3(x - a)^3$$

that satisfies the four matching conditions

$$p_3(a) = f(a), \quad p_3'(a) = f'(a), \quad p_3''(a) = f''(a), \quad \text{and} \quad p_3'''(a) = f'''(a).$$

Because $p_3$ is built "on top of" $p_2$, the first three matching conditions are met. The last condition, $p_3'''(a) = f'''(a)$, is used to determine $c_3$. A short calculation shows that $p_3'''(x) = 3 \cdot 2c_3 = 3!c_3$, so the last matching condition is $p_3'''(a) = 3!c_3 = f'''(a)$.

> Recall that $2! = 2 \cdot 1$, $3! = 3 \cdot 2 \cdot 1$, $k! = k \cdot (k - 1)!$, and by definition, $0! = 1$.

Solving for $c_3$, we have $c_3 = \dfrac{f'''(a)}{3!}$. Therefore, the cubic approximating polynomial is

$$p_3(x) = \underbrace{f(a) + f'(a)(x - a) + \frac{f''(a)}{2!}(x - a)^2}_{p_2(x)} + \frac{f'''(a)}{3!}(x - a)^3.$$

**QUICK CHECK 1** Verify that $p_3$ satisfies $p_3^{(k)}(a) = f^{(k)}(a)$, for $k = 0, 1, 2$, and 3. ◄

Continuing in this fashion (Exercise 66), building each new polynomial on the previous polynomial, the *n*th approximating polynomial for $f$ at $a$ is

$$p_n(x) = \underbrace{f(a) + f'(a)(x - a) + \frac{f''(a)}{2!}(x - a)^2 + \cdots}_{p_{n-1}(x)} + \frac{f^{(n)}(a)}{n!}(x - a)^n.$$

It satisfies the $n + 1$ matching conditions

$$p_n(a) = f(a), \quad p_n'(a) = f'(a), \quad p_n''(a) = f''(a), \dots, \text{ and } p_n^{(n)}(a) = f^{(n)}(a).$$

These conditions ensure that the graph of $p_n$ conforms as closely as possible to the graph of $f$ near $a$ (**Figure 11.4**).

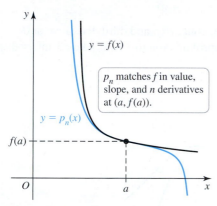

**Figure 11.4**

> Recall that $f^{(n)}$ denotes the *n*th derivative of $f$. By convention, the zeroth derivative $f^{(0)}$ is $f$ itself.

---

**DEFINITION  Taylor Polynomials**

Let $f$ be a function with $f'$, $f''$, ..., and $f^{(n)}$ defined at $a$. The **nth-order Taylor polynomial** for $f$ with its **center** at $a$, denoted $p_n$, has the property that it matches $f$ in value, slope, and all derivatives up to the *n*th derivative at $a$; that is,

$$p_n(a) = f(a), p_n'(a) = f'(a), \dots, \text{ and } p_n^{(n)}(a) = f^{(n)}(a).$$

The *n*th-order Taylor polynomial centered at $a$ is

$$p_n(x) = f(a) + f'(a)(x - a) + \frac{f''(a)}{2!}(x - a)^2 + \cdots + \frac{f^{(n)}(a)}{n!}(x - a)^n.$$

More compactly, $p_n(x) = \displaystyle\sum_{k=0}^{n} c_k(x - a)^k$, where the **coefficients** are

$$c_k = \frac{f^{(k)}(a)}{k!}, \qquad \text{for } k = 0, 1, 2, \dots, n.$$

---

**EXAMPLE 2  Taylor polynomials for $\sin x$** Find the Taylor polynomials $p_1, \dots, p_7$ centered at $x = 0$ for $f(x) = \sin x$.

**SOLUTION** We begin by differentiating $f$ repeatedly and evaluating the derivatives at 0; these calculations enable us to compute $c_k$, for $k = 0, 1, \dots, 7$. Notice that a pattern emerges:

$$\begin{aligned}
f(x) &= \sin x \Rightarrow f(0) = 0 \\
f'(x) &= \cos x \Rightarrow f'(0) = 1 \\
f''(x) &= -\sin x \Rightarrow f''(0) = 0 \\
f'''(x) &= -\cos x \Rightarrow f'''(0) = -1 \\
f^{(4)}(x) &= \sin x \Rightarrow f^{(4)}(0) = 0.
\end{aligned}$$

The derivatives of $\sin x$ at 0 cycle through the values $\{0, 1, 0, -1\}$. Therefore, $f^{(5)}(0) = 1$, $f^{(6)}(0) = 0$, and $f^{(7)}(0) = -1$.

We now construct the Taylor polynomials that approximate $f(x) = \sin x$ near 0, beginning with the linear polynomial. The polynomial of order $n = 1$ is

$$p_1(x) = f(0) + f'(0)(x - 0) = x,$$

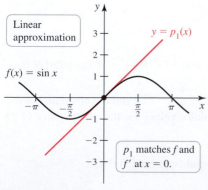

**Figure 11.5**

whose graph is the line through the origin with slope 1 (**Figure 11.5**). Notice that $f$ and $p_1$ agree in value ($f(0) = p_1(0) = 0$) and in slope ($f'(0) = p_1'(0) = 1$) at 0. We see that $p_1$ provides a good fit to $f$ near 0, but the graphs diverge visibly for $|x| > \pi/4$.

► It is worth repeating that the next polynomial in the sequence is obtained by adding one new term to the previous polynomial. For example,

$$p_3(x) = p_2(x) + \frac{f'''(a)}{3!}(x - a)^3.$$

The polynomial of order $n = 2$ is

$$p_2(x) = \underbrace{f(0)}_{0} + \underbrace{f'(0)x}_{1} + \underbrace{\frac{f''(0)}{2!}x^2}_{0} = x,$$

so $p_2$ is the same as $p_1$.

The polynomial of order $n = 3$ is

$$p_3(x) = \underbrace{f(0) + f'(0)x + \frac{f''(0)}{2!}x^2}_{p_2(x) = x} + \underbrace{\frac{f'''(0)}{3!}x^3}_{-1/3!} = x - \frac{x^3}{6}.$$

**QUICK CHECK 2** Verify the following properties for $f(x) = \sin x$ and $p_3(x) = x - x^3/6$:

$$f(0) = p_3(0),$$
$$f'(0) = p_3{}'(0),$$
$$f''(0) = p_3{}''(0), \text{ and}$$
$$f'''(0) = p_3{}'''(0). \blacktriangleleft$$

We have designed $p_3$ to agree with $f$ in value, slope, concavity, and third derivative at 0 (**Figure 11.6**). Consequently, $p_3$ provides a better approximation to $f$ over a larger interval than $p_1$.

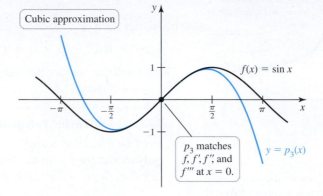

Cubic approximation

$f(x) = \sin x$

$p_3$ matches $f, f', f''$, and $f'''$ at $x = 0$.

$y = p_3(x)$

**Figure 11.6**

The procedure for finding Taylor polynomials may be extended to polynomials of any order. Because the even derivatives of $f(x) = \sin x$ are zero at $x = 0$, $p_4(x) = p_3(x)$. For the same reason, $p_6(x) = p_5(x)$:

$$p_6(x) = p_5(x) = x - \frac{x^3}{3!} + \frac{x^5}{5!}. \qquad c_5 = \frac{f^{(5)}(0)}{5!} = \frac{1}{5!}$$

$y = p_5(x)$

$y = \sin x$

$y = p_7(x)$

$y = p_3(x)$

**Figure 11.7**

**QUICK CHECK 3** Why do the Taylor polynomials for $\sin x$ centered at 0 consist only of odd powers of $x$? ◄

Finally, the Taylor polynomial of order $n = 7$ is

$$p_7(x) = x - \frac{x^3}{3!} + \frac{x^5}{5!} - \frac{x^7}{7!}. \qquad c_7 = \frac{f^{(7)}(0)}{7!} = -\frac{1}{7!}$$

From **Figure 11.7** we see that as the order of the Taylor polynomials increases, more accurate approximations to $f(x) = \sin x$ are obtained over larger intervals centered at 0. For example, $p_7$ is a good fit to $f(x) = \sin x$ over the interval $[-\pi, \pi]$. Notice that $\sin x$ and its Taylor polynomials (centered at 0) are all odd functions.     *Related Exercises 17, 20* ◄

## Approximations with Taylor Polynomials

Taylor polynomials find widespread use in approximating functions, as illustrated in the following examples.

**EXAMPLE 3**   **Taylor polynomials for $e^x$**

**a.** Find the Taylor polynomials of order $n = 0, 1, 2,$ and 3 for $f(x) = e^x$ centered at 0. Graph $f$ and the polynomials.

► Recall that if $c$ is an approximation to $x$, the absolute error in $c$ is $|x - c|$ and the relative error in $c$ is $|x - c|/|x|$. We use *error* to refer to *absolute error*.

**b.** Use the polynomials in part (a) to approximate $e^{0.1}$ and $e^{-0.25}$. Find the absolute errors, $|f(x) - p_n(x)|$, in the approximations. Use calculator values for the exact values of $f$.

**SOLUTION**

**a.** Recall that the coefficients for the Taylor polynomials centered at 0 are

$$c_k = \frac{f^{(k)}(0)}{k!}, \qquad \text{for } k = 0, 1, 2, \ldots, n.$$

With $f(x) = e^x$, we have $f^{(k)}(x) = e^x$, $f^{(k)}(0) = 1$, and $c_k = 1/k!$, for $k = 0, 1, 2, 3, \ldots$. The first four polynomials are

$$p_0(x) = f(0) = 1,$$

$$p_1(x) = \underbrace{f(0)}_{p_0(x) = 1} + \underbrace{f'(0)}_{1}x = 1 + x,$$

$$p_2(x) = \underbrace{f(0) + f'(0)x}_{p_1(x) = 1 + x} + \underbrace{\frac{f''(0)}{2!}}_{1/2}x^2 = 1 + x + \frac{x^2}{2}, \text{ and}$$

$$p_3(x) = \underbrace{f(0) + f'(0)x + \frac{f''(0)}{2!}x^2}_{p_2(x) = 1 + x + x^2/2} + \underbrace{\frac{f^{(3)}(0)}{3!}}_{1/6}x^3 = 1 + x + \frac{x^2}{2} + \frac{x^3}{6}.$$

Notice that each successive polynomial provides a better fit to $f(x) = e^x$ near 0 (**Figure 11.8**). Continuing the pattern in these polynomials, the $n$th-order Taylor polynomial for $e^x$ centered at 0 is

$$p_n(x) = 1 + x + \frac{x^2}{2!} + \frac{x^3}{3!} + \cdots + \frac{x^n}{n!} = \sum_{k=0}^{n} \frac{x^k}{k!}.$$

**b.** We evaluate $p_n(0.1)$ and $p_n(-0.25)$, for $n = 0, 1, 2$, and 3, and compare these values to the calculator values of $e^{0.1} \approx 1.1051709$ and $e^{-0.25} \approx 0.77880078$. The results are shown in Table 11.1. Observe that the errors in the approximations decrease as $n$ increases. In addition, the errors in approximating $e^{0.1}$ are smaller in magnitude than the errors in approximating $e^{-0.25}$ because $x = 0.1$ is closer to the center of the polynomials than $x = -0.25$. Reasonable approximations based on these calculations are $e^{0.1} \approx 1.105$ and $e^{-0.25} \approx 0.78$.

Taylor polynomials for $f(x) = e^x$ centered at 0. Approximations improve as $n$ increases.

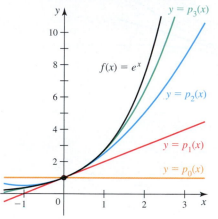

**Figure 11.8**

▶ A rule of thumb in finding estimates based on several approximations: Keep all the digits that are common to the last two approximations after rounding.

**QUICK CHECK 4** Write out the next two Taylor polynomials $p_4$ and $p_5$ for $f(x) = e^x$ in Example 3. ◀

**Table 11.1**

| $n$ | Approximation $p_n(0.1)$ | Absolute Error $\|e^{0.1} - p_n(0.1)\|$ | Approximation $p_n(-0.25)$ | Absolute Error $\|e^{-0.25} - p_n(-0.25)\|$ |
|---|---|---|---|---|
| 0 | 1 | $1.1 \times 10^{-1}$ | 1 | $2.2 \times 10^{-1}$ |
| 1 | 1.1 | $5.2 \times 10^{-3}$ | 0.75 | $2.9 \times 10^{-2}$ |
| 2 | 1.105 | $1.7 \times 10^{-4}$ | 0.78125 | $2.4 \times 10^{-3}$ |
| 3 | 1.105167 | $4.3 \times 10^{-6}$ | 0.778646 | $1.5 \times 10^{-4}$ |

*Related Exercise 31* ◀

**EXAMPLE 4 Approximating a real number using Taylor polynomials** Use polynomials of order $n = 0, 1, 2$, and 3 to approximate $\sqrt{18}$.

**SOLUTION** Letting $f(x) = \sqrt{x}$, we choose the center $a = 16$ because it is near 18, and $f$ and its derivatives are easy to evaluate at 16. The Taylor polynomials have the form

$$p_n(x) = f(16) + f'(16)(x - 16) + \frac{f''(16)}{2!}(x - 16)^2 + \cdots + \frac{f^{(n)}(16)}{n!}(x - 16)^n.$$

We now evaluate the required derivatives:

$$f(x) = \sqrt{x} \Rightarrow f(16) = 4,$$

$$f'(x) = \frac{1}{2}x^{-1/2} \Rightarrow f'(16) = \frac{1}{8},$$

$$f''(x) = -\frac{1}{4}x^{-3/2} \Rightarrow f''(16) = -\frac{1}{256}, \text{ and}$$

$$f'''(x) = \frac{3}{8}x^{-5/2} \Rightarrow f'''(16) = \frac{3}{8192}.$$

Therefore, the polynomial $p_3$ (which includes $p_0$, $p_1$, and $p_2$) is

$$p_3(x) = \underbrace{\underbrace{\underbrace{4}_{p_0(x)} + \frac{1}{8}(x - 16)}_{p_1(x)} - \frac{1}{512}(x - 16)^2 + \frac{1}{16{,}384}(x - 16)^3}_{p_2(x)}.$$

The Taylor polynomials (**Figure 11.9**) give better approximations to $f$ as the order of the approximation increases.

**Figure 11.9**

Letting $x = 18$, we obtain the approximations to $\sqrt{18}$ and the associated absolute errors shown in Table 11.2. (A calculator is used for the value of $\sqrt{18}$.) As expected, the errors decrease as $n$ increases. Based on these calculations, a reasonable approximation is $\sqrt{18} \approx 4.24$.

**Table 11.2**

| $n$ | Approximation $p_n(18)$ | Absolute Error $\lvert \sqrt{18} - p_n(18) \rvert$ |
|---|---|---|
| 0 | 4 | $2.4 \times 10^{-1}$ |
| 1 | 4.25 | $7.4 \times 10^{-3}$ |
| 2 | 4.242188 | $4.5 \times 10^{-4}$ |
| 3 | 4.242676 | $3.5 \times 10^{-5}$ |

*Related Exercises 37–38* ◄

**QUICK CHECK 5** At what point would you center the Taylor polynomials for $\sqrt{x}$ and $\sqrt[4]{x}$ to approximate $\sqrt{51}$ and $\sqrt[4]{15}$, respectively? ◄

## Remainder in a Taylor Polynomial

Taylor polynomials provide good approximations to functions near a specific point. But how accurate are the approximations? To answer this question we define the *remainder* in a Taylor polynomial. If $p_n$ is the Taylor polynomial for $f$ of order $n$, then the remainder at the point $x$ is

$$R_n(x) = f(x) - p_n(x).$$

The absolute value of the remainder is the error made in approximating $f(x)$ by $p_n(x)$. Equivalently, we have $f(x) = p_n(x) + R_n(x)$, which says that $f$ consists of two components: the polynomial approximation and the associated remainder.

> **DEFINITION** **Remainder in a Taylor Polynomial**
>
> Let $p_n$ be the Taylor polynomial of order $n$ for $f$. The **remainder** in using $p_n$ to approximate $f$ at the point $x$ is
>
> $$R_n(x) = f(x) - p_n(x).$$

$y = R_0(x)$

$y = R_1(x)$

$y = R_2(x)$

$y = R_3(x)$

Remainders increase in magnitude as $|x|$ increases. Remainders decrease in magnitude to zero as $n$ increases.

**Figure 11.10**

➤ The remainder $R_n$ for a Taylor polynomial can be expressed in several different forms. The form stated in Theorem 11.1 is called the *Lagrange form* of the remainder.

The idea of a remainder is illustrated in **Figure 11.10**, where we see the remainders associated with various Taylor polynomials for $f(x) = e^x$ centered at 0 (Example 3). For fixed order $n$, the remainders tend to increase in magnitude as $x$ moves farther from the center of the polynomials (in this case 0). And for fixed $x$, remainders decrease in magnitude to zero with increasing $n$.

The remainder for a Taylor polynomial may be written quite concisely, which enables us to estimate remainders. The following result is known as *Taylor's Theorem* (or the *Remainder Theorem*).

---

**THEOREM 11.1    Taylor's Theorem (Remainder Theorem)**

Let $f$ have continuous derivatives up to $f^{(n+1)}$ on an open interval $I$ containing $a$. For all $x$ in $I$,

$$f(x) = p_n(x) + R_n(x),$$

where $p_n$ is the $n$th-order Taylor polynomial for $f$ centered at $a$ and the remainder is

$$R_n(x) = \frac{f^{(n+1)}(c)}{(n+1)!}(x-a)^{n+1},$$

for some point $c$ between $x$ and $a$.

---

**Discussion:** We make two observations about Theorem 11.1 and outline a proof in Exercise 82. First, the case $n = 0$ is the Mean Value Theorem (Section 4.2), which states that

$$\frac{f(x) - f(a)}{x - a} = f'(c),$$

where $c$ is a point between $x$ and $a$. Rearranging this expression, we have

$$f(x) = \underbrace{f(a)}_{p_0(x)} + \underbrace{f'(c)(x-a)}_{R_0(x)}$$
$$= p_0(x) + R_0(x),$$

which is Taylor's Theorem with $n = 0$. Not surprisingly, the term $f^{(n+1)}(c)$ in Taylor's Theorem comes from a Mean Value Theorem argument.

The second observation makes the remainder easier to remember. If you write the $(n + 1)$st Taylor polynomial $p_{n+1}$, the highest-degree term is $\dfrac{f^{(n+1)}(a)}{(n+1)!}(x-a)^{n+1}$. Replacing $f^{(n+1)}(a)$ with $f^{(n+1)}(c)$ results in the remainder for $p_n$.

## Estimating the Remainder

The remainder has both practical and theoretical importance. We deal with practical matters now and theoretical matters in Section 11.3. The remainder is used to estimate errors in approximations and to determine the number of terms of a Taylor polynomial needed to achieve a prescribed accuracy.

Because $c$ is generally unknown, the difficulty in estimating the remainder is finding a bound for $|f^{(n+1)}(c)|$. Assuming this can be done, the following theorem gives a standard estimate for the remainder term.

---

**THEOREM 11.2    Estimate of the Remainder**

Let $n$ be a fixed positive integer. Suppose there exists a number $M$ such that $|f^{(n+1)}(c)| \le M$, for all $c$ between $a$ and $x$ inclusive. The remainder in the $n$th-order Taylor polynomial for $f$ centered at $a$ satisfies

$$|R_n(x)| = |f(x) - p_n(x)| \le M\frac{|x-a|^{n+1}}{(n+1)!}.$$

---

**Proof:** The proof requires taking the absolute value of the remainder in Theorem 11.1, replacing $\left|f^{(n+1)}(c)\right|$ with a larger quantity $M$, and forming an inequality.    ◀

We now give three examples that demonstrate how an upper bound for the remainder is computed and used in different ways.

**EXAMPLE 5   Estimating the remainder for cos x** Find a bound for the magnitude of the remainder for the Taylor polynomials of $f(x) = \cos x$ centered at 0.

**SOLUTION** According to Theorem 11.1 with $a = 0$, we have

$$R_n(x) = \frac{f^{(n+1)}(c)}{(n+1)!} x^{n+1},$$

where $c$ is a point between 0 and $x$. Notice that $f^{(n+1)}(c) = \pm \sin c$, or $f^{(n+1)}(c) = \pm \cos c$, depending on the value of $n$. In all cases, $\left|f^{(n+1)}(c)\right| \leq 1$. Therefore, we take $M = 1$ in Theorem 11.2, and the absolute value of the remainder can be bounded as

$$\left|R_n(x)\right| = \left|\frac{f^{(n+1)}(c)}{(n+1)!} x^{n+1}\right| \leq \frac{|x|^{n+1}}{(n+1)!}.$$

For example, if we approximate $\cos 0.1$ using the Taylor polynomial $p_{10}$, the remainder satisfies

$$\left|R_{10}(0.1)\right| \leq \frac{0.1^{11}}{11!} \approx 2.5 \times 10^{-19}.$$

*Related Exercises 41–42* ◀

**EXAMPLE 6   Estimating a remainder** Consider again Example 4, in which we approximated $\sqrt{18}$ using the Taylor polynomial

$$p_3(x) = 4 + \frac{1}{8}(x - 16) - \frac{1}{512}(x - 16)^2 + \frac{1}{16{,}384}(x - 16)^3.$$

Find an upper bound on the magnitude of the remainder when using $p_3(x)$ to approximate $\sqrt{18}$.

**SOLUTION** In Example 4, we computed the error in the approximation knowing the exact value of $\sqrt{18}$ (obtained with a calculator). In the more realistic case in which we do not know the exact value, Theorem 11.2 allows us to estimate remainders (or errors). Applying this theorem with $n = 3$, $a = 16$, and $x = 18$, we find that the remainder in approximating $\sqrt{18}$ by $p_3(18)$ satisfies the bound

$$\left|R_3(18)\right| \leq M \frac{(18 - 16)^4}{4!} = \frac{2}{3} M,$$

where $M$ is a number that satisfies $\left|f^{(4)}(c)\right| \leq M$, for all $c$ between 16 and 18 inclusive. In this particular problem, we find that $f^{(4)}(c) = -\frac{15}{16} c^{-7/2}$, so $M$ must be chosen (as small as possible) such that $\left|f^{(4)}(c)\right| = \frac{15}{16} c^{-7/2} = \frac{15}{16 c^{7/2}} \leq M$, for $16 \leq c \leq 18$. You can verify that $\frac{15}{16 c^{7/2}}$ is a decreasing function of $c$ on $[16, 18]$ and has a maximum value of approximately $5.7 \times 10^{-5}$ at $c = 16$ (**Figure 11.11**). Therefore, a bound on the remainder is

$$\left|R_3(18)\right| \leq \frac{2}{3} M \approx \frac{2}{3} \cdot 5.7 \times 10^{-5} \approx 3.8 \times 10^{-5}.$$

Notice that the actual error computed in Example 4 (Table 11.2) is $3.5 \times 10^{-5}$, which is less than the bound on the remainder—as it should be.

**Figure 11.11**

*Related Exercise 52* ◄

**EXAMPLE 7**  **Estimating the remainder for $e^x$**  Find a bound on the remainder in approximating $e^{0.45}$ using the Taylor polynomial of order $n = 6$ for $f(x) = e^x$ centered at 0.

**SOLUTION**  Using Theorem 11.2, a bound on the remainder is given by

$$|R_n(x)| \le M\frac{|x - a|^{n+1}}{(n + 1)!},$$

where $M$ is chosen such that $|f^{(n+1)}(c)| \le M$, for all $c$ between $a$ and $x$ inclusive. Notice that $f(x) = e^x$ implies that $f^{(k)}(x) = e^x$, for $k = 0, 1, 2, \ldots$. In this particular problem, we have $n = 6$, $a = 0$, and $x = 0.45$, so the bound on the remainder takes the form

$$|R_6(0.45)| \le M\frac{|0.45 - 0|^7}{7!} \approx 7.4 \times 10^{-7} M,$$

where $M$ is chosen such that $|f^{(7)}(c)| = e^c \le M$, for all $c$ in the interval $[0, 0.45]$. Because $e^c$ is an increasing function of $c$, its maximum value on the interval $[0, 0.45]$ occurs at $c = 0.45$ and is $e^{0.45}$. However, $e^{0.45}$ cannot be evaluated exactly (it is the number we are approximating), so we must find a number $M$ such that $e^{0.45} \le M$. Here is one of many ways to obtain a bound: We observe that $e^{0.45} < e^{1/2} < 4^{1/2} = 2$ and take $M = 2$ (Figure 11.12). Therefore, a bound on the remainder is

$$|R_6(0.45)| \le 7.4 \times 10^{-7} M \approx 1.5 \times 10^{-6}.$$

**Figure 11.12**

► Recall that if $f(x) = e^x$, then

$$p_n(x) = \sum_{k=0}^{n} \frac{x^k}{k!}.$$

Using the Taylor polynomial derived in Example 3 with $n = 6$, the resulting approximation to $e^{0.45}$ is

$$p_6(0.45) = \sum_{k=0}^{6} \frac{0.45^k}{k!} \approx 1.5683114;$$

it has an error that does not exceed $1.5 \times 10^{-6}$.

*Related Exercise 49* ◄

**QUICK CHECK 6** In Example 7, find an approximate upper bound for $R_7(0.45)$. ◄

**EXAMPLE 8**  **Working with the remainder**  In Example 4b of Section 11.2, we show that the $n$th-order Taylor polynomial for $f(x) = \ln(1 - x)$ centered at 0 is

$$p_n(x) = -\sum_{k=1}^{n} \frac{x^k}{k} = -x - \frac{x^2}{2} - \frac{x^3}{3} - \cdots - \frac{x^n}{n}.$$

**a.** Find a bound on the error in approximating $\ln(1 - x)$ by $p_3(x)$ for values of $x$ in the interval $\left[-\frac{1}{2}, \frac{1}{2}\right]$.

**b.** How many terms of the Taylor polynomial are needed to approximate values of $f(x) = \ln(1 - x)$ with an error less than $10^{-3}$ on the interval $\left[-\frac{1}{2}, \frac{1}{2}\right]$?

**SOLUTION**

**a.** The remainder for the Taylor polynomial $p_3$ is $R_3(x) = \dfrac{f^{(4)}(c)}{4!} x^4$, where $c$ is between 0 and $x$. Computing four derivatives of $f$, we find that $f^{(4)}(x) = -\dfrac{6}{(1 - x)^4}$.

On the interval $\left[-\frac{1}{2}, \frac{1}{2}\right]$, the maximum magnitude of this derivative occurs at $x = \frac{1}{2}$ $\left(\text{because the denominator is smallest at } x = \frac{1}{2}\right)$ and is $6 / \left(\frac{1}{2}\right)^4 = 96$. Similarly, the factor $x^4$ has its maximum magnitude at $x = \pm\frac{1}{2}$, and it is $\left(\frac{1}{2}\right)^4 = \frac{1}{16}$. Therefore,

$$|R_3(x)| \le \frac{96}{4!}\left(\frac{1}{16}\right) = 0.25 \text{ on the interval } \left[-\frac{1}{2}, \frac{1}{2}\right]. \text{ The error in approximating } f(x)$$

by $p_3(x)$, for $-\frac{1}{2} \le x \le \frac{1}{2}$, does not exceed 0.25.

**b.** For any positive integer $n$, the remainder is $R_n(x) = \dfrac{f^{(n+1)}(c)}{(n + 1)!} x^{n+1}$. Differentiating $f$ several times reveals that

$$f^{(n+1)}(x) = -\frac{n!}{(1 - x)^{n+1}}.$$

On the interval $\left[-\frac{1}{2}, \frac{1}{2}\right]$, the maximum magnitude of this derivative occurs at $x = \frac{1}{2}$ and is $n! / \left(\frac{1}{2}\right)^{n+1}$. Similarly, $x^{n+1}$ has its maximum magnitude at $x = \pm\frac{1}{2}$, and it is $\left(\frac{1}{2}\right)^{n+1}$. Therefore, a bound on the magnitude of the remainder is

$$|R_n(x)| = \frac{1}{(n + 1)!} \cdot \underbrace{|f^{(n+1)}(c)|}_{\le\, n! 2^{n+1}} \cdot \underbrace{|x|^{n+1}}_{\le\, \left(\frac{1}{2}\right)^{n+1}}$$

$$\le \frac{1}{(n + 1)!} \cdot n! 2^{n+1} \cdot \frac{1}{2^{n+1}}$$

$$= \frac{1}{n + 1}. \qquad\qquad \frac{n!}{(n + 1)!} = \frac{1}{n + 1}$$

To ensure that the error is less than $10^{-3}$ on the entire interval $\left[-\frac{1}{2}, \frac{1}{2}\right]$, $n$ must satisfy $|R_n| \le \dfrac{1}{n + 1} < 10^{-3}$ or $n > 999$. The error is likely to be significantly less than $10^{-3}$ if $x$ is near 0.

*Related Exercises 57, 62* ◄

---

# SECTION 11.1 EXERCISES

## Getting Started

**1.** Suppose you use a second-order Taylor polynomial centered at 0 to approximate a function $f$. What matching conditions are satisfied by the polynomial?

**2.** Does the accuracy of an approximation given by a Taylor polynomial generally increase or decrease with the order of the approximation? Explain.

**3.** The first three Taylor polynomials for $f(x) = \sqrt{1 + x}$ centered at 0 are $p_0 = 1$, $p_1 = 1 + \dfrac{x}{2}$, and $p_2 = 1 + \dfrac{x}{2} - \dfrac{x^2}{8}$. Find three approximations to $\sqrt{1.1}$.

**4.** Suppose $f(0) = 1$, $f'(0) = 2$, and $f''(0) = -1$. Find the quadratic approximating polynomial for $f$ centered at 0 and use it to approximate $f(0.1)$.

**5.** Suppose $f(0) = 1$, $f'(0) = 0$, $f''(0) = 2$, and $f^{(3)}(0) = 6$. Find the third-order Taylor polynomial for $f$ centered at 0 and use it to approximate $f(0.2)$.

**6.** How is the remainder $R_n(x)$ in a Taylor polynomial defined?

**7.** Suppose $f(2) = 1$, $f'(2) = 1$, $f''(2) = 0$, and $f^{(3)}(2) = 12$. Find the third-order Taylor polynomial for $f$ centered at 2 and use this polynomial to estimate $f(1.9)$.

**8.** Suppose you want to estimate $\sqrt{26}$ using a fourth-order Taylor polynomial centered at $x = a$ for $f(x) = \sqrt{x}$. Choose an appropriate value for the center $a$.

## Practice Exercises

**T 9–16. Linear and quadratic approximation**

**a.** Find the linear approximating polynomial for the following functions centered at the given point $a$.

**b.** Find the quadratic approximating polynomial for the following functions centered at $a$.

**c.** Use the polynomials obtained in parts (a) and (b) to approximate the given quantity.

**9.** $f(x) = 8x^{3/2}$, $a = 1$; approximate $8 \cdot 1.1^{3/2}$.

**10.** $f(x) = \dfrac{1}{x}$, $a = 1$; approximate $\dfrac{1}{1.05}$.

**11.** $f(x) = e^{-2x}$, $a = 0$; approximate $e^{-0.2}$.

**12.** $f(x) = \sqrt{x}$, $a = 4$; approximate $\sqrt{3.9}$.

**13.** $f(x) = (1 + x)^{-1}$, $a = 0$; approximate $\dfrac{1}{1.05}$.

**14.** $f(x) = \cos x$, $a = \dfrac{\pi}{4}$; approximate $\cos(0.24\pi)$.

**15.** $f(x) = x^{1/3}$, $a = 8$; approximate $7.5^{1/3}$.

**16.** $f(x) = \tan^{-1} x$, $a = 0$; approximate $\tan^{-1} 0.1$.

**17.** Find the Taylor polynomials $p_1, \ldots, p_4$ centered at $a = 0$ for $f(x) = \cos 6x$.

**18.** Find the Taylor polynomials $p_1, \ldots, p_5$ centered at $a = 0$ for $f(x) = e^{-x}$.

**19.** Find the Taylor polynomials $p_3$ and $p_4$ centered at $a = 0$ for $f(x) = (1 + x)^{-3}$.

**20.** Find the Taylor polynomials $p_4$ and $p_5$ centered at $a = \pi/6$ for $f(x) = \cos x$.

**21.** Find the Taylor polynomials $p_1, p_2$, and $p_3$ centered at $a = 1$ for $f(x) = x^3$.

**22.** Find the Taylor polynomials $p_3$ and $p_4$ centered at $a = 1$ for $f(x) = 8\sqrt{x}$.

**23.** Find the Taylor polynomial $p_3$ centered at $a = e$ for $f(x) = \ln x$.

**24.** Find the Taylor polynomial $p_2$ centered at $a = 8$ for $f(x) = \sqrt[3]{x}$.

**T 25–28. Graphing Taylor polynomials**

**a.** Find the nth-order Taylor polynomials for the following functions centered at the given point $a$, for $n = 1$ and $n = 2$.

**b.** Graph the Taylor polynomials and the function.

**25.** $f(x) = \ln(1 - x)$, $a = 0$    **26.** $f(x) = (1 + x)^{-1/2}$, $a = 0$

**27.** $f(x) = \sin x$, $a = \dfrac{\pi}{4}$    **28.** $f(x) = \sqrt{x}$, $a = 9$

**T 29–32. Approximations with Taylor polynomials**

**a.** Use the given Taylor polynomial $p_2$ to approximate the given quantity.

**b.** Compute the absolute error in the approximation, assuming the exact value is given by a calculator.

**29.** Approximate $\sqrt{1.05}$ using $f(x) = \sqrt{1 + x}$ and $p_2(x) = 1 + x/2 - x^2/8$.

**30.** Approximate $1/\sqrt{1.08}$ using $f(x) = 1/\sqrt{1 + x}$ and $p_2(x) = 1 - x/2 + 3x^2/8$.

**31.** Approximate $e^{-0.15}$ using $f(x) = e^{-x}$ and $p_2(x) = 1 - x + x^2/2$.

**32.** Approximate $\ln 1.06$ using $f(x) = \ln(1 + x)$ and $p_2(x) = x - x^2/2$.

**T 33–40. Approximations with Taylor polynomials**

**a.** Approximate the given quantities using Taylor polynomials with $n = 3$.

**b.** Compute the absolute error in the approximation, assuming the exact value is given by a calculator.

**33.** $e^{0.12}$     **34.** $\cos(-0.2)$     **35.** $\tan(-0.1)$

**36.** $\ln 1.05$     **37.** $\sqrt{1.06}$     **38.** $\sqrt[4]{79}$

**39.** $\sinh 0.5$     **40.** $\tanh 0.5$

**41–46. Remainders** *Find the remainder $R_n$ for the nth-order Taylor polynomial centered at $a$ for the given functions. Express the result for a general value of $n$.*

**41.** $f(x) = \sin x$, $a = 0$     **42.** $f(x) = \cos 2x$, $a = 0$

**43.** $f(x) = e^{-x}$, $a = 0$     **44.** $f(x) = \cos x$, $a = \dfrac{\pi}{2}$

**45.** $f(x) = \sin x$, $a = \dfrac{\pi}{2}$     **46.** $f(x) = \dfrac{1}{1 - x}$, $a = 0$

**T 47–52. Estimating errors** *Use the remainder to find a bound on the error in approximating the following quantities with the nth-order Taylor polynomial centered at 0. Estimates are not unique.*

**47.** $\sin 0.3$, $n = 4$     **48.** $\cos 0.45$, $n = 3$

**49.** $e^{0.25}$, $n = 4$     **50.** $\tan 0.3$, $n = 2$

**51.** $e^{-0.5}$, $n = 4$     **52.** $\ln 1.04$, $n = 3$

**T 53–58. Maximum error** *Use the remainder term to find a bound on the error in the following approximations on the given interval. Error bounds are not unique.*

**53.** $\sin x \approx x - \dfrac{x^3}{6}$ on $\left[-\dfrac{\pi}{4}, \dfrac{\pi}{4}\right]$

**54.** $\cos x \approx 1 - \dfrac{x^2}{2}$ on $\left[-\dfrac{\pi}{4}, \dfrac{\pi}{4}\right]$

**55.** $e^x \approx 1 + x + \dfrac{x^2}{2}$ on $\left[-\dfrac{1}{2}, \dfrac{1}{2}\right]$

**56.** $\tan x \approx x$ on $\left[-\dfrac{\pi}{6}, \dfrac{\pi}{6}\right]$

**57.** $\ln(1 + x) \approx x - \dfrac{x^2}{2}$ on $[-0.2, 0.2]$

**58.** $\sqrt{1 + x} \approx 1 + \dfrac{x}{2}$ on $[-0.1, 0.1]$

**59–64. Number of terms** *What is the minimum order of the Taylor polynomial required to approximate the following quantities with an absolute error no greater than $10^{-3}$? (The answer depends on your choice of a center.)*

**59.** $e^{-0.5}$     **60.** $\sin 0.2$     **61.** $\cos(-0.25)$

**62.** $\ln 0.85$     **63.** $\sqrt{1.06}$     **64.** $\dfrac{1}{\sqrt{0.85}}$

**65. Explain why or why not** Determine whether the following statements are true and give an explanation or counterexample.

   **a.** Only even powers of $x$ appear in the Taylor polynomials for $f(x) = e^{-2x}$ centered at 0.

   **b.** Let $f(x) = x^5 - 1$. The Taylor polynomial for $f$ of order 10 centered at 0 is $f$ itself.

   **c.** Only even powers of $x$ appear in the $n$th-order Taylor polynomial for $f(x) = \sqrt{1 + x^2}$ centered at 0.

   **d.** Suppose $f''$ is continuous on an interval that contains $a$, where $f$ has an inflection point at $a$. Then the second-order Taylor polynomial for $f$ at $a$ is linear.

**66. Taylor coefficients for $x = a$** Follow the procedure in the text to show that the $n$th-order Taylor polynomial that matches $f$ and its derivatives up to order $n$ at $a$ has coefficients

$$c_k = \frac{f^{(k)}(a)}{k!}, \quad \text{for } k = 0, 1, 2, \ldots, n.$$

**67. Matching functions with polynomials** Match functions a–f with Taylor polynomials A–F (all centered at 0). Give reasons for your choices.

   **a.** $\sqrt{1 + 2x}$          **A.** $p_2(x) = 1 + 2x + 2x^2$

   **b.** $\dfrac{1}{\sqrt{1 + 2x}}$      **B.** $p_2(x) = 1 - 6x + 24x^2$

   **c.** $e^{2x}$              **C.** $p_2(x) = 1 + x - \dfrac{x^2}{2}$

   **d.** $\dfrac{1}{1 + 2x}$        **D.** $p_2(x) = 1 - 2x + 4x^2$

   **e.** $\dfrac{1}{(1 + 2x)^3}$    **E.** $p_2(x) = 1 - x + \dfrac{3}{2}x^2$

   **f.** $e^{-2x}$           **F.** $p_2(x) = 1 - 2x + 2x^2$

**T 68. Dependence of errors on $x$** Consider $f(x) = \ln(1 - x)$ and its Taylor polynomials given in Example 8.

   **a.** Graph $y = |f(x) - p_2(x)|$ and $y = |f(x) - p_3(x)|$ on the interval $[-1/2, 1/2]$ (two curves).

   **b.** At what points of $[-1/2, 1/2]$ is the error largest? Smallest?

   **c.** Are these results consistent with the theoretical error bounds obtained in Example 8?

**T 69–76. Small argument approximations** *Consider the following common approximations when $x$ is near zero.*

   **a.** *Estimate $f(0.1)$ and give a bound on the error in the approximation.*

   **b.** *Estimate $f(0.2)$ and give a bound on the error in the approximation.*

**69.** $f(x) = \sin x \approx x$       **70.** $f(x) = \tan x \approx x$

**71.** $f(x) = \cos x \approx 1 - \dfrac{x^2}{2}$    **72.** $f(x) = \tan^{-1}x \approx x$

**73.** $f(x) = \sqrt{1 + x} \approx 1 + \dfrac{x}{2}$   **74.** $f(x) = \ln(1 + x) \approx x - \dfrac{x^2}{2}$

**75.** $f(x) = e^x \approx 1 + x$      **76.** $f(x) = \sin^{-1}x \approx x$

### Explorations and Challenges

**T 77. Errors in approximations** Suppose you approximate $f(x) = \sec x$ at the points $x = -0.2, -0.1, 0.0, 0.1, 0.2$ using the Taylor polynomials $p_2(x) = 1 + x^2/2$ and $p_4(x) = 1 + x^2/2 + 5x^4/24$. Assume the exact value of $\sec x$ is given by a calculator.

   **a.** Complete the table showing the absolute errors in the approximations at each point. Show three significant digits.

**b.** In each error column, how do the errors vary with $x$? For what values of $x$ are the errors largest and smallest in magnitude?

| $x$ | $\|\sec x - p_2(x)\|$ | $\|\sec x - p_4(x)\|$ |
|---|---|---|
| $-0.2$ | | |
| $-0.1$ | | |
| $0.0$ | | |
| $0.1$ | | |
| $0.2$ | | |

**T 78–79. Errors in approximations** *Carry out the procedure described in Exercise 77 with the following functions and Taylor polynomials.*

**78.** $f(x) = \cos x$, $p_2(x) = 1 - \dfrac{x^2}{2}$, $p_4(x) = 1 - \dfrac{x^2}{2} + \dfrac{x^4}{24}$

**79.** $f(x) = e^{-x}$, $p_1(x) = 1 - x$, $p_2(x) = 1 - x + \dfrac{x^2}{2}$

**T 80. Best center point** Suppose you wish to approximate $\cos(\pi/12)$ using Taylor polynomials. Is the approximation more accurate if you use Taylor polynomials centered at 0 or at $\pi/6$? Use a calculator for numerical experiments and check for consistency with Theorem 11.2. Does the answer depend on the order of the polynomial?

**T 81. Best center point** Suppose you wish to approximate $e^{0.35}$ using Taylor polynomials. Is the approximation more accurate if you use Taylor polynomials centered at 0 or at $\ln 2$? Use a calculator for numerical experiments and check for consistency with Theorem 11.2. Does the answer depend on the order of the polynomial?

**82. Proof of Taylor's Theorem** There are several proofs of Taylor's Theorem, which lead to various forms of the remainder. The following proof is instructive because it leads to two different forms of the remainder and it relies on the Fundamental Theorem of Calculus, integration by parts, and the Mean Value Theorem for Integrals. Assume $f$ has at least $n + 1$ continuous derivatives on an interval containing $a$.

   **a.** Show that the Fundamental Theorem of Calculus can be written in the form

$$f(x) = f(a) + \int_a^x f'(t)\, dt.$$

   **b.** Use integration by parts ($u = f'(t)$, $dv = dt$) to show that

$$f(x) = f(a) + (x - a)f'(a) + \int_a^x (x - t)f''(t)\, dt.$$

   **c.** Show that $n$ integrations by parts give

$$f(x) = f(a) + \frac{f'(a)}{1!}(x - a) + \frac{f''(a)}{2!}(x - a)^2 + \cdots$$
$$+ \frac{f^{(n)}(a)}{n!}(x - a)^n + \underbrace{\int_a^x \frac{f^{(n+1)}(t)}{n!}(x - t)^n\, dt}_{R_n(x)}.$$

   **d.** *Challenge:* The result in part (c) has the form $f(x) = p_n(x) + R_n(x)$, where $p_n$ is the $n$th-order Taylor polynomial and $R_n$ is a new form of the remainder, known as the integral form of the remainder. Use the Mean Value

Theorem for Integrals (Section 5.4) to show that $R_n$ can be expressed in the form

$$R_n(x) = \frac{f^{(n+1)}(c)}{(n+1)!}(x-a)^{n+1},$$

where $c$ is between $a$ and $x$.

**83. Tangent line is $p_1$** Let $f$ be differentiable at $x = a$.

  **a.** Find the equation of the line tangent to the curve $y = f(x)$ at $(a, f(a))$.

  **b.** Verify that the Taylor polynomial $p_1$ centered at $a$ describes the tangent line found in part (a).

**84. Local extreme points and inflection points** Suppose $f$ has continuous first and second derivatives at $a$.

  **a.** Show that if $f$ has a local maximum at $a$, then the Taylor polynomial $p_2$ centered at $a$ also has a local maximum at $a$.

  **b.** Show that if $f$ has a local minimum at $a$, then the Taylor polynomial $p_2$ centered at $a$ also has a local minimum at $a$.

  **c.** Is it true that if $f$ has an inflection point at $a$, then the Taylor polynomial $p_2$ centered at $a$ also has an inflection point at $a$?

  **d.** Are the converses in parts (a) and (b) true? If $p_2$ has a local extreme point at $a$, does $f$ have the same type of point at $a$?

**⊤ 85. Approximating $\sin x$** Let $f(x) = \sin x$, and let $p_n$ and $q_n$ be $n$th-order Taylor polynomials for $f$ centered at 0 and $\pi$, respectively.

  **a.** Find $p_5$ and $q_5$.

  **b.** Graph $f$, $p_5$, and $q_5$ on the interval $[-\pi, 2\pi]$. On what interval is $p_5$ a better approximation to $f$ than $q_5$? On what interval is $q_5$ a better approximation to $f$ than $p_5$?

  **c.** Complete the following table showing the errors in the approximations given by $p_5$ and $q_5$ at selected points.

| $x$ | $\lvert \sin x - p_5(x)\rvert$ | $\lvert \sin x - q_5(x)\rvert$ |
|---|---|---|
| $\pi/4$ | | |
| $\pi/2$ | | |
| $3\pi/4$ | | |
| $5\pi/4$ | | |
| $7\pi/4$ | | |

  **d.** At which points in the table is $p_5$ a better approximation to $f$ than $q_5$? At which points do $p_5$ and $q_5$ give equal approximations to $f$? Explain your observations.

**⊤ 86. Approximating $\ln x$** Let $f(x) = \ln x$, and let $p_n$ and $q_n$ be the $n$th-order Taylor polynomials for $f$ centered at 1 and $e$, respectively.

  **a.** Find $p_3$ and $q_3$.

  **b.** Graph $f$, $p_3$, and $q_3$ on the interval $(0, 4]$.

  **c.** Complete the following table showing the errors in the approximations given by $p_3$ and $q_3$ at selected points.

| $x$ | $\lvert \ln x - p_3(x)\rvert$ | $\lvert \ln x - q_3(x)\rvert$ |
|---|---|---|
| 0.5 | | |
| 1.0 | | |
| 1.5 | | |
| 2 | | |
| 2.5 | | |
| 3 | | |
| 3.5 | | |

  **d.** At which points in the table is $p_3$ a better approximation to $f$ than $q_3$? Explain your observations.

**⊤ 87. Approximating square roots** Let $p_1$ and $q_1$ be the first-order Taylor polynomials for $f(x) = \sqrt{x}$, centered at 36 and 49, respectively.

  **a.** Find $p_1$ and $q_1$.

  **b.** Complete the following table showing the errors when using $p_1$ and $q_1$ to approximate $f(x)$ at $x = 37, 39, 41, 43, 45$, and 47. Use a calculator to obtain an exact value of $f(x)$.

| $x$ | $\lvert \sqrt{x} - p_1(x)\rvert$ | $\lvert \sqrt{x} - q_1(x)\rvert$ |
|---|---|---|
| 37 | | |
| 39 | | |
| 41 | | |
| 43 | | |
| 45 | | |
| 47 | | |

  **c.** At which points in the table is $p_1$ a better approximation to $f$ than $q_1$? Explain this result.

**⊤ 88. A different kind of approximation** When approximating a function $f$ using a Taylor polynomial, we use information about $f$ and its derivative at one point. An alternative approach (called *interpolation*) uses information about $f$ at several different points. Suppose we wish to approximate $f(x) = \sin x$ on the interval $[0, \pi]$.

  **a.** Write the (quadratic) Taylor polynomial $p_2$ for $f$ centered at $\pi/2$.

  **b.** Now consider a quadratic interpolating polynomial $q(x) = ax^2 + bx + c$. The coefficients $a$, $b$, and $c$ are chosen such that the following conditions are satisfied:

$$q(0) = f(0), \quad q\left(\frac{\pi}{2}\right) = f\left(\frac{\pi}{2}\right), \text{ and } q(\pi) = f(\pi).$$

Show that $q(x) = -\dfrac{4}{\pi^2}x^2 + \dfrac{4}{\pi}x$.

  **c.** Graph $f$, $p_2$, and $q$ on $[0, \pi]$.

  **d.** Find the error in approximating $f(x) = \sin x$ at the points $\dfrac{\pi}{4}$, $\dfrac{\pi}{2}$, $\dfrac{3\pi}{4}$, and $\pi$ using $p_2$ and $q$.

  **e.** Which function, $p_2$ or $q$, is a better approximation to $f$ on $[0, \pi]$? Explain.

**QUICK CHECK ANSWERS**

**3.** $f(x) = \sin x$ is an odd function, and its even-ordered derivatives are zero at 0, so its Taylor polynomials are also odd functions. **4.** $p_4(x) = p_3(x) + \dfrac{x^4}{4!}$; $p_5(x) = p_4(x) + \dfrac{x^5}{5!}$ **5.** $x = 49$ and $x = 16$ are good choices. **6.** Because $e^{0.45} < 2$, $\lvert R_7(0.45)\rvert < 2\,\dfrac{0.45^8}{8!} \approx 8.3 \times 10^{-8}$. ◄

# 11.2 Properties of Power Series

The preceding section demonstrated that Taylor polynomials provide accurate approximations to many functions and that, in general, the approximations improve as the degree of the polynomials increases. In this section, we take the next step and let the degree of the Taylor polynomials increase without bound to produce a *power series*.

## Geometric Series as Power Series

A good way to become familiar with power series is to return to *geometric series*, first encountered in Section 10.3. Recall that for a fixed number $r$,

$$\sum_{k=0}^{\infty} r^k = 1 + r + r^2 + \cdots = \frac{1}{1-r}, \qquad \text{provided } |r| < 1.$$

It's a small change to replace the real number $r$ with the variable $x$. In doing so, the geometric series becomes a new representation of a familiar function:

$$\sum_{k=0}^{\infty} x^k = 1 + x + x^2 + \cdots = \frac{1}{1-x}, \qquad \text{provided } |x| < 1.$$

This infinite series is a power series and it is a representation of the function $1/(1-x)$ that is valid on the interval $|x| < 1$.

In general, power series are used to represent familiar functions such as trigonometric, exponential, and logarithmic functions. They are also used to define new functions. For example, consider the function defined by

$$g(x) = \sum_{k=1}^{\infty} \frac{(-1)^k k}{4^k} x^{2k}.$$

The term *function* is used advisedly because it's not yet clear whether $g$ really is a function. If so, is it continuous? Does it have a derivative? Judging by its graph (**Figure 11.13**), $g$ appears to be an ordinary continuous and differentiable function on $(-2, 2)$ (which is identified at the end of the chapter). In fact, power series satisfy the defining property of all functions: For each admissible value of $x$, a power series has at most one value. For this reason, we refer to a power series as a function, although the domain, properties, and identity of the function may need to be discovered.

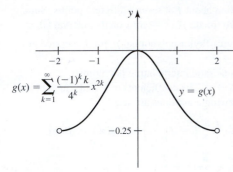

$$g(x) = \sum_{k=1}^{\infty} \frac{(-1)^k k}{4^k} x^{2k}$$

$y = g(x)$

**Figure 11.13**

▶ Figure 11.13 shows an approximation to the graph of $g$ made by summing the first 500 terms of the power series at selected values of $x$ on the interval $(-2, 2)$.

**QUICK CHECK 1** By substituting $x = 0$ in the power series for $g$, evaluate $g(0)$ for the function in Figure 11.13. ◄

Interval of convergence

$a$

Radius of convergence

**Figure 11.14**

## Convergence of Power Series

First let's establish some terminology associated with power series.

---

**DEFINITION  Power Series**

A **power series** has the general form

$$\sum_{k=0}^{\infty} c_k (x - a)^k,$$

where $a$ and $c_k$ are real numbers, and $x$ is a variable. The $c_k$'s are the **coefficients** of the power series, and $a$ is the **center** of the power series. The set of values of $x$ for which the series converges is its **interval of convergence**. The **radius of convergence** of the power series, denoted $R$, is the distance from the center of the series to the boundary of the interval of convergence (**Figure 11.14**).

---

How do we determine the interval of convergence for a given power series? The presence of the term $x^k$ or $(x - a)^k$ in a power series suggests using the Ratio Test or the Root Test to determine the values of $x$ for which the series converges absolutely. By Theorem 10.19, if we determine the interval on which the series converges absolutely, we have a set of values for which the series converges.

Theorem 11.3 spells out the ways in which a power series can converge; its proof is given in Appendix A.

➤ Theorem 11.3 implies that the interval of convergence is symmetric about the center of the series; the radius of convergence $R$ is determined by analyzing $r$ from the Ratio Test (or $\rho$ from the Root Test). The theorem says nothing about convergence at the endpoints. For example, the intervals of convergence $(2, 6)$, $(2, 6]$, $[2, 6)$, and $[2, 6]$ all have a radius of convergence of $R = 2$.

---

**THEOREM 11.3    Convergence of Power Series**

A power series $\sum_{k=0}^{\infty} c_k(x - a)^k$ centered at $a$ converges in one of three ways:

1. The series converges absolutely for all $x$. It follows, by Theorem 10.19, that the series converges for all $x$, in which case the interval of convergence is $(-\infty, \infty)$ and the radius of convergence is $R = \infty$.

2. There is a real number $R > 0$ such that the series converges absolutely (and therefore converges) for $|x - a| < R$ and diverges for $|x - a| > R$, in which case the radius of convergence is $R$.

3. The series converges only at $a$, in which case the radius of convergence is $R = 0$.

---

The following examples illustrate how the Ratio and Root Tests are used to determine the radius and interval of convergence, and how the cases of Theorem 11.3 arise.

**EXAMPLE 1    Radius and interval of convergence**  Find the radius and interval of convergence for each power series.

**a.** $\sum_{k=0}^{\infty} \dfrac{x^k}{k!}$     **b.** $\sum_{k=0}^{\infty} \dfrac{(-1)^k(x - 2)^k}{4^k}$     **c.** $\sum_{k=1}^{\infty} k! \, x^k$

**SOLUTION**

**a.** The center of the power series is 0 and the terms of the series are $x^k/k!$. Due to the presence of the factor $k!$, we test the series for absolute convergence using the Ratio Test:

$$r = \lim_{k \to \infty} \frac{|x^{k+1}/(k+1)!|}{|x^k/k!|} \qquad \text{Ratio Test}$$

$$= \lim_{k \to \infty} \frac{|x|^{k+1}}{|x|^k} \cdot \frac{k!}{(k+1)!} \qquad \text{Invert and multiply.}$$

$$= |x| \lim_{k \to \infty} \frac{1}{k+1} = 0. \qquad \text{Simplify and take the limit with } x \text{ fixed.}$$

Notice that in taking the limit as $k \to \infty$, $x$ is held fixed. Because $r = 0$ for all real numbers $x$, the series converges absolutely for all $x$, which implies that the series converges for all $x$. Therefore, the interval of convergence is $(-\infty, \infty)$ **(Figure 11.15)** and the radius of convergence is $R = \infty$.

Center of
power series

0

Radius of convergence $R = \infty$
Interval of convergence $(-\infty, \infty)$

**Figure 11.15**

**b.** We test for absolute convergence using the Root Test:

$$\rho = \lim_{k \to \infty} \sqrt[k]{\left| \frac{(-1)^k(x - 2)^k}{4^k} \right|} = \frac{|x - 2|}{4}.$$

In this case, $\rho$ depends on the value of $x$. For absolute convergence, $x$ must satisfy

$$\rho = \frac{|x - 2|}{4} < 1,$$

which implies that $|x - 2| < 4$. Using standard techniques for solving inequalities, the solution set is $-4 < x - 2 < 4$, or $-2 < x < 6$. We conclude that the series converges absolutely (and therefore converges) on $(-2, 6)$, which means the radius of convergence is $R = 4$. When $-\infty < x < -2$ or $6 < x < \infty$, we have $\rho > 1$, so the series diverges on these intervals (the terms of the series do not approach 0 as $k \to \infty$ and the Divergence Test applies).

➤ Either the Ratio Test or the Root Test works for the power series in Example 1b.

The Root Test does not give information about convergence at the endpoints $x = -2$ and $x = 6$, because at these points, the Root Test results in $\rho = 1$. To test for convergence at the endpoints, we substitute each endpoint into the series and carry out separate tests. At $x = -2$, the power series becomes

$$\sum_{k=0}^{\infty} \frac{(-1)^k (x-2)^k}{4^k} = \sum_{k=0}^{\infty} \frac{4^k}{4^k} \quad \text{Substitute } x = -2 \text{ and simplify.}$$

$$= \sum_{k=0}^{\infty} 1. \quad \text{Diverges by Divergence Test}$$

The series clearly diverges at the left endpoint. At $x = 6$, the power series is

$$\sum_{k=0}^{\infty} \frac{(-1)^k (x-2)^k}{4^k} = \sum_{k=0}^{\infty} (-1)^k \frac{4^k}{4^k} \quad \text{Substitute } x = 6 \text{ and simplify.}$$

$$= \sum_{k=0}^{\infty} (-1)^k. \quad \text{Diverges by Divergence Test}$$

This series also diverges at the right endpoint. Therefore, the interval of convergence is $(-2, 6)$, excluding the endpoints (**Figure 11.16**).

**c.** In this case, the Ratio Test is preferable:

$$r = \lim_{k \to \infty} \frac{\left| (k+1)!\, x^{k+1} \right|}{\left| k!\, x^k \right|} \quad \text{Ratio Test}$$

$$= |x| \lim_{k \to \infty} \frac{(k+1)!}{k!} \quad \text{Simplify.}$$

$$= |x| \lim_{k \to \infty} (k+1) \quad \text{Simplify.}$$

$$= \infty. \quad \text{If } x \neq 0$$

We see that $r > 1$ for all $x \neq 0$, so the series diverges on $(-\infty, 0)$ and $(0, \infty)$.

The only way to satisfy $r < 1$ is to take $x = 0$, in which case the power series has a value of 0. The interval of convergence of the power series consists of the single point $x = 0$ (**Figure 11.17**), and the radius of convergence is $R = 0$.

*Related Exercises 10–12* ◄

> The Ratio and Root Tests determine the radius of convergence conclusively. However, the interval of convergence is not determined until the endpoints are tested.

Center of power series

Radius of convergence $R = 4$
Interval of convergence $(-2, 6)$

**Figure 11.16**

Center of power series

Radius of convergence $R = 0$
Interval of convergence $\{0\}$

**Figure 11.17**

**QUICK CHECK 2** What are the radius and interval of convergence of the geometric series $\sum x^k$? ◄

> The power series in Example 2 could also be analyzed using the Root Test.

**EXAMPLE 2** **Radius and interval of convergence** Use the Ratio Test to find the radius and interval of convergence of $\sum_{k=1}^{\infty} \frac{(x-2)^k}{\sqrt{k}}$.

**SOLUTION**

$$r = \lim_{k \to \infty} \frac{\left| (x-2)^{k+1} / \sqrt{k+1} \right|}{\left| (x-2)^k / \sqrt{k} \right|} \quad \text{Ratio Test}$$

$$= |x-2| \lim_{k \to \infty} \sqrt{\frac{k}{k+1}} \quad \text{Simplify.}$$

$$= |x-2| \underbrace{\sqrt{\lim_{k \to \infty} \frac{k}{k+1}}}_{1} \quad \text{Limit Law}$$

$$= |x-2| \quad \text{Limit equals 1.}$$

The series converges absolutely (and therefore converges) for all $x$ such that $r < 1$, which implies $|x - 2| < 1$, or $1 < x < 3$; we see that the radius of convergence is $R = 1$. On the intervals $-\infty < x < 1$ and $3 < x < \infty$, we have $r > 1$, so the series diverges.

We now test the endpoints. Substituting $x = 1$ gives the series

$$\sum_{k=1}^{\infty} \frac{(x-2)^k}{\sqrt{k}} = \sum_{k=1}^{\infty} \frac{(-1)^k}{\sqrt{k}}.$$

Center of
power series

$\leftarrow R \rightarrow | \leftarrow R \rightarrow$

0    1    2    3

Radius of convergence $R = 1$
Interval of convergence $[1, 3)$

**Figure 11.18**

This series converges by the Alternating Series Test (the terms of the series decrease in magnitude and approach 0 as $k \to \infty$). Substituting $x = 3$ gives the series

$$\sum_{k=1}^{\infty} \frac{(x-2)^k}{\sqrt{k}} = \sum_{k=1}^{\infty} \frac{1}{\sqrt{k}},$$

which is a divergent $p$-series. We conclude that the interval of convergence is $1 \leq x < 3$ (**Figure 11.18**).

*Related Exercise 18* ◄

The technique used in Examples 1 and 2 to determine the radius and interval of convergence for a given power series is summarized below.

---

**SUMMARY   Determining the Radius and Interval of Convergence of $\sum c_k(x - a)^k$**

1. Use the Ratio Test or the Root Test to find the interval $(a - R, a + R)$ on which the series converges absolutely; the radius of convergence for the series is $R$.

2. Use the *radius* of convergence to find the *interval* of convergence:
   • If $R = \infty$, the interval of convergence is $(-\infty, \infty)$.
   • If $R = 0$, the interval of convergence is the single point $x = a$.
   • If $0 < R < \infty$, the interval of convergence consists of the interval $(a - R, a + R)$ and possibly one or both of its endpoints. Determining whether the series $\sum c_k(x - a)^k$ converges at the endpoints $x = a - R$ and $x = a + R$ amounts to analyzing the series $\sum c_k(-R)^k$ and $\sum c_k R^k$.

---

Although the procedure just outlined can be applied to any power series, it is generally used when nothing is known about the series in question. As we learn shortly in Theorems 11.4 and 11.5, new power series can be built from known power series, and in these cases, other methods may be used to find the interval of convergence.

## Combining Power Series

A power series defines a function on its interval of convergence. When power series are combined algebraically, new functions are defined. The following theorem, stated without proof, gives three common ways to combine power series.

> ➤ New power series can also be defined as the product and quotient of power series. The calculation of the coefficients of such series is more challenging (Exercise 77).

> ➤ Theorem 11.4 also applies to power series centered at points other than $x = 0$. Property 1 applies directly; Properties 2 and 3 apply with slight modifications.

---

**THEOREM 11.4   Combining Power Series**
Suppose the power series $\sum c_k x^k$ and $\sum d_k x^k$ converge to $f(x)$ and $g(x)$, respectively, on an interval $I$.

1. **Sum and difference:** The power series $\sum (c_k \pm d_k)x^k$ converges to $f(x) \pm g(x)$ on $I$.

2. **Multiplication by a power:** Suppose $m$ is an integer such that $k + m \geq 0$, for all terms of the power series $x^m \sum c_k x^k = \sum c_k x^{k+m}$. This series converges to $x^m f(x)$, for all $x \neq 0$ in $I$. When $x = 0$, the series converges to $\lim_{x \to 0} x^m f(x)$.

3. **Composition:** If $h(x) = bx^m$, where $m$ is a positive integer and $b$ is a nonzero real number, the power series $\sum c_k (h(x))^k$ converges to the composite function $f(h(x))$, for all $x$ such that $h(x)$ is in $I$.

---

**EXAMPLE 3   Combining power series**  Given the geometric series

$$\frac{1}{1-x} = \sum_{k=0}^{\infty} x^k = 1 + x + x^2 + x^3 + \cdots, \qquad \text{for } |x| < 1,$$

find the power series and interval of convergence for the following functions.

**a.** $\dfrac{x^5}{1-x}$      **b.** $\dfrac{1}{1-2x}$      **c.** $\dfrac{1}{1+x^2}$

**SOLUTION**

**a.**
$$\frac{x^5}{1-x} = x^5\left(1 + x + x^2 + \cdots\right) \qquad \text{Theorem 11.4, Property 2}$$

$$= x^5 + x^6 + x^7 + \cdots$$

$$= \sum_{k=0}^{\infty} x^{k+5}$$

This geometric series has a ratio $r = x$ and converges when $|r| = |x| < 1$. The interval of convergence is $|x| < 1$.

**b.** We substitute $2x$ for $x$ in the power series for $\dfrac{1}{1-x}$:

$$\frac{1}{1-2x} = 1 + (2x) + (2x)^2 + \cdots \qquad \text{Theorem 11.4, Property 3}$$

$$= \sum_{k=0}^{\infty} (2x)^k.$$

This geometric series has a ratio $r = 2x$ and converges provided $|r| = |2x| < 1$ or $|x| < \frac{1}{2}$. The interval of convergence is $|x| < \frac{1}{2}$.

**c.** We substitute $-x^2$ for $x$ in the power series for $\dfrac{1}{1-x}$:

$$\frac{1}{1+x^2} = 1 + (-x^2) + (-x^2)^2 + \cdots \qquad \text{Theorem 11.4, Property 3}$$

$$= 1 - x^2 + x^4 - \cdots$$

$$= \sum_{k=0}^{\infty} (-1)^k x^{2k}.$$

This geometric series has a ratio of $r = -x^2$ and converges provided $|r| = |-x^2| = |x^2| < 1$ or $|x| < 1$.

*Related Exercises 41–42* ◄

## Differentiating and Integrating Power Series

Some properties of polynomials carry over to power series, but others do not. For example, a polynomial is defined for all values of $x$, whereas a power series is defined only on its interval of convergence. In general, the properties of polynomials carry over to power series when the power series is restricted to its interval of convergence. The following result illustrates this principle.

> ➤ Theorem 11.5 makes no claim about the convergence of the differentiated or integrated series at the endpoints of the interval of convergence.

---

**THEOREM 11.5   Differentiating and Integrating Power Series**
Suppose the power series $\sum c_k(x - a)^k$ converges for $|x - a| < R$ and defines a function $f$ on that interval.

**1.** Then $f$ is differentiable (which implies continuous) for $|x - a| < R$, and $f'$ is found by differentiating the power series for $f$ term by term; that is,

$$f'(x) = \sum k c_k (x - a)^{k-1},$$

for $|x - a| < R$.

**2.** The indefinite integral of $f$ is found by integrating the power series for $f$ term by term; that is,

$$\int f(x)\,dx = \sum c_k \frac{(x - a)^{k+1}}{k + 1} + C,$$

for $|x - a| < R$, where $C$ is an arbitrary constant.

---

The proof of this theorem requires advanced ideas and is omitted. However, some discussion is in order before turning to examples. The statements in Theorem 11.5 about term-by-term differentiation and integration say two things. First, the differentiated and integrated power series converge, provided $x$ belongs to the interior of the interval of convergence. But the theorem claims more than convergence. According to the theorem, the differentiated and integrated power series converge to the derivative and indefinite integral of $f$, respectively, on the interior of the interval of convergence. Let's use this theorem to develop new power series.

**EXAMPLE 4**   **Differentiating and integrating power series**   Consider the geometric series

$$f(x) = \frac{1}{1 - x} = \sum_{k=0}^{\infty} x^k = 1 + x + x^2 + x^3 + \cdots, \quad \text{for } |x| < 1.$$

**a.** Differentiate this series term by term to find the power series for $f'$ and identify the function it represents.

**b.** Integrate this series term by term and identify the function it represents.

**SOLUTION**

**a.** We know that $f'(x) = (1 - x)^{-2}$. Differentiating the series, we find that

$$
\begin{aligned}
f'(x) &= \frac{d}{dx}(1 + x + x^2 + x^3 + \cdots) &&\text{\color{blue}Differentiate the power series for } f. \\
&= 1 + 2x + 3x^2 + \cdots &&\text{\color{blue}Differentiate term by term.} \\
&= \sum_{k=0}^{\infty} (k + 1)\, x^k. &&\text{\color{blue}Summation notation}
\end{aligned}
$$

Therefore, on the interval $|x| < 1$,

$$f'(x) = (1 - x)^{-2} = \sum_{k=0}^{\infty} (k + 1)\, x^k.$$

Theorem 11.5 makes no claim about convergence of the differentiated series to $f'$ at the endpoints. In this case, substituting $x = \pm 1$ into the power series for $f'$ reveals that the series diverges at both endpoints.

**b.** Integrating $f$ and integrating the power series term by term, we have

$$\int \frac{dx}{1 - x} = \int (1 + x + x^2 + x^3 + \cdots)\, dx,$$

which implies that

$$-\ln|1 - x| = x + \frac{x^2}{2} + \frac{x^3}{3} + \frac{x^4}{4} + \cdots + C,$$

where $C$ is an arbitrary constant. Notice that the left side is 0 when $x = 0$. The right side is 0 when $x = 0$ provided we choose $C = 0$. Because $|x| < 1$, the absolute value sign on the left side may be removed. Multiplying both sides by $-1$, we have a series representation for $\ln(1 - x)$:

$$\ln(1 - x) = -x - \frac{x^2}{2} - \frac{x^3}{3} - \frac{x^4}{4} - \cdots = -\sum_{k=1}^{\infty} \frac{x^k}{k}.$$

It is interesting to test the endpoints of the interval $|x| < 1$. When $x = 1$, the series is (a multiple of) the divergent harmonic series, and when $x = -1$, the series is the convergent alternating harmonic series (Section 10.6). So the interval of convergence is $-1 \le x < 1$. Although we know the series converges at $x = -1$, Theorem 11.5 guarantees convergence to $\ln(1 - x)$ only at the interior points. We cannot use Theorem 11.5 to claim that the series converges to $\ln 2$ at $x = -1$. In fact, it does, as shown in Section 11.3.

**QUICK CHECK 3** Use the result of Example 4 to write a series representation for $\ln \frac{1}{2} = -\ln 2$. ◄

*Related Exercises 51, 55* ◄

**EXAMPLE 5**  **Functions to power series**  Find power series representations centered at 0 for the following functions and give their intervals of convergence.

**a.** $\tan^{-1} x$    **b.** $\ln\left(\dfrac{1+x}{1-x}\right)$

**SOLUTION**  In both cases, we work with known power series and use differentiation, integration, and other combinations.

**a.** The key is to recall that

$$\int \frac{dx}{1+x^2} = \tan^{-1} x + C$$

and that, by Example 3c,

$$\frac{1}{1+x^2} = 1 - x^2 + x^4 - \cdots, \qquad \text{provided } |x| < 1.$$

We now integrate both sides of this last expression:

$$\int \frac{dx}{1+x^2} = \int (1 - x^2 + x^4 - \cdots)\, dx,$$

which implies that

$$\tan^{-1} x = x - \frac{x^3}{3} + \frac{x^5}{5} - \cdots + C.$$

Substituting $x = 0$ and noting that $\tan^{-1} 0 = 0$, the two sides of this equation agree provided we choose $C = 0$. Therefore,

$$\tan^{-1} x = x - \frac{x^3}{3} + \frac{x^5}{5} - \cdots = \sum_{k=0}^{\infty} \frac{(-1)^k x^{2k+1}}{2k+1}.$$

> Again, Theorem 11.5 does not guarantee that the power series in Example 5a converges to $\tan^{-1} x$ at $x = \pm 1$. In fact, it does.

By Theorem 11.5, this power series converges to $\tan^{-1} x$ for $|x| < 1$. Testing the endpoints separately, we find that it also converges at $x = \pm 1$. Therefore, the interval of convergence is $[-1, 1]$.

**b.** We have already seen (Example 4) that

$$\ln(1-x) = -x - \frac{x^2}{2} - \frac{x^3}{3} - \cdots.$$

> Nicolaus Mercator (1620–1687) and Sir Isaac Newton (1642–1727) independently derived the power series for $\ln(1+x)$, which is called the *Mercator series*.

Replacing $x$ with $-x$ (Property 3 of Theorem 11.4), we have

$$\ln(1 - (-x)) = \ln(1+x) = x - \frac{x^2}{2} + \frac{x^3}{3} - \cdots.$$

Subtracting these two power series gives

$$\ln\left(\frac{1+x}{1-x}\right) = \ln(1+x) - \ln(1-x) \qquad \color{blue}{\text{Properties of logarithms}}$$

$$= \underbrace{\left(x - \frac{x^2}{2} + \frac{x^3}{3} - \cdots\right)}_{\ln(1+x)} - \underbrace{\left(-x - \frac{x^2}{2} - \frac{x^3}{3} - \cdots\right)}_{\ln(1-x)}, \quad \text{for } |x| < 1$$

$$= 2\left(x + \frac{x^3}{3} + \frac{x^5}{5} + \cdots\right) \qquad \color{blue}{\text{Combine; use Property 1 of Theorem 11.4.}}$$

$$= 2\sum_{k=0}^{\infty} \frac{x^{2k+1}}{2k+1}. \qquad \color{blue}{\text{Summation notation}}$$

**QUICK CHECK 4**  Verify that the power series in Example 5b does not converge at the endpoints $x = \pm 1$. ◄

This power series is the difference of two power series, both of which converge on the interval $|x| < 1$. Therefore, by Theorem 11.4, the new series also converges on $|x| < 1$.

*Related Exercises 57, 62* ◄

If you look carefully, every example in this section is ultimately based on the geometric series. Using this single series, we were able to develop power series for many other functions. Imagine what we could do with a few more basic power series. The following section accomplishes precisely that end. There, we discover power series for many of the standard functions of calculus.

## SECTION 11.2 EXERCISES

### Getting Started

1. Write the first four terms of a power series with coefficients $c_0, c_1, c_2,$ and $c_3$ centered at 0.

2. Is $\sum_{k=0}^{\infty} (5x - 20)^k$ a power series? If so, find the center $a$ of the power series and state a formula for the coefficients $c_k$ of the power series.

3. What tests are used to determine the radius of convergence of a power series?

4. Is $\sum_{k=0}^{\infty} x^{2k}$ a power series? If so, find the center $a$ of the power series and state a formula for the coefficients $c_k$ of the power series.

5. Do the radius and interval of convergence of a power series change when the series is differentiated or integrated? Explain.

6. Suppose a power series converges if $|x - 3| < 4$ and diverges if $|x - 3| \geq 4$. Determine the radius and interval of convergence.

7. Suppose a power series converges if $|4x - 8| \leq 40$ and diverges if $|4x - 8| > 40$. Determine the radius and interval of convergence.

8. Suppose the power series $\sum_{k=0}^{\infty} c_k(x - a)^k$ has an interval of convergence of $(-3, 7]$. Find the center $a$ and the radius of convergence $R$.

### Practice Exercises

**9–36. Radius and interval of convergence** *Determine the radius and interval of convergence of the following power series.*

9. $\sum_{k=0}^{\infty} (2x)^k$

10. $\sum_{k=0}^{\infty} \frac{(x - 1)^k}{k!}$

11. $\sum_{k=1}^{\infty} (kx)^k$

12. $\sum_{k=0}^{\infty} k!(x - 10)^k$

13. $\sum_{k=1}^{\infty} \sin^k\left(\frac{1}{k}\right) x^k$

14. $\sum_{k=2}^{\infty} \frac{2^k(x - 3)^k}{k}$

15. $\sum_{k=0}^{\infty} \left(\frac{x}{3}\right)^k$

16. $\sum_{k=0}^{\infty} (-1)^k \frac{x^k}{5^k}$

17. $\sum_{k=1}^{\infty} \frac{x^k}{k^k}$

18. $\sum_{k=1}^{\infty} \frac{(-1)^k x^k}{\sqrt{k}}$

19. $\sum_{k=0}^{\infty} \frac{x^k}{2^k + 1}$

20. $\sum_{k=0}^{\infty} \frac{(2x)^k}{k!}$

21. $-\frac{x^2}{1!} + \frac{x^4}{2!} - \frac{x^6}{3!} + \frac{x^8}{4!} - \cdots$

22. $x - \frac{x^3}{4} + \frac{x^5}{9} - \frac{x^7}{16} + \cdots$

23. $\sum_{k=1}^{\infty} \frac{(-1)^{k+1}(x - 1)^k}{k}$

24. $\sum_{k=0}^{\infty} (-1)^k \frac{k(x - 4)^k}{2^k}$

25. $\sum_{k=0}^{\infty} \frac{(4x - 1)^k}{k^2 + 4}$

26. $\sum_{k=1}^{\infty} \frac{(3x + 2)^k}{k}$

27. $\sum_{k=0}^{\infty} \frac{k^{10}(2x - 4)^k}{10^k}$

28. $\sum_{k=2}^{\infty} \frac{(x + 3)^k}{k \ln^2 k}$

29. $\sum_{k=0}^{\infty} \frac{k^2 x^{2k}}{k!}$

30. $\sum_{k=0}^{\infty} k(x - 1)^k$

31. $\sum_{k=1}^{\infty} \frac{x^{2k+1}}{3^{k-1}}$

32. $\sum_{k=0}^{\infty} \left(-\frac{x}{10}\right)^{2k}$

33. $\sum_{k=1}^{\infty} \frac{(x - 1)^k k^k}{(k + 1)^k}$

34. $\sum_{k=0}^{\infty} \frac{(-2)^k(x + 3)^k}{3^{k+1}}$

35. $\sum_{k=0}^{\infty} \frac{k^{20} x^k}{(2k + 1)!}$

36. $\sum_{k=0}^{\infty} (-1)^k \frac{x^{3k}}{27^k}$

**37–40. Radius of convergence** *Find the radius of convergence for the following power series.*

37. $\sum_{k=1}^{\infty} \frac{k! x^k}{k^k}$

38. $\sum_{k=1}^{\infty} \left(1 + \frac{1}{k}\right)^{k^2} x^k$

39. $\sum_{k=1}^{\infty} \left(\frac{k}{k + 4}\right)^{k^2} x^k$

40. $\sum_{k=1}^{\infty} \left(1 - \cos \frac{1}{2^k}\right) x^k$

**41–46. Combining power series** *Use the geometric series*

$$f(x) = \frac{1}{1 - x} = \sum_{k=0}^{\infty} x^k, \quad \text{for } |x| < 1,$$

*to find the power series representation for the following functions (centered at 0). Give the interval of convergence of the new series.*

41. $f(3x) = \frac{1}{1 - 3x}$

42. $g(x) = \frac{x^3}{1 - x}$

43. $h(x) = \frac{2x^3}{1 - x}$

44. $f(x^3) = \frac{1}{1 - x^3}$

45. $p(x) = \frac{4x^{12}}{1 - x}$

46. $f(-4x) = \frac{1}{1 + 4x}$

**47–50. Combining power series** *Use the power series representation*

$$f(x) = \ln(1 - x) = -\sum_{k=1}^{\infty} \frac{x^k}{k}, \quad \text{for } -1 \leq x < 1,$$

*to find the power series for the following functions (centered at 0). Give the interval of convergence of the new series.*

47. $f(3x) = \ln(1 - 3x)$

48. $g(x) = x^3 \ln(1 - x)$

49. $p(x) = 2x^6 \ln(1 - x)$

50. $f(x^3) = \ln(1 - x^3)$

**51–56. Differentiating and integrating power series** *Find the power series representation for g centered at 0 by differentiating or integrating the power series for f (perhaps more than once). Give the interval of convergence for the resulting series.*

**51.** $g(x) = \dfrac{2}{(1 - 2x)^2}$ using $f(x) = \dfrac{1}{1 - 2x}$

**52.** $g(x) = \dfrac{1}{(1 - x)^3}$ using $f(x) = \dfrac{1}{1 - x}$

**53.** $g(x) = -\dfrac{1}{(1 + x)^2}$ using $f(x) = \dfrac{1}{1 + x}$

**54.** $g(x) = \dfrac{x}{(1 + x^2)^2}$ using $f(x) = \dfrac{1}{1 + x^2}$

**55.** $g(x) = \ln(1 - 3x)$ using $f(x) = \dfrac{1}{1 - 3x}$

**56.** $g(x) = \ln(1 + x^2)$ using $f(x) = \dfrac{x}{1 + x^2}$

**57–62. Functions to power series** *Find power series representations centered at 0 for the following functions using known power series. Give the interval of convergence for the resulting series.*

**57.** $f(x) = \dfrac{2x}{(1 + x^2)^2}$

**58.** $f(x) = \dfrac{1}{1 - x^4}$

**59.** $f(x) = \dfrac{3}{3 + x}$

**60.** $f(x) = \ln\sqrt{1 - x^2}$

**61.** $f(x) = \ln\sqrt{4 - x^2}$

**62.** $f(x) = \tan^{-1}(4x^2)$

**63. Explain why or why not** Determine whether the following statements are true and give an explanation or counterexample.

   **a.** The interval of convergence of the power series $\sum c_k(x - 3)^k$ could be $(-2, 8)$.

   **b.** The series $\displaystyle\sum_{k=0}^{\infty}(-2x)^k$ converges on the interval $-\frac{1}{2} < x < \frac{1}{2}$.

   **c.** If $f(x) = \displaystyle\sum_{k=0}^{\infty} c_k x^k$ on the interval $|x| < 1$, then $f(x^2) = \sum c_k x^{2k}$ on the interval $|x| < 1$.

   **d.** If $f(x) = \displaystyle\sum_{k=0}^{\infty} c_k x^k = 0$, for all $x$ on an interval $(-a, a)$, then $c_k = 0$, for all $k$.

**64. Scaling power series** If the power series $f(x) = \sum c_k x^k$ has an interval of convergence of $|x| < R$, what is the interval of convergence of the power series for $f(ax)$, where $a \neq 0$ is a real number?

**65. Shifting power series** If the power series $f(x) = \sum c_k x^k$ has an interval of convergence of $|x| < R$, what is the interval of convergence of the power series for $f(x - a)$, where $a \neq 0$ is a real number?

**66. A useful substitution** Replace $x$ with $x - 1$ in the series

$$\ln(1 + x) = \sum_{k=1}^{\infty} \frac{(-1)^{k+1}x^k}{k}$$ to obtain a power series for $\ln x$

centered at $x = 1$. What is the interval of convergence for the new power series?

**67–71. Series to functions** *Find the function represented by the following series, and find the interval of convergence of the series. (Not all these series are power series.)*

**67.** $\displaystyle\sum_{k=0}^{\infty}(\sqrt{x} - 2)^k$

**68.** $\displaystyle\sum_{k=1}^{\infty}\frac{x^{2k}}{4^k}$

**69.** $\displaystyle\sum_{k=0}^{\infty}e^{-kx}$

**70.** $\displaystyle\sum_{k=1}^{\infty}\frac{(x - 2)^k}{3^{2k}}$

**71.** $\displaystyle\sum_{k=0}^{\infty}\left(\frac{x^2 - 1}{3}\right)^k$

**72–74. Exponential function** *In Section 11.3, we show that the power series for the exponential function centered at 0 is*

$$e^x = \sum_{k=0}^{\infty}\frac{x^k}{k!}, \quad \text{for } -\infty < x < \infty.$$

*Use the methods of this section to find the power series centered at 0 for the following functions. Give the interval of convergence for the resulting series.*

**72.** $f(x) = e^{2x}$

**73.** $f(x) = e^{-3x}$

**74.** $f(x) = x^2 e^x$

## Explorations and Challenges

**75. Powers of $x$ multiplied by a power series** Prove that if

$$f(x) = \sum_{k=0}^{\infty}c_k x^k$$ converges with radius of convergence $R$, then the

power series for $x^m f(x)$ also converges with radius of convergence $R$, for positive integers $m$.

**T 76. Remainders** Let

$$f(x) = \sum_{k=0}^{\infty}x^k = \frac{1}{1 - x} \quad \text{and} \quad S_n(x) = \sum_{k=0}^{n-1}x^k.$$

The remainder in truncating the power series after $n$ terms is $R_n = f(x) - S_n(x)$, which depends on $x$.

   **a.** Show that $R_n(x) = x^n/(1 - x)$.

   **b.** Graph the remainder function on the interval $|x| < 1$, for $n = 1, 2,$ and 3. Discuss and interpret the graph. Where on the interval is $|R_n(x)|$ largest? Smallest?

   **c.** For fixed $n$, minimize $|R_n(x)|$ with respect to $x$. Does the result agree with the observations in part (b)?

   **d.** Let $N(x)$ be the number of terms required to reduce $|R_n(x)|$ to less than $10^{-6}$. Graph the function $N(x)$ on the interval $|x| < 1$. Discuss and interpret the graph.

**77. Product of power series** Let

$$f(x) = \sum_{k=0}^{\infty}c_k x^k \quad \text{and} \quad g(x) = \sum_{k=0}^{\infty}d_k x^k.$$

   **a.** Multiply the power series together as if they were polynomials, collecting all terms that are multiples of 1, $x$, and $x^2$. Write the first three terms of the product $f(x)g(x)$.

   **b.** Find a general expression for the coefficient of $x^n$ in the product series, for $n = 0, 1, 2, \ldots$.

**78. Inverse sine** Given the power series

$$\frac{1}{\sqrt{1 - x^2}} = 1 + \frac{1}{2}x^2 + \frac{1\cdot 3}{2\cdot 4}x^4 + \frac{1\cdot 3\cdot 5}{2\cdot 4\cdot 6}x^6 + \cdots,$$

for $-1 < x < 1$, find the power series for $f(x) = \sin^{-1}x$ centered at 0.

**QUICK CHECK ANSWERS**

**1.** $g(0) = 0$  **2.** $|x| < 1, R = 1$  **3.** Substituting

$x = 1/2, \ln(1/2) = -\ln 2 = -\displaystyle\sum_{k=1}^{\infty}\frac{1}{2^k k}$ ◄

# 11.3 Taylor Series

In the preceding section, we saw that a power series represents a function on its interval of convergence. This section explores the opposite question: Given a function, what is its power series representation? We have already made significant progress in answering this question because we know how Taylor polynomials are used to approximate functions. We now extend Taylor polynomials to produce power series—called *Taylor series*—that provide series representations for functions.

## Taylor Series for a Function

Suppose a function $f$ has derivatives $f^{(k)}(a)$ of *all* orders at the point $a$. If we write the $n$th-order Taylor polynomial for $f$ centered at $a$ and allow $n$ to increase indefinitely, a power series is obtained:

$$\underbrace{c_0 + c_1(x - a) + c_2(x - a)^2 + \cdots + c_n(x - a)^n}_{\text{Taylor polynomial of order } n} + \underbrace{\cdots}_{n \to \infty} = \sum_{k=0}^{\infty} c_k(x - a)^k.$$

The coefficients of the Taylor polynomial are given by

$$c_k = \frac{f^{(k)}(a)}{k!}, \quad \text{for } k = 0, 1, 2, \ldots.$$

> ➤ Maclaurin series are named after the Scottish mathematician Colin Maclaurin (1698–1746), who described them (with credit to Taylor) in a textbook in 1742.

These coefficients are also the coefficients of the power series, which is called the *Taylor series for $f$ centered at $a$*. It is the natural extension of the set of Taylor polynomials for $f$ at $a$. The special case of a Taylor series centered at 0 is called a *Maclaurin series*.

---

**DEFINITION** **Taylor / Maclaurin Series for a Function**

Suppose the function $f$ has derivatives of all orders on an interval centered at the point $a$. The **Taylor series for $f$ centered at $a$** is

$$f(a) + f'(a)(x - a) + \frac{f''(a)}{2!}(x - a)^2 + \frac{f^{(3)}(a)}{3!}(x - a)^3 + \cdots$$

$$= \sum_{k=0}^{\infty} \frac{f^{(k)}(a)}{k!}(x - a)^k.$$

A Taylor series centered at 0 is called a **Maclaurin series**.

---

> ➤ There are unusual cases in which the Taylor series for a function converges to a different function (Exercise 90).

For the Taylor series to be useful, we need to know two things:

- the values of $x$ for which the Taylor series converges, and
- the values of $x$ for which the Taylor series for $f$ *equals* $f$.

**QUICK CHECK 1** Verify that if the Taylor series for $f$ centered at $a$ is evaluated at $x = a$, then the Taylor series equals $f(a)$. ◄

The second question is subtle and is postponed for a few pages. For now, we find the Taylor series for $f$ centered at a point, but we refrain from saying $f(x)$ equals the power series.

**EXAMPLE 1** **Maclaurin series and convergence** Find the Maclaurin series (which is the Taylor series centered at 0) for the following functions, and then find the interval of convergence.

**a.** $f(x) = \cos x$  **b.** $f(x) = \dfrac{1}{1 - x}$

**SOLUTION** The procedure for finding the coefficients of a Taylor series is the same as for Taylor polynomials; most of the work is computing the derivatives of $f$.

**a.** The Maclaurin series has the form

$$\sum_{k=0}^{\infty} c_k x^k, \quad \text{where } c_k = \frac{f^{(k)}(0)}{k!}, \quad \text{for } k = 0, 1, 2, \ldots.$$

We evaluate derivatives of $f(x) = \cos x$ at $x = 0$.

$$f(x) = \cos x \implies f(0) = 1$$
$$f'(x) = -\sin x \implies f'(0) = 0$$
$$f''(x) = -\cos x \implies f''(0) = -1$$
$$f'''(x) = \sin x \implies f'''(0) = 0$$
$$f^{(4)}(x) = \cos x \implies f^{(4)}(0) = 1$$
$$\vdots \qquad\qquad\qquad \vdots$$

> ➤ In Example 1a, we note that both $\cos x$ and its Maclaurin series are even functions. Be cautious with this observation. A Taylor series for an even function centered at a point different from 0 may be even, odd, or neither. A similar behavior occurs with odd functions.

Because the odd-order derivatives are zero, $c_k = \dfrac{f^{(k)}(0)}{k!} = 0$ when $k$ is odd. Using the even-order derivatives, we have

$$c_0 = f(0) = 1, \qquad c_2 = \frac{f^{(2)}(0)}{2!} = -\frac{1}{2!},$$

$$c_4 = \frac{f^{(4)}(0)}{4!} = \frac{1}{4!}, \qquad c_6 = \frac{f^{(6)}(0)}{6!} = -\frac{1}{6!},$$

and in general, $c_{2k} = \dfrac{(-1)^k}{(2k)!}$. Therefore, the Maclaurin series for $f$ is

$$1 - \frac{x^2}{2!} + \frac{x^4}{4!} - \frac{x^6}{6!} + \cdots = \sum_{k=0}^{\infty} \frac{(-1)^k}{(2k)!} x^{2k}.$$

Notice that this series contains all the Taylor polynomials. In this case, it consists only of even powers of $x$, reflecting the fact that $\cos x$ is an even function.

For what values of $x$ does the series converge? As discussed in Section 11.2, we apply the Ratio Test:

$$r = \lim_{k \to \infty} \left| \frac{(-1)^{k+1} x^{2(k+1)}/(2(k+1))!}{(-1)^k x^{2k}/(2k)!} \right| \qquad r = \lim_{k \to \infty} \left| \frac{a_{k+1}}{a_k} \right|$$

> ➤ Recall that
> $$(2k+2)! = (2k+2)(2k+1)(2k)!.$$
> Therefore, $\dfrac{(2k)!}{(2k+2)!} = \dfrac{1}{(2k+2)(2k+1)}$.

$$= \lim_{k \to \infty} \left| \frac{x^2}{(2k+2)(2k+1)} \right| = 0. \qquad \text{Simplify and take the limit with } x \text{ fixed.}$$

In this case, $r < 1$ for all $x$, so the Maclaurin series converges absolutely for all $x$, which implies (by Theorem 10.19) that the series converges for all $x$. We conclude that the interval of convergence is $-\infty < x < \infty$.

**b.** We proceed in a similar way with $f(x) = 1/(1-x)$ by evaluating the derivatives of $f$ at 0:

$$f(x) = \frac{1}{1-x} \implies f(0) = 1,$$

$$f'(x) = \frac{1}{(1-x)^2} \implies f'(0) = 1,$$

$$f''(x) = \frac{2}{(1-x)^3} \implies f''(0) = 2!,$$

$$f'''(x) = \frac{3 \cdot 2}{(1-x)^4} \implies f'''(0) = 3!,$$

$$f^{(4)}(x) = \frac{4 \cdot 3 \cdot 2}{(1-x)^5} \implies f^{(4)}(0) = 4!,$$

and in general, $f^{(k)}(0) = k!$. Therefore, the Maclaurin series coefficients are

$$c_k = \frac{f^{(k)}(0)}{k!} = \frac{k!}{k!} = 1, \text{ for } k = 0, 1, 2, \ldots . \text{ The series for } f \text{ centered at 0 is}$$

$$1 + x + x^2 + x^3 + \cdots = \sum_{k=0}^{\infty} x^k.$$

This power series is familiar! The Maclaurin series for $f(x) = 1/(1 - x)$ is a geometric series. We could apply the Ratio Test, but we have already demonstrated that this series converges for $|x| < 1$.

*Related Exercises 12, 18* ◄

**QUICK CHECK 2** Based on Example 1b, what is the Taylor series for $f(x) = (1 + x)^{-1}$? ◄

The preceding example has an important lesson. *There is only one power series representation for a given function about a given point; however, there may be several ways to find it.* Another important point is this: The series in Example 1b represents the function $f(x) = 1/(1 - x)$ only on the interval $(-1, 1)$. If we want a power series representation for $f$ on another interval, a new center must be chosen and the coefficients of the power series must be recomputed, as shown in the next example.

**EXAMPLE 2   Choosing a different center** Find the Taylor series for $f(x) = \dfrac{1}{1 - x}$ centered at 5, and then determine its radius and interval of convergence.

**SOLUTION** Using the pattern discovered in Example 1b, we find that the $k$th derivative of $f$ is $f^{(k)}(x) = \dfrac{k!}{(1 - x)^{k+1}}$. Therefore, the Taylor series coefficients are

$$c_k = \frac{f^{(k)}(5)}{k!} = \frac{k!}{k!(1 - 5)^{k+1}} = \frac{(-1)^{k+1}}{4^{k+1}}, \quad \text{for } k = 0, 1, 2, \ldots,$$

and the series for $f$ centered at 5 is

$$\sum_{k=0}^{\infty} \frac{(-1)^{k+1}(x - 5)^k}{4^{k+1}} = -\frac{1}{4} + \frac{x - 5}{4^2} - \frac{(x - 5)^2}{4^3} + \frac{(x - 5)^3}{4^4} \cdots.$$

To determine the interval of convergence for this series, we apply the Ratio Test:

**QUICK CHECK 3** Verify that the series $\displaystyle\sum_{k=0}^{\infty} \frac{(-1)^{k+1}(x - 5)^k}{4^{k+1}}$ from Example 2 diverges at $x = 1$ and $x = 9$. ◄

$$r = \lim_{k \to \infty} \left| \frac{(-1)^{k+2}(x - 5)^{k+1}/4^{k+2}}{(-1)^{k+1}(x - 5)^k/4^{k+1}} \right| = \lim_{k \to \infty} \frac{|x - 5|}{4} = \frac{|x - 5|}{4}.$$

The series converges absolutely when $r = \dfrac{|x - 5|}{4} < 1$, or when $1 < x < 9$, so the radius of convergence is 4. The series diverges at $x = 1$ and $x = 9$ (Quick Check 3), so we conclude that the interval of convergence is $(1, 9)$.

The graphs of $f$ and the 7th-order Taylor polynomial $p_7$ centered at 5 (blue curve) are shown in **Figure 11.19**, along with the graph of $p_7$ centered at 0 (red curve) found in Example 1b. Notice that in both cases, the interval of convergence extends to the vertical asymptote ($x = 1$) of $f(x) = \dfrac{1}{1 - x}$. In fact, it can be shown that the Taylor series for $f$, with any choice of center $a \neq 1$, has an interval of convergence symmetric about $a$ that extends to $x = 1$. Remarkably, Taylor series for $f$ "know" the location of the discontinuity of $f$.

*Related Exercises 29–30* ◄

**Figure 11.19**

**EXAMPLE 3   Center other than 0** Find the first four nonzero terms of the Taylor series for $f(x) = \sqrt[3]{x}$ centered at 8.

**SOLUTION** Notice that $f$ has derivatives of all orders at $x = 8$. The Taylor series centered at 8 has the form

$$\sum_{k=0}^{\infty} c_k(x - 8)^k, \quad \text{where } c_k = \frac{f^{(k)}(8)}{k!}.$$

Next, we evaluate derivatives:

$$f(x) = x^{1/3} \Rightarrow f(8) = 2,$$

$$f'(x) = \frac{1}{3}x^{-2/3} \Rightarrow f'(8) = \frac{1}{12},$$

$$f''(x) = -\frac{2}{9}x^{-5/3} \Rightarrow f''(8) = -\frac{1}{144}, \text{ and}$$

$$f'''(x) = \frac{10}{27}x^{-8/3} \Rightarrow f'''(8) = \frac{5}{3456}.$$

We now assemble the power series:

$$2 + \frac{1}{12}(x-8) + \frac{1}{2!}\left(-\frac{1}{144}\right)(x-8)^2 + \frac{1}{3!}\left(\frac{5}{3456}\right)(x-8)^3 + \cdots$$

$$= 2 + \frac{1}{12}(x-8) - \frac{1}{288}(x-8)^2 + \frac{5}{20{,}736}(x-8)^3 + \cdots.$$

*Related Exercise 32* ◄

**EXAMPLE 4  Manipulating Maclaurin series**  Let $f(x) = e^x$.

**a.** Find the Maclaurin series for $f$.

**b.** Find its interval of convergence.

**c.** Use the Maclaurin series for $e^x$ to find the Maclaurin series for the functions $x^4 e^x$, $e^{-2x}$, and $e^{-x^2}$.

**SOLUTION**

**a.** The coefficients of the Taylor polynomials for $f(x) = e^x$ centered at 0 are $c_k = 1/k!$ (Example 3, Section 11.1). They are also the coefficients of the Maclaurin series. Therefore, the Maclaurin series for $e^x$ is

$$1 + \frac{x}{1!} + \frac{x^2}{2!} + \cdots + \frac{x^n}{n!} + \cdots = \sum_{k=0}^{\infty} \frac{x^k}{k!}.$$

**b.** By the Ratio Test,

$$r = \lim_{k \to \infty} \left| \frac{x^{k+1}/(k+1)!}{x^k/k!} \right| \qquad \text{Substitute } (k+1)\text{st and } k\text{th terms.}$$

$$= \lim_{k \to \infty} \left| \frac{x}{k+1} \right| = 0. \qquad \text{Simplify; take the limit with } x \text{ fixed.}$$

Because $r < 1$ for all $x$, the interval of convergence is $-\infty < x < \infty$.

**c.** As stated in Theorem 11.4, power series may be added, multiplied by powers of $x$, or composed with functions on their intervals of convergence. Therefore, the Maclaurin series for $x^4 e^x$ is

$$x^4 \sum_{k=0}^{\infty} \frac{x^k}{k!} = \sum_{k=0}^{\infty} \frac{x^{k+4}}{k!} = x^4 + \frac{x^5}{1!} + \frac{x^6}{2!} + \cdots + \frac{x^{k+4}}{k!} + \cdots.$$

Similarly, $e^{-2x}$ is the composition $f(-2x)$. Replacing $x$ with $-2x$ in the Maclaurin series for $f$, the series representation for $e^{-2x}$ is

$$\sum_{k=0}^{\infty} \frac{(-2x)^k}{k!} = \sum_{k=0}^{\infty} \frac{(-1)^k (2x)^k}{k!} = 1 - 2x + 2x^2 - \frac{4}{3}x^3 + \cdots.$$

The Maclaurin series for $e^{-x^2}$ is obtained by replacing $x$ with $-x^2$ in the power series for $f$. The resulting series is

$$\sum_{k=0}^{\infty} \frac{(-x^2)^k}{k!} = \sum_{k=0}^{\infty} \frac{(-1)^k x^{2k}}{k!} = 1 - x^2 + \frac{x^4}{2!} - \frac{x^6}{3!} + \cdots.$$

*Related Exercises 35, 38* ◀

**QUICK CHECK 4** Find the first three terms of the Maclaurin series for $2xe^x$ and $e^{-x}$. ◀

Because the interval of convergence of $f(x) = e^x$ is $-\infty < x < \infty$, the manipulations used to obtain the series for $x^4e^x$, $e^{-2x}$, or $e^{-x^2}$ do not change the interval of convergence. If in doubt about the interval of convergence of a new series, apply the Ratio Test.

## The Binomial Series

We know from algebra that if $p$ is a positive integer, then $(1 + x)^p$ is a polynomial of degree $p$. In fact,

$$(1 + x)^p = \binom{p}{0} + \binom{p}{1}x + \binom{p}{2}x^2 + \cdots + \binom{p}{p}x^p,$$

where the binomial coefficients $\binom{p}{k}$ are defined as follows.

➤ For nonnegative integers $p$ and $k$ with $0 \le k \le p$, the binomial coefficients may also be defined as $\binom{p}{k} = \dfrac{p!}{k!(p-k)!}$, where $0! = 1$. The coefficients form the rows of Pascal's triangle. The coefficients of $(1 + x)^5$ form the sixth row of the triangle.

$$
\begin{array}{ccccccccccc}
 & & & & & 1 & & & & & \\
 & & & & 1 & & 1 & & & & \\
 & & & 1 & & 2 & & 1 & & & \\
 & & 1 & & 3 & & 3 & & 1 & & \\
 & 1 & & 4 & & 6 & & 4 & & 1 & \\
1 & & 5 & & 10 & & 10 & & 5 & & 1
\end{array}
$$

> **DEFINITION** **Binomial Coefficients**
>
> For real numbers $p$ and integers $k \ge 1$,
>
> $$\binom{p}{k} = \frac{p(p-1)(p-2)\cdots(p-k+1)}{k!}, \qquad \binom{p}{0} = 1.$$

**QUICK CHECK 5** Evaluate the binomial coefficients $\binom{-3}{2}$ and $\binom{\frac{1}{2}}{3}$. ◀

For example,

$$(1 + x)^5 = \underbrace{\binom{5}{0}}_{1} + \underbrace{\binom{5}{1}}_{5}x + \underbrace{\binom{5}{2}}_{10}x^2 + \underbrace{\binom{5}{3}}_{10}x^3 + \underbrace{\binom{5}{4}}_{5}x^4 + \underbrace{\binom{5}{5}}_{1}x^5$$

$$= 1 + 5x + 10x^2 + 10x^3 + 5x^4 + x^5.$$

Our goal is to extend this idea to the functions $f(x) = (1 + x)^p$, where $p \neq 0$ is a real number. The result is a Taylor series called the *binomial series*.

> **THEOREM 11.6** **Binomial Series**
>
> For real numbers $p \neq 0$, the Taylor series for $f(x) = (1 + x)^p$ centered at 0 is the **binomial series**
>
> $$\sum_{k=0}^{\infty} \binom{p}{k}x^k = 1 + \sum_{k=1}^{\infty} \frac{p(p-1)(p-2)\cdots(p-k+1)}{k!}x^k$$
>
> $$= 1 + px + \frac{p(p-1)}{2!}x^2 + \frac{p(p-1)(p-2)}{3!}x^3 + \cdots.$$
>
> The series converges for $|x| < 1$ (and possibly at the endpoints, depending on $p$). If $p$ is a nonnegative integer, the series terminates and results in a polynomial of degree $p$.

**Proof:** We seek a power series centered at 0 of the form

$$\sum_{k=0}^{\infty} c_k x^k, \quad \text{where } c_k = \frac{f^{(k)}(0)}{k!}, \quad \text{for } k = 0, 1, 2, \ldots.$$

The job is to evaluate the derivatives of $f$ at 0:

➤ To evaluate $\binom{p}{k}$, start with $p$ and successively subtract 1 until $k$ factors are obtained; then take the product of these $k$ factors and divide by $k!$. Recall that $\binom{p}{0} = 1$.

$$f(x) = (1 + x)^p \Rightarrow f(0) = 1,$$
$$f'(x) = p(1 + x)^{p-1} \Rightarrow f'(0) = p,$$
$$f''(x) = p(p - 1)(1 + x)^{p-2} \Rightarrow f''(0) = p(p - 1), \text{ and}$$
$$f'''(x) = p(p - 1)(p - 2)(1 + x)^{p-3} \Rightarrow f'''(0) = p(p - 1)(p - 2).$$

A pattern emerges: The $k$th derivative $f^{(k)}(0)$ involves the $k$ factors $p(p-1)(p-2)\cdots(p-k+1)$. In general, we have

$$f^{(k)}(0) = p(p-1)(p-2)\cdots(p-k+1).$$

Therefore,

$$c_k = \frac{f^{(k)}(0)}{k!} = \frac{p(p-1)(p-2)\cdots(p-k+1)}{k!} = \binom{p}{k}, \quad \text{for } k = 0, 1, 2, \ldots.$$

The Taylor series for $f(x) = (1+x)^p$ centered at 0 is

$$\binom{p}{0} + \binom{p}{1}x + \binom{p}{2}x^2 + \binom{p}{3}x^3 + \cdots = \sum_{k=0}^{\infty} \binom{p}{k}x^k.$$

This series has the same general form for all values of $p$. When $p$ is a nonnegative integer, the series terminates and it is a polynomial of degree $p$.

The radius of convergence for the binomial series is determined by the Ratio Test. Holding $p$ and $x$ fixed, the relevant limit is

$$r = \lim_{k \to \infty} \left| \frac{x^{k+1}p(p-1)\cdots(p-k+1)(p-k)/(k+1)!}{x^k p(p-1)\cdots(p-k+1)/k!} \right| \qquad \text{Ratio of } (k+1)\text{st to } k\text{th term}$$

$$= |x| \lim_{k \to \infty} \left| \frac{p-k}{k+1} \right| \qquad \text{Cancel factors and simplify.}$$

$$\underbrace{\qquad\qquad}_{\text{approaches } 1}$$

> In Theorem 11.6, it can be shown that the interval of convergence for the binomial series is
>
> • $(-1, 1)$ if $p \le -1$,
> • $(-1, 1]$ if $-1 < p < 0$, and
> • $[-1, 1]$ if $p > 0$ and not an integer.

$$= |x|.$$

With $p$ fixed,
$$\lim_{k \to \infty} \left| \frac{p-k}{k+1} \right| = 1.$$

Absolute convergence requires that $r = |x| < 1$. Therefore, the series converges for $|x| < 1$. Depending on the value of $p$, the interval of convergence may include the endpoints; see margin note. ◄

---

**EXAMPLE 5**   **Binomial series**   Consider the function $f(x) = \sqrt{1 + x}$.

> A binomial series is a Taylor series. Because the series in Example 5 is centered at 0, it is also a Maclaurin series.

**a.** Find the first four terms of the binomial series for $f$ centered at 0.

**b.** Approximate $\sqrt{1.15}$ to three decimal places. Assume the series for $f$ converges to $f$ on its interval of convergence, which is $[-1, 1]$.

**SOLUTION**

**a.** We use the formula for the binomial coefficients with $p = \frac{1}{2}$ to compute the first four coefficients:

**Table 11.3**

| $n$ | Approximation $p_n(0.15)$ |
|-----|---------------------------|
| 0   | 1.0                       |
| 1   | 1.075                     |
| 2   | 1.0721875                 |
| 3   | 1.072398438               |

$$c_0 = 1, \qquad\qquad c_1 = \binom{\frac{1}{2}}{1} = \frac{\frac{1}{2}}{1!} = \frac{1}{2},$$

$$c_2 = \binom{\frac{1}{2}}{2} = \frac{\frac{1}{2}\left(-\frac{1}{2}\right)}{2!} = -\frac{1}{8}, \text{ and } c_3 = \binom{\frac{1}{2}}{3} = \frac{\frac{1}{2}\left(-\frac{1}{2}\right)\left(-\frac{3}{2}\right)}{3!} = \frac{1}{16}.$$

The leading terms of the binomial series are

$$1 + \frac{1}{2}x - \frac{1}{8}x^2 + \frac{1}{16}x^3 - \cdots.$$

> The remainder theorem for alternating series (Section 10.6) could be used in Example 5 to estimate the number of terms of the Maclaurin series needed to achieve a desired accuracy.

**b.** Truncating the binomial series in part (a) produces Taylor polynomials $p_n$ that may be used to approximate $f(0.15) = \sqrt{1.15}$. With $x = 0.15$, we find the polynomial approximations shown in Table 11.3. Four terms of the power series ($n = 3$) give $\sqrt{1.15} \approx 1.072$. Because the approximations with $n = 2$ and $n = 3$ agree to three decimal places, when rounded, the approximation 1.072 is accurate to three decimal places.

**QUICK CHECK 6** Use two and three terms of the binomial series in Example 5 to approximate $\sqrt{1.1}$. ◄

*Related Exercises 46–47* ◄

**EXAMPLE 6  Working with binomial series** Consider the functions

$$f(x) = \sqrt[3]{1 + x} \quad \text{and} \quad g(x) = \sqrt[3]{c + x}, \quad \text{where } c > 0 \text{ is a constant.}$$

**a.** Find the first four terms of the binomial series for $f$ centered at 0.

**b.** Use part (a) to find the first four terms of the binomial series for $g$ centered at 0.

**c.** Use part (b) to approximate $\sqrt[3]{23}, \sqrt[3]{24}, \ldots, \sqrt[3]{31}$. Assume the series for $g$ converges to $g$ on its interval of convergence.

**SOLUTION**

**a.** Because $f(x) = (1 + x)^{1/3}$, we find the binomial coefficients with $p = \frac{1}{3}$:

$$c_0 = \binom{\frac{1}{3}}{0} = 1, \qquad\qquad c_1 = \binom{\frac{1}{3}}{1} = \frac{\frac{1}{3}}{1!} = \frac{1}{3},$$

$$c_2 = \binom{\frac{1}{3}}{2} = \frac{\left(\frac{1}{3}\right)\left(\frac{1}{3} - 1\right)}{2!} = -\frac{1}{9}, \text{ and } c_3 = \binom{\frac{1}{3}}{3} = \frac{\left(\frac{1}{3}\right)\left(\frac{1}{3} - 1\right)\left(\frac{1}{3} - 2\right)}{3!} = \frac{5}{81}.$$

The first four terms of the binomial series are

$$1 + \frac{1}{3}x - \frac{1}{9}x^2 + \frac{5}{81}x^3.$$

**b.** To avoid deriving a new series for $g(x) = \sqrt[3]{c + x}$, a few steps of algebra allow us to use part (a). Note that

$$g(x) = \sqrt[3]{c + x} = \sqrt[3]{c\left(1 + \frac{x}{c}\right)} = \sqrt[3]{c} \cdot \sqrt[3]{1 + \frac{x}{c}} = \sqrt[3]{c} \cdot f\left(\frac{x}{c}\right).$$

In other words, $g$ can be expressed in terms of $f$, for which we already have a binomial series. The binomial series for $g$ is obtained by substituting $x/c$ into the binomial series for $f$ and multiplying by $\sqrt[3]{c}$:

$$g(x) = \sqrt[3]{c}\underbrace{\left(1 + \frac{1}{3}\left(\frac{x}{c}\right) - \frac{1}{9}\left(\frac{x}{c}\right)^2 + \frac{5}{81}\left(\frac{x}{c}\right)^3 - \cdots\right)}_{f(x/c)}.$$

It can be shown that the series for $f$ in part (a) converges to $f(x)$ for $|x| \leq 1$. Therefore, the series for $f(x/c)$ converges to $f(x/c)$ provided $|x/c| \leq 1$, or, equivalently, for $|x| \leq c$.

**c.** The series of part (b) may be truncated after four terms to approximate cube roots. For example, note that $\sqrt[3]{29} = \sqrt[3]{27 + 2}$, so we take $c = 27$ and $x = 2$.

The choice $c = 27$ is made because 29 is near 27 and $\sqrt[3]{c} = \sqrt[3]{27} = 3$ is easy to evaluate. Substituting $c = 27$ and $x = 2$, we find that

$$\sqrt[3]{29} \approx \sqrt[3]{27}\left(1 + \frac{1}{3}\left(\frac{2}{27}\right) - \frac{1}{9}\left(\frac{2}{27}\right)^2 + \frac{5}{81}\left(\frac{2}{27}\right)^3\right) \approx 3.0723.$$

The same method is used to approximate the cube roots of 23, 24, ..., 30, 31 (Table 11.4). The absolute error is the difference between the approximation and the value given by a calculator. Notice that the errors increase as we move away from 27.

*Related Exercises 52–54* ◄

**Table 11.4**

| | Approximation | Absolute Error |
|---|---|---|
| $\sqrt[3]{23}$ | 2.8439 | $6.7 \times 10^{-5}$ |
| $\sqrt[3]{24}$ | 2.8845 | $2.0 \times 10^{-5}$ |
| $\sqrt[3]{25}$ | 2.9240 | $3.9 \times 10^{-6}$ |
| $\sqrt[3]{26}$ | 2.9625 | $2.4 \times 10^{-7}$ |
| $\sqrt[3]{27}$ | 3 | 0 |
| $\sqrt[3]{28}$ | 3.0366 | $2.3 \times 10^{-7}$ |
| $\sqrt[3]{29}$ | 3.0723 | $3.5 \times 10^{-6}$ |
| $\sqrt[3]{30}$ | 3.1072 | $1.7 \times 10^{-5}$ |
| $\sqrt[3]{31}$ | 3.1414 | $5.4 \times 10^{-5}$ |

## Convergence of Taylor Series

It may seem that the story of Taylor series is over. But there is a technical point that is easily overlooked. Given a function $f$, we know how to write its Taylor series centered at a point $a$, and we know how to find its interval of convergence. We still do not know that the series actually converges to $f$. The remaining task is to determine when the Taylor series for $f$ actually converges to $f$ on its interval of convergence. Fortunately, the necessary tools have already been presented in Taylor's Theorem (Theorem 11.1), which gives the remainder for Taylor polynomials.

Assume $f$ has derivatives of *all* orders on an open interval containing the point $a$. Taylor's Theorem tells us that

$$f(x) = p_n(x) + R_n(x),$$

where $p_n$ is the $n$th-order Taylor polynomial for $f$ centered at $a$, that

$$R_n(x) = \frac{f^{(n+1)}(c)}{(n+1)!}(x-a)^{n+1}$$

is the remainder, and that $c$ is a point between $x$ and $a$. We see that the remainder, $R_n(x) = f(x) - p_n(x)$, measures the difference between $f$ and the approximating polynomial $p_n$. When we say the Taylor series converges to $f$ at a point $x$, we mean the value of the Taylor series at $x$ equals $f(x)$; that is, $\lim_{n\to\infty} p_n(x) = f(x)$. The following theorem makes these ideas precise.

---

**THEOREM 11.7   Convergence of Taylor Series**
Let $f$ have derivatives of all orders on an open interval $I$ containing $a$. The Taylor series for $f$ centered at $a$ converges to $f$, for all $x$ in $I$, if and only if $\lim_{n\to\infty} R_n(x) = 0$, for all $x$ in $I$, where

$$R_n(x) = \frac{f^{(n+1)}(c)}{(n+1)!}(x-a)^{n+1}$$

is the remainder at $x$, with $c$ between $x$ and $a$.

---

**Proof:** The theorem requires derivatives of *all* orders. Therefore, by Taylor's Theorem (Theorem 11.1), the remainder exists in the given form for all $n$. Let $p_n$ denote the $n$th-order Taylor polynomial and note that $\lim_{n\to\infty} p_n(x)$ is the Taylor series for $f$ centered at $a$, evaluated at a point $x$ in $I$.

First, assume $\lim_{n\to\infty} R_n(x) = 0$ on the interval $I$ and recall that $p_n(x) = f(x) - R_n(x)$. Taking limits of both sides, we have

$$\underbrace{\lim_{n\to\infty} p_n(x)}_{\text{Taylor series}} = \lim_{n\to\infty} (f(x) - R_n(x)) = \underbrace{\lim_{n\to\infty} f(x)}_{f(x)} - \underbrace{\lim_{n\to\infty} R_n(x)}_{0} = f(x).$$

We conclude that the Taylor series $\lim_{n\to\infty} p_n(x)$ equals $f(x)$, for all $x$ in $I$.

Conversely, if the Taylor series converges to $f$, then $f(x) = \lim_{n\to\infty} p_n(x)$ and

$$0 = f(x) - \lim_{n\to\infty} p_n(x) = \lim_{n\to\infty} \underbrace{(f(x) - p_n(x))}_{R_n(x)} = \lim_{n\to\infty} R_n(x).$$

It follows that $\lim_{n\to\infty} R_n(x) = 0$, for all $x$ in $I$.   ◀

Even with an expression for the remainder, it may be difficult to show that $\lim_{n\to\infty} R_n(x) = 0$. The following examples illustrate cases in which it is possible.

**EXAMPLE 7   Remainder in the Maclaurin series for $e^x$**  Show that the Maclaurin series for $f(x) = e^x$ converges to $f(x)$, for $-\infty < x < \infty$.

**SOLUTION**  As shown in Example 4, the Maclaurin series for $f(x) = e^x$ is

$$\sum_{k=0}^{\infty} \frac{x^k}{k!} = 1 + x + \frac{x^2}{2!} + \cdots + \frac{x^n}{n!} + \cdots,$$

which converges, for $-\infty < x < \infty$. In Example 7 of Section 11.1 it was shown that the remainder is

$$R_n(x) = \frac{e^c}{(n+1)!} x^{n+1},$$

where $c$ is between 0 and $x$. Notice that the intermediate point $c$ varies with $n$, but it is always between 0 and $x$. Therefore, $e^c$ is between $e^0 = 1$ and $e^x$; in fact, $e^c \leq e^{|x|}$, for all $n$. It follows that

$$|R_n(x)| \leq \frac{e^{|x|}}{(n+1)!} |x|^{n+1}.$$

Holding $x$ fixed, we have

$$\lim_{n \to \infty} |R_n(x)| = \lim_{n \to \infty} \frac{e^{|x|}}{(n+1)!} |x|^{n+1} = e^{|x|} \lim_{n \to \infty} \frac{|x|^{n+1}}{(n+1)!} = 0,$$

where we used the fact that $\lim_{n \to \infty} x^n/n! = 0$, for $-\infty < x < \infty$ (Section 10.2). Because $\lim_{n \to \infty} |R_n(x)| = 0$, it follows that for all real numbers $x$, the Taylor series converges to $e^x$, or

$$e^x = \sum_{k=0}^{\infty} \frac{x^k}{k!} = 1 + x + \frac{x^2}{2!} + \cdots + \frac{x^n}{n!} + \cdots.$$

The convergence of the Taylor series to $e^x$ is illustrated in **Figure 11.20**, where Taylor polynomials of increasing degree are graphed together with $e^x$.

*Related Exercise 65* ◄

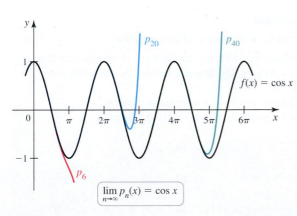

**Figure 11.20**

**EXAMPLE 8   Maclaurin series convergence for cos x**   Show that the Maclaurin series for $\cos x$,

$$1 - \frac{x^2}{2!} + \frac{x^4}{4!} - \frac{x^6}{6!} + \cdots = \sum_{k=0}^{\infty} (-1)^k \frac{x^{2k}}{(2k)!},$$

converges to $f(x) = \cos x$, for $-\infty < x < \infty$.

**SOLUTION**   To show that the power series converges to $f$, we must show that $\lim_{n \to \infty} |R_n(x)| = 0$, for $-\infty < x < \infty$. According to Taylor's Theorem with $a = 0$,

$$R_n(x) = \frac{f^{(n+1)}(c)}{(n+1)!} x^{n+1},$$

where $c$ is between 0 and $x$. Notice that $f^{(n+1)}(c) = \pm \sin c$ or $f^{(n+1)}(c) = \pm \cos c$. In all cases, $|f^{(n+1)}(c)| \leq 1$. Therefore, the absolute value of the remainder term is bounded as

$$|R_n(x)| = \left| \frac{f^{(n+1)}(c)}{(n+1)!} x^{n+1} \right| \leq \frac{|x|^{n+1}}{(n+1)!}.$$

Holding $x$ fixed and using $\lim_{n \to \infty} x^n/n! = 0$, we see that $\lim_{n \to \infty} R_n(x) = 0$ for all $x$. Therefore, the given power series converges to $f(x) = \cos x$, for all $x$; that is, $\cos x = \sum_{k=0}^{\infty} \frac{(-1)^k x^{2k}}{(2k)!}$. The convergence of the Taylor series to $\cos x$ is illustrated in **Figure 11.21**.

*Related Exercise 63* ◄

**Figure 11.21**

The procedure used in Examples 7 and 8 can be carried out for all the Taylor series we have worked with so far (with varying degrees of difficulty). In each case, the Taylor series converges to the function it represents on the interval of convergence. Table 11.5 summarizes commonly used Taylor series centered at 0 and the functions to which they converge.

➤ Table 11.5 asserts, without proof, that in several cases, the Taylor series for $f$ converges to $f$ at the endpoints of the interval of convergence. Proving convergence at the endpoints generally requires advanced techniques. It may also be done using the following theorem:

Suppose the Taylor series for $f$ centered at 0 converges to $f$ on the interval $(-R, R)$. If the series converges at $x = R$, then it converges to $\lim_{x \to R^-} f(x)$. If the series converges at $x = -R$, then it converges to $\lim_{x \to -R^+} f(x)$.

For example, this theorem would allow us to conclude that the series for $\ln(1 + x)$ converges to $\ln 2$ at $x = 1$.

**Table 11.5**

$$\frac{1}{1-x} = 1 + x + x^2 + \cdots + x^k + \cdots = \sum_{k=0}^{\infty} x^k, \quad \text{for } |x| < 1$$

$$\frac{1}{1+x} = 1 - x + x^2 - \cdots + (-1)^k x^k + \cdots = \sum_{k=0}^{\infty} (-1)^k x^k, \quad \text{for } |x| < 1$$

$$e^x = 1 + x + \frac{x^2}{2!} + \cdots + \frac{x^k}{k!} + \cdots = \sum_{k=0}^{\infty} \frac{x^k}{k!}, \quad \text{for } |x| < \infty$$

$$\sin x = x - \frac{x^3}{3!} + \frac{x^5}{5!} - \cdots + \frac{(-1)^k x^{2k+1}}{(2k+1)!} + \cdots = \sum_{k=0}^{\infty} \frac{(-1)^k x^{2k+1}}{(2k+1)!}, \quad \text{for } |x| < \infty$$

$$\cos x = 1 - \frac{x^2}{2!} + \frac{x^4}{4!} - \cdots + \frac{(-1)^k x^{2k}}{(2k)!} + \cdots = \sum_{k=0}^{\infty} \frac{(-1)^k x^{2k}}{(2k)!}, \quad \text{for } |x| < \infty$$

$$\ln(1 + x) = x - \frac{x^2}{2} + \frac{x^3}{3} - \cdots + \frac{(-1)^{k+1} x^k}{k} + \cdots = \sum_{k=1}^{\infty} \frac{(-1)^{k+1} x^k}{k}, \quad \text{for } -1 < x \le 1$$

$$-\ln(1 - x) = x + \frac{x^2}{2} + \frac{x^3}{3} + \cdots + \frac{x^k}{k} + \cdots = \sum_{k=1}^{\infty} \frac{x^k}{k}, \quad \text{for } -1 \le x < 1$$

$$\tan^{-1} x = x - \frac{x^3}{3} + \frac{x^5}{5} - \cdots + \frac{(-1)^k x^{2k+1}}{2k+1} + \cdots = \sum_{k=0}^{\infty} \frac{(-1)^k x^{2k+1}}{2k+1}, \quad \text{for } |x| \le 1$$

$$\sinh x = x + \frac{x^3}{3!} + \frac{x^5}{5!} + \cdots + \frac{x^{2k+1}}{(2k+1)!} + \cdots = \sum_{k=0}^{\infty} \frac{x^{2k+1}}{(2k+1)!}, \quad \text{for } |x| < \infty$$

$$\cosh x = 1 + \frac{x^2}{2!} + \frac{x^4}{4!} + \cdots + \frac{x^{2k}}{(2k)!} + \cdots = \sum_{k=0}^{\infty} \frac{x^{2k}}{(2k)!}, \quad \text{for } |x| < \infty$$

➤ As noted in Theorem 11.6, the binomial series may converge to $(1 + x)^p$ at $x = \pm 1$, depending on the value of $p$.

$$(1 + x)^p = \sum_{k=0}^{\infty} \binom{p}{k} x^k, \text{ for } |x| < 1 \text{ and } \binom{p}{k} = \frac{p(p-1)(p-2)\cdots(p-k+1)}{k!}, \binom{p}{0} = 1$$

---

## SECTION 11.3 EXERCISES

### Getting Started

1. How are the Taylor polynomials for a function $f$ centered at $a$ related to the Taylor series of the function $f$ centered at $a$?

2. What conditions must be satisfied by a function $f$ to have a Taylor series centered at $a$?

3. Find a Taylor series for $f$ centered at 2 given that $f^{(k)}(2) = 1$, for all nonnegative integers $k$.

4. Find a Taylor series for $f$ centered at 0 given that $f^{(k)}(0) = (k + 1)!$, for all nonnegative integers $k$.

5. Suppose you know the Maclaurin series for $f$ and that it converges to $f(x)$ for $|x| < 1$. How do you find the Maclaurin series for $f(x^2)$ and where does it converge?

6. For what values of $p$ does the Taylor series for $f(x) = (1 + x)^p$ centered at 0 terminate?

7. In terms of the remainder, what does it mean for a Taylor series for a function $f$ to converge to $f$?

8. Find the Maclaurin series for $\sin(-x)$ using the definition of a Maclaurin series. Check your answer by finding the Maclaurin series for $\sin(-x)$ using Table 11.5.

### Practice Exercises

**9–26. Taylor series and interval of convergence**

a. *Use the definition of a Taylor/Maclaurin series to find the first four nonzero terms of the Taylor series for the given function centered at a.*
b. *Write the power series using summation notation.*
c. *Determine the interval of convergence of the series.*

9. $f(x) = \dfrac{1}{x^2}, a = 1$

10. $f(x) = \dfrac{1}{x^2}, a = -1$

11. $f(x) = e^{-x}, a = 0$

12. $f(x) = \cos 2x, a = 0$

13. $f(x) = \dfrac{2}{(1 - x)^3}, a = 0$

14. $f(x) = x \sin x, a = 0$

15. $f(x) = (1 + x^2)^{-1}, a = 0$

16. $f(x) = \ln(1 + 4x), a = 0$

17. $f(x) = e^{2x}, a = 0$

18. $f(x) = (1 + 2x)^{-1}, a = 0$

19. $f(x) = \tan^{-1}\dfrac{x}{2}, a = 0$

20. $f(x) = \sin 3x, a = 0$

21. $f(x) = 3^x, a = 0$

22. $f(x) = \log_3(x + 1), a = 0$

**23.** $f(x) = \cosh 3x, a = 0$   **24.** $f(x) = \sinh 2x, a = 0$

**25.** $f(x) = \ln(x - 2), a = 3$   **26.** $f(x) = e^x, a = 2$

**27–34. Taylor series**

a.  Use the definition of a Taylor series to find the first four nonzero terms of the Taylor series for the given function centered at a.
b.  Write the power series using summation notation.

**27.** $f(x) = \sin x, a = \dfrac{\pi}{2}$   **28.** $f(x) = \cos x, a = \pi$

**29.** $f(x) = \dfrac{1}{x}, a = 1$   **30.** $f(x) = \dfrac{1}{x}, a = 2$

**31.** $f(x) = \ln x, a = 3$   **32.** $f(x) = e^x, a = \ln 2$

**33.** $f(x) = 2^x, a = 1$   **34.** $f(x) = x \ln x - x + 1, a = 1$

**35–44. Manipulating Taylor series** *Use the Taylor series in Table 11.5 to find the first four nonzero terms of the Taylor series for the following functions centered at 0.*

**35.** $\ln(1 + x^2)$   **36.** $\sin x^2$

**37.** $\dfrac{1}{1 - 2x}$   **38.** $\ln(1 + 2x)$

**39.** $\begin{cases} \dfrac{e^x - 1}{x} & \text{if } x \neq 1 \\ 1 & \text{if } x = 1 \end{cases}$   **40.** $\cos x^3$

**41.** $(1 + x^4)^{-1}$   **42.** $x \tan^{-1} x^2$

**43.** $\sinh x^2$   **44.** $\cosh(-2x)$

**T 45–50. Binomial series**

a.  Find the first four nonzero terms of the binomial series centered at 0 for the given function.
b.  Use the first four terms of the series to approximate the given quantity.

**45.** $f(x) = (1 + x)^{-2}$; approximate $\dfrac{1}{1.21} = \dfrac{1}{1.1^2}$.

**46.** $f(x) = \sqrt{1 + x}$; approximate $\sqrt{1.06}$.

**47.** $f(x) = \sqrt[4]{1 + x}$; approximate $\sqrt[4]{1.12}$.

**48.** $f(x) = (1 + x)^{-3}$; approximate $\dfrac{1}{1.331} = \dfrac{1}{1.1^3}$.

**49.** $f(x) = (1 + x)^{-2/3}$; approximate $1.18^{-2/3}$.

**50.** $f(x) = (1 + x)^{2/3}$; approximate $1.02^{2/3}$.

**51–56. Working with binomial series** *Use properties of power series, substitution, and factoring to find the first four nonzero terms of the Maclaurin series for the following functions. Give the interval of convergence for the new series (Theorem 11.4 is useful). Use the Maclaurin series*

$$\sqrt{1 + x} = 1 + \frac{x}{2} - \frac{x^2}{8} + \frac{x^3}{16} - \cdots, \quad \text{for } -1 \le x \le 1.$$

**51.** $\sqrt{1 + x^2}$   **52.** $\sqrt{4 + x}$

**53.** $\sqrt{9 - 9x}$   **54.** $\sqrt{1 - 4x}$

**55.** $\sqrt{a^2 + x^2}, a > 0$   **56.** $\sqrt{4 - 16x^2}$

**57–62. Working with binomial series** *Use properties of power series, substitution, and factoring to find the first four nonzero terms of the Maclaurin series for the following functions. Use the Maclaurin series*

$$(1 + x)^{-2} = 1 - 2x + 3x^2 - 4x^3 + \cdots, \quad \text{for } -1 < x < 1.$$

**57.** $(1 + 4x)^{-2}$   **58.** $\dfrac{1}{(1 - 4x)^2}$

**59.** $\dfrac{1}{(4 + x^2)^2}$   **60.** $(x^2 - 4x + 5)^{-2}$

**61.** $\dfrac{1}{(3 + 4x)^2}$   **62.** $\dfrac{1}{(1 + 4x^2)^2}$

**63–66. Remainders** *Find the remainder in the Taylor series centered at the point a for the following functions. Then show that $\lim\limits_{n \to \infty} R_n(x) = 0$, for all x in the interval of convergence.*

**63.** $f(x) = \sin x, a = 0$   **64.** $f(x) = \cos 2x, a = 0$

**65.** $f(x) = e^{-x}, a = 0$   **66.** $f(x) = \cos x, a = \dfrac{\pi}{2}$

**67.  Explain why or why not** Determine whether the following statements are true and give an explanation or counterexample.

a.  The function $f(x) = \sqrt{x}$ has a Taylor series centered at 0.
b.  The function $f(x) = \csc x$ has a Taylor series centered at $\pi/2$.
c.  If $f$ has a Taylor series that converges only on $(-2, 2)$, then $f(x^2)$ has a Taylor series that also converges only on $(-2, 2)$.
d.  If $p(x)$ is the Taylor series for $f$ centered at 0, then $p(x - 1)$ is the Taylor series for $f$ centered at 1.
e.  The Taylor series for an even function centered at 0 has only even powers of $x$.

**68–75. Any method**

a.  Use any analytical method to find the first four nonzero terms of the Taylor series centered at 0 for the following functions. You do not need to use the definition of the Taylor series coefficients.
b.  Determine the radius of convergence of the series.

**68.** $f(x) = \cos 2x + 2 \sin x$   **69.** $f(x) = \dfrac{e^x + e^{-x}}{2}$

**70.** $f(x) = \begin{cases} \dfrac{\sin x}{x} & \text{if } x \neq 0 \\ 1 & \text{if } x = 0 \end{cases}$   **71.** $f(x) = (1 + x^2)^{-2/3}$

**72.** $f(x) = x^2 \cos x^2$   **73.** $f(x) = \sqrt{1 - x^2}$

**74.** $f(x) = b^x$, for $b > 0, b \neq 1$

**75.** $f(x) = \dfrac{1}{x^4 + 2x^2 + 1}$

**T 76–79. Approximating powers** *Compute the coefficients for the Taylor series for the following functions about the given point a, and then use the first four terms of the series to approximate the given number.*

**76.** $f(x) = \sqrt{x}$ with $a = 36$; approximate $\sqrt{39}$.

**77.** $f(x) = \sqrt[3]{x}$ with $a = 64$; approximate $\sqrt[3]{60}$.

**78.** $f(x) = \dfrac{1}{\sqrt{x}}$ with $a = 4$; approximate $\dfrac{1}{\sqrt{3}}$.

**79.** $f(x) = \sqrt[4]{x}$ with $a = 16$; approximate $\sqrt[4]{13}$.

**80. Geometric/binomial series** Recall that the Taylor series for $f(x) = 1/(1 - x)$ centered at 0 is the geometric series $\sum_{k=0}^{\infty} x^k$. Show that this series can also be found as a binomial series.

**81. Integer coefficients** Show that the first five nonzero coefficients of the Taylor series (binomial series) for $f(x) = \sqrt{1 + 4x}$ centered at 0 are integers. (In fact, *all* the coefficients are integers.)

### Explorations and Challenges

**82. Choosing a good center** Suppose you want to approximate $\sqrt{72}$ using four terms of a Taylor series. Compare the accuracy of the approximations obtained using the Taylor series for $\sqrt{x}$ centered at 64 and 81.

**83. Alternative means** By comparing the first four terms, show that the Maclaurin series for $\sin^2 x$ can be found (a) by squaring the Maclaurin series for $\sin x$, (b) by using the identity $\sin^2 x = \dfrac{1 - \cos 2x}{2}$, or (c) by computing the coefficients using the definition.

**84. Alternative means** By comparing the first four terms, show that the Maclaurin series for $\cos^2 x$ can be found (a) by squaring the Maclaurin series for $\cos x$, (b) by using the identity $\cos^2 x = \dfrac{1 + \cos 2x}{2}$, or (c) by computing the coefficients using the definition.

**85. Designer series** Find a power series that has $(2, 6)$ as an interval of convergence.

**86. Composition of series** Use composition of series to find the first three terms of the Maclaurin series for the following functions.

   **a.** $e^{\sin x}$   **b.** $e^{\tan x}$   **c.** $\sqrt{1 + \sin^2 x}$

**T 87–88. Approximations** *Choose a Taylor series and a center point to approximate the following quantities with an error of $10^{-4}$ or less.*

**87.** $\cos 40°$                      **88.** $\sin(0.98\pi)$

**89. Different approximation strategies** Suppose you want to approximate $\sqrt[3]{128}$ to within $10^{-4}$ of the exact value.

   **a.** Use a Taylor polynomial for $f(x) = (125 + x)^{1/3}$ centered at 0.
   **b.** Use a Taylor polynomial for $f(x) = x^{1/3}$ centered at 125.
   **c.** Compare the two approaches. Are they equivalent?

**90. Nonconvergence to $f$** Consider the function

$$f(x) = \begin{cases} e^{-1/x^2} & \text{if } x \neq 0 \\ 0 & \text{if } x = 0. \end{cases}$$

   **a.** Use the definition of the derivative to show that $f'(0) = 0$.
   **b.** Assume the fact that $f^{(k)}(0) = 0$, for $k = 1, 2, 3, \ldots$. (You can write a proof using the definition of the derivative.) Write the Taylor series for $f$ centered at 0.
   **c.** Explain why the Taylor series for $f$ does not converge to $f$ for $x \neq 0$.

**91. Version of the Second Derivative Test** Assume $f$ has at least two continuous derivatives on an interval containing $a$ with $f'(a) = 0$. Use Taylor's Theorem to prove the following version of the Second Derivative Test:

   **a.** If $f''(x) > 0$ on some interval containing $a$, then $f$ has a local minimum at $a$.
   **b.** If $f''(x) < 0$ on some interval containing $a$, then $f$ has a local maximum at $a$.

**QUICK CHECK ANSWERS**

**1.** When evaluated at $x = a$, all terms of the series are zero except the first term, which is $f(a)$. Therefore, the series equals $f(a)$ at this point.   **2.** $1 - x + x^2 - x^3 + x^4 - \cdots$
**3.** The series diverges at both endpoint by the Divergence Test.   **4.** $2x + 2x^2 + x^3$; $1 - x + x^2/2$   **5.** $6, 1/16$
**6.** $1.05, 1.04875$ ◄

# 11.4 Working with Taylor Series

We now know the Taylor series for many familiar functions, and we have tools for working with power series. The goal of this final section is to illustrate additional techniques associated with power series. As you will see, power series cover the entire landscape of calculus from limits and derivatives to integrals and approximation. We present five different topics that you can explore selectively.

## Limits by Taylor Series

An important use of Taylor series is evaluating limits. Two examples illustrate the essential ideas.

**EXAMPLE 1   A limit by Taylor series** Evaluate $\displaystyle \lim_{x \to 0} \frac{x^2 + 2\cos x - 2}{3x^4}$.

> L'Hôpital's Rule may be impractical when it must be used more than once on the same limit or when derivatives are difficult to compute.

**SOLUTION** Because the limit has the indeterminate form $0/0$, l'Hôpital's Rule can be used, which requires four applications of the rule. Alternatively, because the limit involves values of $x$ near 0, we substitute the Maclaurin series for $\cos x$. Recalling that

$$\cos x = 1 - \frac{x^2}{2} + \frac{x^4}{24} - \frac{x^6}{720} + \cdots, \qquad \text{Table 11.5}$$

> In using series to evaluate limits, it is often not obvious how many terms of the Taylor series to use. When in doubt, include extra (higher-order) terms. The dots in the calculation stand for powers of $x$ greater than the last power that appears.

**QUICK CHECK 1** Use the Taylor series $\sin x = x - x^3/6 + \cdots$ to verify that $\lim_{x \to 0} (\sin x)/x = 1$. ◄

we have

$$\lim_{x \to 0} \frac{x^2 + 2\cos x - 2}{3x^4} = \lim_{x \to 0} \frac{x^2 + 2\left(1 - \dfrac{x^2}{2} + \dfrac{x^4}{24} - \dfrac{x^6}{720} + \cdots\right) - 2}{3x^4} \quad \text{Substitute for } \cos x.$$

$$= \lim_{x \to 0} \frac{x^2 + \left(2 - x^2 + \dfrac{x^4}{12} - \dfrac{x^6}{360} + \cdots\right) - 2}{3x^4} \quad \text{Simplify.}$$

$$= \lim_{x \to 0} \frac{\dfrac{x^4}{12} - \dfrac{x^6}{360} + \cdots}{3x^4} \quad \text{Simplify.}$$

$$= \lim_{x \to 0} \left(\frac{1}{36} - \frac{x^2}{1080} + \cdots\right) = \frac{1}{36}. \quad \begin{array}{l}\text{Use Theorem}\\ \text{11.4, Property 2;}\\ \text{evaluate limit.}\end{array}$$

*Related Exercise 13* ◄

**EXAMPLE 2   A limit by Taylor series** Evaluate

$$\lim_{x \to \infty} \left(6x^5 \sin \frac{1}{x} - 6x^4 + x^2\right).$$

**SOLUTION** A Taylor series may be centered at any finite point in the domain of the function, but we don't have the tools needed to expand a function about $x = \infty$. Using a technique introduced earlier, we replace $x$ with $1/t$ and note that as $x \to \infty$, $t \to 0^+$. The new limit becomes

$$\lim_{x \to \infty} \left(6x^5 \sin \frac{1}{x} - 6x^4 + x^2\right) = \lim_{t \to 0^+} \left(\frac{6 \sin t}{t^5} - \frac{6}{t^4} + \frac{1}{t^2}\right) \quad \text{Replace } x \text{ with } 1/t.$$

$$= \lim_{t \to 0^+} \left(\frac{6 \sin t - 6t + t^3}{t^5}\right). \quad \text{Common denominator}$$

This limit has the indeterminate form $0/0$. We now expand $\sin t$ in a Taylor series centered at $t = 0$. Because

$$\sin t = t - \frac{t^3}{6} + \frac{t^5}{120} - \frac{t^7}{5040} + \cdots, \quad \text{Table 11.5}$$

the value of the original limit is

$$\lim_{t \to 0^+} \left(\frac{6 \sin t - 6t + t^3}{t^5}\right)$$

$$= \lim_{t \to 0^+} \left(\frac{6\left(t - \dfrac{t^3}{6} + \dfrac{t^5}{120} - \dfrac{t^7}{5040} + \cdots\right) - 6t + t^3}{t^5}\right) \quad \text{Substitute for } \sin t.$$

$$= \lim_{t \to 0^+} \left(\frac{\dfrac{t^5}{20} - \dfrac{t^7}{840} + \cdots}{t^5}\right) \quad \text{Simplify.}$$

$$= \lim_{t \to 0^+} \left(\frac{1}{20} - \frac{t^2}{840} + \cdots\right) = \frac{1}{20}. \quad \begin{array}{l}\text{Use Theorem 11.4,}\\ \text{Property 2; evaluate limit.}\end{array}$$

*Related Exercise 19* ◄

## Differentiating Power Series

The following examples illustrate ways in which term-by-term differentiation (Theorem 11.5) may be used.

**EXAMPLE 3   Power series for derivatives** Differentiate the Maclaurin series for $f(x) = \sin x$ to verify that $\dfrac{d}{dx}(\sin x) = \cos x$.

**SOLUTION** The Maclaurin series for $f(x) = \sin x$ is

$$\sin x = x - \frac{x^3}{3!} + \frac{x^5}{5!} - \frac{x^7}{7!} + \cdots,$$

and it converges for $-\infty < x < \infty$. By Theorem 11.5, the differentiated series also converges for $-\infty < x < \infty$ and it converges to $f'(x)$. Differentiating, we have

$$\frac{d}{dx}\left( x - \frac{x^3}{3!} + \frac{x^5}{5!} - \frac{x^7}{7!} + \cdots \right) = 1 - \frac{x^2}{2!} + \frac{x^4}{4!} - \frac{x^6}{6!} + \cdots = \cos x.$$

**QUICK CHECK 2** Differentiate the power series for $\cos x$ (given in Example 3) and identify the result. ◄

The differentiated series is the Maclaurin series for $\cos x$, confirming that $f'(x) = \cos x$.

*Related Exercises 25–26* ◄

**EXAMPLE 4   A differential equation** Find a power series solution of the differential equation $y'(t) = y + 2$, subject to the initial condition $y(0) = 6$. Identify the function represented by the power series.

**SOLUTION** Because the initial condition is given at $t = 0$, we assume the solution has a Taylor series centered at 0 of the form $y(t) = \sum_{k=0}^{\infty} c_k t^k$, where the coefficients $c_k$ must be determined. Recall that the coefficients of the Taylor series are given by

$$c_k = \frac{y^{(k)}(0)}{k!}, \quad \text{for } k = 0, 1, 2, \dots.$$

If we can determine $y^{(k)}(0)$, for $k = 0, 1, 2, \dots$, the coefficients of the series are also determined.

Substituting the initial condition $t = 0$ and $y = 6$ into the power series

$$y(t) = c_0 + c_1 t + c_2 t^2 + \cdots,$$

we find that

$$6 = c_0 + c_1 (0) + c_2 (0)^2 + \cdots.$$

It follows that $c_0 = 6$. To determine $y'(0)$, we substitute $t = 0$ into the differential equation; the result is $y'(0) = y(0) + 2 = 6 + 2 = 8$. Therefore, $c_1 = y'(0)/1! = 8$.

The remaining derivatives are obtained by successively differentiating the differential equation and substituting $t = 0$. We find that $y''(0) = y'(0) = 8$, $y'''(0) = y''(0) = 8$, and in general, $y^{(k)}(0) = 8$, for $k = 1, 2, 3, 4, \dots$. Therefore,

$$c_k = \frac{y^{(k)}(0)}{k!} = \frac{8}{k!}, \quad \text{for } k = 1, 2, 3, \dots,$$

and the Taylor series for the solution is

$$y(t) = c_0 + c_1 t + c_2 t^2 + \cdots$$
$$= 6 + \frac{8}{1!} t + \frac{8}{2!} t^2 + \frac{8}{3!} t^3 + \cdots.$$

To identify the function represented by this series, we write

$$y(t) = \underbrace{-2 + 8}_{6} + \frac{8}{1!} t + \frac{8}{2!} t^2 + \frac{8}{3!} t^3 + \cdots$$

$$= -2 + 8 \underbrace{\left( 1 + t + \frac{t^2}{2!} + \frac{t^3}{3!} + \cdots \right)}_{e^t}.$$

➤ You should check that $y(t) = -2 + 8e^t$ satisfies $y'(t) = y + 2$ and $y(0) = 6$.

The power series that appears is the Taylor series for $e^t$. Therefore, the solution is $y = -2 + 8e^t$.

*Related Exercise 36* ◄

## Integrating Power Series

The following example illustrates the use of power series in approximating integrals that cannot be evaluated by analytical methods.

**EXAMPLE 5** **Approximating a definite integral** Approximate the value of the integral $\int_0^1 e^{-x^2}\,dx$ with an error no greater than $5 \times 10^{-4}$.

**SOLUTION** The antiderivative of $e^{-x^2}$ cannot be expressed in terms of familiar functions. The strategy is to write the Maclaurin series for $e^{-x^2}$ and integrate it term by term. Recall that integration of a power series is valid within its interval of convergence (Theorem 11.5). Beginning with the Maclaurin series

$$e^x = 1 + x + \frac{x^2}{2!} + \frac{x^3}{3!} + \cdots + \frac{x^n}{n!} + \cdots,$$

which converges for $-\infty < x < \infty$, we replace $x$ with $-x^2$ to obtain

$$e^{-x^2} = 1 - x^2 + \frac{x^4}{2!} - \frac{x^6}{3!} + \cdots + \frac{(-1)^n x^{2n}}{n!} + \cdots,$$

which also converges for $-\infty < x < \infty$. By the Fundamental Theorem of Calculus,

$$\int_0^1 e^{-x^2}\,dx = \left( x - \frac{x^3}{3} + \frac{x^5}{5 \cdot 2!} - \frac{x^7}{7 \cdot 3!} + \cdots + \frac{(-1)^n x^{2n+1}}{(2n+1)n!} + \cdots \right)\Bigg|_0^1$$

$$= 1 - \frac{1}{3} + \frac{1}{5 \cdot 2!} - \frac{1}{7 \cdot 3!} + \cdots + \frac{(-1)^n}{(2n+1)n!} + \cdots.$$

> ➤ The integral in Example 5 is important in statistics and probability theory because of its relationship to the *normal distribution*.

Because the definite integral is expressed as an alternating series, the magnitude of the remainder in truncating the series after $n$ terms is less than the magnitude of the first neglected term, which is $\left| \frac{(-1)^{n+1}}{(2n+3)(n+1)!} \right|$. By trial and error, we find that the magnitude of this term is less than $5 \times 10^{-4}$ if $n \geq 5$ (with $n = 5$, we have $\frac{1}{13 \cdot 6!} \approx 1.07 \times 10^{-4}$). The sum of the terms of the series up to $n = 5$ gives the approximation

$$\int_0^1 e^{-x^2}\,dx \approx 1 - \frac{1}{3} + \frac{1}{5 \cdot 2!} - \frac{1}{7 \cdot 3!} + \frac{1}{9 \cdot 4!} - \frac{1}{11 \cdot 5!} \approx 0.747.$$

*Related Exercises 37–38* ◀

## Representing Real Numbers

When values of $x$ are substituted into a convergent power series, the result may be a series representation of a familiar real number. The following example illustrates some techniques.

**EXAMPLE 6** **Evaluating infinite series**

**a.** Use the Maclaurin series for $f(x) = \tan^{-1} x$ to evaluate

$$1 - \frac{1}{3} + \frac{1}{5} - \cdots = \sum_{k=0}^{\infty} \frac{(-1)^k}{2k+1}.$$

**b.** Let $f(x) = (e^x - 1)/x$, for $x \neq 0$, and $f(0) = 1$. Use the Maclaurin series for $f$ to evaluate $f'(1)$ and $\sum_{k=1}^{\infty} \frac{k}{(k+1)!}$.

**SOLUTION**

**a.** From Table 11.5, we see that for $|x| \leq 1$,

$$\tan^{-1} x = x - \frac{x^3}{3} + \frac{x^5}{5} - \cdots + \frac{(-1)^k x^{2k+1}}{2k+1} + \cdots = \sum_{k=0}^{\infty} \frac{(-1)^k x^{2k+1}}{2k+1}.$$

▶ The series in Example 6a (known as the *Gregory series*) is one of a multitude of series representations of $\pi$. Because this series converges slowly, it does not provide an efficient way to approximate $\pi$.

Substituting $x = 1$, we have

$$\tan^{-1} 1 = 1 - \frac{1^3}{3} + \frac{1^5}{5} - \cdots = \sum_{k=0}^{\infty} \frac{(-1)^k}{2k + 1}.$$

Because $\tan^{-1} 1 = \pi/4$, the value of the series is $\pi/4$.

**b.** Using the Maclaurin series for $e^x$, the series for $f(x) = (e^x - 1)/x$ is

$$f(x) = \frac{e^x - 1}{x} = \frac{1}{x}\left(\underbrace{\left(1 + x + \frac{x^2}{2!} + \frac{x^3}{3!} + \cdots\right)}_{e^x} - 1\right) \qquad \text{Substitute series for } e^x.$$

$$= 1 + \frac{x}{2!} + \frac{x^2}{3!} + \frac{x^3}{4!} + \cdots = \sum_{k=1}^{\infty} \frac{x^{k-1}}{k!}, \qquad \text{Theorem 11.4, Property 2}$$

which converges for $-\infty < x < \infty$. By the Quotient Rule,

$$f'(x) = \frac{xe^x - (e^x - 1)}{x^2}.$$

Differentiating the series for $f$ term by term (Theorem 11.5), we find that

$$f'(x) = \frac{d}{dx}\left(1 + \frac{x}{2!} + \frac{x^2}{3!} + \frac{x^3}{4!} + \cdots\right)$$

$$= \frac{1}{2!} + \frac{2x}{3!} + \frac{3x^2}{4!} + \cdots = \sum_{k=1}^{\infty} \frac{kx^{k-1}}{(k + 1)!}.$$

**QUICK CHECK 3** What value of $x$ would you substitute into the Maclaurin series for $\tan^{-1} x$ to obtain a series representation for $\pi/6$? ◀

We now have two expressions for $f'$; they are evaluated at $x = 1$ to show that

$$f'(1) = 1 = \sum_{k=1}^{\infty} \frac{k}{(k + 1)!}.$$

*Related Exercises 50–51* ◀

## Representing Functions as Power Series

Power series have a fundamental role in mathematics in defining functions and providing alternative representations of familiar functions. As an overall review, we close this chapter with two examples that use many techniques for working with power series.

**EXAMPLE 7** **Identify the series** Identify the function represented by the power series $\sum_{k=0}^{\infty} \frac{(1 - 2x)^k}{k!}$ and give its interval of convergence.

**SOLUTION** The Maclaurin series for the exponential function,

$$e^x = \sum_{k=0}^{\infty} \frac{x^k}{k!},$$

converges for $-\infty < x < \infty$. Replacing $x$ with $1 - 2x$ produces the given series:

$$\sum_{k=0}^{\infty} \frac{(1 - 2x)^k}{k!} = e^{1 - 2x}.$$

This replacement is allowed because $1 - 2x$ is within the interval of convergence of the series for $e^x$; that is, $-\infty < 1 - 2x < \infty$, for all $x$. Therefore, the given series represents $e^{1 - 2x}$, for $-\infty < x < \infty$.

*Related Exercises 55–56* ◀

**EXAMPLE 8** **Mystery series** The power series $\sum_{k=1}^{\infty} \frac{(-1)^k k}{4^k} x^{2k}$ appeared in the opening of Section 11.2. Determine the interval of convergence of the power series and find the function it represents on this interval.

**SOLUTION** Applying the Ratio Test to the series, we determine that it converges when $|x^2/4| < 1$, which implies that $|x| < 2$. A quick check of the endpoints of the original series confirms that it diverges at $x = \pm 2$. Therefore, the interval of convergence is $|x| < 2$.

To find the function represented by the series, we apply several maneuvers until we obtain a geometric series. First note that

$$\sum_{k=1}^{\infty} \frac{(-1)^k k}{4^k} x^{2k} = \sum_{k=1}^{\infty} k \left(-\frac{1}{4}\right)^k x^{2k}.$$

The series on the right is not a geometric series because of the presence of the factor $k$. The key is to realize that $k$ could appear in this way through differentiation; specifically, something like $\frac{d}{dx}(x^{2k}) = 2kx^{2k-1}$. To achieve terms of this form, we write

$$\underbrace{\sum_{k=1}^{\infty} \frac{(-1)^k k}{4^k} x^{2k}}_{\text{original series}} = \sum_{k=1}^{\infty} k\left(-\frac{1}{4}\right)^k x^{2k}$$

$$= \frac{1}{2}\sum_{k=1}^{\infty} 2k \left(-\frac{1}{4}\right)^k x^{2k} \qquad \text{Multiply and divide by 2.}$$

$$= \frac{x}{2}\sum_{k=1}^{\infty} 2k \left(-\frac{1}{4}\right)^k x^{2k-1}. \qquad \text{Remove } x \text{ from the series.}$$

Now we identify the last series as the derivative of another series:

$$\underbrace{\sum_{k=1}^{\infty} \frac{(-1)^k k}{4^k} x^{2k}}_{\text{original series}} = \frac{x}{2}\sum_{k=1}^{\infty} \left(-\frac{1}{4}\right)^k \underbrace{2kx^{2k-1}}_{\frac{d}{dx}(x^{2k})}$$

$$= \frac{x}{2}\sum_{k=1}^{\infty} \left(-\frac{1}{4}\right)^k \frac{d}{dx}(x^{2k}) \qquad \text{Identify a derivative.}$$

$$= \frac{x}{2}\frac{d}{dx}\left(\sum_{k=1}^{\infty} \left(-\frac{x^2}{4}\right)^k\right). \qquad \text{Combine factors; differentiate term by term.}$$

This last series is a geometric series with a ratio $r = -x^2/4$ and first term $-x^2/4$; therefore, its value is $\dfrac{-x^2/4}{1 + (x^2/4)}$, provided $\left|\dfrac{x^2}{4}\right| < 1$, or $|x| < 2$. We now have

$$\underbrace{\sum_{k=1}^{\infty} \frac{(-1)^k k}{4^k} x^{2k}}_{\text{original series}} = \frac{x}{2}\frac{d}{dx}\left(\sum_{k=1}^{\infty} \left(-\frac{x^2}{4}\right)^k\right)$$

$$= \frac{x}{2}\frac{d}{dx}\left(\frac{-x^2/4}{1 + (x^2/4)}\right) \qquad \text{Sum of geometric series}$$

$$= \frac{x}{2}\frac{d}{dx}\left(\frac{-x^2}{4 + x^2}\right) \qquad \text{Simplify.}$$

$$= -\frac{4x^2}{(4 + x^2)^2}. \qquad \text{Differentiate and simplify.}$$

Therefore, the function represented by the power series on $(-2, 2)$ has been uncovered; it is

$$f(x) = -\frac{4x^2}{(4 + x^2)^2}.$$

Notice that $f$ is defined for $-\infty < x < \infty$ (**Figure 11.22**), but its power series centered at 0 converges to $f$ only on $(-2, 2)$.

*Related Exercises 61, 63* ◀

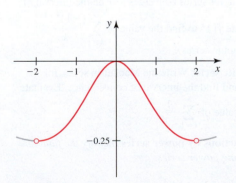

$$\sum_{k=1}^{\infty} \frac{(-1)^k k}{4^k} x^{2k} = -\frac{4x^2}{(4 + x^2)^2} \text{ on } (-2, 2)$$

**Figure 11.22**

## SECTION 11.4 EXERCISES

### Getting Started

1. Explain the strategy presented in this section for evaluating a limit of the form $\lim\limits_{x \to a} \dfrac{f(x)}{g(x)}$, where $f$ and $g$ have Taylor series centered at $a$.

2. Explain the method presented in this section for approximating $\int_a^b f(x)\,dx$, where $f$ has a Taylor series with an interval of convergence centered at $a$ that includes $b$.

3. How would you approximate $e^{-0.6}$ using the Taylor series for $e^x$?

4. Use the Taylor series for $\cos x$ centered at 0 to verify that
$$\lim_{x \to 0} \frac{1 - \cos x}{x} = 0.$$

5. Use the Taylor series for $\sinh x$ and $\cosh x$ to verify that
$$\frac{d}{dx} \sinh x = \cosh x.$$

6. What condition must be met by a function $f$ for it to have a Taylor series centered at $a$?

### Practice Exercises

**7–24. Limits** *Evaluate the following limits using Taylor series.*

7. $\lim\limits_{x \to 0} \dfrac{e^x - 1}{x}$

8. $\lim\limits_{x \to 0} \dfrac{\tan^{-1} x - x}{x^3}$

9. $\lim\limits_{x \to 0} \dfrac{-x - \ln(1 - x)}{x^2}$

10. $\lim\limits_{x \to 0} \dfrac{\sin 2x}{x}$

11. $\lim\limits_{x \to 0} \dfrac{e^x - e^{-x}}{x}$

12. $\lim\limits_{x \to 0} \dfrac{1 + x - e^x}{4x^2}$

13. $\lim\limits_{x \to 0} \dfrac{2 \cos 2x - 2 + 4x^2}{2x^4}$

14. $\lim\limits_{x \to \infty} x \sin \dfrac{1}{x}$

15. $\lim\limits_{x \to 0} \dfrac{\ln(1 + x) - x + x^2/2}{x^3}$

16. $\lim\limits_{x \to 4} \dfrac{x^2 - 16}{\ln(x - 3)}$

17. $\lim\limits_{x \to 0} \dfrac{3 \tan^{-1} x - 3x + x^3}{x^5}$

18. $\lim\limits_{x \to 0} \dfrac{\sqrt{1 + x} - 1 - (x/2)}{4x^2}$

19. $\lim\limits_{x \to 0} \dfrac{12x - 8x^3 - 6 \sin 2x}{x^5}$

20. $\lim\limits_{x \to 1} \dfrac{x - 1}{\ln x}$

21. $\lim\limits_{x \to 2} \dfrac{x - 2}{\ln(x - 1)}$

22. $\lim\limits_{x \to \infty} x(e^{1/x} - 1)$

23. $\lim\limits_{x \to 0} \dfrac{e^{-2x} - 4e^{-x/2} + 3}{2x^2}$

24. $\lim\limits_{x \to 0} \dfrac{(1 - 2x)^{-1/2} - e^x}{8x^2}$

**25–32. Power series for derivatives**

a. *Differentiate the Taylor series centered at 0 for the following functions.*

b. *Identify the function represented by the differentiated series.*

c. *Give the interval of convergence of the power series for the derivative.*

25. $f(x) = e^x$

26. $f(x) = -\cos x$

27. $f(x) = \ln(1 + x)$

28. $f(x) = \sin x^2$

29. $f(x) = e^{-2x}$

30. $f(x) = (1 - x)^{-1}$

31. $f(x) = \tan^{-1} x$

32. $f(x) = -\ln(1 - x)$

### 33–36. Differential equations

a. *Find a power series for the solution of the following differential equations, subject to the given initial condition.*

b. *Identify the function represented by the power series.*

33. $y'(t) - y = 0,\ y(0) = 2$

34. $y'(t) + 4y = 8,\ y(0) = 0$

35. $y'(t) - 3y = 10,\ y(0) = 2$

36. $y'(t) = 6y + 9,\ y(0) = 2$

**T 37–44. Approximating definite integrals** *Use a Taylor series to approximate the following definite integrals. Retain as many terms as needed to ensure the error is less than $10^{-4}$.*

37. $\displaystyle\int_0^{0.25} e^{-x^2}\,dx$

38. $\displaystyle\int_0^{0.2} \sin x^2\,dx$

39. $\displaystyle\int_{-0.35}^{0.35} \cos 2x^2\,dx$

40. $\displaystyle\int_0^{0.2} \sqrt{1 + x^4}\,dx$

41. $\displaystyle\int_0^{0.35} \tan^{-1} x\,dx$

42. $\displaystyle\int_0^{0.4} \ln(1 + x^2)\,dx$

43. $\displaystyle\int_0^{0.5} \dfrac{dx}{\sqrt{1 + x^6}}$

44. $\displaystyle\int_0^{0.2} \dfrac{\ln(1 + t)}{t}\,dt$

**45–50. Approximating real numbers** *Use an appropriate Taylor series to find the first four nonzero terms of an infinite series that is equal to the following numbers.*

45. $e^2$

46. $\sqrt{e}$

47. $\cos 2$

48. $\sin 1$

49. $\ln \dfrac{3}{2}$

50. $\tan^{-1} \dfrac{1}{2}$

51. **Evaluating an infinite series** Let $f(x) = \dfrac{e^x - 1}{x}$, for $x \neq 0$, and $f(0) = 1$. Use the Taylor series for $f$ centered at 0 to evaluate $f(1)$ and to find the value of $\displaystyle\sum_{k=0}^{\infty} \dfrac{1}{(k + 1)!}$.

52. **Evaluating an infinite series** Let $f(x) = \dfrac{e^x - 1}{x}$, for $x \neq 0$, and $f(0) = 1$. Use the Taylor series for $f$ and $f'$ centered at 0 to evaluate $f'(2)$ and to find the value of $\displaystyle\sum_{k=1}^{\infty} \dfrac{k2^{k-1}}{(k + 1)!}$.

53. **Evaluating an infinite series** Write the Taylor series for $f(x) = \ln(1 + x)$ centered at 0 and find its interval of convergence. Assume the Taylor series converges to $f$ on the interval of convergence. Evaluate $f(1)$ to find the value of $\displaystyle\sum_{k=1}^{\infty} \dfrac{(-1)^{k+1}}{k}$ (the alternating harmonic series).

54. **Evaluating an infinite series** Write the Maclaurin series for $f(x) = \ln(1 + x)$ and find the interval of convergence. Evaluate $f\left(-\dfrac{1}{2}\right)$ to find the value of $\displaystyle\sum_{k=1}^{\infty} \dfrac{1}{k2^k}$.

**55–64. Representing functions by power series** *Identify the functions represented by the following power series.*

55. $\displaystyle\sum_{k=0}^{\infty} \dfrac{x^k}{2^k}$

56. $\displaystyle\sum_{k=0}^{\infty} (-1)^k \dfrac{x^k}{3^k}$

57. $\displaystyle\sum_{k=0}^{\infty} (-1)^k \dfrac{x^{2k}}{4^k}$

58. $\displaystyle\sum_{k=0}^{\infty} 2^k x^{2k+1}$

59. $\displaystyle\sum_{k=1}^{\infty} \dfrac{x^k}{k}$

60. $\displaystyle\sum_{k=0}^{\infty} \dfrac{(-1)^k x^{k+1}}{4^k}$

**61.** $\displaystyle\sum_{k=1}^{\infty} (-1)^k \frac{kx^{k+1}}{3^k}$

**62.** $\displaystyle\sum_{k=1}^{\infty} \frac{x^{2k}}{k}$

**63.** $\displaystyle\sum_{k=2}^{\infty} \frac{k(k-1)x^k}{3^k}$

**64.** $\displaystyle\sum_{k=2}^{\infty} \frac{x^k}{k(k-1)}$

**65. Explain why or why not** Determine whether the following statements are true and give an explanation or counterexample.

   **a.** To evaluate $\int_0^2 \frac{dx}{1-x}$, one could expand the integrand in a Taylor series and integrate term by term.

   **b.** To approximate $\pi/3$, one could substitute $x = \sqrt{3}$ into the Taylor series for $\tan^{-1} x$.

   **c.** $\displaystyle\sum_{k=0}^{\infty} \frac{(\ln 2)^k}{k!} = 2.$

**66–68. Limits with a parameter** *Use Taylor series to evaluate the following limits. Express the result in terms of the nonzero real parameter(s).*

**66.** $\displaystyle\lim_{x\to 0} \frac{e^{ax} - 1}{x}$

**67.** $\displaystyle\lim_{x\to 0} \frac{\sin ax}{\sin bx}$

**68.** $\displaystyle\lim_{x\to 0} \frac{\sin ax - \tan^{-1} ax}{bx^3}$

## Explorations and Challenges

**69. A limit by Taylor series** Use Taylor series to evaluate

$$\lim_{x\to 0} \left(\frac{\sin x}{x}\right)^{1/x^2}.$$

**70. Inverse hyperbolic sine** The *inverse of hyperbolic sine* is defined in several ways; among them are

$$\sinh^{-1} x = \ln\left(x + \sqrt{x^2+1}\right) = \int_0^x \frac{dt}{\sqrt{1+t^2}}.$$

Find the first four terms of the Taylor series for $\sinh^{-1} x$ using these two definitions (and be sure they agree).

**71–74. Derivative trick** *Here is an alternative way to evaluate higher derivatives of a function f that may save time. Suppose you can find the Taylor series for f centered at the point a without evaluating derivatives (for example, from a known series). Then $f^{(k)}(a) = k!$ multiplied by the coefficient of $(x - a)^k$. Use this idea to evaluate $f^{(3)}(0)$ and $f^{(4)}(0)$ for the following functions. Use known series and do not evaluate derivatives.*

**71.** $f(x) = e^{\cos x}$

**72.** $f(x) = \dfrac{x^2 + 1}{\sqrt[3]{1+x}}$

**73.** $f(x) = \displaystyle\int_0^x \sin t^2 \, dt$

**74.** $f(x) = \displaystyle\int_0^x \frac{1}{1+t^4}\, dt$

**75. Probability: tossing for a head** The expected (average) number of tosses of a fair coin required to obtain the first head is $\displaystyle\sum_{k=1}^{\infty} k\left(\frac{1}{2}\right)^k$. Evaluate this series and determine the expected number of tosses. (*Hint:* Differentiate a geometric series.)

**76. Probability: sudden-death playoff** Teams A and B go into sudden-death overtime after playing to a tie. The teams alternate possession of the ball, and the first team to score wins. Assume each team has a $1/6$ chance of scoring when it has the ball, and Team A has the ball first.

   **a.** The probability that Team A ultimately wins is $\displaystyle\sum_{k=0}^{\infty} \frac{1}{6}\left(\frac{5}{6}\right)^{2k}$. Evaluate this series.

   **b.** The expected number of rounds (possessions by either team) required for the overtime to end is $\displaystyle\frac{1}{6}\sum_{k=1}^{\infty} k\left(\frac{5}{6}\right)^{k-1}$. Evaluate this series.

**T 77. Elliptic integrals** The period of an undamped pendulum is given by

$$T = 4\sqrt{\frac{\ell}{g}} \int_0^{\pi/2} \frac{d\theta}{\sqrt{1 - k^2 \sin^2\theta}} = 4\sqrt{\frac{\ell}{g}}\, F(k),$$

where $\ell$ is the length of the pendulum, $g = 9.8 \text{ m/s}^2$ is the acceleration due to gravity, $k = \sin\dfrac{\theta_0}{2}$, and $\theta_0$ is the initial angular displacement of the pendulum (in radians). The integral in this formula $F(k)$ is called an *elliptic integral*, and it cannot be evaluated analytically. Approximate $F(0.1)$ by expanding the integrand in a Taylor (binomial) series and integrating term by term.

**78. Sine integral function** The function $\text{Si}(x) = \int_0^x f(t)\, dt$, where

$$f(t) = \begin{cases} \frac{\sin t}{t} & \text{if } t \neq 0 \\ 1 & \text{if } t = 0 \end{cases}, \text{ is called the } \textit{sine integral function}.$$

   **a.** Expand the integrand in a Taylor series centered at 0.

   **b.** Integrate the series to find a Taylor series for Si.

   **c.** Approximate $\text{Si}(0.5)$ and $\text{Si}(1)$. Use enough terms of the series so the error in the approximation does not exceed $10^{-3}$.

**T 79. Fresnel integrals** The theory of optics gives rise to the two *Fresnel integrals*

$$S(x) = \int_0^x \sin t^2 \, dt \quad \text{and} \quad C(x) = \int_0^x \cos t^2 \, dt.$$

   **a.** Compute $S'(x)$ and $C'(x)$.

   **b.** Expand $\sin t^2$ and $\cos t^2$ in a Maclaurin series, and then integrate to find the first four nonzero terms of the Maclaurin series for $S$ and $C$.

   **c.** Use the polynomials in part (b) to approximate $S(0.05)$ and $C(-0.25)$.

   **d.** How many terms of the Maclaurin series are required to approximate $S(0.05)$ with an error no greater than $10^{-4}$?

   **e.** How many terms of the Maclaurin series are required to approximate $C(-0.25)$ with an error no greater than $10^{-6}$?

**T 80. Error function** An essential function in statistics and the study of the normal distribution is the *error function*

$$\text{erf}(x) = \frac{2}{\sqrt{\pi}} \int_0^x e^{-t^2} dt.$$

   **a.** Compute the derivative of $\text{erf}(x)$.

   **b.** Expand $e^{-t^2}$ in a Maclaurin series; then integrate to find the first four nonzero terms of the Maclaurin series for erf.

   **c.** Use the polynomial in part (b) to approximate $\text{erf}(0.15)$ and $\text{erf}(-0.09)$.

   **d.** Estimate the error in the approximations of part (c).

**T 81. Bessel functions** Bessel functions arise in the study of wave propagation in circular geometries (for example, waves on a circular drum head). They are conveniently defined as power series. One of an infinite family of Bessel functions is

$$J_0(x) = \sum_{k=0}^{\infty} \frac{(-1)^k}{2^{2k}(k!)^2} x^{2k}.$$

   **a.** Write out the first four terms of $J_0$.

   **b.** Find the radius and interval of convergence of the power series for $J_0$.

   **c.** Differentiate $J_0$ twice and show (by keeping terms through $x^6$) that $J_0$ satisfies the equation $x^2 y''(x) + xy'(x) + x^2 y(x) = 0$.

**T 82. Newton's derivation of the sine and arcsine series** Newton discovered the binomial series and then used it ingeniously to obtain many more results. Here is a case in point.

a. Referring to the figure, show that $x = \sin s$ or $s = \sin^{-1} x$.
b. The area of a circular sector of radius $r$ subtended by an angle $\theta$ is $1/2\, r^2\theta$. Show that the area of the circular sector $APE$ is $s/2$, which implies that

$$s = 2\int_0^x \sqrt{1 - t^2}\, dt - x\sqrt{1 - x^2}.$$

c. Use the binomial series for $f(x) = \sqrt{1 - x^2}$ to obtain the first few terms of the Taylor series for $s = \sin^{-1} x$.
d. Newton next inverted the series in part (c) to obtain the Taylor series for $x = \sin s$. He did this by assuming $\sin s = \sum a_k s^k$ and solving $x = \sin(\sin^{-1} x)$ for the coefficients $a_k$. Find the first few terms of the Taylor series for $\sin s$ using this idea (a computer algebra system might be helpful as well).

**83. L'Hôpital's Rule by Taylor series** Suppose $f$ and $g$ have Taylor series about the point $a$.

a. If $f(a) = g(a) = 0$ and $g'(a) \neq 0$, evaluate $\lim\limits_{x \to a} \dfrac{f(x)}{g(x)}$ by expanding $f$ and $g$ in their Taylor series. Show that the result is consistent with l'Hôpital's Rule.
b. If $f(a) = g(a) = f'(a) = g'(a) = 0$ and $g''(a) \neq 0$, evaluate $\lim\limits_{x \to a} \dfrac{f(x)}{g(x)}$ by expanding $f$ and $g$ in their Taylor series. Show that the result is consistent with two applications of l'Hôpital's Rule.

**84. Symmetry**

a. Use infinite series to show that $\cos x$ is an even function. That is, show $\cos(-x) = \cos x$.
b. Use infinite series to show that $\sin x$ is an odd function. That is, show $\sin(-x) = -\sin x$.

**QUICK CHECK ANSWERS**

1. $\dfrac{\sin x}{x} = \dfrac{x - x^3/3! + \cdots}{x} = 1 - \dfrac{x^2}{3!} + \cdots \to 1$ as $x \to 0$

2. The result is the power series for $-\sin x$.    3. $x = 1/\sqrt{3}$ (which lies in the interval of convergence) ◄

---

## CHAPTER 11  REVIEW EXERCISES

1. **Explain why or why not** Determine whether the following statements are true and give an explanation or counterexample.

a. Let $p_n$ be the $n$th-order Taylor polynomial for $f$ centered at 2. The approximation $p_3(2.1) \approx f(2.1)$ is likely to be more accurate than the approximation $p_2(2.2) \approx f(2.2)$.
b. If the Taylor series for $f$ centered at 3 has a radius of convergence of 6, then the interval of convergence is $[-3, 9]$.
c. The interval of convergence of the power series $\sum c_k x^k$ could be $(-7/3, 7/3)$.
d. The Maclaurin series for $f(x) = (1 + x)^{12}$ has a finite number of nonzero terms.
e. If the power series $\sum c_k(x - 3)^k$ has a radius of convergence of $R = 4$ and converges at the endpoints of its interval of convergence, then its interval of convergence is $[-1, 7]$.

**2–9. Taylor polynomials** Find the $n$th-order Taylor polynomial for the following functions centered at the given point $a$.

2. $f(x) = \sin 2x, n = 3, a = 0$

3. $f(x) = \cos^3 x, n = 2, a = 0$

4. $f(x) = \cos^{-1} x, n = 2, a = \dfrac{1}{2}$

5. $f(x) = e^{1/x - 1}, n = 2, a = 1$

6. $f(x) = e^{\sin x}, n = 2, a = 0$

7. $f(x) = \cos(\ln x), n = 2, a = 1$

8. $f(x) = \sinh(-3x), n = 3, a = 0$

9. $f(x) = \cosh x, n = 3, a = \ln 2$

**T 10–13. Approximations**

a. Find the Taylor polynomials of order $n = 1$ and $n = 2$ for the given functions centered at the given point $a$.
b. Use the Taylor polynomials to approximate the given expression. Make a table showing the approximations and the absolute error in these approximations using a calculator for the exact function value.

10. $f(x) = \cos x, a = 0; \cos(-0.08)$

11. $f(x) = e^x, a = 0; e^{-0.08}$

12. $f(x) = \sqrt{1 + x}, a = 0; \sqrt{1.08}$

13. $f(x) = \sin x, a = \dfrac{\pi}{4}; \sin\dfrac{\pi}{5}$

**14–16. Estimating remainders** *Find the remainder term $R_n(x)$ for the Taylor series centered at 0 for the following functions. Find an upper bound for the magnitude of the remainder on the given interval for the given value of n. (The bound is not unique.)*

**14.** $f(x) = e^x$; bound $R_3(x)$, for $|x| < 1$

**15.** $f(x) = \sin x$; bound $R_3(x)$, for $|x| < \pi$

**16.** $f(x) = \ln(1 - x)$; bound $R_3(x)$, for $|x| < \dfrac{1}{2}$

**17–26. Radius and interval of convergence** *Use the Ratio Test or the Root Test to determine the radius of convergence of the following power series. Test the endpoints to determine the interval of convergence, when appropriate.*

**17.** $\displaystyle\sum_{k=1}^{\infty} \frac{k^2 x^k}{k!}$

**18.** $\displaystyle\sum_{k=1}^{\infty} \frac{x^{4k}}{k^2}$

**19.** $\displaystyle\sum_{k=1}^{\infty} \frac{(-1)^k (x + 1)^{2k}}{k!}$

**20.** $\displaystyle\sum_{k=1}^{\infty} \frac{(x - 1)^k}{k \, 5^k}$

**21.** $\displaystyle\sum_{k=0}^{\infty} \left(\frac{x}{9}\right)^{3k}$

**22.** $\displaystyle\sum_{k=1}^{\infty} \frac{(x + 2)^k}{\sqrt{k}}$

**23.** $\displaystyle\sum_{k=1}^{\infty} \frac{(x + 2)^k}{2^k \ln k}$

**24.** $\displaystyle\sum_{k=1}^{\infty} \frac{(2k)!(x - 2)^k}{k!}$

**25.** $\displaystyle\sum_{k=1}^{\infty} \frac{(-1)^k (2x + 1)^k}{k^2 3^k}$

**26.** $x + \dfrac{x^3}{3} + \dfrac{x^5}{5} + \dfrac{x^7}{7} + \cdots$

**27–28. Radius of convergence** *Find the radius of convergence of each series.*

**27.** $\displaystyle\sum_{k=1}^{\infty} \frac{(3k)! \, x^k}{(k!)^3}$

**28.** $\displaystyle\sum_{k=1}^{\infty} \frac{2^k k! (x - 5)^k}{k^k}$

**29–34. Power series from the geometric series** *Use the geometric series $\displaystyle\sum_{k=0}^{\infty} x^k = \dfrac{1}{1 - x}$, for $|x| < 1$, to determine the Maclaurin series and the interval of convergence for the following functions.*

**29.** $f(x) = \dfrac{1}{1 - x^2}$

**30.** $f(x) = \dfrac{1}{1 + x^3}$

**31.** $f(x) = \dfrac{1}{1 + 5x}$

**32.** $f(x) = \dfrac{10x}{1 + x}$

**33.** $f(x) = \dfrac{1}{(1 - 10x)^2}$

**34.** $f(x) = \ln(1 - 4x)$

**35–42. Taylor series** *Write out the first three nonzero terms of the Taylor series for the following functions centered at the given point a. Then write the series using summation notation.*

**35.** $f(x) = e^{3x}$, $a = 0$

**36.** $f(x) = \dfrac{1}{2x + 1}$, $a = 1$

**37.** $f(x) = \cos x$, $a = \dfrac{\pi}{2}$

**38.** $f(x) = \dfrac{x^2}{1 + x}$, $a = 0$

**39.** $f(x) = \tan^{-1} 4x$, $a = 0$

**40.** $f(x) = \sin 2x$, $a = -\dfrac{\pi}{2}$

**41.** $f(x) = \cosh(2x - 2)$, $a = 1$

**42.** $f(x) = \dfrac{1}{4 + x^2}$, $a = 0$

**43–46. Binomial series** *Write out the first three terms of the Maclaurin series for the following functions.*

**43.** $f(x) = (1 + x)^{1/3}$

**44.** $f(x) = (1 + x)^{-1/2}$

**45.** $f(x) = \left(1 + \dfrac{x}{2}\right)^{-3}$

**46.** $f(x) = (1 + 2x)^{-5}$

**47–48. Convergence** *Write the remainder term $R_n(x)$ for the Taylor series for the following functions centered at the given point a. Then show that $\lim\limits_{n \to \infty} |R_n(x)| = 0$, for all x in the given interval.*

**47.** $f(x) = \sinh x + \cosh x$, $a = 0$, $-\infty < x < \infty$

**48.** $f(x) = \ln(1 + x)$, $a = 0$, $-\dfrac{1}{2} \le x \le \dfrac{1}{2}$

**49–54. Limits by power series** *Use Taylor series to evaluate the following limits.*

**49.** $\displaystyle\lim_{x \to 0} \frac{x^2/2 - 1 + \cos x}{x^4}$

**50.** $\displaystyle\lim_{x \to 0} \frac{2 \sin x - \tan^{-1} x - x}{2x^5}$

**51.** $\displaystyle\lim_{x \to 4} \frac{\ln(x - 3)}{x^2 - 16}$

**52.** $\displaystyle\lim_{x \to 0} \frac{\sqrt{1 + 2x} - 1 - x}{x^2}$

**53.** $\displaystyle\lim_{x \to 0} \frac{\sec x - \cos x - x^2}{x^4}$ (*Hint:* The Maclaurin series for sec $x$ is $1 + \dfrac{x^2}{2} + \dfrac{5x^4}{24} + \dfrac{61x^6}{720} + \cdots .$)

**54.** $\displaystyle\lim_{x \to 0} \frac{(1 + x)^{-2} - \sqrt[3]{1 - 6x}}{2x^2}$

**T** **55–58. Definite integrals by power series** *Use a Taylor series to approximate the following definite integrals. Retain as many terms as necessary to ensure the error is less than $10^{-3}$.*

**55.** $\displaystyle\int_0^{1/2} e^{-x^2} \, dx$

**56.** $\displaystyle\int_0^{1/2} \tan^{-1} x \, dx$

**57.** $\displaystyle\int_0^1 x \cos x \, dx$

**58.** $\displaystyle\int_0^{1/2} x^2 \tan^{-1} x \, dx$

**T** **59–62. Approximating real numbers** *Use an appropriate Taylor series to find the first four nonzero terms of an infinite series that is equal to the following numbers. There is more than one way to choose the center of the series.*

**59.** $\sqrt{119}$

**60.** $\sin 20°$

**61.** $\tan^{-1}\left(-\dfrac{1}{3}\right)$

**62.** $\sinh(-1)$

**63. A differential equation** Find a power series solution of the differential equation $y'(x) - 4y + 12 = 0$, subject to the condition $y(0) = 4$. Identify the solution in terms of known functions.

**T** **64. Rejected quarters** The probability that a random quarter is *not* rejected by a vending machine is given by the integral $11.4 \int_0^{0.14} e^{-102x^2} dx$ (assuming the weights of quarters are normally distributed with a mean of 5.670 g and a standard deviation of 0.07 g). Estimate the value of the integral by using the first two terms of the Maclaurin series for $e^{-102x^2}$.

**T 65. Approximating ln 2** Consider the following three ways to approximate ln 2.

**a.** Use the Taylor series for $\ln(1 + x)$ centered at 0 and evaluate it at $x = 1$ (convergence was asserted in Table 11.5). Write the resulting infinite series.

**b.** Use the Taylor series for $\ln(1 - x)$ centered at 0 and the identity $\ln 2 = -\ln\dfrac{1}{2}$. Write the resulting infinite series.

**c.** Use the property $\ln\dfrac{a}{b} = \ln a - \ln b$ and the series of parts (a) and (b) to find the Taylor series for $f(x) = \ln\left(\dfrac{1 + x}{1 - x}\right)$ centered at 0.

**d.** At what value of $x$ should the series in part (c) be evaluated to approximate ln 2? Write the resulting infinite series for ln 2.

**e.** Using four terms of the series, which of the three series derived in parts (a)–(d) gives the best approximation to ln 2? Can you explain why?

**T 66. Graphing Taylor polynomials** Consider the function $f(x) = (1 + x)^{-4}$.

**a.** Find the Taylor polynomials $p_0$, $p_1$, $p_2$, and $p_3$ centered at 0.

**b.** Use a graphing utility to plot the Taylor polynomials and $f$, for $-1 < x < 1$.

**c.** For each Taylor polynomial, give the interval on which its graph appears indistinguishable from the graph of $f$.

## Chapter 11 Guided Projects

*Applications of the material in this chapter and related topics can be found in the following Guided Projects. For additional information, see the Preface.*

- Series approximations to $\pi$
- Euler's formula (Taylor series with complex numbers)
- Stirling's formula and $n!$
- Three-sigma quality control
- Fourier series

# 12

# Parametric and Polar Curves

**Chapter Preview**  Up to this point, we have focused our attention on functions of the form $y = f(x)$ and studied their behavior in the setting of the Cartesian coordinate system. There are, however, alternative ways to generate curves and represent functions. We begin by introducing parametric equations, which are featured prominently in upcoming chapters, to represent curves and trajectories in three-dimensional space. When working with objects that have circular, cylindrical, or spherical shapes, other coordinate systems are often advantageous. In this chapter, we introduce the polar coordinate system for circular geometries. Cylindrical and spherical coordinate systems appear in Chapter 16. After working with parametric equations and polar coordinates, the next step is to investigate calculus in these settings. How do we find slopes of tangent lines, rates of change, areas of regions, and arc length when the curves of interest are described with parametric equations or described in polar coordinates? The chapter ends with the related topic of *conic sections*. Ellipses, parabolas, and hyperbolas (all of which are conic sections) can be represented in both Cartesian and polar coordinates. These important families of curves have many fascinating properties and they appear throughout the remainder of the text.

## 12.1 Parametric Equations

So far, we have used functions of the form $y = F(x)$ to describe curves in the $xy$-plane. In this section, we look at another way to define curves, known as *parametric equations*. As you will see, parametric curves enable us to describe both common and exotic curves; they are also indispensable for modeling the trajectories of moving objects.

### Basic Ideas

A motor boat travels counterclockwise around a circular course with a radius of 4 miles, completing one lap every $2\pi$ hours at a constant speed. Suppose we wish to describe the points on the path of the boat $(x(t), y(t))$ at any time $t \geq 0$, where $t$ is measured in hours. We assume the boat starts on the positive $x$-axis at the point $(4, 0)$ (**Figure 12.1**). Note that the angle $\theta$ corresponding to the position of the boat increases by $2\pi$ radians every $2\pi$ hours beginning with $\theta = 0$ when $t = 0$; therefore, $\theta = t$, for $t \geq 0$. It follows that the $x$- and $y$-coordinates of the boat are

$$x = 4 \cos \theta = 4 \cos t \quad \text{and} \quad y = 4 \sin \theta = 4 \sin t,$$

where $t \geq 0$. You can confirm that when $t = 0$, the boat is at the starting point $(4, 0)$; when $t = 2\pi$, it returns to the starting point.

The equations $x = 4 \cos t$ and $y = 4 \sin t$ are examples of **parametric equations**. They specify $x$ and $y$ in terms of a third variable $t$ called a **parameter**, which often represents time.

**Figure 12.1**

In general, parametric equations have the form

$$x = f(t), \quad y = g(t),$$

where $f$ and $g$ are given functions and the parameter $t$ typically varies over a specified interval $a \leq t \leq b$ (**Figure 12.2**). The **parametric curve** described by these equations consists of the points in the plane

$$(x, y) = (f(t), g(t)), \quad \text{for } a \leq t \leq b.$$

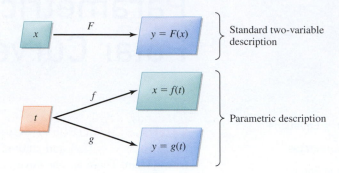

> ➤ With a parametric description, the parameter $t$ is the independent variable. There are two dependent variables, $x$ and $y$.

**Figure 12.2**

**EXAMPLE 1**    **Parametric parabola**    Graph and analyze the parametric equations

$$x = f(t) = 2t, \quad y = g(t) = \frac{1}{2}t^2 - 4, \quad \text{for } 0 \leq t \leq 8.$$

**Table 12.1**

| $t$ | $x$ | $y$ | $(x, y)$ |
|---|---|---|---|
| 0 | 0 | $-4$ | $(0, -4)$ |
| 1 | 2 | $-\frac{7}{2}$ | $\left(2, -\frac{7}{2}\right)$ |
| 2 | 4 | $-2$ | $(4, -2)$ |
| 3 | 6 | $\frac{1}{2}$ | $\left(6, \frac{1}{2}\right)$ |
| 4 | 8 | 4 | $(8, 4)$ |
| 5 | 10 | $\frac{17}{2}$ | $\left(10, \frac{17}{2}\right)$ |
| 6 | 12 | 14 | $(12, 14)$ |
| 7 | 14 | $\frac{41}{2}$ | $\left(14, \frac{41}{2}\right)$ |
| 8 | 16 | 28 | $(16, 28)$ |

**SOLUTION**    Plotting individual points often helps in visualizing a parametric curve. Table 12.1 shows the values of $x$ and $y$ corresponding to several values of $t$ on the interval $[0, 8]$. By plotting the $(x, y)$ pairs in Table 12.1 and connecting them with a smooth curve, we obtain the graph shown in **Figure 12.3**. As $t$ increases from its initial value of $t = 0$ to its final value of $t = 8$, the curve is generated from the initial point $(0, -4)$ to the final point $(16, 28)$. Notice that the values of the parameter do not appear in the graph. The only signature of the parameter is the direction in which the curve is generated: In this case, it unfolds upward and to the right, as indicated by the arrows on the curve.

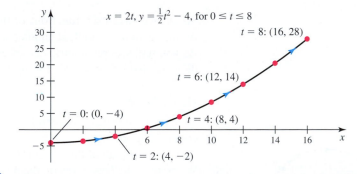

**Figure 12.3**

Sometimes it is possible to eliminate the parameter from a set of parametric equations and obtain a description of the curve in terms of $x$ and $y$. In this case, from the $x$-equation, we have $t = x/2$, which may be substituted into the $y$-equation to give

$$y = \frac{1}{2}t^2 - 4 = \frac{1}{2}\left(\frac{x}{2}\right)^2 - 4 = \frac{x^2}{8} - 4.$$

**QUICK CHECK 1**  Identify the graph generated by the parametric equations $x = t^2$, $y = t$, for $-10 \leq t \leq 10$. ◀

Expressed in this form, we identify the graph as part of a parabola. Because $t$ lies in the interval $0 \leq t \leq 8$ and $x = 2t$, it follows that $x$ lies in the interval $0 \leq x \leq 16$. Therefore, the parametric equations generate the segment of the parabola for $0 \leq x \leq 16$.

*Related Exercise 12* ◀

Given a set of parametric equations, the preceding example shows that as the parameter increases, the corresponding curve unfolds in a particular direction. The following definition captures this fact and is important in upcoming work.

---

**DEFINITION**   **Positive Orientation**

The direction in which a parametric curve is generated as the parameter increases is called the **positive orientation** of the curve (and is indicated by arrows on the curve).

---

**EXAMPLE 2**   **Parametric circle**

**a.** Confirm that the parametric equations

$$x = 4 \cos 2\pi t, \quad y = 4 \sin 2\pi t, \quad \text{for } 0 \le t \le 1$$

describe a circle of radius 4 centered at the origin.

**b.** Suppose a turtle walks with constant speed in the counterclockwise direction on the circular path from part (a). Starting from the point $(4, 0)$, the turtle completes one lap in 30 minutes. Find a parametric description of the path of the turtle at any time $t \ge 0$, where $t$ is measured in minutes.

**SOLUTION**

**a.** For each value of $t$ in Table 12.2, the corresponding ordered pairs $(x, y)$ are recorded. Plotting these points as $t$ increases from $t = 0$ to $t = 1$ results in a graph that appears to be a circle of radius 4; it is generated with positive orientation in the counterclockwise direction, beginning and ending at $(4, 0)$ (**Figure 12.4**). Letting $t$ increase beyond $t = 1$ would simply retrace the same curve.

To identify the curve conclusively, the parameter $t$ is eliminated by observing that

$$x^2 + y^2 = (4 \cos 2\pi t)^2 + (4 \sin 2\pi t)^2$$
$$= 16\underbrace{(\cos^2 2\pi t + \sin^2 2\pi t)}_{1} = 16. \quad \cos^2 \theta + \sin^2 \theta = 1$$

We have confirmed that the graph of the parametric equations is the circle $x^2 + y^2 = 16$.

**Table 12.2**

| $t$ | $(x, y)$ |
|---|---|
| $0$ | $(4, 0)$ |
| $\frac{1}{8}$ | $(2\sqrt{2}, 2\sqrt{2})$ |
| $\frac{1}{4}$ | $(0, 4)$ |
| $\frac{3}{8}$ | $(-2\sqrt{2}, 2\sqrt{2})$ |
| $\frac{1}{2}$ | $(-4, 0)$ |
| $\frac{3}{4}$ | $(0, -4)$ |
| $1$ | $(4, 0)$ |

**Figure 12.4**

**Figure 12.5**

**b.** Duplicating the calculations at the end of part (a) with any nonzero real number $b$, it can be shown that the parametric equations

$$x = 4 \cos bt, \quad y = 4 \sin bt$$

also describe a circle of radius 4. When $b > 0$, the graph is generated in the counterclockwise direction. The *angular frequency* $b$ must be chosen so that as $t$ varies

> ▶ In Example 2b, the constant $|b|$ is called the *angular frequency* because it is the number of radians the object moves per unit time. The turtle travels $2\pi$ rad every 30 min, so the angular frequency is $2\pi/30 = \pi/15$ rad/min. Because radians have no units, the angular frequency in this case has units *per minute*, which is written $\text{min}^{-1}$.

from 0 to 30, the product $bt$ varies from 0 to $2\pi$. Specifically, when $t = 30$, we must have $30b = 2\pi$, or $b = \dfrac{\pi}{15}$ rad/min. Therefore, the parametric equations for the turtle's motion are

$$x = 4 \cos \frac{\pi t}{15}, \quad y = 4 \sin \frac{\pi t}{15}, \quad \text{for } 0 \le t \le 30.$$

You should check that as $t$ varies from 0 to 30, the points $(x, y)$ make one complete circuit of a circle of radius 4 (**Figure 12.5**).

*Related Exercises 20, 37, 55* ◀

Generalizing Example 2 for nonzero real numbers $a$ and $b$ in the parametric equations $x = a \cos bt$, $y = a \sin bt$, we find that

$$
\begin{aligned}
x^2 + y^2 &= (a \cos bt)^2 + (a \sin bt)^2 \\
&= a^2 \underbrace{(\cos^2 bt + \sin^2 bt)}_{1} = a^2.
\end{aligned}
$$

> ▶ For a nonzero constant $b$, the functions $\sin bt$ and $\cos bt$ have period $2\pi/|b|$. The equations $x = a \cos bt$, $y = -a \sin bt$ also describe a circle of radius $|a|$, as do the equations $x = \pm a \sin bt$, $y = \pm a \cos bt$, as $t$ varies over an interval of length $2\pi/|b|$.

Therefore, the parametric equations $x = a \cos bt$, $y = a \sin bt$ describe all or part of the circle $x^2 + y^2 = a^2$, centered at the origin with radius $|a|$, for any nonzero value of $b$. The circle is traversed once as $t$ varies over any interval of length $2\pi/|b|$. If $t$ represents time, the circle is traversed once in $2\pi/|b|$ time units, which means we can vary the speed at which the curve unfolds by varying $b$. If $b > 0$, the positive orientation is in the counterclockwise direction. If $b < 0$, the curve is generated in the clockwise direction.

More generally, the parametric equations

$$x = x_0 + a \cos bt, \quad y = y_0 + a \sin bt$$

**QUICK CHECK 2** Give the center and radius of the circle generated by the equations $x = 3 \sin t$, $y = -3 \cos t$, for $0 \le t \le 2\pi$. Specify the direction of positive orientation. ◀

describe all or part of the circle $(x - x_0)^2 + (y - y_0)^2 = a^2$, centered at $(x_0, y_0)$ with radius $|a|$. If $b > 0$, then the circle is generated in the counterclockwise direction. Example 2 shows that a single curve—for example, a circle of radius 4—may be parameterized in many different ways.

Among the most important of all parametric equations are

$$x = x_0 + at, \, y = y_0 + bt, \quad \text{for } -\infty < t < \infty,$$

where $x_0$, $y_0$, $a$, and $b$ are constants with $a \ne 0$. The curve described by these equations is found by eliminating the parameter. The first step is to solve the $x$-equation for $t$, which gives us $t = \dfrac{x - x_0}{a}$. When $t$ is substituted into the $y$-equation, the result is an equation for $y$ in terms of $x$:

$$y = y_0 + bt = y_0 + b \left( \frac{x - x_0}{a} \right) \quad \text{or} \quad y - y_0 = \frac{b}{a} (x - x_0).$$

This equation describes a line with slope $\dfrac{b}{a}$ passing through the point $(x_0, y_0)$.

---

**SUMMARY  Parametric Equations of a Line**

The equations

$$x = x_0 + at, \, y = y_0 + bt, \quad \text{for } -\infty < t < \infty,$$

where $x_0$, $y_0$, $a$, and $b$ are constants with $a \ne 0$, describe a line with slope $\dfrac{b}{a}$ passing through the point $(x_0, y_0)$. If $a = 0$ and $b \ne 0$, the line is vertical.

---

Notice that the parametric description of a given line is not unique: For example, if $k$ is any nonzero constant, the numbers $a$ and $b$ may be replaced with $ka$ and $kb$, respectively, and the resulting equations describe the same line (although it may be generated in the opposite direction and at a different speed).

### EXAMPLE 3    Parametric equations of lines

**a.** Consider the parametric equations $x = -2 + 3t$, $y = 4 - 6t$, for $-\infty < t < \infty$, which describe a line. Find the slope-intercept form of the line.

**b.** Find two pairs of parametric equations for the line with slope $\frac{1}{3}$ that passes through the point $(2, 1)$.

**c.** Find parametric equations for the line segment starting at $P(4, 7)$ and ending at $Q(2, -3)$.

**SOLUTION**

**a.** To eliminate the parameter, first solve the $x$-equation for $t$ to find that $t = \dfrac{x + 2}{3}$. Replacing $t$ in the $y$-equation yields

$$y = 4 - 6\left(\frac{x + 2}{3}\right) = 4 - 2x - 4 = -2x.$$

The line $y = -2x$ passes through the origin with slope $-2$.

**b.** We use the general parametric equations of a line given in the Summary box. Because the slope of the line is $\dfrac{1}{3}$, we choose $a = 3$ and $b = 1$. Letting $x_0 = 2$ and $y_0 = 1$, parametric equations for the line are $x = 2 + 3t$, $y = 1 + t$, for $-\infty < t < \infty$. The line passes through $(2, 1)$ when $t = 0$ and rises to the right as $t$ increases (**Figure 12.6**). Notice that other choices for $a$ and $b$ also work. For example, with $a = -6$ and $b = -2$, the equations are $x = 2 - 6t$, $y = 1 - 2t$, for $-\infty < t < \infty$. These equations describe the same line, but now, as $t$ increases, the line is generated in the opposite direction (descending to the left).

**c.** The slope of this line is $\dfrac{7 - (-3)}{4 - 2} = 5$. However, notice that when the line segment is traversed from $P$ to $Q$, both $x$ and $y$ are decreasing (**Figure 12.7**). To account for the direction in which the line segment is generated, we let $a = -1$ and $b = -5$. Because $P(4, 7)$ is the starting point of the line segment, we choose $x_0 = 4$ and $y_0 = 7$. The resulting equations are $x = 4 - t$, $y = 7 - 5t$. Notice that $t = 0$ corresponds to the starting point $(4, 7)$. Because the equations describe a line *segment*, the interval for $t$ must be restricted. What value of $t$ corresponds to the endpoint of the line segment $Q(2, -3)$? Setting $x = 4 - t = 2$, we find that $t = 2$. As a check, we set $y = 7 - 5t = -3$, which also implies that $t = 2$. (If these two calculations do not give the same value of $t$, it probably means the slope was not computed correctly.) Therefore, the equations for the line segment are $x = 4 - t$, $y = 7 - 5t$, for $0 \le t \le 2$.

Related Exercises 15, 41–42 ◀

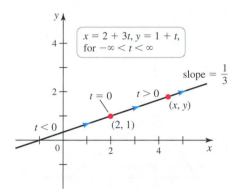

$x = 2 + 3t$, $y = 1 + t$, for $-\infty < t < \infty$

slope $= \dfrac{1}{3}$

$t = 0$    $t > 0$    $(x, y)$

$t < 0$    $(2, 1)$

**Figure 12.6**

➤ The choices $a = -1$ and $b = -5$ result in a slope of $b/a = 5$. These choices also imply that as we move from $P$ to $Q$, a decrease in $x$ corresponds to a decrease in $y$.

➤ Lines and line segments may have unexpected parametric representations. For example, the equations $x = \sin t$, $y = 2 \sin t$ represent the line segment $y = 2x$, where $-1 \le x \le 1$ (recall that the range of $x = \sin t$ is $[-1, 1]$).

$P(4, 7)$

$x = 4 - t$, $y = 7 - 5t$, for $0 \le t \le 2$

$Q(2, -3)$

**Figure 12.7**

**QUICK CHECK 3** Describe the curve generated by $x = 3 + 2t$, $y = -12 - 6t$, for $-\infty < t < \infty$. ◀

### EXAMPLE 4    Parametric equations of curves

A common task (particularly in upcoming chapters) is to parameterize curves given either by Cartesian equations or by graphs. Find a parametric representation of the following curves.

**a.** The segment of the parabola $y = 9 - x^2$, for $-1 \le x \le 3$

**b.** The complete curve $x = (y - 5)^2 + \sqrt{y}$

**c.** The piecewise linear path connecting $P(-2, 0)$ to $Q(0, 3)$ to $R(4, 0)$ (in that order), where the parameter varies over the interval $0 \le t \le 2$

**SOLUTION**

**a.** The simplest way to represent a curve $y = f(x)$ parametrically is to let $x = t$ and $y = f(t)$, where $t$ is the parameter. We must then find the appropriate interval for the parameter. Using this approach, the curve $y = 9 - x^2$ has the parametric representation

$$x = t, \quad y = 9 - t^2, \quad \text{for } -1 \le t \le 3.$$

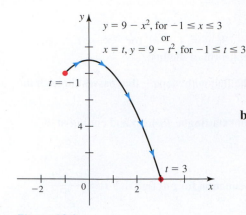

**Figure 12.8**

This representation is not unique. For example, you can verify that the parametric equations

$$x = 1 - t, \quad y = 9 - (1 - t)^2, \quad \text{for} \quad -2 \le t \le 2$$

also do the job, although these equations trace the parabola from right to left, while the original equations trace the curve from left to right (**Figure 12.8**).

**b.** In this case, it is easier to let $y = t$. Then a parametric description of the curve is

$$x = (t - 5)^2 + \sqrt{t}, \quad y = t.$$

Notice that $t$ can take values only in the interval $[0, \infty)$. As $t \to \infty$, we see that $x \to \infty$ and $y \to \infty$ (**Figure 12.9**).

**Figure 12.9**

**Figure 12.10**

**c.** The path consists of two line segments (**Figure 12.10**) that can be parameterized separately in the form $x = x_0 + at$ and $y = y_0 + bt$. The line segment $PQ$ originates at $P(-2, 0)$ and unfolds in the positive $x$-direction with slope $\frac{3}{2}$. It can be represented as

$$x = -2 + 2t, \quad y = 3t, \quad \text{for } 0 \le t \le 1.$$

Finding the parametric equations for the line segment $QR$ requires some ingenuity. We want the line segment to originate at $Q(0, 3)$ when $t = 1$ and end at $R(4, 0)$ when $t = 2$. Observe that when $t = 1, x = 0$ and when $t = 2, x = 4$. Substituting these pairs of values into the general $x$-equation $x = x_0 + at$, we obtain the equations

$$
\begin{aligned}
x_0 + a &= 0 & & x = 0 \text{ when } t = 1 \\
x_0 + 2a &= 4. & & x = 4 \text{ when } t = 2
\end{aligned}
$$

Solving for $x_0$ and $a$, we find that $x_0 = -4$ and $a = 4$. Applying a similar procedure to the general $y$-equation $y = y_0 + bt$, the relevant conditions are

$$
\begin{aligned}
y_0 + b &= 3 & & y = 3 \text{ when } t = 1 \\
y_0 + 2b &= 0. & & y = 0 \text{ when } t = 2
\end{aligned}
$$

Solving for $y_0$ and $b$, we find that $y_0 = 6$ and $b = -3$. Putting it all together, the equations for the line segment $QR$ are

$$x = -4 + 4t, \, y = 6 - 3t, \quad \text{for } 1 \le t \le 2.$$

**QUICK CHECK 4** Find parametric equations for the line segment that goes from $Q(0, 3)$ to $P(-2, 0)$. ◄

You can verify that the points $Q(0, 3)$ and $R(4, 0)$ correspond to $t = 1$ and $t = 2$, respectively. Furthermore, the slope of the line is $\dfrac{b}{a} = -\dfrac{3}{4}$, which is correct.

*Related Exercises 43–44, 47* ◄

Many fascinating curves are generated by points on rolling wheels; Examples 5 and 8 investigate two such curves.

**EXAMPLE 5   Rolling wheels** The path of a light on the rim of a wheel rolling on a flat surface (**Figure 12.11a**) is a **cycloid**, which has the parametric equations

$$x = a(t - \sin t), \quad y = a(1 - \cos t), \quad \text{for } t \geq 0,$$

where $a > 0$. Use a graphing utility to graph the cycloid with $a = 1$. On what interval does the parameter generate the first arch of the cycloid?

**SOLUTION** The graph of the cycloid, for $0 \leq t \leq 3\pi$, is shown in **Figure 12.11b**. The wheel completes one full revolution on the interval $0 \leq t \leq 2\pi$, which gives one arch of the cycloid.

(a)                                                            (b)

**Figure 12.11**

*Related Exercise 58* ◄

## Derivatives and Parametric Equations

Parametric equations express a relationship between the variables $x$ and $y$. Therefore, it makes sense to ask about $dy/dx$, the rate of change of $y$ with respect to $x$ at a point on a parametric curve. Once we know how to compute $dy/dx$, it can be used to determine slopes of lines tangent to parametric curves.

Consider the parametric equations $x = f(t), y = g(t)$ on an interval on which both $f$ and $g$ are differentiable. The Chain Rule relates the derivatives $\dfrac{dy}{dt}, \dfrac{dx}{dt}$, and $\dfrac{dy}{dx}$:

$$\frac{dy}{dt} = \frac{dy}{dx}\frac{dx}{dt}.$$

Provided $\dfrac{dx}{dt} \neq 0$, we divide both sides of this equation by $\dfrac{dx}{dt}$ and solve for $\dfrac{dy}{dx}$ to obtain the following result.

**Figure 12.12**

**QUICK CHECK 5** Use Theorem 12.1 to find the slope of the line $x = 4t$, $y = 2t$, for $-\infty < t < \infty$. ◄

---

**THEOREM 12.1   Derivative for Parametric Curves**

Let $x = f(t)$ and $y = g(t)$, where $f$ and $g$ are differentiable on an interval $[a, b]$. Then the slope of the line tangent to the curve at the point corresponding to $t$ is

$$\frac{dy}{dx} = \frac{dy/dt}{dx/dt} = \frac{g'(t)}{f'(t)},$$

provided $f'(t) \neq 0$.

---

**Figure 12.12** gives a geometric explanation of Theorem 12.1. The slope of the line tangent to a curve at a point is $\dfrac{dy}{dx} = \lim\limits_{\Delta x \to 0} \dfrac{\Delta y}{\Delta x}$. Using linear approximation (Section 4.6), we have $\Delta x \approx f'(t)\Delta t$ and $\Delta y \approx g'(t)\Delta t$, with these approximations improving as $\Delta t \to 0$. Notice also that $\Delta t \to 0$ as $\Delta x \to 0$. Therefore, the slope of the tangent line is

$$\frac{dy}{dx} = \lim_{\Delta x \to 0} \frac{\Delta y}{\Delta x} = \lim_{\Delta t \to 0} \frac{g'(t)\Delta t}{f'(t)\Delta t} = \frac{g'(t)}{f'(t)}. \quad \text{Cancel } \Delta t \neq 0.$$

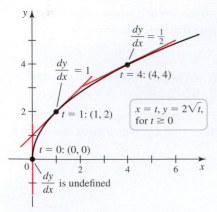

**Figure 12.13**

▶ In general, the equations $x = a \cos t$, $y = b \sin t$, for $0 \le t \le 2\pi$, describe an ellipse. The constants $a$ and $b$ can be seen as horizontal and vertical scalings of the unit circle $x = \cos t$, $y = \sin t$. Ellipses are explored in Exercises 91–94 and in Section 12.4.

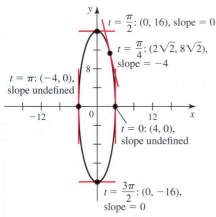

**Figure 12.14**

**EXAMPLE 6   Slopes of tangent lines**  Find $\dfrac{dy}{dx}$ for the following curves. Interpret the result and determine the points (if any) at which the curve has a horizontal or a vertical tangent line.

**a.** $x = f(t) = t$, $y = g(t) = 2\sqrt{t}$, for $t \ge 0$

**b.** $x = f(t) = 4 \cos t$, $y = g(t) = 16 \sin t$, for $0 \le t \le 2\pi$

**SOLUTION**

**a.** We find that $f'(t) = 1$ and $g'(t) = 1/\sqrt{t}$. Therefore,

$$\frac{dy}{dx} = \frac{g'(t)}{f'(t)} = \frac{1/\sqrt{t}}{1} = \frac{1}{\sqrt{t}},$$

provided $t \ne 0$. Notice that $dy/dx \ne 0$ for $t > 0$, so the curve has no horizontal tangent lines. On the other hand, as $t \to 0^+$, we see that $dy/dx \to \infty$. Therefore, the curve has a vertical tangent line at the point $(0, 0)$. To eliminate $t$ from the parametric equations, we substitute $t = x$ into the $y$-equation to find that $y = 2\sqrt{x}$. Because $y \ge 0$, the curve is the upper half of a parabola (**Figure 12.13**). Slopes of tangent lines at other points on the curve are found by substituting the corresponding values of $t$. For example, the point $(4, 4)$ corresponds to $t = 4$ and the slope of the tangent line at that point is $1/\sqrt{4} = \frac{1}{2}$.

**b.** These parametric equations describe an **ellipse** with a major axis of length 32 on the $y$-axis and a minor axis of length 8 on the $x$-axis (**Figure 12.14**). In this case, $f'(t) = -4 \sin t$ and $g'(t) = 16 \cos t$. Therefore,

$$\frac{dy}{dx} = \frac{g'(t)}{f'(t)} = \frac{16 \cos t}{-4 \sin t} = -4 \cot t.$$

At $t = 0$ and $t = \pi$, $\cot t$ is undefined. Notice that

$$\lim_{t \to 0^+} \frac{dy}{dx} = \lim_{t \to 0^+} (-4 \cot t) = -\infty \quad \text{and} \quad \lim_{t \to 0^-} \frac{dy}{dx} = \lim_{t \to 0^-} (-4 \cot t) = \infty.$$

Consequently, a vertical tangent line occurs at the point corresponding to $t = 0$, which is $(4, 0)$ (Figure 12.14). A similar argument shows that a vertical tangent line occurs at the point corresponding to $t = \pi$, which is $(-4, 0)$.

At $t = \pi/2$ and $t = 3\pi/2$, $\cot t = 0$ and the curve has horizontal tangent lines at the corresponding points $(0, \pm 16)$. Slopes of tangent lines at other points on the curve may be found. For example, the point $(2\sqrt{2}, 8\sqrt{2})$ corresponds to $t = \pi/4$; the slope of the tangent line at that point is $-4 \cot \pi/4 = -4$.

*Related Exercises 69–70* ◀

In Section 6.5, we derived the formula for the arc length of curves of the form $y = f(x)$. We conclude this section with a similar derivation for the arc length of curves described parametrically.

## Arc Length

Suppose a curve $C$ is given by the parametric equations $x = f(t)$, $y = g(t)$, for $a \le t \le b$, where $f'$ and $g'$ are continuous on $[a, b]$. To find the length of the curve between $(f(a), g(a))$ and $(f(b), g(b))$, we first subdivide the interval $[a, b]$ into $n$ subintervals using the grid points

$$a = t_0 < t_1 < t_2 < \cdots < t_n = b.$$

The next step is to connect consecutive points on the curve,

$$(f(t_0), g(t_0)), \ldots, (f(t_k), g(t_k)), \ldots, (f(t_n), g(t_n)),$$

with line segments (**Figure 12.15a**).

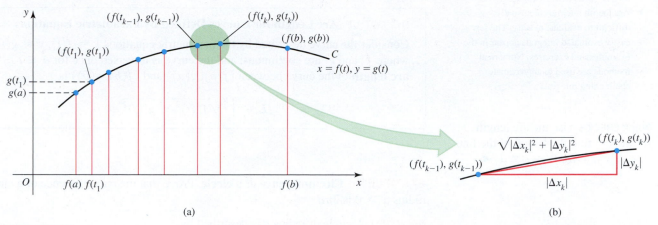

**Figure 12.15**

The $k$th line segment is the hypotenuse of a right triangle, whose legs have lengths $|\Delta x_k|$ and $|\Delta y_k|$, where

$$\Delta x_k = f(t_k) - f(t_{k-1}) \quad \text{and} \quad \Delta y_k = g(t_k) - g(t_{k-1}),$$

for $k = 1, 2, \ldots, n$ (**Figure 12.15b**). Therefore, the length of the $k$th line segment is

$$\sqrt{|\Delta x_k|^2 + |\Delta y_k|^2}.$$

The length of the entire curve $L$ is approximated by the sum of the lengths of the line segments:

$$L \approx \sum_{k=1}^{n} \sqrt{|\Delta x_k|^2 + |\Delta y_k|^2} = \sum_{k=1}^{n} \sqrt{(\Delta x_k)^2 + (\Delta y_k)^2}. \tag{1}$$

The goal is to express this sum as a Riemann sum.

The change in $x = f(t)$ over the $k$th subinterval is $\Delta x_k = f(t_k) - f(t_{k-1})$. By the Mean Value Theorem, there is a point $t_k^*$ in $(t_{k-1}, t_k)$ such that

$$\overbrace{\underbrace{\frac{f(t_k) - f(t_{k-1})}{t_k - t_{k-1}}}_{\Delta t_k}}^{\Delta x_k} = f'(t_k^*).$$

So the change in $x$ as $t$ changes by $\Delta t_k = t_k - t_{k-1}$ is

$$\Delta x_k = f(t_k) - f(t_{k-1}) = f'(t_k^*)\Delta t_k.$$

Similarly, the change in $y$ over the $k$th subinterval is

$$\Delta y_k = g(t_k) - g(t_{k-1}) = g'(\hat{t}_k)\Delta t_k,$$

where $\hat{t}_k$ is also a point in $(t_{k-1}, t_k)$. We substitute these expressions for $\Delta x_k$ and $\Delta y_k$ into equation (1):

$$L \approx \sum_{k=1}^{n} \sqrt{(\Delta x_k)^2 + (\Delta y_k)^2}$$

$$= \sum_{k=1}^{n} \sqrt{(f'(t_k^*)\Delta t_k)^2 + (g'(\hat{t}_k)\Delta t_k)^2} \quad \text{Substitute for } \Delta x_k \text{ and } \Delta y_k.$$

$$= \sum_{k=1}^{n} \sqrt{f'(t_k^*)^2 + g'(\hat{t}_k)^2}\,\Delta t_k. \quad \text{Factor } \Delta t_k \text{ out of square root.}$$

The intermediate points $t_k^*$ and $\hat{t}_k$ both approach $t_k$ as $n$ increases and as $\Delta t_k$ approaches zero. Therefore, given the conditions on $f'$ and $g'$, the limit of this sum as $n \to \infty$ and $\Delta t_k \to 0$, for all $k$, exists and equals a definite integral:

$$L = \lim_{n \to \infty} \sum_{k=1}^{n} \sqrt{f'(t_k^*)^2 + g'(\hat{t}_k)^2}\,\Delta t_k = \int_a^b \sqrt{f'(t)^2 + g'(t)^2}\,dt.$$

We take this integral as the definition of the length of the curve $C$.

➤ Arc length integrals are usually difficult to evaluate exactly. The few easily evaluated integrals appear in the examples and exercises. Numerical methods are used to approximate challenging integrals.

**QUICK CHECK 6** Use the arc length formula to find the length of the line $x = t, y = t$, for $0 \le t \le 1$. ◄

➤ An important fact is that the arc length of a smooth parameterized curve is independent of the choice of parameter (See Section 14.4).

**Figure 12.16**

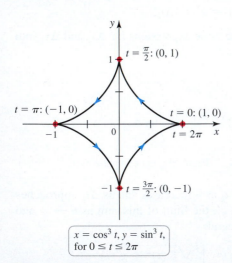

$x = \cos^3 t, y = \sin^3 t,$
for $0 \le t \le 2\pi$

**Figure 12.17**

---

> **DEFINITION  Arc Length for Curves Defined by Parametric Equations**
>
> Consider the curve described by the parametric equations $x = f(t), y = g(t)$, where $f'$ and $g'$ are continuous, and the curve is traversed once for $a \le t \le b$. The **arc length** of the curve between $(f(a), g(a))$ and $(f(b), g(b))$ is
>
> $$L = \int_a^b \sqrt{f'(t)^2 + g'(t)^2}\, dt.$$

**EXAMPLE 7    Circumference of a circle** Prove that the circumference of a circle of radius $a > 0$ is $2\pi a$.

**SOLUTION** A circle of radius $a$ is described by

$$x = f(t) = a \cos t, \; y = g(t) = a \sin t, \quad \text{for } 0 \le t \le 2\pi.$$

Note that $f'(t) = -a \sin t$ and $g'(t) = a \cos t$. The circumference is

$$
\begin{aligned}
L &= \int_0^{2\pi} \sqrt{f'(t)^2 + g'(t)^2}\, dt && \text{Arc length formula} \\[1mm]
&= \int_0^{2\pi} \sqrt{(-a \sin t)^2 + (a \cos t)^2}\, dt && \text{Substitute for } f' \text{ and } g'. \\[1mm]
&= a\int_0^{2\pi} \sqrt{\sin^2 t + \cos^2 t}\, dt && \text{Factor } a > 0 \text{ out of square root.} \\[1mm]
&= a\int_0^{2\pi} 1\, dt && \sin^2 t + \cos^2 t = 1 \\[1mm]
&= 2\pi a. && \text{Integrate a constant.}
\end{aligned}
$$

*Related Exercise 82* ◄

**EXAMPLE 8    Astroid arc length** The path of a point on circle $A$ with radius $a/4$ that rolls on the inside of circle $B$ with radius $a$ (**Figure 12.16**) is an **astroid** or a **hypocycloid**. Its parametric equations are

$$x = a \cos^3 t, \quad y = a \sin^3 t, \quad \text{for } 0 \le t \le 2\pi.$$

**a.** Graph the astroid with $a = 1$ and find its equation in terms of $x$ and $y$.

**b.** Find the arc length of the astroid with $a = 1$.

**SOLUTION**

**a.** Because both $\cos^3 t$ and $\sin^3 t$ have a period of $2\pi$, the complete curve is generated on the interval $0 \le t \le 2\pi$ (**Figure 12.17**). To eliminate $t$ from the parametric equations, note that $x^{2/3} = \cos^2 t$ and $y^{2/3} = \sin^2 t$. Therefore,

$$x^{2/3} + y^{2/3} = \cos^2 t + \sin^2 t = 1,$$

where the Pythagorean identity has been used. We see that an alternative description of the astroid is $x^{2/3} + y^{2/3} = 1$.

**b.** The length of the entire curve is four times the length of the curve in the first quadrant. You should verify that the curve in the first quadrant is generated as the parameter varies from $t = 0$ (corresponding to $(1, 0)$) to $t = \pi/2$ (corresponding to $(0, 1)$; Figure 12.17). Letting $f(t) = \cos^3 t$ and $g(t) = \sin^3 t$, we have

$$f'(t) = -3 \cos^2 t \sin t \quad \text{and} \quad g'(t) = 3 \sin^2 t \cos t.$$

The arc length of the full curve is

$$L = 4 \int_0^{\pi/2} \sqrt{f'(t)^2 + g'(t)^2}\, dt \qquad \text{Factor of 4 by symmetry}$$

$$= 4 \int_0^{\pi/2} \sqrt{(-3 \cos^2 t \sin t)^2 + (3 \sin^2 t \cos t)^2}\, dt \qquad \text{Substitute for } f' \text{ and } g'.$$

$$= 4 \int_0^{\pi/2} \sqrt{9 \cos^4 t \sin^2 t + 9 \cos^2 t \sin^4 t}\, dt \qquad \text{Simplify terms.}$$

$$= 4 \int_0^{\pi/2} 3 \sqrt{\cos^2 t \sin^2 t \underbrace{(\cos^2 t + \sin^2 t)}_{1}}\, dt \qquad \text{Factor.}$$

$$= 12 \int_0^{\pi/2} \cos t \sin t\, dt. \qquad \cos t \sin t \geq 0, \text{ for } 0 \leq t \leq \frac{\pi}{2}$$

> The arc length formula from Section 6.5 could be used to compute the length of the astroid in Example 8 because we found a Cartesian equation for the curve in part (a). However, in cases where it is impossible to eliminate the parameter, the arc length formula derived in this section is required.

Letting $u = \sin t$ with $du = \cos t\, dt$, we have

$$L = 12 \int_0^{\pi/2} \cos t \sin t\, dt = 12 \int_0^1 u\, du = 6.$$

The length of the entire hypocycloid is 6 units.

*Related Exercise 87* ◄

---

## SECTION 12.1 EXERCISES

### Getting Started

1. Explain how a pair of parametric equations generates a curve in the $xy$-plane.

2. Give two pairs of parametric equations that generate a circle centered at the origin with radius 6.

3. Give parametric equations that describe a circle of radius $R$, centered at the origin with clockwise orientation, where the parameter varies over the interval $[0, 10]$.

4. Give parametric equations that generate the line with slope $-2$ passing through $(1, 3)$.

5. Find parametric equations for the complete parabola $x = y^2$. Answers are not unique.

6. Describe the similarities between the graphs of the parametric equations $x = \sin^2 t, y = \sin t$, for $0 \leq t \leq \pi/2$, and $x = \sin^2 t, y = \sin t$, for $\pi/2 \leq t \leq \pi$. Begin by eliminating the parameter $t$ to obtain an equation in $x$ and $y$.

7. Find the slope of the parametric curve $x = -2t^3 + 1, y = 3t^2$, for $-\infty < t < \infty$, at the point corresponding to $t = 2$.

8. In which direction is the curve $x = -2 \sin t, y = 2 \cos t$, for $0 \leq t < 2\pi$, generated?

9. Find three different pairs of parametric equations for the line segment that starts at $(0, 0)$ and ends at $(6, 6)$.

10. Use calculus to find the arc length of the line segment $x = 3t + 1$, $y = 4t$, for $0 \leq t \leq 1$. Check your work by finding the distance between the endpoints of the line segment.

### Practice Exercises

**11–14. Working with parametric equations** *Consider the following parametric equations.*

*a.* *Make a brief table of values of $t$, $x$, and $y$.*
*b.* *Plot the $(x, y)$ pairs in the table and the complete parametric curve, indicating the positive orientation (the direction of increasing $t$).*
*c.* *Eliminate the parameter to obtain an equation in $x$ and $y$.*
*d.* *Describe the curve.*

11. $x = 2t, y = 3t - 4; -10 \leq t \leq 10$

12. $x = t^2 + 2, y = 4t; -4 \leq t \leq 4$

13. $x = -t + 6, y = 3t - 3; -5 \leq t \leq 5$

14. $x = t^3 - 1, y = 5t + 1; -3 \leq t \leq 3$

**15–30. Working with parametric equations** *Consider the following parametric equations.*

*a.* *Eliminate the parameter to obtain an equation in $x$ and $y$.*
*b.* *Describe the curve and indicate the positive orientation.*

15. $x = 3 + t, y = 1 - t; 0 \leq t \leq 1$

16. $x = 4 - 3t, y = -2 + 6t; -1 \leq t \leq 3$

17. $x = \sqrt{t} + 4, y = 3\sqrt{t}; 0 \leq t \leq 16$

18. $x = (t + 1)^2, y = t + 2; -10 \leq t \leq 10$

19. $x = 3 \cos t, y = 3 \sin t; \pi \leq t \leq 2\pi$

20. $x = 3 \cos t, y = 3 \sin t; 0 \leq t \leq \pi/2$

21. $x = \cos t, y = \sin^2 t; 0 \leq t \leq \pi$

22. $x = 1 - \sin^2 s, y = \cos s; \pi \le s \le 2\pi$

23. $x = \cos t, y = 1 + \sin t; 0 \le t \le 2\pi$

24. $x = \sin t, y = 2 \sin t + 1; 0 \le t \le \pi/2$

25. $x = r - 1, y = r^3; -4 \le r \le 4$

26. $x = e^{2t}, y = e^t + 1; 0 \le t \le 25$

27. $x = -7 \cos 2t, y = -7 \sin 2t; 0 \le t \le \pi$

28. $x = 1 - 3 \sin 4\pi t, y = 2 + 3 \cos 4\pi t; 0 \le t \le 1/2$

29. $x = 8 + 2t, y = 1; -\infty < t < \infty$

30. $x = 5, y = 3t; -\infty < t < \infty$

**31–36. Eliminating the parameter** *Eliminate the parameter to express the following parametric equations as a single equation in x and y.*

31. $x = 2 \sin 8t, y = 2 \cos 8t$     32. $x = \sin 8t, y = 2 \cos 8t$

33. $x = t, y = \sqrt{4 - t^2}$     34. $x = \sqrt{t + 1}, y = \dfrac{1}{t + 1}$

35. $x = \tan t, y = \sec^2 t - 1$

36. $x = a \sin^n t, y = b \cos^n t$, where $a$ and $b$ are real numbers and $n$ is a positive integer

**37–52. Curves to parametric equations** *Find parametric equations for the following curves. Include an interval for the parameter values. Answers are not unique.*

37. A circle centered at the origin with radius 4, generated counterclockwise

38. A circle centered at the origin with radius 12, generated clockwise with initial point $(0, 12)$

39. A circle centered at $(2, 3)$ with radius 1, generated counterclockwise

40. A circle centered at $(-2, -3)$ with radius 8, generated clockwise

41. The line segment starting at $P(0, 0)$ and ending at $Q(2, 8)$

42. The line segment starting at $P(-1, -3)$ and ending at $Q(6, -16)$

43. The segment of the parabola $y = 2x^2 - 4$, where $-1 \le x \le 5$

44. The complete curve $x = y^3 - 3y$

45. The vertical line segment starting at $P(2, 3)$ and ending at $Q(2, 9)$

46. The horizontal line segment starting at $P(8, 2)$ and ending at $Q(-2, 2)$

47. The piecewise linear path from $P(-2, 3)$ to $Q(2, -3)$ to $R(3, 5)$, using parameter values $0 \le t \le 2$

48. The path consisting of the line segment from $(-4, 4)$ to $(0, 8)$, followed by the segment of the parabola $y = 8 - 2x^2$ from $(0, 8)$ to $(2, 0)$, using parameter values $0 \le t \le 3$

49. The line that passes through the points $P(1, 1)$ and $Q(3, 5)$, oriented in the direction of increasing $x$

50. The left half of the parabola $y = x^2 + 1$, originating at $(0, 1)$

51. The upper half of the parabola $x = y^2$, originating at $(0, 0)$

52. The lower half of the circle centered at $(-2, 2)$ with radius 6, oriented in the counterclockwise direction

**53–56. Circular motion** *Find parametric equations that describe the circular path of the following objects. For Exercises 53–55, assume $(x, y)$ denotes the position of the object relative to the origin at the center of the circle. Use the units of time specified in the problem. There are many ways to describe any circle.*

53. A go-cart moves counterclockwise with constant speed around a circular track of radius 400 m, completing a lap in 1.5 min.

54. The tip of the 15-inch second hand of a clock completes one revolution in 60 seconds.

55. A bicyclist rides counterclockwise with constant speed around a circular velodrome track with a radius of 50 m, completing one lap in 24 seconds.

56. A Ferris wheel has a radius of 20 m and completes a revolution in the clockwise direction at constant speed in 3 min. Assume $x$ and $y$ measure the horizontal and vertical positions of a seat on the Ferris wheel relative to a coordinate system whose origin is at the low point of the wheel. Assume the seat begins moving at the origin.

**T 57–62. More parametric curves** *Use a graphing utility to graph the following curves. Be sure to choose an interval for the parameter that generates all features of interest.*

57. **Spiral** $x = t \cos t, y = t \sin t; t \ge 0$

58. **Witch of Agnesi** $x = 2 \cot t, y = 1 - \cos 2t$

59. **Folium of Descartes** $x = \dfrac{3t}{1 + t^3}, y = \dfrac{3t^2}{1 + t^3}$

60. **Involute of a circle** $x = \cos t + t \sin t, y = \sin t - t \cos t$

61. **Evolute of an ellipse** $x = \dfrac{a^2 - b^2}{a} \cos^3 t, y = \dfrac{a^2 - b^2}{b} \sin^3 t;$ $a = 4$ and $b = 3$

62. **Cissoid of Diocles** $x = 2 \sin 2t, y = \dfrac{2 \sin^3 t}{\cos t}$

**T 63. Implicit function graph** Explain and carry out a method for graphing the curve $x = 1 + \cos^2 y - \sin^2 y$ using parametric equations and a graphing utility.

64. **Air drop** A plane traveling horizontally at 80 m/s over flat ground at an elevation of 3000 m releases an emergency packet. The trajectory of the packet is given by

$$x = 80t, \quad y = -4.9t^2 + 3000, \quad \text{for } t \ge 0,$$

where the origin is the point on the ground directly beneath the plane at the moment of the release (see figure). Graph the trajectory of the packet and find the coordinates of the point where the packet lands.

**65. Air drop—inverse problem** A plane traveling horizontally at 100 m/s over flat ground at an elevation of 4000 m must drop an emergency packet on a target on the ground. The trajectory of the packet is given by

$$x = 100t, \quad y = -4.9t^2 + 4000, \quad \text{for } t \ge 0,$$

where the origin is the point on the ground directly beneath the plane at the moment of the release. How many horizontal meters before the target should the packet be released in order to hit the target?

**66. Projectile explorations** A projectile launched from the ground with an initial speed of 20 m/s and a launch angle $\theta$ follows a trajectory approximated by

$$x = (20 \cos \theta)t, \quad y = -4.9t^2 + (20 \sin \theta)t,$$

where $x$ and $y$ are the horizontal and vertical positions of the projectile relative to the launch point $(0, 0)$.

**a.** Graph the trajectory for various values of $\theta$ in the range $0 < \theta < \pi/2$.
**b.** Based on your observations, what value of $\theta$ gives the greatest range (the horizontal distance between the launch and landing points)?

**67–72. Derivatives** *Consider the following parametric curves.*

**a.** *Determine $dy/dx$ in terms of t and evaluate it at the given value of t.*
**b.** *Make a sketch of the curve showing the tangent line at the point corresponding to the given value of t.*

**67.** $x = 2 + 4t, y = 4 - 8t; t = 2$

**68.** $x = 3 \sin t, y = 3 \cos t; t = \pi/2$

**69.** $x = \cos t, y = 8 \sin t; t = \pi/2$

**70.** $x = 2t, y = t^3; t = -1$

**T 71.** $x = t + 1/t, y = t - 1/t; t = 1$

**72.** $x = \sqrt{t}, y = 2t; t = 4$

**73–76. Tangent lines** *Find an equation of the line tangent to the curve at the point corresponding to the given value of t.*

**73.** $x = t^2 - 1, y = t^3 + t; t = 2$

**74.** $x = \sin t, y = \cos t; t = \pi/4$

**75.** $x = \cos t + t \sin t, y = \sin t - t \cos t; t = \pi/4$

**76.** $x = e^t, y = \ln(t + 1); t = 0$

**77–80. Slopes of tangent lines** *Find all points at which the following curves have the given slope.*

**77.** $x = 4 \cos t, y = 4 \sin t;$ slope $= 1/2$

**78.** $x = 2 \cos t, y = 8 \sin t;$ slope $= -1$

**79.** $x = t + 1/t, y = t - 1/t;$ slope $= 1$

**80.** $x = 2 + \sqrt{t}, y = 2 - 4t;$ slope $= -8$

**81–88. Arc length** *Find the arc length of the following curves on the given interval.*

**81.** $x = 3t + 1, y = 4t + 2; 0 \le t \le 2$

**82.** $x = 3 \cos t, y = 3 \sin t + 1; 0 \le t \le 2\pi$

**83.** $x = \cos t - \sin t, y = \cos t + \sin t; 0 \le t \le \pi$

**84.** $x = e^t \sin t, y = e^t \cos t; 0 \le t \le 2\pi$

**85.** $x = t^4, y = \dfrac{t^6}{3}; 0 \le t \le 1$

**86.** $x = 2t \sin t - t^2 \cos t, y = 2t \cos t + t^2 \sin t; 0 \le t \le \pi$

**87.** $x = \cos^3 2t, y = \sin^3 2t; 0 \le t \le \pi/4$

**88.** $x = \sin t, y = t - \cos t; 0 \le t \le \pi/2$

**89. Explain why or why not** Determine whether the following statements are true and give an explanation or counterexample.

**a.** The equations $x = -\cos t, y = -\sin t$, for $0 \le t \le 2\pi$, generate a circle in the clockwise direction.
**b.** An object following the parametric curve $x = 2 \cos 2\pi t$, $y = 2 \sin 2\pi t$ circles the origin once every 1 time unit.
**c.** The parametric equations $x = t, y = t^2$, for $t \ge 0$, describe the complete parabola $y = x^2$.
**d.** The parametric equations $x = \cos t, y = \sin t$, for $-\pi/2 \le t \le \pi/2$, describe a semicircle.
**e.** There are two points on the curve $x = -4 \cos t, y = \sin t$, for $0 \le t \le 2\pi$, at which there is a vertical tangent line.

**90. Matching curves and equations** Match equations a–d with graphs A–D. Explain your reasoning.

**a.** $x = t^2 - 2, y = t^3 - t$
**b.** $x = \cos(t + \sin 50t), y = \sin(t + \cos 50t)$
**c.** $x = t + \cos 2t, y = t - \sin 4t$
**d.** $x = 2 \cos t + \cos 20t, y = 2 \sin t + \sin 20t$

(A)                          (B)

(C)                          (D)

**91–92. Ellipses** *An* **ellipse** *(discussed in detail in Section 12.4) is generated by the parametric equations $x = a \cos t, y = b \sin t$. If $0 < a < b$, then the long axis (or **major axis**) lies on the y-axis and the short axis (or **minor axis**) lies on the x-axis. If $0 < b < a$, the axes are reversed. The lengths of the axes in the x- and y-directions are 2a and 2b, respectively. Sketch the graph of the following ellipses. Specify an interval in t over which the entire curve is generated.*

**91.** $x = 4 \cos t, y = 9 \sin t$

**92.** $x = 12 \sin 2t, y = 3 \cos 2t$

**93–94. Parametric equations of ellipses** *Find parametric equations (not unique) of the following ellipses (see Exercises 91–92). Graph the ellipse and find a description in terms of x and y.*

**93.** An ellipse centered at the origin with major axis of length 6 on the $x$-axis and minor axis of length 3 on the $y$-axis, generated counterclockwise

**94.** An ellipse centered at $(-2, -3)$ with major and minor axes of lengths 30 and 20, parallel to the $x$- and $y$-axes, respectively, generated counterclockwise (*Hint:* Shift the parametric equations.)

**95. Intersecting lines** Consider the following pairs of lines. Determine whether the lines are parallel or intersecting. If the lines intersect, then determine the point of intersection.

   **a.** $x = 1 + s, y = 2s$ and $x = 1 + 2t, y = 3t$
   **b.** $x = 2 + 5s, y = 1 + s$ and $x = 4 + 10t, y = 3 + 2t$
   **c.** $x = 1 + 3s, y = 4 + 2s$ and $x = 4 - 3t, y = 6 + 4t$

**96. Multiple descriptions** Which of the following parametric equations describe the same curve?

   **a.** $x = 2t^2, y = 4 + t; -4 \le t \le 4$
   **b.** $x = 2t^4, y = 4 + t^2; -2 \le t \le 2$
   **c.** $x = 2t^{2/3}, y = 4 + t^{1/3}; -64 \le t \le 64$

## Explorations and Challenges

**T 97–100. Beautiful curves** *Consider the family of curves*

$$x = \left(2 + \frac{1}{2}\sin at\right)\cos\left(t + \frac{\sin bt}{c}\right),$$

$$y = \left(2 + \frac{1}{2}\sin at\right)\sin\left(t + \frac{\sin bt}{c}\right).$$

*Plot a graph of the curve for the given values of a, b, and c with $0 \le t \le 2\pi$.*
(*Source: Mathematica in Action*, Stan Wagon, Springer, 2010; created by Norton Starr, Amherst College)

**97.** $a = b = 5, c = 2$     **98.** $a = 6, b = 12, c = 3$

**99.** $a = 18, b = 18, c = 7$     **100.** $a = 7, b = 4, c = 1$

**T 101–102. Lissajous curves** *Consider the following Lissajous curves. Graph the curve, and estimate the coordinates of the points on the curve at which there is (a) a horizontal tangent line or (b) a vertical tangent line. (See the Guided Project* Parametric art *for more on Lissajous curves.)*

**101.** $x = \sin 2t, y = 2\sin t; 0 \le t \le 2\pi$

**102.** $x = \sin 4t, y = \sin 3t; 0 \le t \le 2\pi$

**103–105. Area under a curve** *Suppose the function $y = h(x)$ is nonnegative and continuous on $[\alpha, \beta]$, which implies that the area bounded by the graph of h and the x-axis on $[\alpha, \beta]$ equals $\int_\alpha^\beta h(x)\,dx$ or $\int_\alpha^\beta y\,dx$. If the graph of $y = h(x)$ on $[\alpha, \beta]$ is traced exactly once by the parametric equations $x = f(t), y = g(t)$, for $a \le t \le b$, then it follows by substitution that the area bounded by h is*

$$\int_\alpha^\beta h(x)\,dx = \int_\alpha^\beta y\,dx = \int_a^b g(t)f'(t)\,dt \text{ if } \alpha = f(a) \text{ and } \beta = f(b)$$

$$\left(\text{or } \int_\alpha^\beta h(x)\,dx = \int_b^a g(t)f'(t)\,dt \text{ if } \alpha = f(b) \text{ and } \beta = f(a)\right).$$

**103.** Find the area under one arch of the cycloid $x = 3(t - \sin t)$, $y = 3(1 - \cos t)$ (see Example 5).

**104.** Show that the area of the region bounded by the ellipse $x = 3\cos t, y = 4\sin t$, for $0 \le t \le 2\pi$, equals

$$4\int_{\pi/2}^0 4\sin t(-3\sin t)\,dt.$$ Then evaluate the integral.

**105.** Find the area of the region bounded by the astroid $x = \cos^3 t, y = \sin^3 t$, for $0 \le t \le 2\pi$ (see Example 8, Figure 12.17).

**106–111. Surfaces of revolution** *Let C be the curve $x = f(t)$, $y = g(t)$, for $a \le t \le b$, where $f'$ and $g'$ are continuous on $[a, b]$ and C does not intersect itself, except possibly at its endpoints. If g is nonnegative on $[a, b]$, then the area of the surface obtained by revolving C about the x-axis is*

$$S = \int_a^b 2\pi g(t)\sqrt{f'(t)^2 + g'(t)^2}\,dt.$$

*Likewise, if f is nonnegative on $[a, b]$, then the area of the surface obtained by revolving C about the y-axis is*

$$S = \int_a^b 2\pi f(t)\sqrt{f'(t)^2 + g'(t)^2}\,dt.$$

*(These results can be derived in a manner similar to the derivations given in Section 6.6 for surfaces of revolution generated by the curve $y = f(x)$.)*

**106.** Use the parametric equations of a semicircle of radius 1, $x = \cos t$, $y = \sin t$, for $0 \le t \le \pi$, to verify that surface area of a unit sphere is $4\pi$.

**107.** Consider the curve $x = 3\cos t, y = 3\sin t + 4$, for $0 \le t \le 2\pi$.

   **a.** Describe the curve.
   **b.** If the curve is revolved about the $x$-axis, describe the shape of the surface of revolution and find the area of the surface.

**108.** Find the area of the surface obtained by revolving the curve $x = \cos^3 t, y = \sin^3 t$, for $0 \le t \le \pi/2$, about the $x$-axis.

**109.** Find the area of the surface obtained by revolving one arch of the cycloid $x = t - \sin t, y = 1 - \cos t$, for $0 \le t \le 2\pi$, about the $x$-axis.

**110.** Find the area of the surface obtained by revolving the curve $x = \sqrt{t + 1}, y = t$, for $1 \le t \le 5$, about the $y$-axis.

**T 111.** A surface is obtained by revolving the curve $x = e^{3t} + 1$, $y = e^{2t}$, for $0 \le t \le 1$, about the $y$-axis. Find an integral that gives the area of the surface and approximate the value of the integral.

**T 112. Paths of moons** A idealized model of the path of a moon (relative to the Sun) moving with constant speed in a circular orbit around a planet, where the planet in turn revolves around the Sun, is given by the parametric equations

$$x(\theta) = a\cos\theta + \cos n\theta, \quad y(\theta) = a\sin\theta + \sin n\theta.$$

The distance from the moon to the planet is taken to be 1, the distance from the planet to the Sun is $a$, and $n$ is the number of times the moon orbits the planet for every 1 revolution of the planet around the Sun. Plot the graph of the path of a moon for the given constants; then conjecture which values of $n$ produce loops for a fixed value of $a$.

   **a.** $a = 4, n = 3$     **b.** $a = 4, n = 4$     **c.** $a = 4, n = 5$

**T 113. Paths of the moons of Earth and Jupiter** Use the equations in Exercise 112 to analyze the paths of the following moons in our solar system.

    **a.** The path of Callisto (one of Jupiter's moons) corresponds to values of $a = 727.5$ and $n = 259.6$. Plot a small portion of the graph to see the detailed behavior of the orbit.

    **b.** The path of Io (another of Jupiter's moons) corresponds to values of $a = 1846.2$ and $n = 2448.8$. Plot a small portion of the path of Io to see the loops in its orbit.

(*Source: The Sun, the Moon, and Convexity, The College Mathematics Journal*, 32, Sep 2001)

**114. Second derivative** Assume a curve is given by the parametric equations $x = f(t)$ and $y = g(t)$, where $f$ and $g$ are twice differentiable. Use the Chain Rule to show that

$$y''(x) = \frac{f'(t)g''(t) - g'(t)f''(t)}{(f'(t))^3}.$$

**QUICK CHECK ANSWERS**

**1.** A segment of the parabola $x = y^2$ opening to the right with vertex at the origin   **2.** The circle has center $(0, 0)$ and radius 3; it is generated in the counterclockwise direction (positive orientation) starting at $(0, -3)$.   **3.** The line $y = -3x - 3$ with slope $-3$ passing through $(3, -12)$ (when $t = 0$)   **4.** One possibility is $x = -2t$, $y = 3 - 3t$, for $0 \le t \le 1$.   **5.** $\dfrac{1}{2}$

**6.** $\sqrt{2}$ ◄

# 12.2 Polar Coordinates

Suppose you work for a company that designs heat shields for space vehicles. The shields are thin plates that are either rectangular or circular in shape. To solve the heat transfer equations for these two shields, you must choose a coordinate system that best fits the geometry of the problem. A Cartesian (rectangular) coordinate system is a natural choice for the rectangular shields (**Figure 12.18a**). However, it does not provide a good fit for the circular shields (**Figure 12.18b**). On the other hand, a **polar coordinate** system, in which the coordinates are constant on circles and rays, is better suited for the circular shields (**Figure 12.18c**).

> Recall that the terms *Cartesian* and *rectangular* both describe the usual *xy*-coordinate system.

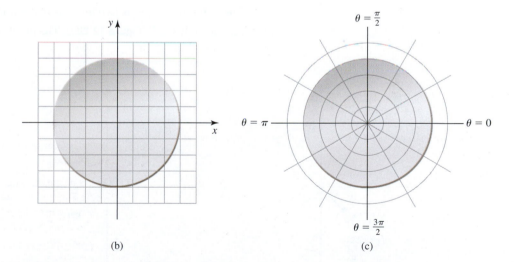

(a)          (b)          (c)

**Figure 12.18**

## Defining Polar Coordinates

> Polar points and curves are plotted on a rectangular coordinate system, with standard "*x*" and "*y*" labels attached to the axes. However, plotting polar points and curves is often easier using polar graph paper, which has concentric circles centered at the origin and rays emanating from the origin (Figure 12.18c).

Like Cartesian coordinates, polar coordinates are used to locate points in the plane. When working in polar coordinates, the origin of the coordinate system is also called the **pole**, and the positive *x*-axis is called the **polar axis**. The polar coordinates for a point $P$ have the form $(r, \theta)$. **The radial coordinate** $r$ describes the *signed* (or *directed*) distance from the origin to $P$. The **angular coordinate** $\theta$ describes an angle whose initial side is the positive *x*-axis and whose terminal side lies on the ray passing through the origin and $P$ (**Figure 12.19a**). Positive angles are measured counterclockwise from the positive *x*-axis.

    With polar coordinates, points have more than one representation for two reasons. First, angles are determined up to multiples of $2\pi$ radians, so the coordinates $(r, \theta)$ and

**QUICK CHECK 1** Which of the following coordinates represent the same point: $(3, \pi/2)$, $(3, 3\pi/2)$, $(3, 5\pi/2)$, $(-3, -\pi/2)$, and $(-3, 3\pi/2)$? ◄

$(r, \theta \pm 2\pi)$ refer to the same point (**Figure 12.19b**). Second, the radial coordinate may be negative, which is interpreted as follows: The points $(r, \theta)$ and $(-r, \theta)$ are reflections of each other through the origin (**Figure 12.19c**). This means that $(r, \theta)$, $(-r, \theta + \pi)$, and $(-r, \theta - \pi)$ all refer to the same point. The origin is specified as $(0, \theta)$ in polar coordinates, where $\theta$ is any angle.

**(a)**   **(b)**   **(c)**

**Figure 12.19**

**EXAMPLE 1    Points in polar coordinates** Graph the following points in polar coordinates: $Q\left(1, \frac{5\pi}{4}\right)$, $R\left(-1, \frac{7\pi}{4}\right)$, and $S\left(2, -\frac{3\pi}{2}\right)$. Give two alternative representations for each point.

**SOLUTION** The point $Q\left(1, \frac{5\pi}{4}\right)$ is one unit from the origin $O$ on a line $OQ$ that makes an angle of $\frac{5\pi}{4}$ with the positive $x$-axis (**Figure 12.20a**). Subtracting $2\pi$ from the angle, the point $Q$ can be represented as $\left(1, -\frac{3\pi}{4}\right)$. Subtracting $\pi$ from the angle and negating the radial coordinate implies that $Q$ also has the coordinates $\left(-1, \frac{\pi}{4}\right)$.

To locate the point $R\left(-1, \frac{7\pi}{4}\right)$, it is easiest first to find the point $R^*\left(1, \frac{7\pi}{4}\right)$ in the fourth quadrant. Then $R\left(-1, \frac{7\pi}{4}\right)$ is the reflection of $R^*$ through the origin (**Figure 12.20b**). Other representations of $R$ include $\left(-1, -\frac{\pi}{4}\right)$ and $\left(1, \frac{3\pi}{4}\right)$.

The point $S\left(2, -\frac{3\pi}{2}\right)$ is two units from the origin, found by rotating *clockwise* through an angle of $\frac{3\pi}{2}$ (**Figure 12.20c**). The point $S$ can also be represented as $\left(2, \frac{\pi}{2}\right)$ or $\left(-2, -\frac{\pi}{2}\right)$.

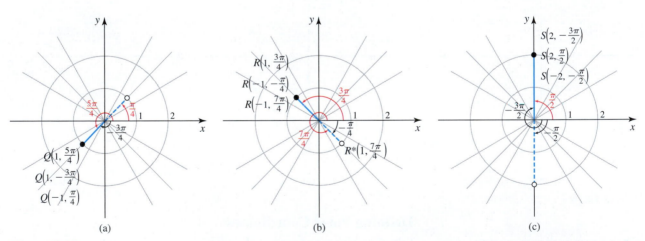

**(a)**   **(b)**   **(c)**

**Figure 12.20**

*Related Exercises 9–13* ◄

## Converting Between Cartesian and Polar Coordinates

We often need to convert between Cartesian and polar coordinates. The conversion equations emerge when we look at a right triangle (**Figure 12.21**) in which

$$\cos \theta = \frac{x}{r} \quad \text{and} \quad \sin \theta = \frac{y}{r}.$$

**Figure 12.21**

**QUICK CHECK 2** Draw versions of Figure 12.21 with $P$ in the second, third, and fourth quadrants. Verify that the same conversion formulas hold in all cases. ◄

➤ To determine $\theta$, you may also use the relationships $\cos \theta = x/r$ and $\sin \theta = y/r$. Either method requires checking the signs of $x$ and $y$ to verify that $\theta$ is in the correct quadrant.

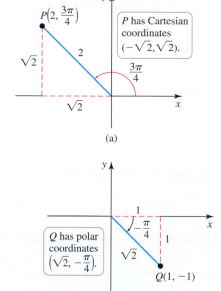

**Figure 12.22**

**QUICK CHECK 3** Give two polar coordinate descriptions of the point with Cartesian coordinates $(1, 0)$. What are the Cartesian coordinates of the point with polar coordinates $\left(2, \frac{\pi}{2}\right)$? ◄

Given a point with polar coordinates $(r, \theta)$, we see that its Cartesian coordinates are $x = r \cos \theta$ and $y = r \sin \theta$. Conversely, given a point with Cartesian coordinates $(x, y)$, its radial polar coordinate satisfies $r^2 = x^2 + y^2$. The coordinate $\theta$ is determined using the relation $\tan \theta = y/x$, where the quadrant in which $\theta$ lies is determined by the signs of $x$ and $y$. Figure 12.21 illustrates the conversion formulas for a point $P$ in the first quadrant. The same relationships hold if $P$ is in any of the other three quadrants.

---

**PROCEDURE** **Converting Coordinates**

A point with polar coordinates $(r, \theta)$ has Cartesian coordinates $(x, y)$, where

$$x = r \cos \theta \quad \text{and} \quad y = r \sin \theta.$$

A point with Cartesian coordinates $(x, y)$ has polar coordinates $(r, \theta)$, where

$$r^2 = x^2 + y^2 \quad \text{and} \quad \tan \theta = \frac{y}{x}.$$

---

**EXAMPLE 2** **Converting coordinates**

**a.** Express the point with polar coordinates $P\left(2, \frac{3\pi}{4}\right)$ in Cartesian coordinates.

**b.** Express the point with Cartesian coordinates $Q(1, -1)$ in polar coordinates.

**SOLUTION**

**a.** The point $P\left(2, \frac{3\pi}{4}\right)$ has Cartesian coordinates

$$x = r \cos \theta = 2 \cos \frac{3\pi}{4} = -\sqrt{2} \quad \text{and}$$
$$y = r \sin \theta = 2 \sin \frac{3\pi}{4} = \sqrt{2}.$$

As shown in **Figure 12.22a**, $P$ is in the second quadrant.

**b.** It's best to locate this point first to be sure that the angle $\theta$ is chosen correctly. As shown in **Figure 12.22b**, the point $Q(1, -1)$ is in the fourth quadrant at a distance $r = \sqrt{1^2 + (-1)^2} = \sqrt{2}$ from the origin. The coordinate $\theta$ satisfies

$$\tan \theta = \frac{y}{x} = \frac{-1}{1} = -1.$$

The angle in the fourth quadrant with $\tan \theta = -1$ is $\theta = -\frac{\pi}{4}$ or $\frac{7\pi}{4}$. Therefore, two (of infinitely many) polar representations of $Q$ are $\left(\sqrt{2}, -\frac{\pi}{4}\right)$ and $\left(\sqrt{2}, \frac{7\pi}{4}\right)$.

*Related Exercises 25, 28, 35* ◄

## Basic Curves in Polar Coordinates

A curve in polar coordinates is the set of points that satisfy an equation in $r$ and $\theta$. Some sets of points are easier to describe in polar coordinates than in Cartesian coordinates. Let's begin by examining polar equations of circles, lines, and spirals.

The polar equation $r = 3$ is satisfied by the set of points whose distance from the origin is 3. The angle $\theta$ is arbitrary because it is not specified by the equation, so the graph of $r = 3$ is the circle of radius 3 centered at the origin. In general, the equation $r = a$ describes a circle of radius $|a|$ centered at the origin (**Figure 12.23a**).

The equation $\theta = \pi/3$ is satisfied by the points whose angle with respect to the positive $x$-axis is $\pi/3$. Because $r$ is unspecified, it is arbitrary (and can be positive or negative). Therefore, $\theta = \pi/3$ describes the line through the origin making an angle of $\pi/3$ with the positive $x$-axis. More generally, $\theta = \theta_0$ describes the line through the origin making an angle of $\theta_0$ with the positive $x$-axis (**Figure 12.23b**).

▶ If the equation $\theta = \theta_0$ is accompanied by the condition $r \geq 0$, the resulting set of points is a *ray* emanating from the origin.

Circle of radius $|a|$ with equation $r = a$

$\theta = \theta_0$

Line through origin with equation $\theta = \theta_0$

(a)                    (b)

**Figure 12.23**

Spiral with equation $r = \theta$

**Figure 12.24**

The simplest polar equation that involves both $r$ and $\theta$ is $r = \theta$. Restricting $\theta$ to the interval $\theta \geq 0$, we see that as $\theta$ increases, $r$ increases. Therefore, as $\theta$ increases, the points on the curve move away from the origin as they circle the origin in a counterclockwise direction, generating a spiral (**Figure 12.24**).

**QUICK CHECK 4** Describe the polar curves $r = 12$, $r = 6\theta$, and $r \sin \theta = 10$. ◄

**EXAMPLE 3** **Polar to Cartesian coordinates** Convert the polar equation $r = 10 \cos \theta + 24 \sin \theta$ to Cartesian coordinates and describe the corresponding graph.

**SOLUTION** Multiplying both sides of the equation by $r$ produces the equation $r^2 = 10r \cos \theta + 24r \sin \theta$. Using the conversion relations $r^2 = x^2 + y^2$, $x = r \cos \theta$, and $y = r \sin \theta$, the equation

$$\underbrace{r^2}_{x^2+y^2} = \underbrace{10r \cos \theta}_{10x} + \underbrace{24r \sin \theta}_{24y}$$

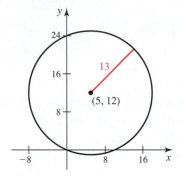

**Figure 12.25**

becomes $x^2 + y^2 - 10x - 24y = 0$. Completing the square in both variables gives the equation

$$\underbrace{x^2 - 10x + 25}_{(x-5)^2} - 25 + \underbrace{y^2 - 24y + 144}_{(y-12)^2} - 144 = 0, \text{ or } (x - 5)^2 + (y - 12)^2 = 169.$$

We recognize $(x - 5)^2 + (y - 12)^2 = 169$ as the equation of a circle of radius 13 centered at $(5, 12)$ (**Figure 12.25**).

*Related Exercises 41, 43* ◄

Calculations similar to those in Example 3 lead to the following equations of circles in polar coordinates.

---

**SUMMARY** **Circles in Polar Coordinates**

The equation $r = a$ describes a circle of radius $|a|$ centered at $(0, 0)$.

The equation $r = 2a \cos \theta + 2b \sin \theta$ describes a circle of radius $\sqrt{a^2 + b^2}$ centered at $(a, b)$.

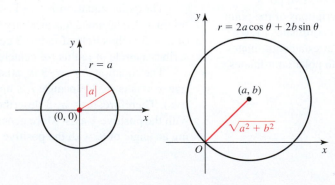

$r = 2a \cos \theta + 2b \sin \theta$

$r = a$

$(0, 0)$

$(a, b)$

$\sqrt{a^2 + b^2}$

$O$

---

▶ Two important cases deserve mention. The equation $r = 2a \cos \theta$ describes a circle of radius $|a|$ centered at $(a, 0)$. The equation $r = 2b \sin \theta$ describes a circle of radius $|b|$ centered at $(0, b)$.

**QUICK CHECK 5** Describe the graph of the equation $r = 2 \cos \theta + 4 \sin \theta$. ◄

▶ When a curve is described as $r = f(\theta)$, it is natural to tabulate points in $\theta$-$r$ format, just as we list points in $x$-$y$ format for $y = f(x)$. Despite this fact, the standard form for writing an ordered pair in polar coordinates is $(r, \theta)$.

## Graphing in Polar Coordinates

Equations in polar coordinates often describe curves that are difficult to represent in Cartesian coordinates. Partly for this reason, curve-sketching methods for polar coordinates differ from those used for curves in Cartesian coordinates. Conceptually, the easiest graphing method is to choose several values of $\theta$, calculate the corresponding $r$-values, and tabulate the coordinates. The points are then plotted and connected with a smooth curve.

**EXAMPLE 4** **Plotting a polar curve** Graph the polar equation $r = f(\theta) = 1 + \sin \theta$.

**SOLUTION** The domain of $f$ consists of all real values of $\theta$; however, the complete curve is generated by letting $\theta$ vary over any interval of length $2\pi$. Table 12.3 shows several $(r, \theta)$ pairs, which are plotted in **Figure 12.26**. The resulting curve, called a **cardioid**, is symmetric about the $y$-axis.

**Table 12.3**

| $\theta$ | $r = 1 + \sin \theta$ |
|---|---|
| 0 | 1 |
| $\pi/6$ | $3/2$ |
| $\pi/2$ | 2 |
| $5\pi/6$ | $3/2$ |
| $\pi$ | 1 |
| $7\pi/6$ | $1/2$ |
| $3\pi/2$ | 0 |
| $11\pi/6$ | $1/2$ |
| $2\pi$ | 1 |

Cardioid $r = 1 + \sin \theta$

**Figure 12.26**

*Related Exercises 53–54* ◀

**Cartesian-to-Polar Method** Plotting polar curves point by point is time-consuming, and important details may not be revealed. Here is an alternative procedure for graphing polar curves that is usually quicker and more reliable.

---

**PROCEDURE** **Cartesian-to-Polar Method for Graphing** $r = f(\theta)$

1. Graph $r = f(\theta)$ *as if r and $\theta$ were Cartesian coordinates* with $\theta$ on the horizontal axis and $r$ on the vertical axis. Be sure to choose an interval for $\theta$ on which the entire polar curve is produced.

2. Use the Cartesian graph that you created in Step 1 as a guide to sketch the points $(r, \theta)$ on the final *polar* curve.

---

**EXAMPLE 5** **Plotting polar graphs** Use the Cartesian-to-polar method to graph the polar equation $r = 1 + \sin \theta$ (Example 4).

**SOLUTION** Viewing $r$ and $\theta$ as Cartesian coordinates, the graph of $r = 1 + \sin \theta$ on the interval $[0, 2\pi]$ is a standard sine curve with amplitude 1 shifted up 1 unit (**Figure 12.27**). Notice that the graph begins with $r = 1$ at $\theta = 0$, increases to $r = 2$ at $\theta = \pi/2$, decreases to $r = 0$ at $\theta = 3\pi/2$ (which indicates an intersection with the origin on the polar graph), and increases to $r = 1$ at $\theta = 2\pi$. The second row of Figure 12.27 shows the final polar curve (a cardioid) as it is transferred from the Cartesian curve.

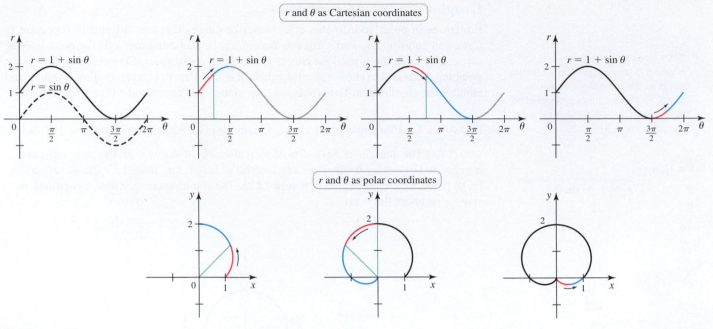

**Figure 12.27**

*Related Exercise 58* ◄

**Symmetry**    Given a polar equation in $r$ and $\theta$, three types of symmetry are easy to spot (**Figure 12.28**).

> ► Any two of these three symmetries implies the third. For example, if a graph is symmetric about both the $x$- and $y$-axes, then it is symmetric about the origin.

---

**SUMMARY    Symmetry in Polar Equations**

**Symmetry about the $x$-axis** occurs if the point $(r, \theta)$ is on the graph whenever $(r, -\theta)$ is on the graph.

**Symmetry about the $y$-axis** occurs if the point $(r, \theta)$ is on the graph whenever $(r, \pi - \theta) = (-r, -\theta)$ is on the graph.

**Symmetry about the origin** occurs if the point $(r, \theta)$ is on the graph whenever $(-r, \theta) = (r, \theta + \pi)$ is on the graph.

---

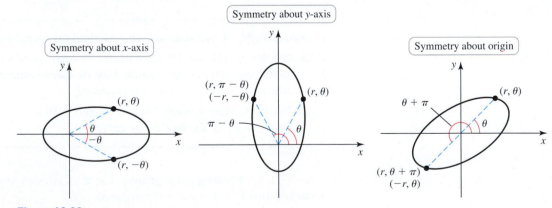

**Figure 12.28**

**QUICK CHECK 6** Identify the symmetry in the graph of (a) $r = 4 + 4 \cos \theta$ and (b) $r = 4 \sin \theta$. ◄

For instance, consider the polar equation $r = 1 + \sin \theta$ in Example 5. If $(r, \theta)$ satisfies the equation, then $(r, \pi - \theta)$ also satisfies the equation because $\sin \theta = \sin (\pi - \theta)$. Therefore, the graph is symmetric about the $y$-axis, as shown in Figure 12.28. Testing for symmetry produces a more accurate graph and often simplifies the task of graphing polar equations.

**EXAMPLE 6**  **Plotting polar graphs**  Graph the polar equation $r = 3 \sin 2\theta$.

**SOLUTION**  The Cartesian graph of $r = 3 \sin 2\theta$ on the interval $[0, 2\pi]$ has amplitude 3 and period $\pi$ (Figure 12.29a). The $\theta$-intercepts occur at $\theta = 0, \pi/2, \pi, 3\pi/2$, and $2\pi$, which correspond to the intersections with the origin on the polar graph. Furthermore, the arches of the Cartesian curve between $\theta$-intercepts correspond to loops in the polar curve. The resulting polar curve is a **four-leaf rose** (Figure 12.29b).

(a)          (b)

**Figure 12.29**

The graph is symmetric about the $x$-axis, the $y$-axis, and the origin. It is instructive to see how these symmetries are justified. To prove symmetry about the $y$-axis, notice that

$$(r, \theta) \text{ on the graph} \Rightarrow r = 3 \sin 2\theta$$
$$\Rightarrow r = -3 \sin 2(-\theta) \qquad \sin(-\theta) = -\sin\theta$$
$$\Rightarrow -r = 3 \sin 2(-\theta) \qquad \text{Simplify.}$$
$$\Rightarrow (-r, -\theta) \text{ on the graph.}$$

We see that if $(r, \theta)$ is on the graph, then $(-r, -\theta)$ is also on the graph, which implies symmetry about the $y$-axis. Similarly, to prove symmetry about the origin, notice that

$$(r, \theta) \text{ on the graph} \Rightarrow r = 3 \sin 2\theta$$
$$\Rightarrow r = 3 \sin(2\theta + 2\pi) \qquad \sin(\theta + 2\pi) = \sin\theta$$
$$\Rightarrow r = 3 \sin(2(\theta + \pi)) \qquad \text{Simplify.}$$
$$\Rightarrow (r, \theta + \pi) \text{ on the graph.}$$

We have shown that if $(r, \theta)$ is on the graph, then $(r, \theta + \pi)$ is also on the graph, which implies symmetry about the origin. Symmetry about the $y$-axis and the origin imply symmetry about the $x$-axis. Had we proved these symmetries in advance, we could have graphed the curve only in the first quadrant—reflections about the $x$- and $y$-axes would produce the full curve.

*Related Exercises 63–64* ◄

> **Subtle Point**
> The fact that one point has infinitely many representations in polar coordinates presents potential pitfalls. In Example 6, you can show that $(-r, \theta)$ does *not* satisfy the equation $r = 3 \sin 2\theta$ when $(r, \theta)$ satisfies the equation. And yet, as shown, the graph is symmetric about the origin because $(r, \theta + \pi)$ satisfies the equation whenever $(r, \theta)$ satisfies the equation. Note that $(-r, \theta)$ and $(r, \theta + \pi)$ are the same point.

**EXAMPLE 7**  **Plotting polar graphs**  Graph the polar equation $r^2 = 9 \cos\theta$. Use a graphing utility to check your work.

**SOLUTION**  The graph of this equation has symmetry about the origin (because of the $r^2$) and about the $x$-axis (because of $\cos\theta$). These two symmetries imply symmetry about the $y$-axis.

A preliminary step is required before using the Cartesian-to-polar method for graphing the curve. Solving the given equation for $r$, we find that $r = \pm 3\sqrt{\cos\theta}$. Notice that $\cos\theta < 0$, for $\pi/2 < \theta < 3\pi/2$, so the curve does not exist on that interval. Therefore, we plot the curve on the intervals $0 \le \theta \le \pi/2$ and $3\pi/2 \le \theta \le 2\pi$ (the interval $\left[-\frac{\pi}{2}, \frac{\pi}{2}\right]$ would also work). Both the positive and negative values of $r$ are included in the Cartesian graph (Figure 12.30a).

Now we are ready to transfer points from the Cartesian graph to the final polar graph (**Figure 12.30b**). The resulting curve is called a **lemniscate**.

Cartesian graph of $r = \pm 3\sqrt{\cos\theta}$, where $\cos\theta \geq 0$

$r = 3\sqrt{\cos\theta}$

$r = -3\sqrt{\cos\theta}$

Lemniscate $r^2 = 9\cos\theta$

(a)

(b)

**Figure 12.30**

*Related Exercise 61* ◄

**EXAMPLE 8    Matching polar and Cartesian graphs**  The butterfly curve is described by the equation

$$r = e^{\sin\theta} - 2\cos 4\theta, \quad \text{for } 0 \leq \theta \leq 2\pi,$$

which is plotted in Cartesian and polar coordinates in **Figure 12.31**. Follow the Cartesian graph through the points $A, B, C, \ldots, N, O$ and mark the corresponding points on the polar curve

▶ See Exercise 105 for a spectacular enhancement of the butterfly curve.

**SOLUTION**  Point $A$ in **Figure 12.31a** has the Cartesian coordinates $(\theta = 0, r = -1)$. The corresponding point in the polar plot (**Figure 12.31b**) with polar coordinates $(-1, 0)$ is marked $A$. Point $B$ in the Cartesian plot is on the $\theta$-axis; therefore, $r = 0$. The corresponding point in the polar plot is the origin. The same argument used to locate $B$ applies to $F, H, J, L,$ and $N$, all of which appear at the origin in the polar plot. In general, the local and endpoint maxima and minima in the Cartesian graph ($A, C, D, E, G, I, K, M,$ and $O$) correspond to the extreme points of the loops of the polar plot and are marked accordingly in Figure 12.31b.

Cartesian graph of $r = e^{\sin\theta} - 2\cos 4\theta$, for $0 \leq \theta \leq 2\pi$

(a)

$B, F, H, J, L,$ and $N$ are at the origin.

Polar graph of $r = e^{\sin\theta} - 2\cos 4\theta$, for $0 \leq \theta \leq 2\pi$

(b)

**Figure 12.31**

(*Source:* T. H. Fay, *Amer. Math. Monthly* 96, 1989, revived in Wagon and Packel, *Animating Calculus*, Freeman, 1994)

*Related Exercises 65–68* ◄

## Using Graphing Utilities

When graphing polar curves that eventually close on themselves, it is useful to specify an interval in $\theta$ that generates the entire curve. In some cases, this problem is a challenge in itself.

➤ **Using a parametric equation plotter to graph polar curves**
To graph $r = f(\theta)$, treat $\theta$ as a parameter and define the parametric equations

$$x = r\cos\theta = \underbrace{f(\theta)}_{r}\cos\theta$$

$$y = r\sin\theta = \underbrace{f(\theta)}_{r}\sin\theta$$

Then graph $(x(\theta), y(\theta))$ as a parametric curve with $\theta$ as the parameter.

➤ The prescription given in Example 9 for finding $P$ when working with functions of the form $f(\theta) = \sin\dfrac{p\theta}{q}$ or $f(\theta) = \cos\dfrac{p\theta}{q}$ ensures that the complete curve is generated. Smaller values of $P$ work in some cases.

**EXAMPLE 9**  **Plotting complete curves**  Consider the closed curve described by $r = \cos(2\theta/5)$. Give an interval in $\theta$ that generates the entire curve and then graph the curve.

**SOLUTION**  Recall that $\cos\theta$ has a period of $2\pi$. Therefore, $\cos(2\theta/5)$ completes one cycle when $2\theta/5$ varies from 0 to $2\pi$, or when $\theta$ varies from 0 to $5\pi$. Therefore, it is tempting to conclude that the complete curve $r = \cos(2\theta/5)$ is generated as $\theta$ varies from 0 to $5\pi$. But you can check that the point corresponding to $\theta = 0$ is *not* the point corresponding to $\theta = 5\pi$, which means the curve does not close on itself over the interval $[0, 5\pi]$ (**Figure 12.32a**).

To graph the *complete* curve $r = \cos(2\theta/5)$, we must find an interval $[0, P]$, where $P$ is an integer multiple of $5\pi$ (so that $f(0) = f(P)$) and an integer multiple of $2\pi$ (so that the points $(0, f(0))$ and $(P, f(P))$ are the same). The smallest number satisfying these conditions is $10\pi$. Graphing $r = \cos(2\theta/5)$ over the interval $[0, 10\pi]$ produces the complete curve (**Figure 12.32b**).

**Figure 12.32**

*Related Exercises 72–73* ◄

---

# SECTION 12.2 EXERCISES

## Getting Started

1. Plot the points with polar coordinates $(2, \pi/6)$ and $(-3, -\pi/2)$. Give two alternative sets of coordinate pairs for both points.

2. Write the equations that are used to express a point with polar coordinates $(r, \theta)$ in Cartesian coordinates.

3. Write the equations that are used to express a point with Cartesian coordinates $(x, y)$ in polar coordinates.

4. What is the polar equation of a circle of radius $\sqrt{a^2 + b^2}$ centered at $(a, b)$?

5. What is the polar equation of the vertical line $x = 5$?

6. What is the polar equation of the horizontal line $y = 5$?

7. Explain three symmetries in polar graphs and how they are detected in equations.

8. Given three polar coordinate representations for the origin.

## Practice Exercises

*9–13. Graph the points with the following polar coordinates. Give two alternative representations of the points in polar coordinates.*

9. $\left(2, \dfrac{\pi}{4}\right)$

10. $\left(3, \dfrac{2\pi}{3}\right)$

11. $\left(-1, -\dfrac{\pi}{3}\right)$

12. $\left(2, \dfrac{7\pi}{4}\right)$

13. $\left(-4, \dfrac{3\pi}{2}\right)$

14. **Points in polar coordinates** Give two sets of polar coordinates for each of the points $A$–$F$ in the figure.

*15–22. Sets in polar coordinates Sketch the following sets of points.*

15. $r \geq 2$

16. $r = 3$

17. $\theta = \dfrac{2\pi}{3}$

18. $2 \leq r \leq 8$

**19.** $\frac{\pi}{2} \le \theta \le \frac{3\pi}{4}$

**20.** $1 < r < 2$ and $\frac{\pi}{6} \le \theta \le \frac{\pi}{3}$

**21.** $|\theta| \le \frac{\pi}{3}$

**22.** $0 < r < 3$ and $0 \le \theta \le \pi$

**23–24. Radar** *Airplanes are equipped with transponders that allow air traffic controllers to see their locations on radar screens. Radar gives the distance of the plane from the radar station (located at the origin) and the angular position of the plane, typically measured in degrees clockwise from north.*

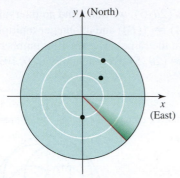

**23.** A plane is 100 miles from a radar station at an angle of 135° clockwise from north. Find polar coordinates for the location of the plane.

**24.** A plane is 50 miles from a radar station at an angle of 210° clockwise from north. Find polar coordinates for the location of the plane.

**25–30. Converting coordinates** *Express the following polar coordinates in Cartesian coordinates.*

**25.** $\left(3, \frac{\pi}{4}\right)$

**26.** $\left(1, \frac{2\pi}{3}\right)$

**27.** $\left(1, -\frac{\pi}{3}\right)$

**28.** $\left(2, \frac{7\pi}{4}\right)$

**29.** $\left(-4, \frac{3\pi}{4}\right)$

**30.** $(4, 5\pi)$

**31–36. Converting coordinates** *Express the following Cartesian coordinates in polar coordinates in at least two different ways.*

**31.** $(2, 2)$

**32.** $(-1, 0)$

**33.** $(1, \sqrt{3})$

**34.** $(0, -9)$

**35.** $(-4, 4\sqrt{3})$

**36.** $(-8\sqrt{3}, -8)$

**37–48. Polar-to-Cartesian coordinates** *Convert the following equations to Cartesian coordinates. Describe the resulting curve.*

**37.** $r \cos \theta = -4$

**38.** $r = \cot \theta \csc \theta$

**39.** $r = 2$

**40.** $r = 3 \csc \theta$

**41.** $r = 2 \sin \theta + 2 \cos \theta$

**42.** $r = -2 \cos \theta$

**43.** $r = 6 \cos \theta + 8 \sin \theta$

**44.** $\sin \theta = |\cos \theta|$

**45.** $r \cos \theta = \sin 2\theta$

**46.** $r = \sin \theta \sec^2 \theta$

**47.** $r = 8 \sin \theta$

**48.** $r = \dfrac{1}{2 \cos \theta + 3 \sin \theta}$

**49–52. Cartesian-to-polar coordinates** *Convert the following equations to polar coordinates.*

**49.** $y = x^2$

**50.** $y = 3$

**51.** $y = \dfrac{1}{x}$

**52.** $(x - 1)^2 + y^2 = 1$

**53–56. Simple curves** *Tabulate and plot enough points to sketch a graph of the following equations.*

**53.** $r = 8 \cos \theta$

**54.** $r = 4 + 4 \cos \theta$

**55.** $r(\sin \theta - 2 \cos \theta) = 0$

**56.** $r = 1 - \cos \theta$

**57–64. Graphing polar curves** *Graph the following equations. Use a graphing utility to check your work and produce a final graph.*

**57.** $r = 2 \sin \theta - 1$

**58.** $r = 2 - 2 \sin \theta$

**59.** $r = \sin^2 (\theta/2)$

**60.** $r^2 = 4 \sin \theta$

**61.** $r^2 = 16 \cos \theta$

**62.** $r^2 = 16 \sin 2\theta$

**63.** $r = \sin 3\theta$

**64.** $r = 2 \sin 5\theta$

**65–68. Matching polar and Cartesian curves** *Cartesian and polar graphs of $r = f(\theta)$ are given in the figures. Mark the points on the polar graph that correspond to the lettered points on the Cartesian graph.*

**65.** $r = 1 - 2 \sin 3\theta$

**66.** $r = \sin (1 + 3 \cos \theta)$

**67.** $r = \dfrac{1}{4} - \cos 4\theta$

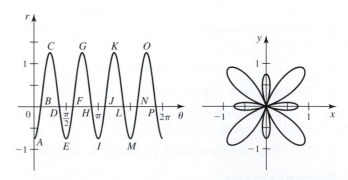

**68.** $r = \cos \theta + \sin 2\theta$

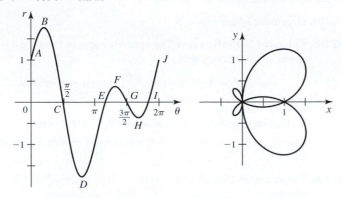

**⊤ 69–76. Using a graphing utility** *Use a graphing utility to graph the following equations. In each case, give the smallest interval $[0, P]$ that generates the entire curve.*

**69.** $r = \sin \dfrac{\theta}{4}$

**70.** $r = 2 - 4 \cos 5\theta$

**71.** $r = \cos 3\theta + \cos^2 2\theta$

**72.** $r = 2 \sin \dfrac{2\theta}{3}$

**73.** $r = \cos \dfrac{3\theta}{5}$

**74.** $r = \sin \dfrac{3\theta}{7}$

**75.** $r = 1 - 3 \cos 2\theta$

**76.** $r = 1 - 2 \sin 5\theta$

**77. Explain why or why not** Determine whether the following statements are true and give an explanation or counterexample.

    **a.** The point with Cartesian coordinates $(-2, 2)$ has polar coordinates $(2\sqrt{2}, 3\pi/4)$, $(2\sqrt{2}, 11\pi/4)$, $(2\sqrt{2}, -5\pi/4)$, and $(-2\sqrt{2}, -\pi/4)$.

    **b.** The graphs of $r \cos \theta = 4$ and $r \sin \theta = -2$ intersect exactly once.

    **c.** The graphs of $r = 2$ and $\theta = \pi/4$ intersect exactly once.

    **d.** The point $(3, \pi/2)$ lies on the graph of $r = 3 \cos 2\theta$.

    **e.** The graphs of $r = 2 \sec \theta$ and $r = 3 \csc \theta$ are lines.

**78. Circles in general** Show that the polar equation

$$r^2 - 2r(a \cos \theta + b \sin \theta) = R^2 - a^2 - b^2$$

describes a circle of radius $R$ centered at $(a, b)$.

**79. Circles in general** Show that the polar equation

$$r^2 - 2rr_0 \cos(\theta - \theta_0) = R^2 - r_0^2$$

describes a circle of radius $R$ whose center has polar coordinates $(r_0, \theta_0)$.

**80–83. Equations of circles** *Use the results of Exercises 78–79 to describe and graph the following circles.*

**80.** $r^2 - 6r \cos \theta = 16$

**81.** $r^2 - 4r \cos(\theta - \pi/3) = 12$

**82.** $r^2 - 8r \cos(\theta - \pi/2) = 9$

**83.** $r^2 - 2r(2 \cos \theta + 3 \sin \theta) = 3$

## Explorations and Challenges

**84. Equations of circles** Find equations of the circles in the figure. Determine whether the combined area of the circles is greater than

or less than the area of the region inside the square but outside the circles.

**85. Navigating** A plane is 150 miles north of a radar station, and 30 minutes later it is 60° east of north at a distance of 100 miles from the radar station. Assume the plane flies on a straight line and maintains constant altitude during this 30-minute period.

    **a.** Find the distance traveled during this 30-minute period.

    **b.** Determine the average velocity of the plane (relative to the ground) during this 30-minute period.

**86. Lines in polar coordinates**

    **a.** Show that an equation of the line $y = mx + b$ in polar coordinates is $r = \dfrac{b}{\sin \theta - m \cos \theta}$.

    **b.** Use the figure to find an alternative polar equation of a line, $r \cos(\theta_0 - \theta) = r_0$. Note that $Q(r_0, \theta_0)$ is a fixed point on the line such that $OQ$ is perpendicular to the line and $r_0 \geq 0$; $P(r, \theta)$ is an arbitrary point on the line.

**87–90. Equations of lines** *Use the result of Exercise 86 to describe and graph the following lines.*

**87.** $r \cos\left(\dfrac{\pi}{3} - \theta\right) = 3$

**88.** $r \cos\left(\theta + \dfrac{\pi}{6}\right) = 4$

**89.** $r(\sin \theta - 4 \cos \theta) - 3 = 0$

**90.** $r(4 \sin \theta - 3 \cos \theta) = 6$

**91. The limaçon family** The equations $r = a + b \cos \theta$ and $r = a + b \sin \theta$ describe curves known as *limaçons* (from Latin for *snail*). We have already encountered cardioids, which occur when $|a| = |b|$. The limaçon has an inner loop if $|a| < |b|$. The limaçon has a dent or dimple if $|b| < |a| < 2|b|$. And the limaçon is oval-shaped if $|a| > 2|b|$. Match equations a–f with the limaçons in figures A–F.

a. $r = -1 + \sin\theta$       b. $r = -1 + 2\cos\theta$

c. $r = 2 + \sin\theta$         d. $r = 1 - 2\cos\theta$

e. $r = 1 + 2\sin\theta$       f. $r = 1 + 2/3\sin\theta$

(A)                                (B)

(C)                                (D)

(E)                                (F)

**92. Limiting limaçon** Consider the family of limaçons $r = 1 + b\cos\theta$. Describe how the curves change as $b \to \infty$.

**93–96. The lemniscate family** *Equations of the form $r^2 = a\sin 2\theta$ and $r^2 = a\cos 2\theta$ describe lemniscates (see Example 7). Graph the following lemniscates.*

**93.** $r^2 = \cos 2\theta$            **94.** $r^2 = 4\sin 2\theta$

**95.** $r^2 = -2\sin 2\theta$         **96.** $r^2 = -8\cos 2\theta$

**97–100. The rose family** *Equations of the form $r = a\sin m\theta$ or $r = a\cos m\theta$, where $a$ is a real number and $m$ is a positive integer, have graphs known as roses (see Example 6). Graph the following roses.*

**97.** $r = \sin 2\theta$            **98.** $r = 4\cos 3\theta$

**99.** $r = 2\sin 4\theta$           **100.** $r = 6\sin 5\theta$

**101. Number of rose petals** Show that the graph of $r = a\sin m\theta$ or $r = a\cos m\theta$ is a rose with $m$ leaves if $m$ is an odd integer and a rose with $2m$ leaves if $m$ is an even integer.

**102–104. Spirals** *Graph the following spirals. Indicate the direction in which the spiral is generated as $\theta$ increases, where $\theta > 0$. Let $a = 1$ and $a = -1$.*

**102.** Spiral of Archimedes: $r = a\theta$

**103.** Logarithmic spiral: $r = e^{a\theta}$

**104.** Hyperbolic spiral: $r = a/\theta$

**T 105. Enhanced butterfly curve** The butterfly curve of Example 8 is enhanced by adding a term:

$$r = e^{\sin\theta} - 2\cos 4\theta + \sin^5\frac{\theta}{12}, \quad \text{for } 0 \le \theta \le 24\pi.$$

a. Graph the curve.

b. Explain why the new term produces the observed effect.

(*Source:* S. Wagon and E. Packel, *Animating Calculus*, Freeman, 1994)

**T 106. Finger curves** Consider the curve $r = f(\theta) = \cos a^\theta - 1.5$, where $a = (1 + 12\pi)^{1/(2\pi)} \approx 1.78933$ (see figure).

a. Show that $f(0) = f(2\pi)$ and find the point on the curve that corresponds to $\theta = 0$ and $\theta = 2\pi$.

b. Is the same curve produced over the intervals $[-\pi, \pi]$ and $[0, 2\pi]$?

c. Let $f(\theta) = \cos(a^\theta) - b$, where $a = (1 + 2k\pi)^{1/(2\pi)}$, $k$ is an integer, and $b$ is a real number. Show that $f(0) = f(2\pi)$ and the curve closes on itself.

d. Plot the curve with various values of $k$. How many fingers can you produce?

**T 107. Earth-Mars system** A simplified model assumes the orbits of Earth and Mars around the Sun are circular with radii of 2 and 3, respectively, and that Earth completes one orbit in one year while Mars takes two years. When $t = 0$, Earth is at $(2, 0)$ and Mars is at $(3, 0)$; both orbit the Sun (at $(0, 0)$) in the counterclockwise direction. The position of Mars relative to Earth is given by the parametric equations

$$x = (3 - 4\cos\pi t)\cos\pi t + 2, \quad y = (3 - 4\cos\pi t)\sin\pi t.$$

a. Graph the parametric equations, for $0 \le t \le 2$.

b. Letting $r = (3 - 4\cos\pi t)$, explain why the path of Mars relative to Earth is a limaçon (Exercise 91).

**108. Channel flow** Water flows in a shallow semicircular channel with inner and outer radii of 1 m and 2 m (see figure). At a point $P(r, \theta)$ in the channel, the flow is in the tangential direction (counterclockwise along circles), and it depends only on $r$, the distance from the center of the semicircles.

a. Express the region formed by the channel as a set in polar coordinates.

b. Express the inflow and outflow regions of the channel as sets in polar coordinates.

c. Suppose the tangential velocity of the water in meters per second is given by $v(r) = 10r$, for $1 \le r \le 2$. Is the velocity greater at $\left(1.5, \dfrac{\pi}{4}\right)$ or $\left(1.2, \dfrac{3\pi}{4}\right)$? Explain.

**d.** Suppose the tangential velocity of the water is given by $v(r) = 20/r$, for $1 \le r \le 2$. Is the velocity greater at $(1.8, \pi/6)$ or $(1.3, 2\pi/3)$? Explain.

**e.** The total amount of water that flows through the channel (across a cross section of the channel $\theta = \theta_0$) is proportional to $\int_1^2 v(r) \, dr$. Is the total flow through the channel greater for the flow in part (c) or part (d)?

Outflow          Inflow

**109. Subtle symmetry** Without using a graphing utility, determine the symmetries (if any) of the curve $r = 4 - \sin(\theta/2)$.

**T 110. Complete curves** Consider the polar curve $r = \cos(n\theta/m)$, where $n$ and $m$ are integers.

   **a.** Graph the complete curve when $n = 2$ and $m = 3$.

   **b.** Graph the complete curve when $n = 3$ and $m = 7$.

   **c.** Find a general rule in terms of $m$ and $n$ (where $m$ and $n$ have no common factors) for determining the least positive number $P$ such that the complete curve is generated over the interval $[0, P]$.

**111. Cartesian lemniscate** Find the equation in Cartesian coordinates of the lemniscate $r^2 = a^2 \cos 2\theta$, where $a$ is a real number.

**QUICK CHECK ANSWERS**

**1.** All the points are the same except $(3, 3\pi/2)$. **3.** Polar coordinates: $(1, 0)$, $(1, 2\pi)$; Cartesian coordinates: $(0, 2)$ **4.** A circle centered at the origin with radius 12; two spirals; the horizontal line $y = 10$ **5.** A circle of radius $\sqrt{5}$ with center $(1, 2)$ **6.** (a) Symmetric about the $x$-axis; (b) symmetric about the $y$-axis ◄

# 12.3 Calculus in Polar Coordinates

Having learned about the *geometry* of polar coordinates, we now have the tools needed to explore *calculus* in polar coordinates. Familiar topics, such as slopes of tangent lines, areas bounded by curves, and arc length, are now revisited in a different setting.

## Slopes of Tangent Lines

Given a function $y = f(x)$, the slope of the line tangent to the graph at a given point is $\frac{dy}{dx}$ or $f'(x)$. So it is tempting to conclude that the slope of a curve described by the polar equation $r = f(\theta)$ is $\frac{dr}{d\theta} = f'(\theta)$. Unfortunately, it's not that simple.

The key observation is that the slope of a tangent line—in any coordinate system—is the rate of change of the vertical coordinate $y$ with respect to the horizontal coordinate $x$, which is $\frac{dy}{dx}$. We begin by writing the polar equation $r = f(\theta)$ in parametric form with $\theta$ as a parameter:

> The slope is the change in the vertical coordinate divided by the change in the horizontal coordinate, independent of the coordinate system. In polar coordinates, neither $r$ nor $\theta$ corresponds to a vertical or horizontal coordinate.

$$x = r \cos \theta = f(\theta) \cos \theta \quad \text{and} \quad y = r \sin \theta = f(\theta) \sin \theta. \tag{1}$$

From Section 12.1, when $x$ and $y$ are defined parametrically as differentiable functions of $\theta$, the derivative is $\frac{dy}{dx} = \frac{dy/d\theta}{dx/d\theta}$. Using the Product Rule to compute $\frac{dy}{d\theta}$ and $\frac{dx}{d\theta}$ in equation (1), we have

$$\frac{dy}{dx} = \frac{\overbrace{f'(\theta) \sin \theta + f(\theta) \cos \theta}^{dy/d\theta}}{\underbrace{f'(\theta) \cos \theta - f(\theta) \sin \theta}_{dx/d\theta}}. \tag{2}$$

If the graph passes through the origin for some angle $\theta_0$, then $f(\theta_0) = 0$, and equation (2) simplifies to

$$\frac{dy}{dx} = \frac{\sin \theta_0}{\cos \theta_0} = \tan \theta_0,$$

provided $f'(\theta_0) \neq 0$. Assuming $\cos \theta_0 \neq 0$, $\tan \theta_0$ is the slope of the line $\theta = \theta_0$, which also passes through the origin. In this case, we conclude that if $f(\theta_0) = 0$, then the tangent

**Figure 12.33**

**QUICK CHECK 1** Verify that if $y = f(\theta) \sin \theta$, then $y'(\theta) = f'(\theta) \sin \theta + f(\theta) \cos \theta$ (which was used earlier to find $dy/dx$). ◀

**Figure 12.34**

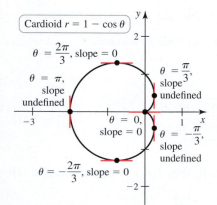

**Figure 12.35**

line at $(0, \theta_0)$ is simply $\theta = \theta_0$ (**Figure 12.33**). If $f(\theta_0) = 0$, $f'(\theta_0) \neq 0$, and $\cos \theta_0 = 0$, then the graph has a vertical tangent line at the origin.

---

**THEOREM 12.2    Slope of a Tangent Line**

Let $f$ be a differentiable function at $\theta_0$. The slope of the line tangent to the curve $r = f(\theta)$ at the point $(f(\theta_0), \theta_0)$ is

$$\frac{dy}{dx} = \frac{f'(\theta_0) \sin \theta_0 + f(\theta_0) \cos \theta_0}{f'(\theta_0) \cos \theta_0 - f(\theta_0) \sin \theta_0},$$

provided the denominator is nonzero at the point. At angles $\theta_0$ for which $f(\theta_0) = 0$, $f'(\theta_0) \neq 0$, and $\cos \theta_0 \neq 0$, the tangent line is $\theta = \theta_0$ with slope $\tan \theta_0$.

---

**EXAMPLE 1    Slopes on a circle**  Find the slopes of the lines tangent to the circle $r = f(\theta) = 10$.

**SOLUTION**  In this case, $f(\theta)$ is constant (independent of $\theta$). Therefore, $f'(\theta) = 0$, $f(\theta) \neq 0$, and the slope formula becomes

$$\frac{dy}{dx} = \frac{f'(\theta) \sin \theta + f(\theta) \cos \theta}{f'(\theta) \cos \theta - f(\theta) \sin \theta} = -\frac{\cos \theta}{\sin \theta} = -\cot \theta.$$

We can check a few points to see that this result makes sense. With $\theta = 0$ and $\theta = \pi$, the slope $\dfrac{dy}{dx} = -\cot \theta$ is undefined, which implies the tangent lines are vertical at these points (**Figure 12.34**). With $\theta = \pi/2$ and $\theta = 3\pi/2$, the slope is zero; with $\theta = 3\pi/4$ and $\theta = 7\pi/4$, the slope is 1; and with $\theta = \pi/4$ and $\theta = 5\pi/4$, the slope is $-1$. At all points $P(r, \theta)$ on the circle, the slope of the line $OP$ from the origin to $P$ is $\tan \theta$, which is the negative reciprocal of $-\cot \theta$. Therefore, $OP$ is perpendicular to the tangent line at all points $P$ on the circle.

*Related Exercises 12–13* ◀

**EXAMPLE 2    Vertical and horizontal tangent lines**  Find the points on the interval $-\pi \leq \theta \leq \pi$ at which the cardioid $r = f(\theta) = 1 - \cos \theta$ has a vertical or horizontal tangent line.

**SOLUTION**  Applying Theorem 12.2, we find that

$$\frac{dy}{dx} = \frac{f'(\theta) \sin \theta + f(\theta) \cos \theta}{f'(\theta) \cos \theta - f(\theta) \sin \theta} \qquad \color{blue}{f'(\theta) = \sin \theta}$$

$$= \frac{\overbrace{\sin \theta \sin \theta}^{\sin^2 \theta = 1 - \cos^2 \theta} + (1 - \cos \theta) \cos \theta}{\sin \theta \cos \theta - (1 - \cos \theta) \sin \theta} \qquad \color{blue}{\text{Substitute for } f(\theta) \text{ and } f'(\theta).}$$

$$= -\frac{(2 \cos^2 \theta - \cos \theta - 1)}{\underbrace{\sin \theta (2 \cos \theta - 1)}_{\sin \theta (2 \cos \theta - 1)}} \qquad \color{blue}{\text{Simplify.}}$$

$$= -\frac{(2 \cos \theta + 1)(\cos \theta - 1)}{\sin \theta (2 \cos \theta - 1)}. \qquad \color{blue}{\text{Factor the numerator.}}$$

The points with a horizontal tangent line satisfy $\dfrac{dy}{dx} = 0$ and occur where the numerator is zero and the denominator is nonzero. The numerator is zero when $\theta = 0$ and $\pm 2\pi/3$. Because the denominator is *not* zero when $\theta = \pm 2\pi/3$, horizontal tangent lines occur at $\theta = \pm 2\pi/3$ (**Figure 12.35**).

Vertical tangent lines occur where the numerator of $\dfrac{dy}{dx}$ is nonzero and the denominator is zero. The denominator is zero when $\theta = 0$, $\pm \pi$, and $\pm \pi/3$, and the numerator is not zero when $\theta = \pm \pi$ and $\pm \pi/3$. Therefore, vertical tangent lines occur at $\theta = \pm \pi$ and $\pm \pi/3$.

The point $(0, 0)$ on the curve must be handled carefully because both the numerator and denominator of $\dfrac{dy}{dx}$ equal 0 at $\theta = 0$. Notice that with $f(\theta) = 1 - \cos\theta$, we have $f(0) = f'(0) = 0$. Therefore, $\dfrac{dy}{dx}$ may be computed as a limit using l'Hôpital's Rule. As $\theta \to 0^+$, we find that

$$
\begin{aligned}
\frac{dy}{dx} &= \lim_{\theta \to 0^+} \left( -\frac{(2\cos\theta + 1)(\cos\theta - 1)}{\sin\theta\,(2\cos\theta - 1)} \right) \\
&= \lim_{\theta \to 0^+} \frac{4\cos\theta\sin\theta - \sin\theta}{-2\sin^2\theta + 2\cos^2\theta - \cos\theta} \qquad \text{L'Hôpital's Rule} \\
&= \frac{0}{1} = 0. \qquad\qquad\qquad\qquad \text{Evaluate the limit.}
\end{aligned}
$$

**QUICK CHECK 2** What is the slope of the line tangent to the cardioid in Example 2 at the point corresponding to $\theta = \pi/4$? ◄

A similar calculation using l'Hôpital's Rule shows that as $\theta \to 0^-$, $\dfrac{dy}{dx} \to 0$. Therefore, the curve has a slope of 0 at $(0, 0)$.

*Related Exercises 25–26* ◄

## Area of Regions Bounded by Polar Curves

The problem of finding the area of a region bounded by polar curves brings us back to the slice-and-sum strategy used extensively in Chapters 5 and 6. The objective is to find the area of the region $R$ bounded by the graph of $r = f(\theta)$ between the two rays $\theta = \alpha$ and $\theta = \beta$ (**Figure 12.36a**). We assume $f$ is continuous and nonnegative on $[\alpha, \beta]$.

The area of $R$ is found by slicing the region in the radial direction, creating wedge-shaped slices. The interval $[\alpha, \beta]$ is partitioned into $n$ subintervals by choosing the grid points

$$\alpha = \theta_0 < \theta_1 < \theta_2 < \cdots < \theta_k < \cdots < \theta_n = \beta.$$

Area of circle $= \pi r^2$
Area of $\Delta\theta/(2\pi)$ of a circle

$$= \left(\frac{\Delta\theta}{2\pi}\right)\pi r^2 = \frac{1}{2} r^2 \Delta\theta$$

We let $\Delta\theta_k = \theta_k - \theta_{k-1}$, for $k = 1, 2, \dots, n$, and we let $\theta_k^*$ be any point of the interval $[\theta_{k-1}, \theta_k]$. The $k$th slice is approximated by the sector of a circle swept out by an angle $\Delta\theta_k$ with radius $f(\theta_k^*)$ (**Figure 12.36b**). Therefore, the area of the $k$th slice is approximately $\frac{1}{2} f(\theta_k^*)^2 \Delta\theta_k$, for $k = 1, 2, \dots, n$ (**Figure 12.36c**). To find the approximate area of $R$, we sum the areas of these slices:

$$\text{area} \approx \sum_{k=1}^{n} \frac{1}{2} f(\theta_k^*)^2\, \Delta\theta_k.$$

(a)

(b)

(c)

**Figure 12.36**

$$\text{Area} = \int_\alpha^\beta \frac{1}{2}(f(\theta)^2 - g(\theta)^2)\,d\theta$$

**Figure 12.37**

➤ If $R$ is bounded by the graph of $r = f(\theta)$ between $\theta = \alpha$ and $\theta = \beta$, then $g(\theta) = 0$ and the area of $R$ is
$$\int_\alpha^\beta \frac{1}{2} f(\theta)^2\,d\theta.$$

➤ Though we assume $r = f(\theta) \geq 0$ when deriving the formula for the area of a region bounded by a polar curve, the formula is valid when $r < 0$ (see, for example, Exercise 54).

➤ The equation $r = 2\cos 2\theta$ is unchanged when $\theta$ is replaced with $-\theta$ (symmetry about the $x$-axis) and when $\theta$ is replaced with $\pi - \theta$ (symmetry about the $y$-axis).

**QUICK CHECK 3** Give an interval over which you could integrate to find the area of one leaf of the rose $r = 2\sin 3\theta$. ◄

This approximation is a Riemann sum, and the approximation improves as we take more sectors ($n \to \infty$) and let $\Delta\theta_k \to 0$, for all $k$. The exact area is given by $\lim_{n\to\infty} \sum_{k=1}^n \frac{1}{2} f(\theta_k^*)^2 \Delta\theta_k$, which we identify as the definite integral $\int_\alpha^\beta \frac{1}{2} f(\theta)^2\,d\theta$.

With a slight modification, a more general result is obtained for the area of a region $R$ bounded by two curves, $r = f(\theta)$ and $r = g(\theta)$, between the rays $\theta = \alpha$ and $\theta = \beta$ (**Figure 12.37**). We assume $f$ and $g$ are continuous and $f(\theta) \geq g(\theta) \geq 0$ on $[\alpha, \beta]$. To find the area of $R$, we subtract the area of the region bounded by $r = g(\theta)$ from the area of the entire region bounded by $r = f(\theta)$ (all between $\theta = \alpha$ and $\theta = \beta$); that is,

$$\text{area} = \int_\alpha^\beta \frac{1}{2} f(\theta)^2\,d\theta - \int_\alpha^\beta \frac{1}{2} g(\theta)^2\,d\theta = \int_\alpha^\beta \frac{1}{2}(f(\theta)^2 - g(\theta)^2)\,d\theta.$$

---

**DEFINITION   Area of Regions in Polar Coordinates**

Let $R$ be the region bounded by the graphs of $r = f(\theta)$ and $r = g(\theta)$, between $\theta = \alpha$ and $\theta = \beta$, where $f$ and $g$ are continuous and $f(\theta) \geq g(\theta) \geq 0$ on $[\alpha, \beta]$. The area of $R$ is

$$\int_\alpha^\beta \frac{1}{2}(f(\theta)^2 - g(\theta)^2)\,d\theta.$$

---

**EXAMPLE 3   Area of a polar region**   Find the area of the four-leaf rose $r = f(\theta) = 2\cos 2\theta$.

**SOLUTION**   The graph of the rose (**Figure 12.38**) *appears* to be symmetric about the $x$- and $y$-axes; in fact, these symmetries can be proved. Appealing to this symmetry, we find the area of one-half of a leaf and then multiply the result by 8 to obtain the area of the full rose. The upper half of the rightmost leaf is generated as $\theta$ increases from $\theta = 0$ (when $r = 2$) to $\theta = \pi/4$ (when $r = 0$). Therefore, the area of the entire rose is

$$8\int_0^{\pi/4} \frac{1}{2} f(\theta)^2\,d\theta = 4\int_0^{\pi/4} (2\cos 2\theta)^2\,d\theta \qquad \textcolor{blue}{f(\theta) = 2\cos 2\theta}$$

$$= 16\int_0^{\pi/4} \cos^2 2\theta\,d\theta \qquad \textcolor{blue}{\text{Simplify.}}$$

$$= 16\int_0^{\pi/4} \frac{1 + \cos 4\theta}{2}\,d\theta \qquad \textcolor{blue}{\text{Half-angle formula}}$$

$$= (8\theta + 2\sin 4\theta)\Big|_0^{\pi/4} \qquad \textcolor{blue}{\text{Fundamental Theorem}}$$

$$= (2\pi + 0) - (0 + 0) = 2\pi. \qquad \textcolor{blue}{\text{Simplify.}}$$

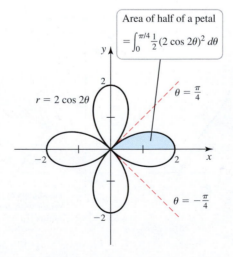

Area of half of a petal
$$= \int_0^{\pi/4} \frac{1}{2}(2\cos 2\theta)^2\,d\theta$$

$r = 2\cos 2\theta$

$\theta = \frac{\pi}{4}$

$\theta = -\frac{\pi}{4}$

**Figure 12.38**

*Related Exercises 38–39* ◄

Crescent-shaped region in Quadrant II is bounded by cardioid from $\theta = \frac{\pi}{2}$ to $\theta = \pi$.

Semicircle has area $= \dfrac{\pi(1)^2}{2} = \dfrac{\pi}{2}$.

**Figure 12.39**

**Figure 12.40**

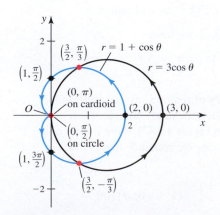

**Figure 12.41**

**EXAMPLE 4** **Areas of polar regions** Consider the circle $r = 1$ and the cardioid $r = 1 + \cos\theta$ (**Figure 12.39**).

**a.** Find the area of the region inside the circle and inside the cardioid.

**b.** Find the area of the region inside the circle and outside the cardioid.

**SOLUTION**

**a.** The points of intersection of the two curves can be found by solving $1 + \cos\theta = 1$, or $\cos\theta = 0$. The solutions are $\theta = \pm\pi/2$. The region inside the circle and inside the cardioid consists of two subregions (Figure 12.39):

- a semicircle with radius 1 in the first and fourth quadrants bounded by the circle $r = 1$, and

- two crescent-shaped regions in the second and third quadrants bounded by the cardioid $r = 1 + \cos\theta$ and the $y$-axis.

The area of the semicircle is $\pi/2$. To find the area of the upper crescent-shaped region in the second quadrant, notice that it is bounded by $r = 1 + \cos\theta$, as $\theta$ varies from $\pi/2$ to $\pi$. Therefore, its area is

$$\int_{\pi/2}^{\pi} \frac{1}{2}(1 + \cos\theta)^2\, d\theta = \int_{\pi/2}^{\pi} \frac{1}{2}(1 + 2\cos\theta + \cos^2\theta)\, d\theta \quad \text{Expand.}$$

$$= \frac{1}{2}\int_{\pi/2}^{\pi}\left(1 + 2\cos\theta + \frac{1 + \cos 2\theta}{2}\right) d\theta \quad \text{Half-angle formula}$$

$$= \frac{1}{2}\left(\theta + 2\sin\theta + \frac{\theta}{2} + \frac{\sin 2\theta}{4}\right)\Big|_{\pi/2}^{\pi} \quad \text{Evaluate integral.}$$

$$= \frac{3\pi}{8} - 1. \quad \text{Simplify.}$$

The area of the entire region (two crescents and a semicircle) is

$$2\left(\frac{3\pi}{8} - 1\right) + \frac{\pi}{2} = \frac{5\pi}{4} - 2.$$

**b.** The region inside the circle and outside the cardioid is bounded by the outer curve $r = 1$ and the inner curve $r = 1 + \cos\theta$ on the interval $[\pi/2, 3\pi/2]$ (**Figure 12.40**). Using the symmetry about the $x$-axis, the area of the region is

$$2\int_{\pi/2}^{\pi} \frac{1}{2}(1^2 - (1 + \cos\theta)^2)\, d\theta = \int_{\pi/2}^{\pi}(-2\cos\theta - \cos^2\theta)\, d\theta \quad \text{Simplify integrand.}$$

$$= 2 - \frac{\pi}{4}. \quad \text{Evaluate integral.}$$

Note that the regions in parts (a) and (b) make up the interior of a circle of radius 1; indeed, their areas have a sum of $\pi$.

*Related Exercise 45* ◄

Part of the challenge in setting up area integrals in polar coordinates is finding the points of intersection of two polar curves. The following example shows some of the subtleties of this process.

**EXAMPLE 5** **Points of intersection** Find the points of intersection of the circle $r = 3\cos\theta$ and the cardioid $r = 1 + \cos\theta$ (**Figure 12.41**).

**SOLUTION** The fact that a point has multiple representations in polar coordinates may lead to subtle difficulties in finding intersection points. We first proceed algebraically. Equating the two expressions for $r$ and solving for $\theta$, we have

$$3\cos\theta = 1 + \cos\theta \quad \text{or} \quad \cos\theta = \frac{1}{2},$$

which has roots $\theta = \pm \pi/3$. Therefore, two intersection points are $(3/2, \pi/3)$ and $(3/2, -\pi/3)$ (Figure 12.41). Without examining graphs of the curves, we might be tempted to stop here. Yet the figure shows another intersection point at the origin $O$ that has not been detected. To find this intersection point, we must investigate the way in which the two curves are generated. As $\theta$ increases from 0 to $2\pi$, the cardioid is generated counterclockwise, beginning at $(2, 0)$. The cardioid passes through $O$ when $\theta = \pi$. As $\theta$ increases from 0 to $\pi$, the circle is generated counterclockwise, beginning at $(3, 0)$. The circle passes through $O$ when $\theta = \pi/2$. Therefore, the intersection point $O$ is $(0, \pi)$ on the cardioid (and these coordinates do not satisfy the equation of the circle), while $O$ is $(0, \pi/2)$ on the circle (and these coordinates do not satisfy the equation of the cardioid). There is no foolproof rule for detecting such "hidden" intersection points. Care must be used.

*Related Exercise 29* ◄

Radial lines enter region A at $r = 1 + \cos \theta$ and exit at $r = 3 \cos \theta$.

**Figure 12.42**

➤ One way to verify that the inner and outer boundaries of a region have been correctly identified is to draw a ray from the origin through the region—the ray should enter the region at the inner boundary and exit the region at the outer boundary. In Example 6a, this is the case for every ray through region A, for $0 \le \theta \le \pi/3$.

**EXAMPLE 6**   **Computing areas** Example 5 discussed the points of intersection of the curves $r = 3 \cos \theta$ (a circle) and $r = 1 + \cos \theta$ (a cardioid). Use those results to compute the areas of the following non-overlapping regions in **Figure 12.42**.

**a.** region A          **b.** region B          **c.** region C

**SOLUTION**

**a.** It is evident that region A is bounded on the inside by the cardioid and on the outside by the circle between the points $Q(\theta = 0)$ and $P(\theta = \pi/3)$. Therefore, the area of region A is

$$\frac{1}{2}\int_0^{\pi/3} ((3 \cos \theta)^2 - (1 + \cos \theta)^2)\, d\theta$$

$$= \frac{1}{2}\int_0^{\pi/3} (8 \cos^2 \theta - 1 - 2 \cos \theta)\, d\theta \qquad \text{Simplify.}$$

$$= \frac{1}{2}\int_0^{\pi/3} (3 + 4 \cos 2\theta - 2 \cos \theta)\, d\theta \qquad \cos^2 \theta = \frac{1 + \cos 2\theta}{2}$$

$$= \frac{1}{2}(3\theta + 2 \sin 2\theta - 2 \sin \theta)\Big|_0^{\pi/3} = \frac{\pi}{2}. \qquad \text{Evaluate integral.}$$

**b.** Examining region B, notice that a ray drawn from the origin enters the region immediately. There is no inner boundary, and the outer boundary is $r = 1 + \cos \theta$ on $0 \le \theta \le \pi/3$ and $r = 3 \cos \theta$ on $\pi/3 \le \theta \le \pi/2$ (recall from Example 5 that $\theta = \pi/2$ is the angle at which the circle intersects the origin). Therefore, we slice the region into two parts at $\theta = \pi/3$ and write two integrals for its area:

$$\text{area of region B} = \frac{1}{2}\int_0^{\pi/3} (1 + \cos \theta)^2\, d\theta + \frac{1}{2}\int_{\pi/3}^{\pi/2} (3 \cos \theta)^2\, d\theta.$$

While these integrals may be evaluated directly, it's easier to notice that

$$\text{area of region B} = \text{area of semicircle } OPQ - \text{area of region A}.$$

Because $r = 3 \cos \theta$ is a circle with a radius of $3/2$, we have

$$\text{area of region B} = \frac{1}{2}\cdot \pi \left(\frac{3}{2}\right)^2 - \frac{\pi}{2} = \frac{5\pi}{8}.$$

**c.** It's easy to *incorrectly* identify the inner boundary of region C as the circle and the outer boundary as the cardioid. While these identifications are true when $\pi/3 \le \theta \le \pi/2$ (notice again the radial lines in Figure 12.42), there is only one boundary curve (the cardioid) when $\pi/2 \le \theta \le \pi$. We conclude that the area of region C is

$$\frac{1}{2}\int_{\pi/3}^{\pi/2} ((1 + \cos \theta)^2 - (3 \cos \theta)^2)\, d\theta + \frac{1}{2}\int_{\pi/2}^{\pi} (1 + \cos \theta)^2\, d\theta = \frac{\pi}{8}.$$

*Related Exercise 61* ◄

## Arc Length of a Polar Curve

> Recall from Section 12.2 that to convert from polar to Cartesian coordinates we use the relations
>
> $$x = r \cos \theta \quad \text{and} \quad y = r \sin \theta.$$

We now answer the arc length question for polar curves: Given the polar equation $r = f(\theta)$, what is the length of the corresponding curve for $\alpha \le \theta \le \beta$ (assuming the curve does not retrace itself on this interval)? The key idea is to express the polar equation as a set of parametric equations in Cartesian coordinates and then use the arc length formula for parametric equations derived in Section 12.1. Letting $\theta$ play the role of a parameter and using $r = f(\theta)$, parametric equations for the polar curve are

$$x = r \cos \theta = f(\theta) \cos \theta \quad \text{and} \quad y = r \sin \theta = f(\theta) \sin \theta,$$

where $\alpha \le \theta \le \beta$. The arc length formula in terms of the parameter $\theta$ is

$$L = \int_{\alpha}^{\beta} \sqrt{\left(\frac{dx}{d\theta}\right)^2 + \left(\frac{dy}{d\theta}\right)^2} \, d\theta,$$

where

$$\frac{dx}{d\theta} = f'(\theta) \cos \theta - f(\theta) \sin \theta \quad \text{and} \quad \frac{dy}{d\theta} = f'(\theta) \sin \theta + f(\theta) \cos \theta.$$

When substituted into the arc length formula and simplified, the result is a new arc length integral (Exercise 84).

**QUICK CHECK 4** Use the arc length formula to verify that the circumference of the circle $r = f(\theta) = 1$, for $0 \le \theta \le 2\pi$, is $2\pi$. ◄

---

**Arc Length of a Polar Curve**

Let $f$ have a continuous derivative on the interval $[\alpha, \beta]$. The **arc length** of the polar curve $r = f(\theta)$ on $[\alpha, \beta]$ is

$$L = \int_{\alpha}^{\beta} \sqrt{f(\theta)^2 + f'(\theta)^2} \, d\theta.$$

---

**EXAMPLE 7   Arc length of polar curves**

**a.** Find the arc length of the spiral $r = f(\theta) = \theta$, for $0 \le \theta \le 2\pi$ (**Figure 12.43**).

**b.** Find the arc length of the cardioid $r = 1 + \cos \theta$ (**Figure 12.44**).

**SOLUTION**

**a.**
$$L = \int_0^{2\pi} \sqrt{\theta^2 + 1} \, d\theta \qquad\qquad f(\theta) = \theta \text{ and } f'(\theta) = 1$$

$$= \left( \frac{\theta}{2} \sqrt{\theta^2 + 1} + \frac{1}{2} \ln\left(\theta + \sqrt{\theta^2 + 1}\right) \right)\Bigg|_0^{2\pi} \qquad \text{Table of integrals or trigonometric substitution}$$

$$= \pi \sqrt{4\pi^2 + 1} + \frac{1}{2} \ln\left(2\pi + \sqrt{4\pi^2 + 1}\right) \qquad \text{Substitute limits of integration.}$$

$$\approx 21.26 \qquad\qquad\qquad\qquad \text{Evaluate.}$$

**b.** The cardioid is symmetric about the $x$-axis, and its upper half is generated for $0 \le \theta \le \pi$. The length of the full curve is twice the length of its upper half:

$$L = 2 \int_0^{\pi} \sqrt{(1 + \cos \theta)^2 + (-\sin \theta)^2} \, d\theta \qquad f(\theta) = 1 + \cos \theta; \ f'(\theta) = -\sin \theta$$

$$= 2 \int_0^{\pi} \sqrt{2 + 2 \cos \theta} \, d\theta \qquad\qquad \text{Simplify.}$$

$$= 2 \int_0^{\pi} \sqrt{4 \cos^2 (\theta/2)} \, d\theta \qquad\qquad 1 + \cos \theta = 2 \cos^2 (\theta/2)$$

$$= 4 \int_0^{\pi} \cos (\theta/2) \, d\theta \qquad\qquad \cos (\theta/2) \ge 0, \text{ for } 0 \le \theta \le \pi$$

$$= 8 \sin (\theta/2) \Big|_0^{\pi} = 8. \qquad\qquad \text{Integrate and simplify.}$$

**Figure 12.43**

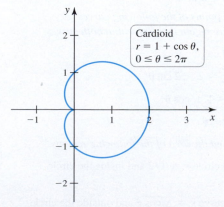

**Figure 12.44**

*Related Exercises 67–68* ◄

## SECTION 12.3 EXERCISES

### Getting started

1. Express the polar equation $r = f(\theta)$ in parametric form in Cartesian coordinates, where $\theta$ is the parameter.

2. Explain why the slope of the line $\theta = \pi/2$ is undefined.

3. Explain why the slope of the line tangent to the polar graph of $r = f(\theta)$ is not $dr/d\theta$.

4. What integral must be evaluated to find the area of the region bounded by the polar graphs of $r = f(\theta)$ and $r = g(\theta)$ on the interval $\alpha \le \theta \le \beta$, where $f(\theta) \ge g(\theta) \ge 0$?

5. What is the slope of the line $\theta = \pi/3$?

6. Find a polar equation of the line with slope $-1$ that passes through the origin.

7. Find the area of the shaded region.

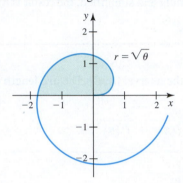

8. Without calculating derivatives, determine the slopes of each of the lines tangent to the curve $r = 8 \cos \theta - 4$ at the origin.

9. Explain why the point with polar coordinates $(0, 0)$ is an intersection point of the curves $r = \cos \theta$ and $r = \sin \theta$ even though $\cos \theta \ne \sin \theta$ when $\theta = 0$.

10. Explain why the point $(-1, 3\pi/2)$ is on the polar graph of $r = 1 + \cos \theta$ even though it does not satisfy the equation $r = 1 + \cos \theta$.

### Practice Exercises

**11–20. Slopes of tangent lines** *Find the slope of the line tangent to the following polar curves at the given points.*

11. $r = 1 - \sin \theta; \left(\dfrac{1}{2}, \dfrac{\pi}{6}\right)$     12. $r = 4 \cos \theta; \left(2, \dfrac{\pi}{3}\right)$

13. $r = 8 \sin \theta; \left(4, \dfrac{5\pi}{6}\right)$

14. $r = 4 + \sin \theta; (4, 0)$ and $\left(3, \dfrac{3\pi}{2}\right)$

15. $r = 6 + 3 \cos \theta; (3, \pi)$ and $(9, 0)$

16. $r = 2 \sin 3\theta;$ at the tips of the leaves

17. $r = 4 \cos 2\theta;$ at the tips of the leaves

18. $r = 1 + 2 \sin 2\theta; \left(3, \dfrac{\pi}{4}\right)$   19. $r^2 = 4 \cos 2\theta; \left(0, \pm\dfrac{\pi}{4}\right)$

20. $r = 2\theta; \left(\dfrac{\pi}{2}, \dfrac{\pi}{4}\right)$

21. **Tangent line at the origin** Find the polar equation of the line tangent to the polar curve $r = \cos \theta + \sin \theta$ at the origin, and then find the slope of this tangent line.

22. **Tangent line at the origin** Find the polar equation of the line tangent to the polar curve $r = 4 \cos \theta$ at the origin. Explain why the slope of this line is undefined.

**T 23–24. Multiple tangent lines at a point**

a. Give the smallest interval $[0, P]$ that generates the entire polar curve and use a graphing utility to graph the curve.

b. Find a polar equation and the slope of each line tangent to the curve at the origin.

23. $r = \cos 3\theta + \sin 3\theta$     24. $r = 1 + 2 \cos 2\theta$

**25–28. Horizontal and vertical tangents** *Find the points at which the following polar curves have horizontal or vertical tangent lines.*

25. $r = 4 \cos \theta$     26. $r = 2 + 2 \sin \theta$

**T 27.** $r = \sin 2\theta$     **T 28.** $r = 3 + 6 \sin \theta$

**29–32. Intersection points** *Use algebraic methods to find as many intersection points of the following curves as possible. Use graphical methods to identify the remaining intersection points.*

29. $r = 2 \cos \theta$ and $r = 1 + \cos \theta$

30. $r = 1 - \sin \theta$ and $r = 1 + \cos \theta$

31. $r = 1$ and $r = 2 \sin 2\theta$

32. $r = \cos 2\theta$ and $r = \sin 2\theta$

**33–40. Areas of regions** *Make a sketch of the region and its bounding curves. Find the area of the region.*

33. The region inside the curve $r = \sqrt{\cos \theta}$

34. The region inside the right lobe of $r = \sqrt{\cos 2\theta}$

35. The region inside the circle $r = 8 \sin \theta$

36. The region inside the cardioid $r = 4 + 4 \sin \theta$

37. The region inside the limaçon $r = 2 + \cos \theta$

38. The region inside all the leaves of the rose $r = 3 \sin 2\theta$

39. The region inside one leaf of $r = \cos 3\theta$

40. The region inside the inner loop of $r = \cos \theta - 1/2$

**41–44. Intersection points and area**

a. Find all the intersection points of the following curves.

b. Find the area of the entire region that lies within both curves.

41. $r = 3 \sin \theta$ and $r = 3 \cos \theta$

42. $r = 2 + 2 \sin \theta$ and $r = 2 - 2 \sin \theta$

43. $r = 1 + \sin \theta$ and $r = 1 + \cos \theta$

44. $r = 1$ and $r = \sqrt{2} \cos 2\theta$

**45–60. Areas of regions** *Find the area of the following regions.*

45. The region outside the circle $r = 1/2$ and inside the circle $r = \cos \theta$

46. The region inside the curve $r = \sqrt{\cos \theta}$ and outside the circle $r = 1/\sqrt{2}$

**47.** The region inside the curve $r = \sqrt{\cos\theta}$ and inside the circle $r = 1/\sqrt{2}$ in the first quadrant

**48.** The region inside the right lobe of $r = \sqrt{\cos 2\theta}$ and inside the circle $r = 1/\sqrt{2}$ in the first quadrant

**49.** The region inside one leaf of the rose $r = \cos 5\theta$

**50.** The region inside the rose $r = 4\cos 2\theta$ and outside the circle $r = 2$

**51.** The region inside the rose $r = 4\sin 2\theta$ and inside the circle $r = 2$

**52.** The region inside the lemniscate $r^2 = 2\sin 2\theta$ and outside the circle $r = 1$

**53.** The region common to the circles $r = 2\sin\theta$ and $r = 1$

**54.** The region inside the inner loop of the limaçon $r = 2 + 4\cos\theta$

**55.** The region inside the outer loop but outside the inner loop of the limaçon $r = 3 - 6\sin\theta$

**56.** The region common to the circle $r = 3\cos\theta$ and the cardioid $r = 1 + \cos\theta$

**57.** The region inside the lemniscate $r^2 = 6\sin 2\theta$

**58.** The region inside the limaçon $r = 2 - 4\sin\theta$

**59.** The region inside the limaçon $r = 4 - 2\cos\theta$

**60.** The region inside the complete three-leaf rose $r = 2\cos 3\theta$

**61. Two curves, three regions** Determine the intersection points of the polar curves $r = 4 - 2\cos\theta$ and $r = 2 + 2\cos\theta$, and then find areas of regions A, B, and C (see figure).

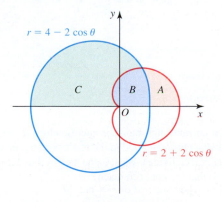

**62.** Find the area in the first quadrant inside the curve $r = 4 + \cos 2\theta$ but outside the curve $r = 3\cos 2\theta$ (see figure), after first finding the intersection points of the two curves.

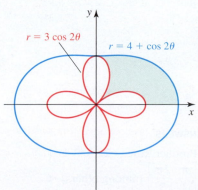

**63–74. Arc length of polar curves** *Find the length of the following polar curves.*

**63.** The complete circle $r = a\sin\theta$, where $a > 0$

**64.** The complete cardioid $r = 2 - 2\sin\theta$

**65.** The spiral $r = \theta^2$, for $0 \le \theta \le 2\pi$

**66.** The spiral $r = e^\theta$, for $0 \le \theta \le 2\pi n$, where $n$ is a positive integer

**67.** The complete cardioid $r = 4 + 4\sin\theta$

**68.** The spiral $r = 4\theta^2$, for $0 \le \theta \le 6$

**69.** The spiral $r = 2e^{2\theta}$, for $0 \le \theta \le \ln 8$

**70.** The parabola $r = \dfrac{\sqrt{2}}{1 + \cos\theta}$, for $0 \le \theta \le \dfrac{\pi}{2}$

**71.** The curve $r = \sin^3\dfrac{\theta}{3}$, for $0 \le \theta \le \dfrac{\pi}{2}$

**T 72.** The three-leaf rose $r = 2\cos 3\theta$

**T 73.** The complete limaçon $r = 4 - 2\cos\theta$

**T 74.** The complete lemniscate $r^2 = 6\sin 2\theta$

**75. Explain why or why not** Determine whether the following statements are true and give an explanation or counterexample.

   **a.** The area of the region bounded by the polar graph of $r = f(\theta)$ on the interval $[\alpha, \beta]$ is $\int_\alpha^\beta f(\theta)\, d\theta$.

   **b.** The slope of the line tangent to the polar curve $r = f(\theta)$ at a point $(r, \theta)$ is $f'(\theta)$.

   **c.** There may be more than one line that is tangent to a polar curve at some points on the curve.

## Explorations and Challenges

**76. Area calculation** Use polar coordinates to determine the area bounded on the right by the unit circle $x^2 + y^2 = 1$ and bounded on the left by the vertical line $x = \sqrt{2}/2$.

**T 77. Spiral tangent lines** Use a graphing utility to determine the first three points with $\theta \ge 0$ at which the spiral $r = 2\theta$ has a horizontal tangent line. Find the first three points with $\theta \ge 0$ at which the spiral $r = 2\theta$ has a vertical tangent line.

**78. Spiral arc length** Consider the spiral $r = 4\theta$, for $\theta \ge 0$.

   **a.** Use a trigonometric substitution to find the length of the spiral, for $0 \le \theta \le \sqrt{8}$.

   **b.** Find $L(\theta)$, the length of the spiral on the interval $[0, \theta]$, for any $\theta \ge 0$.

   **c.** Show that $L'(\theta) > 0$. Is $L''(\theta)$ positive or negative? Interpret your answer.

**79. Spiral arc length** Find the length of the entire spiral $r = e^{-a\theta}$, for $\theta \ge 0$ and $a > 0$.

**80. Area of roses** Assume $m$ is a positive integer.

   **a.** *Even number of leaves:* What is the relationship between the total area enclosed by the $4m$-leaf rose $r = \cos(2m\theta)$ and $m$?

   **b.** *Odd number of leaves:* What is the relationship between the total area enclosed by the $(2m + 1)$-leaf rose $r = \cos((2m + 1)\theta)$ and $m$?

**81. Regions bounded by a spiral** Let $R_n$ be the region bounded by the $n$th turn and the $(n + 1)$st turn of the spiral $r = e^{-\theta}$ in the first and second quadrants, for $\theta \geq 0$ (see figure).

a. Find the area $A_n$ of $R_n$.

b. Evaluate $\lim_{n \to \infty} A_n$.

c. Evaluate $\lim_{n \to \infty} \dfrac{A_{n+1}}{A_n}$.

**82. Tangents and normals** Let a polar curve be described by $r = f(\theta)$, and let $\ell$ be the line tangent to the curve at the point $P(x, y) = P(r, \theta)$ (see figure).

a. Explain why $\tan \alpha = dy/dx$.

b. Explain why $\tan \theta = y/x$.

c. Let $\varphi$ be the angle between $\ell$ and the line through $O$ and $P$. Prove that $\tan \varphi = f(\theta)/f'(\theta)$.

d. Prove that the values of $\theta$ for which $\ell$ is parallel to the $x$-axis satisfy $\tan \theta = -f(\theta)/f'(\theta)$.

e. Prove that the values of $\theta$ for which $\ell$ is parallel to the $y$-axis satisfy $\tan \theta = f(\theta)/f'(\theta)$.

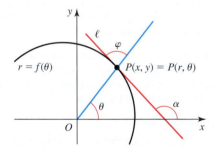

**T 83. Isogonal curves** Let a curve be described by $r = f(\theta)$, where $f(\theta) > 0$ on its domain. Referring to the figure in Exercise 82, a curve is **isogonal** provided the angle $\varphi$ is constant for all $\theta$.

a. Prove that $\varphi$ is constant for all $\theta$ provided $\cot \varphi = \dfrac{f'(\theta)}{f(\theta)}$ is constant, which implies that $\dfrac{d}{d\theta}(\ln f(\theta)) = k$, where $k$ is a constant.

b. Use part (a) to prove that the family of logarithmic spirals $r = Ce^{k\theta}$ consists of isogonal curves, where $C$ and $k$ are constants.

c. Graph the curve $r = 2e^{2\theta}$ and confirm the result of part (b).

**84. Arc length for polar curves** Prove that the length of the curve $r = f(\theta)$, for $\alpha \leq \theta \leq \beta$, is

$$L = \int_\alpha^\beta \sqrt{f(\theta)^2 + f'(\theta)^2}\, d\theta.$$

**85–87. Grazing goat problems** *Consider the following sequence of problems related to grazing goats tied to a rope. (See the Guided Project Grazing goat problems.)*

**85.** A circular corral of unit radius is enclosed by a fence. A goat inside the corral is tied to the fence with a rope of length $0 \leq a \leq 2$ (see figure). What is the area of the region (inside the corral) that the goat can graze? Check your answer with the special cases $a = 0$ and $a = 2$.

**86.** A circular concrete slab of unit radius is surrounded by grass. A goat is tied to the edge of the slab with a rope of length $0 \leq a \leq 2$ (see figure). What is the area of the grassy region that the goat can graze? Note that the rope can extend over the concrete slab. Check your answer with the special cases $a = 0$ and $a = 2$.

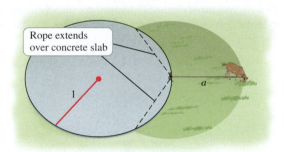

**87.** A circular corral of unit radius is enclosed by a fence. A goat is outside the corral and tied to the fence with a rope of length $0 \leq a \leq \pi$ (see figure). What is the area of the region (outside the corral) that the goat can reach?

**QUICK CHECK ANSWERS**

1. Apply the Product Rule.   2. $\sqrt{2} + 1$

3. $\left[0, \dfrac{\pi}{3}\right]$ or $\left[\dfrac{\pi}{3}, \dfrac{2\pi}{3}\right]$ (among others) ◄

# 12.4  Conic Sections

Conic sections are best visualized as the Greeks did over 2000 years ago by slicing a double cone with a plane (**Figure 12.45**). Three of the seven different sets of points that arise in this way are *ellipses*, *parabolas*, and *hyperbolas*. These curves have practical applications and broad theoretical importance. For example, celestial bodies travel in orbits that are modeled by ellipses and hyperbolas. Mirrors for telescopes are designed using the properties of conic sections. And architectural structures, such as domes and arches, are sometimes based on these curves.

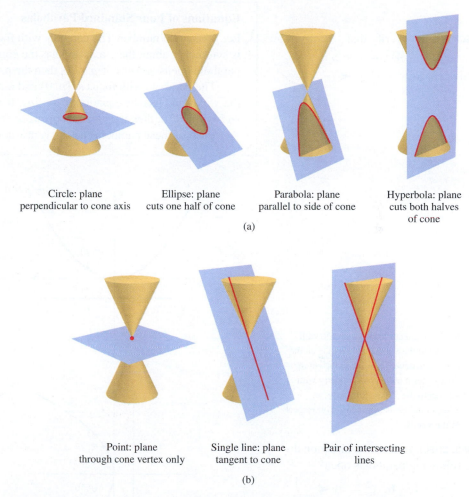

Circle: plane
perpendicular to cone axis

Ellipse: plane
cuts one half of cone

Parabola: plane
parallel to side of cone

Hyperbola: plane
cuts both halves
of cone

(a)

Point: plane
through cone vertex only

Single line: plane
tangent to cone

Pair of intersecting
lines

(b)

**Figure 12.45**  The standard conic sections (a) are the intersection sets of a double cone and a plane that does not pass through the vertex of the cone. Degenerate conic sections (lines and points) are produced when a plane passes through the vertex of the cone (b).

## Parabolas

A **parabola** is the set of points in a plane that are equidistant from a fixed point $F$ (called the **focus**) and a fixed line (called the **directrix**). In the four standard orientations, a parabola may open upward, downward, to the right, or to the left. We derive the equation of the parabola that opens upward.

Suppose the focus $F$ is on the $y$-axis at $(0, p)$ and the directrix is the horizontal line $y = -p$, where $p > 0$. The parabola is the set of points $P$ that satisfy the defining property $|PF| = |PL|$, where $L(x, -p)$ is the point on the directrix closest to $P$

**Figure 12.46**

**QUICK CHECK 1** Verify that $\sqrt{x^2 + (y - p)^2} = y + p$ is equivalent to $x^2 = 4py$. ◄

(Figure 12.46). Consider an arbitrary point $P(x, y)$ that satisfies this condition. Applying the distance formula, we have

$$\underbrace{\sqrt{x^2 + (y - p)^2}}_{|PF|} = \underbrace{y + p}_{|PL|}.$$

Squaring both sides of this equation and simplifying gives the equation $x^2 = 4py$. This is the equation of a parabola that is symmetric about the $y$-axis and opens upward. The **vertex** of the parabola is the point closest to the directrix; in this case, it is $(0, 0)$ (which satisfies $|PF| = |PL| = p$).

The equations of the other three standard parabolas are derived in a similar way.

---

**Equations of Four Standard Parabolas**

Let $p$ be a real number. The parabola with focus at $(0, p)$ and directrix $y = -p$ is symmetric about the $y$-axis and has the equation $x^2 = 4py$. If $p > 0$, then the parabola opens *upward*; if $p < 0$, then the parabola opens *downward*.

The parabola with focus at $(p, 0)$ and directrix $x = -p$ is symmetric about the $x$-axis and has the equation $y^2 = 4px$. If $p > 0$, then the parabola opens *to the right*; if $p < 0$, then the parabola opens *to the left*.

Each of these parabolas has its vertex at the origin (Figure 12.47).

---

▶ Recall that a curve is symmetric with respect to the $x$-axis if $(x, -y)$ is on the curve whenever $(x, y)$ is on the curve. So a $y^2$-term indicates symmetry with respect to the $x$-axis. Similarly, an $x^2$-term indicates symmetry with respect to the $y$-axis.

**QUICK CHECK 2** In which direction do the following parabolas open?

**a.** $y^2 = -4x$   **b.** $x^2 = 4y$ ◄

**Figure 12.47**

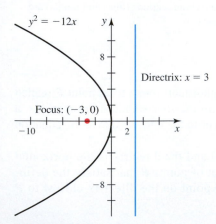

**Figure 12.48**

**EXAMPLE 1   Graphing parabolas** Find the focus and directrix of the parabola $y^2 = -12x$. Sketch its graph.

**SOLUTION** The $y^2$-term indicates that the parabola is symmetric with respect to the $x$-axis. Rewriting the equation as $x = -y^2/12$, we see that $x \le 0$ for all $y$, implying that the parabola opens to the left. Comparing $y^2 = -12x$ to the standard form $y^2 = 4px$, we see that $p = -3$; therefore, the focus is $(-3, 0)$, and the directrix is $x = 3$ (Figure 12.48).

*Related Exercises 13–14* ◄

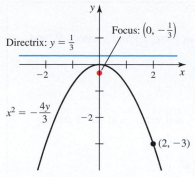

**Figure 12.49**

**EXAMPLE 2** **Equations of parabolas** Find the equation of the parabola with vertex $(0, 0)$ that opens downward and passes through the point $(2, -3)$.

**SOLUTION** The standard parabola that opens downward has the equation $x^2 = 4py$. The point $(2, -3)$ must satisfy this equation. Substituting $x = 2$ and $y = -3$ into $x^2 = 4py$, we find that $p = -\frac{1}{3}$. Therefore, the focus is at $\left(0, -\frac{1}{3}\right)$, the directrix is $y = \frac{1}{3}$, and the equation of the parabola is $x^2 = -4y/3$, or $y = -3x^2/4$ (**Figure 12.49**).

*Related Exercises 35–36* ◄

**Reflection Property** Parabolas have a property that makes them useful in the design of reflectors and transmitters. A particle approaching a parabola on any line parallel to the axis of the parabola is reflected on a line that passes through the focus (**Figure 12.50**); this property is used to focus incoming light by a parabolic mirror on a telescope. Alternatively, signals emanating from the focus are reflected on lines parallel to the axis, a property used to design radio transmitters and headlights (Exercise 83).

**Figure 12.50**

**Figure 12.51**

## Ellipses

An **ellipse** is the set of points in a plane whose distances from two fixed points have a constant sum that we denote $2a$ (**Figure 12.51**). Each of the two fixed points is a **focus** (plural **foci**). The equation of an ellipse is simplest if the foci are on the $x$-axis at $(\pm c, 0)$ or on the $y$-axis at $(0, \pm c)$. In either case, the **center** of the ellipse is $(0, 0)$. If the foci are on the $x$-axis, the points $(\pm a, 0)$ lie on the ellipse and are called **vertices**. If the foci are on the $y$-axis, the vertices are $(0, \pm a)$ (**Figure 12.52**). A short calculation (Exercise 85) using the definition of the ellipse results in the following equations for an ellipse.

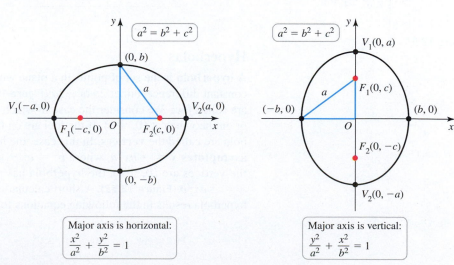

**Figure 12.52**

▶ When necessary, we distinguish between the *major-axis vertices* $(\pm a, 0)$ or $(0, \pm a)$, and the *minor-axis vertices* $(\pm b, 0)$ or $(0, \pm b)$. The word *vertices* (without further description) is understood to mean *major-axis vertices*.

---

**Equations of Standard Ellipses**

An ellipse centered at the origin with foci $F_1$ and $F_2$ at $(\pm c, 0)$ and vertices $V_1$ and $V_2$ at $(\pm a, 0)$ has the equation

$$\frac{x^2}{a^2} + \frac{y^2}{b^2} = 1, \quad \text{where } a^2 = b^2 + c^2.$$

An ellipse centered at the origin with foci at $(0, \pm c)$ and vertices at $(0, \pm a)$ has the equation

$$\frac{y^2}{a^2} + \frac{x^2}{b^2} = 1, \quad \text{where } a^2 = b^2 + c^2.$$

In both cases, $a > b > 0$ and $a > c > 0$, the length of the long axis (called the **major axis**) is $2a$, and the length of the short axis (called the **minor axis**) is $2b$.

---

**QUICK CHECK 3** In the case that the vertices and foci are on the $x$-axis, show that the length of the minor axis of an ellipse is $2b$. ◀

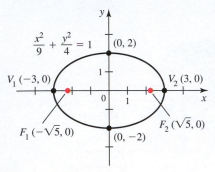

**Figure 12.53**

**EXAMPLE 3    Graphing ellipses** Find the vertices, foci, and length of the major and minor axes of the ellipse $\dfrac{x^2}{9} + \dfrac{y^2}{4} = 1$. Graph the ellipse.

**SOLUTION** Because $9 > 4$, we identify $a^2 = 9$ and $b^2 = 4$. Therefore, $a = 3$ and $b = 2$. The lengths of the major and minor axes are $2a = 6$ and $2b = 4$, respectively. The vertices $V_1$ and $V_2$ are at $(\pm 3, 0)$ and lie on the $x$-axis, as do the foci. The relationship $c^2 = a^2 - b^2$ implies that $c^2 = 5$, or $c = \sqrt{5}$. Therefore, the foci $F_1$ and $F_2$ are at $(\pm\sqrt{5}, 0)$. The graph of the ellipse is shown in **Figure 12.53**.

*Related Exercises 15–16* ◀

**Figure 12.54**

**EXAMPLE 4    Equation of an ellipse** Find the equation of the ellipse centered at the origin with its foci on the $y$-axis, a major axis of length 8, and a minor axis of length 4. Graph the ellipse.

**SOLUTION** Because the length of the major axis is 8, the vertices $V_1$ and $V_2$ are located at $(0, \pm 4)$, and $a = 4$. Because the length of the minor axis is 4, we have $b = 2$. Therefore, the equation of the ellipse is

$$\frac{y^2}{16} + \frac{x^2}{4} = 1.$$

Using the relation $c^2 = a^2 - b^2$, we find that $c = 2\sqrt{3}$ and the foci $F_1$ and $F_2$ are at $(0, \pm 2\sqrt{3})$. The ellipse is shown in **Figure 12.54**.

*Related Exercise 39* ◀

## Hyperbolas

A **hyperbola** is the set of points in a plane whose distances from two fixed points have a constant difference, either $2a$ or $-2a$ (**Figure 12.55**). As with ellipses, the two fixed points are called **foci**. We consider the case in which the foci are either on the $x$-axis at $(\pm c, 0)$ or on the $y$-axis at $(0, \pm c)$. If the foci are on the $x$-axis, the points $(\pm a, 0)$ on the hyperbola are called the **vertices**. In this case, the hyperbola has no $y$-intercepts, but it has the **asymptotes** $y = \pm bx/a$, where $b^2 = c^2 - a^2$. Similarly, if the foci are on the $y$-axis, the vertices are $(0, \pm a)$, the hyperbola has no $x$-intercepts, and it has the asymptotes $y = \pm ax/b$ (**Figure 12.56**). A short calculation (Exercise 86) using the definition of the hyperbola results in the following equations for standard hyperbolas.

➤ Asymptotes that are not parallel to one of the coordinate axes, as in the case of the standard hyperbolas, are called **oblique**, or **slant**, **asymptotes**.

**Figure 12.55**

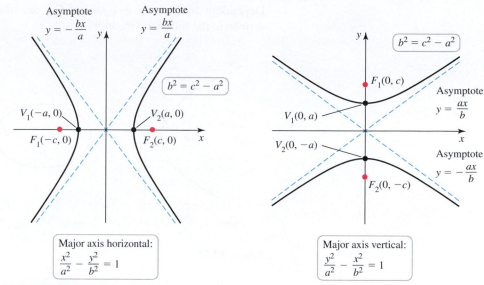

**Figure 12.56**

---

### Equations of Standard Hyperbolas

A hyperbola centered at the origin with foci $F_1$ and $F_2$ at $(\pm c, 0)$ and vertices $V_1$ and $V_2$ at $(\pm a, 0)$ has the equation

$$\frac{x^2}{a^2} - \frac{y^2}{b^2} = 1, \quad \text{where} \quad b^2 = c^2 - a^2.$$

The hyperbola has **asymptotes** $y = \pm bx/a$.

A hyperbola centered at the origin with foci at $(0, \pm c)$ and vertices at $(0, \pm a)$ has the equation

$$\frac{y^2}{a^2} - \frac{x^2}{b^2} = 1, \quad \text{where} \quad b^2 = c^2 - a^2.$$

The hyperbola has **asymptotes** $y = \pm ax/b$.
In both cases, $c > a > 0$ and $c > b > 0$.

---

➤ Notice that the asymptotes for hyperbolas are $y = \pm bx/a$ when the vertices are on the $x$-axis and $y = \pm ax/b$ when the vertices are on the $y$-axis (the roles of $a$ and $b$ are reversed).

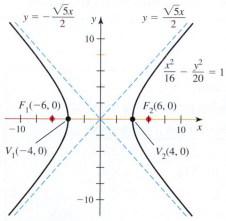

**Figure 12.57**

**QUICK CHECK 4** Identify the vertices and foci of the hyperbola $y^2 - x^2/4 = 1$. ◄

**EXAMPLE 5　Graphing hyperbolas** Find the equation of the hyperbola centered at the origin with vertices $V_1$ and $V_2$ at $(\pm 4, 0)$ and foci $F_1$ and $F_2$ at $(\pm 6, 0)$. Graph the hyperbola.

**SOLUTION** Because the foci are on the $x$-axis, the vertices are also on the $x$-axis, and there are no $y$-intercepts. With $a = 4$ and $c = 6$, we have $b^2 = c^2 - a^2 = 20$, or $b = 2\sqrt{5}$. Therefore, the equation of the hyperbola is

$$\frac{x^2}{16} - \frac{y^2}{20} = 1.$$

The asymptotes are $y = \pm bx/a = \pm \sqrt{5}x/2$ (**Figure 12.57**).

*Related Exercise 41* ◄

## Eccentricity and Directrix

Parabolas, ellipses, and hyperbolas may also be developed in a single unified way called the *eccentricity-directrix* approach. We let $\ell$ be a line called the **directrix** and $F$ be a point not on $\ell$ called a **focus**. The **eccentricity** is a real number $e > 0$. Consider the set $C$ of points $P$ in a plane with the property that the distance $|PF|$ equals $e$ multiplied by the perpendicular distance $|PL|$ from $P$ to $\ell$ (**Figure 12.58**); that is,

➤ The conic section lies in the plane formed by the directrix and the focus.

$$|PF| = e|PL| \quad \text{or} \quad \frac{|PF|}{|PL|} = e = \text{constant}.$$

Depending on the value of $e$, the set $C$ is one of the three standard conic sections, as described in the following theorem.

| Parabola: $\dfrac{\|PF\|}{\|PL\|} = e = 1$ | Ellipse: $\dfrac{\|PF\|}{\|PL\|} = e, 0 < e < 1$ | Hyperbola: $\dfrac{\|PF\|}{\|PL\|} = e > 1$ |

**Figure 12.58**

> Theorem 12.3 for ellipses and hyperbolas describes how the entire curve is generated using just one focus and one directrix. Nevertheless, all ellipses and hyperbolas have two foci and two directrices.

---

**THEOREM 12.3    Eccentricity-Directrix Theorem**

Suppose $\ell$ is a line, $F$ is a point not on $\ell$, and $e$ is a positive real number. Let $C$ be the set of points $P$ in a plane with the property that $\dfrac{\|PF\|}{\|PL\|} = e$, where $\|PL\|$ is the perpendicular distance from $P$ to $\ell$.

1. If $e = 1$, $C$ is a **parabola**.
2. If $0 < e < 1$, $C$ is an **ellipse**.
3. If $e > 1$, $C$ is a **hyperbola**.

---

The proof of this theorem is straightforward; it requires an algebraic calculation that is found in Appendix A. The proof establishes relationships among five parameters $a, b, c, d$, and $e$ that are characteristic of any ellipse or hyperbola. The relationships are given in the following summary.

---

**SUMMARY    Properties of Ellipses and Hyperbolas**

An ellipse or a hyperbola centered at the origin has the following properties.

|  | Foci on $x$-axis | Foci on $y$-axis |
|---|---|---|
| Major-axis vertices: | $(\pm a, 0)$ | $(0, \pm a)$ |
| Minor-axis vertices (for ellipses): | $(0, \pm b)$ | $(\pm b, 0)$ |
| Foci: | $(\pm c, 0)$ | $(0, \pm c)$ |
| Directrices: | $x = \pm d$ | $y = \pm d$ |

Eccentricity: $0 < e < 1$ for ellipses, $e > 1$ for hyperbolas.

Given any two of the five parameters $a, b, c, d$, and $e$, the other three are found using the relations

$$c = ae, \quad d = \frac{a}{e},$$

$$b^2 = a^2 - c^2 \text{ (for ellipses)}, \qquad b^2 = c^2 - a^2 \text{ (for hyperbolas)}.$$

---

**QUICK CHECK 5** Given an ellipse with $a = 3$ and $e = \frac{1}{2}$, what are the values of $b, c$, and $d$? ◄

**EXAMPLE 6    Equation of an ellipse** Find the equation of the ellipse centered at the origin with foci $F_1$ and $F_2$ at $(0, \pm 4)$ and eccentricity $e = \frac{1}{2}$. Give the length of the major and minor axes, the location of the vertices, and the directrices. Graph the ellipse.

$$\frac{y^2}{64} + \frac{x^2}{48} = 1$$

**Figure 12.59**

**SOLUTION** An ellipse with its major axis along the $y$-axis has the equation

$$\frac{y^2}{a^2} + \frac{x^2}{b^2} = 1,$$

where $a$ and $b$ must be determined (with $a > b$). Because the foci are at $(0, \pm 4)$, we have $c = 4$. Using $e = \frac{1}{2}$ and the relation $c = ae$, it follows that $a = c/e = 8$. So the length of the major axis is $2a = 16$, and the major-axis vertices $V_1$ and $V_2$ are $(0, \pm 8)$. Also, $d = a/e = 16$, so the directrices are $y = \pm 16$. Finally, $b^2 = a^2 - c^2 = 48$, or $b = 4\sqrt{3}$. So the length of the minor axis is $2b = 8\sqrt{3}$, and the minor-axis vertices are $(\pm 4\sqrt{3}, 0)$ (**Figure 12.59**). The equation of the ellipse is

$$\frac{y^2}{64} + \frac{x^2}{48} = 1.$$

*Related Exercises 53–54* ◄

## Polar Equations of Conic Sections

It turns out that conic sections have a natural representation in polar coordinates, provided we use the eccentricity-directrix approach given in Theorem 12.3. Furthermore, a single polar equation covers parabolas, ellipses, and hyperbolas.

When working in polar equations, the key is to place a focus of the conic section at the origin of the coordinate system. We begin by placing one focus $F$ at the origin and taking a directrix perpendicular to the $x$-axis through $(d, 0)$, where $d > 0$ (**Figure 12.60**).

We now use the definition $\dfrac{|PF|}{|PL|} = e$, where $P(r, \theta)$ is an arbitrary point on the conic.

As shown in Figure 12.60, $|PF| = r$ and $|PL| = d - r \cos \theta$. The condition $\dfrac{|PF|}{|PL|} = e$

implies that $r = e(d - r \cos \theta)$. Solving for $r$, we have

$$r = \frac{ed}{1 + e \cos \theta}.$$

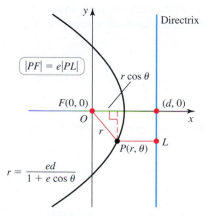

**Figure 12.60**

A similar derivation (Exercise 84) with the directrix at $x = -d$, where $d > 0$, results in the equation

$$r = \frac{ed}{1 - e \cos \theta}.$$

For horizontal directrices at $y = \pm d$ (**Figure 12.61**), a similar argument (Exercise 84) leads to the equations

$$r = \frac{ed}{1 \pm e \sin \theta}.$$

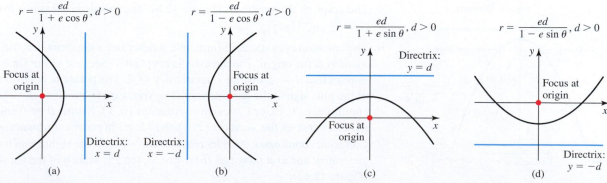

**Figure 12.61**

> **THEOREM 12.4   Polar Equations of Conic Sections**
> Let $d > 0$. The conic section with a focus at the origin and eccentricity $e$ has the polar equation
>
> $$r = \frac{ed}{1 + e \cos \theta} \quad \text{or} \quad r = \frac{ed}{1 - e \cos \theta}.$$
>
> <u>if one directrix is $x = d$</u>     <u>if one directrix is $x = -d$</u>
>
> The conic section with a focus at the origin and eccentricity $e$ has the polar equation
>
> $$r = \frac{ed}{1 + e \sin \theta} \quad \text{or} \quad r = \frac{ed}{1 - e \sin \theta}.$$
>
> <u>if one directrix is $y = d$</u>     <u>if one directrix is $y = -d$</u>
>
> If $0 < e < 1$, the conic section is an ellipse; if $e = 1$, it is a parabola; and if $e > 1$, it is a hyperbola. The curves are defined over any interval in $\theta$ of length $2\pi$.

**QUICK CHECK 6**  On which axis do the vertices and foci of the conic section $r = 2/(1 - 2 \sin \theta)$ lie? ◄

**EXAMPLE 7**   **Conic sections in polar coordinates**  Find the vertices, foci, and directrices of the following conic sections. Graph each curve and check your work with a graphing utility.

**a.** $r = \dfrac{8}{2 + 3 \cos \theta}$     **b.** $r = \dfrac{2}{1 + \sin \theta}$

**SOLUTION**

**a.** The equation must be expressed in standard polar form for a conic section. Dividing numerator and denominator by 2, we have

$$r = \frac{4}{1 + \frac{3}{2} \cos \theta},$$

which allows us to identify $e = \frac{3}{2}$. Therefore, the equation describes a hyperbola (because $e > 1$) with one focus at the origin.

The directrices are vertical (because $\cos \theta$ appears in the equation). Knowing that $ed = 4$, we have $d = 4/e = \frac{8}{3}$, and one directrix is $x = \frac{8}{3}$. Letting $\theta = 0$ and $\theta = \pi$, the polar coordinates of the vertices are $\left(\frac{8}{5}, 0\right)$ and $(-8, \pi)$; equivalently, the vertices are $\left(\frac{8}{5}, 0\right)$ and $(8, 0)$ in Cartesian coordinates (**Figure 12.62**). The center of the hyperbola is halfway between the vertices; therefore, its Cartesian coordinates are $\left(\frac{24}{5}, 0\right)$. The distance between the focus at $(0, 0)$ and the nearest vertex $\left(\frac{8}{5}, 0\right)$ is $\frac{8}{5}$. Therefore, the other focus is $\frac{8}{5}$ units to the right of the vertex $(8, 0)$. So the Cartesian coordinates of the foci are $\left(\frac{48}{5}, 0\right)$ and $(0, 0)$. Because the directrices are symmetric about the center and the left directrix is $x = \frac{8}{3}$, the right directrix is $x = \frac{104}{15} \approx 6.9$. The graph of the hyperbola (Figure 12.62) is generated as $\theta$ varies from 0 to $2\pi$ (with $\theta \neq \pm \cos^{-1}\left(-\frac{2}{3}\right)$).

**b.** The equation is in standard form, and it describes a parabola because $e = 1$. The sole focus is at the origin. The directrix is horizontal (because of the $\sin \theta$ term); $ed = 2$ implies that $d = 2$, and the directrix is $y = 2$. The parabola opens downward because of the plus sign in the denominator. The vertex corresponds to $\theta = \frac{\pi}{2}$ and has polar coordinates $\left(1, \frac{\pi}{2}\right)$, or Cartesian coordinates $(0, 1)$. Setting $\theta = 0$ and $\theta = \pi$, the parabola crosses the $x$-axis at $(2, 0)$ and $(2, \pi)$ in polar coordinates, or $(\pm 2, 0)$ in Cartesian coordinates. As $\theta$ increases from $-\frac{\pi}{2}$ to $\frac{\pi}{2}$, the right branch of the parabola is generated, and as $\theta$ increases from $\frac{\pi}{2}$ to $\frac{3\pi}{2}$, the left branch of the parabola is generated (**Figure 12.63**).

*Related Exercises 57, 60* ◄

**Figure 12.62**

**Figure 12.63**

Directrix:
$x = 1$

$e = 0.4$

$e = 0.6$

$e = 0.8$

$e = 0.2$

**Figure 12.64**

**EXAMPLE 8**   **Conics in polar coordinates**   Use a graphing utility to plot the curves

$$r = \frac{e}{1 + e \cos \theta},$$ with $e = 0.2, 0.4, 0.6,$ and $0.8.$ Comment on the effect of varying the eccentricity, $e.$

**SOLUTION**   Because $0 < e < 1$ in each case, all the curves are ellipses. Notice that the equation is in standard form with $d = 1$; therefore, the curves have the same directrix, $x = d = 1.$ As the eccentricity increases, the ellipses becomes more elongated. Small values of $e$ correspond to more circular ellipses (**Figure 12.64**).

*Related Exercises 65–66* ◄

---

## SECTION 12.4 EXERCISES

### Getting Started

1. Give the property that defines all parabolas.

2. Give the property that defines all ellipses.

3. Give the property that defines all hyperbolas.

4. Sketch the three basic conic sections in standard position with vertices and foci on the $x$-axis.

5. Sketch the three basic conic sections in standard position with vertices and foci on the $y$-axis.

6. What is the equation of the standard parabola with its vertex at the origin that opens downward?

7. What is the equation of the standard ellipse with vertices at $(\pm a, 0)$ and foci at $(\pm c, 0)$?

8. What is the equation of the standard hyperbola with vertices at $(0, \pm a)$ and foci at $(0, \pm c)$?

9. Given vertices $(\pm a, 0)$ and the eccentricity $e$, what are the coordinates of the foci of an ellipse and a hyperbola?

10. Give the equation in polar coordinates of a conic section with a focus at the origin, eccentricity $e$, and a directrix $x = d$, where $d > 0.$

11. What are the equations of the asymptotes of a standard hyperbola with vertices on the $x$-axis?

12. How does the eccentricity determine the type of conic section?

### Practice Exercises

**13–30. Graphing conic sections** *Determine whether the following equations describe a parabola, an ellipse, or a hyperbola, and then sketch a graph of the curve. For each parabola, specify the location of the focus and the equation of the directrix; for each ellipse, label the coordinates of the vertices and foci, and find the lengths of the major and minor axes; for each hyperbola, label the coordinates of the vertices and foci, and find the equations of the asymptotes.*

13. $x^2 = 12y$

14. $y^2 = 20x$

15. $\dfrac{x^2}{4} + y^2 = 1$

16. $\dfrac{x^2}{9} + \dfrac{y^2}{4} = 1$

17. $x = -\dfrac{y^2}{16}$

18. $4x = -y^2$

19. $\dfrac{x^2}{4} - y^2 = 1$

20. $\dfrac{y^2}{16} - \dfrac{x^2}{9} = 1$

21. $4x^2 - y^2 = 16$

22. $25y^2 - 4x^2 = 100$

23. $8y = -3x^2$

24. $12x = 5y^2$

25. $\dfrac{x^2}{4} + \dfrac{y^2}{16} = 1$

26. $x^2 + \dfrac{y^2}{9} = 1$

27. $\dfrac{x^2}{5} + \dfrac{y^2}{7} = 1$

28. $12x^2 + 5y^2 = 60$

29. $\dfrac{x^2}{3} - \dfrac{y^2}{5} = 1$

30. $10x^2 - 7y^2 = 140$

**31–38. Equations of parabolas** *Find an equation of the following parabolas. Unless otherwise specified, assume the vertex is at the origin.*

31. A parabola that opens to the right with directrix $x = -4$

32. A parabola that opens downward with directrix $y = 6$

33. A parabola with focus at $(3, 0)$

34. A parabola with focus at $(-4, 0)$

35. A parabola symmetric about the $y$-axis that passes through the point $(2, -6)$

36. A parabola symmetric about the $x$-axis that passes through the point $(1, -4)$

37.    38.

**39–50. Equations of ellipses and hyperbolas** *Find an equation of the following ellipses and hyperbolas, assuming the center is at the origin.*

39. An ellipse whose major axis is on the $x$-axis with length 8 and whose minor axis has length 6

40. An ellipse with vertices $(\pm 6, 0)$ and foci $(\pm 4, 0)$

**41.** A hyperbola with vertices $(\pm 4, 0)$ and foci $(\pm 6, 0)$

**42.** A hyperbola with vertices $(\pm 1, 0)$ that passes through $(5/3, 8)$

**43.** An ellipse with vertices $(\pm 5, 0)$, passing through the point $(4, 3/5)$

**44.** An ellipse with vertices $(0, \pm 10)$, passing through the point $(\sqrt{3}/2, 5)$

**45.** A hyperbola with vertices $(\pm 2, 0)$ and asymptotes $y = \pm 3x/2$

**46.** A hyperbola with vertices $(0, \pm 4)$ and asymptotes $y = \pm 2x$

**47.**

**48.**

**49.**

**50.**
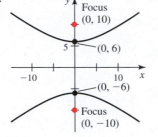

**51. Explain why or why not** Determine whether the following statements are true and give an explanation or counterexample.

  **a.** The hyperbola $\dfrac{x^2}{4} - \dfrac{y^2}{9} = 1$ has no $y$-intercept.

  **b.** On every ellipse, there are exactly two points at which the curve has slope $s$, where $s$ is any real number.

  **c.** Given the directrices and foci of a standard hyperbola, it is possible to find its vertices, eccentricity, and asymptotes.

  **d.** The point on a parabola closest to the focus is the vertex.

**52. Golden Gate Bridge** Completed in 1937, San Francisco's Golden Gate Bridge is 2.7 km long and weighs about 890,000 tons. The length of the span between the two central towers is 1280 m; the towers themselves extend 152 m above the roadway. The cables that support the deck of the bridge between the two towers hang in a parabola (see figure). Assuming the origin is midway between the towers on the deck of the bridge, find an equation that describes the cables. How long is a guy wire that hangs vertically from the cables to the roadway 500 m from the center of the bridge?

**53–56. Eccentricity-directrix approach** *Find an equation of the following curves, assuming the center is at the origin. Sketch a graph labeling the vertices, foci, asymptotes (if they exist), and directrices. Use a graphing utility to check your work.*

**53.** An ellipse with vertices $(\pm 9, 0)$ and eccentricity $1/3$

**54.** An ellipse with vertices $(0, \pm 9)$ and eccentricity $1/4$

**55.** A hyperbola with vertices $(\pm 1, 0)$ and eccentricity $3$

**56.** A hyperbola with vertices $(0, \pm 4)$ and eccentricity $2$

**57–62. Polar equations for conic sections** *Graph the following conic sections, labeling the vertices, foci, directrices, and asymptotes (if they exist). Use a graphing utility to check your work.*

**57.** $r = \dfrac{4}{1 + \cos\theta}$      **58.** $r = \dfrac{3}{2 + \cos\theta}$

**59.** $r = \dfrac{1}{2 - \cos\theta}$      **60.** $r = \dfrac{6}{3 + 2\sin\theta}$

**61.** $r = \dfrac{1}{2 - 2\sin\theta}$      **62.** $r = \dfrac{12}{3 - \cos\theta}$

**63–66. Tracing hyperbolas and parabolas** *Graph the following equations. Then use arrows and labeled points to indicate how the curve is generated as $\theta$ increases from $0$ to $2\pi$.*

**63.** $r = \dfrac{1}{1 + \sin\theta}$      **64.** $r = \dfrac{1}{1 + 2\cos\theta}$

**65.** $r = \dfrac{3}{1 - \cos\theta}$      **66.** $r = \dfrac{1}{1 - 2\cos\theta}$

**T 67. Parabolas with a graphing utility** Use a graphing utility to graph the parabolas $r = \dfrac{d}{1 + \cos\theta}$ for $d = 0.25, 0.5, 1, 2, 3,$ and $4$ on the same set of axes. Explain how the shapes of the curves vary as $d$ changes.

**T 68. Hyperbolas with a graphing utility** Use a graphing utility to graph the hyperbolas $r = \dfrac{e}{1 + e\cos\theta}$ for $e = 1.1, 1.3, 1.5, 1.7,$ and $2$ on the same set of axes. Explain how the shapes of the curves vary as $e$ changes.

**69–72. Tangent lines** *Find an equation of the line tangent to the following curves at the given point.*

**69.** $x^2 = -6y$; $(-6, -6)$      **70.** $y^2 = 8x$; $(8, -8)$

**71.** $y^2 - \dfrac{x^2}{64} = 1$; $\left(6, -\dfrac{5}{4}\right)$      **72.** $r = \dfrac{1}{1 + \sin\theta}$; $\left(\dfrac{2}{3}, \dfrac{\pi}{6}\right)$

**73. Tangent lines for an ellipse** Show that an equation of the line tangent to the ellipse $\dfrac{x^2}{a^2} + \dfrac{y^2}{b^2} = 1$ at the point $(x_0, y_0)$ is

$$\frac{xx_0}{a^2} + \frac{yy_0}{b^2} = 1.$$

**74. Tangent lines for a hyperbola** Find an equation of the line tangent to the hyperbola $\dfrac{x^2}{a^2} - \dfrac{y^2}{b^2} = 1$ at the point $(x_0, y_0)$.

**75–76. Graphs to polar equations** *Find a polar equation for each conic section. Assume one focus is at the origin.*

**75.**

**76.**

**77. Another construction for a hyperbola** Suppose two circles, whose centers are at least $2a$ units apart (see figure), are centered at $F_1$ and $F_2$, respectively. The radius of one circle is $2a + r$ and the radius of the other circle is $r$, where $r \ge 0$. Show that as $r$ increases, the intersection points $P_1$ and $P_2$ of the two circles describes one branch of a hyperbola with foci at $F_1$ and $F_2$.

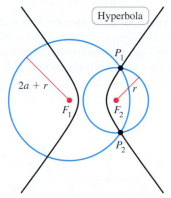

**78. The ellipse and the parabola** Let $R$ be the region bounded by the upper half of the ellipse $\dfrac{x^2}{2} + y^2 = 1$ and the parabola $y = \dfrac{x^2}{\sqrt{2}}$.

**a.** Find the area of $R$.

**b.** Which is greater, the volume of the solid generated when $R$ is revolved about the $x$-axis or the volume of the solid generated when $R$ is revolved about the $y$-axis?

**79. Volume of an ellipsoid** Suppose the ellipse $\dfrac{x^2}{a^2} + \dfrac{y^2}{b^2} = 1$ is revolved about the $x$-axis. What is the volume of the solid enclosed by the *ellipsoid* that is generated? Is the volume different if the same ellipse is revolved about the $y$-axis?

**80. Area of a sector of a hyperbola** Consider the region $R$ bounded by the right branch of the hyperbola $\dfrac{x^2}{a^2} - \dfrac{y^2}{b^2} = 1$ and the vertical line through the right focus.

**a.** What is the area of $R$?

**b.** Sketch a graph that shows how the area of $R$ varies with the eccentricity $e$, for $e > 1$.

**81. Volume of a hyperbolic cap** Consider the region $R$ bounded by the right branch of the hyperbola $\dfrac{x^2}{a^2} - \dfrac{y^2}{b^2} = 1$ and the vertical line through the right focus.

**a.** What is the volume of the solid that is generated when $R$ is revolved about the $x$-axis?

**b.** What is the volume of the solid that is generated when $R$ is revolved about the $y$-axis?

**82. Volume of a paraboloid (Archimedes)** The region bounded by the parabola $y = ax^2$ and the horizontal line $y = h$ is revolved about the $y$-axis to generate a solid bounded by a surface called a *paraboloid* (where $a > 0$ and $h > 0$). Show that the volume of the solid is $3/2$ the volume of the cone with the same base and vertex.

## Explorations and Challenges

**83. Reflection property of parabolas** Consider the parabola $y = \dfrac{x^2}{4p}$ with its focus at $F(0, p)$ (see figure). The goal is to show that the angle of incidence between the ray $\ell$ and the tangent line $L$ ($\alpha$ in the figure) equals the angle of reflection between the line $PF$ and $L$ ($\beta$ in the figure). If these two angles are equal, then the reflection property is proved because $\ell$ is reflected through $F$.

**a.** Let $P(x_0, y_0)$ be a point on the parabola. Show that the slope of the line tangent to the curve at $P$ is $\tan\theta = \dfrac{x_0}{2p}$.

**b.** Show that $\tan\varphi = \dfrac{p - y_0}{x_0}$.

**c.** Show that $\alpha = \dfrac{\pi}{2} - \theta$; therefore, $\tan\alpha = \cot\theta$.

**d.** Note that $\beta = \theta + \varphi$. Use the tangent addition formula $\tan(\theta + \varphi) = \dfrac{\tan\theta + \tan\varphi}{1 - \tan\theta\tan\varphi}$ to show that $\tan\alpha = \tan\beta = \dfrac{2p}{x_0}$.

**e.** Conclude that because $\alpha$ and $\beta$ are acute angles, $\alpha = \beta$.

**84. Deriving polar equations for conics** Use Figures 12.60 and 12.61 to derive the polar equation of a conic section with a focus at the origin in the following three cases.

**a.** Vertical directrix at $x = -d$, where $d > 0$

**b.** Horizontal directrix at $y = d$, where $d > 0$

**c.** Horizontal directrix at $y = -d$, where $d > 0$

**85. Equation of an ellipse** Consider an ellipse to be the set of points in a plane whose distances from two fixed points have a constant sum $2a$. Derive the equation of an ellipse. Assume the two fixed points are on the $x$-axis equidistant from the origin.

**86. Equation of a hyperbola** Consider a hyperbola to be the set of points in a plane whose distances from two fixed points have a constant difference of $2a$ or $-2a$. Derive the equation of a hyperbola. Assume the two fixed points are on the $x$-axis equidistant from the origin.

**87. Equidistant set** Show that the set of points equidistant from a circle and a line not passing through the circle is a parabola. Assume the circle, line, and parabola lie in the same plane.

**88. Polar equation of a conic** Show that the polar equation of an ellipse or hyperbola with one focus at the origin, major axis of length $2a$ on the $x$-axis, and eccentricity $e$ is

$$r = \frac{a(1 - e^2)}{1 + e \cos \theta}.$$

**89. Shared asymptotes** Suppose two hyperbolas with eccentricities $e$ and $E$ have perpendicular major axes and share a set of asymptotes. Show that $e^{-2} + E^{-2} = 1$.

**90–94. Focal chords** A **focal chord** of a conic section is a line through a focus joining two points of the curve. The **latus rectum** is the focal chord perpendicular to the major axis of the conic. Prove the following properties.

**90.** The lines tangent to the endpoints of any focal chord of a parabola $y^2 = 4px$ intersect on the directrix and are perpendicular.

**91.** Let $L$ be the latus rectum of the parabola $y^2 = 4px$ for $p > 0$. Let $F$ be the focus of the parabola, $P$ be any point on the parabola to the left of $L$, and $D$ be the (shortest) distance between $P$ and $L$. Show that for all $P$, $D + |FP|$ is a constant. Find the constant.

**92.** The length of the latus rectum of the parabola $y^2 = 4px$ or $x^2 = 4py$ is $4|p|$.

**93.** The length of the latus rectum of an ellipse centered at the origin is $\dfrac{2b^2}{a} = 2b\sqrt{1 - e^2}$.

**94.** The length of the latus rectum of a hyperbola centered at the origin is $\dfrac{2b^2}{a} = 2b\sqrt{e^2 - 1}$.

**95. Confocal ellipse and hyperbola** Show that an ellipse and a hyperbola that have the same two foci intersect at right angles.

**96. Approach to asymptotes** Show that the vertical distance between a hyperbola $\dfrac{x^2}{a^2} - \dfrac{y^2}{b^2} = 1$ and its asymptote $y = \dfrac{bx}{a}$ approaches zero as $x \to \infty$, where $0 < b < a$.

**97. Sector of a hyperbola** Let $H$ be the right branch of the hyperbola $x^2 - y^2 = 1$ and let $\ell$ be the line $y = m(x - 2)$ that passes through the point $(2, 0)$ with slope $m$, where $-\infty < m < \infty$. Let $R$ be the region in the first quadrant bounded by $H$ and $\ell$ (see figure). Let $A(m)$ be the area of $R$. Note that for some values of $m$, $A(m)$ is not defined.

    **a.** Find the $x$-coordinates of the intersection points between $H$ and $\ell$ as functions of $m$; call them $u(m)$ and $v(m)$, where $v(m) > u(m) > 1$. For what values of $m$ are there two intersection points?

    **b.** Evaluate $\lim\limits_{m \to 1^+} u(m)$ and $\lim\limits_{m \to 1^+} v(m)$.

    **c.** Evaluate $\lim\limits_{m \to \infty} u(m)$ and $\lim\limits_{m \to \infty} v(m)$.

    **d.** Evaluate and interpret $\lim\limits_{m \to \infty} A(m)$.

**T 98. The anvil of a hyperbola** Let $H$ be the hyperbola $x^2 - y^2 = 1$ and let $S$ be the 2-by-2 square bisected by the asymptotes of $H$. Let $R$ be the anvil-shaped region bounded by the hyperbola and the horizontal lines $y = \pm p$ (see figure).

    **a.** For what value of $p$ is the area of $R$ equal to the area of $S$?

    **b.** For what value of $p$ is the area of $R$ twice the area of $S$?

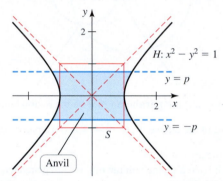

**QUICK CHECK ANSWERS**

**2. a.** Left   **b.** Up   **3.** The minor-axis vertices are $(0, \pm b)$. The distance between them is $2b$, which is the length of the minor axis.   **4.** Vertices: $(0, \pm 1)$; foci: $(0, \pm\sqrt{5})$   **5.** $b = 3\sqrt{3}/2$, $c = 3/2$, $d = 6$   **6.** $y$-axis ◄

---

## CHAPTER 12   REVIEW EXERCISES

**1. Explain why or why not** Determine whether the following statements are true and give an explanation or counterexample.

    **a.** A set of parametric equations for a given curve is always unique.

    **b.** The equations $x = e^t$, $y = 2e^t$, for $-\infty < t < \infty$, describe a line passing through the origin with slope 2.

    **c.** The polar coordinates $\left(3, -\dfrac{3\pi}{4}\right)$ and $\left(-3, \dfrac{\pi}{4}\right)$ describe the same point in the plane.

    **d.** The area of the region between the inner and outer loops of the limaçon $r = f(\theta) = 1 - 4\cos\theta$ is $\dfrac{1}{2}\displaystyle\int_0^{2\pi} f(\theta)^2 \, d\theta$.

    **e.** The hyperbola $\dfrac{y^2}{2} - \dfrac{x^2}{4} = 1$ has no $x$-intercept.

    **f.** The equation $x^2 + 4y^2 - 2x = 3$ describes an ellipse.

**2.** Eliminate the parameter in the parametric equations $x = 1 + \sin t, y = 3 + 2\sin t$, for $0 \le t \le \pi/2$, and describe the curve, indicating its positive orientation. How does this curve differ from the curve $x = 1 + \sin t, y = 3 + 2\sin t$, for $\pi/2 \le t \le \pi$?

**3–6. Eliminating the parameter** *Eliminate the parameter to find a description of the following curves in terms of x and y. Give a geometric description and the positive orientation of the curve.*

**3.** $x = 4\cos t, y = 3\sin t; 0 \le t \le 2\pi$

**4.** $x = \sin t - 3, y = \cos t + 6; 0 \le t \le \pi$

**5.** $x = t + 1, y = \sqrt{t + 1}; 3 \le t \le 8$

**6.** $x = -\cot^2 t, y = 1 + \csc^2 t; \dfrac{\pi}{4} \le t \le \dfrac{\pi}{2}$

**7–8. Parametric curves and tangent lines**

**a.** *Eliminate the parameter to obtain an equation in x and y.*
**b.** *Find the slope of the curve at the given value of t.*
**c.** *Make a sketch of the curve showing the positive orientation of the curve and the tangent line at the point corresponding to the given value of t.*

**7.** $x = 8\cos t + 1, y = 8\sin t + 2$, for $0 \le t \le 2\pi; t = \pi/3$

**8.** $x = 4\sin 2t, y = 3\cos 2t$, for $0 \le t \le \pi; t = \pi/6$

**T 9.** Find parametric equations for the curve $x = 5(y - 1)(y - 2)\sin y$ and then plot the curve using a graphing utility.

**10–12. Parametric curves**

**a.** *Eliminate the parameter to obtain an equation in x and y.*
**b.** *Describe the curve, indicating the positive orientation.*
**c.** *Find the slope of the curve at the specified point.*

**10.** $x = t^2 + 4, y = -t$, for $-2 < t < 0; (5, 1)$

**11.** $x = 3\cos(-t), y = 3\sin(-t) - 1$, for $0 \le t \le \pi; (0, -4)$

**12.** $x = \ln t, y = 8\ln t^2$, for $1 \le t \le e^2; (1, 16)$

**T 13. Tangent lines** Find an equation of the line tangent to the cycloid $x = t - \sin t, y = 1 - \cos t$ at the points corresponding to $t = \pi/6$ and $t = 2\pi/3$.

**14–18. Parametric descriptions** *Write parametric equations for the following curves. Solutions are not unique.*

**14.** The segment of the curve $x = y^3 + y + 1$ that starts at $(1, 0)$ and ends at $(11, 2)$.

**15.** The line segment from $P(-1, 0)$ to $Q(1, 1)$, followed by the line segment from $Q$ to $P$ (two sets of equations are required)

**16.** The segment of the curve $f(x) = x^3 + 2x$ from $(0, 0)$ to $(2, 12)$

**17.** The circle $x^2 + y^2 = 9$, generated clockwise

**18.** The right side of the ellipse $\dfrac{x^2}{9} + \dfrac{y^2}{4} = 1$, generated counterclockwise

**19–20. Area bounded by parametric curves** *Find the area of the following regions. (Hint: See Exercises 103–105 in Section 12.1.)*

**19.** The region bounded by the y-axis and the parametric curve $x = t^2, y = t - t^3$, for $0 \le t \le 1$

**20.** The region bounded by the x-axis and the parametric curve $x = \cos t, y = \sin 2t$, for $0 \le t \le \pi/2$

**T 21. Surface of revolution** Find the area of the surface obtained by revolving the curve $x = \sin t, y = \cos 2t$, for $-\pi/4 \le t \le \pi/4$, about the x-axis. (*Hint:* See Exercises 106–111 in Section 12.1.)

**22–23. Arc length** *Find the length of the following curves.*

**22.** $x = e^{2t}\sin 3t, y = e^{2t}\cos 3t; 0 \le t \le \pi/3$

**23.** $x = \cos 2t, y = 2t - \sin 2t; 0 \le t \le \pi/4$

**24–26. Sets in polar coordinates** *Sketch the following sets of points.*

**24.** $\theta = \dfrac{\pi}{6}$

**25.** $0 \le r \le 4$ and $-\dfrac{\pi}{2} \le \theta \le -\dfrac{\pi}{3}$

**26.** $4 \le r^2 \le 9$

**27–32. Polar curves** *Graph the following equations.*

**27.** $r = 2 - 3\sin\theta$     **28.** $r = 1 + 2\cos\theta$

**29.** $r = 3\cos 3\theta$     **30.** $r = e^{-\theta/6}$

**31.** $r = 3\sin 4\theta$     **32.** $r^2 = 4\cos\theta$

**33. Polar valentine** Liz wants to show her love for Jake by passing him a valentine on her graphing calculator. Sketch each of the following curves and determine which one Liz should use to get a heart-shaped curve.

**a.** $r = 5\cos\theta$     **b.** $r = 1 - \sin\theta$     **c.** $r = \cos 3\theta$

**34. Jake's response** Jake responds to Liz (Exercise 33) with a graph that shows his love for her is infinite. Sketch each of the following curves. Which one should Jake send to Liz to get an infinity symbol?

**a.** $r = \theta$, for $\theta \ge 0$     **b.** $r = \dfrac{1}{2} + \sin\theta$     **c.** $r^2 = \cos 2\theta$

**35. Polar conversion** Write the equation $r^2 + r(2\sin\theta - 6\cos\theta) = 0$ in Cartesian coordinates and identify the corresponding curve.

**36. Polar conversion** Consider the equation $r = \dfrac{4}{\sin\theta + \cos\theta}$.

**a.** Convert the equation to Cartesian coordinates and identify the curve it describes.
**b.** Graph the curve and indicate the points that correspond to $\theta = 0, \dfrac{\pi}{2}$, and $2\pi$.
**c.** Give an interval in $\theta$ on which the entire curve is generated.

**37. Parametric to polar equations** Find an equation of the following curve in polar coordinates and describe the curve.

$$x = (1 + \cos t)\cos t, \quad y = (1 + \cos t)\sin t; \quad 0 \le t \le 2\pi$$

**38. Cartesian conversion** Write the equation $x = y^2$ in polar coordinates and state values of $\theta$ that produce the entire graph of the parabola.

**39. Cartesian conversion** Write the equation $(x - 4)^2 + y^2 = 16$ in polar coordinates and state values of $\theta$ that produce the entire graph of the circle.

**T 40–41. Slopes of tangent lines**

a. *Find all points where the following curves have vertical and hori-zontal tangent lines.*

b. *Find the slope of the lines tangent to the curve at the origin (when relevant).*

c. *Sketch the curve and all the tangent lines identified in parts (a) and (b).*

**40.** $r = 3 - 6\cos\theta$        **41.** $r = 1 - \sin\theta$

**42–43. Intersection points** *Find the intersection points of the follow-ing curves.*

**42.** $r = \sqrt{3}$ and $r = 2\cos 2\theta$    **43.** $r = \sqrt{\cos 3t}$ and $r = \sqrt{\sin 3t}$

**44–49. Areas of regions** *Find the area of the following regions.*

**44.** The region inside one leaf of the rose $r = 3\sin 4\theta$

**45.** The region inside the inner loop of the limaçon $r = 1 + 2\sin\theta$

**46.** The region inside the limaçon $r = 2 + \cos\theta$ and outside the circle $r = 2$

**47.** The region inside the lemniscate $r^2 = 4\cos 2\theta$ and outside the circle $r = \sqrt{2}$

**48.** The region inside $r = \sqrt{2} + \sin\theta$ and inside $r = 3\sin\theta$

**49.** The region inside the cardioid $r = 1 + \cos\theta$ and outside the cardioid $r = 1 - \cos\theta$

**T 50. Intersection points** Consider the polar equations $r = 1$ and $r = 2 - 4\cos\theta$.

    a. Graph the curves. How many intersection points do you observe?

    b. Give approximate polar coordinates of the intersection points.

**T 51–52. Arc length of polar curves** *Find the approximate length of the following curves.*

**51.** The limaçon $r = 3 - 6\cos\theta$

**52.** The limaçon $r = 3 + 2\cos\theta$

**53–57. Conic sections**

a. *Determine whether the following equations describe a parabola, an ellipse, or a hyperbola.*

b. *Use analytical methods to determine the location of the foci, vertices, and directrices.*

c. *Find the eccentricity of the curve.*

d. *Make an accurate graph of the curve.*

**53.** $x^2 - \dfrac{y^2}{2} = 1$        **54.** $x = 16y^2$

**55.** $y^2 - 4x^2 = 16$        **56.** $\dfrac{x^2}{4} + \dfrac{y^2}{25} = 1$

**57.** $4x^2 + 8y^2 = 16$

**58–59. Tangent lines** *Find an equation of the line tangent to the follow-ing curves at the given point. Check your work with a graphing utility.*

**58.** $\dfrac{x^2}{16} - \dfrac{y^2}{9} = 1; \left(\dfrac{20}{3}, -4\right)$    **59.** $y^2 = -12x; \left(-\dfrac{4}{3}, -4\right)$

**60. A polar conic section** Consider the equation $r^2 = \sec 2\theta$.

    a. Convert the equation to Cartesian coordinates and identify the curve.

    b. Find the vertices, foci, directrices, and eccentricity of the curve.

    c. Graph the curve. Explain why the polar equation does not have the form given in the text for conic sections in polar coordinates.

**61–64. Polar equations for conic sections** *Graph the following conic sections, labeling vertices, foci, directrices, and asymptotes (if they ex-ist). Give the eccentricity of the curve. Use a graphing utility to check your work.*

**61.** $r = \dfrac{2}{1 + \sin\theta}$        **62.** $r = \dfrac{3}{1 - 2\cos\theta}$

**63.** $r = \dfrac{4}{2 + \cos\theta}$        **64.** $r = \dfrac{10}{5 + 2\cos\theta}$

**65–68. Eccentricity-directrix approach** *Find an equation of the fol-lowing curves, assuming the center is at the origin. Graph the curve, labeling vertices, foci, asymptotes (if they exist), and directrices.*

**65.** A hyperbola with vertices $(0, \pm 2)$ and directrices $y = \pm 1$

**66.** An ellipse with foci $(\pm 4, 0)$ and directrices $x = \pm 8$

**67.** An ellipse with vertices $(0, \pm 4)$ and directrices $y = \pm 10$

**68.** A hyperbola with vertices $(\pm 4, 0)$ and directrices $x = \pm 2$

**69. Conic parameters** An ellipse has vertices $(0, \pm 6)$ and foci $(0, \pm 4)$. Find the eccentricity, the directrices, and the minor-axis vertices.

**70. Conic parameters** A hyperbola has eccentricity $e = 2$ and foci $(0, \pm 2)$. Find the location of the vertices and directrices.

**71. Bisecting an ellipse** Let $R$ be the region in the first quadrant bounded by the ellipse $\dfrac{x^2}{a^2} + \dfrac{y^2}{b^2} = 1$. Find the value of $m$ (in terms of $a$ and $b$) such that the line $y = mx$ divides $R$ into two subregions of equal area.

**72. Parabola-hyperbola tangency** Let $P$ be the parabola $y = px^2$ and $H$ be the right half of the hyperbola $x^2 - y^2 = 1$.

    a. For what value of $p$ is $P$ tangent to $H$?

    b. At what point does the tangency occur?

    c. Generalize your results for the hyperbola $\dfrac{x^2}{a^2} - \dfrac{y^2}{b^2} = 1$.

**73. Another ellipse construction** Start with two circles centered at the origin with radii $0 < a < b$ (see figure). Assume the line $\ell$ through the origin intersects the smaller circle at $Q$ and the larger circle at $R$. Let $P(x, y)$ have the $y$-coordinate of $Q$ and the $x$-coordinate of $R$. Show that the set of points $P(x, y)$ generated in this way for all lines $\ell$ through the origin is an ellipse.

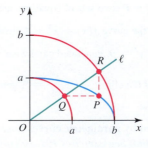

**T 74. Hyperbolas** The graph of the parametric equations $x = a \tan t$, $y = b \sec t$, for $-\pi < t < \pi$ and $|t| \neq \pi/2$, where $a$ and $b$ are nonzero real numbers, is a hyperbola. Graph the hyperbola with $a = b = 1$. Indicate clearly the direction in which the curve is generated as $t$ increases from $t = -\pi$ to $t = \pi$.

**T 75. Lamé curves** The *Lamé curve* described by $\left|\dfrac{x}{a}\right|^n + \left|\dfrac{y}{b}\right|^n = 1$, where $a$, $b$, and $n$ are positive real numbers, is a generalization of an ellipse.

   **a.** Express this equation in parametric form (four pairs of equations are needed).

   **b.** Graph the curve for $a = 4$ and $b = 2$, for various values of $n$.

   **c.** Describe how the curves change as $n$ increases.

**76. General equations for a circle** Prove that the equations

$$x = a \cos t + b \sin t, \quad y = c \cos t + d \sin t$$

where $a$, $b$, $c$, and $d$ are real numbers, describe a circle of radius $R$ provided $a^2 + c^2 = b^2 + d^2 = R^2$ and $ab + cd = 0$.

# Chapter 12 Guided Projects

*Applications of the material in this chapter and related topics can be found in the following Guided Projects. For additional information, see the Preface.*

- The amazing cycloid
- Parametric art
- Polar art
- Grazing goat problems

- Translations and rotations of axes
- Celestial orbits
- Properties of conic sections

# A

# Appendix

## Proofs of Selected Theorems

> **THEOREM 2.3  Limit Laws**
> Assume $\lim\limits_{x \to a} f(x)$ and $\lim\limits_{x \to a} g(x)$ exist. The following properties hold, where $c$ is a real number, and $n > 0$ is an integer.
>
> **1. Sum** $\lim\limits_{x \to a} (f(x) + g(x)) = \lim\limits_{x \to a} f(x) + \lim\limits_{x \to a} g(x)$
>
> **2. Difference** $\lim\limits_{x \to a} (f(x) - g(x)) = \lim\limits_{x \to a} f(x) - \lim\limits_{x \to a} g(x)$
>
> **3. Constant multiple** $\lim\limits_{x \to a} (cf(x)) = c \lim\limits_{x \to a} f(x)$
>
> **4. Product** $\lim\limits_{x \to a} (f(x)g(x)) = \left(\lim\limits_{x \to a} f(x)\right)\left(\lim\limits_{x \to a} g(x)\right)$
>
> **5. Quotient** $\lim\limits_{x \to a} \dfrac{f(x)}{g(x)} = \dfrac{\lim\limits_{x \to a} f(x)}{\lim\limits_{x \to a} g(x)}$, provided $\lim\limits_{x \to a} g(x) \neq 0$
>
> **6. Power** $\lim\limits_{x \to a} (f(x))^n = \left(\lim\limits_{x \to a} f(x)\right)^n$
>
> **7. Root** $\lim\limits_{x \to a} (f(x))^{1/n} = \left(\lim\limits_{x \to a} f(x)\right)^{1/n}$, provided $f(x) > 0$, for $x$ near $a$, if $n$ is even

**Proof:**  The proof of Law 1 is given in Example 6 of Section 2.7. The proof of Law 2 is analogous to that of Law 1; the triangle inequality in the form $|x - y| \le |x| + |y|$ is used (Exercise 43, Section 2.7). The proof of Law 3 is outlined in Exercise 44 of Section 2.7. The proofs of Laws 4 and 5 are given below. The proof of Law 6 involves the repeated use of Law 4. Law 7 is a special case of Theorem 2.12, whose proof is given in this appendix (p. AP-2). ◄

**Proof of Product Law:**  Let $L = \lim\limits_{x \to a} f(x)$ and $M = \lim\limits_{x \to a} g(x)$. Using the definition of a limit, the goal is to show that given any $\varepsilon > 0$, it is possible to specify a $\delta > 0$ such that $|f(x)g(x) - LM| < \varepsilon$ whenever $0 < |x - a| < \delta$. Notice that

$$|f(x)g(x) - LM| = |f(x)g(x) - Lg(x) + Lg(x) - LM| \quad \text{Add and subtract } Lg(x).$$
$$= |(f(x) - L)g(x) + (g(x) - M)L| \quad \text{Group terms.}$$
$$\le |(f(x) - L)g(x)| + |(g(x) - M)L| \quad \text{Triangle inequality}$$
$$= |f(x) - L||g(x)| + |g(x) - M||L|. \quad |xy| = |x||y|$$

▶ Real numbers $x$ and $y$ obey the triangle inequality $|x + y| \le |x| + |y|$.

We now use the definition of the limits of $f$ and $g$, and we note that $L$ and $M$ are fixed real numbers. Given $\varepsilon > 0$, there exist $\delta_1 > 0$ and $\delta_2 > 0$ such that

$$|f(x) - L| < \frac{\varepsilon}{2(|M| + 1)} \quad \text{and} \quad |g(x) - M| < \frac{\varepsilon}{2(|L| + 1)}$$

▶ $|g(x) - M| < 1$ implies that $g(x)$ is less than 1 unit from $M$. Therefore, whether $g(x)$ and $M$ are positive or negative, $|g(x)| < |M| + 1$.

whenever $0 < |x - a| < \delta_1$ and $0 < |x - a| < \delta_2$, respectively. Furthermore, by the definition of the limit of $g$, there exits a $\delta_3 > 0$ such that $|g(x) - M| < 1$ whenever $0 < |x - a| < \delta_3$. It follows that $|g(x)| < |M| + 1$ whenever $0 < |x - a| < \delta_3$. Now take $\delta$ to be the minimum of $\delta_1$, $\delta_2$, and $\delta_3$. Then for $0 < |x - a| < \delta$, we have

$$|f(x)g(x) - LM| \le \underbrace{|f(x) - L|}_{< \frac{\varepsilon}{2(|M| + 1)}} \underbrace{|g(x)|}_{<(|M| + 1)} + \underbrace{|g(x) - M|}_{< \frac{\varepsilon}{2(|L| + 1)}} |L|$$

$$< \frac{\varepsilon}{2} + \frac{\varepsilon}{2} \underbrace{\frac{|L|}{|L| + 1}}_{<1} < \frac{\varepsilon}{2} + \frac{\varepsilon}{2} = \varepsilon.$$

It follows that $\lim\limits_{x \to a} (f(x)g(x)) = LM$.    ◀

**Proof of Quotient Law:** We first prove that if $\lim\limits_{x \to a} g(x) = M$ exists, where $M \ne 0$, then $\lim\limits_{x \to a} \dfrac{1}{g(x)} = \dfrac{1}{M}$. The Quotient Law then follows when we replace $g$ with $1/g$ in the Product Law. Therefore, the goal is to show that given any $\varepsilon > 0$, it is possible to specify a $\delta > 0$ such that $\left| \dfrac{1}{g(x)} - \dfrac{1}{M} \right| < \varepsilon$ whenever $0 < |x - a| < \delta$. First note that $M \ne 0$ and $g(x)$ can be made arbitrarily close to $M$. For this reason, there exists a $\delta_1 > 0$ such that $|g(x)| > |M|/2$, or equivalently, $1/|g(x)| < 2/|M|$, whenever $0 < |x - a| < \delta_1$. Furthermore, using the definition of the limit of $g$, given any $\varepsilon > 0$, there exists a $\delta_2 > 0$ such that $|g(x) - M| < \dfrac{\varepsilon|M|^2}{2}$ whenever $0 < |x - a| < \delta_2$. Now take $\delta$ to be the minimum of $\delta_1$ and $\delta_2$. Then for $0 < |x - a| < \delta$, we have

▶ Note that if $|g(x)| > |M|/2$, then $1/|g(x)| < 2/|M|$.

$$\left| \frac{1}{g(x)} - \frac{1}{M} \right| = \left| \frac{M - g(x)}{Mg(x)} \right| \qquad \text{Common denominator}$$

$$= \frac{1}{|M|} \underbrace{\frac{1}{|g(x)|}}_{< \frac{2}{|M|}} \underbrace{|g(x) - M|}_{< \frac{\varepsilon|M|^2}{2}} \qquad \text{Rewrite.}$$

$$< \frac{1}{|M|} \frac{2}{|M|} \frac{\varepsilon|M|^2}{2} = \varepsilon. \qquad \text{Simplify.}$$

By the definition of a limit, we have $\lim\limits_{x \to a} \dfrac{1}{g(x)} = \dfrac{1}{M}$. The proof can be completed by applying the Product Law with $g$ replaced with $1/g$.

---

**THEOREM 2.12   Limits of Composite Functions**

**1.** If $g$ is continuous at $a$ and $f$ is continuous at $g(a)$, then

$$\lim_{x \to a} f(g(x)) = f(\lim_{x \to a} g(x)).$$

**2.** If $\lim\limits_{x \to a} g(x) = L$ and $f$ is continuous at $L$, then

$$\lim_{x \to a} f(g(x)) = f(\lim_{x \to a} g(x)).$$

**Proof:** Part 1 was proved in Section 2.6 (it is also a direct consequence of Part 2). Let $\varepsilon > 0$ be given. We need to show that there exists $\delta > 0$ such that $|f(g(x)) - f(L)| < \varepsilon$ whenever $0 < |x - a| < \delta$. Because $f$ is continuous at $L$, there exists $\delta_1 > 0$ such that

$$|f(y) - f(L)| < \varepsilon \text{ whenever } |y - L| < \delta_1.$$

We also know that $\lim\limits_{x \to a} g(x) = L$, so there exists $\delta > 0$ such that

$$|g(x) - L| < \delta_1 \text{ whenever } 0 < |x - a| < \delta.$$

Combining these results, if $0 < |x - a| < \delta$, then $|g(x) - L| < \delta_1$, and the inequality $|g(x) - L| < \delta_1$ implies that $|f(g(x)) - f(L)| < \varepsilon$. Therefore, $|f(g(x)) - f(L)| < \varepsilon$ whenever $0 < |x - a| < \delta$. ◄

---

**THEOREM 10.20    Ratio Test**

Let $\sum a_k$ be an infinite series, and let $r = \lim\limits_{k \to \infty} \left| \dfrac{a_{k+1}}{a_k} \right|$.

1. If $r < 1$, the series converges absolutely, and therefore it converges (Theorem 10.19).
2. If $r > 1$ (including $r = \infty$), the series diverges.
3. If $r = 1$, the test is inconclusive.

---

**Proof:** We consider three cases: $r < 1$, $r > 1$, and $r = 1$.

1. Assume $r < 1$ and choose a number $R$ such that $r < R < 1$. Because the sequence $\left\{ \left| \dfrac{a_{k+1}}{a_k} \right| \right\}$ converges to a number $r$ less than $R$, eventually all the terms in the tail of the sequence $\left\{ \left| \dfrac{a_{k+1}}{a_k} \right| \right\}$ are less than $R$. That is, there is a positive integer $N$ such that $\left| \dfrac{a_{k+1}}{a_k} \right| < R$, for all $k > N$. Multiplying both sides of this inequality by $\dfrac{|a_k|}{R^{k+1}}$, we have $\dfrac{|a_{k+1}|}{R^{k+1}} < \dfrac{|a_k|}{R^k}$, for all $k > N$. So the sequence $\left\{ \dfrac{|a_k|}{R^k} \right\}_{k=N+1}^{\infty}$ is decreasing and it follows that $\dfrac{|a_k|}{R^k} < \dfrac{|a_{N+1}|}{R^{N+1}}$, for all $k \geq N + 1$. By letting $c = \dfrac{|a_{N+1}|}{R^{N+1}}$, we have

$$0 < |a_k| \leq cR^k, \text{ for all } k \geq N + 1. \text{ Let } S_n \text{ represent the } n\text{th partial sum of } \sum_{k=N+1}^{\infty} |a_k|;$$

note that the partial sums of this series are bounded by a convergent geometric series:

$$\begin{aligned}
S_n &= |a_{N+1}| + |a_{N+2}| + \cdots + |a_{N+n}| \\
&\leq cR^{N+1} + cR^{N+2} + \cdots + cR^{N+n} \\
&< cR^{N+1} + cR^{N+2} + \cdots + cR^{N+n} + \cdots \\
&= \frac{cR^{N+1}}{1 - R}.
\end{aligned}$$

Because the sequence $\{S_n\}$ is increasing (each partial sum in the sequence consists of positive terms) and is bounded above by $\dfrac{cR^{N+1}}{1 - R}$, it converges by the Bounded Monotonic Sequences Theorem (Theorem 10.5). Therefore, $\sum\limits_{k=N+1}^{\infty} |a_k|$ converges and we conclude that $\sum\limits_{k=1}^{\infty} |a_k|$ converges (Theorem 10.8), which in turn implies that $\sum\limits_{k=1}^{\infty} a_k$ converges (Theorem 10.19).

2. If $r > 1$, there is a positive integer $N$ for which $\left|\dfrac{a_{k+1}}{a_k}\right| > 1$, or equivalently $|a_{k+1}| > |a_k|$, for all $k > N$. So every term in the sequence $\{|a_k|\}_{k=N+1}^{\infty}$ is greater than or equal to the positive number $|a_{N+1}|$, which implies that $\lim\limits_{n\to\infty} a_n \neq 0$. Therefore, the series $\sum\limits_{k=N+1}^{\infty} a_k$ diverges by the Divergence Test, and we conclude that $\sum\limits_{k=1}^{\infty} a_k$ diverges (Theorem 10.8).

3. In the case that $r = 1$, the series $\sum\limits_{k=1}^{\infty} a_k$ may or may not converge. For example, both the divergent harmonic series $\sum\limits_{k=1}^{\infty}\dfrac{1}{k}$ and the convergent $p$-series $\sum\limits_{k=1}^{\infty}\dfrac{1}{k^2}$ produce a value of $r = 1$.    ◄

---

**THEOREM 10.21    Root Test**

Let $\Sigma a_k$ be an infinite series, and let $\rho = \lim\limits_{k\to\infty}\sqrt[k]{|a_k|}$.

1. If $\rho < 1$, the series converges absolutely, and therefore it converges (Theorem 10.19).
2. If $\rho > 1$ (including $\rho = \infty$), the series diverges.
3. If $\rho = 1$, the test is inconclusive.

---

**Proof:** We consider three cases: $\rho < 1$, $\rho > 1$, and $\rho = 1$.

1. Assume $\rho < 1$ and choose a number $R$ such that $\rho < R < 1$. Because the sequence $\{\sqrt[k]{|a_k|}\}$ converges to a number less than $R$, there is a positive integer $N$ such that $\sqrt[k]{|a_k|} < R$, or equivalently $|a_k| < R^k$, for all $k > N$. Let $S_n$ represent the $n$th partial sum of $\sum\limits_{k=N+1}^{\infty} |a_k|$; note that the partial sums of this series are bounded by a convergent geometric series:

$$\begin{aligned}
S_n &= |a_{N+1}| + |a_{N+2}| + \cdots + |a_{N+n}| \\
&\leq R^{N+1} + R^{N+2} + \cdots + R^{N+n} \\
&< R^{N+1} + R^{N+2} + \cdots + R^{N+n} + \cdots \\
&= \frac{R^{N+1}}{1 - R}.
\end{aligned}$$

Because the sequence $\{S_n\}$ is nondecreasing (each partial sum in the sequence consists of nonnegative terms) and is bounded above by $\dfrac{R^{N+1}}{1 - R}$, it converges by the Bounded Monotonic Sequences Theorem (Theorem 10.5). Therefore, $\sum\limits_{k=N+1}^{\infty} |a_k|$ converges, and we conclude that $\sum\limits_{k=1}^{\infty} |a_k|$ converges (Theorem 10.8), which in turn implies that $\sum\limits_{k=1}^{\infty} a_k$ converges (Theorem 10.19).

2. If $\rho > 1$, there is an integer $N$ for which $\sqrt[k]{|a_k|} > 1$, or equivalently $|a_k| > 1$, for all $k > N$. So every term in the sequence $\{|a_k|\}_{k=N+1}^{\infty}$ is greater than or equal to 1, which implies that $\lim\limits_{n\to\infty} a_n \neq 0$. Therefore, the series $\sum\limits_{k=N+1}^{\infty} a_k$ diverges by the Divergence Test, and we conclude that $\sum\limits_{k=1}^{\infty} a_k$ diverges (Theorem 10.8).

3. If $\rho = 1$, the series $\sum\limits_{k=1}^{\infty} a_k$ may or may not converge. For example, both the divergent harmonic series $\sum\limits_{k=1}^{\infty}\dfrac{1}{k}$ and the convergent $p$-series $\sum\limits_{k=1}^{\infty}\dfrac{1}{k^2}$ produce a value of $\rho = 1$.    ◄

---

**THEOREM 11.3    Convergence of Power Series**

A power series $\sum_{k=0}^{\infty} c_k(x - a)^k$ centered at $a$ converges in one of three ways:

1. The series converges absolutely for all $x$. It follows, by Theorem 10.19, that the series converges, in which case the interval of convergence is $(-\infty, \infty)$ and the radius of convergence is $R = \infty$.

2. There is a real number $R > 0$ such that the series converges absolutely (and therefore converges) for $|x - a| < R$ and diverges for $|x - a| > R$, in which case the radius of convergence is $R$.

3. The series converges only at $a$, in which case the radius of convergence is $R = 0$.

---

**Proof:** Without loss of generality, we take $a = 0$. (If $a \neq 0$, the following argument may be shifted so it is centered at $x = a$.) The proof hinges on a preliminary result:

If $\sum_{k=0}^{\infty} c_k x^k$ converges for $x = b \neq 0$, then it converges absolutely, for

$|x| < |b|$. If $\sum_{k=0}^{\infty} c_k x^k$ diverges for $x = d$, then it diverges, for $|x| > |d|$.

To prove these facts, assume that $\sum_{k=0}^{\infty} c_k b^k$ converges, which implies that $\lim_{k \to \infty} c_k b^k = 0$. Then there exists a real number $M > 0$ such that $|c_k b^k| < M$, for $k = 0, 1, 2, 3, \ldots$. It follows that

$$\sum_{k=0}^{\infty} |c_k x^k| = \sum_{k=0}^{\infty} \underbrace{|c_k b^k|}_{< M} \left| \frac{x}{b} \right|^k < M \sum_{k=0}^{\infty} \left| \frac{x}{b} \right|^k.$$

If $|x| < |b|$, then $|x/b| < 1$ and $\sum_{k=0}^{\infty} \left| \frac{x}{b} \right|^k$ is a convergent geometric series. Therefore, $\sum_{k=0}^{\infty} |c_k x^k|$ converges by the comparison test, which implies that $\sum_{k=0}^{\infty} c_k x^k$ converges absolutely for $|x| < |b|$. The second half of the preliminary result is proved by supposing the series diverges at $x = d$. The series cannot converge at a point $x_0$ with $|x_0| > |d|$ because by the preceding argument, it would converge for $|x| < |x_0|$, which includes $x = d$. Therefore, the series diverges for $|x| > |d|$.

Now we may deal with the three cases in the theorem. Let $S$ be the set of real numbers for which the series converges, which always includes 0. If $S = \{0\}$, then we have Case 3. If $S$ consists of all real numbers, then we have Case 1. For Case 2, assume that $d \neq 0$ is a point at which the series diverges. By the preliminary result, the series diverges for $|x| > |d|$. Therefore, if $x$ is in $S$, then $|x| < |d|$, which implies that $S$ is bounded. By the Least Upper Bound Property for real numbers, $S$ has a least upper bound $R$, such that $x \leq R$, for all $x$ in $S$. If $|x| > R$, then $x$ is not in $S$ and the series diverges. If $|x| < R$, then $x$ is not the least upper bound of $S$ and there exists a number $b$ in $S$ with $|x| < b \leq R$.

Because the series converges at $x = b$, by the preliminary result, $\sum_{k=0}^{\infty} |c_k x^k|$ converges for $|x| < |b|$. Therefore, the series $\sum_{k=0}^{\infty} c_k x^k$ converges absolutely (which, by Theorem 10.19,

implies the series converges) for $|x| < R$ and diverges for $|x| > R$.    ◄

▶ The Least Upper Bound Property for real numbers states that if a nonempty set $S$ is bounded (that is, there exists a number $M$, called an *upper bound*, such that $x \leq M$ for all $x$ in $S$), then $S$ has a *least upper bound* $L$, which is the smallest of the upper bounds.

> **THEOREM 12.3**  **Eccentricity-Directrix Theorem**
>
> Suppose $\ell$ is a line, $F$ is a point not on $\ell$, and $e$ is a positive real number. Let $C$ be the set of points $P$ in a plane with the property that $\dfrac{|PF|}{|PL|} = e$, where $|PL|$ is the perpendicular distance from $P$ to $\ell$.
>
> 1. If $e = 1$, $C$ is a **parabola**.
> 2. If $0 < e < 1$, $C$ is an **ellipse**.
> 3. If $e > 1$, $C$ is a **hyperbola**.

$$x = d = \frac{a}{e}$$

$$|PF| = e|PL|$$

**Figure A.1**

**Proof:** If $e = 1$, then the defining property becomes $|PF| = |PL|$, which is the standard definition of a parabola (Section 12.4). We prove the result for ellipses ($0 < e < 1$); a small modification handles the case of hyperbolas ($e > 1$).

Let $E$ be the curve whose points satisfy $|PF| = e\,|PL|$ and suppose $0 < e < 1$; the goal is to show that $E$ is an ellipse. There are two points on the $x$-axis (the *vertices*), call them $V$ and $V'$, that satisfy $|VF| = e|VL|$ and $|V'F| = e|V'L|$. (Note that $|VL|$ is the perpendicular distance from $V$ to the line $\ell$, not the distance from the points labeled $V$ and $L$ in **Figure A.1**. $|V'L|$ should also be interpreted in this way.) We choose the origin such that $V$ and $V'$ have coordinates $(a, 0)$ and $(-a, 0)$, respectively (Figure A.1). We locate the point $F$ (a *focus*) at $(c, 0)$ and let $\ell$ (a *directrix*) be the line $x = d$, where $c > 0$ and $d > 0$. These choices place the center of $E$ at the origin. Notice that we have four parameters ($a$, $c$, $d$, and $e$) that must be related.

Because the vertex $V(a, 0)$ is on $E$, it satisfies the defining property $|PF| = e\,|PL|$, with $P = V$. This condition implies that $a - c = e(d - a)$. Because the vertex $V'(-a, 0)$ is on the curve $E$, it also satisfies the defining property $|PF| = e\,|PL|$, with $P = V'$. This condition implies that $a + c = e(d + a)$. Solving these two equations for $c$ and $d$, we find that $c = ae$ and $d = a/e$. To summarize, the parameters $a$, $c$, $d$, and $e$ are related by the equations

$$c = ae \quad \text{and} \quad a = de.$$

Because $e < 1$, it follows that $c < a < d$.

We now use the property $|PF| = e\,|PL|$ with an arbitrary point $P(x, y)$ on the curve $E$. Figure A.1 shows the geometry with the focus $(c, 0) = (ae, 0)$ and the directrix $x = d = a/e$. The condition $|PF| = e\,|PL|$ becomes

$$\sqrt{(x - ae)^2 + y^2} = e\left(\frac{a}{e} - x\right).$$

The goal is to find the simplest possible relationship between $x$ and $y$. Squaring both sides and collecting terms, we have

$$(1 - e^2)x^2 + y^2 = a^2(1 - e^2).$$

Dividing through by $a^2(1 - e^2)$ gives the equation of the standard ellipse:

$$\frac{x^2}{a^2} + \frac{y^2}{a^2(1 - e^2)} = \frac{x^2}{a^2} + \frac{y^2}{b^2} = 1, \quad \text{where} \quad b^2 = a^2(1 - e^2).$$

This is the equation of an ellipse centered at the origin with vertices and foci on the $x$-axis.

The preceding proof is now applied with $e > 1$. The argument for ellipses with $0 < e < 1$ led to the equation

$$\frac{x^2}{a^2} + \frac{y^2}{a^2(1 - e^2)} = 1.$$

With $e > 1$, we have $1 - e^2 < 0$, so we write $(1 - e^2) = -(e^2 - 1)$. The resulting equation describes a hyperbola centered at the origin with the foci on the $x$-axis:

$$\frac{x^2}{a^2} - \frac{y^2}{b^2} = 1, \quad \text{where} \quad b^2 = a^2(e^2 - 1). \qquad \blacktriangleleft$$

> **THEOREM 15.3**  **Continuity of Composite Functions**
> If $u = g(x, y)$ is continuous at $(a, b)$ and $z = f(u)$ is continuous at $g(a, b)$, then the composite function $z = f(g(x, y))$ is continuous at $(a, b)$.

**Proof:** Let $P$ and $P_0$ represent the points $(x, y)$ and $(a, b)$, respectively. Let $u = g(P)$ and $u_0 = g(P_0)$. The continuity of $f$ at $u_0$ means that $\lim_{u \to u_0} f(u) = f(u_0)$. This limit implies that given any $\varepsilon > 0$, there exists a $\delta^* > 0$ such that

$$|f(u) - f(u_0)| < \varepsilon \quad \text{whenever} \quad 0 < |u - u_0| < \delta^*.$$

The continuity of $g$ at $P_0$ means that $\lim_{P \to P_0} g(P) = g(P_0)$. Letting $|P - P_0|$ denote the distance between $P$ and $P_0$, this limit implies that given any $\delta^* > 0$, there exists a $\delta > 0$ such that

$$|g(P) - g(P_0)| = |u - u_0| < \delta^* \quad \text{whenever} \quad 0 < |P - P_0| < \delta.$$

We now combine these two statements. Given any $\varepsilon > 0$, there exists a $\delta > 0$ such that

$$|f(g(P)) - f(g(P_0))| = |f(u) - f(u_0)| < \varepsilon \quad \text{whenever} \quad 0 < |P - P_0| < \delta.$$

Therefore, $\lim_{(x,y) \to (a,b)} f(g(x, y)) = f(g(a, b))$ and $z = f(g(x, y))$ is continuous at $(a, b)$. ◀

> **THEOREM 15.5**  **Conditions for Differentiability**
> Suppose the function $f$ has partial derivatives $f_x$ and $f_y$ defined on an open set containing $(a, b)$, with $f_x$ and $f_y$ continuous at $(a, b)$. Then $f$ is differentiable at $(a, b)$.

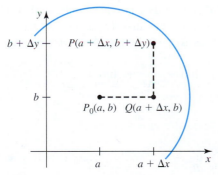

**Figure A.2**

**Proof:** **Figure A.2** shows a region on which the conditions of the theorem are satisfied containing the points $P_0(a, b)$, $Q(a + \Delta x, b)$, and $P(a + \Delta x, b + \Delta y)$. By the definition of differentiability of $f$ at $P_0$, we must show that

$$\Delta z = f(P) - f(P_0) = f_x(a, b)\Delta x + f_y(a, b)\Delta y + \varepsilon_1 \Delta x + \varepsilon_2 \Delta y,$$

where $\varepsilon_1$ and $\varepsilon_2$ depend only on $a, b, \Delta x$, and $\Delta y$, with $(\varepsilon_1, \varepsilon_2) \to (0, 0)$ as $(\Delta x, \Delta y) \to (0, 0)$. We can view the change $\Delta z$ taking place in two stages:

• $\Delta z_1 = f(a + \Delta x, b) - f(a, b)$ is the change in $z$ as $(x, y)$ moves from $P_0$ to $Q$.
• $\Delta z_2 = f(a + \Delta x, b + \Delta y) - f(a + \Delta x, b)$ is the change in $z$ as $(x, y)$ moves from $Q$ to $P$.

Applying the Mean Value Theorem to the first variable and noting that $f$ is differentiable with respect to $x$, we have

$$\Delta z_1 = f(a + \Delta x, b) - f(a, b) = f_x(c, b)\Delta x,$$

where $c$ lies in the interval $(a, a + \Delta x)$. Similarly, applying the Mean Value Theorem to the second variable and noting that $f$ is differentiable with respect to $y$, we have

$$\Delta z_2 = f(a + \Delta x, b + \Delta y) - f(a + \Delta x, b) = f_y(a + \Delta x, d)\Delta y,$$

where $d$ lies in the interval $(b, b + \Delta y)$. We now express $\Delta z$ as the sum of $\Delta z_1$ and $\Delta z_2$:

$$\begin{aligned}
\Delta z &= \Delta z_1 + \Delta z_2 \\
&= f_x(c, b)\Delta x + f_y(a + \Delta x, d)\Delta y \\
&= \underbrace{(f_x(c, b) - f_x(a, b)}_{\varepsilon_1} + f_x(a, b))\Delta x \qquad \text{Add and subtract } f_x(a, b). \\
&\quad + \underbrace{(f_y(a + \Delta x, d) - f_y(a, b)}_{\varepsilon_2} + f_y(a, b))\Delta y \qquad \text{Add and subtract } f_y(a, b). \\
&= (f_x(a, b) + \varepsilon_1)\Delta x + (f_y(a, b) + \varepsilon_2)\Delta y.
\end{aligned}$$

Note that as $\Delta x \to 0$ and $\Delta y \to 0$, we have $c \to a$ and $d \to b$. Because $f_x$ and $f_y$ are continuous at $(a, b)$, it follows that as $\Delta x \to 0$ and $\Delta y \to 0$,

$$\varepsilon_1 = f_x(c, b) - f_x(a, b) \to 0 \quad \text{and} \quad \varepsilon_2 = f_y(a + \Delta x, d) - f_y(a, b) \to 0.$$

Therefore, the condition for differentiability of $f$ at $(a, b)$ has been proved. ◄

---

**THEOREM 15.7    Chain Rule (One Independent Variable)**

Let $z = f(x, y)$ be a differentiable function of $x$ and $y$ on its domain, where $x$ and $y$ are differentiable functions of $t$ on an interval $I$. Then

$$\frac{dz}{dt} = \frac{\partial f}{\partial x}\frac{dx}{dt} + \frac{\partial f}{\partial y}\frac{dy}{dt}.$$

---

**Proof:** Assume $(a, b) = (x(t), y(t))$ is in the domain of $f$, where $t$ is in $I$. Let $\Delta x = x(t + \Delta t) - x(t)$ and $\Delta y = y(t + \Delta t) - y(t)$. Because $f$ is differentiable at $(a, b)$, we know (Section 15.3) that

$$\Delta z = \frac{\partial f}{\partial x}(a, b)\Delta x + \frac{\partial f}{\partial y}(a, b)\Delta y + \varepsilon_1 \Delta x + \varepsilon_2 \Delta y,$$

where $(\varepsilon_1, \varepsilon_2) \to (0, 0)$ as $(\Delta x, \Delta y) \to (0, 0)$. Dividing this equation by $\Delta t$ gives

$$\frac{\Delta z}{\Delta t} = \frac{\partial f}{\partial x}\frac{\Delta x}{\Delta t} + \frac{\partial f}{\partial y}\frac{\Delta y}{\Delta t} + \varepsilon_1 \frac{\Delta x}{\Delta t} + \varepsilon_2 \frac{\Delta y}{\Delta t}.$$

As $\Delta t \to 0$, several things occur. First, because $x = g(t)$ and $y = h(t)$ are differentiable on $I$, $\frac{\Delta x}{\Delta t}$ and $\frac{\Delta y}{\Delta t}$ approach $\frac{dx}{dt}$ and $\frac{dy}{dt}$, respectively. Similarly, $\frac{\Delta z}{\Delta t}$ approaches $\frac{dz}{dt}$ as $\Delta t \to 0$. The fact that $x$ and $y$ are continuous on $I$ (because they are differentiable there) means that $\Delta x \to 0$ and $\Delta y \to 0$ as $\Delta t \to 0$. Therefore, because $(\varepsilon_1, \varepsilon_2) \to (0, 0)$ as $(\Delta x, \Delta y) \to (0, 0)$, it follows that $(\varepsilon_1, \varepsilon_2) \to (0, 0)$ as $\Delta t \to 0$. Letting $\Delta t \to 0$, we have

$$\underbrace{\lim_{\Delta t \to 0}\frac{\Delta z}{\Delta t}}_{\frac{dz}{dt}} = \frac{\partial f}{\partial x}\underbrace{\lim_{\Delta t \to 0}\frac{\Delta x}{\Delta t}}_{\frac{dx}{dt}} + \frac{\partial f}{\partial y}\underbrace{\lim_{\Delta t \to 0}\frac{\Delta y}{\Delta t}}_{\frac{dy}{dt}} + \underbrace{\lim_{\Delta t \to 0}\varepsilon_1}_{\to 0}\underbrace{\frac{\Delta x}{\Delta t}}_{\to \frac{dx}{dt}} + \underbrace{\lim_{\Delta t \to 0}\varepsilon_2}_{\to 0}\underbrace{\frac{\Delta y}{\Delta t}}_{\to \frac{dy}{dt}}$$

or

$$\frac{dz}{dt} = \frac{\partial f}{\partial x}\frac{dx}{dt} + \frac{\partial f}{\partial y}\frac{dy}{dt}.$$  ◄

---

**THEOREM 15.15    Second Derivative Test**

Suppose the second partial derivatives of $f$ are continuous throughout an open disk centered at the point $(a, b)$, where $f_x(a, b) = f_y(a, b) = 0$. Let $D(x, y) = f_{xx}(x, y)\, f_{yy}(x, y) - (f_{xy}(x, y))^2$.

**1.** If $D(a, b) > 0$ and $f_{xx}(a, b) < 0$, then $f$ has a local maximum value at $(a, b)$.

**2.** If $D(a, b) > 0$ and $f_{xx}(a, b) > 0$, then $f$ has a local minimum value at $(a, b)$.

**3.** If $D(a, b) < 0$, then $f$ has a saddle point at $(a, b)$.

**4.** If $D(a, b) = 0$, then the test is inconclusive.

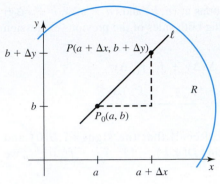

**Figure A.3**

**Proof:** The proof relies on a two-variable version of Taylor's Theorem, which we prove first. **Figure A.3** shows the open disk $R$ on which the conditions of Theorem 15.15 are satisfied; it contains the points $P_0(a, b)$ and $P(a + \Delta x, b + \Delta y)$. The line $\ell$ through $P_0 P$ has a parametric description

$$\langle x(t), y(t) \rangle = \langle a + t\Delta x, b + t\Delta y \rangle,$$

where $t = 0$ corresponds to $P_0$ and $t = 1$ corresponds to $P$.

We now let $F(t) = f(a + t\Delta x, b + t\Delta y)$ be the value of $f$ along that part of $\ell$ that lies in $R$. By the Chain Rule, we have

$$F'(t) = f_x \underbrace{x'(t)}_{\Delta x} + f_y \underbrace{y'(t)}_{\Delta y} = f_x \, \Delta x + f_y \, \Delta y.$$

Differentiating again with respect to $t$ ($f_x$ and $f_y$ are differentiable), we use $f_{xy} = f_{yx}$ to obtain

$$F''(t) = \frac{\partial F'}{\partial x} \underbrace{x'(t)}_{\Delta x} + \frac{\partial F'}{\partial y} \underbrace{y'(t)}_{\Delta y}$$

$$= \frac{\partial}{\partial x}(f_x \, \Delta x + f_y \, \Delta y)\Delta x + \frac{\partial}{\partial y}(f_x \, \Delta x + f_y \, \Delta y)\Delta y$$

$$= f_{xx} \, \Delta x^2 + 2f_{xy} \, \Delta x\Delta y + f_{yy} \, \Delta y^2.$$

Noting that $F$ meets the conditions of Taylor's Theorem for one variable with $n = 1$, we write

$$F(t) = F(0) + F'(0)(t - 0) + \frac{1}{2}F''(c)(t - 0)^2,$$

where $c$ is between $0$ and $t$. Setting $t = 1$, it follows that

$$F(1) = F(0) + F'(0) + \frac{1}{2}F''(c),$$

where $0 < c < 1$. Recalling that $F(t) = f(a + t\Delta x, b + t\Delta y)$ and invoking the condition $f_x(a, b) = f_y(a, b) = 0$, we have

$$F(1) = f(a + \Delta x, b + \Delta y)$$

$$= f(a, b) + \underbrace{f_x(a, b)\Delta x + f_y(a, b)\Delta y}_{F'(0) = 0}$$

$$+ \frac{1}{2}(f_{xx} \, \Delta x^2 + 2f_{xy} \, \Delta x\Delta y + f_{yy} \, \Delta y^2)\Big|_{(a+c\Delta x, \, b+c\Delta y)}$$

$$= f(a, b) + \underbrace{\frac{1}{2}(f_{xx} \, \Delta x^2 + 2f_{xy} \, \Delta x\Delta y + f_{yy} \, \Delta y^2)\Big|_{(a+c\Delta x, \, b+c\Delta y)}}_{H(c)}$$

$$= f(a, b) + \frac{1}{2}H(c).$$

The existence and type of extreme point at $(a, b)$ are determined by the sign of $f(a + \Delta x, b + \Delta y) - f(a, b)$ (for example, if $f(a + \Delta x, b + \Delta y) - f(a, b) \geq 0$ for all $\Delta x$ and $\Delta y$ near $0$, then $f$ has a local minimum at $(a, b)$). Note that $f(a + \Delta x, b + \Delta y) - f(a, b)$ has the same sign as the quantity we have denoted $H(c)$. Assuming $H(0) \neq 0$, for $\Delta x$ and $\Delta y$ sufficiently small and nonzero, the sign of $H(c)$ is the same as the sign of

$$H(0) = f_{xx}(a, b)\Delta x^2 + 2f_{xy}(a, b)\Delta x\Delta y + f_{yy}(a, b)\Delta y^2$$

(because the second partial derivatives are continuous at $(a, b)$ and $(a + c\Delta x, b + c\Delta y)$ can be made arbitrarily close to $(a, b)$). Multiplying both sides of the previous expression by $f_{xx}$ and rearranging terms leads to

$$f_{xx} H(0) = f_{xx}^2 \Delta x^2 + 2f_{xy} f_{xx} \Delta x \Delta y + f_{yy} f_{xx} \Delta y^2$$

$$= \underbrace{(f_{xx}\Delta x + f_{xy}\Delta y)^2}_{\geq 0} + \underbrace{(f_{xx} f_{yy} - f_{xy}^2)}_{D} \Delta y^2,$$

where all derivatives are evaluated at $(a, b)$. Recall that the signs of $H(0)$ and $f(a + \Delta x, b + \Delta y) - f(a, b)$ are the same. Letting $D(a, b) = (f_{xx} f_{yy} - f_{xy}^2)|_{(a, b)}$, we reach the following conclusions:

- If $D(a, b) > 0$ and $f_{xx}(a, b) < 0$, then $H(0) < 0$ (for $\Delta x$ and $\Delta y$ sufficiently close to 0) and $f(a + \Delta x, b + \Delta y) - f(a, b) < 0$. Therefore, $f$ has a local maximum value at $(a, b)$.

- If $D(a, b) > 0$ and $f_{xx}(a, b) > 0$, then $H(0) > 0$ (for $\Delta x$ and $\Delta y$ sufficiently close to 0) and $f(a + \Delta x, b + \Delta y) - f(a, b) > 0$. Therefore, $f$ has a local minimum value at $(a, b)$.

- If $D(a, b) < 0$, then $H(0) > 0$ for some small nonzero values of $\Delta x$ and $\Delta y$ (implying $f(a + \Delta x, b + \Delta y) > f(a, b)$), *and* $H(0) < 0$ for other small nonzero values of $\Delta x$ and $\Delta y$ (implying $f(a + \Delta x, b + \Delta y) < f(a, b)$). (The relative sizes of $(f_{xx} \Delta x + f_{xy} \Delta y)^2$ and $(f_{xx} f_{yy} - f_{xy}^2) \Delta y^2$ can be adjusted by varying $\Delta x$ and $\Delta y$.) Therefore, $f$ has a saddle point at $(a, b)$.

- If $D(a, b) = 0$, then $H(0)$ may be zero, in which case the sign of $H(c)$ cannot be determined. Therefore, the test is inconclusive.    ◄

# Answers

## CHAPTER 1

### Section 1.1 Exercises, pp. 9–13

**1.** A function is a rule that assigns to each value of the independent variable in the domain a unique value of the dependent variable in the range. **3.** $B$ **5.** The first statement **7.** $D = \mathbb{R}, R = [-10, \infty)$ **9.** The independent variable is $h$; the dependent variable is $V$; $D = [0, 50]$. **11.** $-3; 1/8; 1/(2x)$ **13.** The domain of $f \circ g$ consists of all $x$ in the domain of $g$ such that $g(x)$ is in the domain of $f$. **15. a.** 4 **b.** 1 **c.** 3 **d.** 3 **e.** 8 **f.** 1 **17.** 15.4 ft/s; radiosonde rises at an average rate of 15.4 ft/s during the first 5 seconds of its flight. **19.** 2; 2; 2; $-2$ **21.** $A$ is even, $B$ is odd, and $C$ is even. **23.** $D = \{x : x \neq 2\}; R = \{y : y \neq -1\}$ **25.** $D = [-\sqrt{7}, \sqrt{7}]$; $R = [0, \sqrt{7}]$ **27.** $D = \mathbb{R}$ **29.** $D = [-3, 3]$ **31. a.** $[0, 3 + \sqrt{14}]$

**b.**

At time $t = 3$, the maximum height is 224 ft.

**33.** $1/z^3$ **35.** $1/(y^3 - 3)$ **37.** $(u^2 - 4)^3$ **39.** $\dfrac{x - 3}{10 - 3x}$ **41.** $x$

**43.** $g(x) = x^3 - 5, f(x) = x^{10}$ **45.** $g(x) = x^4 + 2, f(x) = \sqrt{x}$

**47.** $|x^2 - 4|; D = \mathbb{R}$ **49.** $\dfrac{1}{|x - 2|}; D = \{x : x \neq 2\}$

**51.** $\dfrac{1}{x^2 - 6}; D = \{x : x \neq \sqrt{6}, -\sqrt{6}\}$

**53.** $x^4 - 8x^2 + 12; D = \mathbb{R}$ **55.** $f(x) = x - 3$
**57.** $f(x) = x^2$ **59.** $f(x) = x^2$ **61. a.** True **b.** False **c.** True
**d.** False **e.** False **f.** True **g.** True **h.** False **i.** True

**63.** 3 **65.** $2x + h$ **67.** $-\dfrac{2}{x(x + h)}$ **69.** $x + a + 1$

**71.** $x^2 + ax + a^2 - 2$ **73.** $\dfrac{4(x + a)}{a^2 x^2}$ **75. a.** 864 ft/hr;

the hiker's elevation increases at an average rate of 864 ft/hr.
**b.** $-487$ ft/hr; the hiker's elevation decreases at an average rate of 487 ft/hr. **c.** The hiker might have stopped to rest during this interval of time and/or the trail was level during this portion of the hike.
**77. a.**

**b.** $m_{\text{sec}} = 112$ ft/s; the object falls at an average rate of 112 ft/s.

**79.** $y$-axis **81.** No symmetry **83.** $x$-axis, $y$-axis, origin
**85.** Origin **87. a.** 4 **b.** 1 **c.** 3 **d.** $-2$ **e.** $-1$ **f.** 7
**89.**

**91.** The equation $y = 2 - \sqrt{-x^2 + 6x + 16}$ can be rewritten as $(x - 3)^2 + (y - 2)^2 = 5^2$. Because $y \leq 2$, the function is the lower half of a circle of radius 5 centered at $(3, 2)$. $D = [-2, 8]; R = [-3, 2]$ **93.** $f(x) = 3x - 2$ or $f(x) = -3x + 4$

**95.** $f(x) = x^2 - 6$ **97.** $\dfrac{1}{\sqrt{x + h} + \sqrt{x}}; \dfrac{1}{\sqrt{x} + \sqrt{a}}$

**99.** $\dfrac{3}{\sqrt{x}(x + h) + x\sqrt{x + h}}; \dfrac{3}{x\sqrt{a} + a\sqrt{x}}$ **101.** None **103.** $y$-axis

### Section 1.2 Exercises, pp. 22–27

**1.** A formula, a graph, a table, words **3.** $y = -\frac{2}{3}x - 1$ **5.** The set of all real numbers for which the denominator does not equal 0

**7.** $y = \begin{cases} x + 3 & \text{if } x < 0 \\ -\frac{1}{2}x + 3 & \text{if } x \geq 0 \end{cases}$ **9.** Shift the graph to the left 2 units.

**11.** Compress the graph horizontally by a factor of $\frac{1}{3}$.
**13.** $f(x) = |x - 2| + 3; g(x) = -|x + 2| - 1$
**15.** $f(x) = 2x + 1$

**17.** $f(x) = 3x - 7$ **19.** $C_s = 5.71; 856.5$ million
**21.** $d = -3p/50 + 27; D = [0, 450]$

**23. a.** $p(t) = 328.3t + 1875$ **b.** 4830

**25.** $f(x) = \begin{cases} 3 & \text{if } x \leq 3 \\ 2x - 3 & \text{if } x > 3 \end{cases}$

**27.** $c(t) = \begin{cases} 0.05t & \text{if } 0 \le t \le 60 \\ 1.2 + 0.03t & \text{if } 60 < t \le 120 \end{cases}$

**29.**   **31.**

**33.**   **35. a.**

**b.** $D = \mathbb{R}$  **c.** One peak near $x = 0$; one valley near $x = 4/3$; $x$-intercept approx. $(-1.3, 0)$, $y$-intercept $(0, 6)$

**37. a.**

**b.** $D = \{x : x \ne -3\}$  **c.** Undefined at $x = -3$; a valley near $x = -5.2$; $x$-intercepts (and valleys) at $(-2, 0)$ and $(2, 0)$; a peak near $x = -0.8$; $y$-intercept $(0, \frac{4}{3})$

**39. a.**

**b.** $D = \mathbb{R}$  **c.** One peak at $x = \frac{1}{2}$; $x$-intercepts $(-1, 0)$ and $(2, 0)$; $y$-intercept $(0, 2)$  **41. a.** $A, D, F, I$
**b.** $E$  **c.** $B, H$  **d.** $I$  **e.** $A$

**43. a.**

**b.** $\theta = 0$; vision is sharpest when we look straight ahead.
**c.** $|\theta| \le 0.19°$ (less than $\frac{1}{5}$ of a degree)  **45.** $S(x) = 2$

**47.** $S(x) = \begin{cases} 1 & \text{if } x < 0 \\ -\frac{1}{2} & \text{if } x > 0 \end{cases}$

**49. a.** 12  **b.** 36  **c.** $A(x) = 6x$  **51. a.** 12  **b.** 21
**c.** $A(x) = \begin{cases} 8x - x^2 & \text{if } 0 \le x \le 3 \\ 2x + 9 & \text{if } x > 3 \end{cases}$
**53. a.** True  **b.** False  **c.** True  **d.** False
**55. a.** Shift 3 units to the right.  **b.** Horizontal compression by a factor of $\frac{1}{2}$, then shift 2 units to the right.

**c.** Shift to the right 2 units, vertical stretch by a factor of 3, reflect across the $x$-axis, and shift up 4 units.  **d.** Horizontal stretch by a factor of 3, horizontal shift right 2 units, vertical stretch by a factor of 6, and vertical shift up 1 unit.

**57.** Shift the graph of $y = x^2$ right 2 units and up 1 unit.  **59.** Stretch the graph of $y = x^2$ vertically by a factor of 3 and reflect across the $x$-axis.

**61.** Shift the graph of $y = x^2$ left 3 units and stretch vertically by a factor of 2.

**63.** Shift the graph of $y = x^2$ to the left $\frac{1}{2}$ unit, stretch vertically by a factor of 4, reflect across the $x$-axis, and then shift up 13 units.

**b.**

**c.** 10

**65.** $(0, 0)$; $(2, 8)$    **67.** $(0, 0)$; $(4, 16)$

**69.**

**71.**

**73.**

**75.**

**77. a.** 0.9; 90% chance that server will win from deuce given that such servers win 75% of their service points  **b.** 0.1; 10% chance that server will win from deuce given that such servers win 25% of their service points  **79. a.** $f(m) = 350m + 1200$  **b.** Buy

**81. a.** $S(x) = x^2 + \dfrac{500}{x}$  **b.** Approx. 6.3 ft

**85. a.**

| $n$ | 1 | 2 | 3 | 4 | 5 |
|---|---|---|---|---|---|
| $n!$ | 1 | 2 | 6 | 24 | 120 |

## Section 1.3 Exercises, pp. 32–34

**1.** $\sin\theta = \text{opp/hyp}$; $\cos\theta = \text{adj/hyp}$; $\tan\theta = \text{opp/adj}$; $\cot\theta = \text{adj/opp}$; $\sec\theta = \text{hyp/adj}$; $\csc\theta = \text{hyp/opp}$  **3.** 3 s
**5.** The radian measure of an angle $\theta$ is the length $s$ of an arc on the unit circle associated with $\theta$.  **7.** $\sin^2\theta + \cos^2\theta = 1$, $1 + \cot^2\theta = \csc^2\theta$, $\tan^2\theta + 1 = \sec^2\theta$  **9.** $\theta = 3\pi/2$
**11.** $\{x : x$ is an odd multiple of $\pi/2\}$
**13.** $-1/2$  **15.** 1  **17.** $-1/\sqrt{3}$  **19.** $1/\sqrt{3}$  **21.** 1  **23.** $-1$
**25.** Undefined  **27.** $\dfrac{\sqrt{2 + \sqrt{3}}}{2}$ or $\dfrac{\sqrt{6} + \sqrt{2}}{4}$
**29.** $\pi/4 + n\pi, n = 0, \pm 1, \pm 2, \ldots$
**31.** $\pi/6, 5\pi/6, 7\pi/6, 11\pi/6$
**33.** $\pi/4 + 2n\pi, 3\pi/4 + 2n\pi, n = 0, \pm 1, \pm 2, \ldots$
**35.** $0, \pi/2, \pi, 3\pi/2$
**37.** $\pi/12, 5\pi/12, 3\pi/4, 13\pi/12, 17\pi/12, 7\pi/4$
**39.** 0.1007; 1.4701  **41.** 17.3°; 72.7°
**43.** $\sec\theta = \dfrac{r}{x} = \dfrac{1}{x/r} = \dfrac{1}{\cos\theta}$
**45.** Dividing both sides of $\cos^2\theta + \sin^2\theta = 1$ by $\cos^2\theta$ gives $1 + \tan^2\theta = \sec^2\theta$.  **47.** Because $\cos(\pi/2 - \theta) = \sin\theta$, for all $\theta$, $1/\cos(\pi/2 - \theta) = 1/\sin\theta$, excluding integer multiples of $\pi$, and $\sec(\pi/2 - \theta) = \csc\theta$.
**49. a.** False  **b.** False  **c.** False  **d.** False  **e.** True
**51.** $\sin\theta = \frac{12}{13}$; $\tan\theta = \frac{12}{5}$; $\sec\theta = \frac{13}{5}$; $\csc\theta = \frac{13}{12}$; $\cot\theta = \frac{5}{12}$  **53.** $\sin\theta = \frac{12}{13}$; $\cos\theta = \frac{5}{13}$; $\tan\theta = \frac{12}{5}$; $\sec\theta = \frac{13}{5}$; $\cot\theta = \frac{5}{12}$  **55.** Amp = 3; period = $6\pi$
**57.** Amp = 3.6; period = 48  **61.** Area of circle is $\pi r^2$; $\theta/(2\pi)$ represents the proportion of area swept out by a central angle $\theta$. Therefore, the area of such a sector is $(\theta/2\pi)\pi r^2 = r^2\theta/2$.
**63.** Stretch the graph of $y = \cos x$ horizontally by a factor of 3, stretch vertically by a factor of 2, and reflect across the $x$-axis.

**65.** Stretch the graph of $y = \cos x$ horizontally by a factor of $24/\pi$; then stretch it vertically by a factor of 3.6 and shift it up 2 units.

**67.** $y = 3 \sin (\pi t/12 - 3\pi/4) + 13$   **69.** About 6 ft
**71.** $d(t) = 10 \cos (4\pi t/3)$   **73.** h

## Chapter 1 Review Exercises, pp. 34–37

**1. a.** True   **b.** False   **c.** False   **d.** True   **e.** False
**3.** $D = \{w: w \neq 2\}; R = \{y: y \neq 5\}$
**5.** $D = (-\infty, -1] \cup [3, \infty); R = [0, \infty)$   **7. a.** 1   **b.** $\sqrt{x^3}$
**c.** $\sin^3 \sqrt{x}$   **d.** $\mathbb{R}$   **e.** $[-1, 1]$   **9.** $2x + h - 2; x + a - 2$

**11.** $3x^2 + 3xh + h^2; x^2 + ax + a^2$   **13. a.** $y = \dfrac{5}{2}x - 8$

**b.** $y = \dfrac{3}{4}x + 3$   **c.** $y = \dfrac{1}{2}x - 2$   **15.** $B = -\dfrac{1}{500}a + 212$

**17. a.**

**b.**

**c.**

**d.**

**19.**

**21. a.**

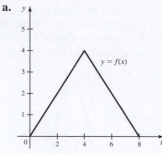

**b.** 2; 14   **c.** $A(x) = \begin{cases} x^2/2 & \text{if } 0 \leq x \leq 4 \\ -x^2/2 + 8x - 16 & \text{if } 4 < x \leq 8 \end{cases}$

**23.** $f(x) = \begin{cases} 4x & \text{if } x < 0 \\ 0 & \text{if } x \geq 0 \end{cases}$

**25.** $D_f = \mathbb{R}, R_f = \mathbb{R}; D_g = [0, \infty), R_g = [0, \infty)$
**27.** Shift $y = x^2$ left 3 units and down 12 units.
**29. a.** $y$-axis   **b.** $y$-axis   **c.** $x$-axis, $y$-axis, origin
**31.** $\theta = \dfrac{\pi}{4}, \dfrac{3\pi}{4}, \dfrac{5\pi}{4}, \dfrac{7\pi}{4}$   **33.** $\theta = \pm\dfrac{\pi}{12}, \pm\dfrac{5\pi}{12}$

**35.** $\theta = 0, \dfrac{\pi}{6}, \dfrac{5\pi}{6}, \pi$

**37. a.** $f(t) = -2 \cos \dfrac{\pi t}{3}$   **b.** $f(t) = 5 \sin \dfrac{\pi t}{12} + 15$

**39. a.** F   **b.** E   **c.** D   **d.** B   **e.** C   **f.** A

**41.** $(7\pi/6, -1/2); (11\pi/6, -1/2)$   **43.** $-\dfrac{\sqrt{2 + \sqrt{2}}}{2}$

**45.** $\tan 2\theta = \dfrac{\sin 2\theta}{\cos 2\theta} = \dfrac{2 \sin \theta \cos \theta}{\cos^2 \theta - \sin^2 \theta}$

$= \dfrac{2 \sin \theta \cos \theta / \cos^2 \theta}{(\cos^2 \theta - \sin^2 \theta)/\cos^2 \theta} = \dfrac{2 \tan \theta}{1 - \tan^2 \theta}$

**47. a.**

| $n$ | 1 | 2 | 3 | 4 | 5 | 6 | 7 | 8 | 9 | 10 |
|---|---|---|---|---|---|---|---|---|---|---|
| $T(n)$ | 1 | 5 | 14 | 30 | 55 | 91 | 140 | 204 | 285 | 385 |

**b.** $D = \{n: n \text{ is a positive integer}\}$   **c.** 14

**49.** $s(t) = 117.5 - 87.5 \sin\left(\dfrac{\pi}{182.5}(t - 95)\right)$

$S(t) = 844.5 + 87.5 \sin\left(\dfrac{\pi}{182.5}(t - 67)\right)$

# CHAPTER 2

## Section 2.1 Exercises, pp. 43–44

**1.** $\dfrac{s(b) - s(a)}{b - a}$  **3.** 20  **5. a.** 36  **b.** 44  **c.** 52  **d.** 60

**7.** 47.84, 47.984, 47.9984; instantaneous velocity appears to be 48

**9.** $\dfrac{f(b) - f(a)}{b - a}$  **11.** The instantaneous velocity at $t = a$ is the slope of the line tangent to the position curve at $t = a$.  **13. a.** 48 **b.** 64  **c.** 80  **d.** $16(6 - h)$  **15.** $m_{\text{sec}} = 60$; the slope is the average velocity of the object over the interval $[0.5, 2]$.

**17.**

| Time interval | Average velocity |
|---|---|
| $[1, 2]$ | 80 |
| $[1, 1.5]$ | 88 |
| $[1, 1.1]$ | 94.4 |
| $[1, 1.01]$ | 95.84 |
| $[1, 1.001]$ | 95.984 |
| $v_{\text{inst}} = 96$ | |

**19.**

| Time interval | Average velocity |
|---|---|
| $[2, 3]$ | 20 |
| $[2.9, 3]$ | 5.60 |
| $[2.99, 3]$ | 4.16 |
| $[2.999, 3]$ | 4.016 |
| $[2.9999, 3]$ | 4.002 |
| $v_{\text{inst}} = 4$ | |

**21.**

| Time interval | Average velocity |
|---|---|
| $[3, 3.5]$ | $-24$ |
| $[3, 3.1]$ | $-17.6$ |
| $[3, 3.01]$ | $-16.16$ |
| $[3, 3.001]$ | $-16.016$ |
| $[3, 3.0001]$ | $-16.002$ |
| $v_{\text{inst}} = -16$ | |

**23.**

| Time interval | Average velocity |
|---|---|
| $[0, 1]$ | 36.372 |
| $[0, 0.5]$ | 67.318 |
| $[0, 0.1]$ | 79.468 |
| $[0, 0.01]$ | 79.995 |
| $[0, 0.001]$ | 80.000 |
| $v_{\text{inst}} = 80$ | |

**25.**

| Interval | Slope of secant line |
|---|---|
| $[1, 2]$ | 6 |
| $[1.5, 2]$ | 7 |
| $[1.9, 2]$ | 7.8 |
| $[1.99, 2]$ | 7.98 |
| $[1.999, 2]$ | 7.998 |
| $m_{\text{tan}} = 8$ | |

**27.**

| Interval | Slope of secant line |
|---|---|
| $[-1.1, -1]$ | 0.475 |
| $[-1.01, -1]$ | 0.4975 |
| $[-1.001, -1]$ | 0.49975 |
| $[-1.0001, -1]$ | 0.499975 |
| $m_{\text{tan}} = \frac{1}{2}$ | |

**29. a.**

**b.** $(2, -1)$

**c.**

| Interval | Slope of secant line |
|---|---|
| $[2, 2.5]$ | 0.5 |
| $[2, 2.1]$ | 0.1 |
| $[2, 2.01]$ | 0.01 |
| $[2, 2.001]$ | 0.001 |
| $[2, 2.0001]$ | 0.0001 |
| $m_{\text{tan}} = 0$ | |

**31. a.**

**b.** $t = 4$

**c.**

| Interval | Average velocity |
|---|---|
| $[4, 4.5]$ | $-8$ |
| $[4, 4.1]$ | $-1.6$ |
| $[4, 4.01]$ | $-0.16$ |
| $[4, 4.001]$ | $-0.016$ |
| $[4, 4.0001]$ | $-0.0016$ |
| $v_{\text{inst}} = 0$ | |

**d.** $0 \le t < 4$  **e.** $4 < t \le 9$  **33.** 0.6366, 0.9589, 0.9996, 1

## Section 2.2 Exercises, pp. 49–53

**1.** As $x$ approaches $a$ from either side, the values of $f(x)$ approach $L$.
**3. a.** 5  **b.** 3  **c.** Does not exist  **d.** 1  **e.** 2
**5. a.** $-1$  **b.** 1  **c.** 2  **d.** 2
**7. a.**

| $x$ | $f(x)$ | $x$ | $f(x)$ |
|---|---|---|---|
| 1.9 | 3.9 | 2.1 | 4.1 |
| 1.99 | 3.99 | 2.01 | 4.01 |
| 1.999 | 3.999 | 2.001 | 4.001 |
| 1.9999 | 3.9999 | 2.0001 | 4.0001 |

**b.** 4

**9. a.**

| $t$ | $g(t)$ | $t$ | $g(t)$ |
|---|---|---|---|
| 8.9 | 5.983287 | 9.1 | 6.016621 |
| 8.99 | 5.998333 | 9.01 | 6.001666 |
| 8.999 | 5.999833 | 9.001 | 6.000167 |

**b.** 6

**11.** As $x$ approaches $a$ from the right, the values of $f(x)$ approach $L$.
**13.** $L = M$  **15. a.** 0  **b.** 1  **c.** 0
**d.** Does not exist; $\lim\limits_{x \to 1^-} f(x) \ne \lim\limits_{x \to 1^+} f(x)$  **17. a.** 3  **b.** 2
**c.** 2  **d.** 2  **e.** 2  **f.** 4  **g.** 1  **h.** Does not exist
**i.** 3  **j.** 3  **k.** 3  **l.** 3

**19.**

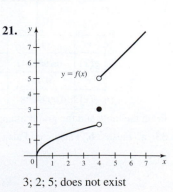

2; 2; 3; does not exist

**21.**

3; 2; 5; does not exist

**23.**

undefined; 10; 10; 10

**25.**

undefined; 3; 3; 3

**27.** From the graph and table, the limit appears to be 1.

| $x$ | 1.5 | 1.9 | 1.99 | 2.01 | 2.1 | 2.5 |
|---|---|---|---|---|---|---|
| $f(x)$ | 1.04291 | 1.00167 | 1.00002 | 1.00002 | 1.00167 | 1.04291 |

**29.** From the graph and table, the limit appears to be 2.

| $x$ | 0.9 | 0.99 | 0.999 |
|---|---|---|---|
| $f(x)$ | 1.993342 | 1.999933 | 1.999999 |
| $x$ | 1.001 | 1.01 | 1.1 |
| $f(x)$ | 1.999999 | 1.999933 | 1.993342 |

**31.**

| $x$ | $-1.1$ | $-1.01$ | $-1.001$ |
|---|---|---|---|
| $g(x)$ | $-0.9983342$ | $-0.9999833$ | $-0.9999998$ |
| $x$ | $-0.999$ | $-0.99$ | $-0.9$ |
| $g(x)$ | 0.9999998 | 0.9999833 | 0.9983342 |

From the table and the graph, it appears that the limit does not exist.

**33. a.** False   **b.** False   **c.** False   **d.** False   **e.** True

**35. a.**

**b.** 0.89   **c.** Because $\lim\limits_{w\to3^-} f(w) = 0.89$ and $\lim\limits_{w\to3^+} f(w) = 1.1$, we know that $\lim\limits_{w\to3^-} f(w) \neq \lim\limits_{w\to3^+} f(w)$. So $\lim\limits_{w\to3} f(w)$ does not exist.

**37.** 3   **39.** 16   **41.** 5

**43. a.**

| $x$ | $\sin(1/x)$ |
|---|---|
| $2/\pi$ | 1 |
| $2/(3\pi)$ | $-1$ |
| $2/(5\pi)$ | 1 |
| $2/(7\pi)$ | $-1$ |
| $2/(9\pi)$ | 1 |
| $2/(11\pi)$ | $-1$ |

The function values alternate between 1 and $-1$.

**b.** The function values alternate between 1 and $-1$ infinitely many times on the interval $(0, h)$ no matter how small $h > 0$ becomes.   **c.** Does not exist

**45.**

**47.**

**49.**

**51. a.** $-2, -1, 1, 2$   **b.** 2, 2, 2
**c.** $\lim\limits_{x\to a^-} \lfloor x \rfloor = a - 1$ and $\lim\limits_{x\to a^+} \lfloor x \rfloor = a$, if $a$ is an integer
**d.** $\lim\limits_{x\to a^-} \lfloor x \rfloor = \lfloor a \rfloor$ and $\lim\limits_{x\to a^+} \lfloor x \rfloor = \lfloor a \rfloor$, if $a$ is not an integer
**e.** Limit exists provided $a$ is not an integer
**53. a.** 8   **b.** 5   **55. a.** 2; 3; 4   **b.** $p$   **57.** $p/q$

## Section 2.3 Exercises, pp. 60–63

**1.** $\lim_{x \to a} p(x) = p(a)$   **3.** Those values of $a$ for which the denominator is not zero   **5.** $\dfrac{x^2 - 7x + 12}{x - 3} = x - 4$, for $x \neq 3$; $-1$

**7.** 32; Constant Multiple Law   **9.** 5; Difference Law
**11.** 8; Quotient and Difference Laws   **13.** 4; Root and Power Laws
**15.** 5; 1   **17.** 20   **19.** 5   **21.** $-45$   **23.** 8   **25.** 3   **27.** 3
**29.** $-5$   **31.** 8   **33.** 2   **35.** $-8$   **37.** $-1$   **39.** $-12$   **41.** $\frac{1}{6}$
**43.** $-\frac{1}{36}$   **45.** $2\sqrt{a}$   **47.** $\frac{1}{8}$   **49.** $-\frac{1}{16}$   **51.** 5   **53.** 10   **55.** 2
**57.** $-54$   **59.** 0   **61.** 1   **63.** 1   **65.** $\frac{1}{2}$   **67.** Does not exist
**69.** Does not exist   **71. a.** False   **b.** False   **c.** False   **d.** False
**e.** False   **73. a.** 2   **b.** 0   **c.** Does not exist   **75. a.** 0
**b.** $\sqrt{x - 2}$ is undefined for $x < 2$.   **77.** 0.0435 N/C
**79.** $\lim_{S \to 0^+} r(S) = 0$; the radius of the cylinder approaches 0 as the surface area of the cylinder approaches 0.

**81. a.** Because $\left| \sin \dfrac{1}{x} \right| \leq 1$, for all $x \neq 0$, we have that

$$\left| x \right| \left| \sin \dfrac{1}{x} \right| \leq \left| x \right|.$$

That is, $\left| x \sin \dfrac{1}{x} \right| \leq |x|$, so $-|x| \leq x \sin \dfrac{1}{x} \leq |x|$, for all $x \neq 0$.

**b.**

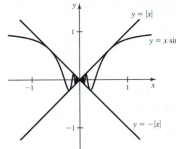

**c.** $\lim_{x \to 0} -|x| = 0$ and $\lim_{x \to 0} |x| = 0$; by part (a) and the Squeeze Theorem, $\lim_{x \to 0} x \sin \dfrac{1}{x} = 0$.

**83. a.**

**b.** $\lim_{x \to 0} \dfrac{\sin x}{x} = 1$

**85.** Because $\lim_{x \to 0^-} |x| = \lim_{x \to 0^-} (-x) = 0$ and $\lim_{x \to 0^+} |x| = \lim_{x \to 0^+} x = 0$, we know that $\lim_{x \to 0} |x| = 0$.   **87.** 1   **89.** $a = -13$; $\lim_{x \to -1} g(x) = 6$

**91.** 6   **93.** $5a^4$   **95.** 6; 5   **97.** $\dfrac{1}{3}$

**99.** $f(x) = x - 1$, $g(x) = \dfrac{5}{x - 1}$

**101.** $b = 2$ and $c = -8$; yes   **103.** 6; 4

## Section 2.4 Exercises, pp. 69–72

**1.** As $x$ approaches $a$ from the right, the values of $f(x)$ are negative and become arbitrarily large in magnitude.
**3.** A vertical asymptote for a function $f$ is a vertical line $x = a$, where one (or more) of the following is true:

$$\lim_{x \to a^-} f(x) = \pm \infty; \quad \lim_{x \to a^+} f(x) = \pm \infty.$$

**5.** $\infty$   **7. a.** $\infty$   **b.** $\infty$   **c.** $\infty$   **d.** $\infty$   **e.** $-\infty$   **f.** Does not exist
**9. a.** $-\infty$   **b.** $-\infty$   **c.** $-\infty$   **d.** $\infty$   **e.** $-\infty$   **f.** Does not exist
**11. a.** $\infty$   **b.** $-\infty$   **c.** $-\infty$   **d.** $\infty$   **13.** $-\infty$   **15.** No; there is a vertical asymptote at $x = 2$ but not at $x = 1$.

**17.**

**19.** a and b are correct.   **21. a.** $\infty$   **b.** $-\infty$   **c.** Does not exist
**23. a.** $-\infty$   **b.** $-\infty$   **c.** $-\infty$   **25. a.** $\infty$   **b.** $-\infty$   **c.** Does not exist   **27. a.** $-\infty$   **b.** $-\infty$   **c.** $-\infty$   **29. a.** $\infty$   **b.** Does not exist   **c.** Does not exist   **31. a.** $\infty$   **b.** $1/54$   **c.** Does not exist
**33.** $-5$   **35.** $\infty$   **37.** $-\infty$   **39.** $\infty$   **41.** $-\infty$   **43.** $\infty$
**45. a.** $1/10$   **b.** $-\infty$   **c.** $\infty$; vertical asymptote: $x = -5$
**47.** $x = 3$; $\lim_{x \to 3^+} f(x) = -\infty$; $\lim_{x \to 3^-} f(x) = \infty$; $\lim_{x \to 3} f(x)$ does not exist   **49.** $x = 0$ and $x = 2$; $\lim_{x \to 0^+} f(x) = \infty$;
$\lim_{x \to 0^-} f(x) = -\infty$; $\lim_{x \to 0} f(x)$ does not exist; $\lim_{x \to 2^+} f(x) = \infty$;
$\lim_{x \to 2^-} f(x) = \infty$; $\lim_{x \to 2} f(x) = \infty$
**51. a.** $-\infty$   **b.** $\infty$   **c.** $-\infty$   **d.** $\infty$   **53. a.** False   **b.** True
**c.** False   **55.** $r(x) = \dfrac{(x - 1)^2}{(x - 1)(x - 2)^2}$   **57.** $f(x) = \dfrac{1}{x - 6}$

**59.** $x = 0$   **61.** $x = -1$   **63.** $\theta = 10k + 5$, for any integer $k$
**65.** $x = 0$   **67. a.** $a = 4$ or $a = 3$   **b.** Either $a > 4$ or $a < 3$
**c.** $3 < a < 4$   **69. a.** $\dfrac{1}{\sqrt[3]{h}}$ regardless of the sign of $h$
**b.** $\lim_{h \to 0^+} \dfrac{1}{\sqrt[3]{h}} = \infty$; $\lim_{h \to 0} \dfrac{1}{\sqrt[3]{h}} = -\infty$; the tangent line at $(0, 0)$ is vertical.

## Section 2.5 Exercises, pp. 79–81

**1.** As $x < 0$ becomes arbitrarily large in magnitude, the corresponding values of $f$ approach 10.   **3.** $\infty$   **5.** 0   **7.** 0   **9.** 3   **11.** 0
**13.** $-\dfrac{1}{2}$; 0; $-\infty$   **15.** 3; 3   **17.** 0   **19.** 0   **21.** $\infty$   **23.** $-\infty$
**25.** $2/3$   **27.** $-\infty$   **29.** 5   **31.** $-4$   **33.** $-3/2$   **35.** 0
**37.** $\lim_{x \to \infty} f(x) = \lim_{x \to -\infty} f(x) = \frac{1}{5}$; $y = \frac{1}{5}$
**39.** $\lim_{x \to \infty} f(x) = \lim_{x \to -\infty} f(x) = 2$; $y = 2$
**41.** $\lim_{x \to \infty} f(x) = \lim_{x \to -\infty} f(x) = 0$; $y = 0$
**43.** $\lim_{x \to \infty} f(x) = \infty$; $\lim_{x \to -\infty} f(x) = -\infty$; none
**45.** $\lim_{x \to \infty} f(x) = \lim_{x \to -\infty} f(x) = \frac{4}{9}$; $y = \frac{4}{9}$
**47.** $\lim_{x \to \infty} f(x) = \frac{2}{3}$; $\lim_{x \to -\infty} f(x) = -2$; $y = \frac{2}{3}$; $y = -2$
**49.** $\lim_{x \to \infty} f(x) = \lim_{x \to -\infty} f(x) = \dfrac{1}{4 + \sqrt{3}}$; $y = \dfrac{1}{4 + \sqrt{3}}$

**51. a.** $y = x - 6$ **b.** $x = -6$
**c.**

**53. a.** $y = \frac{1}{3}x - \frac{4}{9}$ **b.** $x = \frac{2}{3}$
**c.**

**55. a.** $y = 4x + 4$ **b.** No vertical asymptote
**c.**

**57. a.** False **b.** False **c.** True **d.** False **59.** 3500 **61.** 2
**63. a.** $\lim\limits_{x \to \infty} f(x) = \lim\limits_{x \to -\infty} f(x) = 2; y = 2$
**b.** $x = 0; \lim\limits_{x \to 0^+} f(x) = \infty; \lim\limits_{x \to 0^-} f(x) = -\infty$
**65. a.** $\lim\limits_{x \to \infty} f(x) = \lim\limits_{x \to -\infty} f(x) = 3; y = 3$
**b.** $x = -3$ and $x = 4$; $\lim\limits_{x \to -3^-} f(x) = \infty; \lim\limits_{x \to -3^+} f(x) = -\infty$;
$\lim\limits_{x \to 4^-} f(x) = -\infty; \lim\limits_{x \to 4^+} f(x) = \infty$
**67. a.** $\lim\limits_{x \to \infty} f(x) = \lim\limits_{x \to -\infty} f(x) = 1; y = 1$
**b.** $x = 0; \lim\limits_{x \to 0^+} f(x) = \infty; \lim\limits_{x \to 0^-} f(x) = -\infty$
**69. a.** $\lim\limits_{x \to \infty} f(x) = 1; \lim\limits_{x \to -\infty} f(x) = -1; y = 1$ and $y = -1$
**b.** No vertical asymptote
**71. a.** $\lim\limits_{x \to \infty} f(x) = 0; \lim\limits_{x \to -\infty} f(x) = 0; y = 0$
**b.** No vertical asymptote **73. a.** $\lim\limits_{x \to \infty} f(x) = 2; \lim\limits_{x \to -\infty} f(x)$ does not
exist; $y = 2$ **b.** $x = 0; \lim\limits_{x \to 0^+} f(x) = \infty; \lim\limits_{x \to 0^-} f(x)$ does not exist
**75.**

**77.** 1 **79.** 0 **81. a.** No; $f$ has a horizontal asymptote if $m = n$,
and it has a slant asymptote if $m = n + 1$. **b.** Yes;
$f(x) = x^4 / \sqrt{x^6 + 1}$

## Section 2.6 Exercises, pp. 90–93

**1.** a, c **3.** A function is continuous on an interval if it is continuous
at each point of the interval. If the interval contains endpoints, then the
function must be right- or left-continuous at those points. **5.** $a = 1$,
item 1; $a = 2$, item 3; $a = 3$, item 2 **7.** $a = 1$, item 1; $a = 2$, item 2;
$a = 3$, item 1 **9. a.** $\lim\limits_{x \to a^-} f(x) = f(a)$ **b.** $\lim\limits_{x \to a^+} f(x) = f(a)$

**11.** $(0, 1), (1, 2), (2, 3], (3, 5)$; left-continuous at 3
**13.** $(0, 1), (1, 2), [2, 3), (3, 5)$; right-continuous at 2
**15.** $[-1, 1], [-1, 1]$ **17.** No; $f(-5)$ is undefined.
**19.** No; $f(1)$ is undefined. **21.** No; $\lim\limits_{x \to 1} f(x) = 2$ but $f(1) = 3$.
**23.** No; $f(4)$ is undefined. **25.** $(-\infty, \infty)$
**27.** $(-\infty, -3), (-3, 3), (3, \infty)$ **29.** $(-\infty, -2), (-2, 2), (2, \infty)$
**31.** 1 **33.** $2\sqrt{6}$ **35.** 16 **37.** $\dfrac{1}{81}$ **39. a.** $\lim\limits_{x \to 1} f(x)$ does not exist.
**b.** Continuous from the right **c.** $(-\infty, 1), [1, \infty)$ **41.** $(-\infty, 5]$;
left-continuous at 5 **43.** $(-\infty, -2\sqrt{2}], [2\sqrt{2}, \infty)$; left-continuous at
$-2\sqrt{2}$; right-continuous at $2\sqrt{2}$ **45.** $(-\infty, \infty)$ **47.** $(-\infty, \infty)$
**49.** 3 **51.** 1 **53.** 4 **55.** 2 **57.** $-\frac{1}{2}$ **59.** 0
**61.** $(n\pi, (n + 1)\pi)$, where $n$ is an integer; $\sqrt{2}; -\infty$
**63.** $\left(\dfrac{n\pi}{2}, \left(\dfrac{n}{2} + 1\right)\dfrac{\pi}{2}\right)$, where $n$ is an odd integer; $\infty; \sqrt{3} - 2$
**65. b.** $x \approx 0.835$ **67. b.** $x \approx -0.285; x \approx 0.778; x \approx 4.507$
**69. a.** True **b.** True **c.** False **d.** False
**71. a.** $A(r)$ is continuous on $[0, 0.08]$, and 7000 is between
$A(0) = 5000$ and $A(0.08) = 11,098.20$. By the Intermediate Value
Theorem, there is at least one $c$ in $(0, 0.08)$ such that $A(c) = 7000$.
**b.**

$c \approx 0.034$ or $3.4\%$

**73.** $[0, \pi/2]; 0.45$ **75.** $(-\infty, \infty)$ **77.** $[0, 16), (16, \infty)$
**79.** The vertical line segments should not appear.

**81. a, b.**

**83. a.** 2 **b.** 8 **c.** No; $\lim\limits_{x \to 1^-} g(x) = 2$ and $\lim\limits_{x \to 1^+} g(x) = 8$.
**85.** $x_1 = \frac{1}{7}; x_2 = \frac{1}{2}; x_3 = \frac{3}{5}$ **87.** Yes. Imagine there is a clone of the
monk who walks down the path at the same time the monk walks up the
path. The monk and his clone must cross paths at some time between
dawn and dusk. **89.** No; $f$ cannot be made continuous at $x = a$ by
redefining $f(a)$. **91.** Removable discontinuity
**93.** $x = 0$ removable discontinuity; $x = 1$ infinite discontinuity
**95. a.** Yes **b.** No **97. a.** For example, $f(x) = 1/(x - 1)$,
$g(x) = x + 1$ **b.** For continuity, $g$ must be continuous at 0, and $f$
must be continuous at $g(0)$.

## Section 2.7 Exercises, pp. 102–106

**1.** 1   **3.** c   **5.** Given any $\varepsilon > 0$ there exists a $\delta > 0$ such that $|f(x) - L| < \varepsilon$ whenever $0 < |x - a| < \delta$.   **7.** $0 < \delta \le 2$
**9. a.** $\delta = 1$   **b.** $\delta = \frac{1}{2}$   **11. a.** $\delta = 2$   **b.** $\delta = \frac{1}{2}$
**13. a.** $0 < \delta \le 1$   **b.** $0 < \delta \le 0.79$   **15. a.** $0 < \delta \le 1$
**b.** $0 < \delta \le \frac{1}{2}$   **c.** $0 < \delta \le \varepsilon$   **17. a.** $0 < \delta \le 1$   **b.** $0 < \delta \le \frac{1}{2}$
**c.** $0 < \delta \le \frac{\varepsilon}{2}$   **19.** $\delta = \varepsilon/8$   **21.** $\delta = \varepsilon$   **23.** $\delta = \varepsilon$
**25.** $\delta = \varepsilon/3$   **27.** $\delta = \sqrt{\varepsilon}$   **29.** $\delta = \min\{1, \varepsilon/8\}$   **31.** $\delta = \varepsilon/2$
**33.** $\delta = \min\{1, 6\varepsilon\}$   **35.** $\delta = \min\{1/20, \varepsilon/200\}$
**37.** $\delta = \min\{1, \sqrt{\varepsilon/2}\}$   **39.** $\delta = \varepsilon/|m|$ if $m \ne 0$; use any $\delta > 0$ if $m = 0$   **41.** $\delta = \min\{1, 8\varepsilon/15\}$   **45.** $\delta = 1/\sqrt{N}$
**47.** $\delta = 1/\sqrt{N - 1}$   **49. a.** False   **b.** False   **c.** True   **d.** True
**51.** For $x > a$, $|x - a| = x - a$.   **53. a.** $\delta = \varepsilon/2$   **b.** $\delta = \varepsilon/3$
**c.** Because $\lim\limits_{x\to0^+} f(x) = \lim\limits_{x\to0^-} f(x) = -4$, $\lim\limits_{x\to0} f(x) = -4$.
**55.** $\delta = \varepsilon^2$   **57. a.** For each $N > 0$ there exists $\delta > 0$ such that $f(x) > N$ whenever $0 < x - a < \delta$.   **b.** For each $N < 0$ there exists $\delta > 0$ such that $f(x) < N$ whenever $0 < a - x < \delta$.
**c.** For each $N > 0$ there exists $\delta > 0$ such that $f(x) > N$ whenever $0 < a - x < \delta$.   **59.** $\delta = 1/N$   **61.** $\delta = (-10/M)^{1/4}$
**65.** $N = 1/\varepsilon$   **67.** $N = M - 1$

## Chapter 2 Review Exercises, pp. 106–108

**1. a.** False   **b.** False   **c.** False   **d.** True   **e.** False   **f.** False
**g.** False   **h.** True   **3.** 12 ft/s   **5.** $x = -1$; $\lim\limits_{x\to-1} f(x)$ does not exist; $x = 1$; $\lim\limits_{x\to1} f(x) \ne f(1)$; $x = 3$; $f(3)$ is undefined.
**7. a.** 1.414   **b.** $\sqrt{2}$
**9.**

**11.** $\sqrt{11}$   **13.** 13   **15.** 2   **17.** $\frac{1}{3}$   **19.** $-\frac{1}{16}$   **21.** 108   **23.** $\frac{1}{108}$
**25.** 0   **27.** $-\infty$   **29.** $\infty$   **31.** 4   **33.** $-\infty$   **35.** $\frac{1}{2}$   **37.** $-3/\sqrt{a}$
**39.** $2/(1 - a)$   **41.** $-2$   **43.** 1; $\infty$   **45.** 5
**47. a.**

**b.** $\lim\limits_{x\to0} \cos x \le \lim\limits_{x\to0} \dfrac{\sin x}{x} \le \lim\limits_{x\to0} \dfrac{1}{\cos x}$;

$$1 \le \lim\limits_{x\to0} \dfrac{\sin x}{x} \le 1;$$

$$\lim\limits_{x\to0} \dfrac{\sin x}{x} = 1$$

**49.** $\lim\limits_{x\to\infty} f(x) = -4$; $\lim\limits_{x\to-\infty} f(x) = -4$
**51.** $\lim\limits_{x\to\infty} f(x) = 3$; $\lim\limits_{x\to-\infty} f(x) = 3$
**53.** $\lim\limits_{x\to\infty} f(x) = \infty$; $\lim\limits_{x\to-\infty} f(x) = -\infty$
**55. a.** $\infty$; $-\infty$   **b.** $y = 3x + 2$ is the slant asymptote.
**57. a.** $-\infty$; $\infty$   **b.** $y = -x - 2$ is the slant asymptote.

**59. a.** $\infty$, $-\infty$   **b.** $y = 4x + 5$ is the slant asymptote.
**61.** Horizontal asymptotes at $y = 1$ and $y = -1$; vertical asymptote at $x = 1$
**63.** Slant asymptotes are $y = \dfrac{x}{2}$ and $y = -\dfrac{x}{2}$.
**65.** No; $f(5)$ does not exist.   **67.** Yes; $h(5) = \lim\limits_{x\to5} h(x) = 4$
**69.** $(-\infty, -\sqrt{5}]$ and $[\sqrt{5}, \infty)$; left-continuous at $-\sqrt{5}$ and right-continuous at $\sqrt{5}$   **71.** $(-\infty, -5)$, $(-5, 0)$, $(0, 5)$, and $(5, \infty)$
**73.** $a = 3, b = 0$
**75.**

**77. a.** Let $f(x) = x - \cos x$; $f(0) < 0 < f\left(\dfrac{\pi}{2}\right)$   **b.** $x \approx 0.739$
**79. b.** $P(10) < 50 < P(2)$, $P(10) < 50 < P(30)$
**c.** $x = 5, x = 20$   **d.** No; $P(x) > 30$ for all $x > 0$.   **e.** $x = y = 10$
**81.** $\delta = \varepsilon$   **83.** $\delta = \min\left\{1, \dfrac{\varepsilon}{15}\right\}$   **85.** $\delta = 1/\sqrt[4]{N}$

## CHAPTER 3

## Section 3.1 Exercises, pp. 115–118

**1.** Given the point $(a, f(a))$ and any point $(x, f(x))$ near $(a, f(a))$, the slope of the secant line joining these points is $\dfrac{f(x) - f(a)}{x - a}$. The limit of this quotient as $x$ approaches $a$ is the slope of the tangent line at the point.   **3.** The average rate of change over the interval $[a, x]$ is $\dfrac{f(x) - f(a)}{x - a}$. The value of $\lim\limits_{x\to a} \dfrac{f(x) - f(a)}{x - a}$ is the slope of the tangent line; it is also the limit of average rates of change, which is the instantaneous rate of change at $x = a$.   **5.** $f'(a)$ is the slope of the tangent line at $(a, f(a))$ or the instantaneous rate of change in $f$ at $a$.
**7.** $f(2) = 7$; $f'(2) = 4$   **9.** $y = 3x - 1$   **11.** $-5$   **13.** 68 ft/s
**15. a.** 6   **b.** $y = 6x - 14$
**c.**

**17. a.** $-1$   **b.** $y = -x - 2$   **19. a.** $\frac{1}{2}$   **b.** $y = \frac{1}{2}x + 2$
**c.**     **c.**

**21. a.** 2   **b.** $y = 2x + 1$   **23. a.** 2   **b.** $y = 2x - 3$
**25. a.** 4   **b.** $y = 4x - 8$   **27. a.** 3   **b.** $y = 3x - 2$
**29. a.** $\frac{2}{25}$   **b.** $y = \frac{2}{25}x + \frac{7}{25}$   **31. a.** $\frac{1}{4}$   **b.** $y = \frac{1}{4}x + \frac{7}{4}$
**33. a.** 8   **b.** $y = 8x$   **35. a.** $-14$   **b.** $y = -14x - 16$
**37. a.** $-4$   **b.** $y = -4x + 3$   **39. a.** $\frac{1}{3}$   **b.** $y = \frac{1}{3}x + \frac{5}{3}$
**41. a.** $-\frac{1}{100}$   **b.** $y = -\frac{x}{100} + \frac{3}{20}$   **43.** $-\frac{1}{4}$   **45.** $\frac{1}{5}$   **47. a.** True
**b.** False   **c.** True   **49.** $d'(4) = 128$ ft/s; the object falls with an
instantaneous speed of 128 ft/s four seconds after being dropped.
**51.** $v'(3) = -4$ m/s per second; the instantaneous rate of change in
the car's speed is $-4$ m/s² at $t = 3$ s.
**53. a.** $L'(1.5) \approx 4.3$ mm/week; the talon is growing at a
rate of approximately 4.3 mm/week at $t = 1.5$ weeks
(answers will vary).   **b.** $L'(a) \approx 0$, for $a \geq 4$; the talon stops
growing at $t = 4$ weeks.   **55.** $D'(60) \approx 0.05$ hr/day; the number of
daylight hours is increasing at about 0.05 hr/day, 60 days after Jan 1.
$D'(170) \approx 0$ hr/day; the number of daylight hours is neither increasing nor decreasing 170 days after Jan 1.   **57.** $f(x) = 5x^2$; $a = 2$; 20
**59.** $f(x) = x^4$; $a = 2$; 32   **61.** $f(x) = |x|$; $a = -1$; $-1$

**63. a.**

**b.**

| $h$ | Approximation | Error |
|-----|---------------|-------|
| 0.1 | 0.25002 | $2.0 \times 10^{-5}$ |
| 0.01 | 0.25000 | $2.0 \times 10^{-7}$ |
| 0.001 | 0.25000 | $2.0 \times 10^{-9}$ |

**c.** Values of $x$ on both sides of 4 are used in the formula.
**d.** The centered difference approximations are more accurate than the
forward and backward difference approximations.   **65. a.** 0.39470,
0.41545   **b.** 0.02, 0.0003

### Section 3.2 Exercises, pp. 126–130

**1.** $f'$ is the slope function of $f$.   **3.** $\dfrac{dy}{dx}$ is the limit of $\dfrac{\Delta y}{\Delta x}$ as $\Delta x \to 0$.

**5.**

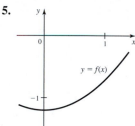

**7.** Yes

**9.** A line with a $y$-intercept of 1 and a slope of 3

**11.** $f'(x) = 7$   **13.** $\dfrac{dy}{dx} = 2x$; $\dfrac{dy}{dx}\Big|_{x=3} = 6$; $\dfrac{dy}{dx}\Big|_{x=-2} = -4$

**15.** a–C; b–C; c–A; d–B

**17.**

**19. a.** Not continuous at $x = 1$   **b.** Not differentiable at $x = 0, 1$
**21. a.** $f'(x) = 5$   **b.** $f'(1) = 5$; $f'(2) = 5$
**23. a.** $f'(x) = 8x$   **b.** $f'(2) = 16$; $f'(4) = 32$

**25. a.** $f'(x) = -\dfrac{1}{(x + 1)^2}$   **b.** $f'\left(-\dfrac{1}{2}\right) = -4$; $f'(5) = -\dfrac{1}{36}$

**27. a.** $f'(t) = -\dfrac{1}{2t^{3/2}}$   **b.** $f'(9) = -\dfrac{1}{54}$; $f'\left(\dfrac{1}{4}\right) = -4$

**29. a.** $f'(s) = 12s^2 + 3$   **b.** $f'(-3) = 111$; $f'(-1) = 15$
**31. a.** $v(t) = -32t + 100$   **b.** $v(1) = 68$ ft/s; $v(2) = 36$ ft/s

**33.** $\dfrac{dy}{dx} = \dfrac{1}{(x + 2)^2}$; $\dfrac{dy}{dx}\Big|_{x=2} = \dfrac{1}{16}$   **35. a.** $6x + 2$

**b.** $y = 8x - 13$   **37. a.** $\dfrac{3}{2\sqrt{3x + 1}}$   **b.** $y = 3x/10 + 13/5$

**39. a.** $\dfrac{6}{(3x + 1)^2}$   **b.** $y = -3x/2 - 5/2$
**41. a.** Approx. 10 kW; approx. $-5$ kW
**b.** $t = 6, 18$   **c.** $t = 12$   **43. a.** $2ax + b$   **b.** $8x - 3$   **c.** 5
**45. a.** $C, D$   **b.** $A, B, E$   **c.** $A, B, E, D, C$   **47.** a–D; b–C; c–B; d–A
**49.**

**51.**

**53. a.** $x = 1$   **b.** $x = 1, x = 2$   **c.**

**55. a.** $t = 0$   **b.** Positive   **c.** Decreasing
**d.**

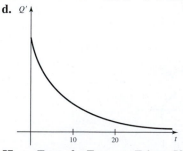

**57. a.** True   **b.** True   **c.** False   **59.** $a = 4$
**61.** Yes

**63.** $y = -\dfrac{x}{3} - \dfrac{2}{3}$   **65.** $y = \dfrac{x}{2} + \dfrac{3}{2}$   **67.** $(1, 2), (5, 26)$

**69.** $(1, 1), \left(-\dfrac{1}{2}, -2\right)$   **71. b.** $f'_+(2) = 1; f'_-(2) = -1$

**c.** $f$ is continuous but not differentiable at $x = 2$.

**73. a.**

Vertical tangent line $x = 2$

**b.**

Vertical tangent line $x = -1$

**c.**

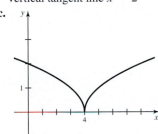

Vertical tangent line $x = 4$

**d.**

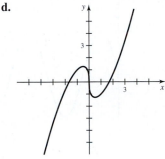

Vertical tangent line $x = 0$

**75.** $f'(x) = \dfrac{1}{3}x^{-2/3}$ and $\lim\limits_{x \to 0^-} |f'(x)| = \lim\limits_{x \to 0^+} |f'(x)| = \infty$

**77. a.**

**b.** 1   **c.** 1   **d.**

**e.** $f$ is not differentiable at 0 because it is not continuous at 0.

## Section 3.3 Exercises, pp. 135–138

**1.** Using the definition can be tedious.   **3.** The slope at every point on the graph of the function is zero.   **5.** Take the product of the constant and the derivative of the function.   **7.** 4   **9.** $-\dfrac{1}{2}$   **11.** $-2$   **13.** 7.5

**15.** $10t^9$; $90t^8$; $720t^7$   **17.** $\dfrac{2}{5}$   **19.** $5x^4$   **21.** 0   **23.** $15x^2$   **25.** $t$

**27.** 8   **29.** $200t$   **31.** $12x^3 + 7$   **33.** $40x^3 - 32$   **35.** $6w^2 + 6w + 10$

**37.** $\begin{cases} 2x & \text{if } x < 0 \\ 4x + 1 & \text{if } x > 0 \end{cases}$   **39. a.** $d'(t) = 32t$; ft/s; the velocity of

the stone   **b.** 576 ft; approx. 131 mi/hr   **41. a.** $A'(t) = -\frac{1}{25}t + 2$

measures the rate at which the city grows in mi$^2$/yr.   **b.** 1.6 mi$^2$/yr

**c.** 1200 people/yr   **43.** $w'(x) = \begin{cases} 0.4 & \text{if } 19 < x < 21 \\ 0.8 & \text{if } 21 < x < 32 \\ 1.5 & \text{if } x > 32 \end{cases}$

**45.** $18x^2 + 6x + 4$   **47.** $2w$, for $w \neq 0$   **49.** $4x^3 + 4x$

**51.** 1, for $x \neq 1$   **53.** $\dfrac{1}{2\sqrt{x}}$, for $x \neq a$

**55. a.** $y = -6x + 5$   **b.**

**57. a.** $y = \dfrac{x}{4} + 1$   **b.**

**59. a.** $x = 3$   **b.** $x = 4$

**61. a.** $(-1, 11), (2, -16)$   **b.** $(-3, -41), (4, 36)$

**63. a.** $(4, 4)$   **b.** $(16, 0)$   **65.** $f'(x) = 20x^3 + 30x^2 + 3$;

$f''(x) = 60x^2 + 60x$; $f'''(x) = 120x + 60$

**67.** $f'(x) = 1$; $f''(x) = f'''(x) = 0$, for $x \neq -1$

**69. a.** False   **b.** True   **c.** False

**71. a.** $y = 7x - 1$   **b.** $y = -2x + 5$   **c.** $y = 16x + 4$

**73.** $b = 2, c = 3$   **75.** $-10$   **77.** 4   **79. a.** $f(x) = x^7 + x^6; a = 1$

**b.** 13   **81. a.** $f(x) = \sqrt{x}; a = 9$   **b.** $\frac{1}{6}$   **85. d.** $\frac{n}{2}x^{n/2 - 1}$

## Section 3.4 Exercises, pp. 144–147

**1.** $\dfrac{d}{dx}(f(x)\, g(x)) = f'(x)\, g(x) + f(x)\, g'(x)$   **3.** $6x + 5$

**5.** $\dfrac{5}{(3x + 2)^2}$   **7. a.** $2x - 1$   **9. a.** $6x + 1$   **11. a.** $2t$, for $t \neq 0$

**13.** 1, for $x \neq a$   **15.** $23; -\dfrac{7}{4}$   **17.** $\dfrac{2}{27}; \dfrac{3}{8}$   **19.** $36x^5 - 12x^3$

**21.** $\dfrac{1}{(x + 1)^2}$   **23.** $\dfrac{11}{2}t^{9/2} + 5t^4$   **25.** $-\dfrac{1}{(t - 1)^2}$

**27.** $3w^2(2w^3 + 3)$   **29.** $4x^3$   **31.** $\dfrac{1}{2\sqrt{w}}$   **33.** $\dfrac{1 - t^2}{(t^2 + 1)^2}$

**35.** $\dfrac{2x(x^4 - 2x^2 - 1)}{(x^2 - 1)^2}$   **37.** $-27x^{-10}$   **39.** $6t - 42/t^8$

**41.** $-\dfrac{3}{t^2} - \dfrac{2}{t^3}$   **43.** $-\dfrac{3}{2x^2}$

**45.** $\dfrac{8x}{(1 - x^2)^2}$   **47.** $\dfrac{\sqrt{w}}{(\sqrt{w} - w)^2}$   **49.** $\dfrac{5w^{2/3}}{3(w^{5/3} + 1)^2}$

**51.** $8x - \dfrac{2}{(5x + 1)^2}$   **53.** $\dfrac{r - 6\sqrt{r} - 1}{2\sqrt{r}(r + 1)^2}$

**55.** $300x^9 + 135x^8 + 105x^6 + 120x^3 + 45x^2 + 15$

**57. a.** $y = -3x/2 + 17/2$   **b.**

**59. a.** $y = -x - 4$   **b.**

**61. a.** $p'(t) = \left(\dfrac{20}{t + 2}\right)^2$   **b.** $p'(5) \approx 8.16$   **c.** $t = 0$

**d.** $\lim\limits_{t \to \infty} p(t) = 200$; the population approaches a steady state.

**e.**

**63. a.** $F'(x) = -\dfrac{1.8 \times 10^{10}\, Qq}{x^3}$ N/m   **b.** $-1.8 \times 10^{19}$ N/m

**c.** $|F'(x)|$ decreases as $x$ increases.   **65. a.** False   **b.** False

**c.** False   **67.** $4x - \dfrac{1}{x^2}$; $2\left(\dfrac{1}{x^3} + 2\right)$; $-\dfrac{6}{x^4}$

**69.** $\dfrac{x^2 + 2x - 7}{(x + 1)^2}$; $\dfrac{16}{(x + 1)^3}$

**71. a.** $y = -\dfrac{108}{169}x + \dfrac{567}{169}$   **b.**

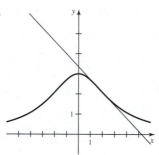

**73.** $-\dfrac{3}{2}$   **75.** $\dfrac{1}{9}$   **77.** $\dfrac{7}{8}$

**79. a.**

**b.** $t \approx 3$

**c.** $f'(3) \approx 0.28 \dfrac{\text{mm/g}}{\text{week}}$; at a young age, the bird's wings are growing quickly relative to its weight.   **d.** $f'(6.5) \approx 0.003 \dfrac{\text{mm/g}}{\text{week}}$; the rate of change of the ratio of wing chord length to mass is nearly 0.   **81.** $\dfrac{15}{2}$

**83.** $-\dfrac{5}{2}$   **85.** 1   **87. a.** $y = -2x + 16$   **b.** $y = -\dfrac{5}{9}x + \dfrac{23}{9}$

**89.** $-90$   **93.** $f''g + 2f'g' + fg''$

**95. a.** $f'gh + fg'h + fgh'$   **b.** $\dfrac{5x^2 + 6x - 3}{2\sqrt{x}}$

## Section 3.5 Exercises, pp. 152–155

**1.** $\dfrac{\sin x}{x}$ is undefined at $x = 0$.   **3.** The tangent and cotangent functions are defined as ratios of the sine and cosine functions.   **5.** $-1$   **7.** $y = x$   **9.** $-\sin x - \cos x$   **11.** 3   **13.** $\dfrac{7}{3}$

**15.** 5   **17.** 7   **19.** $\dfrac{1}{4}$   **21.** $a/b$   **23.** $\cos x - \sin x$

**25.** $3x^3(4 \sin x + x \cos x)$   **27.** $\sin x + x \cos x$   **29.** $-\dfrac{1}{1 + \sin x}$

**31.** $\cos^2 x - \sin^2 x = \cos 2x$   **33.** $-2 \sin x \cos x = -\sin 2x$

**35.** $w^2 \cos w$   **37.** $x \cos 2x + \dfrac{1}{2} \sin 2x$   **39.** $\dfrac{2 \sin x}{(1 + \cos x)^2}$

**41.** $\dfrac{4 \cos 2x}{(4 + \sin x \cos x)^2}$   **43.** $\sec x \tan x - \csc x \cot x$

**45.** $\dfrac{(1 - 2x \cot x) \csc x}{2\sqrt{x}}$   **47.** $-\dfrac{\csc x}{1 + \csc x}$

**49.** $\sin x(1 + \sec^2 x)$   **51.** $2 \sin^2 x$

**55. a.**

**b.** $v(t) = 30 \cos t$

**c.**

**d.** $v(t) = 0$, for $t = (2k + 1)\dfrac{\pi}{2}$, where $k$ is any nonnegative integer; the position is $y\left((2k + 1)\dfrac{\pi}{2}\right) = 0$ if $k$ is even or $y\left((2k + 1)\dfrac{\pi}{2}\right) = -60$ if $k$ is odd.   **e.** $v(t)$ has a maximum at $t = 2k\pi$, where $k$ is a nonnegative integer; the position is $y(2k\pi) = -30$.   **f.** $a(t) = -30 \sin t$

**57.** $2 \cos x - x \sin x$   **59.** $\dfrac{2 \sin x - 2x \cos x - x^2 \sin x}{x^3}$

**61.** $2 \csc^2 x \cot x$   **63.** $2 (\sec^2 x \tan x + \csc^2 x \cot x)$

**65. a.** False   **b.** False   **c.** True   **d.** True   **67.** 2   **69.** $-\dfrac{1}{2}$   **71.** $\dfrac{4}{3}$

**73. a.** $y = \sqrt{3}x + 2 - \dfrac{\pi\sqrt{3}}{6}$  **75. a.** $y = -2\sqrt{3}x + \dfrac{2\sqrt{3}\pi}{3} + 1$

**b.**

**b.**

**c.** $v(1) = -2$ ft/s; $a(1) = 2$ ft/s²  **d.** $a(2) = 2$ ft/s²  **e.** $(2, 5]$

**17. a.**

**b.** $v(t) = 4t - 9$; stationary at $t = \frac{9}{4}$, to the right on $\left(\frac{9}{4}, 3\right]$, to the left on $\left[0, \frac{9}{4}\right)$

**77.** $x = 7\pi/6 + 2k\pi$ and $x = 11\pi/6 + 2k\pi$, where $k$ is an integer
**85.** $a = 0$  **87. a.** $2\sin x \cos x$  **b.** $3\sin^2 x \cos x$  **c.** $4\sin^3 x \cos x$
**d.** $n\sin^{n-1} x \cos x$; the conjecture is true for $n = 1$. If it holds
for $n = k$, then when $n = k + 1$, we have $\dfrac{d}{dx}(\sin^{k+1} x) =$
$\dfrac{d}{dx}(\sin^k x \cdot \sin x) = \sin^k x \cos x + \sin x \cdot k \sin^{k-1} x \cos x =$
$(k + 1) \sin^k x \cos x.$

**c.** $v(1) = -5$ ft/s; $a(1) = 4$ ft/s²
**d.** $a\left(\frac{9}{4}\right) = 4$ ft/s²  **e.** $\left(\frac{9}{4}, 3\right]$
**19. a.**

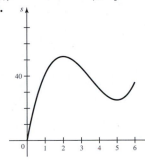

**89.** Because $D$ is a difference quotient for $f$ (and $h = 0.01$ is small), $D$ is a good approximation to $f'$. Therefore, the graph of $D$ is nearly indistinguishable from the graph of $f'(x) = \cos x$.

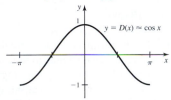

**b.** $v(t) = 6t^2 - 42t + 60$; stationary at $t = 2$ and $t = 5$, to the right on $[0, 2)$ and $(5, 6]$, to the left on $(2, 5)$

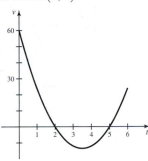

## Section 3.6 Exercises, pp. 163–168

**1.** The average rate of change is $\dfrac{f(x + \Delta x) - f(x)}{\Delta x}$, whereas the instantaneous rate of change is the limit as $\Delta x$ goes to zero in this quotient.  **3.** Small  **5.** At 15 weeks, the puppy grows at a rate of 1.75 lb/week.  **7.** If the position of the object at time $t$ is $s(t)$, then the acceleration at time $t$ is $a(t) = d^2s/dt^2$.  **9.** $v'(T) = 0.6$; the speed of sound increases approximately 0.6 m/s for each increase of 1°C.  **11. a.** 40 mi/hr  **b.** 40 mi/hr; yes  **c.** −60 mi/hr; −60 mi/hr; south  **d.** The patrol car drives away from the station going north until about 10:08, when it turns around and heads south, toward the station. It continues south until it passes the station at about 11:02 and keeps going south until about 11:40, when it turns around and heads north.  **13.** The first 200 stoves cost, on average, $70 to produce. When 200 stoves have already been produced, the 201st stove costs $65 to produce.

**15. a.**

**b.** $v(t) = 2t - 4$; stationary at $t = 2$, to the right on $(2, 5]$, to the left on $[0, 2)$

**c.** $v(1) = 24$ ft/s; $a(1) = -30$ ft/s²  **d.** $a(2) = -18$ ft/s²; $a(5) = 18$ ft/s²  **e.** $\left(2, \frac{7}{2}\right)$, $(5, 6]$  **21.** −64 ft/s; 64 ft/s
**23. a.** $v(t) = -32t + 32$  **b.** At $t = 1$ s  **c.** 64 ft  **d.** At $t = 3$ s  **e.** −64 ft/s  **f.** $(1, 3)$  **25. a.** $v(t) = -32t + 64$  **b.** At $t = 2$ s  **c.** 96 ft  **d.** At $2 + \sqrt{6}$ s  **e.** $-32\sqrt{6}$ ft/s  **f.** $(2, 2 + \sqrt{6})$
**27.** Approx. 90.5 ft/s  **29. a.** $\overline{C}(x) = \dfrac{1000}{x} + 0.1$; $C'(x) = 0.1$
**b.** $\overline{C}(2000) = \$0.60$/item; $C'(2000) = \$0.10$/item
**c.** The average cost per item when 2000 items are produced is $0.60/item. The cost of producing the 2001st item is $0.10.
**31. a.** $\overline{C}(x) = -0.01x + 40 + 100/x$; $C'(x) = -0.02x + 40$
**b.** $\overline{C}(1000) = \$30.10$/item; $C'(1000) = \$20$/item
**c.** The average cost per item is about $30.10 when 1000 items are produced. The cost of producing the 1001st item is $20.  **33. a.** 20
**b.** $20  **c.** $E(p) = \dfrac{p}{p - 20}$  **d.** Elastic for $p > 10$; inelastic for $0 < p < 10$  **e.** 2.5%  **f.** 2.5%  **35. a.** False  **b.** True  **c.** False
**d.** True  **37.** 240 ft  **39.** 64 ft/s  **41. a.** $t = 1, 2, 3$  **b.** It is moving in the positive direction for $t$ in $(0, 1)$ and $(2, 3)$; it is moving in the negative direction for $t$ in $(1, 2)$ and $t > 3$.

**c.**

**d.** $\left(0, \frac{1}{2}\right)$, $\left(1, \frac{3}{2}\right)$, $\left(2, \frac{5}{2}\right)$, $(3, \infty)$

**43. a.** $P(x) = 0.02x^2 + 50x - 100$

**b.** $\dfrac{P(x)}{x} = 0.02x + 50 - \dfrac{100}{x}; \dfrac{dP}{dx} = 0.04x + 50$

**c.** $\dfrac{P(500)}{500} = 59.8; \dfrac{dp}{dx}(500) = 70$

**d.** The profit, on average, for each of the first 500 items produced is 59.8; the profit for the 501st item produced is 70.

**45. a.** $P(x) = 0.04x^2 + 100x - 800$

**b.** $\dfrac{P(x)}{x} = 0.04x + 100 - \dfrac{800}{x}; \dfrac{dp}{dx} = 0.08x + 100$

**c.** $\dfrac{P(1000)}{1000} = 139.2; p'(1000) = 180$

**d.** The average profit per item for each of the first 1000 items produced is \$139.20. The profit for the 1001st item produced is \$180.

**47.** About 1935; approx. 890,000 people/yr (answers will vary)

**49. a.**

**b.** $v = \dfrac{100}{(t+1)^2}$

**c.**

The marble moves fastest at the beginning and slows considerably over the first 5 s. It continues to slow but never actually stops.

**d.** $t = 4$ s    **e.** $t = -1 + \sqrt{2} \approx 0.414$ s

**51. a.** $C'(x) = -\dfrac{125{,}000{,}000}{x^2} + 1.5;$

$\overline{C}(x) = \dfrac{C(x)}{25{,}000} = 50 + \dfrac{5000}{x} + 0.00006x$

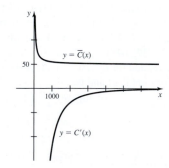

**b.** $C'(5000) = -3.5; \overline{C}(5000) = 51.3$   **c.** Marginal cost: If the batch size is increased from 5000 to 5001, then the cost of producing 25,000 gadgets will *decrease* by about \$3.50. Average cost: When batch size is 5000, it costs \$51.30 *per item* to produce all 25,000 gadgets.

**53. a.** $R(p) = \dfrac{100p}{p^2 + 1}$    **b.** $R'(p) = \dfrac{100(1 - p^2)}{(p^2 + 1)^2}$

**c.** $p = 1$

**55. a.**

**b.** $dx/dt = 10 \cos t + 10 \sin t$

**c.** $t = 3\pi/4 + k\pi$, where $k$ is any positive integer

**d.** The graph implies that the spring never stops oscillating. In reality, the weight would eventually come to rest.

**57. a.** Juan starts faster than Jean and opens up a big lead. Then Juan slows down while Jean speeds up. Jean catches up, and the race finishes in a tie.   **b.** Same average velocity   **c.** Tie   **d.** At $t = 2$, $\theta'(2) = \pi/2$ rad/min; $\theta'(4) = \pi = $ Jean's greatest velocity

**e.** At $t = 2, \varphi'(2) = \pi/2$ rad/min; $\varphi'(0) = \pi = $ Juan's greatest velocity   **59. a.** $-T'(1) = -80, -T'(3) = 80$

**b.** $-T'(x) < 0$ for $0 \le x < 2$; $-T'(x) > 0$ for $2 < x \le 4$   **c.** Near $x = 0$, with $x > 0$, $-T'(x) < 0$, so heat flows toward the end of the rod. Similarly, near $x = 4$, with $x < 4$, $-T'(x) > 0$.

## Section 3.7 Exercises, pp. 173–176

**1.** $\dfrac{dy}{dx} = \dfrac{dy}{du} \cdot \dfrac{du}{dx}; \dfrac{d}{dx}(f(g(x))) = f'(g(x)) \cdot g'(x)$

**3.** $u = x^3 + x + 1; y = u^4; 4(x^3 + x + 1)^3(3x^2 + 1)$

**5.** $u = \cos x, y = u^3, dy/dx = -3 \cos^2 x \sin x;$

$u = x^3, y = \cos u, dy/dx = -3x^2 \sin x^3$   **7.** $g(x), x$   **9.** $\dfrac{2}{\sqrt{4x + 1}}$

**11.** 50   **13.** $u = 3x + 7; f(u) = u^{10}; 30(3x + 7)^9$

**15.** $u = \sin x; f(u) = u^5; 5 \sin^4 x \cos x$

**17.** $u = x^2 + 1; f(u) = \sqrt{u}; \dfrac{x}{\sqrt{x^2 + 1}}$

**19.** $u = 2 + \sin x; f(u) = \sqrt{u}; \dfrac{\cos x}{2\sqrt{2 + \sin x}}$

**21.** $u = 5x^2; f(u) = \tan u; 10x \sec^2 5x^2$   **23. a.** 100   **b.** $-100$

**c.** $-16$   **d.** 40   **e.** 40   **25.** $10(6x + 7)(3x^2 + 7x)^9$

**27.** $\dfrac{5}{\sqrt{10x + 1}}$   **29.** $-\dfrac{315x^2}{(7x^3 + 1)^4}$   **31.** $3 \sec(3x + 1) \tan(3x + 1)$

**33.** $\dfrac{\sec^2 \sqrt{w}}{2\sqrt{w}}$   **35.** $(12x^2 + 3) \cos(4x^3 + 3x + 1)$

**37.** $\dfrac{10}{3(5x + 1)^{1/3}}$   **39.** $-\dfrac{3}{2^{7/4}x^{3/4}(4x - 3)^{5/4}}$

**41.** $5 \sec x (\sec x + \tan x)^5$   **43.** $25(12x^5 - 9x^2)(2x^6 - 3x^3 + 3)^{24}$

**45.** $9(1 + 2 \tan u)^{3.5} \sec^2 u$  **47.** $-\dfrac{\cot x \csc^2 x}{\sqrt{1 + \cot^2 x}}$

**49.** $8 (\sin^2 x + 1)^3 \sin x \cos x$

**51.** $-15 \sin^4 (\cos 3x) (\sin 3x) (\cos (\cos 3x))$

**53.** $\dfrac{1}{2\sqrt{x + \sqrt{x}}} \left(1 + \dfrac{1}{2\sqrt{x}}\right)$

**55.** $f'(g(x^2))g'(x^2) 2x$  **57.** $\dfrac{5x^4}{(x + 1)^6}$

**59.** $(x^2 + 1)^2 (7x^2 + 1)$  **61.** $\theta(2 + 5\theta \tan 5\theta) \sec 5\theta$

**63.** $4((x + 2)(x^2 + 1))^3(3x + 1)(x + 1)$  **65.** $\dfrac{4x^3 - 2 \sin 2x}{5(x^4 + \cos 2x)^{4/5}}$

**67.** $2 (p + 3)(\sin p^2 + p(p + 3) \cos p^2)$

**69.** $f'(x)/(2\sqrt{f(x)})$  **71. a.** True  **b.** True  **c.** True

**d.** False  **73.** $-0.297$ hPa/min  **75.** $2 \cos x^2 - 4x^2 \sin x^2$

**77.** $\dfrac{4(5x^2 - 1)}{(x^2 + 1)^4}$  **79.** $y = 6x - 1$

**81. a.** $h(4) = 9, h'(4) = -6$  **b.** $y = -6x + 33$

**83. a.** $y = 4\sqrt{3} (x - \pi/6) + 2$

**85. a.** $-3\pi$  **b.** $-5\pi$  **87. a.** $\dfrac{d^2 y}{dt^2} = -\dfrac{y_0 k}{m} \cos \left(t\sqrt{\dfrac{k}{m}}\right)$

**89. a.** 10.88 hr  **b.** $D'(t) = \dfrac{6\pi}{365} \sin \left(\dfrac{2\pi(t + 10)}{365}\right)$

**c.** 2.87 min/day; on March 1, the length of day is increasing at a rate of about 2.87 min/day.

**d.**

**e.** Most rapidly: approximately March 22 and September 22; least rapidly: approximately December 21 and June 21

**91. a.** $E'(t) = 400 + 200 \cos \dfrac{\pi t}{12}$ MW

**b.** At noon; $E'(0) = 600$ MW  **c.** At midnight; $E'(12) = 200$ MW

**d.**

**95. a.** $g(x) = (x^2 - 3)^5$; $a = 2$  **b.** 20

**97. a.** $g(x) = \sin x^2$; $a = \pi/2$  **b.** $\pi \cos (\pi^2/4)$  **99.** $10 f'(25)$

## Section 3.8 Exercises, pp. 181–184

**1.** There may be more than one expression for $y$ or $y'$.
**3.** When derived implicitly, $dy/dx$ is usually given in terms
of both $x$ and $y$.  **5.** $\dfrac{1}{2y}$  **7.** $\dfrac{1}{\cos y}$  **9. a.** $(0, 0), (0, -1), (0, 1)$

**c.** Slope at $(0, 0)$ is 2; slope at $(0, -1)$ and $(1, 0)$ is $-1$.

**11.** $\dfrac{d^2 y}{dx^2} = -\dfrac{2}{9y^5}$  **13. a.** $-\dfrac{x^3}{y^3}$  **b.** 1  **15. a.** $\dfrac{2}{y}$  **b.** 1

**17. a.** $\dfrac{20x^3}{\cos y}$  **b.** $-20$  **19. a.** $-\dfrac{1}{\sin y}$  **b.** $-1$  **21. a.** $-\dfrac{y}{x}$  **b.** $-7$

**23. a.** $-\dfrac{1}{4x^{2/3}y^{1/3}}$  **b.** $-\dfrac{1}{4}$  **25. a.** $-\dfrac{3y}{x + 3y^{2/3}}$  **b.** $-\dfrac{24}{13}$

**27.** $\dfrac{\cos x}{1 - \cos y}$  **29.** $-\dfrac{1}{1 + \sin y}$  **31.** $\dfrac{1 - y \cos xy}{x \cos xy - 1}$  **33.** $\dfrac{1}{2y (\sin y^2 + 1)}$

**35.** $\dfrac{3x^2 (x - y)^2 + 2y}{2x}$  **37.** $\dfrac{13y - 18x^2}{21y^2 - 13x}$  **39.** $\dfrac{5\sqrt{x^4 + y^2} - 2x^3}{y - 6y^2\sqrt{x^4 + y^2}}$

**41. a.** $\dfrac{dK}{dL} = -\dfrac{K}{2L}$  **b.** $-4$  **43.** $\dfrac{dr}{dh} = \dfrac{h - 2r}{h}$; $-3$

**45. b.** $y = -5x$  **47. b.** $y = -5x/4 + 7/2$  **49. b.** $y = \dfrac{x}{2}$

**51.** $-\dfrac{1}{4y^3}$  **53.** $\dfrac{\sin y}{(\cos y - 1)^3}$  **55.** $-\dfrac{\cos y}{(1 + \sin y)^3}$  **57. a.** False

**b.** True  **c.** False  **d.** False  **59. a.** $\dfrac{y(3\sqrt{x} + 2y^{3/2})}{x(\sqrt{x} - 2y^{3/2})}$  **b.** $-5$

**61. a.** $y = x - 1$ and $y = -x + 2$

**b.**

**63. a.** $y' = -\dfrac{2xy}{x^2 + 4}$  **b.** $y = \dfrac{1}{2}x + 2, y = -\dfrac{1}{2}x + 2$

**c.** $-\dfrac{16x}{(x^2 + 4)^2}$  **65. a.** $\left(\dfrac{5}{4}, \dfrac{1}{2}\right)$  **b.** No

**67.** Horizontal: $y = -6, y = 0$; vertical: $x = 1, x = 3$

**69. a.** $\dfrac{dy}{dx} = 0$ on the $y = 1$ branch; $\dfrac{dy}{dx} = \dfrac{1}{2y + 1}$ on the other two

branches.  **b.** $f_1(x) = 1, f_2(x) = \dfrac{-1 + \sqrt{4x - 3}}{2}$,

$f_3(x) = \dfrac{-1 - \sqrt{4x - 3}}{2}$  **c.**

**71. a.** $\dfrac{dy}{dx} = \dfrac{x - x^3}{y}$  **b.** $f_1(x) = \sqrt{x^2 - \dfrac{x^4}{2}}$; $f_2(x) = -\sqrt{x^2 - \dfrac{x^4}{2}}$

**c.**

**73.** $y = \dfrac{x}{5}$

**75.** $y = \dfrac{4x}{5} - \dfrac{3}{5}$    **77.** $y = -2x + \dfrac{5\pi}{4}$

**79. a.** Tangent line $y = -\dfrac{9x}{11} + \dfrac{20}{11}$; normal line $y = \dfrac{11x}{9} - \dfrac{2}{9}$

**b.**

**81. a.** Tangent line $y = -\dfrac{x}{3} + \dfrac{8}{3}$;

normal line $y = 3x - 4$

**b.**

**83.** For $y = mx$, $dy/dx = m$; for $x^2 + y^2 = a^2$, $dy/dx = -x/y$.
**85.** For $xy = a$, $dy/dx = -y/x$; for $x^2 - y^2 = b$, $dy/dx = x/y$.
Because $(-y/x) \cdot (x/y) = -1$, the families of curves
form orthogonal trajectories.    **87.** $\dfrac{7y^2 - 3x^2 - 4xy^2 - 4x^3}{2y(2x^2 + 2y^2 - 7x)}$

**89.** $\dfrac{2y^2(5 + 8x\sqrt{y})}{(1 + 2x\sqrt{y})^3}$    **91.** No horizontal tangent line; vertical tangent
lines at $(2, 1)$, $(-2, 1)$

### Section 3.9 Exercises, pp. 188–193

**1.** As the side length $s$ of a cube changes, the surface area $6s^2$ changes
as well.    **3.** The other two opposite sides decrease in length.

**5. a.** $V = 200h$; $\dfrac{dV}{dt} = 200\dfrac{dh}{dt}$   **b.** 50 ft³/min   **c.** $\dfrac{1}{20}$ ft/min

**7. a.** $\dfrac{dV}{dt} = 4\pi r^2 \dfrac{dr}{dt}$   **b.** $128\pi$ in³/min   **c.** $\dfrac{1}{10\pi}$ in/min

**9.** 59   **11. a.** 40 m²/s   **b.** 80 m²/s
**13. a.** 4 m²/s   **b.** $\sqrt{2}$ m²/s   **c.** $2\sqrt{2}$ m/s   **15. a.** $\dfrac{1}{4\pi}$ cm/s
**b.** $\dfrac{1}{2}$ cm/s   **17.** $-40\pi$ ft²/min   **19.** $\dfrac{3}{80\pi}$ in/min
**23.** 720.3 mi/hr   **25.** $\dfrac{3\sqrt{5}}{2}$ ft/s   **27.** 57.9 ft/s   **29.** 4.66 in/s
**31.** $\dfrac{\pi}{2}$ ft³/min   **33.** $-75\pi$ cm³/s   **35.** $2592\pi$ cm³/s
**37.** $9\pi$ ft³/min   **39.** $\dfrac{1}{25\pi}$ m/min   **41.** $\dfrac{5}{24}$ ft/s
**43.** $-\dfrac{8}{3}$ ft/s, $-\dfrac{32}{3}$ ft/s   **45.** $\dfrac{d\theta}{dt} = \dfrac{1}{5}$ rad/s, $\dfrac{d\theta}{dt} = \dfrac{1}{8}$ rad/s
**47.** $-0.0201$ rad/s   **49.** 10 tan 20° km/hr $\approx$ 3.6 km/hr
**51. a.** 187.5 ft/s   **b.** 0.938 rad/s   **53. a.** $P = \dfrac{1}{2}v^2\dfrac{dm}{dt}$
**c.** 17,388.7 W   **d.** 4347.2 W   **55.** 11.06 m/hr
**57.** $\dfrac{1}{500}$ m/min; 2000 min   **59.** 0.54 rad/hr
**61.** $\dfrac{d\theta}{dt} = 0$ rad/s, for all $t \geq 0$   **63. a.** $-\dfrac{\sqrt{3}}{10}$ m/hr   **b.** $-1$ m²/hr

### Chapter 3 Review Exercises, pp. 193–197

**1. a.** False   **b.** False   **c.** False   **d.** False   **e.** True
**3.** $-\dfrac{2x}{(x^2 + 5)^2}$   **9.** $2x^2 + 2\pi x + 7$   **11.** $\dfrac{\cos w - \sin w}{2\sqrt{2} + \sin w + \cos w}$
**13.** $5(\tan w + 10)^4 \sec^2 w$   **15.** $6x^3\sqrt{1 + x^4}$
**17.** $5t^2 \cos t + 10t \sin t$   **19.** $\dfrac{v^{99}(99v + 100)}{(v + 1)^2}$   **21.** $\dfrac{2 \sec 2w \tan 2w}{(\sec 2w + 1)^2}$
**23.** $100t \sin^9(5t^2 + 1) \cos(5t^2 + 1)$   **25.** $1000t(5t^2 + 10)^{99}$
**27.** $(8\theta + 12) \sec^2(\theta^2 + 3\theta + 2)$
**29.** $\dfrac{32u^2 + 8u + 1}{(8u + 1)^2}$   **31.** $(\sec^2 \sin \theta) \cos \theta$
**33.** $-\dfrac{\cos\sqrt{\cos^2 x + 1} \cos x \sin x}{\sqrt{\cos^2 x + 1}}$   **35.** $x^2 \sin x$
**37.** $x(x + 1)^{2/5}$   **39.** $\dfrac{1}{(x + 1)^2} \cos\left(\dfrac{x}{x + 1}\right)$
**41.** $\sec w \tan w$   **43.** $1 + \csc(x - y)$   **45.** $-\dfrac{xy}{x^2 + 2y^2}$
**47.** $\sqrt{2} + \dfrac{\pi\sqrt{2}}{4}$   **49.** $\dfrac{3x + 11}{4(x + 2)^{3/2}}$
**51.** $\dfrac{2(xy + y^2)}{(x + 2y)^3} = \dfrac{2}{(x + 2y)^3}$   **53.** $y = x$   **55.** $y = -\dfrac{4x}{5} + \dfrac{24}{5}$
**57.** $x^2 f'(x) + 2xf(x)$   **59.** $\dfrac{g(x)(xf'(x) + f(x)) - xf(x)g'(x)}{g^2(x)}$
**61.** a–D; b–C; c–B; d–A
**63.**

**65. a.** 27   **b.** $\frac{16}{27}$   **c.** 72   **d.** 1215   **67. a.** $\frac{36}{25}$   **b.** $\frac{1}{2\sqrt{3}}$   **c.** 0

**69.** $y = 24x - 118$  **71. a.** 84 ft/s  **b.** 7 s  **c.** 384 ft
**d.** 96 ft/s  **73. a.** 2.70 million people/yr
**b.** The slope of the secant line through the two points is
approximately equal to the slope of that tangent line at $t = 55$.
**c.** 2.217 million people/yr  **75.** $x = 4; x = 6$
**77.** $f(x) = \tan(\pi\sqrt{3x - 11})$, $a = 5$; $f'(5) = 3\pi/4$
**79. a.** $\bar{C}(3000) = \$341.67$; $C'(3000) = \$280$  **b.** The average cost
of producing the first 3000 lawn mowers is $341.67 per mower. The
cost of producing the 3001st lawn mower is $280.
**81. a.** 6550 people/yr  **b.** $p'(40) = 4800$ people/yr
**83.** 50 mi/hr  **85.** $-5 \sin 65° $ ft/s $\approx -4.5$ ft/s
**87.** $-0.166$ rad/s  **89.** 1.5 ft/s

## CHAPTER 4

### Section 4.1 Exercises, pp. 204–207

**1.** $f$ has an absolute maximum at $c$ in $[a, b]$ if $f(x) \leq f(c)$ for all $x$ in
$[a, b]$. $f$ has an absolute minimum at $c$ in $[a, b]$ if $f(x) \geq f(c)$ for all
$x$ in $[a, b]$.  **3.** The function must be continuous on a closed interval.
**5.**

**7.**

**9.** Evaluate the function at the critical points and at the endpoints of
the interval.  **11.** Abs. min at $x = c_2$; abs. max at $x = b$  **13.** Abs.
min at $x = a$; no abs. max  **15.** Local min at $x = q, s$; local max at
$x = p, r$; abs. min at $x = a$; abs. max at $x = b$  **17.** Local max at
$x = p, r$; local min at $x = q$; abs. max at $x = p$; abs. min at $x = b$
**19.**

**21.**

**23.** $x = \frac{2}{3}$  **25.** $x = \pm 3$  **27.** $x = -\frac{2}{3}, \frac{1}{3}$  **29.** $x = \pm\dfrac{2a}{\sqrt{3}}$

**31.** $t = \frac{1}{2}$  **33.** $x = 1$  **35.** $x = -4, 0$
**37.** If $a \geq 0$, there is no critical point. If $a < 0$, $x = 2a/3$ is
the only critical point.  **39.** $t = \pm a$  **41.** Abs. max: $-1$ at
$x = 3$; abs. min: $-10$ at $x = 0$  **43.** Abs. max: 0 at $x = 0, 3$;
abs. min: $-4$ at $x = -1, 2$  **45.** Abs. max: 234 at $x = 3$;
abs. min: $-38$ at $x = -1$  **47.** Abs. max: 1 at $x = 0, \pi$; abs. min: 0 at
$x = \pi/2$  **49.** Abs. max: 1 at $x = \pi/6$; abs. min: $-1$ at $x = -\pi/6$

**51.** Abs. max: $\dfrac{\pi}{3} + \dfrac{\sqrt{3}}{2}$ at $x = \dfrac{\pi}{3}$; abs. min: 0 at $x = 0$

**53.** Abs. max: 11 at $x = 1$; abs. min: $-16$ at $x = 4$
**55.** Abs. max: 27 at $x = -3$; abs. min: $-\frac{19}{12}$ at $x = \frac{1}{2}$

**57.** Abs. max: $\dfrac{1}{100,000}$ at $x = 1$; abs. min: $-\dfrac{1}{100,000}$ at $x = -1$

**59.** Abs. max: $\sqrt{2}$ at $x = \pm\pi/4$; abs. min: 1 at $x = 0$
**61.** Abs. max: 3 at $x = \pm 1$; abs. min: 0 at $x = -2, 0, 2$
**63. a.** The velocity of the downstream wind $v_2$ is less than or equal to
the velocity of the upstream wind, so $0 \leq v_2 \leq v_1$, or $0 \leq \dfrac{v_2}{v_1} \leq 1$.

**b.** $R(1) = 0$  **c.** $R(0) = \frac{1}{2}$  **d.** 0.593 is the maximum fraction of
power that can be extracted from a wind stream by a wind turbine.
**65.** $t = 2$ s  **67.** $t = 2$ s  **69. a.** 50  **b.** 45  **71. a.** False
**b.** False  **c.** False  **d.** True  **73. a.** $x = 0, 1, 2$
**b.** Abs. max: 8 at $x = -2$; abs. min: 0 at $x = 0, 2$
**75. a.** $x \approx -5.18, -2.03, 1.11, 4.25$
**b.** $x \approx -5.18$ and $x \approx 1.11$ correspond to local max;
$x \approx -2.03$ and $x \approx 4.25$ correspond to local min.
**c.** Abs. max: 2.24; abs. min: $-2.24$  **77. a.** $x = 5 - 4\sqrt{2}$
**b.** $x = 5 - 4\sqrt{2}$ corresponds to a local max.  **c.** No abs. max or min
**79.** Abs. max: 4 at $x = -1$; abs. min: $-8$ at $x = 3$

**81. a.** $T(x) = \dfrac{\sqrt{2500 + x^2}}{2} + \dfrac{50 - x}{4}$  **b.** $x = 50/\sqrt{3}$

**c.** $T(50/\sqrt{3}) \approx 34.15$, $T(0) = 37.50$, $T(50) \approx 35.36$
**d.**

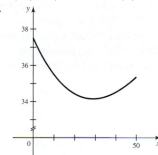

**83. a.** 1, 3, 0, 1  **b.** Because $h'(2) \neq 0$, $h$ does not have a local
extreme value at $x = 2$. However, $g$ may have a local extremum at
$x = 2$ (because $g'(2) = 0$).  **85. a.** $f(x) - f(c) \leq 0$ for all $x$ near $c$

**b.** $\displaystyle\lim_{x \to c^+} \dfrac{f(x) - f(c)}{x - c} \leq 0$  **c.** $\displaystyle\lim_{x \to c^-} \dfrac{f(x) - f(c)}{x - c} \geq 0$

**d.** Because $f'(c)$ exists, $\displaystyle\lim_{x \to c^+} \dfrac{f(x) - f(c)}{x - c} = \lim_{x \to c^-} \dfrac{f(x) - f(c)}{x - c}$. By
parts (b) and (c), we must have that $f'(c) = 0$.

### Section 4.2 Exercises, pp. 211–214

**1.** If $f$ is a continuous function on the closed interval $[a, b]$ and is
differentiable on $(a, b)$, and the slope of the secant line that joins
$(a, f(a))$ to $(b, f(b))$ is zero, then there is at least one value $c$ in $(a, b)$
at which the slope of the line tangent to $f$ at $(c, f(c))$ is also zero.

**3.** $f(x) = |x|$ is not differentiable at 0.  **5. b.** $c = 1$
**7. b.** $c = \pm 2/\sqrt[4]{5} \approx \pm 1.34$
**9.**

**11.** $x = \frac{1}{3}$

**13.** $x = \pi/4$  **15.** Does not apply  **17.** $x = \frac{5}{3}$  **19.** Average lapse rate $= -6.3°/$km. You cannot conclude that the lapse rate at a point exceeds the threshold value.  **21. a.** Yes  **b.** $c = \frac{1}{2}$  **23. a.** Does not apply  **25. a.** Yes  **b.** $c = \frac{9}{4}$  **27. a.** Yes  **b.** $c \approx \pm 0.881$  **29. a.** Yes  **b.** $c = \left(\frac{7}{4}\right)^{3/4} \approx 1.52$  **31. a.** Does not apply  **33. a.** False  **b.** True  **c.** False  **35.** $h$ and $p$

**37.**

**39.** No such point exists; function is not continuous at 2.
**41.** The car's average velocity is $(30 - 0)/(28/60) = 64.3$ mi/hr. By the MVT, the car's instantaneous velocity was 64.3 mi/hr at some time.
**43.** Average speed $= 11.625$ mi/hr. By the MVT, the speed was exactly 11.625 mi/hr at least once. By the Intermediate Value Theorem, all speeds between 0 and 11.625 mi/hr were reached. Because the initial and final speeds were 0 mi/hr, the speed of 11 mi/hr was reached at least twice.  **45.** $\dfrac{f(b) - f(a)}{b - a} = A(a + b) + B$

and $f'(x) = 2Ax + B$; $2Ax + B = A(a + b) + B$ implies that $x = \dfrac{a + b}{2}$, the midpoint of $[a, b]$.  **47.** $\tan^2 x$ and $\sec^2 x$ differ by a constant; in fact, $\tan^2 x - \sec^2 x = -1$.  **49.** Bolt's average speed was 37.58 km/hr, so he exceeded 37 km/hr during the

race.  **51.** *Hint:* By the MVT, there is a value of $c$ in $(2, 4)$ for which $\dfrac{f(4) - f(2)}{4 - 2} = f'(c)$.  **55. b.** $c = \frac{1}{2}$

## Section 4.3 Exercises, pp. 224–227

**1.** $f$ is increasing on $I$ if $f'(x) > 0$ for all $x$ in $I$; $f$ is decreasing on $I$ if $f'(x) < 0$ for all $x$ in $I$.  **3. a.** $x = 3$  **b.** Increasing on $(3, \infty)$; decreasing on $(-\infty, 3)$

**5.**

**7.**

**9.**

**11.**

**13. a.** Concave up on $(-\infty, 2)$; concave down on $(2, \infty)$
**b.** Inflection point at $x = 2$  **15.** Yes; consider the graph of $y = \sqrt{x}$ on $(0, \infty)$.  **17.** $f(x) = x^4$  **19.** Increasing on $(-\infty, 0)$; decreasing on $(0, \infty)$  **21.** Decreasing on $(-\infty, 1)$; increasing on $(1, \infty)$
**23.** Increasing on $(-\infty, 1)$ and $(4, \infty)$; decreasing on $(1, 4)$
**25.** Increasing on $(-\infty, 1/2)$; decreasing on $(1/2, \infty)$
**27.** Increasing on $(-\infty, 0)$, $(1, 2)$; decreasing on $(0, 1)$, $(2, \infty)$
**29.** Increasing on $\left(\dfrac{\pi}{6}, \dfrac{5\pi}{6}\right)$; decreasing on $\left(0, \dfrac{\pi}{6}\right)$, $\left(\dfrac{5\pi}{6}, 2\pi\right)$
**31.** Increasing on $(-\pi, -2\pi/3)$, $(-\pi/3, 0)$, $(\pi/3, 2\pi/3)$; decreasing on $(-2\pi/3, -\pi/3)$, $(0, \pi/3)$, $(2\pi/3, \pi)$
**33.** Increasing on $(-1, 0)$, $(1, \infty)$; decreasing on $(-\infty, -1)$, $(0, 1)$
**35.** Increasing on $(1, 4)$; decreasing on $(-\infty, 1)$, $(4, \infty)$
**37.** Increasing on $\left(-\infty, -\frac{1}{2}\right)$, $\left(0, \frac{1}{2}\right)$; decreasing on $\left(-\frac{1}{2}, 0\right)$, $\left(\frac{1}{2}, \infty\right)$
**39.** Increasing on $(0, \pi)$; decreasing on $(\pi, 2\pi)$  **41. a.** $x = 0$
**b.** Local min at $x = 0$  **c.** Abs. min: 3 at $x = 0$; abs. max: 12 at $x = -3$  **43. a.** $x = \pm\sqrt{2}$  **b.** Local min at $x = -\sqrt{2}$; local max at $x = \sqrt{2}$  **c.** Abs. max: 2 at $x = \sqrt{2}$; abs. min: $-2$ at $x = -\sqrt{2}$
**45. a.** $x = \pm\sqrt{3}$  **b.** Local min at $x = -\sqrt{3}$; local max at $x = \sqrt{3}$  **c.** Abs. max: 28 at $x = -4$; abs. min: $-6\sqrt{3}$ at $x = -\sqrt{3}$  **47. a.** $x = 2$ and $x = 0$  **b.** Local max at $x = 0$; local min at $x = 2$  **c.** Abs. min: $-10\sqrt[3]{25}$ at $x = -5$; abs. max: 0 at $x = 0, 5$  **49.** Abs. max: $-\frac{14}{3}$ at $x = \frac{1}{3}$  **51.** Abs. min: $36\sqrt[3]{\pi/6}$ at $r = \sqrt[3]{6/\pi}$

**53.**

**55.**

**57.** Concave up on $(-\infty, 0)$, $(1, \infty)$; concave down on $(0, 1)$; inflection points at $x = 0$ and $x = 1$  **59.** Concave up on $(-\infty, 0)$, $(2, \infty)$; concave down on $(0, 2)$; inflection points at $x = 0$ and $x = 2$
**61.** Concave up on $(-\infty, -3)$; concave down on $(-3, \infty)$; no inflection point  **63.** Concave down on $(0, \pi)$; concave up on $(\pi, 2\pi)$; inflection point at $\theta = \pi$  **65.** Concave down on $(-1, 0)$; concave up on $(0, \infty)$; inflection point at $x = 0$  **67.** Concave up on $(0, 2)$, $(4, \infty)$; concave down on $(-\infty, 0)$, $(2, 4)$; inflection points at $x = 0, 2, 4$  **69.** Critical pts. $x = 0$ and $x = 2$; local max at $x = 0$, local min at $x = 2$  **71.** Critical pt. $x = 0$; local max at $x = 0$  **73.** Critical pts. $x = 0$ and $x = 1$; local max at $x = 0$; local min at $x = 1$  **75.** Critical pts. $x = 0$ and $x = 1$; local min at $x = 1$; inconclusive at $x = 0$  **77.** Critical pts. $x = 0$ and $x \approx 0.327$; local max at $x = 0$; local min at $x \approx 0.327$
**79.** Critical pts. $t = -3$ and $t = 2$; local max at $t = -3$; local min at $t = 2$  **81.** Critical pt. $x = -a$; inconclusive at $x = -a$
**83.** Critical pts. $x = \pi/9, x = 5\pi/9$, and $x = 7\pi/9$; local max at $x = \pi/9$ and $x = 7\pi/9$; local min at $x = 5\pi/9$
**85. a.** True  **b.** False  **c.** True  **d.** False  **e.** False
**87.**

**89.** a–C–i, b–B–iii, c–A–ii

**91.**

**11.**

**13.**

**93.**

**95.**

**15.**

**17.**

**19.**

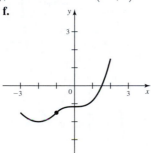

**97. a.** Increasing on $(-2, 2)$; decreasing on $(-3, -2)$
**b.** Critical pts. $x = -2$ and $x = 0$; local min at $x = -2$; neither a local max nor min at $x = 0$ **c.** Inflection pts. at $x = -1$ and $x = 0$
**d.** Concave up on $(-3, -1)$, $(0, 2)$; concave down on $(-1, 0)$
**e.**

**f.**

**21.**

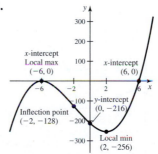

**99. a.** $E = \dfrac{p}{p - 50}$ **b.** $-1.4\%$ **c.** $E'(p) = -\dfrac{ab}{(a - bp)^2} < 0$, for $p \geq 0, p \neq a/b$ **d.** $E(p) = -b$, for $p \geq 0$
**101. a.** 300 **b.** $t = \sqrt{10}$ **c.** $t = \sqrt{b/3}$

### Section 4.4 Exercises, pp. 234–236

**1.** We need to know on which interval(s) to graph $f$. **3.** No; the domain of any polynomial is $(-\infty, \infty)$; there is no vertical asymptote. Also, $\displaystyle\lim_{x \to \pm\infty} p(x) = \pm\infty$, where $p$ is any polynomial; there is no horizontal asymptote. **5.** Evaluate the function at the critical points and at the endpoints. Then find the largest and smallest values among those candidates.

**23.**

**7.**

**9.**

**25.**

**27.**

**29.**

**43.**

**31.**

**33.**

**45.**

**47.** Critical pts. at $x = 1, 3$; local max at $x = 1$; local min at $x = 3$; inflection pt. at $x = 2$; increasing on $(0, 1)$, $(3, 4)$; decreasing on $(1, 3)$; concave up on $(2, 4)$; concave down on $(0, 2)$

**35.**

**49.**

**37.**

**51.**

**39.**

**53.**

**55. a.** False   **b.** False
    **c.** False   **d.** True

**41.**

**57.**

Local max at $x = -\frac{3\pi}{2}$   Local max at $x = -\frac{\pi}{2}$   Local max at $x = \frac{\pi}{2}$   Local max at $x = \frac{3\pi}{2}$

Local min at $x = -\pi$    Local min at $x = 0$   Local min at $x = \pi$

**59.**

Local max at $x = -2$

No local extremum at $x = 0$

Local min at $x = 1$

**61. (A) a.**

**b.** Water is being added at all times.   **c.** No concavity
**d.** $h'$ has an abs. max at all points of $[0, 10]$.

**(B) a.**

**c.** Concave down   **d.** $h'$ has abs. max at $t = 0$.

**(C) a.**

**c.** Concave up   **d.** $h'$ has abs. max at $t = 10$.

**(D) a.**

**c.** Concave down on $(0, 5)$, then concave up on $(5, 10)$; inflection pt. at $t = 5$   **d.** $h'$ has abs. max at $t = 0$ and $t = 10$.

**(E) a.**

**c.** First, no concavity; then concave down, no concavity, concave up, and, finally, no concavity
**d.** $h'$ has abs. max at all points of an interval $[0, a]$ and $[b, 10]$.

**(F) a.**

**c.** Concave down on $(0, 5)$; concave up on $(5, 10)$; inflection pt. at $t = 5$   **d.** $h'$ has abs. max at $t = 0$ and $t = 10$.

**63.**

Local max $(-1.356, 0.467)$   Local max $(1.356, 0.467)$
Inflection point $(-0.494, 0.188)$   Inflection point $(0.494, 0.188)$
Inflection point $(-2.561, 0.186)$   Inflection point $(2.561, 0.186)$
Local min $(0, 0)$
Inflection point $(-5.961, -0.052)$   Inflection point $(5.961, -0.052)$
Local min $(-4.514, -0.207)$   Local min $(4.514, -0.207)$

**65.**

$x, y$-intercept $(0, 0)$   $x$-intercept $(13.15, 0)$
Inflection point $(6.85, -4717)$
Local min $(10, -8000)$

Local max $(1, 19)$
Inflection point $(0.5, 9.3)$
$x, y$-intercept $(0, 0)$   $x$-intercept $(1.5, 0)$

**67.**

$(0, 1)$
$(-1, 0)$   $(1, 0)$
$(0, -1)$

**69.**

$(1.26, 1.59)$   $(1.5, 1.5)$
$(1.59, 1.26)$
$(0, 0)$
$y = -x - 1$

**71.**

$\left(-\frac{\sqrt{2}}{2}, \frac{1}{2}\right)$   $\left(\frac{\sqrt{2}}{2}, \frac{1}{2}\right)$
$(-1, 0)$   $(1, 0)$
$\left(-\frac{\sqrt{2}}{2}, -\frac{1}{2}\right)$   $(0, 0)$   $\left(\frac{\sqrt{2}}{2}, -\frac{1}{2}\right)$

**73.**

$n = 3$
$n = \frac{2}{3}$

## Section 4.5 Exercises, pp. 241–247

**1.** Objective function, constraint(s)   **3.** $Q = x^2(10 - x)$; $Q = (10 - y)^2 y$   **5. a.** $P(x) = 100x - 10x^2$   **b.** Abs. max: 250 at $x = 5$   **7.** $\frac{23}{2}$ and $\frac{23}{2}$   **9.** $5\sqrt{2}$ and $5\sqrt{2}$   **11.** Width = length = $\frac{5}{2}$
**13.** Width = length = 10   **15.** $x = \sqrt{6}, y = 2\sqrt{6}$
**17.** $\dfrac{10}{\sqrt{2}}$ by $\dfrac{5}{\sqrt{2}}$   **19.** Length = width = height = 2

**21.** $\dfrac{4}{\sqrt[3]{5}}$ ft by $\dfrac{4}{\sqrt[3]{5}}$ ft by $5^{2/3}$ ft   **23.** Approx. $(0.59, 0.65)$

**25.** $(5, 15)$, distance $\approx 47.4$   **27. a.** A point $8/\sqrt{5}$ mi from the point on the shore nearest the woman in the direction of the restaurant **b.** $9/\sqrt{13}$ mi/hr   **29.** A point $7\sqrt{3}/6$ mi from the point on shore nearest the island, in the direction of the power station

**31.** 18.2 ft   **33.** $h = \dfrac{20}{\sqrt{3}}; r = 20\sqrt{\tfrac{2}{3}}$

**35. a.** $r = \sqrt[3]{177/\pi} \approx 3.83$ cm; $h = 2\sqrt[3]{177/\pi} \approx 7.67$ cm

**b.** $r = \sqrt[3]{177/2\pi} \approx 3.04$ cm; $h = 2\sqrt[3]{708/\pi} \approx 12.17$ cm; part (b) is closer to the real can.   **37.** $\sqrt{15}$ m by $2\sqrt{15}$ m

**39.** 12″ by 6″ by 3″; 216 in³   **41.** Lower rectangular pane is approximately 5.6 ft wide by 2.8 ft high.   **43.** $r/h = \sqrt{2}$

**45.** $r = \sqrt{2}R/\sqrt{3}; h = 2R/\sqrt{3}$   **47.** 3:1   **49. a.** 0, 30, 25

**b.** 42.5 mi/hr   **c.** The units of $p/g(v)$ are $/mi and so are the units of $w/v$. Therefore, $L\left(\dfrac{p}{g(v)} + \dfrac{w}{v}\right)$ gives the total cost of a trip of $L$ miles.   **d.** Approx. 62.9 mi/hr   **e.** Neither; the zeros of $C'(v)$ are independent of $L$.   **f.** Decreased slightly, to 62.5 mi/hr **g.** Decreased to 60.8 mi/hr   **51.** $\sqrt{30} \approx 5.5$ ft   **53.** The point $12/(\sqrt[3]{2} + 1) \approx 5.3$ m from the weaker source   **55. b.** Because the speed of light is constant, travel time is minimized when distance is minimized.   **57.** $r = h = \sqrt[3]{450/\pi}$ m   **59. a.** $\dfrac{a_1 + a_2}{2}$

**b.** $\dfrac{a_1 + a_2 + a_3}{3}$   **c.** $\dfrac{a_1 + a_2 + \cdots + a_n}{n}$   **61.** $\dfrac{\pi}{3}$

**63.** When the seat is at its lowest point   **65. a.** $P = 2/\sqrt{3}$ units from the midpoint of the base   **67.** You can run 12 mi/hr if you run toward the point $3/16$ mi ahead of the locomotive (when it passes the point nearest you).   **69. a.** $r = 2R/3; h = \tfrac{1}{3}H$   **b.** $r = R/2; h = H/2$

**71.** $(1 + \sqrt{3})$ mi $\approx 2.7$ mi   **73. (i)** $(p - \tfrac{1}{2}, \sqrt{p - \tfrac{1}{2}})$

**(ii)** $(0, 0)$   **75.** Let the angle of the cuts be $\varphi_1$ and $\varphi_2$, where $\varphi_1 + \varphi_2 = \theta$. The volume of the notch is proportional to $\tan \varphi_1 + \tan \varphi_2 = \tan \varphi_1 + \tan(\theta - \varphi_1)$, which is minimized when $\varphi_1 = \varphi_2 = \dfrac{\theta}{2}$.   **77.** $x \approx 38.81, y \approx 55.03$

## Section 4.6 Exercises, pp. 254–256

**1.**

**3.** $f(x) \approx f(a) + f'(a)(x - a)$

**5.** $L(x) = 3x - 1; 2.3$   **7.** 2.7   **9.** $dy = f'(x)\, dx$

**11.** Approx. 25   **13.** 61 mi/hr; 61.02 mi/hr

**15.** $L(x) = T(0) + T'(0)(x - 0) = D - (D/60)x = D(1 - x/60)$

**17.** 84 min; 84.21 min   **19.** $L(x) = 9x - 4$   **21.** $L(t) = t + 5$

**23.** $L(x) = \dfrac{x}{12} + \dfrac{4}{3}$   **25. a.** $L(x) = -4x + 16$   **b.** 7.6

**c.** 0.13% error   **27. a.** $L(x) = 1 - x$   **b.** 1.1   **c.** 1% error

**29. a.** $L(x) = 1$   **b.** 1   **c.** 0.005% error   **31. a.** $L(x) = \dfrac{1}{2} - \dfrac{x}{48}$

**b.** 0.5   **c.** 0.003% error   **33. a.** $L(x) = 1 - x$;   **b.** $1/1.1 \approx 0.9$

**c.** 1% error   **35. a.** $L(x) = 4 + \dfrac{x}{48}$   **b.** $\sqrt[3]{62.5} \approx \tfrac{127}{32} \approx 3.97$

**c.** 0.006% error   **37.** $a = 200; \tfrac{1}{203} \approx 0.004925$

**39.** $a = 144; \sqrt{146} \approx \tfrac{145}{12}$   **41.** $a = 1; 1/1.05 \approx 0.95$

**43.** $a = \tfrac{\pi}{4}; \sin\left(\tfrac{\pi}{4} + 0.1\right) \approx 11\sqrt{2}/20 \approx 0.778$

**45.** $a = 512; \dfrac{1}{\sqrt[3]{510}} \approx 0.125$

**47. a.** $L(x) = -2x + 4$   **b.**

**c.** Underestimates   **d.** $f''(1) = 4 > 0$

**49. a.** $L(x) = -\dfrac{1}{2}x + \dfrac{3}{2}$   **b.**

**c.** Underestimates   **d.** $f''(1) = \dfrac{3}{4} > 0$

**51.** $E(x) \le 1$ when $-7.26 \le x \le 8.26$, which corresponds to driving times for 1 mi from about 53 s to 68 s. Therefore, $L(x)$ gives approximations to $s(x)$ that are within 1 mi/hr of the true value when you drive 1 mi in $t$ seconds, where $53 < t < 68$.

**53. a.** True   **b.** False   **c.** True   **55.** $\Delta V \approx 10\pi$ ft³

**57.** $\Delta S \approx -\dfrac{59\pi}{5\sqrt{34}}$ m²   **59.** $dy = 2\, dx$   **61.** $dy = -\dfrac{3}{x^4}\, dx$

**63.** $dy = a \sin x\, dx$   **65.** $dy = (9x^2 - 4)\, dx$

**67.** $L(x) = 2 + (x - 8)/12$

| $x$ | Linear approx. | Exact value | Percent error |
|-----|----------------|-------------|---------------|
| 8.1 | $2.008\overline{3}$ | 2.00829885 | $1.7 \times 10^{-3}$ |
| 8.01 | $2.00083\overline{3}$ | 2.000832986 | $1.7 \times 10^{-5}$ |
| 8.001 | $2.000083\overline{3}$ | 2.00008333 | $1.7 \times 10^{-7}$ |
| 8.0001 | $2.0000083\overline{3}$ | 2.000008333 | $1.7 \times 10^{-9}$ |
| 7.9999 | $1.9999916\overline{6}$ | 1.999991667 | $1.7 \times 10^{-9}$ |
| 7.999 | $1.999916\overline{6}$ | 1.999916663 | $1.7 \times 10^{-7}$ |
| 7.99 | $1.99916\overline{6}$ | 1.999166319 | $1.7 \times 10^{-5}$ |
| 7.9 | $1.9916\overline{6}$ | 1.991631701 | $1.8 \times 10^{-3}$ |

**69. a.** $f$; the rate at which $f'$ is changing at 1 is smaller than the rate at which $g'$ is changing at 1. The graph of $f$ bends away from the linear function more slowly than the graph of $g$.   **b.** The larger the value of $|f''(a)|$, the greater the deviation of the curve $y = f(x)$ from the tangent line at points near $x = a$.

## Section 4.7 Exercises, pp. 263–264

**1.** If $\lim_{x \to a} f(x) = 0$ and $\lim_{x \to a} g(x) = 0$, then we say $\lim_{x \to a} f(x)/g(x)$ is an indeterminate form $0/0$.   **3.** Take the limit of the quotient of the derivatives of the functions.

**5. a.** $\lim\limits_{x\to 0}\left(x^2\cdot\dfrac{1}{x^2}\right)=1$   **b.** $\lim\limits_{x\to 0}\left(2x^2\cdot\dfrac{1}{x^2}\right)=2$

**7.** If $\lim\limits_{x\to a} f(x)g(x)$ has the indeterminate form $0\cdot\infty$, then

$\lim\limits_{x\to a}\left(\dfrac{f(x)}{1/g(x)}\right)$ has the indeterminate form $0/0$ or $\infty/\infty$.   **9.** $-\dfrac{2}{\pi}$

**11.** $-1$   **13.** $\frac{3}{4}$   **15.** $-4\pi$   **17.** $\frac{1}{2}$   **19.** 2   **21.** $\frac{12}{5}$   **23.** 4   **25.** $\frac{9}{16}$

**27.** $\frac{1}{2}$   **29.** $-\frac{2}{3}$   **31.** $\frac{1}{3}$   **33.** 4   **35.** $-\frac{1}{2}$   **37.** $\frac{1}{\pi^2}$   **39.** 2   **41.** 1

**43.** $\frac{7}{6}$   **45.** 1   **47.** 0   **49.** $-8$   **51.** 0   **53.** $\frac{1}{2}$   **55.** $-\frac{9}{4}$   **57.** $\frac{1}{6}$

**59.** 1   **61. a.** False   **b.** False   **c.** False   **d.** True   **63.** 2

**65.** $-\frac{1}{3}$   **67.** $\sqrt{a/c}$   **69.** $1/3$

## Section 4.8 Exercises, pp. 270–273

**1.** Newton's method generates a sequence of $x$-intercepts of lines tangent to the graph of $f$ to approximate the roots of $f$.
**3.** $x_1=2, x_2=1, x_3=0$   **5.** $x_1=0.75$   **7.** Generally, if two successive Newton approximations agree in their first $p$ digits, then those approximations have $p$ digits of accuracy. The method is terminated when the desired accuracy is reached.

**9.** $x_{n+1}=x_n-\dfrac{x_n^2-6}{2x_n}=\dfrac{x_n^2+6}{2x_n}$; $x_1=2.5, x_2=2.45$

**11.** $x_{n+1}=x_n-(\tan x_n-2)\cos^2 x_n$; $x_1=1.129204, x_2=1.108129$

**13.**

| $n$ | $x_n$ |
|---|---|
| 0 | 3 |
| 1 | 3.1667 |
| 2 | 3.16228 |
| 3 | 3.16228 |

$r\approx 3.16228$

**15.**

| $n$ | $x_n$ |
|---|---|
| 0 | 0.5 |
| 1 | 0.51096 |
| 2 | 0.51097 |
| 3 | 0.51097 |

$r\approx 0.51097$

**17.**

| $n$ | $x_n$ |
|---|---|
| 0 | 1.2 |
| 1 | 1.16935 |
| 2 | 1.16561 |
| 3 | 1.16556 |
| 4 | 1.16556 |

$r\approx 1.16556$

**19.**

| $n$ | $x_n$ |
|---|---|
| 0 | 2 |
| 1 | 1.56511 |
| 2 | 1.4924 |
| 3 | 1.49022 |
| 4 | 1.49022 |

$r\approx 1.49022$

**21.** $x\approx -0.335408, 1.333057$   **23.** $x\approx 1.16717, 2.81286, 5.85098$
**25.** $x\approx -0.161645, 0.675756, 1.19665$
**27.** $x\approx 0, 1.895494, -1.895494$
**29.** $x\approx -2.114908, 0.254102, 1.860806$
**31.** $x\approx 0.062997, 2.230120$   **33. b.** $r\approx 7.3\%$
**35.** $r=0.06$ or $6\%$   **37.** $p(x)=x^4-7$; $r\approx 1.62658$
**39.** $p(x)=x^3+9$; $r\approx -2.08008$   **41.** $x\approx 2.798386$
**43.** $x\approx -0.666667, 1.5, 1.666667$   **45. a.** True   **b.** False
**c.** False   **47.** $x\approx 0.739085$   **49.** $x=0$ and $x\approx 1.047198$

**51. a.**

| $n$ | $x_n$ |
|---|---|
| 0 | 2 |
| 1 | 5.33333 |
| 2 | 11.0553 |
| 3 | 22.2931 |

**b.**

**c.** The tangent lines intersect the $x$-axis farther and farther away from the root $r$.   **53. b.** $x\approx 0.14285714$ is approximately $\frac{1}{7}$.
**55.** $\lambda=1.29011, 2.37305, 3.40918$

## Section 4.9 Exercises, pp. 281–283

**1.** The derivative, an antiderivative   **3.** $x+C$, where $C$ is an arbitrary constant   **5.** $\dfrac{x^{p+1}}{p+1}+C$, where $p\neq -1$   **7.** $-\dfrac{1}{x}+C$

**9.** 0   **11.** $x^5+C$   **13.** $-2\cos x+x+C$   **15.** $3\tan x+C$
**17.** $y^{-2}+C$   **19.** $2\sqrt{x}+C$   **21.** $x^{7/2}+C$   **23.** $\frac{1}{2}x^6-\frac{1}{2}x^{10}+C$
**25.** $\frac{8}{3}x^{3/2}-8x^{1/2}+C$   **27.** $\frac{25}{3}s^3+15s^2+9s+C$
**29.** $\frac{9}{4}x^{4/3}+6x^{2/3}+6x+C$   **31.** $-x^3+\frac{11}{2}x^2+4x+C$
**33.** $-x^{-3}+2x+3x^{-1}+C$   **35.** $x^4-3x^2+C$
**37.** $\dfrac{1}{2}x^2+6x+C$   **39.** $-\cot\theta+2\theta^3/3-3\theta^2/2+C$
**41.** $-2\cot y-3\csc y+C$   **43.** $\tan x-x+C$
**45.** $\tan\theta+\sec\theta+C$   **47.** $t^3-2\cot t+C$
**49.** $\sec\theta+\tan\theta+\theta+C$   **51.** $x-\cos x+C$
**53.** $\frac{4}{15}x^{15/2}-\frac{24}{11}x^{11/6}+C$   **55.** $\dfrac{1}{6}x^6-\dfrac{2}{3}x^3+x+1$
**57.** $2x^4-\cos x+3$   **59.** $\sec v+1, -\pi/2<v<\pi/2$

**61.** $f(x)=x^2-3x+4$   **63.** $g(x)=\dfrac{7}{8}x^8-\dfrac{x^2}{2}+\dfrac{13}{8}$

**65.** $f(u)=4(\sin u+\cos u)-4$

**67.** $y(\theta)=\sqrt{2}\sin\theta+\tan\theta+1, -\pi/2<\theta<\pi/2$

**69.** $f(x)=x^2-5x+4$      **71.** $f(x)=\dfrac{3}{2}x^2-\cos x+4$

      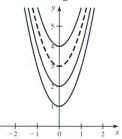

**73.** $s(t)=t^2+4t$   **75.** $s(t)=\frac{4}{3}t^{3/2}+1$
**77.** $s(t)=2t^3+2t^2-10t$   **79.** $s(t)=-16t^2+20t$
**81.** $s(t)=\frac{1}{30}t^3+1$   **83.** $s(t)=t^2+4t-3\sin t+10$
**85.** 200 ft   **87.** Runner A overtakes runner B at $t=\pi/2$
**89. a.** $v(t)=-9.8t+30$   **b.** $s(t)=-4.9t^2+30t$
**c.** Approx. 45.92 m at time $t\approx 3.06$   **d.** $t\approx 6.12$
**91. a.** $v(t)=-9.8t+10$   **b.** $s(t)=-4.9t^2+10t+400$
**c.** Approx. 405.10 m at time $t\approx 1.02$   **d.** $t\approx 10.11$
**93. a.** True   **b.** False   **c.** True   **d.** False
**e.** False   **95.** $F(x)=-\cos x+3x+3-3\pi$
**97.** $F(x)=2x^8+x^4+2x+1$
**99. a.** $Q(t)=10t-t^3/30$ gal

**b.**

**c.** $\dfrac{200}{3}$ gal

## Chapter 4 Review Exercises, pp. 283–286

**1. a.** False   **b.** False   **c.** True   **d.** True   **e.** True   **f.** False
**3.**

**5. a.** $x = 0, \pm 10$; increasing on $(-10, 0)$ and $(10, \infty)$, decreasing on $(-\infty, -10)$ and $(0, 10)$   **b.** $x = \pm 6$; concave up on $(-\infty, -6)$ and $(6, \infty)$, concave down on $(-6, 6)$   **c.** Local min at $x = -10, 10$; local max at $x = 0$   **d.**

**7.** Critical pts. $x = 0, \pm 1$; abs. max: 33 at $x = \pm 2$; abs. min: 6 at $x = \pm 1$   **9.** $x = 3$ and $x = -2$; no abs. max or min

**11.** Critical pts. $x = \dfrac{2\pi}{3}, \dfrac{4\pi}{3}$; abs. max: $\dfrac{3\sqrt{3}}{8}$ at $x = \dfrac{4\pi}{3}$; abs. min: $-\dfrac{3\sqrt{3}}{8}$ at $x = \dfrac{2\pi}{3}$

**13.**

Critical pts.: all $x$ in the interval $[-3, 2]$; abs. max: 9 at $x = 4$; abs. and local min: 5 for $x$ in $[-3, 2]$; local max: 5 for $x$ in $(-3, 2)$

**15. a.** Increasing on $(-\infty, -1)$ and $(1, \infty)$; decreasing on $(-1, 1)$
**b.** Concave up on $(0, \infty)$ and concave down on $(-\infty, 0)$
**17.** Inflection pt. $x = 0$   **19.** Critical pts. $x = -a, a/2$; inflection pts. $x = 0, -a$
**21.**

**23.**

**25.**

**27.**

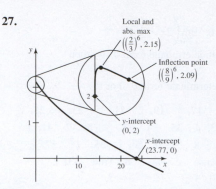

**29.** Approx. 2.5″ by 3.5″ by 9.5″   **31.** Approx. 59 m from the loudest speaker   **33.** $r = 4\sqrt{6}/3; h = 4\sqrt{3}/3$   **35.** $x = 7, y = 14$
**37.** $p = q = 5\sqrt{2}$   **39. a.** $L(x) = \frac{2}{9}x + 3$   **b.** $\frac{85}{9} \approx 9.44$; overestimate   **41.** $f(x) = 1/x^2; a = 4; 1/4.2^2 \approx 0.05625$
**43.** $\Delta h \approx -112$ ft   **45.** $c = 2.5$   **47. a.** $\frac{100}{9}$ cells/week
**b.** $t = 2$ weeks   **49.** $-0.434259, 0.767592, 1$   **51.** 0   **53.** $\frac{3}{4}$
**55.** 0   **57.** 12   **59.** $\frac{2}{3}$   **61.** 0   **63.** $\frac{4}{3}x^3 + 2x^2 + x + C$
**65.** $-\dfrac{1}{x} + \dfrac{4}{3}x^{-3/2} + C$   **67.** $\theta + 3\sin\theta + C$   **69.** $\tan x + C$
**71.** $\frac{4}{7}x^{7/4} + \frac{2}{7}x^{7/2} + C$   **73.** $f(t) = -\cos t + t^2 + 6$
**75.** $f(x) = \dfrac{1}{2}x + \dfrac{1}{2}\sin x + 1$   **77.** $v(t) = -9.8t + 120$;
$s(t) = -4.9t^2 + 120t + 125$. The rocket reaches a height of 859.69 m at time $t \approx 12.24$ s and then falls to the ground, hitting at time $t \approx 25.49$ s.   **79. a.** $v(t) = 64 - 32t$
**b.** $s(t) = -16t^2 + 64t + 128$   **c.** $t = 2$; 192 ft
**d.** $-64\sqrt{3}$ ft/s $\approx -110.9$ ft/s

## CHAPTER 5

### Section 5.1 Exercises, pp. 296–301

**1.** Displacement $= 105$ m

**3. a.** 440 ft   **b.** 400 ft   **5. a.** 340 ft   **b.** 330 ft
**7.** Subdivide the interval $[0, \pi/2]$ into several subintervals, which will be the bases of rectangles that fit under the curve. The heights of the rectangles are computed by taking the value of $\cos x$ at the right-hand endpoint of each base. We calculate the area of each rectangle and add them to get a lower bound on the area.   **9.** Left sum: 34; right sum: 24   **11.** 0.5; 1, 1.5, 2, 2.5, 3; 1, 1.5, 2, 2.5, 3; 1.5, 2, 2.5, 3; 1.25, 1.75, 2.25, 2.75   **13.** Underestimate; the rectangles all fit under the curve.   **15. a.** 67 ft   **b.** 67.75 ft   **17.** 40 m
**19.** 2.78 m   **21.** 148.96 mi   **23.** 20; 25

**25. a. c.**

Left Riemann sum underestimates area.    Right Riemann sum overestimates area.

**b.** $\Delta x = 1; x_0 = 0, x_1 = 1, x_2 = 2, x_3 = 3, x_4 = 4$   **d.** 10, 14

**27. a. c.**

Left Riemann sum overestimates area.    Right Riemann sum underestimates area.

**b.** $\Delta x = \pi/8; 0, \pi/8, \pi/4, 3\pi/8, \pi/2$   **d.** 1.18; 0.79

**29. a. c.**

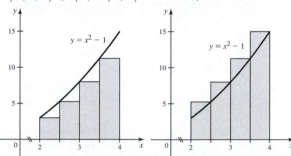

Left Riemann sum underestimates area.    Right Riemann sum overestimates area.

**b.** $\Delta x = 0.5; 2, 2.5, 3, 3.5, 4$   **d.** 13.75; 19.75

**31. a. c.**

Left Riemann sum underestimates area.    Right Riemann sum overestimates area.

**b.** $\Delta x = 0.5; 0, 0.5, 1, 1.5, 2, 2.5, 3$   **d.** 2.9636, 3.8296

**33.** 670   **35. a.** 10,500 m; 10,350 m

**b.** Left Riemann sum   **c.** Increase the number of subintervals in the partition.

**37. a. c.**

**b.** $\Delta x = 1; 0, 1, 2, 3, 4$
**d.** 20

**39. a. c.**

**b.** $\Delta x = \frac{1}{2}; 1, \frac{3}{2}, 2, \frac{5}{2}, 3$
**d.** 2.80

**41. a. c.**

**b.** $\Delta x = 1; 1, 2, 3, 4, 5, 6$
**d.** 1.76

**43.** 5.5, 3.5   **45. b.** 110, 117.5   **47. a.** $\sum_{k=1}^{5} k$   **b.** $\sum_{k=1}^{6}(k+3)$

**c.** $\sum_{k=1}^{4} k^2$   **d.** $\sum_{k=1}^{4} \frac{1}{k}$   **49. a.** 55   **b.** 48   **c.** 30   **d.** 60   **e.** 6

**f.** 6   **g.** 85   **h.** 0   **51. a.** Left: $\frac{3}{10} \sum_{k=1}^{40} \sqrt{\frac{k-1}{10}} \approx 15.6809$;

right: $\frac{3}{10} \sum_{k=1}^{40} \sqrt{\frac{k}{10}} \approx 16.2809$; midpoint: $\frac{3}{10} \sum_{k=1}^{40} \sqrt{\frac{k-0.5}{10}} \approx 16.0055$

**b.** 16   **53. a.** Left: $\frac{1}{25} \sum_{k=1}^{75} \left( \left( 2 + \frac{k-1}{25} \right)^2 - 1 \right) \approx 35.5808$;

right: $\frac{1}{25} \sum_{k=1}^{75} \left( \left( 2 + \frac{k}{25} \right)^2 - 1 \right) \approx 36.4208$;

midpoint: $\frac{1}{25} \sum_{k=1}^{75} \left( \left( 2 + \frac{k-0.5}{25} \right)^2 - 1 \right) \approx 35.9996$   **b.** 36

**55.**

| $n$ | Right Riemann sum |
|-----|-------------------|
| 10  | 21.96             |
| 30  | 21.9956           |
| 60  | 21.9989           |
| 80  | 21.9994           |

The sums approach 22.

**57.**

| $n$ | Right Riemann sum |
|-----|-------------------|
| 10  | 3.14159           |
| 30  | 3.14159           |
| 60  | 3.14159           |
| 80  | 3.14159           |

The sums approach $\pi$.

**59. a.** True   **b.** False   **c.** True   **61.** $\sum_{k=1}^{50} \left( \frac{4k}{50} + 1 \right) \cdot \frac{4}{50} = 12.16$

**63.** $\sum_{k=1}^{32} \left( 3 + \frac{2k-1}{8} \right)^3 \cdot \frac{1}{4} \approx 3639.1$   **65.** $[1, 5]; 4$   **67.** $[2, 6]; 4$

**69. a.**

Left Riemann sum is
$\frac{23}{4} = 5.75$.

**b.**

Midpoint Riemann sum is
$$\frac{53}{8} = 6.625.$$

**c.**

Right Riemann sum is
$$\frac{31}{4} = 7.75.$$

**71. a.** The object is speeding up on the interval $(0, 1)$, moving at a constant rate on $(1, 3)$, slowing down on $(3, 5)$, and moving at a constant rate on $(5, 6)$. **b.** 30 m **c.** 50 m **d.** $s(t) = 80 + 10t$
**73. a.** 14.5 g **b.** 29.5 g **c.** 44 g **d.** $x = \frac{19}{3}$ cm

**75.** $s(t) = \begin{cases} 30t & \text{if } 0 \le t \le 2 \\ 50t - 40 & \text{if } 2 < t \le 2.5 \\ 44t - 25 & \text{if } 2.5 < t \le 3 \end{cases}$

**77.**

| $n$ | Midpoint Riemann sum |
|---|---|
| 16 | 4.7257 |
| 32 | 4.7437 |
| 64 | 4.7485 |

The sums approach 4.75.

**81.** Underestimates for decreasing functions, independent of concavity; overestimates for increasing functions, independent of concavity

## Section 5.2 Exercises, pp. 313–316

**1.** The difference between the area bounded by the curve above the $x$-axis and the area bounded by the curve below the $x$-axis **3.** 60; 0
**5.** $-12; -18; -16$
**7.**

$$\int_{-1}^{4} 2 \, dx = 10$$

**9.**

$$\int_{-1}^{2} 2x \, dx = 3$$

**11.** Both integrals equal 0. **13.** The length of the interval $[a, a]$ is $a - a = 0$, so the net area is 0. **15.** $\dfrac{a^2}{2}$

**17. a.**

**b.** $-16, -24, -20$

**19. a.**

**b.** $-0.948, -0.948, -1.026$

**21. a.**

**b.** $4, -4, 0$ **c.** Positive contributions on $[0, 2)$; negative contributions on $(2, 4]$

**23. a.**

**b.** $0.735, 0.146, 0.530$
**c.** Positive contributions on $(0, \pi/2)$; negative contributions on $(\pi/2, 3\pi/4]$

**25.**

The area is 12; the net area is 0.

**27.**

The area is 2; the net area is 0.

**29. a.**

**b.** $\Delta x = \frac{1}{2}$; 3, 3.5, 4, 4.5, 5, 5.5, 6  **c.** $-22.5$; $-25.5$
**d.** The left Riemann sum overestimates the integral; the right Riemann sum underestimates the integral.

**31. a.**

**b.** $\Delta x = 1$; 1, 2, 3, 4, 5, 6, 7
**c.** $\frac{49}{20}$; $\frac{223}{140}$  **d.** The left Riemann sum overestimates the integral; the right Riemann sum underestimates the integral.

**33.** $\int_0^2 (x^2 + 1)\, dx$  **35.** $\int_1^2 x \cos x\, dx$

**37.** 16

**39.** $-\frac{5}{2}$

**41.** $4\pi$

**43.** 26

**45.** $\pi$  **47.** $-2\pi$  **49. a.** $-32$  **b.** $-\frac{32}{3}$  **c.** $-64$  **d.** Not possible
**51. a.** 10  **b.** $-3$  **c.** $-16$  **d.** 3  **53. a.** 15  **b.** 5  **c.** 3
**d.** $-2$  **e.** 24  **f.** $-10$  **55. a.** $\frac{3}{2}$  **b.** $-\frac{3}{4}$  **57.** 16  **59.** 6

**61.** 32  **63.** $-16$  **65.** $\frac{\pi}{4} + 2$  **67. a.** True  **b.** True  **c.** True
**d.** False  **e.** False

**69. a.** Left: $\sum_{k=1}^{n} \left( \left( \frac{k-1}{n} \right)^2 + 1 \right) \cdot \frac{1}{n}$;

right: $\sum_{k=1}^{n} \left( \left( \frac{k}{n} \right)^2 + 1 \right) \cdot \frac{1}{n}$

**b.**

| $n$ | Left Riemann sum | Right Riemann sum |
|-----|------------------|-------------------|
| 20  | 1.30875          | 1.35875           |
| 50  | 1.3234           | 1.3434            |
| 100 | 1.32835          | 1.33835           |

Estimate: $\frac{4}{3}$

**71. a.** $\sum_{k=1}^{n} 2\sqrt{1 + \left( k - \frac{1}{2} \right)\frac{3}{n}} \cdot \frac{3}{n}$

**b.**

| $n$ | Midpoint Riemann sum |
|-----|----------------------|
| 20  | 9.33380              |
| 50  | 9.33341              |
| 100 | 9.33335              |

Estimate: $\frac{28}{3}$

**73. a.** $\sum_{k=1}^{n} \left( 4\left( k - \frac{1}{2} \right)\frac{4}{n} - \left( \left( k - \frac{1}{2} \right)\frac{4}{n} \right)^2 \right) \cdot \frac{4}{n}$

**b.**

| $n$ | Midpoint Riemann sum |
|-----|----------------------|
| 20  | 10.6800              |
| 50  | 10.6688              |
| 100 | 10.6672              |

Estimate: $\frac{32}{3}$

**75.** 6  **77.** 104  **79.** 18  **81.** 2  **83.** $25\pi/2$  **85.** 25  **87.** 35
**91.** For any such partition on $[0, 1]$, the grid points are $x_k = k/n$, for $k = 0, 1, \ldots, n$. That is, $x_k$ is rational for each $k$ so that $f(x_k) = 1$, for $k = 0, 1, \ldots, n$. Therefore, the left, right, and midpoint Riemann sums are $\sum_{k=1}^{n} 1 \cdot (1/n) = 1$.

## Section 5.3 Exercises, pp. 326–330

**1.** $A$ is an antiderivative of $f$; $A'(x) = f(x)$.
**3.** $\int_a^b f(x)\, dx = F(b) - F(a)$, where $F$ is any antiderivative of $f$.
**5.** Increasing  **7.** The derivative of the integral of $f$ is $f$, or
$$\frac{d}{dx}\left( \int_a^x f(t)\, dt \right) = f(x).$$  **9.** $f(x), 0$  **11.** 16  **13. a.** 0  **b.** $-9$
**c.** 25  **d.** 0  **e.** 16
**15. a.** $A(x) = 5x$   **b.** $A'(x) = 5$

**17. a.** $A(2) = 2$, $A(4) = 8$; $A(x) = \frac{1}{2}x^2$  **b.** $F(4) = 6$, $F(6) = 16$;
$F(x) = \frac{1}{2}x^2 - 2$  **c.** $A(x) - F(x) = \frac{1}{2}x^2 - \left( \frac{1}{2}x^2 - 2 \right) = 2$
**19. a.**

**b.** $A'(x) = \frac{d}{dx}\left( \frac{1}{2}(x + 5)^2 \right) = x + 5 = f(x)$

**21. a.**

$A(x) = \frac{3}{2}x^2 + x - 8$

**b.** $A'(x) = \left(\frac{3}{2}x^2 + x - 8\right)' = 3x + 1 = f(x)$   **23.** $\frac{7}{3}$

**25.** $-\frac{125}{6}$          **27.** $-\frac{10}{3}$

**29.** 16   **31.** 90   **33.** $\frac{7}{6}$   **35.** 8   **37.** $-\frac{32}{3}$   **39.** $-\frac{5}{2}$   **41.** $\frac{9}{2}$   **43.** $-\frac{3}{8}$

**45.** 1   **47.** $\frac{3}{2}$   **49.** $\frac{45}{4}$   **51.** $\frac{2}{3}$   **53.** 1   **55.** 2   **57.** $\sqrt{2} - 1$

**59.** $\frac{3\pi}{2} - 1$

**61. (i)** $\frac{14}{3}$  **(ii)** $\frac{14}{3}$          **63. (i)** $-51.2$  **(ii)** 51.2

**65.** $\frac{94}{3}$   **67.** $\frac{15}{64}$   **69.** 2   **71.** $x^2 + x + 1$   **73.** $-\sqrt{x^4 + 1}$

**75.** $\frac{3}{x^4}$   **77.** $-(\cos^4 x + 6)\sin x$   **79.** $-\frac{\cos z}{\sin^4 z + 1}$

**81.** $\frac{9}{t}$   **83.** $2\sqrt{1 + x^2}$   **85.** a–C, b–B, c–D, d–A

**87. a.** $x = 0, x \approx 3.5$   **b.** Local min at $x \approx 1.5$; local max at $x \approx 8.5$   **c.**

**89. a.** $x = 0, 10$   **b.** Local max at $x = 5$

**c.**

**91.** $-\pi, -\pi + \frac{9}{2}, -\pi + 9, 5 - \pi$

**93. a.** $A(x) = \frac{1}{\pi}\sin \pi x$

**b.**

**c.** $A\left(\frac{1}{2}\right) = \frac{1}{\pi}A(1) = 0$

**95. a.** $A(x) = \sin x$

**b.**

**c.** $A\left(\frac{\pi}{2}\right) = 1; A(\pi) = 0$

**97.** Critical pts. $x = 0, 3,$ and 4; increasing on $(-\infty, 0), (0, 3),$ and $(4, \infty)$; decreasing on $(3, 4)$

**99. a.**

**b.** $g'(x) = \sin^2 x$

**c.**

**101.**

Area = 6

**103.**

Area ≈ 194.05

**105. a.** True   **b.** True   **c.** False   **d.** True
**107.** 3

**109. a.**

**b.** $b = 6$   **c.** $b = \dfrac{3a}{2}$

**111.** $f(x) = -2\sin x + 3$   **113.** $\pi/2 \approx 1.57$

**115.** $(S'(x))^2 + \left(\dfrac{S''(x)}{2x}\right)^2 = (\sin x^2)^2 + \left(\dfrac{2x\cos x^2}{2x}\right)^2$
$$= \sin^2 x^2 + \cos^2 x^2 = 1$$

**117. c.** The summation relationship is a discrete analog of the Fundamental Theorem. Summing the difference quotient and integrating the derivative over the relevant interval give the difference of the function values at the endpoints.

## Section 5.4 Exercises, pp. 334–336

**1.** If $f$ is odd, the regions between $f$ and the positive $x$-axis and between $f$ and the negative $x$-axis are reflections of each other through the origin. Therefore, on $[-a, a]$, the areas cancel each other.
**3. a.** 9   **b.** 0   **5.** $3x^3$ and $x$ are odd functions.   **7.** Even; even
**9.** If $f$ is continuous on $[a, b]$, then there is a $c$ in $(a, b)$ such that
$f(c) = \dfrac{1}{b-a}\displaystyle\int_a^b f(x)\,dx.$   **11.** 0   **13.** $\dfrac{1000}{3}$   **15.** $\dfrac{16}{3}$   **17.** $-\dfrac{88}{3}$
**19.** 0   **21.** 2   **23.** 0

**25.** 0

**27.** $\frac{2}{3}$

**29.** $2/\pi$

**31.** $1/(n+1)$

**33.** 2000   **35.** 21 m/s   **37.** $20/\pi$   **39.** 2   **41.** $a/\sqrt{3}$
**43.** $c = \pm\frac{1}{2}$   **45. a.** True   **b.** True   **c.** True   **d.** False
**47.** 420 ft   **49.** $f(g(-x)) = f(g(x)) \Rightarrow$ the integrand is even;
$\displaystyle\int_{-a}^{a} f(g(x))\,dx = 2\int_0^a f(g(x))\,dx$   **51.** $p(g(-x)) = p(g(x)) \Rightarrow$
the integrand is even; $\displaystyle\int_{-a}^{a} p(g(x))\,dx = 2\int_0^a p(g(x))\,dx$
**53. a.** $a/6$   **b.** $(3 \pm \sqrt{3})/6$, independent of $a$
**57.**

| Even | Even |
|------|------|
| Even | Odd  |

## Section 5.5 Exercises, pp. 343–346

**1.** The Chain Rule   **3.** $u = g(x)$   **5.** The lower bound $a$ becomes $g(a)$ and the upper bound $b$ becomes $g(b)$.   **7.** $\dfrac{(x^2+1)^5}{5} + C$
**9.** $\frac{1}{4}\sin^4 x + C$   **11.** $\dfrac{(x+1)^{13}}{13} + C$   **13.** $\dfrac{(2x+1)^{3/2}}{3} + C$
**15. a.** $-\dfrac{1}{10}\cot x + C$   **b.** $\dfrac{1}{5}\sec 5x + C$
**c.** $-\dfrac{1}{7}\cos 7x + C$   **d.** $7\sin\dfrac{x}{7} + C$
**17.** $\dfrac{(x^2-1)^{100}}{100} + C$   **19.** $-\dfrac{\sqrt{1-4x^3}}{3} + C$   **21.** $\dfrac{(x^2+x)^{11}}{11} + C$
**23.** $\dfrac{(x^4+16)^7}{28} + C$   **25.** $-\dfrac{\sqrt{4-9x^2}}{9} + C$
**27.** $\dfrac{(x^6-3x^2)^5}{30} + C$   **29.** $-\dfrac{1}{2}\csc x^2 + C$   **31.** $\dfrac{1}{10}\tan(10x+7) + C$
**33.** $\dfrac{\sin^2 t^4}{8} + C$   **35.** $\frac{1}{2}\tan^2 x + C$   **37.** $\frac{1}{7}\sec^7 x + C$
**39.** $\dfrac{(\sec w + 3)^{10}}{10} + C$   **41.** $\dfrac{\sqrt{2}}{4}$   **43.** $\frac{7}{2}$   **45.** $4 - 2\sqrt{3}$   **47.** $\frac{1}{3}$
**49.** $(1 + 4\sqrt{2})/6$   **51.** $(\sqrt{3}+3)/6$   **53.** 7/32   **55.** 22/3   **57.** 1
**59.** 10   **61.** $\frac{1}{3}$   **63.** $\dfrac{3}{4}(4 - 3^{2/3})$   **65.** $\frac{32}{3}$   **67.** $\frac{1}{22}$   **69.** $\frac{1}{7}$
**71.** 10 m/s   **73. a.** 160   **b.** $\dfrac{4800}{49} \approx 98$   **c.** $\Delta p = \displaystyle\int_0^T \dfrac{200}{(t+1)^r}\,dt$;
decreases as $r$ increases   **d.** $r \approx 1.28$   **e.** As $t \to \infty$, the
population approaches 100.   **75.** $\frac{2}{3}(x-4)^{1/2}(x+8) + C$
**77.** $\frac{3}{5}(x+4)^{2/3}(x-6) + C$   **79.** $\frac{3}{112}(2x+1)^{4/3}(8x-3) + C$
**81.** $\dfrac{(x+10)^{10}(x-1)}{11} + C$   **83.** $\pi$
**85.** $\dfrac{\theta}{2} - \dfrac{1}{4}\sin\left(\dfrac{60+\pi}{3}\right) + C$   **87.** $\dfrac{\pi}{4}$   **89.** $\dfrac{1}{18}$   **91. a.** True
**b.** True   **c.** False   **d.** False   **e.** False   **93.** 1   **95.** $\dfrac{2}{3}$; constant

**97. a.** $\pi/p$ **b.** 0 **99.** $2/\pi$ **101.** One area is $\int_4^9 \frac{(\sqrt{x}-1)^2}{2\sqrt{x}}\,dx$.

Changing variables by letting $u = \sqrt{x} - 1$ yields $\int_1^2 u^2\,du$, which is the

other area. **103.** $7297/12$ **105.** $\frac{2}{15}(3-2a)(1+a)^{3/2} + \frac{4}{15}a^{5/2}$

**107.** $\frac{1}{3}\sec^3\theta + C$ **109. a.** $I = \frac{1}{8}x - \frac{1}{32}\sin 4x + C$

**b.** $I = \frac{1}{8}x - \frac{1}{32}\sin 4x + C$

**113.** $\frac{4}{3}(-2 + \sqrt{1+x})\sqrt{1+\sqrt{1+x}} + C$ **115.** $-4 + \sqrt{17}$

## Chapter 5 Review Exercises, pp. 346–350

**1. a.** True **b.** False **c.** True **d.** True **e.** False
**f.** True **g.** True
**3. a.**  **b.** 75 **c.** The area is the distance the diver ascends.

**5.** $9.34$; $10.28$; $9.82$
**7.**

| $n$ | Midpoint Riemann sum |
|---|---|
| 10 | 114.167 |
| 30 | 114.022 |
| 60 | 114.006 |

$\int_1^{25}\sqrt{2x-1}\,dx = 114$

**9. a.** $1((3\cdot 2 - 2) + (3\cdot 3 - 2) + (3\cdot 4 - 2)) = 21$

**b.** $\displaystyle\sum_{k=1}^n \frac{3}{n}\left(3\left(1 + \frac{3k}{n}\right) - 2\right)$ **c.** $\frac{33}{2}$ **11.** $-\frac{16}{3}$ **13.** 56

**15. a.** 20 **b.** 0 **c.** 80 **d.** 10 **e.** 0 **17.** 18 **19.** 10
**21.** Not enough information **23. a.** 8.5 **b.** $-4.5$ **c.** 0 **d.** 11.5

**25.** $4\pi$ **27.** $A: \int_0^x f(t)\,dt$; $B: f(x)$; $C: f'(x)$ **29.** $\sqrt{1 + x^4 + x^6}$

**31.** $-\sin x^6$ **33.** $\frac{2}{x^{10} + 1}$ **35.** Increasing on $(3, 6)$; decreasing

on $(-\infty, 3)$ and $(6, \infty)$ **39.** $\frac{212}{5}$ **41.** 0

**43.** $\frac{7}{6}$ **45.** $\frac{4}{\sqrt{3}}$ **47.** 2 **49.** $-\frac{4}{3\sin^{3/4}x} + C$

**51.** $\frac{1}{3}\sin x^3 + C$ **53.** $-\frac{1}{7(16 + \sin^2 7w)} + C$

**55.** $\frac{1}{3}\left(\frac{\sin^4 3x}{4} + \frac{\sin^3 3x}{3}\right) + C$ **57.** 78

**59.** $2\sqrt{3} - 3$ **61.** $-\frac{1}{3}(x^3 + 3x^2 - 6x)^{-1} + C$

**63.** $\pi + \frac{3\sqrt{3}}{4}$ **65.** $\frac{\pi}{2}$ **67.** 0 **69.** $\cos\frac{1}{x} + C$

**71.** $\frac{1}{12}$ **73.** $(x + 3)^{11}\left(\frac{11x - 3}{132}\right) + C$

**75.** 1 **77.** 0 **79.** 48 **81.** $\frac{256}{3}$ **83.** 8 **85.** $-\frac{4}{15}; \frac{4}{15}$

**87. a.** 5 **b.** 20 **89.** Displacement $= 0$; distance $= 20/\pi$

**91.** $\frac{2}{\pi}$ **93. a.** $5/2, c = 3.5$ **b.** $3, c = 3$ and $c = 5$ **95.** 24

**97.** $f(1) = 0$; $f'(x) > 0$ on $[1, \infty)$; $f''(x) < 0$ on $[1, \infty)$

**99. a.** $\frac{3}{2}, \frac{5}{6}$ **b.** $x$ **c.** $\frac{1}{2}x^2$ **d.** $-1, \frac{1}{2}$ **e.** $1, 1$ **f.** $\frac{3}{2}$ **101.** $-1$

**103.** 11 **105.** $\frac{1}{14}\sec^7(\tan x^2) + C$

**107. a.** Increasing on $(-\infty, 1)$ and $(2, \infty)$; decreasing on $(1, 2)$

**b.** Concave up on $\left(\frac{13}{8}, \infty\right)$; concave down on $\left(-\infty, \frac{13}{8}\right)$

**c.** Local max at $x = 1$; local min at $x = 2$ **d.** Inflection point

at $x = \frac{13}{8}$ **109.** Differentiating the first equation gives the

second equation; no.

# CHAPTER 6

## Section 6.1 Exercises, pp. 358–364

**1.** The position $s(t)$ is the location of the object relative to the origin. The displacement is the change in position between time $t = a$ and $t = b$. The distance traveled between $t = a$ and $t = b$ is $\int_a^b |v(t)|\,dt$, where $v(t)$ is the velocity at time $t$. **3.** The displacement between

$t = a$ and $t = b$ is $\int_a^b v(t)\,dt$. **5.** $Q(t) = Q(0) + \int_0^t Q'(x)\,dx$

**7. a.** $[0, 1), (3, 5)$ **b.** $-4$ mi **c.** 26 mi **d.** 6 mi **e.** 6 mi on the positive side of the initial position **9. a.** 3 **b.** $\frac{13}{3}$ **c.** 3

**d.** $s(t) = \begin{cases} -\dfrac{t^2}{2} + 2t & \text{if } 0 \le t \le 3 \\[2mm] \dfrac{3t^2}{2} - 10t + 18 & \text{if } 3 < t \le 4 \\[2mm] -t^2 + 10t - 22 & \text{if } 4 < t \le 5 \end{cases}$

**11. a.** 3 m **b.** 3 m; 0 m; 3 m; 0 m **c.** 12 m
**13. a.** Positive direction for $2 < t \le 3$; negative direction for
$0 < t < 2$ **b.** 0 m **c.** 8 m **15. a.** Positive direction for
$0 \le t < 2$ and $4 < t \le 5$; negative direction for $2 < t < 4$
**b.** 20 m **c.** 28 m **17. a.** $s(t) = 2 - \cos t$ **19. a.** $s(t) = 6t - t^2$

**21. a.** $s(t) = 9t - \dfrac{t^3}{3} - 2$ **23. a.** $s(t) = 2\sin\pi t$

**b.**

$s(t) = 2\sin\pi t$ graph

**c.** $\frac{3}{2}, \frac{7}{2}, \frac{11}{2}$

**d.** $\frac{1}{2}, \frac{5}{2}, \frac{9}{2}$

**25. a.** $s(t) = 10t(48 - t^2)$ **b.** 880 mi **c.** $\frac{2720\sqrt{6}}{9} \approx 740.29$ mi

**27.**

**a.** Velocity is a maximum for $20 \le t \le 45$; $v = 0$ at $t = 0$ and $t = 60$   **b.** 1200 m   **c.** 2550 m   **d.** 2100 m

**29.** $v(t) = -32t + 70$; $s(t) = -16t^2 + 70t + 10$

**31.** $v(t) = -9.8t + 20$; $s(t) = -4.9t^2 + 20t$

**33.** $v(t) = -\frac{1}{200} t^2 + 10$; $s(t) = -\frac{1}{600} t^3 + 10t$

**35. a.** $s(t) = 44t^2$   **b.** 704 ft   **c.** $\sqrt{30} \approx 5.477$ s

**d.** $\frac{5\sqrt{33}}{11} \approx 2.611$ s   **e.** Approx. 180.023 ft

**37.** 6.154 mi; 1.465 mi   **39. a.** 2639 people

**b.** $P(t) = 250 + 20t^{3/2} + 30t$ people   **41. a.** 27,250 barrels

**b.** 31,000 barrels   **c.** 4000 barrels

**43. a.** Alicia: $S(t) = 4 - \dfrac{4}{t + 1}$; Boris: $S(t) = 3 - \dfrac{2}{t + 1}$

**b.** Just after 1 P.M.   **c.** More than 2 mi

**45. a.** $\dfrac{120}{\pi} + 40 \approx 78.20$ m³

**b.** $Q(t) = 20\left( t + \dfrac{12}{\pi} \sin\left( \dfrac{\pi}{12} t \right) \right)$   **c.** Approx. 122.6 hr

**47. a.** $V(t) = 5 + \cos \dfrac{\pi t}{2}$   **b.** 15 breaths/min   **c.** 2 L, 6 L

**49. a.** 7200 MWh or $2.592 \times 10^{13}$ J   **b.** 16,000 kg; 5,840,000 kg
**c.** 450 g; 164,250 g   **d.** About 1500 turbines   **51. a.** $96,875
**b.** $86,875   **53. a.** $69,583.33   **b.** $139,583.33   **55. a.** False
**b.** True   **c.** True   **d.** True   **57.** $\frac{2}{3}$   **59.** $\frac{25}{3}$

**61. a.**

**b.** Theo   **c.** Sasha
**d.** Theo hits the 10-mi mark before Sasha; Sasha and Theo hit the 15-mi mark at the same time; Sasha hits the 20-mi mark before Theo.   **e.** Sasha   **f.** Theo

**63.** $\displaystyle \int_a^b f'(x)\, dx = f(b) - f(a) = g(b) - g(a) = \int_a^b g'(x)\, dx$

## Section 6.2 Exercises, pp. 368–373

**1.** $\displaystyle \int_a^b (f(x) - g(x))\, dx + \int_b^c (g(x) - f(x))\, dx$

**3.**

**5. a.** 1

**7.** $\displaystyle \int_0^1 y\, dy + \int_1^2 (2 - y)\, dy$

**9.** $\dfrac{9}{2}$   **11.** 8   **13.** $\dfrac{7}{6}$

**15.** $2 - \sqrt{2}$

**17.** $\dfrac{4}{3}$   **19.** $\dfrac{32}{3}$

**21.** 9   **23.** $\dfrac{81}{2}$   **25.** 2

**27.** $\dfrac{8\sqrt{2} - 7}{6}$   **29.** $\dfrac{125}{2}$   **31. a.** $\displaystyle \int_0^1 (\sqrt{x} - x^3)\, dx$

**b.** $\displaystyle \int_0^1 (\sqrt[3]{y} - y^2)\, dy$   **33.** $\frac{19}{6} \approx 3.17$ km; the faster runner jogged approximately 3.17 km farther than the slower runner.

**35. a.** 7   **b.** 4   **37.** 25   **39.** $\frac{32}{3}$   **41.** 9   **43.** 1   **45.** $\frac{7}{3}$   **47.** 3

**49.** $\frac{64}{5}$   **51.** 8   **53.** $\frac{5}{24}$   **55.** $\frac{63}{4}$   **57.** $\frac{9}{2}$   **59.** $\frac{32}{3}$   **61.** $2 + \frac{3\pi}{2}$   **63.** $\frac{17}{3}$

**65. a.** False   **b.** False   **c.** True   **67.** $\frac{4}{9}$   **69.** $\dfrac{n - 1}{2(n + 1)}$

**71.** $A_n = \dfrac{n - 1}{n + 1}$; $\lim\limits_{n \to \infty} A_n = 1$; the region approximates a square

with side length of 1.   **73.** $k = 1 - \dfrac{1}{\sqrt{2}}$   **75. a.** The lowest $p\%$

of households own exactly $p\%$ of the wealth for $0 \le p \le 100$.
**b.** The function must be one-to-one and its graph must lie below
$y = x$ because the poorest $p\%$ cannot own more than $p\%$ of the
wealth.   **c.** $p = 1.1$ is most equitable; $p = 4$ is least equitable.

**e.** $G(p) = \dfrac{p - 1}{p + 1}$   **f.** $0 \le G \le 1$ for $p \ge 1$   **g.** $\frac{5}{18}$

**77. a.**

**b.** $A_n(x)$ is the net area of the region between the graphs of $f$ and $g$ from 0 to $x$.   **c.** $x = n^{n/(n^2 - 1)}$; the root decreases with $n$.

## Section 6.3 Exercises, pp. 382–387

**1.** $A(x)$ is the area of the cross section through the solid at the point $x$.

**3. a.** $3 - x$   **b.** $\displaystyle \int_0^2 (3 - x)\, dx$   **5. a.** $\sqrt{\cos x}$   **b.** $\pi \cos x$

**c.** $\displaystyle \int_0^{\pi/2} \pi \cos x\, dx$   **7. a.** $\sqrt{x + 1}$   **b.** 1   **c.** $\pi((\sqrt{x} + 1)^2 - 1)$

**d.** $\displaystyle \int_0^4 \pi((\sqrt{x} + 1)^2 - 1)\, dx$   **9. a.** $\sqrt{x}$   **b.** $\pi x$   **c.** $\displaystyle \int_0^4 \pi x\, dx$

**11.** $\dfrac{4}{3}$   **13.** 1   **15.** $\dfrac{\pi}{3}$   **17.** $36\pi$   **19.** $\dfrac{256\pi}{15}$   **21.** $\dfrac{32\pi}{3}$

**23.** $2\pi(3\sqrt{3} - \pi)/3$   **25.** $\dfrac{2\pi}{5}$   **27.** $\dfrac{\pi^2}{4}$   **29.** $\dfrac{\pi^2}{2}$   **31.** $\dfrac{256\pi}{5}$

**33.** $\dfrac{4\pi - \pi^2}{4}$   **35.** $\dfrac{128\pi}{5}$   **37.** $\dfrac{325\pi}{72}$   **39.** $\dfrac{256\pi}{7}$

**41.** Volumes are equal.   **43.** $x$-axis   **45.** $\dfrac{\pi}{2}$

**47.** $\pi\sqrt{3}$   **49.** $\dfrac{\pi}{6}$   **51.** $2\pi(8 + \pi)$   **53.** $(6\sqrt{3} - 2\pi)\pi$

**55.** $4\pi$   **57. a.** False   **b.** True   **c.** True

**59.** Volume $(S) = 8\pi a^{5/2}/15$; volume $(T) = \pi a^{5/2}/3$

**61. a.**   **b.**

**63.** Left: $166\pi$; right: $309\pi$; midpoint: $219\pi$ **65. a.** $\frac{1}{3}V_C$ **b.** $\frac{2}{3}V_C$
**67.** $24\pi^2$ **69. b.** $V = \pi r^2 h$

## Section 6.4 Exercises, pp. 395–399

**1.** $\int_a^b 2\pi x(f(x) - g(x))\, dx$ **3.** $x; y$ **5. a.** $x$ **b.** $2 - x^2 - x$

**c.** $\int_0^1 2\pi x(2 - x^2 - x)\, dx$ **7. a.** $2 - y$ **b.** $4 - (2 - y)^2 = 4y - y^2$

**c.** $\int_0^2 2\pi(2 - y)(4y - y^2)\, dy$ **9.** $\frac{\pi}{6}$ **11.** $\pi$ **13.** $8\pi$ **15.** $\frac{32\pi}{3}$

**17.** $\pi$ **19.** $\frac{\pi}{2}$ **21.** $\frac{81\pi}{2}$ **23.** $\frac{2\pi}{3}$ **25.** $\frac{3\pi}{10}$ **27.** $90\pi$

**29.** $\frac{\pi}{5}$ **31.** $\frac{4\pi}{15}$ **33.** $500\pi$ **35.** $\frac{11\pi}{6}$ **37.** $\frac{5\pi}{6}$ **39.** $\frac{23\pi}{15}$

**41.** $\frac{52\pi}{15}$ **43.** $\frac{36\pi}{5}$ **47. a.** $4\pi\int_1^5 x\sqrt{4 - (x - 3)^2}\, dx$

**b.** $12\pi\int_{-2}^2 \sqrt{4 - y^2}\, dy$ **c.** $24\pi^2$ **49.** $\frac{\pi}{9}$ **51.** $\frac{16\pi}{3}$ **53.** $\frac{608\pi}{3}$

**55.** $\frac{16\pi}{3}$ **57.** $\frac{5\pi}{6}$ **59. a.** True **b.** False **c.** True **61.** $24\pi$

**63.** $54\pi$ **65. a.** $V_1 = \frac{\pi}{15}(3a^2 + 10a + 15); V_2 = \frac{\pi}{2}(a + 2)$

**b.** $V(S_1) = V(S_2)$, for $a = 0$ and $a = -\frac{5}{6}$ **67.** $10\pi$

**69. a.** $27\sqrt{3}\pi r^3/8$ **b.** $54\sqrt{2}/(3 + \sqrt{2})^3$ **c.** $500\pi/3$

## Section 6.5 Exercises, pp. 402–404

**1.** Determine whether $f$ has a continuous derivative on $[a, b]$.
If so, calculate $f'(x)$ and evaluate the integral $\int_a^b \sqrt{1 + f'(x)^2}\, dx$.

**3.** $\int_{-2}^5 \sqrt{1 + 9x^4}\, dx$ **5.** $\int_1^2 \sqrt{\frac{x^6 + 4}{x^6}}\, dx$ **7.** $4\sqrt{5}$

**9.** $8\sqrt{65}$ **11.** $168$ **13.** $\frac{4}{3}$ **15.** $\frac{123}{32}$ **17.** $60$ **19.** $7\sqrt{5}$

**21. a.** $\int_{-1}^1 \sqrt{1 + 4x^2}\, dx$ **b.** $2.96$ **23. a.** $\int_0^{\pi/4} \sqrt{1 + \sec^4 x}\, dx$

**b.** $1.28$ **25. a.** $\int_3^4 \sqrt{\frac{4y - 7}{4y - 8}}\, dy$ **b.** $1.08$

**27. a.** $\int_0^\pi \sqrt{1 + 4\sin^2 2x}\, dx$ **b.** $5.27$

**29. a.** $\int_1^{10} \sqrt{1 + 1/x^4}\, dx$ **b.** $9.15$

**31.** Approx. $1326$ m **33. a.** False **b.** True **c.** False
**35. a.** $f(x) = \pm 4x^3/3 + C$ **b.** $f(x) = \pm 3\sin 2x + C$
**37.** $y = 1 - x^2$ **39. a.** $L/2$ **b.** $L/c$

## Section 6.6 Exercises, pp. 409–411

**1.** $15\pi$ **3.** Evaluate $\int_a^b 2\pi f(x)\sqrt{1 + f'(x)^2}\, dx$. **5. a.** $4\sqrt{2}\pi$

**7.** $156\sqrt{10}\pi$ **9.** $\frac{2912\pi}{3}$ **11.** $\frac{\pi}{9}(17^{3/2} - 1)$ **13.** $2\pi$

**15.** $15\sqrt{17}\pi$ **17.** $4\pi$ **19.** $96\pi$ **21.** $\frac{9\pi}{125}$ m$^3$

**23. a.** False **b.** False **c.** True **d.** False

**25. a.** $\int_0^{\pi/2} 2\pi(\cos x)\sqrt{1 + \sin^2 x}\, dx$ **b.** Approx. $7.21$

**27. a.** $\int_0^{\pi/4} 2\pi(\tan x)\sqrt{1 + \sec^4 x}\, dx$ **b.** Approx. $3.84$

**29.** $\frac{12\pi a^2}{5}$ **31.** $\frac{53\pi}{9}$ **33.** $\frac{275\pi}{32}$ **35.** $\frac{48,143\pi}{48}$ **39. a.** $c^2 A$ **b.** $A$

**41. a.** $\frac{6}{R}$ **b.** $\frac{3}{R}$ **c.** The ball and ellipsoid each have a volume of $\pi h/3$.

**d.**

| $h$ | SAV ratio of ball | SAV ratio of ellipsoid |
|---|---|---|
| 1.1 | 4.61 | 4.97 |
| 5 | 2.78 | 3.00 |
| 10 | 2.21 | 2.38 |
| 20 | 1.75 | 1.89 |

**e.** A sphere

## Section 6.7 Exercises, pp. 420–424

**1.** 150 g **3.** 25 J **5.** Horizontal cross sections of water at various
locations in the tank are lifted different distances. **7.** 39,200 N/m$^2$

**9.** $\int_5^{10} 25\pi\rho g(15 - y)\, dy$ **11.** $\int_0^{10} 25\pi\rho g(10 - y)\, dy$ **13.** $\pi + 2$

**15.** 3 **17.** $(2\sqrt{2} - 1)/3$ **19.** 10 **21.** 9 J **23. a.** $k = 150$
**b.** 12 J **c.** 6.75 J **d.** 9 J **25. a.** 112.5 J **b.** 12.5 J
**27. a.** 31.25 J **b.** 312.5 J **29. a.** 625 J **b.** 391 J
**31. a.** 22,050 J **b.** 36,750 J **33.** 3675 J **35.** $1.15 \times 10^7$ J
**37** $3.94 \times 10^6$ J **39. a.** $66,150\pi$ J **b.** No **41. a.** $2.10 \times 10^8$ J
**b.** $3.78 \times 10^8$ J **43. a.** 32,667 J **b.** Yes **45.** $7.70 \times 10^3$ J
**47.** $1.47 \times 10^7$ N **49.** $2.94 \times 10^7$ N **51.** 6533 N **53.** 6737.5 N
**55.** $8 \times 10^5$ N **57. a.** True **b.** True **c.** True **d.** False
**59. a.** Compared to a linear spring, $F(x) = 16x$, the restoring force is
less for large displacements. **b.** 17.87 J **c.** 31.6 J **61.** 1,381,800 J
**63.** 0.28 J **65. a.** Yes **b.** 4.296 m **67.** Left: 16,730 N;
right: 14,700 N **69. a.** $8.87 \times 10^9$ J
**b.** $500\, GMx/(R(x + R)) = (2 \times 10^{17})x/(R(x + R))$ J
**c.** $GMm/R$ **d.** $v = \sqrt{2GM/R}$

## Chapter 6 Review Exercises, pp. 425–429

**1. a.** True **b.** True **c.** True **3. a.** Positive direction for
$0 \le t < \frac{1}{2}$ and $2 < t \le 3$; negative direction for $\frac{1}{2} < t < 2$
**b.** 9 m **c.** 22.5 m **d.** $s(t) = 4t^3 - 15t^2 + 12t + 1$
**5.** $s(t) = 20t - 5t^2$; displacement $= 20t - 5t^2$;
$$D(t) = \begin{cases} 20t - 5t^2 & \text{if } 0 \le t < 2 \\ 5t^2 - 20t + 40 & \text{if } 2 \le t \le 4 \end{cases}$$
**7. a.** $v(t) = -\frac{8}{\pi}\cos\frac{\pi t}{4}; s(t) = -\frac{32}{\pi^2}\sin\frac{\pi t}{4}$ **b.** Min value $= -\frac{32}{\pi^2}$;
max value $= \frac{32}{\pi^2}$ **c.** 0; 0 **9. a.** $R(t) = 3t^{4/3}$
**b.** $R(t) = \begin{cases} 3t^{4/3} & \text{if } 0 \le t \le 8 \\ 2t + 32 & \text{if } t > 8 \end{cases}$ **c.** $t = 59$ min
**11.** Both integrals equal the area bounded between $y = x^2$, $x = 2$, and
$y = 0$. **13.** $1 - \frac{\pi}{4}$ **15.** $\frac{8}{3}$ **17.** $\frac{7}{3}$ **19.** 8 **21.** 1
**23.** $\frac{1}{3}$ **25.** $R_1$: $\frac{7}{6}$; $R_2$: $\frac{10}{3}$; $R_3$: $4\sqrt{3} - \frac{10}{3}$ **27.** $\frac{11\pi}{15}$
**29.** $\frac{14\pi}{3}$ **31.** $\int_1^3 2\pi(3 - x)(2\sqrt{x} - 3 + x)\, dx$ **33.** $\frac{7}{3}$ **35.** $\frac{31\pi}{5}$
**37.** $R_1$: $\sqrt{3}$; $R_2$: $\frac{4\pi}{3} - \sqrt{3}$ **39.** $\frac{1}{3}$ **41.** $\frac{5}{6}$ **43.** $\frac{8}{15}$ **45.** $\frac{8\pi}{5}$
**47.** $\frac{\pi}{2}$ **49.** $\pi$ **51.** $\frac{512\pi}{15}$ **53.** About $y = -2$: $80\pi$;
about $x = -2$: $112\pi$ **55.** $\frac{\pi}{3}(4\pi - 3\sqrt{3})$ **57.** $c = 5$
**59.** $2\sqrt{3} - \frac{4}{3}$ **61.** $\int_2^4 \sqrt{4x^2 + 8x + 5}\, dx \approx 16.127$ **63.** 1.74

**65.** 1.28   **67. a.** $9\pi$   **b.** $\dfrac{9\pi}{2}$   **69. a.** $\dfrac{263{,}439\pi}{4096}$   **b.** $\dfrac{483}{64}$

**c.** $\dfrac{\pi}{8}(84 + \ln 2)$   **d.** $\dfrac{264{,}341\pi}{18{,}432}$   **71.** 444 g

**73. a.** 562.5 J   **b.** 56.25 J   **75. a.** 980 J   **b.** 627.2 J

**77. a.** 1,411,200 J   **b.** 940,800 J   **79. a.** 1,477,805 J

**b.** The work required to pump out the top 3 m of water is 1,015,991 J, and the work required to pump out the bottom 3 m of water is 461,814 J. More work is required to pump out the top 3 m of water.

**81.** 4,987,592 J   **83.** 5716.7 N   **85.** $5.2 \times 10^7$ N

# CHAPTER 7

## Section 7.1 Exercises, pp. 439–442

**1.** If a function $f$ is not one-to-one, then there are domain values, $x_1$ and $x_2$, such that $x_1 \neq x_2$ but $f(x_1) = f(x_2)$. If $f^{-1}$ exists, by definition, $f^{-1}(f(x_1)) = x_1$ and $f^{-1}(f(x_2)) = x_2$ so that $f^{-1}$ assigns two different range values to the single domain value of $f(x_1)$.

**3.**

**5.** $(-\infty, -1], [-1, 1], [1, \infty)$   **7.** $f^{-1}(x) = \dfrac{1}{2}x$

**9.**

**11.** $g_1(x) = x^2 + 1; D = [0, \infty); R = [1, \infty); g_1^{-1}(x) = \sqrt{x - 1};$ $D = [1, \infty); R = [0, \infty)$   **13.** 1; 5; 7   **15.** 1/4   **17.** $(-\infty, \infty)$

**19.** $(-\infty, 5) \cup (5, \infty)$   **21.** $(-\infty, 0), (0, \infty)$

**23.** $f^{-1}(x) = -\dfrac{1}{4}x + 2$

**25.** $f^{-1}(x) = x^2 - 2$

**27.** $f^{-1}(x) = 2 + \sqrt{x + 1}$

**29. a.** $f^{-1}(x) = -\dfrac{1}{4}x + \dfrac{3}{2}$   **31. a.** $f^{-1}(x) = x^{3/5} + 7$

**33. a.** $f^{-1}(x) = \dfrac{1}{1 - x}$   **35.** $f^{-1}(x) = \sqrt{6/x + 9}, x > 0$

**37. a.** $f_1(x) = \sqrt{1 - x^2}; 0 \le x \le 1$

$f_2(x) = \sqrt{1 - x^2}; -1 \le x \le 0$

$f_3(x) = -\sqrt{1 - x^2}; -1 \le x \le 0$

$f_4(x) = -\sqrt{1 - x^2}; 0 \le x \le 1$

**b.** $f_1^{-1}(x) = \sqrt{1 - x^2}; 0 \le x \le 1$

$f_2^{-1}(x) = -\sqrt{1 - x^2}; 0 \le x \le 1$

$f_3^{-1}(x) = -\sqrt{1 - x^2}; -1 \le x \le 0$

$f_4^{-1}(x) = \sqrt{1 - x^2}; -1 \le x \le 0$

**39.** $\dfrac{1}{3}$   **41.** $-\dfrac{1}{5}$   **43.** $\dfrac{1}{2}$   **45.** 4   **47.** $\dfrac{1}{12}$   **49.** $\dfrac{1}{4}$   **51.** $\dfrac{5}{4}$

**53. a.** $\dfrac{1}{2}$   **b.** $\dfrac{2}{3}$   **c.** Cannot be determined   **d.** $\dfrac{3}{2}$

**55. a.** False   **b.** True   **c.** False   **d.** True   **e.** True

**57.** $f^{-1}(x) = \sqrt[3]{x} - 1, D = \mathbb{R}$

**59.** $f_1^{-1}(x) = \sqrt{2/x - 2}, D_1 = (0, 1]; f_2^{-1}(x) = -\sqrt{2/x - 2},$

$D_2 = (0, 1]$   **61.** $f^{-1}(x) = \dfrac{x + 4}{3}; (f^{-1})'(x) = \dfrac{1}{3}$

**63.** $f^{-1}(x) = \sqrt{x + 4}, x \ge -4;$

$(f^{-1})'(x) = 1/(2\sqrt{x + 4}), x > -4$

**65.** $f^{-1}(x) = x^2 - 2, x \ge 0; (f^{-1})'(x) = 2x, x \ge 0$

**67.** $f^{-1}(x) = \dfrac{1}{x^2}, x > 0; (f^{-1})'(x) = -\dfrac{2}{x^3}, x > 0$

**69.** $r = \sqrt[3]{\dfrac{3V}{4\pi}}$   **71.** $r = \sqrt{\dfrac{V}{10\pi}}$

**73. a.**

$f$ is one-to-one on the intervals $(-\infty, -1/\sqrt{2}], [-1/\sqrt{2}, 0], [0, 1/\sqrt{2}], [1/\sqrt{2}, \infty)$

**b.** $x = \sqrt{\dfrac{1 \pm \sqrt{4y + 1}}{2}}, -\sqrt{\dfrac{1 \pm \sqrt{4y + 1}}{2}}$

**75.** $f^{-1}(x) = \dfrac{3\left(\sqrt[3]{\dfrac{x}{2} + \dfrac{\sqrt{3}\sqrt{32 + 27x^2}}{18}}\right)^2 - 2}{3\sqrt[3]{\dfrac{x}{2} + \dfrac{\sqrt{3}\sqrt{32 + 27x^2}}{18}}};$

one-to-one on $(-\infty, \infty)$

**77. a.** $a = f'(x_0); b = y_0 - x_0 f'(x_0)$

**b.** $c = \dfrac{1}{f'(x_0)}; \quad d = x_0 - \dfrac{y_0}{f'(x_0)}$

**c.** $L(x) = f'(x_0)x + y_0 - x_0 f'(x_0)$, so

$$L^{-1}(x) = \frac{x - y_0 + x_0 f'(x_0)}{f'(x_0)} = M(x).$$

### Section 7.2 Exercises, pp. 453–456

**1.** $D = (0, \infty), R = (-\infty, \infty)$

**3.** $x = e^y \Rightarrow 1 = e^y y'(x) \Rightarrow y'(x) = 1/e^y = 1/x$

**5.** $\dfrac{d}{dx}(\ln kx) = \dfrac{d}{dx}(\ln k + \ln x) = \dfrac{d}{dx}(\ln x)$   **7.** $f^{-1}(x) = \dfrac{1}{2}\ln x - 3$

**9.** $f^{-1}(x) = \dfrac{e^x - 1}{3}$   **11.** $1/e$   **13.** $\dfrac{\ln 10 - 1}{3}$   **15.** 451 years

**17.** $\dfrac{1}{x}$   **19.** $\dfrac{2}{x}$   **21.** $\cot x$   **23.** $2/(1 - x^2)$   **25.** $(x^2 + 1)/x + 2x \ln x$

**27.** $1/(x \ln x)$   **29.** $\dfrac{1}{x(\ln x + 1)^2}$   **31.** $\dfrac{e^x}{(e^x + 1)^2}$

**33.** $-9e^{-x} - 10e^{2x} - 6e^x$   **35.** $\dfrac{3e^x + 4e^{2x}}{(e^{-x} + 2)^2}$   **37.** 0

**39.** $y = -\dfrac{3x}{4} + \dfrac{1}{4}$   **41.** $3 \ln|x - 10| + C$

**43.** $\ln\left|\dfrac{(x - 4)^2}{(2x + 1)^{3/2}}\right| + C$   **45.** $6(1 - \ln 2)$

**47.** $\dfrac{1}{4}\left(\dfrac{1}{\ln^2(\ln 3)} - \dfrac{1}{\ln^2(\ln 4)}\right)$   **49.** $\ln|\ln(\ln x)| + C$

**51.** $\dfrac{e^{2x} + e^{-2x}}{4} + C$   **53.** $\dfrac{1}{2}e^{2x} + x + C$   **55.** $\dfrac{98}{3}$

**57.** $\ln|e^x - e^{-x}| + C$   **59.** $2e^{\sqrt{x}} + C$

**61.** $65/32$   **63.** $\dfrac{(x + 1)^{10}}{(2x - 4)^8}\left(\dfrac{10}{x + 1} - \dfrac{8}{x - 2}\right)$   **65.** $2x^{\ln x - 1}\ln x$

**67.** $\dfrac{(x + 1)^{3/2}(x - 4)^{5/2}}{(5x + 3)^{2/3}}\left(\dfrac{3}{2(x + 1)} + \dfrac{5}{2(x - 4)} - \dfrac{10}{3(5x + 3)}\right)$

**69.** $(\sin x)^{\tan x}(1 + (\sec^2 x) \ln \sin x); 0 < x < \pi, x \neq \pi/2$

**71. a.** True   **b.** False   **c.** False   **d.** False   **e.** False   **f.** True   **g.** True

**75.** $\dfrac{\ln p}{p - 1}, 0$   **77.** $-20xe^{-10x^2}$   **79.** $\dfrac{1}{2x}$   **81.** $\dfrac{9x^2 - 14x + 11}{(x + 2)^4}$

**83.** $e^x \cos(\sin e^x) \cos e^x$   **85.** $\dfrac{e^x(x^2 + x + 1)}{(x + 1)^2}$   **87.** $\dfrac{1}{3}e^{x^3} + C$

**89.** $2e - 1$   **91.** $\dfrac{32}{3}$   **93.** $x - 2 + 2\ln|x - 2| + C$ or

$x + 2\ln|x - 2| + C$   **95.** $\ln\dfrac{9}{8}$   **97.** $\dfrac{1}{2}\ln|\ln(t^2)| + C$

**99. a.** $x = \pm 1$   **b.** Local min at $x = -1$; local max at $x = 1$

**c.** Concave up on $(-\sqrt{3}, 0), (\sqrt{3}, \infty)$; concave down on $(-\infty, -\sqrt{3}), (0, \sqrt{3})$; inflection points at $x = 0$ and $x = \pm\sqrt{3}$

**101. a.** $L(x) = -\dfrac{1}{2}x + \dfrac{1}{2}(1 + \ln 2)$   **b.**

**c.** $\dfrac{1}{e} \approx \dfrac{1}{2}\ln 2 \approx 0.347$

**103.** $\dfrac{1}{2}(\ln 2 + 1) \approx 0.85$   **105. a.**

**b.** $t = 2\ln 265 \approx 11.2$ years; about 14.5 years

**c.** $P'(0) \approx 25$ fish/year; $P'(5) \approx 264$ fish/year

**d.**

The population is growing fastest after about 10 years.

**107. b.** $r(11) \approx 0.0133; r(21) \approx 0.0118$; the relative growth rate is decreasing.   **c.** $\lim\limits_{t \to \infty} r(t) = 0$; as the population gets close to carrying capacity, the relative growth rate approaches zero.

**109. a.** $\dfrac{10^7(1 - e^{-kt})}{k}$   **b.** $\dfrac{10^7}{k} = $ total number of barrels of oil extracted if the nation extracts the oil indefinitely where it is assumed that the nation has at least $\dfrac{10^7}{k}$ barrels of oil in reserve

**c.** $k = \dfrac{1}{200} = 0.005$   **d.** Approximately 138.6 yr

**113.** $\ln 2 = \displaystyle\int_1^2 \dfrac{dt}{t} < L_2 = \dfrac{5}{6} < 1$

$\ln 3 = \displaystyle\int_1^3 \dfrac{dt}{t} > R_7$

$= 2\left(\dfrac{1}{9} + \dfrac{1}{11} + \dfrac{1}{13} + \dfrac{1}{15} + \dfrac{1}{17} + \dfrac{1}{19} + \dfrac{1}{21}\right) > 1$

### Section 7.3 Exercises, pp. 465–467

**1.** $f'(x) = b^x \ln b$, which is the result for $f(x) = e^x$ when $b = e$.

**3.** $b^x = e^{\ln(b^x)} = e^{x \ln b}$   **5.** $7^x \ln 7$   **7.** $e^{x \ln 3}, e^{\pi \ln x}, e^{(\sin x)(\ln x)}$

**9.** $\dfrac{x^{\pi + 1}}{\pi + 1} + C; \dfrac{\pi^x}{\ln \pi} + C$   **11.** 1000   **13.** 2   **15.** $\ln 21/\ln 7$

**17.** $\ln 5/(3 \ln 3) + 5/3$   **19.** $\pm 2$   **21.** $\pm 4$   **23.** $8^x \ln 8$

**25.** $5 \cdot 4^x \ln 4$   **27.** $3^x \cdot x^2(x \ln 3 + 3)$   **29.** $1000(1.045)^{4t} \ln 1.045$

**31.** $2^{3 + \sin x}(\ln 2)\cos x$   **33.** $\dfrac{2^x \ln 2}{(2^x + 1)^2}$   **35. a.** About 28.7 s

**b.** $-46.512$ s/1000 ft   **c.** $dT/da = -2.74 \cdot 2^{-0.274a} \ln 2$

At $a = 8, \dfrac{dT}{da} = -0.4156$ min/1000 ft

$= -24.938$ s/1000 ft

If a plane travels at 30,000 feet and increases its altitude by 1000 feet, the time of useful consciousness decreases by about 25 seconds.

**37. a.** About 67.19 hr

**b.** $Q'(12) = -9.815 \ \mu\text{Ci/hr}$

$Q'(24) = -5.201 \ \mu\text{Ci/hr}$

$Q'(48) = -1.461 \ \mu\text{Ci/hr}$

The rate at which iodine-123 leaves the body decreases with time.

**39.** $\dfrac{99}{10 \ln 10}$    **41.** 3    **43.** $\dfrac{6^{x^3+8}}{3 \ln 6} + C$

**45.** $e^y y^{e-1}(y + e)$    **47.** $-(\ln 2)2^t \sin 2^t$    **49.** $\dfrac{\sqrt{x}}{2}(10x - 9)$

**51.** $x^{\cos x - 1}(\cos x - x \ln x \sin x); -\ln(\pi/2)$

**53.** $x^{\sqrt{x}}\left(\dfrac{2 + \ln x}{2\sqrt{x}}\right); 4(2 + \ln 4)$

**55.** $\dfrac{(\sin x)^{\ln x}(\ln(\sin x) + x(\ln x)\cot x)}{x}; 0$    **57.** $4(2x)^{4x}(1 + \ln 2x)$

**59.** $(\ln 2)\, 2^{x^2+1} x$    **61.** $2(x + 1)^{2x}\left(\dfrac{x}{x + 1} + \ln(x + 1)\right)$

**63.** $\dfrac{8x}{(x^2 - 1)\ln 3}$    **65.** $-\sin x(\ln(\cos^2 x) + 2)$    **67.** $-\dfrac{\ln 4}{x \ln^2 x}$

**69.** $y = x \sin 1 + 1 - \sin 1$    **71.** $y = e^{2/e}$ and $y = e^{-2/e}$

**73. a.** False   **b.** False   **c.** False   **d.** False   **e.** True

**75.** $A$ is $y = \log_2 x$; $B$ is $y = \log_4 x$; $C$ is $y = \log_{10} x$.

**77.** $-\dfrac{1}{x^2 \ln 10}$    **79.** $\dfrac{2}{x}$    **81.** $1.05^x \ln 1.05$    **83.** $\dfrac{12}{3x + 1}$    **85.** $\dfrac{1}{2x \ln 10}$

**87.** $\dfrac{2}{2x - 1} + \dfrac{3}{x + 2} + \dfrac{8}{1 - 4x}$

**89.** $y = 2$

**91.** $10x^{10x}(1 + \ln x)$    **93.** $x^{\cos x}\left(\dfrac{\cos x}{x} - (\ln x)\sin x\right)$

**95.** $\left(1 + \dfrac{1}{x}\right)^x\left(\ln\left(1 + \dfrac{1}{x}\right) - \dfrac{1}{x + 1}\right)$    **97.** $x^{9+x^{10}}(1 + 10 \ln x)$

**99.** $-\dfrac{1}{9^x \ln 9} + C$    **101.** $\dfrac{10^{x^3}}{3 \ln 10} + C$    **103.** $\dfrac{3 \cdot 3^{\ln 2} - 1}{\ln 3}$

**105.** $\dfrac{1}{e}$    **107.** $27(1 + \ln 3)$

### Section 7.4 Exercises, pp. 475–477

**1.** The relative growth is constant.    **3.** The time it takes a function to double in value    **5.** $T_2 = (\ln 2)/k$    **7.** $\dfrac{\ln 2}{20} \approx 0.03466$

**9.** Compound interest, world population    **11.** $\ln 1.11 \approx 0.1044$

**13.** $\dfrac{df}{dt} = 10.5; \dfrac{dg}{dt} \cdot \dfrac{1}{g} = \dfrac{1}{10}$

**15. a.** $\ln 1.024 \approx 0.02372; y(t) = 90{,}000\, e^{t \ln 1.024}$   **b.** 2028

**17. a.** $\dfrac{\ln 1.1}{10} \approx 0.009531; y(t) = 50{,}000\, e^{t \ln 1.1/10}$   **b.** 60,500

**19. a.** $\ln 1.016 \approx 0.01587; y(t) = 100\, e^{t \ln 1.016}$   **b.** \$126.88

**21.** 3.71%    **23. a.** 88.1 years; 423.4 million
**b.** 99.4 years; 412.2 million    **25.** 28.7 million    **27.** 2026
**29.** $a(t) = 20e^{(t/36)\ln 0.5}$ mg with $t = 0$ at midnight; 15.87 mg;
119.6 hr $\approx$ 5 days    **31.** 1.798 million; the downward turn in the
population size may be temporary.    **33.** 18,928 ft; 125,754 ft
**35.** 1.055 billion yr    **37.** 6.2 hours    **39.** 2 dollars    **41.** 1044 days

**43. a.** False   **b.** False   **c.** True   **d.** True   **e.** True
**45. a.** $V_1(t) = 0.495e^{-0.1216t}$   **b.** $V_2(t) = 0.005e^{0.239t}$
**c.** $V(t) = 0.495e^{-0.1216t} + 0.005e^{0.239t}$
**d.**

The tumor initially shrinks significantly in size but eventually starts growing again.   **e.** 10.9 days; give a second treatment just before the end of the 10th day after the first treatment.

**47. a.** Bob; Abe   **b.** $y = 4 \ln(t + 1)$ and $y = 8 - 8e^{-t/2}$; Bob

**49.** 10.034%; no    **51.** 1.3 s
**53.** $k = \ln(1 + r); r = 2^{1/T_2} - 1; T_2 = (\ln 2)/k$

### Section 7.5 Exercises, pp. 490–492

**1.** Sine is not one-to-one on its domain.    **3.** $3\pi/4$    **5.** Yes; no
**7.** Horizontal asymptotes at $y = \pm\pi/2$

**9.** $\dfrac{d}{dx}(\sin^{-1} x) = \dfrac{1}{\sqrt{1 - x^2}}; \dfrac{d}{dx}(\tan^{-1} x) = \dfrac{1}{1 + x^2};$

$\dfrac{d}{dx}(\sec^{-1} x) = \dfrac{1}{|x|\sqrt{x^2 - 1}}$    **11.** $\dfrac{1}{5}$    **13.** $\dfrac{2}{\sqrt{3}}$    **15.** $a \sin^{-1} x + C$

**17.** $\dfrac{\pi}{2}$    **19.** $-\dfrac{\pi}{6}$    **21.** $\dfrac{2\pi}{3}$    **23.** $-1$    **25.** $\sqrt{1 - x^2}$

**27.** $\dfrac{\sqrt{4 - x^2}}{2}$    **29.** $2x\sqrt{1 - x^2}$

**31.** $\cos^{-1} x + \cos^{-1}(-x) = \theta + (\pi - \theta) = \pi$

**33.** $\dfrac{\pi}{3}$    **35.** $\dfrac{\pi}{3}$    **37.** $\dfrac{\pi}{4}$    **39.** $\dfrac{\pi}{2} - 2$    **41.** $\dfrac{1}{\sqrt{x^2 + 1}}$

**43.** $\sqrt{x^2 + 1}$    **45.** $\dfrac{2x}{\sqrt{4x^2 - 1}}$

**47.** $\theta = \sin^{-1}\dfrac{x}{6} = \tan^{-1}\left(\dfrac{x}{\sqrt{36 - x^2}}\right) = \sec^{-1}\left(\dfrac{6}{\sqrt{36 - x^2}}\right)$

**49.** $\dfrac{2}{\sqrt{1 - 4x^2}}$    **51.** $-\dfrac{4w}{\sqrt{1 - 4w^2}}$    **53.** $-\dfrac{2e^{-2x}}{\sqrt{1 - e^{-4x}}}$

**55.** $\dfrac{4y}{1 + (2y^2 - 4)^2}$   **57.** $-\dfrac{1}{2\sqrt{z}(1 + z)}$   **59.** $\dfrac{1}{|x|\sqrt{x^2 - 1}}$

**61.** $-\dfrac{1}{|2u + 1|\sqrt{u^2 + u}}$   **63.** $\dfrac{2y}{(y^2 + 1)^2 + 1}$

**65.** $\dfrac{1}{x|\ln x|\sqrt{(\ln x)^2 - 1}}$   **67.** $-\dfrac{e^x \sec^2 e^x}{|\tan e^x|\sqrt{\tan^2 e^x - 1}}$   **69.** $-\dfrac{e^s}{1 + e^{2s}}$

**71.** $\dfrac{2w^5}{1 + w^4}$

**73. a.** Approx. $-0.00055$ rad/m

**b.**

The magnitude of the change in angular size, $|d\theta/dx|$, is greatest when the boat is at the skyscraper (i.e., at $x = 0$).

**75. b.** $d\theta/d\ell = -0.0041, -0.0289,$ and $-0.1984$
**c.** $\lim\limits_{\ell \to 10^+} d\theta/d\ell = -\infty$   **d.** The length $\ell$ is decreasing.

**77.** $6 \sin^{-1} \dfrac{x}{5} + C$   **79.** $\dfrac{1}{10}\sec^{-1}\left|\dfrac{x}{10}\right| + C$   **81.** $\dfrac{\pi}{6}$   **83.** $\dfrac{\pi}{12}$

**85.** $\dfrac{\pi}{12}$   **87.** $\tan^{-1}(y - 2) + C$   **89.** $\dfrac{1}{2}\tan^{-1}\dfrac{e^x}{2} + C$   **91. a.** False
**b.** True   **c.** False   **d.** True   **e.** False   **f.** True   **g.** True

**93.** $\sin \theta = \dfrac{12}{13}$; $\tan \theta = \dfrac{12}{5}$; $\sec \theta = \dfrac{13}{5}$; $\csc \theta = \dfrac{13}{12}$; $\cot \theta = \dfrac{5}{12}$

**95.** $\sin \theta = \dfrac{12}{13}$; $\cos \theta = \dfrac{5}{13}$; $\tan \theta = \dfrac{12}{5}$; $\sec \theta = \dfrac{13}{5}$; $\cot \theta = \dfrac{5}{12}$

**97. a.**

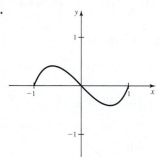

**b.** $f'(x) = 2x \sin^{-1} x + \dfrac{x^2 - 1}{\sqrt{1 - x^2}}$

**99. a.**

**b.** $f'(x) = \dfrac{e^{-x}}{1 + x^2} - e^{-x} \tan^{-1} x$

**101. a.** $d\theta/dc = 1/\sqrt{D^2 - c^2}$   **b.** $1/D$

**103. a.** $\theta = \tan^{-1}\dfrac{h}{80 - h}$   **b.** $\dfrac{1}{50}\dfrac{\text{rad}}{\text{ft}}$

## Section 7.6 Exercises, pp. 498–499

**1.** If $\lim\limits_{x \to a} f(x) = 1$ and $\lim\limits_{x \to a} g(x) = \infty$, then $f(x)^{g(x)} \to 1^\infty$ as $x \to a$, which is meaningless; so direct substitution does not work.

**3.** $\lim\limits_{x \to \infty} \dfrac{g(x)}{f(x)} = 0$   **5.** $\ln x, x^3, 2^x, x^x$   **7.** $\dfrac{1}{2}$   **9.** $\dfrac{1}{e}$   **11.** $\dfrac{1}{24}$   **13.** $\dfrac{\ln 3}{\ln 2}$

**15.** $1$   **17.** $\dfrac{1}{3}$   **19.** $1$   **21.** $e^6$   **23.** $1$   **25.** $e$   **27.** $1$   **29.** $e$

**31.** $e^{0.01x}$   **33.** Comparable growth rates   **35.** $x^x$   **37.** $1.00001^x$
**39.** $x^x$   **41.** $e^{x^2}$   **43. a.** False   **b.** False   **c.** True   **d.** True

**45.** $e^{-1/6}$   **47.** $1$   **49.** $1$   **51.** $-\dfrac{1}{2}$   **53. a.** Approx. $3.44 \times 10^{15}$
**b.** Approx. $3536$   **c.** $e^{100}$   **d.** Approx. $163$   **55.** $1$
**57.** $\ln a - \ln b$   **59. a.** $f(x) = 3^x; a = 0; \ln 3$   **b.** $\ln 3$

**61. b.** $\lim\limits_{m \to \infty}(1 + r/m)^m = \lim\limits_{m \to \infty}\left(1 + \dfrac{1}{(m/r)}\right)^{(m/r)r} = e^r$

**63.** $\lim\limits_{x \to \infty}\dfrac{x^p}{b^x} = \lim\limits_{t \to \infty}\dfrac{\ln^p t}{t \ln^p b} = 0$, where $t = b^x$

**65.** Show $\lim\limits_{x \to \infty}\dfrac{\log_a x}{\log_b x} = \dfrac{\ln b}{\ln a} \ne 0$.   **69. a.** $b > e$
**b.** $e^{ax}$ grows faster than $e^x$ as $x \to \infty$, for $a > 1$; $e^{ax}$ grows slower than $e^x$ as $x \to \infty$, for $0 < a < 1$.

## Section 7.7 Exercises, pp. 511–515

**1.** $\cosh x = \dfrac{e^x + e^{-x}}{2}$; $\sinh x = \dfrac{e^x - e^{-x}}{2}$   **3.** $\cosh^2 x - \sinh^2 x = 1$

**5.** $\sinh^{-1} x = \ln(x + \sqrt{x^2 + 1})$   **7.** Evaluate $\sinh^{-1}\frac{1}{5}$.

**9.** $\displaystyle\int\dfrac{dx}{16 - x^2} = \dfrac{1}{4}\coth^{-1}\dfrac{x}{4} + C$ when $|x| > 4$; the values in the interval of integration $6 \le x \le 8$ satisfy $|x| > 4$.

**23.** $2 \cosh x \sinh x$   **25.** $2 \tanh x \operatorname{sech}^2 x$   **27.** $-2 \tanh 2x$
**29.** $2x (3x \sinh 3x + \cosh 3x)\cosh 3x$   **31.** $4/\sqrt{16x^2 - 1}$
**33.** $2v/\sqrt{v^4 + 1}$   **35.** $\sinh^{-1} x$   **37.** $(\sinh 2x)/2 + C$
**39.** $\ln(1 + \cosh x) + C$   **41.** $x - \tanh x + C$

**43.** $(\cosh^4 3 - 1)/12 \approx 856$   **45.** $\ln(5/4)$

**47.** $\dfrac{1}{2\sqrt{2}}\coth^{-1}\left(\dfrac{x}{2\sqrt{2}}\right) + C$   **49.** $\tanh^{-1}(e^x/6)/6 + C$

**51.** $-\text{sech}^{-1}(x^4/2)/8 + C$   **53.** $-\text{csch}\, z + C$

**55.** $\ln\sqrt{3}\cdot\ln(4/3) \approx 0.158$   **57.** $\dfrac{x^2 + 1}{2x} + C$

**59. a.** The values of $y = \coth x$ are close to 1 on $[5, 10]$.
**b.** $\ln(\sinh 10) - \ln(\sinh 5) \approx 5.0000454$; $|\text{error}| \approx 0.0000454$

**61.**

**a.** $x = \sinh^{-1} 1 = \ln(1 + \sqrt{2})$
**b.** $\pi/4 - \ln\sqrt{2} \approx 0.44$

**63.** $\sinh^{-1} 2 = \ln(2 + \sqrt{5})$   **65.** $-(\ln 5)/3 \approx -0.54$

**67.** $3\ln\left(\dfrac{\sqrt{5} + 2}{\sqrt{2} + 1}\right) = 3(\sinh^{-1} 2 - \sinh^{-1} 1)$

**69.** $\dfrac{1}{15}\left(17 - \dfrac{8}{\ln(5/3)}\right) \approx 0.09$

**71. a.** Sag $= f(50) - f(0) = a(\cosh(50/a) - 1) = 10$;
now divide by $a$.   **b.** $t \approx 0.08$   **c.** $a = 10/t \approx 125$;
$L = 250\sinh(2/5) \approx 102.7$ ft   **73.** $\lambda \approx 32.81$ m

**75. b.** When $d/\lambda < 0.05$, $2\pi d/\lambda$ is small. Because $\tanh x \approx x$ for small values of $x$, $\tanh(2\pi d/\lambda) \approx 2\pi d/\lambda$; therefore,

$$v = \sqrt{\dfrac{g\lambda}{2\pi}\tanh\left(\dfrac{2\pi d}{\lambda}\right)} \approx \sqrt{\dfrac{g\lambda}{2\pi}\cdot\dfrac{2\pi d}{\lambda}} = \sqrt{gd}.$$

**c.** $v = \sqrt{gd}$ is a function of depth alone; when depth $d$ decreases, $v$ also decreases.   **77. a.** False   **b.** False   **c.** True   **d.** False
**79. a.** 1   **b.** 0   **c.** Undefined   **d.** 1   **e.** 13/12   **f.** 40/9
**g.** $\left(\dfrac{e^2 + 1}{2e}\right)^2$   **h.** Undefined   **i.** $\ln 4$   **j.** 1   **81.** $x = 0$

**83.** $x = \pm\tanh^{-1}(1/\sqrt{3}) = \pm\ln(2 + \sqrt{3})/2 \approx \pm 0.658$

**85.** $\tan^{-1}(\sinh 1) - \pi/4 \approx 0.08$   **87.** Applying l'Hôpital's Rule twice brings you back to the initial limit; $\lim\limits_{x\to\infty}\tanh x = 1$.

**89.** $2/\pi$   **91.** 1   **93.** $12(3\ln(3 + \sqrt{8}) - \sqrt{8}) \approx 29.5$
**95. a.** Approx. 360.8 m   **b.** First 100 m: $t \approx 4.72$ s, $v_{av} \approx 21.2$ m/s; second 100 m: $t \approx 2.25$ s, $v_{av} \approx 44.5$ m/s   **97. a.** $\sqrt{mg/k}$

**b.** $35\sqrt{3} \approx 60.6$ m/s   **c.** $t = \sqrt{\dfrac{m}{kg}}\tanh^{-1} 0.95 = \dfrac{\ln 39}{2}\sqrt{\dfrac{m}{kg}}$

**d.** Approx. 736.5 m   **109.** $\ln(21/4) \approx 1.66$

## Chapter 7 Review Exercises, pp. 515–519

**1. a.** True   **b.** False   **c.** True   **d.** False   **e.** True   **f.** False
**g.** False   **h.** False   **i.** True   **3.** $x = 2$; base does not matter
**5.** $(-\infty, 0]$, $[0, 2]$, and $[2, \infty)$
**7.** $f^{-1}(x) = 2 + \sqrt{x - 1}$

**9.** $f^{-1}(x) = \dfrac{5 - 3x}{2}$   **11.** $f^{-1}(x) = -\dfrac{1}{2}\log_{10} x$

**13.** $f^{-1}(x) = \ln\left(\dfrac{2x}{1 - x}\right)$   **15.** $-\sqrt{2}$   **17.** $\dfrac{6}{13}$

**19.** $f^{-1}(x) = -\dfrac{1}{3}x + 2$; $(f^{-1})'(x) = -\dfrac{1}{3}$

**21.** $f^{-1}(x) = \dfrac{x^2 + 1}{4}, x \geq 0$; $(f^{-1})'(x) = \dfrac{x}{2}, x \geq 0$

**23. a.** $(f^{-1})'(1/\sqrt{2}) = \sqrt{2}$   **25.** $(2 + \ln x)\ln x$

**27.** $(2x - 1)2^{x^2 - x}\ln 2$   **29.** $-\dfrac{1}{|x|\sqrt{x^2 - 1}}$   **31.** $\sinh^2 t + \cosh^2 t$

**33.** $\dfrac{e^x}{\sqrt{1 - e^{2x}}}$   **35.** 1   **37.** $\sqrt{3} + \pi/6$   **39.** $x(x\cosh x + 2\sinh x)$

**41.** $\pi/6$   **43.** $-\pi/6$   **45.** $x$, provided $-1 \leq x \leq 1$

**47.** $\cos\theta = \dfrac{5}{13}$; $\tan\theta = \dfrac{12}{5}$; $\cot\theta = \dfrac{5}{12}$; $\sec\theta = \dfrac{13}{5}$; $\csc\theta = \dfrac{13}{12}$

**49.** $\pi/2 - \theta$   **51.** 0   **53.** $\dfrac{2\sqrt{x(1 - x)}}{2x - 1}$   **55.** $\ln 4$

**57.** $\dfrac{1}{2}\ln(x^2 + 8x + 25) + C$   **59.** $\dfrac{3\sqrt{2}}{2}\tan^{-1}(\sqrt{2}x) + C$   **61.** $\pi/6$

**63.** $\pi/6$   **65.** $\cosh^{-1}(x/3) + C = \ln(x + \sqrt{x^2 - 9}) + C$

**67.** $\tanh^{-1}(1/3)/9 = (\ln 2)/18 \approx 0.0385$   **69.** $3\ln 6 + \dfrac{35}{24}$

**71.** $2\pi\ln 17$
**73. a.** $s_{\text{Tom}}(t) = -10e^{-2t} + 10$
$s_{\text{Sue}}(t) = -15e^{-t} + 15$

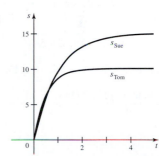

**b.** $t = 0$ and $t = \ln 2$   **c.** Sue
**75.** Approx. 7.3 hours   **77. a.** $y(t) = 29{,}000e^{(t\ln 2)/2}$
**b.** Approx. 41,996,486 transistors (which closely approximates the actual number of transistors)   **79.** 48.37 yr   **81.** Local max at

$x = -\dfrac{1}{2}(\sqrt{5} + 1)$; local min at $x = \dfrac{1}{2}(\sqrt{5} - 1)$;

inflection points at $x = -3$ and $x = 0$;
$\lim\limits_{x\to -8} f(x) = 0$; $\lim\limits_{x\to 8} f(x) = \infty$

**83.** $\sqrt{b^2 + 1} - \sqrt{2} + \ln\left(\dfrac{(\sqrt{b^2 + 1} - 1)(1 + \sqrt{2})}{b}\right)$;

$b \approx 2.715$   **85.** 1   **87.** 0   **89.** $1/e^3$   **91.** $x^{1/2}$   **93.** $\sqrt{x}$   **95.** $3^x$
**97.** Comparable growth rates   **99.** 1; 1
**101.** $\lim\limits_{x\to 0^+} f(x) = 1$; $\lim\limits_{x\to 0^+} g(x) = 0$

**105. a.**

**b.** 0   **d.** $f(x^*) = \dfrac{1}{\sqrt{2\pi}}\dfrac{e^{\sigma^2/2}}{\sigma}$   **e.** $\sigma = 1$

**107. a.** $\theta = \tan^{-1}\dfrac{10}{x} - \tan^{-1}\dfrac{3}{x}$   **b.** $\sqrt{30} \approx 5.5$ ft

**109. a.** $v(t) = -15e^{-t}(\sin t + \cos t)$;
$v(1) \approx -7.6\,\text{m/s}, v(3) \approx 0.63\,\text{m/s}$   **b.** Down $(0, 2.4)$
and $(5.5, 8.6)$; up $(2.4, 5.5)$ and $(8.6, 10)$   **c.** Approx. 0.65 m/s
**111. a.** $\cosh x$   **b.** $(1 - x\tanh x)\,\text{sech}\,x$
**113.** $L(x) = \frac{5}{3} + \frac{4}{3}(x - \ln 3)$; $\cosh 1 \approx 1.535$

# CHAPTER 8

## Section 8.1 Exercises, pp. 523–525

**1.** $u = 4 - 7x$   **3.** $\sin^2 x = \dfrac{1 - \cos 2x}{2}$   **5.** Complete the square in
$x^2 - 4x - 9$.   **7.** $\dfrac{1}{15(3 - 5x)^3} + C$   **9.** $\dfrac{\sqrt{2}}{4}$   **11.** $\dfrac{1}{2}\ln^2 2x + C$

**13.** $\ln(e^x + 1) + C$   **15.** $\dfrac{32}{3}$   **17.** $\dfrac{21}{110}$

**19.** $\dfrac{(\ln w - 1)^9}{9} + \dfrac{(\ln w - 1)^8}{8} + C$

**21.** $\dfrac{1}{2}\ln(x^2 + 4) + \tan^{-1}\dfrac{x}{2} + C$

**23.** $-\dfrac{1}{3}\ln|\csc(3e^x + 4) + \cot(3e^x + 4)| + C$   **25.** 1

**27.** $3\sqrt{1 - x^2} + 2\sin^{-1}x + C$   **29.** $\ln(\sqrt{2} + 1)$

**31.** $\dfrac{1}{3}\tan^{-1}\left(\dfrac{x - 1}{3}\right) + C$   **33.** $\dfrac{x^2}{2} + x + \ln(x^2 + x + 2) + C$

**35.** $\dfrac{3\pi + 10}{12}$   **37.** $\sin^{-1}\left(\dfrac{\theta + 3}{6}\right) + C$   **39.** $\tan\theta - \sec\theta + C$

**41.** $-x - \cot x - \csc x + C$   **43.** $\dfrac{1}{3}\ln(1 + \sinh 3x) + C$

**45.** $\frac{1}{2}\ln|e^{2x} - 2| + C$   **47.** $x - \ln|x + 1| + C$

**49.** $\dfrac{4}{5}(9 + \sqrt{t + 1})^{3/2}(\sqrt{t + 1} - 6) + C$   **51.** $\dfrac{\ln 4 - \pi}{4}$

**53.** $\ln|\sec(e^x + 1) + \tan(e^x + 1)| + C$

**55.** $\dfrac{2\sin^3 x}{3} + C$   **57.** $2\tan^{-1}\sqrt{x} + C$

**59.** $\dfrac{1}{2}\ln(x^2 + 6x + 13) - \dfrac{5}{2}\tan^{-1}\left(\dfrac{x + 3}{2}\right) + C$

**61.** $-\dfrac{1}{e^x + 1} + C$   **63.** $\frac{1}{2}$   **65. a.** False   **b.** False   **c.** False

**d.** False   **69. a.** $\dfrac{\tan^2 x}{2} + C$   **b.** $\dfrac{\sec^2 x}{2} + C$   **c.** The antiderivatives
differ by a constant.   **71. a.** $\frac{1}{2}(x + 1)^2 - 2(x + 1) + \ln|x + 1| + C$

**b.** $\dfrac{x^2}{2} - x + \ln|x + 1| + C$   **c.** The antiderivatives differ by a
constant.   **73.** $\dfrac{\ln 26}{3}$   **75.** $\dfrac{2}{3}(5\sqrt{5} - 1)\pi$

**77.** $\pi\left(\dfrac{9}{2} - \dfrac{5\sqrt{5}}{6}\right)$   **79.** $\dfrac{2048 + 1763\sqrt{41}}{9375}$

## Section 8.2 Exercises, pp. 529–532

**1.** Product Rule   **3.** $\dfrac{x^2(2\ln x - 1)}{4} + C$   **5.** Products for which the
choice for $dv$ is easily integrated and when the resulting new integral is
no more difficult than the original integral
**7.** $(\tan x + 2)\ln(\tan x + 2) - \tan x + C$

**9.** $\dfrac{1}{5}x\sin 5x + \dfrac{1}{25}\cos 5x + C$   **11.** $\dfrac{e^{6t}}{36}(6t - 1) + C$

**13.** $\dfrac{x^2}{4}(2\ln 10x - 1) + C$   **15.** $(w + 2)\sin 2w + \dfrac{1}{2}\cos 2w + C$

**17.** $\dfrac{3^x}{\ln 3}\left(x - \dfrac{1}{\ln 3}\right) + C$   **19.** $-\dfrac{1}{9x^9}\left(\ln x + \dfrac{1}{9}\right) + C$

**21.** $\dfrac{1}{8}\sin 2x - \dfrac{x}{4}\cos 2x + C$   **23.** $\dfrac{1}{4}(1 - 2x^2)\cos 2x + \dfrac{x}{2}\sin 2x + C$

**25.** $-e^{-t}(t^2 + 2t + 2) + C$   **27.** $\dfrac{e^x}{2}(\sin x + \cos x) + C$

**29.** $-\dfrac{e^{-x}}{17}(\sin 4x + 4\cos 4x) + C$

**31.** $-e^{2x}\cos e^x + 2e^x\sin e^x + 2\cos e^x + C$   **33.** $\pi$   **35.** $-\frac{1}{2}$

**37.** $\dfrac{1}{9}(5e^6 + 1)$   **39.** $\dfrac{\pi - 2}{2}$   **41. a.** $x\tan^{-1}x - \dfrac{1}{2}\ln(1 + x^2) + C$

**b.** $\dfrac{1}{2}x^2\tan^{-1}x^2 - \dfrac{1}{4}\ln(1 + x^4) + C$   **43.** $\pi(1 - \ln 2)$   **45.** $\pi$

**47.** $\dfrac{2\pi}{27}(13e^6 - 1)$   **49. a.** False   **b.** True   **c.** True
**51.** Let $u = x^n$ and $dv = \cos ax\,dx$.   **53.** Let $u = \ln^n x$ and $dv = dx$.

**55.** $\dfrac{x^2\sin 5x}{5} + \dfrac{2x\cos 5x}{25} - \dfrac{2\sin 5x}{125} + C$

**57.** $6 - 2e$   **59. a.** $\dfrac{2}{3}(x - 2)\sqrt{x + 1} + C$

**61.** $\displaystyle\int\log_b x\,dx = \int\dfrac{\ln x}{\ln b}\,dx = \dfrac{1}{\ln b}(x\ln x - x) + C$

**63.** $2\sqrt{x}\sin\sqrt{x} + 2\cos\sqrt{x} + C$
**65.** Let $u = x$ and $dv = f''(x)\,dx$.
**67.** $2e^3$   **69.** $x$-axis: $\pi^2/2$; $y$-axis: $2\pi^2$   **71.** $\pi(\pi - 2)$

**75. a.** $t = k\pi$, for $k = 0, 1, 2, \ldots$   **b.** $\dfrac{e^{-\pi} + 1}{2\pi}$

**c.** $(-1)^n\left(\dfrac{e^\pi + 1}{2\pi e^{(n+1)\pi}}\right)$

**d.** $a_n = a_{n-1}\cdot\dfrac{1}{e^\pi}$

**77. c.** $\int f(x)g(x)dx = f(x)G_1(x) - f'(x)G_2(x) + f''(x)G_3(x) - \int f'''(x)G_3(x)dx$

| ƒ and its derivatives | g and its integrals |
|---|---|
| $f(x)$  + | $g(x)$ |
| $f'(x)$  − | $G_1(x)$ |
| $f''(x)$  + | $G_2(x)$ |
| $f'''(x)$  − | $G_3(x)$ |

**d.** $\int x^2 e^{x/2} dx = 2x^2 e^{x/2} - 8xe^{x/2} + 16e^{x/2} + C$

| ƒ and its derivatives | g and its integrals |
|---|---|
| $x^2$  + | $e^{x/2}$ |
| $2x$  − | $2e^{x/2}$ |
| $2$  + | $4e^{x/2}$ |
| $0$  − | $8e^{x/2}$ |

$\dfrac{d^n}{dx^n}(x^2) = 0$, for $n \geq 3$, so all entries in the left column of the table beyond row three are 0, which results in no additional contribution to the antiderivative. **e.** $x^3 \sin x + 3x^2 \cos x - 6x \sin x - 6 \cos x + C$; five rows are needed because $\dfrac{d^n}{dx^n}(x^3) = 0$, for $n \geq 4$.

**f.** $\dfrac{d^k}{dx^k}(p_n(x)) = 0$, for $k \geq n + 1$

**79. a.** $\int e^x \cos x\, dx = e^x \sin x + e^x \cos x - \int e^x \cos x\, dx$

**b.** $\frac{1}{2}(e^x \sin x + e^x \cos x) + C$    **c.** $-\frac{3}{13}e^{-2x} \cos 3x - \frac{2}{13}e^{-2x} \sin 3x + C$

**81. a.** $I_1 = -\frac{1}{2}e^{-x^2} + C$    **b.** $I_3 = -\frac{1}{2}e^{-x^2}(x^2 + 1) + C$

**c.** $I_5 = -\frac{1}{2}e^{-x^2}(x^4 + 2x^2 + 2) + C$

## Section 8.3 Exercises, pp. 536–538

**1.** $\sin^2 x = \frac{1}{2}(1 - \cos 2x)$; $\cos^2 x = \frac{1}{2}(1 + \cos 2x)$    **3.** Rewrite $\sin^3 x$ as $(1 - \cos^2 x) \sin x$.    **5.** A reduction formula expresses an integral with a power in the integrand in terms of another integral with a smaller power in the integrand.    **7.** Let $u = \tan x$.

**9.** $\sin x - \dfrac{1}{3}\sin^3 x + C$    **11.** $\dfrac{x}{2} - \dfrac{1}{12}\sin 6x + C$

**13.** $-\cos x + \frac{2}{3}\cos^3 x - \frac{1}{5}\cos^5 x + C$    **15.** $\frac{1}{5}\cos^5 x - \frac{1}{3}\cos^3 x + C$

**17.** $\dfrac{2}{3}\sin^{3/2} x - \dfrac{2}{7}\sin^{7/2} x + C$    **19.** $\dfrac{7}{24}$    **21.** $\dfrac{8}{45}$

**23.** $\frac{1}{8}x - \frac{1}{32}\sin 4x + C$    **25.** $\frac{1}{48}\sin^3 2x + \frac{1}{16}x - \frac{1}{64}\sin 4x + C$

**27.** $\tan x - x + C$    **29.** $-\frac{1}{3}\cot^3 x + \cot x + x + C$

**31.** $4\tan^5 x - \frac{20}{3}\tan^3 x + 20 \tan x - 20x + C$    **33.** $\tan^{10} x + C$

**35.** $\frac{1}{3}\sec^3 x + C$    **37.** $\frac{1}{3}\tan^3(\ln \theta) + \tan(\ln \theta) + C$    **39.** $\ln 4$

**41.** $\frac{7}{6}$    **43.** $\frac{1}{8}\tan^2 4x + \frac{1}{4}\ln|\cos 4x| + C$    **45.** $\frac{2}{3}\tan^{3/2} x + C$

**47.** $\tan x - \cot x + C$    **49.** $\dfrac{1}{25}$    **51.** $-2 \cot x - \dfrac{\cot^3 x}{3} + C$

**53.** $\frac{4}{3}$    **55.** $\frac{4}{3} - \ln \sqrt{3}$    **57.** $8\sqrt{2}/3$    **59.** $\sqrt{2}$    **61.** $2\sqrt{2}/3$

**63. a.** True   **b.** False    **65.** $\frac{2\pi}{35}$    **67.** $\frac{1}{8}\cos 4x - \frac{1}{20}\cos 10x + C$

**69.** $\frac{1}{2}\sin x - \frac{1}{10}\sin 5x + C$    **73.** $\frac{1}{2} - \ln \sqrt{2}$    **75. a.** $\frac{\pi}{2}; \frac{\pi}{2}$

**b.** $\frac{\pi}{2}$, for all $n$   **d.** Yes   **e.** $\frac{3\pi}{8}$, for all $n$

## Section 8.4 Exercises, pp. 543–546

**1.** $x = 3 \sec \theta$    **3.** $x = 10 \sin \theta$    **5.** $\sqrt{4 - x^2}/x$    **7.** $\pi/6$

**9.** $\dfrac{25\pi}{3}$    **11.** $\dfrac{\pi}{12}$    **13.** $\sin^{-1}\dfrac{x}{4} + C$    **15.** $-\dfrac{\sqrt{x^2 + 9}}{9x} + C$

**17.** $2 - \dfrac{\pi}{2}$    **19.** $\ln\left(\sqrt{x^2 - 81} + x\right) + C$

**21.** $\dfrac{x}{2}\sqrt{64 - x^2} + 32 \sin^{-1}\dfrac{x}{8} + C$    **23.** $\dfrac{x}{25\sqrt{25 - x^2}} + C$

**25.** $-3 \ln\left|\dfrac{\sqrt{9 - x^2} + 3}{x}\right| + \sqrt{9 - x^2} + C$    **27.** $\sqrt{2}/6$

**29.** $\dfrac{1}{16}\left(\tan^{-1}\dfrac{x}{2} + \dfrac{2x}{x^2 + 4}\right) + C$

**31.** $8 \sin^{-1}(x/4) - x\sqrt{16 - x^2}/2 + C$

**33.** $\sqrt{x^2 - 9} - 3 \sec^{-1}(x/3) + C$

**35.** $-1/\sqrt{x^2 - 1} - \sec^{-1} x + C$    **37.** $2 - \sqrt{2}$

**39.** $x/\sqrt{100 - x^2} - \sin^{-1}(x/10) + C$    **41.** $x/\sqrt{1 + 4x^2} + C$

**43.** $\dfrac{\ln 3}{2}$    **45.** $81/(2(81 - x^2)) + \ln \sqrt{81 - x^2} + C$

**47.** $\dfrac{1}{16}(1 - \sqrt{3} - \ln(21 - 12\sqrt{3}))$    **49.** $\dfrac{1}{3} + \dfrac{\ln 3}{4}$

**51.** $\dfrac{x}{2}\sqrt{4 + x^2} - 2 \ln(x + \sqrt{4 + x^2}) + C$

**53.** $\dfrac{9}{10}\cos^{-1}\dfrac{5}{3x} - \dfrac{\sqrt{9x^2 - 25}}{2x^2} + C$

**55.** $\dfrac{\sec^{-1}\frac{x}{10}}{2000} + \dfrac{\sqrt{x^2 - 100}}{200\,x^2} + C$

**57. a.** False   **b.** True   **c.** False   **d.** False

**61.** $\sin^{-1}\left(\dfrac{x + 1}{2}\right) + C$    **63.** $\dfrac{1}{3}\tan^{-1}\left(\dfrac{x + 3}{3}\right) + C$

**65.** $\dfrac{\pi\sqrt{2}}{48}$    **67.** $\dfrac{x - 4}{\sqrt{9 + 8x - x^2}} - \sin^{-1}\left(\dfrac{x - 4}{5}\right) + C$

**69.** $\ln\left((2 + \sqrt{3})(\sqrt{2} - 1)\right)$

**71.** $\dfrac{1}{81} + \dfrac{\ln 3}{108}$      **73.** $3\sqrt{3} - \pi$

**75.** $\dfrac{3}{80}$    **77.** $\dfrac{1}{4a}\left(20a\sqrt{1 + 400a^2} + \ln\left(20a + \sqrt{1 + 400a^2}\right)\right)$

**81. b.** $\lim\limits_{L \to \infty} \dfrac{kQ}{a\sqrt{a^2 + L^2}} = \lim\limits_{L \to \infty} 2\rho k \dfrac{1}{a\sqrt{\left(\dfrac{a}{L}\right)^2 + 1}} = \dfrac{2\rho k}{a}$

**85. a.** $\dfrac{1}{\sqrt{g}}\left(\dfrac{\pi}{2} - \sin^{-1}\left(\dfrac{2 \cos b - \cos a + 1}{\cos a + 1}\right)\right)$

**b.** For $b = \pi$, the descent time is $\dfrac{\pi}{\sqrt{g}}$, a constant.

## Section 8.5 Exercises, pp. 554–556

1. Rational functions   3. a. $\dfrac{A}{x-3}$   b. $\dfrac{A}{x-4}, \dfrac{B}{(x-4)^2}, \dfrac{C}{(x-4)^3}$

c. $\dfrac{Ax+B}{x^2+2x+6}$   5. $\dfrac{A}{x-4}+\dfrac{B}{x-5}$   7. $\dfrac{A}{x-5}+\dfrac{B}{(x-5)^2}$

9. $\dfrac{A}{x}+\dfrac{B}{x+1}+\dfrac{C}{x-1}+\dfrac{D}{x+2}+\dfrac{E}{x-2}$   11. $\dfrac{A}{x}+\dfrac{Bx+C}{x^2+1}$

13. $\dfrac{A}{x-2}+\dfrac{B}{(x-2)^2}+\dfrac{Cx+D}{x^2+x+2}+\dfrac{Ex+F}{(x^2+x+2)^2}$

15. $\dfrac{A}{x-2}+\dfrac{B}{(x-2)^2}+\dfrac{C}{x+2}+\dfrac{D}{(x+2)^2}+\dfrac{Ex+F}{x^2+4}+\dfrac{Gx+H}{(x^2+4)^2}$

17. $\dfrac{2}{x-1}+\dfrac{3}{x-2}$   19. $\dfrac{1}{x-4}-\dfrac{1}{x+2}$   21. $2+\dfrac{3}{x+1}-\dfrac{4}{x+2}$

23. $\ln\left|\dfrac{x-1}{x+2}\right|+C$   25. $3\ln\left|\dfrac{x-1}{x+1}\right|+C$

27. $3\ln|x-1|-\dfrac{1}{3}\ln|3x-2|+C$   29. $-\ln 4$

31. $\ln\left|\dfrac{(x-2)^2(x+1)}{(x+2)^2(x-1)}\right|+C$   33. $3x+\ln\dfrac{(x-2)^{14}}{|x-1|}+C$

35. $\ln\left|\dfrac{x(x-2)^3}{(x+2)^3}\right|+C$   37. $\ln\left|\dfrac{(x-3)^{1/3}(x+1)}{(x+3)^{1/3}(x-1)}\right|^{1/16}+C$

39. $\dfrac{9}{x}+\ln\left|\dfrac{x-9}{x}\right|+C$   41. $\ln 2-\dfrac{3}{4}$   43. $-\dfrac{2}{x}+\ln\left|\dfrac{x+1}{x}\right|^2+C$

45. $\dfrac{5}{x}+\ln\left|\dfrac{x}{x+1}\right|^6+C$   47. $\dfrac{x^2}{2}+2\ln|x-5|-\dfrac{10}{x-5}+C$

49. $\dfrac{3}{x-1}+\ln\left|\dfrac{(x-1)^5}{x^4}\right|+C$   51. $\ln|x+1|+\tan^{-1}x+C$

53. $\ln(x+1)^2+\tan^{-1}(x+1)+C$

55. $\ln\left|\dfrac{(x-1)^2}{x^2+4x+5}\right|+14\tan^{-1}(x+2)+C$

57. $\dfrac{1}{2}\ln|x^2+3|-\dfrac{1}{x^2+3}+C$

59. $\dfrac{1}{2}\ln(x^2+6x+10)-3\tan^{-1}(x+3)-\dfrac{1}{x^2+6x+10}+C$

61. $\ln\left(\dfrac{x^2}{x^2+1}\right)+\dfrac{1}{x^2+1}+C$

63. $\sqrt{\dfrac{3}{7}}\tan^{-1}\left(\sqrt{\dfrac{3}{7}}x\right)-\dfrac{1}{6(3x^2+7)}+C$

65. a. False   b. False   c. False   d. True   67. $\ln 6$

69. $\left(\dfrac{24}{5}-2\ln 5\right)\pi$   71. $\dfrac{2}{3}\pi\ln 2$   73. $\ln\sqrt{\left|\dfrac{x-1}{x+1}\right|}+C$

75. $\dfrac{A}{x}+\dfrac{Bx+C}{x^2+1}+\dfrac{Dx+E}{(x^2+1)^2}+\dfrac{Fx+G}{x^2+x+4}+\dfrac{Hx+I}{(x^2+x+4)^2}$;

$\dfrac{1}{16x}-\dfrac{x+10}{100(x^2+1)}+\dfrac{4x+3}{50(x^2+1)^2}-\dfrac{21x-19}{400(x^2+x+4)}-$

$\dfrac{4x+1}{20(x^2+x+4)^2}$   77. $\ln\left|\dfrac{e^x-1}{e^x+2}\right|^{1/3}+C$

79. $\dfrac{1}{4}\ln\left(\dfrac{1+\sin t}{1-\sin t}-\dfrac{2}{1+\sin t}\right)+C$

81. $\tan^{-1}e^x-\dfrac{1}{2(e^{2x}+1)}+C$   83. $x-\ln(1+e^x)+C$

89. $-\cot x-\csc x+C=-\cot(x/2)+C$

91. $\dfrac{1}{\sqrt{2}}\ln\dfrac{\sqrt{2}+1}{\sqrt{2}-1}$   93. a. Car $A$   b. Car $C$

c. $S_A(t)=88t-88\ln|t+1|$;

$S_B(t)=88\left(t-\ln(t+1)^2-\dfrac{1}{t+1}+1\right)$;

$S_C(t)=88(t-\tan^{-1}t)$

d. Car $C$   95. Because $\dfrac{x^4(1-x)^4}{1+x^2}>0$ on $(0,1)$,

$\displaystyle\int_0^1 \dfrac{x^4(1-x^4)}{1+x^2}\,dx>0$; therefore, $\dfrac{22}{7}>\pi$.

## Section 8.6 Exercises, pp. 560–562

1. Integrate by parts.   3. Let $x=8\sin\theta$.   5. Use the method of partial fractions.   7. $\dfrac{\pi}{4}$   9. $\dfrac{\pi}{6}$   11. $\dfrac{5}{4}-\dfrac{3\pi}{8}$   13. $-\dfrac{\sqrt{1-e^{2x}}}{e^x}+C$

15. $\dfrac{4}{\ln 2}$   17. $\dfrac{3e^4}{2}$   19. $\dfrac{16}{35}$   21. $\dfrac{x^{10}}{10}\ln 3x-\dfrac{x^{10}}{100}+C$

23. $\ln|\cos x+1|-\ln|\cos x|+C$

25. $\ln\left|\dfrac{x}{1+\sqrt{1-x^2}}\right|+C$   27. $\dfrac{3x}{8}-\dfrac{1}{2}\sin x+\dfrac{1}{16}\sin 2x+C$

29. $\ln|\sin x+\sin^2 x|+C$   31. $6\sin^{-1}\dfrac{x}{2}+\dfrac{3}{2}x\sqrt{4-x^2}+C$

33. $\dfrac{1}{a}\tan^{-1}\dfrac{e^x}{a}+C$   35. $\dfrac{11}{6}$   37. $\dfrac{\sqrt{3}+1}{2}$

39. $-\cos x\ln(\sin x)-\ln|\csc x+\cot x|+\cos x+C$

41. $-\dfrac{2}{5}\cot^{5/2}x-\dfrac{2}{9}\cot^{9/2}x+C$   43. $\dfrac{\sin^{-1}x^{10}}{10}+C$   45. $\ln\dfrac{4}{3}-\dfrac{1}{6}$

47. $x^2+3x+4\ln|x-2|+\ln|x+1|+C$

49. $\dfrac{\sec^{11}x}{11}-\dfrac{\sec^9 x}{9}+C$   51. $\dfrac{4}{7}(2^{7/4}-1)$

53. $-\dfrac{\cot^2 e^x}{2}-\ln|\sin e^x|+C$   55. $\ln|x^3+x|+3\tan^{-1}x+C$

57. $-2\sqrt{x}\cos\sqrt{x}+2\sin\sqrt{x}+C$   59. $-\dfrac{1}{x}-\tan^{-1}x+C$

61. $\dfrac{\sqrt{2}e^{\pi/4}-1}{2}$   63. $\dfrac{x^{a+1}}{a+1}\left(\ln x-\dfrac{1}{a+1}\right)+C$

65. $\dfrac{\pi}{18}$   67. $\dfrac{1}{54}\left(\sin^{-1}3x-3x\sqrt{1-9x^2}\right)+C$

69. $\dfrac{1-\sqrt{1-x^2}}{x}+C$   71. $-2\cot x+2\csc x-x+C$

73. $\dfrac{40\sqrt{5}}{3}-\dfrac{224}{15}$   75. $\dfrac{7\pi^2}{144}$   77. $x\cos^{-1}x-\sqrt{1-x^2}+C$

79. $-\dfrac{\sin^{-1}x}{x}+\ln\left|\dfrac{x}{1+\sqrt{1-x^2}}\right|+C$

81. $\ln|x|+2\tan^{-1}x-\dfrac{3}{2(x^2+1)}+C$

83. $\dfrac{\sin^{999}e^x}{999}-\dfrac{\sin^{1001}e^x}{1001}+C$   85. a. True   b. True   c. False

d. False   87. $\pi(\sqrt{2}+\ln(1+\sqrt{2}))\approx 7.212$

89. $\dfrac{\pi(4\sqrt{2}+3)}{3}\approx 9.065$   91. $9800\pi\ln 2\approx 21{,}340.3$ J

93. $4x-2\ln(e^{2x}+2e^x+17)-\tan^{-1}\left(\dfrac{e^x+1}{4}\right)+C$

**95.** $\frac{1}{4}\ln|\tan x + 1| - \frac{1}{4}\ln|\tan x - 1| + \frac{x}{2} + C$

**97.** $x\tan^{-1}\sqrt[3]{x} - \frac{x^{2/3}}{2} + \frac{1}{2}\ln(1 + x^{2/3}) + C$

**99.** $\pi\left(\sqrt{5} - \sqrt{2} + \frac{1}{2}\ln\left(\frac{\sqrt{5} - 1}{\sqrt{5} + 1}\right) - \frac{1}{2}\ln\left(\frac{\sqrt{2} - 1}{\sqrt{2} + 1}\right)\right) \approx 3.839$

## Section 8.7 Exercises, pp. 565–567

**1.** Substitutions, integration by parts, partial fractions **3.** The CAS may not include the constant of integration, and it may use a trigonometric identity or other algebraic simplification.

**5.** $-\frac{1}{3}\sin^3 e^x + \sin e^x + C$ **7.** $x\cos^{-1}x - \sqrt{1 - x^2} + C$

**9.** $\ln(x + \sqrt{16 + x^2}) + C$ **11.** $\frac{3}{4}(2u - 7\ln|7 + 2u|) + C$

**13.** $-\frac{1}{4}\cot 2x + C$ **15.** $\frac{1}{12}(2x - 1)\sqrt{4x + 1} + C$

**17.** $\frac{1}{3}\ln\left|x + \sqrt{x^2 - \left(\frac{10}{3}\right)^2}\right| + C$ **19.** $\ln(e^x + \sqrt{4 + e^{2x}}) + C$

**21.** $-\frac{1}{2}\ln\left|\frac{2 + \sin x}{\sin x}\right| + C$

**23.** $\frac{2\ln^2 x - 1}{4}\sin^{-1}(\ln x) + \frac{\ln x\sqrt{1 - \ln^2 x}}{4} + C$

**25.** $\frac{x}{16\sqrt{16 + 9x^2}} + C$ **27.** $-\frac{1}{12}\ln\left|\frac{12 + \sqrt{144 - x^2}}{x}\right| + C$

**29.** $2x + x\ln^2 x - 2x\ln x + C$

**31.** $\frac{x + 5}{2}\sqrt{x^2 + 10x} - \frac{25}{2}\ln\left|x + 5 + \sqrt{x^2 + 10x}\right| + C$

**33.** $\frac{1}{3}\tan^{-1}\left(\frac{x + 1}{3}\right) + C$ **35.** $\ln x - \frac{1}{10}\ln(x^{10} + 1) + C$

**37.** $2\ln(\sqrt{x - 6} + \sqrt{x}) + C$

**39.** $-\frac{\tan^{-1}x^3}{3x^3} + \ln\left|\frac{x}{(x^6 + 1)^{1/6}}\right| + C$ **41.** $4\sqrt{17} + \ln(4 + \sqrt{17})$

**43.** $\sqrt{5} - \sqrt{2} + \ln\left(\frac{2 + 2\sqrt{2}}{1 + \sqrt{5}}\right)$ **45.** $\frac{128\pi}{3}$ **47.** $\frac{\pi^2}{4}$

**49.** $\frac{(x - 3)\sqrt{3 + 2x}}{3} + C$ **51.** $\frac{1}{3}\tan 3x - x + C$

**53.** $\frac{1540 + 243\ln 3}{8}$

**55.** $\frac{(x^2 - a^2)^{3/2}}{3} - a^2\sqrt{x^2 - a^2} + a^3\cos^{-1}\frac{a}{x} + C$ **57.** $\frac{\pi}{4}$

**59.** $-\frac{x}{8}(2x^2 - 5a^2)\sqrt{a^2 - x^2} + \frac{3a^4}{8}\sin^{-1}\frac{x}{a} + C$

**61.** $2 - \frac{\pi^2}{12} - \ln 4$ **63.** $\frac{27{,}456\sqrt{15}}{7} \approx 15{,}190.9$

**65.** $\frac{1}{8}e^{2x}(4x^3 - 6x^2 + 6x - 3) + C$ **67.** $\frac{\tan^3 3y}{9} - \frac{\tan 3y}{3} + y + C$

**69.** $\frac{1}{24}(128 - 78\sqrt{2} - 3\ln(3 + 2\sqrt{2}))$

**71.** $\frac{1}{a^2}(ax - b\ln|b + ax|) + C$

**73.** $\frac{1}{a^2}\left(\frac{(ax + b)^{n+2}}{n + 2} - \frac{b(ax + b)^{n+1}}{n + 1}\right) + C$

**75. a.** True **b.** True

**79.** $\frac{1}{16}\left((8x^2 - 1)\sin^{-1}2x + 2x\sqrt{1 - 4x^2}\right) + C$

**81.** $-\frac{\tan^{-1}x}{x} + \ln\left(\frac{|x|}{\sqrt{x^2 + 1}}\right) + C$ **83. b.** $\frac{\pi}{8}\ln 2$

**85. a.**

| $\theta_0$ | $T$ |
|---|---|
| 0.10 | 6.27927 |
| 0.20 | 6.26762 |
| 0.30 | 6.24854 |
| 0.40 | 6.22253 |
| 0.50 | 6.19021 |
| 0.60 | 6.15236 |
| 0.70 | 6.10979 |
| 0.80 | 6.06338 |
| 0.90 | 6.01399 |
| 1.00 | 5.96247 |

**b.** All are within 10%.

**87. b.** $\frac{63\pi}{512}$ **c.** Decrease

## Section 8.8 Exercises, pp. 578–582

**1.** $\frac{1}{2}$ **3.** The Trapezoid Rule approximates areas under curves using trapezoids. **5.** 42 **7.** $\frac{112}{3}$ **9.** $-1, 1, 3, 5, 7, 9$

**11.** $1.59 \times 10^{-3}$; $5.04 \times 10^{-4}$ **13.** $1.72 \times 10^{-3}$; $6.32 \times 10^{-4}$

**15.** 576; 640; 656 **17.** 0.643950551 **19.** 704; 672; 664

**21.** 0.622 **23.** 2.28476811; 2.33512377 **25.** 1.76798499

**27.** $M(25) \approx 0.63703884$, $T(25) \approx 0.63578179$; $6.58 \times 10^{-4}$, $1.32 \times 10^{-3}$

**29.**

| $n$ | $M(n)$ | $T(n)$ | Error in $M(n)$ | Error in $T(n)$ |
|---|---|---|---|---|
| 4 | 99 | 102 | 1.00 | 2.00 |
| 8 | 99.75 | 100.5 | 0.250 | 0.500 |
| 16 | 99.9375 | 100.125 | $6.3 \times 10^{-2}$ | 0.125 |
| 32 | 99.984375 | 100.03125 | $1.6 \times 10^{-2}$ | $3.1 \times 10^{-2}$ |

**31.**

| $n$ | $M(n)$ | $T(n)$ | Error in $M(n)$ | Error in $T(n)$ |
|---|---|---|---|---|
| 4 | 1.50968181 | 1.48067370 | $9.7 \times 10^{-3}$ | $1.9 \times 10^{-2}$ |
| 8 | 1.50241228 | 1.49517776 | $2.4 \times 10^{-3}$ | $4.8 \times 10^{-3}$ |
| 16 | 1.50060256 | 1.49879502 | $6.0 \times 10^{-4}$ | $1.2 \times 10^{-3}$ |
| 32 | 1.50015061 | 1.49969879 | $1.5 \times 10^{-4}$ | $3.0 \times 10^{-4}$ |

**33.**

| $n$ | $M(n)$ | $T(n)$ | Error in $M(n)$ | Error in $T(n)$ |
|---|---|---|---|---|
| 4 | $-1.96 \times 10^{-16}$ | 0 | $2.0 \times 10^{-16}$ | 0 |
| 8 | $7.63 \times 10^{-17}$ | $-1.41 \times 10^{-16}$ | $7.6 \times 10^{-17}$ | $1.4 \times 10^{-16}$ |
| 16 | $1.61 \times 10^{-16}$ | $1.09 \times 10^{-17}$ | $1.6 \times 10^{-16}$ | $1.1 \times 10^{-17}$ |
| 32 | $6.27 \times 10^{-17}$ | $-4.77 \times 10^{-17}$ | $6.3 \times 10^{-17}$ | $4.8 \times 10^{-17}$ |

**35.** $T(4) \approx 690.3$ million ft³; $S(4) \approx 692.2$ million ft³ (answers may vary) **37.** 54.5°F, Trapezoid Rule **39.** 35.0°F, Trapezoid Rule **41. a.** Left sum: 204.917; right sum: 261.375; Trapezoid Rule: 233.146; the approximations measure the average temperature of the curling iron on $[0, 120]$. **b.** Left sum: underestimate; right sum: overestimate; Trapezoid Rule: underestimate **c.** 305°F is the change in temperature over $[0, 120]$. **43. a.** 5907.5 **b.** 5965 **c.** 5917

**45. a.** $T(25) \approx 3.19623162$
$T(50) \approx 3.19495398$
**b.** $S(50) \approx 3.19452809$
**c.** $e_T(50) \approx 4.3 \times 10^{-4}$
$e_S(50) \approx 4.5 \times 10^{-8}$

**47. a.** $T(50) \approx 1.00008509$
$T(100) \approx 1.00002127$
**b.** $S(100) \approx 1.00000000$
**c.** $e_T(100) \approx 2.1 \times 10^{-5}$
$e_S(100) \approx 4.6 \times 10^{-9}$

**49.**

| $n$ | $T(n)$ | $S(n)$ | Error in $T(n)$ | Error in $S(n)$ |
|---|---|---|---|---|
| 4 | 1820.0000 | — | 284 | — |
| 8 | 1607.7500 | 1537.0000 | 71.8 | 1 |
| 16 | 1553.9844 | 1536.0625 | 18.0 | $6.3 \times 10^{-2}$ |
| 32 | 1540.4990 | 1536.0039 | 4.50 | $3.9 \times 10^{-3}$ |

**51.**

| $n$ | $T(n)$ | $S(n)$ | Error in $T(n)$ | Error in $S(n)$ |
|---|---|---|---|---|
| 4 | 0.46911538 | — | $5.3 \times 10^{-2}$ | — |
| 8 | 0.50826998 | 0.52132152 | $1.3 \times 10^{-2}$ | $2.9 \times 10^{-4}$ |
| 16 | 0.51825968 | 0.52158957 | $3.4 \times 10^{-3}$ | $1.7 \times 10^{-5}$ |
| 32 | 0.52076933 | 0.52160588 | $8.4 \times 10^{-4}$ | $1.1 \times 10^{-6}$ |

**53. a.** True   **b.** False   **c.** True

**55.**

| $n$ | $M(n)$ | $T(n)$ | Error in $M(n)$ | Error in $T(n)$ |
|---|---|---|---|---|
| 4 | 0.40635058 | 0.40634782 | $1.4 \times 10^{-6}$ | $1.4 \times 10^{-6}$ |
| 8 | 0.40634920 | 0.40634920 | $7.6 \times 10^{-10}$ | $7.6 \times 10^{-10}$ |
| 16 | 0.40634920 | 0.40634920 | $6.6 \times 10^{-13}$ | $6.6 \times 10^{-13}$ |
| 32 | 0.40634920 | 0.40634920 | $8.9 \times 10^{-16}$ | $7.8 \times 10^{-16}$ |

**57.**

| $n$ | $M(n)$ | $T(n)$ | Error in $M(n)$ | Error in $T(n)$ |
|---|---|---|---|---|
| 4 | 4.72531819 | 4.72507878 | $1.2 \times 10^{-4}$ | $1.2 \times 10^{-4}$ |
| 8 | 4.72519850 | 4.72519849 | $9.1 \times 10^{-9}$ | $9.1 \times 10^{-9}$ |
| 16 | 4.72519850 | 4.72519850 | 0 | $8.9 \times 10^{-16}$ |
| 32 | 4.72519850 | 4.72519850 | 0 | $8.9 \times 10^{-16}$ |

**63.** Approximations will vary; exact value is 68.26894921 . . . .
**65. a.** Approx. $1.6 \times 10^{11}$ barrels   **b.** Approx. $6.8 \times 10^{10}$ barrels
**67. a.** $M(50) \approx 34.4345566$

**b.** $f''(x) = \dfrac{3(x^4 + 4x)}{4(x^3 + 1)^{3/2}}$   **d.** $E_M \leq 0.0028$

**69. a.** $T(40) = 0.874799972 \ldots$   **b.** $f''(x) = e^x \cos e^x - e^{2x} \sin e^x$

**d.** $E_T \leq \dfrac{1}{3200}$   **71. a.** $S(20) \approx 0.97774576$

**b.** $E_S \leq 3.5 \times 10^{-8}$   **73.** Approximations will vary; exact value is
38.753792 . . . .   **77.** Overestimate   **79.** $S(20) \approx 1.00000175$

## Section 8.9 Exercises, pp. 590–593

**1.** The interval of integration is infinite or the integrand is unbounded
on the interval of integration.   **3.** $\displaystyle\lim_{b\to\infty}\int_2^b \dfrac{dx}{x^{1/5}}$   **5.** $\displaystyle\int_{-\infty}^{\infty} f(x)\, dx$

**7.** $\dfrac{1}{3}$   **9.** Diverges   **11.** $\dfrac{1}{a}$   **13.** Diverges   **15.** $\dfrac{\pi}{10}$

**17.** Diverges   **19.** Diverges   **21.** $\dfrac{1}{\pi}$   **23.** $\dfrac{\pi}{4}$   **25.** $\dfrac{\pi}{6}$   **27.** 0

**29.** $\dfrac{\pi^3}{12}$   **31.** $\ln 2$   **33.** Diverges   **35.** Diverges   **37.** 6

**39.** Diverges   **41.** Diverges   **43.** $2(e - 1)$   **45.** Diverges
**47.** $4 \cdot 10^{3/4}/3$   **49.** Diverges   **51.** $\pi$   **53.** $-1$
**55.** $\ln(2 + \sqrt{3})$   **57.** 2   **59.** \$41,666.67   **61.** 0.76   **63.** 20,000 hr

**65.** $\dfrac{\pi}{3}$   **67.** $3\pi/2$   **69.** $\pi/\ln 2$   **71.** $2\pi$   **73.** Does not exist

**75.** $\dfrac{72 \cdot 2^{1/3}\,\pi}{5}$   **77.** Converges   **79.** Diverges   **81.** Converges

**83.** Diverges   **85.** Converges   **87. a.** True   **b.** False   **c.** False

**d.** True   **e.** True   **89.** $1/b - 1/a$   **91. a.** $A(a,b) = \dfrac{e^{-ab}}{a}$,   for $a > 0$

**b.** $b = g(a) = -\dfrac{1}{a}\ln 2a$   **c.** $b* = -2/e$   **93.** $\pi$   **107. a.** $\pi$

**b.** $\pi/(4e^2)$   **109. a.** $6.28 \times 10^7\, m\,$J   **b.** 11.2 km/s   **c.** $\leq 9$ mm

## Chapter 8 Review Exercises, pp. 593–596

**1. a.** True   **b.** False   **c.** False   **d.** True   **e.** False

**3.** $2(x - 8)\sqrt{x + 4} + C$   **5.** $\dfrac{1}{3}\sqrt{x + 2}\,(x - 4) + C$   **7.** $\dfrac{\pi}{4}$

**9.** $\dfrac{4}{105}$   **11.** $\sqrt{t - 1} - \tan^{-1}\sqrt{t - 1} + C$   **13.** $\dfrac{2}{15}(1 - e^{3\pi})$

**15.** $7 + \ln 40 - \ln 17$   **17.** $2\ln|x| + 3\tan^{-1}(x + 1) + C$

**19.** $\dfrac{2}{x + 3} - \dfrac{2}{(x + 3)^2} + \ln|x + 3| + C$   **21.** $\sqrt{3} - 1 - \dfrac{\pi}{12}$

**23.** $\dfrac{1}{5}\tan^5 t + C$   **25.** $\dfrac{\pi}{8}$   **27.** $\dfrac{\sqrt{w^2 + 2w - 8}}{9(w + 1)} + C$   **29.** $-\dfrac{\cot^5 x}{5} + C$

**31.** $\dfrac{x\cosh 2x}{2} - \dfrac{\sinh 2x}{4} + C$   **33.** $\dfrac{1}{15}\sec^5 3\theta - \dfrac{1}{9}\sec^3 3\theta + C$

**35.** $\dfrac{1}{6}(x^2 - 8)\sqrt{x^2 + 4} + C$   **37.** $\dfrac{1}{x + 1} + \ln|(x + 1)(x^2 + 4)| + C$

**39.** $\dfrac{t - \ln(2 + e^t)}{2} + C$   **41.** $\dfrac{1}{4}(\csc 4\theta - \cot 4\theta) + C$

**43.** $\dfrac{e^x}{2}(\sin x - \cos x) + C$

**45.** $\ln|x| - \dfrac{1}{x} + \dfrac{1}{2}\ln(x^2 + 4x + 9) - \dfrac{2}{\sqrt{5}}\tan^{-1}\left(\dfrac{x + 2}{\sqrt{5}}\right) + C$

**47.** $\dfrac{\theta}{2} + \dfrac{1}{16}\sin 8\theta + C$   **49.** $\dfrac{\sec^{49} 2z}{98} + C$   **51.** $\dfrac{4}{15}$

**53.** $2\sqrt{x} - 3\sqrt[3]{x} + 6\sqrt[6]{x} - 6\ln\left(\sqrt[6]{x} + 1\right) + C$

**55.** $-\dfrac{\sqrt{9 - y^2}}{9\sqrt{2}\,y} + C$   **57.** $\dfrac{\pi}{9}$   **59.** $-\text{sech}\, x + C$   **61.** $\dfrac{\pi}{3}$

**63.** $\dfrac{1}{8}\ln\left|\dfrac{x - 5}{x + 3}\right| + C$   **65.** $\dfrac{\ln 2}{4} + \dfrac{\pi}{8}$   **67.** 3   **69.** $\dfrac{1}{3}\ln\left|\dfrac{x - 2}{x + 1}\right| + C$

**71.** $2(x - 2\ln|x + 2|) + C$   **73.** $e^{2t}/\left(2\sqrt{1 + e^{4t}}\right) + C$

**75. a.** $\sec e^x + C$   **b.** $e^x \sec e^x - \ln|\sec e^x + \tan e^x| + C$

**77.** $\dfrac{\sqrt{6}}{3}\tan^{-1}\sqrt{\dfrac{2x - 3}{3}} + C$

**79.** $\dfrac{1}{4}\sec^3 x\tan x + \dfrac{3}{8}\sec x\tan x + \dfrac{3}{8}\ln|\sec x + \tan x| + C$

**81.** $2(\ln^3 2 - 3\ln^2 2 + \ln 64 - 3)$   **83.** 1   **85.** $\dfrac{\pi}{2}$

**87.** $\dfrac{2\pi}{\sqrt{3}}$   **89.** Converges   **91.** Diverges   **93.** 1.196288

**95.** $M(4) = 44;\ T(4) = 42;\ S(4) = \dfrac{124}{3}$

**97.** $M(40) \approx 0.398236;\ T(40) \approx 0.398771;\ S(40) \approx 0.398416$

**99.** 0.886227   **101.** $y$-axis   **103.** $\pi(e - 2)$   **105.** $\dfrac{\pi}{2}(e^2 - 3)$

**107. a.** 1.603   **b.** 1.870   **c.** $b \ln b - b = a \ln a - a$
**d.** Decreasing   **109.** $20/(3\pi)$   **111.** 1901 cars

**113. a.** $I(p) = \dfrac{1}{(p - 1)^2}(1 - pe^{1-p})$ if $p \neq 1, I(1) = \dfrac{1}{2}$   **b.** $0, \infty$

**c.** $I(0) = 1$   **115.** 0.4054651   **117.** $n = 2$
**119. a.** $V_1(a) = \pi(a \ln^2 a - 2a \ln a + 2(a - 1))$

**b.** $V_2(a) = \dfrac{\pi}{2}(2a^2 \ln a - a^2 + 1)$

**c.** $V_2(a) > V_1(a)$ for all $a > 1$

**121.** $a = \ln 2/(2b)$   **123.** $\ln(1 + \sqrt{2}/2)$

# CHAPTER 9

## Section 9.1 Exercises, pp. 604–606

**1. a.** 1   **b.** Linear   **3.** Yes   **5.** $\dfrac{\pi}{2} < t < \dfrac{3\pi}{2}$

**21.** $y = 3t - \dfrac{e^{-2t}}{2} + C$   **23.** $y = 2 \ln|\sec 2x| - 3 \sin x + C$

**25.** $y = 2t^6 + 6t^{-1} - 2t^2 + C_1 t + C_2$

**27.** $u = \dfrac{x^{11}}{2} + \dfrac{x^9}{2} - \dfrac{x^7}{2} + \dfrac{5}{x} + C_1 x + C_2$

**29.** $u = \ln(x^2 + 4) - \tan^{-1}\dfrac{x}{2} + C$   **31.** $y = \sin^{-1} x + C_1 x + C_2$

**33.** $y = e^t + t + 3$   **35.** $y = x^3 + x^{-3} - 2, x > 0$
**37.** $y = -t^5 + 2t^3 + 1$   **39.** $y = e^t(t - 2) + 2(t + 1)$

**41.** $u = \dfrac{1}{4} \tan^{-1}\dfrac{x}{4} - 4x + 2$   **43. a.** $v(t) = -9.8t + 29.4$;

$s(t) = -4.9t^2 + 29.4t + 30$; the object is above the ground for
approximately $0 \leq t \leq 6.89$.   **b.** The highest point of 74.1 m is
reached at $t = 3$ s.   **45.** The amount of resource is increasing for
$H < 75$ and is constant if $H = 75$. If $H = 100$, the resource
vanishes at approximately 28 time units.
**47.** $h = (\sqrt{1.96} - 0.1t\sqrt{2g})^2 \approx (1.4 - 0.44t)^2, 0 \leq t \leq 3.16$;
the tank is empty after approximately 3.16 s.

**49. a.** False   **b.** False   **c.** True   **51. c.** $y = C_1 \sin kt + C_2 \cos kt$

**53. b.** $C = \dfrac{K - 50}{50}$

**c.**

**d.** 300

**55. c.** The decay rate is greater
for the $n = 1$ model.

## Section 9.2 Exercises, pp. 611–614

**1.** At selected points $(t_0, y_0)$ in the region of interest draw a short line
segment with slope $f(t_0, y_0)$.   **3.** $y(3.1) \approx 1.6$
**5. a.** D   **b.** B   **c.** A   **d.** C
**7.**

**9.** An initial condition of
$y(0) = -1$ leads to a constant
solution. For any other initial condition, the solutions are increasing
over time.

**11.** An initial condition of
$y(0) = 1$ leads to a constant
solution. Initial conditions
$y(0) = A$ lead to solutions that
are increasing over time if $A > 1$.

**13.**

**15.**

**17. a.** $y = 1, y = -1$
**b.** Solutions are increasing for $|y| > 1$ and decreasing for $|y| < 1$. **c.** Initial conditions $y(0) = A$ lead to increasing solutions if $|A| > 1$ and decreasing solutions if $|A| < 1$.
**d.**

**19. a.** $y = \pi/2, y = -\pi/2$
**b.** Solutions are increasing for $|y| < \pi/2$ and decreasing for $|y| > \pi/2$. **c.** Initial conditions $y(0) = A$ lead to increasing solutions if $|A| < \pi/2$ and decreasing solutions if $\pi/2 < |A| < \pi$.
**d.**

**c.** Time step $\Delta t = 0.025$; smaller time steps generally produce more accurate results. **d.** Halving the time steps results in approximately halving the error. **33. a.** $y(2) \approx 0.00604662$ **b.** 0.012269 **c.** $y(2) \approx 0.0115292$ **d.** Error in part (c) is approximately half of the error in part (b). **35. a.** $y(4) \approx 3.05765$ **b.** 0.0339321 **c.** $y(4) \approx 3.0739$ **d.** Error in part (c) is approximately half of the error in part (b). **37. a.** True **b.** False
**39. a.** $y = 3$ **b, c.**

**21.** The equilibrium solutions are $P = 0$ and $P = 500$.

**23.** The equilibrium solutions are $P = 0$ and $P = 3200$.

**41. a.** $y = 0$ and $y = 3$
**b, c.**

**43. a.** $y = -2, y = 0$, and $y = 3$
**b, c.**

**25.** $y(0.5) \approx u_1 = 4; y(1) \approx u_2 = 8$
**27.** $y(0.1) \approx u_1 = 1.1; y(0.2) \approx u_2 = 1.19$

**29. a.**

| $\Delta t$ | approximation to $y(0.2)$ | approximation to $y(0.4)$ |
|---|---|---|
| 0.20000 | 0.80000 | 0.64000 |
| 0.10000 | 0.81000 | 0.65610 |
| 0.05000 | 0.81451 | 0.66342 |
| 0.02500 | 0.81665 | 0.66692 |

**b.**

| $\Delta t$ | errors for $y(0.2)$ | errors for $y(0.4)$ |
|---|---|---|
| 0.20000 | 0.01873 | 0.03032 |
| 0.10000 | 0.00873 | 0.01422 |
| 0.05000 | 0.00422 | 0.00690 |
| 0.02500 | 0.00208 | 0.00340 |

**c.** Time step $\Delta t = 0.025$; smaller time steps generally produce more accurate results. **d.** Halving the time steps results in approximately halving the error.

**31. a.**

| $\Delta t$ | approximation to $y(0.2)$ | approximation to $y(0.4)$ |
|---|---|---|
| 0.20000 | 3.20000 | 3.36000 |
| 0.10000 | 3.19000 | 3.34390 |
| 0.05000 | 3.18549 | 3.33658 |
| 0.02500 | 3.18335 | 3.33308 |

**b.**

| $\Delta t$ | errors for $y(0.2)$ | errors for $y(0.4)$ |
|---|---|---|
| 0.20000 | 0.01873 | 0.03032 |
| 0.10000 | 0.00873 | 0.01422 |
| 0.05000 | 0.00422 | 0.00690 |
| 0.02500 | 0.00208 | 0.00340 |

**45. a.** $\Delta t = \dfrac{b - a}{N}$ **b.** $u_1 = A + f(a, A)\dfrac{b - a}{N}$
**c.** $u_{k+1} = u_k + f(t_k, u_k)\dfrac{b - a}{N}$, where $u_0 = A$ and $t_k = a + k(b - a)/N$, for $k = 0, 1, 2, \ldots, N - 1$

**47. a.**

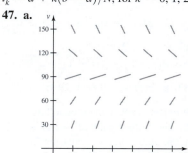

**b.** Increasing for $A < 98$ and decreasing for $A > 98$
**c.** $v(t) = 98$

## Section 9.3 Exercises, pp. 618–620

**1.** A first-order separable differential equation has the form $g(y)\,y'(t) = h(t)$, where the factor $g(y)$ is a function of $y$ and $h(t)$ is a function of $t$. **3.** No **5.** $y = \dfrac{t^4}{4} + C$

**7.** $y = \pm\sqrt{2t^3 + C}$ **9.** $y = -2\ln\left(\dfrac{1}{2}\cos t + C\right)$

**11.** $y = \dfrac{x}{1 + Cx}$ **13.** $y = \pm\dfrac{1}{\sqrt{C - \cos t}}$ **15.** $u = \ln\left(\dfrac{e^{2x}}{2} + C\right)$

**17.** $y = \sqrt{t^3 + 81}$ **19.** Not separable **21.** $y(t) = -e^{e^t - 1}$

**23.** $y = \ln(e^x + 2)$ **25.** $y = \ln\left(\dfrac{\ln^4 t}{4} + 1\right)$

**27.** $y = \sqrt{\tan t}, 0 < t < \pi/2$ **29.** $y = \sqrt{t^2 + 3}$ **31.** $y = \ln t + 2$

**33.** $y^3 - 3y = 2t^3$,
$-1 < t < 1$

**35.** $\cos u = 2 - 2\sin\dfrac{x}{2}, \dfrac{\pi}{3} < x < \dfrac{5\pi}{3}$

**37.**

$(y + 4)^{3/2} = (x + 1)^{3/2} + 19$

**39. a.**

$P = \dfrac{200}{1 + 3e^{-0.08t}}$

**b.** 200

**41. a.** True  **b.** False  **c.** True

**43. a.** $y = -2\ln\left(\dfrac{x^2}{4} + \cos x^2 + C\right)$  **b.** $C = 0, 1, 1 - \dfrac{\pi}{8}$

**c.**

$c = 0$

$c = 1 - \dfrac{\pi}{8}$

$c = 1$

**45.** $y = kx$  **47. b.** $\sqrt{gm/k}$

**c.** $v = \sqrt{\dfrac{g}{a}}\dfrac{Ce^{2\sqrt{ag}\,t} - 1}{Ce^{2\sqrt{ag}\,t} + 1}, t \geq 0,$ where $a = \dfrac{k}{m}$

**49. a.** $h = \left(\sqrt{H} - \dfrac{kt}{2}\right)^2, 0 \leq t \leq \dfrac{2\sqrt{H}}{k}$

**b.** $h = (\sqrt{0.5} - 0.05t)^2, 0 \leq t \leq 14.1$

**c.** Approx. 14.1 s

**51. a.**

$R$ is positive if $0 < M < 4$; $R$ has a maximum value when $M = \dfrac{4}{e}$; $\lim\limits_{M \to 0} R(M) = 0.$

**b.** $M(t) = 4^{1 - e^{-t}}, t \geq 0$; the tumor grows quickly at first and then the rate of growth slows down; the limiting size of the tumor is 4.

**53. a.** $y = \dfrac{1}{1 - t}, t < 1$  **b.** $y = \dfrac{1}{\sqrt{2}\sqrt{1 - t}}, t < 1$

**c.** $y = \dfrac{1}{(n(1 - t))^{1/n}}, t < 1$; as $t \to 1^-, y \to \infty$

## Section 9.4 Exercises, pp. 625–627

**1.** $y = 17e^{-10t} - 13$  **3.** $y = Ce^{-4t} + \dfrac{3}{2}$  **5.** $y = Ce^{3t} + \dfrac{4}{3}$

**7.** $y = Ce^{-2x} - 2$  **9.** $u = Ce^{-12t} + \dfrac{5}{4}$  **11.** $y = 7e^{3t} + 2$

**13.** $y = 4(e^{2t} - 1)$  **15.** $y = 4(2e^{3t-3} - 1)$

**17.** $y = \dfrac{3}{2}$; unstable

**19.** $y = -3$; stable

**21.** $u = -3$; stable

**23.** $B = 100{,}000 - 50{,}000e^{0.005t}$; reaches a balance of zero after approximately 139 months

$B = 100{,}000 - 50{,}000e^{0.005t}$

**25.** $B = 200{,}000 - 100{,}000e^{0.0075t}$; reaches a balance of zero after approximately 93 months

$B = 200{,}000 - 100{,}000e^{0.0075t}$

**27.** Approx. 32 min    **29.** Approx. 14 min

**31. a.**    **b.** 150    **c.** Approx. 115.1 hr

$y(t) = 150 - 150e^{-0.02t}$

**33. a.** $h = 16 \text{ yr}^{-1}$    **b.** 25,000    **35. a.** False    **b.** True    **c.** False
**d.** False    **37. a.** $B = 20,000 + 20,000e^{0.03t}$; the unpaid balance is
growing because the monthly payment of \$600 is less than the interest

on the unpaid balance.    **b.** \$20,000    **c.** $\dfrac{m}{r}$

**39.** $y = 1 + \dfrac{t}{2} + \dfrac{5}{2t}, t > 0$    **41.** $y = \dfrac{1}{2}e^{3t} + \dfrac{7}{2}e^{t}$

**45.** $y(t) = \dfrac{6}{t}, t > 0$    **47.** $y = \dfrac{9t^5 + 20t^3 + 15t + 76}{15(t^2 + 1)}$

## Section 9.5 Exercises, pp. 634–636

**1.** The growth rate function specifies the rate of growth of the
population. The population is increasing when the growth rate function
is positive, and the population is decreasing when the growth rate function is negative.    **3.** If the growth rate function is positive (it does not
matter whether it is increasing or decreasing), then the population is
increasing.    **5.** It is a linear, first-order differential equation.    **7.** The
solution curves in the *FH*-plane are closed curves that circulate around
the equilibrium point.

**9.**

$y = P(t)$

**11.**

$y = P(t)$

**13.**

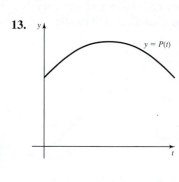

$y = P(t)$

**15.** $P' = 0.2 P\left(1 - \dfrac{P}{300}\right);$

$P = \dfrac{300}{5e^{-0.2t} + 1}, t \geq 0$

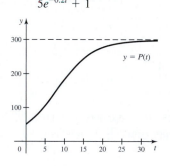

$y = P(t)$

**17.** $P = \dfrac{2000}{9e^{-\ln(27/7)t} + 1}, t \geq 0$    **19.** $M = K\left(\dfrac{M_0}{K}\right)^{e^{-rt}}, t \geq 0$

$y = P(t)$

**21.** $M = 1200 \cdot 0.075^{\exp(-0.05t)},$
$t \geq 0$

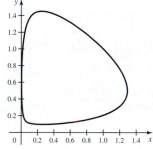

**23. a.** $m'(t) = -0.008t + 80, m(0) = 0$
**b.** $m = 10,000 - 10,000e^{-0.008t}, t \geq 0$
**25. a.** $m'(t) = -0.005\,t + 100, m(0) = 80,000$
**b.** $m = 60,000e^{-0.005t} + 20,000, t \geq 0$
**27. a.** $x$ is the predator population; $y$ is the prey population.
**b.** $x' = 0$ on the lines $x = 0$ and $y = \frac{1}{2}; y' = 0$ on the lines $y = 0$
and $x = \frac{1}{4}$.    **c.** $(0, 0), \left(\frac{1}{4}, \frac{1}{2}\right)$
**d.** $x' > 0$ and $y' > 0$ for $0 < x < \frac{1}{4}, y > \frac{1}{2}$
$\quad x' > 0$ and $y' < 0$ for $x > \frac{1}{4}, y > \frac{1}{2}$
$\quad x' < 0$ and $y' < 0$ for $x > \frac{1}{4}, 0 < y < \frac{1}{2}$
$\quad x' < 0$ and $y' > 0$ for $0 < x < \frac{1}{4}, 0 < y < \frac{1}{2}$
**e.** The solution evolves in the clockwise direction.

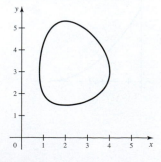

**29. a.** $x$ is the predator population; $y$ is the prey population.
**b.** $x' = 0$ on the lines $x = 0$ and $y = 3; y' = 0$ on the lines $y = 0$
and $x = 2$.    **c.** $(0, 0), (2, 3)$
**d.** $x' > 0$ and $y' > 0$ for $0 < x < 2, y > 3$
$\quad x' > 0$ and $y' < 0$ for $x > 2, y > 3$
$\quad x' < 0$ and $y' < 0$ for $x > 2, 0 < y < 3$
$\quad x' < 0$ and $y' > 0$ for $0 < x < 2, 0 < y < 3$
**e.** The solution evolves in the clockwise direction.

**31. a.** True  **b.** True  **c.** True  **35. c.** $\lim\limits_{t\to\infty} m(t) = C_i V$, which is the amount of substance in the tank when the tank is filled with the inflow solution.  **d.** Increasing $R$ increases the rate at which the solution in the tank reaches the steady-state concentration.

**37. a.** $I = \dfrac{V}{R} e^{-t/(RC)}$  **b.** $Q = VC(1 - e^{-t/(RC)})$

**39. a.** $y'(x) = \dfrac{y(c - dx)}{x(-a + by)}$  **c.**

## Chapter 9 Review Exercises, pp. 636–638

**1. a.** False  **b.** False  **c.** True  **d.** True  **e.** False
**3.** $y = Ce^{-2t} + 3$  **5.** $y = Ce^{t^2}$  **7.** $y = Ce^{\tan^{-1} t}$
**9.** $y = \tan(t^2 + t + C)$  **11.** $y = \sin t + t^2 + 1$
**13.** $Q = 8(1 - e^{t-1})$  **15.** $u = (3 + t^{2/3})^{3/2}, t > 0$

**17.** $s = \dfrac{t\sqrt{2}}{\sqrt{t^2 + 1}}$

**19. a, b.**

**c.** $0 < A < 2$
**d.** $A > 2$ or $A < 0$
**e.** $y = 0$ and $y = 2$

**21. a.** 1.05, 1.09762  **b.** 1.04939, 1.09651  **c.** 0.00217, 0.00106; the error in part (b) is smaller.  **23.** $y = -3$ (unstable), $y = 0$ (stable), $y = 5$ (unstable)  **25.** $y = -1$ (unstable), $y = 0$ (stable), $y = 2$ (unstable)  **27. a.** 0.0713  **b.** $P = \dfrac{1600}{79e^{-0.0713t} + 1}, t \geq 0$

**c.** Approx. 61 hours  **29. a.** $m = 2000(1 - e^{-0.005t})$
**b.** 2000 g  **c.** Approx. 599 minutes  **31. a.** $x$ represents the predator.  **b.** $x'(t) = 0$ when $x = 0$ and $y = 2$. $y'(t) = 0$ when $y = 0$ and $x = 5$.  **c.** $(0, 0)$ and $(5, 2)$  **d.** $x' > 0, y' > 0$ when $0 < x < 5$ and $y > 2$; $x' > 0, y' < 0$ when $x > 5$ and $y > 2$; $x' < 0, y' < 0$ when $x > 5$ and $0 < y < 2$; $x' < 0, y' > 0$ when $0 < x < 5$ and $0 < y < 2$
**e.** Clockwise direction

**33. a.** $p_1 = 3, p_2 = -4$  **b.** $y(t) = t^3 - t^{-4}, t > 0$

## CHAPTER 10

### Section 10.1 Exercises, pp. 647–649

**1.** A sequence is an ordered list of numbers. Example: $1, \frac{1}{3}, \frac{1}{9}, \frac{1}{27}, \ldots$
**3.** 1, 1, 2, 6, 24  **5.** $a_n = (-1)^{n+1} n$, for $n = 1, 2, 3, \ldots$; $a_n = (-1)^n (n + 1)$, for $n = 0, 1, 2, \ldots$ (Answers may vary.)

**7.** $e$  **9.** 1, 5, 14, 30  **11.** $\sum\limits_{k=1}^{\infty} 10$ (Answer is not unique.)

**13.** $\frac{1}{10}, \frac{1}{100}, \frac{1}{1000}, \frac{1}{10,000}$  **15.** $-\frac{1}{2}, \frac{1}{4}, -\frac{1}{8}, \frac{1}{16}$  **17.** $\frac{4}{3}, \frac{8}{5}, \frac{16}{9}, \frac{32}{17}$

**19.** 2, 1, 0, 1  **21.** 2, 4, 8, 16  **23.** 10, 18, 42, 114  **25.** $1, \frac{1}{2}, \frac{2}{3}, \frac{3}{5}$

**27. a.** $\frac{1}{32}, \frac{1}{64}$  **b.** $a_1 = 1, a_{n+1} = \frac{1}{2} a_n$, for $n \geq 1$  **c.** $a_n = \dfrac{1}{2^{n-1}}$, for $n \geq 1$  **29. a.** 32, 64  **b.** $a_1 = 1, a_{n+1} = 2a_n$, for $n \geq 1$  **c.** $a_n = 2^{n-1}$, for $n \geq 1$  **31. a.** 243, 729  **b.** $a_1 = 1, a_{n+1} = 3a_n$, for $n \geq 1$  **c.** $a_n = 3^{n-1}$, for $n \geq 1$  **33. a.** $-5, 5$  **b.** $a_1 = -5$, $a_{n+1} = -a_n$, for $n \geq 1$  **c.** $a_n = (-1)^n \cdot 5$, for $n \geq 1$
**35.** 9, 99, 999, 9999; diverges  **37.** $\frac{1}{10}, \frac{1}{100}, \frac{1}{1000}, \frac{1}{10,000}$; converges to 0  **39.** 2, 4, 2, 4; diverges  **41.** 2, 2, 2, 2; converges to 2
**43.** 54.545, 54.959, 54.996, 55.000; converges to 55

**45.**

| $n$ | $a_n$ |
|---|---|
| 1 | 0.83333333 |
| 2 | 0.96153846 |
| 3 | 0.99206349 |
| 4 | 0.99840256 |
| 5 | 0.99968010 |
| 6 | 0.99993600 |
| 7 | 0.99998720 |
| 8 | 0.99999744 |
| 9 | 0.99999949 |
| 10 | 0.99999990 |

The limit appears to be 1.

**47.**

| $n$ | $a_n$ |
|---|---|
| 1 | 2 |
| 2 | 6 |
| 3 | 12 |
| 4 | 20 |
| 5 | 30 |
| 6 | 42 |
| 7 | 56 |
| 8 | 72 |
| 9 | 90 |
| 10 | 110 |

The sequence appears to diverge.

**49. a.** $\frac{5}{2}, \frac{9}{4}, \frac{17}{8}, \frac{33}{16}$  **b.** 2

**51.**

| $n$ | $a_n$ |
|---|---|
| 1 | 3.00000000 |
| 2 | 3.50000000 |
| 3 | 3.75000000 |
| 4 | 3.87500000 |
| 5 | 3.93750000 |
| 6 | 3.96875000 |
| 7 | 3.98437500 |
| 8 | 3.99218750 |
| 9 | 3.99609375 |
| 10 | 3.99804688 |

The limit appears to be 4.

**53.**

| $n$ | $a_n$ |
|---|---|
| 0 | 1 |
| 1 | 5 |
| 2 | 21 |
| 3 | 85 |
| 4 | 341 |
| 5 | 1365 |
| 6 | 5461 |
| 7 | 21,845 |
| 8 | 87,381 |
| 9 | 349,525 |

The sequence appears to diverge.

**55.**

| $n$ | $a_n$ |
|---|---|
| 1 | 8.00000000 |
| 2 | 4.41421356 |
| 3 | 4.05050150 |
| 4 | 4.00629289 |
| 5 | 4.00078630 |
| 6 | 4.00009828 |
| 7 | 4.00001229 |
| 8 | 4.00000154 |
| 9 | 4.00000019 |
| 10 | 4.00000002 |

The limit appears to be 4.

**57. a.** 20, 10, 5, $\frac{5}{2}$
**b.** $h_n = 20\left(\frac{1}{2}\right)^n$, for $n \geq 0$
**59. a.** 30, $\frac{15}{2}, \frac{15}{8}, \frac{15}{32}$
**b.** $h_n = 30\left(\frac{1}{4}\right)^n$, for $n \geq 0$
**61.** $S_1 = 0.3, S_2 = 0.33, S_3 = 0.333,$
$S_4 = 0.3333; \frac{1}{3}$
**63.** $S_1 = 4, S_2 = 4.9,$
$S_3 = 4.99, S_4 = 4.999; 5$
**65.** $S_1 = 0.6, S_2 = 0.66, S_3 = 0.666,$
$S_4 = 0.6666; \frac{2}{3}$ **67. a.** $\frac{2}{3}, \frac{4}{5}, \frac{6}{7}, \frac{8}{9}$
**b.** $S_n = \frac{2n}{2n+1}; \frac{10}{11}, \frac{12}{13}, \frac{14}{15}, \frac{16}{17}$ **c.** 1

**69. a.** 9, 9.9, 9.99, 9.999 **b.** $S_n = 10 - (0.1)^{n-1}$; 9.9999, 9.99999,
9.999999, 9.9999999 **c.** 10 **71. a.** True **b.** False **c.** True
**73. a.** 20, 10, 5, $\frac{5}{2}, \frac{5}{4}$ **b.** $M_n = 20\left(\frac{1}{2}\right)^n$, for $n \geq 0$ **c.** $M_0 = 20,$
$M_{n+1} = \frac{1}{2}M_n$, for $n \geq 0$ **d.** $\lim_{n\to\infty} M_n = 0$ **75. a.** 200, 190, 180.5,
171.475, 162.90125 **b.** $d_n = 200(0.95)^n$, for $n \geq 0$
**c.** $d_0 = 200, d_{n+1} = (0.95)d_n$, for $n \geq 0$ **d.** $\lim_{n\to\infty} d_n = 0$
**77. a.** 40, 70, 92.5, 109.375 **b.** 160 **79.** 0.739

### Section 10.2 Exercises, pp. 659–662

**1.** $a_n = \frac{1}{n}, n \geq 1$ **3.** $a_n = \frac{n}{n+1}, n \geq 1$ **5.** Converges for
$-1 < r \leq 1$, diverges otherwise **7.** Diverges monotonically
**9.** Converges, oscillates; 0 **11.** $\{e^{n/100}\}$ grows faster then $\{n^{100}\}$.

**13.** 0 **15.** $\frac{3}{2}$ **17.** $\frac{\pi}{4}$ **19.** 2 **21.** 0 **23.** $\frac{1}{4}$ **25.** 2 **27.** 0
**29.** 0 **31.** 3 **33.** Diverges **35.** $\frac{\pi}{2}$ **37.** 0 **39.** $e^2$ **41.** $e^3$
**43.** $e^{1/4}$ **45.** 0 **47.** 1 **49.** 0 **51.** 6
**53.**

**55.** 0 **57.** Diverges
**59.** Diverges **61.** 0 **63.** 0
**65.** 0 **67.** 0 **69.** 0
**71. a.** $d_{n+1} = \frac{1}{2}d_n + 80$, for $n \geq 1$
**b.** 160 mg

Diverges

**73. a.** \$0, \$100, \$200.75, \$302.26, \$404.53
**b.** $B_{n+1} = 1.0075B_n + 100$, for $n \geq 0$ **c.** Approx. 43 months
**75.** 0 **77.** Diverges **79.** 0 **81.** 1 **83. a.** True **b.** False
**c.** True **d.** True **e.** False **f.** True **85. a.** Nondecreasing
**b.** $\frac{1}{2}$ **87. a.** Nonincreasing **b.** 2 **89. a.** $d_{n+1} = 0.4d_n + 75$;
$d_1 = 75$ **c.** 125; in the long run there will be approximately 125 mg
of medication in the blood. **91.** 0.607 **93. b.** 9
**95. a.** $\sqrt{2}, \sqrt{2 + \sqrt{2}}, \sqrt{2 + \sqrt{2 + \sqrt{2}}},$
$\sqrt{2 + \sqrt{2 + \sqrt{2 + \sqrt{2}}}}$, or 1.41421, 1.84776, 1.96157, 1.99037
**c.** 2 **97. a.** 1, 1, 2, 3, 5, 8, 13, 21, 34, 55 **b.** No **99. b.** 1, 2,
1.5, 1.6667, 1.6 **c.** Approx. 1.618 **e.** $\dfrac{a + \sqrt{a^2 + 4b}}{2}$
**101.** Given a tolerance $\varepsilon > 0$, look beyond $a_N$, where $N > 1/\varepsilon$.
**103.** Given a tolerance $\varepsilon > 0$, look beyond $a_N$, where $N > \frac{1}{4}\sqrt{3/\varepsilon}$,
provided $\varepsilon < \frac{3}{4}$. **105.** Given a tolerance $\varepsilon > 0$, look beyond $a_N$,
where $N > c/(\varepsilon b^2)$. **107.** $a < 1$ **109.** $\{n^2 + 2n - 17\}_{n=3}^{\infty}$
**111.** $n = 4, n = 6, n = 25$

### Section 10.3 Exercises, pp. 668–671

**1.** The constant $r$ in the series $\sum_{k=0}^{\infty} ar^k$ **3.** No **5. a.** $a = \frac{2}{3}; r = \frac{1}{5}$
**b.** $a = \frac{1}{27}; r = -\frac{1}{3}$ **7.** $S_n = \frac{1}{4} - \frac{1}{n+4}; S_{36} = \frac{9}{40}$ **9.** 9841
**11.** Approx. 1.1905 **13.** Approx. 0.5392 **15.** $\dfrac{1093}{2916}$
**17.** \$15,920.22 **19. a.** $\frac{7}{9}$ **21.** $\frac{4}{3}$ **23.** $\frac{10}{19}$ **25.** 10 **27.** Diverges
**29.** $\dfrac{1}{e^2 - 1}$ **31.** $\dfrac{1}{7}$ **33.** $\dfrac{1}{500}$ **35.** $\dfrac{3\pi}{\pi + 1}$ **37.** $\dfrac{\pi}{\pi - e}$ **39.** $\dfrac{9}{460}$
**41.** $\dfrac{4}{11}$ **43.** $A_5 = 266.406; A_{10} = 266.666; A_{30} = 266.667;$
$\lim_{n\to\infty} A_n = 266\frac{2}{3}$ mg, which is the steady-state level. **45.** 400 mg
**47.** $0.\overline{3} = \sum_{k=1}^{\infty} 3(0.1)^k = \frac{1}{3}$ **49.** $0.\overline{037} = \sum_{k=1}^{\infty} 37(0.001)^k = \frac{1}{27}$
**51.** $0.\overline{456} = \sum_{k=0}^{\infty} 0.456(0.001)^k = \frac{152}{333}$
**53.** $0.00\overline{952} = \sum_{k=0}^{\infty} 0.00952(0.001)^k = \frac{238}{24,975}$
**55.** $S_n = \frac{n}{2n+4}; \frac{1}{2}$ **57.** $S_n = \frac{1}{7} - \frac{1}{n+7}; \frac{1}{7}$
**59.** $S_n = \frac{1}{9} - \frac{1}{4n+9}; \frac{1}{9}$ **61.** $S_n = \ln(n+1)$; diverges
**63.** $S_n = \frac{1}{p+1} - \frac{1}{n+p+1}; \frac{1}{p+1}$
**65.** $S_n = \left(\frac{1}{\sqrt{2}} + \frac{1}{\sqrt{3}}\right) - \left(\frac{1}{\sqrt{n+2}} + \frac{1}{\sqrt{n+3}}\right); \frac{1}{\sqrt{2}} + \frac{1}{\sqrt{3}}$
**67.** $S_n = \frac{13}{12} - \frac{1}{n+2} - \frac{3}{n+3} - \frac{1}{n+4}; \frac{13}{12}$
**69.** $S_n = \tan^{-1}(n+1) - \tan^{-1}1; \frac{\pi}{4}$ **71. a, b.** $\frac{4}{3}$ **73.** $-\frac{1}{4}$
**75.** $\dfrac{2500}{19}$ **77.** $-\dfrac{2}{15}$ **79.** $\dfrac{1}{\ln 2}$ **81.** $-2$ **83.** $\dfrac{113}{30}$ **85.** $\dfrac{17}{10}$
**87. a.** True **b.** True **c.** False **d.** False **e.** True **f.** False
**g.** True **89. a.** $\frac{1}{5}$ **b.** Approx. 0.19999695 **91.** Approx. 0.96
**95.** 462 months **99.** $\sum_{k=0}^{\infty} \left(\frac{1}{4}\right)^k A_1 = \frac{A_1}{1 - 1/4} = \frac{4}{3} A_1$
**101. a.** $L_n = 3\left(\frac{4}{3}\right)^n$, so $\lim_{n\to\infty} L_n = \infty$ **b.** $\lim_{n\to\infty} A_n = \frac{2\sqrt{3}}{5}$
**103.** $R_n = S - S_n = \frac{1}{1-r} - \left(\frac{1-r^n}{1-r}\right) = \frac{r^n}{1-r}$ **105. a.** 60
**b.** 9 **107. a.** 13 **b.** 15 **109. a.** 1, $\frac{5}{6}, \frac{2}{3}$, undefined, undefined
**b.** $(-1, 1)$ **111.** Converges for $x$ in $(-\infty, -2)$ or $(0, \infty)$; $x = \frac{1}{2}$

### Section 10.4 Exercises, pp. 680–683

**1.** The series diverges. **3.** $\lim_{k\to\infty} a_k = 0$ **5.** Converges for $p > 1$ and
diverges for $p \leq 1$ **7.** $R_n = S - S_n$ **9.** Diverges
**11.** Inconclusive **13.** Diverges **15.** Diverges **17.** Diverges
**19.** Diverges **21.** Converges **23.** Diverges **25.** Converges
**27.** Diverges **29.** Converges **31.** Converges **33.** Converges
**35.** Diverges **37.** Diverges **39. a.** $S \approx S_2 = 1.0078125$
**b.** $R_2 < 0.0026042$ **c.** $L_2 = 1.0080411; U_2 = 1.0104167$

**41. a.** $\dfrac{1}{5n^5}$  **b.** 3  **c.** $L_n = S_n + \dfrac{1}{5(n+1)^5}$; $U_n = S_n + \dfrac{1}{5n^5}$

**d.** $(1.017342754, 1.017343512)$  **43. a.** $\dfrac{3^{-n}}{\ln 3}$  **b.** 7

**c.** $L_n = S_n + \dfrac{3^{-n-1}}{\ln 3}$; $U_n = S_n + \dfrac{3^{-n}}{\ln 3}$

**d.** $(0.499996671, 0.500006947)$  **45.** 1.0083  **47. a.** False
**b.** True  **c.** False  **d.** True  **e.** False  **f.** False  **49.** Converges
**51.** Converges  **53.** Diverges  **55.** Diverges  **57.** Diverges
**59.** Converges  **61.** Converges  **63.** Converges  **65. a.** $p > 1$

**b.** $\displaystyle\sum_{k=2}^{\infty}\dfrac{1}{k(\ln k)^2}$ converges faster.  **67.** $\zeta(3) \approx 1.202, \zeta(5) \approx 1.037$

**69.** $\dfrac{\pi^2}{8}$  **73. a.** $\dfrac{1}{2}, \dfrac{7}{12}, \dfrac{37}{60}$

## Section 10.5 Exercises, pp. 687–688

**1.** Find an appropriate comparison series. Then take the limit of the ratio of the terms of the given series and the comparison series as $n \to \infty$. The value of the limit determines whether the series converges.

**3.** $\displaystyle\sum_{k=1}^{\infty}\dfrac{1}{k^2}$  **5.** $\displaystyle\sum_{k=1}^{\infty}\left(\dfrac{2}{3}\right)^k$  **7.** $\displaystyle\sum_{k=1}^{\infty}\dfrac{1}{k}$  **9.** Converges
**11.** Diverges  **13.** Converges  **15.** Converges  **17.** Converges
**19.** Diverges  **21.** Diverges  **23.** Converges  **25.** Diverges
**27.** Converges  **29.** Diverges  **31.** Diverges  **33.** Diverges
**35.** Converges  **37. a.** False  **b.** True  **c.** True  **d.** True
**39.** Converges  **41.** Diverges  **43.** Diverges  **45.** Diverges
**47.** Converges  **49.** Diverges  **51.** Converges  **53.** Converges
**55.** Diverges  **57.** Converges  **59.** Diverges  **61.** Converges

## Section 10.6 Exercises, pp. 694–696

**1.** Because $S_{n+1} - S_n = (-1)^n a_{n+1}$ alternates sign
**3.** Because the remainder $R_n = S - S_n$ alternates sign
**5.** $|R_n| = |S - S_n| \le |S_{n+1} - S_n| = a_{n+1}$  **7.** No; if a series
of positive terms converges, it does so absolutely and not conditionally.

**9.** Yes, $\displaystyle\sum_{k=1}^{\infty}\dfrac{(-1)^k}{k^2}$ has this property.  **11.** Converges  **13.** Diverges
**15.** Converges  **17.** Converges  **19.** Diverges  **21.** Diverges
**23.** Diverges  **25.** Diverges  **27.** Converges  **29.** $S_4 = -0.945939$;
$|S - S_4| \le 0.0016$  **31.** $S_5 = 0.70696$; $|S - S_5| \le 0.001536$
**33.** 10,000  **35.** 5000  **37.** 10  **39.** $-0.973$  **41.** $-0.269$
**43.** $-0.783$  **45.** Converges conditionally  **47.** Converges absolutely
**49.** Converges absolutely  **51.** Converges absolutely  **53.** Diverges
**55.** Converges conditionally  **57.** Diverges  **59.** Converges absolutely  **61.** Converges conditionally  **63.** Converges absolutely
**65. a.** False  **b.** True  **c.** True  **d.** True  **e.** False  **f.** True
**g.** True  **69.** $x$ and $y$ are divergent series.

**71. b.** $S_{2n} = \displaystyle\sum_{k=1}^{n}\left(\dfrac{1}{k^2} - \dfrac{1}{k}\right)$

## Section 10.7 Exercises, pp. 699–700

**1.** Take the limit of the magnitude of the ratio of consecutive terms of the series as $k \to \infty$. The value of the limit determines whether the series converges absolutely or diverges.  **3.** 999,000

**5.** $\dfrac{1}{(k+2)(k+1)}$  **7.** Ratio Test  **9.** Converges absolutely

**11.** Diverges  **13.** Converges absolutely  **15.** Converges absolutely
**17.** Diverges  **19.** Diverges  **21.** Converges absolutely
**23.** Converges absolutely  **25.** Diverges  **27.** Converges absolutely

**29.** Diverges  **31. a.** False  **b.** True  **c.** True  **d.** True
**33.** Converges absolutely  **35.** Diverges  **37.** Converges absolutely
**39.** Converges conditionally  **41.** Converges absolutely
**43.** Converges absolutely  **45.** Converges conditionally
**47.** Converges absolutely  **49.** Converges conditionally
**51.** $p > 1$  **53.** $p > 1$  **55.** $p < 1$  **57.** Diverges for all $p$
**59.** $-1 < x < 1$  **61.** $-1 \le x \le 1$  **63.** $-2 < x < 2$

## Section 10.8 Exercises, pp. 703–704

**1.** Root Test  **3.** Divergence Test  **5.** $p$-series Test or Limit Comparison Test  **7.** Comparison Test or Limit Comparison Test
**9.** Alternating Series Test  **11.** Diverges  **13.** Diverges
**15.** Converges  **17.** Diverges  **19.** Converges  **21.** Converges
**23.** Converges  **25.** Converges  **27.** Converges  **29.** Diverges
**31.** Converges  **33.** Diverges  **35.** Converges  **37.** Diverges
**39.** Diverges  **41.** Converges  **43.** Diverges  **45.** Converges
**47.** Diverges  **49.** Converges  **51.** Diverges  **53.** Converges
**55.** Diverges  **57.** Converges  **59.** Converges  **61.** Diverges
**63.** Diverges  **65.** Converges  **67.** Converges  **69.** Diverges
**71.** Converges  **73.** Converges  **75.** Converges  **77.** Diverges
**79.** Diverges  **81.** Converges  **83.** Converges  **85.** Converges
**87. a.** False  **b.** True  **c.** True  **d.** False  **89.** Diverges
**91.** Diverges  **93.** Diverges

## Chapter 10 Review Exercises, pp. 704–707

**1. a.** False  **b.** False  **c.** True  **d.** False  **e.** True  **f.** False
**g.** False  **h.** True  **3.** Approx. 1.25; approx. 0.05  **5.** $\lim_{k\to\infty} a_k = 0$,

$\lim_{n\to\infty} S_n = 8$  **7.** $a_k = \dfrac{1}{k}$  **9. a.** 0  **b.** $\dfrac{5}{9}$  **11. a.** Yes; $\lim_{k\to\infty} a_k = 1$

**b.** No; $\lim_{k\to\infty} a_k \ne 0$  **13.** Diverges  **15.** 5  **17.** 0  **19.** 0  **21.** $1/e$

**23.** Diverges  **25. a.** 80, 48, 32, 24, 20  **b.** 16  **27.** Diverges

**29.** Diverges  **31.** Diverges  **33.** $\dfrac{3\pi}{4}$  **35.** 3  **37.** 2/9

**39.** $\dfrac{311}{990}$  **41.** 200 mg  **43.** Diverges  **45.** Diverges  **47.** Converges
**49.** Converges  **51.** Converges  **53.** Converges  **55.** Converges
**57.** Diverges  **59.** Converges  **61.** Converges  **63.** Converges
**65.** Converges  **67.** Converges  **69.** Converges  **71.** Converges
**73.** Diverges  **75.** Diverges  **77.** Converges conditionally
**79.** Converges absolutely  **81.** Diverges  **83.** Converges absolutely
**85.** Converges absolutely  **87.** Diverges  **89. a.** Approx. 1.03666
**b.** 0.0004  **c.** $L_5 = 1.03685$; $U_5 = 1.03706$  **91.** 0.0067
**93.** 100  **95. a.** 803 m, 1283 m, $2000(1 - 0.95^N)$ m  **b.** 2000 m

**97. a.** $\dfrac{\pi}{2^{n-1}}$  **b.** $2\pi$  **99. a.** $T_1 = \dfrac{\sqrt{3}}{16}, T_2 = \dfrac{7\sqrt{3}}{64}$

**b.** $T_n = \dfrac{\sqrt{3}}{4}\left(1 - \left(\dfrac{3}{4}\right)^n\right)$  **c.** $\lim_{n\to\infty} T_n = \dfrac{\sqrt{3}}{4}$  **d.** 0

**101.** $\sqrt{\dfrac{20}{g}\left(\dfrac{1 + \sqrt{p}}{1 - \sqrt{p}}\right)}$ s

# CHAPTER 11

## Section 11.1 Exercises, pp. 718–721

**1.** $f(0) = p_2(0), f'(0) = p_2'(0),$ and $f''(0) = p_2''(0)$
**3.** 1, 1.05, 1.04875  **5.** $p_3(x) = 1 + x^2 + x^3$; 1.048
**7.** $p_3(x) = 1 + (x - 2) + 2(x - 2)^3$; 0.898
**9. a.** $p_1(x) = 8 + 12(x - 1)$
**b.** $p_2(x) = 8 + 12(x - 1) + 3(x - 1)^2$  **c.** 9.2; 9.23

**11. a.** $p_1(x) = 1 - 2x$   **b.** $p_2(x) = 1 - 2x + 2x^2$   **c.** $0.8, 0.82$
**13. a.** $p_1(x) = 1 - x$   **b.** $p_2(x) = 1 - x + x^2$   **c.** $0.95, 0.9525$
**15. a.** $p_1(x) = 2 + \frac{1}{12}(x - 8)$
**b.** $p_2(x) = 2 + \frac{1}{12}(x - 8) - \frac{1}{288}(x - 8)^2$   **c.** $1.958\overline{3}, 1.95747$
**17.** $p_1(x) = 1, p_2(x) = p_3(x) = 1 - 18x^2, p_4(x) = 1 - 18x^2 + 54x^4$
**19.** $p_3(x) = 1 - 3x + 6x^2 - 10x^3$,
$p_4(x) = 1 - 3x + 6x^2 - 10x^3 + 15x^4$
**21.** $p_1(x) = 1 + 3(x - 1), p_2(x) = 1 + 3(x - 1) + 3(x - 1)^2$,
$p_3(x) = 1 + 3(x - 1) + 3(x - 1)^2 + (x - 1)^3$
**23.** $p_3(x) = 1 + \dfrac{1}{e}(x - e) - \dfrac{1}{2e^2}(x - e)^2 + \dfrac{1}{3e^3}(x - e)^3$

**25. a.** $p_1(x) = -x, p_2(x) = -x - \dfrac{x^2}{2}$
**b.**

**27. a.** $p_1(x) = \dfrac{\sqrt{2}}{2} + \dfrac{\sqrt{2}}{2}\left(x - \dfrac{\pi}{4}\right)$,
$p_2(x) = \dfrac{\sqrt{2}}{2} + \dfrac{\sqrt{2}}{2}\left(x - \dfrac{\pi}{4}\right) - \dfrac{\sqrt{2}}{4}\left(x - \dfrac{\pi}{4}\right)^2$
**b.**

**29. a.** $1.0247$   **b.** $7.6 \times 10^{-6}$   **31. a.** $0.8613$   **b.** $5.4 \times 10^{-4}$
**33. a.** $1.1274988$   **b.** Approx. $8.85 \times 10^{-6}$ (Answers may vary if intermediate calculations are rounded.)   **35. a.** Approx. $-0.10033333$
**b.** Approx. $1.34 \times 10^{-6}$ (Answers may vary if intermediate calculations are rounded.)   **37. a.** $1.0295635$   **b.** Approx. $4.86 \times 10^{-7}$
(Answers may vary if intermediate calculations are rounded.)
**39. a.** Approx. $0.52083333$   **b.** Approx. $2.62 \times 10^{-4}$ (Answers may vary if intermediate calculations are rounded.)
**41.** $R_n(x) = \dfrac{\sin^{(n+1)}(c)}{(n + 1)!} x^{n+1}$, for $c$ between $x$ and $0$
**43.** $R_n(x) = \dfrac{(-1)^{n+1}e^{-c}}{(n + 1)!} x^{n+1}$, for $c$ between $x$ and $0$
**45.** $R_n(x) = \dfrac{\sin^{(n+1)}(c)}{(n + 1)!}\left(x - \dfrac{\pi}{2}\right)^{n+1}$, for $c$ between $x$ and $\dfrac{\pi}{2}$
**47.** $2.0 \times 10^{-5}$   **49.** $1.6 \times 10^{-5}$ $(e^{0.25} < 2)$   **51.** $2.6 \times 10^{-4}$
**53.** With $n = 4$, $|\text{error}| \le 2.5 \times 10^{-3}$
**55.** With $n = 2$, $|\text{error}| \le 4.2 \times 10^{-2}$ $(e^{0.5} < 2)$
**57.** With $n = 2$, $|\text{error}| \le 5.4 \times 10^{-3}$   **59.** $4$   **61.** $3$   **63.** $1$
**65. a.** False   **b.** True   **c.** True   **d.** True   **67. a.** C   **b.** E
**c.** A   **d.** D   **e.** B   **f.** F   **69. a.** $0.1; 1.7 \times 10^{-4}$   **b.** $0.2$;
$1.3 \times 10^{-3}$   **71. a.** $0.995; 4.2 \times 10^{-6}$   **b.** $0.98; 6.7 \times 10^{-5}$
**73. a.** $1.05; 1.3 \times 10^{-3}$   **b.** $1.1; 5 \times 10^{-3}$   **75. a.** $1.1; 10^{-2}$
**b.** $1.2; 4 \times 10^{-2}$

**77. a.**

| $x$ | $|\sec x - p_2(x)|$ | $|\sec x - p_4(x)|$ |
|---|---|---|
| $-0.2$ | $3.39 \times 10^{-4}$ | $5.51 \times 10^{-6}$ |
| $-0.1$ | $2.09 \times 10^{-5}$ | $8.51 \times 10^{-8}$ |
| $0.0$ | $0$ | $0$ |
| $0.1$ | $2.09 \times 10^{-5}$ | $8.51 \times 10^{-8}$ |
| $0.2$ | $3.39 \times 10^{-4}$ | $5.51 \times 10^{-6}$ |

**b.** The errors decrease as $|x|$ decreases.

**79. a.**

| $x$ | $|e^{-x} - p_1(x)|$ | $|e^{-x} - p_2(x)|$ |
|---|---|---|
| $-0.2$ | $2.14 \times 10^{-2}$ | $1.40 \times 10^{-3}$ |
| $-0.1$ | $5.17 \times 10^{-3}$ | $1.71 \times 10^{-4}$ |
| $0.0$ | $0$ | $0$ |
| $0.1$ | $4.84 \times 10^{-3}$ | $1.63 \times 10^{-4}$ |
| $0.2$ | $1.87 \times 10^{-2}$ | $1.27 \times 10^{-3}$ |

**b.** The errors decrease as $|x|$ decreases.

**81.** Centered at $x = 0$, for all $n$

**83. a.** $y = f(a) + f'(a)(x - a)$   **85. a.** $p_5(x) = x - \dfrac{x^3}{6} + \dfrac{x^5}{120}$;
$q_5(x) = -(x - \pi) + \dfrac{1}{6}(x - \pi)^3 - \dfrac{1}{120}(x - \pi)^5$

**b.**

$p_5$ is a better approximation on $[-\pi, \pi/2]$; $q_5$ is a better approximation on $(\pi/2, 2\pi]$.

**c.**

| $x$ | $|\sin x - p_5(x)|$ | $|\sin x - q_5(x)|$ |
|---|---|---|
| $\pi/4$ | $3.6 \times 10^{-5}$ | $7.4 \times 10^{-2}$ |
| $\pi/2$ | $4.5 \times 10^{-3}$ | $4.5 \times 10^{-3}$ |
| $3\pi/4$ | $7.4 \times 10^{-2}$ | $3.6 \times 10^{-5}$ |
| $5\pi/4$ | $2.3$ | $3.6 \times 10^{-5}$ |
| $7\pi/4$ | $20$ | $7.4 \times 10^{-2}$ |

**d.** $p_5$ is a better approximation at $x = \pi/4$; at $x = \pi/2$ the errors are equal.

**87. a.** $p_1(x) = 6 + \dfrac{1}{12}(x - 36); q_1(x) = 7 + \dfrac{1}{14}(x - 49)$

**b.**

| $x$ | $|\sqrt{x} - p_1(x)|$ | $|\sqrt{x} - q_1(x)|$ |
|---|---|---|
| $37$ | $5.7 \times 10^{-4}$ | $6.0 \times 10^{-2}$ |
| $39$ | $5.0 \times 10^{-3}$ | $4.1 \times 10^{-2}$ |
| $41$ | $1.4 \times 10^{-2}$ | $2.5 \times 10^{-2}$ |
| $43$ | $2.6 \times 10^{-2}$ | $1.4 \times 10^{-2}$ |
| $45$ | $4.2 \times 10^{-2}$ | $6.1 \times 10^{-3}$ |
| $47$ | $6.1 \times 10^{-2}$ | $1.5 \times 10^{-3}$ |

**c.** $p_1$ is a better approximation at $x = 37, 39$, and $41$.

## Section 11.2 Exercises, pp. 729–730

**1.** $c_0 + c_1 x + c_2 x^2 + c_3 x^3$ **3.** Ratio and Root Tests **5.** The radius of convergence does not change. The interval of convergence may change. **7.** $R = 10; [-8, 12]$ **9.** $R = \frac{1}{2}; \left(-\frac{1}{2}, \frac{1}{2}\right)$

**11.** $R = 0; \{x: x = 0\}$ **13.** $R = \infty; (-\infty, \infty)$ **15.** $R = 3; (-3, 3)$
**17.** $R = \infty; (-\infty, \infty)$ **19.** $R = 2; (-2, 2)$ **21.** $R = \infty; (-\infty, \infty)$
**23.** $R = 1; (0, 2]$ **25.** $R = \frac{1}{4}; \left[0, \frac{1}{2}\right]$ **27.** $R = 5; (-3, 7)$
**29.** $R = \infty; (-\infty, \infty)$ **31.** $R = \sqrt{3}; (-\sqrt{3}, \sqrt{3})$ **33.** $R = 1; (0, 2)$
**35.** $R = \infty; (-\infty, \infty)$ **37.** $R = e$ **39.** $R = e^4$

**41.** $\displaystyle\sum_{k=0}^{\infty} (3x)^k; \left(-\frac{1}{3}, \frac{1}{3}\right)$ **43.** $2\displaystyle\sum_{k=0}^{\infty} x^{k+3}; (-1, 1)$

**45.** $4\displaystyle\sum_{k=0}^{\infty} x^{k+12}; (-1, 1)$ **47.** $-\displaystyle\sum_{k=1}^{\infty} \frac{(3x)^k}{k}; \left[-\frac{1}{3}, \frac{1}{3}\right)$

**49.** $-2\displaystyle\sum_{k=1}^{\infty} \frac{x^{k+6}}{k}; [-1, 1)$ **51.** $g(x) = 2\displaystyle\sum_{k=1}^{\infty} k(2x)^{k-1}; \left(-\frac{1}{2}, \frac{1}{2}\right)$

**53.** $g(x) = \displaystyle\sum_{k=1}^{\infty} (-1)^k k x^{k-1}; (-1, 1)$

**55.** $g(x) = -\displaystyle\sum_{k=1}^{\infty} \frac{3^k x^k}{k}; \left[-\frac{1}{3}, \frac{1}{3}\right)$

**57.** $\displaystyle\sum_{k=1}^{\infty} (-1)^{k+1} 2k x^{2k-1}; (-1, 1)$ **59.** $\displaystyle\sum_{k=0}^{\infty} \left(-\frac{x}{3}\right)^k; (-3, 3)$

**61.** $\ln 2 - \frac{1}{2}\displaystyle\sum_{k=1}^{\infty} \frac{x^{2k}}{k \, 4^k}; (-2, 2)$ **63. a.** True **b.** True **c.** True

**d.** True **65.** $|x - a| < R$ **67.** $f(x) = \dfrac{1}{3 - \sqrt{x}}; 1 < x < 9$

**69.** $f(x) = \dfrac{e^x}{e^x - 1}; 0 < x < \infty$ **71.** $f(x) = \dfrac{3}{4 - x^2}; -2 < x < 2$

**73.** $\displaystyle\sum_{k=0}^{\infty} \frac{(-3x)^k}{k!}; -\infty < x < \infty$

**75.** $\displaystyle\lim_{k \to \infty} \left| \frac{c_{k+1} x^{k+1}}{c_k x^k} \right| = \lim_{k \to \infty} \left| \frac{c_{k+1} x^{k+m+1}}{c_k x^{k+m}} \right|$, so by the Ratio Test, the two series have the same radius of convergence.

**77. a.** $f(x) \, g(x) = c_0 d_0 + (c_0 d_1 + c_1 d_0)x +$
$(c_0 d_2 + c_1 d_1 + c_2 d_0)x^2 +$ **b.** $\displaystyle\sum_{k=0}^{n} c_k d_{n-k}$

## Section 11.3 Exercises, pp. 740–742

**1.** The $n$th Taylor polynomial is the $n$th partial sum of the corresponding Taylor series. **3.** $\displaystyle\sum_{k=0}^{\infty} \frac{(x - 2)^k}{k!}$ **5.** Replace $x$ with $x^2$ in the

Taylor series for $f(x); |x| < 1$. **7.** The Taylor series for a function $f$ converges to $f$ on an interval if, for all $x$ in the interval, $\lim_{n \to \infty} R_n(x) = 0$, where $R_n(x)$ is the remainder at $x$.
**9. a.** $1 - 2(x - 1) + 3(x - 1)^2 - 4(x - 1)^2$

**b.** $\displaystyle\sum_{k=0}^{\infty} (-1)^k (k + 1)(x - 1)^k$ **c.** $(0, 2)$ **11. a.** $1 - x + \dfrac{x^2}{2!} - \dfrac{x^3}{3!}$

**b.** $\displaystyle\sum_{k=0}^{\infty} \frac{(-1)^k x^k}{k!}$ **c.** $(-\infty, \infty)$ **13. a.** $2 + 6x + 12x^2 + 20x^3$

**b.** $\displaystyle\sum_{k=0}^{\infty} (k + 1)(k + 2)x^k$ **c.** $(-1, 1)$ **15. a.** $1 - x^2 + x^4 - x^6$

**b.** $\displaystyle\sum_{k=0}^{n} (-1)^k x^{2k}$ **c.** $(-1, 1)$ **17. a.** $1 + 2x + \dfrac{(2x)^2}{2!} + \dfrac{(2x)^3}{3!}$

**b.** $\displaystyle\sum_{k=0}^{\infty} \frac{(2x)^k}{k!}$ **c.** $(-\infty, \infty)$ **19. a.** $\dfrac{x}{2} - \dfrac{x^3}{3 \cdot 2^3} + \dfrac{x^5}{5 \cdot 2^5} - \dfrac{x^7}{7 \cdot 2^7}$

**b.** $\displaystyle\sum_{k=0}^{\infty} \frac{(-1)^k x^{2k+1}}{(2k + 1)2^{2k+1}}$ **c.** $[-2, 2]$

**21. a.** $1 + (\ln 3)x + \dfrac{\ln^2 3}{2} x^2 + \dfrac{\ln^3 3}{6} x^3$ **b.** $\displaystyle\sum_{k=0}^{\infty} \frac{\ln^k 3}{k!} x^k$ **c.** $(-\infty, \infty)$

**23. a.** $1 + \dfrac{(3x)^2}{2} + \dfrac{(3x)^4}{24} + \dfrac{(3x)^6}{720}$ **b.** $\displaystyle\sum_{k=0}^{\infty} \frac{(3x)^{2k}}{(2k)!}$ **c.** $(-\infty, \infty)$

**25. a.** $(x - 3) - \dfrac{1}{2}(x - 3)^2 + \dfrac{1}{3}(x - 3)^3 - \dfrac{1}{4}(x - 3)^4$

**b.** $\displaystyle\sum_{k=1}^{\infty} \frac{(-1)^{k+1}(x - 3)^k}{k}$ **c.** $(2, 4]$

**27. a.** $1 - \dfrac{(x - \pi/2)^2}{2!} + \dfrac{(x - \pi/2)^4}{4!} - \dfrac{(x - \pi/2)^6}{6!}$

**b.** $\displaystyle\sum_{k=0}^{\infty} \frac{(-1)^k}{(2k)!}(x - \pi/2)^{2k}$

**29. a.** $1 - (x - 1) + (x - 1)^2 - (x - 1)^3$ **b.** $\displaystyle\sum_{k=0}^{\infty} (-1)^k (x - 1)^k$

**31. a.** $\ln 3 + \dfrac{(x - 3)}{3} - \dfrac{(x - 3)^2}{3^2 \cdot 2} + \dfrac{(x - 3)^3}{3^3 \cdot 3}$

**b.** $\ln 3 + \displaystyle\sum_{k=1}^{\infty} \frac{(-1)^{k+1}(x - 3)^k}{k 3^k}$

**33. a.** $2 + 2(\ln 2)(x - 1) + (\ln^2 2)(x - 1)^2 + \dfrac{\ln^3 2}{3}(x - 1)^3$

**b.** $\displaystyle\sum_{k=0}^{\infty} \frac{2(x - 1)^k \ln^k 2}{k!}$ **35.** $x^2 - \dfrac{x^4}{2} + \dfrac{x^6}{3} - \dfrac{x^8}{4}$

**37.** $1 + 2x + 4x^2 + 8x^3$ **39.** $1 + \dfrac{x}{2} + \dfrac{x^2}{6} + \dfrac{x^3}{24}$

**41.** $1 - x^4 + x^8 - x^{12}$ **43.** $x^2 + \dfrac{x^6}{6} + \dfrac{x^{10}}{120} + \dfrac{x^{14}}{5040}$

**45. a.** $1 - 2x + 3x^2 - 4x^3$ **b.** $0.826$
**47. a.** $1 + \frac{1}{4}x - \frac{3}{32}x^2 + \frac{7}{128}x^3$ **b.** $1.029$

**49. a.** $1 - \frac{2}{3}x + \frac{5}{9}x^2 - \frac{40}{81}x^3$ **b.** $0.895$ **51.** $1 + \dfrac{x^2}{2} - \dfrac{x^4}{8} + \dfrac{x^6}{16}$;

$[-1, 1]$ **53.** $3 - \dfrac{3x}{2} - \dfrac{3x^2}{8} - \dfrac{3x^3}{16}; [-1, 1]$

**55.** $a + \dfrac{x^2}{2a} - \dfrac{x^4}{8a^3} + \dfrac{x^6}{16a^5}; |x| \le a$

**57.** $1 - 8x + 48x^2 - 256x^3$ **59.** $\dfrac{1}{16} - \dfrac{x^2}{32} + \dfrac{3x^4}{256} - \dfrac{x^6}{256}$

**61.** $\dfrac{1}{9} - \dfrac{2}{9}\left(\dfrac{4x}{3}\right) + \dfrac{3}{9}\left(\dfrac{4x}{3}\right)^2 - \dfrac{4}{9}\left(\dfrac{4x}{3}\right)^3$

**63.** $R_n(x) = \dfrac{f^{(n+1)}(c)}{(n + 1)!} x^{n+1}$, where $c$ is between 0 and $x$ and

$f^{(n+1)}(c) = \pm \sin c$ or $\pm \cos c$. Therefore, $|R_n(x)| \le \dfrac{|x|^{n+1}}{(n + 1)!} \to 0$

as $n \to \infty$, for $-\infty < x < \infty$. **65.** $R_n(x) = \dfrac{f^{(n+1)}(c)}{(n + 1)!} x^{n+1}$,

where $c$ is between 0 and $x$ and $f^{(n+1)}(c) = (-1)^n e^{-c}$. Therefore,

$|R_n(x)| \le \dfrac{|x|^{n+1}}{e^c (n + 1)!} \to 0$ as $n \to \infty$, for $-\infty < x < \infty$.

**67. a.** False **b.** True **c.** False **d.** False **e.** True

**69. a.** $1 + \dfrac{x^2}{2!} + \dfrac{x^4}{4!} + \dfrac{x^6}{6!}$ **b.** $R = \infty$

**71. a.** $1 - \frac{2}{3}x^2 + \frac{5}{9}x^4 - \frac{40}{81}x^6$ **b.** $R = 1$

**73. a.** $1 - \frac{1}{2}x^2 - \frac{1}{8}x^4 - \frac{1}{16}x^6$ **b.** $R = 1$

**75. a.** $1 - 2x^2 + 3x^4 - 4x^6$  **b.** $R = 1$  **77.** Approx. 3.9149
**79.** Approx. 1.8989  **85.** $\displaystyle\sum_{k=0}^{\infty}\left(\frac{x-4}{2}\right)^k$  **87.** Use three terms of the
Taylor series for $\cos x$ centered at $a = \pi/4$; 0.766  **89. a.** Use
three terms of the Taylor series for $\sqrt[3]{125 + x}$ centered at $a = 0$;
5.03968  **b.** Use three terms of the Taylor series for $\sqrt[3]{x}$ centered at
$a = 125$; 5.03968  **c.** Yes

## Section 11.4 Exercises, pp. 748–750

**1.** Replace $f$ and $g$ with their Taylor series centered at $a$ and evaluate
the limit.  **3.** Substitute $x = -0.6$ into the Taylor series for $e^x$
centered at 0. Because the resulting series is an alternating series, the
error can be estimated easily.  **7.** 1  **9.** $\frac{1}{2}$  **11.** 2  **13.** $\frac{2}{3}$  **15.** $\frac{1}{3}$

**17.** $\frac{3}{5}$  **19.** $-\frac{8}{5}$  **21.** 1  **23.** $\frac{3}{4}$  **25. a.** $1 + x + \frac{x^2}{2!} + \cdots + \frac{x^n}{n!} + \cdots$

**b.** $e^x$  **c.** $-\infty < x < \infty$

**27. a.** $1 - x + x^2 - \cdots (-1)^{n-1}x^{n-1} + \cdots$  **b.** $\frac{1}{1+x}$  **c.** $|x| < 1$

**29. a.** $-2 + 4x - 8\cdot\frac{x^2}{2!} + \cdots + (-2)^n\frac{x^{n-1}}{(n-1)!} + \cdots$

**b.** $-2e^{-2x}$  **c.** $-\infty < x < \infty$  **31. a.** $1 - x^2 + x^4 - \cdots$

**b.** $\frac{1}{1+x^2}$  **c.** $-1 < x < 1$

**33. a.** $2 + 2t + \frac{2t^2}{2!} + \cdots + \frac{2t^n}{n!} + \cdots$  **b.** $y(t) = 2e^t$

**35. a.** $2 + 16t + 24t^2 + 24t^3 + \cdots + \frac{3^{n-1}\cdot 16}{n!}t^n + \cdots$

**b.** $y(t) = \frac{16}{3}e^{3t} - \frac{10}{3}$  **37.** 0.2448  **39.** 0.6958

**41.** 0.0600  **43.** 0.4994  **45.** $1 + 2 + \frac{2^2}{2!} + \frac{2^3}{3!}$

**47.** $1 - 2 + \frac{2}{3} - \frac{4}{45}$  **49.** $\frac{1}{2} - \frac{1}{8} + \frac{1}{24} - \frac{1}{64}$  **51.** $e - 1$

**53.** $\displaystyle\sum_{k=1}^{\infty}\frac{(-1)^{k+1}x^k}{k}$, for $-1 < x \le 1$; $\ln 2$  **55.** $\frac{2}{2-x}$  **57.** $\frac{4}{4+x^2}$

**59.** $-\ln(1-x)$  **61.** $-\frac{3x^2}{(3+x)^2}$  **63.** $\frac{6x^2}{(3-x)^3}$

**65. a.** False  **b.** False  **c.** True  **67.** $\frac{a}{b}$  **69.** $e^{-1/6}$

**71.** $f^{(3)}(0) = 0; f^{(4)}(0) = 4e$  **73.** $f^{(3)}(0) = 2; f^{(4)}(0) = 0$

**75.** 2  **77.** 1.575 using four terms  **79. a.** $S'(x) = \sin x^2$;

$C'(x) = \cos x^2$  **b.** $\frac{x^3}{3} - \frac{x^7}{7\cdot 3!} + \frac{x^{11}}{11\cdot 5!} - \frac{x^{15}}{15\cdot 7!}$;

$x - \frac{x^5}{5\cdot 2!} + \frac{x^9}{9\cdot 4!} - \frac{x^{13}}{13\cdot 6!}$  **c.** $S(0.05) \approx 0.00004166664807$;

$C(-0.25) \approx -0.2499023614$  **d.** 1  **e.** 2

**81. a.** $1 - \frac{x^2}{4} + \frac{x^4}{64} - \frac{x^6}{2304}$  **b.** $R = \infty, -\infty < x < \infty$

## Chapter 11 Review Exercises, pp. 750–752

**1. a.** True  **b.** False  **c.** True  **d.** True  **e.** True

**3.** $p_2(x) = 1 - \frac{3}{2}x^2$  **5.** $p_2(x) = 1 - (x-1) + \frac{3}{2}(x-1)^2$

**7.** $p_2(x) = 1 - \frac{1}{2}(x-1)^2$

**9.** $p_3(x) = \frac{5}{4} + \frac{3}{4}(x - \ln 2) + \frac{5}{8}(x - \ln 2)^2 + \frac{1}{8}(x - \ln 2)^3$

**11. a.** $p_1(x) = 1 + x; p_2(x) = 1 + x + \frac{x^2}{2}$

**b.**

| $n$ | $p_n(x)$ | Error |
|---|---|---|
| 1 | 0.92 | $3.1 \times 10^{-3}$ |
| 2 | 0.9232 | $8.4 \times 10^{-5}$ |

**13. a.** $p_1(x) = \frac{\sqrt{2}}{2} + \frac{\sqrt{2}}{2}\left(x - \frac{\pi}{4}\right)$;

$p_2(x) = \frac{\sqrt{2}}{2} + \frac{\sqrt{2}}{2}\left(x - \frac{\pi}{4}\right) - \frac{\sqrt{2}}{4}\left(x - \frac{\pi}{4}\right)^2$

**b.**

| $n$ | $p_n(x)$ | Error |
|---|---|---|
| 1 | 0.5960 | $8.2 \times 10^{-3}$ |
| 2 | 0.5873 | $4.7 \times 10^{-4}$ |

**15.** $|R_3| < \frac{\pi^4}{4!}$  **17.** $R = \infty, (-\infty,\infty)$  **19.** $R = \infty, (-\infty,\infty)$

**21.** $R = 9, (-9, 9)$  **23.** $R = 2, [-4, 0)$  **25.** $R = \frac{3}{2}, [-2, 1]$

**27.** $R = \frac{1}{27}$  **29.** $\displaystyle\sum_{k=0}^{\infty}x^{2k}; (-1, 1)$  **31.** $\displaystyle\sum_{k=0}^{\infty}(-5x)^k; \left(-\frac{1}{5}, \frac{1}{5}\right)$

**33.** $\displaystyle\sum_{k=1}^{\infty}k(10x)^{k-1}; \left(-\frac{1}{10}, \frac{1}{10}\right)$  **35.** $1 + 3x + \frac{9x^2}{2!}; \displaystyle\sum_{k=0}^{\infty}\frac{(3x)^k}{k!}$

**37.** $-(x - \pi/2) + \frac{(x - \pi/2)^3}{3!} - \frac{(x - \pi/2)^5}{5!}$;

$\displaystyle\sum_{k=0}^{\infty}(-1)^{k+1}\frac{(x - \pi/2)^{2k+1}}{(2k+1)!}$

**39.** $4x - \frac{(4x)^3}{3} + \frac{(4x)^5}{5}; \displaystyle\sum_{k=0}^{\infty}\frac{(-1)^k(4x)^{2k+1}}{2k+1}$

**41.** $1 + 2(x-1)^2 + \frac{2}{3}(x-1)^4; \displaystyle\sum_{k=0}^{\infty}\frac{4^k(x-1)^{2k}}{(2k)!}$

**43.** $1 + \frac{x}{3} - \frac{x^2}{9} + \cdots$  **45.** $1 - \frac{3}{2}x + \frac{3}{2}x^2 - \cdots$

**47.** $R_n(x) = \frac{(\sinh c + \cosh c)\,x^{n+1}}{(n+1)!}$, where $c$ is between 0 and $x$;

$\displaystyle\lim_{n\to\infty}|R_n(x)| = |\sinh c + \cosh c|\lim_{n\to\infty}\frac{|x|^{n+1}}{(n+1)!} = 0$ because

$|x|^{n+1} \ll (n+1)!$ for any fixed value of $x$.

**49.** $\frac{1}{24}$  **51.** $\frac{1}{8}$  **53.** $\frac{1}{6}$  **55.** Approx. 0.4615  **57.** Approx. 0.3819

**59.** $11 - \frac{1}{11} - \frac{1}{2\cdot 11^3} - \frac{1}{2\cdot 11^5}$  **61.** $-\frac{1}{3} + \frac{1}{3\cdot 3^3} - \frac{1}{5\cdot 3^5} + \frac{1}{7\cdot 3^7}$

**63.** $y = 4 + 4x + \frac{4^2}{2!}x^2 + \frac{4^3}{3!}x^3 + \cdots + \frac{4^n}{n!}x^n + \cdots = 3 + e^{4x}$

**65. a.** $\displaystyle\sum_{k=1}^{\infty}\frac{(-1)^{k+1}}{k}$  **b.** $\displaystyle\sum_{k=1}^{\infty}\frac{1}{k2^k}$  **c.** $2\displaystyle\sum_{k=0}^{\infty}\frac{x^{2k+1}}{2k+1}$

**d.** $x = \frac{1}{3}; 2\displaystyle\sum_{k=0}^{\infty}\frac{1}{3^{2k+1}(2k+1)}$  **e.** Series in part (d)

## CHAPTER 12

## Section 12.1 Exercises, pp. 763–767

**1.** Plotting $\{(f(t), g(t)): a \le t \le b\}$ generates a curve in the $xy$-plane.
**3.** $x = R\cos(\pi t/5), y = -R\sin(\pi t/5)$  **5.** $x = t^2, y = t$,

$-\infty < t < \infty$  **7.** $-\frac{1}{2}$  **9.** $x = t, y = t, 0 \le t \le 6; x = 2t, y = 2t$,

$0 \le t \le 3; x = 3t, y = 3t, 0 \le t \le 2$ (answers will vary)

**11. a.**

| t | −10 | −4 | 0 | 4 | 10 |
|---|---|---|---|---|---|
| x | −20 | −8 | 0 | 8 | 20 |
| y | −34 | −16 | −4 | 8 | 26 |

**b.**

**c.** $y = \frac{3}{2}x - 4$
**d.** A line segment rising to the right as $t$ increases

**13. a.**

| t | −5 | −2 | 0 | 2 | 5 |
|---|---|---|---|---|---|
| x | 11 | 8 | 6 | 4 | 1 |
| y | −18 | −9 | −3 | 3 | 12 |

**b.**

**c.** $y = -3x + 15$
**d.** A line segment rising to the left as $t$ increases

**15. a.** $y = -x + 4$ **b.** A line segment starting at $(3, 1)$ and ending at $(4, 0)$ **17. a.** $y = 3x - 12$ **b.** A line segment starting at $(4, 0)$ and ending at $(8, 12)$ **19. a.** $x^2 + y^2 = 9$ **b.** Lower half of a circle of radius 3 centered at $(0, 0)$; starts at $(-3, 0)$ and ends at $(3, 0)$
**21. a.** $y = 1 - x^2, -1 \le x \le 1$ **b.** A parabola opening downward with a vertex at $(0, 1)$ starting at $(1, 0)$ and ending at $(-1, 0)$
**23. a.** $x^2 + (y - 1)^2 = 1$ **b.** A circle of radius 1 centered at $(0, 1)$; generated counterclockwise, starting and ending at $(1, 1)$
**25. a.** $y = (x + 1)^3$ **b.** A cubic curve rising to the right as $r$ increases **27. a.** $x^2 + y^2 = 49$ **b.** A circle of radius 7 centered at $(0, 0)$; generated counterclockwise, starting and ending at $(-7, 0)$
**29. a.** $y = 1, -\infty < x < \infty$ **b.** A horizontal line with $y$-intercept 1, generated from left to right **31.** $x^2 + y^2 = 4$ **33.** $y = \sqrt{4 - x^2}$
**35.** $y = x^2$ **37.** $x = 4 \cos t, y = 4 \sin t, 0 \le t \le 2\pi$
**39.** $x = \cos t + 2, y = \sin t + 3, 0 \le t \le 2\pi$
**41.** $x = 2t, y = 8t; 0 \le t \le 1$
**43.** $x = t, y = 2t^2 - 4; -1 \le t \le 5$ **45.** $x = 2, y = t; 3 \le t \le 9$
**47.** $x = 4t - 2, y = -6t + 3; 0 \le t \le 1$ and $x = t + 1, y = 8t - 11; 1 \le t \le 2$
**49.** $x = 1 + 2t, y = 1 + 4t; -\infty < t < \infty$
**51.** $x = t^2, y = t; t \ge 0$
**53.** $x = 400 \cos\left(\frac{4\pi t}{3}\right), y = 400 \sin\left(\frac{4\pi t}{3}\right); 0 \le t \le 1.5$
**55.** $x = 50 \cos\left(\frac{\pi t}{12}\right), y(t) = 50 \sin\left(\frac{\pi t}{12}\right); 0 \le t \le 24$

**57.**    **59.**

**61.**    **63.** Plot $x = 1 + \cos^2 t - \sin^2 t$, $y = t$.
**65.** Approx. 2857 m

**67. a.** $\frac{dy}{dx} = -2; -2$ **b.**
**69. a.** $\frac{dy}{dx} = -8 \cot t; 0$ **b.**

**71. a.** $\frac{dy}{dx} = \frac{t^2 + 1}{t^2 - 1}, t \ne 0$; undefined **b.**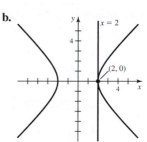

**73.** $y = \frac{13}{4}x + \frac{1}{4}$ **75.** $y = x - \frac{\pi\sqrt{2}}{4}$ **77.** $\left(-\frac{4}{\sqrt{5}}, \frac{8}{\sqrt{5}}\right)$ and $\left(\frac{4}{\sqrt{5}}, -\frac{8}{\sqrt{5}}\right)$ **79.** There is no such point. **81.** 10 **83.** $\pi\sqrt{2}$
**85.** $\frac{1}{3}(5\sqrt{5} - 8)$ **87.** $\frac{3}{2}$ **89. a.** False **b.** True **c.** False **d.** True **e.** True
**91.** $0 \le t \le 2\pi$
**93.** $x = 3 \cos t, y = \frac{3}{2} \sin t, 0 \le t \le 2\pi; \left(\frac{x}{3}\right)^2 + \left(\frac{2y}{3}\right)^2 = 1$; in the counterclockwise direction
**95. a.** Lines intersect at $(1, 0)$. **b.** Lines are parallel. **c.** Lines intersect at $(4, 6)$.

**97.**

**99.**

**13.**

$(4, \pi/2), (4, 5\pi/2)$

**15.**

**101.**

**a.** $(0, 2)$ and $(0, -2)$
**b.** $(1, \sqrt{2}), (1, -\sqrt{2}),$
$(-1, \sqrt{2}), (-1, -\sqrt{2})$

**17.**

**19.**

**103.** $27\pi$   **105.** $\dfrac{3\pi}{8}$   **107. a.** A circle of radius 3 centered at $(0, 4)$

**b.** A torus (doughnut); $48\pi^2$   **109.** $\dfrac{64\pi}{3}$

**111.** $\displaystyle\int_0^1 2\pi(e^{3t} + 1)\sqrt{4e^{4t} + 9e^{6t}}\, dt \approx 1445.9$

**113. a.**

**b.**

**21.**

**23.** $\left(100, -\dfrac{\pi}{4}\right)$

**25.** $(3\sqrt{2}/2, 3\sqrt{2}/2)$
**27.** $(1/2, -\sqrt{3}/2)$
**29.** $(2\sqrt{2}, -2\sqrt{2})$
**31.** $(2\sqrt{2}, \pi/4), (-2\sqrt{2}, 5\pi/4)$
**33.** $(2, \pi/3), (-2, 4\pi/3)$
**35.** $(8, 2\pi/3), (-8, -\pi/3)$

**37.** $x = -4$; vertical line passing through $(-4, 0)$
**39.** $x^2 + y^2 = 4$; circle of radius 2 centered at $(0, 0)$
**41.** $(x - 1)^2 + (y - 1)^2 = 2$; circle of radius $\sqrt{2}$ centered at $(1, 1)$
**43.** $(x - 3)^2 + (y - 4)^2 = 25$; circle of radius 5 centered at $(3, 4)$
**45.** $x^2 + (y - 1)^2 = 1$; circle of radius 1 centered at $(0, 1)$ and
$x = 0$   **47.** $x^2 + (y - 4)^2 = 16$; circle of radius 4 centered at $(0, 4)$
**49.** $r = \tan \theta \sec \theta$   **51.** $r^2 = \sec \theta \csc \theta$ or $r^2 = 2 \csc 2\theta$

### Section 12.2 Exercises, pp. 775–779

**1.**

$(-2, -5\pi/6), (2, 13\pi/6);$
$(3, \pi/2), (3, 5\pi/2)$

**3.** $r^2 = x^2 + y^2, \tan \theta = \dfrac{y}{x}$
**5.** $r \cos \theta = 5$ or $r = 5 \sec \theta$
**7.** $x$-axis symmetry occurs if $(r, \theta)$ on the graph implies $(r, -\theta)$ is on the graph. $y$-axis symmetry occurs if $(r, \theta)$ on the graph implies $(r, \pi - \theta) = (-r, -\theta)$ is on the graph. Symmetry about the origin occurs if $(r, \theta)$ on the graph implies $(-r, \theta) = (r, \theta + \pi)$ is on the graph.

**53.**

**55.**

**9.**

$(-2, -3\pi/4), (2, 9\pi/4)$

**11.**

$(1, 2\pi/3), (1, 8\pi/3)$

**57.**

**59.**

**61.**

**63.**

**91. a.** A   **b.** C   **c.** B   **d.** D   **e.** E   **f.** F

**93.**

**95.**

**65.**

$r = 1 - 2 \sin 3\theta$

Origin:
$B, D, F, H, J, L$

**67.**

$r = \frac{1}{4} - \cos 4\theta$

Origin:
$B, D, F, H, J, L, N, P$

**97.**

**99.**

**69.** $[0, 8\pi]$

**71.** $[0, 2\pi]$

**103.**

For $a = -1$, the spiral winds inward toward the origin.

**73.** $[0, 5\pi]$

**75.** $[0, 2\pi]$

**105. a.**

**107. a.**

**77. a.** True   **b.** True   **c.** False   **d.** True   **e.** True

**81.** A circle of radius 4 and center $(2, \pi/3)$ (polar coordinates)

**83.** A circle of radius 4 centered at $(2, 3)$ (Cartesian coordinates)

**109.** Symmetry about the $x$-axis   **111.** $(x^2 + y^2)^2 = a^2(x^2 - y^2)$

## Section 12.3 Exercises, pp. 786–788

**1.** $x = f(\theta)\cos\theta, y = f(\theta)\sin\theta$   **3.** The slope of the tangent line is the rate of change of the vertical coordinate with respect to the horizontal coordinate.   **5.** $\sqrt{3}$   **7.** $\frac{\pi^2}{4}$   **9.** Both curves pass through the origin, but for different values of $\theta$.   **11.** 0   **13.** $-\sqrt{3}$   **15.** Undefined, undefined   **17.** 0 at $(-4, \pi/2)$ and $(-4, 3\pi/2)$, undefined at $(4, 0)$ and $(4, \pi)$   **19.** $\pm 1$   **21.** $\theta = \frac{3\pi}{4}$; $m = -1$

**23. a.** $[0, \pi]$

**b.** $\theta = \frac{\pi}{4}, m = 1; \theta = \frac{7\pi}{12}, m \approx -3.73$;

$\theta = \frac{11\pi}{12}, m \approx -0.27$

**85. a.** 132.3 miles   **b.** 264.6 mi/hr

**87.**

$y = -\dfrac{x}{\sqrt{3}} + 2\sqrt{3}$

**89.**

$y = 4x + 3$

**25.** Horizontal: $(2\sqrt{2}, \pi/4), (-2\sqrt{2}, 3\pi/4)$; vertical: $(0, \pi/2), (4, 0)$
**27.** Horizontal: $(0, 0)$ $(0.943, 0.955), (-0.943, 2.186), (0.943, 4.097),$
$(-0.943, 5.328)$; vertical: $(0, 0), (0.943, 0.615), (-0.943, 2.526),$
$(0.943, 3.757), (-0.943, 5.668)$   **29.** $(2, 0)$ and $(0, 0)$

**31.** $\left(1, \dfrac{\pi}{12}\right), \left(1, \dfrac{5\pi}{12}\right), \left(1, \dfrac{7\pi}{12}\right), \left(1, \dfrac{11\pi}{12}\right),$

$\left(1, \dfrac{13\pi}{12}\right), \left(1, \dfrac{17\pi}{12}\right), \left(1, \dfrac{19\pi}{12}\right),$ and $\left(1, \dfrac{23\pi}{12}\right)$

**33.** 1

**35.** $16\pi$

**37.** $9\pi/2$

**39.** $\dfrac{\pi}{12}$

**41. a.** $(0, 0), \left(\dfrac{3}{\sqrt{2}}, \dfrac{\pi}{4}\right)$   **b.** $\dfrac{9}{8}(\pi - 2)$

**43. a.** $\left(1 + \dfrac{1}{\sqrt{2}}, \dfrac{\pi}{4}\right), \left(1 - \dfrac{1}{\sqrt{2}}, \dfrac{5\pi}{4}\right), (0, 0)$   **b.** $\dfrac{3\pi}{2} - 2\sqrt{2}$

**45.** $\dfrac{1}{24}(3\sqrt{3} + 2\pi)$   **47.** $\dfrac{1}{4}(2 - \sqrt{3}) + \dfrac{\pi}{12}$   **49.** $\pi/20$

**51.** $4(4\pi/3 - \sqrt{3})$   **53.** $2\pi/3 - \sqrt{3}/2$   **55.** $9\pi + 27\sqrt{3}$

**57.** 6   **59.** $18\pi$   **61.** Intersection points: $\left(3, \pm\dfrac{\pi}{3}\right)$; area of

region A $= 6\sqrt{3} - 2\pi$; area of region B $= 5\pi - 6\sqrt{3}$; area of

region C $= 4\pi + 6\sqrt{3}$   **63.** $\pi a$   **65.** $\dfrac{8}{3}\left((1 + \pi^2)^{3/2} - 1\right)$

**67.** 32   **69.** $63\sqrt{5}$   **71.** $\dfrac{2\pi - 3\sqrt{3}}{8}$   **73.** 26.73

**75. a.** False   **b.** False   **c.** True
**77.** Horizontal: $(0, 0), (4.05, 2.03), (9.83, 4.91)$;

vertical: $(1.72, 0.86), (6.85, 3.43), (12.87, 6.44)$   **79.** $\dfrac{\sqrt{1 + a^2}}{a}$

**81. a.** $A_n = \dfrac{1}{4e^{(4n+2)\pi}} - \dfrac{1}{4e^{4n\pi}} - \dfrac{1}{4e^{(4n-2)\pi}} + \dfrac{1}{4e^{(4n-4)\pi}}$   **b.** 0
**c.** $e^{-4\pi}$   **85.** $(a^2 - 2)\theta^* + \pi - \sin 2\theta^*$, where $\theta^* = \cos^{-1}(a/2)$
**87.** $a^2(\pi/2 + a/3)$

## Section 12.4 Exercises, pp. 797–800

**1.** A parabola is the set of all points in a plane equidistant from a fixed
point and a fixed line.   **3.** A hyperbola is the set of all points in a
plane whose distances from two fixed points have a constant difference.

**5.** Parabola:

Hyperbola:

Ellipse:

**7.** $\left(\dfrac{x}{a}\right)^2 + \dfrac{y^2}{a^2 - c^2} = 1$   **9.** $(\pm ae, 0)$   **11.** $y = \pm\dfrac{b}{a}x$

**13.** Parabola

**15.** Ellipse

Vertices: $(\pm 2, 0)$; foci: $(\pm\sqrt{3}, 0)$;
major axis has length 4; minor axis
has length 2.

**17.** Parabola

**19.** Hyperbola

Vertices: $(\pm 2, 0)$;
foci: $(\pm\sqrt{5}, 0)$;
asymptotes: $y = \pm\dfrac{1}{2}x$

**21.** Hyperbola

**23.** Parabola

**25.** Ellipse

Vertices: $(0, \pm 4)$; foci: $(0, \pm 2\sqrt{3})$; major axis has length 8; minor axis has length 4.

**27.** Ellipse

Vertices: $(0, \pm\sqrt{7})$; foci: $(0, \pm\sqrt{2})$; major axis has length $2\sqrt{7}$; minor axis has length $2\sqrt{5}$.

**29.** Hyperbola

**31.** $y^2 = 16x$  **33.** $y^2 = 12x$

**35.** $x^2 = -\dfrac{2}{3}y$

**37.** $y^2 = 4(x + 1)$

**39.** $\dfrac{x^2}{16} + \dfrac{y^2}{9} = 1$

**41.** $\dfrac{x^2}{16} - \dfrac{y^2}{20} = 1$

**43.** $\dfrac{x^2}{25} + y^2 = 1$  **45.** $\dfrac{x^2}{4} - \dfrac{y^2}{9} = 1$  **47.** $\dfrac{x^2}{4} + \dfrac{y^2}{9} = 1$

**49.** $\dfrac{x^2}{16} - \dfrac{y^2}{9} = 1$  **51. a.** True  **b.** True  **c.** True  **d.** True

**53.** $\dfrac{x^2}{81} + \dfrac{y^2}{72} = 1$

Directrices: $x = \pm 27$

**55.** $x^2 - \dfrac{y^2}{8} = 1$

**57.**

Vertex: $(2, 0)$; focus: $(0, 0)$; directrix: $x = 4$

**59.**

Vertices: $(1, 0)$, $\left(-\tfrac{1}{3}, 0\right)$; center: $\left(\tfrac{1}{3}, 0\right)$; foci: $(0, 0)$, $\left(\tfrac{2}{3}, 0\right)$; directrices: $x = -1$, $x = \tfrac{5}{3}$

**61.**

Vertex: $\left(0, -\tfrac{1}{4}\right)$; focus: $(0, 0)$; directrix: $y = -\tfrac{1}{2}$

**63.**

The parabola starts at $(1, 0)$ and goes through quadrants I, II, and III for $\theta$ in $[0, 3\pi/2]$; then it approaches $(1, 0)$ by traveling through quadrant IV on $(3\pi/2, 2\pi)$.

The parabola begins in the first quadrant and passes through the points $(0, 3)$, $\left(-\tfrac{3}{2}, 0\right)$, and $(0, -3)$ as $\theta$ ranges from 0 to $2\pi$.

**65.**

**67.** The parabolas open to the left due to the presence of a positive $\cos\theta$ term in the denominator. As $d$ increases, the directrix $x = d$ moves to the right, resulting in wider parabolas.

**69.** $y = 2x + 6$  **71.** $y = -\dfrac{3}{40}x - \dfrac{4}{5}$  **73.** $\dfrac{dy}{dx} = -\dfrac{b^2x}{a^2y}$, so $\dfrac{y - y_0}{x - x_0} = -\dfrac{b^2x_0}{a^2y_0}$, which is equivalent to the given equation.

**75.** $r = \dfrac{4}{1 - 2\sin\theta}$  **79.** $\dfrac{4\pi b^2 a}{3}$; $\dfrac{4\pi a^2 b}{3}$; yes, if $a \neq b$

**81. a.** $\dfrac{\pi b^2}{3a^2}(a - c)^2(2a + c)$  **b.** $\dfrac{4\pi b^4}{3a}$  **91.** $2p$

**97. a.** $u(m) = \dfrac{2m^2 - \sqrt{3m^2 + 1}}{m^2 - 1}$; $v(m) = \dfrac{2m^2 + \sqrt{3m^2 + 1}}{m^2 - 1}$; 2 intersection points for $|m| > 1$  **b.** $\dfrac{5}{4}, \infty$  **c.** 2, 2  **d.** $2\sqrt{3} - \ln(\sqrt{3} + 2)$

## Chapter 12 Review Exercises, pp. 800–803

**1. a.** False  **b.** False  **c.** True  **d.** False  **e.** True  **f.** True

**3.** $\dfrac{x^2}{16} + \dfrac{y^2}{9} = 1$; ellipse generated counterclockwise

**5.** Segment of the parabola $y = \sqrt{x}$ starting at $(4, 2)$ and ending at $(9, 3)$

**7. a.** $(x - 1)^2 + (y - 2)^2 = 64$  **b.** $-\dfrac{1}{\sqrt{3}}$  **c.**

**9.** $x = 5(t - 1)(t - 2)\sin t$, $y = t$

**11. a.** $x^2 + (y + 1)^2 = 9$    **b.** Lower half of a circle of radius 3 centered at $(0, -1)$, starting at $(3, -1)$ and ending at $(-3, -1)$    **c.** 0

**13.** At $t = \pi/6$: $y = (2 + \sqrt{3})x + \left(2 - \dfrac{\pi}{3} - \dfrac{\pi\sqrt{3}}{6}\right)$; at

$t = \dfrac{2\pi}{3}$: $y = \dfrac{x}{\sqrt{3}} + 2 - \dfrac{2\pi}{3\sqrt{3}}$    **15.** $x = -1 + 2t, y = t$,

for $0 \le t \le 1$; $x = 1 - 2t, y = 1 - t$, for $0 \le t \le 1$
**17.** $x = 3 \sin t, y = 3 \cos t$, for $0 \le t \le 2\pi$

**19.** $\dfrac{4}{15}$    **21.** 9.1    **23.** $4 - 2\sqrt{2}$

**25.**     **27.**

**29.**     **31.**

**33.** Liz should choose $r = 1 - \sin \theta$.

**35.** $(x - 3)^2 + (y + 1)^2 = 10$; a circle of radius $\sqrt{10}$ centered at $(3, -1)$
**37.** $r = 1 + \cos \theta$; a cardioid
**39.** $r = 8 \cos \theta, 0 \le \theta \le \pi$

**41. a.** Horizontal: $\left(\dfrac{1}{2}, \dfrac{\pi}{6}\right), \left(\dfrac{1}{2}, \dfrac{5\pi}{6}\right), \left(2, \dfrac{3\pi}{2}\right)$;

vertical: $\left(\dfrac{3}{2}, \dfrac{7\pi}{6}\right), \left(\dfrac{3}{2}, \dfrac{11\pi}{6}\right), \left(0, \dfrac{\pi}{2}\right)$    **b.** Undefined

**c.**

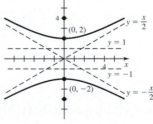

**43.** $\left(\dfrac{\pi}{12}, \dfrac{1}{2^{1/4}}\right), \left(\dfrac{3\pi}{4}, \dfrac{1}{2^{1/4}}\right), \left(\dfrac{17\pi}{12}, \dfrac{1}{2^{1/4}}\right), (0, 0)$

**45.** $\pi - \dfrac{3\sqrt{3}}{2}$    **47.** $2\sqrt{3} - \dfrac{2\pi}{3}$    **49.** 4    **51.** 40.09

**53. a.** Hyperbola
**b.** Foci $(\pm\sqrt{3}, 0)$, vertices $(\pm1, 0)$, directrices $x = \pm\dfrac{1}{\sqrt{3}}$
**c.** $e = \sqrt{3}$

**d.**

**55. a.** Hyperbola
**b.** Foci $(0, \pm2\sqrt{5})$, vertices $(0, \pm4)$, directrices $y = \pm\dfrac{8}{\sqrt{5}}$    **c.** $e = \dfrac{\sqrt{5}}{2}$

**d.**

**57. a.** Ellipse
**b.** Foci $(\pm\sqrt{2}, 0)$, vertices $(\pm2, 0)$, directrices $x = \pm2\sqrt{2}$    **c.** $e = \dfrac{\sqrt{2}}{2}$

**d.**

**59.** $y = \dfrac{3}{2}x - 2$

**61.** $e = 1$

**63.** $e = \dfrac{1}{2}$

**65.** $\dfrac{y^2}{4} - \dfrac{x^2}{12} = 1$

**67.** $\dfrac{y^2}{16} + \dfrac{25x^2}{336} = 1$; foci: $\left(0, \pm\dfrac{8}{5}\right)$

**69.** $e = 2/3, y = \pm9, (\pm2\sqrt{5}, 0)$    **71.** $m = \dfrac{b}{a}$

**75. a.** $x = \pm a \cos^{2/n} t, y = \pm b \sin^{2/n} t$
**c.** The curve becomes more rectangular as $n$ increases.

# Index

# Index of Applications

## Economics and Business

## Engineering and Physical Sciences

# TABLE OF INTEGRALS

| Substitution Rule | Integration by Parts |
|---|---|
| $\int f(g(x))g'(x)\,dx = \int f(u)\,du \quad (u = g(x))$ | $\int u\,dv = uv - \int v\,du$ |
| $\int_a^b f(g(x))g'(x)\,dx = \int_{g(a)}^{g(b)} f(u)\,du$ | $\int_a^b uv'\,dx = uv\Big|_a^b - \int_a^b vu'\,dx$ |

## Basic Integrals

**1.** $\displaystyle\int x^n\,dx = \frac{1}{n+1}x^{n+1} + C;\ n \neq -1$

**2.** $\displaystyle\int \frac{dx}{x} = \ln|x| + C$

**3.** $\displaystyle\int \cos ax\,dx = \frac{1}{a}\sin ax + C$

**4.** $\displaystyle\int \sin ax\,dx = -\frac{1}{a}\cos ax + C$

**5.** $\displaystyle\int \tan x\,dx = \ln|\sec x| + C$

**6.** $\displaystyle\int \cot x\,dx = \ln|\sin x| + C$

**7.** $\displaystyle\int \sec x\,dx = \ln|\sec x + \tan x| + C$

**8.** $\displaystyle\int \csc x\,dx = -\ln|\csc x + \cot x| + C$

**9.** $\displaystyle\int e^{ax}\,dx = \frac{1}{a}e^{ax} + C$

**10.** $\displaystyle\int b^{ax}\,dx = \frac{1}{a\ln b}b^{ax} + C;\ b > 0, b \neq 1$

**11.** $\displaystyle\int \ln x\,dx = x\ln x - x + C$

**12.** $\displaystyle\int \log_b x\,dx = \frac{1}{\ln b}(x\ln x - x) + C$

**13.** $\displaystyle\int \frac{dx}{x^2 + a^2} = \frac{1}{a}\tan^{-1}\frac{x}{a} + C$

**14.** $\displaystyle\int \frac{dx}{\sqrt{a^2 - x^2}} = \sin^{-1}\frac{x}{a} + C, a > 0$

**15.** $\displaystyle\int \frac{dx}{x\sqrt{x^2 - a^2}} = \frac{1}{a}\sec^{-1}\left|\frac{x}{a}\right| + C, a > 0$

**16.** $\displaystyle\int \sin^{-1}x\,dx = x\sin^{-1}x + \sqrt{1 - x^2} + C$

**17.** $\displaystyle\int \cos^{-1}x\,dx = x\cos^{-1}x - \sqrt{1 - x^2} + C$

**18.** $\displaystyle\int \tan^{-1}x\,dx = x\tan^{-1}x - \frac{1}{2}\ln(1 + x^2) + C$

**19.** $\displaystyle\int \sec^{-1}x\,dx = x\sec^{-1}x - \ln(x + \sqrt{x^2 - 1}) + C$

**20.** $\displaystyle\int \sinh x\,dx = \cosh x + C$

**21.** $\displaystyle\int \cosh x\,dx = \sinh x + C$

**22.** $\displaystyle\int \text{sech}^2 x\,dx = \tanh x + C$

**23.** $\displaystyle\int \text{csch}^2 x\,dx = -\coth x + C$

**24.** $\displaystyle\int \text{sech}\,x\tanh x\,dx = -\text{sech}\,x + C$

**25.** $\displaystyle\int \text{csch}\,x\coth x\,dx = -\text{csch}\,x + C$

**26.** $\displaystyle\int \tanh x\,dx = \ln\cosh x + C$

**27.** $\displaystyle\int \coth x\,dx = \ln|\sinh x| + C$

**28.** $\displaystyle\int \text{sech}\,x\,dx = \tan^{-1}\sinh x + C = \sin^{-1}\tanh x + C$

**29.** $\displaystyle\int \text{csch}\,x\,dx = \ln|\tanh(x/2)| + C$

## Trigonometric Integrals

**30.** $\displaystyle\int \cos^2 x\,dx = \frac{x}{2} + \frac{\sin 2x}{4} + C$

**31.** $\displaystyle\int \sin^2 x\,dx = \frac{x}{2} - \frac{\sin 2x}{4} + C$

**32.** $\displaystyle\int \sec^2 ax\,dx = \frac{1}{a}\tan ax + C$

**33.** $\displaystyle\int \csc^2 ax\,dx = -\frac{1}{a}\cot ax + C$

**34.** $\displaystyle\int \tan^2 x\,dx = \tan x - x + C$

**35.** $\displaystyle\int \cot^2 x\,dx = -\cot x - x + C$

**36.** $\displaystyle\int \cos^3 x\,dx = -\frac{1}{3}\sin^3 x + \sin x + C$

**37.** $\displaystyle\int \sin^3 x\,dx = \frac{1}{3}\cos^3 x - \cos x + C$

**38.** $\int \sec^3 x \, dx = \frac{1}{2} \sec x \tan x + \frac{1}{2} \ln |\sec x + \tan x| + C$

**39.** $\int \csc^3 x \, dx = -\frac{1}{2} \csc x \cot x - \frac{1}{2} \ln |\csc x + \cot x| + C$

**40.** $\int \tan^3 x \, dx = \frac{1}{2} \tan^2 x - \ln |\sec x| + C$

**41.** $\int \cot^3 x \, dx = -\frac{1}{2} \cot^2 x - \ln |\sin x| + C$

**42.** $\int \sec^n ax \tan ax \, dx = \frac{1}{na} \sec^n ax + C; \; n \neq 0$

**43.** $\int \csc^n ax \cot ax \, dx = -\frac{1}{na} \csc^n ax + C; \; n \neq 0$

**44.** $\int \frac{dx}{1 + \sin ax} = -\frac{1}{a} \tan \left( \frac{\pi}{4} - \frac{ax}{2} \right) + C$

**45.** $\int \frac{dx}{1 - \sin ax} = \frac{1}{a} \tan \left( \frac{\pi}{4} + \frac{ax}{2} \right) + C$

**46.** $\int \frac{dx}{1 + \cos ax} = \frac{1}{a} \tan \frac{ax}{2} + C$

**47.** $\int \frac{dx}{1 - \cos ax} = -\frac{1}{a} \cot \frac{ax}{2} + C$

**48.** $\int \sin mx \cos nx \, dx = -\frac{\cos (m + n)x}{2(m + n)} - \frac{\cos (m - n)x}{2(m - n)} + C; \; m^2 \neq n^2$

**49.** $\int \sin mx \sin nx \, dx = \frac{\sin (m - n)x}{2(m - n)} - \frac{\sin (m + n)x}{2(m + n)} + C; \; m^2 \neq n^2$

**50.** $\int \cos mx \cos nx \, dx = \frac{\sin (m - n)x}{2(m - n)} + \frac{\sin (m + n)x}{2(m + n)} + C; \; m^2 \neq n^2$

## Reduction Formulas for Trigonometric Functions

**51.** $\int \cos^n x \, dx = \frac{1}{n} \cos^{n-1} x \sin x + \frac{n - 1}{n} \int \cos^{n-2} x \, dx$

**52.** $\int \sin^n x \, dx = -\frac{1}{n} \sin^{n-1} x \cos x + \frac{n - 1}{n} \int \sin^{n-2} x \, dx$

**53.** $\int \tan^n x \, dx = \frac{\tan^{n-1} x}{n - 1} - \int \tan^{n-2} x \, dx; \; n \neq 1$

**54.** $\int \cot^n x \, dx = -\frac{\cot^{n-1} x}{n - 1} - \int \cot^{n-2} x \, dx; \; n \neq 1$

**55.** $\int \sec^n x \, dx = \frac{\sec^{n-2} x \tan x}{n - 1} + \frac{n - 2}{n - 1} \int \sec^{n-2} x \, dx; \; n \neq 1$

**56.** $\int \csc^n x \, dx = -\frac{\csc^{n-2} x \cot x}{n - 1} + \frac{n - 2}{n - 1} \int \csc^{n-2} x \, dx; \; n \neq 1$

**57.** $\int \sin^m x \cos^n x \, dx = -\frac{\sin^{m-1} x \cos^{n+1} x}{m + n} + \frac{m - 1}{m + n} \int \sin^{m-2} x \cos^n x \, dx; \; m \neq -n$

**58.** $\int \sin^m x \cos^n x \, dx = \frac{\sin^{m+1} x \cos^{n-1} x}{m + n} + \frac{n - 1}{m + n} \int \sin^m x \cos^{n-2} x \, dx; \; m \neq -n$

**59.** $\int x^n \sin ax \, dx = -\frac{x^n \cos ax}{a} + \frac{n}{a} \int x^{n-1} \cos ax \, dx; \; a \neq 0$

**60.** $\int x^n \cos ax \, dx = \frac{x^n \sin ax}{a} - \frac{n}{a} \int x^{n-1} \sin ax \, dx; \; a \neq 0$

## Integrals Involving $a^2 - x^2$; $a > 0$

**61.** $\int \sqrt{a^2 - x^2} \, dx = \frac{x}{2} \sqrt{a^2 - x^2} + \frac{a^2}{2} \sin^{-1} \frac{x}{a} + C$

**62.** $\int \frac{dx}{x \sqrt{a^2 - x^2}} = -\frac{1}{a} \ln \left| \frac{a + \sqrt{a^2 - x^2}}{x} \right| + C$

**63.** $\int \frac{dx}{x^2 \sqrt{a^2 - x^2}} = -\frac{\sqrt{a^2 - x^2}}{a^2 x} + C$

**64.** $\int x^2 \sqrt{a^2 - x^2} \, dx = \frac{x}{8} (2x^2 - a^2) \sqrt{a^2 - x^2} + \frac{a^4}{8} \sin^{-1} \frac{x}{a} + C$

**65.** $\int \frac{\sqrt{a^2 - x^2}}{x^2} \, dx = -\frac{1}{x} \sqrt{a^2 - x^2} - \sin^{-1} \frac{x}{a} + C$

**66.** $\int \frac{x^2}{\sqrt{a^2 - x^2}} \, dx = -\frac{x}{2} \sqrt{a^2 - x^2} + \frac{a^2}{2} \sin^{-1} \frac{x}{a} + C$

**67.** $\int \frac{dx}{a^2 - x^2} = \frac{1}{2a} \ln \left| \frac{x + a}{x - a} \right| + C$

## Integrals Involving $x^2 - a^2$; $a > 0$

**68.** $\int \sqrt{x^2 - a^2} \, dx = \frac{x}{2} \sqrt{x^2 - a^2} - \frac{a^2}{2} \ln |x + \sqrt{x^2 - a^2}| + C$

**69.** $\int \frac{dx}{\sqrt{x^2 - a^2}} = \ln |x + \sqrt{x^2 - a^2}| + C$

**70.** $\int \frac{dx}{x^2 \sqrt{x^2 - a^2}} = \frac{\sqrt{x^2 - a^2}}{a^2 x} + C$

**71.** $\int x^2 \sqrt{x^2 - a^2} \, dx = \frac{x}{8} (2x^2 - a^2) \sqrt{x^2 - a^2} - \frac{a^4}{8} \ln |x + \sqrt{x^2 - a^2}| + C$

**72.** $\int \frac{\sqrt{x^2 - a^2}}{x^2} \, dx = \ln |x + \sqrt{x^2 - a^2}| - \frac{\sqrt{x^2 - a^2}}{x} + C$

**73.** $\int \frac{x^2}{\sqrt{x^2 - a^2}} \, dx = \frac{a^2}{2} \ln |x + \sqrt{x^2 - a^2}| + \frac{x}{2} \sqrt{x^2 - a^2} + C$

**74.** $\int \frac{dx}{x^2 - a^2} = \frac{1}{2a} \ln \left| \frac{x - a}{x + a} \right| + C$

**75.** $\int \frac{dx}{x(x^2 - a^2)} = \frac{1}{2a^2} \ln \left| \frac{x^2 - a^2}{x^2} \right| + C$

## Integrals Involving $a^2 + x^2$; $a > 0$

**76.** $\displaystyle\int \sqrt{a^2 + x^2}\, dx = \frac{x}{2}\sqrt{a^2 + x^2} + \frac{a^2}{2}\ln\left(x + \sqrt{a^2 + x^2}\right) + C$

**77.** $\displaystyle\int \frac{dx}{\sqrt{a^2 + x^2}} = \ln\left(x + \sqrt{a^2 + x^2}\right) + C$

**78.** $\displaystyle\int \frac{dx}{x\sqrt{a^2 + x^2}} = \frac{1}{a}\ln\left|\frac{a - \sqrt{a^2 + x^2}}{x}\right| + C$

**79.** $\displaystyle\int \frac{dx}{x^2\sqrt{a^2 + x^2}} = -\frac{\sqrt{a^2 + x^2}}{a^2 x} + C$

**80.** $\displaystyle\int x^2\sqrt{a^2 + x^2}\, dx = \frac{x}{8}(a^2 + 2x^2)\sqrt{a^2 + x^2} - \frac{a^4}{8}\ln\left(x + \sqrt{a^2 + x^2}\right) + C$

**81.** $\displaystyle\int \frac{\sqrt{a^2 + x^2}}{x^2}\, dx = \ln\left|x + \sqrt{a^2 + x^2}\right| - \frac{\sqrt{a^2 + x^2}}{x} + C$

**82.** $\displaystyle\int \frac{x^2}{\sqrt{a^2 + x^2}}\, dx = -\frac{a^2}{2}\ln\left(x + \sqrt{a^2 + x^2}\right) + \frac{x\sqrt{a^2 + x^2}}{2} + C$

**83.** $\displaystyle\int \frac{\sqrt{a^2 + x^2}}{x}\, dx = \sqrt{a^2 + x^2} - a\ln\left|\frac{a + \sqrt{a^2 + x^2}}{x}\right| + C$

**84.** $\displaystyle\int \frac{dx}{(a^2 + x^2)^{3/2}} = \frac{x}{a^2\sqrt{a^2 + x^2}} + C$

**85.** $\displaystyle\int \frac{dx}{x(a^2 + x^2)} = \frac{1}{2a^2}\ln\left(\frac{x^2}{a^2 + x^2}\right) + C$

## Integrals Involving $ax \pm b$; $a \neq 0, b > 0$

**86.** $\displaystyle\int (ax + b)^n\, dx = \frac{(ax + b)^{n+1}}{a(n + 1)} + C;\ n \neq -1$

**87.** $\displaystyle\int (\sqrt{ax + b})^n\, dx = \frac{2}{a}\frac{(\sqrt{ax + b})^{n+2}}{n + 2} + C;\ n \neq -2$

**88.** $\displaystyle\int \frac{dx}{x\sqrt{ax - b}} = \frac{2}{\sqrt{b}}\tan^{-1}\sqrt{\frac{ax - b}{b}} + C$

**89.** $\displaystyle\int \frac{dx}{x\sqrt{ax + b}} = \frac{1}{\sqrt{b}}\ln\left|\frac{\sqrt{ax + b} - \sqrt{b}}{\sqrt{ax + b} + \sqrt{b}}\right| + C$

**90.** $\displaystyle\int \frac{x}{ax + b}\, dx = \frac{x}{a} - \frac{b}{a^2}\ln|ax + b| + C$

**91.** $\displaystyle\int \frac{x^2}{ax + b}\, dx = \frac{1}{2a^3}\left((ax + b)^2 - 4b(ax + b) + 2b^2\ln|ax + b|\right) + C$

**92.** $\displaystyle\int \frac{dx}{x^2(ax + b)} = -\frac{1}{bx} + \frac{a}{b^2}\ln\left|\frac{ax + b}{x}\right| + C$

**93.** $\displaystyle\int x\sqrt{ax + b}\, dx = \frac{2}{15a^2}(3ax - 2b)(ax + b)^{3/2} + C$

**94.** $\displaystyle\int \frac{x}{\sqrt{ax + b}}\, dx = \frac{2}{3a^2}(ax - 2b)\sqrt{ax + b} + C$

**95.** $\displaystyle\int x(ax + b)^n\, dx = \frac{(ax + b)^{n+1}}{a^2}\left(\frac{ax + b}{n + 2} - \frac{b}{n + 1}\right) + C;\ n \neq -1, -2$

**96.** $\displaystyle\int \frac{dx}{x(ax + b)} = \frac{1}{b}\ln\left|\frac{x}{ax + b}\right| + C$

## Integrals with Exponential and Trigonometric Functions

**97.** $\displaystyle\int e^{ax}\sin bx\, dx = \frac{e^{ax}(a\sin bx - b\cos bx)}{a^2 + b^2} + C$

**98.** $\displaystyle\int e^{ax}\cos bx\, dx = \frac{e^{ax}(a\cos bx + b\sin bx)}{a^2 + b^2} + C$

## Integrals with Exponential and Logarithmic Functions

**99.** $\displaystyle\int \frac{dx}{x\ln x} = \ln|\ln x| + C$

**100.** $\displaystyle\int x^n\ln x\, dx = \frac{x^{n+1}}{n + 1}\left(\ln x - \frac{1}{n + 1}\right) + C;\ n \neq -1$

**101.** $\displaystyle\int xe^x\, dx = xe^x - e^x + C$

**102.** $\displaystyle\int x^n e^{ax}\, dx = \frac{1}{a}x^n e^{ax} - \frac{n}{a}\int x^{n-1}e^{ax}\, dx;\ a \neq 0$

**103.** $\displaystyle\int \ln^n x\, dx = x\ln^n x - n\int \ln^{n-1} x\, dx$

## Miscellaneous Formulas

**104.** $\displaystyle\int x^n\cos^{-1} x\, dx = \frac{1}{n + 1}\left(x^{n+1}\cos^{-1} x + \int \frac{x^{n+1}dx}{\sqrt{1 - x^2}}\right);\ n \neq -1$

**105.** $\displaystyle\int x^n\sin^{-1} x\, dx = \frac{1}{n + 1}\left(x^{n+1}\sin^{-1} x - \int \frac{x^{n+1}\, dx}{\sqrt{1 - x^2}}\right);\ n \neq -1$

**106.** $\displaystyle\int x^n\tan^{-1} x\, dx = \frac{1}{n + 1}\left(x^{n+1}\tan^{-1} x - \int \frac{x^{n+1}\, dx}{x^2 + 1}\right);\ n \neq -1$

**107.** $\displaystyle\int \sqrt{2ax - x^2}\, dx = \frac{x - a}{2}\sqrt{2ax - x^2} + \frac{a^2}{2}\sin^{-1}\left(\frac{x - a}{a}\right) + C;\ a > 0$

**108.** $\displaystyle\int \frac{dx}{\sqrt{2ax - x^2}} = \sin^{-1}\left(\frac{x - a}{a}\right) + C;\ a > 0$

# ALGEBRA

## Exponents and Radicals

$$x^a x^b = x^{a+b} \qquad \frac{x^a}{x^b} = x^{a-b} \qquad x^{-a} = \frac{1}{x^a} \qquad (x^a)^b = x^{ab} \qquad \left(\frac{x}{y}\right)^a = \frac{x^a}{y^a}$$

$$x^{1/n} = \sqrt[n]{x} \qquad x^{m/n} = \sqrt[n]{x^m} = (\sqrt[n]{x})^m \qquad \sqrt[n]{xy} = \sqrt[n]{x}\sqrt[n]{y} \qquad \sqrt[n]{x/y} = \sqrt[n]{x}/\sqrt[n]{y}$$

## Factoring Formulas

$$a^2 - b^2 = (a - b)(a + b)$$
$$a^3 - b^3 = (a - b)(a^2 + ab + b^2)$$
$$a^n - b^n = (a - b)(a^{n-1} + a^{n-2}b + a^{n-3}b^2 + \cdots + ab^{n-2} + b^{n-1})$$

$a^2 + b^2$ does not factor over real numbers.
$$a^3 + b^3 = (a + b)(a^2 - ab + b^2)$$

## Binomials

$$(a \pm b)^2 = a^2 \pm 2ab + b^2$$
$$(a \pm b)^3 = a^3 \pm 3a^2b + 3ab^2 \pm b^3$$

## Binomial Theorem

$$(a + b)^n = a^n + \binom{n}{1}a^{n-1}b + \binom{n}{2}a^{n-2}b^2 + \cdots + \binom{n}{n-1}ab^{n-1} + b^n,$$

where $\dbinom{n}{k} = \dfrac{n(n-1)(n-2)\cdots(n-k+1)}{k(k-1)(k-2)\cdots 3 \cdot 2 \cdot 1} = \dfrac{n!}{k!(n-k)!}$

## Quadratic Formula

The solutions of $ax^2 + bx + c = 0$ are

$$x = \frac{-b \pm \sqrt{b^2 - 4ac}}{2a}.$$

# GEOMETRY

**Parallelogram**

$A = bh$

**Triangle**

$A = \frac{1}{2}bh$

**Trapezoid**

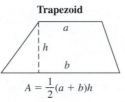

$A = \frac{1}{2}(a + b)h$

**Circle**

$A = \pi r^2$
$C = 2\pi r$

**Sector**

$A = \frac{1}{2}r^2\theta$
$s = r\theta$ ($\theta$ in radians)

**Cylinder**

$V = \pi r^2 h$
$S = 2\pi rh$
(lateral surface area)

**Cone**

$V = \frac{1}{3}\pi r^2 h$
$S = \pi r \ell$
(lateral surface area)

**Sphere**

$V = \frac{4}{3}\pi r^3$
$S = 4\pi r^2$

## Equations of Lines and Circles

$m = \dfrac{y_2 - y_1}{x_2 - x_1}$     slope of line through $(x_1, y_1)$ and $(x_2, y_2)$

$y - y_1 = m(x - x_1)$     point–slope form of line through $(x_1, y_1)$ with slope $m$

$y = mx + b$     slope–intercept form of line with slope $m$ and $y$-intercept $(0, b)$

$(x - h)^2 + (y - k)^2 = r^2$     circle of radius $r$ with center $(h, k)$

# TRIGONOMETRY

$$\cos\theta = \frac{\text{adj}}{\text{hyp}} \qquad \sin\theta = \frac{\text{opp}}{\text{hyp}} \qquad \tan\theta = \frac{\text{opp}}{\text{adj}}$$

$$\sec\theta = \frac{\text{hyp}}{\text{adj}} \qquad \csc\theta = \frac{\text{hyp}}{\text{opp}} \qquad \cot\theta = \frac{\text{adj}}{\text{opp}}$$

$$\cos\theta = \frac{x}{r} \qquad \sec\theta = \frac{r}{x}$$

$$\sin\theta = \frac{y}{r} \qquad \csc\theta = \frac{r}{y}$$

$$\tan\theta = \frac{y}{x} \qquad \cot\theta = \frac{x}{y}$$

*(Continued)*

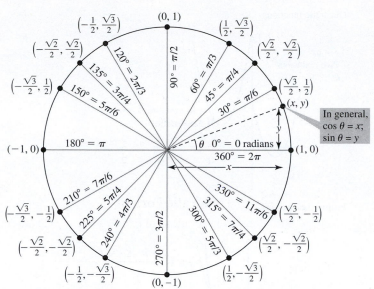

## Reciprocal Identities

$$\tan \theta = \frac{\sin \theta}{\cos \theta} \quad \cot \theta = \frac{\cos \theta}{\sin \theta} \quad \sec \theta = \frac{1}{\cos \theta} \quad \csc \theta = \frac{1}{\sin \theta}$$

## Pythagorean Identities

$$\sin^2 \theta + \cos^2 \theta = 1 \quad \tan^2 \theta + 1 = \sec^2 \theta \quad 1 + \cot^2 \theta = \csc^2 \theta$$

## Sign Identities

$$\sin(-\theta) = -\sin \theta \quad \cos(-\theta) = \cos \theta \quad \tan(-\theta) = -\tan \theta$$
$$\csc(-\theta) = -\csc \theta \quad \sec(-\theta) = \sec \theta \quad \cot(-\theta) = -\cot \theta$$

## Double-Angle Identities

$$\sin 2\theta = 2 \sin \theta \cos \theta \qquad \cos 2\theta = \cos^2 \theta - \sin^2 \theta$$
$$= 2 \cos^2 \theta - 1$$
$$\tan 2\theta = \frac{2 \tan \theta}{1 - \tan^2 \theta} \qquad = 1 - 2 \sin^2 \theta$$

## Half-Angle Identities

$$\cos^2 \theta = \frac{1 + \cos 2\theta}{2} \qquad \sin^2 \theta = \frac{1 - \cos 2\theta}{2}$$

## Addition Formulas

$$\sin(\alpha + \beta) = \sin \alpha \cos \beta + \cos \alpha \sin \beta \qquad \sin(\alpha - \beta) = \sin \alpha \cos \beta - \cos \alpha \sin \beta$$
$$\cos(\alpha + \beta) = \cos \alpha \cos \beta - \sin \alpha \sin \beta \qquad \cos(\alpha - \beta) = \cos \alpha \cos \beta + \sin \alpha \sin \beta$$
$$\tan(\alpha + \beta) = \frac{\tan \alpha + \tan \beta}{1 - \tan \alpha \tan \beta} \qquad \tan(\alpha - \beta) = \frac{\tan \alpha - \tan \beta}{1 + \tan \alpha \tan \beta}$$

## Law of Sines

$$\frac{\sin \alpha}{a} = \frac{\sin \beta}{b} = \frac{\sin \gamma}{c}$$

## Law of Cosines

$$a^2 = b^2 + c^2 - 2bc \cos \alpha$$

## Graphs of Trigonometric Functions and Their Inverses

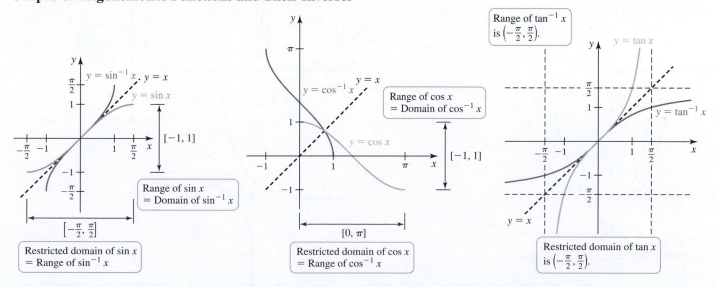

# GRAPHS OF ELEMENTARY FUNCTIONS

Linear functions

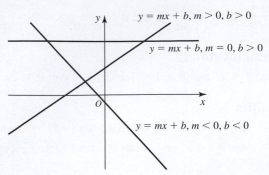

$y = mx + b, m > 0, b > 0$

$y = mx + b, m = 0, b > 0$

$y = mx + b, m < 0, b < 0$

Quadratic functions

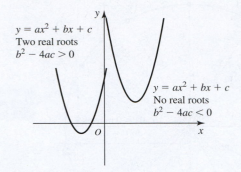

$y = ax^2 + bx + c$
Two real roots
$b^2 - 4ac > 0$

$y = ax^2 + bx + c$
No real roots
$b^2 - 4ac < 0$

Positive even powers

$y = x^6$   $y = x^4$   $y = x^2$

Positive odd powers

$y = x^5$   $y = x^7$   $y = x^3$

Negative even powers

$y = \dfrac{1}{x^6}$   $y = \dfrac{1}{x^4}$   $y = \dfrac{1}{x^2}$

Negative odd powers

$y = \dfrac{1}{x^5}$   $y = \dfrac{1}{x^3}$   $y = \dfrac{1}{x}$

Exponential functions

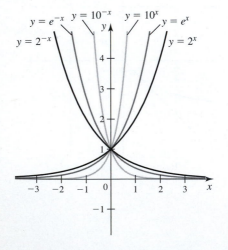

$y = e^{-x}$   $y = 10^{-x}$   $y = 10^x$   $y = e^x$
$y = 2^{-x}$   $y = 2^x$

Natural logarithmic and exponential functions

$y = e^x$   $y = x$   $y = \ln x$

# DERIVATIVES

## General Formulas

$$\frac{d}{dx}(c) = 0$$

$$\frac{d}{dx}(cf(x)) = cf'(x)$$

$$\frac{d}{dx}(f(x) + g(x)) = f'(x) + g'(x)$$

$$\frac{d}{dx}(f(x) - g(x)) = f'(x) - g'(x)$$

$$\frac{d}{dx}(f(x)g(x)) = f'(x)g(x) + f(x)g'(x)$$

$$\frac{d}{dx}\left(\frac{f(x)}{g(x)}\right) = \frac{g(x)f'(x) - f(x)g'(x)}{(g(x))^2}$$

$$\frac{d}{dx}(x^n) = nx^{n-1}, \text{ for real numbers } n$$

$$\frac{d}{dx}\big(f(g(x))\big) = f'(g(x)) \cdot g'(x)$$

## Trigonometric Functions

$$\frac{d}{dx}(\sin x) = \cos x$$

$$\frac{d}{dx}(\cos x) = -\sin x$$

$$\frac{d}{dx}(\tan x) = \sec^2 x$$

$$\frac{d}{dx}(\cot x) = -\csc^2 x$$

$$\frac{d}{dx}(\sec x) = \sec x \tan x$$

$$\frac{d}{dx}(\csc x) = -\csc x \cot x$$

## Inverse Trigonometric Functions

$$\frac{d}{dx}(\sin^{-1} x) = \frac{1}{\sqrt{1 - x^2}}$$

$$\frac{d}{dx}(\cos^{-1} x) = -\frac{1}{\sqrt{1 - x^2}}$$

$$\frac{d}{dx}(\tan^{-1} x) = \frac{1}{1 + x^2}$$

$$\frac{d}{dx}(\cot^{-1} x) = -\frac{1}{1 + x^2}$$

$$\frac{d}{dx}(\sec^{-1} x) = \frac{1}{|x|\sqrt{x^2 - 1}}$$

$$\frac{d}{dx}(\csc^{-1} x) = -\frac{1}{|x|\sqrt{x^2 - 1}}$$

## Exponential and Logarithmic Functions

$$\frac{d}{dx}(e^x) = e^x$$

$$\frac{d}{dx}(b^x) = b^x \ln b$$

$$\frac{d}{dx}(\ln |x|) = \frac{1}{x}$$

$$\frac{d}{dx}(\log_b |x|) = \frac{1}{x \ln b}$$

## Hyperbolic Functions

$$\frac{d}{dx}(\sinh x) = \cosh x$$

$$\frac{d}{dx}(\cosh x) = \sinh x$$

$$\frac{d}{dx}(\tanh x) = \text{sech}^2 x$$

$$\frac{d}{dx}(\coth x) = -\text{csch}^2 x$$

$$\frac{d}{dx}(\text{sech } x) = -\text{sech } x \tanh x$$

$$\frac{d}{dx}(\text{csch } x) = -\text{csch } x \coth x$$

## Inverse Hyperbolic Functions

$$\frac{d}{dx}(\sinh^{-1} x) = \frac{1}{\sqrt{x^2 + 1}}$$

$$\frac{d}{dx}(\cosh^{-1} x) = \frac{1}{\sqrt{x^2 - 1}} \quad (x > 1)$$

$$\frac{d}{dx}(\tanh^{-1} x) = \frac{1}{1 - x^2} \quad (|x| < 1)$$

$$\frac{d}{dx}(\coth^{-1} x) = \frac{1}{1 - x^2} \quad (|x| > 1)$$

$$\frac{d}{dx}(\text{sech}^{-1} x) = -\frac{1}{x\sqrt{1 - x^2}} \quad (0 < x < 1)$$

$$\frac{d}{dx}(\text{csch}^{-1} x) = -\frac{1}{|x|\sqrt{1 + x^2}} \quad (x \neq 0)$$

# FORMS OF THE FUNDAMENTAL THEOREM OF CALCULUS

| Fundamental Theorem of Calculus | $\int_a^b f'(x)\, dx = f(b) - f(a)$ |
|---|---|
| Fundamental Theorem for Line Integrals | $\int_C \nabla f \cdot d\mathbf{r} = f(B) - f(A)$ <br><br> (*A* and *B* are the initial and final points of *C*.) |
| Green's Theorem | $\iint_R (g_x - f_y)\, dA = \oint_C f\, dx + g\, dy$ <br><br> $\iint_R (f_x + g_y)\, dA = \oint_C f\, dy - g\, dx$ |
| Stokes' Theorem | $\iint_S (\nabla \times \mathbf{F}) \cdot \mathbf{n}\, dS = \oint_C \mathbf{F} \cdot d\mathbf{r}$ |
| Divergence Theorem | $\iiint_D \nabla \cdot \mathbf{F}\, dV = \iint_S \mathbf{F} \cdot \mathbf{n}\, dS$ |

# FORMULAS FROM VECTOR CALCULUS

Assume $\mathbf{F}(x, y, z) = f(x, y, z)\mathbf{i} + g(x, y, z)\mathbf{j} + h(x, y, z)\mathbf{k}$, where $f$, $g$, and $h$ are differentiable on a region $D$ of $\mathbb{R}^3$.

Gradient:  $\nabla f(x, y, z) = \dfrac{\partial f}{\partial x}\mathbf{i} + \dfrac{\partial f}{\partial y}\mathbf{j} + \dfrac{\partial f}{\partial z}\mathbf{k}$

Divergence:  $\nabla \cdot \mathbf{F}(x, y, z) = \dfrac{\partial f}{\partial x} + \dfrac{\partial g}{\partial y} + \dfrac{\partial h}{\partial z}$

Curl:  $\nabla \times \mathbf{F}(x, y, z) = \begin{vmatrix} \mathbf{i} & \mathbf{j} & \mathbf{k} \\ \dfrac{\partial}{\partial x} & \dfrac{\partial}{\partial y} & \dfrac{\partial}{\partial z} \\ f & g & h \end{vmatrix}$

$\nabla \times (\nabla f) = \mathbf{0} \qquad \nabla \cdot (\nabla \times \mathbf{F}) = 0$

$\mathbf{F}$ conservative on $D \Leftrightarrow \mathbf{F} = \nabla \varphi$  for some potential function $\varphi$

$\qquad\qquad \Leftrightarrow \oint_C \mathbf{F} \cdot d\mathbf{r} = 0$ over closed paths $C$ in $D$

$\qquad\qquad \Leftrightarrow \int_C \mathbf{F} \cdot d\mathbf{r}$ is independent of path for $C$ in $D$

$\qquad\qquad \Leftrightarrow \nabla \times \mathbf{F} = \mathbf{0}$ on $D$